3rd International School and Conference on Optoelectronics, Photonics, Engineering and Nanostructures (Saint Petersburg OPEN 2016)

Journal of Physics: Conference Series Volume 741

St. Petersburg, Russia
28-30 March 2016

Part 1 of 2

ISBN: 978-1-5108-2995-4
ISSN: 1742-6588

Printed from e-media with permission by:

Curran Associates, Inc.
57 Morehouse Lane
Red Hook, NY 12571

Some format issues inherent in the e-media version may also appear in this print version.

Printed by Curran Associates, Inc. (2016)

For permission requests, please contact the Institute of Physics
at the address below.

Institute of Physics
Dirac House, Temple Back
Bristol BS1 6BE UK

Phone: 44 1 17 929 7481
Fax: 44 1 17 920 0979

techtracking@iop.org

Additional copies of this publication are available from:

Curran Associates, Inc.
57 Morehouse Lane
Red Hook, NY 12571 USA
Phone: 845-758-0400
Fax: 845-758-2633
Email: curran@proceedings.com
Web: www.proceedings.com

TABLE OF CONTENTS

PART 1

CRYSTAL GROWTH AND STRUCTURAL PROPERTIES OF SEMICONDUCTOR MATERIALS AND NANOSTRUCTURES

NANOBIOTECHNOLOGY, BIOPHYSICS AND BIOPHOTONICS

LASERS, SOLAR CELLS AND OTHER OPTOELECTRONIC DEVICES

PART 2

NANOPHOTONICS, SPECTROSCOPY, MICROCAVITIES, OPTICS, PLASMONICS

OTHER ASPECTS OF NANOTECHNOLOGY

3rd International School and Conference

Saint Petersburg OPEN 2016

Optoelectronics, Photonics, Engineering and Nanostructures

Saint Petersburg, March 28–30, 2016

Acknowledgements

RFBR
Russian Foundation for Basic Research

RSF
Russian Science Foundation

SPIE. STUDENT CHAPTER
SAINT PETERSBURG ACADEMIC UNIVERSITY RUSSIAN ACADEMY OF SCIENCES

OSA®
The Optical Society

IOP Institute of Physics

Journal of Physics: Conference Series **741** (2016) 011001 doi:10.1088/1742-6596/741/1/011001

Dear Colleagues,

3rd International School and Conference "Saint Petersburg OPEN 2016" on Optoelectronics, Photonics, Engineering and Nanostructures was held on March 28 – 30, 2016 at St. Petersburg Academic University of the Russian Academy of Sciences. The School and Conference included a series of invited talks given by leading professors with the aim of introducing young scientists to the actual problems and major advances in modern physics and technology. The keynote speakers were

Mircea Guina *(Tampere University of Technology, Finland)*

Evgeny I. Terukov *(Ioffe Institute RAS, Russia)*

Victor M. Ustinov *(Ioffe Institute RAS, Russia)*

Peter G. Kazansky *(University of Southampton, UK)*

Alexander O. Golubok *(ITMO University, Russia)*

Georgy E. Cirlin *(St Petersburg Academic University RAS, Russia)*

Levon V. Asryan *(Virginia Polytechnic Institute and State University, USA)*

Andrey A. Lipovskii *(Peter the Great St.Petersburg Polytechnic University, Russia)*

During the poster session all undergraduate and graduate students attending the conference presented their works. A large number of participants with more than 280 student attendees from all over the world allowed the Conference to provide a fertile ground for the fruitful discussions between the young scientists as well as to become a perfect platform for the valuable discussions between student authors and highly experienced scientists. The best student papers, which were selected by the Program Committee and by the invited speakers based on the theses and their poster presentation, were awarded with diplomas of the conference – see the photos.

This year "**Saint Petersburg OPEN 2016**" is organized by **St. Petersburg Academic University** in cooperation with **Peter the Great St. Petersburg Polytechnic University**. The School and Conference is supported by the **Russian Foundation for Basic Research** (*Project№ 16-32-10060*), **Russian Science Foundation**, **SPIE** (The International Society for Optics and Photonics) and **OSA** (The Optical Society). It is a continuation of the annual schools and seminars for youth on topical problems of physics and technology that is organized by the Academic University since 2009.
We invite all students and young scientists to attend "Saint Petersburg OPEN" in 2017!
Please find details at http://spbopen.spbau.com/

With best wishes, Editorial Board,
Program and Organizing Committees

Journal of Physics: Conference Series **741** (2016) 011001 doi:10.1088/1742-6596/741/1/011001

The head of Program Committee,
rector of St. Petersburg Academic University RAS,
academic Zhores I. Alferov is making opening remarks

The head of Organizing Committee, Prof. Dr. Alexey Zhukov

Poster Session

 doi:10.1088/1742-6596/741/1/011001

ITMO UNIVERSITY

ITMO UNIVERSITY
PREPARATION FIBER BASED AL-DOPED ZINC
FILMS BY ELECTROSPINNING

Journal of Physics: Conference Series **741** (2016) 011001 doi:10.1088/1742-6596/741/1/011001

Award ceremony

Prof. *Alexey Zhukov* and *Andrey Lipovskii* with students

Journal of Physics: Conference Series **741** (2016) 011001 doi:10.1088/1742-6596/741/1/011001

Saint Petersburg OPEN 2016 IOP Publishing
Journal of Physics: Conference Series **741** (2016) 011001 doi:10.1088/1742-6596/741/1/011001

Closing remarks from Prof. Dr. Andrey Lipovskii and Prof. Dr. Alexey Zhukov

Peer review statement

All papers published in this volume of *Journal of Physics: Conference Series* have been peer reviewed through processes administered by the proceedings Editors. Reviews were conducted by expert referees to the professional and scientific standards expected of a proceedings journal published by IOP Publishing.

Journal of Physics: Conference Series **741** (2016) 012001 doi:10.1088/1742-6596/741/1/012001

Grain size and doping effect on structure and electromechanical properties of polycrystalline silicon for MEMS applications

O A Ageev, E Yu Gusev, J Y Jityaeva, M V Ilina, and Al V Bykov
Southern Federal University, Institute of Nanotechnology, Electronics and Electronic Equipment Engineering, Taganrog 347922, Russia

Abstract. Grain size effect on microhardness and Young modulus of polysilicon layers was obtained by nanoindentation experiments: 14.1-16.3 GPa and 240-300 GPa, respectively, for the particle-diameter of 117-430 nm. The range of grain size correlated to mechanical properties exceed theoretical values was defined. The doping effect was considered. The grain size range of PECVD polysilicon with aspects of hardening and suitable conductivity is presented with the objective to formation of micro- and nanomechanical devices.

1. Introduction

Polycrystalline silicon has been widely used in metal-oxide-semiconductor field-effect-transistor, complementary metal–oxide–semiconductor, very-large-scale integration and solar cell processing technologies [1-3]. Doped polysilicon films are also of a great interest for the production of electrostatically movable mechanical elements of micro- and nanoelectromechanical systems (e.g. accelerometers, gyroscopes or pressure sensors) by surface as well as bulk micromachining [1-2,4]. Young's modulus, microhardness, conductivity and mobility of charge carriers are important characteristics of the material. The mechanical properties like Young's modulus and microhardness of these films have a strong impact on the design as well as on the reliability and functionality of these components [5]. Preferably, the material of moving elements should have higher values of the parameters. The questions about possibility to control of electromechanical parameters and monitoring their changes through doping during deposition are important as well as for two steps (deposition and subsequent ex-situ doping) process flow of the films fabrication. Moreover, the concentration of charge carriers of the films should be more 10^{19} cm^{-3} [6].

Intrinsic and doped films and layers of polysilicon mainly formed by low pressure and plasma-enhanced chemical vapor deposition (PECVD) or solid-phase crystallization of amorphous silicon [1,7]. It is known that PECVD allows to precision control properties of deposited films. Its properties are differ from ones formed by other chemical vapor deposition techniques. Growth rate and film uniform were higher for intrinsic films compare with using of doping gases [1,2,7].

The goal of the research is to establish the influence of grain size and ex-situ doping on the mechanical and electrophysical parameters of the plasma-deposited polysilicon films.

2. Experimental details

The plasma-enhanced chemical vapour deposition system was used for deposition of polycrystalline silicon films (PlasmaLab 100). Plasma-deposited polysilicon films, formed from silane and argon [8] were doped using the liquid PCl_3 diffusant (SD.OM-3M). The prediffusion and drive-in steps temperatures were 850°C and 1150°C, respectively. Concentration, mobility of charge carriers and resistivity after doping were measured using Hall/van-der-Pauw and contactless methods (Ecopia HMS-3000/1T and LEI 1510M40). The surface morphology and the mechanical properties were observed with an atomic force microscopy (NTEGRA Vita Probe Nanolaboratory). The mechanical properties like Young's modulus and microhardness were observed by nanoindentation method with used a diamond three sided Berkovich pyramid with an apex angle of $\theta = 70°$ as the indenter. Scratch force was 700-900 μN. The probe was indented into the sample surface at a constant speed; the process was accompanied by recording the values of the load and the appropriate depth of indenter penetration in the material, on the basis of which the resulting dependence (load curve) was plotted.

3. Results and discussion

The uniform polycrystalline silicon films in the thickness range from 0.2 to 2 μm were fabricated. The influence of parameters deposition films on their morphology was investigated early [8].

The experimental research of surface morphology and electro-physical properties of polycrystalline silicon films before and after doping are shown in Figure 1 and listed at Table 1. AFM-images show an increase of grain size. Concentration, mobility of charge carriers and resistivity values were amounted of $2\text{-}5\cdot10^{20}$ cm^{-3}, 12-35 $cm^2/(V\cdot s)$ and $8.6\text{-}9.4\cdot10^{-4}$ $\Omega\cdot cm$.

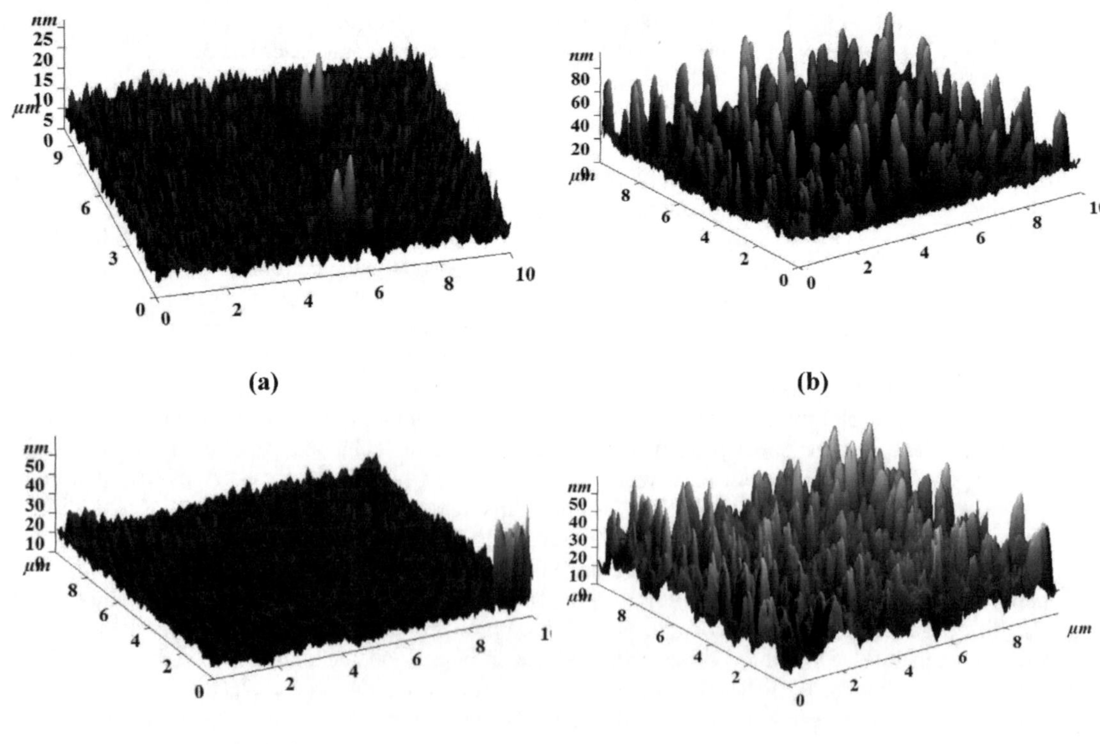

Figure 1(a-d). AFM images of the polycrystalline silicon films before **(a,c)** and after **(b,d)** doping

Root-mean-square (RMS) roughness increases from 2.1 nm and 2.6 nm to 12.1 and 8.9 nm for film thickness of 0.2 μm and 2.0 μm through doping. The average value of the grain size increases from 117 nm to 431 nm for the doped films. According to the well-known Hall-Petch dependence it should effect on microhardness and Young's modulus. However, there is a slightly change for the films with the referred grain size range. It should be mentioned that experimental values defined by nanoindentation are exceed theoretical values for Young's modulus [1,2,9]. It could be concerned with nanoindentation technique, plasma effect, and incorporation of chemicals during deposition, as well as conditions of bottom layer surface.

Table 1. Mechnical and electro-physical properties of the as-deposed and ex-situ doped polysilicon

#	Microhardness, GPa		Young modulus, GPa		Concentration, cm^{-3}, $\times 10^{20}$	Mobility, $cm^2/(V\ s)$	Resistivity, Ω/cm, $\times 10^{-3}$
	undoped	doped	undoped	doped			
17	7.7	23.6	249.5	284.9	3.191	18.78	1.042
18	10.4	18.3	302.3	310.1	3.484	18.55	0.965
19	14.8	13.7	242.7	278.2	5.842	11.35	0.942
20	18.8	21.5	293.2	224.3	7.824	16.53	4.862
21	17.4	18.4	380.4	242.3	1.861	26.42	1.270
22	18.4	17.2	247.1	286.1	3.874	30.69	0.525
23	14.1	16.3	299.3	271.2	2.015	35.98	0.861
24	16.4	18.4	282.9	242.7	2.491	27.74	0.903

It is assumed that, the most suitable polysilicon films for micro- and nanomechanical applications with aspects of high mechanical hardness and elasticity and acceptable conductivity are the films with the grain size within the range 117-430 nm, and the values of electromechanical parameters amounted to: 13-17 GPa, 240-300 GPa, $8.6\text{-}9.4 \cdot 10^{-4}$ Ω/cm, 12-35 $cm^2/(V \cdot s)$. Such films (Figure 2) have been used for fabrication of micro- and nanomechanical gyroscopes and accelerometers (project number 14.575.21.0045) [10,11].

4. Conclusion

Polycrystalline silicon films were deposited by plasma-enhanced chemical vapour deposition. Doping effect on surface morphology was obtained. Root-mean-square roughness increases from 2.1 nm to 12.1 nm for 0.2 μm films thickness and from 2.6 nm to and 8.9 nm for 2.0 μm films. The average value of the grain size increases from 117 nm to 431 nm.

The correlation data of the grain size and root-mean-square roughness to micromechanical properties were presented. According to nanoindentations the microhardness and Young module of doped films were 13.7–23.6 GPa and 224.3–380.4 GPa, respectively. The mechanical parameters of the fabricated films were close to theoretical values or exceed it. Doping effect on mechanical properties was ambiguously and its average values were about ± 15%.

Concentration, mobility of charge carriers and resistivity of doped polysilicon films were (2.0–7.8) $\cdot 10^{20}$ cm^{-3}, 11.35–36.0 $cm^2/(V \cdot s)$ and (0.525–4.87) $\cdot 10^{-3}$ Ω·cm, respectively.

These films were found to be suitable and have been used in fabrication of micro- and nanomechanical gyroscopes and accelerometers (project 14.575.21.0045) to form two conducted layers: 1st bottom layer and 2nd layer of moveable inertial masses.

(a) **(b)**

Figure 2(a,b). SEM images of the bottom **(a)** and top **(b)** polycrystalline silicon layers of the micromechanical accelerometer

Acknowledgment
The equipment of the Centre of Collective Use of Equipment "Nanotechnologies" of Southern Federal University was used for this study. This study was financially supported by The Ministry of Education and Science of Russian Federation within the contract/project 14.575.21.0045 (unique identifier RFMEFI57514X0045).

References

[1] Bhushan B 2010 *Springer Handbook of Nanotechnology* (Heidelberg, Dordrecht, London, New York: Springer)
[2] French P J 2002 *Sensors and actuators A Physical* **99** 3
[3] Gerasimenko N N, Parkhomenko Yu N 2007 *Silicon - a material for nanoelectronics* (Moscow, Tekhnosfera)
[4] Ghodssi R, Lin P 2011 *MEMS Materials and Processes Handbook* (Heidelberg, Dordrecht, London, New York: Springer)
[5] Li X, Bhushan B, Takashima K, Beak C-W, Kim Y-K 2002 *Ultramicroscopy* **97** 481-494
[6] Huikai X, Fedder G K, Sulouff R E 2008 *Comprehensive Microsystems* Ed. B. Yogesh Gianchandani. Amsterdam: Elsevier B V 136–174
[7] Paul K Chu 2009 *Surface and Coatings Technology* **203** 2793-2798
[8] Gusev E Yu, Jityaeva J Y, Bykov Al V, Bespoludin V V 2015 *Izvestiya SFedU. Engineering Sciences* **9** 126
[9] Carlton C E, Ferreira P J 2007 *Acta Materialia* **55** 3749
[10] Ageev O A, Balakirev S V, Bykov Al V, Gusev E Yu, Fedotov A A, Jityaeva J Y, Il'in O I, Il'ina M V, Kolomiytsev A S, Konoplev B G, Krasnoborodko S U, Polyakov V V, Smirnov V A, Solodovnik M S, Zamburg E G 2016 *Springer Proceedings in Physics* (New York: Springer International Publishing) **175** 563
[11] Ageev O A, Gusev E Yu, Jityaeva J Yu, Kolomiytsev A S, Bykov Al V 2016 *CRC Press/Balkema* (London: Taylor & Francis Group) 13

Saint Petersburg OPEN 2016 IOP Publishing
Journal of Physics: Conference Series **741** (2016) 012002 doi:10.1088/1742-6596/741/1/012002

Study of electrical properties of single GaN nanowires grown by molecular beam epitaxy

A M Mozharov[1], F E Komissarenko[1,2], A A Vasiliev[1,5], A D Bolshakov[1], E I Moiseev[1], M S Mukhin[1,2], G E Cirlin[1,3,4,5] and I S Mukhin[1,2]

[1] St. Petersburg Academic University, Khlopina 8/3, 194021 St. Petersburg, Russia
[2] ITMO University, 197101 St. Petersburg, Russia
[3] Institute for Analytical Instrumentation RAS, 190103 St. Petersburg, Russia
[4] St. Petersburg State University, 199034 St. Petersburg, Russia
[5] St. Petersburg State Polytechnical University, 195251 St. Petersburg, Russia

E-mail: mozharov@spbau.ru

Abstract. Electrical properties of single GaN nanowires grown by means of molecular beam epitaxy with N-plasma source were studied. Ohmic contacts connected to single *n*-type GaN wires were produced by the combination of electron beam lithography, metal vacuum evaporation and rapid thermal annealing technique. The optimal annealing temperature to produce ohmic contacts implemented in the form of Ti/Al/Ti/Au stack has been determined. By means of 2-terminal measurement wiring diagram the conductivity of single NW has been obtained for NWs with different growth parameters. The method of MESFET measurement circuit layout of single GaN nanowires (NWs) has been developed. In accordance with performed numerical calculation, free carriers' concentration and mobility of single NWs could be independently estimated using MESFET structure.

1. Intoduction

Semiconductor nitride compounds are very interesting for producing state-of-the-art high efficient semiconductor devices. Having a wide band gap (3.4 eV [1]) with relatively large electron affinity (about 4 eV [1]), GaN is a very promising semiconductor material for optoelectronic devices, in particular for solar cells.

One of the main challenges of 21st century is to develop low cost and high efficient renewable sources of energy including high efficient solar cell (SC). Nowadays the best results in this field have been achieved with tandem SC lattice-matched to GaAs substrate. It's worth noting that the cost of such SC is very high. In order to reduce production costs cheaper Si wafers could be used. The main problem related to production of Si substrate based solar cell is the lack of lattice-matched materials with good opto-electronic properties. One of potential candidates for this purpose could be diluted nitride compounds [2] which are direct bandgap A3B5 semiconductors lattice-matched to Si substrate. Another way is to use nanowire (NW) arrays as top SC junctions [3, 4]. Thus binary nitride materials (AlN, GaN, InN) could be grown on lattice-mismatched Si wafers without dislocations. These structures are very promising due to high crystal quality and excellent anti-reflection properties [5].

In order to achieve device high reproducibility it's important to link the growth parameters with electrical and optical properties of synthesised NWs. In order to independently measure carrier concentration and mobility in thin films, the Hall method is used. But in case of property study of a single NWs, filed-effect transistor (FET) structure is more suitable [6].

2. Experiment

Using molecular beam epitaxy GaN NWs with different n-doping levels have been synthesized. On figure 1 one can see SEM image of GaN NWs array grown on Si substrate using Veeco GEN III MBE machine. The doping level was determined by the temperature value of the Si effusion cell. Two types of NWs samples were studied namely undoped GaN NWs and n-doped NWs with 1050°C temperature of the Si effusion cell. The NWs length of both sample types was about 1 μm. NWs diameter was in the range of (30-40) nm.

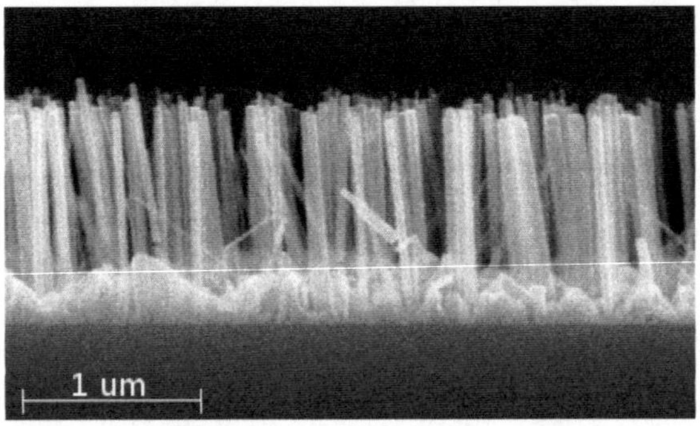

Figure 1. SEM image of GaN NWs array grown on Si substrate by means of molecular beam epitaxy.

With ultrasonic treatment of NW submerged in isopropyl alcohol, the NWs were moved from their native growth Si wafer to the quartz substrate. Using combination of electron beam lithography and metal vacuum evaporation, contacts composed of Ti/Al/Ti/Au stack were attached to single NWs. After contacts formation the samples were annealed in rapid terminal annealing (RTA) machine at 630 – 850°C and 650-700°C for undoped and n-doped NWs, respectively. The ohmic linear behaviour for n-doped NWs was achieved after 7 minutes annealing at 700°C, but for undoped NWs the contacts remained non-ohmic under 800°C. At the temperature exceeding 850°C the contacts were broken. It should be mentioned that in case of non-ohmic behaviour, NWs conductivity could be reasonably well estimated at high voltage zone of volt-ampere characteristic (VAC).

The results of VAC study of both NWs samples in 2-terminal measurement wiring diagram (without gate contact) are indicated on figure 2. Insert on figure 2 (a) shows SEM image of single NW with two Ti/Al/Ti/Au contacts after annealing. Near linear behavior of VAC of both samples confirms the ohmic behavior of the contacts attached to single NWs. The main problem in producing ohmic contact to GaN is necessary to use Al which reacts with any of diffusion barriers as Ti, Ni, Ta, Mo. This leads to formation of intermetallic compounds and breaking the contacts.

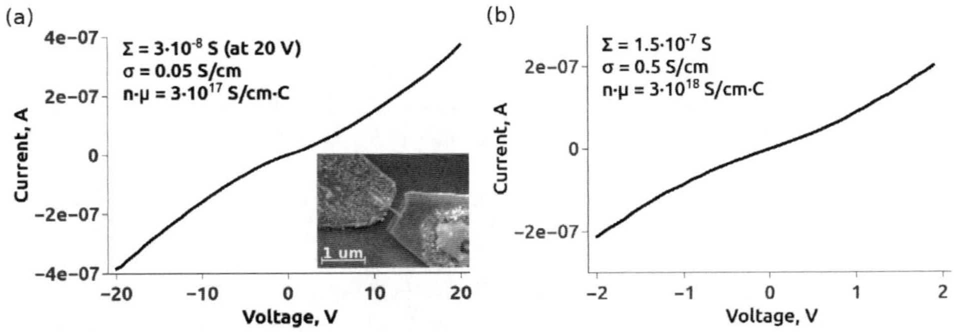

Figure 2 (a, b). Volt-ampere characteristics of undoped (a) and n-doped (b) NWs collected with 2-terminals measurement circuit layout. Inset in (a) shows SEM image of single GaN NW with 2 ohmic Ti/Al/Ti/Au contacts.

According to obtained data, conductivity $\sigma=e\cdot n\cdot\mu$ (where e, n and μ are electron charge, carrier concentration and carrier mobility) of undoped and doped NWs can be estimated to be 0.05 S/cm and 0.5 S/cm, respectively. Thus we can use 2-terminal measurement circuit layout only to obtain the multiplication of the two values $(n \cdot \mu)$, but not any of those values separately.

For independent evaluation of free carriers' concentration and their mobility FET measurement circuit layout could be implemented. In case of FET structure the source and the drain terminals are connected to the semiconductor via ohmic contacts, the channel conductivity (channel width) is a function of the potential applied across the gate and the source terminals. Volume of depletion layer in semiconductor lying under gate contact is mostly determined by free carriers' concentration rather than their mobility. In case of NW, metal-semiconductor FET (MESFET) measurement circuit layout is more suitable for n and μ determination, where gate contact is formed by Schottky contact between NW and metal. For Schottky contact to n-doped GaN, metals with high value of work function, ex. Ni, Pd, Pt, could be used. Figure 3 shows technological process of NW structure fabrication in MESFET structure. Gate Schottky contact could be formed by Pd thermal evaporation.

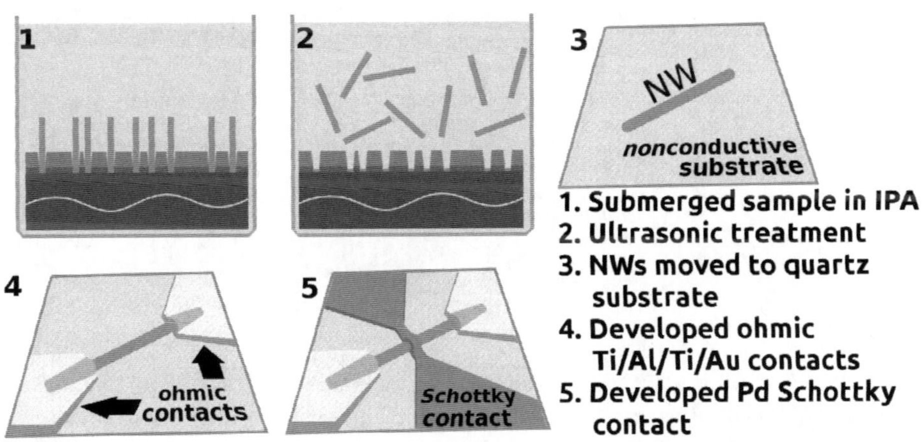

Figure 3. Scheme of technological process of NW structure fabrication in MESFET measurement circuit layout.

3. Modeling

Numerical modeling was conducted to show the possibility of independent estimation of the carriers' concentration and their mobility via studying volt-ampere curves obtained using MESFET structure. In our calculation values of NW conductivity and work function of gate metal were fixed, Pd was considered as a gate contact. Surface states at GaN NW surface and piezo effect contributions were not included in modeling.

Figure 4 shows set of calculated volt-ampere characteristics of 1 μm length and 100 nm diameter NW. Performed calculation showed that gate threshold voltage and current limit values mostly depend on carriers' concentration rather than on their mobility with good approximation. Due to the latter fact these characteristics of semiconductor NWs can be determined independently.

In consideration of fixed NW conductivity σ, figure 4 (a) shows the dependence of drain-source current from potential applied between these contacts in the presence of gate contact. Value of transistor saturation current depends on the combination of carriers' concentration and their mobility. Lowering carriers' concentration increases the volume of depletion region, which in turn leads to increasing of NW electrical resistance and decreasing of drain-source current. Figure 4 (b) indicates the dependence of drain-source current from gate voltage under varying values of carrier concentration and mobility. Figure 4 (c) shows the dependence of drain-source current from gate voltage for different values of NWs mobility in consideration of fixed NW carriers' concentration. It's clear that at fixed value of transistor On/Off ratio, the threshold voltage depends on carriers' concentration rather than on their mobility.

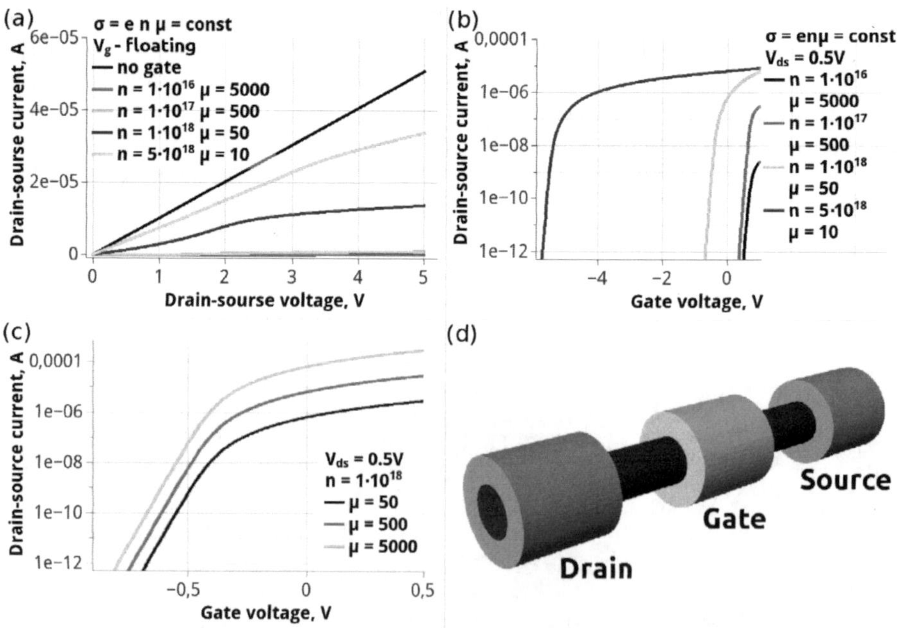

Figure 4 (a, b, c, d). Calculated volt-ampere characteristics of 1 μm length and 100 nm diameter NW. (a) The dependence of drain-source current from drain-source voltage in the presence of gate contact, (b) the dependence of drain-source current from gate voltage at fixed value of conductivity, (c) the dependence of drain-source current from gate voltage at fixed value of carriers' concentration. (d) The symmetrical model of NW with contacts in MESFET structure.

4. Summary

Arrays of GaN NWs with different doping level on Si substrate were synthesized by means of molecular beam epitaxy. Ohmic contacts connected to single NW transferred to fused silica substrate were made by electron beam lithography, metal thermal evaporation and rapid temperature annealing. Conductivity of undoped and n-doped GaN NWs was estimated to be 0.05 S/cm and 0.5 S/cm, respectively. Based on analysis of collected experimental data and results of numerical modelling, realistic estimation of carrier concentration and mobility of single GaN NWs grown under varied conditions could be made using MESFET measurement circuit layout. Obtained data opens perspectives to solar cells with top junctions in the form of nitride nanowire arrays with controllable composition and doping level.

Acknowledgements

This work was carried out with the support of President grant (14.W01.15.7736-MK), the Russian Foundation for Basic Research (15-02-06839 A, 13-02-12031 ofi and 16-32-00560 mol_a), the leading universities of the Russian Federation (grant 074-U01), the Skolkovo Foundation (grant agreement for Russian educational and scientific organisation no.6 dd. 30.12.2015).

References

[1] Sadao Adachi 2009 Properties of Semiconductor Alloys: Group-IV, III–V and II–VI Semiconductors (John Wiley & Sons Ltd)
[2] Henini M 2005 Dilute nitride semiconductors (UK: Elsevier)
[3] Mozharov A, Bolshakov A, Cirlin G and Mukhin I 2015 *Physica Status Solidi RRL* **1–4** 4022
[4] Neplokh V, Ali A, Julien F H, Foldyna M, Mukhin I, Cirlin G, Harmand J-C, Gogneau N and Tchernycheva M 2016 *Materials Science in Semiconductor Processing*
[5] Agarwal R and Lieber C M 2006 *Applied Physics* A **85-3** 209
[6] Blanchard P T, Bertness K A, Harvey T E, Mansfield L M, Sanders A W, Sanford N A 2008 *IEEE of Transactions of nanotechnology* **7-6** 760

High-precision investigation of nanorod and nanosphere topological structures for nanoelectronic issues by means of atomic-force microscopy

M V Zhukov[1,2], V V Lysak[1], I S Mukhin[1,3], A O Golubok[1,2]

[1] Department of material science and nanotechnology, ITMO University, 197101 Kronverkskiy pr., 49, Saint-Petersburg, Russian Federation
[2] Department of scanning probe microscopy and spectroscopy, Institute for Analytical Instrumentation RAS, 198095 Ivan Chernykh, 31-33, Saint-Petersburg, Russian Federation
[3] Laboratory of renewable energy source, St Petersburg Academic University RAS, 194021 Khlopina, 8/3, Saint-Petersburg, Russian Federation

Abstract. Fabrication and study of specialized single nanowhisker probes are performed for high-precision investigation of elements such as nanospheres and nanorods using the atomic force microscopy. It was found that single nanowhisker probe significantly increases the resolution and contrast of images obtained in the semi-contact mode. Furthermore, the roughness analysis and adhesion forces are investigated in contact mode to comprehensively characterize properties of nanospherical and nanorod electronic structures.

1. Introduction

The deviation in properties of nanoelectronic devices greatly depends on the parameter differences of their structure elements. Therefore, it is necessary to ensure high-precision control of topological features for such elements with significant accuracy. Atomic-force microscopy (AFM) is one of the most advanced methods used to perform high-precision study and modification of nanoelectronic structures [1, 2].

The principal element in AFM is a probe, or nanoscale needle tip, that interacts with the surfaces during the scanning process. The geometry and material of the probe can greatly affects on the spatial resolution and contrast of the obtained images. To improve the accuracy of AFM investigations, there is some of research works aimed at development new types of probes [3, 4] at the end of standard tip.

Fabrication of nanowhisker structures has significant advantages, such as the possibility to control composition and orientation of growing structures. It is also should be noted that formation of nanowhisker structures goes directly on tip without any additional manipulation. Nevertheless, the most common method of the single or matrix nanowhisker creating is the molecular-beam epitaxy [5]. This method doesn't allow high-precision orientation of single whisker with controlled geometrical parameters of growth that is very important while scanning complex surface topologies [6]. Such surfaces can be, for instance, structures of nanorods with significant difference in aspect ratio or small nano-objects with smooth walls as nanospheres [7, 8].

In this work, the new method to creating stable single nanowhisker (SNW) probes by electron beam deposition with precursor gases in vacuum chamber [9] was offered at the top of standard probes with control of SNW geometrical parameters.

The aim of this article is to ensure high precision investigation of nanoelectronic elements with complex surface structure like nanorods and nanospheres by means of atomic force microscopy with specialized single nanowhisker probes.

2. Experimental setup

The SNW structures were grown at the tip of the standard probes under the electron beam in a high vacuum and direct injection system of precursor gases [6]. Fabrication of the SNW structures and control their geometrical parameters were performed by means of scanning electron microscope CrossBeam Neon 40 (Carl Zeiss, Germany) with integrated inlet gas $C_9H_{16}Pt$.

Light emitting diodes based on GaN nanorodes and SiO2 nanospheres on their surfaces [10] were selected as nanoelectronic samples with complex topological structure. The investigation of samples was performed by means of scanning probe microscope Ntegra Aura (NT-MDT, Russia) in the semi-contact and contact AFM modes.

3. Results and Discussions

Investigation of SNW probes in the air carried out in tapping mode, which increases the stability of the SNW probes and reducing the damaging effects of the probe to the sample. Figure 1-2 show the distribution of nanospheres and nanorods on the sample surface at the same locations, obtained using standard and SNW probes. It was found that SNW probes allow detect the boundary areas of nanospheres more accurately (diameter ~200-250 nm), whereas applying of standard probes revealed borders thickening of the nanospheres up to diameter ~500 nm (Figure 1). Therefore, the study of nanospheres demonstrated a two-fold decreasing of nanospheres broadening using SNW probes.

(a) (b)

Figure 1. AFM images of nanospheres distribution on the same area of 2x2 mkm, received with standard (a) and SNW probe (b).

Figure 2 shows that the SNW probes area allow visualizing nanorod basis, i.e. penetrating to the basement of the substrate, whereas the standard probes visualize only the upper areas (the tips of nanorods). It was found that SNW probes penetrate to the substrate at the value about 230 nm, whereas the standard probes penetration was only about 150 nm at the same area.

This effect can be explained by the geometry of the SNW probe, which is a cylinder with a radius ~30-40 nm and tip sharpening ~7-12 nm, whereas standard probe is a pyramid, expanding from tip to the basement of cantilever (Figure 3). Thus, the penetration of SNW probes in narrow and deep areas is significantly better because of its high aspect ratio (i.e. the ratio of length to diameter of the whisker). This feature appears most noticeably in the samples having large longitudinal and small transverse dimensions, for instance, nanorods, or areas with sharp height differences and smooth walls, such as nanospheres.

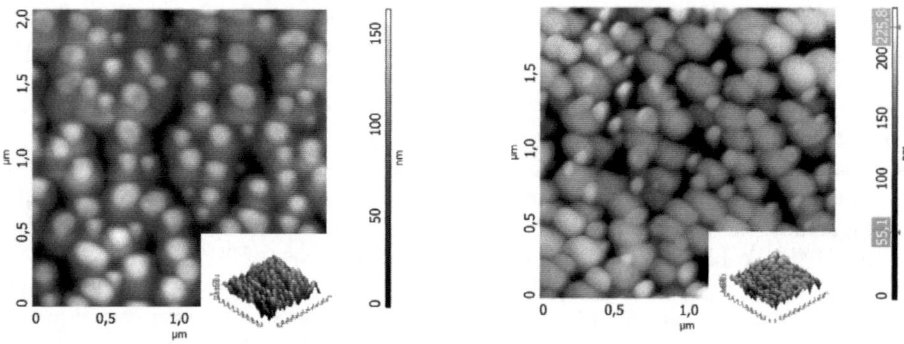

(a) (b)

Figure 2. AFM images of nanorods distribution on the same area of 2x2 mkm, received with standard (a) and SNW probe (b).

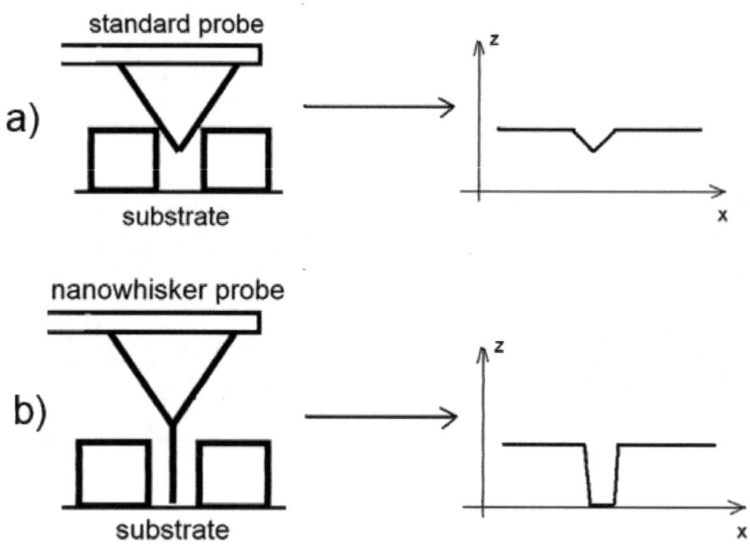

Figure 3. Schematic visualization of a narrow channel using a cone probe (standard Si probe) (a) and cylinder probe (SNW probe) (b).

Results show an overall improvement of SPM resolution and image contrast using SNW probes. This may be explain due to a better penetration ability of SNW probes (contrast height enhancement) and a clear imaging of sharp height difference of objects borders (spatial resolution), correspondingly.

Figure 4 depicts the visualization of the small area of nanorods with cross-sectional view on purpose to measure their spatial parameters. Thus, the single nanorod has a diameter of basement of about 250-300 nm with an altitude of about 100-150 nm. The radius of the tip curvature changes from 20 to 100 nm.

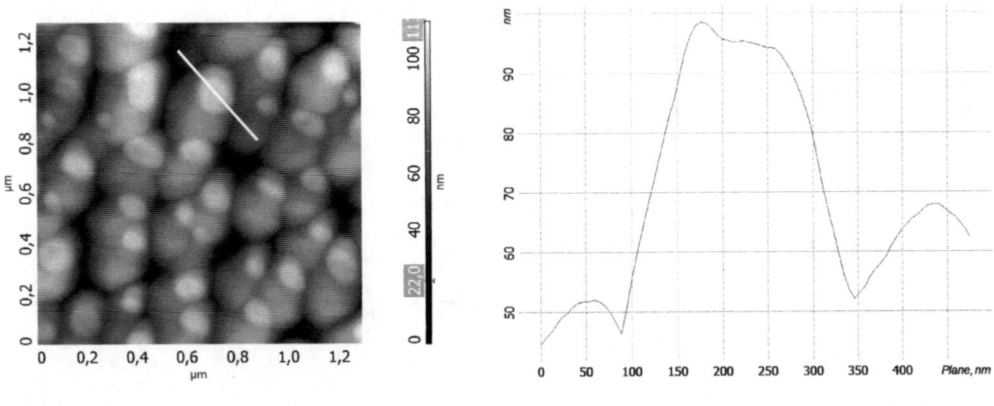

(a) (b)

Figure 4. AFM image of the nanorod distribution (a) and a cross-sectional view of the one nanorod (b), obtained with SNW probe.

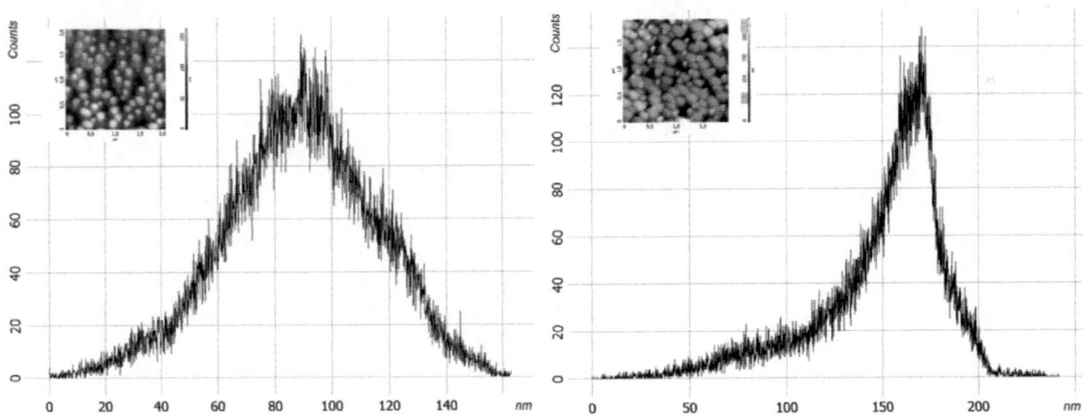

(a) (b)

Figure 5. Histograms of the height distribution on the surface of nanorods on the area about 2x2 mkm, received with standard (a) and SNW probe (b).

Table. 1. Parameters of the nanorods obtained by different types of probes

Parameter	**Standard probe**	**SNW probe**
Height difference, nm	162,917	241,514
Average height, nm	88,736	150,082
Roughness, nm	21,046	24,867
Mean square deviation, nm	26,454	32,879

According to distribution histograms (Figure 5) we identify the parameters of the surface area 2x2 µm of nanorods obtained by standard and SNW probes (Table. 1). Based on this data it is possible to conclude that the roughness of the area has increased values with SNW probes, which means greater penetration capability and, therefore, the reliability of the data obtained using this type of probes.

Analysis of adhesion curves was obtained using standard probe in contact AFM mode (Figure 6). The values of the probe position increment δz (signal Height) corresponding to cantilever deflection (signal DFL) was measured to find the adhesion of probe to the substrate (label A-B and A'-B'). The value of δz calculated as a projection of points A-B (for nanospheres) and A'-B ' (for nanorods) on the X axis. According to the formula (1) it is possible to determine adhesion force as [11]:

$$F = \delta z \cdot k, \quad (1)$$

where δz - deflection of the probe [nm], k - stiffness of the cantilever [N/m].

We use the standard hydrophobic Si contact probes (SiO2 coating on air) with cantilever stiffness of about 0.01 H/m measured by Sader [12].

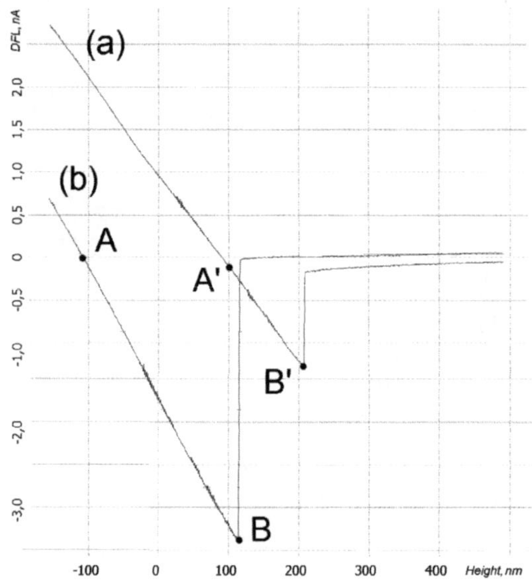

Figure 6. Typical adhesive curves obtained on the surface of nanorods (a) and nanospheres (b) using standard Si probe in contact mode.

The average value of the probe position increment δz was about 120±10 nm for nanorods and about 200±10 nm for nanospheres. This fact suggests a greater accumulation of condensate in the area of nanospheres than in the area of nanorods. That indicates advantage of nanorods as a material for photoelements construction in nanoelectronics, as condensation may interfere with light emission from the surface of the light emitting diodes (LED) and other structures.

The investigation shows that the application of specialized SNW probes has better results than standard silicon probes in the study of complex topological elements with thin structures used in nanoelectronic devises.

4. Conclusions

The method of deposition by a focused electron beam was performed to fabricate specialized probes with nanowhiskers and ensure the precise control of the geometry and composition of the grown structures. It was found that these type of probes may significantly increase the resolution and contrast of images obtained in the study of objects of nanoelectronics, such as nanospheres and nanorods.

Roughness analysis displays a greater degree of penetration ability of specialized whisker probes. In addition, the analysis of adhesion surfaces shows considerable adhesion force in the area of nanospheres than in the area of nanorods that indicates accumulation of condensate in this region.

Acknowledgments
Work carried out with the support of the leading universities of the Russian Federation (grant 074-U01) through ITMO Early Carrier Fellowship scheme, the Russian Foundation for Basic Research (14-02-31703 mol_a), and program U.M.N.I.K.

References

[1] T. Hantschela, A. Schulzea, U. Celanoa, A. Moussaa, K. Arstilaa, P. Eybena, B. Majeeda, D.S. Tezcana, T. Wernera, W. Vandervorsta 2012 TiN scanning probes for electrical profiling of nanoelectronics device structures *Microelectronic Engineering* Vol. 97 pp 255–258.

[2] Takao Inouea, Jun Taniguchia, Toshihiko Ochib 2012 Improving the lifespan of the cantilever during electron assisted AFM lithography *Microelectronic Engineering* Vol. 98 pp 288–292.

[3] J. D. Beard and S. N. Gordeev 2011 Fabrication and buckling dynamics of nanoneedle AFM probes *Nanotechnology* Vol. 22 (17) pp 175303 (8 pp).

[4] Jiang A. N., Gao S., Wei X. L., Liang X. L., and Chen Q. 2008 Amplitude Response of Multiwalled Carbon Nanotube Probe with Controlled Length during Tapping Mode Atomic Force Microscopy *J. Phys. Chem. C* Vol. 112, No. 40 pp 15631-15636.

[5] Dmitry Klinov, Sergei Magonov 2004 True molecular resolution in tapping-mode atomic force microscopy with high-resolution probes *Applied physics letters* Vol. 84 (14) pp 2697-2699.

[6] Levichev V.V., Zhukov M.V., Mukhin I.S, Denisyuk A.I., Golubok A.O. 2013 On the operating stability of a scanning force microscope with a nanowhisker at the top of the probe *Technical Physics* Vol. 58(7) pp 1043-1047.

[7] Héloïse Thérien-Aubin, Ariella Lukach, Natalie Pitch, Eugenia Kumacheva 2015 Coassembly of Nanorods and Nanospheres in Suspensions and in Stratified Films *Angewandte Chemie* Vol. 54(19) pp 5618–5622.

[8] Xingchen Ye, Jaime A. Millan, Michael Engel, Jun Chen, Benjamin T. Diroll, Sharon C. Glotzer, and Christopher B. Murray 2013 Shape Alloys of Nanorods and Nanospheres from Self-Assembly *Nano Lett.* Vol. 13 (10) pp 4980–4988

[9] M. V. Zhukov, I. V. Kukhtevich, V. V. Levichev, I. S. Mukhin and A. O. Golubok 2014 Specialized probes with nanowhisker structures for scanning probe microscopy. *J. Phys.: Conf. Ser.* Vol. 541 (1) pp 012042 (6 pp).

[10] Mustary, M. H., Lysak, V. V. 2014 Fabrication of nanorod light emitting diode by Ni nano-cluster and enhanced extraction efficiency. *IOSR J. of Electrical and Electronics Eng.* Vol. 9 pp 18-22.

[11] Israelachvili J.N. Intermolecular and surface forces 2nd ed. Academic Press 1992 470 p.

[12] Sader, J.E., Larson, I., Mulvaney, P., White, L.R. 1995 Method for calibration of atomic force microscope cantilevers *Rev. Sci. Instrum.* Vol. 66 pp 3789–3498.

The thickness dependence of dielectric permittivity in thin films

Ivan A Starkov[1] and Alexander S Starkov[2]
[1]Nanotechnology Center, St. Petersburg Academic University, RAS,
Khlopin st. 8/3, 194021 St. Petersburg, Russia
[2]National Research University of Information Technologies, Mechanics and Optics,
Kronverksky pr. 49, 197101 St. Petersburg, Russia
starkov@spbau.ru

Abstract. It is well known that the physical properties of thin films depend on their thickness. For a description of such dependences, it is proposed to use a classical model taking into account the presence of film interfaces. A dielectric ball near the half-space was chosen to adopt the approach. The dependence of the effective permittivity of the ball on geometrical and physical parameters of the system is analyzed. It is demonstrated that the dielectric constant of a film can be presented as a sum of the constant of a bulk material and the interface term.

1. Introduction

The classical theory of the dielectric constant (Lorentz-Lorenz, Weiss and others [1]) are based on the expressions for the dipole moment of the sphere (ball) or spheroid placed in a uniform electric field. Obviously, as the sphere is in an infinite space, the effect of boundaries in these models is not considered. In order to account for this issue, it is necessary to examine another model problem – a dielectric ball near the flat surface. This case was repeatedly discussed in the literature (see, e.g., [2]). However, it is possible to obtain an explicit solution only for a particular scenario of the metal sphere [3]. In general, the problem is reduced to a system of difference equations whose solution is given by an infinite product [2]. Due to the complexity of numerical calculations according to the above scheme, it seems convenient to find another solution of the problem. In this study, the approach adopted for this purpose is based on the method of inversion [1].

The method of inversion allows to consistently determine the electric field of a ball or cylinder near the surface. Note that spheroids reproduce the inclusions shape more precisely. Nonetheless, an explicit analytical solution is unknown even for prolate spheroids. In view of this, we consider two bodies simultaneously, as both a ball and cylinder are the extremes of the spheroid. If the dependence of the dielectric constant on the film thickness is known for the both types of inclusions, then for the inclusions of an arbitrary shape this dependence can be found by using the shape factor interpolation, as will be explained below.

2. The statement of the problem

Consider the problem of finding the electrostatic potential of the electric field u from the dielectric sphere located near a semi-infinite dielectric medium in an external electric field $u_{ext} = - E_x x - E_y y$ (see, figure 1). Here $E_{x,y}$ are components of an external electric field. We are interested in the dipole moment p of the described ball, or rather, in the impact of a flat surface on this moment. The potential of the electric field u satisfies the Laplace equation, boundary conditions of continuity of the normal components of the electric displacement at the media interfaces, and tends to u_{ext} for an increasing distance from the origin. The geometry of the problem is greatly simplified if we perform an inversion

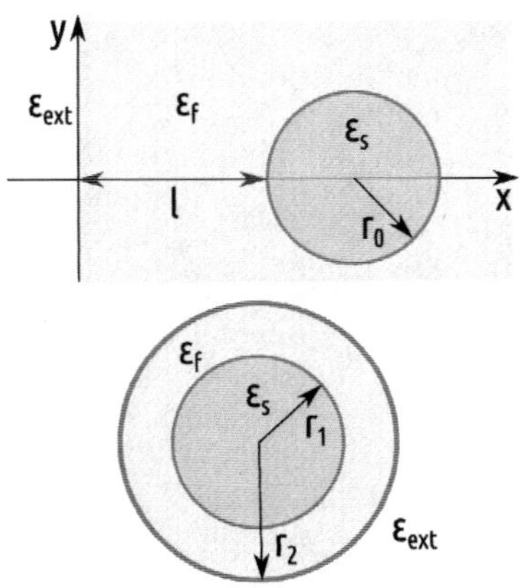

Figure 1. The geometry of the direct (top) and inverse (bottom) problems.

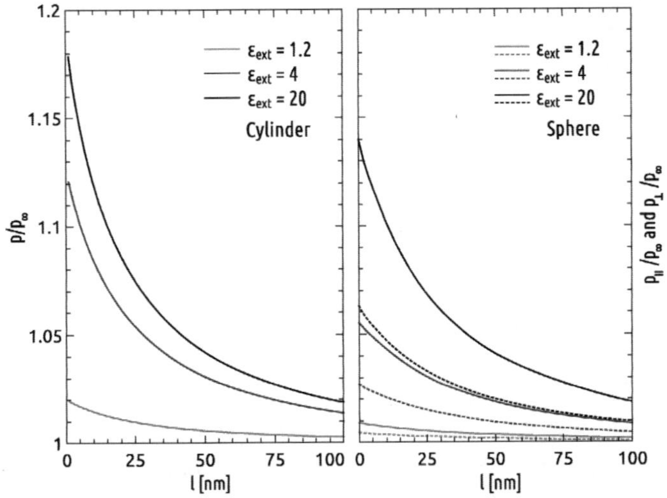

Figure 2. The dependence of the normalized dipole moment on the distance from the interface for the cylinder and sphere. The calculations were performed for a series of ε_{ext} at r_0=60nm, ε_f=1, ε_s=4. The solid and dotted lines reflect the electric field perpendicular and parallel to the x axis.

of the coordinate system [1] with respect to a specially selected point on the symmetry axis and a coordinate system shift. In this way, the eccentric sphere transforms into concentric one, figure 1. Similarly, the eccentric cylinder becomes concentric. Under this transformation the equation and boundary conditions describing the field do not change, and the external uniform field becomes a dipole field. As far as the dipole field may be simply expressed through the field of a point source (charge), the initial problem is reduced to finding a field of a point charge for three-layer media, where the boundaries between the balls are concentric spheres, figure 1 (bottom). Unfortunately, the potential is changed in the case of sphere [1]. After the inversion the Laplace equation is not fulfilled for the potential in the stretched variables, but for a new function that is different from the potential by the factor $F(\rho,\theta)=(\rho^2+\rho_0^2+2\rho\rho_0\cos\theta)^{1/2}$. Here we introduced a spherical coordinate system (ρ,θ,φ) whose origin is located in the centre of the concentric spheres or cylinders, while ρ_0 is the distance to the point at which, after inversion, transforms infinitely distant point. The potential function U is sought as a series in Legendre polynomials

$$U(\rho,\theta,\varphi)=\sum_{n=0}^{\infty}\mathrm{P}_{\mathrm{n}}^{\mathrm{m}}(\cos\theta)\left[\frac{U_n^m}{\rho^{n+1}}+V_n^m\rho^n\right]\delta(\varphi). \qquad (1)$$

In the above expression we have m=0, $\delta(\varphi)$=1 for the field parallel to the z-axis and m=1, $\delta(\varphi)=\sin\varphi$ for the perpendicular one. The series (1) satisfies the Laplace equation with arbitrary coefficients U_n. That is, the problem is reduced to the determination of three pairs of sets (for the three regions under examination) for the coefficients U_n^m and V_n^m. In the outer space $\rho>r_2$, the coefficients V_n^m are known and defined by the external field. In addition, the coefficients U_n^m must be equal to zero in the inner sphere $\rho<r_1$ because of the limitation of the potential at the centre of the coordinate system. There exist four boundary conditions to find the rest four sets of unknown functions. The equality of the potentials at the interfaces $\rho=r_1$ and $\rho=r_2$ makes possible to express two sets of coefficients through the remaining ones. Difficulties arise when considering the second pair of boundary conditions containing derivatives of ρ. Due to the presence of the factor $F(\rho,\theta)$, the boundary conditions become dependent on $\cos\theta$. This leads to the need to solve the system of difference equations. These equations can be solved by performing the Fourier transform. As a result, it manages to obtain explicit equations for the desired potential.

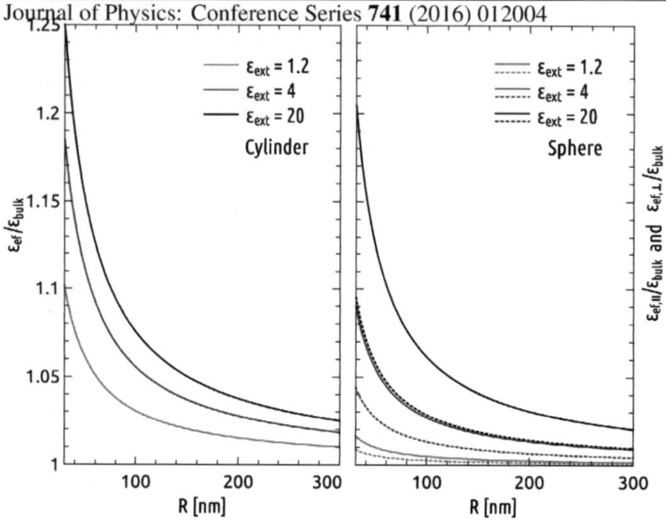

Figure 3. The normalized dielectric permittivity as a function of the film thickness R. Inclusions are modelled by cylinders (left) and spheres (right). The calculations were performed for a series of ε_{ext} at r_0=60nm, ε_f=1, ε_s=4. The solid and dotted lines reflect the electric field perpendicular and parallel to the x axis.

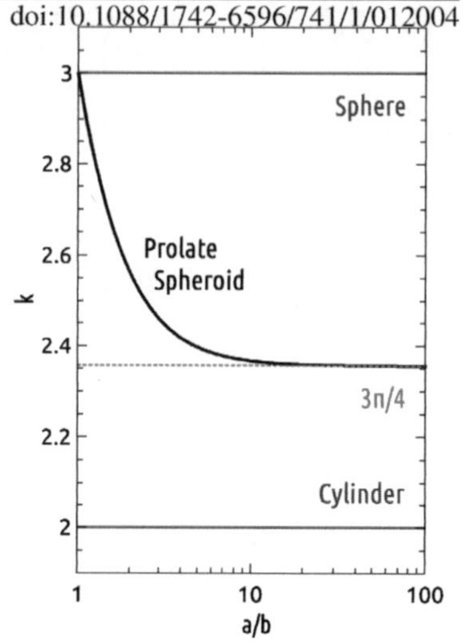

Figure 4. The shape factor for a prolate spheroid depending on the ratio of the minor a and major b axes.

3. Results and discussion

The derived expressions enable us to calculate the required dipole moment of the ball located near the interface, and indicate correction terms to the dielectric constant caused by the presence of the plane. Apparently, the dipole moment p is dependent on the distance from the ball to the plane and the direction of the electric field. Such behaviour is illustrated in figure 2. The dielectric constant of the film ε is completely determined by the dipole moment, which depends on the distance from the film interface, i.e., on the coordinate $p=p(x)$. As a consequence, the film permittivity also should be dependent on the distance to the surface $\varepsilon=\varepsilon(x)$. At large distances from the film interface, the permittivity must tend to a constant value $\varepsilon(\infty)=\varepsilon_{\text{bulk}}$, where $\varepsilon_{\text{bulk}}$ is the dielectric constant of a bulk material. Moreover, because of the corresponding behaviour of the dipole moment, the dielectric constant becomes a tensor. There appear the dielectric permittivities $\varepsilon_\perp$ and $\varepsilon_\parallel$ for the electric field perpendicular and parallel to the x axis. The calculation results are presented in figure 3. The dipole moment of the ball varies largely at short distances from the ball to the dielectric surface. Its maximum value is achieved at the contact of the interface with the ball. The situation is similar for cylinder, but the change of the dielectric constant is more drastic than that of the sphere. This finding is due to the greater area of the interaction between the cylinder and interface.

Such an approach is not always convenient to describe the electrical properties of the film. Therefore, we use the theory of surface phenomena and introduce the Gibbs surface permittivity [4]

$$\varepsilon_{\text{surface}} = \int_0^\infty (\varepsilon(x) - \varepsilon_{\text{bulk}})\, \mathrm{d}x. \tag{2}$$

Hence, the dielectric constant of the film can be replaced by the constant ε_{ef} given by

$$\varepsilon_{\text{ef}} = \varepsilon_{\text{surface}} \frac{S}{V} + \varepsilon_{\text{bulk}} = \varepsilon_{\text{surface}} \frac{k}{R} + \varepsilon_{\text{bulk}}, \tag{3}$$

where $k=S{\cdot}R/V$ is a dimensionless parameter characterizing the shape of an object (shape factor), R typical size of an object (distance from the surface to the most remote interior point of the body). A film of the thickness $2R$, cylinder or sphere of radius R possesses a shape factor 1, 2, and 3, respectively. Thus, the dielectric constant of the film depends on its thickness as $\varepsilon_{ef}=\varepsilon_{bulk}+\varepsilon_{surface}\ k/R$. Naturally, this ratio may be employed for $R>r_0$. The applicability of (3) requires the smallness of the ball size compared to the dimensions of the object. For example, the size of the modelling sphere in the case of crystalline films must be the order of the lattice constant, i.e., several angstroms. Consequently, the film thickness must exceed 2-4nm. Since the shape of inclusions is, as a rule, different from a regular (spherical and cylindrical), it is possible to use an interpolation approach via the shape factor $2< k \leq 3$ for simulation of inclusions of arbitrary shape. The dependence of k on the axis ratio for a prolate spheroid is shown in figure 4. The consideration of the shape factor allows to model various types of inclusions. The inclusions are strictly spherical for k=3. With the reduction of k, their shape becomes more elongated, and with k=2 inclusions transform into cylinders. It is worth to mention that the proposed strategy can be generalized to multilayer cylindrical and spherical inclusions by the matrix homogenization method [5]. Furthermore, the results obtained can be used as a generalization of the Weiss model for ferroelectric thin films [6].

4. Conclusion

The dielectric constant of the film depends on its thickness in a simple way predicted by the developed model. The approach considers polar dielectric regions by inclusions of different shape. The dielectric permittivity varies most strongly (~5%) for cylindrical inclusions. In turn, the most significant variation of the dipole moment is demonstrated by the sphere in contact with the interface. Accounting for the finite film thickness leads to the anisotropy of the dielectric constant. Even if the film is made of a homogeneous material, its dielectric constant depends on the direction of the electric field. For a film thickness of 10nm, the difference of dielectric permittivities $\varepsilon_{\perp,ef}$ and $\varepsilon_{\parallel,ef}$ is about 1%. The presented equations can be used to find the dielectric constant of the mixture and allow to describe the properties of polycrystalline films as a function of the size and shape of the grains. The more exact expressions must take into consideration the influence of the elastic field and the impact of the ball's field on the interface dielectric permittivity.

Acknowledgments

The reported study was funded by RFBR, according to the research project No. 16-32-60189 mol_a_dk. In addition, the research leading to the results exposed has received funding from Government of Russian Federation, Grant 074-U01.

References

[1] Landau L D and Lifshitz E M 1984 *Electrodynamics of Continuous Media*, (Oxford: Pergamon Press), Section 4.
[2] Chaumet P C and Dufour J P 1998 *J. Electrostat.* **43** 145
[3] Starkov A S, Pakhomov O V and Starkov I A 2016, *Tech. Phys.+* **61** 23.
[4] Gibbs J W 1928 *The collected works of J. Willard Gibbs*. eds. Bumstead H A, Longley W R (Longmans, Green and Company), Vol. 1, pp. p-288.
[5] Starkov A S and Starkov I A 2014 *J. Exp. Theor. Phys.+* **119** 861.
[6] Starkov A S, Pakhomov O V and Starkov I A 2013, *J. Exp. Theor. Phys.+* **116** 987.

Raman measurements of dilute nitride alloys GaP(As)N grown on GaP substrates

A Lazarenko, E Pirogov, M Sobolev, A Bukatin, E Nikitina

Nonoelectronics Lab, St. Petersburg Academic University, St. Petersburg 194021, Russia

Abstract. The structural properties of GaP(As)N dilute nitrides alloys grown on GaP substrates by molecular-beam epitaxy are investigated. The samples were studied by Raman scattering and high-resolution X-ray diffraction. In this work the impact of lattice mismatch of GaP(As)N layer and GaP substrate on the form of the spectrum of Raman scattering of samples was detected. It was shown that the addition of arsenic in solid solution GaPAsN can compensate the elastic stresses in the crystal lattice, and we can estimate the lattice mismatch between epitaxial layer GaP(As)N and GaP substrate by the intensity ratio of LO_X/TO_Γ phonon peaks.

1. Introduction

Dilute nitrides materials (InGaPAsN)/GaP with lattice constant close to silicon are very promising for creating lasers, light-emitting diodes (LED), solar cells and detectors for integration on silicon wafers. Recently the yellow-orange light-emitting diodes based on GaP/GaPN heterostructures were demonstrated. Compared with commercial LEDs based on quantum wells AlInGaP, these LEDs have a higher stability of emission wavelength with increasing injection current [1] and a low specific thermal resistance [2].

The electrically pumped diode laser based on Ga(PAsN)/GaP material system which operates at low temperatures [3] and at room temperature [4] was realized.

But GaP(As)N dilute nitrides alloys have unusual properties which are not fully understood. Therefore, this paper investigates structural properties of GaP(As)N/GaP heterostructures.

2. Experiment

The structural properties of GaP(As)N dilute nitrides alloys grown on GaP substrates by molecular-beam epitaxy were investigated. Samples were grown on epi-ready GaP (100) substrates. The concentration of As and N in GaP(As)N layers shown in table 1.

The compositions of the layers were calculated using the BAC-model from the data obtained from the photoluminescence spectra and high-resolution X-ray diffraction. Calculations are shown in previous papers [5,6].

Table 1. The characteristics of the samples 1-3.

№ sample	Concentration N, %	Concentration As, %	LO_X/TO_Γ, a.u.	Lattice mismatch, %
1	0,7	0	0,25	0,23
2	1,6	4,7	0,33	0,14
3	0,5	7,7	0,37	0,11

3. Results and discussion

Structural features of the samples were studied using Raman scattering with excitation energy 2,33 eV (figure 1).

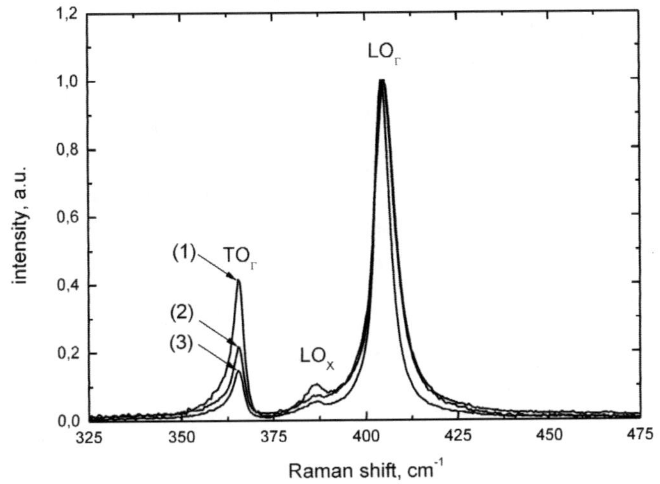

Figure 1. Raman spectra of dilute nitride alloys GaP(As)N grown on GaP substrates

The Raman spectrum is dominated by the longitudinal (LO_Γ) and symmetry-forbidden transverse (TO_Γ) and phonon modes at 404 and 365 cm^{-1}, respectively. Also we observe addition phonon line LO_X (between LO_Γ and TO_Γ modes) at 386 cm^{-1}. Calculation of the intensity ratios of are shown in Table 1

Increasing intensity of additional phonon mode LO_X and symmetry-forbidden phonon line TO_Γ from the third sample to the first may be explained by relaxation of the momentum conservation rules and the symmetry selection rules, respectively. In paper [7] increasing intensity of LO_X mode and LO_X/TO_Γ ratio with increasing of N concentration in $GaP_{1-x}N_x$ layers. But in our case increasing of nitrogen concentration in GaPAsN layers does not lead to increasing of LO_X/TO_Γ ratio.

Therefore samples were studied by high-resolution X-ray diffraction (XRD). Diffraction rocking curves were obtained using a DRON-8 X-ray diffraction system with a BSV 29 highly focused X-ray tube. The anode material was copper with Kα1 radiation line (λ = 1,5405 Å). Characteristic XRD rocking curves near the symmetric GaP (004) reflection are on figure 2.

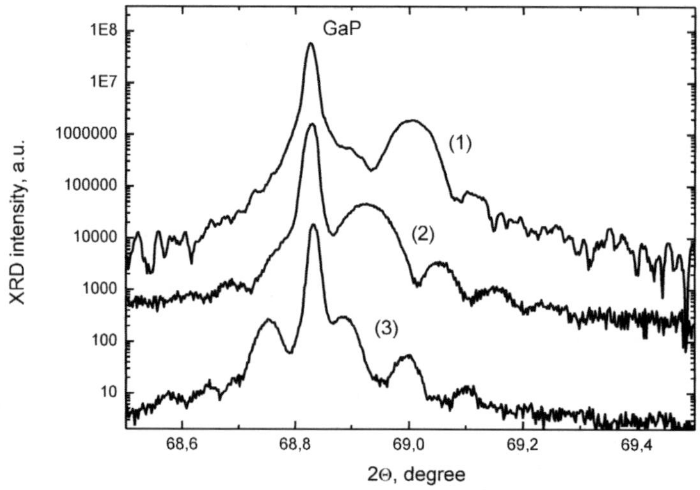

Figure 2. Characteristic XRD rocking curves near the symmetric GaP (004) reflection of epitaxial layers GaP(As)N with different nitrogen concentration

For samples with GaP(As)N layer on GaP substrate, the presence of two diffraction peaks and a series of lines of thickness oscillations are characterized. The maxima correspond to the diffraction on lattice of GaP and the diffraction on elastically strained lattice of GaP(As)N layer. Presence of thickness oscillations demonstrates high quality of heterointerfaces of such layers. Addition of nitrogen in GaP(As)N decreases lattice constant of solid solution and therefore peaks corresponding to XRD of the layers of samples 1 and 2 shifted toward larger angles 2Θ relatively substrate GaP. However for the sample 3 diffraction peak from GaPAsN layer shifted to the left relatively Bragg angle of GaP substrate. This is due to the fact that addition of arsenic in GaPNAs leads to growth of lattice constant and the compensation of elastic stresses caused by the embedding of nitrogen. At certain ratios of nitrogen and arsenic concentrations we can achieve complete matching of the lattice constants of the quaternary solid solution GaPAsN and GaP substrate.

The lattice mismatch of GaP(As)N layers and GaP substrates was calculated by angular distance between diffraction peaks (table 1). We can see that during the growth of ternary solid solution $GaP_{0,993}N_{0,007}$ (sample 1) the mismatch is maximum (0.23%). By adding the arsenic (samples 2 and 3) the mismatch decreases. The smallest mismatch observed in the sample with the highest concentration of arsenic – 7,7% (sample 3). However, it is obvious that a further increasing of arsenic concentration in the solid solution GaPAsN the lattice mismatch will increase.

By XRD and Raman scattering data we can build a dependence between the magnitude of the lattice mismatch of layers GaPAsN and GaP and LO_X/TO_Γ ratio (figure 3).

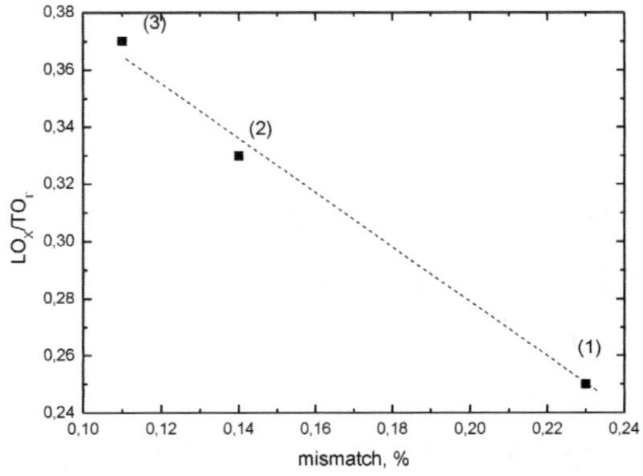

Figure 3. Dependence of lattice mismatch of GaP(As)N layers and GaP substrate on LO_X/TO_Γ phonon peak ratio.

As the increasing the mismatch of lattice constants of growing layer and substrate, the ratio LO_X/TO_Γ decreases linearly and symmetry-forbidden phonon line TO_Γ intensity increases. This may be due to the reducing of crystal symmetry of solid solution GaP(As)N caused by addition of nitrogen. By adding the arsenic in solid solution the ratio of LO_X/TO_Γ increases and intensity of TO_Γ phonon line decreases. This is due to the fact that the addition of arsenic in solid solution GaPAsN can compensate the elastic stresses in the crystal lattice, and by the intensity ratio of LO_X/TO_Γ phonon peaks we can estimate the lattice mismatch between epitaxial layer GaP(As)N and substrate GaP.

Acknowledgments
This work is supported by RFBR Grant № 16-32-00322.

References

[1] M. Kaneko, T. Hashizume, V.A. Odnoblyudov, C.W. Tu. J. Appl. Phys., 101, 103 707 (2007)
[2] S. Adachi, J. Appl. Phys. 54, 1844 (1983)
[3] B. Kunert, S. Reinhard, J. Koch, M. Lampalzer, K. Volz, and W. Stolz, phys. stat. sol. (c) 3, 614–618 (2006)
[4] B. Kunert, A. Klehr, S. Reinhard, K. Volz, and W. Stolz, Electron. Lett. 42, 601 (2006)
[5] A.A.Lazarenko, E.V. Nikitina, E.V. Pirogov, M.S. Sobolev, Semiconductors, Volume 48, Issue 3, pp 392-396 (2014)
[6] A.A.Lazarenko, E.V. Nikitina, M.S. Sobolev, E.V. Pirogov, D.V. Denisov, Semiconductors, Volume 50, Issue 4 (2015)
[7] A. Pulzara-Mora et al. / Vacuum 80 (2006) 468–474

Studying electric field enhancement factor of the nanostructured emission surface

A N Zartdinov[1], **K A Nikiforov**[1]
[1]Saint Petersburg State University, St. Petersburg, 199034, Russia

Abstract. Mathematical model of nanostructured field emission surface is proposed. In order to determine geometrical parameters of the surface structure digital processing of scanning electron microscopy images was used. Effective value of local electrical field enhancement factor is defined and calculated within the Fowler-Nordheim theory. It was found effective enhancement factor decreases as the applied electrical field increases for a fixed geometry.

1. Introduction

The object of this study is field emission array, where microemitters are arranged in a rectangular lattice and have cylindrical shape with sharp top margin (Figure 1) [1-5]. Detailed analysis of the serrated microemitter blade in scanning electron microscope (SEM) revealed a multitude of sharp ledges with curvature radii of less than 10 nm that function as point field emitters with exceptionally large emission surface. This kind of nanolandscape naturally influences the distribution of electrostatic field and defines emission characteristics of the cathode. Example of such a nanolandscape is shown in Figure 1. The goal of this research is assessment of the level of influence of this nanolandscape onto the electric field near the surface. For achieving this goal it is necessary to solve the following problems: define the geometrical parameters of the surface nanolandscape, pick a numerical characteristic describing the influence of nanolandscape onto the fields close to the surface, construct a mathematical model of nanolandscape and compute the coefficient of field strength increase in terms of this model.

2. Model of field enhancement

For finding precise outlines in the images of nanolandscapes, different image processing algorithms were tested. The best results (Figure 1) were demonstrated by the method employing the discrete differential Sobel operator to calculate an approximate value of brightness in each point of the image. The processed images helped us to define the geometrical parameters of nanolandscape structure. The values calculated were the average curvature radius of tip defect of the surface, its height, distance between the tips and the angle at the tip.

The field gain coefficient characterizes the level of influence of the nanolandscape onto the electrostatic field distribution. The field gain coefficient is defined as $\beta(\boldsymbol{s}) = \frac{E_n(s)}{E_0(s)}$, where E_n is the normal component of local field strength of nanostructure surface in point $\boldsymbol{s}$, E_0 is the strength of applied electric field on distance d from the surface without taking the nanolandscape into account.

Knowing the field gain coefficient at given point it is possible to compute the value of current density using the Fowler-Nordheim formula $j\big(\beta(\boldsymbol{s})E_0(\boldsymbol{s})\big) = \frac{A(\beta(s)E_0(s))^2}{\varphi t^2(y)} exp\frac{-B\varphi^{\frac{3}{2}}v(y)}{\beta(s)E_0(s)}$, where φ is the electron work function, $t(y)$, $v(y)$ are elliptic Nordheim functions.

Knowing the current density in area S one can find the value of the current emitted from the surface. The general integral values of the current can be found as $I = \int_S j(\beta(\mathbf{s})E_0(\mathbf{s}))ds$. In this formula, as opposed to "microproblem", the area S is in fact the complex surface of the nanolandscape. In order to simplify the calculations for finding out the emission current we've employed the effective value of field growth coefficient that equates the smooth surface of integrating in "microproblem" to the complex surface of "nanoproblem": $I = \int_{S_0} j(\beta_{эфф}E_0(\mathbf{s}))ds$.

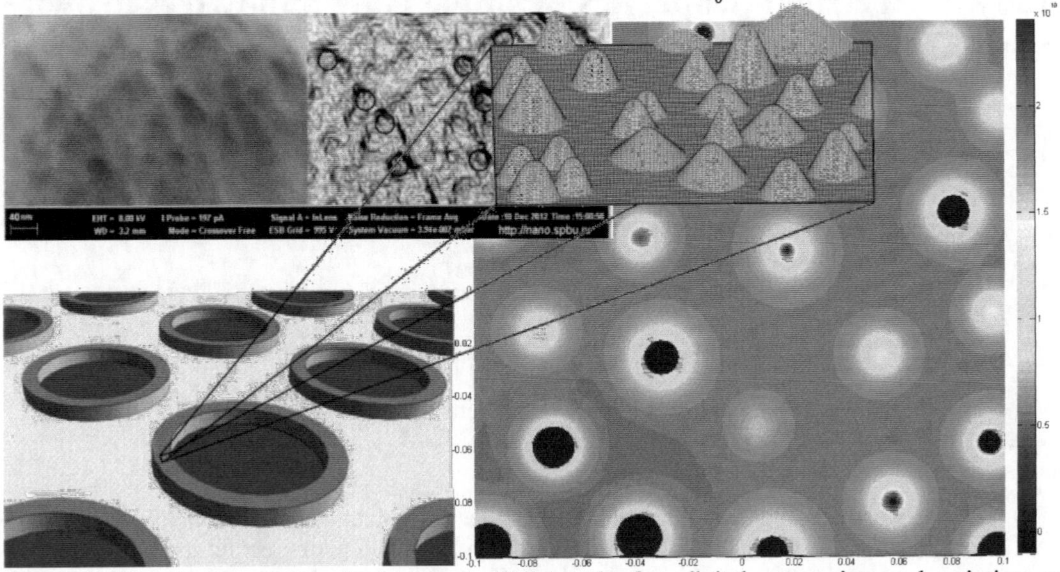

Figure 1. The image of nanolandscape in SEM and after digital processing and pointing out geometrical non-homogeneities. The model of microemitters in the array and an area of emission surface with nanolandscape. Distribution of electrostatic field over the surface.

3. Conclusion

Thus use the effective value of field enhancement coefficient allows to take into account the complex geometry of surface S, but in the same time to simplify the computation by using its flat vertical projection S_0. It is discovered that with increase of field strength the effective growth coefficient decreases. This can be explained thus that the emission surface covers greater part of the nanolandscape surface S while the smooth model surface S_0 remains constant.

4. Acknowledgments

Scientific research were performed at the Interdisciplinary Resource Centre for Nanotechnology and Computing Centre of Research park of St. Petersburg State University.

References

[1] Nikiforov K A, Egorov N V, Saifullin M F 2015 *Technical Physics* **60**, 1626-31
[2] Nikiforov K A, Sayfullin M F 2015 *Journal of Physics: Conference Series* **643**, 012122
[3] Sayfullin M, Nikiforov K Proc. 2014 Int. Vacuum Electron Sources Conf., IVESC 2014 (IEEE, Piscataway), 6892072
[4] Sayfullin M F, Nikiforov K A 2014 *Journal of Physics: Conference Series* **541**, 012020
[5] Nikiforov K A, Antonova L I, Egorov N V et. al. 2012 Proc. RuPAC – 23rd Russian Particle Accelerator Conference (JaCoW, CERN) p. 218-20

Study of electrical properties of single GaN nanowires grown by MOCVD with a Ti mask

A A Vasiliev[1,2], A M Mozharov[2], M M Rozhavskaya[3], V V Lundin[3] and I S Mukhin[2,4]

[1] Experimental physics department, Peter the Great Saint-Petersburg Polytechnic University, Saint Petersburg 194021, Russia
[2] Renewable Energy Lab, St. Petersburg Academic University, St. Petersburg 194021, Russia
[3] Physics of semiconductor heterostructures Lab, Ioffe Institute RAS, Saint Petersburg 194021, Russia
[4] The metamaterial Lab, University ITMO, Saint-Petersburg 197101, Russia

Abstract. We researched electrical characteristics of GaN nanowires (NWs) grown by MOCVD through solid titanium film. The technology of creating the ohmic contacts and MESFET structure on single NWs has been developed. The optimal annealing temperature of contacts has been found and conductivity structure, the free carrier concentration and mobility has been evaluated.

1. Introduction

For the last few years the interest in new growth technologies of III-N structures has increased due to problems in planar technology, where appearing a lot of defects during the epitaxial growth is a main disadvantage. This relates to the large mismatch of the crystal lattice parameters of the substrate and growing layer. Therefore, this leads to researching new ways of organizing material, for example, an organization in the form of nanowires (NWs). The main advantages of the NWs structures is the ability to create an active region on side faces in the non-polar and semi-polar directions and it helps to reduce the stress caused by the mismatch of the crystal lattice parameters of the substrate and a growing nanostructure.

Different structures could be produced using NWs: transistors, light-emitting device with low power consumption, various types of sensors, single electron transistors, solar cells [1,2,3,4]. In this paper we investigate NWs nitride compounds (GaN, InGaN), produced by MOCVD and also consider the possibility of their use.

1.1. Growth technology

At present, the technology of metal-organic chemical vapor deposition (MOCVD) is widely used. MOCVD method has several advantages, such as: the possibility to use a large area substrates, relatively high rate of growth (in compassion it with other methods), good control of the growth parameters, high homogeneity and high crystal quality layers. In this paper GaN NWs were produced by the method of growth in which the sapphire substrate previously was coated with a solid Ti mask of 10-30 nm thickness[5].

Studied structure was synthesized on a sapphire substrate by MOCVD at a temperature of 1040 °C with a precipitated Ti layer (figure 1a). It is a core-shell structure of GaN / InGaN / GaN. The sizes of grown n-type NWs ensemble constituted about 100 microns in length and about 1 micron in diameter at the growth time about 15 minutes. For our research NWs less than 1 micron in diameter were used.

(a) (b)

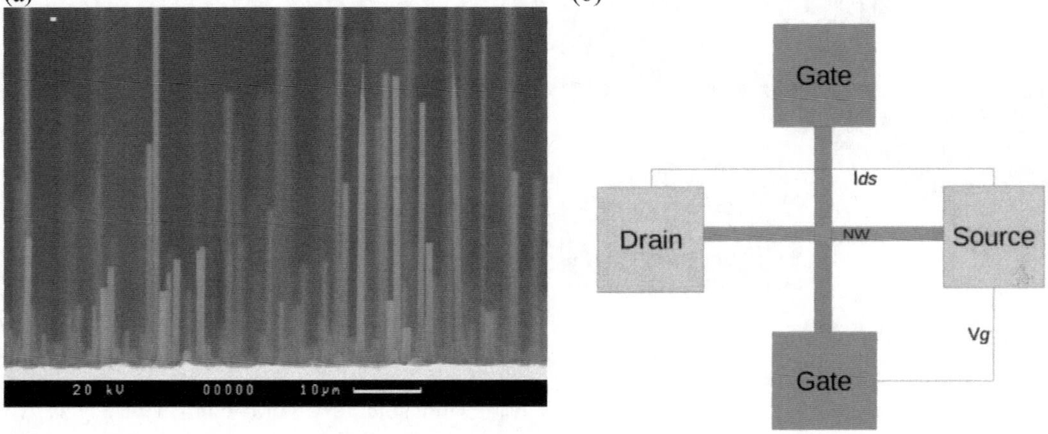

Figure 1(a,b). (a) SEM image of the grown nanowires array; (b) Measurment circuit layout;

1.2. Measurements problem

Creating the devices based on III-nitride NWs there are several problems with the formation of ohmic contacts to the semiconductor structure. In particular, it is known that reduction of contact resistance increases intensity of radiation and efficiency of InGaN of light-emitting diodes. Since the low value of GaN electron affinity (about 3.8-4.1) eV, there are certain difficulties in the selection of materials to create an ohmic contact. One of the appropriate materials for producing ohmic contacts may be a pair of Ti / Al. [6].

Well known method for measurement electrical properties of semiconductors is the Hall method [7]. For using this method we need to make 4 contacts at least and the studied material should be isotropic. In our case we work with structure of InGaN, which is anisotropic material. Implimentation the Hall method to anisotropic material can lead to increase an error.

In order to independently determine the mobility and free carrier concentration in the NWs it is necessary to make FET structure. Knowing the value of the carrier concentration it will be possible to make more accurate estimation of the carrier mobility, using the formula $\rho = 1/en\mu$, where ρ-resistivity, e-electron charge, n-carrier concentration, μ-mobility. Transistor MESFET structure (figure 1b) with Schottky barrier as the gate electrode and good ohmic contacts allows to estimate both parameters. As a material forming a high Schottky barrier to n-type GaN Au, Ni, Pd, Pt may be selected [8,9,10].

2. Experiment

The synthesized NWs with ultrasonic treatment method were separated from the substrate and transferred to the quartz substrate to form a contact to the single NWs. Using methods of the laser lithography and thermal evaporation in vacuum a group of drain / source contacts which based on materials Ti / Al / Ti / Ag (30/150/30/100 nm) was created (figure 2a). Annealing of contacts was performed at 710 ° C. In order to verify the ohmic properties of the produced contacts volt-ampere characteristics (VAC) was measured by using 2-terminal measurement method (figure 2b). From experiment data we calculated conductivity, which equal to 16 S/cm.

(a) (b)

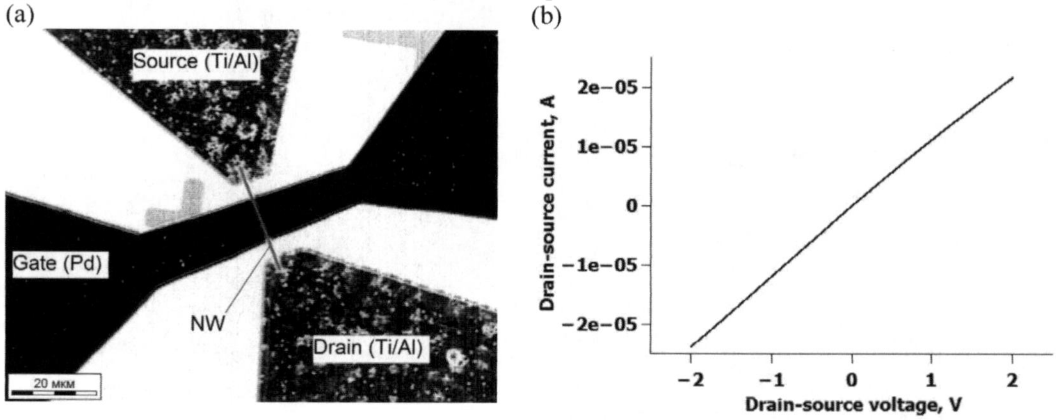

(c)

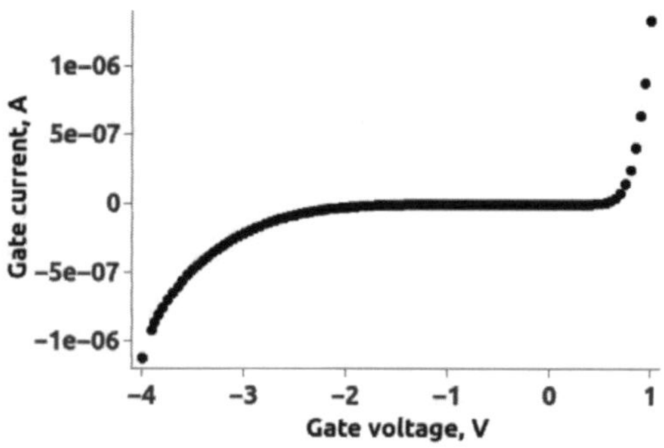

Figure 2(a,b,c). (a) Optical image of the transistor structure. Annealed top and bottom contacts are drain and source. Central contact is a gate; (b) VAC of the drain-source contacts without gate contact; (c) Schottky VAC of the gate contact (Pd contact);

To form MESFET structure gate contact must be Schottky contact. We evaporated Pd (50nm), and after evaporation we measured VAC between source and gate to verify the presence of Schottky barrier (figure 2c).

The results of measurements of the transistor characteristic of our structure is shown on figure 3a. We changed gate voltage and measured drain-source VAC. It can be seen that applying negative gate voltage with higher values leads to increasing the nanowire resistance. It could be explained by the fact that applying the reverse bias leads to increasing the region of the depletion layer in NW. This causes to overlap the conduction channel and the decrease in the current.

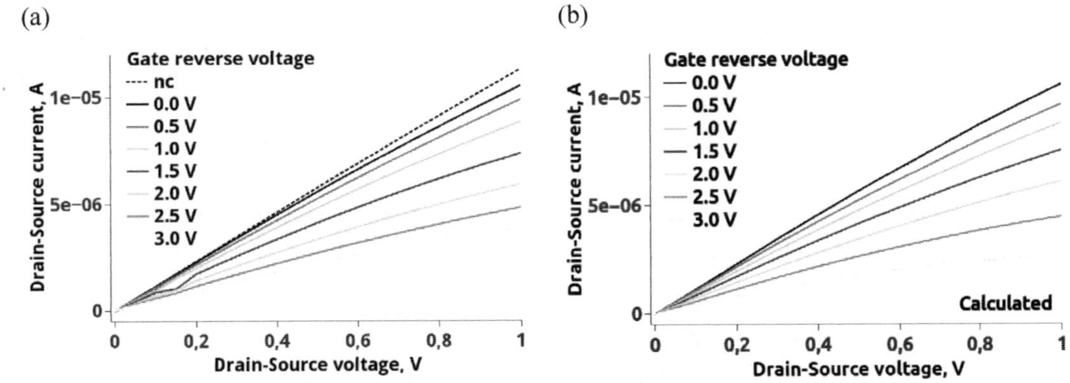

Figure 3(a,b). (a) Transistor's VAC of the manufactured structure (after evaporation Pd contact on the sample); (b) Transistor's VAC of the modeled structure;

3. Modeling

To estimate carrier concentration and mobility we calculated the electrical properties of our NW. The modeling solves symmetric problem, in which the gate contact is around the NW. We chose the such parameters for which the simulation results are similar to the results of the measurements (figure 3). From modeling we estimated that carrier concentration is about (2-3)·$10^{17}cm^{-3}$.·Hence we could estimate a mobility value: 330 cm^2/Vs. We also found that the mobility has no affect the shape of the VAC curves but just scales it.

4. Summary

In this paper we have studied NWs which were grown by MOCVD method with Ti mask. We were able to move nanowires from the growth substrate to the quartz substrate and made ohmic contact to single NW. We found that optimum annealing temperature to form ohmic contacts is 710 °C. Also we formed MESFET structure and measured electrical characteristics of our NW. We found that concentration value is (2-3)·$10^{17}cm^{-3}$ and mobility is 330 cm^2/Vs.

The method of synthesizing the NWs using a solid titanium mask may be perspective due to the simplicity, the quality of the structures and their high growth rate.

References

[1] Dubrovskii V G, Cirlin G E, Ustinov V M 2009 *Semiconductor* **43-12** 1539-84
[2] Li S,Waag A 2012 J. *Appl. Phys.* **111** 071101
[3] Mozharov A , Bolshakov A, Cirlin G, Mukhin I 2015 *Physica status solidi* **9-9** 507-10
[4] Neplokh V, Ali A, Julien F H, Foldyna M, Mukhin I, Cirlin G, Harmand J-C, Gogneau N, Thcernycheva M 2016 *Materia science in Semiconductor Processing*
[5] Rozhavskaya M M, Lundin V V, Lundina E, Davydov S, Troshkov S I, Vasilyev A A, Brunkov P, Baklanov A, Tsatsulnikov A F, Dubrovskii V G 2014 *J. Appl. Phys.* **14** 11399
[6] Blank T V, Gol'dberg Yu A 2007 *Semiconductors* **41-11** 1263-92
[7] Storm K, Halvardsson F, Heurlin M, Lindgren D, Gustafsson A, Wu P M, Monemar Bo, Lars Samuelson 2012 *NATURE NANOTECHNOLOGY* **7** 717-722
[8] Motayed A, Davydov A V, Mohammad S N, Melgailis J 2008 *J. Appl. Phys.* **104** 024302
[9] Kuykendall T, Pauzauskie P, Lee S, Zhang Y, Goldberger J, Yang P 2003 *Nanoletters* **3-8** 1063-66
[10] Blanchard P T, Bertness K A, Harvey T E, Mansfield L M, Sanders A W, Sanford N A 2008 *IEEE of Transactions of nanotechnology* **7-6** 760-65

Sol-gel fabrication of one-dimensional photonic crystals with predicted transmission spectra

V A Ilinykh[1], **L B Matyushkin**[1]
[1]Department of Micro- and Nanoelectronics, St. Petersburg State Electrotechnical University "LETI", St. Petersburg, 197376, Russia

Abstract. One-dimensional multilayer structures of periodically alternating low refractive index (silica) and high refractive index (titania) materials have been deposited by sol–gel spin-coating. Experimental spectra of the structures are in agreement with spectra calculated by transfer matrix technique. As an example, theoretical and experimental spectra with a stop band corresponding 600 nm-reflection are shown.

1. Introduction

Photonic bandgap structures (PBGs), also known as photonic crystals (PCs), are structures whose refractive index is periodic on a length scale of the order of optical wavelengths, which prevents light propagating through the structure due to Bragg reflection. Depending on the dimensionality of this periodicity, PBGs can be one, two or three dimensions (1D, 2D or 3D). In the case of sol-gel derived materials, the main emphasis has been on 1D- and 3D-structures [1]. Distributed Bragg reflectors (DBR), also called Bragg mirrors, are examples of 1D-PBGs. Their overall reflectivity can be very high (higher than that of metallic mirrors) and it's due to Bragg reflection. Sol-gel DBRs can be made, for example, by dip-coating [2] or spin-coating [3]. The refractive index contrast and the number of layers are the most important parameters. Silica (SiO_2) and titanic (TiO_2) are chosen because of the high optical contrast of their refractive indices (~1.46 and ~2.4, respectively). The higher the refractive indices contrast, the wider the 'stop band'. And the higher the number of layers, the higher reflectivity within the stop band. The application of the sol-gel process for the preparation of photonic crystals dealt with a number of issues described in this paper.

2. Math apparatus

Different 1D-PCs composed of SiO_2/TiO_2 alternating layers were analysed. To calculate multilayer structure transmission spectra the transfer matrix method was used. Matrix A describes light propagation between the layers and matrix B describes light transmission through current material layer:

$$A = \frac{1}{2n_2}\begin{pmatrix} n_1 + n_2 & n_2 - n_1 \\ n_2 - n_1 & n_1 + n_2 \end{pmatrix},\quad B = \begin{pmatrix} e^{2\pi i \frac{nL}{\lambda}} & 0 \\ 0 & e^{-2\pi i \frac{nL}{\lambda}} \end{pmatrix},$$

where n_1 and n_2 – refractive indices of alternating materials, n, L – refractive index and thickness of current layer, λ – wavelength of light. Elements of alternate matrix multiplication M ($M = A_1B_1A_2B_2\ldots A_nB_n$) allows to describe the optical properties of the multilayer structures.

3. Experimental section

During the preliminary experiments, the film deposition rate, the annealing temperature and the concentration of the starting sols were optimised. Homogeneous films deposited by spin-coating 100 μL of sol are obtained at speeds of about 3500 rpm. PCs were built by deposition of TiO_2 and SiO_2 on soda-lime glass substrates. The first layer was deposited of TiO_2 sol, because SiO_2 and soda-lime glass refractive indices are almost equal. It has been found that the optimum temperature for annealing of the samples is 550°C, the duration of annealing layers is 60 min for titania, and 30 min for silica. The process is repeated until a total number of layers have been deposited.

4. Results and discussion

To investigate producing of photonic crystals by sol-gel technology it have been researched 1D-PC structures with different numbers and thicknesses of layers. Fig. 1a shows the transmission spectra of the example, the numbers indicate the number of layers. Theoretical transmission spectra (Fig.1b) differs slightly from the experimental spectra due some porosity of layers, but shape and stop-band position are the same.

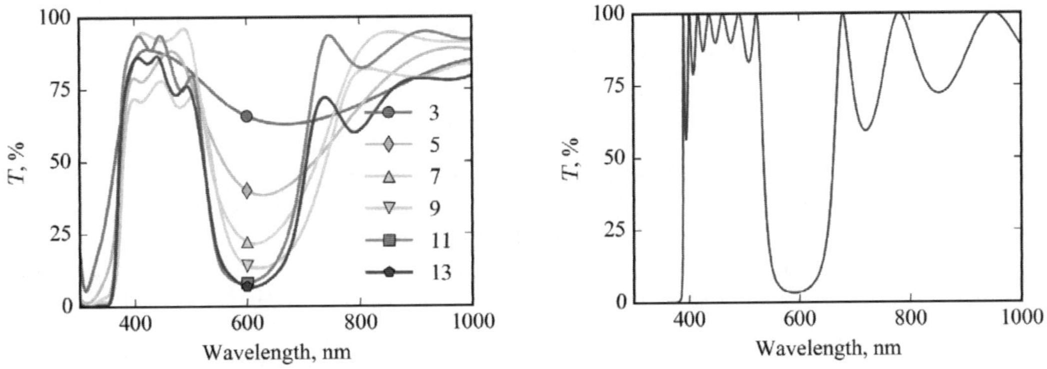

a **b**

Figure 1(a, b). (a) Transmission spectra of photonic crystals with different number of alternating layers and different thicknesses of layers: SiO_2 – 70 nm, TiO_2 – 100 nm; **(b)** corresponding theoretical transmission spectrum for 13-layers structure

5. Conclusion

As a result of experiments, one-dimensional photonic crystals was produced under various process conditions. The possibility of tuning of layer thickness by diluting initial sols is shown. Based on described technique, the photonic crystal structures were made with a minimum of transmittance about 7% at the photonic band gaps (450 and 600 nm). Photonic band gap position was varied by changing the viscosity of the solution and the corresponding changes in the thickness of material layers.

Acknowledgments

This study was supported by the Ministry of Education and Science of the Russian Federation within the framework of the project part of the state assignment for the St. Petersburg Electrotechnical University "LETI" (project 16.2112.2014/K, "Preparation and study of porous systems, functionalized nanomaterials, applications in photonics, sensor technology and medicine").

References

[1] Almeida R M, Sabine P 2003 *Curr. Opin. Solid State Mater. Sci.* **7(2)** 151
[2] Bellessa J, Rabaste S, Plenet J C, Dumas J, Mugnier J, Marty O 2001 *Appl. Phys. Lett.* **79** 2142
[3] Min, Xin, Xavier Orignac, and Rui M. Almeida 1995 *J. Am. Ceram. Soc.* **78(8)** 2254

Field emission spectroscopy of SiC

K A Nikiforov[1], **V V Trofimov**[1], **N V Egorov**[1]
[1]Saint Petersburg Electrotechnical University "LETI", St. Petersburg, 197376, Russia

Abstract. Experimental set up for the natural experiment and measurement model are presented to obtain the feld emission energy distribution spectrum out of silicon carbide in case of the macro-sample having a macroscopic shape of a tip. The prototype of feld emission 6H-SiC monolithic cathode is proposed for spectroscopy measurements, and characterised by current-voltage dependence at macroscale interelectrode distance.

1. Introduction

Energy distribution spectrum of field electrons is directly connected to electron properties of solid bodies. From studying the emitted electron distribution by their total energies one can obtain data on zone structure of the emitter and energy levels of adsorbate.

This article assesses field electron sources based on monolithic silicon carbide, whose mono-crystals have exceptional emission characteristics but, due to their two-element composition, are potentially vulnerable to modifications of emission surface. In particular, Ohmic heating in process of electron field emission can lead to sublimation of silicon and forming of graphene-like carbon coating on emitter surface.

Determining the transitional regimes connected to the silicon carbide surface graphitization in process of field electron emission is an important fundamental problem that also has a broad range of practical applications [1], [2].

The objective of this work is the development of natural model of electrophysical system for energy analysis and study of prototype of field emission 6H-SiC monolythic cathode supposed for spectroscopy measurements. While the potential of the SiC field emission material is recognized, only a few experimental attempts have been made to fabricate and measure emission characteristics of monolithic SiC cathodes. Field emission characteristic of 4H-SiC monolithic structure is reported in [3] and [4] reports strong and stable field emission properties of monolithic 6H-SiC pillar field emission array (FEA). Crystalline p-type 6H-SiC tips in a regular patterned FEA with very noisy emission current (only a few emitters were measured in FEA) is reported in [5] and also [6] reports voltage-current characteristics of monolithic SiC FEA.

While the number of publications on the issue of monolithic SiC field electron emission is scarce, there is no papers at all on field emission energy distribution measurements for both nanostructured or monolithic silicon carbide, as far as we know, i. e. the considered problem is undoubtedly actual.

2. Current-voltage characteristic of 6H-SiC monolithic macro-sample emitter and set up of the natural experiment

Monolithic macro-sample was obtained as a result of split of plane monocrystal 6H-SiC FEA [3], so that tip macroshape of emitter was suitable to current-voltage characteristic measurement at macroscopic interelectrode distance. High voltage up to 10 KV was applied to the anode. The vacuum was at -6 or -7 power of Pa for the duration of the experiment. Figure 1 shows current-voltage characteristic plotted in Fawler-Nordheim coordinates (measured in static mode) and macro-sample view in optical microscope and sample fixation system. It is necessary to note, that there were previously attempts of emission current measurements directly from plane FEA in macroscopic diode configuration (microdiode was studied in [3]), and the emission emerged to the greater part along the margin of the cathode due to electrostatic amplification in macroscopic distribution of the electrical field (as it was calculated in [7], [8]).

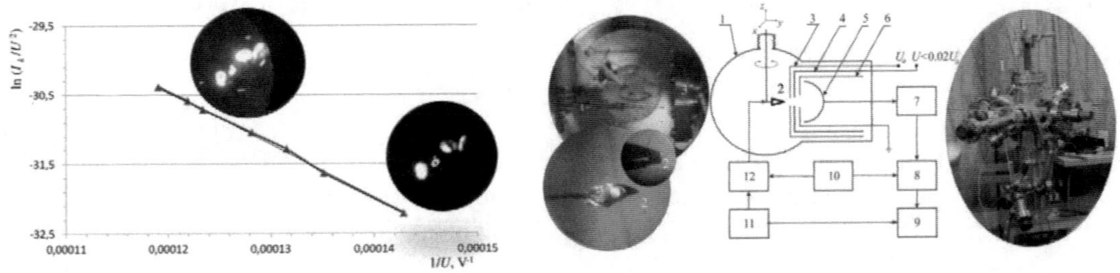

Figure 1(a, b). (a) Current-voltage characteristic and emission images of silicon carbide monolithic macro-sample emitter; **(b)** The principal schematic of the experimental set up, where 1 denotes vacuum chamber, 2 is cathode node, 3 – anode, 4 – electrostatic lens, 5 – hemispherical collector, 6 – protective cylinder, 7 – electrometric amplifier, 8 – synchronous detector, 9 – registering device, 10 – modulation block, 12 – signal adder.

3. Conclusion
This article discussed the problem of measurement of field emission energy distribution from silicon carbide by the retarding potential method. A description of a natural experiment setting is presented as well as a model of its electrophysical system of field emission current differentiation by modulated retarding potential. Prototype of field emission 6H-SiC monolithic cathode is studied and current-voltage characteristic is presented. Future work is characterisation of proposed silicon carbide cathode by field emission spectroscopy measurements in described experimental set up and methodology.

Acknowledgments
The work has been funded by Russian Science Foundation (Grant No. 15-19-30022).

References
[1] Konakova R V, Okhrimenko O B, Svetlichnyi A M, Ageev O A, Volkov E Y, Kolomiytsev A S, Jityaev I L, Spiridonov O B 2015 *Semiconductors* **49**, 1242-5
[2] Volkov E, Jityaev I, Kolomiitsev A 2015 *IOP Conf. Series: Materials Science and Engineering* **93**, 012031
[3] Afanasyev A V, Ivanov B V, Ilyin V A, Kardo-Sysoev A F, Kuznetsova M A, Luchinin V V 2013 *Materials Science Forum* **740-742**, 1010-3
[4] Kang M-G et al. 2013 *Nanotechnology* **24**, 065201
[5] Zuuk A V et al. 2004 *Microelecronic Engineering* **73-74**, 106-10
[6] Kuznetsova M A and Luchinin V V 2012 *Journal of Nano and Microsystem Technique (in Russian)* **12**, 35-40
[7] Nikiforov K A, Egorov N V, Sayfullin M F 2015 *Technical Physics* **60**, 1626-31
[8] Sayfullin M F and Nikiforov K A 2014 *Journal of Physics: Conference Series* **541**, 012020

Chemical potentials and growth rates of gold-catalyzed ternary InGaAs nanowires

J Grecenkov[1], **V G Dubrovskii**[1,2,3]
[1]St. Petersburg Academic University, Khlopina 8/3, 194021, St. Petersburg, Russia
[2]Ioffe Physical Technical Institute RAS, Politekhnicheskaya 26, 194021, St. Petersburg, Russia
[3]ITMO University, Kronverkskiy pr. 49, 197101 St. Petersburg, Russia

Abstract. By applying the Redlich-Kister polynomial model, we calculate chemical potentials of quaternary liquid alloys during the gold-catalyzed vapor-liquid-solid growth of ternary $In_xGa_{1-x}As$ nanowires. We then use these chemical potentials in the Zeldovich nucleation rate to obtain the nanowire axial growth rates versus the deposition conditions and indium compositions.

1. Introduction

Due to their ability to relax elastic strain on free sidewalls, nanowires are promising for monolithic integration of III-V optoelectronic materials with silicon electronic platform [1,2]. III-V nanowires are usually grown by the gold-assisted vapor-liquid-solid (VLS) method wherein the III-V growth species are first transferred from vapor to liquid and then crystallize at the liquid-solid interface under the catalyst droplet [3,4].

Chemical potentials of Au-III-V liquid alloys play a crucial value in understanding the VLS growth and crystal structures of III-V nanowires. Previous works have shown that the analysis of chemical potentials of gold-assisted binary III-V nanowires allows for a delicate control over the VLS growth rates and wurtzite-zincblende polytypism within nanowires [5-12]. In this work, we consider a more complex case of ternary gold-catalyzed InGaAs nanowires and study how chemical potentials affect the VLS growth rates under variable deposition conditions and indium compositions [13].

2. Chemical potential

In the mononuclear growth mode described by classical nucleation theory, the axial growth rate of a nanowire is given by the Zeldovich nucleation rate [10]

$$\frac{dL}{dt} = J_0 \exp\left(-\frac{\Gamma^2}{\Delta\mu k_B T}\right), \tag{1}$$

where J_0 is a prefactor, Γ is the effective surface, or edge energy of a two-dimensional island of a bilayer height, $\Delta\mu$ is the difference of chemical potentials of a III-V pair ($In_xGa_{1-x}As$) in the liquid and solid phases, T is the surface temperature and k_B is the Boltzmann constant. No vapor-solid contributions [14] are considered in Eq. (1). Since the prefactor only weakly depends on the chemical potential and surface energy, the nucleation probability and consequently the nanowire growth rate is mainly determined by the Zeldovich exponent of the nucleation barrier in Eq. (1). The latter is extremely sensitive to the chemical potential value and hence the $\Delta\mu$ needs to be determined as precisely as possible [11].

In the case of gold-catalyzed ternary III-V nanowires, we are dealing with quaternary Au-III-V liquid alloys such as Au-In-Ga-As. These chemical potentials can be calculated by applying the Redlich-Kister polynomial corrections to the Gibbs that take into account the interactions introduced by mixing the components. The Redlich-Kister polynomials have the form

$$G_{R-K} = \frac{1}{2}\sum_{\substack{i,j=1\\ i\neq j}}^{4} c_i c_j \sum_{m=0}^{n} L_{ij}^{m}(c_i - c_j)^m + \frac{1}{3}\sum_{\substack{i,j=1\\ i\neq j\neq k}}^{4} c_i c_j c_k L_{ijk}^{0}, \tag{2}$$

with c_i denoting the concentrations of different elements in the liquid alloy and L_{ij}^{m} are the interaction parameters between the atoms of different elements [15-21]. Once the Gibbs energy is calculated, the chemical potential of a given element is obtained as a derivative of the Gibbs energy with respect to the corresponding concentration

$$\mu_i^l = \frac{\partial G}{\partial c_i}. \tag{3}$$

The liquid-solid chemical potential difference can be written down as

$$\Delta\mu = x(\mu_{In}^l + \mu_{As}^l - \mu_{InAs}^s) + (1-x)(\mu_{Ga}^l + \mu_{As}^l - \mu_{GaAs}^s) \tag{4}$$

under the central simplifying assumption on the identical indium content x in liquid and solid phases. The quantities μ_{In}^l, μ_{Ga}^l and μ_{As}^l denote chemical potentials of the indium, gallium and arsenic atoms in the liquid phase, while μ_{InAs}^s and μ_{GaAs}^s stand for chemical potentials of InAs and GaAs pairs in solid.

3. VLS growth rate

We can now exploit Exp. (1) to estimate the growth rates of gold-catalyzed InGaAs nanowires under different conditions. Assuming, for simplicity, that the nucleation process occurs in the centre of the the liquid-solid interface rather than at the triple phase line, we can use the expression for the surface energy of a newly formed island in the form [10]

$$\Gamma = x\Gamma_{In} + (1-x)\Gamma_{Ga}, \tag{5}$$

with Γ_{In} being the surface energy of an InAs island in an Au-In droplet (the influence of arsenic on the surface energy can be neglected due to its low concentration inside the catalyst droplet) and respectively Γ_{Ga} in the same value for a GaAs island in an Au-Ga droplet . We take $\Gamma_{Ga} = \Gamma_{Ga}(c_{Ga})$ and $\Gamma_{In} = \Gamma_{In}(c_{In})$ as the linear interpolations between pure gold and group III metal droplets [11].
The prefactor in the Zeldovich nucleation rate can be calculated using the expression of Ref. [22]

$$J_0 = \frac{3^{3/4}}{\sqrt{\pi}} D_5 \left(\frac{h}{\Omega_{35}}\right)^2 e^{\mu_{35}^s} c_{As} \Delta\mu^{1/2}. \tag{6}$$

Here, D_5 is the diffusion coefficient of the arsenic adatoms through the liquid, Ω_{35} is the elementary volume of a III-V pair, h is the height of a bilayer and μ_{35}^{s} is the averaged value of chemical potential of a III-V pair in the solid phase. Equations (1) to (6) constitute the complete system for computing the VLS growth rates of gold-catalyzed ternary nanowires versus temperature and droplet composition.

4. Results and discussion

Figure 1 shows the VLs growth rate of gold-catalyzed $In_xGa_{1-x}As$ nanowires versus the indium composition at 750 K, for a fixed arsenic concentration c_5=0.05 and three different total group III concentrations c_3=0.3, 0.5 and 0.7. Quite interestingly, we observe a non-monotonic dependence of dL/dt on x for c_3=0.3, while for larger c_3 the growth rate gradually increases with x. Figure 2 shows the chemical potentials in quaternary Au-In-Ga-As alloy with respect to solid $In_xGa_{1-x}As$, computed for the same parameters as in Fig. 1. Clearly, these chemical potentials increase monotonically with x and hence the non-monotonic behavior of dL/dt on x at c_3=0.3 is due to a combined effect of chemical potential and surface energy.

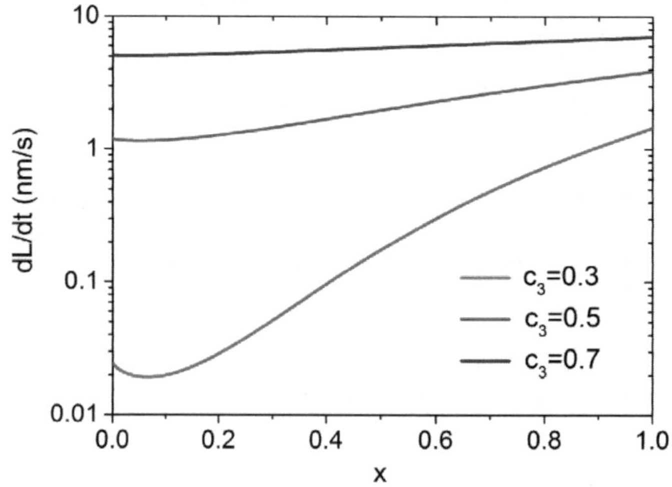

Figure 1. Growth rates of gold-catalyzed $In_xGa_{1-x}As$ nanowires versus x at a fixed temperature of 750 K, arsenic concentration of 0.05 and three different total concentrations of group III atoms in the droplet $c_3 = c_{In} + c_{Ga}$.

Since both chemical potential and surface energy enter the Zeldovich exponent, even the slightest changes in them result in a dramatic effect on the VLS growth rate. In particular, our results show that $In_xGa_{1-x}As$ nanowires of the same composition x grow at very different rates depending on the total group III concentration (determined by the V/III ratio). For example, low total group III flux may stop completely the VLS growth for small enough x, while at higher group III fluxes same nanowires would continue growing normally.

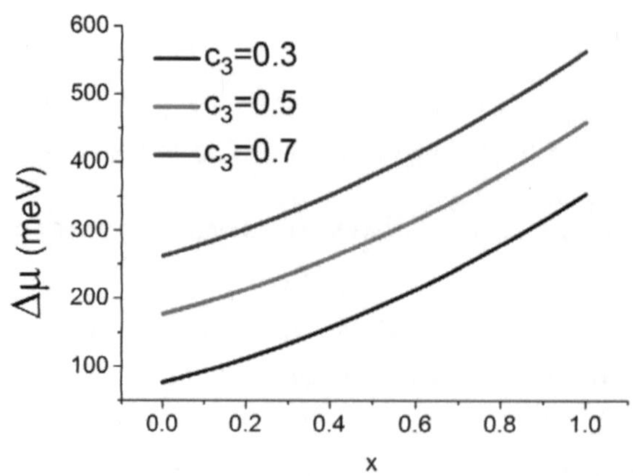

Figure 2. Chemical potentials versus x for the same parameters as in Fig. 1.

5. Conclusion

The model presented is capable of describing the VLS growth rates of gold-catalyzed ternary InGaAs nanowire based on classical nucleation theory involving the macroscopic Zeldovich nucleation rate and quaternary chemical potentials. The latter has been calculated using the Redlich-Kister polynamials. The results show that changing the In/Ga as well as the total V/III flux ratio under otherwise fixed experimental conditions may cause an abrupt growth stop within a plausible range of parameters. We now plan to extend our approach for predicting the crystal structures of InGaAs nanowires and to apply this approach to other ternary nanowires based on both group III and group V intermixing.

Acknowledgments

JG acknowledges the FP7 project NanoEmbrace (Grant Agreement 316751).VGD thanks the Russian Foundation for Basic Research for the financial support under different grants.

References

[1] V.G. Dubrovskii, N.V. Sibirev, X. Zhang, and R. A. Suris, Cryst. Growth Des. **10**, 3949 (2010).
[2] X. Zhang, V. G. Dubrovskii, N. V. Sibirev, and X. Ren, Cryst. Growth Des. **11**, 5441 (2011).
[3] N. Wang, Y. Cai, and R. Zhang, Mater. Sci. Eng. Res. **60**, 1 (2008).
[4] K. A. Dick, Prog. Cryst. Growth Charact. Mater. **54**, 138 (2008).
[5] V. G. Dubrovskii and J. Grecenkov, J. Phys.: Conference Series 541, 012001 (2014).
[6] F. Glas, M. R. Ramdani, G. Patriarche, and J. C. Harmand, Phys. Rev. B **88**, 195304 (2013).
[7] F. Glas, J. Appl. Phys. **108**, 073506 (2010).
[8] C. Colombo, D. Spirkoska, M. Frimmer, G. Abstreiter, and A. Fontcuberta i Morral, Phys. Rev. B **77**, 155326 (2008).
[9] V. G. Dubrovskii, Appl. Phys. Lett. **104**, 053110 (2014)
[10] V. G. Dubrovskii, *Theory of VLS Growth of Compound Semiconductors*. In: Anna Fontcuberta i Morral, Shadi A. Dayeh and Chennupati Jagadish, editors, Semiconductors and Semimetals, v. 93, Burlington: Academic Press, 2015, p. 1-78.
[11] V. G. Dubrovskii, J. Chem. Phys. **142**, 204702 (2015).
[12] E. Gil, V. G. Dubrovskii, G. Avit, Y. André, C. Leroux, K. Lekhal, J. Grecenkov, A. Trassoudaine, D. Castelluci, G. Monier, R. M. Ramdani, C. Robert-Goumet, L. Bideux, J. C. Harmand, and F. Glas., Nano Lett. **14**, 3938 (2014).
[13] A. S. Ameruddin, P. Caroff, H. H. Tan, C. Jagadish, and V. G. Dubrovskii, Nanoscale **7**, 16266 (2015).
[14] V. G. Dubrovskii, V. Consonni, L. Geelhaar, A. Trampert, and H. Riechert, Appl. Phys. Lett. **100**, 153101 (2012).
[15] Calphad, Volume 18, Issue 2, p. 177 (1994).
[16] J. Wang, Y. Liu, L. Liu, H. Zhou, Z. Jin, Calphad **35**, 242 (2011).
[17] H. Liu, Y. Cui, K. Ishida, Z. Jin, Calphad **27**, 27 (2003).
[18] 'T. Anderson, I. Ansara, J. Phase Equilib. **12**, 64, (1991).
[19] M. Ghasemi, B. Sundman, S. G. Fries, and J. Johansson, Journal of Alloys and Compounds **600**, 178 (2014).
[20] A. T. Dinsdale, Calphad **15**, 317 (1991).
[21] I. Ansara, C. Chatillon, H. L. Lukas, T. Nishizawa, H. Ohtani, K. Ishida, M. Hillert, B. Sundman, B. B. Argent, A. Watson, T. G. Chart, and T. Anderson, Calphad **18**, 177 (1994).
[22] V.G. Dubrovskii and J. Grecenkov, Cryst. Growth Des. **15**, 340 (2015).

Planar nanosized field emission cathodes on the basis of graphene/semi-insulating silicon carbide fabricated by focused ion beam

I L Jityaev[1], O A Ageev[1], A M Svetlichnyi[1], A S Kolomiytsev[1] O B Spiridonov[2]
[1]Southern Federal University, Institute of Nanotechnologies, Electronics, and Electronic Equipment Engineering, Taganrog, 347922, Russia
[2] Southern Federal University, Scientific and Design Bureau of Modelling and Controlling Systems, Taganrog, 347928, Russia

Abstract. We investigate the field emission properties of planar graphene structures with nanosized interelectrode distance. The graphene was obtained by thermal decomposition of silicon carbide in vacuum. Planar field emission structures on the basis of graphene on semi-insulating SiC were fabricated by using focused ion beam. We have performed current-voltage measurement on graphene/SiC field emission cathodes. The planar field emission structures showed a threshold voltage less than 1 V.

1. Introduction

Approaching the technological limit of formation possibilities of modern integrated logic components promotes the scientific interest to field emission electronics. The main difficulties of practical realization are related to the large size and low durability of vacuum element, which mainly depends on the field emission cathode material. On-stream a field emission cathode device degrades under a number of destructive environmental influences.

Recently, many studies of field emission cathodes based on carbon materials have been published. The resistance to destructive factors, low threshold voltage and unique geometric dimensions of carbon nanomaterials make them attractive for field emission applications [1-4]. Graphene is one of carbon nanomaterials with high aspect ratio, small thickness and work function, excellent field emission, mechanical and thermal properties [5-10].

Methods of graphene preparation differ in their labor content, film size, imperfection, etc [11]. One of the methods is thermal decomposition of silicon carbide in vacuum [12, 13]. Thermal conductivity, radiation resistance of silicon carbide and opportunity of graphene growth on the entire surface make it progressive substrate for field emission devices, and graphene on silicon carbide is promising material for vacuum micro- and nanoelectronics [14-16].

The object of studying is development of manufacturing technology a planar field emission graphene/semiinsulating silicon carbide structure with nanoscale interelectrode distance and rounding-off radius of the top by gallium focused ion beams and investigation of its emission characteristics.

2. Design

The material and shape are important parameters of the field emission cathodes. Cathodes with high surface curvature are used for field emission. The cathodes in the shape of the tip, the blade and matrix structures are of the greatest interest. Previously, it was found that interelectrode spacing and rounding-off radius of the top having the greatest impact on the electric field strength [17]. It was determined that achievement of a field strength 10^9 V/m in nanoscale field emission structures is possible at a potential difference of less than 10 V. The small size of field emission cathode allows to create a compact matrix structure with a high current density [18].

Since graphene has the atomic thickness, fabrication of planar cathodes in the form of the tip on the basis of graphene proposed in this study. Schematic sketch the planar field emission structure shown on Figure 1.

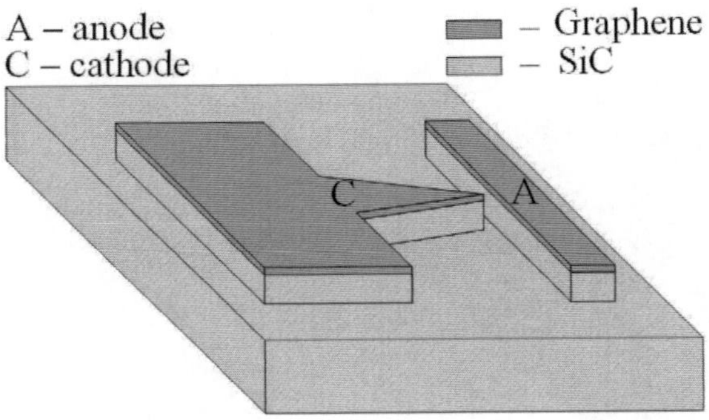

Figure 1. Schematic sketch of planar field emission structure on the basis of graphene on semiinsulating silicon carbide.

3. Experiment and results

Graphene was grown by thermal decomposition of silicon carbide in vacuum at a residual pressure of 10^{-3} Pa [13, 19, 20]. This method is characterized by obtaining high quality graphene large area with good adhesion to the substrate. Only the semiinsulating silicon carbide is required for fabricating planar field emission cathodes on the basis of graphene / SiC, because the cathode and the anode should not be shorted.

Precision maskless etching by gallium focused ion beam (FIB) was used to form the nanosized elements. In this case, FIB allows to form planar anode and cathode in a single technological cycle [21-25]. Other mask technology is more labor intensive, it requires the use of specialized templates, more time, additional photolithography operations and hinders the formation of nanosized elements based on graphene/SiC.

Figure 2. SEM image of planar field emission structure on the basis of graphene/SiC

In this study, the planar field emission structures were fabricated. SEM images of field emission structure (Figure 2) were obtained by using a scanning electron microscope "Nova NanoLab 600". SEM images show a planar field emission structures on the basis of graphene on semiinsulating SiC predetermined shape with 30 nm rounding-off radius of the top and 15 nm distance between the electrodes. Shape of field emission structure was determined by the need for the availability of the cathode and anode contact pads for subsequent measurement of current-voltage characteristics. Interelectrode distance was determined by the resolution of the etching equipment. Total etching time of the planar field emission structure was 2 min. 33 sec.

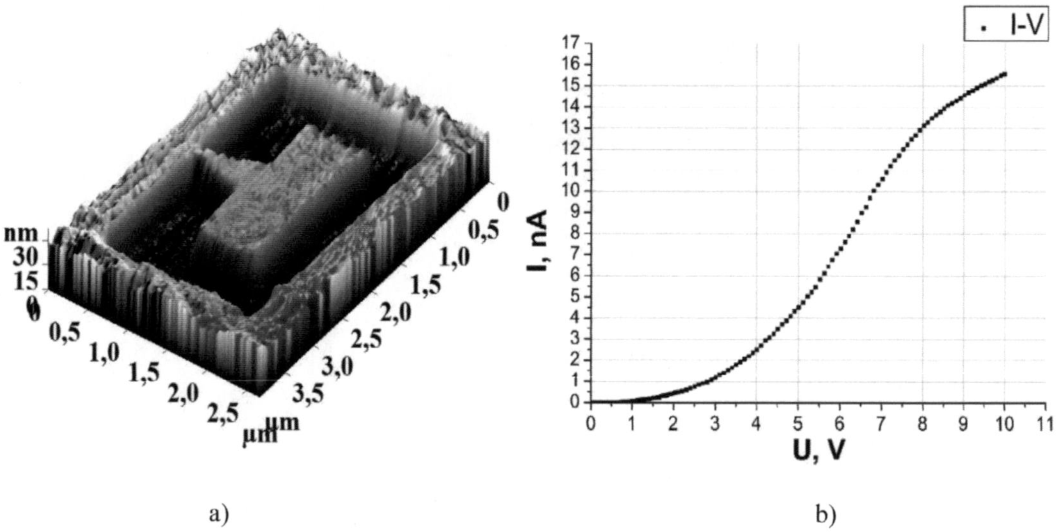

Figure 3. AFM-image of planar field emission graphene/semiinsulating SiC nanostructure (a) and their current-voltage characteristics (b)

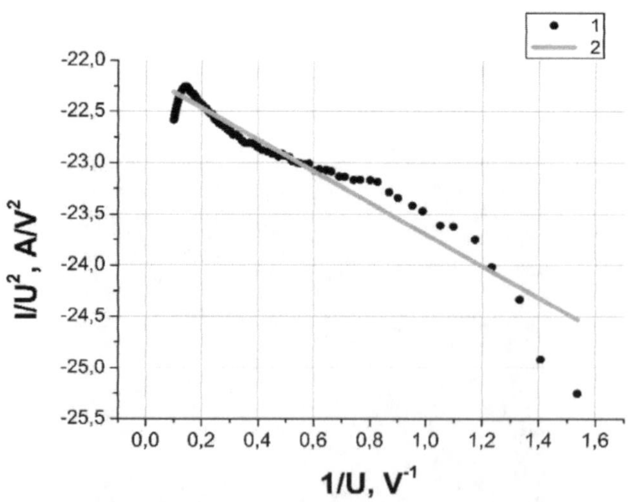

Figure 4. The current-voltage characteristics of planar field emission structure on the basis of graphene/SiC in Fowler–Nordheim coordinates: 1 – experimental data, 2 – approximation.

Structural studies and measurement of current-voltage characteristics was carried out by a scanning probe microscope "Ntegra Vita". AFM images have confirmed parameters of field emission structures and allowed to estimate etch depth, which was 30 nm (Figure 3a).

Measurements of current-voltage characteristics were carried out at standard pressure and a temperature of 23 °C in the voltage range of 0-10 V. The measurement result is shown on Figure 3b.

From the I-V dependence follows that the planar field emission structure is characterized by a threshold voltage of 0.5 V. The maximum current of 20 nA limited by instrumentation capability. I-V dependence in Fowler-Nordheim coordinates was plotted on the basis of experimental data (Figure 4). I-V dependence in Fowler-Nordheim coordinates confirms the field emission nature of planar nanoscale graphene on silicon carbide cathode.

4. Conclusions

The manufacture method of nanoscale planar field emission structures by using a gallium focused ion beam was shown. Forming modes of field emission structures on the basis of graphene on the semiinsulating silicon carbide are determined. The current-voltage characteristics of planar structures were obtained. Using of graphene as a material for planar field emission cathode with nanoscale interelectrode distance and rounding-off radius of the top allowed us to obtain electron emission at a low voltage of less than 1 V. This is significantly better results among nanocarbon field emission cathodes [26-28]. The results of current-voltage characteristics correlate with the results of field emission cathodes on the basis of graphene on SiC vertical type [6]. Low operating voltages can promote the using of nanoscale planar field emitters based on graphene on SiC in modern micro- and nanoelectronic devices.

Acknowledgments

This work was supported by The Ministry of Education and Science of Russian Federation, the State Task in the Sphere of Scientific Activities (project no.16.1154.2014/K). The equipment of the Center for Collective Use "Nanotechnology" and the Research and Education Center "Nanotechnologies" Southern Federal University was used for this study.

References

[1] Le Normand F, Cojocaru C S, Fleaca C, Li J Q, Vincent P, Pirio G, Gangloff L, Nedellec Y, Legagneux P 2007 *Eur. Phys. J. Appl. Phys.* **38** 115

[2] Fursey G N, Petrick V I, Novikov D V 2009 *Technical Physics* **54** 1048

[3] Calderón-Colón X, Geng H, Gao B, An L, Cao G, Zhou O 2009 *Nanotechnology* **20** 325707

[4] Yafyasov A, Bogevolnov V, Fursey G, Pavlova B, Polyakov M, Ibragimov A 2011 *Ultramicroscopy* **111** 409

[5] Novoselov K S, Geim A K, Morozov S V, Jiang D, Zhang Y, Dubonosov S V, Grigorieva I V, Firsov A A 2004 *Science* **306** 666

[6] Konakova R V, Okhrimenko O B, Svetlichnyi A M, Ageev O A, Volkov E Yu, Kolomiytsev A S, Jityaev I L, Spiridonov O B 2015 *Semiconductors* **49** 1242

[7] Eda G, Unalan H E, Rupesinghe N, Amaratunga G A J, Chhowalla M 2008 *Appl. Phys. Lett.* **93** 233502

[8] Kumar S, Duesberg G S, Pratap R, Raghavan S 2014 *Appl. Phys. Lett.* **105** 103107

[9] Zhang Y B, Tan Y W, Stormer H L, Kim P 2005 *Nature* **438** 201

[10] Lee C, Wei X, Kysar J W, Hone J 2008 *Science* **321** 385

[11] Avouris Ph, Dimitrakopoulos Ch 2012 *Materials Today* **15** 86

[12] Kim M, Hwang J, Shields V B, Tiwari S, Spencer M G, Lee J-W 2012 *Nanoscale Research Letters* **7** 186

[13] Liu X, Chen Y, Sun Ch, Guan M, Zhang Y, Zhang F, Sun G, Zeng Y 2015 *Nanomaterials* **5** 1532

[14] Lezhenin F F, Gnesin G G 1967 *Soviet Powder Metallurgy and Metal Ceramics* **6** 114

[15] Lebedev A A, Ivanov A M, Strokan N B 2004 *Semiconductors* **38** 125

[16] Lebedev A A, Kotousova I S, Lavrent'ev A A, Lebedev S P, Makarenko I V, Petrov V N, Titkov A N 2009 *Physics of the Solid State* **51** 829

[17] Svetlichnyi A M, Ageev O A, Volkov E Yu, Jityaev I L and Dem'yanenko M V 2015 *Applied Mechanics and Materials* **752-753** 163

[18] Volkov E, Jityaev I, Kolomiitsev A 2015 *IOP Conf.Series: Materials Science and Engineering* **93** 012031

[19] VanMil B L, Myers-Ward R L, Tedesco J L, Eddy Jr. C R, Jernigan G G, Culbertson J C, Campbell P M, McCrate J M, Kitt S A, Gaskill D K 2009 *Mater Sci Forum*, **615–617** 211

[20] Konakova R V, Kolomys O F, Okhrimenko O B, Strelchuk V V, Volkov E Yu, Grigoriev M N, Svetlichnyi A M and Spiridonov O B 2013 *Semiconductors* **47** 812

[21] Svetlichnyi A M, Kolomiytsev A S, Jityaev I L, Spiridonov O B 2015 *Izvestiya SFedU. Engineering Sciences* **09** 14

[22] Ageev O A, Kolomiitsev A S, Konoplev B G. 2011 *Semiconductors* **45** 89

[23] Ageev O A, Vnukova A V, Gromov A L, Kolomiytsev A S, Konoplev B G, Lisitsyn S A, Alekseev A M 2014 *Nanotechnologies in Russia* **9** 26

[24] Ageev O A, Vnukova A V, Gromov A L, Kolomiytsev A S, Konoplev B G, Alekseev A M 2014 *Nanotechnologies in Russia* **9** 31

[25] Ageev O A, Balakirev S V, Bykov Al V, Gusev E Yu, Fedotov A A, Jityaeva J Y, Il'in O I, Il'ina M V, Kolomiytsev A S, Konoplev B G, Krasnoborodko S U, Polyakov V V, Smirnov V A, Solodovnik M S, Zamburg E G 2016 *Chapter In: Advanced Materials – Manufacturing, Physics, Mechanics and Applications. Parinov, Chang, Topolov (Eds.). Springer International Publishing Switzerland* 563

[26] Deng J-H, Wang F-J, Cheng L, Yu B, Li G-Z, Hou X-G, Li D-J, Cheng G-A 2014 *Materials Letters* **124** 15

[27] Chen L, He H, Yu H, Cao Y, Lei D, Menggen Q, Wu Ch, Hu L 2014 *Journal of Alloys and Compounds* **610** 659

[28] Nguyen D D, Lai Y-T, Tai N-H 2014 *Diamond and Related Materials* **47** 1

Effect of GaAs native oxide upon the surface morphology during GaAs MBE growth

O A Ageev, M S Solodovnik, S V Balakirev, I A Mikhaylin, M M Eremenko
Southern Federal University, Institute of Nanotechnologies, Electronics and Electronic Equipment Engineering, Department of Nanotechnologies and Microsystems
2 Shevchenko Street, Taganrog, 347928, Russia

E-mail: solodovnikms@sfedu.ru

Abstract. The GaAs native oxide effect upon the surface morphology of the GaAs epitaxial layer was studied with taking into account the main growth parameters of MBE technology: substrate temperature, effective As_4/Ga flux ratio and growth rate. The MBE modes of atomically smooth and rough surfaces and surfaces with Ga droplet array formation were determined. The possibility of the obtaining of GaAs nanowires via GaAs native oxide layer was shown.

1. Introduction

The relevance of the experimental studies of the influence of GaAs native oxide on the GaAs epitaxial growth is caused by the need to solve a number of technical problems associated with the formation of nanoscale systems based on A^3B^5 semiconductors using different epitaxial growth techniques on pre-structured surfaces, including those based on the method of molecular beam epitaxy (MBE). However, most of the works are limited to the study of the GaAs oxide interaction with epitaxial layer and its influence on the morphology during the substrate surface treatment by thermal oxide removal in a growth chamber [1-4] in order to obtain an atomically smooth surface, which is very critical for the subsequent formation of self-organized nanostructures [5,6].

This paper presents the results of comprehensive experimental studies of the interaction of the GaAs native oxide with the substrate material, GaAs epitaxial layer, as well as growth components – Ga and As at a stages of heating, oxide removal and epitaxial growth.

2. Experiment

Experimental study of the interaction between GaAs native oxide, the surface of GaAs substrate and growth components was carried out using the MBE system STE35 (SemiTEq) with solid state sources. The deposition of GaAs was carried out from Ga and As_4 fluxes over the GaAs native oxide film on the GaAs(001) substrate without its prior removal by thermal desorption in the growth chamber. The thickness of oxide film was 3-5 nm [7, 8]. Effective deposition thickness of GaAs was 100 nm for all samples. The substrate temperature T during MBE growth was varied from 450 to 650°C, and the effective As_4/Ga flux ratio $J_{As/Ga}$ – from 1 to 4. At low substrate temperatures (below 500°C) growth component feed was started when the substrate temperature reached the desired value. In the regions of the middle (500-600°C) and high (over 600°C) temperatures growth started at the beginning of the structural changes in the GaAs native oxide film, which was observed by RHEED. The set value of flux ratio was maintained in all temperature ranges. We used two growth modes: with Ga pregrowth deposition and without it. After growth, sources of Ga and As_4 were closed simultaneously, and the substrate was rapidly cooled. Then all samples were examined by SEM and AFM-based techniques.

3. Results and discussion

The experiments showed that in the mode without Ga predeposition at $J_{As/Ga} = 4$ and at high and low temperatures (below 550°C) rough surfaces formed, as shown in figure 1*a*, and RHEED dotted pattern was observed. The pits were 30-200 nm in diameter with density in a range $(3\text{-}7)\cdot 10^9$ cm^{-2}. At $T = 550\text{-}600$°C there were observed smooth surfaces with Ga droplets (figure 1*b*), which had a diameter of 50-200 nm, height – 10-20 nm and mean density – up to $(8\text{-}10)\cdot 10^9$ cm^{-2}.

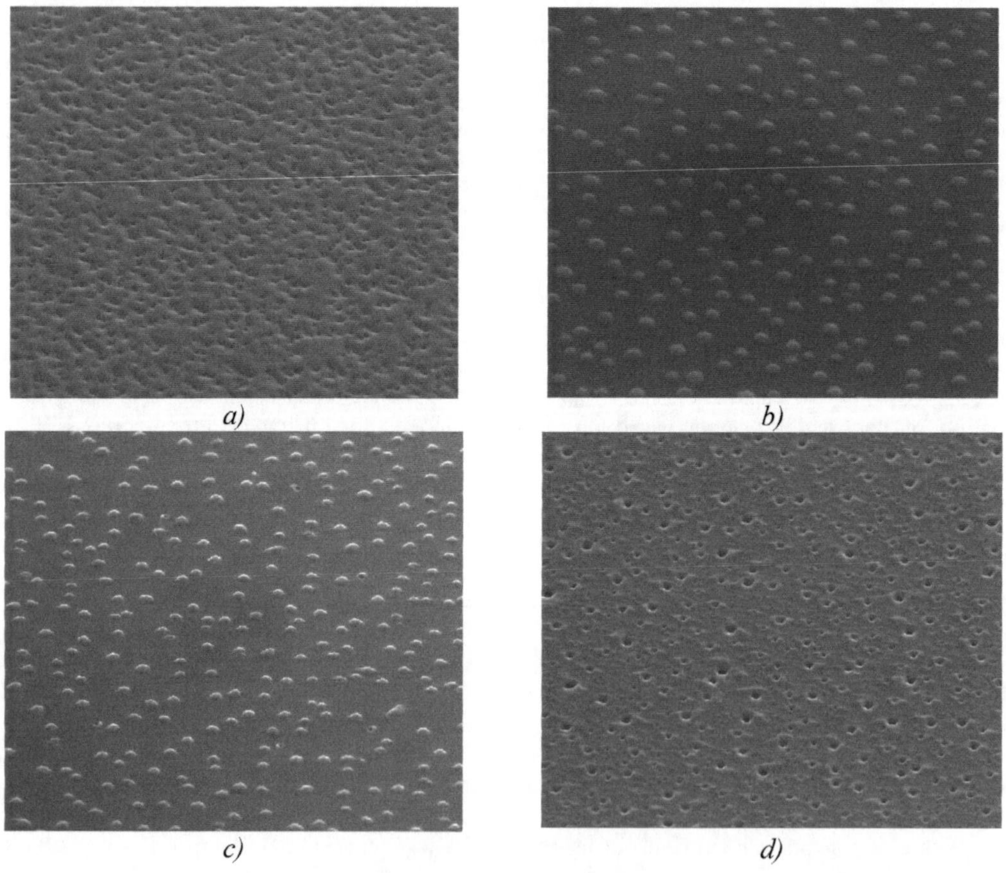

Figure 1. SEM images of GaAs surface after GaAs deposition at: *a)* $J_{As/Ga} = 4$ and $T = 630$°C, *b)* $J_{As/Ga} = 40$ and $T = 590$°C, *c)* $J_{As/Ga} = 1$ and $T = 580$°C, *d)* $J_{As/Ga} = 1$ and $T = 500$°C.

At the same time, reduction of As_4/Ga flux ratio to 1 led to the increase of Ga droplets in diameter and height to 200-500 and 100-120 nm, respectively, and to the decrease of density to $1 \cdot 10^7$ cm^{-2} (figure 1*c*) at the same temperatures. All of these structures had a low crystalline pedestal at its base with diameter of up to 1 μm. At the other temperature modes we obtained results similar to previous case (figure 1*d*).

It should be noted, the droplets did not immerse into growth layer or did not dissolve one and remained on the surface during the GaAs MBE growth. Moreover, the Ga droplets had a liquid state even after cooling of samples to room temperature. This was confirmed by results of surface analysis via AFM techniques [9]. As shown in figure 2, droplets had a sharp phase contrast with the crystalline GaAs surface including its base. The spreading resistance mode showed the significant difference between the droplets and the other surface areas too.

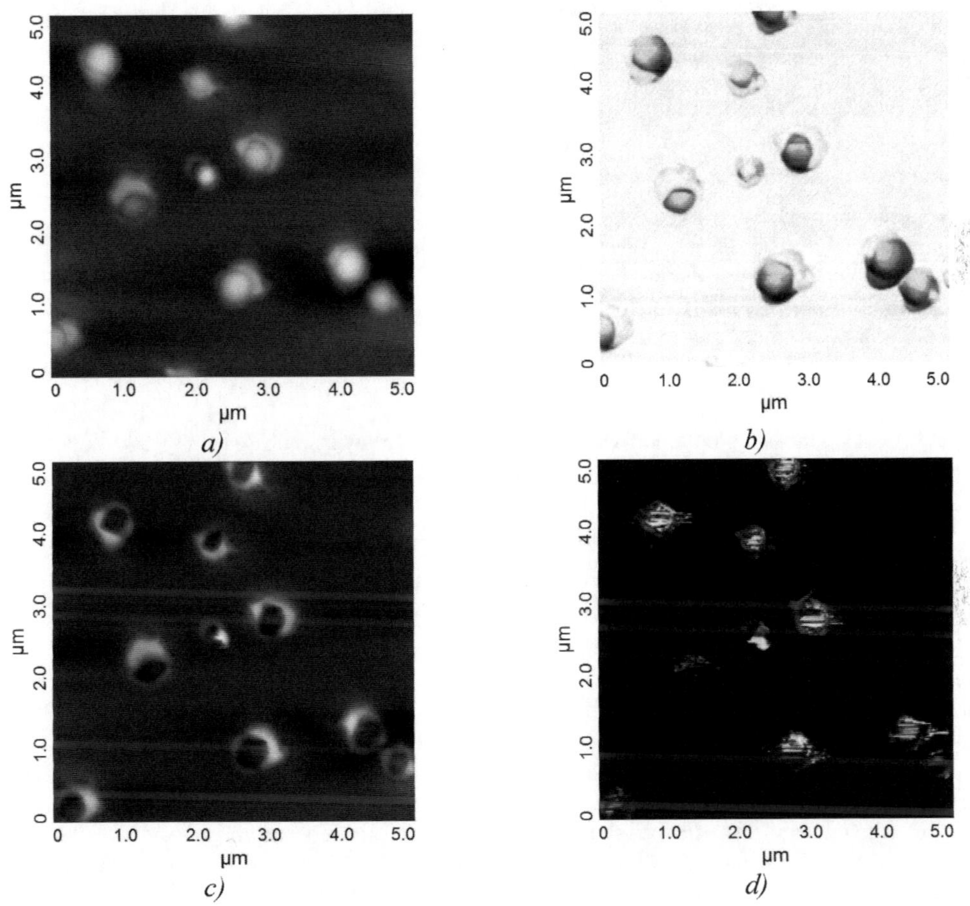

Figure 2. AFM images of surface with Ga droplets in different modes: *a)* non-contact mode, *b)* phase mode, *c)* contact mode, *d)* spreading resistance mode.

Predeposition of 5-10 monolayers of Ga resulted in the smooth surface formation at high substrate temperatures (over 600°C), and in the growth of GaAs layers with pits and particles of metallic Ga at lower temperatures due to the excess Ga accumulation on the surface.

In all cases, the formation of pits or droplets is explained by the interaction between GaAs native oxide film, substrate and growth components [10]. Furthermore, being based on the experimental results obtained, we proposed that at high temperature Ga droplets can act as nucleation centers for nanowire growth or as its initial stage. At some conditions GaAs native oxide film promotes to create a gallium excess and local supersaturation like SiO_x film in [11]. But in the case of GaAs oxide it is difficult to keep these conditions for long time due to the instability of such films at high temperatures.

However, decreasing of the sample heating rate led to the formation of GaAs nanowires which was determined *in situ* by the characteristic RHEED patterns [12]. As shown in figure 3, GaAs crystals had mean density of $5 \cdot 10^8$ cm^{-2}, length – 1-6 μm and diameter – 60-200 nm at an effective deposition thickness of 200 nm. The nanowires tilting relative to substrate caused by surface crystallographic orientation [13].

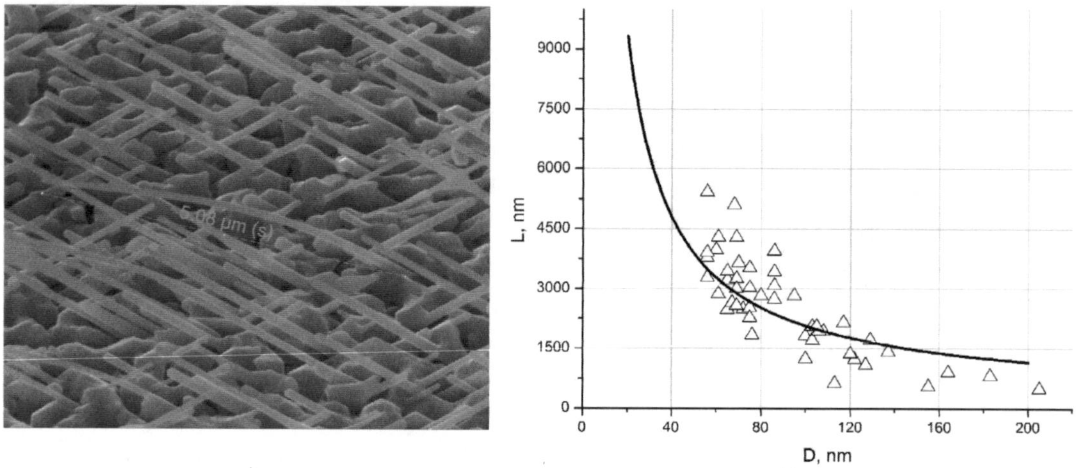

a) *b)*

Figure 3. GaAs nanowires obtained via GaAs native oxide layer: *a)* SEM image, *b)* dependence of the crystal length on its diameter (dots – experiment, line – approximation).

Decreasing dependence of nanowire length on its diameter and comparison of crystal length and deposition thickness suggest the diffusion growth mode [14].

4.Conclusion

In summary, the experimental studies of effect of GaAs native oxide upon the surface morphology of GaAs epitaxial layer was carried out with taking into account the main growth parameters of MBE technology and the features of the interaction between GaAs native oxide, the surface of GaAs substrate and growth components. It was shown that there is a significant influence of the presence of the oxide film during growth on the morphology which depends primarily on the substrate temperature, heating rate and flux ratio. The possibility of the obtaining of different nanostructure types was shown during the MBE growth in the presence of GaAs native oxide film.

Acknowlegments

This work was supported by the Russian Science Foundation Grant No. 15-19-10006. The results were obtained using the equipment of the Research and Education Center and Center for Collective Use "Nanotechnologies" of Southern Federal University.

References

[1] Isomura N, Tsukamoto S, Iizuka K, Arakawa Y 2007 *J. of Crys. Growth* **301-302** 26-29
[2] Guillén-Cervantes A, Rivera-Alvarez Z, López-López M, López-Luna E, Hernández-Calderón I 2000 *Thin Solid Films* **373** 159-163
[3] Asaoka Y 2003 *J. of Crys. Growth* **251** 40-45
[4] Pun A F, Wang X, Durbin S M, Zheng J P 2007 *Thin Solid Films* **515** 4419-4422
[5] Song H Z, Nakata Y, Okada Y 2004 *Physica E* **21** 625-630
[6] Martín-Sánchez J, González Y, González L, Tello M, García R, Granados D, J.M. García J M, Briones F 2005 *J. of Crys. Growth* **284** 313-318
[7] Allwood D A, Carline R T, Mason N J, Pickering C, Tanner B K, Walker P J 2000 *Thin Solid Films* **364** 33-39
[8] Tanner B K, Allwood D A, Mason N J 2001 *Materials Sc. and Eng. B* **80** 99-103
[9] Ageev O A, Balakirev S V, Bykov A V, Gusev E Y, Fedotov A A, Jityaeva J Y, Il'in O I, Il'ina M V, Kolomiytsev A S, Konoplev B G, Krasnoborodko S U, Polyakov V V, Smirnov V A, Solodovnik M S, Zamburg E G 2016 *Springer International Publishing Switzerland* 563-580
[10] Ageev O A, Solodovnik M S, Balakirev S V, Eremenko M M 2016 *Phys. Solid State* **58** 1011-1018
[11] Colombo C, Spirkoska D, Frimmer M, Abstreiter G and Fontcuberta i Morral A 2008 *Phys. Rev. B* **77** 1-5
[12] Shuzhen Y, Guoqing M, Yixin J, Ligong Z, Hang S, Hong J, Zhiming L, Dabing L, Xiaojuan S 2010 *Physica E* **42** 1510-1543
[13] Ageev O A, Konoplev B G, Rubashkina M V, Rukomoikin A V, Smirnov V A, Solodovnik M S 2013 *Nanotechnologies in Russia* **8** 23-28
[14] Johansson J, Wacaser B A, Dick K A and Seifert W 2006 *Nanotechnology* **17** 355-361

The temperature dependence of the electrical conductivity in Cu_2O thin films grown by magnetron sputtering

D Kudryashov[1], A Gudovskikh[1,2] , K Zelentsov[1], A Mozharov[1], A Babichev[3,4,5], A Filimonov[4]

[1]Renewable Energy Lab, St. Petersburg Academic University, St. Petersburg 194021, Russia
[2]St. Petersburg Electrotechnical University "LETI", St. Petersburg 197376, Russia
[3]ITMO University, 197101, St. Petersburg, Russia
[4]Connector Optics LLC, 194292 St. Petersburg, Russia
[5]Ioffe Institute, 194021 St. Petersburg, Russia

Abstract. The temperature dependence of the electrical conductivity in Cu_2O thin films grown by magnetron sputtering at room temperature under different rf-power was investigated. Calculated activation energy of the conductivity for copper oxide (I) films linearly increases with increase in sputtering power reflecting an increasing in concentration of gap states.

1. Introduction

A growing interest in oxide photovoltaics is caused by the low cost, non-toxicity of many metal oxides and scalability at the industrial scale. Due to developing of modern growth technique it is possible to manufacturing oxide solar cells under ambient conditions. Copper oxide (I) is one of the most interesting material in terms of photovoltaic because it is a direct-gap p-type semiconductor with a band gap of 2.2 eV and a high optical absorption coefficient [1]. Numerical simulations of Cu_2O-based solar cell (SC) performance limit show about 12% efficiency [2]. However, the state of the art of recent values for the efficiency of these SC does not exceed 5.38% [3]. The main reason is the low quality of the oxide layer and its interfaces. For poor quality p-type amorphous Cu_2O thin films due to a large concentration of gap states the Fermi level (indicated by the activation energy E_A of the layer) will be close to midgap position. Otherwise the Fermi level will be located near valence band. Its position can be estimated from the measurements of dark conductivity (σ) as a function of temperature [4]:

$$\sigma_{dark}(T) = \sigma_0 \exp(-E_A / kT) \tag{1}$$

The aim of this work is to investigate the temperature dependence of the electrical conductivity in Cu_2O thin films grown by magnetron sputtering under different conditions.

2. Experimental

Deposition of Cu_2O thin films was carried out using rf magnetron sputtering setup (Boc Edwards Auto 500RF). Cu_2O films with thickness of 150 nm were deposited on Si(100) and glass substrates at different rf-power at room temperature. Pure argon (99.999 %) was used as working gas. Control of the oxide layers thickness was carried out using a profilometer Ambios XP-1. SEM images were analyzed using a scanning electron microscope SEM (Zeiss SUPRA 25). XRD spectra were measured using PANalytical X'Pert PRO equipment. Electrical conductivity measurements were carried out with Keithley 2400 and thermo stabilized holder under temperature range of 13-60°C.

3. Results and discussion

SEM image of 150 nm Cu_2O thin films deposited on Si-wafer by magnetron sputtering is shown in figure 1. Cu_2O has a nano-column structure with a sharp interface between Si and copper oxide (I).

Figure 1. Cross-section SEM image of Cu_2O thin film deposited on Si-wafer under different rf-power: (a) - 80 W, (b) - 125 W, (c) - 200 W

The XRD reflex intensity measurements show nonlinear RF power dependent behavior in the range of 80-200 W (Fig. 2). It has a maximum value at the magnetron RF power of 150 W and at power of 200W demonstrates a minimal value.

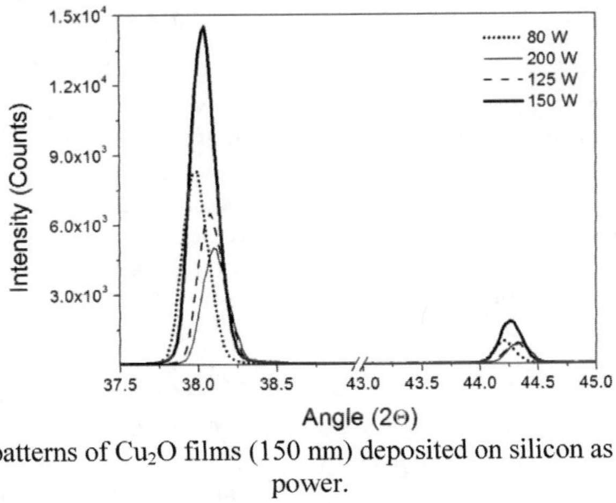

Figure 2. XRD patterns of Cu_2O films (150 nm) deposited on silicon as a function of the RF power.

It should be noted that for stoichiometric and stress-free copper (I) oxide crystal the rocking curve peaks must be at 36.45 for the plane (111) and 42.33 for the plane (200) positions [5]. In our case, these maxima are shifted toward larger angular values, which may indicate a deviation from stoichiometry. Indeed, the formation of Cu_2O was conducted in an argon atmosphere without oxygen addition. A portion of oxygen atoms from Cu_2O target doesn't reach the substrate because of their high reactivity. Under these conditions the oxygen-depleted layer of copper oxide (I) is formed.

Copper (I) oxide grown on glass substrates at rf-magnetron power of 80-200W shows a transmittance of 80% for a wavelength range of 850-900 nm (Fig. 3). Based on transmittance and reflection data for sputtered Cu_2O thin films an absorption spectra was calculated (Fig. 4a). It can be seen that increase in magnetron sputtering power leads to decrease of copper (I) oxide absorption edge (Fig 4b).

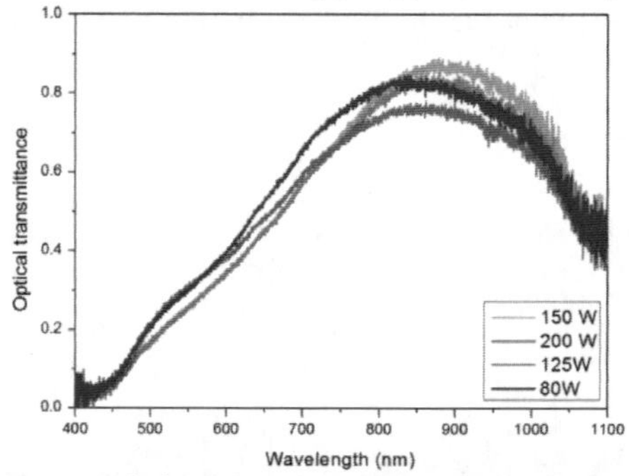

Figure 3. Transmission spectra of rf-magnetron sputtered Cu_2O thin film (150 nm) deposited on glass as a function of the RF power.

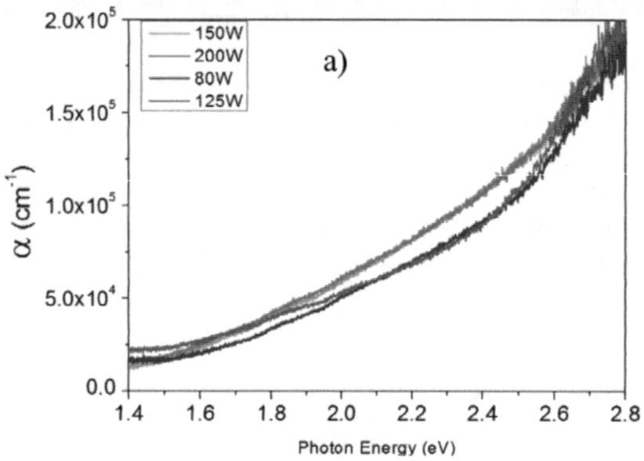

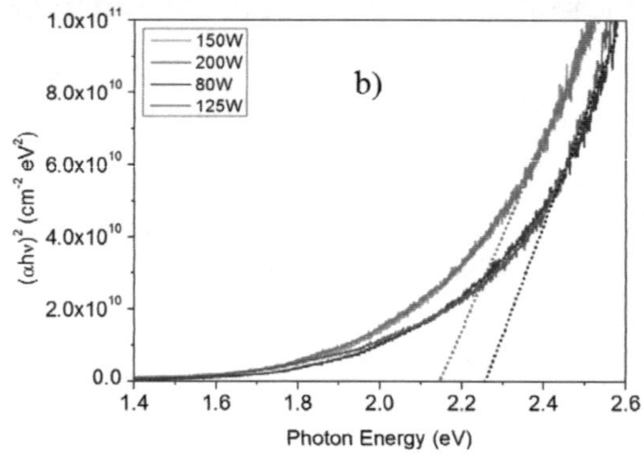

Figure 4. Absorption spectra (a) and a plot of $(\alpha h\nu)^2$ vs. photon energy (b) for Cu_2O thin films deposited on glass as a function of the RF power.

Current-Voltage characteristic curves measured at different temperatures for Cu_2O thin films deposited on glass substrate are shown in figure 5a. All I-V curves exhibit a linear relationship between the applied voltage and the resulting electric current. Arrhenius plot of conductivity is shown in figure 5b where the slope reflects an activation energy for p-type Cu_2O.

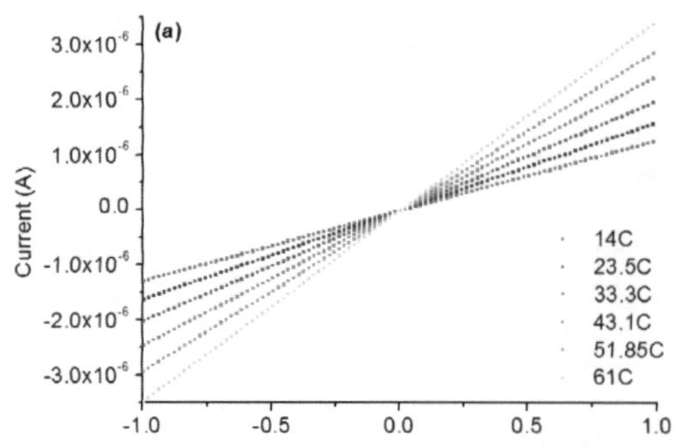

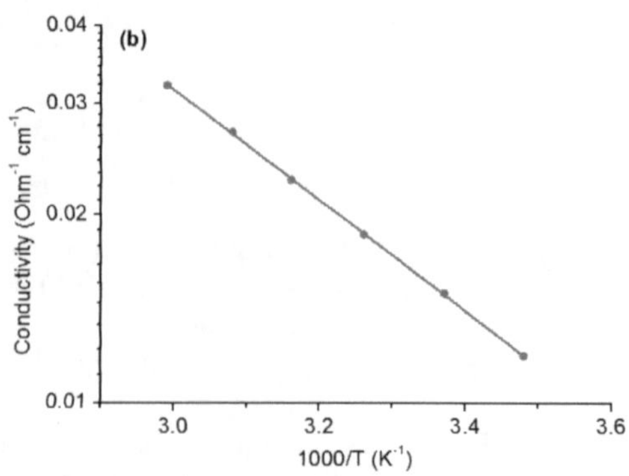

Figure 5(a,b). **(a)** IV curves and **(b)** fitting for Cu_2O thin film deposited at 50 W.

An activation energy values for Cu_2O thin films grown at different sputtering powers at room temperature are shown in figure 6.

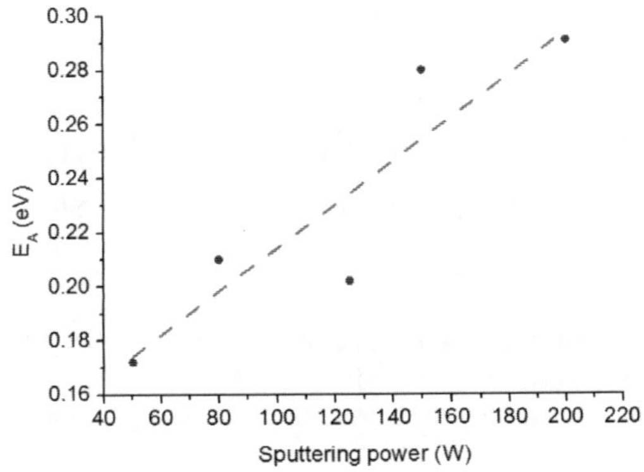

Figure 6. An activation energy of deposited Cu_2O films as a function of the rf-power.

An activation energy of copper oxide (I) films linearly increases with increase in sputtering power. The shift of the Fermi level toward mid gap is associated with increased density of gap states due to the rise of the defect concentration with rf power. Similar behavior was found in [6] where it was shown that increasing of sputtering power results in changing of copper oxide (I) films morphology.

4. Conclusions

The temperature dependence of the electrical conductivity in Cu_2O thin films grown by magnetron sputtering under different rf-power was investigated. Calculated activation energy of copper oxide (I) films linearly increases with increasing of sputtering power reflecting an increasing in concentration of gap states.

Acknowledgment

The reported study was partially supported by RFBR research project № 15-08-06645A

References

[1] Ruhle Sven et al. 2012 *J. Phys. Chem. Lett.* **3** 3755
[2] Kudryashov D, Gudovskikh A 2015 Abstr. E-MRS Spring Meeting (Lille) DP16-7
[3] Minami T, Miyata T, Nishi Y 2014 *Solar Energy* **105** 206
[4] Pavlov P, Khohlov A 2000 *Solid State Physics* (Moscow High School Press) p 497
[5] JCPDS – International Centre for Diffraction Data. Powder Difraction File #05-0667, 1996
[6] Sosnin D, Kudryashov D 2015 Abst. 17th Russ. Youth Conf. on Phys. of Semicond. and Nanostr., Opto- and Nanoelectr. (St. Petersburg) P 41

Influence of modified detonation nanodiamonds on electro-optical properties of nematic liquid crystals

Vera Vashkevich[1], **Valeri Lapanik**[2], **Anatoly Minko**[1]
[1]Belarusian State University, Minsk 220030, Belarus
[2]Institute of Applied Physics Problems, Belarusian State University, Minsk 220045, Belarus
E-mail: vera-live93@mail.ru

Abstract. To modify the structure of detonation nanodiamonds (DNDs) several carboxylate groups were added to DNDs. Activation of COOH-surface functionalized groups allowed attaching of various organic tails to molecules. It was investigated that dielectric and electro-optical properties of nematic liquid crystalline mixtures (LCMs) doped with modified DNDs (MDNDs). It is established that the effect of DNDs on mesomorphic, dielectric and electro-optical properties depends on the size of nanoparticles (NPs) and the type of tail-like organic molecules grafted to DNDs. It is found that NPs of a small size (5–6 nm) do not significantly affect on the parameters of LCMs. At the same time conglomerates of a larger size (50 and 100 nm) depending on the tails polarity can increase or decrease the dielectric anisotropy and response time of LCMs in about 1.2–1.4 times.

1. Introduction

Dispersion of different NPs aimed to improve electro-optical properties of liquid crystals (LCs) have attracted attention of many researchers in recent years [1-4]. The influence of metal particles Ag, MoS and semiconductor particles MnO_2, WS_2 on ferroelectric LCs (FLCs) has been reported lately. According to the research, that was conducted, the rise and decay times of FLCs where Ag metal nanowires had been added did not significantly differ from those of the original FLCs (the nanowire size is about 40-70 nm) [1]. Ag NPs didn't improve the characteristics of FLCs due to the low activity of Ag which is located almost in the end of the series of chemical activity of metals.

The research of Ni particles with a size of 2-8 nm and a concentration of 0.5% showed that the optical transmittance improved and the optical response time of FLCs reduced [2]. The authors suppose that the improvement of properties of FLCs doped with metal NPs is due to the appearance of particles dipole moment, which is explained by their interaction with LC molecules (the dipole moment of Ni NPs is 10 Debye [2]). When voltage was applied, Ni NPs aligned with force lines, attracting LCs, which affected on the temporal characteristics of LCs and promoted better light transmission.

As a result of doping nematic LCs (NLCs) with Ti NPs the optical response time reduction can also be observed [4]. If Ti NPs doping concentration ranges from 0.1 to 1.0 wt% and the size of Ti NPs is 100 nm, then the optical response time reduces twice from 15 ms to 7.5 ms. The authors of this paper propose two mechanisms of Ti NPs interaction with NLC molecules. In the first case though there is no electric field in the conductive metal particle, the electric field density around it increases, which makes NLC molecules move faster. In the second case the screen effect is suppressed due to the presence of ionized impurities in the NLC. Adsorption of impurities around Ti NPs in the absence of the electric field leads to their neutralization, reducing the optical response time.

Nanodiamonds are unique carbon NPs because of their intrinsic hydrophilic surface. The surface contains a complex array of chemical groups, including carboxylic acids, esters, ethers, lactones and amines. Alteration of DNDs surface can lead to a bunch of chemical properties, that can be compared with nanoscale diamond powders, since almost 15% of all atoms in the DND primary particle are located on the surface and, therefore, can dissolve. DNDs show potential for applications in energy storage, composites and catalysis. Besides, DNDs can be used in electroanalysis, electrochemical doublelayer capacitors and batteries. Other potential is biomedical applications including usage of nanodiamonds as supporters for solid-phase peptide synthesis and as sorbents for detoxification and separation [5].

We were interested in characterizing suspensions using a special class of nanodiamond materials having a size of 4–5 nm often called in literature as 'ultradispersed diamonds' or 'detonation nanodiamonds' produced *via* detonation of carbon-containing explosives. Complex research of the effect of modified DNDs on electro-optical characteristics of NLCs was carried out. In the continuation of studies [6] and to test the impact of such NPs on the parameters of compositions based on compounds with different polar groups compositions based on compounds with a isothiocyanates polar group (NCS) and with cyano group (CN) and F atoms were investigated. DNDs of various doping concentrations were dispersed in NLCs cells. The dependence of mesomorphic, dielectric and electro-optic properties of LCs on the NPs size and a type of organic molecules grafted to DNDs was shown.

2. Materials and methods

2.1. Ozonolysis

The DNDs suspension with a concentration of 5 g per liter (water) was poured into the apparatus, producing ozone at a flow rate of 10–15 ml/second, leading to 3–5% of ozone in the gas mixture. Before using, the prepared ozone was purified by means of a concentrated sulphuric acid in the flask and then it was passed through granulated potassium hydroxide. The latter was able to oxidize the intermediate-formed ozonide into a carboxyl group, whereas hydrolysis of the ester bond led to the formation of a carboxylate group. About 0.1 g of caustic soda was added to the water suspension of DNDs under constant stirring. Thereafter, ozone was passed through the suspension for 2 hours (more than 300 times). The final oxidation was reached by adding 30 ml hydrogen peroxide (30%) while incubating the mixture for 1 hour. Then, 10% aqueous hydrochloric acid was added until 2–6 pH acidification was reached. Finally, the upper clear layer was carefully decanted and the precipitate was centrifuged with subsequent washing with water 8–10 times. Residual water was removed by washing the precipitate with dioxane two times. After drying overnight at room temperature, the content of carboxyl groups was determined. This procedure caused 2-3 times increase of the carboxylate groups content. The molecular weight (excluding the sorbed water) was between 1300 and 1560, depending on the kind of experiments

2.2. The activation of COOH-surface functional groups

The activation of COOH-surface functional groups was done by acylchloride obtained by boiling nanodiamonds and $SOCl_2$:

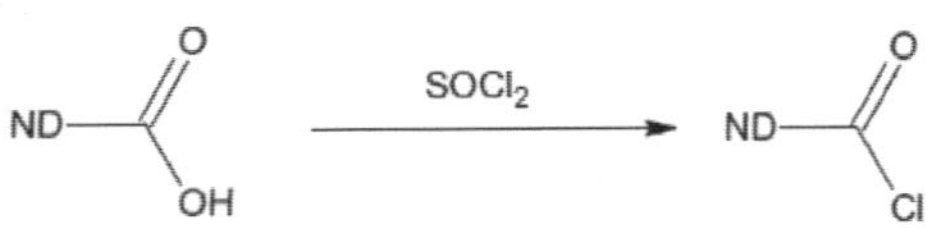

2.3. Acid chloride DND

2 ml thionylchloride was added to 1 g suspension of ozonized DNDs in 30 ml dichloroethaneand it was heated for 1 hour under reflux. After cooling, the product was separated by decantation and washed with dichloromethanetwo times.

2.4. Dicyano-4-4′-diphenyl DND

A mixture of 1 g DND chloride and 0.1 g 4-hydroxy-4′-cyanobiphenyl dissolved in 20 ml pyridine and was heated at 25°C for 3 hours. Then, the ester of the DND was separated and washed with distilled water, with dimethylformamide (DMF) two times and with dichloromethane three times. The resulting ester was dried in the air at 60°C for 3 hours.

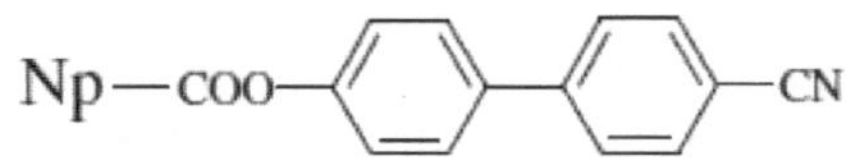

2.5. 4-Octylphenyl DND

A mixture of 1 g chloride of the DND and 0.1 g of 4-octylphenol dissolved in 20 ml of pyridine was heated at 25°C for 3 hours. Then, the ester of DND was separated and washed with distilled water, twice with DMF and three times with dichloromethane. The resulting ester was dried in the air at 60°C for 3 hours.

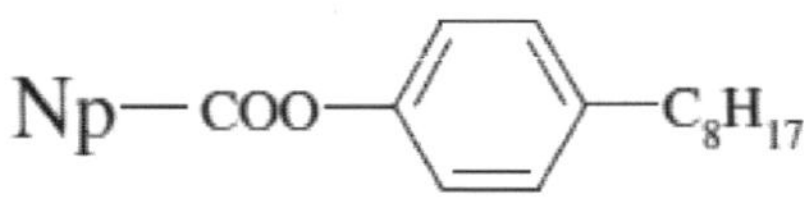

2.6. Preparation of testing mixtures

For the LC base two mixtures with different polar liquid crystalline compounds were prepared. M-1 is the mixture based on the compound with the NCS end group, M-2 is the mixture based on the compound with the CN end group (50%) and F atom (50%). The selection of two mixtures was made in order to verify the effect of modified DNDs (MDNDs) on parameters of LCs with different polar groups. Dicyano-4-4'-diphenyl DNDs having a small size (5-6 nm, N-1), and conglomerates on the basis of dicyano-4-4'-diphenyl DNDs with a size of about 50 nm (N-2) and 100 nm (N-3) were investigated. The conglomerates based on 4-octylphenyl DNDs with a size of about 50 nm led to N-4. DNDs particles and LCs were manually mixed in certain proportions and then subjected to ultrasonic agitation for 20 min in EMMI-20. 0.5% of N-1, N-2, N-3, N-4 were added to the mixturesM-1 and M-2, leading to NLH-1, NLH-2, NLH-3, NLH-4 and NLH-5, NLH-6, NLH-7, NLH-8,respectively. N-4 with concentrations of 0.2%, 0.4%, 0.6%, 0.8%, 1% were added to M-1 for more detailed study of the influence of the MDND conglomerates amount on dielectric properties of LCMs. As a result, NLH-4a, NLH-4b, NLH-4c, NLH-4d, NLH-4e were obtained. Cells with a thickness of 7 μm were filled with mixtures thus prepared. The quality of LCs alignment and sample uniformity were studied using a polarized optical microscope (Olympus BX-51P). Electro-optical parameters were measured using a setup consisting of a He–Ne laser (λ = 632.8 nm), a rotating table, a thermostatic camera (Linkam LTS 350) coupled with a temperature controller (Linkam CI 94, with an accuracy of ±0.1°C), a function generator (HP 33, 120A) and a digital oscilloscope (HP Infinium). The frequency and the voltage of the sinusoidal electric field were 1 kHz and 10 V for all electro-optical measurements.

3. Results and discussion

It was interesting to us to study the impact of different MDNDs on the dielectric characteristics of LCMs. Figure 1 shows the results of measurements of dielectric permittivity for LCMs with MDNDs and conglomerates of MDNDs modified by dicyano-4-4′- diphenyl. It is seen that small MDNDs had no effect on these parameters independent of the LCMs dielectric anisotropy magnitude. Any polar dopant having a concentration of 0.5% cannot change the LC matrix parameters.

Dielectric permittivity of pure LCMs and that of LCs containing conglomerates (50 nm, 100 nm in diameter) of MDNDs were different. The difference is due to long-chain conglomerates of MDNDs that have a permanent dipole moment parallel to the surface. This assumption is based on the fact that functionalized organic groups attached to the surface of DNDs and carrying a permanent dipole moment are parallel to the LCs molecules alignment plane. This alignment is the result of elastic interactions between LC molecules and functionalized organic groups molecules. In this case, long-chain conglomerates of MDNDs are able to change the polarisability anisotropy of mixtures, i.e. the perpendicular component can increase. When the electric field was applied, conglomerates of MDNDs did not reorient like LC molecules, rather their organic tails on the surface became a little bit tilted. Therefore, parallel dielectric permittivity and, consequently, the magnitude of dielectric anisotropy decreased (Figure 1).

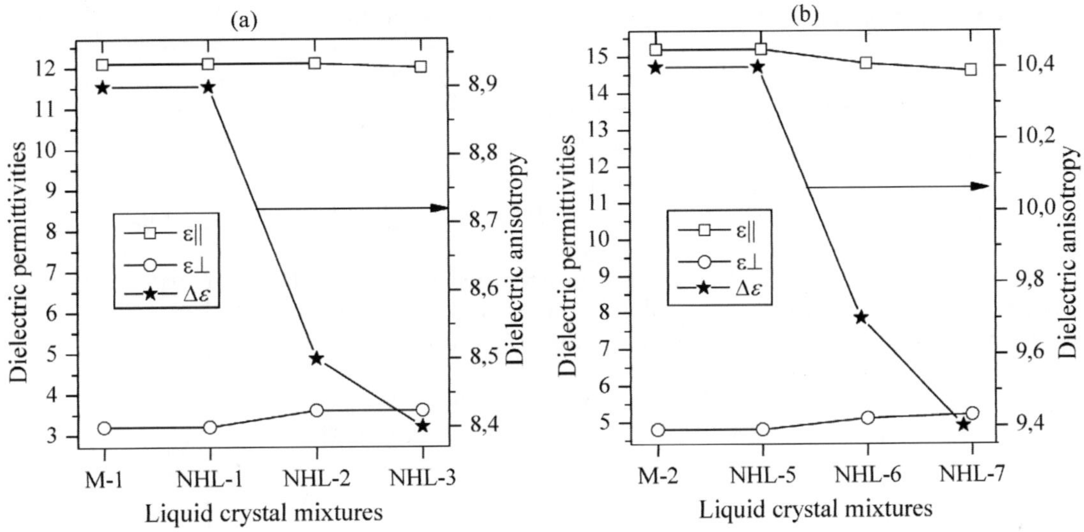

Figure 1. Dependence of dielectric permittivity and dielectric anisotropy of M-1 (a) and M-2 (b) on the size of NPs.

One should note that the influence of MDNDs on the parameters of LCMs with different polar compounds was the same. Naturally, the dielectric anisotropy decrease caused the switching time increase, in particular, the fall time increase. Table 1 shows that the switching time remained virtually unchanged for LCMs with small MDNDs, and it reduced 1.2–1.4 times for LCMs with conglomerates of MDNDs.

Table 1. The switching time of M-1 and M-2.

Mixture	τ_{rize}, ms	τ_{fall}, ms
M-1	1.14	47.87
NHL-1	1.2	50.34
NHL-2	1.32	53.84
NHL-3	1.48	58.62

Mixture	τ_{rize}, ms	τ_{fall}, ms
M-2	1.82	40.3
NHL-5	1.88	43.3
NHL-6	1.98	48.6
NHL-7	2.12	55.7

Similar studies were conducted with LCMs to which conglomerates of MDNDs with non-polar tails had been added. The effect of such conglomerates was opposite compared to the one of conglomerates of MDNDs that had polar tails. Of course, the dielectric permittivity for pure LCMs and LCM nanocolloids with N-4 was also different. This behaviour can be explained by assuming that such particles have the form of long chains. In this case they have a permanent dipole moment which is directed at some angle to the long chain of N-4. According to this assumption, one can explain the reduction of the dielectric anisotropy value of LCMs. Experimental data showed that the polarisability anisotropy of mixtures, i.e. the perpendicular component had decreased. In the case where the electric field was applied, we assumed that N-4 was not totally reoriented together with LC molecules and, therefore, parallel dielectric permittivity also slightly increased. Consequently, the magnitude of dielectric anisotropy also increased (Table 2). The results presented in Table 2 show that the switching time decreased for mixtures with 0.5 % of N-4 as expected.

Table 2. Dependence of dielectric permittivities, dielectric anisotropy and electro-optical parameters of M-1 and M-2 on the conglomerates of N-4.

Mixture	$\varepsilon_{\parallel}$	$\varepsilon_{\perp}$	$\Delta\varepsilon$	V_{10}, V	V_{90}, V	τ_{rize}, ms	τ_{fall}, ms
M-1	12.1	3.2	8.9	1.65	2.49	1.14	47.87
NHL-4	12.6	2.6	10.0	1.46	2.24	0.87	42.04
M-2	15.2	4.8	10.4	1.82	2.44	1.82	40.3
NHL-8	15.7	4.4	11.3	1.64	2.18	1.12	28.3

Finally, the experiments with LCMs having different concentrations of N-4 were performed. Figure 2 shows a strong influence of MDNDs conglomerates on the main parameters of LCMs at concentrations below 0.75 %, i.e. NLH-4a, NLH-4b, NLH-4c.

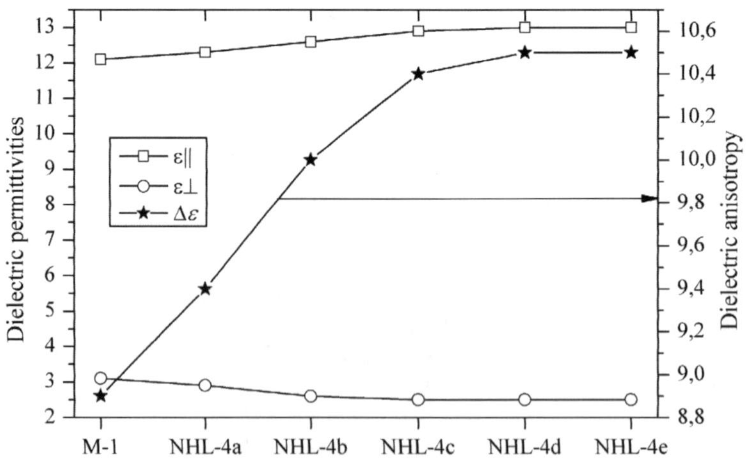

Figure 2. Dependence of dielectric permittivities and dielectric anisotropy of M-1 on the concentration of N-4.

4. Conclusions

It is established that the effect of DNDs on mesomorphic, dielectric and electro-optical properties depends on the size of NPs and a type of tail-like organic molecules. As expected, nps of a small size (4–5nm) don't significantly affect physical properties of LCMs regardless of the value of their dielectric anisotropy. The addition of particles slightly increases the phase transition temperature of

Journal of Physics: Conference Series **741** (2016) 012014 doi:10.1088/1742-6596/741/1/012014

LCMs. At the same time conglomerates of a larger size (50 and 100 nm) depending on the tails polarity can increase or decrease about 1.2–1.4 times the dielectric anisotropy and response time of LCMs. Dicyano-4-4′-diphenyl DNDs and 4-octylphenyl DNDs decrease and increase the dielectric anisotropy and switching time of LCMs, respectively. The change of the dielectric anisotropy and switching time is caused by the increase of the amount of DND particles.

References

[1] Rajh D. 2013 *Physica Status Solidi (A)* **210** 2328
[2] Neeraj K, Raina K 2013 *Optical Materials* **35** 531
[3] Al-Hazmi F, Al-Ghamdi A A, Al-Senany N, Alnowaiser F, Yakuphanoglu F. 2014 *J Mol Liq* **190** 169
[4] Ha Y S, Kim H J, Park H G, Seo D S. 2012 *Opt. Express* **20** 6448
[5] Mochalin V.N., Shenderova O., Ho D., Gogotsi Y. 2011 *Nat. Nanotechnol.* **7** 11
[6] Lapanik V., Lugouskiy A., Timofeev S. and Haase W. 2014 *Liq. Cryst.* **41** 1332

Admittance spectroscopy of Ge/Si p-i-n structures with Ge quantum dots

A A Pishchagin[1], A V Voitsekhovskii[1], A P Kokhanenko[1], V Yu Serokhvostov[1], S M Dzyadukh[1] and A I Nikiforov[1,2]

[1]National Research Tomsk State University, Tomsk, 634050, Russia
[2]Rzanov Institute of Semiconductor Physics SB RAS, Novosibirsk, 630090, Russia

Abstract. The experimental results on synthesis of Si/Ge p-i-n structures with Ge quantum dots in the *i*-region and their investigation by the method of admittance spectroscopy are presented. The activation energies of the emission process from localized states are calculated for two types of structures. Current-voltage characteristics without illumination and under illumination are measured.

1. Introduction

Currently optoelectronics is experiencing rapid development, and the main objects of research are complex heterostructures with nanoscale inclusions. Creating semiconductor structures with new physical properties is the primary goal of nanotechnology, which has the aim of expanding the limits of applicability of semiconductor materials. The discovery of new physical properties in these cases allows creating new devices using advanced technology of silicon microelectronics [1-3]. Also, in recent years the interest in photoelectric properties of Ge/Si heterostructures (primarily in the spectral range of 1.3-1.55 μm) has increased. New types of photodetectors based on silicon-germanium low-dimensional heterostructures using intrasubband and intersubband transitions are intensively being developed. Such devices may be used in optoelectronic communication systems and remote monitoring. [2, 4].

Today a search for new methods and revision of existing traditional methods of diagnostics of nanoelectronic devices and structures takes place. The methods of current-voltage characteristics, capacitance-voltage characteristics, admittance spectroscopy, deep level transient spectroscopy (DLTS) have great potential for characterizing semiconductor structures with quantum inclusions. Admittance measurements can be applied to semiconductors containing a space charge region (it may be a p-n-junction, MIS structure or Schottky barrier). In this case, the reactive component of admittance is formed by barrier capacitance of the space charge region. The source of the active component (conductance) can be both current leakage arising from a junction's imperfection, and deep centers and traps in the band gap of a semiconductor, exciting and emitting charge carriers, depending on the experimental conditions. The second mechanism of conductance is very important for the diagnosis of advanced semiconductor materials and structures and is a subject of dynamic methods of admittance spectroscopy [5].

A large number of studies have been devoted to investigating structures containing Schottky barriers by the methods of admittance spectroscopy [6, 7]. However, for device applications such as solar cells, p-i-n-structures are very important. This work aims to represent the possibility of studying p-i-n-structures based on Si with Ge quantum dots by admittance methods.

2. Experiments

In this paper we present the experimental results on synthesis of Si/Ge p-i-n structures with Ge quantum dots in the *i*-region and their investigation by the method of admittance spectroscopy.

The samples were fabricated by molecular beam epitaxy in an ultra-high vacuum installation "Katun-C" in Institute of Semiconductor Physics. Evaporation of silicon and germanium was carried out by electron beam evaporators, the dopants (Sb and B) were evaporated from effusion cells. The analytical part of the epitaxy chamber consists of a quadrupole mass spectrometer, a quartz thickness meter, and reflection high energy electron diffractometer (RHEED). The growth of Ge quantum dots was carried out on Si(100) substrates with misorientation less than 0.5°. Array of Ge hut-clusters with height 1,5-3 nm, lateral size 10-40 nm, and surface density $\sim 10^{11}$ cm^{-2} was formed on Si surface. In the intrinsic region of the samples multiple layers with Ge quantum dots separated by thin 5 nm layers of silicon are included.

Figure 1 shows the structures of two types of samples The *i*-region of type 1 samples contained 30 layers of 6 monolayer Ge quantum dots separated by 5 nm silicon layers. The *i*-region of type 2 samples contained 30 layers of 6 monolayer Ge quantum dots separated by 5 nm silicon layers, and every 10 layers of Ge quantum dots were additionally separated by 100 nm of Si.

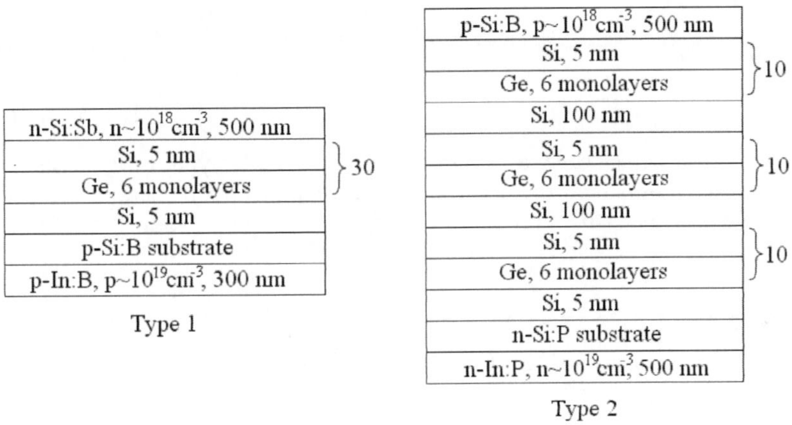

Figure 1. Schematic representation of the structures studied.

Measurements were performed on an automated admittance spectroscopy installation [5]. During one cycle of temperature scanning measurements of the frequency and temperature dependences of capacitance and conductance, as well as measurements of current-voltage characteristics of the studied structures were carried out in the temperature range of 10 300 K.

3. Results

By analogy with deep levels in semiconductors, the principle of admittance spectroscopy of structures with quantum dots is based on measuring the complex conductivity of the system that occurs when discrete energy levels recharge due to emission of charge carriers and their capture by localized states. In the current practice of admittance spectroscopy experimental data processing, the temperature spectra of conductance (*G-T*) at different frequencies of the test signal are often considered the most informative.

In the temperature dependence of conductance of type 1 samples a maximum was observed at low temperatures of 25-40 K. The observed maximum of conductance corresponds to a discrete energy level. The position of this peak is shifted on the temperature scale as the frequency of the applied signal changes (Figure 2 (b)). With fixed bias voltage V_b recharging of the level occurs. The charge carrier emission rate from this level decreases at lower temperatures, so with a decrease in the frequency of the test signal the condition of maximum conductance is achieved at lower temperatures and at different frequencies.

Conductance peak position for the sample remains constant with changes in the applied bias voltage (Figure 2 (a)). Similar results were obtained in studies of type 2 samples. Figure 3 shows the temperature conductance spectra measured at different voltages and different frequencies of type 2 samples.

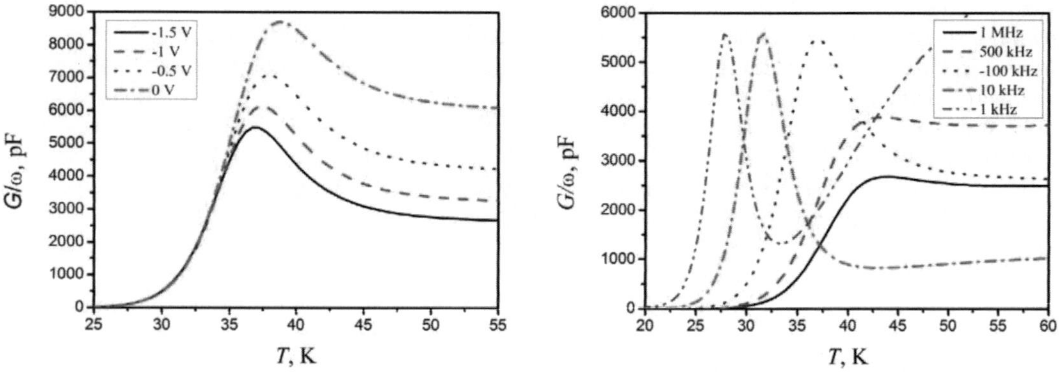

(a) (b)

Figure 2. Temperature dependence of conductance for sample type 1 measured at various bias voltages at the test signal frequency of 100 kHz (a) and at various frequencies of the test signal and the applied bias voltage of –1.5 V (b).

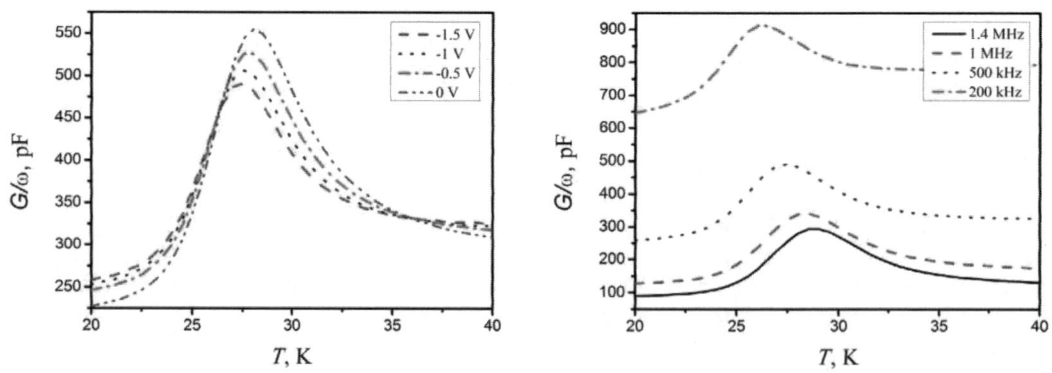

(a) (b)

Figure 3. Temperature dependence of conductance for sample type 2 measured at various bias voltages at the test signal frequency of 100 kHz (a) and at various frequencies of the test signal and the applied bias voltage of –1.5 V (b).

The main characteristic of a discrete energy level is the charge carrier activation energy. To determine the activation energy the experimental points are plotted in coordinates $\ln e = f(1/T)$, called an Arrhenius plot. Processing temperature spectra leads to a typical family of Arrhenius plots for finding activation energies of the emission process. The GT/ω value has a maximum at $\omega = e_p$, where ω is the angular frequency of the test signal, e_p is the charge carrier emission rate from a discrete level. By plotting maxima T_{max} in coordinates $\omega = f(1/T)$ the activation energy characterizing the position of the energy levels is determined. For each frequency a point with coordinates $\ln(e_p/T^2)$, $1/T_{max}$ is plotted and the approximating straight line is built. From the slope of this line the activation energy is calculated. The observed maximum of conductance corresponds to a discrete energy level.

In a further study of type 2 sample a conductance peak at low positive bias was also detected (peak 2 in Figure 4). This maximum is observed only at positive bias at higher temperatures and is most pronounced at low frequencies, while the first maximum is also observed at negative bias.

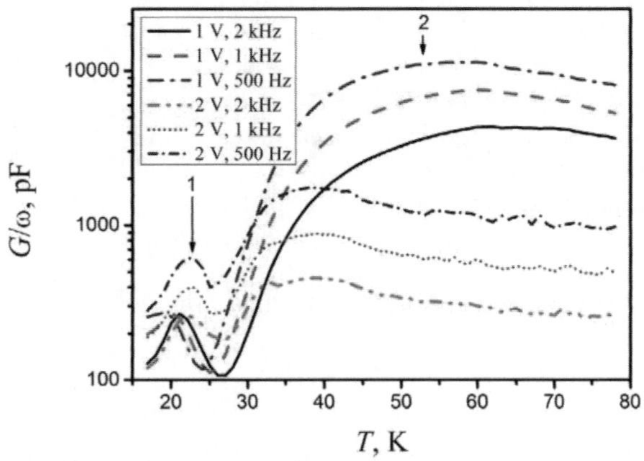

Figure 4. Temperature spectra of conductance of type 2 sample, measured at the voltages of +1 V and +2 V at different frequencies

For both samples activation energies were calculated. For the first peak of conductance calculated activation energies of type 1 and type 2 samples do not depend on the applied bias voltage and are equal to 38±5 meV and 46±4 meV respectively. For the second peak the calculated activation energy at a bias voltage of 1 V is 65±10 meV, at a bias voltage of 2 V it is 165±30 meV. This peak is broadened and probably corresponds not to a single discrete level but to a system of closely lying levels, due to the inhomogeneity of such parameters of quantum dots as their lateral size, height, shape and density in the array.

The first peak on the temperature dependence of conductance may be associated with the impurity level in Si. The second peak is explained by the presence of spatial quantization levels in the system associated with Ge quantum dots. Appearance and modification of peaks can be explained by the fact that with a change in the applied voltage the electrochemical potential occasionally crosses the discrete energy levels, producing oscillations in the charge density distribution. The reason for this is the thermionic emission of charge carriers from a discrete level. Discrete level gives partial charge density increment. This increment of charge leads to an increase in the external circuit current measured as a change in conductance of a sample.

4. Conclusion

This research shows the possibility of studying silicon p-i-n-structures with germanium quantum dots by the method of admittance spectroscopy. Measurements of Si p-i-n-structure samples containing 30 layers of Ge quantum dots in the i-region on an automated admittance spectroscopy installation were conducted. Two peaks are observed on the temperature dependences of conductance of the investigated nanoheterostructures. The first peak is observed at any bias voltage, the second peak is observed in the narrow voltage range. Position of the first peak does not depend on the applied bias voltage, the second peak is shifted on temperature scale with changing the bias voltage from 1 V to

2 V. For both peaks corresponding activation energies were calculated. The first peak of the temperature dependence of conductance may be associated with the impurity level in Si. The second peak is explained by the presence of spatial quantization levels in the system associated with Ge quantum dots.

Acknowledgments
This study was supported by the Ministry of education and science of the Russian Federation, State task No. 16.1032.2014/K.

References

[1] Shklyaev A A and Ichikawa M 2008 *Phys. Usp.* **51** 133-61
[2] Yakimov A I 2013 *Optoelectron., Instrum. and Data Process.* **49** 467-75
[3] Lozovoy K A, Voytsekhovskiy A V, Kokhanenko A P, Satdarov V G, Pchelyakov O P and Nikiforov A I 2014 *Opto-Electron. Rev.* **22** 6-12.
[4] Chen Y, Wu S, Ma Y, Fan Y, Yang X, Zhong Z and Jiang Z 2014 *J. Appl. Phys.* **115** 233502
[5] Zubkov V I 2007 *Automated admittance spectroscopy installation for heterostructures* (St. Petersburg: St. Petersburg Electrotechnical University)
[6] Yakimov A I, Nikiforov A I, Timofeev V A and Dvurechenskii A V 2011 *Semicond. Sci. Technol.* **26** 125002
[7] Asperger T, Miesner C, Brunner K and Abstreiter G 2000 *Thin Solid Films* **380** 227–9

Study of the influence of a spin-orbit exciton on the magnetic ordering in Sr_2IrO_4

E A Dikushina, I L Avvakumov

Department of Physics, Ural Federal University, Yekaterinburg 620002, Russia

eadikushina@gmail.com

Abstract. This study investigates an orbital exciton propagation in quasi-two-dimensional Heisenberg antiferromagnet Sr_2IrO_4 by means of computer simulation. Ising and Heisenberg models are compared, taking in consideration magnetic interactions and exciton hopping. As the result of this simulation the spin structure factor for both models is calculated

1. Introduction

Sr_2IrO_4 is a $5d$ transition metal oxide. It is one of recently discovered Mott insulators and, while other $5d$ transition metal oxides with a similar structure are metallic, this compound does insulate. Sr_2IrO_4 shares several distinctive features with cuprates which are characteristic for superconductivity. It makes Sr_2IrO_4 an object of great interest as a possible high temperature superconductor [1, 2].

Sr_2IrO_4 can be represented as a quasi-two-dimensional Heisenberg antiferromagnet with isospin 1/2. Here isospin is the effective total angular momentum of occupied energy states of $5d$ configuration of Ir as a good quantum number [3, 4].

Unlike the Mott gap in 3d transition metal oxides caused by strong Coulomb repulsion, the Mott gap in Sr_2IrO_4 has a different nature. Under the O_h symmetry $5d$ states of Ir split into the t_{2g} triplet and e_g doublet. In the strong spin-orbit coupling limit the t_{2g} bang splits into a full quartet with isospin 3/2 and a hall-filled doublet with isospin 1/2 (figure 1). Even a relatively small Coulomb repulsion opens up the Mott gap near E_F and splits the $S = 1/2$-manifold into the upper (UHB) and lower Hubbard bands (LHB), shown in figure 2 [3].

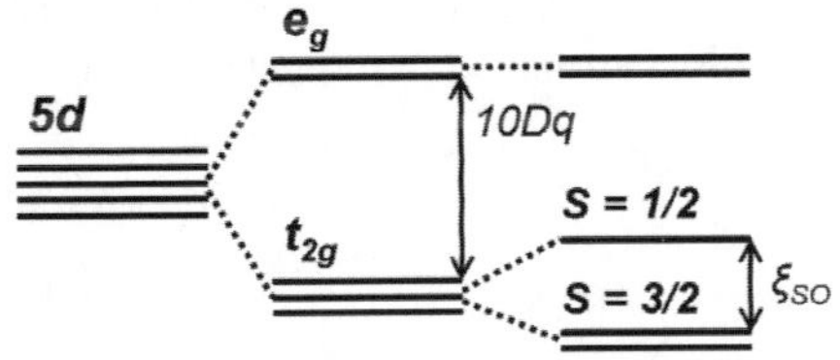

Figure 1. Splitting of $5d$ manifold by the crystal field and spin-orbit coupling [3].

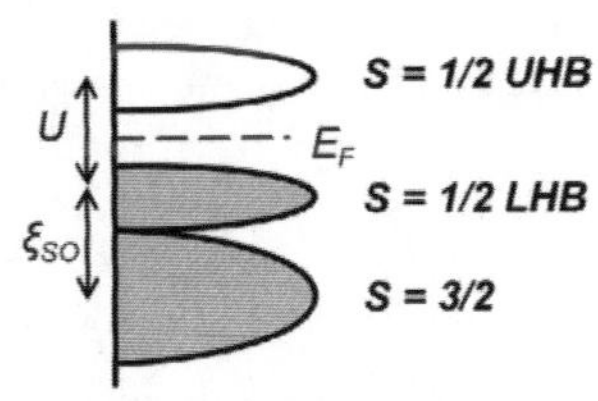

Figure 2. Schematic energy diagram for a $5d$ configuration with the strong spin-orbit coupling and Coulomb repulsion [3].

In experimental spectra of Sr_2IrO_4 one can observe modes of a charge-neutral quasi-particle. This quasi-particle is referred to as 'spin-orbit exciton'. It propagates through the lattice misaligning spins and changing the magnetic ordering [1, 2]. Such an excitation can be induced by X-rays [1] or Rh-doping [5]. Some researchers expect that the spin-orbit exciton propagation in Sr_2IrO_4 can lead to high temperature superconductivity.

In this work we use computer simulation to investigate the propagation of a single spin-orbit exciton through Sr_2IrO_4. The magnon dispersion in this compound is well described by an antiferromagnetic Heisenberg model with isospin one-half moments on a square lattice [2]. For modelling we use a Heisenberg model and a simplier Ising model.

We have developed a program to model a single spin-orbit exciton motion through antiferromagnetically ordered lattice to qualitatively evaluate changes in the system's characteristics. Upon the obtained results we argue that a single spin-orbit exciton propagation through Sr_2IrO_4can be modelled with the Ising model instead of the Heisenberg one.

A rather big value of the spin-orbit coupling constant (ξ_{SO}~0,4 – 0,5 eV [2, 3]) allows to consider ground states in Ir ions as effectively reduced to a half-filled singlet system with isospin 1/2 (figure 3a). For an excited Ir ion we can draw such a scheme with one electron moved from the quartet to the doublet across the spin-orbit split. As now the doublet is filled, we can reduce the system to a quartet with three electrons (figure 3b). Its effective isospin is 3/2 [2].

a) b)

Figure 3. Schematic electronic configurations of a) an Ir ion in a ground state ($S = 1/2$) and b) in an excited state ($S = 3/2$).

The Hamiltonian of this modelling contains longitudinal and transverse exchange coupling and exciton hopping. We take in consideration only nearest neighbours. Any long-range interactions give much smaller value of exchange coupling constants and can be omitted.

The Hamiltonian with exciton hopping and anisotropic exchange coupling can be written as

$$\hat{H} = J_{\parallel} \sum_{\langle ij \rangle} S_i^z S_j^z + J_{\perp} \sum_{\langle ij \rangle} S_i^+ S_j^- - W \sum_{\langle ij \rangle} X_i^\dagger X_j, \quad (1)$$

where $J_{\parallel}$, $J_{\perp}$ are exchange coupling constants, S_i^z is an isospin projection operator, S_i^+, $S_j^-$are raising and lowering operators for isospin projections, $X_i^\dagger$, X_j denote exciton creation and annihilation operators. The hopping parameter $W = 2t^2/U$ is defined by the intraorbital Coulomb repulsion U and hopping integral t. $J_{\perp} = 0$ corresponds to the Ising model, while for the Heisenberg model $J_{\perp} = J_{\parallel}$. The Ising and Heisenberg models are special cases of the anisotropic exchange model.

J, the longitudinal and transverse exchange coupling constant, is the reduced unit of measurement in this modelling. J can be estimated as 60 meV [2], about 650 K. It is omitted in further notation.

2. Method

In order to model such a system we use Stochastic Series Expansion (SSE) [6] – one of quantum Monte-Carlo methods. Its outline is presented here.

The thermal expectation of an operator $\hat{A}$

$$\langle \hat{A} \rangle = \frac{1}{Z} Sp\{Ae^{-\beta\hat{H}}\}, \quad (2)$$

where $Z = Sp\{e^{-\beta\hat{H}}\}$ is the partition function and β is the inverse temperature, is calculated as the expectation value in a combined space of operator sequences and spins. 'Subspace of spins' means that we flip spins in the system. Working in the subspace of operators means changing operator sequences, which are applied to the spin system.

The Hamiltonian could be represented as a sum of M non commutative terms

$$H = \sum_{i=1}^{M} \hat{H}_i, \quad [\hat{H}_i, \hat{H}_j] \neq 0 \tag{3}$$

In this approach we say that β is a small parameter, that is why this method works only for high temperatures. As H-terms do not commutate, the Taylor expansion can be written as a double sum with the product of operators in the sequence

$$e^{-\beta\hat{H}} = 1 - \beta\hat{H} + \frac{1}{2}\left(\beta\hat{H}\right)^2 - \frac{1}{6}\left(\beta\hat{H}\right)^3 + \dots = \sum_{n=0}^{L}\sum_{\{C_n\}}\frac{(-\beta)^n}{n!}\prod_{j=1}^{n}\hat{H}_{l_j}, \tag{4}$$

where $\{C_n\}$is the subspace of operators and $l_j = 1, 2, \dots, M$. Here the expansion is cut up to the Lth term

If we omit zero operators in the sequence, the formula for the expectation value of the operator $\hat{A}$ can be written as

$$\langle\hat{A}\rangle = \frac{1}{Z}\sum_{\alpha}\sum_{\{C_L\}}\frac{(-\beta)^{n(C_L)}[L-n(C_L)]!}{L!}\langle\alpha|\hat{A}\prod_{j=1}^{L}\hat{H}_{l_i}|\alpha\rangle, \tag{5}$$

where $n(C_L)$ denotes the number of nonzero operators in the operator sequence and $\{\alpha\}$ is the subspace of spin configurations.

The straightforward application of the SSE method to the considered system leads to generating a great number of non-physical states, so that many Monte-Carlo iterations are wasted. In order to avoid such configurations some modifications are required, which ensures that hopping operators of a certain sequence make the excitation move along a closed trajectory. According to selection rules we exclude generating those hopping operators which will surely turn the sequence weight into zero.

The algorithm we finally got calculates the spin structure factor

$$S(\boldsymbol{q}) = \frac{1}{N^2}\sum_{i,j} e^{iq(r_i - r_j)}\langle\hat{S}_i^z\hat{S}_j^z\rangle \tag{6}$$

It shows what arrangements of spins are like. $S(\boldsymbol{q})$ is a surface in the reciprocal space. For antiferromagnetically ordered isospins it yields a peak at $\boldsymbol{q} = (\pi, \pi)$, any deviation from the antiferromagnetic ordering makes this peak lower and broader. When spins are ordered ferromagnetically it has a maximum at $\boldsymbol{q} = (0,\ 0)$.

3. Results

First, we have calculated spin structure factors for various temperatures to see what the spin structure factor is like and how the shape of its surface in the reciprocal space changes. For this comparison we consider a system with a localized exciton ($W = 0$) in figure 4 and a system with a propagating exciton (we set $W = 1$, i.e. the hopping parameter is equal to the exchange coupling constant). Here the values of spin structure factor along lines of symmetry in the reciprocal space are depicted.

At first sight it becomes obvious that $S(\boldsymbol{q})$ surfaces and their changes are very similar for both Heisenberg and Ising models. For some values of the system's parameters spin structure factor values for both models coincide within margin of error.

Both models demonstrate a finite peak at $\boldsymbol{q} = (\pi, \pi)$, which corresponds to the antiferromagnetic ordering. It broadens with temperature or hopping parameter. The shape of Heisenberg and Ising models' peaks are almost identical. The peak gets down slightly when the hopping parameter W changes from 0 to 1, but it still corresponds to the antiferromagnetic ordering, i.e. this amount of excitons per site with orbital excitation compared to exchange coupling is not enough to disorder antiferromagnetically ordered spins.

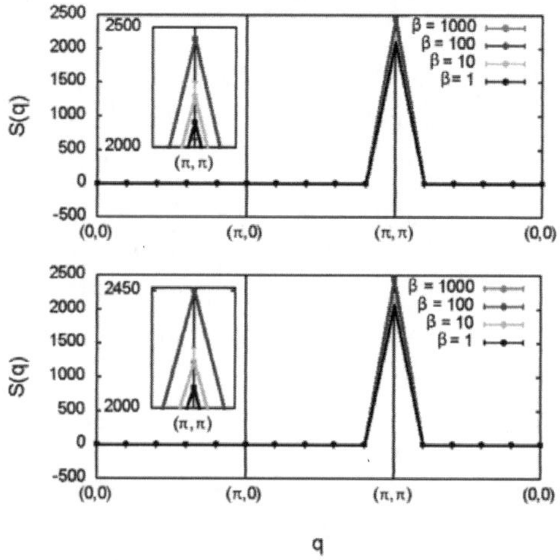

Figure 4. Spin structure factor $S(\boldsymbol{q})$ for the Heisenberg (upper) and Ising models (lower) at various temperatures for the 10x10 lattice at $W = 0$.

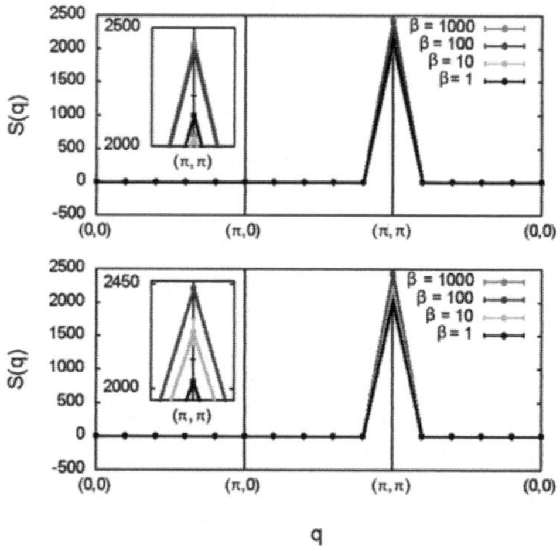

Figure 5. Spin structure factor $S(\boldsymbol{q})$ for the Heisenberg (upper) and Ising models (lower) at various temperatures for the 10x10 lattice at $W = 1$.

In these figures only results for the 10x10 lattice are presented. Other lattice sizes show similar behaviour in general.

Since we have some ideas of how the surface changes with temperature, we can keep an eye only on the point where the peak is expected to be – $\boldsymbol{q} = (\pi, \pi)$ – changes of the value at this point corresponds to the changes in the magnetic ordering. In figures 6 and 7 temperature dependencies for two lattice sizes are presented.

Temperature dependencies of spin factor peak for both models decline in a very similar way. The spin structure factor peak decreases with temperature in the systems for both lattice sizes. For the 4x4 lattice there's obviously a phase transition in the interval $10 < \beta < 40$. For the 10x10 lattice peak height undergoes a sharp decrease as well, but it does not look like the system has switched to the ferromagnetic or paramagnetic order with a peak height around 1000. For both lattice sizes bigger value of the hopping parameter corresponds to the lower temperature of the phase transition.

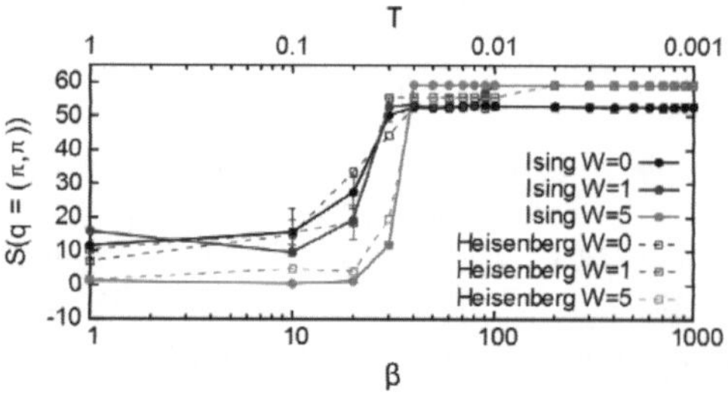

Figure 6. Temperature dependency of spin structure factor peak $S(\boldsymbol{q} = (\pi, \pi))$ for the 4x4 lattice.

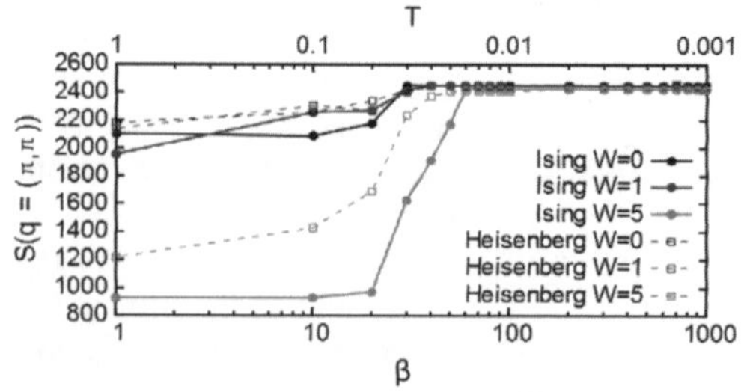

Figure 7. Temperature dependency of spin structure factor peak $S(\boldsymbol{q} = (\pi, \pi))$ for the 10x10 lattice.

Then we tracked how the system evolves when hopping is the varying parameter.

The peak of spin structure factor degrades with hopping for both temperatures in consideration (figures 8 and 9). At a low temperature the Ising and Heisenberg models give almost the same result, except for $W = 2$ (figure 8) – here for the Ising model we get the paramagnetic order while in the Heisenberg model the order is antiferromagnetic.

By the peak's shape and height at a higher temperature (figure 9) antiferromagnetic ordering can be expected at all values of the hopping parameter. In this comparison the Ising and Heisenberg models show themselves very similar – the spin structure factor heights go down with hopping in similar ways.

One can draw an analogy between the influence of spin-orbit exciton propagation (figures 8 and 9) and the influence of heat fluctuations (figures 4 and 5).

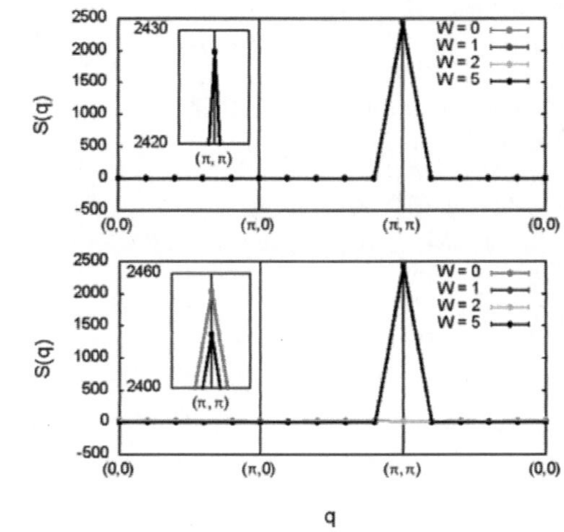

Figure 8. Spin structure factor $S(\boldsymbol{q})$ for the Heisenberg (upper) and Ising models (lower) at various hopping parameters W for the 10x10 lattice at $\beta = 1000$.

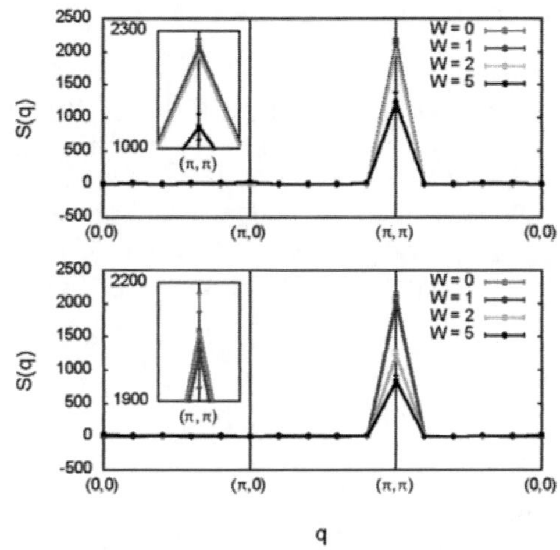

Figure 9. Spin structure factor $S(\boldsymbol{q})$ for the Heisenberg (upper) and Ising models (lower) at various hopping parameters W for the 10x10 lattice at $\beta = 100$.

In figures 10 and 11 peak heights decrease with hopping at high temperatures and remains almost constant at $\beta = 1000$. One can notice that Heisenberg and Ising models yield the same character of losing the antiferromagnetic order with hopping. But at $W = 0$ the Ising model gives peak heights close to 0, though earlier calculations had a high peak (figure 4). It shows that at these parameters ($W = 0$) the Ising model is unstable and can be driven to the paramagnetic ordering.

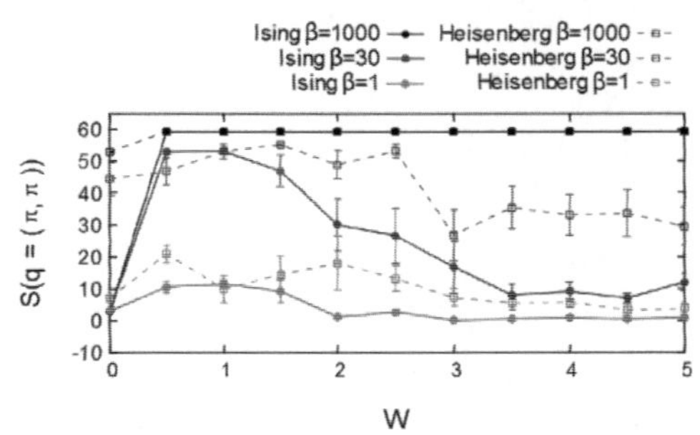

Figure 10. Hopping parameter dependency of spin structure factor peak $S(\boldsymbol{q} = (\pi, \pi))$ for the 4x4 lattice.

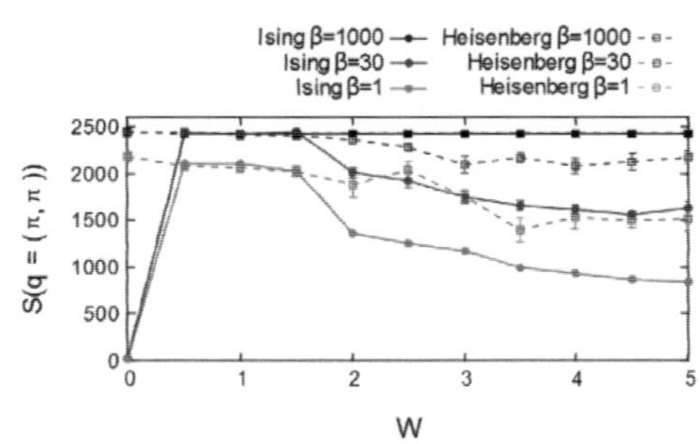

Figure 11. Hopping parameter dependency of spin structure factor peak $S(\boldsymbol{q} = (\pi, \pi))$ for the 10x10 lattice.

4. Conclusion

Both models demonstrate the exciton-induced system getting disordered with spin-orbit exciton hopping and heat fluctuations. Some of results for these two models agree within margin of error. But the Ising model proves to be unstable yielding extremely different results when calculations are repeated. Yet at other values of the system parameters the Ising model qualitatively is rather close to the Heisenberg model's results and experimental and theoretical data.

As it was expected, with exciton hopping an exchange frustration appears. Such frustrations can be observed in both Heisenberg and Ising models, the scales of these frustrations are very much alike. That makes the Heisenberg and Ising models similar for simulation of a spin-orbit exciton propagation.

Thus, for modelling a single spin-orbit exciton propagation in Sr_2IrO_4 the Heisenberg model can be substituted with the Ising model, though with some limitations.

References

[1] Kim J, Daghofer M, Said A H et al. 2014 *Nature Communications* **5** 4453
[2] Kim J, Casa D, Upton M H, Gog T et al. 2012 *Phys. Rev. Lett.* **108** 177003
[3] Kim B J, Jin H, Moon S J et al. 2008 *Phys. Rev. Lett.* **101** 076402
[4] Solovyev I V, Mazurenko V V, Katanin A A 2015 *Phys. Rev.* B **92** 235109
[5] Cao Y, Wang Q, Dessau D S et al. 2014 Hallmarks of the Mott-Metal Crossover in the Hole Doped $J = 1/2$ Mott insulator Sr_2IrO_4 *Preprint* arXiv:1406.4978
[6] Sandvik A W, Kurkijärvi J 1991 *Phys. Rev.* B. **43** 5950

Synthesis of carbon nanostructures on iron nanopowders

A Koshanova[1,2], G Partizan[1,2*], B Mansurov[2], B Medyanova[1,2], M Mansurova[2], B Aliev[1], Xin Jiang[3]

[1]Al-Farabi Kazakh National University, Almaty 050040, Kazakhstan
[2]The Institute of Combustion Problems, Almaty 050012, Kazakhstan
[3]Institute of Materials Engineering, University of Siegen, Siegen 57076, Germany

*E-mail: gulmira.partizan@gmail.com

Abstract. This work presents the results of experiments on synthesis of carbon nanostructures (CNs) by the method of thermal chemical vapor deposition using iron nanopowders obtained by the method of electrical explosion of wires as catalysts. To study the process of nucleation and growth of individual carbon nanostructures, experiments were conducted not only on nanopowders, but also on the separated clusters. To determine the optimum conditions of the carbon nanostructures synthesis and lower temperature limit, experiments were performed at different temperatures (300-700°C) and pressures (100-400 mbar). The experiments have shown that the lower temperature limit for carbon nanostructures synthesis on the iron nanopowders is 350°C and in this process the growth of carbon nanostructures is not so massive. Stable growth of carbon nanostructures for nanopowders began from 400°C during the entire range of pressures. The analysis of Raman spectroscopy showed that the most optimum conditions for obtaining nanotubes of high quality are $P = 100$ mbar and $T = 425$°C.

1. Introduction

Since their discovery carbon nanotubes became the target of numerous experimental and theoretical researches for production and analysis of their unique mechanical, chemical and electrical properties. Today, there are many techniques for synthesis of CNs such as laser ablation, arc discharge and various types of chemical vapor deposition. Thermal chemical vapor deposition (TCVD) unlike other methods of CNs synthesis does not require high energy and complicated equipment. TCVD based on pyrolysis of hydrocarbons such as acetylene, ethylene, methane or ethanol and transition metals Ni, Co, Fe, Cu and others are used as catalysts [1-5]. Prospects of using fine metal powders are related to the possibility of using them as catalysts for synthesis of CNs. One of the methods for production of superfine metal powders is electrical explosion of wires (EEW). EEW nanopowders (NPs) have a number of advantages compared to NPs obtained by other ways: they are resistant to oxidation and sintering at room temperature and show high chemical and diffusion activity during heating [6].

The aims of the study were determination of the lower temperature limit of the growth of carbon nanostructures on EEW Fe NP and search for optimal conditions of low temperature (energetically favorable) synthesis using the most accessible hydrocarbons without expensive additives of inert and other gases.

2. Experimental

NPs were purchased in Tomsk Polytechnic University to be used as catalysts for the synthesis of carbon nanostructures. The procedure of obtaining NPs, experimental details and the results of studies on their morphology and structure are described in detail in works [7-9].

The joint studies on the structure and morphology of EEW NPs, their catalytic activity and the possibility of synthesis of CNs on them by thermal CVD were carried out at The Department of surface and technology of new materials at the Institute of Materials Science of the University of Siegen (Germany) [9, 10].

Experiments were carried out not only on the NPs, but also on the separated clusters to study the process of nucleation and growth of individual CNs. NPs samples were suspended in hexane for separation. Further sonication of solution was carried out (volume of suspension was 30 ml, frequency of ultrasound was 27 kHz, power of generator was 120 W, exposure was conducted for 30 minutes), after which the droplets of suspension with iron particles were deposited on a silicon substrate.

The scheme of technological equipment for synthesis of carbon materials by method of thermal CVD and the procedure of conduction of experiments were described in detail in work [9, 10].

The field emission scanning electron microscope with ultra-high resolution of model Gemini Ultra 55 of the company Zeiss, with a device for X-ray microanalysis of the company «Thermo Scientific» was used to study the morphology of the samples. Investigations were carried out at the Institute of Materials Science of the University of Siegen (Germany).

The samples were investigated by Raman spectroscopy using spectrometer NT-MDT NTegra Spectra (laser wavelength λ = 473 nm) at The National Nanotechnology Laboratory of open type (Almaty, Kazakhstan).

3. Results and discussion

SEM studies

The experiments were performed at different temperatures (200-700°C) and pressures (100-400 mbar) to determine the optimal conditions for CNs synthesis and lower temperature limit. Studies have shown that the lower temperature limit for Fe NPs is 400°C. Only carbonization of clusters is observed for the separated powders at temperatures below 400°C throughout the investigated range of pressure. Thus the results of experiments made it possible to allocate the optimum range for the low-temperature synthesis: temperature is 400-450°C, pressure is mbar 100-300 [11-13].

Further, more detailed studies were carried out in these experimental ranges. Figures 1-3 show SEM images of CNs on EEW Fe NPs obtained at temperatures of 400-450°C and pressures of 100-300 mbar.

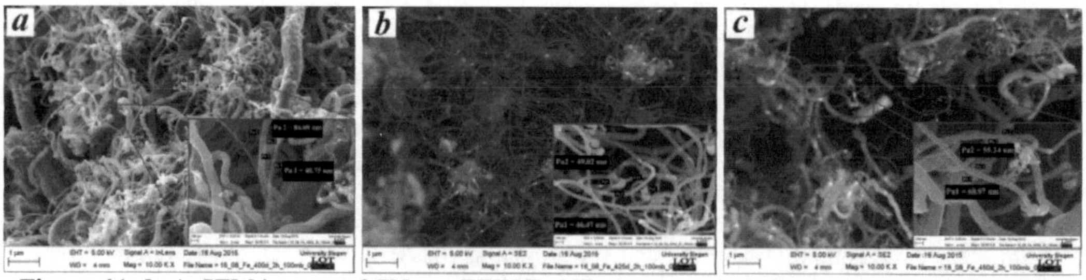

Figure 1(a,b,c). SEM images of CNs on EEW Fe NPs (pressure is 100 mbar): **(a)** – 400°C, **(b)** – 425°C, **(c)** – 450°C.

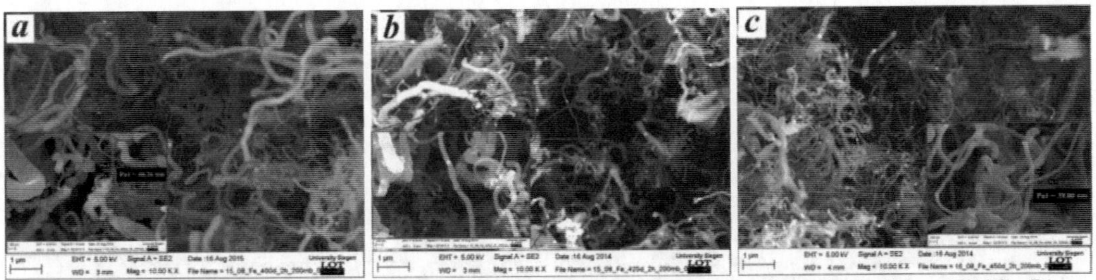

Figure 2(a,b,c). SEM images of CNs on EEW Fe NPs (pressure is 200 mbar): **(a)** – 400°C, **(b)** – 425°C, **(c)** – 450°C.

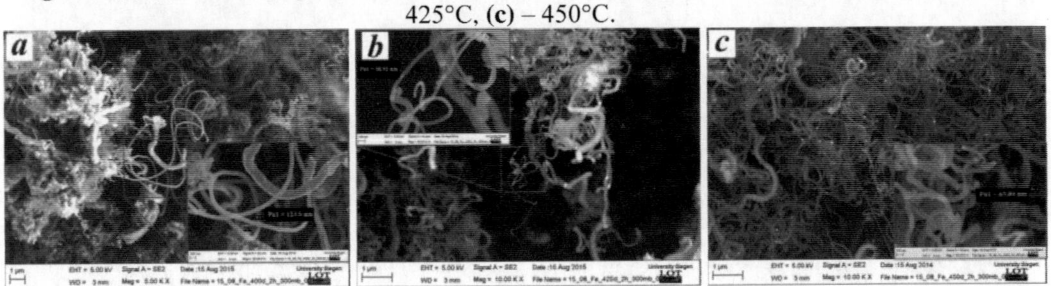

Figure 3(a,b,c). SEM images of CNs on EEW Fe NPs (pressure is 300 mbar): **(a)** – 400°C, **(b)** – 425°C, **(c)** – 450°C.

As seen from SEM images the massive growth of CNs starts for NPs at 400°C during the entire investigated range of pressures. The steady growth of CNs with a fairly large dispersion of diameter from 40 to 100 nm and a different morphology (from spiral to direct) is observed in certain experiments in the temperature range of 400-450°C. Iron clusters are found both on tips and inside of CNs.

Investigation of CNs on the EEW Fe NPs by Raman scattering

Figure 4 presents the Raman spectra of CNs on Fe NPs obtained under optimum temperature and pressure.

All Raman spectra of CNs on Fe NPs grown at 100 mbar predominantly show two groups of first order *D* and *G*. These groups are located in the region of 1330 cm^{-1} and 1576 cm^{-1} for the samples obtained at 400°C, their width at half maximum are 66 cm^{-1} and 77 cm^{-1}, respectively. But intensity of *D* peak is higher compared to the *G* peak, which is a sign of high degree of crystalline disorder of the sample.

Group *G* appears due to interplanar fluctuations of C–C bonds. Also, there is a second order group *2D* at 2648 cm^{-1}, which is an overtone of group *D* [14]. This group is an indicator of the presence of long-range order in the sample and appears due to two-phonon secondary scattering, leading to the formation of inelastic phonon [15]. It may be associated with a level of crystallinity of the carbon nanotubes [16].

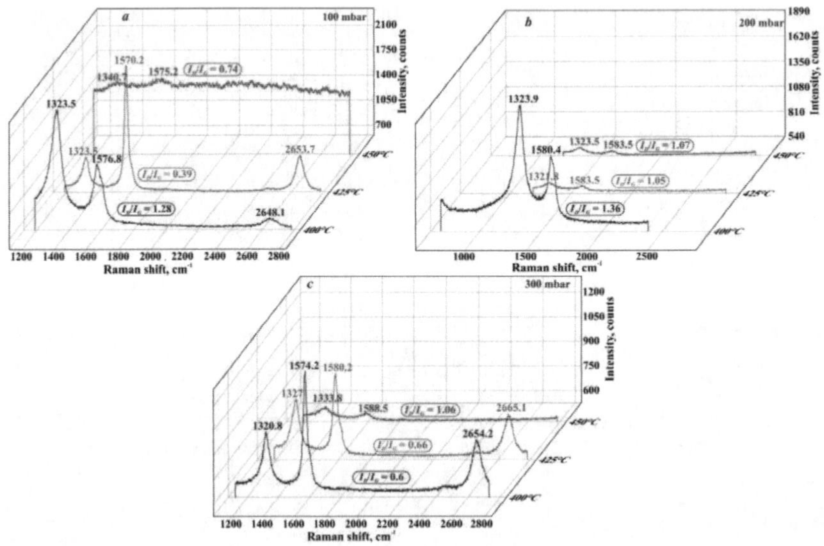

Figure 4(a,b,c). RS spectra of CNs on Fe NPs: **(a)** – 100 mbar; **(b)** – 200 mbar; **(c)** – 300 mbar.

Nanotubes grown at a temperature of 425°C show the good spectra. *D* and *G* peaks are observed at 1323 cm^{-1} and 1570 cm^{-1}, respectively. One can speak about high crystallinity of the sample judging from the width and intensity of the peaks and the presence of second order in the area of 2653 cm^{-1}. In addition, it can be seen that the intensity of *D* peak is quite small, which is also indicative of good quality of nanotubes. Full width at half-maximum (FWHM) of samples are 77 and 34.6 cm^{-1}, respectively. The sample synthesized at 450°C shows a rather blurred spectrum, indicating a defect structure. The shift of group *D* is also observed in the range of 1318 cm^{-1}, and group *G* is located in the range of 1587 cm^{-1}. FWHM of group *D* is much greater than that of the first two samples, indicating the presence of large number of defects and low crystallinity [17].

It can be said that multiwall nanotubes of good quality can be obtained at a pressure of 100 mbar and temperature of 425°C based on the RS results for this series of samples.

The spectra of CNS obtained at 200 mbar show that the intensity of *D* peak is higher than that of *G* peak. This suggests that the nanotubes have a defective structure. Group *G* in the first two samples (400°C and 425°C) is in the range of 1578.5 and 1577 cm^{-1}, but shifted to the high frequency region 1592 cm^{-1} in the sample synthesized at 450°C, indicating a significant graphitisation of sample [15]. Also, this may be indicated by the absence of 2*D* peak, while for the first two samples, it is observed within 2650 and 2662 cm^{-1}, respectively.

The samples obtained at a pressure of 200 mbar and temperatures of 400°C and 425°C show high crystallinity judging from the intensity of 2*D* peak. It can be assumed that they are quite ordered according to the intensity of *D* peak. It is located at 1329 and 1327 cm^{-1}, respectively. Group *G* is observed at 1575 and 1580 cm^{-1}. The third sample shows not smooth spectrum as previous. *D* peak intensity is higher than that of *G* peak, the latter is shifted to 1590 cm^{-1}. This sample otstutstvuet range order as 2*D* peak is not observed.

The lowest value of the ratio of the intensities of *D* and *G* peaks corresponds to nanotubes obtained under the conditions of 100 mbar and 425°C. Also good results were obtained for nanotube grown at 300 mbar and 425°C, 450°C. The samples obtained under the conditions of 100 mbar and 400°C, 200 mbar and 400°C show the lowest quality. One can conclude that the most optimal conditions for obtaining nanotubes are 100 mbar and 425°C after analyzing the Raman spectra. Judging by the width and intensity of the peaks in the Raman spectra, as well as the presence of a second order peak one can talk about the high crystallinity of these samples.

4. Conclusion

The conducted experiments demonstrated the possibility of using Fe NPs, obtained by the EEW as catalysts for growth of CNs. Stable growth of CNs is carried out at temperatures significantly lower than commonly used by thermal CVD on iron catalysts. Studies have shown that the lower temperature limit is 400°C. The massive growth of CNs is observed in the entire investigated range of pressures at the same temperature. SEM studies clearly indicate the existence of a temperature range between the low temperature and high temperature (standard) mode of synthesis in which the growth of CNs occurs.

Raman spectroscopy showed that nanostructures grown on NPs at 100 mbar and 425°C have the highest crystallinity according to intensities of the groups *D* and *G*. The findings of the research results have a high potential for the development of efficient, low-energy, low-cost technology for obtaining CNTs and CNFs, without the use of expensive gas and the ability to control the structure and properties of macroscopic parameters of CNs.

Acknowledgements

This work was performed with partial financial support of grants of Committee of Science of the Ministry of Education and Science No. 3823/ГФ4.

References

[1] Hafner J H, Bronikowski M J, Azamian B R, Nikolaev P, Rinzler A G, Colbert D T, Smith K A, Smalley R E 1998 *Chem. Phys. Lett.* **296** 195–202
[2] Jourdain M, Bichara Ch 2013 *Carbon* **58** 2–39
[3] Wang H, Yuan Y, Wei L, Goh K, Yu D, Chen Y 2015 *Carbon* **81** 1–19
[4] Yan Y, Miao J, Yang Zh, Xiao F, Yang HB, Liu B, Yang Y 2015 *Chem. Soc. Rev.* **44** 3295-3346
[5] Partizan G, Mansurov B Z, Aliyev B, Medyanova B, Tolegen A, Kenzhegulov A, Jiang X, Mansurov Z A 2015 *Proc. World Conf. on Carbon (Dresden)* p 231
[6] Lerner M I, Svarovskaya N V, Psakhie S G, Bakina O V 2009 *Russian Nanotechnologies* **4 (11-12)** 56-68 (in Russian)
[7] Lerner M I 2007 *Doc. Diss. (Tomsk)* 325 p (in Russian)
[8] lyin A P 2003 *Bulletin of the Tomsk Polytechnic University* **306** 133-139 (in Russian)
[9] Partizan G, Mansurov B Z, Medyanova B S, Koshanova A B, Aliyev B A 2016 *Combustion and plasmochemistry* **14** in print (in russian)
[10] Partizan G, Mansurov B Z, Medyanova B S, Aliev B A, Jiang Xin 2015 *Journal of Engineering Physics and Thermophysics* **88** 1151-1158
[11] Partizan G, Yao Ma, Mansurov B Z, Medyanova B S, Xin Jiang, 2014 *Proc. World Conf. on Carbon (Jeju island)* POT2-01
[12] Partizan G, Mansurov B Z, Aliev B A, Xin Jiang 2014 Int. Symp. "Physics and chemistry of carbon materials / nanoengineering (Almaty) p 303-307
[13] Partizan G, Mansurov B Z, Aliyev B A, Medyanova B S, Xin Jiang, Mansurov Z A 2015 *Proc. Annual Int. World Conf. on Carbon* (Dresden) p 232
[14] Xia J H, Jiang X, Jia C L 2009 *Applied Physics Letters* **95** 223110-1-223110-3
[15] Merchan-Merchan W, Saveliev A V, Kennedy L, Jimenez W C 2010 *Progress in Energy and Combustion Science* **36** 696-727
[16] Buranova Yu S 2011 *Proceedings of MIPT* **3** 30-41
[17] Dinadayalane T C, Leszczynski J 2012 Fundamental Structural, Electronic, and Chemical Properties of Carbon Nanostructures: Graphene, Fullerenes, Carbon Nanotubes, and Their Derivatives *Handbook of Computational Chemistry* (vol 2) (London, Springer) pp 793–867

A novel three-jet microreactor for localized metal-organic chemical vapour deposition of gallium arsenide: design and simulation

S A Konakov[1], **V V Krzhizhanovskaya**[1,2,3]
[1] Department of Physics, Chemistry and Microsystems Technology, Peter the Great St. Petersburg Polytechnic University, 195251, St. Petersburg, Russia
[2] University of Amsterdam, 1098 XH Amsterdam, The Netherlands
[3] ITMO University, St. Petersburg, Russia

Abstract. We present a novel three-jet microreactor design for localized deposition of gallium arsenide (GaAs) by low-pressure Metal-Organic Chemical Vapour Deposition (MOCVD) for semiconductor devices, microelectronics and solar cells. Our approach is advantageous compared to the standard lithography and etching technology, since it preserves the nanostructure of the deposited material, it is less time-consuming and less expensive. We designed two versions of reactor geometry with a 10-micron central microchannel for precursor supply and with two side jets of a dilutant to control the deposition area. To aid future experiments, we performed computational modeling of a simplified-geometry (two-dimensional axisymmetric) microreactor, based on Navier-Stokes equations for a laminar flow of chemically reacting gas mixture of $Ga(CH_3)_3$-AsH_3-H_2. Simulation results show that we can achieve a high-rate deposition (over 0.3 μm/min) on a small area (less than 30 μm diameter). This technology can be used in material production for microelectronics, optoelectronics, photovoltaics, solar cells, etc.

1. Introduction

Gallium arsenide (GaAs), with its high electron mobility and direct bandgap, is traditionally used in RF electronics, photonic devices and solar cells [1]. GaAs monocrystals, layers, thin films and nanowires are produced by the liquid-phase, gas-phase, molecular beam epitaxy, or Metal-Organic Chemical Vapour Deposition (MOCVD) [2,3,4]. Electronic circuit patterns are usually shaped by lithography and etching, which often harm material nanostructure. Moreover, this technology is rather expensive and time-consuming. We are developing a novel approach to material patterning by a microreactor-printing technology for localized deposition of thin films, nanoparticles and nanowires. In our earlier work [5], we described the method and its application to a silicon dioxide CVD single-flow microreactor. In this work we extend the technology and investigate GaAs MOCVD in a three-jet microreactor.

To date there are some works dedicated to GaAs MOCVD simulation. Y. C. Chuang in [6] considers the mathematical model and optimal design of a horizontal MOCVD reactor for GaAs film growth. It 3D model showed a good agreement with experimental data and presented an ability in predicting the GaAs film growth rate and uniformity. Interesting paper by H.Song [7] describes a results of two-dimensional numerical simulation on growth rate nonuniformity of selective area GaAs metallorganic vapor phase epitaxy. Described results and applied method reveal surface kinetics and surface reaction mechanism even in the mass-transport limited growth condition. Disadvantages of the selective area growth method is necessity of a mask applying. Our work suggests a solution of the problem. In this paper we firstly describe a microreactor design and then discuss mathematical and computer models which are used in our investigation. Simulations results, discussion and conclusions are presented in the second half of the paper.

2. Microreactor design

To design a microreactor we first choose process parameters. Based on the typical GaAs MOCVD deposition from mixture of $Ga(CH_3)_3$ (a.k.a. TMGa) and AsH_3 diluted by H_2, we set the following process conditions: pressure of 10000 Pa, temperature of 1000 K [8]. The microreactor and substrate wafer are placed in a vacuum chamber with fixed pressure and background gas chemical composition (as shown in Figure 1(a)). The substrate is heated to 1000 K by a heater. Microreactor is situated over the substrate with 10 μm gap. It can be moved by the microreactor X-Y Mobility System in lateral direction to deposit GaAs patterns along the wafer. The vacuum system, heater and X-Y Mobility System are based on standard technical solutions. In the paper we focus only on the microreactor design.

We designed two versions of microreactor geometry:

1. The first simple design (Case 1) is presented in Figure 2(a). It has just one central channel. Inlet gas mixture is injected into the central channel (10 μm diameter) under pressure P_{in}. It then spreads in the gap, reacting with the substrate. Finally, all gases are pumped out from the gap to the vacuum chamber where pressure is P_{ch}.
2. The second microreactor design (Case 2) with three jets is presented in Figure 2(b). It has an additional ring-shaped pumping channel around the central channel, so that the penetration of the inlet mixture is localized in a narrow area of controllable diameter. This microreactor design allows to set three pressure values: P_{in} is the inlet pressure in the central channel, P_{ch} is pressure in the vacuum chamber and P_{out} is outlet pressure in the ring-shaped pumping channel, see Figure 2(b). P_{in} is 10800 Pa, higher than P_{out} and P_{ch}. If the pressure P_{out} is equal or higher than P_{ch}, the gas jets have the same distribution as in the first microreactor design (Case 1). When pressure P_{out} is less than P_{ch}, gas flow in the gap changes direction.

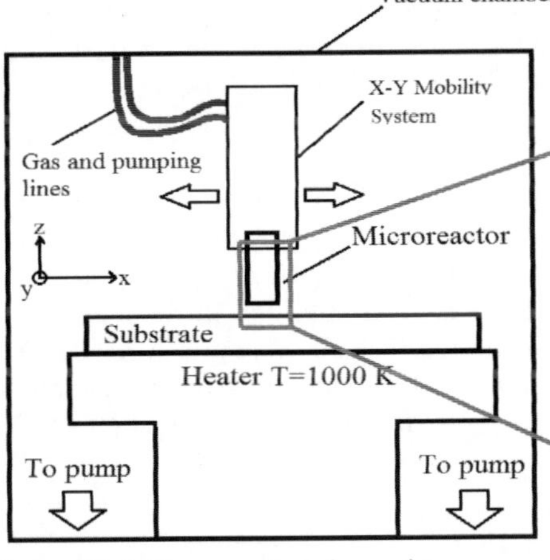

Figure 1(a). Process chamber scheme.

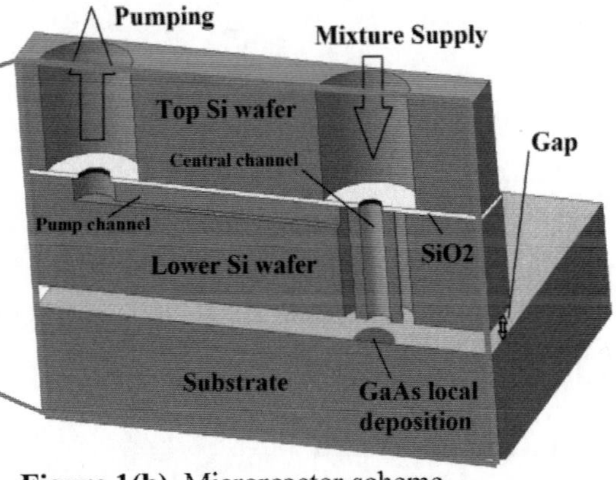

Figure 1(b). Microreactor scheme.

Both microreactor designs can be constructed from silicon chips by microsystem technology [9] . The chip contains two silicon wafers: Top and Lower, see Figure 1(b). Inside the Lower wafer (of 50 μm width) the two channels (Central and pumping) are made by plasma deep silicon etching technology (for instance Bosch-process). Top silicon wafer contains auxiliary microchannels providing interconnections between the channels in the chip and the outer gas supply and pumping lines. The two wafers are direct-bonded to make a microreactor. Silicon wafers were chosen for microreactor construction material because of its selectivity to GaAs. It allows to use wet chemical cleaning processes during a long-term microreactor operation. Some advanced subsystems (for example cooling) and sensors can be included in the microreactor design, but in this work we did not consider it yet.

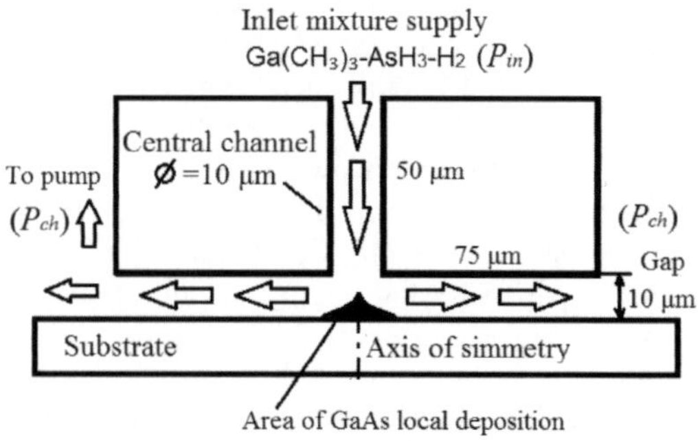

Figure 2(a). The first microreactor design (Case 1).

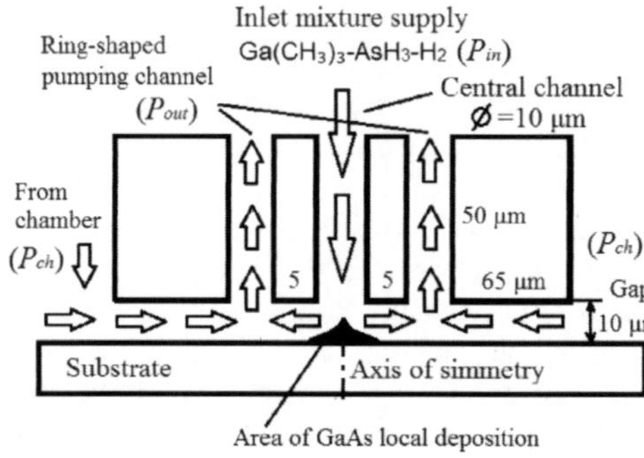

Figure 2(b). The second microreactor design (Case 2).

3. Computer simulation and model parameters

To investigate the properties of microreactors for GaAs MOCVD process we used a computer simulation approach. We modeled a simplified two-dimensional axisymmetric geometry of reactors. The computational model is based on Navier-Stokes equations for a laminar flow (Re=0.02) of chemically reacting gas mixture of $Ga(CH_3)_3$-AsH_3-H_2. A continuum modelling approach is justified at Knudsen numbers $Kn \sim 5\cdot10^{-2}$ calculated for our conditions. Equations of the energy conservation and mass balance for each chemical species were include in the model. Ideal gas low was applied as an equation of state for gas media in the microreactor volume. Model of chemical kinetic was taken from [8]. GaAs deposition reaction is modeled on the substrate and all reactor walls.

For the first microreactor design (Case 1) we simulated MOCVD process and investigated a distribution of GaAs deposition rate along the substrate. TMGa mass concentration in the inlet mixture was 0.00001, AsH_3 was 0.06, and H_2 was 0.93999. Inlet mixture supplying was simulated under constant pressure boundary conditions (P_{in} = 10800 Pa), when the pressure in the vacuum chamber P_{ch} was set at 10100 Pa.

For the second microreactor design (Case 2) we carried out two series of numerical experiments to investigate a GaAs MOCVD process in details. At the first series we change the outlet pressure in the ring-shaped pumping channel P_{out} from 7000 Pa to 10100 Pa. The inlet pressure P_{in} was 10800 Pa and pressure in the vacuum chamber P_{ch} was 10100 Pa. TMGa mass concentration in the inlet mixture was 0.02, AsH_3 was 0.06, and H_2 was 0.96. At the second series we changed a TMGa mass concentration from 0.001 to 0.04. AsH_3 concentration was constant and hydrogen was reduced respectively. P_{in} and P_{ch} were set as in the first series and P_{out} was 7000 Pa. Distribution of GaAs deposition rate along the substrate was investigated in all cases.

4. Results and discussions

4.1 The first microreactor design (Case 1).

Simulation results of the first microreactor design showed an insufficient localization of deposition rate of GaAs, see Figure 3(a). Full Width Half Maximum of deposition rate is 84 μm, eight times larger than the diameter of the central channel. In Figure 3(b) the distribution of TMGa (source gas in inlet mixture) is shown. The TMGa concentration is almost uniformly distributed along the substrate in a large area. It causes the distribution of GaAs deposition rate, shown in Figure 3(a). Therefore

Case 1 microreactor design does not allow to obtain localized deposition under the standard conditions considered.

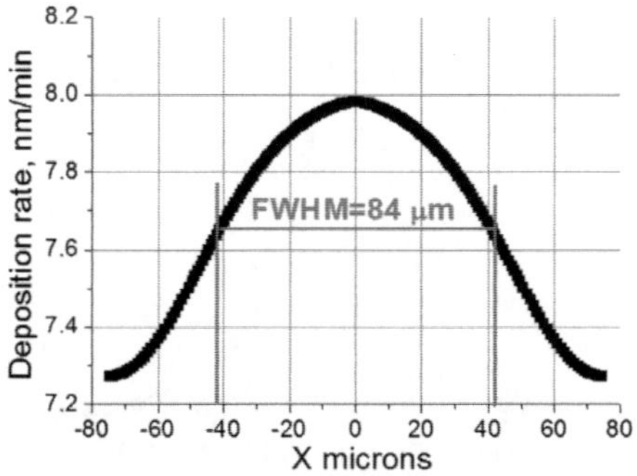

Figure 3(a). Disribution of GaAs deposition rate. The first microreactor design (Case 1).

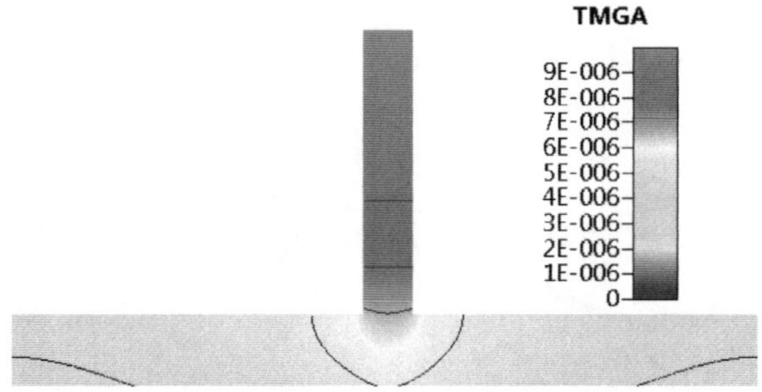

Figure 3(b). Disribution of TMGa. The first microreactor design (Case 1).

4.2 The second three-jet microreactor design (Case 2).

The second microreactor design has an additional ring-shaped pumping channel, which has a strong influence on the flow distribution. In Figure 4(a) stream lines and velocity magnitude distribution in the microreactor are presented.

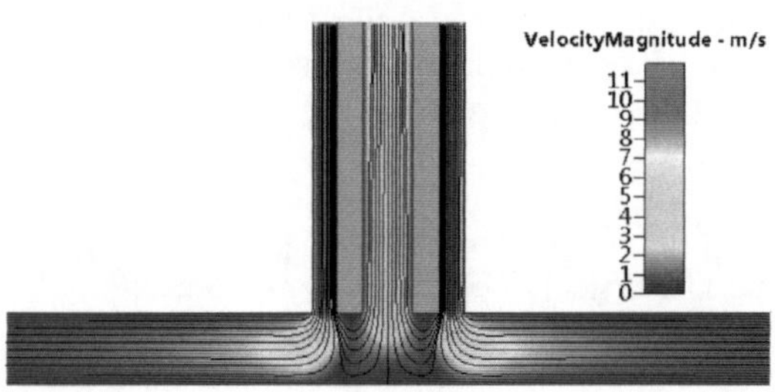

Figure 4(a). Stream lines and velocity magnitude distribution in the microreactor. (Case 2).

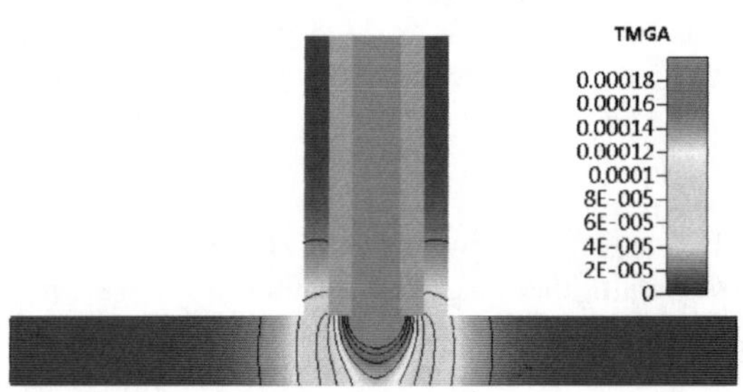

Figure 4(b). Disribution of TMGa. The second microreactor design (Case 2).

The distribution of the TMGa is presented in Figure 4(b). Comparing it to Case 1 distribution (Figure 3(b)), we can note that distribution TMGa is localized in a small volume between the central channel and the ring-shaped pumping channel. The localization of precursor concentration leads to localization of the GaAs deposition rate.

To study the influence of different outlet pressures P_{out} and mass concentration of $Ga(CH_3)_3$ in the inlet mixture, we conducted two series of numerical experiments described in Section 3. In Figure 5 the distribution of normalized (to its maximum) deposition rate for different outlet pressure is presented. In all cases the deposition rate has a maximum in the center. At high outlet pressures P_{out}=10000 Pa and 10100 Pa (close or equal to P_{ch}=10100 Pa), the deposition rate at the edges (x=+/- 75 μm) is very high, meaning poor localization. The gas flow distribution in these two cases is similar to the first microreactor design (Case 1). When the outlet pressure in the ring-shaped microchannel P_{out} is less than 9000 Pa, the distribution of the deposition rate is highly localized. In these cases Full Width Half Maximum is 27 μm. Remarkably, reducing the outlet pressure from 9000 Pa to 7000 Pa does not change the normalized deposition rate distribution. The absolute value of deposition rate in the center slightly changes from 0.1667 μm/ min at 9000 Pa to 0.1655 μm/ min at 7000 Pa.

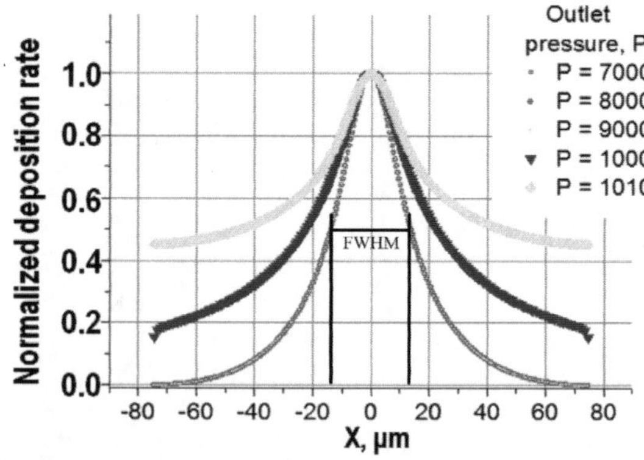

Figure 5. Distribution of normalized deposition rate for different outlet pressure values (P_{out}). P_{in} =10800 Pa, P_{ch}=10100 Pa, mass concentration of $Ga(CH_3)_3$ is 0.02.

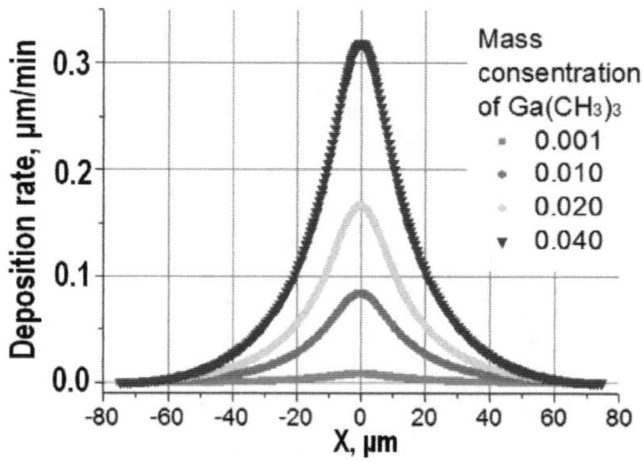

Figure 6. Distribution of deposition rate for different mass concentration of $Ga(CH_3)_3$. P_{in}=10800 Pa, P_{ch}=10100 Pa, P_{out}=7000 Pa.

In Figure 5 distribution of deposition rate for different mass concentration of TMGa is presented. Increasing TMGa concentration leads to the growth of deposition rate. FWHM of the peaks slightly increase from 26.7 μm at 0.001 TMGa concentration to 27.9 μm at 0.04 TMGa concentration. Maximum value of GaAs deposition rate is 0.3188 μm/min for the highest TMGa concentration (0.04). Such deposition rate can provide an efficient "printing" of topological patterns. It is possible to consider a practical application of the microreactor MOCVD process to produce a novel GaAs devices.

5. Discussion and recommendations

To explain the presented results we consider some theoretical topics about microreactor CVD technology. In general the deposition rate is a function of temperature, pressure and precursor concentration. Therefore, to obtain a localized deposition rate we should localize one of the factors limiting the reaction rate. In our process conditions, pressure and temperature are constant, hence we can change only a precursor concentration in the microreactor.

For reactor which is working in ideal displacement regime the precursor concentration reducing along stream lines due to its consumption in chemical reactions, what leads to concentration gradient in the reactor volume. We apply the phenomena in the first microreactor design (Case 1) and observed a reducing TMGa concentration along stream lines in the reactor. The ration between the surface reaction rate and convective mass transport in gas flow and diffusion in the Case is low. TMGa spreads in the gap along large area and provide an almost uniformly deposition.

To provide a high gradient of precursor concentration and localized deposition the ratio between the surface reaction rate and mass transport should be high. To increase the ratio it is possible to rise a temperature (and surface reaction rate) or decrease an initial concentration of precursor in the inlet mixture. It is an universal solution for the first type of CVD microreactor (Case 1). For considered process parameters (fixed temperature) and chemical system, which determines a surface reaction rate, GaAs MOCVD in the first design of microreactor do not provide a localized deposition. To solve the problem we change a scheme of gas flows and design the second version of the microreactor (Case 2). This approach shows perfect results and the localized deposition with 27 μm FWHM was obtained.

Further analyzes of the process and deposited profiles leads to the important suggestion. To narrow down the area where GaAs have been deposited, we suggest to perform an additional etching step after deposition in microreactor. The main idea is uniform etching of GaAs thin film on the whole wafer. In Figure 7 the film profiles of deposited patterns before and after such etching process are shown. Etching process also reduces a FWHM of the peaks. This approach has the most efficiency in case of high selectivity of the etchant to substrate material.

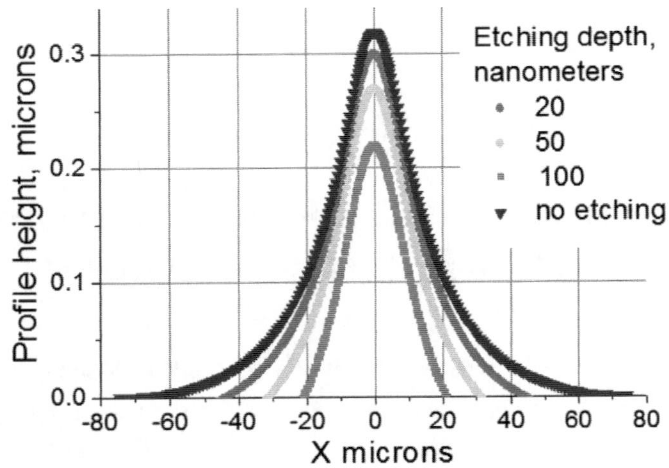

Figure 7. Profiles of deposited film profiles before and after etching process. P_{in}=10800 Pa, P_{ch}=10100 Pa, P_{out}=7000 Pa, mass concentration of TMGa is 0.02. Time of deposition is 1 min.

6. Conclusions and future work

In this work we present a novel three-jet microreactor design for localized deposition of GaAs by low-pressure MOCVD. Two microreactor designs were simulated. Results of simulations shows a possibility to localized deposition of GaAs with 27 μm of FWHM by using the three-jet microreactor design.

We obtain that increasing a pressure (P_{in}) in the ring-shaped pumping channel up to pressure in the chamber (P_{ch}) leads to poor localization of GaAs deposition rate, similar to the first microreactor design. Increasing a TMGa mass concentration in the inlet mixture from 0.001 to 0.04 leads to growth of maximum deposition rate up to 0.3188 μm/min and do not change a deposition distribution. The differences between simulation results in Case 1 and Case 2 were discussed and explained. We suggested addition treatment step to improve a profile after deposition. The next step of this research is the hardware construction of these microreactors and experimental investigations.

References

[1] Einspruch N G, Wisseman W R 2014 *GaAs Microelectronics: VLSI Electronics Microstructure Science Vol. 11* (Academic Press).
[2] Wu D, Tang X, Yoon H S, Wang K, Olivier A, Li X 2015 *Nanoscale research letters* **10**(1) 1-6.
[3] Wu D, Tang X H, Olivier A, Li X Q 2015 *Materials Research Express* **2**(4) 045002.
[4] Jiang N, Gao Q, Parkinson P, Wong-Leung J, Tan H H, & Jagadish C 2013 *Photonics Conference (IPC)*, IEEE pp. 476-477
[5] Konakov S A, Krzhizhanovskaya V V 2015 *JPCS* **574**(1) 012145
[6] Chuang Y C, Chen C T 2014 *Journal of the Taiwan Institute of Chemical Engineers*, **45**(1), 254-267.
[7] Song H, Sugiyama M, Nakano Y, Shimogaki Y 2007 *Journal of the Electrochemical Society*, **154**(2), H91-H96.
[8] Mazumder S, Lowry S A 2001 *J. Cryst. Growth* **224**(1) 165-174.
[9] Wirth T. 2013 *Microreactors in organic chemistry and catalysis* (John Wiley & Sons).

Elongated Quantum Dots of Ge on Si Growth Kinetics Modeling with Respect to the Additional Energy of Edges

K A Lozovoy, A A Pishchagin, A P Kokhanenko, A V Voitsekhovskii

Department of Radiophysics, Tomsk State University, 36 Lenin Av., Tomsk, 634050, Russia

Abstract. In this paper refining of mathematical model for calculation of parameters of self-organised quantum dots (QDs) of Ge on Si grown by the method of molecular beam epitaxy (MBE) is done. Calculations of pyramidal and wedge-like clusters formation energy were conducted with respect to contributions of surface energy, additional edge energy, elastic strain relaxation, and decrease in the atoms attraction to substrate. With the help of well-known model based on the generalization of classical nucleation theory it was shown that elongated islands emerge later than pyramidal clusters. Calculations of QDs surface density and size distribution function for wedge-like clusters with different length to width ratio were performed. The absence of special geometry of islands for which surface density and average size of islands reach points of extremum that was predicted earlier by the model not taking into account energy of edges was revealed when considering the additional contribution of edge formation energy.

1. Introduction

Semiconductor materials containing nanosized Ge clusters are popular for application in optoelectronics since the 1990s. They are used in development of photodetectors, solar cells and light emitting devices that are quite competitive with traditional optoelectronic materials such as AIII-BV compounds.

The one of the most promising Si/Ge heterostructures production methods is molecular-beam epitaxy. The morphology of Ge islands during the growth process can be controlled by changing the temperature of substrate, Ge deposition rate, and the total thickness of Ge layer.

It is known that ensemble of nanoislands of Ge on Si(100) surface consists of pyramidal (hut-) and elongated (wedge-) clusters, which are pyramids with square and rectangle base, respectively. But the differences in the kinetics of formation of these two types of clusters in quantum dots arrays are still studied poorly [1, 2].

In [3, 4] an attempt was made to study kinetics of wedge-like quantum dots formation with the help of the model, based on the classical nucleation theory and proposed in [5]. This model relies on the expression for change of free energy of atoms during transition from the wetting layer to the island taking into account change in free energy due to formation of additional facet surface, elastic strain relaxation, and lessening of atoms attraction to substrate [5]. Meanwhile, the contribution in change of the free energy due to formation of additional edges in the island is not taking into account as it is thought to be negligibly small. But at the same time results obtained in [6, 7] show that taking into account the contribution of edge energy is essential for explanation of some recently discovered effects during growth of quantum dots of germanium on silicon. In [8] an attempt is made to develop kinetic model accounting for the contribution into free energy of island formation due to edge energy.

The aim of this paper is to compare growth kinetics of pyramidal and wedge-like clusters with various length to width ratios with respect to the contribution of additional edge energy into change in free energy during formation of islands.

2. Theoretical model, results and discussion

For calculations we used the kinetic model of Ge QDs on Si growth developed in [8]. This model allows us to define temperature dependencies of QDs surface density N and size distribution function for different growth rates V.

The free energy of pyramidal islands with a rectangular base and constant contact angle φ is a function of the number of atoms i in the island and is represented by

$$\Delta F(i) = Ai^{2/3} - B\zeta i + Ci^{1/3}, \quad (1)$$

where $\zeta = h/h_{eq} - 1$ is the wetting layer superstress, h is the Ge wetting layer thickness, h_{eq} is its equilibrium thickness. In this equation, free energy is expressed in $k_B T$ units, where T is the substrate temperature and k_B is the Boltzmann constant.

With respect to different geometrical shapes of QDs parameters A, B and C will be defined as

$$A = \frac{r[\gamma(\varphi)/\cos\varphi - \gamma(0)]\alpha l_0^2}{k_B T}, \quad (2)$$

$$B = \frac{[1 - Z(\varphi)]\lambda\varepsilon_0^2 l_0^2 d_0}{k_B T} \ln\left\{\frac{\Psi_0}{d_0[1 - Z(\varphi)]\lambda\varepsilon_0^2}\right\}, \quad (3)$$

$$C = \frac{\beta[2 + (r-1)\cos\varphi]\alpha l_0}{k_B T \cos\varphi}, \quad (4)$$

where r is the island's length to width ratio, $\gamma(0)$ and $\gamma(\varphi)$ are the specific surface energies of the base and the lateral faces of pyramid, φ is the angle between the lateral face and the base of the island, l_0 is the mean distance between atoms on the surface, $Z(\varphi)$ is the coefficient of elastic relaxation [9, 10], λ is the material's modulus of elasticity, ε_0 is the lattices' mismatch, d_0 is the height of one monolayer (ML), Ψ_0 is the wetting energy density on the substrate surface, β is the specific energy of edge formation, and α is the geometrical factor depending on the island's shape:

$$\alpha = \left(\frac{12 d_0 ctg\varphi}{(3r-1)l_0}\right)^{1/3}. \quad (5)$$

Parameter α defines the relationship between the lateral width of the island L and the number of atoms in it:

$$i = \left(\frac{L}{\alpha l_0}\right)^3. \quad (6)$$

Using the Zeldovich formula [11], for the nucleation rate elongated islands we will obtain the following dependence on the wetting layer superstress

$$I(\zeta) = \frac{a}{\tau l_0^2}\zeta(\zeta+1)e^{-\Delta F(i_c)}, \quad (7)$$

where i_c is the critical size at which the free energy (1) is at its maximum:

$$i_c = \left(\frac{C}{\sqrt{A^2 + 3CB\zeta} - A} \right)^3, \tag{8}$$

τ is the characteristic time of atoms incorporation processes:

$$\tau = \frac{3l_0^2 \nu}{4(r+1)\alpha BD}, \tag{9}$$

D is the coefficient of diffusion of atoms from the wetting layer to the islands, ν is the cut-off parameter of the elastic strain field [9].

In case of taking into account energy of edges parameter a is defined as

$$a = \frac{3}{2}\sqrt{\frac{|\Delta F^{"}(i_c)|}{2\pi}\frac{i_c^{1/3}}{\zeta}}. \tag{10}$$

Calculations of parameters of pyramidal and wedge-like clusters with different length to width ratio r were carried out. For calculations the following material values were selected [5, 7–9, 12]: $\lambda = 1.27 \cdot 10^{12}$ dyn/cm^2, $\varepsilon_0 = 0.042$, $d_0 = 0.145$ nm, $l_0 = 0.395$ nm, $\Psi_0 = 450$ erg/cm^2, $\gamma(0) \approx \gamma(\varphi) = 800$ erg/cm^2, $\varphi = 20°$, $h_{eq} = 3$ ML, $\nu = 10$, $D = 10^{-4}*\exp(-1.21 / k_BT)$ cm^2/s, $\beta = 1.5*10^{-6}$ erg/cm.

From comparison of the nucleation rates for islands with square and rectangle base it is shown that elongated islands with higher r emerge later but their nucleation doesn't progress more intensively as it was predicted earlier [3, 4] (see figure 1).

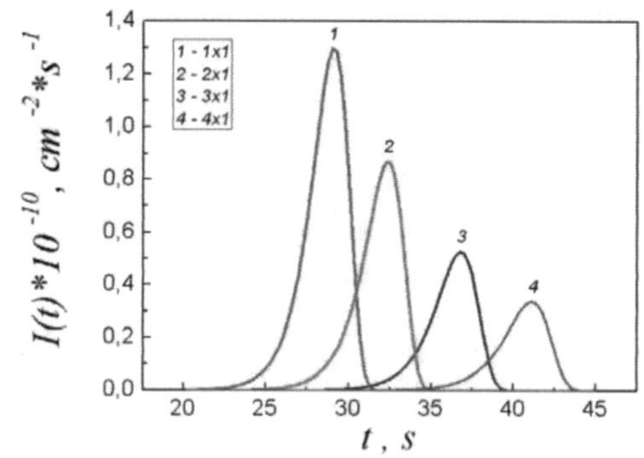

Figure 1. Time dependence of nucleation rate for elongated islands with $r = 1..4$ in Ge/Si(001) system at substrate temperature $T = 470$ °C and deposition rate $V = 0.1$ ML/s.

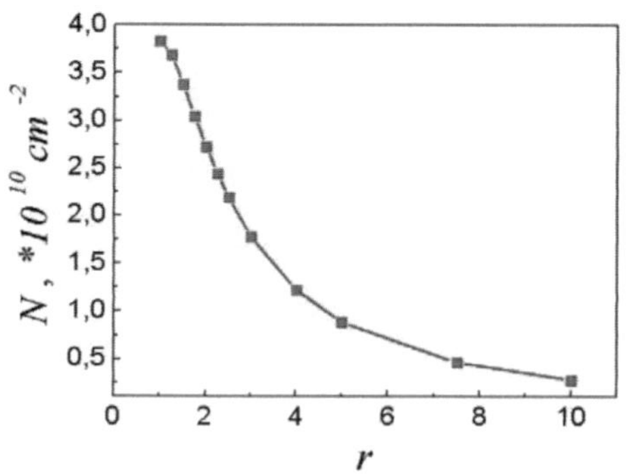

Figure 2. Surface density of wedge-like clusters of Ge on Si(001) surface as a function of r at substrate temperature $T = 470$ °C and deposition rate $V = 0.1$ ML/s.

Modelling of QDs growth dynamics allows us to estimate surface density and size distribution function for the different shaped islands. Results of numerical experiments show that for the same growth conditions average size of elongated islands with small length to width ratio r is larger than for pyramidal clusters when taking into account additional energy of edges (see figures 2 and 3). Their size variation and surface density are smaller.

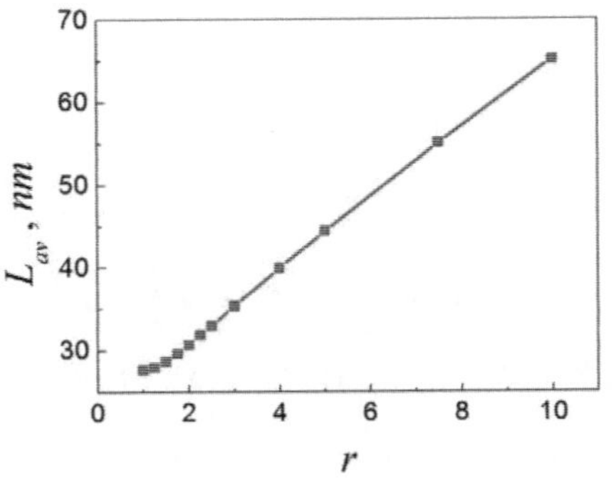

Figure 3. Average size of wedge-like clusters of Ge on Si(001) surface as a function of r at substrate temperature $T = 470$ °C and deposition rate $V = 0.1$ ML/s.

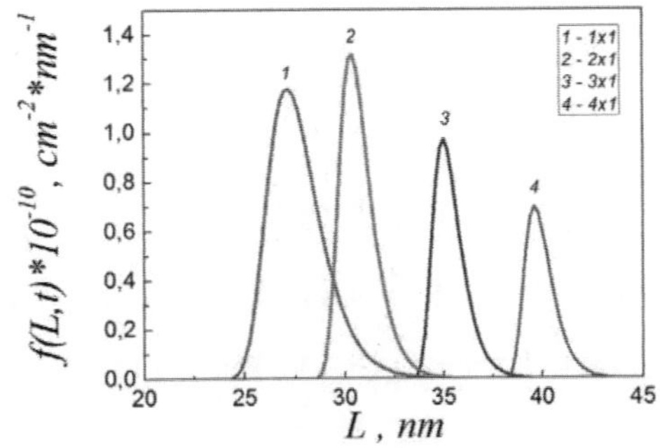

Figure 4. Size distribution function for wedge-clusters of Ge on Si(001) with $r = 1..4$ at substrate temperature $T = 470$ °C and deposition rate $V = 0.1$ ML/s.

It must be noticed, however, that surface density and average size of the elongated islands change monotonically with the ratio increasing as opposed to the results of previous modelling [4]. There is no singular geometry of QDs with $r = 2$ for which average size of the dot reaches its minimum and surface density inversely has its maximum value (for the given growth conditions). With the increase of the length to width ratio from 1 to 10 surface density decreases gradually and the average size increases, both reaching their extrema at $r = 1$ that correspond to pyramidal clusters with square base.

3. Conclusions

Thus, in this paper an attempt is done to apply different shaped QDs of Ge on Si(100) the kinetic model of initial stages of growth of QDs accounting for additional energy of edges. Estimations with the help of this model show that average size and surface density of clusters monotonically depends on length to width ratio r and there is no special geometry of QDs as it was predicted before.

Acknowledgments

This study was supported by the Ministry of education and science of the Russian Federation, State task No. 16.1032.2014/K.

References

[1] Arapkina L V and Yuryev V A 2010 *Physics-Uspekhi* **53** 279
[2] Aqua J, Berbezier I and Favre L 2013 *Phys. Rep.* **522** 59
[3] Lozovoy K A, Voytsekhovskiy A V, Kokhanenko A P, Satdarov V G, Pchelyakov O P and Nikiforov A I 2014 *Opto-Electron. Rev.* **22** 171
[4] Lozovoy K A, Kokhanenko A P and Voitsekhovskiy A V 2014 *Journal of Physics: Conference Series* **541** 012084
[5] Dubrovskii V G 2006 *Semiconductors* **40** 1123
[6] Montalenti F, Scopece D and Miglio L 2013 *Comptes Rendus Physique* **14** 542
[7] Retford C M, Asta M, Miksis M J, Voorhees P W and Webb E B 2007 *Phys. Rev. B* **75** 075311
[8] Lozovoy K A, Kokhanenko A P and Voitsekhovskii A V 2015 *Cryst. Growth Des.* **15** 1055
[9] Osipov A V, Schmitt F, Kukushkin S A and Hess P 2002 *Appl. Surf. Sci.* **188** 156
[10] Ratsch C and Zangwill A 1993 *Surf. Sci.* **293** 123
[11] Kukushkin S A and Osipov A V 1998 *Physics-Uspekhi* **41** 983
[12] Voitsekhovskii A V, Kokhanenko A P, Korotaev A G and Nesmelov S N 2001 *Proceedings of SPIE* **4413** 387

Impact of LT-GaAs layers on crystalline properties of the epitaxial GaAs films grown by MBE on Si substrates

M O Petrushkov, M A Putyato, A K Gutakovsky, V V Preobrazhenskii, I D Loshkarev, E A Emelyanov, B R Semyagin and A V Vasev

Rzhanov Institute of Semiconductor Physics, Siberian Branch of Russian Academy of Sciences, Novosibirsk, 630090, Russia

Email: maikdi@isp.nsc.ru

Abstract. GaAs films with low-temperature GaAs (LT-GaAs) layers were grown by molecular beam epitaxy (MBE) method on vicinal (001) Si substrates oriented 6° off towards [110]. The grown structures were different with the thickness of LT-GaAs layers and its arrangement in the film. The processes of epitaxial layers nucleation and growth were controlled by reflection high energy electron diffraction (RHEED) method. Investigations of crystalline properties of the grown structures were carried out by the methods of X-ray diffraction (XRD) and transmission electron microscopy (TEM). The crystalline perfection of the GaAs films with LT-GaAs layers and the GaAs films without ones was comparable. It was found that in the LT-GaAs/Si layers the arsenic clusters are formed, as it occurs in the LT-GaAs/GaAs system without dislocation. It is shown that large clusters are formed mainly on the dislocations. However, the clusters have practically no effect on the density and the propagation path of threading dislocations. With increasing thickness of LT-GaAs layer the dislocations are partly bent along the LT-GaAs/GaAs interface due to the presence of stresses.

1. Introduction

The silicon and the $A^{III}B^{V}$ compounds are major modern semiconductor materials. Integration of silicon-based electronics and the element base using $A^{III}B^{V}$ compounds on the Si substrates will enable to create the monolithic microwave and optoelectronic devices with silicon control circuits. In addition, the Si substrates are lighter, stronger and less expensive than GaAs and Ge substrates, which are usually used for growing structures based on $A^{III}B^{V}$ compounds.

Practically, the whole spectrum of structures for microwave and optoelectronics based on $A^{III}B^{V}$ compounds and their solid solutions can be obtained by the MBE method. The advantages of the MBE technology consist in the thing that it allows one to grow ultrathin layers with a controllable thickness, form perfect heterointerfaces, study *in situ* processes proceeding on a growing surface. Therefore, the MBE method is prospective regarding the creation of the technology for obtaining perfect $A^{III}B^{V}$/Si heterostructures.

The complexity of device-quality $A^{III}B^{V}$ layers growth on Si is conditioned by a big difference of constant lattice (4.1 % for the GaAs/Si system) and linear thermal extension coefficient (to 50%) of Si with $A^{III}B^{V}$ materials. The differences in the indicated parameters lead to occurrence of threading dislocations in $A^{III}B^{V}$ epitaxial films. To reduce the density of threading dislocations cyclic annealing at various stages of growth and dislocation filters are used [1]. It was established in direct experiments that (for GaAs/Si and GaP/Si [2]) threading dislocations density, directly at the growth temperature, can be decreased to 10^4 - 10^5 cm^{-2}. Nevertheless, after cooling to room temperature, it increases to 10^6- 10^7 cm^{-2}. Thus, for any $A^{III}B^V$ heterosystems on the Si substrate, critical is not the lattice mismatch (e.g. in the GaP/Si, it is tenths of percent), but a big difference in linear thermal extension coefficients of matched materials. Therefore, other ways of reducing the dislocation density are a decreasing the $A^{III}B^V$ layer nucleation and growth temperature and a growing of the intermediate LT-GaAs layers in GaAs/Si(001) films. In this respect, the MBE method that allows one to grow GaAs at temperature 150-200°C is the most fitting epitaxial technology.

Properties of LT-GaAs layers on GaAs substrates are investigated well enough [3,4], but the literature is very little data on the effect of LT-GaAs layers on the quality of GaAs/Si(001) films, containing a large number of threading dislocations.

The aim of the present work was experimental studying the impact of LT-GaAs layers on crystalline properties of the epitaxial GaAs films grown on vicinal (001) Si substrates oriented 6° off towards [110].

2. Experimental methods

The GaAs films with (001) orientation were grown on vicinal (001) Si substrates oriented 6° off towards [110] using the retrofit MBE system "Shtat". To obtain the fluxes of Ga and Si the crucible molecular sources were used, and valve sources with the cracking zone were used to obtain the fluxes of As_4 and P_2 molecules [5]. The surface structure control during growth was realized by the RHEED method.

To achieve this aim it was necessary to solve the following problems. Firstly, we should understand the impact of LT-GaAs layers both formed at the interface of film-substrate and at a distance from it on crystalline perfection of the GaAs/Si films. Second, we should determine the effect of the temperature of growth and annealing of GaAs/Si films containing LT-GaAs layers on its properties. And, thirdly, we should see if arsenic clusters are formed in LT-GaAs layers during annealing, as it occurs in the dislocation-free layers LT-GaAs, grown on GaAs substrates [6,7]. In this regard, GaAs films with LT-GaAs layers both at the film-substrate interface, and at the distance of 800nm from it were grown. The thickness of the LT-GaAs layer at the interface was 170 nm, and the thickness of remote LT-GaAs layer was 700 nm. To understand if the impact is kept from layer to layer the films comprising system of alternating 200nmLT-GaAs/200nmGaAs layers were grown. The profiles of the films are shown in figure 1.

It is known that, during nucleation of GaAs on Si with MBE, applying less than 1 GaAs monolayer leads to the formation of islands with the density to 10^{11} $см^{-2}$. Islands coalescence is accompanied by the appearance of threading dislocations in the film with the density to $10^9 см^{-2}$ [8].To avoid the islands formation and, this way, decrease the threading dislocations density, in the work the nucleation of GaAs layers was realized by atomic-layer epitaxy (ALE) [9,10] through a GaP sublayer at substrate temperature about 250°C. Further the GaAs films thickness of 1 μm, 1.7 μm and 2 μm were grown (figure 1 a, b, c). The growth temperature of GaAs and LT-GaAs layers was 350°C and 170°C, respectively. At a 370nm thickness of the films the 30nm AlAs layers were grown for smoothing of the surface. At a 570nm thickness of the films the cyclic annealing was carried out. The substrate temperature was changing from 200°C to 560°Cduring cycling. In total, 4 cycles for these structures were made. Delta-layer of In (δ-In) was grown in the second structure in 700 nm LT-GaAs layer (figure 1 (b)) to verify the possibility to form ordered arrays of As clusters in the GaAs/Si(001) films, containing a large number of dislocations.

The first sample after the growing of LT-GaAs layer was being annealed at temperature 550°C for 10 min. The second and third samples after growth were cut into four parts and being annealed at temperatures of 500, 550 and 600°C for 10 min. The grown films were investigated by XRD and TEM.

Journal of Physics: Conference Series **741** (2016) 012020 doi:10.1088/1742-6596/741/1/012020

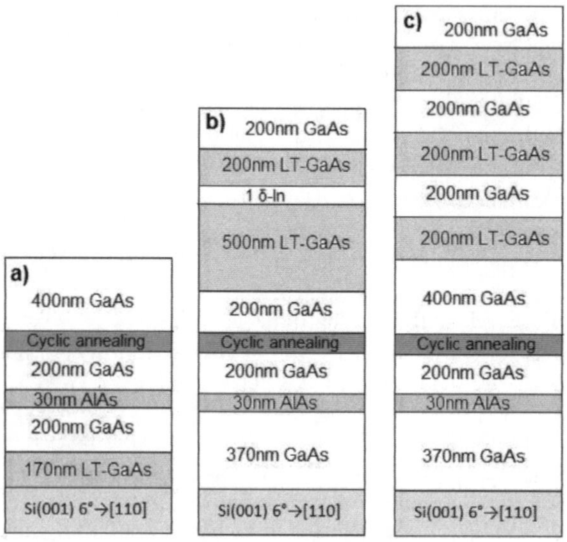

Figure 1 (a, b, c). Profiles of grown GaAs/Si(001) films with LT-GaAs layers:
a) LT-GaAs layer located at the film-substrate interface;
b) LT-GaAs layer located at a distance of 800 nm from the interface;
c) system of alternating LT-GaAs/GaAs layers.

3. Results and discussion

Cross-sectional TEM images of the annealed films are presented in figure 2. The170nm LT-GaAs layer located at the film-substrate interface is shown in figure 2 (a), the 700nm LT-GaAs layer located at a distance of 800 nm from the interface is shown in figure 2 (b), and the system of alternating 200nmLT-GaAs/200nmGaAs layers located at a distance of 1000 nm from the interface is shown in figure 2 (c). It is seen that in all samples the LT-GaAs layers have a darker contrast. This contrast is due to the presence of As clusters [6,7]. After analyzing a number of TEM images, it can be said that large clusters are formed mainly on dislocations (figure 3). This is due to the fact that the As atoms during annealing get into the dislocations and move through it as channel. The higher the annealing temperature, the larger the clusters sizes, and less its concentration.

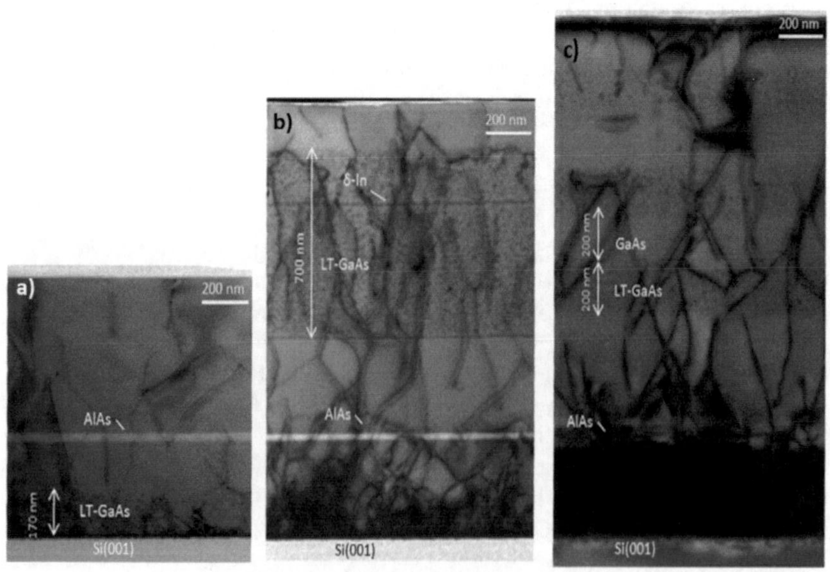

Figure 2 (a, b, c). Cross-sectional TEM images of grown GaAs/Si(001) films with LT-GaAs layers after annealing.

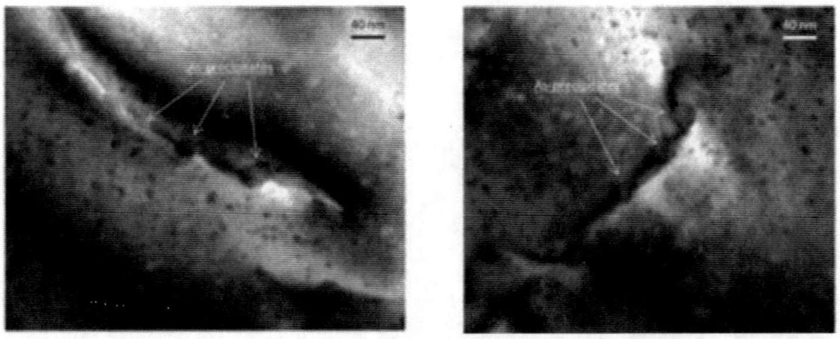

Figure 3. Cross-sectional TEM images of annealed films with As clusters.

The ordered arrays of As clusters is known from the works [11,12] can be formed in the LT-GaAs/GaAs films by δ-layers of In and P. We decided to test if such arrays of As clusters are formed in the GaAs/Si films having threading dislocations. If so, how this array of clusters will effect on the evolution of the threading dislocations system? For this purpose, in the second structure the δ-In was grown in the 700nm LT-GaAs layer. figure 2 (b) shows that the layer of As clusters did formed, but it almost not had affected the dislocation density and direction. The higher the annealing temperature, the more clusters reach to δ-In, and it becomes clearer on the TEM image. As a result, the dislocations affect the position and size of As clusters, and clusters do not affect the evolution of the threading dislocations system.

Low-temperature GaAs is known from the literature [13,14] has a lattice constant larger than the lattice constant of high temperature GaAs. This is due to the adsorption of excess As at low temperatures. There are stresses at the interface of LT-GaAs / GaAs due to the difference in lattice parameters. To reduce the accumulated stress the presence of misfit dislocations lying in the interface is required. The most profitable way to the formation of such misfit dislocations is bending the existing threading dislocations, the so-called process without activation. On the TEM images can be seen that in the annealed samples with a 700 nm LT-GaAs layer the dislocations are partly bent along the interface of LT-GaAs/GaAs (figure 2 (b)) and in the samples without annealing the dislocations are changing its direction of propagation at the interface (figure 4 (a)). However in the samples with 170 nm and 200 nm LT-GaAs layers such the features are observed much less often (figure 2 (c) and 4 (b)). Therefore, with increasing the LT-GaAs layer thickness the stresses at the interface of LT-GaAs/GaAs are increasing and the dislocations are bent more effectively. In addition worth noting that position of LT-GaAs layer in the GaAs/Si(001) film did not play a significant role in changing the density of threading dislocations.

According to XRD analysis of samples can be said that crystalline quality of films is not significantly affected by LT-GaAs layer. On the rocking curves the full width at half maximum (FWHM) of the films peak with LT-GaAs layers are comparable to the FWHM of the films peak without ones. With increasing annealing temperature and film thickness the FWHM value decreases as usual. Also a crystal lattice rotation of the film with respect to the substrate was analyzed [15]. The rotation magnitude is determined by a grid of misfit dislocations in the interface. In the GaAs films with LT-GaAs layers the crystal lattice rotations round the direction <011> the order of several hundred arcsecs (300-700) were detected. These values are comparable to the crystal lattice rotations for films without LT-GaAs layers. Figure 5 shows the (115) reciprocal space map for as-grown GaAs/Si film with alternating LT-GaAs/GaAs layers. The film is in a laterally stretched state due to the difference of the linear thermal expansion coefficients. The (115) GaAs peak has been shifted to the right from a fully relaxation line because the crystal lattice of the film is stretched and has rotation relative to the substrate. The broadening of the (115) GaAs peak indicates the orientation distortion caused mainly misfit dislocations.

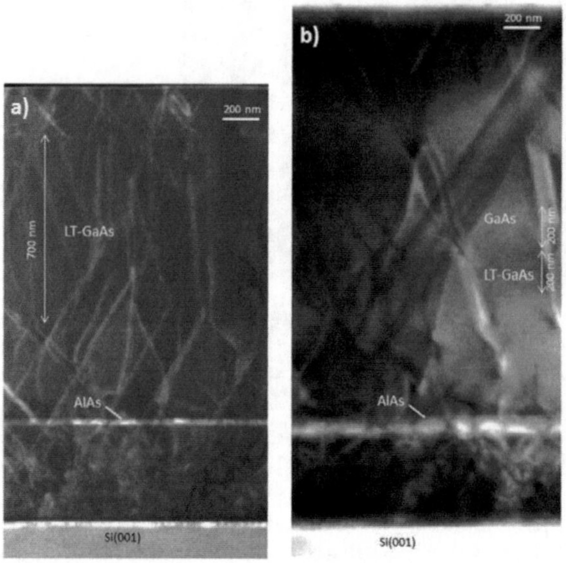

Figure 4 (a, b). Cross-sectional TEM images of as-grown GaAs/Si(001) films with LT-GaAs layers.

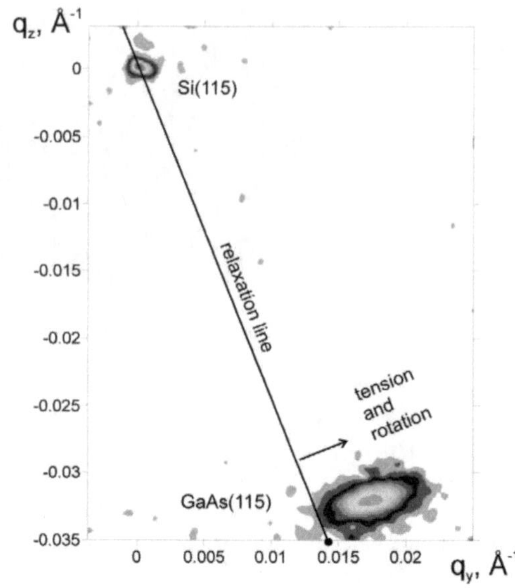

Figure 5. XRD reciprocal space map of the (115) reflections for the GaAs/Si structure.

4. Conclusions

GaAs films with low-temperature GaAs (LT-GaAs) layers were grown by MBE method on vicinal (001) Si substrates oriented 6° off towards [110]. The grown structures were different with the thickness of LT-GaAs layers and its arrangement in the film. Investigations of crystalline properties of the grown structures were carried out by the methods of XRD and TEM.

The crystalline perfection of the GaAs films with LT-GaAs layers and the GaAs films without ones was comparable. In the GaAs/Si structures with LT-GaAs layers the crystal lattice rotation round the direction <011> was detected. It was found that in the LT-GaAs/Si layers the arsenic clusters are formed, as it occurs in the LT-GaAs/GaAs system without dislocation. It is shown that large clusters are formed mainly on the dislocations. It means the dislocations are a kind "channels" for atoms of As. Using δ-In we managed to obtain an ordered array of As clusters. The clusters array proved to have no effect on the density and the propagation path of threading dislocations. Thus, the dislocations affect the position and size of As clusters, and the clusters do not affect the evolution of the threading dislocations system. With increasing LT-GaAs layer thickness the stresses at the interface of LT-GaAs/GaAs are increasing and the dislocations are more bent.

Acknowledgments
This work was supported by Russian Foundation for Basic Research (grants № 16-32-00172 – mol_a and № 16-32-60087 – mol_a_dk).

References

[1] Bolkhovityanov Yu B and Pchelyakov O P – 2008 *J. Advancements in Phys. Sci.*, **178** № 5, p. 459-80
[2] Alferov Zh I, Andreev V M and Rumyantsev V D – 2004 *J. Semiconductors*, **38** №8, p. 937-48
[3] Bert N A, Vainger A I, Vilisova M D, Goloshapov S I, Ivonin I V, Kozyrev S V, Kunitsyn A E, Lavrent'eva L G, Lubyshev D I, Preobrazenskii V V, Semyagin B R, Tret'yakov V V, Chaldyshev V V and Yakubenya M P – 1993 *J. Phys. Solid State*, **35** №10, p.1289-97.
[4] Bobrovnikova I A, Vainger A I, Vilisova M D, Ivonin I V, Lavrent'eva L G, Lubyshev D I, Preobrazenskii V V, Putyato M A, Semyagin B R, Chaldyshev V V and Yakubenya M P – 1998 *Russian Physics Journal*, №9, p. 37-45.
[5] Preobrazhenskii V V, Putyato M A et al. – 2009 *J. Semicond. Sci. Technol.,* **24** p. 055014-20.
[6] Liliental-Weber Z, Cooper G, Mariella R and Kocot C – 1991 *J. Vac. Sci. Technol.*, **9** p.2323-27.
[7] Melloch M R, Otsuka N, Woodall J M, Warrner A C and Freeout J L – 1990 *J. Appl. Phys. Lett.*, **57** №15, p.1531-33.
[8] Yonehara T, Yoshioka S, et al. –1982 *J. Appl. Phys.*, **53** p. 6839-43.
[9] Taylor P J et al. –2001 *J. Appl. Phys.*, **89** № 8. p. 4365-74.
[10] Ohta K, Kojima T and Nakagawa T – 1989 *J. Crystal Growth*, **95** p. 71-74.
[11] Bert N A, Musikhin Yu G, Preobrazenskii V V, Putyato M A, Semyagin B R, Suvorova A A, Chaldyshev V V and Werner P –1998 *J. Semiconductors*, **32** №7, p. 769-74.
[12] Boytsov A V, Bert N A, Chaldyshev V V, Preobrazenskii V V, Putyato M A and Semyagin B R – 2009 *J. Semiconductors*, **43** №12, p. 278
[13] Kin Man Yu, Kaminska M and Liliental-Weber Z – 1992 *J. Appl. Phys.*, **72** №7, p.2850-56.
[14] Baranowsk J M, Liliental-Weber Z, Yau W F and Weber E R – 1991 *J. Phys. Rev. Lett.*, **66** №23, p. 3079-82.
[15] Loshkarev I D, Vasilenko A P, Trukhanov E M, Kolesnikov A V, Putyato M A, Semyagin B R, Preobrazhenskii V V and Pchelyakov O P – *J. Bulletin of the RAS. Physics*, **77** № 3, p. 233-35.

Quantum dot based superstructures: PL decay analysis

D A Volgina[1]**, A V Baranov**[1]**, E V Ushakova**[1]

[1]Department of Optical Physics and Modern Natural Science, ITMO University, Saint Petersburg 197101, Russia

Abstract. The analysis of the optical properties, including transient photoluminescence (PL), of CdSe quantum dots (QDs) self-assembled in supercrystals shows that supercrystals possess efficient QD PL. Although PL band is slightly broadened and blueshifted as compared with QD colloidal solution, PL kinetic is the same showing absence of induced channels of the energy dissipation. The data indicate the opportunity to create quantum dot solids with efficient PL.

1. Introduction

Quantum dots (QDs) are semiconductor nanocrystals which sizes are close or less than the exciton Bohr radius and where movement of charge carriers is space limited in all three directions. This results in size dependent discrete energy spectra of the electronic transitions in QDs often called, therefore, “artificial” atoms. Development of the nanotechnology allows to design novel nanostructured materials based on QDs as building blocks. These materials possess unique optical and electronic properties, which can be tuned by varying the size, shape and chemical composition of the building blocks and, at the same time, by changing the inner morphology of the superstructure. One of the types of nanostructured materials is the supercrystals (SCs) formed from QDs [1]. These materials are very attractive for application in devices with improved performance for photovoltaics, biosensing and elements for telecommunications [2, 3]. However, these materials are far from the mass production, since there are lots of challenges to solve, for instance, there is a lack of information on optical properties, e.g. photoluminescence (PL), of SCs with different chemical composition, morphology and other parameters.

Thus, the investigation of optical properties, including PL kinetics, of SCs formed from CdSe QDs in comparison with that of QDs in colloidal solution is of great interest.

2. Experimental

2.1. Materials. Colloidal 4 nm in diameter CdSe QDs with surface stabilized by trioctylphosphine oxide (TOPO) molecules have been synthesised by standard hot-injection method [4]. Then TOPO molecules have been replaced with oleic acid (OA) ones needed for SCs growth from colloidal QD solution. Absorption (ABS) and PL spectra of QDs in a tetrachlormethane (TCM) solution are shown in figure 1(a). SCs have been obtained by self-assembly of QDs on a glass substrate by slow evaporation of the saturated QD colloidal solution with initial QD concentration of around 10^{-5} M. The obtained SCs have a shape of polyhedra with pronounced edges (see figure 1(b)). The size of typical SC used for investigation is of about 68.5x9.5 um (see figure 1(b)). This QD sample also contains areas with disordered QD ensembles, formed in excess of OA molecules. For comparison a sample with QDs deposited on a glass substrate has been prepared. In this case, a solution with QD concentration around 10^{-7} M has been used. A drop of this colloidal solution has been dripped on the substrate and evaporated in ambient atmosphere leading to QD agglomeration into clusters (hereafter called "QDs in aggregates"). It should be noted that the evaporation process has been much faster than that for the saturated QD solution. Microphotographs of typical SC and of QDs in aggregates on the substrate obtained with a confocal microscope Olympus BX-51 are shown in figure 1(b) and 1(c), respectively.

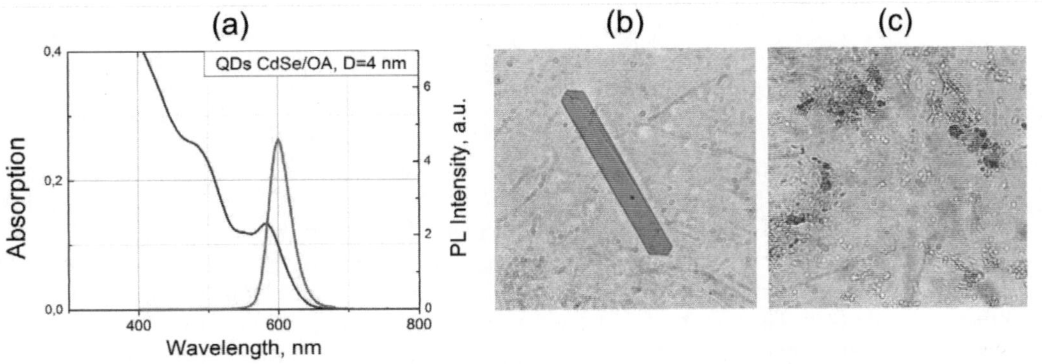

Figure 1. (a) ABS and PL spectra for 4 nm CdSe QDs, covered with OA molecules, in TCM. (b) and (c) 100x100 um microphotographs of QD SC and of QDs in aggregates, respectively.

2.2. Methods. For optical spectral analysis of the samples a spectrophotometer UV-3600 (Shimadzu) and spectrofluorimeter Carry Eclipse (Varian) have been used. PL spectra of QD samples on the substrates have been obtained with a laser scanning microscope LSM-710 (Zeiss). PL decay at room temperature has been investigated by a laser scanning confocal microscope MicroTime 100 (PicoQuant), which implements the method of time-correlated single photon counting. Obtained PL decay curves from the samples have been fitted by biexponential function:

$$I(t) = I_0 + A_1\exp(-t/\tau_1) + A_2\exp(-t/\tau_2). \quad (1)$$

The average PL lifetime has been calculated with:

$$\langle\tau\rangle = \frac{\sum I_i\tau_i^2}{\sum I_i\tau_i}. \quad (2)$$

3. Results

The investigation of the steady-state spectral properties of QD in SCs and in aggregates shows that their ABS and PL bands are broadened and blueshifted in comparison with those in the colloidal solution. Obtained spectra are shown in figure 2.

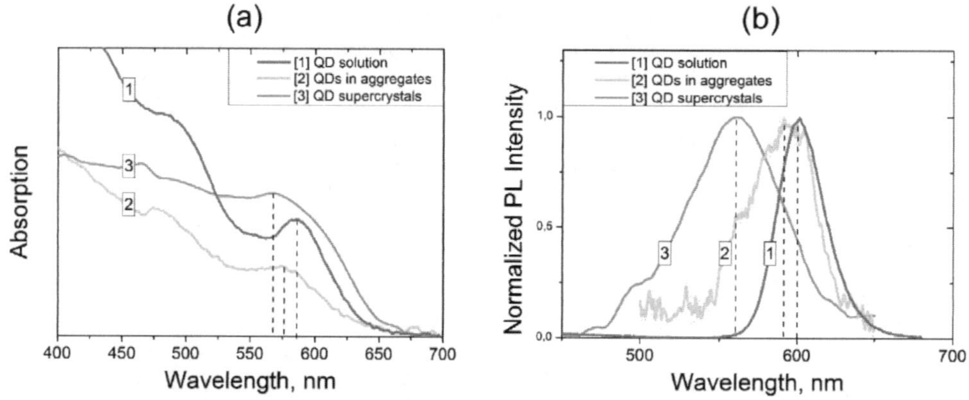

Figure 2. ABS (a) and PL (b) spectra of samples of QDs: (1) in solution, (2) in aggregates, (3) in SCs.

The changing in the optical properties is mainly caused by interactions between the nanocrystals and their chemical environment in solution, in aggregates, and in SCs [5]. The concentration of OA molecules nearby QD is increasing in the series: QDs in solution, QDs in aggregates, disordered QD ensembles and QD SCs. Thus, QDs in SCs and in disordered ensembles are surrounded by a matrix formed from OA molecules.

We have investigated luminescence properties of QD samples on the substrates (QD SCs, QD in aggregates and disordered QD ensembles) by means of confocal microscopy. The obtained PL spectra together with PL images of the samples are shown in figure 3. Comparison of the PL images shows an unusually sharp spatial distribution of PL intensity through SC, where only edges of SC are highly luminescent. As it is seen from the figures 3(b) – 3(d), such PL spatial distribution is not observed for QDs in aggregates and disordered QD ensembles.

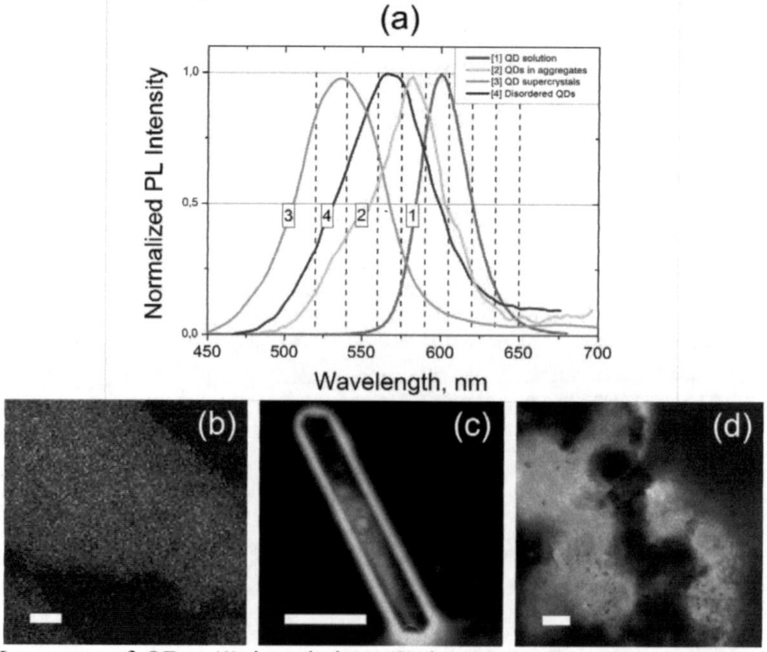

Figure 3. (a) PL spectra of QDs: (1) in solution, (2) in aggregates, (3) in SC, (4) in disordered ensembles. PL images of QDs: (b) in aggregates, (c) in SC, (d) in disordered ensembles.

PL spectral analysis of the samples shows that PL maximum is blueshifted and for SCs this shift is maximal and is of 63 nm compared to QD colloidal solution. This comes, most likely, from differences in QD environment for TCM solution and OA matrix.

A comparison of PL decay curves for QDs in initial colloidal solution and for QD samples on the substrates is presented in figure 4(a).

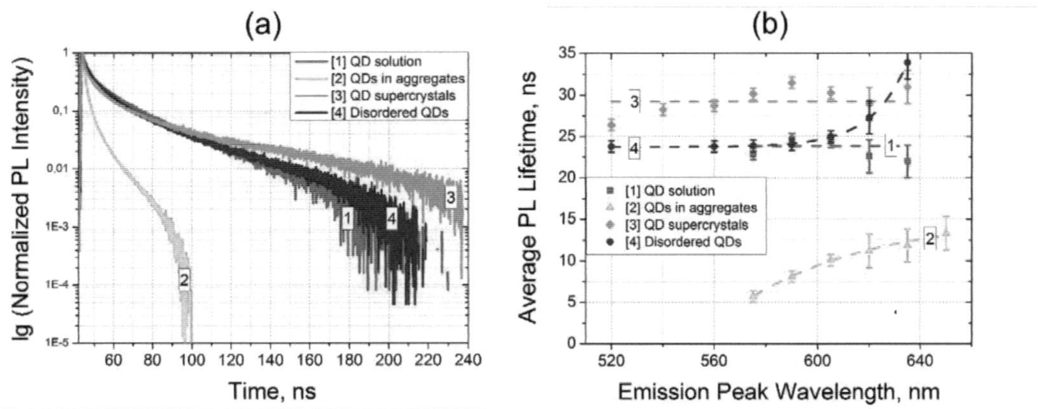

Figure 4. PL decay curves (a) and average PL lifetime dependence on emission peak wavelength (b) for QDs: (1) in solution, (2) in aggregates, (3) in SC, (4) in disordered ensembles.

PL decay analysis shows that the average PL lifetime of QDs in initial colloidal solution is 25.3 ns with 6.7 ns and 30.9 ns of "short" and "long" decay components, which are in agreement with literature [6]. The average PL lifetime of QDs in aggregates is reduced to 9.9 ns that has been expected due to increasing interaction between close packed QDs. At the same time, for disordered QD ensembles in OA the average PL lifetime is close to that in solution and for SCs is as high as 29.9 ns, which is slightly higher that PL decay value for isolated QDs in solution.

For spectrally resolved PL decay analysis interference filters to cut a 5 nm spectral band within PL spectra have been used. The maximum positions of the bands are shown by dashed lines in figure 3(a). In figure 4(b) the average PL lifetimes, estimated by equation (2), are shown. As it was expected, the average PL lifetime of QDs in solution is independent from the emission wavelength. For QDs in aggregates and disordered QD ensembles in OA the increase of PL lifetime with emission wavelength is observed. In case of QDs in aggregates PL lifetime increases from 5.7 to 13.3 ns. It can be explained by taking into account FRET between QDs in aggregates. Contrary to these data PL lifetime of QDs in SCs is independent from the emission wavelength, as it is observed for QDs in solution. This indicates that the interaction of QDs in SCs does not affect PL decay of quasi-isolated QDs.

4. Conclusion

Optical properties of CdSe QDs with diameter of 4 nm self-assembled in supercrystals on the glass substrate have been investigated in comparison with QDs in TCM solution and with QDs in aggregates and disordered QD ensembles on the substrates. The spectral analysis shows that ABS and PL bands of QDs in SCs are broadened and blueshifted in comparison with the colloidal solution. The transient PL analysis shows that QDs in superstructure possess almost the same PL lifetimes comparing to those in colloidal solution, showing absence of induced channels of the energy dissipation. This data indicate the opportunity to create quantum dot solids with efficient PL.

Acknowledgments

The authors thank the Ministry of Education of the Russian Federation (state task № 3.109.2014/K) for financial support.

References

[1] Nie Z, Petukhova A and Kumacheva E 2010 *Nat Nanotechnol* **5**(1) 15-25
[2] Nag A, Zhang H, Janke E and Talapin D V 2015 *Zeitschrift für Physikalische Chemie* **229**(1-2) 85-107 Jiao Y, Han D, Ding Y, Zhang X, Guo G, Hu J, Dong Y and Dong A 2015 *Nature communications* **6** Yu W, Qu L, Guo W and Peng X 2003 *Chemistry of Materials* **15**(14) 2854-2860
[3] Ushakova E, Cherevkov S, Litvin A, Parfenov P, Zakharov V, Dubavik A, Fedorov A and Baranov A 2016 *Opt Express* **24**(2) A58-A64
[4] Wang X, Qu L, Zhang J, Peng X and Xiao M 2003 *Nano Lett* **3**(8) 1103-1106

Numerical simulation of multi-layer graphene structures based on quantum-chemical model

Y Kasper, A Tuchin, A Bokova, L Bityutskaya

Voronezh State University, 394006, Universitetskaya pl.1, Voronezh, Russia

E-mail: solnce-juli_94@mail.ru

Abstract. The electronic structure of the multi-layer graphene has been studied using the density functional theory (DFT). The dependence of the average interlayer distance on the number of layers ($n = 2 \div 6$) has been determined. The analysis of the charge redistribution and the electron density of the bi- and three–layer graphene under the external pressure up to 50 GPa has been performed. The model of the interlayer conductivity of compressed multi-graphene was offered

1. Introduction

A multi-layer graphene (MG) is a two-dimensional allotropic modification of carbon, formed by several layers of the graphene [1–3]. MG combines exceptional properties of graphene, such as low resistivity, high thermal conductivity, transparency and surpasses it in stability and stiffness [4, 5]. The valence band and the conduction band touches or are close to the corners of the hexagonal Brillouin zone, so the MG is a semimetals or narrow-gap semiconductor [5–7]. The control of the band overlap is achieved by the virtue of external influences, in particular by the electric field [8]. The band gap modulation in the range E_g = 0-250 meV in MG with the application of an electric field directed perpendicular to layers have been shown experimentally [9]. Zhang et al. [6] created transistors based on the MG. Scientists are intensively pursuing research and development activities to develop a variety of techniques for syntheses of multi-layer graphene of the large area [4, 9]. The violation of the boundary conditions is the reason of the increase of interlayer distances in MG with compare to the graphite, this is an important practical value [10-12]. Lee et al. [11] received a stable multi-layer graphene suspension in the water and organic solvents without the chemical modification of MG. The precipitate is not dropped within 6 months. This simplifies the deposition on the substrate as opposed to the insoluble graphene substantially. Yoo et al. [13] replaced the graphite to the multi-layer graphene, that allowed to increase the capacity of lithium-ion batteries by 50%. The pressure dependences of the electrical resistance of MG are actively researched. The interlayer distance dependence on the definition of the pressure is an important task in the development of nanoelectromechanical devices, high-capacity batteries, hydrogen cells etc. [14]. For this purpose, pressure dependences of electro physical properties of MG are actively investigated. Interlayer distances and dependences of the effective charge and electron density distribution on layer numbers of the MG under external pressure is the aim of the present work.

2. Computational details

A quantum-chemical modelling is an effective way for the carbon nanomaterials research. All electron simulations of the electronic structure of the multi-layer graphene were performed using the density functional theory (DFT) computational method with a local spin density approximation (LSDA) [15] with the help of the Gaussian09 software in the Supercomputing Centre of Voronezh State University. The initial MG geometry is a ABABA package and the interlayer distance $r_{sh} = c/2 = 3.35$ Å (c - graphite lattice parameter). The interlayer distance was determined after the optimization of the MG geometry. An impact of the external pressure on the electronic structure of MG was determined for bi- and three–layer structures. At the first, the distance between the boundary layers decreases in optimized early structures. Then, the interaction force between layers is defined as $F = -dE_{tot} / dr_{sh}$, where E_{tot} – the total energy of the MG elementary cell. The shift of layers caused by the pressure P defined as $P = F / S$, where S is a twice area of the graphene elementary cell. An upper pressure limit is 50 GPa, that is usual value for experimental works [16].

3. Results and discussion

Let us to present the results of the investigation of the interlayer distance on the number of layer of uncompressed MG. It was found, that the average interlayer distance is a monotonically decreasing function on the number of layers (figure 1). In the range of layer numbers $n = 2, ..., 6$ the interlayer distance decreases from 3.61 Å to 3.49 Å, which is more by 7.8% and 4.2% than in graphite.

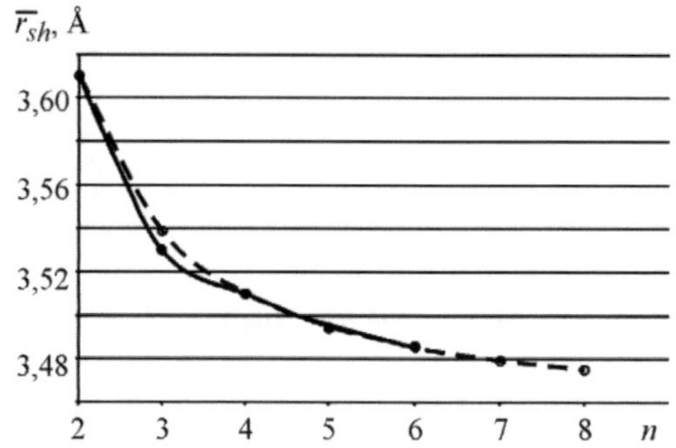

Figure 1. (a) The averaged multi-layer graphene distance, calculated by DFT (solid line) and by the equation (1) (dotted line).

Based on the result of the numerical simulation an analytic dependence of the interlayer distance on the number of layer was installed:

$$r_{sh} = \frac{c}{2} + k\left(\frac{2}{n^2} + \frac{1}{n}\right) + a \quad , \tag{1}$$

where $k = 0.16$ Å; $a = 0.1$ Å.

A maximum absolute error between the numerical calculations and by the equation (1) is 0.01 Å for three–layer MG. Yoo et al. [13] using transmission electron microscopy defined the average interlayer distance $r_{sh} = 3.650$ Å of MG obtained by mechanical graphite cleavage. Agrinskaya et al. [12] experimentally defined the average interlayer distance of eight-layer MG $r_{sh} = 3.39 \pm 0.02$ Å that is less than calculated values according to the equation (1) on 1.9-3.1% $r_{sh} = 3.475$ Å ($n = 8$) (figure 1). It should be noted, that in [12] MG was obtained by the thermal graphitization of SiC. The substrate effect may cause a decrease of the interlayer distance in the relation to simulated free MG.

In this way the interlayer distance is sensitive to the layer number. The layer structure defines the sense to the external pressure. It was found that in the pressure range P = 0–50 GPa the interlayer distance for bi- and three–layer MG is reduced by 47% (from 3.61 to 1.90 Å) and 36% (from 3.56 to 2.28 Å) respectively (figure 2). It is commonly thought that interlayer interaction in the graphite and MG is Van-der-Waals. Consequently, the force and pressure are proportional to r_{sh}^{-7}. With the aim to investigate the features of the interlayer interaction of the MG under the external pressure the dependence of $P(r_{sh}^{-7})$ was calculated. At the pressure P<20 GPa the dependence $P(r_{sh}^{-7})$ of bi-layer MG is close to linear (figure 2a). It was revealed that at the pressure P>20 GPa the dependence becomes steeper, that indicates the interlayer additional attraction. The deviation from the Van-der-Waals interlayer interaction for three–layer MG is occurs at the pressure P>30 GPa. That point out to the stiffness increase of MG with the rice of layer numbers.

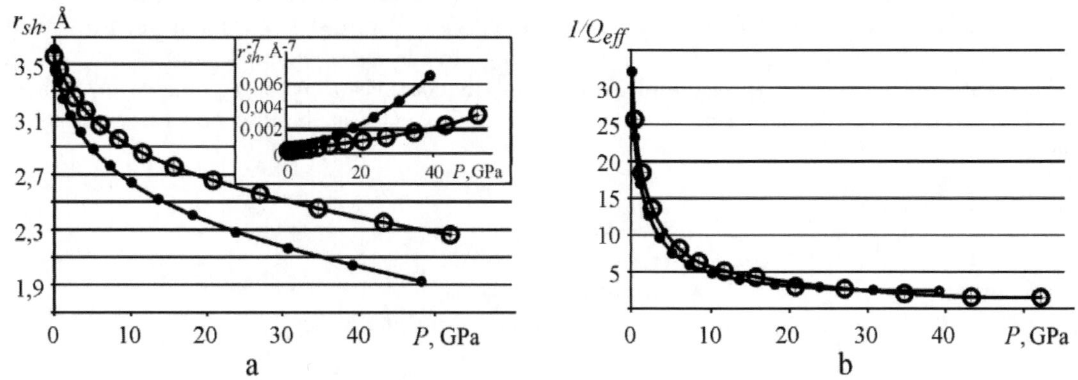

Figure 2(a, b). (a) The interlayer distance versus the pressure value for bi– (○) and three–layer (•) multi-graphene ($r_{sh}^{-7}(P)$ dependence is represented in the insect); **(b)** the reverse module of the induced effective charge versus the pressure value.

The electron density redistribution was analysed to determine the mechanism of the interlayer additional attraction. The evolution of the electron density distribution of the highest occupied crystal orbital (HOCO) and lowest unoccupied crystal orbital (LUCO) of bi- and three–layer MG is shown in the figure 3.

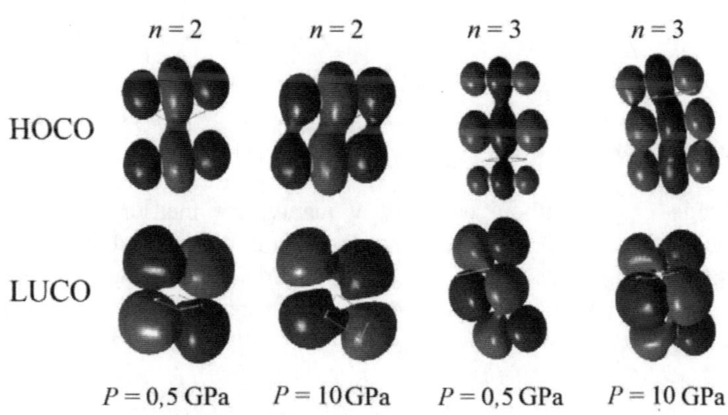

Figure 3. Electron density redistribution for bi- and three–layer MG at 0.5 GPa to 10 GPa pressure.

LUCO is the antibonding orbital, the electron density distribution is weakly depends on the pressure value. At the pressure to P~10 GPa HOCO is weakly binding in the interlayer direction. The electron density is localized in the interatomic and interlayer space. It was found, that at the pressure P ~ 10 GPa interlayer π-bonds are beginning to form, that has two consequences. At the first, the addition binding leads to the deviation of the interlayer interaction from the Van-der Waals. Secondly, in the direction perpendicular to the layers forms the electron transfer channels, this reduce the anisotropy of the MG electrical conductivity. At the pressure P ~ 10 GPa conduction channels are not fully formed, so it can be concluded that the critical pressure of the conductivity channels formation increases with the number of layers.

Based on the results of the visualization of HOCO and LUCO of compressed MG the model of MG conductivity was offered (figure 4). In the case of uncompressed MG the p_z orpitals of MG directed perpendicular to the layers. Probabilities of the finding of electrons above and below of each graphene layer are equal (figure 4). At the external pressure orbitals are deformed as a result of rapprochement layers that lead to the increase of the electrons finding probability on external sides of MG. At the critical value of the pressure (pressure of the conductivity channel formation P_{cf}) interlayer covalent bonds are formed that is the reason of the direct electron transfer in the direction perpendicular to layers.

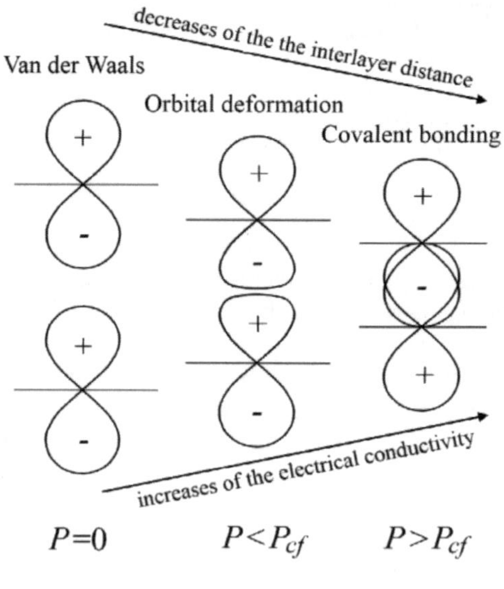

Figure 4. The model of conductivity channels formation for bi- layer MG under an external pressure P (P_{cf}– pressure of the channel forming). A sign plus and minus indicates a positive and negative part of the wave function.

The effective charge Q_{eff} of carbon atoms is a quantitative measure of the electron density redistribution. The ppressure dependence of the effective charge for bi- and three–layer MG is linear. Hence, the hyperbolically dependence of $1/Q_{eff}(P)$ (figure 2b) should diatremes an analogic dependence of the electro resistivity of MG versus the pressure value. This has been demonstrated experimentally by Petrosyan et al [14].

4. Conclusion

The numerical investigation of the electronic structure of MG under external pressure was performed using the density functional theory. For MG with the number of layers n=2–6 was found that the interlayer distance is greater of 4.2–7.8% than the corresponding value in the graphite. The decreasing dependence of the interlayer distance on the layers number n was established $r_{sh} \sim (2+n)/n^2$. The calculated interlayer distances are in a good agreement with experimental works. The π-bonds interlayer formation was detected at the critical pressure value 10 GPa and 30 GPa for bi– and three–layer MG. Interlayer π-bonds serve as channels of the electron transfer in the direction perpendicular to layers, that causes the reduce of the electrical conductivity anisotropy of MG. It was revealed that the critical pressure of the conductivity channel formation increases with the number of graphene layers in the MG. The research results extend the knowledge about the electronic structure and phase transitions in the compressed graphene-like structures and can be used for the development of nanoelectromechanical devices.

Acknowledgments

This work was supported by RFBR, research project No. 16-32-00926 mol_a.

References

[1] Lebedev A A, Agrinskaya N V , Lebedev S P et al. 2011 *Semiconductors* **5** 634
[2] Lebedev A A, Strel'chuk A M, Shamshur D V et al. 2010 *Semiconductors* **10** 1436
[3] Tuchin A V, Bokova A M, Bitytskaya L A, Bormontov E N 2014 *Condensed matter and interfaces* **1** 79
[4] Liu Y et al. 2013 *Nan. Res. Lett.* **8** 335
[5] Wu Y, Wang B, Ma Y et al 2010. *Nano Res* **9** 661
[6] Zhang Y, Tang T, Girit C et al. 2009 *Nature* **459** 820
[7] Craciun M F, Russo S, Yamamoto M et al. 2009 *Nature nanotechnology* **4** 383
[8] McCann E 2006 Phys. Rev. **74** 161403(4)
[9] Russo S, Craciun M F et al. 2011 *Graphene - Synthesis, Characterization, Properties and Applications* (Ed. by J.R. Gong. InTech) 141
[10] Garcia-Sanchez D, Zande A M, Paulo A S et al. 2008 *Nanolett.* **5** 1399
[11] Lee J H, Shin D W, Makotchenko V G et al. 2009 *Adv. Mat* **43** 4383
[12] Agrinskaya N V, Berezovec V A, Kozub V I et ***al*** 2013 *Semiconductors* **2** 267
[13] Yoo E J, Kim J, Zhou H-Sh et al. 2008 *Nano Lett* **8** 2277
[14] Petrosyan T K, Babushkin A N 2013 *Proc. of 15th Vserossiyskoy molodejnoy konferencii po fizike poluprovodnikov i nanostruktur, poluprovodnikovoy opto- i nanoelectronike* (Saint-Petersburg) 90
[15] Kohn W, Sham L J 1965 *Phys. Rev.* **140** 1133
[16] Tihomirova G V, Babushkin A N 2002 *Physics of the Solid State* **4** 618

Synthesis of micro- and nanodiamonds by the method of oxy-acetylene combustion flame

S Sabitov[1,2], B Mansurov[2], B Medyanova[1,2], G Partizan[1,2], A Koshanova[1,2], Ye Merkibayev[2], M Mansurova[1], B Lesbayev[1,2]
[1]Al-Farabi Kazakh National University, Almaty 050040, Kazakhstan
[2]The Institute of Combustion Problems, Almaty 050012, Kazakhstan

Abstract. This work presents the results of experiments on synthesis of micro- and nanodiamonds by the method of oxy-acetylene torch on the surface of pre-deposited copper thin films. The influence of the thickness of the buffer copper film and the concentration ratio of oxygen and acetylene on the structure formation of the deposited samples was investigated during performed experiments. Studies by Raman scattering and scanning electron microscopy showed that the synthesis of micro- and nano-diamonds occurs under certain experimental conditions.

1. Introduction
In recent years carbon allotropes, such as nanodiamonds, have shown promising new applications in many fields due to its physical, chemical and surface characteristics. Their high electron mobility, field electron emission and magnetic properties make them important players in carbon based electronics. Their tribological and mechanical properties give rise to harder coatings, which are biocompatible and can provide improved biological prosthesis joints with decreased wear. This biocompatibility allied to their biosensing, optical and nanoprobing functionalities provide drug delivery and cellular labeling capacity. This wide range of novel applications has fostered the active search for new and more efficient synthesis and production methods of carbon allotropes [1]. The method of oxy-acetylene torch refers to the most simple but efficient methods for producing carbon materials, wherein the deposition occurs at atmospheric pressure, i.e. it does not require complex vacuum and electronic equipment. It has various advantages over other methods, such as high synthesis speed and the safety and low cost of the equipment used [2, 3]. Furthermore, conduction of synthesis does not require expensive precursors, including high purity. The results of studies on micro- and nanodiamonds obtained in the flame of oxy-acetylene torch on the surface of copper film, previously deposited on silicon plates by Raman scattering and scanning electron microscopy are presented in this work.

2. Experimental

Preparation and study of substrates
The monocrystalline silicon plates (analogue of brand KDB-20, manufacturer Siegert Wafer GmbH, Germany) 1×1 cm with orientation [100] and [111] were used as substrates and basis for copper films. The substrates were previously chemically cleaned. The treatment was carried out in a mixed solution of NH_4OH, H_2O_2 and distilled water at a volume ratio of 1:1:6.5, at a temperature of 20°C for 10 minutes, using sound waves with a frequency of 850 kHz, power 250 Watts. Further, washout in

distilled water and drying were performed.

Copper films were deposited on substrates of polished silicon plates by DC magnetron sputtering in equipment VUP-5M. Sputtering was carried out in the flow of working gas Ar at a pressure of 10^{-2} Torr. The flow rate of Ar was 6 cm^3/min and it was controlled by the gas flow controller MCV-500SCCM. Experiments were conducted at a constant voltage on the anode target (740 V), the plasma current of 35 mA. The time of experiments was 30 and 60 minutes.

The obtained samples were investigated by scanning electron microscopy (SEM) in the National nanotechnological laboratory of open type using a microscope Quanta 3D 200i.

Figure 1 shows a cross-sectional SEM image of a thin copper film on the silicon substrate grown for 30 and 60 minutes. SEM studies showed that the thickness of the copper films deposited on silicon wafer for 30 minutes is 437.2 nm, whereas thick layer with thickness of 524.6 nm is observed for deposition time of 60 minutes, regardless of Si orientation.

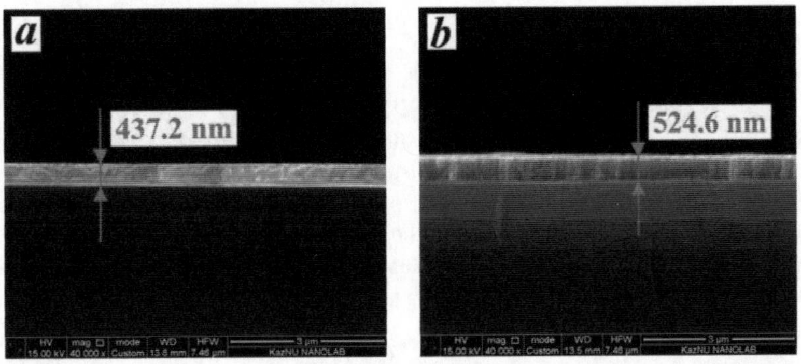

Figure 1(a,b) - Cross SEM image of thin copper film on a silicon substrate: **(a)** - 30 minutes, **(b)** - 60 minutes

Synthesis of micro- and nanodiamonds

The scheme of technological equipment for synthesis of carbon materials in the flame of oxy-acetylene torch and the procedure of conduction of experiments were described in detail in work [3].

A series of experiments in which the distance from the torch nozzle to the substrate (h = 4 mm) and the tilting angle of the flame front (α = 90°) was constant was carried out. The duration of deposition was changed from 15 to 60 minutes, the concentration ratio of oxygen and acetylene (O_2/C_2H_2) was varied from 0.9 to 0.94 in 0.01 increments.

The obtained samples were studied by SEM and Raman scattering (RS). Investigation of the samples was carried out in the National nanotechnological laboratory of open type using spectrometer NT-MDT NTegra Spectra (laser wavelength λ = 473 nm) and Quanta 3D 200i.

The most interesting findings are presented below.

3. Results and discussion

Figure 2 shows the Raman spectra and SEM images of structures synthesized on Cu substrates deposited on Si with orientation [111] for 30 minutes. Raman spectra of the samples show peaks characteristic peak of diamond (sp^3) in the area of 1335.4 cm^{-1}. The shift from the standard value to the 1334-1342 cm^{-1} may take place due to a compressive biaxial stress which occurs because of mismatch of thermal coefficient of expansion between the silicon substrate and the diamond film [7]. A peak corresponding to *G* group is at 1518.3 cm^{-1} in Figure 2*a*. *G* peak is observed in the range of 1511.9 and 1537.3 cm^{-1} for samples 2*b* and 2*c*. The additional peaks can be seen in the spectrum of third sample. The peak at 1406.4 cm^{-1} is a ν_3 mode of transpolyacetylene [8]. The group at 1270.4 cm^{-1} which is a peak in the phonon density of states of diamond can be observed in the case of very small particles [9]. SEM image shows that obtained carbon formations have distinct crystallographic faces. These data are in good agreement with the results of study by Raman scattering.

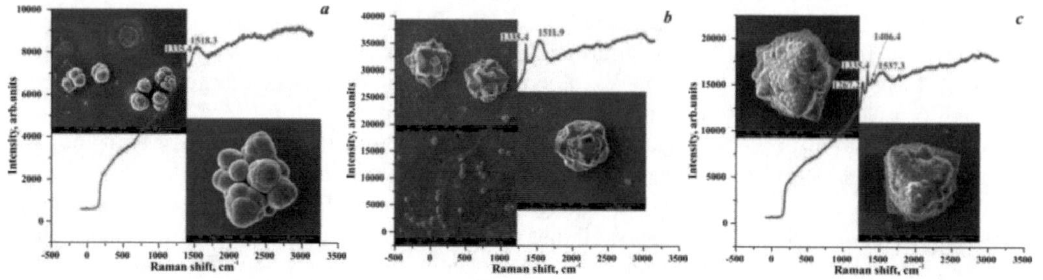

Figure 2(a, b, c). Raman spectra and SEM images of carbon structures obtained on Cu (30 min) deposited on Si (111) for h = 4 mm: **(a)** - $C_{O2}/_{C2H2}$ = 0.93, 30 min, **(b)** - $C_{O2}/_{C2H2}$ = 0.94, 30 min, **(c)** - $C_{O2}/_{C2H2}$ = 0.94, 45 min

Figure 3 presents the Raman spectra and SEM images of carbon structures obtained on Cu films grown on Si (100) for 60 minutes. All spectra show peak at 1331.3 cm^{-1}, which is characteristic of the diamond phase. Location of G group shifts to lower frequencies over time. The second order group $2D$ within 2788.1-2866.1 cm^{-1} can be observed in samples 3*b-d*. The presence of diamond microcrystals is confirmed by SEM images. It is obvious that the smoothing of the crystallite faces (SEM images) and a reduction of the diamond peak intensity (Raman spectra) occurs with an increase in the time of synthesis. This suggests that there is a gradual graphitization of structure.

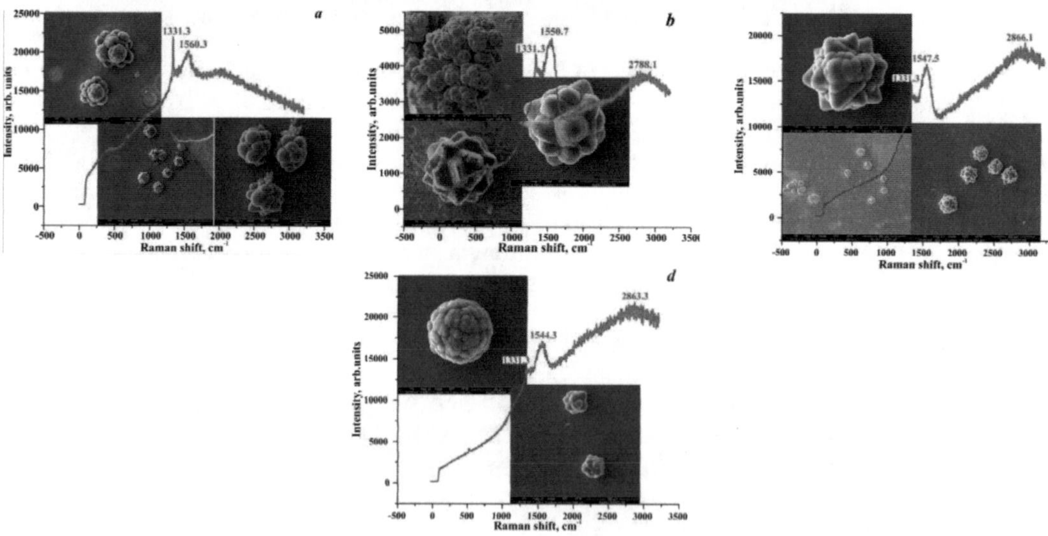

Figure 3(a, b, c, d). Raman spectra and SEM images of carbon structures obtained on Cu (60 min) deposited on Si (100) for $C_{O2}/_{C2H2}$ = 0.94, h = 4mm: **(a)** - 15 min, **(b)** - 30 min, **(c)** - 45 min and **(d)** - 60 min

RS and SEM results show that deposition time of copper film and consequently its thickness mainly effect on the structure formation of diamond crystals. The shift of diamond peak is observed on films deposited for 30 minutes, which is possible due to the presence of stress in the crystals. Whereas, samples synthesized on the copper buffer layer with thickness of 524.6 nm show the peak at 1331.3 cm^{-1}. In addition, it can be seen that the growth of diamond crystals has more massive character in the second case than that of the first.

4. Conclusions

The experiments on synthesis of carbon structures by the method of oxy-acetylene torch on copper films were conducted in the course of studies. According to the performed analysis by RS and SEM some obtained samples have a diamond structure with well-defined crystal faces and edges.

Analysis of the results of the experiments showed that the parameters rendering an important influence on the structure of the samples are:

- The time of deposition of copper film and its thickness;
- The orientation of the silicon substrate;
- The concentration ratio of oxygen and acetylene.

Thus, the process parameters in which occurs the synthesis of micro- and nano-diamonds have been defined. Further studies will be directed at determination of the additional process parameters that effect on the structure formation and the search for their optimal combination.

Acknowledgments

This work is performed with partial financial support of grants of Committee of Science MES the Republic of Kazakhstan №3816/ГФ4, №3823/ГФ4.

References

[1] May P W 2000 *Phil. Trans. R. Soc. Lond.* **358** 473-495
[2] Maia F C B, Samad R E, Bettini J, Freitas R O, Vieira Junior N D, Souza-Neto N M 2015 Scientific Reports 5 11812
[3] Chengzhi L, Xiang Q, Chunxu P, Wenge Y 2015 *Scientific Reports* **5** 13879
[4] Mansurov B Z, Aliyev B A, Medyanova B S, Kenzhegulov A, Tolegen A, Partizan G, Mansurova M E, Mansurov Z A 2015 *Proc. Annual Int. World Conf. on Carbon* 89
[5] Mamoru T, Osamu K 2015 *Advanced Materials Research* **1110** 277-283
[6] Sabitov S, Koshanova A, Medyanova B S, Partizan G, Mansurov B Z, Aliyev B A 2015 Combustion and plasmachemistry 13 47-52 (in Russian)
[7] Ye H, Sun Ch Q, Hing P, Xie H, Zhang S, Wei J 2000 *Surface and Coatings Technology* **123** 129–133
[8] Ferrari A C, Robertson J 2004 *Phil. Trans. R. Soc. Lond.* **362** 2477–2512
[9] Feng X, Matranga Ch, Vidic R, Borguet E 2004 *J. Phys. Chem.* **108** 19949-19954

Evaluation of the influence mode on the CVC GaN HEMT using numerical modeling

Ya M Parnes[1,3], V G Tikhomirov[1,3], V A Petrov[1,3], A G Gudkov[2], I N Marzhanovskiy[2], E S Kukhareva[2], V N Vyuginov[3], V V Volkov[3], A A Zybin[3]

[1]Saint-Petersburg State Electrotechnical University "LETI", St. Petersburg 197376, Russia
[2]Bauman Moscow State Technical University, Moscow 105005, Russia
[3]Closed Joint Stock Company "Svetlana-Electronpribor", St. Petersburg 194156, Russia

Abstract. Done numerically simulated the effects of certain modes of operation on the CVC of field microwave transistors on the basis of heterostructures AlGaN / GaN (HEMT). The results of these studies suggest the possibility of quite efficient use of numerical simulation for the development of HEMT microwave transistors allowing for the real instrument designs.

1. Introduction
The use of the mathematical modeling methods, based on the multidimensional numerical models of calculation of the carrier transport, thermal processes and dynamical characteristics helps to add, check and explain existing experimental data. Also behavior of the device structures with changing parameters can be predicted that is difficult to reproduce experimentally [1-3]. In this work we present the results of numerical modeling of influence of some operating parameters and external impacts on the current-voltage characteristics (CVC) of field effect transistors – high electron mobility transistor (HEMT), based on the AlGaN/GaN heterostructures.

2. Results and discussion
Prediction of field effect transistor output characteristics requires the model which includes the statistics of carrier distribution in the heterostructure layers, peculiarities of carrier transport and electron mobility in the HEMT channel to account the heating up of the carriers and adequate description of current processes under external impacts. A typical AlGaN/GaN HEMT was selected as the object of our modeling. It consists of a 4H semi-insulating silicon carbide (SiC) substrate, a buffer layer of unintentionally doped (UID) 3 μm-thick GaN, 20 nm – thick AlGaN barrier layer. The chematic cross section of the modeled AlGaN/GaN HEMT is shown in Figure 1.

We have optimized the mathematical models used to analyze the characteristics of our heterostructures in accordance with recommendations given in work [4]. A two-dimensional hydrodynamic mathematical model is used in many industrial simulation systems [5], which, combined with the original models of the electrons behavior in material media, gives good results [2]. Similar approaches are described in other works [6-8]. When selecting a model of electron transport, we take into account the balance between the speed of calculations and the required accuracy. The calculation of the drift-diffusion model does not meet the demands of practice in the calculation of sub-micron transistors [9]. We believe that combining of the Monte Carlo method with the well-established and calibrated hydrodynamic model is the best choice in many cases. We have solved the system of four differential equations in partial derivatives: electric field Poisson's equation, current continuity equation, the heat flux (the lattice heat flow equation) and energy balance for the electrons, which are solved self-consistently. This system of equations is supplemented by specific equations for material media (the mobility of the carriers, the electron density, thermal conductivity, etc).

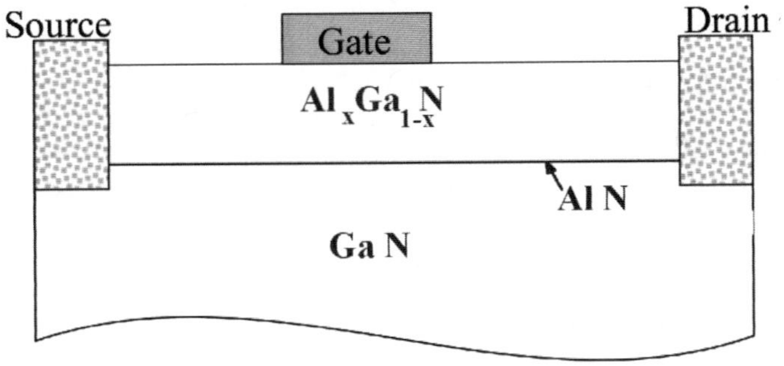

Figure 1. Schematic cross section of the modeled AlGaN/GaN HEMT.

First, we have adapted the numerical models to the features of transistor device design and its technology [2, 3]. Next, the calculation of the static current-voltage and capacitance-voltage characteristics of the HEMT has been done. Figure 2 shows the experimental and calculated static current-voltage characteristics of the modeled AlGaN/GaN HEMT. The very good agreement of modelling results and experimental data were achieved.

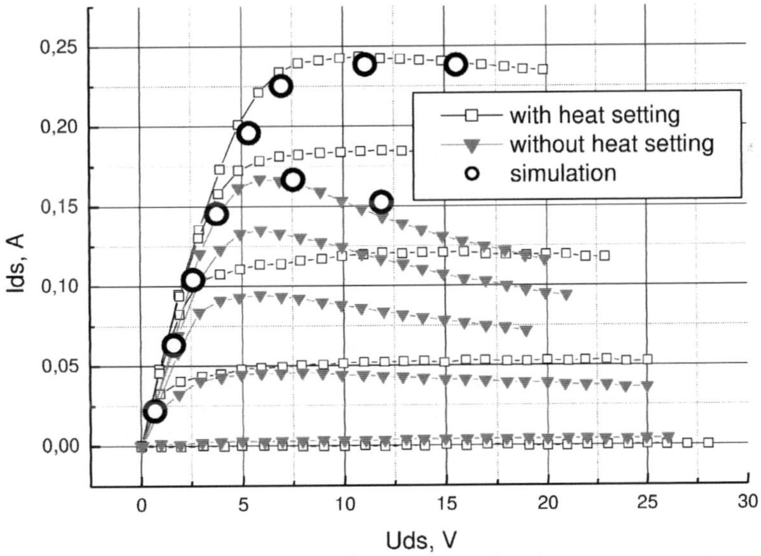

Figure 2. Experimental and calculated in the self-heating of the transistor current-voltage characteristics with and without heat setting.

Next the impact of the temperature on the current-voltage characteristics of AlGaN/GaN HEMT was studied. As described in [10] the heating of the device may results in degradation of the ohmic contacts and in worsening of the CVC of the semiconductor device. Numerical modeling of ohmic contacts degradation under heating was carried out. The dependence of contact resistance (R_C) on the temperature (T) and the time (t) can be written as follows:

$$R_C = R_{C0} + \gamma \cdot \exp\left(\frac{-E_a}{2kT}\right) \cdot \sqrt{t} \, ,$$

where R_{C0} is the contact resistance at the initial time. For contact area S_C = 25 mm^2 and specific contact resistance of 10^{-6} Ohm cm^2 [10], R_{S0} is 4 Ohms. Using the equation and these parameters we calculated CVC of AlGaN/GaN HEMT before and after heating. The effect of the temperature was modeled in five cycles of varying time duration at T = 300 °C. Parameters of the simulated temperature cycles are shown in Table 1.

Table 1. The duration of the temperature effect on transistor.

№ of cycle	time tc, hour	total time of thermal impact tΣ, hour
1	1	1
2	2	3
3	5	8
4	10	18
5	9	27

The simulation results of the temperature impact on the current-voltage characteristics of AlGaN/GaN HEMT with a gate width of 100 μm are shown in Figure 3.

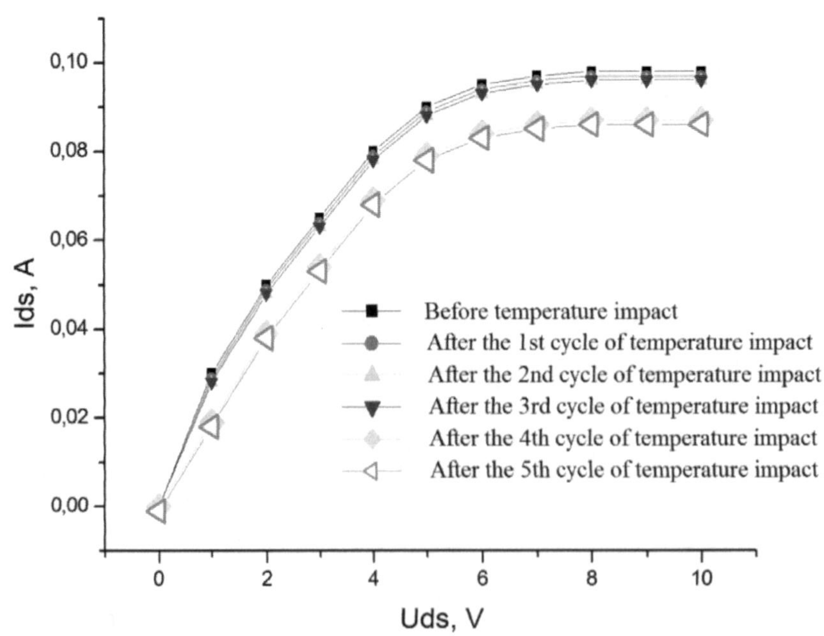

Figure 3. Current-voltage characteristics of AlGaN/GaN HEMT during 5 cycles of heating.

We observe the noticeable change in the CVC behavior after the 4th cycle of thermal impact with duration of 10 hours with and 18 hours of the total heating time. At the subsequent cycle of thermal exposure significant changes of CVC form does not occur compared to previous form of the CVC, measured after the 4th cycle of temperature effects.

3. Conclusion

To conclude, the numerical model of the AlGaN/GaN HEMT is developed. The results of numerical modeling of AlGaN/GaN HEMT current-voltage characteristics and the heating effect are presented. In is shown that the noticeable changes in the shape of CVC begin to appear only after the 4th cycle of thermal impact with duration of 10 hours (total time of thermal impact is 18 hours).

Acknowledgments
This study was supported by the Ministry of Education and Science of the Russian Federation (agreement on grant no. 14.574.21.0116, the unique project identifier is RFMEFI57414XO116).

References

[1] Palankovski V, Quay R, Selberherr S, Industrial Application of Heterostructure Device Simulation, *IEEE J. Solid-State Circuits*, 2001, Vol. 36, N. 9, pp.1365-1370
[2] Tikhomirov V, Zemlyakov V, Volkov V, Parnes Ya, Vyuginov V, Lundin W, Sakharov A, Zavarin E, Tsatsulnikov A, Cherkashin N, Mizerov M, and Ustinov V, *Semiconductors*, 2016, Vol. 50, N. 2, pp. 244–248
[3] Vitanov S, Palankovski V, Murad S, Roedle T, Quay R, Selberherr S. Predictive simulation of AlGaN/GaN HEMTs. In: IEEE compound semiconductor IC symposium; 2007. p.131–4
[4] Vitanov S, et al. Solid-State Electronics 54 (2010) 1105–1112
[5] Grasser T, Ting-Wei T, Kosina H, Selberherr S. A review of hydrodynamic and energy-transport models for semiconductor device simulation. Proc IEEE 2003;91(2):251–74
[6] Faraclas E, Anwar A. AlGaN/GaN HEMTs: experiment and simulation of DC characteristics. Solid-State Electron 2006;50(6):1051–6
[7] Chang Y, Tong K, Surya C. Numerical simulation of current/voltage characteristics of AlGaN/GaN HEMTs at high temperatures. Semicond SciTechnol 2005;20(2):188–92
[8] V. G. Tikhomirov, N. A. Maleev, A. G. Kuzmenkov, Yu. V. Solov'ev, A. G. Gladyshev, M. M. Kulagina, V. E. Zemlyakov, K. V. Dudinov, V. B. Yankevich, A. V. Bobyl', and V. M. Ustinov, Semiconductors 45, 1352 (2011).
[9] Polyakov V, Schwierz F. Influence of electron mobility modeling on DC I–V characteristics of WZ-GaN MESFET. IEEE Trans Electron Dev 2001;48:512–6.
[10] M. Murakami. Development of refractory ohmic contact materials for gallium arsenide compound semiconductors. Science and Technology of Advanced Materials. 2002. V. 3. P. 1-27.

Journal of Physics: Conference Series **741** (2016) 012025 doi:10.1088/1742-6596/741/1/012025

Control of stress and threading dislocation density in the thick GaN/AlN buffer layers grown on Si (111) substrates by low-temperature MBE

D Zolotukhin*[1], D Nechaev[1], N Kuznetsova[1], V Ratnikov[1], S Rouvimov[2], V Jmerik[1], S Ivanov[1]

[1]Ioffe Institute, Polytekhnicheskaya 26, St. Petersburg 194021, Russia
[2]University of Notre Dame, Notre Dame, Indiana 46556, USA

E-mail: zolotukhinspb@gmail.com

Abstract. We report on successful growth by plasma-assisted molecular beam epitaxy on a Si(111) substrate crack-free GaN/AlN buffer layers with a thickness more than 1 μm. The layers fabricated at relatively low growth temperature of 780°C have at room temperature the residual compressive stress of -97 MPa. Intrinsic stress evolution during the GaN growth was monitored *in situ* with a multi-beam optical system. Strong dependence of a stress relaxation ratio in the growing layer vs growth temperature was observed. The best-quality crack-free layers with TDs density of $\sim 10^9$ cm^{-2} and roughly zero bowing were obtained in the sample with sharp 2D-GaN/2D-AlN interface.

1. Introduction

III-Nitrides on Si(111) templates are of great importance for fabrication of high electron mobility transistors (HEMT) [1] and power electronics due to low cost, large available sizes of silicon substrates, and the possibility of integration with the existing Si technologies. However, heteroepitaxy of III-Nitrides on Si(111) suffers from large lattice and thermal expansion coefficients (CTE) mismatches (17% and 54% for GaN on Si(111), respectively [2]) that leads to tensile strain generation during growth and cooling to room temperature (RT) of heterostructures. Conventionally, III-N layers grown on Si substrates by vapour phase epitaxy (VPE) [3] and metal-organic chemical vapour deposition (MOCVD) [4] have better structural quality than layers grown by molecular-beam epitaxy (MBE) [5]. Despite the lower TDs density obtained by MOVPE, the performance of optimized devices grown by MBE is not inferior to that of devices grown by MOVPE [6]. Moreover, VPE and MOCVD require a higher temperature for epitaxial growth of III-N layers than PA MBE. Hence, much more tensile stress generation is observed in samples grown by VPE and MOCVD during cooling down to RT. This tensile stress leads to film cracking [7] and bowing of the heterostructure, which can destroy the device performance or significantly complicate the post growth processing. The solutions for this problem can be reducing the growth temperature and using less lattice-mismatched (~2.4%) [8] AlN buffers. The interface between AlN and growing thick GaN layer is an effective source of compressive stress. Therefore, low growth temperature and *in situ* control of the stress level could make PA MBE suitable for growth of compressively strained GaN/AlN/Si(111) templates for further device fabrication.

The paper reports on fabrication of compressively strained at room temperature GaN/AlN buffer layers on Si(111) substrates by PA MBE In addition, *in situ* analysis of stress evolution and morphology in the growing heterostructure has been presented, which has been done by home-made, multi-beam optical system (MOS), and Reflection High-Energy Electron Diffraction (RHEED)

2. Experimental

The Si substrates with thickness of 275 μm were annealed at 820°C for an hour until the 7×7 surface reconstruction was observed by RHEED. Before AlN growth, 6 MLs of Al were deposited to prevent Si-N interaction. Then, 30-nm-thick AlN:Ga/Si(111) nucleation layer (NL) was grown by migration enhanced epitaxy (MEE) at Al-rich conditions with F_{Al}/F_N=1.02 [9], while substrate temperature simultaneously was increased from T_S=700°C to T_S=780°C, followed by an intermediate 170-nm-thick 2D-AlN buffer layer grown by metal-modulated epitaxy (MME) [10] technique at the flux ratio F_{Al}/F_N=1.6 and the T_S=780°C. Both NL and intermediate AlN layers were grown with the additional Ga-surfactant flux F_{Ga}=0.3 ML/s. The top 1-μm-thick GaN layer was grown at the Ga-rich conditions with the flux ratio F_{Ga}/F_N=1.9 and lower T_S=700°C.

RHEED and laser reflectometry were used for in situ control of the surface morphology and growth rate of the layers. In addition, a home-made multi-beam optical system (MOS) was installed at the standard ellipsometry ports of the setup to measure *in situ* substrate curvature. The surface morphology was investigated *ex situ* by an atomic force microscope (AFM). Crystalline quality of the heterostructures was evaluated by high-angle annular dark-field scanning TEM (HAADF STEM) (FEI Titan 80-300 keV), and X-ray diffraction (XRD) analysis of symmetric (0001) and skew-symmetric (10-15) reflexes. The latter were used to evaluate densities of TDs with Burgers vectors **b=c** or **b=a+c** (N_{screw}) and **b=a** or **b=a+c** (N_{edge}) in approximation of random TDs distribution [11].

2.1. Stress – thickness versus thickness plots

Film stresses can be evaluated by measuring the change in curvature of substrate during stressed film growth by using the modified Stoney equation [12]

$$\langle \sigma_f \rangle h_f = \frac{M_s h_s^2}{6} \int_0^{h_f} \sigma(z) dz = \frac{M_s h_s^2}{6} \Delta\kappa, \tag{1}$$

where subscripts *f* and *s* refer to the film and substrate, respectively, $<\sigma_f>$ is the mean film stress, $\sigma(z)$ is the stress at particular thickness z, κ is the curvature, M_s is the substrate biaxial modulus. Thus, by measuring *in situ* the change in heterostructure curvature, stress-thickness product $<\sigma_f>h_f$ can be determined in real time and converted to film stress if the film thickness is known. For film grown at constant stress, the plot of stress-thickness vs thickness is linear dependence with a slope determining the film stress. For films with varying stress with thickness, stress-thickness vs thickness plot is a non-linear function, where the incremental slope $d(<\sigma_f>h_f)/dh_f$ corresponds to microstructure changes in the layer.

3. Results

3.1. Film characteristics

HAADF STEM images at different magnifications (Fig. 1) illustrate rather planar morphology for all the interfaces in this heterostructure. One can observe also inclination of the TDs above the GaN/AlN interface, followed by their fusion or annihilation during propagation toward the film surface.

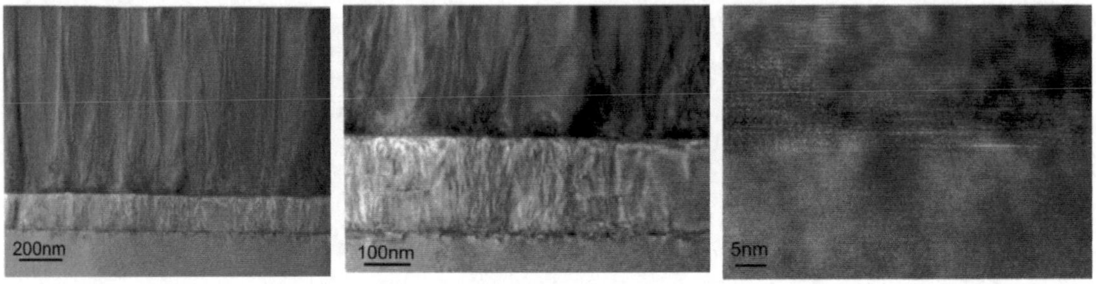

Fig.1 HAADF STEM cross-section images of the GaN/AlN/Si(111) template.

In addition, HAADF STEM revealed a flat interface between the MEE AlN NL and Si(111) substrate without pronounced interaction between Al and Si during first MEE cycles, contrary to step-like interface observed by D. Litvinov *et. al.* [13] in AlN/Si(111) layers grown by PA MBE without special NLs.

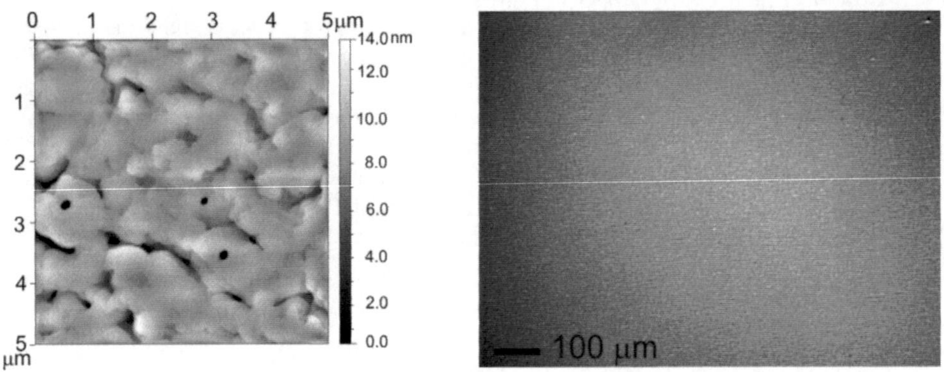

Fig. 2. (left) AFM and (right) OM image of the GaN/AlN/Si(111) PA MBE buffer layer.

The AFM image (figure2. left) of the GaN(1μm)/AlN(170nm)/NL-AlN:Ga(30nm)/Si(111) template showed relatively smooth surface morphology with RMS roughness of 1.7 nm and rather large grain sizes about 2 μm in diameter. The grain boundaries are clearly visible indicating incomplete coalescence, but peak-to-valley height of 14 nm with respect to heterostructure thickness of 1.2 μm corresponds to multi-step coalescence process.

The image of GaN surfaces obtained by optical microscopy (Fig 2. right) revealed crack-free surface of the sample, corresponding to absence of the high level of tensile strain at RT.

XRD θ-2θ scans of the template demonstrated full widths at half maximum (FWHM) of 782 and 1182 for symmetric (0002) and skew-symmetric (10- $\bar{1}$5) reflexes, respectively, that correspond to screw and edge TD densities of $N_{screw}=1.2\times10^9$ cm^{-2} and $N_{edge}=7.4\times10^9$ cm^{-2}

3.2. Intrinsic stress evolution

The stress-thickness versus thickness and stress versus thickness curves for GaN(1 μm)/AlN(200 nm)/Si(111) heterostructure are plotted in figure 3. The stress versus thickness curve that characterize incremental stress in the layer was obtained from polynomial fit and further differentiation, $d(<\sigma_f>h_f)/dh_f$, of stress-thickness vs thickness curve. As seen in figure 3, incremental stress during the growth of AlN layer was tensile and varied from 0.5 to 0.8 GPa. The stress behavior in GaN demonstrates continuous gradual relaxation from the highest observed compressive stress of -3.9 GPa to -0.3 GPa.

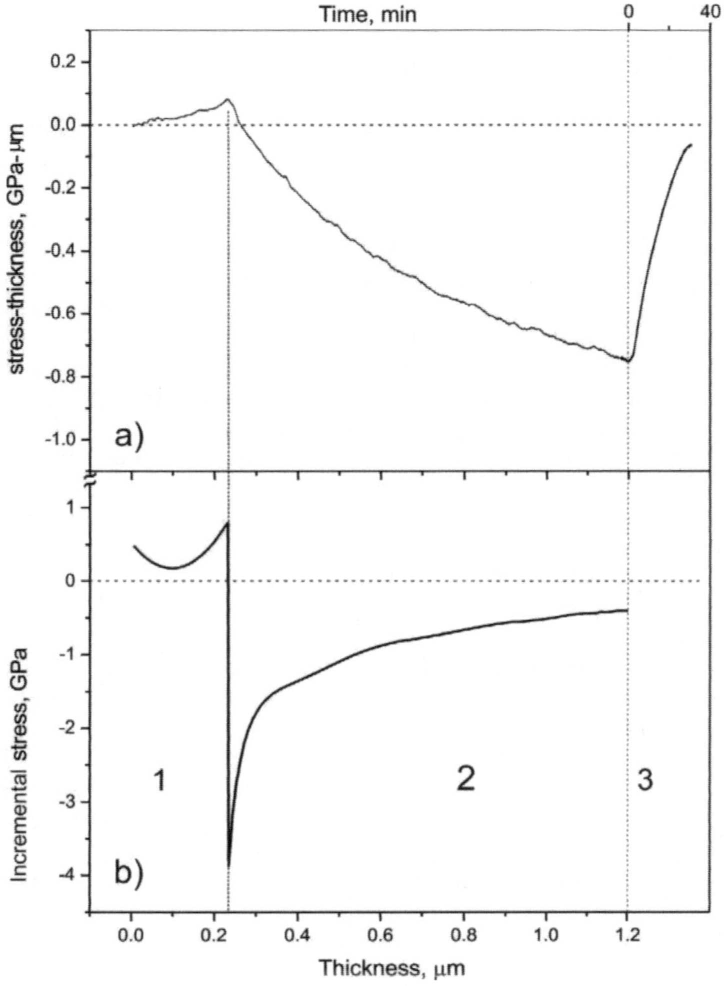

Fig. 3. Stress-thickness vs. thickness (top) and stress vs. thickness (bottom) plots during GaN/AlN/Si(111) heterostructure growth. The digits shows: 1- AlN growth, 2- GaN growth, 3-cooling to RT;

4. Discussion

Theoretically, the stress arising during growth of lattice-mismatched layers should be constant and can be evaluated as:

$$\sigma_f = M_f \varepsilon, \quad (2)$$

wheren M_f is the biaxial modulus of growing layer and ε is the strain. Using M_f=470 GPa for AlN [14] and lattice mismatch between Si(111) and AlN ε=0.19 [8], the tensile stress in AlN layer can be evaluated as σ_f=89 GPa. As shown in figure 3, stress arising during growth AlN layer is varying from initial value of 482 MPa to 800 MPa with the minimum value of 181 MPa and average value of 491 MPa. We speculate that instantaneous relaxation of ~99.5% of tensile stress is related with generation of the misfit dislocation with an edge component and transition to Volmer-Weber (VW) growth mode [15]. Further changes in stress vs thickness behavior are possibly related with multi-step nucleation of grains, their overgrowth, and coalescence processes, which were observed in films grown at VW growth mode [16]. AFM image of overlaying GaN indicates VW mechanism of growth that allows assuming similar mechanism of growth for AlN buffer.

Using biaxial modulus of GaN of 478 GPa [14] and lattice mismatch between GaN and AlN of -2.4% [8], the mismatch stress arising during GaN growth on AlN can be estimated as -11.5 GPa. However, the highest stress observed during GaN on AlN buffer layer deposition is about -3.9 GPa corresponding to 66% relaxation of theoretically calculated stress. The important factor affecting compressive stress formation is the stress level in the AlN buffer layer. As seen in figure 3, at GaN/AlN interface AlN is under a tensile stress of 0.8 GPa reducing the lattice mismatch with GaN to 2.2% and decreasing the theoretical value to -10.7 GPa. Further reduction is expected from generation of the misfit dislocation with an edge component at the interface between two lattice-mismatched layers. As shown in figure 3, incremental stress in GaN layer exhibits two different ratio of relaxation. Starting first monolayer up to thickness of about 100 nm incremental stress in GaN evidences high ratio of stress relaxation, changing from -3.9 to -1.5 GPa. We assume that this stress behavior is similar to behavior observed by N. Baron *et al.* [17], who reported that the main source of stress relaxation in compressively strained layers is reactions between TD's with appropriate burgers vectors. This fact can be confirmed by HAADF STEM image of heterostructure (figure1) which shows that the most intensive interaction between TD's with their further annihilation and TD loop formation was occurred at thickness of up to about 100 nm. We suspect that further gradual stress relaxation is associated with TD's inclination in compressively strained layers, as described in [18]. It should be noted that reactions between TD's at thickness more than 100 nm are also occurred but less intensively due to increasing in distances between TD's. Contrary to S. Raghavan *et al.* [19], who observed transition of incremental stress from compressive to tensile at thickness about 88 nm in samples grown by high-temperature MOCVD, we did not observe in the present study this transition up to thickness of 1 μm in similar heterostructures grown by low-temperature PA MBE. Tensile growth stress along with tensile stresses arising during cooling to RT may lead to film cracking. Since we have not observed the transition of the compressive stress to tensile, we believe that this difference in stress behavior during MOCVD and MBE growth is related to lower growth temperature and metal-rich growth conditions, which imposes certain kinetic barrier for TD's propagation and reactions at early growth stages.

Experimentally measured in this study tensile stress, generated at heterostructure cooling from T_s=700 °C to RT is σ^{CTE}=+0.55 GPa, has been compensated by mean compressive epitaxial stress in the film of σ_f=-0.65 GPa that results in compressively strained at RT GaN(1 μm)/AlN(230 nm)/Si(111) template.

Summarizing, by using low-temperature PA MBE at metal-rich conditions and *in situ* control of substrate curvature, we have obtained compressively strained GaN(1 μm)/AlN(230 nm)/Si(111) template at RT with the smooth crack-free surface with RMS roughness of 1.7 nm. Optimization of NL growth conditions allows us to prevent Al-Si and Ga-Si interaction and achieve flat interfaces between substrate and AlN layer. A comparison of *in situ* measurements of substrate curvature and HAADF STEM of template has confirmed the crucial influence of interplay between TD's and stress behavior in the investigated heterostructures. In addition, a comparison between the high-temperature MOCVD and low-temperature PA MBE growth processes has revealed that the usage of the low growth temperature and the metal-rich conditions is an effective way to suppress relaxation of compressive epitaxial stress arising from the lattice-mismatched AlN/GaN interface.

Acknowledgment

This work has been supported by Russian Science Foundation, grant № 14-22-00107.

References

[1] U. K. Mishra *et al*, 2008 IEEE Proceedings **96**, 287.
[2] M. N. Rahman *et al.* 2016 Appl. Surf. Sci. 362 572.
[3] A. Dadgar et al. 2003 Journal of Crystal Growth **248** 556.

[4] A. Ubukata, K. Ikenaga, N. Akutsu, A. Yamaguchi, K. Matsumoto, T. Yamazaki, T. Egawa 2007 J. Cryst. Growth **298**, 198.
[5] F. Semond, P. Lorenzini, N. Grandjean, J. Massies 2001 Appl. Phys. Lett. **78**, 335
[6] S. W. Kaun, M. H. Wong, U. K. Mishra, J. S. Speck 2013 Semicond. Sci. Technol. **28**, 074001 .
[7] J. Han *et al.* 2001 Appl. Phys. Lett. **78**, 67.
[8] D. Zhu, D.J. Wallis, C.J. Humphreys 2013 Rep. Prog. Phys. **76**, 106501.
[9] D. V. Nechaev, P. A. Aseev, V. N. Jmerik, P. N. Brunkov, Y. V. Kuznetsova, A. A. Sitnikova, V. V. Ratnikov, S. V. Ivanov 2013 J. Crys. Growth **378**, 319.
[10] V. N. Jmerik, A. M. Mizerov, D. V. Nechaev, P. A. Aseev, A. A. Sitnikova, S. I. Troshkov, P. S. Kop'ev, S. V. Ivanov 2012 J. Crys. Growth **354**, 188.
[11] M.A. Moram, M.E. Vickers 2009 Rep. Prog. Phys. **72**, 036502.
[12] J. A. Floro, E. Chason, S. R. Lee, R. D. Twesten, R. Q. Hwang, L. B. Freund 1997 J. Electron. Mater. **26**, 969.
[13] D. Litvinov, D. Gerthsen, R. Vohringer, D.Z. Hu, D.M. Schaadt 2012 J. Cryst. Growth **338**, 283
[14] I. Vurgaftman, J. R. Meyer 2003 J. Appl. Phys. **94**, 3675.
[15] J. Floro, E. Chason, R. Cammarata, D. Srolovitz 2002 MRS. Bull. **27**, 19.
[16] Y. Lua, X. Liua, D. Lua, H. Yuana, G. Hub, X. Wanga, Z. Wanga, X. Duan 2003 J. Cryst. Growth **247**, 91.
[17] N. Baron, Y. Cordier, S. Chenot, P. Vennéguès, O. Tottereau, M. Leroux, F. Semond, and J. Massies 2009 J. Appl. Phys. **105**, 033701.
[18] A. E. Romanov, J. S. Speck // Appl. Phys. Lett. **83**, 2569 (2003)
[19] S. Raghavan, J. M. Redwing // J. Appl. Phys. **98**, 023514 (2005)

Segmented flow reactor for synthesis of quantum dot nanocrystals and plasmonic nanoparticles

R C Mbwahnche[1], L B Matyushkin[1], O A Ryzhov[1], O A Aleksandrova[1], V A Moshnikov[1,2]

[1]St. Petersburg State Electrotechnical University "LETI", Pr. Popova 5, St. Petersburg, 197376 Russia

[2]Peter the Great St. Petersburg Polytechnic University, Polytechnicheskaya, 29, St. Petersburg, 195251, Russia

Abstract. This research presents an automated method of synthesizing semiconductor and metal nanoparticles using flow rector synthesis as a new alternative to the batch method of synthesizing nanoparticles. Experiments were carried out to determining the optimal flow rates of reagents droplets. The reactor was successfully applied to the synthesis of colloidal solutions of semiconductor (CdSe) and metal (Ag) nanoparticles. This instrument is applicable both in material science laboratories and in industry.

1. Introduction

The most popular method for the synthesis of monodisperse semiconductor nanoparticles is the hot injection method, proposed by the Bawendi group in 1993 [1]. However, this technique is performed in a chemical flask, which prevents the technology from scaling to the industrial level and leads to high market price of such nanoparticles. This was the cause for the flow reactor synthesis [2] as a technology that allows large-scale automated production and at the same time offers such benefits as better control of the heat and mass transfer, better precision of the cooling rate, and reproducibility. The flow synthesis allows monitoring parameters and spending reagents more efficiently than the traditional chemical synthesis [3].

There are two major types of flow reactors used in the synthesis of quantum dots; continuous and segmented flow. In continuous flow reactors, reaction and precipitation occur in the same phase. Miscible streams of reagents are injected into the channel where they mix and react. This have been used to produce high quality nanoparticles like metals [4], metal oxides [5] and compound semiconductors [6]. However, the liquid front experience friction with the walls of the channel which induces a parabolic velocity profile across the channel with particles contacting the walls of the channel having greater resident time compared to particles located at the center figure 1. This variation of speed leads to polydisperse size distribution. Also over time, due to contact with the channel walls, precipitating particles begin to accumulate and form a stagnant layer adjacent to the channel walls leading to eventual fouling of the channel.

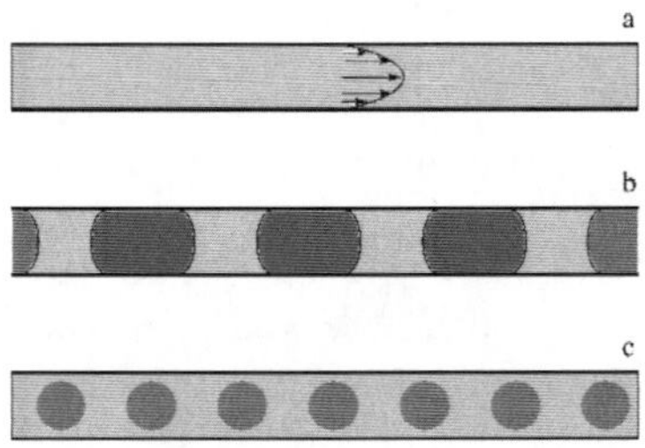

Figure. 1. The diagram illustrating continuous (a) and segmented (b, c) modes of flow in channels with a circular cross section

In contrast to continuous flow microractors, for a segmented flow reactor, an additional immiscible fluid is also injected at the same time with the reagents. This carrier fluid splits and surrounds the reacting mixture forming tiny droplets, which flows through the channel. Inside the droplets, a uniformly circulating flow profile develops providing a completely uniform residence time for the nanoparticles and an enhanced mixing of reagents. [7]. The carrier fluid can be gas as in the case of "gas/liquid" mode or a liquid as in a "liquid/liquid" mode. Although the "gas/liquid" mode prevent the parabolic velocity flow profile, the mixing reagents droplets continue to make contact with the channel walls, eventually leading to channel blockage. This challenges, however, are eliminated by the "liquid/liquid" mode as it not only provides an equal resident time within a droplet, but it also isolates the droplets from the walls of the channel by wetting the full length of the channel wall. This is better achieved when the carrier phase is in excess.

Segmented flow reactor was first used to synthesis nanocrystal at room temperature [8]. However, in order to increase particle crystallinity there is need to synthesis nanocrystal at higher temperature. Chan et al. in 2005 synthesised CdSe quantum dots [9] and after a recorded period of five hours the surface coating of the channel wall began to degrade rapidly resulting to an irregular flow. Later on, using the same reagent and carrier liquid Nightingale et al. developed a more robust segmented flow reactor using polytetrafluoroethylene polymer with the ability to withstand temperature to about 250 °C, beyond which the polymer starts to degrade [10].

Like stated above, flow reactor allows large-scale automated production. Scaling out is rather simpler than in batch reaction since it does not require changing the reactor geometry and also although the material throughput increases, the chemistry is the same. This can be achieved by operating large numbers of identical reaction channel in parallel [11]. Notable success was recorded using five –channel droplet-based microreactor to synthesize CdSe [12]

Recently there have been improvement of flow reactor aiming at widening the range of materials and chemistries to which it can be used. The one-step procedure, in which all reagents are loaded into the droplets at the outset, limits the range of materials that can be used. However, the multistep droplet reactor allows reagents to be sequentially added into the flowing droplets as the reaction proceeds. The report [13] show a three phase system which involves a gas phase and two immiscible liquids.

In our research work, however, we developed a high temperature, two phase segmented flow reactor using polytetrafluoroethylene (PTFE) capillary channel. The reactor is of a modular construction with: isolated temperature zones in order to separate nucleation and growth of particles, a passive generation of droplets and a compact for primary diagnosis of the nanocrystals.

2. Flow reactor design development

The working principle is such that precursors A and B and the carrier liquid C which are contained in three separate syringes are squeezed out at a predetermined speed using three syringe pumps, figure 2. The liquids go into the mixing cell, where a droplet flow of precursor solution in the carrier medium is formed. This total flow moves through the capillary (inner diameter 800 μm), passing through two zones with different temperatures: T_n is the temperature corresponding of nanocrystal nucleation and T_g is the temperature of the nanoparticles growth. The final stage is the selection of colloidal solutions of nanoparticles and their primary diagnosis by the compact spectrophotometer.

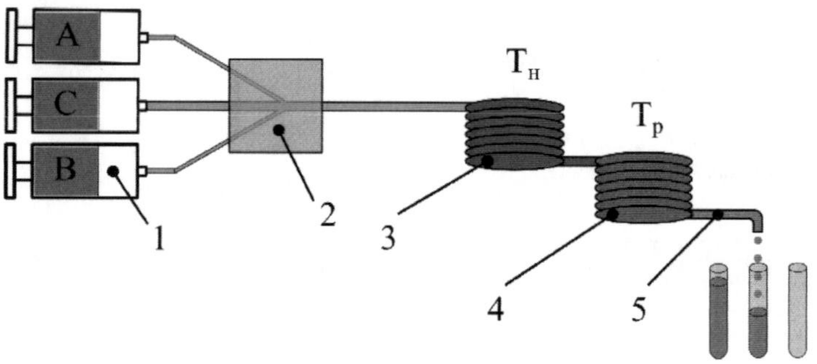

Figure 2. Schematic diagram of the installation: 1 — syringe pumps, 2 — mixing module,3 — nucleation module, 4 — growth module, 5 — selection of nanoparticles

The whole device is operated with the help of a microcontroller Arduino Mega 2560 and a personal computer with special software developed in NI LabVIEW [14]. The program sets the required flow rates and temperatures of nucleation and growth zones. With this design we were able to isolates the various processes; mixing, precipitation process, nucleation and growth. Also we used a transparent PTFE polymer, which enabled us directly observe and precisely control the whole process of the synthesis.

One of the important factors for obtaining the monodisperse nanoparticle in droplet form is a stable flow rate control system. The flow rate control system consists of syringe pumps which are of modular construction, stepper motor and the transmission system which was fabricated using a 3D-printer. The precursors and the carrier liquid are ejected from the syringe pumps using a stepper motor. Due to the need for a relatively low flow rates of the liquid, in our previous work we used a cylindrical gear transmission with a transfer coefficient of six. However, at very low flow rate the motors began to vibrate. The pump design together the software program was therefore modified, replacing the gears with belt having a transfer coefficient of four and the minimum flow rate of 1.5 l / sec.

Although we achieved a minimum flow rate that favour the formation of droplet, there was a need for turbulence at the onset of the reaction before the droplets are formed in order to maximise mixing of the reagent. Three basic structure were proposed; flow-focusing figure 3a, Y-shape, figure 3b and T-type figure 3c microchannel. So as to include the carrier fluid we constructed a device, figure 4, which is a modification of the Y-shape microchannel to enhance mixing, preceding the formation of droplet. This in fact was an important objective of the pump design

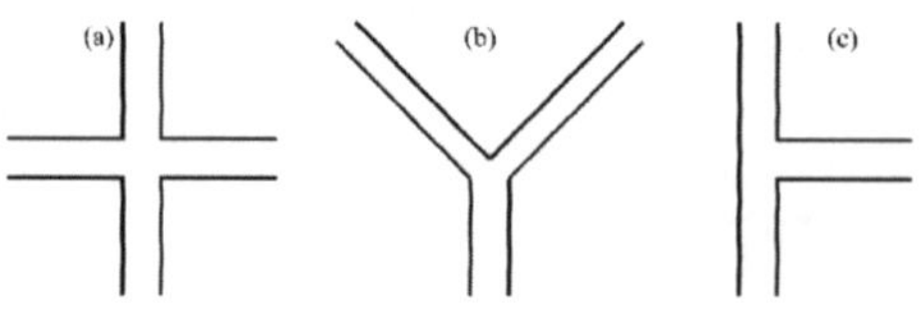

Figure 3. Microchannel for nanoparticle precipitation: (a) flow focusing microchannel; (b) Y-shaped microchannel; and (c) T-type microchannel

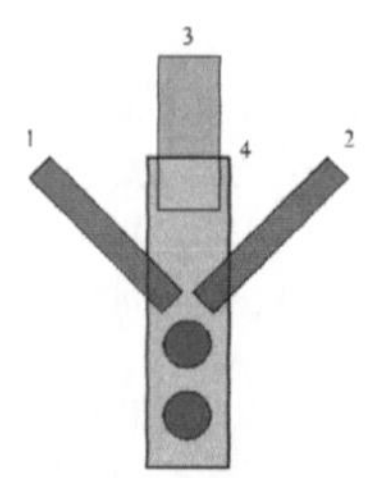

Figure 4. Schematic picture of the device for forming droplets: 1,2– reagents inlet, 3– carrier fluid inlet, 4– microreactor channel

The production of monodisperse colloids requires a temporally discrete nucleation event followed by slower controlled growth on the existing nuclei[15]. While particle size depends on the duration of the growth phase, which is regulated by the flow rate and the number of capillary loops in the growth module. We separated this stages in order to control have a precise control on the synthesis. The main elements for temperature control system are heating element, and oil bath in which is immersed a platinum thermal resistance temperature detector that measures the temperatures T_n and T_g. The heat from the heating element is transferred to a beaker made of heat-resistant glass filled with liquid with a high boiling point, in which a part of the capillary is immersed.

3. Result

To determine the optimum flow rates for precursors mixing and the formation of the "liquid/liquid" flow, we used water-based inks as model precursors and hexane as the carrier liquid as reported in the report [16], figure 5. According to the results, it can be assumed that the optimum flow rate range for three streams, when the precursors are completely mixed and the carrier liquid flow divides into droplets, is in the range 10–50 μl/s. It is important to note that different pairs of liquids require another assessment.

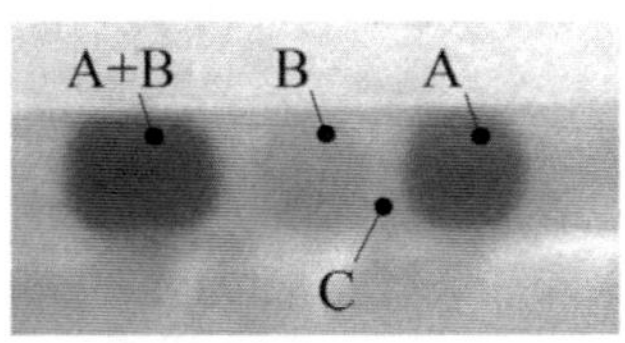

Figure 5. Flow elements in the PTFE capillary: A and B – modelling precursors, C – carrier fluid

Figure 6 shows the results of the experiment, in which all flow rates equally varied vA = vB = vC. The smaller the flow rates, the smaller the droplet size of mixed precursors, but at a rate of 1,2 μl/s the precursor flows do not mix uniformly (color of drops changes). At the same time, too high speed causes the transition of flow shape to laminar. These results suggest that the best speed is in the range 10–20 μl/s

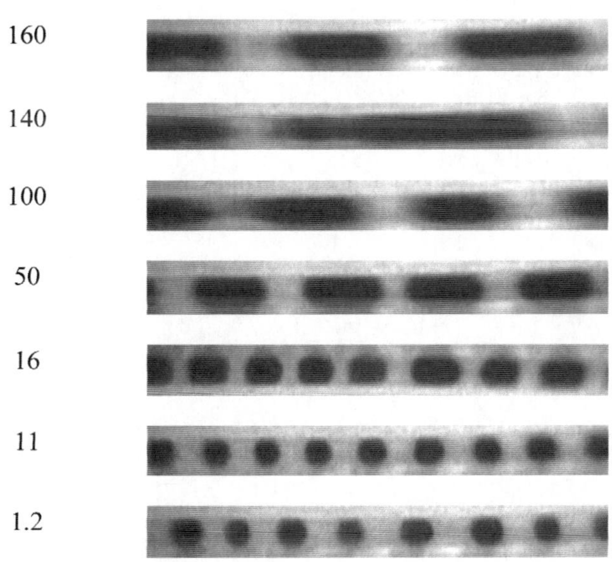

Figure 6. Evolution of the shape of the segmented flow with a simultaneous change of the flow rates (μl/s) precursor A, B and C of the liquid medium

3.1. Metal Nanoparticles Synthesis
The synthesis of metal nanoparticles was carried out using the example of silver nanoparticles. The source of the silver ions (liquid A) is a solution of silver nitrate AgNO3, a solution of ascorbic acid and sodium citrate was used as a reductant and stabilizer (liquid B). The carrier (liquid C) was toluene. The synthesized particles have strongly pronounced plasmon resonance Figure 7. Compared with results in [17], it can be assumed that synthesizing nanoparticles have a diameter of 25–30 nm with the assumption that all particles are spherical.

3.2. Semiconductor Nanoparticles Synthesis
The synthesis of semiconductor nanoparticles was tested on CdSe nanoparticles in aqueous medium, where liquid A was a NaHSe solution (Se2– precursor), B – CdCl2 (Cd2+ precursor) and L-cysteine (a stabilizer of nanoparticles), C – dodecane. Position of the first exciton peak in the absorption spectrum moved to longer wavelengths as the growth of the particles Figure 8. From the absorption spectrum of the synthesis, average diameters of the nanoparticles were found using the formula from [18]: 1 min – 1.4 nm and 15 min – 1.8 nm.

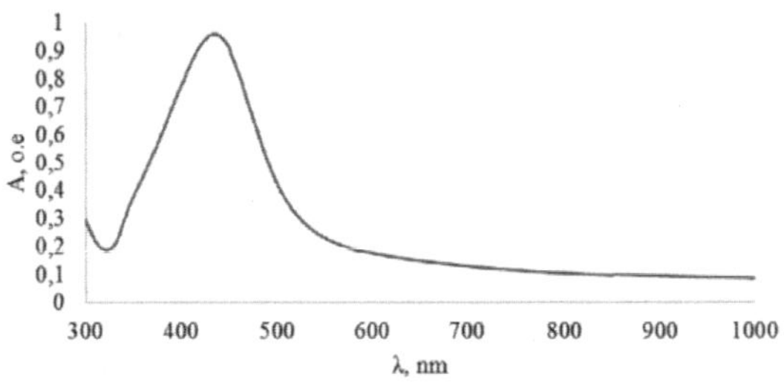

Figure 7. The spectrum of the optical density for silver nanoparticles

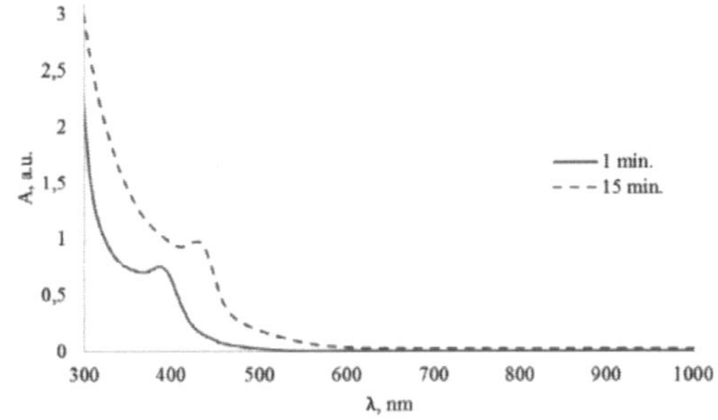

Figure 8. The spectrum of the optical density for CdSe nanoparticles with different growth time

4. Conclusions
The study developed a laboratory model of the flow reactor system for the synthesis of nanoparticles. The proposed technique involving syringe pumps allows creating a segmented stream with precision control. It is shown that using the proposed hardware solution, it is possible to synthesize semiconductor and metal nanoparticles in the aqueous medium. From the absorbance spectra, particle sizes were defined for Ag it is about 25–30 nm, and in the case of CdSe – 1,4 and 1,8 nm for 1 and 15 min. of growth respectively. The obtained nanoparticles were used in studies [20, 21] as a replacement for traditional biomarkers (phosphors) in biology and medicine

Acknowledgments
The research was carried out under a grant from the Russian Research Foundation (project 14–15–00324).

References
[1] Murray C B, Norris D J and Bawendi M G 1993 *J. Am. Chem. Soc.* **115(19)** 8706 reference
[2] Nakamura H, Yamaguchi Y, Miyazaki M, Maeda H, Uehara M and Mulvaney P 2002 *Chem. Commun.* **23** 2844
[3] Pan J, El-Ballouli A O, Rollny L, Voznyy O, Burlakov V M, Goriely A and Bakr O M 2013 *ACS nano* **7(11)** 10158 r reference
[4] Song Y J, Henry L L and Yang W T 2009 *Langmuir* **25** 10209–117
[5] Miyake T, Ueda T, Ikenaga N, Oda H and Sano M 2005 *J. Mater. Sci.* **40** 5011–13
[6] Edel J B, Fortt R, deMello J C and deMello A J 2002 *Chem. Commun.*1136–37

[7] Chan E M, Alivisatos A P and Mathies R A 2005 *J. Am. Chem. Soc.* **127** 13854–61
[8] Shestopalov I, Tice J D and Ismagilov R F 2004 *Lab Chip* **4** 316–21
[9] Chan E M, Alivisatos A P and Mathies R A 2005 *J. Am. Chem. Soc.* **127** 13854–61
[10] Nightingale A, Krishnadasan S H, Berhanu D, Niu X, Drury C, Mcintyre R, Valsami- ones E and de Mello J C 2011 *Lab Chip* **11** 1221–7
[11] Phillips T W, Ioannis G L, and Richard M M 2014 *Lab on a Chip* **14(17)** 3172-80
[12] Nightingale A, Bannock J H, Krishnadasan S H, O'Mahony F T F, Haque S A, Sloan J, Drury C, McIntyre R and de Mello J. C, 2013 J. Mater. Chem. A **1**, 4067
[13] Nightingale A M., et al 2014 *Nature communications* **5**
[14] Ryzhov O A, Matyushkin L B 2015 JPCS **643** 012016
[15] Overbeek JThG 1982 Adv. Coll. Int. Sci. **15** 251–77
[16] Aleksandrova O A, Moshnikov V A, Matyushkin L B, Mbwahnche R, and Ryzhov O A 2015 *Humanities and Science University Journal* **15** 13–21
[17] Pal S, Tak Y K, Song, J M 2007 *Appl. Environ. Microbiol*, **73(6)**, 1712
[18] Yu W W, Qu L, Guo W, Peng X J 2003 *Chem. Mater.* **15(14)**, 2854

The use of SiC/Si(111) hybrid substrate for MBE growth of GaN nanowires

R R Reznik[1-3], K P Kotlyar[2,4], I V Ilkiv[1,2], I P Soshnikov[1,4,5,7], S A Kukushkin[3,6], A V Osipov[3,6], E V Nikitina[1] and G E Cirlin[1-3,5]

[1]St. Petersburg Academic University – Nanotechnology Research and Education Centre RAS, St. Petersburg 194021, Russia
[2]Peter the Great St. Petersburg Polytechnic University, St. Petersburg 195251, Russia
[3]ITMO University, St. Petersburg 197101, Russia
[4]Ioffe Physical Technical Institute RAS, St. Petersburg 194021, Russia
[5]Institute for Analytical Instrumentation RAS, St. Petersburg 190103, Russia
[6]Institute of Problems of Mechanical Engineering Russian Academy of Science, St. Petersburg 199178, Russia
[7]St-Petersburg Electrotechnical University "LETI", St. Petersburg 197376, Russia

Abstract. This work demonstrates the possibility of using a silicon substrate with nanoscale buffer layer of silicon carbide for growth of GaN nanowires by molecular epitaxy on. Morphological and optical properties of the grown arrays are studied. It is shown that the integral intensity of the photoluminescence of such structures is more than 2 times higher than the best NWs GaN structures without buffer layer of silicon carbide.

1. Introduction

The wide bandgap nanoheterostructures based on GaN are of great interest for creating electronic [1] and optoelectronic [2] devices. High mechanical strength combined with chemical resistance and high (2500◦C [3]) melting temperature ensures the stability of the devices. Lack of substrates of gallium nitride encourages researchers to seek suitable substrates for the growth of such structures. The use of sapphire substrates leads to the generation of high dislocation density [4] due to the significant (13%) lattice mismatch. Substrates of silicon carbide, the most suitable in terms of matching the lattice parameters can't be widely used because of the small size and high cost. On the other hand, the successful attempts of growing GaN layers on silicon [5] are very promising, as silicon is a rather cheap material having acceptable thermal conductivity. Moreover, Si substrates can be of a large size and enables the integration of optoelectronic devices based on gallium nitride with the silicon micro- and nanoelectronics base. However, lattice mismatch of Si (111) and GaN (0001) is 17%, and the difference in thermal expansion coefficients is 33% resulting in a high density of different defects suffer its characteristics in created devices. It is known that the optoelectronic GaN based devices can operate for a long time without degrading despite the high density of the linear defect. The dislocation density in the light-emitting diodes based on gallium nitride may exceed the maximum dislocation density in the GaAs based devices to five orders of [6]. The fact that, although dislocations and are recombination centers in GaN, the diffusion length of minority carriers is less than the distance between dislocations (less than 250 nm) [7]. However, to extend the lifetime of GaN based optoelectronic devices an increase their perfection is highly desirable.

In this work to increase the optical quality of the material, we use the nanometer (about 50 ÷ 100 nm) SiC buffer layer which was grown on Si substrate. If we combine two factors (i) the small lattice mismatch between GaN and SiC(111) (only 3%) [2] together with (ii) the nanowires (NWs) growth, the resulting structure can be better in a view of the quality of the structure [8]. Finally, controlled synthesis of GaN NWs allows one to control their physical properties, including the level of doping for both n- and p-type [9], and to create on their basis LED [9] and UV lasers [10].

The aim of this work is to demonstrate the possibility to grow GaN nanowires on the SiC buffer layer on a top of silicon substrates and to compare the optical properties with the GaN NWs grown on silicon without a buffer layer.

2. Experiments

Formation of the buffer layer of silicon carbide on Si (111) substrate was performed by the method described in [11-13]. This method of silicon carbide on silicon films producing is different from all currently existing methods and technologies for growing single crystals, films and nanostructures. Briefly, the method is based on the substitution of silicon atoms on carbon ones on a silicon substrate. At the first stage carbon atom is embedded into interstitial site of silicon. Then nearby silicon atom is removed forming a silicon vacancy. As a result, at the surface of the silicon dilatation dipoles are created (stable complexes consisting of dilatation centers — the silicon vacancy and a carbon atom in interstitial site). These two centers are elastically interacting each other in crystal of cubic symmetry. The rate of the chemical reaction is maximal in the direction along attraction of dilatation dipoles (i.e. along the (111) direction of the silicon substrates). Orientation of the film depends on the crystal structure of the original Si matrix, not only the surface of the substrate, as typically used in conventional film growth techniques. The temperature and pressure of the gas are selected to nucleate the silicon carbide and pores. Formation of the elastic dipoles (carbon atom — silicon vacancy) provides a high quality SiC film.

Growth experiments were carried out using Riber Compact 12 MBE setup equipped with the effusion Ga cell and the nitrogen source. Firstly, the substrate was transferred into the growth chamber, and the substrate temperature was set at 950°C for further cleaning. Then, the substrate temperature was lowered to 600°C and was opened gallium source for 20 seconds to form nanoscale islands on surface to further NWs growth. Finally, and the substrate temperature was raised up to the growth temperature setting at 800°C. Then nitrogen source plasma was ignited and nitrogen flow and the gallium source were opened simultaneously. The growth rate for Ga was set at 0.01 ML/s (according to the previous calibrations). Total growth time of each GaN NWs samples was equal to 16 hours.

3. Results

After the growth, the morphology of grown nanowires was studied by scanning electron microscopy (SEM) SUPRA 25 Zeiss, and their optical properties were investigated by room temperature photoluminescence (PL) and Raman techniques.

Figure 1 shows typical SEM images of GaN nanowires grown on SiC/Si combined substrate. It is seen that on the substrate Si/SiC (111) GaN NWs formed mainly in [111] direction and their average height is about 1.6 mkm. It should be noted that grown structures have high surface density ~ $2 \cdot 10^{10} cm^{-2}$, that's why some of the nanowires became merge. From SEM image (figure 1,a) it is clear that some NWs are inhomogeneous by their diameter - it increases to the top of NWs, possibly due to the rather high substrate temperature and/or relatively low N/Ga fluxes ratio.

Figure 1. SEM images of GaN NWs grown on SiC/Si: (a) tilted by 20 cross section view; (b) top-view.

Figure 2,a shows a comparison of PL spectra of the grown sample and the most successful of GaN NWs sample on silicon measured at room temperature. Maxima of both spectra are coincide in wavelength and conform to experimentally measured radiation value GaN [14]. At the same time, the integral PL intensity of GaN NWs grown on SiC buffer layer more than 2 times higher than the intensity of the best structures of grown on silicon at the same growth conditions.

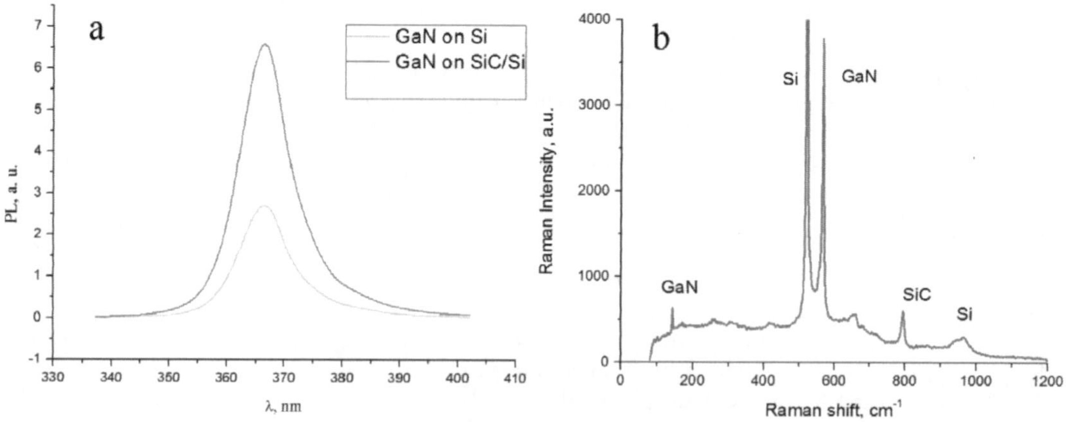

Figure 2. (a) the typical PL spectra of grown samples and the most successful GaN structures grown on silicon; (b) the typical Raman spectrum of grown on SiC/Si GaN NWs.

Figure 2,b shows a typical Raman spectrum of the sample GaN NWs grown on SiC/Si. Despite of the small thickness of the SiC, cubic SiC related-line is clearly visible, along with the lines of the hexagonal GaN and Si. It is worth noting the extremely small width of GaN lines (1.5 cm^{-1}), indicating a high structural perfection GaN NWs.

4. Conclusions

In this paper we demonstrate the possibility of the growth of GaN nanowires by molecular beam epitaxy method on silicon with nanoscale silicon carbide buffer layer for the first time. It was found that the intensity of the photoluminescence spectrum of the GaN NWs on SiC/Si(111) substrate integrally more than 2 times higher than that of the best structures of GaN NWs without a buffer layer of silicon carbide. The extremely small width of GaN lines on Raman spectrum indicates a high crystallographic quality GaN NWs grown on SiC/Si open a new route of the use of these structures in future optoelectronic applications.

Acknowledgments

The nanowire samples were grown under the support of Russian Science Foundation (Project No 14-12-00393). This work was partially supported by RFBR (grant 15-02-06839) and Skolkovo Foundation (6 dd. 30.12.2015). S. A. Kukushkin and A. V. Osipov are grateful for the financial support of the Russian Science Foundation (grant № 14-22-00018).

References

[1] Pearton S J and Ren F 2006 *Adv. Mater.* **12** 1571
[2] Nakamura S and Fasol G 1997 *New York: Springer-Verlag* 368
[3] Ioffe data archive http://www.ioffe.ru/SVA/NSM.
[4] Oliver R A, Kappers M J and McAleese C 2008 *J. Mater. Sci. Mater. Electron.* **19** 208
[5] Aksyanov I G, Bessolov V N, Zhilyaev Y V, Kompan M E, Konenkova E V, Kukushkin S A, Osipov A V, Rodin S N,. Feoktistov N A, Sharofidinov S and Shcheglov M P 2008 *Technical Physics Letters* **34** 479
[6] Cherns D,Young W T, Saunders M A, Ponce F A and Nakamura S 1997 *Microscopy of Semiconducting Materials* **157** 187
[7] Rosner S J, Carr S E, Ludowise M J, Girolami G and Erikson H I 1997 *Appl. Phys. Lett.* **70** 420
[8] Dubrovskii V G, Cirlin G E and Ustinov V M *Semiconductors* **43** 1539
[9] Zhong Z, Qian F, Wang D and Charles M. Lieber 2003 *Nano Letters* **3** 343
[10] Choi H J, Johnson J C and He R 2003 *J. Phys. Chem. B.* **107** 8721
[11] Kukushkin S A and Osipov A V 2008 *Physics Of The Solid State* **50** 1238
[12] Kukushkin S A, Osipov A V, Feoktistov N A 2014 *Physics Of The Solid State* **56** 1507
[13] Kukushkin S A and Osipov A V 2014 *J. of Phys. D: Appl. Phys.* **47** 313001
[14] Tchernycheva M, Sartel C, Cirlin G E, Travers L, Patriarche G, Harmand J-C, Dang Le Si, Renard J, Gayral B, Nevou L and Julien F 2007 *Nanotechnology* **18** 385306

Optical properties of ZnO fabricated by hydrothermal and sonochemical synthesis

M I Zakirov, V V Kuryliuk, O A Korotchenkov

Department of Physics, Taras Shevchenko National University of Kyiv, Kyiv 01601, Ukraine

Abstract. This paper presents the results of experimental study of photoluminescence (PL) and surface photovoltage of ZnO synthesized in this study. Two types of zinc oxide samples, prepared by hydrothermal and sonochemical methods at two different temperatures, are contrasted. The observed differences in their PL spectra are associated with different mechanisms of formation and resulting concentration of point defects. At a high-temperature synthesis, blue luminescence band prevails. Reducing the temperature causes the increase in intensity of the yellow-red luminescence band.

1. Introduction

Zinc oxide (ZnO) is a kind of nontoxic, low cost and abundant material of fundamental importance and great interest to both academic and industrial communities. As a wide-band-gap semiconductor, zinc oxide attracts much attention in current semiconductor research, due to its superior optical properties. Due to its large energy band gap (3.37 eV at room temperature) ZnO becomes an excellent semiconductor material for application in UV devices [1–3]. The unique physical and chemical properties, such as high chemical stability, high electrochemical coupling coefficient, broad range of radiation absorption and high photostability make zinc oxide interesting for study and numerous applications in sensors, photocatalysis, field emission, and solar cells [4, 5].

Solid applied aspects of zinc oxide materials have led to the development of a wide range of techniques for producing ZnO. In these applications of ZnO, the control and design of the point defect structure, as well as the surface and interface defect content, are among the major keys to optimize the device performance [5]. In recent years, many efforts have been made for achieving ZnO nanoparticles via different approaches. Besides gaining information about nanoparticle formation mechanisms, many researches have been focused on different possible methods of obtaining desired particle shapes and different point defect contents. Also, different post-processing techniques capable of varying the defect concentration in ZnO have been given particular attention [6–13].

Room-temperature photoluminescence (PL) of ZnO typically exhibits one emission peak in the near-band-edge ultra-violet (UV) region, which is due to the recombination of free excitons, and one or more peaks in the visible spectral range, which are attributed to defect-related emissions [14]. The latter frequently form at least one broad emission band extending from just above 400 nm to 750 nm, which is referenced to as deep level emission (DLE) band. Different reports have discussed the origin of the deep levels involved in the observed radiative transitions [15].

2. Experiment and Result

Two fabrication methods, namely hydrothermal and sonochemical synthesis, are employed in this work. The detail synthesis procedure has previously been described elsewhere [16]. In both methods, the synthesis temperature is stabilized in the range from 338 to 351 K. The hydrothermally synthesized samples are named below as "HT1" and "HT2" for the growth temperatures of 338 and 351 K, respectively, whereas the ones obtained sonochemically are named as "US1" and "US2". An N_2 laser with the wavelength of λ=337.2 nm (3.67eV) is used as the photogeneration source. Surface photovoltage (SPV) decays were measured in the capacitor arrangement, and details of our setup are given elsewhere [17]. All measurements are performed at room temperature.

The observed changes in the defect-related PL spectra can be associated with the changes in the concentration of point defects. At thermal equilibrium, this concentration is given by

$$c = N_{sites} \exp\left(-\frac{E^f}{kT}\right), \tag{1}$$

where E^f is the formation energy, N_{sites} is the number of sites the defect can reside, k is the Boltzmann constant and T is the temperature [18]. In this framework, the PL intensity is proportional to the defect concentration, provided the involved recombination channels remain unchanged. We therefore give an account of the bulk versus surface concentration N_{sites} in the two fabrication routes, comparing the PL and SPV data, commensurable with the bulk and surface defect concentrations. Using the available reported data, the formation energy E of the different defect types can be related as follows [19]

$$E\,(V_{Zn}) < E\,(V_O) < E\,(Zn_i) < E\,(Zn_O) < E\,(O_i) < E\,(O_{Zn}).$$

PL spectra of our samples are shown in figure 1. Changes in the peak positions and intensities are clearly seen, which are most sufficient in the sonochemically prepared US1 sample.

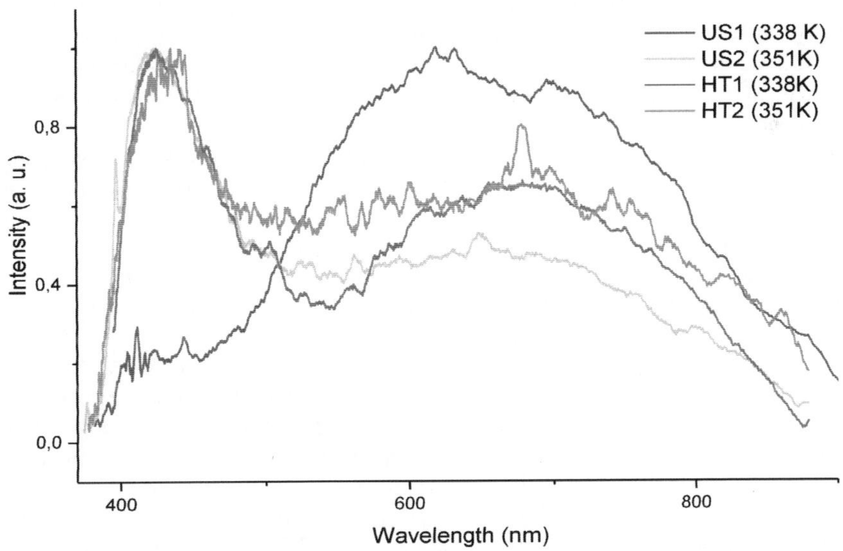

Figure 1. Comparison of the PL spectra taken in ZnO samples fabricated hydrothermally and sonochemically at temperatures of 338 and 351 K.

Of further notice is the fact that, in hydrothermally synthesized samples, PL spectra consist of a strong peak in the blue region and a weak peak in the orange-red region. In sonochemically fabricated samples, strong orange-red peak dominates the emission in US1 prepared at 338 K, whereas two peaks in the blue and orange-red regions are seen in US2 prepared at 351 K, which are quite similar to that observed in the HT samples.

For clarifying the origin of PL spectral bands we can decompose the spectra into individual Gaussian shapes. It is found that each spectrum consists of three elementary bands, which can be associated with three DLE recombination transitions. The data on the featured elementary bands and their relative intensities are given in Table 1. Here we give the peak position, relative intensity of the band (in brackets) with the dominating spectral band marked by asterix, and the likely PL formation mechanism, which is developed using the data of Refs. [20,21].

Table 1. Decomposed elementary spectral bands forming the PL spectra given in figure 1. formation mechanism.

Samples	HT1	HT2	US1	US2
Band 1 (Blue): peak position in nm (relative intensity) likely DLE origin	420 nm (0.63)* CB-V_{Zn}, Zn_i-V_{Zn}	426 nm (0.73)* CB-V_{Zn}	421 nm (0.17) CB-V_{Zn}, Zn_i-V_{Zn}	418 nm (0.63)* CB-V_{Zn}, Zn_i-V_{Zn}
Band 2(Blue-green): peak position in nm (relative intensity) likely DLE origin	456 nm (0.53) Zn_i-V_{Zn}	489 nm (0.34) CB-V_O	571 nm (0.54) Zn_i-V_O, CB-V_O	460 nm (0.49) Zn_i-V_{Zn}
Band 3(Orange-red): peak position in nm (relative intensity) likely DLE origin	673 nm (0.65)* V_O-VB	681 nm (0.67)* CB-V_O	713 nm (0.8)* V_O-VB	646 nm (0.50) CB-O_i

According to the data of Table 1, the HT samples have two elementary bands of nearly equal intensities (two different PL formation mechanisms have the same intensity), while in the US samples one dominant band is observed. Speaking most generally, the observed differences in the PL spectra can be related to different chemistry of defects and their concentrations imposed by different fabrication mechanisms.

At the greater synthesis temperature for hydrothermal method, the long-wavelength shift of the individual band is observed. In clear contrast, sonochemical synthesis produces the short-wavelength shift of the individual band. Significantly, the O_i defect, having rather large formation energy, is realized in the sonochemical route, which can arise in the cavitation process.

The normalized SPV decays for hydrothermally and sonochemically synthesized samples are contrasted in figure 2. It is seen that increasing the synthesis temperature speeds up the SPV decays. Our analysis shows that the decay curves given in figure 1 demonstrate three-component exponential decays. Moore and Thompson [22] observed a two-component exponential decay for ZnO thin films. The fast component in the photocurrent relaxation can be associated with the recombination of free electron-hole pairs [23]. The slow component can be attributed to the lattice relaxation process of a surface state [24]. The SPV decays can also be explained by taking into account the recombination of conduction-band electrons with photogenerated holes captured at the surface [25].

Considering our decays shown in figure 2, we find that the first SPV component, associated with the electron-hole recombination, is present in the inserting. This component is very similar both in the HT and US samples produced at 351 K, whereas it is faster in the sonochemical sample than that in the hydrothermal sample grown at 338 K.

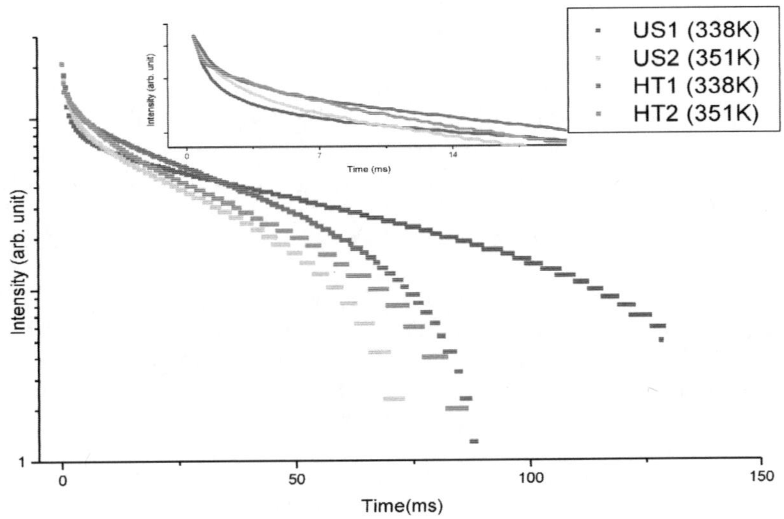

Figure 2. SPV decays in hydrothermally and sonochemically synthesized samples.

The second SPV component, associated with a lattice relaxation process of surface states, exhibit slightly different slopes on a semilog scale (see figure 2) which can be associated with a different concentration of surface states. The data evidence that the greater fabrication temperature causes greater concentration of surface defects.

The third component of the SPV decay is associated with the charge transfer through surfaces defects. There are two causes for observed different density of surface states, including different growing mechanism for both synthesis procedures and temperature.

3. Conclusion

Zinc oxide has been synthesized via hydrothermal and sonochemical procedure at equal temperatures. Both the PL spectra and SPV decays are found to change with the fabrication route. The observed differences are shown to originate from the interplay between bulk and near-surface concentration of point defects. Increasing the synthesis temperature causes the long-wavelength shift of the PL spectrum in hydrothermal samples. On the other hand, for sonochemical synthesis, increasing the temperature causes the short-wavelength shift of the PL band. The increase in the synthesis temperature has no strong effect on the SPV decays in hydrothermal samples. In contrast, SPV decays are sensitive to the growth temperature in sonochemical samples, indicative of varying transited charge transfer processes and surface state densities. The sonochemical route is therefore capable of improving the distribution of defect concentrations, which is important for improving light emission of photonic devices.

References

[1] Gomi M, Oohira N, Ozaki K and Koyano M 2003 *Japanese J. Appl. Physics, Part 1 Regul. Pap. Short Notes Rev. Pap.* **42** 481–5

[2] Li C P, Guo L, Wu Z Y, Ren L R, Ai X C, Zhang J P, Lv Y Z, Xu H B and Yu D P 2006 **139** 355–9

[3] Özgür Ü, Alivov Y I, Liu C, Teke A, Reshchikov M A, Doğan S, Avrutin V, Cho S J and Morkọ H 2005 *J. Appl. Phys.* **98** 041301

[4] Marczak R, Segets D, Voigt M and Peukert W 2010 *Adv. Powder Technol.* **21** 41–9
[5] Oba F, Choi M, Togo A and Tanaka I 2011 *Sci. Technol. Adv. Mater.* **12** 034302
[6] Kwok W M, Djurišić A B, Leung Y H, Chan W K and Phillips D L 2005 *Appl. Phys. Lett.* **87** 16–9
[7] Tam K H, Cheung C K, Leung Y H, Djurišić A B, Ling C C, Beling C D, Fung S, Kwok W M, Chan W K, Phillips D L and Ge W K 2006 J Phys Chem B 110 20865–71
[8] Djuriši A B, Leung Y H and Tam K H 2006 *Appl Phys Lett* **88** 103107
[9] Djurišić A B, Chen X, Leung Y H and Man Ching Ng A 2012 Z *J. Mater. Chem.* **22** 6526
[10] Djurišić A B, Ng A M C and Chen X Y 2010 *Prog. Quantum Electron.* **34** 191–259
[11] Li D, Leung Y H, Djuriši A B, Liu Z T, Xie M H, Shi S L and Xu S J 2004 *Appl Phys Lett* **85** 1601–3
[12] Yang L L, Zhao Q X, Willander M, Yang J H and Ivanov I 2009 *J. Appl. Phys.* **105** 053503
[13] Quang L H, Chua S J, Loh K P and Fitzgerald E 2006 *J. Cryst. Growth* **287** 157–61
[14] Djurišić A B, Leung Y H, Tam K H, Hsu Y F, Ding L, Ge W K, Zhong Y C, Wong K S, Chan W K, Tam H L, Cheah K W, Kwok W M and Phillips D L 2007 *Nanotechnology* **18** 095702
[15] Willander M, Nur O, Sadaf J R, Qadir M I, Zaman S, Zainelabdin A, Bano N and Hussain I 2010 Materials (Basel). 3 2643–67
[16] Zakirov M, Korotchenkov O and Nadtochiy A 2015 PHOTOVOLTAGE DECAY IN SONOCHEMICAL SYNTHESIS ZnO Applied Physics (YSF), 2015 International Young Scientists Forum (Dmipropetrovsk: IEEE) pp 1–4
[17] Podolian A, Kozachenko V, Nadtochiy A, Borovoy N and Korotchenkov O 2010 J. Appl. Phys. 107 1–7
[18] Janotti A and Van De Walle C G 2007 *Phys. Rev. B - Condens. Matter Mater. Phys.* 76 127–32
[19] Janotti A and Walle C G Van De 2009 *Reports Prog. Phys.* **72** 126501 (29)
[20] Zhang L, Yin L, Wang C, Qi Y and Xiang D 2010 *J. Phys. Chem. C* **114** 9651–8
[21] Alvi N H, Ul Hasan K, Nur O and Willander M 2011 *Nanoscale Res. Lett.* **6** 130
[22] Moore J C and Thompson C V 2013 *Sensors* **13** 9921–40
[23] Alivov Y I, Chernykh A V, Chukichev M V and Korotkov R Y 2005 T *Thin Solid Films* **473** 241–6
[24] Reemts J and Kittel A 2007 *J. Appl. Phys.* **101** 013709
[25] Li Q H, Gao T, Wang Y G and Wang T H 2005 *Appl. Phys. Lett.* **86** 123117

Journal of Physics: Conference Series **741** (2016) 012029 doi:10.1088/1742-6596/741/1/012029

The emission characteristics of graphene in different modes of high-voltage power supply

S V Filippov[1], E O Popov[1,2], A G Kolosko[1,3], E I Terukov[1] and P A Romanov[1]

[1]A.F. Ioffe Physico-Technical Institute of Russian Academy of Sciences, ul. Polytechnitscheskaya 26, St.-Petersburg, 194021, Russia
[2]Peter the Great St.-Petersburg State Polytechnical University, ul. Polytechnitscheskaya 29, St.-Petersburg, 195251, Russia
[3]The Bonch-Bruevich St.-Petersburg State University of Telecommunications, pr. Bolshevikov 22, St.-Petersburg, 193232, Russia

E-mail: s.filippov@mail.ioffe.ru

Abstract. We found out the general laws of the current-voltage characteristics hysteresis behavior for emitters of different nanomaterials – graphene, nanotube. Comparison of two power supply modes showed that the IVCs change with growth and decrease of the scanning voltage amplitude in "fast" mode which correlates with current-voltage characteristic hysteresis in "slow" mode. Common patterns forms of current-voltage characteristics in different power supply modes were identified.

1. Introduction

At the recent years due to the progress in the methods of acquisition and processing experimental data the features of the current-voltage characteristics (IVC) of field emitters has been more frequently observed. One of them is so-called hysteresis effect of IVC, which still remains poorly studied. The problems seem to be associated with increasing divergence between developed theory of the single-tip emitter and real multi-tip large area field emitters (LAFE). Such emitters have a potentially huge market of practical applications in the vacuum nanoelectronics.

Before we consider the works on experimental research of the LAFE emitters, it is necessary to introduce some concepts. The emission characteristics can be measured either at increasing voltage (using definition of the threshold field) - a growing branch of the IVC, as well as at a decrease (from the maximum value of the applied voltage to the minimum) – descending branch of the IVC. Let's call *direct* hysteresis situation when the emission current is lower on the descending branch of voltage in comparison with the growing one, and let's call *reverse* hysteresis – situation when descending branch of voltage pulse gives a greater emission current rather than ascending one.

Apparently first who observed the effect of unsymmetrical IVC behavior were Millikan and Eyring in the field emission experiment with the stretched tungsten wire as the cathode and the cylindrical anode [1]. The main source of the features in the IVC is rightly considered the sorption of metal atoms and different molecules. Investigation of the effect of adsorbed particles on the magnitude of the emission current in the single-tip emitters were actively conducted from the middle of the last century. In Gomer's monograph [2] a simple formula connecting the change in the work function and the number of adatoms was proposed. In 1967, Duke and Alferieff shown that resonant tunneling of electrons in a vacuum can occur through energy level of adsorbed particle [3].

The most important achievement of recent years is the question statement about the nature of the adatoms (atoms and molecules which are falling on the surface of the emitter) and the nature of their impact on the emitter work function [4]. For example, Li and Wang in [5] predicted that adsorbed CO and CH_4 increase the work function. Sheng [6] also showed a decrease in current at adsorption CO and CO_2 molecules.

Hysteresis effects of the IVCs (both direct and reverse) on carbon nanotubes of various types were observed in several papers [7-13]. Among the possible causes of the hysteresis most frequently found next ones: the influence of adsorption-desorption processes, Joule heating, the processes of adatom polarization, reversible morphological changes in the emitter surface. In [14-15] authors tried to register type of gases desorbed from the nanotubes surface in the field experiment using a mass spectrometer technique. Desorption of H_2, CO and CO_2 molecules were specified.

The effect of IVC asymmetric behavior was also observed in other LAFEs: porous diamond [16], amorphous Alq3 [17], oriented-AlN thin film [18] and different graphene materials. There are two common experimental approaches to study FE from graphene. The first one consists of local IVC measurements of graphene by single-tip anode in scanning or transmission electron microscopes [19,20]. The second approach utilizes a luminescent screen as anode and provide information about integral IVCs [21,22].

A special place in the studies took the works of Li, Chen, Yan et. al. [23-25]. The authors provided repeatable research of the new graphene materials. They observed direct hysteresis in IVCs of graphene emitters and explained this effect by adsorption-desorption processes. Targeted study of the influence of the inlet gases (oxygen, nitrogen and CO_2) on the field emission of graphene emitter has been provided in [25]. The authors argue that the field emission deteriorates after the adsorption of O_2 and CO_2. In [26,27] on the contrary the reverse hysteresis was observed. In [27] a lateral placement of the graphene film was used and the emission of the side faces was studied (emitter blade length was 1 cm). The hysteresis in this case was explained just by the exfoliation of the graphene film edge from the substrate.

2. Expriment

The study was conducted at a residual pressure in the experimental volume ~10^{-7} torr. The range of the voltage applied to the sample was 0 - 6 kV. The electrode spacing (cathode-anode distance) was 300 µm. As a counter electrode we used tantalum anode. The cathode was a metal tablet with diameter 10 mm coated with nanocomposite film. As nanocomposite we used graphene sheets (G_111, NanoTechCentr TSTU, Tambov, Russia) and polystyrene as a matrix. Prepared graphene-polystyrene suspension was subjected to ultrasonic treatment for 8 hours. Then suspension was deposited onto the substrate by spin coating technique. The technique for characterizing multi-tip field emitters was described in detail in our work [28].

In article [29] we have found out that the prehistory and amplitude of applied voltage pulses define the form and the direction of a hysteresis (direct or reverse). The assumption that the direct hysteresis appears at increasing of amplitude and the reverse - at decreasing of amplitude in comparison with the previous impulse was made.

In this article both direct and reverse hysteresis's of graphene samples have also been received.

3. Results and discussion

3.1. "Slow" mode power supply

In "slow" mode DC voltage levels were applied to the sample. Using power source (FID Technology power supply) controlled by a computer and DAC board NI DAQ PCIe-6351 we able to change voltage level and pulse duration. Figure 1 shows two voltage pulses applied to the sample consequently. One can see from figure 2 that corresponding emission current have a good correlation with pressure changes.

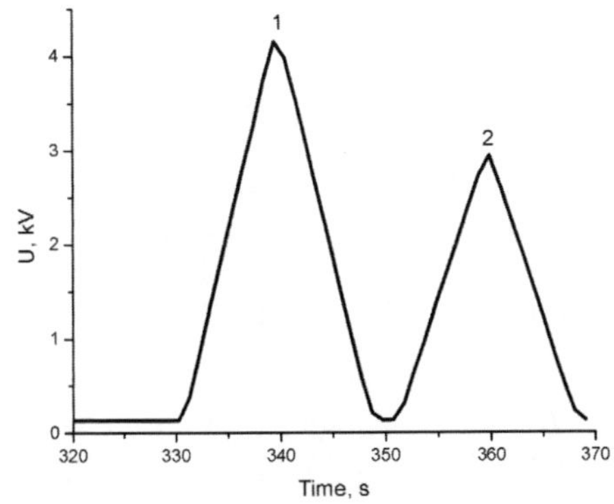

Figure 1. Voltage impulses applied to the sample in "slow" mode

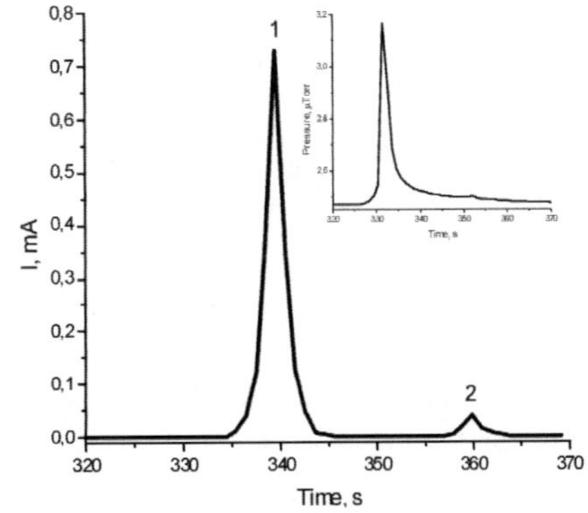

Figure 2. Corresponding current and pressure (on inset) for pulses on figure 1

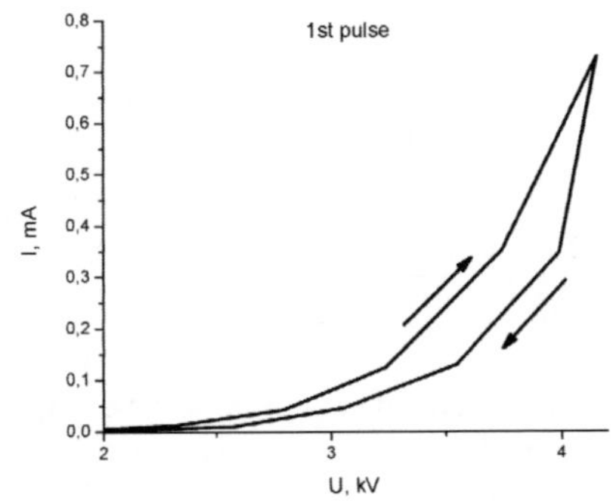

Figure 3. Corresponding IVC for the 1st pulse on figure 1

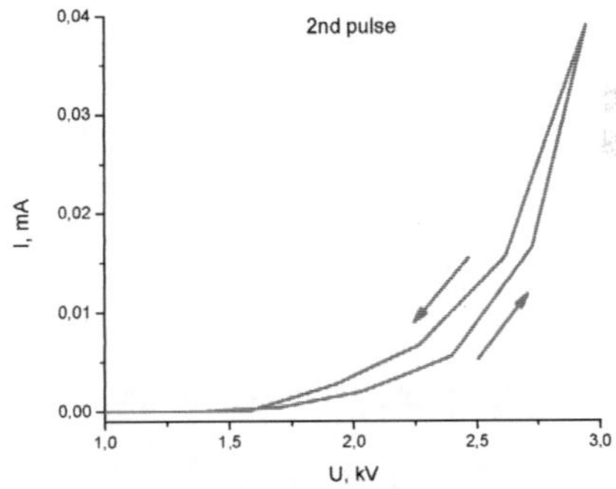

Figure 4. Corresponding IVC for the 2nd pulse on figure 1

Change of pressure in the measuring chamber along with emission current, demonstrates emergence of volatiles products in experimental volume. The source of the pressure increase could be desorption of adatoms from the emitting centers (graphene plates) or from an anode surface. Intensity of streams and speed of a response depend on the obtained emission current value indicate that electron-stimulated desorption takes place to be.

So, using this sequence of pulses we have obtained transition from direct to reverse hysteresis.

Also, we have found such conditions which allow us to get transition from reverse hysteresis to direct hysteresis. From figure 5 it can be seen that two pulses with equal amplitudes can provide this transition. Figure 6 demonstrates the same correlation between emission current and pressure in the experimental volume.

 doi:10.1088/1742-6596/741/1/012029

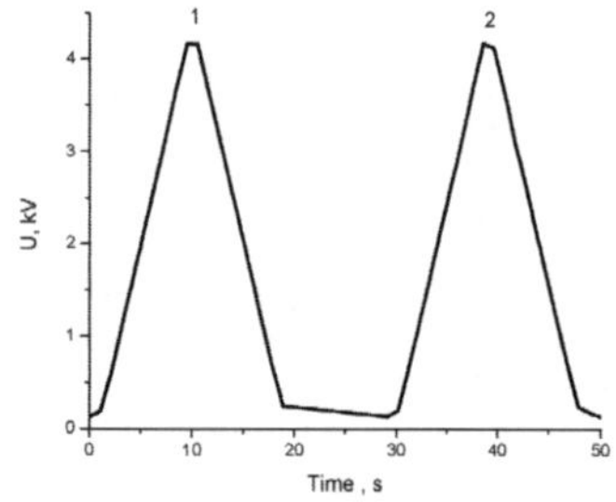

Figure 5. Voltage pulses applied to the sample in "slow" mode

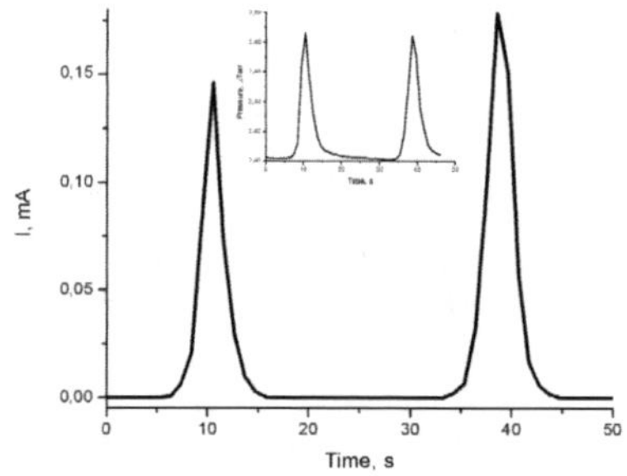

Figure 6. Corresponding current and pressure (on inset) for pulses on figure 5

Probably volatile products dramatically affect on the field emission process. Corresponding IVCs for these pulses are presented on figures 7, 8.

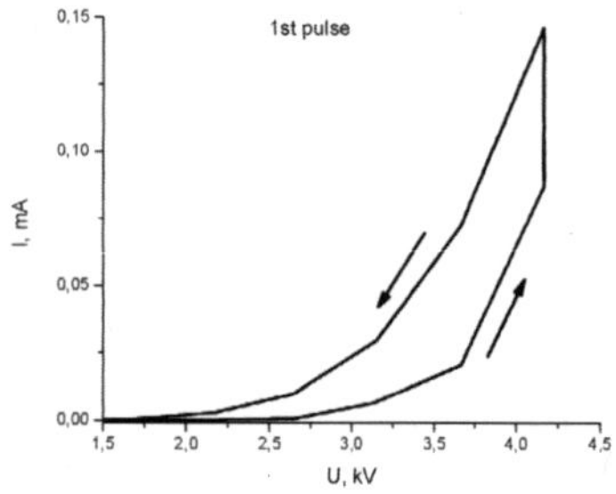

Figure 7. IVC for the 1st pulse on figure 5

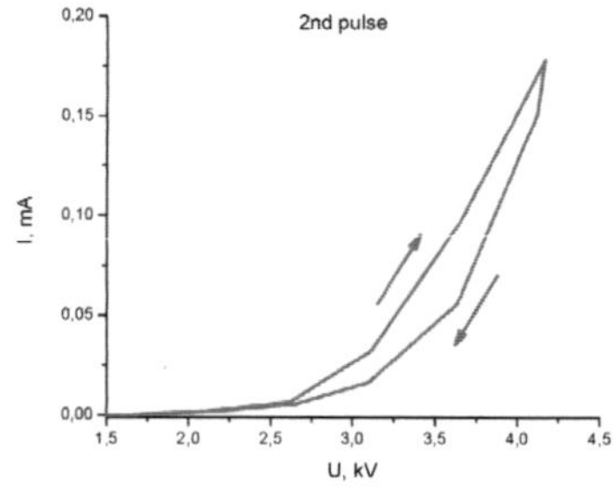

Figure 8. IVC corresponding for the 2nd pulse on figure 5

3.2. "Fast" mode power supply

In fast mode the half-sine pulses with 50 Hz rate and 10 ms duration were applied to sample. The pulse amplitude U_{max} was changed manually. The registration of the corresponding emission current pulses gives about 50 IVCs for 1 second.

The gradually increasing and decreasing of the voltage amplitude leads to hysteresis in the changing of current pulse amplitude I_{max} (see figures 9, 11).

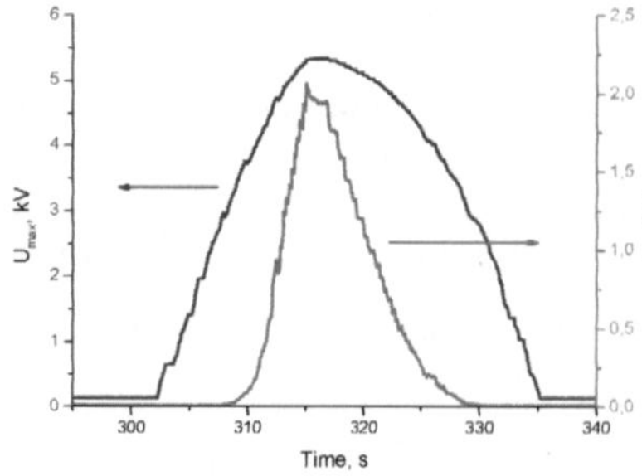

Figure 9. Voltage and emission current time dependences in the "fast" mode

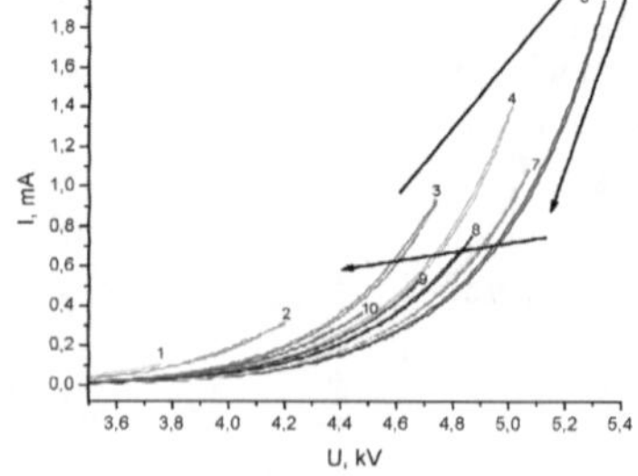

Figure 10. IVCs at different moments for gradually rising and falling of scanning voltage amplitude

The transition from direct hysteresis to reverse hysteresis is observed at the analogue sequences of voltage pulses as for "slow" mode power supply (see figures 10 and 12).

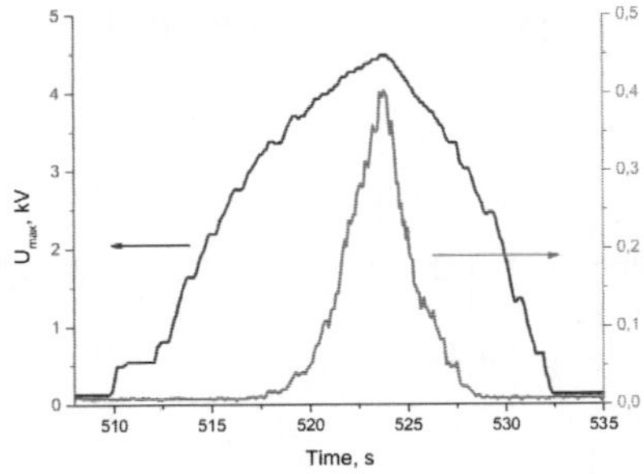

Figure 11. Voltage and emission current time dependences in the "fast" mode

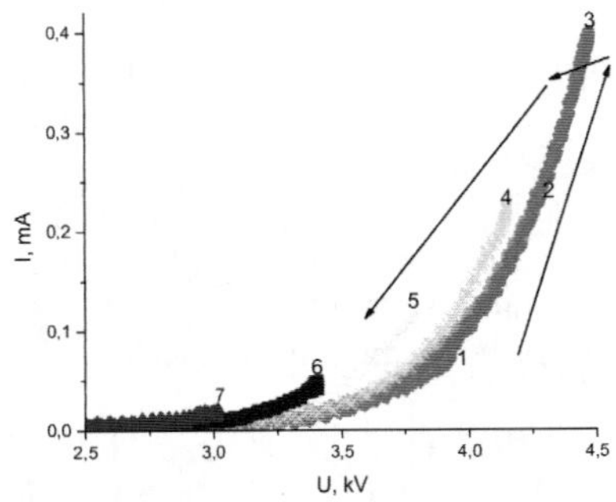

Figure 12. IVCs at different moments for gradually rising and falling of scanning voltage amplitude

3.3. Comparison of two power supply modes

As it can be seen from figure 13 the hysteresis of IVCs obtained in "fast" mode power supply almost have no differences from IVCs registered in "slow" power supply mode. Besides, there is match of hysteresis patterns in both modes of power supply.

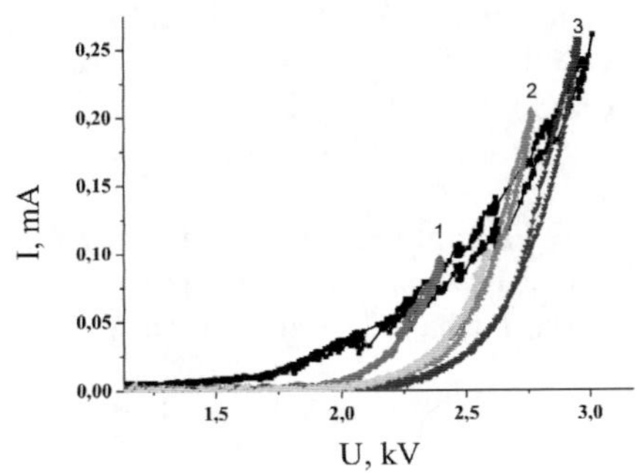

Figure 13. Comparison of IVC hysteresis for two power supply modes.

4. Conclusions

It was found that the effect of a IVC hysteresis can be observed on graphene-polystyrene nanocomposite emitter. We experimentally defined the conditions of emergence of both types of the hysteresis (as direct, and the reverse). Connection of different types of a hysteresis in fast and slow power supply modes has been shown, thereby we have proved reliability of these methods of the IVC studying, and also a community of the nature of this phenomenon irrespective of a power supply mode. Taking into account work [29] we showed that behavior of the hysteresis effect is uniform for emitters of various types, for example, both for nanocomposite with use of nanotubes, and for those containing graphene as a fillers.

According to registration of vacuum and partial pressures, it is possible to take for granted the dependence of vacuum level, namely the partial pressures of vacuum level (CO_2) from the magnitude of emission current and not vice versa. We claim that $P(CO_2) = f(I)$ the defining nature of this dependence is electron-stimulated desorption from an anode surface. In this regard it is possible to claim that the effect of a hysteresis practically doesn't depend on vacuum level, and is exclusive from a condition of surface electrodes purity.

From this follows that the speed of emission area change will be defined by streams of particles as a result of ESD, but not time of sedimentation of residual atmosphere gases (depends on local change of vacuum).

References

[1] Millikan R A and Eyring C F 1926 Physical Review, **27** **1** 51-67
[2] Gomer R. Field Emissions and Field Ionization Cambridge, Harvard University Press, 1961, 195 p
[3] Duke C B and Alferieff M E 1967 *J Chem. Phys.* **46** 923-937
[4] Saito Y Carbon Nanotube and Related Field Emitters: Fundamentals and Applications. Wiley 2010 480 p.
[5] Li Z and Wang C-Y 2006 *Chem. Phys.* **330** 417.
[6] Sheng L M, Liu P, Liu Y M, Qian L, Huang Y S, Liu L and Fan S S 2003 *J. Vac. Sci. Technol.* A **21** 1202.
[7] Yeong K S, Thong J T L 2004 *Appl. Surf. Sci.* **233** 20–23
[8] Lim S C, Lee K, Lee I H, Lee Y H 2007 *NANO: Brief Reports and Reviews.* **2** #2 69–89
[9] Li C, Fang G, Yang X, Liu N, Liu Y, and Zhao X 2008 *J Phys. D: Appl. Phys.* **41** 195401
[10] Okotrub A V, Kurenya A G, Gusel'nikov A V, Kudashov A G, Bulusheva L G, Berdinskii A S, Ivanova Yu A, Ivanov D K, Strel'tsov E A, Fink D, Petrov A V, Belonogov E K 2009 *Nanotechnologies in Russia* **4** **#9–10** 627–633.
[11] Kuznetzov A A, Lee S B, Zhang M, Baughman R H, Zakhidov A A 2010 *Carbon* **48** 41-46
[12] Ageev O A, Blinov Yu F, Il'in O I, Kolomiitsev A S, Konoplev B G, Rubashkina M V, Smirnov V A, and Fedotov A A 2013 *Technical Physics* **58** #12 1831–1836.
[13] D. V. Gorodetskiy, A. V. Gusel'nikov, S. N. Shevchenko, M. A. Kanygin, A.V.Okotrub, Y. V. Pershin. 2016 *J. of Nanophotonics* **10(1)** 012524-1-012524-6
[14] Murray P T, Back T C, Cahay M M, Fairchild S B, Maruyama B, Lockwood N P and Pasquali M 2013 *Appl. Phys. Lett.* **103** 053113
[15] Cahay M, Murray P T, Back T C, Fairchild S, Boeckl J, Bulmer J, Koziol K K K, Gruen G, Sparkes M, Orozco F, and O'Neill W 2014 *Appl. Phys. Lett.* **105** 173107
[16] Mammana V P, Santos T E A, Mammana A P, Baranauskas V, Ceragioli H J and Peterlevitz A C 2002 *Appl. Phys. Lett.* **81** 3470-3472
[17] Wang Z, Zuo Y, Xuemeng H, An Y, Li Y, Guo X, Wang J and Xi L 2013 *J. Phys. D: Appl. Phys.* 46
[18] Yue S L, Gu C Z, Shi C Y, Zhi C Y 2005 *Appl. Surf. Sci.* **251** 215–219
[19] Santandrea S, Giubileo F, Grossi V, Santucci S et al. 2011 *Appl. Phys. Lett.* **98** 163109
[20] Du J, Zhang Y, Deng S, Xu N, Xiao Z, She J et al. 2013 *Carbon* **61** 507–514
[21] Wu Z-S, Pei S, Ren W, Tang D, Gao L, Liu B, et al. 2009 *Adv. Mater.* **21** 1756–1760
[22] Huang C-K, Ou Y, Bie Y, Zhao Q, Yu D 2011 *Appl. Phys. Lett.* **98** 263104
[23] Chen J, Cui L, Sun D, Yang B, Yang J 2014 *Appl. Phys. Lett.* **105** 213111
[24] Chen J, Li J, Yang J, Yan X, Tay B-K and Xue Q 2011 *Appl. Phys. Lett.* **99** 173104
[25] Li J, Yan X, Gou G, Wang Z and Chen J 2014 *Phys. Chem.Chem.Phys.* **16** 1850
[26] Hallam T, Cole M T, Milne W I and Duesberg G S 2013 *Small* **10** 95–99
[27] Kleshch V I, Bandurin D A, Orekhova A S, Purcell S T, Obraztsov A N 2015 *Appl. Surf. Sci.* **357** 1967–1974
[28] Popov E O, Kolosko A G, Filippov S V, Romanov P A, Fedichkin I L 2016 *Nanomaterials and Nanostructures - XXI* **1** 14
[29] Filippov S V, Popov E O, Kolosko, A G, Romanov, P A 2015 J. of Phys.: Conf. Series **643** 1 012101

Growth of vertically aligned multiwalled carbon nanotubes forests on metal alloy Ni-Nb-N with low content of catalyst

S Dubkov[1], A Trifonov[1,2], Yu Shaman[3], A Pavlov[4], A Shulyat'ev[1], S Skorik[3], E P Kirilenko[1], B Rygalin[1]

[1] National research university of electronic technology, Zelenograd 124498, Russia
[2] Federal State Unitary Enterprise Lukin Research Institute of Physical Problems, Zelenograd 124460, Russia
[3] NPK Technological Center, Zelenograd 124498, Russia
[4] Institute of Nanotechnologies of Microelectronics, Russian Academy of Sciences, Moscow 115487, Russia

sv.dubkov@gmail.com

Abstract. This research shows the possibility of carbon nanotubes (CNTs) formation on the surface of low nickel (~ 10 at.%) Ni-Nb-N amorphous metal alloy film by CVD method at 550 °C of the gas mixture based on acetylene. The structure of CNT were studied by transmission and scanning-electron microscopy, energy-dispersive X-ray and the Raman spectroscopy.

1. Introduction

A recent report by ITRS (International Technology Roadmap for Semiconductors) [1] predicts that by 2023 the use of the semiconductor interconnects based on Cu and W will reach it's physical limit in terms of reliability, dimensions, electrical and thermal conductivity. Scaling of VLSI technology increases the interconnect resistance in 5-10 times as compared with the volume value associated with the scattering at the grain boundaries and surfaces [2]. As a result of increased interconnect lines resistance and the number of transistors per unit area will increase the power dissipation, which is a serious problem for modern VLSI technology and future 3d integrated circuits [3].We should not forget about the fact that the production of interconnects of smaller size and higher aspect ratio is a complicated and time-consuming task [4].

One alternative material for semiconductor interconnects is carbon nanotubes are [5] due to their high thermal conductivity [6], low resistance value [7] and the ability to withstand high current densities (up to 10^{12} A/cm^2).

For the synthesis of carbon nanotubes using laser ablation techniques, arc discharge, chemical vapor deposition (CVD) and variations thereof [8,9]. The most preferred method of synthesis of CNTs, embedded in the VLSI technology, considered PECVD-method which allows to obtain the uniform arrays of CNT with desired density on the surface of the solid phase in a suitable temperature range for VLSI technologies [10]. This method is implemented in the presence of catalysts such as Fe, Co, Ni and alloys thereof [11]. Often the catalyst metal is supported by layers of other catalytic metals and oxides (Al_2O_3, SiO_2, MgO), that allows you to control the parameters of the nanotube arrays, up to the cultivation of single-walled nanotubes. [12] There are also reports on the use of these catalysts, alloys with other metals, in which, however, the content of the catalyst is always predominant [13].

In this study we demonstrated CNT growth at 550°C, using a thin Ni-Nb alloy film containing Ni (~ 10 at.%) with addition of nitrogen as a catalyst. The use of Ni-Nb-N alloy as a catalyst will produce reproducible and controlled CNT synthesis on thin catalytic films with different thicknesses, as well as to carry out dry etching operation without the danger of etching of a thin layer of catalyst.

2. Experimental

In work were used Si substrate (100) orientation, coated with SiO_2 layer. The substrates were standard KARO washing solution (H_2SO_4: H_2O_2 = 1:1), followed by rinsing in deionized water and dried in isopropanol vapor.

Formation of a thin layer of Ni-Nb-N alloy was produced on a heated substrate in an atmosphere of Ar + N_2 by magnetron sputtering using a composite target Nb-Ni. The residual pressure in the chamber was about 1×10^{-5} Torr, working gas pressure - 5×10^{-3} Torr.

The CNTs synthesis was carried out by PECVD on the «Plasmalab System 100» made by «Oxford Instruments». CNT formation process occurs in the gas mixture of C_2H_2 + NH_3 at 550°C substrate temperature and a pressure of 3 Torr. CNT growth process occurred within 2 minutes.

CNTs samples were investigated with a scanning electron microscope (SEM) Helios NanoLab 650i, transmission electron microscope (TEM) Tecnai G2 20 S Twin, equipped with EDAX attachment for EDS, by FEI, and Raman spectroscopy on the LabRAM HR Evolution by «Horiba Scientific», with the laser wavelength 514.5 nm.

3. Results and discussion

Investigation by SEM Energy Dispersive X-Ray Analysis (EDX) showed that the film composition of the initial alloy (at.%) was $Ni_{10}Nb_{30}N_{40}O_{20}$.

Investigation by TEM showed the thin alloy film is in an amorphous state, as shown in Figure 1a. This fact confirms the diffraction pattern of the film, having the form of a diffuse halo (inset figure 1a) [14]. It should be noted that at a temperature synthesis of CNTs film is completely crystallized and diffraction pattern takes the form of rings [15]. Crystallization of the thin film layer is accompanied by the appearance of two kinds of catalyst particles of different average size on the substrate surface. As can be seen in Figure 1b average size of large and smaller clusters is about 40 nm and 6 nm, respectively. These two types of catalyst particles, enriched with nickel, are the initiators of the subsequent CNT growth (Figure 1c) [16].

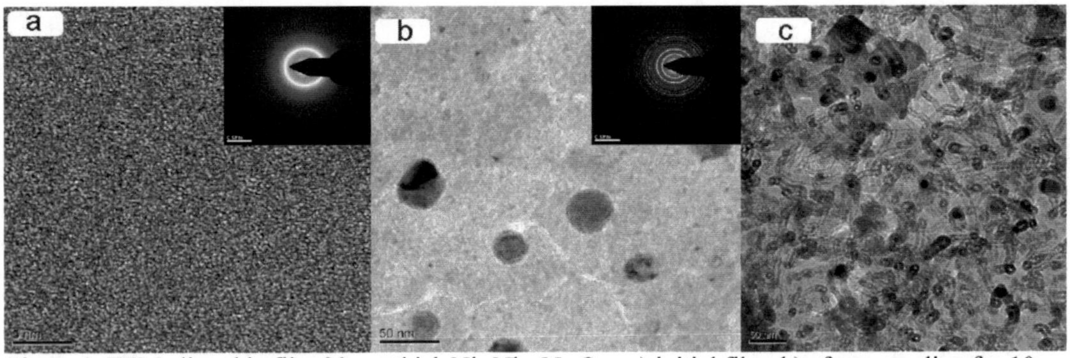

Figure 1. TEM alloy thin film 20 nm thick $Ni_{10}Nb_{30}N_{40}O_{20}$: a) initial film; b) after annealing for 10 minutes at 550°C; c) after the process of synthesis of CNTs at 550°C for 7 seconds.

Figure 2 shows a SEM-picture of a CNT array grown on 20 nm thick film of $Ni_{10}Nb_{30}N_{40}O_{20}$ at substrate temperature of 550°C in the course of 2 min. Analysis of images revealed that there are two kinds of array with different heights and diameters of tubes. CNTs arrays have a height of about 2.6 and 12 μm (Figure 2). The average diameter of the nanotubes: small array ~20nm; high array ~60 nm.

Figure 2. The morphology of CNT sample grown on $Ni_{10}Nb_{30}N_{40}O_{20}$ alloy film at 550°C.

The Raman spectrum of CNTs (Figure 3) shows a well-separated peak D-1350 cm^{-1} and G-peak at 1580 cm^{-1} whose intensity ratio is 0.97, indicating a high quality of the structure [17]. At the same time close the ratio I_D / I_G to one indicates the presence of a large number of defects in the sample of CNTs that TEM microscopy confirms the presence of crystalline structure disorders. Raman-spectrum shows the presence of a sufficiently intense 2D (or G ') peak at 2700 cm^{-1} inherent in high-quality material based on graphite [18].

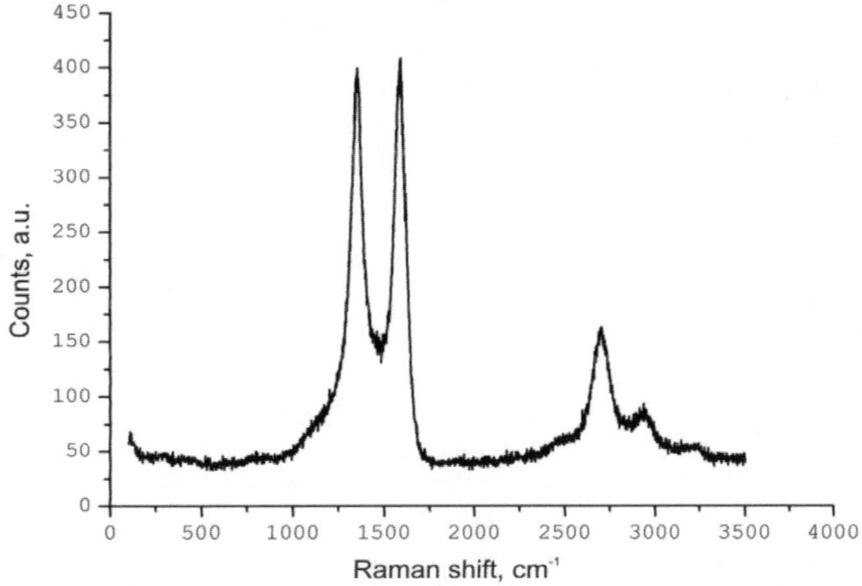

Figure 3. Raman spectra of CNT grown on a thin film of $Ni_{10}Nb_{30}N_{40}O_{20}$ alloy thickness of 20 nm at 550°C.

In the study of individual nanotubes by TEM, it was found that the CNTs are multiwall (inset in Figure 4). It can be seen that the outer diameter of the nanotubes about 20 nm, an inner diameter of ~ 7 nm. It is important to note that the TEM revealed the presence of metal inclusion in the internal channels of the CNT. In these particles revealed the presence of Ni and C, which demonstrates Figure 4, which shows the spectrum of metal cluster within a single tube. The presence of copper peaks in the spectrum related to a noise signal from the sample holder (copper grid).

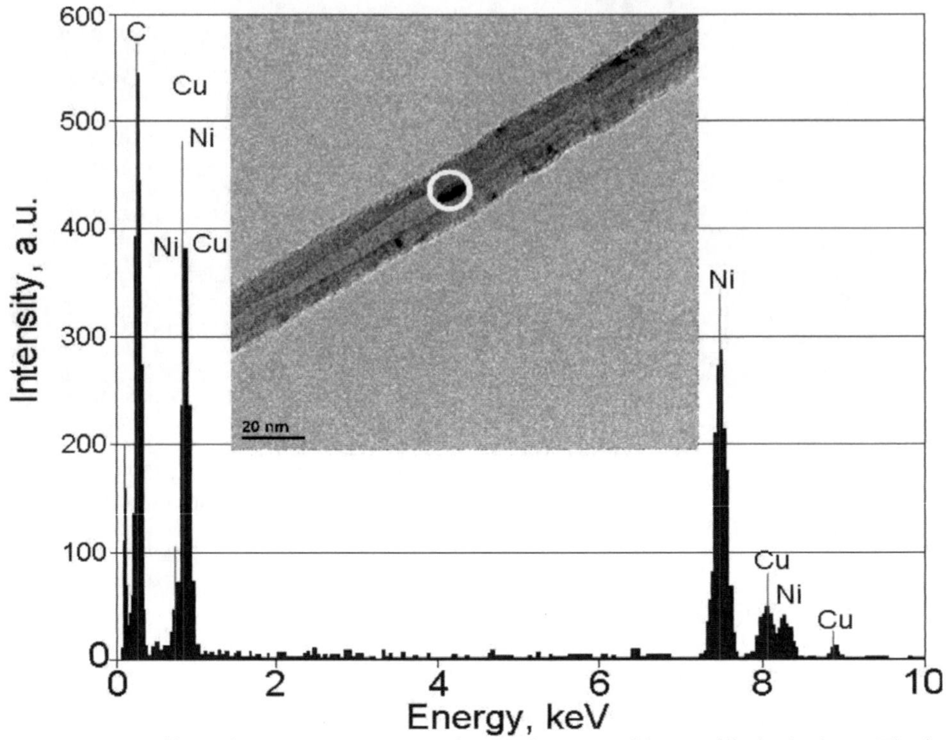

Figure 4. Energy dispersive X-ray spectrum of a single CNT with metallic inclusions. The inset shows a high-resolution TEM image of the object. The location of the analysis is marked by a circle.

4. Conclusions

In this paper demonstrated of CNTs growth by PECVD at 550°C using a catalyst thin film of $Ni_{10}Nb_{30}N_{40}O_{20}$ alloy. It is shown that in one cycle of CVD on the alloy $Ni_{10}Nb_{30}N_{40}O_{20}$ synthesized concurrently two kinds of vertical arrays of CNTs with different average diameter and the height of the tubes. Synthesis of CNTs arrays with different height and diameter due to the presence on the substrate surface the small(~6 nm) and big (~40 nm) clusters of Ni that TEM investigation confirmed. The investigations demonstrated the possibility of using the alloy $Ni_{10}Nb_{30}N_{40}O_{20}$ in integrated circuit technology as a catalyst film for nanotubes synthesis.

Acknowledgments

This work was financially supporte d by the Ministry of Education and Science of the Russian Federation in the framework of public works "Organization of of scientific research" and the agreement of 03.10.2014 № 14.581.21.0007 of (unique identifier RFMEFI58114X0007 project).

References

[1] International technology roadmap for semiconductors, 2012, http://public.itrs.net/
[2] Rossnagel SM, Wisnieff R, Edelstein D, Kuan TS. Interconnect issues post 45 nm. In: IEEE international electron devices meeting; 2005. p. 89–91.
[3] Topol AW et al. 2006 *J. Res. & Dev.* **50** 4.5

[4] Chen JY 2009 IEEE international electron devices meeting p. 1–6

[5] Robertson J 2007 *Materials Today* **10** 1-2

[6] Pop E. et al 2006 *Nano Lett.* **6** 1

[7] Naeemi A, Meindl J D 2008 *IEEE Transactions on Electron Devices* **55** 10

[8] Dresselhaus M S, Dresselhaus G, Avouris P 2001 *Springer-Verlag* 449

[9] Meyyappan M 2009 *J. Phys. D: Appl. Phys.* **42** 21

[10] Vollebregt S et al. 2014 *Carbon* **71**

[11] Liu H, Zhang Y, Li R, Sun X, Abou-Rachid H 2011 *Particuology* **9**

[12] Kitiyanan B, Alvarez W E, Harwell J H and Resasco D E 2000 *Chem. Phys. Lett.* **317** 497–503

[13] Lobiak E V, Shlyakhova E V, Bulusheva L G, Plyusnin P E, Shubin Y V, Okotrub A V 2015 *J. Alloys and Compounds* **621** 351-356

[14] Gromov D G, Dubkov S V, Pavlov A A, Skorik S N, Trifonov A Yu, Kirilenko E P, Shulyat'ev A S, Shaman Y P, Rygalin B N 2016 *Proc. of universities. Electronics* **21** 48-54

[15] Gromov D G, Bulyarskii S, Pavlov A, Scorik S, Shulyat'ev A, Trifonov A Y 2016 *J. Diamond and Related Materials* **64** 97–102

[16] Gromov D G, Dubkov S V, Pavlov A A, Skorik S N, Trifonov A Yu, Shulyatev A S, Shaman Y P, Rygalin B N 2016 *Russian Microelectronics* **45** 98–104

[17] Vollebregt S, Derakhshandeh J, Ishihara R, Wu M, Beenakker C 2010 *J. of electronic materials* **39** 4

[18] Antunes E F, Lobo A O, Corat E J, Trava-Airoldi V J 2007 *Carbon* **45** 5

Application of slope-intercept diagram to determine the parameters of the nanocomposite field emitters in-situ

I U Smirnov[1,2], A G Kolosko[1,2], S V Filippov[1,3], N A Yudkina[1,3] and E O Popov[1,3]

[1]A.F. Ioffe Physico-Technical Institute of Russian Academy of Sciences, ul. Polytechnitscheskaya 26, St.-Petersburg, 194021, Russia
[2]The Bonch-Bruevich St.-Petersburg State University of Telecommunications, pr. Bolshevikov 22, St.-Petersburg, 193232, Russia
[3]Peter the Great St.-Petersburg Polytechnical University, ul. Polytechnitscheskaya 29, St.-Petersburg, 195251, Russia

Abstract. The method of the field emission data online processing of materials, perspective for a nanoelectronics, is developed. The method is based on the slope-intercept diagram analysis of the current voltage characteristics registered during the experiment. It allows to build calibration grids of the microscopic emission parameters, such as work function, effective emission area and field enhancement factor, and and to estimate changes in the sample structure online.

1. Introduction

Quantum field effects - is one of the essential attributes of modern rapidly developing nanoelectronics. One of such effects is field emission of electrons from multi-tip nanostructures. This type of emitters has already established itself as an energetically favorable and reliable source of electrons. However the combination of accompanying effects doesn't allow it to be used in electronics, requiring careful study of fundamental theory of the process.

Two essential characteristics - stability of emission current and its dependence on the applied voltage appear under the influence of several inaccessible to direct monitoring factors: local Joule heating, adsorption-desorption processes, ion bombardment, influence of ponderomotive forces on the surface morphology, the influence of emitting nanocenters on each other, a nonlinear change in the shape of the potential barrier with an applied voltage, and also charge effects on the emitter surface [1-4].

One of the research methods of the field emitter behavior in various operating modes proposed by leading scientists is the so-called SK-analysis. For carrying out this analysis the current voltage characteristics (IVC) or the series of IVCs in Fowler-Nordheim coordinates (IVC-FN) is built. Then its approximation by straight dependence is made. The obtained parameters of this dependence (slope b and intercept a) are put on SK-chart (this chart is in coordinates "katamuki – seppen", i.e. the "slope – intercept").

The analysis of obtained dependence is a rather complex theoretical problem. The main idea of the method is the creation of a calibration grid, intended for microscopic parameters determination of the emitter during its operation. These parameters are the work function, field enhancement factor (FEF), the average curvature radius of nanotips and the total emission area. However for calculation of calibration "equipotential" lines the detailed data on the parameters mutual dependencies is needed.

The change in the work function with the concentration and the kind of atoms adsorbed on the emitter surface was described by R. Gomer in book "Field emission and field ionization" in 1961 [5]. The increment of work function $\Delta\phi$ was estimated with equation:

$$\Delta\phi = 4\pi N\alpha_p F \tag{1}$$

where N is the adatom concentration, α_p – coefficient of the adatom polarization, F - local electric field. By the way it was claimed that electronegative adatoms on the emitter surface act like a barrier which repulses electrons, due to which total emission current decreases, and positive ones act like a window in the potential barrier, due to which the current increases.

The introduction of the dependence from eq. (1) to the standard Fowler-Nordheim equation and the decomposition of subexponential expression in Taylor series for small increment $\Delta\phi$, allowed to obtain a multiplier for emission area correction. So the real area S_0 must be different from the estimated area S_e (calculated without taking into account sorption amendment):

$$S_e = S_0 \exp(-9{,}74 \cdot 10^7 \cdot \gamma \cdot \sqrt{\phi}) \tag{2}$$

where $\gamma = 4\pi N\alpha_P$, coefficient α is connected with deviation of the potential barrier shape from a triangular one under the influence of image forces (Nordheim, 1928).

In 1976 Spindt used this theory to explain fluctuations of the emission current from molybdenum tip and the moving of the emitting centers in the field emission projector picture [6]. Coefficient α was accepted equal to 0,95. The same value was used subsequently by other scientists.

In 1993 the Japanese scientists Gotoh, Ishikawa and Tsuji first tried to build a calibration grid for SK-chart on the basis of Fowler-Nordheim equation without sorption correction [7]. The assessment of the golden microtip work function using this grid showed doubtful results, so it was decided to use an correction from eq.(1). However at the same time in the multiplier for emission area correction was included one more factor – field enhancement factor β:

$$\gamma' = 4\pi N\alpha_p \cdot \beta \tag{3}$$

To determine unknown parameters – S_0 (emission area without adatoms) and γ (proportionality factor $\Delta\phi$ and F from eq.(1)) the equations of slope and intercept of IVC-FN were used:

$$a = \log\left(1{,}4 \cdot 10^{-6} \cdot S_e \cdot \frac{\beta^2}{\phi}\right) + \frac{4{,}26}{\sqrt{\phi}} \qquad b = -2{,}82 \cdot 10^7 \cdot \frac{\phi^{3/2}}{\beta} \tag{4}$$

The resulting dependence was fitted for the golden tip experimental data (ϕ = 5 eV), so that the parameters were equal to: $S_e = 10^{-8}$ cm^2, $\gamma = 10^{-13}$. For this time the experimental data for other materials (nickel, molybdenum and chromium), according to the calibration grid on SK-chart, showed plausible results.

In 1996 the same group of scientists used eq.(2) again for creation of the calibration grid according to the field emission experimental data of tungsten wire with nickel coating [8]. This time they thoroughly showed that the decrease/increase in work function leads to a shift of IVC-FN parameters in SK-chart to the right-upwards/left-down, and decrease/increase of the emitting tip radius – to the left-up/right-down shift.

In 2004 Gotoh et. al. considered the influence of aging in atmosphere of oxygen, hydrogen and carbon dioxide on emission of platinum microtip [9]. In the field emission projector pictures wandering spots were observed. The authors explained its by the adsorption and desorption of individual atoms in different regions of the tip. It was shown that the exposure of the emitter in hydrogen reduces emission current (at the same time the relative flicker noise is reducing), and in oxygen- reduces it even more. Aging in carbon dioxide, on the contrary, increases current level (in this case the noise increased).

SK-chart of IVC noise dependence showed a linear relationship between slope and intercept:

$$b = a \cdot A_c + B_c \quad (5)$$

The observed line was perpendicular to the line of work function change on calibration grid, which allowed to assert that the noise fluctuations are due to a change in FEF. Mathematical analysis of this linear connection indicates that all of noise IVC-FN must converge in one point with coordinates $1/V=-A_c$ and $lg(I/V^2)=-B_c/A_c$.

The application of SK-analysis to arrays of nanotubes (thickness of 10 nm, grown on ITO coating) allowed the scientists from South Korea to evaluate the change of microscopic parameters of the emitter after its operation in the modes of constant (DC) and alternating (AC) voltage [10].

The Chinese scientists also used SK-chart for analyzing of the change of multi-tip emitter made of nanotubes grown by CVD method on a silicon substrate [11]. There was observed a drop in FEF from 5000 to 2000, which the authors associated with the longest burnout of nanotubes under ion bombardment of oxygen atoms.

In 2013 an American scientist A. Persaud simulated a noise dependence of IVC-FN on SK-chart through the introduction of the specified distribution of virtual nanocenters for height, radii, and work function [12]. He showed that the bias of stochastic cloud on SK-chart to the left and down can be caused not only by the change of the work function, but also of the number of emitting nanocenters.

In 2014 we used his assumption of normality of the nanocenters distribution in the nanocomposite emitter (nanotubes in the dielectric polymer matrix) to analyze the dependence of nanocenter distribution by the effective heights from the level of applied voltage [13].

2. Expriment

The experimental installation for the study of the field emitter emission characteristics is described by us in detail in [14]. The installation is computerized and allows to process the registered IVCs and to obtain the microscopic parameters of the emitter online (data processing program was created with the LabView platform). IVCs are registered in a "fast" mode with frequency of 50 Hz by scanning with the half sine pulses. The vacuum in the chamber is maintained at 10^{-7} Torr. The distance between the electrodes d in this work was equal to 300 μm. The sample area S_k was 0,79 cm^2.

The sample for the study was a nanocomposite emitter fabricated on the base of graphene and polystyrene. The substrate was a circular plate of stainless steel. Over it by means of the rotating table was applied the suspension: graphene with polystyrene in the ratio of 1/10 in the organic solvent orthoxylol. For reduction of the size of nanotubes agglomerates and better dissolution of polymer, the suspension before the application was treated in an ultrasound bath for several hours.

Online analysis of the recorded IVCs uses the standard field emission equation in approximation of Elinson [15], written in the coordinates of Fowler Nordheim:

$$\lg\frac{j}{E_0^{\;2}} = \lg(A_\phi) + B_\phi \cdot \frac{1}{E_0} \tag{6}$$

where j is the emission current density, E_0 is the electric field on peaks of the emitter tips, A and B are the coefficients that depend on ϕ - emitter work function (the influence of adatoms on emission current will be incorporated in S_e):

$$A_\phi = \frac{1{,}4\cdot 10^{-6}}{\phi}\cdot 10^{4{,}39/\sqrt{\phi}} \tag{7}$$

$$B_\phi = -2{,}82\cdot 10^{7}\cdot \phi^{3/2} \tag{8}$$

Expressing the microscopic quantities E_0 and j through the macroscopic analogues we can get:

$$\lg\frac{J}{E^2} = \lg\left(\frac{\beta^{2\cdot}\cdot A_\phi \cdot S_e}{S_k}\right) + \frac{B_\phi}{\beta}\cdot\frac{1}{E} = a + \frac{b}{E} \tag{9}$$

where J - the macroscopic current density: $J = I / S_k = j \cdot S_e / S_k$,
E - macroscopic field : $E = E_0 / \beta = U / d$.

The linear approximation of the experimental data plotted in the same coordinates allows us to find the slope b and the intercept a (see Fig.1a,b). The linear dependence of b on a can be analyzed online by obtaining the corresponding coefficients A_C and B_C (see eq.(5)).

During the operation of the emitter its IVC demonstrates a flicker noise, so that slope and intercept of the corresponding approximation have some variations. Received in just a few seconds of the emitter work b-set and a-set were deposited on the SK-chart (see Fig.1c,d).

The coefficients b and a are connected with the parameters S_e, β and ϕ (see eq.(9)):

$$a = \lg\left(\frac{\beta^{2\cdot}\cdot A_\phi \cdot S_e}{S_k}\right) \qquad b = \frac{B_\phi}{\beta} \tag{10}$$

The signal processing program calculates the parameters β и S_e from the experimental a and b automatically (work function ϕ in this calculation is set by the experimenter).

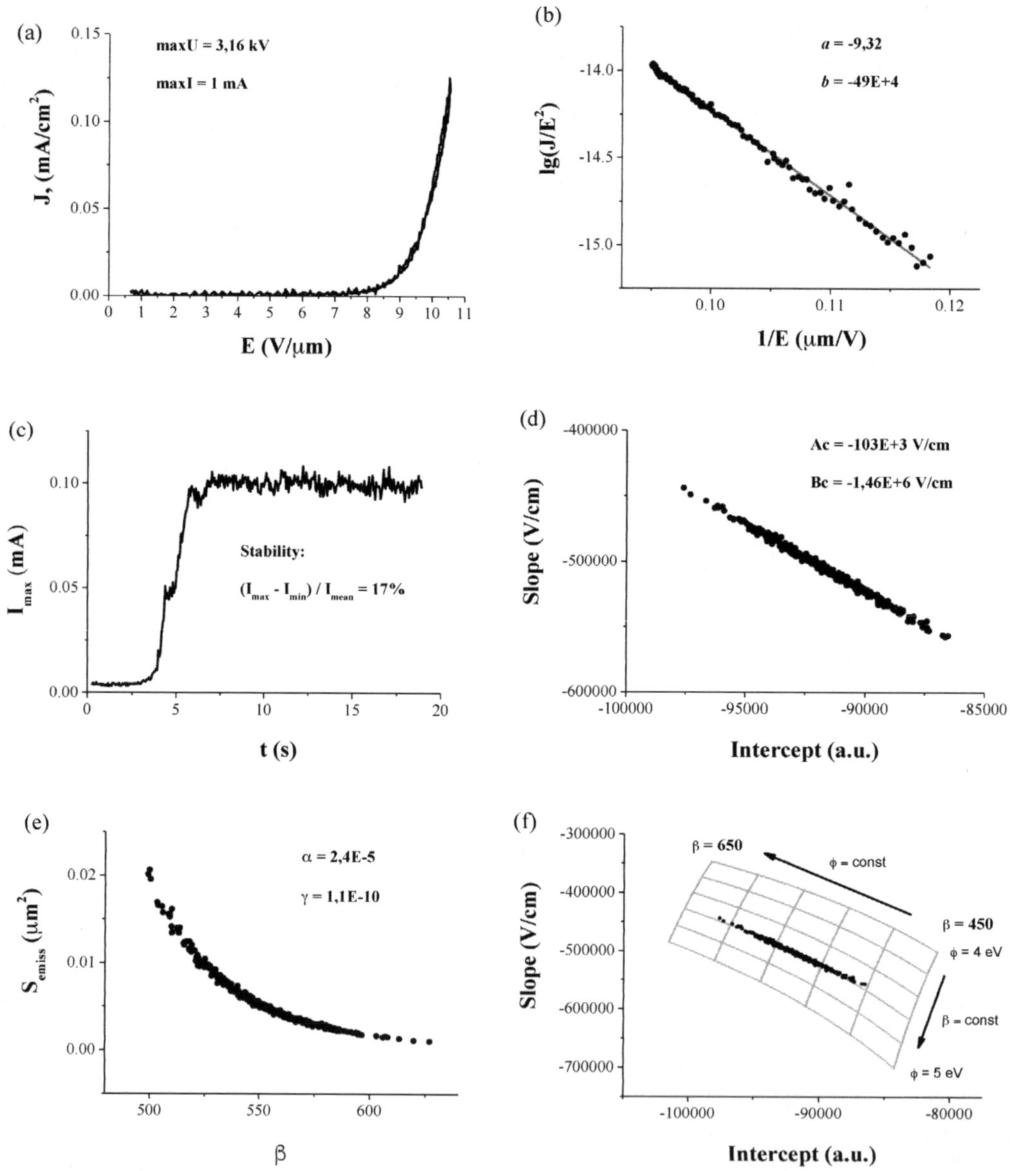

Figure1. The examples of the dependences obtained during field emission experiment with graphene nanocomposite: a) IVC in macroscopic coordinates J and E, b) corresponding IVC-FN, c) noise dependence of amplitude of the emission current pulses from the time in a stable emission mode, d) SK-chart of IVC-FNs obtained for the shown period of time, e) dependence of emission area S_e from the FEF β, calculated for the presented chart, f) calibration grid in the SK-chart which is built with derived coefficients α_0 and γ.

Figure 1e shows the dependence of S_e on β which is obtained by processing noise data from the «cloud» presented in Figure1d. Its exponential shape allows us to apply the approximation in agreement with the dependences in eq.(2) and eq.(3). As a result of approximation of the noise data, adsorption coefficients α_0 and γ are obtained. With the help of these coefficients a separate software module calculates calibration grid and puts it on the experimental SK-chart of the current IVC-FN (see Figure1f).

When the applied voltage amplitude rises the point in the SK-chart, corresponding to the actual IVC-FN, moves across the grid lines. But if the amplitude does not change the noise «cloud» in the SK-chart has almost linear form and seems tangentially to the equipotential line of the grid with corresponding work function ϕ. This fact indicates basic distinction of two processes: fluctuation change of current at the constant level of voltage (U = const) and the change of current due to the change of applied voltage amplitude of ($U \neq$ const).

The first process is obviously connected with wanderings of the emitting areas on the sample surface (we observed this effect in the work [16]). Many authors connect these wanderings with stochastics of the adsorption processes. The second process cardinally changes a condition of fluctuation balance, leading to a change of conditions of nanocenters mutual influence and the speed of adsoption-desorption processes.

Further studying of fundamental dependences in the field emission processes of multi-tip emitters with application of SK-analysis will be continued by us with using computerized field emission projector and also with statistical modeling of multi-tip emission systems.

3. Conclusion

We developed a method for online analysis of SK-charts of the large area field emitters. The calibration grid constructed on the noise dependence of the calculated emission area from the effective FEF, allow to analyze changes in the microscopic parameters of the emitter while changing the scanning voltage amplitude. We consider that the linearity of the noise dependence in the SK-chart at a constant voltage level is due to balancing of the microscopic emitter parameters near the equilibrium point. Correlation of this line with the equipotential curves of the calibration grid requires further research including not only derivation of empirical dependences, but also a modeling of field emission in stochastic multi-tip system.

References

[1] Eletskii A B 2010 *J. Successes of physical sciences* **180** 9 897-930.
[2] Vasilyev E A, Klesh V A, Obrazov A N 2012 *J. Tech. phys. lett.* **82** 7 107-111
[3] Forbes R G 2012 *Nanotechnology* **23** 095706
[4] Yue S L, Gu C Z, Shi C Y, Zhi C Y 2005 *J. Appl. Surface Science* **251** 215
[5] Gomer R, Field emission and field ionization, 1961
[6] Spindt C A, Brodie I, Humphrey L, Westerberg E R 1976 *J. Appl. Phys.* **47** 5248
[7] Ishikawa J, Tsuji H, Gotoh Y, Inoue K, Nagao M, Sasaki T, Kaneko T 1993 *J. Appl. Phys.* **32** 342
[8] Gotoh Y, Nagao M, Matsubara M, Inoue K, Tsuji H, Ishikawa J 1996 *J. Appl. Phys.* **35** 1297
[9] Gotoh Y, Nagao M, Nozaki D and Utsumi K 2004, *J. Appl. Phys.* **95** 1537
[10] Baik C W, Lee J, Choi J H, Jung I, Choi H R, Jin Y W 2010 *Appl. Phys. Lett.* **96** 023105
[11] Zhang Y, Liao M X, Deng S Z, Chen J 2011 *Carbon* **49** 3299
[12] Persaud A, J. 2013 Appl. Phys. **114** 154301
[13] Kolosko A G, Popov E O, Filippov S V, Romanov P A 2015 *JVST B* **33** 03C104-1
[14] Popov E O, Kolosko A G, Filippov S V, Romanov P A, Fedichkin I L 2016 «Nanomaterials and Nanostructures - XXI» **1** 14
[15] Elinson M I, Cold cathode, 1974 Moscow 336 p
[16] Kolosko A G, Popov E O, Filippov S V, Romanov P A, Terukov E I. 2015 IEEE IVNC **1** 40

Self-induced GaN nanowire growth: surface density determination

A A Koryakin[1,2], **L Repetun**[1], **N V Sibirev**[1,2] **and V G Dubrovskii**[1,2]

[1]St. Petersburg Academic University, 8/3 Khlopina, St.Petersburg, 194021, Russia
[2]ITMO University, 49 Kronverkskiy pr., 197101, St. Petersburg, Russia

Abstract. A new numerical approach for the determination of the GaN nanowire surface density on an AlN/Si substrate as a function of the growth time and gallium flux is presented. Within this approach, the GaN island solid-like coalescence and island-nanowire transition are modeled by the Monte-Carlo method. We show the importance of taking into consideration the island coalescence for explaining that the maximum of GaN island surface density is several times larger than the maximum of GaN nanowire surface density. Also, we find that the nanowire surface density decreases with an increase of the gallium flux.

1. Introduction

The wide bandgap and good transport properties make gallium nitride ideal candidates for production of high power electronic and optoelectronic devices [1]. The GaN film growth on substrates with lattice mismatch larger than 2% [2] results in the formation of large amount of dislocations in the GaN layer. The GaN nanowire-based devices can solve this problem. Nanowires (NWs) grow on substrates with a large lattice mismatch without dislocation formation because of a specific elastic stress relaxation mechanism [2]. The present paper is devoted to the modeling of self-induced GaN NW formation (catalyst-free growth) on AlN/Si substrate. The self-induced NW growth enables to avoid an unwonted contamination of the NWs by the catalyst material. Despite the semiconductor NW growth have been studied intensively for the last decade [2, 3] the mechanisms of self-induced NW growth are still far from being fully understood. Consonni [4] distinguishes the three growth stages of self-induced GaN NWs on AlN/Si substrate. The nucleation stage of GaN islands (or the incubation period) is a time interval since the onset of material deposition until the GaN islands appear on the substrate. The incubation period can vary in a typical range of several tens of seconds to several tens of minutes. During the transition stage, the islands grow and change their form (presumably, in the following sequence: cap-shaped islands, truncated pyramid-shaped islands, full pyramid-shaped islands) as a result of elastic stress relaxation. The GaN islands transform into the NW (column-shaped islands) when the island lateral size reaches the critical size that is determined by the onset of plastic relaxation. Finally, an intensive vertical growth of NWs occurs during the NW growth stage. In the work [5], we modeled the nucleation stage of GaN islands and found the size distribution function of islands using the model of the growth via the Stranski-Krastanov mechanism [6, 7]. The island and NW surface density was calculated assuming that any island reached the critical size transforms into NW and neglecting the island coalescence. In this paper, we use a numerical approach wherein the island solid-like coalescence is modeled by the Monte-Carlo method. As a result, the accuracy of the surface density calculation is essentially improved.

2. Model

The initial size distribution function of GaN islands has the form of [5]

$$f(\rho,t) = cn\exp\left(cx(\rho,t) - \exp(cx(\rho,t))\right). \quad (1)$$

Here $\rho = \left(r/\Omega^{1/3}\alpha\right)^2$ is the invariant island size, r is the island base radius, Ω is volume per GaN pair, $\alpha = \left(3/\pi\eta(4\eta^2+3)\right)^{1/3}$ is the geometric factor for the spherical cap-shaped island, η is the island aspect ratio, n is the maximal surface density of island after the nucleation stage, c is a coefficient that characterizes the distribution function dispersion, $x(\rho,t) = \rho_0(t) - \rho$, $\rho_0(t)$ is the island size that corresponds to the distribution function maximum.

We model the island formation on the rectangular substrate with the side $L = (N/n)^{1/2}$, N is the maximal number of islands in our model. To perform the Monte-Carlo modeling, the island size distribution function is discretized and the following rules are used. At time $t=0$, the island distribution function is completely in the region of negative island sizes. At each step of the modeling, the growth rate of the invariant sizes of islands and NWs, $\mathrm{d}\rho/\mathrm{d}t$ and $\mathrm{d}\rho_1/\mathrm{d}t$ respectively, are found from the material balance equation that is written in terms of supersaturation for islands and for NWs [5]

$$\frac{\mathrm{d}\rho}{\mathrm{d}t} = \frac{1}{\tau}\left(\frac{V\chi t}{h_{\mathrm{eq}}} - \frac{S}{L^2 h_{\mathrm{eq}}}\left(N_{\mathrm{ISL}} + N_{\mathrm{NW}}\right)\right), \quad (2)$$

$$\frac{\mathrm{d}\rho_1}{\mathrm{d}t} = \frac{1}{\tau_1}\left(\frac{V\chi t}{h_{\mathrm{eq1}}} - \frac{S}{L^2 h_{\mathrm{eq1}}}\left(N_{\mathrm{ISL}} + N_{\mathrm{NW}}\right)\right). \quad (3)$$

Here ρ_1 is the invariant size for column-shaped islands (i.e. NWs) [5], the first term in the brackets in the right hand side of equations (2) and (3) corresponds to the total amount of material on the substrate, the second term corresponds to the total amount of material in the islands and NWs, V is the Ga flux (in ML/s), χ is the sticking coefficient, t is the growth time (the wetting layer thickness equals to the equilibrium thickness at $t=0$), τ, h_{eq} are the diffusion time and the equilibrium wetting layer thickness for the islands and NWs (index 1 corresponds to NWs), S is the surface area per GaN pair, N_{ISL} and N_{NW} are the total amount of GaN pairs in the islands and NWs. Using the following formulas, we can find the lateral size of islands and the height of NWs: $r = \alpha\Omega^{1/3}\rho^{1/2}$ and $L = \Omega/(\pi r_C^2)\rho_1$, respectively , where r_C is the critical island size. As soon as the island size ρ becomes positive, the island (i.e. a circle with radius r) appears in a random place of the substrate (figure 1). If an island nucleates inside other island or NW this island does not grow. If an island reaches the critical island size r_C before collision with other island or NW the island becomes NW. Otherwise, the island-NW transition does not occur. If two or more islands collide we count these islands as one. When an island collides with other island or a NW the islands and NW continue growth as if there were no collisions. The NW radius is a constant in the model and equals to r_C. The suggested scheme allows the modeling of the GaN NW formation. Figure 1 shows a typical view of the substrate during different growth stages of NWs. One can see in figure 1 multiple collisions of islands that decrease drastically the island surface density. To validate our model, we compare the island and NWs surface density dependence on time with the experimental data [8]. The parameters of the model used in this computation are: $\eta = 0.11$, $\rho_C = 500$ ($r_C \approx 9$ nm that two times smaller than in [4]), $V = 0.09$ ML/s, $\chi = 0.2$, $h_{\mathrm{eq}} = 3$ ML, $\tau = 6.5\times10^{-2}$ s, $h_{\mathrm{eq1}} = 3$ ML, $\tau_1 = 1.0\times10^{5}$ s, $c = 0.013$, $n = 1300\ \mu m^{-2}$. The value of V and h_{eq} is known from the experiment [8], the value of η was

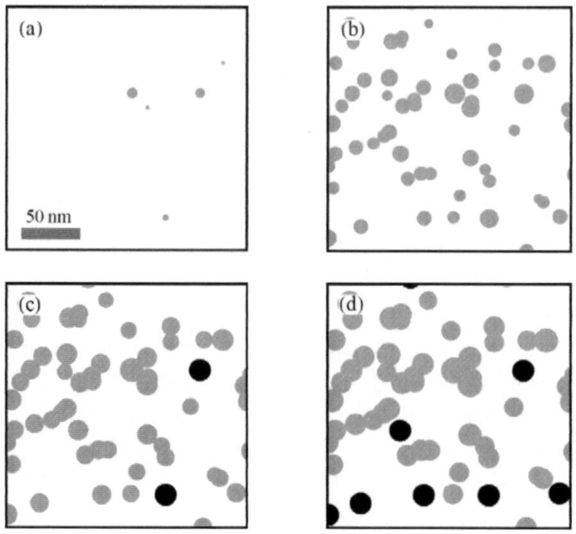

Figure 1. The substrate during modeling at $t = 70$ s (a), $t = 110$ s (b), $t = 150$ s (c), $t = 1300$ s (d). The gray circles are the GaN islands, the black circles are the GaN NWs. The parameters used are described in the text.

calculated in the work [5]. The value of the maximal island density n is higher than in the experiment [8] (about $950\,\mu m^{-2}$) because, as we assume, very small nuclei were not detected in the experiment before they reached some size and collided with each other. In fact, n is a fitting parameter of the model. The parameters χ, τ and c are found using the method [7] and depends on the choice of n. Thus, the fitting parameters of the model are: r_C, n, h_{eq1} and τ_1. Figure 2 shows a good agreement of the theoretical and experimental dependences.

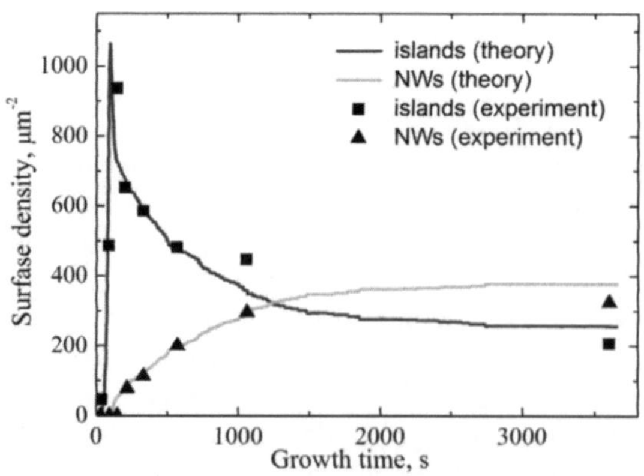

Figure 2. The GaN island and GaN NW surface density versus the growth time. The parameters used to obtain the theoretical curves are described in the text.

On the basis of the modeling, we can make the following notes regarding the NW formation. Firstly, the nucleation stage of GaN islands is short (t = 0-150 s). Secondly, at the very onset of the transition stage (t = 150 s) a large decrease of the island density occurs as a result of the multiple collisions of islands with the size lower than r_C. Therefore the maximal island density is about three times higher than the maximal surface density of NWs. Thirdly, as the transient stage (t = 150-1300 s) starts, the island growth rate decreases drastically and can be even negative (the island evaporation) since the NW growth rate is much higher than the island growth rate ($\tau >> \tau_1$). Fourthly, at t > 1300 s (the NW growth stage), almost all islands either transform into NWs or collide therefore the island and nanowire surface density change insufficiently.

It is noteworthy that the NW coalescence process [9] and the shadowing effect [10] are excluded from our consideration. Although these effects play an important role when the NW height and NW density is large enough. Also, we neglect the island shape change produced by the elastic stress relaxation assuming that the island growth regime changes instantly at the island-NW transition only [5].

Our approach allows the estimation of the NW surface density for different gallium fluxes. For this purpose, we calculate the island size distribution function for different gallium fluxes using the method [7]. Then, we model the NW growth with different initial distribution function. Although the maximal island density increases with an increase of material flux [7], the NW surface density decreases (figure 3) as a result of multiple island collisions that decreases several times the amount of

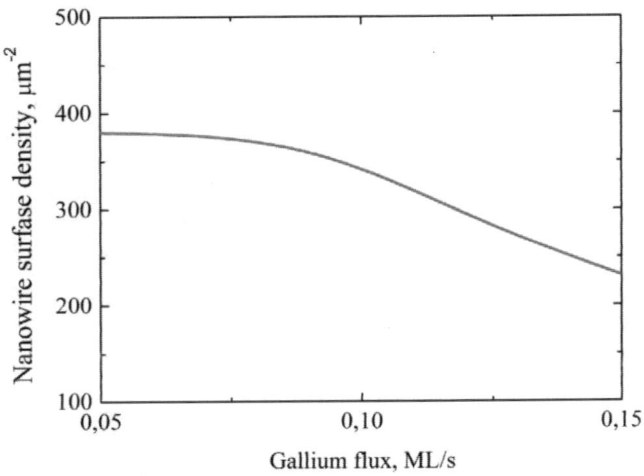

Figure 3. The NW surface density versus the gallium flux.

single islands. It should be noted that the NW surface density is an increasing function of material flux in the model without island collisions [5].

3. Conclusion

A new numerical approach for the determination of the GaN NW surface density on an AlN/Si substrate as a function of the growth time and gallium flux is presented. To model the GaN NW growth, we use the initial size distribution function of GaN islands calculated elsewhere. The island and NW growth rate are found from the material balance equation. The GaN island solid-like coalescence and island-NW transition are modeled by the Monte-Carlo method assuming several rules: if an island reaches the critical island size before collision with other island or NW the island becomes NW, otherwise, the island-NW transition does not occur; if two or more islands collide we count these islands as one; if an island nucleates inside other island or NW this island does not grow;

when an island collides with other island or a NW the islands and NW continue growth as if there were no collisions. We show the importance of taking into consideration the island coalescence for explaining that the maximum of GaN island surface density is several times larger than the maximum of GaN NW surface density. Also, we find that the NW surface density decreases with an increase of the gallium flux as a result of multiple island collisions that decreases several times the amount of single islands.

Acknowledgments
Authors wishing to acknowledge the Russian Foundation for Basic Research (project 14-22-00018).

References
[1] Pearton S J, Ren F 2000 *Adv. Mater.* **12** 1571
[2] Dubrovskii V G, Cirlin G E, Ustinov V M 2009 *Semiconductors* **43** 1539
[3] Dubrovskii V G 2014 *Nucleation Theory and Growth of Nanostructures* (Heidelberg: Springer)
[4] Consonni V, Knelangen M, Geelhaar L *et al* 2010 *Phys. Rev.* B **81** 085310
[5] Koryakin A A, Sibirev N V, Dubrovskii V G 2014 *Technical Physics Letters* **40** 471
[6] Osipov A V, Schmitt F, Kukushkin S A, Hess P 2002 *Appl. Surf. Sci.* **188** 156
[7] Dubrovskii V G, Cirlin G E, Ustinov V M 2003 *Phys. Rev.* B **68** 075409
[8] Landre O, Bougerol C, Renevier H *et al* 2009 *Nanotechnology* **20** 415602
[9] Consonni V, Knelangen M, Trampert A *et al* 2011 *Appl. Phys. Lett.* **98** 071913
[10] Dubrovskii V G, Consonni V, Geelhaar L *et al* 2012 *Appl. Phys. Lett.* **100** 153101

Self-equilibration of the radius distribution in self-catalyzed GaAs nanowires

E D Leshchenko[1,2], **M A Turchina**[2], **V G Dubrovskii**[2,3,4]

[1] St. Petersburg State University, Ulianovskaya st. 3, Petrodvorets, 198504 St. Petersburg, Russia
[2] St. Petersburg Academic University, Khlopina 8/3, 194021 St. Petersburg, Russia
[3] Ioffe Physical Technical Institute of the Russian Academy of Sciences, Politekhnicheskaya 26, 194021 St. Petersburg, Russia
[4] ITMO University, Kronverkskiy pr. 49, 197101 St. Petersburg, Russia

Abstract. This work addresses the evolution of radius distribution function in self-catalyzed vapor-liquid-solid growth of GaAs nanowires from Ga droplets. Different growth regimes are analyzed depending on the V/III flux ratio. In particular, we find a very unusual self-equilibration regime in which the radius distribution narrows up to a certain stationary radius regardless of the initial size distribution of Ga droplets. This requires that the arsenic vapor flux is larger than the gallium one and that the V/III influx imbalance is compensated by a diffusion flux of gallium adatoms. Approximate analytical solution is compared to the numerical radius distribution obtained by solving the corresponding Fokker-Planck equation by the implicit difference scheme.

1. Introduction

Theoretical analysis of self-catalyzed growth of III-V semiconductor nanowires (NWs) fabricated via the vapor-liquid-solid (VLS) method attracts great interest. Due to the absence of unwanted gold contamination, the possibility of integration of NW-based photonic devices on Si and pure zincblende crystal structure, such NWs look very promising for applications. The theoretically predicted [1,2] and experimentally confirmed [3] focusing effect whereby the NW radius converges to a certain stationary one in the course of growth is another important feature of Ga-catalyzed GaAs NWs which enables fabrication of size-uniform NW arrays and may play the key role for their optoelectronic applications. As it is known, the atomic flux of group III elements to the droplet consists of the direct atomic flux, proportional to the droplet surface area, and the diffusion from the NW sidewalls, proportional to the droplet base perimeter. Therefore, the regular growth rate of the NW radius in these conditions can be represented as $dr/dt = A + B/r$. Here, r is the dimensionless NW radius at time t, A and B are the kinetic constants. Depending on the V/III flux ratio, the value of A can be either positive or negative. In the former case, the radial growth is infinite, while in the latter case the radius tends to the stationary value $r_s = -B/A$ regardless of the initial condition. However, accounting for the hyperbolic tangent dependence of the diffusion on the NW length [4], the relationship between the NW radius and the growth time becomes more complicated. Radius dependences at different initial radii of individual NWs are presented in Figure 1.

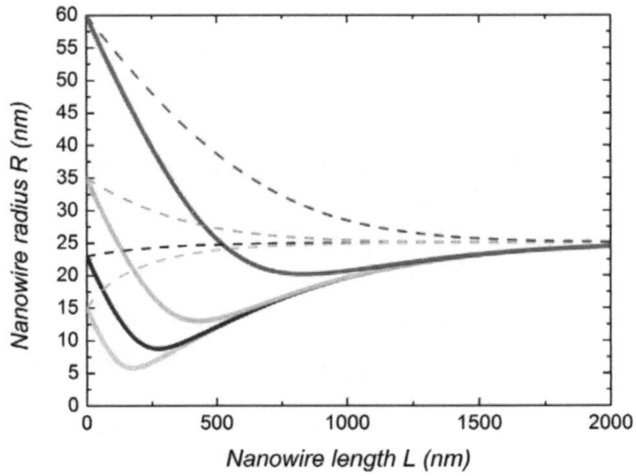

Figure 1. The time-dependent NW radius at different initial radii with (solid curves) and without (dotted curves) accounting for the hyperbolic tangent dependence of the diffusion on the NW length in the self-equilibration regime for $A=-1$ and $B=83.3$.

In the simplified model of Ref. [3] for the radius distribution function, the second derivative in the Fokker-Planck equation was not taken into account. This approximation neglects the role of kinetic fluctuations that should result in the Poissonian broadening of the radius distribution in the infinite growth regime. From an analysis of the radius-time dependence of individual NW, obviously, that in case of self-equilibration growth the radius distribution has to narrow up and tend to the delta function despite the initial droplet size distribution. The purpose of this paper is to describe theoretically the evolution of the NW size distribution.

2. Time evolution of the size spectrum

The After the completion of the nucleation phase, the evolution of the discrete length l or radius r distribution, corresponding to the nuclei growth, is described by the Fokker-Plank type kinetic equation in partial derivatives of the second order [5,6]:

$$\frac{\partial f_s(t)}{\partial t} = -\frac{\partial}{\partial s}\left[(W_s^+ - W_s^-) f_s(t)\right] + \frac{1}{2}\frac{\partial^2}{\partial s^2}\left[(W_s^+ + W_s^-) f_s(t)\right] \tag{1}$$

Here, $s=(r,l)$ is nanowire size, W_s^+ and W_s^- are the attachment and detachment rates. The first derivative describes the drift, whereas the second one describes the pure diffusion effect.

Under steady-state and As limited conditions and neglecting the surface diffusion of group V species, the NW elongation rate equals $W_l^+ - W_l^- = dl/dt = v_5$, where v_5 is the direct atomic flux of group V elements. Then the evolution of the length distribution $f_l(t)$ is given by the Poisson distribution and for large values of l tends to a continuous Gaussian distribution with the dispersion $\sigma^2 = \langle l \rangle$, where $\langle l \rangle = v_5 t$ is the mean nanowire length [7]:

$$f(l,\langle l \rangle) \cong \frac{1}{\sqrt{2\pi\langle l \rangle}} \exp\left(-\frac{(l-\langle l \rangle)^2}{2\langle l \rangle}\right) \tag{2}$$

Now consider the case, when $W_r^+ - W_r^- = W_r^+ + W_r^- = dr/dt$. Using the continuum approximation, that is applicable at $r >> 1$, the evolution of the radius distribution $f(r,t)$ writes:

$$\frac{\partial f(r,t)}{\partial t} = -\frac{\partial}{\partial r}\left[\frac{dr}{dt}f(r,t)\right] + \frac{1}{2}\frac{\partial^2}{\partial r^2}\left[\frac{dr}{dt}f(r,t)\right] \tag{3}$$

Since the length-time dependence is linear, the radial growth rate can be expressed through the mean NW length $dr/d\langle l\rangle = a + b/r$ with constants

$$a = \frac{A}{v_5} = \frac{\Omega_l}{\Omega_{35} f(\beta)}\left(\frac{v_3}{v_5} - 1\right); \qquad b = \frac{B}{v_5} = \frac{\Omega_l}{\Omega_{35} f(\beta)} \frac{2\sin\alpha_3 \lambda_3}{\pi\chi_3}\frac{v_3}{v_5}. \tag{4}$$

Here, $\chi_3 = \chi_5 = 1/(\sin^2\beta)$ are the geometrical functions, $\chi_k V_k$ and $v_k = \chi_k V_k / h_L$ are the direct atomic fluxes ($k = 3,5$) in units of nm/s and ML/s, respectively, $f(\beta)$ is the geometrical function that reflects the relationship between the droplet volume and the contact angle β, α_3 is the incident angle of the Ga beam, Ω_{35} is the volume of a (III,V) pair in the solid, Ω_l is the group III atomic volume in the liquid state, h_L is the monolayer thickness, λ_3 is the diffusion length of the group III adatoms on the NW sidewalls. It can clearly be seen that the radius self-equilibration effect takes place at $v_5 > v_3$.

The continuum second order equation (3) has a simple form in terms of the new distribution defined by $g(\rho,t) = (dr/d\rho) f(r,t)$ with the invariant size ρ according to $d\rho/d\langle l\rangle = 1$:

$$\frac{\partial g(\rho,\langle l\rangle)}{\partial\langle l\rangle} = -\frac{\partial g(\rho,\langle l\rangle)}{\partial\rho} + \frac{1}{2}\frac{d\psi(\langle l\rangle)}{d\rho}\frac{\partial^2 g(\rho,\langle l\rangle)}{\partial^2\rho} \tag{5}$$

$$\frac{d\psi(\langle l\rangle)}{d\rho} = \left(\frac{dr(\rho)}{d\langle l\rangle}\right)^{-1} = \frac{1}{(a+b/r)}, \ \psi(\langle l\rangle = 0) = 0 \tag{6}$$

The invariant size is obtained by integration of the differential equation $d\rho/dr = (a + b/r)^{-1}$ with the boundary condition $\rho(r=0) = 0$:

$$\rho = r/a - b/a^2 \ln(1 + ar/b) \tag{7}$$

The solution for the Green function $g_0(\rho,\langle l\rangle)$ to the equation (5) describing the evolution of the radius distribution from the point source $g(\rho,\langle l\rangle = 0) = \delta(\rho)$, can be approximated as the Gaussian:

$$g_0(\rho,\langle l\rangle) = \frac{1}{\sqrt{2\pi\psi(\langle l\rangle)}}\exp\left(-\frac{(\rho - \langle l\rangle)^2}{2\psi(\langle l\rangle)}\right) \tag{8}$$

Now we introduce the most representative invariant size r_* defined by $dr_*/d\langle l\rangle = 1 + b/r_*$ with $r_*(\langle l\rangle = 0) = 0$, what corresponds to the maximum of the size distribution relating the delta-like initial condition. The value of the most representative invariant size is obtained as the solution to transcendent equation

$$\langle l\rangle = \frac{r_*}{a} - \frac{b}{a^2}\ln\left(1 + \frac{ar_*}{b}\right) \tag{9}$$

The variance can be obtained by integration of the differential equation $d\psi(r_*)/dr_* = (d\psi(r_*)/d\rho)(d\rho/dr_*) = (a+b/r_*)^{-2}$ with the boundary condition $\psi(r_* = 0) = 0$:

$$\psi(r_*) = \frac{r_*}{a^2} + \frac{b}{a^2}\frac{r_*}{ar_*+b} - \frac{2b}{a^3}\ln\left(1+\frac{ar_*}{b}\right) \tag{10}$$

The convolution of the Green function $g_0(\rho,\langle l\rangle)$ with the initial conditions in the terms of invariant size with the main radius r_0 and the variance ψ_0 gives the general solution to the kinetic equation (3):

$$f(r,t) = \frac{1}{|a+b/r|}\frac{1}{\sqrt{2\pi\Psi(r_*)}}\exp\left(-\frac{y^2(r,r_*)}{2\Psi(r_*)}\right) \tag{11}$$

$$y(r,r_*) = \frac{r-(r_*+r_0)}{a} - \frac{b}{a^2}\ln\left(\frac{ar+b}{(ar_*+b)(ar_0+b)}\right) \tag{12}$$

$$\Psi(r_*) = \psi_0 + \frac{r_*}{a^2} + \frac{b}{a^2}\frac{r_*}{ar_*+b} - \frac{2b}{a^3}\ln\left(1+\frac{ar_*}{b}\right) \tag{13}$$

Figure 2 shows the evolution of the radius distribution in the self-equilibration regime in the two cases where the initial mean radius $r_0 << r_s$ and $r_0 >> r_s$, for $a = -0.09$ and $b = 7.5$. As expected, the radius distribution decays at $r_0 >> r_s$ and grows at $r_0 << r_s$. Most importantly, the variance of both distributions gradually decreases so that they acquire a delta-like shape when approaching r_s. For comparison, the numerical solutions of the Fokker-Planck equation by the implicit difference scheme are also shown. The discrepancy between the numerical and analytical solutions is explained by the large step used in numerical calculations and a specific form of the initial condition implied in analytical solution. However, both results match quite well in this case. The analytical scheme does not allow one to consider situations with $r_0 \approx r_s$. The numerical solution of the Fokker-Planck equation by the explicit difference scheme in this case is given in Figure 3. The evolution of the distribution function in the regime of infinite growth is given in Figure 4. It is seen that the distribution spreads up as the NWs grow due to the known effect of Poissonian broadening.

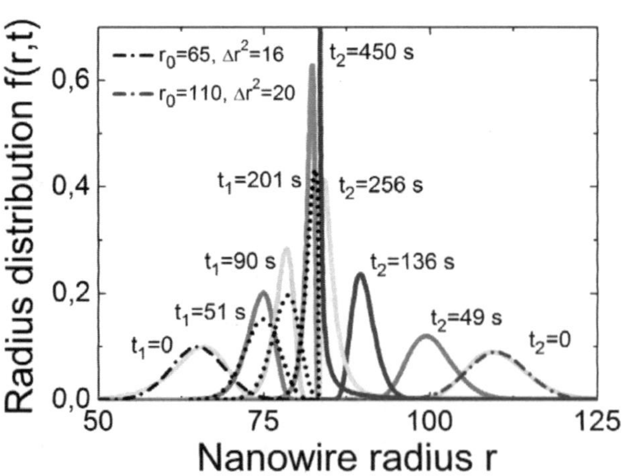

Figure 2. The radius distributions in self-equilibration regime. The solid curves are the analytical solution, the dotted curves are the numerical solution, the dashdotted curves are the initial distributions in Gaussian forms with the expected value r_0 and the variance Δr^2.

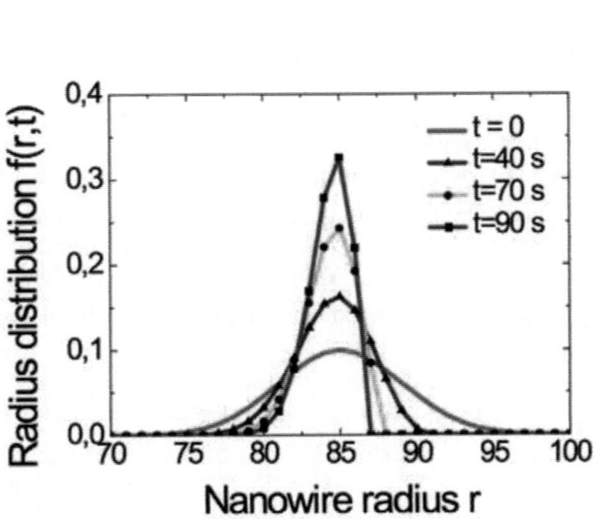

Figure 3. The numerical calculation for the radius distribution for $r_0 = 85$.

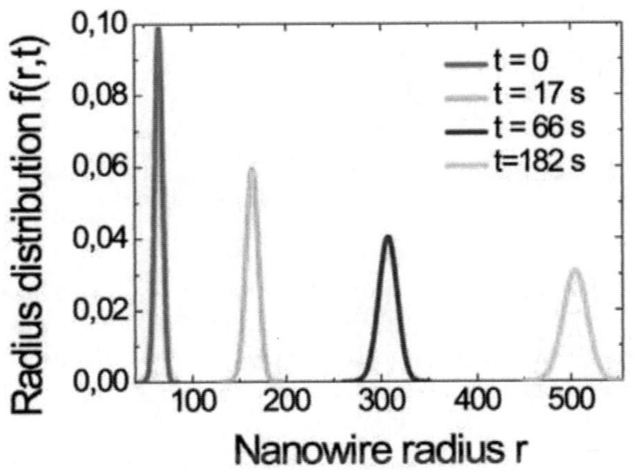

Figure 4. The radius distribution in infinite growth for $r_0 = 65$, $\Delta r^2 = 16$, $a = 0.005$ and $b = 60$.

The Fokker-Planck equation with different kinetic coefficients takes the form

$$\frac{\partial f(r,t)}{\partial t} = -\frac{\partial}{\partial r}\left[\left(a+\frac{b}{r}\right)f(r,t)\right]+\frac{1}{2}\frac{\partial^2}{\partial r^2}\left[\left(c+\frac{b}{r}\right)f(r,t)\right]. \tag{14}$$

Here, constants a and b are the same as in equation (4) and $c = \dfrac{\Omega_l}{\Omega_{35} f(\beta)}\left(\dfrac{v_3}{v_5}+1\right)$.

Analytical solution is the same as in equations (11) and (12) with the variance $\Psi_c(r_*)$

$$\Psi_c(r_*) = \psi_0 + \frac{1}{a^4}\left[acr_* - b(3c-a)\ln\left(1+\frac{ar_*}{b}\right) - b(3c-2a)\left(\frac{1}{1+ar_*/b}-1\right)+\frac{b(c-a)}{2}\left(\frac{1}{(1+ar_*/b)^2}-1\right)\right] \tag{15}$$

The analytical radius distribution and numerical solution are presented in Figure 5. There we can see that under the analytical solution the distribution function does not pass through the stationary radius (what associates with the transition to the invariant size), whereas under the numerical solution it does.

From a physical point of view, due to fluctuations, the distribution function moves to $r > r_s$ region under approaching of the distribution function to the stationary radius, but the function does not have time to return to the stationary value, because of slow growth (which is determined by the first derivative) in $r \approx r_s$ region, and grow further to achieve a stationary form. The form of the distribution function obtained by the numerical method is supported by the experimental data, however, it should be noted that the experimental distribution function is slightly narrower. In the $W_r^+ - W_r^- = W_r^+ + W_r^- = dr/dt$ model both fluctuations and drift speed decrease under approaching of the main radius to the stationary value, what explains the form of the distribution function obtained as a result of the numerical calculations.

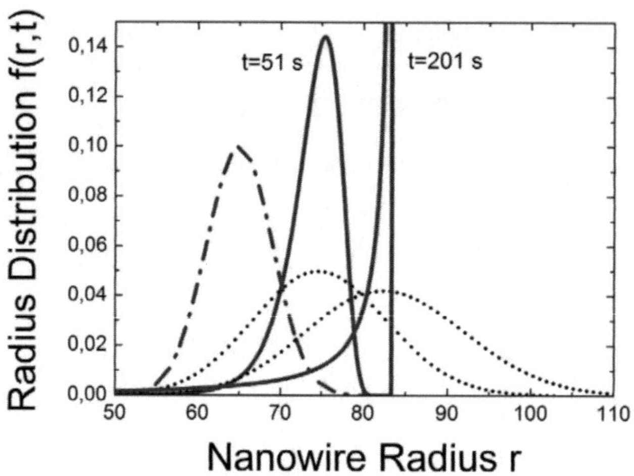

Figure 5. The radius distributions in self-equilibration regime for $a = -0.09$, $b = 7.5$ and $c = 0.116$. The solid curves are the analytical solution, given by equations (11), (12) and (15), the dotted curves are the numerical solution of equation (14), the dashdotted curve is the initial distributions in Gaussian form with $r_0 = 65$ and the variance $\Delta r^2 = 16$.

In conclusion, we demonstrate that the fluctuation-induced broadening of the size distribution can be completely suppressed by appropriate deterministic growth rate that renders the system into the self-equilibration regime. If the absence of radial NW growth, the NW top radius will follow that of the droplet and the NWs will be either tapered or reverse tapered at the beginning to match the stationary radius in the asymptotic stage. Whenever radial growth by a step flow is enabled [8,9], the NWs will have a uniform radius from base to top starting from the time moment at which the droplet reaches the stationary size.

References

[1] Priante G., Ambrosini S., Dubrovskii V. G. *et al.* 2013 *Cryst. Growth Des.* **13** 3976
[2] Tersoff J. 2015 *Nano Lett.* **15** 6609
[3] Dubrovskii V. G., Xu T., Díaz Álvarez A. *et al.* 2015 *Nano Lett.* **15** 5580
[4] Consonni V., Dubrovskii V. G., Trampert A. *et al.* 2012 *Phys. Rev. B* **85** 155313
[5] Dubrovskii V. G. 2009 *J. Chem. Phys.* **131** 164514
[6] Dubrovskii V. G. Nazarenko M. V. 2010 *J. Chem. Phys.* **132** 114507
[7] Dubrovskii V. G. 2013 *Phys. Rev. B* **87** 195426
[8] Dubrovskii V. G., Consonni V.,Geelhaar L. *et al.* 2012 *Appl. Phys. Lett.* **100** 153101
[9] Dubrovskii V. G., Consonni V., Trampert A. *et al.* 2012 Phys. Rev. B **85** 165317

Separation of stress-free AlN/SiC thin films from Si substrate

A V Redkov[1,2], A V Osipov[1-3], I S Mukhin[1,3] and S A Kukushkin[1-4]

[1]St. Petersburg Academic University, St. Petersburg 194021, Russia
[2]Institute of Problems of Mechanical Engineering RAS, St. Petersburg 199178, Russia
[3]ITMO University, St. Petersburg 197101, Russia
[4]Peter the Great St. Petersburg Polytechnic University, St. Petersburg 195251, Russia

Abstract. We separated AlN/SiC film from Si substrate by chemical etching of the AlN/SiC/Si heterostructure. The film fully repeats the size and geometry of the original sample and separated without destroying. It is demonstrated that a buffer layer of silicon carbide grown by a method of substitution of atoms may have an extensive hollow subsurface structure, which makes it easier to overcome the differences in the coefficients of thermal expansion during the growth of thin films. It is shown that after the separation of the film from the silicon substrate, mechanical stresses therein are almost absent.

1. Introduction

Nowadays scientists are actively searching for new ways to obtain high quality semiconductor thin films and heterostructures, in particular, gallium nitride that is used in LEDs and other electronics [1,2] and aluminum nitride. For the growth of such films sapphire and SiC substrates are used mainly [3-5], less often silicon or other substrates, but all of these substrates has some disadvantages. Sapphire has very low thermal conductivity and is dielectric, so the additional conducting contacts should be added. SiC is more suitable, because its thermal conductivity is high and close to the conductivity of the copper. In addition, the lattice parameters of SiC and AlN/GaN are much closer to each other [6]. Since SiC substrates are already widely used in practice, it allows one to suggest, that the use of SiC as a substrate for the growth of AlN/GaN heterostructures looks very promising. Nevertheless, a significant drawback of using single-crystal SiC in the manufacturing of devices is the high price of SiC substrates of good quality and desired polytype. One of the promising types of substrates which solves these problems are substrates of nano-SiC/Si, grown by substitution of silicon atoms to carbon atoms [7] directly in the silicon substrate. In fact, the authors of that work for the first time in world practice implemented method of successive substitution of carbon atoms instead of native silicon atoms right inside the substrate without destroying its crystalline structure. The method resembles the "genetic synthesis" of protein structures in biology. The quality of SiC thin film obtained by this method are far superior than quality of films grown on silicon substrates by conventional methods. The method is simple and cheap, and the presence of pores beneath the surface of SiC allows to overcome the mechanical stresses resulting from the lattice mismatch. Thus the study of the method [7] is prospective for the development of thin film technology.

2. Experiments

For the experiment we have used silicon substrate with orientation <111>, which was subjected to growth process of SiC film [7] at 1250C, CO pressure 2 Torr. The time of the growth was 20 min. According to ellipsometric data, the total thickness of the film was ~100nm. Then in HVPE process on this substrate we have grown AlN film of 18 mkm thickness at 1080C and NH_3/H_2 ratio 1:5. Then, the sample was subjected to chemical etching in a solution containing a mixture of nitric and hydrofluoric acids and water, 4:1:5, respectively. Such a mixture effectively removes silicon [8]. Etching occurred for 15 minutes at 20°C, after which the film has been completely separated from the substrate. After that the resulting film was studied using a Raman microscope Witec Alpha 300 R and the scanning electron microscope.

3. Results and discussion

After 1-5 min of immersion into the etching solution one can note the beginning of the etching with the naked eye. It becomes apparent that the edges of the thin film start to peel off from the substrate. Then over time, the etching front moves towards the center of the substrate, until the film becomes completely separated from the silicon substrate. Probably such behavior is caused by a large number of pores and channels in the subsurface area of silicon, which are quickly filled with acid. This allows acid to penetrate deeper fast, and to etch only small amounts of silicon, which is located between the pores. There is no need to dissolve all of the silicon substrate as a whole, so the separation process takes only a few minutes. Usually this time at room temperature takes 10-20 minutes. The schematic view of this process is presented in Figure 1.

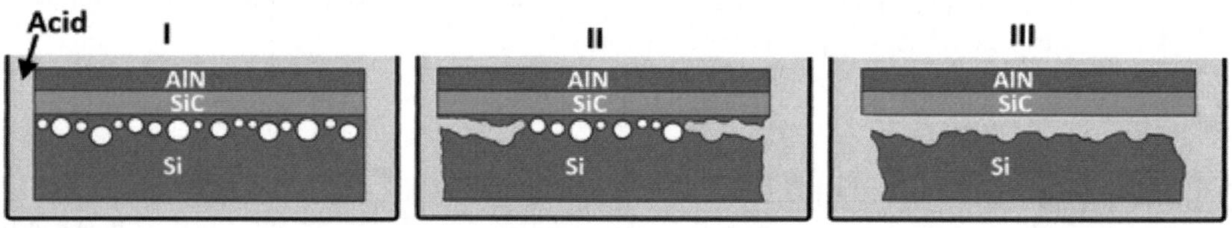

Figure 1. AlN/SiC film delamination mechanism in acid. I - the initial stage. II- etching silicon substrate under the AlN/SiC from the edges towards the center. III - the separation of the film from the substrate

Scanning electron microscope images of AlN/SiC thin film cross-section and "backside" of SiC film are presented in Figure 2. One can clearly see the developed structure of the channels with the thickness of 2 microns remained after etching below the surface of the silicon carbide film. The channels have thin walls about tens of nanometers in thickness and are hollow inside. It can be expected that the ratio of surface area to volume of the material in such a system is quite large. Since the channels have not been dissolved by acid, one can assume that during the growth of SiC film, the surface of the pores formed in the silicon substrate is also being covered by silicon carbide, that was suggested in [7], and channels remaining after etching are the carcass of these pores.

Comparing pores which were visible on cross-section of the SiC films cleaved in [7] and Figure 2, one can conclude that pores in SiC are not enclosed objects, but rather part of the channels which originate at the surface, and have a dendritic structure, which develops deeper into the silicon. The presence of such system of hollow channels can lead to the fact that the mechanical stress arising from the difference in thermal expansion coefficients and lattice parameters mismatch will be much less than in continuous films, because the contact area of the film and the substrate becomes much less. In addition, the presence of pores in silicon subsurface region creates an effective "mechanical decoupling" between the film and the Si substrate.

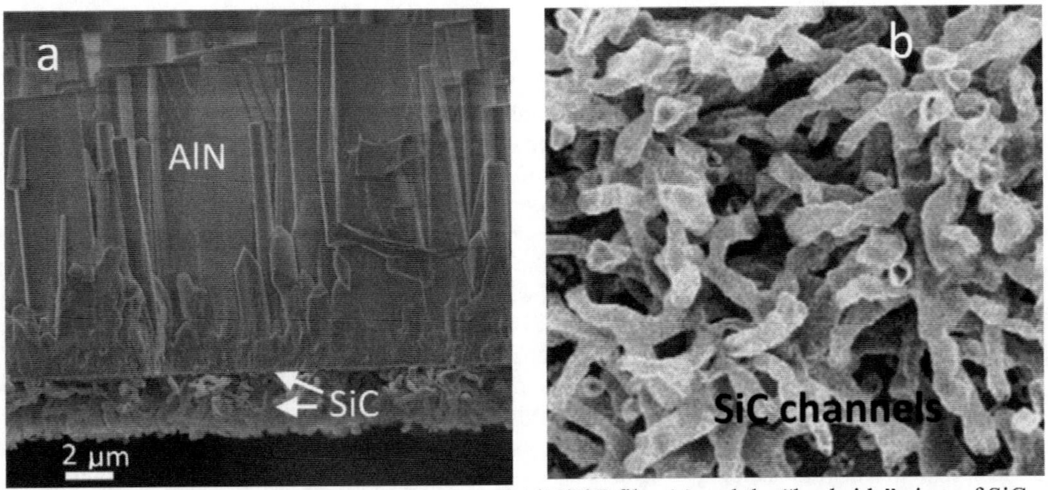

Figure 2. SEM image of the cross-section of AlN/SiC film (a) and the "backside" view of SiC substrate.

The Raman spectra of the initial AlN/SiC/Si structure and AlN/SiC thin film are presented in Figure 3. One can note that the silicon band 521 cm^{-1} after etching totally disappeared, and spectra contains only bands of AlN and SiC. Thus the silicon has been fully dissolved by acid. Inset in Figure 2 demonstrates the position of AlN band 650-665cm^{-1}. The center of this band after etching has been shifted to 657 cm^{-1}, which means that mechanical stress is almost absent in the film [9].

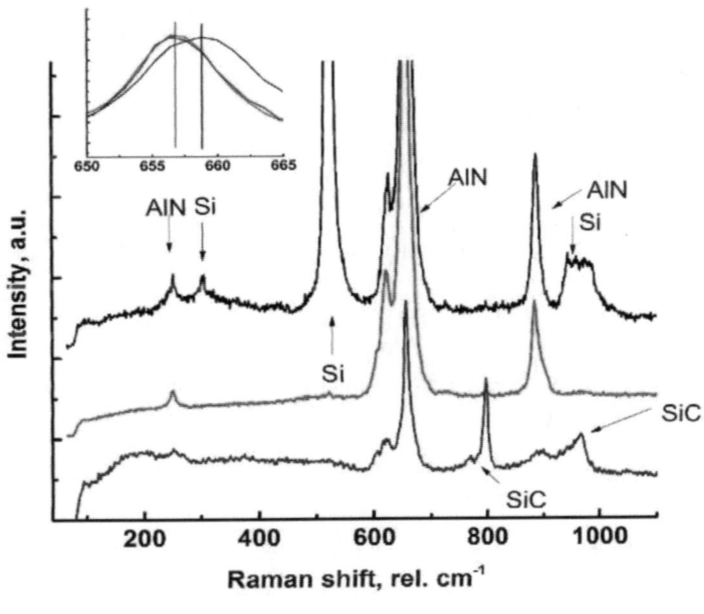

Figure 3. Raman spectra of initial AlN/SiC/Si structure (top graph) and AlN/SiC film from the AlN side (middle) and from SiC side (lowest one).

5. Conclusions

In this work we consider the mechanism of separation of AlN/SiC thin film from Si substrate by chemical etching. It is shown that the presence of SiC buffer which has extensive subsurface porous structure allows to separate thin films or heterostructures from silicon substrates effectively and in a short time without using of sophisticated technical means. This study allowed for the first time to explicitly see the SiC surface grown by substitution of atoms, from the "reverse" side, without the substrate. We identified the structure of pores and channels that appeared in the subsurface area of silicon during the growth of silicon carbide. Such a porous system allows one to "decouple" the mechanical stresses in thin film from the substrate and substantially reduces them, thus leveling the difference in lattice parameters and thermal expansion coefficients. It was shown that after separation from the substrate the mechanical stresses in AlN/SiC thin film are almost completely relaxed. The developed method of the heterostructure separation from Si substrate allows one to transfer it to another substrates, and creates unprecedented conditions for the development of micro and nano-electronics. The proposed method of separation of high-quality epitaxial layers and heterostructures of wide bandgap semiconductors grown on silicon nanocarbide may find applications in the study of various properties of free-standing structures. They can also be used for growth of low-defect crystals. For example, one can use such AlN/SiC/Si substrate as a seed for growth of thick layers of AlN or GaN and then separate them from the silicon substrate.

Acknowledgments

This study was supported by Russian Science Foundation (RNF grant № 14-12-01102). The authors are thankful to V.N. Bessolov for the assistance and to unique scientific facility "Physics, chemistry, and mechanics of crystals and thin films" (IPME RAS, St. Petersburg, Russia).

References

[1] Fujii T, Gao Y, Sharma R, Hu E, DenBaars S and Nakamura S 2004 *Appl. Phys. Lett.* **84** 855-57
[2] Khan M, Kuznia J, Bhattarai A and Olson D 1993 *Appl. Phys. Lett.* **62** 1786
[3] Yoshida S, Misawa S and Gonda S 1983 *Appl. Phys. Lett.* **42** 427-9
[4] Boćkowski M et al 2005 *J. Cryst. Growth* **274**, 55-64
[5] Rizzi A et al 1999 *J. Vac. Sci. Techol. B* **17** 1674-81.
[6] Kukushkin S, Osipov A, Bessolov V, Medvedev B, Nevolin V and Tcarik K 2008 *Rev. Adv. Mater. Sci* **17** 1-32
[7] Kukushkin S and Osipov A 2014 *J. Phys. D: Appl. Phys.* **47** 313001.
[8] Nikolaev A, Nikitina I, Zubrilov A, Mynbaeva M, Melnik Y and Dmitriev V 1999 *MRS Proc.* **595** F99W6-5
[9] Freitas J, Culbertson J, Mastro M, Kumagai Y and Koukitu A 2012 *J. Cryst. Growth* **350** 33-7

A new approach to the high-yield synthesis of nanoparticles by spark discharge

A A Efimov, A A Lizunova, I A Volkov, D A Mylnikov, P V Arsenov, V V Ivanov
Department of Physical and Quantum Electronics, Moscow Institute of Physics and Technology, Dolgoprudny 141700, Russia

E-mail: efimov.aa@mipt.ru

Abstract. A new approach to the high-yield synthesis of nanoparticles in the multi-gap (up to 12) spark discharge generator operating at a repetition rate as high as 2.5 kHz is discussed. This approach allows to increase the production rate by a factor of up to 100 as compared to the conventional approach utilizing single gap generator operating in the self-breakdown mode. In the case of spark erosion of aluminum electrodes in the air atmosphere, the production rate of 300 mg/h has been achieved. From the analysis of transmission electron microscopy images it was found that primary particles of the synthesized material are nearly spherical in shape and their size is 11 nm; these primary particles are united into larger objects (~80 nm) with fractal structure. The fraction of amorphous phase is 54% and the rest 46% is of cubic γ-Al_2O_3 phase with the mean crystallite size of 12 nm as determined from X-ray diffraction analysis. The specific surface area of the nanoparticles determined by BET method is equal 180 m^2/g.

1. Introduction

The synthesis of nanoparticles using spark discharge is a promising method for low-cost industrial-scale nanofabrication of advanced materials [1]. The distinctive features of this method are: a) the possibility of the synthesis of very small primary nanoparticles with a diameter of less than 10 nm from metals, oxides, semiconductors and carbon materials; b) high chemical purity of the synthesized nanoparticles with a low content of impurities; c) the relative simplicity of the implementation of the method. However, low production rate of about 10 mg/h [2] has been a limiting factor for a wide application of nanoparticles produced by spark discharge. In this regard, researchers are developing approaches that will increase the production rate of the synthesis of nanoparticles by spark discharge. The most successful approaches are associated with an increased frequency of discharge impulses [3] or an increased number of consecutive inter-electrode gaps [4].

In this paper, we propose combining two approaches in one setup and investigate the properties of the nanoparticles produced by spark discharge at the maximum production rate. Specifically, the paper presents new results of the synthesis of nanoparticles at high repetition rate of discharge pulses (2.5 kHz) and a large number of series-connected inter-electrode gaps (up to 12) implemented in one setup.

2. Experimental

The multi-spark discharge generator consisted of 12 pairs of aluminium electrodes was used in the experiments (figure 1). The voltage between the electrodes was applied by a 12 nF capacitor. The capacitor is charged from a source of up to 4.5 kV voltage. Initiation of the gas discharge was carried out using the principle of switching multi-spark dischargers by supplying a trigger voltage pulse [5].

The frequency of the discharges of up to 2.5 kHz was regulated by a generator of pulsed currents. The space between the electrodes was constantly purged by the flow of clean air at velocity of 15 m/s. The nanoparticles were deposited onto an aerosol filter made of porous stainless steel.

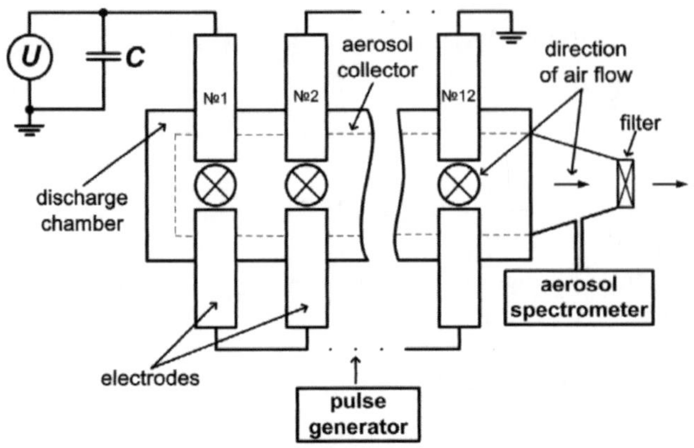

Figure 1. A scheme of the multi-spark discharge generator for the synthesis of nanoparticles.

The particle size distribution, morphological and structural features of the obtained nanoparticles were investigated by a transmission electron microscope (TEM) JEM-2100 (JEOL). The structure and phase composition of the collected nanoparticles were studied by a powder X-ray diffractometer (XRD) D8 DISCOVER (Bruker). Qualitative and quantitative elemental composition of the nanoparticles was determined using a mass spectrometer with inductively coupled plasma (ICP-MS) iCAP Qc (Thermo Scientific). Real-time measurements of size of the nanoparticles in the aerosol flow were performed by an aerosol spectrometer DAS 2702 (AeroNanoTech). The specific surface area of the nanoparticles was analyzed using a gas adsorption analyzer TriStar 3000 (Micromeritics).

3. Results and discussions

The photos of the filter with precipitated particles are shown in figure 2. The area of the filter with precipitated nanoparticles becomes white, which is characteristic of alumina powder.

Figure 2. The photo of the filter before and after the deposition of airborne particles produced by spark erosion of aluminium electrodes in the air.

The mass of the filter was increased by 52 mg within 10 minutes of the generator operation. Thus, the production rate of ~300 mg/h was achieved. This production rate is 100 times higher than the rate of conventional spark discharge generators with single gap [2,6]. The comparison of the production rate of the multi-spark and conventional discharge generators is presented in table 1.

Table 1. Production rate of the new multi-spark discharge generator and a conventional spark discharge for the erosion of aluminium electrodes in the air

Type of generator	Production rate (mg/h)	Number of pairs of electrodes	Frequency of the discharges (Hz)	Voltage (kV)	Flow rate (l/min)	Reference
Multi-spark discharge generator	300	12	2500	4.5	160	This study
Conventional spark discharge generator	0.5	1	30-300	-	0-20	Other [6]

In table 1 the production rate for the conventional spark discharge generator was determined by multiplication of the mass concentration of particles and flow rate [6]. In this study high-yield synthesis of nanoparticles is due to two factors. The first factor is related to the increase in the number of interelectrode discharge impulses by increasing their density in time (i.e. by increasing the repetition rate discharges up to 2.5 kHz). The second factor is the twelve-fold increase in the number of evaporated electrodes, which in turn increases the amount of discharge pulses in space.

A typical TEM-image of the nanoparticles is shown in figure 3a. The sample consists of large agglomerates of spherical primary particles. It is found that the size of the primary nanoparticles remains small despite high production rate. Indeed, the size of the primary particles is in the range of 3 to 30 nm with the mean size of 11±4 nm (figure 3b). The particle size distribution is well described by the log-normal function with the following parameters: median = 10.9 nm and sigma = 0.336.

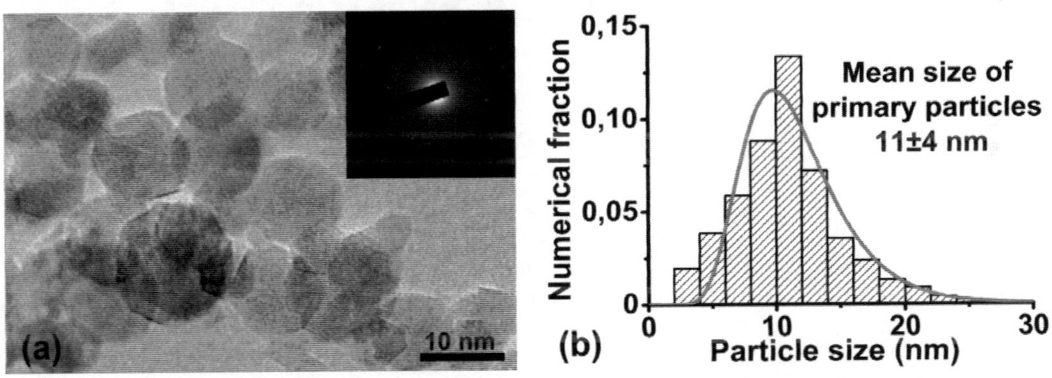

Figure 3. TEM image of nanoparticles and corresponding electron diffraction pattern (inset) **(a)**; Particle size distribution calculated from TEM data **(b)**.

The mean size of the agglomerates is about ~80 nm by TEM-analysis; which is consistent with the data measured by the aerosol spectrometer (figure 4). The large size of the agglomerates is due to the high production rate.

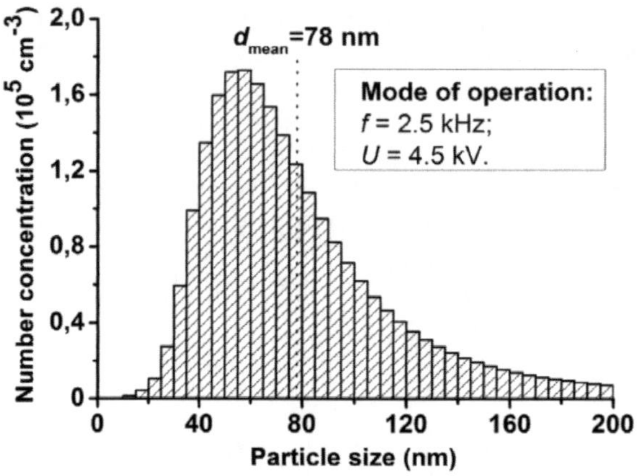

Figure 4. Size distribution of number concentration of as-synthesized airborne nanoparticles measured by diffusional aerosol spectrometer.

Additionally, it is found that the nanoparticles are aluminum oxide with a fraction of amorphous particles of about 54 wt. % and the crystalline fraction of about 46 wt. % with the phase γ-Al_2O_3 according to the results of X-ray diffraction analysis (figure 5). The formation of alumina is due to the fact that erosion of the aluminum electrodes was in the oxygen-containing atmosphere: $4Al + 3O_2 \rightarrow 2Al_2O_3$.

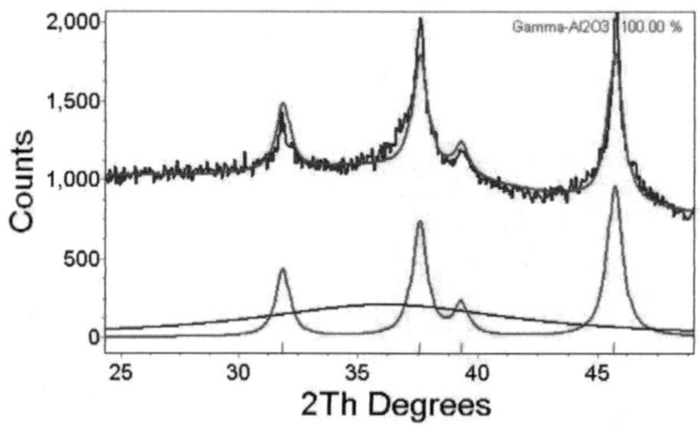

Figure 5. XRD spectrum of nanoparticles produced by spark erosion of aluminum electrodes in the air.

A detailed result of X-ray diffraction analysis of the nanopowder, including lattice parameters and mean size of the crystallite is shown in table 2.

Table 2. Results of XRD phase analysis.

Composition	Parameters of the lattice (Å)	Mean crystallite size (nm)
Al_2O_3 amorphous (54 wt. %)	-	-
γ-Al_2O_3 cubic (46 wt. %)	a = 7.94±0.02	12

It was found that the synthesized nanoparticles contained about 36 wt% of Al according to the results of elemental analysis by ICP-MS. The measured concentration was greater than theoretically-calculated concentration of Al in Al_2O_3 (53 wt. %). This result indicates that the sample is not pure alumina. The specific surface area of the nanoparticles was determined equal to 180 m^2/g. In terms of size this is equivalent to particles with a diameter of 8 nm. Thus, the result of measurement of the specific surface of the particles is in good agreement with TEM-data.

4. Conclusion

New combined approach in the spark discharge allows to increase the production rate of nanoparticles by a factor of up to 100 compared to the conventional approach utilizing single gap generator operating in the self-breakdown mode. The combined approach consists of increasing the repetition rate of discharges up to 2.5 kHz and the number of discharge gaps up to 12, combined in one setup. It was shown to generate nanoparticles of Al_2O_3 at high production rate more than 300 mg/h. The synthesized primary nanoparticles have an mean size of 11 nm, which fractal form agglomerates with size of about 80 nm. The developed method is promising for the application in industrial-scale nanofabrication of advanced materials.

Acknowledgments

This work was supported by the Russian Science Foundation (project № 15-19-00190).

References

[1] Feng J, Biskos G and Schmidt-Ott A 2015 *Sci. Rep.* **5** 15788
[2] Tabrizi N S, Ullmann M, Vons V A, Lafont U and Schmidt-Ott A 2009 *J. Nanoparticle Res.* **11** 315–32
[3] Pfeiffer T V, Feng J and Schmidt-Ott A 2014 *Adv. Powder Technol.* **25** 56–70
[4] Efimov A A, Ivanov V V, Bagazeev A V, Beketov I V, Volkov I A and Shcherbinin S V 2013 *Tech. Phys. Lett.* **39** 1053–6
[5] Mesyats G A 2007 *Pulsed Power* (Springer Science & Business Media)
[6] Bau S, Witschger O, Gensdarmes F and Thomas D 2012 *J. Nanoparticle Res.* **14** 1–17

Preparation of $PbSeO_3$ as a new material, sensitive to the electromagnetic radiation in UV range

V V Tomaev[1], V D Smolyaninov[2], T V Stoyanova[2] and S V Egorov[2,3]

[1] Institute of Chemistry, Saint-Petersburg State University, Saint-Petersburg, 198504, Russia
[2] Department of General and Applied Physics, National Mineral Resources University (Mining University), Saint-Petersburg, 199106, Russia
[3] Ioffe Physical-Technical Institute, Saint-Petersburg, 194021, Russia

Abstract. The new technology of the formation of photoresistive structures sensitive in ultraviolet range of electromagnetic spectrum based on lead selenide and lead selenite composite is discussed. Studies of photosensitivity were carried out using a set of LEDs in the visible and ultraviolet spectral range. Obtained structures show considerable sensitivity in ultraviolet and blue range of radiation, meanwhile that in red and yellow region of light turned to be small. The structures were formed by oxidation of PbSe crystals. Diffusion of the oxygen through the surface layer of PbSe was suggested to be a key mechanism of oxidation. Oxidation kinetics were studied by means of roentgen lines chemical shift and roentgen diffraction.

1. Introduction

The recent years reveal a steady growth of interest aimed at the materials sensitive to ultraviolet (UV) radiation. The reason is not only new scientific results but also increasing need in devices in industry, medicine, ecology and other fields of human activity.

In the last 10 – 15 years the leading scientific centers actively develop semiconducting UV light emitting diodes (LEDs) and photodiodes (PDs) based on AlGaN nanoheterostructures, with a shortest wavelength emitted 210 nm [1], [2]. One can produce detectors of UV radiation as photoresistors, photodiodes based upon p-n transitions, p-i-n structures and Shottky diodes, field transistors with Shottky barrier and so on. Most often used materials for these purposes are A_3B_5 compounds (GaP, GaN, AlN, SiC) and others, A_2B_6 (zinc chalcogenides, cadmium chalcogenides), ZnO and diamond as well [3] – [10].

2. Materials and methods

Materials which do not reveal sensitivity in the visible range, but reveal it in UV are of special interest. They allow one to improve precision of measurements in UV range. As long as energy gap of lead selenite ($PbSeO_3$) is 3.17 eV [11], one can look forward films on its basis to show high sensitivity in UV range of spectrum and no such a one in the visible range.

This work was aimed at production of $PbSeO_3$ as a new material susceptible in UV range and study of its photosensitivity.

As for all optoelectronic devices, for the receivers of UV radiation, it is important to use direct band-gap semiconductors and the fundamental absorption near the edge of the band-gap [12]. The latter condition implies the presence of a red boundary of signal registration. The boundary wavelength λ_{bound}, exceeding which the registration of radiation is absent, may be determined from the following simple relations at interband transfers:

$$h\nu_{bound} = E_g \qquad \lambda_{bound} = \frac{2\pi\hbar c}{E_g} \qquad \lambda_{bound}(\mu m) = \frac{1.22}{E_g(eV)} \tag{1}$$

According to the results presented in [12], Table 1 shows the values of the band gap for the individual components (PbSe and $PbSeO_3$) of a two-phase composite lead selenide – lead selenite and the corresponding boundary wavelengths calculated according to relations (1).

Table 1. Wavelength λ_{bound}, corresponding to the beginning of the interband absorption for the semiconductors PbSe and $PbSeO_3$.

Material	Lead selenide (PbSe)	Lead selenite ($PbSeO_3$)
Bandgap width E_g, eV	0.27	3.17
Wavelength λ_{bound}, µm	4.52	0.39

Thus, for the films of a two-phase composite lead selenide – lead selenite under irradiation in a wide optical spectrum of radiation, it is possible to expect the manifestation of the photosensitivity for the PbSe component in the middle infrared region ($\lambda_{bound\ (IR)} = 4.52$ µm) and for the $PbSeO_3$ component in the near ultraviolet region ($\lambda_{bound\ (UV)} = 0.39$ µm).

Films of lead selenide 1 – 2 µm thick were obtained by vacuum evaporation on a glass substrate. The substrate was rotated during evaporation to improve homogeneity of the film. The subsequent oxidation in a dry air resulted in formation of $PbSeO_3$ phase. It was carried out by means of annealing in electric ovens. Finally compound of two phases PbSe and $PbSeO_3$ was obtained.

The charge state of Se and Pb atoms included in PbSe particles oxidized to a different degree was studied by means of roentgen K-lines chemical shift [13]. Kinetics of oxidation of lead selenide was studied at the temperature equal to 823 K [14]. It was shown that kinetic dependences of oxidation process in lead selenide obtained independently by chemical shift and diffraction corresponds each other quite well. Phase content of reaction products after lead selenide oxidation in a dry air was determined by roentgen diffraction method [15].

Stoichiometrical mass of PbSe was crushed in a gate pounder and subsequently processed in dynamic vacuum to get powder with grains 0.5 – 3 micrometer size. The crystal structure of the powders was studied by roentgen diffraction using K_α line emitted by Cu electrode in continuous regime under voltage of 25 kV and 20 mA electric current. The angle resolution was 0.001°. Oxidation kinetics of PbSe powders were studied by means of roentgen lines chemical shift and roentgen diffraction. Study reveals the following: there is no any other phase inside the sample of PbSe, meanwhile at the surface of it a new structure occurs at the room temperature already. This structure is attributed to $PbSeO_3$; volume of the $PbSeO_3$ at the surface increases proportional to the decrease of PbSe. No any other phase was detected in the samples under the temperatures being used.

Diffusion of the oxygen through the surface layer of oxide film towards the boundary crystal oxide is suggested as a main mechanism of oxidation. The target of further processing is to get oxide film of proper thickness via controlling the value of temperature and time of annealing and thus to get composite materials of two phases with proper electro-physical characteristics.

To simplify further analysis, the distinct grain (cross section of it shown at the Figure 1) of PbSe powder thermally processed in dry oxygen atmosphere could be represented as a sphere or as a cube. Figure 2 shows kinetics of new phase of $PbSeO_3$ formation in the grains of PbSe, thermally processed in dry air at $T = 823$ K.

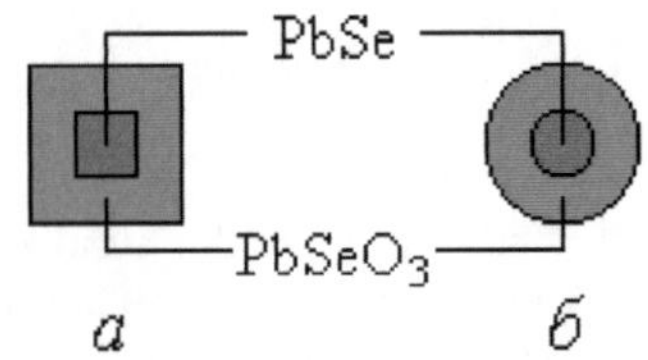

Figure 1. Schematic representation of grain of composite material (1-x)PbSe·(x)$PbSeO_3$ as a cube (a) or a sphere (b).

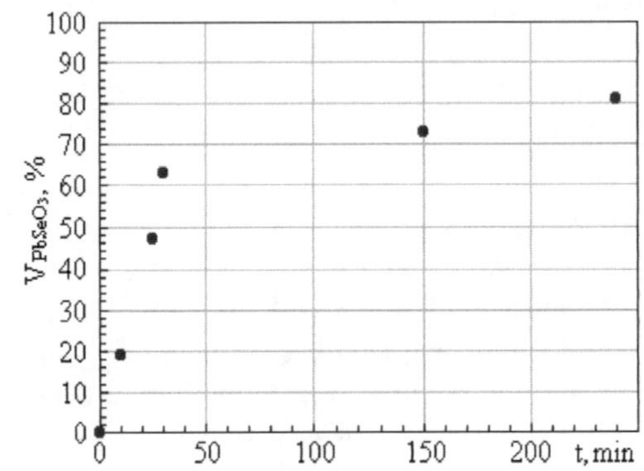

Figure 2. Quantity of $PbSeO_3$ at the grain surfaceof PbSe versus time of oxidation.

One can observe two characteristic ranges: range of a rapid increase of oxidized material and a range of reaction saturation, when oxide phase achieves approximately 80 percent, which is in good agreement with roentgen analysis data. Thus, one can control production of PbSe particles inside the dielectric $PbSeO_3$ coating by means of thermal processing in dry air of various time. Time of the processing governs the size of the particles, which might achieve nanometers.

As an initial material one can choose PbSe polycrystalline films, pressed tablets, powder with grains 0.5 – 3 μm size.

Studies of oxidation of PbSe films in dry air were performed at scanning electron microscope Carl Zeiss SUPRA-40VP [16]. Figure 3(a) shows microphotography of the surface of initial PbSe film, produced by evaporation in a vacuum on a glass. It reveals, that surface consists of separate crystals, which have usual for PbSe cubic structure. One can observe at figure 1(a) dominating growth of (100) sides for the crystals of PbSe with average size ~ 150 – 200 nm. Relatively small variation of the size of the crystals implies homogeneous electrophysical and other properties of the films obtained.

(a) (b)

Figure 3. Microphotographs of the film surface (a) initial PbSe (b) oxidized (1-x)PbSe • x$PbSeO_3$.

Figure 3(b) shows the microphotography of the PbSe surface oxidized at 500 °C during 60 minutes. Average size of oxidized (1-x)PbSe • $xPbSeO_3$ composite particles is of about 50 – 100 nm.

Taking into account, that at 500°C after 60 min ~ 80% of the particle volume is oxidized and oxidation occurs at the surface side, one can figure out that core of each of oxidized particle at figure 3b consists of PbSe and is less than 20 nm size while the coating of the particle occupies the rest of the volume and consists of $PbSeO_3$.

In order to obtain a two-phase composite from lead selenide and selenite, the films of lead selenide were oxidized in dry air atmosphere according to the methods presented in [17 – 23]. In order to determine the region of spectral sensitivity of a receiver of optical radiation based on a composite from lead selenide and selenite, it was proposed to use a chosen set of light-emitting diodes as radiating elements, each of which works in a narrow interval of wavelengths in various regions of the spectrum. Working each in its optical interval of wavelengths, they cover the whole spectral range to be studied. At the same time, an important property of UV receivers is the absence of sensitivity to visible radiation [24]. In connection with this, the investigations were carried out in a wavelength range of 315 – 750 nm (from the short-wavelength radiation of UV to visible radiation). With the effect on a photoresistor of radiation with sufficient photon energy for the generation of pairs of mobile charge carriers (electrons and holes), its resistance should decrease. In our studies, such radiation was provided by a set of light-emitting diodes covering a spectral wavelength range of 315 – 780 nm. Four light-emitting diodes for a visible spectrum of radiation and one diode for UV radiation range were used.

In order to determine photosensitivity of a sample one can measure resistance of it under the illumination. A photoreceiver is a semiconductor film resistor based on a composite from lead selenide and selenite, whose resistance can change under the action of radiation. A layer of semiconductor with the silver contacts at the edges is deposited onto a glass wafer. The semiconductor film has a thickness of 2 μm and a surface area of 2 × 3 mm, the area of the silver contact is equal to 1 × 2 mm.

The scheme of a photoresistor circuit is shown in Figure 4. If a photoresistor is connected in series to a voltage source (Figure 4) and is not subjected to irradiation, then a dark current will flow in its circuit:

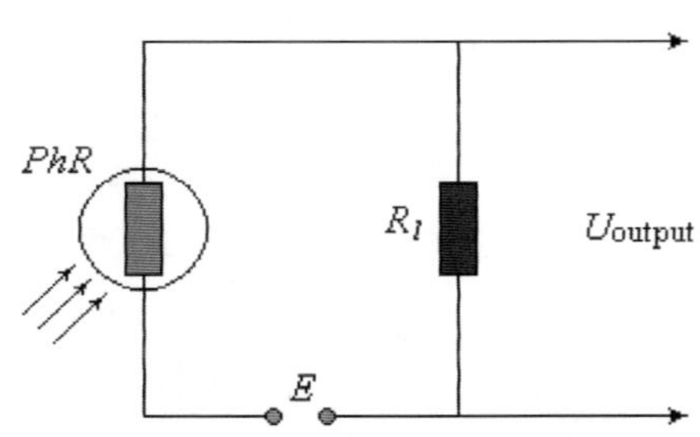

Figure 4. Circuit of the photoresistor

$$I_d = \frac{E}{R_d + R_l} \tag{1}$$

where E is the electromotive force of the supply source; R_d is the value of the electric resistance of the photoresistor in the dark, called dark resistance; and R_l is the resistance of the load. The dark resistance is one of the photoresistor's parameters and has the value ~103 – 105 Ohm. The corresponding current through the photoresistor is called the dark current.

By the illumination of a photoresistor, the energy of the photons is expended to transfer electrons into the conduction band. The number of free electron–hole pairs increases, the resistance of the photoresistor decreases, and a light current, I_i, flows through it:

$$I_i = \frac{E}{R_i + R_l} \tag{2}$$

R_i is resistance of photoresistor under illumination. The difference between the light and dark currents gives the value of current I_{ph}, which is called the primary photocurrent of conductance:

$$I_{ph} = I_i - I_d \tag{3}$$

Another important characteristic of a photoresistor is the value of the specific sensitivity, which is determined by the following expression:

$$\frac{\Delta R}{R} = \frac{R_d - R_i}{R_d} \tag{4}$$

At a certain illumination, the resistance of a photoelement decreases and, hence, the current in the circuit increases, achieving a value sufficient for the operation of the device (it is schematically shown in the form of a load resistance). The useful signal for further amplification or controlling other devices is taken in parallel with R_i.

In our case, the resistance of a photoresistor was determined with the help of measuring bridge AVS-47. The device allows us to measure the resistances of the samples in the range from 2 Ohm to 2 MOhm. The bridge measures the resistance at an alternating current of a frequency of 12.5 Hz. The signal's shape is rectangular pulses. Such a shape allows us to reduce the power released in the studied sample.

3. Experiment and discussion

We have studied resistance of the composite samples $PbSe/PbSeO_3$ illuminated in visible range and UV range of electromagnetic field. We used a set of light emitting diodes as a source of light.

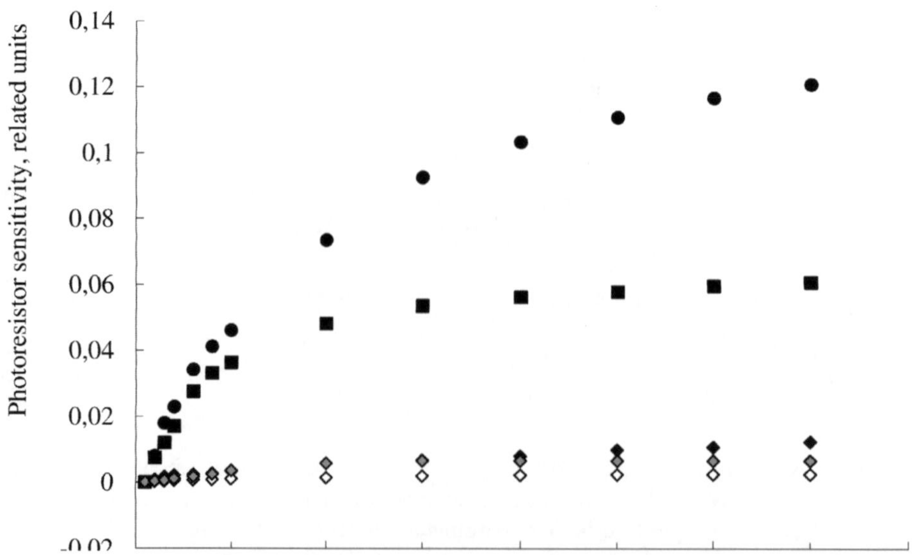

Figure 5. Photosensitivity dependence on LED current.

The diodes were able to emit red, yellow, green and blue light and UV radiation as well. The intensity of diode emission was controlled in accord with its operating current. The samples were illuminated with diodes emitting light of different wavelengths. The sample shows decrease of resistance in UV and blue light region, meanwhile in a red and yellow range of the wavelengths resistance of a sample does not decrease so much. One can calculate sensitivity. This value is shown at the Figure 5.

The energy gap of $PbSeO_3$ is expected to be 3.17 eV [3]. Such energy can be provided by a quantum of radiation with a wavelength about 390 nm, which corresponds to the UV region. We attribute strong susceptibility of resistance under UV light to the layer of $PbSeO_3$ formed over PbSe during the annealing of a sample. Charge carriers are produced by excitation of electrons with a UV energy quanta cross the energy gap. This results into the decrease of resistance.

4. Conclusions

A feature of the proposed material as a new material sensitive to the electromagnetic radiation in UV range is the simplicity of the original film deposition of lead selenide, with its subsequent oxidation to selenite lead and further application of the contact pads on the substrate glass. Photosensitivity of $PbSe/PbSeO_3$ composite films was studied in the visible range and in UV range of spectrum. Maximum sensitivity was observed in the wavelength range 315 – 430 nm. It shows, that $PbSe/PbSeO_3$ demonstrates maximum photosensitivity in UV range of spectrum. Considerable sensitivity of the material exists in blue range, meanwhile it is not sensitive in red and yellow range. It is attributed to the absorption of UV radiation quanta and subsequent excitation of electrons to the conduction band cross the energy gap in $PbSeO_3$ layer formed on surface of PbSe while annealing.

References

[1] Taniyasu Y, Kasu M, Makimoto T 2006 *J. Nature* **441** pp 325–328
[2] Shur M S, Zukauskas A 2004 *UV Solid-State Light Emitters and Detectors* (Proc. NATO ARW Series II 144 Kluwer Dordrecht)
[3] Blank T V and Gol'dberg Yu A 2003 *Semiconductors* **37** pp 999–1030
[4] Razeghi M, Rogalski A 1996 *J. Appl. Phys.* **79** pp 7433–7443
[5] Guo F, Yang B, Yuan Y, Xiao Z, Dong Q and Huang Y 2012 *Nature Nanotechnology* **7** pp 798–802
[6] Sang L, Liao M and Sumiya M A 2013 *Sensors* **13** pp 10482–10518
[7] Soci C, Zhang A, Xiang B, Dayeh S A, Aplin D P R, Park J, Bao X Y, Lo Y H, Wang D 2007 *Nano Lett.* **7** pp 1003–1009
[8] Kind B H, Yan H, Messer B, Law M, Yang P 2002 *Adv. Mater* **14** pp 158–160
[9] Ji L W, Peng S M, Su Y K, Young S J, Wu C Z, Cheng W B 2009 *Appl. Phys. Lett.* **94** 203106
[10] Yan F, Xin X, Aslam S, Zhao Y, Franz D, Zhao, Weiner J H 2004 *IEEE Journal of quantum electronics* **40** pp 1315–1320
[11] Bi G, Zhao F, Ma J, Mukherjee S, Li D, Shi Z 2009 *J Piers Online* **5** pp 61–64
[12] Bube R H 1981 *Photoconductivity of solids* (New York: Wiley)
[13] Tomaev V V, Makarov L L, Tikhonov P A, Popov V P, Rozhdestvenskaya I V 2005 *J Glass Physics and Chemistry* **31** pp 489–498
[14] Tomaev V V, Makarov L L, Tikhonov P A, Solomennikov A A, 2004 *J Glass Physics and Chemistry* **30** pp 349-355
[15] Popov V P, Tikhonov P A, Tomaev V V 2003 *J Glass Physics and Chemistry* **29** pp 494–500
[16] Tomaev V V, Petrov Yu V 2012 *J Glass Physics and Chemistry* **38** pp 240–244
[17] Tomaev V V, Miroshkin V P and Gar'kin L N 2006 *J Glass Physics and Chemistry* **32** pp 579–582

[18] Tomaev V V, Miroshkin V P and Gar'kin L N 2006 *J Glass Physics and Chemistry* **32** pp 583–586
[19] Tomaev V V, Miroshkin V P, Gar'kin L N and Tikhonov P A 2005 *J Glass Physics and Chemistry* **31 pp** 812–819
[20] Tomaev V V and Panov M F 2006 *J Glass Physics and Chemistry* **32** pp 370–373
[21] Tomaev V V, Chernyshova I V and Tikhonov P A 2007 *J Glass Physics and Chemistry* **33** pp 646–651
[22] Tomaev V V 2009 *J Glass Physics and Chemistry* **35** pp 660–667
[23] Panov M F and Tomaev V V 2012 *J Glass Physics and Chemistry* **38** pp 419–426
[24] Diffey B L 2002 *Methods* **28** pp 4–13

Structural changes in high-temperature synthesis of luminescent alumina ceramics

S V Zvonarev, V S Kortov, M V Ryabinina, A N Kiryakov

Ural Federal University, Ekaterinburg, 620002 Russia

Abstract. Scanning electron microscopy was used to study structural changes in luminescent alumina ceramics which was synthesized from nanopowder at high temperatures in reducing environment. An effect of synthesis parameters on size-distribution of grains, their shape and a number of pores in the samples under study was determined. It was found that in a certain temperature range grains are the same ones in the precursor nanopowder, which is associated with the emergence of nanoparticles of lower aluminum oxides.

1. Introduction

Possibilities to use nanosize materials to create luminophors with high light yield and their applications in optical electronics are studied today [1]. Physico-chemical and energetical characteristics of nanomaterials influence greatly their optical properties. Producing luminophors on the base of oxide ceramics synthesized at high temperatures in highly reducing environment is promising [2]. Such ceramics are very strong mechanically and their properties are stable [3]. Luminescence centers caused by oxygen vacancies form in these ceramics. However, high-temperature synthesis of luminescent ceramics is accompanied by changes in structure and growth of the grains [4]. The effects of synthesis conditions on ceramics structure and their luminescent characteristics are little studied. Thus, the aim of the work is to research structural changes in ultrafine alumina ceramics with varying parameters of pressing and sintering.

2. Materials and Methods

Experimental ceramic samples were synthesized from highly pure (99,997%) commercial α-Al_2O_3 nanopowder created by using an alcoholate method. The samples were pressed statically with Specac hydraulic press in metal press moulds under pressure from 450 to 550 MPa. Further, the ceramics were synthesized in a vacuum electric oven (10^{-2} Pa) at the presence and absence of carbon at varying temperatures from 1200 to 1700°C and with annealing for 0.5-5 hours.

The analysis of the particle size distributions of α-Al_2O_3 nanopowder is performed with the Nanophox proton cross correlation spectrometer based on dynamic light scattering. Nanophox provided the twofold detection by 2 laser beams which cross in the measuring volume and 2 detectors which collect the scaterring light. This allows us to select only the single scaterring light information within the total scaterring light under 90^0.

The surface structure and the volume (chip) of the samples were studied by using a scanning electron microscope (SEM) Zeiss Sigma VP. Fragments of ultrafine ceramic structures were analyzed with specialized SIAMS 700 software (SIAMS Photolab complex). This software enabled the quantity characteristics of structural elements of the material to be found. The optical system and computer program of the analyzer allowed us to find all particles in each SEM microphoto and to determine their precise sizes (with tenths of nanometer precision). Particle size distribution for each sample type was defined on the analysis of over 1500 particles.

3. Results and discussion

Particle sizes were analysed by using a method of dynamic light scattering to find the size of structural elements in the precursor α-Al_2O_3 powder. Figure 1 shows SEM image (a) and particle size distribution (b) of α-Al_2O_3 powder. The measurements were repeated several times, so that a set of single measurements was obtained. It can be seen in the figure that a big proportion (> 80 %) in the sample is particles of the sizes 5-50 nm (indicated by arrows). A number of particles with sizes of 100-250 nm is not higher than 20 %. The statistical evaluation contains the absolute standard deviation of the single measurements for each particle size class. In this regard, maximal values up to approximately 20 % are possible and quite normal due to the principle of the applied measurement technique but never reached for any of the analyzed sample. The maximum of the standard deviation calculated for the measurements of this sample reveals a value below 10.9 %.

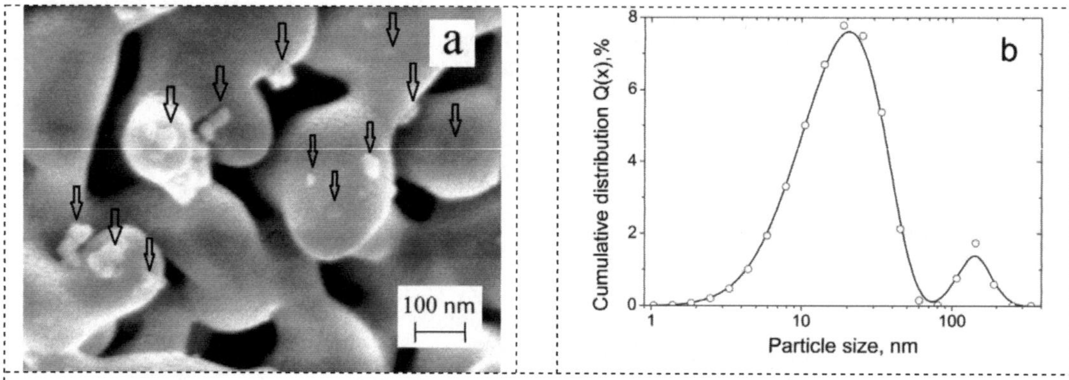

Figure 1. SEM image (a) and particle size distribution (b) of precursor α-Al_2O_3 powder.

To analyze the structure of the ceramic alumina samples synthesized at different heating modes, the obtained SEM microphotos were used. Particle size distributions were found for the samples under study, and the obtained dependences were analyzed. The following parameters were defined: a type and value of the maximum of grain size distribution, mean size and shape of grains. An increasing synthesis temperature leads to a decreasing number of pores and growing density of the samples.

Figure 2 shows SEM images of Al_2O_3 ceramics annealed at 1600 °C for 1 hour with the presence of carbon (a) and with the absence of carbon (b). A highly pure graphite (99.99%) rod with the mass of 20 g was placed 5 mm from the samples in the vacuum chamber of the furnace. It was done to create a strong reducing medium which is needed to synthesize oxygen-deficient samples of alumina ceramics.

Oxygen vacancies create luminescence F-type centers [2]. The absence of carbon in the chamber made the grains grow significantly up to micron sizes. The basic structure was composed of the grains of 1-5 μm. However, the shape of the grains almost did not change. The synthesis of ceramics without carbon was accompanied by a significant decrease in porosity in comparison with the samples synthesized with the presence of graphite. Thus, in the samples annealed at 1700 °C for 1 hour there were no pores at all. With the presence of graphite a layer structure of the sintered particles emerges. The sintered particles were up to 1 μm in size and of polyhedral shape with clearly discernable facets. It is noteworthy that with annealing at 1600 °C for 1 hour there are no particles of the precursor powder (smaller than 250 nm), which indicates uniform sintering throughout bulk of ceramics.

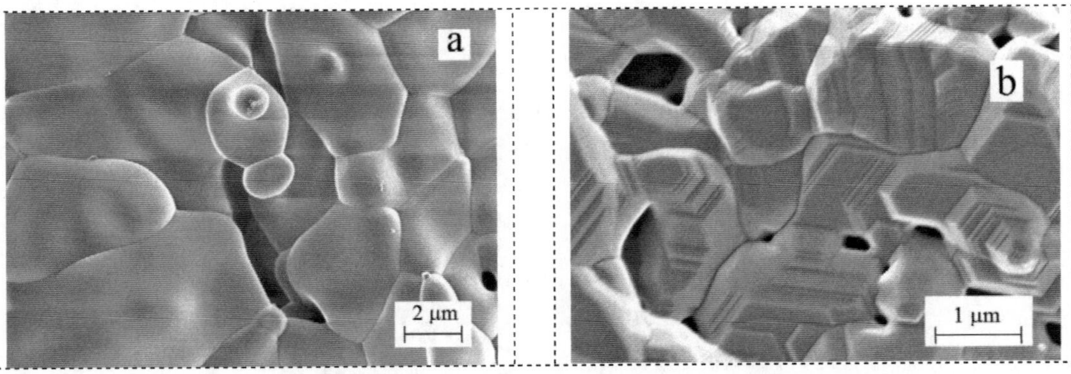

Figure 2. SEM image Al_2O_3 ceramics synthesized at 1600 °C for 1 hour with the absence (a) and presence (b) of carbon.

Figure 3 shows a typical SEM image of alumina ceramics synthesized at 1700°C for 30 minutes with the presence of carbon. The examining of the given microphotos allows us to conclude that as a result of high-temperature ceramics synthesis in vacuum, agglomerates with mean size of about 600-800 nm (Figure 3b) are formed. Figure 3a shows an image of single grains of about 600 nm in size, on the facets of which particles 100-200 nm and 10-40 nm in size were localized. It can be seen that small grains are tightly located along the perimeter of a bigger grain. The sizes of these particles are the same ones in the precursor nanopowder. Such differences in sizes can be explained by the processes of stoichiometric compositional disorder of oxygen at sintering with the presence of carbon. Big grains recrystallized from the precursor α-Al_2O_3 powder, and small grains might emerge from due to lower aluminium oxides (AlO, Al_2O), in which luminescent centers can be formed as well, which is verified by the data in [3]. Further analysis of X-ray diffraction patterns showed that phase ceramic composition does not change and is characterized by 100 % α-phase with the structural changes mentioned above.

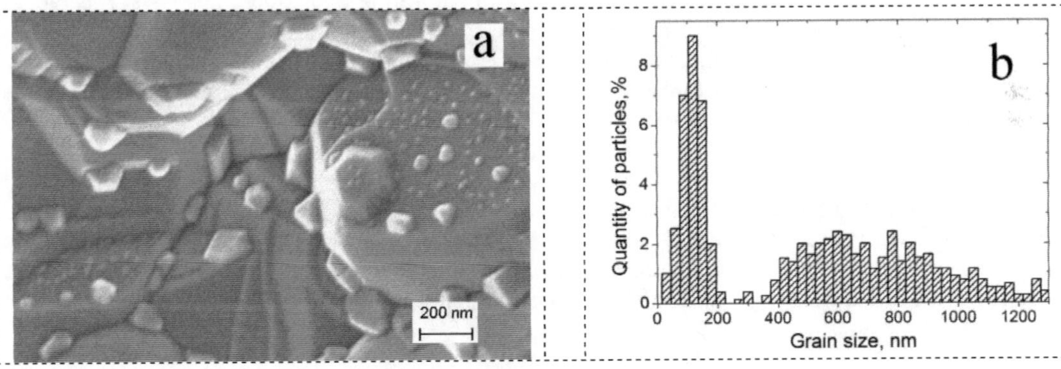

Figure 3. SEM-image (**a**) and grain size distribution (**b**) of ultrafine ceramics Al_2O_3 synthesized at 1700°C for 30 minutes with the presence of carbon.

When the samples are annealed for 1 hour (1700°C) there are no pores in the sintered ceramis including the cases with the presence of carbon (Figure 4). In this situation the surface of the grain facets become uneven. It has hollows with the dimensions comparable with those of the grains found at shorter annealing (Figure 3a). The hollows may be associated with deconvolution of alumina into lower oxides. This is confirmed by the fact that under annealing for over two hours the sample is almost completely decomposed into volatile components including lower oxides which are removed with a vacuum system.

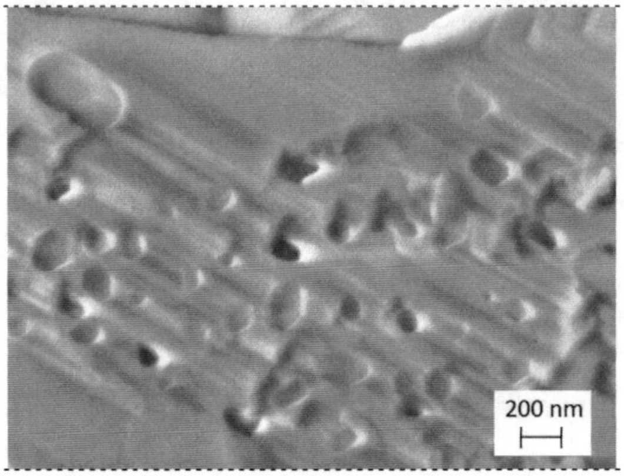

Figure 4. SEM-image ceramics Al_2O_3 synthesized at 1700°C (1 hour).

Photoluminescence (PL) spectra in the emission band of F-centers (420 nm) were analyzed in an express mode to estimate the effect of structure on luminescent yield of the material. It was found that the maximum concentration of luminescence centers is achieved at the highest temperature T=1700°C and annealing time t=1 h. Thus, synthesis of luminescent ceramics at the highest temperature and annealing time leads to two opposite effects: formation of the sample with maximal PL intensity and disorder of its geometrical dimensions down to its complete "burnout". Regarding this, the following optimal parameters for producing luminescent Al_2O_3 ceramics were chosen: synthesis temperature 1600 °C with annealing time over 2 hours.

4. Conclusion

The effect of the synthesis conditions on the structure of ultrafine oxygen-deficient Al_2O_3 ceramics was found. It was shown that with an increasing temperature of ceramics synthesis in vacuum, propotions of particles bigger than 250 nm and new particles of 10-40 nm increase. Luminescent properties also depend on the mode of sample synthesis. Optimal conditions for ceramics synthesis were found. They give a high photon yield of luminescence.

Acknowledgments

This work has been done a part of the government task (3.1016.2014/K) the Ministry of Education and Science of Russian Federation and was supported by the scholarship of the President of the Russian Federation for young scientists and graduate students engaged in advanced research and development in priority areas of modernizing the Russian economy number SP- 983.2015.2.

References

[1] Kortov V S 2010 *Radiat. Meas.* **45** 512
[2] Kortov V, Zvonarev S, Kiryakov A, Ananchenko D 2016 *Mater. Chem. and Phys.* **170** 168
[3] Deryagina I L Khrustov V R Paranin S N Ivanov V V and Beketov I V 2011 *Perspective materials* **1** 123
[4] Barry Carter C Grant Norton M 2007 *Ceramic Materials. Science and Engineering* (Springer Science & Business Media) p. 716

Study of composite MWCNT/pyrolytic Cr interface by NEXAFS spectroscopy

O V Petrova[1], S V Nekipelov[1], A E Mingaleva[1], V N Sivkov[1], A M Obiedkov[2], B S Kaverin[2], K V Kremlev[2], S Yu Ketkov[2], S A Gusev[2], D V Vyalikh[3], S L Molodtsov[4,5]

[1]Komi Science Center Ural Division, RAS, 167982, Syktyvkar, Russia
[2]G.A.Razuvaev Institute of Organometallic Chemistry, RAS, 603950, Nizhny Novgorod, Russia
[3]Technishe Universitat Dresden, D-01062, Dresden, Germany
[4]Institute of Experimental Physiks, TU Bergakademie Freiberg,D-09599, Freiberg, Germany
[5]European XFEL GmbH,Albert-Einstein-Ring 19, 22761 Hamburg, Germany

E-mail: `teiou@mail.ru`

Abstract. A composite material based on MWCNT covered by pyrolytic Cr has been prepared by MOCVD growth technique using bis(arene)chromium compounds as the pyrolytic Cr source. Their structures and morphologies were preliminary studied by X-ray diffraction and scanning and scanning electron microscopy. The atomic and chemical composition of the interface, MWCNT surface and Cr-coating of the composite were studied by total electron yield mode in the range NEXAFS C1s - and Cr2p - absorption edge with use of synchrotron radiation of RGBL at BESSY-II. The study has shown that top layers of the MWCNT in composite have no essential destruction, the coating of the MWCNT surfaces is continuous and consists of Cr_2O_3. The chromium oxide adhesion is provided by chemical binding between the carbon atoms of the MWCNT top layer and the oxygen atoms of the coating.

1. Introduction

Coating the surface of the MWCNT with specific metal films gives them unique physical and chemical properties and leads to the creation of new heterogeneous materials that will have good adhesion to the matrix. Such materials can be used as nanostructured heterogeneous catalysts of various chemical processes, sensors, chemical power sources, elements of various electronic devices and other active elements in many scientific and technical applications. Therefore, the development of methods of synthesis and investigations of nanostructured composite material based on MWCNTs is an actual problem. In our earlier work we have demonstrated for the first time the possibility of synthesis on the MWCNT surfices of the nanosized Fe_3O_4 coating by the metal organic chemical vapour deposition (MOCVD) growth technique [1] and have shown the informativity of the near edge x-ray fine structure (NEXAFS) spectroscopy for the experimental invectigation of the MWCNT/pyrolitic metals geterogenous systems. The present paper is the extension of these researches regarding the development of the metal and metal oxide deposition methods and improvement of the carbon nanotube/matrix interface properties. The MWCNT/pyrolytic Cr composite has been prepared by the MOCVD growth technique using decomposition of the bis(arene)chromium compounds at temperature 400°C. It should be noted that the spectral investigation of the NEXAFS C1s - edge is difficult due to overlaying

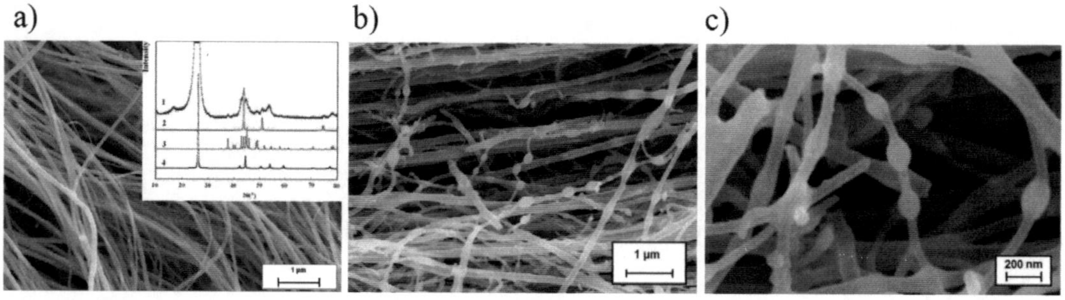

Figure 1. SEM image and XRPD of pristine MWCNTs (a) and SEM image of the MWCNT/chromium composite material prepared by the MOCVD method, using pyrolytic decomposition of chromium-organic liquid (b,c). **1.a** (insert): pristine MWCNT's experimental diffractogram (curve 1) compared with theoretical diffractograms of γ-Fe (curve 2), Fe_3C (curve 3) and graphite (curve 4).

with NEXAFS Cr 2p - edge in the two order diffraction. Therefore, aims of our research were both development of the MOCVD technology for MWCNT coating with pyrolytic chromium and modification of the NEXAFS C1s - edge investigation method. The NEXAFS spectroscopy allows probing the atomic composition and angular momentum character specific densities of unoccupied electronic states and may give information about chemical bonds in the investigated samples. After the absorption of X-rays in the range 250–600 eV the electron escape depths of the primary photoelectrons and Auger electrons in materials are less than 1 nm, in contrast to the secondary electron area occupying a few nm. Therefore NEXAFS spectroscopy by the total electron yield (TEY) method can be used for studies of both the thin coating and coating-substrate interlayer without sample destruction. In case of the composite MWCNT's - pyrolytic Cr the coating and substrate include chromium and carbon atoms, respectively, that enable us to study the state of the MWCNT top layer and to measure the chromium coating effective thickness by the NEXAFS C1s - spectrum, while the coating state can be analyzed from the NEXAFS Cr2p - spectrum.

2. Experimental details

MWCNT's were prepared by the MOCVD method from a toluene-ferrocene mixture in a tubular quartz reactor in the flow of argon at atmospheric pressure. The detailed procedure is described in [2]. The samples produced in the central region of a quartz reactor for MWCNT deposition at 800^oC were selected for the study. The averaged MWCNT diameter was $\sim$ 70 nm, and the MWCNT length was 0.2–2 mm. The laboratory apparatus and the MOCVD technology for the coating of MWCNT's with chromium were developed for the nanostructured composite materials. The chromium-organic liquid (COL) "Barkhos" available from commercial sources was used as an organometallic precursor. COL represents a mixture of alkyl-substituted derivatives of bis(arene)chromium (the average molecular weight is 292, the total chromium content is 15–16 wt %). The technique for deposition of pyrolitic chromium has been described earlier [3]. Briefly, pyrolytic chromium coatings were deposited onto the MWCNT surface as follows. MWCNTs portion of 50 mg was placed into a Pyrex-glass ampoule (10–12 cm^3 in volume) and heated at 200^oC for one hour with continuous vacuum pumping. Then the ampoule was cooled, filled with argon, and 0.1–0.2 cm^3 COL was poured into it. Then the ampoule was sealed using a gas burner and temperature was gradually increased to 400^oC to obtain high-quality pyrolytic chromium coatings. The ampoule was kept at this temperature for two hours, then it was cooled to room temperature, unsealed, and MWCNT samples with deposited pyrolytic chromium coatings were removed.

This method allows controlling the Cr coating thickness from a few nm to tens of nanometers. The MWCNT's and MWCNT's composite samples were characterized by various methods. The X-Ray phase analysis was carried out on a DRON-3M diffractometer with use of Cu-Kα radiation and a graphite monochromator. The thermochemical characteristics of MWCNTs were studied using the Pyris 6TGA thermoscales. The surface morphology of MWCNTs and the composite material was studied on the scanning electron microscope (SEM) ZEISS Supra 50 VP.

The study of the pristine MWCNT's surface, chemical composition, coating thickness and surface coating interaction was carried out by NEXAFS spectroscopy method with using synchrotron radiation of Russian-Germany beamline (RGBL) at BESSY-II [4]. The absorption cross section (CS) spectral dependences in the wide energy interval 250–900 eV and C1s - edge and Cr2p - edge ranges of the pristine MWCNT's and composite MWCNT's/pyrolytic cromium were measured in TEY mode. The samples for studies were prepared by pressing the MWCNT and composite powders into the surface of a pure Cu plate. Additional Ti-film of the 160 nm thickness mounted onto Au grid with small mesh were used in the capacity of absorption (transmisiion) filter for the effective suppression and measurement of the second-order light and the VUV-stray radiation [5]. The incident photon flux was measured using the clean Au-photocathod. The incident monochromatic SR intensity in arbitrary units was obtained by means of the division the TEY monochromatic signal of Au plate by the Au atomic cross section [5]. The CS in arbitrary units was obtained by means of the division the TEY by the SR intensity. In all cases TEY - signal measurements were carried out considering nonmonochromatic stray.

3. Results and discussion

The bottom of figure 2 illustrates the CS measurement problems in the range NEXAFS C1s absorption edge of the chromium-contaned compounds. The composite MWCNT/chromium CS dependence without the Ti-filter (curve 1) shows clearly that NEXAFS Cr 2p - edge and O1s - edge observe in the tow order diffraction. Meanwhile the strong Cr $L_{2,3}$ - lines overlap exactly with NEXAFS C1s - edge. The Ti-absorption filter decreases the Cr $L_{2,3}$ - line and the O1s - fine structure intensities by an order of magnitude. Their intensities were decreased very strongly on the monochromator optical elements, too. This suppression factor is calculated as a ratio of the intensity jumps in the Ti 2p - edge in the first and second orders by the spectral dependence Au (curve 3). This factor is 9.5.

The contribution of the two order diffraction intensity to the range NEXAFS C1s edge (curve 4) can be obtained by dividing the NEXAFS Cr 2p edge intensity by 9.5. As a result the nondistorting TEY-signal in the range NEXAFS C1s absorption edge of composite MWCNT's/pyrolytic chromium was obtained by subtraction of curve 4 from curve 2. The CS of the composite (arb.units) was obtained by dividing the TEY to the SR intensity.

The top of Figure 2 shows the measured CS spectral dependences in the energy interval 250–700eV for the pristine MWCNT's and composite MWCNT/pyrolytic chromium. Examining the CS in Figure 2 one can see, that (i) the pristine MWCNT's dependence has the NEXAFS in the C1s-edge range only and (ii) the composite dependence has the NEXAFS in the C1s-, O1s- and Cr2p-edge ranges.

The first point indicates that the pristine MWCNT surface is clean and the iron compounds detected by x-ray diffractometry method must be contained inside MWCNT's volume. The second point and continuity of the pyrolytic chromium coating suggest that the coating has nanosized thickness and contains chromium oxide. This conclusion based on facts, that the photoelectron's escape depth in metals is about 1 nm only [6] and the NEXAFS Cr2p - spectrum of the composite completely coincides with the Cr2p - spectrum of the pure Cr_2O_3 (in Figure 3 is not shown). Figure 3 shows the decrease of the composite CS dependent area (S2) by a

Journal of Physics: Conference Series **741** (2016) 012038 doi:10.1088/1742-6596/741/1/012038

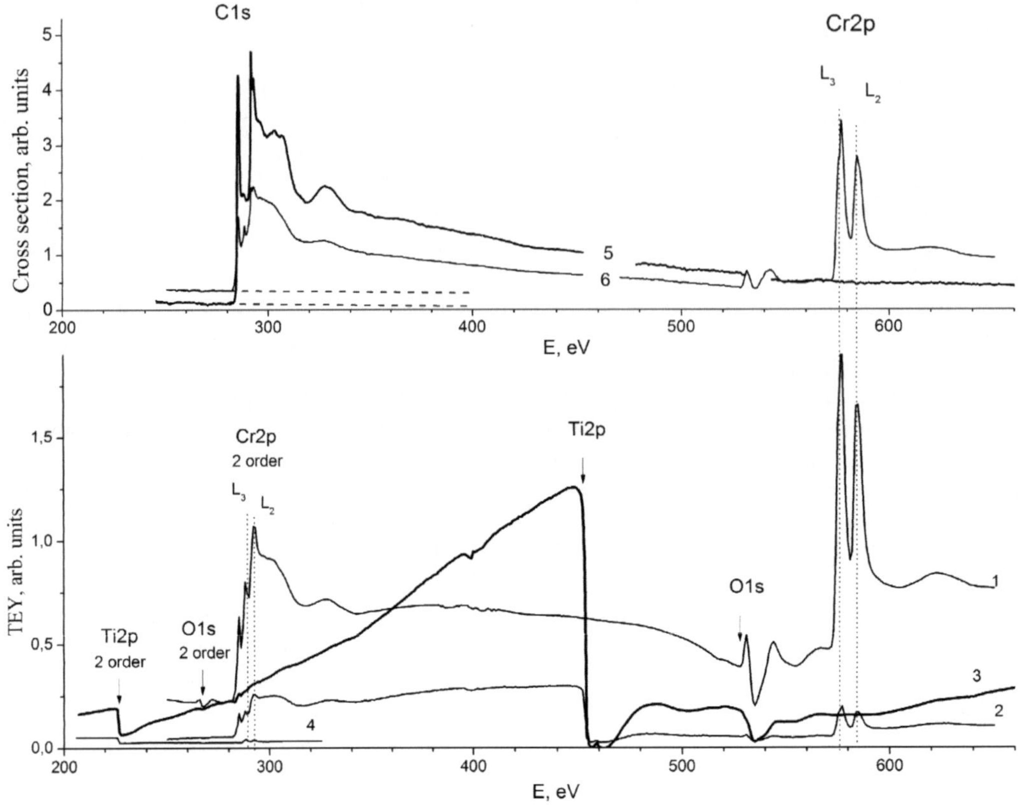

Figure 2. Bottom: TEY of the composite without (curve 1) and with (curve 2) Ti- absorption filter and Au-plate (curve 3) in the range 250–660 eV. Dotted line - interpolation contribution

factor of 2.12 in the NEXAFS C1s-edge range in comparison with the pristine MWCNT area (S1). However, NEXAFS C1s - spectra of pristine MWCNTs and the composite are similar. This decrease is a result of the photoelectron scattering by the atoms of the chromium oxide coating. The photoelectron escape depth in pure Cr_2O_3 is λ=1.0 nm [7]. Using this value we can calculate the minimal effective thickness of Cr_2O_3 coating on the MWCNT's surface as product of λ and ln(S1/S2) and it is equal to 0.75nm. NEXAFS C1s-spectrum of composite has new weak peaks A, B and C with the energy 287.1 eV, 287.8 eV and 288.4eV, respectively. The energies of these peaks coincide with the energies of transitions from C1s - level to π^*-unoccupied orbital of C-O-C, C-O and C=O groups, respectively [4,5]. Taking into account these facts one can believe, that the adhesion of the chromium oxide coating on the MWCNT's surface is a result of double and single bond formation between the surface MWCNT's atoms and oxygen atoms of the Cr_2O_3 coating without destruction of the MWCNT outside surface, and that a small amount of oxygen atoms embedded in the surface of the nanotubes form the epoxy bonds.

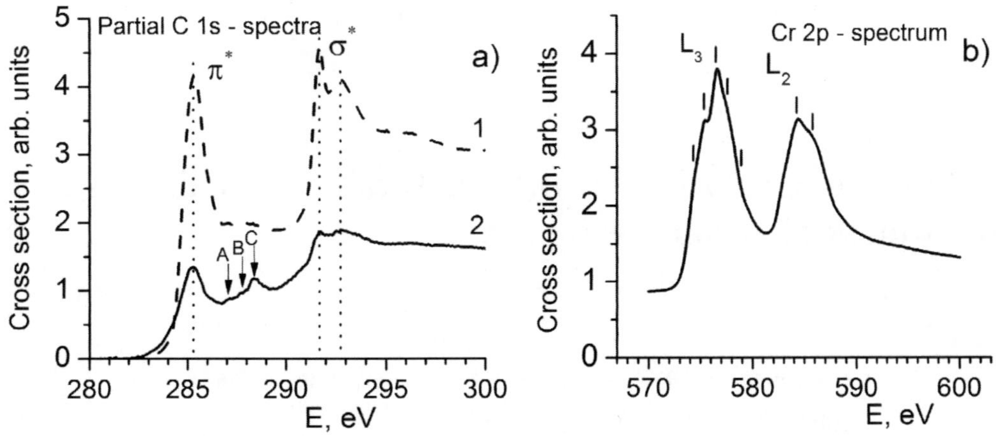

Figure 3. (a) NEXAFS partial C1s - spectra of pristine MWCNTs (curve 1) and MWCNT/chromium composite (curve 2); (b) the NEXAFS Cr2p - spectrum of MWCNT/chromium composite.

4. Conclusion

The study has shown that top layers of the MWCNT's in composite reveal no essential destruction, coating of MWCNT's-surface is Cr_2O_3 and this coating is continuous. The chromium oxide adhesion is provided by chemical binding between the carbon atoms of the MWCNT's top layer and the oxygen atoms of the coating.

Acknowledgments

This work was supported by State Contract П-337, Minfin KR, grants RFBR 16-32-00441мол-а, 16-42-110610 р-а, 16-43-110350 р-а, Basic Research Program of UrD RAS 15-9-1-13 and by the Bilateral Program of the RGBL at BESSY II.

References

[1] Sivkov V N, Petrova O V, Nekipelov S V, Obiedkov A M , Kaverin B S, Kirillov A I, Semenov N M, Domrachev G A, Egorov V A, Gusev S A, Vyalikh D V and Molodtsov S L 2015 *Fullerenes, Nanotubes and Carbon Nanostructures* **23** 17-19.

[2] Korneeva Yu V, Novakova A A, Obiedkov A M, Zaitsev A A, and Domrachev G A 2007 *J. of Surface Investigation* **1** 369-372.

[3] Obiedkov A M, Kaverin B S, Gusev S A, Ezerskii A B , Semenov N M, Zaytsev A A, Egorov V A, and Domrachev G A 2009 *J. of Surface Investigation* **7** 67-72.

[4] Fedoseenko S I, Iossifov I F, Gorovikov S A, Schmidt J, Follath R, Molodtsov S L, Adamchuk V K, Kaindl G 2001 *Nucl. Instr.and Meth. A* **470** 84-88.

[5] Kummer K, Sivkov V N, Vyalikh D V, Maslyuk V V, Bluher A, Nekipelov S V, Bredow T, Mertig I, Mertig M, Molodtsov S L 2009 Physical *Review B* **80** 155433-8.

[6] Stohr J 1992 NEXAFS spectroscopy *Springer: Berlin* 403.

[7] Hollidey P H 1975 *Journal of Vacuum Science and Technology* **12** 1418-1422.

Journal of Physics: Conference Series **741** (2016) 012039 doi:10.1088/1742-6596/741/1/012039

Obtaining solid solutions of InGaAsPInGaAsP solid solutions in the spinoidal decomposition region

A E Marichev[1,2], R V Levin[1,3], G S Gagis[1] and A B Gordeeva[1]

[1] Ioffe Institute, St. Petersburg, Russia
2 Saint-Petersburg State Electrotechnical University "LETI",
St. Petersburg, Russia
[3] Submicron Heterostructures for Microelectronics Research and Engineering Center of the RAS, St. Petersburg, Russia

E-mail: segregate1@yandex.ru

Abstract. Technology for growing layers of InGaAsP solid solutions with Eg ~ 1.0 - 1.2eV on InP substrates in the spinodal decomposition region by the metalorganic vapour phase epitaxy (MOVPE) technique has been developed. Results of investigation of the obtained solid solution layers by the methods of photoluminescence, reflectance anisotropy spectroscopy and fractal analysis are presented.

1.Introduction

The photovoltaic converters for laser radiation with wavelength about 1.06 μm (the local minimum of absorption losses in the Earth's atmosphere), are the most perspective for the free-space power transmission with laser power up to 10 kW. These converters may be used in outer space. The most perspective heterostructure for laser-radiation detector is shown in Figure 1.[1]

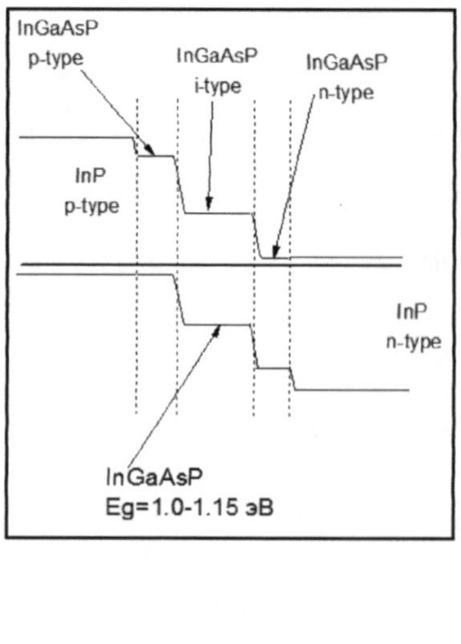

Figure 1. Band diagram of the laser-radiation detector

The photodetectors should be fabricated of the direct band-gap materials with lattice perfection and high mobility of generated charge carriers. The most appropriate material for photodetectors of laser radiation with wavelength $\lambda = 1064$ nm (1,165 eV) is the InGaAsP solid solution, isoperiodic with InP, having the band gap 0.95 – 1.15 eV (300K). However, it is known that these solid solutions grown at 600 - 625°C fall within the spinodal region [2], that complicates obtaining the photoactive layers with the optimal thickness of 1-2 μm.

2. Experimental details

First, the range of spinodal region for InGaAsP solid solutions, isoperiodic with InP, grown at 600^{0}C, has been redefined.

Our calculations were performed within the theory of quasi-regular solutions [3], using the equations and parameters of K.Onabe [2]. According to these calculations, there exists a wide region of spinodal decomposition in the above mentioned system at 600^{0}C. Figure 2 shows that the solid-solution lattice matched to InP with 0.9 - 1.05 eV band gap fall within the spinodal region. [3-7]

Thus, our calculations confirm correctness of location of the boundaries of the spinodal decomposition region.

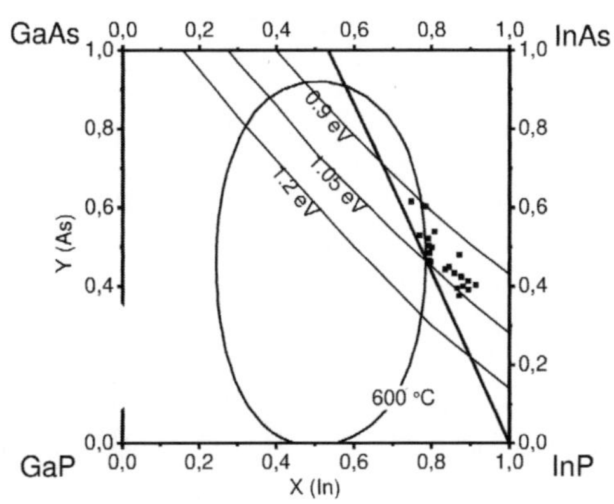

Figure 2. Composition square of $In_{1-x}Ga_xAs_yP_{1-y}$ solid solutions with line of lattice-matched with InP solid solution compositions, spinodal isotherms for 600°C [2] and isobandgap for 0.9 - 1.2 eV lines calculated with patrameters from [3]. Dots indicate compositions of specimens obtained in the present work.

Fabrication of the solid-solution layers and heterostructures

To estimate the appropriate growth parameters for fabrication of solid solutions with required composition, the investigations, including determination of growth temperature, temperature of metal-organic compound precursors, III/V molar ratio, and total hydrogen flow were carried out.

All the epitaxial layers were grown on the AIXTRON AIX-200 installation by the metalorganic vapour-phase epitaxy on n-type InP(001) substrates with misorientation of 4^0 towards (111) at growth temperature 600^{0}C and 100 mbar reactor pressure [8]. Hydrogen, purified through the palladium filter with a dew point of not worse than -100°C was used as a carrier-gas. The total hydrogen flow through the reactor was equal to 5 slpm [8]. The temperature of gallium and indium precursors-triethylgallium (TEGa) and triethylindium (TEIn) – was 17^{0}C. Arsine (AsH_3) and phosphine (PH_3) were used as precursors of V-group elements. The experiments have shown that, to obtain the high-quality samples, the PH_3 flow should be several dozens of times higher than the AsH_3 flow, that is caused by its weak decomposition at 600^{0}C (Figure 3) [9].

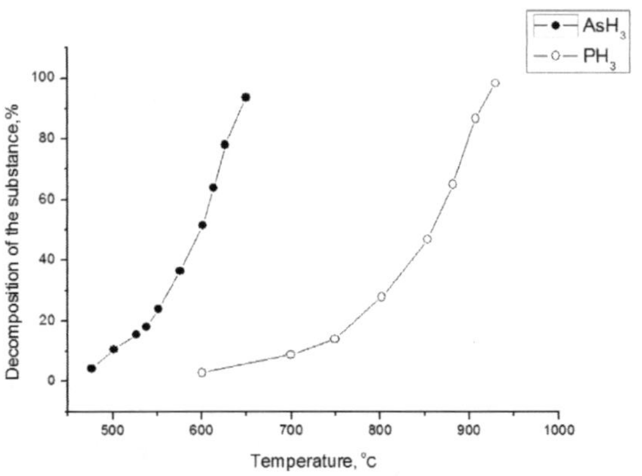

Figure 3. Dependence of AsH_3 and PH_3 pyrolitic decomposition on temperature [9].

The dependence of layer growth rate on III/V molar ratio was studied (see Fig.4). Experimental results show that growth rate of GaInAsP (Eg~1.0-1.15 eV) decreases when III/V molar ratio grows up.

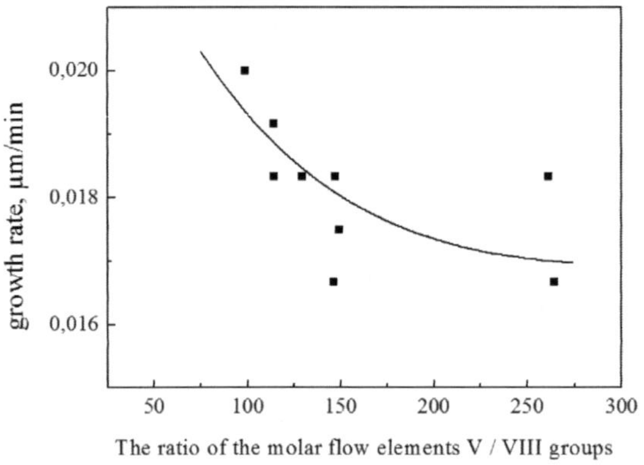

Figure 4. Dependence of GaInAsP layer growth rate (Eg~1.0-1.15 eV) on III/V molar ratio.

Lattice perfection of the obtained samples was controlled by X-ray diffraction and photoluminescence (PL). The X-ray diffraction technique allows to estimate lattice mismatch (peak position on the rocking curve) and lattice perfection (peak half-width). The position and half-width of PL peak allow to determine band gap (Eg) and to characterize material quality respectively.

The X-ray diffraction rocking curve and the PL spectrum of the best GaInAsP sample are shown in Figure 5. The half-width of rocking curve (da) is 0.01Å, maximum PL intensity (see in Figure 6) is 40000 y.e., which is close to the best results obtained for InGaAsP samples with required band gap. Thus, analysis of the experimental data allowed to determine optimal parameters for obtaining GaInAsP solid solutions with Eg~1.0-1.15 eV:
- III/V molar ratio -300;
- Phosphine (PH_3) flow -95 splm;
- Triethylgallium (TEGa) flow- 29 splm;
- triethylindium (TEIn) flow- 113 splm;
- Arsine (AsH_3) flow- 1 splm.

Growth rate - 1.5 μm/hour, layer thickness - 1.5 μm.

The layer material composition was determined by the X-ray spectral analysis, layer thickness and surface defects – by scanning electronic microscopy (SEM), surface roughness – on atomic-force

(AFM) and electronic microscopes. To evaluate the material quality, the following techniques were applied: PL, reflectance anisotropy spectroscopy and multifractal analysis of the surface. This paper summarizes results of these investigations, the details are presented in [8,10].

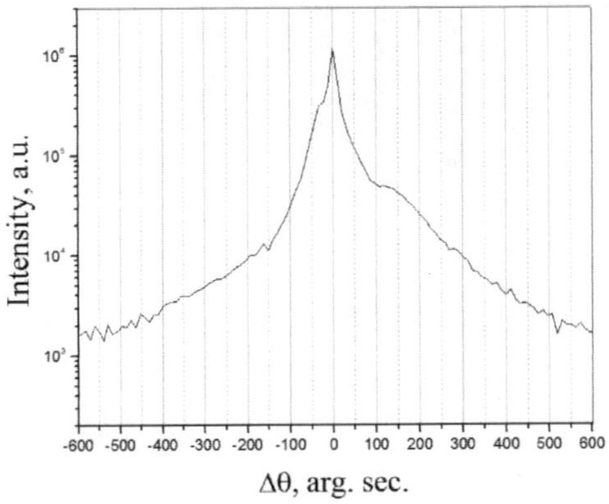

Figure 5. X-ray diffraction rocking curve

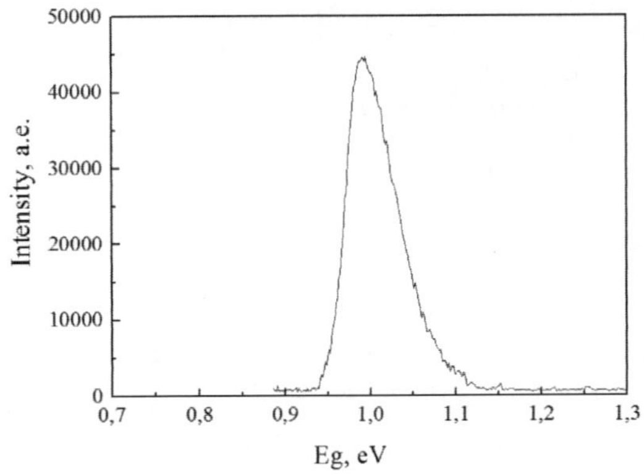

Figure 6. The photoluminescence spectrum (T=300 K) GaInAsP with Eg~1.0-1.15 eV.

3. Results of investigations

Multifractal analysis.

The atomic-force microscope images of samples surface (Figure 7,8) show, that small roughness (Table1) is formed on the surface of all the layers. It could appear when cooling, and is caused by the difference in thermal-expansion coefficients of the layer and the substrate. The tension produced in the layer lead to decomposition of solid solution at the surface and breaking of the top layer part. A high roughness (Table 2) is formed only in thick layers practically in the entire depth and partially in the buffer layer, thus relieving stresses in the bulk.

Table1. Small roughness.

Sample No	d, nm	h, nm	Lo , μm	Ld , μm	σ, din·cm^{-2}
I300	1000	25	0.14	0.145	1.4x 10^{10}
I299	500	17	0.12	0.12	1x 10^{10}
I298	200	20	0.15	0.16	8x10^{9}
I297	50	21	0.15	0.15	1x 10^{10}

Table2. High roughness.

No	d, nm	h, nm	Lo , μm	Ld , μm	σ, din·cm^{-2}
I300	1000	45	0.85	0.9	1.4x 10^{9}
I299	500	31	1.2	1.3	3.6x 10^{8}
I298	200	absent			
I297	50	absent			

where σ - tension, h –height roughness, Ld – difference between roughness, Lo – width at the base of the roughness, d – sample thickness.

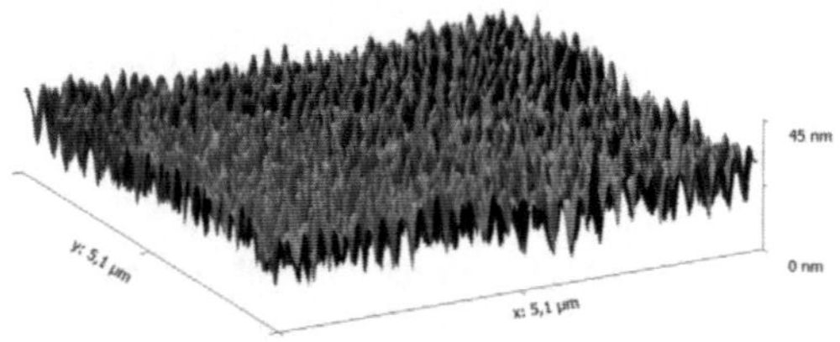

Figure 7. Pictures of the surface of a sample 50nm thick

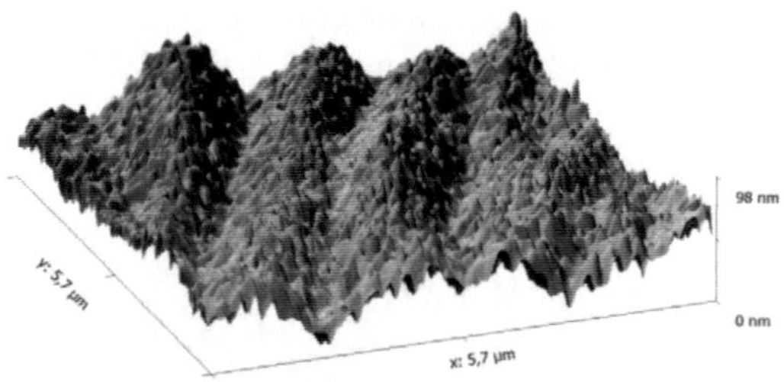

Figure 8. Pictures of the surface of a sample 1000nm thick.

Reflectance anisotropy spectroscopy

To estimate the value of built-in stresses and to define presence of the additional fractions reflectance anisotropy (RA) spectroscopy was used.

Figure 9 shows reflectance anisotropy spectra of surfaces of InGaAsP solid solutions grown on InP(001) substrate for various layer thicknesses. Higher intensity of the InGaAsP spectra, comparing to InP, proves the presence of built-in anisotropic stresses in the layers. The sign of the RA signal indicates that this may be compressive stress in [110] direction or tensile stress in [1-10]. The amplitude of derivative-shape line at approximately 3 eV (optical transitions region E_1-$E_1+\Delta_1$ in InP) is proportional to value of the stress. This amplitude is maximal for the 200 nm layer and decreases for thicker samples. This indicates that the built-in stress is released in layers thicker than 200 nm, that should cause the appearing of anisotropic roughness. This fact was is confirmed by AFM observations.

Presence of small spectral features at energies marked by arrows (the positions of optical transitions in InGaAsP) confirms that the InGaAsP of required composition has been formed.

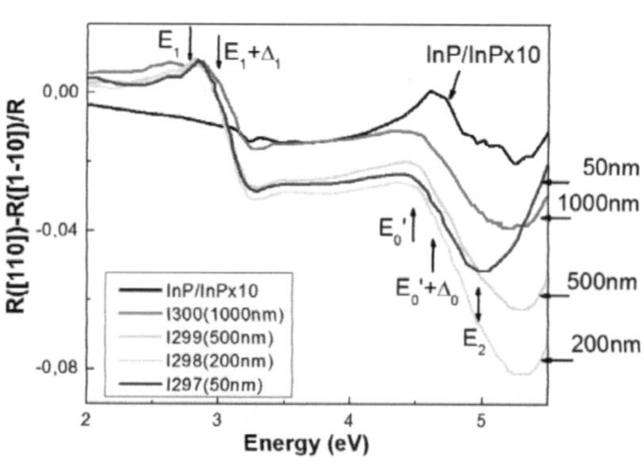

Figure 9. RA spectra of samples I297 - I300 with different thicknesses of InGaAsP layers grown on In(001)

Photoluminescence.

The PL method allows to evaluate material quality – the higher PL signal intensity, the better material: lattice perfection and high mobility of generated charge carriers, long lifetime of carriers.
Analysis of the PL spectra shows that at layer thicknesses under 100-200 nm the PL intensity raises with increasing thickness. That is evidence of improvement of material quality (diffusion length increases, influence of surface recombination reduces). At higher thicknesses, the PL intensity decay

takes place, which confirms the inferences about the spinodal decomposition based on reflectance anisotropy spectroscopy and multifractal analysis data (Figure 10).
Basing on the performed investigations, the 1000 nm ten-layer structure (100 nm each layer) (I313) has been fabricated. The PL intensity magnitude of this structure has increased by 3 orders as compared to that of the 1μm monolithic structure. Figure 11 presents the PL spectrum of this structure.

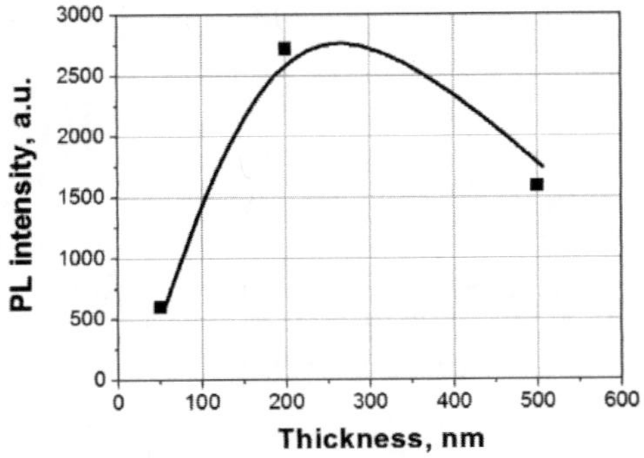

Figure 10. Dependence of the PL intensity on thickness.

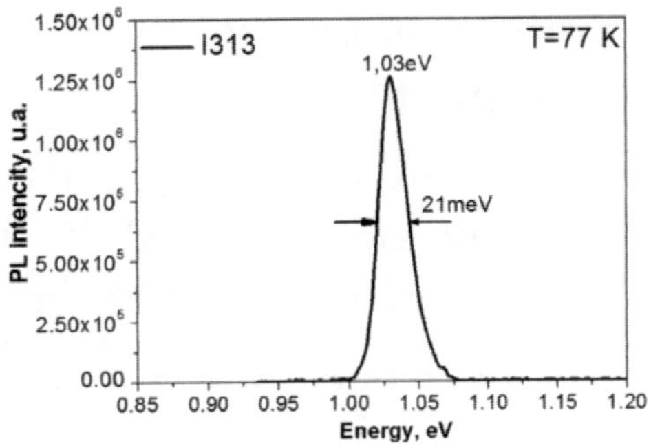

Figure 11. PL spectrum of the best sample I313($In_{0.8}Ga_{0.2}As_{0.46}P_{0.54}$).

Acknowledgements
The authors wish to thank B.V. Pushnyi, N.M. Shmidt, P.N. Brunkow, S.I. Troshkov and A.S. Vlasov for their valuable contribution to the research and discussion of the results.

This work was supported by Ministry of Education and Science of the Russian Federation (Grant Agreement № 14.604.21.0089 dated June 27, 2014, unique identifier of applied research (project) RFMEFI60414X0089).

References
[1] O. Krokhin 1999 *UFN,* **176** 441
[2] Vavilova L S, Kapitonov V A, Murashova A V, Pikhtin N A 1999 *Semiconductors* **33** 1010
[3] Razeghi M, Defour M, Omnes F, Maurel P, Bigan E, Acher O, Nagle J, Brillouet F, Portal J, 1988 *J. Cryst. Growth* **93** 776
[4] Razeghi M, Maure P, Omnes F, Detour M, Acher O 1987 *Appl. Phys. Lett.*, **51** 1821
[5] Onabe K 1982 *Jpn. J. Appl. Phys.* **21** 797
[6] de Cremoux B 1982 *J. Physique* **43** 5
[7] Ipatova I P, Malyshkin V G, Maslov A Yu, Shchukin V A 1993 *Semiconductors* **27** 158
[8] Levin R V, Marichev A E, Shvarts M Z, Marukhina E P, Khvostikov V P, Pushnyi B V, Mizerov M N, Andreev V M 2015 *Semiconductors* **49** 700
[9] William S. and Chang C. 2005 *Principles of Lasers and Optics*.
[10] Marichev A E, Pushnyi B V, Levin R V 2016 *J. of Physics: Conference Series* **690** 012010

Contribution of droplet volume fluctuation to dispersion of nanowire length

Y Berdnikov[1], N V Sibirev[1,2,4] and V G Dubrovskii[1,2,3]

[1]St. Petersburg Academic University, Khlopina 8/3, 194021 St. Petersburg, Russia.

[2]St. Petersburg State University, Physical Faculty, Ulianovskaya Street 3, Petrodvorets, 198504 St. Petersburg, Russia

[3]ITMO University, Kronverkskiy pr. 49, 197101 St. Petersburg, Russia

[4]St. Petersburg State Polytechnical University, 195251 St-Petersburg, Russia

Abstract. In this work we study features of vapor-solid-liquid mechanism of nanowire growth. We consider variation of droplet volume during the growth caused by droplet depletion and refilling. Here we take into account the changes in material transport due to volume variation, which was neglected previously. We also discuss the impact of this effect on nanowire length distributions.

1. Introduction

Size uniformity of III-V nanowires (NWs) is predominant for many applications in optics and optoelectronics [1-3]. For vapor-solid-liquid (VLS) growth mechanism the homogeneity of final structures is primarily determined by features of nucleation process at the liquid-solid interface [3-6].

Commonly assumed approach to VLS mechanism implies layer-by-layer growth of nanowire when 2D island on top facet nucleates and then instantaneously spreads to form a new monolayer [3,6-12]. Most of theoretical models for VLS growth consider steady-state regime where monolayers nucleate randomly and independently from each other which results in Poissonian form for length distributions [9,10]. Meanwhile, there are few recent theories that take into account certain anti-correlation between the successive nucleation events and results in sub-Poissonian distributions of nanowire lengths [11, 12].

However, these models describe oscillations of concentrations in the droplet but neglect the change of radius and volume which, in general, has an impact on the material transport to the droplet. In this work we consider the volume variation due to droplet depletion and refilling after formation of each new monolayer. We study the role of variation of adatoms flux to the top of nanowire and discuss its impact on variance and distribution of nanowire lengths. Here we are not discussing the effect of gradual droplet inflation or deflation which were observed in case of self-catalyzed growth [13, 14], so we assume that volume around constant mean value $\langle V \rangle_t = const$. Also we do not consider the growth kinetics of ensembles of nanowires or effects of diffusion induced growth which can lead to much greater broadening of distributions than the effects considered in the current work [15].

2. Model

In our model we consider direct impingement as the only mechanism of adatoms transport to the droplet. Hence, the material balance implies for variation:

$$\Delta V(t) = J\, \alpha V^{2/3}\, \Delta t - \pi R_{NW}^2\, h\, \Delta L(t), \tag{2}$$

where J is influx rate, R_{NW} is nanowire radius, Δt is the duration of a time step. The geometric factor α depends on droplet contact angle [3]. For following numerical calculations we introduce dimensionless volume $v = V/\pi R_{NW}^2 h$ to Eq. (2)

$$\Delta v(t) = B v^{2/3} - \Delta L(t), \tag{3}$$

with the coefficient $B = \alpha J \Delta t / \pi h R_{NW}^2$. The first term in the right hand side of Eq. (3) corresponds to material influx to the droplet. Further we discuss the effect of its replacement by constant $B\langle v\rangle^{2/3}$.

We model the layer-by-layer growth of nanowire and consider elongation ΔL at each time step as a random variable dependent on droplet volume $v(t)$:

$$\Delta L = \begin{cases} 1 & with\, probability\; p(v) \\ 0 & with\, probability\; 1 - p(v) \end{cases} \tag{4}$$

Common approach assumes a very steep exponential dependence of the nucleation probability on the nucleation barrier through the Zeldovich nucleation rate [3,8-12]. Due to this very steep dependence we consider the nucleation barrier to be linear with the number of atoms in the droplet. Since the droplet volume proportional to the number of atoms inside we obtain exponential dependence for $p(v)$:

$$p(v) = A_{exp}\, e^{\Gamma\, v(t)}, \tag{5}$$

where A_{exp} and Γ are constants. We also consider linear dependence of nucleation probability on the volume of the droplet:

$$p(v) = A_l\, v(t). \tag{6}$$

Next, we apply the condition of constant mean volume $< v >_t = const$ to Eq. (3) and use Eqs. (5) or (6) to obtain $A_l = \frac{2B<v^{2/3}>}{<v>}$ for linear dependence or $A_{exp} = \frac{2B<v^{2/3}>}{<\exp(\Gamma\, v)>}$ for exponential dependence of nucleation probability on droplet volume.

3. Results and discussion

For certain number of time steps t_{max} we repeat 400 times the calculation of nanowire length $L = L(t_{max})$ and evaluate the mean length $\langle L\rangle$ and variance σ^2. For $\langle v\rangle = 100$ we perform two series of simulations with variable and constant material impingement to the droplet to investigate the impact on length homogeneity. In case of constant impingement we substitute the volume by its mean value in the first term in the right hand side of Eq. (3). We obtain much broader distribution with slightly smaller mean length in case of variable adatoms impingement as shown in Figure 1(a). We simulate the growth at different times changing t_{max} and study change of variance with $\langle L\rangle$.We observe saturation of variance in case of constant material influx and almost linear increase for variable rate.

Meanwhile, both cases demonstrate similar dependences in the initial stage. Figure 1(b) summarizes simulation results for $\Gamma = 0.02$ and 0.04.

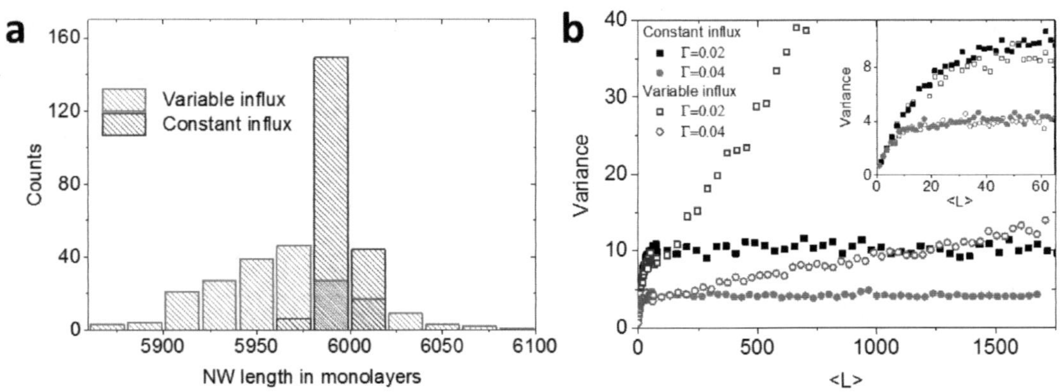

Figure 1. (a) Histograms of nanowire lengths and (b) Variance dependence on mean length at constant and variable material impingement to the droplet. The inset shows the dependences at short lengths.

We repeat the same procedure at different t_{max} for both dependences of nucleation probability (Eq. (5) and (6)) and observe linear increase of NW length variance with $\langle L \rangle$ in both cases. Figure 2(a) shows that calculated $\sigma^2(\langle L \rangle)$ linear dependences are less steep than Poissonian broadening with $\sigma^2 \langle L \rangle$.

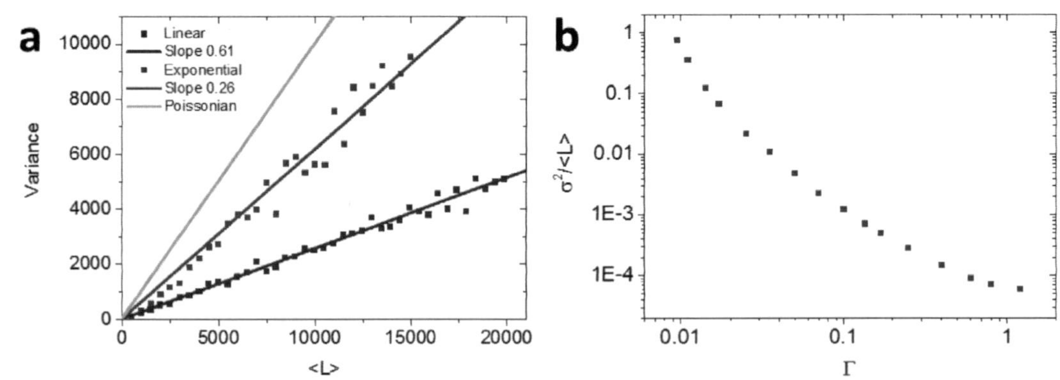

Figure 2. (a) Variance dependence on mean nanowire length measured in number of monolayers; (b) Variance σ^2 to mean length $\langle L \rangle$ratio variation with Γ.

In case of exponential law of nucleation probability the linear slope $\sigma^2/\langle L \rangle$ increases with decrease of parameter Γ and approaches the slope of linear law when Γ is around 0.01.

In conclusion we compare our results with the results of the previous studies with constant material influx to the droplet. Modeling of nucleation process during VLS nanowire growth in references [11, 12] demonstrates saturation of the NW length variance with time in accord with or results for constant impingement to the droplet. Moreover, our model shows that variance increase infinitely and linearly with time or mean length when we take into account the fluctuation volume and its impact on material transport.

Acknowledgements

V.G.D. acknowledges financial support received from Saint-Petersburg State University research grant 11.37.210.2016 . N.V.S. thanks the Russian Foundation for Basic Research for financial support within research grants 15-02-06525 and 16-02-00134. Y.B. acknowledges financial support from FP7 project NanoEmbrace (316751).

References

[1] Yang P, Yan R and Fardy M 2010 *Nano Lett.* 10 1529
[2] Johansson J and Dick K 2011 *CrystEngComm* 13 7175
[3] Dubrovskii V G *Nucleation Theory and Growth of Nanostructures* 2014 (Heidelberg, New York, Dordrecht, London: Springer)
[4] Wagner R S and Ellis W C 1964 Appl. Phys. Lett. 4 89
[5] Givargizov E I 1975 *J. Cryst. Growth* 31 20
[6] Dubrovskii V G, Sibirev N V,Harmand J C, and Glas F 2008 *Phys. Rev. B* 78 235301
[7] Glas F Harmand J C, Patriarche G 2007 *Phys. Rev. E* 99 146101
[8] Sibirev N V, Nazarenko M V, Zeze D A, and Dubrovskii V G 2014 *J. Cryst. Growth* 401 51
[9] F. Glas, J. C. Harmand and G. Patriarche, 2010 *Phys. Rev. Lett.* 104 135501
[10] Dubrovskii V G 2013 *Phys. Rev. B* 87 195426
[11] Glas F 2014 *Phys. Rev. B* 90 125406
[12] Sibirev N V 2013 *Tech. Phys. Lett.* 39 660
[13] Colombo C, Spirkoska D, Frimmer M, Abstreiter G, and Fontcuberta i Morral A 2008 *Phys. Rev. B* 77 155326
[14] Dubrovskii V G, Xu T, Díaz Álvarez A, Plissard S, Caroff P, Glas F, and Grandidier B 2015 *Nano Lett.* 15 5580
[15] Dubrovskii V G, Berdnikov Y, Schmidtbauer J, Borg M, Storm K, Deppert K and Johansson J 2016 *Cryst.& Growth Des.* 16 2167

Features of AlN film grown by ion-plasma sputtering

Ya V Lubyanskiy[1], **A D Bondarev**[1], **I P Soshnikov**[1-3], **K P Kotlyar**[2], **D A Kirilenko**[1], **N A Bert**[1], **K R Ayusheva**[1], **I S Tarasov**[1]

[1] Ioffe Institute of RAS, Saint-Petersburg 194021, Russia
[2] Academic University, Saint-Petersburg 194021 Russia [3] Institute for Analytical Instrumentation of RAS, Saint-Petersburg 198095, Russia

Abstract. The work under consideration presents research of structure, composition and optical properties of aluminium nitride thin films grown by reactive ion plasma sputtering. Aluminium nitride films are shown to contain amorphous and polycrystalline phases. Amorphous phase presence influences on refraction and absorption indexes. Conditions of polycrystalline films with primary (dedicated) orientation synthesis are revealed.

1. Introduction

Aluminium nitride is known to have the next set of the unique characteristics: high thermal conductivity, high strength properties, good insulation properties, piezoproperties, coefficient of thermal expansion close to GaAs and Si, possibility of changing the refractive index in a certain range, etc [1-3]. Above mentioned qualities reveal broad way for the application of this material in the electronics industry, in particular as a high-power semiconductor lasers anti-reflective or/and Bragg mirrors [4-5].The most common methods of the thin films growth are electron beam evaporation and subsequent condensation from the vapor stream onto the substrate under vacuum (EB PVD) and magnetron sputtering. Such techniques are limited to be used in field of high-power diode laser antireflective mirrors process (for instance high heating temperature). The alternative methods based on reactive plasma deposition allow synthesizing the films without such restrictions [6]. Furthermore, such methods are simple, environmentally friendly, economical, and let flexibly vary the manufacturing processes. In addition, these methods can be easily integrated into the production process chains of various electronic devices. Considering the above the presented paper demonstrates the synthesis of aluminum nitride films by reactive ion-plasma sputtering in a glow discharge investigation.

2. Application

AlN films synthesis experiments were carried on a special triode type reactive ion-plasma device based on experimental devise "ТИП УБ 15 1276" with oil-free vacuum pumping system. The residual pressure in the chamber was not more than $2 * 10^{-6}$ torr. During the growth plasma source gases Ar and N_2 with a purity better than 99.998% were used. The target used aluminum target (pure of aluminum – 99.999%). GaAs (100) and Si (111) substrates with different doping types and doping levels are used to be deposition surface. Size of samples was from 1 to 10 sm^2 and thickness is from
300 μm to 500 μm. Influence of the type and level of doping on the basic technological parameters of the process and the refractive index filmt was not observed. Dependence on experiment program preliminary target and samples preparation by Ar were carried. Pressure was set to be $3\text{-}5 \cdot 10^{-3}$ torr. The main process of AlN film synthesizing was carried in the discharge of pure nitrogen at an operating pressure of $3\text{-}5 \cdot 10^{-3}$ Torr. Note that the sample temperature is changed from 30 a process to 100 ^{0}C.

The grown layer structure was investigated by transmission electron microscopy (TEM) and scanning electron microscopy (SEM) (TEM – Joel 2100FE и SEM Supra25 C.Zeiss). The refractive index measurement was performed by ellipsometry (Horiba) on wavelength equal to 633 nm.

3. Results and discussion

AlN film morphology examples are presented on figure 1(a,b). It can be seen that the synthesized film is smooth and uniform in thickness. The structure of the films cross section is tend to be uniform. In some cases, the columnar structures formation was observed figure1(b).

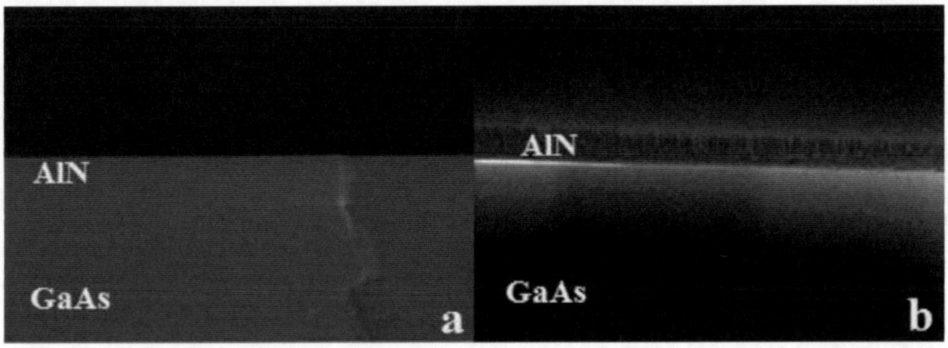

Figure 1(a,b). (a) Uniform structure of AlN film ; (b) Columnar structure of AlN film.

The film thickness dependence on the deposition time increase in the thickness range of 300 nm or more Figure 2. Note that dependence in the field of ultra-thin layers has a threshold with respect to time, i.e., the graph in the low times extrapolation shows that the formation of the layers begins with the deposition time of about 1-2 minutes.

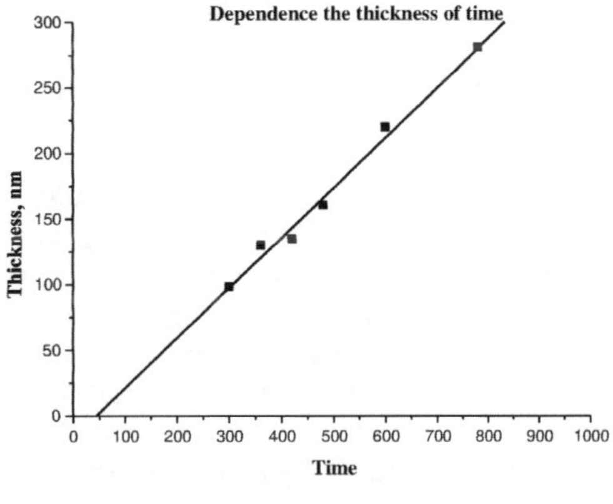

Figure 2. Dependence the thickness of time.

The results of measuring the refractive index depending on the deposition rate (Figure 3) and the films thickness (Figure 4) are showed. The obtained values vary in the range from 1.75 to 2.06. Thus, depending on the thickness and deposition rate explicitly identify problematic.

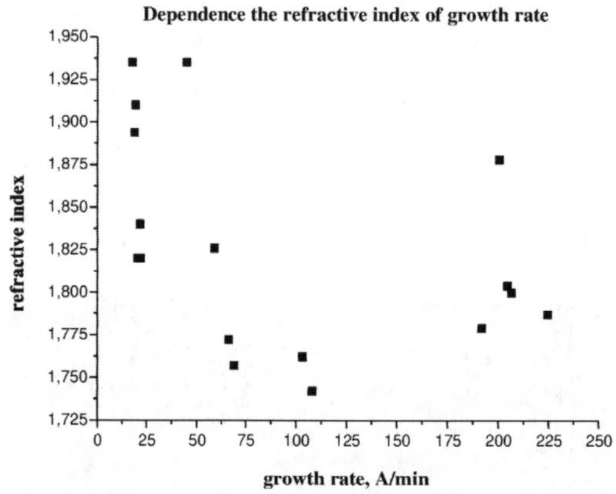

Figure 3. Dependence the refractive index of growth rate.

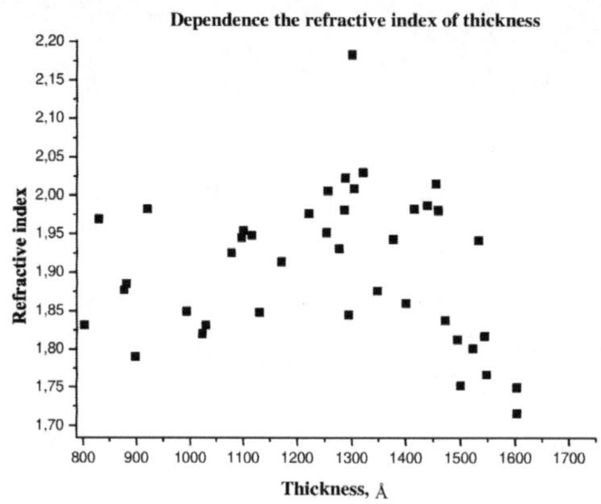

Figure 4. Dependence the refractive index of thickness.

EM results demonstrated that synthesized AlN films may contain an amorphous region adjacent to the substrate, and a polycrystalline region Figure 5(a,b). The thickness of the film amorphous phase varied in the range up to 200 nm depending on the sample preparation, the composition of the plasmaforming gas mixture, temperature and process speed. The conditions of amorphous region absence were revealed. Polycrystalline film region consisted of nanocrystals with typical dimensions on the order of 10 to 30 nm in the lateral direction. The structure corresponds to the wurtzite type of nanocrystals. The preferred orientation directions of single crystals and polycrystalline films in the presence of the amorphous underlayer were not found Figure 5(b). In the case of the AlN films without the amorphous phase synthesis Figure 6(a,b), the formation of columnar crystals with a preferred orientation axis C <0001> was close to the sample surface normal Figure 6(b). The typical dimensions of these crystallites were between 20 to 40 nm in the lateral direction and the film thickness in the order of the normal (300 nm).

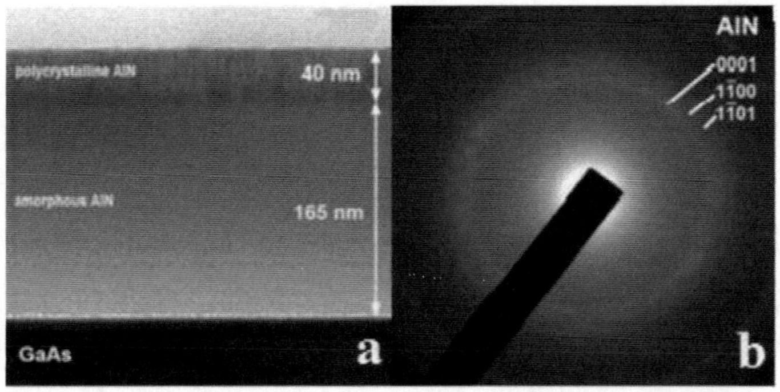

Figure 5(a,b). (a) TEM image of cross-section of AlN amorphous layer grown using ion plasma etched deposition on GaAs(100) ; (b) Selected area microdiffraction.

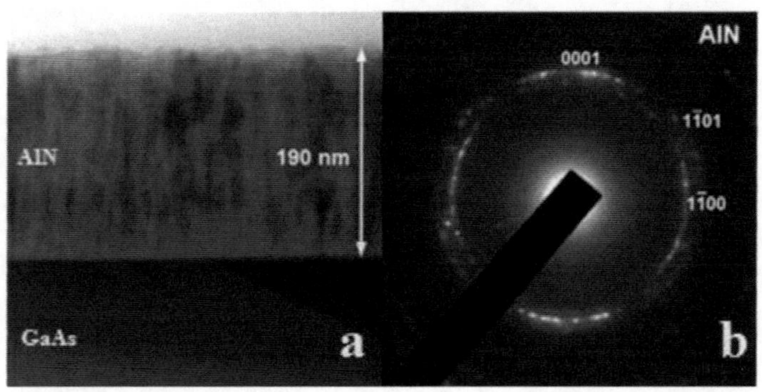

Figure 6(a,b). (a) TEM image of cross-section of AlN textured polycrystal layer grown using ion plasma etched deposition on GaAs(100) ; (b) Selected area microdiffraction.

It is worth to note that the dual structure of the samples (amorphous and polycrystalline) refraction index was failed to measure correctly due probably to the difference in optical properties of amorphous and polycrystalline phase. The obtained opical characteristic were in good agreement with the data obtained in [2] for the synthesis of AlN films by magnetron deposition and ion beam sputtering.

Acknowledgments

Russian Foundation for Basic Research (project# 15-07-06964, 13-02-12031 ofi-m), Ministerium for education and science of Russian Federation (Contract № 02.G25.31.0058 dated 12.02.2013)

References

[1] Zhiliakov L A, Kostanovskii A V 1992 *High Temperature* 30:2 226-229

[2] Kirilenko V V, Zhigarnovski B M, Pavlovski V K, and I P Markelova *Journal of Optiical Technology* 75(12) 800-805

[3] Brien V, Pigeat P 2008 *Journal of Crystal Growth* 310 16 3890-3895

[4] V V Zolotarev, A Yu Leshko, N A Pikhtin, S O Slipchenko, Z N Sokolova, Ya V Lubyanskiy, N V Voronkova, I S Tarasov 2015 *QUANTUM ELECTRON* 45 (12) 1091–1097

[5] V V Zolotarev, A Yu Leshko, N A Pikhtin, A V Lyutetskiy, S O Slipchenko, K V Bakhvalov, Ya V Lubyanskiy, M G Rastegaeva, I S Tarasov 2014 *QUANTUM ELECTRON* 44 (10) 907–911

[6] Berlin E V, Dvinin S A, Seidman L A, 2007 Vacuum technology and equipment (Moscow: Tehnosphera)

Thermal roughening of GaAs surface by dislocation-induced step-flow sublimation

I O Akhundov[1,2], **D M Kazantsev**[1,2], **A S Kozhuhov**[1], **V L Alperovich**[1,2]
[1]Rzhanov Institute of Semiconductor Physics, 630090 Novosibirsk, Russia
[2]Novosibirsk State University, 630090 Novosibirsk, Russia

Abstract. The thermal roughening of epitaxial GaAs film surface is studied under anneals at temperatures 700-775 °C in the presence of a saturated Ga-As melt. Surface roughening consists in the formation of spiral "inverted pyramids" on the initially flat surface due to the step-flow sublimation induced by screw dislocations. The observed roughening indicates that, despite the presence of As and Ga vapors provided by the melt, the annealing conditions are shifted from equilibrium towards sublimation.

1. Introduction

Atomically flat semiconductor surfaces are needed for fundamental surface science, device applications and reproducible fabrication of nanoscale structures [1, 2]. Close to ideal ones, silicon surfaces with atomically flat terraces separated by steps of monatomic height were obtained by annealing chemo-mechanically polished Si substrates in vacuum [3, 4]. The application of this method to III-V semiconductors is problematic because, due to a high evaporation rate of the volatile V component, it is difficult to find the vacuum annealing temperature range, in which the surface diffusion is sufficiently effective for surface smoothing, while the sublimation is still negligible [5]. To avoid the surface morphology deterioration due to depletion with arsenic, GaAs thermal smoothing, which yields GaAs step-terraced surfaces, can be performed under overpressure of arsenic-containing vapors in the MBE or MOCVD growth chambers [6, 7].

Thermal smoothing experiments in the growth chambers of MBE or MOCVD set-ups are expensive and time-consuming. A more efficient technique of GaAs surface thermal smoothing was proposed and developed in Refs. [8-12]. This technique consists in annealing GaAs substrates in the conditions close to equilibrium with Ga and As vapors, provided by the presence of the saturated Ga-As melt. So far, this smoothing technique has been demonstrated only for chemo-mechanically polished epi-ready GaAs substrates with a small root mean square roughness $\rho \leq 0.15$ nm. To apply this method to substrates or epitaxial films with a larger surface roughness and to speed up the smoothing process, one should increase the annealing temperature. However, at an increased temperature $T \geq 700$ °C, GaAs surface smoothing is gradually changed to surface roughening, which reveals itself in step meandering, the formation of deep pits and complete destruction of step-terraced morphology [8]. This roughening may be due to the thermodynamic roughening transition [13, 14], or to kinetic instabilities caused by deviations from equilibrium at high temperatures. On a step-terraced surface, the deviations from equilibrium towards growth or sublimation can be detected by the atomic step motion towards lower of higher lying terraces, respectively. Various origins of atomic steps are possible at a crystal surface. Misorientation of the vicinal crystal surface from a singular face may result in a regular set of straight equidistant atomic steps, which separate terraces with the width determined by the misorientation angle. However, the step motion under step-flow growth or sublimation on an ideal vicinal surface does not change the character of the homogenous step-terrace relief. Therefore, it is difficult to determine the direction and velocity of the step motion by studying the morphology of step-terraced vicinal surfaces because of the problems connected with localization of the same area on the sample with a submicron accuracy in *ex situ* AFM measurements.

A screw dislocation ending at a crystal surface provides an alternative source of a monatomic step. The steps induced by screw dislocations result in the formation of spiral pyramid hills on the surfaces of epitaxially grown films, as it was proposed and theoretically described in the classical paper by Burton, Cabrera and Frank [15] and studied experimentally by means of liquid phase epitaxy on mesa-structured GaAs(001) substrates [16]. In distinction, under sublimation, spiral dips ("inverted pyramids") are formed due to the flow of the atomic steps induced by screw dislocation endings [17]. Thus, unlike vicinal steps, screw dislocation steps lead to the surface morphology that unambiguously indicates the deviation from equilibrium towards growth or sublimation. The present study is aimed at clarifying the reasons for GaAs roughening at high temperatures by observation of the spiral structures induced by screw dislocations under annealing epitaxial GaAs films.

2. Experimental

GaAs films were grown by liquid phase epitaxy (LPE) on conventional flat GaAs(001) substrates and on the substrates with preliminarily prepared lithographic square mesa structures similar to those used in [16, 18]. The size of the mesas ranged from 0.2×0.2 mm^2 to 1×1 mm^2. The morphology of the films grown on flat substrates was usually non-uniform. The surface of such films contained relatively wide (~ 2 – 5 μm) flat terraces with randomly spaced vicinal monatomic steps and high step bunches (up to ~ 40 monolayers), which separated the terraces [18]. The morphology of the films grown on mesa structured substrates was more uniform. Depending on the growth conditions (presumably, on the growth termination conditions, which could not be well-controlled in the LPE set-up), the mesa as-grown surfaces contained spiral pyramid hills [16], or inverted pyramid pits induced by screw dislocations [18]. These morphologies correspond to the dislocation-controlled crystal growth or dissolution, respectively, at the last stage of LPE growth. We also obtained mesa-structured epitaxial film surfaces, containing atomically flat terraces and screw dislocation endings, which induced nearly straight monatomic steps. In the present study these latter mesa-structured films with straight dislocation-induced steps were used in the annealing experiments. The mean concentration of dislocation endings was about 5×10^4 cm^{-2}. The anneals were performed under the molecular hydrogen flow, in a quartz tube of the LPE set-up. The GaAs sample was put in a quasi-closed graphite cassette, which contained the saturated Ga-As melt in order to provide As and Ga vapors. Before annealing the surface oxides were removed in the HCl-isopropanol solution [18]. The morphology of the initial and annealed GaAs surfaces was studied *ex situ* by atomic force microscopy (AFM). The details of the anneals and AFM measurements are described in [8].

3. Results and Discussion

To elucidate the reasons of high-temperature GaAs surface roughening, one hour anneals of the samples cut from the same GaAs epitaxial film were performed at various temperatures. The results are shown in figure 1. The as-grown surface (figure 1a) contains wide (up to ~ 7 μm) atomically flat terraces separated by atomic steps, which are preferentially oriented in the same direction. Marked by the white dashed circles in figures 1a and 1b are the emergence points of "screw dislocations", which are not necessarily "pure screw", but have their Burgers vector component, perpendicular to the surface [15, 16]. The emergence point can be identified by the dislocation-induced step of monatomic height $h \approx 0.3$ nm, which corresponds to the GaAs period in the [001] direction.

Annealing at $T = 700$ °C (figure 1b) did not lead to substantial changes in the surface morphology. Annealing at higher temperature $T = 750$ °C (figure 1c) led to the step displacement by ~ 3 μm with respect to the dislocation ending point. The steps moved towards the higher lying terraces. Consequently, the annealing conditions correspond to crystal sublimation rather than growth. Near the point of dislocation emergence, the step twists into a spiral and forms a distinct spiral pit ("inverted pyramid" hollow), which also corresponds to the step motion towards the higher lying terrace and, thus, to the step-flow sublimation. The increase of the annealing temperature up to 775 °C caused a further spiral spinup (figure 1d). This process is reverse to the dislocation-controlled step-flow crystal growth described in [15]. While the dislocation-controlled step-flow growth was observed experimentally on the surfaces of many crystals, including GaAs [16], we report here on the observation of the dislocation-controlled step-flow sublimation on the GaAs surface.

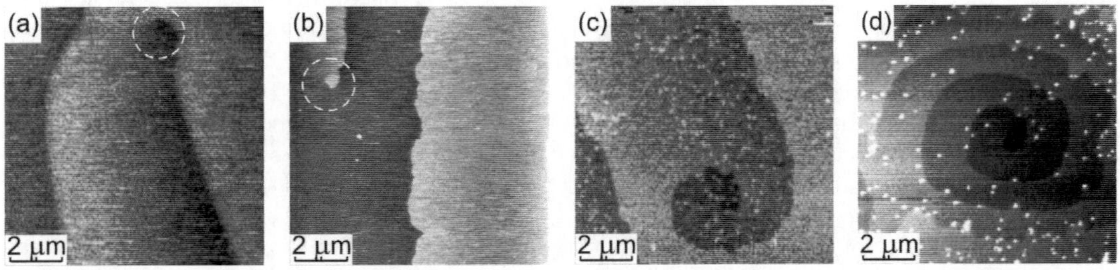

Figure 1. 10×10 µm^2 AFM images of the surface of an epitaxial GaAs film with screw dislocation exits before annealing (a) and after one-hour anneals at T = 700°C (b), 750°C (c) and 775°C (d).

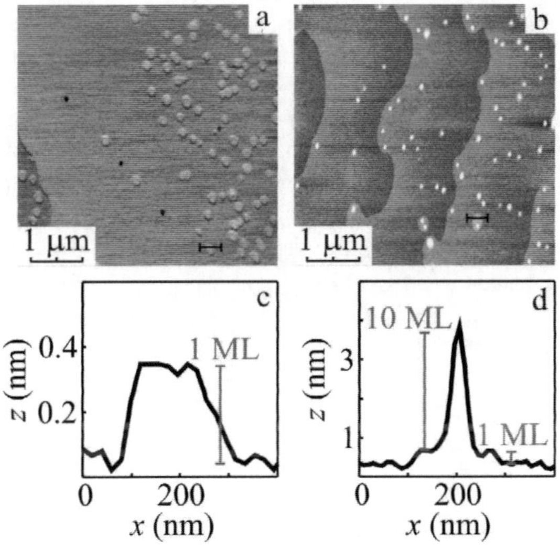

Figure 2. (a, b): Zoomed images of terraces with islands after anneals at T = 750°C (a) and 775°C (b). (c, d): z-x cross sections of selected islands measured along the black segments shown in figures 2a and 2b, respectively.

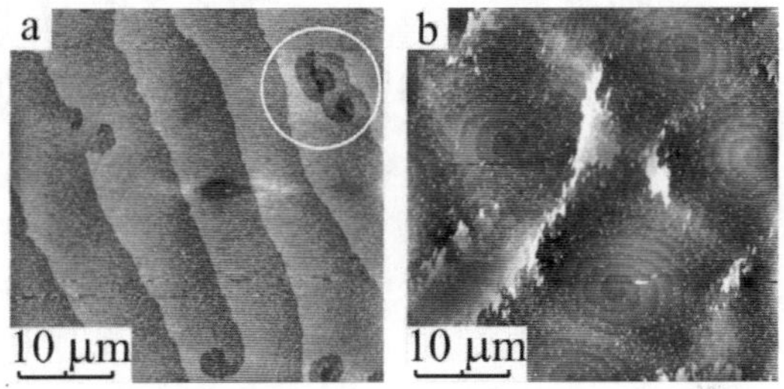

Figure 3. Intersection of spiral structures originating from various points of screw dislocation emergence after annealing at T = 750 °C (a) and 775 °C (b). Indicated by the white circle in (a) is a pair of closely spaced spiral structures.

It is seen from figures 1c and 1d that the spiral step motion is accompanied by the formation of islands on terraces. To consider these islands in more detail, zoomed images of the surface relief after anneals at T = 750 °C and 775 °C are shown in figures 2a and 2b, respectively, together with the z-x cross-sections of typical islands (figures 2c, d). The island lateral sizes are in the range of ~ 100 – 500 nm in diameter. It is seen from figures 2a, and 2c that annealing at T = 750 °C leads to the formation of the islands one or two monolayers (ML) high. At higher temperature T = 775 °C, multilayer islands are formed being from ~ 3 ML to 10 ML high. A possible reason for the formation of these islands is the existence of defect- or contamination-induced surface spots at which the sublimation is inhibited [19]. Under the motion towards higher lying terraces, the steps flow around such spots, leaving the islands behind. Each time a monatomic step passes through the spot, the island height increases by one monolayer with respect to the terrace level. Thus, the formation of a multilayer island at $T = 775$ °C corresponds to the multiple step passages through the island, in accordance with the observed multiple spiral turns (figure 2b). This suggestion is confirmed by the non-uniform monatomic island distribution over the terraces after annealing at $T = 750$ °C (figures 1c and 2a). Indeed, it is clearly seen from figures 1c and 2a that the monatomic islands are observed only on the part of the terrace that has been already passed by the moving dislocation-induced step. It should be also noted that the concentration of the multilayer islands (figure 2b) is smaller than that of the monatomic islands (figure 2a). A possible reason for this fact is that a part of monatomic islands evaporated during high-temperature annealing.

Another peculiarity of the GaAs thermal surface roughening is the intersection of spiral structures at high annealing temperatures. Shown in figure 3 are the 40×40 μm^2 AFM images of the surface relief, which demonstrate this intersection. Very few nucleation centers of the spiral structures are seen after annealing at $T = 750$ °C, with only one pair of closely spaced intersecting spiral structures indicated by the white circle in figure 3a. In distinction, as it is seen from figure 3b, after annealing at 775 °C, the whole surface area consists of mature deep spiral pits separated by high (~ 10 ML or higher) ridges. It is evident that the spiral spin-up is blocked by adjacent spiral structures. Both right-hand and left-hand spirals are observed in figure 3b, indicating different sings of the screw dislocation Burgers vectors.

It should be noted that spiral structure jamming and the respective step motion deceleration hinder the comparison of the experimental morphology with the theory [15] and a correct determination of the parameters, in particular, the undersaturation. However, we can estimate the deviation of the annealing condition from the equilibrium by means of the measured displacement of the dislocation-induced step linear segment after one-hour annealing at $T = 750$ °C. The linear segment step velocity v_∞ and relative Ga vapor supersaturation σ can be expressed as follows [15]:

$$v_\infty = 2\lambda_S \frac{j - j_0}{n_0} , \tag{1}$$

$$\sigma = \frac{j - j_0}{j_0} = \frac{P - P_0}{P_0} , \tag{2}$$

where λ_S is the Ga adatom diffusion length, n_0 is the surface atomic concentration, j and P, j_0 and P_0 are the actual (non-equilibrium) and equilibrium (denoted by subscript 0) atomic flows and pressures, respectively. From measured step velocity $v_\infty = 3.5$ μm/hour and known values $P_0 = 1.3\times10^{-8}$ bar at $T = 750$ °C [20, 21] and $n_0 = 2.5\times10^{15}$ cm^2, we obtain $\sigma = -0.02$. The negative value of σ corresponds to undersaturation of the vapor pressure and, thus, to crystal sublimation.

4. Summary

In conclusion, the surface morphology evolution of epitaxial GaAs films grown on mesa structured substrates is studied under annealing in the presence of the saturated Ga-As melt at temperatures up to 775 °C. Surface roughening is observed at high annealing temperatures T ≥ 750 °C. This roughening consists in the formation of spiral "inverted pyramid" pits on the initially flat surface due to the step-flow sublimation induced by screw dislocations. This observation indicates that, despite the presence of As and Ga vapors provided by the melt, the annealing conditions are shifted from the equilibrium towards sublimation. The observed spiral step spin-up is accompanied by the formation of islands on the terraces. A possible reason for the island formation is that steps flow around the surface spots at which the sublimation is inhibited: each time the step passes through the spot, the island height increases by one monolayer with respect to the terrace level. The velocities of sublimation and relative undersaturation are estimated from the measured step displacement after annealing at $T = 750$ °C. Annealing at higher temperature $T = 775$ resulted in a peculiar "vortex"-like morphology, which consists in deep spiral pits covering the major part of the surface area. To the best of our knowledge, such vortex-like morphology was not observed earlier at semiconductor surfaces.

Acknowledgments
This work was partly supported by RFBR (Grant No. 16-32-00220) and by the Russian Academy of Sciences (Project No. 0306-2015-0020).

References

[1] Crommie M F, Lutz C P, Eigler D M 1993 *Science* **262** 218
[2] Teichert C 2002 *Phys. Rep.* **365** 335
[3] Latyshev A V, Aseev A L, Krasilnikov A B, Stenin S I 1989 *Surf. Sci.* **213** 157
[4] Jeong H-C and Williams E D 1999 *Surface Sci. Rep.* **34** 171
[5] Fan Y, Karpov I, Bratina G, Sorba L, Gladfelter W, Franciosi A 1996 *J. Vac. Sci. Tecnol. B* **14** 623
[6] Ding Z, Bullock D W, Thibado P M, LaBella V P, Mullen K 2003 *Phys. Rev. Lett.* **90** 216109
[7] Epler J E, Jung T A, Schweizer H P 1993 *Appl. Phys. Lett.* **62** 143
[8] Alperovich V L, Akhundov I O, Rudaya N S, Sheglov D V, Rodyakina E E, Latyshev A V, Terekhov A S 2009 *Appl. Phys. Lett.* **94** 101908
[9] Akhundov I O, Alperovich V L, Latyshev A V, Terekhov A S 2013 *Appl. Surf. Sci.* **269** 2
[10] Kazantsev D M, Akhundov I O, Karpov A N, Shwartz N L, Alperovich V L, Terekhov A S, Latyshev A V 2015 *Appl. Surf. Sci.* **333** 141
[11] Kazantsev D M, Akhundov I O, Shwartz N L, Alperovich V L, Latyshev A V 2015 *Appl. Surf. Sci.* **359** 372
[12] Akhundov I O, Kazantsev D M, Alperovich V L, Rudaya N S, Rodyakina E E, Latyshev A V 2016 *Scripta Materialia* **114** 125
[13] Nijs M D and Rommelse K 1989 *Phys. Rev. B* **40** 4709
[14] Lapujoulade J 1994 *Surf. Sci. Rep.* **20** 195
[15] Burton W K, Cabrera N, Frank F C 1951 *Proc. R. Soc. Lond. A* **243** 299
[16] Weishart H, Bauser E, Konuma M, Queisser H-J 1994 *J. Cryst. Growth* **137** 335
[17] Ostadrahimi A H, Dabringhaus H, Wandelt K 2002 *Surf. Sci.* **521** 139
[18] Alperovich V L, Tereshchenko O E, Rudaya N S, Sheglov D V, Latyshev A V, Terekhov A S, 2004 *Appl. Surf. Sci.* **235** 249
[19] Ranganathan M and Weeks J D 2014 *J. Cryst. Growth* **393** 35
[20] Tmar M, Gabriel C, Chatillon C, Ansara J 1984 *J. Cryst. Growth.* **68** 557
[21] Chatillon C, Ansara I, Watson A, Argent B B 1990 *Calphad* **14** 203

Evaluation of nitrogen incorporation into bulk 4H-SiC grown on seeds of different orientation from optical absorption spectra

D D Firsov[1]*, O S Komkov[1], A Yu Fadeev[1] and A O Lebedev[2]

[1]Micro- and Nanoelectronics department, St. Petersburg Electrotechnical University "LETI", 5 Professora Popova Street, St. Petersburg 197376, Russia
[2]Dielectric and Semiconductors Physics department, Ioffe Institute, 26 Polytekhnicheskaya Street, St.-Petersburg 194021, Russia

Abstract. The effectiveness of *n*-type nitrogen doping of bulk 4H-SiC grown on seeds of different orientation is studied by optical absorption measurements. The 4H-SiC ingots have been grown by physical vapour transport (PVT), with nitrogen doping from the SiC source. The nitrogen concentration was determined at room temperature from the absorption peak intensity at 464 nm, with account for the degree of donor ionization. It has been shown that 4H-SiC ingots grown on Si (11-22) faces are significantly less doped by nitrogen than the ones grown on C (11-2-2).

1. Introduction

Silicon carbide is a very significant material for high-power and high-temperature electronics because of its wide energy gap, high breakdown field and thermal conductivity [1]. SiC ingots of large diameter can be grown by physical vapour transport (PVT) on their own seeds. The most prominent *n*-type dopant for silicon carbide is nitrogen, and doping is carried out from the gas phase by adding nitrogen (N_2) into the inert atmosphere (Ar). However, the raw material (SiC powder) appears to be an additional source of nitrogen, leading to unintentional variations in the doping level.

Despite the fact that seeds of basal (0001) orientation are the most popular for PVT growth of SiC, one can also use other orientations, which may be interesting from several points of view. Firstly, some groups have demonstrated the possibility of an increase of charge carrier mobility in the channel of the field effect transistor implemented on a (11-20)-face, compared to that on the basal plane [2]. Secondly, the usage of the substrates of prismatic orientation (11-20) for light-emitting structures can reduce the effect of high piezoelectric and spontaneous polarization along the [0001] axis, and so to increase radiative recombination efficiency [3,4]. Thirdly, because of the strong dependence of the defect structure `on the growth direction, i.e. seed orientation, the ingots grown on the prismatic seed can be described as defect-free, concerning the defects propagating along the [0001] direction, namely threading dislocations and micropipes.

It should be noted that SiC doping levels, even under identical growth parameters, should be different for various seed orientations due to the different incorporation kinetics of nitrogen. The aim of this work is to study the effectiveness of n-type nitrogen doping of bulk 4H-SiC during growth on seeds of various orientations by means of optical absorption measurements.

2. Experimental setup

4H-SiC ingots were grown using the modified Lely (PVT) method in a reactor with resistive heating of the crucible [5]. The growth temperature measured in the bottom of the crucible was 2150°C, and 2060°C at the cap of it. The pressure during growth was 3.5 Torr. The gas atmosphere was argon of high purity, or Ar and N_2 (5% vol.) mixture for intentional nitrogen doping (sample N-2). The nitrogen incorporated in the rest of the grown ingots was from the SiC source. The temperature gradients were in the range of 10-20 K/cm. The duration of the growth was 40 hours. The average growth rate was about 0.5 mm/hour for all the experiments. The seeds fixed at the top of the crucible were 4H-SiC substrates. As a source we used the silicon carbide powder of 100 microns, manufactured by Saint-Gobain (Norway).

Wafers of (0001) and (11-20) orientations were cut from ingots grown on standard off-cut (000-1) C seeds with an inclination of 4 degrees to [11-20]. The (11-22)- and (11-2-2)-oriented wafers were cut from ingots grown on seeds of the same orientations, correspondingly. Initially, (11-22)-oriented 4H-SiC seeds were cut from an (000-1)C ingot at the angle of 73° with respect to the basal plane.

The wafers were faceted by two crystallographically non-equivalent faces, i.e., quasi-polar silicon Si(11-22) and carbon C(11-2-2). The studied samples were approximately rectangular with linear sizes of 51×15 mm. The 4H-SiC single crystals were grown simultaneously on seed faces of both polarities in one experiment.

Prior to the optical measurements, all wafers under consideration were subsequently subjected to mechanical grinding, polishing, and chemical etching in the KOH melt (600°C, 15 min). The parameters of the studied wafers are presented in table 1.

Table 1. Parameters of the studied 4H-SiC samples.

Sample number	Seed orientation	Surface	Thickness, μm	Comments
N-1	(000-1)	C	590	
N-2	(000-1)	C	680	N_2 doping
P-0	(11-20)	no	965	
Si-1	(11-22)	Si	920	
Si-2	(11-22)	Si	920	
Si-3	(11-22)	Si	920	
Si-4	(11-22)	Si	925	
C-1	(11-2-2)	C	705	
C-2	(11-2-2)	C	725	
C-3	(11-2-2)	C	670	
C-4	(11-2-2)	C	630	

The optical setup was based upon a VERTEX 80 Fourier-transform infrared (FTIR) spectrometer equipped for spectra measurements in the 0.4 – 1.1 μm range. This setup enables rapid registration of the spectra with a high signal-to-noise ratio. The samples were placed into a specialized accessory to perform transmission and reflection measurements in the same spot of the sample. The light incidence angle was 11°, corresponding to a ~4° angle inside the SiC crystal – which is rather close to normal. The probe beam was polarized by a linear polarizer, and the studied samples were rotated in order to change the angle between the optical axis of the crystal ***c*** and the electric field vector of the incident electromagnetic wave ***E***. All the measurements were carried out at room temperature.

3. Results and discussion

It has been described in several works [6-8] that nitrogen-doped n-type 4H-SiC exhibits specific optical absorption in the visible range, which is dependent on the polarization of the incident light. In order to evaluate this effect in our case, we have measured the transmission $T(\lambda)$ and reflection $R(\lambda)$

spectra of an (11-20) oriented sample in the extreme cases of the electric field vector $\boldsymbol{E}$ parallel and perpendicular to $\boldsymbol{c}$ axis (figure 1).

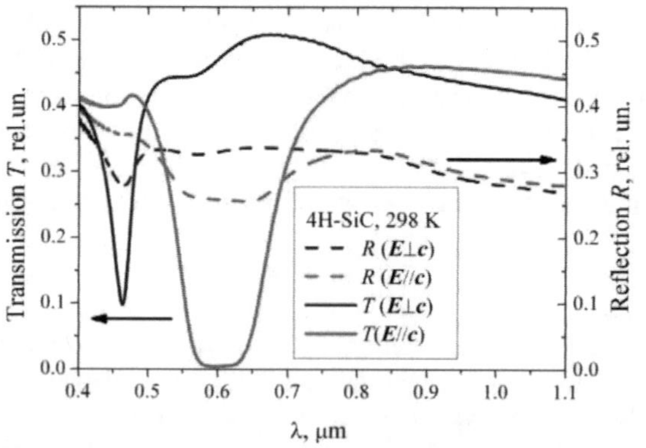

Figure 1. Transmission and reflection spectra of a typical nitrogen-doped 4H-SiC sample of (11-20) orientation for different light polarization.

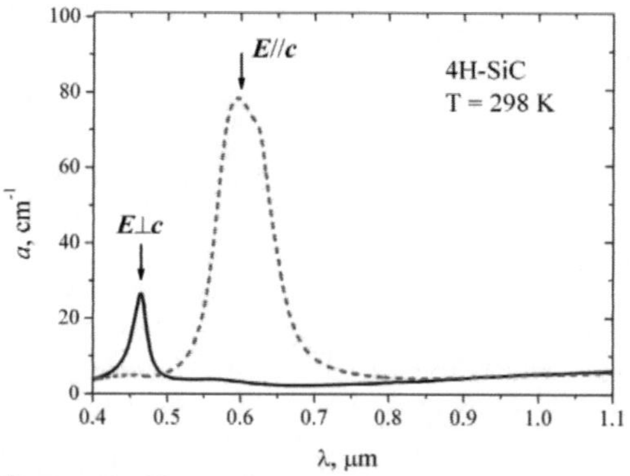

Figure 2. Absorption spectra of the nitrogen-doped 4H-SiC sample (figure 1) in the case of different light polarization.

As can be seen from figure 1, the transmission coefficient T of 4H-SiC decreases significantly at $\lambda \sim 0.46$ μm – or at $\lambda \sim 0.6$ μm, depending on the polarization – which implies the presence of absorption bands. The absorption coefficient of the wafer material a (cm^{-1}) can be determined from the measured transmission T and reflection R coefficients. The transmission coefficient of the plate-parallel sample in the case of normal light incidence is defined by the following expression:

$$T = \frac{(1-R_0)^2 e^{-ad}}{1-R_0^2 e^{-2ad}}, \qquad (1)$$

where d is the sample thickness (cm), and R_0 is the reflection from a single sample-air interface.

The reflection coefficient of the same sample is described as follows:

$$R = R_0 - \frac{(1-R_0)^2}{R_0} + \frac{(1-R_0)^2}{R_0 - R_0^3 e^{-2ad}}. \qquad (2)$$

In the equations (1) and (2), it is taken into account that multiple reflection of light occurs from the front and rear edges of the samples. The solution of these equations for a at each wavelength allows one to determine the absorption spectra $a(\lambda)$ of the sample material without the exact knowledge of $R_0(\lambda)$. The absorption spectra corresponding to the $R(\lambda)$ and $T(\lambda)$ of the n-type 4H-SiC sample from figure 1 are shown on figure 2.

It can be seen that 4H-SiC exhibits strong absorption at $\lambda = 464$ nm, or at $\lambda = 600$ nm, depending on the relative orientation of the $\boldsymbol{c}$ axis and the electric field vector $\boldsymbol{E}$ of the polarized light. The origin of these absorption peaks is conventionally attributed to nitrogen-related optical transitions into the upper sub-bands of the conduction band [6].

In the work [9], an empirical relation between the absorption intensity at 464 nm ($\boldsymbol{E} \perp \boldsymbol{c}$) and the free electron concentration n (cm^{-3}) in nitrogen-doped 4H-SiC has been proposed:

$$a = a_0 + kn, \qquad (3)$$

where $a_0 = 2.4 \pm 1.3$ cm^{-1}, $k = (3.6 \pm 0.1) \cdot 10^{17}$ cm^2.

In order to use the approximation (3) for determining carrier concentration in 4H-SiC, the condition of ($\boldsymbol{E} \perp \boldsymbol{c}$) must be fulfilled during the $R(\lambda)$ and $T(\lambda)$ optical measurements.

The corresponding absorption spectra (***E*** $\perp$ ***c***) obtained by eq. (1) and (2) for two 4H-SiC (0001) wafers are shown on figure 3. It can be seen that the sample N-2, which was grown with additional doping by N_2 from the gas phase, exhibits considerably stronger absorption. The according electron concentration value n obtained with (3) is by an order of magnitude higher than in the sample N-1, as can be seen from the inset of figure 3.

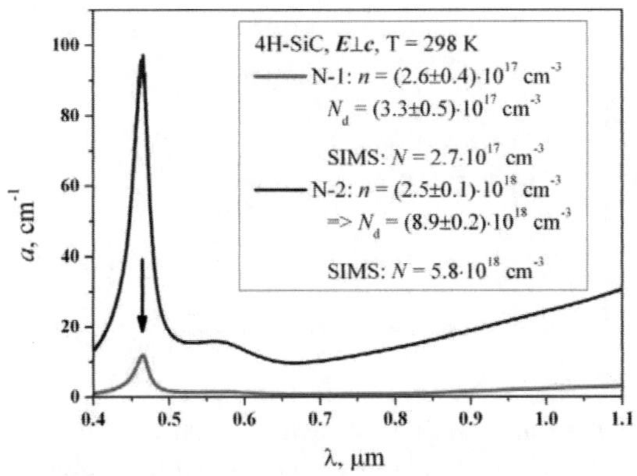

Figure 3. Absorption spectra of 4H-SiC wafers grown on (0001) seeds with (sample N-2) and without (sample N-1) additional nitrogen doping.

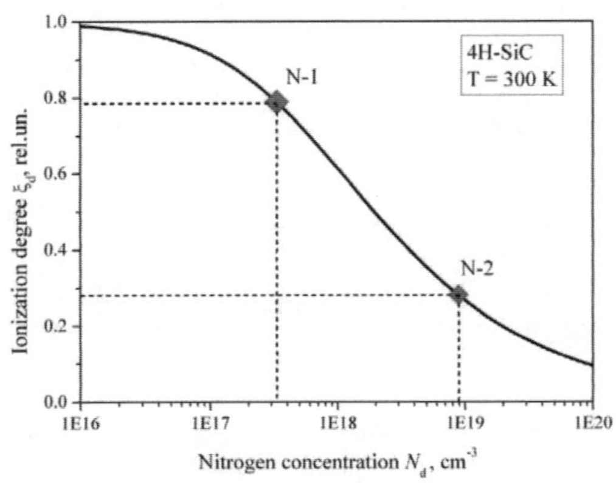

Figure 4. Ionization degree of nitrogen donors in 4H-SiC as a function of nitrogen concentration at room temperature [10]. Dots mark the nitrogen concentration values for the samples from figure 3.

To determine the concentration of nitrogen donors N_d from the free electron concentration values obtained with (3), we have accounted for the degree of donor ionization ξ_d given by the relation [10]:

$$\xi_d = N_d^+ / N_d = \left\{1 + \left[1 + 4g_d \frac{N_d}{N_c} \exp\left(\frac{\Delta E_d}{k_B T}\right)\right]^{1/2}\right\}\left[2g_d \frac{N_d}{N_c} \exp\left(\frac{\Delta E_d}{k_B T}\right)\right]^{-1}, \quad (4)$$

where T is the lattice temperature (K), $g_d = 2$ is the degeneracy factor, k_B is the Boltzmann constant, $N_c = 3.25 \cdot 10^{15}\, T^{3/2}$ cm^{-3} [11] is the effective density of states in the conduction band, and $\Delta E_d = 0.059$ eV [12] is the ionization energy of nitrogen in 4H-SiC. The ionization degree ξ_d decreases with the increase of nitrogen concentration, as plotted in figure 4.

The concentration of the ionized donors can be described by a steady-state Gibbs distribution [10]:

$$N_d^+ = N_d \left[\frac{n g_d}{N_c} \exp\left(\frac{\Delta E_d}{k_B T}\right) + 1\right]^{-1}. \quad (5)$$

Assuming that the equilibrium electron concentration in an extrinsic ($N_d >> N_a$) SiC at room temperature is given by the amount of ionized nitrogen donors ($N_d^+ = n$), we come to the following:

$$N_d = n \left[\frac{n g_d}{N_c} \exp\left(\frac{\Delta E_d}{k_B T}\right) + 1\right]. \quad (6)$$

The nitrogen concentration values N_d obtained with (6) for 4H-SiC samples N-1 and N-2 are shown in an inset of figure 3. The results are in good agreement with the independent data of secondary ion mass spectrometry (SIMS) measurements.

The described method of determining nitrogen concentration in 4H-SiC has been applied to study the crystals grown simultaneously on seeds of various orientations. The absorption spectra $a(\lambda)$ of the sample series grown on Si (11-22) and C (11-2-2) faces are shown on figure 5. The studied samples C-1 and Si-1 were cut from the part of each ingot which was in 4 mm from the seed, C-2 and Si-2 – in

5 mm from the seed and so on. The corresponding values of *n* in the studied wafers are listed on an inset of figure 5.

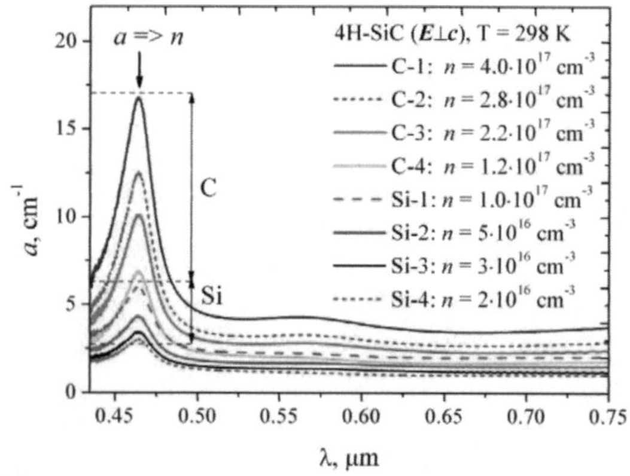

Figure 5. Absorption spectra of 4H-SiC bulk samples grown on Si- and C- faces. The arrow marks the absorption peak used to determine the electron concentration *n*.

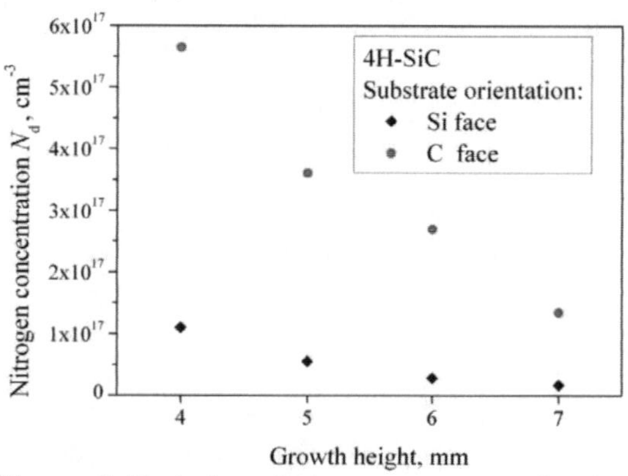

Figure 6. Variation of nitrogen concentration in the (11-22) and (11-2-2) 4H-SiC samples plotted in the ingot growth direction.

The nitrogen concentration values in the studied structures obtained with (6) are plotted in the figure 6. It can be seen that initiating the growth of 4H-SiC on the Si (11-22) faces significantly decreases the injection of nitrogen into the bulk 4H-SiC compared to the C (11-2-2) ones. In addition, a decrease of nitrogen concentration along the growth direction is observed, which is presumably caused by depletion of nitrogen in the SiC source. The obtained results allow us to conclude that growth of 4H-SiC on a Si (11-22) face can be used to obtain lightly doped bulk material.

Acknowledgements
The authors would like to thank Dr. Ber B.Ya. (Ioffe institute of RAS) for the SIMS data. The part of this work accomplished by Firsov D.D. (determination of nitrogen concentration in SiC from the optical absorption spectra) was supported by FASIE (grant "UMNIK 15_12").

References

[1] Casady J B, Johnson R W 1996 *Solid-State Electronics* **39** pp 1409-1422
[2] Yano H, Hirao T, Kimoto T et al. 1999 *IEEE Electr. Dev. Lett.* **20** pp 611-613
[3] Kamiyama S, Iwaya M, Amano H, Akasaki I 2005 *MRS Symp. Proc.* **831** p E10.6.1
[4] Gardner N F, Kim J C, Wierer J J et al. 2005 *Appl. Phys. Lett.* **86** p 111101
[5] Fadeev A Y, Lebedev A O, and Tairov Y M 2012 *Semiconductors*, **46**, pp 1346-1350
[6] Dubrowskii G B, Lepneva A A, and Radovanova E I 1973 *Phys. Status Solidi B* **57**, p 423
[7] Limpijumnong S, Lambrecht W R L, Rashkeev S N, and Segall B 1999 *Phys. Rev. B* **59** pp 12890-12899
[8] Sridhara S G, Bai S, Shigiltchoff O, Devaty RP, and Choyke W J 2000 *Mater. Sci. Forum* **338-342**, pp 551-554
[9] Weingärtner R, Wellmann P J, Bickermann M et al. 2002 *Appl. Phys. Lett.* **80** pp 70-72
[10] Tesfaye A 2004 *SiC semiconductor devices technology, modeling and simulation* (Vienna Univ. Technology) p 3.6
[11] New Semiconductor Materials: http://www.matprop.ru/SiC_bandstr
[12] Choyke W J and Pensl G 1997 *MRS Bulletin* **22** pp 25-29

Synthesis of GaN nanowires on Si (111) substrates by molecular beam epitaxy

A D Bolshakov[1], G A Sapunov[1], A M Mozharov[1], G E Cirlin[1-4], I V Shtrom[1, 3] and I S Mukhin[1,5]

[1] St.Petersburg Academic University, 194021, St. Petersburg, Russia
[2] Institute for Analytical Instrumentation RAS, 190103 St. Petersburg, Russia
[3] St. Petersburg State University, 199034 St. Petersburg, Russia
[4] St. Petersburg State Polytechnical University, 195251 St. Petersburg, Russia
[5] ITMO University, 197101, St. Petersburg, Russia

e-mail: bolshakov@live.com

Abstract. In this work we study growth of semiconductor GaN nanowires (NWs) on Si(111) substrates by means of molecular beam epitaxy. We demonstrate that the substrate temperature affects both the surface density and growth rate of the synthesized NWs. It was determined that at a fixed flux of nitrogen equal to 1.3 cm^3/min the maximum growth rate of NWs is ~ 38 nm/h at a substrate temperature — 800 °C. It was also found that the growth rate of NWs on the substrates treated with the oxide removal procedure is half the growth rate on substrates covered with oxide, while their surface density is twice higher in the first case. In addition we have studied influence of Ga flux on NWs formation.

1. Introduction

For the last few decades low dimensional semiconductor nanostructures such as nanowires, quantum dots, nanoneedles, thin films and monoatomic layers attract a lot of attention because of their unique electronic properties which are believed to give birth for the future generation of opto- and nanoelectronic devices [1-3]. Due to high aspect ratio and large surface area nanowires can be grown epitaxially on highly mismatched substrates (e.g. GaN or GaAs on Si) possessing high crystal quality [4, 5] at the same time. Because of the latter property these structures are considered one of the most promising for the integration of A3B5 optical semiconductors on Si [6, 7].

Semiconductor photovoltaics is one of the implementation field for low dimensional nanostructures such as nanowires [8-11]. First of all, nanowires grown on the face side of the solar cell are known to improve light absorption of the device [12], which especially important if we're considering Si-based cell, because Si itself possess high reflection coefficient close to 30% [13, 14]. Another important property of A3B5 nanowires on Si is their high crystal quality which leads to good carrier transport and make them great conducting channels for photo-generated carriers [6, 15]. As to choice of GaN as a material for nanowires we have to mention its large band gap which makes it ideal semiconductor for the window of the solar cell. Together these features open up broad prospects for their use in photovoltaics, in particular solar cells.

In theoretical work [16] it was shown that the efficiency of the basic solar cell (quite simple in fabrication) with one junction based on GaN NWs can reach 20%. It's been demonstrated theoretically that achievement of the peak efficiency requires optimization of the structure doping levels and its morphology namely surface density, diameter and length. In this work we investigated experimentally the influence of growth parameters on the growth rate and morphology of the synthesized GaN nanowires array on Si (111) substrate by plasma-assisted molecular beam epitaxy technique.

2. NWs synthesis

Before loading n-doped ($5*10^{19}cm^{-3}$) Si (111) substrates into the MBE system they were treated Shiraki cleaning procedure. After preparation every substrate passed degassing procedure and was transferred to the growth chamber, where was annealed at temperatures above 850 °C. The oxide removal was controlled using reflection high-energy electron diffraction (RHEED). The substrate temperature was then lowered to 650 °C and a thin layer of AlN was deposited following the procedure: 1) deposition of 3 nm Al layer, 2) interruption of deposition for 1min, 3) nitrogen deposition for 1.5 minutes at a flux of 1.3 cm^3/min. Formation of AlN layer was also controlled via RHEED. Deposition of AlN in our case played two roles: firstly, it allows formation of GaN nanoislands which during the growth then transform into nanowires and, secondly, it passivates substrate surface which in turn leads to reduction of recombination-induced losses [17-20]. The substrate temperature was then increased to 800±20 °C and Ga and N were deposited simultaneously resulting in formation of NWs. Some of the samples were also doped with Si for the further device application. The experiments were carried out in Veeco GEN III MBE system. Riber plasma cell served as a source of atomic nitrogen.

3. Experimental results

In this work over 10 arrays of GaN NWs on Si (111) substrate were synthesized. In our experiments we have varied annealing temperature of a substrate, a substrate temperature during NWs growth, a Ga deposition flux and a doping Si flux. After the growth, samples were studied by means of scanning electron microscopy (SEM) (Figure 1) in order to obtain morphology of the synthesized array. Results of the experiments are given in the Table 1.

Table 1. Growth parameters and properties of the synthesized GaN NWs array. Here T_{sub} – substrate temperature, T_a – annealing temperature, F_{Ga} – Ga deposition flux, T_{Si} – Si doping effusion cell temperature, t_{gr} – growth time, V_{gr} – average NWs vertical growth rate, D – average NWs top diameter, d – average NW base diameter, σ – average NWs surface density.

sample	T_{sub} (°C)	T_a (°C)	F_{Ga} (Torr)	T_{Si} (°C)	t_{gr} (h)	V_{gr} (nm/h)	D (nm)	d (nm)	σ (mkm^{-2})
1	800	850	$1.5\ 10^{-8}$	-	31	32.3	73	76	31
2	800	850	$1.5\ 10^{-8}$	1050	31.5	38.1	79	75	29
3	800	850	$1.5\ 10^{-8}$	1000	27	38.9	74	71	45
4	800	850	$1.5\ 10^{-8}$	1060	92.5	28.6	153	108	42
5	800	1000	$1.5\ 10^{-8}$	1050	34	20.6	67	61	86
6	800	970	$1.5\ 10^{-8}$	1000	31.6	15.5	62	67	85
7	820	1000	$1.5\ 10^{-8}$	-	18	7.8	32	30	20
8	810	1000	$1.5\ 10^{-8}$	1050	39	10.5	54	48	75
9	790	910	$1.5\ 10^{-8}$	1050	38	13.1	101	75	149
10	800	1000	$2.2\ 10^{-8}$	1000	30	8.8	58	61	159
11	800	1000	$1.5\ 10^{-8}$	1160	31	14.8	87	44	79

4. Discussion

The first few samples were grown on the substrates that didn't pass the oxide removal procedure, as they were annealed at only 850°C. Surprisingly, it turned out that NWs growth rate is twice faster in that case comparing to the samples grown on substrates annealed at 910°C and above. In the latter case the oxide was removed from the substrate which was controlled via observation of changes in RHEED pattern. This effect is probably related to the difference in a mean migration length of the adatoms on these surfaces probably due to the lower sticking coefficient of Ga adatoms on oxide surface. It is expected that the longer adatom motion the faster NWs formation and lower the surface density.

We have also varied the substrate temperature in a small range of 790-820°C. It was found that optimum temperature corresponding to the fastest NWs formation was 800 °C. Lower temperatures correspond to lower growth rate and higher surface density, which is intuitively understandable as adsorption from vapour phase should increase and mean free path decrease. At higher substrate temperatures situation is different: the growth rate decreases again, but also density of the array greatly reduces, which should be caused by reduction of particle adsorption.

We didn't have an opportunity to carry out straight forward study of gallium flux influence on the NWs formation, though one sample was grown using higher flux than the others. It was demonstrated that elevation of the Ga flux leads to fall of the growth rate and rise of the surface density. Again, it should be caused by shift in mean free path related to higher adatom concentration this time.

Another observation concerning shape evolution of the NWs is that they suffer radial extension with time while their surface density doesn't change sufficiently. The latter fact mean that all the sites available for NWs formation (which as far as we know are Si or SiOx sites not covered with AlN) get settled simultaneously at the initial stage of growth. Though NWs do grow radially, no sufficient tapering was observed with one particular exception of growth with high-intensity Si doping flux.

Part of the samples were doped with Si for further device application. The samples were examined by photoluminescence spectroscopic techniques. According to the optical experimental data in the sample number 11 grown at highest intensity doping flux we were able to reach 5 10^{19} cm^{-3} of donor concentration.

a) b)

Figure 1. SEM images of the grown samples: a) sample 1, b) sample 4.

5. Summary

In this work the growth of GaN quasi-onedimensional structures on Si (111) substrate by MBE was studied. We investigated the influence of growth parameters and Si doping on morphology of the synthesized array. Firstly, we have obtained that NWs growth rate is two times faster, while NWs surface density is more than twice smaller on the substrates where an oxide layer wasn't removed by means of annealing, comparing to the substrates treated with oxide removal procedure. It was found that optimum substrate growth temperature corresponding to the fastest NWs formation is 800 °C.

Lower temperatures correspond to lower growth rate and higher surface density, at higher temperatures growth rate decreases, but also density of the array decreases. It was also demonstrated that elevation of the Ga flux leads to fall of the growth rate and rise of the surface density. We speculate that all the mentioned effects related to the change of adatoms mean free path resulting in modification of the growth kinetics.

Acknowledgments
This work was carried out with the support of the Russian Federation President grant (14.W01.15.7736-MK), the Russian Foundation for Basic Research (15-02-06839 A, 16-29-03113 ofi and 16-32-00560 mol_a), the leading universities of the Russian Federation (grant 074-U01). The work is partially supported by the Skolkovo Foundation (grant agreement for Russian educational and scientific organisation no.6 dd. 30.12.2015).

References

[1] Duan X, Huang Yu, Agarwal1 R and Lieber C M 2003 *Nature* **421** 241-5
[2] Garnett E C and Yang P 2008 *J. Am. Chem. Soc.* **130 (29)** 9224–5
[3] Lia S and Waag A 2012 *J. Appl. Phys.* **111** 071101
[4] TangY B et al. 2008 *Nano Lett.* **8(12)** 4191
[5] Cirlin G E et al. 2010 *Nanoscale Res. Lett.* **5(2)** 360
[6] Cirlin G E,et al. 2010 *Phys. Rev.* B **82(3)** 035302
[7] Calarco R et al. 2007 *Nano Lett.* **7(8)** 2248-51
[8] Fan Z et al. 2009 *Nat. Mater.* **8** 648 - 53
[9] Nozik A J 2010 *Nano Lett.* **10 (8)** 2735–41
[10] LongY-Z, Yu M, Sun B, Gud C-Z and Fanc Z 2012 *Chem. Soc. Rev.* **41** 4560-80
[11] Neplokh V, Ali A, Julien F H, Foldyna M, Mukhin I, Cirlin G, Harmand J-C, Gogneau N and Tchernycheva M 2016 *Mater. Sci. Semicond. Process.* In press
[12] Hiralal P 2014 *Nanoscale* **6** 14555-62
[13] Battaglia C et al. 2012 *ACS Nano* **6(3)** 2790 –7
[14] Oh J, Yuan H-C and Branz H M 2012 *Nat. Nanotechnol* **7** 743–8
[15] Thelander C, Björk M T, Larsson M W, Hansen A E, Wallenberg L R and Samuelson L 2004 *Solid State Commun.* **131(9)** 573-9
[16] Mozharov A et al. 2015 *Phys. Stat. Sol. RRL* **9(9)** 507–10
[17] Consonni, V, Dubrovskii V G, Trampert A, Geelhaar L and Riechert H 2012 *Phys. Rev.* B **85** 155313
[18] Chèze C, Geelhaar L, Trampert A and Riechert H 2010 *Appl. Phys. Lett.* **97** 043101
[19] Borysiuk J, Zytkiewicz Z R, Sobanska M, Wierzbicka A, Klosek K, Korona K P, Perkowska P S and Reszka A 2014 *Nanotechnology* **25(13)** 135610
[20] Guo W, Zhang M, Banerjee A and Bhattacharya P 2010 *Nano Lett.* **10 (9)** 3355–9

Interference at the tunnel ionization from double-center potential

P A Golovinski[1,2], A A Drobyshev[2]

[1] Moscow Institute of Physics and Technology, Moscow 141700, Russia
[2] Physics Research Laboratory, Voronezh State University of Architecture and Civil Engineering, Voronezh 394006, Russia

Abstract. The tunnel ionization of an electron bounded by two delta potentials under the influence of a constant electric field is considered. The equations for the electron current density for two different initial states are obtained. The dependence of the emission current on the orientation of the potential with respect to the field direction and the distance between them is studied. Appropriate conditions for observation essential interference effects are defined.

1. Introduction

The model of a short-range potential is widely used in atomic physics, nuclear physics, and in condensed matter physics [1-3]. For the problem of the tunnel ionization of neutral molecules the generalization of asymptotic method [4] for the wave functions in parabolic coordinates in spherically asymmetric states has been developed [5-7]. However, in this case the theory is limited to consideration of the localized states of individual molecular orbitals. The concept of localized states without taken into account possible interference lies at the basis of the description of tunneling effects in nanostructures [8, 9]. At the same time, detecting the possible role of delocalization initial states requires special consideration.

The problem of the particle dynamics in the field of several short-range potentials allows implementation of a complete analytical study [2, 10]. When there is a close located potential centers, the wave functions overlap, and formed a common state. In addition to changing the structure the stationary states itself, such systems have a specific response to the external field [11]. We have considered the interference pattern for electron waves generated by electron tunneling ionization from double-center potential with a different field and potential orientation.

2. Basic equations

The problem of the particle dynamics in a field a few δ-potentials can be formulated in terms of the boundary conditions imposed on the wave function in the potential location points. For a single s-state with $l=0$, the wave function $\psi \sim r^{-1}\exp(-\alpha r)$, and the boundary condition takes the form [1]

$$\left.\frac{d\ln(r\psi)}{dr}\right|_{r=0} = -\alpha , \qquad (1)$$

where $\alpha = \sqrt{-2E_0}$, E_0 is bound energy. We use the atomic system of units in which $|e| = m = \hbar = 1$.

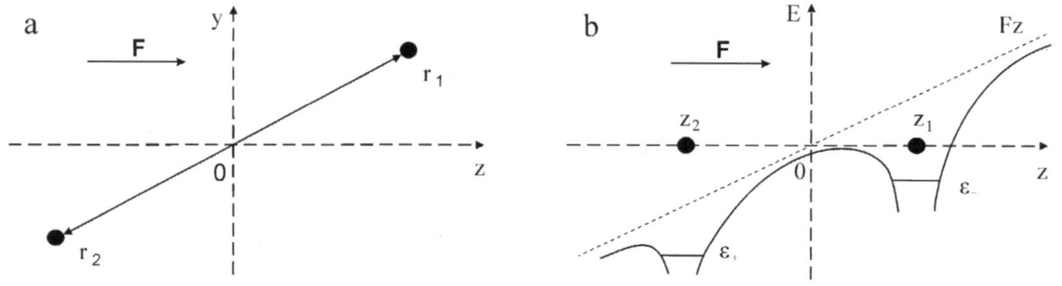

Figure 1. The relative location of the center (a) and the position of the energy levels for the system orientation parallel to the field (b).

For a few potentials their action is provided by the sum of individual potentials. State of a particle in the field of two identical attraction centers and an external electric field of strength $F > 0$ (Figure 1) is described by the stationary Schrödinger equation

$$\left(\varepsilon + \frac{1}{2}\nabla^2 + Fz\right)\psi(\mathbf{r}) = \left(V_1(\mathbf{r} - \mathbf{r}_1) + V_2(\mathbf{r} - \mathbf{r}_2)\right)\psi(\mathbf{r}), \tag{2}$$

where each potential [3]

$$V(\mathbf{r}) = -\frac{2\pi}{\alpha}\delta(\mathbf{r})\frac{\partial}{\partial r}r. \tag{3}$$

The solution, as in the problem without the field [12], is sought in the form of a superposition of the Green's functions

$$\psi(\mathbf{r}) = AG(\mathbf{r}, \mathbf{r}_1, \varepsilon) + BG(\mathbf{r}, \mathbf{r}_2, \varepsilon), \tag{4}$$

where $G(\mathbf{r}, \mathbf{r}_j, \varepsilon)$ is the Green's function with the asymptotic behavior of the outgoing waves, satisfying the Eq. (2) with the replacement of the right side to $\delta(\mathbf{r} - \mathbf{r}_j)$, $\mathbf{r}_1$ is the radius vector of the first center, $\mathbf{r}_2$ is the radius vector of the second center. The Green's function, for a particle moving under the action of a constant force [13], is expressed in terms the Airy function [14]:

$$G(\mathbf{r}, \mathbf{r}', \varepsilon) = \frac{1}{2|\mathbf{r} - \mathbf{r}'|}\left(\mathrm{Ci}(\chi_+)\mathrm{Ai}'(\chi_-) - \mathrm{Ci}'(\chi_+)\mathrm{Ai}(\chi_-)\right), \tag{5}$$

where

$$\chi_\pm = -\frac{1}{(2F)^{2/3}}\left(2\varepsilon + \mathbf{F}(\mathbf{r} + \mathbf{r}') \pm F|\mathbf{r} - \mathbf{r}'|\right), \tag{6}$$

and $\mathrm{Ci}(u) = \mathrm{Bi}(u) + i\mathrm{Ai}(u)$ has the asymptotic behavior of the outgoing wave at $u \to \infty$.

Taking into account the boundary conditions of the form of Eq. (1), we obtain a system of equations for the complex quasi-energy $\varepsilon = \mathrm{Re}\,\varepsilon - i\Gamma/2$ and the coefficients A and B:

$$\frac{1}{\rho_1\psi}\frac{\partial}{\partial\rho_1}(\rho_1\psi)_{\rho_1=0}=-\alpha,$$
$$\frac{1}{\rho_2\psi}\frac{\partial}{\partial\rho_2}(\rho_2\psi)_{\rho_2=0}=-\alpha. \tag{7}$$

Here we use the notation $\rho_1=\mathbf{r}-\mathbf{r}_1$, $\rho_2=\mathbf{r}-\mathbf{r}_2$.

The Eqs. (7) can be written as [15]

$$\begin{pmatrix} b_+ & G(-\mathbf{R}/2,\mathbf{R}/2,\varepsilon) \\ G(\mathbf{R}/2,-\mathbf{R}/2,\varepsilon) & b_- \end{pmatrix}\begin{pmatrix} A \\ B \end{pmatrix}=0, \tag{8}$$

$$b_\pm=\frac{\pi}{(2F)^{1/3}}\left(Ai'(\xi_\pm)Ci'(\xi_\pm)-\xi_\pm Ai(\xi_\pm)Ci(\xi_\pm)\right)+\alpha, \tag{9}$$

$$\xi_\pm=-\frac{2\varepsilon\pm FR\cos\theta}{(2F)^{2/3}},$$

where $\mathbf{r}_1=\mathbf{R}/2$, $\mathbf{r}_2=-\mathbf{R}/2$, $R=|\mathbf{r}_1-\mathbf{r}_2|$. From Eq. (8) follows the coupling of coefficients A and B:

$$\frac{A}{B}=-\left.\frac{2\pi G(-\mathbf{R}/2,\mathbf{R}/2,\varepsilon)+b_\pm}{2\pi G(\mathbf{R}/2,-\mathbf{R}/2,\varepsilon)+b_\mp}\right|_{\varepsilon=\varepsilon_\pm}. \tag{10}$$

Eq. (7) has two solutions, defining two wave function and two energy values [15]

$$\varepsilon_\pm=E_\pm+\Delta\varepsilon_\pm-\frac{i}{2}\Gamma_\pm \tag{11}$$

respective states with their widths, that are found from the condition that the determinant of the system (8) is equal zero. Here, $E_\pm=-k_\pm^2$ are unperturbed energies of the symmetric and antisymmetric states of the system [12] and

$$\kappa_\pm=\alpha\pm\frac{e^{-\kappa_\pm R}}{R}. \tag{12}$$

For a weak electric field ($F/\alpha^3<<1$) the level shifts and width were calculated earlier [11].

3. Numerical simulations

The observed value in the process of ionization is electron current density crossing the plane perpendicular to the z-axis, defined by the direction of the electric field:

$$j_z=\mathrm{Im}\left(\psi^*\frac{\partial\psi}{\partial z}\right). \tag{13}$$

Spatial electron current distribution depends on the angle between the axis connecting the centers, and the direction of the electric field, parameter $\alpha|\mathbf{r}_1-\mathbf{r}_2|$ characterizing the distance between the centers in comparison with attenuation unperturbed wave function localized at a single center, and parameter F/α^3 that defines the magnitude of the force action of the electric field with respect to the characteristic force that bounds the particle in the potential well.

There are two limiting cases of double-center potential orientation with respect to the field: the field parallel and perpendicular to the field. At the orientation centers of attraction perpendicular to the field ($\theta = \pi/2$), the ratio of the coefficients $|A/B| = 1$. This means that the electron wave function is a superposition of the wave functions each δ–potential and an electron is localized on equally both centers, a tunnel current distribution is analogous to the electron current distribution pattern for two coherent point sources.

Figure 2 shows the result of calculating the energy of the ground (excited) state and the ratio of the coefficients $|A/B|$ depending on the angle θ between **R** and **F** for the parameters $E_0 = 1\,\text{eV}$, $F = 10^7\,\text{V/cm}$, $R = 10\,\text{a.u.}$ From the results plotted in the figure it is clear, that for $\theta = \pi/2$ the ratio of the coefficients $|A/B| = 1$ in the ground and excited states, which corresponds to the symmetry.

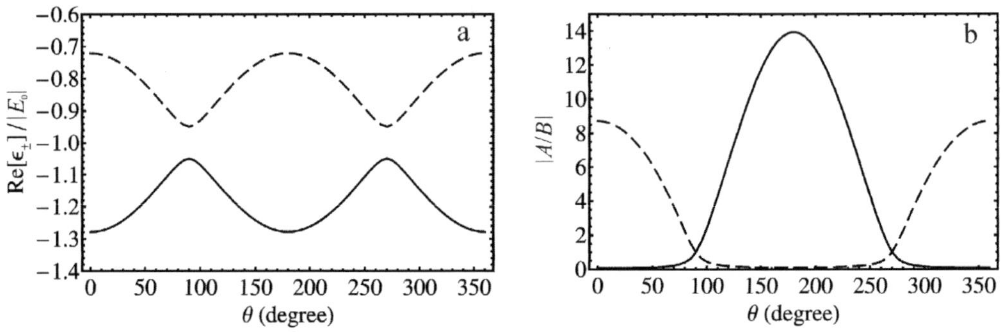

Figure 2. Energy dependence $\varepsilon_\pm$ (a) and the ratio dependence $|A/B|$ of the coefficients (b) on the angle. Solid line is the ground state, dotted is excited state.

When the orientation of the system is parallel to the field ($\theta = 0$), in the ground state $|A/B| = 0.07$, i.e., electron actually is localized on the second center. In the excited state, when $\theta = 0$, the ratio of the coefficients $|A/B| = 9$, corresponding to the electron localization at the first center. Increasing the length and the field strength leads to decrease in the displacement of states, and the electron is localized at one of the centers. In the ground state electron is localized on the second center and tunnels without influence of the first center, and a tunnel current distribution is analogous to the current from a single point source.

4. Conclusions

The considered interference of electron waves generated under the electron tunnel ionization from the double-center potential essentially depends on the orientation of the field and potential due to the degree of electron localization near the different centers. When the orientation of the double-center potential is perpendicular to the field, the distribution of the current is equivalent to the electron current distribution pattern for two coherent point sources. When one have the inclined orientation of the system, the delocalization of the initial states increases with increasing the field strength and the distance between the centers. The calculations indicate the importance of taking into account the type of chemical bonds in molecules in the interpretation of the results of observation of the electron tunnel ionization.

Acknowledgments
This work is supported by the Russian Foundation for Basic Research (grant 16-32-00253).

References
[1] Baz A I, Zel'dovich Ya B, Perelomov A M 1969 *Scattering, Reactions and Decay in Nonrelativistic Quantum Mechanics* (Jerusalem: Israel Program for Scientific Translations)
[2] Demkov Yu N and Ostrovsky V N 1988 *Zero-Range Potentials and Their Application in Atomic Physics* (New York: Plenum Press)
[3] Kirzhnitz D A 1967 *Field Theoretical Methods in Many Body Systems* (Pergamon, Oxford, England)
[4] Smirnov B M and Chibisov M I 1966 *Sov. Phys. JETP* **22** pp 585-592
[5] Batishev P A, Tolstikhin O I, and Morishita T 2010 *Phys. Rev. A* **82** 023416(14)
[6] Jiang W-Ch, Tolstikhin O I, Peng L Y, and Gong Q 2012 *Phys. Rev. A* **85** 023404(15)
[7] Tolstikhin O I, Morishita T, and Madsen L B 2011 *Phys. Rev. A* **84** 053423(17)
[8] Golovinski P A, Drobyshev A A 2014 *JETP* **118**(6) pp 863-868
[9] Golovinski P A, Drobyshev A A 2014 *Journal of Electromagnetic Analysis and Applications* **6** pp 8-14
[10] Demkov Yu N and Subramanian P 1970 *Sov. Phys. JETP* **30**(2) pp 381-383
[11] Dalidchik F I and Slonim V Z 1976 *Sov. Phys. JETP* **43**(1) pp 25-31
[12] Smirnov B M and Firsov O B 1965 *Sov. Phys. JETP* **20**(1) 156-160
[13] Gottlieb D and Kleber M 1992 *Ann. Physic* **1** pp 369-379
[14] Abramowitz M and Stegun I A 1972 *Handbook of mathematical Functions with Formulas, Graphs, and Mathematical Tables* (Washington, National Bureau of Standards: Applied Mathematics Series)
[15] Borzunov S V, Manakov N L, Starace A F, and Frolov M V 2011 *JETP* **112**(5) pp 725-743

Modeling of crystal growth in heteroepitaxial systems

O I Moroz, M V Makarets

Department of Physics, Taras Shevchenko National University of Kyiv, Kyiv 01601, Ukraine

Abstract. This paper investigates the elastic deformation of the structure containing InAs nanoclusters in a pyramid, grown on the substrate GaAs. So far the data have not been grown quantum dots (QDs), one of the reasons is significant difference of periods, which reaches 7%. Ideally atomic plane on the border of QDs and substrate must continuously sews. Due to the difference in lattice periods crosslinking occurring deformations and mechanical stresses, the magnitude of which is proportional to the number of atomic planes, the size of base of the pyramid. Therefore, when you reach a certain size islands (quantum dots), they may experience mechanical stresses p_{cr} sufficient for the appearance of structural defects - dislocations, fractures.

1. Introduction.

One of the methods of forming quantum dots is based on semiconductor nanostructures self organization during its epitaxial growth [1]. Mechanical stresses in epitaxial film of future quantum dot (QD) material and in its islands on the surface of the substrate are critical in the transition from the film's growth to growth of islets (Stranski-Krastanov mechanism [2,3]).These stresses are important in the further growth of QD in size, changing of its shape and its distribution on the substrate. Notable among material for QD occupies the binary semiconductors of the $A^{III}B^{V}$ family such as InAs, InSb [4]. The problem for the cultivation of such QD is the absence of complementary substrates with a close lattice period. Thus, the lattice mismatch periods in QD and in the coming per period substrate for InAs/GaAs system equals to about 7%, and for InSb/GaAs system – 14.5% [4,5]. As a result, in the QD-substrate interface there are occurred the mechanical stresses and deformations, the magnitude of which increases with the number of cross-linked atomic planes of that fundamental size of QD, which is often the pyramid . Therefore, when one reaches a certain size of QD, then the mechanical stresses p_{cr} appear, the value of which is sufficient for the appearance of structural defects namely – fractures, dislocations, etc. [5,6].

In this regard there is still the urgent task to find the mechanical stresses and deformations in InAs/GaAs systems and to estimate the size of QD, the achievement of which in the process of growth process there is running the conditions for the formation of structural defects. It is divided into two tasks: *i)* to find the stresses and deformations occurred in the InAs/GaAs heteroepitaxial system; *ii)* to obtain an estimate of the size of the QD the achievement of which there are appeared the mechanical stresses sufficient for the formation of defects.

2. The calculation of stresses in the InAs/GaAs system.

2.1.Issue formation.

Let's consider the InAs/GaAs heteroepitaxial system of two binary semiconductors in the vicinity of the QD/substrate interface – figure 1. We assume that QD of InAs has the shape of a square pyramid with bases in length l and height h . The QD base's length and height are linked with its growth on a substrate particularly due Wolf-Cashew theorem [7]. InAs and GaAs crystals belong to point group symmetry of $\bar{4}3m$ cubic system (sphalerite-type structure). Also consider that the pieces of QD atomic planes formed one over another from the interface will be a square, their orientation is (001) and the next size is smaller than the previous – see figure 1. If each upper QD atomic plane is shorter than the previous one from both edges (as is when atoms in planes are strictly one above the other), in this case it should be realized the condition $l = na$ where n – pairwise, a – InAs lattice period and then we receive $h = l/2$ where h – the height of the pyramid l – the length of the base of the pyramid. The sides of the base are oriented along the crystallographic axes [010] and in $x = \pm l/2$ and $y = \pm l/2$ [100]. The sides of the pyramid has the orientation (010). The substrate is represented as the parallelepiped of sizes $L_x, L_y, L_z >> h$. The beginning of the coordinate system and its axis were chosen due to figure 1.

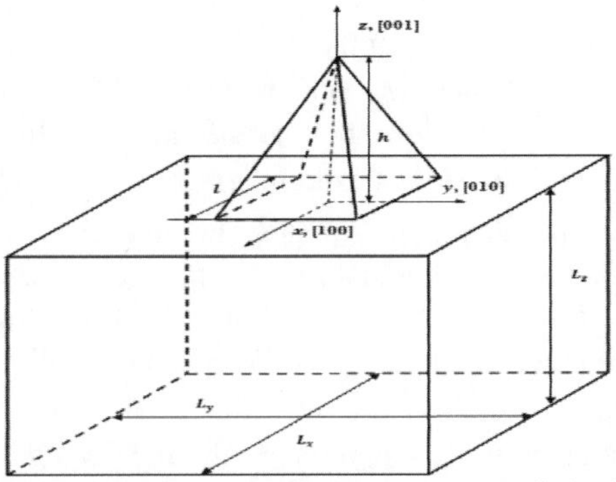

Figure 1. Schematic representation of the piece of substrate (a parallelepiped) and quantum dots (a pyramid) for the system InAs/GaAs

Considering the elastic properties of the system, we will not take its full account of atomic structure, despite the fact that QD as well as substrate contains thousands of atoms. QD for instance shown at figure 1, at $n = 20$ contains 1 771 atoms [8]. This assumption can consider the system as a complete environment. We also believe that as a zero approximation for the deformation description in the system can be applied the linear theory of elasticity.

Within these assumptions the stresses are satisfied with the Lame equation:

$$\frac{\partial p_{ij}}{\partial x_j} = 0 \tag{1}$$

where p_{ij} – (mechanical stress tensor for anisotropic linear environment is given by Hooke's law:

$$p_{ij} = C_{ijkl} u_{kl}, \tag{2}$$

where C_{ijkl} – elastic constants tensor of environment, strain tensor is defined as $2u_{kl} = \partial_l u_k + \partial_k u_l$, deformation vector $\vec{u} = u_x \vec{e}_x + u_y \vec{e}_y + u_z \vec{e}_z$, and ∂_k is partial to coordinate x_k.

GaAs and InAs semiconductor crystals have the same symmetry point group so their elasticity tensors in matrix representation has the same form [9]:

$$C_{ijkl} = \begin{pmatrix} C_{11} & C_{12} & C_{12} & 0 & 0 & 0 \\ C_{12} & C_{11} & C_{12} & 0 & 0 & 0 \\ C_{12} & C_{12} & C_{11} & 0 & 0 & 0 \\ 0 & 0 & 0 & C_{44} & 0 & 0 \\ 0 & 0 & 0 & 0 & C_{44} & 0 \\ 0 & 0 & 0 & 0 & 0 & C_{44} \end{pmatrix} \tag{3}$$

and their value and periods of crystalline structure for GaAs and InAs are shown in Table 1 taken from [9].

Table 1. Modules of elastic constants and lattice period

	InAs	GaAs
$C_{11}(Pa)$	$8,34\times10^{10}$	$1,19\times10^{11}$
$C_{12}(Pa)$	$4,54\times10^{10}$	$5,34\times10^{10}$
$C_{44}(Pa)$	$3,95\times10^{10}$	$5,96\times10^{10}$
$a\ (nm)$	0.605886	0.565321

Data shown shows GaAs period is 7% at the InAs period still the modulus of elasticity is about 15-33% higher. This means that at one and the same deformation tensor u_{kl} on both sides of the interface the QD-substrate in the substrate system there can be more mechanical loads. Similarly, under the same mechanical loads p_{ij} on both sides of the interface in the QD material there can appear larger relative deformation. From a physical point of view, on the both sides of the interface of deformation and mechanical stresses should be continuous. Therefore, detailed answer to the issue concerning deformations and stresses in the system should be due to the equations (1,2) with clear view of the tensor C_{ijkl} due (3) and the values of elasticity coefficients due Table 1.

2.2. Mathematical model

2.2.1.Equation. For isotropic elastic environment the metric representation of tensor has the form [10]:

$$C_{ijkl}^{isotr}=\begin{pmatrix}\lambda+2\mu & \lambda & \lambda & 0 & 0 & 0\\ \lambda & \lambda+2\mu & \lambda & 0 & 0 & 0\\ \lambda & \lambda & \lambda+2\mu & 0 & 0 & 0\\ 0 & 0 & 0 & \mu & 0 & 0\\ 0 & 0 & 0 & 0 & \mu & 0\\ 0 & 0 & 0 & 0 & 0 & \mu\end{pmatrix}, \qquad (4)$$

where λ and μ – Lame parameters while Hooke's law is :

$$p_{ij}=\lambda\delta_{ij}div\vec{u}+2\mu u_{ij}. \qquad (5)$$

The structure of the matrix (3) and (4) is the same that is why for assessing of elastic constants tensor deviations for GaAs and InAs from tensor for isotropic environment we will choose the following value:

$$\delta=\left|\frac{C_{12}+C_{44}-C_{11}}{C_{11}}\right|,$$

which is zero for isotropic environment. Due to Table 1 it is followed that for InAs $\delta\approx0.02$, and for GaAs $\delta\approx0.05$. These values are somewhat smaller than the non-complementarity lattices and significantly less than the relative deviation of the elastic modules of its crystals.

So we make the third step: the deformations in QD-substrate we will describe with the equations of elasticity theory for two isotropic environments. Then the equation (1, 2) takes the form of:

$$(\lambda+\mu)\vec{\nabla}div\vec{u}+\mu\Delta\vec{u}=0, \qquad (6)$$

where Lame parameters are expressed through the matrix coefficients of elastic constants as $\lambda=C_{12}$, $\mu=C_{44}$ on the both parts of the QD-substrate system and its value can be found in the Table 1.

2.2.2. Boundary conditions. As Lame parameters in the equation (6) are piecewise continuous, and their solutions describe continuous deformations than the terms of their will build according to the fact that in a perfect defect-free quantum dot electrons move in a continuous crystalline potential. It is necessary that GaAs and InAs atomic planes continuously without spaces pass one another. This condition is met if the atomic planes of InAs and GaAs crystal lattice are oriented in the same crystallographic direction and deformed without altering the symmetry.

Figure 2 a) it is shown the fragment of QD-substrate geometric contact where the atomic planes [010] in QD and in the substrate (see the figure 1) are shown as the vertical lines. Here due to our choice are stitched only average atomic planes. On the figure 2 b) is shown the same fragment where formally and without gaps all atomic planes are stitched. The average planes of both environments haven't been displaced and the displacement of other planes is increasing together with its distance from the axis of the system. Under such terms the mechanical stresses will increase around the interface. At some distance from the axis of the system they can reach such values when it is energetically favorable to pass the part of atomic plane near the interface. It can lead to structural defect – figure 2c).

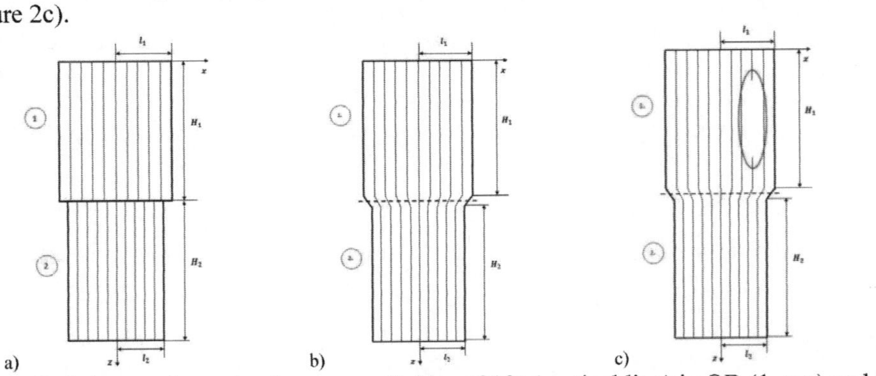

Figure 2. Scheme of atomic planes crosslinking [010] (vertical line) in QD (1, top) and in the substrate (2, below) while its contact in the plane $z=0$: a) geometric contact; b) stitching starting from the central plane; c) stitching of atomic planes with the passing of one of it.

Based on these considerations, we will construct a crosslinking conditions at the interface. The first follows from the z-component conditional continuity of vector deformation under normal passing through the interface's plane along each atomic plane [010] with the same number $n=0,\pm1,\ldots$ that is counted from the average atomic plane in figure 2 a). A similar condition is fair along the atomic planes as well [100]. So $u_z^{(1)}(na_1,ka_1,z=+0)=u_z^{(2)}(na_2,ka_2,z=-0)$ where $k=0,\pm1,\ldots$. If the axis of coordinate system Oz combines with an average atomic plane, axis Ox, Oy direct due to the figure 1, then for the $u_z^{(1,2)}(x,y,z)$ on the verge of a square QD and substrate obtain the following homogeneous boundary conditions:

$$u_z^{(1)}(x,y,0)=u_z^{(2)}(\alpha x,\alpha y,0), \tag{7}$$

where the coordinates $x,y\in[-l_1,l_1]$ are within the QD foundations and multiplier $\alpha=l_2/l_1$ in the arguments of deformations in substrate appears due to the fact that for the same number of atomic planes N in QD and substrate their lateral dimensions $l_{1,2}=Na_{1,2}$ will be different due to different lattice periods $a_1>a_2$,due to the Table 1.

The second and third conditions set continuity of nuclear axes of which are formed the atomic planes in QD and in the substrate while the normal passing through the interface along each of these axes. So we get the following correlation :

$$na_1+u_x^{(1)}(na_1,ka_1,z=+0)=na_2+u_x^{(2)}(na_2,ka_2,z=-0),$$

$$ka_1+u_y^{(1)}(na_1,ka_1,z=+0)=ka_2+u_y^{(2)}(na_2,ka_2,z=-0).$$

If now to move to continuous coordinates we obtain the following inhomogeneous boundary conditions for transverse vector components of deformation in a plane of interface:

$$x + u_x^{(1)}(x, y, z = +0) = \alpha x + u_x^{(2)}(\alpha x, \alpha y, z = -0),$$
$$y + u_y^{(1)}(x, y, z = +0) = \alpha y + u_y^{(2)}(\alpha x, \alpha y, z = -0). \quad (8)$$

The outside sides of QD and the substrate will be assumed as free, so they just of homogeneous boundary conditions

$$p_{ij} n_j \big|_S = 0. \quad (9)$$

We believe that to the QD-substrate system there is not attached the external forces. But the boundary condition (8) is not uniform and it causes the appearance of deformations and stresses in the system with no complementary QD and substrate. It emerged as a result of the conditions of continuity of atomic planes in the system. Provided $a_1 = a_2$ we get $\alpha = 1$ and then conditions (7,8) indicate that the system has the homocrossing. That the substrate and QD grown on it has the same chemical composition or similar symmetry and steel lattice.

2.2.3. Solution of equations. In the QD-substrate system there is no internal sources for the emergence of deformations except the interface so one can introduce the vector displacement such as

$$\vec{u}(x, z) = \vec{\nabla} \psi(x, z), \quad (10)$$

where $\psi(x, y)$ is arbitrary scalar function. Substitute (10) into the equation (6) and obtain the equation of the third order $(\lambda + 2\mu)\vec{\nabla}\Delta\psi(x, z) = 0$. Its solution as a single integral is easy to be found $\Delta\psi(x, z) = \vec{C}$, a solution of this equation is the sum of the general solution of homogeneous equation and the particular solution of inhomogeneous equation.

Since the boundary conditions for ψ-functions arising from (7-10) have the form of conditions for the second derivatives so heterogeneous solution will not give a contribution to the full solution of equations. So put constant $C = 0$. Get the Laplace equation $\Delta\psi(x, z) = 0$.

Using the Lame's equation (6) and above assumptions formulated the components of vector displacement can be found in both environments.

$$u_x^{(1)}(x, z) = \sum_n \tilde{D}_n^{(1)}\left(-k_n^{(1)}\right) \operatorname{sh} k_n^{(1)} z \sin k_n^{(1)} x, \quad (11a)$$

$$u_z^{(1)}(x, z) = \sum_n \tilde{D}_n^{(1)} k_n^{(1)} \operatorname{ch} k_n^{(1)} z \cos k_n^{(1)} x, \quad (11b)$$

$$u_x^{(2)}(x, z) = \sum_n \tilde{D}_n^{(2)}\left(-k_n^{(2)}\right) \times \left(\operatorname{sh} k_n^{(2)} z - \operatorname{th} k_n^{(2)} H_2 \operatorname{ch} k_n^{(2)} z\right) \sin k_n^{(2)} x, \quad (11c)$$

$$u_z^{(2)}(x, z) = \sum_n \tilde{D}_n^{(2)} k_n^{(2)} \times \left(\operatorname{ch} k_n^{(2)} z - \operatorname{th} k_n^{(2)} H_2 \operatorname{sh} k_n^{(2)} z\right) \cos k_n^{(2)} x. \quad (11d)$$

where $u_{x,z}^{(1,2)}(x, z)$ – the components of vector displacement in both environments.

Due to the Hooke's law the components of mechanical stresses will be written through the deformation:

$$p_{xx}^{(1)}(x, z) = C_{11}^{(1)} u_{xx}^{(1)} + C_{12}^{(1)} u_{zz}^{(1)}, \quad (12a)$$
$$p_{zz}^{(1)}(x, z) = C_{11}^{(1)} u_{zz}^{(1)} + C_{12}^{(1)} u_{xx}^{(1)}, \quad (12b)$$
$$p_{xx}^{(2)}(x, z) = C_{11}^{(2)} u_{xx}^{(2)} + C_{12}^{(2)} u_{zz}^{(2)}, \quad (12c)$$
$$p_{zz}^{(2)}(x, z) = C_{11}^{(2)} u_{zz}^{(2)} + C_{12}^{(2)} u_{xx}^{(2)}, \quad (12d)$$

where $u_{xx}^{(1,2)} = \dfrac{\partial u_x^{(1,2)}}{\partial x}$, $u_{zz}^{(1,2)} = \dfrac{\partial u_z^{(1,2)}}{\partial z}$ – the deformation of components.

So we have got the deformations and stresses in the model of QD-substrate that allows us to assess of heir depending on the linear dimensions of quantum dots on the interface.

3. Results and discussion

3.1. Numerical calculations. The analytical estimates and calculations are supplemented with the numerical calculations in the work. We are considering the model of QD-substrate in which the lateral dimensions of substrate are much larger than the QD's dimensions which has the triangular cross section. In the longitudinal direction the system is infinite so the two-dimensional case is considered – figure 3.

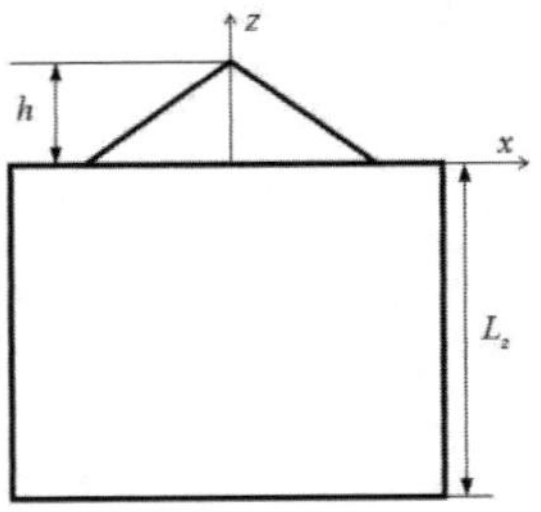

Figure 3. Model quantum dots on a substrate

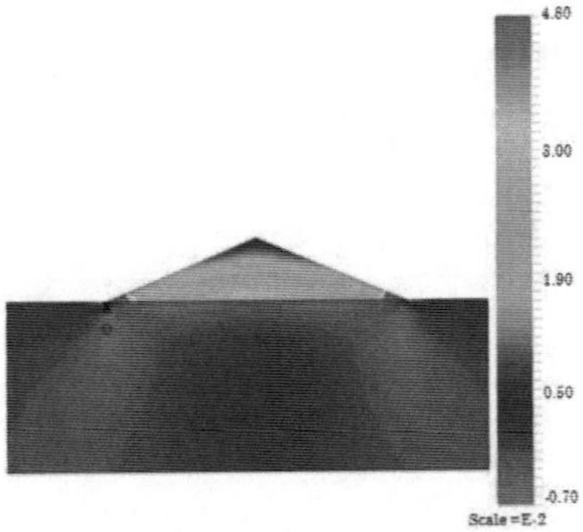

Figure 4. z -component of mechanical stresses

Equation (6) for the deformation together with the boundary conditions on the interface and the outside of the system (7-9) are solved numerically using FlexPDE.

In figure 4 it is shown the z -component of mechanical stresses. We emphasize that within the tops of the foundations the mechanical tensions reach its maximum value. In turn, this will be the most likely QD area for the appearance of dislocations.

4.Conclusion

1. It has been established at the ideal stitching of atomic QD planes and substrates the only source of mechanical stresses and deformations is a mismatch of lattice period.
2. Analytical estimates and numerical calculations show that the maximum stresses in the system occur at the edges of QD-substrate interface, so there will be created conditions for the emergence of defects.
3.The maximum tension of elastic modulus QD order scans occur when its transverse size up to several tens of periods in the event that non-period lattice gratings is about 7%

References

[1] Bimberg D 2008 *Semiconductor Nanostructures* (Springer) p 369
[2] Barnham K and Vvedensky D 2008 *Low-dimensional semiconductor structures* (Cambridge) p 408
[3] Ratsch C 1993 *Surf. Sci.* **293** 123.
[4] Semenov A N, Meltser B Ya, Solov'ev V A, Komissarova T A, Sitnikova A A, Kirylenko D A, Nadtochyi A M, Popova T V, Kop'ev P S and Ivanov S V 2011 *Phys. and Tech. of Sem.* **10** 1379.
[5] Baranov A N, Voronina T I, Gorelenok A A, Lagunova T S, Litvak A M, Sipovska M A, Staroceltseva S P, Tihomirova V A, Sherstnev V V 1992 *Phys. and Tech. of Sem.* **9** 1612.
[6] 1998 *Phys. and Tech. of Semicond.* **4** 385.
Ledentsov N N, Ustinov V M, Shchukin V A, Kop'ev P S, Alferov Zh I, Bimberg D 1998 *Phys. and Tech. of Sem.* **4** 385
[7] Müller P, Kern R 2000 *Surf. Sci.* **457** 229.
[8] Samsonov V M, Karacharov V I 2011 *Vestnyk TGU* **14** 14.
[9] Woong Lee, Jae-Min Myoung, Yo-Han Yoo, Hyunho Shin 2004 *Sol. St. Com.* **132** 135.
[10] Landau L D, Lifshitz E M 1987 *Theory of Elasticity Vol. 7* (Moscow: Science) p 248

Electro-optical properties of the metal oxide-carbon thin film system of CdO-LCC

A V Kokshina[1], A V Smirnov[1], A G Razina[1]

[1]Department of Applied Physics and Nanotechnology, The Chuvash State University, Cheboksary 428015, Russia

Abstract. This article presents the results of a study electrical and optical properties of the thin film system of CdO-LCC. Cadmium oxide films were obtained by method of thermal oxidation. CdO-LCC thin film system was produced by applying on a CdO film a linear chain carbon film in thickness of 100 nm using the ion-plasma method, after which the obtained system was annealed. The studies showed that the obtained CdO-LCC films are quite transparent in the visible region; it has polycrystalline structure, thickness around 300 nm, the band gap to 2.3 eV. The obtained thin film system has photosensitive properties.

1. Introduction

One of the main features of modern high technology is the desire to create and use a new thin film materials, have the ability to actively respond to changing conditions or external influence. Currently the attention of many researchers aimed at studying various photoconductive compounds. This is due to the possibility of the practical application of these compounds as photoresistors, various sensors, solar cells. Photoconduction - increase in conductivity when exposed to light. As is known, photoconductive properties often exhibit cadmium compounds, tellurium, selenium. Binary compounds of these elements is quite well understood. To one of such materials include thin film metal oxide-carbon system comprising carbon various allotropic modifications. However, at present insufficiently known issue is the interaction of carbon in sp-1 state (linear-chain carbon, LCC) with oxides of metals not forming a chemical compound with carbon. In this respect metal oxides thin films which have unique properties such us high transparency in the visible regions and high conductivity, are of outstanding interest. Transparent conductive oxides are used as electrodes in the liquid crystal display, photo diodes, solar cells and other devices. One of them is the cadmium oxide (CdO). CdO is an AIIBVI semiconductor with the band gap of 2.1−3.3 eV, generally with n-type conductivity. CdO thin films have unique properties such us high transparency in the visible regions and high conductivity. It is used in photodiodes, solar batteries, phototransistors and gas sensors [1-3].

Linear chain carbon films create a two-dimensional-ordered structure consisting of carbon chains banded by sp1-hybridization. Linear chain carbon is synthesized using the ion-plasma method in the form of a film. Carbon chains are oriented vertical to the substrate surface and are organized in a close-packed hexagonal lattice. The distance between carbon chains is about 5 Angstroms. The films have one-dimensional conductivity along carbon chains and, according to our research, enhance electrooptical properties of many alloys and elements.

This article presents the results of a study of the influence of films linear-chain carbon (LCC) on the properties of the thin film of CdO.

2. Methodology

Linear chain carbon is synthesized using the ion-plasma method in the form of a film, the model of which is presented in Figure 1.

Cadmium oxide films were obtained by method of thermal oxidation which does not require any special conditions and much time to product an oxide film. In order to select optimum conditions for CdO thin film synthesis Cd films were applied on a silicon or glass substrate by evaporation. Then the films were annealed in air in a furnace at different temperatures (from 250 to 400 ^{0}C) and for different time intervals (20-40 min). CdO-LCC thin film system was produced by applying on a CdO film a linear chain carbon film in thickness of 100-200 nm using the ion-plasma method, after which the obtained system was annealed [4].

For investigation of CdO-LCC thin films structural properties the Raman spectroscopy method was used. The surface's topology and the thickness of the film was investigated by scanning probe microscopy and scanning electron microscopy. Optical properties were investigated by spectrophotometer Lambda 25 in the spectral range 200 - 1100 nm. Electrical properties were investigated using a digital multimeter Source Meter 2400.

3. Results and discussion

Figure 2 illustrates the Raman spectra of CdO-LCC system. From an analysis of the graphs, it can be concluded that the CdO-LCC system going Raman shift wavenumber upwards compared to pure oxide [1]. The presence of G- and D- bands on graph indicates obtaining a composite material in which cadmium oxide and carbon are present.

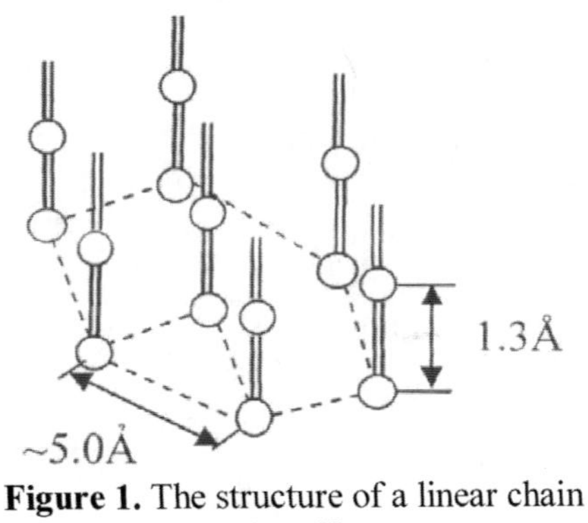

Figure 1. The structure of a linear chain carbon film.

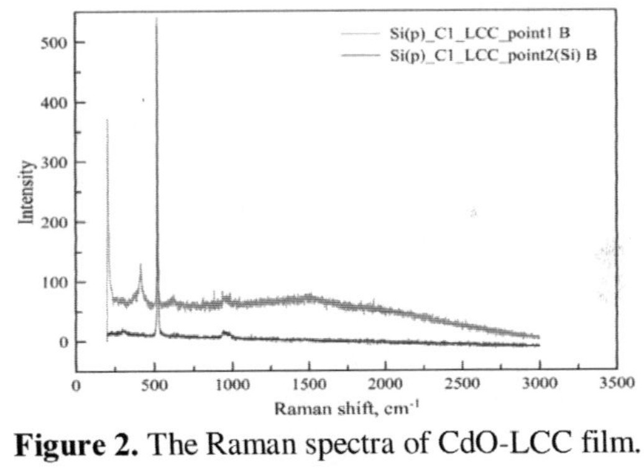

Figure 2. The Raman spectra of CdO-LCC film.

Topology of surfaces were obtained with a scanning probe microscope SolverNext in an atomic force mode. After annealing the film of cadmium surface structure changes, clusters increase in size, the diameter of the cluster is 150-200 nm. CdO film becomes polycrystalline structure. Film thickness was determined by using SPM steps in an atomic force mode, totaled about 200 nm. Figure 3 shows scanning electron micrographs (SEM) of the film CdO-LCC. Implementation in CdO linear-chain carbon also leads to change and streamline the structure of the film surface. The thickness of the resulting CdO-LCC system is approximately 300 nm.

The optical studies (Figure 4) showed that the obtained CdO-LCC films are quite transparent in the visible region, which indicates the perspectives of the application of them as transparent conducting coatings. The obtained cadmium oxide film has photosensitive properties demonstrated by films applied on both silicon and glass substrates. The linear chain carbon film increases photoresponse of the system under research. Figure 5 shows photoresponse of system Si-CdO-LCC.

Figure 3. SEM micrographs of CdO-LCC system.

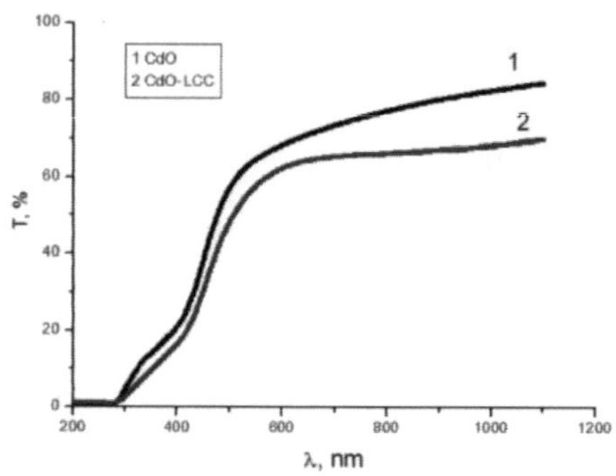

Figure 4. The transmission spectra of films of CdO and CdO-LCC.

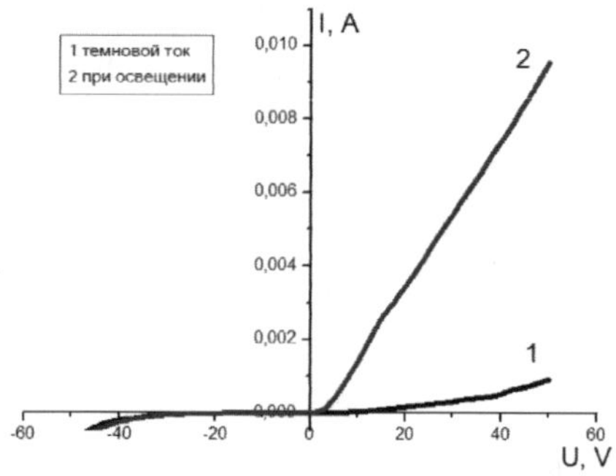

Figure 5. Photoresponse of system Si-CdO-LCC.

The CdO film optical band gap was estimated using the method described in [5] and amounted to 2.5 eV (Figure 6). The optical band gap of the CdO-LCC system amounted to 2.3 eV (Figure 7). In other words, implantation of linear chain carbon into the cadmium oxide film decreases its band gap.

Figure 6. Determination of the width of the bandgap of the film CdO.

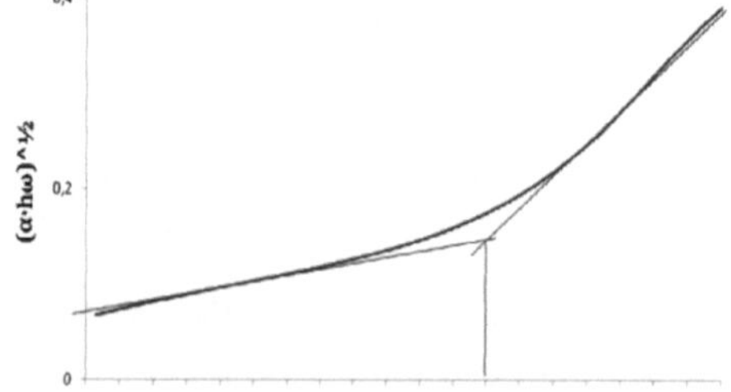

Figure 7. Determination of the width of the bandgap of the film CdO-LCC.

The obtained systems with high transparency and photosensitive properties may be used as transparent electrode in photo cells, phototransistors and optical heaters.

Acknowledgments
The research has been carried out as a part of the scholarship of the President of the Russian Federation for young scientists and postgraduate students СП-2698.2015.1

References
[1] Islam M M, Islam M R, Podder J 2008 *Journal of Bangladesh Academy of Sciences* **32** 1 pp 97-105
[2] Bagheri A, Khatibani, Hallaj Z A, Rozati S M 2015 *The European Physical Journal Plus* **130** 254
[3] Mahaboob Beevi M 2010 *International Journal of Chemical Engineering and Applications* **1**(2) pp 151-154
[4] Kokshina A V *et al.* 2015 *Alternative Energy and Ecology* **19**(183) pp 112-117
[5] Vasilevsky A M *et al.* 2011 Optical and physical methods of research: guidelines for laboratory work

Optical transmission of strained GaN/sapphire structures

S Yu Kurin[1], M V Puzyk[2,3], I A Ermakov[2], A A Antipov[1], I. S. Barash[1], A D Roenkov[1], V V Ratnikov[4], A S Usikov[2,5], B P Papchenko[2], H Helava[5], Yu N Makarov[1,5] and A E Chernyakov[6]

[1]Nitride Crystals Group Ltd., pr. Engel'sa 27, St. Petersburg 194156, Russia
[2]University ITMO, Kronverkskiy pr. 49, St. Petersburg 197101, Russia
[3]Herzen University, Nab. r. Moyki 48, St. Petersburg 194186, Russia
[4]Ioffe Physical Technical Institute, Polytekhnicheskaya 26, St. Petersburg 194021, Russia
[5]Nitride Crystals Inc., 181 E Industry Court, Suite B, Deer Park, NY 11729, USA,
[6]Submicron Heterostructures for Microelectronics Research & Engineering Center, RAS, 26 Polytekhnicheskaya Str., St. Petersburg 194021, Russia

Abstract. In this work we correlated transmission spectra of GaN layers grown on sapphire substrates by hydride vapour phase epitaxy with biaxial stress measured in the layers. It was observed that the sign of stress in the GaN layer is changed by Si doping and growth conditions. Transmission curves are shifted relative to each other depending on the stress in the layer. The cut-off wavelength of the transmission curves has a tendency to shift near parallel to a shorter wavelength range when the GaN layer is under the compression biaxial stress. When the GaN layer is under the tensile biaxial stress the cut off wavelength has a tendency to shift near parallel to a longer wavelength range).

1. Introduction

One of the challenges remaining in the epitaxial growth of III-N materials is the absence of affordable lattice-matched bulk GaN substrate. Sapphire and SiC substrates that are widely used in nitride-based technology are lattice mismatched substrates resulting in threading defects and biaxial strain lowering performance and shortening lifetime of the device structure. A possibility to control the strain at the heteroepitaxial growth of GaN-based structures on foreign substrates is critically important for optimization of the growth conditions to grow strain-less and defect-less layers and structures for understanding the electrical and optical properties of grown structures and, therefore, is of great practical interest. Optical characterizations provide a convenient way to examine the different properties of semiconductors. In particular, measuring the transmittance coefficient of GaN layers grown on sapphire substrates may give information about the strain in the material connected with a specific of the growth process.

In this work we studied transmission spectra of GaN layers grown by hydride vapour phase epitaxy (HVPE) on c-plane sapphire substrates in a horizontal quartz reactor [1]. Typically, HVPE has high growth rate in a range of 0.3-1.2 μm/min. Because of that, HVPE is commonly used for fabricating of GaN quasi-bulk material and GaN templates to use as substrates for III-N device structure growth. The wide range of the growth rate influences on the impurities incorporation that can affect the strain in the grown layers [2]. We correlated transmission spectra of GaN layers grown on sapphire substrates with the biaxial stress measured in the layers. The AlGaN layers (tens of nanometers) were used as thin intermediate layers prior to the GaN layer growth.

2. Experimental details

Several 5-7 μm thick GaN layers were grown by hydride vapour phase epitaxy (HVPE) on 2-inch c-plane sapphire substrates in a horizontal quartz reactor [1]. The growth process was performed at 1040°C and atmospheric pressure with argon as a carrier gas. Ammonia (NH3) and hydrogen chloride (HCl) were used as precursors. For GaN growth, HCl was passed over Ga source. To grow the AlGaN alloy, HCl was passed separately over the Ga and Al sources. Si doping was realized by injection of free HCl gas through a separate quartz tube into a growth zone. Reaction of the HCl gas with the quartz walls of the injection tube and the reactor resulted in the n-type doping by Si. The growth procedure included in-situ sapphire substrate treatment followed by a composite AlGaN buffer layer and GaN layer growth. The growth conditions the AlGaN buffer and the GaN layer were change at the undoped and Si-doped GaN layers growth. Capacitance-voltage (C-V) mercury probe measurements at 1 MHz revealed Nd-Na concentration of (1-4) 10^{16} cm^{-3} and (1-3) 10^{18} cm^{-3} for undoped and Si-doped GaN layers, respectively, depending on growth conditions. All layers were crack free. The biaxial stress in the layers grown at different initial conditions was evaluated based on XRD measurements using Stoney's formula [3].

For transmission spectroscopy we used a commercial spectrophotometer that was capable of recording spectra in the visible range as well as in the near infrared and UV. To compensate for the complicated intensity distribution of the light source the spectrometer did not measure absolute values but compared the signal from the sample to a reference beam. Additionally, a baseline was recorded prior to the actual measurements to calibrate the instrument.

3. Results and discussion

Figure 1 shows X-ray ω-scan rocking curve width (full width at half maximum, FWHM) for symmetric ((00.2) GaN) and asymmetric ((10.2) GaN) reflections with the GaN layers thickness. The FWHM is a figure of merit for crystal quality and usually decreases with the layer thickness (in a wider range of thickness) due to the threading dislocations annihilation for GaN layers grown on c-plane sapphire [4]. No clear dependency of the FWHM values with thickness can be observed in figure 1 and we may consider that difference in layer thickness of GaN samples studies is too small to have an effect on the total stress in the layer. Note also that the (00.2)FWHM and the (10.2)FWHM values for undoped GaN layers are closer to each other than that of Si-doped GaN layers. This is indicative that Si doping disturbs the GaN crystalline cell. Indeed, undoped GaN layers have larger c- and lower a- parameters that Si-doped GaN layers [5].

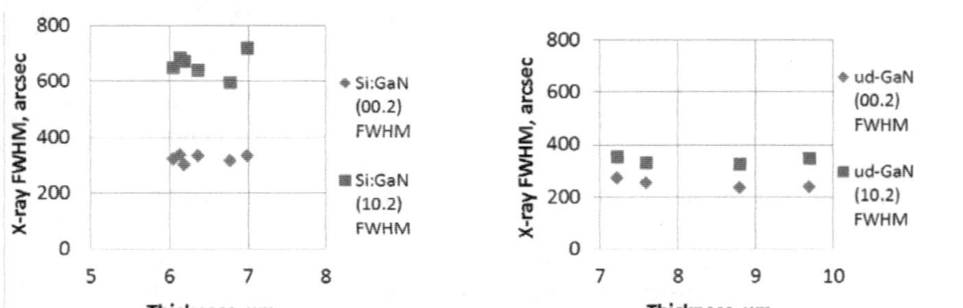

Figure 1. Thickness dependence of X-ray rocking curve width in Si doped GaN layers (a) and undoped GaN layers (b).

Figure 2 shows the radius of curvature evaluated in the same Si-doped undoped GaN layers grown on 2-inch sapphire substrates. All studied undoped GaN samples had concave shape (negative radius of curvature) although growth conditions of the composite AlGaN buffer layer and the GaN layer were different between the samples. The concave shape is attributed to compressive biaxial stress in the sample.

On the other hand, Si-doped samples that were grown in the similar conditions of the AlGaN buffer layer growth had both concave shape (positive radius of curvature) and convex shape (nrgative

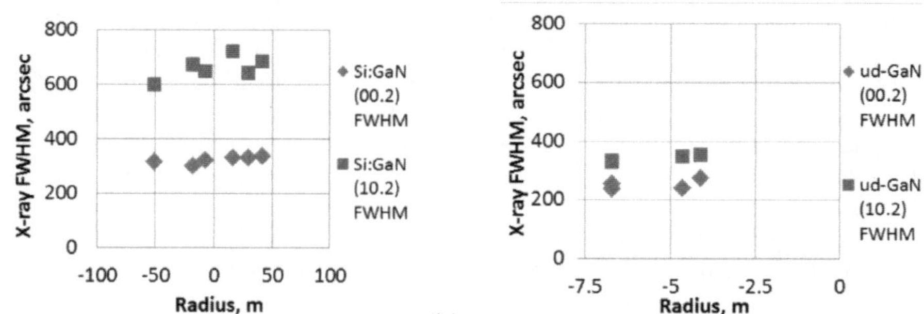

(a) (b)

Figure 2. Radius of curvature evaluated in Si doped GaN layers (a) and undoped GaN layers (b).

radius of curvature). The convex shape is attributed to tensile stress in the sample. Thus, depending on the growth conditions Si-doped GaN layers can be grown on 2-inch sapphire substrates by HVPE having different shape.

Figure 3 shows biaxial stress in Si-doped and undoped GaN samples evaluated based on XRD measurements. It was observed that the stress can reverse sign from σ+ (concave shape of the sample, tensile stress) to σ- (convex shape of the wafer, compressed stress) in Si doped GaN layers depending on the growth conditions of the AlGaN buffer layer mainly. All undoped GaN layers were under compressive stress (σ-) and have convex shape. The sign of stress in the undoped GaN layers did not change. Variation of the layer thicknesses did not influence the X-ray rocking curve width that indicates that the stress changing can be associated with the doping and the growth conditions only.

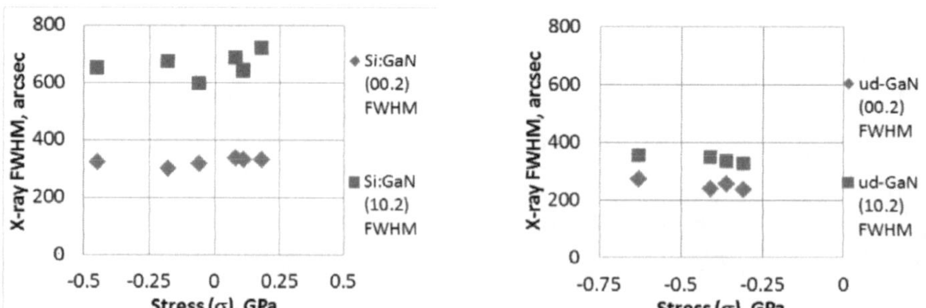

(a) (b)

Figure 3. Biaxial stress evaluated in Si doped GaN layers (a) and undoped GaN layers (b).

The transmission curves of undoped and Si-doped GaN/sapphire samples having different biaxial stress are given in figure 4. The sapphire substrate did not absorb light itself in the visible, near IR and UV spectral ranges. The transmission curves show that for high energies there is no transmission because all the light is absorbed. However, for low energies there are no appropriate electronic transitions so transmission is very high in this range. There is a relatively sharp delimitation (cut-off wavelength) between the areas of high and low absorption. A position of the cut-off wavelength corresponds to the band gap of GaN and strain in the layers grown.

The cut-off wavelength has a tendency to shift near parallel to a shorter wavelength range when the GaN layer is under compression biaxial stress (σ-). The shorter cut-off wavelength means lager effective energy gap associated with the strain. When the GaN layer is under tensile biaxial stress ($\sigma+$) the cut off wavelength has a tendency to shift near parallel to a longer wavelength range. Note that the parallel shifting of the cut-off wavelength in the transmission spectra was usually observed at a fine turning of the growth conditions when the only or a few parameters change a little. If growth conditions disturb crystalline quality (high dislocation density, high background impurity concentration) than the shift of the cut-off wavelength will not be parallel without sharp delimitation between the areas of high and low absorption and no direct correlation with the stress is observed.

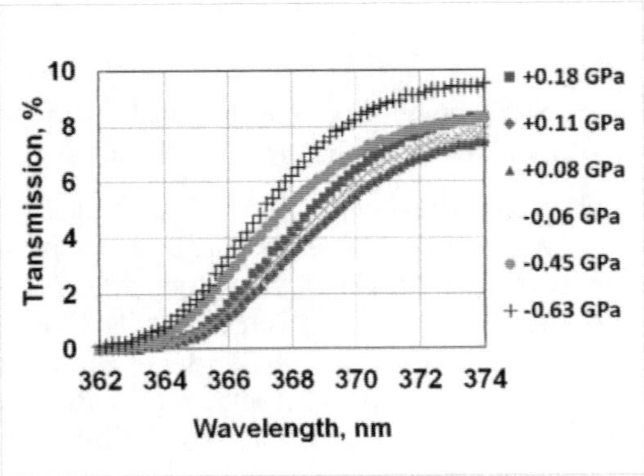

Figure 4. Transmission curves of GaN/sapphire samples. The legend shows values of the biaxial stress measured in the particular sample.

4. Conclusions

Thus, it was observed that shape (convex, concave), radius of curvature and stress (sign of the stress) the sign of stress of the GaN layers grown by HVPE on 2-inch sapphire substrates are changed by Si doping and growth conditions. Transmission curves are shifted relative to each other depending on the stress in the layer. The cut-off wavelength has a tendency to shift near parallel to a shorter wavelength range when the GaN layer is under compression biaxial stress ($\sigma-$).When the GaN layer is under tensile biaxial stress ($\sigma+$) the cut off wavelength has a tendency to shift near parallel to a longer wavelength range. If growth conditions disturb crystalline quality (high dislocation density, high background impurity concentration) than the shift of the cut-off wavelength will not be parallel without sharp delimitation between the areas of high and low absorption and no direct correlation with the stress is observed.

Acknowledgements

Work at University ITMO was supported by the Ministry of Education and Science of Russian Federation (grant agreement 14.575.21.0054, unique identifier of research activities is RFMEFI57514X0054).

References

[1] Kurin S, Antipov A, Barash I, Roenkov A, Usikov A, Helava H, Ratnikov V, Shmidt N, Sakharov A, Tarasov S, Menkovich E, Lamkin I, Papchenko B and Makarov Yu 2014 *Phys. Status Solidi* C**11** 813-816.

[2] A. Usikov, 1, O. Kovalenkov, V. Soukhoveev, V. Ivantsov, A. Syrkin, V. Dmitriev, A. Yu. Nikiforov, S. G. Sundaresan, S. J. Jeliazkov, and A. V. Davydov. phys. stat. sol. (c) 5, No. 6, 1829–1831 (2008).

[3] Stoney G G 1909 *Proc. of Royal Society of London A***82** 172-175

[4] Wagner V, Parillaud A, Buhlmann H J, Illegems M, Gradecak S, Stadelmann P, Riemann T and Christen J 2002 *J. Appl. Phys.* **92** 1307.

[5] Usikov A, Kovalenkov O V, Mastro M M, Tsvetkov D V, Pechnikov A I , Soukhoveev V A, Shapovalova Y V and Gainer G H 2002 *Mat. Res. Soc. Symp. Proc.* **743** L3.41.

Journal of Physics: Conference Series **741** (2016) 012049 doi:10.1088/1742-6596/741/1/012049

Photoelectrochemical corrosion of GaN-based p-n structures

A D Fomichev[1], S Yu Kurin[2], I A Ermakov1, M V Puzyk[1,3], A S Usikov[1,4], H Helava[4], A Nikiforov[5], B P Papchenko[1], Yu N Makarov[2,4] and A E Chernyakov[6]

[1]University ITMO, Kronverkskiy pr. 49, St. Petersburg 197101, Russia
[2]Nitride Crystals Group Ltd., pr. Engel'sa 27, St. Petersburg 194156, Russia
[3]Herzen University, Nab. r. Moyki 48, St. Petersburg 194186, Russia
[4]Nitride Crystals Inc., 181 E Industry Court, Suite B, Deer Park, NY 11729, USA,
[5]Boston University, Photonics Center, 8 St. Mary's St., Boston, MA 02215, USA
[6]Submicron Heterostructures for Microelectronics Research & Engineering Center, RAS, 26 Polytekhnicheskaya Str., St. Petersburg 194021, Russia

Abstract. Direct water photoelectrolysis using III-N materials is a promising way for hydrogen production. GaN/AlGaN based p-n structures were used in a photoelectrochemical process to investigate the material etching (corrosion) in an electrolyte. At the beginning, the corrosion performs through the top p-type layers via channels associated with threading defects and can penetrate deep into the structure. Then, the corrosion process occurs in lateral direction in n-type layers forming voids and cavities in the structure. The lateral etching is due to net positive charges at the AlGaN/GaN interfaces arising because of spontaneous and piezoelectric polarization in the structure and positively charged ionized donors in the space charge region of the p-n junction.

1. Introduction

Hydrogen is considered to be a candidate as the substantial energy carrier. Direct water photoelectrolysis is a promising way for hydrogen production. In this process, a semiconductor material immersed in an aqueous solution of electrolyte allows decomposition of water into H_2 and O_2 gases by solar irradiation of its surface. The required energy to split water molecule is generated by sunlight absorption in the semiconductor. The solar water splitting process may be spontaneous under illumination if electrochemical redox potentials of oxygen evolution reaction (OER) and hydrogen evolution reaction (HER) in an electrolyte are bracketed by the energy gap of the semiconductor. This is fulfilled for GaN and most of InGaN materials [1, 2].

Not all of the electron-hole pairs generated in the semiconductor under illumination in an electrolyte can participate in OER and (or) HER. Carrier recombination at the surface and in the bulk of the layer reduces the gas generation rate. The corrosion or etching of the electrode material is another process having a negative effect on the gas evolution reaction [3]. The corrosion is an issue for n-type materials as photoelectrodes because upward band-bending at the photoelectrode/electrolyte interface promotes photo-generated holes to transfer to the electrolyte. The holes at the interface not only oxidize water (OER) but can oxidize and etch the photoelectrode. In p-type materials the band-bending promotes photo-generated electrons to transfer to the electrolyte for the HER process that almost leaves the surface of the photo-electrode chemically intact.

Journal of Physics: Conference Series **741** (2016) 012049 doi:10.1088/1742-6596/741/1/012049

In this work we use GaN/AlGaN p-n structures as working electrodes with a p-type GaN layer on the surface to investigate the material corrosion of the photoelectrochemical process. Several 8-10 μm-thick AlGaN/GaN p-n structures were grown by chloride HVPE on c-plane 2-inch sapphire substrates.

2. Samples and experimental techniques

Several 8-10 μm-thick AlGaN/GaN p-n structures were grown by chloride HVPE on c-plane 2-inch sapphire substrates in a conventional horizontal-flow reactor. The basic structure included a 50-100 nm-thick GaN active region co-doped with Zn and Si to have emission at 420 nm [4], which was sandwiched between p- and n- AlxGa1-xN barriers (x ~ 0.05–0.12). A 0.5-1 μm-thick p-type GaN contact layer doped with Mg covered the structure. The Mg doping produces hole concentration up to 10^{18} cm^{-3} in the contact layer. Details of the structure growth and characterization can be found elsewhere [5]. Note that it was the p-type GaN layer surface that served as a photoelectrode (working electrode) in this work.

The 2-inch grown samples were immersed in a KOH aqua electrolyte (5.7 weight % of KOH, pH=14) under external electrical bias of +2.5 V and irradiation with a 150 W Xe lamp (AM 1.5 standard spectrum) having a concentration factor of 20x. As a counter electrode a 2-inch diameter Ni plate was used. The electrical contact to the sample (working electrode) was made through a metallic clamp with In pads. The structures were partly immersed in the electrolyte. A segment of the structure, which was used to hang the whole wafer by the clamp, was above the electrolyte to avoid a short circuit. The whole process of photoelectrolysis took about 6 minutes.

After experiments, the samples were cleaved and their surface morphology and luminescence properties were examined in both top-view and cross-sectional configurations by high-resolution scanning electron microscopy (SEM) and cathodoluminescence (CL) spectroscopy and monochromatic CL imaging, respectively.

3. Results and discussion

Figure1 shows SEM images of cross sectional views of the same sample taken at an unetched area on the surface and an area underwent the corrosion (etching). All layers of the structure can be distinguished. The structure total thickness is about 9.5 μm.

(a)
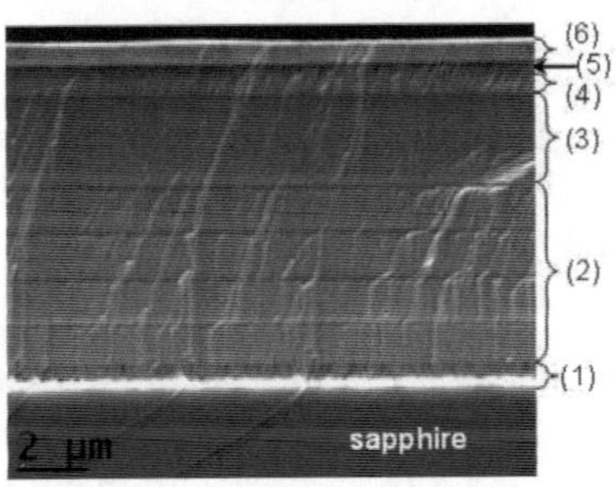

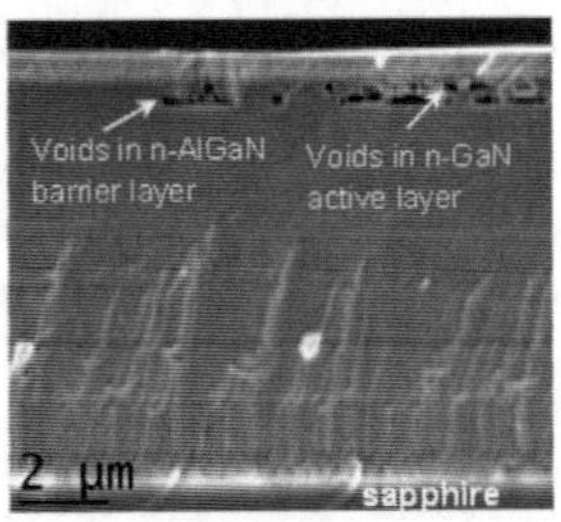

(b)

Figure 1. Cross-sectional SEM image of different areas of the AlGaN/GaN p-n structure after photoelectrolysis. (a) Unetched area on the surface. No peculiarities related to the photoelectrochemical etching are observed. (1) AlN/AlGaN buffer, ~0.65 μm; (2) AlGaN/AlGaN four pairs of stress control layers, 4.97 μm; (3) n-GaN:Si contact layer, ~2.5 μm; (4) n-AlGaN barrier layer, ~0.5 μm; (5) GaN:(Zn+Si) active layer, ~0.2 μm; (6) p-GaN contact +p-AlGaN barrier layers, ~0.66 μm. (b) an area underwent the corrosion (etching) on the surface.

Note that in the experiment a +2.5V bias was applied to the p-type top layer of the structure from external power source, i.e the structure was a photo anode. During the experiment the measured value of photocurrent was found to decrease from 4.3 mA to 2.2 mA (0.5- 0.25 mA/cm^2). The total thickness of the p-type layers is about 0.7 μm. P-type conductivity on the surface was confirmed by C-V measurements. Two rows of micro-voids in Figure 1b can be ascribed to the photoelecrochemical etching. The position of the voids in the stack of layers in the structure makes it possible to assume that these voids are formed in the GaN active layer and in the underlying n-AlGaN barrier layer. The micro-voids can spread laterally and overlap leading to formation of a large-size cavity.

Figure 2 shows the large-size cavity that formed after the photoelctrolysis process performed on another AlGaN/GaN p-n structure but having higher photocurrent of 5.6-9.7 mA (0.65- 1.1 mA/cm^2) under the same illumination. The large-size cavity may result in delamination of the p-layers from the structure. Figure 3 shows a scanning micrograph of surface morphology of the etched area of the sample in figure 1. Pits and large open cavities on the surface are attributed to the photoelectrochemical process. The pits density is about 4-7$\times$10^{8}cm^{-2} and it varies over the sample surface up to (1-2)$\times$10^9 cm^{-2} in the most defective areas near the microcracks. The CL spectra acquired from the sample surface have a weak emission at about 360 nm and a dominant broad peak centred around 420 nm as shown in figure 4a. The strongest luminescence was measured from the area illuminated by the Xe lamp (the etched area) where the density of non-radiative defects was likely decreased as a result of the photoelectrochemical etching. The coarser surface morphology in the etched area may also improve the light extraction efficiency. The dominant emission can be ascribed to the active layer located at a depth of 0.5-0.6 microns from the surface.

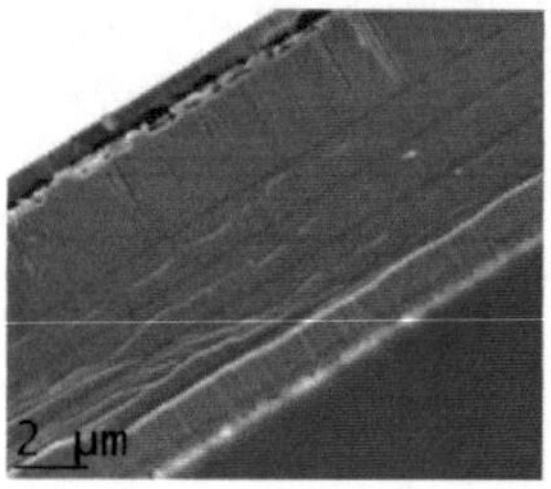

Figure 2. Cross-sectional SEM image of the etched area of another AlGaN/GaN p-n structure. Overlapped micro-voids created large-size cavities that may peel the p-type layer.

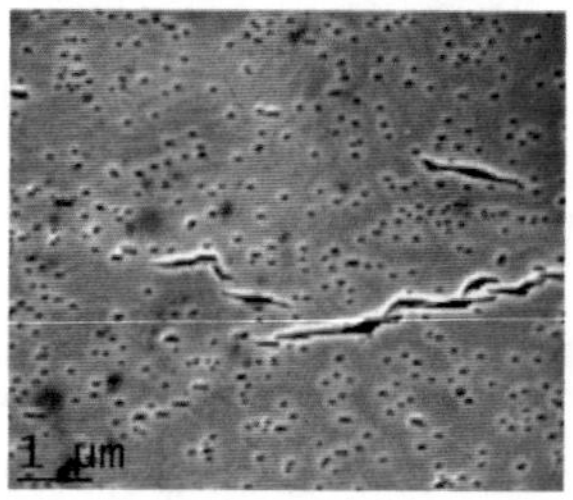

Figure 3. SEM image of surface morphology after the photoelectrochemical etching of the AlGaN/GaN p-n structure. Pits density is about 4-7 $\times$10^8 cm^{-2}.

Obviously, the photoelectrochemical etching follows paths of the photo-current flow. The current flow paths in III-N GaN/AlGaN or GaN/InGaN materials, especially at low current, are non-uniform and can be associated with shunt conductivity in an extended defect system of grain boundaries, threading dislocations, local regions with irregular alloy composition enriched by metallic atoms (Ga or In), and other extended defects [6].

The voids observed in the SEM images in figures 1-2 suggest that the process of their formation proceeds in two stages. During the first stage, the etching process starts via channels (or shunts) in the extended defect system of the structure penetrating the p-type layers all the way into the underlying n-type layers. The shunts could be threading defects, grain boundaries and micro-cracks attributed to mosaic structure of the structures as-grown. During the second stage, the etching on the n-type AlGaN barrier and GaN active layer proceeds in a lateral direction resulting in formation of voids and cavities underneath the p-type layers. Although the p-type layers of the structure remained unetched, the top surface exhibited a lot of pits.

The UV portion of the illumination light gets effectively absorbed in the p-GaN contact layer (at a depth of about 0.2-0.3 µm from the surface) leading to formation of excited electron–hole pairs. These pairs can be separated further by either the internal electric field of the space-charge region in the p-n junction of the structure when they drift towards it or by the specific band-bending at the p-GaN layer/electrolyte interface. The photo-generated carriers may participate in two competitive processes in the structure employed as an anode and immersed into an electrolyte under irradiation: (1) oxygen evolution reaction (OER) and (2) photo-etching of GaN and GaN-based materials. However, because of a too thick p-GaN layer (0.5-0.6 µm thick) the excited electron–hole pairs recombine mostly at the structure surface and in the p-layer before being separated by the p-n junction producing as a result a relatively low photocurrent. It appears that relatively low photocurrent flows non-uniformly through the structure selecting higher conductivity paths that are threading defects and micro cracks.

We speculate that photo-generated holes could assist in oxidative decomposition of p-GaN, primarily in the vicinity of the threading defects forming dangling bonds of Al^{3+} and Ga^{3+} atoms. Hydroxide ions (OH^-) from the electrolyte would then attack the Al^{3+} and Ga^{3+} dangling bonds and form hydroxides $Al(OH)_3$ and $Ga(OH)_3$ [7, 8]. Dissolution of the hydroxides would lead to formation of soluble Al_2O_3 and Ga_2O_3. Continuous formation and dissolution of the oxides in the solution would further simulate the etching process.

The etching process may move deeper into the layers via the threading defects. The Mg-doped p-type GaN and AlGaN layers have ionized Mg- acceptors and may screen partly the positive charge of the dangling bonds preventing the etching of the p-type layers but still allowing the etching to proceed along the threading defects. Indeed we did not observe any appreciable etching of the top p-layers. However, some micro-cracks and numerous pits associated with etching were observed on the top surface as can be seen in figure 3.

The step involving lateral etching in figures 1-2 can be explained on effects resulting in a positive charge at layer interfaces. It is know that spontaneous and piezoelectric polarization between AlGaN/GaN layers in the structure results in formation of a net positive charge at their interface [9]. Figure 4 shows a net positive charge at the AlGaN/GaN interface in the AlGaN layer caused by the sum of the net spontaneous polarization and piezoelectric polarization between the AlGaN and GaN layers in the structure growing along the c-direction [9]. Qπ, AlGaN includes the contribution of spontaneous and piezo-electric charges for the AlGaN layer and Qπ, GaN is the only spontaneous polarization for the GaN layer as it has been assumed to be relaxed due to a much larger thickness than that of the AlGaN layer.

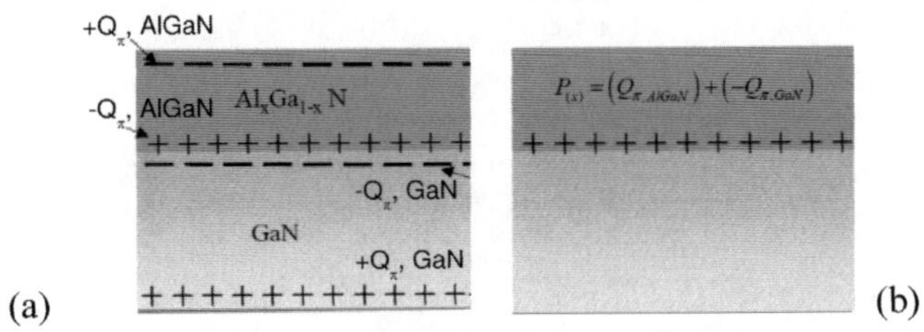

Figure 4. Polarization charges in an AlGaN/GaN structure. (a) Charge polarization in individual layers of the structure (b) a net positive charge at the AlGaN/GaN interface in the AlGaN layer [8].

The structures in this work had a stack of doped GaN and AlGaN layers at the top (i.e the n-GaN active layer was sandwiched between n- and p-AlGaN barrier layers) that influences the net positive charge compared to that depicted in figure 4. First of all, negatively charged ionized acceptors (Mg-) in the p-AlGaN layer may screen positive polarization charge at the p-AlGaN barrier layer/n-GaN active layer interface. In addition, a thin n-GaN active layer is not fully relaxed and may have additional positive piezoelectric polarization charge at the n-GaN active layer/n-AlGaN barrier layer interface.

Then, n-AlGaN barrier also has a positive piezoelectric polarization charge at the interface with n-GaN contact layer beneath (see figure 1a for reference to the structure design). Note also that all thin AlGaN layers having higher composition in the AlGaN/AlGaN four-pair structure, beneath the n-GaN contact layer, have a net positive polarization charge. However these layers are located deep in the structure, further away from the structure surface. Only two thin n-GaN active and n-AlGaN barrier layers are near the p-type layers and have positive piezoelectric polarization charge. An additional positive charge at the layer interfaces can be connected with positively charged ionized donors, which are Si+ presumably, in a space charge region of the p (AlGaN)-n (GaN) junction. This positive charge is located in the n-GaN active region too. Our separate experiments on photoelectrolysis with GaN layers only demonstrated porous-like etching of 5-6 μm thick n-GaN layers through the layer to the substrate whereas in p-n GaN structures having a p-GaN layer on the top the etching process was terminated at the p-n junction.

The net positive charge at the GaN/AlGaN n-n and p-n interfaces promotes OH^- ions from the electrolyte to form $Ga(OH)_3$ and (or) $Al(OH)_3$ hydroxides and the nitrogen vacancy sites and continue the etching process along the positive charge position forming lateral voids and large-size cavities. When the etching process via threading defects reaches a layer with positive net charge it changes the etching direction from vertical to lateral presumably because of a stronger charge at the layer interface than that in the channel associated with threading defects and dangling bonds.

4. Conclusions

The corrosion of GaN/AlGaN p-n structures performs through the top p-type layers via channels associated with threading defects and can penetrate deep into the structure. Then, the corrosion process occurs in lateral direction in n-type layers forming voids and cavities in the structure. The lateral etching is due to net positive charges at the AlGaN/GaN interfaces arising because of spontaneous and piezoelectric polarization in the structure and positively charged ionized donors in the space charge region of the p-n junction.

Acknowledgements
Work at University ITMO was supported by the Ministry of Education and Science of Russian Federation (grant agreement 14.575.21.0054, unique identifier of research activities is RFMEFI57514X0054).

References

[1] Ohkawa K, Ohara W, Uchida D, and Deura M 2013 *Japan. J. Appl. Phys.***52**08JH04.
[2] Aryal K, Pantha B N, Li J, LinJ Y and Jiang H X 2010 *Appl. Phys. Let.* **96** 052110- 1-3.
[3] Wolff N, Rapp M and Rotter T 2005 *Phys. Status Solidi* C2 990–993.
[4] Usikov A S, TsvetkovD V, MastroM A, PechnikovA I, SoukhoveevV A, ShapovalovaY V, Kovalenkov O V, Dmitriev V A, O'Meara B, Gurevich S A, Arakcheeva E M, Zakhgeim A and Helava H 2003 *Phys. Status Solidi* C**0** 2265-2269.
[5] Kurin S, Antipov A, Barash I, Roenkov A, Usikov A, Helava H, Ratnikov V, Shmidt N, Sakharov A, Tarasov S, Menkovich E, Lamkin I, Papchenko B and Makarov Yu 2014 *Phys. Status Solidi* C**11** 813-816.
[6] Shmidt N, Shabunina E, Usikov A, Chernyakov A, Kurin S, Helava H and Makarov Yu 2015 *Phys. Status Solidi* C**12** 1136-1139.
[7] Zhuang D and Edgar J H 2005 *Mat. Sci. and Eng. Rep.* **R48** 1-46.
[8] Macht L, Kelly J J, Weyher J L, Grzegorczyk A and Larsen P K 2005 *J. Cryst. Growth* **273** 347-356.
[9] Mishra U K, Parikh P and Wu Y F 2002 *Proc. IEEE* **90** 1022-103.

Etching of GaN layers at electrolysis under UV-irradiation

T K Zubenko[1], M V Puzyk[1, 2], V M Stozharov[1], I A Ermakov [2], D S Kovalev [2], S A Ivanova [2], A S Usikov[2,3], O S Medvedev[4], B P Papchenko[2], S Yu Kurin[5], A A Antipov[5] and A E Chernyakov[6]

[1]Herzen University, Nab.r.Moyki 48, St.Petersburg 194186, Russia
[2]University ITMO, Kronverkskiy pr. 49, St.Petersburg 197101, Russia
[3]Nitride Crystals Inc., 181 E Industry Court, Suite B, Deer Park, NY 11729, USA
[4]St. Petersburg State University, St. Petersburg, Russia
[5]Nitride Crystals Group, Ltd., pr. Engel'sa 27, St. Petersburg 194156, Russia
[6]Submicron Heterostructures for Microelectronics Research & Engineering Center, RAS, 26 Polytekhnicheskaya Str., St. Petersburg 194021, Russia

E-mail: zubenko_1993@bk.ru

Abstract. Etching of the GaN layers in 1M KOH aqua solution under irradiation was studied by the electro-stimulated photolysis using N_2-laser (337 nm, 60 W/m^2) as a light source. It was observed that the size and the depth of the failure monotonically depend on the optical power and the irradiation time of the N_2 laser and the GaN layer type of conductivity. The GaN layers etching rate was evaluated. A mechanism of the failure in the n-GaN layers is discussed.

1. Introduction

GaN is wide band-gap semiconductor for application in prospective power electronic devices that require rapid charge carrier transport and high breakdown voltage. Because of the chemical resistance to aggressive environment and the large energy gap (E_g = 3.43 eV), GaN is also intensively studied as a base material to be use as working electrode at the water photoelectrolysis to produce molecular hydrogen, a candidate for clean, non-polluting energy storage [1-4].

In the last few years, appreciable rate of hydrogen generation at the water photoelectrolysis (0.37 - 1.44 ml/cm^2 × h) was demonstrated using working anode electrode based on n-GaN and n-GaN/n-InGaN structures [5]. However, systematic studies of the GaN-anode stability were not carried out. In this work we performed investigation of failure in the GaN layers immersed in 1M KOH aqua solution under UV irradiation using N_2-laser (337 nm, 60 W/m^2) as a light source. Portion of the UV light in the solar spectrum is about 3%. However these experiments allow making clear a mechanism of the GaN failure in the electrolyte to prevent the electrode corrosion in real photo-electrochemical cells.

2. Experimental

Hydride vapor phase epitaxy (HVPE) was used to grow 5-10 μm thick GaN layers on c-plane 2-inch sapphire substrates. Undoped and Si-doped GaN layers had net donor concentration (N_D-N_A) measured by C-V method of (1-5) × 10^{16} cm^{-3} and (1-2)×10^{18} cm^{-3}, respectively. The sapphire substrate orientation allows breaking the 2-inch sample into near equal six sectors having an area of about 3 cm^2. The sectors were used for contacts deposition and electrodes producing. The contacts were protected by a chemical resistant varnish.

Diffuse reflectance spectrum of n-GaN is obtained on the Shimadzu UV 2400 spectrophotometer.

The electro-stimulated photolysis (scheme in figure 1) was carried out in a quartz cuvette in 1.0 M aqueous KOH using an external power source (Mastech DS POWER SUPPLY HY 3010E), GaN working electrode (anode) and the Pt plate as the counter electrode (cathode). The current from the power source was 0.5 mA in all experiments. N_2-laser (LGI-21-type, 337 nm, 100 Hz, 60 W/m^2) was used as a light source.

After experiments, the samples were cleaved and their surface morphology were examined in both top-view and cross-sectional configurations by high-resolution scanning electron microscopy (SEM)

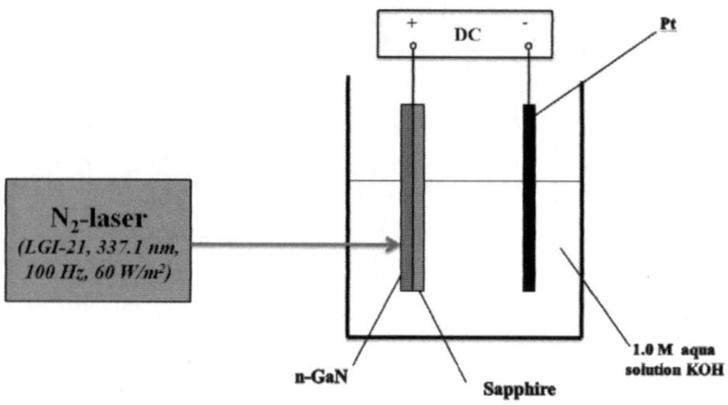

Figure 1. Experimental set-up.

3. Results and discussion

As seen in figure 2, a n-GaN effectively absorbs short-wave radiation up to 352 nm. Therefore, in a nitrogen laser was chosen as the radiation source.

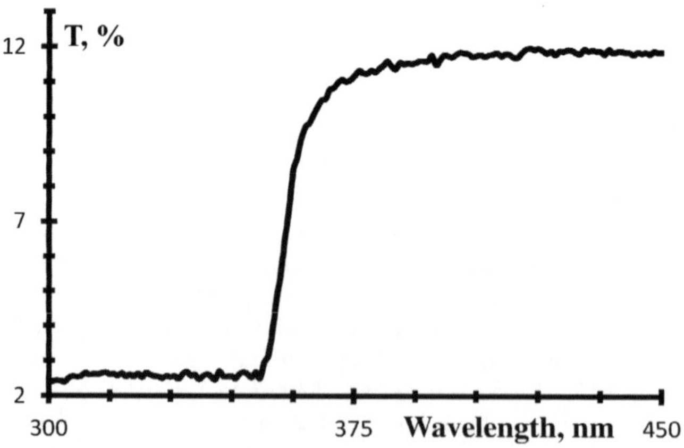

Figure 2. Diffuse reflectance spectrum of n-GaN.

The GaN failures under the laser illumination had a circular spot of a ~2.5 mm in diameter (figure 3a). Honeycombs-like defects of the GaN layer surface in the spot area seems to be due to specific of the etching process in the mosaic (domain) structure attributed to the III-N materials and to HVPE-grown GaN layers in particular (figure 3b).

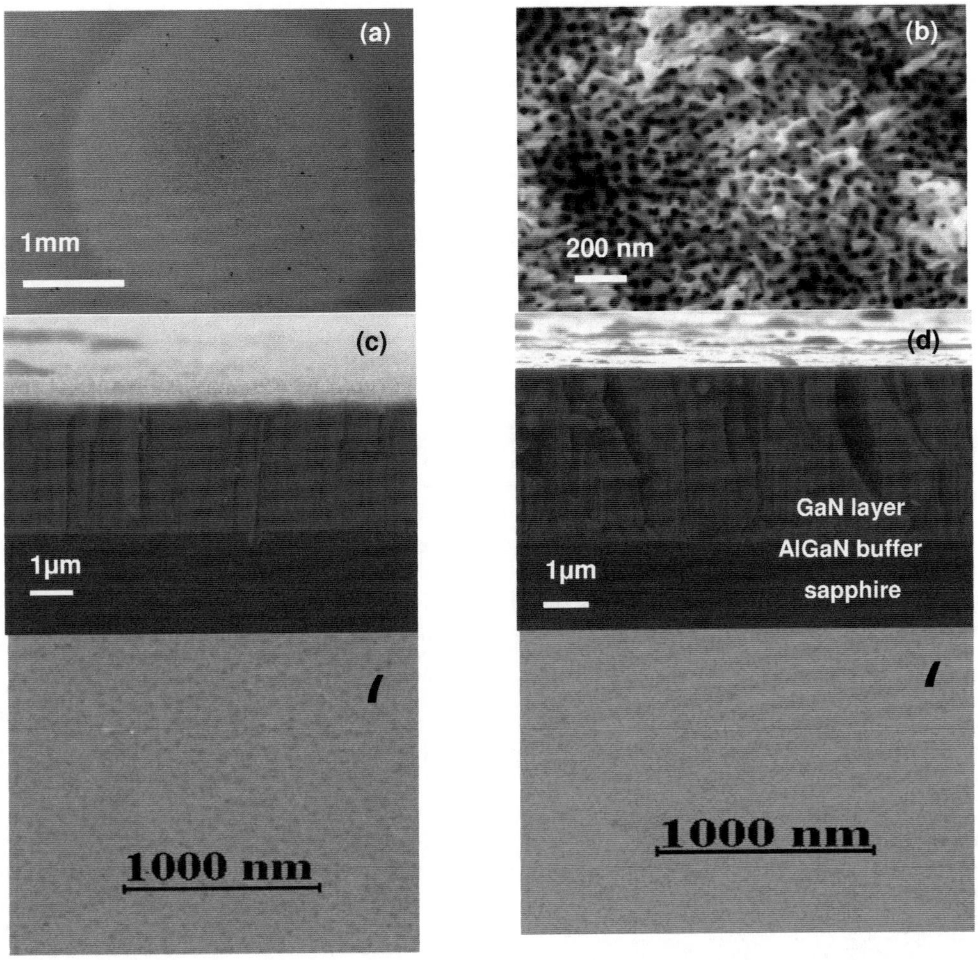

Figure 3 (a-f). SEM images of a GaN sample (N_D-N_A =$1\times10^{18}cm^{-3}$) after the electro-stimulated photolysis. (a) The spot on the GaN surface. (**b**) Over etched GaN surface in the spot under higher magnification after a 40-min experiment. (**c**) Cross-sectional view of the sample in the spot area after a 40-min experiment. The GaN layer thickness is 6.9 µm. (**d**) Cross-sectional view of the sample in unetched area. The GaN layer thickness is 9 µm. (**e-f**) Produced nanodefects on the GaN-surface at a distance of 2 mm (**e**) and 5 mm (**f**) from the spot.

To determine the depth of the photodestruction of the samples were studied from the end (in the area of exposure and beyond figure 3 c-d, respectively).

The etching process may have a preferable way at the domain boundaries. Formation of nanodefects due to the scattered light is observed exposure spot at a distance of 2 mm and 5 mm (figure 3 e-f, respectively). The farther from the spot was, the smaller the formed nanodefects.

In the KOH aqua solution the n- GaN layers underwent etching with an average speed of 46 nm/min (figure 4). The photolysis process with n-GaN electrodes depended on the doping and electrolyte concentration and type and had the max effect at ~0.75M KOH and N_D-N_A = (3-5)$\times10^{18}cm^{-3}$. Photo-destruction of low-alloy specimens n-GaN (N_D-N_A < n × $10^{18}cm^{-3}$) absent. This fact may be useful for creating electrode gallium nitride for splitting water under the influence of UV-light.

The most likely equation describing the processes in the alkaline environment (pH=14) for n-GAN layers may be as follow: 2 GaN + 8 OH^- = 2 $[Ga(OH)_4]^-$ + N_2↑. About excreted nitrogen indicates a small amount of gas bubbles formed on the illuminated part of the anode.

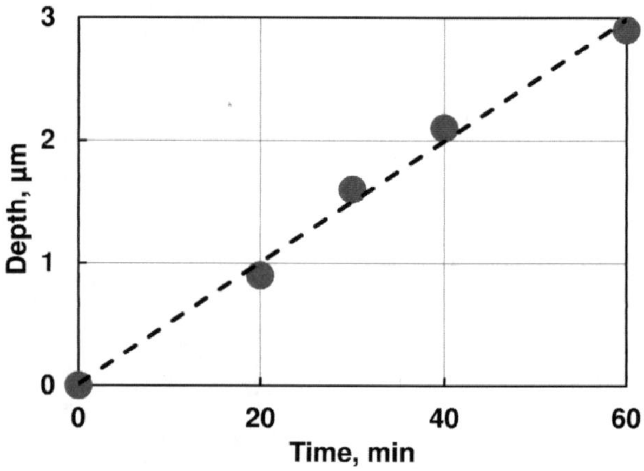

Figure 4. The dependence of the failure depth in n-GaN layer (N_D-N_A=1×10^{18} cm^{-3}) vs. the irradiation time by the N_2 laser. Dashed line is a linear approximation.

4. Conclusions

For the first time held a photo-electro-stimulated destruction of high n-GaN-samples grown by HVPE. Experimentally established linear relationship depth photo-electro-etching n-GaN on the time of UV irradiation.

Acknowledgments

Work at University ITMO was supported by the Ministry of Education and Science of Russian Federation within the grant agreement 14.575.21.0054 (unique identifier of research activities is RFMEF157514X0054).

References

[1] Waki I, Cohen D, Lal R, Mishra U, DenBaars S P and Nakamura S. 2007 *Appl. Phys. Lett.* **91**, 093519

[2] Tseng W J Photoelectrochemical Study of n-GaN for Water Photoelectrolysis. 2014. Dissertation presented in partial fulfilment of the requirements for the degree of Doctor of Science. 3001 Leuven, Belgium

[3] Tseng W J, van Dorp D H, Lieten R R, Vereecken P M and Borghs G. 2014. *J. Phys. Chem.* C **118** (51), 29492-29498

[4] Puzyk M V, Usikov A S, Kurin S Yu, Puzyk A M, Fomichev A D, Ermakov I A, Kovalev D S, Papchenko B P, Helava H and Makarov Yu N 2015 *J. Phys.: Conf. Ser.* **643** 012127

[5] Ohkawa K, Ohara W, Uchida D and Deura M 2013 *Jpn. J. Appl. Phys.* **52** 08JH04

Journal of Physics: Conference Series **741** (2016) 012051 doi:10.1088/1742-6596/741/1/012051

Electronic excitation energy transport in a DNA-Ag cluster complex.

Z Reveguk, E Ikonnikov, A Kononov
Department of Molecular Biophysics and Polymer Physics, St. Petersburg State University, St. Petersburg, Russia

Abstract: We study electronic excitation transfer in the complexes of Ag clusters with oligonucleotides. Steady-state fluorescence excitation spectra show that the excitation energy is transferred to Ag cluster from all the 30 nucleobases of a 15-mer DNA duplex. This is in contrast to DNA-dyes complexes, where the energy transfer to the dye occurs from neighboring DNA bases only. Fluorescence decay curve for the DNA duplex shows that the process of energy transfer occurs within <100 fs. The obtained results suggest coherent excitonic type of the transfer rather than trivial Förster mechanism.

Introduction

Electronically excited states in DNA have been a subject of intense studies for a long time [1] due to their key role in the genetic damage induced by UV sun light. Such damage can lead to mutations, genomic instability, and carcinogenesis [2]. The underlying mechanism of carcinogenicity is associated with the direct absorption of a photon by DNA followed by cyclobutane pyrimidine dimer (CPD) formation [3]. In our recent theoretical study [4] we show that most DNA damages under terrestrial solar radiation occur on the stacked DNA bases with low-lying excitonic states. Such low-lying excited states might act as the traps in energy transfer process. Though energy transfer in DNA is clearly seen from fluorescence anisotropy decay even on femtosecond time scale [5], the range and mechanism of the transfer remain questionable. Dyes bound to DNA and modified nucleobases are often used as the acceptors in the experiments on energy and electron transport in DNA. Most experiments with intercalating dyes and chemically modified bases exhibited short-distance energy transfer only from neighboring DNA bases in the strand [6,7].

In present study, we used a new class of acceptors, namely metallic Ag nanoclusters, to study the process of energy transfer in DNA strand. Fluorescent DNA-bound Ag nanoclusters containing a few metallic atoms attract a lot of attention in last years because of their potential applications in nanotechnology and nanomedicine [8]. It has been shown that Ag clusters on short oligonucleotides were able to be rather efficient acceptors of the electronic excitation in the DNA-cluster complex [9]. This looks promising for developing a new type of ultraviolet to visible light converters, for example, for light emitting diodes [10].

We address the question about the range and mechanism of the excitation energy transfer in the complexes of DNA oligonucleotides with Ag clusters. We synthesized a green-emitting Ag cluster on a self-dimer of 15-mer of DNA 5'- CGCCCCCCTCGGCGT-3' and purified it by high performance liquid chromatography (HPLC) according to the previously described protocol [11]. From comparison of the absorption and fluorescence excitation spectra of the complex, we determined the number of DNA bases from which the energy was transferred to the cluster. The fluorescence life time of the donor, namely DNA duplex, was determined from the fluorescence up-conversion measurements on femtosecond time scale. Based on the observed high efficiency of the transfer on femtosecond time scale, we propose excitonic mechanism of the energy transport in the DNA-Ag cluster complex.

Results.
In Figure 1, shown are fluorescence emission, and excitation spectra of aqueous solution of freshly synthesized Ag clusters-DNA complex along with absorption spectrum of HPLC purified solution of the green-emitting Ag clusters. The fluorescence excitation spectrum of the complex coincides with the absorption spectrum within 10% experimental error. This fact clearly shows that the electronic excitation energy is transferred to the cluster from all 30 nucleobases of the duplex (as has been shown in ref. 11, the Ag cluster was stabilized by two DNA strands). The fluorescence decay curve of the 15-mer DNA duplex recorded at 330 nm in 5 ps time domain is shown in Figure 2. The excited state of major fraction of the DNA decays with a life-time of ca. 300 fs. The energy was shown to be transferred from the whole DNA duplex with possible error of ca. 20%, i.e. with ≥80 % efficiency. At given rate of DNA deactivation of 300 fs it follows that the rate of the transfer should be faster than 100 fs.

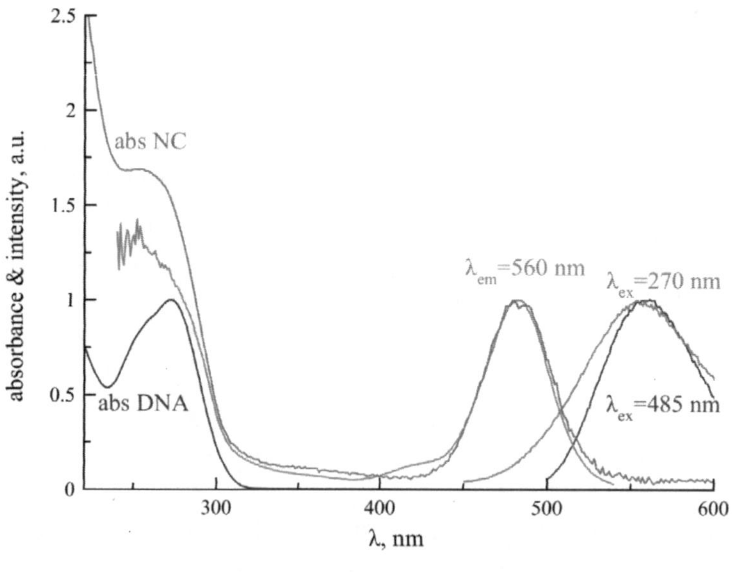

Figure 1. Fluorescence emission and excitation spectra of Ag clusters bound to DNA duplex, and DNA absorption spectrum.

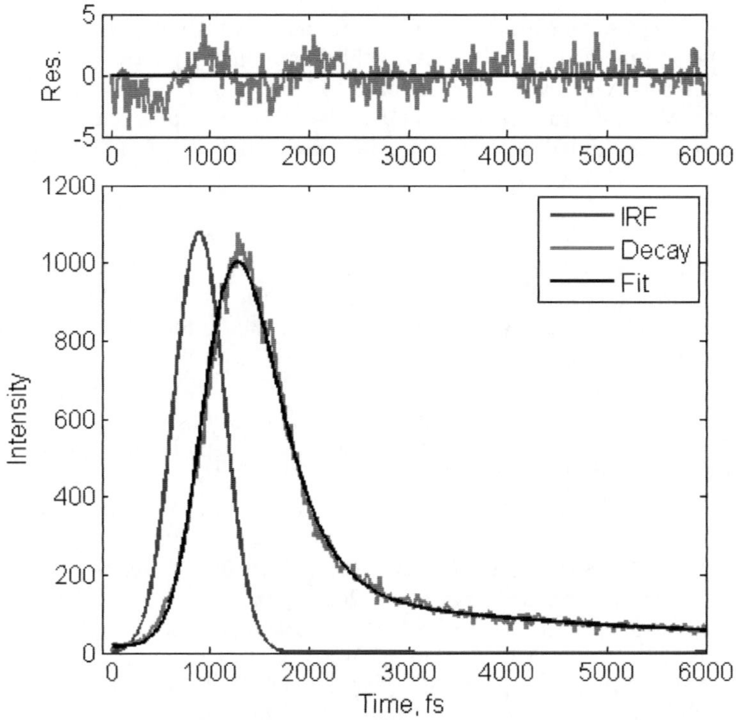

Figure 2. Fluorescence decay curve of the DNA duplex at 330 nm (λex = 266 nm).

Although the detailed structure of the cluster-DNA complex is not known (we are currently working on it), some preliminary conclusions about possible mechanism of the energy transfer can be made. Based on high rigidity of the DNA duplex and small size of Ag cluster consisting of not more than 5 Ag atoms, as it follows from calculations of the excitation spectra of model Ag chains [12], it may be deduced that the distance of the energy transfer between DNA bases and Ag cluster comes to about 20-30 Å. Considering that the transfer occurs much faster than the DNA fluorescence life time of 300 fs, one may rule out trivial Förster mechanism. In this respect, the coherent (exciton) mechanism of the excitation transfer between DNA bases and Ag cluster seems the most probable. The result obtained in this study prompts further efforts in probing the ultrafast dynamics of the excitation in DNA and DNA-Ag cluster complexes in the time domain shorter than 100 fs, which is a challenging task for the deep UV range.

Methods

Synthesis and purification of Ag clusters were performed as describe in ref. 11. Ag clusters were synthesized by mixing DNA buffered solution (ammonium acetate buffer, pH 7) with AgNO3 and by reducing with NaBH4. The Ag/DNA base ratio was 0.4. Ion Pair-Reversed Phase HPLC of the green-emitting Ag cluster was performed on C18 column (Supelco Discovery 250 x 4.6 mm, 5 μm; gradient B in A 1-50% in 30 min; A – 70 mM triethylammonium acetate aqueous buffer (pH 7,5); B – 70 mM triethylammonium acetate solution in methanol).

Fluorescence spectra were obtained on Fluorolog 3 (Horiba Jobin-Yvon) fluorometer at room temperature. Fluorescent emission spectra were corrected for instrument sensitivity as supplied by manufacturers. To eliminate polarization effects, we set a vertically oriented polarizer into the excitation channel and another polarizer was placed at "magic" angle in the emission channel. The fluorescence excitation spectra were corrected for inner filter effect.

The fluorescence time-resolved measurements were made with FOG 100-DX fluorescence up-conversion spectrometer equipped with CDP 2015 third harmonic generator (CDP Corp., Moscow, Russia). The excitation was made by the third harmonic (266 nm) of a mode-locked Ti–sapphire laser (TISSA, CDP Corp., Moscow, Russia) operating at 80 MHz. The apparatus function (IRF) was determined by measuring the signal of Raman line in water. The FWHM of IRF was typically about 450 fs.

Acknowledgments

The authors acknowledge Saint-Petersburg State University for a research grant 11.38.221.2014. Measurements were performed in Resource Center for Optical and Laser Materials Research and Chemical Analysis and Materials Research Centre of St. Petersburg State University. We also thank CDP Systems Corp. for femtosecond measurements.

References

[1] Middleton C , Kimberly L , Su C, Law Y , Carlos E, Crespo-Hernández and Kohler B 2009 *Rev. Phys.Chem.* **60** 217-239
[2] Pfeifer G P, You Y-H and Besaratinia A 2005 *Mutat. Res., Fundam. Mol. Mech. Mutagen* **571** 19-31
[3] Cadet J, Mouret S, Ravanat J-L and Douki T 2012 *Photochem. Photobiol.* **88** 1048-1065
[4] Ramazanov R, Maksimov D and Kononov A 2015 *J. Am. Chem. Soc.* **137** (36) 11656–11665
[5] Vayá I, Gustavsson T, Douki T, Berlin Y and Markovitsi D 2012 *J. Amer. Chem. Soc.* **134** 11366-11368
[6] Nordlund T M 2007 *Photochem. Photobiol.* **83** 625-636
[7] Dumas A and Luedtke N 2011 *Nucleic Acids Res.* **39** 6825-6834
[8] Choi S, Dickson R and Yu J 2012 *Chem. Soc. Rev.* **41** 1867-1891
[9] O'Neill P , Gwinn E and Fygenson D 2011 *J. Phys. Chem. C* **115** 24061-24066
[10] Kamiyama S, Iwaya M, Takeuchi T, Akasaki I, Syväjärvi M and Yakimova R 2011 *J. Semiconductors* **32** 13004
[11] Schultz D and Gwinn E 2012, *Chem. Commun.* **48** 5748–5750
[12] Ramazanov R and Kononov A 2013 *J. Phys. Chem. C* **117** 18681-18687

The study of mixing of reagents within a droplet in various designs of microfluidic chip

N A Filatov[1], K I Belousov[2,3], A S Bukatin[1,3], I V Kukhtevich[3], A A Evstrapov[1,2,4]

[1]Nanobiotech Lab, St. Petersburg Academic University, St. Petersburg 194021, Russia
[2]Department of Nanotechnologies and Materials Science ITMO University, St. Petersburg 197101, Russia
[3]Department of biophysics, Siberian Federal University, Krasnoyarsk 660041, Russia
[4]Laboratory of information and measurement biosensor and chemosensor microsystems, Institute for Analytical Instrumentation RAS, Saint Petersburg 198095, Russia

Abstract. The aim of this work was to determine a design of a microfluidic droplet generator which provides the best mixing of two components within a droplet during its formation. Simulation of droplets' formation in various designs of microfluidic generators was performed. Dependences of a mixing index from a capillary number and a viscosity of continuous phase were obtained for each design. The mixing index was determined directly after the formation of the droplet. It was found that the small capillary numbers correspond to the lower mixing index due to increase of the droplets diameter. The high viscosity of the continuous phase provides vortex flows in the area of droplet formation which leads to an increase in the mixing index for asymmetric designs, but has no significant effect on the mixing in conventional symmetric flow focusing. So for the asymmetric droplet generator designs the mixing index increases up to 1.5 as compared with the conventional flow focusing designs.

1. Introduction

Microfluidic chips are used to solve various analytical tasks in biology, medicine and chemistry, as well as for the synthesis of small quantities of substances. In most of these cases laminar flows of liquids are applied, which allow accurate and reproducible conditions of analysis or synthesis. However, in some tasks the common problem is the efficient mixing of reagents, buffer and analyte, e.g. during an immunoassay, synthesis, etc. Performing mixing in microfluidic devices is not an easy task [1, 2], as in laminar flows (corresponding to low Reynolds numbers $Re < 1$) it is carried out mainly by diffusion. So diffusion time is about 100 s for the path of 100 μm of short proteins with a diffusion coefficient 10^{-10} m^2/s. There are various approaches to accelerate mixing in microfluidics based on different types of passive or active mixers. The first type includes microfluidic chips with a special geometry of channels [3, 4] or specific microstructures in them [5] providing a more efficient interaction of mixed fluid flows. In passive mixers it is difficult to ensure high efficiency of mixing. Active mixers are equipped with various control elements which require additional energy (usually electric [6] or mechanical [7]) for influence on the mixed flow.

Promising devices for mixing substances are droplet on-chip generators [8, 9]. In these systems necessary reactions are conducted in confined volume of droplets with diameter from about 10 to 50 micrometers. Comparing different strategies of droplet formation within microfluidic chips: T-junction, flow focusing and co-flowing, it was found that the best mixing is achieved in the T-junction [10, 11]. However, flow focusing has its own advantages. Its operating principle is based on the fact that the continuous phase flowing through two side channels meets the dispersed phase at a channels' intersection, where the dispersed phase is squeezed by the continuous phase and breaks up into droplets. It allows avoiding a contact of forming droplets with channels' walls and helps prevent potential negative effects on sample's components of the dispersed phase. Therefore, it is reasonable to modify the flow focusing design to achieve better mixing.

2. Methods

Simulation of droplets' formation and distribution of reagents was performed by means of COMSOL Multiphysics which is a software platform based on finite element method for solving physics-based problems.

During the simulation the designs shown in figure 1 were considered.

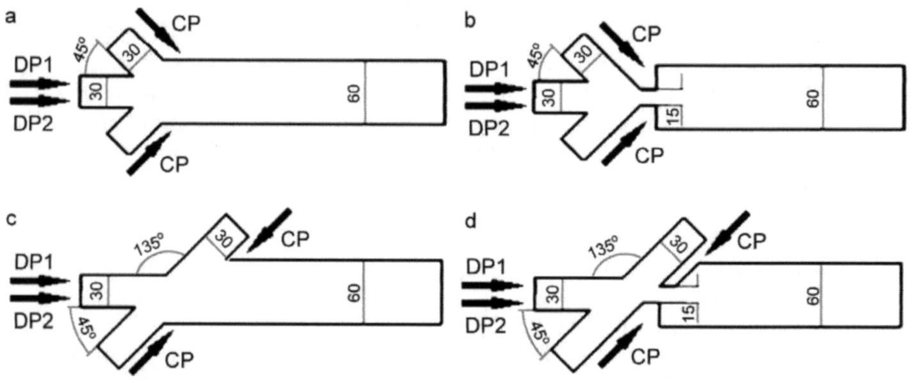

Figure 1. Simulation area for different droplet generator designs: a) conventional flow focusing; b) conventional flow focusing with pinched region; c) asymmetric; d) asymmetric with pinched region. All sizes are in micrometers. DP1 is the disperse phase with reagent, DP2 is the buffer disperse phase, CP is the continuous phase

The Navier-Stokes equations were solved to calculate the velocity profile of the fluid:

$$\rho\left(\frac{\partial \mathbf{u}}{\partial t}+\mathbf{u}\cdot\nabla\mathbf{u}\right)=-\nabla p+\mu\nabla^2\mathbf{u}+\mathbf{F}_s \qquad (1)$$
$$\nabla\cdot\mathbf{u}=0$$

where $\mathbf{u}$ is the flow velocity vector (m/s), ρ is the fluid density (kg/m^3), μ is the fluid viscosity (Pa·s), p is the pressure (Pa), $\mathbf{F}_s$ is the surface tension force (H/m^3).

The Fick's second law with the added convective term was used for modeling the distribution of concentration:

$$\frac{\partial c}{\partial t}+\mathbf{u}\cdot\nabla c=D\nabla^2 c \qquad (2)$$

where D is the diffusion coefficient of reagent (m^2/s), c is the concentration of reagent (mol/m^3).

Displacement of the interface between two immiscible fluids was described by a phase field in which the spatial distribution of the two phases was defined by a phase field variable φ which takes values from -1 to 1. Changing of this variable in time was determined by minimization of the system's free energy [12]:

$$\frac{\partial\varphi}{\partial t}+\mathbf{u}\cdot\nabla\varphi=\nabla\cdot\gamma\nabla\left(\frac{\partial f_{\mathrm{tot}}}{\partial\varphi}-\nabla\cdot\frac{\partial f_{\mathrm{tot}}}{\partial\nabla\varphi}\right) \qquad (3)$$

where f_{tot} is the total free energy density of the system (J/m^3), γ is the relaxation parameter (m^3·s/kg).

Calculation of the mixing index that characterizes the ratio of the variance of reagent to the maximum possible variance was performed using the following expression [13]:

$$I = \left[1 - \left(\frac{\iint (c-\bar{c})^2 \, dA}{A \cdot \bar{c}\left(c_{max} - \bar{c}\right)}\right)^{1/2}\right] \cdot 100\% \quad (4)$$

where A is the area of cross-section of the droplet (m^2), c is the concentration of reagent (mol/m^3), $\bar{c}$ and c_{max} is the average and the maximum concentration of the reagent within the droplet respectively (mol/m^3). Index equals 1 for the complete mixing and 0 when the mixing absent. Mixing index was determined directly after the formation of the droplet.

The dependence of the mixing index on the viscosity of the continuous phase and the capillary number Ca was investigated during the work:

$$Ca = \frac{U \mu_c}{\sigma} \quad (5)$$

where U is the average flow velocity after channel's crossing (m/s), μ_c is the viscosity of the continuous phase (Pa·s), σ is the surface tension between the dispersed and continuous phases (N/m).

The flow velocity was calculated from the volumetric flow rate for channel depth of 30 μm. Flow rate of the continuous phase was varied from 10 μl/min to 1 μl/min and flow rate of dispersed phase was varied from 1 μl/min to 0.1 μl/min to provide a range of capillary number's values from 0.06 to 0.006 at constant viscosity of continuous phase $\mu_c = 0.03$ Pa·s. Diffusion coefficient also was varied to achieve constant Peclet number. It took values from 10^{-10} to 10^{-11} m^2/s corresponded to short proteins. The ratio between continuous and dispersed phases 10:1 was chosen to achieve dripping regime of droplets generator and prevent their contact with the channel's walls to minimize walls' influence. The water with $\rho_d = 1000$ kg/m^3 and dynamic viscosity $\mu_d = 0.001$ Pa·s was considered as the dispersed phase. The oil with density $\rho_c = 880$ kg/m^3 was considered as the continuous phase. When the oil viscosity μ_c was varied from 0.03 to 0.003 Pa·s the flow rate also was changed to achieve constant $Ca = 0.006$. The surface tension σ at the oil-water boundary was equal 0.051 N/m.

3. Results and discussion

During analysis two-dimensional formulation of the problem was used. Since the inertial force was negligible in considered model (Reynolds number equal 0.01–0.1) the process of droplet formation was controlled by the ratio of viscous and surface tension forces that is expressed by the capillary number. Surface tension tries to minimize two-phase interface and interferes with droplet formation. On the other hand viscous stress from the continuous phase elongates disperse phase and breaks up droplets. Thus, droplet diameter L was inversely proportional to the Ca as shown in figure 2a. While droplet size was reducing, the mixing index was increasing (figure 2b) as shorter diffusion length provides enhancing mixing. With increasing of the Ca 10 times the I increased 1.1-1.4 times and the droplet diameter decreased 1.5-1.75 times depended on design. Improvement of mixing ability in droplet generators with pinched region also related to decreasing of droplet diameter. So adding a pinched region improved the mixing for the conventional flow-focusing 1.5 times and for the asymmetric 1.25 times with decreasing of droplet diameter 2.25 and 2.75 times respectively.

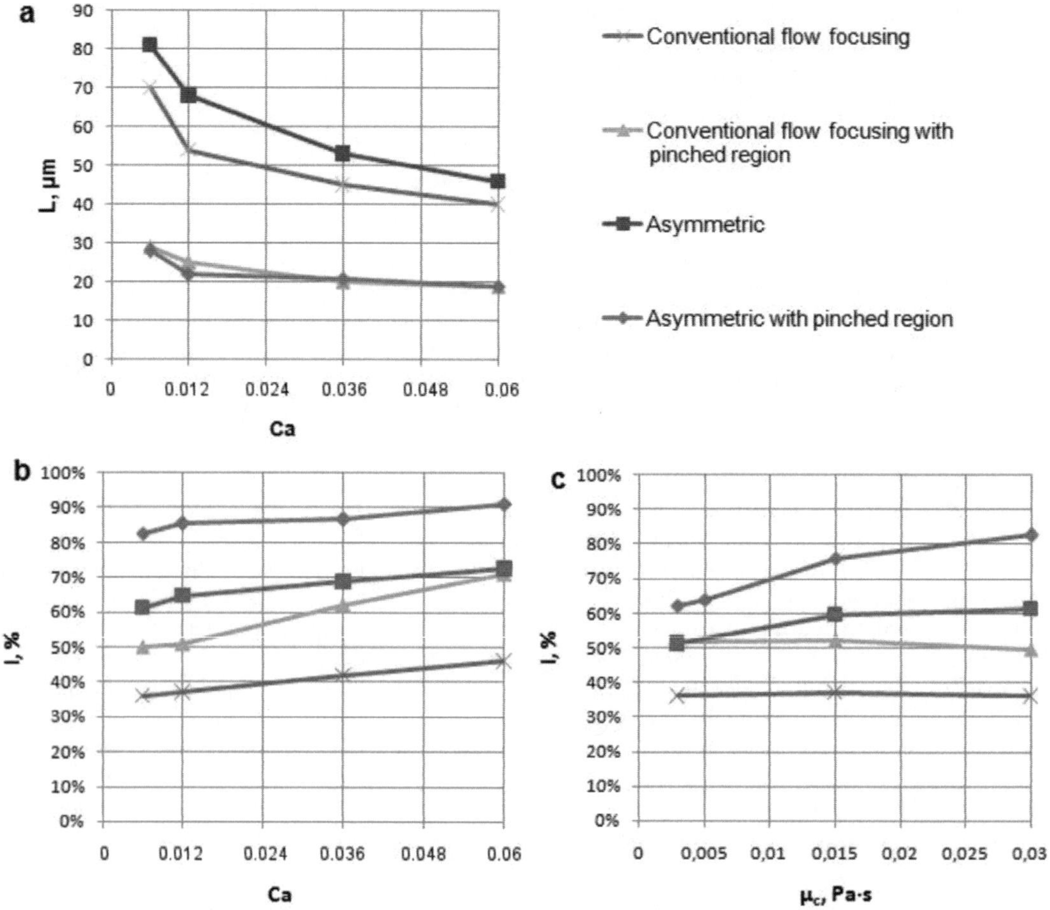

Figure 2. Simulation results: a) dependences of the droplet diameter on the capillary number *Ca*; b) dependences of the mixing index on the capillary number *Ca* with $\mu_c = 0.03$ Pa·s; c) dependences of the mixing index on the viscosity of continuous phase with $Ca = 0.006$

Asymmetric design and asymmetric design with pinched region provided on average up to 1.5 times higher mixing index compared with conventional flow-focusing and conventional flow-focusing with pinched region respectively. The superiority of asymmetric designs related to vortex flows and their asymmetric character in the area of droplet formation. This was evidenced by the results for varying μ_c at constant *Ca* (figure 2c). Decreasing in μ_c leaded to decreasing mixing index by 20% for the asymmetric design with pinched region and by 10% for the asymmetric without it, but had no significant effect on the mixing in conventional flow focusing. Viscosity of fluid relates to momentum transport between its laminae and is proportional to sheer stress, so high viscosity and velocity ratios between continuous and disperse phases induced vortex flows in water which decreased with μ_c reduction (figure 3a, b). In symmetric design there are two vortices occupying left and right sides of forming droplet, such as there is no transfer of reagent from one part of a droplet to another (figure 3c). In asymmetric design most of the two-phase interface is exposed to the flow from the right side channel and any counter vortex isn't created. Thereby, this single vortex flow enhances mixing.

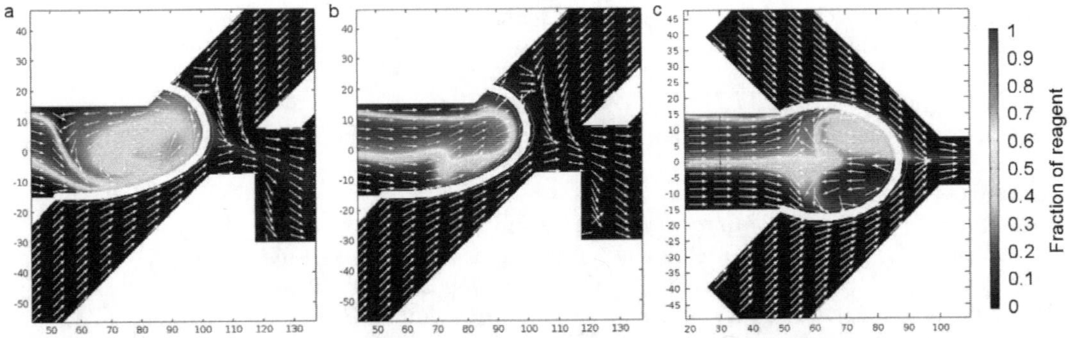

Figure 3. Fraction of reagent in designs with pinched region with the same capillary number $Ca = 0.006$ and different viscosity of the continuous phase μ_c: a) asymmetric with μ_c = 0.03 Pa·s; b) asymmetric with μ_c = 0.003 Pa·s; c) symmetric with μ_c = 0.03 Pa·s. Arrows show the direction of the flow. Solid white contour correspond to the interface between dispersed and continuous phases

In three dimensional case special phenomena might appear, because extra constrains are added at a top and a bottom. Checking the possibility of vortex formation in 3D model the appropriate simulation was conducted at Ca = 0.006 and μ_c = 0.03 Pa·s. As shown in figure 4 the flow at the center of the channels had the same character as during 2D simulation.

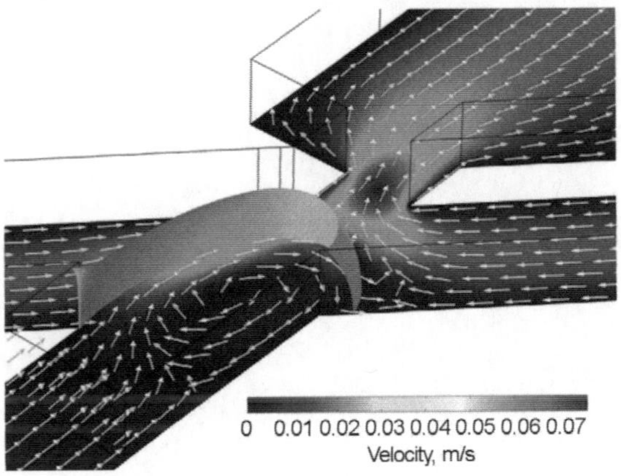

Figure 4. Flow velocity in the center of the channel. Arrows show the direction of the flow. The two-phase interface is marked by gray

4. Conclusions

Dependences of the mixing index on the capillary number Ca and viscosity of the continuous phase μ_c were obtained for four different designs. So for conventional flow-focusing design mixing index was 46%, for conventional flow-focusing with pinched region and asymmetric it was 71% and for asymmetric with pinched region it was 91% when $Ca = 0.06$ and $\mu_c = 0.03$ Pa·s. It was found that with decreasing of the capillary number 10 times the mixing index decreased 1.1-1.4 times associated with increase in droplet diameter. Decreasing of continuous phase viscosity leaded to decreasing of vortex flows and the mixing index for both types of asymmetric designs, but didn't affect on conventional variants, because vortex flows were symmetric in them and didn't give a contribution in mixing.

Thus, the best mixing efficiency was achieved in case of the asymmetric designs. Such designs increased mixing index up to 1.5 compared with conventional flow-focusing. Two-dimensional formulation of the problem and finite element method impose restrictions on the data obtained, but, as three-dimensional simulation showed, they provide an opportunity to get relevant estimates and make principal conclusions. So, we can conclude that proposed in the paper asymmetric design of droplet generator with pinched region allows achieving higher efficiency of mixing in conjunction with the benefits of flow-focusing design.

Acknowledgments
The research was supported by the grant of the Russian Science Foundation (project no.15-19-10041).

References
[1] Suh Y K, Kang S 2010 *Micromachines* **1** 82-111
[2] Capretto L, Cheng W, Hill M, Zhang X 2011 *Topics in Current Chemistry* **304** 27–68
[3] Parsa M K, Hormozi F, Jafari D 2014 *Computers & Fluids* **105** 82–90
[4] Kim D, Lee S, Kwon T, Ahn C 2005 *Lab on a Chip* **5** 739–747
[5] Bhagat A, Papautsky I 2008 *Journal of Micromechanics and Microengineering* **18** 085005
[6] Glasgow I, Batton J, Aubry N 2004 *Lab on a Chip* **4** 558–562
[7] Yaralioglu G, Wygant I, Marentis T, Khuri-Yakub B 2004 *Analytical Chemistry* **76** 3694–3698
[8] Lee C-C, Snyder T M, Quake S R 2010 *Nucleic Acids Research* **38** 2514– 2521
[9] Xu Q B, Hashimoto M, Dang T T, Hoare T, Kohane D S, Whitesides G M, Langer R, Anderson D G 2009 *Small* **5** 1575–1581
[10] Zhao S, Wang W, Zhang M, Shao T, Jin Y, Cheng Y 2012 *Chemical Engineering Journal* **207–208** 267–277
[11] Zhao C X, Middelberg A 2011 *Chemical Engineering Science* **66** 1394–1411
[12] Yue P, Zhou C, Feng J J, Ollivier-Gooch C F, Hu H H 2006 *Journal of Computational Physics* **219** 47–67
[13] Kockmann N 2008 *Transport phenomena in microprocess engineering* (Berlin: Springer)

A numerical model of protein crystallization with counter diffusion method.

A S Sokolovskiy, Yu V Trushin, M N Lubov, I E Eliseev, A N Yudenko and M V Dubina

Nanobiotech Lab., St. Petersburg Academic University, St. Petersburg 194021, Russia

Abstract. A theoretical model is presented to simulate the growth and nucleation of protein lysozyme crystals by counter diffusion method. The comparison of experimental and simulation results shows that the numerical solution accurately describes the initial stages of crystallization. The developed model can be used to describe crystallization processes with various counter diffusion experiment parameters.

1. Introduction

The solution of many biomedical problems requires detailed information about chemical structure of biological molecules. This information can be obtained from X-ray analysis of protein crystal. However, for greater resolution of the protein space structure this crystal should be of sufficient size and quality. One of the widely used methods of protein crystallization is counter diffusion method [1]. In this method, protein crystallization occurs due to diffusion of protein and precipitant solution towards each other through the agar gel layer. This process allows to obtain gradient of concentration from one edge of capillary to another. Diffusion rate of protein solution is much slower than that of precipitant. Thus we get the opposed concentration gradients for protein solution and percipient, which is an initial condition for protein nucleation process. The advantage of this method is that one experiment with counter diffusion technique is equivalent to a huge amount of experiments with "hanging" or "sitting drop" method. Its main characteristic is nonlinear relation between different physical processes: diffusion of protein and precipitant; nucleation and growth of protein crystal. Vacuum silicon grease and wax film were used to seal the crystallization cell.

2. Experiment

2.1. Experiment preparation

In the experiment solubility of lysozyme is controlled by the sodium chloride. Figure 1 illustrates a simplified scheme of counter diffusion experiment.

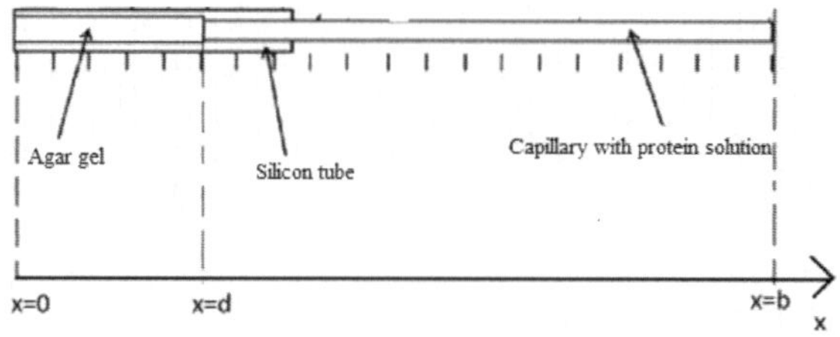

Figure 1. Configuration of the simplified counter-diffusion method

2.2. *Samples preparation*

In the experiment we used 1%(w/v) agar gel with 0.04% NaN_3, obtained in a standard way [2]. Capillary was 200 mkm in diameter. We prepared 30, 65, 100 mg/ml hen-egg lysozyme solution, in 50mM acetate buffer with pH=4.5. Also we prepared 0.5, 0.8, 0.1 M of sodium chloride solution, in 50mM acetate buffer. For one experiment 8 ml of sodium chloride solution is needed. Protein and precipitant solution were degassed. After filling all capillaries, one edge of capillary was sealed with wax with another edge placed in silicon tube filled by agar gel. Finally, samples were placed in precipitant solution.

2.3. *Snapshots of the protein crystal*

The experiment was filmed using Carl Zeiss Axio Observer microscope with 1 hour intervals. To get more accurate results, for every picture along the capillary length (X-axis) different cross section of the capillary were taken (Z-axis) with 1mm step along the X-axis and 50 microns step along the Z-axis. Thus, for a capillary with internal diameter of 200 mm five cross-sections were taken for each point on the X-axis.

3. Theoretical model

3.1. *Model assumptions*

We divide the capillary by N cells. Each i cell (i=1..N) has volume ΔV. Precipitant and protein solution concentrations are considered homogenous, thus dependant only of time.

3.2. *Main physical processes*

Diffusion of Sodium chloride ions through the agar gel from reservoir to capillary. Diffusion of protein molecules (with C_i concentration in i cell) from capillary to reservoir through agar gel. Nucleation of crystals with critical radius when supersaturation S>=1 occurs

$$S(x,t) = \frac{C(x,t)}{C_0(x,t)} \tag{1}$$

Nucleation rate for each cell of capillary at time t is defined by supersaturation function S(x,t).

According to Volmer-Becker-Doring theory, nucleation rate can be described with Arrhenius equation [3]:

$$W(x,t) = k_0 C(x,t) \exp\left(-\frac{\Delta G_{crit}}{(kT)}\right) \tag{2}$$

$$\Delta G_{crit} = \frac{16\pi\Omega^2\gamma^3}{3(kT)^2 \ln^2\left(\frac{C(x,t)}{C_0(x,t)}\right)} \quad (3)$$

where k_0 – empirical constant. C(x,t), C_0(x,t) – concentrations of protein molecules in solution and solubility respectively, Ω – volume of one molecule in crystal, γ - surface free energy of the critical cluster, T – temperature k – Boltzmann constant.

To change solubility of lysozyme sodium chloride was used. When using a simple electrolyte as the precipitant, the protein solubility can be expressed as function of salt concentration [4]

$$C_0(x,t) = C_1(x,t) + C_2(x,t) \quad (4)$$

$$\ln\left(C_1(x,t)\right) = \ln\left(C_{10}\right) - k_{o1}C_S(x,t) \quad (5)$$

$$\ln\left(C_2(x,t)\right) = \ln\left(C_{20}\right) + k_i\left(C_S(x,t)\right)^{1/2} - k_{o2}C_S(x,t) \quad (6)$$

where C_{10},C_{20},k_{o1},k_{o2},k_i empirical constants

Numerical value of function C_S can be obtained from diffusion equation

$$\frac{\partial C_S(x,t)}{\partial t} = \frac{\partial}{\partial x} D_S(x) \frac{\partial C_S(x,t)}{\partial x} \quad (7)$$

Diffusion equation for protein molecules is nonlinear partial differential equation

$$\frac{\partial C_i(x,t)}{\partial t} = \frac{\partial}{\partial x} D(x) \frac{\partial C_i(x,t)}{\partial x} - \sum_j \sum_i \alpha_q D(x) C_i(x,t) R_{q_j}(t) C_{q_j}(t) \quad (8)$$

where C_i - protein molecules concentration in i cell; α_q – empirical constant which depends on interaction between protein molecules and protein crystal; R_{q_j} – radius of j crystal in i cell; C_{q_j} concentration of crystal in j cell

Growth rate of protein crystal in j cell [5]:

$$\frac{dR_{q_j}(t)}{dt} = \frac{\Omega \alpha_q D(x) C_i(x,t)}{4\pi R_{q_j}(t)} \quad (9)$$

D_S(x), D(x) – are piecewise functions, because diffusion coefficient of protein molecules and ions of salt in agar gel and water are different [1,4,6].

Nucleation rate W is number of crystals nucleated per unit volume per unit time:

$$W(x,t) = \frac{\partial^2 N}{\partial V \partial t} \quad (10)$$

The waiting time until the crystal nucleates we can express with τ_0. To get nucleation time for one crystal in i cell we integrate W, with N=1, and set the upper limit of integration to τ_0:

$$\frac{1}{\Delta V} = \int_0^{\tau_0} W(x,t)dt \quad (11)$$

Equations (3,8,9,11) combined give us the final system:

$$\frac{\partial C_i(x,t)}{\partial t} = D(x)\frac{\partial C_i(x,t)}{\partial x} -$$
$$-\sum_j \sum_i \alpha_q D(x) C_i(x,t)\left(R^2_{crit}(x,t) + \frac{\Omega \alpha_q D(x)}{2\pi}\int_0^{\tau_0} C_i(x,t)dt \right)^{1/2} C_{q_j}(t) \quad (12)$$

$$\frac{1}{\Delta V} = \int_0^{\tau_0} k_0 C_i(x,t) \exp\left(-\frac{16\pi\Omega^2\gamma^3}{3(kT)^3 \ln^2\left(\frac{C_i(x,t)}{C_{0i}(x,t)}\right)} \right) dt \quad (13)$$

$$R_{crit}(x,t) = \frac{2\Omega\gamma}{kT \ln\left(\frac{C_i(x,t)}{C_{0i}(x,t)}\right)} \quad (14)$$

The numerical model was made using MATLAB software package. Coordinate Δx and time steps Δt of differencing scheme were selected so that (in capillary and in agar gel)

$$\frac{D\Delta t}{\Delta x^2} = 0.2 \quad (15)$$

4. Results

The experiment was conducted for 24 hours. During this time the crystals appeared at a distance of 10 mm from the beginning of the capillary. Capillary snapshots were taken for analysis of experimental data; the whole capillary was divided into segments of 200 mkm length. To compare the experimental data with numerical simulation, we set step - Δx = 200 mkm in simulation.

Figure 2 shows the theoretical and experimental dependencies of the size and number of crystals. Crystal size: L = 2 * R.

The drop near x=10 mm is due to the features of the experiment, as the outside boundary of the silicone tube is in this area Figure 3.

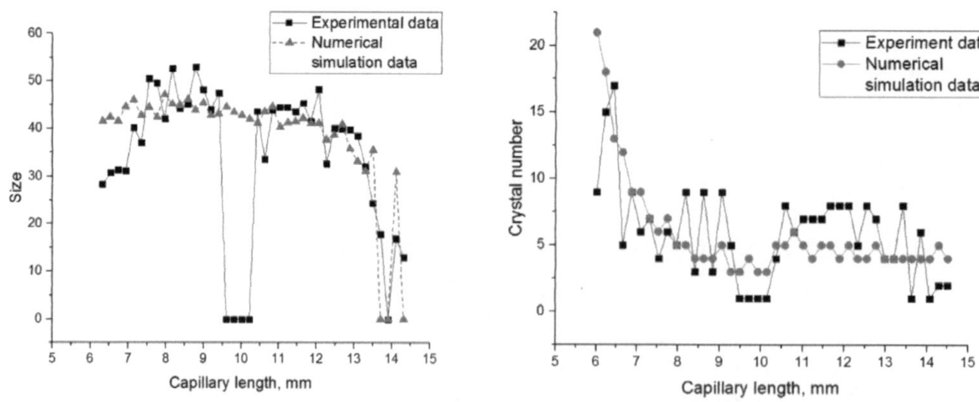

Figure 2. Size and number distribution along the length of capillary.

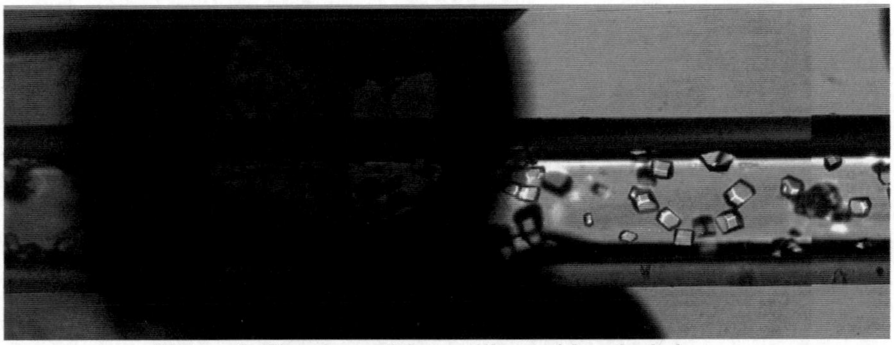

Figure 3. The piece of the capillary with a shaded area.

5. Conclusion

During this work the series of experiments with a model object - lysozyme were conducted and a numerical model for nucleation process was created. The numerical model predicts with high accuracy coordinates of the appearance and size of the crystal, which is confirmed by experiment. This numerical model can be used to determine the optimal crystallization conditions for other macromolecules in the non-ionic composition buffers.

References

[1] García-Ruiz J M 2003, Counterdiffusion Methods for Macromolecular Crystallization, *Methods in Enzymology*, **368** 130-154

[2] Tanaka H, Inaka K, Sugiyama S, Takahashi S, Sano S, Sato M & Yoshitomi S 2010 Optimization of salt concentration in PEG-based crystallization solutions, *J. Synchrotron Rad.,* **18**, 84–87

[3] Galkin O, Vekilov P G 1999, Direct Determination of the Nucleation Rates of Protein Crystals, *J. Phys. Chem.* **103** 10965-71

[4] Otalora F, García-Ruiz J M 1997 Crystal growth studies in microgravity with the APCFI. Computer simulation of transport dynamics, *J. Cryst. Growth*, **182**, 141-154

[5] Trushin Yu V 1996. *Theory of Radiation Processes in Metal Solid Solutions.* (New York, Nova Science Publishers), p 104

[6] Yamanka et al. 2011, Optimization of salt concetration in PEG-based crystalization solutions, *J. Synchrotron Rad.*, **18** 84-87

The study of efficiency of endogenous and exogenous preventive methods of tooth enamel remineralisation by FTIR microscopy using synchrotron radiation

D L Goloshchapov [1], V M Kashkarov [1], P V Seredin[1], Y A Ippolitov [2], Y A Plotnikova[2], K Bambery[3]

[1]Department of Solid State Physics and Nanostructures, Voronezh State University, Voronezh, Russia
[2]Department of Pediatric Dentistry with Orthodontia, Voronezh State Medical University, Voronezh, Russia
[3]Australian Synchrotron (Synchrotron Light Source Australia Pty LTD), Clayton, Victoria, Australia

Abstract. The efficiency carious preventive methods was detected with the use of equipment for IR-spectromicroscopy and high-intensive synchrotron radiation. The results of the experiment are indicative of the use of exogenous caries prevention alone (use of a toothpaste) being inadequate in saturating hard dental tissues by mineral groups and, thus, keeping teeth healthy, as this method is only short-lived. The use of endogenous methods (mineral tablets based on calcium glycerophosphate) in combination with exogenous prevention enhances prevention as part of remineralisation of dental tissues.

1. Introduction

Prevention of carious diseases is of a great importance for every human since any of cariogenic pathologies influence health and aesthetics and as a consequence social and professional activities [1]. Diagnostics of enamel at the early stages of the carious process is still a separate problem [2]. To solve it, new medical devices and techniques of analysis that assist in recognizing this disease at the earlier stages are applied [2–4]. But-it goes without saying that it is better to prevent than to cure tooth decay.

A number of works made in the commercial and state-owned laboratories provide broad and diverse information on the approaches to the prevention of carious diseases. Therefore, all of the investigators assume that there are common grounds in maintaining healthy dental enamel. Their fundamental principle is hygiene of the oral cavity performed with the use of different local (exogenous) methods (tooth brushing, rinsing, applications, etc.). The second ground is based upon maintaining the microelement and organic-mineral composition of the oral fluid at the levels when remineralisation of enamel takes place, i.e. the process of its saturation with various ions and recovery of the lost mineral components [5]. The latter can be achieved by taking in pelleted mineral complex to normalize organic-mineral balance and to saturate oral fluid (endogenous methods). Certainly, oral fluid is a kind of mixed saliva containing organic components as well as inorganic ions coming both from salivary glands and from outside [6,7]. It is supposed that in case when exogenous prophylaxis is adequate and the overall health level is normal, there are no any impediments for the normal organic-mineral metabolism in oral cavity. Studies of the prophylaxis efficiency performed by the exogenous methods, namely, tooth brushing and gels applications have demonstrated efficacy of this trend and the character of activity [8]. But it is suggested that exogenous methods of prevention do not provide the recovery of balance between the natural processes of demineralisation/ remineralisation. Due to the multi-stage character of the ion exchange remineralising agents should be retained for a long time within the oral cavity (in oral fluid) and should be in contact with the dental enamel. But in fact the time of exogenous prophylactics is less than 2 min. The presence of mineral substances in the ionized state in the oral fluid is also a necessary condition of remineralisation. This could be realized only under endogenous supply of the ions in oral fluid. Recovery of the level of natural remineralisation up to the required value depends on the long-term exposure of calcium, phosphate and fluoride ions with the optimal concentrations to the surface of enamel [6]. However, based on the experience of application and comparison of different prevention methods including fluoridation it is safe to say that the use of specimens involving phosphorus- and calcium-containing agents (for example glycerophosphate) provides the best results in the recovery of dental enamel [8–11]. Thus, all changes in the content of oral fluid can be associated with remineralisation/demineralization processes and by estimation of these changes with different kind of methods the best ways for maintaining of enamel surface can be discovered.

In our opinion, the necessity and sufficiency of the different steps in prevention can be determined by studying the changes in the composition of the oral fluid in vivo for the groups of people using different preventive agents of tooth enamel. The changes can be studied by IR-spectroscopy, which is widely used for the analysis of dental materials and for the detection of the nature of pathologies [12–14]. It allows one to obtain an extensive and various information on the phase transformations in different organic and inorganic substance, to detect changes in molecular structures of various materials . That is why this method is suitable for detecting any changes in mineral content of oral fluid. The usage of IR-synchrotron radiation increases the opportunity to detect small changes in IR-spectra with better resolution and to get spectra from micro areas with high intensity of signal.

Thus, the aim of this work is to discover the changes in the oral fluid content through the application of exogenous and endogenous preventive methods of tooth enamel remineralisation by FTIR microscopy using synchrotron radiation.

2. Materials and methods

2.1. Sampling of the materials

10 men and 10 women at the age of 18-25, European, physically healthy, without pernicious habits participated in our experiment. The regulations of making the oral fluid sampling were as follows. During the experiment and a week before the patients ate mainly vegetable food, followed a standard water consumption pattern, did not take any remedies and used no alcohol. On the eighth day after the beginning of the observations without any changes in conditions of the oral cavity hygiene in the morning just before eating, after preliminary oral rinsing patients the oral fluid was collected for the first time. This oral fluid was used as a check reference sample.

On the same day, after check sampling of the oral fluid the patients applied a tooth paste (with multi-mineral complex involving pelleted calcium glycerophosphate) for tooth brushing. Five minutes later after proceeding of the hygienic measures and a **thorough** oral rinsing oral fluid was sampled once again. Additionally, 30 minutes later oral fluid and once again after a **thorough** oral rinsing the oral fluid was sampled off the patients for the third time.

On the next day participants of the experiment just as at the first stage after eating began to take tablets (pelleted vitamin-mineral complex on the basis of calcium glycerophosphate). Participants of the group took one tablet three times a day. Three days after that, in the morning, on an empty stomach the patients thoroughly made oral rinsing and next they were again subjected to oral fluid sampling.

Each time after the sampling oral fluid was immediately centrifuged in order to remove excess water. The residue was dried at 36°C in a desiccator.

2.2. *IR-spectroscopy*

The investigations of molecular structure of the oral fluid and its chemical composition after the described sampling of specimens were performed by IR-spectroscopy. IR microspectroscopy technique was applied utilizing the equipment of the Infrared Microspectroscopy (IRM) beamline at Australian Synchrotron, Victoria, Australia. Bruker V80v Fourier transform infrared (FTIR) spectrometer and a Hyperion 3000 IR microscope (Bruker, Germany) were used to perform the analysis of microprobe of the specimens as well as a diamond high-pressure cell. Transmission IR-spectra were measured within the range of 4000 – 500 cm^{-1}.

3. Results of the experiment

The IR-spectra of the oral fluid collected from the patients, that used endogenous (use of a toothpaste) and exogenous (mineral tablets based on calcium glycerophosphate) preventive methods, are presented in Figure 1 (a),(b).

Figure 1 (a): 1– IR-spectrum of the oral fluid taken before the experiment, 2 – spectrum of the oral fluid taken off the patients five minutes after the use of a tooth paste, 3 – spectrum of the oral fluid taken 30 minutes after the use of a tooth paste, 4 – spectrum of the used tooth paste. Figure 1 (b): 1 – IR-spectrum of the oral fluid taken before the experiment, 2 – IR-spectrum of the oral fluid taken on the fourth day after a three-day taking of the tablets, 3 – spectrum of the tablet on the basis of calcium glycerophosphate.

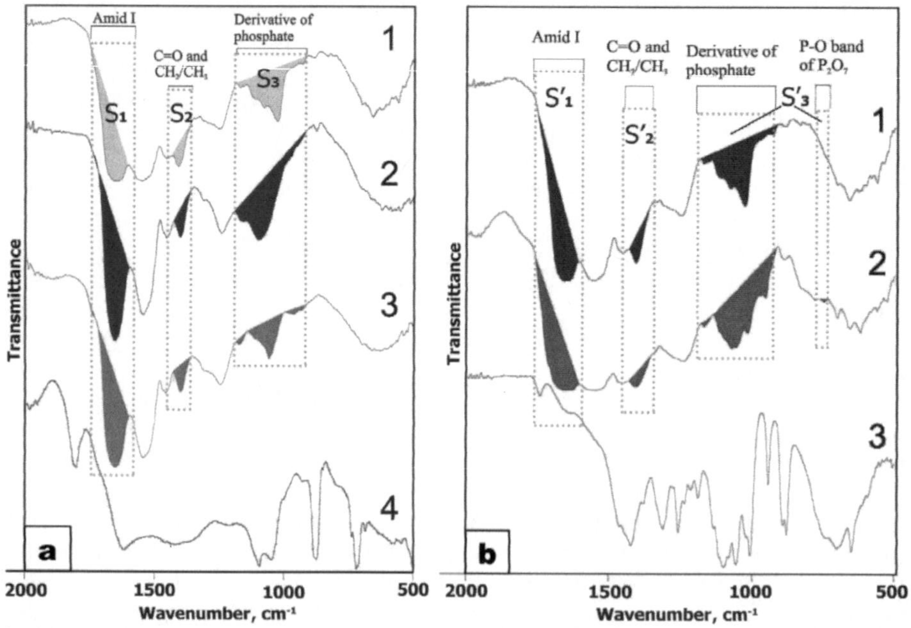

Figure 1. (a) Results from exogenous preventive methods; **(b)** Results from endogenous preventive methods.

From the obtained experimental data (Figure 1 (a),(b)), with comparing of literature it was confirmed that the vibration bands in the transmission IR-spectra of all the samples refer to dry residues of the oral fluids. The group of vibrations arranged in all spectra within 900 – 1200 cm^{-1} is related to the P-O modes. Its appearance is connected with the presence of phosphorus derivatives in the samples: phosphates, glycerophosphates and phospholipids. This conclusion can be made by comparing the data presented in [15] where a detailed analysis of the spectra of oral fluid and human blood was performed. It is possible to confirm that the similarity of the blood and saliva spectra lead to the fact that the modes arranged in the range of 1050 cm^{-1} refer to the organic derivatives of phosphates, glycerophosphates and phosphatase, namely, to the complex of C–O–P–O–C kind.

The next group of the vibration bands localized in the range of 1240 cm^{-1} - 1700 cm^{-1} can be related to the secondary amides: Amid I (80% C=O stretch vibr. in the range of 1615 – 1675 cm^{-1}), Amid II (60% N–H bend and 40% C–N stretch in the range of 1520 – 1575 cm^{-1}) and Amid III (40% C–N stretch, 30% N–H bend in the range of 1270-1315 cm^{-1}). Modes arranged in the experimental spectra in the range of 1400 – 1430 cm $^{-1}$ belong to C=O str (sym) vibrations of COO and CH_2/CH_3 groups. Vibration bands arranged at 500 – 700 cm^{-1} can be associated with the appearance of OCN (Amide IV), NH (Amide V) and C=O (Amide VI) vibrations of proteins.

In the IR-spectra of the oral fluid collected the fourth day (Figure 1 (b)), after three days' taking of the tablets on the basis of a calcium glycerophosphate and the spectrum of the tablet on the basis of calcium glycerophosphate there is one more additional group of vibrations with a peak in the range of 730-770 cm^{-1}. Before proceeding of the endogenous prophylaxis with the use of tablets of the mineral-vitamin complex on the basis of a calcium glycerophosphate as well as at the stage of exogenous prophylaxis using a tooth paste no any vibrations in the above-mentioned range of IR-spectrum. Based on the principles described in [16], it seems impossible to correlate a group of vibrations characterized by the maxima at 730-770 см$^{-1}$ with the vibrations of C-H bonds. However, this vibration mode can be related with the vibrations of P-O bond in P_2O_7 group. Its appearance can be the result of decomposition of the complexes of inorganic phosphates. Arising of these vibration modes was already observed in [17] under thermal decomposition of the inorganic compounds.

The changes in organic-mineral balance of oral cavity can be explored by the calculations and analysis of mineral-organic and carbon-phosphate ratios between the mineral and organic components in the FTIR spectra of the dry residue of the oral fluid (Figure 1 (a),(b)). To calculate the first ratio it is sufficient to consider the ratio of the integral area in the IR-spectra (spectral regions at 900-1200 cm^{-1} and 730-770 cm^{-1}), to the integral area of vibration band 1615 – 1775 cm^{-1}, related to Amid I (Figure 1 (a), S_3/S_1; Figure 1 (b) S'_3/S'_1). A carbon-phosphate ratio can be calculated from the ratio of intensity of vibration bands for C=O and CH_2/CH_3 bonds localized in the range of 1430 - 1400 cm^{-1} to the intensity of phosphate bands in the IR spectrum (900-1200 cm^{-1} and 730-770 cm^{-1}) (Figure 1 (a), S_2/S_3; Figure 1 (b) S'_2/S'_3). These changes in mineral-organic and carbon-phosphate ratios are relative to the values of these ratios calculated for the reference samples at the initial stage of the investigations.

4. Discussion

The analysis of the experimental data and their comparison with the spectra of dry residues of the oral fluid demonstrated that in the IR spectra of the preventive agents one can see the groups of vibrations identical to those observed in the spectra of oral fluids (Figure 1 (a), (b)). It is due to the presence of similar organic-mineral groups and complexes in the composition of these preventive agents. Thus, application of the chosen preventive agents must be revealed in the changes of organic-mineral balance of the oral fluid and it can be observed in the IR-spectra of the investigated samples.

The analyses of the obtained data reveal that a thorough oral rinsing 5 min after tooth brushing of the oral cavity with the use of the tooth paste (involving a calcium glycerophosphate) has completely nullified its positive effect directed at the change of organic-mineral balance of the oral fluid: S_3/S_1= +7.3±2.3 S_2/S_3= -9.4±2.7. Thorough oral rinsing in 30 minutes after the application of the tooth paste for hygienic procedures of the oral cavity showed just the same effect: S_3/S_1= -1.8±2.0 S_2/S_3= -2.0±1.8. In both cases one can observe a negative relative change of the organic-mineral ratio as compared with the reference sample thus meaning washing-out of the mineral groups and complexes remaining in the oral cavity after the application of the tooth paste.

Analyzing the results of the changes of mineral-organic (S'_3/S'_1=+11.8±1.9) and carbon-phosphate (S'_2/S'_3=-53.7±1.5) ratios after taking the tablets on the basis of a calcium glycerophosphate by the patients one can make a conclusion that after taking the tablets for three days an increase in the mineral-organic ratio was observed more than 10% as compared with the reference (check) sample of the oral fluid. All of the above facts mean that organic-mineral balance in the oral fluid was shifted to the increase in the content of the mineral groups and complexes after the application of the tablets comprising of mineral-vitamin complex on the basis of calcium glycerophosphate.

5. Conclusion

Interrelation between the oral fluid composition and efficiency of exogenous and endogenous methods of resistance to caries were shown in the work. Exogenous methods of prevention were found to provide short-term effect in maintaining organic-mineral balance of the oral fluid while endogenous ones resulted in a long-term presence of minerals and organic-mineral groups and complexes required for remineralisation of dental enamel. The organic-mineral balance in the oral fluid was shifted to the increase in the content of the mineral groups and complexes after the application of the tablets comprising of mineral complex on the basis of calcium glycerophosphate. The content of endogenous biologically accessible phosphate groups and complexes in the oral fluid, according to our data, exceeds by more than10% that in the oral fluid before the use of the prevention agents.

Acknowledgments

This work was supported by the grant of Russian Science Foundation, grant number 16-15-00003. The part of this research was undertaken with The Infrared Microspectroscopy (IRM) beamline at the Australian Synchrotron, Victoria, Australia.

References

[1] Fejerskov O, Nyvad B and Kidd E A M 2015 *Dental caries: the disease and its clinical management* (Chichester, West Sussex, UK ; Ames, Iowa: John Wiley & Sons Inc).
[2] Baffi M, Almeida Rodrigues J de and Lussi A 2012 *Contemporary Approach to Dental Caries* ed M-Y Li (InTech)
[3] Carvalho F B de, Barbosa A F S, Zanin F A A, Brugnera Júnior A, Silveira Júnior L and Pinheiro A L B 2013 *Braz. Dent. J.* **24** 59–63
[4] Seredin P, Goloshchapov D, Prutskij T and Ippolitov Y 2015 *PLoS ONE* **10** 1–11
[5] Featherstone J D B 2009 *Adv. Dent. Res.* **21** 4–7
[6] García-Godoy F and Hicks M J 2008 *J. Am. Dent. Assoc.* **139, Supplement 2** 25S – 34S
[7] Hara A T and Zero D T 2014 *Monogr. Oral Sci.* **25** 197–205
[8] Sathe N, Chakradhar Raju R V S and Chandrasekhar V 2014 *Kathmandu Univ. Med. J. KUMJ* **12** 16–20
[9] Memarpour M, Soltanimehr E and Sattarahmady N 2015 *Microsc. Res. Tech.* **78** 801–6
[10] Danelon M, Pessan J P, Neto F N S, de Camargo E R and Delbem A C B 2015 *J. Dent.* **43** 806–13
[11] Damle S G, Deoyani D, Bhattal H, Yadav R and Lomba A 2012 *Dent. Res. J.* **9** 68–73
[12] Reyes-Gasga J, Martínez-Piñeiro E L, Rodríguez-Álvarez G, Tiznado-Orozco G E, García-García R and Brès E F 2013 *Mater. Sci. Eng. C* **33** 4568–74
[13] Seredin P, Kashkarov V, Lukin A, Ippolitov Y, Julian R and Doyle S 2013 *J. Synchrotron Radiat.* **20** 705–10
[14] Siddiqui S, Anderson P and Al-Jawad M 2014 *PloS One* **9** e108879
[15] Elkins K M 2011 *J. Forensic Sci.* **56** 1580–7
[16] Ren Z, Do L D, Bechkoff G, Mebarek S, Keloglu N, Ahamada S, Meena S, Magne D, Pikula S, Wu Y and Buchet R 2015 *PLoS ONE* **10** e0120087
[17] Berzina-Cimdina L and Borodajenko N 2012 Research of Calcium Phosphates Using Fourier Transform Infrared Spectroscopy *Infrared Spectroscopy - Materials Science, Engineering and Technology* ed T Theophile (InTech)

Power law relation between particle concentrations and their sizes in the blood plasma

M N Kirichenko[1], **L L Chaikov**[1,2], **A R Zaritskii**[1]

[1] Lebedev Physical Institute of the Russian Academy of Sciences, Moscow 119991, Russia
[2] National Research Nuclear University "MEPhI", Moscow 115409, Russia

Abstract. This work is devoted to the investigation of sizes and concentrations of particles in blood plasma by dynamic light scattering (DLS). Blood plasma contains many different proteins and their aggregates, microparticles and vesicles. Their sizes, concentrations and shapes can give information about donor's health. Our DLS study of blood plasma reveals unexpected dependence: with increasing of the particle sizes r (from 1 nm up to 1 μm), their concentrations decrease as r^{-4} (almost by 12 orders). We found also that such dependence was repeated for model solution of fibrinogen and thrombin with power coefficient is -3,6. We believe that this relation is a fundamental law of nature that shows interaction of proteins (and other substances) in biological liquids.

1. Introduction

Study of sizes and concentrations of proteins and their aggregates in biological liquids as well as development of methods of their measurement is an important problem of modern analytical biophysics and medicine. Measurement of particle sizes in the range from one to hundreds of nm, in which the traditional microscopy fails, is of great interest. Dynamic light scattering takes an important place among different methods because of its noninvasive nature and the possibility of particle measurements in a wide range of sizes (from 1 to 3000 nm) without disrupting of sample integrity. However, processing of DLS data obtained for complex liquids, which contains more than three groups of scattering particles, sometimes demonstrates non-reproducible results.

The aim of the study was to investigate sizes and concentrations of scattering particles (proteins, their aggregates and vesicles) in blood plasma and model proteins solutions and to adopt DLS method to the case of liquid with wide distribution of particle sizes.

2. Methods and materials

For our study we used the traditional DLS setup with a He-Ne laser (the wavelength of 633 nm, the power of 5 to 10 mV), an optical system ensuring spatial coherence of recorded light, a goniometer allowing measurements at various angles, a special photomultiplier, a Photocor-FCm correlator (in the

«Multiple tau» regime), and a PC with Photocor-FCm and DynaLS programs. For a more detailed description of the setup see paper [1].

Blood plasma and protein solution samples 3 to 4 ml in volume were placed in a dust-free cylindrical cuvette 15 mm in diameter and an immersion cuvette 34 mm in diameter to ease the setup adjustment. Fresh blood was taken from the donor's ulnar vein to 10-ml syringes with heparin (anticoagulant). The blood plasma samples were obtained by centrifuging whole blood for 15 minutes on the lab centrifuge CLC-1 with the rate 3000 rev/min.

For model fibrinogen solution preparation we used fibrinogen from human plasma (Sigma Aldrich, ≥80% of protein is clottable) diluted in Tris-HCl buffer (7.2 pH). Final concentration of fibrinogen was about 2 mg/ml. After sample preparation solution was filtered through 0.2 μm "Superpure PVDF" filter.

Solution of thrombin was prepared using thrombin from human plasma (Sigma Aldrich, lyophilized powder, ≥2,000 NIH units/mg protein) that was diluted in Tris-buffer. Final activity of thrombin after adding it to fibrinogen solution was 0.00125 NIH/ml.

3. Results of plasma study with DLS method.

The DLS study of fresh blood plasma showed that intensity size distributions obtained by processing of autocorrelation functions with conventional DynaLS software were unstable from measurement to measurement. Adaptation of the DLS method is presented in our paper [2] where we demonstrated that the storage of a big array of DLS data (intensity size distributions) with their subsequent information compression allows obtaining stable particle sizes. Figure 1a demonstrates data array where every point corresponds to the particle size obtained at the certain time. The frequencies of registration of each size and their distribution are shown in figure 1b.

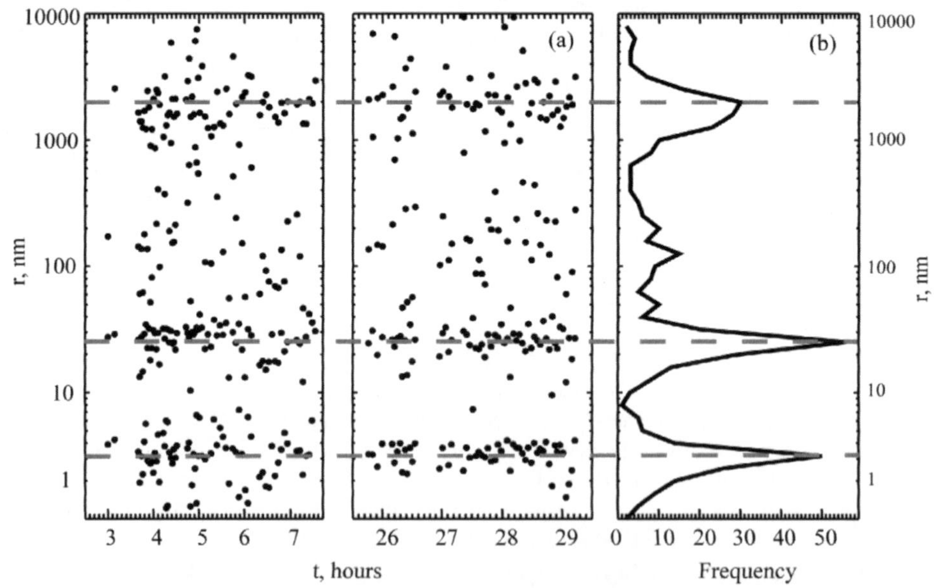

Figure 1 (a, b). (a) Array of data presented as the dependence of particle sizes (from intensity size distribution) on time of measurements. **(b)** Resulting frequency distribution of particle sizes.

3.1. Calculation of particles concentration from scattering intensity in blood plasma.

For each size presented on figure 1a we calculate the particle concentrations N from the Rayleigh-Gans-Debay approximation [3]:

$$N = I_{tot} \cdot A_i \cdot r_i \cdot \left(\sum A_i \cdot r_i\right)^{-1} \cdot \left(\eta_e \frac{V_{sc}}{R^2} \cdot |n_1/n_2 - 1|^2 \cdot G^2(\theta) \cdot r_i^6\right)^{-1} \qquad (1)$$

where I_{tot} is the total light intensity registered in experiment, $A_i \cdot r_i \cdot (\sum A_i \cdot r_i)^{-1}$ is the fraction of intensity of light scattered by particles of hydrodynamic radii r_i (recalculated from the logarithmic scale of sizes), n_1 is the refractive index of particle (for all groups of particle sizes one and the same refractive index of 1.34 was used), n_2 is the refractive index of water; $G^2(\theta)$ is the form-factor of particles for cylindrical shapes [4]. Unknown parameters (such as quantum efficiency of photomultiplier cathode η_e, scattering volume V_{sc}, distance between the point of observation and scattering volume R) we determined by comparing of our signal with light scattering intensity signal of toluene, scattering coefficient of which is known.

The relation between particle concentrations and corresponding sizes in a double logarithmic scale is plotted in figure 2. Approximation of this relation with straight line is presented with red line. Particle concentrations N turned out to depend on the sizes with a power law with an index of -4, i.e. $N=a \cdot r^{-4}$. Dilution of blood plasma in 10 and 100 times does not disturb such power law dependence [5].

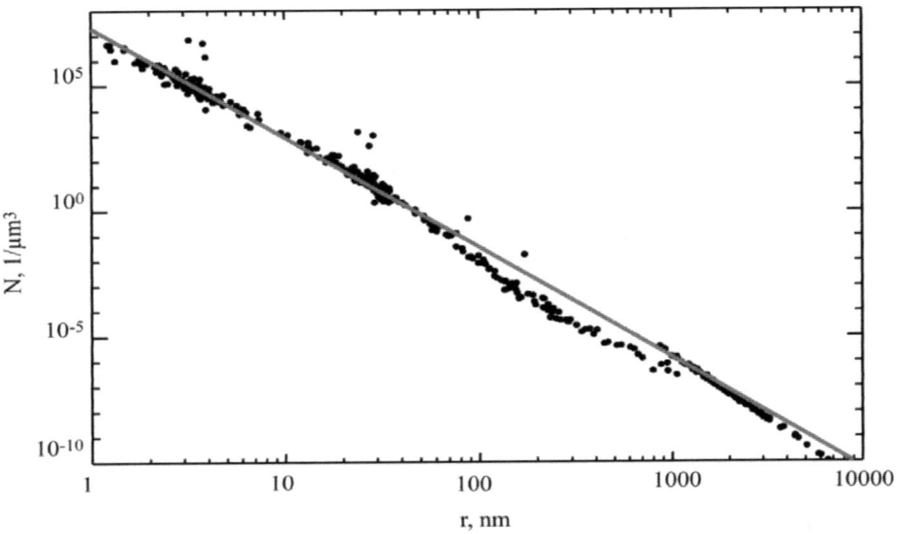

Figure 2. Relation between particles sizes and corresponding concentrations in a double logarithmic scale for nondiluted plasma (black points are concentration of particles with certain sizes and red line is power law approximation of these data).

4. Results of model experiments.

4.1. Model fibrinogen solution.

To understand the nature of the power law relation of concentrations and sizes in blood plasma we carried out the model experiments. It is known that blood plasma consists of different proteins: albumins, globulins, fibrinogen, etc. The processes of enzymatic degradation of proteins and their further aggregation take place there [6]. For our model experiment we chose fibrinogen from blood plasma and thrombin because these proteins are main components of blood cloating system.

We obtained intensity size distribution for fibrinogen solution and found out that the size of main peak that corresponds to fibrinogen monomers was approximately 13 nm (very close to data that were obtained by other authors). Besides the main peak we also identified another peak with bigger size (approximately 200-300 nm) despite of solution filtration before measurements. Apparently, this size coresponds to aggregates of fibrinogen. Calculation of concentrations of the proteins and their aggregates and matching of sizes and concentrations allow us to obtain the correlation size-concentration diagram (figure 3). One can see that the points on the diagram group near the mean value and form the spot (not like in the blood plasma). The diagram is not linear but for simplicity of comparison we fitted this one with linear dependence. Its index is 5.9 that is not very close to that one obtained for blood plasma.

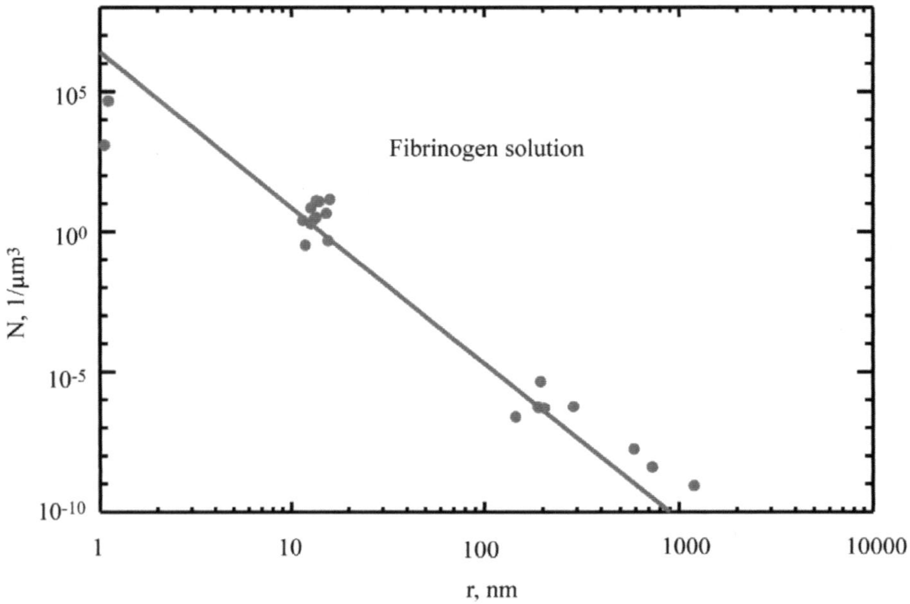

Figure 3. Relation between particles sizes and corresponding concentrations in a double logarithmic scale for model fibrinogen solution.

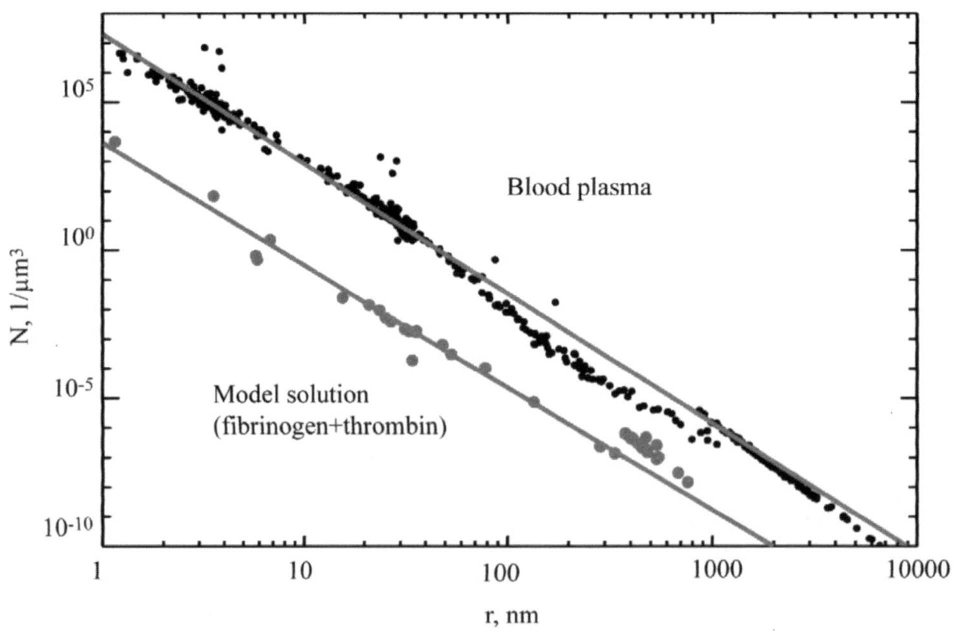

Figure 4. Relation between particles sizes and corresponding concentrations in a double logarithmic scale for nondiluted plasma (black points) and for model solution of fibrinogen with thrombin (red points).

4.2. Model fibrinogen-thrombin solution.

We also investigated intensisty size distribution in the fibrinogen-thrombin solution. All obtained intensity size distributions were treated in the same way like previous data on fibrinogen and blood palsma. Results of calculation of concentrations and matching them to corresponding sizes presented on the figure 4. Red points corresponds to fibrinogen-thrombin system and black points – to blood plasma. One can see that the slope of straight line in a double logarithmic scale is very close to this one in blood plasma (the index is 3.6±0.3), and one can not see the spot of points like in the case of fibrinogen solution.

Conclusions

There are a lot of empirical relations in nature that do not have a rigorous theoretical justification. Widely known Moore's law (Gordon E. Moore), according to which the number of transistors placed on an integrated circuit chip doubles every 24 months. It is also known Damuth's law (John Damuth), which shows that the density of population of mammalian N corresponds with a body mass m as $N = a{\cdot}m^{-3/4}$.We believe that empirical law $N{=}a{\cdot}r^{-4}$, which we found in our investigation, which is also the fundamental law of nature of biological liquids, but now there are no mathematical models to explain such dependence. We believe that relation between particle concentrations and sizes not only shows the distribution of particles in biological solution but also reflects the process of enzymatic degradation of proteins and their further aggregation in biological liquid.

References

[1] Kovalenko K V, Krivokhizha S V, Masalov A V and Chaikov L L 2009 *Bull. Lebedev Phys. Inst.* **36**(4), 65

[2] Chaikov L L, Kirichenko M N, Krivokhizha S V and Zaritskii A R 2015 *J Biomed Opt*, **20** (5), 057003.

[3] Barber P W and Wang D S 1987 *Appl. Opt.* **17**(5), 797.

[4] Neugebauer T 1943 *Ann. Phys.* **42**(7–8), 509.

[5] Kirichenko M N, Masalov A V, Chaikov L L and Zaritskii A R 2015 *Bull. Lebedev Phys. Inst.*, **42**(2), 3.

[6] Yi J, Liu Z, Craft D, O'Mullan P, Ju G and Gelfand C A 2008 *J. Proteome Res.* **7**, 5112.

Saint Petersburg OPEN 2016 IOP Publishing
Journal of Physics: Conference Series **741** (2016) 012056 doi:10.1088/1742-6596/741/1/012056

Activity of site-specific endonucleases on complexes of plasmid DNA with multiwalled carbon nanotubes

V P Egorova[1], H V Krylova[2], I V Lipnevich[2], A A Veligura[1], B G Shulitsky[3], A A Asayonok[1], E V Vaskovtsev[1]

[1]Belarusian State Pedagogical University, 18 Sovetskaya Str, 200005 Minsk, Belarus
[2]Physics Department, Belarusian State University, 4 Nezavisimosti Ave., 200030 Minsk, Belarus
[3]Belarusian State University of Informatics and Radioelectronics, 6 P. Brovki Str, 200013 Minsk, Belarus
E-mail: alla.veligura@mail.ru, grushevskaja@bsu.by

Abstract. We have synthesized and investigated structural and functional properties of plasmid DNA complexes with multi-walled carbon nanotubes (MWCNTs) for detection of changes in structural state of the plasmid DNA at its recognition by site-specific endonuclease. It has been also established that the site-specific endonuclease is functionally active on the surface of MWCNTs.

1. Introduction

Study of nuclease activity is important for understanding reasons of variability of deoxyribonucleic-acid (DNA) structure. Living cells utilize cutting and cleavage of DNA molecules by DNA-hydrolyzing enzymes: DNase, endonuclease, and other various nucleases to recognize and restrict genetic infection. Nuclease activity is proposed to use for revealing, for example, genes with single nucleotide polymorphisms through formation of a three-dimensional structure due to perfect hybridization of DNA with oligonucleotides and then subsequent cutting of the three-dimensional structure by nucleases [1]. Also, development of express methods of estimation of endonuclease activity is an actual biomedical problem of early diagnosis of cardiovascular disease [2]. For its solution the use of bionanosensors is promising [3]. However, the development of nanosensors for biochemical markers is constrained due to low sensitivity of nuclease activity detection. One solution to this problem is the use of multi-walled carbon nanotubes (MWCNTs), possessing unique electrophysical properties and biocompatibility [4]. Therefore, carbon nanotube-based sensorics of structural changes in DNA under the action of nucleases is perspective to diagnose and monitor diseases [5-7]. In connection with this, development of DNA-nanosensors based on complexes of DNA/carbon nanotube (DNA/CNT), which are designed for express DNA-diagnosis by changing of the structural state of nucleic acids under the action of nucleases, is relevant.

In this paper, to explore nuclease activity on a surface of MWCNTs the simplest model system is used, namely the site-specific restriction endonuclease (restrictase) – plasmid. In such a system, restrictase cleaves small covalently-closed circular double-stranded DNA molecules (plasmids) that coexist with a chromosome of the host bacterium.

A goal of this work is the development, synthesis and investigation of structural and functional properties of DNA/MWCNT complexes for detection of changes in structural state of plasmid DNA at its recognition by site-specific nucleases.

2. Materials and Methods

MWCNTs with diameters ranging from 2.5 to 7 nm and length of ~5–10 μm were obtained by the method of chemical vapor deposition (CVD-method) [8]. MWCNTs were covalently modified by carboxyl groups.

As a model plasmid DNA, the pBR322 plasmid (Fermentas, Lithuania) has been used, the DNA of which is characterized by homogeneity and existence of different structural forms. Plasmid pBR322 represents itself a covalently closed circular DNA molecule of 4363 bp in length. The pBR322 plasmid has several unique restriction sites indicated on the genetic map. If supercoiled pBR322 plasmid DNA was treated with a restriction enzyme which cleaves it at a single site located in one of genes for resistant to one or another antibiotics, then depending on time of restriction transition from the supercoiled to circular and then to linear double-stranded DNA takes place. Linearization (restriction) of plasmid DNA has been carried out by site-specific endonuclease EcoRI (Thermo Fisher Scientific, Germany).

2.5 μg of plasmid DNA pBR322 has been hydrolyzed by site-specific restriction endonuclease EcoRI (3 activity units) during appropriate time interval at 37 °C in 50 μl of Tango buffer, containing 33 mM Tris-HCl pH 7.9; 10 mM $MgCl_2$; 66 mM NaCl; 0.1 mg/ml BSA. One unit of activity has been defined as the amount of enzyme required to hydrolyze 1 μg of plasmid DNA in one hour at 37 °C in a 50 μl reaction mixture containing 1 × Tango buffer. At appropriate intervals, aliquots have been taken from the reaction mixture and the reaction was stopped by addition of EDTA into reaction mixture to a final concentration of 2 mM at the temperature 2 °C. Restriction products were analyzed by electrophoresis in agarose gel. Enzyme activity was estimated by electron microscopic analysis and electrophoresis of complexes plasmid DNA/MWCNT with and without the endonuclease in agarose gel.

DNA gel-electrophoresis was performed in 1% gel in 0.5 × TBE buffer 0.09 M (pH 8.0) at a voltage of 75 V and a temperature of 10-12 °C for 4 hours. As the DNA molecular weight marker, the "DNA molecular weight marker, M50Kb" (Praymteh, Belarus) has been used. The gel was stained in ethidium bromide solution (0.5 μg/ml) for 7 minutes, then washed three times in TE buffer (10 mM Tris-HCl; 1 mM EDTA, pH 8.0) with subsequent review and documentation of gels on the system ImageQuant 300 (GE Healthcare, USA).

Transmission electron microscopic (TEM) images were obtained by means of transmission electron microscope JEM-100CX (JEOL, Japan) at accelerating voltage of 100 kV. After the objects were previously deposited on a copper grid with a formvar polymer coating, structural analysis has been performed.

3. Structural Analysis

Let us denote an open circular DNA with cleavage in one of two strands and, respectively, with one sticky end by ocDNA. A supercoiled DNA, representing itself a closed double chain, is denoted by cDNA. The linear double-stranded helical DNA and single-stranded DNA are denoted by dsDNA and ssDNA, respectively.

TEM-image of a native plasmid DNA is shown in figure 1a. A supercoiled cDNA in MWCNT-surroundings is shown in figure 1b. As one can see, cDNA molecules do not form a complex with MWCNTs. Single-stranded ssDNA, obtained by dsDNA denaturation, forms a complex ssDNA/MWCNT, shown in figure 1c. Adhesion of ssDNA nucleotides on CNT surface proceeds due to π-π stacking interactions. Denaturated dsDNA molecules link closely situated MWCNTs in figure 1c. Figure 1c demonstrates also a helicoidal part of denaturated dsDNA, which links two CNTs. At that, unwound ends of this dsDNA are fixed on the MWCNT surface, as one of arrows in figure 1c indicates. ocDNA, which has only one closed chain, can adhere by sticky end to MWCNT, as figure 1d demonstrates.

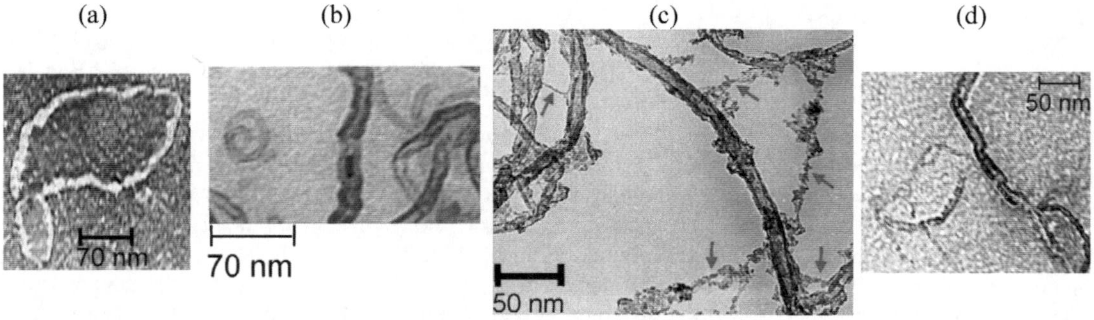

Figure 1. TEM-images of plasmid DNA molecules with different structure: (a) a native plasmid DNA; (b) a supercoiled cDNA in MWCNT-surroundings; (c) complexes ssDNA/MWCNT, arrows indicate denaturated dsDNA molecules linking closely situated MWCNTs; (d) open circular ocDNA adhered to MWCNT.

4. Electrophoretic Analysis

Electrophoresis in figure 2 allows to estimate time for which the restrictase EcoRI cleaves completely rings in closed double chain of plasmid DNA. This time is about of 1 hour. We observe that EcoRI cleaves the chemical bonds just in the ring as the concentration of linear dsDNA increases uniformly in time. The intensity of luminescence of dye-labeled cDNA and, respectively, its amount decreases uniformly in time. The intensity of luminescence of dye-labelled ocDNA and, respectively, its amount initially increases with time t, reaches a maximum at t = 15 min, and then decreases to near zero at t = 60 min. Therefore, we can assume that the strands of native plasmid DNA are cleaved one by one.

Electrophoretic lanes 3-6 in figure 3 show that the localization site of the native plasmid cDNA is narrowed in mixture of DNA with MWCNT. Since the intensity of this dye-labelled DNA from the mixture is practically independent on MWCNT concentration, it can be assumed that carbon nanotubes appear to orient DNA molecules as their localization site in lanes 4-6 is narrowed compared to lane 3. Interaction of MWCNT with the plasmid DNA affected by restrictases for 40 minutes results in narrowing of the lines and sharp decrease in luminosity of ocDNA localization cites in lane 8 in comparison with lane 7 for ethidium bromide stained gel electrophoresis in figure 3. π-π stacking interaction of ocDNA sticky end with MWCNT should lead to fixing whole dye-labelled ocDNA on the surface of the nanotube with subsequent fluorescence quenching. Decreasing the intensity of luminescence of dye-labelled ocDNA in lane 8 in comparison with lane 7 for electrophoresis in figure 3 confirms this assumption.

The electrophoresis lanes 3-5 in figure 4 show mobility of DNA subjected to restrictase digestion and then treated with increased amounts of MWCNTs. According to these lanes, the probability of binding with MWCNT increases not only for ocDNA, but for the cDNA and dsDNA as well, since the phenomenon of their fluorescence quenching by MWCNTs is observed. dsDNA has sticky ends that bind it to MWCNT. This explains the fluorescence quenching of dye-labelled dsDNA MWCNTs. The fluorescence of dye-labelled cDNA, not cleaved with a restrictase after incubation, is quenched by MWCNTs. It allows to suggest changes in the structure of supercoiled DNA due to its local destabilization by the enzyme. This destabilization facilitates the exit of the nucleotide bases and, respectively, leads to the formation of a complex between the cDNA and MWCNT. To prove the latter assumption, a short-term thermal destabilization of the native plasmid DNA, which was previously subjected to restrictase digestion, has been carried out.

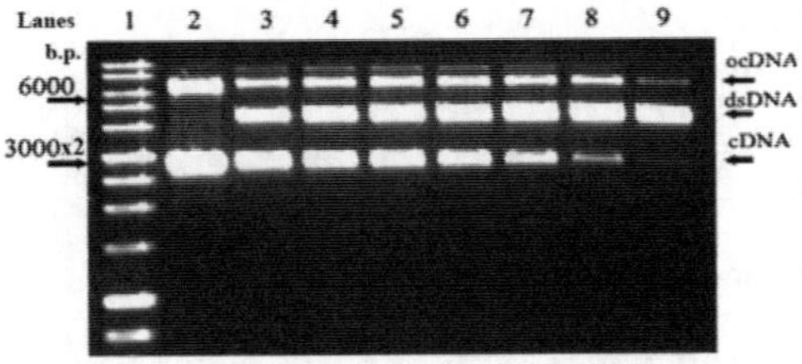

Figure 2. Electrophoregram of the plasmid-DNA hydrolysis with site-specific endonuclease EcoRI for different restriction time. Lanes: 1 – M50kb marker; 2 – native plasmid DNA (control); 3-9 – mixture of ocDNA, cDNA, and dsDNA, the restriction time equals to 5, 10, 15, 20, 30, 40, and 60 minutes, respectively.

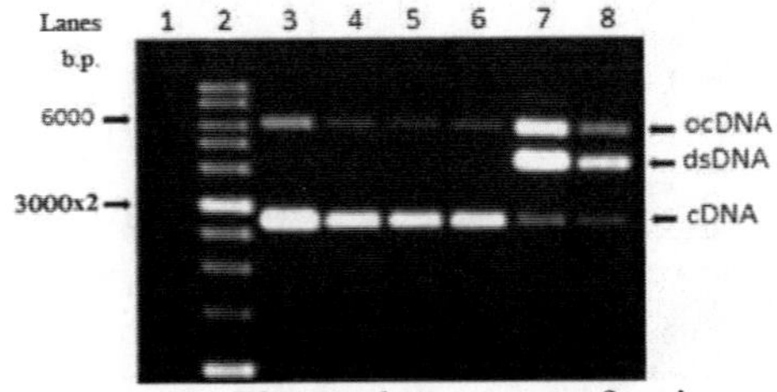

Figure 3. Electrophoregram of mixture from MWCNTs and native plasmid DNA. Lanes: 1 – MWCNTs; 2 - M50kb marker; 3 - native plasmid DNA (control 1); 4-6 – cDNA + MWCNT for nanotube concentration C_{MWCNT} 50, 100, and 200 μg/ml, respectively; 7 – mixture of cDNA, ocDNA, and dsDNA, the restriction time is 40 min (control 2); 8 – mixture of cDNA, ocDNA, and dsDNA+ MWCNTs , C_{MWCNT} =50 μg/ml.

Lanes 6-8 of electrophoresis in figure 4 gives information about the adsorption ability of thermally destabilized mixture of plasmid DNA with the endonuclease EcoRI on the surface of MWCNT. Comparison of lane 6 with lanes 7 and 8 in figure 4 gives the following. Denatured (single-stranded) DNA molecules in the free state have been not detected in the mixture of fermented DNA + MWCNTs already at concentration MWCNT C_{MWCNT} equal to 50 μg/ml (lane 7, figure 4). Thermally destabilized ocDNA is practically all linked at C_{MWCNT} = 100 μg/ml (lane 8, figure 4). Similarly, since the emission intensity I_{cDNA} for the free cDNA (lane 8, figure 4) at C_{MWCNT} = 100 μg/ml is significantly less I_{cDNA} (lane 7, figure 4) at C_{MWCNT} = 50 μg / ml, the thermal destabilization of cDNA leads to increase in the probability of the cDNA/MWCNT complex formation. Comparison of lanes 4 and 5 with lanes 7 and 8 in figure 4 for dsDNA shows that the thermal destabilization does not practically change the probability of dsDNA/MWCNT complex formation. Therefore, the electrophoretic analysis shows that the lifetime of thermally destabilized state of dsDNA is much smaller than for ocDNA and cDNA due restrictase linked to the latter.

Thus, the electrophoretic analysis shows, on the one hand, a large value of the formation probability for complex of carbon nanotubes with denatured DNA compared to other types of DNA-structures and, on the other hand, thermal enhancement of restrictase destabilization of cDNA and ocDNA.

Further, we examine the effect of MWCNTs on enzymatic activity of site-specific endonucleases and show that dsDNA is formed by transition of double destabilized (unwound) chain of dsDNA-molecule in a low-energy state of the double helix. According to comparison of lanes 7-9 in figure 2 with 3-4 in figure 5, restriction-nuclease concentration increase from 3 to 6 activity units is equivalent to increasing of restrictase-digestion time from 30 to 60 minutes. Since the fluorescence intensity for free dye-labelled cDNA from the mixture of cDNA + MWCNTs subjected to restrictase digestion (lane 8, figure 5) is decreased compared to the control (lane 5, figure 5), the restrictase functionality is retained in the mixture of cDNA + MWCNTs. Electrophoresis lanes 5-7 and 8-10 in figure 5 show a dependence of motilities on amounts of MWCNTs for complexes of cDNA with MWCNT and for complexes between MWCNT and cDNA subjected to restrictase digestion, respectively.

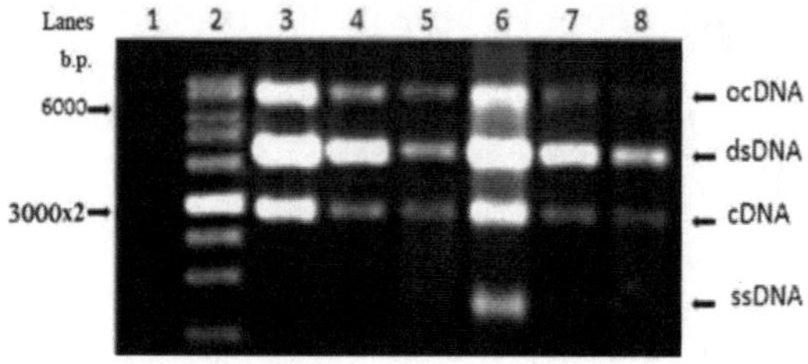

Figure 4. Electrophoregram of complexes ssDNA/MWCNT. Lanes: 1 – MWCNTs; 2 – M50kb marker; 3 – mixture of cDNA, ocDNA, and dsDNA, the restriction time is 30 min (control 1); 4 and 5 – mixture of cDNA, ocDNA, and dsDNA + MWCNT, C_{MWCNT} = 50, 100 μg/ml (control 2); 6 – mixture of sDNA, ocDNA, dsDNA, and ssDNA, the restriction time is 30 min; 7 and 8 – mixture of cDNA, ocDNA, dsDNA, and ssDNA + MWCNT, C_{MWCNT} = 50, 100 μg/ml.

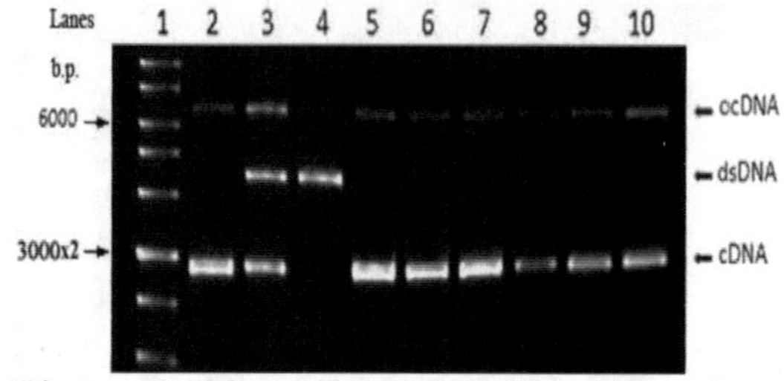

Figure 5. Electrophoregram of mixture from MWCNTs and native plasmid DNA after restriction enzyme treatment. Lanes: 1 – M50kb marker; 2 – native plasmid DNA (control 1); 3, 4 – mixture of cDNA, ocDNA, and dsDNA, restriction-enzyme concentration 3 and 6 activity unit respectively, the restriction time is 30 min (control 2); 5-7 – native plasmid DNA + MWCNT at C_{MWCNT} equal to 50, 100, and 200 μg/ml, respectively; 8-10 – mixture from native plasmid DNA and MWCNTs after treatment with a restriction enzyme at C_{MWCNT} = 50, 100 and 200 μg/ml, respectively; the restriction time is 45 min.

We observe an increase in the emission intensity of free dye-labelled cDNA and ocDNA in the mixture of cDNA + MWCNT subjected to restrictase digestion with an increase of MWCNT concentration (lanes 8-10, figure 5). It can be explained by an inactivation of the restrictases during adhesion to the surface of the nanotubes.

The observed absence of free dsDNA in the mixture of cDNA + MWCNTs which is subjected to restrictase digestion, regardless of MWCNT concentration (lanes 8-10, figure 5) is explained by sufficiently long lifetime of unstable DNA-structure formed immediately after the cleavage by restrictase of the second ringed DNA- chain. This time is sufficient for formation of the complex of destabilized linear dsDNA with MWCNT.

5. Discussion and Conclusions

So, the investigation performed demonstrates that supercoiled DNA molecules do not form a complex with MWCNTs. Endonucleases are functionally active on the surface of MWCNTs. An interaction of plasmid DNA with restrictase on nanotube surface amplifies the dsDNA binding with MWCNT. The changes in the structure of dsDNA are due to its local destabilization by the enzyme and adsorption of "sticky" ends of the dsDNA on the surface of MWCNTs on account of π–π-stacking interaction. However, increasing of MWCNT concentration leads to decrease of luminescence intensity of free open-circular and supercoiled DNA molecules in the gel. It can be explained in the following way. Restrictases change a conformation on the surface of MWCNTs and, therefore, are inactivated. Thus, plasmid DNA in various conformational forms (supercoiled, open circular, double-stranded and single-stranded linear (denatured)) can form thermodynamically stable complexes with MWCNTs in the presence of endonucleases only.

References

[1] Sobrino B, Brion M, Carracedo A 2005 *Forensic Sci Int.* **154** 181
[2] Cherepanova A V, Tamkovich S N and Vlasov V V 2007 *Biomedical chemistry* **53** 488
[3] Chen R J, Bangsaruntip S, Drouvalakis K A, Shi Kash N W, Shim M, Li Y, Kim Woong, Utz P

J and Dai H 2003 *PNAS* **100** 4984
[4] Egorova V P, Krylova H V, Lipnevich I V, Veligura A A, Shulitsky B G and Fedotencova L Y 2015 *J. Phys*. CS **643** 012023
[5] Kerman K, Kobayashi M, Tamiya E 2004 *Meas. Sci. Technol.* **15** Rl
[6] Iost M R, da Silva W C, Madurro J M, Madurro A G, Ferreira L F, Crespilho F N 2011 *Front Biosci (Elite Ed.)* **1(3)** 663
[7] Abdullin T I, Bondar O V, Rizvanov A, Nikitina I I 2009 *Applied Biochem. Microbiol.* **45** 229
[8] Labunov V, Shulitski B, Prudnikava A, Shaman Y P, Basaev A S Quant 2010 *Electron. Optoelectron.* **13** 137

Neural signal registration and analysis of axons grown in microchannels

Y Pigareva[1*], E Malishev[2], A Gladkov[1, 3], V Kolpakov[1], A Bukatin[2], I Mukhina[1,3], V Kazantsev[1] and A Pimashkin[1]

[1] Department of Neuroengineering, Center of Translational Technologies, Lobachevsky State University of Nizhny Novgorod, Nizhny Novgorod, Russia
[2] Saint-Petersburg Academic University, Nanotechnology Research and Education Centre of the RAS, Saint-Petersburg, Russia
[3] Central Research Laboratory, Cell Technology Department, Nizhny Novgorod State Medical Academy, Nizhny Novgorod, Russia

[*] corresponding author. E-mail: pigareva@neuro.nnov.ru

Abstract. Registration of neuronal bioelectrical signals remains one of the main physical tools to study fundamental mechanisms of signal processing in the brain. Neurons generate spiking patterns which propagate through complex map of neural network connectivity. Extracellular recording of isolated axons grown in microchannels provides amplification of the signal for detailed study of spike propagation. In this study we used neuronal hippocampal cultures grown in microfluidic devices combined with microelectrode arrays to investigate a changes of electrical activity during neural network development. We found that after 5 days *in vitro* after culture plating the spiking activity appears first in microchannels and on the next 2-3 days appears on the electrodes of overall neural network. We conclude that such approach provides a convenient method to study neural signal processing and functional structure development on a single cell and network level of the neuronal culture.

1. Introduction

An Action Potential (AP) generated in neuron propagates through an axon with amplitude about 100 mV. Such signal mostly occurs inside cell body while extracellular measurements record AP as 20-100 μV spikes. It was shown that extracellular voltage potentials of axons proportionally depend on the resistance of extracellular medium [1]. The resistance can be significantly increased by reducing extracellular medium volume around axon near recording electrode which can be done by growing axons in microfluidic microchannel with 5-10 μm diameter. AP signals of isolated axons in microchannels combined with extracellular recording on microelectrode arrays (MEA) can be amplified up to 4.5 mV [2]. Such method allows to study axonal signal transmission for various medical applications [3,4], implantable neurointerfaces [1] and information processing in axons. Microfluidic methods allow to control morphology of the neural culture of dissociated neurons to study functional characteristics of realistic unidirected synaptic pathways in the neural tissue. Specific form of the microchannels (figure 1c) can be used to guide axon outgrowth in preferred direction between chambers [5, 6]. Such directed connectivity provides a directed spike propagations and localization of pre- and postsynaptic neurons in the network for fundamental studies of synaptic plasticity, neural information coding/decoding, memory detalization and learning on the network level of neural tissue.

However axonal signals in asymmetric microchannels with specific geometry [6] has not been yet studied. Axonal signals in microchannels recorded on early stages of the neural culture development during axon growth in specified direction can be used to uncover mechanisms of neuronal connection appearance. We used the method to investigate the relations between electrical activity and morphology on the cellular and network wide level of the neural culture organization.

2. Materials and methods

2.1. Microfluidic device fabrication

General design of the microfluidic device is shown in figure 1 (a). Microelectrode array (MEA) was precisely positioned and connected with poly-dimethylsiloxane (PDMS) microchannels, connecting two chambers of the device. The mcrofluidic chip was fabricated by pouring PDMS (Sylgard 184, Dow Corning) onto the structured molds and curing at 70°C for at least 4 h. Then PDMS chips were removed from molds and chamber structures were punched. Finally, each PDMS chip was positioned and mounted onto the surface of a planar microelectrode array (MEA). For mold fabrication, standard two layer lithography was used: silicon wafers were spin-coated with SU-8 2025 negative photoresist (MicroChem, USA), diluted by the SU-8 Thinner (MicroChem, USA) for the first layer (4.5 μm thick) and SU-8 2075 for the second layer (~200 μm). Exposure was performed through a chromium photomask on MJB4 (SUSS Microtec, Germany) with an UV-filter PL-360-LP (Omega Optics, USA). Patterns were developed by using SU-8 Developer (MicroChem, USA).

2.1. Cell culture and electrophysiology recording

Hippocampal neuronal cells were dissociated from embryonic mice (E18) and plated on microelectrode arrays (MEAs) (Multichannel systems, Germany) at a final density of approximately 15,000–20,000 cells/mm2 [6]. The MEAs consisted of 60 electrodes with the diameter of 30 μm and 200 μm interelectrode space. Electrophysiological signals of spontaneous spiking activity were recorded with USB-MEA120-Inv-2-BC-System (Multichannel systems, Germany) and analyzed with custom made software Meaman developed in Matlab. The experiments were performed from 5 to 12 days in vitro (DIV) after culture plating. For more details of the experimental procedures see [6,7].

3. Results and Conclusions

3.1. Results

In our previous study [6] we developed the microfluidic devices which consisted of two chambers with neuronal subcultures connected by microchannels for axons and dendrites outgrowth (figure 1b, c). The device was placed on microelectrode array to record neuronal bioelectrical activity in microchannels and chambers independently (figure 1a). Because of specific form of microchannels axons grew mostly in one direction: from presynaptic subculture (left chamber) to postsynaptic (right chamber). We investigated electrophysiological signals recorded from the electrodes in microchannels (axonal spikes) and chambers (somata, dendritic and axonal signals).

The signal from the microelectrode composed of extracellular potentials from axon, dendrites and neuron's somata. Amplitude of the potentials proportionally depended on extracellular resistance that mostly depended on extracellular medium volume around axon near the recording electrode. The structure of the microfluidic microchannels limited a volume of extracellular medium around axon increasing its resistance and hence extracellular potential. Example of signal amplitude amplification in the microchannel is shown on figure 1 (d, right), in contrast to regular spiking activity recorded from neurons on the electrode where the signal is lower (figure 1d, left).

Microscopic image (20x) of electrodes in the microchannel and the chamber are shown on figure 1 (e) on the right and left respectively.

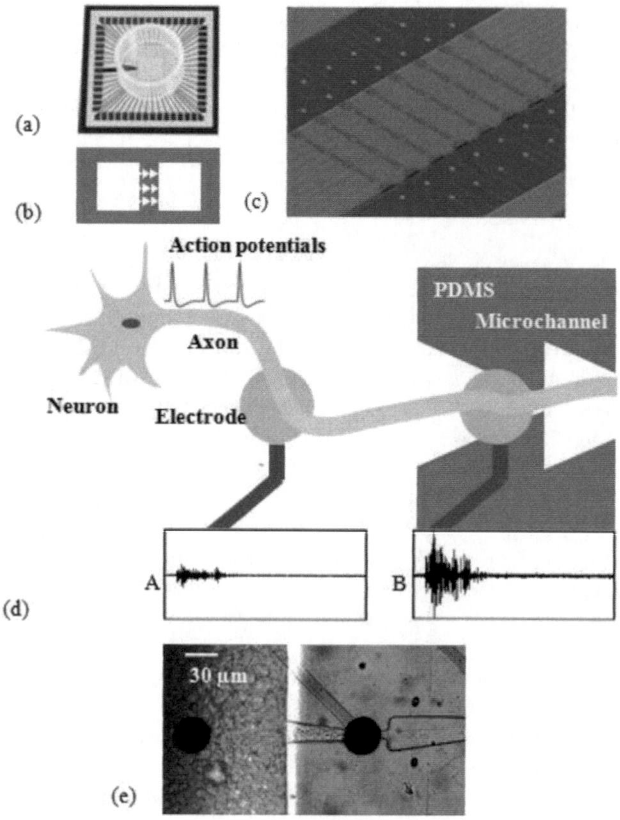

Figure 1. Hippocampal culture grown on MEA in microfluidic chip with two chambers connected by microchannels. (a) Microelectrode array with 60 electrodes combined with microfluidic device (b) with 8 microchannels. (c) Scheme of microfluidic device with two chambers connected by microchannels with length of 600 µm. (d) Schematic representation of signal registration in the microfluidic structure and examples of recorded signals. (e) Hippocampal cells grown on microelectrode array in microchannels at the places illustrated in (d). Microscope photo was made with 10x objective on 5 DIV.

We recorded the spiking activity on the MEA starting from 5th day of development *in vitro* (DIV) every two days until DIV 12. In one culture on 5 DIV the spikes were obtained in all electrodes of the MEA, but a firing rate in the microchannels was greater (40-300 Hz) than in the chambers (20-50 Hz). On 6 DIV spontaneous bursting activity appeared in the microchannels. The bursts were characterized by rapid spiking periods about 200-500 ms that followed by silence phase. The bursts in the chambers appeared on 8 DIV. Example of the recorded signal from electrodes in chamber and microchannels are shown in figure 2 (a). Note that due to the amplification inside the microchannel the noise level of the signal was also higher than in the chamber (figure 2b). Higher noise level leads to increase in absolute value of spike detection threshold estimation. Distributions of spike amplitudes observed for one electrode in the microchannel and one in the chamber were presented in figure 2 (c). The activity from microelectrodes in chambers and microchannels on different days of development is presented on figure 2 (e), (f). Note that even in case where the electrode has a minimal coverage area with axon in the microchannel bursts were still registered (figure 2d).

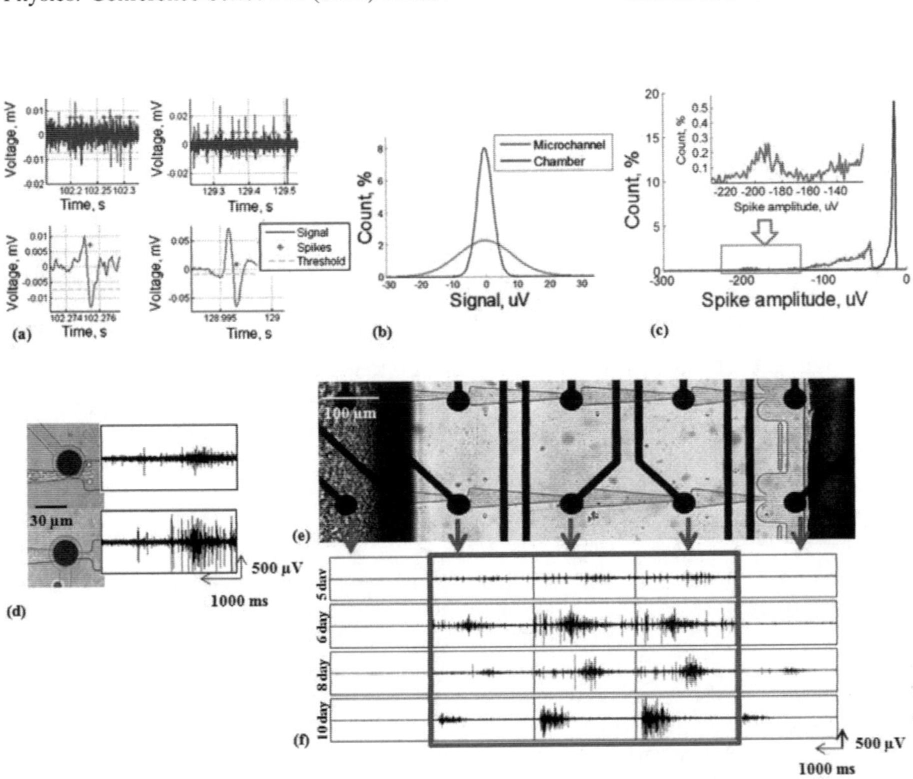

Figure 2. Neuronal signal amplification in microchannels. (a) Example of signal and one spike recorded in chamber (left) and microchannel (right). Spikes marked by asterisks. (b) Distribution of 60 seconds of the signal recorded from microchannel (red) and chamber electrodes (blue). (c) Distribution of spike amplitudes recorded from microchannel (red) and chamber electrodes (blue). (e) Microscope photo of microchannel with axons and two chambers with neurons. (f) Activity recorded from microelectrodes in chambers and microchannels (red box) on different days of development. Bursts in microchannels appeared earlier than in chambers and demonstrated greater amplitude.

Next we investigated electrophysiological characteristics of the 5 neuronal cultures during culture development. Mean firing rate, number of bursts in microchannels and chambers and amplitude amplification during each 2 days are presented on the figure 3 (n=5 cultures). Firing rate increased during culture development and was always greater in microchannels (figure 3a). Number of bursts per minute in microchannels and chambers increased until 8-10 DIV and then decreased. Number of bursts dominated in microchannels. Importantly that the bursting activity appeared in microchannels two days before it was obtained in chambers (figure 3b). We estimated a ratio between the spike amplitude recorded in the microchannels and chambers during each burst. The amplitude ratio in microchannels was about 1.4 in three experiments, 3 and 4.7 in other two cultures (figure 3c). Note that in average the ratio was relatively stable during the culture development.

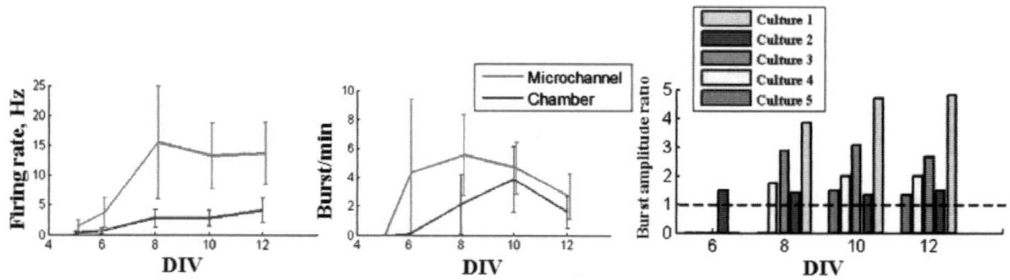

Figure 3. Changes of neuronal activity in microchannels during early culture development. (a) Mean firing rate measured per electrode in micrichannels (red) and chambers (blue). (b) Bursting frequency per minute obtained in microchannels (red) and chambers (blue). (c) Ratio of spike amplitudes registered in microchannels and chambers during bursts.

3.2. Conclusion

In this study we investigated the spiking activity of hippocampal cultures grown on microelectrode arrays combined with microfluidic device with microchannels to guide axon growth in specific direction. Our previously designed microchannels consisted of asymmetric blocks with specific shape in order to maximize probability of axon growth in one direction [6]. We used amplification feature of the extracellular signal in microchannels to obtain single cell activity on early stage of the culture development. Similar to symmetric shaped microchannels [2,3] our structure was also capable to enhance axonal signal due to small amount of the medium around axons (figure 2). Axonal spikes were found to be in the range of 30-300 μV while the spikes from neurons in chambers were 20-50 μV. We obtained a high variability across different microchannels within the single culture. We suggested that the density of survived cells on the 1st and 2nd day after culture plating may vary which contributes to the number of axons growing inside the channel. The signal also depended on the overlap area between the electrode and the microchannel. These questions will be addressed in further study.

The bursting activity in the microchannels was obtained earlier (6-8 days in vitro) than bursting activity of the whole culture network (starting from 8-10 days in vitro). Thus the synchronized bursting activity in the network started to develop in certain fraction of the culture and could not be observed in neuron's somata (chambers). Note that axonal bursts were synchronous in all 8 microchannels which suggest that all neurons in the network contribute to global burst generation. We also obtained the activity in microchannels when it was undetectable in the rest of the culture. The mechanisms of the effect will be explored in future studies and may uncover fundamental features of functional connectivity development in neural networks. The results suggest that presented microfluidic platform provides new physical methods to study spike pattern generation and transmission in neuronal cultures. The amplification effect in microchannels can be also used to develop bioartificial neurointerfaces and neuroimplants. High signal-to-ratio characteristics and spikes in range of millivolts will simplify signal registration methods and reduce applied electronics requirements.

Acknowledgments
This research was supported by the Russian Science Foundation (№ 14-19-01381).

References

[1] FitzGerald J, Lacour S, McMahon S, Fawcett J. 2008 Microchannels as axonal amplifiers. IEEE Trans Biomed Eng. 55: 1136–1146. doi: 10.1109/TBME.2007.909533 PMID: 18334406
[2] Pan L, Alagapan S, Franca E, DeMarse T, Brewer G, Wheeler B 2013. Large Extracellular Spikes Record- able From Axons in Microtunnels. IEEE Trans Neural Syst Rehabil Eng

22(3): 453–459 doi: 10. 1109/TNSRE.2013.2289911 PMID: 24240004

[3] Lewandowska M, Bakkum D, Rompani S, Hierlemann A. 2015 Recording large extracellular spikes in microchannels along many axonal sites from individual neurons. PLoS ONE, 10(3), 1–24

[4] Habibey R, Golabchi A, Blau A. 2015. Neurotechnology, Electronics, and Informatics, 13.

[5] Le Feber J, Postma W, de Weerd E, Weusthof M., Rutten W 2015 Barbed channels enhance unidirectional connectivity between neuronal networks cultured on multi electrode arrays. Frontiers in Neuroscience, 1–10

[6] Malishev E, Pimashkin A, Gladkov A, Pigareva Y, Bukatin A, Kazantsev V, Mukhina I, Dubina M 2015 Microfluidic device for unidirectional axon growth. Journal of Physics: Conference Series, 643, 012025

[7] Pimashkin A, Gladkov A, Mukhina I, Kazantsev V 2013 Adaptive enhancement of learning protocol in hippocampal cultured networks grown on multielectrode arrays. Front. Neural Circuits 7:87. doi: 10.3389/fncir.2013.00087

Dissolution and mixing of flavin mononucleotide in microfluidic chips for bioassay

K I Belousov[1,2], I A Denisov[2], K A Lukyanenko[2], A S Yakimov[2], A S Bukatin[2,3], I V Kukhtevich[2], V V Sorokin[2], E N Esimbekova[2,4], P I Belobrov[2,4] and A A Evstrapov[1,3,5]

[1] Department of material science and nanotechnology, ITMO University, St. Petersburg 197101, Russia
[2] Department of biophysics, Siberian Federal University, Krasnoyarsk 660041, Russia
[3] Nanobiotech Lab, St. Petersburg Academic University, St. Petersburg 194021, Russia
[4] Laboratory of Photobiology, Institute of Biophysics SB RAS, Krasnoyarsk 660041, Russia
[5] Laboratory of information and measurement biosensor and chemosensor microsystems, Institute for Analytical Instrumentation RAS, St. Petersburg 198095, Russia

E-mail: belousov_k.i@mail.ru

Abstract. Dissolution and mixing of flavin mononucleotide (FMN), which activates a luminescent reaction, were considered in various designs of microfluidic chip for pollution analysis of liquid samples. The aim was to determine the velocity mode of fluid flow ensured the uniform distribution of the FMN in the reaction chamber. Simulation of concentration distribution of FMN in various designs of microfluidic chips was conducted. It was shown that the passive mixing techniques based on the constant flow rate didn't provide mixing of FMN in acceptable time (3 seconds). The most efficient mixing was achieved using variable flow rate with a gradually increasing frequency of oscillation.

1. Introduction

The use of microfluidic technique and bioluminescent method increases an accuracy and sensitivity of a biological assay of liquid samples on toxicity, reduces the cost of analysis and its time compared to traditional methods using experimental animals and protozoa [1, 2]. One can make conclusions about the degree of a pollution of a sample by measuring a change in luminescence intensity associated with a reduction in the catalytic activity of enzymes in the presence of pollutants [3-5]. In the considered method the bioluminescent reaction starts when flavin mononucleotide (FMN) mixed with other components of the reaction stored in gel film solved in a reaction chamber. As FMN activates the reaction of light emission [6] it is important to have an uniform concentration of FMN in all parts of the reaction chamber. For example, poor and slow mixing can lead to that the reaction won't come to its peak, and the analysis won't be possible. FMN is initially immobilized in dried form in a microfluidic chip, so various designs of microfluidic chips and velocity regimes of fluid flow were considered using a numerical simulation to provide fast dissolution and efficient mixing of FMN.

2. Methods

Simulation was performed by means of COMSOL Multiphysics software for solving differential equations based on finite element method. The Navier-Stokes equations were solved to calculate the velocity profile of the fluid:

$$\rho\left(\frac{\partial \mathbf{u}}{\partial t}+\mathbf{u}\cdot\nabla\mathbf{u}\right)=-\nabla p+\mu\nabla^2\mathbf{u} \tag{1}$$

$$\nabla\cdot\mathbf{u}=0$$

where $\mathbf{u}$ is the flow velocity vector (m/s), ρ is the fluid density (kg/m^3), μ is the fluid viscosity (Pa·s), p is the pressure (Pa).The Fick's second law with the added convective term was used for modeling the concentration distribution:

$$\frac{\partial c}{\partial t}+\mathbf{u}\cdot\nabla c=D\nabla^2 c \tag{2}$$

where c is the bulk concentration of FMN (mol/m^3), $D = 4.8 \cdot 10^{-10}$ m^2/s is its diffusion coefficient [7]. The simulation of dissolution process was based on Noyes–Whitney equation. On the boundary related to the placement of dissolved reagent the equations for the rate of dissolution process, the change of FMN boundary concentration c_s and reagent flux N were applied with the following conditions:

$$\text{if } c_s > c, \text{ then } \frac{dQ_h}{dt}=(c_s-c), \text{ else } \frac{dQ_h}{dt}=0 \tag{3}$$

$$\text{if } Q_h > Q_s, \text{ then } \frac{dc_s}{dt}=0, \text{ else } \frac{dc_s}{dt}=-1000\cdot c_s \tag{4}$$

$$N=\frac{dQ_h}{dt} \tag{5}$$

where Q_h is the quantity of dissolved FMN from unit area (mol/m^2), Q_s = 0.005 mol/m^2 is initial quantity of dried FMN per unit area. As initial condition c_s is equal to saturated concentration c_{sat} = 200 mol/m^3 [8].

The quality of mixing was determined by calculating the variance of concentration normalized by average concentration c_{av}:

$$\sigma^2 = A^{-1}\iint\left(\frac{c}{c_{av}}-1\right)^2 dA \tag{6}$$

where A is a total area of cross-section (m^2) and dA is its element.

To characterize the flow the Reynolds number Re and Womersley number α were evaluated:

$$Re=\frac{U\rho L}{\mu} \tag{7}$$

$$\alpha=L\left(\frac{2\pi f\rho}{\mu}\right)^{1/2} \tag{8}$$

where L is an appropriate length scale (m), U is average velocity (m/s), f is frequency of the oscillations (Hz). The Reynolds number describes the transition between laminar and turbulent flow regimes. The Womersley number characterizes the changes of velocity pattern at pulsatile flow.

During the work the designs shown in figure 1 were considered. The depth of their structures was 0.5 mm.

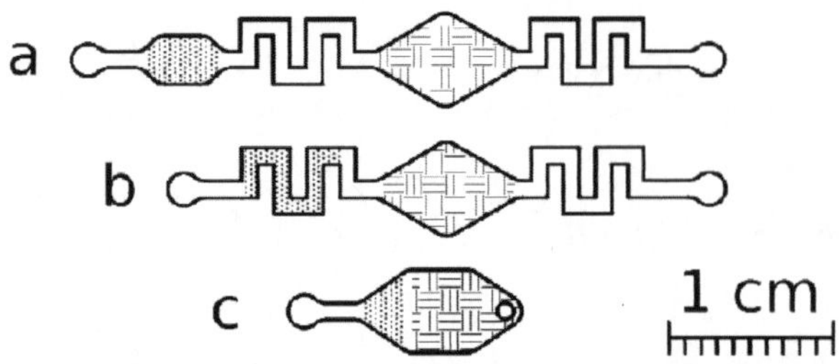

Figure 1. The designs of microfluidic chips: (a), (b) for passive mixing; (c) for active mixing. Stripes indicate the place of immobilization of reagents for the bioluminescent reaction, the points indicate immobilized FMN

3. Results and discussion

The distribution of FMN concentration was obtained for designs shown in figure 1 (a, b) during its dissolution by exposure constant fluid flow with characteristic download velocities (0.04 and 0.08 m/s). It was found that dissolution process was going rather fast and dissolution time didn't exceed 1 second. Immobilization of FMN in serpentine microchannel provided the fastest dissolution. So with the average flow velocity 0.08 m/s the 99% of reagent dissolved in 0.67 s in the channel, while in the chamber it took 0.78 s. With the velocity decreasing to 0.4 m/s the dissolution time increased by 0.11 s in serpentine channel and by 0.05 s in the chamber. The acceleration of dissolution in the serpentine channel mainly related to Dean vortices, which provided vertical convection of reagent. For this reason, the growth of flow velocity had a stronger effect on the dissolution time in the serpentine channel: it not only reduced the diffusion layer thickness, but also actively produced vortices. Checking this statement the study of dissolution in the straight channel with the same length was conducted. The dissolution time in this case was close to that for the chamber.

Although this scheme provided a rather rapid dissolution, efficient mixing was not achieved and reagent mainly located at the bottom of the channel with the variance down to 6.5. This was due to the initial very uneven distribution of the reagent and its position at the bottom of the channel, whereby Dean vortices drove reagent only at down part of the channel. In addition, the FMN flux through the cross-section of the channel showed significant time dependence (figure 2). There was increase in flux associated with the beginning of the dissolution and decrease caused by the completion of the dissolution process and the start of reagent washout. Thus, the passive mixing in suggested designs hasn't provided the uniformity of sample injection.

The experimental studies confirmed the simulation results that passive dissolution and mixing in considered designs don't provide required quality of reaction: kinetics of the reaction had no maximum, which made quantitative analysis impossible.

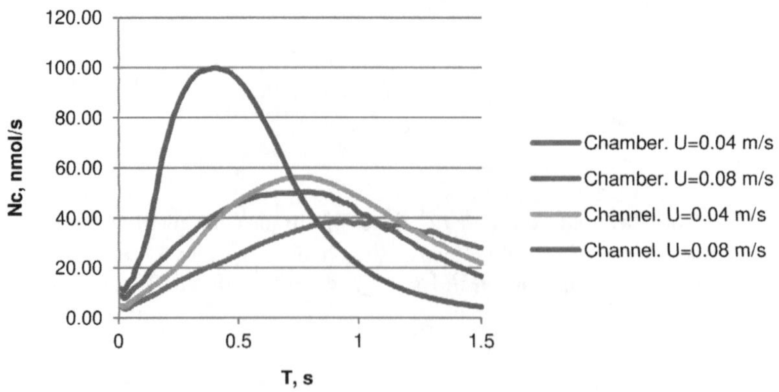

Figure 2. The flux of FMN N_c through the cross section of the channel during the time T for FMN immobilization in the chamber and in the serpentine channel for different average fluid velocities U

The active influence on FMN mixing was conducted by using alternating flow velocity in the design shown in figure 1 (c). During analysis two-dimensional formulation was used and only FMN mixing was considered. Its initial distribution was set at FMN immobilization area marked by points in figure 1 (c). The simulation area was changed as shown in figure 3 to provide the same velocity pattern as at the 3D version.

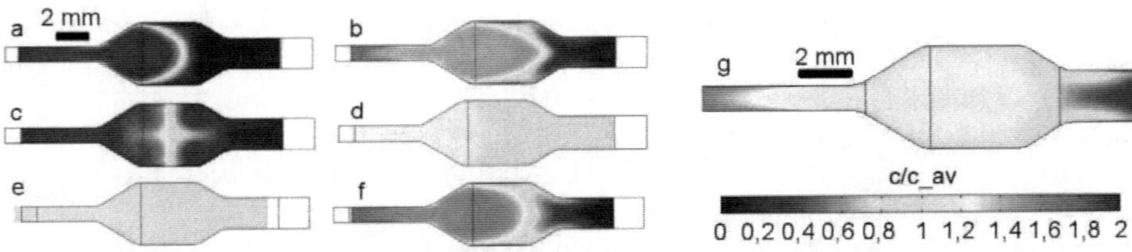

Figure 3. The distribution of FMN concentration after 3 seconds sinusoidal wave oscillations with frequency: (a) 5 Hz; (b) 10 Hz; (c) 20 Hz; (d) 50 Hz; (e) 75 Hz; the distribution of FMN concentration after 3 seconds rectangular wave oscillations with uniformly increasing frequency from 1 to 10 Hz and peak-to-peak amplitude: (f) 3 ul; (g) 4 ul.

Variance of concentration distribution of FMN was calculated for sinusoidal flow rate oscillations with peak-to-peak amplitude 3 ul and frequencies changed from 5 to 75 Hz (figure 3 (a-e)). As the varying volume was kept constant with the change of frequency, the average velocity was increased and hence the *Re*, which varied from 60 to 900 proportional to frequency. These values correspond to the transitional regime between laminar and turbulent flow when vortexes are actively produced, but don't have chaotic nature. The results showed that variance value after 3 second mixing decreased from 0.81 to 0.0042 with the change of frequency from 5 to 75 Hz (figure 4). As the variance exponentially depended on time the mixing efficiency could be expressed as the time constant of the process divided by period of oscillations (or multiplied by their frequency) that shows the number of periods needing to decrease the variance of FMN concentration e times (figure 5). Thus, the maximum mixing efficiency per cycle was achieved for frequencies not exceeding 10 Hz. As can be seen, a significant deterioration in the mixing efficiency occurred when the Womersley number exceeded the threshold quantity of 10. In this case, a flow velocity profile became flatter and the vortex formed as a result of the fluid velocity direction change localized only in an expanding part of the design in contrast to the case with a low frequency when it covered the entire reaction chamber. As a result, instead of distribution all over the chamber reagent occupied only half of it. Separating the impact of α and *Re* the additional studies were conducted. Thus, maintaining the average velocity (keeping *Re* constant) and increasing frequency from 10 to 20 Hz the variance was found to be 1.22 up to 3 seconds. Such reduction in mixing efficiency, which is more significant compared to constant varying volume condition, shows that Womersley number has the major impact on mixing efficiency.

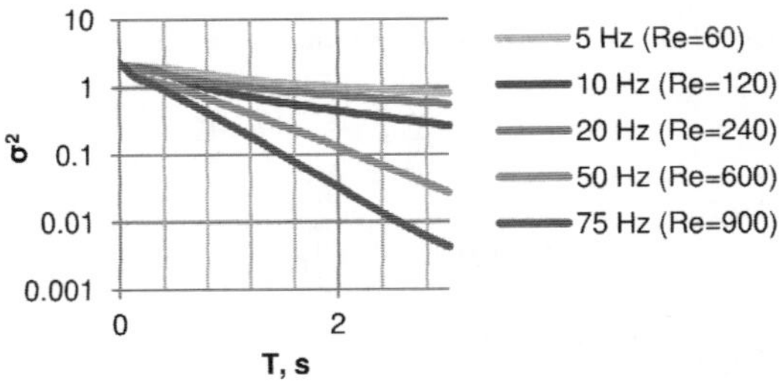

Figure 4. The variance of FMN concentration during sinusoidal oscillations of fluid velocity and peak-to-peak amplitude 3 ul

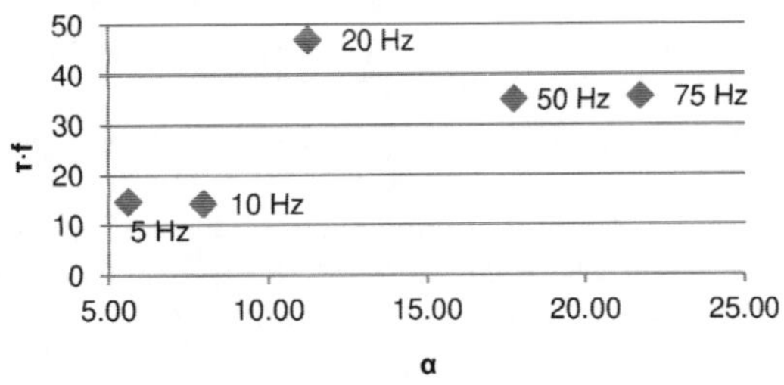

Figure 5. The number of periods needed to decrease the variance of FMN concentration in e times (time constant τ multiplied on oscillations frequency f)

Besides sinusoidal wave flow rates a square wave with a frequency of 10 Hz and rectangular wave with uniformly increasing frequency from 1 to 10 Hz were tested. Square wave signal with a constant frequency showed worse mixing compared to sinusoidal wave with the same period. However, the use of variable frequency provided the same variance as sinusoidal wave (figure 3 (f)). Thus, the variable frequency signal allowed increasing the mixing efficiency per one period of oscillation, since the same results was achieved with lower average frequency equal 5 Hz. This was because at low frequencies the reagent was spread over the chamber and then at higher rates was finally mixed. The variance of concentration after 3 seconds in the reaction chamber equal 0.011 was achieved using this mode with peak-to-peak amplitude 4 ul (figure 3 (g)).

Experimental results for the dye mixing have confirmed the efficiency of selected mode with varying frequency as well as its deterioration at 20 Hz. Mixing efficiency has been also confirmed by the growth of luminescent kinetics quality. As shown in figure 6, after the active mixing was started at 10 s the reaction accelerated and light intensity reached the maximum.

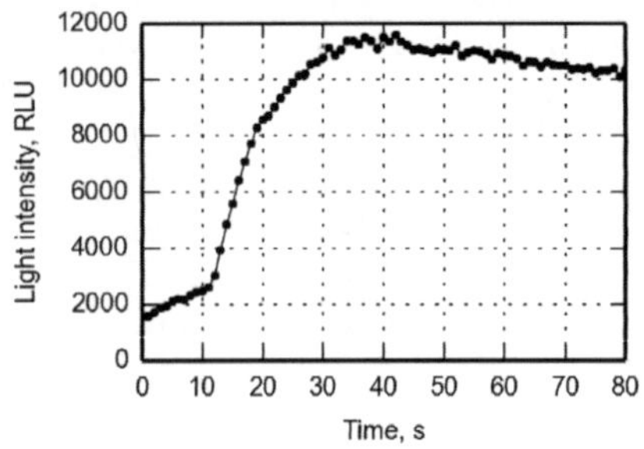

Figure 6. Bioluminescent reaction rate before mixing at 10 s and after it

4. Conclusions

The passive mixing didn't provide the uniform distribution of dissolved FMN. On the other hand, active mixing will be able to provide constant amount of FMN in the reaction chamber and a suitable mixing time less than 3 seconds. The highest mixing efficiency per cycle was achieved by varying frequency of rectangular flow rate oscillations from 1 to 10 Hz in 3 seconds which provided variance of FMN equal 0.011.

Acknowledgments

The research was supported by the grant of the Russian Science Foundation (project no.15-19-10041).

References

[1] Sackmann E, Fulton A, Beebe D 2014 *Nature* **507** 181-189
[2] Girotti S, Ferri E N, Fumo M G, Maiolini E 2008 *Analytica Chimica Acta* **608** (1) 2-29
[3] Hastings J, Nealson K 1977 *Annual Reviews in Microbiology* **31** 549–595
[4] Vetrova E, Esimbekova E, Remmel N, Kotova S, Beloskov N, Kratasyuk V, Gitelson I 2007 *Luminescence* **22** (3) 206–14
[5] Esimbekova E, Kratasyuk V, Shimomura O 2014 *Adv Biochem Eng Biotechnol* **144** 67-109
[6] Petushkov V N, Kratasyuk G A, Rodionova N S, Fish A M, Belobrov P I 1984 *Biochemistry. Academy of Sciences of the USSR* **49** (4) 593-604
[7] Nguyen H D, Renslow R, Babauta J, Ahmed B, Beyenal H 2012 *Sensors and Actuators B: Chemical* **161** (1) 929-937
[8] http://www.ymdb.ca/compounds/YMDB00085

View of the bacterial strains of Escherichia coli M-17 and its interaction with the nanoparticles of zinc oxide by means of atomic force microscopy

A Sagitova[1] I Yaminsky[1,2] G Meshkov[1]
[1]Faculty of Physics Lomonosov Moscow State University
[2]Advanced Technologies Center, Moscow

Abstract. Visualization of the structure of biological objects plays a key role in medicine, biotechnology, nanotechnology and IT-technology. Atomic force microscopy (AFM) is a promising method of studying of objects' morphology and structure. In this work, AFM was used to determine the size and shape of the bacterial strains of Escherichia coli M-17 and visualization its interaction with the nanoparticles of zinc oxide. The suspension of E.coli bacteria was applied to natural mica and studied by contact mode using the FemtoScan multifunctional scanning probe microscope.

1. Introduction

Bacteria are among the simplest structures of living organisms. It is believed that they are the first organisms that appeared on Earth. In extreme conditions, unsuitable for the existence of other organisms, bacteria can be the only form of life. In this regard, the study of bacteria is of great interest. A promising area of application of the bacteria is the IT-technology. Using the bacteria as a storage device is not a new idea: it was proposed by a group of Bancroft in 2001, and a team of scientists involved in Yachi information storage in the genome of the bacteria since 2007. Bacteria provide secure storage of information. Since it multiplies and makes additional copies of the data. Some microorganisms can survive the radiation of a nuclear explosion preserving while the recorded data. One species of bacterium E.coli can save only one kilobyte of information. One gram bacterium E.coli produces the number of individuals of this species equal to the total amount of information 931322 GB (450 hard disk memory to 2 TB). Bacterial cells have an ability to change the genome in the process of adaptation to changing external conditions. Due to high speed of division and possible changes in genome they can withstand antibiotics, chemical agents, nanoparticles with functional properties. Here we study the morphology of Escherichia coli M-17 bacteria and their interaction with zinc oxide nanoparticles.

2. AFM of the bacterial strains of *Escherichia coli M-17*

This work demonstrates the application of AFM to the study of bacteria. AFM uses a cantilever that measures the force of van der Waals interactions, or electrostatic attraction, repulsion forces. It's a fine needle fixed vertically on the elastic console. The method helps to measure the attraction or repulsion of the atoms in the sample and the atoms in the cantilever. The laser beam is directed at the reflective surface of the free end of the console, while the tip is placed on the opposite side of the console. The beam is reflected by the surface of console and then is directed to the photodiode. Depending on the changes of the cantilever-surface interaction force the cantilever console bending occurs and the laser beam is deflected from the central position on the photodiode. The feedback system changes the position of the sample, so that the reflected laser beam returns it to the "zero" central position on the photodiode. Thus, by registering the sample shift required to return the laser beam to the "zero" point the system measures a surface topography. FemtoScan scanning probe microscope was used in the experiments. Images were processed using FemtoScan Online software [3].

Figure 1,2 shows images of bacteria. Bacteria have oval flattened shape. The size of bacteria is of 1 to 2 micron.

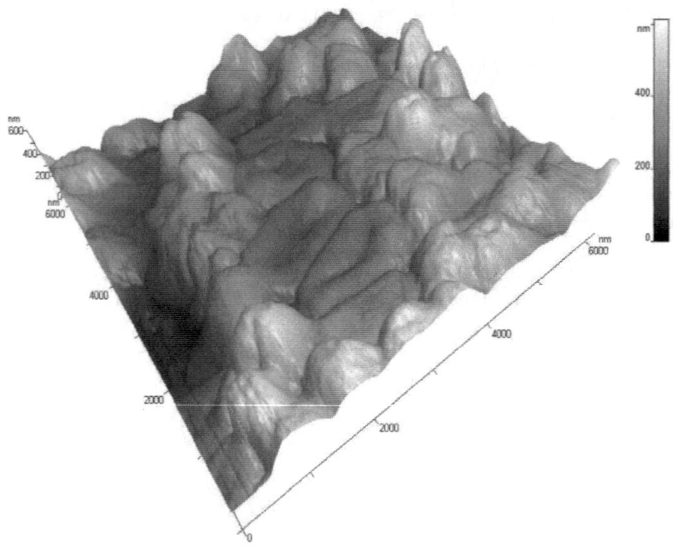

Figure 1. AFM 3D image of strains of Escherichia coli M-17.

Figure 2. AFM top image of strains of Escherichia coli M-17.

3. Zinc oxide nanoparticles

Liquid colloidal solutions of metals and metal oxides were obtained using a condensing-disperazione electropulse installation, in the low-temperature plasma area. Synthesized nanoparticles of zinc oxide have a high zeta potential, greater than 20 mV by absolute value. This indicates on some kinetic stability of colloidal solutions. The presence of negative surface charge on the diffusion mobility boundary of zinc oxide nanoparticles, is caused by the concentrating of negatively charged particles on the surface of the solid phase particles. By adding quaternary ammonium compounds in a zinc oxide colloidal solution the adsorption of cation surfactant takes place on the surface of the nanoparticles.

4. The interaction of bacteria with nanoparticles of zinc oxide

Figure 3, 4 shows images of the bacterial strains of Escherichia coli M-17 after adding nanoparticles of zinc oxide. Bacteria form agglomerates and retain their shape.

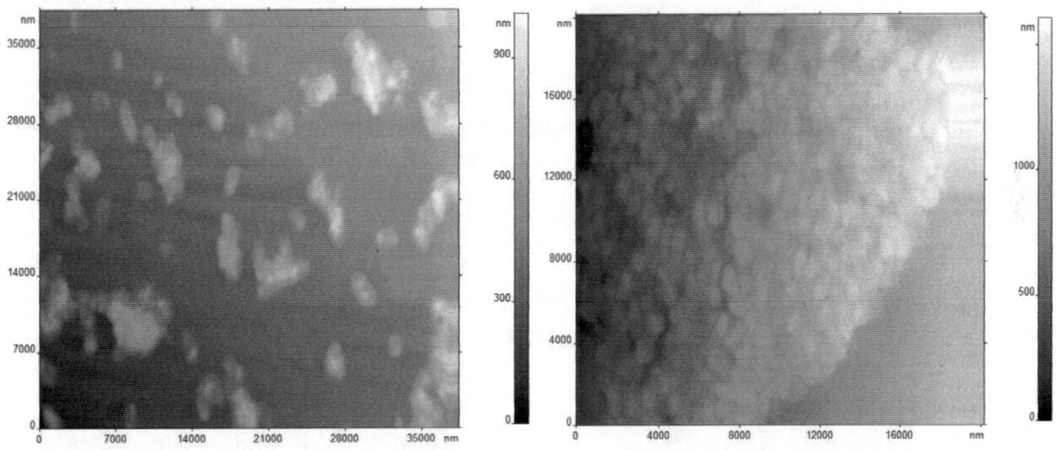

Figure 3 (a,b). AFM top image of strains of Escherichia coli M-17 with nanoparticles.

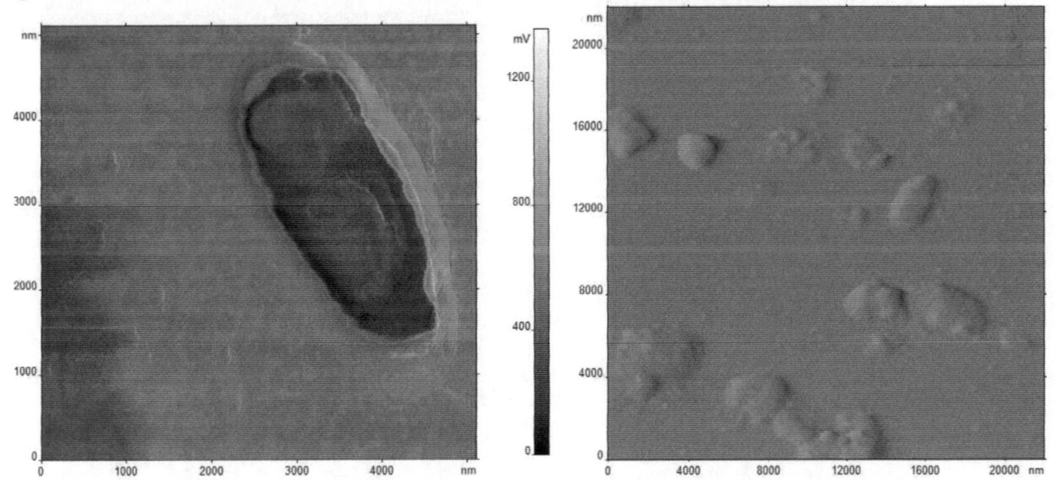

Figure 4 (a,b). AFM top image of strains of Escherichia coli M-17 with nanoparticles, friction mode (left image) and deflection mode (right image).

During our preliminary observations we did not see the visible changes of the morphology of individual cells (fig 4a). Still there are seen some difference in bacteria aggregation (fig 3). The bacteria in the presence of Zinc oxide nanoparticles do not produce dense films of closely packed bacterial cells as it is seen in the fig 1 and 2 for the samples with no addition of nanoparticles. Probably the intercellular interaction is affected by Zinc oxide nanoparticles.

Acknowledgments
The authors are grateful for the support of Russian Foundation of Basic Research (project N 15-04-07678) and Ministry of Science and Education (project N 02.G25.31.0135). The authors are thankful to Pogorelsky I.P. (Microbiology institute, Kirov) and Frolov A.G. (MISIS, Moscow) for the provided samples of the Escherichia coli M-17 bacteria.

References
[1] Yaminsky I., Filonov A., Sinitsyna O., Meshkov G. FemtoScan Online software. Nanoindustry, N2 (64), 42-46 (2016).
[2] Butt, H.J., Cappella, B., Kappl, M. Surf. Force measurements with the atomic force microscope: technique, interpretation, applications Sci. Rep., 59, 1–152 (2005).
[3] Gross L., Mohn F., Moll N., Liljeroth P., Meyer G. The chemical structure of a molecule resolved by atomic force microscopy. Science 325, 1110–1114 (2009).

Continuous agglomerate model for identifying the solute-indifferent part of colloid nanoparticle's surface charge

A V Alfimov, E M Aryslanova and S A Chivilikhin

ITMO University, Kronverkskiy, 49, St. Petersburg, 197101, Russia

Abstract. This work proposes an explicit analytical model for the surface potential of a colloidal nano-agglomerate. The model predicts that when an agglomerate reaches a certain critical size, its surface potential becomes independent of the agglomerate radius. The model also provides a method for identifying and quantifying the solute-indifferent charge in nanocolloids, that allows to assess the stability of toxicologically significant parameters of the system.

1. Introduction

Biocompatibility of inorganic colloidal nanoparticles (NPs) depends greatly on their surface charge. This charge governs the state of agglomeration in nanocolloids, which, in turn, affects internalization, distribution, metabolism, and excretion of NPs in biological systems [1-6]. It was recently reported that the surface charge itself also has an impact on the toxicity of NPs [1,7-9].

Synthesized metal oxide NPs in aqueous solutions can acquire surface charge through a number of mechanisms: adsorption and desorption of protons on surface hydroxyl groups; physical adsorption of ions; reversible and irreversible chemical adsorption of ions; isomorphic substitution [10,11]. Most of these processes are dependent on the pH of the solution. Therefore, the effect of the medium on the surface charge is usually studied by measuring the surface potential of agglomerated NPs in a relevant pH range.

However, the part of the surface charge arising due to irreversible chemical adsorption and/or isomorphic substitution can be indifferent to the solution used for titration. When NPs are released into a solution of unknown chemistry, the species responsible for this constant charge may react with the medium, leading to neutralisation of their charge. As a result, the surface potential can significantly deviate from the expected values, presenting a considerable toxicological threat.

During experimental studies of colloidal systems, one often observes agglomerates rather than individual particles. We assume that, while for solid particles charging occurs only on the surface, agglomerates, like porous particles, can exhibit volumetric charging. Consequently, it is important to focus theoretical studies of the colloidal charging phenomena not just on individual NPs, but on their agglomerates as well.

In the present work, we propose an analytical model of a colloidal nano-agglomerate that allows to identify the presence of the constant solute-indifferent charge (SIC) and study its effect on the surface potential.

2. Methods

We consider a spherical agglomerate of metal oxide NPs. The agglomerate is immersed in a solution of a strong inert monovalent electrolyte in deionized water. The solution is isolated from atmospheric CO_2. The pH of this colloidal solution can be varied using acid-base titration at constant ionic strength $I = const$.

The agglomerate is assumed to be a porous particle. The volume fraction of void space in the agglomerate (porosity Ω) depends on the ordering of constituent NPs and their hydration.

Nanoparticles inside the agglomerate acquire surface charge by protonation and deprotonation of surface hydroxyl groups. They can also carry a SIC originating from isomorphic substitution and/or irreversible chemisorption. Due to the porosity of the agglomerate we assume that the adsorption and desorption of protons on NP surface occurs in the entire volume of the agglomerate.

It is now generally agreed that the adsorption and desorption of protons happens on two types of surface hydroxyl groups, basic and acidic in nature respectively [12-16]:

$$\mathrm{M-OH_b + H^+ \overset{K_b}{\rightleftarrows} M-OH_2^+} \ , \tag{1}$$

$$\mathrm{M-OH_a \overset{K_a}{\rightleftarrows} M-O^- + H^+} \ , \tag{2}$$

where $\mathrm{M-OH_b}$, $\mathrm{M-OH_a}$ - basic and acidic surface OH groups; K_b, K_a - equilibrium constants for protonation (1) and deprotonation (2) reactions. We consider the reactive surface hydroxyls to be evenly split between acidic and basic type [13,14].

Surface density of hydroxyl groups for NPs usually lies in the range of $\gamma_S = 1-20$ nm^{-2} [10,12-14,17]. It should be noted that the average density of reactive sites on NPs comprising the agglomerate could be lower than the values for individual NPs.

When NPs do not posses the constant charge (SIC), the pH at which the particle's surface charge is zero (*point of zero charge*) can be found as:

$$\mathrm{PZC_0} = \frac{\mathrm{p}K_b + \mathrm{p}K_a}{2} , \tag{3}$$

where $\mathrm{PZC_0}$ - point of zero charge in absence of SIC; $\mathrm{p}K_b = -\log_{10}(K_b)$ and $\mathrm{p}K_a = -\log_{10}(K_a)$. Parameters $\mathrm{p}K_b$ and $\mathrm{p}K_a$ have the physical meaning of pH values at which half of the basic (acidic) sites are protonated (deprotonated). We can now express:

$$\mathrm{p}K_b = \mathrm{PZC_0} - \Delta\mathrm{p}K \quad \text{and} \quad \mathrm{p}K_a = \mathrm{PZC_0} + \Delta\mathrm{p}K \ , \tag{4}$$

where $\Delta\mathrm{p}K = \mathrm{p}K_a - \mathrm{p}K_b$. Values of $\Delta\mathrm{p}K$ for metal oxide NPs can span over a wide range of $\Delta\mathrm{p}K = 1-14$ [18].

Electrostatic potential φ in the described system is governed by the Poisson-Boltzmann equation [10]. A number of assumptions were made to derive an analytical solution to the Poisson-Boltzmann problem for the potential inside and outside of the agglomerate. First, the Debye-Huckel approximation for low potentials in the system is used. According to this approximation, the maximum potential in the system should satisfy the condition: $\max(\varphi) < kT/e$ (k - Boltzmann constant; T - temperature; e - elementary charge). Second, we consider only the non-saturated regime of protonation/deprotonation: $\mathrm{p}K_b < \mathrm{pH} < \mathrm{p}K_a$. Finally, the internal space of the agglomerate is approximated with continuous distributions of effective volume charge density ρ and dielectric permittivity ε_A:

$$\rho(\mathbf{r}) = e\frac{1-\Omega}{V_{NP}}\left(Z_C + \frac{P}{2}\left[10^{-\Delta\mathrm{pH}}\left\{1-\frac{e\varphi(\mathbf{r})}{kT}\right\} - 10^{\Delta\mathrm{pH}}\left\{1+\frac{e\varphi(\mathbf{r})}{kT}\right\}\right]\right) + \varepsilon_A\varepsilon_0\cdot\Omega\kappa^2\cdot\varphi(\mathbf{r}) , \tag{5}$$

$$\varepsilon_A = \varepsilon_{NP}(1-\Omega) + \varepsilon_M\Omega , \tag{6}$$

where Z_C - average charge number of SIC per NP, $P = S_{NP} \cdot \gamma_S \cdot 10^{-\frac{1}{2}\Delta pK}$ - charge regulation parameter; V_{NP}, S_{NP} - volume and surface area of primary NPs; $\kappa^2 = 2I \cdot e^2 / \varepsilon_M \varepsilon_0 kT$ - Debye-Huckel screening parameter of the solution; ε_{NP}, ε_M - dielectric permittivities of NPs and the medium (solution); ε_0 - dielectric constant; $\mathbf{r}$ - position vector.

By solving the resulting Debye-Huckel problem we obtained the following expression for the surface potential of the agglomerate φ_S:

$$\varphi_S = \frac{e}{\varepsilon_A \varepsilon_0} \cdot \frac{1-\Omega}{V_{NP}} \cdot \frac{Z_C + P \cdot \sinh[-\ln(10) \cdot \Delta\mathrm{pH}]}{\alpha^2 \cdot \left(1 + \frac{\varepsilon_M}{\varepsilon_A} \cdot \frac{\kappa R + 1}{\alpha R \cdot \coth(\alpha R) - 1}\right)}, \quad (7)$$

where R - agglomerate radius; α - effective Debye-Huckel screening parameter of the agglomerate's interior that is given by:

$$\alpha^2 = \left(\frac{e^2}{\varepsilon_A \varepsilon_0 kT} \cdot \frac{1-\Omega}{V_{NP}} \cdot P \cdot \cosh[-\ln(10) \cdot \Delta\mathrm{pH}]\right) + \left(\Omega \kappa^2\right). \quad (8)$$

As follows from (8), the parameter α consists of contributions from the surface charge of constituent NPs (first term) and from ions penetrating the void spaces between the NPs (second term).

3. Results and discussion

3.1. Surface potential as a function of agglomerate size

From expressions (7) and (8) it follows that when $\kappa R \gg 1$, the surface potential becomes independent of agglomerate radius. This behavior is illustrated in figure 1A. The saturated size-independent surface potential of the agglomerate is given by:

$$\varphi_S^{\mathrm{Sat}} = \frac{e}{\varepsilon_A \varepsilon_0} \cdot \frac{1-\Omega}{V_{NP}} \cdot \frac{Z_C + P \cdot \sinh[-\ln(10) \cdot \Delta\mathrm{pH}]}{\alpha^2 \cdot \left(1 + \frac{\varepsilon_M}{\varepsilon_A} \cdot \frac{\kappa}{\alpha}\right)} \quad (9)$$

The critical radius above which the agglomerate's surface potential becomes size-independent is defined only by the ionic strength of the solution. This relation is shown in figure 1B.

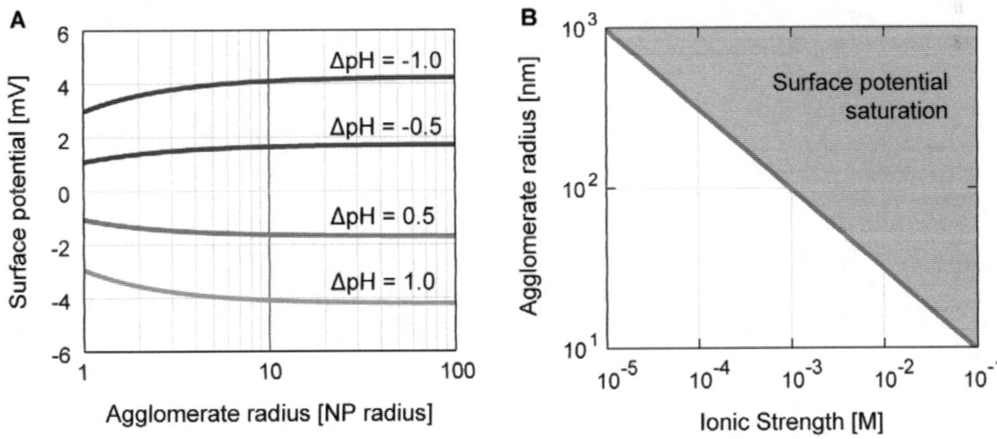

Figure 1. A – surface potential as a function of agglomerate radius (normalized to the primary NP radius) at different values of $\Delta\mathrm{pH}$; B – dependency of critical agglomerate radius on ionic strength of the solution; filled area corresponds to agglomerate radii above the critical value for which surface potential becomes independent of agglomerate size.

3.2. *Effect of solute-indifferent charge on the agglomerate's point of zero charge*

The model of the agglomerate's surface potential was developed under the Debye-Huckel approximation. Therefore, it is quantitatively valid only at low potentials. This condition is generally not satisfied throughout entire experimental surface potential curves. However, expression (7) is always applicable around the point of zero charge of the agglomerate. Consequently, we can use it to find the condition for zero agglomerate charge in case of non-zero SIC:

$$\frac{Z_C}{P} = \sinh\left[\ln(10)\cdot\Delta\mathrm{PZC}\right], \tag{10}$$

where $\Delta\mathrm{PZC} = \mathrm{PZC} - \mathrm{PZC}_0$ - shift of point of zero charge in relation to its value in absence of SIC.

Equation (10) shows that the presence of SIC shifts the agglomerate's PZC in relation to PZC_0 (point of zero charge when SIC is zero):

$$\Delta\mathrm{PZC} = \log_{10}\left(\frac{Z_C}{P} + \left[\left\{\frac{Z_C}{P}\right\}^2 + 1\right]^{\frac{1}{2}}\right); \tag{11}$$

this dependency is illustrated in figure 2A. As a result, it is possible to identify the presence of SIC by detecting such PZC shift.

As shown by (3), PZC_0 lies exactly in the middle between $\mathrm{p}K_b$ and $\mathrm{p}K_a$, which signify the onset of protonation and deprotonation saturation correspondingly. We use a numerical non-linear Poisson-Boltzmann model of charged flat surface to demonstrate that non-zero SIC does not alter pH values at which protonation and deprotonation saturate (figure 2B). It is thus possible to assess the position of PZC_0 as a mid-point between the points of saturation onset.

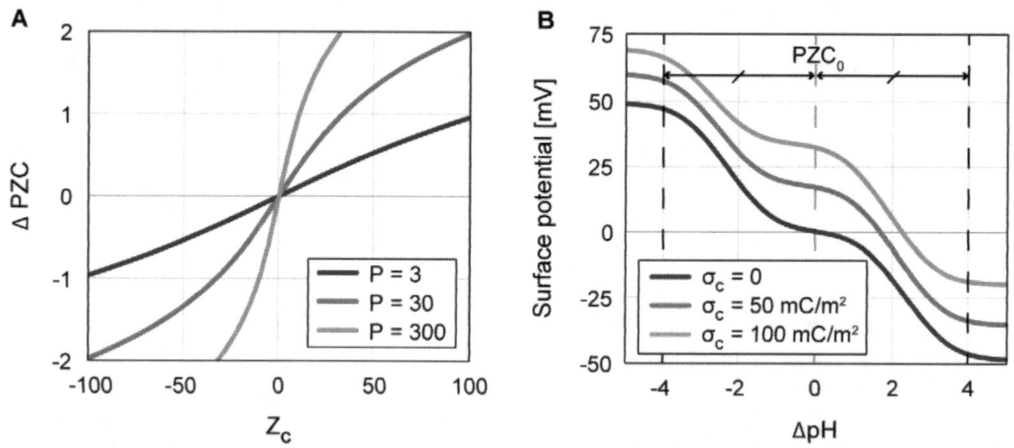

Figure 2. A – point of zero charge shift ($\Delta\mathrm{PZC}$) as a function of average charge number of SIC per NP (Z_C) at different values of charge regulation parameter P; B – surface potential as a function of $\Delta\mathrm{pH}$ predicted by the numerical model of flat surface at different values of surface charge density of SIC.

3.3. *Limiting cases of continuous agglomerate model*

We now consider the interior Debye-Huckel screening parameter α. According to (8), this parameter consists of contributions from NPs' surface charge and ions in the inter-particle cavities. In a limiting case when one of these contributions significantly outweighs the other, expression (9) allows for further simplification.

The interior Debye-Huckel parameter α reduces to just the NP surface charge contribution when the following condition is met:

$$\frac{e^2}{\varepsilon_0 kT}\cdot\frac{S_{\mathrm{NP}}}{V_{\mathrm{NP}}}\cdot\gamma_{\mathrm{S}}\cdot 10^{-\frac{1}{2}\Delta pK}\cdot\left(\varepsilon_{\mathrm{NP}}+\varepsilon_{\mathrm{M}}\frac{\Omega}{1-\Omega}\right)^{-1} \gg \Omega\kappa^2 . \quad (12)$$

We denote this limit as *Case A*. For closed-packed agglomerates of NPs with site density $\gamma_{\mathrm{S}}=1$ nm^{-2} immersed in electrolyte solution with ionic strength $I=0.001$ M, condition (12) is met when primary NPs radius is less or equal to 25 nm and $\Delta pK \le 4$.

The simplified agglomerate's surface potential expression for *Case A* is given by:

$$\varphi_{\mathrm{S}}^{\mathrm{Case\,A}}=\frac{kT}{e}\cdot\frac{Z_{\mathrm{C}}+P\cdot\sinh\left[-\ln(10)\cdot\Delta\mathrm{pH}\right]}{P\cdot\cosh\left[-\ln(10)\cdot\Delta\mathrm{pH}\right]\cdot\left(1+\frac{\varepsilon_{\mathrm{M}}}{\varepsilon_{\mathrm{A}}}\cdot\frac{\kappa}{\alpha}\right)} . \quad (13)$$

For highly porous agglomerates/particles with large ΔpK in solutions with high ionic strength, the opposite limiting case can be relevant:

$$\frac{e^2}{\varepsilon_0 kT}\cdot\frac{S_{\mathrm{NP}}}{V_{\mathrm{NP}}}\cdot\gamma_{\mathrm{S}}\cdot 10^{-\frac{1}{2}\Delta pK}\cdot\left(\varepsilon_{\mathrm{NP}}+\varepsilon_{\mathrm{M}}\frac{\Omega}{1-\Omega}\right)^{-1} \ll \Omega\kappa^2 . \quad (14)$$

The interior Debye-Huckel parameter alpha then reduces to just the contribution from ions in agglomerate's cavities. We denote this limit as *Case B*. For agglomerates with porosity $\Omega\ge 0.5$ composed from NPs with site density $\gamma_{\mathrm{S}}=1$ nm^{-2} this condition is met in solutions with ionic strength $I\ge 0.001$ M when primary NP radius is greater or equal to 25 nm and $\Delta pK \ge 6$.

The simplified agglomerate's surface potential expression for *Case B* is given by:

$$\varphi_{\mathrm{S}}^{\mathrm{Case\,B}}=\frac{e\cdot(1-\Omega)}{\varepsilon_0\kappa^2 V_{\mathrm{NP}}}\cdot\frac{Z_{\mathrm{C}}+P\cdot\sinh\left[-\ln(10)\cdot\Delta\mathrm{pH}\right]}{\varepsilon_{\mathrm{NP}}\Omega(1-\Omega)+\varepsilon_{\mathrm{M}}\left(\Omega^2+\sqrt{\Omega}\right)} \quad (15)$$

Expressions (13) and (15) explicitly show the dependency of agglomerate's surface potential on the ionic strength of the solution. In both cases the effect of SIC on the potential increases at lower ionic strengths. Moreover, this dependency is suggested to be stronger for the case of highly porous agglomerates/particles (15).

3.4. Solute-indifferent charge assessment

Simplified expressions (13) and (15) are always valid near to the agglomerate's point of zero charge for the same reason as the original formula (7). Therefore, we can use these expressions to find the relation between the model parameters (Z_{C} and P) and the first derivative of the surface potential dependency on pH at the point of zero charge.

The derivative of the surface potential curve at PZC can be determined experimentally. Then, by differentiating expression (13) or (15) and using the relation (10), we can find the estimate for the SIC charge number per NP Z_{C} and the charge regulation parameter P:

$$Z_{\mathrm{C}}=\tanh\left[\ln(10)\cdot\Delta\mathrm{PZC}\right]\cdot Y , \quad (16)$$

$$P=\mathrm{sech}\left[\ln(10)\cdot\Delta\mathrm{PZC}\right]\cdot Y , \quad (17)$$

where the Y function is specific to the employed limiting case. For the *Case A* we have:

$$Y_{\mathrm{Case\,A}}=\left(\frac{\varphi_{\mathrm{S}}'\big|_{\Delta\mathrm{PZC}}\cdot\varepsilon_{\mathrm{M}}\cdot\kappa}{\ln(10)\cdot\frac{kT}{e}+\varphi_{\mathrm{S}}'\big|_{\Delta\mathrm{PZC}}}\right)^2\cdot\left(e^2\frac{\varepsilon_{\mathrm{NP}}\left[1-\Omega\right]^2+\varepsilon_{\mathrm{M}}\Omega\left[1-\Omega\right]}{V_{\mathrm{NP}}\varepsilon_0 kT}\right)^{-1} , \quad (18)$$

and for the *Case B*:

$$Y_{\text{Case B}} = \left| \frac{\varphi'_{\text{S}}\Big|_{\Delta\text{PZC}}}{\ln(10)} \right| \cdot \frac{\varepsilon_0 \kappa^2 V_{\text{NP}}}{e} \cdot \left(\varepsilon_{\text{NP}} \Omega + \varepsilon_{\text{M}} \frac{\Omega^2 + \sqrt{\Omega}}{1-\Omega} \right) \quad (19)$$

where $\varphi'_{\text{S}}\Big|_{\Delta\text{PZC}}$ – experimentally determined first derivative of the surface potential dependency on pH at the agglomerate's point of zero charge. Note that this assessment requires the additional knowledge of estimates for the agglomerates porosity Ω and primary NP's volume V_{NP} and dielectric permittivity ε_{NP} .

4. Conclusions

We developed an explicit analytical model for the surface potential of a colloidal nano-agglomerate. Analysis of this model suggests that

1) after an agglomerate reaches a certain critical size, its surface potential becomes independent of the agglomerate radius. The value of this critical size is defined solely by the ionic strength of the solution.
2) Presence of solute-indifferent charge (SIC) in the agglomerate causes the point of zero charge to shift in relation to its value in absence of SIC. The sign of this shift coincides with the sign of SIC.
3) The effect of SIC on the surface potential increases at lower ionic strength. This correlation is stronger for highly porous agglomerates/particles.
4) It is possible to identify the presence of SIC by comparing the point of zero charge and the mid-point between the pH values corresponding to protonation and deprotonation saturation onsets. The developed agglomerate model provides a method for quantitative assessment of the average amount of SIC per primary nanoparticle.

Acknowledgements
This work was financially supported by the Government of Russian Federation, Grant 074-U01.

References

[1] Abdelmonem A M, Pelaz B, Kantner K, Bigall N C, Del Pino P and Parak W J 2015 *Journal of Inorganic Biochemistry* **153** 334
[2] Cupi D, Hartmann N B and Baun A 2016 *Ecotoxicology and Environmental Safety* **127** 144
[3] Jiang J, Oberdörster G and Biswas P 2009 *J Nanopart Res* **11** 77
[4] Kim K-M *et al* 2012 *Toxicol. Environ. Health Sci.* **4** 121
[5] Berg J M, Romoser A, Banerjee N, Zebda R and Sayes C M 2009 *Nanotoxicology* **3** 276
[6] Oberdörster G, Oberdörster E and Oberdörster J 2005 *Environmental Health Perspectives* **113** 823
[7] Barisik M, Atalay S, Beskok A and Qian S 2014 *J. Phys. Chem. C* **118** 1836
[8] Chusuei C C *et al* 2013 *Chemico-Biological Interactions* **206** 319
[9] Schwegmann H, Feitz A J and Frimmel F H 2010 *J. Colloid Interface Sci.* **347** 43
[10] Stumm W 1992 *Chemistry of the solid-water interface* (John Wiley & Sons)
[11] Cosgrove T 2010 *Colloid Science* (John Wiley & Sons)
[12] Fierro J L G 2005 *Metal Oxides* (CRC Press)
[13] Tamura H, Mita K, Tanaka A and Ito M 2001 *J. Colloid Interface Sci.* **243** 202
[14] Tamura H 1999 *J. Colloid Interface Sci.* **209** 225
[15] Hiemstra T, Van Riemsdijk W H and Bolt G H 1989 *J. Colloid Interface Sci.* **133** 91
[16] Hiemstra T and Van Riemsdijk W H 1991 *Colloids and Surfaces* **59** 7
[17] McCafferty E and Wightman J P 1998 *Surface and Interface Analysis* **26** 549
[18] Hiemstra T 1989 *J. Colloid Interface Sci.* **133** 105

Study of improving the quality of bread and wheat-aegilops hybrids with the biotechnological ways

Aygun Ganbarzada, Sudaba Hasanova

Sumgait, Azerbaijan

Abstract. The great need of the people to bread demands to increase high qualitative grain plants. At present time for solving these problem different methods of biochemistry, genetics and molecular biology are widely used in the process of selection. To investigate biochemical peculiarities of wheat-aegilops hybrids and to define the correlative relation between these characteristics. To investigate the technological peculiarities of wheat-aegilops hybrids and to define the relation between their main biochemical and technological characteristics. The conclusion of this investigation showed the followings- the wheat-aegilops hybrids according to their morphological and biochemical characteristics have approached to wheats. The electrophoretic spectres of the wheat-aegilops hybrids which have stable for their morphological characteristics are homogeny and heterogenic. Hereditarily some group protein components have passed to their tribes from their parents. But spontaneous hybridisation results in taking part the components of other unknown wheats in these electrophoretic spectres. There is a relation between the electrophoretic spectres and the indications of the grain quality.

1. General characteristics of investigation

The study of theoretical and practical importance of biochemical part of wheat - aegilops hybrids that is the product of gender distant hybridization, including its biotechnological peculiarities, also the relations of protein components and grain's quality is very necessary.

Purpose of the work - is the study of peculiarities of Triticum and Aegilops intergeneric hybridization with the biotechnological and biochemical ways and perfection of bread baking technique.

General amount of nitrogen in grain was according to the Keldal methods, amount of starch was according to the Evers method, amount of lysine and tryptophan were according to the N.P.Yarosh [1], the amount of protein fractions were according to the Osborn method [2], electrophoresis of spare proteins and enzymes were investigated according to the methods developed in All-Union Institute of Genetics and Selection, in the laboratory of Genetic Biochemistry, physical and technological indicators were carried out according to the standard methods. [3]

2. Progress of research work

During the course of experiment has determined that the main studying aspects are the following. (Table 1)

2.1. Amount of protein in the grain

Protein is one of the main qualities of grain. In the external environment affects to the amount of protein. However, protein is a sign transferring along with hereditary.

In the T. dicoccum v. atratum x Ae.ovata hybrids, the price of protein has changed between 15, 29-16, 38% during 3 years. İn the hybrids' line of number 5 and 6 the amount of protein was higher than the standard varieties. Number 1 and 5 hybrid lines, has changed little under the influence of climatic conditions. The coefficient of variation of the hybrid combination (CV %) were between 2,25-8,18%.

In the T.durum v.leucurum x Ae.ventricosa hybrid combination the average price of protein was between 15,36-19,13%. Number 11 hybrid line was having the highest protein (19,13%). Coefficient of variation among the hybrids changed between 2,09-10,5%.

To the highest protein of Teyakan 60 x Ae.ovata hybrid combination includes the 13, 14 and 15 of the hybrid lines. The average price of 16, 88-18, 98%, of protein hybrids within three years were between 3,8-9,

94%. Number 14 hybrid line has been more resistant to the effects of climatic conditions.

In the line of hybrid number 16, in the Teyakan 60 x Ae.ventricosa hybrid combination the standard amount of protein was higher than the standard sorts (16.18%), and the coefficient of variation was 1.7%.

In the whole hybrid line, which includes to Ae.ventricosa x Teyakan 60 hybrid combinations, the average price for a standard sort was higher within three years. The amount of protein in these hybrids changed between 15, 01-17, 64%, and coefficient of variation changed between 4, 26-9,93%.

Number 11, from hybrid combination of T.durum v.leucurum x Ae.ventricosa, and the number 14 from the Teyakan 60 x Ae.ovata hybrid combination in the durum wheat the quality of the bread is near the soft wheat in the hybrid lines. In the T.durum v.leucurum x Ae.ventricosa hybrid combination between the bread volume and protein correlation coefficient have been positive during two years (2014, 2015), (r = + 0,955, r = + 0,920). The coefficient of correlation between the amount of bread and gluten was positive (r = + 0,948, r = + 0,588), but between the amount starch in bread volume was negative (r = -0.921, r = - 0,696). [4]

2.2. *Physical indicators*

In the studied wheat-aegilops hybrids, in the hybrid lines weight of 1000 grains that looks like durum wheat vitreous and soft wheat grain was relatively higher in comparison with similar lines.

In the hybrid line of number 3, including to the T.dicoccum v.atratum x Ae.ovata hybrid combination the weight of 1000 grains reached to 64.9%. Most hybrids like hard wheat, soft wheat hybrids, in the hybrid lines of number 21, 22 and 23, weight of 1000 grains was higher than standard sorts.

Between the amount of protein of vitreous and weight of 1000 grains has observed the correlation. In the hybrid combinations of Teyakan 60 x Ae.ovata (r=+0,635) və Ae.ventricosa x Teyakan 60 (r=+0,538) the proteins and weight of 1000 grains was positive, and in the other combinations the correlation observed negative. The majority of wheat hybrids like the sort of vitreous is close to the sort of East, from the hybrids like soft wheat number 19, 20, 21, 22 and 23, including to the Ae.ventricosa x Teyakan 60 hybrid combination the vitreous was higher than the sort of standart Bezostaya I. Hybrid combinations in the T.durum v.leucurum x Ae.ventricosa (r=+0,368) and Teyakan 60 x Ae.ovata (r=-0,215) the amount of protein between the vitreous observed with the weak positive correlation.

2.3. *Amount of ash element*

The main mineral elements of grain are potassium and phosphorus. Amount of mineral elements of the grain depending on growing years and geographic factors are different.

Amount of ash elements was 1.84 - 2.48%, in a solid wheat forms, and in soft wheat form amount was 1.92 - 2.25% in a hybrid lines in studied wheat - Aegilops hybrids. Amount of ash changes under the influence of climate. Only in the combination T.dicoccum v.atratum x Ae.ovata hybrids amount of ash elements were less, because of standard sorts, but in other combinations were more.

2.4. Amount of starch

Technological quality of grains directly depends on amount of starch. In the hybrid lines that are similar to the soft wheat in the investigated hybrids the starch was 55.13% - 61.4%, in the hybrid lines to be similar to solid wheat was 55.63 - 65.36%. In the solid wheat hybrids has observed negative correlation in the amount of bread and starch, but in the soft wheats' hybrids observed positive correlation.

2.5. Amount of gluten

It is the most important biotechnological indicator of wheat quality Beside 4 No. of the T. dicoccum v. atratum x Ae. ventricosa x Teyakan, and 19 No. hyper lines of 60 combination the amount of gluten was higher than the standards sorts. In the hybrid line No. 22 the amount of gluten reached to 52.96% in Ae.ventricosa x the Teyakan 60 hybrid combination.

In the hybrid combination of Teyakan 60 Ae.ovata the amount of gluten was 43, 83-49, 43%, and the coefficient of variation was small too.

In the hybrid combinations T.durum v.leucurum x Ae.ventricosa (r=+0,886) and Ae.ventricosa x Teyakan 60 (r=+0,946) were more connection. The amount of gluten determines with the IDK-I device. According to this device gluten's quality has been "solid" and "good". [5-6]

2.6. Bread quality

The volume of the bread is an important biotechnology. The larger volume of bread would be more porous and can be digesting better by the body. [7]

In studied hybrids, both hard and soft wheat hybrids lines the qualities of bread were near and higher than the standard sort. In the hybrid lines that is similar to the soft wheat the quality of bread was better than the lines of heavy wheat line. No. 5 including to the hybrid combination of T.dicoccum v.atratum x Ae.ovata, No. 16, including to the hybrid combination Teyakan 60 x Ae.ventricosa, No. 22 hybrid lines the average quality of bread and volume of bread was higher than the standard sort Bezostaya I. (Table 2)

Table 1. In the wheat-aegilops hybrids the amount of bread and coefficient correlation between some biochemical indicators

Names of hybrid combinations	2014			2015		
	Gluten	Starch	Ashes' elements	Gluten	Starch	Ashes' elements
Tt.dicoccum v.atratum x Ae.ovata	-0.187	-0.0082	-0.192	+0.255	-0.407	+0.416
T.durum v.leucurum x Ae.ventricosa	+0.948	-0.921	-0.979	-0.699	-0.699	+0.956
Teyakan 60 x Ae.ovata	-0.807	+0.840	-0.199	+0.193	+0.193	-0.188
Ae.ventricosa x Teyakan 60	-0.378	-0.497	+0.082	+0.378	+0.378	+0.182

Table 2. Bread quality of the wheat-aegilops hybrids

№	Years				
	2014		2015		The correlation coefficient for bread amount
	Bread amount sm^3	Per price	Bread amount sm^3	Per price	
T.dicoccum v.atratum x Ae.ovata					
1.	470	3,4	430	2,8	6,29
2.	410	3,1	470	3,8	8,13
3.	400	2,7	470	3,6	16,8
4.	420	2,7	460	4	6,43
5.	460	3,3	600	4,8	18,6
6.	440	4	500	4	9,02
7.	400	3,63	460	4	9,87
T.durum v.leucurum x Ae.ventricosa					
8.	400	3	410	3,7	1,7
9.	420	3	410	3	1,7
10.	390	3	430	2,4	6,89
11.	520	4,4	480	4	5,56
12.	410	2,9	440	4,1	4,99
Teyakan 60 x Ae.ovata					
13.	500	3,4	470	3	4,37
14.	480	4,4	470	4,1	1,49
15.	500	3,4	400	4	15,71
Teyakan 60 x Ae. Ventricosa					
16.	500	5	540	4,6	5,43
Ae.ventricosa x Teyakan 60					
17.	480	4,4	530	4,3	7,0
18.	460	4,3	420	3,1	6,43
19.	590	4,9	430	4,4	22,18
20.	430	3,1	530	4,4	20,16
21.	460	3	480	4,3	3,0
22.	540	3,6	600	4,9	7,44
23.	500	3	460	4,5	9,87
Bezostaya I	450	2,8	510	4,6	8,84
East	420	3	430	3,4	1,66

References

[1]Ermakov V.I. Methods of biochemical research of plants. *L.: Kolos*.: 1972, 224 p.
[2]Pleshkou B.P. Workshop on biochemistry plant. *M.: Kolos*, 1976, 254 p.
[3]Dospexov B.A. Planning of field experience and static processing of its data. *M .: Kolos*, 1972, p.8.
[4]Akhmetov M.G. Polymorphism and genetic analysis of replacement varieties of wheat zoned in Azerbaijan- Abstract dis. *Ph.D.* in *Biological Sciences,* Tank, 1992, 21, p.
[5]Konarev V.G. Plant proteins as genetic markers. *M .:Kolos*, 1983, 320 p.
[6]Sozinov A.A. Polymorphism of proteins and its significance for genetics and breeding. *M.: Nauka*, 1985, 272p.
[7]Kudryavtsev A.M. Metakovsky E.V., Upelniek V.P. Catalog of components blocks of gliadin chromosome 6A summer durum wheat // *Genetics*, 1987, t.23, №8, p. 1465-1479

Switching assay as a novel approach for specific antigen-antibody interaction analysis using magnetic nanoparticles

M Parr[1], R Illarionov[2], Y Marchenko[3], L Yakovleva[3], B Nikolaev[3], A Ischenko[3], M Shevtsov[4]

[1]Department of Physics, Saint-Petersburg State University, Saint-Petersburg 199034, Russia

[2]Department of Chemical and Biotechnology, Saint-Petersburg State Institute of Technology, Saint-Petersburg 190013, Russia

[3]Research Institute of Highly Pure Biopreparations, Saint-Petersburg 197110, Russia

[4]Institute of Cytology of the Russian Academy of Sciences (RAS), Saint-Petersburg 194064, Russia

Abstract. Switching assay was applied for the detection of antigen-antibody interaction between 70-kDa heat shock protein (Hsp70) and anti-Hsp70 monoclonal antibodies in water solutions using conjugates with magnetic iron oxide nanoparticles (MNPs). Hsp70 is a ubiquitous intracellular protein that plays a crucial role in cancerogenesis and many other pathologies. Detection of the Hsp70 level in the biological fluids might have a prognostic and diagnostic value in clinic. The developed switch assay for the detection of Hsp70 demonstrated high sensitivity for antigen-antibody interaction analysis thus proving its potential for further preclinical and clinical studies.

1. Introduction

Switching assay [1] is an accurate and robust technique for detection of water soluble biomarkers in various biological fluids (e.g., blood, urine, etc). Due to magnetic field local inhomogeneities induced by MNPs in water suspensions the proton's transverse relaxation time (T_2) is significantly decreased. The interaction between magnetic conjugates and molecules of the biomarker dramatically affects the relaxation behavior of water protons, thus giving the possibility to detect the biomarker in the solution and measure its concentration. In the presented study the feasibility of the Switching assay for the detection of antigen-antibody interaction between Hsp70 and anti-Hsp70 monoclonal antibody in water solutions was assessed. Molecular chaperone Hsp70 is involved in pathogenesis of many diseases (e.g., cancer, stroke) and its detection represents a high diagnostic and prognostic potential for management of the patients.

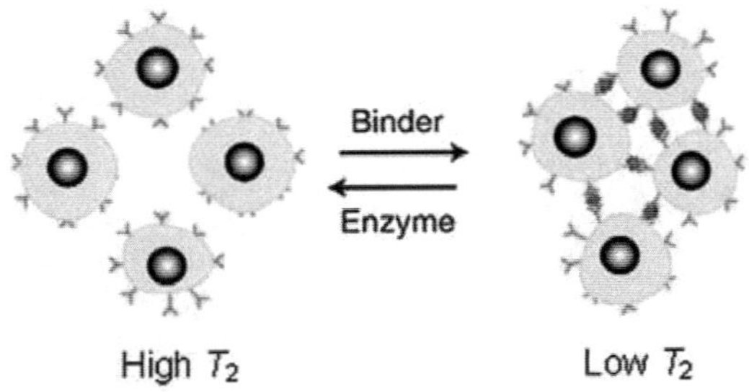

Figure 1. Scheme of magnetic switching assay [2].

2. Materials and methods

Iron oxide MNPs were prepared by co-precipitation of water dissolved Fe(II) and Fe(III) salts and covered with dextran (10kD, Sigma-Aldrich, USA)[3]. Magnetic conjugates of dextran coated MNPs with proteins (Hsp70 or anti-Hsp70 monoclonal antibody) were synthesized by coupling of COO-protein groups to carbodiimide activated surface dextran. The prepared non-conjugated MNPs and protein-decorated nanoparticles were assessed with a high sensitive method of longitudinal nonlinear response of magnetic nanoparticles to a weak *ac* magnetic field with measurements of second harmonic of magnetization (NLR-M_2). The obtained results demonstrated the superparamagentic properties of the MNPs. The size distribution of the MNPs in suspension was measured by dynamic light scatering (DLS). The distribution for both MNPs coated with dextran and MNPs conjugated with antibodies was bimodal with average hydrodynamic radii of modes equal to 34 nm and 159 nm and 34 nm and 240 nm respectively (Figure 2). Recombinant human Hsp70 was prepared from *E. coli* transformed with a pMSHsp70 plasmid. Anti-Hsp70 monoclonal antibody was produced by hybridoma Mus musculus 2E4 [4]. The prepared suspensions of magnetic conjugates were mixed with Hsp70 (or anti-Hsp70 antibody) and the dynamics of the immune reaction was studied by measuring T_2 time of relaxation using CXP-300 (Bruker) 7.1 T for 90 min. For estimation of proton magnetic relaxation times the inversion recovery and Carr-Purcell-Meiboom-Gill impulse sequences were employed. The magnetic relaxation rate R_2* was measured from half width of resonance line.

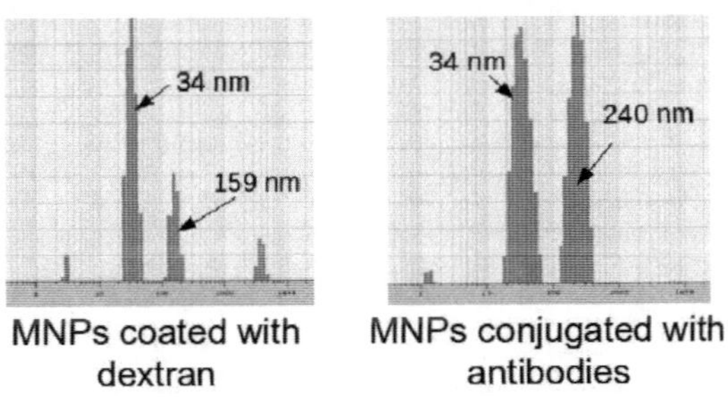

Figure 2. Size distribution of the synthesized MNPs.

3. Results and discussion

In experiment the suspension of MNPs was located in highly homogeneous field B_0 of superconducting magnet of NMR spectrometer. In dipole approximation the MNPs induce additional inhomogeneous field which z component is described by relation

$$B = \frac{\mu M}{3}\left(\frac{R}{r}\right)^3 \left(3\cos^2\theta - 1\right) \tag{1}$$

where B – magnetic induction, μ = 4π 10^{-7} H/m, M – magnetization, r – distance from center to reference point, θ – the angle between r vector and direction of constant magnetic field of spectrometer [5],[6]. The water protons in vicinity of MNPs experience the dephasing of spin precision in magnetic field Bo. The spin-lattice and spin-spin relaxation times T_1, T_2 and their inverse values R_1, R_2 are related with static and fast diffusional mechanisms of outer-sphere relaxation theory which leads to formula:

$$R_1 = \frac{1}{T_1} = \frac{32\pi}{135}\gamma_I^2 \frac{n}{RD}\left[3J^A\left(\sqrt{2\omega_I\tau_D}\right)\right]\langle\mu_z\rangle^2$$

$$R_2 = \frac{1}{T_2} = \frac{32\pi}{135}\gamma_I^2 \frac{n}{RD}\left[\frac{3}{2}J^A\left(\sqrt{2\omega_I\tau_D}\right) + 2J^A(0)\right]\langle\mu_z\rangle^2 \tag{2}$$

where γ_I is the gyromagnetic ratio of the proton, D is the self-diffusion constant of the water, n is the number of MNPs per unit volume, μ is the magnetic moment, R is the effective radius associated with a magnetic nanoparticle, $\tau_D = R^2/D$, and $J^A(z)$ is the spectral density of the correlation function related to molecular diffusion in the inhomogeneous local magnetic field produced by the magnetic particles. The increased relaxivity r_2 of water protons arises from the time fluctuation of the dipolar interaction between the magnetic moment of superparamagnetic domain and proton magnetic moment of the H_2O. The fluctuating local magnetic field in the vicinity of a magnetic core is governed by outer sphere relaxation mechanism of nonhomogeneous broadening in magnetic suspensions. The proton relaxation times T_1, T_2, T_2^* of water protons were found to depend on the concentration MNPs in suspension. The magnetic relaxation rates R_1, R_2, R_2^* as inverse time relaxation values have shown a linear dependence on total Fe content. The experimental results for R_2 concentration dependence are shown in Figure 3. The concentration plots followed to a linear function

$$R_i = r_i C + A \tag{3}$$

where $R_i = 1/T_i$, C – MNPs concentration in mM, r_i – molar relaxivity, and A – a constant determined by the rate of relaxation of water protons in the absence of paramagnetic entity in solution, i = 1, 2.

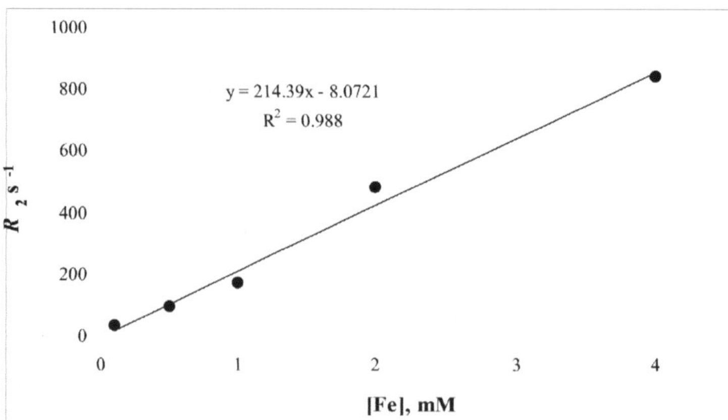

Figure 3. Magnetic relaxation rate R_2 of magnetic conjugate in dependence on the Fe content in MNPs.

According to the relations of outer-sphere magnetic relaxation theory (2, 3) the magnetic relaxation rates increase with growth of MNPs diameter in initial stages of aggregation provoked by antibody-antigen interaction. Magnetic field enhanced targeted aggregation is the key principle of swiching assay. The decreasing number of MNPs due to the formation of magnetic clusters leads to creation of motional averaging of local magnetic fields with subsequent increase of estimated relaxation time T_2. Addition of biomarker to the suspension of magnetic conjugates led to the formation of MNPs clusters that further precipitated from the detection area of NMR spectrometer and thus increased T_2 relaxation time as shown in Figure 4 and Figure 5.

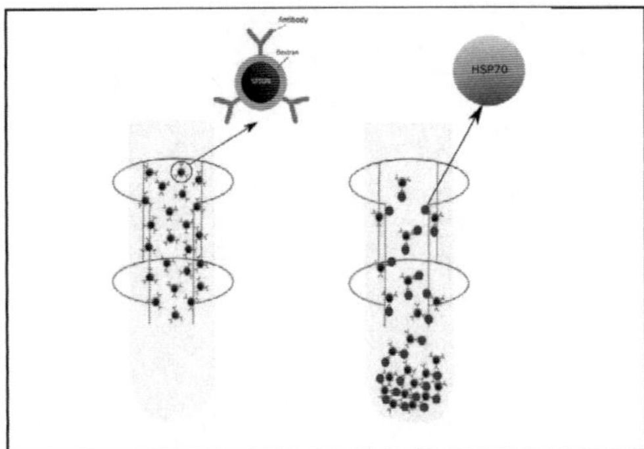

Figure 4. Scheme of MNPs clusters formation following biomarker addition to the nanoparticles suspension.

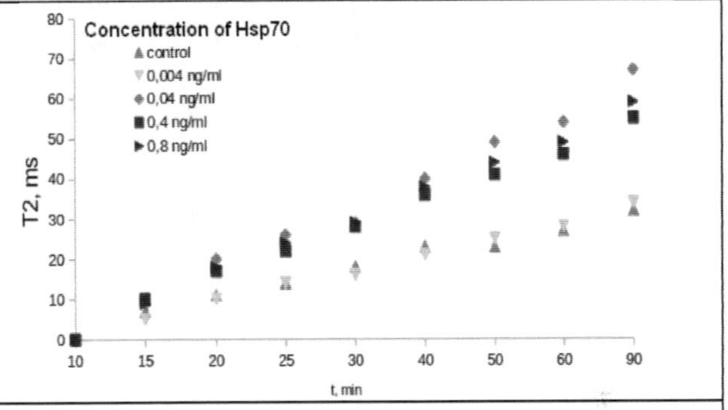

Figure 5. The dynamics of the reaction between Hsp70 and anti-Hsp70 antibodies conjugated with MNPs.

The Figure 4 in schematic form illustrates the influence of antigen-induced aggregation on detected T_2 relaxation of water protons in NMR sample. The observed tendency of increased T_2 relaxation time (Figure 5) indicates the predominant mechanism of motional averaging at analyte concentrations in the analyzed samples. The inhomogeneous contribution of large clusters is suppressed by their sedimentation in the suspension and leaving from the detection coil. The biosensing reaction was observed in the wide range of biomarker concentrations (0.004 - 0.8 ng/ml of Hsp70, 0.4 - 4 ng/ml of antibodies) that is significantly higher than that of the widely applied enzyme-linked immunosorbent assay (ELISA). It is worth noting the fact that standard ELISA is based on the absorption spectroscopy that limits it's accuracy in biological fluids, while switching assay allows to overcome this limit due to independency on the scattering centers in media.

4. Conclusions

The obtained data clearly demonstrate the feasibility to apply MNPs conjugated with anti-Hsp70 antibodies for Hsp70 detection in the switching assay. Formation of MNPs clusters due to the interaction with the soluble protein significantly affects T_2 relaxation time of water protons thus enabling to detect the Hsp70 with high sensitivity in comparison to standard diagnostic tools (i.e., ELISA).

Acknowledgments
Authors acknowledge O. Zlobina for preparation of Hsp70. The study was supported by the Russian Foundation for Basic Research 15-08-08148A.

References

[1] Min C, Shao H, Liong M, Yoon T, Weissleder R, Lee H 2012 *ACS Nano* **6** 6821
[2] Swierczewska M, Liu G, Lee S, Chen X 2012 *Chem. Soc. Rev.* **41** 2641
[3] Shevtsov M, Yakovleva L, Nikolaev B, Marchenko Y, Dobrodumov A, Onokhin K, Selkov S, Mikhrina A, Guzhova I, Martynova M, Bystrova O, Ischenko A, Margulis B 2014 *Neuro-Oncology* **16** 38
[4] Vetchinin S, Belova E, Baranov A, Sapozhnikov A, Margulis B Patent 2 381 271 C1 Russian Federation 2010
[5] Haan H 2011 *Magnetic Resonance in Medicine* **66** 1748
[6] Roch A, Gossuin Y, Muller R, Gillis P 2005 *J Magn Magn Mater* **293** 532

Photodamage of the cells in culture sensitized with bilirubin

O A Kozlenkova, L G Plavskaya, A V Mikulich, I A Leusenko, A I Tretyakova, V Yu Plavskii.
Laboratory of organic heterogeneous media, B.I. Stepanov Institute of Physics of NAS of Belarus, Minsk, Republic of Belarus

Abstract. It has been shown that exposure to radiation of LED sources of light with an emission band maximum at about 465 and 520 nm having substantially identical damaging effects on animal cells in culture, that are in a logarithmic growth phase and preincubated with pigment. Photobiological effect is caused by photodynamic processes involving singlet oxygen generated by triplet excited sensitizer. Mono-exponential type dependence of cell survival on the energy dose indicates that it is bilirubin that acts as a sensitizer but not its photoproducts. The inclusion of bilirubin in the cells, where it is primarily localized in the mitochondria cells, it is accompanied by multiple amplification photochemical stability compared to pigment molecules bound with albumin

1. Introduction

The development of syndrome of neonatal hyperbilirubinemia (jaundice) is caused by excessive accumulation in the blood, as well as in subcutaneous fat of gall pigment (breakdown product of normal heme catabolism) – Z, Z-bilirubin IXα, due to its overproduction and / or low excretion rate from the body. Phototherapy - the main and most common method for treating neonatal hyperbilirubinemia, which lies in exposure of body surface to radiation with spectral range, which corresponds to the long-wavelength absorption band of bilirubin. It is believed [7-11] that the determining role in reducing the level of bilirubin in newborns under irradiation play photoisomerization processes of the pigment - the formation of its configurational and structural isomers that are more hydrophilic compounds than the native bilirubin, and quickly eliminated from the body. In addition to photoisomerization self-sensitized photolysis of bilirubin involving singlet oxygen also contributes to the reduction of bilirubin level [12]. At the same time, one cannot exclude that the singlet oxygen generated by triplet excited bilirubin, will cause photodynamic damage to other vital structures of molecular and cell organelles. This indicates the need to investigate sensitizing effect of bilirubin and its photoproducts on important biological structures to develop measures for reducing the possible side effects of the action of light during phototherapy. In recent years, the urgency this problem is even more acute in connection with the application for the treatment of neonatal hyperbilirubinemia of new light sources based on high-brightness LEDs, allowing vary over a wide range both the intensity and the wavelength of incident radiation within the absorption band of the pigment [7 - 11].

The aim of present work – comparable studies of photostability of bilirubin in cells and in a complex with albumin as well as study of mechanism of sensitized by pigment damage of cells in culture upon irradiation with LED sources of light of blue and green spectral range.

2. Materials and methods

Photostability of bilirubin was investigated in MEM medium (with 10% bovine serum), in which bilirubin is in a complex with albumin, as well as in kidney cells of an African green monkey BGM in a logarithmic growth phase. The cells were grown in disposable Petri dishes in a nutrient medium MEM (minimal essential medium) with 10% bovine serum at 37 °C and a 5% CO_2 incubator. The cellular monolayers were preincubated with bilirubin in a concentration of 40 mM for 2 hours, and then exposed to radiation of LED sources with emission band maximum at about 465 nm or 520 nm. Bilirubin photostability was evaluated spectrophotometrically by comparing the absorbance of the bilirubin solution extracted from the cells using DMSO immediately after irradiation of the LED source and unirradiated cells. We use an MTT-test as a criterion for the metabolic cells activity assessment.

3. Results and discussion

Studies have shown that bilirubin can have a sensitizing effect on the cells (as manifested in the reduction of viability) when excited by optical radiation with wavelength $\lambda = 465$ nm and $\lambda = 520$ nm. It is found that the biological effect induced by the light in the presence of bilirubin depends on the physiological state of the cells, the dose of incident radiation and the concentration of the photosensitizer. Changing the power density of light in 3-4 times with appropriate compensation of dose due to the irradiation time has practically no influence on the photobiological effect (table 1). Adding in cell culture together with the sensitizer a quencher of singlet oxygen - sodium azide - significantly reduces the damaging effect of light towards the cells. This is indicated by the MTT-test data presented in Table 1. One can see that fraction of viable cells (relative to control) in the absence of other additives upon irradiation of the cells ($P = 20$ mW / cm^2, $t = 5$ min) sensitized with bilirubin is $\gamma = (45{,}5 \pm 1{,}6)\%$. Under the same conditions of irradiation for cells in the presence of 5 mM of sodium azide fraction of viable cells is significantly increased and $\gamma = (97{,}9 \pm 5{,}7)\%$. Thus, sodium azide actually blocks the destruction of the cells sensitized with bilirubin. This fact is evidence of the decisive role of singlet oxygen in the mechanism of photodegradation of cells sensitized with bilirubin. Also the addition of sodium azide in the culture medium prior to irradiation reduces in some degree bilirubin discoloration. Thus, singlet oxygen plays a determining role in the study of photochemical processes. In the absence of bilirubin, as well as in case of incubation of the cells with bilirubin ($C_{BR} = 40$ μM) without light exposure, the effect is extremely low.

Table 1. Effect of sodium azide on photosensitized bilirubin cell damage in the conditions of constant light exposure dose of energy ($\lambda = 465$ nm) by varying the time and power density of incident radiation.

Parameters	Control group	Experimental group		
The power density of incident radiation, mW/sm^2	0	20	10	5
Exposure time, min	0	5	10	20
Energy dose, J / cm2	0	6	6	6
Percentage of viable cells, incubated for 2 hours with 40 μM bilirubin,%	100,0±7,5	45,5±1,6	50,0±6,4	48,7±12,5
Percentage of viable cells, incubated for 2 hours with 40 μM bilirubin and 5 mM sodium azide,%	100,0±13,5	97,9±5,7	91,2±10,4	90,2±1,0

It is well known [13 - 15] that the lipophilic bilirubin being readily soluble in lipids of cell membranes, can penetrates and accumulates in the mitochondria. For this reason, the high concentration of bilirubin in the mitochondria, even in dark conditions may alienate respiration and oxidative phosphorylation in them, induce apoptosis, disrupt protein synthesis, and have other toxic effects. As already mentioned, at the chosen concentrations (40 μM in growth medium) bilirubin in dark conditions has no appreciable effect on the survival of kidney cells of African green monkey BGM. Nevertheless, these results indirectly confirm the localization of bilirubin in mitochondrial structures, since preincubation of cells with bilirubin greatly affects the structure of crystals of formazan formed in the mitochondria of living cells for the MTT-test. Pictures (figure1) obtained at the same magnification (40^X) clearly show that after incubation of cells with bilirubin formazan crystals become significantly larger (figure 1B) compared with the control intact cells (figure 1A). Irradiation of cells in the presence of bilirubin significantly affect cell viability, as evidenced by low packing density of crystals of formazan formed in the mitochondria (figure 1D). Thus in the absence of irradiation bilirubin cells (figure 1C) has practically no effect on the arrangement density of mentioned crystals.

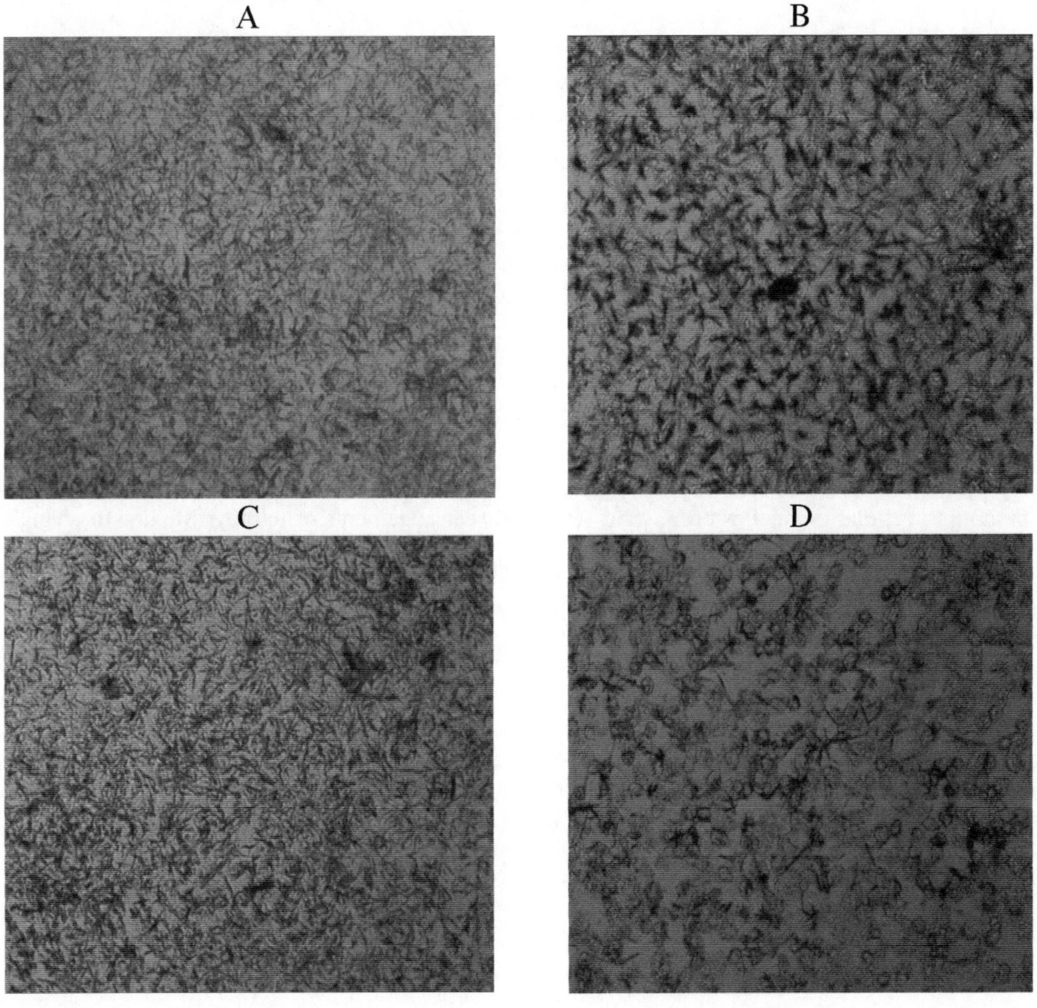

Figure 1. Effect of bilirubin and light exposure (λ_{ex} = 465 nm, P = 10 mW / cm^2, t = 10 min) on the structure of crystals of formazan formed in the mitochondria of metabolically active cells A - cells in the absence of bilirubin and without irradiation (control); B - cells with bilirubin without irradiation; C - irradiated cells in the absence of bilirubin; T - cells are irradiated with bilirubin.

Dependence of cell survival on energy dose upon irradiation of cells in the presence of bilirubin with LED sources of light with an emission band maximum at about 465 and 520 nm is mono-exponential function (figure 2). This indicates that it is bilirubin that acts as a sensitizer but not its photoproducts. Another unique feature of these curves - practically identical photobiological effect for radiation with λ = 465 nm, corresponding to the maximum of the absorption spectrum of bilirubin-albumin complex, and for radiation with λ = 520 nm corresponding to the long-wavelength slope of the specified spectrum.

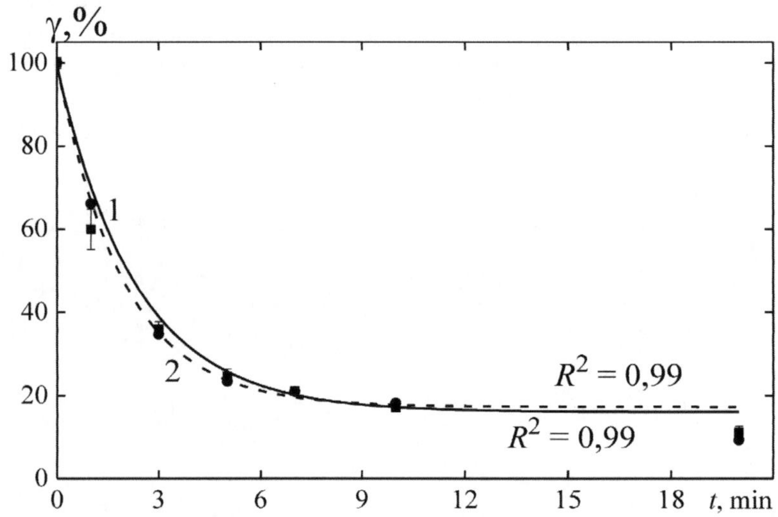

Figure 2. The dependence of viable cells percentage sensitized with bilirubin on the irradiation time by LED sources of light with emission band maximum at λ_{em} = 465 nm (curve 1) and λ = 520 nm (curve 2), power density P = 20 mW / cm^2.

Apparently, it should be stated about sharp changes in the spectral characteristics of bilirubin upon its binding to cellular organelles because of structural rearrangements of tetrapyrrole due to changing microenvironment.

It was shown that exposure to radiation of LED sources of light with emission band (maximum at about 465 nm), corresponding to the absorption spectrum of bilirubin, in the conditions causing almost complete bilirubin photodestruction in the nutrient media practically does not cause photolysis of bilirubin localized in the mitochondria (Figure 3). The findings suggest that bilirubin has a high photochemical stability in the lipid environment. Among the possible causes of increased bilirubin photostability at its intracellular localization are considered: a) the formation, along with the monomers, of tetrapyrrole dimeric forms; b) the presence of closely located antioxidants; c) slow diffusion of oxygen involved in the reaction self-sensitized discoloration of the pigment; g) quenching of the triplet state of bilirubin by other biomolecules.

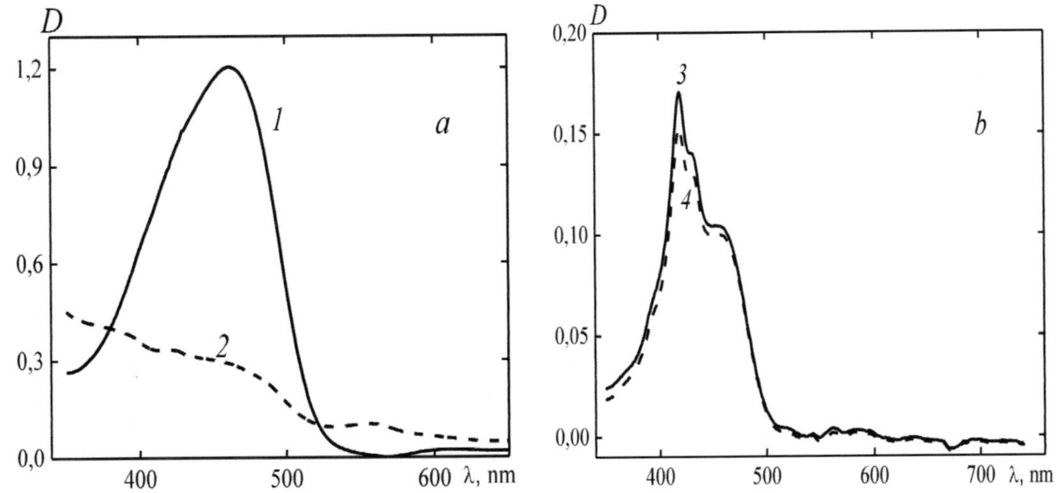

Figure 3. Absorption spectra of bilirubin in medium (*a*) and a DMSO solution (*b*) after extraction from the cells without prior exposure (curves 1 and 3) and after exposure to radiation λ = 465 nm, power density P = 20 mW / cm^2 for t = 5 min (curves 2 and 4).

4. Conclusions

It is shown that bilirubin located within cells is characterized by high photochemical stability and can act as a photosensitizer, causing photodamaging lethal effect on animal cells in culture. The exposure to radiation of LED sources of light with an emission band maximum at about 465 and 520 nm causes substantially identical damaging effects on animal cells in culture that may be due to by a significant change in the spectral characteristics of bilirubin upon entering into the cells. The addition of sodium azide to the system being irradiated, the quencher of singlet oxygen, reduces the biological effect that points to the decisive role of singlet oxygen in the mechanism of photodynamic action sensitized with bilirubin.

References

[1] Xiong T, Qu Y, Cambier S, Mu D 2011. *Eur J Pediatr.* **170** p 1247–1255.
[2] Плавский В Ю, Мостовников В А, Третьякова А И, Мостовникова Г Р 2008. *Журн. прикл. спектр.* **75** с 383–394.
[3] Böhm F,.Drygalla F, Charlesworth P, Böhm K, Truscott T G, Jokiel K 1995. *Photochem. Photobiol.* **62** p 980–983.
[4] Oláh J, Tóth-Molnár E, Kemény L, Csoma Z 2013. *Brit J Dermatol.* **169** p 243-249.
[5] Tanaka K, Hashimoto H, Tachibana T, Ishikawa H, Ohki T 2008 *Pediatr Surg Int.* **24** p 837–842.
[6] Liu G S, Wu H, Wu B Q, Huang R Z, Zhao L H, Wen Y 2008. *World J Pediatr.* **4** p 31–35.
[7] Plavskii V Yu 2012 *Nova Science Publishers Inc* (New York). ed. J F Novotny and F Sedlacek p 1–65.
[8] Maisels M J, McDonagh A F 2008. *N. Engl. J. Med.* **358** p 920–928.
[9] Stokowski L A 2006. *Adv. Neonatal Care.*p 303—312.
[10] Плавский В Ю, Третьякова А И, Мостовникова Г Р 2014. *Опт.журн.* **81** с 51–62.
[11] Плавский В Ю, Третьякова А И, Микулич А В, Плавская Л Г 2014. *Инновационные технологии в медицине* **2** с 84-98.
[12] Landen G L, Park Y T , Lightner D A 1983 *Tetrahedron* **39** p 1893–19071.

[13] Chuniaud L, Dessante M, Chantoux F, Blondeau J P , Francon J, Trivin F 1996. *Clin Chim Acta.* **256** p 103–114.
[14] Akin E, Clower B, Tibbs R, Tang J, Zhang J 2002. *Brain Res* **931** p 168–175.
[15] Keshavan P, Schwemberger S. J, Smith D L, Babcock G F, Zucker S D 2004 *Int J Cancer.* **112** p 433–445.

Properties of calcium-containing microparticles formed in the process of biomineralization of the human aortic wall

O Yu Zlatoustova[1,2], S V Vasilev[1], A S Rudy[1,2]

[1] Diagnostics of micro- and nanostructures Laboratory, Yaroslavl Branch of the Physics and Technology Institute RAS, Yaroslavl 150007, Russia
[2] Microelectronics and general physics department, P.G. Demidov Yaroslavl State University, Yaroslavl 150000, Russia

Abstract. The results of the study of human aortic walls, subjected to different stages of mineralization are presented. By means of scanning electron microscopy and X-ray microanalysis the morphology, elemental composition and characteristics of the mineral component localization were investigated. The key differences in the initial, intermediate and final stages of pathological mineralization of the aorta wall were identified. The data obtained may be useful in describing the mechanism of biomineral deposits formation in human body.

1. Introduction

The pathological mineralization in human body can usually be the sight of such diseases as: kidney stone disease, arthritis, arthrosis, atherosclerosis, calcium thesaurismosis. Mineralization phenomenon is the deposition of minerals in the organs and tissues. The development of this process is facilitated by the alteration in the ratio between specific types of proteins, the activity of hydrogen ions in biological fluids, disorders of calcium metabolism, enzymatic and other factors.

It is commonly considered that first stages of atherosclerosis are associated with the deterioration of the vascular wall [1-7]. For example, immune complexes may cause such deterioration. The immune complexes are formed at the time when there is an excess of body antigens (infections, viruses, bacteria) and cause local inflammation of soft tissues in the body [1-2]. Hypercholesterolemia leads to disruption of condition of the vascular wall also. It is accompanied by increased levels low-density lipoprotein (LDL) in the blood. Oxidized LDL exerts toxic effect on endothelial cells. In response, the monocyte clusters are trying to eliminate inflammation, but in complex with LDL, are transformed into foam cells, subsequently they accumulate in the vascular wall [1,3-4]. The toxic factors such as smoking, radiation, foreign chemicals may result in oxidative stress for endothelial cells. Oxidant molecules have one or more unpaired electrons, and aspire compensate their quantity by means of endothelial cells. As a result, the endothelial cells become unstable and endothelial layer of the aortic wall structure is damaged. Toxic factors cause disorder in the metabolism of cells. It may be accompanied by a mutation, apoptosis or cell proliferation [2,5]. Hemodynamic loads lead to wear, disruption properties and elasticity of the vascular wall tissue. The sudden change in blood pressure causes the vascular spasm. This results in nutrition disturbance of the large arteries as well as tissues of vital organs [6]. Vascular spasm occurs also with stress, causing disruption of the receptor system and subsequent failure in the transmission of nerve impulses and the protective functions of the organism [7].

Journal of Physics: Conference Series **741** (2016) 012064 doi:10.1088/1742-6596/741/1/012064

The only violation of the state of the vascular wall, of course, is unlikely to lead to development of its atherosclerotic lesions. Atherosclerosis can spread like an avalanche, as a result of chronic inflammation of soft tissues and organs, as well as influenced of promotor factor. The most important one of such factors is adverse genetic code. Thus, for example, may be deteriorated production of some enzymes, which should participate in the removing died off cells. As a result these cells accumulation in the body and their subsequent mineralization takes place. Metabolic disorders and acid-base balance, increased concentrations mineral ions in the blood and, as a result, elution of ions from the bone may occur if we have unbalanced diet and lack of physical activity. Under conditions of high acidity the one type of mineral deposits is formed, while in alkaline environment are formed rather different deposits [1,6-7].

Final stages of the disease are related with deposition of calcium salts in the vascular wall, resulting in aorta elasticity loss in fully mineralized tissues. The areas of calcification may constitute lesions, clearly visible to the naked eye. However, these lesions are generally composed of minute particles which are detectable by means of electron microscopy only.

The main role in mineralization process in a human body and particularly in soft tissues belongs to calcium containing nanoparticles [8-10]. Sizes of the particles varies in 50-400 nm range. But till now the mechanism of their nucleation, their properties and characteristics remain completely unclear.

To clarify the role of these particles in the development of cardiovascular disease the careful analysis of their morphological features, the elemental and phase composition as well as the characteristics of their localization at different stages of formation of mineral deposits in the soft tissues of the human cardiovascular system is required.

This work presents the results on study of morphology, location and composition of the mineral deposits in the specimens of human aortic walls, obtained by the methods of electron microscopy and X-ray microanalysis at various stages of their damage by mineral compounds.

2. Materials and methods

As a samples of tissue of the human cardiovascular system were used aorta sections, exposed to various degrees of damage by mineral compounds and with varying degree of the aortic wall integrity. Tissue samples have been stored in formalin and transported within one day. Further the samples were washed by an isotonic solution pH = 7,4 and were stored in the same solution within a few days in the refrigerator at a temperature between 1-5 .

For samples study by means of electron microscopy and X-ray microanalysis they were frozen in liquid nitrogen for about an hour, followed by preparation of several cleavages of the samples. Obtained samples were fixed by a glue on a silicon substrate cleaved surface up. Finaly the samples were washed in 0.1 M citrate buffer pH = 2 for 15 minutes.

All of the samples were examined at electron-ion microscope Quanta 3D 200i in a low-vacuum mode. Morphological characteristics of mineralized sections of the samples, a feature of the location and composition of the mineral components in the aortic wall were thoroughly studied. To produce the image both the secondary electron detector and the detector of the backscattered electrons were used.

3. Results and Discussion

The results obtained evidence that most probably the biomineral formation of deposits in the soft tissues of the human cardiovascular system is a multistage process. The initial stage of biomineralization features the presence of solitary mineral spherical particles of 50 nm to 2 mm in size in the media layer of the vessel wall (Figure 1).

Journal of Physics: Conference Series **741** (2016) 012064 doi:10.1088/1742-6596/741/1/012064

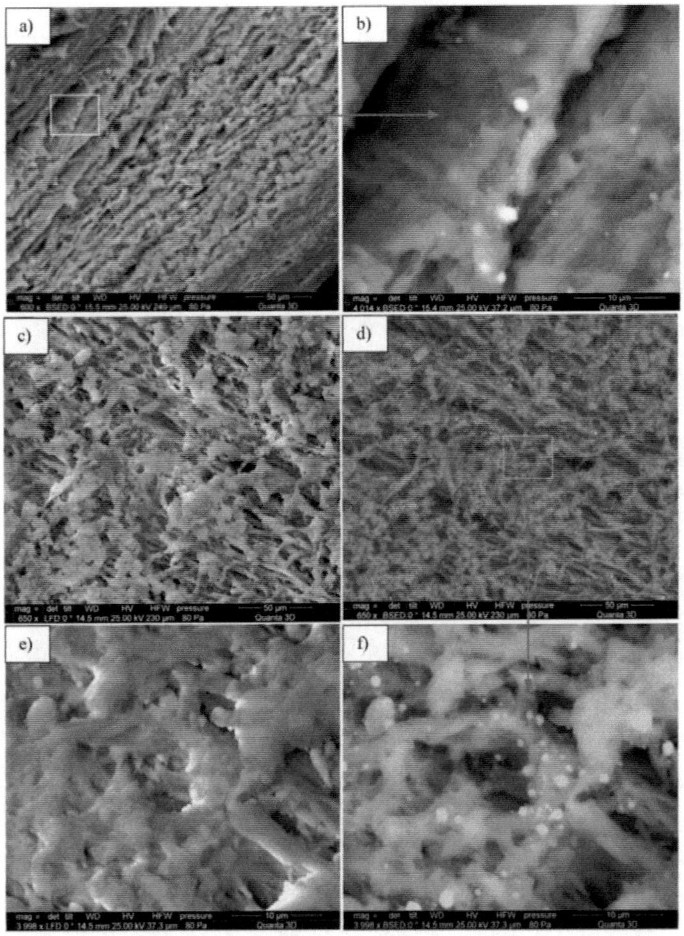

Figure 1. SEM microphotographs of human aortic walls at initial stages of pathological mineralization:
a) - transverse cleavage of aortic wall with the initial symptoms of pathological mineralization. Magnification - 600. (BSED - back-scattered electrons detector allows to reveal the area of mineralization at low magnifications);
b) - micro and nanoparticles in the media layers of the aortic wall (this is the first sign of pathological mineralization) BSED;
c-d) - the same domain of the media layer of aortic wall. Magnification - 650 (c - LFD - Live fiber detector, d) - BSED);
e-f) - the same domain (the rectangular area in Figure 1d) of the media layer of aortic wall obtained at magnification - 4000 (e - LFD, f - BSED)

The next stage is associated with appearing of particle clusters localized in aorta wall and increase of particles volume. The typical size of the mineral particles at this stage reaches several micrometers. The clusters have a size varying from 5 micrometers to hundreds of micrometers (Figure 3). Outbreaks of extensive mineralization are predominantly arranged along the fibers (Figure 3b), between them (Figure 2, 3g-h) and inside them (Figure 3c-f). Mineralization at intermediate stages takes the form of thin layers, located between the fibers of the media layer. (Figure 2a).

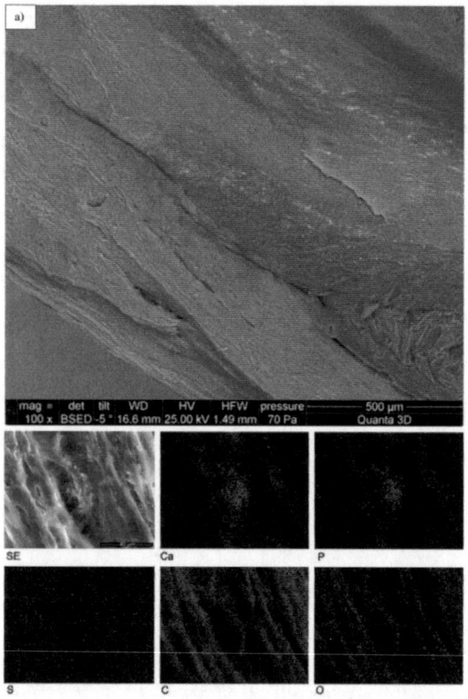

Figure 2. SEM microphotograph and X-ray map of elements distribution in human aortic walls at intermediate stages of the mineralization:
a) - transverse cleavage of aortic wall with the sign of pathological mineralization. Magnification - 100.

SE - electron image of a media domain of aortic wall and map of calcium (Ca), phosphorus (P), sulfur (S), carbon (C) and oxygen (O). The bright domain on the map is a localization place of the corresponding element in the sample.

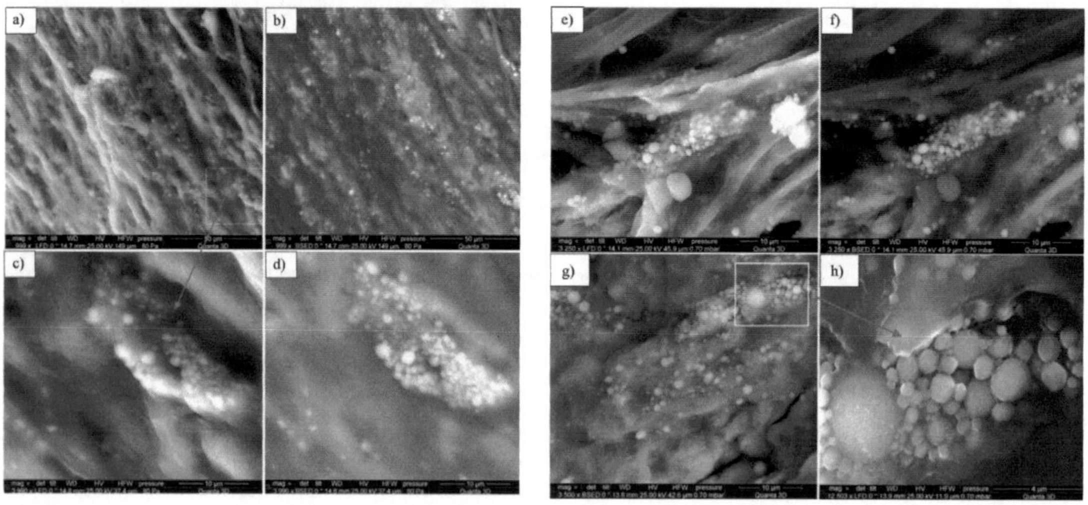

Figure 3. SEM microphotographs of human aortic walls at intermediate stages of the mineralization. (a-b) The same domain of the media layer of aortic wall with magnification 1000 (a-LFD, b-BSED); (c-d) The rectangular area highlighted in Figure 3a with magnification 4000 (c - LFD, d - BSED); (e-f) The same domain of the media layer of aortic wall at magnification 3250 (e- LFD, f- BSED); (g) The domain of the media layer of aortic wall at magnification 3250 (BSED); (h) The rectangular area highlighted in Figure 3g at magnification 12500.

The final stages are characterized by increasing volume of mineralized areas and the dense packs of mineral particles, measuring several micrometers. Subsequently, the mineralized areas are identified as whole mineral agglomerates (Figure 4b). It can also be noted, that the initial stages of mineralization are observed in the media layer, but at a later stage the shift of the extensive mineralization towards the aortic intima is observed (Figure 4c).

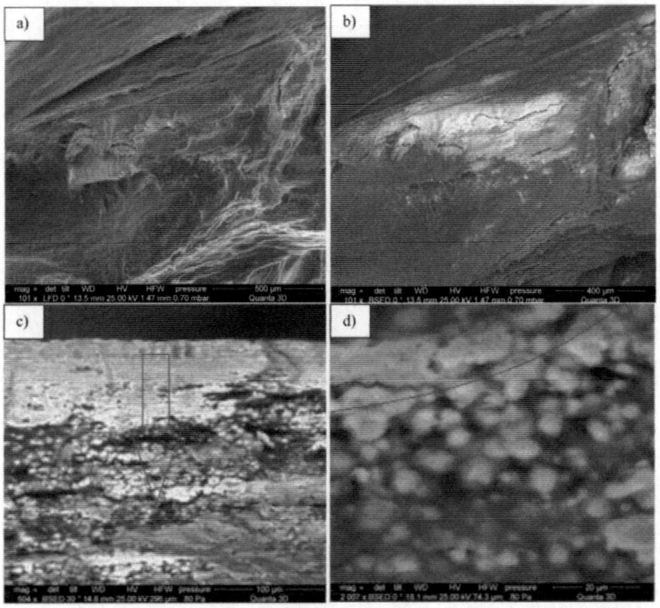

Figure 4. SEM microphotographs of human aortic walls at final stages of pathological mineralization:
a-b) - the same domain of the media layer of aortic wall at magnification 100 (intima is at the top of the image, below is a layer of the media, and at the bottom of image is adventatsiya aortic wall) a) - LFD, b) - BSED (more light domains are the places of mineralization localization).
c) – a cleavage of the aortic wall at the site of mineral aggregate at magnification 500. The arrow indicates the direction from the intima to the media (LFD).
d) - the conventional boundary of transition from the intima to the media at the late stage of mineralization (LFD).

It remains unclear why the mineralization is initiated and proceeds along the muscle fibers of the media, and is often observed in the form of layers disposed between them? Media is composed of aortic smooth muscle cells, and extracellular matrix proteins, first of all of collagen and elastin. Mineral micro- and nanoparticles, ranging in size from 50 nm to 2 microns were detected on the surface of the protein fibers (Figure 1b, 1f); between them (Figure 2, 3g-h); and also in the volume of fibers (Figure 3c-f).

The hotbeds of the initial stages of mineralization of aortic wall lie in a media layer. Media of aortic wall has a high density distribution of smooth-muscle cells as compared with the intima [11]. In turn, in the processes of cellular metabolism the regulation of calcium level is the responsibility of Ca-binding proteins such as calmodulin, troponin C and parvalbumin [12]. Violation of secretion of Ca-binding proteins or of its structure inevitably results in violation of cell homeostasis. The cell consists of various organelles (nucleus, nucleolus, vesicles, and lysosomes) which have a spherical shape. As a result of calcium excess in the cell, these organelles may serve as the matrix for the deposition of mineral components. The proteins, which make up the cells, in this process are most probably degrading. During the protein molecules destruction, the sulfide bonds, that support the protein structure, are torn in the first turn. Apparently, at destruction of sulfide bonds sulfur somehow is deduced from affected area when the sulfide bonds are broken, because it is not detected at the later stages of mineralization.

4. Conclusion

To summarize let us mention that the key differences in the initial, intermediate and final stages of pathological mineralization of the aorta wall were revealed. Combining the obtained results, it can be argued that the aortic mineralization indicated in the layer media. The size of the mineral particles detected in the local conglomeration is different. Dimensions of local conglomerations correlate well with the size of the smooth muscle cells. Dimensions of single mineral particles correlate well with the size of the cell organelles. In this regard, one can assume that the mineralization begins with the pathology of the homeostasis of a smooth muscle cells. The presence of extensive areas of mineralization can be explained, for example, by the disturbances in several adjacent cells. The data obtained may be useful in reveling the mechanism of biomineral deposits formation in human body.

Acknowledgments
The authors would like to extend special thanks to Pukhov D.E. for valuable advice and comments. This work was performed at the equipment of the facilities sharing centre "Micro- and Nanostructures Diagnostics" and was supported by the Ministry of Education and Science of Russian Federation.

References

[1] Lutai M I. 2004 *Ucr J Cardiol* **1** 22
[2] Avaliani V M, Popov V A, Martushov S I. 2005 *Human ecology* **4** 24
[3] Cormode D P, Roessl E, Thran A. et al. 2010 *Radiology* **256(3)** 774782.
[4] Abela G S, Aziz K., Vedre A. et al. 2009 *The American Journal of Cardiology* **103** 959
[5] Benditt E P 1977 *Sci Am* 74
[6] Ermoshin V I 2012 *Health and education in XXI century* **14(1)** 65
[7] Kiechl Stefan et al. 2001 *Circulation* **103** 1064.
[8] John C. Lieske 2008 *Int. J. Nanomedicine* **3(2)** 24324
[9] Martel J, Young J D. 2008 *Proc. Natl. Acad. Sci.* **105** 5549
[10] Miller V M et al. 2004 *Am. J. of Physiology-Heart and Circulatory Physiology* **287 (3)** 1115
[11] Wolinsky H, Glagov S. 1964 *Circulation Research* **14** 400-413.
[12] Leninger A. 1985 *Fundamentals of biochemistry* **3** 794-807

Chirality as a physical aspect of structure formation in biological macromolecular systems

E V Malyshko, V A Tverdislov

Faculty of Physics, Lomonosov Moscow State University, Moscow, 119991 Russia

E-mail: katyamalyshko@mail.ru, tverdislov@mail.ru

Abstract. A novel regularity of hierarchical structures is found in the formation of chiral biological macromolecular systems. The formation of structures with alternating chirality (helical structures) serves as an instrument of stratification. The ability of a carbon atom to form chiral compounds is an important factor that determined the carbon basis of living systems on the Earth as well as their development through a series of chiral bifurcations. In the course of biological evolution, the helical structures became basic elements of the molecular machines in the cell. The discreteness of structural levels allowed the mechanical degrees of freedom formation in the molecular machines in the cell.

1. Structure formation in chiral systems

One of the efficient methods of theoretical biology is a geometrization of the problem. It is natural to assume that the world of nucleic acids and the world of proteins should operate in the rank-one symmetric space, but with a certain quantitative difference in tangible media. We are talking about symmetry relations in the primary, secondary, etc. structural levels of nucleic acids and proteins. However, both systems must be constructed hierarchically to have executive and regulating subsystems. Chiral duality becomes a natural tool in combinatorics of structural correlations in these subsystems at all levels of structural organization. The purpose of the biological hierarchies is the ability to match the processes of different time and space scales.

Finding the criteria of discreteness and mechanisms of stratification in the hierarchy of primary, secondary, tertiary and quaternary structures of biological macromolecules and, as a result, their functional characteristics is one of the most important physical problems of the molecular biology [1]. The work is aimed at developing a model of the formation of the hierarchical structures in biological macromolecules. We proposed to formulate a new synergetic law that, in our view, is of common physicochemical and biological nature: in the process of self-organization an evolving system, possessing free energy and elements with chiral asymmetry, may change the type of symmetry inside one hierarchical level, thereby increasing its «complexity», but with transition to a higher level the system may also change its scale and chirality sign together with a change in the functionality of an enantiomorph. The change of chirality sign provides for the evolutionary irreversibility of stratification (stratification is understood as the splitting of a hierarchical level of a system into two levels with different chirality signs, i.e. formation of chiral layers in a system). Indeed, to switch its chirality sign a chiral object, which does not possess a symmetry axis, must be disassembled and then reassembled from the same elements. The chirality of biological structures of different levels ensures that the process of L/D stratification is universal and hierarchical levels are deterministic and stable.

An enantiomorph of a higher level, while preserving its mirror equivalence, acquires a broader spectrum of functionality. Earlier we detected the succession of the hierarchical levels with an alternating chirality sign in the structural hierarchy of DNA and proteins: from the "lower" asymmetric carbon atom to the helices, superhelices and supramolecular structures [2]. The regular alternations of the chirality sign D-L-D-L for DNA and L-D-L-D for protein structures (figure 1) were observed during the transition to a higher level of the structural and functional organization [3].

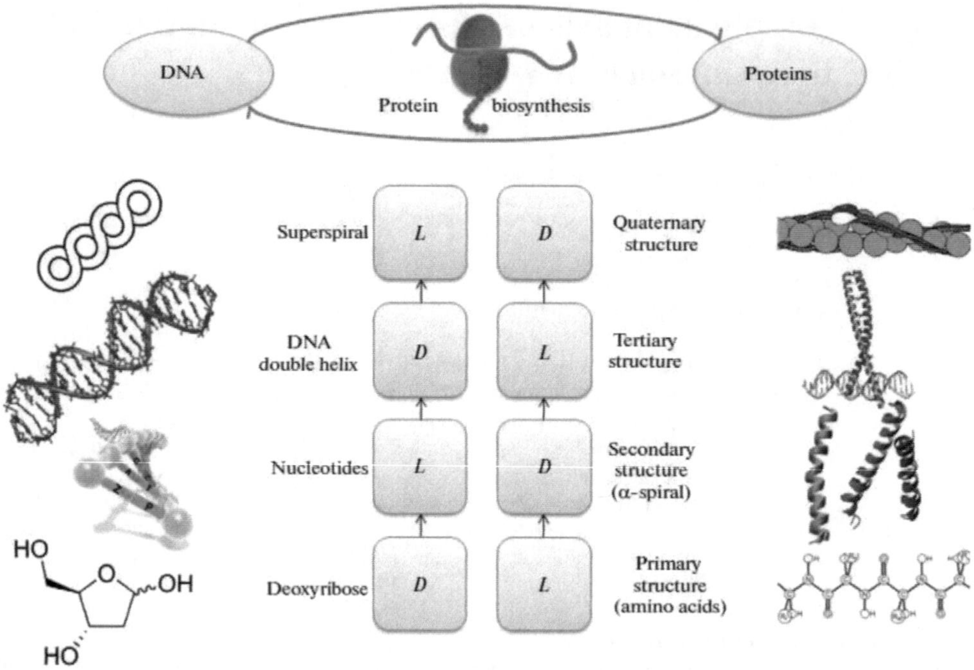

Figure 1. The sign-alternating hierarchies from primary to quaternary chiral (helical) structures for DNA (left column) and proteins (right column). L denotes the left enantiomer or the helical configuration; D, the right enantiomer or the helical configuration [3].

In principle, the premises of the proposed paradigm prompt an answer to the question about the basis of precisely this relation between chiral hierarchies of nucleic acids and proteins. Probably the answer is in the complemental principles of chirality and complementarity in the chain DNA-RNA-protein. It appears that the mechanism of complementary interactions that connects DNA and proteins at the level of RNA and ribosomes provides for another level in the chiral hierarchy of informationally determined macromolecular systems (D/L-sequences of DNA and proteins are mirrored). As a matter of fact nucleic acids and proteins were formed in nature as linear polymers for the purpose «to build», in the course of early evolutionary adjustment, a «sensible» chiral pair «DNA - protein».

2. Structural and functional hierarchy of proteins

The protein primary structure is a sequence of L-amino acids residues. The polypeptide chain is laid in the helix or in the pleated sheet. Generally, α-helix is D-enantiomer, because the right-handed α-helix is more stable than the left-handed one [4]. Interacting right-handed α-helices tend to form left superhelices. The quaternary structure of proteins is represented by the supramolecular structures, mainly formed with the left-handed superhelices interacting with the right-coiling structures. A phase shift is obvious because the protein hierarchy starts from L-amino acids, and the nucleotide hierarchy starts from D-carbohydrates deoxyriboses.

While the chirality signs are determined for the first two levels of the hierarchy of the protein structures, other levels of more complex construction of tertiary and quaternary structures are not that easy to determine. Soon after the Pauling's α-helix model, studying complicated structures with several α-helices became interesting topic for researchers. Significant research progress was made by Pauling and Crick independently of each other in the early 1950s. (see [5]). Crick noted that if two a-helices twist around each other at an angle of about 20°, their side-chains would interlock systematically along the core of the structure, repeating the same interactions every 7 residues (or two turns of the a-helix). He called such a bundle of the supercoiled helices the "coiled coil" and such type of interaction – the "knobs into holes" (KIH).

Later Crick's assumptions were confirmed: a typical sequence of the coiled coils therefore shows a "heptad" repeat of 7 amino acid residues (see [6].). The seven structural positions are labeled *a-b-c-d-e-f-g*. Usually there are hydrophobic residues in positions *a* and *d* (most frequently - Leu, Ile and Val), polar amino acid residues in positions *g* and *e* (most frequently - Lys and Glu). Such distribution of amino acid residues makes the helices amphipathic with residues at *a* and *d* forming the inter-helical hydrophobic core and residues at *e* and *g* forming inter-helical ionic interactions.

The authors of the paper [7] have introduced systematical classification of superhelices by the "periodic table". Protein structures were divided into two groups: classical (structure with one coiled coil and consequently one hydrophobic core) and complex (protein structures that contain two or more classical coiled coils that are connected in some way). Classical coiled coils with two, three, four or five helices form the first row of the table and the heads of the columns, whereas complex coiled-coil structures are separated into classes according to the highest-order classical coiled coil. Each protein containing superhelix can be studied separately using the source [8].

To prove the dominance of the left-handed superhelices in all types of configurations, we analyzed all available protein structures from the "periodic table" [7, 9]. In almost all the cases where intermolecular overlap of α-helices was visually detected there was a tendency for the secondary left superhelix formation. The structure analysis gives quantitative evaluations of the presence of left-handed superhelical motif in each class selected from the table [7]. In the simplest case - for superhelix consisted of two α-helices, - left twisting tendency was observed almost in all structures in this class. Increasing a number of α-helices, number of structures with a similar trend is reduced, but in particular classes, which are presented by more complex superhelices, containing 5-9 α-helices, the percentage of structures with a tendency to left twisting increases sharply. We pay special attention to the fact that the right-handed twisting was observed only in two cases among the variety of structures [7]: in tetrabrahione of *Staphylothermus marinus* (1FE6 in [8]) and in artificially synthesized tetramer (1RH4 in [8]). In some cases it was difficult to classify the twisting due to accumulations of α-helices and β-sheets near the coiled coil. Sometimes it was difficult to observe the tendency because of too short fragments of α-helices in a superhelix. In addition in some cases (most often in superhelices based on 4 α-helices) there was almost a parallel layer of α-helices with a light left-handed twisting tendency, and we did not rank these structures as having left-handed motif (as an "incomplete criterion").

Thus, we consider as proven our earlier statement regarding the characteristic chirality sign change, which is a general physical principle during transition from the secondary to the tertiary structure. It is worthy to note that the trend of the chirality sign change can be observed in protein quaternary structure (in fibrous proteins: actin protofibril, myosin, α-keratin, tropomyosin, paramyosin and the light chain of meromyosin) [2].

In our paper [3] there is a description of possible solution of Levinthal paradox: complementary sign-changing chiral sequences D–L–D–L and L–D–L–D for DNA and proteins form an Ariadne's thread that directs macromolecular folding along the necessary trajectory toward Levinthal's trap. It is known that a homochiral molecular substance (e.g., an amino acid or carbohydrate solution) undergoes racemization to equalize the concentration of enantiomers, raising the system's entropy to its maximum and lowering its free energy. Meanwhile, a linear homochiral polymer is able to reduce the free energy not only through monomer racemization ("horizontal racemization") but through "vertical racemization" as well (i.e., by creating higher level structures with the different chirality sign). In this case the "right-handed" and the "left-handed" structural units of different structural levels are the particles of the system. The system thus distributes its homochirality, lowering its free energy and leading to the development, among some (or all) macromolecules, of a more stable (longer lived, harder) shell, as opposed to the initial state.

3. Conclusions

A model for the chiral periodicity with an alternating chiral sign in hierarchies of protein and nucleic acid structures is proposed and substantiated. The model also describes chiral conformities in interactions of macromolecules of the same or different types. Regular alternation of the chirality sign is revealed in transitions from the lowest to the higher levels of structural-functional organization both in DNA molecules where it is D-L-D-L (the structure of the deoxyribose polymeric chain is taken to be the primary structure), and in proteins where it is shifted in chiral phase to L-D-L-D. Such cross-chiral (antiphase) correspondence between proteins and DNAs forms the basic structural unit of the molecular biology as an achiral invariant allowing to realize chiral motifs in the interactions of molecules of the same or different classes through interactions of specific chemical groups. There are reasons to believe that the helical structures are required not only as rigid constructions in order to provide specific mechanical translational degrees of freedom for the molecular machines, but also as gating devices ensuring unidirectional cyclical motion of the molecular machines that create rotational degrees of freedom. Prospects of practical application of the approach developed here is related to the modeling of the interaction of chiral drugs with macromolecular targets as well as with the selection of chiral cryoprotectants in the methods associated with the cryopreservation of biomaterials.

To summarize, any chiral system, possessing free energy, tends to spontaneous formation of a new, higher structural level with the same type of symmetry but with an opposite sign of chirality and on an enlarged scale. The resulting hierarchy of conjugated sign-alternating chiral structures kinetically stabilizes them by inhibiting spontaneous racemization; forms a conjugated system with selected degrees of freedom, which makes the work of biological machines possible; in living systems macromolecular and autowave chiral structures are conjugated as a consequence of their channeling through the structures of chiral biological machines; defines the vector of the general development of a system in the direction of an upper, "open-ended" hierarchical level [10].

Acknowledgements

The authors would like to thank Yakovenko L V, Sidorova A E, Ilchenko S A and Skurida V D for their contribution to this research.

This work was supported by the grant of the Russian Scientific Fund (project №14-50-0029).

References

[1] Tverdislov V A, Sidorova A E and Yakovenko L V 2012 From symmetries to the laws of evolution. I. Chirality as a means of active media stratification *Biophysics* **57(1)** 120

[2] Tverdislov V A 2013 Chirality as a primary switch of hierarchical levels in molecular biological systems *Biophysics* **58(1)** 128–32

[3] Tverdislov V A, Malyshko E V and Ilchenko S A 2015 From autowave mechanisms of self-assembly to molecular machines *Bulletin of the Russian Academy of Sciences. Physics* **79(12)** 1516–20

[4] Finkelshtein A V, Ptitsyn O B 2005 *Protein Physics: A Course of Lectures* (Academic Press, An Imprint of Elsevier Science; Amsterdam - Boston - London - New York - Oxford - Paris - San Diego - San Francisco - Singapore - Sydney - Tokyo)

[5] Lupas A and Gruber M 2005 The structure of a-helical coiled coils *Adv Protein Chem* **70** 37-78

[6] Yu Y B 2002 Coiled-coils: stability, specificity, and drug delivery potential *Advanced Drug Delivery Reviews* **54** 1113–29

[7] Moutevelis E and Woolfson D 2009 A Periodic Table of Coiled-Coil Protein Structures *J. Mol. Biol.* **385** 726–32

[8] http://www.rcsb.org/; Berman H M, Westbrook J, Feng Z, Gilliland G, Bhat T N, Weissig H, Shindyalov I N and Bourne P E 2000 The Protein Data Bank *Nucleic Acids Research* **28** 235-42

[9] Testa O D, Moutevelis E and Woolfson D N 2009 CC+: a relational database of coiled-coil structures *Nucleic Acids Res*. **37** D315-D322

[10] Tverdislov V A Chirality as an Instrument of Stratification of Hierarchical Systems in Animate and Inanimate Nature *arXiv*:1212.1677

3D super-resolution microscopy of bacterial division machinery

A D Vedyaykin[1, a], A V Sabantsev[1], I E Vishnyakov[2], N E Morozova[1], V S Polinovskaya[1] and M A Khodorkovskii[1]

[1] Research Institute of Nanobiotechnologies, Peter the Great St.Petersburg Polytechnic University, Saint-Petersburg, 195251, Russia

[2] Institute of Cytology, Russian Academy of Sciences, Tikhoretsky av. 4, Saint Petersburg 194064, Russia

E-mail: misterkotlin@gmail.com

Abstract. Super-resolution microscopy is a promising tool for the field of microbiology, as bacteria sizes are comparable to the resolution limit of light microscopy. Bacterial division machinery and FtsZ protein in particular attract much attention of scientists who use different super-resolution microscopy techniques, but most of the available data on FtsZ structures was obtained using two-dimensional (2D) super-resolution microscopy. Using 3D single-molecule localization microscopy (SMLM, namely dSTORM) to visualize FtsZ, we demonstrate that this approach allows more accurate interpretation of super-resolution images and provides new opportunities for the study of complex structures like bacterial divisome.

1. Introduction

Single-molecule localization microscopy (SMLM) is a powerful fluorescence microscopy technique, surpassing the diffraction limit of resolution by an order of magnitude [1]. SMLM appears to be a flexible tool to study bacterial division machinery and provides possibilities to investigate complex mechanisms, including arrangement of proteins which are involved in bacterial division.

Cytokinesis in bacteria is based on formation of a ring-like structure (Z-ring) in the middle of dividing cell which is composed of several proteins. One of these proteins – FtsZ – is a prokaryotic tubulin homologue that plays a key role in cell division [2, 3]. The Z-ring guides septum formation and thereby allows to segregate future daughter cells, thus it is crucial to ensure proper Z-ring localization. There are several regulation systems that provide correct Z-ring positioning in the cell [4]. In *Escherichia coli* two partially overlapping systems are involved in this process – Min-system and nucleoid occlusion that restrict Z-ring formation to the middle of the cell by direct interactions with FtsZ.

FtsZ was the first identified cytoskeletal protein among prokaryotes and is considered to be an ancestor of tubulin, but unlike tubulin, FtsZ does not form hollow tubes [5]. FtsZ is a highly conserved protein and is present in most bacteria and archaea. It binds to inner plasma membrane via accessory proteins (FtsA, ZipA in *E.coli*) and attracts several downstream proteins which are involved in cell

---

[a] To whom any correspondence should be addressed

wall synthesis and carry out septum formation, thus Z-ring is considered to be a kind of scaffold in this process.

FtsZ functions have been extensively investigated both *in vitro* and *in vivo* [4, 6]. Using *in vitro* approach it was shown that FtsZ in physiological conditions is able to polymerize into filaments with length up to 200 nm and thickness of 1 subunit [6]. FtsZ bundling ability varies from one species to another and depends on conditions. In appropriate conditions FtsZ is capable of forming longer and thicker filaments, including bundles, tubules, ribbons and other higher-order structures, although usually these conditions are far from those present in cells [7]. To explain how FtsZ which forms short filaments *in vitro* could form relatively large Z-ring, two alternative models were proposed. The first model suggests that the Z-ring is a relatively disordered array of short FtsZ filaments that weakly interact with each other and possibly form bundles. The second model depicts the Z-ring as an ordered structure (continuous ring or helix), formed by annealed short filaments [6].

To investigate FtsZ organization in the Z-ring and understand which of two models better fits *in vivo* situation, two main visualization approaches were used, first based on fluorescence microscopy (FM), including super-resolution, and second – cryo-electron tomography (CryoET). Super-resolution FM (SMLM, SIM, STED) allowed to obtain images of the Z-ring which seems to demonstrate uneven or discontinuous structure in several species [8-11]. At the same time data obtained using CryoET supports continuous Z-ring arrangement and thereby is in clear contradiction with FM data [12]. It seems to be important to gather more data on the Z-ring structure under different conditions and using different methods to resolve this controversy.

In the current work 3D SMLM in combination with immunofluorescence labeling was utilized to obtain images of Z-rings during exponential growth in *E.coli.*

2. Materials and methods

All experiments were conducted using Top10 *E. coli* strain transformed with pGEX-4T-2 plasmid. Cells were grown overnight in LB medium supplemented with ampicillin (100 μg/ml) at 37°C, then passaged to fresh medium and grown until $OD_{600}\approx0.5$. Then cells were fixed directly in culture by addition of sodium-phosphate buffer (pH 7.4, 30 mM), formaldehyde (2.6%) and glutaraldehyde (0.04%) for 10 minutes at room temperature followed by 50 minutes on ice.

Cells were harvested by centrifugation, washed in PBS three times and immobilized on poly-L-lysine coated coverslips. Then cells were permeabilized using 0.1% Triton X-100 in PBS, followed by 10 μg/ml lysozyme in GTE buffer (glucose 50 mM, Tris 32.5 mM pH=7.5, EDTA 10 mM), each for 5 minutes.

Next cells were blocked using 2% w/v BSA in PBS (PBS-BSA) for 1 hour and then incubated overnight with rabbit anti-FtsZ polyclonal antibodies (Agrisera) diluted 1:200 in PBS-BSA. Cells were washed 5 times with 0.01% Tween-20 in PBS, incubated with goat anti-rabbit secondary antibodies (Alexa 647 conjugated $F(ab')_2$ fragments of goat anti-rabbit antibodies (LifeTechnologies)) diluted 1:100 in PBS-BSA for 1 hour at room temperature and washed another 5 times as described above.

To prevent sample drift during data acquisition, custom active stabilization system was used. To track sample position, polystyrene microspheres (2.1 μm, Spherotech) were immobilized on the coverglass surface.

All images were obtained using custom set-up based on AxioImager.Z1 (Carl Zeiss) motorized microscope which was described previously [13]. Images were acquired by EM-CCD camera (Andor iXon 897) at 100× total magnification. One CCD pixel corresponded to 108 nm in the focal plane. LF635/LP-B-000 (Semrock) filter set was used to visualize Alexa 647 fluorescence (excitation filter was removed to enable dye activation using a 405 nm laser.

Raw images were acquired using MicroManager [14] in PBS-Tris buffer with pH 7.5, containing 10% w/v glucose, 10 mM 2-mercaptoethylamine combined with 50 mM 2-mercaptoethanol, 2 mM cyclooctatetraene and an oxygen scavenging system (2.5 mM protocatechuic acid and 50 nM protocatechuate 3,4-dioxygenase) [15]. Alexa 647 fluorescence was excited using 635 nm diode laser with a power density of approximately 1 kW/cm^2 at the focal plane and 2000-10000 frames containing individual Alexa 647 molecule images were acquired. To measure axial positions of individual molecules, a weak cylindrical lens (ESCO, 8 m focal length) was introduced to the optical path between the objective and the tube lens.

Data processing was performed using ImageJ, SMLM reconstruction was carried out using ThunderSTORM plugin for ImageJ [16].

3. Results and discussion

SMLM previously allowed us to obtain two-dimensional (2D) images of Z-ring on different division stages and to demonstrate that Z-ring thickens during constriction [17]. We were not able to detect Z-ring thickening using conventional fluorescence microscopy, and this fact emphasizes the usefulness of SMLM for the study of bacterial division. However, images obtained using 2D SMLM represent projections, and this fact substantially complicates interpretation of visualized structures. To implement 3D SMLM, a weak cylindrical lens was introduced into the optical path between the objective and the tube lens.

SMLM with artificial astigmatism allowed us to obtain 3D images of Z-rings in *E.coli* cells. The method used in this work allows to visualize area about 1 μm in depth what is comparable with bacterium outer diameter. Obtained images demonstrate hollow ring-like structure in the middle of some cells. Z-rings on these images demonstrate uneven structure reminiscent of beads on a string (Fig. 1). These structures are in strong agreement with data obtained using 2D-SMLM and most other methods [8-12, 17]. However, we were not able to show that observed discontinuity is inherent to actual FtsZ distribution since several stochastic effects of the method used could sufficiently influence obtained images.

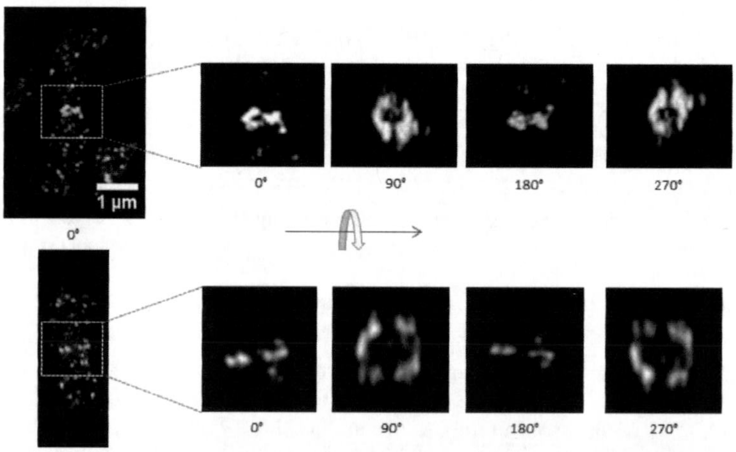

Figure 1. 3D reconstruction of FtsZ ring structure (Z-ring) in *E. coli* dividing cells. 2D image of cell (left panel) and a fragment of the same cell image (in yellow rectangle) obtained using rendering with four different angles (from 0 to 270 degrees).

To rule out the possibility that method used could distort actual FtsZ distribution in Z-ring, further work is needed. For instance, to decrease effects of labeling stochasticity, other labeling techniques which provide one label to one target molecule ratio should be used (e.g. fluorescent fusion proteins). Utilization of photoactivatable or photoconvertible dyes may allow to reduce probability of multiple detections of a single molecule which also affect obtained images.

4. Conclusions

In this work 3D-SMLM in combination with immunofluorescence labeling was successfully used to visualize Z-ring in fixed *E.coli* cells. Obtained images demonstrate uneven FtsZ distribution in the Z-ring which is in agreement with data obtained using other FM methods, but contradicts the results of cryo-electron tomography. To verify if the Z-ring is actually organized in that way, further study is needed.

3D-SMLM expands opportunities of localization microscopy and allows to investigate complicated structures like bacterial divisome. SMLM In combination with immunofluorecent staining can be especially useful in the cases of microorganisms for which effective genetic manipulation protocols haven't yet been developed (e.g. Mycoplasma).

Acknowledgements

The reported study was supported by Russian Science Foundation, research project No. 14-34-00023. The work was carried out using the unique installation «Laser tweezers».

References

1. Rust, M.J., M. Bates, and X. Zhuang, *Sub-diffraction-limit imaging by stochastic optical reconstruction microscopy (STORM).* Nat Methods, 2006. **3**(10): p. 793-5.
2. Addinall, S.G., E. Bi, and J. Lutkenhaus, *FtsZ ring formation in fts mutants.* J Bacteriol, 1996. **178**(13): p. 3877-84.
3. de Boer, P., R. Crossley, and L. Rothfield, *The essential bacterial cell-division protein FtsZ is a GTPase.* Nature, 1992. **359**(6392): p. 254-6.
4. Haeusser, D.P. and W. Margolin, *Splitsville: structural and functional insights into the dynamic bacterial Z ring.* Nat Rev Microbiol, 2016. **14**(5): p. 305-19.
5. Bi, E.F. and J. Lutkenhaus, *FtsZ ring structure associated with division in Escherichia coli.* Nature, 1991. **354**(6349): p. 161-4.
6. Erickson, H.P., D.E. Anderson, and M. Osawa, *FtsZ in bacterial cytokinesis: cytoskeleton and force generator all in one.* Microbiol Mol Biol Rev, 2010. **74**(4): p. 504-28.
7. Huang, K.H., J. Durand-Heredia, and A. Janakiraman, *FtsZ ring stability: of bundles, tubules, crosslinks, and curves.* J Bacteriol, 2013. **195**(9): p. 1859-68.
8. Holden, S.J., et al., *High throughput 3D super-resolution microscopy reveals Caulobacter crescentus in vivo Z-ring organization.* Proc Natl Acad Sci U S A, 2014. **111**(12): p. 4566-71.
9. Jacq, M., et al., *Remodeling of the Z-Ring Nanostructure during the Streptococcus pneumoniae Cell Cycle Revealed by Photoactivated Localization Microscopy.* MBio, 2015. **6**(4).
10. Strauss, M.P., et al., *3D-SIM super resolution microscopy reveals a bead-like arrangement for FtsZ and the division machinery: implications for triggering cytokinesis.* PLoS Biol, 2012. **10**(9): p. e1001389.
11. Rowlett, V.W. and W. Margolin, *3D-SIM super-resolution of FtsZ and its membrane tethers in Escherichia coli cells.* Biophys J, 2014. **107**(8): p. L17-20.
12. Szwedziak, P., et al., *Architecture of the ring formed by the tubulin homologue FtsZ in bacterial cell division.* Elife, 2014. **3**: p. e04601.
13. Vedyaykin, A.D., et al., *Localization microscopy study of FtsZ structures in E. coli cells during SOS-response.* 1st International School and Conference Saint-Petersburg Open 2014 on Optoelectronics, Photonics, Engineering and Nanostructures, 2014. **541**.
14. Edelstein, A., et al., *Computer control of microscopes using microManager.* Curr Protoc Mol Biol, 2010. **Chapter 14**: p. Unit14 20.
15. Olivier, N., et al., *Resolution doubling in 3D-STORM imaging through improved buffers.* PLoS ONE, 2013. **8**(7): p. e69004.

16. Ovesny, M., et al., *ThunderSTORM: a comprehensive ImageJ plug-in for PALM and STORM data analysis and super-resolution imaging.* Bioinformatics, 2014. **30**(16): p. 2389-90.
17. Vedyaykin, A.D., et al., *New insights into FtsZ rearrangements during the cell division of Escherichia coli from single-molecule localization microscopy of fixed cells.* Microbiologyopen, 2016.

Use of optogenetic technology in cell culture models

A I Erofeev[1], O A Zakharova[1], M V Matveev[1], S G Terekhin[1], I B Bezprozvanny[1,2], O L Vlasova[1]

[1]Laboratory of Molecular Neurodegeneration, Peter the Great Polytechnic University, Saint-Petersburg, 195251, Russia
[2]Department of Physiology, University of Texas Southwestern Medical Center, Dallas, TX, USA

E-mail: alexandr.erofeew@gmail.com, leshazakharova@mail.ru, olvlasova@yandex.ru

Abstract. Optogenetic is a powerful method that allows to modulate cellular physiological properties. In our article, we demonstrate changes of electrical properties of cellular membranes on HEK-293T and hippocampal neurons transfected with channelrhodopsins and halorhodopsins induced by blue and orange light stimulation.

1. Introduction
Brain is one of the most complicated and poorly understood parts of the human body. There is a large group of disorders related to abnormalities in brain activity called neurodegenerative diseases. At the moment etiology and pathological basis of these diseases are unknown. That's why the fundamental assays in neurobiological field become more and more important. New technology that allows scientists to solve these biomolecular problems is called optogenetics. Optogenetics is a modern approach to modulate physiological status of excitable cells, including neurons. This modulation is achieved by combining the techniques of genetic engineering and photonics [1, 2, 3].

2. Materials and methods

2.1. Human embryonic kidney cell line 293T
HEK-293T cells were cultured in DMEM with addition of 10% FBS, 1% L-glutamine and 1% PEST, after 3 days of culturing, medium was refreshed. Transfection with FCK-ChR_2-HALO plasmid using turbofect transfection reagent was performed in order to perform optogenetic experiments.

2.2. Whole-cell patch recordings in HEK-293T cells
Whole-cell recordings in external solution (140mM NaCl, 5mM KCl, 2mM $CaCl_2$, 1mM $MgCl_2$, 10mM HEPES) were performed in voltage-clamp mode (MultiClamp 700B amplifier) unsing 5–10 MΩ pipettes filled with internal solution (35mM NaCl, 2mM $MgCl_2$, 10mM EGTA, 10mM HEPES). Following establishment of whole-cell configuration, the depolarizing current steps 1 sec in duration from 20 pA to 50 pA in amplitude were injected and the corresponding potential changes were recorded.

2.3. Primary hippocampal neuron cultures

The hippocampal cultures of mice were established from postnatal day 0–1 pups and maintained in culture as we described previously [4, 5]. Briefly, after dissection and dissociation, neurons were plated on coverslips (pre-treated with poly-lysine) and cultured in neurobasal A medium with addition of 1% FBS and 2% B27. At third day in vitro (DIV3), Ara-C (4 _M) was added to prevent glial cell growth. At DIV7 and DIV14, 50% of medium was exchanged with fresh neurobasal A medium containing 2% B27 without FBS. In these culture conditions, the astrocytes constitute about 10–20% in total cells in our cultures at DIV15 as determined by GFAP staining (data not shown). For assessment of synapse morphology, hippocampal cultures were transfected with TD-tomato for easily detection of transfected cells and for optogenetics experiments -FCK-CHR$_2$ plasmids at DIV7 using the calcium phosphate method.

2.4. Whole cell patch recordings and loose patch recordings in hippocampal cultures

Whole cell recordings in ACSF external solution (124mM NaCl, 26mM $NaHCO_3$, 10mM glucose, 5mM KCl, 2.5mM $CaCl_2$, 1.3mM $MgCl_2$, 1mM NaH_2PO_4) were performed in a current-clamp mode (Axopatch-200B amplifier) using 5–10 MΩ pipettes filled with internal solution (K-Gluconate 140mM, MgCl2 2mM, NaCl 2mM, ATP-Na2 2mM, GTP Mg 0.3mM,HEPES 10mM). Following establishment of whole-cell configuration, the depolarizing current steps 1 sec in duration from 10 pA to 100 pA in amplitude were injected and the corresponding potential changes were recorded.

Loose patch recordings in Hibernate A solution with B27 and glutamine (Life Technologies) were performed in a voltage-clamp mode (Axopatch-200B amplifier) held at 0mV using 5–10 MΩ pipettes filled with ACSF external solution. A loose patch (>100MΩ) was generated at the neuron soma close to the axon hillock. Spontaneous action potential currents were recorded 10min from each cell.

3. Optogenetic approach in HEK cells

Cells of human embrionic kidney (line HEK-293T) are easily transfected so they are often used as an object of study [6]. In our preliminary experiments HEK-293T cells were transfected with Channelrhodopsin and Halorhodopsin constructs. The responses were recorded using the patch-clamp technique in voltage clamp whole-cell mode using blue (depolarisation, Fig. 1,A) and orange (hyperpolarisation, Fig. 1, B) light stimulation.

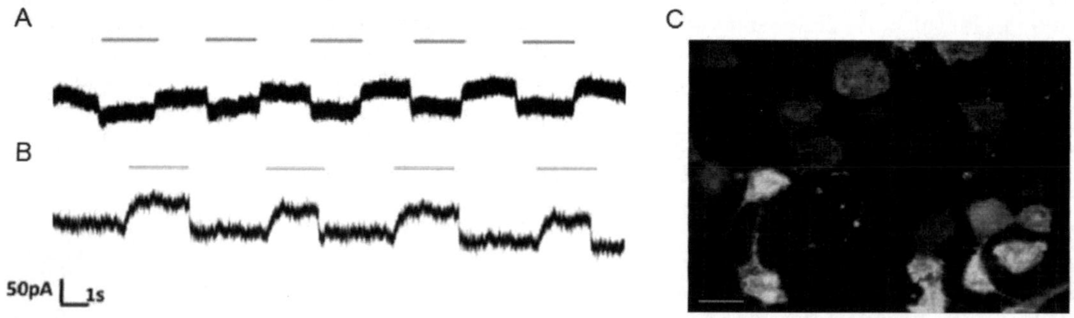

Figure 1. (A, B, C). (A) Depolarization of the cell membrane during the optogenetic stimulation of transfected by ChR$_2$-GFP plasmids cells lines HEK-293 (λ_{ex} = 470 nm, whole-cell, voltage-clamp mode); (B) Hyperpolarization of the cell membrane during the optogenetic stimulation of transfected by HR-GFP plasmids cells lines HEK-293 (λ_{ex} = 590 nm, whole-cell, voltage-clamp mode); (C) confocal image of transfected cell lines HEK-293T.

4. Optogenetic approach in hippocampal neurons

After successful approbation on HEK cells, optogenetic experiments were conducted in mouse primary hippocampal neuron cultures. Neurons were transfected with ChR$_2$-GFP plasmid (Fig. 2) and stimulated by blue (470 nm) light. Traces of neurons activity were recorded in whole-cell and loose-patch, voltage and current clamp modes (Fig. 3, 4).

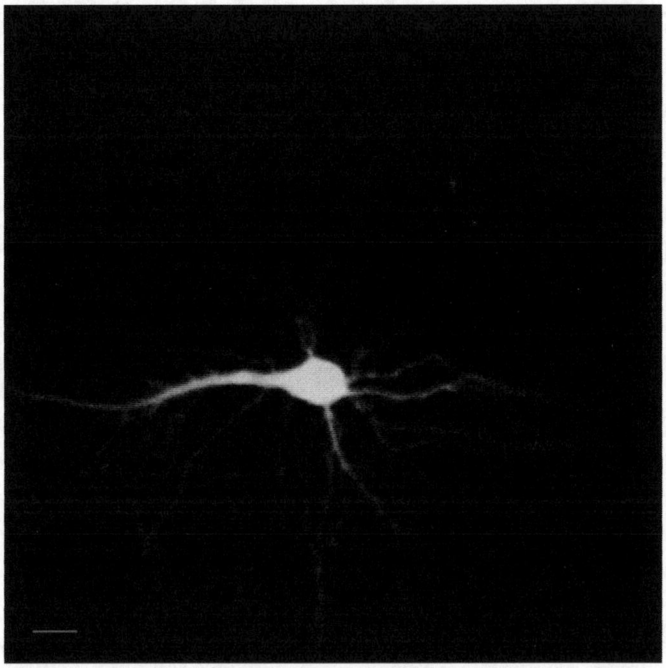

Figure 2. Confocal image of transfected hippocampal neuron expressing ChR$_2$-GFP (scale bar 20 µm, λ_{ex} = 470 nm)

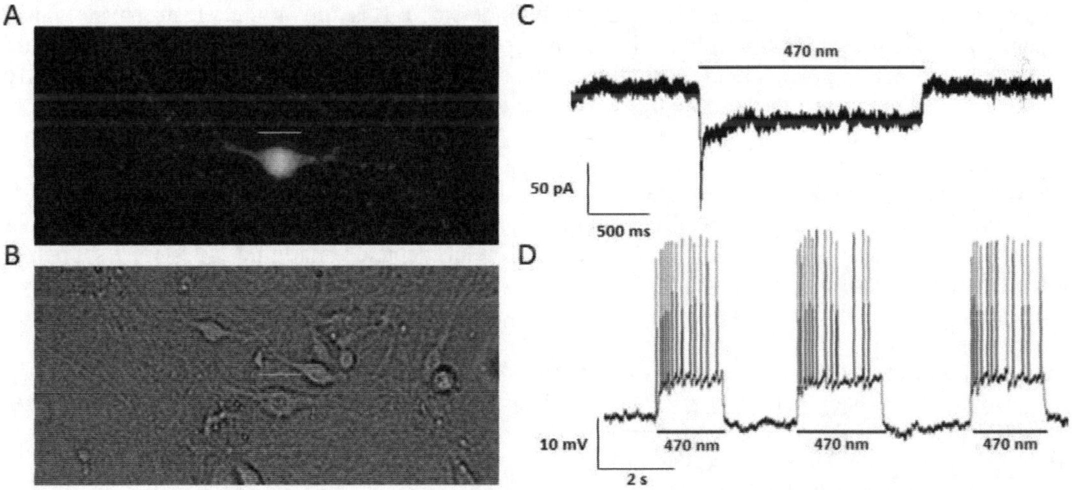

Figure 3. (A, B, C, D). (A) Hippocampal neurons expressing dTomato and ChR$_2$-GFP (scale bar 20 µm); (B) Hippocampal neuron culture (DIV15); (C) Inward current in voltage-clamped neuron evoked by 470 nm light (indicated by blue bar, whole-cell, voltage-clamp mode); (D) Voltage traces showing response to light stimulation (indicated by blue bar, whole-cell, current-clamp mode).

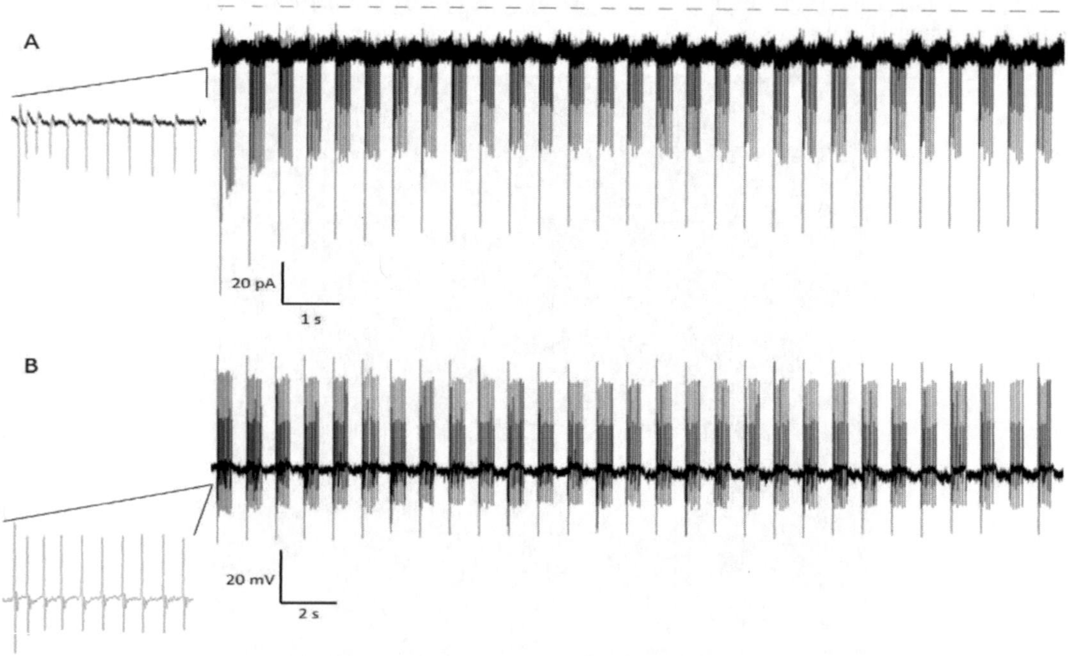

Figure 4. (A, B). Neuron activity evoked by blue light 470 nm (indicated by blue bar); (A) Inward current in voltage-clamped neuron (loose-patch, voltage-clamp mode) (B) Voltage traces showing response to light stimulation (loose-patch, current-clamp mode).

5. Future plans and conclusions

Alzheimer's disease (AD) and aging are resulting in impaired ability to store memories, but the mechanisms responsible for these defects are poorly understood. It is known that electrophysiological response of mutant mouse neuron cultures in case of electrical stimulation results in significant decrease in frequency of action potentials [7]. Our future plans include use of optogenetics in slices and live animals from AD models. For these studies we expect to use optoelectronic implant.

Acknowledgments

This work was supported by the state grant 17.1360.2014/K, by the Russian Scientific Fund grant 14-25-00024: The financial support was divided in the following way: experiments depicted on Figure 1 was supported by the state grant 17.1360.2014/K, experiments depicted on Figure 2, 3, 4 were supported by the Russian Scientific Fund grant.

References

[1] Deisseroth K 2010 *Scientific American* **303(5)** 48
[2] Boyden ES, Zhang F, Bamberg E, Nagel G, Deisseroth K 2005 *Nature Neuroscience* **8(9)** 1263
[3] Windhorst U, Johansson H 1999 *Springer* 190
[4] Zhang H, Sun S, Herreman A, De Strooper B, Bezprozvanny I 2010 *The Journal of Neuroscience* **30** 8566-8580
[5] Sun S, Zhang H, Liu J, Popugaeva E, Xu N-J, Feske S, White L, 3rd, Bezprozvanny I 2014 *Neuron* **82** 79-93
[6] He B, Soderlund D M 2010 *Neuroscience Letters* **469(2)** 268
[7] Zhang H, Jie Liu J, Sun S, Pchitskaya E, Popugaeva E and Bezprozvanny I 2015 *Journal of Alzheimer's Disease* (**45**) 561

3D-printed scaffolds based on PLA/HA nanocomposites for trabecular bone reconstruction

K V Niaza, F S Senatov, S D Kaloshkin, A V Maksimkin, D I Chukov

National University of Science and Technology "MISIS"

Abstract. In the present work porous PLA scaffolds filled with micro- and nano- HA were studied. Both composites with micro- and nano-HA were obtained by extrusion in the same conditions. Scaffolds were obtained by 3D-printing by fused filament fabrication method. Structure of porous scaffolds was pre-modeled by computer software. Compression and three - point flexural tests were used to study mechanical properties of the scaffolds.

1. Introduction

Reconstruction of the structural integrity of the damaged bone tissue is an urgent problem for people with traumas or bone diseases. Polylactide (PLA) is a bioresorbable polymer which is widely used for medical applications, especially bone implants [1]. Some works [2, 3] revealed the need to reinforce PLA matrix to reach mechanical properties of human bone tissue. Bioactive ceramics like hydroxyapatite (HA) may be used to increase mechanical properties and osseointegration [4]. In case of increasing mechanical properties in compression it is more preferable to use nanoparticles as it was observed by *Senatov et al* [5]. Nanocomposite system with nano-HA in the form of nanorods is able to mimic a natural morphology of bone apatite. Due to the fact that HA is identical to the bone tissue and is actually a source of minerals for bone cells, it leads to the regeneration approach. In case of nano-HA such a bioactivity increases. Composite materials with the addition of ceramic nanoparticles showed a significant increase in polymer degradation rate compared with the unfilled polymer [6]. This indicates the improvement in osseoconductive properties of nanocomposites. Rapid prototyping structures shows mechanical properties significantly higher than those structures fabricated by other well – known techniques such as solvent – casting and particle leaching, thermal-induced phase separation and gas foaming, among others [4-8]. In this work 3D-printing as an actively developing method for formation of polymer implants was used to produce highly-porous scaffold [1].

2. Materials and methods

Pellets of polylactide (PLA) (Mw = 60 000 g/mol, Aldrich) were used as a material for bioresorbable matrix. Hydroxyapatite (HA) powders with the average particle size of 90 nm and 1 micron produced by JSC "Polystom" (Russia) were used as the bioactive fillers.

Mixing of PLA and 15% weight HA powder was carried out in a screw extruder HAAKE MiniLab II Micro Compounder (Thermo Fisher Scientific, USA) for 15 min at 30 rev / min and temperature of 185 °C. Structure of 3D scaffold (Figure 1,2) was modeled using SolidWorks 2015 software (SolidWorks Corporation,USA). 3D-printing of porous scaffolds was performed using Picaso Designer Pro-250 (Picaso 3D, Russia) by fused filament fabrication (FFF) at nozzle temperature of 220 °C.

Studies of PLA/HA nanocomposites and scaffolds were conducted by scanning electron microscopy (SEM). SEM was performed with Hitachi TM-1000 (accelerating voltage 5kV). To investigate non-conductive polymer samples, the surface of the samples was covered with a layer of platinum (10 – 20 nm) by Auto Fine Coater JFC-1600 (Jeol, USA). Structural characterization of HA nanoparticles was performed by transmission electron microscopy (TEM). TEM was performed with JEM 1400 (accelerating voltage 120kV). Mechanical testing was studied in compression and three-point flexural test. Mechanical testing was performed in a universal testing machine Zwick/Roell Z 020, (Zwick GmbH & Co.KG, Germany) using rectangular samples of scaffolds with the cross-sectional dimensions of 24 x 12.5 mm for compression tests (ASTM D695) and 80 x 10 x 4 mm for three point flexural test (ASTM D790).

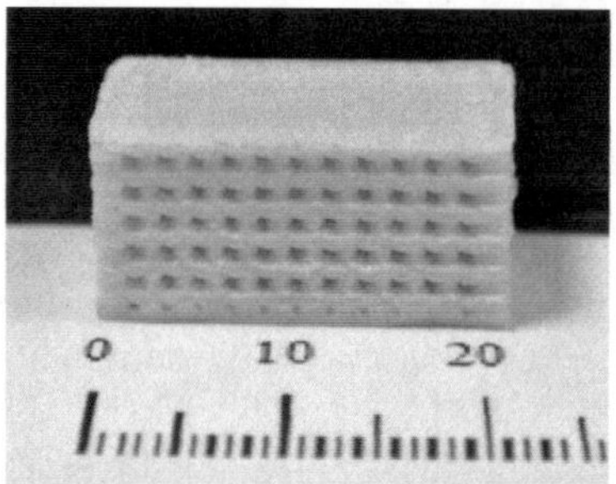

Figure 1. 3D-printed scaffold based on bioactive nanocomposite PLA/HA

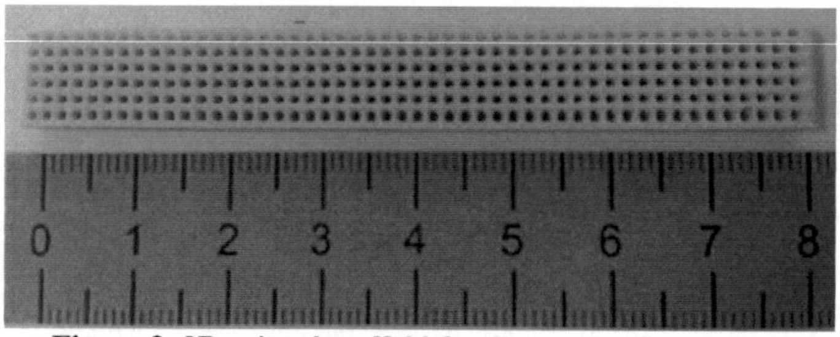

Figure 2. 3D-printed scaffold for three-point flexural test

3. Results and Discussion

Figure 3 shows TEM image of HA nanoparticles. The particles have an elongated non-equiaxed shape. The average length is 90 nm, the average diameter – 20 nm. Nanoparticles have a porous structure, which is shown in Figure 3a. The pore size is less than 5 nm. All pores are mainly open and belong to meso- and micropores. This indicates an increased specific surface of nanoparticles.

It was shown, that nanocomposites had higher mechanical properties such as Young's modulus, ultimate strength than composites with micro-HA in compression test, as shown on Figure 4. The mean value of Young's modulus was 4.0±0.2 GPa and 2.8 ±0.2 GPa for PLA/nano-HA and PLA/micro-HA, respectively. Young's modulus of biocomposite with nano-HA is closer to the Young's modulus of human bone tissue, than modulus of composite with micro HA. It is known, that modulus of trabecular bone is in the range from 3 to 5 GPa [12]. Incorporation of HA nanoparticles into PLA matrix led to higher mechanical properties, than in case of micro-HA. Therefore, PLA/nano-HA scaffolds demonstrated mechanical properties of a trabecular bone tissue preserving highly porous structure.

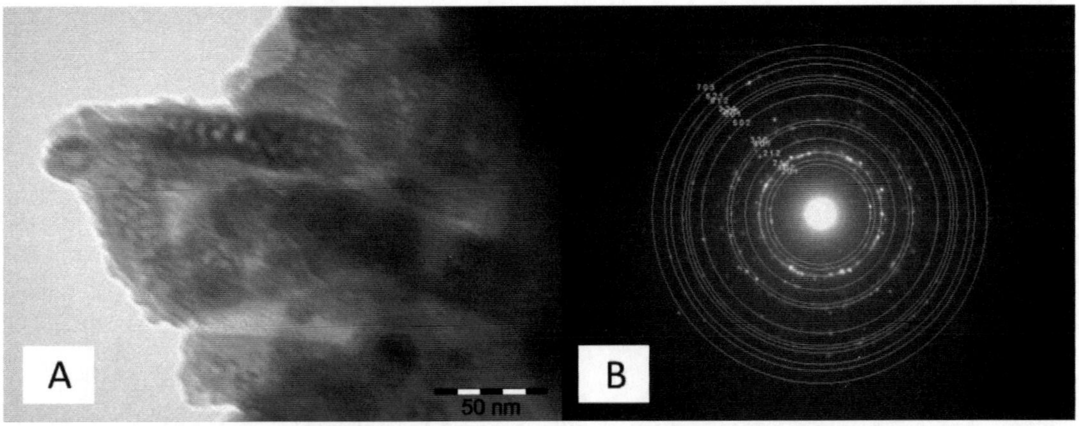

Figure 3. TEM image (A) and electron diffraction pattern (B) of HA nanoparticles

Three-point flexural test confirmed previous results. Flexural modulus is 1.93±0.11 and 1.57±0.11 GPa for PLA/nano-HA and PLA/microHA, respectively (Figure 5). This is the result of greater surface interaction between nanoparticles and polymer matrix. Porous scaffold is a construction, that can withstands not only compressive loadings, but also flexural up to 30 MPa, which are inherent in the trabecular bone.

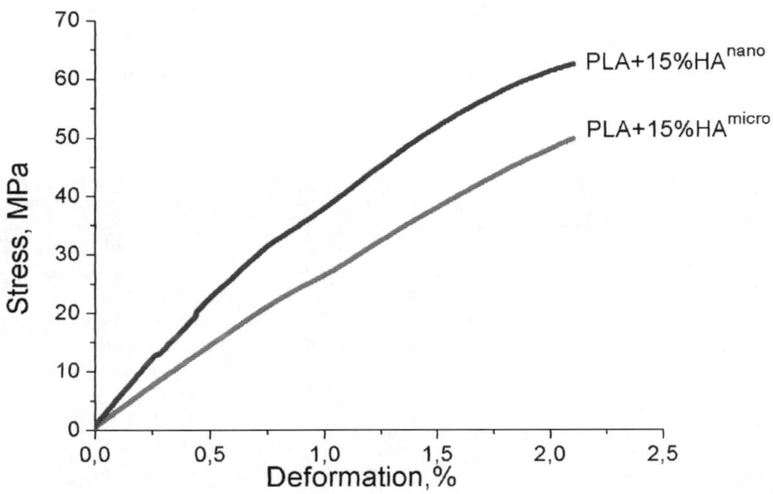

Figure 4. Typical stress-deformation curves of compression test for PLA-based composites

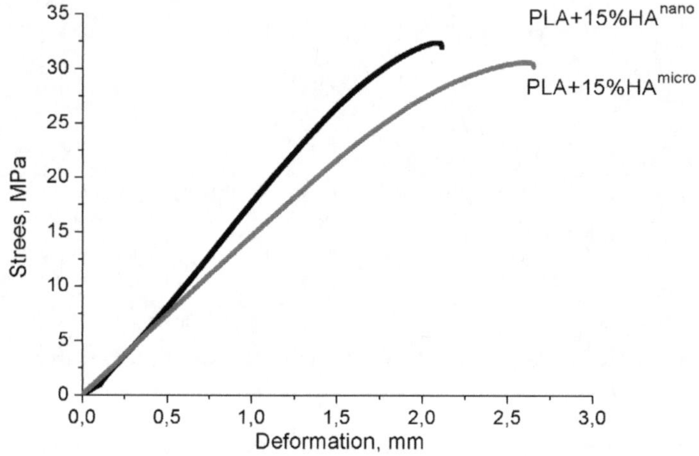

Figure 5.Typical three-point flexural curves for PLA-based composites

Figure 6 shows SEM image of fracture surface after three-point flexural test for scaffold with micro- and nano-HA. The brittle fracture in all cases was observed. The main difference in the microstructure lies in the character of the particle distribution. Nanoparticles did not form agglomerates more than 10 μm in diameter and showed better distribution in the polymer matrix in contradistinction to micro-HA particles, which formed agglomerates more than 100 μm. The presence of the plastic flow regions (Figure 5b) after destruction of the nanocomposites indicates better adhesion of nano-HA to the polymer matrix.

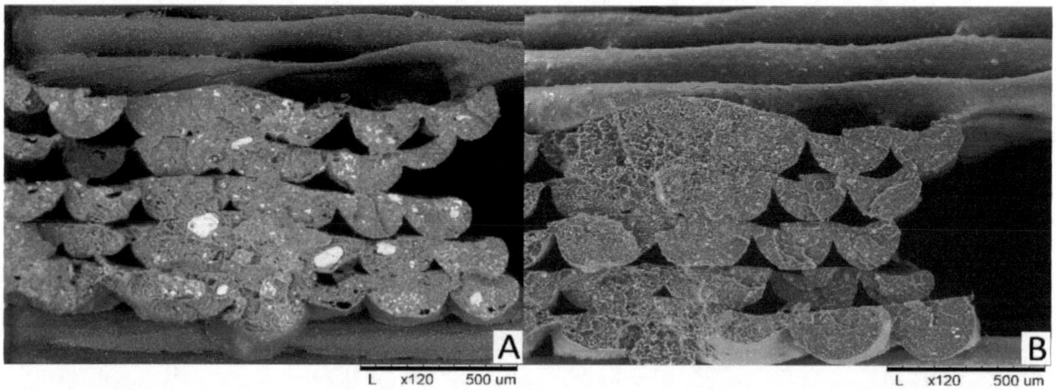

Figure 6 –SEM imager of fracture surface of PLA-based scaffold after three-point flexural test
A - PLA / micro-HA
B - PLA / nano-HA

Incorporation of the nano-HA into the polymer matrix instead of micro-HA led to a hardening of the material and more regular microarchitecture because of greater interfacial surface by formation of ultrafine structure.

Theoretical porosity of scaffolds is 30 vol. %, however, real porosity is higher, because of pore formation during 3D-printing. This is the result of special character of sintering between the layers. This character is single-point and the size of such pores is about 100 μm.

The strength of construction is also achieved by solid sintering of perpendicular layers.

Conclusions

PLA-based porous scaffolds with porosity of 30 vol. % for bone implants were prepared by 3D-printing based on fused filament fabrication. Hydroxyapatite nanorods with the average length of 90 nm and the average diameter of 20 nm increased mechanical properties of the scaffolds due to the higher Young's modulus similar to the trabecular human bone tissue. The architecture of porous scaffolds mimics structure of trabecular bone. Mechanical properties of scaffolds are in range of mechanical properties of bone tissue. 3D-printed scaffolds based on PLA/HA nanocomposites may be perspective constructions for trabecular bone reconstruction.

Acknowledgments

This work was supported by the Federal Targeted Program "Research and development in priority directions of development of scientific technological complex of Russia in 2014–2020", with the funding from the Ministry of Education and Science of Russian Federation: agreement 14.575.21.0088, 21 October 2014, RFMEFI57514X0088

References

[1] Senatov F S, Niaza K V, Zadorozhnyy M Yu, Maksimkin A V, Kaloshkin S D, Estrin Y Z, 2016 *J. Mech. Behav. Biomed.* **57** 139
[2] Zhongkui Hong, Peibiao Zhang, Chaoliang He, Xuyeyu Qiu,Aixue Liu, Li Chen, Xuesi Chen, Xiabin Jing,2015 *J. Biomater.* **26** 6296
[3] Jean-Marie Raquez J-M, Habibi Y, Murariu M, Dubois P, 2013 *J Prog Polym Sci* **38** 1504
[4] Kikuchi M, Suetsugu Y, Tanaka J, Akao M 1997 *J Mater Sci Mater Med* **8** 361
[5] Senatov F S, Baranov A A, Muratov D S, Gorshenkov M V, Kaloshkin S D, TcherdyntsevV V 2014 *J. Alloy Compd* **615** 573
[6] Loher S, Reboul V, Brunner TJ, Simonet M, Dora C, Neuenschwander P, 2006 *Nanotechnology.* **17** 2054
[7] Serra T, Planell J A, Navarro M, 2013 *Acta Biomater.* **9** 5521
[8] Hutmacher D W, 2000 *J. Biomater.* **21** 2529
[9] Hutmacher D W, Schantz J T, Lam C X F, Tan K C, Lim T C. 2007 *J. Tissue Eng. Regen. Med.* **1** 245
[10] Hollister S J, 2005 *J. Nat. Mater.* **4** *518*
[11] Moroni L, Elisseeff J, 2008 J. Mater. Today **11** 44
[12] Goldstein S A, 1987 *J Biomechanics* **11** 1055

Structure of fluorescent metal clusters on a DNA template.

A A Vdovichev, T S Sych, Z V Reveguk, A A Smirnova, D A Maksimov, R R Ramazanov, A I Kononov

Department of Molecular Biophysics and Polymer Physics, St. Petersburg State University, St. Petersburg, Russia
E-mail:artvd.jr@gmail.com

Abstract. Luminescent metal clusters are a subject of growing interest in recent years due to their bright emission from visible to near infrared range. Detailed structure of the fluorescent complexes of Ag and other metal clusters with ligands still remains a challenging task. In this joint experimental and theoretical study we synthesized Ag-DNA complexes on a DNA oligonucleotide emitting in violet-green spectral range. The structure of DNA template was determined by means of various spectral measurements (CD, MS, XPS). Comparison of the experimental fluorescent excitation spectra and calculated absorption spectra for different QM/MM optimized structures allowed us to determine the detailed structure of the green cluster containing three silver atoms in the stem of the DNA hairpin structure stabilized by cytosine-Ag+-cytosine bonds.

Introduction

A few-atomic metal clusters are a subject of increasing interest in the past decade due to their bright emission from visible to near infra-red range [1,2]. For example, DNA-stabilized Ag clusters, exhibiting excellent brightness and photostability, are considered as a new promising type of emitters for various applications in photonic devices, bioassays, chemical sensors, biosensors [3-10]. Detailed structure of the fluorescent complexes of Ag and other metal clusters with ligands still remains a challenging task. In this joint experimental and theoretical study we synthesized Ag-DNA complexes on a DNA oligonucleotide emitting in violet-green spectral range. The structure of DNA template was determined by means of various spectral measurements (CD, MS, XPS). Comparison of the experimental fluorescent excitation spectra and calculated absorption spectra for different QM/MM optimized structures allowed us to determine the detailed structure of the green-emitting cluster-DNA complex consisting of three silver atoms in the stem of the DNA hairpin structure stabilized by cytosine-Ag^+-cytosine bonds.

Methods.

Cluster synthesis. Oligonucleotide strand 5'-CCTCCTTCCTCC-3' (BioBeagle ltd.) and $AgNO_3$ aqueous solutions were mixed and stored for 15 min at room temperature. After that, $NaBH_4$ aqueous solution was added followed by vigorous stirring. The final concentrations were C_{DNA} = 20 μM, C_{AgNO3} = 120 μM, and C_{NaBH4} = 30 μM. The sample was kept in the dark about 1 month at 21 °C to reach maximum fluorescence.

Theoretical methods. DNA-cluster complex equilibrium structures were obtained with use of CP2K[11] program by utilizing computational protocol comprising of several steps. Initially from a nucleotide sequence described above was obtained the equilibrium DNA-hairpin structure stabilized by 4 silver ions in explicit solvent environment (Figure 1A). For this purpose we performed geometry optimization and dynamic equilibration of suggested structure by use of QM/MM scheme. The QM part is comprised of orthogonal box 19.73x18.12x33.16 Å, all nitrogenous bases and 4 silver ions linking of 4 cytosine pairs, while the MM part is comprised of orthogonal water box 3x3x4 nm and sugar-phosphate backbone of DNA. The full charge of the QM/MM system was set to zero by adding of corresponded number of counterions Na^+. QM/MM calculations where performed by use of pseudopotential plane-wave density functional theory in reciprocal space, namely pseudopotentials of Goedecker, Teter and Hutter (GTH)[12] for core electrons, molecularly optimized DZVP-MOLOPT-GTH basis set for valence electrons and hybrid functional PBE[13]. The model of DNA in MM part utilized periodic boundary conditions, TIP3P water solvent and parmbsc0 forcefield[14]. Subsequent MD equilibration performed in the NPT ensemble using a Nosé-Hoover thermostat with a 1 fs integration step. The obtained hairpin structure further was used as a template for Ag cluster formation.

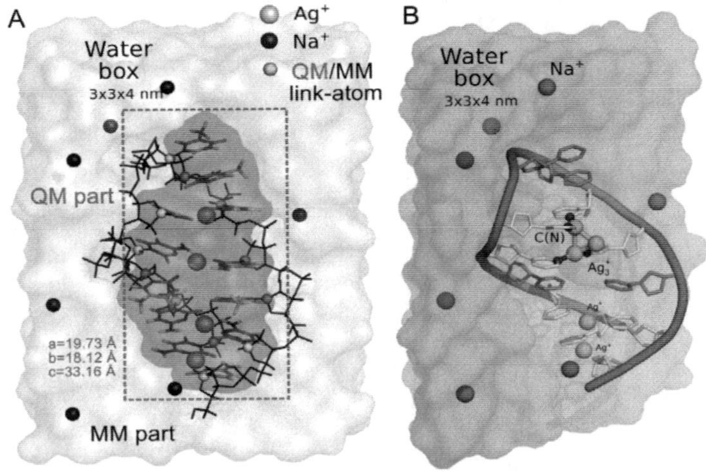

Figure 1. A) DNA-Ag^+ equilibrium hairpin model structure obtained by QM/MM optimization protocol; B) DNA-silver green-emitting model structure obtained on the hairpin template after adding of one metal silver atom.

Based on the obtained hairpin structure we constructed a QM part comprising of 4 silver ions, 1 metal silver atom and all of the nucleobases (Figure 1B). After QM/MM optimization procedure and addition dynamic equilibration the equilibrium cluster-nucleotide complexes where extracted for gas phase electronic spectra calculations. Calculations of absorption spectra were performed by use of the second order algebraic-diagrammatic construction scheme (ADC(2)[15]) method and def2-SVP basis set with Turbomole program[17].

Results

A palindrome 12-mer DNA sequence 5'-CCTCCTTCCTCC-3' was used as a template for Ag cluster formation and theoretical investigations. The corresponding fluorescent emission and excitation spectra are shown in Figure 2.

The major fluorescent fraction of the Ag-DNA complex has a maximum at about 520 nm. As has been shown previously [17], CD spectrum of the complex of DNA with Ag cations indicates formation of C-C pairs mediated by cytosine-Ag^+-cytosine bonds. Since MS spectra show single-stranded DNA-Ag^+ adducts only, it means formation of hairpin rather than a self-dimer structure. XPS analysis gives about 0.5 Ag/P ratio, suggesting that all silver cations are indeed bound with DNA matrix.

Comparison of experimental luminescent excitation spectra and preliminary calculated excitation spectra of different QM optimized isolated clusters and cytosine-cluster complexes consisting up to 3 silver atoms showed that green fluorescent silver clusters included not more than 3 silver atoms. We obtained an equilibrium structure of the DNA-cluster complex consisting of 3 silver atoms in fluorescent site in the stem of the DNA hairpin structure stabilized by cytosine-Ag^+-cytosine bonds. This structure obtained with QM/MM approach is shown in Figure 3.

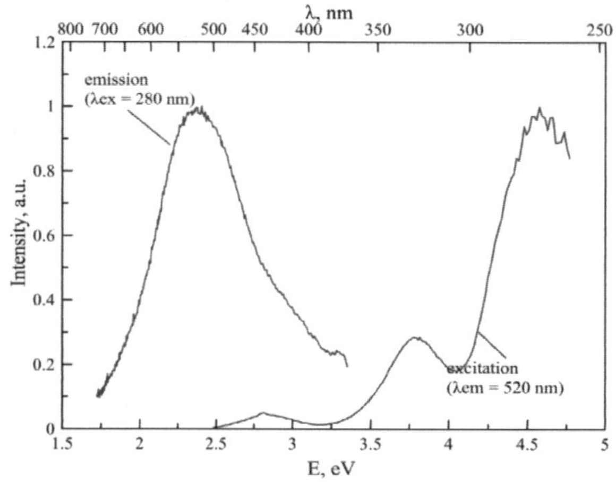

Figure 2. Experimental emission and excitation spectra of green-emitting Ag clusters-DNA complex.

A reasonable agreement between the electronic spectrum of the complex, calculated by the ADC(2), and the experimental fluorescence excitation spectrum supports the model of the fluorescent DNA-cluster complex. The proposed approach can be further used for determining the structure of other ligand-stabilized fluorescent metal clusters.

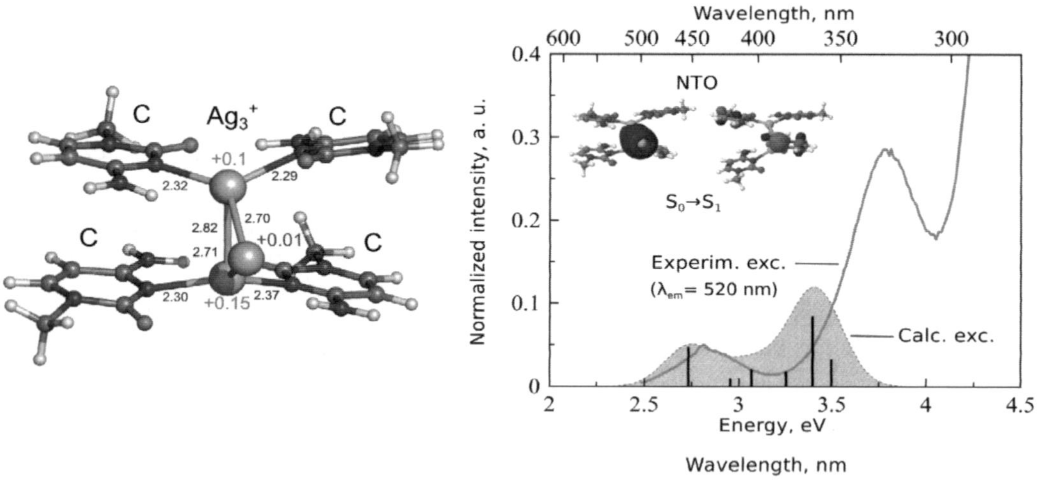

Figure 3. Equilibrium structure of DNA and Ag^{3+} green-emitting cluster and corresponded experimental and calculated excitation spectra.

Acknowledgments

The authors acknowledge Saint-Petersburg State University for a research grant 11.38.221.2014. The reported theoretical study was also funded by RFBR according to the research project No 16-32-00293. The reported calculations were performed in the Supercomputing Center of Lomonosov Moscow State University [18]. Experimental data presented in this work were obtained using the equipment of Center for Optical and Laser Materials Research, Chemical Analysis and Materials Research Centre, and Centre for Physical Methods of Surface Investigation of St. Petersburg State University, Russia.

References

[1] Yizhong Luab and Wei Chen. Sub-nanometre sized metal clusters: from synthetic challenges to the unique property discoveries. *Chem. Soc. Rev.*, 2012, **41**, 3594–3623
[2] Sungmoon Choi, Robert M. Dickson and Junhua Yu. Developing luminescent silver nanodots for biological applications. *Chem. Soc. Rev.*, 2012, **41**, 1867–1891

[3] W. Guo, J. Yuan, E. Wang, Oligonucleotide-stabilized Ag nanoclusters as novel fluorescence probes for the highly selective and sensitive detection of the Hg^{2+} ion, *Chem. Commun. (Cambridge, U. K.)* (2009) 3395–3397.

[4] W. Guo, J. Yuan, Q. Dong, E. Wang, Highly sequence-dependent formation of fluorescent silver nanoclusters in hybridized DNA duplexes for single nucleotide mutation identification, *J. Amer. Chem. Soc.* **132** (2010) 932–934.

[5] J. T. Petty, B. Sengupta, S. P. Story, N. N. Degtyareva, DNA sensing by amplifying the number of near-field emitting, oligonucleotide-encapsulated silver clusters, *Anal. Chem.* **83** (2011) 5957–5964.

[6] P. Shah, A. Rørvig-Lund, S. B. Chaabane, P. W. Thulstrup, H. G. Kjaergaard, E. Fron, J. Hofkens, S. W. Yang, T. Vosch, Design aspects of bright red emissive silver nanoclusters/DNA probes for microRNA detection, *ACS Nano* **6** (2012) 8803–8814.

[7] H.-C. Yeh, J. Sharma, I.-M. Shih, D. M. Vu, J. S. Martinez, J. H. Werner, A fluorescence light-up Ag nanocluster probe that discriminates single-nucleotide variants by emission color, *J. Amer. Chem. Soc.* **134** (2012) 11550–11558.

[8] J.M. Obliosca, C. Liu, R.A. Batson, M.C. Babin, J.H. Werner, H.-C. Yeh, DNA/RNA detection using DNA-templated few-atom silver nanoclusters, *Biosensors* **3** (2) (2013) 185–200.

[9] S. M. Copp, D. E. Schultz, S. Swasey, E. G. Gwinn, Atomically precise arrays of fluorescent silver clusters: a modular approach for metal cluster photonics on DNA nanoclusters, *ACS Nano* **9** (3) (2015) 2303–2310.

[10] Schultz D., Gwinn E., Silver atom and strand numbers in fluorescent and dark Ag:DNAs, *Chem. Commun.*, 2012, **48**, 5748–5750.

[11] J. VandeVondele, M. Krack, F. Mohamed, M. Parrinello, T. Chassaing, J. Hutter, Comput. Phys. Commun. 2005, 167, 103–128.

[12] S. Goedecker, M. Teter, J. Hutter, *Phys. Rev. B* 1996, **54**, 1703–1710

[13] J. P. Perdew, K. Burke, M. Ernzerhof, *Phys. Rev. Lett.* 1996, **77**, 3865–3868.

[14] Pérez, A.; Marchán, I.; Svozil, D.; Sponer, J.; Cheatham, T. E.; Laughton, C. A.; Orozco, M.; Pe, A.; Cheatham, T. E., III *Biophys. J.* 2007, **92**, 3817–3829

[15] Trofimov, A. B.; Schirmer, J. J. Phys. B: At., *Mol. Opt. Phys.* 1995, **28**, 2299–2324

[16] Ahlrichs, R.; Bär, M.; Häser, M.; Horn, H.; Kölmel, C. *Chem. Phys. Lett.* 1989, **162**, 165–169

[17] I. Volkov, T. Sych, P. Serdobintsev, Z. Reveguk, A. Kononov, Fluorescence saturation spectroscopy in probing electronically excited states of silver nanoclusters, *Journal of Luminescence* 2016, **172**, 175-179.

[18] Voevodin, Vl. V.; Zhumatiy, S. A.; Sobolev, S. I.; Antonov, A. S.; Bryzgalov, P. A.; Nikitenko, D. A.; Stefanov, K. S.;Voevodin, Vad. V. *Open Syst. J. - Moscow Open Syst. Publ.* 2012, **7**.

Development and design of advanced two-photon microscope used in neuroscience

M S Doronin[1], **A V Popov**[1,2]
[1]Institute of Neuroscience, Lobachevsky State University of Nizhny Novgorod, Nizhny Novgorod, 603950, Russia
[2] Department of computer control systems and telecommunications, Volga State University of Water Transport, Nizhny Novgorod, Russia, 603950

Abstract. This work represents the real steps to development and design advanced two-photon microscope by efforts of laboratory staff. Self-developed microscopy system provides possibility to service it and modify the structure of microscope depending on highly specialized experimental design and scientific goals. We are presenting here module-based microscopy system which provides an opportunity to looking for new applications of this setup depending on laboratories needs using with galvo and resonant scanners.

1. Introduction

Application of modern techniques offer high resolution properties for living systems research is significant key for neurobiology progress. Often researchers face the necessity both of enhancing of research techniques and using several techniques based on one system. Working with selected methods require both of high-technology equipment availability and knowledge and skills to applying non-trivial mathematical algorithms for control two-photon microscopy system [1] and for well-formed data analysis [2,3]. This work may be regarded as a quick source to laboratory staff who wish to develop and design their own microscope system for self-service it and modify the structure of microscope in order to study highly specialized tasks.

2. Optical scheme

Proposed type of microscope is module-based (Figure 1). Main module of microscope is body represented as metal base to which other modules are mounted. It could be chamber for fix the sample (stereotaxis in case of *in vivo* experiments), moving platform with fixed objectives nosepiece, scanners, optical elements designed for work of differential interference contrast (DIC) [4] and multi-photon mode. DIC produces contrast by visually displaying the refractive index gradients of different areas of a specimen. Galvo [5] and resonant [6] scanners are presented on Figure 1. The first one may be used for rectangular area (line-by-line) scanning, line scanning or point scanning and for uncaging some drug (e.g. glutamate, GABA), the second one – for fast frame rectangular area scanning. Maximal speed for galvo scanner is up to 2 fps (for 512x512 pixels frame), and for resonant scanner is up to 30 fps (for 512x512 pixels frame). IR femtosecond pulsed lasers (tuning wavelength range is 690–1080 nm) are using for fluorophore activation. Non-descanned detectors (NDDs) are using for detecting of fluorescence. Each NDD based on pair of bandpass filter to extract dye absorption wavelength and photomultiplier to detect fluorescent signal. NDDs are mounted exactly above the objectives aperture which is useful for low level fluorescence detecting in living tissues. Often researchers are using at list two detectors to split fluorescence produced by different dyes (e.g.,

Journal of Physics: Conference Series **741** (2016) 012070 doi:10.1088/1742-6596/741/1/012070

morphology and calcium indicators). Modern microscopy systems allow to mount up to 4–6 detectors depending on researcher goals.

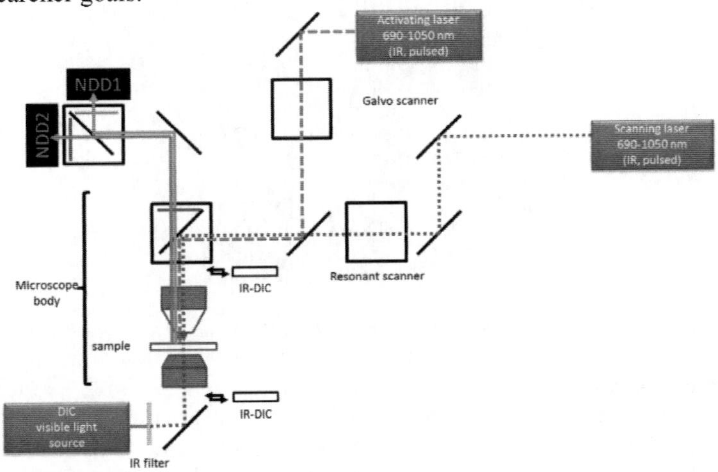

Figure 1. Optical scheme of two-photon microscope with galvo and resonant scanners.

3. Development and design

Researchers should develop and design high-technology platform for realisation module-based microscope (e.g. using Kompas software). Microscope consists of high-technology alloy allowed both to follow requirements of structure rigidity and minimal mass of microscope. All the components and elements of microscope should be developed in such a way as to have possibilities for receiving all the performance goals on the one hand and provide to evolution and addition techniques which could be used in that unique microscopy system on the other. For example if researchers are planning to have an *in vivo* experiments than working space (Figure 2) and, thus, the body of microscope should be higher than systems without *in vivo* approaches. Moving platform located above DIC elements and working space is using for mounting of objective nosepiece, filter cube(s) and NDDs. Moving platform should be constructed as handy element in the context of ergonomics and functional. Moreover, developer should turn attention to minimal approach increment (stepper motor with at list 500 steps per rev should be used) of moving platform on the one hand, and to high-speed displacement and no slop (free motion) on the other hand. Other important aspect is accurate positioning of chamber for brain tissue (or alive animals) as well as micromanipulators for using electrophysiological and other (e.g., bolus loading) methods. U-shaping table can be adapted for specified type of microscope. It is possible to to set up this table independently from the microscope. For maintenance purposes it is possible to remove this table without disassembling the microscope. Motorized axes could be controlled by independent controller (control box and joystick or keyboard) or PC using especial software which could be integrated in mail control software for microscopy system.

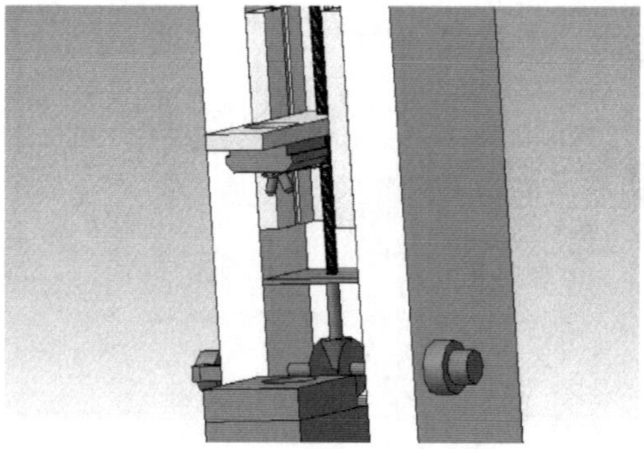

Figure 2. Working space of the microscope model developed using Kompas software.

4. Microscope system control

Control of system is realized either exactly by using DA/AD converter(s) or by using control element of microscope components (e.g. USB, COM connection). However development of cross functional and convenient software to control the system is very important object for researchers. For achieving these goals developers could use commercially available software or create their own software using Matlab, LabView, Python or others. In case of software development researcher could create few sub-programs for each of elements controlled by software (e.g. manipulators, focus, lasers, scanners, acousto-optic modulator (AOM), detectors and others). After this step sub-programs could be combine to one main program which will control all of them and synchronize their integrated work. Software also should implement algorithms to scan region(s) of interest. These algorithms will depend on scanner type. Galvo scanner could be used for rectangular area (line-by-line, few options are presented in Figure 3) scanning, line scanning or point scanning and for uncaging some drug (e.g. glutamate, GABA) in sequence of points, resonant scanner could be used for fast frame rectangular area scanning in relation to feature of construction. Working cooperatively, these scanners could fast scan rectangular area (resonant scanner) and activate caged drugs (galvo scanner) using different wavelength from different laser sources. Important aspect during any type of scanning is synchronizing between movement of laser beam by scanner and input signal from detector(s). Input signal received in one moment of time is detected fluorescent signal from the point where was directed laser beam by scanner at the same moment of time. Thus, software could construct fluorescent signal frames and simply analyze them (overlapping to each other, scale bar, construct 3D images and others). Moreover, researchers could realize more complicated data analysis allow represent data correctly. Additionally all the experimental data (include parameters of hardware such as laser intensity and wavelength, photomultipliers sensitivity and others) could be saved in especial file which could be opened both in that software to repeat experimental parameters and special software to represent data and analyze it (e.g. Matlab and others).

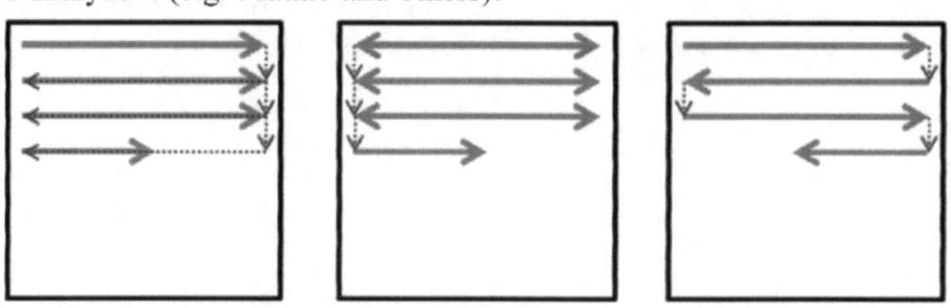

Figure 3. Few options of laser beam movement in case of rectangular area (line-by-line) scanning using by galvo scanner.

Acknowledgments
The work was performed at the Lobachevsky State University of Nizhny Novgorod with financial support of the government represented by the Ministry of Education and Science of the Russian Federation, the unique identifier of the project is RFMEFI57814X0079, the agreement on granting a subsidy №14.578.21.0079 dated by 28.11.2014.

References

[1.] Svoboda K., Yasuda R. Principles of two-photon excitation microscopy and its applications to neuroscience. // Neuron. 2006. Vol. 50, № 6. P. 823–839.
[2.] Danielyan A., Katkovnik V., Egiazarian K. BM3D frames and variational image deblurring // IEEE Trans. Image Process. 2012. Vol. 21, № 4. P. 1715–1728.
[3.] Danielyan A. et al. Cross-color BM3D filtering of noisy raw data // 2009 Int. Work. Local Non-Local Approx. Image Process. LNLA 2009. 2009. № 118312. P. 125–129.
[4.] Nomarski G. From phase contrast to contrast by interference // Rev Hematol. 1957. Vol. 12, № 4. P. 439–442.
[5.] Aylward R.P. Advanced galvanometer-based optical scanner design // Sens. Rev. 2003. Vol. 23, № 3. P. 216–222.
[6.] Sheppard C.J.R., Kompfner R. Resonant scanning optical microscope // Appl. Opt. 1978. Vol. 17, № 18. P. 2879–2882.

Implantable optical-electrode device for stimulation of spinal motoneurons

M V Matveev[1], A I Erofeev[1,2], O A Zakharova[1,2], E N Pyatyshev[3], A N Kazakin[3] and O L Vlasova[1,2]

[1] Department of Medical Physics, Peter the Great St. Petersburg Polytechnic University, Saint-Petersburg, 195251, Russia

[2] Molecular Neurodegeneration Lab, Peter the Great Polytechnic University, Saint-Petersburg, 195251, Russia

[3] Nano and Microsystem Technology Lab, Peter the Great Polytechnic University, Saint-Petersburg, 195251, Russia

E-mail: m.v.matveev@bk.ru, alexandr.erofeew@gmail.com, olvlasova@yandex.ru

Abstract. Recent years, optogenetic method of scientific research has proved its effectiveness in the nerve cell stimulation tasks. In our article we demonstrate an implanted device for the spinal optogenetic motoneurons activation. This work is carried out in the Laboratory of Molecular Neurodegeneration of the Peter the Great St. Petersburg Polytechnic University, together with Nano and Microsystem Technology Laboratory. The work of the developed device is based on the principle of combining fiber optic light stimulation of genetically modified cells with the microelectrode multichannel recording of neurons biopotentials. The paper presents a part of the electrode implant manufacturing technique, combined with the optical waveguide of ThorLabs (USA).

1. Introduction

It is proved that each separate group of brain neurons is responsible for performing a specific elementary function. And it is the interplay between these systemic units that ensures the whole neural network operation. Until recently obtaining of experimental data was mostly limited to the studying of the consequences resulted from damages to the certain parts of brain or to the record of the brain activity during animal's execution of stereotype tasks [1].

In terms of the therapeutic effect, researchers used electrical stimulation and pharmacological products most frequently. In order to activate different parts of brain they implanted electrodes into them. In this case, however, electrical current affects almost all the undifferentiated neuron groups (electrode stimulates all neural tissues) and, as a result, it was very difficult to localize a function's generator. When they used pharmacological products that could selectively inhibit nerve cells of a specific group, the effect of the chemical substance determined the time delay compared with the natural neural stimulation [2].

2. Optogenetic method

The application of a fundamentally new approach to the studying of brain neurons processing is optogenetics. An important advantage of the optogenetic research method is selectivity in terms of affecting the specific part of brain or the neuron group. Milliseconds matching data allow researchers to carry out experiments with the speed of the living cells biological response when determining the significance of research models specific actions in neurons. [2]

For combined optogenetic experiments with the use of electrophysiological stimulation of freely behaving animals, we use implantable cannulas that unite optical fiber and implanted electrode. This configuration offers real-time and accurate probe of the influencing organs to the experiment area.

3. Optical-electrode interface

Brain function studying requires neuron interface that could record parameters and stimulate brain with high time-space accuracy. Most researchers who use optogenetic method in laboratory conditions on *in vivo* animals now use optical fiber that is sent through the implantable cannula [3]. Parameters of the pulses sequence and their generation is controlled by computer graphical interface or manual switch of modes.

Programmed LED control drivers ensure the setting of direct current values for one or several separate LEDs or a cluster that consists of several diodes united in one output fiber. Each channel is controlled autonomously (manually in modes of CW, external TTL or analog modulation types) or by software installed on computer.

At the department of Medical Physics and in the Laboratory of Molecular Neurodegeneration we are working on the development and testing of implantable device for monitoring of brain neurons physiological parameters (action potential).

Together with "Nano and Microsystem Technology" research laboratory we are developing a combined optical-electrode device that allows you to carry out combined research with the use of intravital microelectrode stimulation and optogenetic activation of genetically differentiated neurons (figure 1).

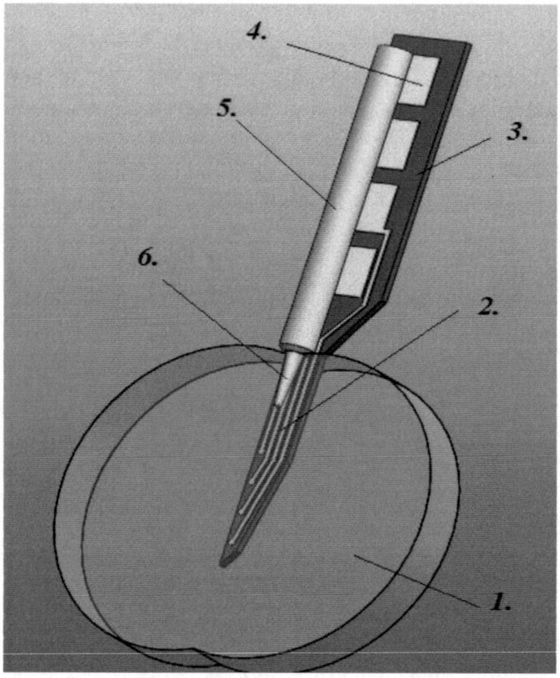

Figure 1. Implantable device layout: 1 – biopsy sample containing modified photosensitive motoneurons of the spinal cord; 2 – Submersible microneedles tip with electrodes outputs; 3 – silicon substrate of the implant; 4 – contact pad of output cable; 5 – light guide conductor; 6 – light guide tip.

4. Microneedle manufacturing technology

The first experiments on registration electrical activity from cells in vitro with the electrodes of 30-element microelectrode arrays were performed over 40 years ago. The multi-electrode array was fabricated by etching thin metal films deposited on glass cover slips; the fabrication employs techniques developed by the microelectronics industry [5].

To date, a plenty of probes have been used in neuroscience research based on multi-electrode arrays fabricated by Micro Electro Mechanical Systems (MEMS) [6].

The following section describes the development of base silicon chip design and technology containing needles with metal electrodes and contact pads for acquisition system.

Many research groups investigating neural probe technologies are faced with different challenges including non-standard and unconventional fabrication processes leading to low yield and high cost, lack of on-site and monolithic Integrated Circuit (IC) integration leading to high noise and reduced sensitivity, shorter sized probes mainly limited by fabrication technology, low design flexibility, and limited selection of materials having the mechanical properties that fulfill both the implantation application requirements and being compatible with standard micro fabrication processing [6].

Preliminary design is based on the composite scheme, combining micro fabricated silicon chip with microelectrode array and optical fiber. Chip design is shown in figure 2.

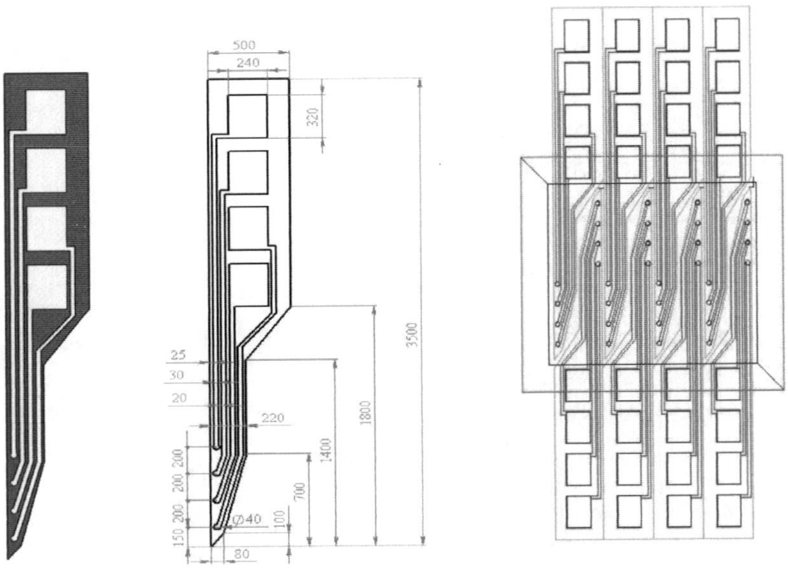

Figure 2. The appearance of a single needle chip, mutual arrangement of needles on photo mask.

Dimensions – 3500x500x400 μm, electrode diameter – 40 μm, number of electrodes – 4.

A given shape needles are made by step-by-step etching a standard silicon wafer in wet etching and dry plasma etching by Bosch-process. For pattern of metallization, dielectric isolation and contact pads we are used standard microelectronic photolithography and vacuum deposition techniques.

The standard silicon wafer (100) with dual side alignment signs is processed by wet etching at a back side depth of 350 μm under nitride mask. After metal film deposition, deposition of dielectric isolation and autopsy contact pads, perform deep silicon etching by Bosch-process. As result wafer is divided into individual chips - needles.

Silicon chip technological sequence schematically shown in figure 3.

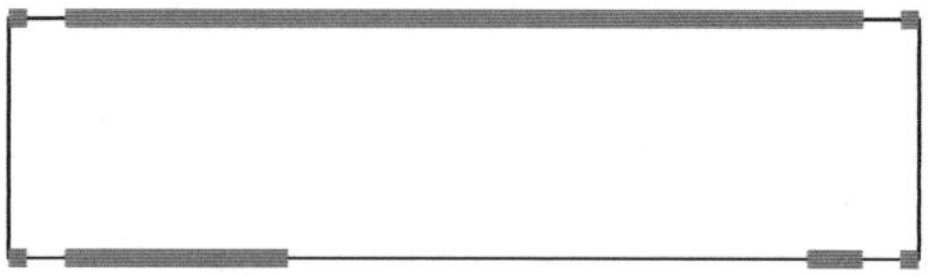

a. Silicon nitride deposition and photolithography (for back side hollow and alignment signs)

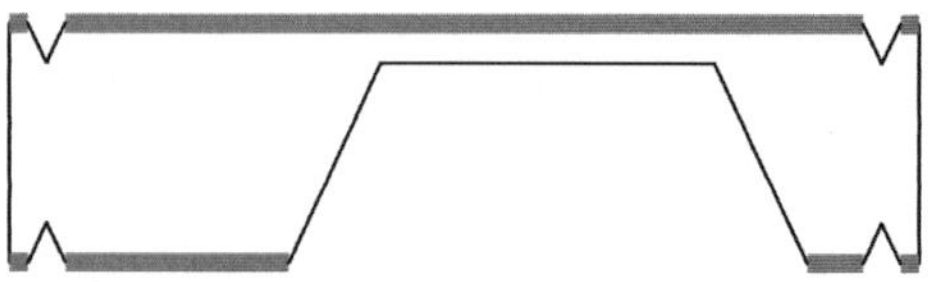

b. Silicon wet etching (50 μm membrane thickness formation)

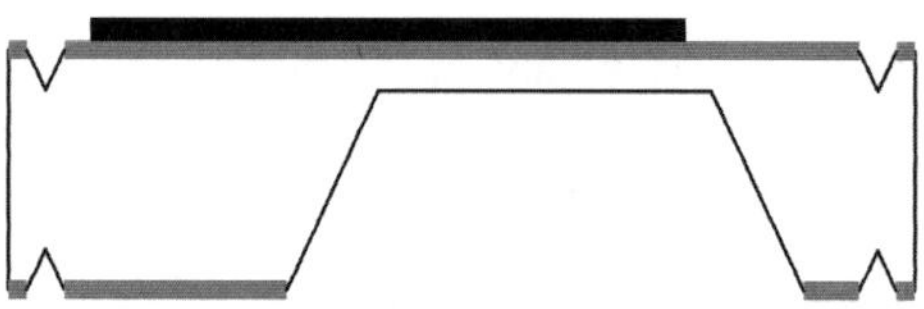

c. Metal film deposition and photolithography (electrode formation)

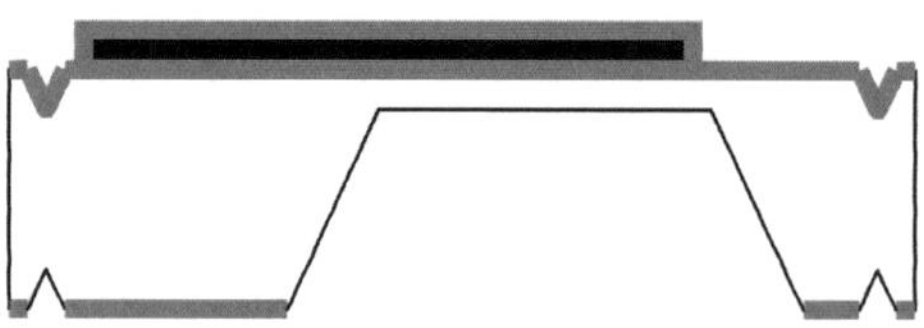

d. Gas phase deposition of silicon oxide

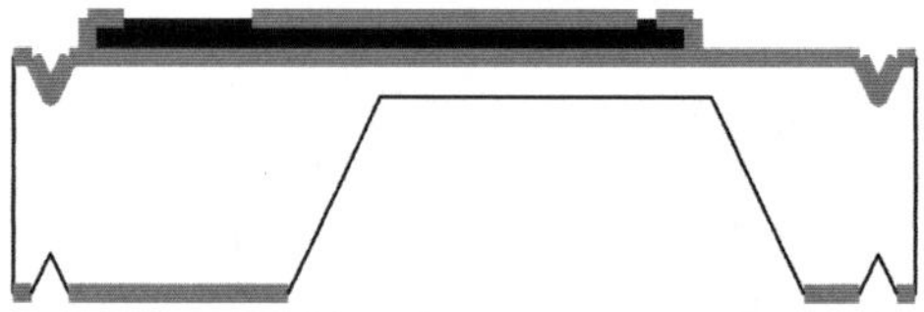

e. Silicon oxide window autopsy (contact formation)

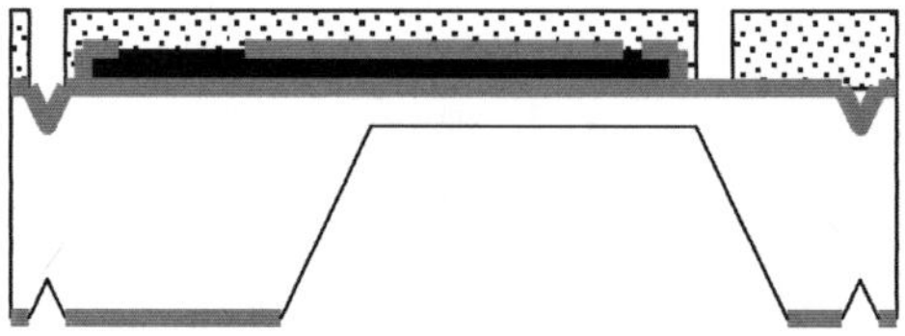

f. Photo resist deposition and photolithography (plasma etching mask formation)

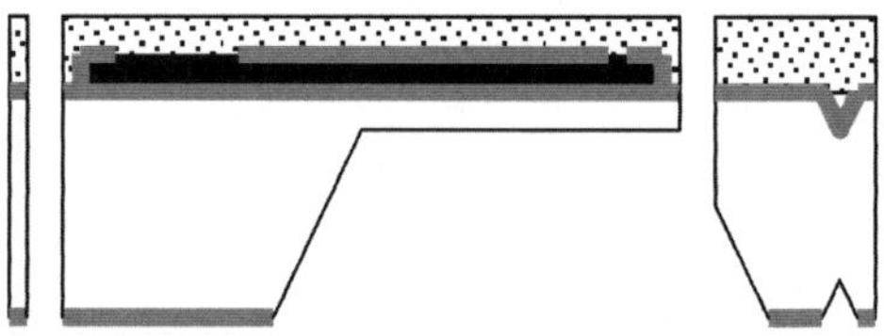

g. Silicon end-to-end plasma etching (silicon formation needles and divide into individual chips)

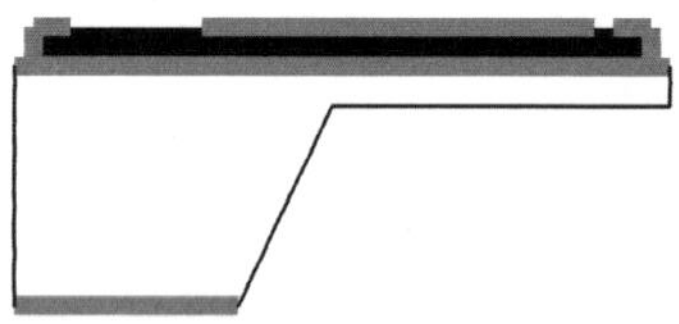

e. Photo resist mask delete

Figure 3. A schematic representation of the process of silicon chip technological sequence

The next step of experimental studies will be the production of microelectrode implants consisting of the micro needles and fiber group represented in figure 1. It is located in the central part of the device to obtain a three-dimensional data array of the distribution of neurons excitation at different distances from the radiation source (figure 4).

An implant consists of the coaxial conical optical waveguide (optrode) integrated inside the implantable electrode array (multi-electrode array-MEA) for recording the experimental data.

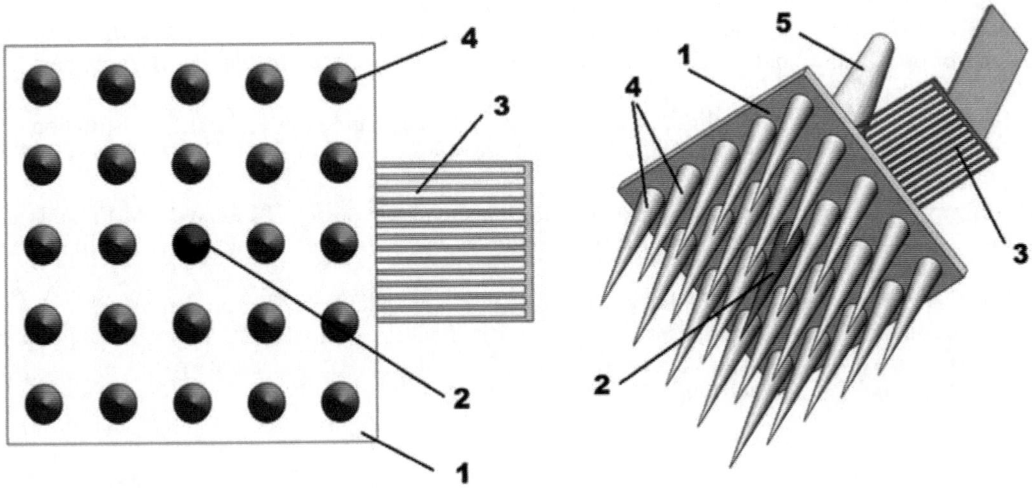

Figure 4. Schematic representation of an implantable optical-electronic array: 1 – the body of the array; 2 – integrated optrode; 3 – multichannel electrical data recording; 4 – electrode needle; 5 – optogenetic control signal

Optrode allows to record electrical activity during optogenetic experiments. This combination of several microelectrodes allows to record the activity of several neurons in light affected areas. It minimizes the effects of light diffusion inside the tissue and the mismatch of positions of the light source and the detector that records neurons excitation / inhibition parameters. [4]

5. Future plans and conclusions

Neurointerface is used in problems of activation and one-time registration of neuronal activity and is constantly being improved. Our future plans include using the device in long-term experiments on the spinal cord motor neurons stimulation using optogenetic techniques. The use of the implant is planned in slices and live animals like AD models. Simultaneously, the recording system, the filtering and processing of data received from the device are improved.

Acknowledgments

The authors express their gratitude to I.B. Bezprozvanny, the head of the Laboratory of Molecular Neurodegeneration, doctor of biological sciences and Carl J. and Hortense M. Thomsen Chair in Alzheimer's Disease Research, and to the whole team of the laboratory for consultations and assistance in research.

The part of the study dedicated to staging and testing optogenetic methods is supported by the Russian Scientific Fund Grant no. 14-25-0024.

References

[1] Deisseroth K 2011 Optogenetics *Nat. Meth.* **8**(1) pp 26-29
[2] Matveev M, Erofeev A, Terekhin S, Plotnikova P, Vorobyov K and Vlasova O 2015 Implantable devices for optogenetic studies and stimulation of excitable tissue *St. P. St. Pol. Un. J. Phys. and Math.* **3**(225) pp 75-83
[3] Sparta D, Stamatakis A, Phillips J, Hovelsø N, van Zessen R and Stuber G 2012 Construction of implantable optical fibers for long-term optogenetic manipulation of neural circuits *Nat. Protoc.* **7**(1) pp 12-23
[4] Anikeeva P, Andalman A, Witten I, Warden M, Goshen I, Grosenick L, Gunaydin L, Frank L and Deisseroth K 2011 Optetrode: a multichannel readout for optogenetic control in freely moving mice *Nat. Neurosci.* **15**(1) pp 163-70
[5] Thomas C, Springer P, Loeb G, Berwald Y, Netter Y and Okun L 1972 A miniature microelectrode array to monitor the bioelectric activity of cultured cells *Exp. Cell Res.* **74**(1) pp 61-66
[6] HajjHassan M, Chodavarapu V and Musallam S. 2008 NeuroMEMS: Neural Probe Microtechnologies *Sensors* **8**(10) pp 6704-26

Development of the technique of terahertz pulse spectroscopy for diagnostic malignant tumors during gastrointestinal surgeries

A A Goryachuk[1], M K Khodzitsky[1], M A Borovkova[1], A K Khamid[2], P S Dutkinskii[2] and D A Shishlo[2]
[1]Photonics and Optical information technology Department, ITMO University, Saint Petersburg, Russia
[2]Pavlov First Saint Petersburg State Medical University, Saint Petersburg, Russia

E-mail: goryachuk-anna@yandex.ru

Abstract. Samples of fresh excised tissues obtained from patients who had undergone gastric cancer have been investigated . Samples were consisted of cancer zone, normal zone and zone mixed of normal and cancer tissues. Their optical properties and spectral features were investigated by terahertz time-domain spectroscopy (TDS) in reflection mode. It was found that waveforms of reflected signals from normal and cancer tissues were well distinguished so it can be concluded that it is easy to discriminate gastric cancer tissue from normal by using THz TDS.

1. Introduction

Terahertz (THz) reflectometry has been applied to a lot of biomedical research and cancer diagnosis has become the main course of this researches. As soon as endoscope was invented the potential of THz spectroscopy increased and during the past few years, the basic tissue characteristics of each digestive organ in THz regime have been investigated. Although advanced gastric cancers have shown poor prognosis, the early staged gastric cancers can be cured by surgery in patients with good operator-independent technique of cancer tissues diagnostics around tumors during surgeries.

The main advantage of the THz spectroscopy which is really important for biophotonics is that the characteristic energies of molecule's rotational and vibrational motions lay in THz frequency region, so many chemical and biological molecules can be identified by their characteristic resonant peaks. In additional THz radiation is non-ionizing and might be applied in vivo diagnostics that is the big advantage for biological structures, because there are not appeared any destructions during the procedure.

The aim of the study was to evaluate the feasibility of THz time-domain reflectometry for the discrimination of gastrointestinal cancer tissues from normal tissues and searching of appropriate method for discrimination unique spectral features and optical properties of gastrointestinal tract cancer tumors for better it distinguishing during surgeries. In this study samples of fresh excised tissues obtained from five patients who had undergone gastric cancer have been investigated. Samples were consisted of cancer zone and normal zone.

Journal of Physics: Conference Series **741** (2016) 012072 doi:10.1088/1742-6596/741/1/012072

Waveforms of signals reflected from the fresh tissues were obtained by the THz time-domain spectroscopy system in reflection mode. In addition, several optical properties extracted from the THz reflection signals were analyzed to quantitatively discriminate gastric cancer tissues from normal regions.

2. Description of the samples and experimental setup

Subject of study were fresh excised tissue specimens obtained from patients who had undergone gastric cancer. Samples were consisted of cancer zone, normal zone and zone mixed of normal and cancer tissues (figure 1).

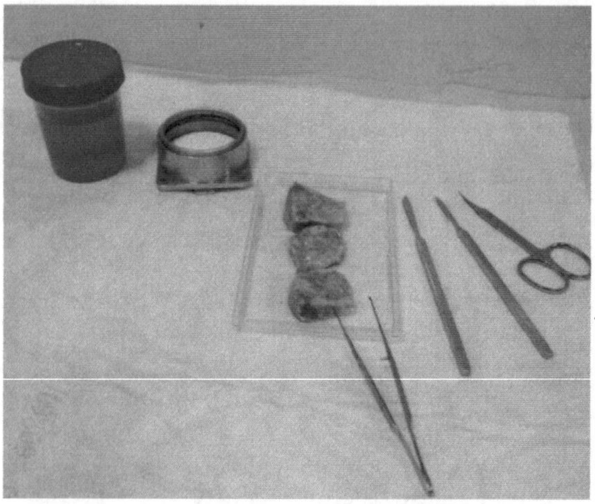

Figure 1. Photo of sample preparation. All of the specimens were consisted of cancer zone, normal zone and zone mixed of normal and cancer tissues.

The specimens were investigated within 4 hours after their extraction by the THz time domain (TD) spectrometer in the reflection mode (figure 2). The characteristics of the spectrometer are following: the spectral resolution of 15GHz, the sensitivity of lock-in amplifier of 1mV, the lock-in amplifier time constant of 1s. The THz broadband pulsed radiation had the following parameters: the spectral range of 0.1–1.5THz, the pulse duration of 2.7ps, the average power of 30 mW. THz radiation was generated by femtosecond laser (Yb:KYW) irradiation of undoped indium arsenide crystal. The femtosecond laser parameters are following: the wavelength of 1040nm, the pulse duration of 120fs, the pulse repetition rate of 75MHz, the power of 1W.

The aim of the study was to evaluate the feasibility of THz time-domain reflectometry for the discrimination of gastrointestinal cancer tissues from normal tissues. The spectra in frequency domain were obtained by the Fourier transform of the measured time domain waveforms of signals reflected from the fresh tissues and then were analyzed to quantitatively discriminate gastric cancer tissues from normal regions by their optical properties.

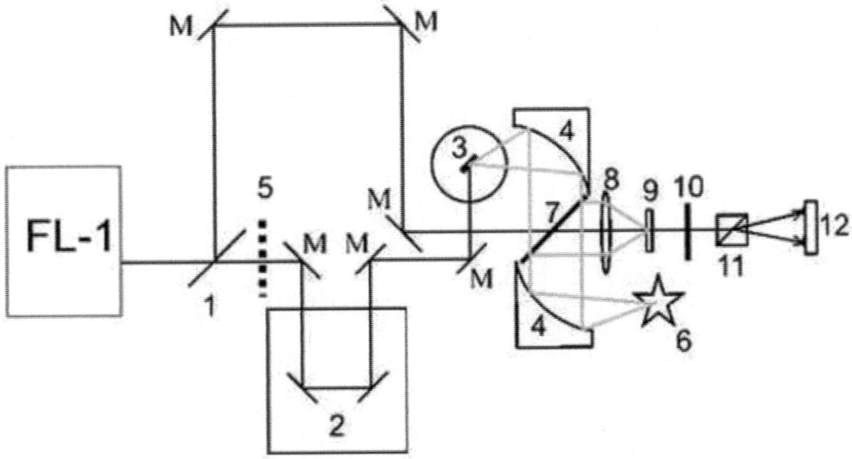

Figure 2. Scheme of the THz pulsed spectrometer in the reflection mode. FL-1 — infrared laser; M — mirrors; 1 — beam splitter; 2 — optical delay line; 3 — THz radiation generator: crystal of InAs in a magnetic system; 4 — parabolic mirrors; 5 — chopper; 6 — biological object; 7 — THz radiation beam splitter on the basis of high-resistivity silicon wafer; 8 — TPX polymer lens; 9 — electro optical crystal of CdTe; 10 — achromatic quarter-wave plate; 11 — Wollaston prism; 12 — balanced photodiodes.

3. Results and Discussions

In our work the optical properties and spectral features of gastric cancer tissue and normal tissue were obtained. Averaging was made by five measurements. Due to the fact that THz radiation is strongly absorbed by water and oncological tissues have increased water content compared with normal tissues it is easy to discriminate these two tissues by intensity of the reflected signals. This study have shown such differentiation that you can see in figure 3.

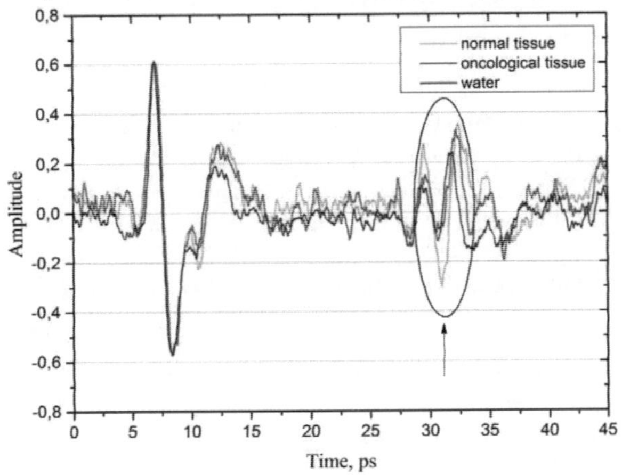

Figure 3. Differentiating of reflected signals intensity of cancer and normal tissues occurred because of different water content in these tissues. Also the reflected signal of pure water is shown so it is clear that absorption coefficient of cancer tissue and water pretty similar.

For more accurate results it is necessary to calculate the optical properties of normal and oncological tissues and quantitatively discriminate these parameters. The basic optical properties such as refractive index and dielectric permittivity were calculated (figure 4 – 5). The dispersion of refraction index is about 1.4 – 1.6 for normal tissues, 2.15 – 2.55 for cancer tissues and 2.2 – 2.7 for water in the frequency range of 0.25 – 0.6 THz. The comparison of refractive indices could be an additional approach in the diagnostics of gastrointestinal cancer.

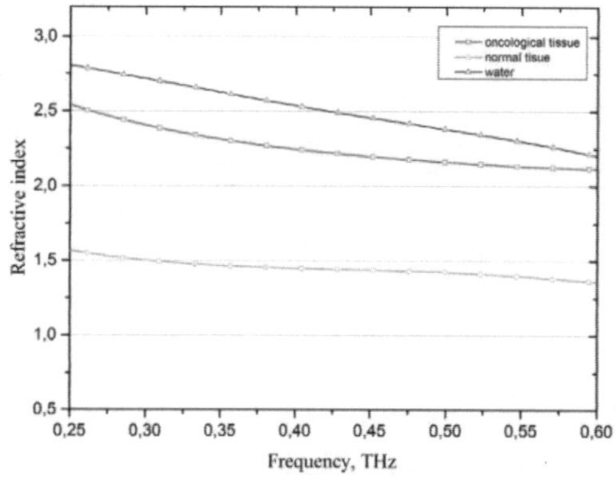

Figure 4. The dispersion of real part of refractive indices of water, cancer and normal tissues in frequency range of 0,25 – 0,60 THz.

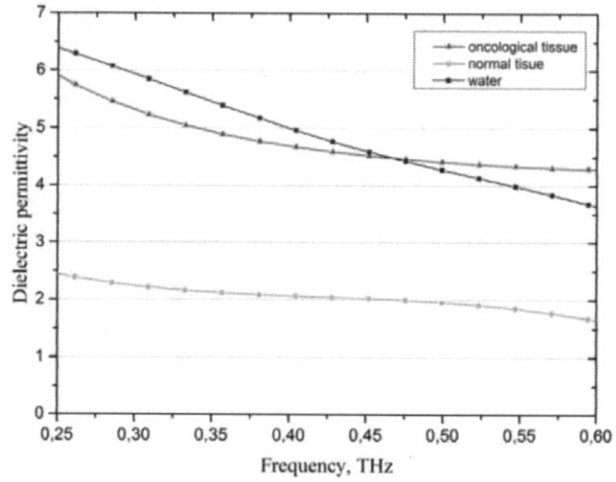

Figure 5. The dispersion of real part of dielectric permittivities of water, cancer and normal tissues in frequency range of 0,25 – 0,60 THz.

4. Conclusions

Thus, the possibility of THz reflectometry for discrimination of cancer tissues from normal tissues has been investigated. It was shown that it is easy to discriminate such tissues by using THz TDS because of their differences of water content. Due to this fact it is enough to get waveforms of reflected signals from different tissues for further its comparison. And to reach more accurate results the comparison of the optical properties might be used. Therefore, this method of investigation might be easily applied in clinical in vivo gastric cancer diagnostics.

Acknowledgments
This work was financially supported by the Government of Russian Federation, Grant 074-U01.

References

[1] Ji Y B, Kim S-H, Jeong K, Choi Y, Son J-H, Park D W, Noh S K, Jeon T-I, Huh Y-M, Haam S, Lee S K, Oh S J and Suh J-S 2014 *Biomed. Opt. Express* **6** 1398–1406
[2] Taylor Z D, Singh R S, Bennett D B, Tewari P, Kealey C P, Bajwa N, Culjat M O, Stojadinovic A, Lee H, Hubschman J-P, Brown E R and Grundfest W S 2011 *IEEE Trans. Terahertz Sci. Technol.* **1** 201–219
[3] Soon J-H 2014 Terahertz Biomedical Science and Technology (*New York: CRC Press*)

Cross MFF–analysis in studying the obsessive-compulsive disorder

S A Demin, O Yu Panischev and N Yu Demina

Kazan Federal University, Institute of Physics, Kazan, 420008 Russia

Kazan Federal University, Engelhardt Astronomical observatory, Zelenodolski region, st. Observatory, 422526 Russia

Abstract. There were demonstrated capabilities of the Memory Function Formalism (MFF) in analyzing cross correlations in human brain bioelectric activity at obsessive–compulsive disorder (OCD). To extract the information about collective phenomena in (electroencephalogram) EEG brain activity we use the power spectra of memory functions and the memory quantifiers. We discover the pairs of the electrodes with the greatest differences in dynamic and stochastic parameters for patients with the different condition. The high OCD condition is characterized by the influence of the memory effects. The MFF cross correlation analysis allow to describe the collective phenomena in EEG dynamics at OCD including the dynamic, spectral and stochastic behavior.

1. Introduction

The objective diagnostics of mental disorders at early developmental stages is obstructed due to absence of instrumental method of fixation, though relation of such disease to inevitable changes in the activity of separate areas of the brain cortex of a man seems evident. The most disorders haven't the objective diagnostic criteria. At this rate, schizophrenia, bipolar affective or obsessive-compulsive disorders (BAD and OCD) are diagnosed based on criteria in either the American Psychiatric Association's fifth edition of the Diagnostic and Statistical Manual of Mental Disorders (DSM 5), or the World Health Organization's International Statistical Classification of Diseases and Related Health Problems (ICD-10). One of the limitations of instrumental diagnosis of psychiatric disturbances is a study of the signals of human brain activity: electroencephalogram (EEG), magnetoencephalograms (MEG), reflecting functional activity of different brain regions. However in papers [1, 2] it is shown that BAD and epilepsy are manifested in the specific behavior of synchronization effects and in spectral properties of alpha rhythm. In [3] on the basis of the flicker-noise spectroscopy authors have formulated the criteria of degree of the susceptibility to schizophrenia in children/adolescents. In [4] it is demonstrated that the method based on the memory function formalism (MFF) can discover the susceptibility to schizophrenia by analyzing the collective phenomena in the cerebral cortex bioelectric activity. Within the framework of MFF [5, 6, 7] the present work analyzes the spectral properties and effects of statistic memory of bioelectric activity of the cerebral cortex of healthy subjects (with low level of OCD) and patients with high level of OC-symptoms for the purpose of determination of diagnostic criteria. MFF is a method that allows to bring in a complex of information measures for quantitative comparison of correlation fall time and existence of statistic memory in the dynamics under. The method contains equations for describing the frequency-phase synchronization.

2. Theory and method

We consider the stochastic dynamics of the studying signals, registered in two different brain regions as the sequences x_j and y_j of random values X, Y [4]:

$$X = \{x_0, x_1, x_2, \ldots, x_{N-1}\}, \quad Y = \{y_0, y_1, y_2, \ldots, y_{N-1}\}. \tag{1}$$

Mean values and fluctuations are determined as follows:

$$\langle X \rangle = \frac{1}{N}\sum_{j=0}^{N-1} x_j, \quad \delta x_j = x_j - \langle X \rangle; \quad \langle Y \rangle = \frac{1}{N}\sum_{j=0}^{N-1} y_j, \quad \delta y_j = y_j - \langle Y \rangle.$$

Here τ is the time interval of signal discretisation.

To describe the probabilistic relation between the sequences of random variables X and Y we use the normalized time dependent cross-correlation function (CCF):

$$c(t = m\tau) = \frac{1}{(N-m)\sigma_x \sigma_y} \sum_{j=0}^{N-m-1} \delta x_j \delta y_{j+m}, \tag{2}$$

$$t = m\tau, \qquad 1 \le m \le N-1.$$

Parameters σ_x and σ_y – are the mean-square derivations of random variables X and Y:

$$\sigma_x = \sqrt{\frac{1}{N}\sum_{j=0}^{N-1} \delta x_j^2}, \quad \sigma_y = \sqrt{\frac{1}{N}\sum_{j=0}^{N-1} \delta y_j^2}.$$

Further we represent the CCF through the vectors of system state: initial

$$\mathbf{A}_k^0 = \mathbf{A}_k^0(0) = \{\delta x_0, \delta x_1, \dots, \delta x_{k-1}\}, \quad \mathbf{B}_k^0 = \mathbf{B}_k^0(0) = \{\delta y_0, \delta y_1, \dots, \delta y_{k-1}\}; \tag{3}$$

and in time moment $t = m\tau$:

$$\mathbf{A}_{m+k}^m(t) = \{\delta x_m, \delta x_{m+1}, \dots, \delta x_{m+k-1}\}, \quad \mathbf{B}_{m+k}^m(t) = \{\delta y_m, \delta y_{m+1}, \dots, \delta y_{m+k-1}\}; \tag{4}$$

$$c(t) = \frac{\langle \mathbf{A}_k^0(0)\ \mathbf{B}_{m+k}^m(t)\rangle}{\langle \mathbf{A}_k^0(0)\ \mathbf{B}_k^0(0)\rangle}. \tag{5}$$

Then, using the technique of projector operators we derive the dynamic orthogonal variables:

$$\begin{aligned} \mathbf{W}_0^X = \mathbf{A}_k^0(0),\ \mathbf{W}_1^X = (i\hat{L} - \lambda_1^{XY})\mathbf{W}_0^X,\ \mathbf{W}_2^X = (i\hat{L} - \lambda_2^{XY})\mathbf{W}_1^X - \Lambda_1^{XY}\mathbf{W}_0^X - \dots, \\ \mathbf{W}_0^Y = \mathbf{B}_k^0(0),\ \mathbf{W}_1^Y = (i\hat{L} - \lambda_1^{XY})\mathbf{W}_0^Y,\ \mathbf{W}_2^Y = (i\hat{L} - \lambda_2^{XY})\mathbf{W}_1^Y - \Lambda_1^{XY}\mathbf{W}_0^Y - \dots, \end{aligned} \tag{6}$$

kinetic and relaxation parameters, memory functions (CCFs of the higher order)

$$\lambda_n^{XY} = \frac{\langle \mathbf{W}_{n-1}^X \hat{L} \mathbf{W}_{n-1}^Y \rangle}{\langle \mathbf{W}_{n-1}^X \mathbf{W}_{n-1}^Y \rangle},\ \Lambda_n^{XY} = i\frac{\langle \mathbf{W}_n^X \mathbf{W}_n^Y \rangle}{\langle \mathbf{W}_{n-1}^X \mathbf{W}_{n-1}^Y \rangle}, \tag{7}$$

$$M_{n-1}^{XY}(t) = \frac{\langle \mathbf{W}_{n-1}^X \{1 + i\tau \hat{L}_{22}\}^m \mathbf{W}_{n-1}^Y \rangle}{\langle \mathbf{W}_n^X \mathbf{W}_n^Y \rangle}. \tag{8}$$

To analyze the phase synchronization we use the power spectra of memory functions:

$$\mu_i^{XY}(\nu) = \left| \tau \sum_{j=0}^{N-1} M_i^{XY}(j\tau) \cos 2\pi\nu j\tau \right|^2. \tag{9}$$

To analyze the stochastic and memory effects in the studied collective dynamics we use the frequency-dependent non-Markovian parameter:

$$\varepsilon_i^{XY} = \left\{ \frac{\mu_{i-1}^{XY}(\nu)}{\mu_i^{XY}(\nu)} \right\}^{\frac{1}{2}}. \tag{10}$$

3. Experimental Data

Obsessive-compulsive disorder is a mental disorder where people feel the need to check things repeatedly, perform certain routines repeatedly, or have certain thoughts repeatedly [8]. People are unable to control either the thoughts or the activities. The EEG data obtained from a group of ninety eight subjects including fifty patients with low OCD symptoms and forty eight patients with high OCD symptoms. Sixty-four channel EEG signals were continuously recorded by active scalp electrodes according to the extended 10-20 International system of electrode placement (figure 1). Additional electrodes were placed above and below each eye, and at the outer can thus of each eye, to record vertical and horizontal eye movements, respectively. The sampling rate was 512 Hz. EEG data during the 2 min eyes closed resting state were re-referenced offline to the average of the earlobe electrodes and data were notch-filtered between 49 and 51 Hz to remove line-noise. The data was then separated into 2 sec epochs, and bad channels were replaced by means of nearest neighbor interpolation [9, 10].

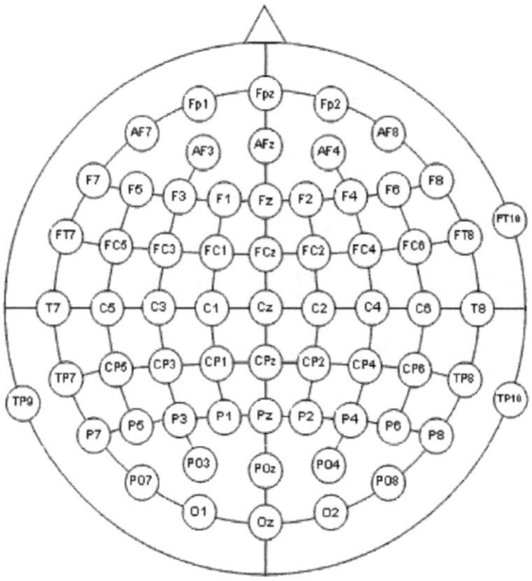

Figure 1. Positions of the 64 electrodes including their number and their designations. The schemata are based on the internationally established 10-20 system.

4. Discussion

Studying the frequency-phase synchronization using the power spectra of memory functions allows to determine the spatial connection distribution of cerebral cortex areas. It is found that the low OCD condition is characterized by long range phase synchronization in comparing with high OCD where the strong synchronization is observed in closely adjacent cortex areas (figure 2). To study the different type of synchronization we use the cross non-Markovian parameter (10) which can describes the synchronization of memory effects. Figure 3 demonstrates the areas with high level of the memory effect synchronization in patients with different OCD conditions. The finding are derived for all subjects. Using the window-time representation [11] we can study the local properties of phase synchronization. Figure 4 demonstrates the local deviations of main frequency peak at high OCD symptoms and it stability at the low OCD condition.

5. Conclusions

It is shown that the cerebral cortex signals from healthy people are characterized by high degree of long-range synchronization, whereas EEG of OCD patients has a strongly pronounced

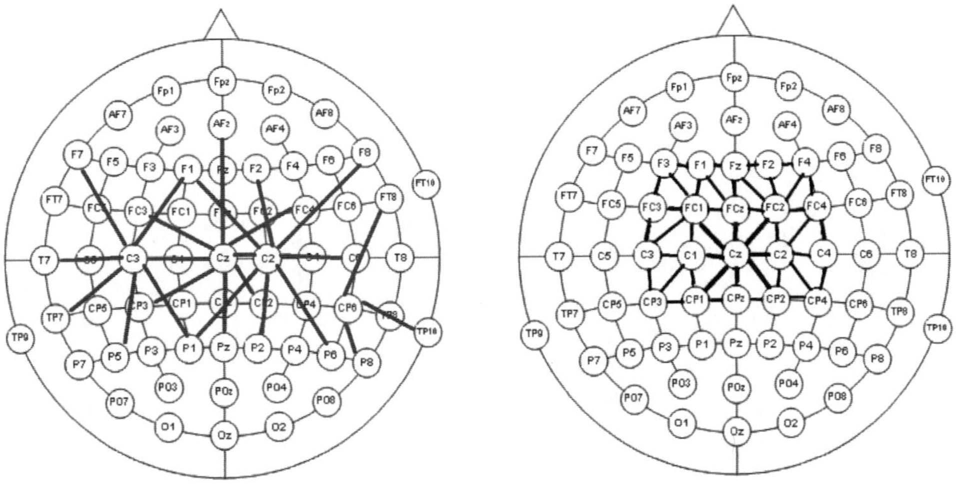

Figure 2. Schematic representation of the ranges of strong phase synchronization at low OCD and high OCD symptoms.

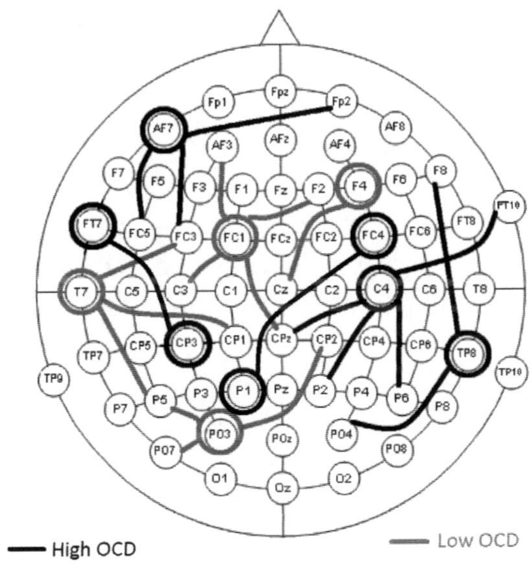

Figure 3. Synchronization of memory effects at low OCD and high OCD symptoms.

short-range synchronization. The obtained results allow to make progress in understanding the physical mechanisms of pathological changes in functioning of the brain cortex of a man, occurring as a result of OCD, and also to control the effectiveness of therapy in prospect. The work observes the possibilities of using EEG records in the development of objective methods of detection and identification of neuropsychopathy.

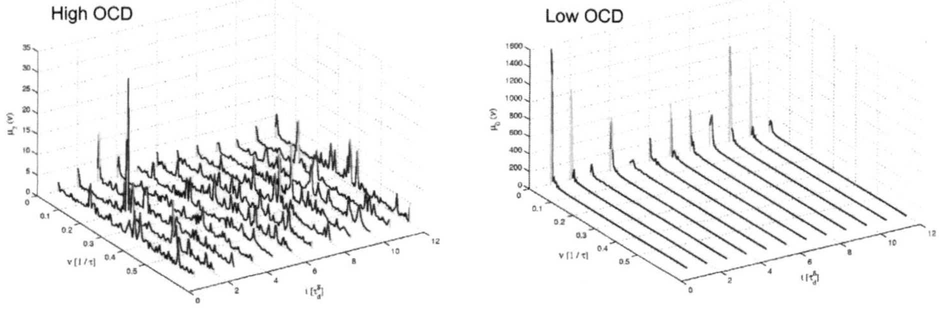

Figure 4. Intermittency in phase synchronization at low and high OCD symptoms.

Acknowledgments

The work is performed according to the Russian Government Program of Competitive Growth of Kazan Federal University. This work was supported by the Russian Foundation for Basic Research, grant No 14-02-31385_mol_a.

References

[1] Bhattacharya J, Kanjilal P P and Nizamie S H 2000 *IEEE Trans. Biomed. Engin.* **46** 738

[2] Bhattacharya J 2001 *Acta Neurobiol. Exp.* **61** 309

[3] Timashev S F, Panischev O Yu, Polyakov Yu S, Demin S A and Kaplan A Ya 2012 *Physica A* **391** 1179

[4] Panischev O Yu, Demin S A, Kaplan A Ya and Varaksina N Yu 2013 *Biomedical Engineering* **47** 153

[5] Yulmetyev R M, Emelyanova N A, Demin S A, Gafarov F M, Hänggi P and Yulmetyeva D G 2004 *Physica A* **331** 300

[6] Demin S A, Yulmetyev R M, Panischev O Yu and Hänggi P 2008 *Physica A* **387** 2100

[7] Panischev O Y, Demin S A, Panischeva S N and Bhattacharya J 2015 *Nonlinear Phenomena in Complex Systems* **18** 230

[8] Yaryura-Tobias J and Neziroglu F A 1997 *Obsessive-compulsive disorder spectrum: pathogenesis, diagnosis, and treatment* (Berlin: American Psychiatric Publishing)

[9] Jones R and Bhattacharya J 2014 *NeuroImage Clinical.* **4** 112

[10] Jones R and Bhattacharya J 2012 *J. Behav. Addit.* **1** 96

[11] Yulmetyev R M, Demin S A, Panischev O Yu and Hänggi P 2005 *Physica A* **353** 336

Transparent Nd doped YAG ceramics

Catalina-Andreea Stanciu, Traian Dascalu, George Stanciu, Nicolaie Pavel
Laboratory of Solid-State Quantum Electronics, *National Institute for Laser, Plasma and Radiation Physics, Bucharest R-077125, Romania*

Abstract. The reasearch main objective is to obtain ceramic laser materials based on pure YAG ($Y_3Al_5O_{12}$) and Nd doped YAG ($Y_{3-x}Nd_xAl_5O_{12}$, with x = 0.5 and 1.0 at. %), by conventional solid state reaction method. Stoichiometric compositions of $Y_3Al_5O_{12}$ (YAG), $Y_{2.985}Nd_{0.015}Al_5O_{12}$ (0.5 at.% Nd:YAG) and $Y_{2.97}Nd_{0.03}Al_5O_{12}$ (1.0 at.% Nd:YAG) were prepared using high purity Y_2O_3 (99.999%), Al_2O_3 (99.999%) and Nd_2O_3 (99.999%) nanopowders. Green bodies were sintered at 1750 °C for 16 h under vacuum (1.0×10^{-3} Pa) and then annealed at 1450 °C for 10 h in the air.

Ceramic laser materials are now use as alternative solutions to replace single-crystals for realizing lasers with high output performances. Large volume of transparent ceramics doped with rare earth ions are desirable for high power lasers. The development of short pulse lasers and their applications fueled the research in this field, especially for Yb and Nd doped $Y_3Al_5O_{12}$ material that allows ceramic engineering for new laser architectures [1-3].

In this work we report recent results on fabrication of transparent pure YAG ($Y_3Al_5O_{12}$) and Nd doped YAG ceramic media by solid state reaction method. Nd:YAG nanopowders were prepared by solid-state reaction method using high purity Y_2O_3 (99.999%), Al_2O_3 (99.999%) and Nd_2O_3 (99.999%) nanopowders. Polyethylene glycol (PEG-400) was added during the mixing process in order to avoid clustering; also, tetraethyl orthosilicate (TEOS) and MgO were used as sintering additives, in order to obtain dense ceramic pellets by decreasing the inter- and intra- particle porosity. The powders were molded into cylindrical pallets of 4.0-mm thickness and 12.0-mm diameter using a steel die at 20 MPa. To increase the density of the samples they were cold isostatic pressed for 10 minutes at 245 MPa.

The structure of each sample was investigated with an X-ray diffractometer (PANalytical X'Pert PRO MRD, -θ2θ geometry, Bragg Brentano). All the diffraction peaks were attributed to the cubic symmetry of YAG (*Ia3d* spatial group); the results are similar for all samples, indicating a single phase of $Y_3Al_5O_{12}$.

An electronic microscope (SEM, Quanta Inspect F) was used to investigate the morphology (in fracture) of sintered samples. The SEM micrographs allow us to appreciate the homogeneous and dense microstructure with small pores for the transparent ceramics (Fig.1).

Fig. 1 Photos of two 1.0-at.% Nd:YAG.ceramic media obtained.

Acknowledgments
Authors wishing to acknowledge Romanian National Authority for Scientific Research, project NUCLEU for financial support assistance.

References

[1] Wei GC 2005 *J. Phys. D: Appl. Phys.* **38** 3057-3065
[2] Li Y, Zhou S, Lin H, Hou X, Li W, Teng H, Jia T 2010 *J. Alloys Comp.* **502** 225-230
[3] Krell A, Hutzler T, Kinke J 2009 *J. Eur. Ceram. Soc.* **29** 207-221

Quantum dot laser optimization: selectively doped layers

Vladimir V Korenev, Sergey S Konoplev, Artem V Savelyev, Yurii M Shernyakov, Mikhail V Maximov, Alexey E Zhukov

Saint-Petersburg Academic University, Hlopina Str. 8/3A, Saint-Petersburg, 194021, Russian Federation

E-mail: `korenev@spbau.ru`

Abstract. Edge emitting quantum dot (QD) lasers are discussed. It has been recently proposed to use modulation p-doping of the layers that are adjacent to QD layers in order to control QD's charge state. Experimentally it has been proven useful to enhance ground state lasing and suppress the onset of excited state lasing at high injection. These results have been also confirmed with numerical calculations involving solution of drift-diffusion equations. However, deep understanding of physical reasons for such behavior and laser optimization requires analytical approaches to the problem. In this paper, under a set of assumptions we provide an analytical model that explains major effects of selective p-doping. Capture rates of elections and holes can be calculated by solving Poisson equations for electrons and holes around the charged QD layer. The charge itself is ruled by capture rates and selective doping concentration. We analyzed this self-consistent set of equations and showed that it can be used to optimize QD laser performance and to explain underlying physics.

1. Introduction

Compact semiconductor lasers emitting near 1.2-1.3 μm are important light sources for a wide range of practical applications, from optical coherence tomography and aesthetic surgery to ultrafast data transmission and Raman amplifiers [1, 2]. The usage of InAs/InGaAs quantum dot (QD) lasers emitting via the ground-state (GS) optical transition allows one to overlap the required wavelength range with small room temperature threshold currents [3 – 4]. However, as the injection current increases, an additional short-wavelength spectral line associated with the first excited-state (ES) optical transition of QDs appears in the lasing spectra [5 – 18]. For example, if the GS transition is centered at 1.28 μm, the ES one appears near 1.2 μm. Further increase of injection current usually results in the GS-lasing quenching [11 – 17]. Because simultaneous lasing via the GS and ES transitions (so-called multi-state or two-state lasing) put obstacles in practical usage of such QD lasers, deep understanding of this phenomenon as well as finding its key parameters and possible methods of its suppression becomes important from both fundamental and practical points of view.

It has been found that p-doping of a thin, typically 10 nm, layer that is adjacent to QD layer enhance laser performance in a sense pointed above. Namely, p-doping enhances maximum power from GS and increases threshold current of ES lasing [13]. Therefore an explanation of this effect is required to understand experimental data and provide a basis for laser optimization. A few earlier works concerning GS/ES lasing were not quite consistent with details of the experiment and/or could not explain a serious impact of p-doping on competition between GS and ES lasing. For example the [5 – 8] papers are based on the system of rate equations formulated in terms of excitonic energy levels and excitonic exchange processes. However, this model is capable of explaining a saturation of the GS lasing transition rather than its quenching. Also it was suggested that laser self-heating in CW regime is the reason for the GS-lasing quenching [11, 12]. However, it has been experimentally demonstrated [13] that the GS-lasing quenching is observed even in pulsed regime, when a temperature increment is as small as 1K.

In our previous work [15], it has been demonstrated that the hole-to-electron capture rate ratio, which we denoted as the h-factor has a serious impact on a QD laser operation regime. In particular, if the h-factor is below a certain critical value, decline and complete suppression of the GS lasing is expected. Obviously, the capture rates are determined by the carrier concentration in the GaAs matrix surrounding QD layer. Concentration itself is depending on QD layer charge. This self-consistent problem could be solved numerically as we presented earlier [19]. The results obtained confirm the idea that p-doping change the hole capture rate (increase h-factor) and favors to GS lasing by this means. Therefore, up to date, phenomenological theory of QD laser operation with different capture rates of electrons and holes has been presented together with numerical modeling of h-factor. In this paper we would like to present an analytical approach to calculation of h-factor at high injection currents. This approach could be useful for laser optimization as well as for a deeper view on underlying physics. Though suggested approach can be applied to QD lasers based on various material systems, for the sake of simplicity, conventional InAs/InGaAs/GaAs QD laser [17] structure details and material parameters will be used in this paper.

2. Analytical approach

In this part we discuss GaAs volume that surrounds layer of quantum dots. A few assumptions have to be made in order to construct simple and clear theoretical model. Let us list them:

- The "flat band" assumption. The electric field in the volume of GaAs waveguide is zero except for the areas near QD layers. That means that all charges including the charge of matrix/emitter interface (GaAs/AlGaAs) are effectively screened. This screening is due to free carriers in the matrix. Strong screening could be achieved only when injection and carrier concentration is high enough. That is why only the case of high injection and high output power is discussed here.
- Carrier mobility in matrix is high enough so we can neglect the gradient of chemical potentials F_n and F_p inside the matrix. The steps of their spatial dependence are located near matrix/emitter barrier.
- The screening is strong enough so neighboring layers of QDs do not affect each other. In the case of single layer QD laser that assumption is the same as the first one.
- The doping layer and QD layer can be modeled like a zero width layers.

One may notice here that the first assumption is an exact opposite to one in [20] where screening is completely neglected. The "no screening" should work well at low injection. Actually, first three assumptions has been checked in numerical simulations (see [19]) and have been proven increasingly realistic with the rise of the injection current.

Due to screening there are no charges far from QD layer. That means electron concentration equals hole concentration that we denote as n_0. Let us note as z the direction along a growth access and let $z = 0$ be the position of QD layer. Then we may write down the following:

$$n(z) = N_C \exp\left(\frac{F_n - E_C(z)}{T}\right) = N_C \exp\left(\frac{F_n - E_{C0}}{T}\right) \exp\left(\frac{q\phi(z)}{T}\right) , \quad (1)$$

$$p(z) = N_V \exp\left(\frac{E_V(z) - F_p}{T}\right) = N_V \exp\left(\frac{E_{V0} - F_p}{T}\right) \exp\left(-\frac{q\phi(z)}{T}\right) , \quad (2)$$

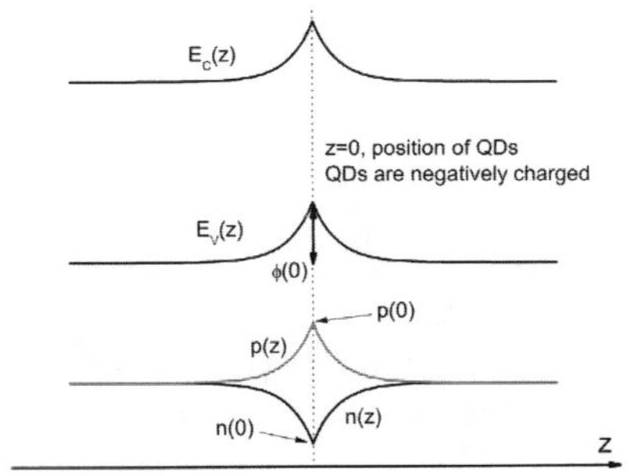

Figure 1. Scheme of energy bands and carrier concentration: screening of negatively charged QD layer

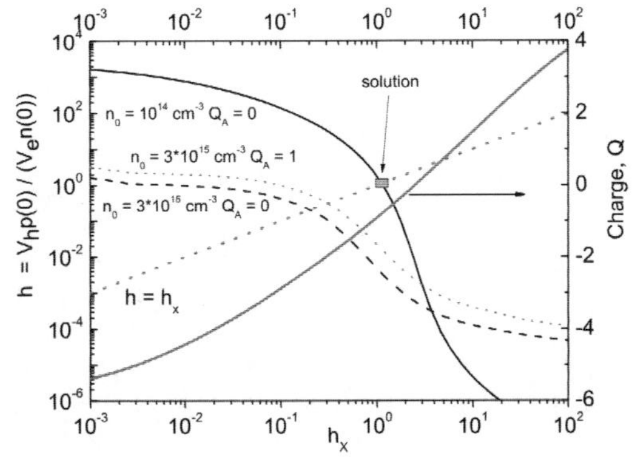

Figure 2. Calculated h - factor versus g_{hc}/g_{ec} ration. See details in text. The square at the intersection of $h(h_x)$ dependence and $h = h_x$ line marks the solution of equation set. The solid gray line shows Q dependence of h_x for $Q_A = 0$ and $n_0 = 10^{14}$ cm^{-3}

where ϕ is electrostatic potential, F_n, F_p are constant Fermi levels for electrons and holes, N_C, N_V are effective densities of states, $E_C(z)$, $E_V(z)$ are z-dependent edges of conduction and valence bands, E_{C0}, E_{V0} are their values far from QD layer, q is an absolute value of electron charge, T is a temperature. Let us assume that $\phi = 0$ far from QD layer. Then one can note that

$$N_C \exp\left(\frac{F_n - E_{C0}}{T}\right) = N_V \exp\left(\frac{E_{V0} - F_p}{T}\right) = n_0 \; . \tag{3}$$

After that we can write down carriers density in the matrix $\rho(z)$ and Poisson equation for potential ϕ:

$$\rho(z) = -qn_0 \mathrm{sh}\left(\frac{q\phi(z)}{T}\right) \; ; \tag{4}$$

$$\phi(z)'' = \frac{2qn_0}{\epsilon\epsilon_0}\mathrm{sh}\left(\frac{q\phi(z)}{T}\right) \; , \tag{5}$$

where ϵ is matrix dielectric permittivity and ϵ_0 is a vacuum permittivity. If we denote electric field as $F \equiv -\phi'$ then $F(0)$ is connected with the total density charge σ of the layer with quantum dots and acceptors:

$$|F(0)| = \frac{\sigma}{2\epsilon\epsilon_0} \; . \tag{6}$$

After integrating (5) we may achieve the relation between potential near the QD layer and its charge:

$$\frac{1}{2}F(z)^2 = \frac{2Tn_0}{\epsilon\epsilon_0}\mathrm{ch}\left(\frac{q\phi(z)}{T}\right) + \mathrm{const} \; ; \; F(-\infty) = 0 \; ; \; \phi(-\infty) = 0 \; ; \tag{7}$$

$$|F(0)|^2 = \frac{4n_0T}{\epsilon\epsilon_0}\left(\mathrm{ch}\left(\frac{q\phi(0)}{T}\right) - 1\right) = \frac{\sigma^2}{4\epsilon^2\epsilon_0^2} . \tag{8}$$

Let us introduce parameter x

$$x = 1 + \frac{\sigma^2}{16n_0T\epsilon\epsilon_0} \; , \tag{9}$$

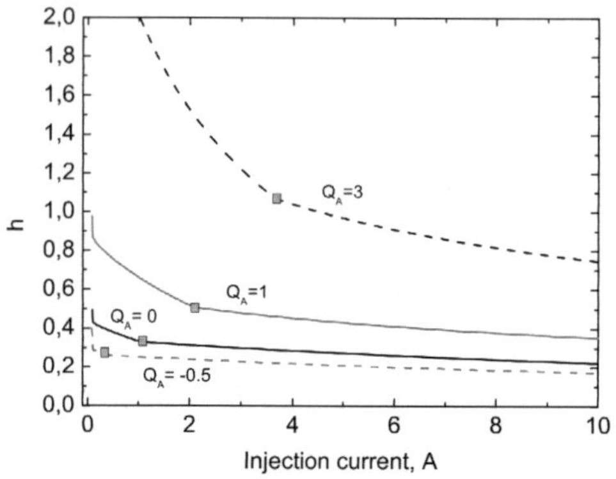

Figure 3. Dependence of h-factor on the injection current at different levels of doping. $Q_A < 0$ or > 0 means selective doping with donors and acceptors respectively. Square marks on the plot show the onset of ES lasing.

then carrier concentration near QD layer can be expressed as

$$n(0) = n_0 \left(x \pm \sqrt{x^2 - 1} \right) \ , \ p(0) = p_0 / \left(x \pm \sqrt{x^2 - 1} \right) \ , \tag{10}$$

where sign '+' is for $\sigma > 0$ and vice versa.

The suggested approach can be adopted to different models of quantum dots. Here, for example, we will use model that was successfully exploited in order to explain interplay between GS and ES lasing [15, 17]. In that model one need to know just two parameters about the environment of QD those are electrons and holes capture rates: g_{ec} and g_{hc}. Those are the rates of carrier capture at the highest energy level of quantum dot when it is empty. All other parameters of quantum dot in a given laser structure can be calculated as a function of g_{ec}, g_{hc} (see details in [17]). For our consideration the charge of quantum dot is of major importance. So let us introduce function $Q(g_{ec}, g_{hc})$ that is the charge of QD at a given capture rates. The actual captures to ES for example are likely to come from 3rd excites state or 4th excited states. However, under an assumption that these (3rd or 4th) excited states are weakly populated it is easy to show that their population is proportional to the carrier concentration in the matrix. Finally we can write down:

$$g_{ec} = V_e n(0) \ , \ g_{hc} = V_h p(0) \ , \ \sigma = N_S Q(g_{ec}, g_{hc}) - N_A h \ , \tag{11}$$

where V_e, V_h are capture constants (measured in $\mathrm{s}^{-1}\mathrm{cm}^3$), N_S is surface QD density in one layer, N_A and h are concentration and thickness of an acceptor layer respectively. The set of equations (9 - 11) together with $Q(g_{ec}, g_{hc})$ function describe laser operation point. Below in the paper we will specify doping by convenient parameter $Q_A = N_A h/(qN_S)$ that equals number of acceptors per one QD.

The parameters V_e and V_h could be calculated from the experimental data, for example, from carrier escape times. The function Q depends on the parameters of laser structure such as ratio of losses to maximal gain and could be non-trivial [17]. In order to clarify nature of the solution of (9 - 11) set we provide here figure 2. At the fixed value of n_0 one can easily see that

$$n(0)p(0) = n_0^2 \ , \ \text{and} \ g_{ec} g_{hc} = V_e V_h n_0^2 \equiv G^2 \ . \tag{12}$$

The last value is constant at a given n_0. Let us vary g_{ec} and g_{hc} keeping their product constant: $g_{hc} = G\sqrt{h_x}$, $g_{hc} = G/\sqrt{h_x}$. For any given h_x the charge Q can be calculated and then we get carriers concentration $n(0)$ and $p(0)$ (n_0 is already fixed). At figure 2 we plot

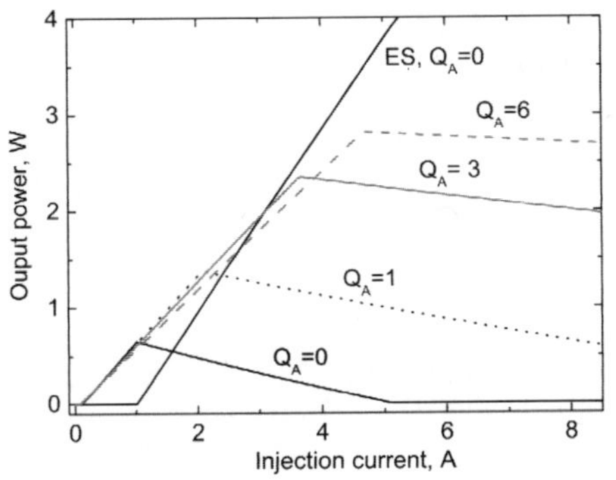

Figure 4. Dependence of output power on the injection current for different levels of doping.

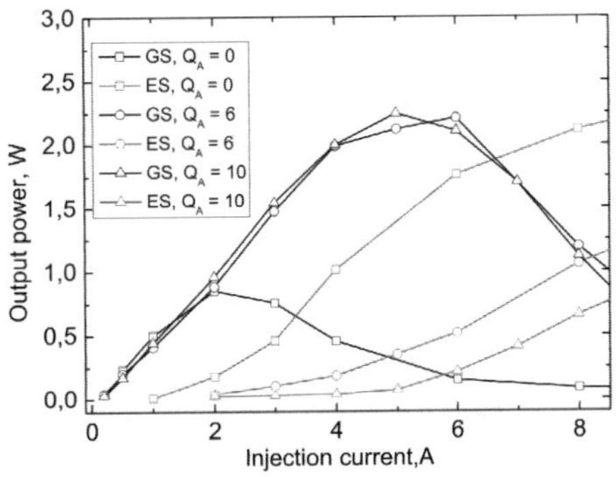

Figure 5. Experimental data for the structure modeled at figures 2 – 4. Stripe size 600x50 μm, 10 QD layers.

$V_h p(0)/(V_e n(0))$ versus h_x. Obviously, the equality of these two values at certain $h_x = h$ means the solution of the (9 - 11) system and determines the h-factor in the operation laser at a given injection. The QD parameters used in the calculations were generally taken from [15, 17]. The laser under consideration has 10 QD layers and stripe size 600x50 μm. The laser operates at room temperature, measured internal losses in the laser are 1.6 cm^{-1}. Below in the paper all calculations are made with the same parameters. The rising dependence of $Q(g_{ec}, g_{hc})$ on h_x is also shown in figure 2. The gray dotted line shows h dependence, when there is selective doping of one acceptor per one quantum dot (Q_A =1). In that case due to negative charge of "empty" acceptor the electron concentration decreases and hole concentration increases. Both these factors leads to increasing of h-factor that favors GS lasing.

3. Results and discussions

First of all we would like to discuss dependence of h-factor on the injection and output power for the different doping levels. Such dependencies can be seen in figure 3. The h - factor is decreasing while injection rises. That could be easily explained with more effective screening by the higher carrier concentration in the matrix. In fact, the QD layer in conventional laser is negatively charged due to the slow hole captures. Therefore, screening would reduce hole's flow towards QD layer and decrease h-factor. Square marks in the figure 3 show the onset of ES lasing. At higher powers when ES lasing is on the growth of n_0 with current become more steady due to more efficient captures in the ES state comparing with GS. That explains less decrease rate of h at higher currents. Just for illustration, the unpractical case of n-doping is also shown (Q_A = - 0.5). In that case h-factor is so low that GS lasing is almost totally suppressed.

Now we can directly discuss the interplay between GS and ES lasing. In the figure 4 laser output power from GS and ES are shown as a function of injection current. The power and injection current were calculated in the same way as in [17]. In figure 4 the data are shown for Q_A that equals 0, 1, 3 and 6. One can see that the laser characteristics are rather sensitive to Q_A and even moderate Q_A values significantly enhances the maximum power that can be generated at GS. Further increase of Q_A leads to higher internal losses that are of order of 0.06 cm$^{-1} \cdot Q_A$ for each QD layer. Obviously that suppress output power and overall laser efficiency. In this way the optimization could be performed depending on the preferred laser properties: Q_A should be high enough to take effects and low enough not to significantly rise the losses.

The modeling could be compared with experimental data shown on figure 5. The laser sample parameters were as in models. Despite we omit some processes in the laser such as non-radiative recombination there is fair agreement between experiment and theoretical modeling.

In conclusion, the analytical approach was suggested in order to describe effect of screening and selective doping in QD lasers. The results allow to optimize doping level according to preferable laser properties and clearly demonstrate positive effect of selective doping on laser performance.

Acknowledgments
This work was supported by the Russian Sceince Foundation, project 14-42-00006 "A novel type of diode lasers with characteristics improved by using asymmetric barriers".

References

[1] Zhukov A E, Kovsh A R 2008 *Quantum Electronics* **38** 409
[2] Zhukov A E, Kovsh A R, Nikitina E V, Ustinov V M, Alferov Zh I 2007 *Semicond.* **41** 606
[3] Savelyev A V, Novikov I I, Maximov M V, Shernyakov Yu M, Zhukov A E 2009 *Semicond.* **43** 1597
[4] Korenev V V, Savelyev A V, Zhukov A E, Maximov M V, Omelchenko A V, 2015 *Proc. of SPIE* **9503** 950305
[5] Markus A, Chen J X, Paranthon C, Fiore A, Platz C, Gauthier-Lafaye O 2003 *Appl. Phys. Lett.* **82** 1818
[6] Markus A, Fiore A 2004 *Phys. Stat. Sol. (a)* **201** 338
[7] Fiore A, Markus A, Rossetti M 2005 *Proc. of SPIE* **5840** 464
[8] Shi L W, Chen Y H, Xu B, Wang Z C, Wang Z G 2007 *Physica E* **39** 203
[9] Jiang L, Asryan L V 2006 *IEEE Phot. Tech. Lett.* **18** 2611
[10] Lee J, Lee D 2011 *J. Korean Phys. Society* **58** 239
[11] Y. J. Kim, Y. K. Joshi, A. G. Fedorov 2010 *J. Appl. Phys.* 107 073104
[12] Ji H M, Yang T, Cao Y L, Xu P F, Gu Y X, Wang Z G 2010 *Jpn. J. Appl. Phys.* **49** 072103
[13] Zhukov A E, Maximov M V, Shernyakov Yu M, Livshits D A, Savelyev A V, Zubov F I, Klimenko V V 2012 *Semicond.* **46** 231
[14] Viktorov E A, Mandel P, Tanguy Y, Houlihan J, Huyet G 2005 *Appl. Phys. Lett.* **87** 053113
[15] Korenev V V, Savelyev A V, Zhukov A E, Omelchenko A V, Maximov M V 2013 *Appl. Phys. Lett.* **102** 112101
[16] Wu Y, Asryan L V 2014 *J. of Appl. Phys.* **115** 103105
[17] Korenev V V, Savelyev A V, Zhukov A E, Omelchenko A V, Maximov M V 2013 *Semicond.* **47** 1397
[18] Maximov M V, Shernyakov Yu M, Zubov F I, Zhukov A E, Gordeev N Yu, Korenev V V, Savelyev A V, Livshits D A 2013 *Semicond. Sci. Technol.* **28** 105016
[19] Konoplev S S, Savelyev A V, Korenev V V, Maimov M V, Zhukov A E 2015 *J. Phys.: Conf. Series* **643** 012039
[20] Asryan L V, Suris R A 1997 *IEEE J. of Select. Top. in Quant. Electron.* **3** 148

Electrically pumped microdisk lasers with semitransparent conducting pyrolytic carbon film

E I Moiseev[1], Yu S Polubavkina[1], N V Kryzhanovskaya[1,2], M M Kulagina[3], Yu M Zadiranov[3], M V Maximov[1,2,3], F E Komissarenko[1,4], T Kaplas[5], Yu P Svirko[5], M Silvennoinen[5], A A Lipovskii[1,2], F I Zubov[1,2], A E Zhukov[1,2]

[1] St Petersburg Academic University, Khlopina 8(3), St Petersburg 194021, Russia
[2] Peter the Great St Petersburg Polytechnic University, St Petersburg 195251, Russia
[3] Ioffe Institute, St Petersburg, Russia
[4] ITMO University, 197101 St. Petersburg, Russia
[5] University of Eastern Finland, Joensuu, Finland

Abstract. Electrically driven microdisk lasers with top contacts made of a semitransparent conducting pyrolytic carbon film are developed. Electrical properties of the pyrolytic carbon contact to a *p*-type doped GaAs epitaxial layer are studied. Room temperature electroluminescence spectra from an array of the microdisk lasers and a single 27 µm in diameter microdisk laser are demonstrated.

1. Introduction

Last decade the essential efforts were aimed at the miniaturization of semiconductor lasers. Recently, whispering gallery mode (WGM) semiconductor microlasers based on ring/disk cavities with low threshold (2 uW) [1] and high, up to $7x10^5$, quality factor (Q) [2] have been demonstrated. The lasing operation at 100°C of microring laser as small as 2 µm under optical pumping has been recently reported [1]; room temperature lasing was reported in 1.16 µm GaInAsP microdisk laser [3] and in 1µm InAs/GaInAs quantum dot (QD) microdisk laser [4]. At the same time, the smallest reported diameter of the microdisk laser with current injection operating at room temperature is 6.5 µm [5]. Further reduction of the injection microdisk laser size requires electron beam lithography, planarization stage or air-bridge technology to form an electrical contact to upper (typically, *p*-doped) layers. In this work we study a possibility to use the pyrolytic carbon (PyC) film to make an electric contact to a *p*-doped GaAs cap layer. Room temperature electroluminescence spectra of the microdisk lasers with diameters 15-30 µm, which are electrically pumped through the semitransparent PyC layer, are demonstrated.

2. Experiment

Epitaxial structures were grown with molecular beam epitaxy on an n+ GaAs(100) substrate. Ten layers of $InAs/In_{0.15}Ga_{0.85}As$ QDs were deposited in the middle of a 0.44 µm thick GaAs waveguiding layer confined with $Al_{0.25}Ga_{0.75}As$ claddings. Microdisk resonators with diameters of 15…30 µm were formed by means of chemical plasma etching. Etch depth was about 7µm. AgMn/NiAu (AuGe/Ni/Au) metallization was used to form ohmic contacts to p+ GaAs cap layer (n+ substrate, respectively). A schematic illustration of the injection microdisk laser structure is shown in Figure 1. 20-nm PyC film was synthesized using chemical vapor deposition on quartz substrate using CH_4:H_2 gas mixture and then transferred onto PMMA layer. PyC film transmittance was about 50% at λ = 1.3 μm. To make an electrode to the *p*-type contact, the PyC film was attached to the top of an array of the microdisk lasers. Owing to mechanical strength the film did not show any sagging. Current was injected through the PyC film to the array of the microdisk lasers using a copper tape. Schematic of the microdisk lasers with PyC contact is shown in figure 2. Needle-shaped metal probe was used for current pumping a single microdisk. Lasers were mounted on a copper heatsink and tested at room temperature without external cooling. Emitted light was collected with a piezoelectrically adjustable x10 Mitutoyo MPlan NIR objective.

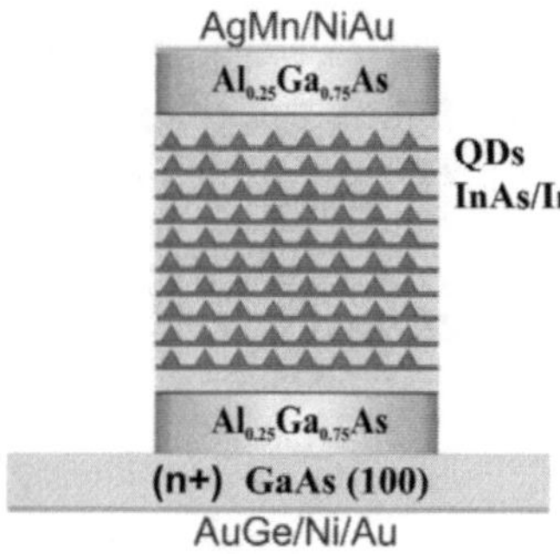

Figure 1. Schematics of the injection microdisk laser epitaxial structure.

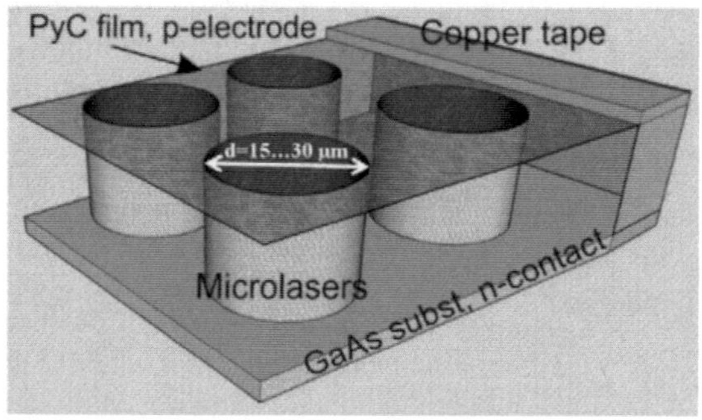

Figure 2. Schematic illustration of the array of microdisk lasers with top PyC electrode.

3. Results

The microlasers demonstrate well pronounced turn-on behavior of current-voltage characteristic with opening voltage under forward bias of U_0 ~1 V and series resistance R_S =1kOhm (fig.3). Figure 4 shows electroluminescence spectra of the microdisk array with diameters of 15-30 μm recorded at different currents. The emission spectrum has a form typical for spontaneous emission of self-organized InAs/InGaAs QDs with the ground-state peak located at 1.26 μm (@ 1 mA). We did not obtain any WGM lasing because all lasers in tha array were excited simultaneously via semitransparent . With increasing pump current, the electroluminescence maxima demonstrates red shifts 0.8nm/mA caused by the heating of the structure. From this shift we can estimate the structure overheating to be 60°C at 15mA, which is nearly the same as was obtained for a single microlaser without the PyC film.

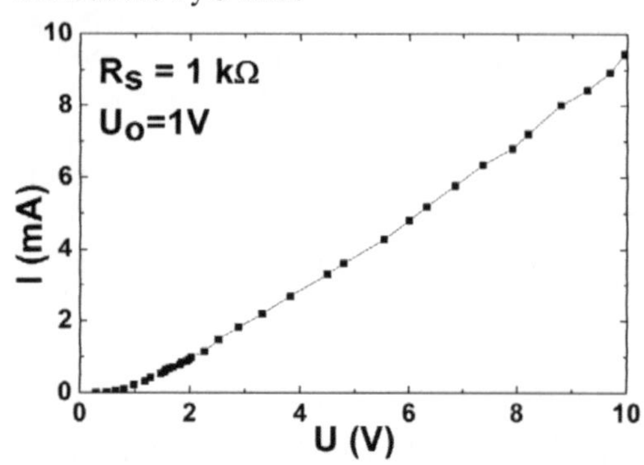

Figure 3. Current–voltage characteristics of the microlasers with PyC electrode.

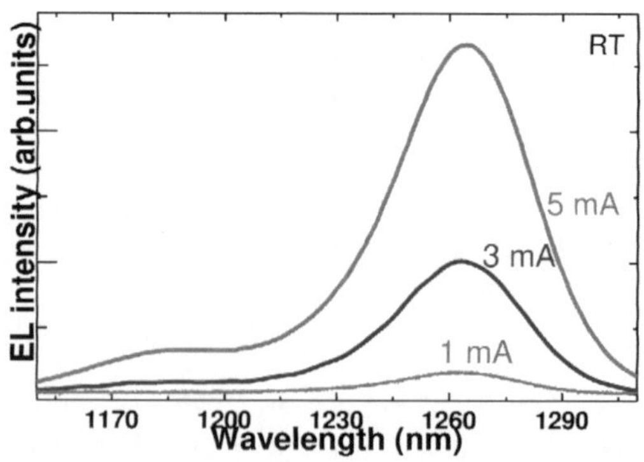

Figure 4. Room temperature electroluminescence from an array of the microdisk lasers with PyC electrode.

Needle-shaped metal probe was used to pump a single microdisk with diameter D = 27 μm (a video camera image is shown in inset to figure 6). Electroluminescence spectra of this microdisk obtained at room temperature with PyC film and without PyC film are compared in figure 5. The emission intensity of the microdisk with PyC film is approximately 35% of that without the PyC film in the whole spectral range. We attribute the losses to the absorption and reflection of the PyC film. The sharp lines observed in the spectra correspond to WGMs of the disk microcavity. Lasing in the PyC-capped microlaser was observed at the same resonance line of 1266 nm as before PyC film placing. Figure 6 shows a dependence of integral intensity of the resonant line on pump current. The threshold current is 5.5 mA in both cases.

To conclude, we demonstrate room temperature lasing of the 27 μm quantum dot microdisk laser with top contact made of a semitransparent conducting pyrolytic carbon film. Electrical properties of the contact between PyC film and p-type doped GaAs epitaxial layer are studied. Room temperature electroluminescence spectra from the array of the microdisk lasers and a single microdisk laser with D = 27 μm are demonstrated.

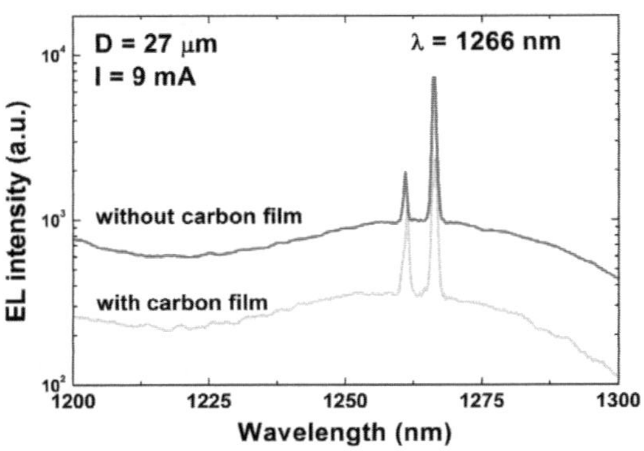

Figure 5. Electroluminescence spectra of the microdisk laser with D = 27 μm obtained at room temperature.

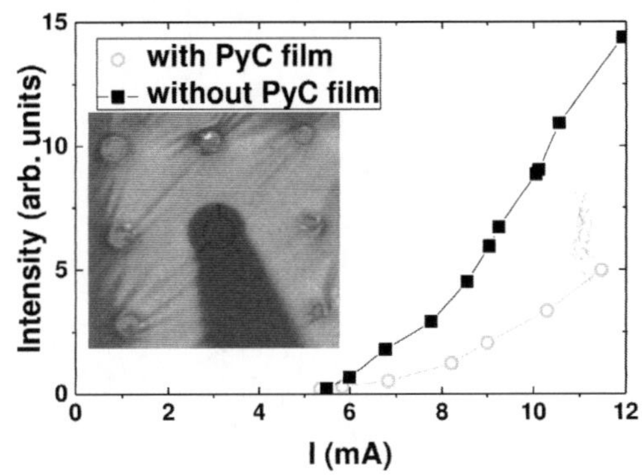

Figure 6. Integrated intensity of the WGM line (λ = 1266 nm) as a functions of the pump current. Inset shows the camera-image of the single microdisk with the needle probe used for electric pumping.

Acknowledgments
The work is supported in different parts by RFBR projects 15-32-20238 and 15-02-03624. E.I. Moiseev acknowledges the support by Program "U.M.N.I.K." and Government of St Petersburg and the Committee on Science and Higher Education. Y.P. Svirko and T.Kaplas acknowledge support from the H2020 Graphene Flagship Core 1 program

References

[1] Maximov M V, Kryzhanovskay N V, Nadtochiy A M, Moiseev E I, Shostak I I, Bogdanov A A, Sadrieva Z F, Zhukov A E, Lipovskii A A, Karpov D V, Laukkanen J and Tommila J 2014 *Nanoscale Research Lett.* **9** 657.

[2] C.P.Michael, K.Srinivasan, T.J.Johson, O.Painter, K.H. Lewe, K.henessy, H.Kim, E.Hu, *Appl.Phys. Lett.* 90, 051108 (2007).

[3] R.Perahia, T.P.Mayer Alegre, A.H.Savafi-Naeini, O.Painter, *Appl.Phys. Lett,* 95, 201114 (2009).

[4] N.V. Kryzhanovskaya, A.E. Zhukov, M.V. Maximov, E.I. Moiseev, I.I. Shostak, A. M. Nadtochiy, Yu.V. Kudashova, A.A. Lipovskii, M.M. Kulagina, S.I. Troshkov, *IEEE JSTQE,* 21, 1-5 (2015).

[5] Mao M-H, Chien H-C, Hong J-Z and Cheng C-Y 2011 *Opt. Exp.* **19** 14145.

[6] T.Kaplas, Yu.Svirko, *Journal of Nanophotonics*, **6**, 061703, 2012.

Journal of Physics: Conference Series **741** (2016) 012077 doi:10.1088/1742-6596/741/1/012077

Deep-level study of Ga(In)P(NAs) alloys grown on Si substrates

A I Baranov[1,2], J P Kleider[2], A S Gudovskikh[1,3], A Darga[2,4], E V Nikitina[1], A Yu Egorov[5]

[1] St Petersburg Academic University, 194021 St Petersburg, Russia

[2] Group of electrical engineering of Paris (GeePs), 91192 Gif sur Yvette Cedex, France

[3] St Petersburg Electrotechnical University "LETI", 197376 St Petersburg, Russia

[4] Sorbonne Universités, UPMC Univ Paris 06, UMR 8507, Paris, France.

[5] ITMO University, 197101 St Petersburg, Russia

Artem.Baranov@supelec.fr

Abstract. Defect properties of Ga(In)P(NAs) layers with different composition were studied by admittance spectroscopy. For nitrogen content layers the defect level with energy of 0.44-0.47 eV, which related to nitrogen incorporation into GaP, was observed. Its concentration is lower for GaPNAs layers compared to GaPN/InP due to better compensation by arsenic than by indium in lattice of GaP. Other defect level with energy of 0.30 eV was detected in GaPAs and GaPN/InP layers. Likely, the both observed defects in GaPAs and GaPN/InP have the same nature.

1. Introduction

Dilute nitrides III V-N form a new family of perspective materials for optoelectronic devices such as light emitting diodes, lasers and high-efficiency multijunction solar cells, and open the way to novel integrated optoelectronics combining III-V compound semiconductors with silicon [1]. The incorporation of just 0.43 % atomic fraction of nitrogen into the gallium phosphide (GaP) lattice leads to direct band transition while the band gap decreases drastically with small (few percents) N content [2,3]. Furthermore, addition of indium and (or) arsenic atoms allows one to vary the band gap in a wide range of 1.5 to 2.0 eV while remaining lattice-matched to Si substrates. The latter opens a way for the development of high efficiency lattice-matched multijunction solar cells on Si substrates [4]. However, experimental studies demonstrate low minority carriers' lifetime in active GaPNAs layers grown on GaP and Si substrates by MOCVD [5] and by MBE [6]. This is associated with defect formation and appearance of a significant concentration of non-radiative recombination centers during the growth. Therefore, the study of defects in Ga(In)PN(As) is an important issue for further material development. However the electronic properties of point defects in III-V-N alloys grown on Si are still not enough studied. Most of defect studies were performed for alloys grown on GaP substrates [7-11] which seems necessary as a first step. But there could be a strong difference in defect formation mechanisms for the growth on GaP and Si substrates. For example, formerly it was shown that GaPNAs grown on GaP has a defect level with a characteristic activation energy of 0.18 eV, which is not observed in GaPNAs grown on Si [12]. Another issue is that a significant part of the experimental study for (Al)Ga(In)PNAs layers was performed using the optically detected magnetic resonance (ODMR) technique [13], which does not allow one to determine the activation energy and capture cross sections of defects that are the most important parameters for recombination mechanisms analysis. To determine these parameters the capacitance techniques like deep-level transient spectroscopy (DLTS) and admittance spectroscopy should be rather used. However the only n-type GaP:N layers were studied by DLTS [7-10]. In addition, according to references [5, 6] the lifetime of minority charge carriers is extremely low in doped GaPNAs layers. Thus p-i-n structures using the i (undoped) layer as the absorber should have the highest interest for optoelectronic applications. Therefore, the properties of undoped III-V-N layers are one of the key issues for optoelectronic devices technology development. To our knowledge no defect studies of undoped Ga(In)PN(As) layers grown on silicon substrates have been reported. Here we use the admittance spectroscopy technique to investigate the defect properties of undoped III V-N layers grown on Si.

2. Experiment and samples

Three different p-i-n structures with active III-V-N i-layers of different composition were grown on p-type Si substrates (concentration of acceptors $10^{16}cm^{-3}$) by molecular-beam epitaxy using Veeco GEN III setup with a plasma nitrogen source. The parameters of intrinsic (not intentionally doped) layers are presented in Table 1. The layers were lattice-matched to silicon. Sample #1 based on $GaP_{0.70}As_{0.30}$ was grown as a reference and to define the nitrogen role in defect formation. Sample # 2 has the undoped region based on a supperlattice consisting of a sequence of 10 nm $GaP_{0.99}N_{0.01}$ layers separated by monolayers of InP to improve indium incorporation into the alloy. Sample # 3 was grown with 10 atomic percent of As for a better incorporation of nitrogen.

Table 1. Parameters of III-V-N i-layers.

Sample #	Composition	Width, nm	Lattice mismatch to Si at 300 K, %	Growth temperature, °C
1	$GaP_{0.70}As_{0.30}$	400	1.48	610
2	$GaP_{0.99}N_{0.01}$/InP	200 (10.0/0.3)×20	0.19	490
3	$GaP_{0.882}N_{0.018}As_{0.10}$	200	0.42	530

The admittance spectroscopy technique allows one to determine the activation energy (E_A), i.e. the energy location of defects with reference to the majority carrier band, and the corresponding capture cross section (σ_A) from the analysis of the temperature and frequency dependence of the capacitance C(T,f) and conductance G(T,f) [14]. The measurements were performed using a liquid nitrogen cryostat in the temperature range of 80…380 K and at frequencies from 20 Hz to 1MHz.

3. Result and discussion

The measured temperature dependences of the capacitance C(T, f) are shown in Figures 1a—1c. The steps in the capacitance (which are accompanied by conductance peaks) observed for the samples #1, #2, #3 are shifted toward higher frequency when the measurement temperature is increased. Such steps are characteristic for the response of gap states and may be caused either by bulk defect levels in the undoped layer or by interface states at the III-V-N/GaP(N) heterojunctions. The measurements of C(T, f) at different bias voltages (not shown here) demonstrate that the step positions are independent of the applied bias voltage indicating that the response originates from bulk defects rather than from the interface.

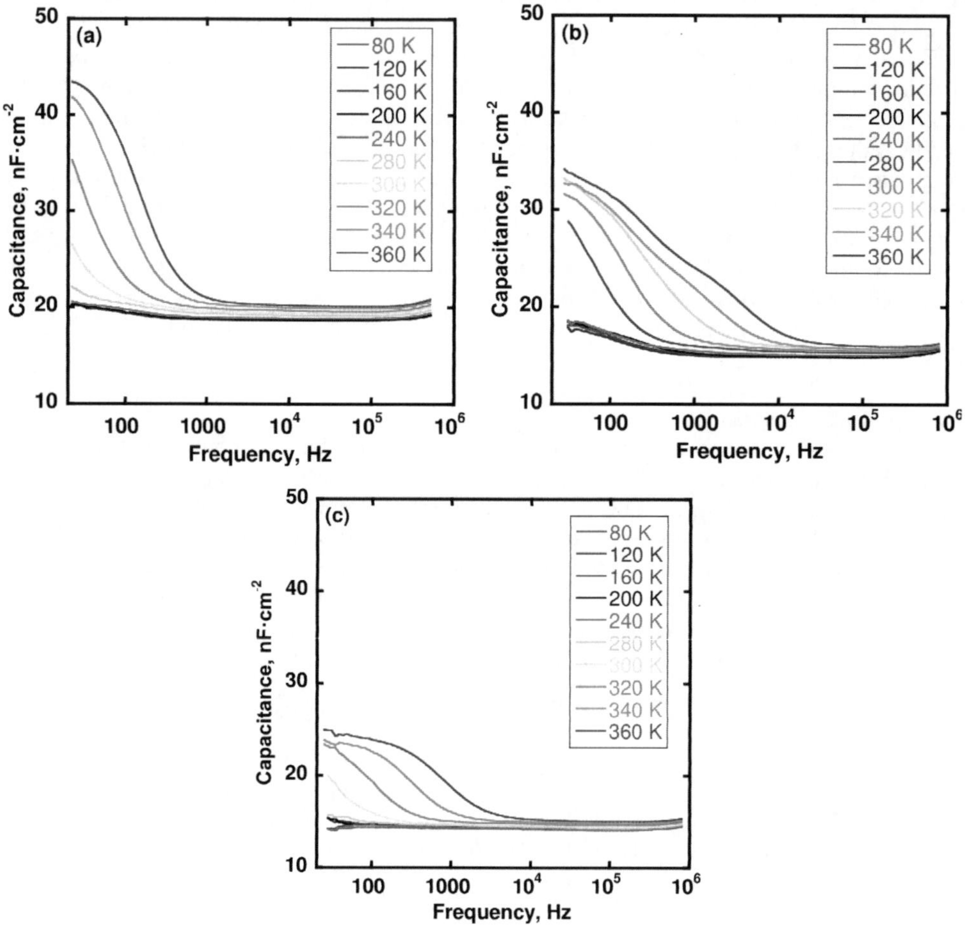

Figure 1. Capacitance spectrum C(T, *f*) for a- GaPAs, b- GaPN/InP, c- GaPNAs.

Sample # 1 with an active layer of GaPAs exhibit a pronounced step in the capacitance curves at high temperatures 320-360 K (Figure 1a). The Arrhenius plots of f/T^2 (ω being the turn-on frequency at the capacitance step or maximum of the related conductance peak) are used to derive the activation energy and capture cross section (from the extrapolation to infinite temperature) of the defect responsible for the capacitance step. The Arrhenius plots are shown in Figure 2 for all structures. For GaPAs layer only one defect level with activation energy of 0.30 eV and low capture cross-section $\sigma=3.8\cdot10^{-19}cm^2$ was observed. For the sample #2 C(T, *f*) curve (Figure 1b) is not monotonic meaning overlap of few defect responses for GaPN/InP layer. Detailed analysis of the C(T, *f*) derivative and G(T, *f*) curves (not presented) has allowed us to distinguish two capacitance steps (conductance peaks) that correspond to the response from two different defect levels, which are presented in Figure 2. While exact determination of the activation energy and capture cross section is difficult for this structure, the first defect level seems to have the similar parameters to the one observed for GaPAs sample. The second defect level with $E_a=0.44$ eV and $\sigma=1.3\cdot10^{-16}cm^2$ has similar parameters to the sample#3 (GaPNAs), which according to Figure 1c exhibits only one defect response with $E_a=0.47$ eV and $\sigma=8.8\cdot10^{-17}cm^2$.

The last defect with activation energy of 0.44-0.47 eV was detected only in GaPN/InP and GaPNAs layers, while it was not observed for GaPAs layer. It means that nitrogen plays key role in defect origin. Also, we suppose it is response from the defect B2 which we detected in GaPNAs in our previous work

[12]. According to the literature, the similar defect was described in Ref. [9] and [10] for GaP doped with nitrogen where it was associated with formation of pair N_P-N_P and V_{Ga}.

The second defect in GaPN/InP (Figure 2b) with lower activation energy was not detected in GaPNAs. Moreover the total defect concentration is higher for GaPN/InP compared to GaPNAs. This fact is in the agreement with the photoluminescence measurements performed in Ref. [15] where indium incorporation into III-V-N was demonstrated to lead to higher non-radiative recombination losses compared to As contain III-V-N alloys. Indium being a III-group element is less efficient to compensate elastic stresses in sub-lattice of V-group compared to arsenic in phosphorus sites (both being V-group elements). We also experimentally demonstrate lower external quantum efficiency for p-i-n structures based on GaPN/InP active layer compared to GaPNAs (Figure 3) indicating lower charge carrier lifetime in GaPN/InP. However we suggest that this defect (0.44-0.47 eV) is not single and has energy distribution due to varying composition of alloy in local area. Thus it is very complicated to obtain its exact parameters values.

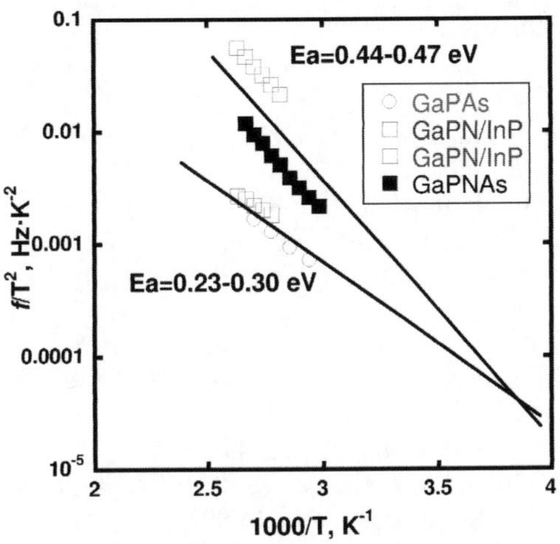

Figure 2. Arrhenius plot of f/T^2 for the defects identified by admittance spectroscopy for the GaPAs, GaPN/InP and GaPNAs structures.

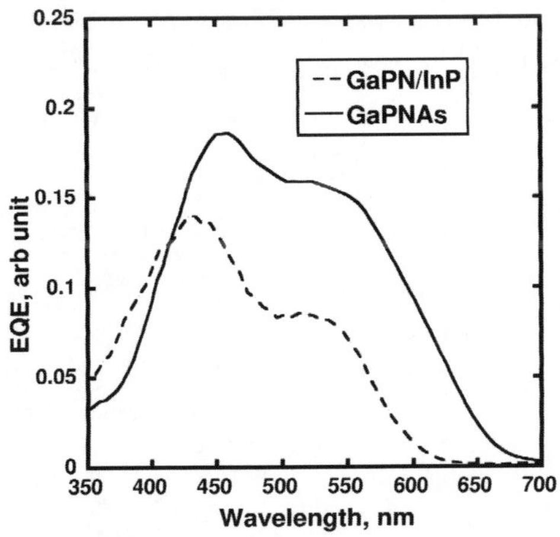

Figure 3. External quantum efficiency of p-i-n structures with GaPN/InP and GaPNAs i-layer.

Finally, only one defect level with activation energy of 0.30 eV was detected in GaPAs layers (Figure 1a). The defect concentration of GaPAs layer is higher compared to GaPN/InP and GaPNAs. We suggest, it is the same defect with low activation energy as in GaPN/InP. Further studies will be done applying deep-level transient spectroscopy (DLTS) to get inside in the origin of this defect.

4. Conclusion

Defect properties of Ga(In)P(NAs) layers with different composition were studied by admittance spectroscopy. For nitrogen content layers the defect level with energy of 0.44-0.47 eV, which related to nitrogen incorporation into GaP, was observed. Its concentration is lower for GaPNAs layers compared to GaPN/InP due to better compensation by arsenic than by indium in lattice of GaP. Other defect level with energy of 0.30 eV was detected in GaPAs and GaPN/InP layers. Likely, the both observed defects in GaPAs and GaPN/InP have the same nature. Further studies using DLTS will be carried out.

Acknowledgments
This work was carried out under financial support by Ministry of Science and Education of Russian Federation (RFMEFI61615X0040).

References

[1] Yonezu H 2005 *Dilute Nitride Semiconductors*, ed. M. Henini (Elsevier, Amsterdam) p.451
[2] Shan W et al. 2000 *Appl. Phys. Lett.* **76** 3251
[3] Buyanowa I A et al. 2002 *Appl. Phys. Lett.* **81** 52
[4] Geisz J F and Friedman D J 2002 *Semicond. Sci. Technol.* **17** 769–777
[5] Geisz J F et al. 2005 *Proc. of the 31st IEEE Photovoltaic Specialists Conference PVSC2005* (Orlando, USA) p. 695
[6] Baranov A I, Gudovskikh A S, Nikitina E V and Egorov A Yu 2013 *Tech. Phys. Lett.* **39** 1117
[7] K. Zdansky, J. Zavadil, Nohavica D and Kugleret S 1970 *J. Appl. Phys.* **83** 7678
[8] Kol'tsov G I et al. *Sov. Phys. Semicond.* **24** 494
[9] Tell B and Kuijpers F P J 1978 *J. Appl. Phys.* **49** 5938
[10] Kaminski P, Strupinski W and Roszkiewicz K 1991 *J. Cryst. Growth* **108** 699
[11] Rumyantsev O I et al. 2010 *Semiconductors* **44** 893
[12] Baranov A I, Gudovskikh A S, Zelentsov K S, Nikitina E V and Egorov A Yu 2015 *Semiconductors* **49** 524-528
[13] Dagnelund D et al. 2013 *Appl. Phys. Lett.* **102** 021910
[14] Loose D L 1975 J. Appl. Phys. **46** 2204
[15] Lazarenko A A, Nikitina E V, Pirogov E V, Sobolev M S, and Egorov A Yu 2014 Semiconductors **48** 392

Saint Petersburg OPEN 2016 IOP Publishing
Journal of Physics: Conference Series **741** (2016) 012078 doi:10.1088/1742-6596/741/1/012078

Mechanism of the polarization control in intracavity-contacted VCSEL with rhomboidal oxide current aperture

M A Bobrov[1], N A Maleev[1], S A Blokhin[1], A G Kuzmenkov[2,1], A P Vasil'ev[2,1], A A Blokhin[1], M M Kulagina[1], Yu A Guseva[1], S I Troshkov[1], V M Ustinov[1,2,3]

[1] Ioffe Institute, Politekhnicheskaya 26, St. Petersburg, 194021 Russia
[2]Submicron Heterostructures for Microelectronics Research and Engineering Center of the RAS, St Petersburg, Russia
[3]Peter the Great St. Petersburg Polytechnic University, St Petersburg, Russia
E-mail: *Bobrov.mikh@gmail.com*

Abstract. The possible mechanisms of the polarization control in single-mode intracavity-contacted vertical-cavity surface-emitting lasers (IC-VCSELs) with the rhomboidal selectively-oxidized current aperture were investigated. It was found that the lasing emission polarization of all single-mode VCSELs is fixed along the minor diagonal of the rhomboidal-shape aperture (the $[\bar{1}10]$ direction). Numerical modelling of carrier transport did not reveal any sufficient injection anisotropy in the laser active region, while the transverse optical confinement factors calculated for the fundamental mode with two orthogonal polarizations are identical. Optical loss anisotropy and/or gain anisotropy are the most likely mechanisms of inducing the polarization fixation.

1. Introduction

A GaAs-based vertical-cavity surface-emitting lasers (VCSELs) are attractive low-cost laser source for high-speed data communication systems, optical interconnects and different sensors [1]. New VCSEL applications (compact atomic clock and magnetometer) require single-mode temperature-stable operation with fixed linear polarization [2]. While the conventional VCSELs with small oxide aperture diameters demonstrate stable fundamental transverse mode operation, the combination of cylindrical symmetry and isotropic gain leads to polarization uncertainty and instability [3]. Despite a number of efforts (polarization-dependent gain, asymmetric resonators, polarization-dependent mirrors, external optical feedback) this issue is still under investigation [4-7].
In this work we discuss possible mechanisms for polarization control in single-mode VCSELs containing rhomboidal selectively-oxidized current aperture and intracavity contacts (IC).

2. VCSEL design

Since it is challenging to achieve both high conductivity and low loss using fully-doped AlGaAs distributed Bragg reflectors (DBRs) grown by molecular beam epitaxy (MBE), VCSEL design with two relatively high-doped contact layers sandwiched between undoped DBRs and optical microcavity was widely used [8]. To improve the conventional short-wavelength IC-VCSEL design we propose to embed a few pairs of doped $Al_{0.15}Ga_{0.85}As/Al_{0.9}Ga_{0.1}As$ DBR (so called composite lattice) between InGaAlAs-based active region and the buried AlGaAs contact layers to decrease the optical field amplitude in high-doped layers, to employ a top dielectric SiO_2/TiO_2 DBR for better optical confinement and to insert p^+-GaAs contact layer directly between the electrical p-contact and the p-

$Al_{0.15}Ga_{0.85}As$ contact layer to decrease contact resistance and free carrier absorption [9]. Selective wet oxidation of thin AlAs/AlGaAs aperture layer embedded into the p-doped composite lattice was used to create lateral confinement for both the injected current and the optical field. Device fabrication steps include mesa etching, selective oxidation, p- and n-contact formation, top dielectric DBR deposited after local window opening in p^+-GaAs contact layer, planarization and contact pads deposition. The details of the VCSEL epitaxial structure and processing can be found in [9]. The schematic cross-section of the processed 850 nm IC-VCSEL is shown in figure 1.

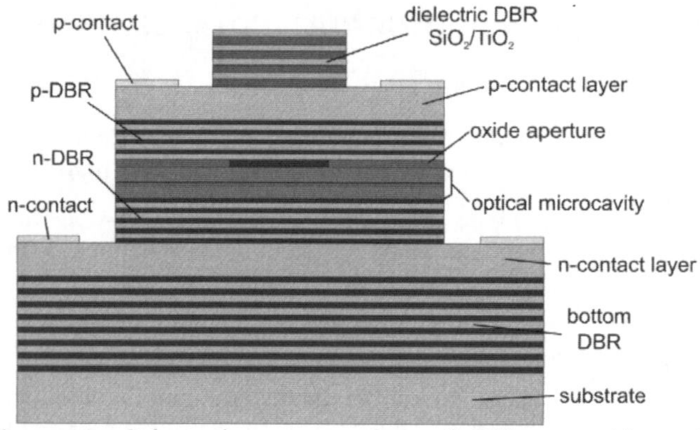

Figure 1. Schematic cross-section of 850 nm oxide-confined VCSEL with dielectric DBR and intra-cavity contacts.

3. Results and discussion

Figure 2 shows the light output power-current characteristics and the typical lasing spectra at 20°C and 80°C for the InGaAlAs IC-VCSELs with aperture size of about 2.5 μm. The devices demonstrate continuous-wave single-mode lasing at 845-852 nm with side-mode suppression ratio (SMSR) higher than 30 dB with output power more than 1.5 mW, threshold current less than 0.9 mA and orthogonal polarization suppression ratio (OPSR) higher than 20 dB for operation temperature of 20-80°C. No polarization switching was observed through the entire current and temperature range. According to the near field studies (see inset in Figure 2.b), the IC-VCSELs with current aperture size less than 10 μm have a rhomboidal shape of oxide aperture with the major diagonal along the [110] direction, which is caused by crystallographic sensitivity of oxidation speed of AlAs layers [10]. Moreover the emission polarization is always fixed along the minor diagonal of the rhomboidal-shape aperture (along the $[\bar{1}10]$ direction).

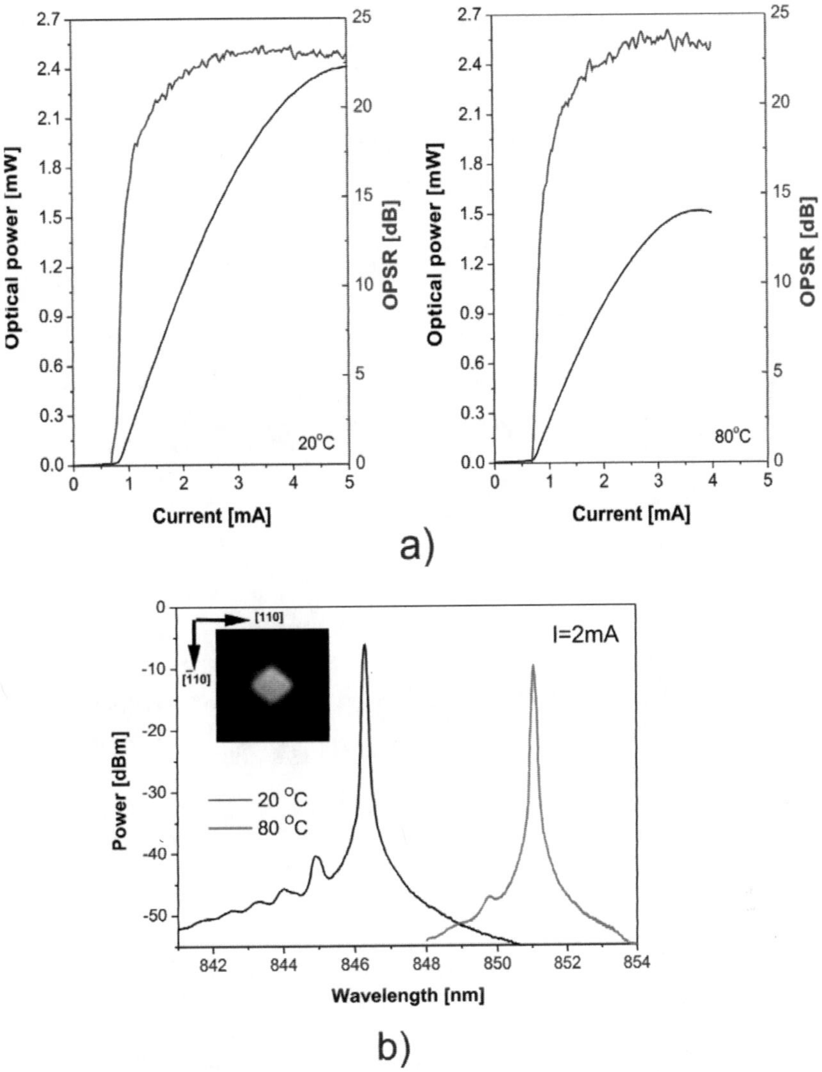

Figure 2. 850 nm IC-VCSEL with aperture size of about 2.5 μm: (a) Output power and OPSR as function of current at different temperatures; (b) Lasing spectra at 2 mA for different temperatures. Inset: near-field image below threshold current.

The observed polarization behavior can be attributed to the electro-optic effect [11], which caused a refractive index ellipsoid with [110] and [$\bar{1}$10] as the principal axes, however this effect is inherently unstable (rotation or switching of the polarization with current or temperature). To clarify the possible mechanism of the strong polarization stability, the simplified numerical modeling of optical mode and carrier distributions were done.

The 2D microcavity optical field was calculated using effective index method [12]. Figure 3.a shows the refractive index and the corresponding electric field intensity profile for the investigated 850 nm IC-VCSEL. For the rhomboidal microcavity the fixation of the longitudinal component of electric field E_z along one of two possible directions (along the major or minor diagonals) was observed (see

Figure 3.b). According to the calculated distribution of the transverse component of electric field E_{xy}, the transverse optical confinement factor for two orthogonal polarizations of the fundamental mode differs less than 3%, which is too small for the stable polarization control.

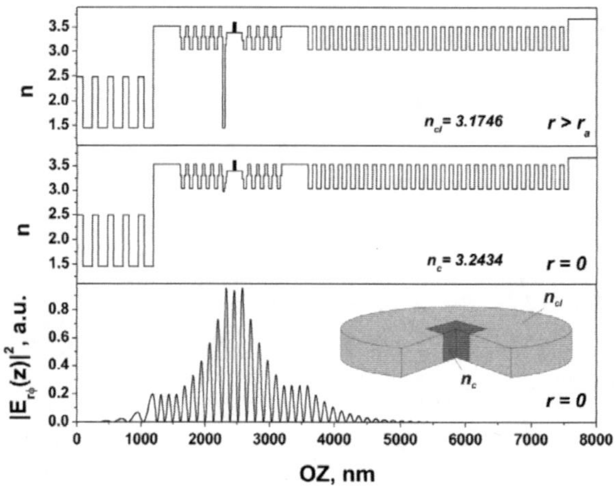

a)

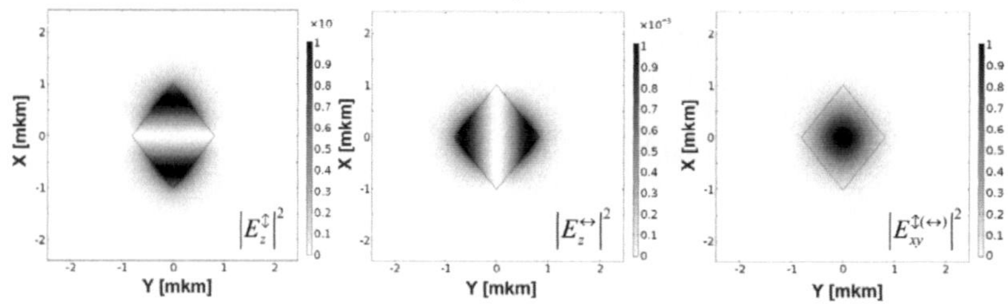

b)

Figure 3. Optical calculation for 850 nm IC-VCSEL: (a) Refractive index profile and normalized electric field intensity versus distance from the top IC-VCSEL surface. Inset: the schematic effective waveguide with the rhomboidal selectively-oxidized current aperture; (b) The fundamental mode distribution of longitudinal (E_z) and transverse (E_{xy}) components of electric field for two possible polarizations.

To perform 3D carrier transport numerical simulation, the composite lattices were replaced by the single layer with the anisotropic carrier mobility (see inset in Fig.4.a). The carrier spatial distributions below lasing threshold for the simplified VCSEL structure revealed the carrier concentration near the rhombus acute angles for multimode devices with aperture size more than 5 μm (see inset in Figure 4.b). Above lasing threshold the anisotropic carrier distribution on the oxide aperture area can be enhanced due to the spatial hole-burning effect. However, the carrier distributions for the aperture size less than 5 μm are more uniform into the aperture and, hence, the observed polarization control cannot be explained by the injection anisotropy in the laser active region.

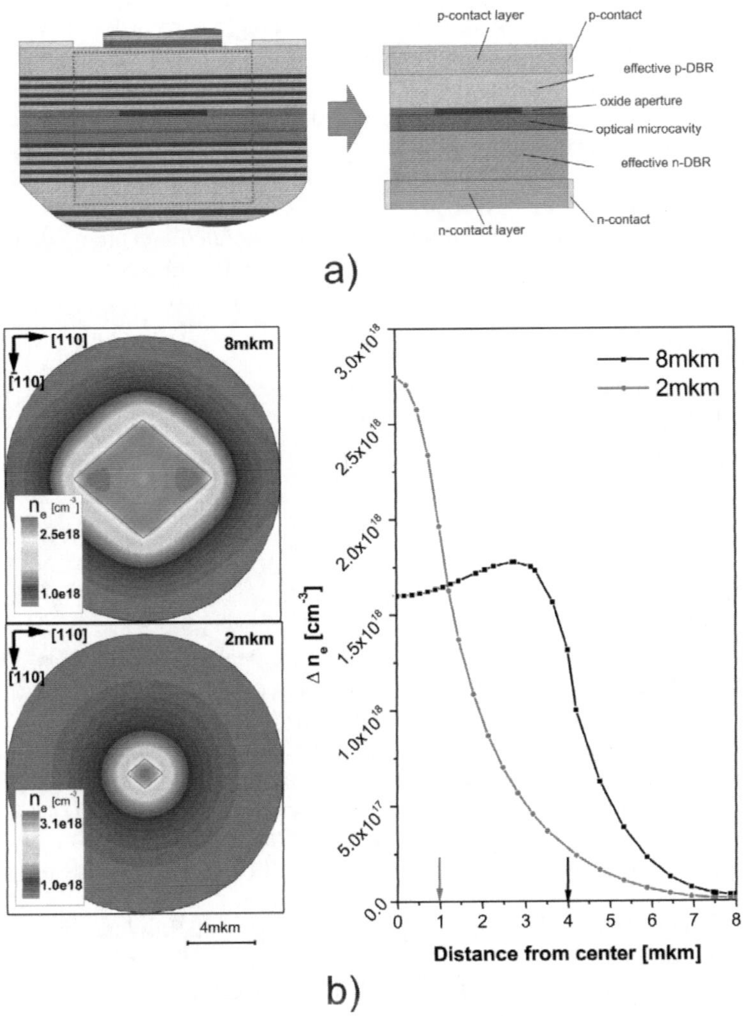

Figure 4. Carrier calculation for 850 nm IC-VCSEL: (a) Scheme of the simplified IC-VCSEL with the rhomboidal-shape current aperture; (b) Carrier distributions and corresponding cross-section profiles for two aperture size about 2 μm and 8 μm.

We attribute the polarization fixation to the polarization-dependent scattering losses inside the asymmetric VCSEL cavity, for example, the light diffraction on the rhomboidal-shape oxide aperture. On the other hand, the spontaneous emission of such IC-VCSELs is partially polarized along the $[\bar{1}10]$ direction, which can be associated with the gain anisotropy of InGaAlAs-based active region. More detailed modelling based on the real design of the optical microcavity and/or the investigation of the spontaneous emission via the oxide aperture shape for IC-VCSELs is necessary to clarify the origin for the polarization fixation along the $[\bar{1}10]$ direction.

4. Conclusion

The alternative approach for single-mode polarization-stable VCSELs based on rhomboidal selectively-oxidized current aperture combined with intracavity contacts was discussed. The single-mode devices have output power more than 1.5 mW and threshold current less than 0.9 mA into the temperature range of 20-80°C. The lasing emission is polarized along the minor diagonal of the rhomboidal selectively-oxidized current aperture (oriented along the $[\bar{1}10]$ direction). Optical simulation of the effective waveguide with the rhomboidal-shape core results in the similar transverse optical confinement factor for two orthogonal polarizations of the fundamental mode. On the other hand, the carrier transport numerical simulation of single-mode devices did not reveal the significant anisotropic carrier distribution on the oxide aperture area. The anisotropy of optical losses in the rhomboidal microcavity and/or the gain anisotropy in the strained quantum wells seems to be responsible for the observed polarization fixation along the $[\bar{1}10]$ direction.

Acknowledgments
This work was supported in part by the Ministry of Education and Science of the Russian Federation (grant № 14.607.21.0001, unique ID RFMEFI60714X0001).

References
[1] R. Michalzik, VCSELs: Fundamentals, Technology and Applications of Vertical-Cavity Surface-Emitting Lasers, Berlin: Springer-Verlag, 2013.
[2] D.K. Serkland, G.M. Peake, K.M. Geib, R. Lutwak, R.M. Garvey, M. Varghese, M. Mescher, "VCSELs for atomic clocks", *Proc. SPIE* **6132** 613208, 2006.
[3] G. Verschaffelt, K. Panajotov, J. Albert, B. Nagler, M. Peelers, J. Danckaert, I. Veretennicoff and H. Thienpont, "Polarisation switching in vertical-cavity surface-emitting lasers: From experimental observations to applications," *Opto-Electron. Rev.* **9** 257–268, 2001.
[4] N. Nishiyama, M. Arai, S. Shinada, M. Azuchi, T. Miyamato, F. Koyama, K. Iga, "Highly strained GaInAs/GaAs quantum well vertical-cavity surface-emitting laser on GaAs (311)B substrate for stable polarization operation, " *IEEE J. Select. Topics Quantum Electron.* **7** 242, 2001.
[5] Y. Sato, K. Furuta, T. Katayama, H. Kawaguchi, "Polarization switching in vertical-cavity surface-emitting lasers by asymmetrical current injection," *IEEE Photon. Technol. Lett.* 20, 1446, 2008.
[6] S. Riyopoulos and E. Nhan, "Polarization selection due to scattering losses in nonaxisymmetric vertical cavity surface emitting laser cavities," *Appl. Phys. Lett.* **85** 3038, 2004.
[7] J.M. Ostermann, P. Debernardi, R. Michalzik, "Optimized integrated surface grating design for polarization-stable VCSELs," *IEEE J. Quantum Electron.* **42** 690, 2006.
[8] D.L. Huffaker, D.G. Deppe, "Intracavity contacts for low-threshold oxide-confined vertical-cavity surface-emitting lasers," *IEEE Photon. Technol. Lett.* **11** 934–936, 1999.
[9] N.A. Maleev, A.G. Kuz'menkov, M.M. Kulagina, Yu.M. Zadiranov, A.P. Vasil'ev, S.A. Blokhin, A.S. Shulenkov, S.I. Troshkov, A.G. Gladyshev, A.M. Nadtochiy, M.M. Pavlov, M.A. Bobrov, D.E. Nazaruk, V.M. Ustinov, "Single-spatial-mode semiconductor VCSELs with a nonplanar upper dielectric DBR," *Semiconductors* **47** 993, 2013.
[10] K.D. Choquette, K.M. Geib, C.I.H. Ashby, R.D. Twesten, O. Blum, H.Q. Hou, D.M. Follstaedt, B.E. Hammons, D. Mathes, R. Hull, "Advances in selective wet oxidation of AlGaAs alloys," *IEEE J. Select. Topics Quantum Electron.* **3** 916, 1997.
[11] M.P. Van Exter, A.K. Jansen van Doorn and J.P. Woerdman, "Electrooptic effect and birefringence in semiconductor vertical-cavity lasers," *Phys. Rev. A, Gen. Phys.* **56** 845, 1999.
[12] G.R. Hadley, "Effective index model for vertical-cavity surface-emitting lasers", *Opt. Lett.* **20** 1483, 1995.

Model of mode-locked quantum-well semiconductor laser based on InGaAs/InGaAlAs/InP heterostructure

D A Rybalko[1], I S Polukhin[1], Y V Solov'ev[1], G A Mikhailovskiy[1], M A Odnoblyudov[1], A.E. Gubenko[1], D.A. Livshits[1], A.N. Firsov[1], A.N. Kirsyaev[1], A.A. Efremov[1], V E Bougrov[2]
[1]Peter the Great St. Petersburg Polytechnic University, Saint Petersburg 195251, Russia
[2]ITMO University, Saint Petersburg 197101, Russia

Abstract. We propose a model for operation of mode-locked (ML) quantum-well semiconductor laser consisting of a reverse biased saturable absorber and a forward biased amplifying section. To describe the dynamics of this laser we use the traveling wave model. Numerical simulations performed for the InGaAs/InGaAlAs laser structure emitting at 1,55 um.

1. Introduction
ML semiconductor lasers could play a significant role as microwave photonic components at telecom wavelengths [1,2]. Such devices can be used for determination of a sampling rate in microwave photonic analog-to-digital converters [3]. Passive ML lasers based on semiconductors materials have many benefits in comparison with liquid organic dyes and solid crystal passive ML lasers, such a small size, a low power consumption and an ability to plug in a monolith integrated circuit. Today the passive mode-locked lasers are successfully deployed in quantum dots and wells for 1.3-1.55 μm wave length range on GaAs and InP substrates [4-6]. In this work, we consider a model of a two-section quantum-well ML laser consisting of a reverse biased saturable absorber and a forward biased amplifying (gain) section. We perform a comprehensive numerical study of the operation regimes of this laser and estimate main characteristics of ML pulses such as the repetition rate, width and shape of the pulse depending on parameters of laser structure. One of the most important parts of the ML model is a function of the spontaneous noise, which affects the initial emission, width of the pulse and the time jitter. In our model, we propose to use the Rayleigh distribution to describe the spontaneous processes. The numerical calculations performed for the novel laser structure emitting at 1.55 um wavelength based on InGaAs/InGaAlAs quantum well structure grown on InP substrate.

2. Model
This model allows to define the main laser specifications, such as pulse repetition rate, pulse shape/width and power. Passive mode-locking models can be divided into two groups as follows: electromagnetic field equation depending on time and electromagnetic field equation depending on frequency [7-9]. Models with frequency-based equations can be advantageous in terms of time for numerical calculations [7]. In this model, we consider the spatio-temporal evolution of the two counter-propagating optical fields $E^+(z,t)$ and $E^-(z,t)$ and carrier density $n(z,t)$. After fixing wavelength λ_0, the field in the laser cavity can be represented as a superposition of two slowly varying complex amplitudes ($\left|\frac{1}{E^\pm}\frac{\partial E^\pm}{\partial z}\right| \ll k_0$, $\left|\frac{1}{E^\pm}\frac{\partial E^\pm}{\partial z}\right| \ll \omega_0$) $E^+(z,t)$ and $E^-(z,t)$ of forward and backward propagating waves. Here, t denotes time and z is the coordinate along the cavity axis ($0<z<L$), k_0 is wave vector, ω_0 is carrier wave frequency. The amplitudes $E^\pm(z, t)$ of the two waves counter-propagating in the cavity satisfy the traveling wave equations:

$$\frac{\partial E^+}{\partial t} + v_g \frac{\partial E^+}{\partial z} = v_g \frac{\Gamma\sigma(n-n_t)}{2}(1 - i\alpha_H)E^+ + \Gamma \frac{v_g \sigma n_t}{2\omega_g{}^2} - v_g \frac{\alpha_i}{2} E^+ + F_{sp}^+ , \quad (1)$$

$$\frac{\partial E^-}{\partial t} - v_g \frac{\partial E^-}{\partial z} = v_g \frac{\Gamma\sigma(n-n_t)}{2}(1 - i\alpha_H)E^- + \Gamma \frac{v_g \sigma n_t}{2\omega_g{}^2} - v_g \frac{\alpha_i}{2} E^- + F_{sp}^- , \quad (2)$$

where v_g is group velocity, σ is differential amplification in the gain section or absorption in the saturable absorber, n is the carrier concentration in the quantum-well ground state, n_t is transparency concentration, α_H is the Henry-factor, α_i is the internal loss, Γ is the optical confinement factor, ω_g is the frequency dispersion coefficient, F_{sp} is the spontaneous emission noise. The measurement unit of the field amplitude is chosen so that the sum $|E^+|^2 + |E^-|^2$ corresponds to the local photons concentration. The rate equation in this system is

$$\frac{\partial n}{\partial t} = \frac{I(z)}{eV} - R - v_g \sigma(n - n_t)(|E^+|^2 + |E^-|^2) , \quad (3)$$

where e is the elementary charge, $I(z)$ is the injection current. $I(z)=0$ in the saturable absorber and $I(z)=I_G$ in the gain section. V is volume of active area, R described all recombination mechanisms in structure. For absorber section we consider R as:

(4)

where D is the absorber saturation rate. For amplifier section R was:

(5)

where A is the nonradiative recombination coefficient, B is the radiative recombination coefficient, C is the Auger recombination coefficient. Boundary and initial condition for (1)-(3):

$$E^+(0,t) = r_L E^-(0,t); E^-(L,t) = r_0 E^+(L,t);$$
$$E^+(z,0) = E^-(z,0); \frac{\partial E^{\pm}(z,0)}{\partial t} = 0; \quad (6)$$
$$n(z,0) = n_t ,$$

where $r_{0,L}$ are factors describing a reflectivity of the laser facets. Equations (1)-(3) with the initial and boundary conditions (6) make a closed equation system that enables to determine repetition rate, as well as width and shape of the pulse. The structure considered in this model consisted of five InGaAs/InGaAlAs quantum wells on InP substrate. The structure parameters present in table 1.

Table 1. Simulated structure parameters

No.	Parameter name	Symbol	Value	Unit
1	Group velocity	v_g	0,91 10^{10}	cm/s
2	Resonator overall length	L	0.455	cm
3	Absorber length	l_a	0.0455	cm
4	Internal loss rate	α_i	5	cm^{-1}
5	Optical confinement factor	Γ	0.025	
6	Differential gain	σ_g	4 10^{-16}	cm^2
7	Differential absorption	σ_a	2 10^{-15}	cm^2
8	Henry-factor [10]	α_H	2	
9	Transparency threshold concentration	n_t	10^{18}	cm^{-3}
10	Frequency gain dispersion rate	ω_g	7.5 10^{13}	s^{-1}
11	Nonradiative recombination coefficient [11]	A	2 10^8	s^{-1}
12	Radiative recombination coefficient [11]	B	0.96 10^{-10}	cm^3/s

13	Auger recombination coefficient [11]	C	7 10^{-29}	cm^6/s
14	Absorber relaxation rate	D	10^{11}	s^{-1}
15	Active region volume	V	1.7 10^{-10}	cm^{-3}
16	Waveguide spontaneous emission contribution [12]	β_{sp}	10^{-4}	
17	Amplitude reflection coefficient at 0 boundary	r_0	1	
18	Aamplitude reflection coefficient at L-boundary	r_L	0.9	
19	Pumping current	I_G	300-900	mA

The output power can be calculated using the following formula:

$$Power[W] = \hbar w_0 \left|E^+\right|^2 v_g \frac{V}{L \cdot \Gamma}(1 - R_L), \quad (7)$$

where $\hbar$ – the Planck constant, R_L – reflection factor of one of the mirrors.

3. Experimental Results and Discussion

The dynamics of the transition to the operation regime for a simulated laser was acquired. Transition to the one pulse generation mode is possible for parameters listed in the table and provided the voltage at the absorber is more than 4V. Figure 1 shows the result. At the initial moment of time there is the generation of multiple longitudinal modes in the amplifier area (figure 1a). Their intensity is not sufficient to illuminate the absorber, but they gradually start its saturation. Further, there is mode discrimination when the weakest modes are attenuated and the strong modes start coming to the absorber simultaneously with each other (figure 1b, c). As a result by the moment of time of 30 ns modes are locked and there is only one large pulse left. With voltage at the absorber of less than 2V the situation is slightly different. Thus, by the moment of time of 30 ns, there are two pulses instead of a single one formed in the resonator which are rather powerful to pass through the saturable absorber (figure 2b). The mechanism of ultrafast pulse formation is very sensitive to laser initial conditions. In case of low losses in resonator, the two or more pulses can be obtained. A similar result was shown in the work [12,13].

Journal of Physics: Conference Series **741** (2016) 012079 doi:10.1088/1742-6596/741/1/012079

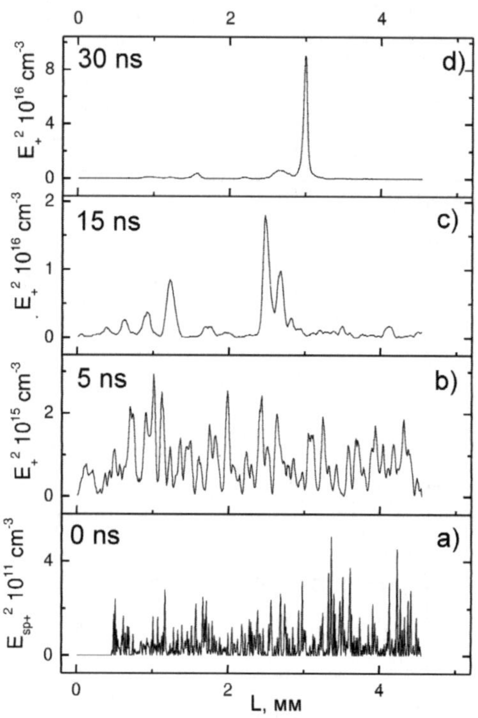

Figure 1. Distribution of radiation intensity along the laser resonator in second harmonic mode-locking.

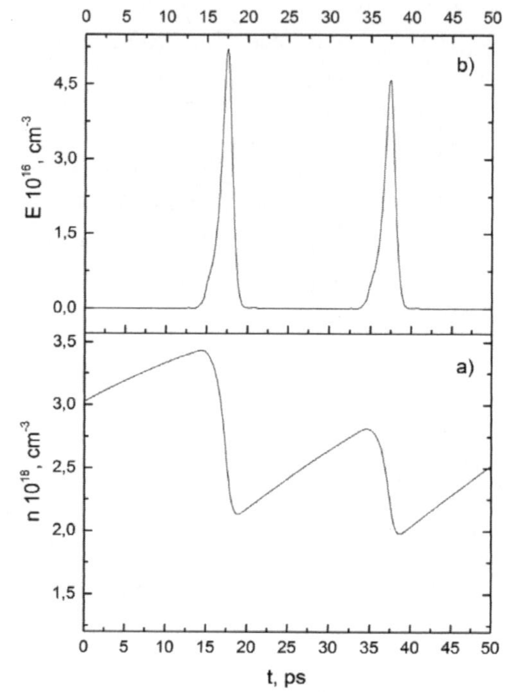

Figure 2. Changes in concentration of electrons and photons in the resonator depending on a time

The model provides an opportunity to track the changes in the concentration of both photons and electrons. Figure 2a shows the dependence of the electrons concentration on a coordinate. One can see that maximum concentration is about to coincide with the pulse shown in the figure 2b. After the pulse has passed, the electron concentration decreases which corresponds to the radiating interband transitions of electrons.

We have also investigated the influence of absorber section length on the formation of ultra-short pulses. Figure 3 shows the results. With reduction of the absorber length absorption in the resonator decreases and spurious pulses occur. Application of a laser with such radiation regime poses some difficulties: pulses have relatively small power, their position in time regarding one another may vary. Accordingly, generation of spurious pulses can be regarded as a negative process while the construction with such kind of operation is deemed as non-optimal one. Increase of absorber length is accompanied by increased losses in the resonator. This fact adds up to the mode discrimination and allows to cut off spurious modes. As a result a single pulse is formed (figure 3d, e). Optimum is found with absorber length equal to 14% of resonator length (637 μm). With further increase in absorber length no transition to mode-locking occurs since dramatically increased losses do not allow any mode amplification. One should note that these results are valid for the system with specific parameters, the optimal length of the absorbing section will be different provided the different voltage value at the absorber, as well as gain and absorption value of the medium.

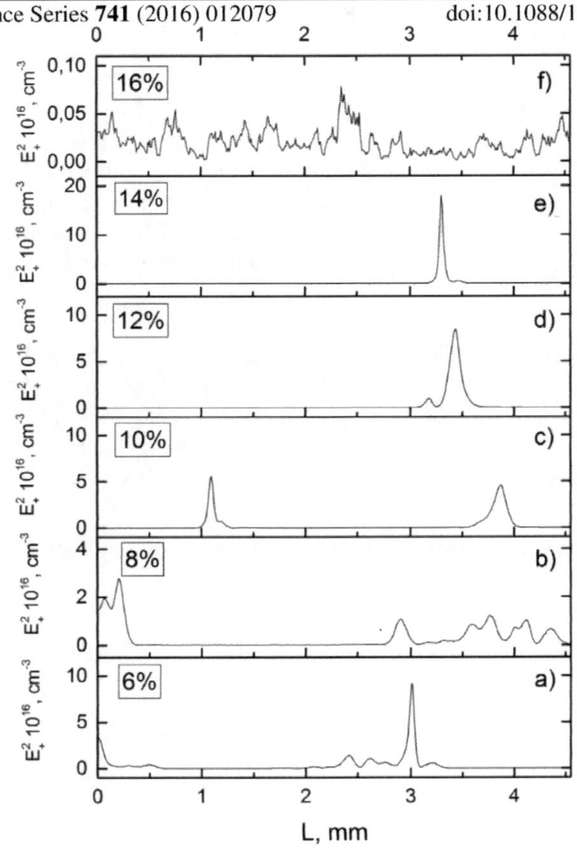

Figure 3. Intensity distribution in the resonator for different lengths of the absorber.

4. Conclusion

Represented model allows to perform numerical simulation of processes in the resonator of the semiconductor laser on quantum structures with passive mode-locking. We have received pulses of 2 ps long and 10 HHz repetition frequency for a structure with four InGaAs/InGaAlAs quantum wells and the resonator length equal to 4.55 mm. Power of the received pulses was 9.4 mW. Varying the length of the absorber has shown that the simulation predicts the possibility of optimizing laser geometric dimensions for the most effective transition to the passive mode locking. In this case the optimal length was equal to 637 μm. Further we plan to include the refractive index in the dispersion model, the effect of gain saturation and nonlinear coherent effects.

Acknowledgments

This work supported by the Ministry of Education and Science of the Russian Federation as a part of federal purpose-oriented program "Researches and development works on priority direction of Russian scientific and technological complex for 2014-2020 years", code number 2015-14-579-0015, agreement № 14.578.21.0100, unique identifier RFMEFI57814X0100.

Journal of Physics: Conference Series **741** (2016) 012079 doi:10.1088/1742-6596/741/1/012079

References

[1] G.C. Valley, Optics Express, **15** (2007), p. 1955.
[2] A. Khilo, S.J. Spector, E. Matthew, et. al., Optics Express, **20** (2012), p. 4454
[3] R.S. Staricov, Uspekhi sovremennoi radioelektroniki (in Russian). **2** (2015), p. 3.
[4] K. Merghem, A. Akrout, A. Martinez et al., Optics Express, **16** 14 (2008) p. 10675
[5] Xiaodong Huang, A. Stintz, Hua Li, Appl. Phys. Lett. **78**, 2825 (2001) p. 2825.
[6] R. Rosales, K. Merghem, A. Martinez, IEEE Journal of Selected Topics in Quantum Electronics, **17**, 5, (2011), p. 1292
[7] E.A Avrutin, J.H. Marsh, E.L. Portnoi, Proceedings-Optoelectronics, **147** (2000), p. 251.
[8] M. Radziunas, A. G. Vladimirov, E.A. Viktorov et al., J. of Quant. El., **47** (2011), p. 935.
[9] A. G. Vladimirov, D. Turaev, Physical Review A **72** (2005), p. 033808.
[10] G.A. Mikhailovskiy, I.S. Polukhin, D.A. Rybalko et. al., Tech. Ph. Lett., vol **42** (2016)
[11] A. Fali, E. Rajaei, and Z. Kaftroudi, J. of the Korean Physical Society, **64** (2014), p. 16.
[12] M. Radziunas, A. G. Vladimirov and E. A. Viktorov, Proceedings of SPIE, **7720** (2010), p. 77200X-1.
[13] A. G. Vladimirov, D. Turaev, Physical Review A **72**, 033808 2005

Transport level in disordered organics: correlated energetic disorder in dipole glass model

V R Nikitenko and V M Sukharev
National Research Nuclear University MEPhI (Moscow Engineering Physics Institute), Moscow, 115409, Russia

E-mail: VMSukharev@mephi.ru

Abstract. In this paper Monte-Carlo modeling is in use in order to further approve and quantify the concept of effective transport level in the respect to organic materials with correlated disorder. We consider a model of dipole glass (simple cubic lattice, which sites are occupied by randomly oriented dipoles). Both the absolute values and dependence of the effective transport level on the disorder and temperature, in the limit of low electric field, is very similar to the same in the case of uncorrelated disorder

1. Introduction

Organic optical-electronics devices, such as light-emitting diodes, photovoltaic cells, consist of thin active layers. Predictive modeling of operation of such devices unavoidably needs correct characterization of charge transport in organic layers. The concept of transport level [1] (or transport energy) is known to be a useful tool for analytic modeling, providing uncorrelated Gaussain disorder [2], because it can simplify the description of hopping transport greatly. By the use of transport level concept, which is an analog of mobility edge, mobility of charge carriers can be calculated in analogy with multiple trapping model.

Transport level is neither an energy of final state after the first jump from a deep state nor a most visited energy in course of transport. So one needs to consider random walk of a carrier inside a region of some radius around initial state [1]. Model of dipole glass is the simplest model of disordered organic materials that have polar molecules and where energetic disorder is spatially correlated [2]. Previously, an applicability of effective transport level concept [1] to disordered organics with correlated disorder was shown qualitatively by Monte-Carlo modelling in the framework of the dipole glass model [3]. The purpose of this paper is to test this applicability again by the means of more effective algorithm, quantify the temperature dependence of the effective transport level and compare it with analogous results, which were obtained previously [1] in the framework of uncorrelated Gaussian disorder model [2].

2. Monte-Carlo procedure and scheme of modeling

In MC simulations we use the same approach to the modeling of random walk in disorder media [4], that was applied in our previous works [[5], [6]]. Hopping centers are located at sites of a simple cubic lattice, all sites occupied by dipoles. The whole volume is divided by 3 sub-volumes, see Figure. 1a. A carrier allowed to walk inside the first (the inner) sub volume; energies are calculated also for any state of the second sub volume, which is necessary to evaluate probabilities of hopping at the outer boundary of the first sub volume correctly. The 3th (rather large) sub volume is necessary, in order to provide correct (isotropic) energy distribution of energies of any state inside the 1st and 2nd sub-volumes.

One can calculate the energy of a carrier, which is initially located in a center of a volume (at the "initial" site), as a sum of charge-dipole interaction energies,

$$E_{central} = \frac{e}{\kappa}\sum_m \frac{\left(\vec{r}_m, \vec{p}_m\right)}{r_m^3} = E_0 \sum_{i,j,k} \frac{i\sin\theta^{ijk}\cos\varphi^{ijk} + j\sin\theta^{ijk}\sin\varphi^{ijk} + k\cos\theta^{ijk}}{\left(i^2+j^2+k^2\right)} \quad (1)$$

$$E_0 = \frac{ep}{\kappa a_0^2}, \kappa = 4\pi\varepsilon_0\varepsilon, \tag{2}$$

p - the dipole moment, ε – dielectric constant of the medium; i, j, k - the specified coordinates of the cubic lattice sites (arbitrary integers) and a_0 – the cubic lattice constant (distance between the nearest lattice points), θ, φ – angles that define the spatial orientation of the dipoles at lattice sites. Dipole orientation in space is defined by the random distribution in space (see Figure. 1a).

In analogy, one can calculate energy of a carrier in any site of the first and second sub volumes. Miller–Abrahams hopping rates from site i to site j are presumed [3]:

$$\omega_{i\to j} = \omega_0 \exp\left(-2\gamma r_{ij} - \frac{E_j - E_i + |E_j - E_i|}{2kT}\right), \tag{3}$$

where ω_0 is the attempt-to-escape frequency, $1/\gamma$ is the decay length of the localized wave function, E_i is the energy of the carrier at the site i, r_{ij} is the distance between sites i and j. The values $T = 295$ K, $2\gamma a_0 = 10$ and $\varepsilon = 3$ are used in calculations. Sizes of the sub volumes are the following: $5a_0, 8a_0$ and $15a_0$(by "size" we mean a half of square side of the volume). If the electron appears in the nearest position of the second sub-volume, event is over. The value $15a_0$ seems to be minimal sufficient to provide size-independence of node energies inside the 1st and 2nd sub volumes with appropriate accuracy.

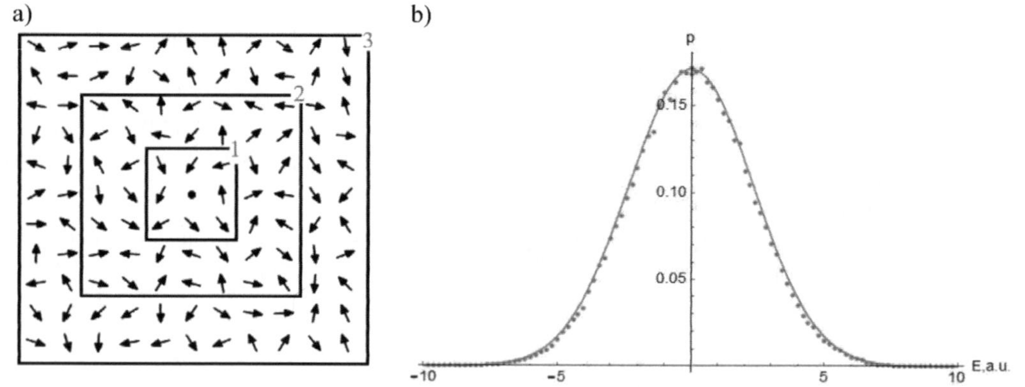

Figureure 1. a) Randomly orientated dipoles in the volume and, as the result, b) Gaussian distribution for the energy of the central node.

In order to fix the energy of the central node, $E_{central}$, we choose the interval from $E_{central} - kT/2$ to $E_{central} + kT/2$. If the energy $E_{central}$ is beyond this interval , the procedure of random dipoles setting starts again, until the energy $E_{central}$ falls into the given interval.

To make sure that the energy of the central node has Gaussian distribution, we plot the distribution of energies of the central node for more than 100 000 events (Figure. 1b, E is given in arbitrary units). It is shown that the distribution is well fitted by Gaussian function, and it's mean-square variation corresponds to well- known expression [2]:

$$\sigma = E_0\sqrt{\frac{1}{3}\sum_{i,j,k}\left(i^2 + j^2 + k^2\right)^{-4}} \approx 2.35E_0. \tag{4}$$

The correlated character of site energies is demonstrated in Figure. 2, where typical distributions of energies of nearest neighbors of an initial (central) nodes are shown for several events. Points with coordinates (r, E) are shown, where r is the distance from the central node to the current node, and E is the energy of the current node. As one can see, there are only several states that are deeper than the central one.

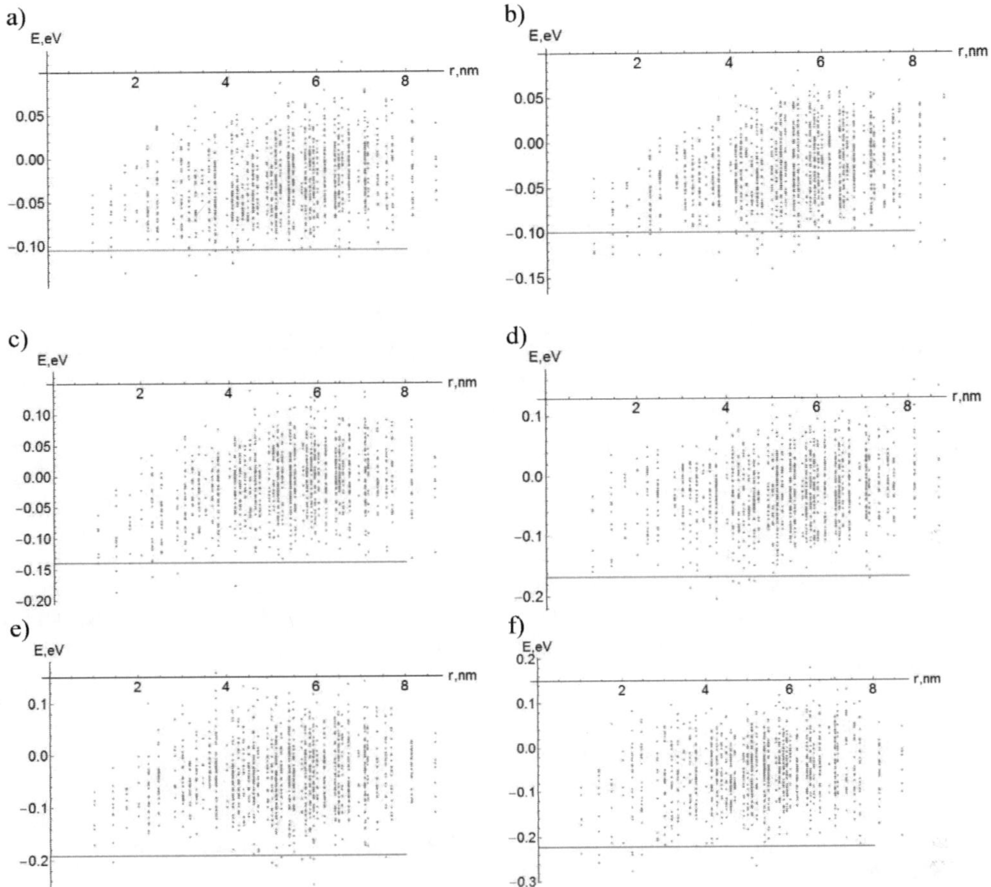

Figureure 2. 2D distribution of a node energy versus distance between this node and the central node that shows correlated nature of disorder of the considered system. Parameters are: $E_0 = \sqrt{2}\sigma = 0.05\ eV, E_0/kT = 2, E_{central} = -2E_0(a), -2.2E_0(b); E_0/kT = 3, E_{central} = -2E_0(c)$, $-2.2E_0$ $(d); E_0/kT = 4,\ E_{central} = -2E_0(e), -2.2E_0(f)$. The solid line is the energy of the central node calculated according to current dipole orientation.

3. Results and Discussion

The following dependence of mean release time of a carrier from a rather deep initial state on the energy of this state, E_i, should be valid, if transport level concept is applicable:

$$< t_{esc} > (E_i) = t_0 \exp\left[(E_{tr} - E_i)/kT\right], \tag{5}$$

where $t_0 = \omega_0^{-1}\exp(2\gamma a_0)$ and the effective transport level E_{tr} are constants [1].

Figure. 3 shows the results for mean escape time vs initial energy that were obtained earlier in [1] for the case of non-correlated disorder, plus brand new results for correlated energetic disorder in dipole glass model at various values of the disorder parameter, σ / kT. One has to conclude that new data are in very proximity with the results that were obtained for non-correlated disorder. However, it is clear, that stars, representing the result for correlated disorder, are located a little bit higher that the results of [1]. So there is the difference, but transport levels for both cases are close enough to each other (within the interval *kT*).

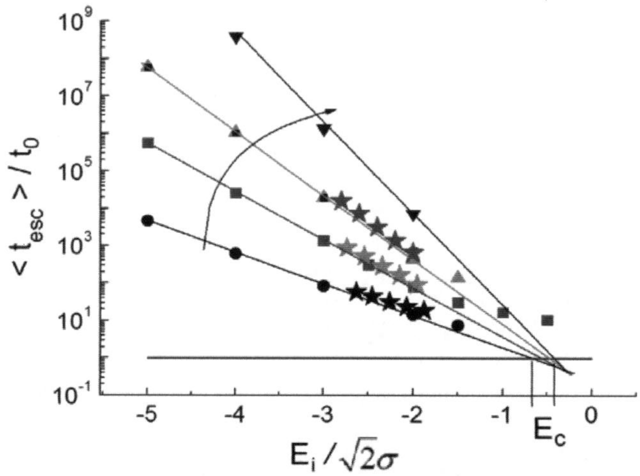

Figureure 3. Mean escape time vs initial energy, parametric in disorder parameter, σ / kT: 1.414, 2.12, 2.83 , 4.0 (it increases as shown by arrow), $T =$ 295K , providing uncorrelated disorder [1]. Stars are brand new results for correlated disorder for $\sigma / kT = 1.414,\ 2.12,\ 2.83$. Time and initial energy are normalized by the values $t_0 = \omega_0^{-1} exp(2\gamma a)$, and $\sqrt{2}\sigma$, respectively. Horizontal line shows the level $\langle t_{esc} \rangle\ = t_0$, and vertical dashed lines shows the values of E_c .

4. Conclusion

Following the work [3], one can conclude, that the concept of effective transport level is applicable to the organic materials with correlated disorder. Moreover, both the absolute values and dependence of this energy level on the disorder parameter, σ/ kT , is very similar to the same in the case of uncorrelated disorder [1].

References

[1] Nikitenko V R, Strikhanov M N, Transport level in disordered organics: An analytic model and Monte-Carlo simulations 2014 *Journal of Applied Physics* **115** (7) 073704

[2] Novikov SV, Dunlap D H, Kenkre V M, Parris P E, and Vannikov AV, Essential role of correlations in governing charge transport in disordered organic materials 1998 *Physical Review Letters* **81** 4472

[3] Porvatkina O V, Nikitenko V R and Strikhanov M N, Analysis of applicability of the transport level concept in organic materials, providing correlated disorder 2014 *Journal of Physics: Conference Series* **541** 012102

[4] Bässler H Charge transport in disordered organic photoconductors 1993 *Phys. Status Solidi B* **175** 15-56

[5] Sukharev V M, Nikitenko V R, Monte-Carlo modeling of drift mobility in thin organic layers: lattice type influence 2015 *Journal of Physics: Conference Series* **643** (1) 012036

[6] Nikitenko V R, Sukharev V M, Sannikova N A, Drift mobility in thin organic layers: joint Monte-Carlo and analytic modeling 2014 *Journal of Physics: Conference Series* **541** 012028

Saint Petersburg OPEN 2016 IOP Publishing
Journal of Physics: Conference Series **741** (2016) 012081 doi:10.1088/1742-6596/741/1/012081

Post-processing procedure for industrial quantum key distribution systems

Evgeny Kiktenko[1,2], **Anton Trushechkin**[1,3], **Yury Kurochkin**[4], **and Aleksey Fedorov**[1,4]

[1]Theoretical Department, DEPHAN, 100A Novaya St., Skolkovo, Moscow 143025, Russia
[2]Bauman Moscow State Technical University, 5 2nd Baumanskaya St., Moscow 105005, Russia
[3]Steklov Mathematical Institute of Russian Academy of Sciences, 8 Gubkina St., Moscow 119991, Russia
[4]Russian Quantum Center, 100A Novaya St., Skolkovo, Moscow 143025, Russia

E-mail: akf@rqc.ru

Abstract. We present algorithmic solutions aimed on post-processing procedure for industrial quantum key distribution systems with hardware sifting. The main steps of the procedure are error correction, parameter estimation, and privacy amplification. Authentication of classical public communication channel is also considered.

1. Introduction

Significant attention to quantum key distribution [1] is related to the fact of breaking of public-key encryption algorithms using quantum computing. Security of public-key exchange schemes can be justified on the basis of the complexity of several mathematical problems. Nevertheless, the Shor's algorithm [2] allows solving these problems in a polynomial time. Absence of efficient classical (non-quantum) algorithms breaking public-key cryptosystems still remains unproved.

In view of possibility to establish a shared private key with unconditional security between two users (Alice and Bob) via quantum key distribution [1], using of information-theoretically secure one-time-pad encryption technique becomes a practical tool. Privacy of quantum keys is guaranteed by the laws of quantum physics [3]. Quantum key distribution has been realized in experiments [5–9]. Devices for quantum key distribution are available on the market [10].

On the top of using of single quantum objects (photons) as information carriers, in quantum key distribution protocols such as the seminal BB84 protocol [4] classical communications and post-processing are required [4]. A key requirement for quantum key distribution is that classical communications are not distorted. In other words, an eavesdropper is able to obtain information from this channel, but cannot to change it. To this end, one can use authentication schemes. Due to the technological limitations of quantum key distribution, keys of Alice and Bob are initially different even in the absence of eavesdropping. Excluding of this effect can be realized by using error correction procedures. On the basis of compassion of keys after error corrections, one can estimate the quantum bit error rate (QBER). If QBER exceeds a critical value, the parties receive warning message about possible eavesdropping. In other cases, they use privacy amplification methods to exclude announced information on previous stages from keys.

In this contribution, we report about the joint research project, which is aimed on a design of an industrial fiber based quantum key distribution system in Russia in collaboration between four teams. The quantum key distribution engine is based on the decoy states BB84 protocol [11]. We present the developed post-processing procedure for sifted quantum keys (*i.e.*, keys after basis and intensity reconciliations), which consists of the following steps: error correction, parameter estimation, and privacy amplification. Communications over public channel are authenticated.

In Sec. 2, we briefly describe the hardware engine of our system. In Sec. 3, we consider post-processing algorithms for error correction, parameter estimation, and privacy amplification. We discuss authentication of public channel in Sec. 4. The workflow of the post-processing procedure is presented in Sec. 5. We summarize our results in Sec. 6.

2. Quantum key distribution setup

Our setup for quantum key distribution consist of two modules on the each side (Alice and Bob). First, control units, that are connected via the optical fiber and perform all operations with photons. Second, conjugation units, that are connected with the classical public channel, perform all post-processing, and export final keys to external applications.

Operation of the control units is based on the "plug and play" [12] principle of realization of the BB84 protocol with decoy states [11]. Control units work as follows. Bob sends a sequence ("train") of strong coherent pulses to Alice. On her side these pulses are (i) reflected by a Faraday mirror, (ii) phase modulated with four possible values according to random basis and bit value, (iii) attenuated to the one of three intensities (vacuum, signal, or decoy), and (iv) sent back. The scheme passively compensates slowly varying polarization fluctuations in the fiber. In the Bob's control unit, the returned attenuated pulses are phase modulated with two possible values that correspond to random choice of basis, and measured by pair of single photon detectors.

After an each "train" of pulses, Bob's conjugation unit obtains time indices of detected signals and corresponding bit and basis values. These time indices and values of basis choices are sent to Alice's conjugation unit via classical channel. Alice compares the measurement bases with preparation bases and sends Bob time indices of signal bits, which have been measured and prepared in the same basis. This procedure is know as sifting. The values of signal qubits that were prepared and measured in the same basis give so-called sifted keys, that we denote as $K_{\rm sift}^{A}$ and $K_{\rm sift}^{B}$. We note that Alice's conjugation unit also obtains rate of detection events for all three intensities that are used for revealing of photon-number splitting attack. However, in this work we restrict ourselves to post-processing without paying attention to analysis of decoy states.

We also note that random choices of bit values and preparation basis in Alice's control unit and random choices of measurement bases in Bob's one are performed using random number generators. Random number generators are important and often forgotten ingredient of quantum key distribution systems [9]. Their development is a part of the project [13]. Random number generator is used several times during the protocol: (i) to generate random photon states, (ii) to choose random measurement bases, (iii) to generate random information specifying the verification hash function (part of error correction) and the privacy amplification procedure.

3. Algorithms for the post-processing procedure

In this section, we consider processing procedures of two sifted quantum keys $K_{\rm sift}^{A}$ and $K_{\rm sift}^{B}$.

3.1. Error correction

The error correction algorithm is applied for making the sifted quantum keys $K_{\rm sift}^{A}$ and $K_{\rm sift}^{B}$ to be identical on the both sides. We employ the following error correction algorithm with two basic stages. The first is to use the low-density parity-check (LDPC) syndrome coding/decoding [14] to correct discrepancies between keys. The universal polynomial hashing [15] to verify an identity between keys after previous step is used as the second stage.

Alice and Bob share a pool of LDPC parity-matrices with code-rates R from 0.9 up to 0.5 (with step being equal to 0.05) and the frame size (length of processed strings) being equal to $n = 4096$. These matrices are constructed with progressive edge-growth algorithm [16] using polynomial (degree distributions) from Ref. [17]. For each coding and decoding process parties employ code with the minimal rate R, which satisfies the following condition:

$$\frac{1-R}{h_{\rm b}(\mathrm{QBER}_{\rm est})} \leq f_{\rm crit} \tag{1}$$

where $h_{\rm b}$ is the standard binary entropy function, $\mathrm{QBER}_{\rm est}$ is the estimated level of QBER (see below), and $f_{\rm crit} = 1.22$ is the critical efficiency parameter in our setup, that is the tolerable ratio between level of disclosed information about sifted key and theoretical limit for successive error correction, which is predicted by the classical information theory.

To decrease a frame error rate (probability of unsuccessive decoding) in the constraint that the resulting efficiency is not greater than $f_{\rm crit}$, Alice and Bob use the shortening technique [18]. The number of shortened bits $n_{\rm s}$ is obtained from the following expression

$$n_{\rm s} = \lfloor n - \frac{m}{f_{\rm crit} h_{\rm b}(\mathrm{QBER}_{\rm est})} \rfloor, \tag{2}$$

where $\lfloor x \rfloor$ stands for the maximal integer less than x. Alice and Bob construct $\mathcal{N} = 256$ strings of length n possessing $n - n_{\rm s}$ bits of their sifted keys $K^A_{\rm sift}$ and $K^B_{\rm sift}$ and $n_{\rm s}$ shortened bits, whose positions and values come from synchronised pseudo-random generator. Bob multiplies the chosen parity-check matrix on the constructed strings to obtain syndromes that are sent to Alice. For the LDPC syndrome decoding, Alice applies iterative sum-product algorithm [18], which uses log-likelihood ratios for messages between symbol and parity-check nodes of the corresponding Tanner graph. To avoid costly calculations, we employ optimization techniques considered in Ref. [19]. Then Alice removes shortened bits obtaining corrected key.

If the algorithm of decoding does not converge for particular block in a specified number of iterations (we use about 60 iterations in our setup), then this block is considered as unverified. However, rarely decoding process converges to a wrong result, *i.e.*, to a incorrect key but still with proper syndrome. To avoid such situations, the second step of verification is used for blocks successively completed decoding.

In our setup, we employ comparison of hash-tags constructed with ε-universal polynomial hashing [15]. In particular, we use modified 50-bit variant of PolyR hash function that provide collision probability for a $\mathcal{N}$ blocks of $n - n_{\rm s}$ bits on the level of $\varepsilon_{\rm ver} < 2 \times 10^{-12}$. If the hash tags on the both sides match, then the corresponding blocks are concerned to be identical on the both sides and are added to verified keys $K_{\rm ver}$. We note that $K_{\rm ver}$ is a part of $K^B_{\rm sift}$ as all modifications of the key are performed on the Alice's side.

3.2. Parameter estimation

The purpose of the parameter estimation procedure is to determine QBER that is a probability of bit-flipping in a quantum channel. This problem could be resolved by comparison of input $K^A_{\rm sift}$ and output $K_{\rm ver}$ keys of the error correction, because as it was noted all changes of the key were performed exclusively by Alice. The ratio of corrected bits is averaged over a set of $\mathcal{N}$ the error correction blocks, and for unverified blocks a conservative maximal value 1/2 is assumed. If we denote all verified blocks in the set as $\mathcal{V}$ than the estimated QBER reads

$$\mathrm{QBER}_{\rm est} = \mathcal{N}^{-1} \left(\sum\nolimits_{i \in \mathcal{V}} \mathrm{QBER}_i + \left|\overline{\mathcal{V}}\right|/2 \right), \tag{3}$$

where QBER_i is the ratio of bit-flips in ith block and $|\overline{\mathcal{V}}|$ is number on unverified blocks.

3.3. Privacy amplification

After these procedures, both sides have identical bit strings. Nevertheless, Eve may have some amount of information about them. The privacy amplification procedure is used to reduce this potential information of an adversary to a negligible quantity. This is achieved by a contraction of the input bit string into a shorter string. The output shorter string is a final private key $K_{\rm sec}$ of length $l_{\rm sec}$. Our algorithm of the privacy amplification firstly checks (according to Ref. [20]) whether it is possible to distill private key with the given length and security parameter $\varepsilon_{\rm pa}$ (which is set to $\varepsilon_{\rm pa} = 10^{-12}$). Namely, define the quantity

$$\nu = \sqrt{\frac{2(l_{\rm ver} + k)(k+1)\ln\frac{1}{\varepsilon_{\rm pa}}}{l_{\rm ver}k^2}}, \tag{4}$$

where k in the number of bits used to estimate the QBER. Then if the inequality

$$2^{-\frac{1}{5}(l_{\rm ver}(1-h_{\rm b}(\delta+\nu))-r-t-l_{\rm sec})} \leq \varepsilon_{\rm pa} \tag{5}$$

is satisfied, then the generation of a private key of length $l_{\rm sec}$ and security parameter $\varepsilon_{\rm pa}$ is possible. Here r is the total length of the syndromes in the error correction procedure, t is the total length of the verification hashes and δ is a critical level of QBER.

If formula (5) gives the positive answer on the question about the possibility of key generation with the desired parameters, then the final private key is computed as a hash function of the input bit string. The family of hash functions is required to be 2-universal [20]. The generalization to almost universal families of hash functions is also possible.

A hash function is chosen randomly from the Toeplitz universal family of hash functions [21]. A matrix T of dimensionality $l_{\rm sec} \times l_{\rm ver}$ is a Toeplitz matrix if $T_{ij} = T_{i+1,j+1} = s_{j-i}$ for all $i = 1, \dots, l_{\rm sec} - 1$ and $j = 1, \dots, l_{\rm ver} - 1$. Thus, to generate a Toeplitz matrix, we need a random bit string $S = (s_{1-l_{\rm sec}}, s_{1-l_{\rm sec}+1}, \dots, s_{l_{\rm ver}-1})$ of length $l_{\rm ver} + l_{\rm sec} - 1$. The string S is generated randomly by one side (say, Alice) and sent to another side (Bob) by public channel. Let us denote the corresponding Toeplitz matrix as T_S. Then the final private key is computed as $K_{\rm sec} = T_S K_{\rm ver}$ (with multiplication and addition modulo 2).

It is possible to use not a random, but a pseudo-random bit string, which uses a shorter random seed, to specify the Toeplitz matrix [21, 22]. In this case, the family of hash functions is not universal, but almost universal, which is also acceptable for privacy amplification with some modifications in formula (5). The security of privacy amplification is based on the Leftover hash lemma, which can be proved for almost universal hash functions as well [23]. However, neither the rate of random number generator, nor the amount of publicly amount information are critical parameters of our setup. Thus, we adopt the standard family of Topelitz functions.

4. Authentication

The purpose of the authentication is to ensure that messages received by each side via public channel were sent by the other legitimate side (not by the adversary) and were not changed during the transmission.

Usual way to deal with this problem is hashing of the messages by hash functions dependent on a private key $K_{\rm aut}$ known only to legitimate parties. In general, the procedure is as follows: Alice sends to Bob a message with the hash tag. After receiving the message, Bob also computes the hash tag of the message and compare it to that received from Alice. If the hash tags coincide, Bob acknowledge the message as sent by Alice, otherwise he breaks the protocol. The hash function requires to assure that, whenever an eavesdropper does not know the private key, he cannot modify the message of send his own message and guess the correct hash tag of the message except for negligible probability (we require it not to exceed $\varepsilon_{\rm aut} = 10^{-12}$).

For unconditional security, the hash function must be chosen from some universal family: almost strongly universal [24,25], almost xor-universal [21,26], or almost Δ-universal family [27].

After consideration of many universal families of hash functions, we have decided in favour of the Toeplitz hashing (see Sec. 3.3 above) due to its computational simplicity. In the privacy amplification procedure, we exploit that it is a 2-universal family. Here we exploit its xor-universality [21]. Let the lengths of the authenticated messages and their hash tags be l_M and l_h respectively. The hash tag of the ith message M_i is calculated as

$$h(M_i) = T_S M_i \oplus r_i, \tag{6}$$

where T_S is a $l_h \times l_M$ Toeplitz matrix generated by a string S of length $l_h + l_M - 1$, r_i is a bit string of length l_h, and $\oplus$ is the bitwise xor. Both S and r_i are private and taken from the common private key K_{aut}. Then, the probability that an eavesdropper will guess the hash tag of a modified message is not more than 2^{-l_h}. The demand $\varepsilon_{\text{aut}} = 10^{-12}$ gives $l_h = 40$.

As in privacy amplification, we do not adopt a pseudo-random generation of a string S from a shorter string. This reduces the consumption rate of the private key. However, formula (6) allows to generate the bit string S only once, for the first message. For further messages, the private key is consumed in the rate l_h bits per message (for strings r_i). Thus, large initial consumption of the private is not critical. The string S may be used for many sessions of quantum key distribution.

Also this consideration suggests that it is advantageous to authenticate several messages at once: except for the first message, the consumption of the private key is l_h bit per message and does not depend on the length of the message. However, the reduction of the number of authentications raises the risk of denial-of-service attacks from an eavesdropper: he is able to simulate messages from legitimate parties and force them to do calculations before his interference will be disclosed (in the nearest authentication stage). Thus, there is a trade-off between the private key consumption rate and the risk of denial-of-service attacks.

During the post-processing, the public classical channel is used twice: (i) to send the syndrome in the error correction stage along with the verification hash tag and (ii) to send the decision about possibility of key generation and (if the answer is positive) estimated level of QBER and bit string used to generate the Toeplitz matrix in the privacy amplification algorithm.

Another possibility is to use the hash function based on the known GOST cipher [28], though it is not unconditionally secure.

5. Workflow of the post-processing procedure

The workflow of the post-processing procedure is as follows. Sifted keys go through the error correction that is adjusted on the current value of QBER. After accumulation of necessary number of blocks they input to the parameter estimation (together with their versions before the error correction). If an estimated value of QBER given by (3) is higher than the critical value needed for efficient privacy amplification, the parties receive warning message about possible of eavesdropping. Otherwise, verified blocks input privacy amplification and estimated QBER is used in next round of the error correction algorithm.

The overall (in)security parameter of the quantum key distribution system is

$$\varepsilon_{\text{qkd}} = \varepsilon_{\text{ver}} + \varepsilon_{\text{pa}} + \varepsilon_{\text{aut}} = 2 \times 10^{-11} + 10^{-12} + 10^{-12} < 3 \times 10^{-11}. \tag{7}$$

This parameter majorizes both the probability that the keys of Alice and Bob do not coincide and the probability of guessing the common key by Eve. If this parameter exceeds a critical value, then the protocol is terminated.

After the privacy amplification procedure, a fraction of the key is used for authentication in the next rounds. In our setup, the fraction of generated private key consumed by authentication procedure does not exceed 15%.

6. Conclusion

We have present post-processing procedure for industrial quantum key distribution systems. The post-processing procedure consisting of error correction, parameter estimation, and privacy amplification has been described. Also authentication of classical communications over a public channel has been considered.

Acknowledgments

We thank Andrey Fedchenko for useful comments. The support from Ministry of Education and Science of the Russian Federation in the framework of the Federal Program (Agreement 14.579.21.0104, ID RFMEFI57915X0104) is acknowledged. We thank the organizers of the 3rd International School and Conference Saint-Petersburg OPEN 2016 for kind hospitality.

References

[1] For a review, see Gisin N, Ribordy G, Tittel W, and Zbinden H 2002 *Rev. Mod. Phys.* **74** 145; Scarani V, Bechmann-Pasquinucci H, Cerf N J, Dusek M, Lütkenhaus N, and Peev M 2009 *Rev. Mod. Phys.* **81** 1301

[2] Shor P W 1997 *SIAM J. Comput.* **26** 1484

[3] Wootters W K and Zurek W H 1982 *Nature* **299** 802

[4] Bennet C H and Brassard G 1984 *Proceedings of IEEE International Conference on Computers, Systems and Signal Processing (Bangalore, India, 1984)* p. 175 (New York: IEEE)

[5] Bennett C H, Bessette F, Salvail L, Brassard G, and Smolin J 1992 *J. Cryptol.* **5** 3

[6] Gobby C, Yuan Z L, and Shields A J 2004 *Appl. Phys. Lett.* **84** 3762

[7] Schmitt-Manderbach T, Weier H, Fürst M, Ursin R, Tiefenbacher F, Scheidl T, Perdigues J, Sodnik Z, Kurtsiefer C, Rarity J G, Zeilinger A, and Weinfurter H 2007 *Phys. Rev. Lett.* **98** 010504

[8] Stucki D, Walenta N, Vannel F, Thew R T, Gisin N, Zbinden H, Gray S, Towery C R, and Ten S 2009 *New J. Phys.* **11** 075003

[9] Walenta N, Burg A, Caselunghe D, Constantin J, Gisin N, Guinnard O, Houlmann R, Junod P, Korzh B, Kulesza N, Legré M, Lim C W, Lunghi T, Monat L, Portmann C, Soucarros M, Thew R T, Trinkler P, Trolliet G, Vannel F, and Zbinden H 2014 *New J. Phys.* **16** 013047

[10] Characteristics of devices from ID Quantique (Switzerland), SeQureNet (France), and Austrian Institute of Technology (Austria) are available on their websites.

[11] Lo H-K, Ma X, and Chen K 2005 *Phys. Rev. Lett.* **94** 230504; Zhao Y, Qi B, Ma X, Lo H-K, and Qian L 2006 *Phys. Rev. Lett.* **96** 070502

[12] Muller A, Herzog T, Huttner B, Tittel W, Zbinden H, and Gisin N 1997 *Appl. Phys. Lett.* **70** 793-5

[13] Chizhevsky V N 2010 *Phys. Rev. E* **82** 050101

[14] Gallager R 1962 *Inf. Theory, IRE Trans.* **8** 21; MacKay D 1999 *IEEE Trans. Inf. Theory.* **45** 399

[15] Krovetz T and Rogaway P 2000 *Lect. Notes Comp. Sci.* **2015** 73

[16] Martinez-Mateo J, Elkouss D, and Martin V 2010 *IEEE Commun. Lett.* **14** 1155

[17] Elkouss D, Leverrier A, Alléaume R, and Boutros J 2009 *Proceedings of IEEE International Symposium Information Theory (Seoul, South Korea, 2009)* p. 1879 (New York: IEEE)

[18] Elkouss D, Martinez-Mateo J, and Martin V 2011 *Quant. Inf. Comput.* **11** 0226

[19] Hu X-Y, Eleftheriou E, Arnold D M, and Dholakia A 2001 *Proceedings of IEEE Global Telecommunications Conference (San Antonio, USA, 2001)* **2** 1036 (New York: IEEE)

[20] Tomamichel M and Leverrier A 2015 *arXiv:1506.08458*

[21] Krawczyk H 1994 *Lect. Notes Comp. Sci.* **839** 129

[22] Krawczyk H 1995 *Lect. Notes Comp. Sci.* **921** 301

[23] Tomamichel M, Schaffner C, Smith A, and Renner R 2011 *IEEE Trans. Inf. Theor.* **57** 5524

[24] Wegman M N and Carter J L 1981 *J. Comp. Syst. Sci.* **22** 265

[25] Stinson D R 1992 *Lect. Notes Comp. Sci.* **576** 74

[26] Shoup V 1996 *Lect. Notes Comp. Sci.* **1109** 313

[27] Halevi S and Krawczyk H 1997 *Lect. Notes Comp. Sci.* **1267** 172

[28] Schneier B 1996 *Applied Cryptography* (John Wiley & Sons, Inc.)

Photoluminescence and electrical properties of silicon oxide and silicon nitride superlattices containing silicon nanocrystals

D V Shuleiko[1] and A S Ilin[1,2]

[1]Faculty of Physics, Lomonosov Moscow State University, 1/2 Leninskie Gory, Moscow, 119991, Russia

[2]National Research Centre "Kurchatov Institute", 1 Kurchatov Sq., Moscow, 123182, Russia

Abstract

Photoluminescence and electrical properties of superlattices with thin (1 to 5 nm) alternating silicon-rich silicon oxide or silicon-rich silicon nitride, and silicon oxide or silicon nitride layers containing silicon nanocrystals prepared by plasma-enhanced chemical vapor deposition with subsequent annealing were investigated. The entirely silicon oxide based superlattices demonstrated photoluminescence peak shift due to quantum confinement effect. Electrical measurements showed the hysteresis effect in the vicinity of zero voltage due to structural features of the superlattices from $SiO_{0.93}/Si_3N_4$ and $SiN_{0.8}/Si_3N_4$ layers. The entirely silicon nitride based samples demonstrated resistive switching effect, comprising an abrupt conductivity change at about 5 to 6 V with current-voltage characteristic hysteresis. The samples also demonstrated efficient photoluminescence with maximum at ~1.4 eV, due to exiton recombination in silicon nanocrystals.

1. Introduction

Current nanostructuring methods provide an opportunity for controllable varying of optical and electrical properties of siliceous films, by forming areas with silicon nanocrystals in the bulk of the film. These structures can be used to produce new types of photovoltaic and optoelectronic devices, such as all-silicon light emitting devices and third-generation tandem solar cells [1]. Moreover, silicon oxide-based films [2, 3] or silicon nitride-based films [4] with silicon nanocrystals can be used to produce logical elements based on memristive effect [5] (the dependence of the circuit element conductivity on the current passed). For practical applications in all cases precise control of the shape, size and concentration of silicon nanocrystals is required.

Silicon nanocrystals embedded in siliceous films can be produced by several techniques such as chemical vapor deposition (CVD) [6], reactive evaporation [7], and magnetron sputtering [8]. Control over all three parameters mentioned above (size, density, spherical shape) can be achieved by depositing layers with alternating material stoichiometry by plasma-enhanced chemical vapor deposition (PECVD) with subsequent high temperature annealing [6, 9]. In this case silicon-rich oxide or nitride layers with thickness of several nanometers are separated with barrier layers of stoichiometric silicon oxide or silicon nitride. These barrier layers effectively constrain the growth of quasi-ordered silicon nanocrystal arrays within the silicon-rich layers during thermal annealing [9, 10].

The purpose of this work is an investigation of nonlinear electrical properties of such silicon nanocrystal arrays in the silicon oxide and silicon nitride superlatices, as these structures may potentially find application in microelectronics, and detailed study of such structures electrical properties has not been yet carried out. Photoluminescence (PL) properties of the superlattices also were examined to confirm the formation of silicon nanocrystals.

2. Experiment

2.1 Sample fabrication

The superlattices consisting of 20 pairs of thin (1 to 5 nm) alternating $SiO_{0.93}$ or $SiN_{0.8}$, and SiO_2 or Si_3N_4 layers (40 layers in total) were deposited by plasma-enhanced chemical vapor deposition. The samples were deposited at n-type Si substrate with crystallographic surface orientation (100). High temperature annealing during 1 hour at 1150 °C in nitrogen atmosphere caused nanocrystal formation in the silicon-rich layers. Three series of the samples were fabricated: consisting of alternating $SiO_{0.93}/SiO_2$ (SO series), $SiO_{0.93}/Si_3N_4$ (SO/SN series) and $SiN_{0.8}/Si_3N_4$ (SN series) layers. A more detailed description of the sample preparation procedure can be found in works [10, 11]. The samples properties are given in table 1.

Table 1. Structural parameters of the samples.

Sample name	Silicon-rich layer (with nanocrystals)		Barrier layer	
	Material	Thickness (nm)	Material	Thickness (nm)
SO-1	$SiO_{0.93}$	1.5	SiO_2	1
SO-2		3		
SO-3		5		
SO/SN-1		1.5	Si_3N_4	1.5
SO/SN-2		3		
SO/SN-3		5		
SN-1	$SiN_{0.8}$	1.5		
SN-2		3		
SN-3		5		

2.2 Optical measurements

The PL spectra were measured at room temperature using a spectral complex based on the monochromator/spectrograph SOLAR TII MS3504i, which allows to record spectra in the visible and near-infrared (up to 1100 nm) range with CCD digital camera Proscan HS 101H. A HeCd laser with 325 nm wavelength in pulsed mode was used as the excitation source. Output power was 30 mW.

2.3 Electrical measurements

To measure the current-voltage characteristics (CVC), the aluminum contacts were deposited. One common contact was deposited on the entire bottom surface of the substrate, and the group of contacts with 600 μm diameter was deposited on the top of the superlattice. During the measurements the voltage was applied between the common bottom contact, and one of the top contacts.

CVC were measured using picoammeter Keithley 6487. During the measurements voltage was applied cyclically: at first positive bias from 0 to +10 V and back to 0 V was applied to the top contact and then negative bias from 0V to –10V and back to 0 V was applied. Measurement step was 0.2 V and frequency was 1 measurement per second. To control the reproducibility, the measurements were carried out on several contacts for each sample.

3. Experimental results and discussion

According to the measurements carried out, the superlattices possess nonlinear CVCs due to potential barriers between the superlattice layers, at the silicon nanocrystal surfaces, as well as on the film-substrate and film-contact interfaces. All the samples also possess effective PL.

3.1 PL properties

The PL spectra of the samples from SO series are presented in figure 1. Maximum of PL was observed at 1.40 eV for SO-1, 1.50 eV for SO-2 and 1.63 eV for SO-3. The PL reason in this spectral range can only be exciton recombination in the silicon nanocrystals [9]. The PL peak shift can be explained by the quantum confinement effect in the silicon nanocrystals. Due to the quantum confinement effect, the size of the nanocrystals can be evaluated using calibration graph from [12], or the equation:

$$E_{gap}(D) = E_{gap}(\infty) + \frac{2\hbar^2 \xi_{n,l}^2}{m_r D^2} = E_{gap}(\infty) + \frac{A}{D^2} \qquad (1)$$

where $E_{gap}(D)$ – band gap in silicon nanocrystal with diameter D, $E_{gap}(\infty)$ – band gap in bulk Si, $\xi_{n,l}^2$ – Bessel spherical function root of half-integer argument l+1/2 (l=0,1,2...); for l = 0, $\xi = \pi n$ (n=1,2,3...). The second term in the equation can be simplified using constant for spherical nanocrystals in the strong confinement mode A=3.57 eV·nm^2 [13]. Calculated silicon nanocrystal sizes are in a good agreement with Si-rich layers thickness in the superlattices (Table 2). Minimal crystallization diameter expected in previous works was 1.8 nm [10], hence in 1.5 nm Si-rich layers silicon nanocrystals are formed with larger diameter for all structures.

Maximum of PL was observed at approximately 1.40 eV for all samples from SO/SN and SN series also. The PL reason is a same exiton recombination in the silicon nanocrystals correspondingly. However, the PL peak shift absent in these cases. Therefore, there is no silicon nanocrystal size control effect in the superlattices with Si_3N_4 barrier layers.

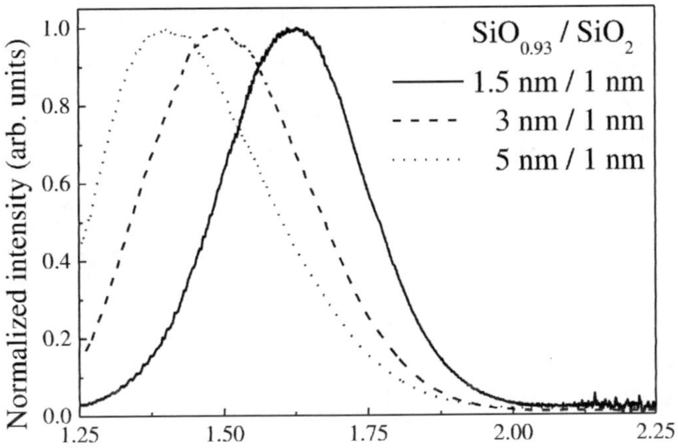

Figure 1. PL spectra of the $SiO_{0.93}/SiO_2$ superlattices.

Table 2. Silicon nanocrystal size compared to thickness of Si-rich layers in the superlattice

Sample name	Thickness of the layers with nanocrystals (nm)	Silicon nanocrystals mean size calculated using:	
		Calibration graph [12] (nm)	Equation (1) (nm)
SO-1	1.5	2.0	2.6
SO-2	3	3.0	3.1
SO-3	5	4.5	3.6

3.2 Current-voltage characteristic hysteresis at zero voltage

For all samples of the SO/SN and SN series, a symmetrical CVC hysteresis effect at 0 V was observed. Figure 2 shows the CVC of the sample SO/SN-1 compared to the CVC of a 100 GΩ test impedance. This comparison confirms that the observed effect is not associated with the capacity of the experimental setup parts. It should also be noted that the hysteresis effect at 0 V was not observed in the SO series as the samples from SO series have resistance by 2 orders of magnitude higher compared to the samples from other series. And as will be shown below, the resistance of the multilayer structure is important in the explanation of the observed hysteresis.

The effect of hysteresis can be explained by the relatively high capacity of the superlattices constituting a structure of alternating layers with different dielectric constants. This structure can be considered as serially connected flat capacitors that cause charge retention with certain impedance. Thus the equivalent circuit can be represented with parallel-connected resistor and capacitor, and is characterized by charge relaxation time $\tau = RC_\Sigma$. The capacity of each layer is

$$C_i = \frac{\varepsilon\varepsilon_0 S}{d}, \quad (2)$$

where S – capacitor plate area, d – layer thickness, ε – dielectric constant of the material layer. The table values $\varepsilon(SiO_{0.93}) = 4.8$, $\varepsilon(Si_3N_4) = 9$ were used. The total capacity C_Σ of the samples from SO/SN series is calculated by the formula for the series-connected capacitors:

$$C_\Sigma = \left(\sum_{i=1}^{40}\left(\frac{1}{C_i}\right)\right)^{-1}, \quad (3)$$

where 40 is the total number of layers.

The resistance R of the samples at $U = 1$ V, and the relaxation time $\tau = RC_\Sigma$ were calculated using the experimentally measured CVC. The data are given in table 3. Also, time-resolved charge relaxation of the superlattices from the SO/SN series was measured. Charge relaxation behavior of the structures is biexponential: figure 3 shows electric current relaxation for the sample SO/SN-3 and its approximation. Measured decay times are also presented in table 3.

Table 3. Resistance, capacity and charge relaxation time for the sampler from SO/SN series

Sample name	Resistance R at U=1V (GΩ)	Capacity C_Σ (pF)	Relaxation time $\tau=RC$ (s)	Experimental relaxation time τ_1 (s)	Experimental relaxation time τ_2 (s)
SO/SN-1	58	260	15	6.7±1.9	0.9±0.5
SO/SN-2	24	160	4	6.0±0.6	0.8±0.1
SO/SN-3	88	100	9	7.8±0.7	1.0±0.1

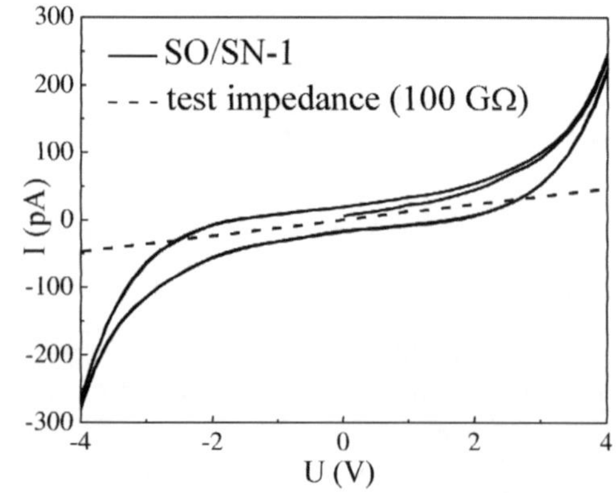

Figure 2. CVC hysteresis of the SO/SN-1 superlattice compared to test impedance

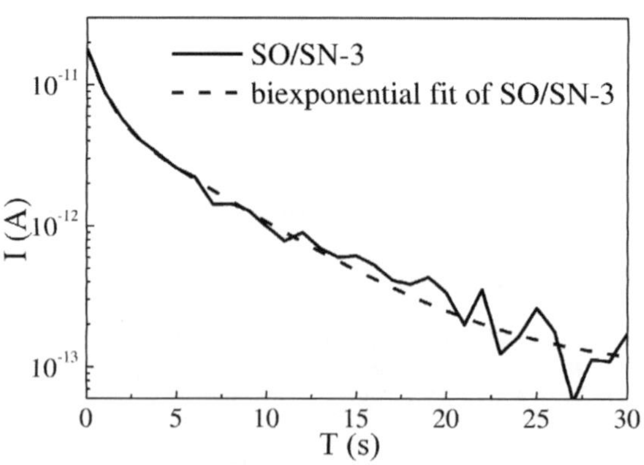

Figure 3. Electrical current relaxation in the sample SO/SN-3

The binexponential nature of charge relaxation can be attributed to presence of the silicon nanocrystals in the volume of the superlattice, and in the model used above these nanocrystals were not taken into account (see below). Thus capacity of the interfaces between matrix and nanocrystal should be taken into account [14]. Also, the contact–superlattice and superlattice–substrate interfaces can contribute to the sample capacity. It is also important to note that the presence of the silicon nanocrystals leads to a change in the dielectric constant of the layers.

In addition, silicon nanocrystals embedded in the dielectric matrix should be considered as quantum dots (QDs) with discrete electron spectrum, separated by potential barriers. The arrays of coupled QDs can accumulate electrons by trapping them, and current flow in such system occurs by electron tunneling. Electron tunneling is governed by the Coulomb interaction between localized electrons and is highly dependent on the morphology of the QDs (size, shape and spacing). The QDs morphology determines the energy level intervals and transition rates of electrons between the QDs [15]. Similar effect of charge retention due to accumulation of electrons into the silicon nanocrystals that act as QDs was observed in [16]. In that work charge retention occurred in the silicon nitride films with silicon nanocrystals, with relaxation time of several hundred of microseconds.

As biexponential charge relaxation behavior was observed in our work, the presence of an additional charge relaxation channel can be assumed; and this channel is most likely associated with electron tunneling between the silicon nanocrystals acting as QDs. Due to the presence of such supplemental channel the relaxation rate in the initial stage is fast, while the amount of electrons trapped on silicon QDs is large.

3.3 Resistive switching with hysteresis (memristive effect)

Resistive switching effect, comprising the abrupt conductivity change with CVC hysteresis was demonstrated by the sample SN-2. At the voltage of 5 to 6 V, the conductivity increased by almost two orders of magnitude (Figure 4). The sample withstood at least 10 measurement cycles without substantial change in the switching parameters. This hysteresis effect is similar to the memristive switching observed in SiO_2 films containing silicon nanocrystals [2, 3] or Si_3N_4 films with thickness of 4 to 13 nm [4]. However, the hysteresis effect was observed only at positive voltage bias, and no hysteresis was observed at negative bias.

This conductivity switching most likely occurs due to formation of silicon filaments in the matrix between the nanocrystals [2–4]. In the low conductivity ("off") state electrons are trapped in the silicon nanocrystals and conductive path is formed by electron tunneling between the nanocrystals. At certain voltage (from 5 to 6 V) a conductive path is formed from the silicon filaments in the spaces between the silicon nanocrystals due to nitride ion migration or generation of nitride vacancies [4] by current-induced heating [3]. The local current density in the regions between closely spaced nanocrystals is higher than in the surrounding matrix, and current-induced local heating is sufficient for such structural changes and formation of silicon filaments. When conductive path is formed, the sample is switched into high conductivity state ("on").

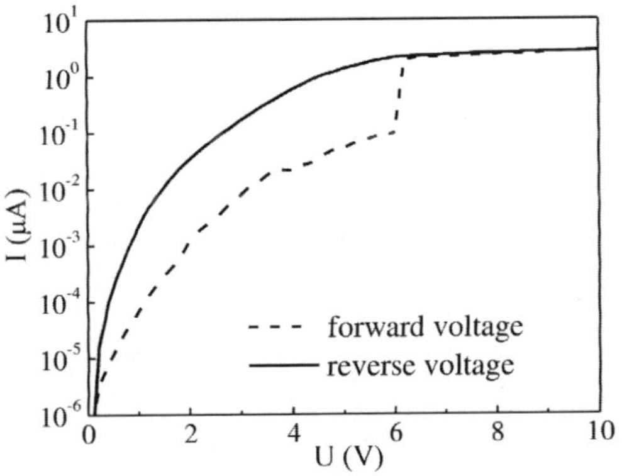

Figure 4. Current-voltage characteristic of the sample SN-2.

An important feature of the observed resistive switching is distinct CVC hysteresis at the switching voltage, formed by CVC curves of the sample in the "off" and "on" states (Figure 4). The hysteresis occurs due to the fact that after the conductive path of silicon filaments is formed, it exists as long as the electric current flows, and degrades after the voltage is turned off.

4. Conclusions

The silicon oxide and silicon nitride superlattices with embedded silicon nanocrystals demonstrate PL properties in the red and near-infrared band. PL maximum was observed at ~1.4 eV for all $SiO_{0.93}/Si_3N_4$ and $SiN_{0.8}/Si_3N_4$ samples while PL peak position for $SiO_{0.93}/SiO_2$ samples depended on size of the nanocrystals and shifted from 1.40 to 1.63 eV. For all samples from SO/SN and SN series PL maximum was observed at approximately 1.40 eV and had no shift.

Superlattices of $SiO_{0.93}/Si_3N_4$ and $SiN_{0.8}/Si_3N_4$ layers, demonstrated current-voltage characteristics hysteresis at zero voltage. This effect is most likely due to resistive and capacitive parameters of the structures representing a multilayer RC circuit, as well as presence of silicon nanocrystals acting as QDs in the bulk of the superlattice.

The memristive effect was detected in the $SiN_{0.8}/Si_3N_4$ superlattice with silicon nanocrystals. The conductivity abruptly increased by almost two orders of magnitude at 5 to 6 V during forward voltage sweep, followed by CVC hysteresis with a smooth voltage decrease at reverse voltage sweep. This effect is most likely associated with silicon filament formation between silicon nanocrystals at switching voltage and gradual degradation of the filaments after the voltage is turned off.

Acknowledgments

This work was financially supported by the Russian Foundation for Basic Research (project 15-32-21153, photoluminescent study, and project 16-32-80066, electrical study). Authors wishing to acknowledge Albert-Ludwig University of Freiburg, Germany. Authors also wishing to acknowledge S V Zabotnov for fruitful discussions.

References

[1] Cheng K–Yu, Anthony R, Kortshagen U R and Holmes R J 2011 *Nano Lett.*, **11(5)** 1952; Green M A 2001 *Prog. Photovoltaics* **9** 123
[2] Yao J, Sun Z, Zhong L, Natelson D and Tour J M 2010 *Nano Lett.* **10** 4105
[3] Wang Y, Chen Y–T, Xue F, Zhou F, Chang Y–F, Fowler B and Lee J C 2012 *Appl. Phys. Lett.* **100** 083502
[4] Kim H–D, An H–M, Hong S M and Kim T G 2013 *Phys. Status Solidi* A **210(9)** 1822
[5] Strukov D B, Snider G S, Stewart D R and Williams R S 2008 *Nature* **453** 80
[6] Creazzo T, Redding B, Marchena E, Murakowski J and Prather D W 2010 *J. Lumin.* **130** 631
[7] Cheylan S, Elliman R G, Gaff K and Durandet A 2001 *Appl. Phys. Lett.* **78** 1670
[8] So Y–H, Huang S, Conibeer G and Green M A 2011 *Thin Solid Films*, **519** 5408
[9] Zacharias M, Heitmann J, Scholz R, Kahler U, Schmidt M and Blaesing J 2002 *Appl. Phys. Lett.* **80** 661
[10] Zelenina A, Dyakov S A, Hiller D, Gutsch S, Trouillet V, Bruns M, Mirabella S, Löper P, López–Conesa L, López–Vidrier J, Estradé S, Peiró F, Garrido B, Bläsing J, Krost A, Zhigunov D M and Zacharias M 2013 *J. Appl. Phys.* **114** 184311
[11] Zelenina A, Sarikov A, Zhigunov D M, Weiss C, Zakharov N, Werner P, López–Conesa L, Estradé S, Peiró F, Dyakov S A and Zacharias M 2014 *J. Appl. Phys.* **115** 244304
[12] Takeoka S, Fujii M and Hayashi S 2000 *Phys. Rev.* B **62(24)** 16820
[13] Goncharova L V, Nguyen P H, Karner V L, D'Ortenzio R, Chaudhary S, Mokry C R and Simpson P J 2015 *J. Appl. Phys.* **118** 224302.
[14] Forsh F A, Martyshov M N, Timoshenko V Yu and Kashkarov P K 2006 *Semiconductors* **40(9)** 476
[15] Mantsevich V N, Maslova N S and Arseyev P I 2013 *Solid State Commun.* **168** 36
[16] Liao W, Zeng X, Wen X, Zheng W, Wen Y and Yao W 2015 *J. Electron. Mater.* **44(3)** 1015

InGaN/GaN heterostructures with lateral confinement for light emitting diodes

K P Kotlyar[1], B I Soshnikov[2], I A Morozov[1], D A Kudryashov[1], K S Zelentsov[1], V V Lysak[2], I P Soshnikov[1,3-5]
[1]Nanotechnology Research and Education Center, Academic University, RAS, St. Petersburg 194021, Russia
[2]Department of Nanotechnology and Material Science, ITMO University, St Petersburg 192000, Russia.
[3]Ioffe Physical-Technical Institute of the RAS, St. Petersburg 194021, Russia
[4]Institute for Analytical Instrumentation of RAS, St. Petersburg 198095, Russia
[5]Saint Petersburg Electrotechnical University "LETI", St. Petersburg 197376, Russia

E-mail: konstantin-kt21@rambler.ru

Abstract. InGaN/GaN nanorod structure for light emission diode fabricated by the reactive plasma etching through the self-assembled Ni nano-cluster mask is presented. Fabricated array structure, presented in the form of truncated cones with average diameter height and period of nanorods is 250±50 nm, 430nm and 650±50 nm, respectively. The side angle of single structure about 80°. EL spectrum has maximum at 460 nm.

1. Introduction

Fabrication of high-efficient optoelectronic devices (LEDs, PDs, lasers) is the key to the development of modern nanoelectronics. Further way of increasing efficiency is connected with using 3-dimentional structures, such as nanorods (NRs) and nanowire [1]. The formation of lateral confinement allows to decrease the transportation of the carriers in the dimension of the layers of heterostructures (to defects) and also to concentrate current in the light generating area where is performed (quantum dots). Furthermore, nanowires and NRs have practically non-defective structures [2]. One of the problems that can be solved with nanowires is increasing the light extraction efficiency from the structure by formation of effective resonator [3, 4]. With increasing of the efficient area, heating removing is also increased. There are different ways to fabricate such structures: catalytic and non-catalytic growth [5], growth with using electron-beam lithography [6]. Judging by technological process the easiest way to transfer to mass production is the technology of plasma chemical etching of planar LED structures with Ni nano-particles mask [7-9]. The aim of this study is development the technology of fabrication and investigation of properties of InGaN/GaN heterostructures with lateral confinement.

2. Experimental

The lateral confinement structure is fabricated using reactive plasma etching of planar LED heterostructures through the self-assembled Ni nano-cluster mask. In the work we use planar QD heterostructures (p-GaN/InGaN QD/n-GaN) grown by MOCVD on a-sapphire substrate. Ni mask is

formed by a deposition of ultrathin Ni film (about 5 nm) on the planar InGaN QD/GaN LED samples and followed by rapid thermal annealing (RTA). The samples through the self-assembled Ni nano-cluster mask are etched using ICP-DEP in gas mixture Cl_2 (20sccm) and BCl_3 (10sccm). Afterwards, the residual Ni nanoclusters is removed in solution of HCl:HNO_3 3:1. The contact formation is realized by polymer (SU-8) and ITO coating.
The surface morphology is studied using electron microscopy (SEM).

3. Results and discussions

3.1. Formation of Ni nano-masks

Preliminary, we investigated the formation of nanoparticles of Ni on Si(111) surface. Figure 1(a) shows SEM image of sample with a thickness of Ni layer: 2 nm, at the temperature of 850°C and annealing time of 120 sec. The size distributions of Ni particles on the sample are given on the figure 1(b). Optimal RTA temperature of 850°C is found for Ni nanoparticle (NP) formation.

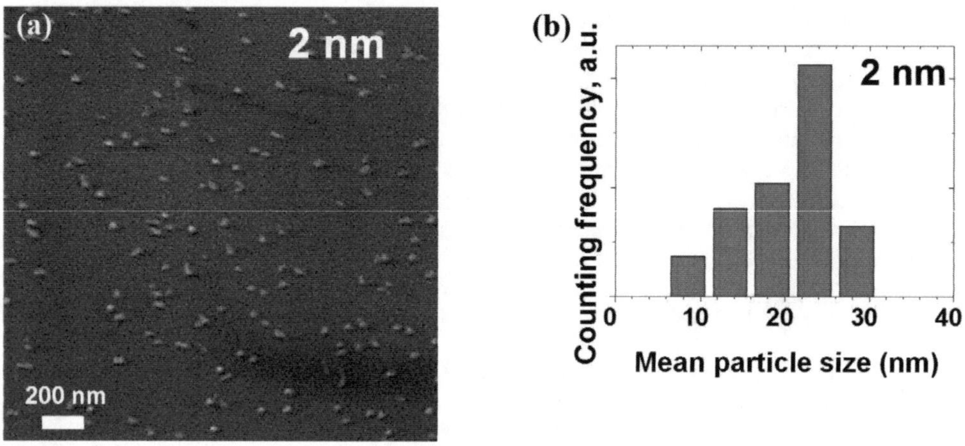

Figure 1(a, b). (a) SEM image of Ni particles on the Si(111) surface after annealing of Ni films thickness 5 nm, at 850°C and 120 sec; **(b)** Size distributions of Ni.

When increasing the temperature of annealing to 900°C the evaporation of the film is possible. With lower temperature the fabrication of particles is not noticed. During investigation this effect of the annealing time, we observe of increasing size particles by diffusion atomic and Osvald process [10]. On sample with efficient thickness of film 5 nm, we saw displacement islands as a result of the Brown's motion of the island as a whole. As time of annealing was increased, the density of the islands decreased and their series increased. Figure 2 shows that averaged size of Ni particles and their density dependent on the thickness of film by the matter conservation law in the system. The studying of Ni nano-mask formation on GaN based heterostructures (figure 3) demonstrates results near for Ni on Si. Notice the Ni islands have a hexagonal shape that is likely result from the crystal structure of the substrate.

3.2. Formation of nanorod structure

The average diameter of the NRs is ranging from 250±50 nm, height of the rod is about 430nm and the period is about 650±50 nm. Different etching angles (77±2° and 80±2° for p- and n-GaN layer, respectively) show clearly the active layer position in the structure (figure 4). Most probably, every nanorod acquired at least one quantum dot.

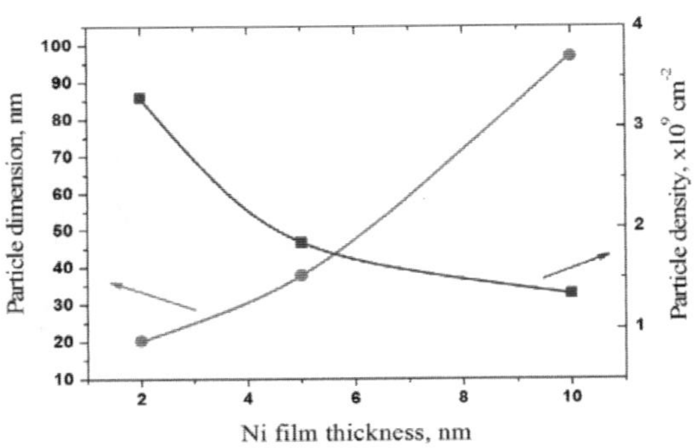

Figure 2. Average size and density of the Ni particles as a function of the initial Ni thickness for RTA at 850◦C, 120 sec.

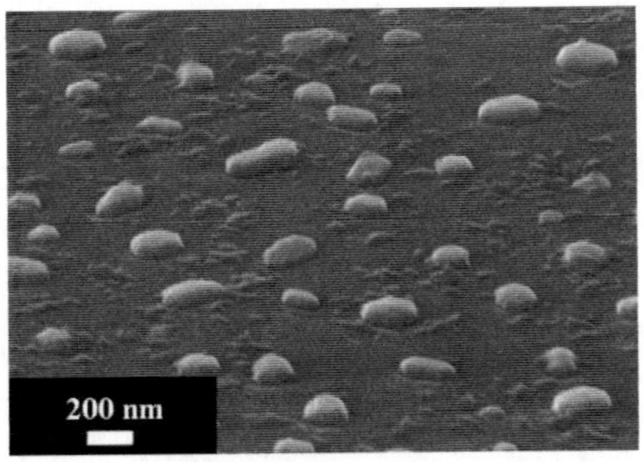

Figure 3. Ni particles on GaN surface fabricated by 5 nm Ni film deposition and RTA at 850◦C, 120 sec.

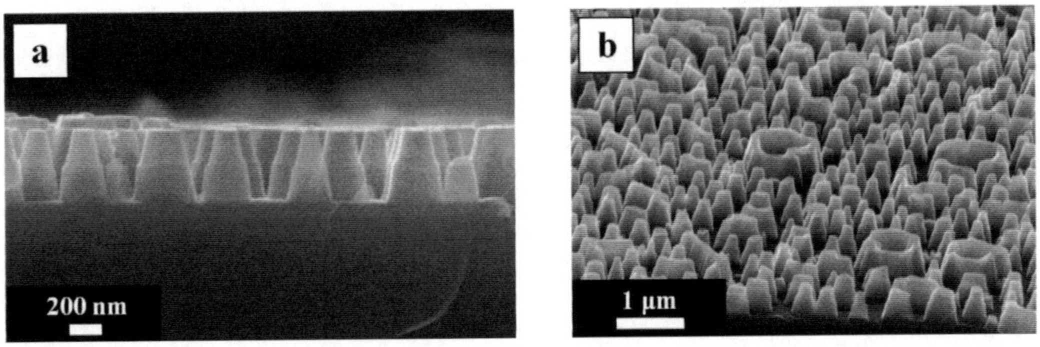

Figure 4(a, b). Cross section **(a)** and isometric **(b)** view of the GaN sample after ICP etching.

3.3. Optoelectronic properties

Figure 5(a) shows PL intensity of the NR structure with measured at temperatures of 10 K and 300 K (blue and black lines, respectively) comparing to planar one at room temperature (green line). The short wavelength peak can be described by the luminescence of the bulk GaN layer. The peak shifts expectedly to the longer wavelengths with rising of the temperature. Long-wavelength peak describes optical answer of the InGaN active layer and leads himself in a more complex manner. The peak hasn't red-shift nature and increasing of spectral width. Such behavior can be explained by indium surface segregation (ISS) effect. After the formation of NR, these islands can act as separated quantum dots with superior enhancing of the optical gain properties and temperature stability. Increasing of the indium content decreases band gap energy, while quantum dots naturally spread energy [11]. Both effects compensate each other and keep wavelength stable. While the PL intensity of the single quantum dot is naturally very narrow, we can suggest that the broad width of our NRS spectrum at low temperature causing by the total sum of different quantum dots [12] and random values of NR diameters. In addition, it is well known that there exists strong strain-induced piezoelectric fields exerted across InGaN/ GaN MQWs, causing a quantum confined Stark effect (QCSE) [13] and thus generating a red shift in emission peak. The optical pumping induced blue shift in emission peak, which occurs to our InGaN/GaN NRS structure, confirms that the strain has been relaxed. The peaks of the spectra are between 450-470 nm. Presented on figure 5(b) electroluminescence (EL) spectrum has similar peak in the same area. Light extracted from NR structure is shown in the inset of figure 5(b). One of the problems during etching is the evenness of the height of NR that decreases the light extraction efficiency. *VI* characteristic of the fabricated nanorod LED shows 2.2 V of the cutoff voltage.

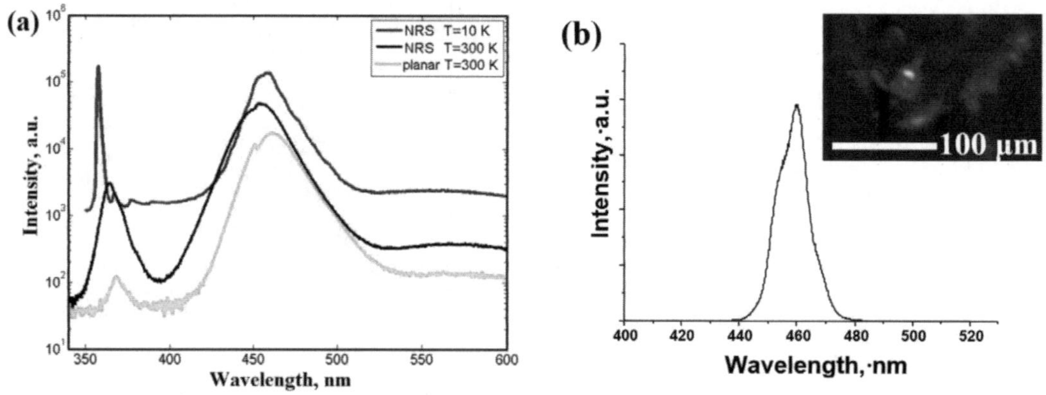

Figure 5(a, b). (a) PL spectra of initial and nanorod structures; **(b)** EL spectrum nanorod structures.

4. Conclusion

Lateral confined InGaN/GaN heterostructures were fabricated. We investigated the main processes of the formation of Ni – mask. PL intensity behaviour shows high temperature stability over a wide range of temperatures. The peak of improved structure showed 2.8 times enhancement and 9nm (54meV) blue-shifting comparing to the planar one at room temperature. After the nanorod formation, quantum dots, formed by the indium surface segregation, can act as separated quantum dots with superior enhancing of the optical gain properties and temperature stability. Optoelectronic properties of the nanorod LEDs were demonstrated in the range of 450-470 nm.

Acknowledgments

We are greatly M.H. Mustary for provided sample of LED structure. This work is partly supported by the Government of the Russian Federation (Grant 074-U01) through ITMO Early Carrier Fellowship scheme, Russian Foundation for Basic Research (project# 15-07-06964, 13-02-12031 ofi-m), Ministry for education and science of Russian Federation (Contract № 02.G25.31.0058 dated 12.02.2013).

References

[1] Li S, Waang A 2012 *J. Appl. Phys.* **111** 071101
[2] Glas F 2006 *Phys. Rev. B* **74** 121302(R)
[3] Wang R, Liu D, Zuo Z, Yu Q, Feng Z, Xu X 2012 *AIP Adv.* **2** 012109
[4] Lin C F, Zheng J H, Yang Z J, Dai J J, Lin D Y, Chang C Y, Z. X. Lai Z X, Hong C S 2006 *Appl. Phys. Lett.* **88** 083121
[5] Dubrovskii V G , Cirlin G E, Ustinov V M 2009 *Semiconductors* **43** 1585
[6] Soshnikov I P, Afanas'ev D E, Cirlin G E, Petrov V A, Tanklevskay E M, Yu. B. Samsonenko Yu B, Bouravlev A D, Khrebtov A I, Ustinov V M 2011 *Semiconductors* **45** 840
[7] Mustary M H, Lysak V V 2014 *IOSR JEEE* **9** 18
[8] Yang G, Xie F, Tong Y, Chen P, Yu Z, Yan D, Xue J, Zhu H, Guo Y, Li G, Gao S 2015 *Materials Science in Semiconductor Processing* **30** 694
[9] Yang G, Guo Y, Zhu H, Yan D, Li G, Gao S, Dong K 2013 *Appl. Surf. Sci.* **285**(Part B) 772
[10] Kukushkin S A, Slyozov V V 1996 *Disperse systems on the surface of solids. (Evolution approach): Mechanisms of thin film formation* (St. Petersburg: Nauka) p. 309

[11] Muth J F, Lee J H, Shmagin I K, Kolbas R M, Casey J H C, Keller B P, Mishra U K, DenBaars S P 1997 *Appl. Phys. Lett.* **71** 2572
[12] Jarjour A F, Green A M, Parker T J, Taylor R A, Oliver R A, Briggs G A D, Kappers M J, Humphreys C J, Martin R W, Watson I M 2006 *Physica E* **32** 119
[13] Bae S Y, Kong D J, Lee J Y, Seo D J, Lee D S 2013 *Optics Express* **21** 16854

Journal of Physics: Conference Series **741** (2016) 012084 doi:10.1088/1742-6596/741/1/012084

Ultrahigh precision synchronization of optical and microwave frequency sources

A Kalaydzhyan[1], M Y Peng[2], F X Kärtner[1,2,3]

[1]Center for Free-Electron Laser Science, Deutsches Elektronen-Synchrotron DESY, D-22607 Hamburg, Germany

[2]Research Laboratory of Electronics, Massachusetts Institute of Technology, Cambridge, MA 02139, USA

[3]Department of Physics, University of Hamburg and the Hamburg Center for Ultrafast Imaging, D-22761 Hamburg, Germany

E-mail: aram.kalaydzhyan@cfel.de

Abstract. In this paper we demonstrate that balanced optical-microwave phase detectors (BOMPD) are able to provide a robust long-term optical-RF synchronization with sub-femtosecond residual timing drift over 24 hours in laboratory conditions without active temperature control of optical and electronic paths. Moreover, 10.833 GHz Sapphire-loaded cavity oscillator (SLCO) was successfully disciplined by 216.66 MHz laser oscillator using the BOMPD which resulted in a sub-femtosecond RMS jitter integrated from 1 Hz to 1 MHz.

1. Introduction

Optical frequency combs have become a part of everyday high-accuracy frequency and phase measurements by the outstanding spectral resolution and possibility to reference to both optical and radio-frequency (RF) standards [1]. In particular, they find a place in ultra-low phase noise microwave generation based on direct detection of the pulse train emitted by a mode-locked laser (MLL). Such systems can achieve sub-femtosecond short-term stability of the generated multi-GHz signals [2]. However, nonlinear effects by the photodetection degrade long-term phase stability of the extracted microwave signals by making it sensitive to optical power fluctuations and environmental drifts [3]. This fact forces to study the electro-optical features of each particular photodetector used in an experiment to find its optimal operation conditions and minimize the influence of non-linear effects.

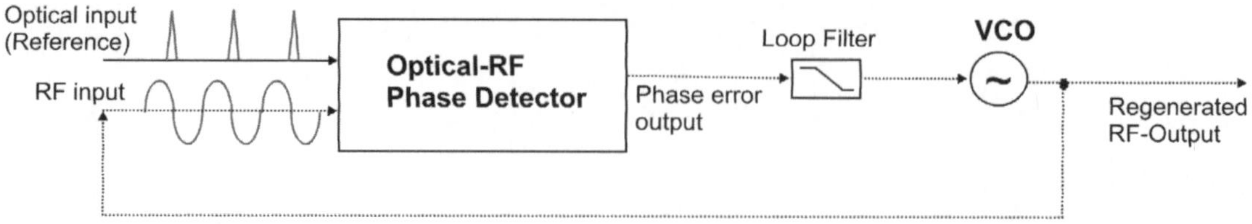

Figure 1. Schematic diagram of an opto-electronic phase-locked loop.

There is a workaround for these issues - one should utilize the detected signal as a frequency reference in a RF phase-locked loop (PLL) that would be insensitive to the input optical power fluctuations. Narrowband voltage-controlled oscillators (VCOs) can be disciplined by an optical pulse train generated by a femtosecond MLL to significantly improve its long-term stability. Such a perfect marriage requires a matchmaker who translates frequency and phase stability from optical domain to electronic domain, a hybrid optical-microwave phase detector (see Figure 1). Various schemes of such phase detectors were successfully implemented to the hybrid PLLs [4, 5], however their high sensitivity to temperature and humidity drifts put obstacles in the way to achieve sub-1-fs RMS stability over several hours of operation.

2. Experiment

Here, we are using a hybrid balanced optical-microwave phase detector (BOMPD) to connect a 216 MHz MLL (OneFive) and a 10.8 GHz Sapphire-loaded cavity oscillator (PSI) together in the optoelectronic PLL. The BOMPD operation principle is based on balanced optical heterodyne detection with use of the fiber Sagnac interferometer [6]. In addition to the last modifications of the BOMPD scheme [4] with multi-GHz and, therefore, unidirectional phase modulation of the optical pulses in the Sagnac loop, we have implemented an independent RF demodulation arm for the error signal (see Figure 2). The signal demodulation is performed at the lowest possible frequency (half of the MLL repetition rate) to maximize SNR at photodetection and to minimize thermally- and humidity-induced phase drifts in the electronic and optical paths for long-term stability.

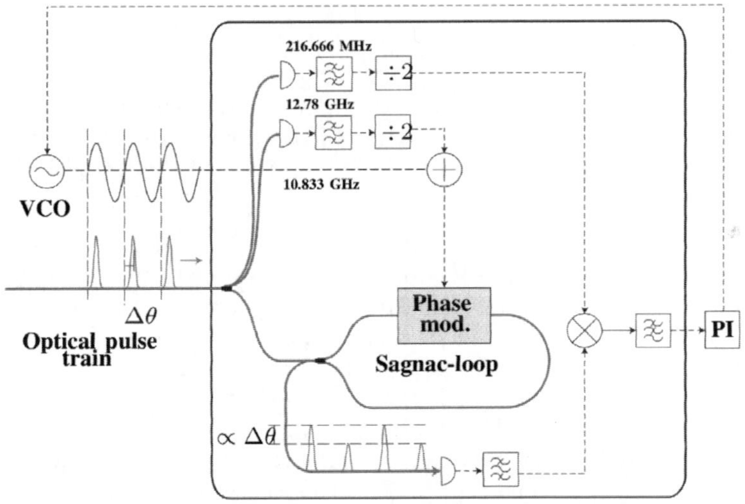

Figure 2. Scheme of the BOMPD with an additional independent reference path down-mixing of the modulated pulse train at the output of Sagnac-loop interferometer. Much lower operating frequency of this reference path makes possible further suppression of the photodetection noise floor without limiting of the PLL's operating bandwidth.

Microwave phase trimmers in the RF paths were replaced by free-space optical delay lines to eliminate the spurious losses and improve precision of phase tuning. Low power consumption low noise RF amplifiers (Micran) have eliminated the need of active cooling of electronics and, as consequence, improved the thermal stability of the setup.

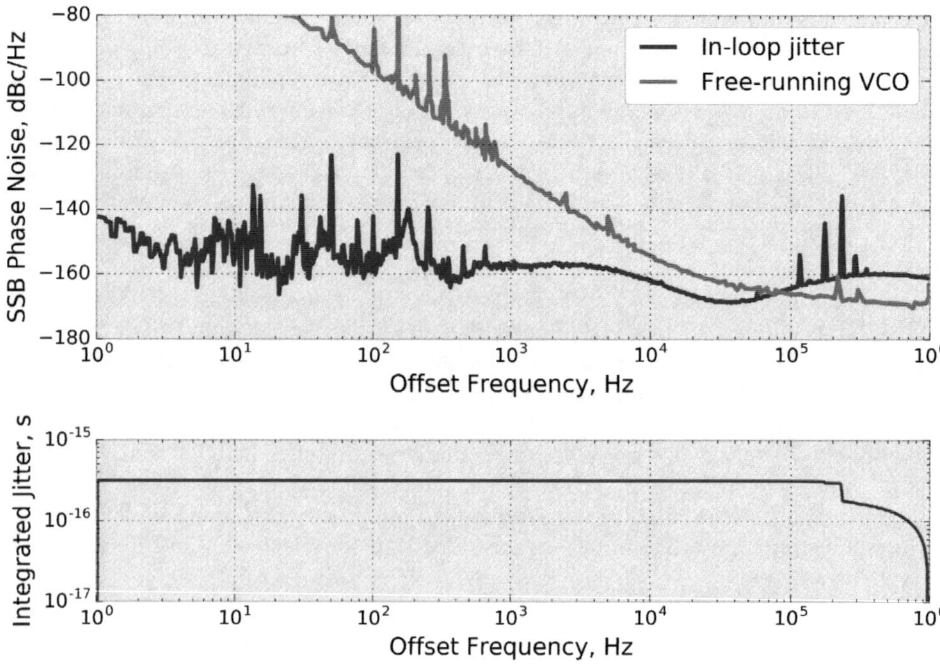

Figure 3. *Top:* Variation of the single-sideband (SSB) phase noise of the free-running SLCO (red) and residual in-loop phase noise of the locked PLL (blue) at 10.833 GHz carrier frequency with offset frequency. *Bottom:* Integrated RMS in-loop jitter of the locked PLL.

As one would expect, phase noise performance of the VCO improves dramatically after switching on of the PLL (see Figure 3). The SSB phase noise curve of the 10.833 GHz carrier stays mainly below -140 dBc/Hz level except several sharp spikes at 50 Hz, 150 Hz and others which are induced by external noise sources, such as 50 Hz power lines, air conditioning system, EMI and others. However, they don't influence much on the value of RMS jitter integrated from 1 Hz to 1 MHz which remains on the level of about 500 as.

Thanks to the optimized power consumption of the RF paths of the BOMPD and, therefore, reduced heat production, it became possible to place the whole setup into thermo-insulating housing without a danger of running out of operating temperature ranges of power RF amplifiers. After a warming-up process the temperature of the most of system components is passively stabilized within a range of 0.1 K, while the value of relative humidity could drift by about 2% per day. The result of the long-term out-of-loop measurement captured by the second identical BOMPD one can observe on Fig. 4. One can see a day-night cycle oscillation, because some components of the PLL are still placed outside of thermo-insulating housing. However, the RMS drift of the regenerated 10.833 GHz signal stays below 1 fs for one day of operation.

The measurement result (Figure 4) shows long-term stable drift with sub-femtosecond RMS deviation for 24 hours of optical to microwave synchronization.

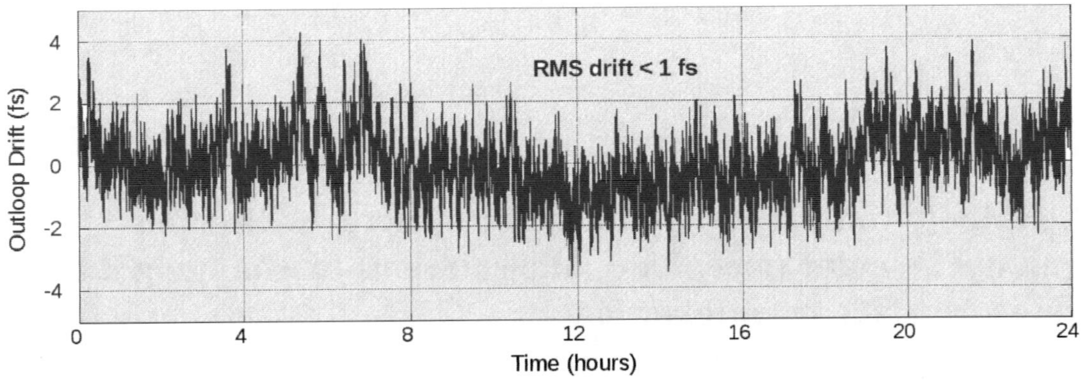

Figure 4. Long-term drift of the system. The RMS value stays slightly below 1 fs during the 24 hour experiment (ADEV $< 1.15 * 10^{-20}$ for 1 day).

3. Conclusion

We have achieved optical-to-RF synchronization of a 10.833 GHz SLCO with a 216.66 MHz laser frequency comb with residual RMS jitter of about 0.5 fs integrated from 1 Hz to 1 MHz and sub-femtosecond daily drift. The new scheme of the hybrid phase detector will be used in one of the future FEL timing distribution systems. In order to make the design of the BOMPD more compact and, therefore, even more stable all RF paths will be placed on a single ceramic PCB (see Figure 5).

Figure 5. Microwave components of one of the BOMPD path's placed on a customized ceramic PCB.

References

[1] Ye J, Schnatz H and Hollberg L W 2003 *IEEE J. Sel. Top. Quantum Electron.* **9**(4) 1041
[2] Fortier T, Quinlan F, Hati A, Nelson C, Taylor J, Fu Y, Campbell J and Diddams S 2013 *Opt. Lett.* **38** 1712
[3] Williams K J, Esman R D and Dagenais M 1996 *J.Lightwave Technol.* **14**(1) 84
[4] Peng M Y, Kalaydzhyan A and Kärtner F X 2014 *Opt. Express* **22** 27102
[5] Jung K and Kim J 2012 *IEEE Int. Frequency Control Symp.* (Baltimore, USA) 768
[6] Kim J and Kärtner F X 2010 *Laser Photon. Rev.* **4**(3) 432

Journal of Physics: Conference Series **741** (2016) 012085 doi:10.1088/1742-6596/741/1/012085

Selectivity control of photosensitive structures based on gallium arsenide phosphide solid solutions by changing the rate of surface recombination

S A Tarasov, M Y Andreev, I A Lamkin, A V Solomonov
Saint-Petersburg Electrotechnical University "LETI", Prof. Popova 5, St. Petersburg 197376, Russia

Abstract. In this paper, we demonstrate the effect of surface recombination on spectral sensitivity of structures based on gallium arsenide phosphide solid solutions. Simulation of the effect for structures based on a p-n junction and a Schottky barrier was carried out. Photodetectors with different rates of surface recombination were fabricated by using different methods of preliminary treatment of the semiconductor surface. We experimentally demonstrated the possibility to control photodetector selectivity by altering the rate of surface recombination. The full width at half maximum was reduced by almost 4 times, while a relatively small decrease in sensitivity at the maximum was observed.

1. Introduction

Photosensitive structures and devices based on such structures are widely used in modern optoelectronics. An important characteristic of such structures and devices is the spectral sensitivity to photon flux radiation. Generally, the range is broad and greater than 300 nm. The benefit of such broadband photodetectors is versatility for recording signals from sources of various types. At the same time, the broadbandness is a drawback for some applications, because in addition to the desired signal, background signals from the Sun, illuminating lamps and other interference units can be recorded. To solve this problem, selective photodetectors with narrow spectral response should be employed [1], [2]. Besides, selective photodetectors should be capable of recording signals from a specified irradiator under strong background noise within other wavelength ranges. High selectivity can be achieved by using optical filters. Unfortunately, it is often the case that filters with necessary selectivity parameters cannot be found, or tend to degradation (solarization). Moreover, their implementation increases structural complexity, dimensions and device cost. Therefore, the optimum is to achieve the high selectivity by modifying operational principle of the photosensitive structures [3].

2. Experimental results and discussions

Surface and interface make significant contribution to operation of photosensitive structures due to a large number of defects and impurities, which appear as deep-level states in the band diagram, called surface states. In most cases, surface states have a negative impact on device operation, because their presence increases the recombination probability and as a consequence the photocurrent value decreases. However, it should be taken into account that contribution from surface recombination differs depending on sensitivity subranges, i.e. the difference in contribution is due to absorption coefficient dependence of the photon energy. Thus, surface recombination is particularly evident in short-wavelength region of the spectral sensitivity.

In this work, we demonstrate the effect of surface recombination on spectral sensitivity of structures based on the p-n junction and the Schottky barrier. Based on solutions of the continuity equation we estimated the current flow through different parts of the structure and obtained expressions which describe transfer of photogenerated charge carriers. The calculations take into account effects of surface recombination on the illuminated side of the structure. Expressions for the quantum efficiency and the structure photosensitivity were obtained. Spectral sensitivity at various rates of surface recombination was calculated.

Consider an n-type semiconductor where electron-hole pair generation is caused by incident light with frequency hν. The number of recombining minority carriers on the crystal surface per surface unit area is proportional to the deviation of the minority carrier concentration from the equilibrium value [80]. The distribution of the minority carrier concentration is given by the continuity equation.

Figure 1 shows the depth distribution of the generated carriers. The minority carrier concentration at the surface recombination rates s=0 cm/s and s=10^5 cm/s are marked by the solid and dashed curves correspondingly. The absorption coefficient is equal to 10^2 cm^{-1}, the quantum efficiency is determined by the area under the curve.

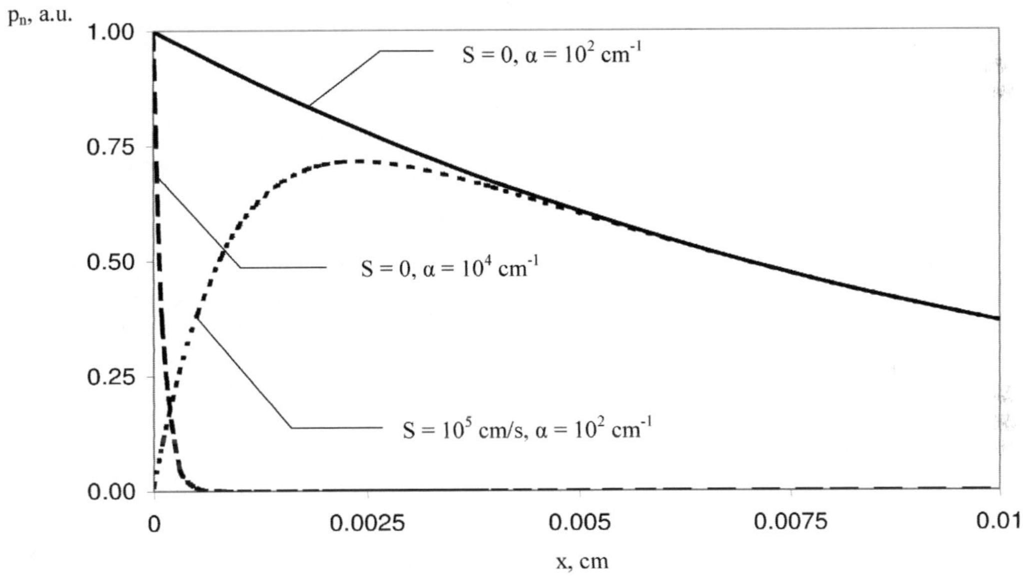

Figure 1. The distribution of carriers generated in the sample

As shown in the Figure 1, when increasing the surface recombination rate the carrier concentration at x = 0 begins to fall, and the area under the curve as well as the quantum efficiency are reduced. However, at low values of the absorption coefficient the incident light deeply penetrates into the semiconductor layer, and a small fraction of carriers recombines on the surface. The more the absorption coefficient is, the more carriers are generated in the surficial region, the less carriers contribute to the photocurrent. Since α increases with increasing the incident radiation energy (i.e. the illumination wavelength decreases), then we can control the photodetector sensitivity by varying the surface recombination rate in the short-wavelength range.

We have analysed two types of selective photodetectors based on A3B5 solid solutions. Figure 2 shows the result of calculation for a p-n junction with n-zone as a substrate.

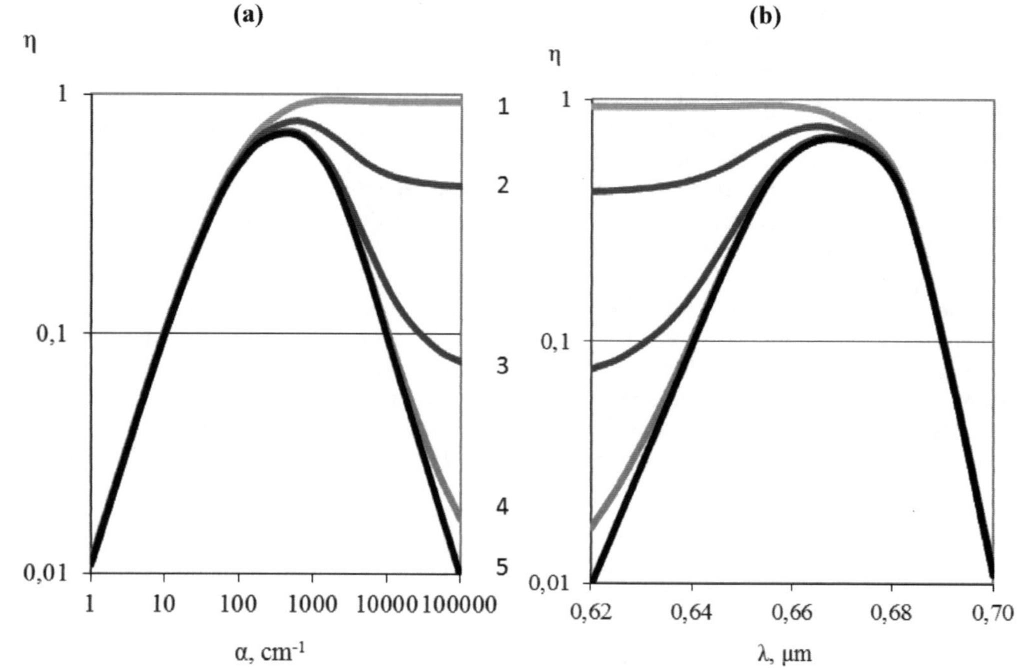

Figure 2 (a, b). The quantum efficiency as a function of the absorption coefficient **(a)** and the incident light wavelength **(b)** for a p-n-junction to $GaAs_{0,7}P_{0,3}$ $L_n = L_p = w = 5 \cdot 10^{-4}$cm; $d=5 \cdot 10^{-5}$cm: **1** - s=0 cm/s; **2** - s=10^3cm/s; **3** - s=10^4cm/s; **4** - s=10^5cm/s; **5** - s=10^7cm/s.

Figure 3 shows the result of calculation for the metal-semiconductor contact illuminated from the metal film side. While the quantum efficiency significantly reduces at larger absorption coefficients, it remains almost unchanged at the maximum. At $\alpha = 10^5$ cm^{-1} the sensitivity decreases by almost an order of magnitude, whereas at the maximum it remains at the level of 80 % of the quantum efficiency at s = 0 cm/s

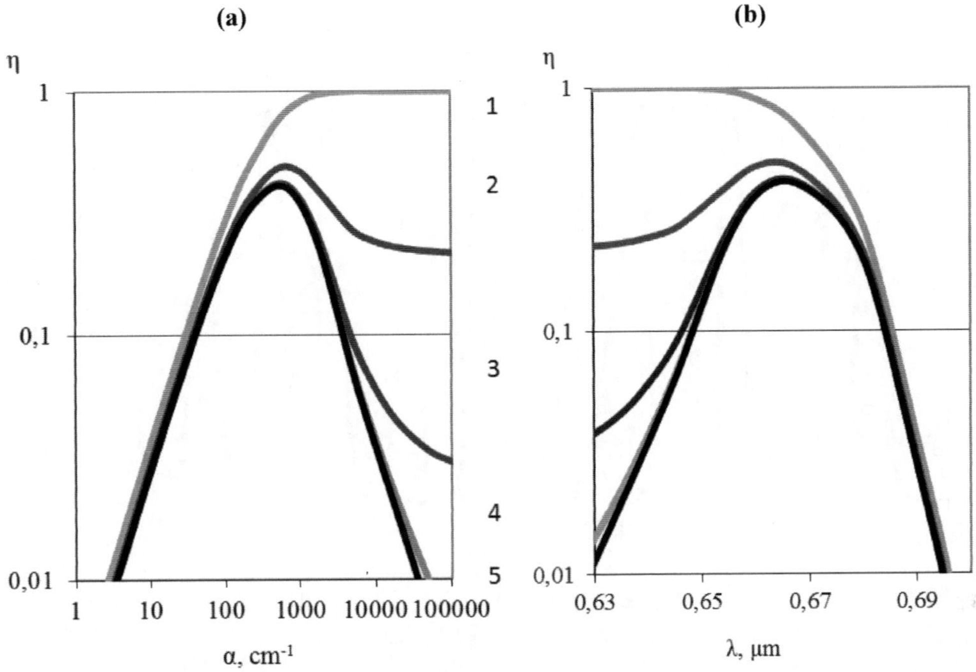

Figure 3 (a, b). The quantum efficiency as a function of the absorption coefficient **(a)** and the incident light wavelength **(b)** for a p-n-junction to $GaAs_{0,7}P_{0,3}$ $L_n = L_p = w = 5 \cdot 10^{-4}$cm; $d = 5 \cdot 10^{-5}$cm: **1** - s=0 cm/c; **2** - s=10^3cm/c; **3** - s=10^4cm/c; **4** - s=10^5cm/c; **5** - s=10^7cm/c.

The calculations demonstrate the significant difference between the surface recombination effects on the sensitivity at the maximum and in the short-wavelength range. In general, the increase of surface recombination rate decreases the sensitivity in the photodetector operating range. Nevertheless, while the decrease at the maximum is relatively small and equal to 20-50 % depending on the structure type at high rates of surface recombination, the sensitivity decreases by 2-3 orders of magnitude in the short-wavelength range. Such effect can be used to achieve high selectivity without extra filtering (Figure 4).

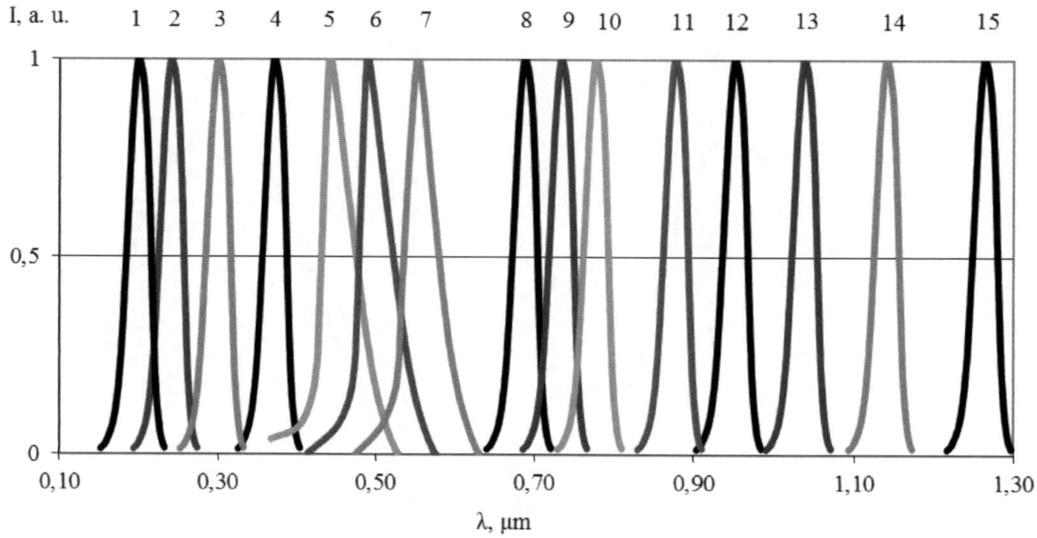

Figure 4. The spectral characteristics of the structures based on a p-n-junction at s=10^7cm/s: **1-3 - $Al_xGa_{1-x}N$**: 1-x=0,7, 2-x=0,4, 3-x=0,1; **4-SiC; 5-10 - $GaAs_xP_{1-x}$**: 5-x=0, 6-x=0,2, 7-x=0,4, 8-x=0,6, 9-x=0,7, 10-x=0,8; **11-15 - $InAs_xP_{1-x}$**: 11-x=0, 12-x=0,1, 13-x=0,2, 14-x=0,4, 15-x=0,5.

3. Conclusion

The rate of surface recombination can be set by proper choice of preliminary treatment of the semiconductor surface. For this purpose, various etching reagents such as acids (hydrochloric acid, nitric acid, aqua regia etc.), alkali (KOH etc.) and their combinations with other reagents were used. Samples treated with abovementioned methods as well as non-treated samples were fabricated. The structures were sensitive in the visible and ultraviolet spectrum ranges. While the long- wavelength edge was set by the solid solution composition and the Schottky barrier height, the short-wavelength edge was primarily dependent on the rate of surface recombination. The structures based on epitaxial gallium phosphide layers had the following parameters: responsivity of 0.19 A/W at the maximum, full width at half maximum (FWHM) of 200 nm, dark current of -0,02pA at a reverse bias of V = -1 V. Surface treatment reduced FWHM by almost 4 times. At the same time, the sensitivity at the maximum was reduced by less than 50 %, whereas in short-wavelength region the sensitivity decreases by almost two orders of magnitude. Currently, we work on further improving the selectivity and increasing the sensitivity of the photodetectors.

Acknowledgments

The work was supported by the Ministry of Education of the Russian Federation in the framework of the project of the state task in the field of scientific activity, project number 16.1307.2014K.

References

[1] Lamkin I A, Tarasov S A, Petrov A A, Menkovich E A, Solomonov A V, Kurin S Yu *2014 Journal of Physics: Conference Series* **572** 012063
[2] Lamkin I A, Andreev M Y, Tarasov S A, Solomonov A V, Kurin S Yu 2015 *Proceedings of the 2015 IEEE North West Russia Section Young Researchers in Electrical and Electronic Engineering Conference* p 30
[3] Lamkin I, Tarasov S 2013 *Journal of Physics: Conference Series* **461** 012025

Optimization of structural and growth parameters of metamorphic InGaAs/GaAs photoconverters grown by MOCVD

D V Rybalchenko[1], S A Mintairov[1], M Z Shvarts[1], and N A Kalyuzhnyy[1]
[1]Photovoltaic Laboratory, Ioffe Institute, Saint Petersburg, Russia

Abstract. Metamorphic $Ga_{0.76}In_{0.24}As$ heterostructures for PV converters of 1064 nm laser radiation have been grown by the MOCVD. Parameters of the GaInAs metamorphic buffer layer with a stepwise profile of In composition variation were calculated. Its epitaxial growth conditions have been optimized, which allowed improving collection of charge carriers from the n-GaInAs base region and obtaining the photo-response quantum yield of 83% at 1064 nm wavelength. It has been found that, due to discontinuity of valence bands at the $In_{0.24}Al_{0.76}As$-p/$Ga_{0.76}In_{0.24}As$-p heterointerface (window/emitter) a potential barrier for holes arises as a result of low carrier concentration in the wide-band-gap material. The use of InAlGaAs solid solution with Al concentration of < 40% has allowed raising the holes concentration in the wide-band-gap window, eliminating completely the potential barrier and reducing the device series resistance. Optimization of the PV converter metamorphic heterostructure has resulted in obtaining 1064 nm laser radiation conversion efficiency at the level of 38.5%.

1. Introduction

Development of photovoltaic (PV) converters of laser radiation (LR) with wavelength of the atmosphere transparency for wireless transmission of energy and information is an important task for elaborating cellular communication in remote and hard to reach places. In conditions, when construction of communication systems is difficult to perform or is not profitable, it is possible to apply unmanned aerial vehicles. They can play a role of base stations or Internet access points in communicating with terrestrial networks by means of a laser – PV converter optical pair [1, 2]. For such an application, a conventional Nd:YAG laser with a powerful emission mode at 1064nm is the best suited and also has a low absorption factor in the atmosphere. This leads to only 8% difference at 1064nm wavelength between space and terrestrial solar spectra. Currently, important is development of heterostructures for PV converters of the 1064nm LR energy and having high efficiency.

A record 1064 nm laser radiation conversion efficiency for silicon-based structures is close to 40% [3]. However, this result was achieved by using a very complicated procedure for the surface texturing. Furthermore, the maximum efficiency values were achieved at a LR concentration less than 1.5 W/cm^2, which is not enough for wireless power transmission networks, and the low carrier mobility makes Si unreasonable for the manufacture of high-frequency PV converters.

In the present work, PV converters of 1064 nm LR based on metamorphic $In_xGa_{1-x}As$ heterostructures grown by the MOCVD technique on cheap and conventional GaAs substrates have been developed.

2. Experimental details

Metamorphic PV structures were grown in a low pressure MOCVD reactor under 100 mbar on n–GaAs (100) substrates misoriented by 6° towards the [111] direction. Metal alkyls were used as sources of the III-group atoms: threemethylgallium, threemethylaluminum, threemethylindium. Arsine was used as a source of As atoms. Monosilane and diethylzinc were used to dope layers with n–type impurity p–type impurity respectively.

The PV structure represents a *p-n* diode with photoactive base (of n-type) and emitter (of p-type) regions, as well as functional layers: a back surface field (BSF) and wide-band-gap window layer (figure 1). Base and emitter thicknesses have been chosen 3.5 μm and 500 nm respectively from consideration of total absorption for GaAs PV converter. As a window layer ternary $Al_{1-x}In_xAs$ and quaternary $In_x(Al_{0.5}Ga_{0.5})_{1-x}As$ solid solutions were used. A metamorphic buffer (MB) was grown for the compensation of the lattice mismatch between $Ga_{1-x}In_xAs$ layer and GaAs substrate. Variation of In concentration *"x"* in $Ga_{1-x}In_xAs$ layers of the metamorphic buffer was performed in "steps" with a step of about 2.5%, each 120 nm thick, to create the sufficient amount of interfaces, at which dislocations can be bent.

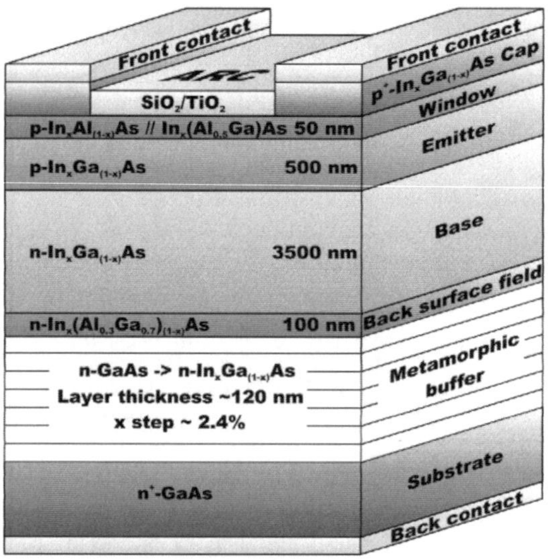

Figure1. Schematic of the metamorphic $In_xGa_{1-x}As$ PV converter cross-section.

Using the grown heterostructures, 3x3.4 mm PV converters were fabricated using photolithography, and then their spectral and photovoltaic characteristics were measured. Modeling of a band diagrams was carried out by means of the AFORS-HET software [4].

3. Results and discussion

Internal quantum efficiency (IQE) at 1064 nm wavelength for initial $Ga_{0.76}$ $In_{0.24}As$ PV converter was approximately 77%. Incomplete charge carrier collection from the n-GaInAs base region was due to small diffusion lengths of charge carriers [5], that is common for lattice mismatched heterostructures. One of the factors which might decrease diffusion lengths is a formation of threading dislocations in MB region.

Threading dislocation density in the base layer depends on parameters of MB – thickness and number of its sublayers with constant composition, as well as its growth temperature and growth rate. Thickness of each sublayer was approximately 120 nm, which is 1.5 times more than a critical thickness for $GaAs/Ga_{0.975}In_{0.025}As$ layer calculated by means of the Matthews-Blakeslee theory. In order to reduce threading dislocation density in the base layer and increase carrier's diffusion lengths, optimization of the growth parameters of MB was carried out: buffer layer growth temperature was varied within 550 – 700 °C, and the growth rate – within 1.5 – 6 μm/h. Decrease of MB growth temperature to 550 °C at a constant growth rate (3 μm/h) resulted in decreasing the IQE value (figure 2 (a)). At this temperature, adatom migration rate is dramatically reduced, therefore, at a constant delivery rate of precursors to the growth zone, the morphology of the structure becomes worse, leading to wave-like surface. Optimal growth parameters of forming MB have been found to be – 3 μm/h at 600 °C.

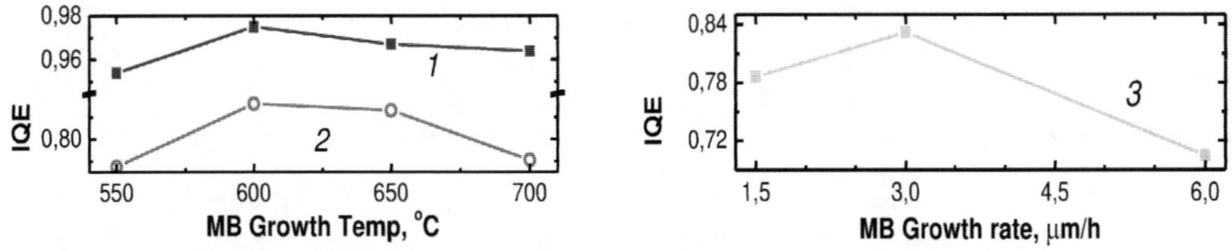

Figure2. *(a)* – Internal quantum efficiency (IQE) vs. MB growth temperature: 1 – maximal IQE, 2 – IQE at 1064 nm wavelength; *(b)* – IQE at 1064 nm wavelength (MB growth temp – 600 °C) vs. MB growth rate.

As a result, charge carrier collection from the n-GaInAs base region was improved, which allowed achieving the photo-response quantum efficiency at 1064 nm wavelength of 83% (figure 3).

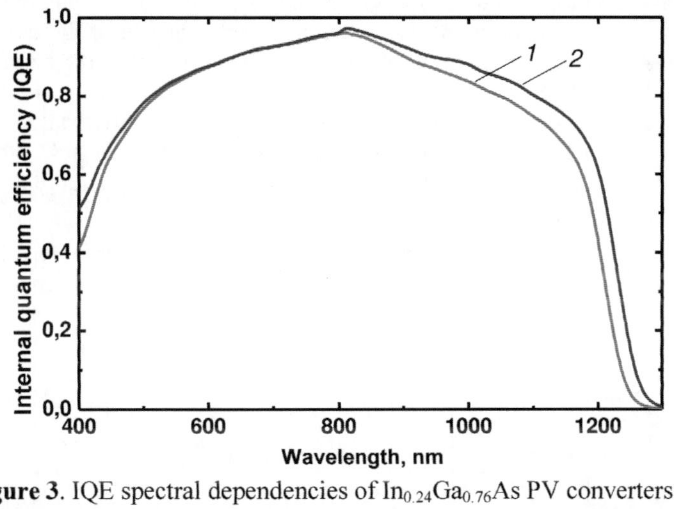

Figure 3. IQE spectral dependencies of $In_{0.24}Ga_{0.76}As$ PV converters with $In_{0.24}(Al_{0.5}Ga_{0.5})_{0.76}As$ window layer and: 1 – non-optimized MB, 2 – with optimized MB.

However, the light IV curves of such PV converters were characterized by a large series resistance and low fill factor (FF) (figure 4, blue curve). It may be caused by a potential barrier in PV structure for majority charge carriers.

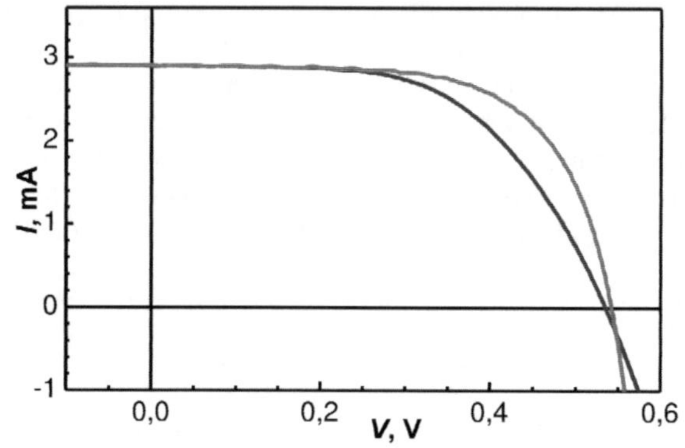

Figure4. Light IV curves of $Ga_{0.76}In_{0.24}As$ PV converter with window layer composition: 1 – $Al_{0.76}In_{0.24}As$, 2 – $In_{0.24}(Al_{0.5}Ga_{0.5})_{0.76}As$.

Simulation of the $Ga_{0.76}In_{0.24}As$ PV converter heterostructure band diagram was carried out by means of the AFORS-HET software. It has been found that a potential barrier for holes is present on the window/emitter heterointerface caused by discontinuity of the valence bands at the $In_{0.24}Al_{0.76}As$-p/$Ga_{0.76}In_{0.24}As$-p interface (figure 5). In case of $In_{0.24}Al_{0.76}As$ wide-band-gap window layer, the Fermi level departs substantially from the valence band. Such a shape of the band diagram is also associated with a low level of the wide-band-gap window carrier concentration compared with that of the emitter and contact layers. To eliminate the potential barrier introduction of Ga atoms into sublattice of the third group, at a concentration equal to Al, has been proposed. The use of the quaternary InAlGaAs solid solution (with Al concentration reduced down to 38%) as the wide-band-gap window has permitted to raise the holes concentration of up to about ~ 9×10^{18} cm^{-3}. Analysis of the band diagram has shown that application of the narrow-band $In_{0.24}(Al_{0.5}Ga_{0.5})_{0.76}As$-p material with a high acceptor concentration allows eliminating completely the observed potential barrier, which is supported experimentally by a significant reduction of the PV converter resistance (figure 5) and by fill factor increasing (figure 4, red curve).

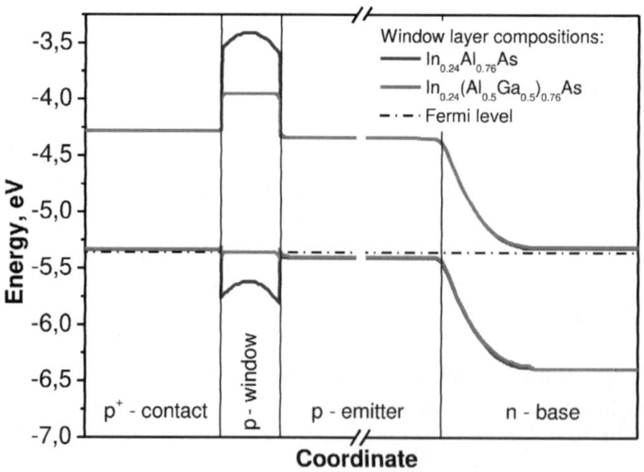

Figure5. Band diagram of the $Ga_{0.76}In_{0.24}As$ PV converter heterostructures: with $Al_{0.76}In_{0.24}As$ window (blue curve) and $In_{0.24}(Al_{0.5}Ga_{0.5})_{0.76}As$ window layer (red curve).

The photovoltaic converters fabricated on the base of metamorphic heterostructures with optimized wide-band window layers and metamorphic buffer layer have demonstrated the 1064 nm LR conversion efficiency of about 38.5% (figure 6, curve 2).

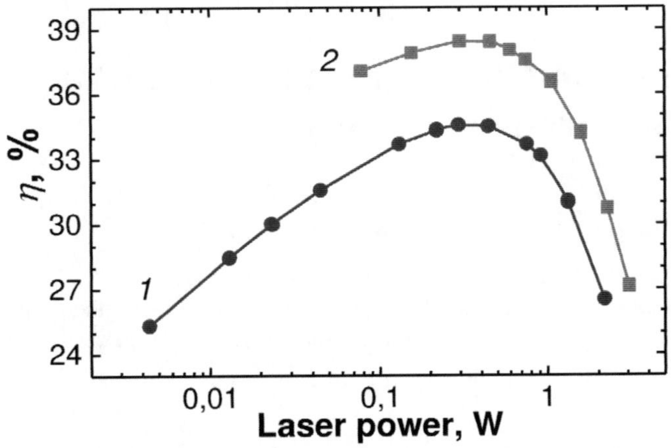

Figure6. Dependence of the 1064 nm LR conversion efficiency on laser power of $In_{0.24}Ga_{0.76}As$ PV converter with $In_{0.24}(Al_{0.5}Ga_{0.5})_{0.76}As$ window layer and: 1 – non-optimized MB, 2 – optimized MB.

4. Conclusion

Optimization of structural and growth parameters of the $In_{0.24}Ga_{0.76}As$ metamorphic PV converters consisted in analysis of the band diagram and development of growth parameters of stepwise MB has allowed reaching 1064 nm radiation conversion efficiency of 38.5% (at laser radiation power density ~ 5 W/cm^2).

Acknowledgments

The study was supported by the Russian Foundation for Basic Research (grant #. 16-08-00742 a) in part of structure epitaxial growth.

References

[1] https://www.nasa.gov/centers/armstrong/news/FactSheets/FS-087-DFRC.html
[2] Landis G, Stavnes M, Oleson S, Bozek S 1992 *Proc. of the 28th SAE, ASME, and ASEE Joint Propulsion Conference and Exhibit*
[3] Green M, Zhao J, Wang A, Wenham S 1992 *IEEE Electron Device Lett.* **13** 317
[4] Stangl R, Kriegel M, Schmidt M 2006 *Proc. of the 4th World Conf. on Photovoltaic Energy Conversion* 1350
[5] Mintairov S A, Emelyanov V M, Rybalchenko D V, Salii R A, Timoshina N K, Shvarts M Z, Kalyuzhnyy N A 2016 *Semiconductors* **50** (4)

Evaluation of the quality of green InGaN LEDs by values of the threshold current

O A Radaev[1], V A Sergeev[1,2] and I V Frolov[2]

[1]Ulyanovsk State Technical University, Ulyanovsk, 432027, Russia
[2]Ulyanovsk Branch of Kotel'nikov Institute of Radio-Engineering and Electronics of RAS, 48/2 Goncharov Street, Ulyanovsk 432071, Russia

Abstract. The article shows the possibility of using the threshold current for evaluating the quality of green InGaN/GaN LEDs. It was determined that the current threshold correlated with the position of the maximum of the current dependence of external quantum efficiency, and a concentration gradient of charge carriers in the heterostructure.

1. Introduction

LEDs based on InGaN/GaN heterostructures are currently widely used in various branches of engineering. In order to improve the reliability of technical devices and systems based on InGaN/GaN LEDs devices and systems of nondestructive testing their quality is actively being developed [1, 2]. One of the least understood, but quite informative parameter, which gives representation about the quality of the LED heterostructure is the threshold current. For lasers this term refers to the current at which the optical radiation from spontaneous enters induced. For LEDs this term refers to the minimum current at which the optical radiation can be registered [3, 4].

It is known that luminescence of an ideal LED comes with an external voltage, compensating the contact potential difference $U_D = E_g/e$, where E_g – band gap, e – electron charge [3]. However, the voltage drop across the base resistance and the ohmic contacts of the LED, the presence of discontinuities of the conduction band and valence band of a semiconductor heterostructure leads to a deviation of the threshold voltage from the U_D value. Thus the current corresponding to the threshold voltage, characterized non-radiative recombination losses in the system of defects penetrating the LED active region [4], therefore, the characteristics of the sampling distributions of the LEDs on the threshold current values can be used to evaluation the quality of the light emitting heterostructure.

2. Experimental details

This article presents the results of a study of green InGaN/GaN LEDs type ARL-5215PGC produced by Arlight in the number of N = 100 pieces, with the following parameters: the wavelength of the maximum of emission spectrum λ_{max} = 525 nm, the maximum constant operating current I_{max} = 30 mA, the crystal size 270x320 μm.

Measurement of threshold current of LEDs made on the automated installation [5], the operation algorithm which is to measure of the LED emission power depending from the current $P(I)$ by highly sensitive photodetector in the current range, corresponding to the beginning of the LED illumination, and subsequent approximation of this dependence by an exponential function of the form $P_0 \cdot (I/I_0)^a$

(figure 1). The current I_{th} at which the approximating function crosses the level of the photodetector dark signal, is taken as the numerical value of the threshold current.

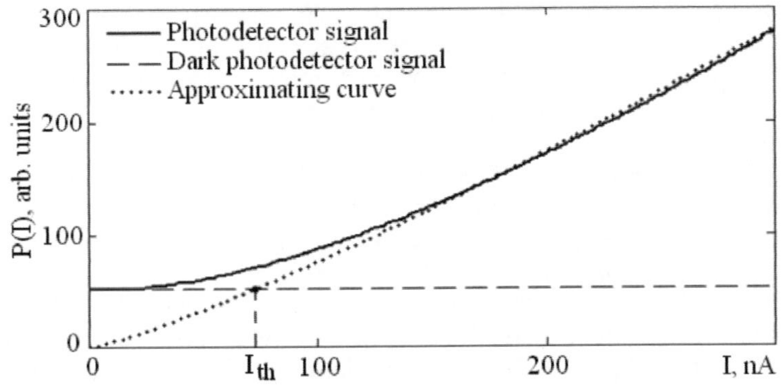

Figure 1. Graphs illustrating the principle of determining the value of the LED threshold current.

Measurement of the external quantum efficiency (EQE) of LEDs was performed using an integrating sphere TKA-KK1.

Measurement of the capacitance-voltage characteristics of LEDs performed in an automated installation [6] in a range of bias voltages +2...–45 V at room temperature. According to the measured capacitance-voltage characteristics was performed calculation of the profile of concentration charge carriers distribution in the heterostructure $n(w)$.

3. Results and discussion

The measurement results showed that the current threshold I_{th} of the researched type LEDs (at an ambient temperature T = 25 °C) are in the range 21...607 nA. 72 LEDs in sampleare the values of threshold current in the range 21...90 nA, while the remaining 28 samples are the values of I_{th} in the range of 90...607 nA. Histogram of LEDs distribution by threshold current values is shown in figure 2. The average value of the threshold current in the sample is 110 nA, the standard deviation is 113 nA. The proximity of the estimation of mathematical expectation value and the estimation of standard deviation indicates the exponential law of distribution of LEDs by threshold current values. We assume that the LEDs are characterized by lower concentrations of defects in the active region, are in the first interval of the distribution histogram, which is marked in figure 2 a gray background, that is, have a threshold current value in the range of 21...90 nA.

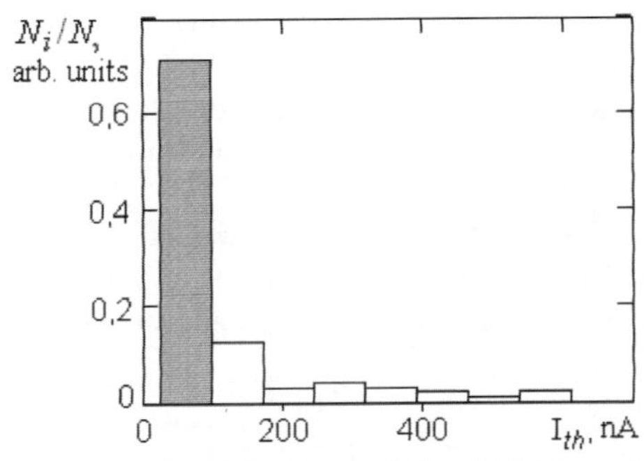

Figure 2. Histogram of distribution of LEDs group of N = 100 pieces by threshold current values.

To confirm this statement, we investigated the relationship between the values of the threshold current and character of the EQE versus current density depending $\eta(j)$ of the LED. Figure 3 shows plots of EQE versus current density for three investigated LEDs samples having different values of the threshold current I_{th}. It can be seen that the most noticeable variation of the EQE values is reveals at the low current density range, corresponding to rise range of function. At current densities corresponding to range of decrease of EQE, LEDs have similar external quantum efficiency values. According to the model described in [7], it is explained as follows. At low current densities the non-radiative recombination centers present in the LED active region can efficiently trap the injected carriers, thus reducing the probability of radiative recombination, and consequently, the radiation power. At high current densities the injection of a large number of charge carriers leads to a saturation of non-radiative centers, so differences in the radiation power and therefore the EQE of LEDs at high measurement currents is much smaller than for low current.

Statistical analysis of the sample measurement results showed that the value of the threshold current is correlated with the value of the EQE of LEDs at low current densities, corresponding to rising of the current dependence of EQE. The dependence of the external quantum efficiency $\eta(j)$ from the LEDs with a low threshold current value I_{th} = 22 nA reaches a maximum at 0.1 A/cm^2 current density, while the LEDs with a high threshold current value I_{th} = 607nA – at 0.6A/cm^2, while absolute value of EQE according to maximum of $\eta(j)$ is the greater, the smaller the value I_{th} (figure 3).

These results suggest that the greater the threshold current of the LED, the higher the probability of tunneling non-radiative recombination, and hence the higher the concentration of defects in the active region of the LED.

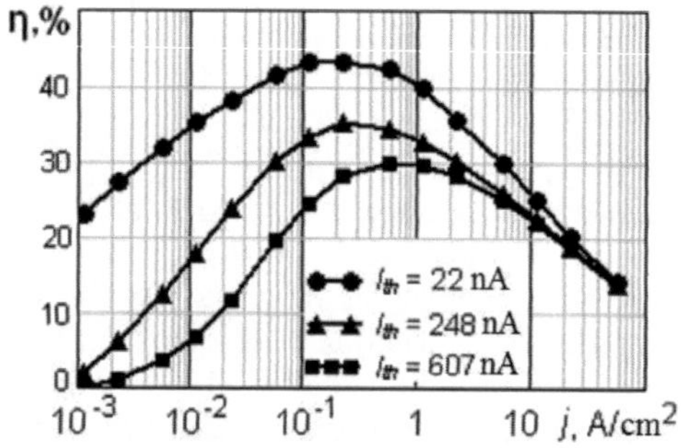

Figure 3. The current density dependence of the external quantum efficiency for three LEDs with different values of the threshold current.

It is known that the defect concentration is largely determined by the degree of ordering of the structure [4]. Evaluation of ordering of the structure can be performed by profile of concentration charge carriers distribution in the heterostructure is obtained by measuring the capacitance-voltage characteristics of the LED.

In [8] the correlation between the value of the LED light power and the character of the concentration profile is shown: the wider the area of increase of charge carriers concentration from the minimum to the maximum, the higher the power of the LED light. The authors explain this effect to the improvement of injection and distribution of charge carriers in the active region of the LED.

We propose to use to assess the of ordering of the active region of LEDs concentration gradient of charge carriers in the heterostructure [9], defined according to the formula:

$$\frac{\Delta n}{\Delta w} = \frac{\int_{w_{min}}^{w_{max}} n(w)dw}{(w_{max} - w_{min})^2}, \quad (1)$$

where w_{min} and w_{max} – the borders of measured profile concentrations.

As aresult of research of LEDs sample we have found that LEDs with a low threshold current value (about 21...50 nA) have a more uniform concentration profile (figure 4, curve 1) than the LEDs with a high threshold current (of the order of 200...607 nA) (figure 4, curve 2). The correlation coefficient between the values of the threshold current I_{th} and the concentration gradient of charge carriers $\Delta n/\Delta w$ is 0.7. Thus, LEDs with a more uniform concentration profile have less defects in the active region and are characterized by small values of the threshold current.

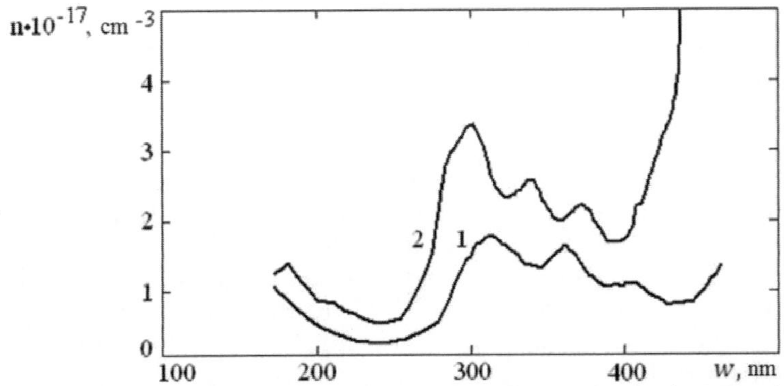

Figure 4. Profiles profile of concentration charge carriers distribution in the LEDs heterostructure with a small (1), and large (2) value of the threshold current.

4. Conclusions

The results of the studies showed that the threshold current can be used to assess the quality of the LEDs based on InGaN/GaN nanoheterostructure. The distribution of LEDs by values of threshold current has a form close to exponential, indicating the correlation between this parameter and defect density in the active region of the heterostructure [10]. The threshold current values is correlated with the value of the EQE of the LED at low current densities range corresponding increase external quantum efficiency with increasing current, and with a charge carrier concentration gradient value. The more uniform concentration profile and the higher the quantum efficiency of the LED, measured at low currents, the less defects in the active region of LEDs, which are non-radiative recombination centers, and thus, the lower the threshold current. LEDs that are less than the threshold current average value in the sample, characterized by lower concentrations of defects in the active region, thus their quality is higher. Thus, the threshold current of the LED is an important informative parameter that can be used to evaluation of the light-emitting heterostructures quality.

Acknowledgments

The reported study was funded by RFBR, according to the research project No. 16-32-60051mol_a_dk.

References

[1] Nippert F, Karpov S, Pietzonka I, Galler B, Wilm A, Kure T,Nenstiel C, Callsen G, Straburg M, Lugauer H-J, and Hoffmann A 2016 *Japanese Journal of Applied Physics* **55** 05FJ01
[2] Meneghini M, Grassa M la, Vaccari S, Galler B, Zeisel R, Drechsel P, Hahn B, Meneghesso G, Zanoni E 2014 *Applied Physics Letters* **104** 113505
[3] Schubert E F 2006 *Light Emitting Diodes* (Cambridge University Press)
[4] Averkiev N S, Levinshtein M E, Petrov P V, Chernyakov A E, Shabunina E I, Shmidt N M 2009 *Technical Physics Letters* **35** 922
[5] Radaev O A, Sergeev V A 2015 *Radioelectronic technique***2** 249 (in Russian)
[6] Sergeev V A, Frolov I V, Shirokov A A 2014 *Instruments and Experimental Techniques* **1** 137
[7] Meneghesso G, Meneghini M, Zanoni E 2010 *Journal of Physics D: Applied Physics* **43** 354007
[8] Kim C S, Choi R J, Youn H S, Bae S J, Cho H K, Lee B K, Kang D S, HahnY B, Lee H J, and Hong C-H 2003 *Journal of the Korean Physical Society* **42** S367
[9] Sergeev V A, Frolov I V, Shirokov A A, Nizametdinov A M 2013 *Nonlinear world* **7** 493 (in Russian)
[10] Gurvich A K, Ermolov I N, Sazhin S G 1992 *Nondestructive control* (Moscow: High school) (in Russian)

Thin film GaP for solar cell application

I A Morozov [1], A S Gudovskikh [1,2], D A Kudryashov[1], E V Nikitina[1], J-P Kleider[3], A V Myasoedov[4], V Levitskiy[5]

[1]Saint-Petersburg Academic University RAS, 194021, St.-Petersburg, Russia
[2]Saint-Petersburg Electrotechnical University "LETI", 197376, St.-Petersburg, Russia
[3]GeePs, Group of electrical engineering - Paris, CNRS, CentraleSupélec, Univ. Paris-Sud, Université Paris-Saclay Sorbonne Universités, UPMC Univ Paris 063&11 rue Joliot Curie, Plateau de Moulon, 91192 Gif sur Yvette Cedex, France
[4]Ioffe Physical-Technical Institute, 26 Polytechnicheskaya str., 194021 Saint-Petersburg, Russia
[5]Research and development center for thin-film technologies in energetics 26 Polytechnicheskaya str., 194021 Saint-Petersburg, Russia

Abstract. A new approach to the silicon based heterostructures technology consisting of the growth of III-V compounds (GaP) on a silicon substrate by low-temperature plasma enhanced atomic layer deposition (PE-ALD) is proposed. The basic idea of the method is to use a time modulation of the growth process, i.e. time separated stages of atoms or precursors transport to the growing surface, migration over the surface, and crystal lattice relaxation for each monolayer. The GaP layers were grown on Si substrates by PE-ALD at 350°C with phosphine (PH3) and trimethylgallium (TMG) as sources of III and V atoms. Scanning and transmission electron microscopy demonstrate that the grown GaP films have homogeneous amorphous structure, smooth surface and a sharp GaP/Si interface. The GaP/Si heterostructures obtained by PE-ALD compare favourably to that conventionally grown by molecular beam epitaxy (MBE). Indeed, spectroscopic ellipsometry measurements indicate similar interband optical absorption while photoluminescence measurements indicate higher charge carrier effective lifetime. The better passivation properties of GaP layers grown by PE-ALD demonstrate a potential of this technology for new silicon based photovoltaic heterostructure

1. Introduction

Silicon based solar cells are by far dominating the PV market due to the availability and relatively low price of Si substrates. One of the most efficient ways to increase the Si solar cell performance is to use heterojunction designs. Thus heterojunctions combining hydrogenated amorphous silicon (a-Si:H) and crystalline silicon (c-Si) have reached 24.7% conversion efficiency for n-type Si wafers with conventional architecture, and 25.6% when combined with interdigitated back contacts [1]. Thin layers of a-Si:H deposited by PECVD at temperature below 250 °C with a gap of 1.65-1.75 eV provide an excellent passivation of the c-Si surface with relatively low light absorption in a-Si:H. An increase of the Si heterojunction solar cell efficiency could be achieved by further reducing the absorption in the emitter layer using a material with a higher band gap. On the other hand a-Si:H/c-Si heterojunctions are not so successful for p-type Si wafers, which are still of interest for low orbit space applications due to better radiation hardness [2].

GaP could fit both requirements as a wide gap emitter. GaP has a gap of 2.26 eV and has the smallest lattice mismatch beyond III-V binary compounds (less 0.4 %) providing conditions for low defect density at the interface with the Si substrate [3]. Therefore, GaP/Si heterojunctions are of great interest for photovoltaic applications, which according to simulations give an advantage of more than 1% of efficiency compared to passivated emitter and rear cells made of p-type Si wafers [4]. However, the commonly used techniques for the growth of GaP layers such as molecular beam epitaxy (MBE) and metal organic vapor-phase epitaxy (MOVPE) require high temperatures of 800-900°C at least at the initial stage for silicon oxide removing and surface reconstruction [5-9]. High temperatures affect the Si wafers quality and the properties of III V/Si interface leading to low photovoltaic performance of GaP/Si heterojunctions obtained by epitaxy compared to the high efficiency a-Si:H/c-Si solar cells [10, 11]. The problem can be related to elastic stresses in the growing layer, which lead to the appearance of dislocations. This problem is associated primarily with the high temperatures required for the growth due to different coefficients of thermal expansion in silicon and GaP. More-over, the use of high temperatures leads to inter-diffusion of group-III and -V atoms into Si and opposite, which act as dopants affecting the electrical properties of heterojunctions, and it can also promote the fast diffusion of species that can degrade the carrier lifetime in c-Si [12].In this paper we propose to use a new technological approach for the growth of III-V compounds on Si substrates using low temperatures (less than 400°C) and plasma-enhanced atomic layer deposition (PE-ALD). This technique consists of alternatively changing the phosphorus and gallium atom source flows providing the growth of one monolayer by cycle.

2. Experiment

3 inch p-type silicon substrates with a doping level of 10^{16} cm^{-3} and (100) orientation (4° cut-off towards (110)) were used for GaP deposition. At the initial stage the native oxide was removed from the Si surface by HF-dip (10%) immediately before the growth of the GaP layer. The time between HF dip and deposition start does not exceed 30 min.

The PE-ALD deposition was performed using the Oxford Plasmalab 100 PECVD (13.56 MHz) setup supplied with phosphine (PH_3) and trimethylgallium (TMG) lines for sources of phosphorus and gallium, respectively. Hydrogen was used as a gas carrier for TMG. The hydrogen pressure was 3 bar while the temperature of the TMG cylinder was kept at 3°C.

Three different deposition modes were used: i) continuous PECVD mode (sample # 1) with PH_3 and TMG/H_2 flows of 50 sccm, RF power of 5 W, gas pressure P = 250 mTorr and deposition time of 7.5 min with growth rate of about 8 nm/min; ii) ALD mode where PH_3 and TMG/H_2 flows were alternatively changed with continuous plasma discharge due to Ar flow addition during the purge step (samples #2, 3); iii) ALD mode with plasma interruption, evacuation and purge steps (sample # 4). In this case in order to provide reliable plasma ignition at each growth step the higher RF power (50 W) was used and Ar was added to the TMG/H_2 flow. The deposition parameters for ALD modes are given in table 1. All ALD processes consist of 50 cycles. The substrate temperature of 350°C was the same for all the processes.

Epitaxial single-crystal GaP layers grown by MBE [10] on Si substrates from the same batch were used as reference samples.

Table 1 Deposition parameters of the four samples.

Step	**ALD mode with continues plasma**		**ALD mode with plasma interruption**
	Sample # 2	Sample # 3	Sample # 4
P step	PH_3=50 sccm t = 10 s P = 250 mTorr RF=5 W		PH_3=50 sccm t = 5 s P = 250 mTorr RF=50 W
evacuation	None		t = 20 s P < 0.1 mTorr
purge	Ar =50 sccm t = **10** s P = 250 mTorr RF=5 W	Ar =50 sccm t = **20** s P = 250 mTorr RF=5 W	Ar =30 sccm t = 10 s P = 250 mTorr
Ga step	TMG/H_2=50sccm t= 10s P =250mTorr RF=5W		TMG/H_2= 50 sccm Ar =30 sccm t = 5 s P = 250 mTorr RF=50 W
Thickness per cycle	~1.6 nm	~1 nm	~1.3 nm

3. Results

3.1. *Structural properties*

The layer morphology and thickness were determined using scanning electron microscopy (SEM) (SUPRA 25 Zeiss). The grown GaP films have homogeneous structure and smooth surface. The RMS roughness measured by AFM is in the range of 0.2-0.3 nm. The structural properties were studied by transmission electron microscopy (TEM) (Philips EM420). An example of TEM images for the PE-ALD GaP layer is presented in Figure 1. TEM demonstrates a sharp GaP/Si interface. The selected area electron diffraction (SAED) pattern of the GaP film obtained by TEM (inset of Figure 1) shows a dominant amorphous phase.

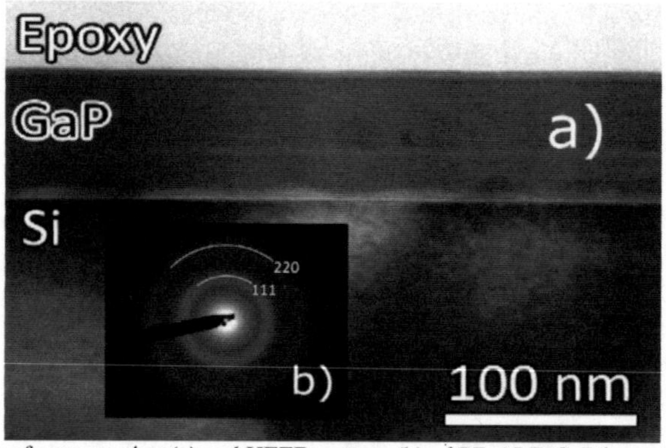

Figure 1 TEM photography of cross section (a) and HEED pattern (b) of PE-ALD GaP layer (sample #2).

3.2. *Optical properties*

Ellipsometric angles Ψ and Δ were measured using spectral ellipsometer Uvisell 2 (Horiba Scientific). The measurement range was from 0.6 to 6.5 eV with a step of 0.05 eV. An estimation of refractive index (n) and extinction coefficient (k) was performed in terms of Bruggeman effective medium approximation "BEMA" [13]. The spectra were simulated in approximation of isotropic substrate/isotropic layer obtaining a good agreement with experimental data. GaP dispersion was set in the approximation of four harmonic oscillators (quatre amorphous) [14]. Optical properties of GaP layers grown by PE-ALD were compared with ones of GaP grown on Si by MBE [10] (Figure 2). Also n and k spectra of bulk GaP [15] are presented in Figure 2 for the reference. The general behaviour of n and k spectra for MBE GaP is similar to that of bulk GaP with the same peaks positions. The difference in absolute values could be caused by the low thickness (<200 nm) of the MBE GaP layer. On the contrary the behaviour of PE-ALD GaP n and k spectra differs from that of the bulk GaP. The peaks are unclear and their positions are shifted. This difference is obviously due to band structure changes of amorphous GaP. However the absorption of PE-ALD GaP does not exceed that of MBE GaP for photon energy less than 2.3 eV indicating low interband absorption of PE-ALD GaP. This is an essential issue for the photovoltaic structures which require to minimize absorption in the wide gap emitter GaP layer.

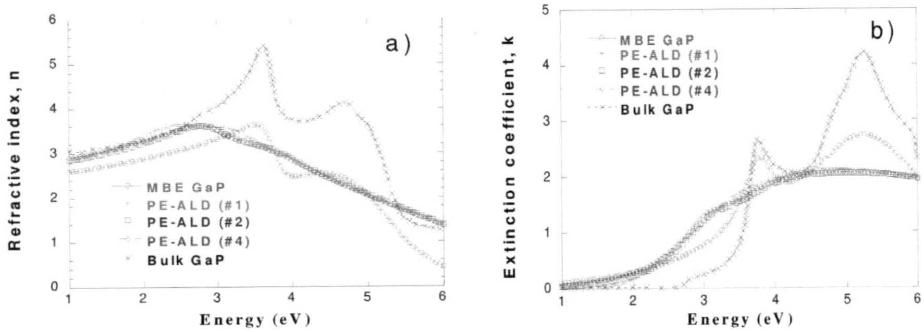

Figure 2 Experimental refractive index (a) and extinction coefficient (b) determined from spectroscopic ellipsometry for GaP layers on Si grown by MBE and PE-ALD.

3.3. Passivation properties

Photoluminescence is known to be a powerful tool for the characterization of interface properties in Si based heterojunctions [16,17]. The radiative recombination of excess carriers in the volume depends on the effective lifetime, which is affected by surface recombination. The signal of the PL from Si (at λ=1150 nm) decreases with increasing recombination rate at the GaP/Si interface. The photoluminescence (PL) measurements were performed at room temperature with excitation at 778 nm wavelength (minimum absorption in GaP) using Accent RPM Sigma PL.

The PL spectra for the series of the GaP/Si samples fabricated using PECVD technology are presented in Figure 3a. Also the PL spectra for the MBE layers are shown. The back side of all the samples was passivated by a thin a-Si:H layer in the same process to avoid any influence of the back interface. The intensity of PL signal for the samples #2 and #4 obtained in ALD mode is significantly higher compared to epitaxial GaP/Si structures indicating considerable improvement of the Si surface passivation. Low PL signal for the sample # 3 is obviously caused by double Ar plasma time. Ar plasma, which was used to keep the discharge during gas exchange, is known to create radiation defects and to reduce charge carrier lifetime in Si. The sample #1 obtained in continuous mode exhibits low PL intensity.

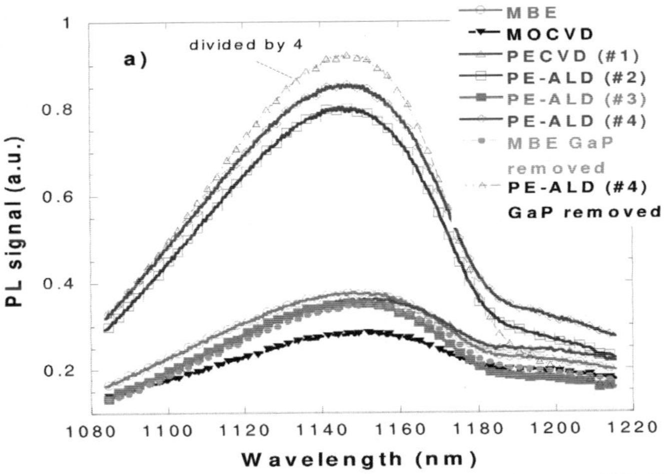

Figure 3 Photoluminescence spectra for MBE, MOCVD, PECVD and PE-ALD GaP/Si heterostructures

4. Discussion

The fact that PE-ALD provides better passivation of Si wafers compared to MBE technology is encouraging. We should stress that initially all the Si substrates were similar (with the same charge carrier lifetime). However high temperatures of MBE processes could lead to higher defect density in Si substrates, due for instance to rapidly diffusing defects into the Si wafer [12]. On the contrary low temperature PE-ALD process should not affect the properties of Si wafers. To verify this difference the GaP layer was removed by wet etching (acid based etchant) for MBE and PE-ALD (#4) grown samples followed by silicon surface passivation. The intensity of PL signal (Figure 3a) for MBE sample remained at the same level while for the PE-ALD sample the PL intensity increased by factor of four being 70% of the PL signal from initial Si wafer passivated by a-Si:H (not presented). This fact clearly indicates the silicon degradation during the high temperature MBE growth. The 30% difference between PL signal for Si wafer at initial state and after PE-ALD GaP layer removing is associated with radiation defects. The decrease of PL signal with Ar plasma time demonstrates that Ar addition should be rather avoided. The hydrogen will be used in further technology development.

The difference for the samples obtained in continuous and ALD modes indicates the sensitivity of Si passivation quality to growth conditions and GaP layer properties. Therefore there is still a room for improvement when further detailed study using in particular high resolution TEM will be performed.

5. Conclusion

GaP layers were grown on Si substrates by PE-ALD at 350°C. The grown GaP films have homogeneous amorphous structure, smooth surface and a sharp GaP/Si interface. The proposed PE-ALD provides better passivation properties of GaP/Si heterostructures, which does not deteriorate Si wafer, compared to MBE demonstrating a high potential of this technology for new silicon based photovoltaic heterostructures that has to be confirmed by further development of the proposed technology (including a doping facility) and layer properties study.

Acknowledgements

This work was carried out in the framework of the joint Russian-French project LAMSOL supported by Ministry of Science and Education of Russian Federation (RFMEFI61615X0040) and by PHC Kolmogorov Program (35522TL).

References

[1] K. Masuko, M. Shigematsu, T. Hashiguchi, D. Fujishima, M. Kai, N. Yoshimura, T.Yamaguchi,Y. Ichihashi, T. Yamanishi, T. Takahama, M. Taguchi, E. Maruyama, S. Okamoto, IEEE J. of Photovoltaics **4**, 1433 (2014).

[2] William C. Cooley and Robert J. Janda, Handbook of Space-Radiation Effects on Solar-Cell Power Systems, published by NASA, Washington, D.C., 1963

[3] S. M. Sze, Physics of Semiconductor Devices, 2nd Edition (John Wiley & Sons, New-York, 1981).

[4] H. Wagner, T. Ohrdes, A. Dastgheib-Shirazi, B. Puthen Veettil, D. König, P. P. Altermatt, J. Appl. Phys. **115**, 044508 (2014)

[5] Mahdad Sadeghi, Shumin Wang, J. Cryst. Growth **227–228**, 279 (2001)

[6] Steven L. Wright, Herbert Kroemer and Masanori Inada, J. Appl. Phys. **55**, 2916 (1984)

[7] J.P. André, J. Hallais, C. Schiller, Journal of Crystal Growth **31**, 147 (1975)

[8] L. Samuelson, P. Omling, H.G. Grimmeiss, J. Cryst. Growth **68**, 340 (1984)

[9] J.M. Olson, M.M. Al-Jassim, A. Kibbler, K.M. Jones, J. Cryst. Growth **77**, 515 (1986)

[10] M. S. Sobolev, A. A. Lazarenko, E. V. Nikitina, E. V. Pirogov, A. S. Gudovskikh, A. Yu. Egorov, Semiconductors **49**, 559 (2015)

[11] J.F. Geisz, J.M. Olson, D.J. Friedman et al. // 31th IEEE PVSC, 2005, p. 695.

[12] R. Varache, M. Darnon, M. Descazeaux, M. Martin, T. Baron, D. Muñoz, Energy Procedia **77**, 493 (2015)

[13] D. A. G. Bruggeman, Annalen der Physik, 24, 637 (1935)

[14] A. R. Forouhi and I. Bloomer, Phys. Rev. B 38, 1865

[15] D.E. Aspnes and A. A. Studna, Phys. Rev. **B27**, 985 (1983).
[16] S. Tardon, M. RoËsch, R. Brüggemann, T. Unold, G.H. Bauer, J. Non-Cryst. Solids **338-340**, 444 (2004).
[17] G.H. Bauer, S. Tardon, M. Rösch, T. Unold, Phys. Status Solidi (c) **1**, 1308 (2004).

Luminescence of rare-earth ions and intrinsic defects in Gd_2O_3 matrix

Yu A Kuznetsova, A F Zatsepin, R A Tselybeev, V N Rychkov, V A Pustovarov
Ural Federal University, Yekaterinburg, Russia

Abstract. The X-ray luminescence and photoluminescence of the ultradispersed Gd_2O_3 powders with different purity were investigated in 90-360 K temperature range. Both impurity and intrinsic optical active centers were detected. The effect of energy transfer from Gd^{3+} to RE^{3+} ions was observed.

1. Introduction

Gadolinium oxide is of interest as a wide-band gap insulator (E_g = 5.6 eV) for optoelectronics, photonics, fluorescence sensing, etc., because of its optical and chemical properties, in particular, the ability of being easily doped with rare-earth (RE) ions in a wide concentrations range. The trivalent RE ions doping such as Pr, Dy, Tb ions or Yb-Er, Yb-Ho, Yb-Tm pairs determines prospects of Gd_2O_3 as down- and up-conversion material for energy transformation devices, including solar cells [1].

Energy and quantum efficiency of oxide matrixes doped with RE^{3+} is mainly determined by dynamic of excitation states of material electronic subsystem [2]. Processes of charge carrier transport, especially the probabilities of non-radiative relaxations, depend on crystal structure and defectiveness of the host lattice. In general, the low-symmetric modifications of host matrixes along with the intrinsic and impurity-related defects lead to the creation of additional luminescence quenching channels [2, 3]. On the other hand, there is information about enhancement the luminescence efficiency of RE-doped phosphors after co-doping that distorts the stoichiometry of lattice [4]. Thus, the investigation of matrix role in the spectral-optical properties of RE-doped materials will make it possible to find ways for improving of their functional characteristics.

In the present work we have studied the features of different excitation channels of RE luminescence in ultradispersed Gd_2O_3 powders and revealed the influence of intrinsic defects on energy transfer processes.

2. Samples and Methods

The objects of this research were microstructured Gd_2O_3 powders prepared by aqueous solution deposition method. The samples had different degree of purity and corresponded to the GdO-Zh (99.99 %) and GdO-G (99.9999 %) technical marks. The XRD profiles of these samples showed single cubic phase (Ia-3 space group) with lattice constants a = 10.81 and 10.79 Å respectively. The average particle sizes of powders were estimated to be 8-10 μm by means of FRITSH laser particle size analyzer.

The X-ray excited luminescence (XRL) spectra were recorded by experimental setup equipped with BSV-2 X-ray tube with Mo anode at a voltage of 30 kV and current of 10 mA. The photoluminescence (PL) and PL excitation studies were performed on a setup assembled using 400 W deuterium lamp, R6358-10 Hamamatsu photomultiplier and two prismatic DMR-4 computer-controlled monochromators. Low-temperature measurements were carried out by using vacuum cryostat (P = 10^{-4} Pa). The temperature of the sample was controlled by platinum resistance thermometer HRTS-5760-B-T-1-12 with accuracy of 1% and changed by the external heater inserted into a sample holder.

3. Results

In nominally pure Gd_2O_3 powders under X-ray excitation at room temperature a number of narrow lines of luminescence in blue (489 nm), green (550 nm), yellow (572-594 nm) and red (612-628 nm) regions of the spectrum has been observed (figure 1). Based on the Dieke diagram these emission bands were attributed to the intraconfiguration *f-f* transitions of Pr^{3+}, Tb^{3+}, Er^{3+} and Eu^{3+} ions existing in matrix as micro impurities. It should be noted that in the high-purity sample (GdO-G) quantum yield of impurity RE ions luminescence considerable decreases while $^6P_J \rightarrow {}^8S_{7/2}$ radiative transition of Gd^{3+} ions with 314 nm characteristic line is firstly detected.

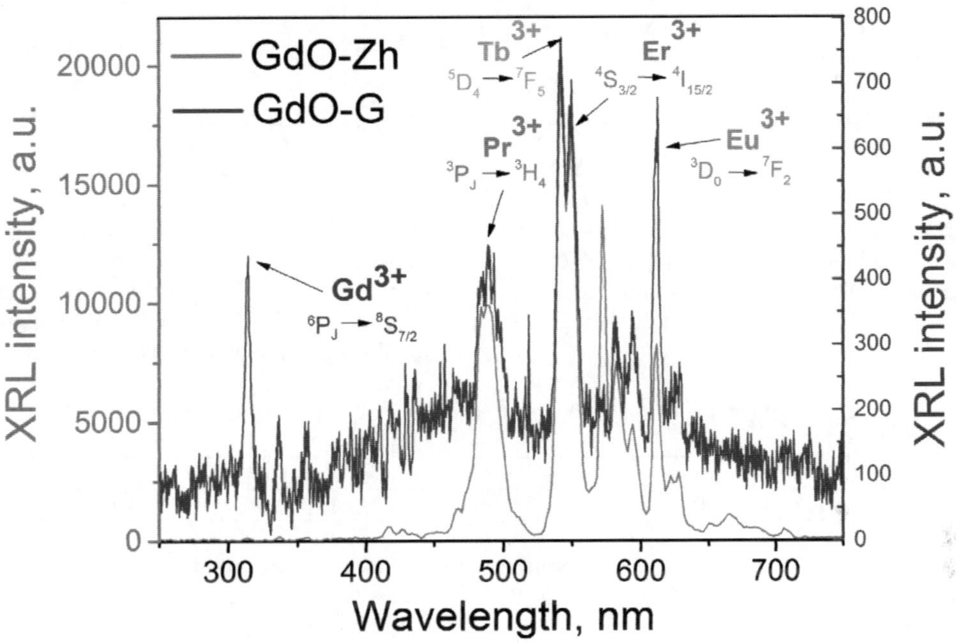

Figure 1. XRL spectra of GdO-Zh (99.99 % purity) and GdO-G (99.9999 % purity) samples at room temperature. Electron transitions corresponding to *f-f* emission of RE^{3+} impurity ions are shown by narrows.

In low-temperature XRL spectrum of the GdO-G sample the narrow emission lines of RE^{3+} ions also appear but the most intense luminescence corresponds to the broad band centered at 430 nm with a full-width-half-maximum of about 130 nm (figure 2). The temperature dependence of the intensity of this band in the range 110-350 K is plotted in figure 3. The noticeable luminescence quenching starts at the 125 K approximately. At first, we tried to describe the obtained experimental data by the classic Mott's law that is valid for a single activation model. However, such approximation is satisfactory only at low temperatures up to 200 K (red line in figure 3). Therefore we have used the modified Mott's law expression that accounts for the realization of two-activated processes with different activation energies [5-7]:

$$I(T) = \frac{I_0}{1 + A_a \exp(-E_a/kT) + A_b \exp(-E_b/kT)}. \quad (1)$$

I_0 is the intensity in the limit of low temperature; k is the Boltzmann constant; A_a and A_b are the pre-exponent factors reflected the non-radiative over radiative probability ratios; E_a and E_b are the activation energies, which characterize the height of the berries of non-radiative relaxation channels. Using the equation 1 allows adequately describing the $I(T)$ dependence over the whole investigated temperature range (blue line in figure 3). The parameters of fit are listed in table 1.

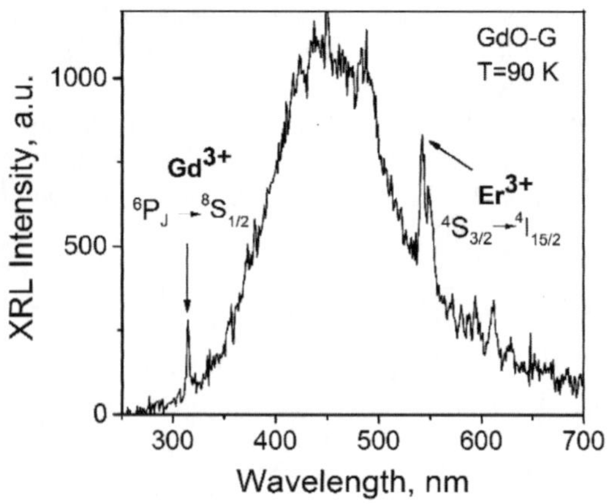

Figure 2. XRL spectra of GdO-G (99.9999 % purity) sample at 90 K temperature. The narrows represents the transitions of Gd^{3+} host ions and Er^{3+} impurity ions.

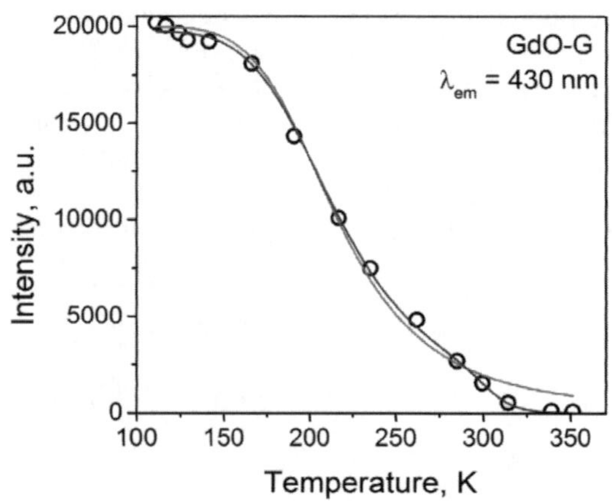

Figure 3. Temperature quenching of luminescence at 430 nm under X-ray excitation. The red line represents the approximation by classic Mott's law. The blue line corresponds to a double-activation Mott's law (equation (1)).

Table 1. Parameters of the quenching of luminescence at 430 nm for Gd_2O_3 with 99.9999 % purity

I_0	A_a	A_b	E_a (eV)	E_b (eV)
19760 ± 113	1700 ± 56	1.2E18 ± 4E17	0.141 ± 0.005	1.033 ± 0.08

The PL spectra recorded at room temperature under excitation corresponding to interband Gd_2O_3 transitions (5.6 eV) demonstrated the lines at the same wavelengths as in XRL spectra (not shown here). The PL excitation spectra of Pr^{3+}, Er^{3+} and Eu^{3+} ions are represented in figure 4. It was found that *f-f* emission of RE^{3+} can be efficiently excited by different ways: (*i*) – under intracenter excitation; (*ii*) – at energies corresponding to ^6P_J, ^6I_J, ^6D_J excited states of Gd^{3+} ions; (*iii*) – at energies more than 5.6 eV in the region of Gd_2O_3 fundamental absorption edge [8].

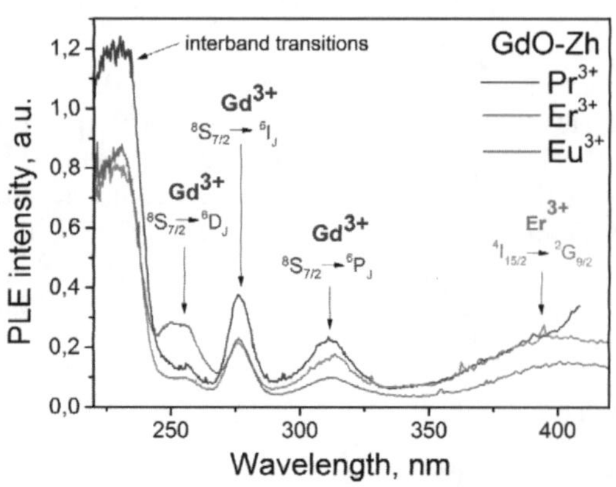

Figure 4. PL excitation spectra of *f*→*f* emission at 486 nm (Pr^{3+}), 550 nm (Er^{3+}) and 612 nm (Eu^{3+}) for GdO-Zh (99.9999 % purity) sample at 295 K.

4. Discussion

The obtained results clearly indicate that, firstly, there is efficient energy transfer by band charge carriers to impurity centers as evidenced by high X-ray luminescence output (figure 1). Secondly, the presence in excitation spectra the intense Gd^{3+} transition lines (figure 4) points out to non-radiative energy transfer from gadolinium to Pr^{3+}, Er^{3+} and Eu^{3+} ions.

The reduction of impurity ions concentration leads to decreasing of $Gd^{3+} \rightarrow RE^{3+}$ energy transfer efficiency and simultaneous increasing the probability of radiative relaxation in Gd^{3+} ion (see figure 1). This fact is confirmed by the appearance of $^6P_J \rightarrow {}^8S_{7/2}$ emission band in the XRL spectrum of greater purity sample (GdO-G) against the background of dramatically decreasing of luminescence intensity related to impurity RE^{3+} ions. It denotes that the radiative $^6P_J \rightarrow {}^8S_{7/2}$ transition of Gd^{3+} and energy transfer to impurity RE^{3+} ions are competing processes.

The presence of local energy levels of Gd^{3+} ions in the region of Gd_2O_3 optical transparency indicates the realization of the self-activation effect. In other words, matrix Gd^{3+} ions behave as extrinsic activators that act as energy donors. In different host lattices the nature of the self-activation centers has been related to the point distortions of lattice caused by deviations from stoichiometry [9, 10]. In our case, the most possible intrinsic defects involved in absorption and luminescence processes are either interstitial Gd^{3+} ions or Gd^{3+} at normal lattice sites associated with other nearest point defects. In any instances the local environment of the "defective" Gd^{3+} ions is distorted by internal crystal field. It results in partly weakening of the parity selection rule for *f-f* transitions of Gd^{3+} as confirmed by the appearance of corresponding bands in XRL and PL excitation spectra [11].

Low-temperature band at 430 nm manifested in XRL spectra of the high-purity Gd_2O_3 (figure 2), obviously, cannot be related to transition of any RE ions. The trivalent lanthanides are characterized by narrow emission lines originating from 4*f* intraconfiguration transitions while the divalent rare-earth ions (for example, Eu^{2+}) show broad 5*d*-4*f* luminescence but in another spectrum region [12, 13]. It is most likely that the nature of this band is associated with intrinsic defects of Gd_2O_3 lattice acting as optically active centers. In cognate La_2O_3 matrix the broad emission peak in the deep blue region has been attributed to single ionized oxygen vacancy [14]. Therefore we suggest that low-temperature luminescence at 430 nm also can be related with the presence of intrinsic lattice defects like *F*-centers.

The analysis of the temperature quenching curve of luminescence at 430 nm (figure 3) revealed that there are at least two non-radiative relaxation channels. The first activation process is realized at low temperatures (up to 200 K) and characterizes by E_a energy barrier value of 0.141 eV. The low activation energy along with a small value of probability A_a (table 1) indicates that the most likely this quenching process is intracenter. At the same time, at higher temperatures the another quenching channel with significantly higher barrier (1.033 eV) is dominant. For this case the non-radiative relaxation can also occurs within the center provided the complexity of configuration curves. However, it is possible that the second channel is due to the external quenching process with the transfer of excitation to other optically active centers. For detail information about luminescence quenching mechanisms the additional spectroscopic investigations are required.

5. Conclusion

The luminescence of impurity RE^{3+} ions in Gd_2O_3 powders has been detected. We have found that so called "defective" Gd^{3+} ions act as donors of excitation energy and presumably associated with intrinsic lattice defects. With decrease of impurity concentration the self-activated luminescence of Gd^{3+} has been observed. The low-temperature luminescence at 430 nm was assigned to the host lattice defects like *F*-centers. The analysis of temperature quenching revealed two non-radiative relaxation channels: in temperature region up to 200 K the intracenter quenching process is realized while at higher temperatures the second activation channel with a higher energy barrier prevails.

The obtained results indicate that the processes of energy transport and efficiency of radiative transitions in ultradispersed Gd_2O_3 powders largely depend on intrinsic defects. The control of impurity composition and lattice disordering allows modifying the optical properties to improve the functional parameters of Gd_2O_3-based energy converters.

Acknowledgments
This work was supported by the Ministry of Education and Science of Russia, the Grain-in-Aid Agreement no. 14.581.21.0002, September 29, 2014 (the unique identifier RFMEFI58114X0002) in the frame of the Federal Aim Program «Research and Development in Priority Directions of the Science and Technology Complex of Russia for 2014-2020». The part of results was obtained within the framework of the Act 211 Government of the Russian Federation (contract № 02.A03.21.0006).

References
[1] Strumpel C, McCann M, Beaucarne G et all 2007 *Sol. Energ. Mat. Sol. C.* **91** 238–249
[2] Case M A, Bradford C Stout 2012 *Photoluminescence: Applications, types and efficacy* (New York: Nova Science Publishers, Inc.) p 55
[3] Zhang J Z 2009 *Optical Properties and Spectroscopy of Nanomaterials* (Hackensack: World Scientific Publishing Co Pte. Ltd.) p 271
[4] Liu B, Gu M, Liu X at all 2006 *J. Alloy. Compd.* **440** 341-345
[5] Bieber H, Gilliot P, Gallart M et all 2007 *Catal. Today* **122** 101-108
[6] Hirai T, Asahi M, Iwasaki Y et all 2009 *J. Lumin.* **129** 406-409
[7] Zatsepin A F, Biryukov D Yu, Kortov V S, Cholakh S O 2002 *Phys. Solid State* **44** 1671–75
[8] Dhananjaya N, Nagabhushana H, Nagabhushana B M et all 2012 *Spectrochim. Acta A* **96** 532-540
[9] Chen Y N, Xu S J, Zheng C C et all 2014 *Appl. Phys. Lett.* **105** 041912
[10] Wang T, Gou J, Xu X et all 2015 *Optics Express* **10** 12595-604
[11] Kasap S, Capper P 2007 *Springer Handbook of Electronic and Photonic Materials* (New York: Springer Science-Business Media Inc.)
[12] Biswas K, Sontakke A, Sen R, Annapurna K 2012 *J. Fluoresc.***22** 745-752
[13] Liu Z, Nengli D, Yang L, Li J 2015 *Appl. Phys. A* **119** 553-557
[14] Hu C, Liu H, Dong W 2007 *Adv. Mater.***19** 470-474

Implementation of decoy states in a subcarrier wave quantum key distribution system

A Gaidash[1], A Kozubov[1], V Egorov[1], A Gleim[1,2]
[1]ITMO University, 199034, Russia
[2]Kazan National Research Technical University KAI, 420111, Russia

Abstract. Subcarrier wave quantum key distribution systems demonstrate promising capabilities for secure quantum networking. However for this class of devices no implementation of secure decoy states protocol was developed. It leaves them potentially vulnerable to photon-number splitting attacks on quantum channel and limiting the key distribution distance. We propose a practical solution to this problem by calculating the required parameters of light source and modulation indices for signal and decoy states in a subcarrier wave system and describing the corresponding experimental scheme.

1. Introduction

Quantum key distribution (QKD) systems allow secretly sharing identical binary sequences that can possibly be used as symmetric keys, between authorized parties [1]. Subcarrier wave QKD [2 - 5] is a promising approach in terms of compatibility with optical network infrastructure. In these systems the quantum signal is generated at subcarrier frequencies (sidebands) by phase modulation of classical optical pulse at some central frequency. The advantages of this approach include polarization independence, long range and high bitrates at a relatively low clock frequency [5] and potentially remarkably high spectral efficiency: up to 40%, as discussed in [3]. Moreover, most QKD phase protocols can be easily implemented in this system by altering the modulation parameters accordingly. Unfortunately, the quantum channel generation method in SCW systems also leads to complications in maintaining security of the system against photon-number-splitting (PNS) attacks [1, 6]. There are several ways of dealing with this problem. Monitoring the multiphoton statistics, as suggested in [6], requires an advanced type of detector which is currently not widely available. Using the strong reference protocol [7] may reduce the secure key distribution distance and complicates the filtering system in receiver module [5]. Implementing the widely used decoy states method [8] in SCW system is not as straightforward as in other systems, because switching between signal and decoy photons results in energy fluctuations at the classical carrier, and therefore Eve can easily monitor and distinguish them, as discussed in [9]. In this work we for the first time propose a practical solution for implementing decoy states in the SCW QKD system.

2. Results and discussion

Carrier ratio (CR) represents a relation of decoy power at central frequency to the signal one. According to [9], a direct approach to generating undistinguishable decoys in SCW is to keep CR as close as possible to unity. However, solving the equations given in [9] leads to non-quantum signal at sidebands or alternatively non-classical signal at central frequency due to modulation index is higher than unity. To avoid this problem, the value of modulation indices chosen should be less than unity.

Therefore, one needs different initial amplitudes of the carrier wave before modulation for decoy and signal pulses.

Let us consider a practical case of SCW system with a low-intensity carrier, with its energy corresponding to photon number before modulation $\mu_L \sim 20$. It lowers the requirements for filtration and at the same time allows for a sufficiently high modulation index. Subcarrier wave generation can be represented by first terms of Jacobi-Anger expansion:

$$A \cdot \exp(i\omega t) \rightarrow A \cdot \exp[i\omega t + im \cdot \sin(\Omega t)] =$$
$$= A \cdot \exp(i\omega t) \cdot \left[J_0(m) + \sum_{k=1}^{\infty} J_k(m) \cdot \exp(ik\Omega t) + \sum_{k=1}^{\infty} (-1)^k \cdot J_k(m) \cdot \exp(-ik\Omega t) \right] \approx \quad (2)$$
$$\approx A \cdot \exp(i\omega t) \cdot [J_0(m) + J_1(m) \cdot \exp(i\Omega t) - J_1(m) \cdot \exp(-i\Omega t)],$$

where A is amplitude of initial pulse, ω is central frequency, m is modulation index, Ω is RF frequency, $J_k(m)$ is k-th order Bessel function. Then optical power (intensity) W at the carrier with frequency ω and the quantum sidebands with frequencies $\omega \pm \Omega$ can be expressed as follows:

$$W_0 = A^2 \cdot J_0^2(m) = \mu_L \cdot J_0^2(m) = \mu_C,$$
$$W_{\pm 1} = A^2 \cdot J_1^2(m) = \mu_L \cdot J_1^2(m) = \mu_Q, \quad (3)$$

where μ_Q is the average photon number per pulse in the quantum channel, μ_C is the average photon number per pulse at the central frequency after modulation.

Thereby our goal is to calculate m and μ_L for both signal (s) and decoy (d) pulses since other parameters are known from experimental setup. CR needs to be equal to unity, so $\mu_{Cs} = \mu_{Cd} = \mu_C$. Quantum ratio (QR), which gives a relation between intensities of signal and decoy states at subcarrier frequencies ($\mu_{Qs} \cdot QR = \mu_{Qd}$), can be chosen arbitrarily between 0 and 1 [8]; in this work we shall use QR = 0.05. The optimal value of μ_{Qs} can be found according to formulae given in [8]:

$$(1 - \mu_{Qs}) \cdot \exp(-\mu_{Qs}) = \frac{f \cdot H_2(e_{\det})}{1 - H_2(e_{\det})}, \quad (4)$$

where f is bidirectional error correction efficiency, $H_2(x)$ is binary entropy of x, $e_{\det}$ is detection error. Using values $f = 1.08$ and $e_{\det} = 0.0075$ from experimental data in [5] we finally calculate $\mu_{Qs} \approx 0.831$, $\mu_{Qd} = 0.041$. For $\mu_C = 20$, one can derive set of equations for both signal and decoy pulses as following:

$$\begin{cases} \mu_L \cdot J_0^2(m) = \mu_C \\ \mu_L \cdot J_1^2(m) = \mu_Q \end{cases} \Rightarrow \frac{J_1^2(m)}{J_0^2(m)} = \frac{\mu_Q}{\mu_C}, 0 < m < 1 \Rightarrow \frac{m^2}{4} + \frac{m^4}{16} + \frac{11 \cdot m^6}{768} + O(m^7) = \frac{\mu_Q}{\mu_C} \quad (5)$$

We get modulation indices for generating signal and decoy states are $m_s = 0.4$ and $m_d = 0.091$. The respective the initial carrier photon numbers $\mu_{Ls} = 21.68$ and $\mu_{Ld} = 20.08$.

The last parameters that need to be considered are the values of losses induced by the programmed attenuator that regulates the light source power. In [7] a 10 mW semiconductor laser emitting at 1550 nm with repetition rate 100 MHz is used. For this configuration one can calculate the values of attenuation in dB to achieve μ_L power for signal and decoy as following:

$$A = 10 \log_{10} \left(\frac{h \cdot c \cdot v \cdot \mu_L}{\lambda \cdot W_0} \right) \quad (6)$$

where h is Plank's constant, c is speed of light, v is repetition rates, λ is laser wavelength, W_0 is initial laser power. Values of attenuation for signal and decoy are found to be $A_s = -75.6$ dB and $A_d = -75.9$ dB correspondingly. The proposed solution can be implemented in an experimental setup with controlled attenuator and phase-modulator.

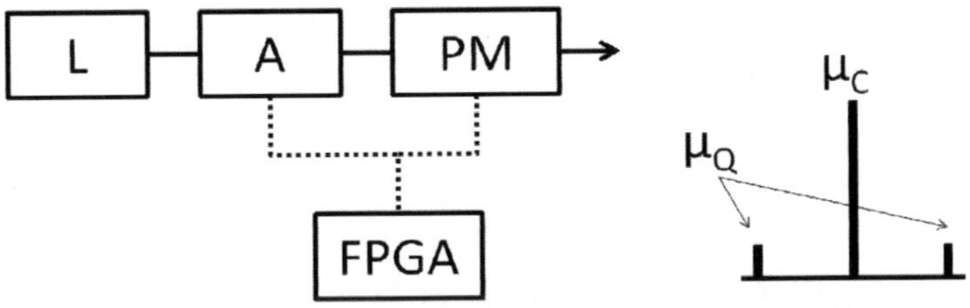

Figure 1. Proposed experimental scheme, where L denotes laser, A is controllable attenuator, PM is controllable phase modulator, FPGA is field-programmable gate array, which switches between two states to achieve calculated values of attenuation and modulation index

In order to generate the states with parameters defined in this work, an FPGA must be added to the SCW scheme, which randomly chooses the attenuation loss and phase modulation index between two pairs of values A_s, m_s and A_d, m_d. Therefore, in this setup Eve cannot distinguish the signal and decoy states since both have same mean power at carrier frequency μ_C.

3. Conclusion

We propose a practical solution to implementing decoy states protocol in a SCW QKD system by calculating the required parameters of light source and modulation indices for signal and decoy states and describing the corresponding experimental scheme. We show that a simple subsystem consisting of a controllable attenuator and phase modulator can generate signals with required parameters.

Acknowledgments

This work was financially supported by Government of Russian Federation, Grant 074-U01 and by the Ministry of Education and Science of Russian Federation (project № 14.578.21.0112).

References

[1] V Scarani, H Bechmann-Pasquinucci, N J Cerf et al. 2009 *Rev. Mod. Phys.* **81** 1301
[2] J-M Merolla, Y Mazurenko, J-P Goedgebuer, W T Rhodes 1999 *Phys. Rev. Lett.* **82** 8
[3] J Mora, A Ruiz-Alba, J Capmany et al. 2012 *Opt. Lett.* **37** 11
[4] A Gleim, V Egorov et al. 2014 *Bull. Russ. Acad. Sci., Physics* **78** 3
[5] A Gleim,V Egorov et al. 2016 *Optics Express* **24** 3
[6] A Gaidash, V Egorov, A Gleim 2014 *J. Phys.: Conf. Ser.* **541** 012062
[7] O Guerreau, F J Malassenet, J.-M. Merolla et al. 2005 *IEEE Photonics Technol. Lett.* **17** 8
[8] X Ma, B Qi, Y Zhao, H-K Lo 2005 Phys. Rev. A 72 012326
[9] S Bhattacharya, P Kumar Krishnamurthy 2013 *J. Opt. Soc. Am. B* **30** 4

Current relaxation in MJ SCs and its influence on IV curve formation in presence of light coupling

E D Filimonov, S A Kozhukhovskaia, V M Emelyanov, M A Mintairov and M Z Shvarts

Ioffe Institute, St. Petersburg, 194021, Russia

E-mail: dexterfree1@gmail.com

Abstract. In the work, a procedure for simulating I-V characteristics of multijunction solar cells (MJ SCs) with accounting for the processes of current relaxation determined by the luminescent coupling between separate p-n junctions has been considered. The effect of the mentioned above processes on the I-V characteristic shape was investigated. It has been shown that, in the case of well pronounced reemission processes in high-efficient MJ SCs, in recording their I-V characteristics, variation of fill factor and rise of the open circuit voltage values being registered take place.

1. Introduction

Gains in the sunlight conversion efficiency being demonstrated in the last two decades in developing III-V MJ SCs have, on the one hand, outlined the practical feasibility of record SC efficiency values at the level of more than 45% [1-3] and, on the other hand, - a set of problems in determining these values. First of all, the problems are associated with transfer to using high-quality semiconductor materials characterized by a high quantum yield of charge carrier radiative recombination [4, 5]. The luminescent coupling determined by these processes is expressed in the appearance of an induced current in the narrow-bandgap (NB) p-n junction due to absorption of recombination radiation from an optically adjacent wide-bandgap (WB) subcell. The effects of emission-absorption of luminescent photons will affect the photocurrent redistribution between photoactive p-n junctions, on the dark current values and on the MJ SC open circuit voltage, which complicates the simulation procedure and investigation of MJ SC characteristics and affects the accuracy of results being obtained. However, the widely applied methods for simulating of MJ SC I-V characteristic do not always account fully for the pointed out optical interaction between separate p-n junctions (subcells), which results in an inconsistency between the I-V characteristic and values of photovoltaic parameters being obtained, first of all evaluated values of the open circuit voltage V_{oc}, and data experimentally registered.

2. The model

In the work, the SPICE method for simulation is considered. Method is based on a standard equivalent circuit of a MJ SC modified for I-V curve tailoring with allowing for the luminescent coupling between subcells. The equivalent circuit (Figure.1) includes three photovoltaic p-n junctions, each of which consists of the following components: J_{ph}– generator setting a subcell photocurrent value; J_d– diode allowing for the p-n junction diffusion component; J_r– p-n junction current recombination component;

J_{rev}– diode for forming an I-V characteristic reverse branch; R_{sh}– shunt resistance simulating various current leakages and losses on internal resistances. Additionally, resistance R_s, which plays a role of a linear series resistance and electronic load (source of a constant voltage V) are included into the scheme. Parameters of the equivalent circuit used in SPICE simulation are presented in Table 1 [6]:

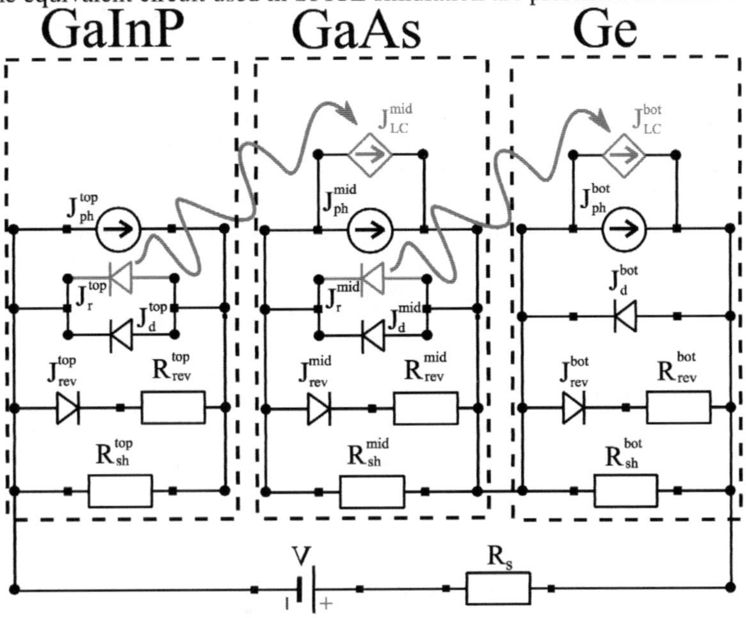

Figure 1. Equivalent circuit of a triple-junction GaInP/GaAs/Ge SC.

Table 1. Parameters of the equivalent circuit.

Parameter	Subcell		
	GaInP	GaAs	Ge
$R_s, ohm/cm^2$	0.023		
$J_{pn}, mA/cm^2$	100	30	15
$J_r, A/cm^2$	$8 \cdot 10^{-15}$	$1.85 \cdot 10^{-11}$	–
$J_d, A/cm^2$	$9.5 \cdot 10^{-27}$	$1.05 \cdot 10^{-20}$	$1.8 \cdot 10^{-6}$
γ	0.07		0.5

Optical interaction between p-n junctions is ensured by adding emitting diodes to the WB subcells circuits. The J_{pn}^{top} and J_{pn}^{mid} recombination currents produce luminescent light and as a result the corresponding induced currents (J_{LC}^{mid} or J_{LC}^{bot}) in an adjacent NB p-n junctions:

$$\begin{cases} J_{LC}^{mid} = \gamma_1 \cdot J_{pn}^{top} = \gamma_1 (J_{ph}^{top} - (J_{ph}^{mid} + J_{LC}^{mid}); \\ J_{LC}^{bot} = \gamma_2 \cdot J_{pn}^{mid} = \gamma_2 (J_{ph}^{mid} + J_{LC}^{mid} - (J_{ph}^{bot} + J_{LC}^{bot})), \end{cases} \quad (1)$$

where J_{ph}^{top}, J_{ph}^{mid}, J_{ph}^{bot} – photocurrents generated in top GaInP, middle GaAs and bottom Ge subcells owing to external light; γ_k $(k = 1, 2)$ - effectiveness of the luminescent coupling for subcell pairs GaInP-GaAs (*k=1*) and GaAs-Ge (*k=2*), γ_s - the limiting (saturated) value of the luminescent coupling effectiveness for corresponding pair of subcells [7]:

$$\gamma_k = \gamma_s \frac{J_{rd}}{4J_{pn}}\left(1 - \sqrt{\left(1 + \frac{4J_{pn}}{J_{rd}}\right)}\right);\ \gamma_s = \lim_{J_{pn}\to\infty} \gamma_k\left(J_{pn}\right), \quad (2)$$

J_{pn} - current flowing through a p-n junction of the emitting WB subcell; J_{rd} - a conditional current boundary between the recombination and diffusion sections of the dark I-V characteristic for the radiating p-n junction

It is obvious (formula 1) that the greater the coupling effectiveness γ_k , the greater is the effect of the WB subcell on the adjacent NB one. It is known [7] that, in a triple-junction GaInP/GaAs/Ge SC $\gamma_1 \ll \gamma_2$. For this reason, from the point of view of studying optical interaction of subcells, the greatest interest is the GaAs-Ge pair, which is the object for simulation in the present work.

In modelling a MJ SC I-V characteristic, the optical interaction between subcells should be taken into account by a process describing the current relaxation [8]. In the initial moment of time, the subcells are in conditions of current equilibrium ($i = 0$) and the photocurrent in the WB subcell is $J^{mid} = J_{ph}^{mid} + J_{LC}^{mid}$ corresponding to a definite level of external illumination, and in the NB one is the induced by luminescence current J_{LC}^{bot} , which is determined by the current flowing through the p-n junction of WB subcell. Any variation in illumination on the emitting or absorbing p-n junctions results in arising of uncontrolled change of conditions for photocurrent matching and actuates the relaxation process (Figure.2). After its completion, a new steady state mode $J_{LC\,i}^{bot}(J_{pn\,i}^{mid})$ comes at $i \approx 7$. As shown on figure.2, b for the presented practical cases the difference between J_{LC1}^{bot} and J_{LC7}^{bot} ($J_{LC1}^{bot} > J_{LC7}^{bot}$) is $\Delta J \approx 5\,mA/cm^2$. So, in simulating real output characteristics of a MJ SC, it is necessary to take into account the given process to avoid the use of a too high value of induced current in NB subcell.

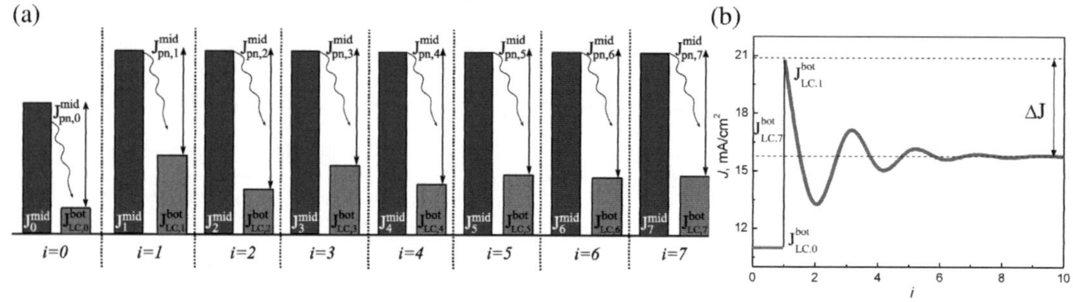

Figure 2. Relaxation of luminescent current. Insert "a" - bar chart (i – is the iteration number). Insert "b" - graph chart. *ΔJ* – difference between expected and recorded current values.

3. Results and discussion

Transfer to a new equilibrium state $J_{LC,i}^{bot}(J_{pn,i}^{mid})$ can be initiated also in varying the current through the emitting p-n junction, as, for example, in recording the I-V characteristic: the value of current in the external circuit is set by applying a bias voltage V_{bias}, and, in "moving" along the I-V characteristic from J_{sc} to V_{oc}, the current flowing through the emitting WB p-n junction is thus increased ($J_{pn,0}^{mid} < \cdots < J_{pn,4}^{mid}$) (Figure.3). In turn, in the optical coupling conditions, this current causes the rise of the current induced in the NB subcell: $J_{LC,1}^{bot} <\ldots< J_{LC,4}^{bot}$. Thus, as a result, new conditions for the photocurrent matching arise, for which a new "grown" up value of the NB subcell current and, correspondingly, a new voltage value at this current are typical. Hence, for every preset current value there is an increased registered voltage magnitude (including the open circuit point on the I-V characteristic), i.e., we have a shift of points on the MJ SC I-V characteristic towards the region of great voltages with the ΔV_{oc} values of about 20mV (Figure.3).

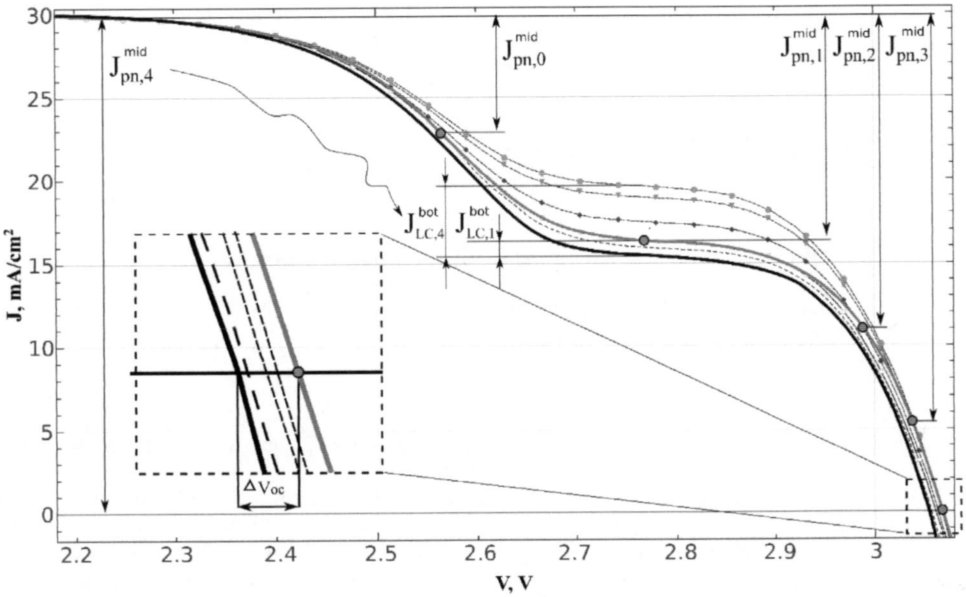

Figure 3. Effect of the luminescent coupling on shaping of an I-V characteristic leading to the open circuit voltage increase.

4. Conclusion

The reemission effects and the processes of current relaxation in a MJ SC associated with them are capable to affect substantially the NB subcell photocurrent and, as a result, the open circuit voltage values being registered in recording the I-V characteristic. It is obvious that, in interpreting experimental data and in obtaining adequate results of simulation of MJ SC characteristics, it is extremely important to take into account the considered mechanism of optical interaction between subcells in a multijunction structure.

References

[1] Green M A, Emery K, Hishikawa Y, Warta and Dunlop E D 2016 Solar cell efficiency tables *Progr. Photovoltaic: Res. Appl.* **24,** Issue 1, 3–11

[2] Aiken D, Dons E, Je S.-S; Miller N, Newman F, Patel P, Spann J 2013 Lattice-Matched Solar Cells With 40% Average Efficiency in Pilot Production and a Roadmap to 50% *IEEE J. Photovoltaics* **3(1)** 542-547

[3] Wiemer M, Sabnis V, Yuen H 2011 43.5% efficient lattice matched solar cells *Proc. of SPIE* **8108** 810804-1–810804-5

[4] Yoon H, King R R, Kinsey G S, Kurtz S, Krut D D 2003 Radiative coupling effects in GaInP/GaAs/Ge multijunction solar cells *3rd WCPEC* **1** 745–748

[5] Baur C, Hermle M, Dimroth F and Bett A W 2007 Effects of optical coupling in III-V multilayer systems Appl. Phys. Lett. **90** 192109-1–192109-3

[6] Mintairov M A, Evstropov V V, Mintairov S A, Shvarts M Z, Timoshina N Kh and Kalyuzhnyy N A 2015 Estimation of the Potential Efficiency of a Multijunction Solar Cell at a Limit Balance of Photogenerated Currents *Semiconductors* **49(5)** 668–673

[7] Shvarts M Z, Mintairov M A, Emelyanov V M, Evstropov V V, Lantratov V M and Timoshina N Kh 2013 Method for direct measurements of luminescent coupling efficiency in concentrator MJ SCs *AIP Conf. Proc.* **1556** 147

[8] Sogabe T, Ogura A, Hung C, Evstropov V V, Mintairov M A, Shvarts M Z and Okada Y 2013 Experimental characterization and self-consistent modeling of luminescence coupling effect in III-V multijunction solar cells *Appl. Phys. Lett.* **103** 263907

Visualizer of magnetic fields

S E Logunov, A Yu Koshkin, V V Davydov, A A Petrov

Department of Quantum Electronics, Peter the Great Saint-Petersburg Polytechnic University, Saint Petersburg 195251, Russia

Abstract. The method of constructing of the optical image of the magnetic field intensity spatial structure distribution with the help of the ferrofluid cell is considered. The experimental research results showed, that the suggested method allows to determine in real time besides the intensity lines structure also the irregularity of the magnetic field and its direction. The possibility of the magnetic system tuning with the help of intensity field lines optical image was examined having in view the realization of necessary field parameters in the course of the experiment. The experimental research results of the magnetic system constructing and tuning quality at different temperatures are represented.

1. Introduction

In order to solve many problems of applied and theoretical physics, as well as to improve the operation efficiency of different devices and instruments (for instance, thermomagnetic engines, spectrometers, magnetic sorting systems, etc.) one must create between the magnetic system pole tips the fields with the assigned values of magnetic induction and its irregularity, as well as with the intensity field lines structure [1, 2]. In order to measure the magnetic field induction B and it is irregularity ΔB were developed and successfully applied methods based on the Hall's sensor and nuclear magnetic resonance [3]. But difficulties appeared while investigating the intensity field lines spatial structure between the poles in of the system. The measured values of B and ΔB in the different points of the gap between the magnetic tips don't allow the construct the adequate pictures of the intensity field lines.

The traditional methods of intensity field lines image construction with the help of magnetic particles or needles either can't be applied or give the image of lines only in one plane. One can use a vessel filled with the liquid having magnetic particles and immerse the magnetic system into it to get the spatial image of lines but it is not rationally, because the size and weight of magnetic systems differ greatly. Besides that, the tuning of the magnetic system placed in liquid is impossible.

Besides that, in magnetic system where "shimaes " [4] are used in order to improve the uniformity of the magnetic field in the gap between the magnetic poles tips, the magnetic particles form the united structure, because the force lines are located too close. In case the width of the gap between the poles is smaller than 10 cm it is rather difficult to observe the spatial structure of the force lines, because the most of particles are attracted towards the magnetic poles and distort the field distribution with the increase of B (magnetic induction) between the poles the number of attracted particles increases, and the spatial structure of the force lines is destroyed. And the application of this lines method becomes non-effective. The application of the methods on the basis of the Hall probe line can be effective only in case, if the distance between the measuring elements matches the distance between the force lines. This correspondence it rather difficult to realize in the fields with the high homogeneity of the magnetic intensity. That is why the development of the new methods of the magnetic field force lines spatial images construction represents an actual problem.

2. The method and experimental technique

One of the possible solution of this problem is the use (the application) of the colloidal solution of the ferromagnetic nanoparticles placed in the glass vacuum-sealed volume of the rectangular form (ferrofluid cell) having different size. This ferrofluid cell can be placed either between the magnetic system poles, or near the magnetic system of different geometry. Under the action of the magnetic field ferrofluid ferromagnetic nanoparticles (in the colloidal solutions the magnetite particles with the size from 30 to 120 mm are used) are magnetized and orientated along the field force lines. The smaller the ferromagnetic particles size, the more precise is the reproduction of the magnetic field force lines structure. Besides that, in case of the extremely high force lines density, the application of such a small particles allows to avoid the effect of the particles fusion with the formation of big clumps in comparison with the case when the iron filings are used. The magnetic field induction B it's the gap between the poles in connected with the magnetic flux Φ by the formula

$$B = \Phi/S \quad (1)$$

where S - is the area of one of the poles tip. But this relation is not always suitable, that is why the induction is expressed in terms of the function the force lines density by distribution j_B over the cross section of the magnet between the poles tips. In this case the induction B can be represented in the following way

$$B = \int j_B(S)dS \quad (2)$$

In this case due to the fact that $j_B(S) \sim N$ (proportional to the number of force lines over the area element), appears the necessity to estimate the quantity N in different points of the space, as well as the distance between the force lines.

If the cell, placed in the magnetic field, is illuminated by the laser light, ferromagnetic particles, located at the force lines, will form a kind of the grating with the spacing equal to d_p. The rating spacing is equal to $d_p = a + c$, where a is the thickness of the force lines, on which the ferromagnetic nanoparticles are located and c - is the distance between lines. The constancy of the grating period, for instance, along the length of the magnetic system will depend upon the irregularity of the magnetic field. For the solution of different problems, the direction of the magnetic field irregularity investigation is chosen depending upon the configuration of the magnetic system or magnet. For example, while investigating the inter polar space of the magnetic system; in which the magnetic induction B is orientated along the axis z, the density of the force lines distribution must be checked (controlled) both along the x and y directions

In the figure 1 the experimental set-up for the investigation of the magnetic field uniformity represented. The laser radiation is focused by the lens 3 onto the assigned space region of the ferrofluid cell 4, placed in the magnetic system 5. The ferrofluid cell is installed on the special non-magnetic construction so that it can be moved over the magnetic field area, where one must know the field force lines distribution for carrying out different experiments [1,2,5-7]. While performing experiments the ferrofluid cells with different thickness of the colloidal solution layer d_k in the direction of the laser radiation propagation are used, that is why for the regulation of the dissipated laser radiation (diffraction picture), received by the camera, the polarizer 6 is used. The diffraction pictures images are processed with far help of special programs.

Journal of Physics: Conference Series **741** (2016) 012092 doi:10.1088/1742-6596/741/1/012092

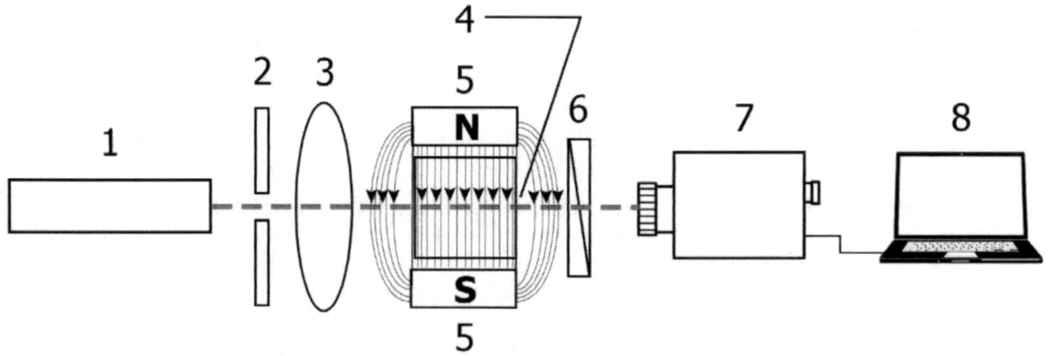

Figure 1. The arrangement of the experimental installation. 1 – semiconductor laser; 2 – diaphragm; 3 – lens; 4 – ferrofluid cell; 5 – magnetic system; 6 – polarizer; 7 – camera; 8 – computer.

The diffraction picture of the laser radiation passed through the ferrofluid cell, that is registrated by the camera, represents a number of alternating maxima and minima. If the period d_p is constant in the area of the laser radiation influence on the ferrofluid cell (i.e. the field homogeneity is high), the then diffraction picture is symmetric with respect to the central maximum. In this case for maxima and minima for following relations are valid

$$d_p \sin \varphi_k = k\lambda. \ d_p \sin \theta_k = (2k + 1) \lambda/2 \qquad (3)$$

where φ_k and θ_k - are angles at which the number "k" maxima and minima are registrated.

But before the registration device the radiation passes interfaces of different mediums with different reflection indies, that is why an additional displacement of the diffraction maxima and minima by angles $\Delta\varphi_k$ and $\Delta\theta_k$, takes place, that can be calculated for each maximum and minimum. By measuring the angle φ_1' between the zero and first maxima, that is equal:

$$\varphi_1' = \varphi_1 + \Delta\varphi_1 \qquad (4)$$

It is possible to calculation the value of a + c with the help of the relation (3). The similar measurements and calculations can be carried out for the angle θ between the zero maximum and zero minimum. As a result, we get two equations with two unknowns, that allow to determine the values of a and c. The received data give a possibility to estimate the distance between the magnetic field force lines (the density of magnetic field force lines). Displacing by the fixed distance the ferrofluid cell over the gap of the magnetic field in the xy plain we investigation the force lines structure in the magnetic system "operational sector" [6-8].

In case the period d_p in the ferrofluid cell changes due to the non-homogeneity of the magnetic field in the region of the laser radiation influence on the cell, the structure and position of diffraction maxima and minima along the x and y axes will change. Comparing received diffraction pictures it is possible to estimate the change of the irregularity of the magnetic field along the ferrofluid cell displacement direction.

It is necessary to mention also, that during these experiments the medium optical density (the spectral transparent coefficient) value appeared to be a very important characteristics. That is why before the magnetic field irregularity investigation the dependence of the spectral transparent coefficient of colloidal solution upon the different parameters (volume concentrations, nanoparticles size and so on) is determined. At permits selecting the colloidal solution parameters and d_k, exclude the polarizer 6 from the experimental installation. This simplifies the calculation of $\Delta\varphi_k$ and $\Delta\theta_k$ values and allows to decrease the errors in calculating a and c.

3. Results and discussion

As an example in the figure 2 (a, b) the curves of the dependence of the colloid solutions transmission coefficient (the intensity relation of the transmitted I_p and falling I_0 laser radiation). Upon the light wave length are represented. As the colloid solution the water solution of unidomain magnetite nanoparticles (the volume concentration of 0.027) with kerosene as the surface active substance is used. Before the colloid solution investigation, the cell where this solution is placed must be tested the curve 1 in figure 2a, 2b.

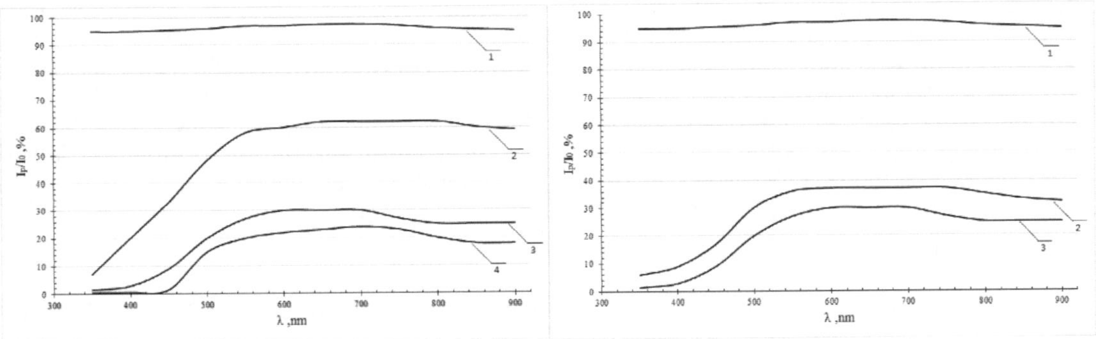

Figure 2 (a, b). The spectral transmittance of the colloidal solution of **(a)** – graphs 2, 3, 4 correspond to the volume concentration of 0.008; 0.027; 0.054, **(b)** – graphs 2,3 corresponds to a particle size in mm: 60; 80. The thickness d_k = 1 mm, T = 293.1 K.

The received results analyses shows that during the magnetic field irregularity investigation by colloid solution application one must use the laser radiation of the visible range with λ from 500 to 600 nm. It was determined that besides the absorption process in the colloid solution the intensity of the transmitted radiation depends also upon the dissipation process on the ferromagnetic particles (figure 2b).

In the figure 3 (as an example) are represented the registrated by the camera images of the laser radiation, transmitted through the ferrofluid cell, placed in the field of the laboratory and in the disk magnet field. The environment temperature was equal to T = 291.4 K.

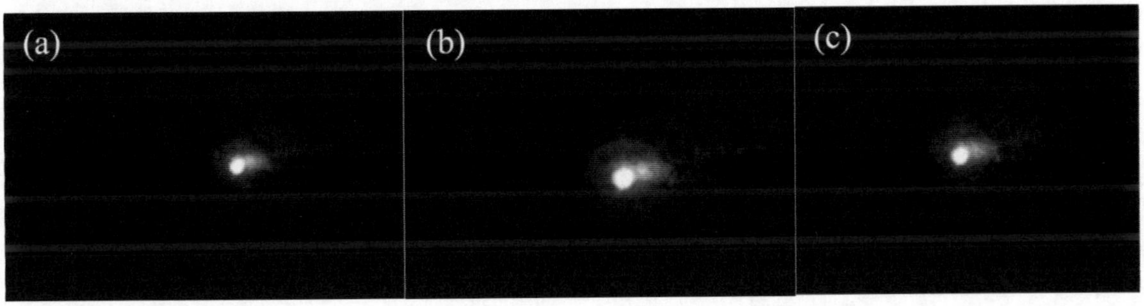

Figure 3 (a, b, c). The diffraction pattern of the laser radiation: **(a)** - magnetic field in the laboratory, **(b)** and **(c)** - for different directions in of the external magnetic field acting on the cell.

The received results show, that the magnetic field, where the ferrofluid cell was located is non-uniform both in the laboratory and in the area of the disc magnet action. After the transformation of the received images into the pseudo-colors MRI with the help of them it is possible to calculate the integral static-moments, to determine the spot area, the mass center position and orientation of the magnetic field in the area of the radiation influence area for a) – 33.095^0, b) – 22.094^0, c) – 26.056^0.

As a results of the carried out investigation of it was started that for the determination magnetic field force lines structure between the poles and the degree of their irregularity the most effective way is to use the ferrofluid cell with thickness d_k equal from 1.5 to 2 mm and with the volume concentration of particles not less than 0,02. For dissipated magnet fields structure investigation, the value of d_k must be not less than 3mm, and the particle volume concentration must not exceed 0,03. The ferromagnetic cells constructed to day allow to determine the magnetic field direction change of 1 degree with the accuracy of 1 percent.

The experimental results analysis shows the possibility of the magnetic field intensity lines investigation with the help of the suggested method, as well as the field parameters determination by the registered image. Besides that, it is possible to tune more precisely the magnetic system using the image of the scattered laser radiation in order to solve different problems. On the other hand, the experiments carried out, showed that the necessity of the ferrofluidic cell construction improvement, as well as the development of the new colloidal solutions. It is especially actual in case of the visualizer application for the investigation of the geomagnetic field intensity lines in the magnetosphere during the influence of the solar wind upon it. Processes taking peace as a result of that influence are the reason of the Earth magnetic storm, northern lights, etc.

Reference

[1] Andreevskiy K N, Mandjavidze A G, Margvelashvili I G, Sobolevskaya S V 1998 *Technical Physics Journal* **68** 119

[2] Davydov V V, Velichko E N, Dudkin V I, Karseev A Yu 2015 *Instruments and Experimental Techniques* **58** 62

[3] Davydov V V, Dudkin V I, Karseev A Yu 2015 *Technical Physics Journal* **85** 138

[4] Vasiliev P A, Saykin K S 1993 *Instruments and Experimental Techniques* **36** 211

[5] Timofeev S S, Turubanov M A 2016 *Sensors and systems* **1** 45

[6] Marusina M Y, Bazarov B A, Galaidin P A, Silaev A A, Marusin M P, Zakemovskay E Yu, Gilev A G, Alekseeva A V 2014 *Measurement Techniques* **57** 217

[7] Mironov E A, Voitovich A V, Palashov O V 2011 *Quantum Electronic* **41** 71

[8] Baranov V E, Baranov P F, Muraviev S V, Uchaikin S V 2015 *Measurement Techniques* **58** 317

Polarization-sensitive photocurrent in the resistive Ag/Pd films

A S Saushin, R G Zonov, K G Mikheev, R R Shamshetdinov and G M Mikheev

Institute of Mechanics of Ural Branch of RAS, 34 ul. T. Baramzinoy, Izhevsk, Russia

Abstract. We report on the observation of helicity dependent photocurrent of the 20 μm thick silver–palladium (Ag/Pd) films manufactured by the thick-film technology. The transverse photocurrent is observed at oblique incidence of laser radiation with different wavelengths in the spectral range of 266 – 2100 nm. At the wavelength range of 532 - 2100 nm the polarity of the transverse photocurrent is positive (negative) for the left- (right-) circular polarized beam. We show that action of high temperature on the films in vacuum results in the decrease of longitudinal photocurrent due to the reduction of PdO content. The photon drag effect is suggested to be the origin of the polarization-sensitive photocurrent in the Ag/Pd films. The obtained results show that the Ag/Pd resistive films may be of interest for polarization–sensitive measurements.

1. Introduction

Interaction of pulsed laser radiation with matter can lead to the interesting phenomena. One of such phenomena is helicity dependent transverse photocurrent induced by the circular polarized laser beam. The transverse photocurrent is perpendicular to the plane of the incidence. The origin of helicity dependent photocurrent can be the circular photogalvanic effect (CPGE) [1] and the photon drag effect [2].

The CPGE arises in gyrotropic media where the symmetry with respect to mirror transformation is broken, and is conditioned by the peculiarities in the band structure of gyrotropic crystals [3]. It should be noted that depending on the type of symmetry, the CPGE can be observed both at oblique and normal incidence of elliptically polarized light onto the sample surface [4].

The photon drag effect (PDE), first discovered in [5, 6], leads to the generation of the photo-electromotive force (photocurrent) by transferring the momentum of photons of incident radiation to the charge carriers in the intraband or interband energy transitions. Unlike the CPGE, the PDE can be observed in centrosymmetric media [2]. In the presence of circular polarization of exciting radiation, the PDE can be accompanied by the helicity dependent photocurrent.

The present work is aimed at studying the spectral and temperature dependences of a polarization–sensitive photocurrent in the silver-palladium (Ag/Pd) films at the wavelength range of 266 - 2100 nm.

2. Experimental

2.1. Ag/Pd films

Ag/Pd resistive films have stable electrical characteristics and have long been used in electronics. They are widely applied for obtaining hybrid microcircuits, multicrystal modules, integrated circuits assemblies and passive electronic components such as resistors, inductive elements, multilayer capacitors. The Ag/Pd films are produced using technologies [7] based on burning of a special paste, which contains silver oxide (Ag_2O) and palladium (Pd), at high temperatures on the insulating substrate surface. The optical and electronic properties of the Ag/Pd films are varied in a wide range depending on a number of factors including thermodynamics and kinetics of Pd oxidation during burnout and firing, Ag diffusion and migration, properties and concentration of inorganic and organic additives. The film studied in our experiments was fabricated at temperature of 878 K using a well-known technology of the thick-film resistors on ceramic substrates. The size of the film obtained was 20 × 20 mm (figure 1a). The thickness of the film was about 20 μm. To measure the photocurrent the film was provided with two parallel silver film electrodes which were arranged along the opposite sides of the film between the dielectric substrate and the Ag/Pd film. Scanning electron microscopy showed that the Ag/Pd film is porous with pore size ranging from 25 to 500 nm (figure 1b).

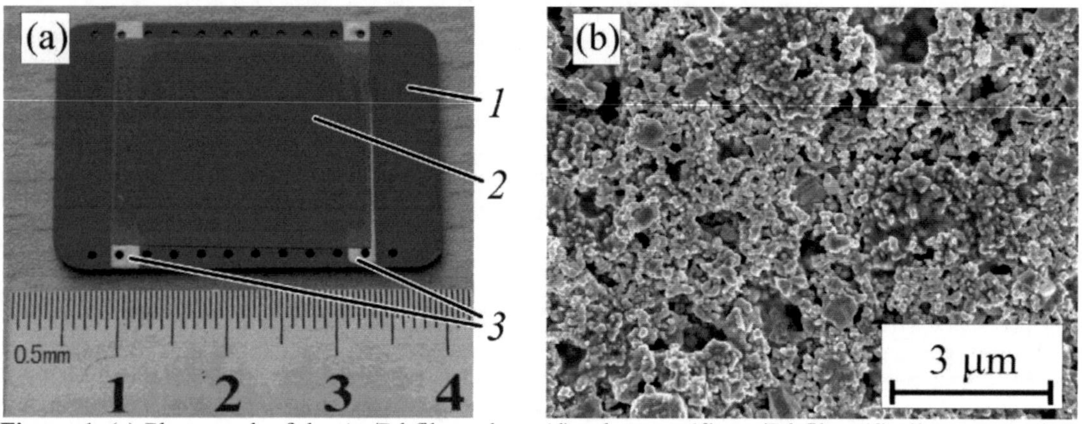

Figure 1. (a) Photograph of the Ag/Pd film, where (*1*) substrate; (*2*) Ag/Pd film; (*3*) silver measuring electrodes located between the substrate and the film, and (b) the scanning electron microscope image of a film surface section.

2.2. Experimental setup

The transverse photocurrent in the resistive Ag/Pd film was induced by laser radiation at different wavelengths in the range of 266 – 2100 nm. In the range of 266 – 1064 nm we used a passive Q-switched single-mode pulsed YAG:Nd^{3+} laser (λ = 1064 nm) with frequency converters to second (λ = 532 nm), third (λ = 355 nm), and fourth (λ = 266 nm) harmonics. To carry out the experiments in the range of 1350 – 2100 nm we have used an automated frequency tunable infrared laser source (Laser Vision). The linearly polarized radiation of laser was converted into circularly polarized radiation by zero order quartz quarter-wave plates (figure 2a). These quarter-wave plates were orientated with their planes perpendicular to the direction of laser radiation propagation. The ellipticity of the laser beam was determined by the angle φ between the vertical axis x' and the slow axis n_e (figure 2a). The radiation fell onto the Ag/Pd film at an angle $\alpha = 45°$. The measuring electrodes *A* and *B* were orientated in such way that they were parallel to the incidence plane σ (the transverse configuration). The laser radiation did not fall on these electrodes during the experiments. The pulsed voltage induced by polarized laser radiation was measured using a TDS7704B digital oscilloscope (Tektronix) with an impedance of $r = 50\ \Omega$ connected directly to the sample electrodes.

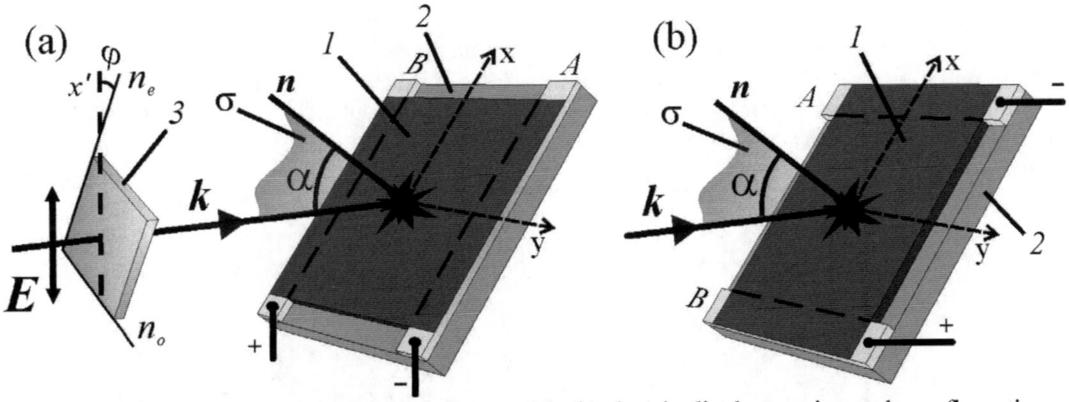

Figure 2. The sketch of (a) the transverse and (b) longitudinal experimental configurations: (*1*) Ag/Pd-film, (*2*) ceramic substrate, (*3*) quarter-wave plate; (*A*, *B*) electrodes, (***n***) normal to the film surface, (***k***) wave vector, (σ) plane of incidence, (α) angle of incidence, (n_o) and (n_e) – fast and slow axis of the quarter-wave plate respectively ($n_e \perp n_o$), (*x*') vertical axis, and (***E***) electric field vector (***E***, *x*', *A*, *B* || σ).

To study the temperature influence on the photocurrent in Ag/Pd films we used *p*-polarized laser pulses of 1064 nm and the longitudinal configuration, where the electrodes were oriented perpendicular to the incidence plane and the photocurrent along *x* axis was measured (figure 2b). The Ag/Pd film was cyclically heated in vacuum. The film was placed in a vacuum chamber, where atmosphere pressure was reduced upon $1.3{\cdot}10^{-3}$ Pa. Heating was performed with the special 15 Ω ceramic isolated wolfram heater. The same oscilloscope was used for measuring photocurrent. Temperature T was changed cyclical: first the Ag/Pd film was heated to $T = 270°$ C and this temperature was carried awhile, then the film cooled down. During the process of heating and cooling the film was irradiated by the laser and the photocurrent was registered. The film resistance was also registered by a digital ohmmeter. After cooling film upon the room temperature the heating and cooling process was repeated.

To estimate the changing composition of the Ag/Pd film the X-ray diffraction (XRD) analysis was performed using a D2 PHASER diffractometer with monochromatized CuK_α radiation.

3. Experimental results

A nanosecond optical pulse produces an electric current directed perpendicular to the σ in the film that manifests itself as a unipolar nanosecond electric pulse of voltage U_y. The photocurrent can be determined as $i_y = U_y/r$.

Experimental investigation of the value i_y as function of the incidence angle α show that the relation $i_y(\varphi, \alpha) = -i_y(\varphi, -\alpha)$, where $-90° < \alpha < 90°$, holds true in the entire range of wavelengths (266 – 2100 nm) and for any fixed angle of φ. That means that the signal polarity depends on the direction of the wave vector ***k*** of incident radiation. According to [2, 8], this allows us to consider the observed effect as PDE.

The dependences of the photocurrent on the angle φ have been obtained for the wavelengths of 266, 355, 532, 1064, 1350, 1450, 1650, 1750, 1850, 1950 and 2100 nm. For example insets in figure 3 shows the normalized photocurrent dependences of the peak amplitude of the measured signal on φ for λ = 355 nm and 1750 nm. The normalized photocurrent is determined as $\overline{i_y}(\varphi) = i_y(\varphi)/i_y^{max}$, where i_y^{max} is the maximum absolute value of i_y in the range $0 < \varphi < 180°$. Similar experimental dependences of $\overline{i_y}(\varphi)$ have been obtained for other wavelengths. For all wavelengths the signal is absent at φ = 0, 90°, 180° (*p*-polarized light). The photocurrent is also absent in the case of *s*-polarization, which has been proven by additional experiments. For λ = 1750 nm (figure 3, right inset) the signal is positive for circular polarization with the negative sign ($0 < \varphi < 90°$) and negative for circular polarization with the positive sign ($90° < \varphi < 180°$), the photocurrent depending considerably on the degree of ellipticity of light polarization, i.e. on the angle φ. However for $\lambda = 355$ nm (figure 3, left inset) on the experimental dependence obtained with both left- and right-hand polarized light and the phase angle varied within $0 < \varphi < 90°$and $90° < \varphi < 180°$, respectively, the signal polarity takes both negative and positive values.

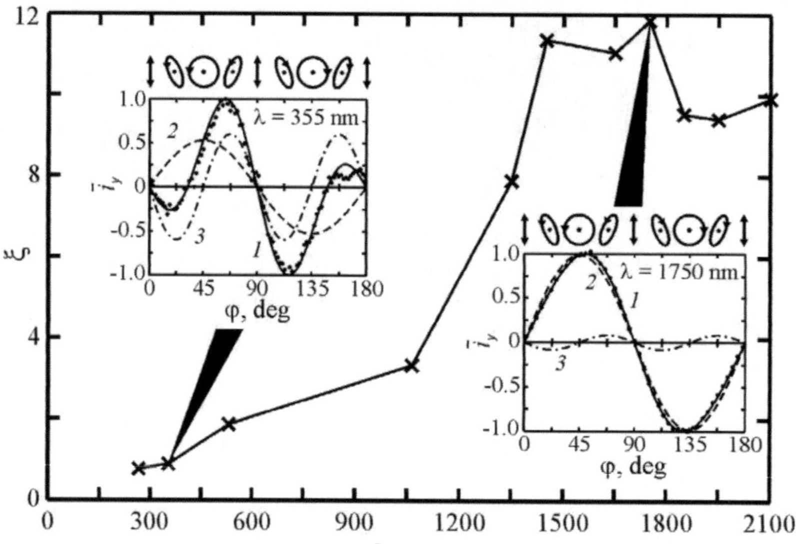

Figure 3. Ratio ξ of the amplitudes of photocurrent circular and linear contributions vs. the incident radiation wavelength λ. Insets show the normalised photocurrent as a function of the angle φ (dots), obtained for the excitation wavelengths of 355 (left inset) and 1750 nm (right inset). Curves (*1*) show approximating dependences, curves (*2*) and (*3*) – dependences of circular and linear contributions, respectively. Polarization ellipses of radiation for different angles φ are shown at the top of insets.

For all wavelengths the data obtained is well approximated with relation:

$$\overline{i_y} = i_1(\lambda)\sin(2\varphi) - i_2(\lambda)\sin(4\varphi), \quad (1)$$

where $i_1(\lambda)$ and $i_2(\lambda)$ represent the magnitudes of the helicity-sensitive (circular) and insensitive (linear) contributions of normalized photocurrent, respectively. The magnitudes of both contributions change with changing the wavelength independently. For $\lambda = 355$ nm $i_1 = 0.527$, $i_2 = 0.602$ and for $\lambda = 1750$ nm $i_1 = 0.986$, $i_2 = 0.083$. The dependence $\xi = i_1(\lambda)/i_2(\lambda)$ has been plotted in figure 3. It can be seen that, in the ultraviolet spectral region, the magnitude ξ is comparable with unity, while in the infrared region for the wavelengths of 1350 – 2100 nm the amplitude $i_1(\lambda)$ of the circular contribution is many times greater than the amplitude $i_2(\lambda)$ of the linear contribution. Thus, it may be assumed that virtually "pure" circular PDE is observed in the wavelength range of 1350 – 2100 nm. Expression (1) indicates that, for the counterclockwise-polarized light $0 < \varphi < 90°$ the photocurrent $i_y(\varphi)$ is strictly positive for $\xi > 2$. It follows from figure 4 that for $\lambda = 532$ nm the ratio $\xi \approx 2$. This means that in the spectral range of $532 < \lambda < 2100$ nm, for transverse experimental geometry, the photovoltaic signal has positive polarity in the case of the negative sign of circular polarization and negative polarity in the case of the positive sign. Thereby the pronounced polarization dependence of the photocurrent makes the Ag/Pd films an interesting material for fabrication of helicity sensitive photon drag photodetectors.

The results of experimental investigations of the temperature influence on the longitudinal photocurrent $i_x=U_x/r$ are presented in the figure 4, where U_x is longitudinal photo-induced voltage. Here, inset demonstrates the example of the changing longitudinal photocurrent during a cycle of heating and cooling. One can see that during heating i_x increases. When the film cools down, i_x reduces and becomes lower than initial value. Figure 4 shows the plots of i_x vs. quantity of heating and cooling cycles. It can be seen that every cycle of heating and cooling leads to decrease of longitudinal photocurrent. The behaviour of the interelectrode resistance of the film is similar. All these results obtained can be explained by the varying of the content of Ag/Pd film during of it heating. It is well known that PdO is unstable at high temperature [9, 10]. In our experiments decomposition of PdO occurs in vacuum. For this reason oxygen is removed from the films and, therefore, Pd is not oxidized backward.

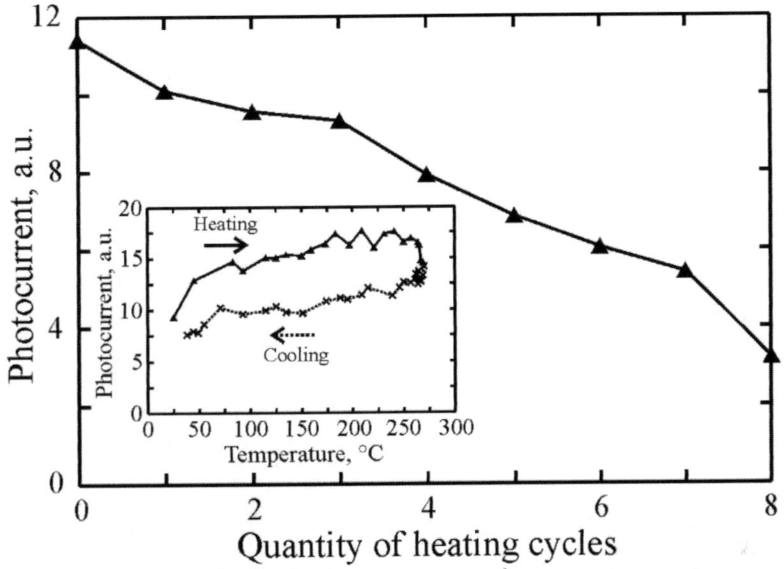

Figure 4. The longitudinal photocurrent at the room temperature as a function of the quantity of heating and cooling cycles. Inset shows the longitudinal photocurrent as a function of temperature during a cycle of heating and cooling.

XRD patterns of the film is shown in figure 5. We can see that XRD pattern of the film before heating contains the Ag-Pd, PdO, and Al_2O_3 reflections (figure 5, curve *1*). The Ag-Pd and PdO reflections are in agreement with composition of initial resistive paste, therefore corresponding solid solutions are present in the film. Al_2O_3 reflection belongs to substrate on which the film was produced. After cyclical heating and cooling the intensity of the PdO reflections have been reduced with respect to Ag-Pd reflections (figure 5, curve *2*). Therefore the content of PdO in the film after heating is lower than before. Moreover new little reflections of Pd appear in the XRD pattern. Thus PdO content is reduced, that results in the photocurrent decrease. These results indicate that PdO content influence on polarization-sensitive photocurrent in Ag/Pd films.

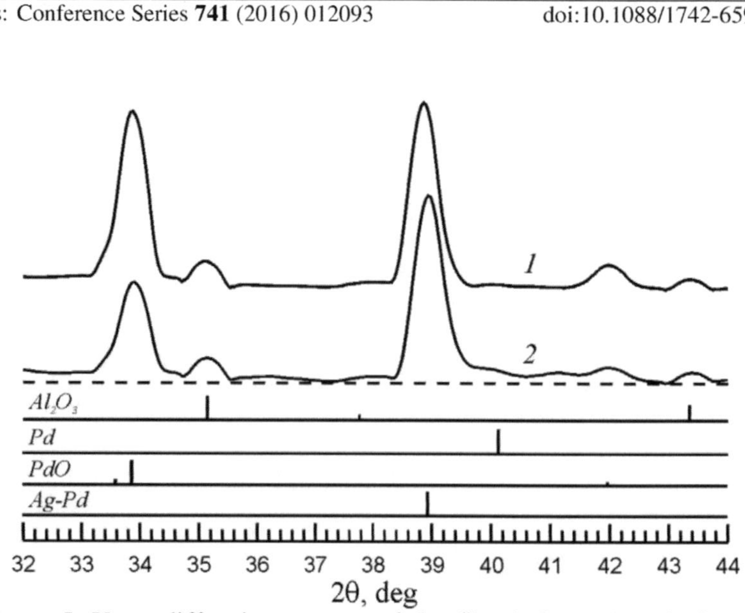

Figure 5. X-ray diffraction patterns of the film before (*1*) and after (*2*) heating and bar diffraction patterns of detected phases (CuK_α).

4. Conclusion

Thus, the present work demonstrates that helicity dependent photocurrent is generated in Ag/Pd resistive films at broad wavelength range from 266 to 2100 nm. In the wavelengths range of 532 - 2100 nm, the polarity of the photocurrent is unambiguously determined by the sign of circular polarization. These results demonstrate the possibility of employing resistive Ag/Pd films for the creation of high-performance sensors of the circular polarization sign of the incident light in a wide spectral range.

The action of high temperature on the films in vacuum results in the photocurrent decrease due to the reduction of PdO content.

The PDE is suggested to be the origin of the helicity dependent photocurrent.

Acknowledgments
This study was supported by the RFBR (project no. 16-38-00552).

References

[1] Belinicher V I and Sturman B I 1980 *Sov. Phys. Usp.* **23** 199–223
[2] Glazov M M and Ganichev S D 2014 *Phys. Rep.* **535**(3) 101–138
[3] Asnin V M, Bakun A A, Danishevskii A M, Ivchenko E L, Pikus G E and Rogachev A A 1978 *JETP Lett.* **28**, 74-77
[4] Ivchenko E L 2002 *Phys. Usp.* **45** 1299–1303
[5] Danishevskii A M, Kastal'skii A A, Ryvkin S M and Yaroshetskii I D 1970 *Sov. Phys. JETP* **58**(2) 544–550
[6] Gibson A F, Kimmitt M F and Walker A C 1970 *Appl. Phys. Lett.* **17**(2). 75–77
[7] Larry J, Rosenberg R and Uhler R 1980 *IEEE Trans. Components, Hybrids, Manuf. Technol.* **3**(2) 211–225
[8] Karch J, Olbrich P, Schmalzbauer M, Zoth C, Brinsteiner C, Fehrenbacher M, Wurstbauer U, Glazov M M, Tarasenko S A, Ivchenko E L, Weiss D, Eroms J, Yakimova R, Lara-Avila S, Kubatkin S and Ganichev S D 2010 *Phys. Rev. Lett.* **105**(22) 227402-6
[9] Mikheev G M, Saushin A S, Goncharov O Yu, Dorofeev G A, Gil'mutdinov F Z and Zonov R G 2014 *Phys. Solid State* **56**(11) 2286–93
[10] Wang S F and Huebner W 1991 *J. Am. Ceram. Soc.* **74**(6) 1349–53

Betavoltaic p^{-}-n^{+}-structure simulation

S U Urchuk, V N Murashev, S A Legotin, A A Krasnov, O I Rabinovich, K A Kuzmina, Y K Omel'chenko, U V Osipov and S I Didenko

National university of science and technology "MISiS" 119049, Leninskiy pr., bl.4, Moscow, Russian Federation

Abstract. In order to increase the betavoltaic batteries efficiency output characteristics of the p^{-}-n^{+} (n^{-}-p^{+}) - structures were simulated. Replacing the p^{+}-n-structures on the p-n^{+} and n-p^{+} -structures enables the space-charge expansion to the crystal surface and thus to reduce the recombination loss in the heavy doped p^{+}-layer and improve conversion of betavoltaic elements efficiency.

1. Introduction

Modern electronic systems more and more need power batteries for electrical and mechanical devices (MEMS). Traditional chemical power batteries are not suitable for power supplying MEMS devices due to their low power density and short lifetime [1]. The need for low output power, long lifetime and no need to repair inherent of devices that are used in electronic equipment, space satellites, sensors in remote locations (mines, far north, etc.), as well as in implantable medical devices.

Betavoltaic batteries are promising power sources with a long lifetime, which are suitable for a variety of low-power devices. Work of betavoltaic sources is based on the principle of converting the β-radiation emitted by the decay of a radioactive isotope, into electrical energy in the semiconductor body. The vast betavoltaic power supply consists of a beta-emitting material connected with p-n-junction [2-6].

Betavoltaic cells produce the power like photovoltaic devices and solar cells. One of the main processes taking place in the semiconductor when an electron contacts with it with energy 1-100 keV, is the ionization of atoms along the electron trajectory. As a result, in the semiconductor forming a great number of electron-hole pairs along the trajectory of high power electron. With a relatively low energy of the outer electron (up to 100 keV), knocking out electrons from the semiconductor atoms and the formation of pairs are the dominant processes, so the number of pairs is proportional to the energy of the outer electron. Under the action of the static electric field these pairs form the current having a value proportional to the product of the electrons flow and their energy.

2. Results and discussion

Optimizing the performance of radiation-stimulated power sources by selecting the doping level of the various regions and their sizes has limitations, so further improvement of device structures requires the selection of new betavoltaic element structures. Simulating of the formation of the power batteries output parameters allows to optimize the design of the device, without the need for expensive experiments. Simulation of the characteristics was carried out using a numerical solution of the basic system of equations given in [7, 8]. As the initial structure for simulation was taken p^{-}-n^{+}-structure shown in figure 1.

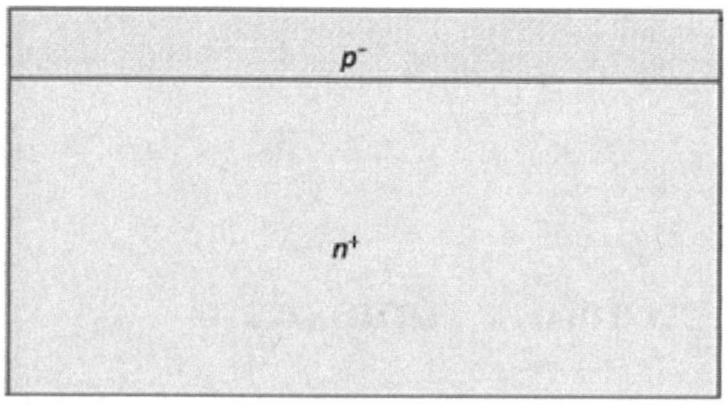

Figure 1. Structure for simulating, the structure square - 1 cm^2.

Previous simulation results obtained showed definite increase efficiency while reducing the depth of the p-n-junction, which is caused by the approach of the space charge region to the area with higher velocity of the generation radiation-stimulated of the mobile charge carriers [8]. Another way to increase the power efficiency can be changes in the level of doping regions so that the surface was light doped region, i.e. replacing p^+n-structure by p^--n^+ structure. This achieves expansion space charge region not into the semiconductor but to the surface, that reduces recombination losses in heavy doped p^+-layer.

Figure 2 shows the influence of the p^--n^+-junction depth on the output parameters betavoltaic elements at different levels of doping the upper light doped p^--layer.

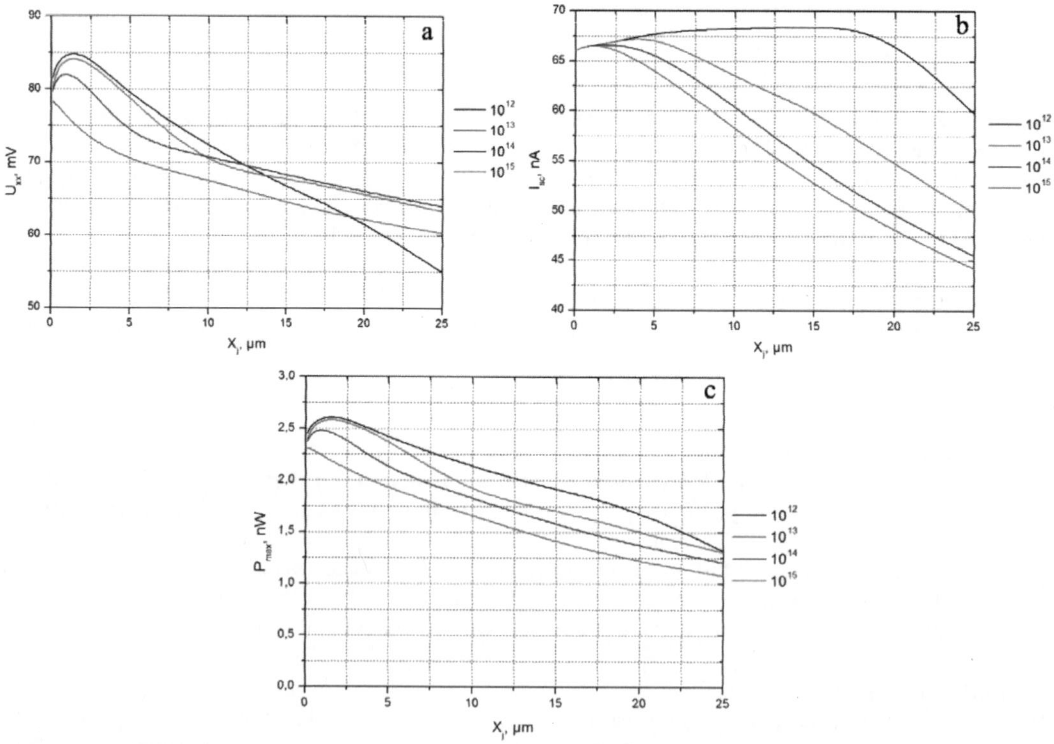

Figure 2. Influence of the upper light doped p^--layer depth to a) open circuit voltage, b) short-circuit current, c) output power (Source activity ^{63}Ni – 2.7 mCi / cm^2).

The released electric power increases with a decrease of the doping level in the light doped region and reducing the p-n-junction depth up to 1 micron, i.e., when approaching depletion region to the surface structure. When the impurity concentration in the p^--layer is 10^{15} cm^{-3}, the space charge region reaches the surface structure with a thickness of a doped region light less than 1 micron. By reducing the doping level less than 10^{12} cm^{-3}, the space charge region doesn't reach the surface only at a thickness more than 25 microns. The increase in the width of the depletion region at a concentration of 10^{12} cm^{-3}and a thickness more 25 microns causes to a significant decrease in efficiency compared to the structures with a higher doping level. This is due to a significant increase of the space charge region and, consequently, an increase of the current generation, which reduces the floating voltage, and thus efficiency. It is interesting to consider how the structure various areas influence on the generation current. Figure 3 shows the change of the different areas contribution at the doping level of 10^{12} cm^{-3}.

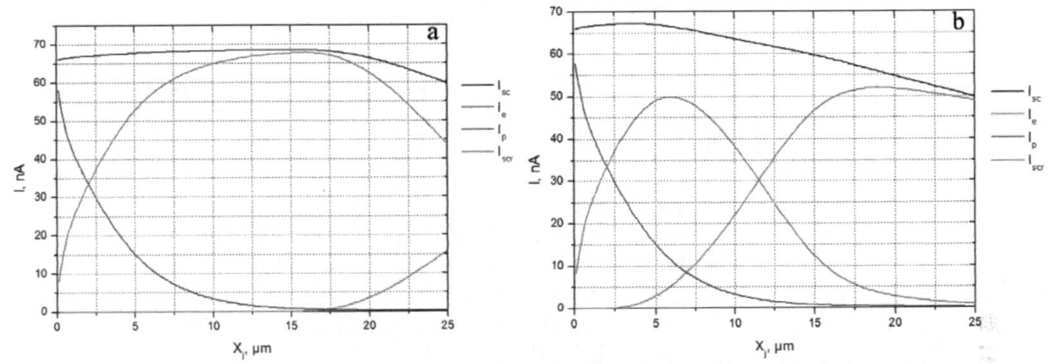

Figure 3. Changing the different regions contribution in p-n $^+$-structure: a) the doping level of p-type region 10^{12} cm^{-3}, b) the doping level of p-type region 10^{13} cm^{-3}.
(Source activity ^{63}Ni – 2.7 mCi / cm^2).

At the light doped region thickness is 25 microns, when the space charge region does not reach the surface the significant contribution to the electron stimulated current the p-type region electrons contribute. By reducing the light doped region thickness, when the space charge region reaches the surface the p-field electron current decreases to zero, the short circuit current is determined only by the space-charge region current. With a further decrease in the p-type region thickness it is reduced space charge region contribution, and simultaneously the total current is reduced. Then the contribution of the depletion region continues to decline, but the area of charge carriers generation reaches the n-region and the contribution of holes begins to increase.

By increasing the doping level of the p-region to 10^{13} cm^{-3} the change in the areas contribution is another trend (figure 3b). At the big thickness of the light doped region the main contribution to the short-circuit current of the p-region electrons give and at light doped region thickness less than 10 microns the significant contribution depletion area provides.

Another problem to be considered is the type of conductivity of light doped p region or n-type. For the p$^+$n-junction answer is obvious because at the required large depletion region the main contribution comes from the space charge region and p$^+$-region electrons. The electrons mobility is higher, so p$^+$n-structures are preferable. Simulation results presented in figure 4 show that the doping level of the light doped region 10^{12} cm^{-3} and the light doped region thickness less than 20 microns greater efficiency is obtained for n$^-$-p$^+$-structures.

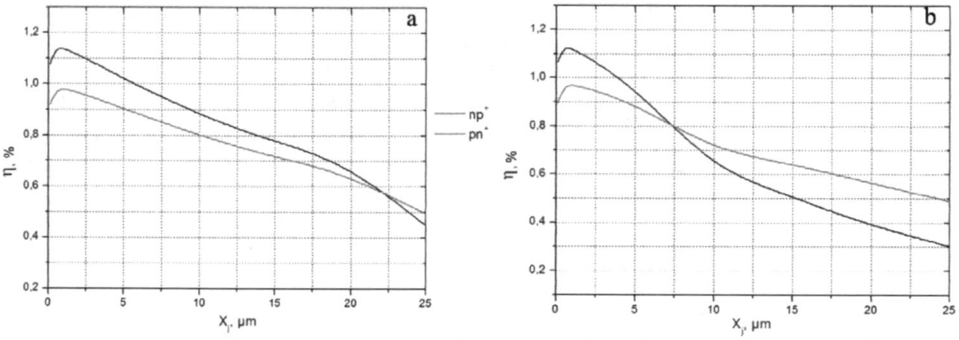

Figure 4. The thickness influence of the light doped region for p^--n^+ - and n^--p^+ structures on betavoltaic elements efficiency: a) the doping level of the light doped region -10^{12} cm^{-3}, b) doping level of the light doped region - 10^{13} cm^{-3} (Source activity ^{63}Ni - 2.7 mCi / cm^2).

It is caused by the fact that at small thickness of the light doped region, when the depletion region reaches the surface, the mobile carriers of quasi-neutral part in the light doped region do not contribute, and an additional contribution to the radiation-induced signal the mobile heavily doped region carriers provide. In the case of n^--p^+-structure these are electrons, which mobility is higher. By increasing the doping level to 10^{13} cm^{-3}, the efficiency is higher for p^--n^+-structure at light doped region thickness of more than 10 microns (figure 4b).

3. Conclusions

Usage as upper layer the light doped semiconductor (p-n^+ - and n-p^+ -structure) gives the possibility for extension of the space charge region to the direction of the surface and allows theoretically to increase the efficiency more than 1 %.

However, it requires manufacturing high quality surface structures to provide a low surface recombination velocity, i.e. the lifetime of mobile charge carriers close to the bulk one. At the doping level of the light doped region of 10^{12} cm^{-3} and a small thickness of the upper i-layer (when the depletion region reaches the surface) the n^--p^+-structure efficiency is higher than p^--n^+-structure.

Acknowledgements

The current study was supported by the Federal Targeted Program "Research and development on priority directions of scientific-technological complex of Russia for 2014-2020", state contract № 14.575.21.0051. (Unique identifier for applied scientific research (project) RFMEFI57514X0051).

References

[1] Sun L, Yuan W and Qiao D 2006 *Microfabric Techno.* **3** 32.
[2] Liu Yu, Tang X and Jang L 2014 *Science China* **57(1)** 14.
[3] Li Da-R, Jiang L and Angfjin L 2012 *CHIN. PHYS. LETT.* **29(7)** 078102.
[4] Krasnov A A, Legotin S A, Omel'chenko Yu K, Didenko S I, Murashev V N, Rabinovich O I, Yurchuk S Yu, Yaromsky V P and Popkova A V 2015 *J. Nano- Electron. Phys.* **7(4)** 04004.
[5] Krasnov A A, Legotin S A, Murashev V N, Didenko S I, Rabinovich O I, Yurchuk S Yu, Omelchenko Yu K, Yakimov E B and Starkov V V 2016 *IOP Conf. Series: Materials Science and Engineering* **110** 012029.
[6] Guo H and Lal A 2003 *IEEE Transducer* **1** 36.
[7] Yurchuk S Yu, Legotin S A, Murashev V N, Krasnov A A, Omel'chenko Yu K, Osipov Yu V, Didenko S I, Rabinovich O I 2015 *J. Nano- Electron. Phys.* **7(3)** 03014.
[8] Yurchuk S Yu, Krasnov A A, Legotin S A, Didenko S I, Murashev V N and Rabinovich O I 2015 *J. Nano- Electron. Phys.* **7(4)** 04005.

On specific features of investigation of fluid flows by photometric techniques

V A Vologdin[1], V V Davydov[1], E N Velichko[1,2]
[1]Department of Quantum Electronics, Peter the Great Saint Petersburg Polytechnic University, Saint Petersburg 195251, Russia
[2]Chair of Computer System Design and Security, Saint Petersburg National Research University of Information Technologies, Mechanics and Optics, St. Petersburg, 197101, Russia

Abstract. Specific features of investigation of the fluid flow structure in a pipeline by photometric techniques are considered. The applicability of the photometric techniques based on the Doppler effect to such studies is discussed. A new method for detecting defects on inner walls of a pipeline that involves the use of the laser radiation scattered from particles in a flowing fluid is suggested.

1. Introduction

Experimental and theoretical studies of flows of fluids and suspensions are among the most urgent problems of fundamental physics [1-3]. Such investigations are necessary for improvement of such devices as flowmeters, spectrometers, pulseoxymeters, life support systems, water-based cooling systems, etc. [3-4]. The data obtained in such studies can make important contributions into development of advanced devices and methods for measurements of physical parameters of flows [2, 5-7].

Of major complexity are the studies of flows of aggressive and hazardous media (e.g., benzene, heptane, or concentrated sulfuric acid) and also of biological solutions and medical suspensions where sterility conditions should be satisfied. Such measurements must be carried out by non-contact methods. For example, high-speed video cameras are successfully used for temperature measurements in plasma [8]; the techniques based on scattered laser radiation enable investigation of surface inhomogeneity [1, 2], etc.

The fluid flow velocity in a transparent pipeline can be measured by the technique based on the Doppler effect. From the measured flow velocity one can find the shape and nature of the flow (laminar or turbulent), but this method has not found a wide use for studying fluid flows and measuring the flow rate. One of specific features of this photometric technique is that in order to measure a fluid flow velocity with an error of not more than 1%, the sizes of the particles from which the laser radiation is scattered must be not less than λ [1, 3, 5-7]. To achieve this, various chemical components or particle are added to water and other media in many experiments. However, such additives cannot be used in studies of hazardous liquids or suspensions in chemical or pharmaceutical industries, and for this reason the structure of the flows of these media in pipes is examined by NMR spectrometers [3, 4].

At present, the problem of detecting internal defects in pipelines by a non-contact method is of vital importance for life-support systems and production of artificial blood vessels. This is also significant for manufacturing biological solutions and medical suspensions because an accumulation of a substance and a subsequent formation of a foreign particle occur in the zone of the defect. The detachment of this particle can lead to undesirable consequences. NMR techniques are unable to reveal defects in a flowing liquid. One of possible solutions to this problem may be the use of photometric techniques.

2. The method

The experimental setup which was used for investigation of the fluid flow velocity in the pipeline with circular cross-section is presented in figure 1.

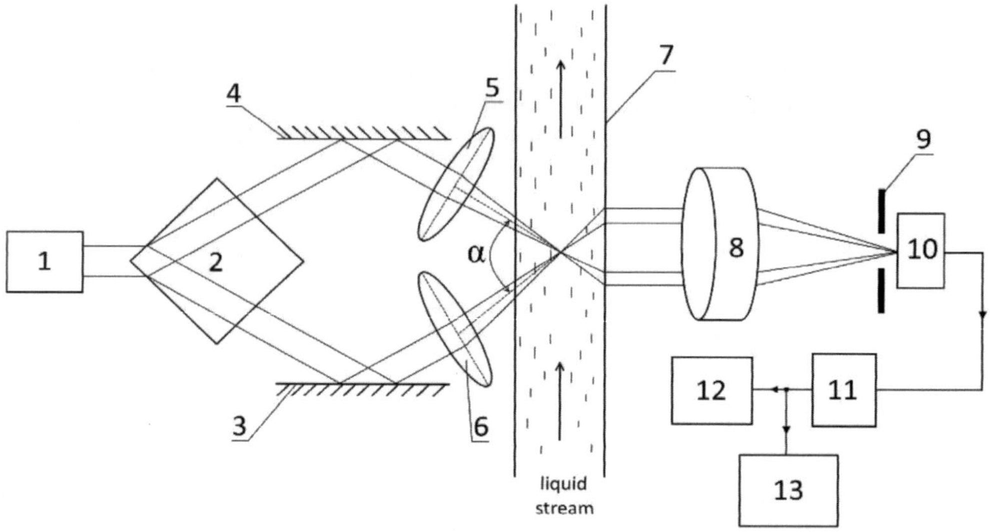

Figure 1. Scheme of differential optical Doppler velocity meter: 1 – optical quantum generator (laser), 2 – rhomb-shaped prism, 3 and 4 – mirrors, 5 and 6 – focusing lenses, 7 – glass pipeline, 8 – object glass, 9 –diaphragm with adjustable aperture, 10 – photoreceiver, 11 – wideband amplifier, 12 – spectrum analyzer, 13 – oscilloscope.

A differential type measuring scheme allowed us to exclude the condition of coincidence of directions of the wave number vectors of laser radiation k_0 and k_s in measurements of velocity with an accuracy of no more than 1% [7]. After laser beam splitting into two uniform beams (by a beam splitter rhomb-shaped prism (2) in figure 1) both coherent beams are passed through the optical elements and are focused on a flowing fluid in the glass pipeline (7) to be studied. The scattered light passes through an object glass (8) and is detected by a photomultiplier (10). The wave fronts of the scattered laser beams in our method are always parallel.

One more feature of the differential circuit presented here is the independence of the spectral width of beats on the object glass aperture. The lower limit of the measuring range is determined by the bandwidth of laser radiation and the upper limit is dictated by the photoreceiver bandwidth (if we use a photomultiplier, the bandwidth is up to 300 MHz)

To calculate the flow velocity υ by the Doppler method, classical equations may be used [6, 7]:

$$\Delta f_D = \upsilon \cdot \frac{2n}{\lambda} \cdot \sin\left(\frac{\beta}{2}\right) \cdot \cos\varphi \ , \quad (1)$$

where β is the angle between the directions of incident and scattered waves, φ is the angle between the difference vector k_s and velocity vector, and Δf_D is the Doppler frequency shift.

In our case the wave number vectors of scattered light which are detected by the photoreceiver are coplanar (a specific feature of the differential circuit). Therefore, equation (1) may be written as

$$\Delta f_D = \frac{2\upsilon}{\lambda} \cdot \sin\left(\frac{\alpha}{2}\right) \qquad (2)$$

where υ is the flow velocity, and α is the angle between laser incident beams.

It is evident from (2) that the upper measuring range boundary of υ for a specified λ is determined by angle α. For example, the experimental results showed that in the case λ = 630 nm and α = 20^0, the maximal flow velocity which can be measured was 5 m/s

In addition, we used the transfer matrix method for calculating the transformation of the laser Gaussian beam and the beam waste shift ΔX_p during the beam successive passage through two different interfaces (air - glass and glass - fluid) and also the coordinate of the beam intersection point Xp. In the meridional plane ΔXp is given by the following relationship:

$$\Delta X_p = \frac{(n_2^2-1)\sin^2\alpha/2}{n_2\,(1-\sin^2\alpha/2\,)^{3/2}} \cdot X_k + \frac{(n_1^2-n_2^2)\sin^2\alpha/2}{n_2\left(n_1^2-\sin^2\alpha/2\right)^{3/2}} \cdot L_p \qquad (3)$$

where n_1 and n_2 are the refractive indexes of the material of pipeline and flowing fluid, X_k is the distance from the outer pipeline wall in the laser radiation propagation direction to the point where the beams would intersect after the lenses (Figure 1) if no pipe with a flowing fluid were present in their path, and L_p is the pipe wall thickness.

Expression (3) was calculated for the following method of setting of the measuring optical system. The setting was carried out by the maximum value of Δf_D which corresponded to the highest velocity in the pipeline. Regardless of the regime of fluid flow, the flow velocity is maximum at the center of the pipeline. The point at which the beams would intersect after the lenses was determined in the absence of the pipeline with the flowing fluid. The point of location of the pipeline was found and X_k was determined.

Then ΔX_p was calculated. ΔX_p had to be lower than d/4 (where d is the inner pipe diameter), otherwise it was difficult to set the system to measuring the maximum υ. After this the pipeline moved along the X axis which coincided with the propagation direction of radiation emerging from 1 (see Fig. 1). The distance X_k in this case increased or decreased until the maximum in Δf_D appeared. After setting a new value of X_k was determined and ΔX_p which was compared with the difference (L_p + d/2) – X_k was calculated. The discrepancy between these values was less than 2%, all the inaccuracies in measuring the distances being taken into account. The result obtained allowed us to accurately determine the position of the point in the pipeline at which the measurements of υ were performed. The error in measurements of flow velocity υ was less than 1%.

In the case a defect was formed on the inner surface of the pipeline, the flow velocity profile changed in the pipeline cross-section (above the defect formation zone) as compared with the profile of υ in other sections of the pipeline. In the case of laminar fluid flow, the flow velocity profile remained unvaried along the entire length of the pipeline [9, 10]. Figure 2 shows a section of the pipeline with the defect in the form of protrusion (the most commonly encountered defects on the inner walls of pipelines). Comparison of the measured velocity profiles were performed in two sections 1 and 2.

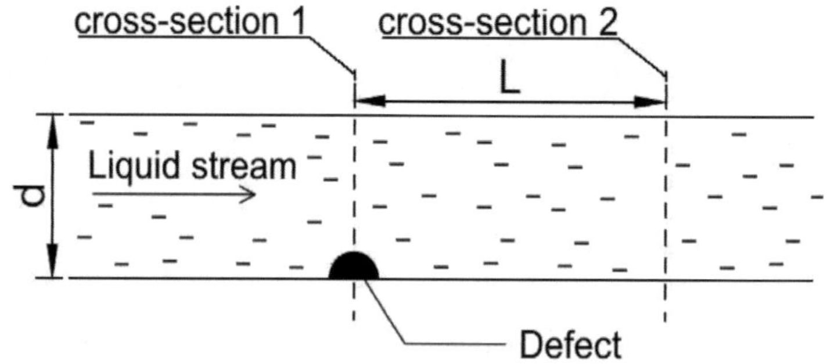

Figure 2 Pipeline with a flowing fluid

3. Results and discussion

Figure 3 shows, as an example, the measured flowing fluid velocities in two sections of the pipeline spaced at L = 20 cm (Figure 1 (a)). In the experiment, the tap water with an additive in the form of iron-3 oxidized in water (particle size of 1 - 3 microns) was used.

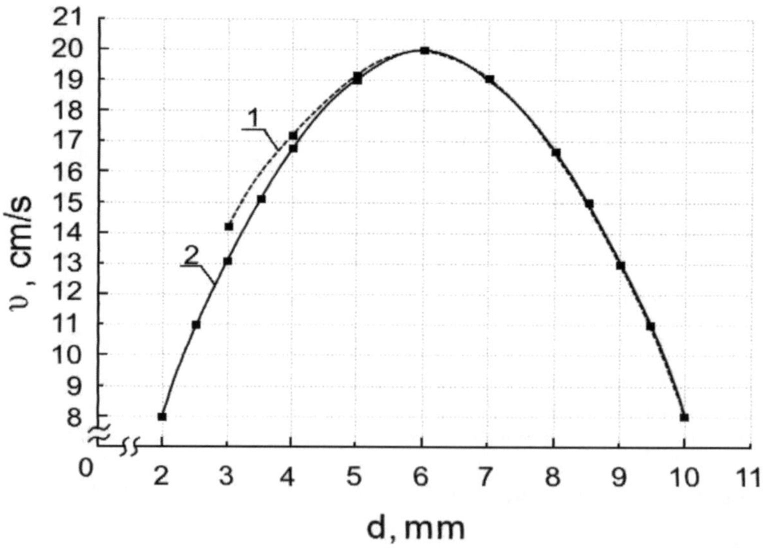

Figure 3 Flow velocity υ distribution over the pipeline cross section (d – diameter). Curve 1 is for cross-section 1, curve 2 is for cross section 2.

The experimental results confirmed that it was possible to establish the presence of internal defects in the form of protrusions in the pipelines with the diameters d more than 4 mm by the method involving measurements of velocity changes. If the internal defect in the pipeline was in the form of a crack, it was difficult to detect it by measuring the flow velocity because a vortex zone was formed in the vicinity of the crack near the pipeline wall.

The flow velocity changed only slightly in this zone. In order to reveal defects of this type in pipelines, we developed the following method. The camera recorded a speckle pattern of the scattered laser radiation from the fluid flow. The changes in the speckle pattern shape arising due to eddy currents in the defect zone during the flow of the fluid had specific features from which the presence of the defects could be revealed.

Conclusion

The experimental results have shown that the method we propose is much more efficient and reliable than the detection of defects on the inner pipeline walls by optical techniques without the use of a flowing fluid.

References

[1] Rozanov N N 2012 *Optics and Spectroscopy* **113** 613
[2] Popovac M, Hanjalic K 2007 *Flow, Turbulence and Combustion* **78** 177
[3] Davydov V V, Dudkin V I, Vologdin V A 2015 *Journal of Applied Spectroscopy* **82** 936
[4] Karseev A Yu, Vologdin V A, Davydov V V 2015 *Journal of Physics: Conference Series* **643** 012108
[5] Sobolev V S, Prokopenko M N 2000 *Quantum Electronics* **30** 1109
[6] Borozdova M A, Fedosov I V, Tuchin V V 2015 *Quantum Electronics* **45** 275
[7] Zybov V A, Rinkevichus B S 1997 *Quantum Electronics* **24** 1161
[8] Bronenko M P, Gulyaev I P, Gulyaev P Yu, Seregin A E 2014 *Russian Physics Journal* **57** 279
[9] Alekseeva K G, Borzenko E I 2012 *Russian Physics Journal* **55** 684
[10] Girgidov A D 2007 *Mechanics of fluid and gas* (Saint – Petersburg: Publisher Polytechnic University)

GaP/Si anisotype heterojunction solar cells

K S Zelentsov[1], A S Gudovskikh[1,2]

[1] St. Petersburg Academic University, St. Petersburg 194021, Russia
[2] St. Petersburg Electrotechnical University "LETI", 197376, St.-Petersburg, Russia

e-mail: veorg00@gmail.com

Abstract. Multijunction solar cells grown on Si substrates using III-V compounds are very promising due to potentially high efficiency (over 40%). In that case homojunction in silicon can be replaced with GaP heterojunctions which can be more effective in terms of lowering surface recombination. Silicon substrates of both n- and p-type doping were used to form GaP/Si anisotype heterostructures. It was demonstrated that for p-GaP/n-Si structures charge carrier transport is mainly blocked due to large value of valence band offset at the heterointerface. For n-GaP/p-Si solar cells optimal thickness and doping level were calculated using AFORS-HET software. The influence of interface states density and GaP layer parameters was investigated.

1. Introduction

Further development of silicon based multijunction solar cells requires new concepts to be applied. The use of anisotype heterojunctions may potentially increase the efficiency by lowering surface recombination. This approach was successfully used for GaAs based solar cells as reported here [1]. However, in case of GaP/Si heterojunctions the polar/nonpolar nature and band offsets at the interface must be taken into account.

Numeric simulations of band diagram (figure 1) for both p-GaP/n-Si and n-GaP/p-Si heterostructures were performed using AFORS-HET software [2]. According to data on band offsets [3, 4], high value of ΔE_v (0.8-0.95 eV) results in significant barrier for majority carriers (holes) at the heterointerface for p-GaP/n-Si structure. This barrier may affect charge carrier transport even at room temperature. On the other hand, in this structure minority carriers (electrons) are effectively blocked off the surface due to band bending in GaP layer and ΔE_c value, thus surface recombination is suppressed.

In case of n-GaP/p-Si structure, barrier height for majority carrier (electrons) is significantly lower due to conductance band offset ΔE_c = 0.19-0.32 eV. However, potential barrier at the interface for minority carrier, which is mostly because of valence band offset, still effectively blocks them off the surface and thereby lowers recombination rate.

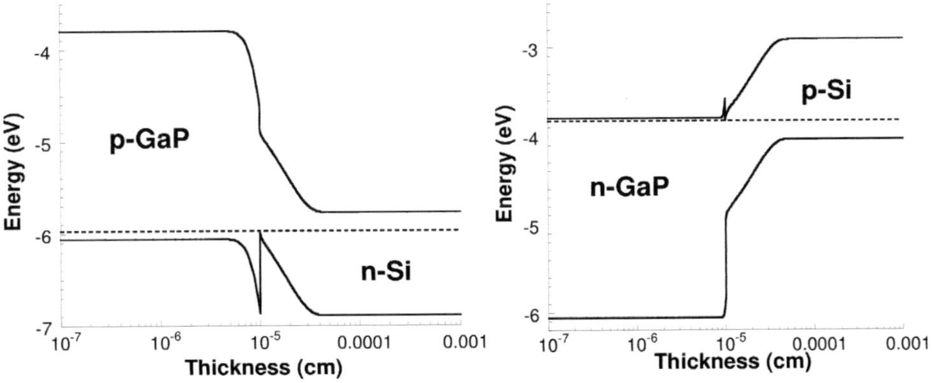

Figure 1. Simulated band diagrams for p-GaP/n-Si and n-GaP/p-Si heterostructures.

2. Experiment

Both p-GaP/n-Si and n-GaP/p-Si heterostructures were fabricated using molecular-beam epitaxy. Silicon substrates of <100> orientation with 4° offcut towards <110> with doping level of 10^{16} cm^{-3} were used. Thickness of GaP layer was 100 nm for both structures and the doping level was more than 10^{18} cm^{-3}. Al and In were used to form back contact to p-type and n-type substrates respectively. Top contact was In for both structures.

Admittance spectroscopy technique [5,6] was used to investigate charge carrier transport in GaP/Si structures. Potential barriers at the interface result in step-like increase in admittance signal as shown in figure 2. Obtained value of activation energy E_a, which is in fact an effective barrier height, is 0.7 eV and 0.12 eV for p-GaP/n-Si and n-GaP/p-Si structures respectively. This is less than values of band offsets reported in literature because of tunneling effect in highly doped GaP layer.

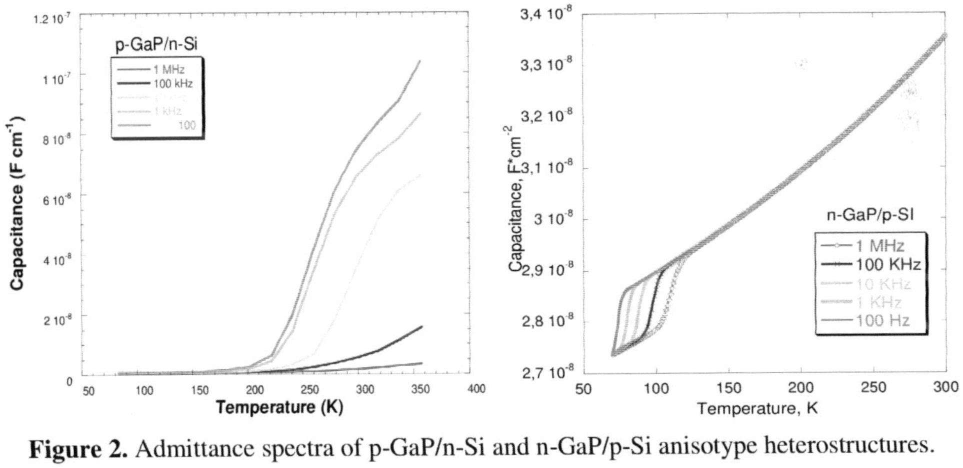

Figure 2. Admittance spectra of p-GaP/n-Si and n-GaP/p-Si anisotype heterostructures.

Current-voltage characteristics under illumination flux of ~1 sun, measured for these structures are presented in figure 3a. The potential barrier of 0.12 eV is not sufficient enough to affect charge carrier transport over the interface, so n-GaP/p-Si structure demonstrates decent values of short circuit current, J_{sc}= 38 mA/cm^2, and fill factor, FF=84%. Short circuit current for p-GaP/n-Si is about two orders of magnitude lower that that for n-GaP/p-Si due to potential barrier of 0.7 eV which blocks charge carrier transport over the interface.

The result above was also confirmed by spectral measurements of EQE for these structures (figure 3b). N-GaP/p-Si demonstrates noticeable EQE in the range of 300-1200 nm due to absorption in silicon substrate. However, for p-GaP/n-Si structure potential barrier for holes blocks transport over heterointerface, so this structure exhibits non-zero EQE only for wavelengths less than 400 nm and only charge carriers generated in GaP layer contribute to photocurrent.

Thus, there are fundamental problems, limiting the use of p-GaP/n-Si heterostructures for photovoltaic applications. However, GaP/Si structures grown on p-type substrates have no such limitations and can be used for high efficiency solar cells.

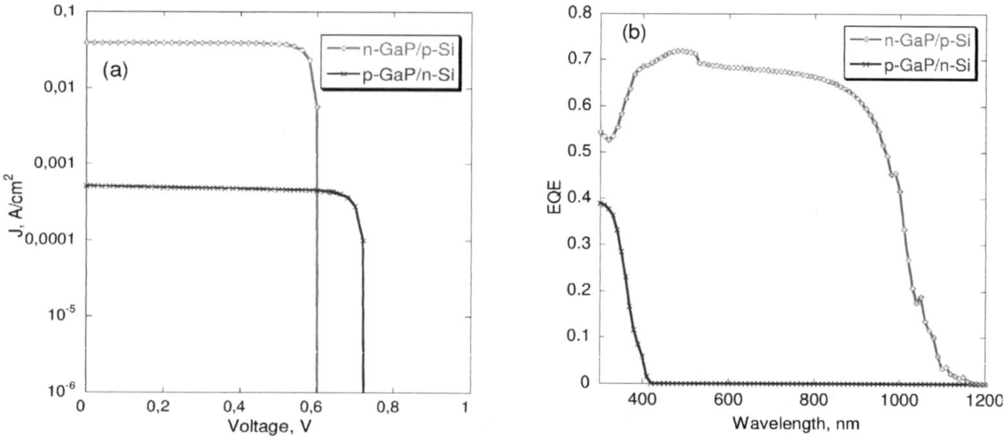

Figure 3(a, b). (a) I-V curves for GaP/Si heterostructures; **(b)** Quantum efficiency spectra (no ARC).

3. Simulation

It was demonstrated in previous section that p-GaP/n-Si heretostructure can be used as an effective photoconverter. In this section the influence of GaP thickness, doping level and interface states density on device properties is considered using numeric simulation.

Effective lifetime of charge carrier in GaP epi-layer is substantially less than that of substrate, so photocurrent through GaP/Si structure mostly depends on absorption in Si. Absorption in GaP layer result in short circuit current decrease when layer thickness is more than 50 nm as shown in figure 4a. However, V_{oc} is only slightly depends on GaP thickness, but largely on interface trap density D_{it}. In order to reduce effect of D_{it} on V_{oc}, the minimum doping level of GaP layer should be not less than $N_d > 10^{16}$ cm^{-3} or else $(E_C - E_F) \leq 0{,}2$ eV. On the other hand, the surface state density should be less than $D_{it} < 10^{12}$ $cm^{-2}eV^{-1}$ (see figure 4b).

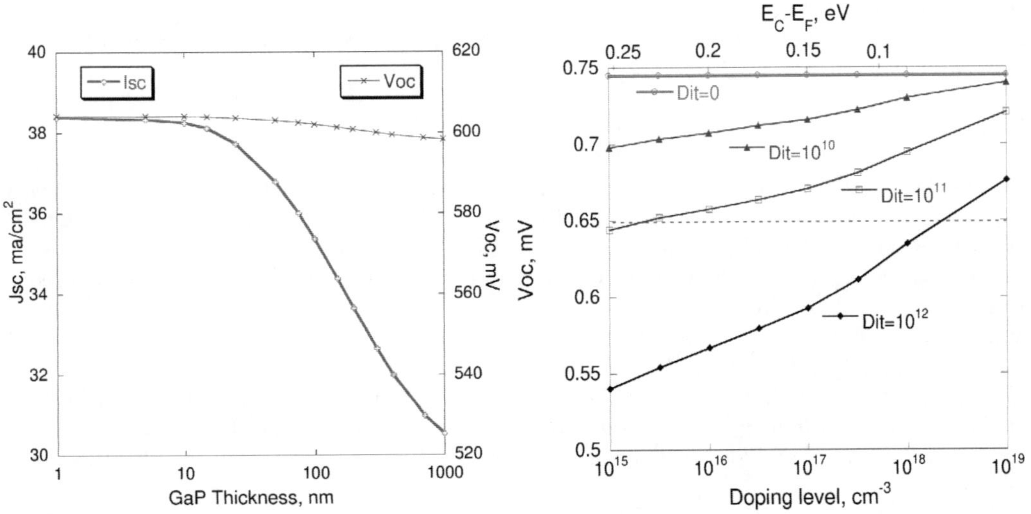

Figure 4(a, b). (a) Effect of GaP layer thickness on I_{sc}; **(b)** V_{oc} at different Dit and GaP doping level.

In figure 5 anisotype heterojunction n-GaP/p-Si is compared to conventional n-GaP/n-Si/p-Si solar cells (isotype). I-V curves demonstrate that both structures are sensitive to heterointerface trap density, but there is almost no difference between them in terms of open voltage circuit or fill factor at equal D_{it}. However, isotype structure is commonly formed by diffusion of phosphorus into substrate during GaP growth. This requires the use of high temperatures for any growth technique used. Anisotype heterostructures doesn't rely on diffusion processes, so it can be possible to grow this structure at lower temperatures using ALD instead of MBE.

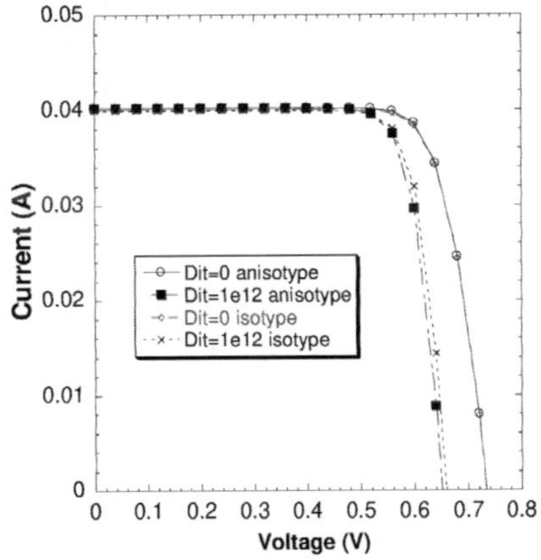

Figure 5. Simulated I-V characteristics for isotype and anisotype GaP/Si heterostructures.

4. Conclusion

In this paper new approach to high efficient Si based solar cell was discussed. It was demonstrated that charge carrier transport is mainly blocked in p-GaP/n-Si heterostructures, while n-GaP/p-Si does not concede to conventional isotype heterostructures, but can be grown at lower temperatures. The requirements to doping type, doping level and layer thickness for GaP/Si solar cells were investigated.

Acknowlegements
This work was carried out under financial support by Ministry of Science and Education of Russian Federation (RFMEFI61615X0040).

References

[1] A.S. Gudovskikh, K.S. Zelentsov, N.A. Kalyuzhnyy, V.M. Lantratov, S.A. Mintairov, *Anisotype GaAs based heterojunctions for III-V multijunction solar cells*, Proc. of 25th European Photovoltaics Solar Energy Conference and Exhibition, 472-476, 2010
[2] R.Stangl, M.Kriegel, M.Schmidt, Proc WCPEC-4, 4th World Conference on Photovoltaic Energy Conversion, Hawaii, USA, May 2006
[3] Wagner, H *J. Appl. Phys.*115, 044508. – 2014
[4] Sakata I. *Appl. Phys. Express* vol 1, Nr 9 2008
[5] J. Li, R.S. Crandall, I. Repins, A. Nardes, D. Levi, O. Sulima, *Applications of admittance spectroscopy in photovoltaic devices beyond majority-carrier trapping defects*, PVSC 37 2011
[6] Zelentsov K.S. *Study of diffusion processes at the III-V/IV interface in MJ SC* 2014 E-MRS Spring Meeting

Effect of a boron implantation on the electrical properties of epitaxial HgCdTe with different material composition

D V Lyapunov[1], A A Pishchagin[1], D V Grigoryev[1], A G Korotaev[1], A V Voitsekhovskii[1], A P Kokhanenko[1], I I Iznin[1,2], H V Savytskyy[3], A U Bonchik[3], S A Dvoretskii[1,4], N N Mikhailov[4]

[1] Tomsk State University, 36 Lenin Av., Tomsk, 634050, Russia
[2] Scientific Research Company "Caratt", L'viv, Ukraine
[3] Institute for Applied Problems in Mechanics and Mathematics, L'viv, Ukraine
[4] Institute of Semiconductor Physics SB RAS, Novosibirsk, Russia

Abstract. In this work the experimental results of investigations of the dynamics of accumulation and spatial distribution of electrically active radiation defects when irradiating epitaxial films of $Hg_{1-x}Cd_xTe$ (MCT) with different material composition (x). The films, grown by molecular beam epitaxy (MBE) were irradiated by B ions at room temperature in the radiation dose range 10^{12} –10^{15} ions/cm^2 and with ion energy 100 keV. The results give the differences in implantation profiles, damage accumulation and electrical properties as a function of the material composition of the films.

1. Introduction

The semiconductor narrow-band material $Hg_{1-x}Cd_xTe$ (MCT) where the material composition (x) is the molar content of Cd is one of the main materials used for manufacturing inherent IR-photodetectors for the wavelength ranges of 3-5 and 8-14 micrometers [1]. At present the epitaxial MCT films grown by the method of molecular-beam epitaxy (MBE) present a promising material for creating multi-element semiconductor detectors of IR radiation providing signal processing directly in the focal region. One of the basic technological methods used to create photodiode structures is boron implantation [2 - 4]. For a precise control of the implantation, a strong insight is necessary into the processes of formation and evolution of radiation defects after ion implantation, which in the case of MCT determines all the electro-physical properties in the irradiated region [2 - 10]. It should be noted that in recent years to improve the operational characteristics of the device structures, the epitaxial material is used with different graded-band gap layers, in particular, in the surface region [11, 12]. Thus the study of the process of radiation defect formation in MCT, with a variable composition x, is important.

In the previous works of the authors concerning experiments about the effect of radiation using molecular nitrogen, Ar and boron ions on the electro-physical parameters of MCT epitaxial films [13, 14] showed that the observed differences due to ion implantation in heteroepitaxial MBE MCT as compared to the implantation in volume crystals of MCT can be explained by a different varied layer in the surface region. The authors make the assumption that the observed differences of ion implantation in the epitaxial films of MBE MCT with a variable composition in the region of implantation and in the volume material of MCT are mainly conditioned by various dynamics of accumulation of electrically active defects and the dependence of electro-physical characteristics on the MCT composition. But proof of this assumption requires more research.

The aim of this paper is to investigate the specific properties of the formation and accumulation of the spatial distribution of electrically active radiation defects after B ion implantation in MCT epitaxial films of different material composition in the region of introduction of the implant.

2. Samples and measurement procedure

The epitaxial films were grown at the Semiconductor Physics Institute of the Siberian Branch of the Russian Academy of Sciences in Novosibirsk, specifically with different mercury to cadmium ratios (x = 0.22, 0.3, 0.38, 057). After growth, the films had n-type conductivity, for conversion to p-type conductivity, the films were annealed in a neutral atmosphere of hydrogen or helium. Irradiation by B ions was made at room temperature for B ion radiation dose in the range 10^{12} –10^{15} ions/cm^2 and with ion energy 100 keV.

Measurements of the electro-physical parameters of the samples before and after irradiation were made at the temperature of liquid nitrogen using the Hall Electromotive Force (EMF) method in the Van-der-Pau configuration. The electron concentration distribution as a function of semiconductor depth was determined by the method of differential Hall measurements using an etching process. Thin layers of material were removed by chemical etching in a 0.02% Br solution in dimethylformamide. The rate of etching was determined using a test sample with a step height measured by an electron microscope MII-4 after multiple cycles of material removal. The error in the Hall and conductivity measurements did not exceed 2-3%. The error in the volume concentration and mobility of charge carrier measurements is mainly determined by the accuracy in the measurement of the thickness of the removed layer and did not exceed 10-12%.

3. Experimental results and discussion

Measurement of the electro-physical parameters of the samples after implantation shows that B ion irradiation resulted in a conversion of the conduction type, with the formation of a highly alloyed n$^+$-layer in the surface of the material due to the generation of radiation defects. The dependence of the layer charge carrier concentration N_S on the radiation dose Φ is shown in Figure 1.

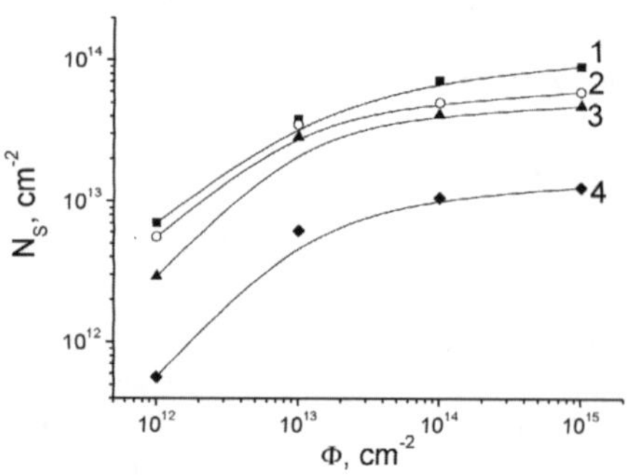

Figure 1. The dependence of the electron concentration N_S plotted against radiation dose. The material composition (x): 1 – 0.22, 2 – 0.3, 3 – 0.38, 4 – 0.57.

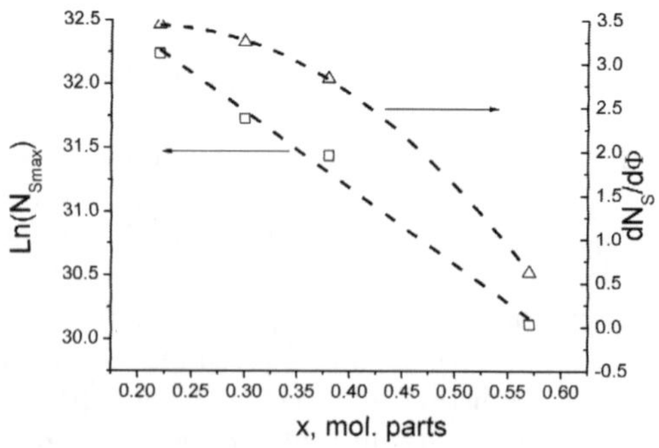

Figure 2. The dependence of the natural logarithm of the maximum layer charge carrier concentration N_{Smax} and rate of introduction of electrically active radiation defects dNS/dΦ on the composition x of the MCT film

From the figures we can see that for all the samples, a monotonic increase of N_S with radiation dose is observed. However, for the same radiation dose the value of N_S depends strongly on the material composition of the films. The dependence of N_S on radiation dose shows a behavior typical of what would be expected in bulk MCT crystals. When attaining the radiation dose of 10^{14} ion/cm^2, the layer concentration is maximal N_{Smax} and saturation takes place.

However, the magnitude of the maximum value N_{Smax} in the saturation region dose depends strongly on the material composition of the films. With the increase in the value of the composition of the epitaxial film (x) the value of the N_{Smax} decreases (Figure 2). The different behavior of N_S on dose for the four series samples occurs because implantation take place in the region, where the material composition differs, the generated primary radiation defects in the metal sublattices of CdTe and HgTe will also differ. Using the software package SRIM2003 [15] a calculation was made of the total number of Hg interstitials that would form for a given number of collision cascades. The calculation shows that for sample 4 the number of primary shifts is 60% less than for sample 1. Because the number of generated interstitial Hg atoms decreases, we might expect a delay in the dynamics of formation of the secondary electrically active defects whose kinetics of accumulation is proportional to the primary radiation defect concentration. Therefore for samples of different series at doses $\Phi < \Phi sat$ the rate of introduction of electrically active radiation defects $dN_S/d\Phi$ was varied. With increasing composition x decreases the rate of defect introduction. Figure 2 shows the dependence of the rate of introduction of electrically active radiation defects on the composition of the epitaxial film.

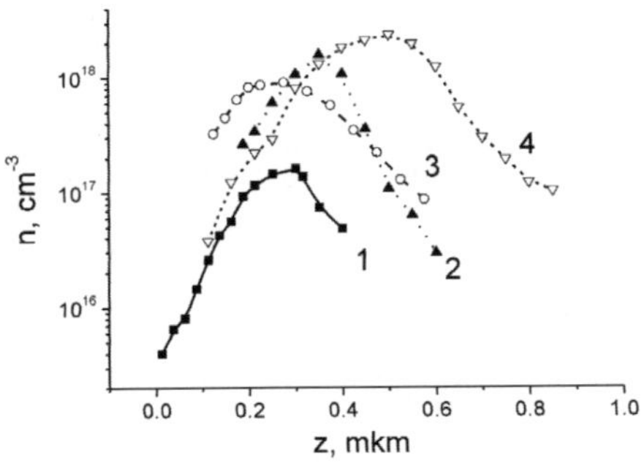

Figure 3. The electron concentration n plotted against depth z for the samples with material composition x = 0.22 after implantation by B ions. The radiation doses are : 1 – 10^{12} ion/cm^{-2}, 2 – 10^{13} ion/cm^{-2}, 3 – 10^{14} ion/cm^{-2}, 4 - 10^{15} ion/cm^{-2}.

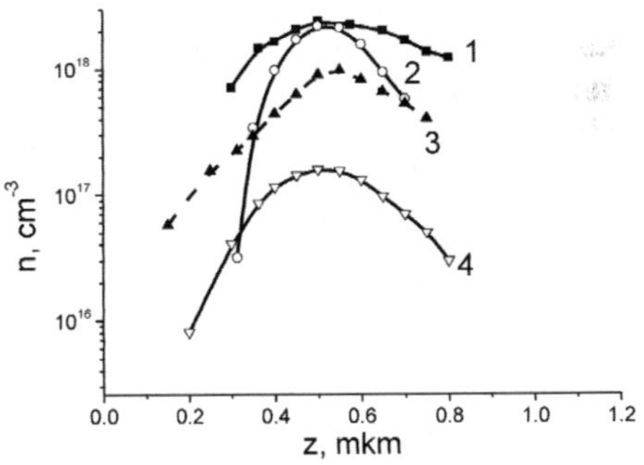

Figure 4. The electron concentration n plotted against depth z after implantation by B ions. The radiation doses are 10^{15} ion/cm^{-2}. The material composition (x): 1 – 0.22, 2 – 0.3, 3 – 0.38, 4 – 0.57.

The experimentally determined profiles of the volume electron concentration n as a function of the depth z show that at an irradiation dose of 10^{12} ion/cm^2 the maximum electron concentration is detected in the region of the mean free projection path of B ions. Increasing the irradiation dose up to 10^{14} ion/cm^2 results in an increase in the maximum value n_{max}. A further increase of irradiation dose results in a shift of n_{max} deeper into the sample but the value of n_{max} itself does not change, so electron concentration saturation takes place (Fig. 3). For a fixed dose the peak values of n occur at the same depth for all the samples considered. However the value of the volume electron concentration at the profile peak is determined by material composition (Fig. 2). The value of the n_{max} is reduced by increasing the material composition.

According to experimental data the high-energy irradiation of the semiconductor (independence of irradiation condition and a previous history of a material) with increase of radiation dose lead to pinning of the electrochemical potential (Fermi level) in limiting state (F_{lim}). In the case of the material HgCdTe the pinning of a Fermi level is explained by coincidence of it energy with energy of radiation-induced defects (E_R), generating by irradiation. With increase of radiation dose the donor type defect concentration increase and accordingly an electron concentration increase down to coincidence of Fermi level energy and a defect energy level ($E_F=E_{RD}$) is observed. At the further increase of radiation dose the electron concentration is invariable value that is shown by our experiments for irradiation HgCdTe by ions, protons, electrons and the relevant theoretical model is suggested [14, 16]. The energy position of the level E_R is estimated to be 0.15 eV higher than the bottom of the conduction band.

In reference [18] it is shown that high energy radiation of semiconductors in the diamond structure and binary mixtures of group $A^{III}B^{V}$ leads to fixing the Fermi level in the limiting state F_{lim} independent of the irradiation conditions and the previous history of the material. It was also shown that the stationary location of the Fermi level in an irradiated semiconductor coincided with the level of local electro-neutrality of a crystal, whose position can be calculated theoretically. Reference [19] uses such an approach applied for narrow-gap solid solutions of MCT.

A comparison of experimental data [2 - 14] and theoretical calculations [19] enables one to assume that the energy position of the level of radiation defects E_R coincides with the position of local electro-neutrality of a crystal lattice of the material relative to the bottom of the conductivity band E_C. Within the limits this approach we have made the calculation of the energy position of the Fermi stationary level F_{lim} and the corresponding limiting concentration of electrons n_{lim} depending on the MCT composition. When calculating we took into account the results of reference [19] and the following.

1. The energy position of the level of local electroneutrality of the material relative to the level E_C is linearly dependent on the MCT composition;
2. For the composition x = 0.5 the energy position of the level coincides with the edge of the bottom of the conductivity band;
3. For the composition x = 0.2 the level is located 0.15 eV higher than the bottom of the conductivity band E_C, i.e., it coincides with the energy position of the level E_R determined from the experimental data [2 - 14].

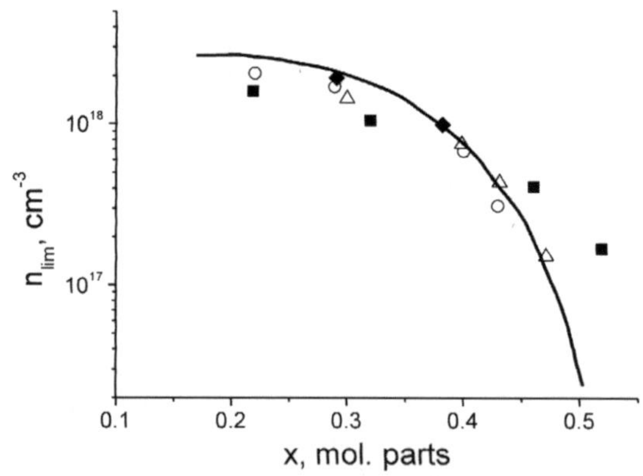

Figure 5. The maximum electron concentration n_{lim} plotted as a function of the material composition x. The solid curve gives the calculated values and the symbols refer to experimental points obtained from figures 3-4.

The results of the calculation of the limiting (i.e. maximum) electron concentration n_{lim} as a function of composition are shown in Figure 5. The symbols refer to the experimental values of the maximum electron concentration as determined from Figures 3-4. As seen from the figure the theoretical curve describes satisfactorily the experimental results.

4. Conclusion

Thus, in this work it is clearly shown that the material composition (x) of an epitaxial film significantly effect on boron implantation results and determine both the electrical parameters of the implanted layer, and the spatial distribution of the donor type radiation defects. Practically the obtained results can be used for the optimization of ion implantation regimes in MBE MCT for producing diode structures.

Acknowledgments
This research was supported by grant (№ 8.2.10.2015) from "The Tomsk State University Academic D.I. Mendeleev Fund Program".

References

[1] Rogalski A 2011 *Infrared Detectors* (CRC Press, Boca Raton)
[2] Ryssel H, Lang G, Biersack J P, Muller K and Kruger W 1980 *IEEE Transaction on electron devices* **ED-27** 58
[3] Destefanis G L 1983 *Nucl. Instr. Meth.* **209/210** 567
[4] Sidorov Yu G, Dvoretskii S A, Mikhailov N N, Yakushev M V, Varavin V S, Vasil'ev V V, Syslyakov A O and Ovsyuk V U 2000 *Applied Physics* **5** 108
[5] Ovsyuk V N, Kuryshev G L and Sidorov Yu G 2001 *Matrix photo-receiving devices of IR range* (Novosibirsk: Nauka)
[6] Margalit S, Nemirovsky Y and Rotstein I 1979 *J. Appl. Phys.* **50** 6386
[7] Vodopyanov L K and Kozyrev S P 1982 *Phys. Stat. Sol.* **72** 734
[8] Bahir C and Kalish R 1983 *J. Appl. Phys.* **54** 3129
[9] Bubulac L O, Tennant W E, Shin S H, Wang C C, Lanir M, Gerther E R and Marshall E D 1980 *Japan Appl. Phys.* **19** 495
[10] Voitsekhovskii A V and Kokhanenko A P 1998 *Russian Physical Journal* **1** 101
[11] Voitsekhovskii A V and Kokhanenko A P 1998 *Russian Physical Journal* **41** 44
[12] Voitsekhovskii A V, Denisov Yu A, Kokhanenko A P, Varavin V S, Dvoretskii S A, Mikhailov N N, Sidorov Yu G and Yakushev M V 1998 *Autometry (Novosibirsk: Russian Academy of Sciences)* **4** 47
[13] Voitsekhovskii A V, Grigoriev D V, Korotaev A G, Kokhanenko A P, Leontiev D V and Kul'chitskii N A 2004 *Izv. Vyssh. Uchebn. Zaved. Materials of electron techniques* **2** 60
[14] Voitsekhovskii A V, Grigoryev D V and Smith R *Semiconductor Science and Technology* **23** 055020
[15] Ziegler J F, Biersack J P and Littmark U 1985 *The Stopping and Range of Ions in Solids*, (New York: Pergamon Press)
[16] Voitsekhovskii A V, Kokhanenko A P, Shulga S A and Smith R 2004 *Nucl. Instrum. and Meth.* **B 215** 109
[17] Monastyrskii L S and Sokolovskii B S 1992 *Fiz. Tech. Poluprovodn.* **26** 2143
[18] Brudnyi V N, Grinyaev S N and Stepanov V E 1995 *Physica B: Cond. Matter* **212** 429
[19] Brudnyi V N and Grinyaev S N 2001 *Semiconductors* **35** 784
[20] Mikhailov N N, Sidorov Yu G, Vasiliev V V 2008 *Semiconductors* **42** 1298

Electro-physical characteristics of a HgCdTe epitaxial films upon exposure by a volume discharge in air at atmospheric pressure

A V Barko[1], A A Pishchagin[1], D V Grigoryev[1,2], V F Tarasenko[2], M A Shulepov[2]

[1] Tomsk State University, 36 Lenin Av., Tomsk, 634050, Russia
[2] High Current Electronics SB RAS, 2/3 Akademichesky, Tomsk, 634055, Russia

Abstract. In this paper the influence of the volume discharge of nanosecond duration formed in a non-uniform electric field at atmospheric pressure on samples of epitaxial films HgCdTe (MCT) of p-type conductivity. It is suggested that after exposure on film surface oxide layer was formed. This layer has a built positive charge that leads to the formation of an inversion layer which "shunts" the rest of the sample so that the measured field dependence of Hall coefficient corresponds to the material of n-type of conductivity.

1. Introduction

Electric discharges of different types as well as electron beams are now widely used for the modification of near-surface layers of various materials. It was shown in [1, 2] that the metal surface can be modified and cleaned by a volume discharge induced by a runaway electrons preionized diffuse discharge (REP DD). At higher pressures of different gases, including air at atmospheric pressure, a volume (diffusion) discharge is formed in the inhomogeneous electrical field due to runaway electrons and X-ray radiation. A special feature of a REP DD is a possibility of realizing high specific power of energy contribution (up to 800 MW/cm3) [3]. In so doing, beams of runaway electrons with the current amplitude of tens-to-hundreds amperes are generated from the discharge plasma, and the half-amplitude pulse duration of the beam current is in no excess of 100 ps [4]. Thus, during formation of a nanosecond volume discharge in the air, the anode is acted upon by a combination of a dense nanosecond-discharge plasma with the specific power of energy contribution of hundreds of megawatts per cubic centimeter and a supershort electron beam with a wide energy spectrum. In addition, the anode is affected by a shock wave as well as UV- and VUV radiation from the discharge plasma.

The solid solution $Hg_{1-x}Cd_xTe$ (HgCdTe, MCT) is a widely used material for preparation of intrinsic detectors of infrared (IR) radiation [5]. HgCdTe has proved to be a versatile semiconductor material for IR detection, as its tunable band gap offers the flexibility in the choice of wavelength band as well as the temperature of operation. For temperature imaging the most commonly used region of wavelengths is 8-14 μm, which corresponds to the $Hg_{1-x}Cd_xTe$ composition with x = 0.22. A wide variety of modern industrial, strategic, defense, and space applications require IR focal plane array detectors in large pixel formats. Therefore of particular interest is the characterization and study of properties of an epitaxial MCT material grown by molecular-beam epitaxy (MBE). Along with the study of the initial properties of epitaxial films of MCT grown by MBE a very urgent task is controlled

IOP Publishing
doi:10.1088/1742-6596/741/1/012098

changing of the electro-physical parameters of the material in order to obtain the desired semiconductor structures.

The aim of this work was to study the effect of volume nanosecond discharge in air at atmospheric pressure on the electro-physical properties of epitaxial HgCdTe grown by the method of molecular beam epitaxy.

2. Samples and measurement procedure

For experiments the samples of epitaxial MCT films of the n-type conductivity grown by MBE on GaAs (013) substrates with ZnTe and CdTe buffer layers were prepared at the Institute for Semiconductor Physics of the Siberian Branch of the Russian Academy of Sciences (Novosibirsk). The content of CdTe in the working layer of epitaxial films was x = 0.22. The width of the upper variband layer was close to 0.4 micrometers while the content of CdTe on the surface was 0.44. The value of the content x and the width of the epitaxial film were controlled by in situ ellipsometric measurements.

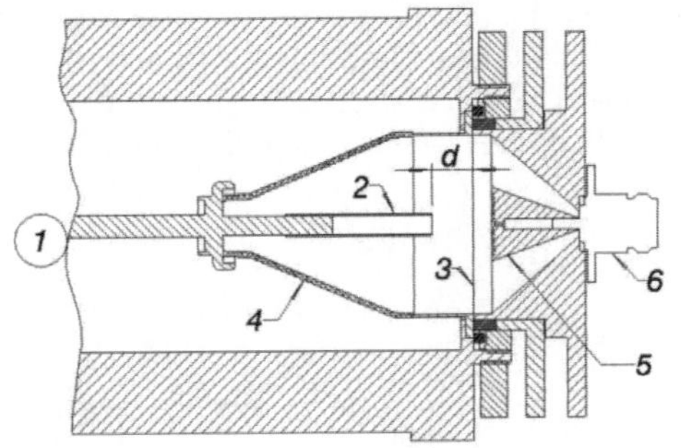

Figure 1. Schematic diagram of the gas-filled diode and collector assembly: (1) connection to a high-voltage pulse generator; (2) cathode; (3) foil anode; (4) insulator; (5) cone-shaped collector; (6) connecting socket.

Figure 2. The image of the diffuse discharge in atmospheric pressure air.

The irradiation of the MCT epifilms samples by REP DD was carried out with specially designed setups. The irradiations were performed using a discharge chamber, which is schematically depicted in figure 1. The as-grown samples were placed in a gas diode on a copper anode. A distance from a flat copper to a tubular electrode could be varied within 8–16 mm. The interelectrode voltage was supplied from a pulser of the RADAN-220 type, which generates voltage pulses with an amplitude of ~230 kV (in the open-circuit regime), a FWHM of ~ 2 ns (on a matched load), and a leading front width of ~ 0.5 ns. The discharge current was measured using a shunt composed of chip resistors connected between the foil anode and the discharge chamber housing. The results of measurements have shown that the current pulse amplitude for both polarities of the applied voltage was ~3 kA and the total duration of the discharge current pulse was ~30 ns (the first half-period of the discharge current pulse had a duration of ~8 ns). The generator RADAN-220 provides specific power input in gas discharge plasma above 0.8 GW/cm3 under atmospheric pressure in air at electrodes spacing of 8 mm. Figure 2 shows a photograph of the discharge. The samples were irradiated in the repetitively pulsed mode with a pulse repetition rate of 1 Hz. Exposure was carried out in the range of 100 - 1200 pulses. The surface of the irradiated samples was studied using atomic force microscope (AFM) "Ntegra Prima" (by NT-MDT) and optical profilometer MicroXAM-100 at ambient conditions.

Electro-physical parameters of the MCT samples before and after exposure were determined from measurements of the Hall effect by van der Pauw method. The measurements were performed at a constant current flowing through the sample (I = 1 mA) for the two directions of the current and two directions of a constant magnetic field. Removal of thin layers from irradiated material's surface was carried out in a 0.2% solution of bromine in dimethylformamide.

3. Results and discussion

The results of measurements of electro-physical parameters of the irradiated samples of epitaxial MCT have shown that after irradiation in the range of 100 - 1200 pulses an increase in conductivity is visible for all the samples. Moreover, for samples exposed to pulses in the range from 100 to 400, a decrease in the value of the Hall coefficient is observed. In this case, a shift of the inversion point for the sign of the Hall coefficient in the region of higher magnetic fields (from 0.17 T to 0.28 T) at the field dependence of the Hall coefficient is observed. Increasing the number of pulses of impacts of volume discharge to 600 pulses results in a reversal of the sign of the Hall coefficient (Figure 3, curve 2). Further increasing of the number of pulses reduces the Hall coefficient values (Figure 3, curves 2, 3, 4), wherein the epitaxial HgCdTe samples are characterized by low electron mobility values ~ (2 - 3)·10^3 $cm^2V^{-1}s^{-1}$, which is two orders of magnitude lower than the corresponding values for the epitaxial n-type material of high quality. The results obtained made it possible to suggest that in the process of impact of volume discharge on the samples of epitaxial films a layer with high electron concentration is formed at the surface or in the surface region of the material. The conductivity of this layer is such that it bypasses the bulk of the epitaxial film in the measurements of the Hall effect.

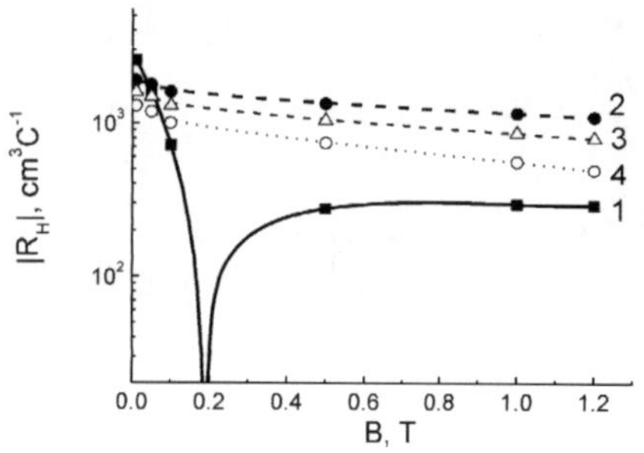

Figure 3. The field dependence of the Hall coefficient for samples of HgCdTe epitaxial films before (1) and after exposure to a volume discharge. The number of exposure pulses: (2) - 600, (3) – 800, (4) – 1000.

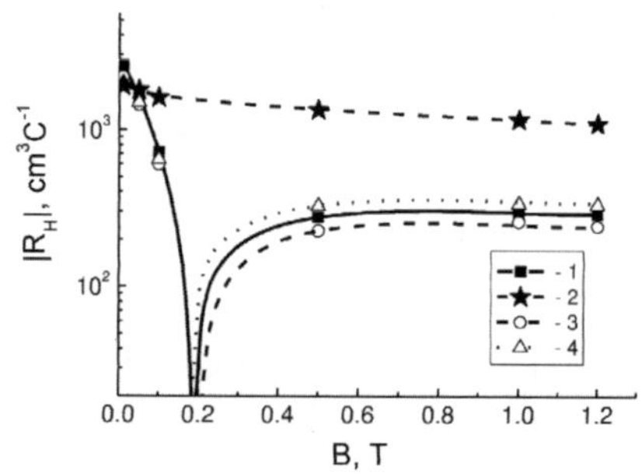

Figure 4. The field dependence of the Hall coefficient for samples of HgCdTe epitaxial films. (1) – before exposure; (2) – 600 volume discharge pulses; after etching of the irradiated sample in a solution of bromine dimethylformamide (3) and hydrochloric acid (4).

The study of the surface structure of initial and irradiated epitaxial films with atomic force microscope has revealed that the surface quality of the samples after exposure to a volume discharge does not change. Surface roughness slightly increases from 1.6 to 2.2 nm. Additionally, the surface distribution profile of the sample for which part of the surface was closed with the dielectric plate during irradiation was investigated. Surface profile measurements were carried out with an optical profilometer MicroXAM-100. Analysis of measurement results have shown that the interface irradiated / unirradiated sample has no characteristic step that leads to the conclusion about the absence of an alien film on the surface of the exposed part of the test sample.

After etching of the surface of the irradiated material by 0.1 mm electro-physical parameters of the irradiated samples return to baseline values (Figure 4, curve 3). The obtained results allow concluding that during the impact of volume discharge on the samples of epitaxial films high conductivity layer forms in the surface region of the material. Furthermore, it was found that after treatment of the irradiated samples in a solution of concentrated hydrochloric acid the restoration of the original values of electro-physical parameters of the material takes place (Figure 4, curve 4).

However, hydrochloric acid does not chemically react with initial MCT or with a material subjected to a radiation treatment. In the gas diode during a volume nanosecond discharge in air after a short voltage pulse in the discharge gap plasma consisting of an electron beam of positively and negatively charged ions is formed and the current flows through the diode. In this case, the sample of epitaxial MCT film located at the anode is exposed to complex effect consisting of a beam of electrons and negative ions, which cause the formation of different chemical MCT compounds after the influence on the sample surface.

The obtained data allow concluding that in the surface region of irradiated material the formation of chemical compounds of MCT with oxygen and nitrogen atoms occurs (such as anodic oxide which dissolve in hydrochloric acid). Analysis of published data shows that such chemical compounds contain significant concentration of positively charged centers that lead to the formation of an inversion layer at the interface with the epitaxial MCT film of p–type of conductivity. The formation of such an inversion layer is shown during the study of the properties of the interface between the oxide film and MCT [6].

Modeling for this assumption was conducted by growing oxide on the surface of the MCT epitaxial layer by the method of anodic oxidation. After growth of anodic oxide electro-physical parameters of obtained multilayer structure were measured. The measurement results have shown that the observed changes in the field dependence of the Hall coefficient before and after the deposition of anodic oxide are similar to the results observed after irradiating the samples of MCT by volume discharge (Figure 5).

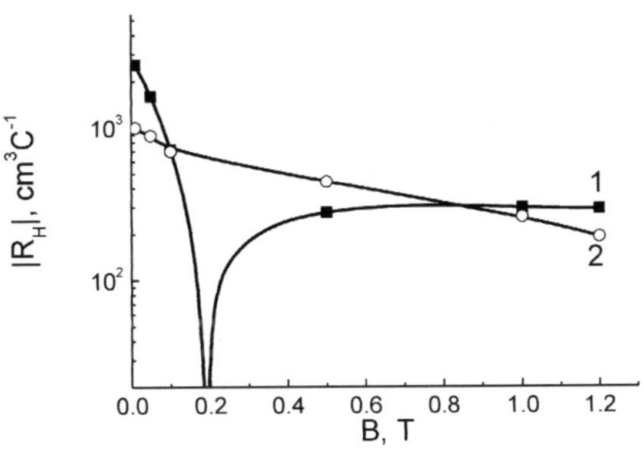

Figure 5. The field dependence of the Hall coefficient for samples of HgCdTe epitaxial films before (1) and after (2) growing of anodic oxide on the surface.

In the paper [7] the volume discharge impact on the electrical parameters HgCdTe MIS structures are investigated. It shown, that irradiated sample has a greater density of positive fixed charge. The density of the mobile charge in this sample is also great. The differential resistance of the space-charge region in the strong inversion for the MIS-structure was calculated. The value of RSCR at 200 kHz and 77 K for initial sample was about 1800 Ohm. For irradiated structure the value of the RSCR was so small that it does not exceed the error in determining RSCR in a wide temperature range. This result explains the low-frequency type of CV characteristics at sufficiently high frequencies for irradiated sample and indicates a high flow rate of minority carriers (electrons) in the inversion layer. The most likely mechanism for the emergence of additional minority carriers in the inversion layer is exchange of electrons with the inversion layer outside the field electrode. This layer is induced by a significant positive fixed charge. One possible reason for the presence of a positive fixed charge can be the appearance of a thin dielectric film on the surface of HgCdTe. It is connected with the formation of oxides by stimulated discharge. It is known that a positive fixed charge is typical for the anodic oxide film in HgCdTe [5, 6]. It is usually associated with the presence of oxygen vacancies in the anodic oxide [8]. For the studied structures the thickness of the dielectric layer must be several nanometers. At larger thicknesses of the dielectric layer reduction in capacitance will be observed due to the increase in passivating coating thickness. Therefore, a dielectric layer with a small thickness has a higher density of fixed positive charge than the typical anodic oxide. It is possible that after exposure in discharge a layer with a high density of embedded positive charges appears near the surface of HgCdTe. These charges cannot change its state of charge due to the potential barriers surrounding the charges.

4. Conclusion

Thus, the experimental data have shown that the exposure of MCT epitaxial films to the pulse volume nanosecond discharge in air at atmospheric pressure leads to formation of a dielectric layer with a small thickness in the surface layer of material. The formed layer has a built-in positive charge that leads to the formation of an inversion layer at the oxide / MCT interface which "shunts" the rest of the sample so that the measured field dependence of Hall coefficient corresponds to the material of n-type of conductivity. Our results show that it is possible to use the volume discharge induced by an avalanche electron beam for the development of technologies for the controlled change of the properties of MCT narrow-band solid solutions and production of structures which are heterogeneous with respect to conduction.

Acknowledgments

The authors gratefully acknowledge N.N. Mikhailov and V.S. Varavin for providing the MBE HgCdTe epitaxial films.

This research was supported by grant (№ 8.2.10.2015) from "The Tomsk State University Academic D.I. Mendeleev Fund Program" and by grant (№14-29-00052) from Russian Science Foundation.

References

[1] Shulepov M A, Akhmadeev Y K, Tarasenko V F, Kolubaeva Y A, Krysina O V, Kostyrya I D 2011 *Russian Physics Journal* **53** 1290
[2] Shulepov M A, Tarasenko V F, Goncharenko I M, Koval' N N, Kostyrya I D 2008 *Technical Physics Letters* **34** 296
[3] Alekseev S B, Gubanov V P, Kostyrya I D, Orlovskii V M, Skakun V S, Tarasenko V F 2004 *Quantum Electronics* **34** 1007
[4] Tarasenko V F 2001 *Plasma Physics Reports* **37** 409
[5] Rogalski A 2011 *Infrared Detectors* (CRC Press, Boca Raton)
[6] Brogowski P, Piotrowski J 1990 *Semiconductor Science and Technology* **5** 530
[7] Grigoryev D V, Voitsekhovskii A V, Lozovoy K A, Nesmelov S N, Dzyadukh S M, Tarasenko V F, Shulepov M A, Dvoretskii S A 2015 *Proceedings of SPIE* **9810** 98100U
[8] Voitsekhovskii A V, Nesmelov S N, Dzyadukh S M, Varavin V S, Dvoretskii S A, Mikhailov N N, Sidorov Yu G, Vasiliev V V 2008 *Semiconductors* **42** 1298

The properties of photoconductivity of the IIa-type diamond related to the band gap structure

A A Altukhov[1,2], V S Feshchenko[1,3], V A Shepelev[1,3], A V Popov[1]

[1] Ltd "Industrial-Technological Center "UralAlmazInvest", 4 I. Franko st., Moscow 121108, Russia
[2] Federal State Budget Educational Institution of Higher Education "Moscow Technological University", 78 Vernadsky Avenue, Moscow 119454, Russia
[3] JSC "Central Scientific-Research Institute of Technology "Tekhnomash", 4 I. Franko st., Moscow 121108, Russia

E-mail: valq2006@rambler.ru

Abstract. We investigate the properties of the photosensitivity spectra of the UV photodetectors based on natural diamond. The effect of the structural defects associated with nitrogen impurities to the photosensitivity is analyzed. We confirm that the polychrome light bias application enhances the photosensitivity of these detectors in the spectral range 240–340 nm due to the quasi-two-photon absorption which originates due to the complicated structure of the band gap impurity states of a natural diamond. The possibility to influence the photosensitivity spectra in the λ<220 nm spectral range of these detectors by the polychrome light bias application is revealed.

1. Introduction

Research and development of UV photodetectors and photosensitive materials plays an important role in the growing importance of UV photometry and spectroscopy for a lot of applications in ecology, medicine and others. The important photosensitive material for these applications is the diamond due to the unique semiconducting and physical properties [1, 2]. However the feature of a diamond is high concentration of structural defects associated with nitrogen. These defects have high density of band gap states ($N_g \leq 10^{18}$ cm^{-3}) [1]. The structure of the band gap states is complicated (Figure 1) [1]. The density of the conduction band states and the valence band states are $N_c = 1.1 \cdot 10^{19}$ cm^{-3} and $N_v = 3.3 \cdot 10^{19}$ cm^{-3} [1]. Therefore IIa-type diamond [1] has density of the conduction band states only 10 – 100 times higher than the density of the band gap impurity states.

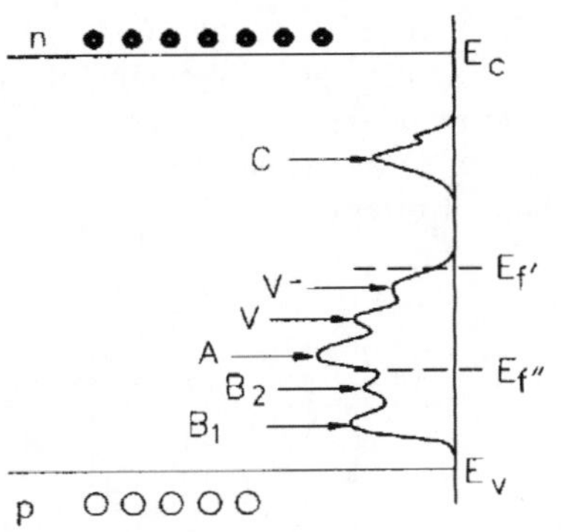

Figure 1. The structure of the band gap states of IIa type diamond: A, B_1, B_2, C, V, V^- – are the legends of the energy states of various nitrogen defects. [1].

The complicated structure of the band gap states makes possible the effects of polarization [3, 4] and quasi-two-photon absorption [5]. The last one is appeared when a photon of one wavelength throws a charge to the trap in the band gap, and another photon throws it into the conduction band. Moreover, each of the two photons has energy smaller than the band gap. It is appeared as sensitizing of the diamond photodetector to the emission with photon energy lesser than the band gap [5]. The relaxation time is tens of hours because the energy traps of nitrogen impurities are located deep ($E_N \sim$ 2 eV [5]) in the band gap. These trapped states were pumped at room temperature for several tens of pulses of UV laser with a wavelength of 266 nm and intensity of 250 mW/cm^2. This diamond detector demonstrated a time-dependent photosensitivity to emission of an optical parametric oscillator, which generated visible light for two days.

The last mentioned photosensitive feature of diamond provides both negative (the accuracy of the photometric measurements of the UV solar-blind range is strongly reduced) and positive effects on the diamond optoelectronic devices [6, 7]. The question of the pumping source power is issue of the day as well. The known studies of diamond sensitizing [5] were performed using a laser. We investigate the possibility of diamond detector to detect the low-power polychrome visible and near IR light.

2. Objects of the Research

We selected two diamond plates with different concentrations of nitrogen impurities as objects of the investigation. The diamond plate #1 has size of 3.2×2.5×0.3 mm and nitrogen concentration of $C_N = 9.0 \cdot 10^{18}$ cm^{-3} (IIa type diamond). The diamond plate #2 has size of 2.9×2.3×0.3 mm and nitrogen concentration of $C_N = 3.6 \cdot 10^{19}$ cm^{-3} ("transitional" type diamond). We produced two photodetectors using these diamond plates correspondingly. Figure 2 demonstrates the pattern of the detector (Figure 2).

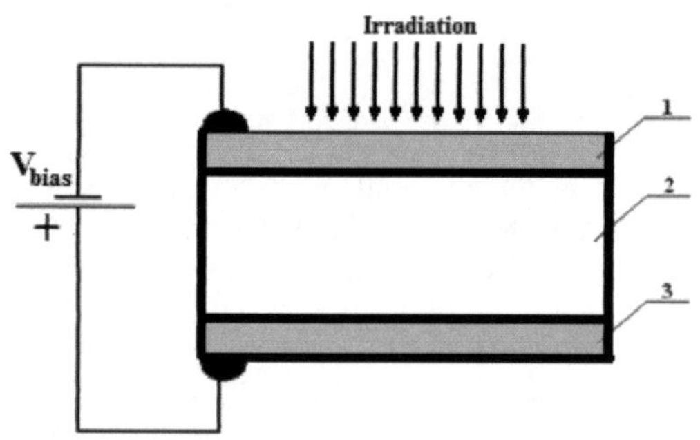

Figure 2. The pattern of the diamond photodetector: 1 – semitransparent platinum layer, 2 – diamond plate; 3 – aluminum electrical contact.

3. The Experimental Setup

The experimental setup is demonstrated on the Figure 3. The optical condenser (2) directs UV light of the deuterium lamp (1) into monochromator (3) which directs the dispersed light to the diamond photodetector (4) (Figure 3). The preamplifier (8) amplifies and converts the photocurrent into a voltage which is fed into voltmeter (9). The setup allows illuminating of the diamond photodetector active area using a halogen lamp (5) broadband non-coherent radiation passing through the aperture plate (6) and the optical filter (7). By monochromator scanning of the deuterium lamp radiation we obtain the spectral dependencies of the diamond detector photosensitivity.

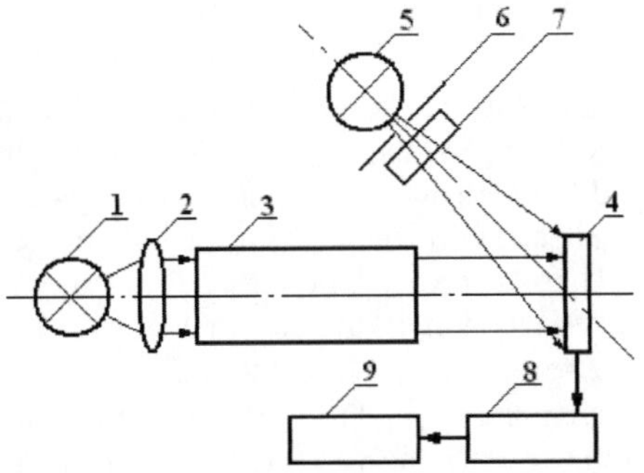

Figure 3. The experimental setup: 1 – deuterium lamp, 2 – optical condenser, 3 – monochromator, 4 – diamond photodetector, 5 – halogen lamp, 6 – aperture plate, 7 – optical filter, 8 – preamplifier, 9 – voltmeter.

4. Experimental details and Discussion

We apply bias voltage of $V_{bias} = -100$ B to the diamond photodetectors #1 and #2 (Figure 2). Then we shortly time apply the light bias illuminating by the halogen lamp to the detector active area. The UV deuterium lamp is not used in this step of the experiment. As expected, the photocurrent is not generated when illuminating of the active detector area by the halogen lamp emission during both short time and several days as the lamp emits the light with wavelengths of more than 380 nm though the "red" optical limit of the type IIa diamond photosensitivity is about 230 nm. Probably, the relaxation rate of charge carriers at the impurity energy states equals to the generation rate of them using the halogen lamp emission.

We turn on the UV deuterium lamp and detect the spectral dependencies of the photosensitivity of the diamond detectors without the bias light application and with one. We obtain the dependencies of the photoresponse multiplicity (I_{photo}/I_{dark}) from the UV wavelengths (Figure 4). The photocurrent is intensively amplified when applied the light bias of the halogen lamp emission to the photodetectors (Figure 4, curves 2, 4). For the diamond detector #1 the photocurrent is amplified twice in the spectral range of the maximum photosensitivity (λ = 220 – 227 nm) and the increase is much smaller in the range (240 – 340 nm) which corresponds the photosensitive impurity energy states (Figure 4, curves 1 and 2). While, for the diamond detector #2 the photocurrent is amplified only 1.5 times in the $\lambda = 220 - 227$ nm spectral range and 2.5 times in the $\lambda = 240 - 340$ nm spectral range (Figure 4, curves 3 and 4). Such a different behavior of the photosensitivity spectral dependencies of two types of the diamond detectors is due to the difference in the mechanisms of the photocurrent amplifying in these spectral ranges as well as the different structure of the band gap impurity states of diamond.

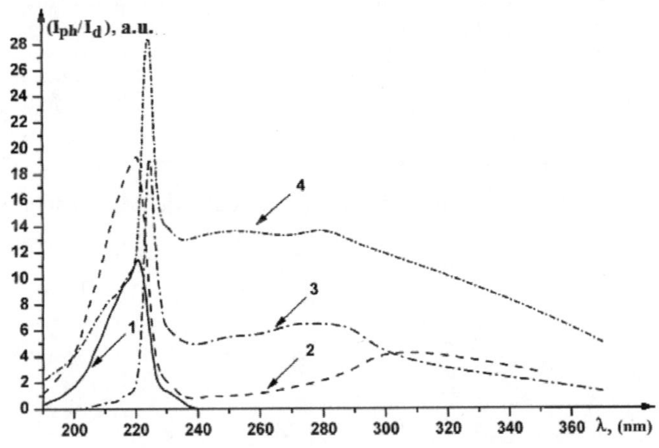

Figure 4. The dependencies of the photoresponse multiplicity from the UV wavelength for the diamond photodetectors:
the detector #1 – curves 1 and 2 correspond to the dependencies without application of the light bias broadband radiation of the halogen lamp and with one;
the detector #2 – curves 3 and 4 correspond to the dependencies without application of the light bias radiation and with one.

We assume that amplifying of the photocurrent in the spectral range of the maximum photosensitivity of a diamond is due to the charge carriers which are generated by the UV radiation in the valence band or the conduction band then partially trapped by the band gap impurity states. Some trapped charge carriers are thrown back from the band gap into the conduction band or the valence band when the broadband spectral light bias application. These charges generates additional the impurity photocurrent. The charge carriers throwing to the conduction band and valence band is possible due to the broadband spectral light bias and a lot of the band gap impurity states.

The mechanism of the photocurrent generation in the spectral range (λ = 240 – 340 nm) of the band gap impurity states is different. In this case the UV radiation throws the charge carriers into the band gap states and then the light bias throws the charge carriers to the conduction band or the valence band. Therefore the differences become clear in the photoconductivity spectra of the IIa type diamond (Figure 4, curves 1 and 2) and the "transitional" type diamond (Figure 4, curves 3 and 4).

We assume that a lesser amplifying of the photocurrent at the maximum photosensitivity of the "transitional" diamond is caused by a decrease the mobility and lifetime of nonequilibrium charge carriers [1] due to a lot of the energy traps in the band gap. In addition there are a lot of the deep energy traps in the band gap of IIa diamond. Therefore the narrow UV radiation, which selected by the monochromator, does not throw sufficient amount of the charge carriers to the band gap states. The small amplifying of the photocurrent in the 300 – 340 nm spectral range (Fig 4, curve 2) indirectly confirms that the band gap impurity states are deeper in the IIa diamond in comparison with the "transitional" diamond.

We set up an experiment to determine the effect of the band gap states for the photosensitivity of these diamond detectors. Using the set of the light filters (Figure 5) which are established between the halogen lamp and the diamond detector we obtain the dependencies of photoresponse multiplicity (I_{photo}/I_{dark}) from the UV wavelengths (Figure 6) for the detector #2. We apply the broadband spectrum light bias of the halogen lamp to the diamond detector at the same time. The set of the SZS25 and SZS21 light filters passes visible light of the 380 – 560 nm range (Figure 5a). The light filter KC19 passes near infra-red spectral range of the halogen lamp radiation (Figure 5b).

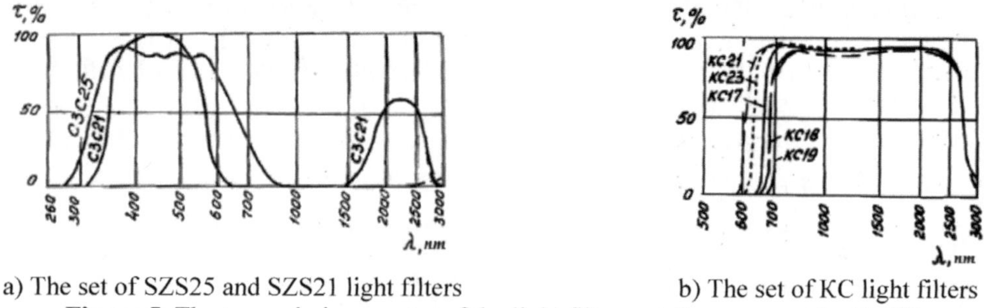

a) The set of SZS25 and SZS21 light filters b) The set of KC light filters

Figure 5. The transmission spectra of the light filters set (in correspond with [8])

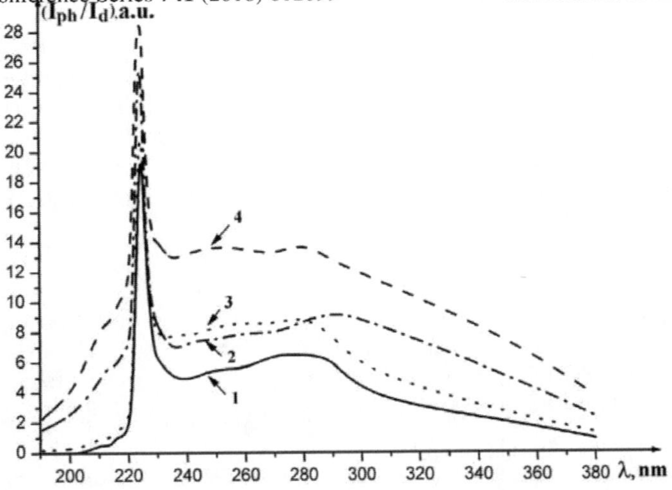

Figure 6. The dependencies of photoresponse multiplicity from the UV wavelength for the diamond detector #2: 1 – without the light bias application, 2 – with application of the light bias and the set of SZS25 and SZS21 light filters, 3 – with application of the light bias and the light filter of KC19; 4 – with application of the light bias without light filters.

The amplifying in the photocurrent in the spectral range of 240 – 380 nm indicates the distribution of the band gap impurity states. The "transitional" diamond has equally smeared-out density of the gap states (Figure 1). Therefore the amplifying in the photocurrent for application both the visible (Figure 6, curve 2) and near IR (Figure 6, curve 3) light bias is about the same and depends on both the filtered spectral bands (Figure 5) and the halogen lamp spectrum. Though there are some differences of the photoresponse spectra in the 280 – 380 nm range (Figure 6, curves 2 and 3) at application both the near IR and the visible light bias which are determined by density of the gap states. In order to study in detail the band gap states, it is necessary detect the spectral photosensitivity depending on the UV wavelength and at the same time scan the light bias emission of the halogen lamp by second monochromator. However, in the production of devices based on diamond, we need the knowledge of the integral dependencies of the influence of background illumination on the spectral photosensitivity of a diamond detector.

There is the interesting spectral range of 190 – 220 nm of the photosensitivity spectrum of the "transitional" diamond. While the detector #1 at the light bias application demonstrates a proportional amplifying in the actual photocurrent (Figure 4, curves 1 and 2), the diamond #2 does not demonstrate the photosensitivity in this spectral range without the light bias application (Figure 4, curve 3). Therefore, it is unclear, why is the photosensitivity take place at application of the safe light bias? The light bias application of the different spectral ranges (Figure 5), which are selected from the broadband spectral radiation of the halogen lamp, allows you to answer it.

Well known for semiconductors [9], there is a spectral flat-topped dependence of photocurrent in the range of wavelengths higher than the band gap absorption. This dependence is related to the fact that the charge carriers are thrown into the conduction band by a radiation with a higher energy than the band gap. Then the charge carriers rapidly relax to the bottom of the conduction band or to the top of the valence band and then generate a photocurrent. This dependence usually has a smooth slope [9, 10] from the band gap absorption toward shorter wavelengths. This slope is associated with increased optical absorption of the semiconductor with decreasing wavelength. Therefore penetration of the light emission into the semiconductor depth is reduced and hence photocurrent is weakening.

However, a diamond semiconductor has a strong slope of the photocurrent in the range of shorter wavelengths ($\lambda < 220$ nm) near the maximum photoresponse (in addition, the slope is slower for the diamond is impurities-free – the comparison of the curves 1 and 3 at Figure 5). This is due to a lot of nitrogen associated defects are displaced to the diamond surface. These defects are traps for charge carriers generated by light emission. Therefore the photocurrent is completely quenched for the "transitional" diamond into $\lambda < 220$ nm the spectral range (Figure 5, curve 3 and Figure 6, curve 1).

The application of the light bias and light filters set allows splitting the charge carriers which are trapped by the different energy states of the band gap. In the spectral range of $\lambda < 220$ nm the photocurrent is only amplified with the visible light bias application (Figure 6, curves 2 and 4) while as the near IR light bias application does not amplify the photocurrent into the spectral range (Figure 6, curve 3). Consequently there are not shallow states of the band gap on the surface of the "transitional" diamond. Therefore the charge recombination takes place at deeper states of the band gap.

5. Conclusion

To produce high stability operational devices based on diamond photodetectors it is necessary to apply best possible of clean impurities diamonds. During operation of such devices must be considered the effect of background lighting on the photosensitivity of the detector, in spite of the diamond photodetector is a solar-blind (insensitive to visible and infrared light emission) without UV illumination as well.

Acknowledgments

The research is supported by Ministry of Education and Science of Russian Federation (government contract №14.579.21.0047, unique identifier for Applied Scientific Research RFMEFI57914X0047).

References

[1] Kvaskov V B 1997 *Prirodnye almazy Rossii* (Moskva: Polyaron) (In Russian)
[2] Mityagin A Yu and Feshchenko V S 2013 UV Photodetectors Based on Natural Diamond (Ukraine, Odessa: Politehperiodika)
[3] Mukhachev Yu S et al 1983 *Soviet Journal Fizika I Tekhnika Poluprovodnikov* **17** 1308 (In Russian)
[4] Kato K et al 1987 *Jap. J. Appl. Phys.* **26** 70
[5] Klimentov S M et al 2000 *Quantum Electronics* **30** 459
[6] Milns A G 1973 *Deep Impurities in Semiconductors* (New York: Wiley)
[7] Sze S M 2007 *Physics of Semiconductor Devices* (third edition; Hoboken, New Jersey: Wiley Interscience)
[8] Petrovskii G T 1990 *Cvetnoe opticheskoe steklo I osobye steykla* (Catalog; Moskva: Dom Optiki) (In Russian)
[9] Ryivkin S M 1963 *Fotoelektricheskie yavleniya v poluprovodnikakh* (Moskva: FIZMALIT) (In Russian)
[10] Tauc J 1965 *Optical Properties of Semiconductors in the Visible and Ultraviolet Ranges* Progr. Semiconductors **9** 89

Improved photoanode structure based on anodic titania nanotube array covered by TiO_2-NPs/nanographite composite layer for ETA-cells

I M Gavrilin[1], A A Dronov[1], Yu I Shilyaeva[1], E A Lebedev[1], M S Kuzmicheva[1], T P Savchuk[1] and S A Gavrilov[1]

[1]Department of Materials for Functional Electronics, National Research University of Electronic Technology, Zelenograd 124498, Russia

Abstract. This work is devoted to the morphology and electrical properties optimization of flexible photoanodes based on anodic titanium oxide nanotubular arrays (TiO_2-NTAs) for solar cells with extremely thin absorbing layer (ETA-cells) by TiO_2-nanographite thin composite layer formation on the TiO_2-NTAs surface. First, the carbon doped TiO_2-NTAs were synthesized by annealing of the as-anodized TiO_2-NTAs in argon without foreign carbonaceous precursor. The residual ethylene glycol absorbed on the nanotube wall during anodization serves as the carbon source and the C species are uniformly distributed along the entire nanotube to form the C-TiO_2 NTAs. Further decorating of C-TiO_2-NTAs surface by TiO_2 nanoparticles to form the TiO_2-nanographite (NG) composite layer with high conductivity and increased photoanode effective area showed improved ETA-cells performance.

1. Introduction

Due to structural, optical, and chemical properties, anodic titanium oxide nanotubular arrays (TiO_2-NTAs) is a promising material for photoanodes in lightweight and flexible dye-sensitized solar cells (DSSCs) and solid-state solar cells with extremely thin absorbing layer (ETA-cells) [1]. However, one of the disadvantages of using titanium dioxide layers in the photoelectric converters is its low electrical conductivity, which reduces the charge carriers collecting efficiency. To reduce the TiO_2-NTAs resistance various conductive particles commonly are used to introduce them to TiO_2-NT structure. For example, modification TiO_2-NTAs with carbon by thermal annealing in acetylene produces TiO_2-NT arrays with semimetallic conductivity [2]. But using carbon-containing precursor gases requires high temperatures (600 °C or above) and does not allow obtaining a uniform coating along the tube depth.

Previously, Schmuki and al. [3] has been found that walls of TiO_2 nanotubes formed in ethylene glycol based electrolytes have a multilayered structure and one of these layers, located on the TiO_2 nanotube inner surface, is saturated by carbonaceous products of electrochemical reactions.

In this paper, we use the presence of the inner layer as a carbon source for its further conversion to covering nanographite ultrafine layer, by heat treatment in an argon atmosphere. Further decoration of modified TiO_2-NTAs surface by TiO_2 nanoparticles forms the TiO_2-nanographite (NG) composite layer with improved conductivity and increased photoanode effective area.

2. Experiment

2.1. TiO_2-NTAs formation

For photoanode structures the TiO_2-NTAs were prepared by electrochemical anodization of Ti metal foils in an ammonium fluoride (NH_4F) - ethylene glycol (EG) solution. The electrolyte was EG containing 0.3 wt.% NH_4F and 2 vol.% H_2O. Anodization was performed at 60V for 1 hour. The anodized samples were rinsed in deionized water, dried in air stream, and then annealed for an hour at 500°C in a tube furnace in air ambient (TiO_2-NTAs/air) and in Ar atmosphere (TiO_2-NTAs/Ar). The morphology, structure, and composition of the samples were characterized by field emission scanning electron microscopy (FE-SEM), X-ray diffraction (XRD), energy-dispersive x-ray (EDS) and Raman spectroscopy. Processes occurring during samples heat treatment were studied by differential scanning calorimetry (DSC) and thermogravimetric analysis (TG). Studies carried out at a heating rate of 5°C/min in air and argon streams.

2.2. TiO_2-NG composite layer and ETA-cell formation

For the TiO_2-NG composite layer formation, annealed TiO_2 nanotubes were immersed in 0.2M $TiCl_4$ solution for 1 hour and then annealed in air at 350°C for 30 min.

For ETA-cell formation, In_2S_3/InxPb$_{1-x}$S light absorber layer was deposited on TiO_2-NTAs/air and TiO_2-NTAs/Ar by SILAR method [4]. First, the substrates were immersed sequentially in the 0.01M $InCl_3$ and 0.01M Na_2S*9H_2O water solutions for 35 cycles. Then 10 ml of the 0.1M $Pb(CH_3COO)_2$ was added to indium chloride solution and additional 15 cycles were done.

After SILAR process the samples were dried in airflow and annealed in air at 130°C for 30 min. On the top of $In_xPb_{1-x}S$ absorber layer the CuSCN holes conductor was deposited on heated up to 80°C substrates by impregnation from a 0.05M CuSCN solution in dipropylsulfide. For CuSCN layer deposition, custom-made capillary system with thermo-stabilized table was used to improve the drop contact between solution and the substrates.

Finally, ZnO:Ga transparent conductive electrodes with the thickness of 500 nm were formed on the top of the solar cell structure by magnetron sputtering.

3. Results

3.1. TiO_2-NTAs Thermal analysis

By thermogravimetry (TG) and differential scanning calorimetry (DSC) methods it has been found that during obtained structures heating in the air stream, at the temperature range of 300-400°C two exothermic peaks are observed (figure 1a), accompanied by a large loss of weight. At a temperature about 300°C exothermic peak should correspond to oxidation of carbon to CO and CO_2 with heat release due to the combustion reaction. Second exothermic peak at 380°C corresponds to crystallization of the amorphous structure of TiO_2-NTAs.

On DSC curve for TiO_2-NTAs heated in argon stream a sloping hump at the temperature range of 200-300°C and sharp peak near 350°C (figure 1b) are observed. First small low-temperature exothermal peak arise most probably from dehydration and the small degree of crystallization from anodized hydrated titania to anatase TiO_2 [5] as well as partial decomposition of EG. Larger exothermal peak can be contributed to TiO_2 crystallization, total EG decomposition and graphite phase formation. Observed exothermal peak appeared at temperature lower than at the same experiment but in the air stream due to overheating of the system during combustion reaction. These results are confirmed by obtained Raman and XRD spectra (figure 2).

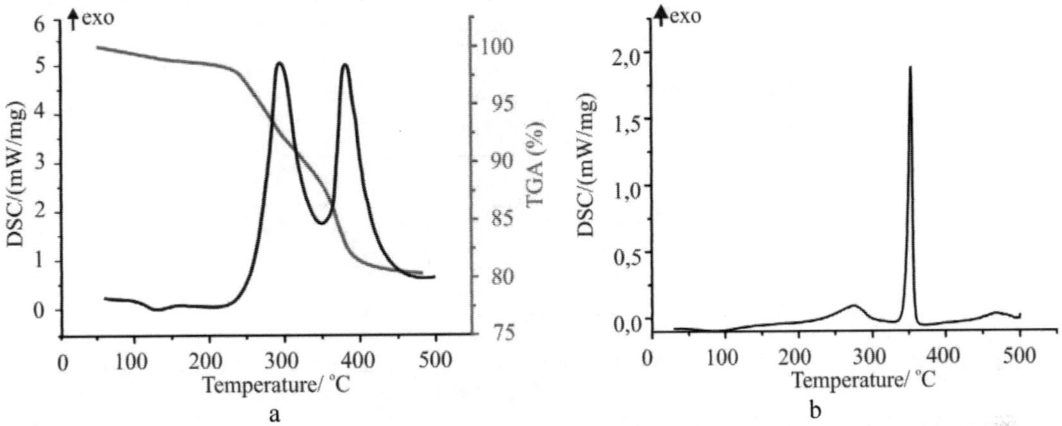

Figure 1. (a) DSC and TGA curves for TiO_2-NTAs heated in air stream; (b) DSC curve for TiO_2-NTAs heated in argon stream.

3.2. Chemical and structural analysis

Phase composition of the samples after thermal post-treatment by XRD showed the strong presence of crystalline phase of anatase type. That has also been confirmed by Raman spectroscopy (figure 2b). However, both methods show the presence of carbon in the graphite phase on samples annealed in argon.

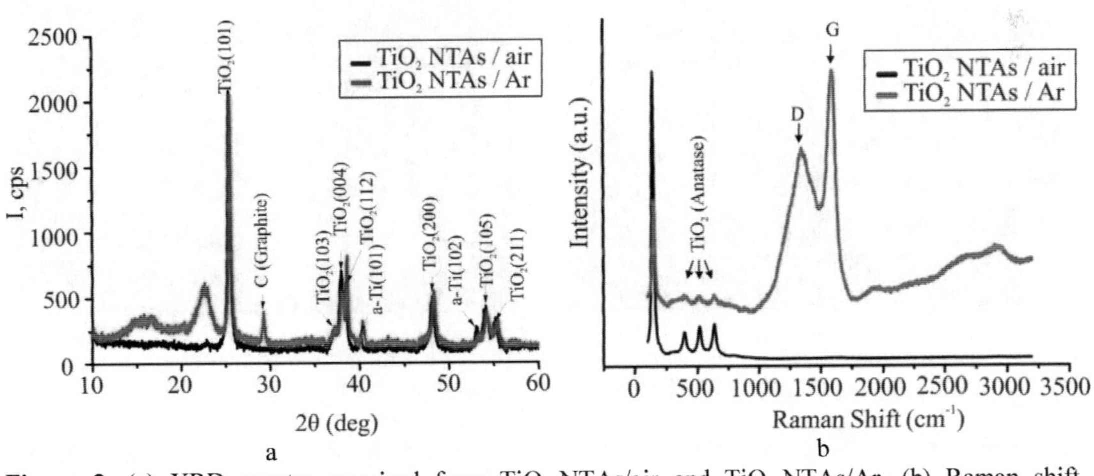

Figure 2. (a) XRD spectra acquired from TiO_2-NTAs/air and TiO_2-NTAs/Ar. (b) Raman shift acquired from TiO_2-NTAs/air and TiO_2-NTAs/Ar.

For Raman spectra the peaks positions 393 cm^{-1}, 504 cm^{-1}, 615 cm^{-1} which correspond to anatase vibronic states were found in both samples. Along with this, in TiO_2-NTAs/Ar the sample peaks at

1198 cm^{-1}, 1378 cm^{-1} and 1579 cm^{-1} appeared. They are attributed to carbon modifications (D and G - lines). Such modes perfectly describe the graphite type compounds and confirmed by peak position 27 (2θ, deg.) on XRD spectra.

From EDS results (table 1) it can be seen that almost all carbon removed from the TiO_2-NTA after annealing in air stream. The opposite situation for TiO_2-NTA annealed in argon atmosphere is observed. In this case, the carbon percentage content is practically unchanged as compared with not annealed sample, what also clearly coincides with other research methods.

Table 1. Relative values of the content of elements atoms on the surface of TiO_2-NTAs obtained by EDS

Atomic, %	C	O	F	Ti
TiO_2-NTA/as is	8,18	45,92	10,18	35,72
TiO_2-NTAs/air	1,20	68,22	0	30,58
TiO_2-NTAs/Ar	15,21	55,90	0	28,61

In addition, TiO_2-NTAs electrical properties studies have been performed. It has been found that the electrical conductivity of the sample TiO_2-NTAs/Ar increases more than an order than TiO_2-NTAs/air that is consistent with the results described in [6].

3.3. Morphology

FE-SEM images of TiO_2-NTAs annealed in air and argon are presented in figure 3.

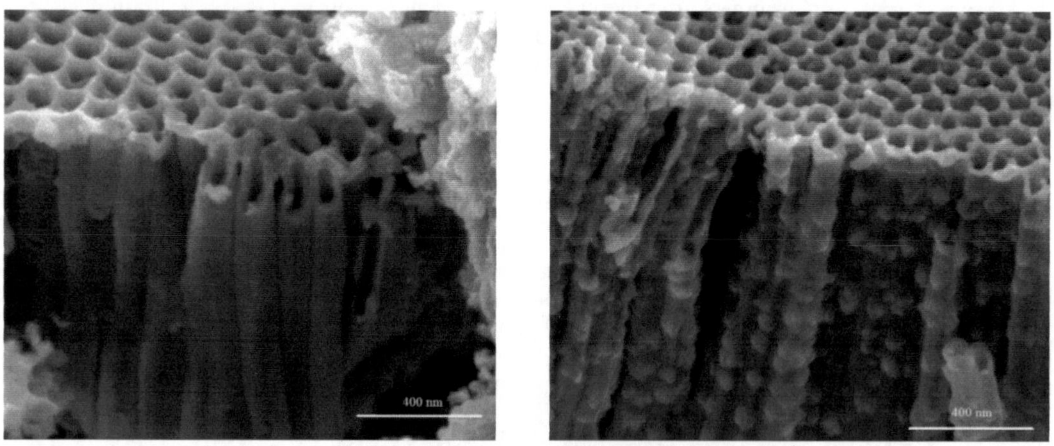

a b

Figure 3. Morphology of TiO_2-NT layers after thermal post-treatment in different ambient: (a) TiO_2-NTAs heated in air stream; (b) TiO_2-NTAs heated in argon stream.

By comparing the titanium oxide nanotubular layers annealed in inert and oxidizing atmospheres strong differences in morphology are observed. Thus, after thermal treatment in argon stream, nanoparticles are observed on the outer and inner titania nanotubes wall surfaces. Presumably these nanoparticles formed from carbon rich layer appeared at inner surface of TiO_2 nanotube walls during the TiO_2-NTAs formation by anodizing in organic-based electrolytes. In this case, as opposed to calcination in air stream, carbon not oxidizes to CO_2, but constitutes graphite phase nanoparticles at TiO_2-NTAs walls surfaces.

3.4. Characterization of the ETA-cells

Test structures of solar cells (ETA-cells) have been formed on the obtained samples and their current-voltage characteristics (CVC) and main electrical characteristics were measured.

CVCs (figure 4) were obtained with custom-made test bench. As the light source Newport solar simulator with irradiation power of light that incident on the sample - 100mW/cm^2 was used.

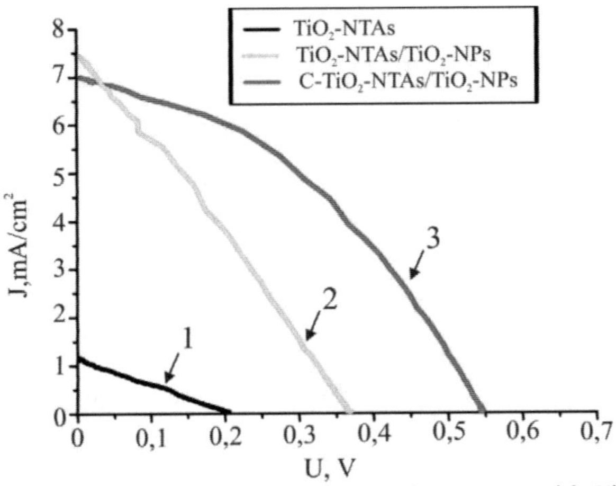

Figure 4. C-V curves of ETA-cells based on TiO_2-NTAs with different post treatment procedure (1) TiO_2-NTAs; (2) TiO_2-NTAs/TiO_2-NPs; (3) C-TiO_2-NTAs/TiO_2-NPs.

Analysis of the CVCs shows that TiO_2-NTAs with $TiCl_4$ treatment enhances efficiency of ETA-cell dramatically due to the healing of structural defects occurring during the formation and annealing of TNTs, such as cracks, fractures, nanotubes detached from the substrate, etc. Also, this treatment procedure increases the effective light absorption area by forming TiO_2 nanoparticles assemblies on the walls of TiO_2-NTs.

At the same time, combining the treatment of the TiO_2-NTAs with $TiCl_4$ and the heat treatment in argon atmosphere, which allows to reduce the structure series resistance significantly by increasing the photoanode conductivity due to graphite nanoparticles formation, increases the ETA-cell efficiency in 2 times (table 2) in comparison with ETA-cell with photoanode, which was annealed in air.

Table 2. Main characteristics of obtained ETA-cells based on TiO_2-NTAs with different post treatment procedure

Structure	U_{oc}, V	J_{sc}, μA/cm^2	η,%	FF,%
TiO_2-NTAs	0,211	116	0,06	25,5
TiO_2-NTAs/TiO_2-NPs	0,371	735	0,76	27,8
C-TiO_2-NTAs/TiO_2-NPs	0,547	697	1,52	39,8

As a result, ETA-cells based on TiO_2-NTAs photoanodes with TiO_2-NG composite layer showed improved efficiency compared to solar cells with TiO_2-NTAs photoanodes annealed on air.

Conclusions

In the present work a comprehensive comparative analysis of the TiO_2-NTAs samples annealed in oxidizing and inert atmospheres was performed. The calcination process features of nanotubular titania containing a carbon-rich layer in different environments were investigated. It is found that carbon located in the body of titanium oxide nanotubes during annealing in an inert atmosphere is not oxidized to carbon dioxide, as in the case of air ambient or other oxidizing atmospheres, but forms crystalline graphite nanoparticles on the inner and outer nanotube wall surfaces. The contribution of the modified photoanode based on TiO_2-NTAs with increased conductivity on the solar cells with extremely thin absorbing layer efficiency was rated. Measurements of current-voltage characteristics of the obtained ETA-cell experimental samples showed efficiency increase in two times with carbon doped modified nanocomposite photoelectrode.

Acknowledgments

This work was supported by RFBR grant № 16-33-60217 mol_a_dk.

References

[1] Belov A, Gavrilin I, Gavrilov S, Dronov A and Shulyatyev A 2011 *Izv. Vyssh. Uchebn. Zaved., Elektron.* **88** 38 (in Russian)
[2] Hahn R, Schmidt-Stein F, Salonen J, Thiemann S, Song Y, Kunze J, Lehto V, and Schmuki P 2001 *Angew. Chem. Int. Ed.* **48** 1
[3] Albu P, Ghicov A, Aldabergenova S, Drechsel P, LeClere D, Thompson G, Macak J and Schmuki P 2008 *J. Adv. Mater.* **20** 4135
[4] Gavrilov S, Dronov A, Shevyakov V, Belov A, Poltoratskii E 2009 *Nanotechnol Russia* **4** 237
[5] Grimes, C.A., Mor, G., 2009 *TiO_2 Nanotube Arrays: Synthesis, Properties, and Applications* (New York: Springer)
[6] Tighineanu A, Ruff T, Albu S, Hahn R, and Schmuki P, 2010 *Chemical Physics Letters*, **494** 260

The contribution of 3D nanoscale indium fluctuations into the efficiency droop in the green and blue InGaN/GaN LEDs

N A Talnishnikh[1], A E Chernyakov[1], E I Shabunina[2], N M Shmidt[2]
[1] Submicron Heterostructures for Microelectronics Research and Engineering Center RAS, Polytekhnicheskaya 26, St. Petersburg 194021, Russia.
[2] Ioffe Physical Technical Institute, Polytekhnicheskaya 26, St. Petersburg 194021, Russia.

Abstract. It was shown that the contribution of 3D nanoscale indium fluctuations into the efficiency droop at $j < 50$ A/cm^2 can be very high and results in different shapes of the efficiency dependency on current $\eta(j)$ in blue and green LEDs. Reducing 3D nanoscale indium fluctuations can decrease the efficiency droop in these LEDs. The contribution of delocalized carriers is predominant at $j > 50$ A/cm^2 and the current dependences of efficiency, approximated by as $\eta(j) \propto j^{-b}$ where $0.2 < b < 0.3$, are alike in both green and blue LEDs.

1. Introduction

The efficiency droop is currently one of the most popular research topics for GaN-based light-emitting diodes (LEDs). The reduction in the external quantum efficiency in green and blue InGaN/GaN light emitting diodes (LEDs) with increasing current injection, the so-called droop, is a definite limitation for the future improvement of the light output in these LEDs including the white lamps [1]. The physical mechanisms behind the droop have been discussed intensely during the last decade, as reviewed by several authors [1-3]. In the majority of papers, the analysis of the efficiency droop is based on the assumption that these LEDs have homogeneous nanomaterial properties [2, 3]. However, it is known that the complicated nanomaterial arrangement of nitrides [4-6] includes 3D nanoscale indium fluctuations and extended defects, piercing device active layers, that are both incorporated into InGaN/GaN heterostructure. This leads to the existence of the vast variety of nanomaterial arrangements in nitride based compounds, the inhomogeneity of properties and the current crowding effect. These peculiarities result in numerous NA forms with different forms of nanomaterial disordering which are governed by crystallites' coalescence of near 3D growth mode to near 2D one. It is known that nanomaterial disordering causes crowding current effect, the local Joule heating and degradation effect in light emitting devices based on III-N nano materials [7], the percolation paths in InGaN layers [5] and the conductive paths localized in extended defects [8], Yet, not many papers consider the contribution of extended defects [4, 6] and 3D nanoscale indium fluctuations [5] into the efficiency droop in the blue and green InGaN/GaN LEDs.

The aim is to clarify the contribution of 3D nanoscale indium fluctuations into the external quantum efficiency (η) droop in blue and green commercial LEDs, using the approach based on the concept that the nitrides have the complicated NA.

2. Experimental Detailes

Commercially available LEDs from several leading manufactures had maximum values of external quantum efficiency (η) in the range of 40 – 50 % for blue and 15 – 30 % for green LEDs. The maximum of radiation emission was observed at wavelength 450 – 460 nm for blue LEDs and 527 – 530 nm for green ones. The current–voltage (I–V) characteristics in the region of 10^{-13} – 1 A were determined. I–V characteristics at the voltage less than 1.5 V were used to reveal the conductive paths (and identify the values of their conductivity) localized at the extended defects. The shape of the η (j) dependences was used to separate the contribution in radiative recombination of localized and delocalized carriers and to determine the external quantum efficiency droop. The peak wavelength at each voltage step (at fixed the current value) during the η (j) dependences measure of LEDs was determined carefully. The distribution of carriers in MQW and the MQW position in depleted layer of p-n junction were clarified from C-V profiling.

3. Results and Discussion

I–V characteristics of blue and green InGaN/GaN LEDs at the voltage lower than 1.5 V allow us to distinguish two groups of LEDs differing by a several order of magnitude in the conductivity values of shunts (conductive paths) and the shape of the η (j) dependences Figure 1a. The η values of blue (Figure 1a, curves 1, 2) and green (Figure 1a, curves 3, 4) InGaN/GaN LEDs differ noticeably. However, the shape of the η (j) dependences of all LEDs Figure 1a have common properties are determined by the competitive contribution of localized and delocalized carriers into radiative recombination for both green and blue LEDs with low value of the paths' conductivity ($j < 10^{-9}$ A/cm^2 at the voltage less than 1.5 V) Figure 1a, curves 1, 3. The contribution of localized carriers is predominant at $j < 50$ A/cm^2. Both green and blue LEDs with high value of paths' conductivity ($j > 10^{-6}$ A/cm^2 at < 1.5 V) Fig.1a, curves 2 demonstrate a shift of maximal η values to the higher injection current region, which is observed on the η (j) Fig.1a, curves 1, 3. The shift is caused by the carrier transport along the conductive paths localized at the extended defect. This process prevents the carriers localization in MQW and leads to a decrease in η values. This, however, is not the only mechanism behind the droop.

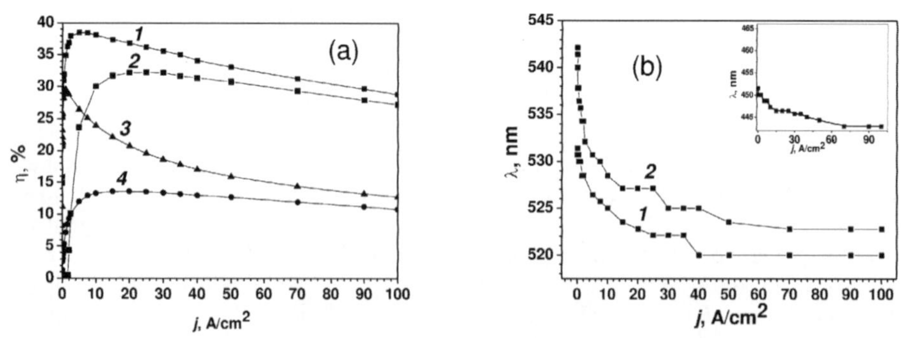

Figure 1 (a, b). The external quantum efficiency dependences on current density of blue (curves 1, 2) and green (curve 3, 4) LEDs with different values of path conductivity at the voltage less than 1.5 V: $j < 10^{-9}$ A/cm^2 – 1, 3, 4; $j > 10^{-6}$ A/cm^2 – 2 (a). The distribution of peak wave lengths on current density for green LEDs curves 1 and 2 for LEDs with the η (j) dependences are presented at Fig. 1a, accordingly curves 3 and 4 (b). The insert shows the distribution of peak wave lengths on current density for blue LEDs.

The study of the peak wavelength distribution on current densities Fig. 1b allows us to reveal another cause of low η values in the green LEDs without prominent conductivity paths (Fig. 1a, curve 4). It is shown that the low η values in LEDs Fig.1a curve 4 is due to the losses related to the radiative recombination at local regions enriched by In. According to the peak wavelength distribution on current densities Fig.1b curve 2 the radiative recombination begins approximately at 0.02 A/cm^2 at 542.2 nm, while the peak wavelength at current densities corresponding to the maximum value of quantum efficiency (15 A/cm^2) is 527.1 nm. The low η values at 542.2 nm allow us to assume that the InGaN composition in these In enriched local regions is highly irregular. Moreover, another feature in 3D nanoscale indium fluctuations leading to a large η droop (~ 50 %) in green LEDs is observed (Fig.1a, curve 3). The similar peculiarities were observed in blue LEDs earlier, but now the more typical η(j) dependence is shown at figure 1a, curve 1 [8].

In green LEDs (Figure 1a, curve 3) the radiative recombination of localized carriers begins at low current density j< 0.03 A/cm^2 until 0.1 A/cm^2 at the peak wavelength 532.8nm, at j > 0.15A/cm^2 the continuous change of the peak wavelength until 520 nm (Figure 1b, curve 1) correlating with the huge η droop by approximately 50% is observed (Figure 1a, curve 3). The 3D nanoscale indium fluctuations in such LEDs is also confirmed by C-V profiling (Figure 2a).The distribution of carriers in MQW in green LED differs drastically from that in blue LED (Figure 2b) as well as the shape of the peak wavelength distribution (the insert on Figure 1b).

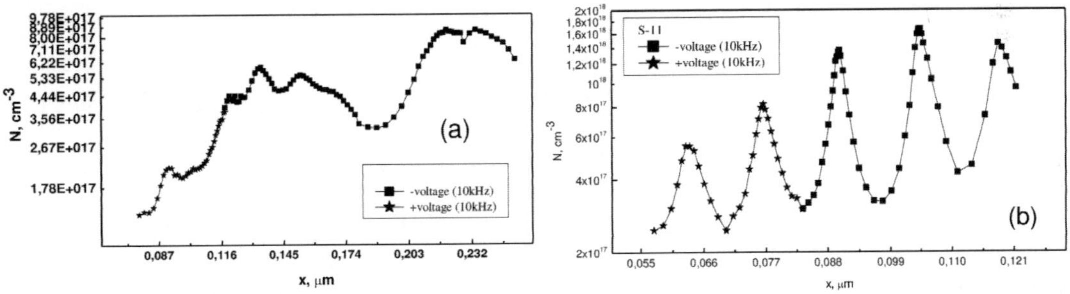

Figure 2. The distribution of carriers in MQW of green (a) and blue (b) LEDs. The MQW position in depleted region under zero bias in green (a) and blue (b) LEDs.

For blue LEDs in the region of the maximal efficiency droop (at j < 15 A/cm^2) the indium fluctuations are significantly lower according to the peak wavelength distribution (Figure 1b, on the insert). The position of first well at the depletion region of this green LED provides a good localization for carriers. However, other wells are distorted by In fluctuation and disposed outside the depleted region. Thus, the both phenomena leads to the huge η droop. The common feature for all investigated LEDs is the predominant contribution of delocalized carriers at j > 50 A/cm^2 and the current dependences of efficiency approximated by as $\eta(j) \propto j^{-b}$ where 0.2 < b < 0.3. The η droop values are about 20 %.

4. Conclusions

The shape of external quantum efficiency dependences on current density in green and blue LEDs is determined by radiative recombination of localized and delocalized carriers. The contribution of localized carriers predominates at j < 50 A/cm^2. In this range the most prominent variety of dependences' shapes is observed. The impact of 3D nanoscale indium fluctuations and conductive paths on external quantum efficiency values and their droop at j > 10 A/cm^2 as well as the shape of η (j) dependences in green and blue LEDs is of high variety and is sometimes contradictory. The conductive paths localized at extended defects reduce the radiative recombination of localized carriers and the values of the relative efficiency droop. 3D nanoscale indium fluctuations resulting in the formation of irregular InGaN composition containing In enriched local regions also reduce the radiative recombination of localized carriers and the values of relative efficiency droop. The continuous 3D nanoscale indium fluctuations resulting in distorted MQWs and the thin depleted layer leads to the huge external quantum efficiency droop (~ 50 %). The common feature for all investigated LEDs is the predominant contribution of delocalized carriers at j > 50 A/cm^2 and the current dependences of efficiency, approximated by as $\eta\ (j) \propto j^{-b}$ where $0.2 < b < 0.3$.

References

[1] Piprek J et al *Appl. Phys. Letters* 2015 **106** 101101
[2] Verzellesi G et al *Journal of Applied Physics* 2013 **114** 071101
[3] Weisbuch C et al *Physyca Status Solidi A* 2015 **212** 899
[4] Shmidt N M et al *Physyca Status Solidi C* 2015 **42** 1-4
[5] Wu Yu-R et al *Appl. Phys. Letters* 2012 **101** 083505
[6] Pozina G et al *Appl. Phys. Letters* 2015 **107** 251106
[7] A.E.Chernyakov et al.*Microelectronics Reliability*2012 **52** 2180
[8] Ber B Ya et al *Semiconducters* 2011 **45** 425

Axicons for power conversion efficiency enhancement in solar cells for the visible spectrum

V V Podlipnov[1,2], A P Porfirev[1,2], S N Khonina[1,2]

[1] Samara National Research University, Moskovskoe shosse 34, Samara, 443086, Russia
[2]Image Processing Systems Institute of RAS, Molodogvardeyskaya 151, Samara, 443001, Russia

Abstract. We investigate the possibility of using diffractive microaxicons with different periods for power conversion efficiency enhancement in solar cells. The microaxicons were manufactured by using electron beam lithography. The parameters of the manufactured microaxicons were measured using scanning electron microscopy (SEM). For imitation of solar light, we utilised a tunable laser (the used wavelength range is from 400 nm to 800 nm). Experimentally measured dependence of solar cell efficiency for the case of a combination of a solar cell and microaxicons of various types demonstrates a power conversion efficiency enhancement in the case of using such structures.

1. Introduction

The primary goal of manufacturing solar cells (SC) is to enhance their energy efficiency. The problems of increasing the conversion efficiency of solar light into electricity could be conditionally divided into two classes: the problems associated with an increase in the efficiency of conversion of the incident light into electricity (quantum efficiency), and the problems associated with an increase in the rate of optical energy reaching the absorber semiconductor (optical efficiency).

In order to enhance quantum efficiency, reliability and durability, and to compete in the market of photovoltaics, the scientists extend the research area of photovoltaic materials from crystalline silicon to polycrystalline and amorphous silicon, as well as one- and multi-junction SC based on thin film of CdS, CdTe, GaAs, InP, GaInAsP, and more complex compositions of Cu(In, Ga) Se2-KCC. It allowed using new types of substrates, such as organic, polyimide films for flexible solar cells [1]. Another promising direction in the area of photovoltaics is to create a photovoltaic SC in which the photo-generating carriers are the organic and polymeric materials with various impurities [2-3] and organo-metallic perovskite materials [4]. Such approaches allow you to vary the properties of the material in a wide range and to increase solar cell efficiency. The last mentioned materials are particularly interesting. Solar cells on the basis of their demonstrated unprecedented progress and increase of conversion efficiency to 21%.

However, the world-record efficiency of 46% belongs to the solar cell based on inorganic materials, developed at the Fraunhofer Institute for Solar Energy Systems ISE. This result could not have been achieved without the use of methods to increase optical efficiency. In this solar cell, multiple active junctions were used. Each of them converts light of a specific spectral range into electricity with the greatest efficiency. The light is the first incident through the lens, focusing it on a very small element [5].

Basic principles of the methods for improving optical efficiency are described in [6]. Problems associated with the use of additional optical elements, mirrors, and refractive or diffractive secondary optics to concentrate the light and to direct the light to the solar cell with high efficiency are of great interest. The solution to these problems allows reducing the cost of electricity generation through the use of solar cells of a significantly smaller area [7]. To increase the efficiency, texturing of the solar cell surfaces is widely used [8]. Structures based on diffraction gratings [9], luminescent structures based on quantum dots [10], randomly arranged textures that increase the diffuse scattering of light [11], plasmonic structures on the surface [12], photonic crystal structures [13], pyramidal surface structures [14], the structure in the form of nanocones [15], and silicon nanowires [16] are used to increase the capture of light and reduce reflection by texturing.

Thus, it can be concluded that current manufacturing elements with structures based on diffraction gratings, which may be created using, for example, inexpensive nanoimprinted lithography [17] and further plasma-chemical etching, are an inexpensive and appropriate approach for industrial production of solar cells. This technology is well combined with the technologies of microelectronic production and manufacturing of the classic silicon solar cells. Diffraction gratings may be used for texturing the front-side of the solar element [18], forming the back-side of the solar element [19], and for their combination in a single solar cell [20]. This allows you to maximise the light trapping and power conversion efficiency. However, the choice of the most effective types of diffractive gratings is an actual problem. To date, the results of the use of one-dimensional arrays, two-dimensional arrays, and solar cells with two filling-factor gratings [21] have been described in many works. Despite this, the use of diffraction and Fresnel diffraction axicon lenses as light-trapping structures has been investigated poorly. In most papers, their use is limited to use as secondary optics thin-film elements, namely a light concentrator [22]. In this paper, we demonstrate investigation of axicon applications to improve the efficiency of silicon solar cells.

2. The principle of a diffractive axicon

As is known, axicons are widely used for generating ring light distribution in a far-field region and for generating axial intensity distribution in a near-field region (see figure 1). Thus, they can be used for concentrating energy on the optical axis. However, if the period of the diffraction axicon is close to the wavelength of the illuminating radiation, most of the energy is scattered in the near field of the optical element at high diffraction orders [23, 24].

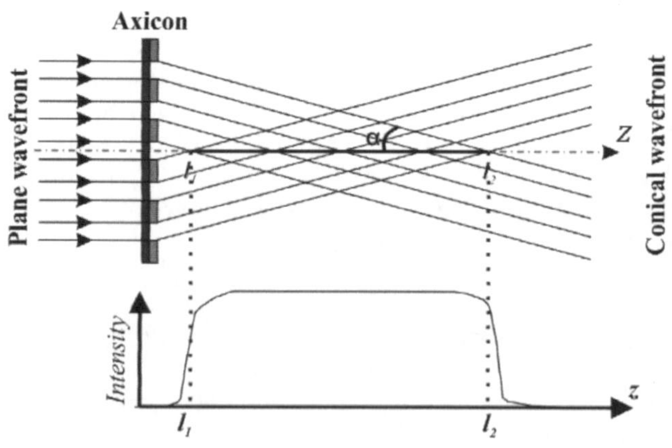

Figure 1. The principle of a diffractive axicon.

Let us consider a linearly polarized field in x-direction when flat beam bounded in circle of radius R illuminates the diffraction axicon:

$$\tau(r) = \exp(ik\alpha_0 r), \quad (1)$$

where α_0 is an axicon parameter that determines its numerical aperture.

Then the spectral distribution of the x-component will be obtained as

$$P_x(\sigma)=\int_0^R \exp\left(ik\alpha_0 r\right)J_0(kr\sigma)r\,\mathrm{d}r\,. \tag{2}$$

Then the field at distance z from the axicon takes form [25-27]:

$$\mathbf{E}(\rho,\theta,z)=k^2\int_{\sigma_1}^{\sigma_2}\begin{bmatrix} t_{TE}(\sigma)B_0^{SS}(k\sigma\rho,\theta)+t_{TM}(\sigma)\sqrt{1-\sigma^2}\,B_0^{CC}(k\sigma\rho,\theta) \\ B_0^{SC}(k\sigma\rho,\theta)\left(t_{TM}(\sigma)\sqrt{1-\sigma^2}-t_{TE}(\sigma)\right) \\ -t_{TM}(\sigma)\sigma B_0^{C}(k\sigma\rho,\theta)\end{bmatrix} P_x(\sigma)\exp\left[ikz\sqrt{1-\sigma^2}\right]\sigma\,\mathrm{d}\sigma, \tag{3}$$

where $B_0^{C}(k\sigma\rho,\theta)=iJ_1(k\sigma\rho)\cos\theta$, $B_0^{CC}(k\sigma\rho,\theta)=0.5\left[J_0(k\sigma\rho)-J_2(k\sigma\rho)\cos 2\theta\right]$, $B_0^{SS}(k\sigma\rho,\theta)=0.5\left[J_0(k\sigma\rho)+J_2(k\sigma\rho)\cos 2\theta\right]$, $B_0^{SC}(k\sigma\rho,\theta)=-0.5J_2(k\sigma\rho)\sin 2\theta$.

Now the x- and z-components of the electric field make the main contribution to the intensity near by the optical axis (z-axis):

$$E_x(\rho,\theta,z)=\frac{k^2}{2}\int_{\sigma_1}^{\sigma_2}\left[J_0(k\sigma\rho)\left(t_{TE}(\sigma)+t_{TM}(\sigma)\sqrt{1-\sigma^2}\right)+J_2(k\sigma\rho)\cos 2\theta\left(t_{TE}(\sigma)-t_{TM}(\sigma)\sqrt{1-\sigma^2}\right)\right]\times \\ \times P_x(\sigma)\exp\left[ikz\sqrt{1-\sigma^2}\right]\sigma\,\mathrm{d}\sigma, \tag{4a}$$

$$E_z(\rho,\theta,z)=ik^2\int_{\sigma_1}^{\sigma_2}J_1(k\sigma\rho)t_{TM}(\sigma)P_x(\sigma)\exp\left[ikz\sqrt{1-\sigma^2}\right]\cos\theta\sigma^2\,\mathrm{d}\sigma. \tag{4b}$$

The value of the equation (2) in neighborhood of $\sigma=\sigma_0$ is significantly higher than the values in other regions, so the integrals of equations (4a) and (4b) can be approximated as their integrands. The electric field x-component reaches its maximum intensity value at $\rho=0$ and z-component at $\rho=1.84\lambda/(2\pi\alpha_0)$. The ratio of the maximum values of the longitudinal and transverse components in the interval $\rho_\lambda\in[0,\lambda]$ along the polarization direction $\theta=0$ is then

$$\eta_\lambda=\frac{\left|E_z^{\max}(\rho_\lambda,0,z)\right|^2}{\left|E_x^{\max}(\rho_\lambda,0,z)\right|^2}\approx\frac{\left(1.16\cdot\alpha_0 t_{TM}(\alpha_0)\right)^2}{\left\|\left(t_{TE}(\alpha_0)+t_{TM}(\alpha_0)\sqrt{1-\alpha_0^2}\right)\right\|^2}.. \tag{5}$$

If we denote $\zeta(\alpha_0)=t_{TE}(\sigma)/t_{TM}(\sigma)$ we obtain

$$\eta_\lambda\approx\frac{1.34\alpha_0^2}{\left(\varsigma(\alpha_0)+\sqrt{1-\alpha_0^2}\right)^2}\underset{\alpha_0\to 1}{=}\frac{1.34}{\varsigma^2(1)}.. \tag{6}$$

In addition, if we denote the refractive index of the axicon by n_1 and the refractive index of the environment by n_2, assume that $n_1>n_2$, and write $\gamma=\sqrt{1-\alpha_0^2}$, then for the real values of $\gamma>n_2/n_1$ we obtain

$$\varsigma(\alpha_0)=\frac{2n_1n_2-\gamma\left(n_1^2+n_2^2\right)}{n_1^2+n_2^2-2\gamma n_1 n_2}. \tag{7}$$

If $\alpha_0=1$ and $\gamma=0$ we get

$$\varsigma(1) = \frac{2n_1 n_2}{n_1^2 + n_2^2}. \tag{8}$$

As a result, equation (6) will increase by a decrement of equation (8), i. e. with an increment of the relative refractive index $n=n_1/n_2$. If $n_1=1.5$ and If $n_2=1$, then $\zeta(1)\approx0.857$ and the maximum of equation (6) is $\eta_{max}(n=1.5)\approx1.83$.

3. Experimental measurements

In this article, we propose using a combination of solar cells and an array of microaxicons. Such a structure is shown in figure 2. An array of microaxicons is placed on the front-side of the solar cells. In this case, each of the microaxicons concentrates energy on a specific area of the solar cell. Thus, the microaxicon array formed a high-energy density region on the surface of the solar cell. It can be used to enhance the absorption of light energy in the upper layer of n-type silicon. In addition, each of the microaxicons refracts light rays normally incidental on the surface of each of the microaxicons. This leads to the fact that the refracted light beams propagate at an angle normal of the surface of the solar cell; thus, the probability of light trapping increases. As a result, light absorption efficiency must be increased. This must lead to an increase in the efficiency of conversion of light energy into electricity.

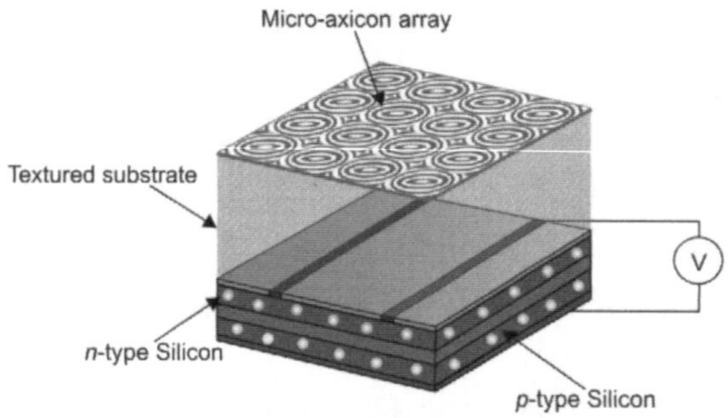

Figure 2. The scheme of a combination of a solar cell and a microaxicon array for enhancement of the light absorption efficiency.

For experimental investigation of the above-discussed combination of solar cells and microaxicons of different types, we use the scheme shown in figure 3. In the experiments, we used a tunable laser (EKSPLA NT200). This allowed us to change the laser wavelength of the generated laser radiation in the range from 400 to 800 nm. The output laser beam is directed by means of a mirror to a structure composed of a photocell and an axicon. We used three different microaxicons with periods of 645 nm, 1000 nm and 1500 nm. These micro-electronic lithography axicons were manufactured with a scanning electron microscope, Carl Zeiss Supra 25, and a Xenos XeDraw 2 high-speed writer. Images of the manufactured axicon are shown in figure 4.

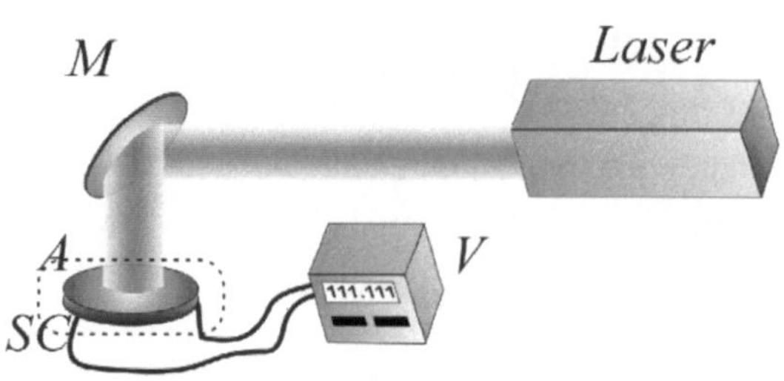

Figure 3. Experimental optical setup: *Laser* is a tunable EKSPLA NT200, *M* is a mirror, *A* is an axicon, *SC* is a solar cell, *V* is a multimeter.

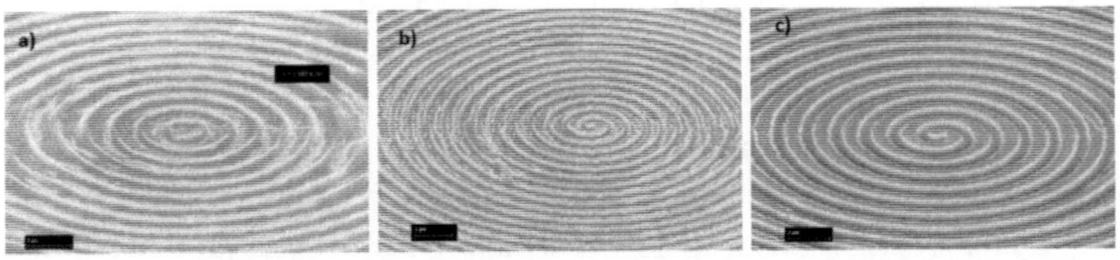

Figure 4. SEM images of microaxicons with periods of 645 nm (a), 1000 nm (b) and 1500 nm (c).

Figure 5 shows the experimentally measured dependence of the voltage on the wavelength. From this graph it is seen that for each of the three cases, using a combination of solar cells with different types of microaxicons in the wavelength range of 450 to 700 volts causes a distinct increase compared with the case of using only one solar cell without axicons. In the case of the combination of solar cells and axicons with a period of 645 nm, we achieved the maximum voltage increase — a relative increase compared to the combination of a solar cell and a quartz substrate was about 7% (the peak corresponds to a wavelength of 650 nm). This effect can be explained by the fact that the axicons with a large period should work efficiently for radiation with larger wavelengths in comparison with the wavelengths in the visible spectrum.

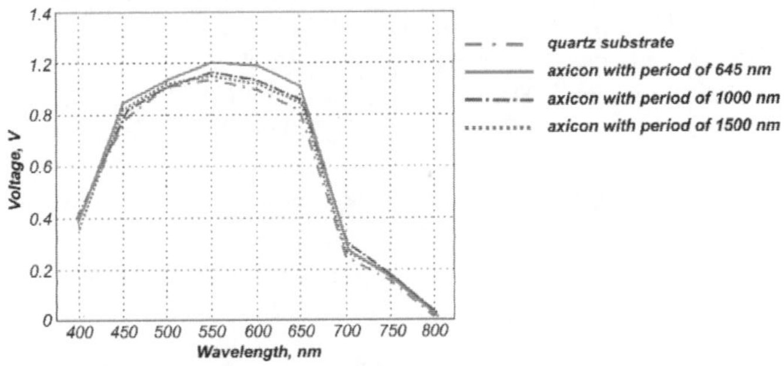

Figure 5. The dependence of the voltage on the wavelength.

4. Conclusion

We have investigated the possibility of using diffractive microaxicons with different periods for power conversion efficiency enhancement in solar cell. In the case of the combination of solar cells and axicons with a period of 645 nm, we achieved the maximum voltage increase — a relative increase compared to the combination of a solar cell and a quartz substrate was about 7% (the peak corresponds to a wavelength of 650 nm).

Acknowledgments

This work was financially supported by the grant of the President of the Russian Federation (project no. MD-5205.2016.9) and the Russian Foundation for Basic Research (grant No. 16-07-00494 A, 16-37-00241).

References

[1] Khrypunov G S, Chernykh E P, Kovtun N A, Belonogov E K 2009 *Semiconductors* **43** 1084
[2] Thool G S, Narayanaswamy K, Venkateswararao A, Singh S P 2016 *Langmuir*
[3] Singh A, Dey A, Das D, and Iyer P K 2016 *ACS Appl. Mater. Interfaces*
[4] Song Z, Watthage S C, Phillips A B, and Heben M J 2016 *J. Photon. Energy* **6** 022001
[5] Dimroth F, Tibbits T N D, Niemeyer M, Predan F, Beutel P, Karcher C, Oliva E, Siefer G, Lackner D, Fuß-Kailuweit P, Bett A W, Krause R, Drazek C, Guiot E, Wasselin J, Tauzin A, Signamarcheix T 2015 *IEEE J. Photovolt.* **6** 343

[6] Wehrspohn R B, Rau U, Gombert A 2015 *Photon Management in Solar Cells* (New York: John Wiley & Sons)
[7] Andreev V M, Davidyuk N Y, Ionova E A, Rumyantsev V D 2013 *Tech. Phys.* **83** 1323
[8] Polman A, Atwater H A 2012 *Nature Mater.* **11** 174
[9] Dewan R and Knipp D 2009 *J. Appl. Phys.* **106** 074901
[10] Gupta V, Chaudhary N, Srivastava R, Sharma G D, Bhardwaj R, and Chand S 2011 *J. Am. Chem. Soc.* **133** 9960
[11] Fahr S, Rockstuhl C, and Lederer F 2010 *Appl. Phys. Lett.* **97** 173510
[12] Atwater H A, Polman A 2010 *Nature Mater.* **9** 205
[13] Wehrspohn R B, Üpping J 2012 *J. Opt.* **14** 024003
[14] Schneider B W, Lal N N, Baker-Finch S, and White T P 2014 *Opt. Express* **22** A1422
[15] Wang K X, Yu Z, Liu V, Cui Y, and Fan S 2012 *Nano Lett.* **12** 1616
[16] Garnett E and Yang P 2010 *Nano Lett.* **10** 1082
[17] Hauser H, Mellor A, Guttowski A, Wellens C, Benick J, Müller C, Hermle M, and Bläsi B 2012 *Energy Procedia* **27** 337
[18] Nguyen-Huu N, Cada M, and Pištora J 2014 *Opt. Express* **22** A68
[19] Wellenzohn M and Hainberger R 2012 *Opt. Express* **20** A20
[20] Meng X, Drouard E, Gomard G, Peretti R, Fave A, and Seassal C 2012 *Opt. Express* **20** A560
[21] Lee Y C, Huang C F, Chang J Y, and Wu M L 2008 *Opt. Express* **16** 7969
[22] Chemisana D 2011 *Renew. Sustainable Energy Rev.* **15** 603
[23] Khonina S N, Serafimovich P G, Savelyev D A, Pustovoi I A 2012 *J. Opt. Technol.* **79** 626
[24] Khonina S N, Nesterenko D V, Morozov A A, Skidanov R V, Pustovoy I A 2011 *Computer Optics* **35** 11
[25] Luneburg R K 1966 *Mathematical Theory of Optics* (Berkeley: University of California Press)
[26] Mansuripur M 1989 *J. Opt. Soc. Am. A* **6** 786
[27] Khonina S N, Nesterenko D V, Morozov A A, Skidanov R V, Soifer V A 2012 *Opt. Mem. Neur. Net.* **21** 17

Optical parameters of ITO/TPD/Alq3/Al luminescent structures, containing arrays of CdSe/ZnS colloidal quantum dots

I I Mikhailov, S A Tarasov, I A Lamkin, P O Tadtaev, L I Kozlovich, A V Solomonov, E M Stepanov

Saint Petersburg Electrotechnical University «LETI» (ETU)

E-mail: iimihalov@gmail.com

Abstract. The luminescent organic ITO/TPD/Alq3/Al structures and CdSe/ZnS quantum dots (QD) arrays were created. Electrical and optical properties of the samples were examined. The luminescence of the layers and QD arrays was shown in the range of wavelengths from 400 to 680 nm. Luminescent structures with phosphors corresponding to the emission standards with CRI>98 and with color temperature of 5500 K and 6504 K were created.

1. Introduction
For the last decade colloidal quantum dots (CQDs) have become an object of particularly intensive studying [1]. CQDs are semiconductor particles of nanometer scale synthesized in liquid media and characterized by unique properties which substantially differ from bulk material properties. One of the most important advantages of CQDs is a facility of geometric size tuning through technological parameters variation (time, temperature etc.). The procedures of their synthesis do not require expensive equipment and materials. CQDs are a promising material [2] for the creation of high-performance LED and laser structures, as well as photosensitizers in solar cells or phosphors in micro-devices for optical tomography of biological tissues. The application of CQDs in the creation of the organic light emitting structures such as OLEDs is of special interest. Insertion of CQDs into OLED structures provides the substantial increase in luminescence efficiency without seriously complicating the techniques of creating light-emitting structures.

2. Materials and methods
In this article, attention is paid to the study of CdSe/ZnS and CdSeZnS/ZnS colloidal quantum dots with peaks of photoluminescence in the range of wavelength from 500 to 680 nm. Surface shell of quantum dots by wide bandgap shell of ZnS significantly improves their properties. Concentration of the surface defects in such particles is greatly reduced. It leads to noticeable increase in luminescence quantum yield. In addition, the oxidation resistance of shell CQDs increases. Nevertheless, were also investigated nonenveloped CQDs CdS. This provided an opportunity to compare their basic optical parameters. It is

also allowed to extend the range of luminescence of produced arrays of CQDs toward shorter wavelengths to 400 nm.

CQDs deposition techniques on the substrate have been tested. These techniques allow retaining high emission properties without significant loss in the luminescence efficiency for all types of particles. As substrates used were different plates of crystalline and amorphous materials [3], promising for creating organic light emitting diodes (OLED) and other luminescent devices. The large number of samples has been created based on glass and quartz substrates without the use of additional layers. The deposition of CQDs was carried out by centrifugation or by applying microdroplets using for a part of structures subsequent annealing in an inert atmosphere [4].

Essential optical and luminescent properties of CQDs were measured and carefully analyzed [5-7]. Experiments were performed on the installation, based on the double-grating monochromator SDL-1, which enabled the optical resolution of 0.01 nm. Ocean Optics USB-4000 и HR4000 fast scanning luminescence spectrometers were applied also. 445-nm semiconductor laser, 290-nm optical emitter, mercury vapor and other gas lamps were used to excite the emission. The lock-in amplifier Stanford SR-810 was applied in order to increase the sensibility of the installation.

3. Results

Figure 1 shows the luminescence spectra of colloidal quantum dots of various composition and size, deposited on the substrate. CQDs deposited arrays were obtained, luminescence parameters, which almost did not differ from the characteristics of the particles present in the solution. It was shown that photoluminescence of deposited CQDs based on CdSe with shell still was mainly caused by excitonic transitions. It was confirmed by the study of the absorption spectra. It is shown that the first exciton absorption peak is close to the maximum of the luminescence spectrum of such dots. In addition, on the graph is no long-wave radiation, which is characteristic for transitions involving surface states. It indicates the high level of surficial defect passivation due to the application of wide-bandgap ZnS shell.

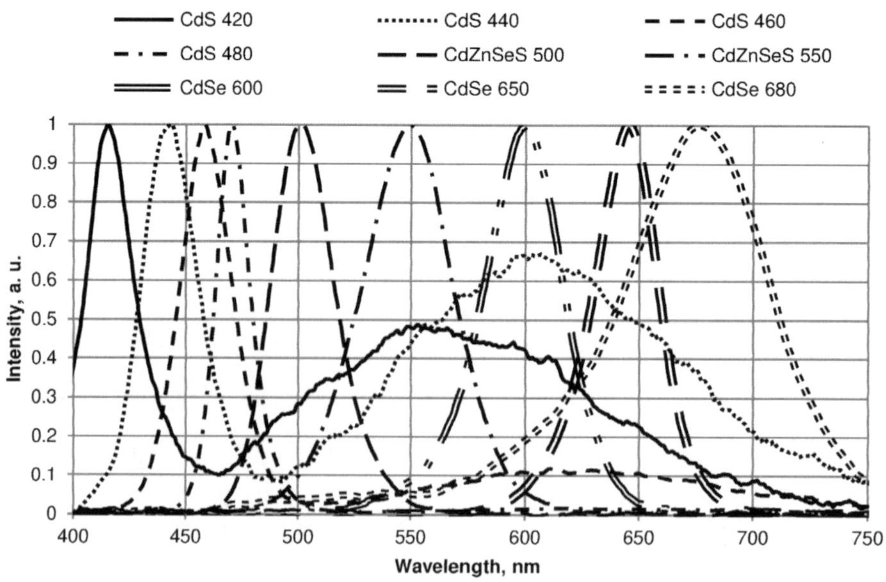

Figure 1. The luminescence spectra of CQD's.

For samples based on non-enveloped CQDs based on CdS is observed the broadened line at wavelengths of 500-750 nm (Figure 1). This is explained by the influence of transitions involving surface states. Radiation in this range is observed in nanoparticles, which are in the colloidal solution.

Although in some cases, the existence of such a wide range of radiation is necessary to recognize drawback investigated material. In the article, this effect is found useful application. Using such CQDs white phosphorus coatings were created based on a small number of components. Color parameters of such coatings are similar to the characteristics of white LEDs based on InGaN. CdSe 500 an additional component has been applied to improve the parameters of these phosphor coatings (Figure 2a). This is allowed to smooth cavity in the green region and achieve CRI> 90.

To further improve the color characteristics and the luminescence efficiency of the phosphor layers properties of combined arrays CQDs containing nanoparticles of various compositions and sizes was simulated. Calculations have shown that it is possible to create phosphors, CRI which is practically equal to 100. To create coatings with high CRI were conducted experiments to study the deposition techniques of CQDs arrays with different composition and their luminescence properties. We were determined coefficients depending on the initial concentration of particles in the solution and on the efficiency of luminescence. It was also determined the effect of particles arrays deposition techniques, substrate and annealing. The processes of interaction of nanoparticles of various types that make up the array are also considered. It is allowed to create phosphor coating, CRI that exceeded> 98. The structures corresponding emission standards with the color temperatures of 6504 K (Figure 2a) and 5500 K (Figure 2b) and solar radiation on the ground have been developed.

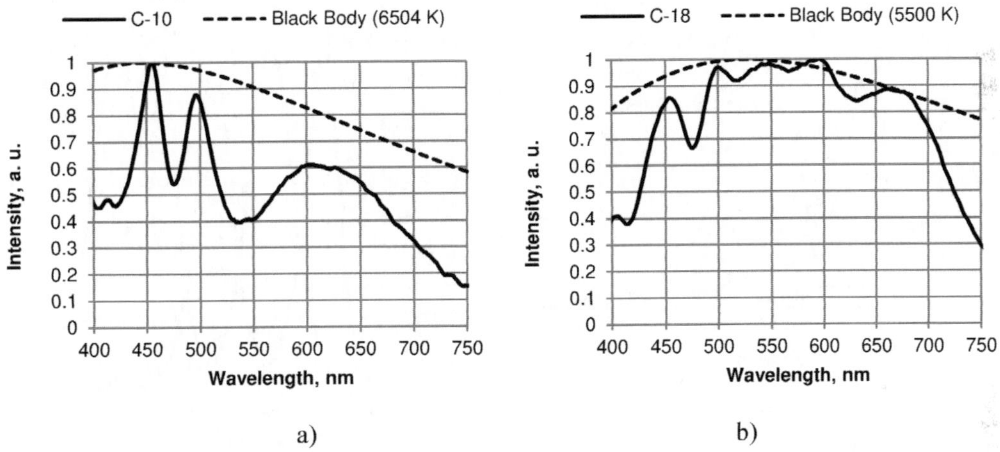

Figure 2. White phosphor coatings based on colloidal quantum dots CdSe (a) and structures with improved CRI (b).

The calculation of the CQDs' dimensions, based on their absorption and photoluminescence spectra, was run and it was shown that the diameter of the core was to range from 3 to 6 nm. The substantial divergence between the SEM data and results of the modelling might be explained by the fact that the calculation didn't take wide-bandgap shell's and surfactant's dimensional parameters into account. Therefore, the thickness of the outer shell might be estimated at 5-7 nm.

Previously examined CQDs were inserted into the active layer of the OLED as the deposited array of nanoparticles. The general OLED structure is a complex system of several thin layers between two electrodes, one of which is transparent to allow the emittance the light. The simplest structure is a single-layer device architecture, where the organic emitter is deposited between anode and cathode. More complex structures have a total of 7-9 active materials: an anode; anode buffer layer (ABL), hole injecting layer (HIL) or electron blocking layer (EBL); hole transporting layer (HTL); emissive layer (EML); electron transporting layer (ETL) or hole blocking layer (HBL), electron injecting layer (EIL); cathode buffer layer (CBL), a cathode and a protective barrier layer. Excitonic radiative transitions provide the growth of luminescence efficiency. Insertion of the CQDs also provides allowance for the OLED spectrum tuning by controlling QDs' sizes, structure and composition.

ITO/TPD/Alq3/Al OLED structures were created (Fig. 3a): TPD (N,N'-bis(3-methylphenyl)-N,N'-diphenylbenzidine) was a hole transporting layer, Alq3 (Tris(8-hydroxyquinolinato)aluminum) was an emissive/electron transporting layer, aluminum and ITO (indium tin oxide) were used to create electrodes. All the layers were deposited by means of spin-coating or vacuum thermal evaporation on the glass substrate. The process of deposition lasted no more than several seconds which positively affected the quality of the layers. The upper contact was created of the magnesium or aluminum by vacuum resistive coating.

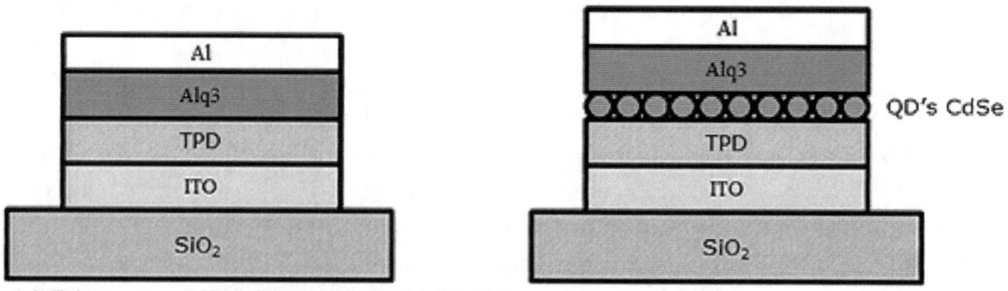

Figure 3. (a) Diagram of ITO/TPD/Alq3/Al OLED structure; (b) Diagram of ITO/TPD/Alq3/Al OLED structure, containing CQDs.

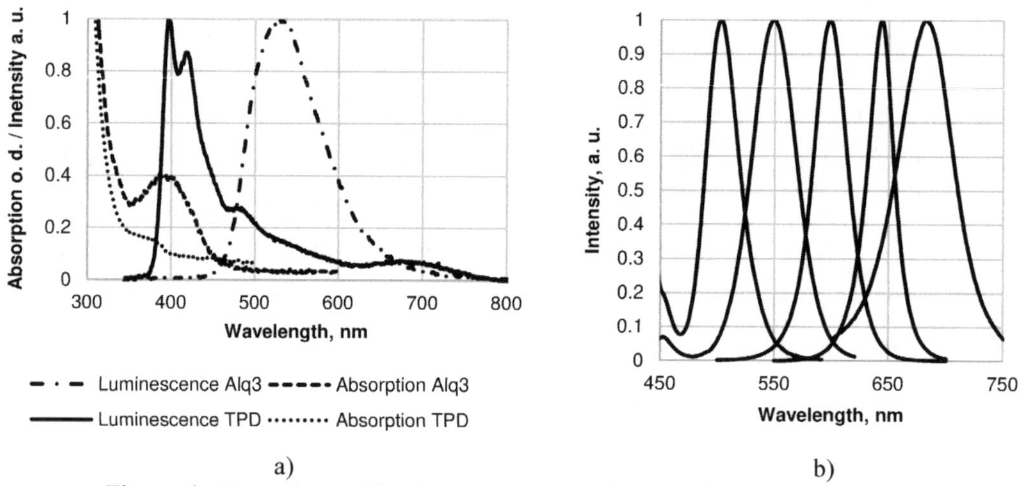

Figure 4. Absorption and luminescence spectra for organic layers in structure ITO/TPD/CQD/Alq3/Me (a) and luminescence spectra of arrays of colloidal quantum dots of various sizes (b)

In this paper we studied essential optical and electrical parameters of the created structures; thus, absorption and luminescence spectra of the structures and separate layers of TPD and Alq3 were examined. Absorption and luminescence structure layers charts shown in figure 4 a.

The luminescence spectrum of Alq3 was almost symmetrical with a strongly marked maxima at 530 nm and half-width of the spectral curve around 90 nm. There was also a peak of the absorption spectra at 390 nm. The luminescence spectrum of TPD was not symmetrical and had two local maxima at 400 and 420 nm. These spectra were in a good correlation with the data from third-party sources which allows the application of the vacuum thermal evaporation with a high speed of layer formation in the OLED technology. Adding to the array of colloidal quantum dots (Fig. 3b) changed band diagram of the structure, improving the collectability of the charge carriers in its active region. Current-voltage curves of the samples were obtained; rectifying effects were observed. Addition of the array of CdSe/ZnS CQDs is to allow the broadening of the luminescence spectrum to the wavelength of 680 nm, as well as the control over emission range of the structure (Fig. 4b).

Thus, arrays of CdS, CdSe/ZnS and CdSeZnS/ZnS colloidal quantum dots with peaks of photoluminescence in the range of wavelength from 400 to 680 nm were investigated. The luminescent organic ITO/TPD/Alq3/Al structures and CdSe/ZnS quantum dots (QD) arrays were created. Electrical and optical properties of the samples were investigated. Fuorescent nanostructures with phosphors corresponding emission standards with color temperature of 5500 K and 6504 K and a CRI> 98 have been created.

4. Acknowledgements

The work was supported by the Ministry of Education of the Russian Federation in the framework of the project of the state task in the field of scientific activity, project number 16.1307.2014K.

5. References

[1] Y Shirasaki, G Supran, M Bawendi. Nature Photonics, 2012, 7.
[2] C Burda, S Link, M. Mohamed. 2001, J. Phes. Chem. B 105, 12286.
[3] S Tarasov, O Aleksandrova, I Lamkin, A Maksimov, E Maraeva, I Mikhailov, V Moshnikov, S.Musikhin, S. Nalimova. Semiconductors, 2015, 49, 1710.
[4] S. Tarasov, O. Aleksandrova, A. Maksimov, E. Maraeva, L. Matyushkin, V. Moshnikov, S. Musikhin. Semiconductors, 2014, 48, 13, 1729–1731.
[5] I Mikhailov, STarasov, A Solomonov, L Matyushkin, D Mazing. Functional materials, 2014, 21, 2, 142-145.
[6] I Mikhailov, S Tarasov, A Solomonov, O Aleksandrova, L Matyushkin, D Mazing. Journal of Physics: Conference Series, 2014, 572, 012029.
[7] I Mikhailov, S Tarasov, I Lamkin, M Andreev, A Solomonov. Journal of Physics: Conference Series, 2015, 643, 012080.

On-chip realization of quantum circuits by using waveguides on Si_3N_4

A Vasilev[1], A Kozubov[1], A Gaidash[1], S Chivilikhin[1]
[1]ITMO University, 199034, Russia

Abstract. In this paper we consider the dependence of the supported modes and the type of its polarization on the profile of the waveguide. Investigated possibility of production of beam splitters designed on the conventional waveguide.

1. Introduction
In the present the issue of creating quantum chip [1-3] is very important, in view of the fact that all possible technological solutions are moving in the direction of reducing the size, which means that we need to find ways to implement it. In particular, quantum chips are essentially nanoscale objects, which gives a considerable gain in size. Thus, the possible transfer of various linear quantum optical systems is a very important technological problem. In this work will be considered acceptable and optimal size to create waveguides in Si_3N_4, and the possibility of making structures based on waveguides and optical modes overflow between them.

2. Method
In the first part of this work we will be consider how different modes in different waveguides are supported, and why in different sizes dominant modes of the waveguides can be displayed. Also we will give an explanation how to reduce the size of the waveguide for solving this problem. On the Figure 1 you can see the waveguide's profile without cladding.

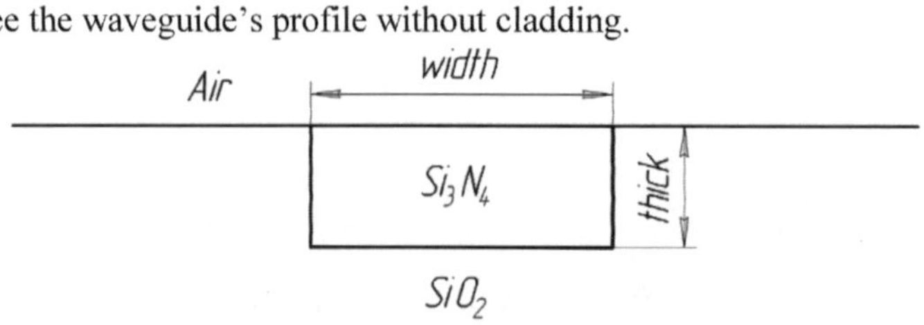

Figure 1.Waveguide without cladding.Refractive indexes are:n(Air)=1,n(Si3N4)=2.02, n(SiO2)=1.553

We assume that E_y and E_x - it's vertical and horizontal polarization, respectively, if we look at the profile of the waveguide. To find the number of supported modes used Mode Solver tool from OptiFDTD package. In the second part we will explain the possibility of constructing some linear optical elements such as beam splitters (Figure 2) in Si_3N_4 using only waveguides. We obtained optimal sizes for these elements and investigated any arbitrary deviation from this values. Furthermore we will show which different circuits from different beam splitters can be realized this way and show all optimal values for its sizes.

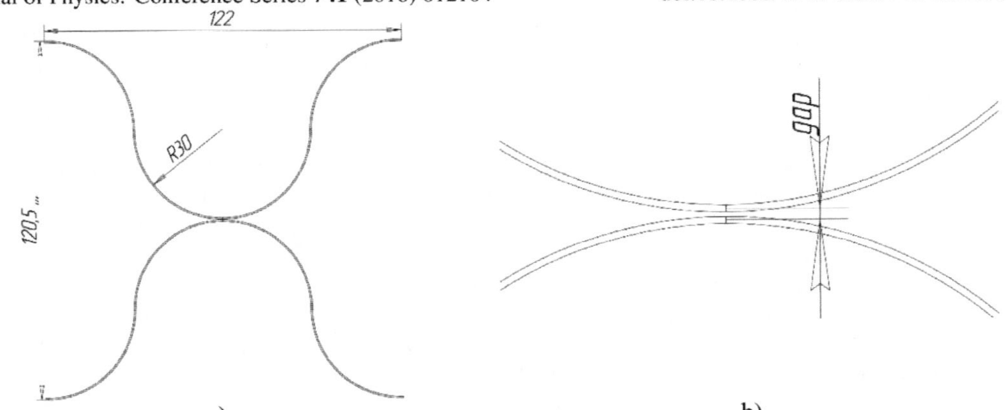

a) b)

Figure 2. a) The general scales of the coupler. b) The gap between waveguides.

For modeling the process of the light propagation we use the next method, which is known as FDTD. In the 2D TE case, only H_x, E_y and H_z, and have nonzero components. In lossless media, Maxwell's equations take the following form:

$$\frac{\partial E_y}{\partial t} = \frac{1}{\epsilon}\left(\frac{\partial H_x}{\partial z} - \frac{\partial H_z}{\partial z}\right)$$
$$\frac{\partial H_x}{\partial t} = \frac{1}{\mu_0}\frac{\partial E_y}{\partial z} \qquad (1)$$
$$\frac{\partial H_z}{\partial t} = -\frac{1}{\mu_0}\frac{\partial E_y}{\partial x},$$

where $\epsilon = \epsilon_0 \epsilon_r$ is the dielectric permittivity of the material, μ_0 is the magnetic permeability of free-space.
The refractive index of the material is defined by:

$$n = \sqrt{\epsilon_r}. \qquad (2)$$

Each field is represented by a 2D array: `Ey(i,k)`, `Hx(i,k)` and `Hz(i, k)`. The indices `i` and `k` account for the number of space steps in the X and Z direction, respectively. In the case of TE (Figure 3), the location of the fields in the mesh is shown below.

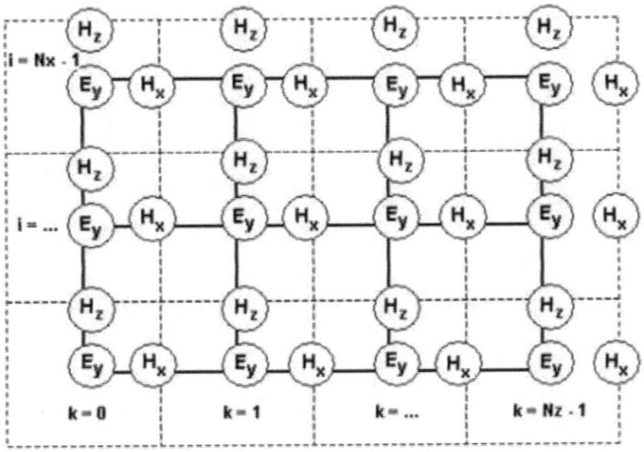

Figure 3. Location of the TE fields components in the computational domain

3. Results and conclusion

The number and polarization of modes, supported by the waveguide, depends on its profile and materials, surround the waveguide. In the Table 1 dimensions of the waveguide are given in micrometres. A dash indicates the absence of a supported modes. Stacked Recording format: <Ex>/<Ey>.

The following table indicates the kind of polarization is maintained waveguide. Presented that thin and wide waveguide supports horizontal polarization, but narrow and deep - vertical. At the same waveguide width of 0.5 microns and 0.2 microns reliably does not support any mode – for the light there is no waveguide structure.

The smallest waveguide observed in the symmetric case - when one is surrounded by single material on all sides. Also supported type of polarization depends on the proportions. It makes possible to make polarizers only with conventional waveguide. We have also demonstrated the possibility of constructing some linear optical elements such as beam splitters based on waveguides on Si_3N_4. We modeled directed beam splitters with different values of beamsplitting ratio (Figure 4), which are based on optical tunneling. Obtained parameters for the beam splitters with a few basic values: ½, ⅔, ¾. As an example you can see the distribution of the electric field in 50:50 beam splitter on the Figure 5. This work is extremely important because we can use such circuits to make controlled quantum networks.

		width						
		0.1	0.2	0.3	0.4	0.5	0.6	0.7
thick	0.2						-/-	-/-
	0.25						-/-	-/-
	0.3					-/-	Ex/-	Ex/-
	0.4		-/-	-/-	Ex/-	Ex/-		
	0.5		-/-	-/-	Ex/Ey	Ex/Ey		
	0.6	-/-	-/Ey	-/Ey	Ex/Ey	Ex/Ey	1Ex/2Ey	
	0.7	-/-	-/Ey	-/Ey	Ex/Ey			

Table 1. Dependence of supported modes from waveguide's scales.

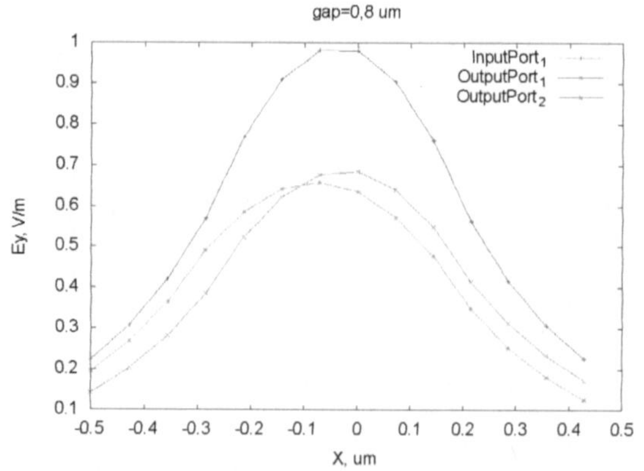

Figure 4. Figure illustrating the amplitude on one input port at two outputs of beam splitter circuit modeled in Si_3N_4.This figure shows the amplitude distribution along cross-section for 50:50 beam splitter.

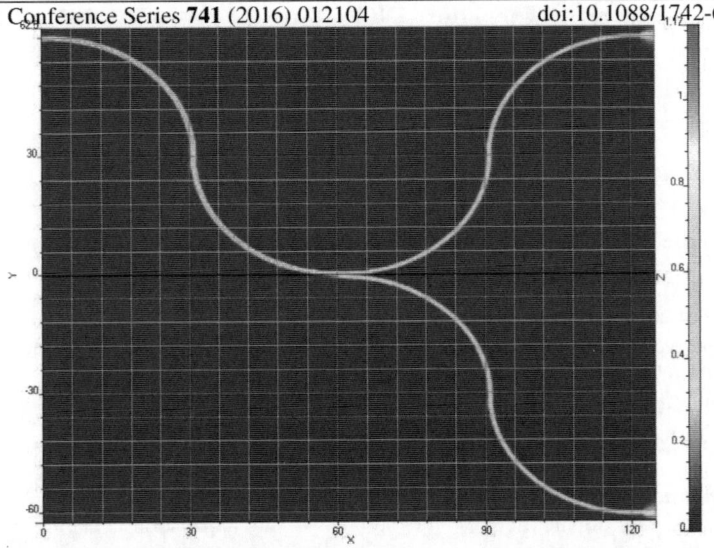

Figure 5. Figure illustrating the distribution of the electric field in 50:50 beam splitter. Waveguides presented by Si_3N_4 on SiO_2 substrate.

Acknowledgments
This work was financially supported by Government of Russian Federation, Grant 074-U01 and by the Ministry of Education and Science of Russian Federation (project № 14.578.21.0112).

References

[1] M. A. Foster, A. C. Turner, J. E. Sharping, B. S. Schmidt, M. Lipson, and A. L. Gaeta, Broad-band optical parametric gain on a silicon photonic chip, Nature 441, 960–963 (2006)

[2] V T. Gerrits, N. Thomas-Peter, J. C. Gates, A. E. Lita, B. J. Metcalf, B. Calkins, N. A. Tomlin, A. E. Fox, A. Lamas Linares, J. B. Spring, N. K. Langford, R. P. Mirin, P. G. R. Smith, I. A. Walmsley, S. W. Nam, On-chip, photon-number-resolving, telecommunication-band detectors for scalable photonic information processing, Phys. Rev. A 84, 060301(R) (2011)

[3] B. Desiatov, I. Goykhman and U. Levy, Nanoscale mode selector in silicon waveguide for on-chip nanofocusing applications, Nano Letters 9, 3381–3386 (2009).

Reactive ion etching of indium-tin oxide films by CCl_4-based Inductivity Coupled Plasma

Sucheta Juneja [1,3], Sergey D Poletayev[1,2], Sergey Fomchenkov[1,2], Svetlana N Khonina[1,2], Roman V Skidanov[1,2], Nikolay L Kazanskiy[1,2]

[1] Samara State Aerospace University, 443086 Samara, Russia.
[2] Image Processing Systems Institute of the Russian Academy of Sciences, 443086 Samara, Russia.
[3]Physics of Energy Harvesting Division, CSIR – National Physical Laboratory, Dr. K.S. Krishnan Marg, New Delhi, India, 110012.

Abstract. Indium tin oxide (ITO) films have been a subject of extensive studies in fabrication of micro-electronic devices for opto-electronic applications ranging from anti-reflection coatings to transparent contacts in photovoltaic devices. In this paper, a new and effective way of reactive ion etching of a conducting indium-tin oxide (ITO) film with Carbon tetrachloride (CCl_4) has been investigated. CCl_4 plasma containing an addition of gases mixture of dissociated argon and oxygen were used. Oxygen is added to increase the etchant percentage whereas argon was used for stabilization of plasma. The etching characteristics obtained with these gaseous mixtures were explained based on plasma etch chemistry and etching regime of ITO films. An etch rate as high as ~20 nm/min can be achieved with a controlled process parameter such as power density, total flow rate, composition of reactive gases gas and pressure. Our Investigation represents some of the extensive work in this area.

1. Introduction

Thin films with high electrical conductivity and optical transparency have been the subject of extensive studies by material scientists and engineers for many applications, such as transparent electrodes in liquid crystal displays or electroluminescent displays, [1] Photovoltaic devices [2]. The most employed thin films are Flat Panel Displays (FPDs) are made from transparent conducting oxide (TCO) materials [3-4]. Among various materials, Indium tin oxide (ITO, 90% wt.% In_2O_3/10Wt. % SnO_2) is one of the most widely investigated and useful transparent conducting oxide, because of its relatively low resistivity and high visible transmittance as compared with other (TCO) materials such as SnO_2 or ZnO [5-10]. It has been known that ITO has both excellent and optical properties [11, 12]. The theoretical understanding of ITO has been limited because of the complex crystal structure of ITO thin film. Indium Tin Oxide is essentially formed by subsititutional doping of In_2O_3 with Sn which replaces the In^{3+} atoms from the cubic bixbyte structure with a unit cell containing 40 atoms and two non-equivalent cation sites [13] .For the applications mentioned above, their etching characteristics and the availability of patterning for the films become an important factor and are directly linked with the productivity. The growths of anisotropic dry etch process for ITO films are an important task to be resolved. Conventionally, wet chemical etching process is used. However, this process requires multiple process steps, large expensive equipment's and geometry patterning [14]. To overcome these problems, dry etching processes have been developed using fluorine–based and hydrogen based chemistries [15-16]. The main concern with the etching of ITO film using fluorine and hydrogen-based gases (such as CF_4 in plasma methods) has the disadvantage of inducing contamination due to polymerization. Whereas the etching of ITO films by plasma, containing Cl_2 and BCl_3 lead to fast etch rate, films with smooth surface without polymerization [17-19]. The etch selectivity of ITO over photoresist films is also an important factor in many applications [20].

In this work, dry etching characteristics and surface structure of ITO films using $CCl_4/Ar/O_2$ in inductively coupled plasmas (ICP) were studied. At the same time, plasma-etching phenomenon were investigated.

2. Experimental

Research on structures to the substrate was made by ion-reactive etching in an inductively coupled plasma installation (RIE-ICP) "Caroline PE-15". Figure1 shows schematic of the ICP etching system. Ion energy to the substrate, may influence the physical (ion) etching component controls the RF bias generator with a frequency of 13.56 MHz, is fed through to the chuck power matching device. To generate plasma using RF generator matching network was used to convert the loaded impedance in the plasma chamber. The chamber is evacuated with a turbo molecular pump of 800 L / s capacity.

We investigated the effect of bias power on the rate of etching of ITO glass. Etching was carried out in an environment $CCl_4/O_2/Ar$ under the following conditions: power of coil 500 W, current of electromagnet 2A, flow rate CCl_4 7 sccm, flow rate O_2 3 sccm and flow rate Ar 20 sccm . Power of bias took different modes values from 180 to 250 W. The time of etching for all was only 3 min. Aluminium masks was used with approximately 40 nm thickness deposited by magnetron sputtering method on market TCO glass with surface roughness of rms = 1 nm. The choice of aluminum masks determines their relative chemical inertness with respect to the chlorine-containing plasma. Carbon tetrachloride was poured into the flask and connected to a channel flow rate of gases. When the channel is turned on by the action of the vacuum chamber is a partial evaporation of CCl_4 and the formation of the gas stream of vapors and etching was done with these CCl_4 vapors. After etching mask residues are removed by chemical means and then etching rate was determined using a profilometer (KLA-Tencor P16+ model). The 3-D images was taken with Interferometer (ZYGO 7600 model), Field emission scanning electron microscopy (ZEISS SUPRA 25) was used to take surface images.

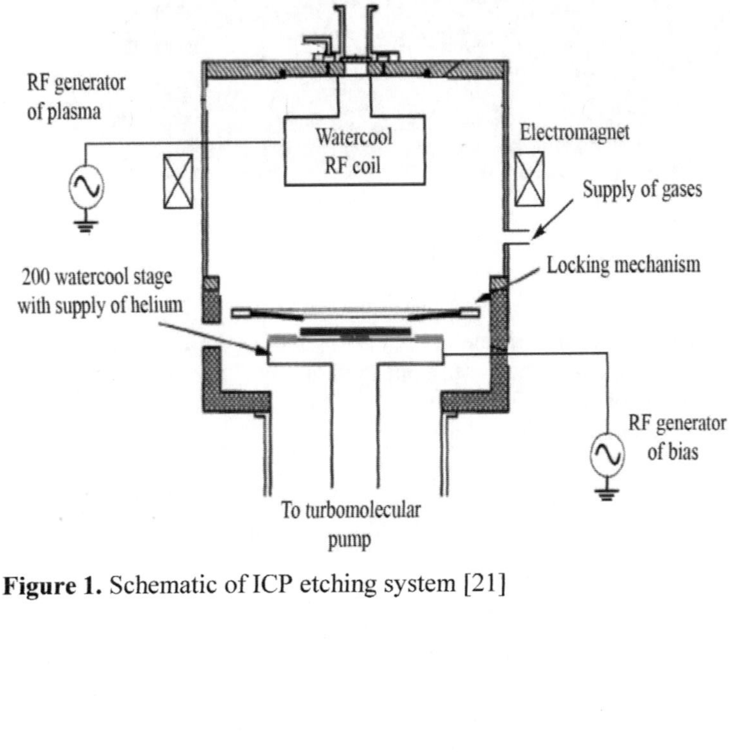

Figure 1. Schematic of ICP etching system [21]

3. Results and discussion

3.1 Etching characteristics of ITO films

Figure 2 shows the etch rate of ITO thin films with CCl_4 /Ar/O_2 gas mixing ratio at bias power of 250W. The etch rate were found to be 5nm/min, 12nm/min and ~20nm/min respectively with different bias power applied.

We may consider that the maximum etch rate is related to the effects of surface chemistry and that can be explained by chemical processes. This etching behavior shows that the inert gas (Ar) results in ion assisted chemical etching and consequently the etch rate is increased. The argon species are ionized in the plasma and accelerate towards the surface of the ITO owing to the DC bias, which could provide energy for the reaction to take place between the etchant and ITO films. The etch reaction in plasma with argon take via sputter deposition mechanism due to energetic ion bombardment [22]. The decreased etch rate obtained at low power is due to the lowering of physical bombardment effect and rate of chemical reaction. Oxygen was added to plasma to increase the etchant percentage by converting unsaturates to etchants. There was lesser reaction between the In and oxygen atoms due to the chemical reactions of In and Sn atoms. It is assume that the plasma containing argon or inert gas broke the chlorine bonding with In_2O_3 in the tetrahedral interstices of the face-centered-cubic In^{3+} lattice. Therefore, the inert gas transported the reaction products such as $InCl_x$ [23,24].

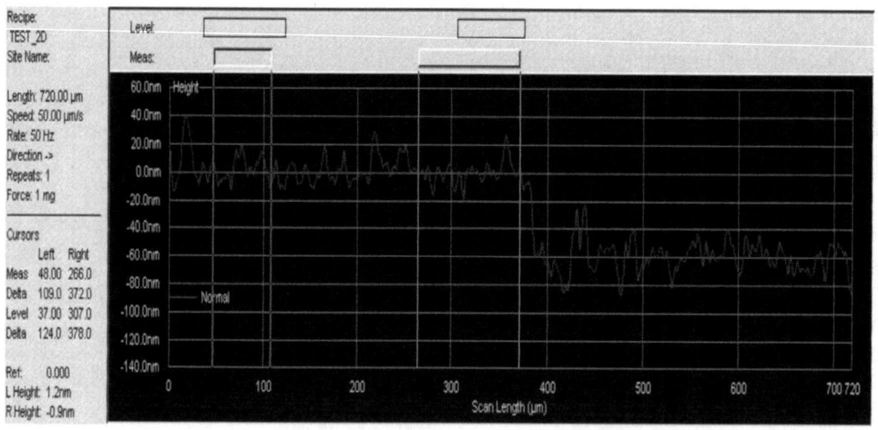

Figure 2. Etch rate of ITO films at bias power of 250W

3.2 Interferometer result

Figure 3 shows the 3-D images obtained from Interferometer. The surface of ITO film is change with etching. It is observed that the average roughness increases after etching unsignificantly. Root mean square remains the same. The smoothest surface of ITO film obtained using CCl_4/Ar/O_2 plasma with high etch rate at bias power of 250W was obtained.

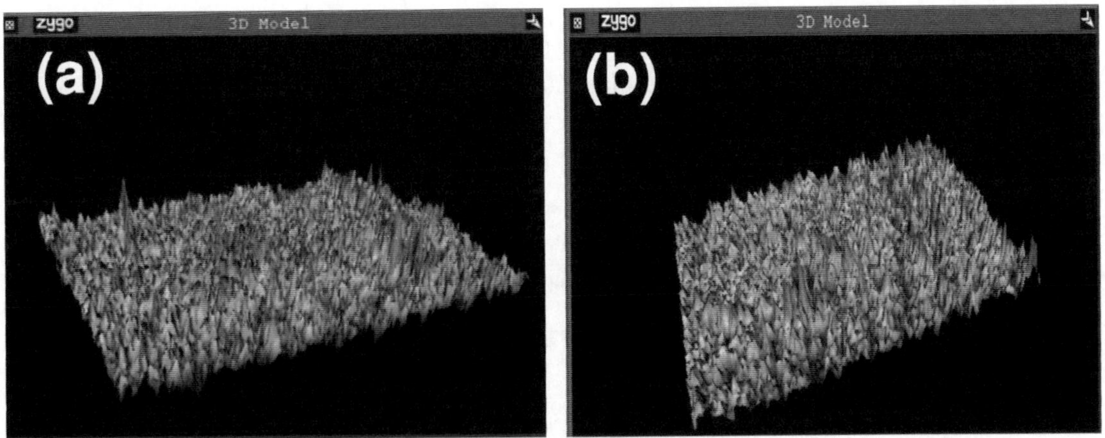

Figure 3. 3-D images of ITO films a) Before etching b) after etching.

3.3 Field emission scanning electron microscopy (FESEM) studies

Figure 4 shows the FESEM images of ITO film at bis power of 250W. The surface of ITO film is change with etching as observed from the images (Figuge 4(a) &(b)). The etching height is also mentioned. This clearly shows the etching of ITO films in gaseous mixture of $CCl_4/Ar/O_2$.

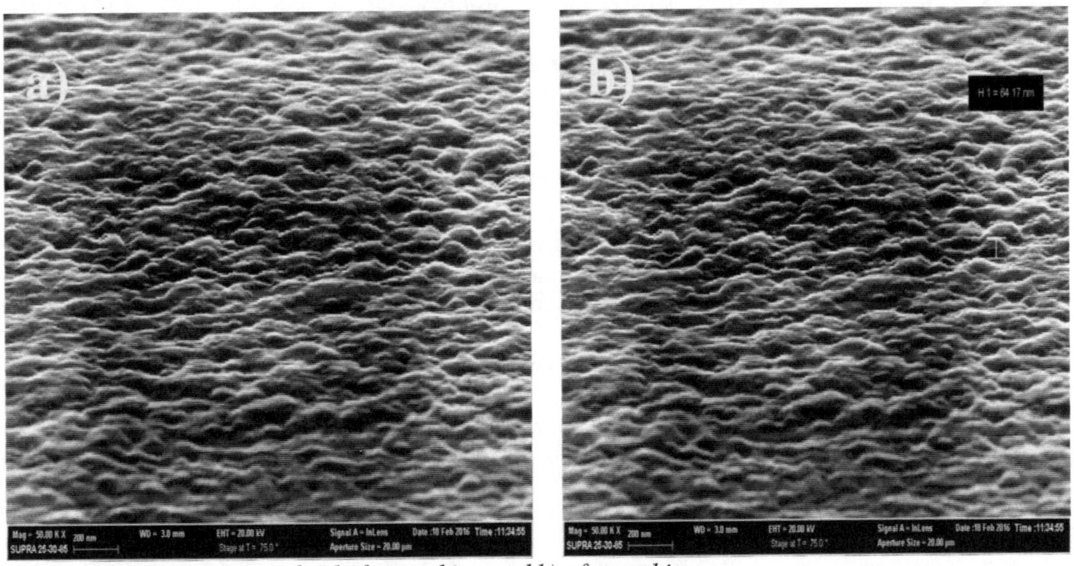

Figure 4. FESEM images of a) before etching and b) after etching

4. Conclusion

ICP etching of ITO has been studied in $CCl_4/Ar/O_2$ plasma. We vary bias power to increase the etch rate in the present investigation. The maximum etch rate of ITO films was ~20 nm/min. When CCl_4 was added to argon, the reaction between ITO and CCl_4 may enhanced by sufficient ion bombardment resulting increase in ITO etch rate. The etching of the ITO thin films in gaseous mixture CCl_4/ O_2 containing inert gas (argon) result in high etch rate and films with smooth surface without any polymerization. Further, more Investigation are needed in detail.

Acknowledgements

This work is financially supported by the Ministry of Education and Science of the Russian Federation.

References

1. Kaneko E 1987. Liquid Crystal TV Displays, chapter 7 (Tokyo: KTK Scientific)
2. Herrero J, Guillen C 2002 Vacuum **67** 611
3. Chou T H, Cheng K Y, Chang T L, Ting C J, Hsu H C, Wu C J, Tsai J H, Huang T Y 2009 Microelec. Eng. **86** 628
4. Lee CS, Cui ZY, Jin HF, Sung SW, Lee HG, Kim NS 2011 Trans. Electr. Electron. Mater. **12** 35
5. Kim YS, Park YC, Ansari SG, Lee JY, Lee BS, Shin HS 2003 Surface and Coatings Technology **173** 299
6. Hwang MS, Lee HJ, Jeong HS, Seo YW, Kwon SJ 2003 Surface and coatings technology **171** 29
7. Keshmiri SH, Rezaee-Roknabadi M, Ashok S 2002 Thin solid films **413** 167
. Chandrasekhar R, Choy KL 2001 Thin solid films **398** 59
9. Fahland M, Karlsson P, Charton C 2001 Thin Solid Films **392** 334
10. Yang ZW, Han SH, Yang TL, Ye L, Zhang DH, Ma HL, Cheng CF 2000 Thin Solid Films **366** 4
11. K.H. Choi, J.Y. Kim, Y.S. Lee, H.J. Kim, Thin Solid Films 341 (1999) 152.
12. A. Klo¨ppel, W. Kriegseis, B.K. Meyer, A. Scharmann, C. Daube, J. Stollenwerk, J. Trube, Thin Solid Films 365 (2000) 139.
13. Ho-Chul Lee Applied Surface Science 252 (2006) 3428–3435
14. Kim JH, Her JH, Lim YJ, Kumar P, Lee SH, Park KH, Lee JH, Kim BK 2010 Transactions on Electrical and Electronic Materials **11**134
15. Shin H, Kim C, Bae C, Lee JS, Lee J, Kim S 2007 Applied Surface Science **253** 8928
16. Kuo Y, Tai TL 1998 Journal of The Electrochemical Society **145** 4313
17 Kuo Y 1997 Japan. J. Appl. Phys. **36** L629
18. Grigonis A, Knizikevičius R, Rutkūnien Ž, Tribandis D 2003 Vacuum **70** 319
19. Andagana HB, Cao XA 2010 Journal of Vacuum Science & Technology A **28** 189
20. S. W. Na, M. H. Shin, Y. M. Jung, J. G. Han, and N.-E. Lee, J. Vac. Sci. Technol. A, **23**, 898 (2005).
21. Berlin, S.Dvinin, L.Seidman , E.Vaccum technology and equipment for coating and etching of thin films M.: Technosphere,2007-172p.
22. Young-Hee Joo, Jong-Chang Woo, Kyung-Rok Choi, Han-Soo Kim, Jae-Hyung Wi, and Chang-Il Kim, TRANSACTIONS ON ELECTRICAL AND ELECTRONIC MATERIALS Vol. 13, No. 3, pp. 157-161, June 25, 2012
23. D. Y. Kim, J. H. Ko, M. S. Park, N. E. Lee, Thin Solid Films 516, 3512 (2008) [DOI:10.1016/j.tsf.2007.08.021].
24. W. R. Salaneck, N. Johansson, K. Z. Xing, F. Cacialli, R. H. Friend, G. Beamson, D. T. Clark, Synth. Met. 92, 207 (1998) [DOI:10.1016/S0379-6779(98)80088-X].

Red photoluminescence of living systems at the room temperature : measurements and results

I S Kudryashova[1], V Yu Rud[1,3], V Ch Shpunt[2], Yu V Rud'[2], A P Glinushkin

[1]Department of Quantum Electronics, Peter the Great Saint Petersburg Polytechnic University, Saint Petersburg 195251, Russia
[2]Department of Solid State Physics, Ioffe Physico-Technical Institute, Russian Academy of Sciences 26 Politekhnicheskaya, St Petersburg 194021, Russia
[3]Department of Ecology, Russian Research Institute for Phytopathology, 143050, Moscow Region, Odintsovo district, B.Vyazyomy, Russia

Abstract. Presents results of a study of the red luminescence of living plants at room temperature. The analysis of obtained results allows to conclude that the photoluminescence spectra for green leaves in all cases represent the two closely spaced bands.

1. Introduction

The rapid development of biological and agricultural sciences, due to the need to feed the growing world population dictates the importance of in-depth study of structure and functioning of living systems. Knowledge of these issues can improve the situation of plant breeding and the development of sustainable agriculture, improve the comfort of human life.

In science a long time it turned out that for the breakthrough in what is new fields tend to use well-established in other scientific questions methods and experimental techniques. Despite greater complexity, to date there are already enough successful attempts of the physical description of basic processes occurring in the interaction of light with the sheet and aimed at the creation of a generalized optical model sheet. In this case, is also widely known technique, when light of different spectral composition, or microwave signals treat seeds to increase the reproducibility of the plants. With all of this virtually unknown experiments in which green plant objects were investigated as would a conventional solid state semiconductor samples. The analysis results can be performed based on the established solid objects for views and to get new information about plant objects. That is why the use of well-developed in application to solid-state physics methods for the investigation of photoluminescence for the study of living green plants seems to us entirely innovative and promising demand.

Therefore, the use of a well-developed as applied to solid-state physics photoluminescence research methods [1] for the study of living green plants seems to us entirely innovative and promising demand [2,3,4,5].

2. The method

Photoluminescence studies were conducted on green leaves and flower elements of several types of plants. As a source for photoluminescence excitation used ILA 120-1 Argon laser with photon energies in the range of 2.41, 2.50, 2.54, 2.60 and 2.71 eV with power density is 50 - 100 mW / cm^2

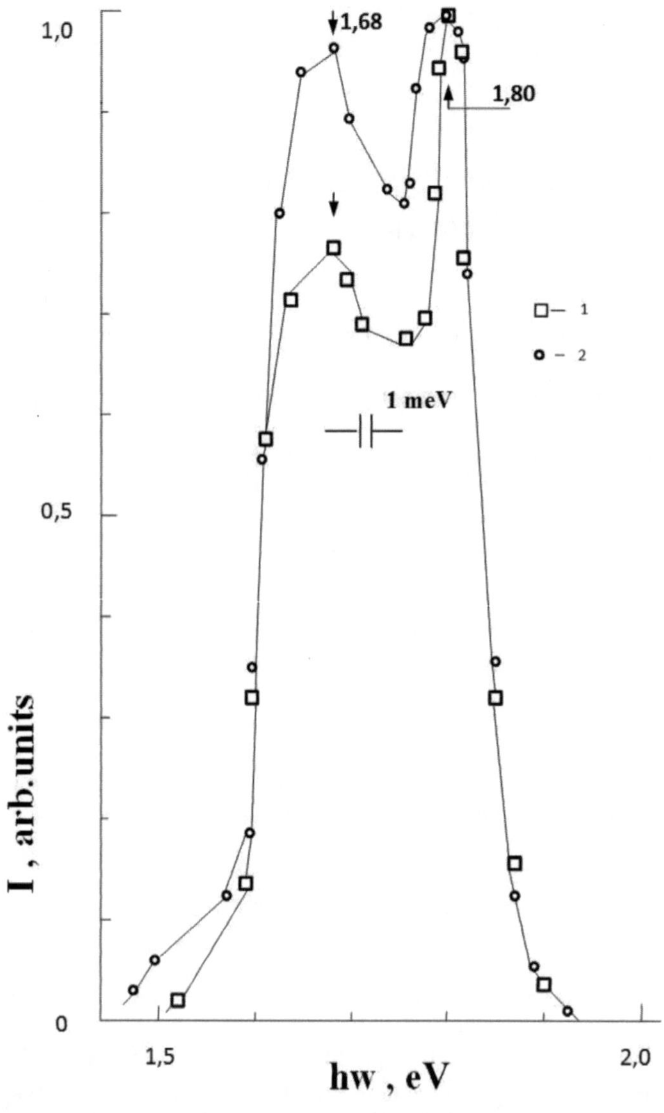

Figure 1. The spectral dependences of photoluminescence of leave Calendula officinalis L. at T=300K ($hw_{exc.}$, eV : for curves 1- 2.705, for curve 2- 2.410) each of the previously mentioned energy. Radiation photoluminescence taken monochromator MDR-3 with a grating 600 lines / and detected by a photomultiplier.The studies were conducted at T = 300 K, with a spectral resolution of the installation better than 1 meV.

3. Results and discussion

The results showed the presence of bright red photoluminescence for the green leaves when they are excited by radiation energies mentioned above.

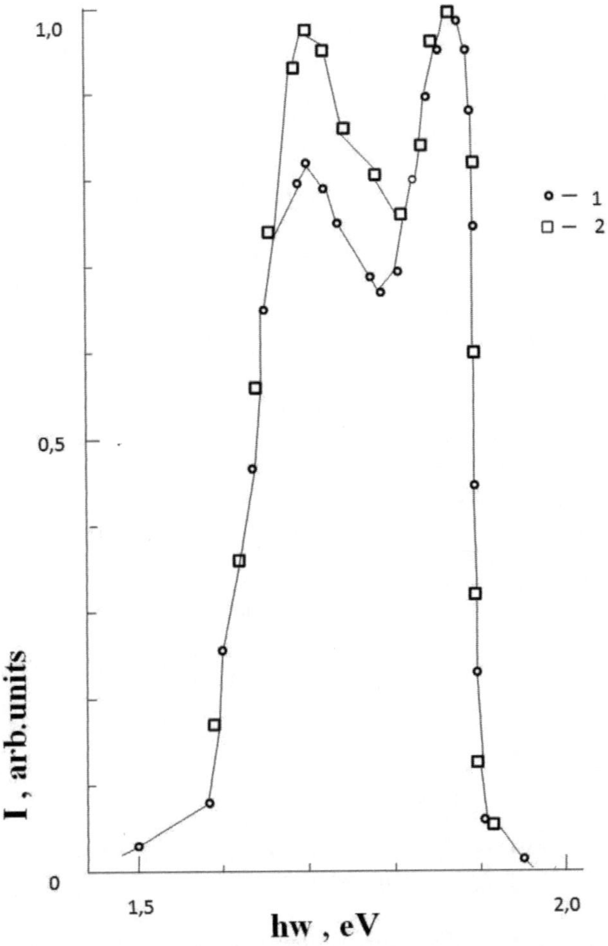

Figure 2. The spectral dependences of photoluminescence of leave Allium officinalis L. at T=300 K (hw_{exc}. , eV : for curve 1- 2.705, for curve 2- 2.410)

The photoluminescence red visually observed at all of these energies, photoluminescence excitation spectra for green leaves in all cases represent two closely spaced bands (Figures 1, 2, 3). In this regard, it is important to note that from fluorescence studies it is known that regardless of the wavelength of the incident light , the chlorophyll fluoresces in the red part [6] of the spectrum . This indicates that red luminescence can be associated with chlorophyll associated with the life cycle of the plant world. Thus both energy bands for different types of plants did not differ from each other. It is seen that the values of half-widths of the component ($_{lv}$ and $_{sv}$) and the ratio of their intensities I_2/I_1 is close in leaves of different plants. When you change the flux density of the exciting radiation the intensity of the photoluminescence field detection of peaks is described by a linear dependence of I~L . It is important to note that the important thing for photoluminescence analysis in physics, the ratio of the photoluminescence peak of the solid and half-width at half-maximum values were also close to all types of plants. This is an important result. When changing the exciting flux density observed linear law as a function of the photoluminescence intensity of the flux density. Thus the spectral contour of the luminescence and the energy position of peaks and ratio of their value remained unchanged (pic.3).

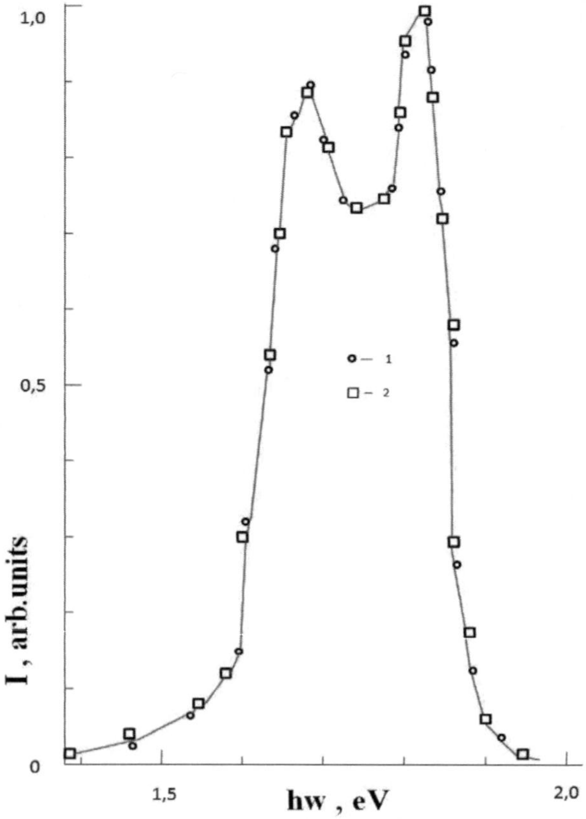

Figure 3. The spectral dependences of photoluminescence of leave Calendula officinalis L. at T=300 K (L/L_0: for curve 1 – 0.5, for curve 2 is 0.07 when $hw_{exc.}$ = 2.705 eV).

These patterns in semiconductor physics is typical for radiative transitions between free zones and for transitions from levels in the forbidden gap. At lower energies of excitation, however, increases the contribution of long wavelength components of the luminescence. The observed feature can be related to the fact that the centres can be localized in different positions in the tissue. That is responsible for the photoluminescence band with a peak at energies of hw_2 , the centres are localized more deeply in the layers of fabric, than those centres that are associated with hw_1. To assess the importance of quantum efficiency visible photoluminescence of live leaves gives a value of a few percent. Analysis of the survey results for the flower fragments shows that the efficiency drops by 2-3 orders of magnitude compared with a green plant with a maximum markedly shifted to shorter wavelengths. and reaches hw_1= 2.34 eV. It was found that the photoluminescence band is broadened in comparison with those obtained for the leaves, and as much extends into the long wavelength region of the spectrum. Based on these data, it is possible to draw a conclusion about the complex structure of the centers responsible for radiation. It should also be noted that the emission peak of the flower according to their colour.

4. Conclusion

The observed features of the photoluminescence may be of interest to biologists, physicians [2, 7], and agricultural workers.

Acknowledgement

The work was supported by the Russian Foundation for Basic Research under the project 15-29-05893.

References

[1] Rud' V Yu 2005 *Thesis for the degree of doctor of physico-mathematical sciences* Ulyanovsk UlSU 374
[2] Shpunt V Ch, Rud' Yu V, Tsendin K D 1993 *Journal Technical Physics* **19** 13 41 (in Russian)
[3] Samoilov V O 2004 *Medical Biophysics* SPb. 479
[4] Pryakhin Yu Yu, Kartabaeva B B, Sidorova, O P I, Glinushkin A P 2015 *Russian Agricultural Science Review* **6** 6-2 29
[5] Surov N V, Bikteev R Sh, Ragulin V S, Glinushkin A P, Duskaev G K 2015 *Biotics.* **3** 2. 15
[6] Dubessy J, Lhomme T, Boiron M C, Rull F 2002 *Applied Spectroscopy.* **56** 99
[7] Glinushkin A P, Rud' Yu V, Shpunt V Ch, Rud' V Yu, Kudryashova I S, Mozhaiko A A 2015 The use of proven solid state physical experimental techniques for the study of plants Biotics **6** 7 29

The efficiency of GaN/AlGaN p-n heterostructures in UV spectral range

S Yu Kurin[1], A S Usikov[2,3], B P Papchenko[2], H Helava[3], Yu N Makarov[1,3], A S Evseenkov[4], S A Tarasov[4] and A V Solomonov[4]

[1]Nitride Crystals –AlN Ltd., pr. Engel'sa 27, St. Petersburg 194156, Russia
[2]University ITMO, Kronverkskiy pr. 49, St. Petersburg 197101, Russia
[3]Nitride Crystals Inc., 181 E Industry Court, Suite B, Deer Park, NY 11729, USA
[4]Saint-Petersburg Electrotechnical University "LETI", 5 Prof. Popov Str., 197376 St. Petersburg, Russia

E-mail: sergey.kurin@nitride-crystals.com

Abstract. GaN/AlGaN p-n heterostructures emitting in UV spectral range obtained by HVPE approach were investigated. It was shown that the peak wavelength of UV LEDs was in the range of 360-380 nm with FWHM of 10-13 nm. At operating current of 20 mA, the active region temperature T_j was 43°C, the output optical power and efficiency - 1.14 mW and 1.46%, respectively. The model based on corpuscular Monte Carlo method for calculation of the light extraction index was presented. The simulation results allow us to propose the ways to increase the efficiency of UV LEDs: surface interfaces texturing, optimization of the design of heterostructures, and the use of lenses.

1. Introduction

In modern optoelectronic industry in order to create LED structures, the compositions based on solid solutions of semiconductor nitrides are used. Their most important characteristic is the conversion efficiency of electrical energy into optical radiation. The main problems of UV LEDs are low quality of epitaxial heterostructures [1, 2], low light extraction index [3], and self-heating effect. Addressing these problems will allow us to create high-quality GaN/AlGaN heterostructures for such promising areas as photoelectrochemical hydrogen generation, photopolymerization, medicine. Note also that GaN/AlGaN heterostructures can be used for other application like transistor modules and water electrolysis systems.

The main factor limiting the light extraction from the chip and reducing the efficiency of LEDs is total internal reflection [4, 5] at the interface of materials with high and low optical density. The effective way to solve this issue is introducing into the structure the surfaces that scatter light. When creating a specific shape roughness the light extraction index increases due to a significant increase in the critical angle. At present time, the texturing of the substrates and epitaxial layers is widely used [6]. The texturing can be created using a variety of methods: the technique of wet etching in combination with the Laser-lift-off technology [7], creation of microoptical elements on a sapphire substrate [8], micromachining [9] etc.

The experimental study of the optical and electrical properties of UV LEDs based on GaN/AlGaN [10-12] heterostructures has been carried out. Special attention was paid to investigation of the electroluminescence spectra at different forward currents and temperatures and determination of optical power and luminescence efficiency. In addition, in order to optimize parameters of the emitters a method based on corpuscular Monte Carlo method to simulate light extraction from the LED structures had developed. This approach allows us to not only predict the optical properties of the test structures and describe in detail the processes that lead to the loss of the output radiation, but also to make recommendations in order to improve the luminescence efficiency of LED heterostructures.

2. Experimental results

The main objective of the experiment was to investigate the structural perfection and features of the distribution of impurities in the GaN/AlGaN p-n heterostructures grown by HVPE [13], and determine their most important operating parameters, including output optical power, the temperature of the active region and efficiency [14, 15].

The design of GaN/AlGaN p-n heterostructures emitting in UV spectral range is shown in figure 1. Counting from the surface of the sapphire substrate (0001), the heterostructure includes the following epitaxial layers: nucleation layer, stress control layers, electron emitter, undoped active region, electron-blocking layer, p-contact layer. The total thickness of the above-mentioned layers is about 5 μm. Planar chips were fabricated with dimensions of 0.31×0.31 mm^2 and then packaged. The Ti/Al and Ni/Au metal compositions were used as n-and p-type electrodes, respectively.

Good quality of the structures was confirmed [16] by XRD method. The FWHM of the active layer peak on XRD rocking curve obtained in a symmetric reflection (0002) was less than 400 arcsec. The threading dislocation density measured by AFM varied from $8 \cdot 10^7$ to $9 \cdot 10^8$ cm^{-2}, which is typical for thin epitaxial nitride layers grown on sapphire substrates. The wavelength of PL maximum was in the range of 359.5−362 nm, the average PL FWHM was 11 nm. The position of the PL peak wavelength and FWHM of the main peak were almost identical for different areas on the wafer surface. The results showed that the use of rapid and low-cost HVPE method allowed us to achieve a high degree of structural perfection of epitaxial structures.

When studying the operating parameters of LEDs we focused on the investigation of spectral (figure 2), power-current (P-I) (figure 3) and I-V characteristics, the measurement of power and efficiency of the luminescence, as well as the study of the impact on them of the direct current and ambient temperature. The investigated structures demonstrated high enough output optical power. At 20 mA current it was 1.15 mW and reached a maximum of 4.2 mW at 120 mA. A deviation of the P-I characteristics from the linear dependence was observed at current values above 30 mA. The LEDs showed a working capacity in the continuous operation mode up to the values of the direct current of 140 mA.

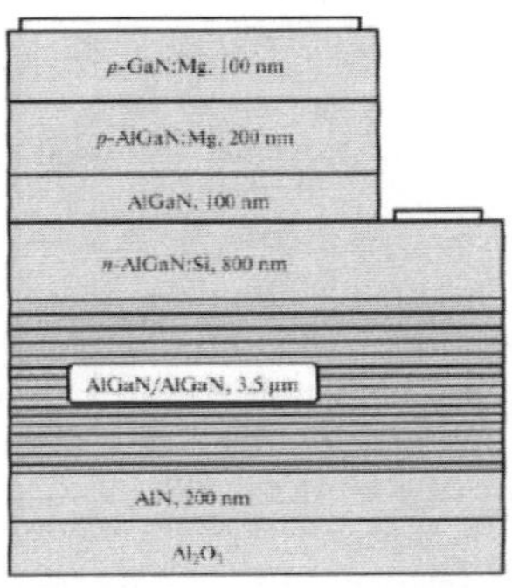

Figure 1. The design of GaN/AlGaN p-n heterostructure.

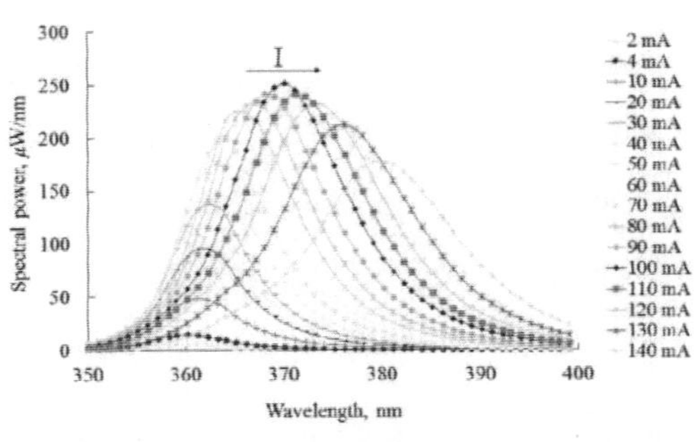

Figure 2. Electroluminescence spectra of UV LEDs at different values of direct current.

Depending on the wavelength of the maximum of the luminescence spectrum on the current value (figure 2) one can see that the shift of the spectral characteristics to long-wave region accelerates with an increase in current. This effect is due to a rapid increase in the temperature of the crystal. In general, the peak wavelength is seen to shift by 20 nm with increasing of the current from 2 mA to 140 mA (figure 4). Special attention was paid to the influence of direct current on the temperature of the active region of the structure (figure 5). The experimental data indicate that the p-n-junction temperature T_j at 20 mA is 43°C, which is higher than the values of blue emitters based on InGaN. At the current of 60 mA T_j exceeded 100°C and then the temperature continued to increase significantly. A comparison dependencies of wavelength and temperature on the current leads to a conclusion that self-heating is a one of the main process that reduces the output optical power of the structures. It was determined that the thermal resistance of the samples at I_f was equal to 250°C/W.

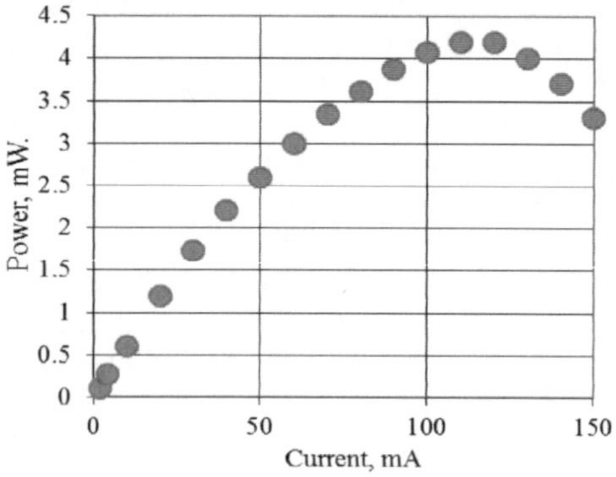

Figure 3. The power-current characteristics of UV LEDs.

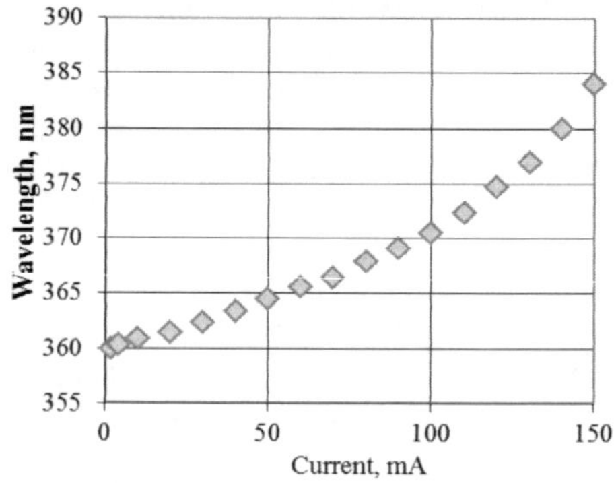

Figure 4. The dependence of the wavelength at the maximum of the spectral characteristics on the direct current.

Maximum wall-plug efficiency of packaged UV LED chips was equal to 1.5% at I_f (figure 6). At 100 mA the efficiency decreased by about 2 times.

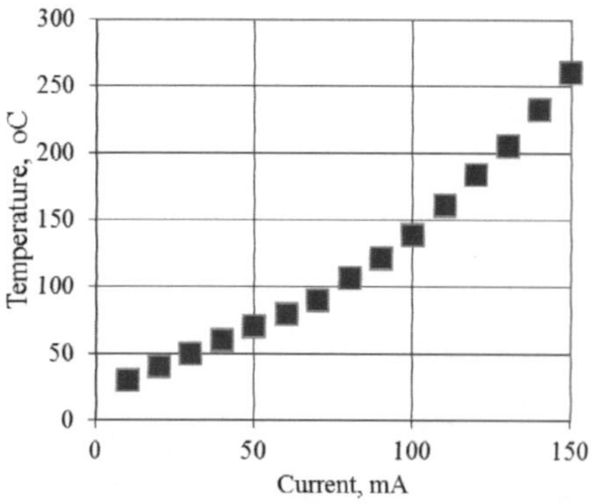

Figure 5. The dependence of the temperature of the UV LED active region on the direct current.

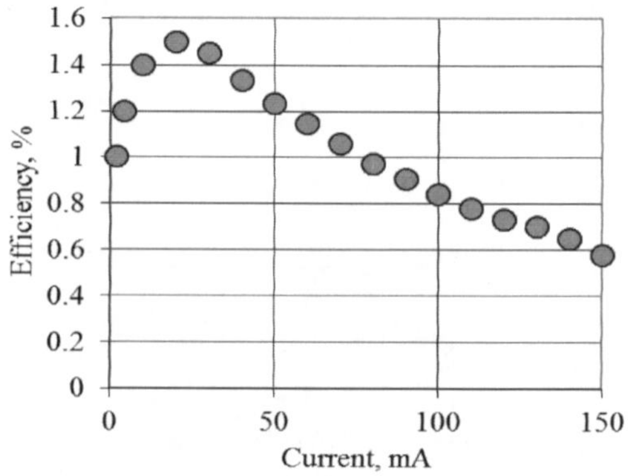

Figure 6. The dependence of the efficiency of the UV LED on the direct current active region on the direct current.

3. Numerical simulation by Monte Carlo method

In order to calculate the efficiency of transmission by Monte Carlo method [17] it is necessary to find the ratio of photons passing through the structure to their total number, i.e. one should take into account all possible variants of interaction of photons with the structure through randomly distributed parameters.

The random variables are the initial position of a photon generated in the active layer, its propagation vector (it sets the initial solid angle), as well as the factor that characterizes the probability of absorption of a photon. When photon passes through the structure its incidence angle on the surface of the phase boundary can be found. Depending on this angle either a rotation of the vector of a photon or its reflection from the interface occurs. As a result of consecutive reflections and refractions photon can be either absorbed by multilayer structure or passed through it. For a sufficiently large number of photons one can determine the transmittance of the structure that was actually done in the framework of the established model.

For simulation, the grid step was chosen to be 10 nm, the number of cross-sectional nodes was $8.25 \cdot 10^{14}$, the number of generated photons – 10^6 (per one point). The choice of these parameters is explained by a compromise between accuracy and computational speed, as well as finding the optimal interaction of the photon with the structure (with too small step a photon will "feel" too small irregularities, with too large step – not adequately describe the interaction with the roughness).

The light extraction index was calculated for structures with different thicknesses of the GaN layers and sapphire. The dependencies obtained indicate significant improvement of transmission of the structure with decreasing the thickness of semiconductor layer and small influence of varying the thickness of the sapphire substrate. The first effect is explained by a decrease in the average path length of photons, which accounts for the layers with the highest light absorption. As a result, a decrease in the thickness of the GaN layer from 5.5 to 1.5 μm causes an increase in transmittance more than two times (figure 7). In the case of pyramidal roughness aspect ratio is the ratio between the height of the pyramid and its base. The improvement is observed for all the aspect ratios of texturing, because absorption occurs for all photon trajectories. The decrease in the thickness of the sapphire has very little effect on the transmission of the structure (figure 8), since the absorption therein is very small. This indicates a decisive impact of the thickness of the semiconductor layer on the efficiency of the light extraction from LED.

Adding a lens allows us to achieve maximum improvement of transmission at low aspect ratios of texturing (figure 8), since in this case the influence of texturing is weak. In this case, one can double the transmittance. For higher aspect ratio the effect is minimized.

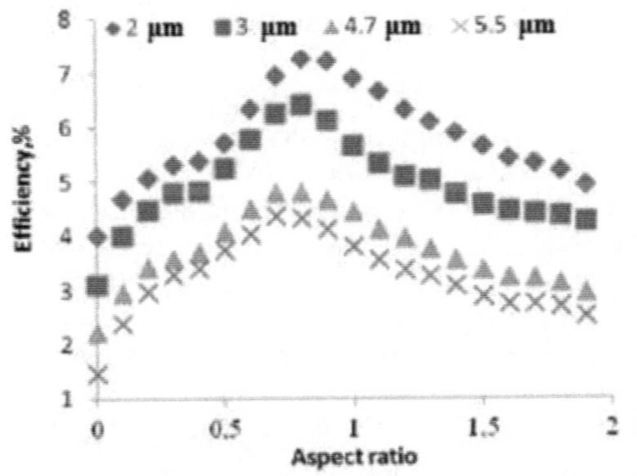

Figure 7. The dependence of transmittance on the aspect ratio for different thicknesses of the semiconductor layer.

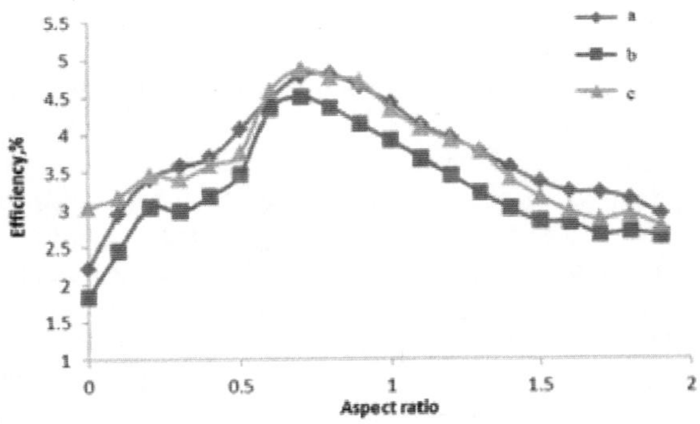

Figure 8. The dependence of transmittance on the aspect ratio for samples with various modifications: (a) thinned sapphire layer, (b) without modifications, (c) use of the lens.

Figure 9 shows the experimental dependence of the efficiency on the direct current in comparison with the simulation results for different aspect ratios of texturing relief – 0.5, 1 and 1.5. The semiconductor/sapphire interface has the greatest impact, since the radiation passing through the first interface through roughness comes in sapphire layer at small angles to the normal. Maximum efficiency is observed when the aspect ratio corresponds to the critical angle of total internal reflection for the GaN/sapphire interface – in the studied structure it was equal to one, i.e. the faces of the pyramid should be placed at the angle of 45° to the interface.

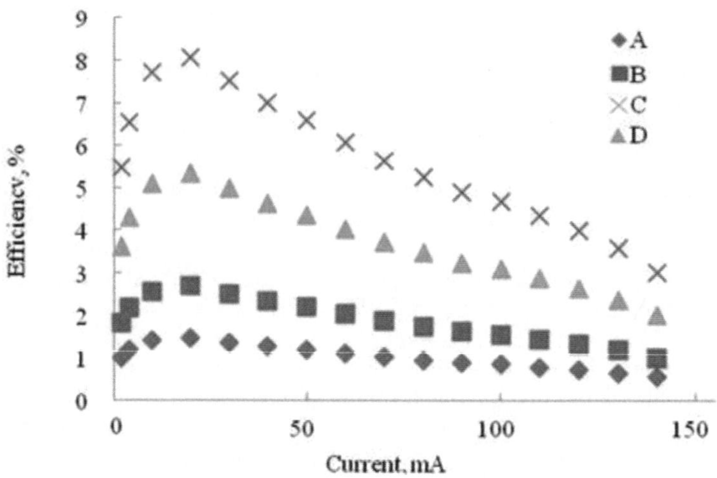

Figure 9. The dependence of the efficiency on the current. A - experimental data, B - theoretical dependence with aspect ratio equal to 0.5, C - 1, D - 1.5.

4. Conclusions

The GaN/AlGaN p-n heterostructures emitting in UV spectral range were grown by HVPE approach. The FWHM of the active layer peak on XRD rocking curve obtained in a symmetric reflection (0002) was less than 400 arcsec. The threading dislocation density measured by AFM varied from $8 \cdot 10^7$ to 9 10^8 cm^{-2}. PL measurements revealed a uniform distribution of peak wavelength and FWHM across the wafer. The results showed that the use of HVPE method allowed us to achieve a high degree of structural perfection of epitaxial structures.

The experiment showed that the peak wavelength of UV LEDs was in the range of 360-380 nm with FWHM of 10-13 nm. At operating current of 20 mA the active region temperature Tj is 43°C, the output optical power - 1.14 mW. At 100 mA the efficiency decreased by about 2 times.

It was developed a model based on Monte Carlo numerical method to calculate the light extraction index. The calculated data suggest the possibility of increasing the efficiency of appliances by 3-4 times due to texturing surface interfaces, optimization of the design of heterostructures, and the use of lenses. According to the simulation, the interface that mainly influences the light extraction index is the first interface – GaN/sapphire. This is due to the fact that the main absorption layer is GaN and the rays that have passed through the first layer, leave the roughness at right angles (i.e. almost all the photons reaching the second layer leave the structure). In order to increase the efficiency of light extraction the texturing parameters at the interfaces GaN/Al_2O_3 and Al_2O_3/air need to be different. It is shown that the aspect ratio for the texturing relief should tend to cotangents of critical angles of total internal reflection at the interface of respective environments.

Acknowledgments
Work at University ITMO was supported by the Ministry of Education and Science of Russian Federation (grant agreement 14.575.21.0054, unique identifier of research activities is RFMEFI57514X0054).

References
[1] JE.F. Schubert 2006 Light Emitting Diodes New York, Cambridge University Press
[2] M Kneissl et al 2011. *Semicond. Sci. Technol* **26** 01403
[3] F I Manyakhin et al, 2010 *Izv. Vyshh. Uchebn. Zaved* **50** 54
[4] J K Kim et al 2008 *Appl Phys Lett* **93** 221111
[5] J H Son et al 2012 *Advanced Materials* **24** 2259-2262
[6] C Rooman et al 2005 *Photonics Technology Letters* 2649-2651
[7] A Deinega et al 2011 *J.Opt.Soc. Am.* A **28** 5
[8] V Osincky et al 2003 *FIP* **1** 94.
[9] K D Lee et al 2010 *Semiconductor lighting* **5** 22.
[10] E A Menkovich, S A Tarasov, I A Lamkin 2012 *Functional Materials* **19** 233-237.
[11] I Lamkin, S Tarasov 2013 *Journal of Physics: Conference Series* **461** 012025
[12] E A Menkovich, A V Solomonov, S A Tarasov, P A Yurgin 2014 *Functional materials* **21** 186-189
[13] S Kurin, A Antipov, I Barash, A Roenkov, H Helava, S Tarasov, E Menkovich, I Lamkin, Yu Makarov 2013 *Physica status solidi (c)* **10** 289–293
[14] E A Menkovich, S A Tarasov, I A Lamkin, S Yu Kurin, A A Antipov, et al 2013 *Journal of Physics: Conference Series* **461** 012027
[15] A V Solomonov, S A Tarasov, E A Men'kovich, I A Lamkin, S Yu Kurin, A A Antipov et al 2014 *Semiconductors* **48** 245–250
[16] E A Menkovich, S A Tarasov, I A Lamkin, A V Solomonov, S Yu Kurin, A A Antipov, I S Barash, A D Roenkov, A S Usikov, H I Helava and Yu N Makarov 2014 *Journal of Physics: Conference Series* **541** 012054
[17] V Kandidatov et al 2006 Quantum Electronics **36** 1003-1008

AUTHOR INDEX

AUTHOR INDEX

AUTHOR INDEX

AUTHOR INDEX

AUTHOR INDEX

AUTHOR INDEX

AUTHOR INDEX

AUTHOR INDEX

Institute of Physics
Dirac House, Temple Back
Bristol BS1 6BE UK

ISSN: 1742-6588
ISBN 978-1-5108-2995-4

CURRAN ASSOCIATES INC.
proceedings.com

3rd International School and Conference on Optoelectronics, Photonics, Engineering and Nanostructures (Saint Petersburg OPEN 2016)

Journal of Physics: Conference Series Volume 741

St. Petersburg, Russia
28-30 March 2016

Part 2 of 2

IOP
Institute of Physics

3rd International School and Conference on Optoelectronics, Photonics, Engineering and Nanostructures (Saint Petersburg OPEN 2016)

Journal of Physics: Conference Series Volume 741

St. Petersburg, Russia
28-30 March 2016

Part 2 of 2

ISBN: 978-1-5108-2995-4
ISSN: 1742-6588

Printed from e-media with permission by:

Curran Associates, Inc.
57 Morehouse Lane
Red Hook, NY 12571

CURRAN ASSOCIATES INC. proceedings.com

Some format issues inherent in the e-media version may also appear in this print version.

Printed by Curran Associates, Inc. (2016)

For permission requests, please contact the Institute of Physics at the address below.

Institute of Physics
Dirac House, Temple Back
Bristol BS1 6BE UK

Phone: 44 1 17 929 7481
Fax: 44 1 17 920 0979

techtracking@iop.org

Additional copies of this publication are available from:

Curran Associates, Inc.
57 Morehouse Lane
Red Hook, NY 12571 USA
Phone: 845-758-0400
Fax: 845-758-2633
Email: curran@proceedings.com
Web: www.proceedings.com

TABLE OF CONTENTS

PART 1

NANOBIOTECHNOLOGY, BIOPHYSICS AND BIOPHOTONICS

LASERS, SOLAR CELLS AND OTHER OPTOELECTRONIC DEVICES

PART 2

NANOPHOTONICS, SPECTROSCOPY, MICROCAVITIES, OPTICS, PLASMONICS

ELECTRIC, MAGNETIC AND MICROWAVE DEVICES

OTHER ASPECTS OF NANOTECHNOLOGY

Specific of a photocurrent in GaN-based photoelectrochemical cell

O N Marchenko[1], I A Ermakov [1], M V Puzyk[1,2], D S Kovalev [1], S A Ivanova [1], B P Papchenko[1], A S Usikov [1,3] and A E Chernyakov[4]

[1]University ITMO, Kronverkskiy pr. 49, St. Petersburg 197101, Russia
[2]Herzen University, Nab. r. Moyki 48, St. Petersburg 194186, Russia
[3]Nitride Crystals Inc., 181 E Industry Court, Suite B, Deer Park, NY 11729, USA,
[4]Submicron Heterostructures for Microelectronics Research & Engineering Center, RAS, 26 Polytekhnicheskaya Str., St. Petersburg 194021, Russia

Abstract. An influence of various parameters of a photoelectrochemical cell (PECC) having a GaN working electrode on the photocurrent was studied. Type of the aqua electrolyte (alkaline (KOH)-, neutral salt (Na_2SO_4)- and acid (H_2SO_4)- based electrolytes) influences on transient time for the photocurrent stabilization. A transient time for the photo current stabilization was observed under illumination by the UV LED light sources. The shortest transient time and the highest photocurrent were observed in the alkaline-based electrolyte (~0.5M KOH) with n-GaN working electrodes (N_D-N_A =(3-5)×10^{16} cm^{-3}). PECC with electrolytes based on sodium sulfate and sulfuric acid demonstrated longer transient time (up to ten minutes) for the photocurrent stabilization and smaller photocurrent.

1. Introduction

Since the end of the 20th century, investigations to make a photoelectrochemical cell (PECC) based on wide band-gap semiconductor materials for hydrogen production under the solar illumination is still prospective and challenge topic for researchers. The reason is well known: hydrogen is considered to be a candidate as the substantial energy carrier and the Sun is an inexhaustible source of energy that provides to the Earth's surface more than 1000- 800 W/m^2 in the summer time and 500-300 W/m^2 in the winter time and water resources are enormous. GaN-based semiconductors are promising materials for not only light-emitting devices, but also for splitting water into H_2 and O_2 photoelectrochemically being immersed into an electrolyte aqua solution [1] although GaN itself can effectively absorb only the UV portion (about 5%) of the solar spectrum.

At the direct water splitting process, the GaN-based material as a photoelectrode in an aqueous solution of electrolyte may spontaneously decompose water into H_2 and O_2 gases by sunlight absorption. The sunlight absorption creates the electron-hole pairs (carriers) that being separated can participate in the gases evolution reaction at the semiconductor/electrolyte interface. However, the carrier recombination and the corrosion (or etching) of the photelectrode material substantially reduce the electron-hole pairs population and the gas production. The corrosion is an issue for n-type materials because upward band-bending at the photoelectrode/electrolyte interface promotes to separate carriers and to transfer photo-generated holes to the electrolyte. The holes not only oxidize water producing oxygen (hydrogen is produced on the counter electrode of PECC in this case) but can oxidize and etch the photoelectrode. This process is usually accomplished by dropping the photocurrent with time indicative non stability of the electrode. In this work we studied an influence of various parameters of PECC having a GaN layer as a working electrode on the photocurrent.

2. Experimental details

Several 5-8 μm thick GaN layers were grown by hydride vapour phase epitaxy (HVPE) on c-plane 2-inch sapphire substrates. The growth procedure included in-situ sapphire substrate treatment and an AlGaN buffer layer deposition followed by GaN layers growth. Details of the structure growth and characterization can be found elsewhere [2]. For n-GaN layers, the N_D-N_A concentration measured by C-V characterization was from of 3×10^{16} cm^{-3} to $4\times10^{18}cm^{-3}$. The N_A-N_D concentration for p-GAN layers was near $(1\text{-}2)\times10^{18}cm^{-3}$. Figure 1a shows a view of a cross section of one of the layer taken from scanning electron microscope (SEM).

To make electrodes for PECC, the intra-volume laser scribing was applied to cut 15 mm × 15 mm chips from 2-inch GaN/sapphire wafers. This scribing technique is associated with a plasma breakdown in sapphire substrate under irradiation by a short pulse of high power density laser beam [3]. The procedure did not pollute the chip surface. After that, a standard photolithography procedure was applied to deposit a 2-mm wide Ni/Au contacts on the chip edges. Figure 1b shows a view of the electrode made. The irradiated area was adjusted to 1 cm^2. The contact was protected by fluorine-based varnish to prevent its etching in the electrolyte.

A digital multi-meter and an UV-radiometer were used to measure the photocurrent and the luminous flux (the optical power of the light) of the incident light from the light sources on the GaN-electrode surface. Typically, PECC has GaN working electrode and Pt-plate as counter electrode. The parameters under consideration were (1) electrolytes based on KOH, Na_2SO_4, and H_2SO_4 water solutions with a concentration of 0.25-0.75M, (2) n-type and p-type GAN layers, and (3) three types of light sources used for samples illumination: a solar simulator (Sun Simulator Newport), a 345-nm light emitting diode (LED), and a 365-nm LED, hereinafter SSN, LED-345, and LED-365, respectively. The initial optical power of SSN, LED-345, and LED-355 was 2.5 W/m^2, 40 W/m^2, and 80 W/m^2, respectively.

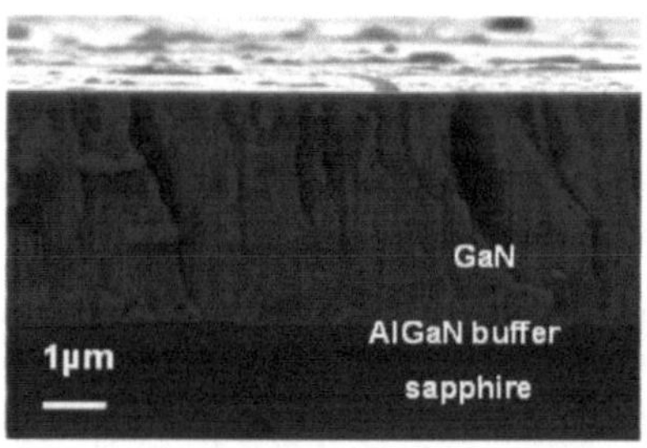

(a)

(b)

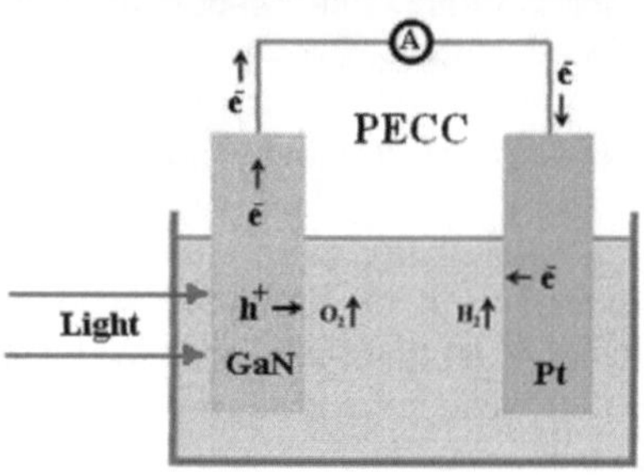

(c)

Figure 1. Schematic of the photoelectrochemical cell (PECC) with GaN photoelectrode.
(a) SEM cross sectional view of GaN layer on sapphire;
(b) Top view of 15 mm×15 mm GaN photoelectrode. Irradiated area is ~1 cm^2. Contact bars on the edge is covered by fluorine-based varnish;
(c) schematic of photoelectrochemical cell.

Journal of Physics: Conference Series **741** (2016) 012108 doi:10.1088/1742-6596/741/1/012108

3. Results and discussion

Figure 2 shows results of photocurrent measurements as a function of time using three light sources and KOH aqua electrolyte with concentration of 0.75M. The experiments were performed with the same sample. A position of the light sources was adjusted to have the same power of the incident light on the electrode surface of 2.5 W/m^2. Any serious corrosion of the sample surface was not observed because of the relatively low photocurrent. A transient time for the photo current stabilization was observed under illumination by the LED-based light sources. The most stable over time (although smaller in magnitude) photocurrent was measured under illumination by SSN. One of the reasons for that is the SSN has smaller intensity of the UV portion in the total spectrum than UV LED sources.

A specific band-bending at the n-GaN electrode (anode)/electrolyte interface forms the extended space charge region in the semiconductor, which separates the photo carriers and creates photocurrent. Illumination by LED-365 gave the highest photocurrent. LED-365 has relatively narrow peak emission that is pretty close to the GaN energy gap and effectively absorbs mostly on a distance from the surface lowering the carriers' recombination before they can be split by the space charge region having an associated electric field. However estimated efficiency of the process is near 0,5% or lower.

Experiments with a neutral salt (Na_2SO_4) and acid (H_2SO_4) aqua electrolytes in PECC demonstrated longer transient time (up to ten minutes) for the photocurrent stabilization and smaller photocurrent. Obviously the n-GaN electrode (anode)/electrolyte space charge region (potential barrier) depends on the electrolyte type. Narrower space charge region (lower potential barrier) results in lower photocurrent (lesser electron-hole pairs can be separated). It looks that among the neutral salt and the acid aqua electrolytes alkaline (KOH-based) electrolyte has higher barrier at the n-GaN/electrilyte interface. Further experiments were performed using KOH-based aqua electrolyte and SSN light source because of stable photocurrent.

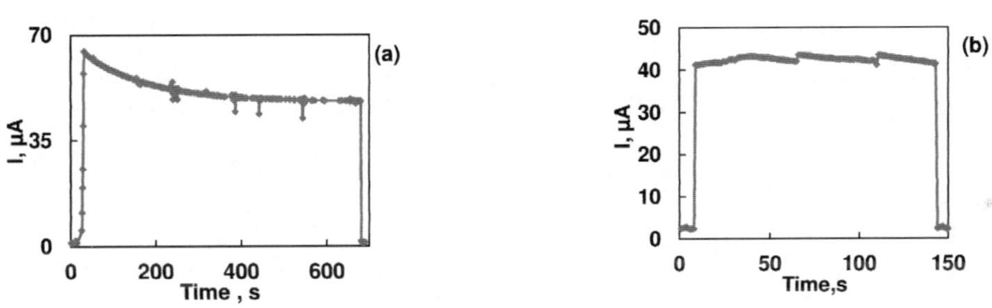

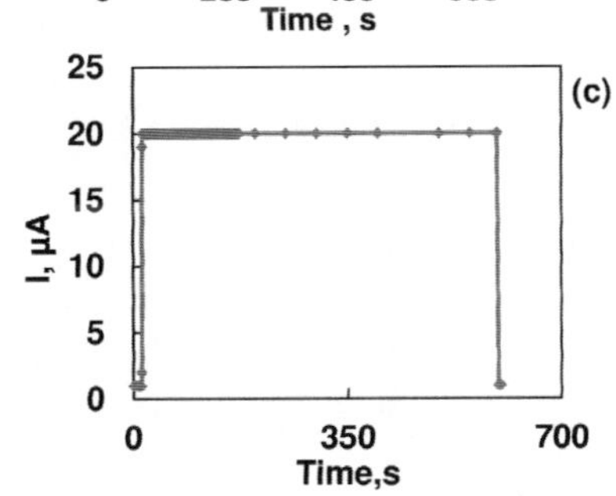

Figure 2. The photocurrent as a function of time in 0.75 mol/l KOH (n-GaN, Nd-Na~ 5×10^{16}cm^{-3}) under illumination from different light sources.
(a) 365-nm LED;
(b) 345-nm LED;
(c) SSN

Figure 3 summarizes some results of the photocurrent measurements under SSN illumination in relation with the KOH electrolyte concentration and the GaN layer doping. The max photocurrent was measured in n-GaN (N_D-N_A = (3-5)$\times10^{16}$cm^{-3}) immersed in 0.5M KOH. KOH concentration has minor effect on the photocurrent.

Note, n-GaN layer under illumination became the anode, i.e. holes generated at the surface go to the electrolyte and excited electrons go through the external surface to the Pt counter electrode. At the same conditions, p-GaN layers became the cathode with much lower photocurrent.

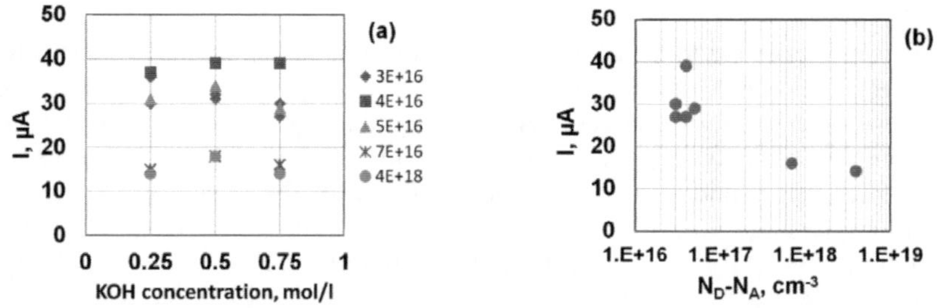

Figure 3 The influence of the electrolyte concentration and the GaN electrode doping on the photocurrent. **(a)** KOH aqua solution. Legend shows N_D-N_A concentration in GaN layers used; **(b)** N_D-N_A concentration in n-GaN layers.

4. Conclusions

Type of the aqua electrolyte (alkaline (KOH)-, neutral salt (Na_2SO_4)- and acid (H_2SO4)- based electrolytes) in PECC influences on transient time for the photocurrent stabilization. The shortest transient time and the highest photocurrent were observed in the alkaline-based electrolyte.

The max photocurrent was measured in a low-doping n-GaN layers (N_D-N_A =(3-5)×10^{16}cm-3) in 0.5M KOH. KOH concentration has minor effect on the photocurrent. n-GaN layer under illumination became the anode, i.e. holes generated at the surface go to the electrolyte and excited electrons go through the external surface to the Pt counter electrode. At the same conditions, p-GaN layers became the cathode with much lower photocurrent.

Acknowledgments

Work at University ITMO was supported by the Ministry of Education and Science of Russian Federation within the grant agreement 14.575.21.0054 (unique identifier of research activities is RFMEFI57514X0054).

References

[1] Ohkawa K, Ohara W, Uchida D, Deura M 2013 *Jpn. J. Appl. Phys.* **52** 08JH04.
[2] Kurin S, Antipov A, Barash I, Roenkov A, Usikov A, Helava H, Ratnikov V, Shmidt N, Sakharov A, Tarasov S, Menkovich E, Lamkin I, Papchenko B and Makarov Yu 2014 *Phys. Status Solidi* C**11** 813-816.
[3] Medvedev O S, Khait O, Kurin S Yu, Usikov A S, Papchenko B P, Helava H and Makarov Yu N 2015 J. of Phys.: Conf. Ser. **643** 012129.

1.5 µm InAs/InGaAsP/InP quantum dot laser with improved temperature stability

F I Zubov[1,2], S P Gladii[2], Yu M Shernyakov[3,1], M V Maximov[1,2,3], E S Semenova[4], I V Kulkova[5,6,*], K Yvind[4], A E Zhukov[1,2]
[1]St Petersburg Academic University, St Petersburg 194021, Russia
[2]Peter the Great St.Petersburg Polytechnic University, St Petersburg 195251, Russia
[3]Ioffe Institute, St Petersburg 194021, Russia
[4]DTU Fotonik, Technical University of Denmark, DK-2800 Kgs. Lyngby, Denmark
[5]IMEC, Kapeldreef 75, BE-3001 Leuven, Belgium
[6]KU Leuven, Department of Physics and Astronomy, Celestijnenlaan 200d, BE-3001 Leuven, Belgium

Abstract. Temperature characteristics of InAs/InGaAsP quantum dot (QD) lasers synthesized on InP (001) substrate are presented. The lasers demonstrate high temperature stability: a threshold current characteristic temperature as high as 205 K in the temperature range between 20 to 50°C was measured. Lasing wavelength of 1.5 µm was achieved by covering QDs with 1.7 monolayers of GaAs.

1. Introduction

Quantum dot (QD) lasers are of great interest owing to 3D confinement leading to a number of device improvements [1]. 1.3 µm InAs/GaAs QD lasers are already close to their maturity. For example, temperature insensitive threshold current at room temperature (RT) has been demonstrated for such lasers [2]. A significant interest exists in studying devices with 1.5 µm lasing emission, which is the key wavelength for telecom applications. However, this is beyond the ability of GaAs-based laser diodes with pseudomorphic QDs. At the same time, InAs QDs formed on an InP substrate are typically characterized by much longer wavelength emission (see e.g. [3]) because of twice lower lattice mismatch in this material system compared to the InAs/GaAs combination. Recent advances in synthesis of InAs QDs on InP substrates allowed to reach 1.5 µm laser generation. However, so far these lasers have poor temperature stability. To the best of our knowledge, the highest characteristic temperature of threshold current was reported to be 135 K in the 200-300K temperature interval [4]. The characteristic temperature estimated above room temperature is 106K [5] or even lower.

In the present work, we report on 1.5 µm InAs/InGaAsP/InP QD lasers that have improved temperature stability of threshold current (and external differential efficiency) compared to previously reported results.

2. Experiment details and results

A laser heterostructure was grown by metal organic vapour-phase epitaxy (MOVPE) on an InP (001) substrate. An active region comprised 5 layers of 1.65 monolayer (ML) InAs QDs, which surface density was estimated as $4.2\cdot10^{10}$ cm^{-2} per layer [6]. Each QD layer was covered with 1.7 ML GaAs

[*] on leave from DTU Fotonik, Technical University of Denmark, DK-2800 Kgs. Lyngby, Denmark

and separated from each other by 60 nm $In_{0.83}Ga_{0.17}As_{0.38}P_{0.62}$ (Q1.08) spacers. The active region was positioned in the centre of a 450 nm thick Q1.08 waveguide, confined by *n*- and *p*-type InP claddings.

Using standard post-growth techniques, the structure was processed into 2 μm wide ridge waveguides of 1-4 mm lengths. The devices were soldered *p*-side up onto heatsinks. No facet coatings were deposited. Measurements were performed in a pulsed pumping regime (5 μs and 150 Hz) in 20-50°C temperature range.

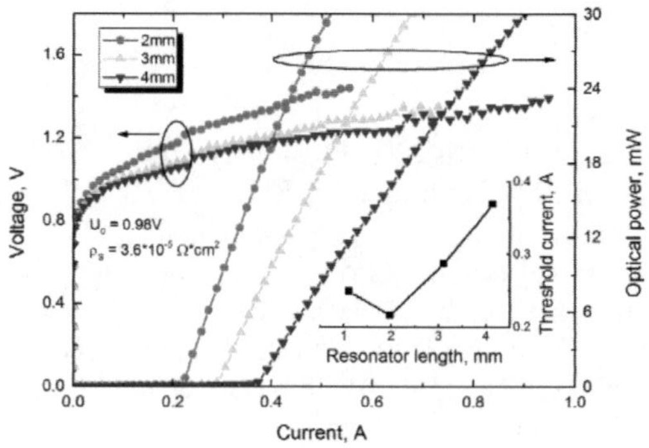

Figure 1. Voltage and optical power versus bias current at 20°C for 2, 3 and 4-mm-long QD lasers. Output power corresponds to both facets. Inset: threshold current *vs.* resonator length.

In Figure 1, current- voltage and light-current characteristics of 2, 3 and 4 mm-long samples measured at 20°C are presented. Their turn-on voltage and specific series resistance were estimated as 0.98 V and $3.6 \cdot 10^{-5}$ $\Omega \cdot cm^2$, respectively. The threshold current (I_{th}) versus resonator length dependence is presented in inset of Figure 1. The smallest I_{th} of 215 mA was found in 2-mm-long device. The lasers with resonator length in the 2-4 mm interval demonstrate lasing wavelength close to 1.5 μm, which corresponds to the QD ground state transition. At the same time, the lasing line is blue-shifted to 1.45 μm in 1 mm-long devices. This is accompanied by a strong rise in threshold current density from 5.4 kA/cm^2 (in 2-mm device) to 11.2 kA/cm^2 in 1 mm device.

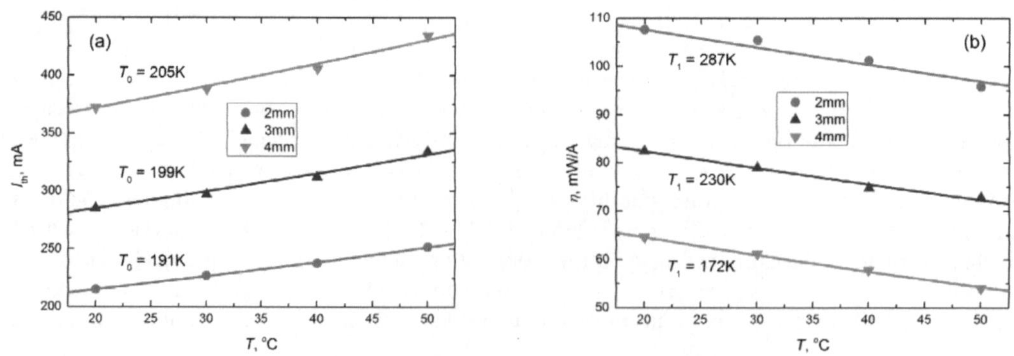

Figure 2. Temperature dependencies of threshold current I_{th} (a) and external differential efficiency η (b) of 2, 3 and 4 mm long lasers. Solid lines are fittings to exponential function (see description in the text). External differential efficiency corresponds to both facets.

Figure 2 shows temperature dependencies of the threshold current I_{th} (a) and the external differential efficiency η (b) of the lasers under study. Characteristic temperatures T_0 of the threshold current and T_1

of the external differential efficiency were extracted from fitting of the experimental data to exponent functions $I_{th} = I_0\exp(T/T_0)$ and $\eta = \eta_0\exp(-T/T_1)$, correspondingly. High values of T_0 and T_1 were reached for this type of lasers. For example, 4 mm long device demonstrates T_0 of 205 K and T_1 of 172 K in the 20-50°C temperature interval. We attribute this improvement of the temperature characteristics to reduced aspect ratio of grown QDs [6] as well as high carrier localization energy, which is achieved by using InGaAsP (Q1.08) barriers.

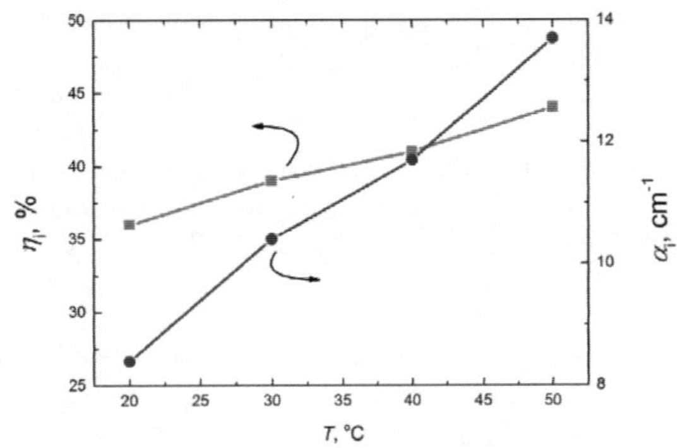

Figure 3. Temperature dependencies of internal differential quantum efficiency η_i and internal losses α_i.

Temperature dependence of internal differential quantum efficiency η_i and internal losse α_i was also extracted (see Figure 3). A standard procedure was used to determine η_i and α_i from inverse external efficiency versus cavity length dependence. The samples are characterized by a moderate value of internal losses (about 8 cm^{-1}) at room temperature. However, the α_i rapidly grows with temperature by one quarter per 10 centigrade. The value of η_i is rather low at room temperature (~ 36%). This can probably be attributed to a large inhomogeneous broadening of QDs, i.e. only part of QDs contributes to lasing. As temperature rises, η_i slightly increases to 44% at 50°C.

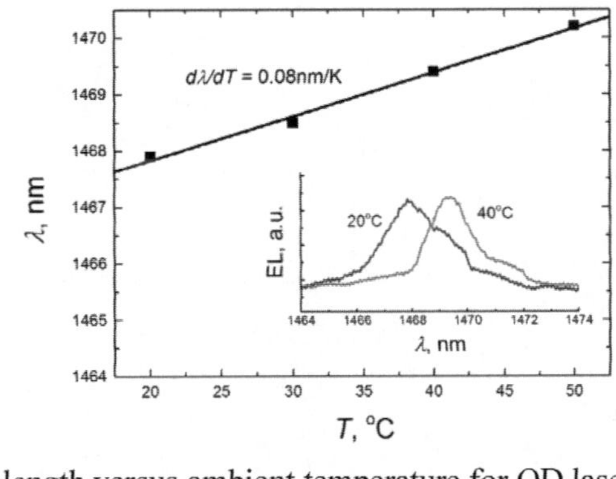

Figure 4. Emission wavelength versus ambient temperature for QD laser having length of 1 mm. Solid line represents a linear fit to the data. Inset shows emission spectra slightly above the threshold at 20 and 40°C.

For 1-mm-long device a low temperature-induced shift of lasing line was observed, see Figure 4. The temperature coefficient was estimated to be 0.08 nm/K, whereas it typical values in InP-based QD lasers is about 0.4 nm/K (e.g. see [7]). This can be explained by proximity of the ground state optical gain to its saturated value in this device that leads to a strong counteraction between a blue shift of lasing line caused by thermally activated escape of carriers and its red shift due to band gap shrinkage with temperature.

In conclusion, we have studied 1.5 μm InAs/InGaAsP/InP QD lasers grown by MOVPE. Owing to reduced QDs aspect ratio, as a result of capping QDs with thin GaAs layer, and considerably high localization energy, we have achieved high temperature stability of threshold current and differential efficiency in pulsed regime: in the 20-50°C interval, the 4 mm-long device is characterized by T_0 of 205 K (to the best of our knowledge, this is the highest characteristic temperature ever reported for 1.5 μm QD laser grown on InP) and T_1 of 172 K.

Acknowledgments
The work was supported by the Russian Science Foundation (project 14-42-00006 «A novel type of diode lasers with characteristics improved by using asymmetric barriers»). ESS, IVK and KY wish to thank the the Villum Foundation for the support via YIP QUEENs.

References

[1] Zhukov A E 2011 *Quantum dot diode lasers, In: "Landolt-Börnstein: Numerical Data and Functional Relationships in Science and Technology - New Series"* (Berlin: Springer-Verlag) p 95
[2] Fathpour S, Mi Z, Bhattacharya P, Kovsh A R, Mikhrin S S, Krestnikov I L, Kozhukhov A V, Ledentsov N N 2004 *Appl. Phys. Lett* **85** 5164
[3] Zaitsev S V, Gordeev N Yu, Kopchatov V I, Ustinov V M, Zhukov A E, Egorov A Yu, Kovsh A R, Kop'ev P S 1999 *Jpn. J. Appl. Phys* **38** 601
[4] Kim J S, Lee J H, Hong S U, Han W S, Kwack H-S, Lee C W, Oh D K 2004 *IEEE Photon. Technol. Lett* **16** 1607
[5] Lelarge F, Dagens B, Renaudier J, Brenot J, Accard A, Dijk F, Make D, Gouezigou O L, Provost J-G, Poingt F, Landreau J, Drisse O, Derouin E, Rousseau B, Pommereau F, Duan G-H 2007 *IEEE J. Select. Topics Quantum Electron* **13** 111
[6] Semenova E S, Kulkova I V, Kadkhodazadeh S, Schubert M, Yvind K 2011 *Appl. Phys. Lett* **99** 101106
[7] Gilfert C, Ivanov V, Oehl N, Yacob M, Reithmaier J P 2011 *Appl. Phys. Lett* **98** 201102

Compact microdisk cavity laser with GaInNAs/GaAs quantum well

N V Kryzhanovskaya[1,3], A E Zhukov[1,3], M V Maximov[1,2], E I Moiseev[1], Yu S Polubavkina[1], M M Kulagina[2], S I Troshokov[2], Yu A Guseva[2], A A Lipovskii[1,3], R Isoaho [4], T Niemi[4], M Guina[4]

[1] St Petersburg Academic University of RAS, Khlopina 8(3), St Petersburg 194021, Russia
[2] Ioffe Institute, St Petersburg, Russia
[3] Peter the Great St Petersburg Polytechnic University, St Petersburg, Russia
[4] Optoelectronics Research Centre, Tampere University of Technology, Tampere, Finland

Abstract. Optically pumped 3-6μm in diameters microdisk lasers with InGaAsN/GaAs quantum well active region has been studied. Single-mode CW lasing at 78K temperature in microdisk laser with 3 μm diameter is demonstrated.

Introduction

Since the first demonstration of the semiconductor microdisk laser in the early 1990s [1] a large number of papers devoted to this type of light sources, reporting various active regions, have been published. Most of the results were demonstrated on (GaInAsP)/InP microdisk/microring lasers including successful transfer to the Si substrate using wafer bonding technology [2]. The main advantage of the InP-based material system is its low surface recombination rate which is highly desirable in case of compact microdisk cavity with sidewalls etched through the active region. Alternatively, quantum dots have also emerged as capable of providing suppression of the surface recombination at the disk sidewall even for AlGaInAs-based laser materials, owing to suppression of the carrier diffusion. Moreover, microlasers with InAs/InGaAs quantum dots have demonstrated high temperature stability of the operation parameters with the maximal operation temperature of 107°C under optical pump [3] and 100°C in injection structures [4]. However, quantum dot-based lasers suffer from the saturation of the ground state gain due to the finite numbers of the quantum dots. To overcome the problem of the gain saturation and still take advantages of GaAs-based materials (such as larger band offsets, high refractive index contrast and better thermal conductivity) we propose in this work to use InGaAsN/GaAs quantum well as an active region in microdisk lasers. High-performance InGaAsN quantum well edge-emitting lasers have been realized by both metalorganic chemical vapor deposition and molecular beam epitaxy [5-9] enabling lasing in a broad spectral range from 1.1 μm to 1.6 μm.

In this paper we demonstrate compact microdisk laser with novel active region based on an InGaAsN/GaAs quantum well. Lasing characteristics of the microdisk lasers with 3-6μm diameters are studied at different temperatures (78-300 K). We present single-mode CW lasing at 78K temperature in such microdisk laser with 3 μm diameter.

Experiment

The epitaxial structure was grown by molecular-beam epitaxy on *n*+-doped GaAs(100) substrate. Three layers of InGaAsN quantum wells (QWs) separated with 10-nm-thick GaAs layers were grown as an active region. This QW material has a ground-state emission peak around 1.2 μm at room temperature. The active region was inserted within a 0.2-μm thick GaAs waveguide layer confined by 10-nm-thick $Al_{0.3}Ga_{0.7}As$ barriers. A 1-μm-thick $Al_{0.8}Ga_{0.2}As$ cladding layer was grown beneath the waveguide layer serving as a support pedestal for the microdisk. Microdisks were defined by photolithography and subsequent circular mesa etching in the etching solution, based on a K2Cr2O7/HBr composition through the waveguide layer and the $Al_{0.8}Ga_{0.2}As$ cladding layer. The outer diameter was varied in different structures from D =3 μm to 6 μm. Next, anisotropic HF etching was used to etch $Al_{0.8}Ga_{0.2}As$ layer and to form a desired diameter of the pedestal underneath the GaAs disk.

For optical investigations, a CW-operating YAG:Nd laser ($\lambda = 532$nm) was used. For low temperature measurements the samples were mounted in a flow cryostat. A piezoelectrically adjustable Olympus LMPlan IR objective x10 used to focus the incident laser beam to pump the microresonator homogeneously and to collect micro-photoluminescence (μPL) signal from a microresonator. The collected μPL was then dispersed via a 1000-mm monochromator Horiba FHR and measured by a cooled InGaAs single-channel detector. The overall spectral resolution was 0.1 nm for 900 l/mm grating. Scanning electron micrograph (SEM) of the 3 μm diameter microdisk laser is shown in figure 1.

Figure 1. SEM images of the microdisk laser with GaInNAs/GaAs quantum wells active region.

Results

Optical properties of the QW media were first investigated in an unprocessed (as-grown) sample in the temperature range 78-300 K under low excitation density (200 mW/cm^2). We observe insignificant decrease (by ~8 times) of QW photoluminescence intensity as temperature increases from 78 to 300 K evidencing high optical quality of the InGaAsN/GaAs QWs (figure 2, a). At low temperature (78K) full-width at half maximum (FWHM) of the spectrum is ~ 18 meV revealing reasonable homogeneity of the QW thickness and composition. Temperature dependence of FWHM demonstrates a monotonic increase to 30meV at 300 K.

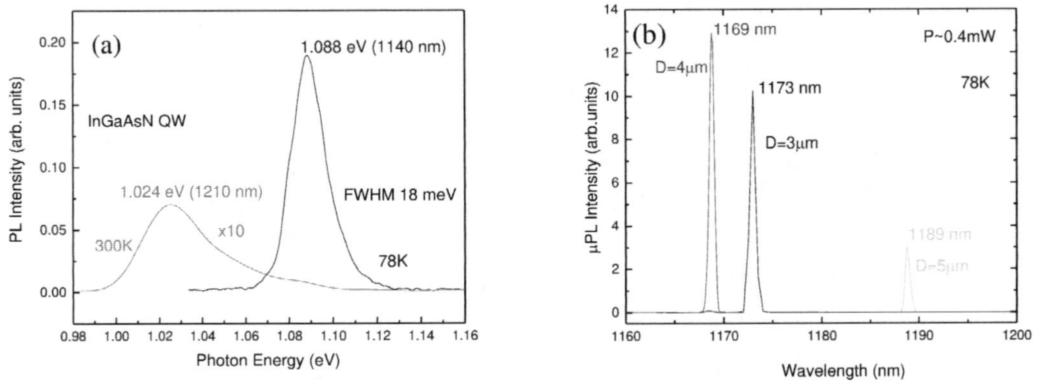

Figure 2 . Photoluminescence spectra of the unprocessed GaInNAs/GaAs QWs obtained at 78K and 300K (a) and microphotoluminescence spectra of the microdisks based on GaInNAs/GaAs QWs obtained at 78K with 3, 4 and 5 μm cavity diameters (b).

Emission spectra taken at 78 K for the microdisks of different diameter (3…6 μm) are shown in figure 2, b. A single narrow line corresponding to the one of the whispering gallery mode of the resonator is observed on the long-wavelength side of the QW spectra for all the structures. The output intensity-input pump curves (L-L curves) for these lines are shown in figure 3. Each curve has a pronounced kink indicating the lasing threshold of the microdisk (marked by arrows). The smallest threshold pump power at 78K is 0.17 mW in a microlaser with diameter of 3 μm. Threshold pump power is scaled as D^{-2} *vs* diameter of the microlasers. This indicates that the threshold power density remains nearly constant and, hence, no significant growth of optical loss and/or non-radiative recombination takes place with decreasing the disk diameter (see inset in figure 3). The emission is TE-polarized and has narrow far-field diagram in vertical direction.

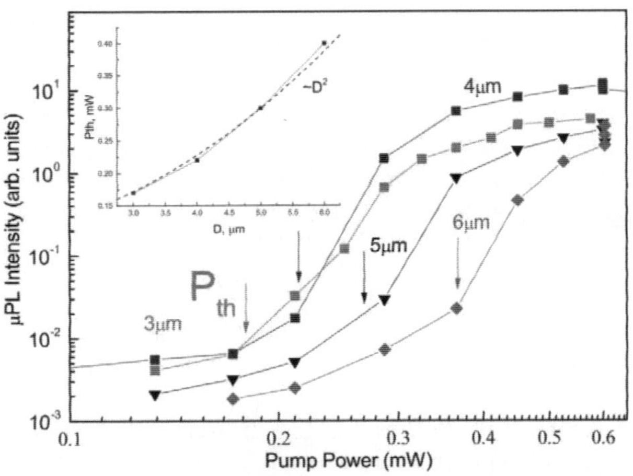

Figure 3. Plot of the WGM mode intensity against excitation power for the microdisks of 3-6 μm cavity diameters. Inset: the threshold pump power against microdisk diameter.

Next, we studied temperature dependence of the lasing threshold for 3 μm in diameter microdisk laser. The threshold power increases from 0.17 to 0.3 mW in the 78-130 K interval. Further increase of the temperature or pump power results in overheating of the microdisk laser. The temperature dependence of the lasing threshold can be described by fitting the experimental points with T_0=100 K. Such a high threshold characteristic temperature indicates high thermal stability of the microdisk laser based on GaInNAs/GaAs QW active region. Evidently, surface passivation should be applied for further improvement of the laser characteristics.

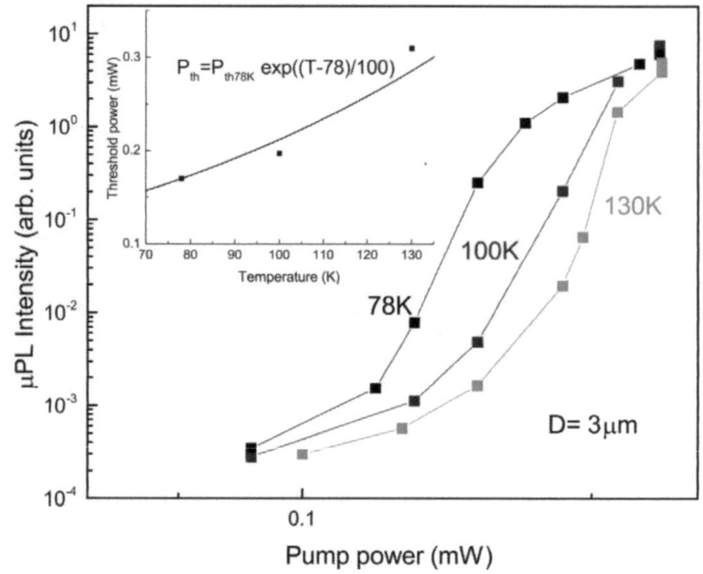

Figure 4. The L-L curves of the 3µm microdisk laser obtained at different temperatures. Insert: the measured threshold power versus operating temperature.

In summary, GaInNAs/GaAs quantum well microdisk lasers with 3-6µm diameter were studied. The lasing at 1173 nm wavelength with a low threshold (170µW) was achieved in 3 µm in diameter microdisk at 78K. The lasing was observed up to 130K. Passivation of the surface is required to increase the operation temperature.

Acknowledgments
This work was supported by the project the Ministry of Education of the Russian Federation RFMEFI61615X0052 "Development of microlasers with disk geometry based on quantum-sized active region for integrated photonics" (agreement # 14.616.21.0052). R.I., M.G. and T.N. acknowledge European Commission FP7 project RAPIDO (www.rapido-project.eu) and Dr. Arto Aho for support concerning epitaxy of GaInAsN.

References

[1] S.L. McCall, A.F. Levi, R.E. Slusher, S.J. Pearton, R.A. Logan, Appl.Phys.Lett., 60, 289-291, (1992).

[2] Di Liang, M. Fiorentino, S. Srinivasan, J.E. Bowers, R. G. Beausoleil «Low threshold electrically-pumped hybrid silicon microring lasers», IEEE Jour. of Sel. Topics in Quant. El., 17 (6), 1528 (2011).

[3] N.V. Kryzhanovskaya, A.E. Zhukov, A.M. Nadtochy, I.A. Slovinsky, M.V. Maximov, M.M. Kulagina, A.V.Savelev, E.M. Arakcheeva, Yu.M. Zadiranov, S.I. Troshkov, A.A. Lipovskii «High-temperature lasing in a microring laser with an active region based on InAs/InGaAs quantum dots», Semiconductors, 46, 1040-1043, (2012).

Journal of Physics: Conference Series **741** (2016) 012110 doi:10.1088/1742-6596/741/1/012110

[4] N.V. Kryzhanovskaya, E.I. Moiseev, Yu. V. Kudashova, F.I. Zubov, A.A. Lipovskii, M.M. Kulagina, S.I. Troshkov, Yu. M. Zadiranov, D.A. Livshits, M.V. Maximov, A.E. Zhukov, «Continuous-wave lasing at 100°C in 1.3 μm quantum dot microdisk diode laser», Electron. Lett. 51(17), 1354 – 1355 (2015).
[5] D. A. Livshits, A. Yu. Egorov, and H. Riechert, Electron. Lett. 36, 1381 (2000).
[6]M.Guina, V.-M. Korpijärvi, J. Rautiainen, P. Tuomisto, J.Puustinen, A.Härönen, O. Okhotnikov, «3.5 W GaInNAs disk laser operating at 1220 nm», Proc. SPIE 6997, Semiconductor Lasers and Laser Dynamics III, 69970Q (2008).
[7] V.-M.Korpijärvi, J.Viheriälä, M. Koskinen, Antti T. Aho, and M. Guina, "High-power temperature-stable GaInNAs distributed Bragg reflector laser emitting at 1180 nm," Opt. Lett. 41, 657-660 (2016).
[8] J. Wei, F. Xia, C. Li, and S. R. Forrest, IEEE Photonics Technol. Lett. 14, 597 (2002).
[9] N.Tansua, L.J. Mawst, «Current injection efficiency of InGaAsN quantum-well lasers», Jour. of Appl. Phys. 97, 054502 (2005).

Lasers with asymmetric barrier layers: A promising type of injection lasers

Levon V Asryan[1], Fedor I Zubov[2], Natalia V Kryzhanovskaya[2], Mikhail V Maximov[2], and Alexey E Zhukov[2]

[1] Virginia Polytechnic Institute and State University, Blacksburg, VA 24061, USA

[2] St. Petersburg National Research Academic University of Russian Academy of Sciences, Saint Petersburg 194021, Russia

E-mail: asryan@vt.edu

Abstract
We present an overview of our theoretical and experimental work on a novel type of semiconductor lasers – quantum well (QW) lasers with asymmetric barrier layers (ABLs). Our experimental work supports our theoretical derivations — ABL QW lasers demonstrate superior operating characteristics as compared to conventional QW lasers. In particular, the threshold current is lower and more temperature-stable, the light-current characteristic is more linear, and the wall-plug efficiency is higher in ABL lasers.

1. Introduction

Low lasing threshold, as well as temperature-stable and high-power operation, have always been desirable in semiconductor lasers [1]–[13]. In conventional diode lasers with a quantum-confined active region, a significant fraction of electrons and holes does not enter into the active region and thus does not contribute to stimulated recombination therein. Instead, this fraction is consumed by spontaneous recombination in the waveguide region [optical confinement layer (OCL)], wherein the carriers are initially injected from the cladding layers and which contains (in its central part) the active region [Figure 1(a)]. The parasitic electron-hole recombination outside a low-dimensional active region presents a major challenge in conventional injection lasers. Due to this recombination, the laser characteristics are adversely affected – the threshold current is increased and more temperature-sensitive and the light-current characteristic (LCC) is sublinear, even in the absence of heating effects.

To overcome the limitations placed on the operating characteristics by recombination outside the active region, novel designs of semiconductor lasers were proposed – one using double tunnelling-injection (injection of both electrons and holes) into the active region [14]–[22] and the other using asymmetric barrier layers (ABLs) [15]–[17] [one on each side of the active region – see Figure 1(b, c)]. The active region in these novel lasers can be in the form of either a single quantum well (QW) or a single layer with quantum dots.

Here we present an overview of our theoretical [23]–[27] and experimental [28]–[32] work on ABL QW lasers. Our experimental work supports our theoretical derivations — ABL QW lasers demonstrate superior operating characteristics as compared to conventional QW lasers. In particular, the threshold current is lower and more temperature-stable, the LCC is more linear, and the wall-plug efficiency is higher in ABL lasers.

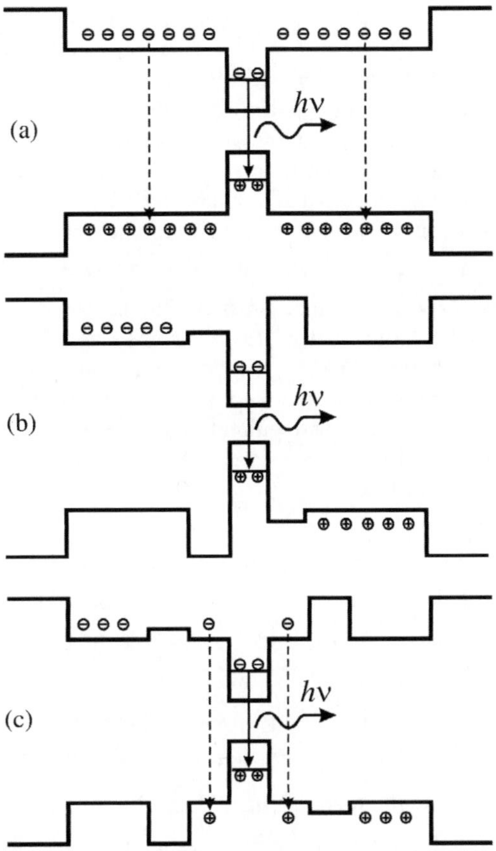

Figure 1. Schematic energy band diagrams of a conventional laser (a) and ABL lasers without (b) and with (c) intermediate layers. The vertical arrows show the electron-hole recombination.

2. ABL laser structures

Figures 1(b, c) show the schematic energy band diagrams of ABL lasers. The barrier layers are asymmetric in that they have considerably different heights for the carriers of opposite signs. The layer located on the electron- (hole-) injecting side of the structure [left- (right-) hand side in Figs. 1(b, c)] provides a low barrier (ideally no barrier) for electrons (holes) [so that it does not prevent electrons (holes) from easily approaching the active region] and a high barrier for holes (electrons) [so that holes (electrons) injected from the opposite side of the structure do not overcome it].

The use of ABLs in the structure of Figure 1(b) will thus secure that there will be no electrons and holes simultaneously (and hence no parasitic electron-hole recombination) outside the active region. In the structure of Figure 1(c), there will however be both electrons and holes (and hence some electron-hole recombination) not only in the active region but also in two intermediate layers located between the active region and each of the ABLs. The presence of these thin intermediate layers may be required in order to facilitate the flux switches during epitaxial growth process and to prevent from the active region re-evaporation [33, 34].

We discuss below the calculated and experimental characteristics of ABL QW lasers.

3. Threshold characteristics

The threshold current density of a QW laser is given as the sum of the current densities of spontaneous radiative recombination in and outside the QW,

$$j_{\rm th} = j_{\rm spon}^{\rm QW} + j_{\rm spon,th}^{\rm outside} \,. \tag{1}$$

The current densities of spontaneous radiative recombination in and outside the QW are

$$j_{\rm spon}^{\rm QW} = eB_{\rm 2D}n^{\rm QW}p^{\rm QW}\,, \tag{2}$$

$$j_{\rm spon,\,th}^{\rm outside} = eb^{\rm outside}B_{\rm 3D}n_{\rm th}^{\rm outside}p_{\rm th}^{\rm outside}\,, \tag{3}$$

where e is the electron charge, B_{2D} is the spontaneous radiative recombination coefficient for a two-dimensional region (QW) (see [35] for the expression for B_{2D}), n^{QW} and p^{QW} are the two-dimensional electron and hole densities in the QW, $b^{\rm outside}$ is the thickness of the region outside the QW wherein the parasitic electron-hole recombination occurs, B_{3D} is the spontaneous radiative recombination coefficient for the material of that region (see [5, 35] for the expression for B_{3D}), and $n_{th}^{outside}$ and $p_{th}^{outside}$ are the free electron and hole densities in that region at the lasing threshold.

In conventional QW lasers (CQWLs), $b^{outside}$ is the thickness of the entire OCL and hence is large [300 nm in the structure of Figure 5(a)]. In the ABL QW laser with intermediate layers, $b^{outside}$ is the sum of the thicknesses of the intermediate layers, which are very thin [5 nm each in the structure of Figure 5(b)], and hence is very small. In the ABL QW laser without intermediate layers, $b^{outside}$ is simply zero since there is no region outside the QW wherein the electron-hole recombination occurs. Consequently, the parasitic recombination current density $j_{spon,th}^{outside}$ is high in CQWLs, low in the ABL QW laser with intermediate layers, and zero in the ABL QW laser without intermediate layers. This is illustrated in figures 2 and 3, which show the calculated threshold current density against cavity length and temperature, respectively. As seen from the figures, the threshold current density is considerably higher in the reference CQWL as compared to both ABL QW lasers and there is no much difference between the threshold current density in the ABL QW lasers without and with intermediate layers.

The temperature dependence of threshold current in semiconductor lasers is described by the parameter T_0 termed as 'characteristic temperature' and defined as

$$T_0 = \left(\frac{\partial \ln j_{th}}{\partial T} \right)^{-1} . \tag{4}$$

As clear from (4), the higher T_0, the more temperature-stable is j_{th}.

In CQWLs, the temperature dependence of j_{th} is primarily due to such dependence of the parasitic recombination current density in the entire OCL, $j_{spon,th}^{outside}$, and hence T_0 is low. In the ABL QW laser without intermediate layers, the temperature dependence of j_{th} is merely due to such dependence of the recombination current density in the QW, j_{spon}^{QW}, and hence T_0 is very high (ideally, T_0 calculated for j_{spon}^{QW} is equal to T – see [23]). In the ABL QW laser with intermediate layers, the temperature dependence of j_{th} is due to such dependence of both recombination current densities in and outside the QW. For the cavity lengths considerably exceeding the shortest cavity length (the minimum tolerable cavity length below which the lasing is impossible to attain – see [36]), the contribution of the parasitic recombination in the intermediate layers is small and hence T_0 is high also in this ABL QW laser [23]. This is illustrated in Figure 4, which shows the calculated T_0 against cavity.

Figures 6 and 7 show the threshold characteristics measured in our experimental ABL QW laser structure with intermediate layers [the structure of Figure 5(b)] – T_0 against reciprocal cavity length, and T_0 and j_{th} against temperature. For comparison, the characteristics measured for the reference CQWL structure [the structure of Figure 5(a)] are also presented. As seen from the figures, T_0 is considerably higher in the ABL QW laser as compared to the reference CQWL. In particular, at the operating temperature 20 °C, T_0 is 143 K in the ABL QW laser while it is 99 K in the reference CQWL.

4. Power characteristics

The LCC of a diode laser (the output optical power P versus injection current density j) is given by

$$P(j) = \frac{\hbar\omega}{e} S\,(j - j_{th})\,\eta_{int}(j), \tag{5}$$

where $\hbar\omega$ is the photon energy, $S = WL$ is the cross-section of the junction, W is the lateral size of the device, L is the cavity length, and $\eta_{int}(j)$ is the internal differential quantum efficiency (efficiency of stimulated recombination).

In the CQWL, the internal quantum efficiency is a decreasing function of the injection density. The following expression was derived for η_{int} in [37]–[39]:

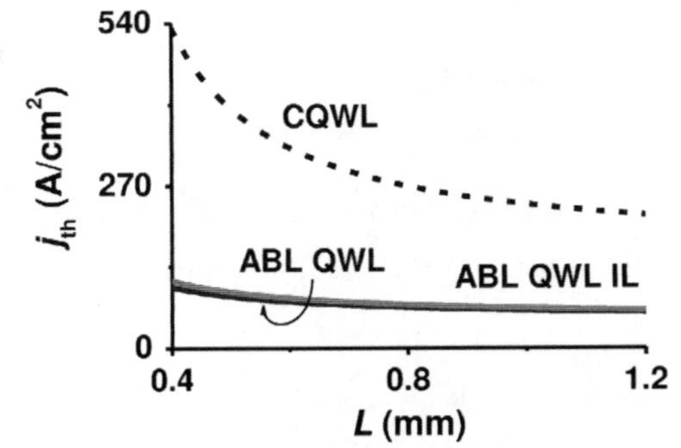

Figure 2. Threshold current density vs. cavity length in the ABL QW lasers without and with intermediate layers and reference conventional quantum well laser (CQWL).

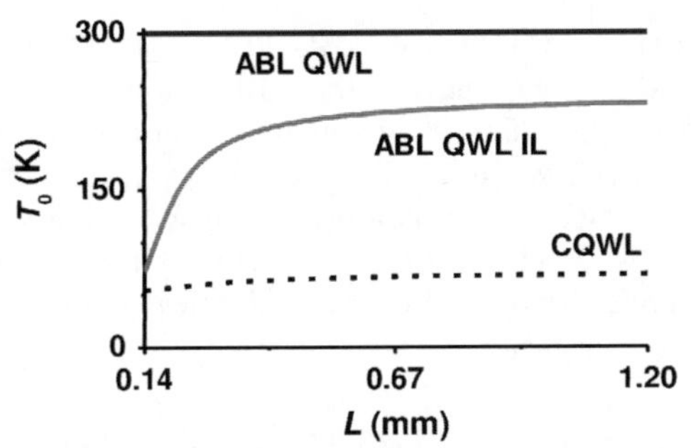

Figure 4. Characteristic temperature vs. cavity length in the ABL QW lasers without and with intermediate layers and reference CQWL.

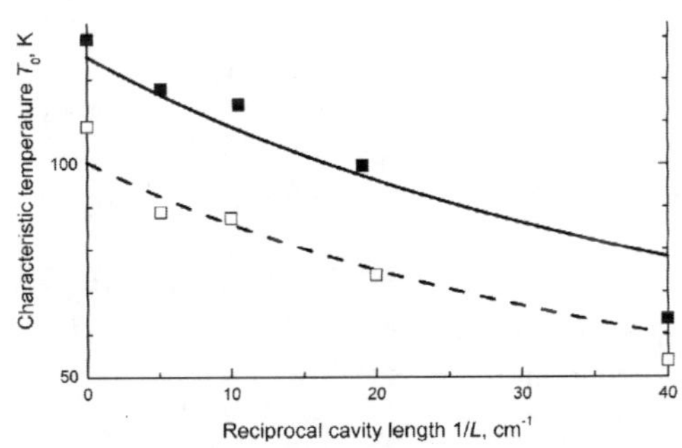

Figure 6. Experimental (symbols) and calculated (curves) dependences of the characteristic temperature on the reciprocal cavity length in the ABL QW laser with intermediate layers (solid curve, dark squares) and reference CQWL (dashed curve, open squares).

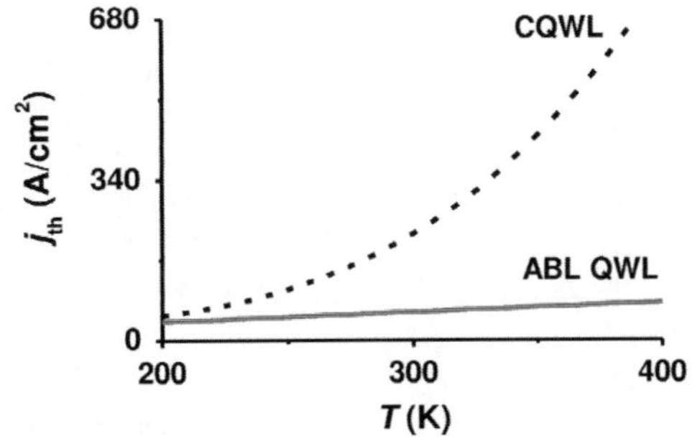

Figure 3. Threshold current density vs. temperature in the ABL QW laser without intermediate layers and reference CQWL.

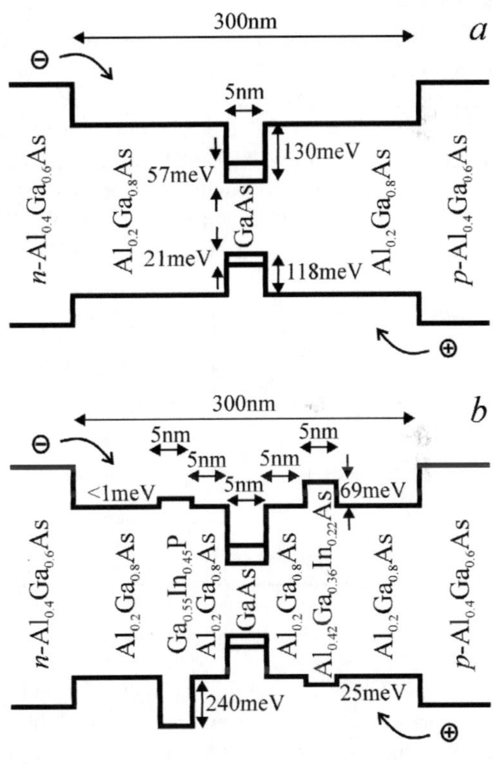

Figure 5. Energy band diagrams of the experimental structures: (a) reference CQWL and (b) ABL QW laser with intermediate layers. The lasing wavelength at 20 °C is 833.8 nm in the CQWL and 835.6 nm in the ABL QW laser.

$$\eta_{\mathrm{int}}(j)=\left[\frac{1}{2}+\frac{j_{\mathrm{spon,\,th}}^{\mathrm{outside}}}{\langle j_{\mathrm{capt,\,th}}\rangle_{\mathrm{harmon}}}+\sqrt{\left(\frac{1}{2}+\frac{j_{\mathrm{spon,\,th}}^{\mathrm{outside}}}{\langle j_{\mathrm{capt,\,th}}\rangle_{\mathrm{harmon}}}\right)^2+\frac{j_{\mathrm{spon,\,th}}^{\mathrm{outside}}}{\langle j_{\mathrm{capt,\,th}}\rangle_{\mathrm{geom}}}\frac{j-j_{\mathrm{th}}}{\langle j_{\mathrm{capt,\,th}}\rangle_{\mathrm{geom}}}}\,\right]^{-1}, \quad (6)$$

where $\langle j_{\mathrm{capt,th}}\rangle_{\mathrm{harmon}}$ and $\langle j_{\mathrm{capt,th}}\rangle_{\mathrm{geom}}$ are the harmonic and geometric means of the current densities $j_{\mathrm{capt,\,n,\,th}}$ and $j_{\mathrm{capt,\,p,\,th}}$ of electron and hole capture into the QW at the lasing threshold,

$$\frac{1}{\langle j_{\mathrm{capt,\,th}}\rangle_{\mathrm{harmon}}}=\frac{1}{2}\left(\frac{1}{j_{\mathrm{capt,\,n,\,th}}}+\frac{1}{j_{\mathrm{capt,\,p,\,th}}}\right), \quad (7)$$

$$\langle j_{\mathrm{capt,\,th}}\rangle_{\mathrm{geom}}=\sqrt{j_{\mathrm{capt,\,n,\,th}}\,j_{\mathrm{capt,\,p,\,th}}}\,. \quad (8)$$

As seen from (6), the higher the parasitic recombination current density at the lasing threshold, $j_{\mathrm{spon,th}}^{\mathrm{outside}}$, the stronger the decrease in η_{int} with increasing j. Hence, η_{int} decreases considerably with j and the LCC is strongly sublinear in the CQWL. As already mentioned above, $j_{\mathrm{spon,th}}^{\mathrm{outside}}$ is very low in the ABL QW laser with intermediate layers and simply zero in the ABL QW laser without intermediate layers. Hence, η_{int} decreases only slightly with increasing j in the ABL QW laser with intermediate layers and constant (unity) in the ABL QW laser without intermediate layers; consequently, the LCC is just slightly sublinear in the ABL QW laser with intermediate layers and virtually linear in the ABL QW laser without intermediate layers. This is illustrated in Figures 8-10, which present the following characteristics calculated as functions of the injection current: the parasitic recombination current outside the QW, the internal differential quantum efficiency, and the LCC.

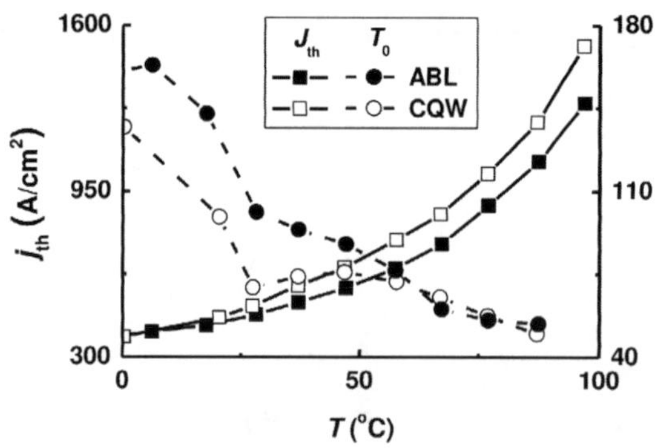

Figure 7. Experimental threshold current density (squares, left axis) and characteristic temperature (circles, right axis) vs. temperature: dark symbols – ABL QW laser with intermediate layers, open symbols – reference CQWL. The cavity length is 0.5 mm.

Figure 12 shows the experimental LCC in the ABL QW laser and reference CQWL of Figure 11. As seen from Figure 12, at both operating temperatures 20 and 60 °C, the LCC is more linear and the output power is higher in the ABL laser.

Figure 13 shows the experimental wall-plug efficiency in the ABL QW laser with intermediate layers and reference CQWL of Figure 5. As seen from Figure 13, at both operating temperatures 20 and 75 °C, the efficiency is distinctly higher in the ABL laser.

5. Conclusions

We have presented an overview of our theoretical and experimental work on a novel type of semiconductor lasers – quantum well (QW) lasers with asymmetric barrier layers (ABLs). Our experimental work supports our theoretical derivations — ABL QW lasers demonstrate superior operating characteristics as compared to conventional QW lasers. In particular, the threshold current is lower and more temperature-stable, the LCC is more linear, and the wall-plug efficiency is higher in ABL lasers.

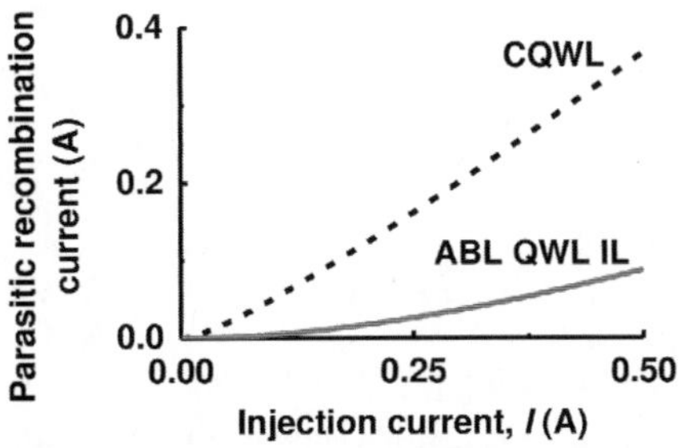

Figure 8. Parasitic recombination current vs. injection current in the ABL QW laser with intermediate layers and reference CQWL.

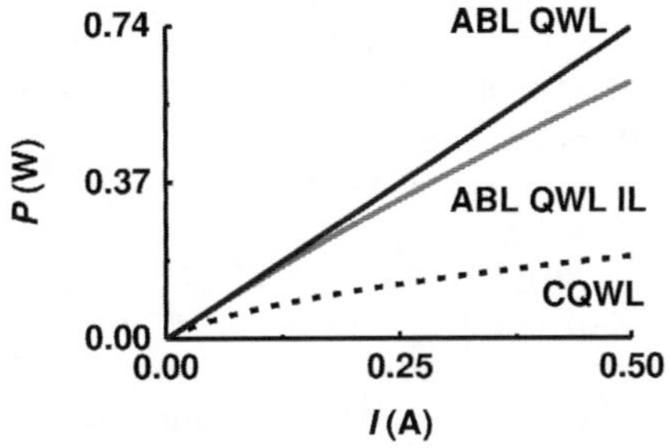

Figure 10. Light-current characteristic in the ABL QW lasers without and with intermediate layers and reference CQWL.

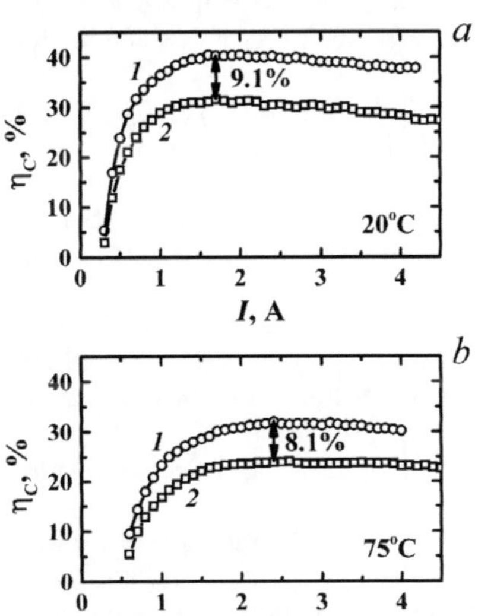

Figure 13. Experimental wall-plug efficiency vs. injection current in (1) the ABL QW laser with intermediate layers and (2) reference CQWL of Figure 5 at (a) 20 and (b) 75°C.

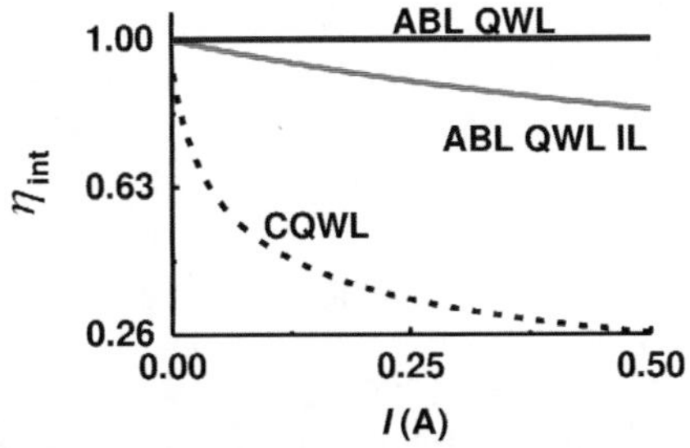

Figure 9. Internal differential quantum efficiency vs. injection current in the ABL QW lasers without and with intermediate layers and reference CQWL.

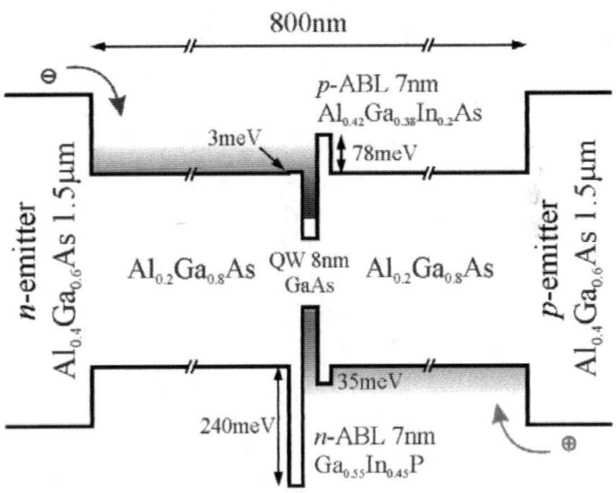

Figure 11. Energy band diagram of the experimental ABL QW laser used for the measurements in Figure 12.

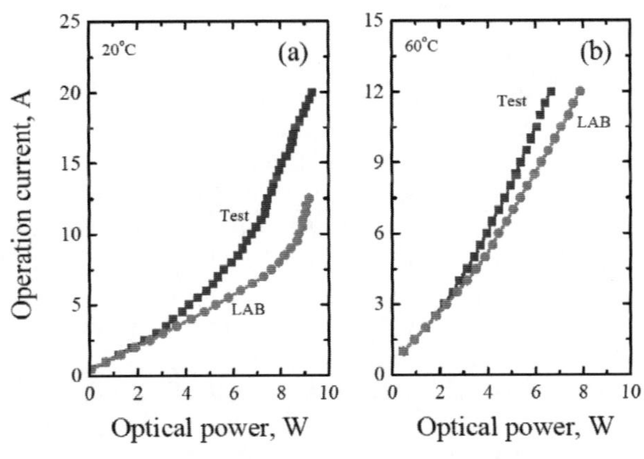

Figure 12. Operating current vs. output optical power measured in the ABL QW laser of Figure 11 and CQWL at 20°C (a) and 60°C (b).

Acknowledgment
This work was supported by the Russian Science Foundation (project 14-42-00006 "A novel type of diode lasers with characteristics improved by using asymmetric barriers").

References

[1] Agrawal G P and Dutta N K 1986 *Long-Wavelength Semiconductor Lasers* (New York: Van Nostrand)
[2] Zory P S Jr ed 1993 *Quantum Well Lasers* (Boston: Academic)
[3] Kapon E ed 1999 *Semiconductor Lasers I: Fundamentals*, 1st Edition (New York: Academic)
[4] Temkin H, Coblentz D, Logan R A, Vandenberg J M, Yadvish R D and Sergent A M 1994 Strained quaternary quantum well lasers for high temperature operation *Appl. Phys. Lett.* **63** 2321-3
[5] Asryan L V and Suris R A 1996 Inhomogeneous line broadening and the threshold current density of a semiconductor quantum dot laser *Semicond. Sci. Technol.* **11** 554-67
[6] Seki S, Oohashi H, Sugiura H, Hirono T and Yokoyama K 1996 Study on the dominant mechanisms for the temperature sensitivity of threshold current in 1.3 μm InP-based strained-layer quantum-well lasers *IEEE J. Quantum Electron.* **32** 1478-95
[7] Kazarinov R F and Shtengel G E 1997 Effect of carrier heating on static linearity of MQW InGaAsP/InP lasers *J. Lightw. Technol.* **15** 2284-6
[8] Maximov M V, Asryan L V, Shernyakov Y M, Tsatsul'nikov A F, Kaiander I N, Nikolaev V V, Kovsh A R, Mikhrin S S, Ustinov V M, Zhukov A E, Alferov Z I, Ledentsov N N and Bimberg D 2001 Gain and threshold characteristics of long wavelength lasers based on InAs/GaAs quantum dots formed by activated alloy phase separation *IEEE J. Quantum Electron.* **37** 676–83
[9] Mawst L J, Bhattacharya A, Lopez J, Botez D, Garbuzov D Z, DeMarco L, Connolly J C, Jansen M, Fang F and Nabiev R F 1996 8 W continuous wave front-facet power from broad-waveguide Al- free 980 nm diode lasers *Appl. Phys. Lett.* **69** 1532-4
[10] Garbuzov D, Xu L, Forrest S R, Martinelli R and Connolly J C 1996 1.5 μm wavelength, SCH-MQW InGaAsP/InP broadened-waveguide laser diodes with low internal loss and high output power *Electron. Lett.* **32** 1717-9
[11] Al-Muhanna A, Mawst L J, Botez D, Garbuzov D Z, Martinelli R U and Connolly J C 1998 High-power (>10 W) continuous-wave operation from 100-μm-aperture 0.97-μm-emitting Al-free diode lasers *Appl. Phys. Lett.* **73** 1182-4
[12] Livshits D A, Kochnev I V, Lantratov V M, Ledentsov N N, Nalyot T A, Tarasov I S and Alferov Zh I 2000 Improved catastrophic optical mirror damage level in InGaAs/AlGaAs laser diodes *Electron. Lett.* **36** 1848-9
[13] Vinokurov D A, Zorina S A, Kapitonov V A, Murashova A V, Nikolaev D N, Stankevich A L, Khomylev M A, Shamakhov V V, Leshko A Yu, Lyutetskiy A V, Nalyot T A, Pikhtin N A, Slipchenko S O, Sokolova Z N, Fetisova N V and Tarasov I S 2005 High-power laser diodes based on asymmetric separate-confinement heterostructures *Semicond.*, **39** 370-3
[14] Asryan L V and Luryi S 2001 Tunneling-injection quantum-dot laser: Ultrahigh temperature stability *IEEE J. Quantum Electron.* **37** 905-10
[15] Asryan L V and Luryi S 2003 Temperature-insensitive semiconductor quantum dot laser *Solid-State Electron.* **47** 205-12
[16] Asryan L V and Luryi S 2004 Quantum dot lasers: Theoretical overview *Semiconductor Nanostructures for Optoelectronic Applications* (Boston: Artech House Publishers) (pp 113–58), chapter 4
[17] Asryan L V and Luryi S 2005 Semiconductor laser with reduced temperature sensitivity *U.S. Patent 6 870 178 B2*
[18] Han D-S and Asryan L V 2008 Tunneling-injection of electrons and holes into quantum dots: A tool for high-power lasing *Appl. Phys. Lett.* **92** 251113

[19] Han D-S and Asryan L V 2008 Characteristic temperature of a tunneling-injection quantum dot laser: Effect of out-tunneling from quantum dots *Solid-State Electron.* **52** 1674-9

[20] Han D-S and Asryan L V 2009 Effect of the wetting layer on the output power of a double tunneling-injection quantum-dot laser *J. Lightw. Technol.* **27** 5775-82

[21] Han D-S and Asryan L V 2010 Output power of a double tunneling-injection quantum dot laser *Nanotechnology* **21** 015201

[22] Asryan L V 2015 Modulation bandwidth of a double tunnelling-injection quantum dot laser *Semicond. Sci. Technol.* **30** 035022

[23] Asryan L V, Kryzhanovskaya N V, Maximov M V, Egorov A Yu and Zhukov A E 2011 Bandedge-engineered quantum well laser *Semicond. Sci. Technol.* **26** 055025

[24] Zhukov A E, Kryzhanovskaya N V, Maximov M V, Egorov A Yu, Pavlov M M, Zubov F I and Asryan L V 2011 Semiconductor lasers with asymmetric barrier layers: An approach to high temperature stability *Semicond.* **45** 530-5

[25] Asryan L V, Kryzhanovskaya N V, Maximov M V, Zubov F I and Zhukov A E 2013 Light-current characteristic of a quantum well laser with asymmetric barrier layers *J. Appl. Phys.* **114** 143103

[26] Zhukov A E, Asryan L V, Semenova E S, Zubov F I, Kryzhanovskaya N V and Maximov M V 2015 On the optimization of asymmetric barrier layers in InAlGaAs/AlGaAs laser heterostructures on GaAs substrates *Semicond.* **49** 935-8

[27] Asryan L V, Zubov F I, Kryzhanovskaya N V, Maximov M V and Zhukov A E 2016 Theory of power characteristics of lasers with asymmetric barrier layers: Inclusion of asymmetry in electron and hole states filling *Semicond.* **50**

[28] Zhukov A E, Kryzhanovskaya N V, Zubov F I, Shernyakov Y M, Maximov M V, Semenova E S, Yvind K and Asryan L V 2012 Improvement of temperature-stability in a quantum well laser with asymmetric barrier layers *Appl. Phys. Lett.* **100** 021107

[29] Zhukov A E, Asryan L V, Shernyakov Yu M, Maximov M V, Zubov F I, Kryzhanovskaya N V, Yvind K and Semenova E S 2012 Effect of asymmetric barrier layers in the waveguide region on the temperature characteristics of quantum-well lasers *Semicond.* **46** 1027-31

[30] Zubov F I, Zhukov A E, Shernyakov Yu M, Maximov M V, Kryzhanovskaya N V, Yvind K, Semenova E S and Asryan L V 2015 The effect of asymmetric barrier layers in the waveguide region on power characteristics of QW lasers *Technical Phys. Lett.* **41** 439-42

[31] Zubov F I, Maximov M V, Shernyakov Yu M, Kryzhanovskaya N V, Semenova E S, Yvind K, Asryan L V and Zhukov A E 2015 Suppression of sublinearity of light–current curve in 850 nm quantum well laser with asymmetric barrier layers *Electron. Lett.* **51** 1106-8

[32] Zubov F I, Zhukov A E, Shernyakov Yu M, Maximov M V, Semenova E S and Asryan L V 2015 Diode lasers with asymmetric barriers for 850 nm spectral range: experimental studies of power characteristics *J. Phys. Conf. Ser.* **643** 012042

[33] Zhukov A E, Kovsh A R, Egorov A Yu, Maleev N A, Ustinov V M, Volovik B V, Maximov M V, Tsatsul'nikov A F, Ledentsov N N, Shernyakov Yu M, Lunev A V, Musikhin Yu G, Bert N A, Kop'ev P S and Alferov Zh I 1999 Photo- and electroluminescence in the 1.3-μm range from quantum-dot structures grown on GaAs substrates *Semicond.* **33** 153-6

[34] Ledentsov N N, Kovsh A R, Shchukin V A, Mikhrin S S, Krestnikov I L, Kozhukhov A V, Karachinsky L Y, Maximov M V, Novikov I I, Shernyakov Yu M, Soshnikov I P, Zhukov A E, Musikhin Yu G, Ustinov V M, Zakharov N D, Werner P, Kettler T, Posilovic K, Bimberg D, Hu M, Nguyen H K, Song K, and Zah C-E 2006 1.3-1.5 μm quantum dot lasers on foreign substrates: growth using defect reduction technique, high-power CW operation, and degradation resistance *Proc. SPIE* **6133** 61330S-1–12

[35] Asryan L V 2005 Spontaneous radiative recombination and nonradiative Auger recombination in quantum-confined heterostructures *Quantum Electron.* **35** 1117-20

[36] Asryan L V and Suris R A 1999 Spatial hole burning and multimode generation threshold in quantum-dot lasers *Appl. Phys. Lett.* **74** 1215-7

[37] Asryan L V, Luryi S and Suris R A 2002 Intrinsic nonlinearity of the light-current characteristic of semiconductor lasers with a quantum-confined active region *Appl. Phys. Lett.* **81** 2154-6
[38] Asryan L V, Luryi S and Suris R A 2003 Internal efficiency of semiconductor lasers with a quantum-confined active region *IEEE J. Quantum Electron.* **39** 404-18
[39] Asryan L V and Sokolova Z N 2014 Optical power of semiconductor lasers with a low-dimensional active region *J. Appl. Phys.* **115** 023107

Laser-Induced Periodical Structures Fabrication for Third Harmonic Generation

T A Voytova, S V Makarov, A N Tsypkin, V A Milichko, I S Mukhin, A V Yulin, S E Putilin, M A Baranov, A E Krasnok, P A Belov

ITMO University, 49 Kronverksky Pr., St. Petersburg, 197101, Russia

Abstract. We propose a novel strategy for self-adjusted fabrication of large-scale array of resonant silicon nanoparticles (metasurface) on a thin silicon film. The self-adjusting mechanism is based on the effect of resonant nanogratings formation under intense multishot femtosecond irradiation of a thin silicon film. The resulting metasurfaces allow for generation of ultraviolet laser pulses at a wavelength of 270 nm with conversion efficiency up to 10^{-6} and high peak (≈100 kW/sm^2) and average power (≈1.5μW). Such high peak power from ultrathin metasurface makes the generated UV pulses applicable in a wide range of applications: precise nanolithography, ultrafast photoexcitation etc.

1. Introduction

All-dielectric nanostructures and metasurfaces based on high refractive index dielectric nanoparticles have emerged as promising objects for nonlinear nanophotonics [1-3], competing with conventional dielectric microdevices [4]. Indeed, the low-loss all-dielectric nanoresonators, possessing electric and magnetic resonances in the visible range, provide efficient control of light scattering [5-9], strong light localization [10, 11], and effective wave front engineering [12], whereas nonlinear response of high-index dielectrics is very high [13]. For such high-intensity applications as optical harmonics generation from isotropic materials, e.g. silicon, it is important to have an effective thermal sink, in order to increase damage threshold and achieve higher output power of nonlinear signal. For this purpose, it is more reasonable to use homogeneous materials rather than nanoparticle-based metamaterials. Non-lithographical methods [14-17] imply formation of separate nanoparticles rather than surface nanostructuring preserving a good thermal contact with the remained film. On the other hand, femtosecond laser surface nanostructuring is proven to be high-efficient tool for fabrication of different single subwavelength nanoantennae [18-21], and for large-scale periodical surface nanostructuring [22-26]. In this paper, we apply self-organized laser-induced nanostructuring of thin *a*-Si:H film to fabricate array of periodically arranged nanoparticles (metasurface) placed on the homogeneous film, which provide resonance optical response at a pumping wavelength for enhanced third-harmonic generation (THG). The self-organized metasurfaces demonstrate up to 30-fold increase of the efficiency as compared with arbitrary nonresonant *a*-Si films and 2-fold improving as compared with an initial film supporting a Fabry-Perot resonance.

2. Samples fabrication

In order to enhance TH, we fabricate metasurface by irradiation of *a*-Si film with the thickness (100 nm) corresponding to the most effective THG. *a*-Si:H films with thicknesses were deposited on a substrate of fused silica by plasma enhanced chemical vapor deposition from SiH_3 precursor gas (initial hydrogen concentration ~10%), were used. The film irradiation is carried out by focusing of fs-laser pulses (100 fs) into 0.9 μm spot at 1/e level through a NA=0.75 objective. Surface scanning by the focused beam with 0.1 J/cm2, repetition rate 10 kHz, and scan speed 20 μm/s provides ≈450 absorbed laser pulses per spot. The resulting metasurface represents periodical array of nanobumps with mean periods ≈260 nm along one axis and ≈700 nm along the perpendicular axis, the width of the bumps is ≈130 nm and the length is ≈330 nm (figure 1a). The 700-nm period is equal to scanning period of the laser nanostructuring. The all dimensions of the metasurface are smaller than the wavelength of fs-laser (800 nm) and all visible spectrum. Therefore, the structure can be considered as metasurface. Optimization of surface nanostructuring is represented in figure 2.

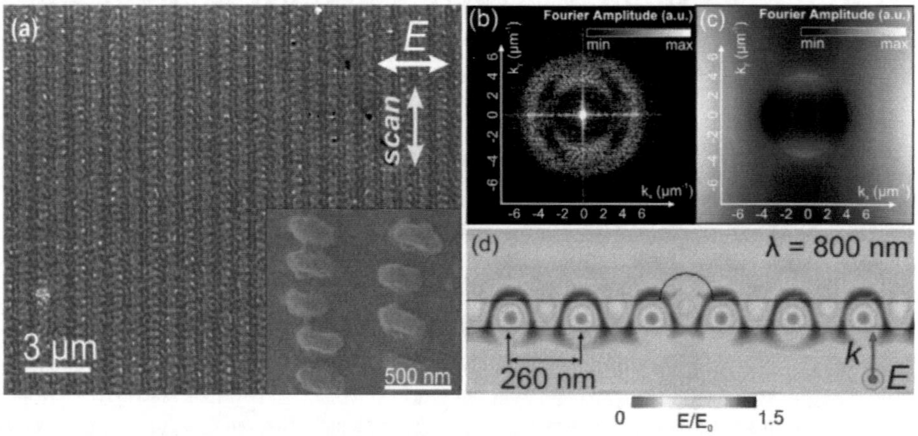

Figure 1. (a) SEM image of self-organized nc-Si metasurface. Inset: zoomed part of the metasurface. Experimental (b) and theoretical (c) spatial 2D Fourier spectra of the fabricated metasurface. (d) Calculated value of total electric field (averaged over time) near subwavelength hemispherical feature on 100-nm *a*-Si film after incidence of a plane electromagnetic wave.

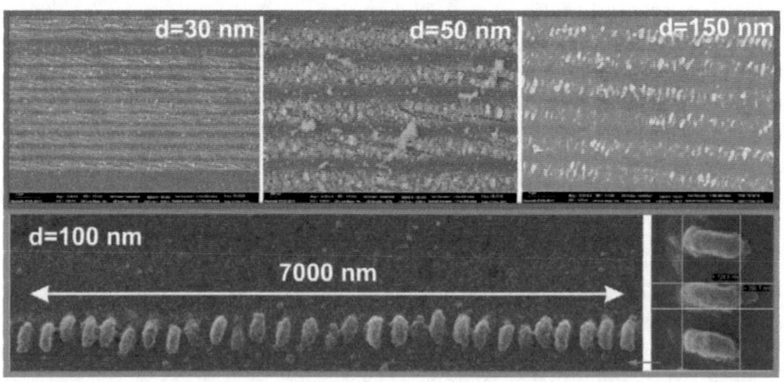

Figure 2. SEM images of nanostructured films by femtosecond laser irradiation with different thicknesses d.

3. Model for laser-induced metasurface formation

The most widely accepted theory of laser-induced periodical surface structures (LIPSS) is based on the interference of the incident laser beam with some form of a surface-scattered electromagnetic wave. The work of Sipe et al. [27] represents a first-principle theory which takes into consideration the interaction of an electromagnetic wave with a microscopically rough surface. This theory predicts possible LIPSS wave vectors $\mathbf{k}$ of the surface [with $|\mathbf{k}| = 2\pi/\Lambda$] as a function of the laser parameters angle of incidence θ, polarization direction, and wave vector of the incident radiation $\boldsymbol{k}_L$ (with $|\mathbf{k}_L| = 2\pi/\lambda$) which has a component $\mathbf{k}_i$ in the surface plane and surface parameters (dielectric constant and surface roughness). It leads to an expression for the inhomogeneous energy deposition into the irradiated material which is proportional to $\eta(\mathbf{k}, \mathbf{k}_i) \times |b|$, where η is a response function describing the efficacy with which the surface roughness at **k** leads to inhomogeneous absorption of radiation. The second factor b represents a measure of amplitude of the surface roughness at **k** which is a slowly varying function for a surface with a homogeneously distributed roughness. In contrast to the behavior of b, the efficacy factor η can exhibit sharp peaks at certain **k** values which then determine the spatial ripple periods. It worth noting that once the LIPSS are formed, the Fourier spectrum of the surface $b(\mathbf{k})$ can also exhibit sharp peaks coincident with the peaks in $\eta(\mathbf{k})$ which were initially responsible for the surface damage.

In the theory of Sipe et al., the efficacy factor is defined as $\eta(\mathbf{k},\mathbf{k}_i) = 2\pi|\nu(\mathbf{k}_+) + \nu^*(\mathbf{k}_-)|$. For the two cases of s- or p-polarized light, incident under an angle of θ and having polarization vectors and a wave vector component $\mathbf{k}_i$ the complex function ν is given by

$$\nu(\mathbf{k}_\pm, \mathrm{s-pol.}) = \left[h_{ss}(k_\pm)\left(\hat{\mathbf{k}}\pm\cdot\hat{\mathbf{y}}\right)^2 + h_{kk}(k_\pm)\left(\hat{\mathbf{k}}\pm\cdot\hat{\mathbf{x}}\right)^2\right]\gamma_t|t_s(\mathbf{k}_i)|^2, \quad (1)$$

$$\nu(\mathbf{k}_\pm, \mathrm{p-pol.}) = \left[h_{ss}(k_\pm)\left(\hat{\mathbf{k}}\pm\cdot\hat{\mathbf{x}}\right)^2 + h_{kk}(k_\pm)\left(\hat{\mathbf{k}}\pm\cdot\hat{\mathbf{y}}\right)^2\right]\gamma_t|t_s(\mathbf{k}_i)|^2 +$$
$$h_{kz}(k_\pm)\left(\hat{\mathbf{k}}\pm\cdot\hat{\mathbf{y}}\right)\gamma_z\varepsilon t_x^*(\mathbf{k}_i)t_z(\mathbf{k}_i) + h_{zk}(k_\pm)\left(\hat{\mathbf{k}}\pm\cdot\hat{\mathbf{y}}\right)\gamma_t t_x(k_i)t_z^*(\mathbf{k}_i) +$$
$$h_{zz}(k_\pm)\gamma_z\varepsilon|t_z(\mathbf{k}_i)|^2, \quad (2)$$

with inner products $\left(\hat{\mathbf{k}}\pm\cdot\hat{\mathbf{y}}\right) = (\sin\theta \pm \kappa_y)/\kappa_\pm$ and $\left(\hat{\mathbf{k}}\pm\cdot\hat{\mathbf{x}}\right) = \kappa_x/\kappa_\pm$. Here, the definition $\kappa_\pm = [\kappa x^2 + (\sin\theta \pm k_y)^2]^{1/2}$ has been used and all lengths have been normalized with the factor $\lambda/(2\pi)$. Hence, the dimensionless LIPSS wave vectors $\kappa = k\cdot\lambda(2\pi) \equiv \lambda/\Lambda$ are used in the following. With these definitions and with ε being the complex dielectric function of the material at the irradiation wavelength, the functions h_{ss}, h_{zz}, h_{kk}, h_{kz}, h_{zk} used in equations (2) and (3) can be expressed as

$$h_{ss}(k_\pm) = 2i\left(\sqrt{1-\kappa_\pm^2} + \sqrt{\varepsilon-\kappa_\pm^2}\right)^{-1}, \quad h_{zz}(k_\pm) = 2i\kappa_\pm^2\left(\sqrt{1-\kappa_\pm^2} + \sqrt{\varepsilon-\kappa_\pm^2}\right)^{-1}, \quad (3)$$

$$h_{kk}(k_\pm) = \frac{2i}{\varepsilon}\sqrt{(\varepsilon-\kappa_\pm^2)(1-\kappa_\pm^2)}\left(\sqrt{1-\kappa_\pm^2} + \sqrt{\varepsilon-\kappa_\pm^2}\right)^{-1}, \quad (4)$$

$$h_{kz}(k_\pm) = \frac{2i\kappa_\pm}{\varepsilon}\sqrt{(\varepsilon-\kappa_\pm^2)}\left(\sqrt{1-\kappa_\pm^2} + \sqrt{\varepsilon-\kappa_\pm^2}\right)^{-1}, \quad (5)$$

$$h_{zk}(k_\pm) = \frac{2i\kappa_\pm}{\varepsilon}\sqrt{(1-\kappa_\pm^2)}\left(\sqrt{1-\kappa_\pm^2} + \sqrt{\varepsilon-\kappa_\pm^2}\right)^{-1}. \quad (6)$$

The complex functions t_s, t_z, and t_x are given by

$$t_s(\mathbf{k}_i) = 2|\cos\theta|\left(|\cos\theta| + \sqrt{\varepsilon-(\sin\theta)^2}\right)^{-1}, \quad t_z(\mathbf{k}_i) = 2\sin\theta\left(\varepsilon|\cos\theta| + \sqrt{\varepsilon-(\sin\theta)^2}\right)^{-1}, \quad (7)$$

$$t_x(\mathbf{k}_i) = 2\sqrt{\varepsilon-(\sin\theta)^2}\left(\varepsilon|\cos\theta| + \sqrt{\varepsilon-(\sin\theta)^2}\right)^{-1}. \quad (8)$$

The surface roughness is included in the theory in the factors γ_t and γ_z via two numerical factors, s (shape factor) and f (filling factor), by

$$\gamma_t = \frac{\varepsilon-1}{4\pi(1+0.5(1-f)(\varepsilon-1)(F(s)-R\cdot G(s)))}, \quad \gamma_z = \frac{\varepsilon-1}{4\pi(\varepsilon-(1-f)(\varepsilon-1)(F(s)+R\cdot G(s)))}, \quad (9)$$

with $R = (\varepsilon-1)/(\varepsilon+1)$ and the scalar functions $F(s) = \sqrt{s^2+1} - s$ and $Q(s) = 0.5\left(\sqrt{s^2+4} + s\right) - \sqrt{s^2+1}$.

Following [27], the surface roughness was modeled with the values s=0.4, f=0.1, which represent the assumption of spherically shaped islands. On the basis of equations (1) - (9) it is possible to calculate numerically the efficacy factor η as a function of the normalized LIPSS wave vector components κ_x, κ_y at given values for the irradiation parameters (θ, λ and polarization direction) and for parameters characterizing the optical (ε) and the surface roughness properties (s and f). The calculated 2D Fourier spectrum of laser energy deposition (for real and imagine parts or complex refractive index, respectively, n=3.90 and K=0.11 at λ=800 nm [28]) exhibits pronounced maxima for structures with period around $\lambda/n \approx$205 nm and with wavevectors perpendicular to the laser polarization (figure 1b,c).

Despite excellent qualitative agreement, analytical modeling based on the widely acknowledged Sipe's model [27] does not provide quantitative prediction of the periods between nanoparticles within the metasurface, because it does not take into account film thickness. In numerical simulations we observe coupling of normally incident laser wave to a planar waveguide mode within the Si film. This coupled waveguide mode interferes with the incident field, providing the formation of a standing wave within the film and modulated heating with the period about 260 nm. Figure 1b supports the interference mechanism, where the scattering of the incident electromagnetic wave with 800 nm wavelength on a single subwavelength particle on Si 100-nm. The electric field enhancement within each maximum is about 1.5, resulting in ultrafast non uniform heating and deformation of the film due to thermal expansion and fast resolidification.

It is worth noting, that the surface plasmon-polaritons excitation [20, 24, 25] is unlikely to be the origin of the metasurface formation in our case, because the wavevector of the periodical arrays is perpendicular to polarization of the incident light (Fig.1a), whereas plasmon-polaritons are mostly longitudinal waves [29]. Moreover, quality of nanostructuring is the best for 100-nm film as compared to the other *a*-Si:H films, revealing an important role of the waveguiding mode.

4. Third harmonics generation from the nanostructure

For THG generation a commercial femtosecond laser system (based on a Ti:Sapphire regenerative amplifier Regulus 35F1K, Avesta Poject) was used. Laser pulses have 800 nm central wavelength, with maximum pulse energy of 2 mJ, and pulse duration of 40 fs at the repetition rate of 1 kHz. Laser energy was varied and controlled by an polarising filter and a power meter (Nova II, Ophir), respectively, while the pulse duration was measured by an autocorrelator ASF-20 (Avesta Poject). The THG signal was separated from the fundamental wavelength by a set of filters and measured by the power meter. Measured UV signal can be clearly seen by a spectrometer, excluding the influence of possible one-photon or multiphoton luminescence.

Importantly, a smooth Si film, supporting Fabry-Perot resonance at 800 nm and formation of resonant metasurface, shows a sharp maximum of UV signal at 100—120 nm thicknesses (figure 3a). The largest difference in THG signal is observed between 170-nm and 100-nm films, giving 15-fold enhancement [30]. In figure 3b the comparison of UV signal from the metasurface with UV signal from the thin 100-nm film is presented, revealing 2.5-fold enhancement of third-harmonics conversion efficiency and the 30-fold enhancement as compared to the nonresonant 170-nm film. The observed maximum TH conversion efficiency is about $0.9 \cdot 10^{-6}$ for the 100-nm film and incident pulse energy 1.5 mJ. In our experiments, maximum laser intensity is much smaller than the damage threshold of a Si film (100 mJ/cm^2) [17] and the level of dense electron-hole plasma generation (F=1 mJ/cm^2) [1, 2, 31, 32]. Despite the conversion efficiency for our system is not a record among previously reported values achieved in high-Q resonantors [4, 32], the measured output energies and average power of UV femtosecond laser pulses are much larger.

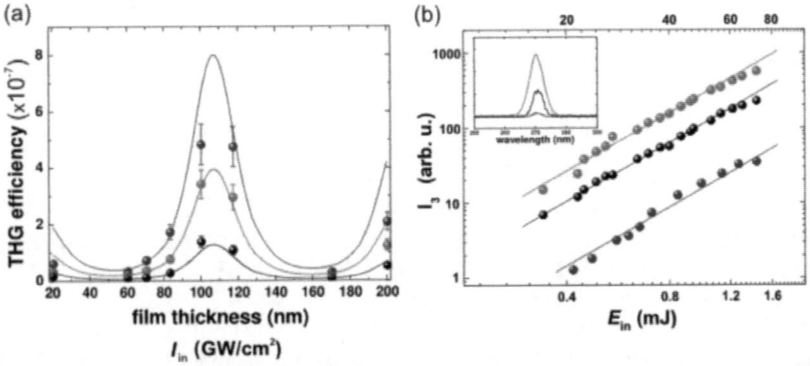

Figure 3. (a) Experimental (circles) and theoretical (lines) dependencies of third harmonic generation (THG) efficiency on Si film thickness at various pump intensities: 20 GW/cm^2 (black), 35 GW/cm^2 (red), 55 GW/cm^2 (blue). (b) Experimental dependencies of third harmonic intensity from initial (black dots) and nanostructured Si films with 100-nm thickness on fused silica substrate with resonant (red dots). Third harmonics signal from smooth 170-nm Si film (blue dots). All approximation lines correspond to 2.8 slope. Inset: spectral range in the vicinity of the third harmonic signal.

Indeed, the advantage of our systems is that they allow working with much more powerful pump and to generate UV femtosecond pulses with the energy as high as W=1.3 nJ per pulse. This value is at least 5 orders of magnitude higher than that reported and yields high average power of femtosecond UV beam about P_{av}==1.3 μW, being high enough to provide visible fluorescence of strontium aluminate doped with europium and dysprosium in quartz cuvette. The maximum estimated peak power of the UV pulse is up to P_{peak}~10^5 W at the λ =270 nm. The shortest pulse duration τ_3 of the UV pulse at such parameters is estimated to be 15 fs.

The experimental results can be described in terms of excitation of modes within a film or nanostructure. The intensity of the emitted third harmonic I_3 from the unit square of the surface can be expressed through the integral of the product of the Green function $G(x)$ and the nonlinear current $j_n(x)$ over film or nanostructure thickness d as $I_3 \sim |E_3|^2 = \left|\int_0^d G(x) j_n(x) dx\right|^2$. In turn, the nonlinear current can be expressed as $j_n = \chi^{(3)} \cdot E_1^3$, where E_1 is electric field at the fundamental frequency and $\chi^{(3)}$ is nonlinear susceptibility. It is clearly seen that the efficiency of the third harmonics depends not only on the intensity distribution of the nonlinear current $j_n(x)$ but also on the Green function $G(x)$, which characterize resonances of the structure. Thus the resonances on the frequency of the third harmonics can also enhance the emission of the radiation. Figure 3a shows the dependencies of THG efficiency η on the film thickness calculated using the Green function rigorously calculated for the planar structure. The fundamental field distribution cannot be found analytically for real geometry. However, for the experimental conditions the losses are high at the frequency of the third harmonic and thus only the radiation from a thin layer close to the interface is important.

5. Conclusion

In summary, we have applied self-organized laser-induced nanostructuring of thin Si film to fabricate all-dielectric metasurface, which provides resonance optical response at a pumping wavelength for third-harmonic generation. The proposed method is extremely convenient and fast, because one does not need to develop a specific design of a metasurface for the strongest field enhancement. The self-adjusted metasurface has demonstrated 2.5-fold improvement of THG as compared with an initial film with the highest conversion efficiency. This allows to achieve UV radiation at the wavelength 270 nm with high peak and average power (10^5 kW and 1.5 μW, correspondingly). Moreover, according to spectral width of the UV pulses, their duration could be as short as 15 fs.

Acknowledgments
This work was financially supported by Government of Russian Federation (RFMEFI58414X0009).

References

[1] Shcherbakov M R 2014 *Nano Letters* **14** 6488
[2] Makarov S, Kudryashov S, Mukhin I, Mozharov A, Milichko V, Krasnok A and Belov P 2015 *Nano Letters* **15** 6187
[3] Carletti L, Locatelli A, Stepanenko O, Leo G and De Angelis C 2015 *Opt. Express* **23** 26544
[4] Leuthold J, Koos C and Freude W 2010 *Nature Photonics* **4** 535
[5] Popa B-I and Cummer S A 2008 *Physical Review Letters* **100** 207401
[6] Zhao Q, Zhou J, Zhang F and Lippens D 2009 *Materials Today* **12** 60
[7] Evlyukhin A B, Reinhardt C, Seidel A, Lukyanchuk B S and Chichkov B N 2010 *Physical Review* B **82** 045404
[8] Garcia-Etxarri A, Gomez-Medina R, Froufe-Perez L S, Lopez C, Chantada L, Scheffold F, Aizpurua J, Nieto-Vesperinas M and Saenz J J 2011 *Opt. Express.* **19** 4815
[9] Miroshnichenko A E, Evlyukhin A B, Yu Y F, Bakker R M, Chipouline A, Kuznetsov A I, Lukyanchuk B, Chichkov B N and Kivshar Y S 2015 *Nature Communication* **6** 8069
[10] Bakker R M et al. 2015 *Nano Letters* **15** 2137
[11] Caldarola M, Albella P, Cortes E, Rahmani M, Roschuk T, Grinblat G, Oulton R F, Bragas A V, and Maier S A 2015 *Nature Communication* **6** 7915
[12] Decker M, Staude I, Falkner M, Dominguez J, Neshev D N, Brener I, Pertsch T and Kivshar Y S 2015 *Advanced Optical Materials* **3** 813
[13] Boyd R W 2003 *Nonlinear optics* (Academic Press)
[14] Shi L, Tuzer T U, Fenollosa R and Meseguer F 2012 *Advanced Materials* **24** 5934
[15] Moitra P, Slovick B A, Li W, Kravchenko I I, Briggs D P, Krishnamurthy S and Valentine J 2015 *ACS Photonics* **2** 692
[16] Abbarchi M et al. 2014 *ACS Nano* **6** 11181
[17] Dmitriev P, Makarov S V, Milichko V, Mukhin I, Gudovskikh A, Sitnikova A, Samusev A, Krasnok A and Belov P 2015 *Nanoscale* **8** 5043
[18] Kuznetsov A I, Evlyukhin A B, Goncalves M R, Reinhardt C, Koroleva A, Arnedillo M L, Kiyan R, Marti O and Chichkov B N 2011 *ACS Nano* **5** 4843
[19] Reininghaus M, Wortmann D, Cao Z, Hoffmann J M and Taubner T 2013 *Opt. Express* **21** 32176
[20] Kuchmizhak A, Ionin A, Kudryashov S, Makarov S, Rudenko A, Kulchin Y N, Vitrik O and Efimov T 2015 *Opt. Lett.* **40** 1678
[21] Makarov S V, Milichko V A, Mukhin I S, Shishkin I I, Zuev D A, Mozharov A M, Krasnok A E and Belov P A 2014 *Laser Photonics Reviews* **10** 91
[22] Le Harzic R, Schuck H, Sauer D, Anhut T, Riemann I and Konig K 2005 *Opt. Expess* **13** 6651
[23] Tang M, Zhang H and Her T-H 2007 *Nanotechnology* **18** 485304
[24] Huang M, Zhao F, Cheng Y, Xu N and Xu Z 2009 *ACS Nano* **3** 4062
[25] Bonse J, Rosenfeld A and Kruger J 2009 *Journal of Applied Physics* **106** 104910
[26] Vorobyev A, Makin V and Guo C 2009 *Physical Review Letters* **102** 234301
[27] Sipe J, Young J F, Preston J and Van Driel H 1983 *Physical Review* B **27** 1141
[28] Palik E D 1998 *Handbook of optical constants of solids* (Academic Press)
[29] Raether H 1998 *Surface plasmons on smooth surfaces* (Springer)
[30] Saeta P N and Miller N 2001 *Applied Physics Letters* **79** 2704
[31] Shcherbakov M R et al. 2015 *Nano Letters* **15** 6985
[32] Yang Y, Wang W, Boulesbaa A, Kravchenko I I, Briggs D P, Puretzky A, Geohegan D and Valentine J 2015 *Nano Letters* **15** 7388

Ultrathin and Nanostructured Au Films with Gradient of Effective Thickness. Optical and Plasmonic Properties

S V Tomilin[1], V N Berzhansky[1,2], A N Shaposhnikov[1], A R Prokopov[1], E T Milyukova[1], A V Karavaynikov[1], O A Tomilina[2]

[1]Research Centre of Functional Materials and Nanotechnologies, Physical-Technical Institute, Vernadsky Crimean Federal University, Crimea, Simferopol 295007, Russian Federation
[2]Experimental Physics Department, Physical-Technical Institute, Vernadsky Crimean Federal University, Crimea, Simferopol 295007, Russian Federation

Abstract. In present work the results of investigation of optical (transmission spectra) and plasmonic (surface plasmon-polariton resonance) properties of ultrathin and nanostructured Au films are presents. Methods and techniques for the syntheses of samples of ultrathin and nanostructured metallic films, and for the experimental studies of optical and plasmonic properties are representative. Au films on SiO_2 (optic glass) substrates were investigated.

1. Introduction
The interaction of light radiation with the metallic low-dimensional objects (nanofilms, nanoparticles) is determined by the structure of the electronic spectrum of these objects. For the films and nanoparticles with the band type of the energy spectrum, the conduction electrons are the plasma with a specific resonant frequency of its oscillations. The effect of plasmon resonance excitation in nanoislet metallic films can be observed in the optical transmission spectra as the spectral dips at the resonant wavelength. Another method of resonance surface plasmons excitation based on the phenomenon of plasmon and polariton waves phase matching when the effect of total internal reflection of plane-polarized coherent light beam is executed. In this case, the plasmon resonance is observed as an intensity reduction of the reflected light beam when the light is incident under the "resonance angle" [1,2].

2. Experimental technique
The samples of ultrathin and nanostructured Au films were prepared by thermal evaporation and condensation in vacuum. As the substrates were used the plates of optical glass (SiO_2) with a refractive index n = 1.545 and the plates of gallium-gadolinium garnet (GGG). Metal evaporation was carried out from molybdenum (Mo) boats at a residual pressure of atmospheric gases not more than $5\cdot10^{-4}$ Pa. Before deposition the substrates were annealed in vacuum at T = 450°C during 30 min. The deposition of Au was carried out on substrates at 150°C. At the deposition process the original technique for obtaining the films with a gradient of thickness along the sample plane was used. [3]. This technique allows one to research films with different thicknesses, which are obtained in a single technological cycle.

So in one cycle the four samples (3 on GGG substrate and 1 on SiO_2 substrate) were obtained. The maximum thickness of the coating was 15 nm, and the minimum approached to zero. The sample length was about 15 mm, so the gradient of effective thickness grad $h \approx 1$ nm / mm.

Then, for the nanoislet structures formation the samples of Au films on GGG substrates were annealed in air at atmospheric pressure at temperatures 750, 850 and 950°C during 10 min. The sample of Au film on the SiO_2 substrate didn't annealed and was used for investigations of plasmon-polariton resonance by Kretchmann's method.

The analysis of surface morphology of nanoislet Au films was maked using Scanning Electron Microscope REM-106 (SELMI).

The investigation of transmission spectra of ultrathin and nanostructured Au films was carried out using an automated spectrophotometer KFK-3. Spectral range was 400–970 nm.

Surface plasmon resonance (SPR) was excited by the Kretschmann's method using a triangular prism (optical glass, n = 1,514). For plasmon excitation normally-polarized light beam (TM mode) with a wavelength $\lambda = 632.8$ nm (He-Ne laser) was used.

3. Results and discussion

Figure 1 shows the results of investigation of the surface morphology of nanoislet Au film on a GGG substrate, annealed at 850°C. The studies were carried in areas with different original thickness of coating, for islet films it is the effective thickness h_{eff}. The step value along the thickness gradient was 1 mm. As it was found the annealing temperature significantly affects to the morphology parameters of nanoislet coatings and, as result, its optical properties. But, the limited scope of this article does not allow us to show the results of morphology investigation of films, who was annealed at temperatures of 750°C and 950°C.

The present images (Figure 1) clearly show that the parameters of nanoislet (size, surface concentration, etc.) significantly depend on the effective thickness of the annealed coating. Table 1 shows the parameters of morphology nanoislet coating in investigated areas.

Table 1. Morphology parameters of nanoislet Au film in different investigated areas.

Step (mm)	Effective thickness h_{eff} (nm)	Islands concentration (μm^{-2})	Islands size dispersion (nm)	Maximum of dispersion (nm)
1	14.4	3.6	100–450	230
2	11.5	4	100–400	225
3	10.3	4.4	100–300	215
4	9.1	5.4	50–300	200
5	7.6	6.5	50–300	185
6	5.9	7.1	50–200	148
7	5.2	9	50–200	127
8	4.6	8.5	50–200	125
9	4.2	7.9	50–200	122
10	3.9	8	50–200	120
11	3.1	7.2	50–200	105
12	2.7	8.3	50–200	100
13	2	10.3	50–200	90
14	1.6	12	50–150	84

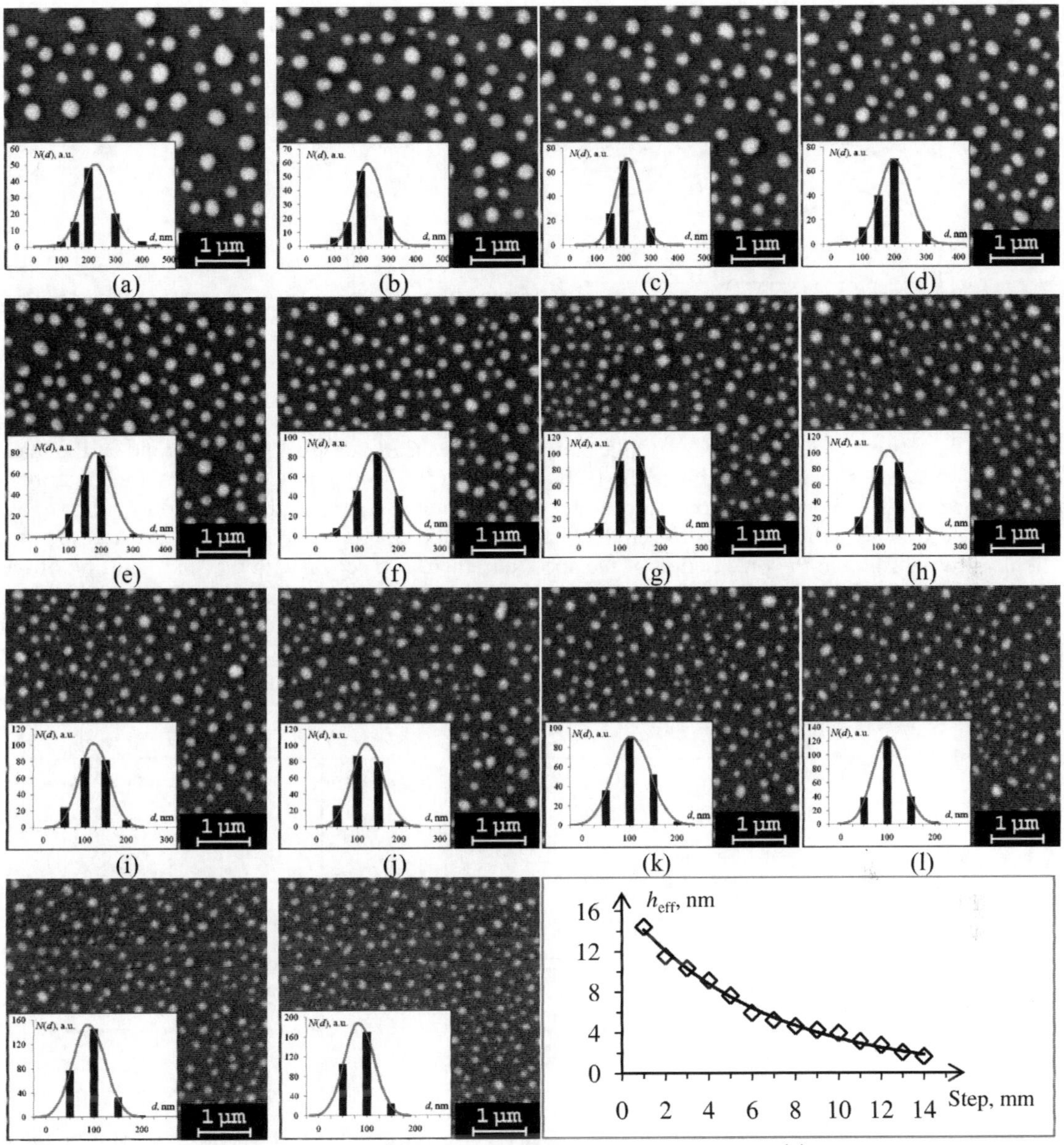

Figure 1. SEM-study of surface morphology in different areas of nanostructured Au film with gradient of the effective thickness. Film was annealed at 850ºC, 10 min. Insets shows the dispersion of islands size: columns are experimental data, line is Gaussian approximation. (a–n) – the different areas of thickness gradient, step is 1 mm; (o) – change of h_{eff} along thickness gradient)

The researches of transmission spectra of Au gradient films in areas with different effective thickness are gave the next results. Figure 2 shows the transmission spectra of unannealed Au film on GGG substrate. The legend shows the values of the film effective thickness in the investigated areas. As one can see, the transmission spectra correspond to the volume absorption of Au, so films are solids [4].

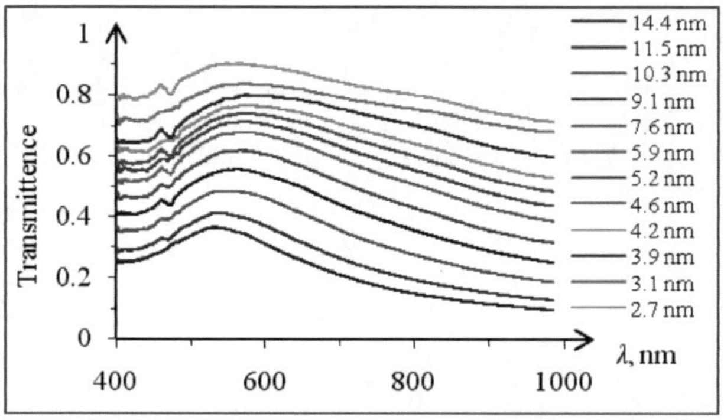

Figure 2. Transmission spectra of unannealed Au nanofilm on GGG substrate, which are measured in different areas along the gradient of thickness.

Figure 3 shows the transmission spectra of the Au film after annealing at 750°C, 10 min in areas with different effective thickness h_{eff}. The spectrum analysis has led to the following conclusions. At first, at the present spectra the peaks of plasmon resonance clearly defined [5]. At second, at the thick coating two peaks are observed: the first "blue" peak (500–540 nm) corresponds to quadrupole resonance, the second "red" peak – to dipole [6]. It is noted that the peak position of the dipole resonance strongly depends to h_{eff}. When the thickness of the coating is decrease the dipole resonance is not always shifts to the blue region of the spectrum, there is a sector with the "red" shifting of the resonance peak (figure 3(b)). This effect may be related to a change of islands size in the normal direction to a plane of the substrate, namely with the formation of islands with spheroidal form and with a different aspect index. The position of the quadrupole resonance peak of almost does not change with h_{eff}, and for the thinner areas of nanoislands Au film the quadrupole resonance is not observed.

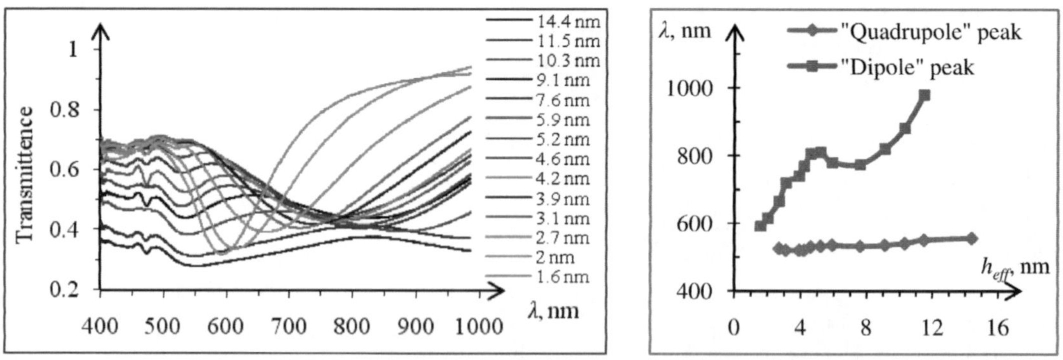

(a) (b)

Figure 3. (a)Transmission spectra in different areas of Au nanofilm on GGG substrate with gradient thickness. Film was annealed at 750°C during 10 min. (b) Shift of plasmon peaks for different effective thicknesses of Au nanoislet film.

The transmission spectra of Au film on GGG substrate, annealed at a temperature of 850°C, are shown in Figure 4. The difference from the previous sample is the changes in position of the dipole resonance peak in the thicker areas of coating. (figure 4(a)), and the relocation of sector of "red" shifting of dipole resonance into a region with smaller thickness (figure 4(b)).

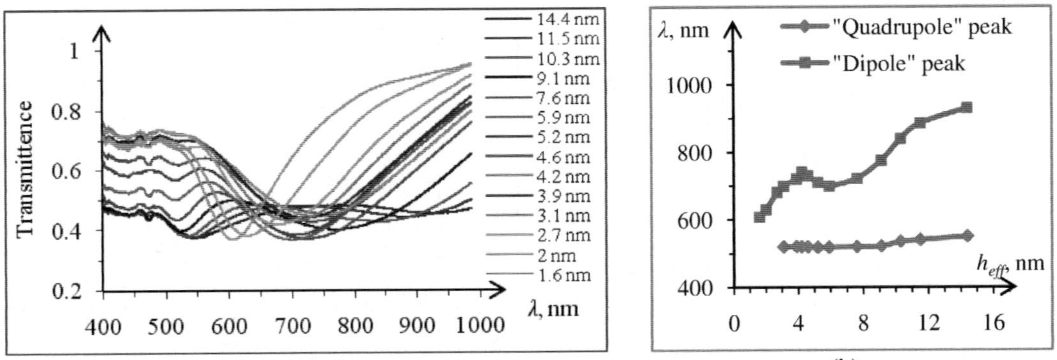

(a) (b)

Figure 4. (a)Transmission spectra in different areas of Au nanofilm on GGG substrate with gradient thickness. Film was annealed at 850°C during 10 min. (b) Shift of plasmon peaks for different effective thicknesses of Au nanoislet film.

Figure 5 shows the transmission spectra in different areas of the effective thickness gradient of the nanoislet Au film after annealing at 950°C. Compared to the samples annealed at 750°C and 850°C for this film a "blue" shift of the dipole plasmon resonance peaks is observed, and a small decreasing in the region where for the low annealing temperature "red" shifts are observed (figure 5(b)).

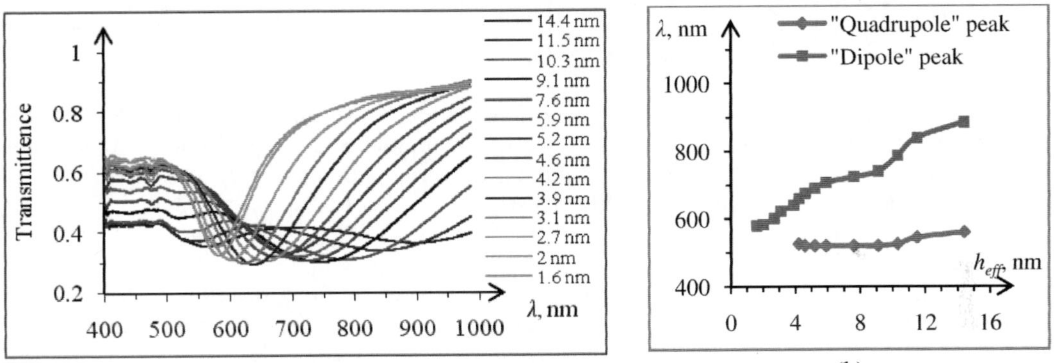

(a) (b)

Figure 5. (a)Transmission spectra in different areas of Au nanofilm on GGG substrate with gradient thickness. Film was annealed at 950°C during 10 min. (b) Shift of plasmon peaks for different effective thicknesses of Au nanoislet film.

The investigations of surface plasmon-polariton resonance (SPR) in a gradient Au film on SiO_2 substrate, excited by Kretchmann's method gave the next results. Figure 6 (a) presents the normalized resonance curves of the reflected beam intensity in dependence to incidence angle (in the legend the relocation of investigated area along h_{eff} gradient are shown, the values of the thickness see in table 1). It can be seen that for all studied areas the SPR peak is observed, besides the last three areas [7]. The absence of the resonant peak for the most thin parts of Au film may be indicative about island structure of coating. The structural heterogeneity of the coating does not allow realizing the SPR [8]. Figure 6 (b-d) shows the main parameters of SPR peaks. It has been shown, that when h_{eff} is decreased, the quality coefficient of the resonant peak is decreased, the resonance angle is increased and the height of peak is decreases (increases the intensity of the reflected beam).

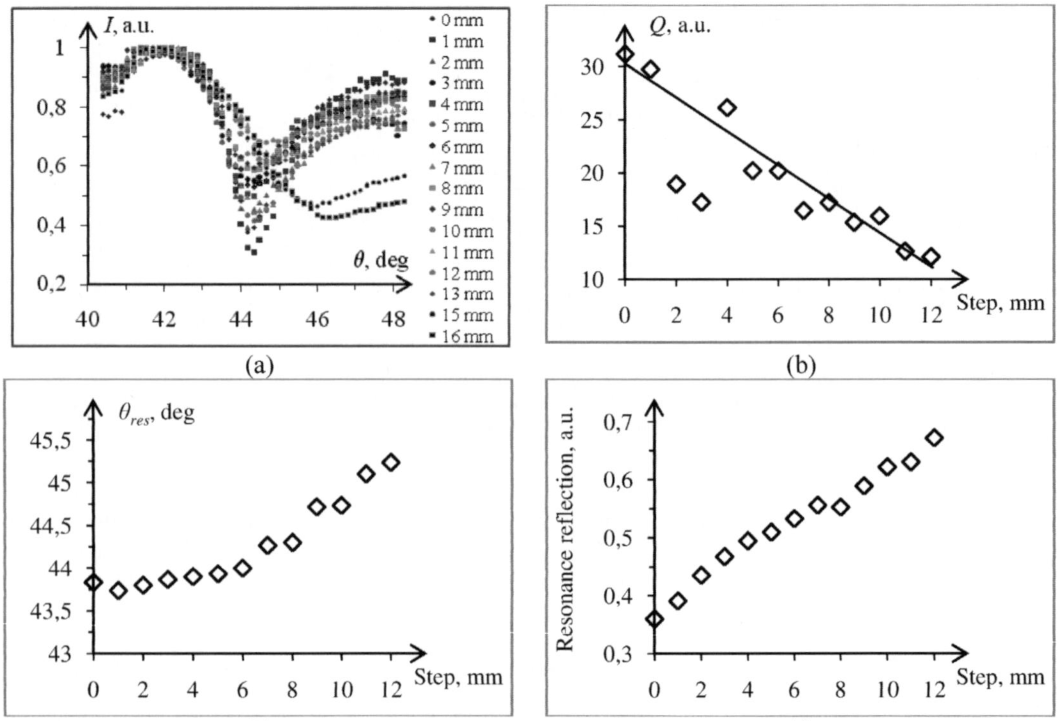

Figure 6. (a) SPR resonance curves for Au nanofilms with different thicknesses (in different areas along thickness gradient) on SiO_2 substrates. In legend the translocation (step in mm) of studied area from maximum film thickness (15 nm) to minimum are specified. Parameters of SPR: (b) quality factor, (c) resonance angle, (d) resonance reflection.

Conclusions

It is shown that the optical and plasmonic properties of ultrathin Au films are strongly depend on the coating structure. So for "thick" nanofilms with continuous structure of the coating the peak of surface plasmon-polariton resonance is observed, while in transmittance spectra the resonance absorption is absent. For ultra-thin films with nanoislet structure of coating the resonance absorption in the transmission spectra is observed, but clearly defined SPR peak is absent. For islet films the presence of a discrete quasi-regular structure leads to the beam energy absorption at incidence angles higher than the resonance SPR angle.

Acknowledgments

This work is supported by the RF Ministry of Education and Science in the framework of the base part of the state task №2015 / 701 (project 3879) and the State Council of Republic Crimea in the framework of the grant for young scientists of the Crimea (resolution № п170-1/16 from 02.02.2016).

References

[1] Chentsova E I, Dubovik V I, Khovivchak V S. 2012 *Bulletin of Omsk Univ.* **2** 110.
[2] Bremer J, Vaicikauskas V, Hansteen F, Hunderi O. 2001 *J. of Applied Physics*. vol. **89 11** 6177.
[3] Tomilin S V, Yanovsky A S. 2013 *J. of Nano- and Electronic Physics*. vol. **5 3** 03014.
[4] Axelevitch A, Apter B, Golan G. 2013 *Optics Express*. vol. **21 4** 4126.
[5] Kossoy A, Merk V, Simakov D, Leosson K. 2015 *Adv. Optic. Mater*. vol. **3 1** 71.
[6] Rodrıguez-Fernandez J, Perez-Juste J, Garcıa de Abajo F J. 2006 *Langmuir* **22** 7007.
[7] Kostukevich E V, Kostukevich S A. 2014 *Optoelectronics and Semicond. Tech.* **49** 60.
[8] Frumin L L, Nemykin A V, Perminov S V, Shapiro D A. 2013 *J. Opt.* **15** 085002.

Morphology and optical properties of aluminum oxide formed into oxalic electrolyte with addition surface active agents

B Kazarkin[1], A Stsiapanau[1], V Zhilinski[2], A Chernik[2], V Bezborodov[2], G Kozak[3], S Danilovich[3] and A Smirnov[1]

[1] Department of Micro- and Nanoelectronics, Belorussian State University of Informatics and Radioelectronics , Brovki Str., 6, Minsk, 220013, Belarus
[2] Belarusian State Technological University, Sverdlova Str., 13a, Minsk, 220006, Belarus
[3]Scientific Production Unitary Enterprise "Display", Gikalo Str., 5, Minsk, 220005, Belarus.

Abstract. The article discusses the results of investigations of porous films of alumina, formed into oxalic electrolyte with addition surface active agents, in particular, ordering structure, roughness of a surface, the optical transparency of the electrolyte concentration and surface active agents. Also discusses the features of the formation of porous films of temperature and IR radiation.

1. Introduction

Presently the quality of image, formed by devices on the basis of liquid-crystal materials became much higher and these devices fully forced out cathode-ray tubes. The best combination of electrical conductivity, optical transmission and processability processing using physicochemical methods etching demonstrates film of mixed oxide In_2O_3-SnO_2. However, the world's proven reserves of indium is very limited, and in the next few years the industry will inevitably face its significant deficit. Therefore, the search for new alternative materials with improved performance and processability, as well as economic parameters, is relevant and practically important problem [1-4]. The purpose of this research is to investigate the process of formation and growth of electrochemical-based porous structural alumina in the presence of additives and anisotropic surfactants.

2. Methodology

Nanoporous alumina formed into oxalic electrolyte with addition surface active agents. As additives used anisotropic ammonium salt 3- (4-hexylphenyl) -N, N-dimethyl-3-oksopropanammoniy chloride (A1) and 3- (4-methoxyphenyl) -N, N, -dimethyl-3-oksopropanammoniy chloride (A2) trimetildodetsilammony bromide (A3) and dimetilbenzildodetsilammony bromide (A4) in an amount of 0.1 g/l to 0.7 g/l. Material of cathode was stainless steel.

The samples were a glass substrate with a layer of aluminum 1 micron thick, deposited by vacuum evaporation. The dimensions of substrates was 20 x 10 mm. A layer of chemically resistant lacquer, which separated the direct surface anodizing and place of fixation plates, was deposited on the substrate.

For anodised samples used DC power supply brand MNIPI B5-78/7. To control the temperature values using a magnetic stirrer with heating DRAGONLAB MS7-H550-Pro. Digital registration of chronoamperogrammes made with a multimeter APPA 107. Microphotographs of the surface of anodized aluminum made on AFFRI-DM8 device at magnifications of X400. The morphology of the coatings were examined using a scanning electron microscope JSM-5610 LV chemical analysis system EDX JED-2201 JEOL. Optical transmittance measurements were carried out using a spectrophotometer SF-26.

Chronoamperogrammes of aluminum anodizing are shown in Figure 1.

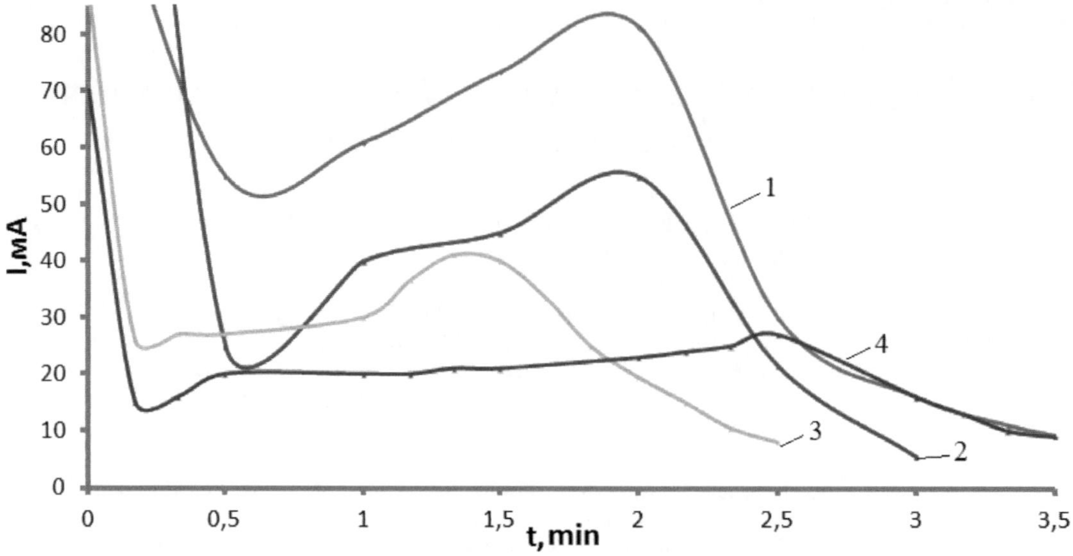

1 – A1; 2 – A2; 3 – A3; 4 – A4

Figure 1. Chronoamperogrammes of anodizing aluminum at U = 50 V in a 0.5 M oxalic acid with the addition of additives.

Presented chronoamperogrammes have a classical form. When applying a DC voltage begins the growth of the barrier layer of the anode, creating significant resistance, which leads to a drop in the current start time of anodization.

3. Results

Due to the presence in the electrolyte of oxalic acid happens loosening dense layer and forming a porous nucleation structure, that is shown on an increasing anode current. Further current drop indicate complete transformation aluminum in anodic alumina.

Pore diameter in the investigated electrolytes on the surface of aluminum oxide was in the range of from 0.4 to 0.9 microns. With increasing concentration of additives increase the pore size is observed.

The optically transparent aluminum oxide film with a transmission coefficient of 50-55% in the wavelength range 450-650 nm obtained by introducing additives A1 into the electrolyte. Use A2 reduces transmittance by 10%.

Using A3 and A4 additives allows to obtain alumina having a degree of transmittance of 40 and 35% respectively. It should be noted that increasing the concentration of additives in the electrolyte is in the range of 0.1 - 0.5 g/l leads to a decrease in the degree of light transmittance alumina with the additive A3 and increase with the additive A4.

Furthermore, when the anodizing of aluminum in solution with 0.3 M oxalic acid at a temperature of from 4 to 20 °C and a voltage of 30-80V for 3 - 420 minutes under the influence of infrared radiation, we found that the formation of the small pore structure of the film is observed only at elevated temperatures, which is associated with an increase in the chemical reaction of aluminum oxide etch rate when heated.

The research data of the surface microprofile shows that an average of the samples with small pores roughness parameters Ra = 0,015 microns, Rz = 0,0706 microns, Rmax = 0,1009 microns. The maximum depth of the pores of 0.03 microns.

The dependence of the optical density of the wavelength for the sample (anodization with 60V, 10 °C) after the infrared radiation 10 000 lm/m^2 are shown in Figure 2.

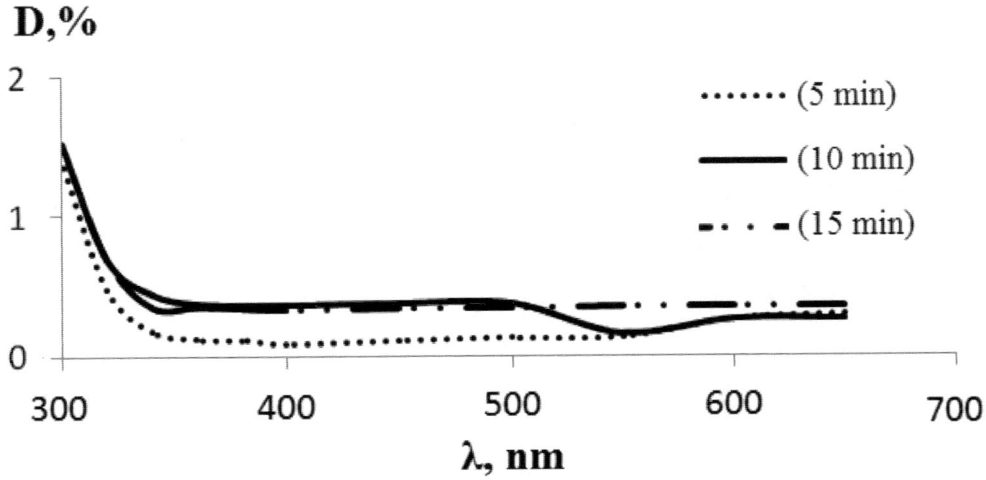

Figure 2. The optical density of the wavelength dependence of the sample anodized aluminum (anodization at 60 V, 10 ° C) after irradiation of infrared 10000 lm/m^2 for 5, 10 and 15min.

Figure 2 shows that the longer the sample is exposed to infrared emission, the greater its increased absorbance in the visible light spectrum. To restore the ability of transmitting the sample was kept in the dark for 5 minutes and recovered to a normal state. This fact indicates that the flow of the remaining adverse oxidation of aluminum with thermal stimulation.

4. Conclusions

Thus, researches have shown that the use of oxalic acid allows to obtain the most ordered, repeatable and uniform across the square anodized aluminum films. In this case variation of the additives and their concentrations allows you to change the ratio and pore sizes of aluminum films; that transparent nanostructured aluminum layers are obtained by self-organized growth of porous anodic alumina have a roughness parameters Ra = 0,015 microns, Rz = 0,0706 microns, Rmax = 0,1009 microns and are characterized by high chemical activity in IR heating. These films can be used as a dimming glass IR optics, display technology for various purposes.

References

[1] Fu-Lai Y T 2014 Technological catching-up in East Asia (*International Economic Development: Leading Issues and Challenges*) pp 78-92.

[2] Hubarevich A, Marus M, Stsiapanau A, Smirnov A, Zhao J, Fan W, Wang H, Sun X W 2015 Transparent conductive nanoporous aluminium mesh prepared by electrochemical anodizing *Physica Status Solidi* A **212** pp 2174-2178.

[3] Самарин А 2002 Жидкокристаллические дисплеи: схемотехника, конструкция, применение *М.: Наука* p 310.

[4] Carter C 2006 Modification of indium-tin oxide surfaces: enhancement of solution electron transfer rates and efficiencies of organic thin-layer devices (*The University of Arizona: dis. degree of doc. of phil*) p 324.

Dry e-beam etching of resist for optics

A Rogozhin[1], M Bruk[1, 2], E Zhikharev[1], D Streltsov[3], A Spirin[2], J Hramchihina[1, 4]

[1] Institute of Physics and Technology of RAS, Moscow 117218, Russia
[2] L.Ya. Karpov Institute of Physical Chemistry, Moscow 105064, Russia
[3] Enikolopov Institute of Synthetic Polymer Materials of RAS, Moscow 117393, Russia
[4] Moscow Institute of Physics and Technology, Moscow, 141700, Russia

Abstract. Method of dry e-beam etching of resist (DEBER) is described. It appears that the method could be extremely useful for formation of wide range of structures for optics and optoelectronics. It is relatively simple to form diffraction or binary gratings, some diffractive optical elements (DOE), 3D structures or planar photonic crystals. Method could be realised in any focused e-beam induced process (FEBIP) system or in e-beam lithographer with minor modifications. DEBER method is significantly more productive than standard or grayscale e-beam lithography. Typical exposure time for 3x3.9 mm^2 area is about 10-100 s. Examples of structures formed by the DEBER method that could be used in optoelectronics are presented.

1. Introduction

Dry electron beam etching of resist (DEBER) proposed by Bruk et. al. [1, 2] could be used for formation of wide range of optical or optoelectronic structures. The method is based on the chain depolymerization reaction, which takes place in the polymer resists during e-beam exposure at the glass-transition or higher temperatures. The volatile reaction products (monomers) are pumped out during exposure.

The method provides quite simple way for formation of well-rounded or 3D structures. In some cases it could be much more flexible than usual methods, in others it could be more productive or convenient.

In this paper we compare DEBER method with usual formation techniques for diffraction or binary gratings, some diffractive optical elements (DOE), 3D structures or planar photonic crystals. Benefits and drawbacks are analyzed.

2. Dry e-beam etching of resist

The DEBER method is based on the chain depolymerization reaction which takes place in the polymer resists during e-beam exposure at the glass-transition or higher temperatures. Volatile reaction

Journal of Physics: Conference Series **741** (2016) 012115 doi:10.1088/1742-6596/741/1/012115

products (monomers) are pumped out during exposure (figure 1). Various resists that could be effectively decomposed to monomer under these conditions can be used in the method (poly(methyl methacrylate), poly-α-methylstyrene, polymethyl isopropenyl ketone etc.).

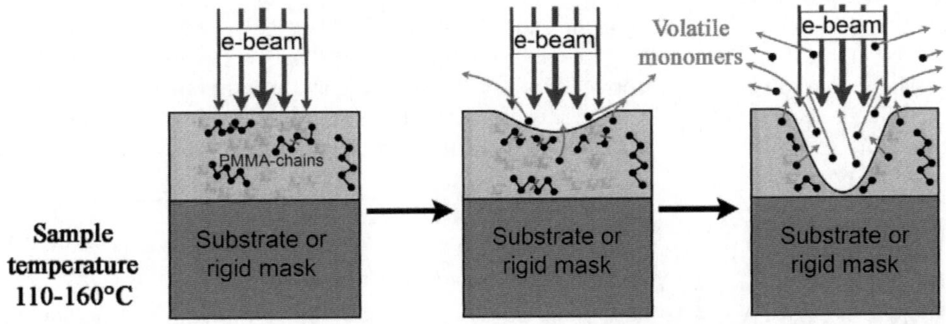

Figure 1. Dry e-beam etching of resist process scheme.

In the DEBER method at a temperature higher than glass-transition temperatures e-beam stimulated chain depolymerization reaction takes place. In this process polymer bonds are broken during e-beam exposure. As a result molecules of terminal macroradicals are formed. These macroradicals at increased temperatures split off monomer molecules one by one with "zipper" mechanism [2, 3]. Lots of monomer molecules are evacuated during exposure. The process is faster at higher temperatures. The relief (trenches or holes) formation is defined by appearance in the exposed region free space due to evacuation of volatile monomers and polymer relaxation as affected by surface tension. Well-rounded shape of the structures is determined by a specific form of the etching kinetic curves of DEBER method [4].

PMMA sensitivity to e-beam in the DEBER method is about 100 times higher than that in the standard "wet" e-beam lithography process. Because of high vertical resolution (about 1 nm) the metod could be used for high-precision 3D structuring. On the other hand DEBER lateral resolution (about 100 nm) and contrast (0.7 – 1.5) are rather low.

DEBER process could be implemented in some scanning electron microscopes (SEM), e-beam lithography or focused e-beam induced process (FEBIP) systems. Most of the systems require minor modifications for it.

It is possible to transfer relief after DEBER process from the resist to the silicon or fused silica substrate or to the metal mask [4].

3. Diffraction gratings

Three main technologies are used for diffraction grating formation: ruling, holographic technology and replication [5]. Master grating for replication could be obtained by ruling, holographic or DEBER method. Ruling is a complex and low-productive process but it provides the grating of the highest efficiency. Moreover very large grating can be obtained by ruling. On the other hand light scattering could be relatively high for ruled gratings.

Holographic gratings have sinusoidal shape. Their efficiency could be high for the radiation of some wavelength range. Holographic gratings may offer advantages to spectroscopic systems in which light scattered from the grating surface is performance-limiting. Holographic method is high-productive. Only some special resists could be used for holographic method.

Simple gratings obtained by DEBER method are quite similar to holographic gratings (figure 2). They also have sinusoidal shape and limited efficiency. Throughput for DEBER method is lower than ruling or holographic method but it is high enough for small gratings production. DEBER exposure time of 3x3.9 mm^2 is about 10-100 s (exposure dose 0.1-1 $\mu C/cm^2$). DEBER method is extremely accurate due to e-beam system exploitation. Moreover the shape of the grating produced by DEBER method could be modified. The shape of the grooves could be skewed, for example. Also complex grating like binary gratings could be produced by DEBER method if e-beam lithography or FEBIP system is used. Large set of positive resists could be used with DEBER method.

4. Diffractive optical elements

Diffraction optical elements (DOE) contain staircase structures. Usually these structures are formed by grayscale e-beam lithography or multiple e-beam lithography (EL) processes followed by plasma etching. The main problems are incorrect depth of the structure and misalignment. These effects lead to the significant decrease of the DOE efficiency. Grayscale e-beam lithography provides low vertical resolution. And for multiple EL processes it is difficult to align them properly. Both processes are extremely low-productive.

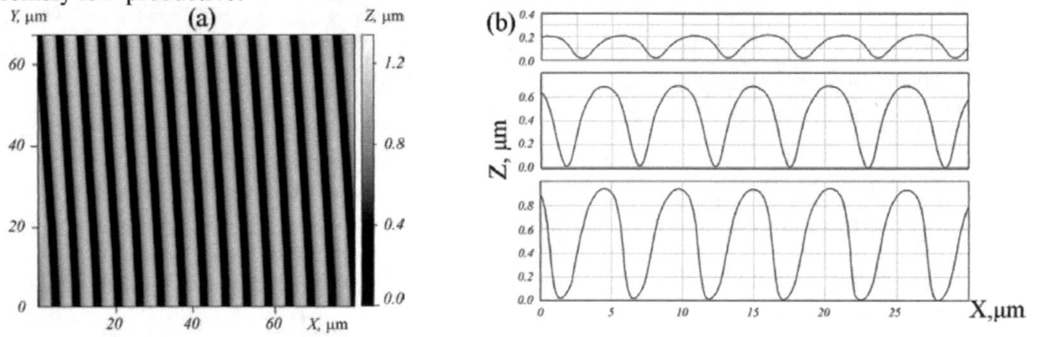

Figure 2. AFM image (a) and profiles (b) of diffraction gratings obtained by DEBER method.

Formation of staircase structures by DEBER process is quite simple (Figure 3). The technique is similar to grayscale e-beam lithography. But due to specific etching mechanism vertical resolution of the DEBER method is 1-2 nm. So incorrect depth is not a problem for the method. DEBER is a one-step process and misalignment also does not take place for the method. The method also is high-productive.

Drawbacks of DEBER method here are edge rounding and low lateral resolution. Edge rounding leads to insignificant efficiency decrease [6]. It appears that if the DEBER method is implemented in the modern e-beam lithography or FEBIP system lateral resolution will be enough for DOE fabrication.

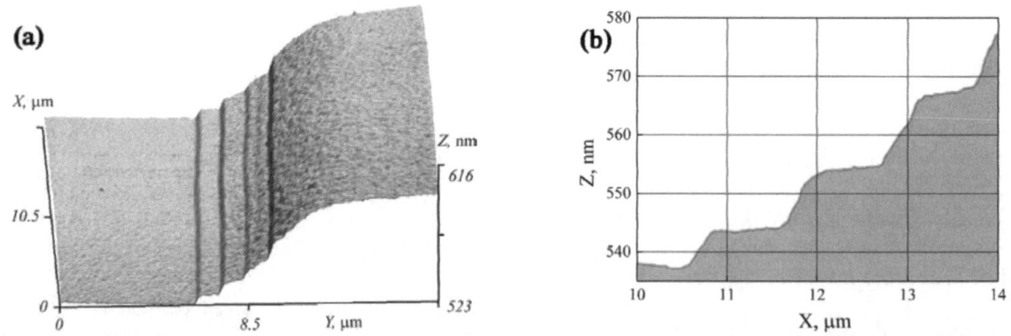

Figure 3. 3D AFM image (a) and profile (b) of staircase structure obtained by DEBER method; the structure was obtained in the Zeiss Ultra-55 SEM by the sequential exposure of overlapped square areas.

5. Other 3D structures

Staircase structures could be also used for angle resolved spectroscopy. DEBER method could be useful for formation of different 3D structures due to high productivity. Microlens arrays, focusators, waveguide couplers, reflecting devices could be fabricated by the DEBER method if it is implemented in e-beam lithography or FEBIP system. The method works here like grayscale e-beam lithography with high vertical resolution, accurate lateral positioning but limited lateral resolution.

6. Binary gratings and planar photonic crystals

DEBER method could be used not just for simple diffraction grating formation but also for formation of complex binary gratings or planar photonic crystals (figure 4). In this case DEBER works like standard e-beam lithography with limited lateral resolution. On the other hand it seems that edge rounding could somewhat decrease efficiency of these structures.

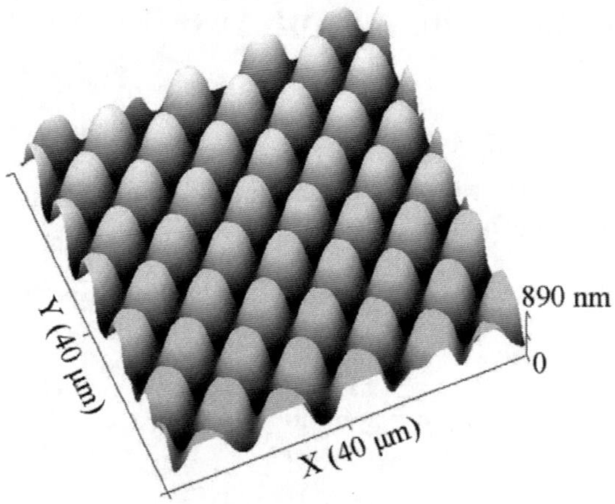

Figure 4. 3D AFM image of structure obtained by DEBER method during exposure of two sets of lines.

7. Conclusion

It seems that DEBER method can be used for formation of wide range of structures and devices for optics and optoelectronics (diffraction and binary gratings, DOE, planar photonic crystals, microlens arrays etc.). If the resist is an appropriate material for application, DEBER method could be used without any additional procedures. In other cases the pattern could be transferred in the substrate material by anisotropic etching. Also the pattern could be used as a model for replication, stamps and matrices.

References

[1] Bruk M A, Zhikharev E N, Kalnov V A, Spirin A V and Streltsov D R, Method of masking pattern formation in positive e-beam resist layers 2013 *Russian Bulletin of Inventions* **9** patent 2478226.

[2] Bruk M A, Zhikharev E N, Streltsov D R, Kalnov V A and Spirin A V 2013 *Microelectronic Eng.* **112** 1.

[3] Bruk M A, Kondrat'eva M V, Baranov A A, Pebalk K V, Sergeev A M, Kozlova N V 1999 *Polymer Science. Series A.* **41** 159.

[4] Bruk M A, Zhikharev E N, Rogozhin A E, Streltsov D R, Kalnov V A, Averkin S.N. and Spirin A V, Formation of micro- and nanostructures with well-rounded profile by new e-beam lithography principle 2016 *Microelectronic Eng.* **155** 92.

[5] Loewen E 2005 *Diffraction Grating Handbook* (New-York:Newport corp.).

[6] Kemme S 2010 *Microoptics and Nanooptics Fabrication* (Miami: CRC Press).

Design and compressibility of Langmuir monolayers from organometallic nanocyclic complexes

H V Grushevskaya, N G Krylova, I V Lipnevich

Belarusian State University, 4 Nezavisimosti Ave., Minsk 220030, Belarus
E-mail: krylovang@bsu.by

Abstract. We propose to control design of organometallic conducting Langmuir monolayer by utilizing geometrodynamic approach. By means of this approach the compressibility of the monolayer, which consists of molecules of thiophene-pyrrole series oligomer with covalently bound hydrophobic alkyl chain, has been analyzed for different subphase with iron salts. It has been discovered that the monolayers are polymerized and turn into monolayers of nanocyclic organometallic coordination compounds at a two-dimensional phase transition of the first order.

1. Introduction

At present, conducting polymers are perspective materials for effective optimization of electrical and electro-chemical characteristics of nanosensors [1-3]. One of the promising directions to design ultra-thin films of metal-polymer compounds is their self-organized formation from metal complexes of macrocyclic coordination compounds [4]. The Langmuir – Blodgett (LB) technique is a nanothechnology to fabricate monolayers self-assembling on liquid-subphase surface in the process of a two-dimensional (2D) first-order phase transition from 2D gas up to condense state of monolayer.

The consideration of monolayer compressibility κ can provide new information about thermodynamics and structural characteristics of the monolayers. The experimental determination of the surface compressional modulus (K) in dependence on the area per molecule (A) shows a nonlinear dependence K on A [5]. The K value reaches a maximum in two phase transition points and then decreases nearly to zero while the area per molecule decreases within the phase transition region named isotherm plateau.

However, the known procedures (e.g., [6]), which handle the statistical analysis of experimental data set for the monolayer compressibility κ in the whole region of surface pressure π, allow us to determine only the mean value of the compressibility coefficient – the "apparent" compressibility C' equals to $1/K$ in relatively linear sections of the isotherms.

Besides this, as known [7-11], in the Langmuir monolayer one finds two-dimensional lattices of metal ions underneath the air–water interface and these lattices evolve in structure with pH, ion concentration, and time. Therefore, new methods for control of these monolayers states during time have to be developed. In this paper we use a geometrodynamic approach [12,13] to perform statistical analysis of compression isotherms for metastable Langmuir monolayers. It allows us to discover features of the compressibility coefficient κ based on experimental data concerning the 2D phase transitions of first order at different compression rates.

A goal of this paper is to study the compressibility of monolayer from coordination complexes of thiophene-pyrrole series oligomer on surface of different iron-salt subphases and propose a method for design of LB-films according to features of monolayer-compressibility behavior at 2D phase transition of the first order.

2. Materials and methods

Salts $Fe(NO_3)_3 \cdot 9H_2O$ и $FeCl_3$ (Sigma-Aldric, USA), hydrochloric acid, deionized water were used to preparate subphases. Monolayers and Langmuir – Blodgett films was formed from amphiphilic oligomer of thiophene derivatives: 3-hexadecyl-2,5-di(thiophen-2-yl)-1*H*-pyrrole (H-DTP, H-dithionilepyrrole). H-dithionilepyrrole was synthesized by a method proposed in [14].

All salt solutions have been prepared with deionized water with resistivity of 18.2 MΩ·cm.

Iron-containing Langmuir monolayers are fabricated on an automated home-built Langmuir trough with controlled deposition on a substrate [15]. Hexan solution of DTP was dripped with the help of a micropipette on a liquid subphase surface. Langmuir monolayers were obtained by compressing of DTP molecules on air / aqueous subphase interface. Control of the surface tension was performed by a highly sensitive resonant inductive sensor. The Y-type transposition of monolayers on supports was performed by their vertical dipping. The complexes $Fe(II)DTP_3$ of high-spin Fe(II) with thiophene-pyrrole ligands were synthesized by LB-technique at compression of H-dithionilepyrrole molecules on the surface of subphase with salts of three-valence Fe [15]. $Fe(NO_3)_3$ и $FeCl_3$ solutions were used as subphases for the monolayer formation.

Spectral studies in visible range were carried out using a confocal micro-Raman spectrometer Nanofinder HE ("LOTIS-TII", Tokyo, Japan–Belarus) by laser excitation at wavelength 532 nm and spectral resolution equals 0.1 nm.

Valent states of transition metal centers in coordination-complexes, from which iron-containing LB-monolayers are formed, have been detected by Mössbauer spectroscopy. Mössbauer spectra were registered in transmission geometry at room temperature utilizing spectrometer MS2000 with source of $^{57}Co/Rh$ (40 mCu). All isomer shifts are represented relative to α-Fe.

3. Thermodynamical analysis

Typical experimental dependences of the surface pressure π upon the area per molecule A (compression $\pi - A$-isotherm) for DTP-monolayer fabricated on surfaces of $Fe(NO_3)_3$ - and $FeCl_3$ - containing subphases are shown in figure 1. As one can see, the phase transition occurs in a monolayer formed on $FeCl_3$ subphase at less surface pressure value than in a monolayer formed on subphase with $Fe(NO_3)_3$. The corresponding dependences $\kappa(\pi)$ at different compression rates V have been approximated by theoretical curves, as represented in figure 1 (red dashed lines). The phase-transition width $\Delta\pi_{pht}$ is determined as a difference between two singular points on a theoretical curve. In figure 2 dependences of $\Delta\pi_{pht}$ on compression rates V are shown for monolayers fabricated on $FeCl_3$ - and $Fe(NO_3)_3$ -containing subphases. $\Delta\pi_{pht}$ is seen to decrease with increase of V, at that, this dependence behaves as $1/V^3$. $\Delta\pi_{pht}$ was established to be larger for the monolayer formed on subphase with $Fe(NO_3)_3$ than for the monolayer formed on subphase with $FeCl_3$. In contrast to a model system of Van der Waals gas type, the theoretical dependence of compressibility on π changes a sign twice and, consequently, describes metastable state well. Thus, a range of values of surface pressure, at which phases coexist due to their instability, diminishes with rise of the compression rate V and, consequently, the plateau degenerates into a tricritical point. Respectively, increase of the rate V results in the monolayer turns into a glass-like state of a supercooled liquid state type.

In Table 1 values of the monolayer compressibility, which were determined according data in figure 1, are represented for the Langmuir monolayers before the phase transition, in the phase transition state and for the LB-monolayer after the phase transition. As one can see, though before the phase transition the compressibility values for monolayers fabricated on subphases with $Fe(NO_3)_3$ are lower than for $FeCl_3$ at all V, κ values after the phase transition are lager for $Fe(NO_3)_3$ than for $FeCl_3$. The compressibility κ of monolayers after the phase transition is 3-5 times lower than before phase transition. Thus, a structure of the monolayer formed on the subphase with $FeCl_3$ is the more rigid one.

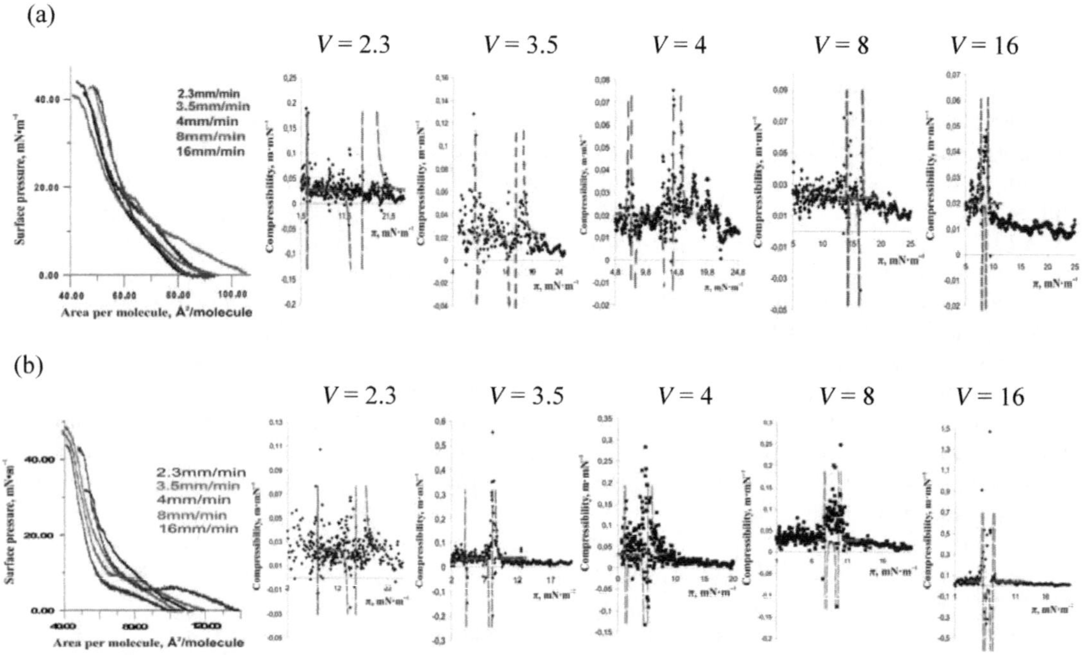

Figure 1. $\pi - A$-isotherms (left) and dependences of compressibility κ on surface pressure π at different compression rates V =2.3, 3.5, 4.0, 8.0, 16.0 mm/min (right) in first-order 2D-phase transition for monolayers fabricated on subphases with $Fe(NO_3)_3$ (a) and $FeCl_3$ (b). Symbols denote experimental data processing results, dashed red curves are the theoretical predictions.

Table 1. The value of Langmuir-monolayer compressibility formed on $Fe(NO_3)_3$ - and $FeCl_3$-containing subphases at different compression rate before, at, and after first-order phase transition (PT).

Compression rate, mm/min	Compressibility, m/N					
	before PT		at PT[a]		after PT	
	$Fe(NO_3)_3$	$FeCl_3$	$Fe(NO_3)_3$	$FeCl_3$	$Fe(NO_3)_3$	$FeCl_3$
2.3	33±10	45±16	24±11	25±15	8.5±1.0	9.0±1.9
3.5	34±10	47±16	24±10	30±15	8.6±1.0	7.4±0.9
4.0	20±3	50±20	20±4	30±19	8.0±0.4	7.6±1.9
8.0	23±3	35±7	22±3	25±9	11.0±1.5	8.8±1.4
16.0	19±5	40±20	19±4	30±15	6.7±0.4	6.2±1.1

[a] Values of compressibility are determined in the middle between the two singular points of the theoretical curves

4. Raman spectroscopy of Fe-containing LB-films

In figure 3 Raman spectra of 5 LB-DTP-monolayers fabricated on $Fe(NO_3)_3$ - and $FeCl_3$ -containing subphases are shown. In [3] characteristic molecular vibrations observed in the Raman spectrum of 5-, 7-monolayer LB-DTP-film formed on subphase with $Fe(NO_3)_3$ have been examined. At fabrication of

LB-monolayer on subphase with $FeCl_3$ narrowing of all Raman lines is observed that evidence the weaker interaction of the metal centers with surroundings. Intensities of Raman bands 556 cm^{-1}, 653 cm^{-1} (ring in-plane deformation) and 1235 cm^{-1} (CH-bending) for LB-films formed on $FeCl_3$ - containing subphase increse. Besides this, intensification of spectral band 1158 cm^{-1}, being characteristic for Cl–O vibration, is registered. Shift of 2-subst pyrrole Quadrant stretch Raman band into high-frequency region (from 1462 to 1467 cm^{-1}) is observed.

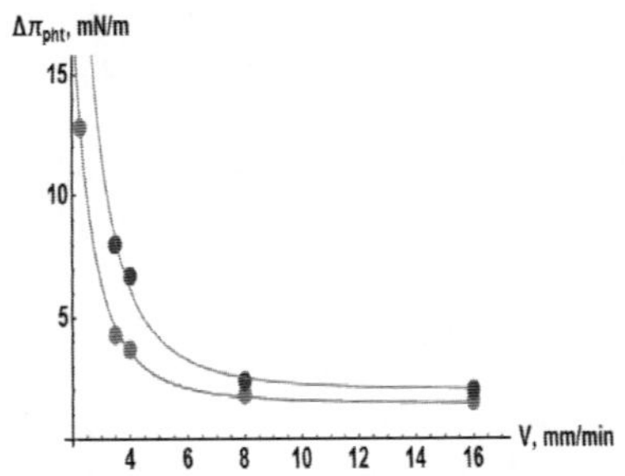

Figure 2. The dependences of phase transition width value $\Delta\pi$ on compression rate V for Langmuir monolayers formed on $Fe(NO_3)_3$ - (blue) and $FeCl_3$ - containing (red) subphases.

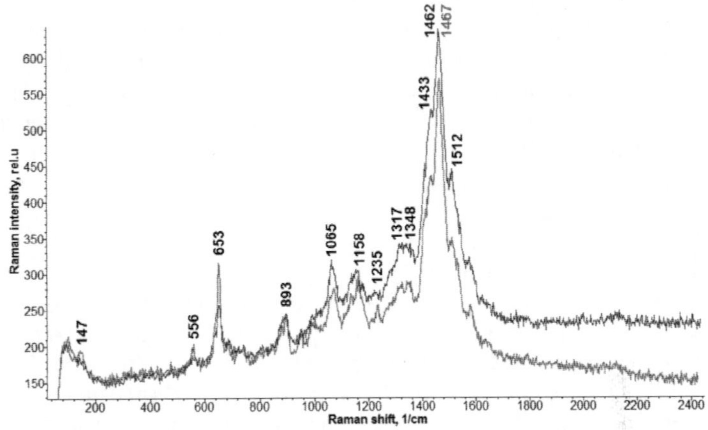

Figure 3. Raman spectra of 5 monolayer LB-DTP-films formed on $Fe(NO_3)_3$- (blue) and $FeCl_3$-containing (red) subphases at compression rate 4 mm/min. Pressure of LB-film deposition π was 25 mN/m. Excitation by green laser at wavelength 532 nm and laser power 5.4 mW. Accumulation time was 1 s.

5. Mössbauer spectroscopy

To establish a structure of studied supramolecular compound, it is important to find out a valent state of metal center. Mössbauer spectrum of a sample of nanoporous anodic alumina (nanopore diameter is 10 nm) coated by iron-containing 13-monolayer LB-DTP-film formed on subphase with $Fe(NO_3)_3$ is shown in figure 4.

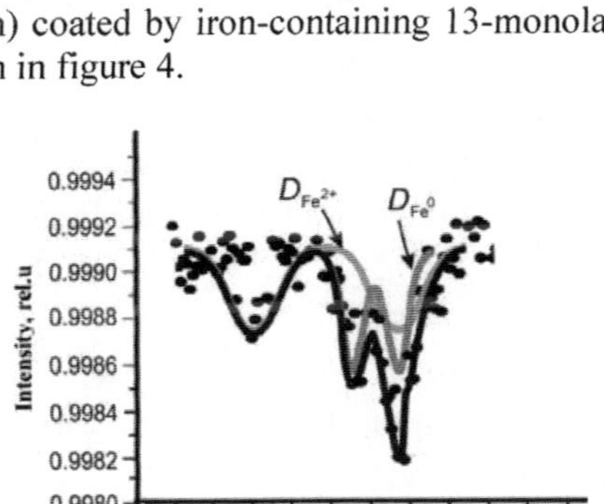

Figure 4. Mössbauer spectrum of iron-containing 13-monolayer LB-film of dithionilepyrrole.

As one can see, two doublets are observed in Mössbauer spectrum of 13-monolayer LB-film of dithionilepyrrole. One of them $D_{Fe^{2+}}$ is registered at rate v about 1.6 mm/s, and another doublet $D_{Fe^{0}}$ – at rate v about 0.3 mm/s. These values of v is close to v equals 1.7 mm/s and 0.4 mm/s, which are characteristic for doublets of paramagnetic ion Fe^{2+} in oxide of metal alloy FeCoZr and superparamagnetic Fe^0 in non-oxidized FeCoZr, respectively [16]. The broadened lines $D_{Fe^{2+}}$ in figure 4 evidence interaction of iron ions with dielectric 2D matrix. Vice versa, the narrowness of D_{Fe^0} doublet lines in figure 4 means that reduction of the Fe atoms is accompanied by the exit of these atoms from LB-monolayer plane.

Organometallic complexes, which contain S, can reduce Fe^{3+} up to Fe^0 [17-19]. Due to this Mössbauer spectrum indicates that complexification of high-spin ion Fe^{2+} with S in quasi-two-dimentional parts of LB-film (with 7 monolayers and less) proceeds. According to the Mössbauer spectrum in figure 4, an interlayer interaction in thick LB-film (with 9 monolayers and more) leads to partial reduction of Fe^{2+} up to Fe^0. It allows to assume that an iron ion coordinated by six sulphur atoms of three from six H-dithionilepyrrole residues to form a hexagon.

Thus, the Mössbauer spectroscopy analysis of the iron-containing DTP-LB-film consisted of 13 LB-monolayers shows that ions Fe^{3+} in compressed metal-aromatic complexes of DTP can turn into high-spin bivalent state Fe(II) and be reduced up to Fe^0.

6. Discussion and conclusion

According to foregoing data, the monolayers are polymerized and turn into monolayers of nanocyclic organometallic coordination compounds with a mixing valent state. Figure 5 shows schemes of the LB-monolayers of Fe-DTP-complexes in high-spin Fe(II) and low-spin Fe(III) states. As one can see, Fe(III)DTP_3 complexes have planar structure in opposite to dome-shaped structure of Fe(II)DTP_3 complexes. As provided by the compressibility analysis and the data of Raman spectroscopy, one can conclude that the value of mixing valent state of iron in LB-monolayers formed on $FeCl_3$-containing subphase tends to one of high-spin state, while a mixing valent state of iron in monolayers formed on $Fe(NO_3)_3$-containing subphase is closer to one of low -spin state.

Figure 5. Schemes of LB-monolayer of Fe(II)DTP_3 (a) and Fe(III)DTP_3 (b) complexes.

So, the method of experimental isotherm analysis utilizing geometrodynamic approach has been proposed to estimate the compressibility of metastable Langmuir monolayers. This method allows to find out the monolayer features, which determine the design of ultrathin LB-film after its deposition on a support.

References

[1] Rahman Md M, Li X-B, Lopa N S, Ahn S J, Lee J-J 2015 *Sensors* **15** 3801
[2] Egorov A S, Egorova V P, Grushevskaya H V, Krylova N G, Lipnevich I V, Orekhovskaya T I, Shulitsky B G 2016 *Letters in Applied NanoBioScience* **5(1)** 343
[3] Grushevskaya H V, Krylova N G, Lipnevich IV 2016 *Int. J. Mod. Phys.* **30** 1642018
[4] Pfeil A, Lehn J-M 1992 *J.Chem.Soc. Chem. Comun.* **11** 838
[5] Vollhardt D and Fainerman V B 2009 New aspects in the consideration of compressibility of Langmuir monolayers *Interfacial reology* ed R. Miller and L. Liggieri (Brill, Boston) pp 429–475
[6] Toma L M, Gengler R Y N, Prinsen E B, Gournis D, Rudolf P 2010 *Phys. Chem. Chem. Phys.* **12** 12188
[7] Leveiller F, Jacquemain D, Lahav M, Leiserowitz L, Deutsch M, Kjaer K, Als-Nielsen J 1991

Science **252** 1532
[8] Datta A, Kmetko J, Yu C-J, Richter A G, Chung K-S, Bai J-M, Dutta P 2000 *J. Phys. Chem. B* **104** 5797
[9] V Dupres, S Cantin, F Benhabib, F Perrot, P Fontaine, M Goldmann, J Daillant, O Konovalov 2003 *Langmuir* **19** 10808
[10] Cantin S, Pignat J, Perrot F, Fontaine P, Goldmann M 2004 *Phys. Rev. E* **70** 050601
[11] Kundu S, Datta 2006 *A Colloids and Surfaces A: Physicochem. Eng. Aspects* **289** 250
[12] Grushevskaya H V, Krylov G G, Krylova N G, Lipnevich I V 2015 *J. Phys. CS* **643** 012015
[13] Neagu M, Krylova N G, Grushevskaya H V 2013 *J. Math. Phys.* **54** 031508
[14] Kel'in A, Kulinkovich O 1995 *Folia pharm. Univ. Carol.* **18**(supplementum) 96
[15] Grushevskaya H V, Lipnevich I V, Orekhovskaya T I 2013 *J. Mod. Phys*. **4** 7
[16] Kasuk Y V, Larkin A V, Fedotova Y A 2011 *Herald of BSU Ser. 1*. **2** 52
[17] Brintzinger H, Palmer G, Sands R 1966 *J. Am. Chem. Soc.* **88** 623
[18] Alexandrov I A, Karmilov A Y, Shevchenko V G, Obolonkova E S, Aleksandrov A I, Solodovnikov S P 2009 *High molecular compounds. Ser.B* **51** 1573
[19] Rakhimov R R, Arrington S A, Hwang J S, Prokof'ev A I,· Alexandrov I A, 2004 *J. Appl. Phys.* **95** 7133

Nonlinear dynamics of nonequilibrium exciton-polaritons in a periodic potential

M V Charukhchyan[1], I Yu Chestnov[1], A P Alodjants[1,2], X Ma[3] and O A Egorov[3]

[1]Department of Physics and Applied Mathematics, Vladimir State University named after A. G. and N. G. Stoletovs, 87 Gorkii St., Vladimir 600000, Russia
[2]National Research University of Information Technology, Mechanics and Optics, (ITMO), Saint Petersburg 197101, Russia
[3]Institute of Condensed Matter Theory and Solid State Optics, Abbe Center of Photonics, Friedrich-Schiller-Universität Jena, Max-Wien-Platz 1, 07743 Jena, Germany

Abstract. An exciton-polariton condensate formed in a semiconductor microcavity coupled to an exciton reservoir in the strong coupling regime is studied. The condensate is trapped in one-dimensional periodic potential, and we work in the centre of Brillouin zone. We develop a model for coupled three spatial harmonics of mean field. Using the simplified model we get important analytical relations for polaritonic eigenstates and band-structure. The analytical results are supported by numerical analysis. The strong influence of external potential and nonlinearity is discussed and the feedback induced by the inhomogeneity of the incoherent reservoir on the dynamics of coherent polaritons.

1. Introduction

Strongly correlated bosonic particles placed in lattices represent an indispensable tool for fundamental studies of quantum phenomena in modern condensed matter and solid state physics [1]. During the past decade there was a remarkable progress in this area, which was related to exciton polaritons occurring inside a high quality semiconductor microcavity due to the strong light-mater coupling [2–4]. Extremely small effective masses of these composite bosons enable the observation of high temperature nonequilibrium BEC with exciton polaritons [5–7]. An interesting direction is connected with study of polariton condensate placed in a periodic potential [8,9].

In this paper an exciton-polariton nonequilibrium condensate is described confined in a weak-contrast periodic potential embedded into a planar microcavity driven by a homogeneous incoherent pumping [see Figure 1(a)]. Polaritons are subject to rapid radiative decay and their population is maintained due to optical pumping, thus, the condensate of exciton-polaritons demonstrate a nonequilibrium open-dissipative behavior.

The main goal of this work is to explore the macroscopic coherent oscillations of the exciton polaritons (Figure 1) coupled with the exciton reservoir.

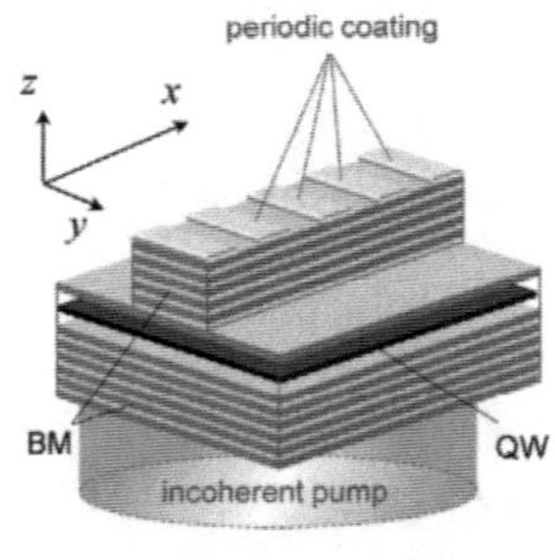

Figure 1. (Color online) Sketch of the considered system. A one-dimensional microcavity with periodic coating is driven by an incoherent optical pump [9]. Semiconductor quantum well (QW) is placed between two Bragg mirrors (BM).

2. The model

2.1 Gross-Pitaevskii dissipative model.
Here it is considered an open-dissipative mean-field Gross-Pitaevskii (GP) model that describes the incoherently pumped condensate coupled with an exciton reservoir. The considered system is described by GP-type equation for polaritonic order parameter ψ and by rate equation for the reservoir density n [6]

$$i\hbar\frac{\partial\psi}{\partial t}=\left[-\frac{\hbar^2}{2m}\frac{\partial^2}{\partial x^2}+V(x)+g_c|\psi|^2+i(Rn-\gamma_c)/2+g_r n\right]\psi, \quad (1)$$

$$\frac{\partial n}{\partial t}=-\left(\gamma_r+R|\psi|^2\right)n+P_0, \quad (2)$$

where m is the effective mass of the polariton, which is much smaller than the free electron mass m_e ($m\approx m_e\times10^{-4}$), $R=0.01\,ps^{-1}\mu m^2$ defines the condensation rate, $\gamma_c=0.33\,ps^{-1}$ and $\gamma_r=0.495\,ps^{-1}$ represent the decay rates of polaritons and reservoir excitons, respectively. $g_c=6\times10^{-3}meV\,\mu m^2$ and $g_r=2g_c$ characterize the strengths of polariton-polariton and polariton-reservoir interactions, respectively, P_0 characterizes incoherent (nonresonant) homogeneous pumping. Here it is discussed a quasi-one-dimensional polariton condensate [8] in the periodic potential $V(x)=V_0cos(\beta x)$, where $\beta=2\pi/l$ and l is the period of modulation [see Figure 1(a)].

A typical attribute of exciton-polariton condensation dynamics is the presence of the threshold. Above some critical value of pumping, when all the losses are compensated, a non-trivial solutions of the system (1), (2) appears. In the case of homogeneous system ($V_0=0$) they are traveling waves

$$\psi^{HS}(x,t)=\psi_0^{HS}e^{-i\mu_0 t+ikx}, \quad (3)$$

where $\hbar\mu_0=\left(\hbar^2/2m\right)k^2+g_c|\psi_0|^2+g_r n_0$. The threshold value is determined by $P_{th}=\gamma_c\gamma_r/R$. The coherent polariton and incoherent reservoir densities are given by $\left|\psi_0^{HS}\right|^2=(P_0-P_{th})/\gamma_c$ and $n_0^{HS}=\gamma_c/R$, respectively [6]. It is justified to suggest that in the presence of weak periodic potential the threshold value of pumping P_{th} remains the same as in homogeneous system. Numerical simulations of system (1) and (2) confirm this estimate.

2.2 Numerical simulations

It is well known that a rigorous analytical description of the periodic nonlinear system is possible only in the case of a specific shape of potential. Thus, firstly we solved the system (1) numerically. In this paper we are mostly interested in the states of the condensate with a wavevector close to zero. In the case of homogeneous pumping, which corresponds to the wide pumping laser field, and periodical boundary conditions the formation of the state with $k=0$ from the weak initial noise is the most probable. Simulating the formation of polariton condensate from the small initial noise we observed that the final state of the polariton condensate depends on the system parameters. Two typical scenarios of the evolution of the condensate density $|\psi|^2$ are presented in Figure 2.

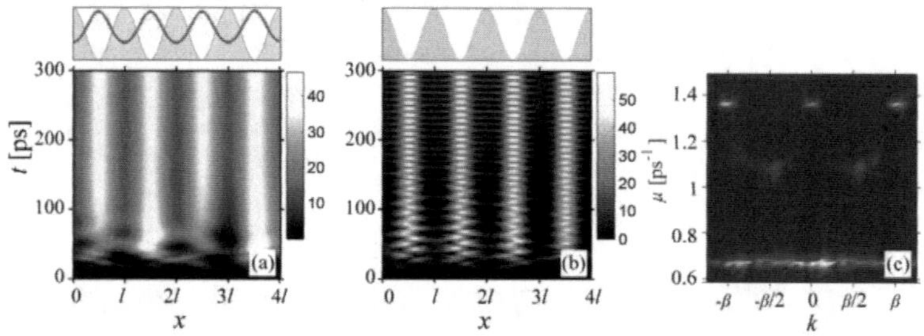

Figure 2. (Color online) (a) Formation of the ground state of polariton condensate from a noisy initial conditions under $V_0/\hbar = 0.1\,\text{ps}^{-1}$ and $P_0 = 24\ \mu\text{m}^{-2}\text{ps}^{-1}$. An upper panel shows the distribution of the condensate mode (red curve) and the shape of periodic potential (shaded region) (b) formation of the oscillating state under the same value of pumping rate but for $V_0/\hbar = 0.4\,\text{ps}^{-1}$. Spatio-temporal spectrum of this dynamical solution is shown on the panel (c).

In the first case which is shown in Figure 2(a) the condensate forms in the steady state with a wavevector close to zero. The maxima of the condensate distribution are located at the minima of the potential in this case [see the upper bar of panel (a)]. It means that the condensate is in the state with minimal energy, i.e. in ground state. The analysis of the spatio-temporal spectra of such solutions confirms it. This scenario is realized mostly when the potential depth V_0 is small [for instance $V_0 = 0.1\,ps^{-1}$ in Figure 2(a)].

In the case of large modulation the dynamical solution forms which has a form of metastable oscillations of polariton density. After long time (more than several tens of nanoseconds) such a state breaks spontaneously. This scenario is realized for all the values of V_0 greater than some threshold value which is determined by the pumping rate P. The oscillating nature of the observed solution clearly indicates that two states of the condensate are excited. From the spectrum of the solution, Figure 2(c), it is seen that one of these states is a ground state with zero wavevector and another one is located at the upper band.

To understand the reasons of the formation of such oscillations we use a simplified model which allows determining the spatial structure of the observed states.

2.3 Three spatial harmonics model

We search for the solutions in form of Bloch functions $\psi = \psi(x)e^{ikx}$, where k has a sense of quasi-impulse. Since we consider only low contrast lattices the solution $\psi(x)$ can be constructed as a sum of plane waves. This is equivalent to Fourier expansion of condensate order parameter ψ and reservoir density n on the spatial harmonics of the periodic potential. According to numerical simulations the condensate forms only on the lowest energy bands, thus we can use only first three spatial harmonics. In this case an approximate solution of Eqs. (1), (2) is

$$\psi(x,t) \approx \left(a_0 + a_- e^{-i\beta x} + a_+ e^{i\beta x}\right)e^{ikx}, \tag{4a}$$

$$n(x,t) \approx n_0 + n_+ e^{i\beta x} + n_- e^{-i\beta x}, \tag{4b}$$

where a_0, a_-, and a_+ are time-dependent amplitudes of the main spatial harmonics of coherent polariton condensate. The modulation of the coherent exciton polaritons evokes a spatial modulation of the reservoir described by the terms with $n_\pm$ in Eqs. (4b). Inserting (4) into the system (1), (2) and neglecting the higher spatial harmonics arising from the nonlinear terms we obtain the set of equations for three condensate components (a_0, a_-, and a_+), for the three components of the reservoir (n_0, n_+ and n_-).

3. Steady-state solutions in the frame of three spatial harmonics model

The energy band-structure of the condensate and the spatial structure of the specific states can be obtained numerically by the solution of the steady-stated equations in the frame of the three spatial harmonics model (4). The analytical description is reasonable only in the vicinity of the condensation threshold ($P_0 \simeq P_{th}$) where the nonlinear effects are small and can be neglected. In this case the condensate band structure, i.e. the dependence of the chemical potential μ on the wavevector k, resembles the dispersion of a single particle in the periodic potential, see Figure 3(a). Since pump is homogeneous the reservoir modulation is small just above the threshold, i.e. $n_0 \simeq n_0^{HS}$. Thus, the average density of the condensate can be approximated by the value $|\psi_0|^2$ characterizing the homogeneous system.

Searching for the steady-state solutions of a linear problem in the form $a_0(t) = a_0' e^{-i\mu t}$, $a_-(t) = a_-' e^{-i\mu t}$, and $a_+(t) = a_+' e^{-i\mu t}$ one can determine the spatial shapes and the frequencies of the states with $k = 0$.

The first solution possessing minimal energy is the *ground state*

$$\begin{pmatrix} a_0' \\ a_-' \\ a_+' \end{pmatrix} = \sqrt{\frac{|\psi_0|^2}{2H(H+4K)}} \begin{pmatrix} H+4K \\ -1 \\ -1 \end{pmatrix}, \tag{5}$$

where $K = \hbar^2 \beta^2 / 8mV_0$ and $H \equiv \sqrt{16K^2 + 2}$. In the case of weak modulation and relatively large lattice period we have $K > 1$. It means that zero-momentum component a_0' makes the main contribution in the solution. In this case condensate is periodically modulated on the zero-momentum background.

Solutions corresponding to the excited states have a symmetric and antisymmetric eigenvectors in respect to the signs of the components a_-' and a_+'. The lower *antisymmetric state* has a perfect sine shape with $a_-' = -a_+'$ and vanishing zero-momentum component $a_0' = 0$.

$$\begin{pmatrix} a_0' \\ a_-' \\ a_+' \end{pmatrix} = \frac{|\psi_0|}{\sqrt{2}} \begin{pmatrix} 0 \\ 1 \\ -1 \end{pmatrix}. \tag{6}$$

In this case the maxima of the condensate density is located at the maxima of the potential.

An upper *symmetric state* has a symmetric structure ($a_-' = a_+'$)

$$\begin{pmatrix} a_0' \\ a_-' \\ a_+' \end{pmatrix} = \sqrt{\frac{|\psi_0|^2}{2H(H-4K)}} \begin{pmatrix} H-4K \\ -1 \\ -1 \end{pmatrix}. \tag{7}$$

The above solutions are precise only in the vicinity of the condensation threshold. Under larger pumpings the condensate modulation initiates modulation of the reservoir which results in a spatially distributed gain proportional to $Rn(x)$. In this case steady-state solutions can be obtained numerically. Nevertheless, our simulations show that the states keep their symmetry even in this case.

The typical profiles of polariton condensate at the ground (GR), antisymmetric (AS), and symmetric (SY) states are shown on the panels (b), (c) and (d) of Figure. 3. Corresponding distributions of the reservoir are shown on the panels (e), (f) and (g).

In the next section we show how the spatial structure of the condensate and especially spatially inhomogeneity of the reservoir influence on the condensate dynamics triggering the oscillatory dynamics.

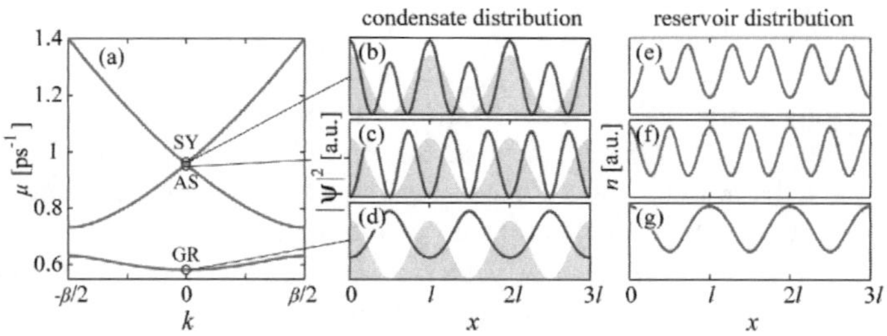

Figure 3. (Color online) (a) Energy band structure of the polariton condensate in the vicinity of the condensation threshold P_{th}. Eigenstates GR, AS and SY are ground, antisymmetric, and symmetric states of polaritons with $k = 0$. Other parameters are $V_0 / \hbar = 0.1$ ps^{-1} and $l = 2\pi / \beta = 8\mu$m. Panels (b) – (d) show the distribution of condensate in three steady states marked by blue circles in panel (a). The corresponding reservoir profiles are shown on the panels (e)-(g).

4. Oscillation and relaxation dynamics of the condensate

As it was mentioned above in the case of a weak potential the condensate forms in the ground state [compare Figure 3(d) and an upper panel of Figure 2(a)]. Filling of the upper state in the more contrast lattices can be explained by the selective saturation of the gain arising from the spatial modulation of reservoir. With the increase of V_0 the spatial modulation of the condensate becomes deeper since zero-momentum component a_0' reduces – see Eq. (5). In this case due to the spatial dependency of the reservoir density the gain is saturated mostly in the vicinity of the condensate maxima, i.e., in the minima of the periodic potential. It initiates the growth of the symmetric mode, whose peaks are located near the maxima of the periodic potential [see panel (b)]. Such a phenomenon resembles hole burning effect know from laser physics, but the saturation of the gain occurs in a real space domain and, thus, this effect can be called *spatial hole burning*.

Since oscillations appear due to the temporal beating between the ground and the excited symmetric modes, the oscillation frequency ω can be approximated by the difference of their eigenfrequencies μ_{GR} and μ_{SYM}. The latter can be determined analytically in the frame of the three spatial harmonics model (4) in the linear limit which is valid in the vicinity of the condensation threshold. Using the same approach as in Section 3 one obtains

$$\hbar\omega \approx V_0\sqrt{4K^2 + 2}\,. \tag{8}$$

Numerical modeling shows that Eq. (8) accurately predicts the oscillation frequency in the vicinity of the condensation threshold ($P_0 \approx P_{th}$) [Figure 2(c)]. For a stronger pumping the influence of nonlinear effects and reservoir inhomogeneities become significant. As a consequence the oscillation frequency increases with the pump amplitude P_0 until the critical value where the oscillations disappear. The value of oscillation frequencies versus pumping rate for different values of modulational depth is shown on Figure 4. The recovery to the steady state solution occurs because the ground mode of the condensate stabilizes. This threshold is determined by both the potential depth V_0 and the pumping rate P_0 and is shown on the panel (b) of Figure 4.

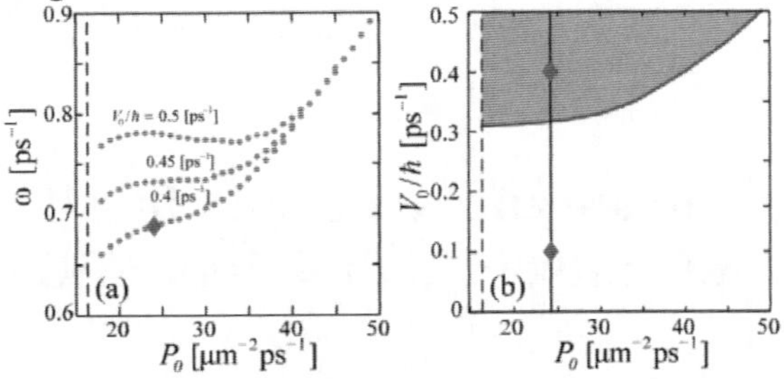

Figure 4. (Color online) (a) numerically calculated frequency of oscillation for different values of modulation depth. (b) Existence domain of "zero momentum" oscillations as a function of the pump P_0 and the modulation depth V_0 obtained both in the GP model (green shaded area). Red diamonds indicates the points corresponding to Figure 1 (a) (lower) and Figure 1(b) (upper).

5. Conclusion

We considered nonlinear dynamics of coherent exciton-polaritons in weak-contrast lattices embedded into a planar microresonator under homogeneous incoherent pumping. Within this approach we developed a simplified mean-field model for three spatial harmonics and found analytical expressions for the relevant eigenstates of the condensate. We have observed that the metastable oscillations between the states locating in the center of the Brillouin zone can form from the weak noise. The numerical results are supported by the analytical analysis. The strong influence of the dissipative effects and the incoherent reservoir on the dynamics of coherent polaritons was also discussed.

Acknowledgments

The financial support from the EU project (FP7, PIRSES-GA-2013-612600) LIMACONA is acknowledged. Russian authors (I.Yu.C., M.V.C., A.P.A.) acknowledge financial support from Russian Foundation for Basic Research (RFBR) and the Russian Ministry of Education and Science, state task No. 2014/13. The reported study was also supported by the Ministry of Education and Science of the Russian Federation (state project no. №16.440.2014/K), and the RFBR grants #16-32-60102 and #15-59-30406.

References

[1] Kramer M, Menotti C, Pitaevskii,L and Stringari S 2003 Eur. Phys. J. D **27** 247; Lewenstein M, Sanpera A, Ahufinger V, Damski B, De A S and Sen U 2007 *Adv. Phys.* **56** 243; Kantian A, Daley J, Torma M and Zoller P 2007 *New J. Phys.* **9** 407
[2] Weisbuch C, Nishioka M, Ishikawa A and Arakawa Y 1992 *Phys. Rev.* Lett. **69** 3314
[3] Houdre R, Stanley R P, Oesterle U, Ilegems M and Weisbuch C 1994 *Phys. Rev.* B **49** 16761
[4] Sanvitto D and Timofeev V 2012 *Exciton polaritons in Microcavities* (Berlin: Springer-Verlag)
[5] Kasprzak J and *et al.* 2006 *Nature* (London) **443** 409
[6] Wouters M and Carusotto I 2007 *Phys. Rev.* Lett. **99** 140402
[7] Deveaud-Pl´edran B 2012 *J. Opt. Soc. Am.* B **29** A138
[8] Lai C and *et al.* 2007 *Nature* (London) **450** 529
[9] Cerda-M´endez E A, Krizhanovskii D N, Wouters M, Bradley R, Biermann K, Guda K, Hey R,Santos P V, Sarkar D and Skolnick M S 2010 *Phys. Rev.* Lett. **105**, 116402

Enhanced photoluminescence efficiency in AlGaN quantum wells with gradient-composition AlGaN barriers

E A Shevchenko, D V Nechaev, V N Jmerik, V Kh Kaibyshev, S V Ivanov and A A Toropov
Ioffe Institute RAS, Saint Petersburg 194021, Russia

Abstract. We present photoluminescence studies of $Al_xGa_{1-x}N/Al_yGa_{1-y}N$ ($y = x+0.3$) quantum well (QW) heterostructures with graded Al content in barrier layers, emitting in the range 285–315 nm. The best-established internal quantum efficiency of the QW emission is as high as 81% at 300 K, owing to enhanced activation energy of charge carriers and exciton binding energy in the QW heterostructure with optimized design.

1. Introduction

AlGaN alloys are widely considered as the most promising materials for design and manufacturing of optoelectronic devices for deep ultraviolet spectral range. The emission efficiency in III-N films grown on commercially available lattice-mismatched sapphire substrates is, however, limited by the emergence of high density of extended and point defects. Under the circumstances, fabrication of the $Al_xGa_{1-x}N$-based heterostructures with high internal quantum efficiency essentially relies on the realization of enhanced fluctuating localization potential, which suppresses charge carriers transport towards non-radiative recombination centers [1]. Provided the activation energy of localized carriers is big enough (e.g. $> 2k_BT$ at 300 K), a relatively high internal quantum efficiency can be achieved even in the structures with the typical defect density exceeding 10^9 cm^{-2}. A convenient measure of the structure internal quantum yield at room temperature is the relative internal quantum efficiency (RIQE) defined as the ratio of photoluminescence (PL) integrated intensities at room and low temperatures I^{RT}/I^{LT}. In particular, Shatalov *et al.* [2] reported the best-known RIQE value of about 70%, evaluated under above-barrier excitation conditions in AlGaN multi-QW structures emitting below 300 nm.

In this paper, we demonstrate an optimized AlGaN single-QW structure grown on *c*-sapphire, emitting near 300 nm with the RIQE of at least 81% under resonant (below-barrier) excitation conditions. Theoretical and technological optimization was performed in order to achieve enhanced electron and hole activation energies both in the vertical direction (along the c-axis) and in the QW plane. Gradient-composition $Al_yGa_{1-y}N$ barriers were introduced in the structures in order to improve collection in the $Al_xGa_{1-x}N$ QW ($x<y$) of the carriers photoexcited within the barriers. Nevertheless, a pronounced difference in the RIQE measured under either above- or below- barrier excitation indicates emergence of noticeable nonradiative losses, which accompany carrier transport within the barrier layers.

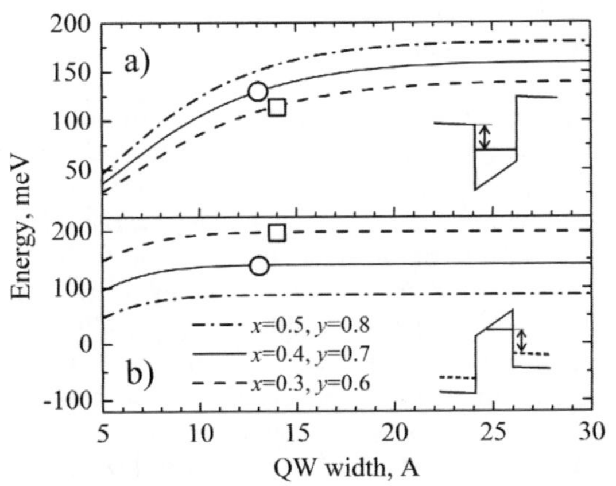

Figure 1. Electron (a) and hole (b) activation energies, calculated as a function of QW width for $Al_xGa_{1-x}N/Al_yGa_{1-y}N$ ($y = x+0.3$) QWs with various combinations of the alloy composition in the QW and barrier layers. Square and circle marks, respectively, indicate parameters of sample A and sample B. The inserts illustrate QW band line-ups.

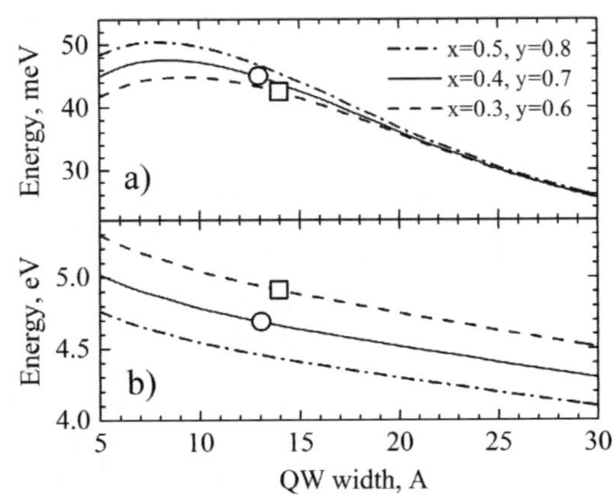

Figure 2. Exciton binding energy (a) and exciton resonance energy (b), calculated as a function of QW width for $Al_xGa_{1-x}N/Al_yGa_{1-y}N$ ($y = x+0.3$) QWs with various combinations of the alloy composition in the QW and barrier layers. Square and circle marks, respectively, indicate parameters of sample A and sample B.

2. Samples design and experiment

2.1. Theoretical simulations

In order to optimize design of the QW structures we estimated the activation energy of carriers as well as parameters of the QW exciton as a function of the QW width. For this aim, band structure and hole effective masses in AlGaN alloys were obtained using a 6-band *kp* model [3], implying pseudomorphic growth with the in-plane lattice constant of an AlN buffer. The QW potential profile was obtained in the approximation of a uniform electric field induced by both spontaneous and piezoelectric polarizations in the pyroelectric compounds with wurtzite crystal structure. We used a transfer matrix method to calculate single-particle energy states and wave functions of electrons and holes, while exciton parameters were evaluated variationally in the approximation of independent excitons by means of a two-parameter trial wave function [4]. All material parameters were taken from [5] besides valence band deformation potentials of AlGaN, which were linearly extrapolated from those of GaN and AlN reported in Refs. [6,7].

Figure 1 shows the activation energies of electrons and heavy holes calculated for three combinations of the alloy composition in the QW and barrier layers in the $Al_xGa_{1-x}N/Al_yGa_{1-y}N$ QW, namely (x=0.3, y=0.6), (x=0.4, y=0.7), and (x=0.5, y=0.8). The difference $y - x = 0.3$ is chosen to achieve reasonably large band offsets at the QW interfaces, while keeping as low Al content in the barrier layers as possible that is a necessary condition of the reliable *p*-doping in respective light emitting diode (LED) structures. The activation energies are obtained as the energy gap between the ground electron (hole) QW energy level and the bottom (top) of the lowest (topmost) band in the barriers (see insertions in Figure 1). For the considered elastic strain conditions, crossing of the heavy-hole and split-off valence bands occurs in $Al_xGa_{1-x}N$ slightly above $x = 0.6$. Therefore, in all three cases, the upper valence band in the QW alloy is the heavy-hole one. The upper valence band in the $Al_yGa_{1-y}N$ barriers is the heavy-hole band for $y = 0.6$, while for two other composition combinations the topmost valence band is the split-off one. The former situation always corresponds to TE polarized emission needed for efficient operation of surface-emitting LEDs [8]. For $y > 0.6$, the TM polarized emission can take place for thin enough QWs when the heavy-hole quantum confined energy level is pushed down below the top of the split-off band in the barriers. This corresponds to type-II band line-ups and negative hole activation energy. For the chosen parameters, however, this situation does not come true even in the thinnest QWs.

The activation energy of electrons increases with the increase of both QW width and Al content. Even in case of (x=0.3, y=0.6) the activation energy is larger than 60 meV at the QW width exceeding 10 Å, which is twice greater than k_BT at room temperature. The activation energy of holes also increases with the increase of the QW width but decreases with the increase of the Al content. For thick enough QWs (thicker than 10 Å), it exceeds 60 meV even for the combination (x=0.5, y=0.8) with the largest chosen Al content. Note that a further increase of the Al content results in a fast reduction of the hole activation energy and for $x > 0.6$ it becomes negative for any QW width. Figure 2 shows the exciton binding energy (Figure 2 (a)) and exciton resonance energy (Figure 2 (b)) calculated versus QW width for the same composition combinations. To evaluate emission energy of the localized QW excitons one should take into account the Stokes shift, which owes to the exciton localization by fluctuations of the QW width and composition. Even for an ideal QW the Stokes shift should be larger than the energy relevant to a single-monolayer (~2.5 Å) thickness fluctuation, which can be estimated from Figure 2 (b) as 100–200 meV depending on the QW width. Then, one can expect the exciton emission wavelength to be below 300 nm (above 4.13 eV) for all considered composition combinations, provided the QW width is below 15–20 Å. Note that the exciton binding energy strongly increases in the thinnest QWs (see Figure 2 (a)) that is also favorable for light emission due to the additional enhancement of the carriers activation energy [2].

To sum up the above analysis, we declare the optimized parameters limits as 10–15 Å for the QW width and x = 0.3–0.4 for the Al content in the QW. Within these limits, both electron and hole activation energies exceed 100 meV, exciton binding energy exceeds 40 meV, the emission wavelength is below 300 nm, and the Al content in the barriers does not exceed y = 0.6–0.7.

2.2. *Experimental techniques*

Following the performed analysis, we have grown two optimized QW structures by plasma-assisted molecular beam epitaxy on a c-sapphire substrate. Intended parameters of these samples (named further as A and B) are Al content of (x=0.3, y=0.6) and (x=0.4, y=0.7) and QW widths of 14 and 13 Å, respectively. Sample A represents the design with the smallest possible Al content in the barrier layers, which nevertheless secures reasonably high band offsets at the QW interfaces and the exciton emission energy below 300 nm (see Figs. 1 and 2). Sample B is designed to provide a shorter emission wavelength at the expense of larger Al content in both QW and barriers. Both samples possess a 2 μm thick AlN buffer containing six 3.5 nm thick GaN layers aimed to reduce threading dislocations density in the active region [9]. The density of threading dislocations was estimated from XRD measurements as 6.9×10^9 cm^{-2} and 4×10^9 cm^{-2} of edge dislocations, and 9×10^8 cm^{-2} and 7×10^8 cm^{-2} of screw dislocations for samples A and B, respectively. The active region itself consists of a 32 nm thick $Al_yGa_{1-y}N$ barrier, $Al_xGa_{1-x}N$ QW, a 65 nm thick $Al_zGa_{1-z}N$ graded barrier layer ($z = y \rightarrow y+0.3$), and a 10 nm thick AlN cap layer. Digital sub-monolayer epitaxy technique was used to form the QW as a superlattice composed of several sub-monolayer thick sheets [10,11]. The rotation period of substrates was set to 3 seconds for the growth of barriers and to 7 seconds during the QW growth. Due to the ununiformed distribution of metal and nitrogen fluxes across the substrate the slower rotation speed leads to additional alloy composition disordering within the QW layer [12].

Time-integrated PL spectra were measured in the grown samples in a wide temperature range between 10 and 300 K, using harmonics of a femtosecond Ti-sapphire laser (Coherent MIRA 900) as excitation and an OceanOptics QE65Pro spectrometer for registration. We used the fourth and the

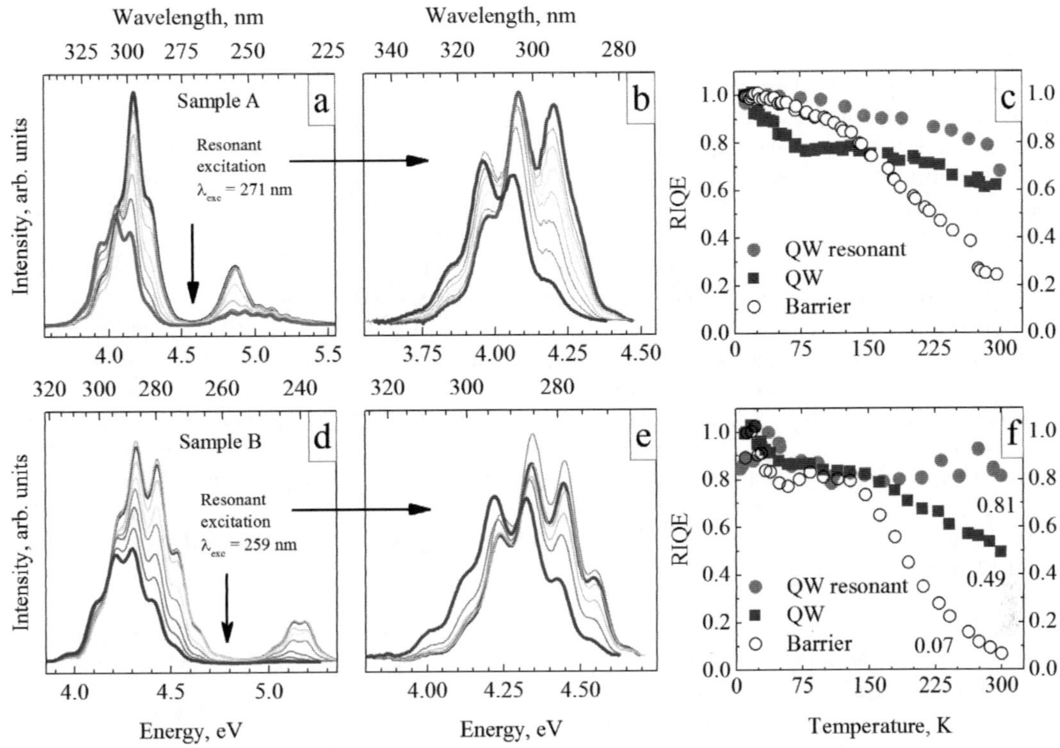

Figure 3. PL spectra of samples A (a, b, c) and B (d, e, f) measured at different temperatures under above-barriers (a, d) and below-barriers (b,e) excitation. RIQE of samples A (c) and B (f) obtained under both excitation modes.

third harmonics of the laser in separate measurement series to excite either barriers or QW, respectively.

3. Results and discussion

The PL spectra measured in both samples are qualitatively similar. Under the excitation by a 210 nm laser line (forth harmonic) the spectra demonstrate two relatively wide separate peaks assigned to the emission of localized excitons in QWs and barriers (Figs. 3 (a) and 3 (d)). In agreement with the intended design, the QW PL peak emerges in Sample A near 300 nm, while in Sample B it shifts toward shorter wavelengths. The whole PL signal is periodically disturbed by light interference defined by reflections from the structure surface and from the interface with the sapphire substrate. Both PL peaks survive up to 300 K, however, the temperature-induced quenching of the QW PL intensity is weaker than that of the barriers. In particular, the ratio of the spectrally-integrated intensities of the QW and barrier peaks (I_{QW}/I_B) decreases with the temperature increase from 28% and 11% to 7% and 1% for samples A and B, respectively (see Figs. 3 (c) and (f)). Since the AlN cap layer and the gradient in the barrier potential profile prevent transport of the photoexcited carriers toward the structure surface and their subsequent surface-induced recombination, we conclude that the drop of the I_{QW}/I_B ratio should be due to the temperature-activated transport of the photoexcited carriers toward QW and, probably, due to their enhanced rate of non-radiative recombination inside barriers as compared with that in QW. In particular, this difference might owe to the enhanced carriers localization within the QW plane, originating from the QW interface roughness and compositional modulations.

Figures 2 (b) and (e) show the series of the PL spectra measured under the resonant (below-barrier) QW excitation by a third laser harmonic. For both samples, the RIQE value of the QW PL for this excitation regime is higher than that under the above-barrier excitation (Figures 2 (c) and (f)). In sample A, the difference between the RIQE curves obtained under different excitation modes first increases with the temperature rise, reaches ~20% at 75 K, and is roughly constant at higher temperatures. Generally, this behavior indicates temperature independence of the additional losses in the barrier layers above 75 K. In Sample B, the behavior is qualitatively different: The difference between the RIQE curves is negligible below ~175 K and drastically increases above this temperature. The distinction between the two samples relies, probably, on varies particular distribution of defects there and is not completely understood at present. Nevertheless, the higher RIQE values observed under the resonant excitation at elevated temperatures clearly indicate emergence of additional non-radiative losses during the carriers transport within the barrier layers.

This observation implies that the true RIQE value of the QW PL is that measured under the resonant excitation. Compare Figs. 3 (c) and (f), one can conclude that the highest RIQE is observed in Sample B. Exact defining of this value is hampered by certain oscillations of the RIQE temperature dependence, originating, most probably, from interplay of the temperature-induced red shift of the whole PL line and the superimposed oscillating interference pattern. Still in case of the worst estimation the RIQE in Sample B is as high as 81%.

4. Conclusions

We have demonstrated the successful design optimization and technological implementation of improved AlGaN based QW structures emitting below 300 nm. The worst-case estimation of RIQE of the QW PL in the best-optimized sample amounts to 81% that is the best-known result to date.

Acknowledgements

This work was supported by Russian Science Foundation (Project #14-22-00107).

References

[1] Chichibu S F *et al.* 2006 *Nat. Mater.* **5**, 810–816

[2] Shatalov M, Yang J, Sun W, Kennedy R, Gaska R, Liu K, Shur M and Tamulaitis G 2009 *J. Appl. Phys.* **105** 073103

[3] Bir G I and Pikus G E 1974 *Symmetry and Strain-induced Effects in Semiconductors* (New York: Wiley)

[4] Toropov A A, Shevchenko E A, Shubina T V, Jmerik V N, Nechaev D V, Yagovkina M A, Sitnikova A A, Ivanov S V, Pozina G, Bergman J P and Monemar B 2013 *J. Appl. Phys.* **114** 124306

[5] Vurgaftman I and Meyer J R 2003 *J. Appl. Phys.* **94** 3675

[6] Ishii R, Kaneta A, Funato M, Kawakami Y and Yamaguchi A A 2010 *Phys. Rev. B* **81** 155202

[7] Ishii R, Kaneta A, Funato M and Kawakami Y 2013 *Phys. Rev. B* **87** 235201

[8] Northrup J E, Chua C L, Yang Z, Wunderer T, Kneissl M, Johnson N M and Kolbe T 2012 *Appl. Phys. Lett.* **100** 021101

[9] Jmerik V N, Lutsenko E V and Ivanov S V 2013 *Phys. Status Solidi* A **210** 439–50

[10] Zhmerik V N, Mizerov A M, Shubina T V, Sakharov A V, Sitnikova A A, Kop'ev P S, Ivanov S V, Lutsenko E V, Danil'chik A V, Rzheutskii N V and Yablonskii G P 2008 *Semiconductors* **42** 1420

[11] Ivanov S V, Nechaev D V, Sitnikova A A, Ratnikov V V, Yagovkina M A, Rzheutskii N V, Lutsenko E V and Jmerik V N 2014 *Semicond. Sci. Technol.* **29** 084008

[12] Toropov A A, Shevchenko E A, Shubina T V, Jmerik V N, Nechaev D V, Pozina G, Bergman J P, Monemar B, Rouvimov S and Ivanov S V 2016 *Physica Status Solidi* C accepted for publication

Journal of Physics: Conference Series **741** (2016) 012119 doi:10.1088/1742-6596/741/1/012119

Laser printing of Au/Si core-shell nanoparticles

G P Zograf[1,*], **D A Zuev**[1], **V A Milichko**[1], **I S Mukhin**[1,2], **M A Baranov**[1], **E V Ubyivovk**[1,3], **S V Makarov**[1,*] and **P A Belov**[1]
[1]ITMO University, St. Petersburg 197101, Russia
[2]St. Petersburg Academic University, St. Petersburg 194021, Russia
[3]St. Petersburg State University, St. Petersburg, 199034 Russia

*g.zograf@metalab.ifmo.ru
*s.makarov@metalab.ifmo.ru

Abstract. We develop a novel method of core-shell nanoparticles fabrication based on laser ablation of multilayer thin films using femtosecond laser pulses. Transmission electron microscopy proves that the obtained structures have Si and Au separated parts. We theoretically show that the combination of high refractive index dielectric and plasmonic nanoparticles shows possibility of interplay between magnetic optical responses and plasmon resonances . This opens a possibility to manipulate by both scattering power pattern and local optical field enhancement via precise engineering of the core/shell nanoparticles.

1. Introduction

Plasmonic nanoparticles actively studied in the last two decades because of their ability to localize and amplify light at the nanoscale [1]. However, metal nanoparticles of simple shapes (nanospheres, nanocubes, nanorods, etc.) have only resonances of electrical nature [2]. For magnetic response in the optical range it is necessary to create nanoparticles of more complex shapes applying lithographic techniques [3-5]. More recently, magnetic optical response has been observed experimentally in dielectric nanoparticles with a high refractive index (for example, silicon spheres) [6]. Indeed, according to the classical Mie theory, dielectric nanostructure can exhibit optically induced resonant magnetic response in the optical range. This opens up opportunities for controlling radiation pattern of such nanoparticles due to the interference of electric and magnetic resonances [7]. The combination of dielectric and plasmonic nanoparticles shows magnetic optical response due to dielectric part together with a high degree of localization by metal one.

There are fundamentally different configurations of metal-dielectric (hybrid) nanoparticles. The first type corresponds to a case where metal and dielectric parts are separate closely spaced particles. Such particles can be created using nanolithography techniques, laser ablation or colloidal chemistry [8-10]. Completely different type of a structure such as "core-shell", where one component is integrated into another one. Such structure is mechanically much more stable and compact. However, until now state-of-the-art technologies allows to create only core-shell particles of low-index dielectrics (oxides of metals and semiconductors), which hinders the induction of magnetic resonances in the visible range [11]. Indeed, covering a semiconductor nanoparticle by a thin metal layer faces the problem of shell dewetting [12], whereas fabrication of a crystalline semiconductor shell on a metallic core is still challenge for chemical and lithographical approaches.

2. Results and discussion

2.1 Fabrication

In our work, we developed, for the first time, a method to create core-shell nanoparticles made of gold and silicon. Our novel method based on laser ablation of multilayer thin films using femtosecond laser pulses (figure 1).

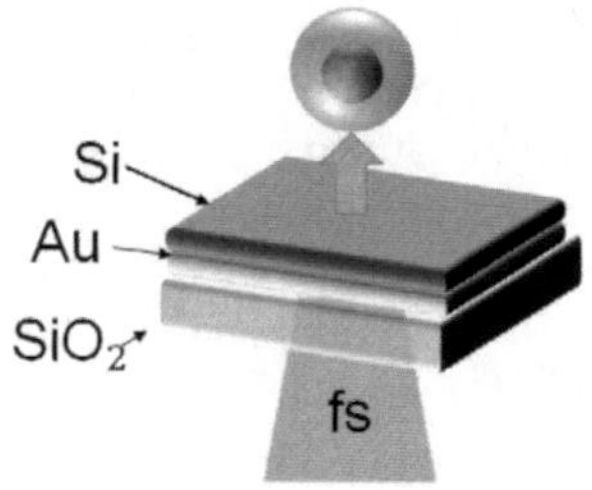

Figure 1. Femtosecond laser ablation method scheme.

Ablation is a method of removing material from the surface by a laser pulse. At low laser powers vaporizes or sublimates the substance in the form of free atoms and clusters. When the intensity of the laser pulse exceeds the ablation threshold, there is a micro-explosion with the formation of a hole in a thin film together with liquid particles erupting off the surface. In our case, we place a receiving substrate to catch the ablated particles above the films. Gold and amorphous silicon thin films (different thickness) were deposited on a glass substrate. Firstly, a gold layer was deposited by thermal deposition. Then, a-Si:H films with thicknesses in the range of 15-100 nm were deposited on a substrate of fused silica with the 15-nm gold layer by plasma enhanced chemical vapor deposition from SiH_3 precursor gas (initial hydrogen concentration 10%), were used. Energy regime was chosen to ablate both films but to keep glass substrate unaffected. As simulations and Raman scattering demonstrated that in the process of ablation amorphous silicon is restructuring into crystalline silicone [13]. The crystalline silicon is known to be better material for nanoptical applications as compared with amorphous one.

Our ablation setup (figure 2) consisted of commercial femtosecond laser system (TeMa, Avesta Project), which has central wavelength of $\lambda \approx 1050$ nm (monitored by a spectrometer) and a pulse duration of 150 fs (monitored by an autocorrelator) with 80 MHz repetition rate. Laser pulses energy is varied by two attenuators and selected by a Pockels cell-based pulse picker (Avesta Project). Pulses were tightly focused on films with microscope objective (Olympus 40x) with numerical aperture of 0.75.

The resulting nanoparticles fabricated from 15/15 nm Au/Si films were studied experimentally by transmission electron microscopy. The electron microscopy shows that the fabricated nanoparticles have Au-core and Si-shell structure with an average size of less than 200 nm (figure 4-a).

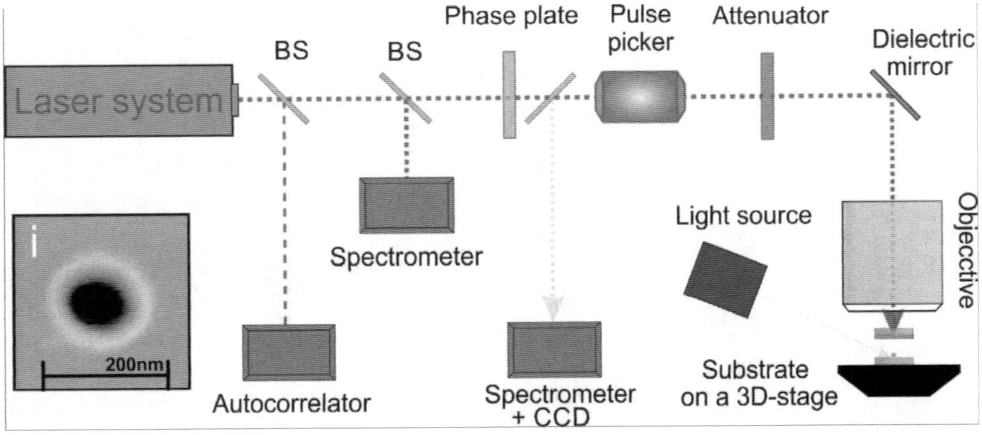

Figure 2. Optical setup for laser printing of core-shell nanoparticles. Inset shows a fabricated core-shell nanoparticle with Si core and Au shell. Inset: TEM image of a core-shell nanoparticle.

2.2 Optical properties

Despite we obtained experimentally only Si-core/Au-shell nanoparticles, the two main scenario are possible. In our case, Si part is crystallized much earlier than Au one, yielding a solid Si core covered by a thin Au layer, which is dewetted at large sizes of the Si core. However, in the case of formation of core-shell nanoparticle by liquid Au and Si parts, the Si is covered by Au with much higher surface energy. These two possible regimes are governed by kinetics of cooling and solidification, which depends on laser intensity (i.e. on initial temperature) and time of flight before deposition on a substrate with relatively high thermal conductivity (i.e. on time of slow cooling rate). Therefore, we consider two possible cases in our theoretical analysis of optical properties of Au/Si core-shell nanoparticles.

The numerical analysis in commercial software CST Microwave Studio of electromagnetic field distribution around a Si-core/Au-shell nanoparticle reveals three peaks corresponding to three resonances when the inner and outer diameters are 80 nm and 140 nm, respectively (figure 3a). The numerical simulations of near fields at these resonances are shown in figure 3, confirming their origin. As shown in figure 3b the high-frequency resonance at λ=410 nm corresponds to magnetic dipolar response of Si core, whereas the other two resonances at λ=550 nm and λ=760 nm are hybridized plasmon resonances of the Au shell. According to previous studies [14], the low-frequency plasmon resonance is symmetric mode, whereas high-frequency plasmon resonance is asymmetric mode of the shell. The low-frequency plasmon (symmetric coupling of the cavity and sphere plasmon), couples to light much more strongly than the high-frequency plasmon. The near-field enhancement is up to 3 times. This is because the dipole moment of the symmetric plasmon is much larger than the antisymmetric plasmon, where the dipole moment of the bare cavity and sphere plasmons oppose each other. Therefore, the resulting electric field inside the Si-core Au-shell is much stronger for the low-frequency plasmon (figures 3b,c). We also studied numerically power patterns at the resonances, which show the possibility of manipulation by scattering properties only around magnetic dipolar resonance owing to its interference with high-frequency plasmon in shell.

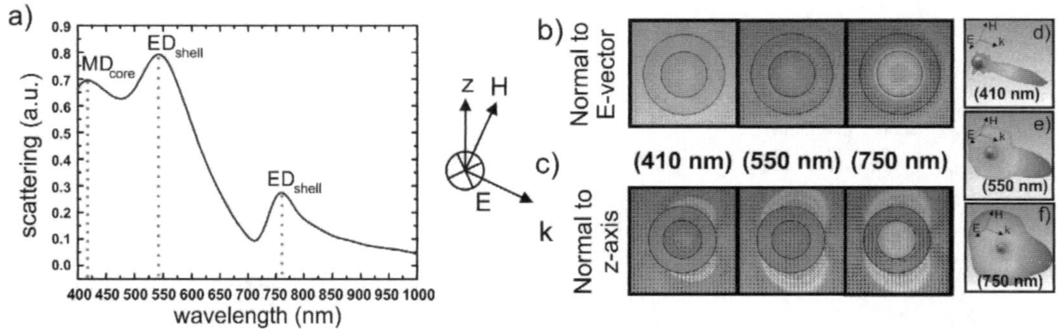

Figure 3. a) Theoretical scattering cross-sections for Si-core/Au-shell with diameters of 80/140 nm. Electric near-fields at ED, MD, MQ resonances b) cross-section is normal to E-vector direction. c) cross-section is normal to Z-axis through the core-shell center. Power patterns for d) MD_{core}, e) antisymmetric ED_{shell}, and f) symmetric ED_{shell}.

In turn, calculated Au-core/Si-shell nanoparticle scattering spectrum (figure 4-a) indicates the presence of three types of resonances - the electric dipole (ED), magnetic dipole (MD) and magnetic quadrupole (MQ) - in the visible range. These results were confirmed by the numerical calculations of the total scattering cross section and vector maps of electric field of hybrid core-shell nanoparticles with a geometry corresponding to the dimensions defined by the electron microscopy. Moreover, numerical calculations show the ability to effectively control the radiation pattern of the scattered light by nanoparticles. In figures 4-d,e we show power patterns of electric and magnetic dipole resonances, respectively. It is inherent for these types of resonances the absence of radiation in the far field along the E-vector direction for electric and along H-vector for MD resonance, as well as equal scattering to front and back directions. The power-pattern at MQ resonance is given in figure 4-f. In comparison with MD, the side radiation is decreased. In case of overlapping MD and ED resonances, there is, so-called, Huygens-source with almost complete scattering in forward direction (figure 4-g). In addition, core-shell nanoparticles allow for strong field localization and enhancement up to 5 times at ED as well as high Purcell factor in the dielectric shell due to the presence of the resonance of the gold core.

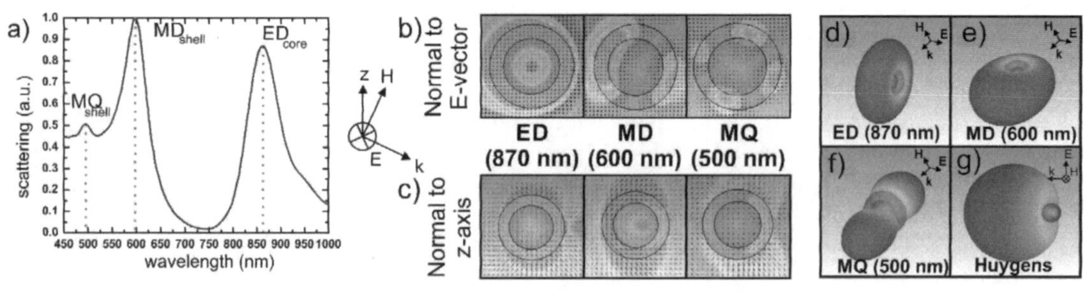

Figure 4. a) Theoretical scattering cross-sections for Au-core/Si-shell with diameters of 126/174 nm. Electric near-fields at ED, MD, MQ resonances b) cross-section is normal to E-vector direction. c) cross-section is normal to Z-axis through the core-shell center. Power patterns for d)-ED, e) MD, f) MQ and g) Huygens source at 750 nm.

3. Conclusions

In conclusion, we have shown for the first time that the laser printing allows to achieve fabrication of nanoparticles made of noble metal shell and high-refractive index dielectric core. Our numerical simulations provided insight into mode structure of the obtained core-shell nanoparticles, revealing plasmons excitation and magnetic optical response in the visible range. Such type of nanoparticles is a promising platform for creation of directional nanolasers, all-optical switches and biosensors.

Acknowledgements

This work was supported by the Ministry of Education and Science of the Russian Federation (project №14.584.21.0009 with unique identificator RFMEFI58414X0009).

References

[1] S.A.Maier, 2007 Plasmonics: Fundamentals and Applications.
[2] V.V.Klimov 2010 Nanoplasmonics.
[3] A.E.Krasnok 2015 *Proc. SPIE* **950203**.
[4] J.Zhou et al. 2005 *Phys. Rev. Lett.* **95** 22390
[5] S.V.Makarov et al. 2015 *Nano Lett.* **15(9)** 6187
[6] A.I.Kuznetsov et al. 2012 *Sci. Rep* **2** 492
[7] A.E.Krasnok et al. 2014 *Nanoscale* **6** 7354
[8] L.Shi et al. 2012 *Adv. Mater.* **24** 5934
[9] U.Zywietz et al. 2014 *Nature Comm.* **5** 3402
[10] D.A. Zuev et al. 2016 *Advanced Materials* **28** 3087
[11] G.Chen et al. 2012 *ACS Nano* **6(9)** 8280
[12] Liu P et al. 2015 *J. Phys. Chem. C* **119(2)** 1234
[13] P.A.Dmitriev et al. 2015 *Nanoscale* **8** 5043
[14] E. Prodan and P. Nordlander 2004 *J. Chem. Phys.* **120(11)** 5448

Plasmonic properties of metal nanoislands: surface integral equations approach

S A Scherbak[1,2] and A A Lipovskii[1,2]

[1]Department of Physics and Technology of Nanoheterostructures, St. Petersburg Academic University RAS, St. Petersburg 194021, Russia
[2]Institute of Physics, Nanotechnology and Telecommunications, Peter the Great St. Petersburg Polytechnic University, St. Petersburg 195251, Russia

Abstract. The surface integral equations method is used to analyse the surface plasmon resonance position in a metal island film formed by non-interacting axisymmetrical prolate/oblate hemispheroids placed on a dielectric substrate. The approach is verified via the comparison of results obtained for a hemisphere on a substrate with the ones obtained using the multipole expansion method. The preference of the integral equations method is in obtaining a simple final analytical expression for a particle polarizability in which any dielectric function of a metal can be substituted. Such simple formulae for the hemispherical particle on the substrate and calculated dependences of the hemispheroid resonant wavelength on its aspect ratio are presented.

1. Introduction

The surface plasmon resonance (SPR) phenomenon in metal nanoparticles [1] is of a particular interest for both applied optics/plasmonics and fundamental physics. A special and interesting case of metal nanoparticles are metal island films (MIF) which can be used as polarizers and spectral filters [2] and, because of the high enhancement of the local electric field, in Raman spectroscopy [3]. The spectral position of the SPR in MIF is usually calculated for metal nanoislands of a hemispherical shape observed in experiments [4]. Using the multipole expansion method [5, 6] the dependences of the SPR wavelength on the dielectric permittivity of the substrate and on the dielectric constant and the thickness of a layer covering the nanoparticles were calculated [7]. However, not hemispherical but hemispheroidal shape of metal nanoislands has recently also been reported [8], the nanoislands being axisymmetric prolate/oblate hemispheroids (see Figure 1). This makes important the understanding of plasmonic characteristics of such nanoparticles and, respectively, of MIF formed by metal hemispheroids.

To describe plasmonic properties of a particle it is usually enough to find its polarizability dispersion. In the case of free standing spherical and spheroidal nanoparticles there are the analytical solutions allowing one to find the SPR frequency and the absorption spectrum [9, 10]. However, the majority of even relatively simple shapes of metal nanoparticles require applying numerical techniques to describe their polarizability. Diverse numerical and semi-numerical methods allow this, and below we present the application of the surface integral equations method [11, 12] to the semi-numerical analysis of the SPR in a metal hemisphere and an axisymmetrical metal hemispheroid placed on a dielectric substrate.

2. The approach

We consider a plasmonic particle with somehow induced surface charges of a density $\sigma(\mathbf{r})$ and surrounded by a dielectric medium with permittivity ε_{out}. General relations of the Potential Theory [13] and the standard boundary conditions lead to the equation [12]:

$$\sigma_k(\mathbf{r}) = \frac{\lambda_k}{2\pi}\oint_S \sigma_k(\mathbf{s})\frac{\mathbf{n}(\mathbf{r})(\mathbf{r}-\mathbf{s})}{|\mathbf{r}-\mathbf{s}|^3}dS. \quad (1)$$

Here the index k corresponds to a particular plasmonic mode, $\lambda_k = (\varepsilon_k - \varepsilon_{out})/(\varepsilon_k + \varepsilon_{out})$, ε_k is the particle permittivity at which the kth plasmonic mode can be excited, $\mathbf{r}$ and $\mathbf{s}$ are position vectors of points on the nanoparticle surface, $\mathbf{n}(\mathbf{r})$ is a unit vector normal to the surface (see Figure 2). Eq. 1 is an eigenvalue problem which determines all possible surface plasmon modes in the particle. Since the plasmonic resonant frequency is to be found, it is convenient to consider the permittivity of the surrounding dielectric, ε_{out}, as a parameter. Note that the particle permittivity ε_k defines an eigenvalue λ_k and corresponds to the kth distribution of surface charge density $\sigma_k(\mathbf{r})$. These values ε_k are called resonant permittivities.

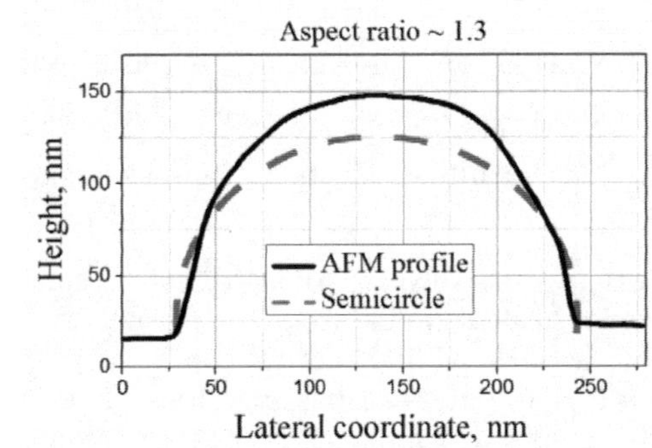

Figure 1. The example of an atomic force microscope (AFM) measured profile of a particle. The nanoisland is a prolate hemispheroid with the aspect ratio ~ 1.3.

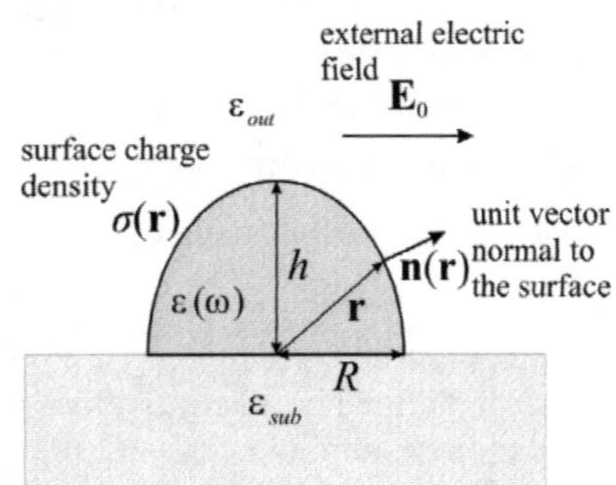

Figure 2. The scheme of the problem

If an external electric field $\mathbf{E}_0(\mathbf{r})$ is applied to the nanoparticle, Eq. 1 transforms to:

$$\sigma(\mathbf{r}) = \frac{\lambda}{2\pi}\oint_S \sigma(\mathbf{s})\frac{\mathbf{n}(\mathbf{r})(\mathbf{r}-\mathbf{s})}{|\mathbf{r}-\mathbf{s}|^3}dS + \frac{\lambda}{2\pi}\mathbf{n}(\mathbf{r})\mathbf{E}_0(\mathbf{r}). \quad (2)$$

Note that Eq. 2, as well as Eq. 1, can be written in a more general form using the Green function $G(\mathbf{r},\mathbf{s})$:

$$\sigma(\mathbf{r}) = \frac{\lambda}{2\pi}\oint_S \sigma(\mathbf{s})\mathbf{n}(\mathbf{r})\nabla_{\mathbf{r}}G(\mathbf{r},\mathbf{s})dS + \frac{\lambda}{2\pi}\mathbf{n}(\mathbf{r})\mathbf{E}_0(\mathbf{r}).$$

Here $G(\mathbf{r},\mathbf{s}) = |\mathbf{r}-\mathbf{s}|^{-1}$ for charges distributed in a homogeneous medium. In Eq. 2, λ is no longer an eigenvalue, and it is defined by the permittivity $\varepsilon(\omega)$ of the particle at a certain frequency ω: $\lambda = (\varepsilon(\omega) - \varepsilon_{out})/(\varepsilon(\omega) + \varepsilon_{out})$. The solution of Eq. 2 can be presented through the expansion by the eigenfunctions given by Eq. 1:

$$\sigma(\mathbf{r}) = \frac{\lambda}{2\pi}\sum_k \frac{\lambda_k}{\lambda_k - \lambda}\langle \mathbf{n}(\mathbf{r}')\mathbf{E}_0(\mathbf{r}');\tau_k(\mathbf{r}')\rangle \sigma_k(\mathbf{r}), \tag{3}$$

where τ_k is a corresponding eigenfunction of the equation adjoint to Eq. 1, λ_k is a corresponding eigenvalue, and $\langle \mathbf{n}(\mathbf{r}')\mathbf{E}_0(\mathbf{r}');\tau_k(\mathbf{r}')\rangle = \oint_S \mathbf{n}(\mathbf{r}')\mathbf{E}_0(\mathbf{r}')\tau_k(\mathbf{r}')dS'$. According to Ref. 12, the solutions of Eq. 1 and the adjoint one form two biorthogonal sets: $\oint_S \tau_i(\mathbf{s})\sigma_j(\mathbf{s})dS = \delta_{ij}$, where δ_{ij} is the Kroneckers delta. Since the eigenfunctions of Eq. (1) do not form an orthogonal set because the kernel of the equation is not symmetrical, it is convenient to use the eigenfunctions of the adjoint equation in the expansion (3). Note that the adjoint equation describes the surface dipole density which provides the same electric field as a single charged layer [12].

The particles dipole momentum, **p**, and, respectively, the polarizability, α ($\mathbf{p} = \alpha\mathbf{E}_0$), are given by the formula:

$$\mathbf{p} = \oint_S \mathbf{r}\sigma(\mathbf{r})dS = \frac{\varepsilon(\omega) - \varepsilon_{out}}{4\pi\varepsilon_{out}}\sum_k \frac{\varepsilon_{out} - \varepsilon_k}{\varepsilon(\omega) - \varepsilon_k}\left(\oint_S \mathbf{r}\langle \mathbf{n}(\mathbf{r}')\mathbf{E}_0(\mathbf{r}');\tau_k(\mathbf{r}')\rangle \sigma_k(\mathbf{r})dS\right). \tag{4}$$

The values ε_k are called the resonant permittivities since the polarizability increases greatly when $\varepsilon(\omega) \approx \varepsilon_k$. Hereinafter, we use notification ε_{res} for ε_k.

The substrate influence can be considered [12] using the method of mirror images. In this case the Green function transforms from $G(\mathbf{r},\mathbf{s}) = |\mathbf{r} - \mathbf{s}|^{-1}$ to $G_{sub}(\mathbf{r},\mathbf{s}) = |\mathbf{r} - \mathbf{s}|^{-1} - |\mathbf{r} - \mathbf{s}'|^{-1}(\varepsilon_{sub} - \varepsilon_{out})/(\varepsilon_{sub} + \varepsilon_{out})$ [12], where $\mathbf{s}'$ is the mirror image of $\mathbf{s}$ in the substrate. This Green function should be substituted in all the equations written above instead of $G(\mathbf{r},\mathbf{s}) = |\mathbf{r} - \mathbf{s}|^{-1}$.

The formula (4) allows calculating the polarizability and, respectively, the SPR position. The most time-consuming routine here – eigenvalues and eigenfunctions calculation – should be run only twice (for Eq. 1 and for the adjoint one), and the calculation of the polarizability spectral dependence does not require direct numerical study that considerably simplifies the analysis. Contrary, other techniques, like the multipole expansion method [5], the discrete dipole approximation (DDA) [14] or the finite elements methods (FEM) [15], require full calculation cycle for each spectral point.

3. Polarizability of hemispherical nanoparticles

Below we present the numerical solution of Eq. 1 and the adjoint one using the surface discretization. To be more specific we consider the external electric field $\mathbf{E}_0$ directed along the substrate surface; this corresponds to an incident s-polarized electromagnetic wave. We also assume that metal particles under consideration are much smaller than the light wavelength [4]. This allows neglecting retardation of the electric field phase that is to consider the problem within the frames of quasistatic (dipole) approximation [16]. In these assumptions we analyse the polarizability dispersion, $\alpha(\omega)$, of a metal hemisphere placed on a dielectric substrate.

Keeping in (4) the resonant term only, allows obtaining a simple formula for the polarizability spectral dependence in general form:

$$\alpha(\omega) = AR^3 \frac{\varepsilon(\omega)/\varepsilon_{out} - 1}{\varepsilon(\omega)/\varepsilon_{out} - \varepsilon_{res}}, \tag{5}$$

where scale coefficient A and resonant permittivity ε_{res} depend on specific problem geometry and substrate permittivity ε_{sub}; R^3 is a dimensional factor.

In case of a hemispherical particle in a homogeneous medium ($\varepsilon_{sub} = \varepsilon_{out}$) and on a substrate with $\varepsilon_{sub}/\varepsilon_{out} = 2.25$ we, respectively, obtain:

$$\alpha(\omega) = 0.63R^3 \frac{\varepsilon(\omega)/\varepsilon_{out} - 1}{\varepsilon(\omega)/\varepsilon_{out} + 3.95}, \tag{6}$$

$$\alpha(\omega) = 0.52R^3 \frac{\varepsilon(\omega)/\varepsilon_{out} - 1}{\varepsilon(\omega)/\varepsilon_{out} + 6.61}, \tag{7}$$

where the resonant permittivities 3.95; 6.61 are the direct result of the eigenvalues calculations and the scale coefficients 0.63; 0.52 are defined by the integrals in Eq. 4. All these values depend on the substrate permittivity because it appears in the kernels of the integral equations. The results presented by formulae like (6, 7) are the indisputable advantage of the used approach: the analysis of the polarizability spectral dependence does not require direct numerical study - it is enough to substitute specific dielectric function $\varepsilon(\omega)$ in Eqs. 6, 7. In Figure 3 we plot the dependences of the magnitude coefficient, A, and the resonant permittivity, ε_{res}, on the substrate dielectric constant. The plots in Figure 3 allow one to find the polarizability of a metal hemispherical particle with a dielectric function $\varepsilon(\omega)$ for given permittivities of the substrate, ε_{sub}, and the surrounding media, ε_{out}.

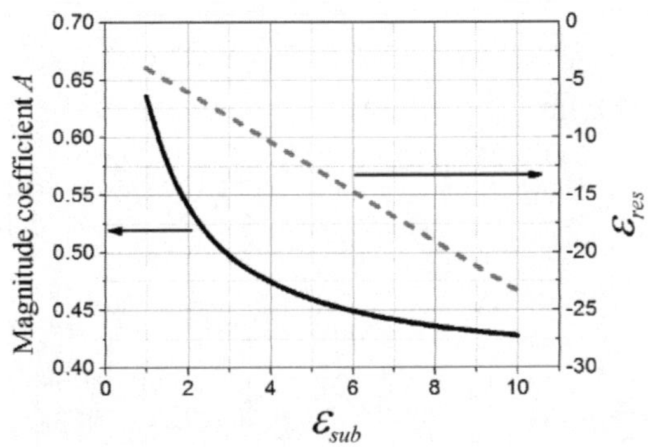

Figure 3. The dependences of the magnitude coefficient, A, and the resonant permittivity, ε_{res}, on the substrate dielectric constant, ε_{sub}.

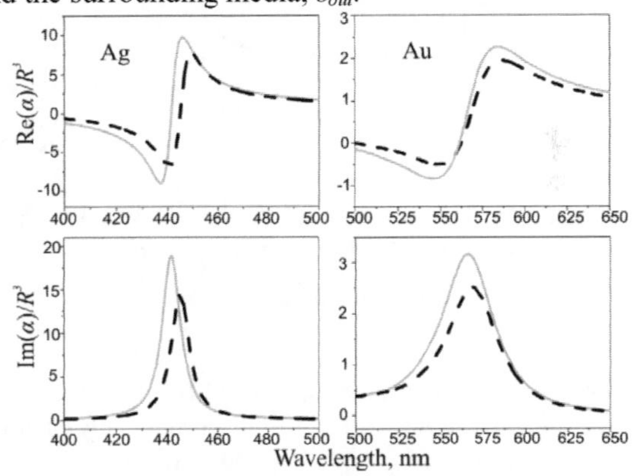

Figure 4. The dispersion of the real (upper row) and the imaginary (lower row) parts of polarizability of silver and gold hemispheres on a glass substrate (ε_{sub} = 2.25) in the air (ε_{out} = 1). Solid lines correspond to the results of the surface integral equations method, dash lines – multipole expansion method [6, 7]. The polarizability is normalized to dimensional factor R^3.

4. Surface integral equation method vs multipole expansion method

To verify the approach under discussion, we calculate the polarizability dispersion, $\alpha(\omega)$, of silver and gold hemispheres on a glass substrate with ε_{sub}=2.25 in the air, ε_{out}=1, and compare these with the results of multipole expansion method [6, 7], see Figure 4. The metals permittivities are taken from [17]. The calculated polarizability is normalized to dimensional factor R^3. Thus, the dependences in Figure 4 are dimensionless.

There is a slight discrepancy between the results given by these two approaches: 3-5 nm difference in the resonant wavelengths and about 20% difference in magnitudes. However, inasmuch as both methods are approximate this discrepancy does not seem to be crucial.

The result for a hemisphere in a homogeneous medium also coincides well with the outcome of the multipole expansion method. According to [18], the SPR in a metal hemisphere in a homogeneous medium occurs if $\varepsilon(\omega)/\varepsilon_{out} \approx -4$. The surface integral equations method gives the value -3.95 (see Eq. 6).

5. Polarizability of prolate/oblate hemispheroids

We use the surface integral equations approach to find the polarizability spectra and the SPR wavelengths of axisymmetrical hemispheroids differing in aspect ratio (Figures 5, 6). Figure 5 illustrates the polarizability dispersion of silver and gold hemispheroids with aspect ratio *h/R*=0.7 and aspect ratio *h/R*=1.5.

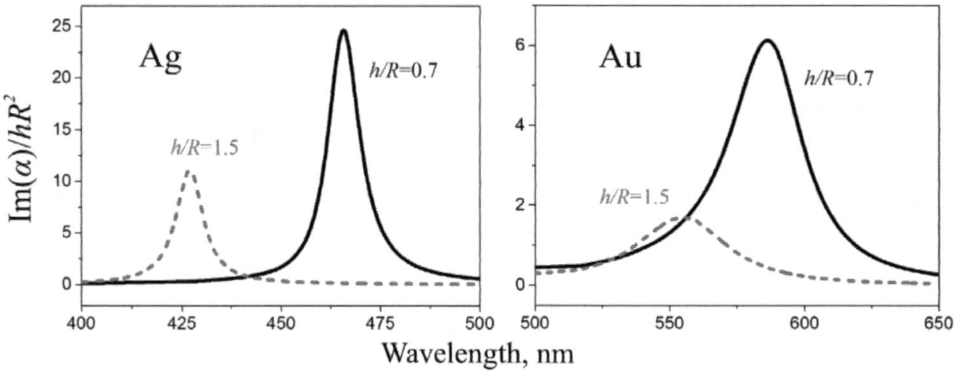

Figure 5. The dispersion of the imaginary part of polarizability of silver (left) and gold (right) hemispheroids with different aspect ratios (*h/R*) on a glass substrate ($\varepsilon_{sub} = 2.25$). The polarizability is normalized to the dimensional factor hR^2.

As expected, the resonant polarizability for s-polarization grows with the redistribution of metal from height to lateral direction. SPR wavelength also grows tending to approach to the polarizability of a thin disk. One can see that the red shift of the SPR about 50 nm occurs after the decrease in the aspect ratio from 1.5 to 0.7, about two times.

In Figure 6 we present dependences of hemispheroids resonant wavelength on their aspect ratio. The curves are calculated for silver and gold particles and for different substrate dielectric constants.

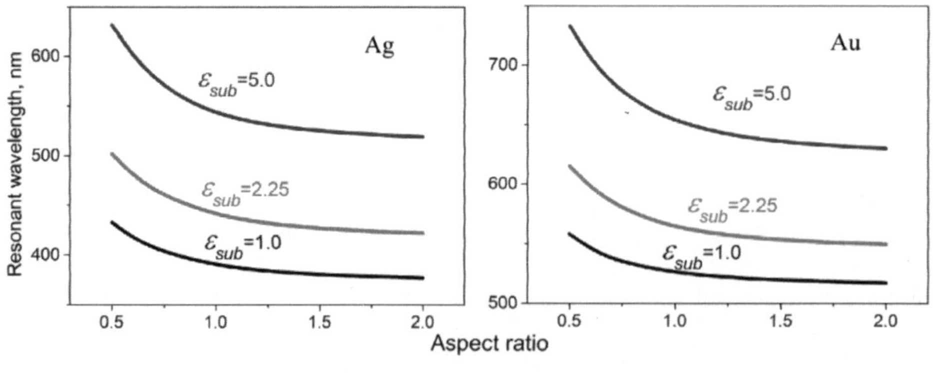

Figure 6. The dependence of the resonant wavelength on the aspect ratio of silver (left) and gold (right) hemispheroids for different substrate permittivities.

Note that the higher substrate permittivity corresponds to the wider range in which the resonant wavelength varies. The behavior of the calculated dependences qualitatively corresponds to the trend well known from the analytical expressions obtained for spheroids in a homogeneous medium [9, 10].

6. Conclusion

We used the surface integral equations method to calculate the polarizability dispersion of a metal hemisphere placed on a dielectric substrate and compared the result with the one obtained using the multipole expansion method. The comparison evidenced reasonable coincidence of the results obtained with these techniques. The preference of the surface integral equations method is in the simple analytical expression for polarizability spectrum derived. The dependences of numerical parameters in the expression on a substrate dielectric constant were presented. The importance of this result is in avoiding numerical calculations for estimating a particle polarizability dispersion and the SPR wavelength. Used approach allowed us to analyze the influence of an axisymmetrical hemispheroid aspect ratio on the polarizability spectrum. Finally we presented the dependences of the SRP wavelength on silver and gold nanoislands aspect ratio. All the results presented are applicable to sparse MIF covered with dielectrics and can be used to estimate their SPR position and the polarizability.

Acknowledgements
This work was supported by the Ministry of Education and Science of the Russian Federation (project #16.1233.2014/K).

References

[1] Kreibig U and Vollmer M 1995 *Optical Properties of Metal Clusters* (Berlin: Springer)
[2] Heger P, Stenzel O and Kaiser N 2004 *Proc. of SPIE* **5250**, 21.
[3] Dieringer J, McFarland A, Shah N, Stuart D, Whitney A, Yonzon C, Young M, Xiaoyu Z and Van Duyne R 2006 *Faraday Discuss.* **132**, 9
[4] Zhurikhina V, Brunkov P, Melehin V, Kaplas T, Svirko Y, Rutckaia V and Lipovskii A 2012 *Nanoscale Res. Lett.* **7**, 676
[5] Hafner C 1960 *Generalized Multipole Thechnique for Computational Electromagnetics* (Boston: Artech House)
[6] Wind M, Vlieger J and Bedeaux D 1987 *Physica* **141A**, 33-57.
[7] Scherbak S, Shustova O, Zhurikhina V and Lipovskii A 2015 *Plasmonics* **10**, 519
[8] Heisler F, Babich E, Scherbak S, Chervinsky S, Hasan M, Samusev A and Lipovskii A 2015 *J. Phys. Chem. C* **119**, 26692–97
[9] Noguez C 2007 *J. Phys. Chem. C* **111**, 3806-19
[10] Landau L and Lifshitz E 1960 *Electrodynamics of Continuous Media, Volume 8 of "A Course of Theoretical Physics"* (Pergamon Press)
[11] Fuchs R 1974 *Phys. Rev. B* **11**, 1732
[12] Mayergoyz I, Fredkin D and Zhang Z 2005 *Phys. Rev. B* **72**, 155412
[13] Kellogg O 1929 *Foundations of Potential Theory* (NY: McGraw-Hill)
[14] Draine B and Goodman J 1993 *Astrophys. J. 405*, 685
[15] Morgan M and Mei K 1979 *IEEE Trans. Antennas Propag.* 27, 202–14
[16] Mayer S 2007 *Plasmonics: Fundamentals and Applications* (NY: Springer Science)
[17] Johnson P and Christy R 1972 *Phys. Rev B* **6**, 4370
[18] Kettunen H, Wallén H and Sihvola A 2008 *J. Appl. Phys.* **102**, 094112

One-dimensional photonic crystals with highly Bi-substituted iron garnet defect in reflection polar geometry

T V Mikhailova, V N Berzhansky, A V Karavainikov, A N Shaposhnikov, A R Prokopov and S D Lyashko

V.I. Vernadsky Crimean Federal University, 4 Vernadskogo Prospekt, Simferopol, 295007, Russia

Abstract. It is represented the results of modelling of magnetooptical properties in reflection polar geometry of one-dimensional photonic crystal, in which highly Bi-substituted iron garnet defect of composition $Bi_{1.0}Y_{0.5}Gd_{1.5}Fe_{4.2}Al_{0.8}O_{12}$ / $Bi_{2.8}Y_{0.2}Fe_5O_{12}$ is located between the dielectric Bragg mirrors $(SiO_2 / TiO_2)^m$ (were *m* is number of layer pairs) and buffer SiO_2 and gold top layers of different thicknesses is placed on structure. The modification of spectral line-shapes of microcavity and Tamm plasmon-polariton modes depending on *m* is found.

1. Introduction

One-dimensional photonic crystals (1D-PCs) [1] are artificial optical-order periodic structures, that can affect the propagation of light and actively used for micro- and nano- optoelectronic devices, displays, information transmitting devices, storage, magnetic holography and sensors. The functionality of such devices may be based on the magnetooptical (MO) effects, Faraday and polar Kerr effects, that occur in included in 1D-PC structure magnetoactive (MA) defect. The transparent Bi-substituted iron garnet (Bi:IG) films may be taken as a defect. Enhancing of MO effects defect due to the phenomena of interference and diffraction in 1D-PC reaches record values of 20 times at visible [2] and 100 times at near infrared [3] ranges of wavelengths. Recently research of MO effects in 1D-PCs, in which the simultaneous excitation of several resonance modes (microcavity or defect, Tamm plasmon-polariton, surface plasmon-polariton, exciton) exist, also became relevant.

This work predominantly focuses on the investigation of microcavity and Tamm plasmon-polariton modes in modelled polar Kerr rotation (KR) spectra of 1D-PCs $[TiO_2/SiO_2]^m$ / $Bi_{1.0}Y_{0.5}Gd_{1.5}Fe_{4.2}Al_{0.8}O_{12}$ / $Bi_{2.8}Y_{0.2}Fe_5O_{12}$ / $[SiO_2/TiO_2]^m$ with addition of buffer SiO_2 and gold top layers of different thicknesses. The model is based on the fitting parameters of the synthesized 1D-PCs without buffer SiO_2 and gold top layers. Optimality of multiresonance 1D-PCs for application depends on the Bi:IG layers absorption, that are most strongly manifested in the visible range of wavelengths and that contribution was not previously observed in the polar Kerr effect modelling or experimental study [4].

2. Synthesis of 1D-PCs with highly Bi:IG defect. Experimental reflectance and polar Kerr rotation spectra

As was found previously in our investigations to create nanostructures of 1D-PCs with highly Bi:IG defect and high optical quality necessary to satisfy the optimal synthesis conditions. Thus, the double MA defect of composition $Bi_{1.0}Y_{0.5}Gd_{1.5}Fe_{4.2}Al_{0.8}O_{12}$ / $Bi_{2.8}Y_{0.2}Fe_5O_{12}$ (the buffer layer $Bi_{1.0}Y_{0.5}Gd_{1.5}Fe_{4.2}Al_{0.8}O_{12}$ hereinafter is M1 and the main layer $Bi_{2.8}Y_{0.2}Fe_5O_{12}$ hereinafter is M2) in considered 1D-PCs caused by the technological difficulties of synthesis of garnet films with a high Bi content on SiO_2 layers. Bi:IG layers were fabricated by reactive ion beam sputtering of corresponding ceramic targets in argon-oxygen mixture on "cold" fused quartz substrate KU-1 and crystallized in the annealing process at the air. High (up to 80%) oxygen content in the sputtering gas mixture and low heating rates during annealing crystallization used to achieve low surface roughness of the Bi:IG layers. M1 and M2 layers were formed and crystallized separately at optimal annealing temperature of 680 °C. The first layer M1 was deposited on SiO_2 and after its crystallization the layer M2 was deposited and crystallized. The thickness of MA layers was controlled by the sputtering rates and the deposition times. Necessary MO and the crystalline quality of the double layer is achieved at the minimum optimal thickness of the buffer layer, which approximately corresponds to the optical thickness $\lambda_0/4$, where λ_0 is the wavelength of photonic band gap (PBG) center. TiO_2 and SiO_2 layers were fabricated by electron beam evaporation with an optical thickness control by monitor in-situ. To minimize the size and influence of transition layer "substrate – structure" ion-beam treatment of the substrates (low-energy plasma of Ar^+, ion current density 2.5 $mA{\cdot}cm^{-2}$) was applied.

The experimental and theoretical spectra of polar KR and reflectance of synthesized 1D-PCs with MA defect of $(2.5{\cdot}\lambda_0/2)$-optical thickness and repetition number of layer pairs $m = 4$ and $m = 7$ are shown in Figure 1 (data taken from [2]). MA defect of thickness $h_M = 317$ nm includes a buffer layer M1 of thickness $h_{M1} = 68$ nm. PBG center is $\lambda_0 = 685$ nm. The thicknesses of SiO_2 and TiO_2 layers are $h_{SiO_2} = 116\,\text{nm}$ and $h_{TiO_2} = 75\,\text{nm}$, respectively. For these structures the maximum Faraday rotation –13.6 and –20.6 deg and MO quality factor 15.1 and 8.0 deg have been achieved for resonance at 625 nm for $m = 4$ and $m = 7$, respectively.

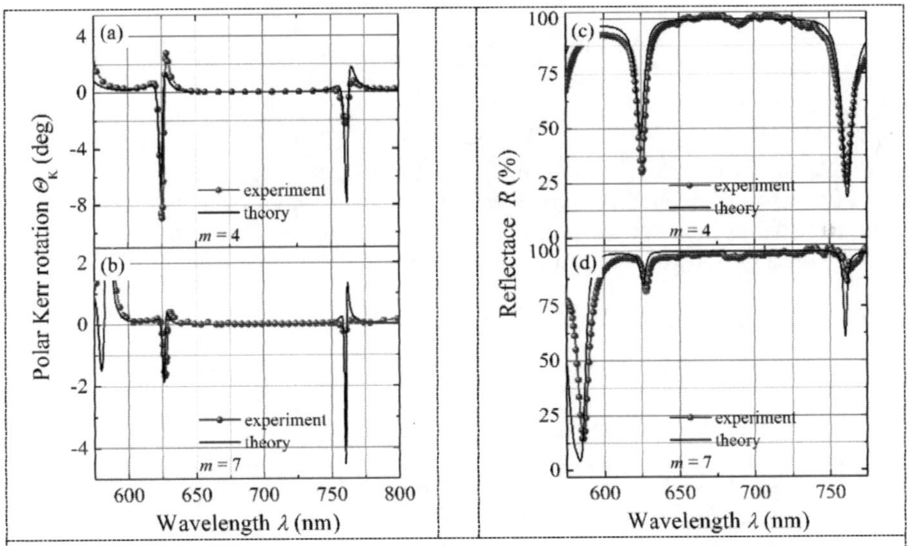

Figure 1. Spectra of polar Kerr rotation Θ_K (a, b) and reflectance R (c, d) of synthesized 1D-PCs with highly Bi:IG defect of $(2.5{\cdot}\lambda_0/2)$-optical thickness and different number m [2].

As will be shown in subsection 3.1, the number of resonance peaks (one or two) in PBG corresponding to first I, second II or third III microcavity modes (with resonance conditions $(\lambda_0/2)$, (λ_0) or $(3{\cdot}\lambda_0/2)$, respectively) and their spectral position depends on the thickness of MA defect (or its optical thickness l_M). In Figure 1 the spectra include II and III microcavity modes at 760 and 625 nm, accordingly. The features of polar KR spectra of considered 1D-PCs without gold top layer are a large positive peak and two minor negative asymmetric peaks in the vicinity of the resonance wavelengths.

In contrast to the Faraday rotation values the maximum polar KR observed at 625 nm for the structures with $m = 4$ ($\Theta_K = -8.9$ deg), but not with $m = 7$ ($\Theta_K = -1.7$ deg).

3. Polar Kerr rotation spectra modelling

The light propagation in 1D-PCs were modelling by computational solution of Maxwell's equation

$$\nabla \times \mathbf{E}(\mathbf{r},t) = i\omega\mu_0 \mathbf{H}(\mathbf{r},t),\ \nabla \times \mathbf{H}(\mathbf{r},t) = -i\omega\varepsilon_0 \hat{\varepsilon}\,\mathbf{E}(\mathbf{r},t) \qquad (1)$$

by conventional 4×4 transfer matrix method [5]. Here $\mathbf{E}(\mathbf{r},t)$ and $\mathbf{H}(\mathbf{r},t)$ are magnetic and electric fields vectors of light wave respectively, $\mathbf{r}$ is radius-vector, t is time, μ_0 and ε_0 are vacuum permittivity and permeability respectively, $\hat{\varepsilon}$ is the permittivity tensor of corresponding layer. For optical frequencies $\mu = 1$. General view of $\hat{\varepsilon}$ is

$$\hat{\varepsilon} = \begin{pmatrix} \varepsilon_{xx} & -i\varepsilon_{xy} & 0 \\ i\varepsilon_{xy} & \varepsilon_{xx} & 0 \\ 0 & 0 & \varepsilon_{zz} \end{pmatrix}. \qquad (2)$$

Components ε_{xx} and ε_{xy} for MA layers are complex, $\varepsilon_{xx} = \varepsilon'_{xx} + i\varepsilon''_{xx}$, $\varepsilon_{xy} = \varepsilon'_{xy} + i\varepsilon''_{xy}$. The value of ε_{zz} is very close to ε_{xx} below Curie temperature T_C and is equal to ε_{xx} above T_C. Component ε_{xx} determines by refractive index n and extinction coefficient k: $\varepsilon_{xx} = (n^2 - k^2) + i \cdot 2 \cdot n \cdot k$. Component ε_{xy} relates to Faraday rotation Θ_F, magnetic circular dichroism $\Delta\alpha$, refractive index n, extinction coefficient k of film and wavelength of incident light λ as

$$\varepsilon'_{xy} = -\left(\frac{\lambda}{\pi}\right)\left(n \cdot \Theta_F + \frac{k \cdot \Delta\alpha}{4}\right),\ \varepsilon''_{xy} = -\left(\frac{\lambda}{\pi}\right)\left(\Theta_F \cdot k - \frac{n \cdot \Delta\alpha}{4}\right) \qquad (3)$$

For nonmagnetic layers of BM $\varepsilon_{xy} = 0$ and $\varepsilon_{zz} = \varepsilon_{xx}$. It is considered that TM polarized light falls to the structure. The fused quartz substrate KU-1 and the spectral dependence of permittivity tensor components of layers taken into account. Components of permittivity tensor of 1D-PC layers is defined using the experimental data of single films and synthesized 1D-PCs and presented in Table 1. These values are used for shown in Figure 1 calculations and below.

Table 1. Permittivity tensor components for layers of model 1D-PCs

Wavelength λ (nm)	M1 $\varepsilon^1_{xx}, \varepsilon^1_{xy}$	M2 $\varepsilon^2_{xx}, \varepsilon^2_{xy}$	SiO_2 and KU-1 ε_{SiO_2}	TiO_2 ε_{TiO_2}	Au ε_{Au}
600	$7.039 + 0.121 \cdot i$, $-0.014 + 2.656 \cdot 10^{-3} \cdot i$	$8.35 + 0.177 \cdot i$ $-0.072 + 0.019 \cdot i$	2.125	5.409	$-9.34 + 1.47 \cdot i$
650	$6.591 + 0.066 \cdot i$, $-9.187 \cdot 10^{-3} + 2.702 \cdot 10^{-3} \cdot i$	$7.863 + 0.063 \cdot i$ $-0.037 + 0.01 \cdot i$	2.12	5.282	$-12.864 + 1.198 \cdot i$
700	$6.277 + 0.038 \cdot i$, $-7.37 \cdot 10^{-3} + 2.813 \cdot 10^{-3} \cdot i$	$7.518 + 0.044 \cdot i$ $-0.024 + 7.801 \cdot 10^{-3} \cdot i$	2.117	5.182	$-16.519 + 1.113 \cdot i$
750	$6.06 + 0.021 \cdot i$, $-6.619 \cdot 10^{-3} + 2.95 \cdot 10^{-3} \cdot i$	$7.279 + 0.041 \cdot i$ $-0.018 + 7.886 \cdot 10^{-3} \cdot i$	2.114	5.102	$-20.22 + 1.122 \cdot i$

Thus, the following structure is taken as a model

KU-1 / $[TiO_2/SiO_2]^m$ / M1 / M2 / $[SiO_2/TiO_2]^m$ / SiO_2 / Au with thicknesses of M1 layer $h_{M1} = 68$ nm, SiO_2 and TiO_2 layers of 1D-PC $h_{SiO_2} = 116\,nm$ and $h_{TiO_2} = 75\,nm$, respectively. M2 h_{M2}, top gold h_{Au} and top buffer SiO_2 h_{bSiO_2} layers thicknesses and repetition number of layer pairs m are varied depending on the calculations purpose.

3.1. 1D-PC with highly Bi:IG defect of different optical thickness and without top SiO_2 and gold layers

Figure 2 shows the spectral displacement of resonance peaks corresponding to I, II and III microcavity modes occurring with increasing of optical thickness of MA defect. The thickness of MA defect is augmented by change of M2 layer thickness h_{M2}. For $(\lambda_0/2)$-, $(1.5\cdot\lambda_0/2)$-, (λ_0)-, $(2.5\cdot\lambda_0/2)$- and $(3\cdot\lambda_0/2)$-structures the values of h_{M2} are 64, 126, 193, 244 and 315 nm, respectively. The modification of KR peaks is absent. Dependence of polar KR on the number m for I microcavity mode in Figure 3 (a) demonstrates the achievement of maximum KR for 1D-PC with optimum repetition number of layer pairs $m_{opt} = 4$, as the experiment show earlier (see section 2). It is caused the growing influence of significant absorption of highly Bi:IG during the amplification of the light localization inside the defect with increasing of number m. It is evident from the transmittance and reflectance shown here. The optimum value of m_{opt} is different for each resonance modes and changes with the its shifting in PBG. Figure 3 (b) shows the polar Kerr rotation as a function of the number m for I and II microcavity modes. KR can be increased by shifting of cavity mode from the PBG center to longer wavelengths with increasing of m_{opt} (curves 2 and 3) or by a change of the microcavity mode at the BPG center wavelength with increasing of h_{M2} and decreasing of m_{opt} (curves 1 and 4).

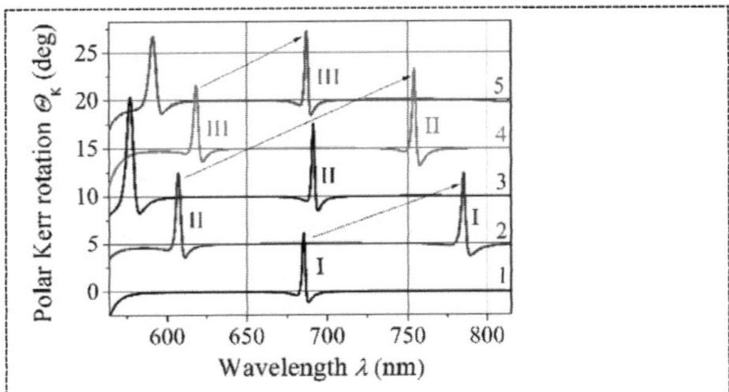

Figure 2. Polar Kerr rotation spectra of model 1D-PC with MA defect of $(\lambda_0/2)$-, $(1.5\cdot\lambda_0/2)$-, (λ_0)-, $(2.5\cdot\lambda_0/2)$- and $(3\cdot\lambda_0/2)$-optical thickness (curves 1, 2, 3, 4 and 5, respectively). Repetition number of layer pairs is $m = 4$. The arrows indicate the displacement of microcavity modes. Curves is shifted up to 5 degrees relative to the previous curve on the scale of polar Kerr rotation.

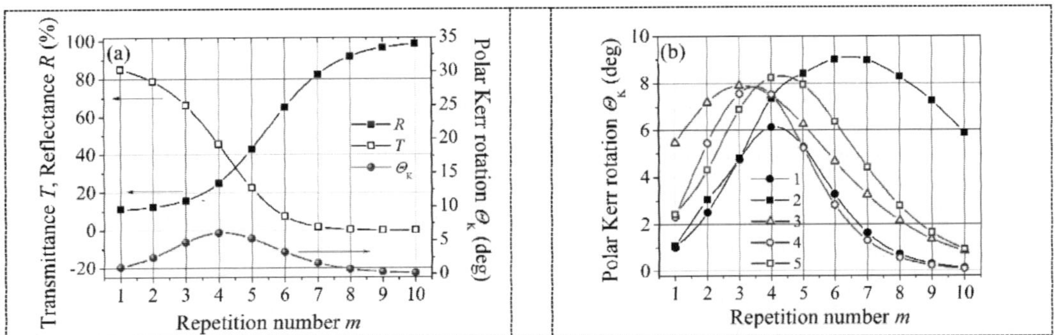

Figure 3. (a) Reflectance R, transmittance T and polar Kerr rotation Θ_K as a function of the number m of model 1D-PC with MA defect of $(\lambda_0/2)$-optical thickness. (b) Polar Kerr rotation Θ_K as a function of the number m for I microcavity mode of model 1D-PC with MA defect of $(\lambda_0/2)$- and $(1.5\cdot\lambda_0/2)$-optical thickness (curves 1 and 2, respectively) and for II microcavity mode of model 1D-PC with MA defect of $(1.5\cdot\lambda_0/2)$-, (λ_0)- and $(2.5\cdot\lambda_0/2)$-optical thickness (curves 3, 4 and 5, respectively).

3.2. 1D-PC with highly Bi:IG defect of (λ_0/2)-optical thickness and top gold layer and without top SiO_2 buffer layer

For the 1D-PC structure with the top gold layer the modification of line-shape of microcavity mode takes place. The Tamm plasmon-polariton mode is observed in the wavelength range from 740 to 850 nm and its behaviour is considered in subsection 3.3. The inversion of KR peak of microcavity mode occurs for 1D-PC with $m = 4$ at increasing of gold layer thickness h_{Au} from 0 to 200 nm, that is shown in Figure 4. For structures with $m < 4$ the line-shape and sign of microcavity mode are similar to the line-shape and sign of Faraday rotation peaks. For the structure with $m > 4$ the line-shape of KR peak is similar to that shown in Figures 1 and 2 for 1D-PC without the top gold layer. The spectral position of the resonance peak is strongly dependent on the number m. The maximum KR is achieved at "inversion region" for $m = 4$ $\Theta_K = -17.7$ deg and $m = 5$ $\Theta_K = 21.3$ deg. So, for the structures with $m = 5$ the presence of a top gold layer increases the KR by 4 times at preservation the line-shape of the resonance peak. The inversion of KR peak is caused by the competition of "passing through" (absorbed) and "reflected by" MA layer contributions to the polarization rotation angle of 1D-PC.

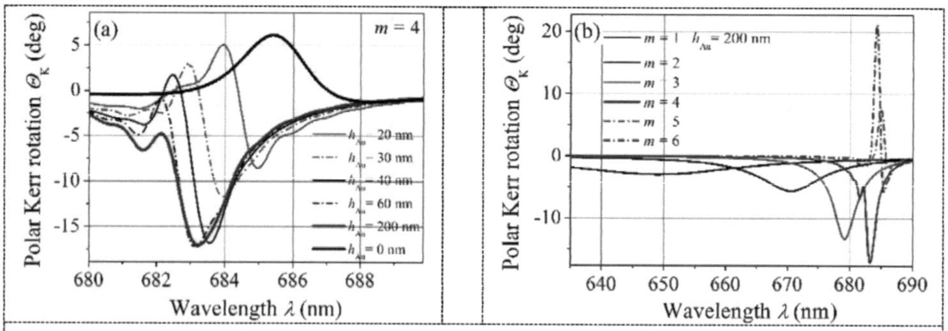

Figure 4. Spectra of polar Kerr rotation Θ_K of model 1D-PC with MA defect of (λ_0/2)-optical thickness for I microcavity mode as a function of the thickness h_{Au} for the number $m = 4$ (a) and as a function of the number m at unchanged thickness $h_{Au} = 200$ nm (b).

3.3. 1D-PC with highly Bi:IG defect of (λ_0/2)-optical thickness and top buffer SiO_2 and gold layers

Spectral position and line-shape of Tamm plasmon-polariton mode depend on the introduction of the buffer layer in structure before top gold layer. To reduce the absorption of used materials as the buffer layer the SiO_2 layer of thickness h_{bSiO_2} has been selected, but the MA layer can also be applied. Figure 5 shows the modification of KR spectra of considered 1D-PC with unchanged top gold layer thickness $h_{Au} = 200$ nm versus number m and the thickness h_{SiO_2}, which is varied from 174 to 232 nm.

When a buffer layer is absent, the spectral position of Tamm plasmon-polariton mode corresponds to PBG long-wave edge, similarly to that shown in Figure 5 (e) and (f). When m changes its spectral position and line-shape also change. S-shape peak is observed for structures with $m > 4$. When the buffer layer is present, in a some buffer thickness range the Tamm plasmon-polariton mode arises at the short-wave edge (Figure 5 a, b) and shifts to the long-wavelength edge of PBG. At the same time the alignment of microcavity and Tamm plasmon-polariton modes does not occur. But at the closest approach these modes are spectral dependent. That is, the modes have a similar line-shape and behave the same way with the change of m. When $m = 4$ for both modes the sign inversion of peaks takes place (Figure 5 c, d). Thus, as shown by the results, to achieve the highest angles of rotation of light polarization plane in reflection the using of 1D-PC with Tamm plasmon-polariton mode is not profitable.

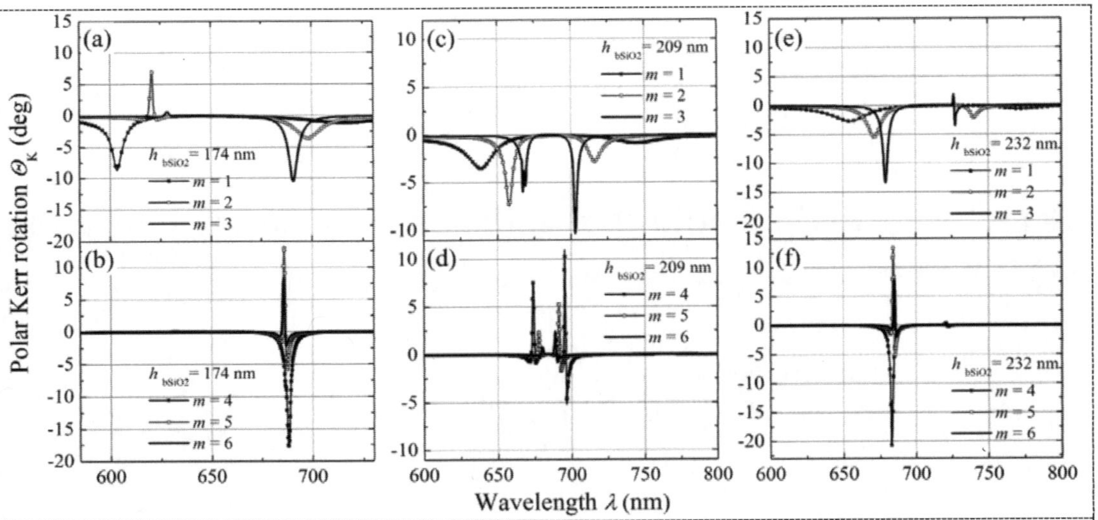

Figure 5. Spectra of polar Kerr rotation Θ_K of model 1D-PC with MA defect of $(\lambda_0/2)$-optical thickness for I microcavity mode and controlled by the thickness of buffer layer h_{bSiO_2} Tamm plasmon-polariton mode as a function of the number m at unchanged thickness of top gold layer h_{Au} = 200 nm.

4. Summary

The modelling of polar Kerr rotation spectra of one-dimensional photonic crystal with highly Bi-substituted iron garnet defect and top buffer SiO_2 and gold layers of different thicknesses demonstrates a strong dependence of spectral line-shapes of microcavity and Tamm plasmon-polariton modes on number of layer pairs m. In the vicinity of $m = 4$ or $m = 5$ the signs of the peaks of both modes inverse as mode approach to each other. Similar occurs for only one microcavity mode, if modes spectrally separated. The modification of the spectral dependences caused by the large absorption of highly Bi-substituted iron garnet defect. In order to enhance the magnetooptical response in reflection geometry is better to use only one top gold layer without a buffer layer. The maximum of polar Kerr rotation $\Theta_K = 21.3$ deg have the structure with $m = 5$ and amplification effect is 4 times as compared with the structure without gold.

Acknowledgments

This work is supported by the RF Ministry of Education and Science in the framework of the base part of the state task №2015 / 701 (project 3879).

References

[1] Inoue M, Baryshev A V, Goto T, Baek S M, Mito S, Takagi H and Lim P B 2013 *Magnetophotonics* (Springer-Verlag Berlin Heidelberg)
[2] Berzhansky V N, Mikhailova T V, Karavainikov A V, Prokopov A R, Shaposhnikov A N, Kharchenko Y M, Lukienko I M, Miloslavskaya O V, Kharchenko M F, Belotelov V I and Golub V O 2015 *Solid State Phenomena* **230** 241
[3] Kato H, Matsushita T, Takayama A et al. 2004 *J. Magn. Magn. Mater.* **272–276** e1305
[4] Kato H and Inoue M 1997 *J. Appl. Phys.* **81** 5659
[5] Yin C P, Wang T B and Wang H Z 2012 *Eur. Phys. J. B* **85** 104

Journal of Physics: Conference Series **741** (2016) 012122 doi:10.1088/1742-6596/741/1/012122

Bound state in the continuum in the one-dimensional photonic crystal slab

Z F Sadrieva and A A Bogdanov
ITMO University, St. Petersburg, 197101,Russia

zfsadrieva@corp.ifmo.ru

Abstract. In this work we developed a design of one-dimensional Si/SiO2 photonic crystal slab supporting so called optical bound states in the continuum – infinitely high-Q optical states with energies lying above the light line of the surrounding space. Such high-Q states are very perspective for many potential applications ranging from on-chip photonics and optical communications to biological sensing and photovoltaics.

1. Introduction

Localization of light in cavities (waveguides) implies that there is no coupling of cavity modes with modes of free space forming a continuum spectrum. Eigenstates of optical systems is localized only if their energies lie under the light line of the surrounding space. Any modes with energies lying in the continuum spectrum (leaky modes) have low quality factor because of the radiation losses. As a result, they have not found wide implementation. However, in 1929 von Neumann and Wigner proposed that the single-particle Schrodinger equation may possesses spatially localized states with energy lying above the asymptotic limit of the potential [1]. For optics, it means existence of localized states with energies from the continuum spectrum of leaky modes. Such states are called bound states in the continuum (BIC).

Marinica et al, following the results by von Neumann and Wigner, proposed the existence of infinitely high-Q (non-radiative) optical states with energies lying above the light line of the surrounding space [2]. As well as in the case considered by von Neumann and Wigner, radiation of optical BIC into the surrounding space disappeared because of the destructive interference. Theory of BIC is actively developed now. Nature of BIC can be explained by interference effects [3], in terms of topological charge [4], or by symmetry properties of eigenmodes and photonic structures [4,5]. First experimental observation of optical BIC was carried out by Plotnik et al in 2011 [6]. Nowadays a lot of scientific groups worldwide investigate the potential applications of BIC in various fields of science and technology [5,7]. Nevertheless, the majority works are theoretical [2-4, 8, 9].

In this work we theoretically analyzed CMOS-compatible one-dimensional photonic structure based on silicon-on-insulator wafer with optical BIC at telecommunication wavelengths (1.5-1.6 microns).

2. Results and discussion

The photonic structure under consideration is shown in the figure 1(a). It consists of rectangular bars with a period *a*, width *w* and height *t*. We propose that the bars are made of crystalline silicon with refractive index about 3.47 surrounded by fused silica with refractive index 1.44. The periodic geometry results in photonic band structure. The photonic slab supports guided and leaky modes with frequencies embedded in the continuum modes of free space. Leaky resonances have a finite lifetime, however, in the vicinity of specific points the lifetime tends to infinity. These points correspond to BIC.

We consider two types of slabs with fixed ratio of the width w to the period a: w=0.45a and w=0.5a. Here we focus on the lowest TE-like mode (electric field lies along bars). The dependence of frequency and Q-factor for this mode on the Bloch wavenumber are shown in figure 1(b, c). Mode structure of the photonic crystal slab was simulated using the finite-element method in Comsol Multiphysics software. The eigenvalue solver returns complex eigenfrequencies f, so it is possible to calculate radiative Q-factor as

$$Q_r = Re(f)/Im(f). \quad (1)$$

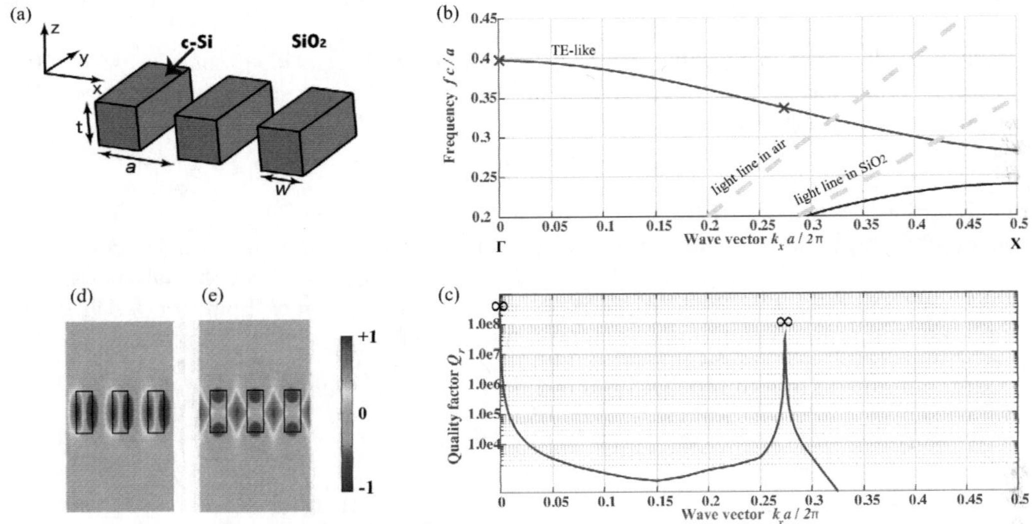

Figure 1. (a) Design of a photonic crystal slab with one-dimensional periodicity in x-direction and is infinite along y-direction. (b) Calculated band structure for photonic slab with width w=0.45a and thickness t=1.2a . Cross symbols indicate embedded eigenstates (BIC). (c) Calculated radiative lifetime, that is quality factor of photonic slab. At the certain k points leaky resonance does not decay and light becomes perfectly confined in the slab. (d), (e) Electric field profile E_y of the trapped state at zero and nonzero wave vector k_x respectively, plotted on the $y=0$ slice.

One can see that at some k points lifetime tends to infinity and light becomes perfectly confined in the slab. At these points leaky resonances turn into localized eigenmodes that do not decay. Sometimes eigenvalues existed within light cone, where there is continuum of radiation modes, are called embedded eigenvalues [3].

We discovered that for the photonic crystal slabs with w=0.45a two BIC may occur (see figure 1). The first one has zero wave vector (Γ-point) that corresponds to the normal incidence of the exciting wave, and the second one has a nonzero wave vector that corresponds to an oblique incidence of the exciting wave. Figure 2 shows the dependence of eigenfrequency at both zero and nonzero x-component of wave vector from photonic structure design. The second BIC emerges at certain t/a value. As t/a increases, the incident angle becomes bigger till it reaches value 90° from air, see figure 3.

In contrast to structures discussed above, photonic crystal slabs with $w=0.5a$ have embedded states at the Γ point only.

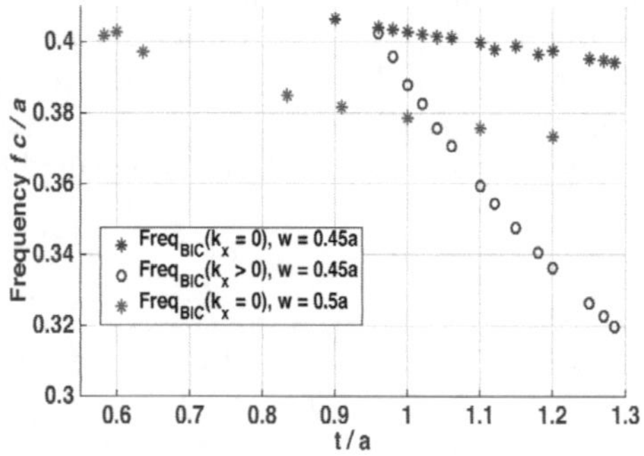

Figure 2. BIC position at both zero and nonzero x-component of wave vector versus photonic structure design.

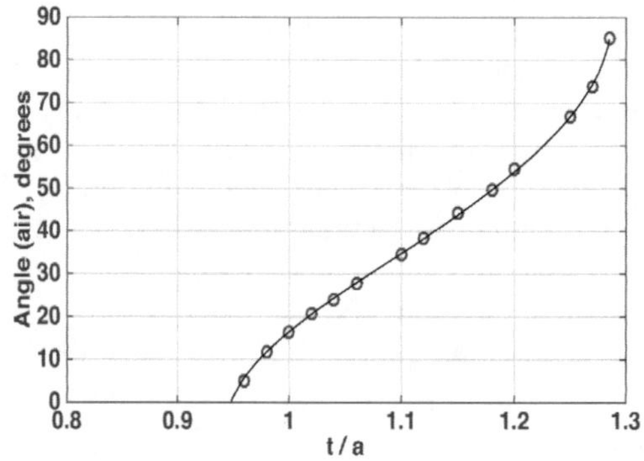

Figure 3. The incident angle at which BIC occurs in photonic crystal slab with $w=0.45a$.

Both resonances (Γ and off-Γ BICs) can be tuned to the telecommunication frequency range (1.5 to 1.6 microns) by proper choice of the geometrical parameters of the sample (w, t, a). As soon as figure 2 is plotted in terms of dimensionless parameters, it describes any photonic crustal slab specified design. As an example, let us consider the structures with thickness t=600 nm, for which figure 4 illustrates results. The frequency range of interest determines the rest geometric parameters (w, a).

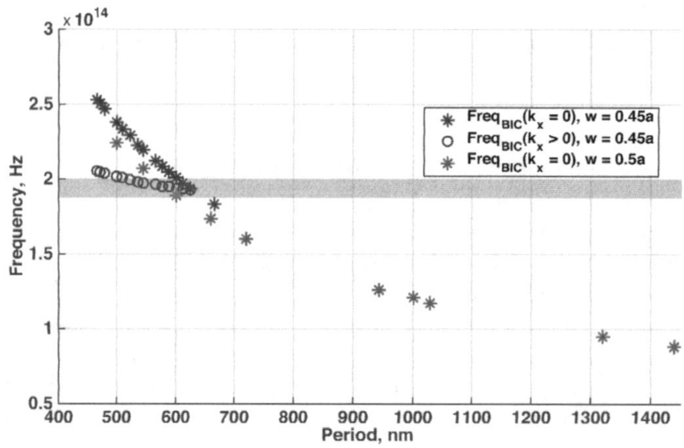

Figure 4. BIC position at both zero and nonzero x-component of wave vector versus photonic structure period a. The thickness t is fixed on 600 nm. The shaded area corresponds to telecommunication frequency range.

As an example, we consider the structures with thickness t=600 nm. Figure 4 shows the dependence of BIC frequency on the period of the structure for different ratio w/a. So, one can see from figure 4 that in order to BIC frequency lies in the telecommunication frequency range the rest parameters of the structure could be taken, for example, as follows: period a=500 nm and width w=225 nm.

For these parameters, the first resonance (at the normal incidence) corresponds to 238 THz (1.260 microns) and the second one (at the angle of incidence about 34.3 degrees from air) corresponds to 201 THz (1.491 microns). One can see that the second resonance is close enough to telecommunication wavelength range.

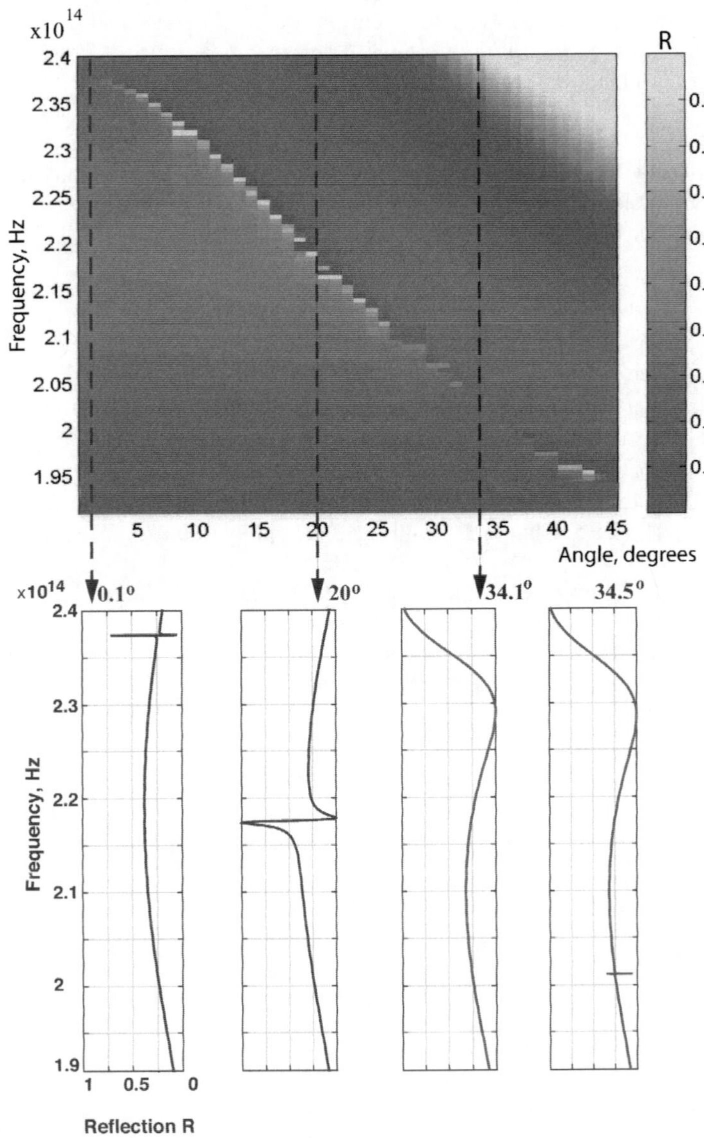

Figure 5. Calculated reflectance spectrum: map and cross sections. An incident angle is measured in SiO_2 medium. Disappearance of the response around 0° and 34.1° indicates a trapped state with no leakage.

To confirm existence of BIC we simulated the reflection spectra from the structure under consideration for different angles of incidence. The interference of incident light with leaky modes results in appearance of Fano-type resonances in the reflectance spectra [10]. We observed them nearby the frequencies corresponding to position BIC obtained within the simulation of the mode structure (figure 5). In contrast, perfect BIC has no response in reflection spectrum, because it is decoupled from free space waves. So, disappearance of the response around 0° and 34.1° indicates a trapped state with no leakage.

To summarize, we observed two types of photonic crystal slabs and found out that there are two BICs for slabs with width w=0.45a only. The eigenfrequencies, Q-factors and reflection spectra were calculated. The photonic structures with the proposed design can be fabricated using standard CMOS-compatible silicon-on-insulator wafers.

Acknowledgments
This work is supported by the Ministry of Education and Science of the Russian Federation (Grant of the President of the Russian Federation (MK-6462.2016.2)) and the Russian Foundation for Basic Research (#16-37-60064).

References

[1] von Neumann J, Wigner E 1929 *Phys. Z.,* **30** 465 – 467
[2] Marinica D C, Borisov A G, and Shabanov S V 2008 *J. Phys. Rev. Lett.* **100** 183902
[3] Hsu C W, Zhen B, Lee J, Chua S L, Johnson S G, Joannopoulos J D, Soljacic M 2013 *Light: Science & Applications* **2** e84
[4] Zhen B, Hsu C W, Ling L, Stone A D, Soljacic M 2014 *J. Phys. Rev. Lett.* **113** 257401
[5] Hsu C W, Zhen B, Lee J, Chua S L, Johnson S G, Joannopoulos J D, Soljacic M 2013 *Nature* **499** 188
[6] Plotnik Y, Peleg O, Dreisow F, Heinrich M, Nolte S, Szameit A, Segev M 2011 *Phys. Rev. Lett.* **107** 183901
[7] Lepetit T and Kante B 2014 *Phys. Rev. B* **90** 241103(R)
[8] Bulgakov E N, Sadreev A F 2001 *J. Tech. Phys.* **46** 10
[9] Zou C L, Cui J M, Sun F W, Xiong X, Zou X B, Han Z F and Guo G C 2015 *Laser Photonics Rev.* **9** 114
[10] Fan S, Joannopoulos J D 2002 *Phys. Rev. B* **65** 235112

Influence of AlN/GaN superlattice period on frequency of polar optical modes

D V Pankin[1,2], M B Smirnov[1]

[1] Physical faculty, St.Petersburg State University, 199034 St. Petersburg, Russia.
[2] Resource Center for Optical and Laser Materials Research, St. Petersburg State University, St. Petersburg, 199034 St. Petersburg, Russia.

E-mail: dima-pankin@mail.ru

Abstract. Polar optical phonons in the binary AlN/GaN superlattices (SL) were studied in the framework of dielectric continuum model (DCM). It is shown that the modes propagating in the interface plane are delocalized over all layers and the modes propagating along the SL axis are delocalized in certain layers. Frequencies of former strongly depend on the layer thickness ratio and frequencies of the latter do not depend on the SL structure. Such behavior is typical for the short-period SL. We show that a strong coupling between two types of the modes takes place at increasing SL period. The phenomenon is described mathematically by the frequency-structure relations and its physical meaning is discussed.

1. Introduction

The nanostructures based on the $A^{III}N$ nitrides attract much attention owing their great potentiality for modern optoelectronics. Particularly, the AlN/GaN superlattices (SL) are used as active material in various devices operating in ultraviolet and infrared spectral regions. They are used as a key element in deep-ultraviolet light emitting diodes and in high-speed infrared intersubband optical devices such as intensity modulators, optically pumped light emitters, and photodetectors [1, 2]. Due to great importance of these systems for applications, the structure control of the grown SL is an imperative issue. Raman spectroscopy is a rather efficient technique for a fast nondestructive sample characterization. It was found that some of the Raman-active phonon modes are rather sensitive to particularities of the SL's structure: frequencies of the delocalized polar modes with wave vectors parallel to the interface plane strongly depend on the layer thickness ratio [3]. This phenomenon is due to the strong long-range dipole-dipole interactions and is specific feature of the nitride-based SLs.

Strong correlations between phonon frequencies and SL's structure opens possibility to use Raman spectroscopy for the structure characterization. In this connection it is important to established precise quantitative relations between phonon frequencies and structural parameters. Structure of a binary SL is described by thickness of the constituent layers d_1 and d_2 (here and below we use '1' for GaN and '2' for AlN) or alternatively by total period $d=d_1+d_2$ and the layer thickness ratio $f=d_1/(d_2+d_1)$. Another key parameter for the Raman scattering process is the wavelength of the scattered light whose wave length λ customary is about 500 nm. Depending on the d value all studied SL can be divided in two groups: the short-period SLs ($d \ll \lambda$) and long-period SLs ($d \sim \lambda$). Most attention was paid to the short-period SL owing their great potentiality for the optoelectronic applications [4]. The long-period SL also attract some interests due to their applications in the distributed Bragg reflectors [5] and in the ultraviolet light-emitting diodes [6]. First thorough discussion of the phonon-structure relations was focused on the dependence of the phonon frequencies on the SL period, or rather on the ratio d/λ

referred as *s*-parameter [7]. Later, dependence of the phonon frequencies on the layer thickness ratio (i.e. on the f-parameter) also was considered [3]. However, this analysis dealt only with the short-period SL (for $s<<1$). A complex analysis which considered both these impacts was not done. This study is aimed to fill this gap.

2. Method

The long-range dipole-dipole interactions can be taken into account in the frame of the dielectric continuum model (DCM) [8]. Such approach does not deal with atomistic structure of the materials and consider SL as sandwich-like infinite periodic solid matter. Dielectric matrix is periodic function which takes is any layer the values of corresponding constituent crystals. Both AlN and GaN are hexagonal wurtzite-like crystals, and the AlN/GaN SL grows along the hexagonal axes. Thus, the SL is a uniaxial system and dielectric constant matrix has the form

$$\boldsymbol{\varepsilon} = \begin{pmatrix} \varepsilon_{xx} & & \\ & \varepsilon_{xx} & \\ & & \varepsilon_{zz} \end{pmatrix} \quad (1)$$

in all layers. Polar phonon modes in the short-period nitride-based SL can divided A and E symmetry representations [7,9]. The A-modes are polarized along z-axis(axis of structure growth), and E-modes are polarized in perpendicular direction (in *xy*-plane corresponding to the interface plane). Hence the phonon contributions to the dielectric function can be written as follows [10]:

$$\varepsilon_{xx}(\omega) = \varepsilon^{\infty}\frac{\omega_{E(LO)}^2 - \omega^2}{\omega_{E(TO)}^2 - \omega^2}, \quad \varepsilon_{zz}(\omega) = \varepsilon^{\infty}\frac{\omega_{A(LO)}^2 - \omega^2}{\omega_{A(TO)}^2 - \omega^2} \quad (2)$$

Dielectric parameters for AlN and GaN crystals are listed in Table 1.

Table 1. Dielectric parameters for AlN and GaN crystals.

	Unit	AlN	GaN
$\omega_{A(TO)}$	cm^{-1}	610 [11]	531 [18]
$\omega_{E(TO)}$	cm^{-1}	669 [11]	559 [18]
$\omega_{A(LO)}$	cm^{-1}	890 [11]	734 [18]
$\omega_{E(LO)}$	cm^{-1}	911 [11]	742 [18]
ε^{∞}		4.84 [13]	5.35 [14]

Essence of the DCM approach is solving the Maxwell equations

$$\mathrm{rot}\mathbf{E} = 0, \quad \mathrm{div}\boldsymbol{\varepsilon}\mathbf{E} = 0 \quad (3)$$

for the plane-wave field

$$\mathbf{E}(\mathbf{r}) = \mathbf{E}_0 \exp(\mathrm{i}\mathbf{qr}) \quad (4)$$

and the frequency dependent dielectric matrix (1-2).

The *z*-axis is directed along the periodicity direction, *x* and *y*-axes lay in the interface plane. We consider the waves with wave vector **q** directed along *x*-axis. Hence, the electric field components can be written as

$$E_x = E_x(z)\exp(\mathrm{i}qx), \quad E_y = 0, \quad E_z = E_z(z)\exp(\mathrm{i}qx). \quad (5)$$

Using the equations (3) one obtains

$$\frac{\mathrm{d}}{\mathrm{d}z}E_x(z) - \mathrm{i}qE_z(z) = 0, \quad \mathrm{i}q\varepsilon_{xx}E_x(z) + \varepsilon_{zz}\frac{\mathrm{d}}{\mathrm{d}z}E_z(z) = 0. \quad (6)$$

Solutions of these linear differential equations can be found by substituting $E_x(z) = E_x^0 \exp(gz)$ and $E_z(z) = E_z^0 \exp(gz)$. This leads to the relations:

$$gE_x^0 - iqE_z^0 = 0, \quad iq\varepsilon_{xx}E_x^0 + g\varepsilon_{zz}E_z^0 = 0 \tag{7}$$

from which one can conclude:

$$-q^2\varepsilon_{xx} + g^2\varepsilon_{zz} = 0. \tag{8}$$

Using equation (8) one can determine *g*-index for any given *q*-index. Recall the latter defines wavelength of oscillations along *x*-direction and the former describe exponential decay along *z*-axis. Value of the *g*-index depends on the dielectric constant ratio:

$$g = \pm q\sqrt{\frac{\varepsilon_{xx}}{\varepsilon_{zz}}} = \pm qr \tag{9}$$

Recall that in the DCM approach the dielectric constants are frequency-dependent and can take positive or negative values. Thus, depending on sign of the $\varepsilon_{xx}/\varepsilon_{zz}$ ratio *g* can be real or imaginary. The first type of solutions corresponds to the exponential decay *z*-dependence, and the second one is related to the oscillatory *z*-dependence. On analyzing the ε_{xx} and ε_{zz} values for the GaN and AlN crystals in the wide frequency interval one can determine the frequency regions in which different types of DCM solutions are expected.

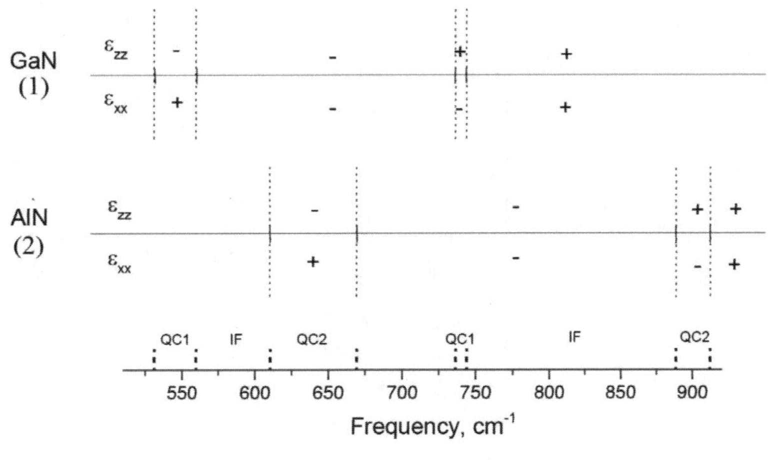

Figure 1. Frequency dependence of sign of the $\varepsilon_{xx}/\varepsilon_{zz}$ ratio for AlN and GaN crystals.

One can see that there are two intervals in which the $\varepsilon_{xx}/\varepsilon_{zz}$ ratios are positive for both crystals. They correspond to the so-called interface (IF) phonons. Besides there are intervals in which the $\varepsilon_{xx}/\varepsilon_{zz}$ ratio is positive for one crystal and negative for another. They correspond to the so-called quasi-confined (QC) phonons which have oscillatory character in one material and the exponential decay character in another. We shall discriminate them by index '1' (oscillations in GaN) and '2' (oscillations in AlN). Solutions of equations (8-9) for all the layers 1 (GaN) and 2 (AlN) must be supplemented by the boundary conditions:

$$E_{1,x} = E_{2,x}, \; \varepsilon_{zz}^1 E_{1,z} = \varepsilon_{zz}^2 E_{2,z} \tag{10}$$

The considered SL structure is symmetric with respect to plane passing through the middle of any layer. Hence, all solutions must be invariant or anti-invariant with respect to the mirror-plane transformation. Besides, the solution invariant with respect to E_x must be anti-invariant with respect to E_z and vice versa (see equation 6). We shall refer the solutions invariant with respect to E_z as IFZ and QCZ, and the solutions invariant with respect to E_x as IFX and QCX.

Journal of Physics: Conference Series **741** (2016) 012123 doi:10.1088/1742-6596/741/1/012123

Summarizing, the phonon frequency region can be divided in six intervals (see figure 1). Within every interval, the DCM allows determining two phonon states: either IFZ and IFX, or QCZ and QCX. Frequencies of all phonons depend on two structural parameters: the total period described by dimensionless parameter $s=q(d_1+d_2)$ and the layer thickness ratio which we shall characterize by parameter $f=d_1/(d_2+d_1)$.

It was shown that these dependencies can be easily established for the short-period SL, i.e. in the $s\rightarrow 0$ limit. In this case one can assume q=0 and g=0. This means that the electric fields are homogeneous in all layers, the quasi-confined (QCX and QCZ) modes are fully confined in the certain layers, and the IFZ and IFX modes are delocalized over all layers. Moreover, in the $s\rightarrow 0$ limit, all modes are strongly polarized: the solutions IFZ and QCZ correspond to E_x=0, and the solutions IFX and QCX correspond to E_z=0.

Evidently, frequencies of the completely confined QCX and QCZ modes correspond to corresponding parameters of the constituent bulk material. Frequencies of the completely delocalized IFX and IFZ modes can be determined by simple calculations based on the equations (2, 10). These frequencies ω depend on the layer ratio parameter f as it is shown in figure 2. Now we turn to the central question – how do these dependencies change along with increasing s value?

Analytic equations for solving this problem were presented in [7]. Solving these equations with various s and f values we have calculated The results are shown in figure 2(b) as three sets of the $\omega(f)$ dependencies corresponding the for s= 0.1, 1, 10. In this paper we focus our attention on the frequency of A_1(TO)-mode.

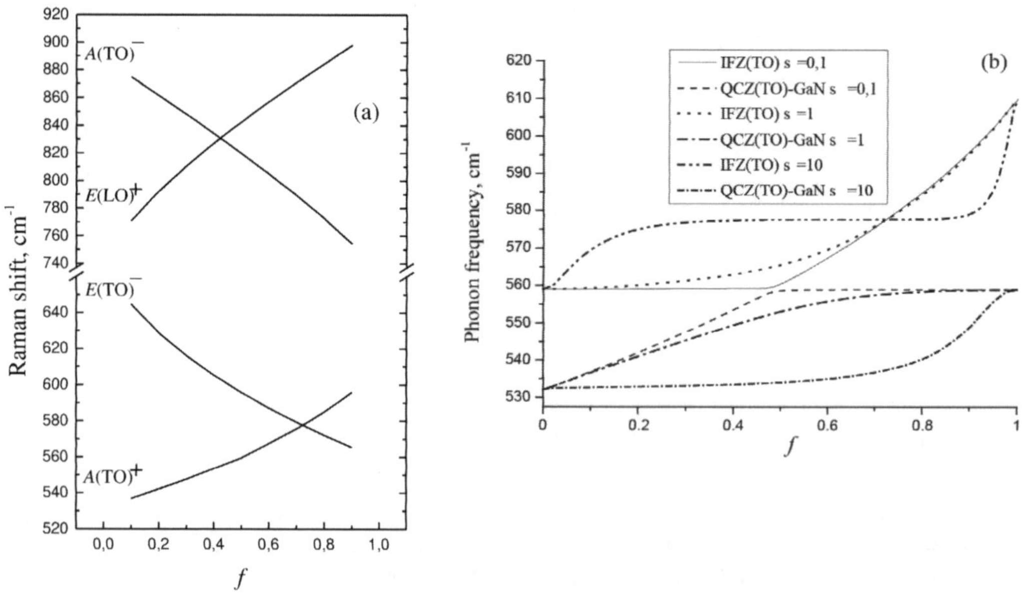

Figure 2. Frequencies of polar modes of $(GaN)_m(AlN)_n$ in dependence on the layer thickness parameter *f*. (a): s=0 (b): IFZ(TO) and QCZ(TO)-GaN modes at different s values.

3. Discussion

In the case of s=0.1 it is seen that IFZ(TO) and QCZ(TO)-GaN smoothly merge together at $\omega=\omega_{E(TO)1}$ forming the curve which is very close to that representing the dependence for the $A(TO)^+$(or A_1(TO)) mode in figure 2a (i.e. in the s = 0 limit).

In the case of s = 1 the curves have no common points: they diverge on approaching to the line $\omega = \omega_{E(TO)1}$. Such behaviour indicates a strong coupling between IFZ(TO) and QCZ(TO)-GaN modes. The latter is no longer confined at s ≠ 0. The nature of this coupling is quite evident: at finite wave vector directed along interface plane the atomic oscillations in z and x directions are dynamically coupled. The dependencies presented in figure 2b show that at f < 0.47 QCZ(TO)-GaN curve at s=1 deviates from that at s=0.1 not far than 8 cm^{-1}. The same is true in respect IFZ(TO) curves at s=1 and s=0.1 at f > 0.47. This allows an estimation of accuracy of the long wave approximation as 8 cm^{-1} at s< 1.

At thicker layers, divergence of the QCZ(TO)-GaN and IFZ(TO) curves becomes more considerable. This is illustrated by calculation at s= 10. This evidences essential difference between IF and QC modes in a long- period SL. At s=10 the curves exhibit a large plateau in the large interval 0.2 <f< 0.8. Physically, this means that frequencies of these modes do not depend on the layer thickness. Frequencies of QCZ(TO)-GaN modes take the values of the bulk A(TO)-GaN phonons, and frequency of the IFZ(TO) mode takes the value of $\omega_{IF(TO)}$=578.5 cm^{-1}. Note, that this value exactly corresponds to the IF mode localized on the single free-standing heterojunction [15].

The structure related phonon mode splitting can be illustrated by experimental data reported in [3]. Raman spectra of the $GaN/Ga_{0.28}Al_{0.72}N$ SLs with d_1 = d_2 and with d varying from 2.5 up to 1500 nm are shown in figure 9. They clearly manifest that at increasing d the following transformations occur: the single intense line corresponding to IFZ mode is clearly seen in the spectrum of the short-period SLs (up to s = 1); then it becomes to enlarge and at s≈10 evidently splits into two components which gradually diverge and at s≈100 achieve the frequencies typical for A(TO) modes in bulk crystals. This phenomenon is in line with the transformations shown in figure 2b: the single line observed in short period SL (curves IFZ(TO) and QCZ(TO)-GaN at s=0.1) splits in a doublet (curves IFZ(TO) and QCZ(TO)-GaN at s=1) at intermediate s values, and the two spectral lines can be observed in the long-period SLs (curves IFZ(TO) and QCZ(TO)-GaN at s=10). The calculated splitting value is shown in figure 3b.

4. Conclusions

It was found that magnitude of frequency splitting between two A_1(TO) phonon components in the equi-thick AlN/GaN SL depends on SL period. It is equal to zero in the case of short-period SL and reaches maximal value in the case of long-period SL (the terms 'short' and 'long' imply the ratio between the SL period and the wavelength of scattered light). It is shown that in the case of long-period SL there are two A_1(TO) modes: one with frequency equal to the A_1(TO) phonon in bulk GaN and another one with frequency equal to the interface phonon on the isolated AlN/GaN heterojunction. The established dependence between frequency splitting and the SL period can be well approximated by exponential function.

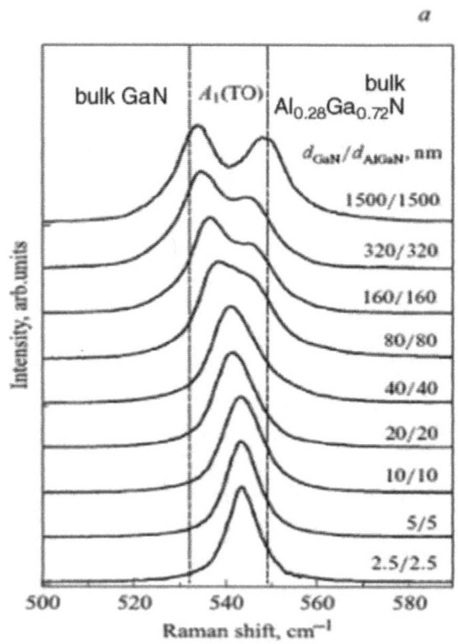

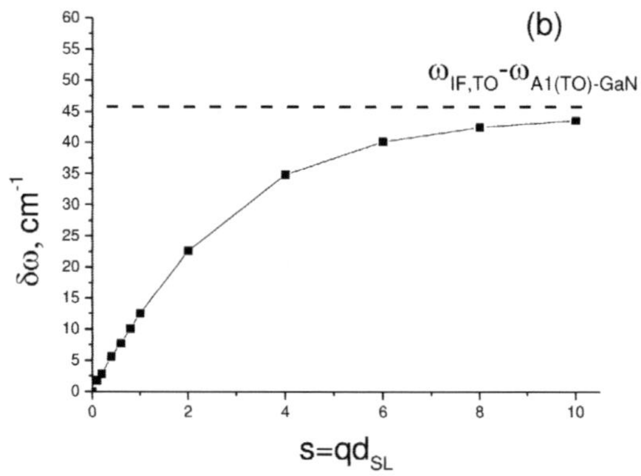

Figure 3. (a): A1(TO) Raman spectra of the GaN/$Al_{0.28}Ga_{0.72}$N SLs with d_1 = d_2 and different d=d_1+d_2. Corresponding s values are as follows: s = 0.25 (2.5/2.5), 0.5 (5/5), 1 (10/10), 2 (20/20), 4 (40/40), 8 (80/80), 16 (160/160), 32 (320/320), 150 (1500/1500). Positions of the A(TO) lines in bulk constituents are shown by dotted vertical lines. (b): calculated frequency splitting values in dependence on the s-parameter.

Referencies

[1] Nikishin S, Holtz M and Temkin H 2005 *Jpn. J. Appl. Phys.*, **44**, 7221
[2] Hofstetter D, Baumann E, Giorgetta F R, Dawlaty J, George P A, Rana F, Guillot F and Monroy E 2008 *Appl. Phys. Lett.*, **92**, 231104
[3] Smirnov M B, Karpov S V, Davydov V Yu, Smirnov A N, Zavarin E E, and Lundin V V 2005 *Phys. Solid State* **47**, 742
[4] Nikishin S, Holtz M, and Temkin H 2005 *Jpn. J. Appl. Phys.* **44**,7221
[5] Nakada N, Ishikawa H, Egawa T and Jimbo T 2003 *Jpn. J. Appl. Phys.* **42**, L144
[6] Khan A, Balakrishnan K and Katona T 2008 *Nature Photonics* **2**, 77
[7] Gleize J, Renucci M A, Frandon J and Demangeot F 1999 *Phys. Rev. B*, **60**, 15985
[8] Rytov S M 1956 *Sov. Phys. JETP*, **2**, 466
[9] Kitaev Yu E, Limonov M F, Tronc P and Yushin G N 1998 *Phys. Rev. B*, **57**, 14209
[10] Loudon R 1964 *Adv. Phys.*, **13**, 423
[11] Kuball M, Hayes J M, Prins A D, van Uden N W A, Dunstan D J, Shi Y and Edgar J H 2001 *Appl. Phys. Lett.*, **78**, 724
[12] Davydov V Yu, Kitaev Yu E, Goncharuk I E, Smirnov A N, Graul J, Semchinova O, Uffmann D, Smirnov M B, Mirgorodsky A P and Evarestov R A 1998 *Phys. Rev. B*, **58**, 12899
[13] Collins A T, Leitowlers E C and Dean P J 1967 *Phys. Rev. B*, **158**, 833
[14] Misek J, and Srobar F 1979 *Elektrotech. Cas.*, **30**, 690
[15] Lee B C, Kim K W, Stroscio M A, and Dutta M 1998 *Phys. Rev. B*, **58**, 4860

Electric-field-induced mid-infrared birefringence of the double quantum wells

M Ya Vinnichenko, V M Babich, R M Balagula, A N Sofronov, D A Firsov, L E Vorobjev

Peter the Great St. Petersburg Polytechnic University, 29 Polytechnicheskaya str., St. Petersburg195251, Russia

Abstract. Birefringence in double tunnel-coupled GaAs/AlGaAs quantum wells was studied in the mid-infrared spectral range close to the intersubband resonance. Phase-sensitive optical studies allowed us to deduce simultaneously the differences of the refraction index and absorption coefficient for the normal waves polarized in the plane of the structure and along the structure growth, including electric-field induced effects. The optical absorption data are in a good agreement with the direct optical transmission measurements.

1. Introduction

The semiconductor structures with multiple quantum wells (QW) represent the uniaxial media with respect to the optical properties. The permittivity tensor $\hat{\varepsilon}$ has the non-zero components $\varepsilon_{xx} = \varepsilon_{yy}$ and ε_{zz}, where z is the growth direction, and x and y are in-plane directions. Microscopically, the difference in the permittivity components is related to the contribution of the electron intersubband optical transitions, which are allowed only for the light with non-zero z-component of polarization vector. In general, application of the lateral electric field (for example, along the x axis) to the quantum well structure reduces the symmetry, so all three components of the permittivity tensor become unequal and the structure becomes optically biaxial. However, under practical conditions, not so strong electric field affects only the values of the permittivity tensor components due to electron heating, and its symmetry is conserved. The variation of the $\hat{\varepsilon}$ under the electric field opens the way for both amplitude and phase modulation of the infrared radiation with photon energies close to intersubband resonances, since both the absorption coefficient and refraction index are influenced.

The modulation of the mid-infrared absorption under the strong lateral electric field is well studied in the quantum well structures. For example, in the recent works [1, 2] the intersubband absorption modulation was measured by means of the Fourier transform transmission spectroscopy in the tunnel-coupled quantum well structures. The modification of the absorption coefficient was associated firstly, with the redistribution of the charge carriers between the QW states, and, secondly, with the modification of distribution function of the hot electrons, which becomes slightly anisotropic. In the present work, we focus on the phase-sensitive optical studies of the tunnel-coupled quantum well structures.

2. Experimental details

Research was carried out on the MBE-grown multiple GaAs/AlGaAs QW structure. One period of the structure contains a pair of narrow and wide GaAs layers separated with the tunnel-transparent AlGaAs barrier. The electronic structure of the one pair of QWs contains three electronic subbands *e*1, *e*2, and *e*3. The first electronic state is localized mostly in the wide well, while the second one is localized in the narrow well.

The sample was prepared in multi-pass geometry. The sidewall of the sample was illuminated with the linearly polarized mid-infrared light. The angle between the polarization vector and the *x*-*y* plane was 45°, so the incident beam created both *s*- and *p*- normal waves in the sample. The polarization of the output beam was analyzed with the metal grid polarizer. Electric field was applied through indium contacts made on the surface of the sample.

In this work, the Fourier transform spectroscopy was used to measure the spectral dependence of the both equilibrium and electric field induced birefringence. The globar was used as a source of mid-IR radiation. We recorded spectra of the light intensity passed through the system "polarizer – sample – analyzer" at different analyzer angles. Spectra were measured by photovoltaic liquid-nitrogen-cooled HgCdTe photodetector in a time resolved step-scan mode in order to detect the response on the electric field pulse applied to the sample.

3. Theory

In contrast to the birefringence in transparent media, in our case the resulting elliptical polarization of output beam is caused by the difference of both refraction index and absorption coefficient for the normal waves. Moreover, the absorption contribution should be rather significant as the photon energy is close to the intersubband resonance, which takes place only for *p*-polarized light.

In order to resolve carefully the phase-related effect on the emergent polarization ellipse against the strong intersubband absorption we obtain analytical expressions for the dependence of the light intensity I after the analyzer upon its azimuth angle α in the framework of the Jones matrix formalism [3]. In order to take into account difference in the both phase and amplitude for the *s*- and *p*-light polarizations we used a complex refraction index $n^* = n + ik$, where n and k are real refraction and extinction coefficients, respectively. The resulting expression for $I(\alpha)$ is

$$I(\alpha) \sim \cos^2(\alpha)(1 + 2\beta\gamma tg(\alpha) + \beta^2 tg^2(\alpha)), \tag{1}$$

where

$$\beta = \exp\left(-\Delta k L \frac{2\pi}{\lambda}\right), \quad \gamma = \cos\left(\Delta n L \frac{2\pi}{\lambda}\right), \tag{2}$$

and $\Delta n = n_p - n_s$, $\Delta k = k_p - k_s$, L – optical path length, λ – radiation wavelength.

To obtain a spectral dependence of β and γ (and, respectively, Δn and Δk) one can measure a spectra with three analyzer angles $\alpha = \pm 45°, 0°$. In this case

$$\beta^2 = \frac{I(45^\circ) + I(-45^\circ)}{I(0^\circ)} - 1, \quad \gamma = \frac{I(45^\circ) - I(-45^\circ)}{\beta I(0^\circ)}. \tag{3}$$

4. Results and discussion

Figure 1 shows the sample response on the electric filed pulse as an integral photodetector AC-coupled signal change with the analyzer angle $\alpha = +45°$. The optical and electric filed pulse shapes are very close so we can expect the effect to be related to the redistribution of charge carriers but not to the sample heating.

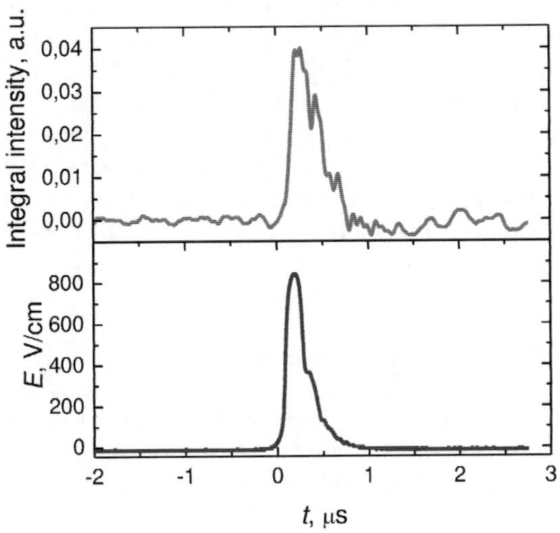

Figure 1. Optical response (red line) of the sample with double GaAs/AlGaAs QWs on the electric filed pulse (blue line) as an integral photodecetor AC-coupled signal change with the analyser azimuth angle $\alpha = +45°$.

We measured spectra of the DC-coupled photodetector output at the different time moments with respect to the electric field pulse with different analyzer angles (α = ±45° and 0°) under different lateral electric field at the temperature of the liquid nitrogen. Thus, we studied both equilibrium and field-induced birefringence in one experiment series. Figure 2 shows the equilibrium spectra ΔnL and ΔkL at $T = 80$ K calculated using the equations 3 and 2 from the measured data before the field pulse. Absorption peak corresponds to the $e1$-$e3$ transitions measured elsewhere with traditional optical transmission spectroscopy [2].

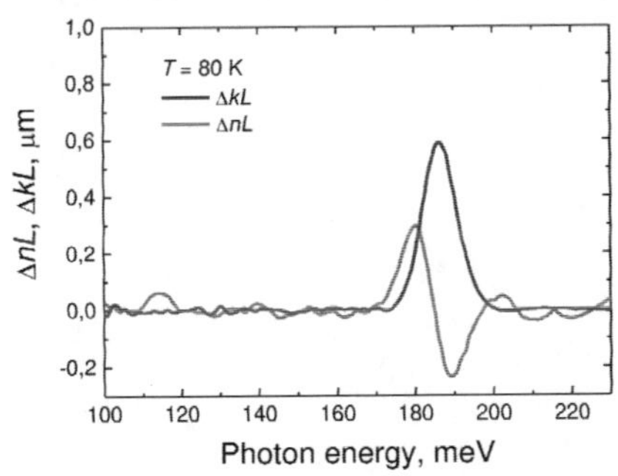

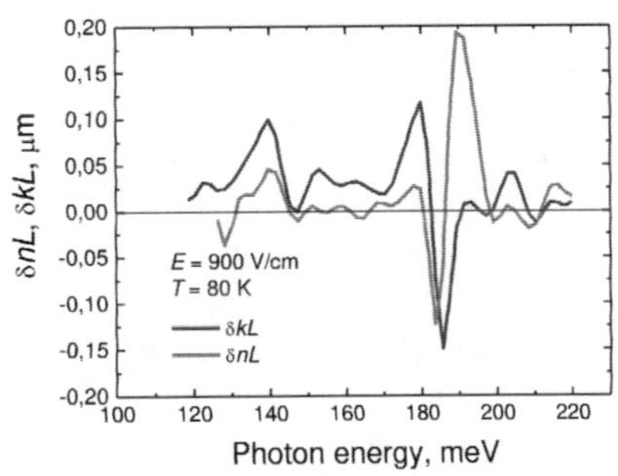

Figure 2. Equilibrium spectra of the differences in the real refraction Δn and extinction Δk coefficients for p- and s- normal waves. L is the optical path length.

Figure 3. Modification spectra of the refraction δn and extinction δk coefficients under lateral electric field. L is the optical path length.

Taking the measured data at the time moment when the applied electric filed is maximal, we can obtain the modifications δn, δk of the "natural" refraction and extinction differences in the electric filed: $\delta n(E) = \Delta n(E) - \Delta n(E=0)$, $\delta k(E) = \Delta k(E) - \Delta k(E=0)$. The corresponding spectra are shown in figure 3. One can see the decrease and slight red shift of absorption peak $e1$-$e3$.

Comparison of the modification spectra of real refractive index δn (figure 3) under lateral electric field and the spectrum of refractive index variation using Kramers–Kronig relations showed the similar result. The variation in refractive index was determined from the spectra of intersubband absorption change under lateral electric field [1] using the Kramers–Kronig relation and method described in [2]. Observed modification spectra of real refractive index δn (figure 3) under lateral electric field could be explained by redistribution of hot charge carriers between the two lowest electronic states belonging to different quantum wells of the pair and by subsequent variation of space charge in the structure. Further microscopic description of the electric field induced birefringence, including the contribution of the anisotropy of the hot electron distribution function, is under consideration.

5. Conclusion

In summary, the analysis of the state of the elliptical polarization of the output beam allowed us to determine both the refraction index and the absorption coefficient differences for the *s*- and *p*- waves in the tunnel-coupled QW structure, including the electric field induced differences. The absorption data are consistent with the direct optical transmission measurements [1]. The observed electric field induced birefringence can be used for phase modulation of the mid-infrared radiation.

Acknowledgments

The reported study was funded by RFBR project No. 14-02-00336, Russian Federation President Grant for young scientists No MK-4616.2016.2 and the Ministry of Education and Science of the Russian Federation (state assignment).

References

[1] Shumilov A A, Vinnichenko M Ya, Balagula R M, Makhov I S, Firsov D A and Vorobjev L E 2016 *Journal of Physics: Conference Series* **690** 012017

[2] Firsov D A, Vorobjev L E, Vinnichenko M Ya, Balagula R M, Kulagina M M, Vasil'iev A P. 2015 *Semiconductors* **49(11)** 1425

[3] Jones R C 1941 *Journal of the Optical Society of America (JOSA)* **31(7)** 488

X-ray-induced fluorescent centers formation in zinc-phosphate glasses doped with Ag and Cu ions

D A Klyukin, A S Pshenova, A I Sidorov, M V Stolyarchuk
Department of Optical Informational Technologies and Materials, ITMO University, Saint-Petersburg 199034, Russia

Abstract. Fluorescent properties of silver and copper doped zinc-phosphate glasses were studied. By X-ray irradiation of silver and copper co-doped glasses we could create and identify new emission centers which do not exist in single-doped samples. Doping of the glass with both silver and copper ions leads to the increase of quantum yield by 2.7 times. The study was complemented by quantum chemical calculations using the time-dependent density functional theory. It was shown that fluorescence may be attributed to the formation of mixed Ag-Cu molecular clusters.

1. Introduction

Over the last decades, fluorescent molecular clusters (MCs) consisted of a few metal atoms have attracted tremendous interest since their unique electronic structure and the optical properties [1]. Doping MCs with different transition metals may strongly affect their physical and chemical properties. Inorganic oxide glasses provide a solid matrix that stabilizes clusters and makes them promising polyfunctional materials for use in various fields including photonics, solar power engineering and optical data storage [2].

Here, we report a result of combined experimental and theoretical investigation of the luminescent properties of silver- and copper-containing zinc-phosphate glasses. In particular, we have studied the effect of the formation of new emission centers by subsequent X-ray irradiation and thermal treatment.

2. Experimental section

Several types of zinc-phosphate glasses composed of Na_2O-ZnO-P_2O_5 (ZPG) with additives of $AgNO_3$ and CuO_2 as activators were synthesized in the ITMO University by melt-quenching approach. The glasses are further denoted by ZPG-X, where X refers to a dopant of the following elements: Ag and/or Cu. The prepared glass samples were irradiated with the radiation of a CuKα (λ=1.54Å) sealed X-ray tube working at 50 kV and 50 mA. After the irradiation, the samples were heat-treated for 1 hour at 200 °C.

Figure 1a shows photographs of the samples and their luminescence after X-ray irradiation and thermal treatment. The excitation spectrum of ZPG-Ag-Cu differs significantly from the samples with one type of activator (Figure 1b). Its maximum is red-shifted by 15 nm and 45 nm in comparison with single doped ZPG-Ag and ZPG-Cu respectively. The silver and copper co-doped glasses (ZPG-Ag-Cu) with concentrations of $AgNO_3$ 0.1 mol.% and Cu_2O 0.1 mol.% have a very intense emission with the excitation peak at 365 nm and in the nearly whole visible spectral range (Figure 2). The difference in excitation spectra can be caused by the appearance of new emission centers in ZPG-Ag-Cu. This is

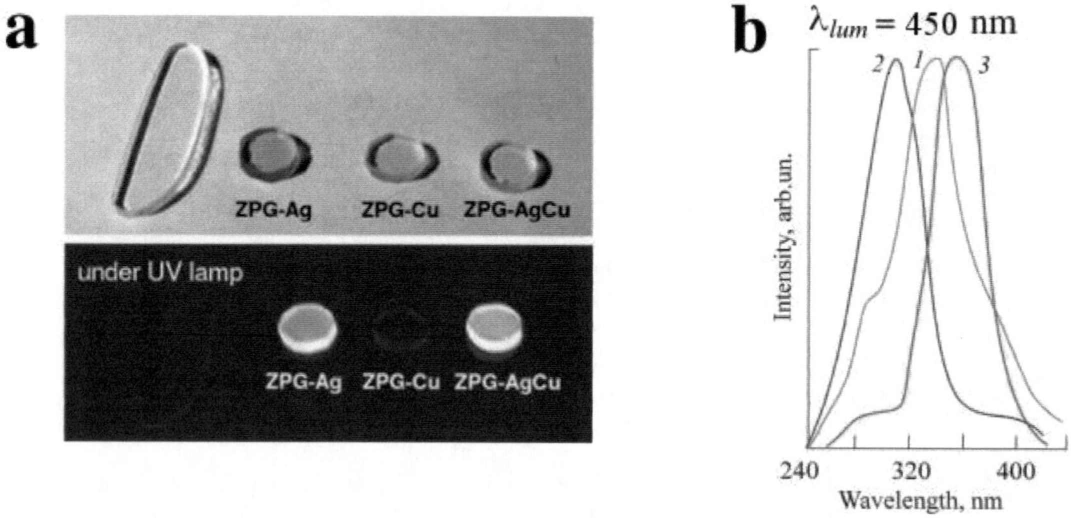

Figure 1. **(a)** Photographs (top) and luminescence (bottom) of the samples after X-ray irradiation and thermal treatment at 200 °C. Excitation wavelength – 365 nm; **(b)** Normalized excitation spectra of ZPG-Ag (1), ZPG-Cu (2), and ZPG-Ag-Cu (3)

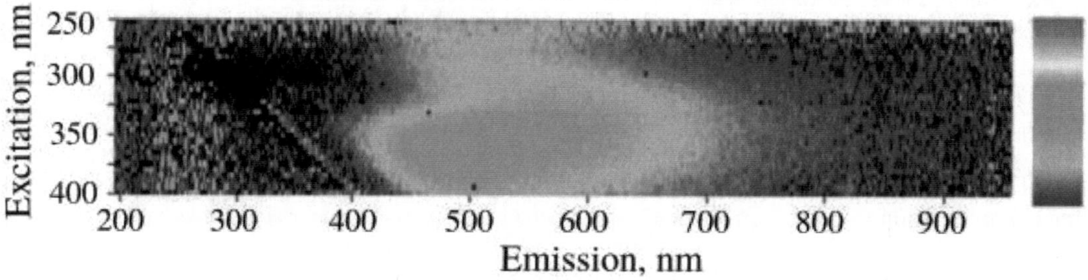

Figure 2. Fluorescence excitation-emission map of ZPG-Ag-Cu

confirmed also by the increase of luminescence intensity in this glass with the comparison with other two glass types: the measured absolute quantum yields are equal 9.6 % for ZPG-Ag, 0.5 % for ZPG-Cu, and 27 % for ZPG-Ag-Cu.

3. Computation

In order to gain insight into the structure of possible species, the effect of Cu doping of neutral homonuclear Ag_n MCs have been investigated with the linear response formalism of time-dependent density functional theory (TDDFT). The silver tetramer (Ag_4) doping was chosen as an example. The calculations were performed using the Amsterdam Density Functional program package at the CAM-B3LYP/TZ2P level. The long-range corrected exchange correlation functional CAM-B3LYP with parameters recommended by Rabilloud [4] was used. It has been shown to perform well for the calculation of ground and excited state properties of silver MCs [5].

Calculated optical absorption spectra for the lowest energy 4-atom Ag-Cu MCs in the photon energy range $\hbar\omega$ = 1.5–5.0 eV are given in Figure 3. Before reporting on the results, it should be noted that our calculations on monoatomic MCs show remarkably good agreement with previously reported theoretical data as well as with experimental results [6]. To best of our knowledge, there were no experimental studies on the optical absorption of such small mixed Ag-Cu MCs. The overall structure of the spectra is the same. All calculated spectra are characterized by one dominant band between 2.5 and 4.0 eV, except for the copper tetramer (Cu_4) which possesses three pronounced lines. It can be seen that for silver-containing MCs the main absorption peak continuously red-shifted from 2.9 eV to 2.7 eV with increasing number of copper atoms. The copper tetramer (Cu_4) exhibits a dominant peak located at 3.75 eV. This behaviour is also observed in the experiments (see Figure 1b). In addition, the oscillator strength of the main transitions of the lowest energy MCs gradually become weak with increasing Cu atoms. Weak oscillator strength of copper tetramer (Cu_4) can explain the weak luminescence of copper-containing zinc-phosphate glass.

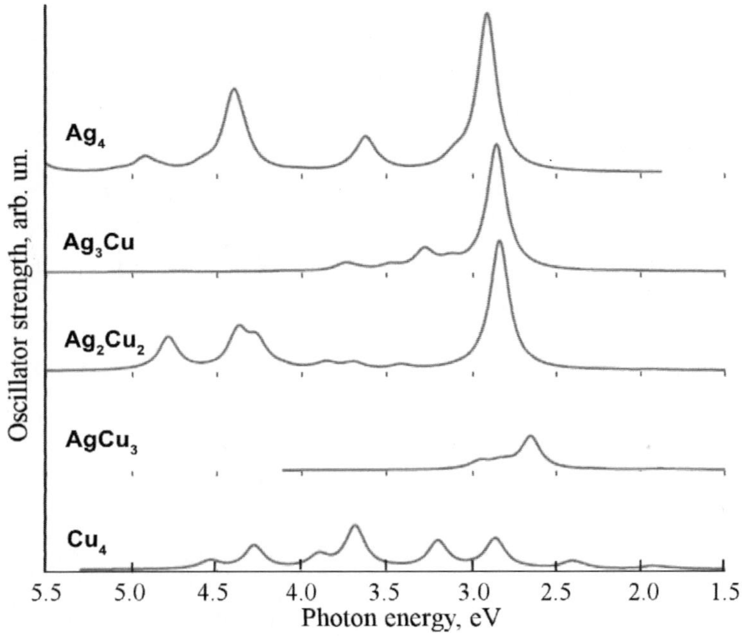

Figure 3. Calculated normalized vertical excitation spectra for Ag_nCu_{4-n} ($n < 4$) MCs

4. Conclusions

In summary, we have examined the luminescent properties of silver- and copper-containing zinc-phosphate glasses after subsequent X-ray irradiation and thermal treatment. The obtained results of a joint experimental and first-principles investigations demonstrate that the addition of Ag and Cu ions could lead to the formation of emission centers most likely attributed to bimetallic Ag-Cu MCs. The unique tunable optical properties of MCs in a broad spectral range make the studied glasses a promising candidates for photonic and sensing applications including wavelength converters and radio-photoluminescent dosimeters.

Acknowledgement

This work has been supported by the Ministry of Education and Science of Russian Federation (Project No.RFMEFI58114X0006).

References

[1] Zhang L, Wang E 2014 *Nano Today* **9** 132
[2] Dubrovin V D, Ignatiev A I, Nikonorov N V, Sidorov A I, Shakhverdov T A. Agafonova D S 2014 *Optical Materials* **4** 753
[3] Rabilloud F 2013 *J. Phys. Chem. A* **117** 4267
[4] Anak B, Bencharif M, Rabilloud F 2014 *RSC Adv.* **4** 13001

[5] Lecoultre S, Rydlo A, Félix C, Buttet J, Gilb S, Harbich W 2011 *J. Chem. Phys.* **134** 84504; Lecoultre S, Rydlo A, Félix C, Buttet J, Gilb S, Harbich W 2011 *J. Chem. Phys.* **134** 074303

Mode adjustment in hexagonal microresonators with an active region

D R Kazanov*, N Kuznetsova, D S Zolotukhin, V N Jmerik, V Yu Davydov, A N Smirnov, T V Shubina
Ioffe Physical-Technical Institute of the RAS, St.-Petersburg, 194021, Russia
* E-mail: kazanovdr@gmail.com

Abstract. We present original types of III-nitride monocrystal microresonators with an inserted active quantum-sized region. Modelling of microresonator's modes allows us to select right parameters of a polarized quantum well (width, composition) in the way that excitation of its optical transitions would be selectively amplified by interaction with the resonator modes. Adjustment of a GaN nanocolumn resonator to an InGaN quantum well is performed.

1. Introduction
Tunable microresonators, which can be fabricated by molecular-beam epitaxy (MBE) in a single-growth-run process, attract attention as promising elements for nanophotonics [1]. Especially an active region can be inserted inside of them immediately during growth without any additional cleanroom processing. Such active structures allow to amplify radiation of active layers selectively using the Purcell effect [2]. These structures can be used as a source of quantum or coherent light [3]. In this paper, we demonstrate III-nitride microresonators with inserted InGaN quantum wells (QWs), which can exhibit bright photoluminescence [4], and show how to adjust the resonator modes energy to the transition energy of such active region.

2. Samples and characterization
In our experiments GaN nanocolumns were grown on cone-shaped patterned c-sapphire substrates by plasma-assisted (PA) MBE. The PAMBE process has been developed to control the monocrystals shape [5]. The variation of substrate temperature, flux ratios $F_{Ga,In}/F_N$, and polarity, N or Ga(In), can change its shape from nanocolumn with cups on the top to tapered nanocolumns (Figure.1). The fabrication process includes the consecutive growth of GaN nucleation layer (Ga flux 5.6 nm/min) with flux ratio F_{III}/F_N=0.86 at T_s=760°C. Then the temperature of Ga source was decreased to maintain flux ratio F_{III}/F_N=0.2 (Ga flux 1.32 nm/min) to grow GaN nanocolumns while keeping always constant the active N flux and the growth temperature at 760°C. InGaN QWs of varying thicknesses of 3-10 nm were grown at the substrate temperature T_s=600°C and flux ratio F_{III}/F_N=0.22. High-spatial-resolution technique such as micro-photoluminescence (μPL) is widely used to determine experimentally the spectral characteristics of GaN nanocolumn and InGaN active layer inside of it. The μPL measurements were done with cw excitation by a 325nm laser line at room temperature. We exploited the objective Mitutoyo 40x NUV and liquid nitrogen-cooled charge-coupled detector (CCD). The beam impinging normally to the surface was focused by the objective into a spot with FWHM of 1 μm that is enough to measure separately the monocrystals. The same objective collected the PL signal from sample surface.

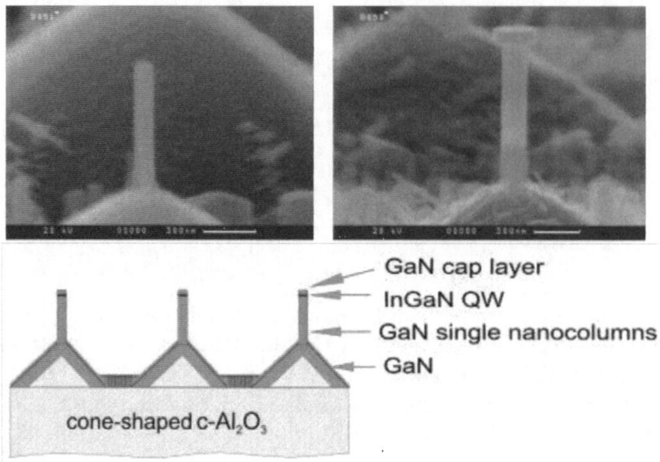

Figure 1. Different shapes of GaN nanocolumns grown by PA MBE (top) and design of GaN nanocolumn (bottom).

3. Modelling of InGaN quantum well in GaN barriers

Optical transitions of InGaN QWs depend not only on the band gap values, but also on the large internal electric fields arising due to piezoelectric and spontaneous polarizations. The former depends on the lattice mismatch between the InGaN well and the GaN barriers, whereas the latter one is a material constant due to dipole moment of ions of wurtzite crystal [5]. Figure 2 shows an example of the band profile and electron and hole ground state energies in $In_{0.1}Ga_{0.9}N$/GaN QW of 3 nm in width.

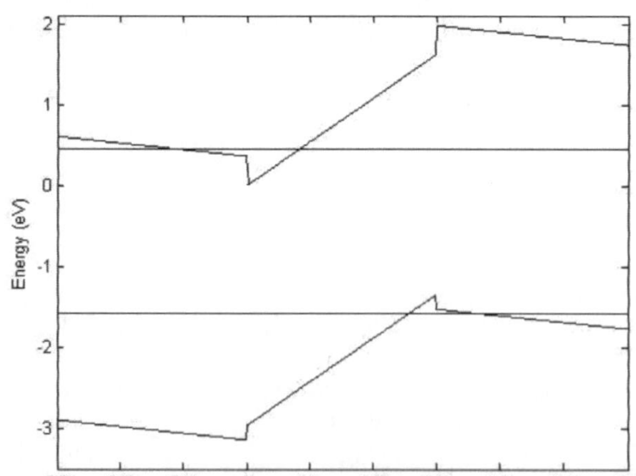

Figure 2. Example of band profile, electron and hole ground state energies in an $In_{0.1}Ga_{0.9}N$/GaN with a well width of 3 nm.

In this paper, optical transition energy is calculated using an effective-mass approximation. The single-particle ground-state electron and hole energies were found numerically by solving Harrison's finite difference version of the one-dimensional Schrödinger equation that reads as

$$\frac{\psi(z+h)}{m^*(z+{}^h/_2)} \simeq \left\{\frac{2h^2}{\hbar^2}[V(z)-E]+\frac{1}{m^*\left(z+{}^h/_2\right)}+\frac{1}{m^*\left(z-{}^h/_2\right)}\right\}\psi(z)-\frac{\psi(z-h)}{m^*\left(z-{}^h/_2\right)},$$

where ψ is the wave function, m^* is the effective mass, z is the distance in the growth direction, V is the band potential, and E is the energy. The results of calculations are presented in Figure. 3 for two different temperatures of 300 K and 77 K as dependence on the In content in the $In_xGa_{1-x}N$ QW.

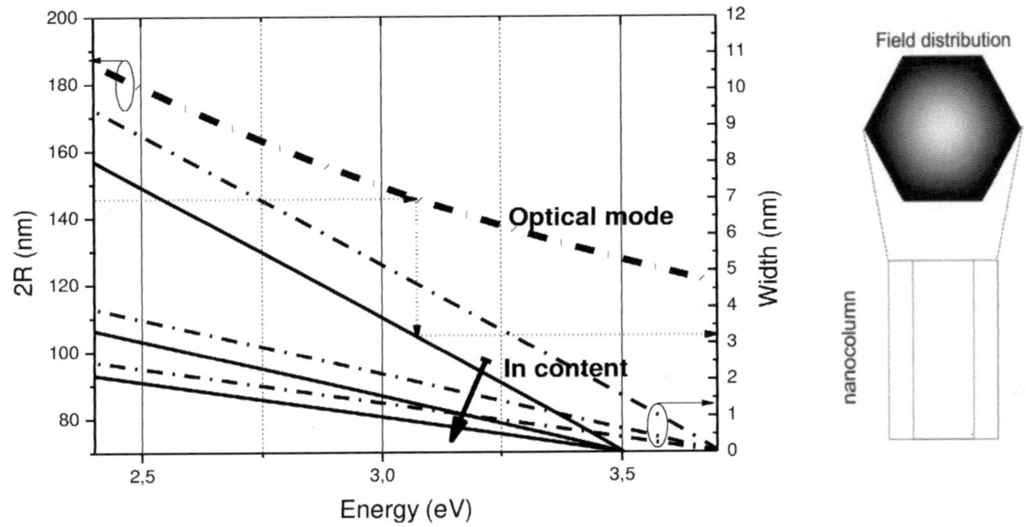

Figure 3. (left) Ground state transition energies in an $In_xGa_{1-x}N$/GaN QW as a function of a well width for various In content (x equals to 0.1, 0.2 and 0.3) at temperatures of 300 K (solid lines) and 77 K (dot-dashed lines). The mode energy of the first order is shown as a function of nanocolumn diameter 2R (bold dot-dashed line). Fine dashed lines with arrows show how to choose the diameter of nanocolumn and the width of QW to fit the energy of optical mode to the energy of optical transition. (right) Typical field distribution of the low order optical mode in a nanocolumn (shown from its top).

In Figure. 3, there is an example of how we can adjust the InGaN QW to GaN nanocolumn mode of the first order, while keeping the diameter of nanocolumn. We should consider its mode energy and then adjust QW emission to be in resonance with a microresonator mode by varying its width and In content. The energies of optical transitions in the InGaN/GaN QWs are calculated taking into account internal electric fields. The nanocolumn optical modes energy depends only on the geometry of resonator and its refractive index. Regarding to all the aforesaid we are able to resonantly adjust the transitions in the active layer with the microresonator mode. This approach can be used to analyse emission spectra.

4. Experimental results

The InGaN QW layer is not homogeneous because it is grown in three possible places: inside GaN nanocolumn, on an unperturbed layer and on cone-shaped patterned substrate. Therefore, to prove that the Purcell effect in resonator and respective amplification exist, we perform μPL measurements. In Figure. 4, we present μPL spectra of 150-nmwide GaN nanocolumns with the 8-nm InGaN QW inside. Measurements were done in two different points between nanocolumns and on a nanocolumn. The spectra measured at different excitation power densities are normalized to the intensity of GaN emission to control the relative enhancement of InGaN QW emission from its different parts. The analysis shows that the enhancement really exists with increasing laser power density from 1.1 kW/cm^2 to 11 kW/cm^2 on the nanocolumns (3.1 eV energy), while between them the increase of signal intensity has not been detected (Figure. 4).

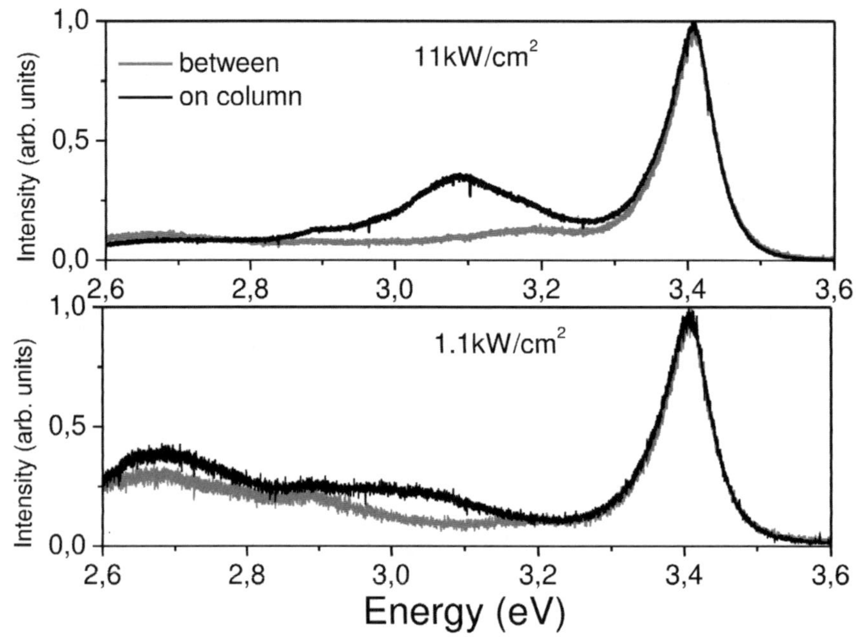

Figure 4. Room temperature μPL spectra of a measured nanocolumn (black solid line) and signal between nanocolumns (red solid line) with varying laser power.

5. Summary

We have presented monocrystal GaN nanocolumns on cone-shaped patterned c-sapphire substrates grown by PA MBE. These nanocolumn cavities contain InGaN QW inside. We have shown that it is possible to adjust the quantum level energies in the QW to optical mode in nanocolumn by solving one-dimensional Schrödinger equation and calculating numerically optical mode energy of nanocolumn. Analyzing the μPL spectra in different points of our structures, we have proved approaching the enhancement of QW emission. Therefore, this method could be used to make precise adjustment of the resonances of different types in microcavities with QW active regions.

Acknowledgments

These studies are supported by Russian Science Foundation (Project #14-22-00107).

References

[1] Shubina T V et al. *Sci. Rep.* **5**, 17970 (2016).
[2] Purcell E M *Phys. Rev.* **69**, 681 (1946).
[3] Holmes M *Nano Lett.* **14,** 982 (2014).
[4] Ivanov S V et al. *J Crystal Growth*, Vol. 301-302, 465-468 (2007).
[5] Shubina T V et al. *Phys. Stat. Sol. B* (2016).
[6] Christmas Ursula M. E. et al *J. Appl. Phys.* 98, 073522 (2005).

Photonics and application of dipyrrinates in the optical devices

Iu Aksenova[1], D Bashkirtsev[1], A Prokopenko[1], R Kuznetsova[1], N Dudina[2], M Berezin[2]
[1] Laboratory of Photophysics and Photochemistry of Molecules, Tomsk State University, Tomsk 634050, Russia
[2] Laboratory of Physical Chemistry of Solutions of Macrocyclic Compound, G.A. Krestov Institute of Solution Chemistry RAS, Ivanovo 153045, Russia

Abstract. In this paper spectral–luminescent, lasing, photochemical, and sensory characteristics of a number of Zn(II) and B(III) coordination complexes with dipyrrinates with different structures are presented. We have discussed relations of the structure of investigated compounds and formed solvates with their optical characteristics. The results showed that alkyl substituted dipyrrinates derivatives have excellent luminescent characteristics and demonstrated effective lasing upon excitation of Nd:YAG-laser. They can be used as active media for liquid tunable lasers. Zinc and boron fluoride complexes of dipyrrinates with heavy atoms in structure don't have fluorescence but have long-lived emission due to increased nonradiative intersystem processes in the excited state by the mechanism of a heavy atom. For solid samples based on halogenated complexes was found dependency of the long-lived emission intensity of the oxygen concentration in gas flow. The presence of line segment indicates the possibility of the use of these complexes as a basis for creation of optical sensors for oxygen. Moreover, results of a study of halogen-substituted aza-complexes under irradiation are presented. Such complexes are promising for the creating media for generation of singlet oxygen ($^1O_2$), which is important for photodynamic therapy in medicine and photocatalytic reactions in the industry.

1. Introduction

To date, the study of complexes of dipyrrinates with different structure is one of the most successful developing areas of modern chemistry. The increased interest to this class of compounds is associated with widespread complexes of dipyrrinates in many branches of science, technology and medicine. The most perspective members of this family are boron fluorinated complexes of dipyrrinates (BODIPY). Depending on the structure of the ligand and a solvent, they have optimal characteristics for the fluorophores (high fluorescence quantum yield, a good photostability). BODIPYs can be widely applied as active media for tunable dye lasers with high energy conversion outputs, as well as light harvesting arrays to develop antenna systems [1, 2]. Furthermore, dipyrrinates can form stable complexes with ions of d-elements, such as Zn(II). In comparison with BODIPY, advantage complexes of d-metals with dipyrrinates is easy to "self-assembly" in the "soft" conditions on complexing ions in solutions and in biological systems, as well as high sensitivity spectral-luminescence characteristics to changes in the structure of the chromophore and the nature of the solvent, which makes them very promising to fluorescent probes [3]. To successfully apply dipyrrinates complexes and creation of various optical devices based on them must be systematic study of photochemical and photophysical properties and the establishment of their relationship from the structural features of the complexes. Therefore, the purpose of the work is to study the spectral–luminescent, lasing, photochemical, and sensory properties of a number of Zn(II) and B(III) coordination complexes with dipyrrinates, the optimal combination of which will indicate the direction of the most effective use of these dyes.

2. Experimental details

In this work we studied the photophysical and photochemical characteristics of Zn(II) and B(III) complexes with dipyrrinates with different alkyl and halogen substituted which were synthesized at the Institute of Solution Chemistry RAS. The structural formulas and denotations of the investigated compounds are given in Figure 1. The individuality and structure of synthesized compounds were verified using the thin layer chromatography, NMR, and IR spectroscopy methods in accordance with [4]. As solvents, we used ethanol, ethyl acetate, and cyclohexane (reagent grade).

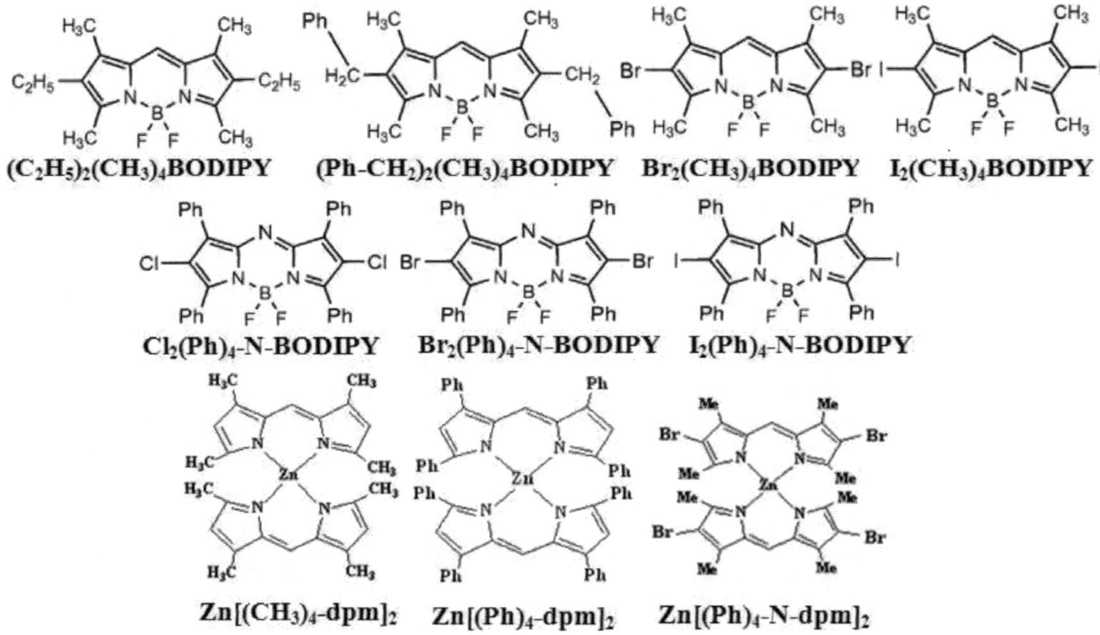

Figure 1. Structural formulas and denotation of investigated of dipyrrinates.

The spectral and fluorescence characteristics of solutions were measured at room temperature on a SM2203 spectrometer (SOLAR, Belarus). Fluorescence, phosphorescence, and luminescence excitation spectra and lifetimes of the long-lived emission in frozen (77 K) non-degassed ethanol were measured accurate to 10% on a Cary Eclipse spectrometer (Varian) with an OptistatDN-100 cryostat (Oxford Instruments). The luminescence quantum yield was determined with an error of 10% by the reference method.

Lasing and photochemical characteristics of the dipyrrinates complexes were studied with excitation by the second harmonics of Nd-YAG laser (SOLAR, Belarus). Lasing and pumping energies were measured accurate within 3% with an OPHIR-NOVA (Israel) and GentecE-100 (Canada) instruments. Generation spectra were measured by an AVANTES laser spectrometer (Netherlands) with an accuracy of 0.5 nm.

For the study the sensory properties the fibers of methylcellulose (MC-100, ReaChem) were pressed in tablets with thickness of about 1 mm and diameter of about 7 mm, which were impregnated by dyes solution in ethanol with different concentrations and dried under standard conditions. Polymer swelling in ethanol was not observed because of small quantities of solvent: volume of impregnating solution was 0.15 ml. After impregnating the volume of tablet increased by 20%. The sensor properties were investigated on the setup built around a CM2203 spectrometer consisted of a UFPGS-4 device intended for generating gas mixture flow, a gas cell with the examined sample put in the cell block of the spectrometer, and a personal computer. Pure oxygen was used as a gas-analyte, and argon was used as a gas-diluent. A more detailed description of the experimental procedure has been described in [5].

3. Results and disacussion

Results of our investigations of spectral-luminescent, lasing, and photochemical characteristics are presented in the Table 1. Alkyl substituted BODIPY show excellent luminescent characteristics. Since the fluorescence quantum yield of such complexes in solutions is close to unity, stimulated emission was generated by pumping with the second harmonics of the Nd-YAG laser (532 nm). The solutions of $(C_2H_5)_2(CH_3)_4$BODIPY and (Ph-$CH_2)_2(CH_3)_4$BODIPY generates in the range of 557-560 nm and exceeds well known commercial laser dye in efficiency.

Table 1. Spectral luminescent, lasing, and photochemical properties of alkyl dipyrrinates.

Compound, solvent	λ_{abs}^{max}, nm	λ_{fl}^{max}, nm (λ_{ex}, nm)	γ_{fl} (λ_{ex}, nm)	λ_{las}, nm (λ_{ex}, nm, W_{pump}, MW/cm^2)	Eff. $_{las}$, % (λ_{ex}, nm, W_{pump}, MW/cm^2)	$\phi_{phot}\times10^5$ (λ_{ex}, nm)
$(C_2H_5)_2(CH_3)_4$ BODIPY, ethanol	528	545 (457)	0.8 (475)	560 (532, 25)	74 (532, 25)	7 (532)
$(Ph\text{-}CH_2)_2(CH_3)_4$ BODIPY, cyclohexane	531	539 (500)	0.9 (500)	557 (532, 25)	76 (532, 25)	4 (532)

The solution of alkyl BODIPY generates with low threshold $W_{pump}<1$ MW/cm^2, and, even at a laser-pump density of 1 MW/cm^2, the efficiency achieves 43%. With increasing excitation intensity, the maximum of lasing is shifted toward the long wavelength range by 1.5 nm (Figure 2). High generation efficiency and low yield of phototransformation point to the possibility of creating active media for tunable lasers based on alkyl derivatives of boron fluorinated complexes of dipyrrinates.

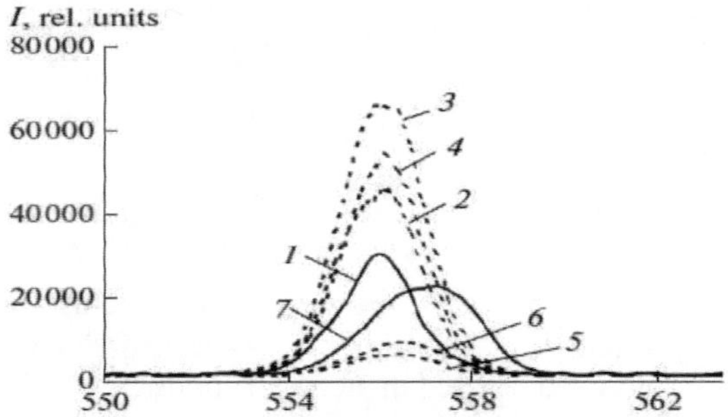

Figure 2. Lasing spectra of $(Ph\text{-}CH_2)_2(CH_3)_4$BODIPY upon excitation by second harmonic of a Nd:YAG laser in cyclohexane (5×10^{-4} M, $\Delta\lambda_{1/2}$=2 nm). Curves in: 1 – λ_{max}=556 nm, W=0.5 MW/cm^2; 2 – λ_{max}=556 nm, W=1.1 MW/cm^2; 3 – λ_{max}=556 nm, W=1.3 MW/cm^2; 4 – λ_{max}=556 nm, W=3.4 MW/cm^2; 5 – λ_{max}=556.3 nm, W=6 MW/cm^2; 6 – λ_{max}=556.5 nm, W=11 MW/cm^2; 7 – λ_{max}=557.5 nm, W=46 MW/cm^2.

Introduction bromine or iodine into the dipyrrinates core enhances intersystem crossing, which leads to decrease of the fluorescence yield and to emergence a long-lived emission (Table 2). The replacement of the central atom boron(III) by zinc(II) adds to non-planarity of the chelates and contribution of non radiative processes in excitation energy deactivation, resulting in a decrease in the fluorescence yield. The higher efficiency of intersystem crossing and the higher triplet population in the zinc(II) dipyrrinates complexes in comparison with BODIPY can be due to both the heavy atom effect and, possibly, an increased fraction of the unpaired "promoted" electrons in the non-planar transition metal chelates. This increase results in an appearance of the long lived emission of the zinc chelates in the form of phosphorescence and delayed fluorescence generated via the T-T annihilation mechanism.

Table 2. Spectral luminescent properties of halogenated dipyrrinates of B(III) and Zn(II).

Compound in ethanol	λ_{abs}^{max}, nm	λ_{fl}^{max}, nm (λ_{ex}, nm)	γ_{fl} (λ_{ex}, nm)	λ_{ph}, nm *77 K	τ_{ph}, ms *77 K	$\phi_{phot}\times10^5$ (λ_{ex}, nm)
$Br_2(CH_3)_4$BODIPY	528	545 (470)	0.4 (480)	795	3.7	12 (532)
$I_2(CH_3)_4$BODIPY	535	550 (500)	0.04 (500)	792	0.6	35 (532)
$Zn[(CH_3)_4\text{-dpm}]_2$	485	492 (460)	0.008 (460)	740+820	11+4.2	7.5 (355)
$Zn[(Ph)_4\text{-dpm}]_2$	525	602 (510)	0.01 (510)	600+830	0.7+17.7	1.2 (532)
$Zn[(Ph)_4\text{-N-dpm}]_2$	588	350 (209)		760+810	0.2+2.2	5 (532)

The obtained characteristics of long-lived emission from complexes allow us to recommend this compound for study of his interaction with oxygen with aim of creation oxygen sensor. Because sensory properties required evaluation at room temperature were prepared solid samples (matrices based on methyl cellulose, solid-state films based on polyvinyl butyral) to minimize nonradiative processes of the T-state characteristic of liquid solutions. For studied solid samples (matrices and films) noted the change in intensity of luminescence dipyrrinates complexes under various conditions, i.e., quenching of phosphorescence in an oxygen atmosphere and its increase in argon (Figure 3). Measuring dependencies of the intensity of the phosphorescence of the oxygen concentration and plotting the calibration curve (Stern-Volmer dependences have linear portion (Figure 4)) can be used to determine the unknown concentration of oxygen. Thus, the obtained results point to the possibility of the use of dipyrrinates complexes as a basis for creation of optical sensors for oxygen.

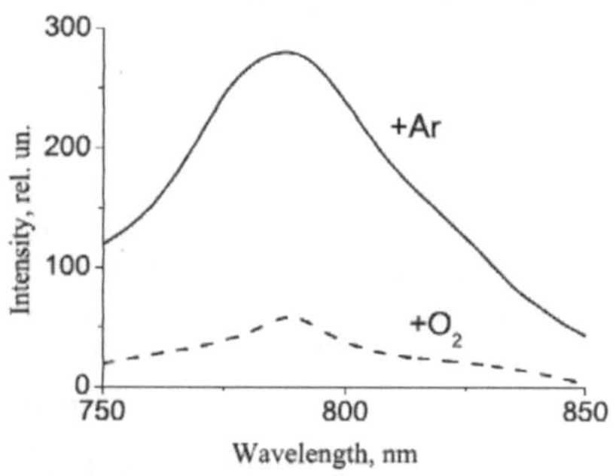

Figure 3. Luminescence spectra of $I_2(CH_3)_4$BODIPY in methyl cellulose matrix in the argon and oxygen atmospheres at λ_{ex}=520 nm

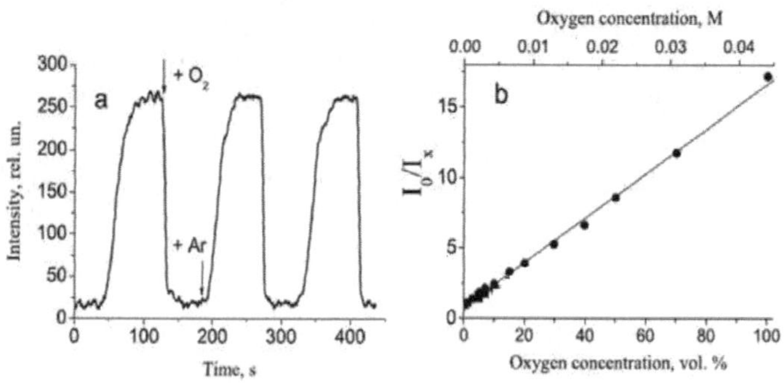

Figure 4. (a) Typical change in phosphorescence intensity (λ_{ex}=520 nm, λ_{em}=790 nm) of $I_2(CH_3)_4$BODIPY; (b) Stern–Volmer plots sequestered in methyl cellulose.

We also studied the spectral properties of halogenated tetraphenyl-aza-dipyrrinates. They have a low fluorescence quantum yields due to increased intersystem processes by the mechanism of "heavy" atoms (Table 3). The absence of phosphorescence may be due to high efficient T-S_0 intersystem crossing and quenching triplet oxygen. According to the literature, there is evidence of generation singlet oxygen by similar structures [6]. Therefore, the efficiency of the formation of singlet oxygen when transferring the excitation energy from the BODIPY complexes to 1,3-diphenylisobenzofuran (DPBF well known as a "trap" of singlet oxygen) was measured. Under irradiation mixture a dye with DBPF was found decrease concentration of DPBF (Figure 5). This indicates the formation of singlet oxygen in solution by reacting triplet excited BODIPY (λ_{ex}=640 nm) with molecular O_2. So, halogenated aza-dipyrrinates is promising for the creating media for generation of singlet oxygen ($^1O_2$), which is important for photodynamic therapy in medicine and photocatalytic reactions in the industry.

Table 3. Spectral luminescent properties of halogenated aza-dipyrrinates.

Compound, solvent	λ_{abs}^{max}, nm	λ_{fl}^{max}, nm (λ_{ex}, nm)	γ_{fl} (λ_{ex}, nm)	$\varphi_{phot}\times10^5$ (λ_{ex}, nm)
$Cl_2(Ph)_4$-N-BODIPY, ethyl acetate	644	674 (590)	0.26 (590)	1.3 (532)
$Br_2(Ph)_4$-N-BODIPY, ethyl acetate	643	674 (560)	0.02 (560)	0.13 (532)
$I_2(Ph)_4$-N-BODIPY, ethanol	647	674 (600)	0.32 (600)	0.18 (532)

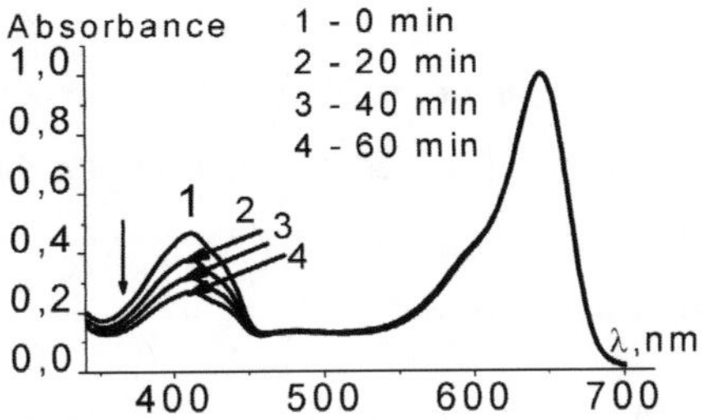

Figure 5. The change in the absorption spectra of the mixture $I_2(Ph)_4$-N-BODIPY (3×10^{-5} M) and 1,3-DPBF (3×10^{-5} M) upon irradiation (λ_{ex}=640 nm) in ethanol in oxygen atmosphere

Substance used in optical devices must have a high photostability under regular excitation. And so the stability of the complexes in the ground and excited states isthe most important characteristic for practical applications. To study this, photostability of dipyrrinates in ethanol solutions was investigated under 2nd and 3rd harmonic of Nd:YAG-laser irradiation. Alkyl and halogen dipyrrinates complexes exhibit high photostability (Table 1, 2, 3). Thereby this value of phototransformations quantum yield is quite enough for successful use in optoelectronic devices.

4. Conclusion

We have investigated the spectral–luminescent properties and lasing characteristics in relation to the excitation parameters (intensity and excitation wavelength) for new dipyrrinates complexes. We have ascertained and analyzed relations of obtained characteristics with the molecular structure and properties of the solvent and its phase state (liquid, solid). Our results showed that dipyrrinates derivatives can be used as active media for different optical devices: tunable lasers, sensors for determining the oxygen in the gas phase and so on.

References

[1] Loudet A, Burgess K 2007 *Chem. Rev.* **107** 4891
[2] Harriman A 2015 *Chem. Commun.* **51** 11745
[3] Kuznetsova R T, Aksenova Yu V, Bashkirtsev D E, Prokopenko A A, Tel'minov E N, Mayer G V, Dudina N A, Antina E V, Nikonova A Yu, Berezin M B, Semeikin A S 2015 *High Energy Chem.* **49** 16
[4] Dudina N A, Nikonova A Yu, Antina E V, Berezin M B, V'yugin A I 2013 *Khim. Geterotsikl. Soedin.* **12** 1878
[5] Ermolina E G, Kuznetsova R T, Aksenova Yu V, Gadirov R M, Kopylova T N, Antina E V, Berezin M B, Semeikin A S 2014 *Sensors and Actuat. B* **197** 206
[6] Zhang X, Yang X 2013 *J. Phys. Chem. B* **117** 5533

Local field effects in periodic metamaterials

O V Porvatkina, A A Tishchenko and M N Strikhanov

National Research Nuclear University MEPhI (Moscow Engineering Physics Institute), Moscow, Russia

E-mail: OVPorvatkina@mephi.ru, Tishchenko@mephi.ru

Abstract. In this article we investigate dielectric and magnetic properties of periodic metamaterials taking into account the so-called local field effect, caused by interaction between single particles the material consists of. We also consider the spatial dispersion effects. As a result, generalized Clausius-Mossotti techniques have been extended to the case of periodic metamaterials; permittivity tensor and permeability tensor were obtained.

1. Introduction

Metamaterials are artificial materials, which attract much attention because of their exotic properties [1]-[3] and promising applications ranging from biosensors to photonic devices [4],[5]. These materials may consist of periodic or random collections of, for example, metallic nanoparticles, metallic split-ring resonators, I-shaped particles and other magnetodielectric objects. The unit elements of metamaterials are usually designed to provide a resonant response to the electric and magnetic fields [6].

In this work we will be focusing on periodic metamaterials taking into consideration the local-field effects.

2. Local field effects in periodic metamaterials

We investigate periodic metamaterial consisting of N anisotropic particles with dielectric polarizability tensor $\alpha_{ij}^{e}(\omega)$ and magnetic polarizability tensor $\alpha_{ij}^{m}(\omega)$:

$$\alpha_{ij}^{e}(\omega)=\alpha_{\perp}^{e}(\omega)\left(\delta_{ij}-e_ie_j\right)+\alpha_{l}^{e}(\omega)e_ie_j, \tag{1}$$

$$\alpha_{ij}^{m}(\omega)=\alpha_{\perp}^{m}(\omega)\left(\delta_{ij}-e_ie_j\right)+\alpha_{l}^{m}(\omega)e_ie_j. \tag{2}$$

The microscopic field in medium is the sum of the primary field $(\mathbf{E}_0,\mathbf{B}_0)$ generated by external charges and secondary fields produced by all particles of the medium. Maxwell's equations for the Fourier transform of the microscopic fields acting at a point $\mathbf{r}$ give [7]:

$$\mathbf{E}^{mic}(\mathbf{r},\omega)=\mathbf{E}^{0}(\mathbf{r},\omega)+\frac{4\pi i}{\omega}\int d^3q e^{i\mathbf{qr}}\frac{k^2\mathbf{j}^{mic}(\mathbf{q},\omega)-\mathbf{q}\left(\mathbf{q},\mathbf{j}^{mic}(\mathbf{q},\omega)\right)}{q^2-k^2}, \tag{3}$$

$$\mathbf{H}^{mic}(\mathbf{r},\omega)=\mathbf{H}^{0}(\mathbf{r},\omega)+\frac{4\pi i}{c}\int d^3q e^{i\mathbf{qr}}\frac{[\mathbf{q},\mathbf{j}^{mic}(\mathbf{q},\omega)]}{q^2-k^2}, \tag{4}$$

where

$$j_i^{mic}(\mathbf{r}',\omega) = -i\omega\alpha_{ij}^{e}(\omega)\sum_{b=1}^{N} E_j^{mic}(\mathbf{R}_b,\omega)\delta(\mathbf{r}'-\mathbf{R}_b) + c\alpha_{ij}^{m}(\omega)\sum_{b}\left(rot\mathbf{H}^{mic}(\mathbf{R}_b,\omega)\right)_j\delta(\mathbf{r}'-\mathbf{R}_b). \quad (5)$$

One can solve this system only approximately because $N \gg 1$. For $\mathbf{r} = \mathbf{R_a}$ one can easily obtain from Eqs.(3),(4) a system of N equations for acting on the a –th particle field with use so-called the local fields. The local field is an effective field can be obtained by averaging over the positions of all particles of the medium except the given one [8]:

$$E_i^{loc}(\mathbf{R}_a,\omega) = E_i^0(\mathbf{R}_a,\omega) + \frac{1}{2\omega\pi^2}\int d^3p \frac{k^2\delta_{ij} - p_i p_j}{p^2 - k^2}\int d^3R_{ba} e^{-i\mathbf{pR}_{ba}} N w(\mathbf{R}_{ba})$$
$$\times\left(\omega\alpha_{jk}^{e}(\omega)E_k^{loc}(\mathbf{R}_a + \mathbf{R}_{ba},\omega) - c\alpha_{jk}^{m}(\omega)\left[\mathbf{p},\mathbf{B}^{loc}(\mathbf{R}_a + \mathbf{R}_{ba},\omega)\right]_k\right), \quad (6)$$

$$B_i^{loc}(\mathbf{R}_a,\omega) = B_i^0(\mathbf{R}_a,\omega) + \frac{1}{2c\pi^2}\int d^3p \frac{1}{p^2 - k^2}\int d^3R_{ba} e^{-i\mathbf{pR}_{ba}} N w(\mathbf{R}_{ba})$$
$$\times\left(\omega\alpha_{ij}^{e}(\omega)\left[\mathbf{p},\mathbf{E}^{loc}(\mathbf{R}_a + \mathbf{R}_{ba},\omega)\right]_j - c\alpha_{ij}^{m}(\omega)\left[\mathbf{p},\left[\mathbf{p},\mathbf{B}^{loc}(\mathbf{R}_a + \mathbf{R}_{ba},\omega)\right]\right]_j\right). \quad (7)$$

On the other hand, dependence of the local field on the structure of the medium is provided by the function of distribution $w(\mathbf{R}_{an})$. For periodic material one can write:

$$w(\mathbf{R}_{an}) = \frac{1}{N}\sum_{m=1}^{N}\delta(\mathbf{R}_{an} - \mathbf{R}_m). \quad (8)$$

At the same time, averaging equations Eqs. (3),(4) over the positions of all the particles one gets equations for usual average macroscopic fields. Combining these expressions, we obtain equations connecting the Fourier transforms of the local and macroscopic fields:

$$G_{ij}(\mathbf{q},\omega)E_j^{loc}(\mathbf{q},\omega) = E_i(\mathbf{q},\omega), \quad (9)$$

$$F_{ij}(\mathbf{q},\omega)B_j^{loc}(\mathbf{q},\omega) = B_i(\mathbf{q},\omega), \quad (10)$$

where

$$G_{ik}(\mathbf{q},\omega) = \delta_{ik} - \frac{1}{2\pi^2}\sum_{m=1}^{N} S_{ij}(-\mathbf{R}_m,\omega)e^{i\mathbf{qR}_m}\alpha_{jk}^{e}(\omega) + 4\pi n S_{ij}(\mathbf{q},\omega)\alpha_{jk}^{e}(\omega), \quad (11)$$

$$F_{ik}(\mathbf{q},\omega) = \delta_{ik} - \frac{1}{2\pi^2}\sum_{m=1}^{N} F_{ij}(-\mathbf{R}_m,\omega)e^{i\mathbf{qR}_m}\alpha_{jk}^{m} + 4\pi n F_{ij}(\mathbf{q},\omega)\alpha_{jk}^{m}(\omega), \quad (12)$$

$$S_{ij}(-\mathbf{R}_m,\omega) = \int d^3p \frac{k^2\delta_{ij} - p_i p_j}{p^2 - k^2}e^{-i\mathbf{pR}_m}, \quad (13)$$

$$F_{ij}(-\mathbf{R}_m,\omega) = \int d^3p \frac{p^2\delta_{ij} - p_i p_j}{p^2 - k^2}e^{-i\mathbf{pR}_m}. \quad (14)$$

We do our calculations in a long-wave limit. According to this limit, if

$$d << L << c/\omega, \quad (15)$$

the effective field acting on a molecule is formed by adding the fields of many molecules lying in some volume of the crystal [9]. In (15) d is the lattice constant, L is the linear dimension of the region responsible for the formation of the local field.

Neglecting the anisotropy of individual particles

$$\alpha_l(\omega) \equiv \alpha_\perp(\omega) \equiv \alpha(\omega) \quad (16)$$

we can rewrite Eqs. (13),(14) in the following form:

$$\int d^3p \frac{k^2\delta_{ij} - p_i p_j}{p^2 - k^2}e^{-i\mathbf{pR}_m} = a(R_m)\delta_{ij} + b(R_m)\frac{R_i^m R_j^m}{R^2}, \quad (17)$$

$$\int d^3 p \frac{p^2\delta_{ij} - p_i p_j}{p^2 - k^2} e^{-i\mathbf{pR}_m} = c(R_m)\delta_{ij} + d(R_m)\frac{R_i^m R_j^m}{R^2}. \quad (18)$$

It is well known that the probability of unlimited approaching of molecules is negligibly small [10]. For this reason, convolutions of Eqs. (17),(18) with δ_{kj} and $\frac{R_j^m R_i^m}{R^2}$ allow us to obtain coefficients $a(R_m), b(R_m), c(R_m), d(R_m)$:

$$a(R_m) = -2\frac{\pi^2 k}{R_m^2}\sin(kR_m) - 2\frac{\pi^2}{R_m^3}\cos(kR_m), \quad (19)$$

$$b(R_m) = 6\frac{\pi^2 k}{R_m^2}\sin(kR_m) + 6\frac{\pi^2}{R_m^3}\cos(kR_m) + 4\frac{\pi^2 k^2}{R_m}\cos(kR_m), \quad (20)$$

$$c(R_m) = -2\frac{\pi^2 k}{\mathrm{R}_m^2}\sin(k\mathrm{R}^m) - 2\frac{\pi^2}{\mathrm{R}_m^3}\cos(k\mathrm{R}^m) + 2\frac{\pi^2 k^2}{\mathrm{R}_m}\cos(\mathrm{kR}^m), \quad (21)$$

$$d(R_m) = 6\frac{\pi^2}{\mathrm{R}_m^3}\cos(k\mathrm{R}^m) + 6\frac{\pi^2 k}{\mathrm{R}_m^2}\sin(k\mathrm{R}^m) - 2\frac{\pi^2 k^2}{\mathrm{R}_m}\cos(k\mathrm{R}^m). \quad (22)$$

So, we get tensors $G_{ik}(\mathbf{q},\omega), F_{ik}(\mathbf{q},\omega)$ in the form

$$G_{ik}(\mathbf{q},\omega) = \delta_{ik} - \frac{1}{2\pi^2}\sum_{m=1}^{N}\left(a(\mathbf{R}_m)\delta_{ik} + b(\mathbf{R}_m)\frac{R_i^m R_k^m}{R_m^2}\right)e^{i\mathbf{qR}_m}\alpha^e(\omega) + 4\pi n\frac{k^2\delta_{ij} - p_i p_j}{p^2 - k^2}\alpha^e(\omega), \quad (23)$$

$$F_{ik}(\mathbf{q},\omega) = \delta_{ik} - \frac{1}{2\pi^2}\sum_{m=1}^{N}\left(c(\mathbf{R}_m)\delta_{ik} + d(\mathbf{R}_m)\frac{R_i^m R_k^m}{R_m^2}\right)e^{i\mathbf{qR}_m}\alpha^m(\omega) + 4\pi n\frac{p^2\delta_{ij} - p_i p_j}{p^2 - k^2}\alpha^m(\omega). \quad (24)$$

Considering that polarization and magnetization of the medium can be expressed in terms of both the macroscopic fields and the local fields, one can obtain equations for permittivity and permeability of periodic metamaterials:

$$\varepsilon_{ij}(\mathbf{q},\omega) = \delta_{ij} + 4\pi n\alpha^e(\omega)G_{ij}^{-1}(\mathbf{q},\omega), \quad (25)$$

$$\mu_{ij}(\mathbf{q},\omega) = \delta_{ij} + 4\pi n\alpha^m(\omega)F_{ij}^{-1}(\mathbf{q},\omega). \quad (26)$$

Eqs. (25),(26) are analogues of the well-known Clausius-Mossotti relation, but for periodic metamaterial. In case of non-magnetic media our results coincide with those from work [11].

3. Discussion

In this paper dielectric and magnetic properties of periodic metamaterials were investigated taking into account the local field effects. We have obtained permittivity tensor (25) and permeability tensor (26). Our results describe relation between macroscopic properties of periodic metamaterial and microscopic properties of the particles it consists of. This fact is important, because it allows the potential features of metamaterials to be incorporated into a design and analysis of new devices, such as transmission devices, electromagnetic absorbers, *etc* [12],[13].

Acknowledgments

This work was supported by the Ministry of Education and Science of the Russian Federation, the project 3.1110.2014/K.

References

[1] Pendry J B 2000 Negative refraction makes a perfect lens *Phys. Rev. Lett.* **85** 3966-69
[2] Wells B M, Zayats A V, and Podolskiy V A 2014 Nonlocal optics of plasmonic nanowire metamaterials *Physical Review B* **89** 035111
[3] Smith D R, Padilla W J, Vier D C, Nemat-Nasser S C and Schultz S 2000 Composite Medium

with Simultaneously Negative Permeability and Permittivity *Phys. Rev. Lett.* **84** 4184-87
[4] Linden S, Enkrich C, Wegener M, Zhou J, Koschny T, and Soukoulis C 2004 Magnetic response of metamaterials at 100 Terahertz *Science* **306** 1351-53
[5] Alù A 2011 First-principles homogenization theory for periodic metamaterials *Phys. Rev. B* **84** 075153
[6] Airoldi L, Senesi M, Ruzzene M 2012 Piezoelectric Superlattices and Shunted Periodic Arrays as Tunable Periodic Structures and Metamaterials *Wave Propagation in Linear and Nonlinear Periodic Media* **540** 33-108
[7] Ryazanov M I and Tishchenko A A 2006 Clausius-Mossotti-type relation for planar monolayers *JETP* **103** 539-545 [ZhETF 130 621]
[8] Ryazanov M I 1984 Electrodynamics of Condensed Matter (Moscow: Nauka) (in Russian)
[9] Gorkunov M V and Ryazanov M I 1997 The effect of a local field on Raman scattering in a uniaxial crystal *JETP* **85** 97-103 [ZhETF 112 180]
[10] Porvatkina O V, Tishchenko A A, and M. N. Strikhanov 2015 Local Field Effects for Left-handed Planar Metamaterials *PIERS Proceeding* 1689-1692
[11] Anokhin M N, Tishchenko A A, Ryazanov M I and Strikhanov M N 2015 *J. Phys.: Conf. Ser.* **643** 012066
[12] Gardner D F, Evans J S and Smalyukh I I 2011 *Molecular Crystals and Liquid Crystals* **545** 1221-45
[13] Zhou S, Li W and Li Q 2010 Design of 3-D Periodic Metamaterials for Electromagnetic Properties *IEEE Transactions On Microwave Theory And Techniques* **58** 910-916

Luminescence and structural properties of germanium nanocrystals formed by annealing multilayer GeO_x/Al_2O_3 nanostructures

D A Grachev[1], S A Garakhin[1], A V Belolipetsky[2], A V Nezhdanov[1], A V Ershov[1]

[1] Department of Semiconductor Physics and Optoelectronics,
Lobachevsky University, Nizhny Novgorod 603950, Russia
[2] Sector of Theory of Optical and Electrical Phenomena in Semiconductors,
Ioffe Institute RAS, Saint Petersburg 194021, Russia

E-mail: grachov@phys.unn.ru

Abstract. By Raman scattering, luminescence, and IR-absorption spectroscopy multilayer nanoperiodic structures Ge/Al_2O_3 & GeO_x/Al_2O_3 have been investigated. The samples have been obtained by the physical evaporation; their properties have been varied by changing the layer thicknesses (2–20 nm) and annealing temperature (500–1000 °C). It is found that germanium nanocrystals are formed in the temperature range of 500–800 °C and exhibit intense size-depend photoluminescence at 1.2 eV and 1.8–2.0 eV.

1. Introduction

In recent years there has been a significant growth in interest in the study of germanium nanocrystals (Ge-NCs) in a dielectric, the prospective composite material for LED components and long-term charge storage cells. Compared with silicon nanocrystals (Si-NCs), which are used in the modern microelectronics, Ge-NCs have some important advantages. The process of Ge NC forming is realized at lower temperatures than for Si-NCs that extends the capability of band engineering. Moreover germanium has a low reactivity that improves the stability of the composite [1]. Differences in the band structures of silicon and germanium makes research of systems with Ge-NCs in oxide matrix fundamentally important for better understanding of the quantum size effects in indirect-band-gap materials in general.

One way to obtain Ge-NCs is to anneal nanoperiodic multilayer sandwich structure GeO_x/oxide. This approach is to make possible the control of nanocrystal sizes and their density. The purpose of the study was to form multilayer nanoperiodic samples Ge/Al_2O_3 & $GeO_{x\approx 1}/Al_2O_3$, containing germanium nanocrystals, and to investigate their structural and luminescent properties as well.

2. Preparation of samples and measurements procedure

The nanoperiodic multilayer Ge/Al_2O_3 & GeO_x/Al_2O_3 samples (Figure 1) have been formed on silicon substrates by vacuum physical vapor deposition. As to the Ge/Al_2O_3 structure design, thickness of the Al_2O_3 layers was 5 nm, and the Ge layer thickness was varied between 3 and 7 nm that depended on particular sample. For the GeO_x/Al_2O_3 structure thickness of the Al_2O_3 layers was 8 nm; size of the Ge layers was a value in the range of 2–20 nm. The number of layers was 19. After deposition the

Journal of Physics: Conference Series **741** (2016) 012129 doi:10.1088/1742-6596/741/1/012129

structures have been divided into chips, which were annealed in nitrogen atmosphere at the temperature from 400 °C to 1000 °C.

Raman spectra have been carried out in the range of 400–600 cm^{-1}. IR-transmission spectra in the range of 400–1400 cm^{-1} have been obtained. The photoluminescence (PL) spectra have been measured at room temperature in the wavelength range of 580–1100 nm (Nd:YAG laser: 5.29 mW, 532 nm).

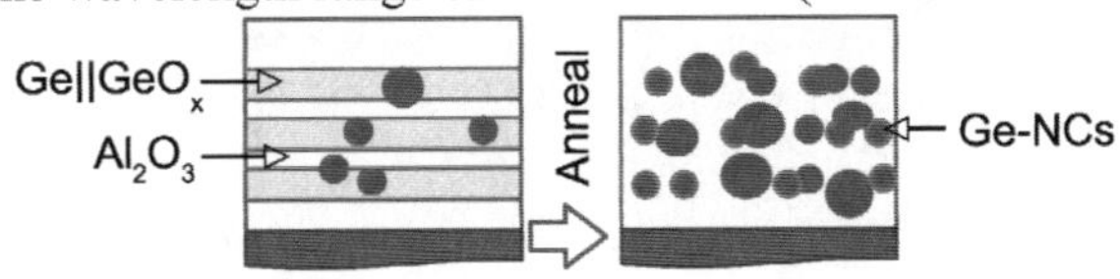

Figure 1. Schematic representation of the GeO_x/Al_2O_3 nanostructure modification.

3. Experimental results and discussion

Figure 2 shows the typical Raman spectra of the Ge/Al_2O_3 nanostructures. Prior to annealing germanium is amorphous, since there is a band of 260–280 cm^{-1} [2,3]. When the temperature is over 600 °C the sharp peak of 299 cm-1 appears for the reason that germanium crystallizes. As to the luminescence spectra (Figure 3), annealing at 600 °C gives rise to intense PL-band in the range of 1.6–2.0 eV, peak position of which is about 1.9 eV. It can be inferred from papers [4,5] that origin of the band should be associated with nothing but crystalline germanium phase.

Maximum volume fraction of a crystalline germanium phase corresponding to the higher scattering peak is observed at the temperature of 800 °C. In this case, the PL counter plot (Figure 3) shows two luminescence bands at 1.2 eV and 1.8–2.0 eV. The first may be also associated with Ge-NCs, but of smaller size [2,6,7]. That small particle formation seems to be caused by the oxidation process, which confirms the second luminescence band of the oxygen-deficient centers in GeO_2 [2]. At the temperatures above 900 °C germanium metal phase on the Raman spectra as well as Ge-NCS luminescence bands are not observed. We attribute this to total oxidation of germanium and formation of a volatile compound of germanium monoxide [1].

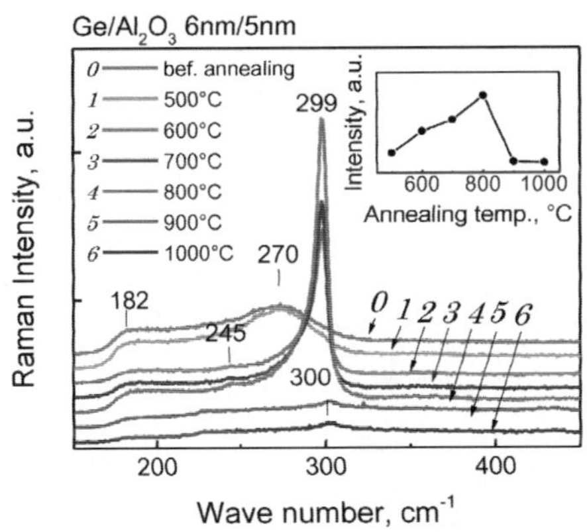

Figure 2. Raman spectra of the Ge/Al_2O_3 nanostructure (3 nm/ 5 nm bilayer) annealed in the temperature range of 500-1000 °C.

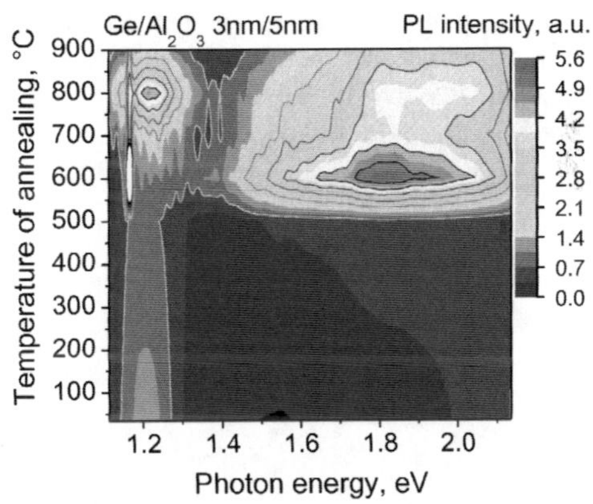

Figure 3. PL counterplot of the Ge/Al_2O_3 nanostructure (3 nm/ 5 nm bilayer) annealed in the temperature range of 500-1000 °C.

The non-annealed structures also exhibit slight luminescence at 1.2 eV. This may indicate that the deposited amorphous germanium layers are inhomogeneous. Presumably, the structures before annealing are consisted of small amount of Ge-NCs, that volume concentration is not enough for detecting by the Raman spectrometer.

As far as the thin GeO_x/Al_2O_3 system is concerned, it is observed the similar problem of the Raman spectra interpretation. The maximum intensity of the 1.2 eV PL peak corresponds to the annealing temperature of 500 °C, but in this case the sharp of the Raman curve (Figure 4) doesn't have the features peculiar to the crystalline phase, there is a domed peak instead. According to the Raman spectra crystallization of germanium starts at 600 °C. It is quite possible that effect stems from extremely small sizes of the Ge NCs, 3 nm.

Notwithstanding that further growth of the annealing temperature leads to drastic decrease in the volume fraction of germanium (Figure 4) and its PL intensity (Figure 5), the 1.2 eV band appears in the second time when the temperature reaches 800 °C, as to the Ge/Al_2O_3 system (Figure 3). The phenomenon is assumed to be caused by rising diffusivity.

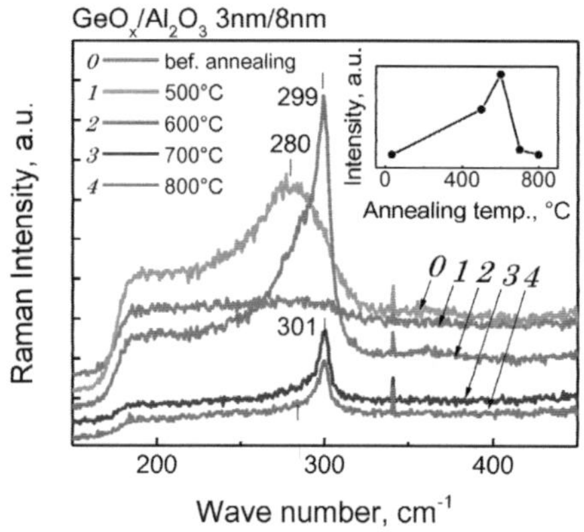

Figure 4. Raman spectra of the GeO_x/Al_2O_3 nanostructure (3 nm/ 8 nm bilayer) annealed in the temperature range of 500-1000 °C.

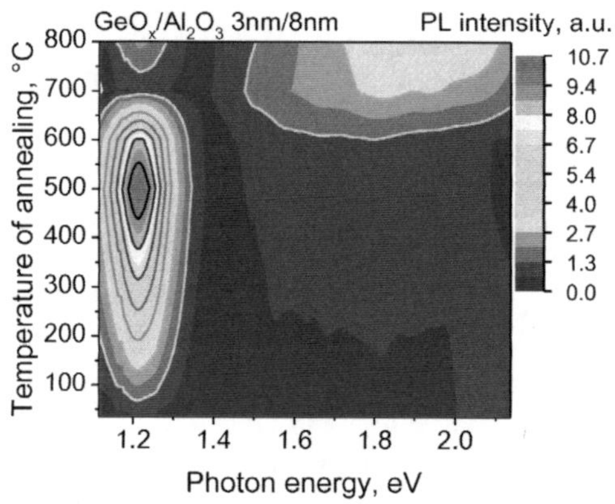

Figure 5. PL counterplot of the GeO_x/Al_2O_3 nanostructure (3 nm/ 8 nm bilayer) annealed in the temperature range of 500-1000 °C.

Figures 6 shows FTIR spectra of the GeO_x/Al_2O_3 systems where thickness of the GeO_x layers is more than 6 nm. For the non-annealed sample weak-intensity bending band of GeO_x (780 cm^{-1}) is observed [8]. After temperature treatment of 500 °C band at 800 cm^{-1}, GeO_2 characteristic [2], clearly appears. This confirms the fact [9] that the process of splitting GeO_x into the Ge and GeO_2 occurs at a temperature not more than 500 °C.

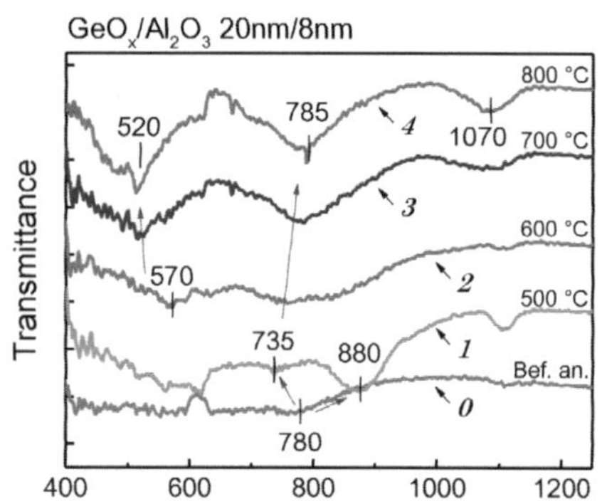

Figure 6. FTIR spectra of the GeO_x/Al_2O_3 nanostructure (20 nm/ 8 nm bilayer) annealed in the temperature range of 500-1000 °C.

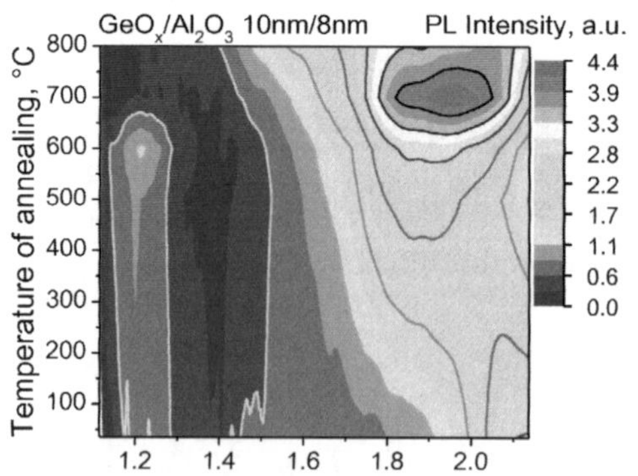

Figure 7. PL counterplot of the GeO_x/Al_2O_3 nanostructure (10 nm/ 8 nm bilayer) annealed in the temperature range of 500-1000 °C.

Besides the GeO_2 band, at 735 cm^{-1} it is observed light absorption by GeO_x molecules, $x < 1$. As the temperature increases the position of the band shifts from 735 to 785 cm^{-1} indicating the growth of *x* [2], likewise the 550 cm^{-1} stretching GeO_x band occurs [10]. That points to the strong process of germanium oxidation for temperatures of 600–800 °C. These data are in good agreement with the PL spectra (Figure 7) in particular with the 1.8–2.0 eV GeO_x band. In particular GeO emits at 1.6eV, and GeO_2 mentioned above in region of 1.8-2.0 eV.

In discussing the PL spectra of the thick-layer systems, we have to note that the temperature of the 1.2 eV band observing is 600 °C which agrees well with the Raman data (Figure 4). We can conclude that the most suitable temperature to form small Ge-NCs is between 500 °C and 600 °C.

Apart from the 1.2 eV band, in the Figure 7 there is an intensive band at 1.9 eV [5], which is also observed for the Ge/Al_2O_3 system (Figure 3). As it is detected only for the thick-layer structures, we associate it with large Ge-NCs, since GeO_x layers contain enough Ge-fraction to obtain large quantum dots. Presumably their size is about of 4–20 nm [4].

As evidence of whether the band was caused by Ge-NCs luminescence, it was shown how the peak position and its intensity depend on the thickness of the germanium-rich layers (Figure 8). The photon energy decreases with increasing layer size. In the case of the GeO_x/Al_2O_3 systems the 1.9 eV band intensity has maximum when thickness of the GeO_x layers is approximately 10 nm. It should be noted that in Figure 8, as well as in Figure 9, the abscissa shows a value of the thickness of the germanium-rich layers. This value differs from the size of the Ge-NCs. Thus, due to diffusion the average diameter of the Ge-NCs is large than the Ge layer thicknesses [3] and is smaller than the GeO_x layer thicknesses.

As to the 1.2 eV band we have not observed the quantum size effect (Figure 9). It could be explained by prolonged time of annealing (1 hour), during which the nanocrystals are taking thermodynamic equilibrium size about 3 nm [11]. We would like to mention the fact that the intensity of the 1.2 eV band increases exponentially with a decrease in the size of the Ge-NCs [12].

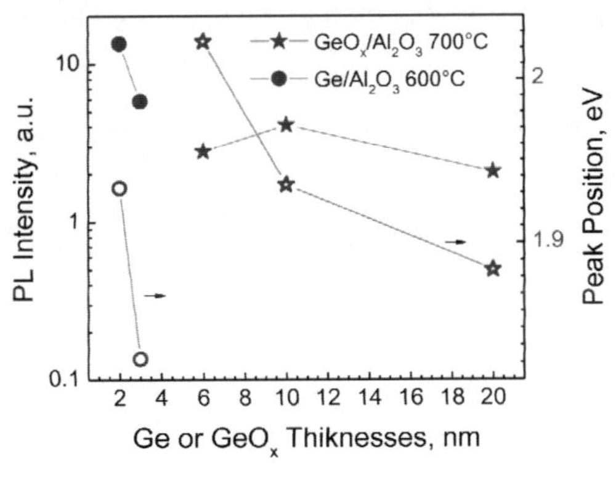

Figure 8. The 1.9–2.0eV L peak properties.

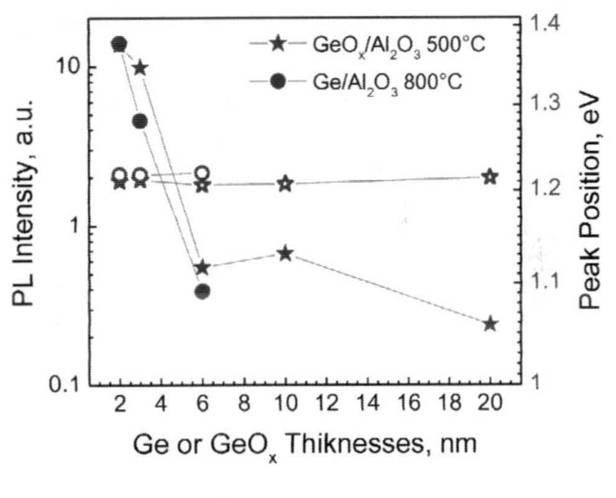

Figure 9. The 1.2eV PL peak properties.

4. Summary

Annealing the Ge/Al_2O_3 nanoperiodic multilayer structures at temperature of 800 °C as far as the GeO_x/Al_2O_3 of 500–600 °C leads to the formation of 3 nm Ge-NCs that exhibit intense 1.2 eV photoluminescence. In case of annealing at 600–700 °C it is observed size-depended 1.8-2.0 eV band for both systems. We associate it with luminescence of large germanium nanocrystals, sizes of which are about 4-20 nm.

Acknowledgments
The study is partially supported by RFBR (grant 14-02-00119-a).

References

[1] Tananaev I V and Shpirt M I 1967 *Khimii͡a germanii͡a* (Moscow: Khimia)

[2] Ardyanian M, Rinnert H and Vergnat M 2006 Structure and photoluminescence properties of evaporated GeOx/SiO2 multilayers *Journal of Applied Physics* **100** 113106

[3] Lepadatu A-M, Stoica T, Stavarache I, Teodorescu V S, Buca D and Ciurea M L 2013 Dense Ge nanocrystal layers embedded in oxide obtained by controlling the diffusion–crystallization process *J Nanopart Res* **15** 1–12

[4] Maeda Y 1995 Visible photoluminescence from nanocrystallite Ge embedded in a glassy SiO2 matrix: evidence in support of the quantum-confinement mechanism *Phys. Rev. B* **51** 1658–70

[5] Buljan M, Radić N, Ivanda M, Bogdanović-Radović I, Karlušić M, Grenzer J, Prucnal S, Dražić G, Pletikapić G, Svetličić V, Jerčinović M, Bernstorff S and Holý V 2013 Ge quantum dot lattices in Al2O3 multilayers *J Nanopart Res* **15** 1–13

[6] Peibst R, de Sousa J S and Hofmann K R 2010 Determination of the Ge-nanocrystal / SiO 2 matrix interface trap density from the small signal response of charge stored in the nanocrystals *Physical Review B* **82**

[7] Kanno T, Fujii M, Sugimoto H and Imakita K 2014 Colloidal hydrophilic silicon germanium alloy nanocrystals with a high boron and phosphorus concentration shell *Journal of Materials Chemistry C* **2** 5644

[8] Wu X L, Gao T, Siu G G, Tong S and Bao X M 1999 Defect-related infrared photoluminescence in Ge+implanted SiO2 films *Applied Physics Letters* **74** 2420

[9] Sahle C J, Sternemann C, Conrad H, Herdt A, Feroughi O M, Tolan M, Hohl A, Wagner R, Lützenkirchen–Hecht D, Frahm R, Sakko A and Hämäläinen K 2009 Phase separation and nanocrystal formation in GeO *Applied Physics Letters* **95** 021910

[10] Shen J K, Wu X L, Yuan R K, Tang N, Zou J P, Mei Y F, Tan C, Bao X M and Siu G G 2000 Enhanced ultraviolet photoluminescence from SiO2/Ge:SiO2/SiO2 sandwiched structure *Applied Physics Letters* **77** 3134

[11] Peibst R, Dürkop T, Bugiel E, Fissel A, Costina I and Hofmann K R 2009 Driving mechanisms for the formation of nanocrystals by annealing of ultrathin Ge layers in SiO2 *Phys. Rev. B* **79** 195316

[12] Takeoka S, Fujii M, Hayashi S and Yamamoto K 1998 Size-dependent near-infrared photoluminescence from Ge nanocrystals embedded in SiO2 matrices *Physical Review B* **58** 7921–5

Calculation of Electronic Absorption Spectra with Account of Thermal Geometry Fluctuations

Maris V Guzha[1], **Andrew I Svitenkov**[1]

[1]eScience Research Institute, ITMO University, St. Petersburg 197101, Russia

Abstract. An influence of thermal fluctuations of molecule's geometry on calculated electronic-absorption Vis/Uv spectra is considered. Paper presents the quantum chemical modeling of the electronic-absorption spectra for the collection of graphene samples (44, 56, 60, 68 atoms). The calculations were performed by time dependent density functional theory (TDDFT) method in combination with molecular dynamics (MD) simulation at T=300 K. The noticeable changing of spectra relative to single point TDDFT calculation was discovered for two of four structures. We associate achieved results with perturbation of hydrogen and carbon atoms on the edges of the structures. We believe that suggested methodology will be useful in application engineering researches of novel molecules and molecular complexes.

1. Introduction

Quantum chemistry (QC) calculations is a powerful tool in basic researches and applications engineering of novel molecules and complexes. Many distinctive features in electronic structure and optical properties of graphene nanostructures were explored by first-principles or semi-emperical QC calculations. There is significant research interest on the optical properties of graphene related to highly selective adsorption spectrum and its dependences on size, shape and structure of the edges of the graphene samples.

Band gap engineering by varying structure of graphene edge has been an exciting area of research and was studied extensively [1, 2]. Strong adsorption and high oscillator strengths of electronic transitions are as well of paramount importance in such studies. Osella *et al.* [3] described it in respect with the optical properties of graphenenanoribbons (GNR). Denk *et al.* [4] studied the exictone optical response of GNRs with zigzag edge and its precursor. Chopra *et al.* [5] investigated small graphene samples of various form and found the highest oscillator strength for the rectangle one. Cocchi [6] denoted the growth of oscillator strength with the length of GNRs.

At the present time graphene structures attract a lot of interest for practical application; educed features were explored properly. The great interest is arisen to the research of particular engineering applications of graphene in photonics and optoelectronics. In spite of augmentation of experimental studies, method of QC is still high evaluated, though some auxiliary aspects have to be taken into account regarding practical side of investigation.

We consider an influence of thermal fluctuations on the QC calculated graphene absorption spectra. Most part of the recent QC studies are usually based on single point calculations for ground state optimized geometries [4-7]. At the same time the relation between geometrical deformations of graphene samples and their electronic properties was shown previously [7]. Thus we suppose that single-point methodology of studies is challenging for the practical aspect of graphene engineering.

In this study the methodology of *ab-initio* calculations of absorption spectra with account of thermal fluctuations based on combination of time dependent density functional theory (TDDFT) and molecular dynamics (MD) is presented. The set of graphene samples described in [5] was considered.

We believe that suggested methodology will be advantageous in application engineering researches of novel molecules and molecular complexes and will provide better agreement between the calculated and measured spectra.

2. Methodology

It should be noted that the similar issues were discussed before [8, 9], however an account of only metastable geometries of molecules was considered. The question about an influence of thermal geometry fluctuations of graphene samples has not been systematically examined previously.

The range of considered graphene samples was taken from the study [5] by S. Chopra and L. Maidich. Primary parameters of QC calculations are the same as in [5]. The similarity of objects and parameters of calculations was kept of the verification considerations. The proposed graphene structures are also very useful for discussion of noticed hypothesis about influence of fluctuation of graphene edge on the obtained spectra. Two triangle structures, depicted on the figure 1 (b, d) produce larger contribution of edge that rectangle (a, b) samples. The highest oscillator strengths were obtained by Chopra for rectangle samples as well, moreover the question about an influence of thermal fluctuations on the strongest electronic transition is of special interest.

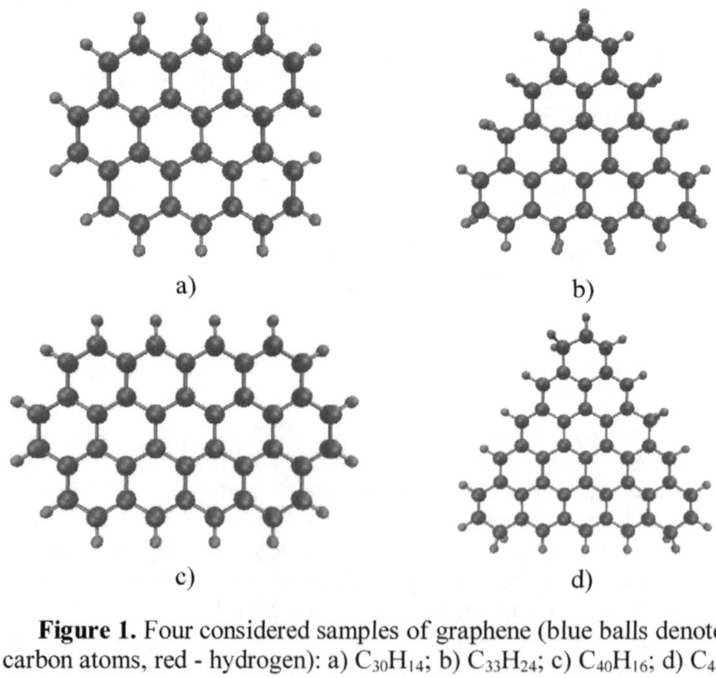

Figure 1. Four considered samples of graphene (blue balls denote carbon atoms, red - hydrogen): a) $C_{30}H_{14}$; b) $C_{33}H_{24}$; c) $C_{40}H_{16}$; d) $C_{46}H_{22}$

We considered the temperature T = 300K. Suggested method of account of thermal fluctuations is implemented by four-step workflow:

1. Obtaining of optimized geometry of a molecule;
2. Producing the set of fluctuated geometries by MD simulation;
3. QC calculations of spectrum of electronic transitions for each geometry form the set;
4. Averaging of spectrum with Boltzmann weights $\exp[(U_0 - U)/kT]$ (k – is a Boltzmann constant). Estimation of distribution and etc.

The workflow was implemented based on CLAVIRE program platform [10].

The optimized geometries of graphene samples were obtained by HF/6-31G(d) calculations. Energies U_0 and U were estimated as QC total energy values and thus the details of MD simulation are not so important. It was performed by NAMD package with CHARMM22 force field [11] during 100ps with 1ps output period that is enough to avoid correlation among the states. Remarkable revelation is that we had to decrease the temperature of MD ensemble to 200K since the value of 300K provided many low probability states. The UV-visible absorption spectrum were calculated at TDDFT/B3LYP approach using 6-31G (d) basis set with 50 singlet excited states. All calculations have been performed with the Firefly package, v.8.1.1 [12].

Though we believe that this QC method is substantially preferable for molecules of carbon and hydrogen atoms [13], we note that the same workflow can be implemented with another QC approaches.

3. Results and discussion

The passivation of edges by hydrogen structure as it is depicted at the figure 1 provides good convergence of SCF procedure. We have not observed any discrepancies in rate of convergence for optimized and fluctuated geometries.

3.1.1. Rectangle samples

The calculated spectrum of electronic transitions for rectangle samples (figure 1) are depicted at the figure 2. At the upper two charts the transparency of red bars is referred to probability of the related fluctuated geometry, so the distribution is visible.

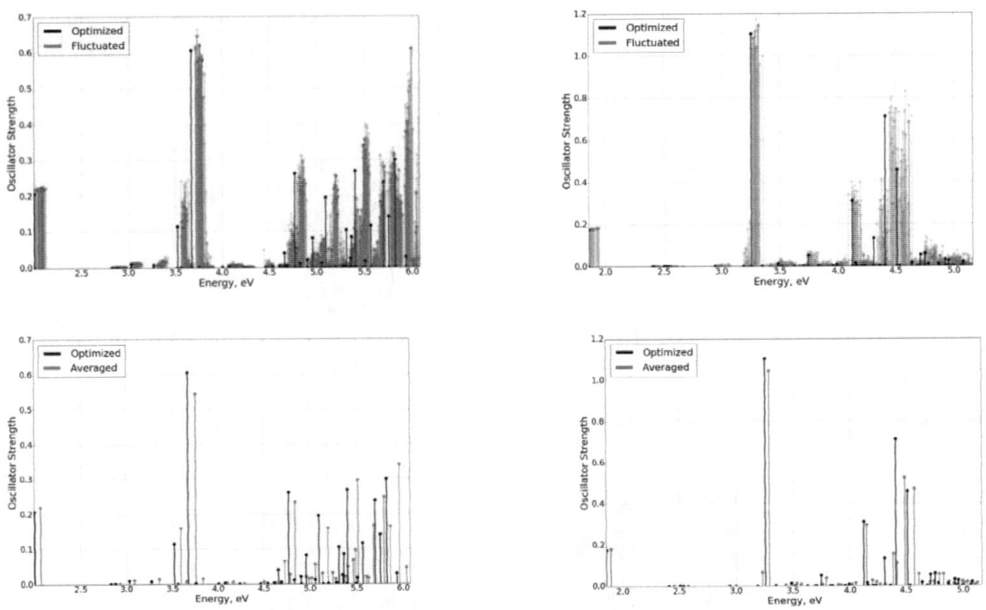

a) b)

Figure 2. The spectrum of electronic transitions for a) $C_{30}H_{14}$ and b) $C_{40}H_{16}$ graphene samples. The transitions for optimised geometries are depicted by black bars, the transitions for fluctuated geometries, T=300K (upper), and average spectra (lower) are depicted by red bars

We distinguished restricted shift of spectrum and small variations of oscillator strengths due to thermal fluctuations. The intensity of strongest electronic transitions did not change considerably and look even more stable for the larger molecule.

3.1.2. Triangle samples

The different results were obtained for triangle graphene samples (figure 3). We can see the significant shift of transition energies (for the $C_{46}H_{22}$ especially) and that is more interesting – appearance of more intensive transitions during the fluctuations (however they are still lower than for the rectangle samples).

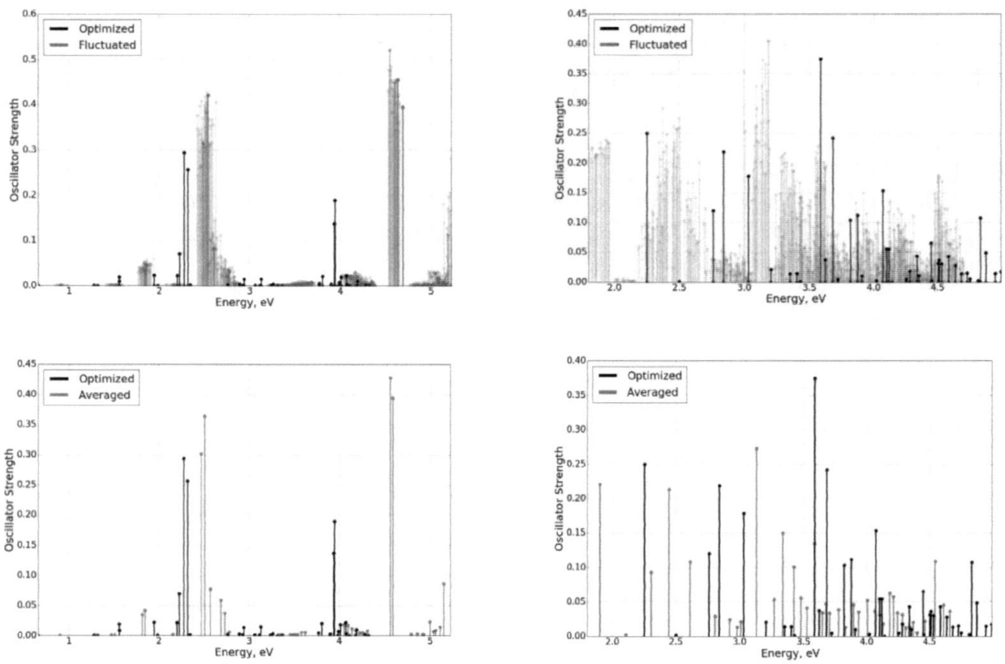

a) b)

Figure 3. The spectrum of electronic transitions for a) $C_{33}H_{24}$ and b) $C_{46}H_{22}$ graphene samples. The transitions for optimised geometries are depicted by black bars, the transitions for fluctuated geometries, T=300K (upper), and average spectra (lower) are depicted by red bars

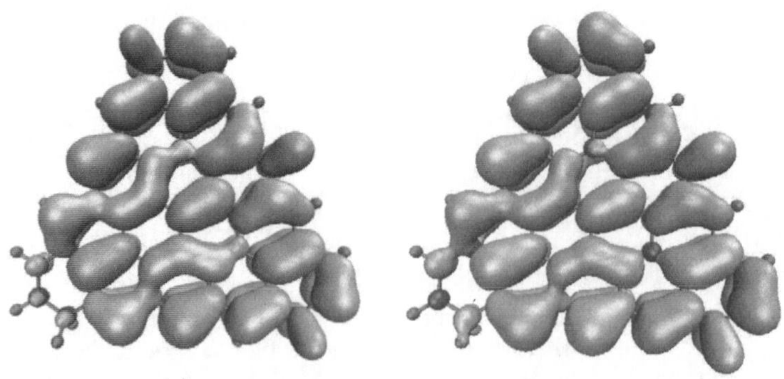

a) b)

Figure 4. LUMO for a) optimized and b) fluctuated geometries; cyan – isovalue 0.005, orange – isovalue -0.005

This is in agreement with our hypothesis of fluctuation of edge atoms influence since one is considerable for triangle structures then for the rectangle. The figure 4 illustrates the altering in electronic structure (LUMO) and the localization of electron density with fluctuations can be perceived.

4. Conclusion

Two main purposes were fulfilled in the present study: to show the necessity of account of thermal fluctuation of the geometry in QC modelling and a methodology for it; to examine the stability of most intensive electronic transitions in graphene samples. Study presents high tolerance of the spectrum of rectangle graphene samples and its growth with the size of the structure. It allows to assume the stability of the optical response of large GNRs, engineering application of which is implied. On the other hand QC calculation for triangle samples demonstrates the case that the thermal fluctuations have an essential influence. This indicates the necessity in account of thermal fluctuations and supports the suggested combination of MD and QC for modeling in engineer application.

Acknowledgments
This paper is financially supported by Ministry of Education and Science of the Russian Federation, Agreement #14.587.21.0024 (18.11.2015).

This work was partly performed by the Master student Maris V Guzha of the Master's Programme in Computational Science [14].

References

[1] Allen M J, Tung V C, Kaner R B 2010 *Chem. Rev.* **110** 132–45
[2] Mohanty N, Moore D, Xu Z, Sreeprasad T S, Nagaraja A, Rodriguez A A, and Berry V 2012 *Nature communications* **3** 844
[3] Osella S, Narita A, Schwab M G, Hernandez Y, Feng X, Müllen K and Beljonne D 2012 *ACS Nano* **6** 5539–48
[4] Denk R, Hohage M, Zeppenfeld P, Cai J, Pignedoli C A, Söde H, Fasel R, Feng X, Müllen K, Wang S, Prezzi D, Ferretti A, Ruini A, Molinari E, Ruffieux P 2014 *Nat. Commun.* **5**
[5] Chopra S, Maidich L 2014 *RSC Adv.* **4** 50606–13
[6] Cocchi C, Prezzi D, Ruini A, Benassi E, Caldas M J, Corni S, and Molinari E 2012 *The journal of physical chemistry letters* **3**(7) 924-929
[7] Liu F, Ming P, Li J 2007 *Physical Review B* **76**(6) 064120
[8] Koścień E, Gondek E, Kityk A V 2008 *Spectrochimica Acta Part A: Molecular and Biomolecular Spectroscopy* **70**(5) 1097-1103
[9] Mandal T K, Dutta S, Pati S K 2009 *Journal of Chemical Sciences*, **121**(5), 873-880
[10] Knyazkov K V, Nasonov D A, Tchurov T N and Boukhanovsky A V 2013 *Procedia Computer Science* **18** 2223-2232
[11] Phillips J C, Braun R, Wang W, Gumbart J, Tajkhorshid E, Villa E, ... and Schulten K 2005 *Journal of computational chemistry* **26**(16) 1781-1802
[12] Granovsky A A 2014 Firefly ver. 8.1.1.
[13] Suendo V and Viridi S 2012 *ITB J. Sci* **44**(2) 93-112
[14] Krzhizhanovskaya V V, Dukhanov A V, Bilyatdinova A, Boukhanovsky A V and Sloot P M A 2015 *Journal of Computational Science* **10** 288-298

Raman spectroscopy for identification of wood species

V A Gerasimov[1], A M Gurovich[2], D K Kostrin[1], L M Selivanov[1], V A Simon[1], A B Stuchenkov[2], A V Paltcev[2] and A A Uhov[1]
[1] Department of electronic instruments and devices, Saint Petersburg Electrotechnical University "LETI", 197376, Saint Petersburg, Russia
[2] Research and production center "Innovative engineering and technology", 196084, Saint Petersburg, Russia

Abstract. This article discusses the application of Raman spectroscopy for identification of wood species. Use of Raman spectroscopy allows increasing the certainty of determining the type of wood compared to the analysis of spectra of diffuse reflectance. Raman spectrums of different wood samples when irradiated by laser radiation are shown. Ways to improve the determination reliability of wood species due to the modernization of the identification technique are discussed. The stages of data processing, allowing carrying out correct further analysis are described.

1. Introduction

A number of studies in the field of identification of various natural and artificial objects by spectral methods were conducted in the past years. Instruments for spectral analysis have been successfully used in many areas of scientific research and technical control. However, there are a number of tasks that could be solved by spectral methods, but only in the field, because not every object of study can be delivered to the laboratory. One such area is the identification of wood species for customs control.

One of the first methods proposed for identification of wood species was based on diffuse reflection. The operating experience of this method showed that to improve the accuracy of identification the increase in the signal-to-noise ratio of the photodetector is required, because the resultant spectrums of different wood samples differ very slightly [1]. However, the use of more advanced sensor greatly increases the cost of research equipment. Thus, it is highly relevant to develop new methods to identify wood species in particular with the use of Raman spectrums.

2. Identification of wood species by spectrums of diffuse reflectance

The diffuse reflectance spectrophotometer for the identification of wood species contains a continuous spectrum source (halogen incandescent lamp), diffuse radiation of which falls onto the investigated sample using a photometric sphere. Reflected light through the slit illuminates the dispersing element (diffraction grating) that splits the incident radiation by wavelengths into the spectrum on the photodetector surface. To obtain the background (dark) signal in the scheme a controlled optical shutter is used [2]. To minimize measurement error of the spectral coefficient of diffuse reflectance algorithmic (numerical) compensation that is focused on the readings of temperature sensors located in different parts of the device is implemented [3]. Fragments of the diffuse reflectance spectrums of some wood species are shown in Figure 1.

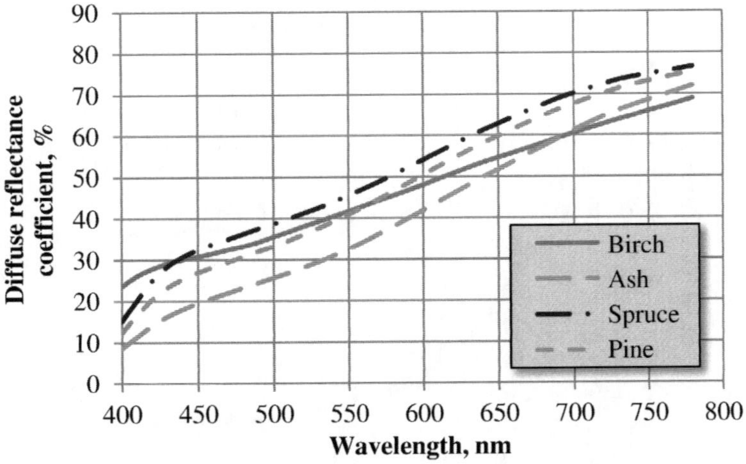

Figure 1. Fragments of the diffuse reflectance spectra of some woods species.

As can be seen from the figure, spectra of diffuse reflectance of wood are continuous. For different species it is impossible to distinguish the individual absorption bands; however the overall level of reflectance and steepness of the spectral functions make it possible to build identification algorithm [4].

The use of this method also imposes a number of restrictions on sample preparation. For confident identification preparation of the log cut ends by machining to achieve low roughness is required. Also for reliable operation of the algorithm the existence and regular updating of a library of spectral information is requested. These factors do not allow using devices based on method of diffuse reflectance effectively in the field.

3. Identification of wood species using Raman spectrums

When using Raman spectroscopy the investigated substance is irradiated with monochromatic radiation. A small portion of the transferred energy is converted into radiation, the characteristics of which are associated with the chemical composition of a substance and the parameters of the incident radiation; meanwhile most of the energy is absorbed or dissipated.

For the laser with a wavelength of 532 nm for the registration of Raman spectrums it is necessary to use a device with the spectral sensitivity region from 532 to 650 nm (this corresponds to Raman shift from 0 to 3400 cm^{-1} relative to the incident radiation). In this work AvaSpec-ULS2048 spectrometer was used. For the laser with a wavelength of 1064 or 785 nm it is possible to examine the spectrums in the direction of decreasing wavelengths from 1064 to 750 nm (corresponding to Raman shift from 0 to 3900 cm^{-1}). In our case Nicolet NXR Raman spectrometer was used. With increasing the wavenumber the spectral sensitivity of the photodetector also increases that enables the registration of the Raman spectrum over a wider range and this corresponds to the greater number of different substances which can be identified.

To separate the reflected radiation in the case of a laser with a wavelength of 532 nm it is possible to use filter glasses OS-13 or OS-14. In the case of the 1064 nm laser it is efficient to use the Semrock notch filter NF03-532/1064E-25.

Instead of the incandescent lamp (as in the devices based on diffuse reflectance method) a laser diode, which is included into the scheme of constant current switching converters, is used. The current value is programmed using the corresponding commands from PC. The current control circuit is required in cases where the laser radiation is able to heat the test samples to temperatures of combustion.

The need for the use of different lasers is related to the ability of substances to fluorescamine under the influence of powerful monochromatic radiation. If the substance is irradiated in the spectral region close to the absorption maximum, then fluorescence will be at its maximum and its value will be several orders of magnitude greater than the magnitude of the Raman radiation. Figure 2 shows spectra of different wood species when irradiated with the 785 nm laser.

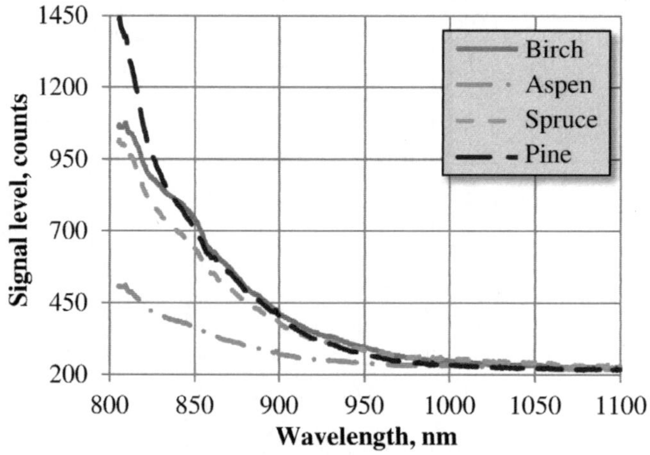

Figure 2. Spectrums of different wood species achieved using laser with a wavelength of 785 nm.

The figure shows that the spectra are also close to continuous, however small local extremes are present. If for each of the spectra the polynomial trend-line of low-order is found and then normalized by the maximum value difference between the spectrum and the trend-line is build [5], extrema can be considered (Figure 3).

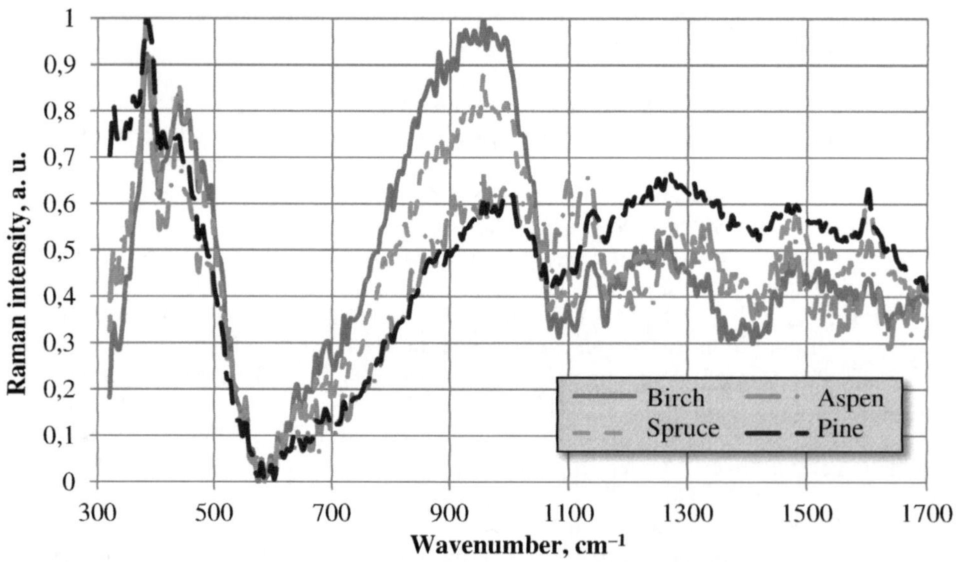

Figure 3. The difference between the Raman spectra and the trend-lines for different wood species.

The figure shows that the differences between the spectra of different wood species are more distinct than in the analysis of spectra of diffuse reflectance.

Spectrum obtained when irradiated by a 1064 nm laser is the most informative as it contains a large number of extrema, characteristic for certain substances. Figure 4 shows a comparison of Raman spectra using irradiation of the object with different lasers. Figure 5 shows spectrums of different wood species under laser irradiation with a wavelength of 1064 nm.

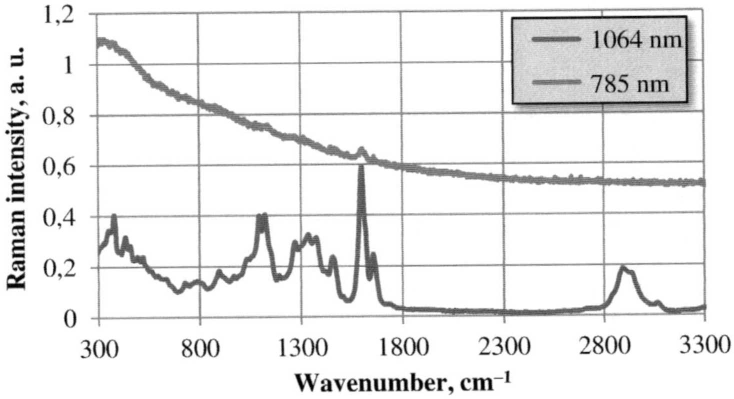

Figure 4. Raman spectra of pine under laser irradiation with different wavelength.

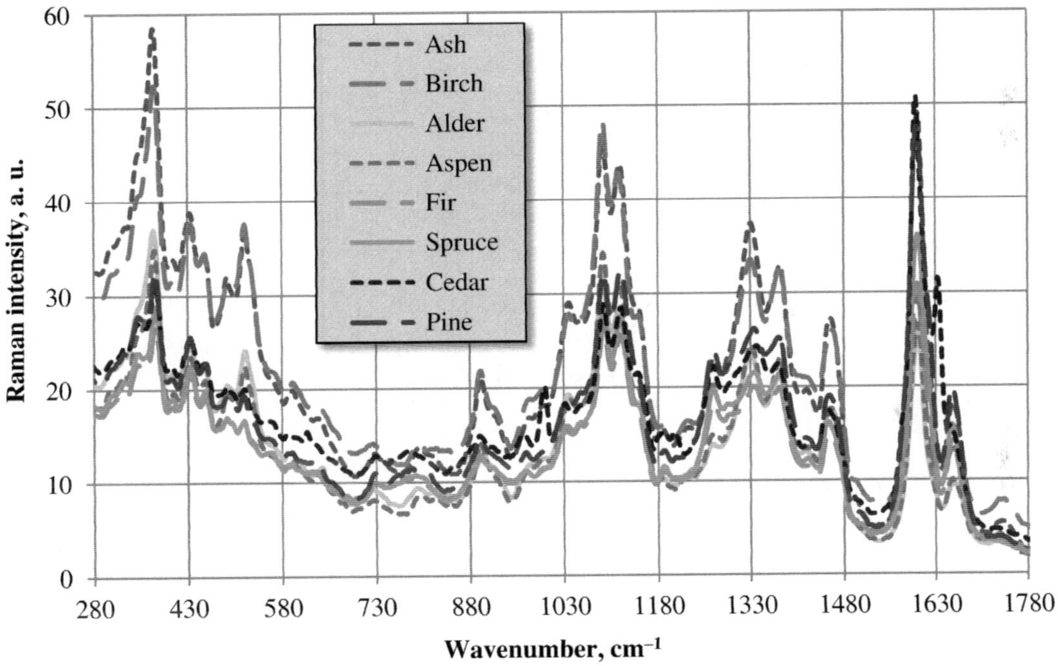

Figure 5. Raman spectrums of wood species achieved using laser with a wavelength of 1064 nm.

The figure shows that the structure of spectrums for different wood species is practically the same. However different concentrations of individual substances in these samples make identification of specific wood species possible.

4. Processing of the obtained spectral data

After receiving the spectrum it is necessary to process the data using the following three stages:

- smoothing;
- subtraction of the baseline;
- normalization.

The source data is always noisy. To reduce the noise effect one of the algorithms of smoothing or their combination should be used. At the moment, authors implemented four smoothing algorithms:

- moving average;
- exponential smoothing;
- median filter;
- Savitzky–Golay smoothing algorithm.

The last type of smoothing is particularly relevant for spectral data, because it carefully processes spectral peaks, not aligning them.

Obtained spectrums almost always have the effect of luminescence, which does not allow direct estimation of the Raman responses. Furthermore, luminescence is different in various samples (and sometimes in different regions of the same sample). This effect greatly complicates the analysis and requires a software solution. The problem is solved by finding the underlying background (base line) on the graph of the spectrum with its subsequent subtraction from the main spectrum. The analysis of the current works in this area has shown the perspectivity of the two approaches:

- modeling of the baseline by a polynomial of fixed order, with evaluation of the quality of approximation by minimizing the assigned cost function [6];
- modeling of the baseline with wavelet functions of different types. In practice two algorithms were implemented:
  - a simple algorithm based on the selection of the wavelet function parameters that use as input the Daubechies 10 wavelet or discrete Meyer wavelet [7];
  - more complex and high-quality algorithm, based on preliminary peak detection, using as basic a mexican hat wavelet [8].

After subtracting the baseline spectra need to be normalized, so that in future, while solving the identification problem, to estimate relative (and not absolute) influence of individual features (the intensity of the spectra at different wavelengths) in the common spectral measurements.

Three methods of normalization were used:

- normalization on the maximum;
- normalization by the area under the spectrum curve;
- normalization on the Euclidean norm (spectrum is considered as a normalized feature vector).

Formally, the problem of identification of the spectra is a problem of classification, that is, it requires preliminary training on pre-prepared sample spectra. The classification is based on a small set (3…5) of the spectra corresponding to multiple measurements of a single substance. An important condition for correct identification is matching of methods and parameters of pre-treatment for the identifiable spectra and spectra of the training samples.

The training sample should include an equal number of spectra of each class and obtained in the same conditions. The order in which the spectra are acquired should not depend on the class (the spectra of one class should not be measured consistently), so that the real differences between classes are not replaced by the possible variation of the measurement conditions.

The analysis is based on applying the principal component analysis (PCA) [9], which allows using the methods of linear algebra to identify in a large data set a relatively small number of parameters, which accurately characterize each of the elements of the set. Thus, reduction of the dimensionality of the processed data (and therefore processing time) can be achieved; also this solution allows to get rid of the slight "noise" effect present in the original data. Modern algorithms for calculating PCA method are based on obtaining singular value decomposition (SVD) of the matrix of original data, which in turn is associated with obtaining the eigenvalues. The quality of the method depends on the number of the selected components. The optimal number of principal components in our case ranges from 11 to 14 (instead of the original 512 points).

After applying the method of principal components and reducing the dimensionality of the original data the method of classification based on linear discriminant analysis (LDA) [10] is used. This method allows constructing linear combinations of available features, which allow maximizing differences between objects from different groups (classes) and minimizing difference between objects from one group. Mathematically the idea of the method is based on selection coefficients, which maximize the dispersion of the average members of the classes and minimize the variance within classes (Fisher criterion). The adjustment of coefficients for LDA decomposition is performed on the matrix, rows of which are the measurements (spectra) and columns are the main components of each measurement obtained by PCA analysis. The overall results of the LDA analysis are:

- matrix of coefficients that projects the original (training) data into the space with a maximum ratio between the differences amongst classes to intraclass differences;
- set of similar linear functions with different coefficients (linear discriminant functions), each designed for a specific class and depending on the average value of the class and the covariance matrix of its elements.

A necessary step prior to the identification of a new spectrum is to perform preliminary processing identical to that carried out on the spectra of training samples. The task of classification is implemented by the following sequence of steps:

- after preprocessing the spectrum is transformed into the principal component space for which the vector corresponding to the spectrum is multiplied by the matrix of loadings in PCA decomposition of the training sample;
- resulting vector is projected into the principal component space of LDA decomposition for which the vector is multiplied by the projecting matrix, obtained during the LDA analysis;
- for a decision about belonging of the constructed LDA vector to a particular class this vector should be substituted into the linear discriminant function, corresponding to this class. The result of the lookup is a scalar value associated with the probability of belonging of the element to the given class (the larger the function value, the higher the probability that the analyzed spectrum relates to the corresponding class).

Calculating the values of the discriminant functions for all classes, the class for which the result of the function is maximal is selected. When performing classification of a set of spectra evaluation results for each spectrum are summarized. At present, discussed method and implemented system of processing and analysis of the spectra allows to obtain Raman spectra of wood and its identification with a probability of about 90 %.

5. Conclusion

Application of the principles of Raman spectroscopy allows increasing the reliability of the identification method compared to the spectral analysis of diffuse reflectance due to the registration of more informative spectral data. The conducted research gives opportunity to conclude that Raman spectroscopy is a promising method of identifying wood species.

References

[1] Woronin A A, Gerasimov W A, Kostrin D K, Smirnov P E, Turubarov A W, Oukhov A A 2013 *Biotechnosphere* **3** 16–20
[2] Uhov A A, Gerasimov V A, Kostrin D K, Selivanov L M 2014 *Journal of Physics: Conference Series* **567** 012039
[3] Voronin A A, Smirnova E V, Faizov I N 2011 *Journal of Instrument Engineering* **1** 74–78
[4] Voronin A A, Smirnova E V, Smirnov A P 2010 *Scientific and Technical Journal of Information Technologies, Mechanics and Optics* **2** 5–11
[5] Voronin A A, Smirnov P E, Turubarov A V, Gerasimov V A, Kostrin D K, Uhov A A 2014 *Testing. Diagnostics* **5** 35–38
[6] Mazet V, Carteret C, Brie D, Idier J, Humbert B 2005 *Chemometrics and intelligent laboratory systems* **2** 121–133
[7] Galloway C M, Le Ru E C, Etchegoin P G 2009 *Applied Spectroscopy* **12** 1370–1376
[8] Zhang Z-M, Chen S, Liang Y-Z, Liu Z-X, Zhang Q-M, Ding L-X, Yec Fei, Zhou H 2010 *Journal of Raman Spectroscopy* **6** 659–669
[9] Jolliffe I T 2002 Principal Component Analysis (New York: Springer-Verlag)
[10] Perriere G, Thioulouse J 2003 *Computer Methods and Programs in Biomedicine* **70** 99–105

Scanning near-field microscopy of microdisk resonator with InP/GaInP quantum dots using cantilever-based probes

A V Shelaev[1], A M Mintairov[2,3], P S Dorozhkin[1,4], V A Bykov[1]

[1] NT-MDT Co., 100 Zelenograd, Moscow, 124482 Russia
[2] Ioffe Institute, 194021 Saint Petersburg, Russia
[3] University Notre Dame, Notre Dame, 46556 IN, USA
[4] Skolkovo Institute of Science and Technology, Skolkovo Innovation Center, Moscow, 143026, Russia

Abstract. We present cantilever-probe based scanning near-field microscopy (SNOM) studies of GaInP microdisks resonators (radii R=2 um and quality factors Q~1000) with embedded InP quantum dots (QDs) emitting at ~750 nm. Near-field photoluminescence spectroscopy in collection regime, using side excitation from micro-objective, was used for imaging of whispering-gallery modes (WGMs) with a spatial resolution below the light diffraction limit. Using collection-illumination regime we imaged the position of single InP/GaInP QDs in microdisk.

Introduction

Using scanning near-field optical microscopy (SNOM) allows to measure optical fields in passive [1,2] and active [3,4] nano-photonic cavity structures with spatial resolution beyond light diffraction limit under optical pumping. While in passive structures the field is excited by external waveguide coupled to the cavity, in active structures the emission of active element is detected. In the last case the effects of interaction of emitting media with cavity modes can be probed [4], which is important for optimization of nano-lasers and single photon sources. So far most of measurements of optical fields in active cavity structures have been performed using tapered fiber metal coated aperture probes [5] or uncoated fiber apertureless probes [2–4]. Recently cantilever-based probes were used for measurements of the field of lasing mode [6] and plasmonic waveguide in stripe laser structures under electrical pumping [7]. In the present paper we used cantilever-probe SNOM for photoluminescence (PL) spectroscopy measurements of GaInP microdisks with InP quantum dots (QDs) embedded. Using collection and collection-illumination regimes we demonstrate using cantilever-probe SNOM to measure optical fields of cavity whispering-gallery modes (WGMs) and location of individual QDs in the cavity.

1. Experimental details

Cantilever-based SNOM setup consists of AFM head with build in 6-mm working distance 100$^{\times}$ objective (0.7NA, Mitutoyo), light input-output module connected to AFM base unit (NTEGRA Spectra, NT-MDT) and confocal spectrometer (OMU34, NT-MDT) with a 473nm laser (Cobolt Blues, Cobolt) and a CCD (iDus, Andor Technology) coupled to the AFM system. AFM registration system consists of 1064nm laser which goes through high NA 100$^{\times}$ objective and 4-section InGaP photodiode which detects tip-sample interaction. AFM tapping mode applied keep intraction between tip and surface on the same level and to collect SNOM images. Sample is mounted onto XYZ piezotube scanner with build-in capacitance sensors.

Hollow-pyramid near-field cantilever probe (SNOM_NC, NT-MDT) with aperture size of about 100 nm was used. It has 70° internal angle coated by 100 nm Al layer. Probe is mounted in the AFM tip holder with the tilt angle 11° to sample plane and positioned in the center of high NA objective with the help of viewing microscope. 100$^{\times}$ objective with 0.7NA collects most of the signal collected by aperture as objective cone angle to the axis is 45°.

Two excitation SNOM configurations were applied: side illumination and illumination-collection through the aperture (figure 1). In side illumination configuration laser beam coupled to optical fiber is focusing to the sample surface by 0.28NA lens. Angle of incidence between beam optical axis and sample plane is equal to 17°. Side illumination optics is positioned on micrometer manual positioner (Newport) allowing bringing laser spot under the hollow-pyramid cantilever. Focused laser spot has 1.7μm×5.5μm dimensions. Taking into account 10mW of laser intensity focused to the sample and 85% power focused into the central Eire spot, an average power density on the sample surface is 1.15mW/μm^2. Polarization in the focused laser spot is aligned along X axis.

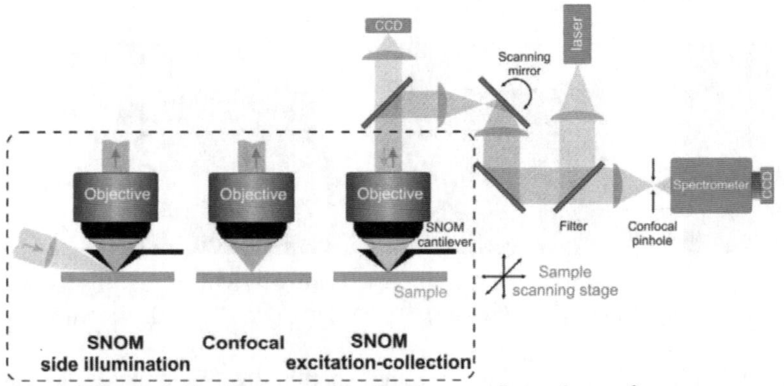

Figure 1. Excitation and collection configurations of setup.

In illumination-collection SNOM configuration 7mW laser is focused to hollow-pyramid apex. Same objective is using to collect luminescence collected through aperture. Precise control of laser beam position is performed by mirror-scanner which build-in into light input-output module. Tightly focused laser spot scans around the aperture by mirror-scanner. Collected light is directed from input-output module to confocal microscope with two 473 nm edge filters (Semrock) and cross slit. In addition confocal upright scheme applied. In this case no SNOM cantilever used. Sample excitation and collection were performed via the same objective with 0.07 mW laser power.Confocal detection scheme allows minimize background singal and therefore improve resolution and SNOM signal/background signal ratio. 520 mm focal length spectrometer equipped by 150 lines/mm grating and 1024 pixel CCD provides spectrum in each point. Acquisition time varies from 0.1 s/point in confocal mode or side illumination SNOM mode to 3 s/point in illumination-collection SNOM mode. Spectrum and data analysis was performed using the image analisys Nova software (NT-MDT).

2. Sample description

Microdisk resonators – were prepared from a 150 nm thick GaInP waveguide structure with embedded self-organized InP quantum dots grown by Metal-Organic Chemical Vapor Deposition (MOCVD). For uncapped structure, used to estimate dot density, the upper part of $Ga_{0.52}In_{0.48}P$ waveguide was not grown. The structures have been grown in a horizontal AIX200/4 reactor under pressure of 100 mbar. Initially a 700 nm thick AlAs layer was deposited on the GaAs wafer. Then a 150 nm thick $Ga_{0.52}In_{0.48}P$ waveguide layer lattice-matched to GaAs was grown with InP QDs at the center. The QDs were grown at 725 °C by depositing 7 monolayers of InP. Microdisks having D = 2 μm were fabricated using electron beam lithography (EBL) and inductively-coupled plasma reactive ion etching (ICP-RIE) by using either a wafer bonding (WB). The waveguide was first wafer-bonded to a Si wafer using a spinon-glass (SOG) precursor (Filmtronics, Inc.) solution, then the GaAs substrate and the AlAs sacrificial layer were removed by mechanical polishing and chemical etching. Finally the MDs were prepared from this structure using EBL and ICP-RIE.

3. Results and discussions

The emission spectrum of the ensemble of QDs, measured in the confocal configuration from an area of about square micrometer, contain broadband peak ~750 nm. To obtain spectra from separate QDs in the near-field configuration, aperture cantilevers with improved transmission (~10^{-2}) was selected. The transmission of light by the aperture was measured at the excitation wavelength of 473 nm. Same 3×3 μm^2 area was investigated by AFM and SNOM providing an AFM surface topography (figure 2) and the confocal (figure 3) and SNOM (figure 4) luminescence intensity map from InP QDs in the spectral range 750-780 nm. Topography and SNOM image were obtained in tapping mode to avoid any damage to the sample.

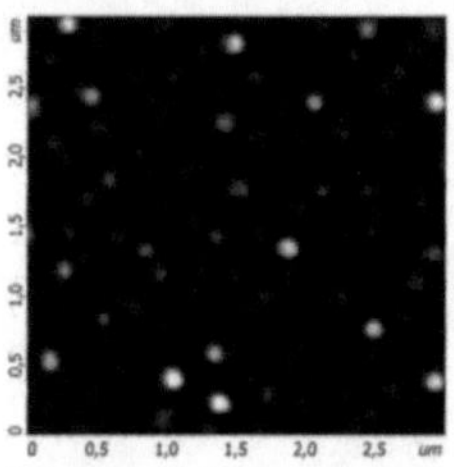

Figure 2. Topography.

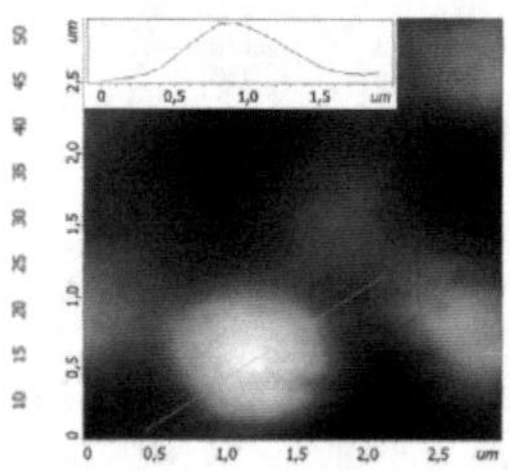

Figure 3. confocal map (750-780 nm).

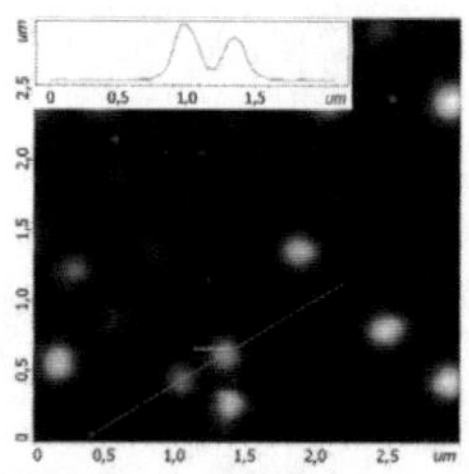

Figure 4. SNOM map (750-780 nm).

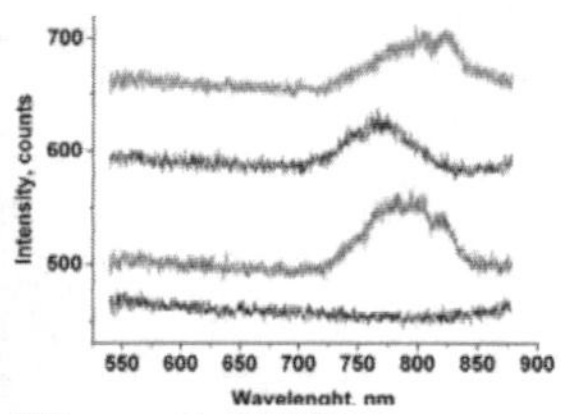

Figure 5. SNOM spectra of QDs and background (bottom).

SNOM photoluminescence was excited through aperture by 473 nm laser and collected by SNOM aperture with 0.5 s exposure time in each point. Spectrum was recorded for each point of the scanning area. It can be seen that individual QDs are resolved in the SNOM luminescence intensity maps. The full width at half-maximum (FWHM) of the optical signal profile taken via individual small QDs is approximately 350 nm (see the inset of figure 4). Using the Rayleigh criterion and taking into account the real sizes of the 200-250 nm emission sources, we obtain for the SNOM with the subwavelength aperture used in the study an estimated resolution of about 100 nm.

Individual QD spectra are clearly resolved and destiguished from each other (figure 5) even for QDs with less than 500 nm distance. The main advantages of near-field spectroscopy through nanoapertures arise from the inherent and absolute stability concerning the arrangement of the aperture relative to the QD. We should pointed out that previously spectra of individual semiconductor QDs were recorded using metalized fiber probes at cryogenic temperatures [8] and there was only one publication on imaging of individual QDs at room temperature [9]

AFM topography of the selected microdisk is shown on the figure 6. Spatial distribution of the luminescence from QDs in the microdisk resonator (figure 7) was obtained in SNOM illumination-collection configuration. It represents a location of individual dots. Spectral distribution of selected QDs in microdisk is presented on the figure 8. It was shown that light emitted from QDs in the central region of resonator is involved in Fabry-Perot mode oscillations only [4]. Whispering-gallery modes appears only when the reflection angle equils to total internal reflection angle and therefore QDs which are located close to the edge are involved into whispering-gallery mode oscillations. Optical pumping of the resonator was performed by side illumination and covers the entire disk area.

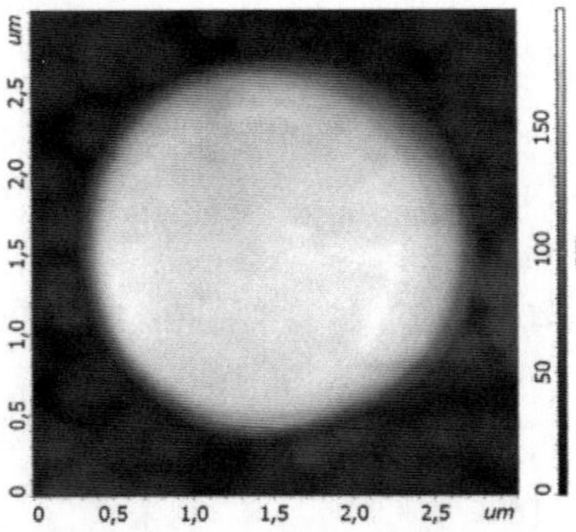

Figure 6. Topography, obtained by SNOM cantilever probe.

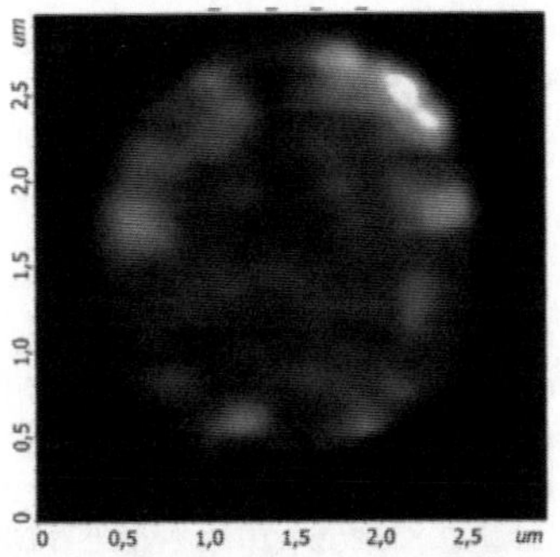

Figure 7. SNOM map (720-730 nm) with 0.5 s per point.

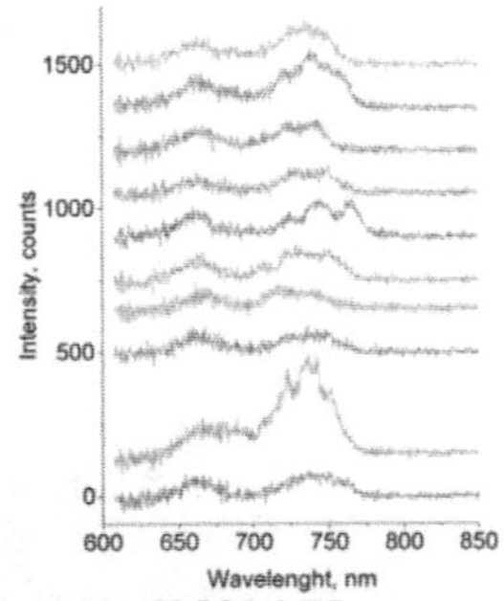

Figure 8. SNOM PL spectra of single QDs in excitation-collection mode.

Figure 9 presents the experimental confocal and SNOM spectrum from microdisk collected from the edge of the disk. Both Fabry-Perot and whispering-gallery modes are detected, but in contrast to confocal detection scheme Fabry-Perot modes are much weaker compare to whispering-galleries modes. Most likely this is due to smaller area from which spectra acquired. SNOM maps of 733.4 nm peak and 755.3 nm peak which corresponds to TE_{20} WGM and TE_{18} WGM respectively are shown on the figure 10 and figure 11.

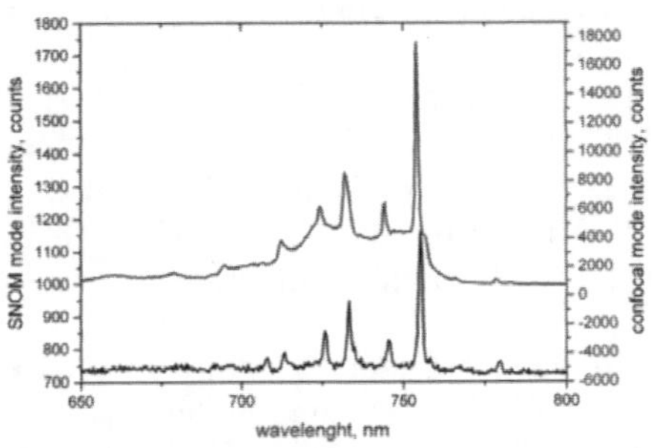

Figure 9. Confocal WGMs (top) and SNOM (bottom) spectra.

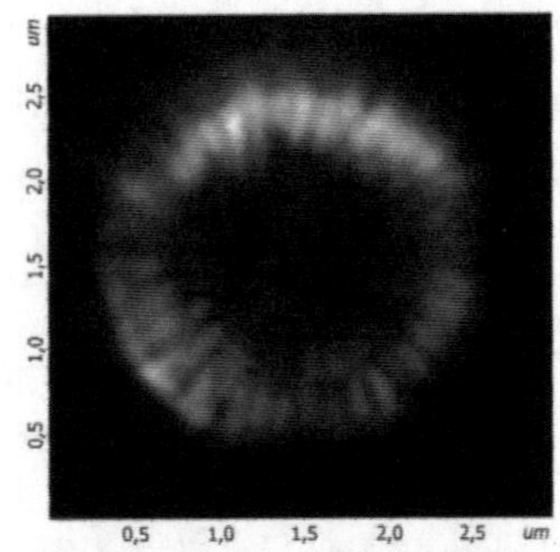

Figure 10. SNOM WGM TE_{20} mode, 732-735 nm.

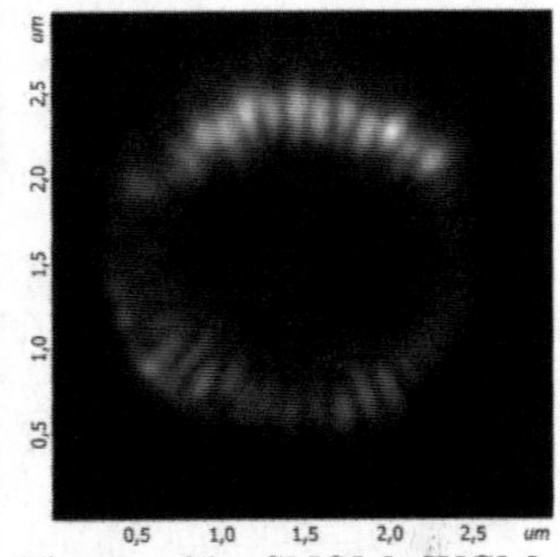

Figure 11. SNOM WGM TE_{18} mode, 753-757 nm.

4. Conclusions

Photoluminescence of self assembled InP/GaInP QDs with the size of 250 nm was studied. Illumination-collection regime of cantilver-based SNOM was used (excitation and emission collection through the same SNOM aperture). Hyperspectral SNOM photoluminescence maps resolving individual QDs were obtained. Luminescence spectra of individual QDs (with emission at 700-780 nm) were measured; spectral difference between single QDs was observed. Corellated images of topography and SNOM photoluminescence were obtained. Both tapping and contact feed back mechanisms were applicable in cantilever based SNOM. Optical resolution of photoluminescence maps was 100 nm, which is about $\lambda/7$.

Configuration with far field side excitation under the SNOM cantilever probe and collection through SNOM aperture was applied to investigate whispering-gallery modes distribution in a microdisk resonator of 2 μm diameter containing InP/GaInP QDs. Distribution of individual QDs in the microdisk resonator was obtained. Fine structure of optical whispering-gallery modes was identified both spectrally and spatially. Q-factor of selected whispering gallery mode was of order of 1000.

Acknowledgement
A.M.M acknowledge a support by Ministry of the Education and Science of the Russian Federation (contract № 14.Z50.31.0021, 7th April 2014).

References

[1] Quidant R, Weeber J-C, Dereux A, Lévêque G, Weiner J and Girard C 2004 Addressing and imaging microring resonators with optical evanescent light *Phys. Rev.* B **69** 081402
[2] Skacel M, Pagliano F, Hoang T, Midolo L, Fattahpoor S, Li L, Linfield E H and Fiore A 2013 Coupling of single quantum dots to photonic crystal cavities investigated by low-temperature scanning near-field optical microscopy *Phys. Rev.* B **88** 035416
[3] Louvion N, Gérard D, Mouette J, de Fornel F, Seassal C, Letartre X, Rahmani A and Callard S 2005 Local Observation and Spectroscopy of Optical Modes in an Active Photonic-Crystal Microcavity *Phys. Rev. Lett.* **94** 113907
[4] Mintairov A M, Chu Y, He Y, Blokhin S, Nadtochy A, Maximov M, Tokranov V, Oktyabrsky S and Merz J L 2008 High-spatial-resolution near-field photoluminescence and imaging of whispering-gallery modes in semiconductor microdisks with embedded quantum dots *Phys. Rev.* B **77** 195322
[5] Okamoto K, Loncar M, Yoshie T, Scherer A, Qiu Y and Gogna P 2003 Near-field scanning optical microscopy of photonic crystal nanocavities *Appl. Phys. Lett.* **82** 1676
[6] Ankudinov A V, Yanul M L, Slipchenko S O, Shelaev A V, Dorozhkin P S, Podoskin A A and Tarasov I S 2014 Investigation of the light field of a semiconductor diode laser *Opt. Express* **22** 26438
[7] Costantini D, Greusard L, Bousseksou A, Rungsawang R, Zhang T P, Callard S, Decobert J, Lelarge F, Duan G-H, De Wilde Y and Colombelli R 2012 In Situ Generation of Surface Plasmon Polaritons Using a Near-Infrared Laser Diode *Nano Lett.* **12** 4693
[8] Mintairov A M, Sun K, Merz J L, Li C, Vlasov A S, Vinokurov D A, Kovalenkov O V, Tokranov V and Oktyabrsky S 2004 Nanoindentation and near-field spectroscopy of single semiconductor quantum dots *Phys. Rev.* B **69** 155306
[9] Matsuda K, Saiki T, Saito H and Nishi K 2000 Room-temperature photoluminescence spectroscopy of self-assembled $In_{0.5}Ga_{0.5}As$ single quantum dots by using highly sensitive near-field scanning optical microscope *Appl. Phys. Lett.* **76** 73

Generation of photon pairs through parametric processes in nonlinear waveguides with the account of losses

D N Vavulin[1], **A A Sukhorukov**[2]

[1] ITMO University, 49 Kronverksky Ave., St.Petersburg, 197101, Russia

[2] Nonlinear Physics Centre, Research School of Physics and Engineering, Australian National University, Canberra 2601, Australia

E-mail: dima-vavulin@mail.ru, Andrey.Sukhorukov@anu.edu.au

Abstract. We present an analytical description of the process of spontaneous four-wave mixing in a cubic nonlinear fiber with linear losses. We consider the generation of photon pairs in the fiber when in the input of fiber is fed the pumping wave and single signal photon. The focus of attention is on three cases: when the signal photon propagates in the fiber without generating of biphotons; when the photon pair is generated; and when the photon is lost in the fiber. We also consider the cascade processes, but do not give them an analytical description because of their smallness. Description of the biphotons generation process we provide using the Schrödinger-type equation, and take into account the losses in the fiber through the introduction of the virtual beam splitters. We demonstrate the effectiveness of the generation of photon pairs through parametric processes.

1. Introduction

A lot of quantum communication systems, including quantum cryptography [1, 2] and quantum teleportation [3, 4] schemes, are based on the generation of biphoton pairs for the transmission of quantum information. Most modern quantum communication systems are using the effect of spontaneous parametric down-conversion (SPDC) as a source of biphotons [5, 6]. The generation of biphotons through effect of SPDC in $\chi^{(2)}$-nonlinear fibers has been described in [7]. However, this effect requires the fibers with the $\chi^{(2)}$-nonlinearity, whose crystal lattice has no inversion center [8]. Thus, the production of such fibers is entailed with some difficulties. Therefore, in this paper we consider the generation of photon pairs through the effect of spontaneous four-wave mixing (SFWM), which occurs in standard $\chi^{(3)}$-nonlinear fiber-optic communication lines.

For efficient flow of the SFWM process it is necessary that, along with the pump signal at frequency ω_p, in the fiber present single photon of frequency ω_s at which we want to generate the signal photons of photon pairs. The frequences of idler photons ω_i of biphotons pairs will be determined from the condition $2\omega_p = \omega_s + \omega_i$.

This paper is organized as follows. In Section 2 gives a description of the process of biphotons generation, possible variants of photon propagation in the fiber and scheme of losses. In Section 3 provides an analysis of the obtained equations. Also in this section we describe the mostly effective modes of biphotons generation for different coefficients of losses in the fiber. Summary and conclusions of the work we present in Section 4.

2. Process description

The focus of our attention is on three cases (Figure 1): 1. The signal photon propagates in the fiber without generating of biphotons; 2. The photon pair is generated; 3. The photon is lost in the fiber.

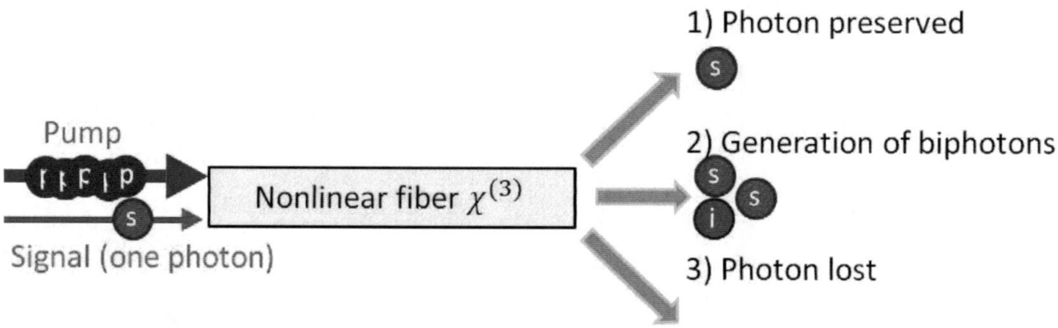

Figure 1. Possible cases of signal photon propagation through $\chi^{(3)}$-nonlinear fiber with losses in presence of pumping.

We consider the following state of a single photon propagating within the fiber:

$$|1\rangle = a_s^\dagger(\omega_s)|0\rangle. \tag{1}$$

We write the Hamiltonians for the case when it is 1 photon in the input, before their action on that photon:

$$\begin{aligned}\hat{H}_{nl}(z) &= \int d\omega_s \beta_s^{(0)}(\omega_s) a_s^\dagger(\omega_s) a_s(\omega_s) \\ &+ \int d\omega_i \beta_i^{(0)}(\omega_i) a_i^\dagger(\omega_i) a_i(\omega_i) \\ &+ \int d\omega_s \int d\omega_i [\chi^{(3)} E_p^2(z, \omega_s + \omega_i) a_s^\dagger(\omega_s) a_i^\dagger(\omega_i) \\ &\quad + \chi^{(3)} E_p^{*2}(z, \omega_s + \omega_i) a_s(\omega_s) a_i(\omega_i)],\end{aligned} \tag{2}$$

To avoid difficult calculations, we will hold the renormalization: $\chi(z) \ll 1$ to $\chi(z) = 1$, and $A \ll 1$ to $A = 1$. This renormalization will not affect the final results and the dynamics of the process. In fact, we consider a weak-pump regime $A \ll 1$, because we are interested in the generation of a single photon pair rather than cascading processes that we neglect due to their smallness.

$$\begin{aligned}\hat{H}_{bs}(z) &= \int d\omega_s \sqrt{2\gamma_s(\omega_s)} [a_s(\omega_s) b_s^\dagger(z, \omega_s) \\ &\quad + a_s^\dagger(\omega_s) b_s(z, \omega_s)] \\ &+ \int d\omega_i \sqrt{2\gamma_i(\omega_i)} [a_i(\omega_i) b_i^\dagger(z, \omega_i) \\ &\quad + a_i^\dagger(\omega_i) b_i(z, \omega_i)],\end{aligned} \tag{3}$$

$$\hat{H}(z) = \hat{H}_{nl}(z) + \hat{H}_{bs}(z). \quad (4)$$

Now we write the Hamiltonians after their action on the signal photon:

$$\begin{aligned}\hat{H}_{nl}(z)|1\rangle &= \int d\omega_s \beta_s^{(0)}(\omega_s) a_s^\dagger(\omega_s) a_s(\omega_s)|1\rangle \\ &+ \int d\omega_i \beta_i^{(0)}(\omega_i) a_i^\dagger(\omega_i) a_i(\omega_i)|1\rangle \\ &+ \int d\omega_s \int d\omega_i [\chi^{(3)} E_p^2(z, \omega_s + \omega_i) a_s^\dagger(\omega_s) a_i^\dagger(\omega_i)|1\rangle \\ &\quad + \chi^{(3)} E_p^{*2}(z, \omega_s + \omega_i) a_s(\omega_s) a_i(\omega_i)|1\rangle], \end{aligned} \quad (5)$$

$$\begin{aligned}\hat{H}_{bs}(z)|1\rangle &= \int d\omega_s \sqrt{2\gamma_s(\omega_s)} [a_s(\omega_s) b_s^\dagger(z, \omega_s)|1\rangle \\ &+ a_s^\dagger(\omega_s) b_s(z, \omega_s)|1\rangle] \\ &+ \int d\omega_i \sqrt{2\gamma_i(\omega_i)} [a_i(\omega_i) b_i^\dagger(z, \omega_i)|1\rangle \\ &+ a_i^\dagger(\omega_i) b_i(z, \omega_i)|1\rangle], \end{aligned} \quad (6)$$

$$\hat{H}(z)|1\rangle = \hat{H}_{nl}(z)|1\rangle + \hat{H}_{bs}(z)|1\rangle. \quad (7)$$

Now we write the Eqs.(5)-(7) in the simplified form, taking into account the standart properties of the operators ($a^\dagger a|n\rangle = n|n\rangle; a|0\rangle = 0; a|n\rangle = n^{1/2}|n-1\rangle; a^\dagger|n\rangle = (n+1)^{1/2}|n+1\rangle$) [9]:

$$\begin{aligned}\hat{H}_{nl}(z) \quad |1\rangle &= \int d\omega_s \beta_s^{(0)}(\omega_s)|1\rangle \\ + \quad & \int d\omega_s \int d\omega_i [\chi^{(3)} E_p^2(z, \omega_s + \omega_i) a_s^\dagger(\omega_s) a_i^\dagger(\omega_i)|1\rangle], \end{aligned} \quad (8)$$

$$\begin{aligned}\hat{H}_{bs}(z)|1\rangle &= \int d\omega_s \sqrt{2\gamma_s(\omega_s)} [a_s(\omega_s) b_s^\dagger(z, \omega_s)|1\rangle \\ &+ a_s^\dagger(\omega_s) b_s(z, \omega_s)|1\rangle] \\ &+ \int d\omega_i \sqrt{2\gamma_i(\omega_i)} [a_i(\omega_i) b_i^\dagger(z, \omega_i)|1\rangle], \end{aligned} \quad (9)$$

We seek a solution for a given state at a distance of z as (we assume that the losses are weak):

$$\begin{aligned}|\Psi(z)\rangle &= \Phi_1(z) a_s^\dagger|0\rangle + \Phi_2(z) a_i^\dagger a_s^\dagger a_s^\dagger|0\rangle \\ &+ \int_0^z dz_l \tilde{\Phi}_3^{(i)}(z, z_l) b_s^\dagger(z_l)|0\rangle \\ &+ \int_0^z dz_l \tilde{\Phi}_4^{(i)}(z, z_l) a_i^\dagger b_s^\dagger(z_l) a_s^\dagger|0\rangle \\ &+ \int_0^z dz_l \tilde{\Phi}_5^{(s)}(z, z_l) b_i^\dagger(z_l) a_s^\dagger a_s^\dagger|0\rangle \\ &+ \cdots \end{aligned} \quad (10)$$

Here we neglect terms of higher order, to avoid cumbersome writing formulas. However, we take into account that in the fiber both photons may be lost.

In our paper $E_p(z, \omega_p)$ - is a function proportional to the pump amplitude A at the frequency ω_p and inversely proportional to the cubic nonlinearity $\chi^{(3)}$. Also, $E_p(z, \omega_p)$ is normalized to the attenuation with the loss coefficient γ_p:

$$\chi^{(3)} E_p^2(z, \omega_p) = A^2 e^{-2\gamma_p z}. \tag{11}$$

According to our notation all the propagation constants β is calculated with respect to the propagation constant of the pump β_p. Therefore, the physical meaning of $\Delta\beta = \beta_s + \beta_i$ - is a mismatch of nonlinear SFWM and term $-2\beta_p$ is contained in $\Delta\beta$.

In our case the evolution of the state vector in assumptions describes by approximate Schrödinger equation for the biphoton wavefunction: $d\Psi(z)/dz \approx -i\hat{H}(z)(|0\rangle + |\Psi(z)\rangle)$.

After the necessary calculations, we obtain the following equations for the evolution of biphoton wave functions with account of losses:

$$\begin{aligned}
\partial\Phi_1(z)\partial z &= -\left(i\beta_s + \gamma_s\right)\Phi_1(z) \\
&- i2A^2 e^{-2\gamma_p z}\Phi_2(z), \Phi_1(0) = 1, && (12)\\
\partial\Phi_2(z)\partial z &= -\left(i(2\beta_s + \beta_i) + 2\gamma_s + \gamma_i\right)\Phi_2(z) \\
&- iA^2 e^{-2\gamma_p z}\Phi_1(z), \Phi_2(0) = 0, && (13)\\
\partial\tilde{\Phi}_3^{(i)}(z, z_l)\partial z &= 0, z \geq z_l, \\
&\quad \tilde{\Phi}_3^{(i)}(z_l, z_l) = -i\sqrt{2\gamma_s}\Phi_1(z_l), && (14)\\
\partial\tilde{\Phi}_4^{(i)}(z, z_l)\partial z &= -(i(\beta_s + \beta_i) + \gamma_s + \gamma_i)\tilde{\Phi}_4^{(i)}(z, z_l), \\
z &\geq z_l, 8mu\tilde{\Phi}_4^{(i)}(z_l, z_l) = -i\sqrt{2\gamma_s}\Phi_2(z_l), && (15)\\
\partial\tilde{\Phi}_5^{(s)}(z, z_l)\partial z &= -(i2\beta_s + 2\gamma_s)\tilde{\Phi}_5^{(s)}(z, z_l), \\
z &\geq z_l, 8mu\tilde{\Phi}_5^{(s)}(z_l, z_l) = -i\sqrt{2\gamma_i}\Phi_2(z_l). && (16)
\end{aligned}$$

Here we consider the pump in the form of Eq.(11).

3. An analysis of equations

Of particular interest to us are the Eqs.(12)-(14), because values $|\Phi_{1,2}(z)|^2$ and $\int\limits_0^z |\Phi_3(z)|^2 dz$ represents a probability of density of three considered cases.

Equations (12)-(13) can be solved analytically when $\gamma_p = 0$ and $\Delta\beta^2 \neq -8A^4$, where $\Delta\beta = \beta_s + \beta_i$:

$$\begin{aligned}
\Phi_1(z) &= e^{ik_0 z}\left[\cos(\Delta k z) + \frac{\tilde{\beta}_s + \tilde{\beta}_i}{2\Delta k} i \sin(\Delta k z)\right], && (17)\\
\Phi_2(z) &= -i\frac{A^2}{\Delta k} e^{ik_0 z}\sin(\Delta k z), && (18)\\
k_{1,2} &= k_0 \pm \Delta k, && (19)\\
k_0 &= \frac{-(3\tilde{\beta}_s + \tilde{\beta}_i)}{2}, && (20)\\
\tilde{\beta}_s &= \beta_s - i\gamma_s, \tilde{\beta}_i = \beta_i - i\gamma_i, && (21)\\
\Delta k &= \frac{1}{2}\sqrt{(\tilde{\beta}_s + \tilde{\beta}_i)^2 + 8A^4}. && (22)
\end{aligned}$$

In the absence of losses, when $\gamma_s = \gamma_i = 0$, we obtain $k_{1,2} = \frac{1}{2}(-\beta_3 \pm \sqrt{8A^4 + \Delta\beta^2})$, where $\beta_3 = 3\beta_s + \beta_i, \Delta\beta = \beta_s + \beta_i$. Thus, if the condition $\beta_3^2 = 8A^4 + \Delta\beta^2$ is satisfied, propagation constant k_1 is zero.

When $\gamma_p \neq 0$ the equations (12)-(13) can not be solved analytically in terms of elementary functions, so we will solve them by perturbation theory, assuming that A^2 is small:

$$\Phi_1(z) = e^{-i\tilde{\beta}_s z}, \tag{23}$$

$$\Phi_2(z) = \frac{-2iA^2}{\tilde{\beta}_s + \tilde{\beta}_i + i2\gamma_p} \sin\left[\frac{z}{2}(\tilde{\beta}_s + \tilde{\beta}_i + i2\gamma_p)\right] \times \times e^{ik_0 z} e^{-\gamma_p z}, \tag{24}$$

$$\tilde{\beta}_s = \beta_s - i\gamma_s, \tag{25}$$

$$\tilde{\beta}_i = \beta_i - i\gamma_i, \tag{26}$$

$$k_0 = \frac{-(3\tilde{\beta}_s + \tilde{\beta}_i)}{2}. \tag{27}$$

In creating the figures, we calculate the normalized intensity of the photons generated by the SFWM effect. The intensity $I_1(z) = |\Phi_1(z)|^2$ is equal to the average number of signal photons without generating photon pairs that can be detected at the coordinate z, per unit time. The intensity $I_2(z) = |\Phi_2(z)|^2$ is equal to the average number of sets of three photons (signal photon + photon pair) at coordinate z per unit time.

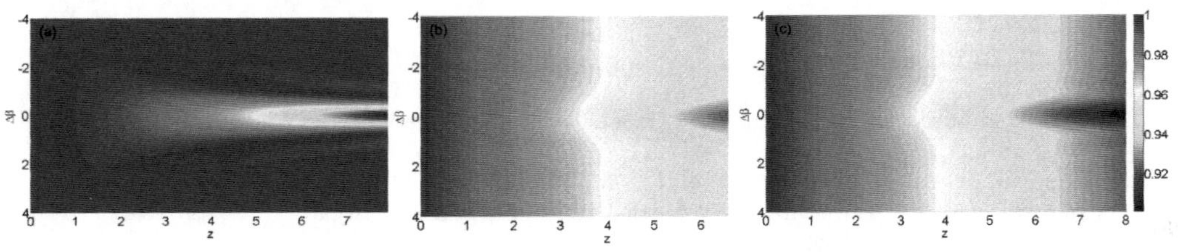

Figure 2. Normalized number of signal photons $I_1(z)$ without generating of biphotons in a single waveguide vs the phase mismatch $\Delta\beta$ and propagation distance z for $A = 0.12$, and different losses: (a) $\gamma_p = \gamma_s = \gamma_i = 0$, (b) $\gamma_p = 0, \gamma_s = \gamma_i = 0.005$, (c) $\gamma_p = \gamma_s = \gamma_i = 0.005$.

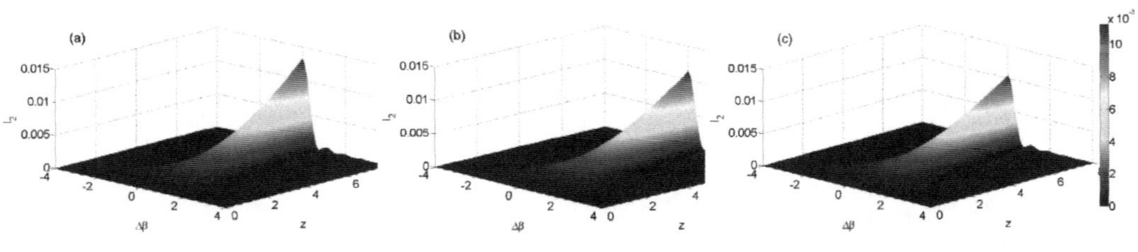

Figure 3. Normalized number of signal photons with biphotons generated through SFWM $I_2(z)$ in a single waveguide vs the phase mismatch $\Delta\beta$ and propagation distance z for $A = 0.12$, and different losses: (a) $\gamma_p = \gamma_s = \gamma_i = 0$, (b) $\gamma_p = 0, \gamma_s = \gamma_i = 0.005$, (c) $\gamma_p = \gamma_s = \gamma_i = 0.005$.

First of all, we have to provide general analyze the dependence of the intensities I_1 and I_2 of coordinate z and phase mismatch $\Delta\beta$. This dependences are presented at [Figure 2 and Figure 3]: for different values of the loss coefficients. On both figures: a). $\gamma_p = \gamma_s = \gamma_i = 0$

b) $\gamma_p = 0, \gamma_s = \gamma_i = 0.005$, a) $\gamma_p = \gamma_s = \gamma_i = 0.005$. Figure 2 is the above view to $I_1(z, \Delta\beta)$. Figure 3 is the side view to $I_2(z, \Delta\beta)$.

Note that the expressions for $\tilde{\Phi}_3^{(i)}(z, z_l)$, $\tilde{\Phi}_4^{(i)}(z, z_l)$ and $\tilde{\Phi}_5^{(s)}(z, z_l)$ can be easily obtained from the Eqs.(14)-(16).

4. Conclusion
In this paper we have performed the analysis of the effect of linear losses on SFWM in cubic nonlinear fiber, considering in detail states of single-photon, triple-photon (single-photon + biphoton) and without photon outputs under a different distances and phase mismatching. In particulary, we provide a detail analysis of a case with small losses by perturbation theory. We also demonstrate the effectiveness of the generation of photon pairs through parametric processes.

Acknowledgments
This work was financially supported by the Government of Russian Federation, Grant 074-U01.

References
[1] A. Glejm *et al.*, "Quantum key distribution in an optical fiber at distances of up to 200 km and a bit rate of 180 bit/s," Bulletin of the Russian Academy of Sciences. Physics 78 (2014).
[2] A. Gleim *et al.*, "Secure polarization-independent subcarrier quantum key distribution in optical fiber channel using BB84 protocol with a strong reference," Optics Express **24,** 2619–2633 (2016).
[3] D. Bouwmeester, J. W. Pan, K. Mattle, M. Eibl, H. Weinfurter, and A. Zeilinger, "Experimental quantum teleportation," **390,** 575–579 (1997).
[4] D. Boschi, S. Branca, F. De Martini, L. Hardy, and S. Popescu, "Experimental realization of teleporting an unknown pure quantum state via dual classical and Einstein-Podolsky-Rosen channels," Phys. Rev. Lett. **80,** 1121–1125 (1998).
[5] S. Tanzilli, H. De Riedmatten, W. Tittel, H. Zbinden, P. Baldi, M. De Micheli, D. B. Ostrowsky, and N. Gisin, "Highly efficient photon-pair source using periodically poled lithium niobate waveguide," **37,** 26–28 (2001).
[6] Q. Zhang, H. Takesue, C. Langrock, X. P. Xie, M. M. Fejer, and Y. Yamamoto, "Hong-Ou-Mandel Dip Using Degenerate Photon Pairs from a Single Periodically Poled Lithium Niobate Waveguide with Integrated Mode Demultiplexer," **49,** 064401–4 (2010).
[7] D. A. Antonosyan, A. S. Solntsev, and A. A. Sukhorukov, "Effect of loss on photon-pair generation in nonlinear waveguide arrays," Physical Review A **90,** 043845 (2014).
[8] G. P. Agrawal, *Nonlinear fiber optics* (Academic press, 2007).
[9] R. Loudon, *The Quantum Theory of Light*, 3rd ed. (Oxford University Press, New York, 2000).

Particles of porous silicon formed from silica powders of plant origin and their structural properties

I Kashko[1], K Girel[1], K Yanushkevich[2] and H Bandarenka[1]

[1] Department of Micro- and Nanoelectronics, Belarusian State University of Informatics and Radioelectronics, Minsk, 220013, Belarus
[2] Scientific and Practical Materials Research Center of Belarusian Academy of Sciences, Minsk, 220072, Belarus

Abstract. In this paper, we present the results on study of formation regularities, morphology, composition and photoluminescence of porous silicon particles fabricated by the magnesiothermal reduction of different samples of biogenic silicon dioxide based on rice husk, bamboo husk and bamboo joints at 650 °C at argon atmosphere.

1. Introduction

Porous silicon has been one of the most actively studied nanomaterials since the very beginning of its discovery [1]. PS presents the bulk silicon permeated with pores of micro- or nanometer diameters. However, the properties of PS differ a lot from those of the bulk silicon. In particular, it can behave like a direct-gap semiconductor, showing intense luminescence [2], which is due to quantum effects in nanoscale silicon structures [3]. Using PS can solve the problem of creating light-emitting diodes integrated with other electronic devices on a silicon substrate [4]. PS presents a favourable template for deposition of other semiconductors [5] and metals [6, 7] to fabricate nanomaterials with novel properties. The other outstanding feature of PS is in its biocompatibility, which is not typical for the single crystal silicon [8, 9]. This combined with PS high adsorption and biodegradation has been widely used in biomedicine to deliver nutrients and drugs into living tissue [10, 11].

Biomedical application of PS requires producing large volumes of this porous material. At the same time, the production process has to be cost-effective and simple [12]. Despite the traditional fabrication of PS by an electrochemical anodization is easily realized, it does not allow meeting these requirements because the price of silicon wafers is high enough to ensure the profitability of the production. To solve this problem, some researches have turned their attention to formation of PS by thermal reduction of silicon oxide, released from plants [12 – 14]. Many plants have a high silicon content, due to absorption of liquids from the soil containing silicic acids. The prospective plants for the PS mass production have to be fast-growing variety, which are adapted to the conditions of acid soils. Some agricultural wastes (rice, bamboo husks) and weeds (horsetail) can be used for this purpose.

At present, thermal reduction of SiO_2 is predominantly carried out in the furnaces filled with carbonaceous medium [15]. However, such a method requires high temperatures (> 2000 °C), which cause silicon sintering and pore closing. In addition, the side product of the reactions is carbon dioxide that leads to undesirable "greenhouse" effect. From the point of view of ecology, the two-step method using solar energy is very attractive [16]. The first step is a thermal treatment of SiO_2 in the presence of nitrogen to form Si_3N_4 while the second one is the solar dissociation of the nitride resulting in silicon fabrication. Nevertheless, this technique requires long processing time and thus is not effective. Alternative method is reduction of silicon dioxide in the presence of active metals such as aluminium and magnesium. The temperature in this case is much lower than at carbonaceous thermal reduction. It should be noted that using aluminium results in formation of the chemically stable aluminium oxide, which then is hardly removed from the final product. That is why more attention has been paid to the thermal reduction of SiO_2 in magnesium vapour [12 – 14]. The temperature of magnesiothermal treatment is varied in the range from 630 to 650 °C. The family of morphological types of PS particles formed by this method is very rich. Such PS has been actively studied for application in the oral delivery of drugs and nutrients, cosmetics, orthopaedics, tissue engineering, etc. Paper [17] reported development of coating which contains PS particles filled with medicines to disinfect and cure open wounds. It is very important to manage the concentration of medicines and to control pH of the covered area, which is different for injured and healthy living tissue. The PS particles can play a role of both drug carriers and sensors due to their specific optical properties [18]. However, papers on spectral sensitivity of the biogenic PS have not been published yet. It has been found that biogenic PS is a prospective candidate for use in biomedicine and cosmetics as a body powder [1]. In this case, an important requirement to the PS powder is photoluminescence in the visible range but by nowadays it has not been studied yet.

In this paper, we studied the regularities of the magnesiothermal reduction of SiO_2 based on bamboo and rice grass, as well as morphology, elemental and phase composition of the resulting PS.

2. Experimental details

Powders of SiO_2 based on rice husk, bamboo husk and joints were used to form PS. Silica powders were mixed with magnesium (Mg) powder and salt (NaCl) in the molar ratio 1:2.5:10. The salt was added to the reagents to prevent silicon sintering and thus pore closing. The mass of the mixture components was measured using balance Sartorius-CP225D with the 0.01 mg accuracy. The mixtures were poured to ceramic boat for the following thermal treatment. After that, ceramic boats were fixed in the quartz tube, which then was placed in the furnace Carbolite HST 1200/600 and filled with the inert gas (argon). Thermal treatment was performed at 650 °C for 6 h. The final powder samples cooled to the room temperature were mixed with the distilled water to the formation of suspension to which concentrated hydrochloric acid (HCl) was slowly added. The distilled water allowed to dissolute NaCl in the reaction product while HCl was used to remove magnesium oxide (MgO). The samples were kept in such a solution during 4 h. To provide complete interaction of the reagents and removal of the gas bubbles we used magnetic stirrer. Finally, we rinsed samples with the distilled water 3 times, filtered them and dried in the air at room temperature.

Morphology of the experimental samples was studied with scanning electron microscope (SEM) Hitachi S-4800 that provided resolution of 1 nm. Energy-dispersive X-ray analyser (EDX) Link Analytical AN 10000 built in electron microscope Cambridge Instruments Stereoscan-360 was used to reveal elemental composition of the samples. The phase analysis was performed with X-ray diffractometer DRON-3 equipped with copper cathode (Kα-radiation) at wavelength $\lambda = 0.5406$ nm. Voltage and anode current were 30 kV and 20 A respectively. Photoluminescence spectra of the PS were registered using complex included Xe Newport 6271 (excitation at 375 nm), focusing optical system, monochromators SOLAR DM 160 and MS 7504i. Pore size and porosity of PS was estimated with adsorption analyser of porosity and specific surface area ASAP 2020.

3. Results and discussion

Thermal reduction of silica to silicon in presence of magnesium occurs according to the following reaction:

$$SiO_2 + 2Mg \rightarrow Si + 2MgO.$$

In this work to form PS we mixed powders of biogenic silica, magnesium and salt and annealed them in the argon atmosphere at 650 oC during 6 h. It led to the formation of the composite PS/MgO, which was released from the magnesium oxide by the treatment in HCl. As a result, we obtained the powder composed of the PS particles.

The study of the structural properties of the PS powders we performed comprehensive analysis using various methods, which allowed revealing phase and elemental composition as well as morphology of the samples.

Figure 1 shows the results of the XRD-analysis of the powders formed by magnesiothermal reduction of the SiO2 powders made of rice husk, bamboo husk and bamboo joints. It is well seen, that the whole powders consist of silicon polycrystals as the spectra have the characteristic silicon peaks. There are the silicon crystallites with the hexagonal and cubic cells in the powder based on the rice husk (Figure 1, a). At the same time, we observe just the cubic cell crystallites in the powders based on the bamboo (Figure 1, b, c). The wide band of the weak intensity in the range of $2\Theta = 20 - 25$ degree typical for all the spectra evidences about minor oxidation of the silicon crystallites.

XRD-analysis gave us the information on the crystalline parts of the samples while the elemental composition of the components, which atoms are not organized in the crystalline lattice were still hidden. To solve the problem, we used EDX-analysis that allows identifying all the atoms in the sample. Electron beam of the 1 μm in diameter at the energy of 20 keV was able to penetrate to 1.3 – 1.5 μm depth of the samples. Figure 3 shows EDX-spectra of the PS powder samples based both on the rice and bamboo plants. Prevalent content of silicon and oxygen atoms is typical for all the samples. Low peaks of magnesium, natrium and chorine are also observed on the spectra. The Mg presence can be caused by the two reasons. Firstly, magnesium did not fully react with silica and was partially left in the samples. Secondly, magnesium oxide was nor completely removed from the samples during the HCl rinsing. High oxygen content can be explained by the oxidation of the silicon nanocrystals. PS is characterized by the vast specific surface area that is covered with the thin oxide film in the air. Natrium and chlorine were probably left in the samples due to poor rinsing.

It should be noticed that the greatest content of the silicon atoms (65,76 %) is typical for the PS sample based on bamboo husk while the less silicon atom percentage (38,44 %) – for the product of the bamboo joints. The initial silica powder from the bamboo joints consists of particles, which sizes are much larger than those of powders made of husks. Therefore, the specific surface area of the SiO_2 particles based on bamboo joints was not enough to provide complete interaction between the reagents during the thermal treatment and the reduction of silica to silicon was not fully realized.

SEM-analysis was used to study the morphology of the samples. Figure 4 shows SEM images of the PS powders formed by the magnesiothermal reduction of the biogenic silica samples. The samples are composed of particles riddled with pores which diameter varies in the range from 10 to 1000 nm. Comparing the results of XRD and SEM analyzes allows to conclude that the powders present a large set of the PS particles. Dimensions of the PS particles alter from 1 to more than 200 μm.

The sample of PS made of rice husk silica has the morphology slightly different from the other samples. Its silicon nanocrystals look like fibers (Figure 3, b) while the samples based on bamboo are similar to the sponge (Figure 3, d, f). Such effect takes a place as the PS samples inherit the morphology of the initial biogenic silica.

Gas adsorption analysis revealed that specific surface and average pore diameter of PS based on rice husk were 9 m^2/g and 14.4 nm respectively. The same parameters of the PS made of bamboo husk were 44 m^2/g and 7.2 nm while for the sample based on bamboo joints – 12 m^2/g and 18.5 nm. Thus, the obtained materials are of mesoporous media.

It was found that pure PS powders based on the bamboo silica (both husk and joints) demonstrated weak photoluminescence in the near-IR range. It was also possible to achieve some photoluminescence of the bamboo husk PS in the range of 400 – 550 nm by its mixing with skin sebum. The PS powder based on rice husk did not show any photoluminescence at all.

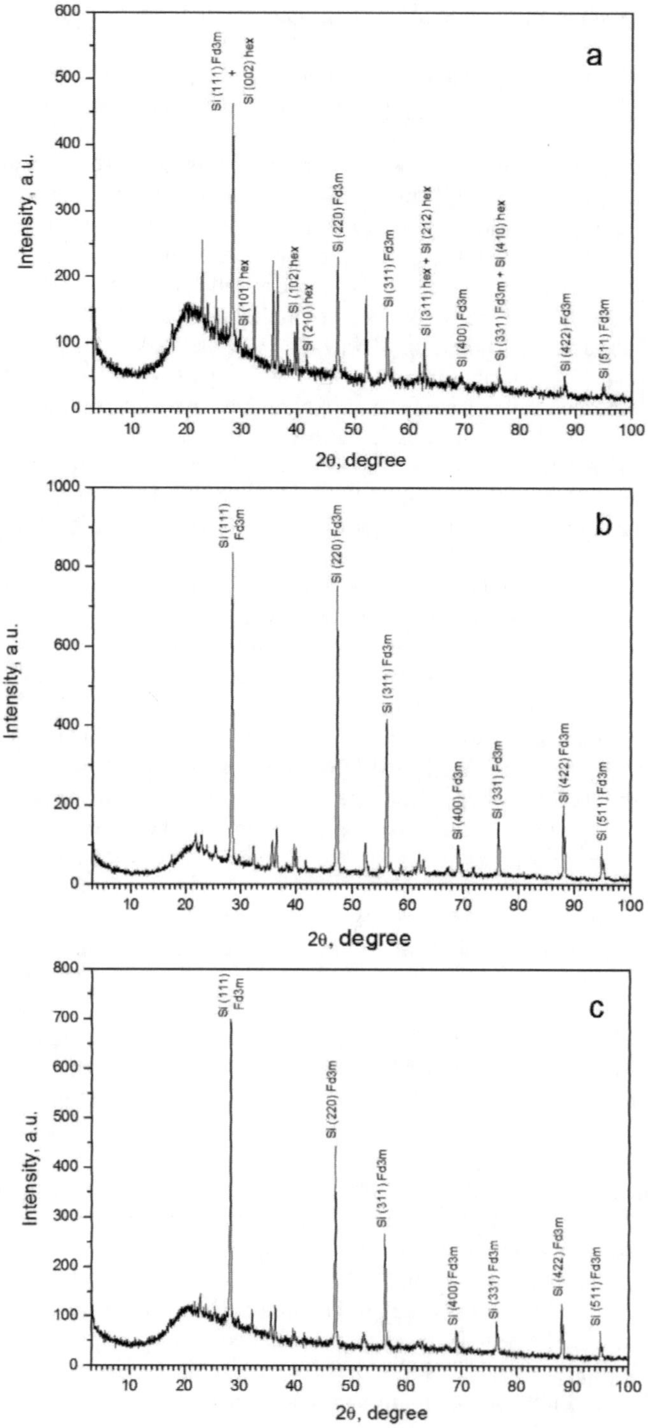

Figure 1 (a, b, c). X-ray diffractograms of the PS powder samples formed from the silica powders based on rice husk (a), bamboo husk (b) and bamboo joints (c)

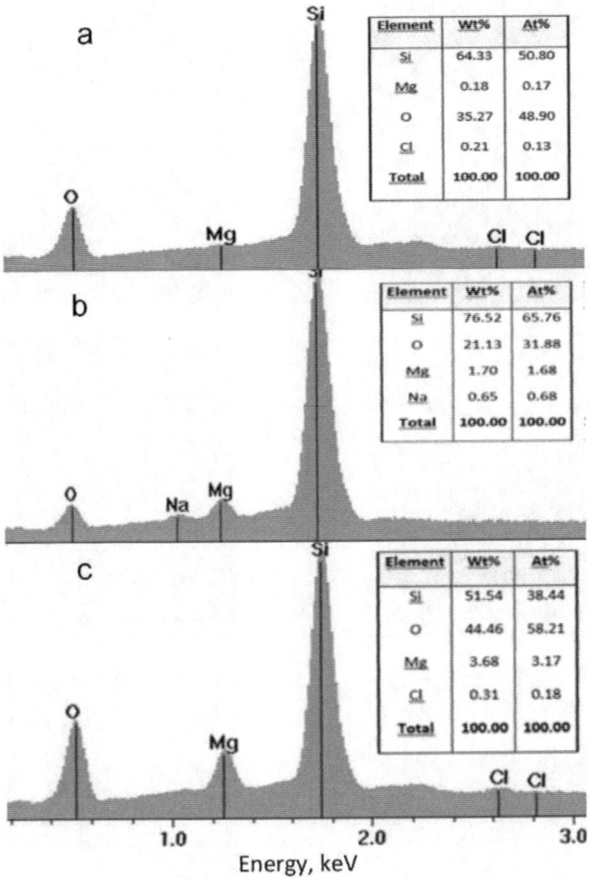

Element	Wt%	At%
Si	64.33	50.80
Mg	0.18	0.17
O	35.27	48.90
Cl	0.21	0.13
Total	100.00	100.00

Element	Wt%	At%
Si	76.52	65.76
O	21.13	31.88
Mg	1.70	1.68
Na	0.65	0.68
Total	100.00	100.00

Element	Wt%	At%
Si	51.54	38.44
O	44.46	58.21
Mg	3.68	3.17
Cl	0.31	0.18
Total	100.00	100.00

Figure 2 (a, b, c). EDX spectra of the PS powder samples formed from the silica powders based on rice husk (a), bamboo husk (b) and bamboo joints (c)

3. Conclusion
In present work, the mesoporous silicon powders were fabricated by the magnesiothermal reduction of biogenic silica powders based on the rice husk, bamboo husk and bamboo joints. It was found that the silicon skeleton of PS samples has polycrystalline nature. We suppose that the surface of the silicon skeleton of PS is highly oxidized in the air. The PS powder based on bamboo joints contains the smallest amount of the silicon atoms comparing to other PS powders due to limited access of magnesium to the whole volume of the silica particles upon thermal reduction. It was shown that the morphology of the PS particles is defined by the morphology of the initial biogenic silica. This fact opens an opportunity to vary morphology of the porous material by the choosing plant type. The possibility to form PS powders based on the biogenic silica, which can demonstrate photoluminescence, was revealed. Additional research on optimization of the reduction regimes are carried out now to achieve greater specific surface area of the PS samples and increase effectiveness of their photoluminescence.

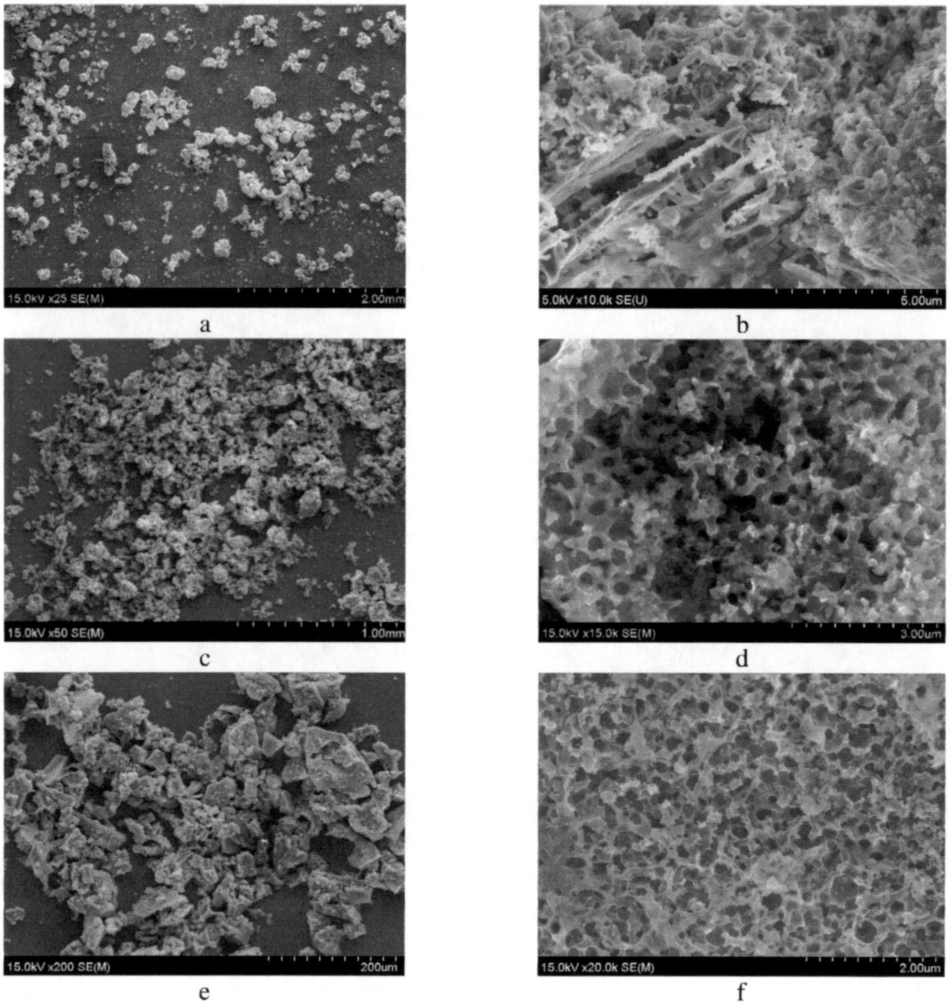

Figure 3 (a, b, c, d, e, f). SEM images of the PS powder samples formed from the silica powders based on rice husk (a, b), bamboo husk (c, d) and bamboo joints (e, f)

Acknowledgments
Authors would like to thank very much Prof. Leigh Canham for fruitful discussions on the topic. This work has been supported by the Fund for Fundamental Research and Ministry of Education of Belarus.

References

[1] Canham L 2014 *Handbook of Porous Silicon* (Springer International Publishing Switzerland)
[2] Lazarouk S, Bondarenko V, Pershukevich P, La Monica S, Maiello G, Ferrari A 1995 *Proc. MRS Symp. (Boston, MA, USA)* vol 358 pp 659-664
[3] Cullis A G, Canham L T 1991 *Nature* vol 353 pp 335-338
[4] Yakovtseva V, Dolgyi L, Vorozov N, Kazuchits N, Bondarenko V, Balucani M, Lamedica G, Franchina L, Ferrari A 2000 *J Por Mat* vol 7 (1) pp 215-222
[5] Chubenko E B, Klyshko A A, Petrovich V A, Bondarenko V P 2009 Thin Solid Films vol 517(21) pp 5981-5987

[6] Bandarenka H, Redko S, Smirnov A, Panarin A, Terekhov S, Nenzi P, Balucani M, Bondarenko V 2012 *Nanoscale Res Lett* vol 7(1) 477
[7] Dolgyi A, Bandarenka H, Prischepa S, Yanushkevich K, Nenzi P, Balucani M, Bondarenko V 2012 ECS Trans 41(35) pp 111-118
[8] Canham L T 1995 *Adv Mater* vol 7(12) pp 1033-1036
[9] Low S P, Voelcker N H, Canham L T, Williams K A 2009 *Biomaterials* vol 30 pp 2873-2880
[10] Anglin E J, Cheng L, Freeman W R, Sailor M J 2008 *Adv Drug Deliv Rev* vol 60 pp 1266-1277
[11] Salonen J, Laitinen L, Kaukonen A M, Tuura J, Björkqvist M, Heikkilä T, Vähä-Heikkilä K, Hirvonen J, Lehto V-P 2005 *J Control Release* vol 108 pp 362-374
[12] Batchelor L, Loni A, Canham L T, Hasan M, Coffer J L 2012 *Silicon* vol 4 pp 259-266
[13] Bao Z, Weatherspoon M R, Shian S, Cai Y, Graham P D, Allan S M, Ahmad G, Dickerson M B, Church B C, Kang Z, Abernathy III H W, Summers C J, Liu M, Sandhage K H 2007 *Nature* vol 446(8) pp 172-175.
[14] Xing A, Tian S, Tang H, Losic D, Bao Z 2013 *RSC Adv* vol 3 pp 10145-10149
[15] Koch E C, Clement D 2007 *Propell Explos Pyrotech* vol 32(3) pp 205-212
[16] Murray J P, Flamant G, Roos C J 2006 *Solar Energy* vol 80(10) pp 1349-1354.
[17] Pace S, Vasani R B, Zhao W, Perrier S, Voelcker N H 2014 *Nanoscale Res Lett* vol 9(1) 420
[18] McInnes S J P, Szili E J, Al-Bataineh S A, Xu J, Alf M E, Gleason K K, Short R D, Voelcker N H 2012 *Appl Mater Interfaces* vol 4 pp 3566-3574

Indium phosphide all air-gap Fabry-Pérot filters for near-infrared spectroscopic applications

A Ullah[1], M A Butt[2], S A Fomchenkov[2,3] and S N Khonina[2,3]

[1]Department of Electronic Engineering, Balochistan University of Information Technology, Engineering and Management Sciences (BUITEMS), Takatu Campus, Airport Road, Quetta, 87300, Pakistan
[2] Samara State Aerospace University, Moskovskoye Shosse 34, Samara, 443086, Russia
[3]Image Processing Systems Institute of the Russian Academy of Sciences, Molodogvardeyshaya 151, Samara, 443001, Russia

Abstract. Food quality can be characterized by noninvasive techniques such as spectroscopy in the Near Infrared wavelength range. For example, 930 –1450 nm wavelength range can be used to detect diseases and differentiate between meat samples. Miniaturization of such NIR spectrometers is useful for quick and mobile characterization of food samples. Spectrometers can be miniaturized, without compromising the spectral resolution, using Fabry-Pérot (FP) filters consisting of two highly reflecting mirrors with a central cavity in between. The most commonly used mirrors in the design of FP filters are Distributed Bragg Reflections (DBRs) consisting of alternating high and low refractive index material pairs, due to their high reflectivity compared to metal mirrors. However, DBRs have high reflectivity for a selected range of wavelengths known as the stopband of the DBR. This range is usually much smaller than the sensitivity range of the spectrometer detector. Therefore, a bandpass filter is usually required to restrict wavelengths outside the stopband of the FP DBRs. Such bandpass filters are difficult to design and implement. Alternatively, high index contrast materials must be can be used to broaden the stopband width of the FP DBRs. In this work, Indium phosphide all air-gap filters are proposed in conjunction with InGaAs based detectors. The designed filter has a wide stopband covering the entire InGaAs detector sensitivity range. The filter can be tuned in the 950–1450 nm with single mode operation. The designed filter can hence be used for non-invasive meat quality control.

1. Introduction

Near Infrared (NIR) Reflectance Spectroscopy in the ~930–1450 nm wavelength range, can be used for both, detection of diseases and noninvasive quality control, of food samples, such as meat, vegetables and wheat [1], [2], [3], [4], [5]. Miniaturized of such spectrometers will provide quick and mobile characterization of food samples. This can help health inspectors to detect the food quality on the spot rather than taking samples back to the laboratory, thus saving time and resources. Such miniaturized spectrometers can be built using Fabry-Pérot (FP) filters. Such filters consist of two highly reflecting mirrors with a central cavity between them, figure 1. The central cavity governs the output wavelength of the FP filter. Distributed Bragg Reflectors (DBRs) are the most commonly used mirrors in FP filters, due to their high reflectivity. DBRs consist of alternating layers of high and low refractive index materials. DBRs have high reflectivity for wavelengths around a central wavelength, which is governed by the optical thickness (refractive index × physical thickness) of the constituent layers, and is four times their optical thickness. DBR layers are therefore a quarter–wave thick of the center wavelength. The high reflection region of a DBR is known as the DBR stopband, and is governed by the refractive index contrast between the constituent layers. A broad stopband is desired, which can be achieved by a high index contrast between the DBR materials. Combining two DBRs to form a FP filter will result in a sharp transmission line inside the stopband of the DBRs. For FP filters with air-gap cavities, the transmission line is twice the cavity thickness. The transmission line can be tuned within the DBR stopbands by varying the cavity thickness. This can be achieved by electrostatic actuation of two electrodes placed at the two DBRs of the filter. An increased actuation voltage will result in an increased attractive force between the two DBRs. This will result in a reduced cavity and hence a shifted transmission line.

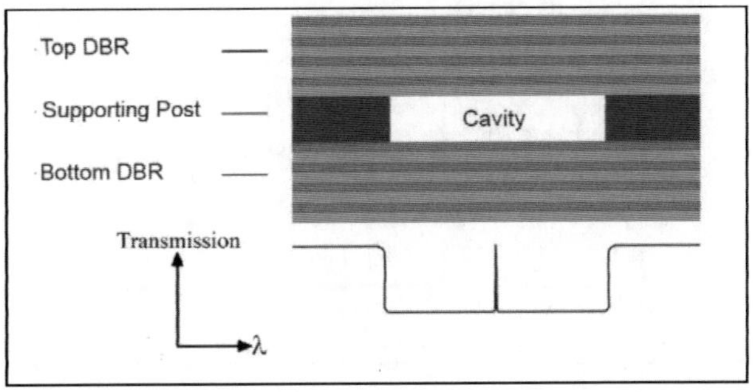

Figure 1. A FP filter with two DBRs and a cavity between them. A DBR consists of alternating pairs of high and low refractive index materials (shown in grey and green colors). Shown is also the transmission spectrum of the filter.

An exemplary NIR spectrometer is shown in figure 2. It consists of a NIR source, a FP filter and a detector.

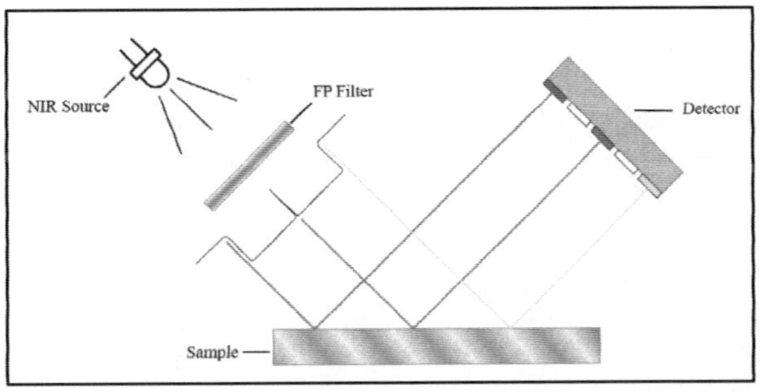

Figure 2. An exemplary NIR spectrometer. It consists of an NIR source, a FP filter and a detector. Wavelengths outside the stopband of the FP DBRs are also detected in this setup.

A detector usually has a much wider detection range compared to the stopband of the FP filter DBRs, figure 3. Indium gallium arsenide (InGaAs) based detectors provide the ideal detection range for the specified wavelength range. Indium phosphide (InP) and silicon dioxide (SiO_2) based DBRs provide a wide stopband at NIR wavelengths. However, the detection range of the InGaAs detector is still wider than the stopband of the FP filter DBRs, figure. 3. This will result in unwanted wavelengths reaching the detector, causing erroneous measurements. Therefore, the wavelengths outside the FP filter DBRs must be stopped from reaching the detector. However, InP all air-gap filters have a much wider DBR stopband. A carefully selected central wavelength could result in covering the whole range of the detector, therefore completely eliminating the need of any bandpass filter.

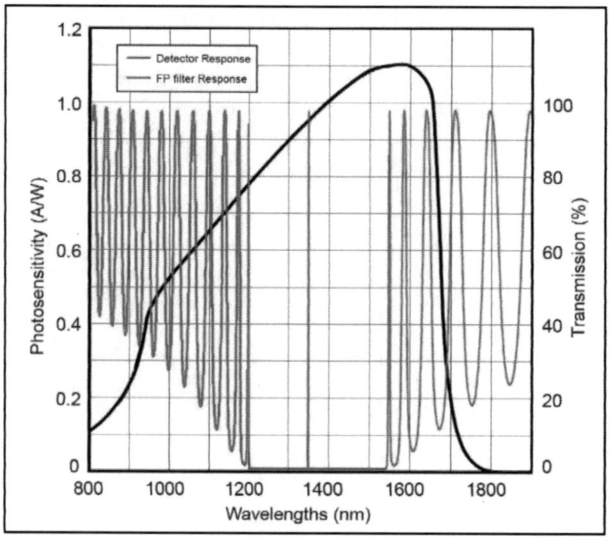

Figure 3. Photosensitivity of an InGaAs based detector (black) [6] and simulated transmission response of a FP filter (red) composed of InP and SiO_2. Wavelengths outside the stopband of the FP DBRs are also detected.

Although InP all air-gap filters have been developed for telecom applications [7], agri-food and environmental processing [8], no work has been reported for the above mentioned wavelength range (~930–1450 nm) to the best of our knowledge.

2. Design Requirements

As discussed earlier, the wavelengths of interest for classification and defect detection of meat samples are in the range ~930–1450 nm, while InGaAs based detectors are highly sensitive in the ~900–1700 nm range. Since, InP is highly absorbing below 950 nm, therefore only filters with passband above 950 nm are possible. Therefore, a FP filter with a stopband of 900–1700 nm is required which can be easily tuned in the 930–1450 nm range. The transmission line with in this range should ideally be more than 90%. Another requirement is that the filter must have single mode of operation for the entire tuning range (930–1450 nm). The line-width which is measured at half the maximum transmission should be as low as possible, ideally less as 2 nm.

3. Filter Design and Discussion

As discussed earlier, the central wavelength of the FP DBRs defines the stopband and the tuning range of the filter. Therefore, choosing the right central wavelength is essential to the filter design. An open source software *OpenFilters* is used to vary the central wavelength of the FP DBRs and to measure the output spectrum. The refractive index of InP in the wavelengths of interest used for the design of the filter is as in figure 4.

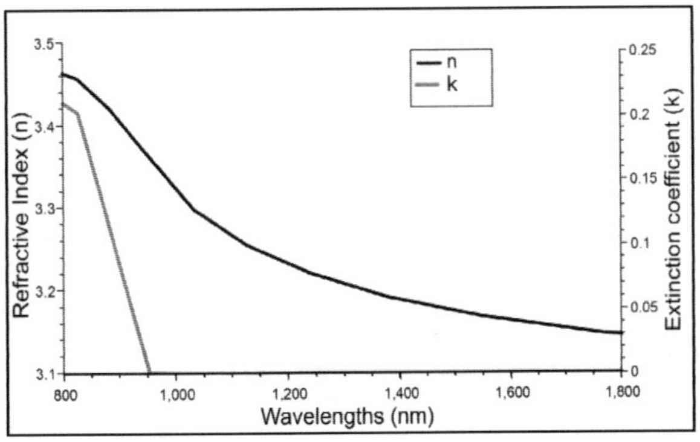

Figure 4. Refractive indexes n (black) and extinction coefficient k (red) of InP.

The advantage of using InP all air-gap filters is that they require only a few alternating layers to implement, due to the large refractive index contrast between the constituent layers. The filter with a DBR central wavelength of 1200 nm resulted in a stopband covering the whole range of the detector sensitivity, figure 5. The filter consists of only 5 InP/air-gap layers for each DBR, and therefore is easy to implement.

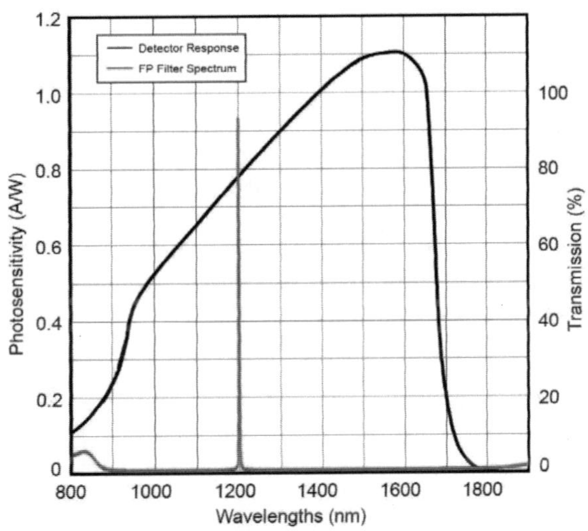

Figure 5. Photosensitivity of InGaAs based detector [6] and simulated transmission response of a FP filter (red) with a DBR central wavelength of 1200 nm. The stopband of the FP DBRs is wide enough to block the entire sensitivity range of the detector.

The InGaAs detector has very low sensitivity below 900 nm. Therefore, the small transmission of the filter below 900 nm can be ignored. For highly sensitive applications, a color filter (absorption filter) can be used as a substrate to eliminate those wavelengths. The filter is easily tuned between 950–1450 nm and beyond while maintaining single mode of operation, figure 6. The maximum and minimum transmission within in this range is 99% and 97% respectively. While the minimum and maximum line-widths within this range are 3.1 nm and 6 nm respectively. The minimum line-width occurs at a cavity thickness equal to the DBRs central wavelength, while the maximum line-width occurs at highest measured wavelength, i.e. 1450 nm.

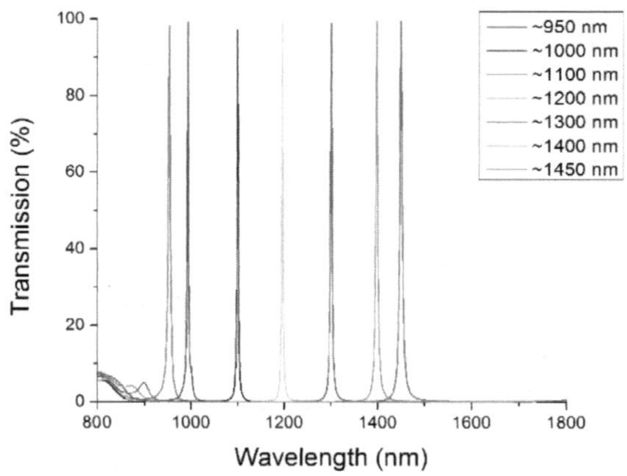

Figure 6. Transmission spectrum of an InP all air-gap FP filter (having DBR central wavelength of 1200 nm) with different cavity thicknesses.

4. Conclusions

In this work an InP all air-gap FP filter is presented to be used as an NIR spectrometer for food quality control. The filter has only 5 InP and air-gap layers for each DBR of the FP filter and hence can be implemented economically. The filter DBRs have a stopband covering the whole range of an InGaAs detector (~900–1700 nm), eliminating the need for an extra bandpass filter. The filter can be easily tuned in the 950–1450 nm wavelength range with more than 97% of transmission and a line-width not more than 6 nm. The best values of transmission and line-width, within the mentioned range, being 99% and 3.1nm respectively. The designed filter can easily be used for non-invasive quality control of food items, namely meat samples.

Acknowledgements

The authors would like to acknowledge Balochistan University of Information Technology, Engineering and Management Sciences (BUITEMS), Quetta, Pakistan for the research environment provided. The authors would also like to thank the colleagues at Samara State University, Russia and RED (Electronic Devices Research group) BUITEMS for the fruitful discussions and encouragement.

References

[1] Huang H, Liu L and Ngadi M 2014 *Sensors* **14** 7248
[2] Kamruzzaman M, Elmasry G, Sun D W and Allen P 2011 *J. Food Eng.* **104** 332
[3] Pedersen D K, S. Morel, H. J. Andersen, and S. B. Engelsen 2003 *Meat Sci.* **65** 581
[4] ElMasry G, Sun D W and Allen P 2011 *Food Res. Int.* **44** 2624
[5] ElMasry G, Sun D W and Allen P 2012 *J. Food Eng.* **110** 127
[6] Hamamatsu 2016 *InGaAs PIN photodiodes* pp 9–11
[7] Kusserow T, Irmer S, Nethaji D and Hillmer H 2009 *Photonic International* **1**, pp 14–16
[8] Garrigues M, Leclrecq J, Gil-sobraqués R, Parillaud O, Crochon M, et al. 2007 Proc. of MOEMS and Miniaturized Systems vol 6 (California: SPIE) p 646607

Biomedical bandpass filter for fluorescence microscopy imaging based on TiO_2/SiO_2 and TiO_2/MgF_2 dielectric multilayers

M A Butt[1], S A Fomchenkov[1,2], A Ullah[3], P Verma[1] and S N Khonina[1,2]

[1]Samara State Aerospace University, Moskovskoye Shosse 34, Samara, 443086, Russia

[2]Image Processing Systems Institute of the Russian Academy of Sciences, Molodogvardeyshaya 151, Samara, 443001, Russia

[3]Department of Electronic Engineering, Balochistan University of Information Technology, Engineering and Management Sciences (BUITEMS) Quetta, Pakistan

Abstract. We report a design for creating a multilayer dielectric optical filters based on TiO_2 and SiO_2/MgF_2 alternating layers. We have selected Titanium dioxide (TiO_2) for high refractive index (2.5), Silicon dioxide (SiO_2) and Magnesium fluoride (MgF_2) as a low refractive index layer (1.45 & 1.37) respectively. Miniaturized visible spectrometers are useful for quick and mobile characterization of biological samples. Such devices can be fabricated by using Fabry-Perot (FP) filters consisting of two highly reflecting mirrors with a central cavity in between. Distributed Bragg Reflectors (DBRs) consisting of alternating high and low refractive index material pairs are the most commonly used mirrors in FP filters, due to their high reflectivity. However, DBRs have high reflectivity for a selected range of wavelengths known as the stopband of the DBR. This range is usually much smaller than the sensitivity range of the spectrometer range. Therefore a bandpass filters are required to restrict wavelength outside the stopband of the FP DBRs. The proposed filter shows a high quality with average transmission of 97.4 % within the passbands and the transmission outside the passband is around 4 %. Special attention has been given to keep the thickness of the filters within the economic limits. It can be suggested that these filters are exceptional choice for florescence imaging and Endoscope narrow band imaging.

1. Introduction

Thin film optics is well developed technology. Therefore, many devices such as passband filters, stopband filters, polarizers and reflectors are realized with the help of multilayer dielectric thin films [1]. These devices consist of alternating layers of high and low refractive index materials with particular thicknesses with good knowledge of their refractive index and absorption. These devices work on the principle of multiple reflections between high and low index materials interface. Distributed Bragg Reflectors (DBRs) are quarter wave thick of the center wavelength. The high reflection region of a DBR is known as the DBR stopband, and can be obtained by the refractive index contrast between the constituent layers. A broad stopband can be realized by using high index contrast thin films. By inserting a half wave cavity between two highly reflecting mirrors governs an output at the design wavelength [1]. This kind of device is known as Fabry-Perot (FP) filter. DBRs are the most commonly used mirrors in FP filters, due to their high reflectivity. FP has numerous applications in laser line filtering, astronomical observations, fluorescence microscope imaging and spectroscopic instrumentation.

In two photon fluorescence microscopy imaging, a fluorophore absorbs light at its excitation wavelength, and typically emits light at a longer wavelength. The emission spectra of the biological samples has high emission peak at 511 nm which is easy to detect [2, 3, 4]. Therefore, there is a need to develop such filter which can selectively transmit the emitted wavelength and block the excitation wavelength, thus improving the contrast for both sensing and imaging these fluorophores. DBRs consist of alternating layers of high and low refractive index materials. DRBs have high reflectivity for wavelengths around a central wavelength, which is governed by the optical thickness (refractive index x physical thickness) of the constituent layers, and in four times their optical thickness. DBR layers are therefore a quarter-wave thick of the central wavelength.

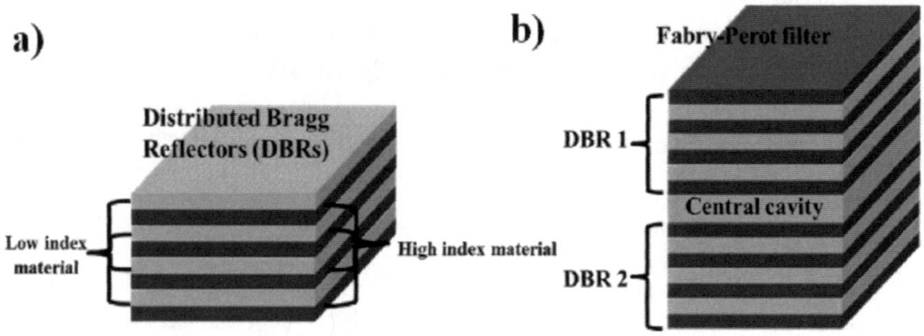

Figure 1. Schematic diagram of (a) Distributed Bragg Reflectors, (b) FabryPerot filter

2. Filter design and discussion

In this work, the designs of Fabry-perot filters based on TiO_2/SiO_2 and TiO_2/MgF_2 are proposed at a central wavelength of 511 nm for biomedical bandpass filter for fluorescence microscopy imaging. We tried to design the filters with less number of layers and high transmission peak at central wavelength with narrow width (FWHM). Open-source software, OpenFilters, is used in this work to design and optimize the required filter.

2.1. Filter 1: Fabry perot filter design based on TiO_2/SiO_2

TiO_2 and SiO_2 are chosen as high and low refractive index materials, respectively. The choice of materials is made on the basis of low absorption and high index contrast in the wavelengths of interest. The FP filter is designed for visible spectrum ranges from 390-700 nm. In our first design, we proposed a spacer (SiO_2) embedded between two DBRs consisting of 7 layers each. The optimized thickness of the layers is shown in table I. The total thickness of the filter is 1117 nm.

Table I. Layer thickness of TiO_2/SiO_2 based Fabry Perot filter with a passband of 511 nm.

Layer no.	Materials	Thickness (nm)	Layer no.	Materials	Thickness (nm)
1	TiO_2	53	9	TiO_2	53
2	SiO_2	85	10	SiO_2	85
3	TiO_2	53	11	TiO_2	53
4	SiO_2	85	12	SiO_2	85
5	TiO_2	53	13	TiO_2	53
6	SiO_2	90	14	SiO_2	85
7	TiO_2	53	15	TiO_2	53
8	SiO_2	178		-	-

2.2. Filter 2: Fabry perot filter design based on TiO_2/MgF_2

TiO_2 and MgF_2 (1.37) are chosen as high and low refractive index materials, respectively. The choice of materials is made on the basis of low absorption and high index contrast in the wavelengths of interest. The FP filter is designed for visible spectrum ranges from 390-700 nm. In our second design, we proposed a spacer (MgF_2) embedded between two DBRs consisting of 7 layers each. The optimized thickness of the layers is shown in table II. The total thickness of the filter is 1171 nm. The transmission spectra in the visible range of the filter shows a characteristic peak with a maximum of 99.4 % at the reference wavelength of 511 nm. The width (FWHM) of the transmission spectra is observed at 3.5 nm.

Table II. Layer thickness of TiO_2/SiO_2 based Fabry Perot filter with a passband of 511 nm.

Layer no.	Materials	Thickness (nm)	Layer no.	Materials	Thickness (nm)
1	TiO_2	53	9	TiO_2	53
2	MgF_2	90	10	MgF_2	90
3	TiO_2	53	11	TiO_2	53
4	MgF_2	85	12	MgF_2	90
5	TiO_2	53	13	TiO_2	53
6	MgF_2	90	14	MgF_2	60
7	TiO_2	53	15	TiO_2	53
8	MgF_2	189		-	-

The transmission spectra of both designs of FP filters are shown in figure 2. Black and red color represents the transmission response of the design 1 design 2 respectively. It can be observed that design 2 provides much broader stopband region in visible region and high transmission peak at 511 nm as compared to design 1 due to high refractive index contrast. The transmission spectra in the visible range of the filter shows a characteristic peak with a maximum of 94.4 % and 99.4 % with width (FWHM) of 4 and 3.5 nm for design 1 and 2 respectively.

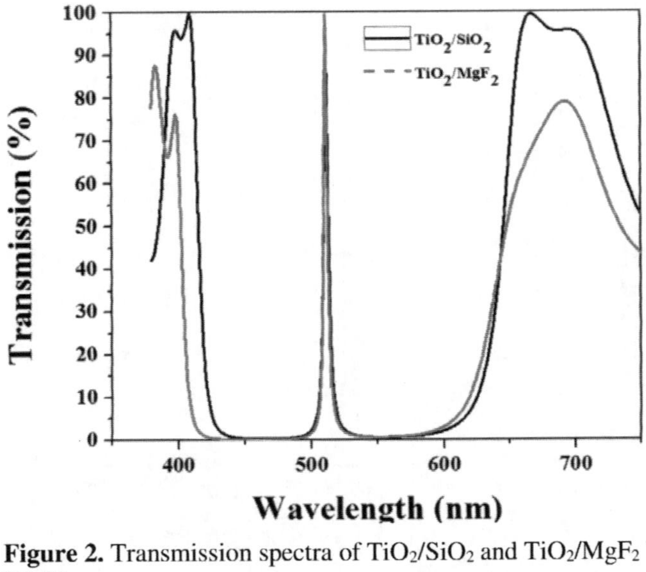

Figure 2. Transmission spectra of TiO_2/SiO_2 and TiO_2/MgF_2 based FP filters.

2.3. Filter 3: DBR filters design based on TiO_2/SiO_2

In order to obtain the desired specific wavelength in full visible region, two DBRs with 9 layers each were designed to stop the undesired band of wavelengths. TiO_2 and SiO_2 are chosen as high and low refractive index materials, respectively. The total thickness of DBR 1 and DBR 2 is 455 and 832 nm respectively. The combined effect of all three filters fulfills the requirement of fluorescence microscopic imaging and can transmit the excitation wavelength while suppressing all the wavelengths outside the stopband of the FP DBRs. The optimized layer thicknesses of both DBRs are shown in table III.

Table III. Layer thickness of bandpass filters based on TiO_2/SiO_2.

DBR1: Stopband below 450 nm			**DBR2: Stopband above 550 nm**		
Layer no.	Materials	Thickness (nm)	Layer no.	Materials	Thickness (nm)
1	TiO_2	25	1	TiO_2	83
2	SiO_2	70	2	SiO_2	86
3	TiO_2	35	3	TiO_2	93
4	SiO_2	70	4	SiO_2	103
5	TiO_2	45	5	TiO_2	93
6	SiO_2	70	6	SiO_2	98
7	TiO_2	30	7	TiO_2	85
8	SiO_2	90	8	SiO_2	103
9	TiO_2	20	9	TiO_2	88

The transmission spectrums of the combined filters (filter 1 + filter 3) and (filter 2+filter 3) are shown in figure 3. Black and red color represents the transmission response of (filter 1+ filter 3) and (filter 2 + filter 3) respectively. It is well noted that that the overall transmission spectra in the visible range of the spectrum shows a characteristic peak at 511 nm with a maximum of 92.4 % and 97.4 % .These combined filters reduced the transmission outside the passband up to 4%.

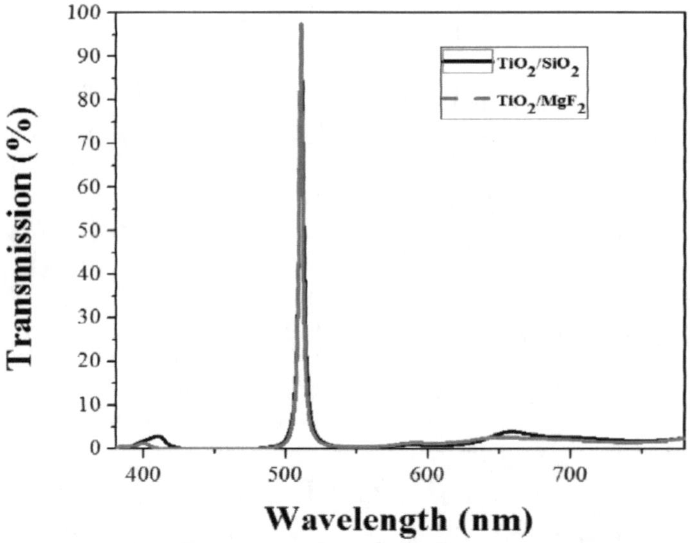

Figure 3. Over all transmission spectra of (filter 1+ filter 3) and (filter 2 + filter 3) indicating a peak at 511 nm in full visible spectrum.

3. Conclusions

In this work, we presented two FP filter designs based on TiO_2/SiO_2 and TiO_2/ MgF_2 combined with two DBR filters based on TiO_2/SiO_2. These filters provide a peak transmission of 92.4 and 97.4 % at 511 nm. By combining two DBRs with FP filter, the transmission outside the passband reduced to 4 %. The combined effect of all three filters fulfils the requirement of fluorescence microscopic imaging and can transmit the excitation wavelength while suppressing all the wavelengths outside the stopband of the FP DBRs.

Acknowledgments

This work was supported by the Ministry of Education and Science of the Russian Federation and the Russian Foundation for Basic Research (grant No. 14-07-00177). The authors would also like to thank the colleagues at BUITEMS, Quetta, Pakistan for the fruitful discussions and encouragement.

References

[1] Macleod 1989 H A Thin film Optical filters (McGraw-Hill)

[2] Nirmala R 2000 Neoplasia **2** 89

[3] Grazyna P, Zhiqian D, Marcin G, Jennifer J H, David R W, Nathan S A and Krzsztof P 2014 Nat Med **20** 785

[4] Spiess E, Bestvater F, Heckel P A, Toth K, Hacker M, Stobrawa G, Feurer T, Wotzlaw C, Berchner P U, Porwol T and Acker H 2004 *J. of Mic.* **217** 200

Inversion of the topological charge of optical vortices in a coil fibre resonator

C N Alexeyev, M C Alexeyeva, B P Lapin, A V Milodan and O. V. Vishnevskaya

V.I. Vernadsky Crimean Federal University, Vernadsky Prospekt, 4, Simferopol, 295007, Russian Federation.

Abstract. We have analytically studied the evolution of optical vortices in a fibre resonator with two coils in the presence of an inter-coil coupling. We have shown that such a system can invert the orbital number of the incoming optical vortices if the resonance condition is fulfilled.

1. Introduction

As is known, in optical fibres light can be made to interact with itself in certain regions. This phenomenon is present in directional couplers as well as fibre loop resonators [1]. If a microfibre is wound onto a coil, light in adjacent coils gets coupled through decaying fields, which leads to specific properties of coil resonators [2] relevant for telecommunications, for designing delay lines [3]. The study of fibre-based resonators can be relevant in connection with nonlinear optics [4]. They can also prove useful while creating optical gyroscopes [5]. Coil resonators can serve as a platform for demonstration of Berry's phase [6]. These research, however, are connected with propagation of regular lower-order modes in fibre resonators. Recently, a large amount of interest has been evoked owing to communications on the base of the signals, which carry orbital angular momentum (OAM) [7]. In this connection, it is interesting to study propagation of modes with higher orbital numbers in coil resonators. In this paper we study the evolution of optical vortices (OVs) in a two-coil microfibre resonator. We show that such a coil resonator can to invert the sign of the topological charge of the incident OV if the resonance condition is fulfilled. We also demonstrate that such a system is able to change the OAM of the transmitted field.

2. Conversion of OVs in a coil resonator

Consider a two-coil resonator, as shown in Figure 1. Instead of building the equations in slow amplitudes, which are essential in coupled mode theory,we use the technique based on expressions for coupled modes of parallelfibres. To this endweplot an equivalent scheme of the system in study (see Figure 2). The two-coil optical snake can be regarded as two parallel opticalfibres. To achieve the correspondence with our system we connect the output end of one of such fibres with the input end of the other fibre. In this configuration the output field of the first fibre provides the boundary condition to the second fibre.

In this way, we bring down the problem of a two-coil optical snake to the problem of coupled fibres. The expressions forhigher-order normal modes of parallelfibres, which allow for the spin-orbit interaction (SOI),were obtained in [8]. In the scalar approximation one should set the SOI constants A_l, B_l to zero. It is convenient to introduce the following combinations:

$$|l,L/R\rangle+|-l,L/R\rangle\equiv|E,L/R\rangle,\quad |l,L/R\rangle-|-l,L/R\rangle\equiv|O,L/R\rangle, \tag{1}$$

where $|l, L/R\rangle = \exp(il\varphi_{L/R})F_l(r_{L/R})$ and the subscripts L/R denote that axial-polar coordinates (r,φ) are measured with respect to centres of left (*L*) or right (*R*) fibres. Here F_l is the radial function of the ideal fibre mode of order *l*, where $l = 0,1,2...$. In terms of these combinations the expressions for normal modes of parallel fibres can be expressed in the scalar approximation as follows:

$$|\psi_{1,3}\rangle = (|E,L\rangle \pm |E,R\rangle),\ |\psi_{2,4}\rangle = (|O,L\rangle \pm |O,R\rangle). \tag{2}$$

The propagation constants for these modes look like:

$$\beta_1 = \tilde{\beta}_l + (C_l + D_l)/2\tilde{\beta}_l,\ \beta_3 = \tilde{\beta}_l - (C_l + D_l)/2\tilde{\beta}_l,$$
$$\beta_2 = \tilde{\beta}_l + (C_l - D_l)/2\tilde{\beta}_l,\ \beta_4 = \tilde{\beta}_l + (-C_l + D_l)/2\tilde{\beta}_l, \tag{3}$$

where $\tilde{\beta}_l$ is the scalar propagation constant, C_l and D_l are corresponding exchange integrals determined in [8]. Now we are in a position to study the evolution of arbitrary fields in a two-coil optical snake.

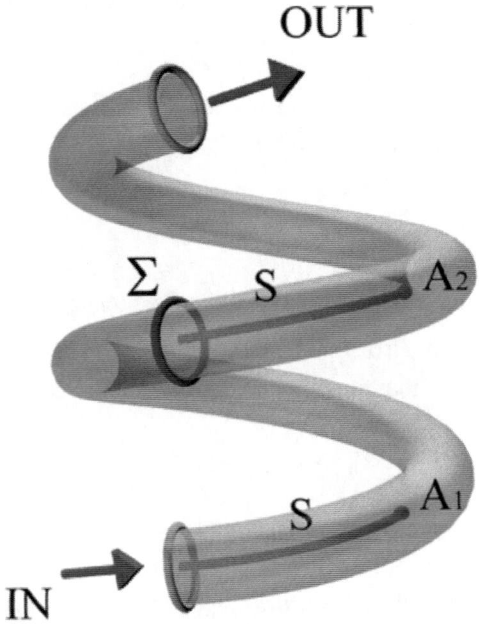

Figure 1. The scheme of a two-coil optical snake. Σ marks the boundary between the coils. Coordinates s of the points A_1 and A_2 are the same

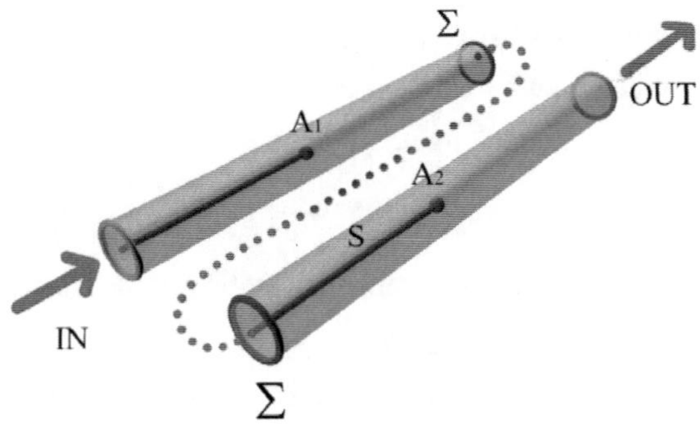

Figure 2. An equivalent scheme of a two-coil optical snake. Red dotted line shows actual path of light from input to output ends.

Following [2,3] we define in each of coilsthe length of the fibre's central line*s*(Figure 1). The value of *s* is chosen to beidentical for the points with the same angular coordinates measured in the frame connected with the optical snake's axis. In this manner, *s*is the position in an individual coil andsatisfies $0 \le s \le s_0$. Here s_0 is the coil's length. Then we use the expressions for normal modes (2). We note that the field in the *m*-th coil ($m = 0,1$)is given by the combination:

$$\Psi_m = \sum_{i=1}^{4} C_i \left|\psi_i^{(m)}\right\rangle \exp(i\beta_i s). \tag{4}$$

Here the upper index in $\left|\psi_i^{(m)}\right\rangle$ specifies that in the corresponding formula for the normal mode $|\psi_i\rangle$ one should retain only the terms, which are related to the coil with index *m*. In this way, for example, the expressions for modes $\psi_3^{(1)}$ and $\psi_3^{(2)}$ reduce as: $\psi_3^{(1)} = (|l,L\rangle + |-l,L\rangle) \equiv |l\rangle + |-l\rangle$ and $\psi_3^{(2)} = (-|l,R\rangle - |-l,R\rangle) \equiv -|l\rangle - |-l\rangle$. Note that in the final expressions we omit the marker $L(R)$ since it is redundant in a chosen coil (1) or (2).

Let us study the evolution of an OV with topological charge l excited at the input end of the two-coil optical snake. In terms of the introduced combinations $|E/O\rangle$ the expression for such an OV reads as:

$$\Psi_{in} = |l\rangle = \frac{1}{2}(|E\rangle + |O\rangle). \tag{5}$$

The boundary condition at the input end leads to the following identity:

$$\frac{1}{2}(|E\rangle + |O\rangle) \equiv (C_1 + C_3)|E\rangle + (C_2 + C_4)|O\rangle. \tag{6}$$

One should take into consideration that the field at the end of the first coil gives the input field for the second coil. This gives the second identity:

$$(C_1 e^{i\beta_1 s_0} + C_3 e^{i\beta_3 s_0})|E\rangle + (C_2 e^{i\beta_2 s_0} + C_4 e^{i\beta_4 s_0})|O\rangle \equiv (C_1 + C_3)|E\rangle + (C_2 + C_4)|O\rangle. \tag{7}$$

Equations (6), (7) enables us to obtain the equations in the unknown decomposition coefficients C_i. The outcoming field then can beobtained as decomposition overvortices $|\pm l\rangle$ as:

$$\begin{aligned}\Psi_{out} &= |l\rangle\left\{C_1 e^{i\beta_1 s_0} - C_3 e^{i\beta_3 s_0} + C_2 e^{i\beta_2 s_0} - C_4 e^{i\beta_4 s_0}\right\} + \\ &+ |-l\rangle\left\{C_1 e^{i\beta_1 s_0} - C_3 e^{i\beta_3 s_0} - C_2 e^{i\beta_2 s_0} + C_4 e^{i\beta_4 s_0}\right\} \equiv K_+|l\rangle + K_-|-l\rangle\end{aligned} \tag{8}$$

Note that this is essentially the field in the second coil taken at $s = s_0$. The powers associated with the constituent OVs $|\pm l\rangle$ are given by: $N_\pm = |K_\pm|^2$. One can also show that $N_+ + N_- = 1$, which expresses conservation of energy in the optical snake. For N_+ the following expression can be readily obtained:

$$N_+ = \frac{1}{2}\left\{1 + \cos\left[(\Delta_+ - \Delta_-)s_0 + 2(\gamma_+ - \gamma_-)\right]\right\}. \tag{9}$$

Here $\Delta_\pm = (C_l \pm D_l)/2\tilde{\beta}_l$, $\tan\gamma_\pm = \dfrac{\sin\Phi_\pm \cos 2\Phi_\pm}{\cos\Phi_\pm + \sin\Phi_\pm \sin 2\Phi_\pm}$ and $\Phi_\pm = \Delta_\pm s_0 / 2$.

As is shown in Figure 3, at specific values of s_0 the optical snake at the output end changes the sign of OAM of the input vortex. At these points one has $N_+ = 0$, that is the energy of the field is associated with the vortex $|-l\rangle$. As the distance between the coils increases, the scale of spatial variations also increases (see Figure 3b). At $s \neq s_0$ the OAM of the output field can assume values from l to $-l$. This property can be used for an all-fibre inversion of OAM and,in general, for control of OAM. Indeed, the time-averaged OAM $\langle L\rangle$ can be calculated through technique described in [9]:

$$\langle L\rangle = l(2N_+ - 1). \tag{10}$$

Figure 4 shows evolution of OAM for the input OV $|1,1\rangle$. Values of s_0 where $\langle L\rangle = \pm 1$ stand to the fields with well-defined OAM, or OVs. For other s_0 the field possesses of non-integer OAM.

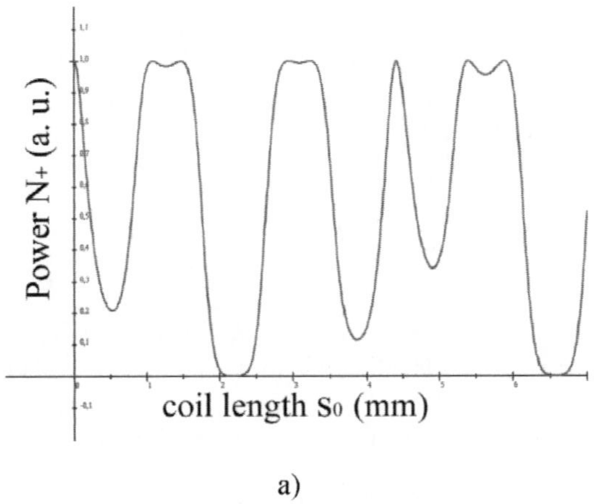

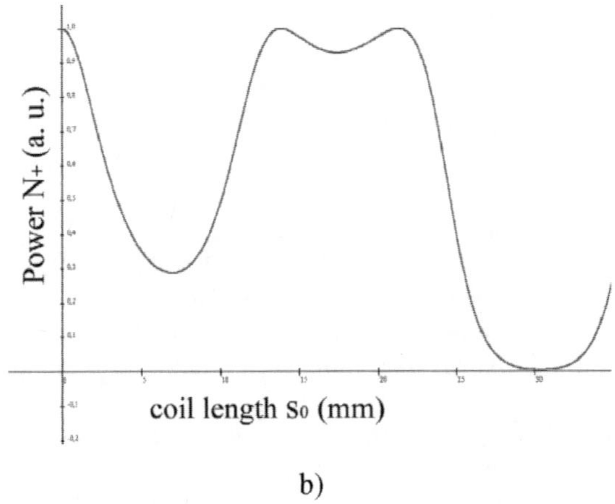

Figure 3. Plot of power N_+ (in relative units)associated with optical vortex $|l\rangle$ as function of the coil's length s_0 (in millimeters) in the two-coil optical snake excited with the vortex $|l\rangle$. Fibre parameters: refractive index of the core $n_{co} = 1.5$, the height of the refractive index profile $\Delta = 5 \cdot 10^{-3}$, core's radius $r_0 = 4\lambda_0$, wavelength $\lambda_0 = 6.33 \cdot 10^{-7}$ m. The topological charge of the incoming OV equals 1. The distance between the centres of adjacent fibres in panels (a), (b) are $2r_0$ and $3r_0$, respectively.

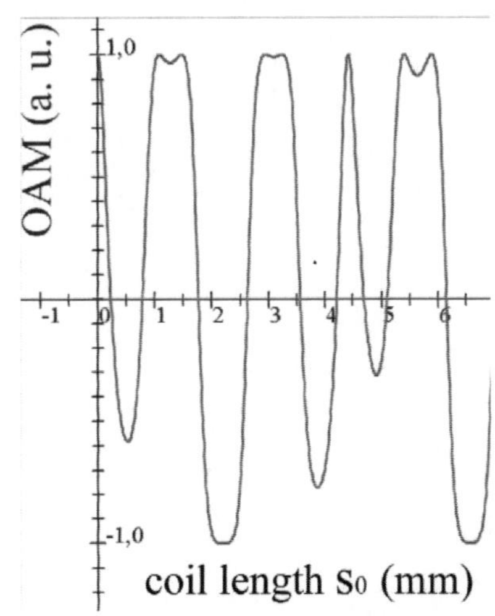

Figure 4. OAM vs coil length at the fibre's output for excitation with unity-charged OV. Fibre parameters as in Figure 3a.

3. Conclusion

In conclusion, in this paper we have studied the evolution of optical vortices in a two-coil optical snake. We have demonstrated that such a system can change the sign of the topological charge of the incident optical vortex at certain resonance of the coil's length.

References

[1] SumetskyM, DulashkoY, FiniJ, HaleA and DiGiovanniD 2006*J. Lightwave Technol.* The

Microfiber Loop Resonator: Theory, Experiment, and Application **24** 242-250.
[2] SumetskyM2005*Opt. Express*Uniform coil optical resonator and waveguide: transmission spectrum, eigenmodes, and dispersion relatio, **13**, 4331-40.
[3] SumetskyM2009*Opt. Express*Optical microfiber coil delay line **17** 7196-7205.
[4] BroderickN 2008*Opt. Express*Optical Snakes and Ladders: Dispersion and nonlinearity in microcoil resonators**16** 16247-254.
[5] ScheuerJ 2009*Opt. Lett.*Fiber microcoil optical gyroscope**34** 1630-32.
[6] LeeT, BroderickN and BrambillaG2011*Opt. Lett.* Berry phase magnification in optical microcoil resonators**36** 2839-41.
[7] BozinovicN, YueY, RenY, TurM, KristensenP, HuangH, WillnerA and RamachandranS2013 *Science*Terabit-scale orbital angular momentum mode division multiplexing in fibers **340**1545-8.
[8] AlexeyevC, BoklagN and YavorskyM2010*J. Opt.*Higher order modes of coupled optical fibres **12**, 115704.
[9] AlexeyevC, VolyarA YavorskyM2007 *Lasers,Optics and Electro-Optics Research Trends*, edL I Chen (Nova Publishers, New York), pp. 131–223.

The compact converter of Bessel beams of zero and second orders on the basis of z-cut lithium niobate

V D Paranin[1], S V Karpeev[1,2], S N Khonina[1,2], K N Tukmakov[1]
[1] Samara State Aerospace University, 34, Moskovskoe shosse, Samara, 443086, Russian Federation
[2] Image Processing Systems Institute of the Russian Academy of Sciences, 151, Molodgvardeyskaya Street, Samara, 443001, Russian Federation

Abstract. Transformation of zero-order Bessel beams into a second-order vortex beam in the process of propagation in a c-cut of lithium niobate $LiNbO_3$ crystal has been investigated experimentally. The possibility of controlling beam transformation by means of changing the curve radius of the illuminating beam is shown. The possibility of Bessel beam transforming by compact devices on the basis of thin c-cuts of uniaxial crystals with a diffraction mask formed on their surface is proved.

1. Introduction

Laser Bessel beams are widely used in scientific research and optical systems of advanced technological systems. This is explained by high energy density in the central light spot of the Bessel beam and the long distance of maintenance comparing with Gaussian beams. Zero-order beams can be used for imaging of extended or moving objects, especially in medical science [1], non-destructive testing [2], scanning metrology [3], high-resolution lithography [4]. Formation of higher-order Bessel beams, namely, vortex Bessel beams, is a topical problem for a number of applications, including tasks of optical manipulation by micro-objects [5, 6], free space communications [7] and displacement measurements in a range from a few up to tens microns [8]. Certain problems are encountered in the formation of such beams with the help of diffractive optical elements (DOE) due to the formation of a vortex phase. In this case DOEs with multilevel microrelief [9] need to be produced, which cannot be easily implemented with high precision [10]. Besides, a changing of illumination beam wavelength leads to a decomposition of the higher order vortex beam into several off-axis vortices of the first order [11]. Low-level quantized frequency-carrier DOEs make it possible to form vortex beams with the highest-order diffraction but limited energy efficiency [12, 13]. Meanwhile, binary ring arrays (diffractive axicons) are simple and convenient elements of generating zero-order Bessel beams. The subject of our research is the development of methods of efficient transformation of zero-order Bessel beams into higher-order vortex Bessel beams.

The use of anisotropic crystals is one of the ways of developing devices and methods of zero-order Bessel beam transformation. Theoretical research [14-17] shows that in uniaxial crystals zero-order Bessel beams propagating non-paraxially along the c-axis are transformed periodically into second-order vortex Bessel beams. To obtain complete transformation the crystal thickness must be a multiple of the transformation period. As it is shown in [18-20] the beam form recurrence interval in the crystal depends on the crystal refraction indices and the numerical aperture of the axicon, for axicons with the numerical aperture of 0.15 – 0.5 and a lithium niobate crystal the period being equal to tens of micrometers. Up-to-date methods of making DOEs make it possible to produce photomasks for even greater values of the numerical aperture, therefore, it is quite realistic to produce compact Bessel beam transforming devices less than 1 mm thick. In this case the task of integrating a diffractive axicon and an anisotropic crystal comes to the forefront. The matter in question is, essentially, the formation of diffractive microrelief directly on the surface of a uniaxial crystal. Thus, the aim of this work is to develop and investigate experimentally a small-dimensioned integrated zero-order Bessel beam transforming device based on a uniaxial crystal.

2. Experimental investigation

The following design parameters of the beam-transforming device have been determined on the basis of the mathematical model developed in [18]: the period of the diffractive axicon is 3.5 μm, the thickness of the crystal (lithium niobate $LiNbO_3$) is 500 μm. The choice of the crystal thickness is also conditioned by sufficient mechanical strength for 10x10-15x15 mm dimensions and high chemical stability. Besides, a lithium niobate crystal is characterized by high operational performance (electrooptical, thermooptical, acoustooptic effects etc.) that allow trimming of its thickness and the refraction index. A nominally pure crystal of congruent lithium niobate ($LiNbO_3$) of a Z-cut 514±2 μm thick was used as the substrate. The surface condition conformed to PIII, the surface flatness problem did not exceed 10 arcsec, the roughness did not exceed 2 nm. $LiNbO_3$ crystals were purified by washing in an alkaline cleaner, soaking in chromium solution H_2SO_4:K_2CrO_7 for 5 min and rinsing in deionized water. A 100- nm thick aluminum layer was deposited with the help of magnetron sputtering with additional heating of the crystals up to 150 °C. A layer of positive electronic resist PMMA 950 was spin-coated at 5000 rpm on the metal layer obtained. The obtained PMMA 950 layer was soaked at room temperature for 2 minutes and then dried on a warm table at 160 °C for 10 minutes. The exposure dose of the electron resist was determined experimentally and is equal to 200 $\mu C/cm^2$. The structure recorded was developed in the mixture MIBK:IPA 1:3 for 1 minute. Immediately prior to mask etching of the metal layer with the use of a PMMA 950 mask the crystal was quickly heated to the temperature of 80 °C on the warm table. Aluminum was etched with the help of a weak 0.01 % alkali solution. This resulted in the formation of an amplitude diffractive axicon with a period of 3.5 μm and a diameter of 400 μm. The width of the groove was equal to the width of the chromium band. The transmission of the chromium mask at the wavelength of λ=632.8 nm did not exceed 0.2%. The photo of the central part of the diffractive axicon is presented in figure 1.

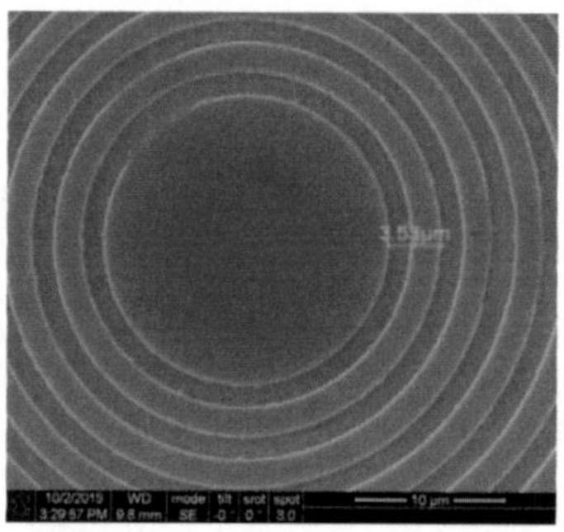

Figure 1. Diffraction axicon.

Bessel beam transformation was investigated using an optical setup (figure 2) including a helium-neon laser LGN-208A, a spatial filter that acts as a beam expander, an axicon on a substrate of a lithium niobate z-cut, an image-forming lens 20x, a CCD-camera DCM310 with the resolution of 2048x1536.

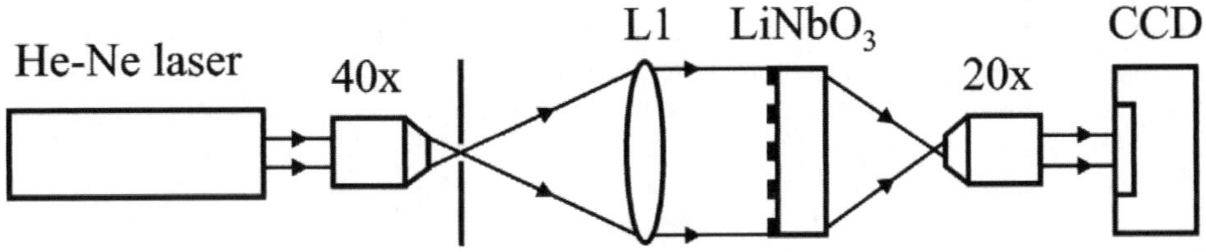

Figure 2. Scheme of the experimental setup.

Apart from the axicon with a period of 3.5 μm axicons with periods of 2 μm and 4 μm and a diameter of 20 mm were also used which were mechanically pressed to the crystal. These axicons were produced by the method of thermochemical oxidation of 100 nm- thick chromium deposited on the surface of a quartz substrate in vacuum. The photos of Bessel beams formed by the axicons with periods 2 μm, 3.5 μm, 4 μm and transformed in the crystal are presented in figures 3-5.

Figure 3(a, b). (a) The beam at the axicon output;**(b)** The beam at the crystal output.

Figure 4(a, b). (a) The beam at the axicon output;**(b)** The beam at the crystal output.

Figure 5(a, b). (a) The beam at the axicon output;**(b)** The beam at the crystal output.

The analysis carried out showed that the most complete transformation of beams takes place if the axicon period is 3.5 μm (figure 4). Thus, the crystal thickness is a multiple of the recurrence interval of the beam shape for the specified axicon period. In the other cases shown in figures 3, 5 the peak intensity in the central part of the beam is small. This maximum is indicative of the presence of a residual zero-order Bessel beam. All this confirms the validity of the mathematical model used that describes non-paraxial propagation of a Bessel beam in a uniaxial crystal and high instrument accuracy in the production of beam-transforming devices.

3. Conclusions
The results obtained demonstrate the possibilities of producing small-dimensioned devices of high-order laser beam transformation on the basis of uniaxial crystals and diffractive optical elements. In order to reduce the dimensions of the beam-transforming devices diffractive microrelief can be formed on the crystal surface in the form of an amplitude or phase mask. This will predetermine high mechanical strength and reliability of the optical element operation.

Acknowledgments
This work was supported by the Russian Foundation for Basic Research (grant 16-07-00825) and by the Ministry of Education and Science of Russian Federation.

References
[1] Lee K S, Rolland J P 2008 *Opt. Lett.* **33**(15) 1696.
[2] Matsuoka Y, Kizuka Y, Inoue T 2006 *Appl. Phys. A.* **84** 423.
[3] Fortin M, Piché M, Borra E F 2004 *Opt. Express* **2**(24) 5887.
[4] Cagniot E et al. 2011 *J. Opt. Soc. Am. A.* **28** 1709.
[5] Hegner M 2002 *Nature* **419** 125.
[6] Khonina S N, Kotlyar V V, Skidanov R V, Soifer V A 2004 *J. Mod. Opt.* **51**(14) 2167.
[7] Zhu K, Zhou G, Li X, Zheng X, Tang H 2008 *Opt. Express* **16**(26) 21315.
[8] Kotlyar V V, Skidanov R V, Khonina S N 2001 *Computer Optics* **21** 102. (in Russian).
[9] Kotlyar V V, Almazov A A, Khonina S N, Soifer V A, Elfstrom H, Turunen J 2005 *J. of Opt. Soc. Am. A.* **22**(5) 849.
[10] Tsai H Y, Smith H I, Menon R J 2007 *J. Vac. Sci. Technol. B.* **25**(6) 2068.
[11] Khonina S N, Porfirev A P, Ustinov A V 2015 *Journal of Optics* **17** 125607.
[12] Vasara A, Turunen J, Friberg A T 1989 *J. of Opt. Soc. Am A.* **6** 1748.
[13] Soifer V A 2012 *Computer Design of Diffractive Optics* (Cambridge: Cambridge Inter. Scien. Pub. Ltd.& Woodhead Pub. Ltd.).
[14] Khilo N A, Petrova E S, Ryzhevich A A 2001 *Quantum Electron* **31**(1) 85.
[15] Stepanov M A 2002 *Optics Communications* **212** 11.
[16] Zusin D H et al. 2010 *J. Opt. Soc. Am. A.* **27**(8) 1828.
[17] Khonina S N, Volotovsky S G, Kharitonov S I 2013 *Computer Optics* **37**(3) 297. (In Russian).
[18] Khonina S N, Morozov A A, Karpeev S V 2014 *Laser Phys.* **24** 056101.
[19] Khonina S N et al. 2014 *Computer Optics* **38**(4) 598. (In Russian).
[20] Paranin V D, Karpeev S V, Khonina S N 2016 *Quantum Electron.* **46**(2) 163.

Per-pixel adjustment of the afterglow effect EOC screen.

K A Borodina[1], S A Sorokina[1], A G Bebiya[1], M P Boronenko[1], P Yu Gulyaev[1]

[1]Ugra State University, Chehov st. 16, Khanty-Mansiysk, 628012 Russia

Abstract. Information capacity of optoelectronic measuring systems significantly exceeds the volume of the others. The residual afterglow EOC creates additional uncertainty. There is a partial solution to the problem [1-2]. The work is dedicated to the enhancement of existing methods to fight the afterglow.

1. Introduction

The main achievements of modern science and technological processes are closely connected to improved methods of obtaining and processing of measurement information, diagnostics of parameters studied or controlled processes. Optoelectronic system distinguishes fast entry of registered images to the computer, processing, extraction of image quantitative characteristics of the process being studied, data analysis and visualization. In such systems used a photodiode array made by CMOS technology. The main indicator of the quality of the measuring device is obtained from him the amount of information about the value of the measured value The optoelectronic system has a large information capacity in comparison with other devices. The residual illumination of the electro-optical converter creates difficulties in measurement. The work is dedicated to the refinement of existing techniques to combat afterglow[1-2].

2. Experimental setup

We shall determine thermal characteristics of high-speed processes. For this camcorder VideoSprint with nanosecond electro-optical shutter it has been used. Experimental stand presented in figure 1.

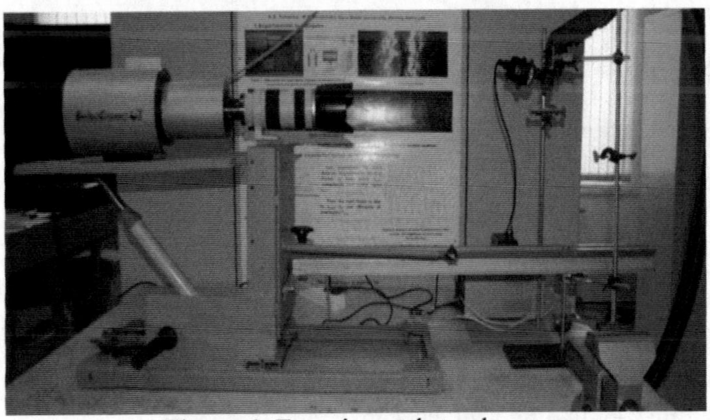

Figure 1. Experimental stand.

Performance provides additional gate - pulse, which is carried out in the interval photocathode-MCP. Solid-state image detectors convert light into a two-dimensional image brightness. For information processing and has been selected ImageJ - a freely available program. ImageJ written in Java, so it is one of the world's fastest image processing programs. In ImageJ presented all the necessary tools for image analysis. ImageJ has the ability to segment automation actions by writing macros. Discovered the afterglow of the phosphor is forcing us to process the image data pixel by pixel.

3. The algorithm refines the technique

Let the amplitude of the observed signal exceeds in I_{nois} N times background. The signal can be represented as exhibitors:

$$I(N) = m \cdot I_{nois} \cdot \exp(-a \cdot N).$$

The signal decreases e times during 1/0.36. Therefore, the lifetime of the afterglow three frames. The main contribution to the afterglow of investing 3 frames. Figure 2 illustrates the luminous particle by 4 frames, N is 0 to 3.

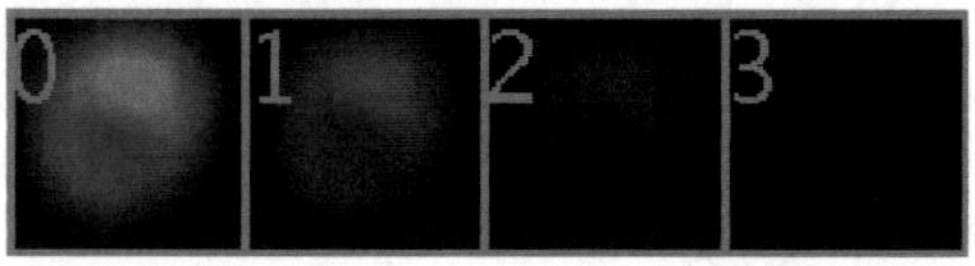

Figure 2. The image of the glowing particles in 4 consecutive frames.

Accordingly, the frame number, the minimum residual glow corresponds to the third frame:

$$I_{PS\min} = m \cdot I_{nois} \cdot \exp(-0,36 \cdot 3).$$

It was established experimentally that the appearance of the attenuation law does not depend on the initial intensity of the signal. Let the amplitude of the new signal I_2 exceeds a background in k times:

$$I_2(N) = k \cdot I_{nois} \cdot \exp(-0,36 \cdot 0).$$

If the signal traces overlap, recorded the amount of:

$$I(N) = k \cdot I_{nois} \cdot \exp(-0,36 \cdot 0) + m \cdot I_{nois} \cdot \exp(-0,36 \cdot N).$$

Where the value N depends on the time from the start of the afterglow of the previous signal. The signal-to-noise:

$$SNR(dB) = 20 \cdot \lg\left(\frac{A_{signal}}{A_{nois}}\right).$$

The background adds afterglow. For simplicity, we assume that the amplitude across the region of signal are the same. Then the signal/noise:

$$SNR(dB) = 20 \cdot \lg\left(\frac{k \cdot I_{nois} \cdot \exp(-0,36 \cdot 0) + m \cdot I_{nois} \cdot \exp(-0,36 \cdot N)}{I_{nois} + m \cdot I_{nois} \cdot \exp(-0,36 \cdot N)}\right).$$

Calculation of signal-noise with minimal afterglow:

$$SNR(dB) = 20 \cdot \lg\left(\frac{k \cdot I_{nois} \cdot \exp(-0,36 \cdot 0) + m \cdot I_{nois} \cdot \exp(-0,36 \cdot 3)}{I_{nois} + m \cdot I_{nois} \cdot \exp(-0,36 \cdot 3)}\right),$$

$$SNR(dB) = 20 \cdot \lg\left(\frac{k + 0,34 \cdot m}{1 + 0,34 \cdot m}\right).$$

If the images of the objects do not overlap and are joined along the border:

$$SNR(dB) = 20 \cdot \lg\ k + 0,34 \cdot m\ .$$

If the new signal appeared earlier, for example N = 1:

$$SNR(dB) = 20 \cdot \lg\ k + 0,7 \cdot m\ .$$

Therefore, when using optoelectronic systems for the study of SHS is important to eliminate afterglow. If we study one moving object, we use the algorithm [1-2] and there is no problem. If during the observation appear and disappear a lot of new objects, the difficulties arise. Then it is not enough to apply for all image pixels the same algorithm. You must use feedback.

Any image can be described by a set I_{nois}, I, $I_{PS}(N)$. Feedback based on the estimation of the ratio of the amplitude of the signal I_n the pixel to the amplitude of the signal in the same pixel in the previous frame I_{n-1}. Based on the comparison of selected factor b residual glow.

$$I_{PS}(N) = I\cdot\exp(-a\cdot N) = I\cdot b.$$

The value of the amplitude of the signal of the next frame in the same pixel is calculated by the formula:

$$I(N) = I_n - b\cdot I_{n-1}.$$

Let the currently registered signal In the pixel with coordinates (x,y). If the ratio is greater than one, then this new signal. Then the coefficient is b=1, N=0. When you move to the next frame in the memory cell "0" signal is automatically reset into the cell "1". The ratio of the signals stored as "d".

Table 1. Part of the algorithm, pixel-by-pixel processing

	The memory	Calculation in the pixel	Comparison		
			If	Then	
				b	N
RAM memory	«0» a readable signal	$\frac{I_i}{I_{i-1}}$	>1 (this is a new signal)	1	0
	«1» storage I_{n-1}		=0,7 (this afterglow)	0,7	1
	«2» storage $d = \frac{I_n}{I_{n-1}}$		=0,5 (this afterglow)	0,5	2
			=0,3 (this afterglow)	0,3	3
	«3» (background)		<1 (this background)	1	

Sometimes a very weak signal. Then the ratio of the signal more than the previous, but less than 1.

If you need high accuracy, supplemented by comparison algorithm. The resulting ratio should not exceed d.

4. The application of optoelectronic systems for studies of SHS

The use of optoelectronic systems to study the SHS can be measured: the time of heat generation and heat, the speed of propagation of the combustion front, etc. There is also an opportunity to assess the size and times of the phases. Figure 3 shows the SHS [3-5] combustion process.

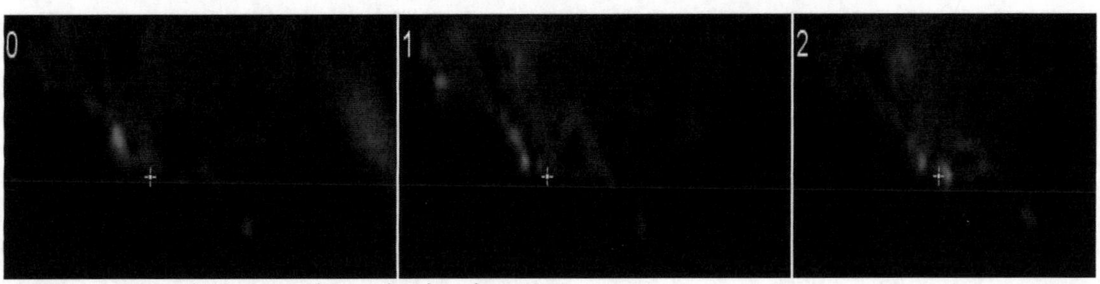

Figure 3. SHS obtain potassium - titanium bronze.

The frames obtained using the optoelectronic system. Pixel-by-pixel processing of data will make the results more accurate. The application of optoelectronic systems to measure the rate of heating will help to solve important tasks of nanotechnology.

5. Conclusions

The application of optoelectronic systems for research, very informative. However, existing methods of removing residual emission electron-optical converter contain flaws. In order to more accurately measure the need to: per-pixel processing and feedback. The proposed algorithm helps to select the correction factor for each pixel.

Acknowledgments

The study was partially supported by RFBR in the framework of a research project № 15-42-00106.

References

[1] Boronenko M P, Gulyaev P Yu, Seregin A E, Poluhina K G 2015 *Journal of Physics Conference Series* **643** 012028

[2] Boronenko M P, Gulyaev P Yu, Seregin A E, Bebiya A G 2015 *IOP Conference Series: Materials Science and Engineering* **93** 012021

[3] Lisitsyn V M, Valiev D T, Oleshko V I, Tupitsyna I A, Polisadova E F, Andryuschenko L A, Yakubovskaya A G, Lisitsyna L A, Vovk O M 2014 *Journal of Luminescence* **153** 130

[4] Gulyaev P Yu, Kotvanova M K, Pavlova S S, Sobol E N, Omelchenko A I 2012 *Nanotechnologies in Russia* **7** 127

[5] Gulyaev P, Cui H, Gulyaev I, Milyukova I 2015 *High Temperatures* **44** 83

[6] Loktev M J, Abanin V A, Savin V F, Ermolaev D A, Suranov A Y 2010 *IEEE 2nd Russia school and seminar on fundamental problems of micro/nanosystems technologies (MNST)* 26

[7] Dolmatov A V, Gulyaev I P, Jordan V I, *IOP Conference Series: Materials Science and Engineering* 2015 **81** 012041

Manipulating Fano resonance via fs–laser melting of hybrid oligomers at nanoscale

S I Lepeshov[1], D A Zuev[1], S V Makarov[1], V A Milichko[1], I S Mukhin[1,2], A E Krasnok[1] and P A Belov[1]
[1]ITMO University, Saint Petersburg 197101, Russia
[2]St. Petersburg Academic University, St. Petersburg 194021, Russia

E-mail: s.lepeshov@gmail.com

Abstract. Here, the novel concept of asymmetric metal–dielectric (hybrid) nanoparticles is proposed. The experimental data and the results of numerical simulation of the optical properties of hybrid nanostructures are presented. The change of their optical response after fs-laser modification is shown. The possibility of manipulating Fano resonance in hybrid oligomers by the gold nanoparticles reshaping is demonstrated.

1. Introduction

Metallic nanostructures have proved to be an efficient tool for light manipulation at nanoscale. While these structures have some well-known advantages, e.g. very strong field localization, subwavelength dimensions, etc. [1], the main problem that prevents realization of practical applications on their basis is severe losses, especially at optical frequencies [2]. On the contrary, low dissipation losses in dielectric nanostructures make them promising materials for light manipulation [3–5]. Moreover, optical nanostructures based on nanoparticles with high-refractive index exhibit both electric and magnetic responses, which makes it possible to control both electric and magnetic components, simultaneously. [6–7]. The use of metal and dielectric materials together in hybrid nanostructures and metasurfaces opens a wide range of perspective applications for medicine, photonics, photovoltaics, etc [4, 8–9]. Initially plasmonic and all-dielectric nanostructures have been studied separately and only recently the idea of hybrid nanostructures with advantages from both of them has been introduced [10–12].

The novel concept of asymmetric hybrid nanodimer and fabrication methods of such nanostructures have been recently developed [13]. Hybrid nanodimer includes silicon (Si) nanocone and gold (Au) nanodisc. Important feature of such nanostructure is the possibility of modification of Au nanodisc to nanosphere via fs-laser irradiation of nanodimer. At the same time Si nanocone is almost not exposed to any impacts of the fs-laser radiation. It opens great opportunities for the control of the properties of such hybrid nanodimers.

Here, we study optical properties of modified and unmodified asymmetric hybrid nanodimers. Moreover, we demonstrate numerically how fs-laser modification of metal nanoparticle affects the Fano resonance in oligomers consisting of hybrid nanodimers.

2. Optical characterization of hybrid nanodimer

First, we fabricated an array of hybrid nanodimers by the method proposed in Ref. [13]. We used the same geometrical parameters of nanoparticles: the bottom base diameter of the nanocone and the diameter of the nanodisc are d = 190 nm, the height of the nanocone is h = 200 nm, the diameter of the top nanocone base is a = 50 nm and the thickness of the nanodisc is l = 20 nm. Then, we performed fs-laser post-processing of the array in order to modify the structures. In both cases (reshaped and unreshaped arrays) we applied the method of dark-field microscopy to obtain the scattering spectra of a single hybrid nanoparticle (**Figure 1**(a,d)). We used the software package CST Microwave Studio for numerical calculations of reflection spectra, power patterns (**Figure 1**(b,e)) and electric field distributions (**Figure 1**(c,f)).

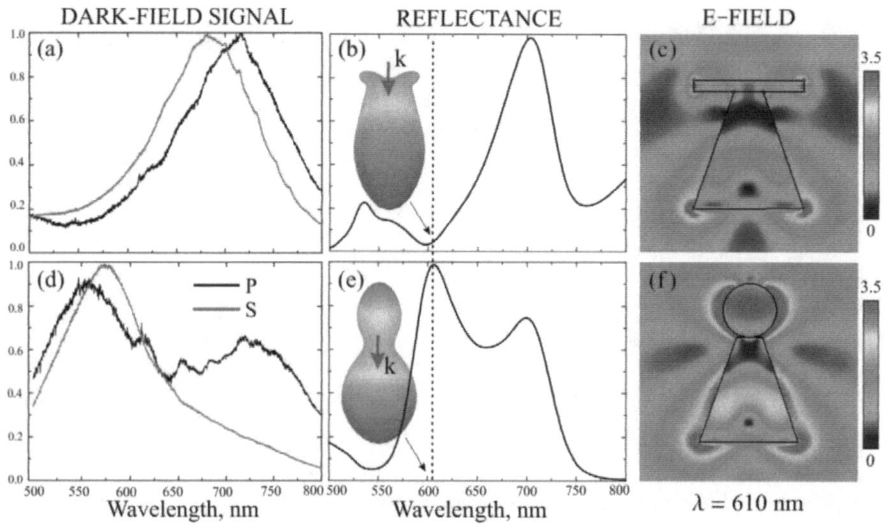

Figure 1. (a) Dark-field optical microscopy spectra for p- (black line) and s- (red line) polarized light; **(b)** Numerically calculated reflectance spectrum and **(c)** electric field distribution of the unreshaped hybrid nanodimer. **(d)** Dark-field spectra; **(e)** Reflectance spectrum and **(f)** electric field distribution of the reshaped hybrid nanodimer.

In case of unreshaped hybrid nanostructures, magnetic dipole resonance of silicon nanocone overlaps with electric dipole resonance of Au disk (**Figure 1**(a)). The light scattering by the Au nanodisc is dominant. The results of numerical modeling demonstrate that it appears in suppression of back reflectance from the structures at the wavelength of 610 nm (**Figure 1**(b)). Analysis of the electric field distribution of the unreshaped nanoparticle shows the absence of resonances at 610 nm (**Figure 1**(c)).In turn, the fs-laser reshaping of the gold nanodisc to nanosphere causes a shift of the electric dipole resonance of Au nanoparticle to the shorter wavelength region (**Figure 1**(d)). It changes the power pattern, which results in the apperance of the reflectance peak at the wavelength of 610 nm (**Figure 1**(e)). The electric field distribution at 610 nm confirms that the peak at this wavelength is caused by localized plasmonic resonance of nanosphere (**Figure 1**(f)).

Thus, we can tune the spectral position of the electric dipole resonance of Au nanoparticle as well as the power pattern of the hybrid structures. These results open the possibility of tuning the optical properties of hybride nanodimers and nanostructures.

Journal of Physics: Conference Series **741** (2016) 012140 doi:10.1088/1742-6596/741/1/012140

3. Manipulating Fano resonance in hybrid oligomers

Then, we studied the optical properties of hybrid oligomers composed of reshaped and unreshaped nanodimers. The importance of all-dielectric and plasmonic oligomers has been recently discussed in works [14,15].

We considered six unreshaped hybrid nanodimers arranged in hexamer. The distance between the centers of the bottom bases of nanocones in this regular hexagonal structure is 200 nm. We carried out numerical simulation of the scattering spectrum of the hexamer by using CST Microwave Studio. The structure was excited by a plane wave along the axis of symmetry. The scattering spectrum of hybrid hexamer is shown in **Figure 2**(a). It is seen that magnetic dipole mode of Si nanocones is splitted into two broad low-Q magnetic modes as a result of interaction between nanocones.

Then, we simulated the hybrid nanodimer with smaller diameter of the bottom base of Si nanocone (d = 150 nm). The resonant wavelength of the dielectric nanoparticles is directly proportional to its size. Therefore, the scattering spectrum of the magnetic dipole resonance of a single nanoparticle is shifted to the shorter wavelength region relatively to the collective magnetic modes (**Figure 2**(b)).

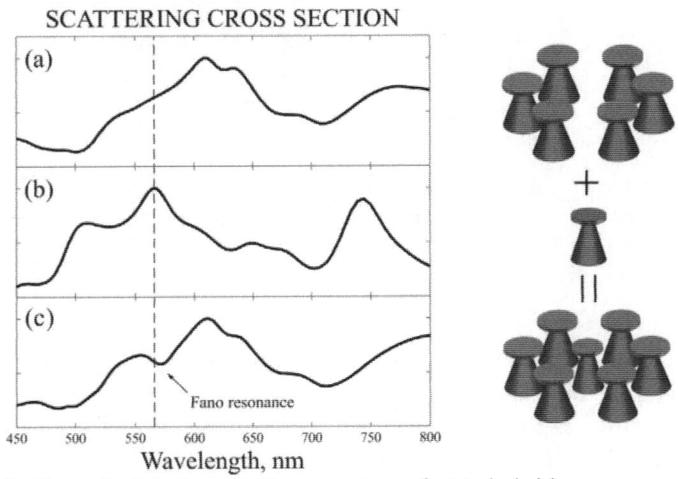

Figure 2. Numerically calculated scattering spectra of **(a)** hybrid gexamer, **(b)** single hybrid nanodimer and **(c)** hybrid oligomer.

Finally, we put nanodimer with a smaller diameter of the bottom base in the center of the hexamer. Numerically calculated scattering spectrum of such oligomer is presented in **Figure 2**(c). We observe that there is an asymmetrical Fano type resonance at the wavelength of the magnetic dipole resonance of the single nanodimer. Fano resonance in this nanostructure is caused by the existence of two scattering channels: scattering by the broad collective magnetic mode of hexamer and scattering by the narrow resonant magnetic mode of single nanodimer. Because of the spectral shift between the resonances of hexamer and nanodimer excited modes oscillate at the different wavelengths in the phase and antiphase. The minimum of Fano resonance in scattering spectrum at 572 nm is due to the destructive interference (**Figure 3**(b)), while the maximum at 553 nm is due to the constructive interference.

However, after reshaping of Au nanodiscs to nanospheres Fano resonance cannot be observed in the scattering spectrum (**Figure 3**(c)). It is caused by a shift of the electric dipole resonance of Au nanoparticles to the short wavelength region (**Figure 3**(d)).

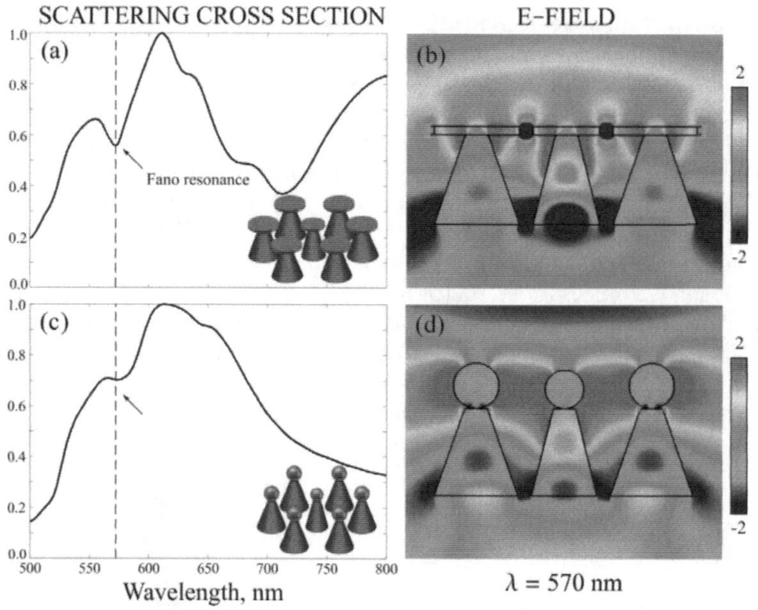

Figure 3. **(a)** Numerically calculated scattering spectrum and **(b)** electric field distribution of unreshaped hybrid heptamer. **(c)** Numerical calculated scattering spectrum and **(d)** electric field distribution of reshaped hybrid heptamer.

4. Conclusion

In summary, we have demonstrated the opportunity of Fano resonance tuning via fs-laser reshaping of Au nanoparticles in hybrid oligomers. Our results open a way for experimental realization of hybrid oligomers with specially engineered Fano resonances and for developing new switchable nanophotonic devices.

Acknowledgements
Authors are thankful to Dr. R.S. Savelev for useful comments and corrections. This work was supported by the Ministry of Education and Science of the Russian Federation (project №14.584.21.0009 with unique identificator RFMEFI58414X0009).

References

[1] Giannini V, Fernández-Domínguez A I, Heck S C and Maier S A 2011 Plasmonic nanoantennas: fundamentals and their use in controlling the radiative properties of nanoemitters *Chemical Reviews* **111** 3888-3912.

[2] Maier S A 2007 *Plasmonics: fundamentals and applications* Springer Science & Business Media.

[3] Kuznetsov A I, Miroshnichenko A E, Fu Y H, Zhang J and Luk'yanchuk B 2012 Magnetic light *Scientific Reports* **2**.

[4] Krasnok A E, Miroshnichenko A E, Belov P A and Kivshar Y S 2012. All-dielectric optical nanoantennas *Optics Express* **20** 20599-604.

[5] Shcherbakov M R, Neshev D N, Hopkins B, Shorokhov A S, Staude I, Melik-Gaykazyan E V, Decker M, Ezhov A A, Miroshnichenko A E, Brener I, Fedyanin A A 2014 Enhanced third-harmonic generation in silicon nanoparticles driven by magnetic response *Nano Letters* **14** 6488-92.

[6] Evlyukhin A B, Novikov S M, Zywietz U, Eriksen R L, Reinhardt C, Bozhevolnyi S I and Chichkov B N 2012 Demonstration of magnetic dipole resonances of dielectric nanospheres in the visible region *Nano Letters* **12** 3749-55.

[7] Schuller J A, Zia R, Taubner T and Brongersma M L 2007 Dielectric metamaterials based on electric and magnetic resonances of silicon carbide particles *Physical Review Letters* **99** 107401.

[8] Lapotko D 2009 Therapy with gold nanoparticles and lasers: what really kills the cells? *Future Medicine* **4** 253-6.

[9] Brongersma M L, Cui Y and Fan S 2014 Light management for photovoltaics using high-index nanostructures *Nature Materials* **13** 451-460.

[10] Joshi B P and Wei Q H 2008 Cavity resonances of metal-dielectric-metal nanoantennas *Optics Express* **16** 10315-22.

[11] Boriskina S V and Reinhard B M 2012 Molding the flow of light on the nanoscale: from vortex nanogears to phase-operated plasmonic machinery *Nanoscale* **4** 76-90.

[12] Krasnok A E, Maksymov I S, Denisyuk A I, Belov P A, Miroshnichenko A E, Simovski C R and Kivshar Y S 2013 Optical nanoantennas *Physics-Uspekhi* **56** 539.

[13] Zuev D A, Makarov S V, Milichko V A, Starikov S V, Mukhin I S, Morozov I A, Shishkin I I, Krasnok A E and Belov P A 2016 Fabrication of Hybrid Nanostructures via Nanoscale Laser-Induced Reshaping for Advanced Light Manipulation *Preprint* arXiv:1601.02013.

[14] Miroshnichenko A E and Kivshar Y S 2012 Fano resonances in all-dielectric oligomers *Nano Letters* **12** 6459-63.

[15] Kuznetsov A I, Evlyukhin A B, Gonçalves M R, Reinhardt C, Koroleva A, Arnedillo M L, Kiyan R. Marti O and Chichkov B (2011). Laser fabrication of large-scale nanoparticle arrays for sensing applications *ACS Nano* **5** 4843-9.

Simulation of Ti-indiffused lithium niobate waveguides and analysis of their mode structure

M Parfenov[1], P Agruzov[2], I Il'ichev[2], A Shamray[1,2]
[1]Peter the Great St. Petersburg Polytechnic University, Saint Petersburg 195251, Russia
[2]Quantum Electronics Laboratory, Ioffe Institute, Saint Petersburg 194021, Russia

Abstract. Simulation of Ti-indiffused waveguides on $LiNbO_3$ is described. The influence of diffusion conditions and technological parameters on the waveguide mode structure is considered. Boundaries of multimode, single-mode, and single-mode polarizing regimes of operation are defined.

1. Introduction

In integrated optics, where optical and optoelectronic devices are fabricated on a single wafer, optical waves propagate in the waveguides formed in the surfactant area of the wafer material. Since waveguides are basic elements of every integrated optical device, their properties are of a great importance. In order to fabricate an integrated optical device, it is necessary to know, for example, the waveguide size, its optical losses per unit length, and its eigen mode structure. These properties depend on the method of fabrication and, hence, on the technological parameters used in the fabrication. The waveguide properties can be tailored to specific applications by varying technological parameters. In addition, since production of devices demands reproducible results, it is important to not only find the fabrication parameters under which the desired waveguide properties can be obtained, but also estimate the allowable deviations from these parameters which will not lead to a deterioration of necessary waveguide properties.

The influence of technological parameters on the waveguide properties can be studied experimentally or theoretically (analytically or numerically). However, the first approach encounters difficulties because of a long production cycle of experimental waveguides. It is much more suitable to define the necessary range of technological parameters and only then to test experimentally whether the calculations have given a correct result.

Nowadays lithium niobate ($LiNbO_3$) is a widely spread material for integrated optics [1, 2]. This crystal is very suitable as a wafer because of the presence of electrooptical properties (which is important for fabrication of modulators) [3], its transparency in the visible and near infrared spectral ranges, piezoelectric effect [4], and applicability in nonlinear optics [5]. There are two basic technologies for fabrication of optical waveguides on lithium niobate with low losses (<0.1 dB/cm) [1]. The first method, i.e., titan indiffusion, leads to an increase in both the ordinary and extraordinary refractive indexes in this birefringent crystal. The second method, i.e., the so-called proton exchange, increases the extraordinary refractive index alone, and thus allows one to create polarizing waveguides [2, 6].

Both methods produce channel waveguides with a gradient refractive index distribution and, therefore, the waveguide parameters cannot be analyzed analytically. To get a proper understanding of the influence of technological parameters on the characteristics of the waveguide being fabricated, numerical simulation is needed. This report is devoted to the simulation of Ti-indiffused waveguides.

2. Ti:LiNbO$_3$ waveguides

Indiffusion of Ti is a traditional method of fabrication of LiNbO$_3$ waveguides [2]. In this technique, a metal strip is sputtered on a crystal surface and then is annealed at temperatures near 1000°C (figure 1).

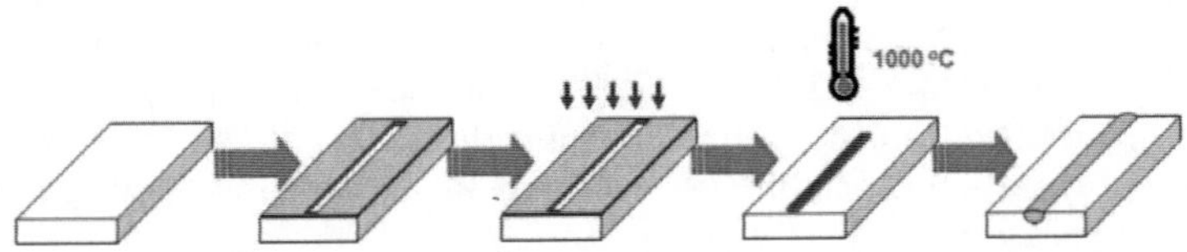

Figure 1. Fabrication of Ti:LiNbO$_3$ waveguides

The result of the diffusion process for the x-cut crystal orientation can be described analytically in terms of error functions as [7]

$$C(x,z) = C_0\left[\mathrm{erf}\left\{\frac{h+x}{\sqrt{4D_X t}}\right\} + \mathrm{erf}\left\{\frac{h-x}{\sqrt{4D_X t}}\right\}\right] \times \left[\mathrm{erf}\left\{\frac{W/2+z}{\sqrt{4D_Z t}}\right\} + \mathrm{erf}\left\{\frac{W/2-z}{\sqrt{4D_Z t}}\right\}\right], \quad (1)$$

where C is the Ti concentration, D_Z and D_X are the diffusion coefficients along the Z and X axes, respectively, t is the annealing time, and h and W are the thickness and width of the sputtered Ti strip, respectively (figure 2).

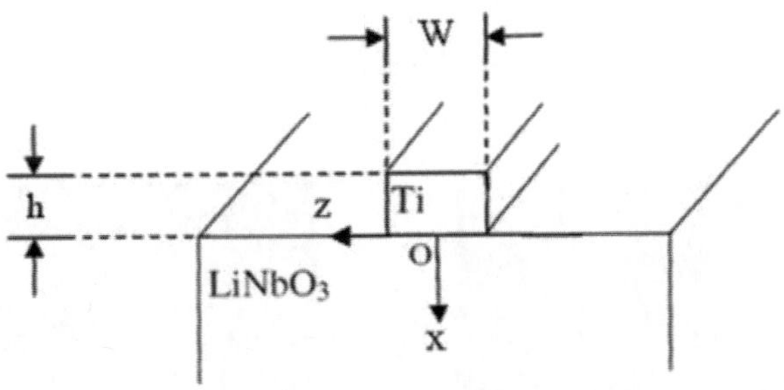

Figure 2. Crystal orientation (x-cut) and parameters of the sputtered Ti strip

The change in the refractive index caused by the diffusion process [2] can be described by semi-empirical equations [7]

$$\Delta n_{o,e}(x,z,\lambda) = A_{o,e}(\lambda)[C(x,z)]^{\alpha_{o,e}}, \quad (2)$$

where $A_{o,e}$ and $\alpha_{o,e}$ are unknown constants at a fixed wavelength.

Since the Ti indiffusion increases both the ordinary and extraordinary refractive indexes, it allows propagation of the orthogonally polarized TM and TE modes through the waveguide.

3. Simulation

To perform the simulation, the numerical finite element method in the COMSOL Multiphysics package was used.

The refractive index distribution which was necessary for the mode structure analysis could not be defined without knowing the numerical values of the constants used in expressions (1) and (2) (D_x, D_z, α_e, α_o, A_e, A_o, C_0). In order to find them, the experimentally captured mode field distributions were used (figure 3). They were obtained for a single-mode waveguide fabricated under technological parameters W=6 μm, h=100 nm, t=17 h, and temperature T=1000°C, which yielded the waveguide sizes which were most suitable for optimum coupling to standard single-mode optical fibers. The constants were found through an iterative algorithm, so that the computed intensity distributions of the mode field (figure 4) matched the experimentally obtained ones (figure 3). The search and fitting were performed for the constants similar by the order of magnitude to those reported in the literature for other crystal orientations and annealing times [7].

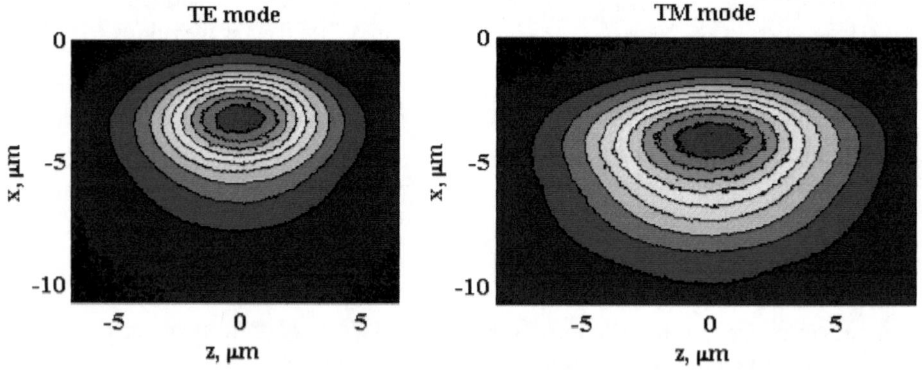

Figure 3. Experimentally captured intensity distributions for TE and TM modes

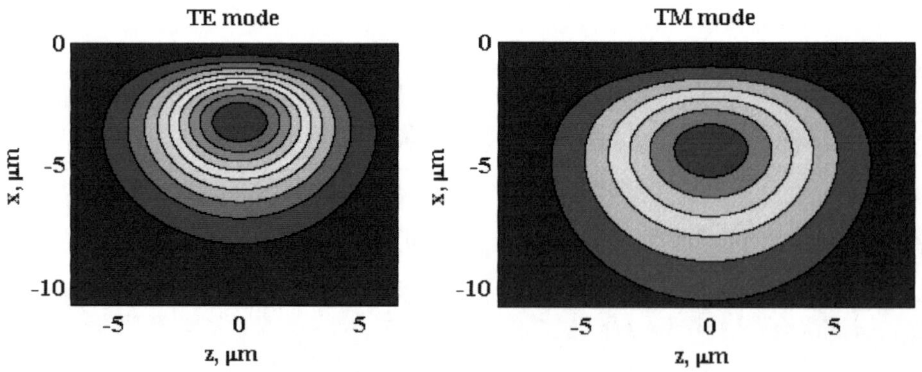

Figure 4. Computed intensity distributions for TE and TM modes

Then the technological parameters were varied at a fixed wavelength (λ=1550 nm) and at the constants derived above (D_x=D_z=0.19 um^2/h, α_e=0.69, α_o=0.46, $A_e(C0)^{\alpha e}$=0.0121, $A_o(C0)^{\alpha o}$=0.0419), and the resulting mode composition was analyzed. To simplify the analysis we varied only two parameters, i.e., the Ti strip width and annealing time. The Ti strip height was fixed at a value of

100 nm, at which the experimentally fabricated waveguides had the lowest losses. The strip width and time of annealing were varied in the vicinity of 6 μm and 17 h, respectively. Results of simulation are presented in figure 5.

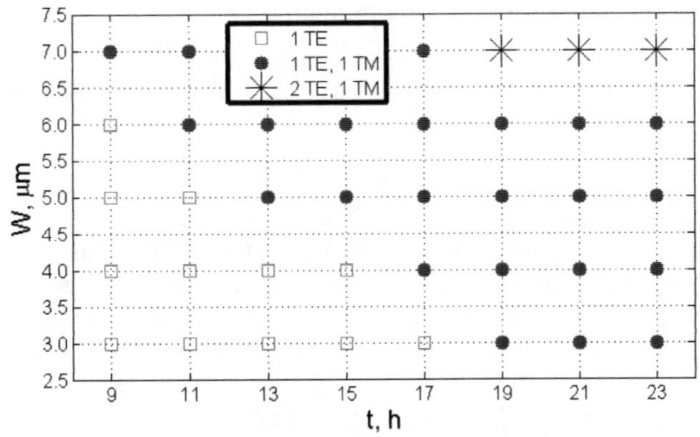

Figure 5. Results of simulation: dependence of supported modes on technological parameters (Ti strip width and annealing time). The blue circles correspond to the single-mode regime, the red squares mark the polarizing regime, and the asterisks are for the multimode regime.

4. Results

The analysis revealed how technological parameters affected the mode structure of waveguides. It was found that large widths of a sputtered metal strip and long annealing times resulted in multimode waveguides. If both parameters were decreased, the waveguides supported only one TE and one TM mode (a single-mode regime). A further decrease in the annealing time and strip width caused the TM mode cutoff, and the waveguides became polarizing. Such an influence of the technological parameters on the mode composition was caused by different ordinary and extraordinary refractive index distributions (figure 6) which were formed during the Ti indiffusion.

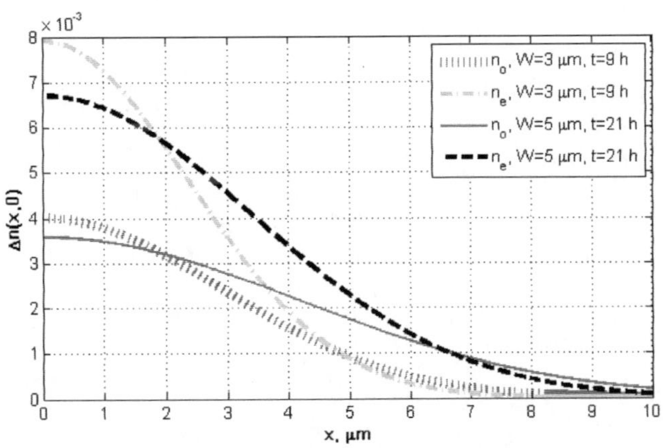

Figure 6. Computed distributions of refractive index increase for n_e and n_o at z=0 for a polarizing waveguide (W=3 μm, t=9 h) and for a single-mode waveguide with both the TE and TM modes (W=5 μm, t=21 h).

Therefore, it is possible to fabricate polarizing waveguides by using indiffusion of titan. This result is highly important because a single-mode polarizing regime of operation is necessary for production of such devices as amplitude modulators with a high extinction ratio and others. As mentioned in the Introduction, the polarizing properties are generally associated with proton exchanged waveguides.

To suppress the TM mode in single-mode Ti:$LiNbO_3$ waveguides, plasmon-polariton polarizers are employed at present, but such polarizers cause optical losses and a special technological step is needed to obtain polarizing properties [8]. The use of new Ti-induffused waveguides that are originally polarizing is much more promising.

5. Conclusions

The simulation of Ti-indiffused waveguides on $LiNbO_3$ was described. The analysis of the effect of technological parameters on waveguide eigen modes was carried out. As an example, the influence of the Ti strip width and annealing time on the mode structure was investigated. The boundaries of multimode, single-mode, and single-mode polarizing regimes of operation were defined. The technological parameters for production of polarizing waveguides by Ti-indiffusion were estimated. The results obtained were found to be in good agreement with the experimental results. Our findings can prove useful for the development of integrated optical devices.

Acknowledgement

This work was funded by the Ministry of Education and Science of the Russian Federation, project RFMEFI60414X0055.

References

[1] Bazzan M, Sada C 2015 *Appl. Phys. Rev.* **2(4)** 040603
[2] Wooten Ed L, Kissa K M, Yi-Yan Al, Murphy Ed J et al. 2000 *IEEE J. Sel. Top. Quantum Electron.* **6(1)** 69-82
[3] Toney J E 2015 *Lithium Niobate Photonics* (Norwood: Artech House)
[4] Arizmendi L 2004 *Phys. Status Solidi (a)* **201(2)** 253-283
[5] Weis R S, Gaylord T K 1985 *Appl. Phys. A* **37** 191-203
[6] Il`ichev I, Kozlov A, Gaenko P, Shamray A 2009 Quantum Electron. **39(1)** 98-104
[7] Ganguly P, Biswas J C, Lahiri S K 2010 *J. Opt.* **39(4)** 175-180
[8] Il'ichev I, Toguzov N, Shamray A 2009 *Tech. Phys. Lett.* **35(9)** 831-833

Simulation of vortex laser beams propagation in parabolic index media based on fractional Fourier transform

O A Mossoulina[1], **M S Kirilenko**[1,2] **and S N Khonina**[1,2]
[1]Samara National Research University, 34, Moskovskoye Shosse, Samara, Russia, 443086
[2]Image Processing Systems Institute of the Russian Academy of Sciences, 151, Molodogvardeyskaya, Samara, Russia, 443001

Abstract. We use radial Fractional Fourier transform to model vortex laser beams propagation in optical waveguides with parabolic dependence of the refractive index. To overcome calculation difficulties at distances proportional to a quarter of the period we use varied calculation step. Numerical results for vortex modes superposition propagation in a parabolic optical fiber show that the transverse beam structure can be changed significantly during the propagation. To provide stable transverse distribution input scale modes should be in accordance with fiber parameters.

1. Introduction

Beams with helical phase singularity are very perspective in telecommunication applications as new information channels as mode multiplexing [1, 2]. They can be used for nano-imaging [3]. We use radial Fractional Fourier transform (FrFT) to model vortex laser beams propagation in optical waveguides with parabolic dependence of the refractive index [4]. The canonical FrFT was considered [5] as the Fourier transform of p-order, where p is the real value. Currently the FrFT is actively used in optical image processing [6] and can be optically implemented by means of several spherical and/or cylindrical lenses [7-9]. Note, cylindrical lenses are used for astigmatic transformation in order to form vortex beams [9-14]. For real fibers a bounded FrFT should be used [15, 16] and problems of beams propagation are solved numerically. Application of FrFT for laser beam propagation in gradient index media has calculation problems at distances proportional to a quarter of the period [17, 18]. We have overcome these difficulties using the varied calculation step. Several examples of vortex beams propagation are demonstrated.

2. Theoretical background

In a gradient parabolic fiber the refractive index is given by $n^2(r)=n_0^2\left(1-\alpha^2 r^2\right)$, where n_0 is the refractive index on the fiber's optical axis and α is a constant that defines the curvature of the refractive index profile. Based on the paraxial approximation for the parabolic medium $\alpha r \ll 1$ we can get the integral propagation operator:

$$E(r,\varphi,z)=-\frac{ik\alpha}{2\pi\sin(\alpha z)}\exp(ikz)\exp\left[\frac{ik\alpha}{2\tan(\alpha z)}r^2\right]$$
$$\times\int_0^{2\pi}\int_0^{\infty}E_0(\rho,\theta)\exp\left[\frac{ik\alpha}{2\tan(\alpha z)}\rho^2\right]\exp\left[-\frac{ik\alpha}{\sin(\alpha z)}\rho r\cos(\theta-\varphi)\right]\rho d\rho d\theta, \tag{1}$$

which is FrFT in polar system with order $p=\alpha z$; $k=2\pi/\lambda$ is the wavenumber, λ is the wavelength of an illuminating light.

Consider a laser beam with a singular component:

$$E_0(\rho,\theta)=E_m(\rho)\exp(im\theta), \tag{2}$$

where $E_m(\rho)$ is the radial component of the beam, $\exp(im\theta)$ is the optical vortex of order m.

Substituting the expression (2) into the formula (1), we obtain:

$$E(r,\varphi,z)=-\frac{ik\alpha}{2\pi\sin(\alpha z)}\exp(ikz)\exp\left[\frac{ik\alpha}{2\tan(\alpha z)}r^2\right]\exp(im\varphi)$$
$$\times\int_0^{\infty}E_m(\rho)\exp\left[\frac{ik\alpha}{2\tan(\alpha z)}\rho^2\right]\int_0^{2\pi}\exp\left[im(\theta-\varphi)\right]\exp\left[-\frac{ik\alpha}{\sin(\alpha z)}\rho r\cos(\theta-\varphi)\right]d\theta\rho d\rho. \tag{3}$$

In order to calculate the integral over θ, we apply this relation:

$$J_m(x)=\frac{i^m}{2\pi}\int_0^{2\pi}\exp(imt)\exp(-ix\cos t)dt, \tag{4}$$

where $J_m(x)$ is Bessel function of order m.

Using the expression (4) instead of equation (3), we obtain:

$$E_m(r,\varphi,z)=\mathrm{FrFT}\left[E_m(\rho)\exp(im\theta)\right]=-\frac{ik\alpha}{\sin(\alpha z)}\exp(ikz)\exp\left[\frac{ik\alpha}{2\tan(\alpha z)}r^2\right]$$
$$\times i^{-m}\exp(im\varphi)\int_0^{\infty}E_m(\rho)\exp\left[\frac{ik\alpha}{2\tan(\alpha z)}\rho^2\right]J_m\left[\frac{k\alpha\rho r}{\sin(\alpha z)}\right]\rho d\rho. \tag{5}$$

We can make use formula (5) as the sum to modulate the propagation of the superposition of vortex functions. For example, let the superposition is situated at the entrance:

$$E_0(\rho,\theta)=\sum_q C_{m_q}E_{m_q}(\rho)\exp(im_q\theta), \tag{6}$$

where C_{m_q} are the superposition coefficients of functions as in equation (2). Then the result of the transformation is easily obtained from the relation:

$$E(r,\varphi,z)=\sum_q C_{m_q}\,\mathrm{FrFT}\left[E_{m_q}(\rho)\exp(im_q\theta)\right]. \tag{7}$$

3. Numerical simulations

The following parameters were used in the calculations: radius of the parabolic fiber is $R=10\lambda$, parameter $\alpha=0.0628\lambda^{-1}$. Laguerre-Gaussian modes (LG) [19] were considered in the calculations:

$$GL_{nm}(r,\phi,z)=\frac{\sigma_0}{\sigma(z)}\left[\frac{2n!}{\pi\left(n+|m|\right)!}\right]^{1/2}\exp\left[\mathrm{i}\left(2n+m+1\right)\eta(z)\right]\exp\left[\mathrm{i}kz+\mathrm{i}\frac{kr^2}{2R(z)}\right]\times$$
$$\times\exp\left[-\frac{r^2}{\sigma^2(z)}\right]\cdot\left[\frac{\sqrt{2}r}{\sigma(z)}\right]^{|m|}L_n^{|m|}\left[\frac{2r^2}{\sigma^2(z)}\right]\exp\left(\mathrm{i}m\phi\right), \tag{8}$$

where $L_n^m(x)=\frac{1}{n!}e^x x^{-m}\frac{\mathrm{d}^n}{\mathrm{d}x^n}\left(e^{-x}x^{n+m}\right)$ is the generalized Laguerre polynomial [20], $k=2\pi/\lambda$ is the light wavenumber for the wavelength λ, $\sigma(z)=\sigma_0\left[1+\left(z/z_R\right)^2\right]^{1/2}$ is the beam effective radius, σ_0 is the waist radius, $z_R=k\sigma_0^2/2$ is the confocal parameter, $R(z)=z\left[1+\left(z_R/z\right)^2\right]$ is the radius of curvature of the parabolic front of light field, $\eta(z)=\arctan\left(z/z_R\right)$ is Gouy phase, $r^2=x^2+y^2$, $\phi=\arctan(y/x)$.

By using the diffraction optical elements [21-22], we can effectively generate separate LG modes (8) and their superposition (6) with given complex coefficients.

The propagation of LG modes with indexes $(n,m)=(3,2)$ and $(n,m)=(0,5)$, respectively, in a parabolic fiber at the distance of $z\in[0,50\lambda]$ is shown in figures 1 and 2. The modes sizes were matched with the fiber parameters to avoid large-scale changes during the propagation. For this purpose the waist radius (8) was chosen as $\sigma=2\lambda$. If we increase or decrease the waist radius, the transverse distribution will change its scale during propagation as shown in figure 3.

Although the idea of mode multiplexing in the fiber communication channels has been known for a long time [23-24], practical interest in this subject has been renewed lately due to vortex laser beams [1-3], possessing orbital angular momentum. «Single-ring» LG modes with a zero radial index $n=0$ and different azimuthal indexes m (figures 2, 3) are of particular interest.

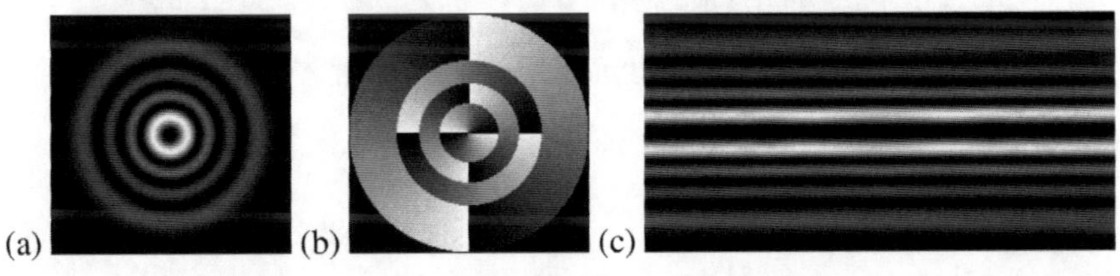

Figure 1. The propagation of LG modes with indexes $(n,m)=(3,2)$ in a parabolic fiber, $\sigma=2\lambda$: the distribution at the entrance (a) the intensity, (b) the phase and (c) the longitudinal intensity distribution $z\in[0,50\lambda]$.

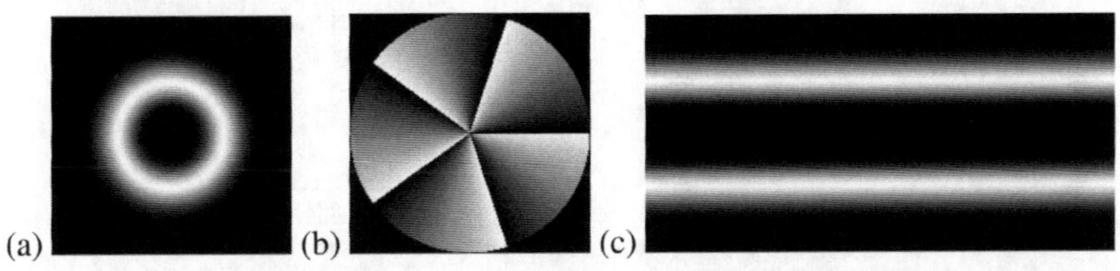

Figure 2. The propagation of LG modes with indexes $(n,m)=(0,5)$ in a parabolic fiber, $\sigma=2\lambda$: the distribution at the entrance (a) the intensity, (b) the phase and (c) the longitudinal intensity distribution $z\in[0,50\lambda]$.

(a) (b)
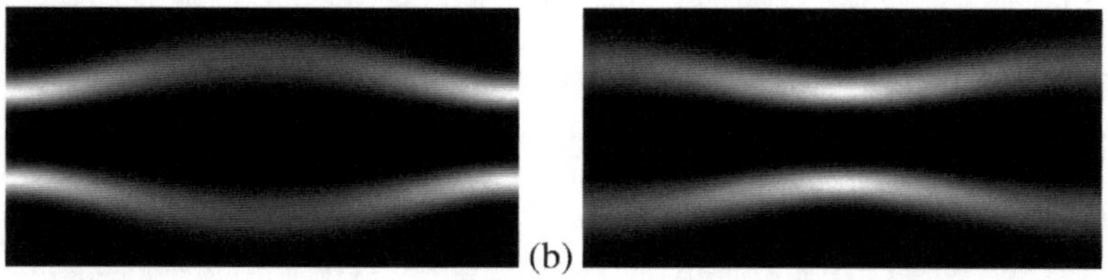

Figure 3. The propagation of LG modes with indexes (*n*,*m*)=(0,5) is the longitudinal intensity distribution where (a) $\sigma = 1.5\lambda$ and (b) $\sigma = 2.5\lambda$.

The propagation of the «single-ring» LG modes superposition with different vortex of order *m* and the waist radius σ in a single physical carrier (an optical fiber) is shown in Tables 1-3.
In particular, the propagation of the two LG modes superposition with parameters $m_1 = 1$, $m_2 = 7$, $\sigma_1 = \sigma_2 = 2\lambda$ was shown in Table 1.

Table 1. The propagation of the two LG modes superposition with parameters $m_1 = 1$, $m_2 = 7$, $\sigma_1 = \sigma_2 = 2\lambda$.

The transverse distributions on a period (intensity and phase)				
z=0	z=12.5λ	z=25λ	z=37.5λ	z=50λ

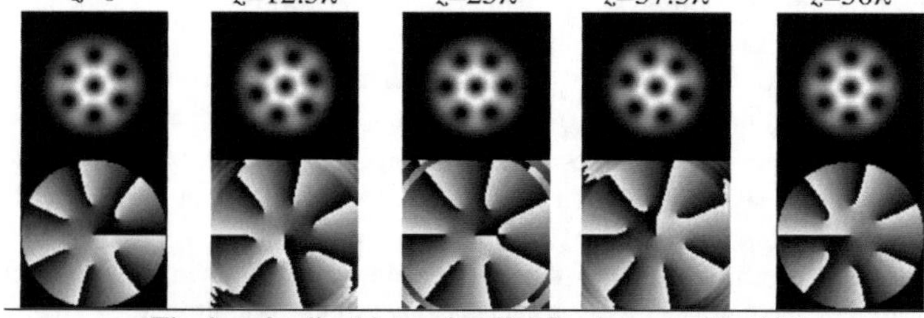

The longitudinal intensity distribution, $z \in [0, 50\lambda]$

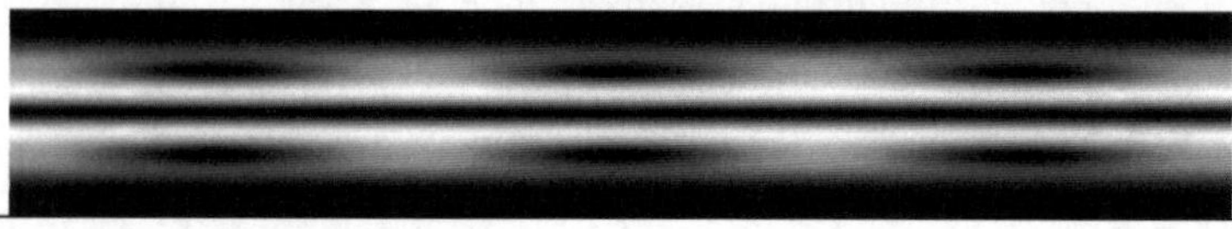

Table 2. The propagation of the two LG modes superposition with parameters $m_1 = 2$, $m_2 = -5$, $\sigma_1 = \sigma_2 = 2\lambda$.

The transverse distributions on a period (intensity and phase)				
z=0	z=12.5λ	z=25λ	z=37.5λ	z=50λ

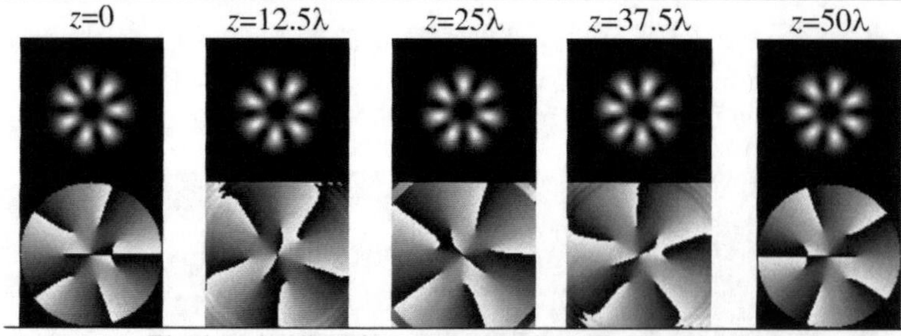

The longitudinal intensity distribution, $z \in [0, 50\lambda]$

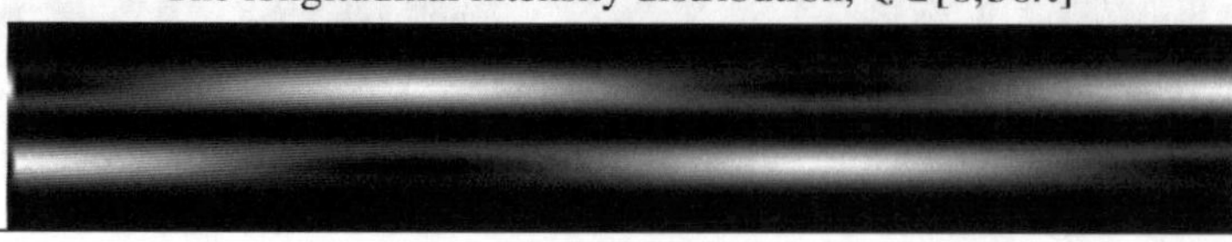

As can be seen from Tables 1 and 2, the transverse pattern in an ideal fiber does not change by using of the harmonized scales of the two modes superpositions, and it only rotates [26]. When there are mismatched scales and when a larger number of modes are used in the superposition (6), more complex patterns appears (Tables 3 and 4).

Table 3. The propagation of the two LG modes superposition with parameters $m_1 = 2,\ m_2 = -5,\ \sigma_1 = 2\lambda,\ \sigma_2 = 3\lambda$.

The transverse distributions on a period (intensity and phase)				
z=0	z=12.5λ	z=25λ	z=37.5λ	z=50λ

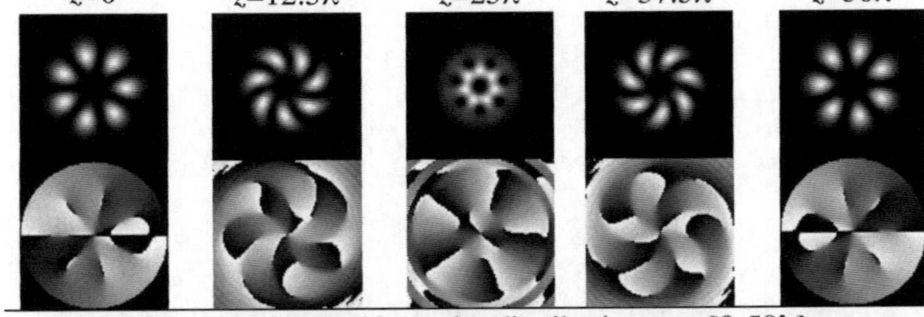

The longitudinal intensity distribution, $z \in [0, 50\lambda]$

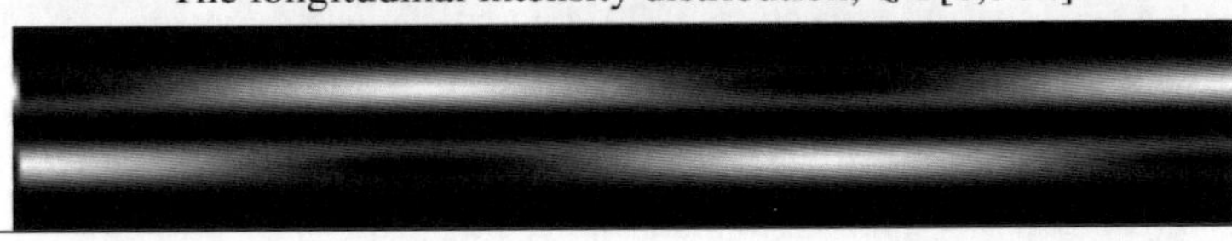

Table 4. The propagation of the three LG modes superposition with parameters $m_1 = 2,\ m_2 = 5,\ m_3 = 7,\ \sigma_1 = 3\lambda,\ \sigma_2 = 2\lambda,\ \sigma_2 = \lambda$.

The transverse distributions on a period (intensity and phase)				
z=0	z=12.5λ	z=25λ	z=37.5λ	z=50λ

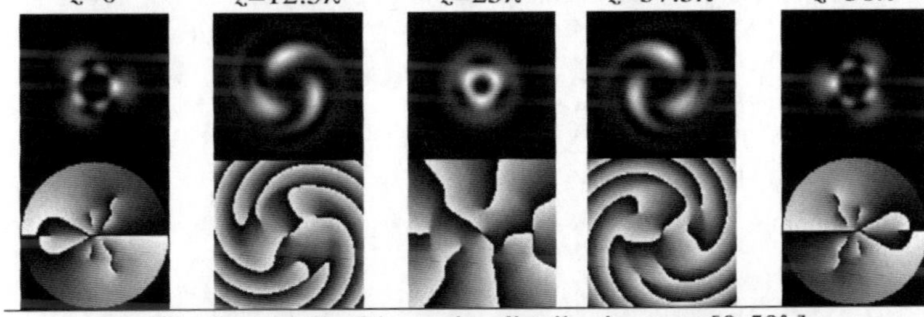

The longitudinal intensity distribution, $z \in [0, 50\lambda]$

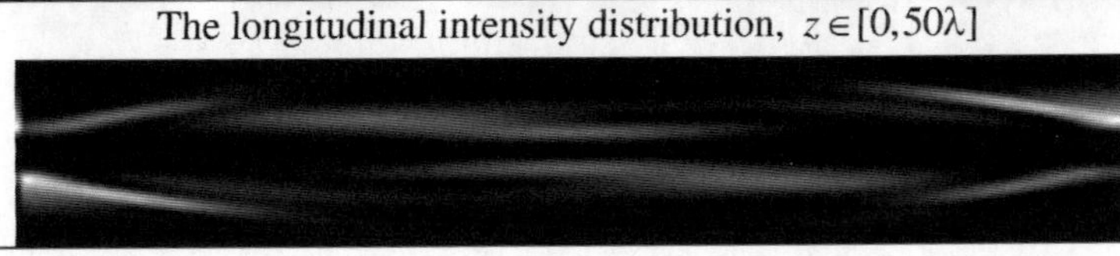

4. Summary and conclusions

To sum up, might be said that the propagation of the vortex modes superpositions in an optical fiber with a parabolic refractive index based on the FrFT was considered. It was presented that the transverse beam structure can be changed significantly during the propagation, especially when there is the inconsistency of the scales included in the superposition of modes with fiber parameters.

Acknowledgments

Authors wish to acknowledge the Russian Foundation for Basic Research (grants 16-07-00825, 16-29-11698) for financial support.

References

[1] Bozinovic N, Yue Y, Ren Y, Tur M, Kristensen P, Huang H, Willner A E and Ramachandran S 2013 *Science* **340** 1545–8

[2] Khonina S N, Kazanskiy N L and Soifer V A 2012 Optical vortices in a fiber: mode division multiplexing and multimode self-imaging *Recent progress in optical fiber research* ed M Yasin *et al* (Croatia: INTECH Open Access Publisher) chapter 15 pp 327 352

[3] Ramachandran S and Kristensen P 2013 *Nanophotonics* **2** 455-74

[4] Feit M D and Fleck J A 1978 *Appl. Opt.* **17** 3990-8

[5] Namias V 1980 *J. Inst. Math. Appl.* **25** 241-65

[6] Alieva T, Bastiaans M J and Calvo M L 2005 *J. Appl. Signal Processing* **10** 1-22

[7] Dorsch R G and Lohmann A W 1995 *Appl. Opt.* **34** 4111-2

[8] Cai L Z and Wang Y Q 2002 *Opt. Laser Technol.* **34** 249-52

[9] Malutin A A 2004 *Quantum Electronics* **2** 165-71

[10] Abramochkin E and Volostnikov V 1991 *Opt. Commun.* **83** 123-35

[11] Beijersbergen M W, Allen L, Van der Veen H E L O and Woerdman J P 1993 *Opt. Commun.* **96** 123-32

[12] Khonina S N, Karpeev S V and Ustinov A V 2009 *News of the Samara Science Center of the RAS* **11** 13-3

[13] Zoteeva O V and Khonina S N 2010 *Vestnik of the Samara State Aerospace University* **24** 184-92

[14] Khonina S N, Zoteeva O V and Karpeev S V 2011 *Pacific Science Review* **13** 233-41

[15] Kirilenko M S, Zubtsov R O and Khonina S N 2015 Computer Optics **39** 332-8

[16] Zubtsov R O and Kirilenko M S 2015 *Proc. of Information Technology and Nanotechnology (ITNT-2015), CEUR Workshop Proc.* **1490** 105-11

[17] Striletz A S and Khonina S N 2008 Computer Optics **32** 33-8

[18] Khonina S N, Striletz A S, Kovalev A A and Kotlyar V V *Proc. SPIE* **7523** 75230B

[19] Yariv A 1985 *Optical Electronics* (NewYork: Holt, Rinehart and Winston)

[20] Prudnikov A P, Bychkov Yu A and Marychev O I 1983 *Integrals and Series: Volume 2: Special Functions* (Moscow: Nauka)

[21] Khonina S N, Kotlyar V V and Soifer V A 1998 *Optics and Spectroscopy* **85** 636-44

[22] Khonina S N, Kotlyar V V, Skidanov R V, Soifer V A, Laakkonen P and Turunen J 2000 *Optics Communications* **175** 301-8

[23] Berdague S and Facq P 1982 *Appl. Optics* **21** 1950-5

[24] Thornburg W Q, Corrado B J and Zhu X D 1994 *Opt. Lett.* **19** 454-6

[25] Dubois F, Emplit Ph and Hugon O 1994 *Opt. Lett.* **19** 433-5

[26] Kotlyar V V, Soifer V A and Khonina S N *Optics and Lasers in Engineering* **29** 343-50

Asymmetric devices based on carbon nanotubes as detectors of sub-THz radiation

I A Gayduchenko[1], G E Fedorov[1], T S Stepanova[1], N Titova[1], B M Voronov[1], D But[2], D Coquillat[2], N Diakonova[2], W Knap[2] and G N Goltsman[1]

[1]Moscow State Pedagogical University, Moscow, 119991, Russia
[2]Laboratoire Charles Coulomb (L2C), UMR 5221 CNRS - Univ. Montpellier, Montpellier, France

Abstract. Demand for efficient terahertz (THz) radiation detectors resulted in intensive study of the asymmetric carbon nanostructures as a possible solution for that problem. In this work, we systematically investigate the response of asymmetric carbon nanodevices to sub-terahertz radiation using different sensing elements: from dense carbon nanotube (CNT) network to individual CNT. We conclude that the detectors based on individual CNTs both semiconducting and quasi-metallic demonstrate much stronger response in sub-THz region than detectors based on disordered CNT networks at room temperature. We also demonstrate the possibility of using asymmetric detectors based on CNT for imaging in the THz range at room temperature. Further optimization of the device configuration may result in appearance of novel terahertz radiation detectors.

1. Introduction

Terahertz (THz) radiation is used in applications ranging from security to medicine. However, sensitive room-temperature detection of terahertz radiation is notoriously difficult. The use of nano-scale objects is a promising route for obtaining cost-effective solutions for new THz detectors. One particular route is the use of carbon nanomaterials: single-wall carbon nanotubes (SWCNTs) and graphene and it already brought very encouraging results even at room temperature [1-7].

This work is a continuation of the author's investigation of the response of asymmetric devices based on CNTs to THz radiation [6-7]. It was previously shown that under certain conditions such devices demonstrate strong response to terahertz radiation even at room temperature. Analysis of the experimental data shows that the response contains two components. The first is the thermal reflecting different increase in temperature in the areas of the nanotube/metal interfaces. The second is the response due to the non-linearity of the current-voltage characteristic of the device at zero bias. Here, we systematically investigate the response of asymmetric carbon nanodevices to sub-THz radiation using different sensing element in order to determine the optimal morphology of it.

2. Experimental

The schematic device configuration is illustrated in Figure 1 (a). Our devices are made in a configuration of a field-effect transistor. The conduction channel between source and drain electrodes is formed by CNT networks of different density: disordered CNT networks, consisting of a mixture of the single-wall semiconducting CNTs (sm-CNTs) and quasi-metallic CNTs (qm-CNTs) of the p-type (Figure 1 (c)) or individual CNTs (Figure 1 (b)) both semiconducting and quasi-metallic. The carbon nanotubes are grown by a CVD method on Si/SiO_2 substrates using methane as a carbon feedstock. Room temperature resistivity of the silicon substrate is 10 Ω·cm and is high enough to use it as a backgate. The devices are coupled to the radiation with a logarithmic spiral antenna which also serves for DC contacts. The devices are fabricated in such way that the contact material is different for source and drain electrodes: vanadium and gold (Figure 1 (a)).

The experimental set-up is shown on Figure 1 (d). The sub-THz radiation was provided by a multiplier chain based on Schottky diodes allowing to measure spectral characteristics in the region from 140 GHz to 220 GHz. The DC voltage signal appearing at the asymmetric structures under the radiation was measured using lock-in amplifier. In all measurements gold electrode was grounded. The gate voltage between the silicon substrate and the electrode was supplied by the source - meter Keithley 2600.

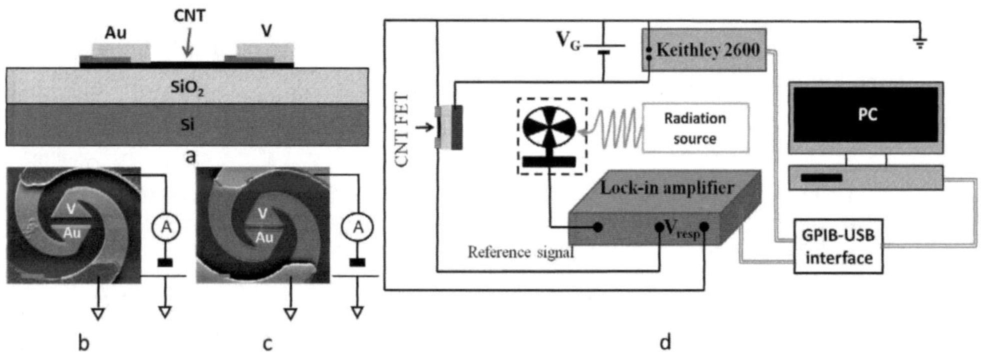

Figure 1. (a): Device configuration: schematic representation of the sensing element of the device illustrating the current path between gold and vanadium electrodes; (b-c): SEM images of the devices based on individual CNT (b) and disordered CNT networks (c). Scale bars are 5 μm; (d): the schematic representation of the experimental setup

3. Results and discussion

The transfer characteristics of our devices are shown on figure 2. As seen on Figure 2 (a), in the case of semiconductor CNT conductivity of the device is suppressed when gate voltage is positive. This is because that the Fermi level is in the band gap. When we apply negative voltage on the gate the Fermi level moves in the valence band and the conductivity of the device increases. Note that the conductivity of quasi-metallic CNTs is almost insensitive to the gate voltage (Figure 2 (b)). The asymmetry of the transistor characteristics is due to the fact that we use the different metals for source and drain electrodes. In the case of disordered CNT network ON/OFF ratio is about 2.4:1 (Figure 2 (c)), which corresponds to the ratio of average number of semiconductor nanotubes in the array to the amount of metal [8].

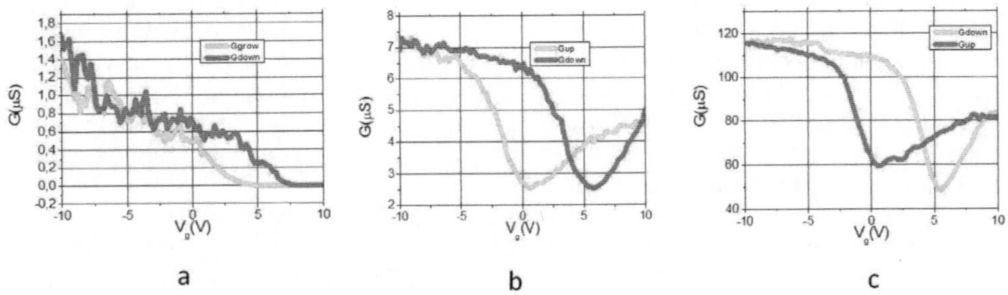

Figure 2. The transfer characteristics of the devices based on individual sm-CNT (a), individual qm-CNT (b) , mixed CNT network (c)

As it was shown in our previous works the asymmetry incorporated into our devices should result in DC voltage response to incident radiation at room temperature [6-7]. There are two effects that in turn may lead to observed DC voltage response to incident radiation at room temperature. One is related to the nonlinearity of the IV curve – so called diode type response. Second, the thermal effect is related to the temperature gradient building up across the devices under the action of the radiation and resulting thermal electromotive force [7]. In order to illustrate the contribution of this effect into the response we measured the gate voltage dependence of the response voltage in case of disordered CNT film (see Figure 3). We compare the observed signal with the gate voltage dependence of the Seebeck coefficient calculated using the Mott formula [9]:

$$S \equiv \frac{\Delta V}{\Delta T} = -\frac{\pi^2 k_B^2 T}{3|e|}\frac{1}{G}\frac{dG}{dE}\bigg|_{E_F} = -\frac{\pi^2 k_B^2 T}{3|e|}\frac{1}{G}\frac{dG}{dV_G}\frac{dV_G}{dE}\bigg|_{E_F} \tag{1}$$

where k_B is the Boltzmann constant and e is the electron charge. We use the measured $G(V_G)$ curves to calculate the $S(V_G)$ dependence using the formula (1).

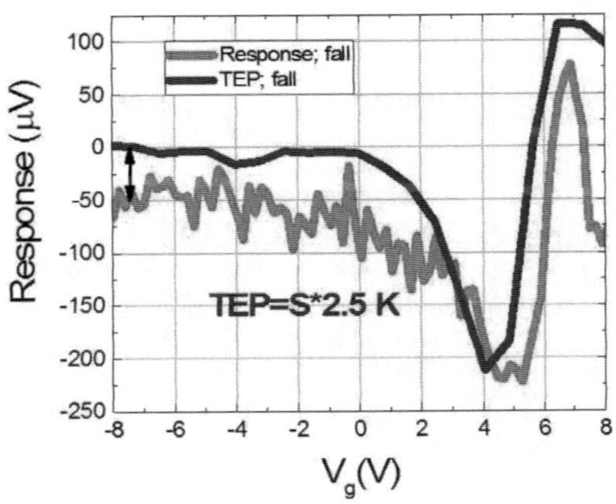

Figure 3. Room temperature response of device based on disordered CNT network as a function of gate voltage (magenta) compared to the TEP (blue) calculated using Eq. (1).

The most important feature of the calculated S(VG) dependence is that it tends to zero as the gate voltage goes to more negative values. Same should apply to the thermoelectric power (TEP) across the device. At the same time the measured response is not going to zero at negative gate voltages. This additional component of response (so called diode response [7]) should be almost the same at negative gate voltage for a given type of sensing element. This operating mode of the detector is the most promising from a practical point of view. In particular, the operation of the detectors at the same gate voltage is necessary to unite them in a matrix to create imaging systems. In this regard, we compared the response of our devices at negative gate voltage.

Figure 4 shows the spectral response of our devices to sub-THz radiation in the region from 140 GHz to 220 GHz. As we can see the detectors based on individual CNTs both semiconducting and quasi-metallic demonstrate much stronger response in sub-THz region than detectors based on disordered CNT networks. The weaker response of the network is most probably to extra scattering in the channel introduced by the CNT crossing points. At the same the shape of spectrum does not depend on sensitive element and is determined by characteristics of antenna. The explanation of this dependence is beyond the scope of this work and is the subject of further study. The response time of the devices is determined by device resistance and geometric capacitance between the electrodes and the gate [6], indicating that the detectors based on CNT networks should be faster due to the higher conductivity. The presence of a large number of CNTs in the conduction channel should lead to a more stable operation of the detector.

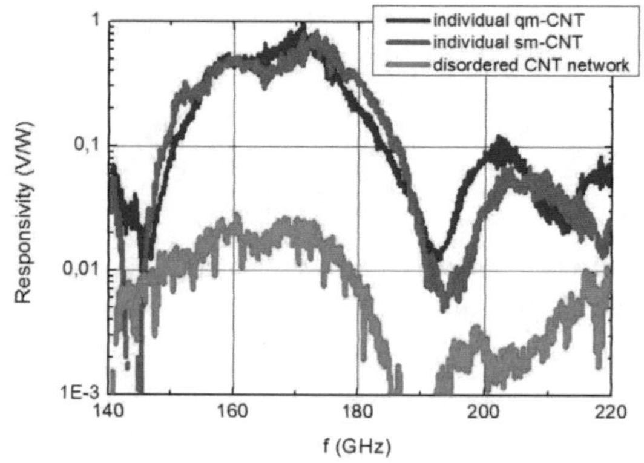

Figure 4. The spectral response of the devices with different numbers of CNTs in the channel at negative gate voltage

In order to demonstrate prospects of the use of the asymmetric carbon nanostructures as room temperature THz detectors we make a THz image of object hidden in carton box (Figure 5). The visualized object is placed inside carton box and fixed on the motorized translation stage allowing to move automatically in horizontal and vertical direction. The radiation source and the detector are fixedly mounted, so that the radiation is focused on the detector by a lens system. The carton box with the object inside it moves so that the entire surface of the box is scanned by radiation. Figure 5 (c) shows the resulting THz image of the hidden object.

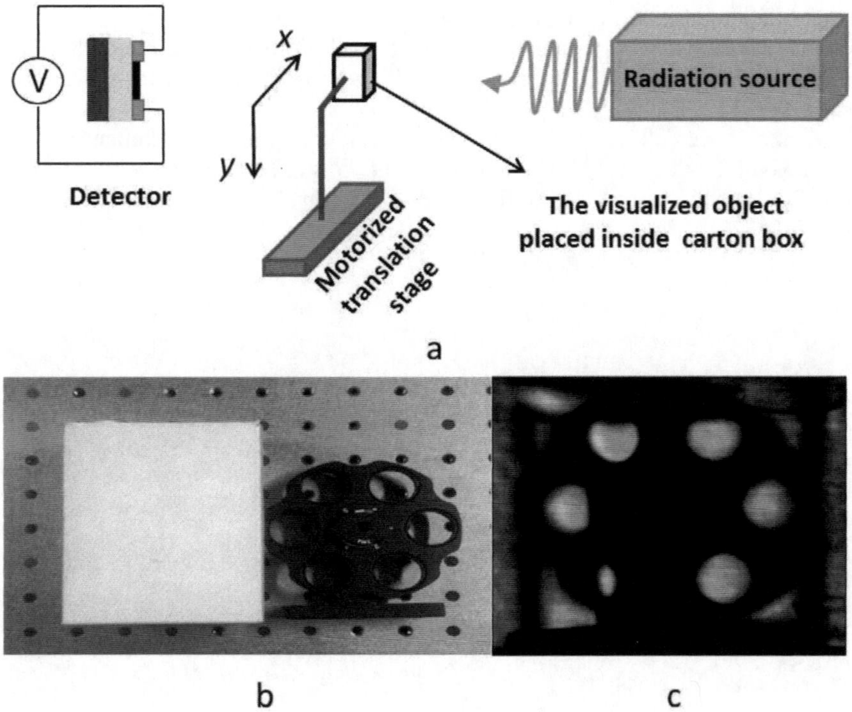

Figure 5. The schematic representation of the experimental setup for THz imaging (a). Optical (b) and THz (c) image of the rotator placed inside carton box

4. Conclusion

We conclude that the detectors based on individual CNTs both semiconducting and quasi-metallic demonstrate much stronger response in sub-THz region than detectors based on disordered CNT networks at room temperatureWe also demonstrate the possibility of using asymmetric detectors based on CNT for imaging in the THz range at room temperature. Further improvement of the network morphology, e.g. aligning the CNTs along the same direction will make it possible to make a detector that is as sensitive as one based on an individual CNT, but with better stability.

Acknowledgments

This work was supported by the Ministry of Education and Science of Russian Federation under Contract No. 14.586.21.0003 (project ID RFMEFI58614X0003)..

References

[1] Cai X, Sushkov A B, Suess R J, Jadidi M M, Jenkins G S, Nyakiti L O, Myers-Ward R L, Li S, Yan J, Gaskill D K, Murphy T E, Drew H D and Fuhrer M S 2014 *Nature Nanotechnology* **9** 814

[2] Vicarelli L, Vitiello MS, Coquillat D, Lombardo A, Ferrari AC, Knap W, Polini M, Pellegrini V, Tredicucci A 2012 *Nature Materials* **11** 865

[3] Fu K, Zannoni R, Chan C, Adams S H, Nicholson J, Polizzi J E and Yngvesson K S 2008 *Appl. Phys. Lett.* **92** 033105

[4] Rinzan M, Jenkins G, Drew H D, Shafranjuk S and Barbara P 2012 *Nano Lett.* **12 (6)** 3097

[5] He X, Fujimura N, Lloyd J M, Erickson K J, Talin A A, Zhang Q, Gao W, Jiang Q, Kawano Y, Hauge R H, Léonard F, and Kono J 2014 *Nano Letters* **4 (7)** 3953,

[6] Fedorov G, Kardakova A, Gayduchenko I, Charayev I, Voronov B M, Finkel M, Klapwijk T M, Morozov S, Presniakov M, Bobrinetskiy I, Ibragimov R and Goltsman G 2013 *Appl. Phys. Lett.* **103** 181121

[7] Gayduchenko I, Kardakova A, Fedorov G, Voronov B, Finkel M, Jimenez D, Morozov S, Presniakov M, and Goltzman G N 2015 *J. Appl. Phys*. **118** 194303

[8] Saito R, Dresselhaus G, and Dresselhaus M S 1998 *Physical Properties of Carbon Nanotubes Imperial College Press* 259

[9] Cutler M and Mott N F 1969 *Phys Rev* **181** 1336

Structure and spectral properties of the silver-containing high-silica glasses

M A Girsova[1], G F Golovina[1], I N Anfimova[1], M Yu Arsent'ev[2], T V Antropova[1]
[1]Laboratory of Physical Chemistry of Glass, Grebenshchikov Institute of Silicate Chemistry RAS, Saint-Petersburg 199034, Russia
[2]Laboratory of Research of Nanostructures, Grebenshchikov Institute of Silicate Chemistry RAS, Saint-Petersburg 199034, Russia

Abstract. Silver-containing high-silica glasses were synthesized by an impregnation of the silica porous glasses (PGs) first with $AgNO_3$ aqueous solution (with or without the presence of the sensitizers, such as $Cu(NO_3)_2$ or $Ce(NO_3)_3$), next in the mixed halide salt (NH_4Cl, KBr, KI) solution. Then some part of the samples was sintered at the temperatures from 850 to 900°C up to closing of the pores. The structure of glasses was studied by UV-VIS-NIR and IR spectroscopy and X-ray diffraction (XRD) techniques. According to XRD data the silver-containing high-silica glasses contain the AgBr, AgI, Ag_3PO_4, $(CuBr)_{0.75}(CuI)_{0.25}$ phases. IR spectra confirmed B–O–B, Si–O–Si, P–O–P, O–P–O, O–B–O bonds, $(PO_4)^{3-}$ and $P–O^-$ groups in glasses.

1. Introduction

Silver-containing materials and glasses are used as materials for creation of the plasmonic waveguides, 3D laser modified areas, in superlenses and sensors [1–6].

2. Experimental

Synthesis of the silver-containing high-silica glasses was carried out according to the procedure [1, 7–9], which consist in following. At first, PG NFF matrixes (in the form of the rectangular plane-parallel plates in size (1.5±0.03) x (5–15) x (5–25) mm^3) were placed in 100 mg/ml $AgNO_3$ aqueous solution (with or without the presence of sensitizers, such as 10 mg/ml $Cu(NO_3)_2$ or 10 mg/ml $Ce(NO_3)_3$). Next the samples were immersed in a mixed halide salt (NH_4Cl, KBr, KI) solution. As the result the samples of photochromic porous glasses (PPGs) with silver halides / with silver and copper halides / with silver and cerium halides were obtained. PPGs were sintered at temperatures from 850 to 900°C for 10–40 minutes which leads to closing of the pores and the photochromic quartz-like glasses (PQGs), i.e. a monolitic high silica material is formed. Depending on the concentration of the impregnating solutions containing silver and sensitizer, in this paper we made the following notation of PPGs and PQGs: 100Ag, 100Ag / 10Cu, 100Ag / 10Ce. The pure quartz-like glasses (QGs) were prepared from PGs without impregnation under the same sintering conditions. The glasses under investigation have following compositions (as analyzed, wt. %): 0.55 Na_2O, 6.85 B_2O_3, 92.56 SiO_2, 0.04 |F|, traces P_2O_5 (PGs), 0.46 Na_2O, 6.42 B_2O_3, 93.09 SiO_2, 0.03 |F|, traces P_2O_5 (QGs); (0.43–0.49) Na_2O, (0.52–0.77) K_2O, (5.83–6.39) B_2O_3, (91.00–91.89) SiO_2, (0.02–0.13) |F|, (0.08–0.10) P_2O_5, (0.89–1.60) Ag_2O, and in the presence or without the sensitizer (0.02 CuO / 0.04 Ce_2O_3) (PPGs and PQGs).

The synthesized glasses were investigated by various techniques. IR transmission spectra were obtained wih SPECORD M-80 spectrophotometer in the range of 4000 to 300 cm^{-1} at the spectral resolution of 4 cm^{-1}. For measurements of IR spectra the glasses were powdered and mixed with KBr in order to obtain thin pellets by vacuum pressing. Transmission spectra were measured by SF-2000 UV / VIS spectrophotometer in the wavelength range of 190–1100 nm with a step of 0.115 nm. Spectral measurements were conducted at room temperature. X-ray diffraction (XRD) studies were carried out with the D8-Advance "Bruker" powder diffractometer with monochromatic CuK_{α}-radiation (operating voltage 40 kV, current 40 mA) from the sample surface. Phases were identified using the International Centre for Diffraction Data (ICDD-2006) database.

3. Results and discussion

According to XRD data (figure 1), in all the glasses only phases with cubic structure were found. The **PPGs 100Ag** contain AgBr (Pattern no. 01-071-3754, the unit cell parameter is a = 5.780 Å), Ag_3PO_4

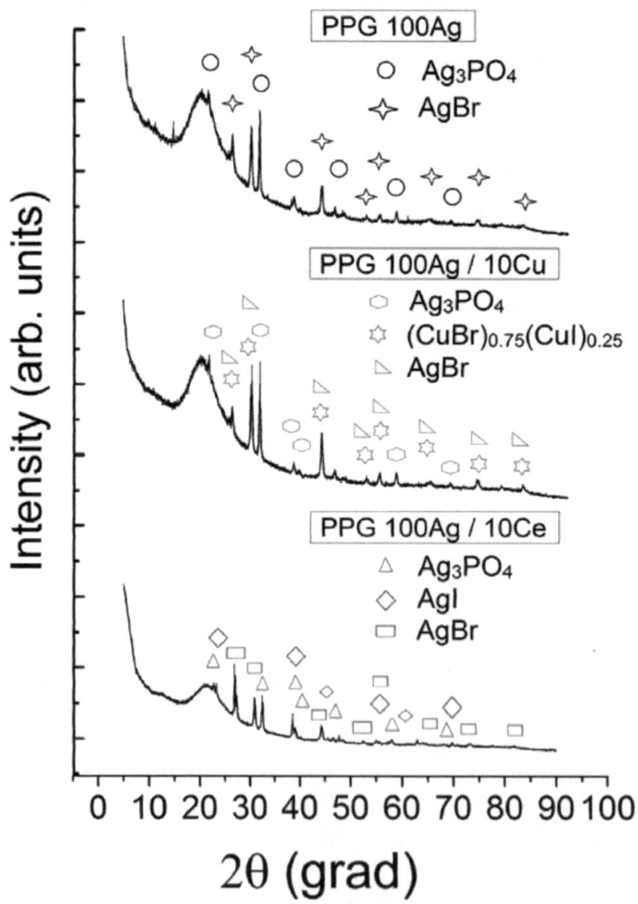

Figure 1. X-ray diffraction spectra of the photochromic porous glasses.

(Pattern no. 00-030-1137, a = 7.720 Å) phases, the **PPGs 100Ag / 10Cu** – AgBr (Pattern no. 01-079-0148, a = 5.769 Å), Ag_3PO_4 (Pattern no. 00-030-1137, a = 7.752 Å) and $(CuBr)_{0.75}(CuI)_{0.25}$ (Pattern no. 01-077-2388, a = 5.772 Å) phases, **PPGs 100Ag / 10Ce** – AgBr (Pattern no. 01-071-3754, a = 5.766 Å), Ag_3PO_4 (Pattern no. 00-030-1137, a = 7.768 Å), and AgI (Pattern no. 01-071-3829, a =

6.606 Å) phases. The glasses additionally doped with copper or cerium have smaller unit cell parameter of AgBr and larger unit cell parameter of Ag_3PO_4, which may be due to the influence of sensitizers. Such reduction of the unit cell parameters can be explained by the fact that for mixed crystals of the silver halides the superimposition and shift of the diffraction peaks are characteristic [8].

IR and UV-VIS-NIR transmission spectra of the glasses are shown in figure 2. In the IR spectra (figure 2a) of all types of glasses (PG and PPGs) the absorption bands are observed at: 3676–3672,

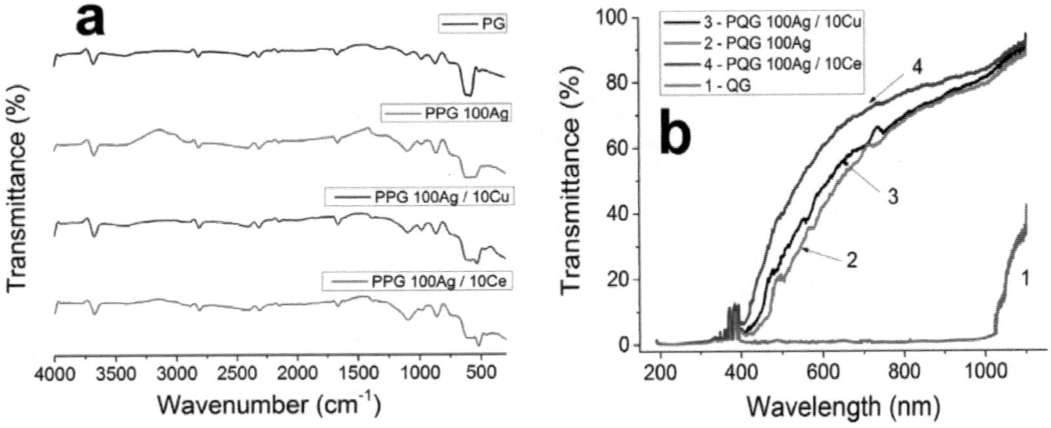

Figure 2 (a, b). (a) IR transmission spectra of the glasses; **(b)** UV-VIS-NIR transmission spectra of the glasses.

3500–3476, 3428–3404, 2896–2888, 2824–2812, 1664–1660, 1388–1368, 1100–1096, 984–972, 864–856, 732–724, 692–684, 476–472, 464–456 cm^{-1}. Broad bands placed at 3676–3672, 3500–3476, 3428–3404, 1664–1660 cm^{-1} are assigned to stretching vibrations of hydroxyl and water groups [10–11]. Two bands at 2896–2888 and 2824–2812 cm^{-1} are associated with hydrogen bonds [10]. The band at 3428–3404 cm^{-1} is due to Si–OH stretching and also due to vibrational structure of Si–O–Si [12]. The band located at 1388–1368 cm^{-1} is assigned to B–O asymmetric stretching vibrations in BO_3 units [13]. The absorption band at 1100–1096 cm^{-1} is connected with Si–O–Si asymmetric stretching and assigned to $P–O^-$ groups, the phosphate non-bridging oxygen portion of PO_4 tetrahedra in a chain structure [14–16]. Bands have been found in the range 984–972 cm^{-1} assigned to the stretching and deformation modes of $(PO_4)^{3-}$ groups and confirming the presence of $P–O^-$ groups in the glasses [15–16]. Stretching vibrations of B–O bonds in BO_4 units are responsible for band at 864–856 cm^{-1} [13]. The band observed at 732–724 cm^{-1} suggest the presence of symmetric stretching modes of the linear P–O–P linkages, ν_{sym} (P–O–P) [15, 17]. The absorption band at 692–684 cm^{-1} is assigned to ν_{sym} (P–O–P) and B–O–B bond bending vibrations of the borate network [17–19]. The two weak bands appearing at 472 cm^{-1} and in the range 464–456 cm^{-1}, which may be attributed to bending vibrations δ (Si–O–Si), δ (O–P–O) and δ (O–B–O) linkages [12, 15, 17–18].

All samples of PQGs are highly transparent in the range 370–1100 nm (figure 2b). QG has low transparency (strong absorbance) from 190 nm to 370 nm and from 395 nm to 980 nm. Other character of light transmittance of the QG plates may be caused by a formation scattering and absorbing structure (possible microcrystalline inhomogeneities) [20]. The broad absorption band from 190 to 322 nm and the maxima at 344, 352, 364, 379, 392–393 nm (in all glasses) are associated with the glass matrix. The absorption bands at 395 nm (curve 4), 410 (curves 2, 3), 505, 577, 722 nm (curve 2) are conditioned by silver. Several absorption bands around 481, 559, 712, and 748 nm (curve 3) are caused by the presence of copper. The spectra showed three surface plasmon resonance peaks two of which centered at 395 and 410 nm corresponding to the silver and the another centred at 559 nm to that of copper in the glasses [21–23].

It should be noted that the given results (figure 2) obtained at the study of the PPGs and PQGs on the basis of PG NFF matrixes differ from data for the similar samples received on the basis PG 8V-NT [8].

The optical absorption spectra of PPGs are shown in figure 3. A plurality of absorption bands were

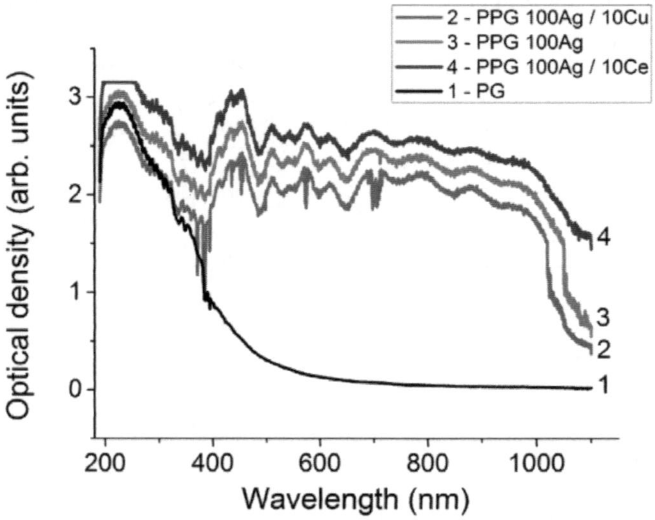

Figure 3. Absorption spectra of the photochromic porous glasses in comparison with porous glass.

detected on absorbance spectra. The maxima at 364, 376–379, 396, 410–411, 433–434, 454, 511–514, 539–544, 570, 617–620, 693–698, 779–780, 875–876, 966–971 nm (curves 2–4) can be connected with presence of silver inside matrix of PPGs. The peaks observed at 581, 689, 720, 789 nm (curve 2) could be corresponding to the presence in the glasses of divalent copper ions [23]. Absorption band centered 789 nm which is due to the characteristic $^2B_{1g} \rightarrow {}^2B_{2g}$ transition of Cu^{2+} ions in a distorted octahedral site [19].

4. Conclusion

X-ray diffraction spectra of PPGs show that they contain AgBr, AgI, Ag_3PO_4, $(CuBr)_{0.75}(CuI)_{0.25}$ phases in the cubic system. According to IR transmission spectra, PPGs contain B–O–B, Si–O–Si, P–O–P, O–P–O, O–B–O bonds, $(PO_4)^{3-}$ and P–O$^-$ groups in glasses. UV–VIS–NIR transmission spectra have shown that PQGs are transparent throughout the visible and near infrared spectral range. Optical absorption spectrum in PPG 100Ag / 10Cu shows broad band at 789 nm which is due to the characteristic $^2B_{1g} \rightarrow {}^2B_{2g}$ transition of Cu^{2+} ions.

Acknowledgments

Work is carried out within the limits of the State Assignment under the Program of Fundamental Scientific Researches of the State Academies of Sciences for 2013-2020 (a theme N 0097-2015-0021). The authors thank L. F. Dikaya, L. A. Doronina, L. N. Kurilenko, T. G. Kostyreva for chemical analysis of the glasses.

References

[1] Kostyuk G K, Sergeev M M, Girsova M A, Yakovlev E B, Anfimova I N, Antropova T V 2014 *Glass Physics and Chemistry* **40** 415.

[2] Veronis G, Fan S 2005 *Applied Physics Letters* **87** 131102.

[3] Veiko V, Sergeev M, Kostyuk G 2015 *Journal of Laser Micro/Nanoengineering* **10** 20.

[4] Moore C P, Blaikie R J, Arnold M D 2009 *Optics Express* **17** 14260.
[5] Jeppesen C, Nielsen R B, Boltasseva A, Xiao S, Mortensen N A, Kristensen A 2009 *Optics Express* **17** 22543.
[6] Zhao J, Zhang X, Yonzon C R, Haes A J, Van Duyne R P 2006 *Nanomedicine* **1** 219.
[7] Girsova M A, Drozdova I A, Antropova T V 2014 *Glass Physics and Chemistry* **40** 162.
[8] Girsova M A, Golovina G F, Drozdova I A, Polyakova I G, Antropova T V 2014 *Optica Applicata* **44** 337.
[9] Antropova T, Girsova M, Anfimova I, Drozdova I, Polyakova I, Vedishcheva N 2014 *Journal of Non-Crystalline Solids* **401** 139.
[10] Ramadevudu G, Rao S L S, Shareefuddin MD A H, Chary M N 2011 *International Journal of Engineering Science and Technology* **3** 6998.
[11] Kundu V, Dhiman R L, Maan A S, Goyal D R 2009 *Journal of optoelectronics and advanced materials* **11** 1595.
[12] Jeon H-J, Yi S-C, Oh S-G 2003 *Biomaterials* **24** 4921.
[13] Ardelean I, Cora S 2008 *J. Mater. Science: Materials in Electronics* **19** 584.
[14] Patra A, De G, Kundu D, Ganguli D 2000 *Materials Letters* **42** 200.
[15] Das S S, Srivastava P K, Singh N P, Srivastava V 2010 *Indian Journal of Engineering & Materials Sciences* **17** 123.
[16] Okura T, Yoshida N 2012 *World Academy of Science, Engineering and Technology* **6** 186.
[17] Das S S, Singh N P, Srivastava V, Srivastava P K 2008 *Indian Journal of Engineering & Materials Sciences* **15** 256.
[18] Ardelean I, Timar V 2008 *Journal of optoelectronics and advanced materials* **10** 246.
[19] Bhogi A, Kistaiah P 2015 *Phys. Chem. Glasses: Eur. J. Glass Sci. Technol. B* **56** 197.
[20] Drozdova I, Antropova T 2008 *Optica Applicata* **38** 17.
[21] Manikandan D, Mohan S, Nair K G M 2003 *Physica B* **337** 64.
[22] Zamkovets A D, Ponyavina A N, Baran L V 2010 *Journal of Optical Technology* **77** 456.
[23] Podsvirov O A, Sidorov A I, Tsekhomskiĭ, Vostokov A V 2010 *Physics of the Solid State* **52** 1906.

Plasmon coupled nanoparticle arrays for fluorescence, photoluminescence and Raman scattering enhancement.

V E Kaydashev, N Lyanguzov, D Zhilin, A Tsaturyan, E A Raspopova and E M Kaidashev

Southern Federal University, Rostov-on-Don, 344090, Russia

Abstract. We study the enhancement of the fluorescence an photoluminescence by large homogeneous arrays of plasmon coupled 5-8 nm Au and Ag nanoparticle separated by distances less than 10 nm. A red shift of the major "symmetric" plasmon mode near 780 nm, and additional "anti-symmetric" plasmon mode centered near 310 nm evidence the plasmon coupling in Au particle arrays. The systems were found to be effective in enhancement of the fluorescence/photoluminescence processes.

1. Introduction

The ability of plasmon nanoparticles to concentrate electromagnetic radiation in the near field make them powerful probe to study ultralow concentrations of chemical and biological species by surface enhanced Raman scattering (SERS), fluorescence or chemiluminescence imaging [1-4]. Electromagnetic filed redistribution in vicinity of a single plasmon nanoparticle may cause a quenching or enhancement of molecule fluorescence [5]. The metal nanoparticle may change i) a local field distribution and ii) increase the intrinsic decay rate of the fluorophore. Thus, the total effect of particle on the fluorescence is a superposition of two factors, notably, stronger excitation rate and the increase of the quantum yield (QY) / decrease the fluorescence lifetime [6-8]. When the electric field is M times enhanced near the isolated nanoparticle, which is not plasmon coupled with others, a Raman cross section of scattering molecule is enhanced by a factor of ~$|M|^4$, whereas the fluorescence cross-section of the species is increased ~$|M|^4/|M_d|^2$ times. Here $|M_d|$ denotes an amplification factor of the excited state decay rate for a molecule in vicinity of nanoparticle. More specifically, when for instance, a phosphor molecule is in the immediate vicinity of 40 nm silver particle, i.e. the distance is ~5Å, $|M|^2$ factor reaches value of 20-40, whereas $|M_d|^2 \approx 10^6$. Thus, one can expect a fluorescence quenching by a factor of ~10^3 compared with the free-molecule case [9,10]. With increasing the particle-molecule distance to ~8-10 nm the fluorescence cross section is increased and passes through the maximum because $|M|^4$ decreases more rapidly that $|M_d|^2$ [6,9]. Still, M is not too high at any distance from a single spherical nanoparticle and, thus, large Raman enhancement does not occur and one can expect only a moderate fluorescence enhancement at distance of ~5-10 nm from the metal particle [5]. The local electromagnetic fields are completely changed when the distance between adjacent particles becomes smaller then ~10 nm and the plasmon modes of single nanoparticles become coupled [11]. So called "hot spots" in percolation films and "self-similar" nanoparticle arrays are discussed since pioneer paper by Li et.al.[12], further theoretically characterized in [13,14] and are summarized in review [15]. The considerable shift of the "symmetric" plasmon resonance band from 500 nm to 1500 nm and appearance of another band in UV-green of an "anti-symmetric" plasmon mode in strongly coupled ~80 nm particle assemblies with inter-particle spacing of ~1-2 nm was recently demonstrated by Hanske et.al [16]. Gaint SERS and fluorescence enhancement induced by strongly coupled nanoparticles using precisely controlled nano-engineering were demonstrated only a few years ago [8,17]. More specifically, a 473-fold fluorescence enhancement for fluorophore with QY=0.3 was reported for Au dimers with gap distance of ~6 nm [17] and more than 1000-fold fluorescence enhancement near nanorod for low QY (1.9-2.5%) [18]. Electron beam lithography is normally used to design plasmon nanoobjects in precisely controlled way but on small area. In particular, an array of bowtie nanoantennas with a gap down to 6 nm was presented recently [11]. Self-assembling techniques namely nanosphere lithography and template assisted assembling [16], drying of colloids [19], use of ligands to produce nanogaps [20] and other deposition techniques allow design quasi-homogeneous arrays of nanoparticles on the large areas. We study the capability of vacuum based methods such as pulsed laser deposition (PLD) and magnetron sputtering to design large homogeneous arrays of plasmon coupled Ag and Au nanoparticles for fluorescence enhancement and SERS.

2. Experimental

The special feature of nanoparticle arrays morphology produced by depositing a metal film in vacuum and Ar gas were characterized using FE-SEM Zeiss SUPRA 25. Optical absorption spectra were studied using UV/Vis Spectrophotometer Varian 5000. The influence of Au particle arrays on the fluorescence of Rhodamine B dye and on the ZnO photoluminescence were studied under excitation at 514 nm (Ar+ laser) and at 325 nm (He-Cd laser), correspondingly. The spectra were detected by using Renishaw inVia Reflex Raman spectrometer with spectral resolution better than 1 cm^{-1}. All the samples were excited at normal angle to the substrate with the fluorescence/PL emission being collected in the backward direction. As a metal-enhanced fluorescence has a radiation pattern with maxima in forward and backward directions one may expect that ~50% of emitted light was collected. The areas with/without nanoparticles were prepared on the same substrate and the identical measurement conditions were used to minimize an uncertainty.

3. Results and discussion

To illustrate the influence of the ambient gas on the nucleation of a metal film we studied in-situ the PLD growth of semi-continuous Au film in vacuum (10^{-5} mbar) and in Ar (0.7 mbar) atmosphere at room temperature. Laser ablation was done by focusing a KrF (248 nm, 10Hz) laser beam on a rotating target to give a fluence of ~2 J cm^{-2} on a 10 mm^2 rectangular spot. For deposition in vacuum a target to substrate distance was 5 cm. The DC resistance of the Au film growing in vacuum on ZnO/a-Al_2O_3 substrate was measured in situ as a function of laser pulse number (Figure 1). With increasing deposited material and fractional coverage f, nanoparticles grow to reach a percolation threshold at $f=f_c$. As the film keeps growing the isolated nanoparticles ((1) in Figure 1) transform to a semi-continuous ($f>f_c$) ((2, 3) in Figure 1) and, finally, a continuous ($f=1$) film at an equivalent mass thickness more than 10 nm. The percolation threshold for Au films in the discussed deposition regime corresponds to ~100-200 laser pulses.

When ablating Au in the heavy ambient gas such as Ar, one may easier obtain more homogeneous arrays of ~5-8 nm Au nanoparticles separated by distances less than 10 nm. The number of laser shots corresponding to percolation threshold is ~600-800 pulses in this case. The vacuum chamber was evacuated down to 2×10^{-4} mbar and an Ar flow was introduced to maintain a pressure of 0.7 mbar. Au particles were deposited at room temperature. A reduced target to substrate distance of ~3.5 cm was used, which roughly corresponds to stopping distance of a laser plume at this pressure. An array of "small" Au nanoparticles less than 10 nm separated by gaps less than 10 nm is obtained for 500 laser shots (Figure 2a).

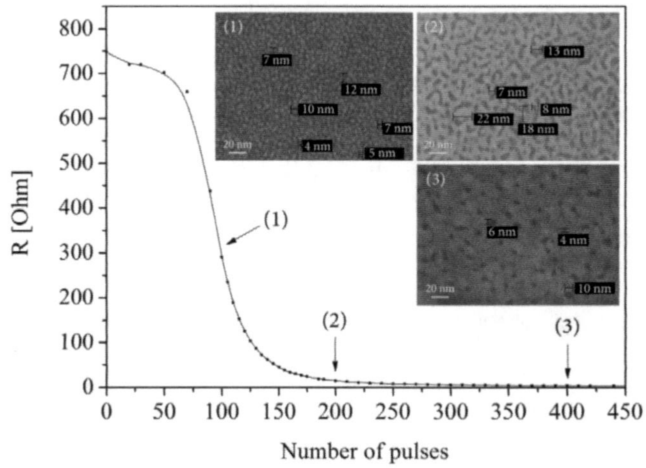

Figure 1. The DC resistance of the growing Au film on $ZnO/a\text{-}Al_2O_3$ substrate measured in situ as a function of laser pulse number. SEM images (1)-(3) of the samples deposited for respective number of laser shots are shown in the inset.

Ag nanoparticle arrays were deposited by DC-magnetron sputtering at room temperature (Figure 3a). A rotating substrate was maintained at 8 cm from the Ag (99.8%) target. Ag films were deposited at $3.2*10^{-2}$ mbar of Ar ambient pressure. The discharge voltage and current were maintained at to be 100 V and 35 mA, respectively.

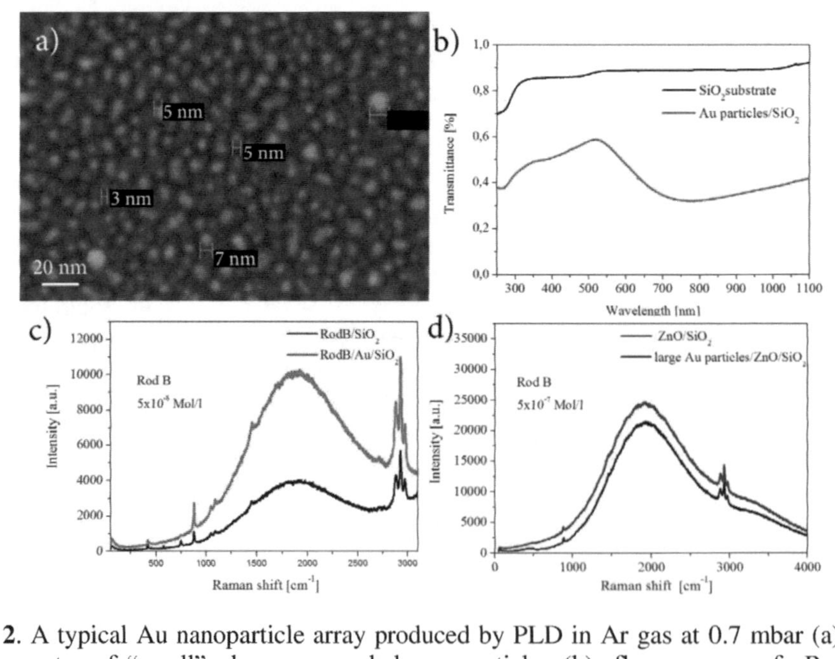

Figure 2. A typical Au nanoparticle array produced by PLD in Ar gas at 0.7 mbar (a); optical absorption spectra of "small" plasmon coupled nanoparticles (b); fluorescence of Rod B and Raman scattering enhanced by the array of "small" nanoparticles; minor effect of "large" not coupled nanoparticle array on fluorescence (d)

Optical absorption spectra of isolated small Au nanoparticles normally show a single major plasmon resonance peak near 520 nm. Higher energy shoulder is often attributed to d→ sp transitions of gold. Still, additional plasmon mode caused by plasmon coupling may also appear in this range when the inter-particle distance is reduced to nanometers [13]. Absorption spectra of Au nanoparticle array onto SiO_2 substrates deposited in Ar gas revealed two plasmon bands evidencing that plasmon coupling occurs (Figure 2b). More specifically, a major "symmetric" plasmon mode is shifted to red and appears near 780 nm, whereas additional "anti-symmetric" plasmon mode was centered near 310 nm [13,14]. Similar increase of optical absorption in the UV range can be followed in spectra presented in several previous studies of strongly plasmon coupled nanoparticles, namely in [18,16], but the origin of such behavior was not discussed there.

Similar, Ag nanoparticle arrays deposited onto ZnO films show plasmon coupling behavior. Apart from the major plasmon resonance near 500-530 nm an additional broad band in UV/blue region with photon energy higher the band gap of ZnO was observed (Figure 3b, c). In the case of silver the d-band of the ground state is completely separated from the sp-band, thus the d→ sp transitions show higher photon energies than those of Au. Thus, the observed increase of the optical absorption in the UV region is likely originated from the "beginning" of the plasmon coupling rather than from the d→ sp transitions [14]. For Ag particle array a "symmetric" mode is shifted to lower photon energy for ~57 nm and ~35 nm with increase of Ag content in the case of particle arrays on SiO_2 snd ZnO surfaces, respectively (Figure 3b, c). The UV broad band corresponding to "anti-symmetric" plasmon mode becomes more pronounced with increase of Ag content for both cases (Figure 3b, c).

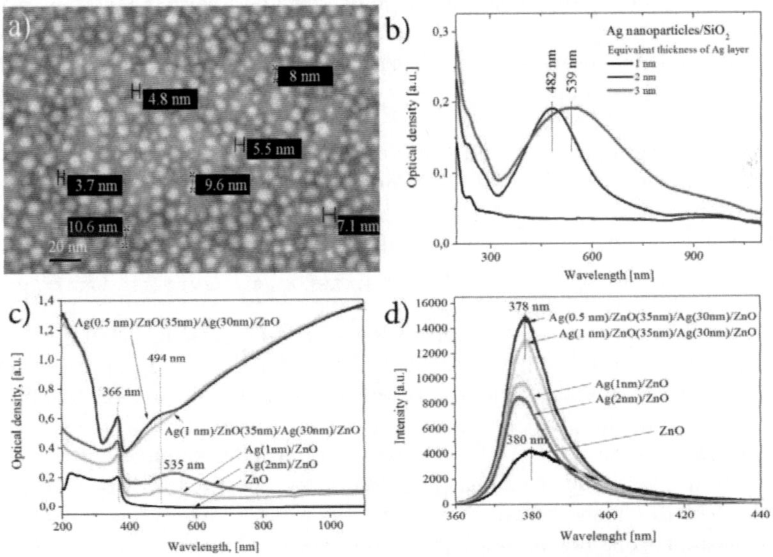

Figure 3. SEM image of Ag nanoparticle array (a), optical density of Ag particles on SiO_2 (b), optical density of Ag particles onto ZnO (substrate subtracted) (c), and ZnO photoluminescence enhanced by Ag particles (d).

We examined the effect of "small" nanoparticle arrays on the fluorescence of Rhodamine B (Rod B) dye dissolved in the ethanol with concentrations of 5×10^{-5}-5×10^{-8} Mol/l. The fluorescence self-quenching is insignificant at such Rod B content [21]. Rod B molecules have an optical absorption in the range of 480-560 nm with a maximum near 550 nm and a fluorescence in the range of 530-630 nm with maximum near 569 nm. Thus, we use Ar^+ laser light (514 nm) to excite fluorescence of Rod B dye.

Both Ag and Au nanoparticle arrays were found to be capable to enhance fluorescence/ photoluminescence cross-section concentrating the electric field near the surface in the gaps between the particles. In particular, the detected fluorescence of a Rod B (5×10^{-8} Mol/l) was ~2.55 fold and Raman modes of dye and ethanol were more than ~2.3 fold enhanced by "small" Au nanoparticle array (Figure 2c), whereas a homogeneous array of "large" (~40 nm) well separated (not coupled) nanoparticles does not show or show minor (~1.4 fold) enhancement (Figure 2d). This supports our assumption that our "small" particle arrays are plasmon coupled systems, though the coupling is not too strong yet.

As noted above the plasmon coupled particles enhance fluorescence enhancement by i) concentrating the local electric field intensity in gaps, i.e. by increase of the excitation rate of a fluorophore ii) by increasing the quantum yield/decreasing the fluorescence lifetime. The first effect is optimal when the absorption of fluorophore is maximal at the wavelength of the plasmon resonance. On the other hand spectral overlap between plasmon mode and fluorophore emission spectrum should also have overlap. The most fluorescence enhancement is expected to the red of the plasmon resonance [22]. The efficiency of the two effects is defined by geometry of an array, position of fluorescent molecule, by the overlap between plasmon modes and absorption/emission spectra of fluorophore and by its initial quantum efficiency. We believe that the geometry on an array may be further optimized by adjustment the deposition conditions to diminish the inter-particle gaps. Note, that in present study optical absorption of a "small" Au nanoparticle array has a minimum near ~520 nm and the "symmetric"/"anti-symmetric" plasmon modes have lower/higher photon energy, which is not optimal to enhance a fluorescence of Rod B. Rod B has high quantum yield of ~65%, which is already high, and this value can`t higher than 100%. Thus, we expect that the discussed nanoparticle arrays would be more effective to enhance the fluorescence of chemical and biological species with better overlap of particle plasmon bands and absorption/emission spectra of fluorophore. Fluorophores with low QY may also help better characterize the ability of the arrays and is the subject of further study.

To further enhance the efficiency to detect fluorescence and Raman response we designed a two sandwich-like structures of Ag particles/ ZnO(35nm)film/Ag(30nm)/ZnO(200nm)/a-Al_2O_3 by DC-magnetron sputtering and PLD methods. The Ag "mirror" was separated from the particle array to effectively collect a light emitted in backward direction. The 35 nm thick ZnO spacer was deposited at room temperature. The equivalent thickness of the top Ag layers was calculated to be 1 and 0.5 nm.

Ag nanoparticle arrays without "mirror" show ~2 fold enhancement of ZnO exciton photoluminescence (Figure 3d). Using a sandwich-like structure with Ag "mirror" under the Ag particle array the ZnO photoluminescence was further enhanced and the PL response was ~3.5 times higher in comparison with pure ZnO layer.

In summary we studied the possibilities to produce a plasmon coupled nanoparticle arrays by PLD and magnetron sputtering techniques. Absorption spectra of arrays of "small" Au nanoparticles less then 10 nm and separated by less than 10 nm gaps revealed two plasmon bands evidencing that plasmon coupling occurs. The fluorescence of a Rod B dye molecules (5×10^{-8} Mol/l) is ~2.55 fold enhanced by the "small" particle array. The 1.5 fold higher efficiency to collect the fluorescence response by use of Ag "mirror" under the particle array is demonstrated.

Acknowledgments

Fluorescence enhancement study was supported by President Grant № MK-8369.2016.2 "A study of an optical absorption and electromagnetic field concentration in strongly coupled plasmon nanosystems". Study of the nanowire based systems growth was supported by Project No.

16.219.214/K "Development of ZnO nanowire based systems synthesis methods for photo-detectors, optical antennas and piezo- and chemisensors elements".

References

[1] Kneipp K, Wang Y, Kneipp H, Perelman L T, Itzkan I, Dasari R R, and Feld M S, 1997 *Phys. Rev. Lett.* **78** 1667
[2] Feng A L, You M L, Tian L, Singamaneni S, Liu M, Duan Z, Lu T J, Xu F, and Lin M 2015 *Sci. Reports* **5** 7779
[3] Chowdhury M H, Aslan K and Malyn S N, Lakowicz J R, Geddes C D, 2006 *Appl. Phys. Lett.* **88**, 173104
[4] Aslan K, Huang J, Wilson G M, Geddes C D, 2006 *J. Am. Chem. Soc.* **128** 4206
[5] Anger P, Bharadwaj P, Novotny L, 2006 Phys. Rev. Lett. **96** 113002
[6] Johansson P, Xu H, Käll M, 2005 Phys. Rev. B. **72** 035427
[7] Lakowicz J R, Geddes C D, Gryczynski I, Malicka J, Gryczynski Z, Aslan K, Lukomska J, Matveeva E, Zhang J, Badugu R, Huang J, 2004 *J. Fluorescence,* **14** 425
[8] Ray K, Lakowicz J R, 2013 *J. Phys. Chem C* **117** 15790
[9] Vasilev K, Knoll W, Kreiter M, 2004 *J. Chem. Phys.* **120** 3439
[10] Xue C, Xue Y , Dai L, Urbas A, Li Q, 2013 *Adv. Opt. Mater.* **1** 581
[11] Zhang J, Irannejad M, Cui B. 2015 *Plasmonics* **10** 831
[12] Li K, Stockman M I, Bergman D J 2003, Phys. Rev. Lett., 91 227402
[13] Toroghi S, Kik P G, 2011 *Proc. SPIE* **8054** 80540E
[14] Bakhti S, Destouches N, Tishchenko A V, 2015, Plasmonics **10**, 1391
[15] Halas N J, Lal S, Chang W S, Link S, Nordlander P, 2011 *Chem. Rev.* **111** 3913
[16] Hanske C, Tebbe M, Kuttner C, Bieber V, Tsukruk V V, Chanana M, König T A F, and Fery A 2014 *Nano Lett.* **14** 6863
[17] Zhang T, Gao N, Li S, Lang M J, Xu Q H 2015, J. Phys. Chem. Lett. **6** 2043
[18] Yuan H, Khatua S, Zijlstra P, Yorulmaz M, and Orrit M, 2013 *Angew. Chem. Int. Ed.* **52** 1217
[19] Pazos-Perez N, Ni W, Schweikart A, Alvarez-Puebla R A, Fery A, Liz-Marzan L M, 2010 *Chem. Sci.* **1** 174
[20] Lim D K, Jeon K S, Kim H M, Nam J M, and Suh Y D 2010 *Nat. Mater.* **9** 60
[21] Kristoffersen A S, Erga S R, Hamre B, Frette Ø 2014, J. Fluoresc **24** 1015
[22] Gedded C D ed. 2010 *Metal-Enhanced Fluorescence* (New Jersey: John Wiley & Sons)

COANP-fullerenes system for optical modulation

S V Likhomanova[1,2,*], N V Kamanina[1,3,]**

[1]Department for "Photophysics of media with nanoobject", Vavilov State Optical Institute, St. Petersburg, 199053, Russia
[2]ITMO University, St. Petersburg, 197101, Russia
[3]Saint Petersburg Electrotechnical University "LETI", St. Petersburg, 197376, Russia

Abstract. The advanced investigations of π-conjugated organic molecule COANP sensitized with fullerenes have been revealed to consider this system as an affective medium for optical limiting and phase modulation. The special accent has been given to influence of the nanostructured relief at the interface on the spectral and photoconductive features.

1. Introduction

Materials with π-conjugated organic molecules [1-4] are ones of the perspective media used for optical modulators, in particular for optical limiter. These materials can be polarized at a distance more than molecules dimension that leads to the appearance of nonlinear properties of the medium.

In our previous works [5, 6] we have considered the optical limiting properties of 2-cyclooctylamino-5-nitropyridine (COANP) solution doped with fullerenes C_{70}. Experiments demonstrated the limiting of the laser radiation with energy density from 0.1 J/cm^2 up to 0.6 J/cm^2. We have obtained the maximum attenuation of the laser radiation with energy density of 0.35 J/cm^2 for the solution COANP-C_{70} with fullerene content of 20 wt.%. The energy value passed through solution has been decreased up to 18 times.

Reverse saturable absorption (RSA) and the complex formation process have been considered as the main mechanisms responsible for the optical limiting (OL) of COANP-C_{70} system. The RSA mechanism, as the first one, is based on a difference in cross section for excited state of the fullerene molecule and unexcited ones. The cross section of singlet-triplet excited states in the fullerene is larger than the one of the unexcited molecule. The population of excitation levels increases with laser energy increase, so absorption increases as well.

The complex formation process, as the second OL mechanism, is corresponded to the following reason. Intermolecular charge transfer complex with large absorption cross section can be formed in π-conjugated organic system doped with C_{70} due to the large electron affinity energy of fullerene. The electron affinity of acceptor fragment NO_2 in COANP is close to 0.45 eV. This parameter for fullerene is 2.65 eV that leads fullerene attracts weakly bound electrons from COANP donor part. Moreover, the distance for the electron pathway is larger for the *inter*molecular charge transfer than the one for the *intra*moleculat transfer process. These two facts provoke the formation of the large dipole moment correlated with the increased absorption cross section under the laser irradiations. The model of the

Journal of Physics: Conference Series **741** (2016) 012146 doi:10.1088/1742-6596/741/1/012146

intermolecular charge transfer process which can dominate on the intramolecular one is shown in figure 1.

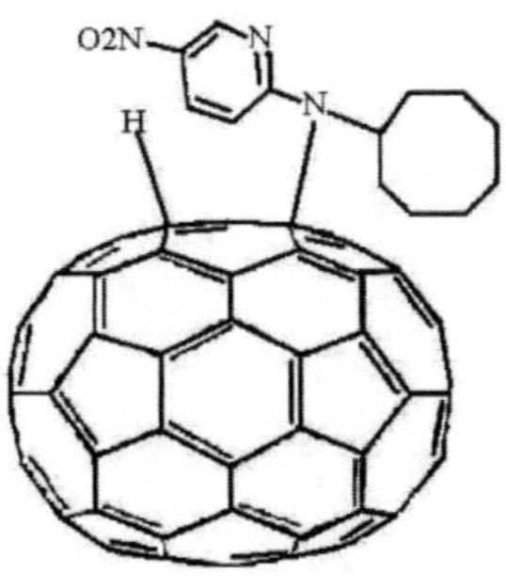

Figure 1. Possible interpretation of the *inter*molecular electron transfer in the COANP-fullerene system

The formation of the intermolecular charge transfer complex C_{70}- NO_2 has been supported by the results of DSC analysis (figure 2). The results of DSC demonstrate shifting of the transition temperature values (such as a melting point, a point of crystallization and a glass transition temperature) in the case of adding some amount of fullerene C_{70} in COANP.

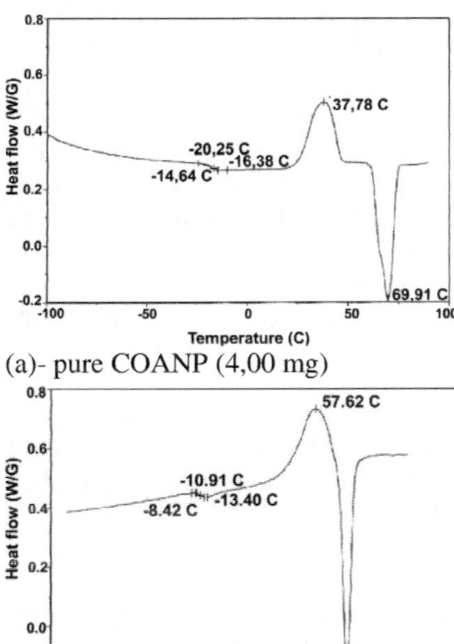

(a)- pure COANP (4,00 mg)

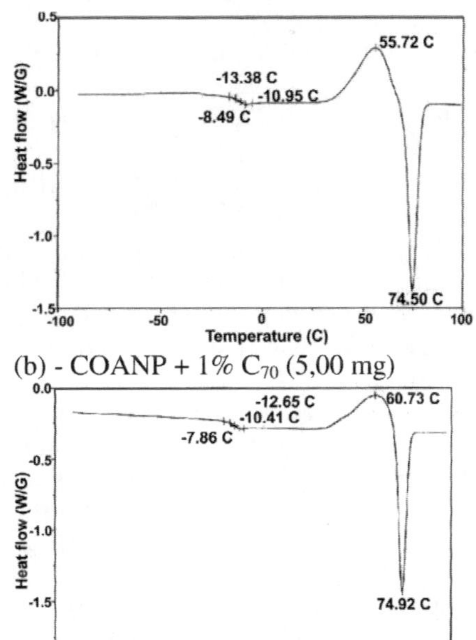

(b) - COANP + 1% C_{70} (5,00 mg)

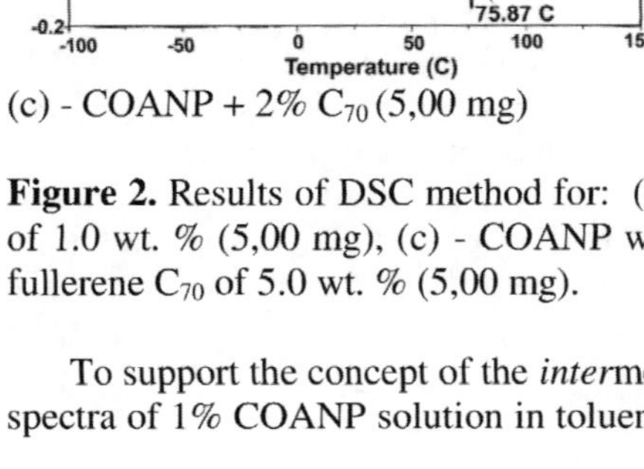

(c) - COANP + 2% C_{70} (5,00 mg)

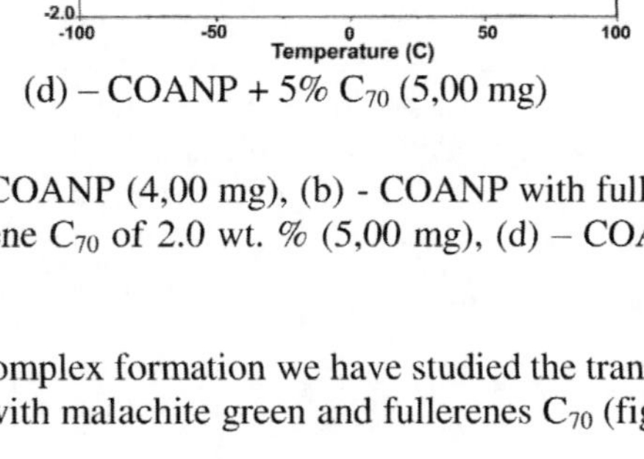

(d) – COANP + 5% C_{70} (5,00 mg)

Figure 2. Results of DSC method for: (a) - pure COANP (4,00 mg), (b) - COANP with fullerene C_{70} of 1.0 wt. % (5,00 mg), (c) - COANP with fullerene C_{70} of 2.0 wt. % (5,00 mg), (d) – COANP with fullerene C_{70} of 5.0 wt. % (5,00 mg).

To support the concept of the *inter*molecular complex formation we have studied the transmittance spectra of 1% COANP solution in toluene doped with malachite green and fullerenes C_{70} (figure 3) as well. The bathochromic shift for COANP solution doped with fullerenes C_{70} has been obtained. There has been no additive effect of malachite green dye as a sensitizer for COANP solution.

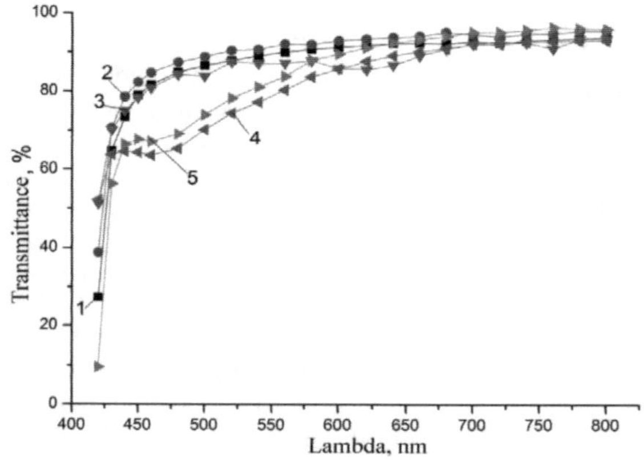

Figure 3. The transmittance spectra of 1% COANP solution in toluene: 1 – pure mol. COANP, 2 – COANP + malachite green of 0.1 wt. %, 3 – COANP + malachite green of 1.0 wt. %, 4 – COANP + C_{70} of 0.1 wt. %, 5 – COANP + C_{70} of 0.5 wt. %

The complex based on the COANP-C_{70} has been included into the liquid crystal (LC) cells and it has been considered for the increase of the materials applicable in optoelectronics devices. For example, the calculated nonlinear coefficients for the LC system doped with the COANP-C_{70} have been found as: n_2= 0,77 × 10^{-10} cm^2/W and $\chi^{(3)}$ = 2.4 × 10^{-9} cm^3/erg [7], which coincided with the comparatively observed for the LC doped with the PANI-C_{60} with n_2= 0,12 × 10^{-11} cm^2/W and $\chi^{(3)}$ = 3.42 × 10^{-10} cm^3/erg. Nonlinear properties of the LC-fullerene system via increase of the system dipole polarizability has been obtained analogous the procedure shown in work [8]. The influence of the refractivity change due to the modification of the substrate relief has been discussed too [9,10].

2. Experimental

In the current study we have explored the mutual effect the sensitization of the nematic LC with COANP – fullerenes complex and formation of the carbon nanostructured substrate relief on the doped LC properties. The LC cell has been consisted from two glass substrates with indium tin oxide (ITO) coating. The carbon nanotubes structure relief is created on the ITO under the electric field of 100 and 200 V/cm. The mixture LC-COANP-C_{70} has been used as a filling for cell.

3. Results and discussion

Transmittance spectra measurements of empty LC cell with carbon nanotubes relief and filling with LC-COANP-C_{70} complex are shown at the figure 4. One can see that the transparency of the doped LC structure can be increased drastically via ITO coating modified with the oriented in the electric field carbon nanotubes (CNTs) use. It can be explained via decrease of the Fresnel losses due to the reason that the refractive index of the CNTs is close to 1.1. Moreover, we can conclude for the filled LC cell that the self-arrangement of LC molecules with the transition from the nematic phase to quasi-smectic one can be provoked. The LC phase order parameter change leads to the transmittance increases as well. Furthermore, under these conditions the decrease of the level of the bias voltage has been obtained.

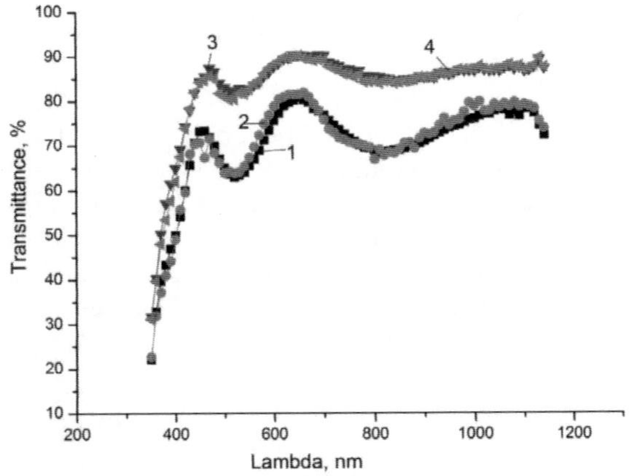

Figure 4. The transmittance spectra of the full and empty LC cell: 1 – LC cell empty, relief ITO+CNTs (100 V/cm) + SEW, 2 - LC cell empty, relief ITO+CNTs (200 V/cm) + SEW, 3 – LC+COANP+C_{70}, relief ITO+CNTs (100 V/cm) + SEW, 4 – LC+COANP+C_{70}, relief ITO+CNTs (200 V/cm) + SEW

Quantum chemical calculations with LAMMPS program have been done, which demonstrate the conjunction of carbon nanotubes with ITO substrate (figure 5).

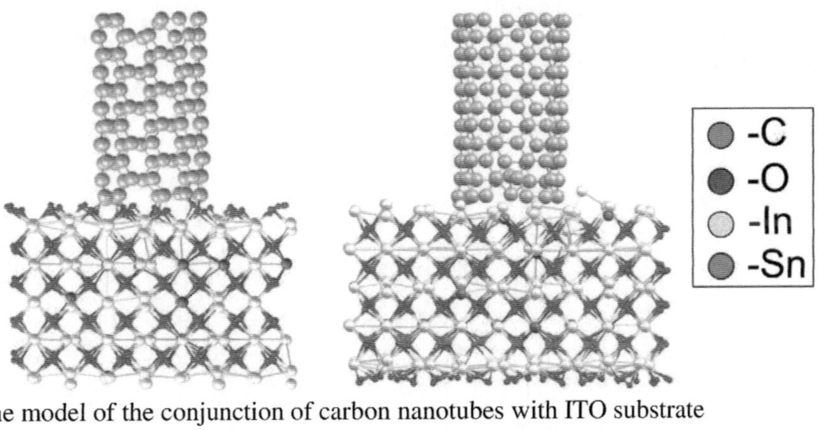

Figure 5. The model of the conjunction of carbon nanotubes with ITO substrate

One can see that the formation of a chemical bond between carbon nanotubes and ITO layer exists. The connection between two conductivity materials, such as ITO and carbon nanotubes leads to the conductivity increase and consequently to change the temporal characteristics (switch-on and switch-off time) of LC cell.

4. Conclusion

To summarize the results, it can be concluded that both the LC mesophase doping procedure as well as the interface relief nanostructuring process provoke the following features:

1). Observed transmittance increase permits to use the optoelectronic devices, for example the spatial light modulators and optical limiters, with good advantage.

2). Decreased level of the bias voltage which activates the LC cell and prognoses to improve the LC speed permit to consider the COANP-fullerene structures as the systems applied for the display and biomedicine technique.

3). Knowledge and experience used to study the intermolecular charge transfer process testify that the LC-COANP-fullerene materials can be proposed to be used not only in the visible spectral range but in the near IR-spectral range too.

Acknowledgments
The obtained results have been supported by RFBR grant #13-03-00044 (2013-2015), project "Nanocoating-GOI" (2012-2015) and international FP7 Program, Marie Curie Action, Project "BIOMOLEC" (2011-2015). Authors wish to thank their colleagues Dmitry Kvashnin (Emanuel Institute of biochemical physics RAS, Institute of Science and Technology "MISiS") and Pavel Sorokin (Technological Institute for Superhard and Novel Carbon Materialstheir, Institute of Science and Technology "MISiS") for quantum chemical calculations. The authors appreciate their colleagues from the department for "Photophysics of media with nanoobject", Vavilov State Optical Institute and abroad too for the helpful discussion.

References

[1]. Sreeja S, Sreedhanya S, Smijesh N, Reji P, Muneera C. I. 2013 *J. Mater. Chem.* C **1** 3851
[2]. Rao S V, Anusha, P T, Prashant T S, Swain D, Tewari S P 2011 *Materials Sciences and Applications* **2** 299
[3]. Carvalho C M B, Brocksom T J, de Oliveira K T 2013 *Chem. Soc. Rev.* **42** 3302
[4]. Jyothi L, Kuladeep R, Rao D N 2015 *Journal of Nanophotonics* 9(1) doi: 10.1117/1.JNP.9.093088
[5]. Likhomanova S.V., Kamanina N.V. 2011 *Processing and Application of Ceramics* **5** 425
[6]. Likhomanova S.V., Kamanina N.V. 2012 *Technical Physics Letters* **38** 425
[7]. Kamanina N.V. 2002 *Journal of Optics A: Pure and Applied Optics* **4(4)** 571
[8]. Kamanina N V 2005 *Physics-Uspekhi* **48** 419
[9]. Kamanina N., Kukharchik A., Kuzhakov P., Vasilyev P. 2015 *Proc. of the Fourteenth Israeli – Russian Bi-National Workshop 2015* p. 115.
[10]. Kamanina N.V., Likhomanova S.V., Zubtcoval Yu.A., Kamanin A.A., Pawlicka A. 2016 *Journal of Nanomaterials* – has been accepted for the publication

Terahertz absorption in GaN epitaxial layers under lateral electric field

R M Balagula[1], M Ya Vinnichenko[1], G A Melentev[1], M D Moldavskaya[1], V A Shalygin[1], L E Vorobjev[1], D A Firsov[1], S N Danilov[2], S Suihkonen[3]

[1] Peter the Great St. Petersburg Polytechnic University, St. Petersburg, Russia
[2] University of Regensburg, 93040 Regensburg, Germany
[3] School of Electrical Engineering, Aalto University, Otakaari 1, Espoo, Finland

Abstract. Variation of absorption of terahertz radiation in lateral electric field was investigated in GaN epitaxial layers. Different behaviour of the absorption modulation in electric field was observed for radiation polarized along electric field and perpendicular to it. Joint analysis of optical and transport measurements let us obtain field dependencies of mobility, electron concentration and absorption cross-section. For terahertz radiation polarized perpendicular to the electric field, results are in accordance with Drude model of free electron absorption. Another polarization demonstrates significant deviation that is yet to be studied more thoroughly.

1. Introduction

Gallium nitride is a promising material for high-speed electronics and photonics. One of possible implementations for this material is terahertz (THz) emitters with electrical pumping [1]. In this respect investigation of GaN optical properties in THz spectral range are of a great importance. Nevertheless, until now there were no works devoted to the investigation of terahertz absorption by free electrons in GaN. The reason for this lack of knowledge is the small value of equilibrium absorption coefficient in GaN which makes direct measurements of this value in epitaxial GaN layers quite problematic. However, modulation technique can provide the possibility to measure nonequilibrium absorbance. In particular, application of electric field provides the means to detect rather small changes of the absorption coefficient.

A lot of electronics devices, in particular, high frequency transistors, operate in highly nonequilibrium conditions of carrier heating. The investigation of the hot electron distribution function as a very important characteristic of electron gas seems to be of interest. Particularly, anisotropy of distribution function is a necessary condition for the stimulated generation of millimeter radiation in semiconductors with negative effective mass [2] and under the condition of electric-field-induced electron-transit-time resonance [3]. The anisotropy of carrier distribution function in momentum space can manifest itself in the polarization dependence of terahertz radiation absorption by free electrons in high electric field.

In the present paper we publish the results of the investigation of terahertz radiation absorption modulation in GaN epitaxial layers under lateral electric field.

2. Experimental technique

Wurtzite gallium nitride epitaxial layers were grown by metal-organic vapor phase epitaxy (MOVPE) on a *c*-plane sapphire covered by 2 μm undoped GaN buffer layer. The epitaxial layer had thickness $L = 4.3$ μm. In accordance with electro-physical measurements in low electric fields at room temperature electron concentration is $1.7 \cdot 10^{17}$ cm^{-3}, and electron mobility is 212 cm^2/V·s. At the liquid nitrogen temperature these parameters have values of $5.1 \cdot 10^{16}$ cm^{-3} and 329 cm^2/V·s, respectively. Two Ti/Au electrical contacts with 4 mm length were deposited on the surface of the structure for the application of lateral electric field. Distance between the contacts was 1 mm. Gallium nitride layers were covered with protective SiO_2 layer in order to prevent surface electrical breakdown under high electric field.

The optical experimental setup is shown in figure 1. Samples were mounted into liquid nitrogen cryostat ($T_0 = 77$ K) with TPX (polymethylpentene) windows. Terahertz radiation was generated by Edinburgh Instruments CW molecular gas laser with optical pumping. Absorption studies were carried out at a fixed laser wavelength $\lambda = 118$ μm ($\nu = 2.54$ THz). A Ge:Ga photodetector at liquid helium temperature was used to detect the intensity of terahertz radiation passed through the sample. The photoresponse I without applied electric field was measured using mechanical chopper. Photoresponse change ΔI was measured without the chopper during electric field pulse. In order to prevent the sample heating and destruction, high lateral electric field **E** was applied in pulsed mode with 1 μs pulse duration and repetition rate less than 1 Hz using a generator based on IGBT (isolated gate bipolar transistor). Attenuator was used to reduce THz laser radiation intensity. Half-wave plate was used to rotate the polarization plane of radiation. The resulting modulation M was obtained as a ratio of photoresponse change signal to the equilibrium signal:

$$M = \frac{\Delta I}{I} = \frac{I(E) - I}{I} = \frac{I_0 \cdot \exp(-(\alpha + \Delta\alpha)L) - I_0 \cdot \exp(-\alpha L)}{I_0 \cdot \exp(-\alpha L)} = \exp(-\Delta\alpha L) - 1,$$

where I_0 is the input beam intensity, α is the absorption coefficient, $\Delta\alpha$ is the change of absorption coefficient in electric field, L is the thickness of the GaN epitaxial layer. In case of a small absorption change, the modulation is proportional to $\Delta\alpha$: $M = -\Delta\alpha L$.

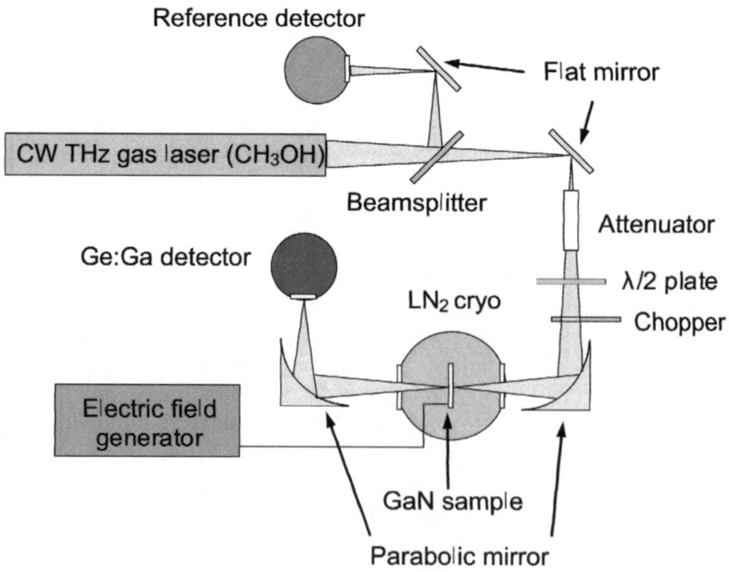

Figure 1. Optical experimental setup scheme.

3. Results and discussion

The dependencies of the absorption modulation on the lateral electric field **E** at $\lambda = 118$ μm are shown in figure 2. Solid red circles correspond to the absorption modulation for radiation polarized perpendicular to the electric field ($\mathbf{e}_\omega \perp \mathbf{E}$, $\mathbf{e}_\omega$ is radiation polarization vector). Empty black circles correspond to the absorption modulation for radiation polarized along the electric field ($\mathbf{e}_\omega || \mathbf{E}$). The negative modulation M (which corresponds to the decrease of radiation intensity detected by a photodetector) means an increase of absorption ($\Delta\alpha > 0$) . Modulation for both radiation polarizations demonstrates similar behavior in electric fields up to $E = 3000$ V/cm. There is a significant difference in the absorption modulation for different polarizations at high lateral electric fields ($E > 4000$ V/cm). In this range for polarization $\mathbf{e}_\omega \perp \mathbf{E}$, the absorption modulation monotonically increases with electric field. Absorption modulation field dependence drastically differs for polarization $\mathbf{e}_\omega || \mathbf{E}$. In the fields exceeding 4000 V/cm modulation is almost constant.

In order to understand the peculiarities of optical modulation we additionally measured current-voltage characteristic (CVC) in wide range of the electric fields (see figure 3). The general behavior of the obtained CVC can be explained qualitatively in the following way. The superlinear increase of the current in the field range $E = 200 - 2000$ V/cm can be attributed to the impurity breakdown and the sublinear increase of current in the fields exceeding 4000 V/cm is the result of mobility decrease due to the electron heating [4].

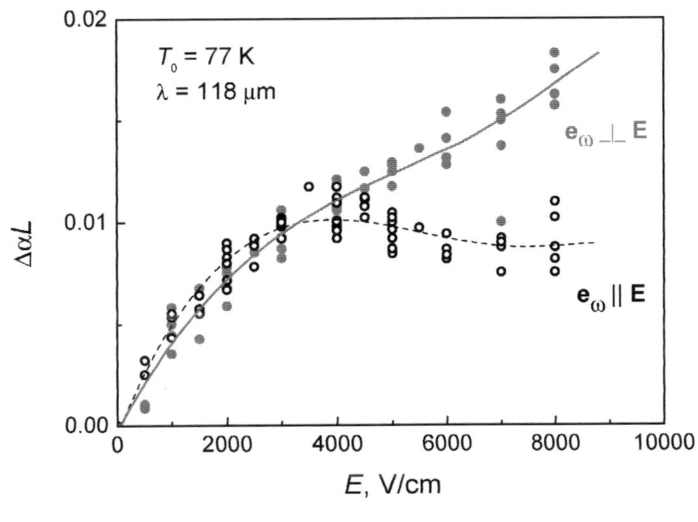

Figure 2. Dependences of the absorption modulation on the lateral electric field E for radiation polarized parallel (empty black circles) and perpendicular (solid red circles) to the electric field. $T_0 = 77$ K, $\lambda = 118$ μm. The solid and dashed lines are guides for the eyes.

The following model was considered to describe the abovementioned experimental results on absorption modulation and CVC. Variation of the absorption coefficient of GaN epitaxial layer in electric field can be described as $\Delta\alpha(E) = \alpha(E) - \alpha_0$, where α_0 and $\alpha(E)$ are equilibrium and nonequilibrium absorption coefficients, respectively. The latter quantity can be expressed as $\alpha(E) = s(E)\cdot n(E)$ where $s(E)$ and $n(E)$ are nonequilibrium absorption cross-section and electron concentration in electric field E. The equilibrium absorption coefficient $\alpha_0 = s_0 \cdot n_0$ where s_0 and n_0 are equilibrium absorption cross-section and equilibrium electron concentration at $T_0 = 77$ K. CVC allows us to find field dependence of conductivity $\sigma(E) = e \cdot \mu(E) \cdot n(E)$ which is determined by nonequilibrium electron concentration as well as electron mobility in electric field $\mu(E)$.

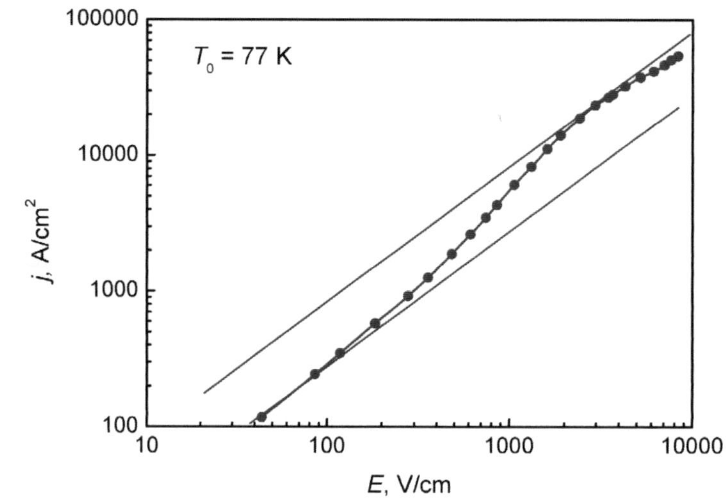

Figure 3. Current-voltage characteristic at temperature $T_0 = 77$ K. Straight lines correspond to linear dependence.

The field dependencies of μ, *s* and *n* uniquely determine the behavior of the dependencies $\Delta\alpha(E)$ and $\sigma(E)$ but not vice versa. Nevertheless the problem of determination of $\mu(E)$, $s(E)$ and $n(E)$ can be solved on the base of experimental results on $\Delta\alpha(E)$ and $\sigma(E)$ if there is some additional coupling equation. Such an equation can be derived from the Drude model of free carrier absorption. In the framework of the model, electron absorption cross-section and mobility are coupled by inverse proportion: $s(E) \propto \lambda^2/\mu(E)$.

Using abovementioned relations, we performed joint analysis of the experimental data on absorption modulation for polarization $\mathbf{e}_\omega \perp \mathbf{E}$ and CVC. The results are presented in figures 4 and 5. Free electron concentration monotonically increases with electric field that can be attributed to the process of impurity breakdown [4]. Total increase of the electron concentration is rather large (3.6 times) because in equilibrium at $T_0 = 77$ K the major part of electrons is frozen-out. At electric fields exceeding 4000 V/cm, one can notice simultaneous significant decrease of electron mobility (see figure 4) and increase of absorption cross-section (see figure 5). Maximum deviation of these quantities from their equilibrium values is close to 40%. The decrease of the electron mobility is caused by electron heating under applied electric field, and it results in an increase of the absorption cross-section in accordance with Drude model. It should be noted that equilibrium value of the absorption cross-section $s_0 = (2.0 \pm 0.15) \cdot 10^{16}$ cm^{-3} was found from the experimental data on absorption modulation under relatively low fields (~ 1000 V/cm) when decrease of the mobility due to electron heating is negligibly small (see figure 4). Consequently, $\alpha_0 \approx 10$ cm^{-1} and the equilibrium optical transmission of the GaN epitaxial layer $\exp(-\alpha_0 L) \approx 0.996$. This result proves the statement from Introduction concerning the impossibility of the direct measurements of equilibrium absorption coefficient.

After that using the obtained dependencies $n(E)$ and $\mu(E)$, we analyzed the experimental data on absorption modulation for polarization $\mathbf{e}_\omega || \mathbf{E}$ and found corresponding absorption cross-section. Dependence $s(E)$ for this polarization (see dashed curve in figure 5) demonstrates non-monotonic behaviour and can not be understood in the framework of Drude model.

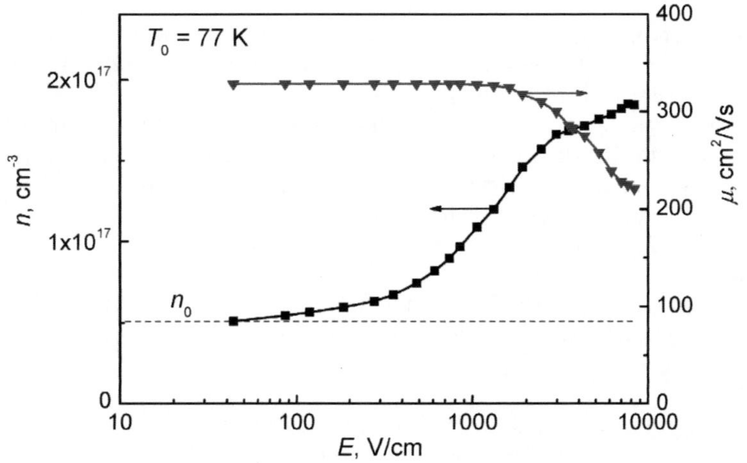

Figure 4. Field dependencies of electron concentration (left axis) and electron mobility (right axis) at $T_0 = 77$ K. Dashed line shows the equilibrium electron concentration value at the named temperature.

Microscopic mechanism of the observed polarization dependence of absorption modulation remains to be discovered. It can be related to anisotropy of the hot electron distribution function which arises in high electric fields [5, 6]. In particular, scattering on nonequilibrium optical phonons can have a significant influence on electron distribution.

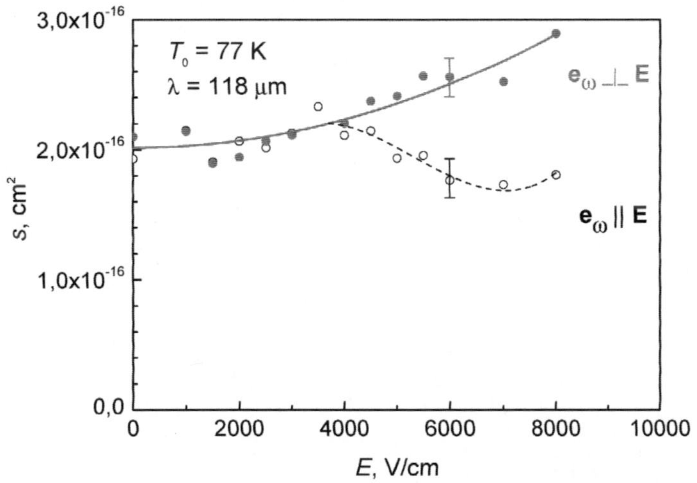

Figure 5. Absorption cross-section field dependence. Terahertz radiation is polarized parallel (empty black circles) and perpendicular (solid red circles) to the electric field. $T_0 = 77$ K, $\lambda = 118$ μm. The solid and dashed lines are guides for the eyes.

4. Summary
In this work, modulation of absorption in GaN epitaxial layers under electric field was investigated in terahertz spectral range. Joint analysis of optical and electrical measurements resulted in a determination of field dependencies of mobility, electron concentration and absorption cross-section. Experimental and analytical results in case of the terahertz radiation polarized perpendicular to electric field are consistent with Drude model of free electron absorption. A surprising polarization dependence was observed for absorption cross-section. The microscopic mechanism of this phenomenon for now is unclear and will be further studied.

Acknowledgments
The reported study was funded by RFBR (research projects 16-02-00863 and 16-32-80036 mol_ev_a) and the Ministry of Education and Science of the Russian Federation (state assignment).

References
[1] Hao Y, Yang L A and Zhang J-C 2008 *Terahertz Science and Technology* **1** 51
[2] Andronov A A, Belyantsev A M, Gavrilenko V I, Dodin E P, Krasil'nik Z F, Nikonorov V V and Pavlov S A 1984 *JETP Lett.* **40** 989
[3] Vorob'ev L E, Danilov S N, Tulupenko V N and Firsov D A 2001 *JETP Lett.* **73** 219
[4] Shalygin V A *et al* 2009 *Journ. Appl. Phys.* **106** 123523
[5] Bray R and Pinson W E 1963 *Phys. Rev. Lett.* **11** 268
[6] Syngayivska G I and Korotyeyev V V 2007 *Semiconductor Physics, Quantum Electronics and Optoelectronics* **10** 54

Photo- and electroluminescence in strong electric fields in Sb-containing narrow gap semiconductor materials

M Ya Vinnichenko[1], I S Makhov[1], V Yu Panevin[1], A V Selivanov[1], D A Firsov[1], L E Vorobjev[1], N A Pikhtin[2], K V Bakhvalov[2], L Shterengas[3], G Belenky[3], and G Kipshidze[3]
[1]Peter the Great St. Petersburg Polytechnic University, St. Petersburg, Russia
[2]Ioffe Physical Technical Institute, St. Petersburg, Russia
[3]State University of New York at Stony Brook, New York, USA

Abstract. Spectra of the mid-infrared interband luminescence under interband optical pumping and impact ionization in strong electric fields are experimentally studied in the InAsSb epilayer and in the monocrystalline InSb in the temperature range from 10 K to 85 K. The recombination radiation anisotropy in InSb arising due to electron heating and drift in strong electric fields is observed.

1. Introduction

The narrow band semiconductors, semiconductor alloys and superlattices allow to develop mid-infrared photodetectors, detector arrays and light sources [1, 2]. Such materials as a HgCdTe and Sb-based A3B5 alloys are rivals in this field. On the characterization stage of the semiconductor material, the information about its fundamental properties can be obtained from a luminescence study. In the present work we report on the spectral study of interband photoluminescence (PL) under interband optical pumping and electroluminescence (EL) under strong electric fields due to impact ionization in the InAsSb epilayer and in the InSb monocrystalline samples. The luminescence spectra obtained under optical excitation gave information about fundamental bandgap value. Electroluminescence spectra allow us to study the nonequilibrium electron heating phenomena in strong electric field as well as electric field induced luminescence polarization anisotropy observed earlier in ref. [3] on integrated over the spectrum luminescence.

2. Samples and experimental details

Our first sample is bulk 1 μm $InAs_{0.6}Sb_{0.4}$ MBE-grown, unintentionally doped epilayer. Details of the sample structure could be found in ref. [4]. The second sample is monocrystalline InSb with low donor impurity concentration of about $6\cdot 10^{13}$ cm^{-3} and electron mobility of about $6\cdot 10^{5}$ cm^2/(V·s) at the lattice temperature $T = 77$ K. Electric field was applied using contacts created on the surface of the samples: the $InAs_{0.6}Sb_{0.4}$ epilayer sample had a gold contacts; the InSb sample had an indium contacts annealed at 300°C. In order to increase the quantum efficiency of luminescence we have performed optical and photoelectrical studies at low lattice temperatures (from 10 K to 85 K) using the closed cycle cryostat. To prevent sample overheating, the electric field was applied in short pulse regime (250 ns / 10 Hz). For PL studies, we used interband optical pumping of the sample surface with radiation wavelength of 1064 nm (peak power up to 30 W in 100 ns pulse regime [5]). The mid-infrared PL and EL spectra were studied using FTIR-based setup described earlier in ref. [6] supplemented with lock-in amplifier.

3. Experimental results and discussion

The PL spectrum of the $InAs_{0.6}Sb_{0.4}$ epilayer obtained at the lattice temperature $T = 10$ K is presented in figure 1. It shows luminescence band lying in the spectral range 100 - 240 meV with the maximum near the photon energy of about 160 meV. The EL spectra of the sample studied at the $T = 10$ K at different applied electric fields F are presented in figure 2. Long-wavelength edges of the PL and EL spectra are similar and correspond well to the calculated $InAs_{0.6}Sb_{0.4}$ bandgap value $Eg = 136$ meV [1] (this value is marked in figures 1 and 2 with arrows). Thus, we suppose, that the emission bands observed on the EL and PL spectra are related to a direct interband electron-hole recombination. In case of EL the nonequilibrium electrons and holes were generated due to an impact ionization in strong electric field. Probable due to low carrier mobility, the intensive impact ionization was observed on measured current-voltage characteristics of the $InAs_{0.6}Sb_{0.4}$ sample in strong electric fields near 1000 V/cm. The EL spectrum measured at the electric field $F = 600$ V/cm shows a luminescence band lying in the spectral range 100 - 190 meV with the maximum near 145 meV (see figure 2). The increase of the electric field up to 1000 V/cm leads to broadening of the EL spectrum till 300 meV and to the change in a slope of the high-energy part of the spectrum. The luminescence band broadening could be connected with the increase of the nonequilibrium carrier concentration due to impact ionization with the increase of electric field. A slope change could be a result of the electron temperature increase in electric field.

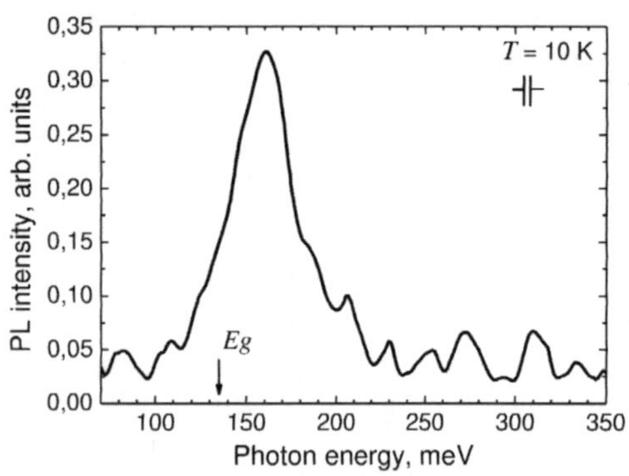

Figure 1. Interband PL spectrum of the $InAs_{0.6}Sb_{0.4}$ sample.

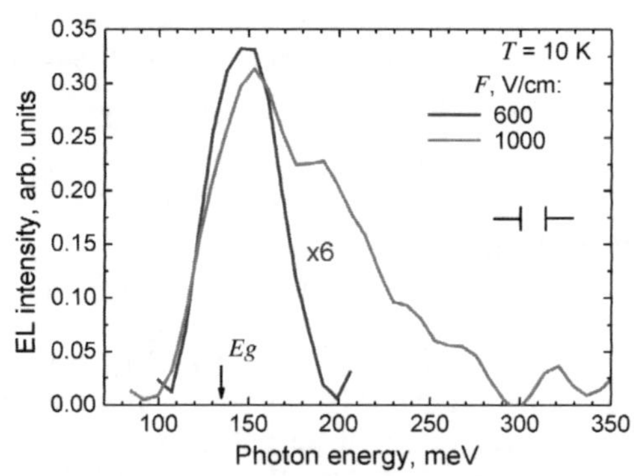

Figure 2. Interband EL spectra of the $InAs_{0.6}Sb_{0.4}$ epilayer under different applied electric fields (please, see the legend on the plot). The EL intensity at the $F = 600$ V/cm is multiplied by 6.

The n-InSb sample current-voltage characteristic (CVC) studied at $T = 77$ K is presented in figure 3. In figure 3, one can observe the impact ionization threshold (marked with arrow) at the electric field F of about 200 V/cm when the Ohmic behaviour of the CVC changes to the sharp current density increase with F (like F^3). In the electric fields stronger than 600 V/cm, the current increase is slowing down because of the mobility decrease due to rise of polar optical phonon scattering.

The EL spectra of the InSb sample were measured in the range of electric fields from 300 V/cm to 1125 V/cm ($T = 85$ K). They also shows the emission band related to direct interband electron-hole recombination (see figure 4). As the electric field rises, the luminescence band of the InSb sample is broadening and a change in a slope of the high-energy part of the EL spectrum can be observed as well as for the $InAs_{0.6}Sb_{0.4}$ sample. Analysis of this shortwave slope can give information about the hot carrier temperature.

Journal of Physics: Conference Series **741** (2016) 012148 doi:10.1088/1742-6596/741/1/012148

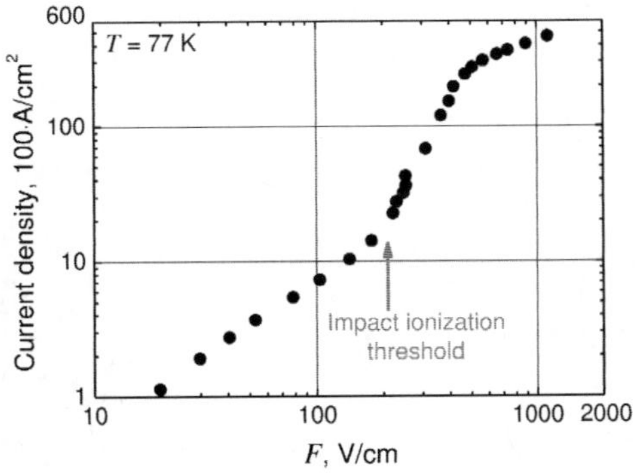

Figure 3. Current-voltage characteristic of the *n*-InSb sample.

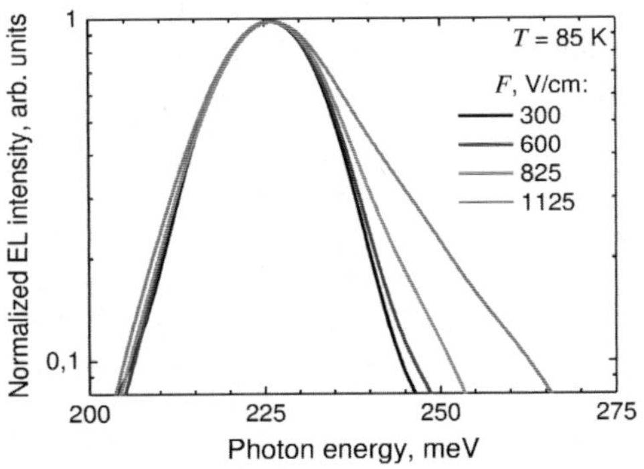

Figure 4. Interband EL spectra of the *n*-InSb sample under different applied electric fields *F* (please, see the legend on the plot).

Dependence of the spontaneous emission intensity I on the photon energy $\hbar\omega$ is determined by the following expression [7]:

$$I(\hbar\omega) \sim (\hbar\omega)^2 \cdot \alpha(\hbar\omega) \cdot f_e(\hbar\omega) \cdot f_h(\hbar\omega)\,, \tag{1}$$

where $\alpha(\hbar\omega)$ is the absorption coefficient, f_e and f_h are electron and hole distribution functions. For large photon energies $\hbar\omega$ one can simplify expression (1) using Maxwell-Boltzmann electron and hole distribution functions:

$$I(\hbar\omega) \sim \frac{(\hbar\omega)^2 \cdot \alpha(\hbar\omega)}{\exp\left(\hbar\omega/kT_c\right)}, \tag{2}$$

where k is the Boltzmann constant, T_c is the effective carrier temperature [7]:

$$\frac{1}{T_c} = \frac{m_h}{m_e + m_h}\frac{1}{T_e} + \frac{m_e}{m_e + m_h}\frac{1}{T_h}, \tag{3}$$

where m_e and m_h are the effective masses of the electron and hole, T_e and T_h are the electron and hole temperatures, respectively. As far as m_e is much smaller than m_h, the equation (3) can be simplified to

$$T_c = T_e \tag{4}$$

Thus, electrons temperature T_e could be determined from fitting the high-energy tail of luminescence spectra (figure 4) with expression (2). The dependence of the hot electron temperature change $\Delta T = T_e - T$ (where T is the lattice temperature) on the applied electric field F determined from the EL spectra for the *n*-InSb sample at the $T = 85$ K is presented in figure 5 by black squares. The values of ΔT obtained for the different electric fields are much lower than ones calculated from ref. [8]. This could be explained with a distortion of the EL spectra by self-absorption. Self-absorption occurs when we collected the emission from deep layers of the sample (deep in comparison with penetration depths of light with photon energy comparable with the InSb bandgap). The distortion is as large, as sample thicker, if we take into account the pinch effect [9]. To demonstrate the significant influence of self-absorption on measured EL spectrum, we plot in figure 6 both corrected spectrum taking into account self-absorption for 60 μm thick sample and original measured spectrum. We have performed as well the self-absorption correction for all curves from figure 4 and have determined the corresponding electron temperatures. Dependence of the hot electron temperature change for self-absorption corrected spectra is plotted by red squares in figure 5. Finally, after self-absorption correction for 60 μm thick sample we have obtained significantly higher values of hot electron temperature.

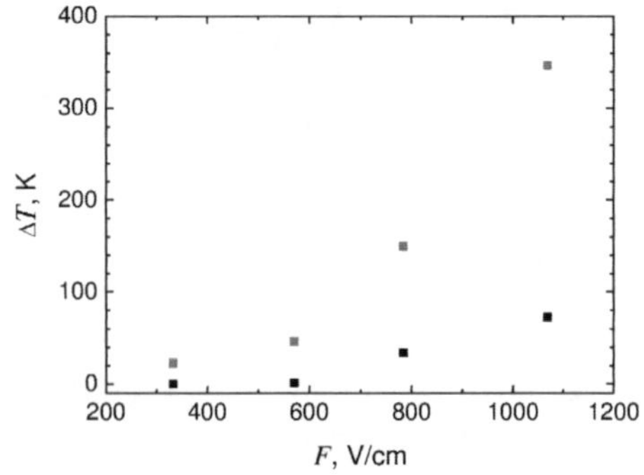

Figure 5. Dependence of the hot electron temperature change ΔT on the applied electric fields F for the n-InSb sample at the $T = 85$ K without self-absorption correction (black squares) and with correction for 60 μm thick sample (red squares).

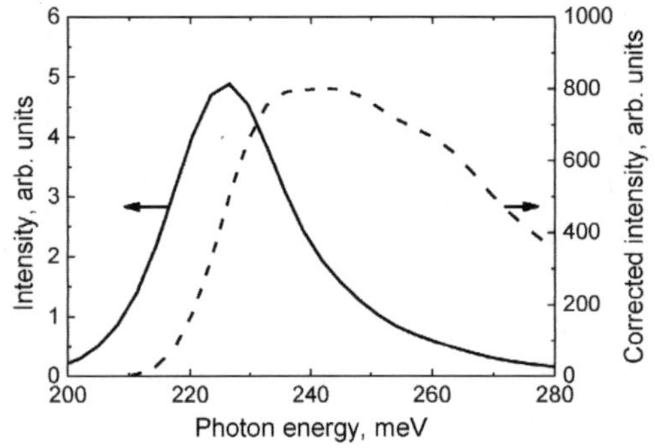

Figure 6. The comparison of the original measured spectrum of the n-InSb sample at the $T = 85$ K and $F = 1125$ V/cm (solid curve) and the results of self-absorption correction of the measured spectrum (dashed curve).

It should be mentioned that in our conditions the main mechanism of charge carrier energy loss in InSb is the polar optical phonon scattering (electron-electron scattering and impurity scattering are negligible unless the impurity or carrier concentrations are more than 10^{16} cm^{-3}). Acoustic phonon scattering times are much longer (approximately two orders of magnitude) than those for optical phonon scattering [10]. The probability for the polar optical phonon scattering $W_{kk'}$ is given by

$$W_{kk'} \sim \frac{1}{|\boldsymbol{k}-\boldsymbol{k}'|^2}, \tag{5}$$

where $\boldsymbol{k}$ and $\boldsymbol{k}'$ are the wave vectors of the initial and final carrier states, respectively. From this equation, one can conclude that small-angle scattering is more likely than large-angle scattering. In strong electric field the energy gained by electron from the electric field will be partially dissipated with the emission of polar optical phonons, but after each collision the electron will continue to move close to the direction of electric field. Therefore, the distribution of electrons in $\boldsymbol{k}$-space will be focused along the direction of the electric field. The squared module of the optical matrix element for electron transitions from conduction band to heavy holes subband $|M_{e\to hh}|^2$ depends on the angle between polarization vector of the electromagnetic wave $\boldsymbol{e}_\omega$ and electron wave vector [11]:

$$|M_{e\to hh}|^2 \sim sin^2(\widehat{\boldsymbol{e}_\omega, \boldsymbol{k}}). \tag{6}$$

Consequently, the difference between interband luminescence intensity for light polarized along applied electric field and perpendicular to it should appear.

The EL spectra for the n-InSb sample measured for light polarized along applied electric field ($I_\parallel$) and perpendicular to it ($I_\perp$) are presented in figure 7. Comparing these spectra measured at strong electric field ($F = 1000$ V/cm) when electron heating and drift are significant (see figure 5), one can observe a difference between them. This polarization anisotropy is related to nonequilibrium electron distribution function anisotropy and angle dependence of the interband optical matrix element in $\boldsymbol{k}$-space discussed above. For greater clarity we also presented in figure 7 the interband luminescence anisotropy as the difference $\Delta I = I_{\perp} - I_{\|}$ related to the $I_{\perp}$ (black curve). In accordance with the model of luminescence anisotropy discussed in ref. [3], anisotropy increases with photon energy increase.

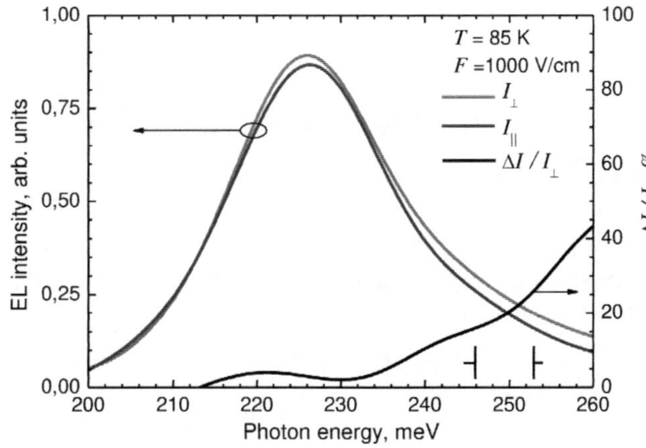

Figure 7. Polarized EL spectra of the *n*-InSb sample ($I_{\perp}$ and $I_{\|}$) and relative change in the EL intensity $\Delta I/I_{\perp}$ (please, see the legend on the plot).

4. Summary

It is shown that $InAs_{0.6}Sb_{0.4}$ epilayer can be used as mid-infrared (~10 μm) light sources with electrical pumping. Significant influence of the self-absorption on interband luminescence spectra excited by a strong electric field, inherent to the "thick" samples, is demonstrated. The polarization anisotropy of electroluminescence spectra in strong electric fields was studied in InSb monocrystalline sample.

Acknowledgments

The reported study was funded by the Russian Foundation for Basic Research (grant 16-02-00863), Russian Federation President grant for young scientists No MK-4616.2016.2 and the Ministry of Education and Science of the Russian Federation (state assignment).

References

[1] Lin Y, Donetsky D, Wang D, Westerfeld D, Kipshidze G, Shterengas L, Sarney W L, Svensson S P and Belenky G 2015 *Journal of Electronic Materials* **44**(10) 3360
[2] Morozov S V, Rumyantsev V V, Dubinov A A, Antonov A V, Kadykov A M, Kudryavtsev K E, Kuritsin D I, Mikhailov N N, Dvoretskii S A and Gavrilenko V I 2015 *Applied Physics Letters* **107** 042105
[3] Vasil'eva M A, Vorobjev L E and Stafeev V I 1969 *Sov. Phys. Semicond.* **3**(9) 1374
[4] Svensson S P, Sarney W L, Hier H, Lin Y, Wang D, Donetsky D, Shterengas L, Kipshidze G and Belenky G 2012 *Physical Review* B **86** 245205
[5] Petrunov A N *et al* 2010 *Semiconductors* **44** 789
[6] Selivanov A V, Makhov I S, Panevin V Yu, Sofronov A N, Firsov D A, Vorobjev L E and Shterengas L 2015 *Journal of Physics: Conference Series* **643** 012078
[7] Shah J *Spectroscopy of nonequilibrium electrons and phonons*, ed. by C.V. Shank, B.P. Zakharchenya (Amsterdam–London–N.Y.–Tokyo, Elsevier, 1992) 2, pp. 57 – 112
[8] Glicksman M and Higinbothem W A 1963 *Physical Review* **129**(4) 1572
[9] Chynoweth A G and Murray A A 1961 *Physical Review* **123**(2) 515
[10] Dumke W P 1968 *Physical Review* **167**(3) 783
[11] Baumgardner C A and Woodruff T O 1968 *Physical Review* **173**(3) 746.

Optical properties of GaAs/AlGaAs double quantum wells in lateral electric field

M Ya Vinnichenko, R M Balagula, I S Makhov, A A Shumilov, D A Firsov, L E Vorobjev

Peter the Great St. Petersburg Polytechnic University, 29 Polytechnicheskaya str., St. Petersburg 195251, Russia

Abstract. The influence of the lateral electric field on the mid-infrared intersubband light absorption and near-infrared interband photoluminescence is experimentally investigated in tunnel-coupled GaAs/AlGaAs quantum wells. Absorption and emission modulation in lateral electric field are related to the electron heating and redistribution of hot electrons between the quantum well states resulting in variation of the space charge in the structure.

1. Introduction

Electric field influence on the optical properties of semiconductor nanostructures is investigated for a long time. Transverse electric field applied along the structure growth axis changes the energy spectrum of the structure (quantum confined Stark effect). Rather strongly optical properties are influenced by the transverse electric field in the structures with tunnel-coupled quantum wells (QW) [1 - 3]. Effects of lateral electric field in quantum wells are not studied thoroughly enough. In the structures with a single rectangular quantum well, the application of lateral electric field can lead to a slight change in carrier concentration on the quantum-size levels related to carrier heating [4]. More complicated band structure of tunnel-coupled quantum wells facilitates the observation of the influence of carrier heating on the QW optical properties. The present paper is devoted to the results of electro-optical investigations in tunnel-coupled QWs under the lateral electric field. Spectra of intersubband light absorption and interband photoluminescence affected by the electron heating and redistribution of hot electrons between the quantum well states were studied. Earlier, the influence of lateral electric field on the photoluminescence (PL) spectrum was investigated in [5]. In contrast to our work, the modulation of PL in [5] was not directly connected with electron heating. Carrier heating effect on the intersubband absorption in tunnel-coupled quantum wells was studied in [6] at a fixed radiation wavelength. Comparing the effect of an electric field and lattice heating on the spectra of intersubband absorption and interband PL, we were able to estimate the temperature of the hot electrons.

2. Sample and experiment

The sample was MBE-grown on a semi-insulating GaAs substrate. The investigated structure consisted of 100 pairs of GaAs/AlGaAs tunnel-coupled QWs. Tunnel-transparent barrier (12 Å wide) separated 56 and 40 Å wide quantum wells. The barrier between pairs of QWs had a width of 120 Å. The quantum well layers were uniformly doped with Si to the donor concentration $N_d = 6\cdot 10^{11}$ cm^{-2}. The calculated energy of $e1$ - $e3$ and $e2$ - $e3$ transitions at $T = 77$ K are 187 and 135 meV,respectively (see inset in figure 1). We used material parameters taken from the review paper [7]. In order to characterize the structure, interband photoluminescence spectra without electric field were measured with Horiba FHR-640 grating monochromator paired with a Symphony II CCD-camera and YAG laser as a source of interband optical pumping. Spectra of PL modulation in electric field were studied with another grating monochromator and Si detector.

The intersubband absorption was measured in a multi-pass sample geometry where two 45 degrees beveled facets and the substrate side of a sample were polished to allow multiple total internal reflections. Electric field was applied using indium contacts created on the surface of the sample. In order to avoid sample overheating, the lateral electric field was applied to the sample in pulsed mode (electric field pulse duration was 250 ns with a duty cycle from 50 000 to 250 000).

The absorption spectra were obtained using Bruker Vertex 80v vacuum Fourier transform infrared spectrometer. Globar was used as a source of broadband infrared radiation. A photovoltaic HgCdTe photodetector was chosen as a detector of radiation. Infrared absorption coefficient α was determined by taking the ratio of the transmission spectra T of TM and TE-polarized light: $\alpha L = -\ln(T_{TM}/T_{TE})$, were L is the optical path length. According to the polarization selection rules intersubband optical electron transitions can occur only upon exposure to the light with a polarization vector component directed along the structure growth axis, i.e., upon exposure to TM-polarized light. No significant absorption was observed in mid-infrared spectral range for TE-polarized light. The sample was placed in the closed cycle cryostat which allows the temperature to be maintained in the range of 4–320 K with an accuracy of 0.1°. Spectra of absorption modulation under lateral electric field were measured with Fourier transform spectrometer in step-scan mode.

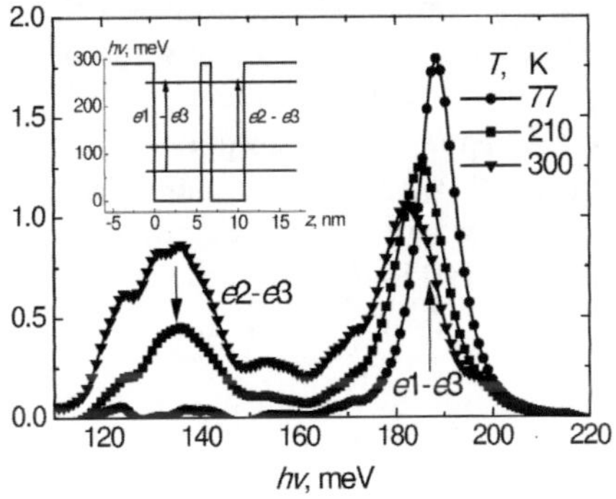

Figure 1. Intersubband absorption spectra at different lattice temperatures, L is the optical path length. The arrows indicate the calculated electron transition energies at $T = 77$ K. Inset shows the potential profile of one GaAs/AlGaAs QW pair at $T = 77$ K.

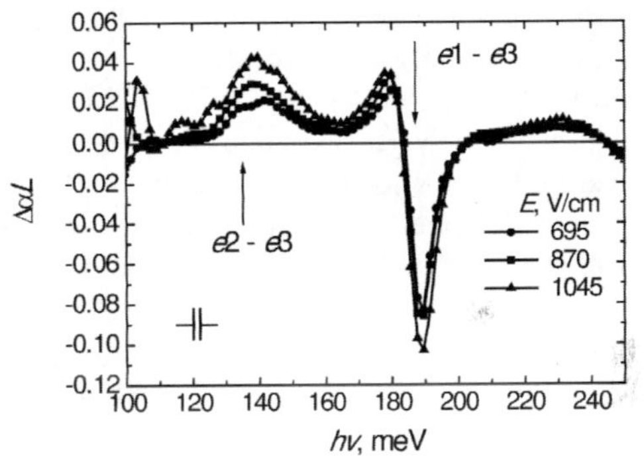

Figure 2. Spectra of intersubband absorption modulation under different lateral electric fields E at $T = 77$ K. The arrows indicate the calculated electron transition energies.

3. Results and discussions

The interband photoluminescence spectra without application of electric field were measured at different lattice temperatures. Spectral peaks demonstrate a temperature shift corresponding to the bandgap change. At room temperature, three peaks corresponding to the electron transitions *e*1-*hh*1, *e*2-*hh*2 and *e*1-*lh*1 were observed. At liquid nitrogen temperature only one peak associated with transitions *e*1-*hh*1 was registered due to the lack of electrons on the *e*2 level and holes on the *lh*1 and *hh*2 levels. The spectra are in a satisfactory agreement with the calculated energy spectrum.

The spectra of intersubband light absorption for different lattice temperatures without applied electric field are shown in figure 1. At low temperatures, there is only one peak associated with the optical transitions $e1$-$e3$, which means that the carriers are localized mainly at the level $e1$. When the temperature rises, carriers begin to fill the overlying energy level. The observed longwavelength shift of the $e1$-$e3$ absorption peak with the temperature increase is primarily associated with temperature dependence of the conduction band potential profile of QWs. In addition, the space charge changes the structure potential due to the redistribution of carriers between QWs with the temperature increase.

The modulation of absorption under lateral electric field was measured at $T = 77$ K using two different experimental methods: i) direct measurement of only the transmission variation signal and ii) measurement of signals under lateral electric field and without it with subsequent comparison of these signals. The transmission change signal can be of both signs. Application of the fast Fourier transform technique assumes the positivity of the resulting spectrum, therefore in case of the first method the spectra of resulting signal change do not contain the information about the sign of absorption variation. Thus, this technique requires phase correction procedure to obtain the real modulation spectra. The procedure of phase correction is described more specifically in [1, 8]. The second method is accurate in the sense of signal sign but is not so reliable in case of small signal values. The three-dimensional time and photon energy dependences of the TM light transmission were obtained as a result of the second method implementation. The transmission change spectra of the structure under electric field obtained by both methods are identical which proves the feasibility of the chosen phase correction procedure. Modulation of intersubband absorption coefficient $\Delta\alpha$ in electric field was determined by taking the ratio of the TM light transmission spectra change measured under the electric field and in equilibrium conditions: $\Delta\alpha \cdot L = - \ln (\Delta T_{TM} (E) / T_{TM} (E{=}0) + 1)$. The absorption associated with optical transitions $e2$-$e3$ increases in electric field (see figure 2). This can be explained with the increase of the carrier concentration at the electron level $e2$. Absorption related to $e1$-$e3$ transitions changes due to the decrease of $e1$ concentration and due to the shift of the $e1$-$e3$ absorption peak to longer wavelengths under lateral electric field. Nonequilibrium hot electrons are being redistributed in the real space between the states of two quantum wells. Comparing the absorption variation $\Delta\alpha L$ induced by temperature change (see figure 1) and $\Delta\alpha L$ induced by lateral electric field (see figure 2) we were able to estimate the level of electron heating. The experimentally determined electron temperature was $T_e = 97$ K at $E = 695$ V/cm and $T_e = 103$ K at $E = 1045$ V/cm.

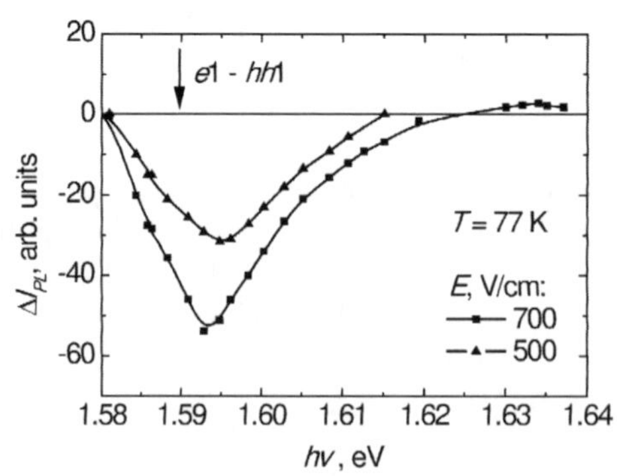

Figure 3. Modulation of interband photoluminescence at $T = 77$ K under lateral electric field E. Arrow shows the calculated energy of interband $e1$-$hh1$ transitions.

The interband photoluminescence variation spectra were measured at different lateral electric fields at $T = 77$ K. The hot carrier temperature was not high enough to significantly change the concentration of carriers at hole and electron levels $lh1$, $hh2$ and $e2$. Only modulation of interband photoluminescence associated with the $e1$-$hh1$ optical transitions was observed (see figure 3). It should be noted that interband PL modulation was observed at high level of optical pumping. High nonequilibrium carrier concentration is created in such conditions, therefore we can consider the equality of electron and hole temperatures due to strong electron-hole scattering. The electron temperature T_e was estimated by means of the shortwave spectra shoulder analysis. To derive T_e value one should obtain the energy value $h\nu_0$ which corresponds to the zero photoluminescense modulation $\Delta I_{PL} = 0$ or, in other terms, $dI_{PL}/dT_e = 0$. Photoluminscense intensity I_{PL} is proportional to the number of spontaneous interband radiation quanta originating from elementary volume during elementary time period at the frequency range from ν to $\nu + \Delta\nu$. Therefore one can obtain nonequilibrium charge carrier temperature using Boltzmann distribution function: $T_e = (h\nu_0 - \varepsilon_g)/2k$, were k is Boltzmann constant, ε_g is effective band gap. The relative increase of electron temperature due to the heating was determined to be about 30 K. This temperature was close to the temperature obtained from the analysis of intersubband absorption modulation spectra.

4. Conclusion

The influence of the lateral electric field on the mid-infrared intersubband light absorption and near-infrared interband photoluminescence in tunnel-coupled GaAs/AlGaAs quantum wells is experimentally observed. The temperature modification of the intersubband absorption spectra was studied as well. Modulation of intersubband absorption and interband photoluminescence under lateral electric field is explained by redistribution of hot charge carriers between the two lowest electronic states belonging to different quantum wells off the pair and by subsequent variation of space charge in the structure. The electron temperature T_e was estimated from the comparison of equilibrium intersubband absorption spectra and modulation of absorption spectra under lateral electric field. The analysis of interband photoluminescence modulation spectra allowed us to derive electron temperature and compare it with temperature value obtained from the absorption spectra analysis. These temperature values were close. The value of the observed absorption variation suggests the application prospects of tunnel-coupled quantum wells as the basis for fast infrared light modulators.

Acknowledgments

The authors are grateful to A.N. Sofronov and V.Yu. Panevin for their help with the experiment preparation. This work has been supported by the Russian Foundation for Basic Research (Projects №№ 14-02-00336, 16-32-60085) and the Ministry of Education and Science of the Russian Federation (state assignment).

References

[1] Firsov D A, Vorobjev L E, Vinnichenko M Ya, Balagula R M, Kulagina M M and Vasil'iev A P 2015 *Semiconductors* **49(11)** 1425
[2] Dupont E, Delacourt D, Berger V, Vodjdani N and Papuchon M 1993 *Appl. Phys. Lett.* **62** 1907
[3] Rybalko D A, Vinnichenko M Ya, Vorobjev L E, Firsov D A, Balagula R M, Panevin V Yu, Kulagina M M and Vasil'iev A P 2014 *J. Phys.: Conf. Ser.* **541** 012081
[4] Vorobjev L E, Danilov S N, Zerova V L and Firsov D A 2003 *Semiconductors* **37(5)** 586
[5] Vorobjev L E, Vinnichenko M Ya, Firsov D A, Zerova V L, Panevin V Yu, Sofronov A N, Thumrongsilapa P, Ustinov V M, Zhukov A E, Vasil'iev A P, Shterengas L, Kipshidze G, Hosoda T and Belenky G 2010 *Semiconductors* **44(11),** 1402-1405
[6] Zerova V L, Vorob'ev L E, Firsov D A, Towe E 2007 Semiconductors **41(5)** 596
[7] Vurgaftman I, Meyer J R and Ram-Mohan L R 2001 *Journal of Applied Physics* **89(11)** 5815-5875
[8] Firsov D D and Komkov O S 2013 *Tech. Phys. Lett.* **39(12)** 1071–1073

E-beam lithography exposure conditions for the fabrication of RGB filter based on metal/dielectric subwavelength grating

S A Fomchenkov[1,2], M A Butt[1], V V Podlipnov[1,2], S D Poletaev[1,2], R V Skidanov[1,2], N L Kazanskiy[1,2]
[1]Samara State Aerospace University, Moskovskoye Shosse 34, Samara, 443086, Russia
[2]Image Processing Systems Institute of the Russian Academy of Sciences, Molodogvardeyshaya 151, Samara, 443001, Russia

Abstract. The main idea of this work was to determine the optimized parameters of E-beam lithography to obtain metal grating over dielectric thin film. This combination of metal/dielectric can provide high transmission spectrum of RGB colors. Different electric flux densities were used during E-beam writing and the best resolution and symmetric periodicity was obtained at 53 μC/cm^2 dose.

1. Introduction

Colour filters are playing important role as a key element for wide area of applications such as light emitting diodes (LED), liquid crystal displays (LCD) and antireflection filters [1,2]. In order to meet the demanding market requirements for large display panels, it is mandatory to make the color filter thinner, larger and easy to fabricate. In this work, we have presented the steps of fabrication of the RBG color filter and studied the effect of various power intensities of e-beam writing system on E-resist at nanometer scale resolution.

2. Experimental details

In this section, the fabrication steps of RGB filter are presented. The schematic of the total fabrication steps are shown in figure 1.

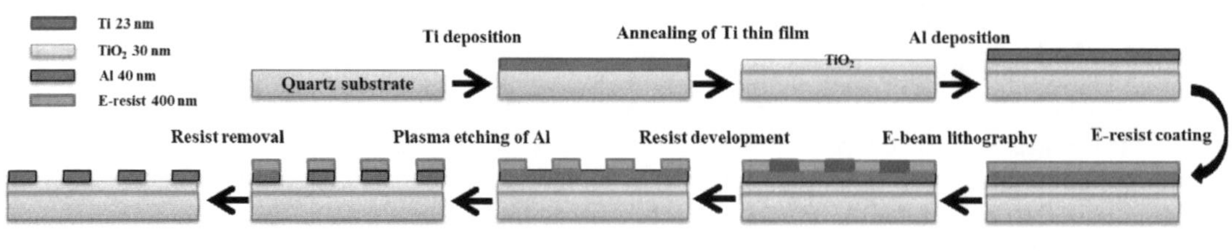

Figure 1. Fabrication steps of RGB filter.

2.1. Deposition and oxidation of Ti layer

Ti thin film was deposited on quartz with the help of magnetron deposition system Caroline D12A. For all deposition experiments, we used DC power of 300 W, 2x10^{-5} Torr and the gap between substrate/target was 100 mm. The adhesion of Ti films was improved by heating the substrate but the film morphology strongly depends on the substrate temperature during the deposition process. Ti films of 80-200 nm was deposited for morphological and optical characterization.

Usually, Ti thin films deposited below150 °C are amorphous and attain crystallinity after annealing process. TiO_2 was obtained by high temperature annealing (oxidation) of Ti films. Annealing process was carried out in muffler furnace by using Conventional Thermal Annealing (CTA). The samples were loaded in the muffler at room temperature and slowly heated till 700 °C and then maintain the temperature for 1 hr. Once the annealing was complete, the temperature was reduced till room temperature with a slow ramp. In order to obtain rutile phase of TiO2, the annealing was done between 500-700 °C for 1-2 h. After annealing, the obtained TiO_2 films were fully transparent which can be observed visually.

We tried to investigate the effect of substrate temperature during deposition of Ti layer. Substrate temperature of 50 and 150 °C were used during Ti metal deposition to test the adhesion and scratch of the layer. After several attempts of tape test, it was confirmed that, Ti layer deposited at both 50 and 150 °C showed good adhesion on quartz.

Ti thin film of 120 nm was deposited on quartz substrate at 50 and 150 °C and annealed at 600 °C for 1 h. From SEM images, it can be observed that TiO_2 layer obtained after annealing at 600 °C (substrate temperature of 150 °C) has comparable less surface roughness as compared to the TiO_2 film obtained after annealing (substrate temperature 50 °C) as shown in figure 2 (a) and (b). These films are smooth and can be used in the fabrication of various optical components.

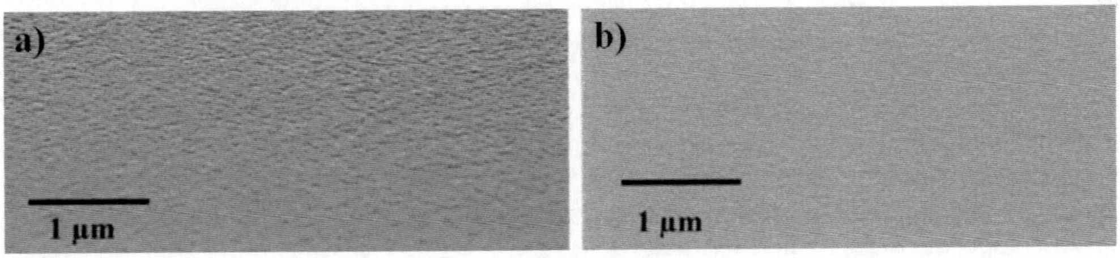

Figure 2. SEM images of the surface morphology of TiO_2 on quartz substrate obtained after annealing at 600 °C (a) substrate temperature 50 °C, (b) 150 °C.

2.2. Optical characterization

To verify the phase of annealed thin TiO2 films deposited on quartz substrate, we carried out Raman spectroscopy and obtained one weak Raman shift peaks at 143 and two strong peaks at 447 and 612 cm^{-1} which indicates rutile phase as shown in figure 3.

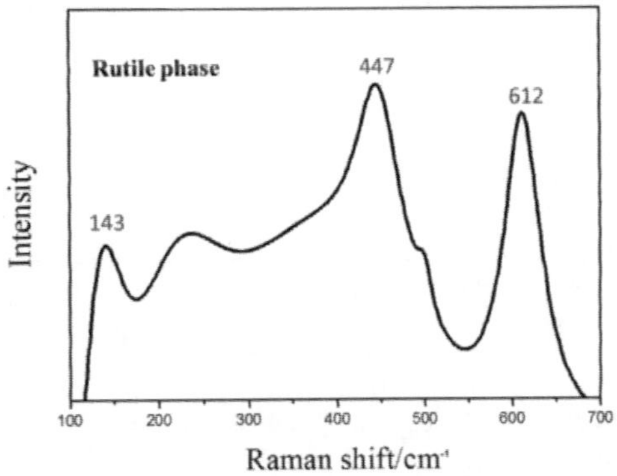

Figure 3. Raman spectra of TiO_2 thin film converted by annealing Ti layer deposited on quartz substrate (substrate temperature 150 °C).

Refractive index of TiO_2 layer of same thickness deposited on quartz substrate annealed at different temperatures (500 and 700 °C) was measured.

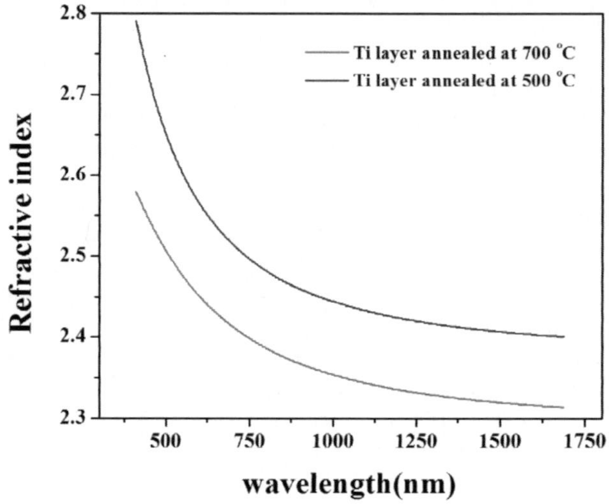

Figure 4. Refractive index measurement of TiO_2 layer on quartz substrate.

From figure 4, it can be observe that, the TiO_2 films on quartz substrate annealed at 500 °C has higher refractive index range as compared to sample annealed at 700 °C. The gradient in refractive index can be due to the expansion of layers at different ratio during annealing at different temperature. Transmittance spectra were measured in the wavelength range of 350-1600 nm. For the visible wavelength range, the transmittance is around 75%.

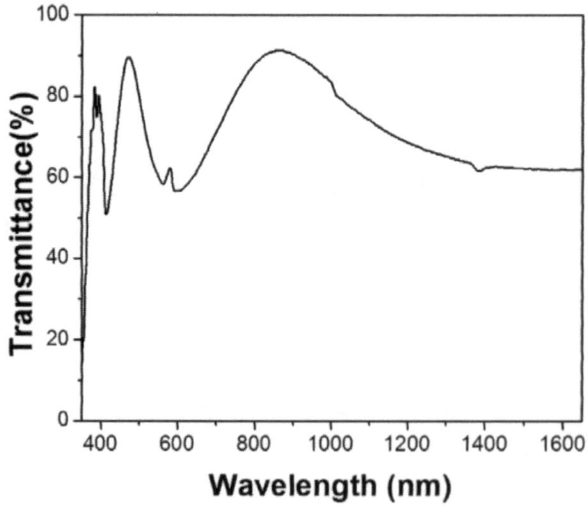

Figure 5. Transmittance spectrum of TiO_2 deposited on quartz.

The optical properties of TiO_2 film deposited on quartz substrate were measured with the help of Ellipsometer M2000DI.

2.3. E-beam lithography
The sample was coated with electron positive resist (EPR-40) at 5000 rpm to obtain 400 nm layer. The grating design was written on resist at different electric flux density as mentioned in table 1.

Table I. E-beam dose and actual size of RGB gratings.

E-beam dose ($\mu C/cm^2$)	**Actual size(nm) Period/Duty cycle**		
	Red	**Green**	**Blue**
	420/231	350/193	260/143
47	Underexposed	Underexposed	Underexposed
50	430/234	360/188	260/137
53	434/215	362/165	259/139

From the table, it can be observe that the resist is underexposed at doses less than 50 $\mu C/cm^2$. The design closest to the actual parameters was obtained at dose level of 50 $\mu C/cm^2$. The SEM images of the grating fabricated on resist after development is shown in figure 6.

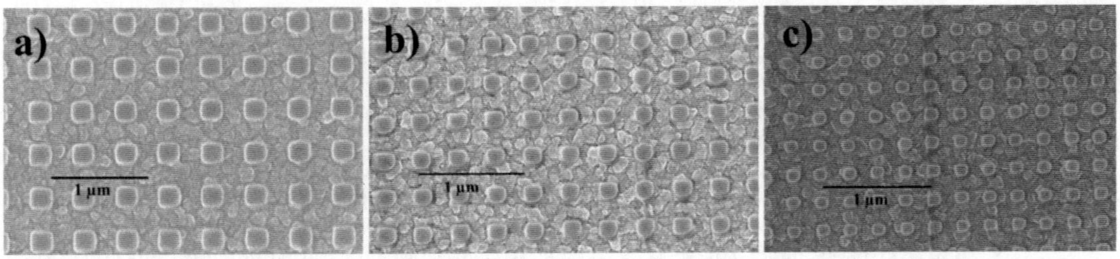

Figure 6(a, b, c). SEM images of the RGB grating fabricated at 50 µC/cm2 dose on resist **(a)** Red, **(b)** Green, **(c)** Blue.

3. Conclusion
In this work, we obtained transparent oxide layer by annealing Ti layer which is the fundamental basis to increase the transmission efficiency of RGB filter. Titanium metal layers deposited by magnetron sputtering on different substrates were converted into titanium dioxide rutile phase at 500-700 °C. Polycrystalline phase was determined by Raman spectroscopy. Surface morphology of the dielectric film deposited on different substrates was observed by SEM and found the low surface roughness on quartz substrate and the maximum transmittance when the deposition temperature of Ti metal layer was 150 °C. Transmittance could be enhanced substantially through the surface plasmon resonance induced at the dielectric-metal interfaces. This filter was fabricated on EPR with E-beam lithography at different electric flux densities with an accuracy of 10 nm. The best resolution and symmetric periodicity was obtained at 53 $\mu C/cm^2$ dose.

Acknowledgments
This work was supported by the Ministry of Education and Science of the Russian Federation and the Russian Foundation for Basic Research (grant No. 14-19-00114).

References
[1] Yan Y, Yun Z, Linsen C 2009 *Appl. Opt.* **48** 5035
[2] Kanamori Y, Kikuta H, Hane K 2000 *Jpn. J. Appl. Phys.* **39** L735

Optical properties of InP/ZnS quantum dots deposited into nanoporous anodic alumina

S S Savchenko, A S Vokhmintsev and I A Weinstein

NANOTECH Centre, Ural Federal University, Mira str., 19, Yekaterinburg 620002, Russia

Abstract. Spectral characteristics of InP/ZnS core/shell colloidal quantum dots of two different sizes (QD-1 and QD-2) were investigated. Absorption and luminescence spectra were analyzed for a series of solutions with a concentration range from 0.04 to 40 g/l. Energies of the optical transitions are evaluated. The obtained values of 2.60 eV (QD-1) and 2.38 eV (QD-2) correspond to the InP first excitonic transitions while 4.06 (QD-2) and 4.70 eV (QD-1, QD-2) are assumed to be caused by the ZnS shell absorption. Structures based on nanoporous anodic aluminum oxide (AAO) with the QDs were synthesized via an electrochemical oxidation and ultrasonic-assisted deposition. Chromaticity coordinates and correlated color temperatures for all phosphors under study were calculated. The fabrication possibilities of InP/ZnS@AAO nanostructures with tunable emission color (including the border of white region) were shown.

1. Introduction

Colloidal quantum dots (QDs) attract much attention and research efforts over the past decades [1,2,3]. Generally, a size of this semiconductor nanocrystals should be comparable with the exciton Bohr radius or the electron and hole de Broglie wavelength of the material they are made of. Such condition leads to the manifestation of the quantum-confinement and therefore to a collapse of the continuous energy bands of a bulk material into discrete, atomic-like energy levels [4]. Moreover, the electronic structure strongly depends on the dot size providing these objects with tunable physical and chemical properties. At present, possible applications of QDs in many fields are intensively studied, in particular for creating light-emitting diodes, lasers, solar cells, phosphors, biosensors and biomarkers, etc [5-9].

QDs based on III-V semiconductor compounds are of special interest, because they do not contain toxic components such as Cd, Pb and Hg, which have low acceptability for technological applications. Indium phosphide is the most studied compound, as the emission of InP nanocrystals can be tuned throughout the visible and near infrared range by changing their size [10]. Their overgrowth with wider band gap ZnS shell is used to improve rather poor optical properties of as-prepared InP nanocrystals [10]. However, synthesis of the InP-based QDs emitting in the blue region with a high quantum yield (QY) is the challenging task. In this regard, the use of widegap solids is promising to create luminescent composites with adjustable emission based on discrete color mixing. Nanoporous anodic alumina characterized by a large surface area and intrinsic blue emission [11] can be used as functional substrates for constructing such phosphors. Thus, this work deals with optical and luminescence study of the InP/ZnS colloidal QDs deposited into anodic aluminum oxide (AAO) matrices.

2. Experimental

InP/ZnS colloidal QDs of two different sizes marked DS-72a (QD-1) and DS-72b (QD-2) were synthesized by FSUE "NIIPA" (Dubna, Russia). They have core/shell structure coated with modified polyacrylic acid. Semiconductor nanocrystals are dispersed in water with initial concentrations (c) of 40 g/l for QD-1 and 48.4 g/l for QD-2. OA measurements of the samples were carried out at room temperature in a range of 190 to 900 nm by means of a Shimadzu UV-2450 spectrophotometer in a 1 cm quartz cuvette. A slit width and sampling interval were set to the 2 and 0.1 nm, respectively. Absorption spectra of the QDs were recorded for a series of solutions with a concentration range from 40 g/l to 0.04 g/l prepared by two-fold dilutions with distilled water for 10 times. Photoluminescence (PL) of the QDs was analyzed for the colloidal solutions by means of a Perkin Elmer LS 55 luminescent spectrometer in a fluorescence mode at room temperature.

QYs of the semiconductor nanocrystals under investigation were determined relative to the standard according to the procedure reported in [12]. Rhodamine 6G (R6G) was chosen as a reference since it is considered as one of the well-established standards [13] and its emission region is close to that of the QDs. First of all water solutions with absorbance A ≈ 0.5 at a 370 nm wavelength were prepared. Afterward they were diluted in a volumetric ratio of 1:9 in order to avoid inner filter effects [14]. Finally, their fluorescence spectra were detected, areas under emission curves calculated and QYs evaluated.

Nanoporous matrices of Al_2O_3 were grown by two-step Al foil anodizing at the constant current density using oxalic acid as an electrolyte [10]. The synthesized AAO structures with the average pore size of ≈ 80 nm were annealed at 500 and 700 °C for 5 hours in ambient atmosphere. All the samples had amorphous structure in accordance with the X-ray diffraction analysis [10]. To deposit QDs the AAO substrates were placed in the InP/ZnS colloidal solutions and treated in an Eurosonic 4D ultrasonic bath for 60 minutes. Four structures were obtained this way. Their PL measurements under 270 nm excitation in a range of 300 to 700 nm were performed in a fluorescence mode at room temperature using LS55. A sampling interval, excitation and emission slit widths were 0.5, 10 and 5 nm, respectively.

3. Results and discussion

3.1. Optical absorption and photoluminescence of InP/ZnS quantum dots

Electronic energy structure of semiconductor nanocrystals appears in the OA spectra. They are presented in figure 1 for colloidal QD-1 and QD-2 solutions of various concentrations. It can be seen that the QDs begin to absorb light in different spectral ranges. Namely, QD-1 has an A = 0.1 at a wavelength of 560 nm and the QD-2 – at a wavelength of 600 nm for c=40 g/l. When diluting different regions of the spectrum are spanned by absorbance range of A = 0.1 – 3, which allows absorption of samples to be studied in a wide diapason of wavelengths revealing its features. The QD-1 OA exhibits a distinct peak in the region of 450-480 nm after several dilutions. For QD-2 a shoulder in the range of 490-570 nm is observed. This fact can be connected with the wider particle size distribution for the second type colloidal nanocrystals or with different quantum confinement regimes in the samples. Experimental OA spectra are also influenced by light scattering at nanocrystals, which should be taken into account when evaluating optical transition energies. In addition, the concentration dependencies were analyzed at various wavelength. Observed linear character indicates the absence of deviations from the Beer–Lambert–Bouguer law for c = 0.04 - 40 g/l.

A known method for distinguishing of the hidden peaks is derivative spectrophotometry [3,15]. For the all experimental data $d^2A/d\lambda^2$ spectra were calculated. For instance, figure 2 shows some OA curves and their second derivatives whose minima correspond to the hidden absorption maxima. The technique was used to evaluate energies of the optical transitions. Obtained values are 2.60, 4.70 ± 0.02 eV for the QD-1 and 2.38, 4.06, 4.68 ± 0.02 eV for the QD-2 and do not depend on the solution concentration. The energies of 2.60 and 2.38 eV correspond to the first excitonic absorption peaks of the InP core and the shift between them results from size-dependent effect. Noteworthy that the values

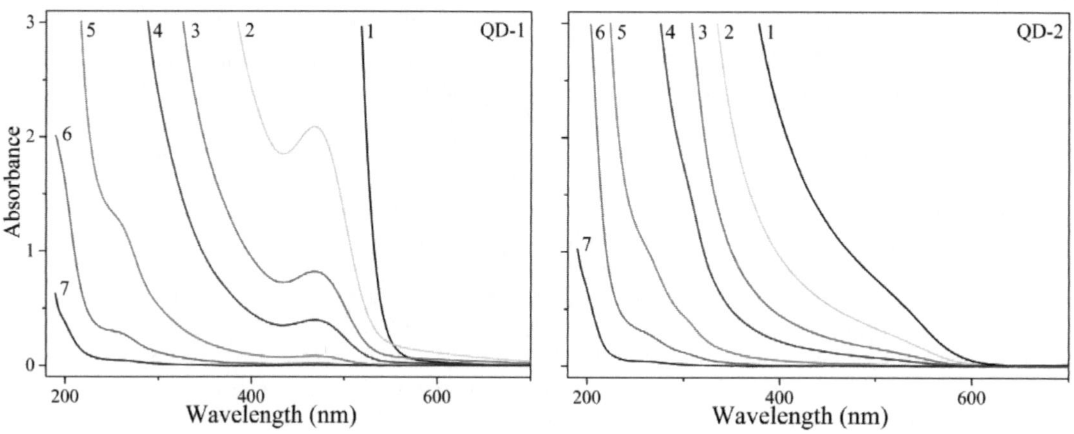

Figure 1. OA spectra of the InP/ZnS solutions of various concentrations.
QD-1: 1 – 40 g/l, 2 – 10 g/l, 3 – 5 g/l, 4 – 2.5 g/l, 5 – 0.63 g/l, 6 – 0.16 g/l, 7– 0.04 g/l;
QD-2: 1 – 40 g/l, 2 – 20 g/l, 3 – 10 g/l,4 – 5 g/l, 5 – 2.5 g/l, 6 – 0.31 g/l, 7– 0.04 g/l.

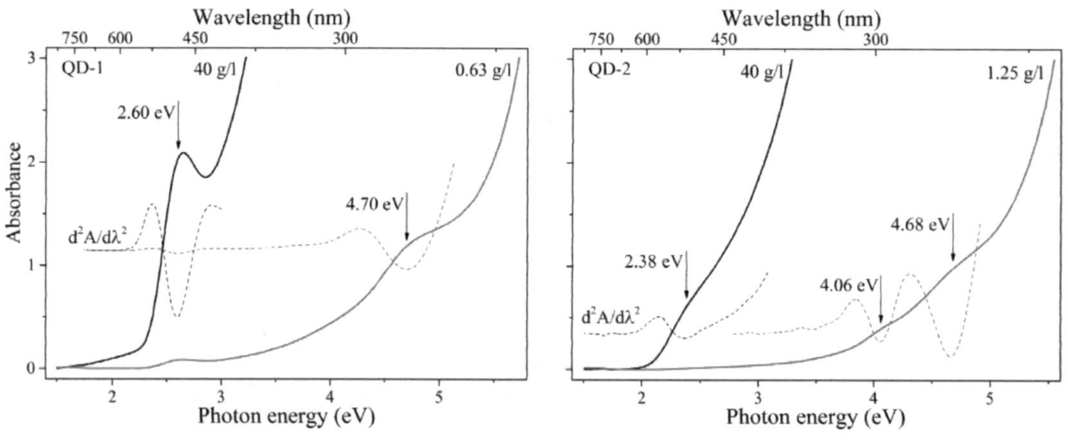

Figure 2.OA spectra for QDs of the marked concentrations (solid lines) and their second derivatives (dash lines). The arrows indicate minima of the $d^2A/d\lambda^2$ curves.

of 4.68 and 4.70 eV are equal within the experimental error and fall in the range of band gap energies from 4.82 eV to 4.47 eV observed for nanocrystalline zinc sulfide powders [16]. In [17] the absorption edge of the ZnS colloidal nanocrystals is determined to be 4.07 eV, which matches with our data for the QD-2 samples. Thereby these optical transitions might be ascribed to the ZnS shell, however clarifying their origin requires additional researches.

PL measurements were carried out for the InP/ZnS solutions of the same concentration range as for OA. The QD-1 has $h\nu_{max}$ = 2.38 eV (525 nm), FWHM = 0.25 eV and QD-2 exhibits $h\nu_{max}$ = 2.18 eV (570 nm), FWHM = 0.33 eV (green and orange dash lines in figure 3). The analysis of the data registered indicates that emission maxima is not affected by the concentration. QYs under excitation of 370 nm are 27% for the QD-1 and 10% for the QD-2 solutions.

3.2. Photoluminescence of InP/ZnS@AAO nanostructures

PL spectra under the 270 nm excitation of the synthesized structures based on AAO membranes with the deposited QDs are demonstrated in figure 3. Cyan and blue dash lines display the emission of aluminum oxide matrices annealed at 500 and 700 °C. Black solid lines illustrate PL of the corresponding fabricated structure. The spectra of the synthesized phosphors contain luminescence bands of the QD-1 and QS-2. This result agrees with our previous work [8] where another type of InP/ZnS semiconductor nanocrystals ($h\nu_{max}$ = 2.26 eV, FWHM = 0.23 eV) was deposited into AAO matrices. Preservation of the fluorescence emission properties of the QDs after the deposition suggests a reliable passivation of the its surface.

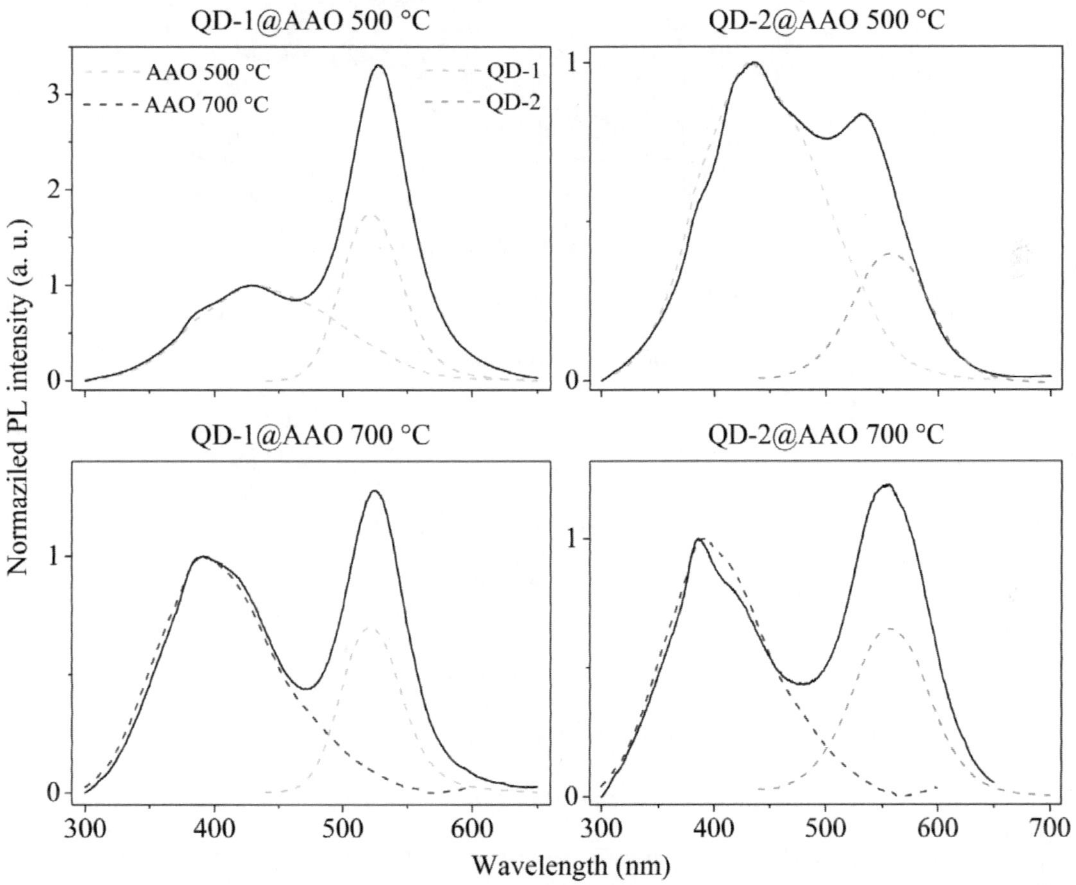

Figure 3. PL spectra of synthesized InP/ZnS@AAO structures.

Figure 4 shows chromaticity diagram where appropriate calculated coordinates for the samples under study are marked and photos of the fabricated phosphors are presented. Circles indicate objects synthesized here and triangles refer to the [8]. It is seen that phosphors with the tunable emission color can be designed by filling AAO membranes with InP/ZnS QDs of various sizes. AAO matrices annealed at 500 and 700 °C with deposited QD-2 (6 and 8 diagram points) exhibit luminescence on the border of the white region. Correlated color temperatures for these phosphors are 4068 K and 3232 K, respectively. Obtained data for all samples are summarized in table 1.

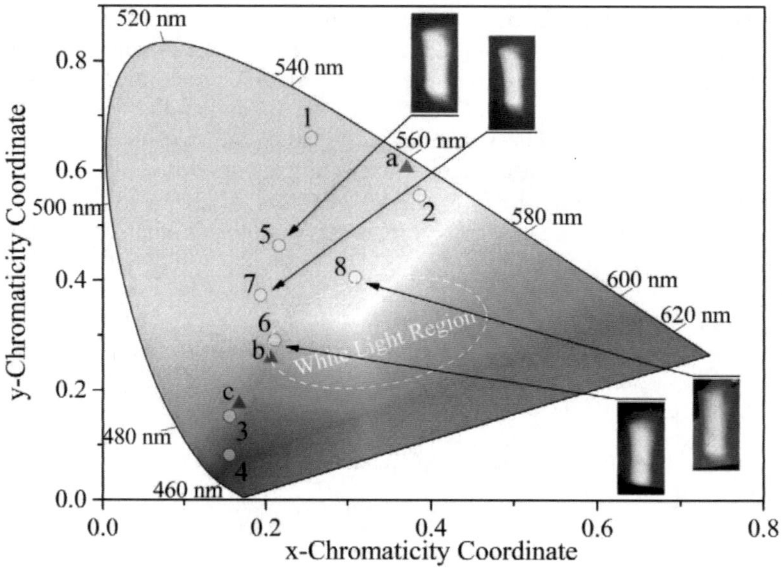

Figure 4. Chromaticity diagram for the samples.

Table 1. Calculated chromaticity coordinates.

Diagram point	Sample	Coordinates
1	QD-1	(0.25, 0.66)
2	QD-2	(0.39, 0.56)
3	AAO 500 °C	(0.16, 0.15)
4	AAO 700 °C	(0.15, 0.08)
5	QD-1@ AAO 500 °C	(0.22, 0.46)
6	QD-2@ AAO 500 °C	(0.21, 0.29)
7	QD-1@ AAO 700 °C	(0.19, 0.37)
8	QD-2@ AAO 700 °C	(0.31, 0.41)
a	InP/ZnS [8]	(0.37, 0.61)
b	InP/ZnS @ AAO 500 °C [8]	(0.21, 0.26)
c	InP/ZnS @ AAO 700 °C [8]	(0.17, 0.17)

4. Conclusion

InP/ZnS core/shell colloidal QDs of two different sizes were investigated. OA and PL spectra were studied for a series of the QD-1 ($h\nu_{max}$ = 2.38 eV) and QD-2 ($h\nu_{max}$ = 2.18 eV) solutions within a concentration range from 0.04 to 40 g/l. The OA experimental data were analyzed by techniques of derivative spectrophotometry characterize hidden absorptions features. Energies of the optical transitions are evaluated to be 2.60, 4.70 ± 0.02 eV for the QD-1 and 2.38, 4.06, 4.68 ± 0.02 eV for the QD-2. The values of 2.60 eV and 2.38 eV correspond to the InP excitonic absorption peaks. Other transition are assumed to be caused by the ZnS shell. The calculated values as well as measured maxima of the QD emission bands do not depend on the solution concentration. Quantum yields of the samples evaluated relative to the R6G aqueous solution are 27% for the QD-1 and 10% for the QD-2.

Phosphors based on nanoporous aluminum oxide membranes with core/shell InP/ZnS semiconductor nanocrystals were synthesized via an anodizing process and ultrasonic-assisted deposition. The fabricated InP/ZnS@AAO samples with QD-2 exhibit emission on the border of the white region. The nanostructures are characterized by chromaticity coordinates of (0.21, 0.29), (0.31, 0.41) and correlated color temperatures of 4068 K, 3232 K.

Acknowledgments

The work was supported by Act 211 Government of the Russian Federation, contract № 02.A03.21.0006.

References

[1] Rossetti R, Nakahara S and Brus L E 1983 *J. Chem. Phys.* **79** 1086-8
[2] Spanhel L, Haase M, Weller H and Henglein A 1987 *J. Am. Chem. Soc.* **109** 5649-55
[3] Weller H 1993 *Angew. Chem. (Int. Ed. Engl.)* **32** 41-53
[4] Gaponenko S V 1998 *Optical Properties of Semiconductor Nanocrystals* (Cambridge: Cambridge University Press)
[5] Zhang Q, Wang C-F, Ling L-T and Chen S *J. Mater. Chem.* C 2014 **2** 4358-73
[6] Klimov V I, Mikhailovsky A A, Xu S, Malko A, Hollingsworth J A, Leatherdale C A, Eisler H-J and Bawendi M G 2000 *Science* **290** 314-7
[7] Kamat P V 2008 *J. Phys. Chem.* C **112** 18737-53
[8] Savchenko S S, Vokhmintsev A S and Weinstein I A 2016 *AIP Conf. Proc.* **1717** 040028
[9] Chatterjee K, Sarkar S, Jagajjanani Rao K and Paria S 2014 *Adv. Colloid Interface Sci.* **209** 8-39
[10] Reiss P, Protiere M and Li L 2009 *Small* **5** 154-68
[11] Ilin D O, Vokhmintsev A S, Chukin A V and Weinstein I A 2015 *Physics. Technologies. Innovations: Conf. Proc.* (Yekaterinburg: UrFU) pp. 96 – 100 (in Russian)
[12] Grabolle M, Spieles M, Lesnyak V, Gaponik N, Eychmüller A and Resch-Genger U 2009 *Anal. Chem.* **81** 6285-94
[13] Brouwer A M 2011 *Pure. Appl. Chem.* **83** 2213-28
[14] Fery-Forgues S and Lavabre D 1999 *J. Chem. Educ.* **76** 1260-4
[15] Talsky G 1994 *Derivative Spectrophotometry: Low and Higher Order* (Weinheim: VCH Verlagsgesellschaft mbH)
[16] Kho R, Torres-Mart´ınez C L and Mehra R K 2000 *J. Colloid and Interface Sci.* **227** 561-6
[17] Zhang Y, Ma M, Wang X, Fu D, Gu N, Liu J, Lu Z, Ma Y, Xu L and Chen K 2002 *J. Phys. Chem. Solids* **63** 2115-8

Optical tuning of near and far fields form hybrid dimer nanoantennas via laser-induced melting

Stanislav A Kolodny[1], Yali Sun[2], Dmitry A Zuev[1], Sergey V Makarov[1], Alexander E Krasnok[1], and Pavel A Belov[1]

[1]ITMO University, St. Petersburg 197101, Russia

[2]Huazhong University of Science and Technology, Wuhan, Hubei, 430074, China

Abstract. Hybrid nanophotonics based on metal-dielectric nanostructures unifies the advantages of plasmonics and all-dielectric nanophotonics providing strong localization of light, magnetic optical response and specifically designed scattering properties. Here, we propose a new method for optical properties tuning of hybrid dimer nanoantenas via laser-induced melting at the nanoscale. We demonstrate numerically that near- and farfield properties of a hybrid nanoantenna dramatically changes with fs-laser modification of Au particle. The results lay the groundwork for the fine-tuning of hybrid nanoantennas and can be applied for effective light manipulation at the nanoscale, as well as biomedical and energy applications.

1. Introduction

The concept of hybrid nanostructure, utilizing the advantages of both plasmonics and all-dielectric nanophotonics, has recived heightened interest because of unique optical properties of sych type nanostructures [1-3]. These properties strongly depend on the geometrical parameters of each constituent part and their relative position. Therefore, one of the most important research directions here is the development of fabrication methods which are still poorly developed [3]. The most common methods of hybrid nanostructures fabrication are the high-throughput wet chemistry preparation [3, 4] and combination of conventional methods of lithography [5, 6]. However, wet-chemistry methods suffer on the lack of spatial arrangement, whereas designs of highly ordered lithographically produced nanostructures usually are limited by geometrical properties of masks. Such limitations can be avoided by means of additional cost-effective postprocessing of lithographically produced hybrid nanostructures. One of the most promising post-processing technique is femtosecond (fs) laser heating by tightly focused beam, which has been successfully implemented for plasmonic nanoparticles reshaping accompanied by a strong change in their optical properties. [7].

Recently, we have developed a new method of tuning of non-symmetrical hybrid nanostructures [8]. The method is based on fs-laser melting of gold nanodisc placed on the top of truncated silicon nanocone. Silicon nanocone has electric and magnetic dipole resonances, whereas gold nanodisc has only electric one in visible range. Laser melting makes it possible to change the position of the plasmon resonance and adjust the properties of the whole hybrid particle.

Here, we apply this method for tuning of hybrid dimer nanoantenas. We demonstrate numerically that near- and far-field properties of a hybrid nanoantenna dramatically changes with fs-laser modification of Au particle.

2. Tuning of optical properties of hybrid nanostructures via fs-laser reshaping

For numerical simulations, we have choosen the assymetrical (not a core-shell structure) hybrid nanoantenna proposed and fabricated in [7]. This concept (gold nanodisk placed on the tip of silicon nanocone) has a several advantages. At first, these nanostructures can be fabricated via the sequence of fabrication steps including standart methods of litography: e-beam lithography, metal evaporation, lift-off procedure, gas phase chemical etching. The selective fs-laser modification (reshaping) of the metal components of the hybrid nanoparticles without affecting dielectric ones is applied to complete the fabrication process and precisely tailoring magnetic and electric optical resonances of the nanoparticles. So, the final geometry of the hybrid nanodimer (and optical properties as well) after modification is generally determined by the laser fluence and can be varied considerably [6]. Therefore, such nanostructures and metasurfaces are very perspective for application in nanofotonics.

To understand influence of fs-laser reshaping on the coupling effects taking place in array of hybrid nanostructures, the system including a pair of hybrid nanostructures (dimer) was choosen for numerical simulation.

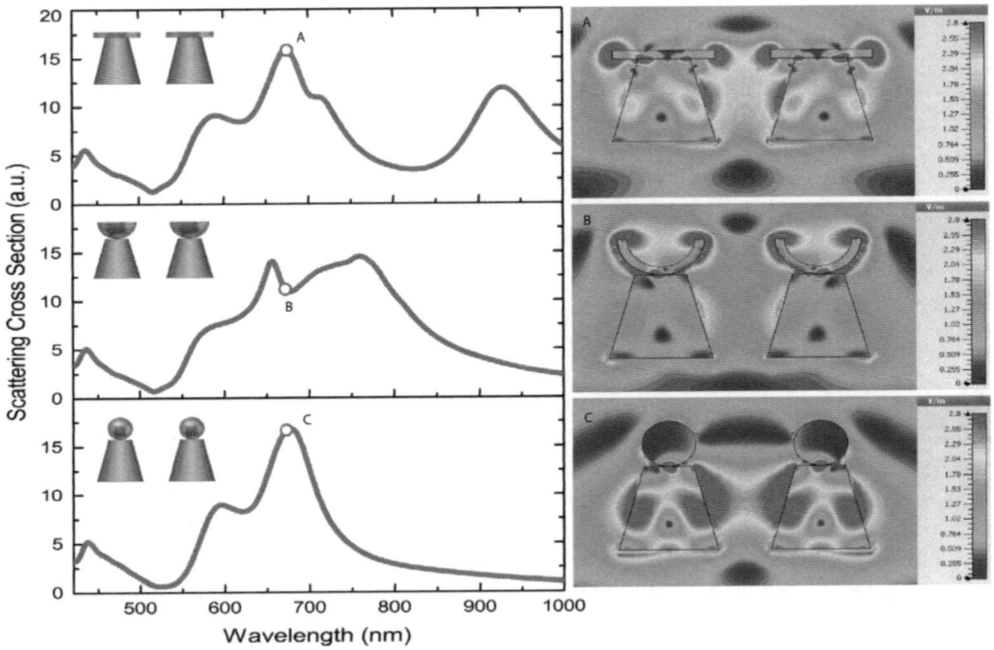

Figure 1. The scattering spectra of hybrid dimer nanoantennas with different fs-laser reshaping degree. On the right, the distributions of local E-field for wavelength of 670 nm.

Scattering spectra of hybrid dimer nanoantennas have been calculated in range of wavelength 420-1000 nm for different values of gold particles forms (see **Figure.1**). The distance between the axes of the hybrid nanoparticles is 100 nm in all cases. The first plot on the **Figure.1** is the scattering spectrum of hybrid dimer nanoantanna without reshaping (gold nanoparticle has a disk shape). The second plot is the scattering spectrum of hybrid dimer nanoantenna with gold particles in the form of nanocup. The third plot is the scattering spectrum of hybrid dimer nanoantenna in the final stage of reshaping. It is seen that the scattering spectral dramatically changes with reshaping of gold particle in the hybrid nanoantenna. Under reshaping, the electric dipole resonance of Au nanoparticle (point A on the **Figure.1**, 1st plot) shifts to magnetic dipole resonance of silicone one. The scattering spectrum of the hybrid dimer nanoantenna with gold nanocups (see **Figure.1**, 2nd plot) is more difficult because of complex form of Au nanoparticles. In the **Figure.1** on the right the electric field distribution for the hybrid dimer nanoantennas with varying degrees of reshaping are presented. We have observed an increasing of a local electric field in the gap of the hybrid nanoantanna. Thus we have shown that the fs-laser melting of hybrid nanoantenna makes it possible to tune its farfield and nearfield properties.

Let's study the changing of local density of states (LDOS) caused by fs-laser melting of hybrid dimer nanoantenna. The normalized LDOS describes by Purcell factor which is a measure of so-called Purcell effect. The Purcell effect is defined as a modification of the spontaneous emission lifetime of a quantum source induced by its interaction with environment. For calculation of Purcell factor we have used the method of its extraction from an input impedance of small dipole antenna (1):

$$F = \frac{R_{in}}{R_{0,in}} \equiv \frac{\mathrm{Re}Z_{in}}{\mathrm{Re}Z_{0,in}}, \tag{1}$$

where R_{in} is the resistive part of the input impedance of the small dipole in presence of hybride nanoantenna, $R_{0,in}$ is the same in free space [9].

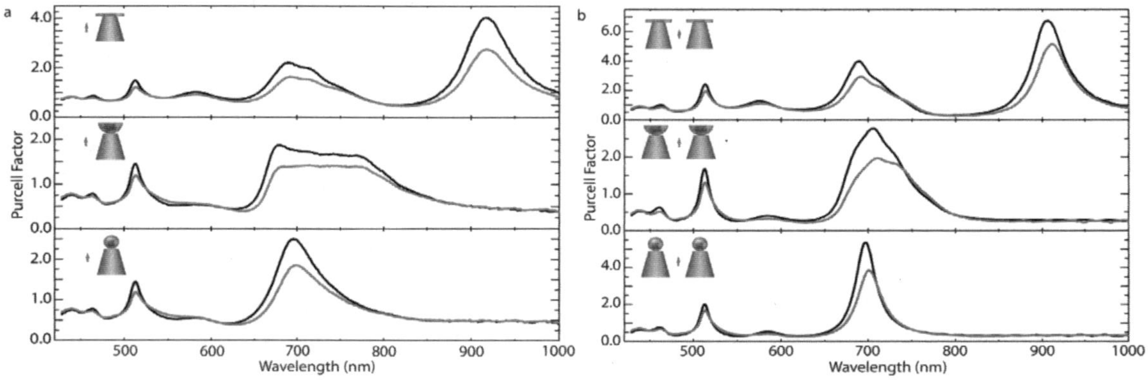

Figure 2. The Purcell factor of hybrid nanoantennas with different fs-laser reshaping degree: a – single nanoparticles, b – hybrid dimer nanoantennas.

We have calculated the Purcell factor in the wavelength range of 420-1000 nm for two cases of the small dipole located close to (i) a single hybrid nanoparticle (**Figure.2a**), and (ii) to the hybrid dimer nanoantenna (**Figure.2b**). The results are presented for two distances between the dipole and nanostructure of 30 nm and 50 nm (black and red curve, respectively). We have observed that the position of Purcell factor resonances changes after the modification of gold nanoparticles. The value of Purcell factor strongly depends on the distance between dipole and nanoantenna and the gold nanoparticle shape. Thus, we have shown that the fs-laser melting of hybrid nanoantenna makes it possible to tune its Purcell factor.

3. Conclusions

In conclusion, we have proposed a new method of optical properties tuning of hybrid dimer nanoantenas via laser-induced melting at the nanoscale. We have demonstrated numerically that near- and far-field properties of a hybrid nanoantenna dramatically changes with fs-laser modification of Au particle. We belive that the results lay the groundwork for the fine-tuning of hybrid nanoantennas arrays and can be applied for effective light manipulation, as well as biomedical and energy applications.

Acknowledgements

This work was supported by the Ministry of Education and Science of the Russian Federation (project №14.584.21.0009 with unique identificator RFMEFI58414X0009).

References

[1] G. Grinblat and M. Rahmani and E. Cortes and M. Caldarola and D. Comedi and S.A. Maier and A.V. Bragas, Nano Letters, vol. 14, p. 6660-6665 (2014).
[2] O.Bar-Elli and E. Grinvald and N. Meir and L. Neeman and D. Oron, ACS Nano, vol. 9(8), p. 80648069 (2015).
[3] R. Jiang, B. Li, C. Fang, and J. Wang, Advanced Materials 26, 5274–5309 (2014).
[4] B. Khanal, A. Pandey, L. Li, Q. Lin, W. K. Bae, H. Luo, V. Klimov, and J. Pietryga, ACS Nano 6, 3832–3840 (2012).
[5] H. Aouani, M. Rahmani, M. Navarro-Cia, and S. A. Maier, Nature Nanotechnology 9, 290–294 (2014).
[6] N. Hatab, C.-H. Hsueh, A. Gaddis, S. Retterer, J.-H. Li, G. Eres, Z. Zhang, and B. Gu, Nano Lett 10, 4952–4955 (2010).
[7] P. Zijlstra, J. W. Chon, and M. Gu, Nature 459, 410 (2009).
[8] Zuev, D. A., Makarov, S. V., Mukhin, I. S., Milichko, V. A., Starikov, S. V., Morozov, I. A.,

Shishkin, I. I., Krasnok, A. E. and Belov, P. A. (2016), Fabrication of Hybrid Nanostructures via Nanoscale Laser-Induced Reshaping for Advanced Light Manipulation. Adv. Mater., 28: 3087–3093. doi:10.1002/adma.201505346 (Accepted).
[9] A. Krasnok et al, An antenna model for the Purcell effect., Scientific Reports, 5: 12956 (2015).

Recombination dynamics in heterostructures with two planar arrays of II-VI quantum dots

T N Mikhailov[1], K G Belyaev[1], A A Toropov[1], S V Sorokin[1], G Pozina[2], and T V Shubina[1]

[1]Ioffe Institute, 26 Politekhnicheskaya, St Petersburg 194021, Russia
[2]Linköping University, S-581 83 Linköping, Sweden

Abstract. We present time-resolved photoluminescence studies of epitaxial heterostructures with two arrays of Cd(Zn)Se/ZnSe quantum dots (QDs), which are formed by the successive insertion of CdSe fractional monolayers of different nominal thicknesses into a ZnSe matrix. Our data are suggestive of the appearance of effective channels of the energy transfer from the insertion comprising the array with smaller QDs, emitting at higher energy, towards the array with larger QDs, emitting at lower energy. The effect of dark excitons on characteristic times of radiative recombination is discussed.

1. Introduction

Fractional monolayer (ML) insertions of CdSe within a ZnSe matrix have been successfully used as an active region in green lasers pumped optically or by an electron beam [1]. During molecular beam epitaxy (MBE) growth these insertions are transformed into the planar arrays of quantum dots (QDs). As a rule, the lasers developed for the visible range contain several sheets of such QD arrays. The nominal thickness of each CdSe insertion, *d*, is in the 1.4-2.7 ML range. As an example, the design of an optically-pumped laser developed for violet–green optical converters includes an active region comprising 5-9 CdSe sheets with d=2.7 ML placed in the center of ZnSe wells [2,3]. These QD arrays are formed at slightly different growth conditions depending on their order in a growth run. The respective variation of the QD sizes and Cd content in each sheet can, in turn, affect the QD transition energies and efficiency of pumping.

On the other hand, it has been recently demonstrated that effective Förster resonance energy transfer (FRET) [4] is possible between subsystems of small and large QDs within an inhomogeneous CdSe/ZnSe QD array [5], where the density of states can have the maximum at higher energy than the photoluminescence (PL) peak [6]. This FRET takes place mainly between the ground states of smaller QDs and the excited states of larger QD, followed by the fast energy relaxation to the ground level of the large QDs. We assume that similar process can be between two arrays comprising small and large QDs when the separation between them does not markedly exceed in order of magnitude the critical Förster radius when this mechanism is still efficient (~10-13 nm). This process can affect the recombination dynamics of the QD arrays [7].

In this paper, we focus on exciton recombination dynamics in structures with two fractional ML insertions separated by a ZnSe barrier. These insertions were intentionally made different by the variation of a nominal CdSe layer thickness to provide the inhomogeneous QD arrays with predominantly either small or large QDs. We demonstrate that there is the energy transfer for separations between the QD arrays, which are typical for green lasers. Our data are also suggestive of the influence of dark excitons in small QDs on this process.

1. Samples and characterization

Two types of samples were grown by MBE on GaAs substrates. Both comprise two sheets of QDs, which in the first approximation can be considered as the arrays of "small" or "large" QDs, depending on the nominal thickness of the CdSe insertion. In the first-type structures, the insertion of smaller thickness (d=1.4-2.1 ML) was grown first, followed by the insertion with d= 2.7 ML, while in the second type this order was reversed. The basic results of optical studies of these structures turned out to be similar. Here, we present mostly data obtained using the first-type structures, where the array exhibiting PL at lower energy E_2 is located above the array emitting at higher energy E_1 (Figure 1, a). We believe that this arrangement minimizes the energy transfer between the neighboring arrays via radiation. Two samples of this type were thoroughly studied: Sample A (d values are 2.1 and 2.7 ML) and sample B (d values are 1.4 and 2.7 ML). Thus, in each structure a PL band contains the lower energy line (lower line) and the higher energy one (upper line). The QD sheets in these two samples are separated by a barrier with a thickness w=12-14 nm. Note that such QD arrays were considered in previous studies as efficiently separated [8].

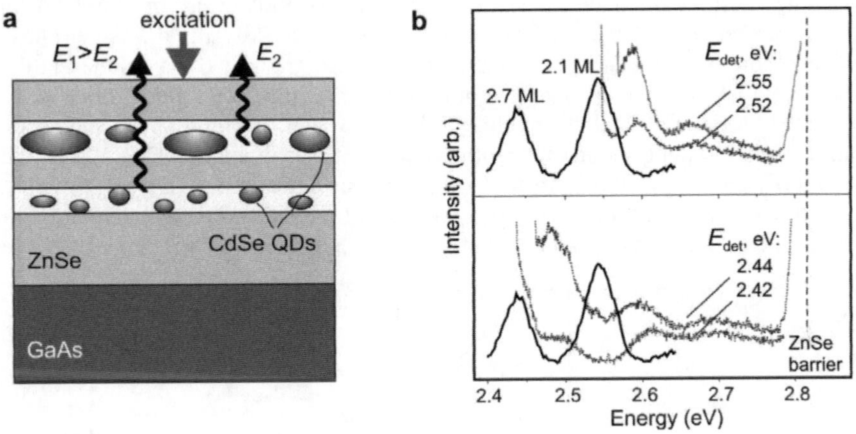

Figure 1. (a) Schematic drawing of the first-type structures with two different arrays of QDs. The inferior array radiates at higher energy E_1. (b) PL (461-nm excitation) and PLE spectra measured at 28 K in sample A with 2.1 and 2.7 ML CdSe insertions. The PLE spectra detected from respective PL lines contain similar peaks at ~2.6 eV energy.

The structures were first characterized by cw PL and PL excitation (PLE) spectroscopies in a closed-cycle helium cryostat at low temperature. PL measurements performed using cw lasers show that the ratio of the PL intensities of the two lines depends on the excitation energy. PLE spectra were measured with the excitation by emission of a tungsten lamp, dispersed by a monochromator. In sample B, LO phonon replicas are well pronounced in the PLE spectra recorded from the upper 2.62-eV PL line that indicates phonon-assisted energy transfer to nano-islands. The energy transfer has more complicated character for the lines originated from the arrays with the higher CdSe content (2.1 and 2.7 MLs). Their PLE spectra contain two or three smooth peaks (figure 1, b). Surprisingly, the energy position for some of them is similar in the PLE spectra recorded from upper and lower PL lines in the same structure (compare the peaks near 2.6 eV in the top and bottom panels of figure 1, b). It may indicate that these QD arrays are not completely separated, as assumed previously, but linked via some kind of energy transfer.

3. Recombination dynamics

To elucidate the mechanism of excitation transfer we have performed time-resolved PL studies. Temporal behavior of PL intensity in the structures was investigated at low temperatures (2 K) using a Hamamatsu streak-camera with temporal resolution of ~15 ps. A pulsed femtosecond laser with tunable wavelength was used for excitation. Measured PL decay curves are presented in figure 2 (a, b). To fit PL intensity decay with time, *t*, we use the following expression

$$I(t) = A_1\left(\exp\left(\frac{-t}{t_1}\right) - \exp\left(\frac{-t}{t_{rise}}\right)\right) + A_2\exp\left(\frac{-t}{t_2}\right),$$

where t_1 and t_2 are the decay time constants characterizing, respectively, fast and slow decaying PL components; A_i (i=1,2) is a respective amplitude; t_{rise} is the rise time constant (~15-25 ps). The PL decay curves turn out to be almost similar for both the upper and lower emission lines at the above-barrier excitation by a 266-nm (4.66 eV) laser line. Their decay can be fitted by a single exponent with the time constant t_1=280-300 ps, while the contribution of the other (slower) component is negligible.

With under-barrier excitation by a 461-nm line (2.69 eV), the decay curves are strongly modified. In both structures, we observe shortening of the PL decay time of the upper lines. It is especially strong in sample B, where t_1 of the 2.62-eV line decreases down to 130 ps; the t_1 variation for the 2.54-eV line of sample A is much smaller. We assume that the fast PL decay in sample B indicates the efficient energy transfer from the small QDs in the 1.4-ML array towards the larger QDs in the 2.7-ML array which exhibits lower-energy emission. This energy transfer can be done via dipole-dipole interaction between QDs of different sizes, in other words, via the FRET [5]. It takes place when the ground and excited levels of these dots are close to be in resonance (Figure 2, c). Such energy transfer (and acceptable detuning) depends on the exciton radiative time. When its value is less than 1 ns, this process can be efficient at a distance up to 13 nm between interacting QDs [7]. Such a distance is comparable with the barrier width in our structures. The PL decay variation is less pronounced in the sample A, because the excited and ground QD levels in the neighbouring arrays are out of resonance.

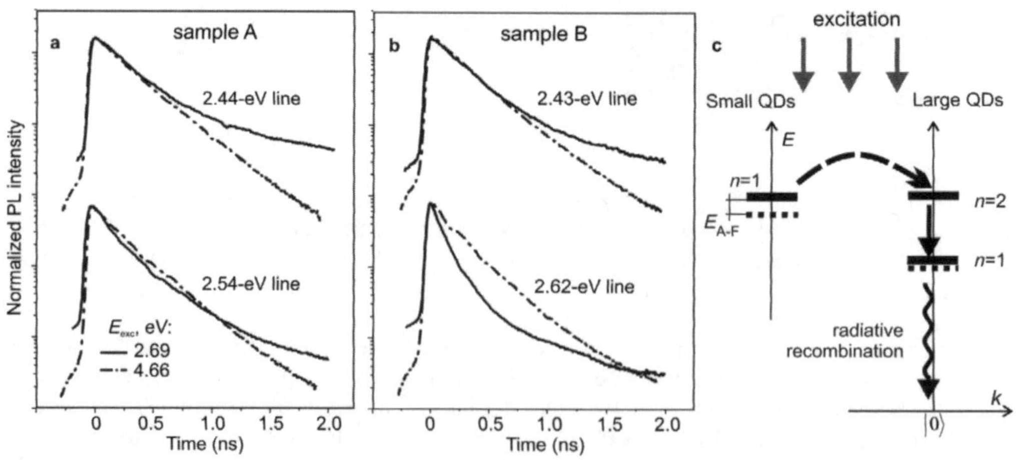

Figure 2. PL decay curves measured in samples A (a) and B (b) in the maxima of the upper and lower PL lines at different energies of excitation: 2.69 eV (solid lines) and 4.66 eV (dash-dotted lines). (c) Schematic drawing of possible resonance energy transfer (shown by dashed arrow) between excitonic levels in the small and large QDs. $\Delta E_{A\text{-}F}$ is the splitting of allowed-forbidden exciton states.

Other intricate finding is the enhancement of the slow decaying components with under-barrier excitation in both the lower and upper PL lines. These slow components are characterized by t_2=1.0-1.9 ns. To understand this phenomenon, let us remind that the fine structure of exciton ground states comprise optically allowed (AX) and forbidden (FX) dark excitons (see figure 2, c), separated by the energy $\Delta E_{A\text{-}F}$ defined by the electron-hole exchange interaction [9]. The FX radiative recombination reveals itself as an appearance of slowly decaying component in time-resolved PL spectra [10-12]. Actually, such FX contribution is observed due to its mixing with the AX state. The efficiency of the mixing depends on the $\Delta E_{A\text{-}F}$ value, which is strongly dependent on a QD size. For bulk CdSe, this value is 0.12 meV; for QDs, it ranges between 1.3 and 3.8 meV, increasing with the decrease of the QDs size [10]. When $\Delta E_{A\text{-}F}$ is ≤ 1 meV, as in the large QDs, the mixing is effective and the recombination rate is high. In the small QDs with large $\Delta E_{A\text{-}F}$ separation, the mixing is hampered and the rate of recombination goes down. This explains the appearance of the slow component in the PL band originating from the small QDs. The recombination rate in the large QDs is decreased as well, because it is controlled by the resonance energy transfer via the dipole-dipole interaction with the allowed states of the small QDs, which are slowly and insufficiently populated with the large $\Delta E_{A\text{-}F}$ value.

Summarizing, we demonstrate that QD planar arrays formed by CdSe fractional monolayer insertions into a ZnSe matrix cannot be considered as isolated with barrier widths typical for lasers. Effective energy transfer to one of these arrays can be realized if the ground level of small QDs is in resonance with the excited level of large QDs. This conclusion is supported by the results of time-resolved studies demonstrating the acceleration of PL decay in small QDs which transfer the energy to the large ones. The enhancement of slow decaying components evidences the influence of dark excitons on the recombination process. Our finding can be used for optimization of II-VI laser design.

Acknowledgement

This work has been supported by Russian Science Foundation, grant № 14-22-00107.

References

[1] Ivanov S V *et al.*, "Molecular beam epitaxy of wide-gap II-VI laser heterostructures" in Molecular Beam Epitaxy, 611, Elsevier.
[2] Lutsenko E V *et al.*, *Phys. Stat. Sol.* B 247, No. 6, 1557 (2010).
[3] Yablonskii G P *et al.*, *J. Non-Crystalline Solids* **356**, 1928 (2010).
[4] Förster T, *Ann. Phys.* (Leipzig) **437**, 55 (1948)
[5] Shubina T V *et al. Phys. Stat. Sol. B*, (2016) / DOI 10.1002/pssb.201600095.
[6] Golovatenko A A *et al. Acta Physica Polonica A* **129**, 107 (2016).
[7] Poddubny A N and Rodina A V, *JETP* **149**, 614 (2016).
[8] A. Reznitsky *et al.*, *Phys. Stat. Sol.* B **252**, 1717 (2015).
[9] Nirmal M *et al.*, *Phys. Rev. Lett.* **75**, 3728 (1995).
[10] Labeau O *et al.*, *Phys. Rev. Lett.* **90**, 257404 (2003).
[11] Biadala L *et al.*, *Phys. Rev. Lett.* **103**, 037404 (2009).
[12] Patton B *et al.*, *Phys. Rev. B* **68**, 125316 (2003).

Optical field distribution in quasy-1D nanostructures

B I Soshnikov[1], V V Lysak[1], K P Kotlyar[2], I P Soshnikov[2,3-5]

[1] Department of Nanotechnology and Material Science, ITMO University, St Petersburg 192000, Russia.
[2] Nanotechnology Research and Education Center, Academic University, RAS, St. Petersburg 194021, Russia
[3]Ioffe Physico-Technical Institute of the RAS, St. Petersburg 194021, Russia
[4]Institute for Analytical Instrumentation of RAS, St. Petersburg 198095, Russia
[5]Saint Petersburg Electrotechnical University "LETI", St. Petersburg 197376, Russia

Abstract. The FDTD optical distribution model in the quasi single dimensional nanostructures is presented. The polar diagram is analyzed for nanowire structure with various diameters. The main conditions for the maximum output in vertical (0° and 180°), horizontal (90°) and leaky mode (30°, 60° and 165°) directions are discussed. Theoretical data show good agreement with experiments.

1. Introduction

One dimensional (1-D) nanostructures such as nanowires (NW), nanotubes, nanobelts and nanosprings are fast becoming a key instrument in intensive research due to their unique potential for fabrication into high density optoelectronic nano-scale devices [1]. In fact, 1-D nanostructures are the smallest dimension systems that can be used for efficient transport of electrons and are thus critical to the function and integration. Their electrical properties are strongly influenced by minor perturbations due to high surface-to-volume ratio and tunable electron transport properties. Compared with the thin-film/thin-film LEDs [2-4], which may suffer from the total internal reflection, nanowire/thin-film heterostructures are utilized in order to increase the extraction efficiency of the LEDs by virtue of the waveguiding properties of the nanowires and decreasing the defect density of the carrier's nonradiative recombination [5-8]. Growth technique is developed to create the real optoelectronic devices based on nanowires [9-12]. Optical field distribution in such structures depends on geometrical and material parameters and can be easily analysed using numerical simulation. In this work, the structure is modelled using using finite-difference time-domain (FDTD) principle [13]. In this method, space is divided into small blocks representing the materials in the region of interest. Then by taking small steps in time, the propagation of the electromagnetic field is followed through space as it enters the region and interacts with the structure. The method allows for the effective and powerful simulation and analysis of sub-micron devices with very fine structural details. A sub-micron scale implies a high degree of light confinement and correspondingly, the large refractive index difference of the materials (mostly semiconductors) to be used in a typical device design.

Some modal properties of circular section nanowires have already been described by Maslov and Ning [14–16] and others. These properties include the dispersion for guided modes, the end-facet reflectivity [14], the mode quality factor [17], the far-field emission [15], and the distribution of the emitted power between the free-space modes and the guided modes for an axial heterostructure configuration [16].

In this work we analyze angular optical field distribution in the single NW with different diameters.

2. Theoretical simulation

The finite-difference time-domain (FDTD) method for electromagnetic fields is a computational method for solving Maxwell's equations on a discrete grid in space and time [18]. With it, the real-valued time-dependent electric and magnetic field for a point in time and space are deduced from the surrounding electromagnetic fields a short time interval earlier. In this manner, the evolution of the fields can be followed. The concept of the FDTD method is thus very intuitive as it is closely related to how we experience the physical universe.

Solutions calculated with a basic FDTD method converge with second order with respect to the number of the grid cells per unit volume, but only if there are no material interfaces present. All simulations are performed as if the scatterers are contained in a perfectly conductive box. To obtain a unique solution to a given problem, in addition to the governing differential equations, boundary conditions have to be specified. An efficient implementation of an absorbing layer, called the perfectly matched layer (PML) is used in this calculation [19].

2.1. Model description

The model consists dot optical source inside single NW. A quantum dot can be modeled by a source dipole with in-plane orientation if we consider that the lateral confinement is not important (same behavior as an axial quantum well) [20]. NW is characterized by fixed normalized length ($L = 13\lambda/n$), optical source position ($h = 6.5\lambda/n$), period ($T = 4\lambda$) and variable diameter ($D = k\lambda/n$) where n is the refractive index, λ is wavelength and k=0.05-2 is variable parameter (Figure 1).

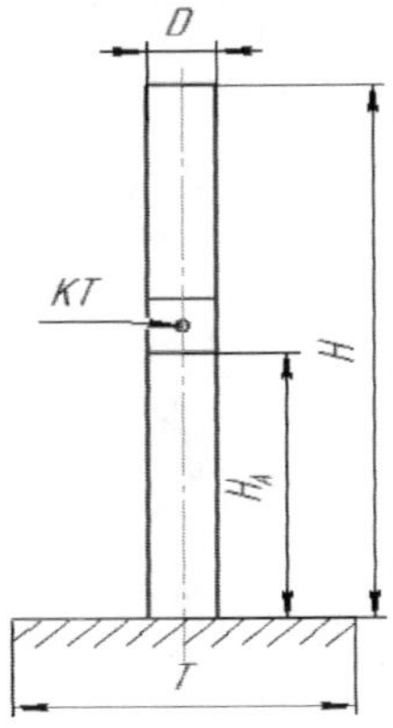

Figure 1. NW structure used in simulation

2.2. Optical field distribution in nanowire

Calculated results of E field distribution in the NW for different k are presented in Figure 2. Results show that perfect confinement of the optical field is possible for structure with k is less than quarter wave.

2.3. Polar diagrams and modes in the structure

Results presented on Figure 2 were transferred in the form of polar diagram for analysis of the NW optical properties for different normalized diameters ($D = k\lambda/n$) (Figure 3). Polar diagram shows five main directions of the light distribution: vertical (0° and 180°), horizontal (90°) and leaky mode (30°, 60° and 165°). For thin structures with k region 0.1-0.35 optical field transfers mostly in the vertical direction. Horizontal distribution prevails for NW with k=1, 2, … that is in good agreement with interference law and measured PL intensity [21].

We also found that at k=1.25 and 1.6 strong and narrow leaky modes are transferred.

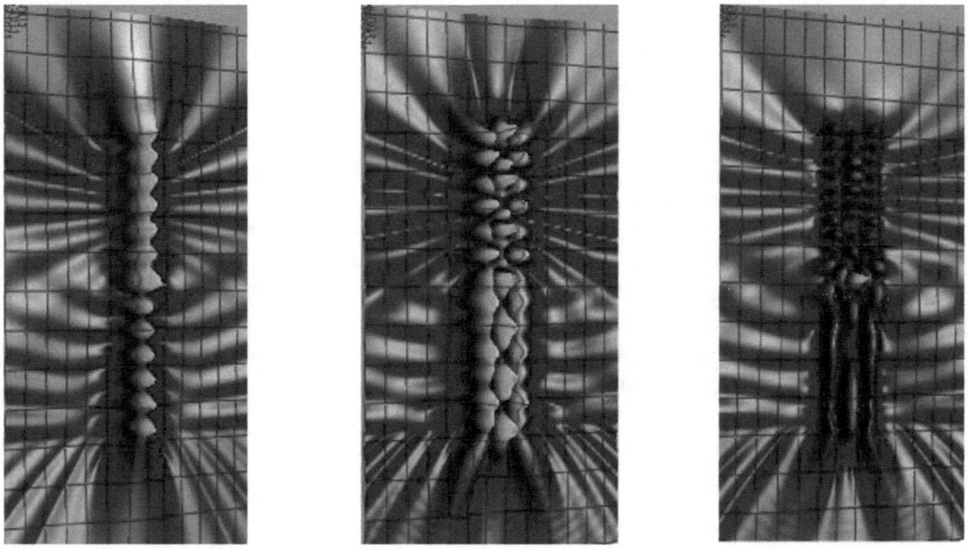

a) b) c)

Figure 2. E field distribution in nanowire for k equal a) 0.1; b) 1.25; c) 2

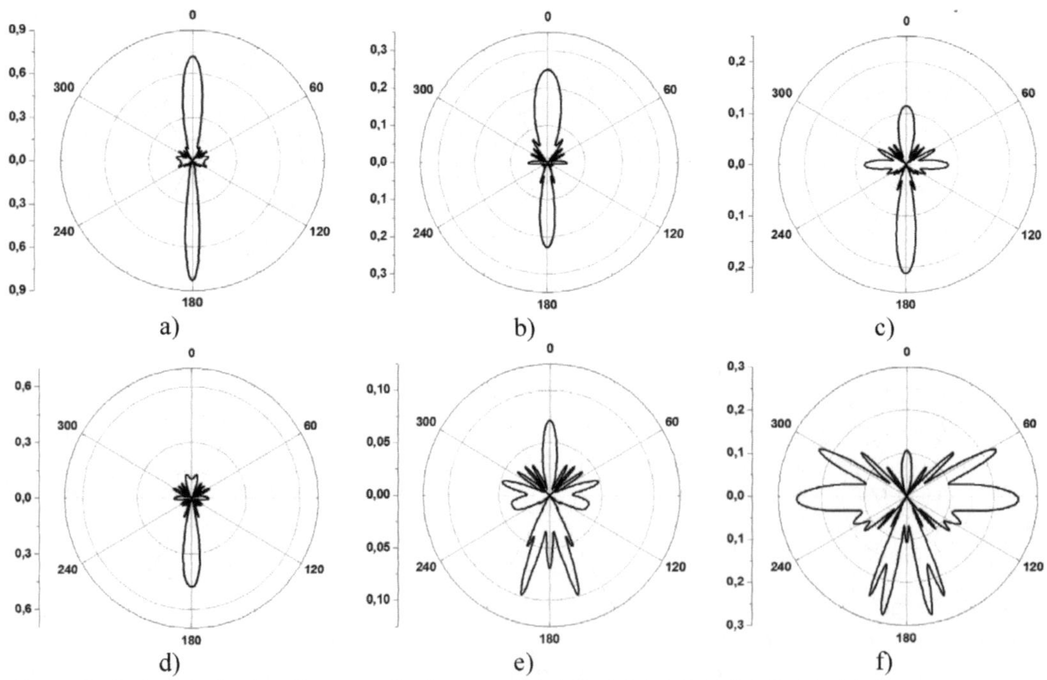

Figure 3. Calculated polar diagrams for k equal a) 0.1 b) 0.35 c) 0.8 d) 1.2 e) 1.6 f) 2

Figure 4a shows dependence of the normalized diameter on the amplitude in each direction. For k = 0.05-0.85 light radiates mainly in vertical direction (up to 90%). The relation between up and down directions changes non-monotonically and quasi-periodically. At k=1 and 2 light distributes laterally that in agreement with field interference effect. At k=1.25 and 1.6 field redistributes in the leaky mode (30°, 60° and 165°) with narrow bandwidth. Calculated results agree well with different experimental data presented by plus markers (Figure 4b) [22-25].

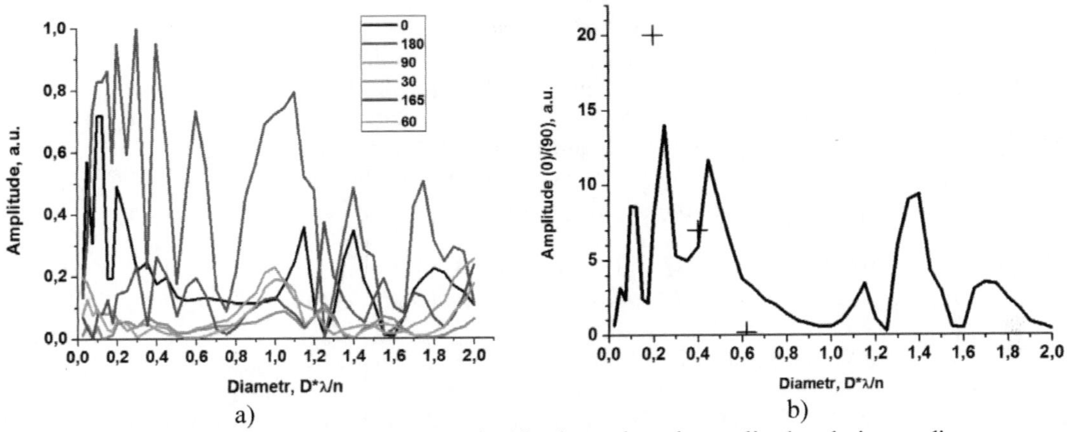

Figure 4. a) Calculated angular modes b) Vertical/horizontal mode amplitude relation vs diameter

3. Conclusions

In this paper we have investigated the optical properties of the near fields generated by a nanowire laser. The polar diagram of the light distribution in the linear nanostructures is analyzed for different structure diameters. For diameter lower than optical wavelength light mainly distributes in vertical directions. For structure with whole wavelength diameters light prevails in horizontal direction due to interference. We found conditions at k=1.25 and 1.6 with strong and narrow leaky modes.

Acknowledgments

This work is partly supported by the Government of the Russian Federation (Grant 074-U01) through ITMO Early Carrier Fellowship scheme, Russian Foundation for Basic Research (project# 15-07-06964, 13-02-12031 ofi-m), Ministerium for education and science of Russian Federation (Contract № 02.G25.31.0058 dated 12.02.2013).

References

1. Tilke A, Simmel F, Lorenz H, Blick R, Kotthaus J. 2003 *Phys. Rev. B* **68** 075311
2. Alivov Y, van Nostrand J, Look D, Chukichev M, Ataev B. 2003 *Appl. Phys. Lett.* **83** 2943
3. Vispute R, Talyansky V, Choopun S, Sharma R, T. Venkatesan , M. He , X. Tang , J. B. Halpern , M. G. Spencer , Y. X. Li , L. G. Salamanca-Riba , A. A. Iliadis , K. A. Jones 1998 *Appl. Phys. Lett.* , 73 , 348
4. Asil H, Gur E, Cinar K, Coskun C. 2009 *Appl. Phys. Lett.* **94** 253501 .
5. Zhong J, Chen H, Saraf G, Lu Y, Choi C, Song J, Mackie D, Shen H. 2007 *Appl. Phys. Lett.* **90** 203515
6. An S, Chae J, Yi G, Park G. 2008 *Appl. Phys. Lett.* **92** 121108
7. Kim K, Kim S, Jeong H, Jeong M, Jung G. 2010 *Adv. Funct. Mater.* **20** 1076
8. Xu S, Xu C, Liu Y, Hu Y, Yang R, Yang Q, Ryou J, Kim H, Lochner Z, Choi S, Dupuis R, Wang Z. 2010 *Adv. Mater.* **22** 4749

9. Soshnikov I, Cirlin G, Tonkikh A, Samsonenko Y, Dubrovskii V, Ustinov V, Gorbenko O, Litvinov D, Gerthsen D. 2005 *Phys. Solid State* **47** 2213
10. Dubrovskii V, Cirlin G, Soshnikov I, Tonkikh A, Samsonenko Y, Ustinov V, Sibirev N. 2005 *Phys. Rev. B* **71** 205325
11. Samuelson L, 2014 Proc. Conf. on Las. and Electro-optics (CLEO) # SM2J.5
12. Harmand J, Glas F, Patriarche G. 2010 *Phys. Rev B* **81** 235436
13. www.Optiwave.com
14. Maslov A, Ning C. 2003 *Appl. Phys. Lett.* **83** 1237
15. Maslov A and Ning C. 2004 *Opt. Lett.* **29** 572
16. Maslov A, Bakunov M, and Ning C. 2006 *J. Appl. Phys.* **99** 024314
17. Wang M, Huang Y, Chen Q, Cai Z, 2006 *IEEE J. Quantum Electron.* **42** 146
18. Yee K. 1966 *IEEE Transactions on Antennas and Propagation* **14** 302
19. Berenger J. 1994 *Journal of Computational Physics*, **114** 185
20. Henneghien A, Gayral B, Désières Y, Gérard J. 2009 *J. Opt. Soc. Am. B* **26** 2396
21. Lazić S, Chernysheva E, Gačević Ž, García-Lepetit N, van der Meulen H, Müller M, Bertram F, Veit P, Christen J, Torres-Pardo A, Calbet J, Calleja E, Calleja J, 2015 *Proc. of SPIE* **9363** 0U-1
22. Talalaev V. 2016 *Thech. Phys. Lett.* (to be publish)
23. Mustary M , Lysak V. 2014 *IOSR Journal of Electrical and Electronics Engineering* **9** 18
24. Yang G, Xie F, Tong Y, Chen P, Yu Z, Yan D, Xue J, Zhu H, Guo Y, Li G, Gao S. 2015 *Materials Science in Semiconductor Processing* **30** 694
25. Yang G, Guo Y, Zhu H, Yan D, Li G, Gao S, Dong K. 2013 *Applied Surface Science B* **285** 772

Switching of resonance energy transfer mechanism in a dense array of II-VI quantum dots

K G Belyaev, T V Shubina, M A Semina, A V Rodina, A A Toropov, and S V Ivanov

Ioffe Institute, 26 Polytekhnicheskaya, St. Petersburg 194021, Russia

Abstract. We report on Förster resonance energy transfer (FRET) in a dense array of II-VI epitaxial quantum dots (QDs). Besides FRET between the ground states of QDs of different sizes, we observe the energy transfer via the excited levels of large QDs which are in resonance with the ground levels of small QDs. The switching of dominant energy transfer mechanism, revealed by photoluminescence excitation spectroscopy, takes place at the energy controlled by the architecture of quantum levels in these arrays.

Introduction

Epitaxial CdSe/ZnSe II-VI quantum dots (QDs) have great potential for nanolaser application. One of main features of the arrays of such QDs, which have high lateral densities $>10^{10}$ cm^{-2}, is a relatively narrow width of photoluminescence (PL) band (40-50 meV) that contradicts the broad distribution of QD sizes, revealed by transmission electron microscopy (TEM) studies [1,2]. Besides, our recent micro-PL spectroscopy studies exhibited narrow exciton lines of single QD emission instead of an expected smooth PL band [3]. These findings evidence the spectral selection of a limited number of radiating QDs. To explain such features it has been proposed that Förster resonance energy transfer (FRET) [4] between the smaller QDs ("donors") and larger ones ("acceptors") takes place in such QD ensembles [5]. The FRET mechanism between QDs can be realized in different ways. First, it can take place between the ground states of the "donor"-QDs and "acceptor"-QDs, provided their absorption and emission contours are overlapped ($FRET^1$). This mechanism is realized in an ensemble of colloidal QDs [6], between sheets of vertically coupled direct and indirect bandgap QDs [7], and in Si nanocrystals [8]. In the dense and inhomogeneous array of epitaxial II-VI QDs, the energy transfer can be performed by other way, namely, from the ground levels of smaller QDs towards the excited levels of larger QDs, followed by the fast excitation relaxation to their ground levels ($FRET^2$) [5]. The coincidence of the ground and excited levels in the QDs of different sizes is the necessary prerequisite for such mechanism. It can provide emission from a limited number of selected large QDs, while the emission from a huge number of small QDs will be completely suppressed [9]. Note that both FRET processes can act in competition and coexistence with radiative, tunneling, and hopping mechanisms of energy transfer.

2. Experimental

To prove this concept, we have performed the comprehensive studies of structures containing CdSe fractional monolayer (ML) insertions in a ZnSe matrix, which are spontaneously transformed into the QD arrays during growth by molecular beam epitaxy [10]. The structure A contains one CdSe insertion with a 2.7-ML nominal thickness. Structures with two insertions (B and C) contain 1.4 and

2.7 ML sheets. In these structures, the sheet with lower Cd content, and hence smaller QDs, can act as a donating array, while the sheet with higher Cd content will accept the energy. These two insertions are separated by a ZnSe barrier of a different width to control the FRET efficiency. The nominal barrier widths were 7 nm in the structure B and 15 nm in the structure C; their gradient along a substrate allows us to elucidate precisely the effect of width variation on the efficiency of energy transfer.

PL and PL excitation (PLE) measurements were performed at low temperature (20 K) using detection by a nitrogen-cooled CCD camera and excitation by a halogen lamp dispersed by a monochromator. The hyper-PLE spectra comprise the PL spectra measured at continuously changed excitation energy. Thus, the tracks of PL in the coordinates "excitation energy – detection energy" are obtained synchronously with the excitation energy variation. The TEM studies have shown that the structure A consists of distinct nano-islands with the lateral density of 10^{11}-10^{12} cm^{-2} and average distance between them about 5-10 nm (sufficiently small for FRET). The QD characteristic lateral sizes are distributed within the 2-12 nm range with maximum at 2-4 nm [3,5]. In contrast to that, the PL maximum position is near 2.42 eV (Figure 1, a). In accordance with preliminary modeling [5,9], it corresponds to the emission from the QDs with the lateral sizes of 5-7 nm and Cd content about 85%. Thus, a significant part of QDs is out of radiation process.

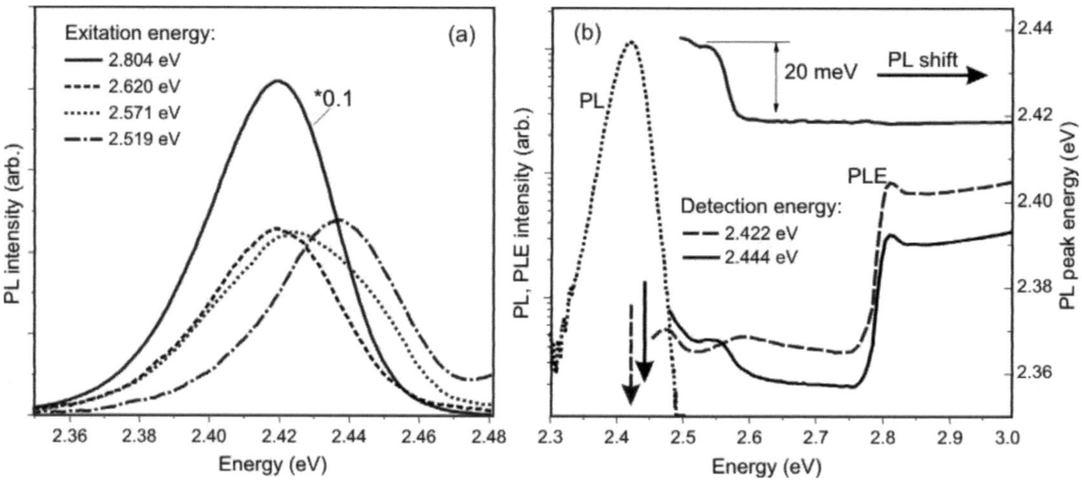

Figure 1. (a) PL spectra measured in the structure A at different excitation energies. (b) PLE spectra recorded in the same structure at two detection energies corresponding to the PL maximum positions with above-barrier (dashed line) and under-barrier (solid line) excitations. On the top, the solid line shows the shift of PL maximum.

In the studied structures, we observe an unusual feature – the modification of PL spectra when the excitation energy changes. Namely, the PL maximum shifts toward higher energies by 10-25 meV at ~2.55-2.6 eV energy with respect to the position corresponding to the excitation above the ZnSe barrier energy gap (2.8 eV). In particular, this shift approaches 20 meV in the structure A (Figure 1, a). The PLE spectra recorded from the shifted PL maximum differ from those registered at the initial PL peak position. Namely, no features are observed above the 2.6-eV boundary. Thus, the states in the 2.6-2.8 range do not supply any excitation for this shifted PL peak. These peculiarities look even more pronounced in the hyper-PLE spectra (Figure 2). Besides, the hyper-PLE spectroscopy shows that the value of the shift depends on the distance between two QD sheets (the closer sheets, the stronger shift). Moreover, in the case of the small barrier width between these sheets (7-10 nm) the "donor" QD array has dramatically decreased PL intensity (Figure 2, a). Apperantly, this QD array transfers the most part of excitation to the lower-energy states in the "acceptor" sheets. We highlight that the insertions with small Cd content do not exhibit such blue shift at all (see PL tracks of the 1.4 ML CdSe insertions in the hyper-PLE spectra shown in Figure 3). It is also essential that the blue shift of PL peaks takes place at the same energy around 2.6 eV in different structures.

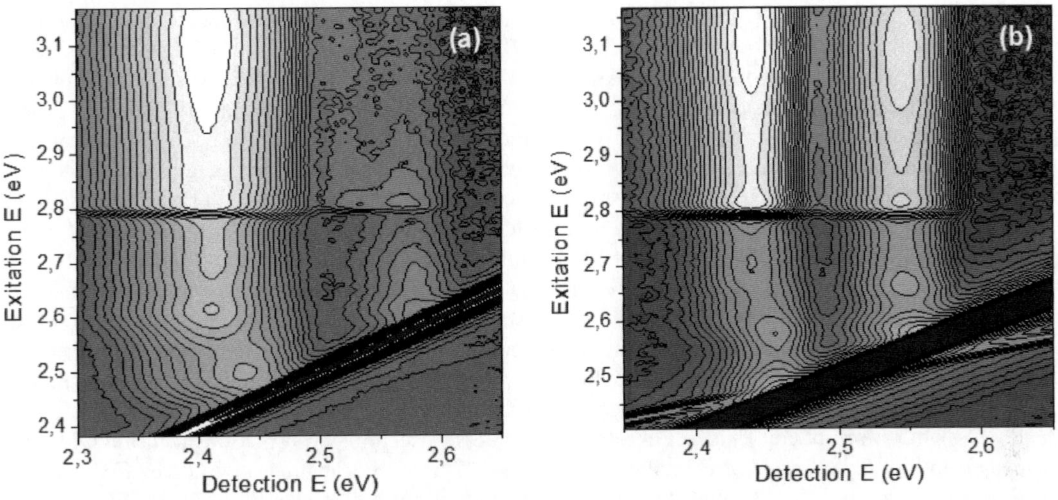

Figure 2. Hyper-PL excitation spectra measured at 20 K in the B (a) and C (b) structures with two insertions. The brightness represents the PL intensity, decreasing from white to black, shown in logarithmic scale. Contours connect the points of equal intensity. Inclined black line in bottom part of the images is the track of excitation with continuously changed energy.

3. Modelling and discussion

Theoretical modeling of excitonic levels, considering the QDs as spheroids with a Gaussian potential profile, is performed as described in Refs. [7,8].

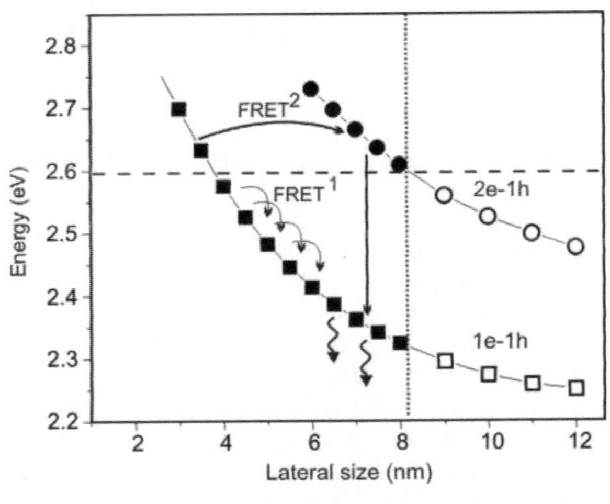

Figure 3. Dependences of exciton energies on QD lateral sizes calculated for CdSe QDs with 85% Cd content in their center. Possible energy transfer processes are marked by arrows. The white marks denote data for largest QDs which are presumably defective; they are separated from radiating QDs by a vertical dotted line. Horizontal dashed line separates the energy ranges where different FRET mechanisms dominate.

The calculated electron (hole) levels in the QDs of different lateral sizes are used to compose the excitonic levels, whose symmetry allows their participation in the process of excitation transfer. Figure 3 presents the simplified scheme of possible energy transfer channels, which involve only 1e-1h and 2e-1h excitonic levels. Here, 1e (1h) corresponds to the first electrons (holes) level, 2e denotes excited electron level. The existence of other excitonic states such as 1e-2h is neglected. We consider the $FRET^2$ via the excitonic states 1e-1h of the small QDs (lateral size of 2-4 nm) and 2e-1h states of the larger ones (lateral size of 6-8 nm). The 1e-1h states of these 6-8-nm-sized dots are radiating, while the CdSe islands of larger sizes are usually defective and cannot participate in emission process [11,12]. Therefore, the $FRET^2$ can exist only until the energy is higher than 2.6 eV, since below there are no suitable states. Underneath this 2.6-eV boundary, the $FRET^1$ can dominate. This mechanism acts via the ground states 1e-1h in two closely situated QDs which have small energy detuning. In our inhomogeneous systems, the $FRET^1$ can be a multistep process, which includes several steps of energy transfer from a "donor" QD to a next "acceptor" dot, which plays a role of the "donor" for the next step. The required closeness of QD energies for such mechanism results in the blue shift of the PL band as a whole. Note that involvement of 1e-2h states to the energy transfer process will provides some blue shift of PL maximum as well. In any case, we assume that the basic reason of observed PL modification near 2.6 eV in the studied structures is the change of the energy transfer mechanism.

The modeling also shows that in the case of low Cd content the excited levels exist only in the too large QDs which are optically inert. Therefore, the blue shift cannot be pronounced in the QD arrays formed from thin CdSe insertions. Indeed, the 1.4 ML CdSe sheets, which have lower Cd content in the dots, do not exhibit the PL blue shift at all. The modeling predicts also the sharpest $FRET^2$ cutoff for bimodal QDs size distribution. On the contrary, with large dispersion of QD sizes the cutoff must be more smooth. This is also consistent with experimental data obtained from structures with different QD size statistics (compare 2.7-ML insertion PL tracks in Figure 3, (a) and (b)).

In conclusion, we demonstrate that both $FRET^1$ and $FRET^2$ can be dominating energy transfer mechanisms in CdSe/ZnSe QDs at helium temperatures. Switching between these resonance mechanisms takes place at certain energy of excitation, which depends on the architecture of quantum levels in an array. Our data provide a new insight into unusual optical properties of fractional monolayer CdSe/ZnSe insertions studied from the end of the last century. The $FRET^2$ can be used for the spectral selection of single QDs in combination with spatial isolation to fabricate efficient nano-emitters.

Acknowledgments
These studies are supported by Russian Science Foundation (Project #14-22-00107).

References

[1] Peranio N et al., *Phys. Rev. B* **61**, 16015 (2000).
[2] Litvinov D et al., *Phys. Status Solidi A* **205,** 2892 (2008).
[3] Shubina T V et al. 23 Int. Symp. "Nanostructures: Physics and Technology" Saint-Petersburg, Russia, 86 (2015).
[4] Förster T., *Discuss. Faraday Soc.* **27**, 7 (1959).
[5] Shubina T V et al. *Phys. Stat. Sol. B*, (2016) / DOI 10.1002/pssb.201600095.
[6] Crooker et al. S A *Phys. Rev. Lett.* **89**, 186802 (2002).
[7] Shamirzaev T S et al. *Appl. Phys. Lett.* **97**, 263102 (2010).
[8] Limpens R et al. *J. Phys. Chem. C* **119**, 19565–19570 (2015).
[9] Golovatenko et al. *Acta Phys. Polon. A* **129**, 107 (2016).
[10] Ivanov S V et al. *J. Appl. Phys.* **83**, 3168 (1998).
[11] Sorokin S V et al., *J. Crystal Growth*, **201**, 461 (1999).
[12] Toropov A A et al., *Jpn. J. Appl. Phys. Part 1*, **38**, 566 (1999).

Magnetic hot-spots in hollow silicon cylinders

K V Baryshnikova[1], A B Evlyukhin[1,2] and A S Shalin[1]

[1] ITMO University, 49 Kronverksky Ave., St. Petersburg, 197101, Russia
[2] Laser Zentrum Hannover e.V., Hollerithallee 8, D-30419 Hannover, Germany

k.baryshnikova@metalab.ifmo.ru

Abstract. Silicon nanoparticles can possess magnetic Mie-resonant response in the visible and near infrared wavelength ranges. In this paper, we consider numerically the features of magnetic hot-spots realized inside silicon nanocylinders at the conditions of the optical magnetic resonances, and show that the intensity of the magnetic field inside nanoparticles with a coaxial through hole can be much stronger than the intensity of incident light waves.

1. Introductions

Last decades all-dielectric high-index nanostructures (consisting of, e.g., silicon or germanium) attract great attention in the field of nanophotonics and related areas [1, 2]. Despite the non-magnetic nature, nanostructures composed of silicon, germanium etc. can possess magnetic response [3, 4], which is following from the Mie theory [5]. Conductive currents inside such high-index semiconductor nanoparticles are almost zero [6] while the displacement currents can have special circular distribution, which corresponds to the excitation of magnetic dipole and high-order magnetic multipoles [7, 8]. Magnetic dipole mode in the high-index structures was considered previously [8, 9] with respect to the light scattering problem, while magnetic field distributions and magnetic field localization effects were not studied in details. At the same time, the resonant enhancement of magnetic field or, by other words, the magnetic hot-spots (MHSs) in the nanoparticles can be used in large amount of applications, such as, e.g., optomechanics, optical trapping, sensing of magnetic particles etc [8]. In this paper we investigate in details resonant MHSs inside silicon nanocylinders. Moreover, as it is shown that a coaxial through cavity in the nanoparticle (such holes can be produced by focused ion-beam milling or electron beam lithography and reactive ion etching [10]) does not lead to the total destroying of the MHSs, providing possibility for practical applications of MHSs.

Note that several approaches to get the MHSs in free space have been proposed in literature including the hot-spots in the gap between silicon nanoparticles [11, 12] and indirect magnetic field enhancement in plasmonic nanostructures [13, 14]. However, there are principal limitations, for example, very small region of magnetic field enhancement between two nanoparticles and strong electromagnetic field absorption inside metal material. Scattering of light by high-index dielectric nanoparticles with cavities has been considered previously [10, 15], however, without investigation of the MHSs.

2. Eigen modes of nanocylinders

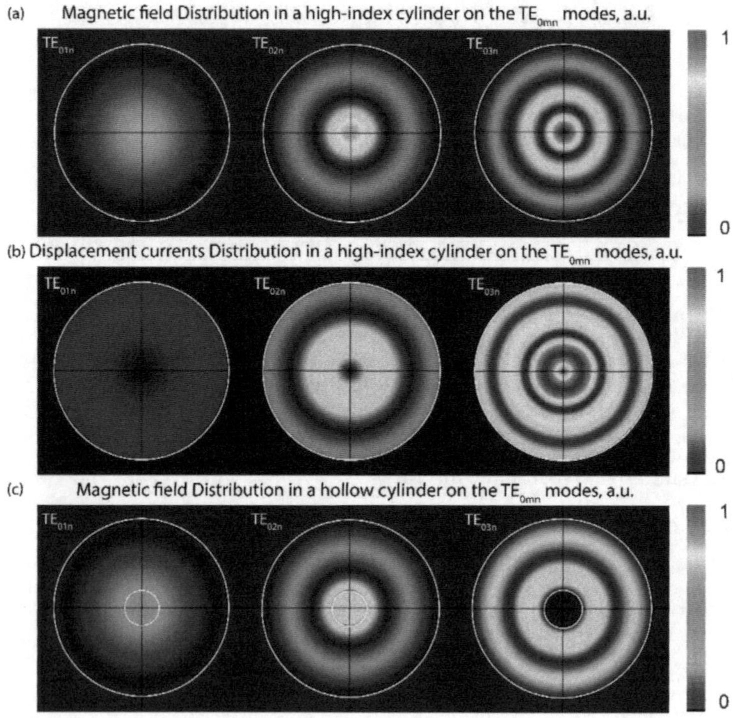

Figure 1 (a) Magnetic field distribution (a), and displacement currents distribution (b) in the dielectric cylinder on the TE_{0mn} eigen modes, m=1, 2, 3. (c) Magnetic field distribution in the hollow dielectric cylinder on the TE_{0mn} eigen modes, m=1, 2, 3. White contours highlight boundaries of cylinder (a-c) and hollow (c).

In order to design electromagnetic response of silicon nanoparticles, their modal structure should be analyzed. Among the number of high-index cylinder eigen modes there are some of them which correspond to the magnetic field concentration inside the cylinder. At the figure 1a magnetic field distributions on the TE_{0mn} modes, m=1, 2, 3, n=1, are shown. These distributions, for cylinder with height of 100 nm and radius of 200 nm and dielectric permittivity equals to 16, were obtained numerically (with help of Comsol Multiphysics [16]). Here modes with n>1 are not shown because for n>1 magnetic field distribution at the frontal cross section is repeated manually this for TE_{0m1} mode but number of magnetic field maxima on the cylinder axis are increasing (1 for TE_{0m1} mode, 2 for TE_{0m2} mode and so on). TE_{0mn} modes are of highest interest, because special circular distribution of displacement currents (see figure 1b), slightly depends on the defects on the cylinder axis.

Actually, in coaxial cylinders with a cavity the magnetic field concentration stays quite high for low-order modes but disappears for high-order modes (see figure 1c) because of overlapping between the cavity and the area of displacement currents maxima. In order to estimate the magnetic field enhancement in the cavity a full-wave simulations should be done. Results are presented in the next section.

3. Magnetic hot-spots in hollow cylinders under plane wave irradiation

In figure 2 results of numerical simulations for cylinder (radius is 200 nm, height is 100 nm) irradiated by linearly polarized plane wave are presented. Here we take into account dispersion of silicon dielectric function [3]. We consider two different propagation directions: along and perpendicular to the cylinder axis (see figure 2a, b). Results for third case, when propagation vector is perpendicular to the cylinder axis, and electric field oscillations are along cylinder axis, are similar to the case of frontal excitation and, therefore, are omitted.

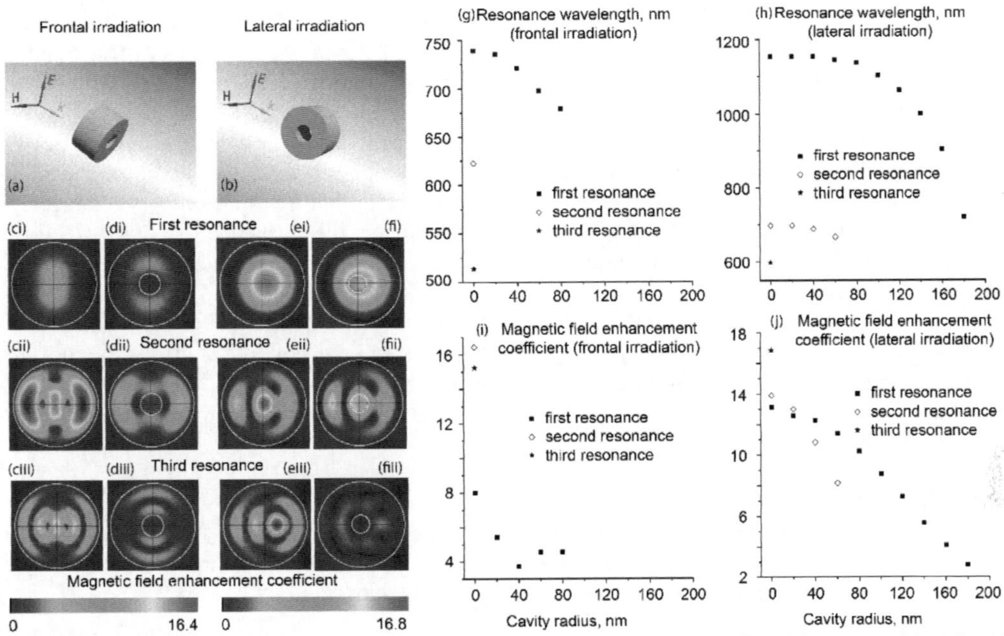

Figure 2 (a, b) Frontal (a) and lateral (b) excitation of cylinder with radius 200 nm and height 100 nm with a coaxial through hole. (ci-fi) Distributions of normalized (with respect to the magnetic field of incident light waves) magnetic field for the first resonance: (ci) cylinder without cavity, frontal irradiation, wavelength 739 nm; (di) with coaxial cavity of 40 nm radius, frontal irradiation, wavelength 721 nm; (ei) cylinder without cavity, lateral irradiation, wavelength 1154 nm; (fi) with coaxial cavity of 40 nm radius, lateral irradiation, wavelength 1154 nm. (cii-fii) Distributions of normalized magnetic field on the second resonance: (cii) cylinder without cavity, frontal irradiation, wavelength 622 nm; (dii) with coaxial cavity of 40 nm radius, frontal irradiation, wavelength 622 nm; (eii) cylinder without cavity, lateral irradiation, wavelength 698 nm; (fii) with coaxial cavity of 40 nm radius, lateral irradiation, wavelength 688 nm. (ciii-fiii) Distributions of normalized magnetic field on the third resonance: (ciii) cylinder without cavity, frontal irradiation, wavelength 514 nm; (diii) with coaxial cavity of 40 nm radius, frontal irradiation, wavelength 514 nm; (eiii) cylinder without cavity, lateral irradiation, wavelength 598 nm; (fiii) with coaxial cavity of 40 nm radius, lateral irradiation, wavelength 598 nm. (giii) Resonance wavelength as a function of cavity radius, frontal irradiation. (hiii) Resonance wavelength as a function of cavity radius, lateral irradiation. (i) Maximal normalized magnetic field in the cavity as a function of cavity radius, frontal irradiation. (j) Maximal normalized magnetic field in the cavity as a function of cavity radius, lateral irradiation.

Distribution of electromagnetic field for the first resonance (corresponding to the TE_{011} mode) in particles with relatively small cavity and without the cavity is similar (figure 2ci-fi). Inside the cylinders without cavity magnetic field is quite homogeneous and have a Lorentz-like maximum near the middle of cylinder axis, and for cylinders with void core maximum is shifted to the cavity boundary. Maximal magnetic field in the cavity decreases when cavity size enlarges. Interestingly, for lateral irradiation case (figure 2b) the magnetic field in the cavity is 2 times higher than the magnetic field of the incident wave even in the case of very thin silicon shell. For all the resonances the lateral excitation of cylinder is more prospective than the frontal one from the application point of view. This is a consequence of the circular character of displacement currents, which are not changed significantly for the lateral irradiation. The wavelength of resonance decreases monotonically when the size of cavity goes up in the both cases of excitation (figure 2g, h). Note the following important feature of the MHSs for the first resonance: it is a large region (up to 80% of overall cylinder radius) where the enhancement coefficient exceeds the factor 4. The obtained dependencies of frequency and enhancement factor on the cavity size can be used allow for the construction of systems with required level of magnetic field enhancement at the certain frequency.

Higher-order hybrid modes excited in the cylinder can also provide the magnetic field enhancement. In figure 2cii-fii, ciii-fiii distributions of the magnetic field for the second and third resonance are shown for both irradiation cases, respectively. Note that the high-order resonances give stronger magnetic field enhancement (see figure 2i, j). However, the influence of cavity size on the high-order resonances is more considerable because the regions of magnetic field concentration decrease with the increase of the resonant order.

As a conclusion, we investigated the magnetic field distributions inside the silicon cylinders with and without coaxial through holes (cavities) at the resonant optical conditions using the full-wave numerical simulations. Obtained results can be used for development of new optical components and devices for management, trapping and detection of magnetic nanoparticles and molecules with magnetic transitions.

Acknowledgements
This work has been supported by the Russian Science Foundation Grant No. 16-12-10287. A.S. Shalin acknowledges the support of the President of Russian Federation in the frame of Scholarship SP-4248.2016.1. The research of K.V. Baryshnikova was partially supported by FASIE.

References

[1] Jahani S and Jacob Z 2016 *Nat. Nanotechnol.* **11** 23
[2] Krasnok A E, Miroshnichenko A E, Belov P A and Kivshar Y S 2012 *Opt. Express* **20** 20599
[3] Evlyukhin A B, Reinhardt C, Seidel A, Luk'yanchuk B S and Chichkov B N 2010 *Phys. Rev. B - Condens. Matter Mater. Phys.* **82** 1
[4] Garcia-Camara B, Gomes-Medina R, Saenz J J and Sepulveda B 2013 *Opt. Express* **21** 23008
[5] Bohren C F and Huffman D R 1998 *Absorption and Scattering of Light by Small Particles* Wiley
[6] Vuye G, Fisson S, Van N, Wang Y, Rivory J and Abeles F 1993 *Thin Solid Films* **233** 166
[7] Evlyukhin A B, Eriksen R L, Cheng W, Beermann J, Reinhardt C, Petrov A, Prorok S, Eich M, Chichkov B N and Bozhevolnyi S I 2014 *Sci. Rep.* **4** 4126
[8] Evlyukhin A B, Novikov S M, Zywietz U, Eriksen R L, Reinhardt C, Bozhevolnyi S I, Chichkov B N 2012 *Nano Lett.* **12** 3749
[9] Fu Y H, Kuznetsov A I, Miroshnichenko A E, Yu Y F and Luk'yanchuk B *Nat. Commun.* **4** 1527
[10] van de Haar M A, van de Groep J, Brenny B J M, Polman A 2016 *Optics Express* **24** 191
[11] Bakker R M, Permyakov D, Yu Y F, Markovich D, Paniagua-Domínguez R, Gonzaga L, Samusev A, Kivshar Y, Luk'yanchuk B and Kuznetsov A I 2015 *Nano Lett.* **15** 2137
[12] Mirzaei A and Miroshnichenko A 2015 *PIERS Proceedings* Prague 218
[13] Grosjean T, Mivelle M, Baida F I, Burr G W and Fischer U C 2011 *Nano Lett.* **11** 1009
[14] Nazir A, Panaro S, Proietti Zaccaria R, Liberale C, De Angelis F, Toma A 2014 *Nano Lett* **14** 3166
[15] Bakunov M I, Maslov A V, Kuznetsova S M and Zhukov S N 2014 *Photonics Nanostructures - Fundam. Appl.* **12** 114
[16] Available on www.comsol.com

Peculiarities of the exposure of actinic radiation on polymeric holographic recording media

B G Manukhin, N V Andreeva and O V Andreeva
ITMO University, Saint Petersburg 197101, Russia

manukhin.ifmo@gmail.com

Abstract. The results of experiments that allow to evaluate changes of optical parameters of polymeric recording medium with diffusional amplification occurring during recording of information are presented. It is shown that phase characteristics of the sample compared to its initial state are observed during recording of information and in the post-exposure period, i.e. in a stable condition of the finished element. Quantitative estimates which can be used for planning conditions of holographic experiment during creating highly selective holographic optical elements (HOE) with given parameters are obtained.

1. Introduction

Polymeric recording media with the thickness of the order of millimeter are used at the present time for scientific purposes and for the creation of highly selective holographic optical elements (HOE), for recording and archival storage of information. [1-3]. One of the most promising practically realizable materials for recording volume holograms are polymeric media with diffusion enhancement based on polymethyl methacrylate (PMMA) and phenanthrenequinone (PQ). Such materials are characterized by high resolution capability and information capacity of the samples in combination with available technology of their production. Their disadvantages include the fact that the change of physico-mechanical characteristics of polymer samples and their optical parameters during changing external conditions (including temperature and humidity) is significantly higher than other optical materials such as silicate glass or optical crystals.

Changes of parameters of the sample due to increase of its temperature should be taken into account when considering the recording process of the hologram, because the effect of actinic radiation (radiation with a wavelength in the spectral region of absorption of the sample) is accompanied by heating which leads to the change of the average refractive index and the thickness of the sample.

Thus, during recording of the hologram under the influence of the interferential field on the spatial-modulated changes of the optical characteristics of the sample (photoinduced changes) are superimposed spatial-unmodulated changes of its optical characteristics caused by heating.

If temperature changes of the refractive index of the sample are comparable with the amplitude of the photoinduced changes, this can lead to the fact that the spatial structure of recorded hologram will be different from the structure of the recorded interferential field. The accounting of such deformations, as well as optimization of the recording conditions to reduce thermal effects is particularly important in obtaining volume holograms designed for the creation of highly selective HOE. The available literature information about the effect of influence of temperature changes on parameters of recorded polymeric holograms is ambiguous, and often contradictory, as it refers to samples obtained in different conditions [1].

In this paper the spatial-temporal changes of phase characteristics of the polymer samples during action of actinic radiation, the presence of which may lead to deformation of recorded highly selective holographic structures, are considered and their quantitative evaluations are given.

2. Object and methods

Object of research: sample of material "Difphen" - one of the modifications of polymeric volume recording medium with diffusion enhancement based on PMMA/PQ [4]. On figure 1a are shown spectral characteristics of a typical sample of this material in the initial state (curve 1 – absorption spectrum due to the presence of PQ molecules in the sample of PMMA) and after complete bleaching of the sample (curve 2), when under the action of actinic radiation, there was a complete transformation of the light-sensitive molecules of PQ into not light-sensitive molecules of photoproduct (PP), which don't have absorption in the visible region of the spectrum.

For research of changes of optical characteristics of the sample in the process of radiation exposure and post-exposure period was used experimental stand designed for the research of phase transformations in transparent objects by method of digital holographic interferometry (DHI) [5], block-diagram of which is shown on figure 1b. On this stand an object node (3) placed outside the main scheme of the interferometer (1) which allows monitoring the process of external influence on the object by beam of radiation 473 nm DPSS-laser (4). The sample (disk with diameter 40 mm) is rigidly mounted in a special cassette and installed in the object beam in accordance with the optical scheme of interferometric experiment. The object beam of interferometer is a non-actinic test beam for this sample of radiation of 532 nm, which tests a workspace of the sample with the size not less than 25x25 mm. The beam of actinic radiation, extended by lens, passes through the sample in the center of workspace. This stand has a computer terminal to control registration of sequence of digital holograms and for their further processing.

The methodology of each experiment includes the following main stages [5]:
- recording of the sequence of digital holograms that show the current state of the sample during the experiment: before radiation exposure (the initial state), during the period of exposure (exposure) and in post-exposure period (relaxation);
- receiving a series of digital interferograms by processing sampling of recorded holograms;
- digital processing of interferograms and presentation of results in a convenient way.

The digital interferogram of the current state of the object represents by itself a spatial distribution of phase difference of the testing radiation in the plane of observation of object relevant to the change of the sample at the moment of time t_i compared to initial (at the moment of time t_0 before exposure) – $\varphi_{i0}(x,y) = \varphi_i(x,y) - \varphi_0(x,y)$. A typical view of interferogram of the sample status during the period of radiation exposure is shown in figure 1c. Interference pattern in this case takes the form of concentric rings with the center in the center of the exposure area. Outside of interference pattern ("pertubation" area of the sample) there is a uniform background. In further calculations it is assumed that in background area the phase difference between rays that have passed through the sample at time t_i and t_0 is equal to zero, i.e. $\varphi_{i0}(x,y)_{bg} = \varphi_i(x,y)_{bg} - \varphi_0(x,y) = 0$. Thus, $\varphi_i(x,y)_{bg} = \varphi_0(x,y)$ and the spatial distribution of the phase difference $\varphi_{i0}(x,y)$ will represent the value of $\varphi_i(x,y)$, measured from $\varphi_i(x,y)_{bg}$. The results of digital processing of interferogram is shown on figure 1d in the form of "detailed" phase difference along the x axis, $\varphi_{i0}(x)$, obtained by summing the phase incursions along the line A-A from background to the left from the center of exposure, where $\varphi_{i0} = 0$, through the center, where φ_{i0} is maximum, to the background right from the center; there are also the limits of beam of exposed radiation Δx_{exp} that gives an ability to compare their limits and the area of "phase perturbations" caused by this exposure.

By each current interferogram during processing were obtained the following quantitative data:
- change of the phase difference in the centre of exposure area $(\varphi_{i0})_c$ in the moment of time t_i;
- the sample area, Δx_{pert}, in which at the current time $\varphi_{i0} \neq 0$.

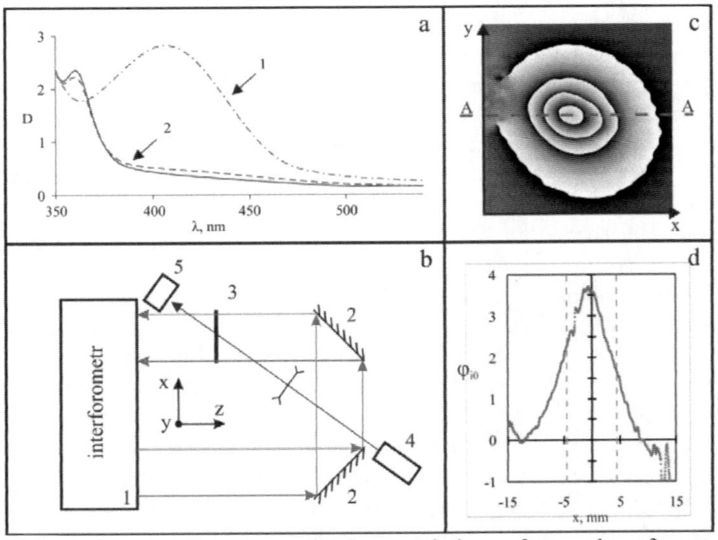

Figure 1. a – the spectral characteristics of sample of material "Difphen" before radiation exposure (curve 1) and after complete cycle of processing (curves 2); b – the block-diagram of the stand for experiments: 1 – interferometer, 2 – mirrors, 3 – object, 4 – laser 473 nm, 5 - detector; c – digital interferogram that shows the phase change of the test radiation (532 nm) during actinic radiation exposure (473 nm), d – detailed distribution of the phase difference, $\varphi_{i0}(x)$, along the line A-A and the limits of beam of exposed radiation Δx_{exp} (dashed lines).

3. Experimental results and analysis

The results of experimental data processing are presented on figure 2 and figure 3 for two sequences, in each of them holograms were recorded with an interval of one second and with 1/1024 s exposure: during exposure of radiation of 473 nm - during 10 minutes and after its ending – during 5 minutes. The power of radiation beam of 473 nm in experiment 1 was 20 mW, in experiment 2 it was 36 mW.

On figure 2 results of processing of mostly typical holograms are presented in the form of detailed distribution of phase difference on x-axis, $\varphi_{i0}(x)$, (centre of radiation exposure of 473 nm is located at x = 0). In both experiments during the period of radiation exposure $\varphi_{i0}(x) \geq 0$, and "perturbed" area considerably exceeds the area of exposure, i.e. $\Delta x_{pert} > \Delta x_{exp}$. In stable post-exposure condition of sample $\varphi_{i0}(x) \leq 0$ и $\Delta x_{pert} \approx \Delta x_{exp}$. This situation can be explained by the fact that the phase changes of the sample, that are diagnosed in the post-exposure period, due to photochemical transformations (conversion of PQ to PP), that occurred in the exposed area of the sample (Δx_{exp}) during the influence of actinic radiation. At the same time, the phase changes of the sample at the stage of exposure radiation is caused by two processes: phototransformation of PQ and heating of the sample. In the post-exposure period heating of the sample is not presented, and the fact that $\Delta x_{pert} \approx \Delta x_{exp}$ also suggests that in this case, the change of phase, $\varphi_{i0}(x)$, is determined only by phototransformation of PQ.

Exactly this process provides recording of information in given recording medium. Phototransformation of PQ leads to the change of the refractive index of the medium (Δn_{PQ}) ,thus, it is assumed that spatial changes of polymeric framework in this process are not happening - $\Delta l_{PQ} = 0$. The process of transformation of PQ into PP in the experiments is monitored by measuring the transmittance of the sample, τ, at 473 nm (curve 3, figure 3a). During full transformation of PQ transmittance of the sample reaches maximum value which doesn't change in the future. In the experiments it happens during first 200 seconds of exposure of the sample. Thus, $\Delta n_{PQ}(t_i) \approx$ const at t_i

> 200 seconds and change $\varphi_{i0}(t_i)$ is only due to heating by absorption of radiation energy. Heating of the sample leads to change of its refractive index and to change of its geometric dimensions and temperature changes of optical thickness of the sample $\Delta(nl)_{term}$ in this case determined by the properties of the polymeric framework and specific modification of the polymer.

Dependences 1 and 2 on figure 3a characterize phase changes of the sample $\varphi_{i0}(t_i)$ during radiation expose of 20 mW (1) and 36 mW (2). In the period of exposure the value $\varphi_{i0}(t_i)$ becomes higher after power increase; while during relaxation the difference between values $\varphi_{i0}(t_i)$ in experiments 1 and 2 are within the measurement error. The difference of values $\Delta x(t_i)$ (figure 3b) in both experiments is also within measurement error.

The obtained experimental results allow us making the following assessments:
- Changes of the sample during the period of exposure of actinic radiation include irreversible changes due to phototransformation of PQ, $\{\varphi_{i0}(x,y)\}_{PQ}$, and reversible due to heating $\{\varphi_{i0}(x,y)\}_{term}$.
- During exposure of radiation of $200 < t_i < 600$ seconds phase characteristics of the sample don't change significantly. During this period irreversible changes of the sample that determine the process of information recording, with the full phototransformation of PQ make the quantity of $\{\varphi_{i0}(x,y)\}_{PQ} = -(1,0\pm0,1)\pi$ rad, and reversible changes of the sample due to the heating make the quantity of $\{\varphi_{i0}(x,y)\}_{term} = \{\varphi_{i0}(x,y)\} - \{\varphi_{i0}(x,y)\}_{PQ}$, equal to $\approx 3\pi$ rad for experiment 1, which significantly exceeds the quantity of the irreversible changes which determine information recording.

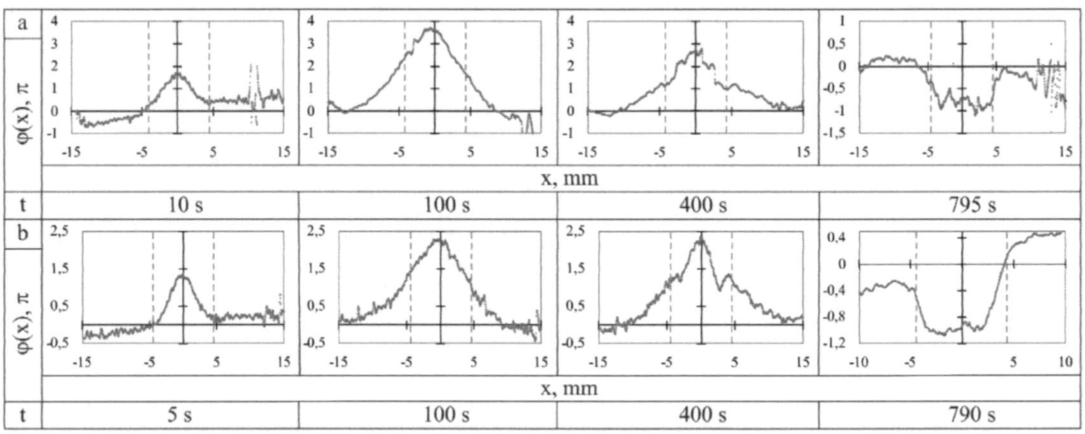

Figure 2. The change of phase $\varphi_{i0}(x)$ along the line A-A in experiments 1 (a) and 2 (b) - exposure of radiation of 473 nm in the period from 0 to 600 s. Vertical dashed lines determine Δx_{exp}.

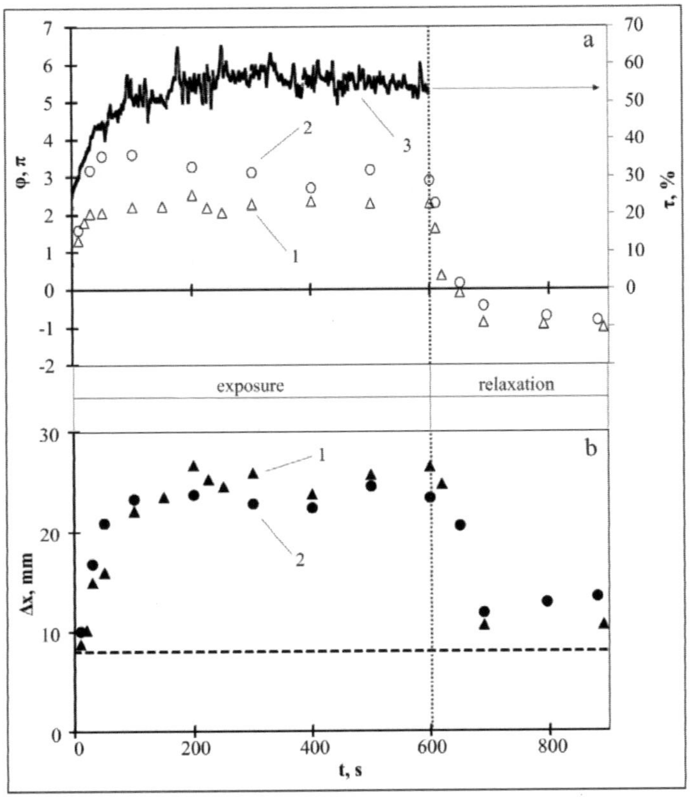

Figure 3. Dynamics of change of characteristics of polymeric sample during exposure of actinic radiation of 473 nm (exposure) and in post-exposure period (relaxation): a - phase change of testing radiation, φ_{i0}, in the center of the area of exposure of actinic radiation with power of 20 mW (experiment 1) and 36 mW (experiment 2, change of transmittance of the sample, τ, at wavelength of 473 nm (curve 3); b – change of the value of "phase perturbation" area (Δx_{pert}) in experiments 1 and 2; dashed line is the exposed sample area, Δx_{exp}.

4. Conclusions

The experiments with the use of the method developed by authors allowed quantifying the phase change of the sample of polymeric recording medium "Difphen" during exposure of actinic radiation of 473nm.

It is shown that irreversible changes of the sample with the thickness of 2.7 mm under exposure of radiation of 473 nm with power (20÷36) mW at full transformation of PQ into PP is characterized by the negative phase difference of testing radiation of 532 nm with a value of (1,0±0,1)π radian. The value of reversible changes, which due to thermal effects, is characterized by the positive phase difference of testing radiation of 532 nm which is estimated be the value ≈ 3π radian.

Thus, at these experimental conditions reversible changes of optical characteristics of a polymeric sample, that accompany the process of exposure of actinic radiation, greatly exceed irreversible changes that provide recording holographic information.

Perhaps these facts explain the failure of some experiments carried out before, including the authors, on recording of highly selective reflective holograms on samples of polymeric medium with diffusional amplification.

To reduce distorting influence of thermal effects during holographic recording it is necessary to choose conditions of the experiment associated with the increase of influence of phototransformation of PQ (selection of optimal working wavelength) and for reduce the influence of thermal effects (wavelength and power of exposed radiation, duration of exposure and etc.)

Acknowledgments

The authors thank O.V. Bandyuk for production of samples and discussion of the results.

References

[1] Veniaminov A V, Mahilny U V 2013 *Opt. Spectros.* **115**(6) 906

[2] Shelby R M 2002 *Proc. of SPIE* **4659** 344

[3] Hemmer P, Shahriar S, Ludman J, Caulfield H J 2002 *Holography for the New Millenium* (New York: Springer) p 179

[4] Andreeva O V, Bandyuk O V, Paramonov A A, Cherkasov A S, Gavrilyuk E R, Andreev P V 2000 J. Opt. Tech. **67**(12) 1043

[5] Manukhin B G, Gusev M E, Kucher D A, Chivilikhin S A, Andreeva O V 2015 *Opt. Spectros.* **119**(3) 392

Improved emission outcoupling from microdisk laser by Si nanospheres

Yu S Polubavkina[1], N V Kryzhanovskaya[1], E I Moiseev[1], M M Kulagina[3], I S Mukhin[1,2], F E Komissarenko[1,2], Yu M Zadiranov[3], M V Maximov[1], A E Krasnok[2], A A Bogdanov[2], A E Zhukov[1], A V Shelaev[4]

[1] St Petersburg Academic University of RAS, Khlopina 8(3), St Petersburg 194021, Russia
[2] ITMO University, St Petersburg, Russia
[3] Ioffe Institute, St Petersburg, Russia
[4] NT-MDT Co., 100 Zelenograd, Moscow, 124482 Russia

Abstract. Far field emission of optically pumped QD microdisk lasers with spherical Si nanoantenna placed on the top surface of microdisk is studied theoretically and experimentally. The interaction of the whispering gallery mode microdisk resonator with properly designed and positioned Si nanospheres results in improvement of light outcoupling and directed emission.

1. Introduction

Optical whispering-gallery mode (WGM) microcavities having ultrahigh Q factor (up to 10^{11} [1]) and small mode volumes enhance significantly the interaction between light and matter, becoming an excellent platform for achieving ultralow-threshold microlasers [2]. However, the emission of WGM microlasers is isotropic in plane due to the rotational symmetry, whereas for practical applications the directional emission is required. To increase the collection efficiency and improve the directivity of the emitted light several types of defects such as a point scatterer [3], linear defect [4], air hole [5] have been proposed but they introduce Q spoiling accompanied by essential increase of the threshold characteristics. In this work we used a subwavelength silicon scatterer placed on the top surface of the microdisk, to obtain the improved emission outcoupling due to the effect of the Si nanosphere-enhanced scattering. Optimization of the Si nanosphere size and location results in increased intensity of the outcoupled WGM and its directional far field emission without Q/threshold spoiling.

2. Experiment and calculation details

The epitaxial structures used for fabrication of the microdisk lasers were grown by molecular beam epitaxy on a n-doped GaAs substrate. An active region represents 5 layers of $InAs/In_xGa_{1-x}As$ quantum dots inserted into a 0.35-μm-thick GaAs waveguiding layer cladded with 400-nm-thick $Al_{0.98}Ga_{0.02}As$ layer from the substrate side. Spectral position of quantum dot ground-state transition was located around 1.28 μm at room temperature. Microdisk resonators were fabricated using photolithography and Ar+ ion beam etching. Outer diameter of the microdisks was 6 μm. The $Al_{0.98}Ga_{0.02}As$ bottom cladding layer was selectively oxidized to be transformed into an AlGaO oxide.

Spectrum and field structure of the eigenmodes in the microdisk cavity were calculated by use COMSOL Multiphysics software using the approach described in [6-7]. Si nanospheres with a different diameters were made by laser ablation of thick a-Si:H layers on quartz substrates and placed on the top surface of the microdisk by a microtip using pick-and-place technique. The scanning electron micrograph (SEM) of the microdisk laser with nanosphere is shown in figure1. The spectra and distribution of the emitted light of the microdisk laser were studied by confocal scanning optical spectroscopy (Integra Spectra, NT MDT). The microdisk was uniformly optically pumped by YAG:Nd laser (λ=532nm) using lensed fiber.

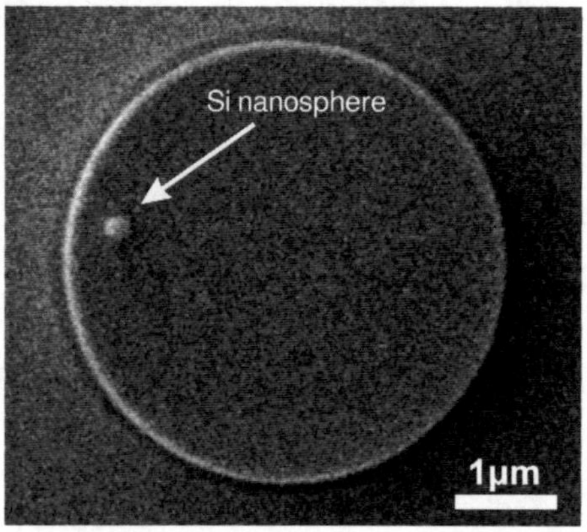

Figure 1. Scanning electron micrograph of the 6-μm microdisk with a nanosphere (the position of sphere is shown with mark).

3. Results

First, optimization of the Si nanosphere size and location at the microdisk surface for directional light outcopling was done. We studied the dependence of the microdisk laser dominant modes intensities, azimuthal distributions, thresholds and widths on the size and position of the Si nanosphere. If the nanosphere diameter was less than 200nm and/or its position did not coincide with the maximum of electric field of the microdisk resonator mode we did not observe any difference in the microlaser spectra, threshold power, and quality factor of the modes. Figure 2 shows SEM image of the 6-μm microdisk laser with nanosphere placed at the position of the maximum of electric field of the microdisk resonator mode but with diameter 170 nm that is smaller than optimal. The room temperature spectra of the microlaser without and with 170 nm Si nanospere are shown in figure 3. We also do not observe any difference in the threshold power, and quality factor of the modes.

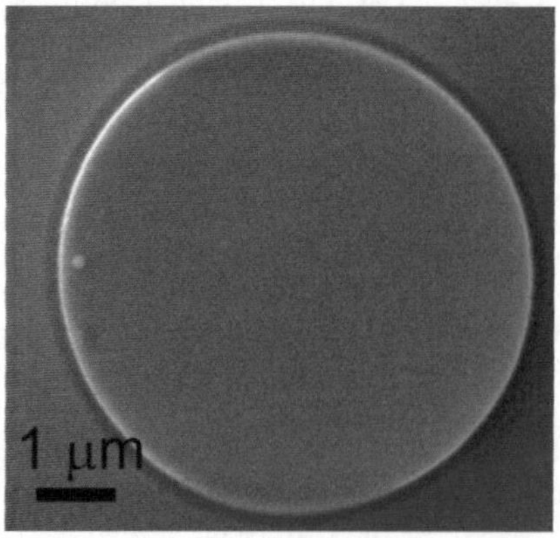

Figure 2. SEM of the 6-μm microdisk with a170 nm in diameter nanosphere.

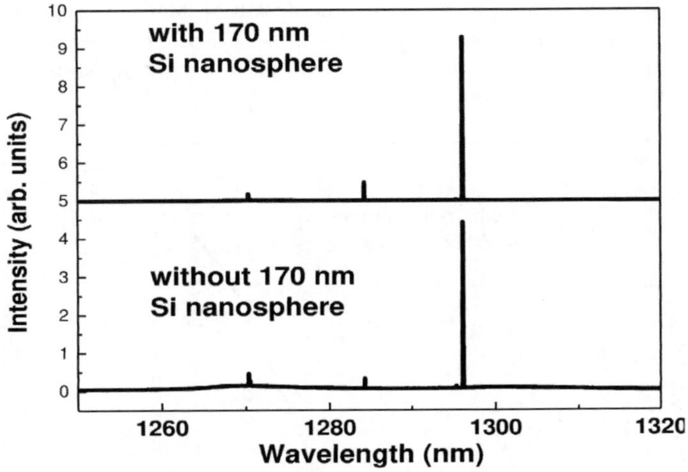

Figure 3. The room temperature μPL spectra of the microlaser without and with 170 nm Si nanosphere obtained at optical power $P \sim 1.5P_{th}$. The spectra are vertically shifted for clarity.

From the calculation of the coupling parameters of the Si nanosphere and 6-μm microdisk resonator optimal diameter of the single Si nanosphere was found to be 300nm. In order to influence the high quality $TE_{39,1}$ WGM with low radial index the nanosphere was placed near the maximum of its electric field and the distance between the Si nanosphere center and the microdisk edge was 480 nm. Figure 4 demonstrates the room temperature spectra of the 6-μm microlaser without and with 300 nm Si nanosphere. After the placing of the Si nanosphere with optimal size the enhancement of the $TE_{39,1}$ WGM line at 1294 nm intensity is observed.

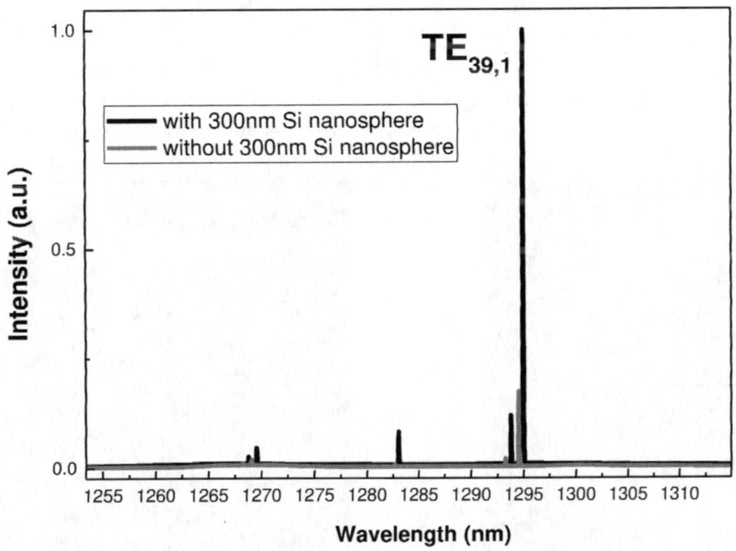

Figure 4. The room temperature μPL spectra of the microlaser without and with 300 nm Si nanosphere obtained at optical power $P \sim 2P_{th}$.

We studied the dependence of the intensity of the WGM line $TE_{39,1}$ at λ=1294 nm on the pump power of the same microdisk laser without Si nanosphere and with Si nanosphere (figure 5). More than 20-times increase of the line intensity (at the pump power ~$2P_{th}$) provided by Si microsphere-enhanced scattering is observed. The threshold pump power and line full-width remains the same P_{th}~0.3 mW, FWHM~ 30 pm (within the setup resolution).

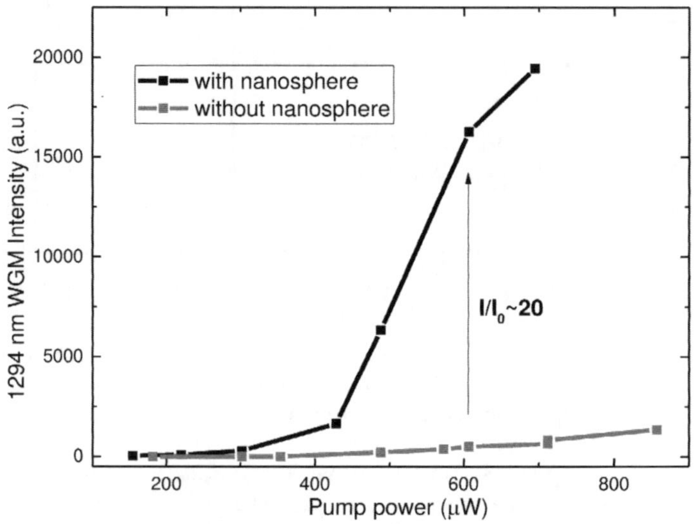

Figure 5. The dependence of the intensity of the WGM line at λ=1294 nm on the pump power.

Figure6 (a) and (b) demonstrates the spatial distributions of the lasing WGM $TE_{39,1}$ line intensity of the initial microdisk and microdisk with nanosphere. The initial microdisk demonstrates nearly isotropic emission distribution with some modulation in azimuthal direction due to the light scattering at the microdisk surface roughness due to technology imperfections. Light distribution from the microdisk with nano-antenna is highly directional. The maximum of the intensity is observed at the Si microsphere position resulting in a highly directional far field emission (~38°). The other modes of the microdisk resonated that are not coupled to the Si microsphere demonstrate the random spatial distribution of the intensity mainly governed by the scattering on the surface roughness.

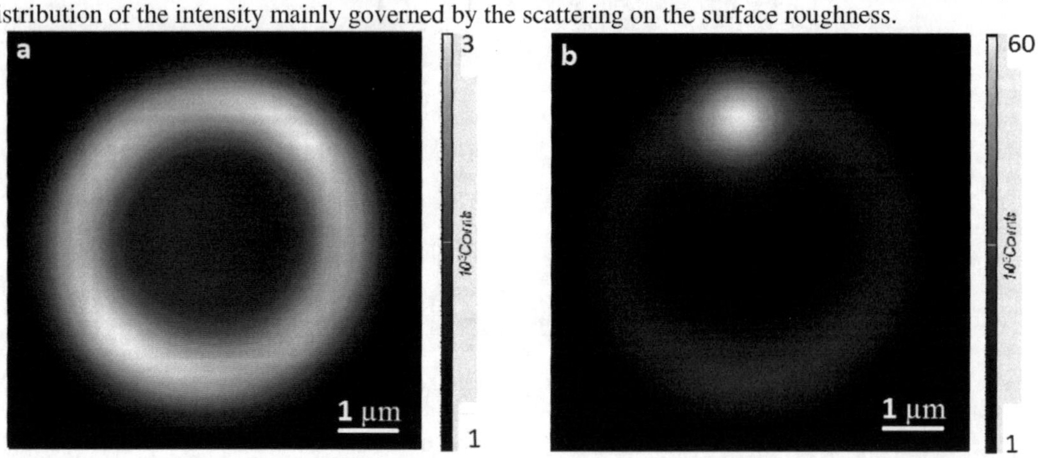

Figure 6. The distribution of the WGM line intensity (λ=1294 nm) in microlaser without (a) and with Si nanosphere on the top surface (b).

To conclude we demonstrated the method of enhancement of a dominant mode intensity by means of a Si nanosphere placed at the top of the microdisk laser. Using Si nanosphere with optimized size of 300 nm we achieved more than 20-times increase of the WGM line intensity enhancement by 20 times in room-temperature spectra of 6-μm-diameter QD microdisk. The azimuthal emission intensity distribution for microdisk cavity with Si nanospere is built and divergence angle at half height of maximum intensity is about 38^0, which corresponds to a highly directional far field emission.

Acknowledgments
The work is supported by the Skolkovo Foundation (grant agreement for Russian educational and scientific organization no.6 dd. 30.12.2015). YSP acknowledges Government of St. Petersburg and The Committee on Science and Higher Education. AAB acknowledges RFBR (16-37-60064) and the President of Russian Federation (MK- 6462.2016.2).

References

[1] Foreman M R, Swaim J D, Vollmer F 2015 *Whispering gallery mode sensors* (Adv. Opt. Photon. 7) pp 168-240
[2] McCall S L, Levi A F J, Slusher R E, Pearton S J, Logan R A 1992 *Whispering-gallery mode microdisk lasers* (Appl. Phys. Lett. 60) pp 289-291
[3] Dettmann C P, Morozov G V, Sieber M, Waalkens H 2009 *Unidirectional Emission from Circular Dielectric Microresonators with a Point Scatterer* (Physical Review A 80) 063813
[4] Apalkov V M and Raikh M E 2004 *Directional emission from a microdisk resonator with a linear defect* (Physical Review B 70) 195317
[5] Wiersig J, Hentschel M 2006 *Unidirectional light emission from high-Q modes in optical microcavities* (Physical Review A 73) 031802(R)
[6] Oxborrow M 2007 *Traceable 2-d finite-element simulation of the whispering gallery modes of axisymmetric electromagnetic resonators* (IEEE Transactions on Microwave Theory and Techniques) vol 55 pp 1209–1218
[7] Bogdanov A A, Mukhin I S, Kryzhanovskaya N V, Maximov M V, Sadrieva Z F, Kulagina M M, Zadiranov Y M, Lipovskii A A, Moiseev E I, Kudashova Y V, Zhukov A E 2015 *Mode selection in InAs quantum dot microdisk lasers using focused ion beam technique* (Opt. Lett. 40(17) pp 4022-4025

Optical Nanowaveguides Based on Zinc Oxide Plasmonic Materials

A M Lerer[1], **P E Timoshenko**[1], **T Yu Chernikova**[1], **E M Kaidashev**[1,2], **A S Puzanov**[1]

[1]Southern Federal University, Faculty of Physics, Rostov-on-Don, Russia
[2]Southern Federal University, Nanomaterials Lab., Rostov-on-Don, Russia

lerer@sfedu.ru

Abstract. The numerical study of nanowaveguides containing ZO, AZO, GZO and other materials are studied in infrared range. The dispersion relation for the waveguides contained an arbitrary number of layers are solved numerically. The integral-differential equations in vector form of the electromagnetic waves propagation in nanowaveguides containing a finite width plasmon films are calculated by the Galerkin' method and the effective dielectric constants method. The opportunity of surface plasmon-polariton propagation with a large deceleration ratio are shown.

Introduction

The surface TM-wave (surface plasmon-polariton [1], SPP) with a large deceleration ratio [2] can propagate at the interface of plasma-insulator or metal-dielectric in optical range. The phenomenon allows to create plasmonic waveguides conducting the light signals in the volumes smaller than the diffraction limit. It can significantly reduce the size and increase the efficiency of optical integrated circuits. The main disadvantage of plasmon waveguides are strong losses due to the large absorption in metals. The metal waveguides provide a large deceleration at wavelengths less than micrometer. The main aims of the nanophotonic circuits developers [3] are to extend the operating band of plasmon waveguides and to achieve an optimal balance of SPP localization and waveguide losses.

Recently, the new materials based on zinc oxide (ZO) such as AZO, GZO and other were created [4, 5]. So the goal of this paper is a numerical study properties of nanowaveguides containing such materials in the infrared range. Subject matter of the study is the following nanowaveguides: 1) semi-infinite dielectric structures, the new material, dielectric, 2) planar nanowaveguides formed by plane-parallel dielectric layers, the new material, and 3) planar nanowaveguides formed by the finite thickness films of the new material lying on the multilayered dielectric substrate.

The well-known analytical expressions [6] are used to calculate the wavelength and the decay constant of SPP and to study the first type waveguides. The optimal wavelength ranges for spread of SSP are found. The ranges are used to investigate the second type nanowaveguides.

The dispersion relation for the second type waveguide contained an arbitrary number of layers are obtained and solved numerically. The waveguide is a prototype of the third type nanowaveguide that can be practically implemented.

The way to find the complex propagation constant in waveguide structures with small losses are proposed in this article. The properties of SPP propagating in offered structures are investigated and applied to study the third type nanowaveguides. The solution of the integral-differential equations in vector form of the electromagnetic waves propagation in nanowaveguides containing a finite width plasmon films is given by the Galerkin' method [7]. The possibility of SPP propagation with a large deceleration ratio is also shown.

1. The wave propagated on "dielectric – plasmonic material" boundary

The expression [6] are used to calculate the dispersion in case of "dielectric – plasmonic material" boundary:

$$n = n' - in'' = \frac{\beta}{k} = \left(\frac{\varepsilon\varepsilon_{pl}}{\varepsilon + \varepsilon_{pl}}\right)^{1/2} \quad (1)$$

where β is the complex propagation constant, k is the wave number in vacuum, ε and ε_{pl} are the dielectric constants of dielectric and plasmonic material, respectively. The length L of wave propagation can be expressed as the distance at which one is attenuated *e* times.

The numerical results of SPP propagated on "dielectric – plasmonic material" boundary are presented in figure 1: solid curves are dielectric with $n = 1.77$, dashed curves are ZnO dielectric, curves 1 are HfN material, 2 – TiN and 3 – ZrN. The refractive index is normalized $n'_{\text{norm}} = n'/(\text{Re}\,\varepsilon)^{1/2}$, where ε is the permittivity of dielectric semi-infinite layer.

The figure shows that the losses of SPP propagation at this boundary are high and the wave propagation length is also comparable to the wavelength. Moreover, the lower wavelength is, the higher losses are and so shorter the propagation distance is. Compared with the silver and argentum metalic layers, the high deceleration ration can be obtained at $\lambda \geq 800\text{nm}$.

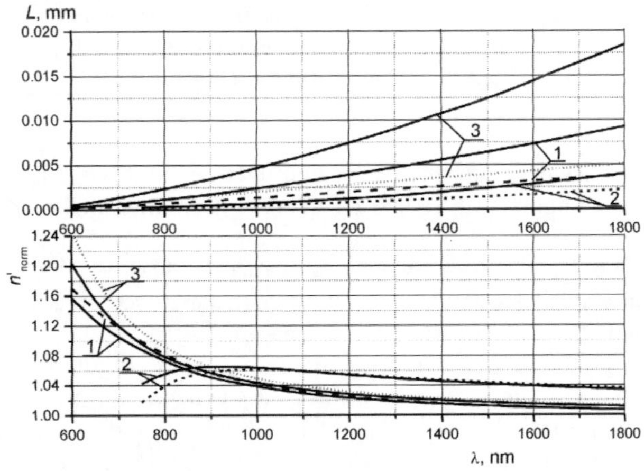

Figure 1. Dispersion curves of the SPP propagated on "dielectric – plasmonic material" boundary. Solid curves are dielectric with refractive index $n = 1.77$, dashed curves are ZnO dielectric, curves 1 are HfN plasmonic material, 2 – TiN and 3 – ZrN.

2. Planar waveguides

The dispersion relation in case of arbitrary number of dielectric layers having complex permittivity is obtained in recurrent form. The approximate method for numerical searching the complex roots of a complex function used during the finding of dispersion solutions is described in the article [6].

The calculated layered structure consists of the substrate with refractive index n = 1.77, the ZnO layer having thickness h_2 and the ZrN layer with thickness h_1. As shown in figure 2, the deceleration ratio n' at wavelengths far from the critical wavelength weakly depends on the layer ZnO and ZrN thickness. It confirms the existence of SPP near the interface of layers. When the wavelength is close to critical wavelength n' and the thickness of ZnO layer decreased, the losses are reduced.

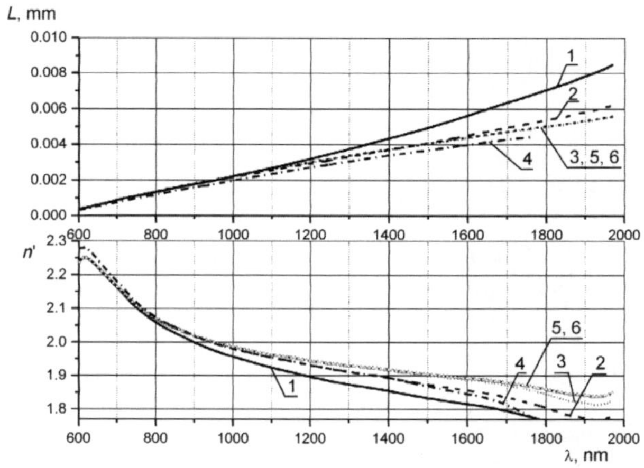

Figure 2. Dispersion curves of the planar waveguide: 1 – h_1 = 100nm, h_2 = 200nm; 2 – h_1 = 100nm, h_2 = 400nm; 3 – h_1 = 100nm, h_2 = 1000nm; 4 – h_1 = 50nm, h_2 = 1000nm; 5 – h_1 = 200nm, h_2 = 1000nm; 6 – h_1 = 400nm, h_2 = 1000nm.

3. Rectangular waveguides

The investigated structure consists of the substrate with refractive index n = 1.77, the ZnO layer having thickness 400nm and the rectangular waveguide filled with plasmonic material placed on ZnO layer with thickness 100nm and width w.

The two different methods are used to calculate the structure. The first one is the volume integral method [6]. The second method is the effective dielectric constants method (EDCM). The EDCM has simple numerical implementation [7]. It comprises two stages.

At first step, the waveguide having complex geometric shape divides by the vertical lines of multilayer waveguide structures (uniform waveguides). In general, the investigated model has an arbitrary number of horizontal layers. Each of these layered waveguides is planar in the horizontal direction. In case of the model in the vertical direction is uniform, the deceleration rate of the waveguide is equal to the refractive index. If the structure can't propagate some Eigen modes (e.g., structure formed by two semi-infinite vertical dielectric layers having same signs of permittivity), the deceleration ration is equal to the refractive index of the waveguide layer adjacent to the studied waveguide. So the deceleration ratio can be calculated, if the Eigen waves exists.

A planar waveguide formed by multiple parallel vertical layers are considered at the second step. Because the refractive index of each layer and the deceleration ratio has already calculated previously, the deceleration ratio of waves propagated in studied waveguide can be found considering it as an original waveguide deceleration rates with complex cross-sectional shape.

The EDCM provides sufficient accuracy and high speed to calculate the nanoplasmonic waveguides parameters in wavelength range in which the surface waves can propagate. The rigorous volume integral equations method is used to verify the EDCM.

The numerical results of rectangular nanowaveguides are presented in figure 3 and 4. If the width of the waveguide reduces, the operating wavelength range becomes narrower and the deceleration ratio reduces too, but at same time the losses reduces significantly. The dispersion characteristics of the ZnN rectangular waveguide shown in figure 3. The figure 4 contains the comparison of waveguides parameters formed by different plasmon materials. It's seen that the losses are rather high and comparable with optical waveguides formed by thin metal films. The losses could be reduced by using of active substrates or layers.

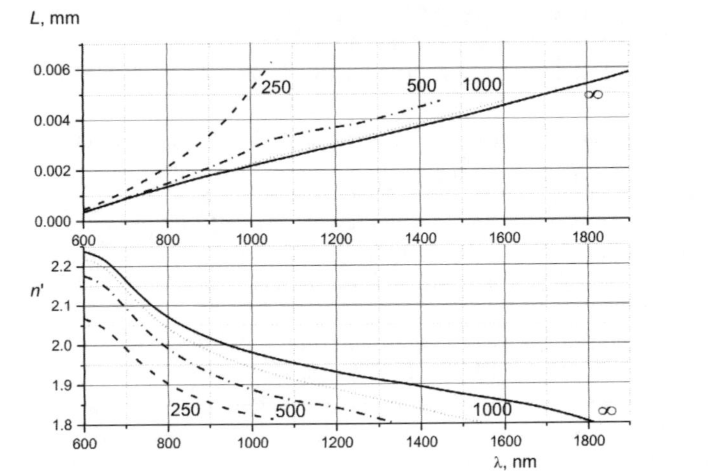

Figure 3. The dispersion curves of ZrN rectangular waveguide with various width *w*.

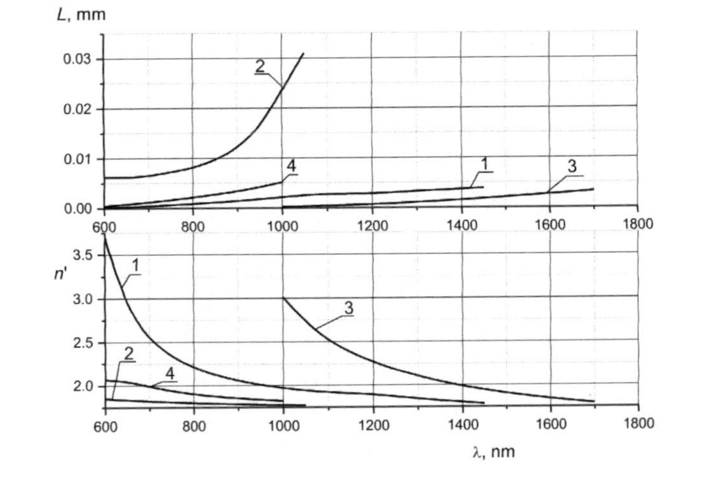

Figure 4. Dispersion curves of the rectangular waveguide with *w* = 250nm: 1 – HfN, 2 – ITO, 3 – TiN, 4 –ZrN.

Summaries

The Galerkin' method and the effective dielectric constants method are used to find the solution of a vector integral-differential equation to describe electromagnetic wave diffraction on three-dimensional bodies with complex dielectric permeability. Three types of nanowaveguide were calculated: 1) semi-infinite dielectric structures, the new material, dielectric, 2) planar nanowaveguides formed by plane-parallel dielectric layers, the new material, and 3) planar nanowaveguides formed by the finite thickness films of the new material lying on the multilayered dielectric substrate.

Acknowledgement

The authors acknowledge the support of the Russian Ministry of Education and Science, provided by grant 16.219.2014/K "The synthesis methods development, the theoretical and experimental investigation of the nanostructures based on ZnO performs to create photodetectors, optical nanoantennas, piezo- and chemosensors", and the part of the Internal Grant project of Southern Federal University in 2014 – 2016y., provided by grant 213.01.-07.2014/08PCHVG, for carrying out this research study.

References

[1] Klimov V *Nanoplasmonics*. 2014 CRC Press. 598

[2] Golovacheva E V, Lerer A M and Parkhomenko N G *Diffraction of electromagnetic waves of optical range on a metallic nanovibrator*. 2011 Moscow Univ. Phys. Bulletin, **66**(1):5–11.

[3] Stockman M I *Nanoplasmonics: past, present, and glimpse into future*. 2011 Opt. Express. **19** 22029–22106.

[4] Naik G V, Kim J and Boltasseva A *Oxides and nitrides as alternative plasmonic materials in the optical range*. 2011 Optical Materials Express, **1**(6):1090.

[5] Naik G V, Shalaev V M and Boltasseva A. *Semiconductor plasmonic metamaterials for near-infrared and telecommunication wavelength* 2010 Proc. SPIE 7754, Metamaterials: Fundamentals and Applications III, 77540M.

[6] Lerer A M, Donets I V, Kalinchenko G A and Makhno P V *Volume integral method for investigation of plasmonic nanowaveguide structures and photonic crystals*. 2014 Photon. Res. **2** 31–37

[7] Lerer A M *Theoretical investigation of 2D periodic nanoplasmon structures*. 2012 J. Commun. Technol. Electron. **57** 1151–1159

RTD application in low power UHF rectifiers

V Yu Sinyakin, M O Makeev, S A Meshkov

Bauman Moscow State Technical University, 2-nd Baumanskaya, 5, Moscow 105005, Russia

Abstract. In the current work, the problem of UHF RFID passive tag sensitivity increase is considered. Tag sensitivity depends on HF signal rectifier efficiency and antenna-rectifier impedance matching. Possibility of RFID passive tag sensitivity increase up to 10 times by means of RTD use in HF signal rectifier in comparison with tags based on Schottky barrier diode is shown.

1. Introduction

Low power UHF rectifiers are used in UHF RFID passive tags. A passive tag consists of antenna and micro-chip. Chip includes detector (demodulator), controller, nonvolatile memory, modulator and UHF rectifier with voltage limiter those represent passive tag power supply.

Passive tag sensitivity increase is the important problem in UHF RFID system design. It gives a possibility to improve system's range and to enhance electromagnetic compatibility and ecological situation in the work area of RFID system.

2. Passive tag sensitivity increase

UHF RFID passive tag sensitivity is defined, as a rule, by the ability of tag to transform the reader's electromagnetic radiation energy into the energy of DC used for tag supply [1]. This function is performed by rectenna that includes antenna and UHF low power rectifier.

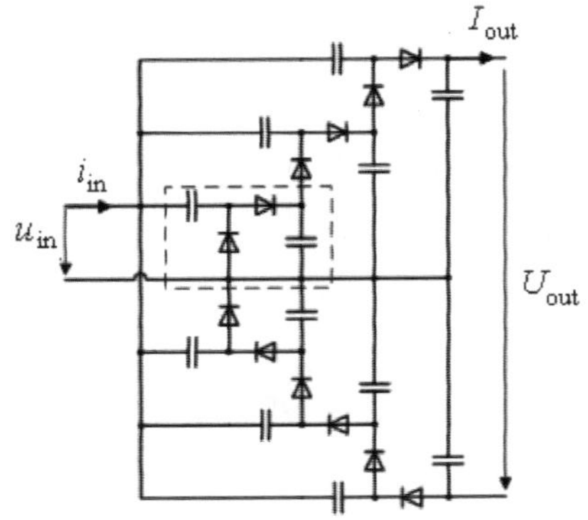

Figure 1. Low power UHF rectifier scheme.

Rectifier is a voltage multiplying scheme consisting of N nonlinear elements (Schottky barrier diode or MOS-transistors) and N capacitors (Figure 1). The multiplying factor is defined by the number of elemental cells. One cell is a voltage doubler (it is limited by dotted line on Figure 1). In the ideal case, the output voltage of the cell is

$$U_1 = 2 \cdot U_{in} \tag{1}$$

where U_{in} is the amplitude of input alternating voltage. Serial connection of N_C ideal cells allows to achieve output DC voltage

$$U_{out} = 2N_C U_{in} = NU_{in}. \tag{2}$$

The model of multiplying rectifier is described in [2]. One can calculate input impedance, required input RF power and efficiency of the rectifier if specify output voltage, output current and IV-curve of nonlinear elements using this model.

Rectenna performance is described by efficiency coefficient:

$$K_{eff} = \eta_R \cdot 4R_R R_A |Z_R + Z_A|^{-2} \tag{3}$$

where $Z_A = R_A + iX_A$ – input antenna impedance, $Z_R = R_R + iX_R$ – input rectifier impedance, η_R – rectifier efficiency. Last one is the rectifier characteristic, and it shows which part of input RF power P_{RF} is transformed into DC power P_{DC}:

$$\eta_R = P_{DC} / P_{RF} \tag{4}$$

The feature of low power UHF rectifiers is the necessity of low-amplitude alternating voltage rectification. Nonlinear elements such as Schottky barrier diodes have IV-curve with exponential character. It is described by the equation [3]

$$I(U) = I_S \cdot \left[\exp\left(\frac{U - IR_S}{0,026} \right) - 1 \right] \tag{5}$$

where I_S – saturation current, R_S – serial resistivity. IV-curve of Schottky barrier diode has nonzero threshold voltage that value is about 0.2 V (see Figure 2, dotted curve). This fact means that rectification of alternating voltage with amplitude less than 0.2 V is complicated. Thus, the efficiency of low power UHF rectifier based on Schottky barrier diode will extremely decrease with input power reduction.

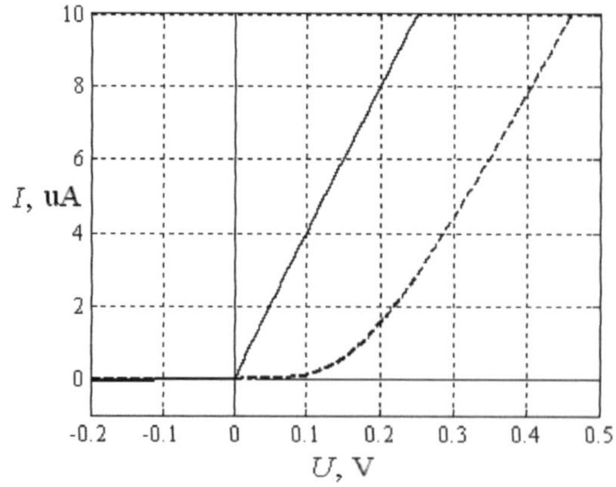

Figure 2. IV-curve of RTD (solid) and Schottky diode (dotted).

In [1],[4]-[6], it is shown that application of resonant-tunneling diodes (RTD) based on A_3B_5 semiconductor heterostructure that have optimized for low power signal rectification IV-curve allows to increase η_R up to 10 times. This IV-curve has zero threshold voltage (see Figure 2, solid curve) and it is described by the system of equations:

$$I(U)=\begin{cases} S_{fw}U & \text{if} \quad U \geq 0; \\ S_{bw}U & \text{if} \quad U < 0 \end{cases} \tag{6}$$

where S_{fw} and S_{bw} – conductivity of RTD in forward and backward direction respectively. Changing heterostructure layers parameters, it is possible to obtain RTD with different S_{fw} and S_{bw}. These values may vary from $1\cdot10^{-7}$ to $1\cdot10^{-1}$ Ohm^{-1}. Thus, the efficiency of low power UHF rectifier based on RTD is close to 100% even if the amplitude of input voltage is less than 0.2 V. Besides, it is possible to design rectifiers with input impedance from several Ohms to dozens of kilo-Ohms. Thanks to this, a perfect impedance matching can be achieved with any type of antenna without additional transformers.

Investigating the problem of antenna and rectifier impedance matching, it is necessary to take into account parasitic capacitance of nonlinear elements. That means that the rectifier impedance is capacitive. Typical values of high-speed Schottky diode or MOS-transistors parasitic capacitance are hundredth and tenth of pico-Farad [2], [3]. The input capacitance of the multiplying rectifier based on such nonlinear elements may achieve ones of pico-Farad. On UHF, it will lead to the antenna and rectifier inputimpedance mismatch because the antenna impedance has purely active component on resonant frequency. Regarding this fact, one should use a matching transformer in passive tag design in this case. Its dimensions may be compatible with dimensions of tag's antenna. In [2], it is shown that the absence of such transformer between antenna and rectifier leads to rectenna efficiency reduction up to 4 times or RFID system range decrease in 1.5-2 times.

RTD based on nanoscale A_3B_5 semiconductor heterostructures has noticeably less parasitic capacitance that can be equal to ones of femto-Farad [7]. Input capacitance of the multiplying rectifier based on RTD should be less than 1 pF even if multiplying factor N is equal to several dozens. At such a value of input capacitance of the rectifier, the need in matching transformer disappears or its dimensions are negligible comparing to antenna dimensions.

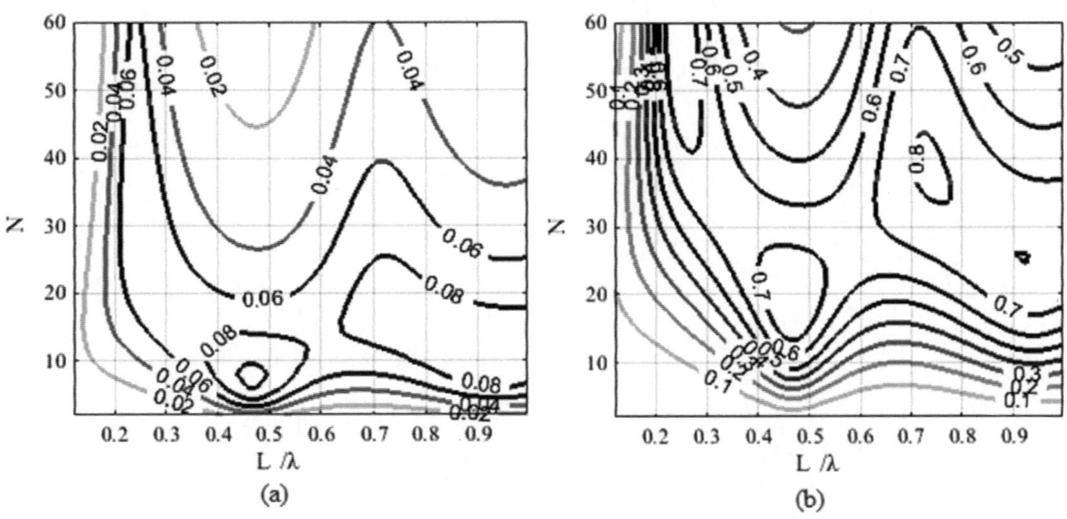

Figure 3. Efficiency coefficient of Shottky diode (a) and RTD (b) rectenna.

In this work, rectennas consisting of electric dipole antenna and rectifiers based on RTD and Schottky barrier diode were investigated. Figure 3 shows the efficiency coefficient as a function of the rectifier multiplying factor N and the length of antenna L normalized to wavelength λ of HF signal with frequency 910 MHz when DC power is 3 μW. Calculation shows that RTD rectenna's efficiency coefficient is up to 10 times greater than Schottky diode rectenna's one when DC power is 3 μW, 4.5 times greater when DC power is 10 μW and 4 times greater when DC power is 20 μW.

3. Conclusion

It is possible to achieve an advantage in rectenna efficiency up to 10 times by means of RTD use instead of Shottky barrier diode in UHF signal rectifier. Thus, the range of RFID system can be increased up to 3 times or radio emission of reader can be decreased up to 10 times that is crucial considering electromagnetic compatibility and ecological situation in the RFID system area of action. Besides, dimensions of a tag can be reduced because of additional impedance matching components absence. These advantages can extend the sphere of RFID application.

Acknowledgments

The research work was supported by Ministry of Education and Science of the Russian Federation under state task № 16.1116.2014/K.

References

[1] Ivanov Yu A, Meshkov S A, Sinyakin V Yu, Fedorenko I A, Fedorkova N V, Fedorov I B, Shashurin V D 2011 *Nanoengineering* **2** 20
[2] Curty JP, Declercq M, Dehollain C, Joehl N2007*Design and Optimization of Passive UHF RFID Systems*(New York:Springer Science+Business Media)
[3] *HSMS-285x Series. Surface Mount Zero Bias Schottky Detector Diodes. Datasheet.* Avago Technologies. Available at: http://www.avagotech.com/docs/AV02-1377EN (accessed 29.03.2016)
[4] Ivanov Yu A, Meshkov S A, Sinyakin V Yu, Fedorenko I A, Fedorkova N V, Shashurin V D 2010 *20th Int. Crimean Conf. CriMiCo'2010* (Sevastopol) p. 859
[5] Ivanov Yu A, Gudkov A G, Agasieva S V, Meshkov S A, Sinyakin V Yu, Makeev M O 2014 *24th Int. Crimean Conf. CriMiCo'2014* (Sevastopol) p. 1063
[6] Gudkov A G, Ivanov Yu A, Meshkov S A, Agasieva S V, Petrov V I, Sinyakin V Yu, Shchukin S I 2015 *Biomedical Engineering* **2** 26
[7] Sollner T CLG, Goodhue W D, Tannenwald PE, Parker C D, Peck DD 1983*Appl. Phys. Lett.* **43** 588

Adaptive behaviour of silicon oxide memristive nanostructures

D S Korolev, A N Mikhaylov, A I Belov, V A Sergeev, I N Antonov, O N Gorshkov and D I Tetelbaum

Lobachevsky University, 23/3 Gagarin prospect, Nizhny Novgorod, 603950, Russia

Abstract. The response to electrical pulses of various parameters has been studied for the CMOS-compatible memristive nanostructures on the basis of silicon oxide demonstrating reproducible resistive switching. It is established that an increase in the amplitude or width of a single programming pulse is followed by the gradual decrease in the device resistivity. By applying periodic pulse sequences of different polarity it is possible to obtain both lower and higher resistance states. This adaptive behavior is analogous to synaptic plasticity and considered as one of the main conditions for the application of memristive devices in neuromorphic systems and synaptic electronics.

1. Introduction

Memristor or a similar memristive device can be defined as a nonlinear resistor with memory and is usually implemented in a simple capacitor-like nanostructure of the 'metal-insulator-metal' (MIM) type. The key advantages of memristive devices are simple structure (the change of electric resistance with applied voltage occurs in thin nanometer film of a material placed between two electrodes), small size (down to 10 nm and smaller) and low power consumption. Because of the ability (by analogy with the biological synapse) to continuously change the resistance in dependence on the input electrical signal, the memristive device is considered as a basis of the next-generation synaptic electronics intended to mimic the adaptive behavior of biological systems [1-2]. The realization of adaptive behavior is based on a gradual character of switching between the different resistive states. It has been already shown that such memristive devices can mimic the plasticity of biological synapses, namely, the short-term and long-term facilitation/depression and spike-timing-dependent plasticity (STDP) [3-5]. Together with the already developed CMOS-integrable neuron-like generators, the memristive devices can be used for the development of artificial networks approaching to the capabilities of brain neural systems.

A challenge is to develop such synaptic electronics on the basis of memristive devices, the technology and materials of which are fully compatible with the traditional CMOS technology. The usage of silicon oxide as a dielectric layer in memristive devices fits organically the technology requirements [6-7]. However, the demonstration of ability to mimic the functions of biological synapse needs special investigation of the device response to the parameters of an individual electrical pulse or their periodic sequences analogous to the bioelectrical activity patterns in a living neural system.

In the present work, the dependence of resistance on the amplitude and width of a single switching pulse as well as on the amplitude, polarity and frequency of the repeated identical pulses is studied for the silicon oxide based memristive nanostructures compatible with the conventional microelectronic technology.

2. Experimental

The capacitor-like MIM nanostructure was deposited on the CMOS-compatible TiN (25 nm) / Ti (25 nm) / SiO_2 (500 nm) / Si substrate by magnetron sputtering using the Torr International® MagSputt™ 3G2 and 2G1-1G2-EB4-TH1 vacuum thin-film deposition systems. The SiO_x film (40 nm) was deposited by RF-magnetron sputtering of fused silica in the argon-oxygen gas mixture (50% oxygen content) at the substrate temperature of 300 °C. Top Au electrodes (40 nm) with the adhesive Zr underlayer (3 nm) and area of $1.2 \cdot 10^{-3}$ cm^{-2} were deposited through the mask by the method of DC-magnetron sputtering in an argon atmosphere and substrate temperature of 200°C. The electrical characteristics in the continuous and pulse regimes were measured by the Agilent B1500A semiconductor device analyzer. The sign of bias on the device corresponds to the potential of top electrode (Au) relative to the grounded bottom electrode (TiN).

3. Results and discussion

To obtain the memristive behavior, the as-deposited MIM devices were subjected to electroforming by applying a negative voltage in the range of 4-8 V. Typical current-voltage (*I-V*) hysteresis of a memristive device after electroforming is shown in Figure 1. The device can switch repeatedly between the two states – high resistance state (HRS) and low resistance state (LRS) that differ in resistance by more than one order of magnitude. The observed bipolar resistive switching is originated from the partial oxidation and recovery (RESET and SET processes, respectively) of conducting filaments grown in silicon oxide film during the electroforming [7].

The measurements of *I-V* characteristics in a sweeping regime show the dependence of current values and switching voltages on the sweeping amplitude (Figure 1). As can be also seen, the SET transition from HRS to LRS occurring at negative voltages is not abrupt and represented by a gradual stepwise change in resistance. This gradual behavior is potentially important from the viewpoint of multi-bit information storage and can provide the continuous spectrum of final resistive states of memristive nanostructures, the transition between which can be controlled by changing the electrical pulse parameters.

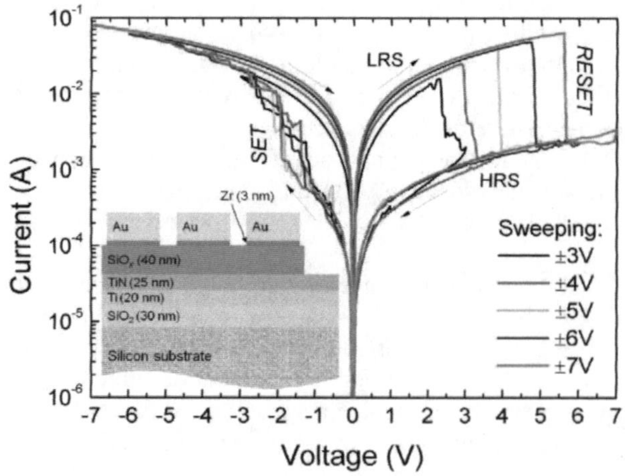

Figure 1. *I-V* characteristics measured in the cyclic sweeping regime with different sweeping amplitudes. The inset shows schematic cross-section of the Au/Zr/SiO_x/TiN/Ti device.

For the establishment of regularities of adaptive behavior of the same memristive devices, their resistance was measured by applying reading voltage pulses of +0.5 V depending on the parameters (amplitude and width) of programming pulses applied in the SET transition region. The results are shown in Figure 2. An increase in the negative pulse width at a constant amplitude of 7 V leads to the gradual decrease in resistance due to the partial SET transition. After each applied pulse memristive structure was switched back to the HRS by applying a long positive pulse (2 s) with amplitude of 7 V. A similar regularity can be observed in the case of increasing voltage amplitude at a constant pulse width (5 ms) – an increase in the pulse amplitude leads to the decrease in the final LRS resistance. Switching back to HRS in this case occurs by applying long positive voltage pulse (2 s) with the amplitude corresponding to that for the SET transition.

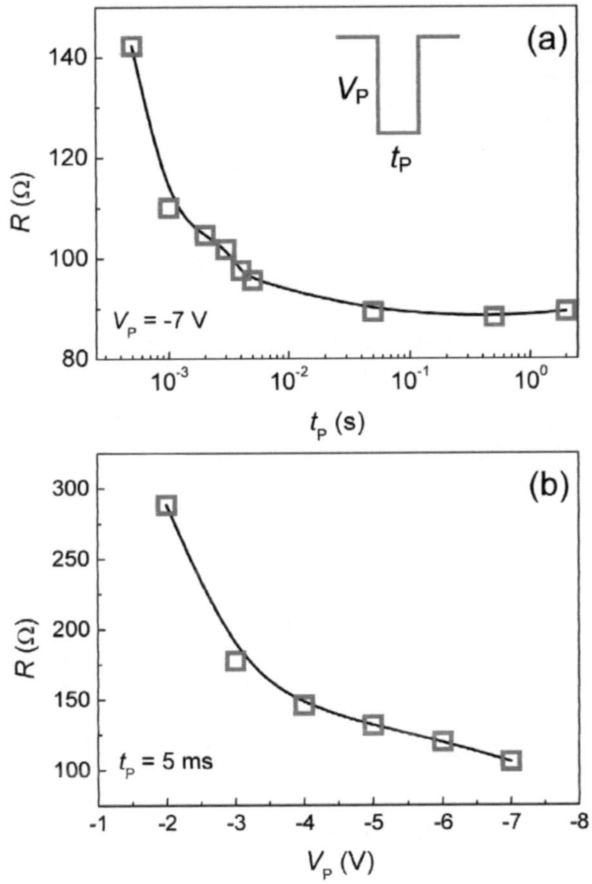

Figure 2. The resistance of memristive device measured at +0.5 V vs. the width (a) and amplitude (b) of a single pulse applied to the device.

The response of memristive device to the voltage pulse amplitude and width can be explained as related to the physical-chemical processes occurring under the recovery of conductive filament in an oxide film [7]. During the application of a low-voltage pulse, the filament recovers only partially, and the degree of the filament ripening depends on the pulse amplitude. The dependence of resistance on the pulse width goes into saturation in the case of long-time pulses. This may be due to the fact that the reduction (SET) process has some finite duration at a fixed voltage.

The fabricated memristive devices were also investigated by applying the sequences of identical voltage pulses with fixed width and amplitude (see typical current response in Figure 3). The ability of device to change the resistance is found to depend on the polarity and frequency of a periodic signal. The observed response of memristive devices is functionally similar to the facilitation and depression properties of a biological synapse.

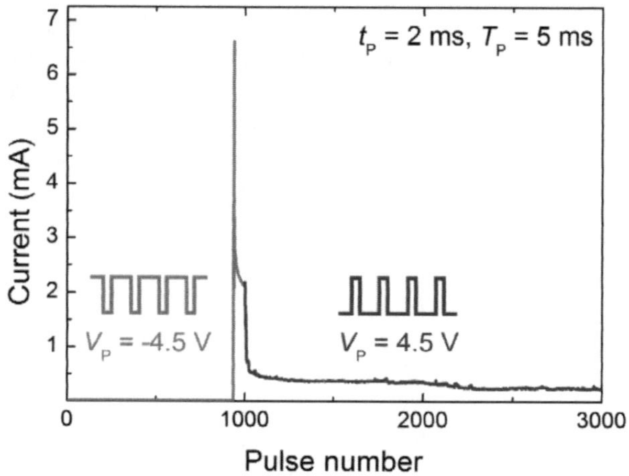

Figure 3. The dependence of current through the memristive device on the number of repeated identical pulses of different polarity.

Further experiments will be focused on the experimental realization of plasticity phenomena using the FitzHugh-Nagumo neuron-like spike generators and modelling the artificial neuronal network based on memristive devices.

4. Conclusions

In summary, the measured dependencies of resistance of SiO_2-based memristive devices on the amplitude and duration of applied voltage pulses show that a continuous spectrum of resistive states is realized and the device resistance depends on the parameters of periodic pulse sequences. The studied memristive devices on the basis of CMOS-integrable materials and technology reveal the synaptic (adaptive) behavior and can be implemented in the development of new-generation artificial neural networks and neuroprocessors.

Acknowledgments
The study is supported by the grant of Russian Science Foundation (project № 16-19-00144).

References

[1] Kuzum D, Yu S and Philip Wong H-S 2013 *Nanotechnology* **24** 382001
[2] Thomas A 2013 *Journal of Physics D: Applied Physics* **46** 93001
[3] Choi S-J, Kim G-B, Lee K, Kim K-H, Yang W-Y, Cho S, Bae H-J, Seo D-S, Kim S-I and Lee K-J 2011 *Applied Physics A* **102** 1019–25
[4] Tsuruoka T, Hasegawa T, Terabe K and Aono M 2012 *Nanotechnology* **23** 435705
[5] Ohno T, Hasegawa T, Tsuruoka T, Terabe K, Gimzewski J K and Aono M 2011 *Nature Materials* **10** 591
[6] Mehonic A, Cueff S, Wojdak M, Hudziak S, Jambois O, Labbé C, Garrido B, Rizk R and Kenyon A J 2012 *Journal of Applied Physics* **111** 74507
[7] Mikhaylov A N, Belov A I, Guseinov D V, Korolev D S, Antonov I N, Efimovykh D V, Tikhov S V, Kasatkin A P, Gorshkov O N, Tetelbaum D I, Gryaznov E G and Yatmanov A P 2015 *Mat. Sci. Eng. B* **194** 48

The influence of signals correlation on the long-term stability of a tandem of quantum magnetometers with laser pumping

M I Fedorov[1], S V Ermak[1], M V Petrenko[2], V V Semenov[1]

[1] Institute of Physics, Nanotechnology and Telecommunications, Peter the Great Saint Petersburg Polytechnic University, St. Petersburg 195251, Russia
[2] Ioffe Institute RAS, St. Petersburg 194021, Russia

E-mail: qfs-lab@mail.ru

Abstract. The results of studies of the long-term frequency stability as a function of the correlation coefficient for a tandem of two quantum magnetometers with laser pumping of ^{87}Rb in wall-coated vapour cell are represented. Measurement scheme includes a low-frequency self-generating magnetometer and a quantum microwave discriminator working at magnetic dipole transitions of radio-optical end state resonance. The difference of synchronously detected signals is processed to determine the Allan variance as a function of averaging time and correlation coefficient of signals. These parameters are essentially dependent both on the pumping light intensity and polarization and the intensity of radio fields that are produced in the working cell.

1. Introduction

The long-term frequency stability of alkali vapor quantum discriminators with optical pumping, which are used in quantum magnetometers and frequency standards, is known to be dependent on technical (flicker) noises caused by slow variations of parameters of such devices (for example, a temperature of the working cell, an alkali vapor pressure, a variation of intensity and spectral composition of pumping light and etc.). The total influence of these noises on the frequency dynamics of self-oscillating device does not allow determining the destabilizing role of different flicker components, which makes it difficult to find new ways of improvement of the metrological of measurement scheme. Nevertheless, numerous studies in this field allow to state that the most important destabilizing factor for such devices are so-called light shifts of the measured frequency, due to the influence of non-resonant components in pumping light spectrum.

2. Frequency stability investigations

The experimental basis of the studies in present work is the magnetic induction meter [1] that consists of a tandem of optically pumped quantum magnetometers (OPQMs), one of which is a low-frequency spin generator; another is a radio-spectrometer with frequency lock loop to the microwave resonance line. The functional scheme of a two-channel system of quantum magnetometers with optical pumping is presented on figure 1.

The external cavity laser Toptica DL-100-L, tuned to the D_2 line of the head doublet of ^{87}Rb atoms, was used as a pumping light source. Figure 2 shows part of the energy spectrum of ^{87}Rb atoms, which corresponds to the D_2 line of the head doublet, where solid arrows in the $S_{1/2}$ ground state indicate magnetic dipole transitions corresponding to the frequency of the self-oscillating magnetometer and dashed arrows indicate magnetic-field-dependent transitions corresponding to the microwave signal of radio-optical resonance at frequency ν_1 or ν_2, depending on the sign of the circular polarization of pumping light ($\sigma-$ versus $\sigma+$).

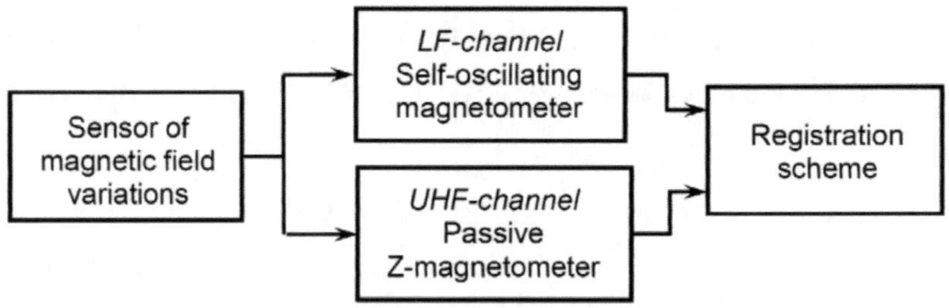

Figure 1. Functional scheme of a two-channel system of quantum magnetometers with optical pumping.

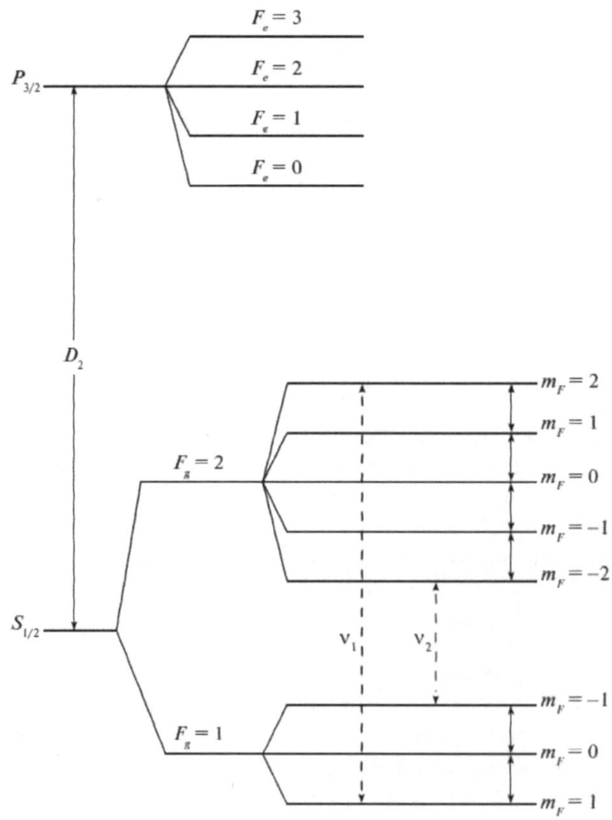

Figure 2. Energy spectrum of alkali metal atoms for the electric-dipole transition of the D_2 line of the head doublet.

The experiments were performed on the laboratory setup of quantum discriminator, based on the set of elements which are commonly used in vapor cell based frequency standards [2]. The laboratory setup is described in detail in [3]. We used the wall-coated 1 cm^3 working cell. The single optical path for the OPQMs tandem was placed in the center of coils, generating the magnetic field with 45° orientation relative to the optical axis. The whole system was placed into two-layer magnetic screen with shielding factor of 10^2 to reduce external magnetic field fluctuations. Resonant microwave field, inducing the transitions in hyperfine structure of rubidium atoms, was created by the traditional scheme of the frequency multiplying of reference quartz oscillator and subsequent amplification of the signal.

The resonance frequency of the self-oscillating magnetometer was determined by a frequency meter with a built-in rubidium frequency standard and processed by a computer, to which the error signal of the microwave channel (proportional to residual variations of magnetic field inside the screen) was also fed, as it was described in [3]. Figure 3 shows the signals from two channels (low-frequency spin generator and UHF radio-spectrometer) and the difference of magnetic field variations detected on the magnetic-field-dependent microwave transition by a self-oscillating magnetometer. The difference of these synchronous variations was reduced to the same scale in magnetic field units and processed to determine the Allan variance depending on the averaging time.

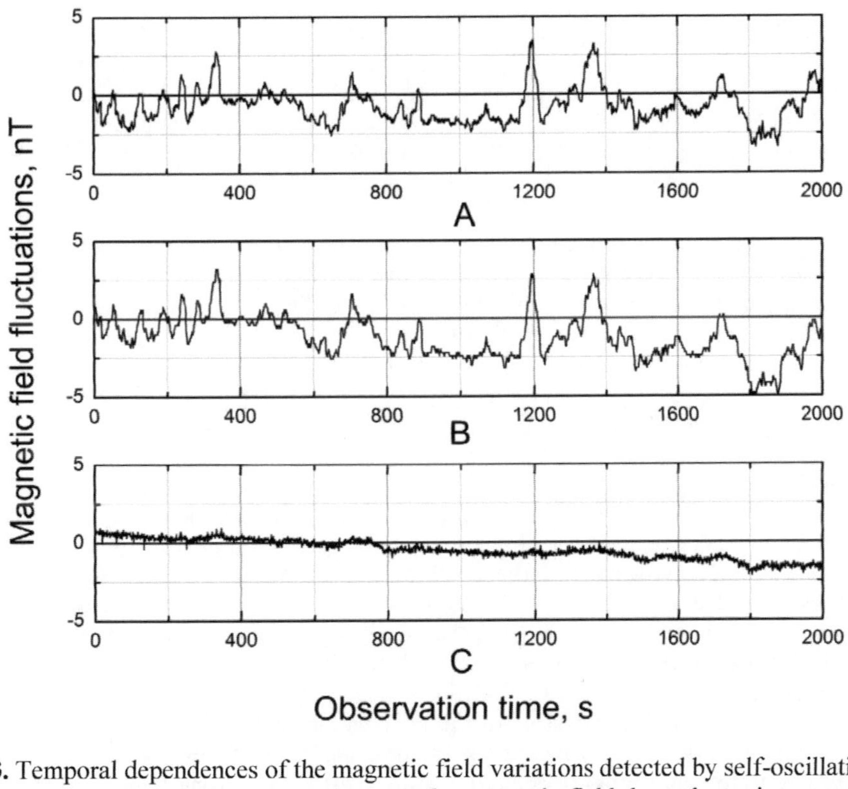

Figure 3. Temporal dependences of the magnetic field variations detected by self-oscillating magnetometer (A) and by radio-spectrometer on the magnetic-field-dependent microwave transition (B). Temporal dependence of the difference of magnetic field variations (C) registered by LF and UHF channels for 0.88 – correlation coefficient, σ+ circular polarization of microwave pumping light.

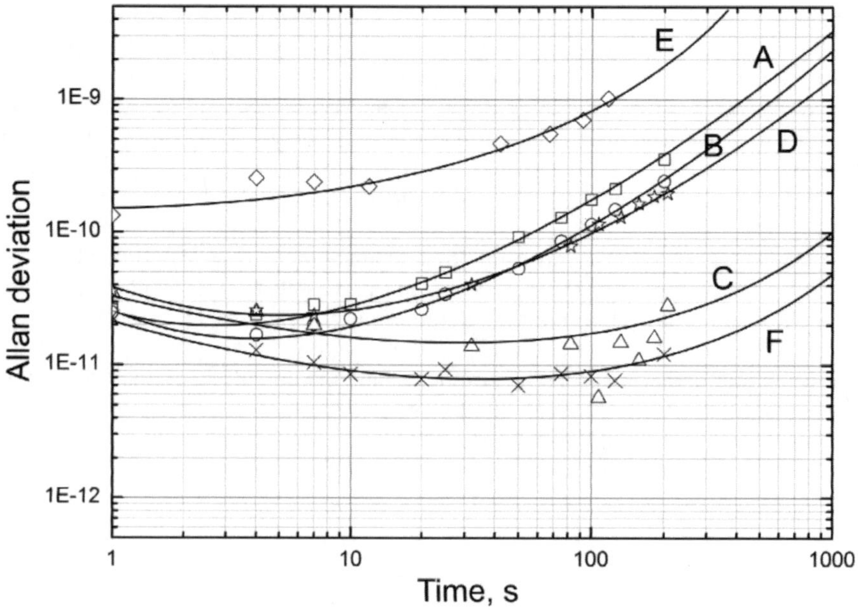

Figure 4. The time dependence of Allan variance for the difference of OPQMs signals. Curves A – F correspond to different signs of the circular polarization of microwave pumping light and various values of correlation coefficient: A – σ+, 0.88; B – σ+, 0.94; C – σ–, 0.99; D – σ–, 0.93; E – σ–, 0.65; F – σ+, 0.99.

During the processing of experimental data there was performed the computation of correlation coefficient, defining the relation between microwave and low-frequency signals, detected in a tandem, which depends on the pumping rate and the intensity of radio fields in the working cell. The value of correlation coefficient was varied in the range between 0.3 and 0.99 depending on the sign of circular polarization of the pumping light, the intensities of applied fields and the power of spin oscillator signal. Figure 4 represents the Allan variance for the difference of frequencies of quantum magnetometers as a function of averaging time.

3. Conclusions

The obtained data lead to the conclusion that mutual compensation of low-frequency flicker noise components in the optically pumped alkali discriminator is possible provided by a correct choice of the direction of circular polarization of the pumping light. This conclusion allows such OPQMs tandem to be considered as an alternative to an atomic clock with a low level of long-term instability. The use of miniature quantum magnetometers based on vertical cavity surface emitted lasers and small-sized alkali vapor cells in OPQMs tandem allows it to be considered as an alternative to miniature atomic clock based on coherent population trapping [4].

Such OPQMs tandem may also be of interest in the development of the magnetic induction meters with laser pumping on D_1 line, where the light shift component, which depends on the magnetic-field orientation relative to the optical axis, becomes significant [5]. The strong light shifts correlation on the microwave and low-frequency transitions, which is implemented in the tandem on the atomic level, gives an opportunity to reduce systematic and orientation errors in quantum magnetometers in conditions of varying intensity and spectral composition of the pump light.

References

[1] Baranov A A, Ermak S V, Sagitov E A, Smolin R V and Semenov V V 2015 *JETP* **121** 393
[2] Riehle F 2004 *Frequency Standards: Basics and Applications* (Weinheim, Germany: Wiley)
[3] Baranov A A, Ermak S V, Sagitov E A, Smolin R V and Semenov V V 2016 *Tech. Phys. Lett.* **42** 186
[4] Ermak S V, Petrenko M V, and Semenov V V 2016 *Tech. Phys. Lett.* **42** 127
[5] Semenov V V 1999 *Russian Physics Journal* **42** 226

Experimental investigation of the electrocaloric response and simulation of solid-state ferroelectric cooler

A V Es'kov[1,2], P Yu Belyavskiy[1,2], A S Anokhin[2], A A Semenov[1], O V Pakhomov[2], I L Mylnikov[1,2], A A Nikitin[1,2], V A Krylov[2]

[1] St. Petersburg Electrotechnical University, St. Petersburg, 197376, Russia
[2] ITMO University, St. Petersburg, 197101, Russia

E-mail: pbeliavskiy@gmail.com

Abstract. The electrocaloric response in ferroelectrics was experimentally investigated in nonequilibrium thermal conditions. The electrocaloric response reached 6 mK per one cycle polarization-depolarization of the ferroelectric sample. Computer modeling based on experimental data demonstrated the cooling capacity of multilayered structure at 25 W/cm^3.

The finding of new cooling principles for the development of compact, ecologically friendly, solid-state heat pumps and coolers with low power consumption, working around room and cryogenic temperatures, are extremely important for modern society. The effective heat transformation could be implemented in the solid state structures using electrocaloric effect [1, 2]. The main technical difficulty standing in the way of developing a small-sized high-performance solid-state coolers on electrocaloric effect, is the need to use heat switches for heat removal process from the cooling object [3-5]. Earlier, in works [6-9], principle and thermodynamic cycle of the electrocaloric cooler without the use of thermal switches have been proposed. The principle of operation is based on the effect of the difference values of electrocaloric effect in ferroelectric materials with polarization and depolarization of the sample [10].

In this paper electrocaloric response in bulk ferroelectric samples based on a solid solution of barium strontium titanate ($Ba_{0.65}Sr_{0.35}TiO_3$) was experimentally investigated. The temperature dependence of the electrocaloric response was investigated for ferroelectric samples placed in special thermal conditions. The plane-parallel ceramic capacitors with a diameter of 10 mm and a thickness of 1 mm were used as a measured samples. The copper electrodes were created by vacuum evaporation. A thickness of electrodes was about 1 μm.

Figure. 1 shows the view of the electric field E time dependence in the plane-parallel ferroelectric capacitor and temperature variation due to the electrocaloric effect. At the initial time the sample is in contact with an environment at the temperature T_0. At time t_1 occurs polarization and temperature of the sample changes from T_0 to T_1 due to electrocaloric effect. The time interval from t_1 to t_2 corresponds to a temperature relaxation due to heat exchange with the environment. The time of changing of electric field in the sample from E_1 to E_2 and from E_2 to E_1 must be much less than the time of thermal relaxation. At time t_2 occurs depolarization of ferroelectric sample and its temperature decrees below T_0. In the time interval from t_2 to t_3 the sample temperature increase occurs due to heat exchange with the environment. Then, the cycle is repeated with a different initial temperatures. The measurement of electrocaloric response was carried out in the thermostatic bath of FP40-MA Refrigerated/Heating Circulator with using of arbitrary waveform generator Agilent 33522A, a high-voltage amplifier Trek 609E-6 and nanovoltmeter Agilent 34420A. Direct measurement of the temperature change due to polarization and depolarization of the ferroelectric sample produced with calibrated miniature platinum thermistor. On the sample were applied voltage pulses with amplitude of 2 kV which led in the sample the electric field 2 V/μm. Duration of periods t_1 - t_2 and t_2 - t_3 were 100 s. The durations of the leading and trailing edge of the pulse were 200 μs. Figure. 2 shows the dependence of the electrocaloric response due to polarization and depolarization of the ceramic ferroelectric sample based on $Ba_{0.65}Sr_{0.35}TiO_3$.

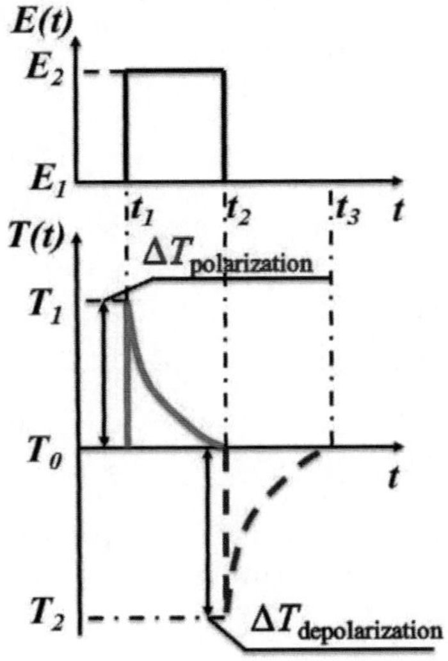

Figure 1. The time dependence of the electric field E in the plane parallel ferroelectric capacitor and temperature changes due electrocaloric effect.

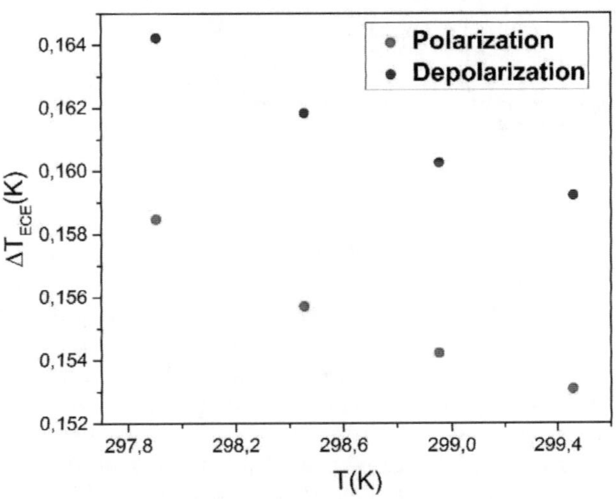

Figure 2. The dependence of the electrocaloric response due to polarization and depolarization of the ceramic ferroelectric sample based on $Ba_{0.65}Sr_{0.35}TiO_3$.

The results of the electrocaloric response experimental measurements have been used for computer simulation of active element of electrocaloric cooler without using of heat switches. The active element of cooler is a MLC capacitor with ferroelectric layers based on $Ba_{0.65}Sr_{0.35}TiO_3$. Figure. 3 shows a simple model of the cooling capacitor. It consists of ten ferroelectric layers separated from each other by nickel electrodes. The thickness of ferroelectric layer is 10 μm, thickness of nickel electrode is 1 μm. We will investigate a one-dimensional model under the assumption that the temperature changes only along the x axis. One boundary ($x = 0$) is thermally insulated, whereas the other boundary ($x = l$) is held under slow convection by air with temperature T_0, with equal to the initial temperature T_0. The heat capacity $C(x)$ and thermal conductivity $\lambda(x)$ are assumed to be constant for each segment of the line according to the materials of layers. In this case, the temperature distribution $T(x, t)$ along the line can be found by solving the heat conduction equation:

$$C(x)\frac{\partial T}{\partial x} = \frac{\partial}{\partial x}\lambda(x)\frac{\partial T}{\partial x} + Q(x,t,T) \tag{1}$$

which satisfies the initial and boundary conditions.

$$T(x,\ 0) = T_0; -\lambda\frac{\partial T}{\partial x}\bigg|_{x=0} = 0,\ \lambda\frac{\partial T}{\partial x}\bigg|_{x=L} = h(T - T_0) \tag{2}$$

where h is the heat transfer coefficient. The function $Q(x, t, T)$ determines the quantity of heat released (or absorbed) by a thermal electrocaloric source and is expressed in the following form:

$$Q(x,t,T) = -T\frac{\partial P}{\partial T}\frac{dE}{dt} \tag{3}$$

where E is the electric field strength of the electrocaloric element and P is the polarization of the ferroelectric. The function $Q(x, t, T)$ is nonzero only within the ferroelectric layers.

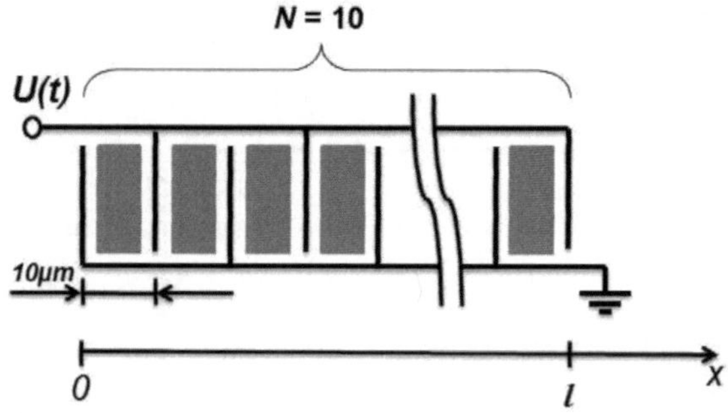

Figure 3. Schematic model of multilayered capacitor with 10 ferroelectric layers.

The initial temperature 298.5 K has been selected for the simulation procedure, in this case the temperature variation due to polarization and depolarization was about 0.006 K. Periodic square pulses of the electric field were supplied to the ferroelectric elements with different frequencies. The pulses were cause periodic heating (cooling) of the electrocaloric elements and a redistribution of the temperature field along the structure. After a series of cycles of the polarization and depolarization of the solid—state structure, we determined the temperature distribution on the free end of the cooling line ($x = 0$). We consider two different situation. In the first case, we have considered a simulation with in-phase pulses supplied to the ferroelectric elements. Figure 4 presents the results of the computer simulation of the temperature variation for in-phase pulses for all layers with frequency 1 kHz. In the second case, we have considered a simulation with pulses shifted by a half period for odd and even layers. Figure. 5 shows the schematic diagram of a such connection. Figure 6 presents the results of the computer simulation of the temperature variation on the free end of the cooling line ($x = 0$) for out-of-phase pulses with frequencies 500 Hz, 1k Hz and 2 kHz. Volumetric cooling capacity in case of phase shift was about 6 W/cm^3 for frequency 500 Hz, 9 W/cm^3 for 1 kHz and and correspondingly 25 W/cm^3 for 2 kHz. Thus, the cooler may show considerable cooling capacity when used the multilayer elements with thin layers and high-frequency out-of-phase pulses, even if there is a small difference between temperature changes due to polarization and depolarization.

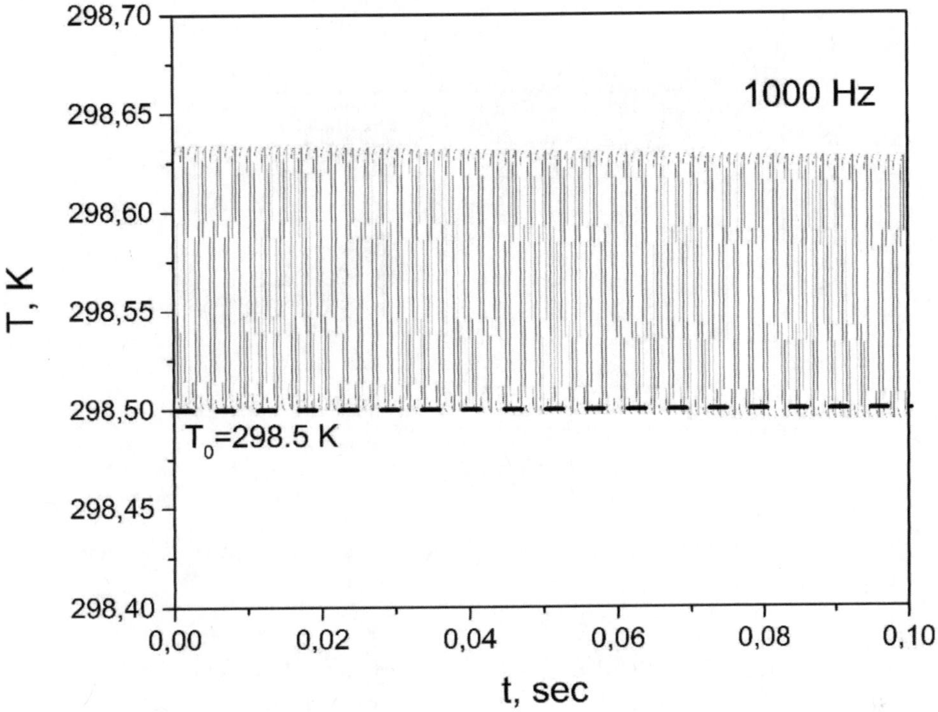

Figure 4. Dynamics of change in the temperature at the left end ($x = 0$) of the cooling elementin the case of in-phase square pulses.

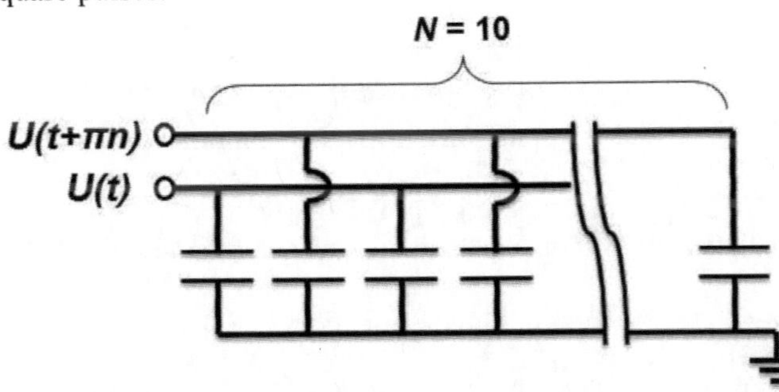

Figure 5. Schematic diagram of electric connection of layers in the case of out of phase pulses.

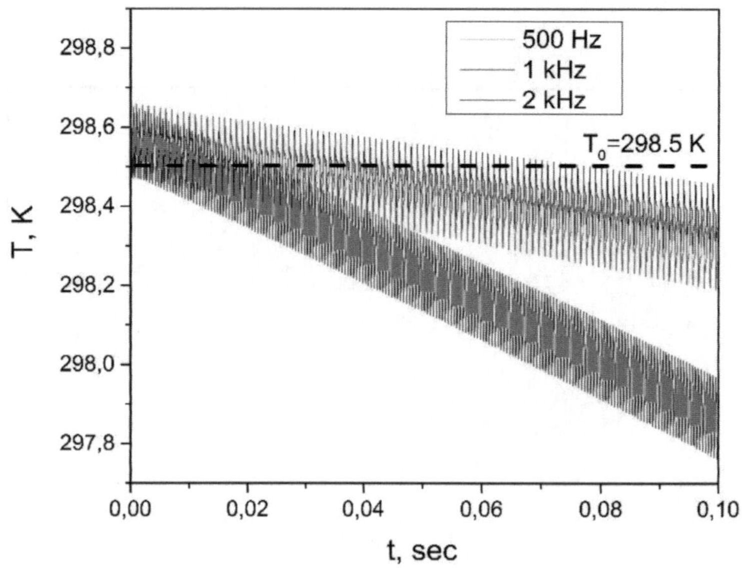

Figure 6. Dynamics of change in the temperature at the left end ($x = 0$) of the cooling element in the case of out of phase square pulses.

Acknowledgements
The samples were prepared at the St. Petersburg Electrotechnical University with support of the Russian Science Foundation (Grant 14-12-01296). The measurements and investigations were supported in part by grant of the President of the Russian Federation for young scientists and PhDs MK- 6229.2015.8, the Government of the Russian Federation (Grant 074-U01), and the Russian Foundation of Basic Research (project № 16-32-00715 mol_a).

References

[1] Valant M 2012 *Progress in Materials Science* **57**(6) 980-1009
[2] Ožbolt M, Kitanovski A, Tušek J and Poredoš A 2014 International journal of refrigeration 40 174-188
[3] Jia Y and Ju Y S 2012 Appl. Phys. Lett. 100(24) 242901
[4] Gu H, Qian X, Li X, Craven B, Zhu W, Cheng A and Zhang Q M 2013 Appl. Phys. Lett. 102(12) 122904
[5] Guo D, Gao J, Yu Y J, Santhanam S, Slippey A, Fedder G K and Yao S C 2014 Int. J. Heat. Mass. Tran. 72 559-564
[6] Karmanenko S F, Pakhomov O V, Prudan A M, Starkov A S and Es'kov A V 2007 J. Europ. Ceramic Soc. 27 3109
[7] Es'kov A V, Karmanenko S F, Pakhomov O V and Starkov A S 2009 Phys. of the Solid State 51(7) 1574-1577
[8] Starkov A S, Karmanenko S F, Pakhomov O V, Es'kov A V, Semikin D and Hagberg J 2009 Phys. of the Solid State 51(8) 1510-1514
[9] Pakhomov O V, Karmanenko S F, Semenov A A, Starkov A S and Es'kov A V 2010 Technical Physics 55(8) 1155-1160
[10] Karmanenko S, Semenov A, Dedyk A, Es'kov A, Ivanov A, Beliavskiy and Pakhomov O 2014 Springer 34 183-223

InAlN/AlN/GaN heterostructures for high electron mobility transistors

S O Usov[1,2], A V Sakharov[1,2], A F Tsatsulnikov[1,2], V W Lundin[1,2], E E Zavarin[1,2], A E Nikolaev[1,2], M A Yagovkina[1,2], V E Zemlyakov[3], V I Egorkin[3], V M Ustinov[1,2]

[1]Submicron Heterostructures for Microelectronics, Research & Engineering Center, RAS, 26 Politechnicheskaya, Saint Petersburg 194021, Russia
[2]Ioffe Institute, 26 Politechnicheskaya, Saint Petersburg 194021, Russia
[3]National Research University of Electronic Technology - MIET, Shokin Square, Bld. 1, Zelenograd, Moscow 124498, Russia

Abstract. The results of development of InAlN/AlN/GaN heterostructures, grown on sapphire substrates by metal-organic chemical vapour deposition, and high electron mobility transistors (HEMTs) based on them are presented. The dependencies of the InAlN/AlN/GaN heterostructure properties on epitaxial growth conditions were investigated. The optimal indium content and InAlN barrier layer thicknesses of the heterostructures for HEMTs were determined. The possibility to improve the characteristics of HEMTs by *in-situ* passivation by Si_3N_4 thin protective layer deposited in the same epitaxial process was demonstrated. The InAlN/AlN/GaN heterostructure grown on sapphire substrate with diameter of 100 mm were obtained with sufficiently uniform distribution of sheet resistance. The HEMTs with saturation current of 1600 mA/mm and transconductance of 230 mS/mm are demonstrated.

1. Introduction

Currently the high electron mobility transistors (HEMTs) based on InAlGaN material system are most promising for production of electronic components for high-power amplifiers operating at high frequencies in extreme conditions at high temperatures, in the near-Earth space environment, at exposure to ionizing radiation [1]. High values of the band gaps, conduction band discontinuities at heterojunctions of AlGaN/GaN or InAlN/GaN, high electron saturation velocities and breakdown voltages and other features of the InAlGaN material system allow to produce HEMTs with the parameters exceeding achievable for transistors based on III-As or Si material systems.

The lattice-mismatched heterostructures AlGaN/GaN are traditionally used for HEMTs. Two dimensional channel is formed in the heterostructure on the interface between AlGaN barrier layer and GaN buffer layer both by difference in the coefficients of spontaneous polarization and piezoelectric polarization caused by strain. The increase of the mobility and charge carrier concentration in HEMT AlGaN/GaN heterostructures to achieve maximal output characteristics requires the increase of the Al content and variation of the barrier layer thickness. The increase of the Al content and AlGaN layer thickness leads to increase of the elastic strain, which reduces the material quality and in turn causes deterioration of the transistors parameters and lifetime.

The implementation of InAlN layer as a barrier in HEMT structures instead of AlGaN allows to control the strain in the InAlN/GaN heterostructures. Since $In_xAl_{1-x}N$ layer with indium content of about 18 % is lattice matched to GaN it allows to realize strain-free InAlN/GaN heterostructures [2, 3]. The formation of the two-dimensional channel at the interface is caused not only by difference of band gap energies, but also by electric field induced by difference of the coefficients of spontaneous polarization and piezoelectric polarization caused by strain in lattice mismatched case. The variation of indium content below/above 18 % allows to obtain tensile/compressive strained barrier layers which in turn permits to control charge carrier concentration.

It should be noted that the InAlN alloy is the most technologically complicated material among the III-nitrides due to difference in the optimal growth conditions for the AlN (growth temperature over 1100 °C) and InN (~600 °C) which is related to the low mobility of Al adatoms and desorption of In atoms. The optimization of the relatively thick epitaxial layer growth regimes, carried out previously [4], allows us to determine growth parameters and to obtain high-quality InAlN layers, which permits to produce InAlN/GaN distributed Bragg reflectors with reflectivity more than 99 % in a wide range from blue (460 nm) to red (610 nm) spectral regions [5]. Implementation of the InAlN layers for modification of the active region based on InGaN/GaN quantum wells allows to expand the range of emission of light emitting diodes [6].

2. Experiment

The InAlN/AlN/GaN heterostructures were grown on (0001) oriented sapphire substrates by metal-organic chemical vapour deposition (MOCVD) in Dragon-125 setup with horizontal reactor equipped with induction-heated graphite susceptor [7]. Nitrogen, hydrogen and nitrogen-hydrogen mixtures were used as carrier gases and ammonia, trimethylgallium (TMGa), trimethylindium (TMIn), trimethylaluminum (TMAl) and monosilane (SiH_4) - as precursors.

The heterostructures (Table 1) consist of GaN nucleation layer, GaN insulating buffer layer of 2 μm thick, AlN layer of ~1 nm thick and undoped $In_xAl_{1-x}N$ barrier layer with varied thickness and indium content. InAlN layers were grown under reduced reactor pressure of 200 mbar without implementation of the intentional doping by silicon. A detailed description of the epitaxial growth conditions of the InAlN and GaN layers by MOCVD method was described previously in the studies [4 - 6]. The use of the nanometer-thick AlN layer at the interface allows to increase the mobility of electrons in the channel by improving the interface quality and by increasing of the charge carriers wave function localization in the channel region [8, 9]. The growth conditions of the GaN and AlN layers for all the heterostructures were selected the same and growth temperature for InAlN varied (Table 1) from 820°C (structure H1) to 880°C (structure H5).

Table 1. Parameters of the $In_xAl_{1-x}N/AlN/GaN$ heterostructures.

Structure	T_{gr}, °C	d_{InAlN}, nm	T= 300 K		
			σ, $(\text{Ohm cm})^{-1}$	μ, $cm^2/(V\ s)$	n_s, 10^{13} cm^{-2}
H1	820	19	24.3	746	2.03
H2	835	18	30.6	837	2.28
H3	850	17	36.5	747	3.05
H4	865	16	30	559	3.35
H5	880	15	20.3	372	3.41
H6	850	12	36.6	795	2.88
H7	850	12	38	840	2.82

The structural properties of the grown samples were investigated by the scanning electron microscopy and high-resolution X-ray diffraction (HRXRD) using Bruker D8 Discover (Bruker AXS,

Germany) diffractometer. The ω-2θ rocking curves for symmetric Bragg reflection (0002) were analyzed [10]. Approximation of experimental data was carried out using LEPTOS software [10]. Measurements of the electrical parameters of conductivity σ, charge carrier concentration n_s and carrier drift mobility μ in the channel were performed by Hall effect measurements by van der Pauw method. The sheet resistance measurements and maps of resistance distribution on the wafer were carried out by LEI 1510A SA (Lehighton Electronics) contactless measurement system. Structural and electrical parameters of the heterostructures versus growth temperature (T_{gr}) of $In_xAl_{1-x}N$ are summarized in Table 1.

3. Results and Discussion

3.1. *InAlN/AlN/GaN heterostructures for high electron mobility transistors*

Set of samples was grown on 2 inch wafers with variation of growth temperature of the $In_xAl_{1-x}N$ barrier layer from 820 to 880 °C (structures H1 - H5), maintaining the same growth time. It was observed that the increase of growth temperature from 820° to 880° C leads to simultaneous reduction of thickness from 20 to 15 nm and to nearly twice decrease of indium content from 18 % to 9 % (Figure 1 (a)). No evidence of phase separation in the InAlN layers was revealed for all investigated heterostructures from the high-resolution X-ray diffraction studies as shown in Figure 1 (b). The comparison of the variation of the InAlN and partial AlN thicknesses, calculated as the product of layer thickness on Al content, with increasing of T_{gr} shows that the reduction of the InAlN layer thickness occurs mostly due to decrease of the indium incorporation.

The highest values of conductivity 36.5 $(\text{Ohm cm})^{-1}$ and mobility of 837 $cm^2/(V\ s)$ were observed for heterostructures with InAlN layer grown at temperatures of 835 (H2) and 850 °C (H3) with indium content of about 14-16 %. Increase of the barrier layer growth temperature up to 880 °C (H5) leads to significant increase of the charge carrier concentration in the channel and to decrease of the mobility more than twice to 372 $cm^2/(V\ s)$, resulting in decrease of conductivity. Despite being strained at room temperature the $In_xAl_{1-x}N$ layers of the heterostructures H2 and H3 contain less defects, originating during epitaxy due to higher temperature and during cooling due to the difference of thermal expansion coefficients of the $In_xAl_{1-x}N$, AlN and GaN.

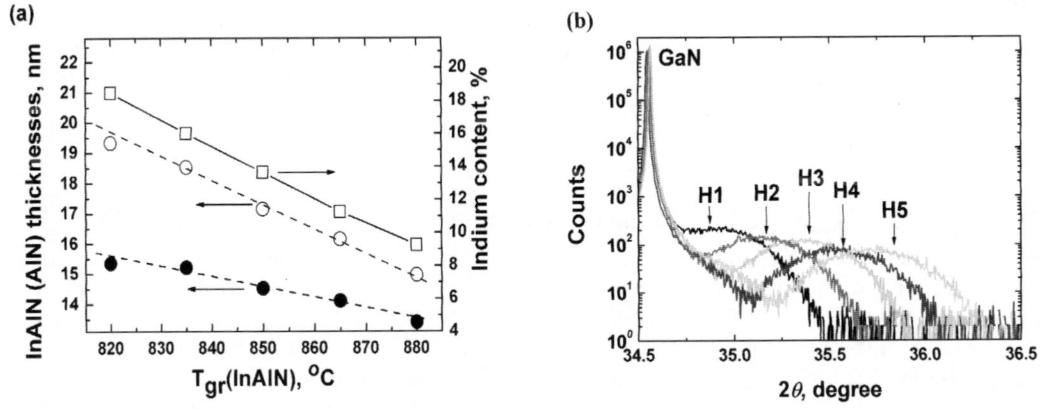

Figure 1. HRXRD rocking curves (a), InAlN barrier layer (open circles) and AlN layer (solid circles) thicknesses and indium content (open squares) (b) of the InAlN/AlN/GaN heterostructures with barrier layer grown at different temperatures from 820° (structure H1) to 880° C (H5).

Hence, $In_xAl_{1-x}N$ barrier layers with indium content of about 14-16 % are optimal for HEMT based on InAlN/AlN/GaN heterostructures. The structure H6, grown at T_{gr} = 850 °C with reduced InAlN barrier layer thickness (12 nm), demonstrates characteristics similar to those of H3 structure. So the variation of InAlN barrier layer thickness from 12 to 20 nm does not significantly affect the performance of InAlN/AlN/GaN HEMT heterostructures.

To study the effect of surface passivation, a pair of identical structures were grown with *in-situ* deposited 5 nm thick Si_3N_4 layer (H7) and without one (H6). The protective passivation layer was grown *in-situ* in the same epitaxial process using ammonia and SiH_4 after the deposition of the $In_xAl_{1-x}N$ barrier layer [11, 12]. The values of charge carrier concentration obtained for the structures H6 and H7 were found to be very close, some variation may be due to surface modification of the InAlN layer or by the addition strain caused by deposition of Si_3N_4 layer.

In order to analyze the possibility of growth of InAlN/AlN/GaN heterostructures on larger diameter substrates, the structure with identical to H7 growth regimes was grown on (0001) oriented sapphire substrate with diameter of 100 mm. The distribution of sheet resistance R_s over the wafer is shown on Figure 2. The sheet resistance R_s measured in different regions of the wafer is found to be uniform and varied in the range from 211 to 216 Ohms/square with mean value of 213 ± 1.5 Ohms/square.

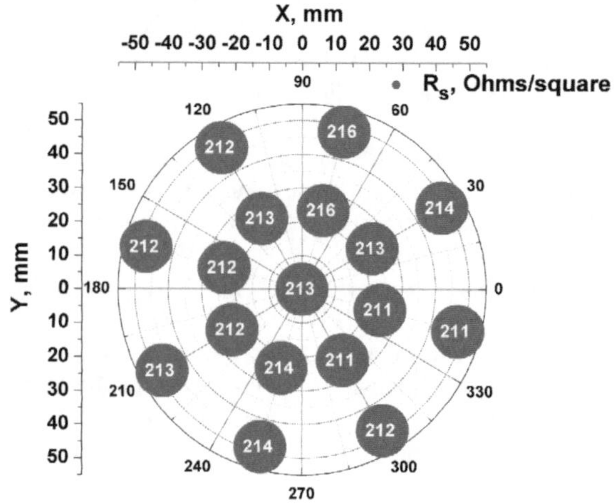

Figure 2. Distribution map of sheet resistance R_s over the wafer of the InAlN/AlN/GaN heterostructure, grown on sapphire substrate with diameter of 100 mm.

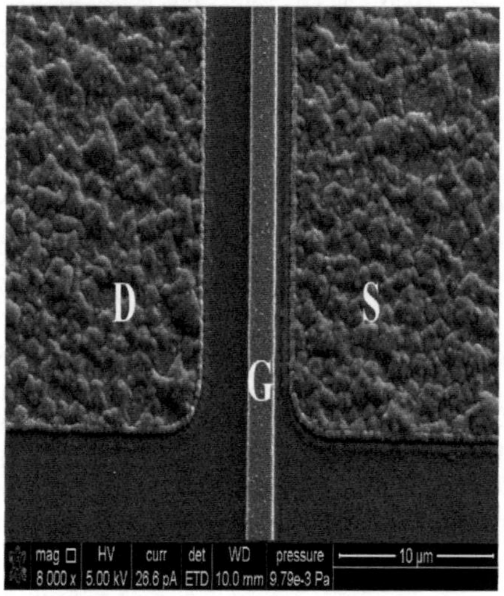

Figure 3. Scanning electron microscopy image of the surface of HEMT test structure with drain (D), gate (G) and source (S) regions indicated.

3.2. InAlN/AlN/GaN transistor test structures

Heterostructures with the highest electrical parameters (H2, H6 and H7) were processed using optical photolithography (Figure 3). Standard metallization sequence of Ti/Al/Ni/Au was used for ohmic and Ni/Au for barrier contacts. Gate length (L) and gate width (W) were 1 and 90 microns, respectively. The gates of the transistors were passivated by Si_3N_4 layer, deposited by plasma-enhanced chemical vapor deposition method. In order to increase the drain saturation current (I_{dss}), transconductance (g_m), and breakdown voltage (U_{br}) of HEMT structures the distance between the drain (D) and the gate (G) were selected larger (3.5 μm) than the distance (1.5 μm) between the gate and the source (S) [13].

The output characteristics of the test transistor structures T1 (Figure 4 (a)), T2 (Figure 4 (b)) and T3 (Figure 4 (c)) are shown in Figure 4. The parameter determined from the analysis of the current-voltage characteristics are summarized in Table 2. Transistor structures are opened at zero gate-to-source voltage (U_{gs}) and almost open at $U_{gs}>0$ V. The increase of the negative U_{gs} leads to decrease of the drain-to-source current (I_{ds}) until the total cut-off of the transistor channel.

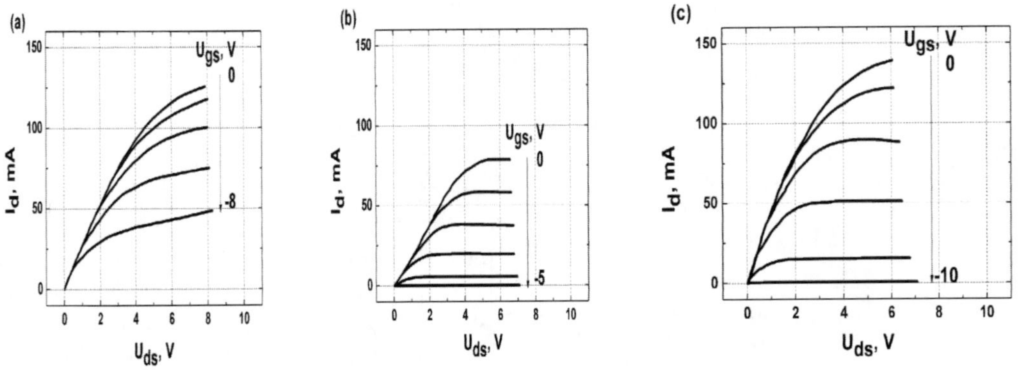

Figure 4. The output current-voltage characteristics drain current versus drain-to-source voltage of the transistor structures T1 (a), T2 (b) and T3 (c).

It should be noted that the channel of the structure T1 does not close completely at $U_{gs}<0$ possibly due to inhomogeneous indium distribution in the InAlN barrier layer. Barrier heights φ_b, determined from the analysis of the current-voltage characteristics of the gate current (I_{gs}) versus the gate-to-source voltage (U_{gs}), were 1.7, 1.5 and 2.5 eV for the structures T1, T2 and T3, respectively. The breakdown voltages (U_{br}) were determined from the forward source-gate and reverse gate-drain diode characteristics (Table 2).

Table 2. Parameters of the HEMT transistors based on $In_xAl_{1-x}N/AlN/GaN$ heterostructures.

Transistor structure	Structure	d_{Si3N4}, nm	d_{InAlN}, nm	φ_b, eV	U_{br}, V	I_{dss}, A/mm	g_m, mS/mm
T1	H2	-	16	1.7	10	1.35	100
T2	H6	-	12	1.5	40	1.2	230
T3	H7	5	12	2.5	80	1.6	217

The transconductance (g_m) and saturation current (I_{dss}) for the HEMT structure T1 are 100 mS/mm and 1350 mA/mm, respectively. The comparison of the HEMT structures T2 and T3 allows to reveal that the structure T3, grown with *in-situ* Si_3N_4 passivation layer, shows higher electron mobility than structure without Si_3N_4 layer with nearly the same values of carrier concentration. Also despite the high barrier height of 2.5 eV structure T3 demonstrates better output characteristics and maximal saturation current reaching 1600 mA/mm (Figure 5). Breakdown voltage U_{br} significantly increases from ~40 V to ~80 V. The maximal value of g_m slightly decreases from 230 to 217 mS/mm. As shown in Figure 5, cut-off voltage and the region of the maximal transconductance shifts towards the negative U_{gs}. The fact that deposition of the ~5 nm thick *in-situ* Si_3N_4 layer leads to increase of the distance between the 2D channel and the gate and to passivation of the InAlN/AlN/GaN heterostructure surface, which reduces the density of surface states.

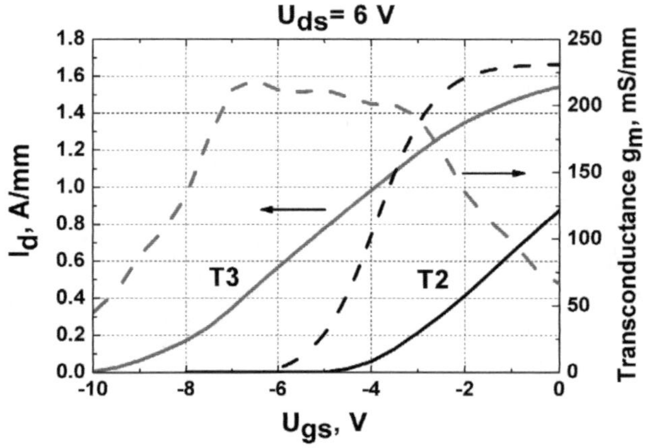

Figure 5. Dependencies of the drain current (I_d) (solid line) and transconductance (g_m) (dashed line) versus gate-to-source voltage (U_{gs}) of the InAlN/AlN/GaN transistor structures, grown with (T3) and without (T2) passivation layer.

4. Conclusions

The optimal design of the InAlN/AlN/GaN heterostructures: indium content and InAlN barrier layer thicknesses were determined. InAlN/AlN/GaN heterostructure grown on sapphire substrate with diameter of 100 mm were obtained with very high uniformity of sheet resistance distribution with 3 % standard deviation. The InAlN/AlN/GaN heterostructures with high electron mobility up to 837 $cm^2/(V s)$ and high charge carrier concentration up to 3.05 10^{13} cm^{-2} were used for HEMT fabrication. The optimal design for InAlN/AlN/GaN heterostructures and HEMTs was found to be slightly different. High electron mobility transistors with transconductance over 200 mS/mm were demonstrated. Transistor structure, grown with *in-situ* Si_3N_4 passivation layer, demonstrates improved output characteristics and maximal saturation current of 1600 mA/mm.

Acknowledgments

This study was supported by the Ministry of Education and Science of the Russian Federation (agreement on grant no. 14.607.21.0003, 05.06.2014, the unique project identifier is RFMEFI60714X0003). XRD structural characterization was made on the equipment of the Federal Joint Research Centre «Material science and characterization in advanced technology» (Ioffe Institute, St.Petersburg, Russia).

References

[1] Kuzmik J 2001 *Electron Device Letters* **22** 510-512
[2] Neuburger M, et al. 2004 *Int. J. Hi. Spe. Ele. Syst.* **14** 785-790
[3] Vurgaftman I, Meyer J R, Ram-Mohan L R 2001 *J. Appl. Phys.* **89** 5815-75
[4] Lundin W V, Sakharov A V, Tsatsulnikov A F, Ustinov V M 2011 *Semicond. Sci. Tech.* **26** 014039
[5] Usov S O, et al. 2010 *Semiconductors* **44** 949-953
[6] TsatsulnikovA F, et al. 2012 *Semiconductors* **46** 1281-85
[7] Lundin W V, et al. 2015 Technical Physics Letters **41** 213-216
[8] Lundin W V, Nikolaev A E, Sakharov A V, Brunkov P N, Zavarin E E, Tsatsulnikov A F 2010 Technical Physics Letters **36** 1133-35
[9] Fang T, Tahy K, Jena D 2009 *Proc. Wide Bandgap Device Physics (*Notre Dame*, ND)* Electrical Engineering **EE 87024**
[10] Bruker Corporation: D8 DISCOVER - Overview https://www.bruker.com/products/x-ray-diffraction-and-elemental-analysis/x-ray-diffraction/d8-discover/overview.html
[11] Derluyn J, et al. 2005 *J. Appl. Phys.* **98** 054501
[12] Yunin P A, Drozdov Yu N, Drozdov M N, Korolev S A, Okhapkin A I, Khrykin O. I, Shashkin V I 2015 *Semiconductors* **49** 1421-24
[13] Russo S, Di Carlo A 2005 *arXiv:cond-mat/0510049*

Journal of Physics: Conference Series **741** (2016) 012165 doi:10.1088/1742-6596/741/1/012165

Attojoule energy resolution of direct detector based on hot electron bolometer

S V Seliverstov[1], A A Rusova[1], N S Kaurova[1], B M Voronov[1] and G N Goltsman[1,2]

[1]Physics Department, Moscow State Pedagogical University, 29 Malaya Pirogovskaya St.,Moscow, 119992, Russia

[2]Moscow Institute of Electronics and Mathematics, National Research University Higher School of Economics, 34 Tallinskaja St., Moscow, 123458, Russia

Abstract. We characterize superconducting antenna-coupled NbN hot-electron bolometer (HEB) for direct detection of THz radiation operating at a temperature of 9.0 K. At signal frequency of 2.5 THz, the measured value of the optical noise equivalent power is 2.0×10^{-13} W·Hz$^{-0.5}$. The estimated value of the energy resolution is about 1.5 aJ. This value was confirmed in the experiment with pulsed 1.55-μm laser employed as a radiation source. The directly measured detector energy resolution is 2 aJ. The obtained risetime of pulses from the detector is 130 ps. This value was determined by the properties of the RF line. These characteristics make our detector a device-of-choice for a number of practical applications associated with detection of short THz pulses.

1. Introduction

Terahertz (THz) frequency range is quite attractive for a number of important practical applications, such as security systems [1] and medicine [2]. Unfortunately, employing of THz radiation in real practical applications associates with a number of difficulties related to the CW sources of THz radiation such as quantum-cascade lasers, gas-discharge lasers, frequency multipliers based on semiconductor superlattices and others. These sources are complex and not quite stable in operation in comparison with the pulsed sources.

For this reason, the design of devices based on effective methods of generating pulsed THz radiation is the most promising way of further development in this area. The optical rectification [3] and deference-frequency generation in nonlinear optical crystals [4] are among these methods. Pulsed THz sources are more reliable, so they are good candidates for using in the practical THz systems. At the same time, these systems require fast and sensitive detectors of THz radiation. That would provide for obtaining THz pulses without distortion of their characteristics.

Both features are combined in a direct detector based on the HEB that demonstrates a relatively good sensitivity with a short response time and operates at convenient liquid-helium temperatures. The HEB sensitive element is integrated with the logarithmic-spiral antenna for better coupling of the input radiation [5]. The measured value of the optical noise equivalent power of the superconducting NbN HEB direct detector is 2.0×10^{-13} W·Hz$^{-0.5}$ at 2.5 THz [6]. The time constant can be calculated from the measurements of the HEB gain bandwidth established at the critical temperature [7]. The corresponding value is about 50 ps. The calculated value of the energy resolution is about 1.5 aJ.

This value was confirmed in additional experiment. In this paper, we report on the direct measurements of the HEB energy resolution at 1.55-μm wavelength. The obtained value of the energy resolution is also consistent with the THz range, because of the frequency independence of detector response in a wide range.

2. Device and setup

The HEB detectors were made from NbN films with thickness of about 4 nm on quartz substrates. We used both side polished substrates with roughness much less that the thickness of the NbN layer. Typical value of the sheet resistance was about 500 Ω. The critical temperature was about 9 K. The NbN deposition was followed by the *in situ* deposition of a 30 nm Au layer. Another Au layer on top formed a contact pads. The detectors were fabricated using DC magnetron sputtering, ion milling, lift-off patterning, and ion and chemical etching techniques. More details about the fabrication process one can find elsewhere [5].

The calculated value of the critical current density at 4.2 K was about $5 \cdot 10^6$ A·cm^{-2}. The detector length-to-width ratio was 0.1 for better rf matching between the bolometer and the 50 Ω of employed coplanar line. Figure 1 shows the optical and atomic-force microscope (AFM) images of the bolometer sensitive element. In order to obtain the AFM photos, the upper dielectric SiO_2 layer was removed using the chemical etching, and thus the quartz substrate was also partially etched during this process.

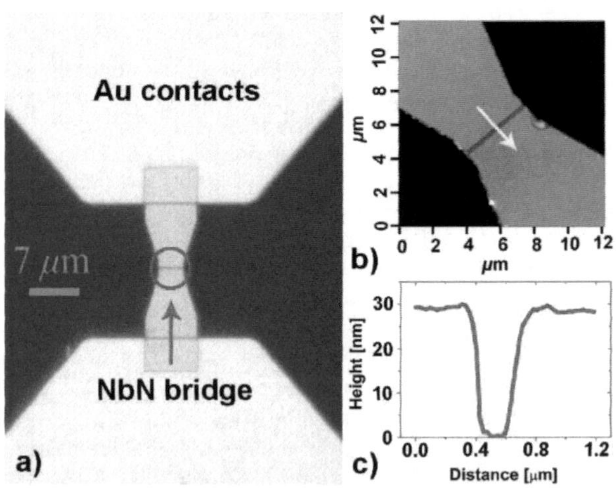

Figure 1. Photos of the central path of the bolometer: a – obtained from the optical microscope, b – obtained from the atomic-force microscope, c – the profile of the bolometer sensitive element along the direction marked with the yellow arrow.

Figure 2 shows the schematic of the experimental setup. Pulsed 1.55-μm laser with pulse duration of less than 50 ps and repetition rate of 10 MHz was employed as a radiation source. The HEB was wire-bonded to a microstrip transmission line, and connected to a dc bias source, and low-noise amplifier (LNA) via the broadband bias-tee. The LNA gain was about 30 dB through the bandwidth of 0.2-4 GHz. The whole HEB arrangement was placed on special holder on a deepstick and maintained at the temperature close to the temperature of liquid-helium.

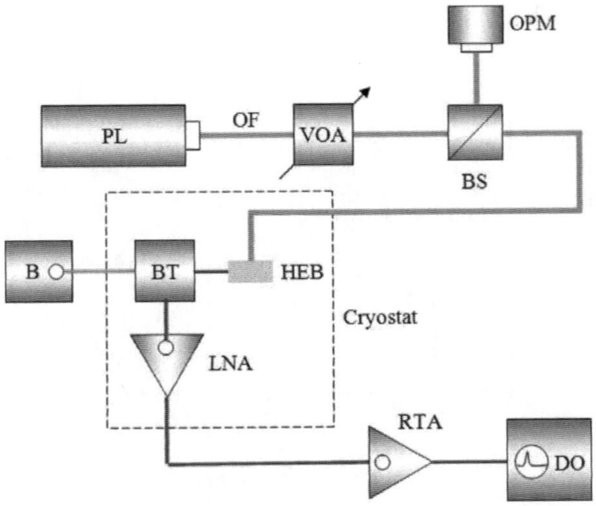

Figure 2. The schematic representation of the experimental setup. PL – 1.55-μm laser with pulse duration of less than 50 ps, OF – optical fiber, VOA – variable optical attenuator, BS – beam splitter, OPM – optical power meter, B – bias source, BT – bias-tee, LAN – low-noise amplifier with the bandwidth of 0.2-4.0 GHz, RTA – room-temperature amplifier with a bandwidth of 0.7-18.0 GHz, DO – digital oscilloscope with the upper frequency limit of 4.0 GHz.

A signal from LNA was transmitted to the room-temperature amplifier with the gain of 24 dB across the bandwidth of 0.7-18 GHz. The amplified signal was measured via the digital oscilloscope with the upper cut-off frequency of 4 GHz. We used the optical fiber to couple the input power to the bolometer. The power adjustment of the incident radiation was carried out using the optical tuneable attenuator. The laser beam was divided into two optical fibers, one of which was connected to the optical power meter. The radiation power in both fibers was approximately the same. The polarization control was carried out via special polarization rotator.

3. Experimental results and discussion

Temperature dependence of the detector resistance together with the current-voltage characteristic at the operating temperature is shown in the figure 3. There are two superconducting transitions. First one at T_{c1} = 8.3 K corresponds to the transition of the detector sensitive element. The second one at T_{c1} = 6.5 K occurs due to the proximity effect in the NbN film under the Au contacts.

Figure 4 shows the dependence of the detector response on the absorbed power of the laser. The absorbed power was estimated using the well-known technique described in [8]. The planar sizes of the detector sensitive element were $0.15\times1.5\ \mu m^2$. The upper limit of the 3-dB detector dynamic range corresponds to the absorbed power of 220 nW. The minimum detected signal determined by the intrinsic device noise corresponds to the absorbed power of 8.5 nW, thus the detector dynamic range is about 14 dB.

The inset in figure 4 shows the time transient of the device response. The risetime of the pulse is about 130 ps determined by properties of the RF line. The falltime is about 0.4 ns. The pulse duration is τ_0 = 250 ps at half maximum. The obtained detector intrinsic noise equivalent power is $NEP_{Rx} = 1.1\times10^{-13}\ W\cdot Hz^{-0.5}$. This result is in good agreement with that obtained in [6]. The detector energy resolution is about $\delta E = NEP_{Rx}\cdot\tau_0^{0.5}$ = 2 aJ.

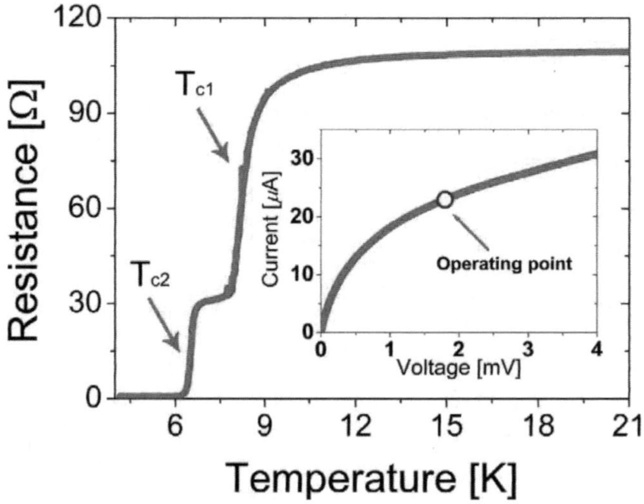

Figure 3. The detector resistance-temperature curve. Inset shows the current-voltage characteristic of the detector at the operating temperature.

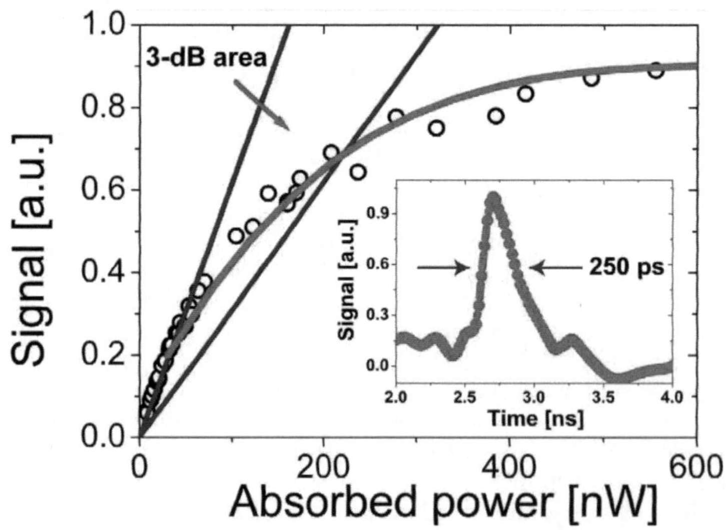

Figure 4. The output of the detector versus the absorbed power. Detector 3-dB dynamic range is marked with the blue lines. Inset shows the time transient of the device response from a 1.55-μm laser pump with pulse duration of less than 50 ps.

4. Conclusion

Finally, performance of the fast and sensitive bolometric detector based on hot-electron effect in ultra-thin superconducting films has been demonstrated. The directly measured energy resolution of this detector is about 2 aJ and the internal response time is less than 130 ps. This detector is a device-of-choice for a number of important practical applications associated with pulsed THz radiation.

Acknowledgments
This work was supported in part by the Russian Foundation for Basic Research (RFBR) Grant No. 16-32-00622, and by the Ministry of Education and Science of the Russian Federation, Grant No. 14.B25.31.0007 and Grant No. 3.2575.2014/K.

References

[1] Shen Y C, Lo T, Taday P F, Cole B E, Tribe W R, Kemp M C 2005 Detection and identification of explosives using terahertz pulsed spectroscopic imaging *Applied Physics Letters* **86**(24) 241116
[2] Siegel P H 2004 Terahertz technology in biology and medicine. In Microwave Symposium Digest *IEEE MTT-S International* **3** 1575-1578
[3] Tuchak A N, Gol'tsman G N, Kitaeva G K, Penin A N, Seliverstov S V, Finkel M I, Yakunin P V 2012 Generation of nanosecond terahertz pulses by the optical rectification method *JETP letters* **96**(2) 94-97
[4] Angeluts A A, Bezotosnyi V V, Cheshev E A, Goltsman G N, Finkel M I, Seliverstov S V, Kostryukov P V 2013 Compact 1.64 THz source based on a dual-wavelength diode end-pumped Nd: YLF laser with a nearly semiconfocal cavity *Laser Physics Letters* **11**(1) 015004
[5] Shurakov A, Seliverstov S, Kaurova N, Finkel M, Voronov B, Goltsman G 2012 Input bandwidth of hot electron bolometer with spiral antenna *Terahertz Science and Technology, IEEE Transactions on* **2**(4) 400-405
[6] Seliverstov S, Maslennikov S, Ryabchun S, Finkel M, Klapwijk T M, Kaurova N, Goltsman G 2015 Fast and Sensitive Terahertz Direct Detector Based on Superconducting Antenna-Coupled Hot Electron Bolometer *Applied Superconductivity, IEEE Transactions on* **25**(3) 1-4
[7] Tretyakov I, Ryabchun S, Finkel M, Maslennikova A, Kaurova N, Lobastova A, Gol'tsman G 2011 Low noise and wide bandwidth of NbN hot-electron bolometer mixers *Applied Physics Letters* **98**(3) 033507
[8] Elant'ev A I and Karasik B S 1989 Effect of high-frequency current on Nb superconductive film in resistive state *Fizika Nizkikh Temperatur* **15**(7) 675-683

Laser formation of tip emitting structures with high aspect ratio on glass-carbon field-emission cathodes

D A Bessonov[1], T N Sokolova[1], V I Shesterkin[2], E L Surmenko[1], I A Popov[1] and Yu V Chebotarevsky[1]

[1] Ed.-Res. Lab. of Laser Technics & Technology, Yuri Gagarin State Technical University of Saratov, Saratov, 410054, Russia
[2] JSC "Research & Production Enterprise "Almaz", Saratov, 410033, Russia

Abstract. Paper describes the method of creating the sharp emitting tips on the surface of a field-emission cathode. The needle-shaped structures with a high aspect ratio on the glass-carbon plate are produced by means of nanosecond laser micromachining. The operations of laser scribing, milling and cleaning were applied for fabrication of the single- and multi-tip field-emission cathode. The special technique of laser rough and fine milling provided high tips with sharp and smooth apexes. As a result we obtained the tips with aspect ratio up to 600. The tests of cathodes showed that high density of current emission and a shorter technological route of production to be reached.

1. Introduction

Field-emission cathodes (FEC) are an attractive alternative to mainstream thermionic cathodes in microelectronics [1], which are power hungry and require high vacuum and high temperature to operate. One of the promising material for FEC is glass-carbon. The unique properties of this material make it suitable for manufacturing of FEC of different configurations [2,3].

With a number of advantages over thermionic cathodes (lack of energy costs for heating the cathode, microsecond time availability), nevertheless, FEC of carbon materials have a low value of the field-emission current density divided by the square of the cathode disk. The maximum value of the electron beam current density in triode electron gun with spike-shaped FEC made of glass-carbon is about 1.5 A/cm^2, which is approximately 30 times lower than for modern metal porous cathodes. This is due to lack of planarity of the emitting surface of the cathode, and is determined by the method of tips forming, namely thin-film technology. For modern vacuum electronic devices the cathode current density of 30 A/cm^2 and above is required.

It was found out that the emission area of the tip and the effective field on the tip apex are key design parameters of an optimized tip structure and depends mainly on the apex radius and tip height. Since the effective field on the tip surface has been proven to have an intimate relation with the ratio of tip height to apex radius (aspect ratio), many research groups over the world have long devoted huge efforts to developing emitter tips with longer lengths but smaller radii for lower voltage operation [4]. For example, carbon nanotubes were considered appropriate emitter structure due to their extremely high aspect ratio (up to 270 [5]) of length in micrometer to radius in nanometer [6]. But such cathodes based on nanotubes with a small radius often show distinct emission instability and fast and nonhomogeneous destruction [5]. So, the requirements of stability demanded less fragile structures to be applied, such as needle-shaped tips with sharp apexes of micrometer dimensions.

A method of glass-carbon microtip FEC producing by plasma chemical etching [7] is time-consuming and includes several different processes (metallization, creating a mask, etching, removing the mask, thermochemical sharpening) that require complex and expensive equipment. Samples, obtained with traditional methods, contain inclusions of the catalysts (Fe, Ni), and the structures themselves have many sources of secondary emissions that negatively affect at work of devices based on them (rapid destruction of the emitting structure, poor performance of the electron beam). In addition, the aspect ratio is low (about 3). It weakens the current emission. Thus, we faced with the task of creating the emitting structure of glass-carbon using a complex of laser processes.

2. Method

Fabrication of the glass-carbon field-emission cathode by laser radiation comprised the operations of laser scribing, milling, cutting and cleaning. The operations were carried out with D-Mark laser commercial installation with parameters: Q-switched Nd:YAG diode-pumped laser, λ=1.06 μm, mean power P=6.7 W, pulse duration 70 ns. Visual investigation was carried out with MIRA/TESCAN electron microscope. Research on the plate surface cleanliness was carried out by laser induced breakdown spectroscopy (LIBS).

As the most developers of field-emission cathodes, we chose the commercial brand of glass-carbon SU-2000. This material is treated well by laser radiation. It has excellent absorption capacity at 1064 nm wavelength, which is the most characteristic of the process equipment.

Fabrication of FEC includes several stages, each of which used a separate driving computer program (figure 1). The first operation (figure 1a) is scribing and subsequent fracture of the glass-carbon preform of the needed dimensions. (This operation can be implemented by any suitable method, as it is preliminary). Then the array of pillars with flat tops, or one pillar in the case of the single-tip cathode, was formed by laser milling (figure 1b). At the next stage the tips were sharpened to a needle-shape form (figure 1c and 1d) by laser milling with other parameters. After that the technological ledge was formed (figure 1e) and the cathode disc itself was cut (figure 1f). Finally, the laser cleaning of the cathode and the tips took place.

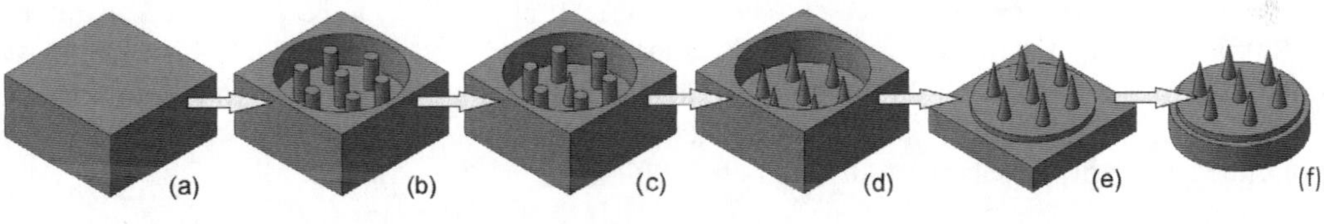

Figure 1. Schematic technological process of producing glass-carbon tip structure.

Laser shaping and cleaning of emitting structures is possible in a single process, using a single laser facility. The milling of pillars is considered to be a stage of "rough" milling. The height and diameter of the created straight cylinder was up 700 μm and about 200 μm respectively. Repeatable results were obtained with the following, found empirically, parameters: focal length of the lens 50 mm, radiation power 3.9 W, speed of the laser beam 343 mm/s, pulse repetition rate 30 kHz.

After the formation of pillars, the process of "fine" milling started for the individual forming of each needle-shaped tip. According to the developed algorithm, the laser beam moved radially from the periphery of each pillar towards its center. Each subsequent movement of the beam was carried out at an angle 30' to the previous. The complete number of lines was 720. The movement of the laser beam for each pillar carried out in such a way that the final point of the movement of the center of the laser beam was on circle D=5 μm, centered on the axis of the pillar. Since the energy of the laser radiation is distributed by the Gauss law, the removal of material in the center of the pillar was minimal. As a result, in the center of the pillar an intact spot of 2-5 μm diameter appeared. At each subsequent azimuth moving the value D became 5 μm more. Thus, after each full turn, the needle-shaped tip was formed with an increasing diameter to its base. The programmable image of the laser beam movement during a single tip formation is shown on Figure 2.

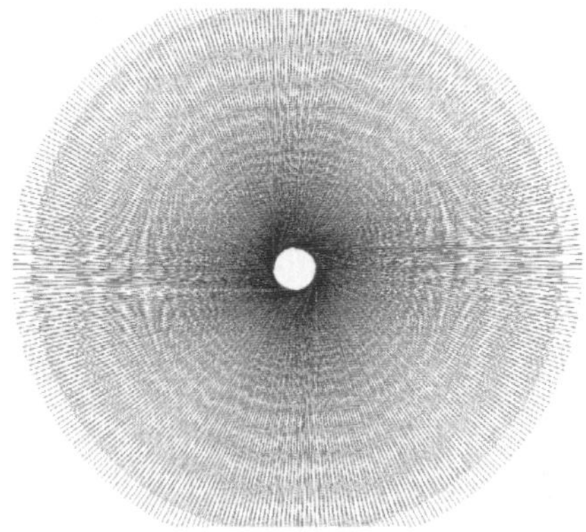

Figure 2. Image of paths of movement of the laser beam on one of the tip.

The following parameters were applied for fine milling: the focal length of the lens 50 mm, radiation power 1.15 W, speed of the laser beam 171 mm/s, pulse repetition rate 8 kHz.

After both milling processes, the FEC surface was cleaned with the following parameters: radiation power 1.3 W, speed of the laser beam 1150 mm/s, pulse repetition rate 30 kHz. Cleaning the surface of the needle-shaped tips is analogical to fine milling with some changes: radiation power 0.65 W, speed of the laser beam 630 mm/s, pulse repetition rate 8 kHz.

3. Results and discussion

Figure 3-5 represents the FEC and, separately, one tip and its apex. The diameter of the cathode is 3.1 mm. 19 needle-shaped tips are located on two concentric circles: 6 tips on the 1.25 mm diameter circle and 12 tips on 2.5 mm diameter circle. One tip is located in the center. The height of the tips on some samples of cathodes reached 700 μm. For each cathode, the tops of all tips are supposed to be lying on the same plane. The radii of curvature of the apexes on some cathode samples were 0.8-3 μm. The aspect ratio of the geometric dimensions of the tips was 500-600.

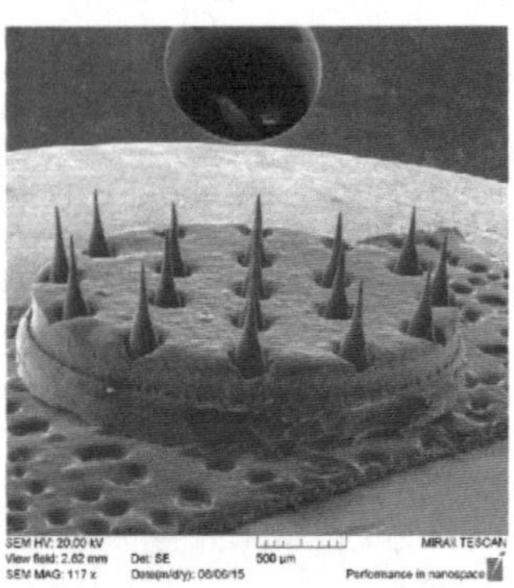

Figure 3. Image of paths of movement of the laser beam on one of the tip.

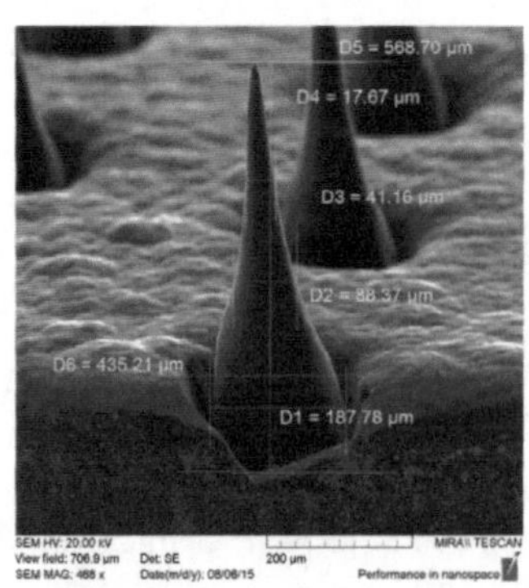

Figure 4. Image of paths of movement of the laser beam on one of the tip.

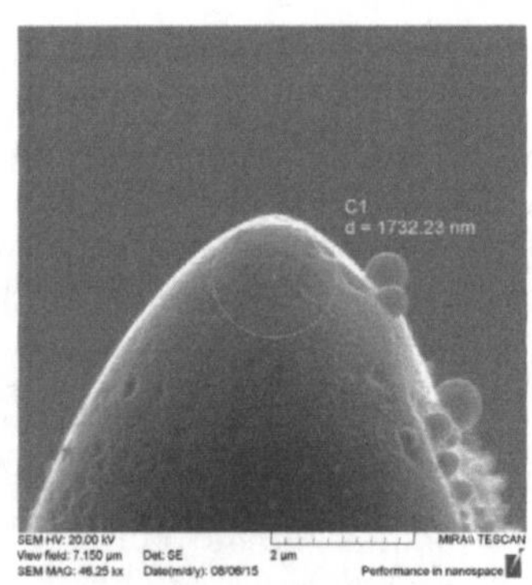

Figure 5. Image of paths of movement of the laser beam on one of the tip.

The figures 6 and 7 show the tips before and after cleaning respectively. As seen from the photos the surface of tips is smooth. As glass-carbon consists of volume cavities (50 nm globules), divided by 5 nm partitions [7], obtaining such a surface indicates the presence of the liquid phase under the influence of laser radiation, which is consistent with observations [8].

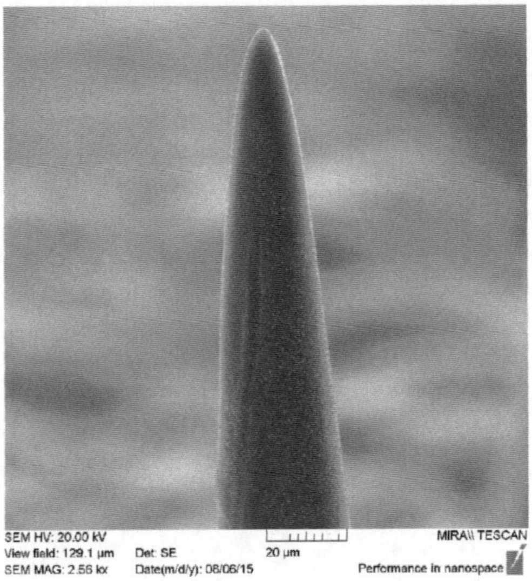

Figure 6. The apex before laser cleaning.

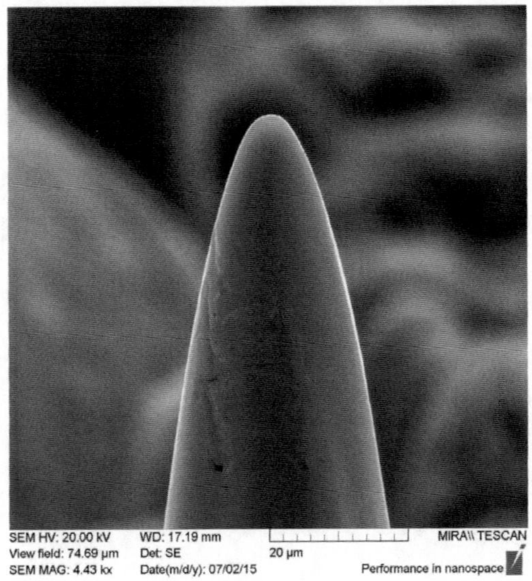

Figure 7. The apex after laser cleaning.

As the structure is cut from a single piece, the resulting matrix has a high strength against degradation by ponderomotive force. The manufacture time of one tip out of the forth-mentioned preform (disc diameter 3.1 mm, thickness 1.3 mm) did not exceed 35 minutes.

4. Conclusions

The laser processing technique including scribing, milling and cleaning on the same laser set-up allows the glass-carbon FEC with high, more than 500, aspect ratio to be created. Due to the special technique of rough and fine laser milling with programmable azimuth movement the tips have a needle-shape. The surface of the tips is smooth, which is a prerequisite for stability in time of field-emission.

The advantage of this technology is its high performance. The manufacture of a 19-tip cathode wastes about 45 minutes. Laser manufacturing, unlike the other methods, performed on the same equipment and the same production site. The technological process is ecologically clean, requires minimal energy and labor costs.

Acknowledgements

The work was implemented on the equipment of the regional branch of CCU "Laser and Optical Technologies", contract 02.552.11.7021 (08.05.2007) in the framework of the basic part of the State Task № 732.

References

[1] Fursey G N 2005 *Field Emission in Vacuum Microelectronics* (New York: Kluwer Academic / Plenum Publishers) p 205

[2] Darmaev A N, Komarov D A, Morev S P, Shesterkin V I and Shalaev P D Experimental investigations and numerical simulation of the electron flow shaped by Grigorev-Shesterkin matrix carbon field-emission cells 2014 *J. of Communications Techn. and Electronics* **59(8)**

pp 820-2

[3] Chebotarevsky Yu V, Sokolova T N, Surmenko E L, Popov I A, Bessonov D A 2012 Laser structuring of glass-carbon for improvement of its emitting properties *Proc. Int. Conf. on Advanced Laser Technologies (Thun).* Laser – matter interaction and processing technologies series vol 1 (Bern: Bern Open Publishing) 44-270-2-PB

[4] Smith R C, Carey J D, Forrest R D and Silva S. R. P. 2005 Effect of aspect ratio and anode location on the field emission properties of a single tip based emitter *J. Vac. Sci. Technol.* B **23(2)** pp 632-35

[5] Lee Myoung-Bok Aspect ratio of nanosized field emitters for optimized emission current 2004 *J. of the Korean Phys. Soc.* **44(4)** pp L800-03

[6] Baylor L R, Merkulov V I, Ellis E D et al. 2002 *J. Appl. Phys.* **91** 4602

[7] Grigorjev J A, Vasilkovskij S V, Shesterkin V I, Jartseva Z A 1993 A method of topology forming of the preferentially multipeak cathode *Patent RF N1738013. H01J 9/02*

[8] Abramov D V, Gerke M N, Kucherik et al. 2007 Formation of nanostructures at the glass-carbon surface exposed to laser radiation *Quantum Electronics* **37(11)** pp 1051-54

A simple electrochemical micropump: Design and fabrication

I V Uvarov[1], S S Lemekhov[1], A E Melenev[1], V V Naumov[1], O M Koroleva[1], M O Izyumov[1] and V B Svetovoy[1,2]

[1]Yaroslavl Branch of the Institute of Physics and Technology RAS, 150007, Universitetskaya Street 21, Yaroslavl, Russia
[2]MESA+ Institute for Nanotechnology, University of Twente, PO 217, Enschede 7500 AE, The Netherlands

Abstract. A micropump based on water electrolysis with a fast change of voltage polarity is presented. It is designed to demonstrate a new pumping principle with the gas termination time as short as 100 microseconds. The device consists of a working chamber with metallic electrodes, inlet and outlet diffusers, and channels for liquid. The chamber and the channels are filled with the electrolyte, which is the pumped liquid. The pump is fabricated on a glass substrate with the deposited metallic electrodes. The substrate is bonded with a polydimethylsiloxane structure containing the channels and the chamber. Design, fabrication procedure and preliminary testing of the device are described.

1. Introduction

Microfluidic devices are widely used in chemical, biological and medical applications including microreactors, implantable drug delivery systems, micro total analysis systems (μTAS), lab-on-a-chip [1]. Micropump is an essential component of microfluidic devices, it is responsible for generation of the fluid movement. Several actuation mechanisms of the micropump are known: pneumatic, piezoelectric, electromagnetic, and others [2]. Electrochemical actuation is based on reversible electrochemical reactions that enable gas bubble expansion and reduction by electrolysis in the aqueous electrolyte solution. In comparison with the other mechanisms, the electrochemical actuation provides large driving force, accurate flow control, low power consumption, low heat generation and compliance with lab-on-a-chip technology. The main drawback is a long response time due to slow recombination of the gas inside of the working chamber [3]. Recently a short-time electrolysis using alternating polarity voltage pulses was demonstrated, which allows fast generation and termination of the gas in the chamber [4]. Here we describe design and fabrication of a simplest micropump, which will be able to demonstrate the new pumping principle.

2. Design of the micropump and operation principle

Design of the pump is schematically shown in figure 1. The pump consists of a chamber with metallic electrodes inside it, inlet and outlet diffusers, and channels for liquid. The pump has a planar configuration, all its elements have a rectangular cross-section with the same height. Working fluid of the pump is the electrolyte that is molar solution of Na_2SO_4 in distilled water.

Operation principle of the pump is schematically shown in figure 2. The pump operates in a cyclic mode. Each cycle consists of pumping and supplying phases. In the pumping phase a series of short (~10 μs) voltage pulses of alternating polarity is applied to the electrodes. These pulses produce hydrogen and oxygen due to water electrolysis but all the gas is collected in nanobubbles [4]. The pressure increases in the chamber and pushes the fluid through the diffusers. More liquid is going via the outlet diffuser because it has smaller flow resistance. In the supplying phase the pulses are switched off and the gas in nanobubbles recombine into water in 100 μs or so releasing the pressure. This time more liquid is going via the inlet diffuser. Thus, the liquid is pumped from the inlet to the outlet. Then the cycle is repeated. The pump has no movable parts such as passive check valves. Flow direction is determined by the geometrical design of the diffusers. Instead of a working fluid the device is pumping the electrolyte. This simplified design is used only for the purpose of demonstration.

Valveless design of microfluidic pumps is well known [5-7]. It has a number of advantages in comparison with the design including check valves: high reliability due to absence of mechanical fatigue, low sensitivity to contamination, fabrication simplicity. Performance of the pump is determined by the dimensions of the diffusers. It has been shown that the optimal performance is achieved when the diffuser angle is about 10° and the ratio between the diffuser length and the neck width is about 10 [6]. The dimensions of diffusers were selected using these considerations. The main geometrical parameters of the pump are demonstrated in figure 3. Channels and chamber of the pump have a height of 6 μm.

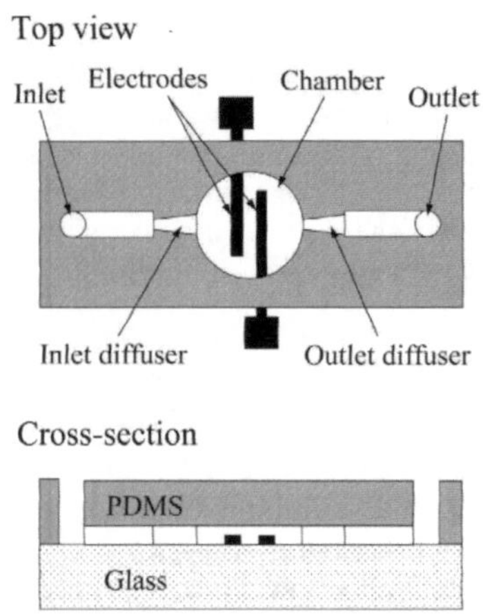

Figure 1. Design of the micropump.

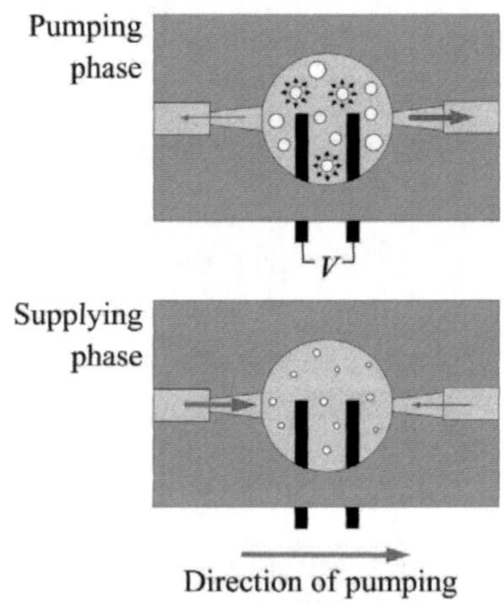

Figure 2. Operation principle of the pump.

3. Fabrication

The pump was constructed from two parts: one is a glass substrate (76×26×1 mm^3) with deposited metallic electrodes and the other one is a polydimethylsiloxane (PDMS) structure containing channels for the electrolyte. These parts were fabricated separately and then were bonded together.

The main steps of the glass substrate fabrication are shown in figure 4. At the first step the substrate was covered by the positive photoresist layer (Shipley S1813) and the UV contact photolithography was performed. Then an adhesion layer (Ti, 10 nm) and 100 nm thick Pt layer were deposited by magnetron sputtering. At the final step the lift-off process was performed. Photograph of the glass substrate with close-up view of platinum electrodes is shown in figure 5. Electrodes have interdigitated structure to increase the efficiency of the electrolytic bubble generation and pumping. Each finger have a width of 20 μm, the spacing between fingers is also 20 μm. The diameter of the overall electrode area is 450 μm which is close to the diameter of the working chamber. This is done in order to saturate the entire volume of the chamber evenly with the gas bubbles.

PDMS structure was fabricated by soft lithography technique [8]. Fabrication steps are shown in figure 6. First, a silicon mold containing the chamber, diffusers and channels was fabricated using deep reactive ion etching. SEM image of the mold is shown in figure 7. The height of the mold features was 6 µm. Then liquid PDMS pre-polymer (Sylgard 184, a mixture of a base and a curing agent at 10:1 ratio) was poured onto the mold, degassed in vacuum and cured at the temperature of 100 °C during 45 min. After the curing PDMS replica was peeled off from the mold. At the last step the inlet and outlet holes with a diameter of 0.5 mm were made in the PDMS structure by a biopsy puncher.

Glass substrate with the metallic electrodes and PDMS structure were assembled together using irreversible bonding of PDMS to glass. To achieve this both structures were exposed to an oxygen plasma with the pressure of 0.5 mbar and 300 W power during 5 s. Exposed surfaces were brought into contact quickly after the plasma treatment aligning the electrodes and the chamber manually. After 15 min samples were ready for filling with the electrolyte and testing. Cross-section of the assembled micropump is schematically shown in figure 1. Photograph of the fabricated pump with close-up view of the working chamber is shown in figure 8. For clarity channels were filled with water colored by food dye.

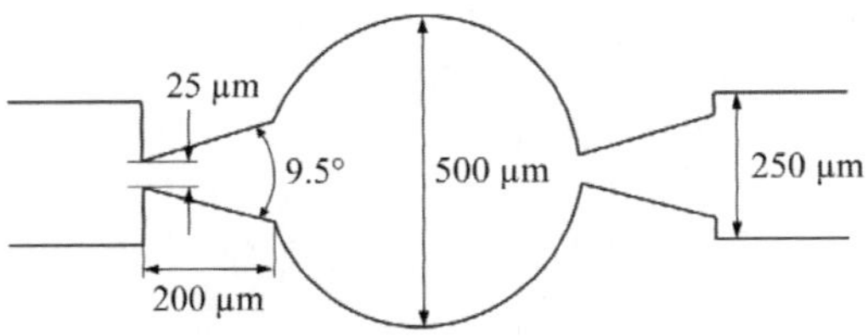

Figure 3. Dimensions of the main structural elements of the pump.

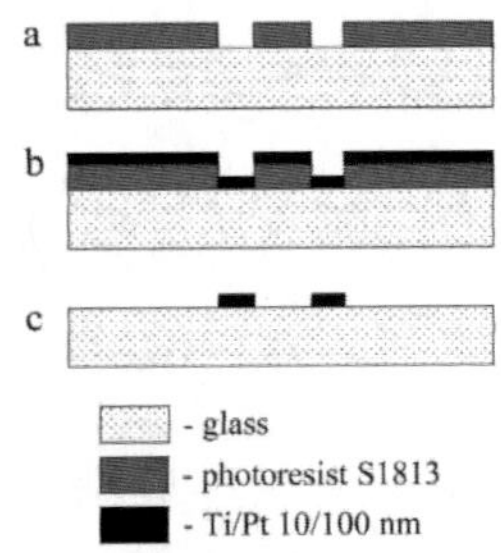

Figure 4. Preparation of the glass substrate.

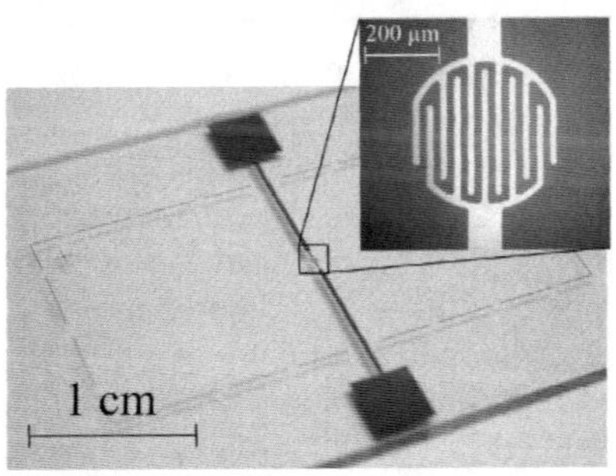

Figure 5. Glass substrate with metallic electrodes.

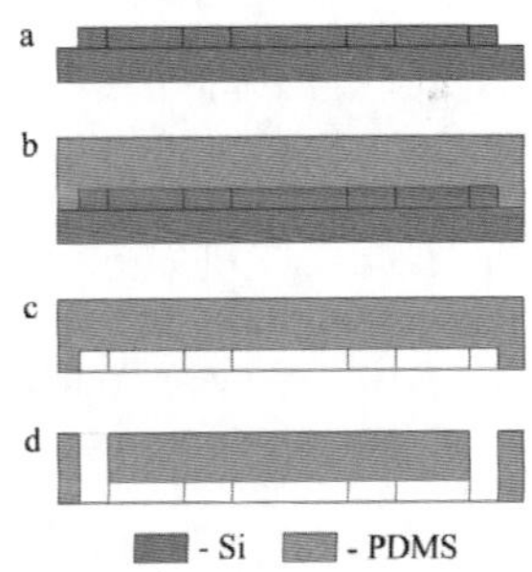

Figure 6. Preparation of the PDMS structure.

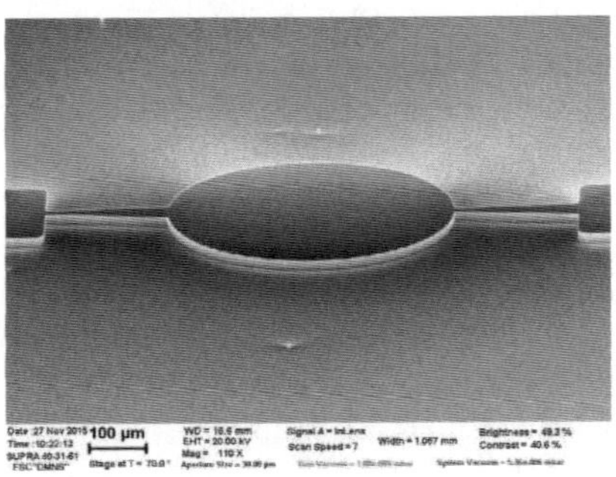

Figure 7. Silicon mold.

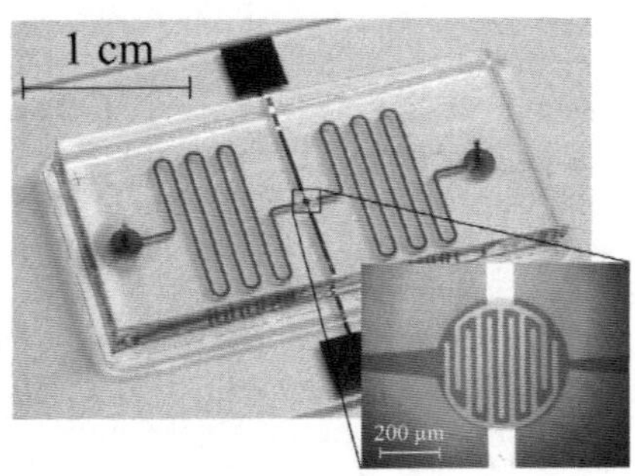

Figure 8. Fabricated micropump.

4. Preliminary testing of the pump

Standard operation mode of the electrochemical pump is the single polarity mode. In the pumping phase a voltage pulse is applied to one of the electrodes while the other electrode is grounded. Hydrogen is produced above the negative electrode and oxygen is produced above the positive one (figure 9). The hydrogen and oxygen bubbles are spatially separated. H_2 and O_2 gases do not mix and have no opportunity to react. In the supplying phase the voltage is removed and the bubbles recombine into water during a relatively long time.

Fabricated micropump was tested in the single polarity mode. In the pumping phase the voltage pulse with the amplitude of 5 V and duration of 1 ms was applied to the sample. Gas bubbles were produced above the electrodes. Photograph of the working chamber made at 200 ms after the pumping phase is shown in figure 10. Gas bubbles are clear visible in the chamber. The time of complete disappearance of bubbles, and therefore, the duration of the supplying phase, was about 5 s. Such a long duration did not allow to achieve a high pumping speed, operation frequency of the pump was less than 0.2 Hz.

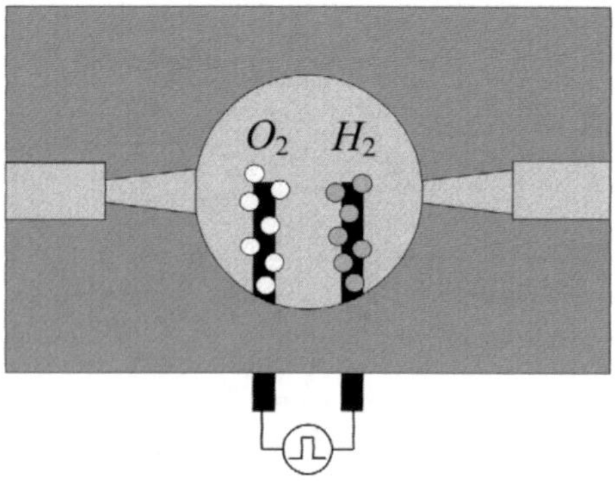

Figure 9. Gas production in the single polarity mode.

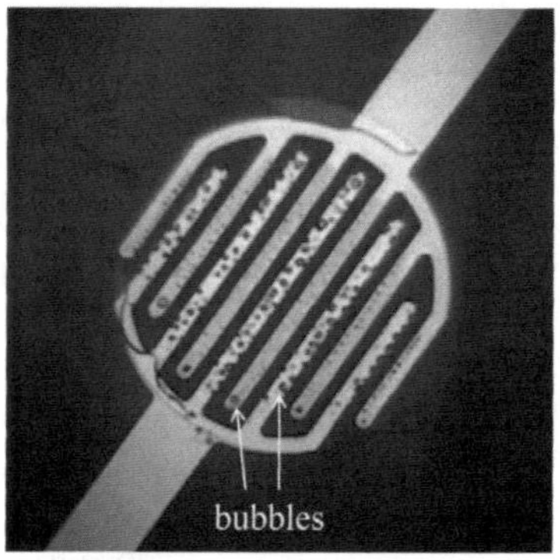

Figure 10. Working chamber in the single polarity mode.

Recently a short-time electrolysis by alternating polarity voltage pulses was demonstrated, which allows fast generation and termination of the gas in the chamber [4]. This regime can be used for fast pumping of liquids. Voltage applied to the working electrode (the other one is grounded) is shown in figure 11. In the pumping phase it is a series of N rectangular pulses of alternating polarity with duration t_0 and amplitude U. These pulses produce hydrogen and oxygen above each electrode (figure 12). For short pulses ($t_0 \sim 10$ μs) all the gas is collected in nanobubbles, which are invisible because they do not scatter light [4]. Due to gas production the pressure increases in the chamber and pushes the fluid through the diffusers. In the supplying phase the pulses are switched off. Unlike the single polarity mode, hydrogen and oxygen molecules are spatially not separated and form not only H_2 and O_2 bubbles, but also bubbles containing mixtures of H_2 and O_2 [9]. The latter disappear in the reaction, which is ignited in nanobubbles spontaneously. The mechanism of the reaction is still not completely clear but the reaction happens with the assistance of the gas-liquid interface [9]. When the pulses are switched off the pressure is reduced in about 100 μs. Short recombination time makes it possible to significantly reduce the duration of the supplying phase.

To test the micropump in the alternating polarity mode a series of $N = 6000$ pulses was applied to the electrode. The pulses were repeated with a frequency of $f = 100$ kHz ($t_0 = 5$ μs) at the amplitude $U = 12.5$ V. Duration of the pumping phase was $t_p = Nt_0 = 30$ ms. For this test duration of the supplying phase was chosen to be $t_s = 220$ ms to make the precess observable with the naked eye but much shorter t_s can be used. Total duration of the pumping cycle was $t_p + t_s = 250$ ms.

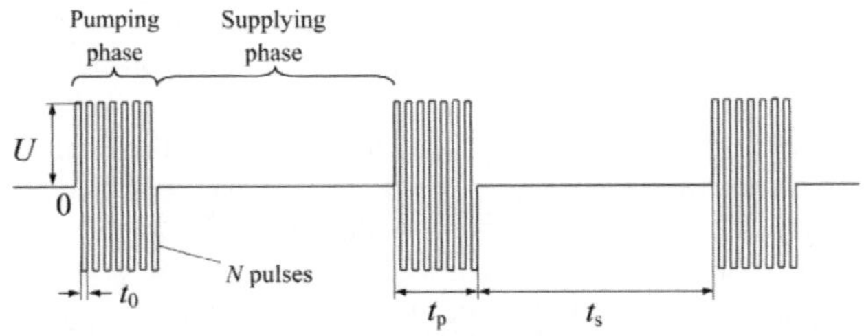

Figure 11. Signal applied to the electrode in the alternating polarity mode.

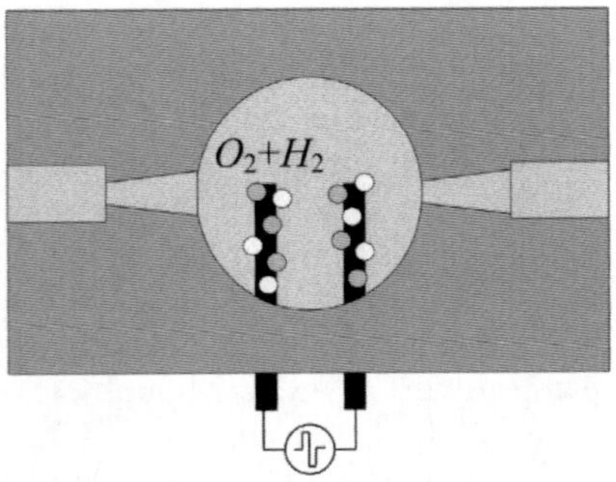

Figure 12. Gas production in the alternating polarity mode.

Figure 13. Working chamber in the alternating polarity mode.

Photograph of the working chamber at the pumping phase made after several hundreds of cycles is shown in figure 13. There were no visible bubbles inside the chamber because the gas was collected in nanobubbles with the size ~ 100 nm. A strong damage of Pt electrodes was observed. To all appearance the damage was caused by the significant energy release in the close proximity of the electrode surface as a result of combustion reaction between hydrogen and oxygen in nanobubbles [10].

Expected flow rate produced by the micropump can be estimated as follows. Total number of gas molecules M produced during the pumping phase is

$$M = \frac{3I_F t_p}{4e} \approx 1.4\times 10^{15},$$

where e is the absolute value of the electron charge and $I_F \approx 10$ mA is the amplitude of the Faraday current. All these molecules are densely packed in nanobubbles with a size of $2r \sim 100$ nm. Pressure in these bubbles is dominated by a high Laplace pressure, $P \approx 2\gamma/r \approx 28$ bar, where $\gamma \approx 0.072$ J/m^2 is the surface tension of water. Significant part of the produced molecules recombines back to water without producing a useful work. As a conservative estimate one can take that only 10% of gas molecules [9] takes part in the pumping. Assuming also that the design of the diffusers provides a medium level of rectification $\varepsilon = 0.1$ we come to the estimate of the flow rate

$$R \sim 0.1\varepsilon\, f_p \frac{MkT}{P} \approx 5 \text{ nl/min},$$

where $f_p = (t_p+t_s)^{-1} = 4$ Hz is the frequency of pumping and the thermal energy kT originates from the equation of state. This value is three orders of magnitude smaller than the flow rate $R \sim 10$ µl/min in micropumps with much larger volume of the chamber [3,12,13]. The pump described in this paper allows much higher pumping frequency of about 1 kHz or, probably, even more. The use of higher f_p can compensate for the low pumping rate.

The flow rate was determined experimentally observing the movement of the meniscus in the channel. The result $R_{exp} \approx 0.15$ nl/min is significantly smaller than predicted. It happens because PDMS is a very soft material with Young's modulus as low as $E = 1.3$ MPa. Since the meniscus is moving the overpressure in liquid P_l is estimated somewhat larger than the Laplace pressure produced by the meniscus $P_l \geq \gamma/h = 12$ kPa, where $h = 6$ µm is the height of the channel. FEM modelling of the pump using COMSOL Multiphysics has shown that for this pressure the maximal deflection of the upper wall will be about $d_0 = 2.2$ µm. With this deflection the increment of the chamber volume $\Delta V \approx 0.14$ nl is close to the volume of gas produced in one cycle $\delta V = 0.1MkT/P \approx 0.21$ nl. Therefore, instead of pumping a significant part of the produced gas is used to inflate the chamber. It is clear that such a soft material cannot be used for fabrication of chamber and channels. One has to use instead more stiff material such as SU-8, PMMA, glass, or Si.

5. Conclusions

An electrochemical valveless micropump was presented. It was designed to demonstrate a new pumping principle based on water electrolysis with a fast change of voltage polarity. The device consisted of the glass substrate with metallic electrodes and PDMS structure with the chamber and channels for liquid. Preliminary testing of the micropump in the standard single polarity mode and recently proposed alternating polarity mode was performed. Alternating polarity mode allowed to achieve sufficiently higher operation frequency of the pump due to very short gas termination time. However, the flow rate in this regime was lower than expected. Two main problems were encountered. One is that PDMS is too soft material for the chamber, which increases its volume during pumping. The chamber of the micropump has to be fabricated from a more stiff material. The second problem is the fast degradation of the electrodes in the alternating polarity regime. An additional work has to be done to choose an appropriate material for the electrodes.

Acknowledgments

This work is supported by the Russian Science Foundation (Grant No. 15-19-20003).

References

[1] Whitesides G M 2006 *Nature* **442** 368-73
[2] Au A K, Lai H, Utela B R and Folch A 2011 *Micromachines* **2** 179-220
[3] Yi Y, Buttner U, Carreno A A A, Conchouso D and Foulds I G 2015 *J. Micromech. Microeng.* **25** 105011
[4] Svetovoy V B, Sanders R G P, Ma K and Elwenspoek M C 2014 *Sci. Rep.* **4** 4296
[5] Stemme E and Stemme G 1993 *Sens. Actuators* A **39** 159-67
[6] Cui Q, Liu C and Zha X F 2008 *Int. J. Adv. Manuf. Technol.* **36** 516-24
[7] Komatsuzaki H, Suzuki K, Liu Y, Kosugi T, Ikoma R, Youn S-W, Takahashi M, Maeda R and Nishioka Y 2011 *Jpn. J. Appl. Phys.* **50** 06GM09
[8] McDonald J C and Whitesides G M 2002 *Acc. Chem. Res.* **35** 491-9
[9] Svetovoy V, Postnikov A, Uvarov I, Sanders R and Krijnen G 2016 *Energies* **9** 94
[10] Svetovoy V B, Sanders R G P, Lammerink T S J and Elwenspoek M C 2011 *Phys. Rev.* E **84** 035302(R)
[11] Singhal V, Garimella S V and Murthy J Y 2004 *Sens. Actuators* A **113** 226-35
[12] Lui C, Stelick S, Cady N and Batt C 2010 *Lab Chip* **10** 74-9
[13] Li P-Y, Sheybani R, Gutierrez C A, Kuo J T W and Meng E 2010 *J. Microelectromech. Syst.* **19** 215-28

Modeling and experimental study of resistive switching in vertically aligned carbon nanotubes

O A Ageev, Yu F Blinov, M V Ilina, O I Ilin, V A Smirnov
Southern Federal University, Institute of Nanotechnologies, Electronics and Electronic Equipment Engineering, Taganrog, st. Shevchenko, 2, 347928, Russia

ageev@sfedu.ru, mailina@sfedu.ru

Abstract. Model of the resistive switching in vertically aligned carbon nanotube (VA CNT) taking into account the processes of deformation, polarization and piezoelectric charge accumulation have been developed. Origin of hysteresis in VA CNT-based structure is described. Based on modeling results the VACNTs-based structure has been created. The ration resistance of high-resistance to low-resistance states of the VACNTs-based structure amounts 48. The correlation the modeling results with experimental studies is shown. The results can be used in the development nanoelectronics devices based on VA CNTs, including the nonvolatile resistive random-access memory.

1. Introduction

The development of nonvolatile high-access resistive random-access memory (RRAM) is one of the priority trends in the development of present nanoelectronics [1-4].Vertically aligned carbon nanotubes (VA CNTs) are of special interest for advanced memory applications due to high scalability and possibility the locally growth nanotubes by plasma-enhanced chemical vapor deposition (PECVD) compatible with the silicon technology [5-7].

We have previously established by scanning tunnel microscopy (STM) that structures based on VA CNTs exhibit resistive switching with a resistance ratio in high-resistance (R_{HR}) to low-resistance (R_{LR}) states more 25 [8, 9]. Experimental studies have shown that the ratio of R_{HR}/R_{LR} increases with increasing applied between the STM probe and VA CNT voltage and depends on the form of the voltage pulse [9]. The mechanism of resistive switching based on deformation, polarization and piezoelectric charge accumulation of the VA CNT was proposed [9-11]. However there is no theoretical description of the resistive switching in VA CNTs. This fact limits application of the VA CNTs for the design and creation of RRAM based on them.

The aim of this work is to develop the resistive switching model of VA CNT-based structure taking into account the processes of deformation, polarization and the piezoelectric charge accumulation of nanotube.

2. Modeling description

The voltage pulse $U(t)$ applied to the VA CNT-based structure results in the polarization of the nanotube [12, 13], which gives rise to the surface attractive force $F_{at}(x,t)$ occurring between the VACNT and the upper electrode (Figure 1). This force is directed at the area of maximum field intensity $E(x,t)$ [9, 12]:

$$F_{at}(x,t) = 0.5\varepsilon\varepsilon_0 E(x,t)^2 S, \tag{1}$$

where $0 \le x \le L$; ε_0 is the electric constant; ε is the relative permittivity of the medium between the top of the VACNT and the upper electrode; S is the cross-section of the VA CNT.

The attraction force causes the elongating of the VACNT by $\Delta L(x,t)$, which gives rise to the elastic force $F_{el}(x,t)$. The elongation of the VACNT $\Delta L(x,t)$ can be determined by the solution of a hyperbolic differential equation with the following initial and boundary conditions:

$$\rho\frac{\partial^2 \Delta L(x,t)}{\partial t^2} = \frac{F_{at}(x,t)}{V} - Y\frac{\partial^2 \Delta L(x,t)}{\partial x^2}, \tag{2}$$

where $0 \le x \le L$; ρ is the density of the VA CNT; V is volume of the VACNT; Y is Young's modulus of VA CNT determined using the technique in [14]. Figure 2 shows the results of the solution of (2) at $x = L$ and $t = t_0$ (t_0 is a point in time corresponding the maximum amplitude of the sawtooth voltage pulse) for VACNT with diameter D = 90 nm, length L = 2 μm and Y = 1 TPa and the following distances between the tip of the nanotube and the upper electrode d: 0.5, 1.0, 1.5 and 2.0 nm.

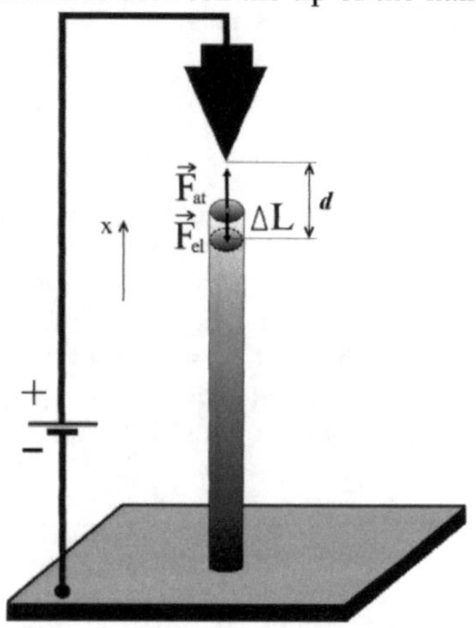

Figure 1. The representation of VA CNT-based structure by applying a local external electric field.

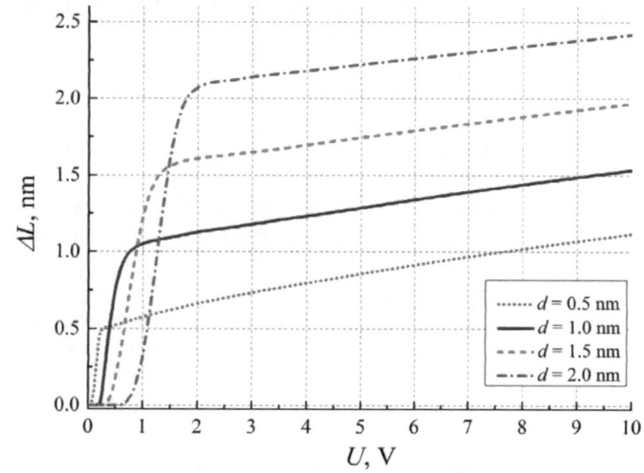

Figure 2. The deformation of VACNT (D = 90 nm, L = 2 μm, Y = 1 TPa) by applying the voltage.

Figure 2 shows that at the initial segment of $\Delta L(U)$ dependency, where $E(x,t) < 10^9$ V/m, the deformation of nanotubes is close to zero. When the threshold value ($E(x,t) = 10^9$ V/m) is achieved, there is a jump of the deformation curve and the nanotube touches the upper electrode. Then the deformation curves $\Delta L(U)$ go into saturation due to the fact that the maximum value of the external electric field is concentrated in the tip area of the upper electrode and the nanotube comes into contact with the electrode within the bounds of this area. The jump of $\Delta L(U)$ dependencies is connected with a non-linear increase in the external electric field $E(x,t) = U(t)/(d-\Delta L(x,t))$ which takes place in the process of VACNT elongation.

According to the power balance equation, the energy W, which is expended on the processes of deformation $\Delta L(x,t)$, polarization, accumulation of piezoelectric charge Q_{cnt} and conduction current creation in the VA CNT-based structure, is equal to the consumed energy of the external electric field:

$$\frac{dW(x,t)}{dt} = vF_{at}(x,t),\ v = \frac{\tau}{m}F_0(x,t), \tag{3}$$

where v is the charge carrier transport speed; τ is the relaxation time of the charge in the VACNT; m is the mass of the charge carriers; $F_0(x,t)$ is the resultant force arising in the nanotube under the influence of the external electric field $E(x,t)$ and equal to the vector sum of the forces of surface attraction $F_{at}(x,t)$ and elasticity $F_{el}(x,t)$.

When $W=U(t)\cdot Q(t)$, the current which flows in the VACNT-based structure (prior to contact with the upper electrode) is equal to:

$$\frac{dQ_1}{dt} = \frac{\tau \varepsilon_0 SE^2(x,t)}{2mU(t)} F_0(x,t) - \frac{Q_{cnt}}{U(t)} \frac{dU}{dt}. \tag{4}$$

When the VA CNT touches the upper electrode conductivity current is inversely proportional to the resistance $R(t)$ of the VACNT arises in the memristor structure. The $R(t)$ is the sum of the intrinsic resistance of the nanotube R_0 which depends on its resistivity and the additional resistance $R_{def}(t)$ which is associated with the deformation and piezoelectric charge:

$$\frac{dQ_2}{dt} = U(t)/(R_0 + R_{def}(t)). \tag{5}$$

According to the power balance and the equations of the piezoelectric theory the $R_{def}(t)$ of the VACNT with length L and longitudinal permittivity $\varepsilon_{||}$ is:

$$\frac{1}{R_{def}(t)} = \frac{dQ_{cnt}(t)}{dt} \frac{1}{U(t)} + \frac{Q_{cnt}(t)}{U^2(t)} \frac{dU}{dt}, \tag{6}$$

$$\frac{dQ_{cnt}(t)}{dt} \frac{1}{U(t)} = \left[\frac{dE(L,t)}{dt} \Delta L(t) - \frac{d\Delta L(t)}{dt} E(L,t) \right] \frac{SL\varepsilon_0 \varepsilon_{||}}{\Delta L^2(t) U(t)}.$$

Despite the fact that the VACNT is held on the upper electrode and cannot be deformed, the external electric field $E(L,t)=U(t)/L$ creates an additional mechanical tension $\sigma'(t)\cdot= bE(L,t)$, where b is the piezoelectric coefficient of VACNT. Then the total deformation of the VACNT $\Delta L(t)$ will be the vector sum of the elongation of the nanotube $\Delta L_0(t)$ which corresponds to the recording voltage and the deformation $\Delta L'(t)$ which corresponds to the additional mechanical tension $\sigma'(t)$. At $dU/dt > 0$ we have the additional nanotube tensile stress $\sigma'(t)$ with $\Delta L'(t) > 0$; at $dU/dt < 0$ – additional compressive stress $\sigma'(t)$ with $\Delta L' < 0$. Thus, if we apply sawtooth voltage pulses to the VACNT-based structure, there will be a hysteresis in the current-voltage characteristic (CVC), which is associated with redistribution of stress and their corresponding piezoelectric charge of the VACNT.

3. Experimental studies

As the experimental sample we used a VA CNTs array grown by PECVD at the multifunctional complex NANOFAB NTC-9 (NT-MDT, Russia). As the substrate, we used a silicon plate with a bilayer structure consisting of a 20-nm-thick titanium film and a 10-nm-thick nickel film formed on its surface. The reaction gas was acetylene. The studies of the array of the VA CNTs on a Nova NanoLab 600 scanning electron microscope (SEM) (FEI, the Netherlands) and on Ntegra probe nanolaboratory (NT-MDT, Russia) made it possible to estimate $D = 108\pm39$ nm, $L = 2.23\pm0.37$ μm and $Y = 1.05\pm0.21$ of VA CNTs in the array.

For study of individual VA CNTs resistive switching the two nanotubes were extracted by mechanical atomic force lithography from the experimental sample using a Solver P47 Pro scanning probe microscope (NT-MDT, Russia). Diameters of the VA CNTs are 88 and 76 nm, length is 2.1 μm (Figure 3). The role of the lower electrode in the VA CNTs-based structure was played by the conducting layer formed on the silicon substrate surface after the growth of the carbon nanotubes, while the upper electrode was the tungsten STM probe 146 nm in radius sharpened by electrochemical etching (Figure 4). The resistive switching of the vertically aligned carbon nanotubes were determined by scanning tunneling microscopy (STM) in spectroscopy mode at $d = 1$ nm using the Solver P47 Pro.

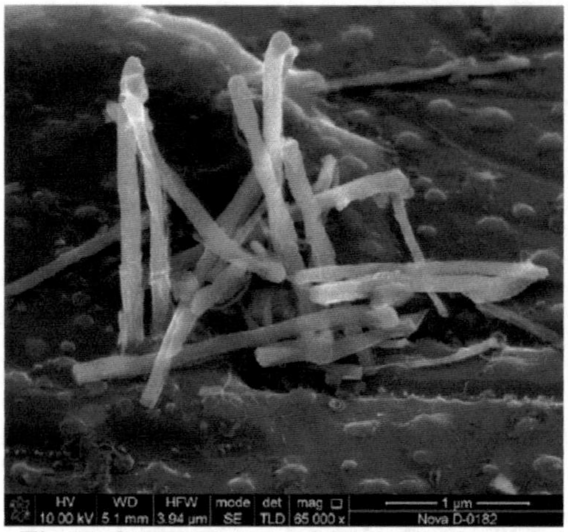

Figure 3. SEM image of the vertically aligned carbon nanotubes.

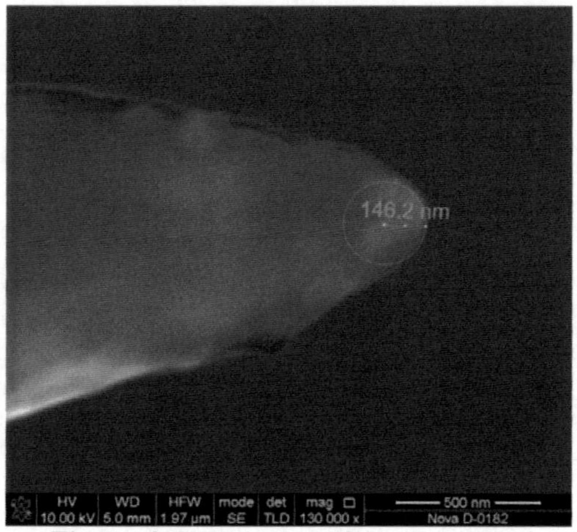

Figure 4. SEM image of the the upper electrode.

The CVC of the VA CNTs-based structure measured in the case of applying the sawtooth voltage pulse of 1.5 V shown in Figure 5. The CVC shows that the structure resistance is changed from the high-resistance to low-resistance state under action of an external electric field; i.e., the resistive switching effect is observed. The ratio $R_{HR}/R_{LR} = 48$ at reading voltage of ±0.2 V, recoding voltage of 1.5 V and erasing voltage of -1 V. Thus the results of experimental studies show the correlation with the modeling.

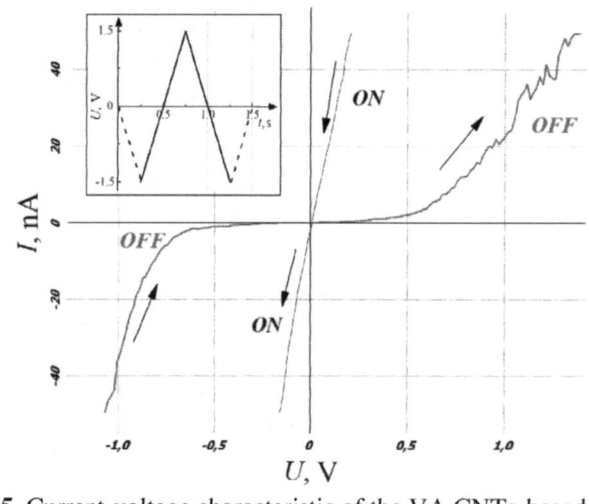

Figure 5. Current-voltage characteristic of the VA CNTs-based structure. In the inset is shown the applied voltage pulse.

Switching time of the VACNT-based structure is estimated based on the analysis of the CVC at U < 1 V and equation (4) and amounts 13 ps. The number of switching cycles of this structure theoretically has no limit, as operating principle associated with low elastic deformation (less than 1%) and the electronic polarization of VACNT that does not lead to its degradation.

4. Conclusion

In summary, the developed model taking into account the processes of deformation, polarization and piezoelectric charge accumulation makes it possible to explain the mechanism of the resistive switching in vertically aligned carbon nanotubes. Origin of hysteresis in VA CNT-based structure is redistribution of stress and their corresponding piezoelectric charge of the nanotube under the influence of an external electric field.

The results of experimental study of VA CNTs-based structure show that the ratio of resistance of the high-resistance to low-resistance states amounts 48 at reading voltage of ±0.2 V, recoding voltage of 1.5 V and erasing voltage of -1 V. Switching time of this is 13 ps. The obtained results can be used in the development nanoelectronics devices based on VA CNTs, including the nonvolatile resistive random-access memory with ultra-high speed and density.

The results were obtained on the equipment of the Collective Usage Center "Nanotechnologies" and the Research and Education Center "Nanotechnologies" of the Southern Federal University.

Acknowledgments

The study was financially supported by the Russian Foundation for Basic Research (project nos. 16-37-00101 mol_a).

References

[1] Akinaga H, Shima H 2010 *Proc. IEEE* **98** (12) 2237.
[2] Chua L 2011 *Appl. Phys. A Mater. Sci. Proces.* **102** 765.
[3] Vasu K S, Sampath S, Sood A K 2011 *Solid State Commun.* **151** 1084
[4] Ageev O A, Balakirev S V, Bykov Al V, Gusev E Yu, Fedotov A A, Jityaeva J Y, Il'in O I, Il'ina M V, Kolomiytsev A S, Konoplev B G,. Krasnoborodko S U Polyakov V V, Smirnov V A, Solodovnik M S, Zamburg E G 2016 Development of new metamaterials for advanced element base of micro- and nanoelectronics, and microsystem devices *Advanced Materials – Manufacturing, Physics, Mechanics and Applications* (*Springer Proceedings in Physics* vol 175) eds I A Parinov, Sh-H Chang, V Yu Topolov (Springer International Publishing Switzerland) chapter 40 pp. 563-580
[5] Chen H, Roy A, Baek J - B, Zhu L, Qua J, Dai L 2010 *Mater. Sci. Eng. R* **70** 63
[6] Jang J E, Cha S N, Choi Y, Amaratunga G A J 2005 *App. Phys. Lett.* **87** 163114
[7] Bichoutskaia E, Popov A M, Lozovik Yu E 2008 *Mater.Today* **11**(6) 38
[8] Ageev O A, Blinov Yu F, Il'in O I, Konoplev B G, Rubashkina M V, Smirnov V A and Fedotov A A 2015 *Phys. Solid State* **57** (4) 825
[9] Ageev O A, Blinov Yu F, Il'in O I, Kolomiitsev A S, Konoplev B G, Rubashkina M V, Smirnov V A and Fedotov A A 2013 *Tech. Phys.* **58** (12) 1831
[10] Michalski P J, Mele E J 2007 *Phys. Rev. B* **76** 205419
[11] Gao Y, Zhai Q, Barrett R, Dalal N S, Kroto H W, Acquah S F A 2013 *Carbon* **64** 537
[12] Lu W, Wang D, Chen L 2007 *Nano Lett,* **7** (9) 2729
[13] Mayer A, 2005 *Phys. Rev. B* **71** 235333
[14] Ageev O A, Il'in O I, Kolomiitsev A S, Konoplev B G, Rubashkina M V, Smirnov V A and Fedotov A A 2012 *Nanotechnol. Russ.* **7** (1–2) 47

Saint Petersburg OPEN 2016 IOP Publishing
Journal of Physics: Conference Series **741** (2016) 012169 doi:10.1088/1742-6596/741/1/012169

Superconducting integrated terahertz receiver for spectral analysis of gas compounds

N V Kinev[1], L V Filippenko[1], K V Kalashnikov[1], O S Kiselev[1], V L Vaks[2], E G Domracheva[2] and V P Koshelets[1]

[1]Kotel'nikov Institute of Radio Engineering and Electronics RAS, Moscow 125009, Russia
[2]Lobachevsky State University, 603950 Nizhni Novgorod, Russia

Abstract. A new highly sensitive device for analysis of gas compounds in terahertz frequency range based on the superconducting integrated receiver is being developed. Such receiver for spectral research of Earth atmosphere from balloon-borne instrument was developed earlier in Kotel'nikov Institute of Radio Engineering and Electronics and successfully operated during several flight missions. In this work, the laboratory setup for gas spectroscopy in the range of 450-700 GHz with the noise temperature below 150 K and spectral resolution better than 0.5 MHz is presented. First results of measurements of NH_3 and H_2O absorption spectra are obtained.

1. Introduction

Spectrum analysis of gas compounds in terahertz region (0.1-10 THz) is well known to be very promising method for different applications [1-3]. This is because of huge number of substances having strong absorption lines at THz frequencies. For instance, molecules of H_2O, CO, NO, NH_3, NO_2, CS_2, H_2O_2, which are very important for biology, chemistry and medicine, can be detected by THz spectral analysis. Furthermore, many explosives have «partner» compounds, such as NO, NO_2, NH_3, formaldehyde, acetaldehyde and others, which also have absorption lines in THz range. This makes measurements of absorption lines for detecting gases quite promising technique in the following fields: in medicine for analysis of exhaled air and diagnostics of various diseases indicated by substances-markers, in security systems for detecting dangerous (explosive or poisonous) substances in situ, and others. In addition, high-technology processes in manufacturing fields require controlling the purity of gases and impurities structure and hence can efficiently use spectroscopic method for these aims.

Therefore, development of sensitive devices for gas detecting is actual problem nowadays. We are developing the spectrometer for gas analysis in THz range with high sensitivity, such spectrometer could be used for medical diagnostics of exhaled air and for detecting dangerous substances in security systems. This spectrometer is based on superconducting integrated receiver (SIR) developed in recent 15 years by our team in collaboration with Netherlands Institute for Space Research [4-5]. The SIR was built up in scope of TELIS project for monitoring the atmospheric compounds (ClO, HCl, H_2O isotopes, NH, O_3, BrO etc) at high altitude in the frequency range 490 - 630 GHz and demonstrated perfect operation at extreme flight conditions. While operating at altitude 20-35 km the absorption pass length (effective path of the beam at which the radiation is absorbed) reaches tens of kilometers, but in laboratory conditions we are strongly limited by dimensions of gas cell. The main purpose of current work is to develop the compact laboratory spectrometer which can provide high sensitivity of detecting important gases.

2. Experimental setup & technique

The spectrometer under development uses the SIR as the receiver in 450-700 GHz range. The SIR is heterodyne receiver that utilizes extremely high non-linearity of tunnel Josephson junction "superconductor-insulator-superconductor" (SIS). The SIS-mixer is based on high-quality trilayer Nb/AlOx/Nb or Nb/AlN/NbN with critical current density 5-10 kA/cm^2 fabricated on Si substrate and integrated with planar quasioptical dipole antenna, it's dimensions are about 0,8 x 0,8 μm^2. The superconducting flux-flow oscillator (FFO) is long Josephson junction is used as the on-chip local oscillator, SIR intermediate frequency (IF) range is 4-8 GHz. For FFO stabilization an additional harmonic SIS-mixer (HM) is used in the feedback phase-lock loop system. All the elements needed for heterodyne detection are located on the 4x4 mm^2 microchip placed on the flat back surface of the silicon lens. The SIR block diagram is shown in figure 1. The concept, fabrication and experimental technique of SIR with more detailes are discussed in [4-6], the most important results of TELIS flight missions are presented in [6-8]. As for the cooling system for SIR microchip and other elements operating at low temperatures, two different systems are used. The first one is vacuum cryostat filled with liquid helium (helium volume is about 3.5 l) having the additional compartment for liquid nitrogen (about 2.5 l). Another option is to use closed cycle cooling system based on cryogenic refrigerator, which is more complicated but consumes only power supply and needs no liquid helium. In both cases, operating SIR temperature is about 4.2-5 K. By now the best double sideband noise temperature of receiver as low as 120 K is obtained, operating frequency range is from 450 to 700 GHz with continuous frequency tuning of the phase-locked local oscillator having free-running linewidth between 0.3 and 5 MHz, the spectral resolution is better than 0.5 MHz.

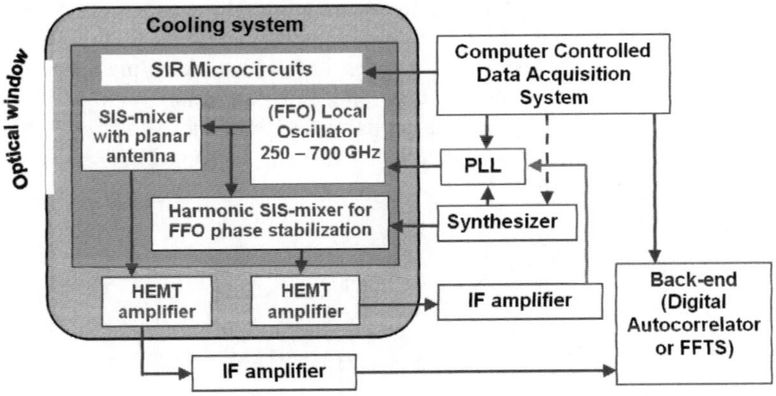

Figure 1. Block diagram of the superconducting integrated receiver (SIR) [6].

The simplified diagram of spectrometer setup for measurements of absorption lines is shown in figure 2. To measure the absorption spectral lines of gas samples by SIR we use the backward-wave oscillator (BWO) with a frequency multiplier and gas cell. The BWO has the operating frequency range 120-160 GHz, it was developed and fabricated in Institute for Physics of Microstructures. The BWO output signal is multiplied by quantum semiconductor superlattice (QSSL) multiplier by factor of 3-5 forming the output signal in the range 0.36-0.8 THz [9] passing then through the gas cell with the gas sample. The gas cell is a 60 cm length tube made of stainless steel, the diameter of cell of about 10 cm is large enough to reduce significantly the influence of absorbed molecules to the side surface of the cell on the results of measurement. The cell has the windows made of polyethylene

2 mm thick. To set and control the gas pressure in the range from atmospheric pressure to $\sim 10^{-6}$ mbar the turbo-molecular pump and the pressure gauge are used.

The measurement technique of the absorption lines is as described below. The output operating frequency of BWO with QSSL multiplier is tuned to the precise frequency (say, f_0) of the specified gas in THz range, the SIR detecting frequency is also tuned to the same frequency f_0 according to $f_{FFO} \pm f_{IF}$ (plus or minus depends on choice of upper or lower sideband operation). Then the multiplied frequency of BWO is slightly changed in the range $f_0 \pm \Delta$, where Δ should be deliberately larger than the width of absorption line. If necessary, the SIR operating frequency is also tuned in a simple way to search the spectrum or/and record the spectrum wider than IF range. Thus, using described technique absorption line is directly seen on the screen of spectrum analyzer and recorded by computer.

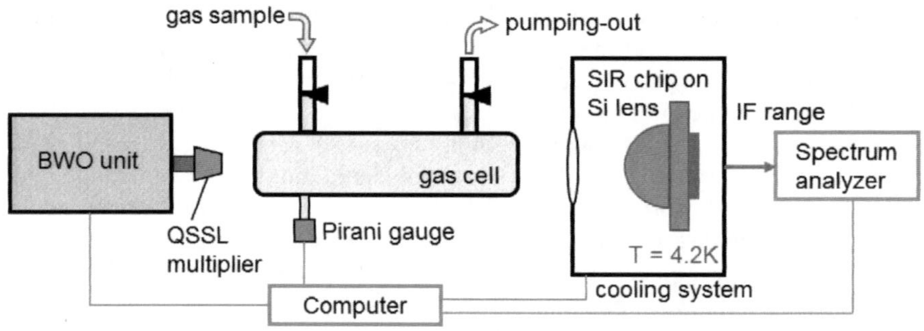

Figure 2. Experimental setup scheme for gas absorption spectra research at 450-700 GHz by SIR. Cooling setup based on liquid helium cryostat or closed cycle system can be used.

Another measurement technique that allows to improve significantly signal-to-noise ratio is using the synchronous detection. This methodic utilizes the frequency modulation of BWO output signal by frequency 0.2-1.5 MHz (with frequency deviation up to 5 MHz) and detecting the SIR output signal in a narrower range of 0.1-0.8 GHz by another back-end detector setup. In figure 3a the scheme of frequency modulation of BWO is shown, the signal from frequency modulator then comes as a reference signal to lock-in amplifier which is connected to SIR output. In figure 3b the scheme of signal storage and recording is presented.

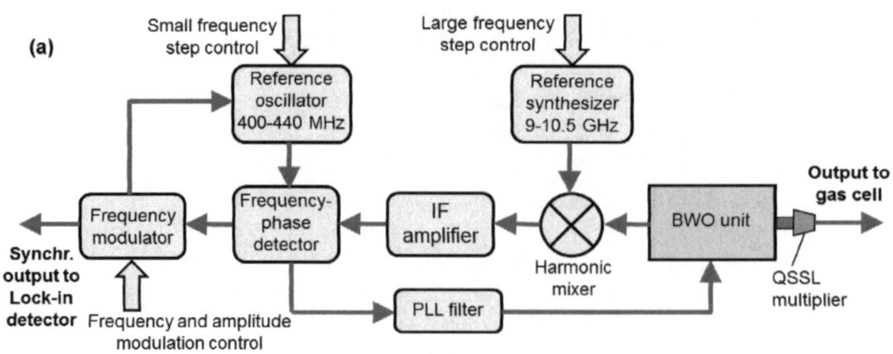

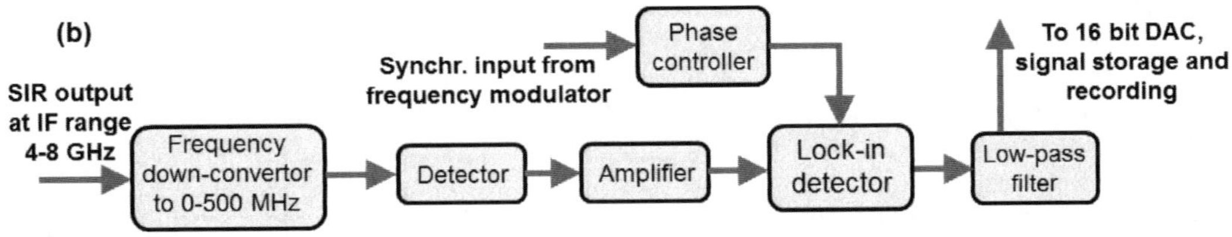

Figure 3. (a) Block diagram of BWO signal frequency and amplitude modulation. (b) Block diagram of detecting the output SIR signal by lock-in detector and its storage and recording.

3. First results and discussion

For the first testing measurements, the most common and simple in preparation and handling gases were chosen: ammonia NH_3 and water vapour. Their strong absorption lines in SIR operating range are at the frequencies 572.5 GHz and 556.9 GHz accordingly. The measured NH_3 absorption lines at different pressures with using spectrum analyzer as the back-end and without BWO frequency/amplitude modulation are presented in figure 4. It is clearly visible by our setup that the intensity of absorption depends on the gas pressure; the pressure limit for NH_3 to distinguish it's presence turned out to be about 10^{-3} mbar. The absorption line of water vapour is stronger, hence the pressure limit for water in our measurements turned out to be better than 10^{-5} mbar. We assume this resolution limit can be considerably improved by using multi-pass gas cell and measuring technique with frequency modulation and lock-in detector as described in figure 3. Nevertheless, these results demonstrate the applicability of such technique and show the spectrometer under development to be an actual device for spectral analysis of gases.

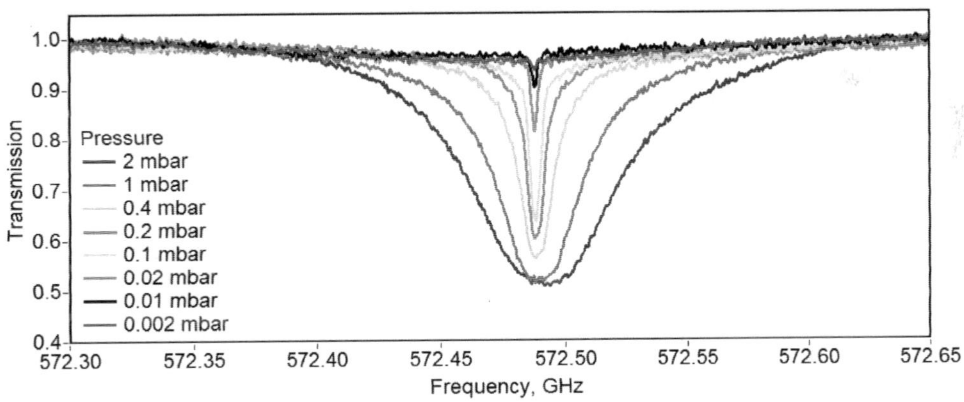

Figure 4. Absorption lines of ammonia NH_3 measured by SIR. Transmission level is calculated from the SIR output IF signal power measured by spectrum analyzer as back-end.

4. Conclusion

The new spectrometer with high sensitivity for gas analysis in 450-700 GHz range is being developed and presented in this work. The spectrometer is based on superconducting receiver with a noise temperature as low as 120 K and having SIS-mixer made of Nb/AlOx/Nb or Nb/AlN/NbN junctions. The introduced technique and the spectrometer under development can have many important practical applications. We concentrate primarily on medicine application, such as spectral non-invasive analysis of exhaled air. By detecting marker-substances this diagnostics can efficiently determine many different human diseases such as oncology, asthma, oxidative stress, schizophrenia, radiation injury, diseases of blood, respiratory tract, gastro-enteric tract, liver, kidney etc at different stages. Another field of great importance nowadays that can require such devices is development of new security systems for detecting different explosive or poisonous substances, or, for instance, leakage of some gases in high-technology processes or on manufacture. Finally, such a device can be used for monitoring of atmosphere and ecological processed, climate changed and others.

Acknowledgments
The work is supported by Ministry of Education and Science of the Russian Federation (project No. 14.607.21.0100, ID RFMEFI60714X0100).

References

[1] Haddad J E, Bousquet B, Canioni L, Mounaix P 2013 *Trends Anal. Chem.* **44** 98
[2] Fischer B M, Walther M, Epsen P U 2002 *Phys. Med. Biol.* **47** 3807
[3] Kawase K, Shibuya T, Hayashi D, Suizu K 2010 *C. R. Physique* **11** 510
[4] Koshelets V P, Shitov S V, Filippenko L V, Baryshev A M, Luinge W, Golstein H, Stadt H, Gao J R, Graauw T et al 1997 *IEEE Trans. on Appl. Supercond.* **7** 2905
[5] Koshelets V P, Shitov S V 2000 *Supercond. Sci. Technol.* **13** R53
[6] Kiselev O, Birk M, Ermakov A, Filippenko L, Golstein H, Hoogeveen R, Kinev N, Kuik B, Lange A, Lange G, Yagoubov P, Koshelets V 2011 *IEEE Trans. on Appl. Supercond* **21** 612
[7] Lange G, Boersma D, Dercksen J, Dmitriev P, Ermakov A B, Filippenko L V, Golstein H, Hoogeveen R, Jong L, Khudchenko A V, Kinev N V, Kiselev O S et al 2010 *Supercond. Sci. Technol.* **23** 045016
[8] Koshelets V P, Dmitriev P N, Faley M I, Filippenko L V, Kalashnikov K V, Kinev N V, Kiselev O S, Artanov A A, Rudakov K I, Lange A, Lange G, Vaks V L, Li M Y, Wang H B 2015 *IEEE Trans. Terahertz Sci. Technol.* **5** 687
[9] Vaks V L, Paveliev D G, Koshurinov I Yu, Koshelets V P, Panin A N 2005 *Proc. The 16th Int. Symp. on Space Terahertz Technol.* (Gothenburg: Curran Associates) p 317

Experimental data and numerical modelling of heating of titanium implants using high-frequency currents

A Fomin[1], M Fomina[1], V Koshuro[1], A Voyko[1], I Rodionov[1]
[1]Yuri Gagarin State Technical University of Saratov, Saratov 410054, Russia

Abstract. High temperature heating of experimental titanium samples using high frequency currents is studied. The velocity of heating is determined depending on the current value on the inductor. Experimental results are compared to the data of numerical modeling of heat transfer in metallic products. The peculiarities of the heating of cylindrical designs of two types: I (diameter - 3.75 mm, length - 10 mm) and II (diameter - 3.95 mm, length - 62 mm) implanted into the bone tissue were studied. The temperature of the constructional elements of different implant parts was calculated.

1. Introduction

Heating with high-frequency currents (HFC) is widely used in the treatment of metals, e.g. in heat treatment of steel. However, this technology is not practically used in the treatment of titanium products. There is experimental data on strengthening and modifying effect of HFC on titanium and its medical alloys [1]. Various intraosseous and transosseous implants, e.g. dental implant, osteofixation devices and more complex designs, such as components of endoprostheses of large joints, are fabricated from titanium and its alloys.

In order to develop technological recommendations on heat modifying treatment with HFC different geometries of the system “inductor – sample” should be considered. As the choice of treatment parameters for the products with complex geometry appears to be difficult numerical modeling of physical distribution of AC magnetic field and non-stationary heat transfer can be applied to make it easier. Thus, the aim of this study is to establish the dependency of heating of commercially pure titanium samples and implants using HFC on the current values of the inductor and to compare the experimental results with numerical modeling data.

2. Methodology

2.1. Titanium samples

Titanium samples are have the disc shape, the diameter of 14 mm, thickness of 2 mm and are fabricated from commercially pure (CP) titanium. The samples served as control ones and were used to build the temperature curves of heating and cooling using the optical imaging techniques and infrared pyrometer. The samples corresponding to the geometry of the actual implant designs were cylinders of two types: I (diameter - 3.75 mm, length - 10 mm) and II (diameter 3.95 mm, length - 62.5 mm). There were threaded elements of complete and incomplete profiles, conical and wedge elements on the cylindrical parts.

2.2. HFC heating device
The laboratory apparatus for induction heat treatment includes power supplies with rectified high (up to +300 V) and stable low (+15 V) voltage, a generator unit and an induction heating unit. It includes an filters, an AC-DC and DC-AC converters, and load – the system "inductor – sample". The generating unit includes an oscillator with adjustable frequency in the range from 50 to 150 kHz. During the heating process power consumption did not exceed 550 W and changed by autoconnected transformer. The heating efficiency is determined by two factors: the amount of voltage supply and resonance. The peculiarities of the heating device performance were related to the geometry of both the product and the inductor. The inductor is a tool for power supply of AC magnetic field and has a shape of a single-layer coil (sometimes it has two or more layers), spiral or more complex geometry.

2.3. Temperature measurement
The temperature of titanium samples with the diameter of 14 mm and length of 2 mm was measured using a non-contact method (Figure 1). For this purpose we used infrared pyrometer "DT-8828" with the limit of -50…1100 °C. The figure shows: a copper inductor (1), a ceramic chamber (2), a control titanium sample (3), PTFE insulation (4), a sample holder (5), a water cooling tube of the inductor (6) and an IR pyrometer (7).

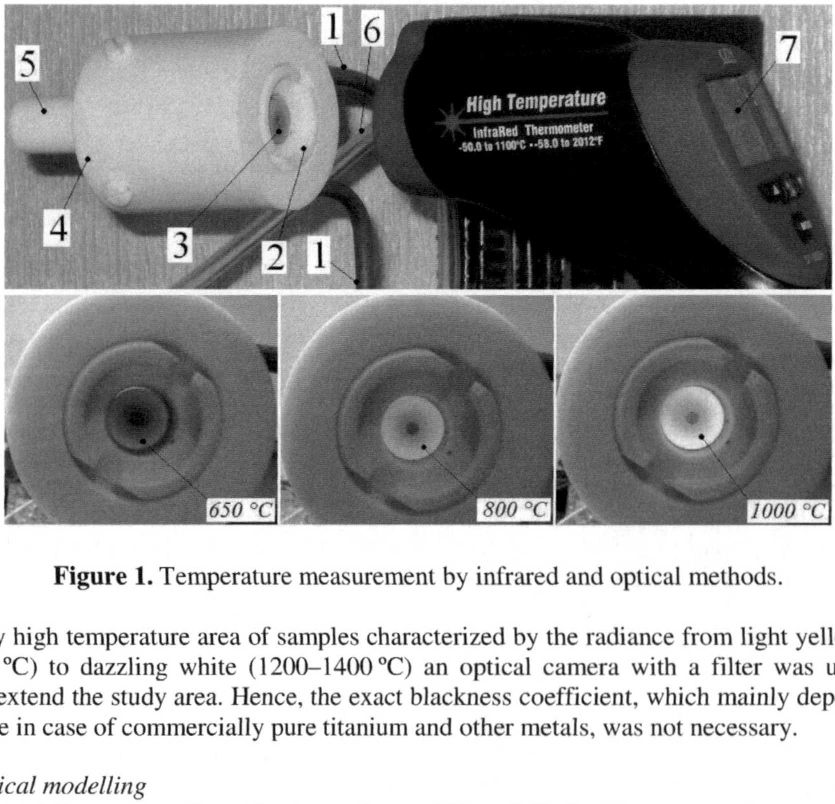

Figure 1. Temperature measurement by infrared and optical methods.

To study high temperature area of samples characterized by the radiance from light yellow (around 1000–1050 °C) to dazzling white (1200–1400 °C) an optical camera with a filter was used, which enabled to extend the study area. Hence, the exact blackness coefficient, which mainly depends on the surface state in case of commercially pure titanium and other metals, was not necessary.

2.4. *Numerical modelling*
Numerical modeling was performed using software "Elcut 6.1". In this program the geometry of the system "inductor – sample" was drawn, further a finite element mesh was generated and AC magnetic field distribution and non-stationary heat transfer were modeled depending on the peculiarities of heat source and all types of heat loss (Figure 2a). Main elements of the studied system are: inductor (1); water (2); quartz camera (3); titanium sample (4) with its central part and periphery; ceramic holder (5) and locking (6) of the sample; lines of magnetic field potential (7).

3. Results

3.1. Heating of control titanium samples

Field images are the result of modeling, e.g. current density *j* and magnetic field strength *H* (Figure 2a). When the heat transfer problem was solved the curves of heating of a titanium sample in its central part and periphery – the hottest area of the skin-layer were built (Figure 2b).

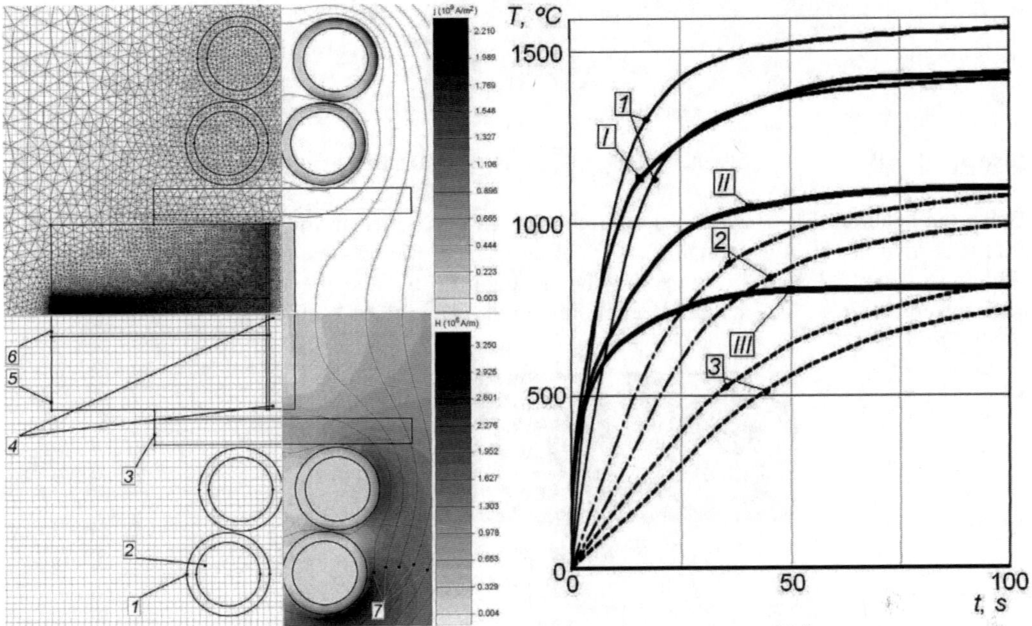

Figure 2(a, b). System “inductor – sample” and modelling results **(a)**; heating curves **(b)**.

The calculation results *1-3* were compared to the experimental data *I-III* at current values on the inductor 2000, 1700, and 1100 A, respectively. The differences between the experimental results and the calculation values were observed at low heat duration of 60±30 s.

3.2. Peculiarities of heating of titanium implant designs

The peculiarity of heating of implant samples with more complex shape was the presence of internal cylindrical and conical holes (Figure 3a). At the end part with the hole there was an area, which has a small thickness of about 0.2-0.3 mm. These areas were unevenly situated with regard to the inductor windings, which affected their bakeout temperature. The presence of the thread on the surface had no substantial effect on the temperature field inside the implant. The most significant temperature nonuniformity across the section was caused by the presence of internal cylindrical and conical holes, which lead to the temperature difference in areas II and III of about 200-250 °C (Figure 3b). Therefore, the hole must be insulated with metal (titanium) plug, which would "level" the picture of the magnetic field potential and its other characteristics.

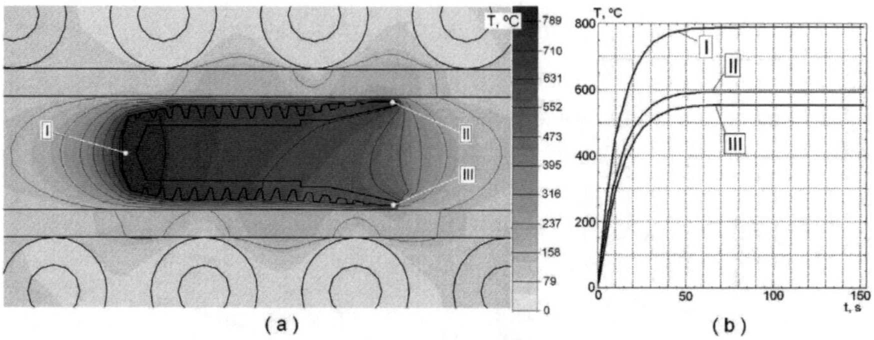

Figure 3(a, b). The picture of the temperature field at t = 150 s **(a)**; heating curves at points I-III **(b)**.

When the heating of the 62.5 mm long implant was performed in the inductor of the same design and at the current of 180 A several areas were observed in it. The hottest area was in the central part of the inductor at coils 3 and 4 (Figure 4). The temperature outside the inductor fell by 250-400 °C towards the end with the metric thread depending on the duration of heating.

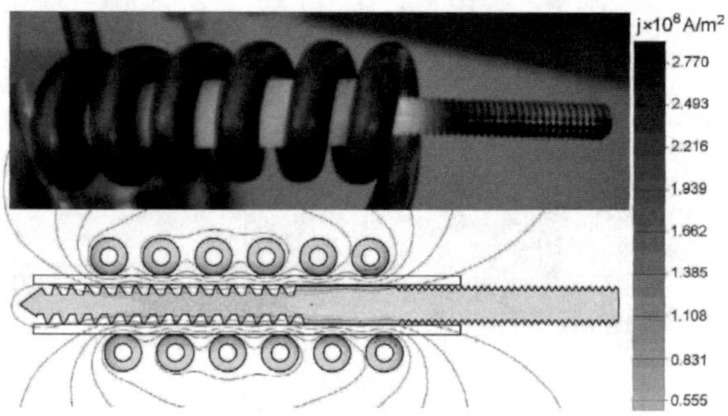

Figure 4. Heating of the osteofix device and picture of the current density field in it.

4. Conclusions

Thus, experimental data on heating at 2200 A correspond to the modeling results. At lower current the results start to converge in the quasi-stationary area of heat treatment when the treatment duration is at least 100 s. According to the numerical calculation and experimental results, the temperature uniformity for the implants can be achieved if the functional elements are located in the central part of the inductor and if construction parts (plugs) are used to minimize the magnetic field inhomogeneity. The results are necessary to determine the inductor current and the treatment duration as these parameters define the biomechanical qualities of implants surface.

Acknowledgments

The research was carried out with financial support of grant 11.687.2016/DAAD.

References

[1] Fomin A A, Rodionov I V, Fomina M A, Petrova N V 2015 *IOP Conf. Ser.: Mater. Sci. Eng.* **77** 012015

[2] Fomin A, Dorozhkin S, Fomina M, Koshuro V, Rodionov I, Zakharevich A, Petrova N and Skaptsov A 2016 *Ceramics International* **42** 10838

Compensation of chromatic and polarization mode dispersion in fiber-optic communication lines in microwave signals transmittion.

A N Ermolaev, G P Krishpents, V V Davydov, M G Vysoczkiy
Department of Quantum Electronics,Peter the Great Saint – Petersburg Polytechnic University, Saint Petersburg 195251, Russia

Abstract. Methods of dispersion compensation in fiber-optic communication lines. A new proposed method of electronic dispersion compensation in the transmission of microwave signals through fiber-optic lines . Represents is proposed the results of experimental studies of this method.

1. Introduction

In fiber-optic communication lines (FOCL) with a single-mode optical fiber there are a number of factors that limit the speed of its transmission are considered. Attenuation, chromatic and polarization mode dispersion and nonlinear effects are among them. These phenomena give rise to the inter symbol interference (ISI) because adjacent pulses begin to overlap are their broadening. This leads to errors during information processing.

By increasing the speed of information transmission impulses (figure 1) start to "deliquesce" in time due to intersymbol interference (figure 2). This leads to overlap of adjacent pulses on one another. In this regard, it is increasing the probability of incorrect signal detection. And at speeds of 10 Gbit/s pulse distortion is not as critical as at speeds of 40 Gbit/s or 100 Gbit/s, which pass all the modern company.

To compensate the chromatic dispersion (CD) and polarization mode dispersion (PMD) different methods are used to-day [1]. In optical data transmission systems optical methods are traditionally used to compensate the chromatic dispersion and polarization mode dispersion. But in the transmission of microwave signals in the range from 10 GHz to 33GHz, the use of optical compensation techniques becomes unacceptable[2, 3].Therefore methods of the electronic compensation of dispersion were developed [1-5], which allow to make the signal processing in digital form in the receiving device.

Nowadays there are a large number of algorithms for compensation of chromatic and polarization mode dispersion. But these algorithms have several drawbacks that in case of a wrong choice of parameters don't allow to decipher the transmitted message without errors.An most cases it leads to the loss of information.

2. Proposed method of compensation of dispertion

To exclude errors in the decoding of the received signals associated with chromatic and polarization mode dispersion, we developed a new algorithm of compensation of these electronic dispersions and the method of determining values according to characteristics of transmitted signals.

For experimental verification of our proposed new method compensation chromatic and polarization mode dispersion was designed and assembled experimental setup (figure 3).

Journal of Physics: Conference Series **741** (2016) 012171 doi:10.1088/1742-6596/741/1/012171

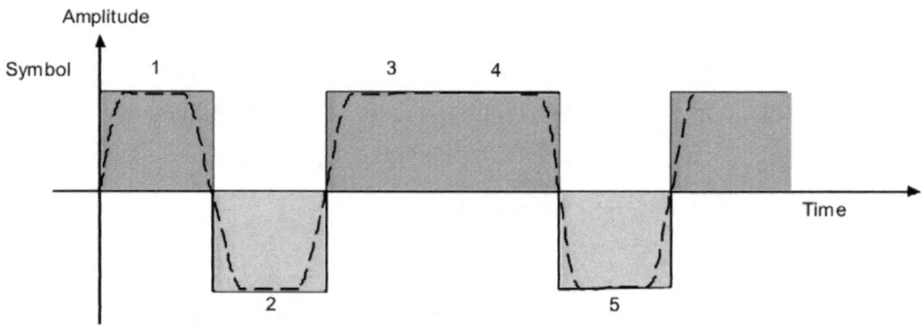

Figure 1. The signal without the intersymbol interference.

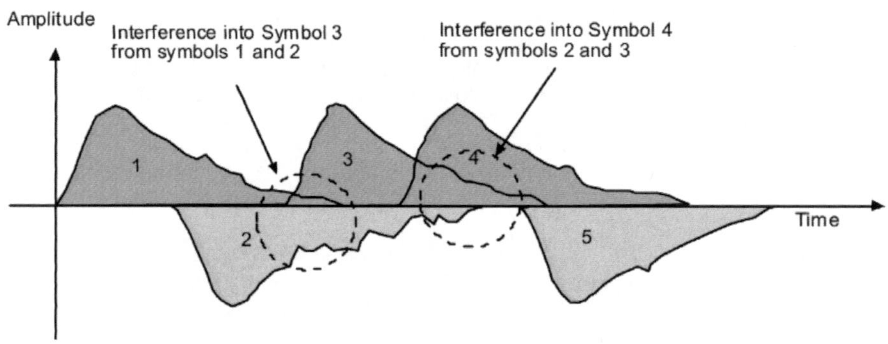

Figure 2. The signal with the intersymbol interference.

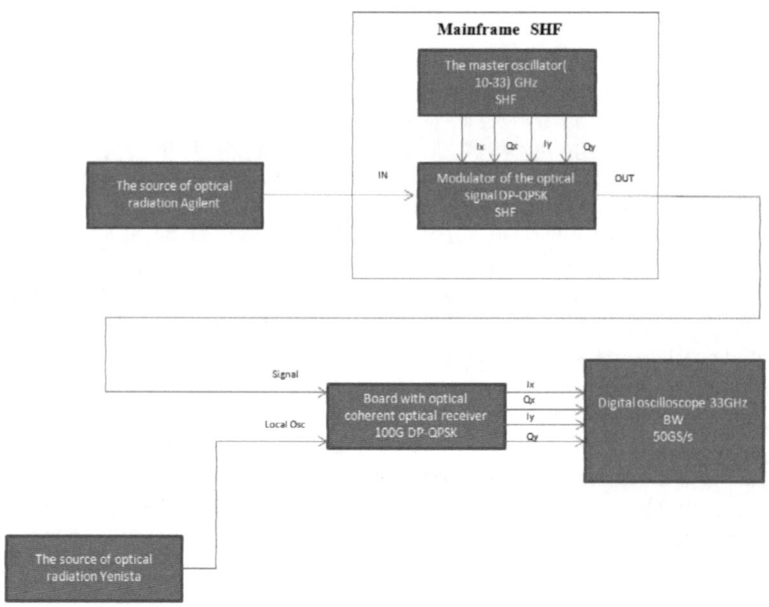

Figure 3. The scheme of experimental installation.

For researches using optical signal (λ = 1510 nm) DP-QPSK format (figure 4) (modulation frequency 25 GHz) is transmitted at a rate of 100 Gbit/s. The structure of the signal in the DP-QPSK format is illustrated in figure 4. The signal includes two information components for formatted QPSK orthogonal polarizations - which means that spectral multiplexing is carried out. In turn, each of the two orthogonally polarized signals in the QPSK format can be represented as the union of two binary phase signals BPSK, phase-shifted by $\pi/2$. Each QPSK signal symbol can be one of the four phase values: 0, $\pi/2$, π, $3\pi/4$. Consequently, each signal QPSK symbol carries 2 bits of information. Each symbol of DP QPSK signal comprises two symbols of QPSK signal, hence it carries 4 bits of information.

In the experimental setup reference radiation with a wavelength of 1550 nm is modulated by the optical modulator, with the frequency of modulation 25 GHz/s. The radiation is directed to a single mode fiber. The fiber length is 20 m. The optical dispersion compensator is built on the basis of the Bragg grating. Dispersion values are compensated by this device lie in the range from -1200 to 1200 ps/nm. Since at such a considerable distance chromatic dispersion does not have time "to express itself" - we artificially distort using of an optical compensator. With a step of 100 ps/nm, we have changed the setting of the optical dispersion compensator. The optical signal was fed to the optical coherent receiver [4], the output signals of which are phase and quadrature components of each of the two polarizations. In figure 2 shows the block diagram of our proposed method for dispersion compensation. Numbers 1 – 6 marked charts of a signal in a phase plane at different compensation stages of the dispersion. In figure 2 shows the block diagram of our proposed method for dispersion compensation. Numbers 1 – 6 marked charts of a signal in a phase plane at different compensation stages of the dispersion.

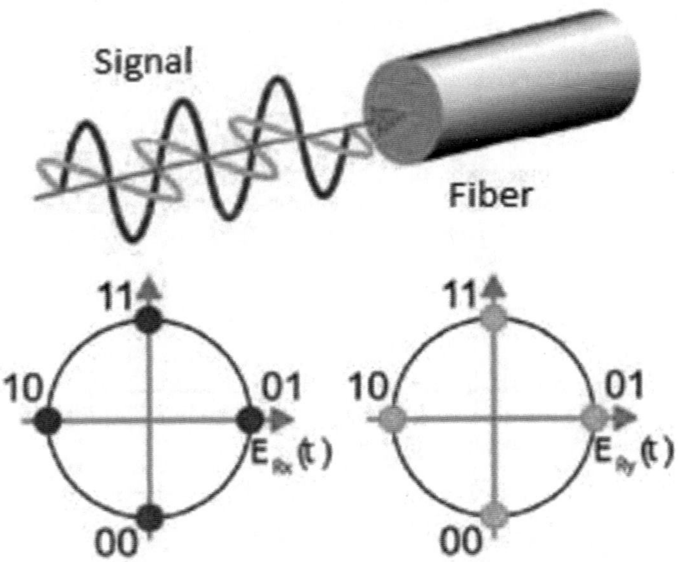

Figure 4. The structure of DP-QPSK signal.

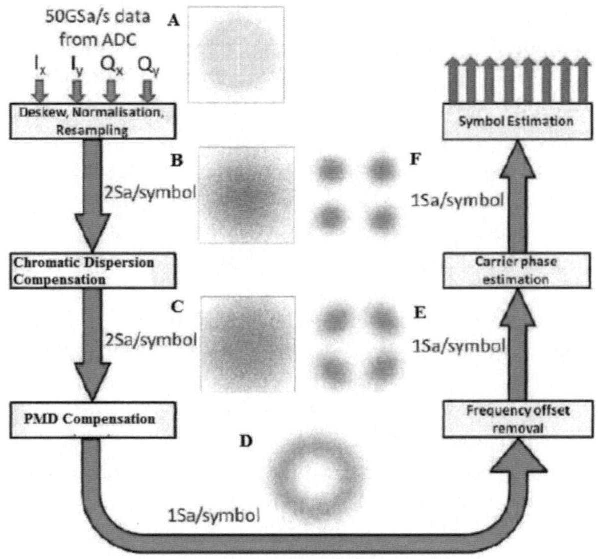

Figure 5. The algorithm of digital signal processing.

To compensate for chromatic dispersion, we use a digital filter with a transfer function like this:

$$_{CD} = \exp(jf^2\pi\frac{\lambda^2}{c}CD),$$

where f is the clock frequency, λ the signal wavelength, c the speed of light and CD the value of CD.

The chromatic dispersion compensation unit executes static compensation of chromatic dispersion independently for two polarizations (chart C), and then to the two polarizations simultaneously apply rapid adaptive distortion compensation methods. The next step DP QPSK signal processing is used constant value signal module search algorithm (CMA - constant modulus algorithm). This algorithm minimizes the signal amplitude deviations from a desired mean value (chart D).The next step is estimated and eliminates the difference frequency between the reference laser and the optical carrier of the received signal, which is determined by the laser transmitter (chart E).The calculated value of the difference frequency is used to eliminate the rotation of the graph in phase plane. In the following steps is estimated and compensated for phase noise (chart F), then determine the values of received symbols.

3. Conclusion

Based on the obtained results we can conclude that if we use the signal chart F (figure 5) after CD and PMD in the signal processing system, bit error rate becomes less than 10^{-12}. This confirms that our proposed method allows us to use of CD and PMD in long communication lines (up to 500 km) without using of optics and other dispersion compensators.

References

[1] Friman R K 2012 *Fiber-optic communication systems* 496
[2] Gurkin N, Nanii O, Novikov A, Plaksin S, Treshchikov V,Ubaydullaev R 2013 *Quantum Electronics* **43** 550
[3] Poggiolini P 2012 *J. Lightw. Technol.* **30** 3857
[4] Le Nguyen B 2014 *Digital Processing. Optical Transmission and Coherent Receiving Techniques* 510
[5] Rice M 2009 *Digital communications: A Discrete - Time Approach* 796

Impact of nanomaterial arrangement on the reliability and the electron mobility in AlGaN/GaN HEMTs

S I Vidyakin[1], A G Gudkov[1], G A Oganesyan[2], V N Petrov[2], A V Sakharov [2], E I Shabunina[2], A A Zybin[3], Ya M Parnes[3]

[1] Bauman Moscow State Technical University, Moscow 105005, Russia
[2] Ioffe Institute, St. Petersburg 194021, Russia
[3] JSC «Svetlana-Electronpribor», St. Petersburg 194156, Russia

Abstract. The obtained results demonstrate that the improvement of nanomaterial arrangement in AlGaN/GaN HEMT structures quantitatively characterized with the use of a multifractal parameter (the degree of disorder) results in the increase at several times in the electron mobility values at 2DEG channel in HEMT structures and the reliability of HEMT parameters.

1. Introduction

AlGaN/GaN high electron mobility transistors (HEMTs) have received increased attention because of their prospective use in microwave power devices. Recently, the impressive progress in improving the performance of such devices for the use in RF-power amplifiers has been achieved. However, open points related to the output power/current collapse and the gate-drain leakage current and its relation to reliability issues still remain [1,2]. The origin of both phenomena limiting AlGaN/GaN HEMTs reliability is now under the study. Inverse piezo-electric effect, hot carriers, the creation of stress-induced percolation path across the AlGaN barrier by defect formation, the increase in material traps concentration due to high electric fields have been considered as the possible causes of HEMT reliability deterioration [2-4]. It seems that all causes have a common source related to the complicated nanomaterial arrangement (NA) of AlGaN/GaN HEMT structures. The complicated NA in nitrides is determined by the presence of extended defects system (EDS) piercing the device's active layers and nano scale phase separation in AlGaN and InGaN alloys [5-7]. These peculiarities result in numerous NA forms with different nanomaterial disordering which are governed by crystallite coalescence of near 3D growth mode to near 2D one. It is well known that nanomaterial disordering causes current crowding effect, the local Joule overheating and degradation in light emitting devices based on III-N nano materials [8], the percolation paths in AlGaN and InGaN layers [6,7]. It also has an impact on the electron mobility values and electron transport mechanism in GaN layers [9]. However, the impact of nanomaterial disordering on AlGaN/GaN HEMT reliability and the electron mobility in two-dimensional gas (2DEG) channels has not been considered yet.

In this paper we will show some experimental results allowing us to clarify the correlation of the electron mobility in 2DEG channel and AlGaN/GaN HEMT's reliability with the nanomaterial disordering, using the approach based on the concept that the nitrides have the complicated NA.

2. Experimental

The plates of HEMT structures were grown on (0001) sapphire substrates using a low-temperature GaN nucleation layer in the standard AIX2000HT MOVPE reactor equipped with an in situ optical reflectance monitoring the system. Ammonia, TMGa, TMAl, and SiH_4 were used as precursors, and H_2, N_2, and H_2N_2 mixtures were used as carrier gases for various stages of the growth. The heterostructure represents a standard HEMT structure consisting of 4 nm insulating GaN buffer, 1 nm AlN interface layer, 7 nm undoped $Al_{0.3}Ga_{0.7}N$ spacer, 13 nm silicon doped $Al_{0.3}Ga_{0.7}N$ layer (the Si concentration is approximately $1 \cdot 10^{19}$ cm^{-3}) and 5 nm undoped GaN cap layer. To suppress the formation of a conductive layer at GaN/sapphire interface, GaN nucleation layer was annealed in a hydrogen-free ambient, which resulted in quasi-2D growth mode of high-temperature GaN buffer from the very beginning of the growth. The growth regimes for the top of GaN buffer and AlN interface layer were optimized for suppression of interface erosion due to the GaN hydrogen interaction [10]. The change in the buffer layer growth conditions enables one to obtain AlGaN/GaN HEMT structures with different NA. The plates of AlGaN/GaN HEMT structures were grown in Ioffe Institute.

It was shown that the complicated NA is reflected at the surface morphology [9]. Surface monitoring by using atomic force microscopy (AFM) and the data processing by the multifractal analysis allow one to quantitatively characterize nanomaterial desodering [9]. The AFM investigation was carried out by using the SOLVER Pro set up produced by NT-MDT (Russia). The surface topography was recorded in the tapping mode using silicon cantilevers with a 10 nm radius tip. The procedure of image processing and of multifractal parameters' extraction, such as the Renyi dimension (D) and the degree of disorder (Δ_p) at local level, includes several steps: the AFM images are converted to high-contrast rough images by using standard computer software for the image processing, these images are approximated by a binary matrixes containing "0" for bright spots and "1" for dark spots, the matrixes are processed by using the mathematical procedure based on multifractal analysis (MFA) in the MFRDrom program developed by Vstovsky [11]. In contrast with the conventional fractal analysis which uses only one level of self-similarity and only one correlation function of digit ensemble density for an entire system, MFA uses several levels of self-similarity and several correlation functions. As a result, the obtained multifractal parameters can characterize the complicated structure of nanomaterial more precisely. The lower Δ_p value, the more ordered nanomaterial. The experimental application of MFA to study the properties of nitride based materials and the devices on their base has shown the effectiveness of the use of the degree of disorder [8,9]. The relative error in Δ_p value determining is $\pm$ 0.002 at simultaneous determination of Δ_p in investigated materials with the use of AFM images of the same scale. HEMTs devices and the samples for the determination of the electron mobility values by the Van-der-Paw method were obtained from each plate. HEMTs devices were fabricated at JSC «Svetlana-Electronpribor». The reliability of HEMTs devices was controlled by step-stress test [12] which was performed at 300 K by sweeping the gate voltage from -20 to -100 V in steps of 5 V with the source and drain voltage set to 0 V. Step time is value 100 s.

3. Results and Discussion

The examples of the surface topology for two HEMT structure plates differing strongly by the nanomaterial arrangement are presented at figure 1 (a,b). The strong violation of layer by layer growth is typical for the HEMT structure with the higher value of the degree of disorder which is presented at figure 1 (a). It results in the heterogeneity of nanomaterial and the decrease in the values of electron mobility in 2DEG channels of HEMT structures to 500 $cm^2 V^{-1}s^{-1}$ at 300 K and 800-1000 $cm^2 V^{-1} s^{-1}$ at 77 K (Figure 2, curves 1,2). Moreover, the uniformity of electron transport at the same sample was revealed by the discrepancy in the electron mobility values for the same sample with the change of the current and potential contacts (Figure 2, curves 1,2).

(a) (b)

Figure 1 (a,b). AFM views of AlGaN/GaN HEMT structures having different the degree of disorder - Δ_p: (a) – 0.357; (b) - 0.345.

The improvement of nanomaterial arrangement (the decrease in Δ_p values) by the realization of nearly 2D growth mode (Figure 1, b) is accompanied by an increase in the electron mobilily values in 2 DEG until 10 000 cm^2 V^{-1} s^{-1} at 77 K and 2000 cm^2 V^{-1} s^{-1} at 300 K. The density of 2DEG is 1.2·1013 cm^{-2}. HEMTs were fabricated from these structures. The dispersion of the leakage current of the gate values and drain current near one order is typical for HEMTs fabricated from structure with Δ_p = 0.357. Moreover, the output power/current collapse and the increase of gate-drain leakage current were observed for many parts of devices after several switching without the reliability test. The reliability of HEMTs fabricated from the structure with Δ_p = 0.345 is significantly higher. The degradation in these HEMTs occurs only after 250 hours working under the reliability test.

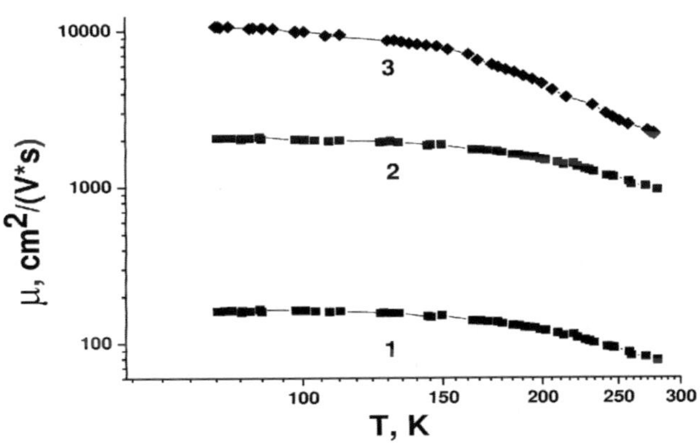

Figure 2. The temperature dependences of electron mobility in 2DEG channels of HEMT structures with different the degree of disorder (Δ_p): 1 and 2 (at different couples of contacts) - 0.357; 3 - 0.345.

4. Conclusion

The obtained results demonstrate that the improvement of nanomaterial arrangement of AlGaN/GaN HEMT structures, quantitative characterized by the use of multifractal parameter, the degree of disorder, results in the increase by several orders of magnitude in the electron mobility values at 2DEG channel of HEMT structures and the reliability of HEMTs parameters. It seems that quantitative control of the nanomaterial disordering is one of the ways to optimize the grow process of AlGaN/GaN HEMT structures and HEMTs reliability.

Acknowledgments
This work was supported by the Ministry of Education and Science of Russian Federation on behalf of the Government of Russian Federation (the contract № 14.574.21.0116, the project identification number is RFMEFI57414XO116).

References
[1] Kim H, Tilak V, Green V M, Cha H, Smart J A 2001 *Phys. stat. sol. A* **188** 203-206
[2] Marko P, Menegini M, Bychikhin S, Meneghesso G, Pogany D 2012 *Microelectronics Reliability* **52** 2194-2199
[3] Lossy R et al 2012 *Microelectronics Reliability* **52** 2144-2148
[4] Lambert B et al 2012 *Microelectronics Reliability* **52** 2184-2187
[5] Lymperakis L et al 2004 *Physical Review Letters* **93** 196401
[6] Nath D N et al 2013 *Applied Physics Letters* **103** 022102
[7] Wu Yu-R et al 2012 *Applied Physics Letters* **101** 083505
[8] Shabunina E et al 2013 *Phys. stat. sol. C* **10** 335-337
[9] Shmidt N M et al 2001 *Nanotechnology* **12** 471-474
[10] Shalygin V A et al 2011 *Journal of Applied Physics* **109** 073108
[11] Vstovsky G V, Kolmakov A G, Bunin I Zh 2001 *Introduction to maltifructal characterisation of material structures* (Izhevsk: Science Publishing Center: Regular and Chaotic Dynamics) p 116 c.
[12] Marcon D et al 2012 Microelectronics Reliability **52** 2188-2193

The research of the temperature difference effect on the sensitivity of the LNA parameters

A A Zybin[1], V N Vyuginov[1], V G Tikhomirov[1], P A Shaganov[1], S I Vidyakin[2], E S Kukhareva[2]

[1] "Svetlana-Electronpribor", CJSC, St. Petersburg 194156, Russia
[2] Bauman Moscow State Technical University, Moscow 105005, Russia

Abstract. There is an investigation of the functional characteristics changes of a low noise amplifier under the influence of temperature changes with helping of a mathematical modelling in particular CAD AWRDE Microwave Office. The relative changes of the functional characteristics of electrical parameters of a low noise amplifier at the fixed frequency as a result of influence of the environment temperature are researching. As a result, temperature changes have a greater impact on the reflection coefficient than on the noise figure and gain.

Microwave solid-state amplifiers are the most important semiconductor devices, determining the main characteristics of the modern radar systems. The sensitivity of the microwave receivers in the modern communication systems is determined by the LNA. The ambient noise level and natural interference in the microwave range is significantly lower than in the HF and VHF ranges. In this case, the noise parameters of receivers determine the quality of reflection low power signals.

For the successful RF module design and operating in the conditions of the external factors effect it is necessary to determine kinetics of the low-noise amplifiers functional characteristics.

The research of the low-noise amplifier functional characteristics under the temperature difference effect carried out by the mathematical modeling.

To research the low-noise amplifier characteristics sensitivity the model of the Ka-Band MMIC LNA was designed in Microwave Office. To research the kinetics of the functional characteristics under the influence of temperature changes the standard model of the non-linear element pHEMT was used.

The developed LNA model has the following parameters: operating frequency:

- f_0 = 30 GHz;
- gain: G≥20 dB;
- noise figure : NF ≤ 5 dB;
- the input reflection coefficient: S_{11}≤ -10 dB;
- the output reflection coefficient: S_{22}≤ -5 dB;
- substrate material: GaAs; nominal relative dielectric constant: ε = 12,9;
- dielectric loss tangent: tgδ = $1 \cdot 10^{-3}$;
- substrate thickness : H = 100 mm;
- plating thickness: T = 3 microns.

Researching the sensitivity of the functionally complete devices electrical parameters to external effects (e.g. ambient temperature) analysis should begun from the research of the sensitivity of its elements (mainly active). We investigate the effect of ambient temperature on the transistor electrical parameters.

There are dependences of the saturation current (IDSS – triangular markers) and maximum drain current (I_{dmax} – square markers) of 100 microns gate width pHEMT from ambient temperature in Figure 1. The graphs show that the maximum current with temperature rising from -80 ° C to + 115 ° C decreases by 10 % and the saturation current – 3 %. Herewith the essential changes in the cutoff voltage are not detected. The change in I-V characteristics leads to a change in the bias condition of the transistor, consequently, the changing of the functional parameters such as gain and noise figure.

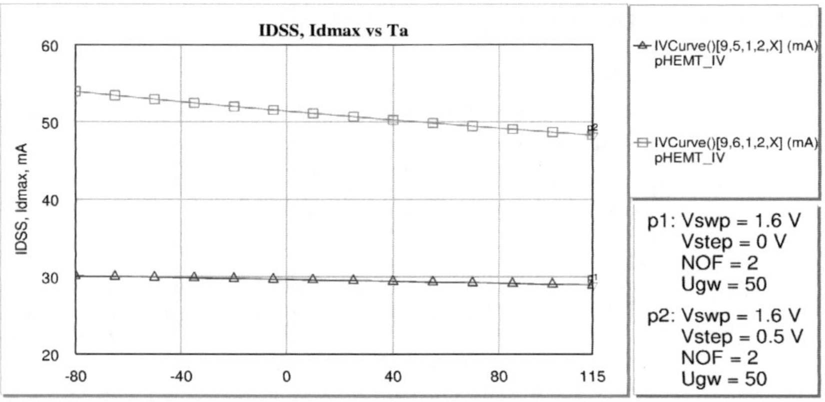

Figure 1 The dependence of the maximum current and pHEMT transistor saturation current on the temperature

It should be noted that aside from choosing of the transistor bias condition the designer use transistor parameters such as F_{min}, R_n, G_{opt}, G_a, where F_{min} – minimum noise factor at the optimum input reflection coefficient; R_n – noise temperature; G_{opt} – optimum reflection coefficient of the input corresponding to the minimum of the noise factor; G_a – maximum available gain of the transistor. The first three parameters are included in the noise matrix of the transistor. Following the definitions, it can be stated that any changings of the transistor parameters will also change the noise parameters of the LNA.

The ambient temperature dependences of F_{min}, R_n, G_a, as a result of mathematical modeling are presented in Figure 1, "b". With the increasing of the temperature from -80 ° C to +115 ° C F_{min} and R_n monotonically increased by 34 % and G_a decreased by 14 %.

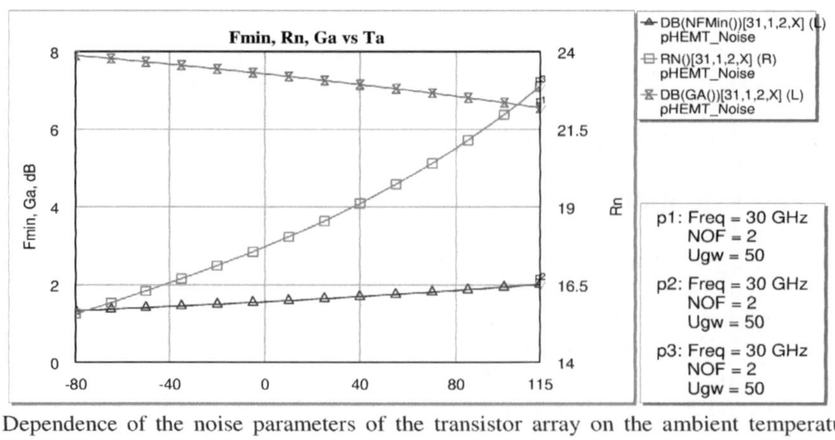

Figure 2. Dependence of the noise parameters of the transistor array on the ambient temperature at a fixed frequency of 30 GHz

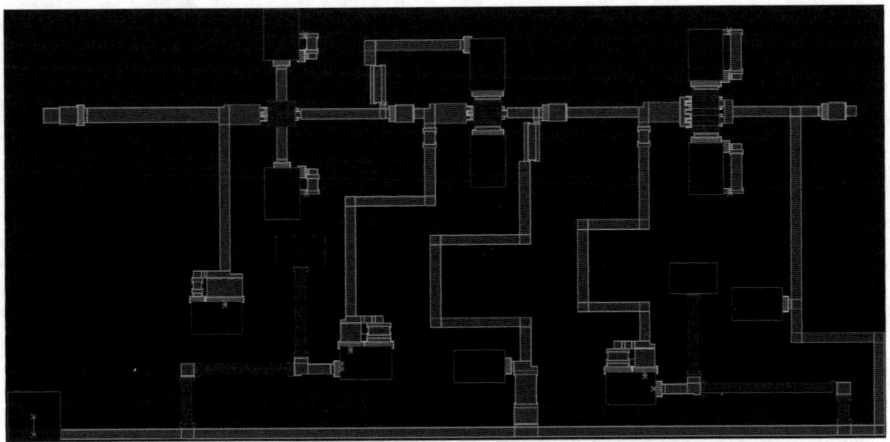

Figure 3. The layout of monolithic integrated circuit of the three-stage LNA

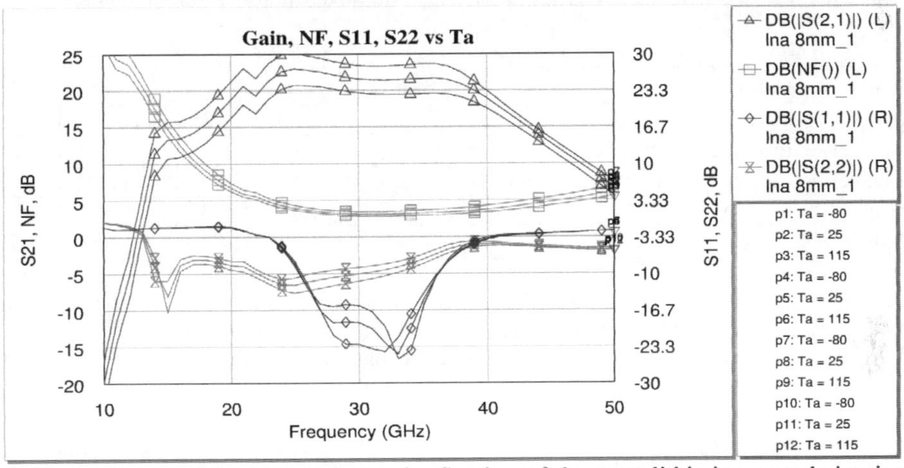

Figure 4. Dependences of the gain, noise and reflection of the monolithic integrated circuit at ambient temperatures ranging from -80 °C to 115 °C

Further it was designed monolithic integrated circuit of the LNA, which layout is presented on the Figure 3. The results of the electromagnetic simulations in the temperature range from -80 °C to + 115 °C across the all frequency range and on the fixed frequency are shown in Figures 4, 5.

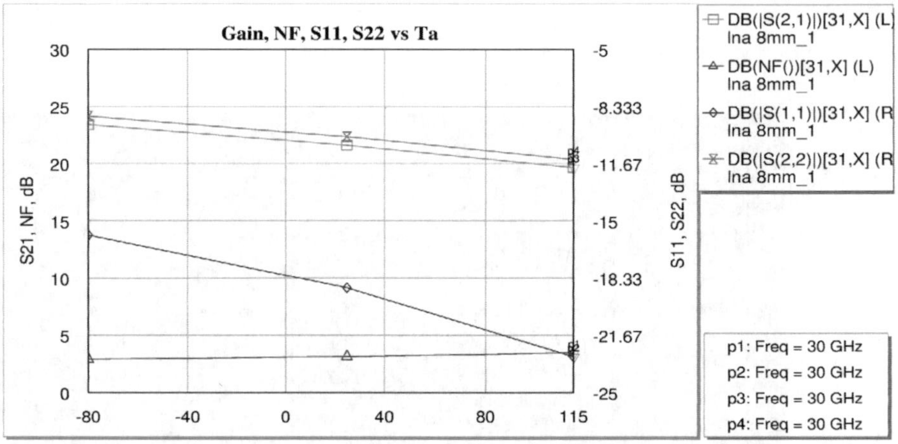

Figure 5. Dependence of the gain, noise and reflection of the monolithic integrated circuit of LNA on the ambient temperatures on the fixed frequency of 30GHz

Analyzing the obtained results on the fixed frequency, we can conclude about the meaning of the scattering of the main electrical parameters of LNA as a result of entertainment temperature influence. There are a data about relative errors of the electrical parameters LNA changes.

Table 1

Name of parameter	Symbol	min	nom	max	R	δR
Environment temperature, °C	T_A	-80	25	115		
The reflection coefficient on the input, dB	S_{11}	-15,83	-18,9	-22,91	7,08	0,37
The reflection coefficient at the output, dB	S_{22}	-8,896	-10,09	-11,42	2,524	0,25
Gain, dB	G	2,899	3,136	3,467	0,568	0,18
Noise figure, dB	NF	23,43	21,6	19,66	3,77	0,17

Finally it is shown a bar chart (Figure 6) of relative errors of the electrical parameters LNA changes under the expose of temperature changes. Based on this graph we can conclude that temperature changes have stronger influence on the LNA adjustment than on the gain and noise figure. Analogically it can be shown the influence of the other factors on the amplifier characteristics.

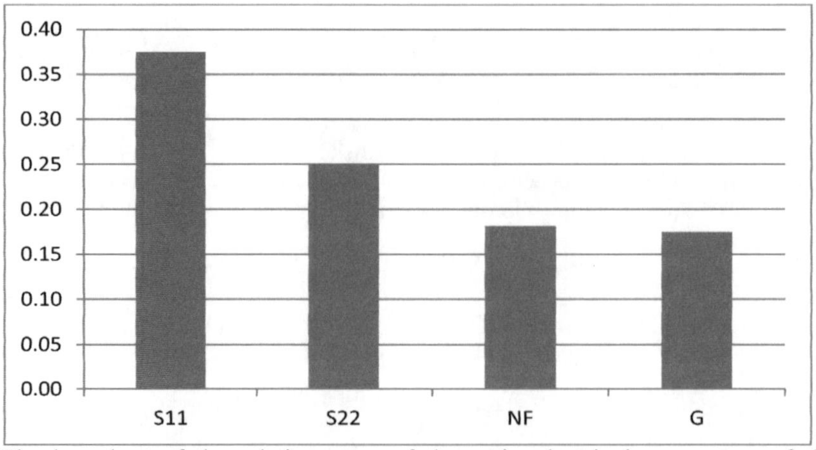

Figure 6. The bar chart of the relative error of the main electrical parameters of the monolithic integrated circuits of LNA under the environment temperature changes.

Acknowledgments
This study was supported by the Ministry of Education and Science of the Russian Federation (agreement on grant no. 14.574.21.0116, the unique project identifier is RFMEFI57414XO116).

References

1. Amplifier Design. Low Noise Amplifier. UMS, sept., 2011/
2. Friis H. T. Noise Figures of Radio Receivers. Proc. of the IRE,July 1944,pp.419-422.
3. Gemener J. J., Oliva J. A., Septier A. Very low noise cryodenic. MESFET amplifier. «Evropean Macrowave Conf.», 1973, v. l,pp. 123-126.
4. Weinreb S. Low-noise cooled GASFET amplifier. «IEEE Trans», 1980, v. MTT-28, N10, pp. 1041-1054.

Resistive switching in the Au/Zr/ZrO_2–Y_2O_3/TiN/Ti memristive devices deposited by magnetron sputtering

O N Gorshkov, A N Mikhaylov, A P Kasatkin, S V Tikhov, D O Filatov, D A Pavlov, A I Belov, M N Koryazhkina, A I Bobrov, N V Malekhonova, E G Gryaznov, I N Antonov and M E Shenina

Lobachevsky University, Nizhny Novgorod, 603950 Russia

Abstract. Bipolar resistive switching phenomenon in the Au/Zr/ZrO_2–Y_2O_3/TiN/Ti memristive devices deposited by magnetron sputtering has been studied. The structure of devices and electrical measurements data for the temperature range from 77 to 490 K are analyzed. The stable switching is demonstrated at room temperature, but the decrease in the resistive switching performance at elevated temperatures is observed.

1. Introduction

The effects of memristive (resistive) switching and memory in the metal–insulator–metal (MIM) devices have been studied extensively in recent years because of the fundamental interest to the structural and electrical phenomena taking place in memristive nanostructures as well as of the potential applications of these phenomena. The nonvolatile memory of new generation – Resistive Random Access Memory (ReRAM) is one of such applications [1-3].

Due to the ability to change the resistance continuously subject to applied voltage (analogously to the biological synapse), the memristive devices are considered now as the basic elements of artificial neural networks of new generation designed to imitate the adaptive behavior of living systems [4, 5]. Transition metal oxides are considered now to be the most promising dielectric layer materials from the technological viewpoint. The bipolar resistive switching (RS) in these oxides is currently attributed to the formation and destruction of conductive filaments as a result of reduction/oxidation (redox) electrochemical reactions under the application of electric bias.

Zirconium oxide (ZrO_x) is one of the most extensively studied oxide materials for the memristive devices [2, 3], since it is well compatible with the modern complimentary-symmetry metal-oxide-semiconductor (CMOS) integrated circuit technology. It is worth noting that the application of yttria-stabilized zirconia ZrO_2–Y_2O_3 (YSZ) along with the stabilized hafnia in the memristive devices has the following advantages. The change of stabilizing oxide (Y_2O_3 in the case of YSZ) fraction allows tailoring the concentration of oxygen vacancies, which determines their mobility, the rate of redox reactions responsible for the electric field distribution in a dielectric film and their electric conductivity, which, in turn, determines the RS parameters.

The goal of the present work was to investigate the elementary processes responsible for the bipolar RS in the Au/Zr/YSZ/TiN/Ti memristive devices prepared by magnetron sputtering compatible with the modern Si-based CMOS technology. The data of structural investigation and electrical measurements for the memristive devices in the temperature range of 77 ÷ 490 K are discussed.

2. Materials and Methods

The Au/Zr/YSZ/TiN/Ti devices were deposited by magnetron sputtering using the Torr International® MagSputt™ 3G2 and 2G1-1G2-EB4-TH1 vacuum thin-film deposition systems on the SiO_2/Si substrates. The YSZ layers (12 mol.% of Y_2O_3) were deposited by radio-frequency (RF) magnetron sputtering at the substrate temperature of 300°C. The Au top electrodes with the area of 8.2×10^{-3} cm^2 were deposited by direct-current (DC) magnetron sputtering. The intermediate Zr layer was introduced between the YSZ film and top Au electrode to improve the adhesion of Au to YSZ and to reduce the work function of the top electrode material. The nominal layer thicknesses were 500 nm for SiO_2, 25 nm for Ti, 25 nm for TiN, 40 nm for YSZ, 3 nm for Zr, and 40 nm for Au.

The structural investigation was carried out by high-resolution cross-sectional transmission electron microscopy (XTEM) using the Jeol® JEM-2100F microscope operated at acceleration voltage of 200 kV (a part of shared research facilities of Research and Educational Center for Physics of Solid State Nanostructures at Lobachevsky University). Such electrical parameters of the memristive devices as the current-voltage (*I–V*) characteristics within the temperature range $T = 77 \div 490$ K and the admittance were measured using the Agilent® B1500A semiconductor device analyzer. Particularly, the frequency dependencies of the differential capacitance $C_{p(s)}$ and conductance $G_{p(s)}$ of the memristive devices in the parallel (p) and serial (s) equivalent circuits of a capacitor [6] at zero voltage bias were measured in the frequency range $f = 10^3 \div 10^6$ Hz. Then, the frequency dependencies of the resistance $R_{p(s)}$ were derived from the measured data. The endurance characteristics were measured by reading current values at +0.5 V between the 120 ms write and erase pulses with amplitudes of +5 and –5 V, respectively. The voltage bias was applied to the top electrode. The measurements were carried out in air ambient unless otherwise specified.

3. Results and Discussion

The XTEM images of a memristive device are presented in Figure 1. The YSZ film is polycrystalline (similar to the literature data [7]). It is worth noting that the interface of the YSZ film with top metal electrode is rough that originates from the polycrystalline nature of the YSZ film.

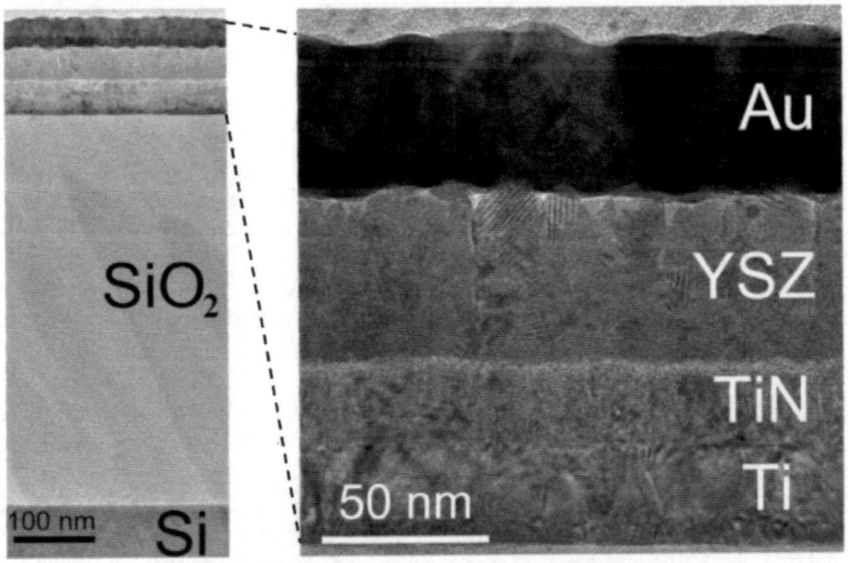

Figure 1. XTEM images of the memristive device.

Typical *I–V* curves of the studied memristive device measured at room temperature are shown in Figure 2. The curves corresponds to the forming process and the bipolar RS from the low-resistive ("ON") state to the high-resistive ("OFF") state and back to the "ON" state.

The electron work function is 4.05 eV for Zr, 5.1 eV for Au [8], and 4.5 eV for TiN [9]. Therefore, the intermediate Zr layer reduces the electron work function of the top electrode material essentially. Since the electron work function of Zr is lower than that of TiN, the electron injection from the top electrode into YSZ should be more intensive than that from the base electrode at the same voltage bias. Therefore, the forming (without the current compliance) was performed at negative voltage bias. As a result of forming, the memristive device is switched to the "ON" state. The subsequent application of a positive voltage bias switches the device into the "OFF" state. The corresponding RESET process takes place at $V = V_{reset}$. The reverse switching from the "OFF" state to the "ON" state takes place at a SET voltage bias $V_{set} < 0$. Thus, the bipolar RS in the devices under study is of the type classified in [10] as a "counter figure of eight".

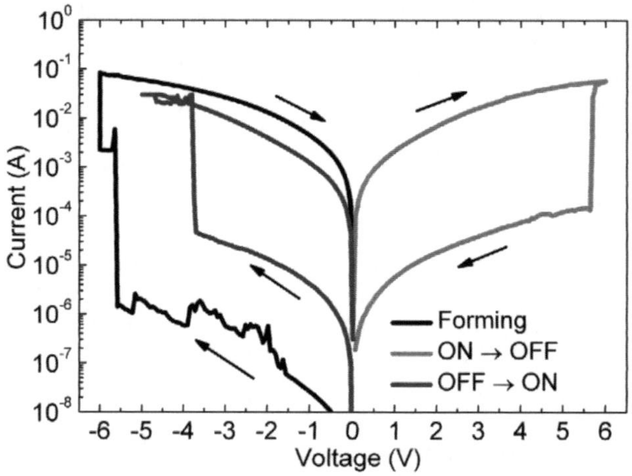

Figure 2. The *I–V* curves of the memristive device measured at $T = 300$ K and voltage sweep rate $u = 1.4$ V/s.

In order to clarify the conduction mechanisms in YSZ in the "ON" and "OFF" states, the *I–V* curves presented in Figure 2 are plotted in Figure 3 in double logarithmic scales. These dependencies demonstrate the same slope equal to ≈ 1.0 at both $V < 0$ and $V > 0$ in the low-voltage range. In the high voltage range, the slope of curves increases up to 1.9 in the "OFF" state and up to 2.1 in the "ON" state. The transition between the low-voltage and high-voltage ranges takes place at nearly the same voltage values at $V < 0$ and $V > 0$ (at $|V| \approx 0.7$ V in the "OFF" state and at $|V| \approx 1.1$ V in the "ON" state). So far, the conductivity of the memristive device appears to be the ohmic one in the low-voltage range for both voltage polarities. However, the conduction mechanism changes with increasing the voltage magnitude, and the *I–V* curves take the form of $I \sim V^n$ with $n \approx 2$ pointing to the space charge limited current (SCLC) conduction mechanism [11]. Thus, the SET process is preceded by trapping of electrons at the trap sites in YSZ associated with the T and C centers [12], whereas the RESET process leaves the traps empty.

The admittance measurements [6, 13] at zero voltage bias are used to determine the values of dielectric constant (ε_1, ε_2) of a dielectric layer determined from the capacitance of the memristive devices measured in the parallel capacitor equivalent circuit, as well as the dielectric losses angle tangent (tgδ_1, tgδ_2) and the resistances (R_{p1}, R_{p2}, R_{s1}, R_{s2},). Here, the index 1 denotes the measurement frequency $f = 1$ kHz and the index 2 corresponds to $f = 100$ kHz. The parameters of parallel equivalent scheme are determined by the electronic phenomena in YSZ, while the parameters featuring the serial equivalent scheme are determined by the resistance of metal electrodes and the one of the transition layer between the metal contacts and YSZ film. The values of these parameters for the studied memristive devices in the initial state as well as in the "ON" and "OFF" states at room temperature are given in Table 1.

The data presented in Table 1 demonstrate that the measured values of dielectric constant for YSZ coincide approximately with the values reported in the literature [14]. This means that the electrode resistance in the studied devices is low enough. Note that the values of dielectric constant of the YSZ films almost doesn't change when the transition comes from the initial state to the "ON" state that evidences the formation of filaments in the YSZ films during forming, which determines the high dielectric losses. The values of dielectric losses in the "ON" state decrease by almost two orders of magnitude with increasing the measurement frequency from 1 kHz up to 100 kHz that can be related to the hopping conductivity in the conductive filaments.

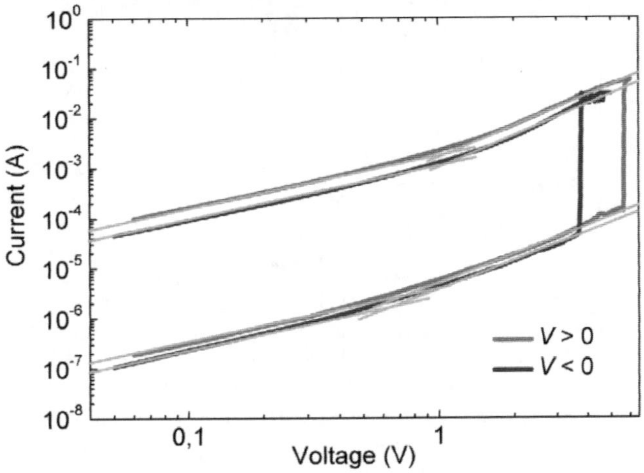

Figure 3. The *I–V* curves of the memristive device in the "ON" and "OFF" states (Figure 2) plotted in double logarithmic scales.

Table 1. The electrical parameters of the memristive devices in the initial state, in the "ON" state, and in the "OFF" state obtained by admittance measurements in the parallel and serial capacitor equivalent circuits.

State	ε_1	ε_2	tgδ_1	tgδ_2	R_{p1} (Ω)	R_{p2} (Ω)	R_{s1} (Ω)	R_{s2} (Ω)
initial	23	22	$2.64\cdot10^{-2}$	$8.87\cdot10^{-2}$	$1.46\cdot10^{6}$	$4.58\cdot10^{3}$	$1.02\cdot10^{3}$	35.8
ON	21	19	70.0	$8.21\cdot10^{-1}$	611	552	611	222
OFF	22	21	$3.47\cdot10^{-1}$	$9.09\cdot10^{-2}$	$1.13\cdot10^{5}$	$4.53\cdot10^{3}$	$1.21\cdot10^{4}$	37.1

According to the XTEM data, the roughness of the YSZ film interface with the top Au/Zr electrode correlates with the grain boundaries in the oxide film and is greater than that of the base TiN electrode. The top interface inhomogeneities can play a role of electric field concentrators, which favor the formation of filaments during the forming process at negative voltage bias applied to the top electrode. A relatively low electron work function of Zr intermediate layer material (≈ 4.05 eV) also favors the growth of filaments from the top interface. Besides, one should take into account that the electron affinity of YSZ is ≈ 2.65 eV [14]. Therefore, the α-band lying ≈ 1.5 eV below the conduction band edge in YSZ [12] is ≈ 4.15 eV below the vacuum level, i.e. below the Fermi level in Zr. This means, that electrons from Zr can be injected into the α-band of YSZ in the thermodynamic equilibrium that should favor the formation of conductive filaments at negative voltage bias applied to the top electrode. It should be noted that along with the preferential formation of the filaments at negative voltage bias, the SET process taking place at negative voltage bias occurs at lower voltage magnitude as compared to the RESET process (see Figure 2).

Figure 4 shows a dependence of the current through the Au/Zr/YSZ/TiN/Ti memristive device in the "ON" and "OFF" states at room temperature on the number of pulse-driven switching cycles. As it follows from Figure 4, the memristive devices demonstrate a high switching endurance. The ratio of the resistances in the "OFF" and "ON" states R_{ON}/R_{OFF} doesn't change essentially after 10^3 switching cycles. The variation of the current through the memristor device in the "ON" and "OFF" states is determined by the stochastic nature of the filament formation and destruction. It is worth noting that the switching between the "ON" and "OFF" states is preserved for the duration of voltage pulses as low as 100 ns, what is important for the fast write/erase operation of memory devices.

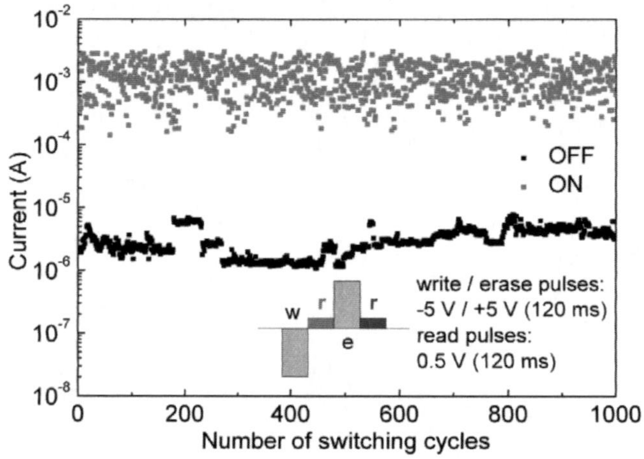

Figure 4. The dependence of current through the memristive device in the "ON" and "OFF" states at room temperature on the number of the pulse-driven switching cycles.

It is worth noting that the RS phenomenon in the investigated Au/Zr/YSZ/TiN/Ti memristive devices depends weakly on the ambient (vacuum, air) unlike the analogous SiO_x-based devices [15]. In particular, the character of *I–V* hysteresis almost doesn't change when the memristive device is measured at room temperature in ambient air, in low vacuum (3-5·10^{-2} Torr), and after letting the ambient air back in the vacuum chamber.

The temperature stability of the RS parameters was studied by measuring the cyclic *I–V* curves at the temperatures in the range $T = 77 \div 490$ K (Figure 5).

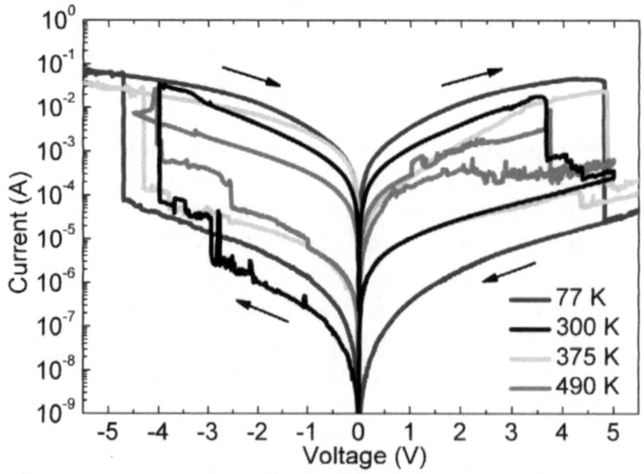

Figure 5. The *I–V* curves of the memristive device measured at various temperatures. The voltage sweep rate $u = 3.0$ V/s.

The switching voltages V_{set} and V_{reset} in the studied devices depend on the measurement temperature and fall into the ranges –2.5 ÷ –5.0 V and +3.5 ÷ +5.0 V, respectively (see Figure 5). The resistive switching performance decreases with increasing temperature starting from about 313 K: the values of current through the memristive device in the "ON" and "OFF" states become closer to each other, i.e. the R_{OFF}/R_{ON} ratio decreases at all values of the voltage V considered. It is important to note that the increase in current through the memristive device in the "OFF" state with increasing temperature obeys the Arrhenius law: $I(T) \sim \exp(-E_a/kT)$, where k is the Boltzmann's constant and E_a is the activation energy. The value of E_a at $T > 313$ K obtained from the $I(T)$ dependencies measured at $V = +0.5$ V is ≈ 80 meV. The mechanism of the observed degradation of the bipolar RS performance in the Au/Zr/YSZ/TiN/Ti memristive devices with increasing the temperature is not clear at the moment and requires further investigation.

4. Conclusion

It can be concluded from the present study that the bipolar RS in the studied Au/Zr/YSZ/TiN/Ti devices originates from the formation and destruction of conductive filaments. According to the XTEM data, the inhomogeneities of the YSZ film interface with the top Au/Zr electrode can act as electric field concentrators, which favor the growth of filaments during the forming process at negative voltage bias applied to the top electrode. A relatively low electron work function of Zr intermediate layer also favors the electron injection in YSZ from the top electrode. The decrease of dielectric losses in the "ON" state with increasing measurement frequency evidences the hopping mechanism of the filament conductivity. However, the conduction mechanism changes to the SCLC at the increased voltage magnitude due to the participation of YSZ defect traps in the electron transport. The bipolar RS in the studied devices is found to be affected weakly by the ambient (air, vacuum) that originates from the fact that the oxygen vacancy concentration in YSZ is determined by the Y_2O_3 molar fraction and, therefore, independent of the oxygen content in the ambient. This is an important advantage of YSZ as a material for the memristive devices as compared to other oxides. The YSZ-based MIM nanostructures fabricated by magnetron sputtering demonstrate high enough endurance, fast switching and are promising for the ReRAM applications.

Acknowledgements

The study was supported by the Ministry of Education and Science of the Russian Federation (applied research project identifier RFMEFI57514X0029).

References

[1] Mickel P R, Lohn A J, Marinella M J, 2014 *Mod. Phys. Lett. B* **28** (no. 10) 1430003-1
[2] Panda D and Tseng T-Y, 2013 *Thin Solid Films* **531** 1
[3] Karkkanen I, Shkabko A, Heikkila M et al., 2015 *Phys. Stat. Sol. A* **212** 751
[4] Kuzum D, Yu S, Wong H, 2013 *Nanotechnology* **24** (no.38) 382001-1
[5] Thomas A, 2013 *J. Phys. D: Appl. Phys.* **46** (no. 9) 093001-1
[6] Epstein S L, Measurement of capacitor characteristics (1965, Moscow and Leningrad, Energy) [in Russian].
[7] Tarancón A, Morata A, Pla D et al., 2015 ECS Transactions, **69** (16) 11
[8] Michaelson H B, 1977 *J. Appl. Phys.* **48** 4729
[9] Westlinder J, Sjoblom G, Olsson J, 2004 *Microelectronic Eng.* **75** 389
[10] Lee J S, Lee S, Noh T W, 2015 *Appl. Phys. Rev.* **2** 031303-1
[11] Lampert M A, Mark P, Current Injection in Solids (1970, NewYork/London, Academic Press)
[12] Orera V M, Merino R I, Chen Y et al., 1990 *Phys. Rev. B* **42** (no. 16) 9782
[13] Oreshkin P T, Tutorial on physics of semiconductors and dielectrics (1977, Moscow and Leningrad, Higher School) [in Russian]
[14] Wilk G D, Wallace R M, Anthony J M, 2001 *J. Appl. Phys.* **89** (no. 10) 5243
[15] Mikhaylov A N, Belov A I, Guseinov D V et al., 2015 *Mat. Sci. Eng. B* **195** 48

Temperature dependence of silicon carbide drift step recovery diodes injection electroluminescence

A V Afanasyev[1], B V Ivanov[1], V A Ilyin[1], V V Luchinin[1], K A Serguschichev[1], A A Smirnov[1], A F Kardo-Sysoev[2]
[1] Center of diagnosis and microtechnology, Saint Petersburg Electrotechnical University "LETI", Saint Petersburg 197376, Russia
[2] Ioffe Institute RAS, Saint Petersburg 194021, Russia

Abstract. The experimental results of silicon carbide (SiC) drift step recovery diodes (DSRDs) temperature dependence of injection electroluminescence (IEL) spectra were presented. It was shown that in the forward current range I_f = 0,1…1 A the DSRD-dies temperature was raised from 327 K to 546 K correspondingly. While the short-wavelength maximum of IEL spectra - λ_{max1} shifts from 392.4 to 402.1 nm and possesses dependence close to linear. On the basis of obtained calibration curves it is possible the non-contact temperature measuring of SiC-DSRDs by electroluminescence spectra at their operation in the generator of high voltage pulses.

1. Introduction

Among the diode opening switches which are used in the voltage pulse generators with the inductive energy storage, the drift step recovery diodes (DSRDs) occupy particular place [1]. In [2,3] are presented the results of development of the voltage pulse generators with use of silicon carbide DSRDs which form the voltage impulses of nano- and subnanosecond duration with a peak power of hundreds of kilowatt up to megawatt units and pulse repetition rate of tens to hundreds of kHz in continuous operation. For designing of the optimal structure of 4H-SiC-DSRDs, evaluating of the effectiveness of the cooling system and to prevent the overheating and selection of the optimal functioning mode of SiC switches at their operation as a part of the voltage pulse generator it is necessary to have accurate information on the device temperature. The application of optical temperature control methods allows measuring the DSRD temperature in operation mode of high-voltage pulse generator. In this case, the additional temperature sensors are not necessary in generator scheme.

The aim of this work is the experimental determination of the temperature dependence of the injection electroluminescence (IEL) spectra of 4H-SiC-DSRDs at various values of direct forward currents.

2. Experimental Details

4H-SiC DSRD samples were fabricated on 4° off *n*-type 4H-SiC wafers with p^+-p-n^+ CVD grown epitaxial stack. The n^+-emitter layer of 3μm with net donor concentration of $3\cdot10^{18}$cm^{-3} was grown directly on the substrate. The *p*-drift layer of 15μm with Al concentration is of $5\cdot10^{15}$ cm^{-3}. The p^+-emitter of 1.5 μm with Al concentration of $1\cdot10^{19}$cm^{-3} forms the structure's top layer. This p^+-p-n^+ stack has been optimized using Synopsys Sentaurus TCAD software for switching time of a DSRD with a maximum reverse voltage of 1800V. Based on the data obtained in the work [4,5] the effects of high doping and high injection levels, cascade generation and incomplete dopants ionization were accounted in the simulation. DSRD samples were fabricated in form of mesa-diodes with active area $2mm^2$ and die-size 2,5x2,5 mm^2. Top view of fabricated 4H-SiC-DSRDs is presented in Figure 1.

Figure 1. Top view of fabricated 4H-SiC-DSRDs.

A direct current has been transmitted through 4H-SiC-DSRDs which value have been changed within the range from 0.1...1 A in increments of 100 mA. Diodes` temperature has been measured by a chromel-alumel thermocouple. Additional control of diodes temperature has been carried out by thermal imaging infrared camera FLIR A315. In preliminary experiments divergence diode`s temperature measurements using thermocouples and an infrared camera does not exceed a few percent. The spectral characteristics of the 4H-SiC-DSRDs IEL have been recorded by compact optical spectrometer ISM3600 developed by St. Petersburg Electrotechnical University [6]. With precision positioner Cascade Microtech DCM210 fiber optic line installed at a distance of 100...200 μm from the end of the test sample 4H-SiC-DSRDs.

3. Experimental Results

The spectral characteristics of 4H-SiC-DSRDs IEL for the respective temperature values are shown in Figure 2. From the presented dependence is seen that at the diode temperature T = 327 K short wavelength (λ_{max1} = 392.4 nm) and long wavelength (λ_{max2} = 425.1 nm) maximums are equal by value. With increasing of forward current (and diode temperature respectively), the short wavelength maximum (λ_{max1} = 392.4 nm) is shifted to longwave. The inset to Figure 2 shows the temperature dependence of the energy position of short wavelength maximum $E_{\lambda max1}$ spectral dependences of the electroluminescence DSRDs. A certain tilt of the approximation straight line to the experimental data of the temperature coefficient β of $E_{IEL}(T)$ has a value $3.2 \cdot 10^{-4}$ eV/K. The resulting value is close to the temperature coefficient of silicon carbide bandgap obtained in [7,8]. Thus, we can say that the change in the maximum spectral characteristics DSRDs electroluminescence associated with the temperature dependence of the band gap of silicon carbide.

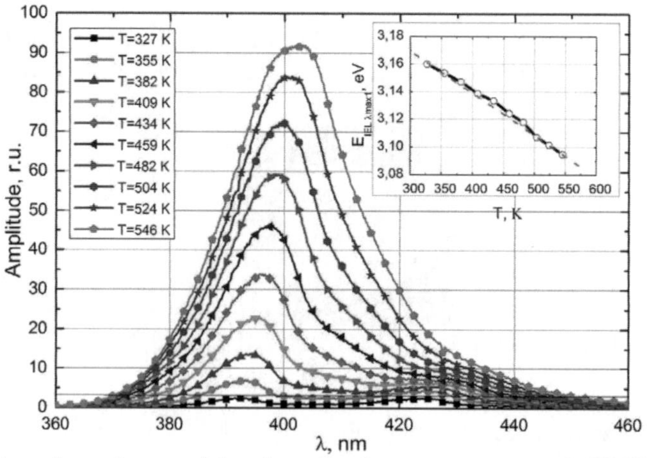

Figure 2. Temperature dependences of the electroluminescence spectra of 4H-SiC-DSRDs.

Experimental verification of the possibility of non-contact measurement of the silicon carbide DSRDs temperature was carried out at his work as an opening switch in the voltage pulse generator, which scheme is shown in Figure 3. The operating principle of scheme is as follows. From the external clock oscillator enters the rectangular control pulses by duration of 50...100 ns, performing the activation of the switch S1, which acts as a MOSFET-transistor IXZ631DF18N50 (IXYS Corp.), whose maximum pulse current is of 95 A. After opening of the switch S1 through the diode D1 (4H-SiC-DSRD) is passed the direct current pulse which carries out injection of the electron-hole plasma to the diode *p*-base. At the end of the control pulse from an external clock oscillator, the voltage at the DSRD abruptly changes its polarity, wherein the reverse current flows through the diode D1. When all the nonequilibrium electron-hole plasma is derived from the diode by the reverse current, the diode is changed over to the locked condition and at the Rload is formed fast rising voltage pulse which amplitude and timing parameters are determined both by DSRDs and a scheme in which it operates.

Figure 3 represents the photo of the generator output part which is responsible for operation of silicon carbide DSRDs and formation of high-voltage pulses. In Figure 3 are pointed: 1 – 4H-SiC-DSRD, 2 – the clamping contact to the p^+-electrode of the diode, 3 – the optical fiber cable.

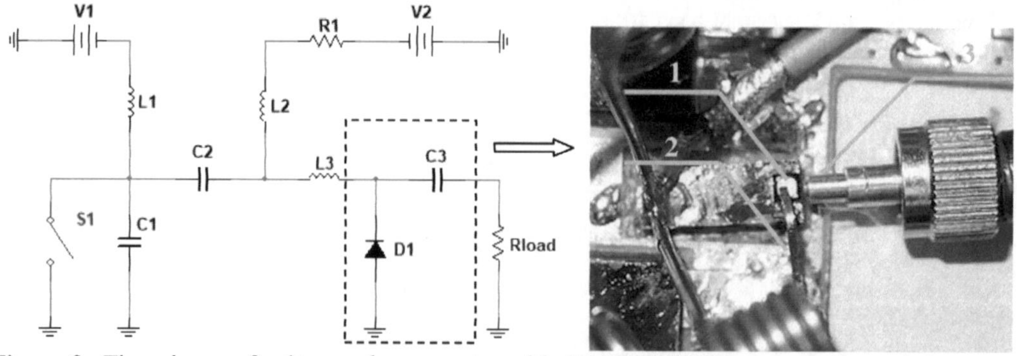

Figure 3. The scheme of voltage pulse generator with 4H-SiC-DSRD as an opening switch.

In Figure 4 is shown the oscillogram of the voltage pulses which are formed by the generator with the developed 4H-SiC-DSRDs as an opening switch. The amplitude of the pulses is equal to 1800 V, the duration at full width at half maximum (FWHM) is of 2.3 ns, the duration of the leading edge is

1050 ps. The pulse repetition frequency can vary from a few to hundreds of kHz in continuous mode and is limited to the maximum operating temperature of the diode.

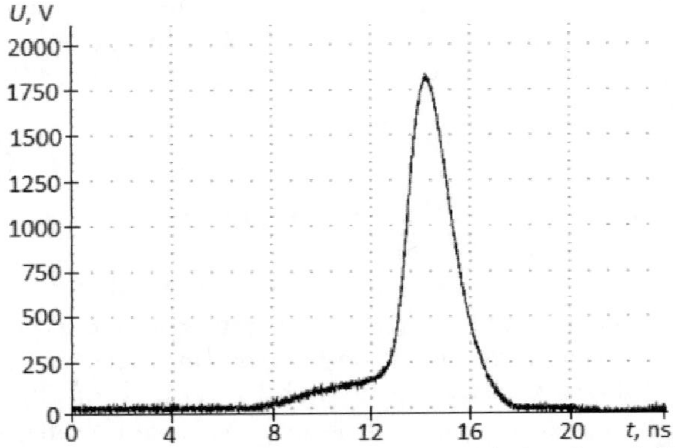

Figure 4. The oscillogram of formed pulses by generator with developed 4H-SiC-DSRDs.

For experimental verification of the possibility of non-contact measurement of silicon carbide DSRD temperature at switching of high-voltage pulses was done as follows. Sequentially change the pulse repetition rate of output voltage pulses, which led to an increase in the average switching power, and as a consequence, to the heating of the diode structure. Using measurements of the electroluminescence spectra of 4H-SiC-DSRDs, it was found that an increase in the frequency of the output voltage pulses corresponding to the diode heating occurs.

Described in this paper the non-contact temperature measurement method can be used to determine the temperature not only of individual diodes, but diodes assembled in a high-voltage stack that is capable of switching the value of units-hundreds of kilowatts, and consisting of tens of diodes. Figure 5 shows a high-voltage assembly, consisting of four silicon carbide DSRDs and capable of switching voltage of about 7 kV value.

Figure 5. High-voltage diode assembly, consisting of four 4HiSiC-DSRDs.

4. Conclusion

As a result of experimental studies of the temperature dependence of the spectra of the injection electroluminescence developed SiC drift step recovery diodes it was found that by increasing the diode temperature from 327 to 546 K, the maximum λ_{max} of the spectral lines shift (λ_{max1} = 392.4 nm, λ_{max2} = 425.1 nm) in long-wave part of the spectrum and is associated with the heating of the DSRDs, which is explained by the temperature dependence of the band gap. Definite of the received dependencies the temperature coefficient β of E_{IEL}(T) has a value of $3.2 \cdot 10^{-4}$ eV/K.

On the basis of experimental data on the temperature of silicon carbide diode structure when working with direct current and temperature dependences of the spectrum maximum injection electroluminescence it is possible to determine with high accuracy (a few Kelvin) in a contactless manner the temperature of 4H-SiC-DSRDs at their work in the structure of the high-voltage generator of ultrashort voltage pulses.

References

[1] Grekhov I V and Mesyats G A 2005 "Nanosecond semiconductor diodes for pulsed power switching" *Phys.-Usp.* vol. **48** no. 7 pp 703-712

[2] Afanasyev A V, Demin Y A, Ivanov B V, Ilyin V A, Kardo-Sysoev A F, Luchinin V V, Smirnov A A 2013 Miniature high-voltage silicon carbide source of power nanosecond pulses to generate the X-ray and microwave radiation *Nano and microsystem technique* **2** p 30

[3] Afanasyev A V, Demin Y A, Ivanov B V, Ilyin V A, Kardo-Sysoev A F, Luchinin V V, Smirnov A A 2015 Megawatt generator of nanosecond pulses on the basis of silicon carbide drift step recovery diodes *Proceedings of the Russian Universities: Radioelectronics* **3** p 21

[4] Ilyin V A, Afanasyev A V, Ivanov B V, Kardo-Sysoev A F, Luchinin V V, Reshanov S A, Schöner A, Sergushichev K A, Smirnov A A 2016 High-Voltage Ultra-Fast Pulse Diode Stack Based on 4H-SiC *Materials Science Forum* Vol. **858** pp 786-789

[5] Afanasyev A V, Ivanov B V, Ilyin V A, Kardo-Sysoev A F, Luchinin V V, Smirnov A A, Shevchenko S A 2015 Study the process of switching silicon carbide drift step recovery diodes *All-Russian Scientific and Technical Conference "Electronics and Microelectronics Microwave"* pp 215-219

[6] Kostrin D K, Ukhov A A 2013 Hardware-software complex for spectrometric investigation of parameters of light-emitting diodes *Biotechnosphera* **3** pp 21-25

[7] Grivickas P, Grivickas V, Linnros J, Galeckas A 2007 Fundamental band edge absorption in nominally undoped and doped 4H-SiC *Appl. Phys.* **101** p 123510

[8] Galeckas A, Grivickas P, Grivickas V, Bikbajevas V, Linnros J 2002 Temperature Dependence of the Absorption Coefficient in 4H- and 6H-Silicon Carbide at 355 nm Laser Pumping Wavelength *Phys. Status Solidi* (*A*) **191** p 613

State diagnostics of RTD based on nanoscale multilayered AlGaAs heterostructures

M O Makeev, S A Meshkov, V Yu Sinyakin
Bauman Moscow State Technical University, Moscow, Russia

Abstract. In the present work the problems of technical diagnostics of RTD based on nanoscale multilayered AlGaAs heterostructures are being solved. The technique and the algorithms of RTD functionality region developing are being considered.

1. Introduction

One of the most promising elements of SHF and EHF radioelectronics is a resonant tunneling diode (RTD). RTDs can be regarded as nanoobjects due to the physical phenomenon of electron resonant tunneling occurring in resonant tunneling structure, which consists of the layers with the thicknesses from one to dozens nanometers [1-6].

The prospects of RTD applications for modern radioelectronics are caused by the following aspects:

- RTD operates in a temperature range and under other external effects required for technical applications;
- technologies and facilities for RTD production are validated in the framework of micro and nanoelectronics technologies;
- limiting operating frequency of RTD reaches up to THz units;
- changing the heterostructure layers parameters (thickness, chemical composition), one can manage the IV-curve shape and create diode with optimal IV-curve for the exact type of nonlinear transformation.

In [1-6] it is shown, in particular, that the possibility of IV-curve optimization as well as declining section presence in it allow creating a number of nonlinear SHF radiosignal transformers with improved characteristics based on RTD. A set of possible nonlinear transformations with RTD usage is very wide: radiosignals generation, frequency modulation, mixing of radiosignals, amplitude detection, rectification, frequency tags grid generation etc.

The task of assignment parameters improvement by usage of a new domestic hardware components is particularly true for space radioelectronic systems, for which an emergence of qualitatively new functions, increased requirements to accuracy and effectiveness of a task performance, intensification of operational modes are characteristic. Especially rigid demands are made to the reliability of space radioelectronic systems. On the one hand, they are dictated by hard exploitation conditions, on the other hand - by stepped-up active shelf periods of space objects. Maximum consistent and complete reliability parameters of radioelectronic systems are usually obtained as a result of exploitation. However, obtaining of such information takes time. In many cases, traditional test methods do not allow confirmation of radioelectronic systems reliability target due to hidden defects presence as well.

One of the most promising direction in the development of reliability and quality estimation methods of space-related application radioelectronic systems is prediction of their further condition.

The problem of the technical system reliability increase with reliability prediction development on specified exploitation period in specified conditions is solved by technical diagnostics (TD), the scientific base of which are outlined particularly in [7].

2. Results

Technical diagnostics solves many different problems, which are intersected with the problems of other scientific disciplines. One of the main problems of TD is the identification of the object state under conditions of limited information. Object state identification is based on identification algorithms, which are usually considered as classification problems.

Identification algorithms in technical diagnostics are based on the diagnostic models, which establish relation between object state and its representation in the diagnostic signals dimension. An important part of identification problem is the rules of decision making (decision rules).

Enumerated elements of the TD structure need to be developed regarding to the specific object of diagnostics. In this work, the major elements of TD structure regarding to RTD as a part of SHF radio signal mixer are considered. Herewith, two hypotheses are accepted:

- Reliability of RTD and radio signal mixer are based on gradual failure model.
- Degradation of radio signal mixer parameters is determined by degradation of RTD. Degradation of RTD leads to the changes of its IV-curve shape, resulting in mixer parameters changes. The failure of mixer and RTD occurs, when at least one of the mixer assignment parameters falls outside of set limits range (region of functionality).

2.1. Decision making rules

The rules of decision-making are meant for decision making of RTD IV-curve location in the region of functionality ***F***. Figure 1 shows in a simplified manner IV-curves of RTD used in mixers and conditionally marked regions of functionality ***F***, correct operation ***CO***, catastrophic failure ***CF***.

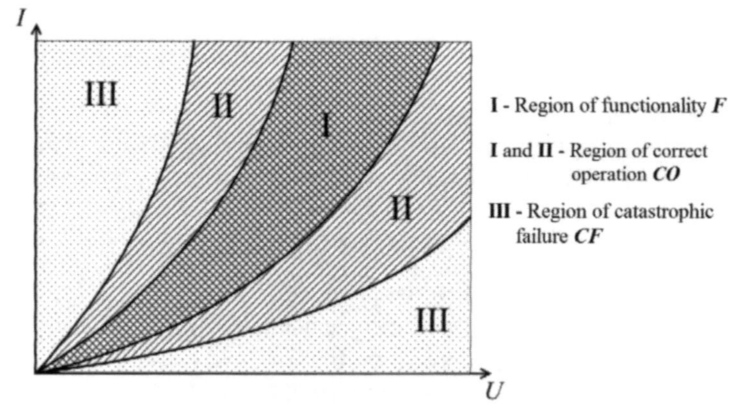

Figure 1. IV-curves range of RTD used in mixers.

RTD IV-curves rating depends on radio signal mixer characteristics based on it. Mixer functionality is determined by the following main parameters [8] conversion loss, noise factor, service band, SWR of input and output, suppression of signal on image channel, suppression of intermodulation components, signal-heterodyne channels isolation, dynamic range, and required power of heterodyne.

In this paper, conversion loss L_s, dynamic range P_{IP3}, input signal reflection factor K_{ref} are considered as the appointment parameters of a mixer. Consider that the region of mixer functionality is determined by following conditions:

$$L_s \leq L_{s\ tol}, \quad (1)$$

$$P_{IP3} \geq P_{IP3\ tol}, \quad (2)$$

$$K_{ref} \leq K_{ref\ tol}. \quad (3)$$

Decision making algorithm of relating RTD IV-curve to a certain category is shown in figure 2.

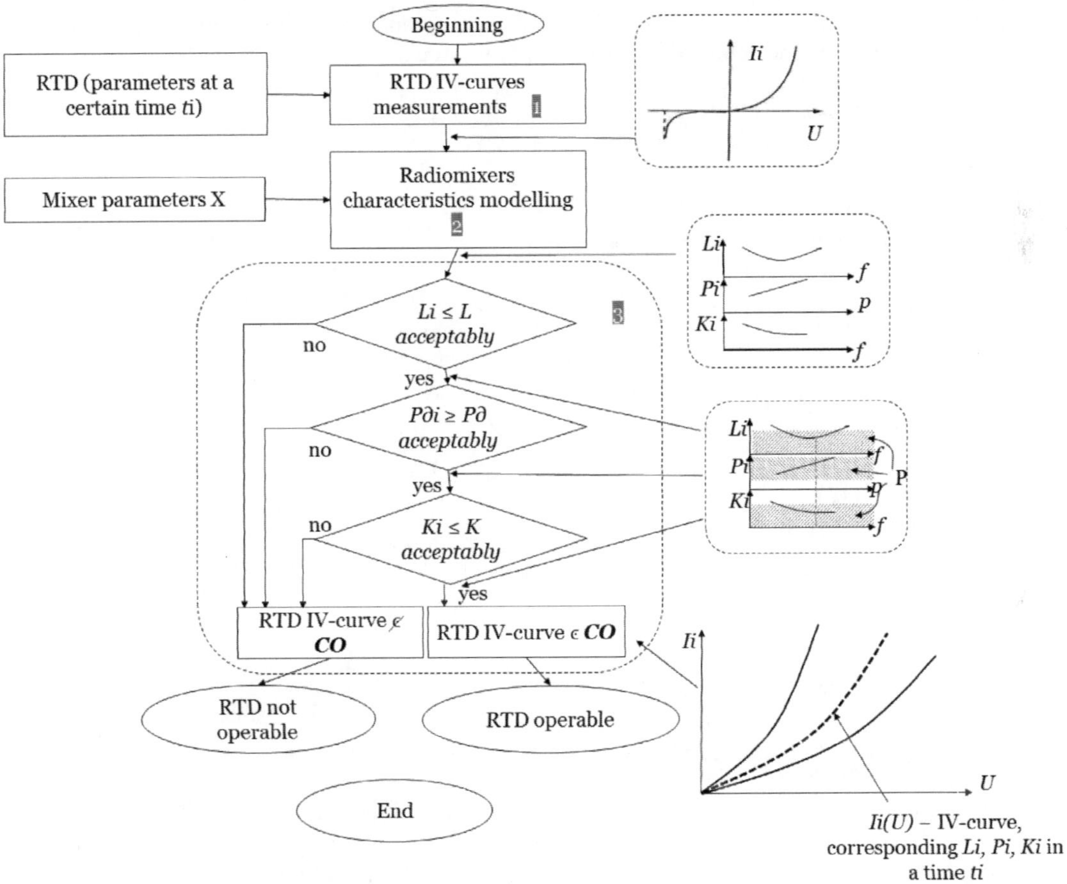

Figure 2. Decision making algorithm.

Special feature of decision making of the RTD IV-curve classification involves region of functionality F transformation from mixer parameters dimension into RTD IV-curves plan, so that one point in mixer parameter dimension $\{L_{si}, P_{IP3i}, K_{refi}\}$ corresponds to one RTD IV-curve $I_i(U_i)$.

On the first stage, a RTD IV-curve is measured, which is used in the program module, where mixer characteristics are calculated. Particularly, MWO system, widely used among the developers of SHF hardware, could be used for this particular task [9].

The electrical parameters (1-3) are the mixer modelling results.

In the following algorithm module calculated mixer characteristics are compared with the limitations (1-3). If the parameters (1-3) are fallen within corresponding limitations, it is concluded that the mixer parameters and RTD IV-curves are laid in functionality region *P*; if any of mixer parameters is out of the limitations framework, it is concluded that mixer parametric failure is detected and RTD IV-curve is related to ***CO*** or ***CF*** region.

In general, the program module of a mixer characteristics modelling can be replaced by measuring module, however such a replacement deals with necessity of a mixer mocking-up and RTD integration in it.

2.2. *Detection algorithm*

The algorithm is designed to forming RTD functionality region. Above mentioned decision making rules are the integral part of detection algorithm shown in figure 3.

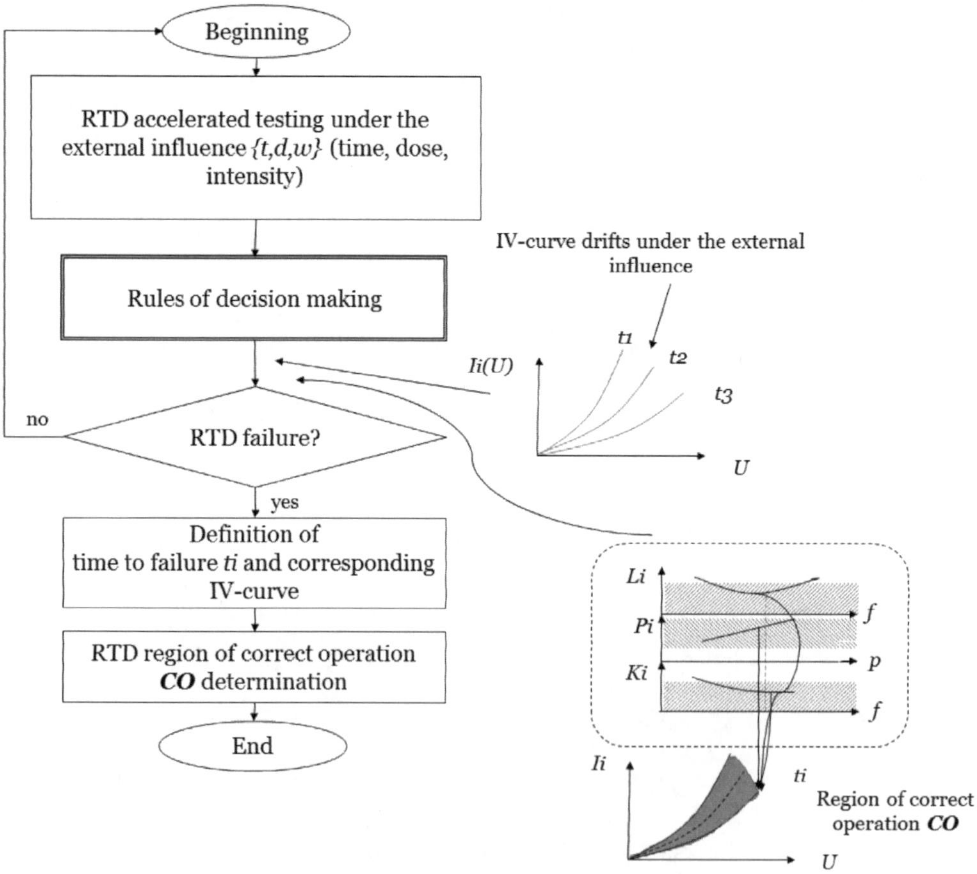

Figure 3. Detection algorithm.

The algorithm assumes carrying of RTD accelerated testing, during which the destabilizing impacts and accelerating degradation processes are applied to RTD. While choosing the type, intensity and impact duration only the characteristic of normal exploitation conditions which may cause mechanism failure of the device are considered [10]. Practically, it is high temperature tests that are frequently used for degradation process of semiconductor devices acceleration. The number of the impact cycles is chosen to be no less than 4-5 in order to estimate the RTD IV-curve kinetics. Minimal intensity and impact duration are set to such a value, so that the variation of IV-curve induced by the impact is no less than the measuring instrument error rate; maximum intensity and impact duration are limited by the risk of IV-curve overranging of ***F*** region for one impact cycle. Thus, at the input of the «decisive rules» unit goes RTD IV-curve after one cycle of destabilizing impacts, at the output of the unit–classification of RTD as failed or working component.

Termination of the cycle «destabilizing impact – decision making rule» is implemented as soon as at the time moment t_i the failure of mixer and RTD is detected. IV-curve, corresponding to the time moment t_i is a boundary of RTD functionality region.

3. Conclusions

Consideration of the proposed technique and the algorithms in relation to mixer RTD does not limit their generality. Considered technique and algorithms may be used in developing RTD's region of functionality as a part of any nonlinear device, which characteristics depend on the shape of RTD IV-curve.

The considered technique and algorithms may be used in manufacturing process while planning of control technological operations for hardware components.

Acknowledgments

The research work was supported by Ministry of Education and Science of the Russian Federation (grant № 14.574.21.0116, unique identifier RFMEFI57414X0116).

References

[1] Ivanov Y A, Meshkov S A, Fedorenko I A, Fedorkova N V and Shashurin V D 2010 *J. Commun. Technol. El.* 55 982
[2] Ivanov Y A, Meshkov S A, Sinjakin V Y, Fedorenko I A, Fedorkova N V, Fedorov I B and Shashurin V D 2011 *Nanoengineering* 1 34
[3] Ivanov Yu A, Meshkov S A, Sinyakin V Yu, Fedorenko I A, Fedorkova N V and Shashurin V D 2010 *20th International Crimean Conference "Microwave & Telecommunication Technology"* 859
[4] Ivanov Yu A, Meshkov S A, Fedorenko I A, Fedorkova N V and Shashurin V D 2011 *Microwave and Telecommunication Technology (CriMiCo), 2011 21th International Crimean Conference* 181
[5] Makeev M O, Ivanov Y A and Meshkov S A 2015 *J. Phys.: Conf. Ser.* **584** 012014
[6] Makeev M O, Ivanov Y A and Meshkov 2016 *Semiconductors* **50** 83
[7] Birger I A 1978 Tehnicheskaja diagnostika *Moscow: Mashinostroenie* 240
[8] Bova N T, Efremov Y G and Konin V V 1984 Microwave microelectronic devices Kiev: Technika 184
[9] Razevig V D, Potapov Y V and Kurushkin A A 2003 *Designing of microwave devices using Microwave Office* Moscow: SOLON-press 492
[10] Gludkin O P 1991 Metody i ustrojstva ispytanij RES i EVS *Moscow: Vysshaja shkola* 336

Journal of Physics: Conference Series **741** (2016) 012177 doi:10.1088/1742-6596/741/1/012177

Fabrication of tunnelling gap of nanomechanicalaccelerometer by focused ion beam

O A Ageev[1], E Yu Gusev[2], A S Kolomiytsev[3], S A Lisitsyn[4], and Al V Bykov[5]
Southern Federal University, Institute of Nanotechnology, Electronics and Electronic Equipment Engineering, Taganrog, 347922, Russia

[1]ageev@sfedu.ru, [2]eyugusev@sfedu.ru, [3]askolomiytsev@sfedu.ru , [4]lisitsynsa@sfedu.ru, [5]alvbykov@nt.ru

Abstract. The two-layer polysilicon surface micromachining process flow of nanomechanical accelerometerincluded a high aspect-ratio etch step was presented.In the experiments we defined modes, and developed the technology of tunnelling gap formation using focused ion beam. The nanomechanical accelerometer crystals were fabricated by surface micromachining and focused ion beams.

1. Introduction

Many micromechanical gyroscopes and accelerometers were fabricated by bulk and surface micromachining [1-6]. Its process flows mainly based on optical lithography and dry/wet etching of substrate or sacrificial layer. Typically the substrate was silicon, polysilicon and silicon dioxide used as structural and sacrificial layers, respectively. The process flow of nanomechanical system could be also based on microelectronic IC technology and materials but need to be added special steps. For high-sensitivityaccelerometers operating on tunnelling effectit could a high aspect-ratio etch [1,7-9].

The tunnelling gap of nanomechanical accelerometer can be manufactured using Focused Ion Beam (FIB) technology[1,7-13]. FIB technology enables reproducible and precise material processing with high accuracy [14]. The key feature of FIB is the high spatial resolution which is provided by the application of a gallium ion beam 7 nm in diameter, as well as by the possibility of varying the impact of the parameters over wide limits [15]. In contrast to conventional fabrication techniques based on optical lithography with the application of photoresist and material processing a direct writing mode by FIB allows precise nanopatterning even on sample with advanced topography [16-19].

Focused ion beam allows creating nanosized structures and nanodevices with different shapes and parameters. Using the polycrystalline silicon allows minimizing the negative impact of the implanted gallium ions. Combination of FIB nanopatterning and FIB-induced local deposition of materials allows precise nanofabrication of tunnelling gaps with sub 5 nm width [20].

The objective of this work is the fabrication and study of tunnelling gap of nanomechanical accelerometer using the combination of FIB technology and atomic force microscopy measurements.

2. Formation of nanomechanical accelerometer samples for high aspect-ratio etch step

We propose integrated two-layer polysilicon surface micromachining process flow of micro- and nanomechanical gyroscope and accelerometers [21]. The process flow of nanomechanical accelerometer need to be include a high aspect-ratio etch step by focused ion beams.

We fabricate wafer of non-finished nanomechanical accelerometer for latest high aspect-ratio etch step (Figure 1) according following process flow. The substrate was 100 mm n-type silicon wafer Si (100). Insulator film of silicon nitride of 0.6 μm thickness was grown on the substrate using inductively coupled plasma CVD (SemiTEq ICPd81). Than sandwich-like structures of bottom polycrystalline silicon, sacrificial silicon oxide and top polysilicon layers of 0.3, 1.5 and 2.0 μm were prepared by plasma enhanced (Oxford Instruments PlasmaLab 100) and ICP CVD. The polycrystalline silicon films were doped using the liquid PCl_3diffusant (SD.OM-3M). Polysilicon inertial masses were fabricated by optical lithography (SUSS MJB4), dry etching under masks of photoresist, aluminum and nickel (SemiTEq ICPe68), and wet etching of sacrificial layer [21,22].

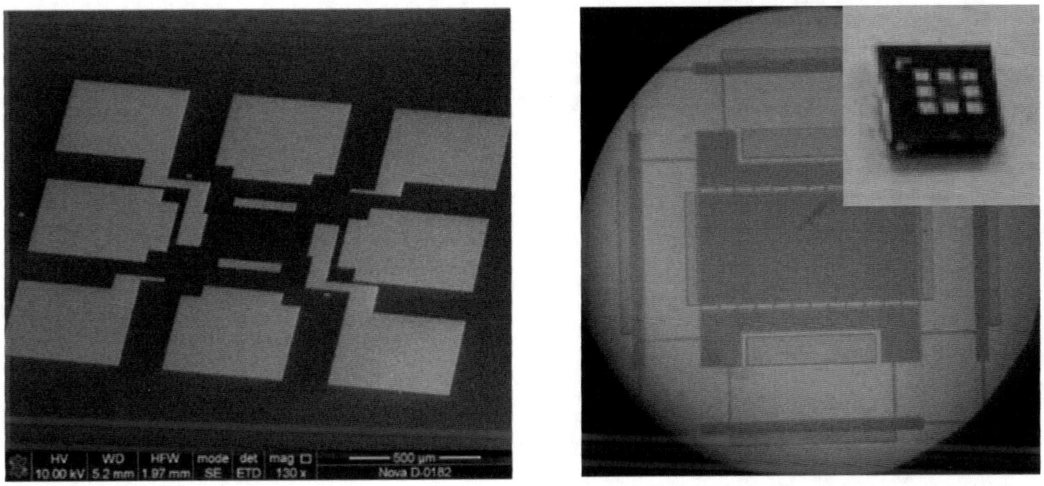

(a) **(b)**

Figure 1. Optical image of first (**a**), secondpolysilicon layers and nanomechanical accelerometer crystal (**b**).

3. Fabrication of tunneling gap using focused ion beam

At the first step of our research we investigated the dependence between the ion milling depth and FIB current for doped and undopedpolysilicon.The experimental studywas carried outby the following procedure: an array of 4 by 2 micron squares was produced by FIB local milling on a polysiliconsubstrate at various ion-beam-treatment parameters: 1)FIB current I = 10, 30, 50, 100, 300 nA; 2) Beam dwell timeDt = 50 microsecond; 3) Total FIB milling time at square structure T = 60 sec.; 4) FIB overlap OL = 50%. Thenumberofion-beampasses (#)for a given graphic pattern was chosen so that the total time of FIB exposure at each structure is 1 minute for each value of the ion beam current.

The process of ion-beam etching was monitored byscanning electron microscopy. The morphology and geometric characteristics of the structureswere analyzed using an Ntegra Vita scanningprobenanolaboratory in the intermittent mode of atomic force microscopy (Figure 2).

The results allow to calculate the dependence between FIB milling depth and ion beam current (Figure 3). The analysis of curves gives an indication of the ion-beam milling speed for different values of ion-beam current. The values of the etching rates were identified during the research. It is found that the formation of tunneling gap must apply the minimum FIB current values (approx. 1-10 pA) that corresponds to the milling rate of about 10 nm/min.

The fabrication of tunneling gap was performed with a Unique scientific equipment "Electron beam microscope with electron lithography system" Nova Nanolab 600, combining a Ga^+ FIB and a field emission scanning electron microscope [10].

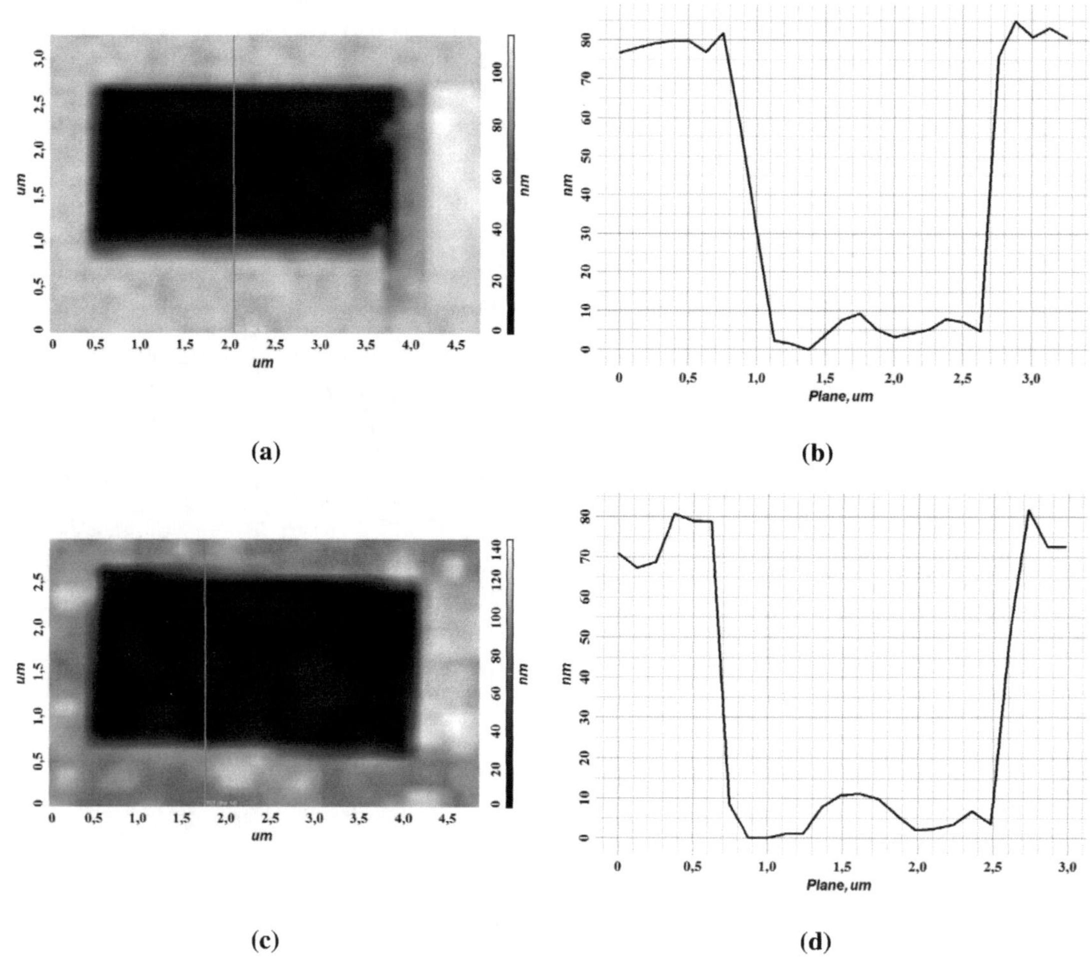

Figure 2 (a-d). Profiling of substrate surface using focused ion beam. AFM-images and profiles of undoped**(a,b)** and doped **(c,d)** samples after FIB local milling.

For the tunneling gap fabrication the following FIB parameters were used: the accelerating voltage of the ion beam – 30 keV; the ion beam current – 1.0 pA; and the dwell time of the ion beam - 1.0 μs. Figure 4(a) shows the SEM image of FIB-fabricated tunnelling gap on polysilicon structure. The total width of the gap is approximately 22.0 nm which is not suitable for tunneling devices. To reduce the width of the gap the method of FIB-induced deposition of tungsten was used. After the ion-induced deposition, the gap width was reduced to approximately 4.0 nm as shown at Figure 4(b).

4. Conclusion

Previouslywe fabricate wafer of half-finished nanomechanical accelerometer by surface micromachining for latest high aspect-ratio etch step.

Then we have illustrated the procedure, based on FIB milling and deposition, to create of tunneling gap of nanomechanical accelerometer.It is shown that the use of FIB technology allows forming tunneling gaps with different values of the electrode spacing.It is shown that at minimum FIB current values the polysilicon milling rate is about 10 nm/min.

Finally the nanomechanical accelerometer with tunneling gap were finished by FIB.

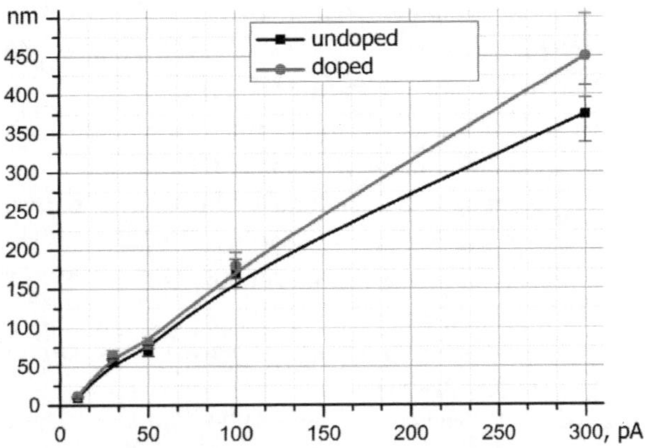

Figure 3. FIB milling depth vs. ion beam current (exposure time of 1 minute).

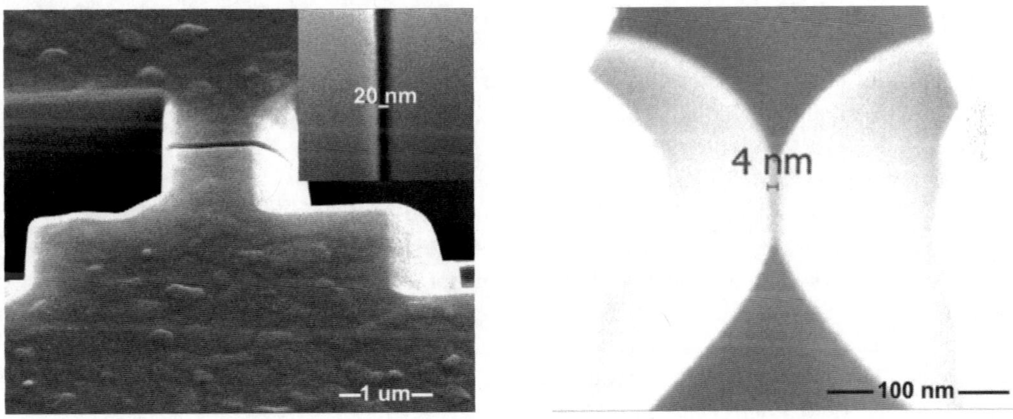

(a) **(b)**

Figure 4 (a, b). SEM image of nanoaccelerometer gap: etched (**a**)and deposited (**b**).

Combining scanning probe technology and bitmap-based FIB technology will open a new direction of development of MEMS and NEMS. The obtained results can be used to develop the technological processes of the production of nanomechanicalaccelerometers.

Acknowledgment

The results were obtained by using the equipment of the Centre of Collective Use of Equipment "Nanotechnologies" of Southern Federal University. This research was financially supported by the

Ministry of Education and Science of Russian Federation within the contract 14.575.21.0045 (unique identifier RFMEFI57514X0045).

References

[1] Bhushan B 2010 *Springer Handbook of Nanotechnology* (New York: Springer)
[2] Narasimhan V, Li, H, Jianmin M 2015 *Journal of Micromechanics and Microengineering***25(3)**
[3] Biswas K, Sen S, Dutta P K 2007 *Sensor Letters***5(3-4)** 471
[4] Zhanshe G, Fucheng C, Boyu L, Le C, Chao L, Ke S 2015 *Microsystem Technologies***21(10)** 2053
[5] Maenaka K 2014 *IEEJ Transactions on Sensors and Micromachines***134(7)** 175
[6] Xia D, Yu C, Kong L 2014 *Sensors (Switzerland)***14(1)** 1394
[7] Patra S, Bhattacharyya T K 2010 *Proceedings of IEEE Sensors***5690642** 1192
[8] Daniel J H, Moore D F 1999 *Sensors and Actuators***73**201
[9] Moore D F, Burgess S C, Chiang H-S, Klaubert H, Shibaike N, Kiriyama, T 1995 *Proceedings of SPIE - The International Society for Optical Engineering***2639** 253
[10] Ageev O A, Kolomiytsev A S, Bykov Al V, Smirnov V A, Kots I N 2015 *Microelectronics Reliability* **55** 2131
[11] Giannuzzi L A, Stevie F A 2004 *Introduction to Focused Ion Beams: Instrumentation, Theory, Techniques and Practice* (New York: Springer)
[12] Schmidt M, Johari Z, Ismail R, Mizuta H, Chong H 2012 *Microelectron. Eng.***98** 313
[13] Tseng A 2004 *J. Micromech. Microeng.***14** 15
[14] Han J, Lee H, Min B, Lee S 2010 *Microelectron. Eng.***87** 1
[15] Konoplev B G, Ageev O A, Smirnov V A, Kolomiitsev A S, Serbu N I 2012 *Russ. Microelectron.***41(1)** 41
[16] Orloff J, Swanson L W, Utlaut M 2003 Springer New York
[17] NagaseM, NakamatsuK, Matsui S, Namatsu H 2005 *Japanese J. of Appl. Phys.***44(7)** 5409
[18] Orloff J 2009 CRC Press New York
[19] SchafferM, SchafferB, Ramasse Q 2012 *Ultramicroscopy* **114** 62
[20] Ageev OA, Kolomiytsev A S, Konoplev B G 2011*Semiconductors***45(13)** 89
[21] Ageev O A, Balakirev S V, Bykov Al V, Gusev E Yu, Fedotov A A, Jityaeva J Y, Il'in O I, Il'ina M V, Kolomiytsev A S, Konoplev B G, Krasnoborodko S U, Polyakov V V, Smirnov V A, Solodovnik M S, Zamburg E G 2016 *Springer Proceedings in Physics* (New York: Springer International Publishing)**175** 563
[22] Ageev O A, Gusev E Yu, Jityaeva J Yu, Kolomiytsev A S, Bykov Al V 2016*CRC Press/Balkema*Taylor & Francis GroupLondon13

The study of influence of the gas flow rate to etched layer thickness, and roughness of the anisotropy field of gallium arsenide is etched in the plasma chemical etching process

O A Ageev, V S Klimin, M S Solodovnik, A V Eskov, S Y Krasnoborodko
Department of Nanotechnologies and Microsystems, Institute of Nanotechnologies, Electronics and Electronic Equipment Engineering, Southern Federal University, 2 Shevchenko Street, Taganrog, 347928, Russia

E-mail: ageev@sfedu.ru, kliminv.s@mail.ru, solodovnik@sfedu.ru, eskov@sfedu.ru, krasnoborodko@sfedu.ru.

Abstract. In the experiments on the etched surface of gallium arsenide were performed. We studied the effect of BCl_3 gas flow rate on the thickness of the etched layer. GaAs etching rate was: 537,4 nm/min 28,7 nm/min 2,6 nm/min, the values of the flow rate of BCl_3 N_{BCl3} - 15, 10, 5 cc/min, respectively. The effect of BCl_3 gas flow rate to the mean-square roughness of the etched surface. The influence of the anisotropy of the process on the geometry of the etched area. Revealed that the deflection angle for the samples treated with the working gas flow rate N_{BCl3} - 15 cc/min in the [110] direction was $\alpha_{[110]}$ = 65,5° in direction [111] was $\alpha_{[111]}$ = 45,58°. For samples treated with the working gas flow rate N_{BCl3} - 10 cc/min in the [110] direction was $\alpha_{[110]}$ = 20,94° in direction [111] was $\alpha_{[111]}$ = 11,37°. For samples treated with the working gas flow rate N_{BCl3} - 5 cc/min in the [110] was $\alpha_{[110]}$ = 0,32° in direction [111] was $\alpha_{[111]}$ = 0,21°. The results can be used to produce discrete diodes, heterojunction devices, and others.results.

1. Introduction

The widespread use of plasma chemical etching processes in micro- and nanoelectronics technology represents an important development trend of technology, is the transition from the liquid to the processes of "dry" gas-phase processes [1-6]. For plasma BCl_3 characterized by significantly lower concentrations of chlorine atoms, but is often seen planting solids plasma-chemical reactions on the walls of the reactor and the surface of the treated material, but its benefits are not fully realized due to poor knowledge of the mechanisms of active plasma particles interact with GaAs[7-8]. The lack of information about the types of active particles, interoperable, limiting stage of the process, their kinetic characteristics (probability, rate constants) does not control the result of the processing efficiently, optimize the process mode, as well as to carry out an adequate comparison of the output parameters of the process with a studied system [9- 12].
Atomic force microscopy is one of the most promising methods for studying the surface topography, the past treatment. The big advantage of this method is the possibility of obtaining nanometric resolution without the use of vacuum [13- 17].

The aim of this work was to study by atomic force microscopy etched layer thickness, quality of the etched surface of the structure and anisotropy profile gallium arsenide after plasma chemical etching in combination plasma capacitive and inductive discharge of the mixture at changing the working gas flow [18- 21].

2. Description of the experiment and research methods

In this paper, the method of plasma chemical etching in combination with perchloric capacitive and inductive plasma discharge. This method makes it possible to combine the two modes of plasma chemical etching: reactive ion etching and etching in inductively coupled plasma. Installation STE ICPe68 is used for plasma chemical etching. The diameter of the processed wafers - 150 mm, with the ability to use samples of arbitrary shape, it can be used as a platform for application development and research and for industrial production. The unit is equipped with a reactor gas supply system, to optimize the uniformity of processes and reduce pumping time is also provided for installation in helium cooling for long etching processes [22].

As experimental samples fragments own GaAs wafers were used having a standard liquid surface after polishing. Preparation of gallium arsenide substrate was the fact that the surface of the deposited protective mask pattern plasma resistant photoresist.

Control surfaces after plasma treatment was carried out by means of nano-lab «NTegra» probe using a "semi-contact" mode. For samples treated with plasma AFM images were obtained which was built profillogramma on which determines the depth of etched layer and the etching angle. So with the help of a specialized package «Image Analysis» estimated etched surface roughness. Prior to the measurements on the microscope photoresist mask is removed.

In this paper, experimental studies were conducted to identify the etching rate, the roughness of the etched area and the process of the anisotropy of BCl_3 working gas flow rate in the combined plasma. Samples were processed at a pressure of 2 Pa, a capacitive plasma source power W_{RIE} - 35 W, bias voltage U_{bias} = 102 V, the power source is inductively coupled plasma W_{ICP} - 300 W, the buffer gas flow rate was N_{Ar} - 100 cc/min speed working gas flow changed N_{BCl3} - 5, 10, 15 cc/min, the etching time at all stages was t = 1,2,3 minutes.

3. Results and discussion

Experimental samples obtained during the first phase, were investigated by atomic force microscopy. Figure 1 shows the SEM image obtained by scanning and profillogramma samples after the plasma etching process gas flow rate was N_{BCl3} - 15 cc/min.

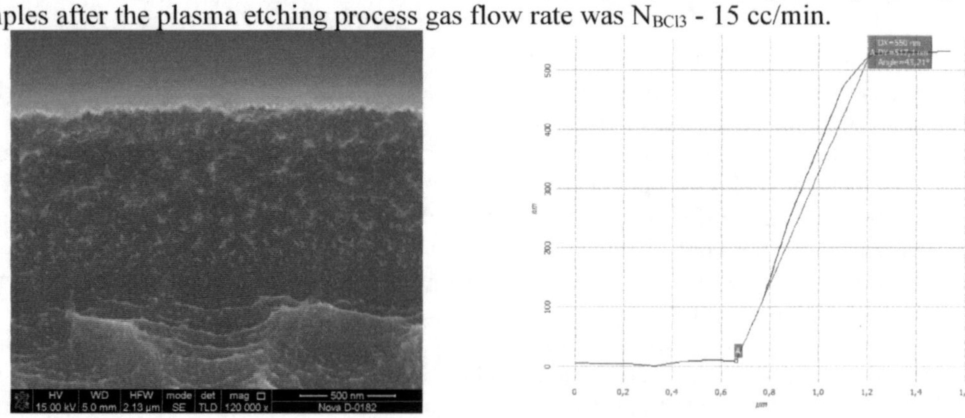

Figure 1. SEM of etched layer and profillogramma GaAs at BCl3 - 15 cc/min, the etching time of 1 minute.

According to the results of this stage of experimental studies have been constructed according to the thickness of the layer etched by the etching time for various values of working gas flow rates shown in Figure 2.

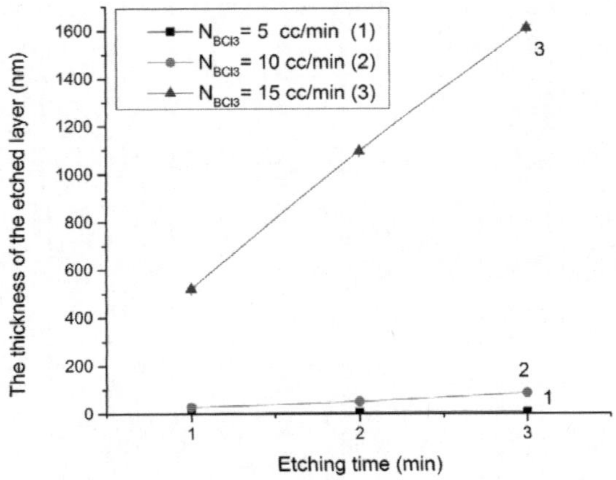

Figure 2. The dependence of the thickness of the etched GaAs times, for different values of the working gas flow rate.

The results of the research of the dependencies, presented in Figure 3, we can estimate the rate of etching. GaAs etching rate was: 537,4 nm/min; 28,7 nm/min; 2.6 nm/min, the values N_{BCl3} working gas flow rate - 15, 10, 5 cc/min, respectively.

As was the dependence of the mean square roughness of the etched area of the etching time for various values of the flow rate of the working gas, which is shown in Figure 3.

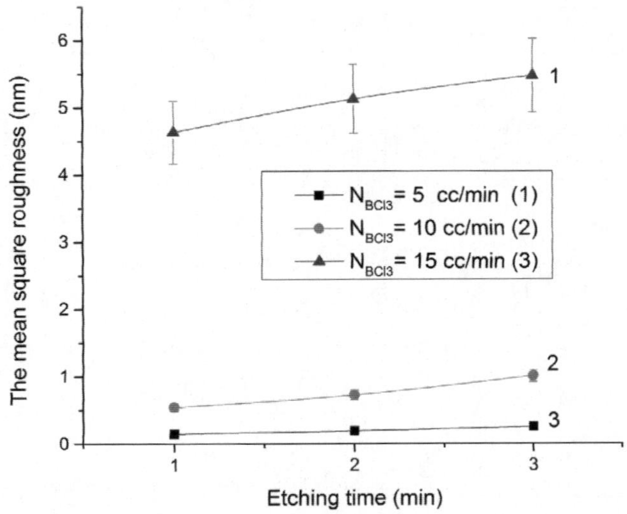

Figure 3. The dependence of the mean square roughness of the etched GaAs layer from time to time, for various values of the working gas flow rate.

Journal of Physics: Conference Series **741** (2016) 012178 doi:10.1088/1742-6596/741/1/012178

Because of dependencies that at least the values of the working gas flow rate are reduced etch rate and, accordingly, the root mean square roughness of the etched surface.
As it is known, in plasma-chemical etching of the etched area has a trapezoidal shape and anisotropy etched surface can be estimated by the vertical wall angle different directions. Figure 4 shows two lateral profillogrammy etched surfaces.

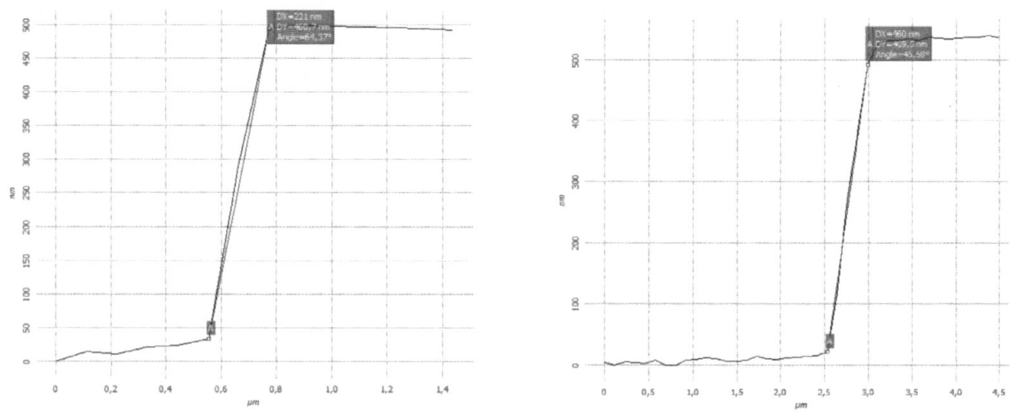

Figure 4. Profilogram etched sides along different crystallographic directions:

a) crystallographic direction [110];

b) crystallographic direction [111].

Revealed that the deflection angle for the samples treated with the working gas flow rate N_{BCl3} - 15 cc/min in the [110] direction was $\alpha_{[110]}$ = 65,5° in direction [111] was $\alpha_{[111]}$ = 45,58°. For samples treated with the working gas flow rate N_{BCl3} - 10 cc/min in the [110] direction was $\alpha_{[110]}$ = 20,94° in direction [111] was $\alpha_{[111]}$ = 11,37°. For samples treated with the working gas flow rate N_{BCl3} - 5 cc/min in the [110] was $\alpha_{[110]}$ = 0,32° in direction [111] was $\alpha_{[111]}$ = 0,21°.
The results of the experiments were built according to the angle of the anisotropy of the working gas flow rates N_{BCl3} (Figure 5).

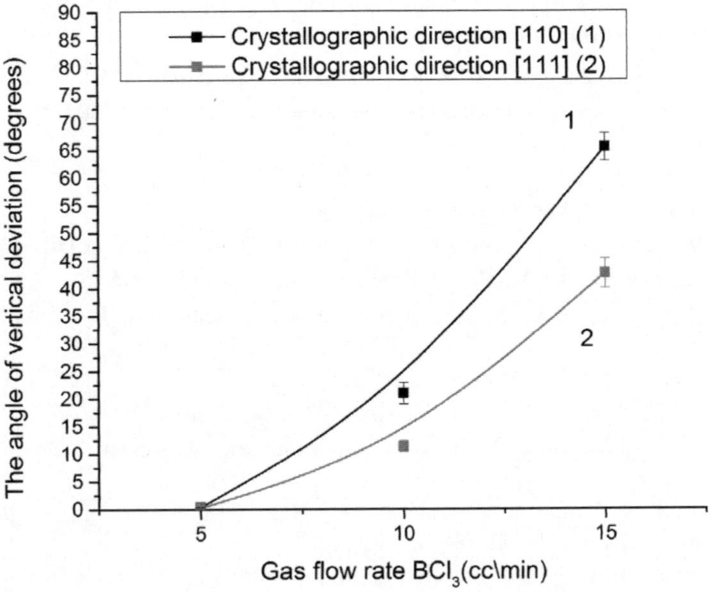

Figure 5. The dependence of the deflection angle from the vertical of the working gas flow rate.

The dependence can be seen that at lower flow of the working gas and, respectively, and the angle of anisotropy etching is reduced at lower speeds, in different crystallographic areas seeking to vertical.

4. Conclusion

In this paper we studied the influence of working gas flow rate of plasma-chemical etching process in a combined plasma etched layer thickness, roughness of the etched surface and the anisotropy of the process in an environment of technological gases BCl3 and Ar.

The rates of the etching, the value of the mean square roughness and anisotropy angles for modes shown in the work. It was also found that when the power source is inductively coupled plaza burr remains relatively constant, while reducing capacitive plasma source power and bias voltage decreases accordingly decreases the rms roughness of the etched surface.

Using the experimental results, we can find the optimal etching modes, which allow the most precise control over the thickness of the etched layer is etched surface roughness based on the anisotropy of the process.

Acknowledgments

This work was supported by the Russian Science Foundation Grant No. 15-19-10006. The results were obtained using the equipment of Common Use Center and Education and Research Center "Nano-technologies" of Southern Federal University.

References

[1] LaBella V P, Bullock D W, Ding Z, Emery C, Harter W G and Thibado P M 2000 *J. Vac. Sci. Tech.* A **18** 1526

[2] Ledentsov N N *et al.* 1996 *Sol. St. Electron.* **40** 785

[3] Riel B J, Hinzer K, Moisa S, Fraser J, Finnie P, Piercy P, Fafard S and Wasilewski Z R 2002 *J. Cryst. Growth* **236** 145

[4] Morgan C G, Kratzer P and Scheffler M 1999 *Phys. Rev. Lett.* **82** 4886

[5] Murdick D A, Wadley H N G and Zhou X W 2007 *Phys. Rev.* B **75** 125318
[6] Shiraishi K and Ito T 1998 *Phys. Rev.* B **57** 6301
[7] Amrani A, Djafari Rouhani M and Mraoufel A 2011 *Appl. Nanosci.* **1** 59
[8] Kangawa Y, Ito T, Taguchi A, Shiraishi K, Irisawa T and Ohachi T 2002 *Appl. Surf. Sci.* **190** 517
[9] Daweritz L and Ploog K 1994 *Semicond. Sci. Tech.* **9** 123
[10] Foxon C T and Joyce B A 1977 *Surf. Sci.* **64** 293
[11] Tok E S, Neave J H, Zhang J, Joyce B A and Jones T S 1997 *Surf. Sci.* **374** 397
[12] Kley A, Ruggerone P and Scheffler M 1997 *Phys. Rev. Lett.* **79** 5278
[13] Avery A R, Dobbs H T, Holmes D M, Joyce B A and Vvedensky D D 1997 *Phys. Rev. Lett.* **79** 3938
[14] Nishinaga T and Shen X Q 1994 *Appl. Surf. Sci.* **82-83** 141
[15] Nishinaga T 2004 *Prog. Cryst. Growth Charact. Mater.* **48-49** 104
[16] Higuchi Y, Uemura M, Masui Y, Kitada T, Shimomura S, Hiyamizu S 2003 *J. Cryst. Growth* **251** 80
[12] Schmidt M, Johari Z, Ismail R, Mizuta H, Chong H 2012 Microelectron. Eng. 98 313
[13] Tseng A 2004 J. Micromech. Microeng. 14 15
[14] Han J, Lee H, Min B, Lee S 2010 Microelectron. Eng. 87 1
[15] Konoplev B G, Ageev O A, Smirnov V A, Kolomiitsev A S, Serbu N I 2012 Russ. Microelectron. 41(1) 41
[16] Orloff J, Swanson L W, Utlaut M 2003 Springer New York
[17] Nagase M, Nakamatsu K, Matsui S, Namatsu H 2005 Japanese J. of Appl. Phys. 44(7) 5409
[18] Orloff J 2009 CRC Press New York
[19] Schaffer M, Schaffer B, Ramasse Q 2012 Ultramicroscopy 114 62
[20] Ageev O A, Kolomiytsev A S, Konoplev B G 2011 Semiconductors 45(13) 89
[22] http://www.semiteq.ru

Catalytic activity of metallic nanoisland coatings. The influence of size effects on the recombination properties

O A Tomilina[1], V N Berzhansky[1,2], S V Tomilin[2], A N Shaposhnikov[1]

[1]Research Centre of Functional Materials and Nanotechnologies, Physical-Technical Institute, Vernadsky Crimean Federal University, Crimea, Simferopol 295007, Russian Federation
[2]Experimental Physics Department, Physical-Technical Institute, Vernadsky Crimean Federal University, Crimea, Simferopol 295007, Russian Federation

Abstract. The results of investigations of the quantum-size effects influence on selective properties of heterogeneous nanocatalysts are presents. As etalon exothermic reaction was used the reaction of atomic hydrogen recombination. The nanostructured Pd and Pt films on Teflon substrate were used as a samples of heterogeneous nanocatalysts. It was shown that for nanoparticles with various sizes the catalytic activity has the periodic dependence. It has been found that for certain sizes of nanoparticles their catalytic activity is less than that of Teflon substrate.

1. Introduction

In ultra-small nanoparticles (< 10 nm) the structure of the electron and phonon spectrum is discrete due to the small number of atoms and the effect of spatial quantization of energy levels. These nanoparticles can't be clearly identified as metals or semiconductors. They are called quantum dots [1,2]. For these nanoobjects the energy absorption for the excitation of new phonons or electron transitions occurs not in continuous range, but discrete – with certain portions (quants). In the case of optical radiation absorption this effect appeared in the form of the depression and peaks in the transmission (absorption) spectra. In the study of heterogeneous catalysis, this effect leads to a selectivity properties of the solid state catalysts in relation to reactions with various releases of energy. The goal of this work was to make comparative researches of catalytic properties of nanostructured metal films with different effective thicknesses which have obtained by thermal deposition.

2. Experimental technique

The samples were produced in the form of Pt and Pd nanostructured films, deposited on a teflon (poly-tetra-fluor-ethylene, PTFE) substrate. Before the deposition of the catalytically active coating, the substrates were treated with acetone, ethanol and were sustained at least a day in a vacuum (pressure of 10^{-5} Torr). Samples were obtained by method of thermal vacuum deposition of metals. To make the nanostructured coatings the original deposition technique of the films with a gradient of thickness along its surface was used. Conditions and parameters of deposition are chosen in such a way to deliberately get the nanoislet coating (ultrasmall thickness of film, low deposition rate, non-wetting substrate). This technique allows one to obtain and research films with different effective thicknesses h_{eff} (and structure) in a single technological cycle [3].

After deposition the sample in the area of thickness gradient was cut into 6 pieces of 20 × 6 mm. The catalytic activity of nanostructured metallic films to the reaction of atomic hydrogen recombination was determined using a method of recombination sensor (figure 1). In this method the atomic hydrogen, which is produced by HF glow discharge, flows in the PTFE atom-pipes through catalyst into recombination sensor. As sensor a serial thermo-manometer transducer PMT-2 was used, in which the concentration of free H-radicals was determined calorimetrically.

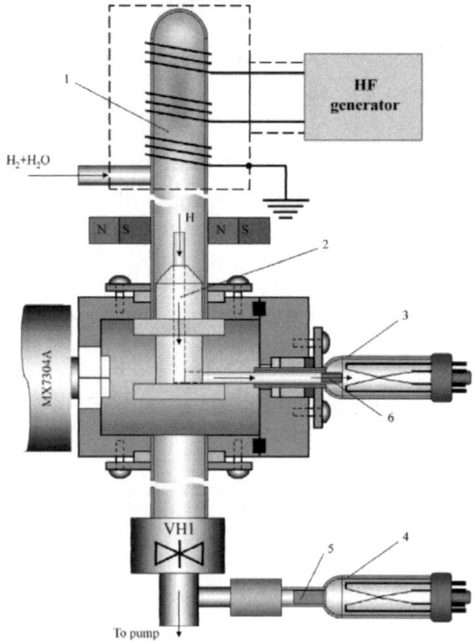

Figure 1. Scheme of catalytic properties experimental investigation by method of recombination sensor. 1 – discharge tube, 2 – PTFE atom-pipe, 3 – thermo-manometer transducer PMT-2 (recombination sensor), 4 – etalon sensor, 5 – sample of bulk catalyst, 6 –sample of investigated nanocatalyst.

3. Results and discussion

Since the catalytic activity of nanostructured coating determines as the number of free radicals, entering to the sensor, should consider how the output data of sensor are associated with radicals concentration. Based on the experiment methodology, we can assume that the system at the moment of maximum heating of the sensors catalytically active Pt wire at small but finite times was in quasi-equilibrium thermodynamic state. On the basis of this assumption consider the following model.

The output data of the sensor is electromotive force (EMF) of thermocouple, which is connected to a platinum wire. Therefore, when the temperature of the wire is changed the changing of thermal electromotive force can be written as

$$dU = \gamma(T)dT, \quad (1)$$

where $\gamma(T)$ – the proportionality factor of a specific sensor (not confused with thermal EMF coefficient), T – the temperature.

Then the total potential difference can be found by integrating the equation (1). The heat flow radiated from the heated wire is determined by the equation of heat transfer

$$\vec{q} = -\partial(T) \cdot \overrightarrow{gradT}, \quad (2)$$

where $\partial(T)$– the coefficient of thermal conductivity.

At equilibrium conditions the heat flow from wire equals to the heat flow transferred to it. So accounted that the heat is transferred to wire by portions from radical recombination we can write

$$q' = N_r E_0 \mu(T), \tag{3}$$

where N_r – number of the recombination acts on unit of surface area per one unit time, E_0 – reaction energy in one recombination act, $\mu\,(T)$– coefficient of energy accommodation.

From [4] it is known that the surface recombination can be realized through two mechanisms: shock impact (by Ridil-Ili) and diffusion (by Langmuir-Hinshelwood). Then the total N_r can be defined as

$$N_r = \sigma(T)\, jN + k(T)N^2, \tag{4}$$

where $\sigma(t)$ – the effective cross section of the recombination reaction by shock impact, j – density of free atoms flow, $k(t)$ – the reaction rate constant by the Langmuir-Hinshelwood mechanism, N – the concentration of atoms adsorbed on the surface.

Let's take assume about the small changes of wire temperature (the coefficients which depends on the temperature let's considered as constant). Then from (1) - (4) by solving a quadratic equation we can obtain the approximate formula of the surface radicals concentration depending to sensor's EMF

$$N = -\frac{\sigma j}{2} + \frac{1}{2}\left(\sigma^2 j^2 + 4\frac{\partial \cdot k}{\gamma E_0 \mu}\Delta U\right)^{1/2}, \tag{5}$$

where ΔU – the difference of output signals of the recombination and etalon sensors.

The surface concentration of radicals may also be determined from classical Langmuir adsorption theory. Therefore on condition of the thermodynamic equilibrium we have the equation for the partial pressure of radicals

$$P = \frac{\beta(T)N}{\alpha(T)(N^* - N)}, \tag{6}$$

where N^* – the surface concentration of adsorption centers, P – partial pressure of the radicals, $\alpha(T)$ – adsorption coefficient, $\beta(T)$ – desorption probability, and N described by (5)

Given that at relatively low partial pressure of the radicals $N >> N^*$ and concentration $n \sim P$, it can be shown that

$$n \sim \Delta U^{1/2} \tag{7}$$

Therefore for the possibility of carrying out a comparative analysis of the experimental results obtained for Pd and Pt the dimensionless relative measure of the catalytic activity was introduced

$$\xi = \frac{n_k}{n} = \frac{n - n_d}{n} = \frac{\sqrt{\Delta U_0} - \sqrt{\Delta U_i}}{\sqrt{\Delta U_0}}, \tag{8}$$

where n_k – the concentration of free radicals, which are lost during the recombination on the investigated nanocatalysts; n_d – the concentration of radicals, which determined by sensor; $n = n_k + n_d$ – normalizing constant, defined as the concentration of radicals which get to the sensor after recombination on clean PTFE substrate without catalyst; ΔU_0 – sensor's EMF without catalyst; ΔU_i – sensor's EMF for i-th sample of nanocatalyst.

Figure 1 shows the dependence of the relative index of the catalytic activity from the effective thickness of the nanostructured films of Pd and Pt.

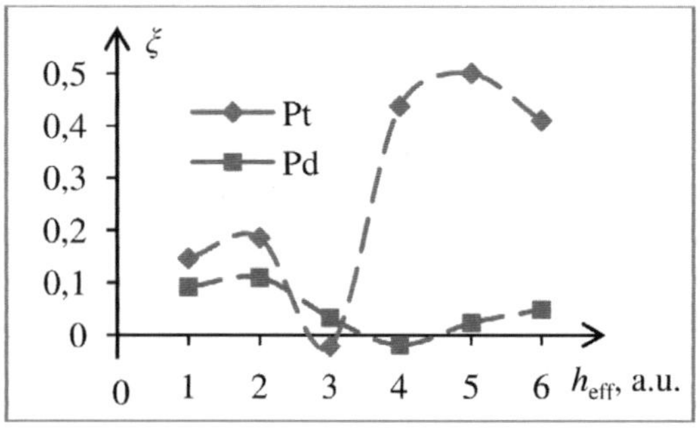

Figure 2. The dependence of the relative index of the catalytic activity from the effective thickness h_{eff} of the Pd and Pt nanostructured films.

As the figure shows the catalytic activity of the investigated coatings for different values of the effective thickness has an oscillation character. This dependence can be explained in the scopes of the band theory and the theory of lattice dynamics in stipulates of discrete type of electrons and phonons energy-spectra in the nanoparticles (quantum dots). Since the size of nanoislands is changing with h_{eff} increase, respectively it energy spectra are changed to. This leads to oscillating changes in catalytic activity. This result is observed in the experiment.

Analysis of the dependencies in Figure 1 reveals the following patterns. Firstly, both graphics have a characteristic minimum, which indicates a low catalytic activity of nanostructured coatings. Secondly, both minima lie lower than the catalytic activity level of pure PTFE. Those catalytic inactive nanoparticles are filling the structural defects on the PTFE surface and it's modified surface with decreasing of its catalytic activity.

Also the figure shows that the essential difference of Pt and Pd catalytic activity observed only at relatively high values of h_{eff} (about 45%). For small h_{eff} the difference of Pt and Pd catalytic activity is decreasing (about 7.5%).

Conclusions

In was shown that the catalytic activity of nanostructured Pd and Pt films has an oscillatory character in depending on the effective thickness of coating (parameters of nanostructures). The catalytic activity of the individual nanostructures in some areas of gradient coatings is lower than the pure PTFE substrate. For large values of h_{eff} the Pt catalytic activity (to the reaction of atomic H recombination) is almost an order of magnitude greater than the Pd catalytic activity. When the effective thickness is reduced the Pt catalytic activity is also significantly reduced and doesn't exceed the activity of Pd more than 50%.

Acknowledgments

This work is supported by the RF Ministry of Education and Science in the framework of the base part of the state task №2015 / 701 (project 3879) and the State Council of Republic Crimea in the framework of the grant for young scientists of the Crimea (resolution № п170-1/16 from 02.02.2016).

References

[1] Dubrovskii V G. 2006 *Semiconductors* **40 10** 1153
[2] Eliseev A A, Lukashin A V. Functional materials / Ed. by Tretyakov Yu.D. Fizmatlit, 2010. 456p.
[3] Tomilin S V, Yanovsky A S. 2013 J. of Nano- and Electronic Physics **5 3** 03014.
[4] Kharlamov V F. 1987. J. of Applied Spectroscopy **46 3** 273.

Journal of Physics: Conference Series **741** (2016) 012180 doi:10.1088/1742-6596/741/1/012180

Silicon-based porous nanocomposite thin-films as an active anode material for lithium-ion batteries

L A Mazaletskiy[1,2], A S Rudy[2,1], A V Metlitskaya[2,1]

[1] Department of Nanotechnology in Electronics, Yaroslavl State University, Sovetskaya Street 14, Yaroslavl 150003, Russia

[2] Yaroslavl Branch of the Institute of Physics and Technology, Russian Academy of Sciences, Universitetskaya Street 21, Yaroslavl 150007, Russia

E-mail: boolvinkl@yandex.ru

Abstract. The results of experimental studies of porous silicon nanocomposite materials for future usage as an anode material of lithium-ion batteries are presented. Comparison between original and porous structures in terms of their qualitative and quantitative characteristics is given.

One of the most promising secondary chemical current sources for energy storage and savings is a lithium-ion battery (LIB). In the last decade the researchers have been actively working to improve both the cathode and anode materials for LIB. Now as the basis for cathodes manufacturing still remains lithium cobalt $LiCoO_2$. As an anode material the graphite C_6 is used.

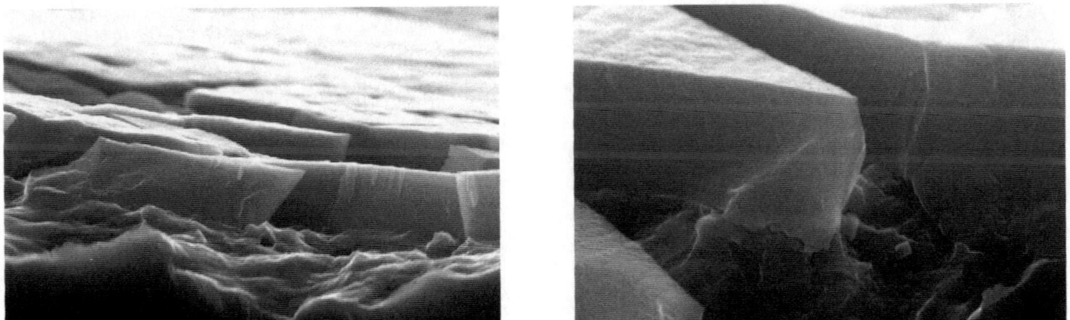

Figure 1. The side view of the nanocomposite silicon material on a titanium foil before processing. Images obtained by the scanning electron microscope Supra 40 of facilities sharing center "Diagnostics of micro- and nanostructures"

In this paper it is proposed to use as the anode material the thin film of silicon, which is the traditional material of microelectronics. Its specific charging capacitance with respect to lithium intercalation is 4211 mA·h/g [1,2]. That more than 10 times exceeds he charging capacity of graphite (372 mA·h/g [3,4]). However, silicon is unstable material and degrades during the process of lithium intercalation. This is due to the fact that the introduction of 4 lithium atoms per one silicon atom leads to rapid destruction of the silicon and anode materials on its basis. Therefore, the anode material is performed as a composite (Si-O-Al) and its capacitance is artificially limited up to 2000 - 3000 mA·h/g by partial oxidation of the silicon during the sputtering.

One of the factors that determine the stability, capacitance and charging characteristics of silicon-based composite materials is their porosity. Therefore, the main goal is developing of methods to control the porosity of the composite films Si-O-Al, as well as study of the effect of porosity on the charge-discharge characteristics.

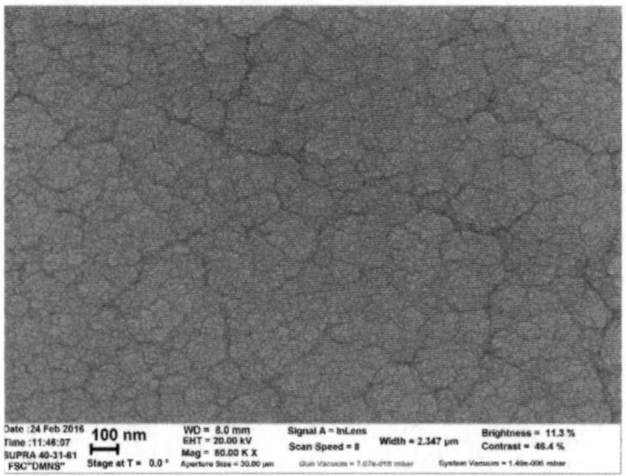

Figure 2. Initial sample, magnification 50k.

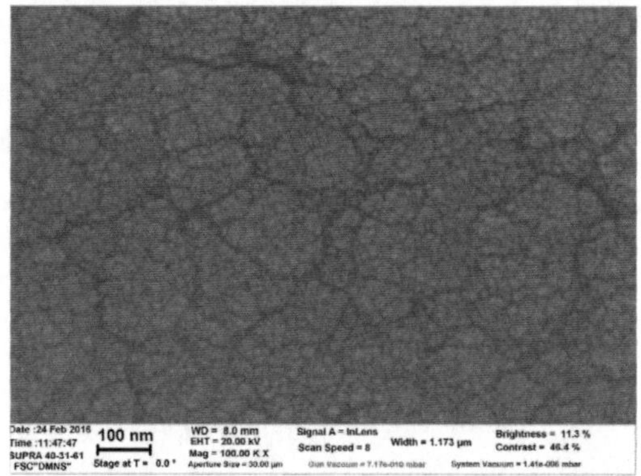

Figure 3. Initial sample, magnification 100k.

Samples of thin films of Si-O-Al composition were fabricated by magnetron deposition on the installation Magna TMR. On figure 1 one can see side view of one of these samples. Thickness of active material is about 600 nm. Using the technique developed by a team was made a series of samples of different porosity. In figure 2 and 3 the morphology of initial sample is shown. Figure 4, 5, 6, and 7 shows morphology of two other already processed samples. The porosity of the samples was monitored by scanning electron microscopy and was quantitatively estimated by the degree of grey by a specially developed technique.

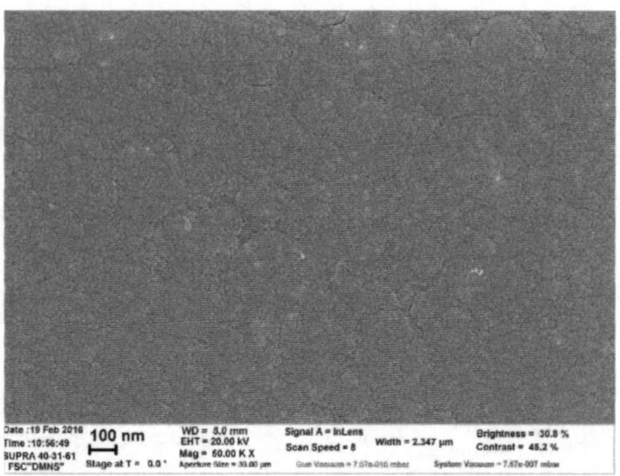

Figure 4. Sample №2, magnification 50k.

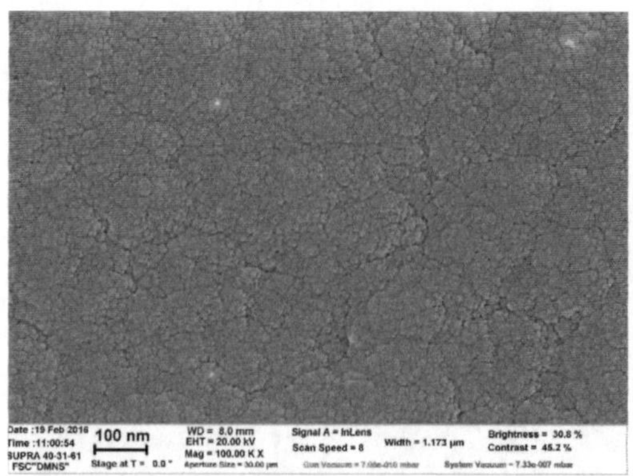

Figure 5. Sample №2, magnification 100k.

To confirm the changes in the chemical composition of the composite material the EDX analysis on the attachment INCA (Oxford Instruments) for SEM Supra 40 was carried out. Analytical data are presented in the table below. As one can see from the data in the table the changes are insignificant. For more detailed investigation the EDX-spectra were compared. Comparison of the spectra is presented on the figure 8.

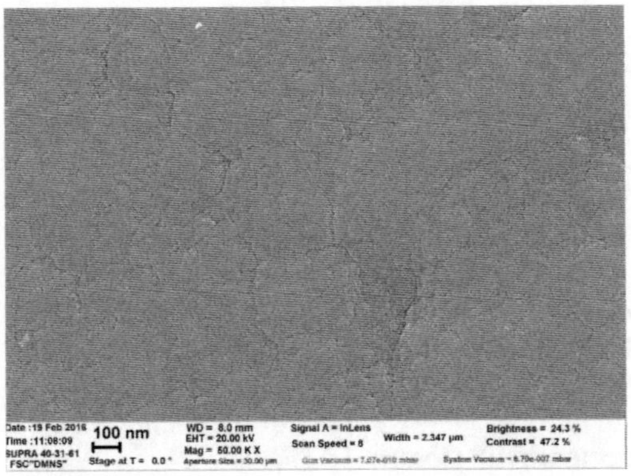

Figure 6. Sample №3, magnification 50k.

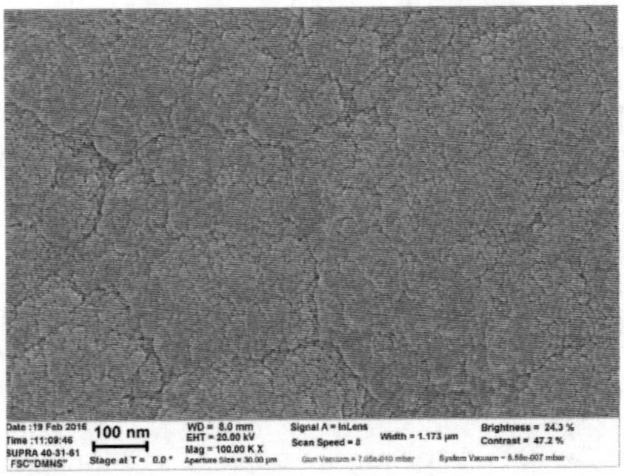

Figure 7. Sample №3, magnification 100k.

Table. The energy dispersive microanalysis data on initial sample and 6 different samples with a developed porous structure

Sample № / Element (at. %)	Initial	№1	№2	№3	№4	№5	№6
Si	81,57	81,58	81,38	81,46	81,98	82,16	81,81
O	10,85	10,41	10,77	10,61	10,12	10,30	10,46
Al	7,58	8,01	7,85	7,93	7,90	7,54	7,73

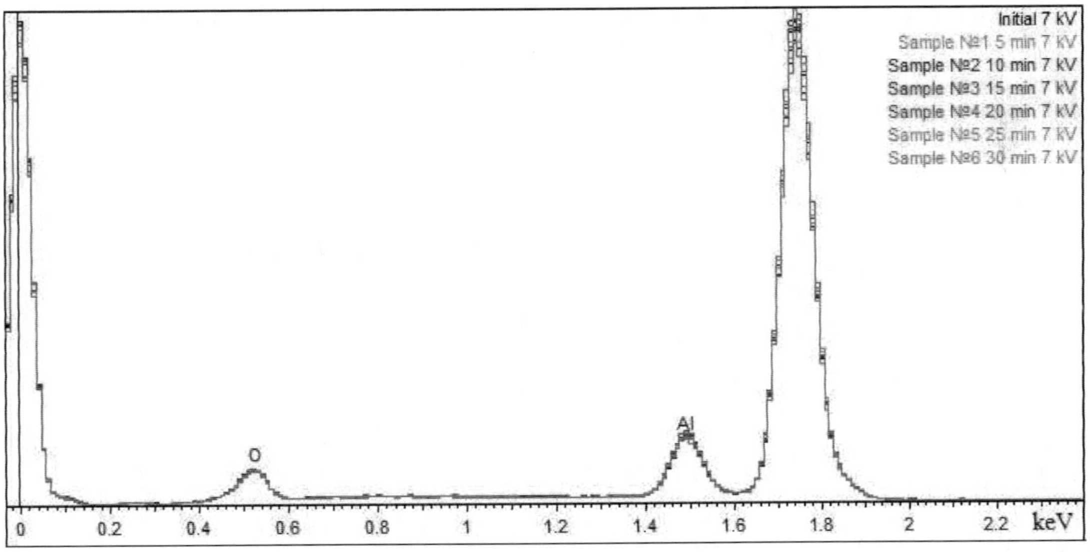

Figure 8. EDX-spectra comparison.

All spectra are virtually indistinguishable, except the silicon concentration, which reveals small deviations. Nevertheless according the data, obtained during films' charge-discharge characteristics testing, the samples with developed pore structure exceeds the other ones in storage capacity (up to 3000 mA·h/g) and are more stable. Samples survive the continuous testing within 10 C mode and then recover their initial capacity.

Acknowledgments
This work was carried out at the equipment of the Facilities sharing centre "Diagnostics of Micro- and Nanostructures" and was supported by the Ministry of Education and Science of the Russian Federation (project no. 14.574.21.0099; unique project ID RFMEFI57414X0099).

References

[1] Sharma R A, Seefurth R N 1976 *J. Electrochem. Soc.* **123** 1763–1768
[2] Wen C J, Huggins R A 1981 *J. Solid State Chem.* **37** 271-278
[3] Yoon S H, Park C W, Yang H, Korai Y, Mochida I, Baker R T K, Rodriguez N M 2004 *Carbon* 42 21-32
[4] Liu B, Yu Y, Chang J, Yang X, Wu D, Yang X 2011 Electrochem. Comm. 13 558–561

The influence of non-spherical nanoparticles' shape on sedimentation process

Ksenia D Mukhina, Sergey A Chivilikhin

ITMO University, 49 Kronverkskiy pr., Saint-Petersburg, 197101, Russia

Abstract. The influence of non-spherical nanoparticles' shape on sedimentation process was studied. Six different shape of particles were considered. From proposed mathematical model it was shown that for non-spherical nanoparticles the perturbation of Z-component of velocity is slightly lower comparing to spherical particles.

1. Introduction

There is no doubt that the process of sedimentation has important technical applications. Since 1950 when J. Taylor first draw attention to the instability of liquid surface [1], the Rayleigh-Taylor instability (or RT instability) was observed in a various kind of situations including electrodynamics, magnetodynamics and hydrodynamics, nanohydrodynamics [2] in particular. In Glowinski, Pan and Joseph's study [3] two-dimensional simulation of this phenomena in suspension was created and few years later the three-dimensional simulation was considered in Guda, Bukharina and Mucha's work [4].

The RT-instability induced by particles themselves was studied in work [5]. The single-phase model and Euler-Lagrange two-phase model were used. The impact of Coriolis force for Rayleigh-Taylor instability under rotation was investigated in [6], the critical rotation rate was found for stabilizing the system. Influence of other factors, such as pH, ionic strength, and adsorption of humic acid on the sedimentation process of TiO2 nanoparticles was studied by J. Lu et al. in 2015 [7].

The Rayleigh-Taylor instability of viscous fluids in spherical geometry was first analyzed in 1955 by Chandrasekhar [8] and was later improved by Mikaelian [9].

In several cases the occurring instability is quite similar to RT instability, thus someone can consider it identical. For example, gravitational instability of thin gas layer in liquid was solved for inviscid fluid and viscous gas [10]. In 2008, Saveliev and Rozanov [11] proposed the mathematical model for nanoparticles with fixed size in moving gas. However, there is a lack of research considering the particles with different sizes and shapes.

2. Problem statement

The main goal of this work is to study the influence of nanoparticles' shape on the sedimentation process of its statistical ensemble. To measure the shape difference from the sphere the coefficient c was used, which is called particle circularity (or surface sphericity). The circularity gives exact the same value in case of two-dimensional and three-dimensional objects, and it ranges between 0 and 1, where 1 corresponds to a perfect circle or sphere.

Journal of Physics: Conference Series **741** (2016) 012181 doi:10.1088/1742-6596/741/1/012181

In this study we are considering a six simple shapes with different value of circularity [12]. The examined shapes are obtained by combining several spherical particles together. The sphericity coefficients are presented in Table 1.

Table 1. Particle circularity for different shapes

Particle shape		Particle circularity, *c*
Close-to-sphere		$\frac{\sqrt{2}}{3}\left(2+\frac{3\sqrt{3}}{\pi}\right)^{1/2} \approx 0.9011$
Pyramids		$\frac{1}{5}\left(5+\frac{16}{\pi}\right)^{1/2} \approx 0.6354$
Stars	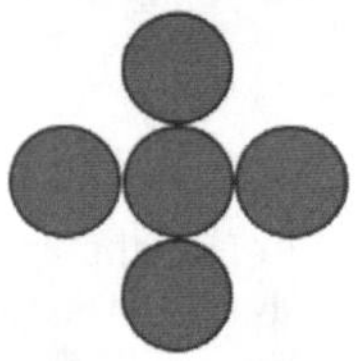	$\frac{1}{\sqrt{5}} \approx 0.4472$
H-shaped particle	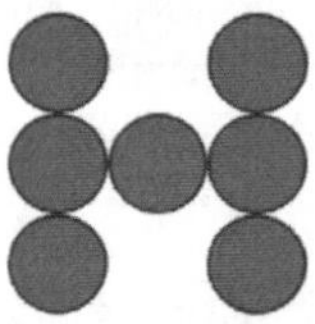	$\frac{1}{\sqrt{7}} \approx 0.3779$
Cross-shape particle	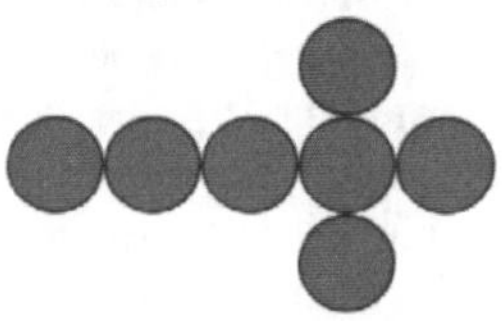	$\frac{1}{\sqrt{7}} \approx 0.3779$
Cylindrical bar	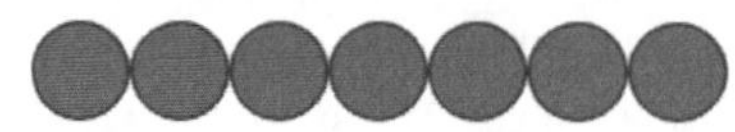	$\frac{1}{\sqrt{7}} \approx 0.3779$

2.1. Mathematical model

The aforementioned process in a gravity field (see the scheme on Figure 1) was modeled, where the upper-half area is the gas dusted with nanoparticles and other half in zero moment of time contains the gas only. System is unbounded in horizontal direction; the nanoparticles are considered as nanosized balls with distribution in sizes. The friction force according to Stokes' law is $F = 6\pi\mu R\gamma R^2$.

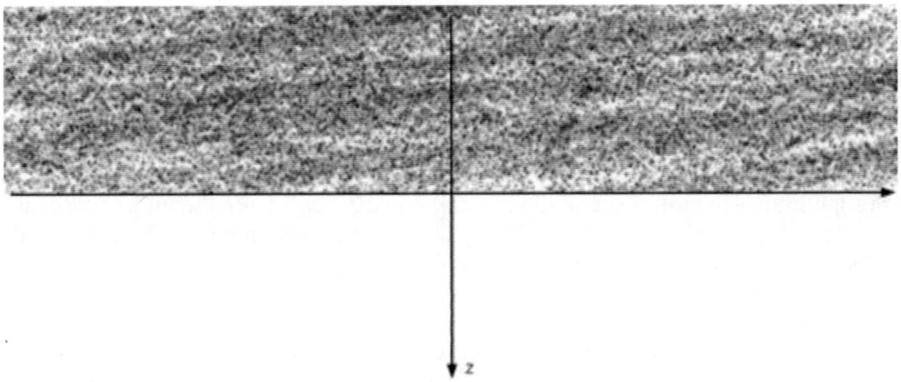

Figure 1. The scheme of the problem. Z-axis is directed as a gravity force.

The Navier–Stokes equations and kinetic equation were used to describe the system of spherical nanoparticles:

$$\begin{cases} \frac{\partial \hat{f}}{\partial t} + \gamma R^2 \frac{\partial \hat{f}}{\partial z} + \hat{v}_z \frac{\partial f^{(0)}}{\partial z} = 0 \\ \rho \frac{\partial \hat{v}_z}{\partial t} = \mu(\frac{\partial^2 \hat{v}_z}{\partial z^2} + \frac{\partial^2 \hat{v}_z}{\partial y^2} + \frac{\partial^2 \hat{v}_z}{\partial x^2}) - \frac{\partial \hat{p}}{\partial z} + \int_0^\infty 6\pi\mu R\gamma R^2 \hat{f} dR \\ \frac{\partial^2 \hat{p}}{\partial x^2} + \frac{\partial^2 \hat{p}}{\partial y^2} + \frac{\partial^2 \hat{p}}{\partial z^2} = \int_0^\infty 6\pi\mu R\gamma R^2 \frac{\partial \hat{f}_k}{\partial z} dR \end{cases}$$

After solving this system, the mathematical expression for Z-component of velocity was obtained previously in case of spherical nanoparticles [13]. Its numerical solution represented below:

$$V_{j,n+1} = V_{j,n} + \nu\Delta t\left(\frac{V_{j+1,n} + V_{j-1,n} - 2V_{j,n}}{(\Delta z)^2} - k^2 V_{j,n}\right) + \frac{k\Delta t}{2} e^{k\cdot j\Delta z} \sum_{j'=j}^{\infty} e^{-k\cdot j'\Delta z} \frac{J(j')}{\rho} \Delta z + \frac{k\Delta t}{2} e^{-k\cdot j\Delta z} \sum_{j'=0}^{j} e^{k\cdot j'\Delta z} \frac{J(j')}{\rho} \Delta z$$

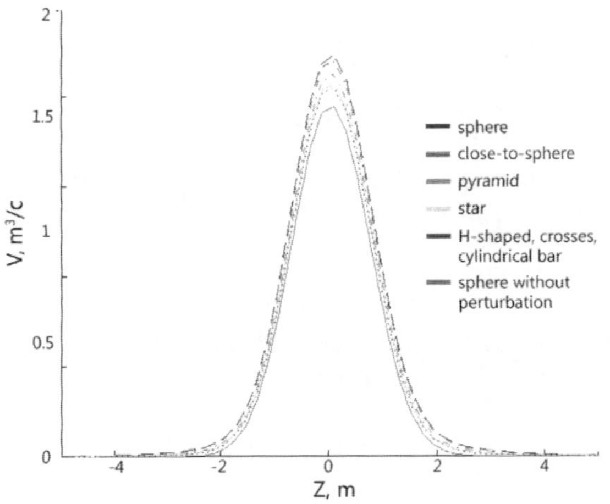

Figure 2. Comparing Fourier transform of velocity's Z-component with and without perturbation depending on different shape of particles; $\nu = 10^{-6}$; $\gamma = 2.173 \cdot 10^6$; $n_0 = 10^{16}$

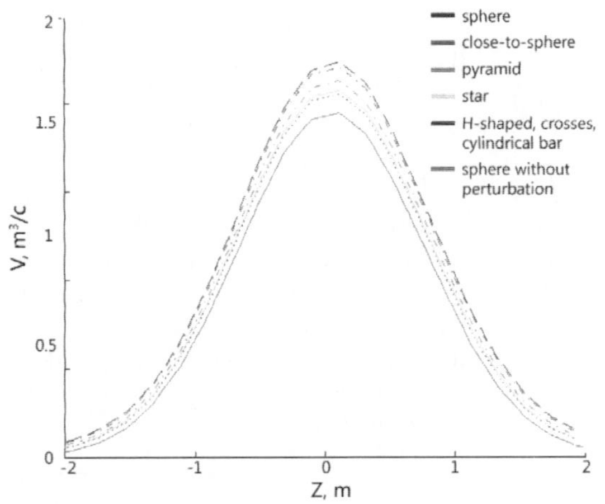

Figure 3. Comparing Fourier transform of velocity's Z-component with and without perturbation with different value of *c*; $\nu = 10^{-6}$; $\gamma = 2.173 \cdot 10^6$; $n_0 = 10^{16}$

3. Results and Discussions

The influence of particle shape on its velocity is presented on figures 2 and 3. As we can see from the charts with growing of coefficient c the velocity of nanoparticle increases. Comparing given values with obtained velocity without perturbation (Figure 3) we can conclude that the shape of particle has significant influence on the value of Z-component of velocity. Thus, the highest value of velocity corresponds to shape which is the closest to sphere. It should be noticed that for spherical nanoparticles the limit value of concentration is $n_0 = 3 \cdot 10^{10}\ m^{-3}$, for the higher concentration the instability is appearing. However, the influence of shape on concentration for non-spherical particles requires further investigations.

Acknowledgments

This paper is financially supported by Ministry of Education and Science of the Russian Federation, Agreement #14.587.21.0024 (18.11.2015), RFMEFI58715X0024.

References

[1] Taylor G.I. 1950 The instability of liquid surfaces when accelerated in a direction perpendicular to their planes *Proc. Roy. Soc. A.* **201** p 192–196

[2] Chandrasekhar S. 1981 Hydrodynamic and Hydromagnetic Stability NY.: Dover **652**

[3] Völtz C, Pesch W and I Rehberg 2001 Rayleigh-Taylor instability in a sedimenting suspension Phys. Rev. E. **65** p 1–7.

[4] Mucha, P., & Guda, S. 2007 Rayleigh-Taylor Instability in a Sedimenting Suspension. In APS Division of Fluid Dynamics Meeting Abstracts **1**

[5] Chou, Y. J., & Shao, Y. C. 2016 Numerical study of particle-induced Rayleigh-Taylor instability: Effects of particle settling and entrainment. *Physics of Fluids (1994-present)* **28(4)** p 043302.

[6] Baldwin, K. A., Scase, M. M., & Hill, R. J. 2015 The inhibition of the Rayleigh-Taylor instability by rotation. *Scientific reports*, **5**.

[7] Lu, J. et al. 2015 Sedimentation of TiO2 nanoparticles in aqueous solutions: influence of pH, ionic strength, and adsorption of humic acid. *Desalination and Water Treatment*, p 1-8.

[8] Chandrasekhar, S. 1955 The character of the equilibrium of an incompressible fluid sphere of variable density and viscosity subject to radial acceleration. *The Quarterly Journal of Mechanics and Applied Mathematics*, **8(1)** p 1-21.

[9] Mikaelian, K. O. 2016 Viscous Rayleigh-Taylor instability in spherical geometry.*Physical Review E*, **93(2)** p 023104.

[10] Pimenova A.V. and Goldobin D.S. 2015 Gravitational instability of thin gas layer between two thick liquid layers

[11] Saveliev R.S., Rozanov N.N., Sochilin G.B., Chivilikhin S.A. Rayleigh-taylor instability of a dusty gas (in Russian) *Scientific and Technical Journal of Information Technologies, Mechanics and Optics* **3(73)** p 18-21

[12] Tran-Cong, S., Gay, M., & Michaelides, E. E. 2004 Drag coefficients of irregularly shaped particles. *Powder Technology*, **139(1)** p 21-32.

[13] Mukhina, K., & Chivilikhin, S. 2015 Theoretical investigation of sedimentation process for nanoparticles statistical ensemble. In *Journal of Physics: Conference Series* **643(1)** p 012124 IOP Publishing.

Journal of Physics: Conference Series **741** (2016) 012182 doi:10.1088/1742-6596/741/1/012182

Rate of thermal transitions in kagome spin ice

S Y Liashko[1,2], **V M Uzdin**[2,3], **H Jónsson**[1,4]
[1]Faculty of Physical Sciences, VR-III, University of Iceland, Reykjavik, Iceland
[2]Department of Natural Sciences, University ITMO, St. Petersburg, 197101 Russia
[3]Department of Physics, St. Petersburg State University, St. Petersburg, 198504 Russia
[4]Department of Applied Physics, Aalto University, FIN-00076 Espoo, Finland

Abstract. The rate of thermal transitions in a kagome spin ice element is calculated using harmonic transition state theory for magnetic systems. Each element consists of six prolate magnetic islands. Minimum energy paths on the multidimensional energy surface are found to estimate activation energy. Vibrational frequencies are also calculated to estimate the rate of the various transitions. An overall transition rate between equivalent ground states is calculated by using the stationary state approximation including all possible transition paths. The resulting transition rate is in a good agreement with experimentally measured lifetime.

1. Introduction

Artificial spin ice is a frustrated system of magnets with ground state entropy. Such systems are made to study various new physical effects such as magnetic monopoles [1], temperature induced disordering (melting) [2] and new kinds of thermodynamic phase transitions [3]. In particular, experimental measurements of temperature induced magnetic switching [4] with timescale ranging from days to seconds depending on temperatures have been made for kagome spin ice. This research is dedicated to theoretical study of such transitions.

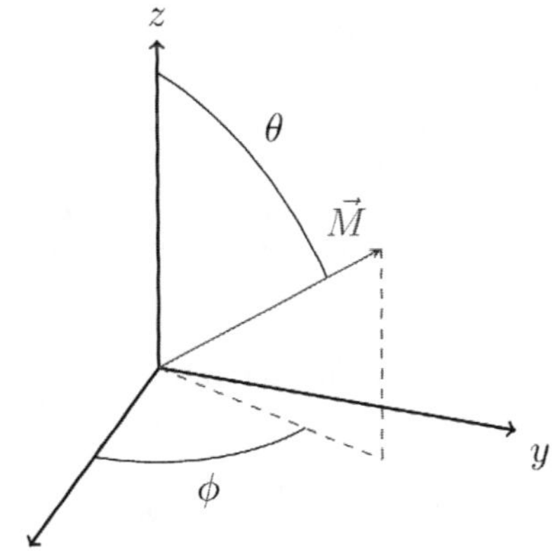

Figure 1. Schematic representation of the magnetic island in 3 dimensions, where $\vec{M}$ is a magnetic moment of the island with respected angles ϕ and θ. Axis 0Z is aligned with easy anisotropy direction $\vec{K}$.

We focus here on a single kagome ring formed by six islands and study how the interaction between them affects the activation energy and mechanism of the transitions.
A single prolate magnetic island is described as a single-domain ferromagnetic particle with uniaxial anisotropy (Figure 1). Such approach assumes that the re-magnetization process involves simultaneous rotation e.g. at any moment of time all magnetic moments of the island are parallel. In such cases, the total energy of the islands includes an anisotropy term of the form:

$$E_{islad} = -\left(\vec{M}\vec{K}\right)^2 \tag{1}$$

In order to form the kagome spin ice element we combine six islands and include the interaction between them as described by equation (2). Since the size of a typical island in spin ice is a few nanometers, the interaction between islands is dominated by the dipole-dipole interaction. Therefore, the energy of the system becomes

$$E = \sum_i^N\left(\vec{M}_i\vec{K}_{out}\right)^2 - \sum_i^N\left(\vec{M}_i\vec{K}_{in}\right)^2 - \frac{1}{2}\sum_{i\neq j}^N\left(\frac{(\vec{M}_i\vec{M}_j)}{|\vec{r}_{ij}|^3} - \frac{3(\vec{M}_i\vec{r}_{ij})(\vec{M}_j\vec{r}_{ij})}{|\vec{r}_{ij}|^5}\right) \tag{2}$$

Where $\vec{K}_{out}$ and $\vec{K}_{in}$ are respectively the in-plane shape anisotropy and out-of-plane shape anisotropy. The vector $\vec{r}_{ij}$ is the distance vector between centres of i-th and j-th islands. The summation limit N is the number of islands.

2. Simulations

The kagome spin ice element consists of 6 islands (N = 6) placed on the kagome lattice. Given the geometry, the distance vectors $\vec{r}_{ij}$ can easily be found. The calculations are based mainly on experimentally determined values of the parameters e.g. the volume of an island and saturation magnetization of an island (V = 470 × 170 ×3 nm^3, M_s = 166 × 103 A/m). The in-plane shape anisotropy parameter K_{in} and out of plane shape anisotropy parameter K_{out} were calculated by using micro-magnetic simulation from the volume and saturation magnetization. This is in an effort to simulate as closely as possible the experiment and to avoid any fitting parameters. The ratio between out-of-plane and in-plane anisotropy is about a factor of 4 (K_{in} = 4550 erg/cm^3 and K_{out} = 17991 erg/cm^3), so only rotation in plane is possible and this approximation implied in equation (2) is fully valid for the simulations.

The nudged elastic method (NEB) [5] was applied to find minimum energy paths (MEPs) on the energy surface determined by our model. The MEP represents the most probable, from a statistical point of view, path for the remagnetization process between the two ground states. Such paths show the mechanism of the transition and identify intermediate states. Maxima along the MEP give estimates of the activation energy for each of the elementary steps in the transition. There are, in fact, multiple different minimum energy paths and they have all been calculated (an example of such a path is shown in Figure 2). The two ground states of the kagome spin ice element correspond to clockwise and anticlockwise orientation of the magnetic moments of the islands (number I and VII in Figure 2).

3. Rate calculation

The transition rates are calculated using harmonic transition state theory (HTST) for magnetic systems [6]. This gives an Arrhenius law for the transition rate of each elementary step:

$$k = A\, exp(-\frac{E_a}{k_bT}) \tag{3}$$

Here, $E_a$is the activation energy which in HTST is given by the difference in energy between the maximum along the MEP and the initial state, $k_b$is the Boltzmann constant and T is temperature. The pre-factor A is in many cases taken from the literature or is considered to be a fitting parameter. but we have applied the HTST to calculate the pre-exponent factor for our

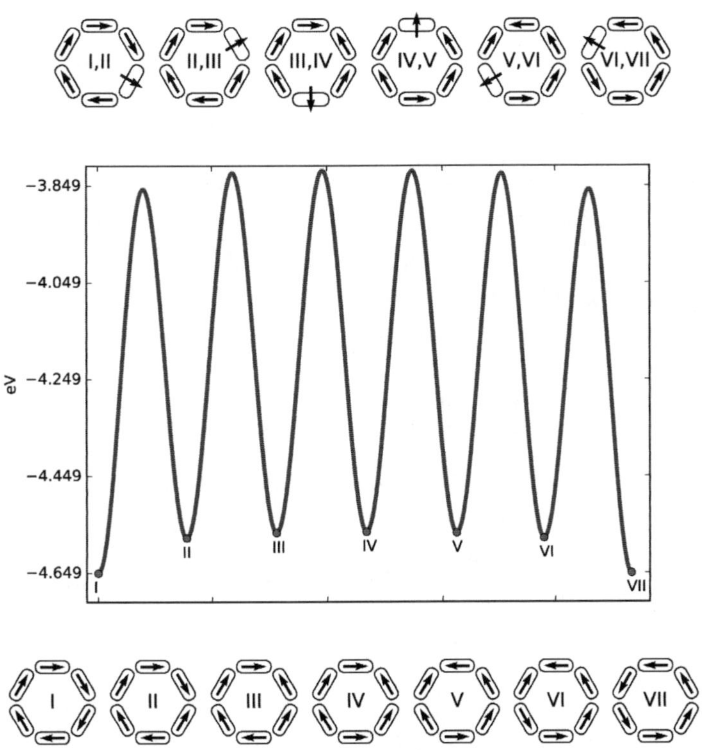

Figure 2. Minimal energy path(MEP) between ground states with respected configurations in the minima and maxima.

particular system:

$$k_{htst} = \frac{1}{2\pi}\frac{J_s}{J_m}\sqrt{\sum_{j=2}\frac{a_j^2}{\lambda_{s,j}}}\sqrt{\frac{\det H_{min}}{\det^* H_{sad}}} \quad (4)$$

Here, J_s and J_m are Jacobians at the saddle point and at the initial state minimum. $H_{min}$and H_{sad} are the Hessian matrices at the minimum and at the saddle point. The $\lambda_{s,j}$ are the eigenvalues of the Hessian at the minimum. $\det^*$ is the determinant of the matrix excluding the row with the negative eigenvalue. This expression makes it possible to calculate the transition rate as well as the lifetime of a given state by using only observable parameters of the material as mentioned earlier, without any fitting to the measured rates.

As can be seen from Figure 2, the transition between the two degenerate ground states occurs via a set of intermediate minima. So, there is not just one activation barrier but rather a set of activation barriers. In such a case HTST can be used to find the rate of each elementary step and the overall rate of the sequence of steps estimated using stationary state approximation including all possible transition paths. This is based on a set of master equations, differential equations for the variation over time of the probabilities that the system occupies each of the different possible states

$$\begin{cases} \frac{dn_0}{dt} = -W_{0,1}n_0 + W_{1,0}n_1 \\ \dots \\ \frac{dn_i}{dt} = -(W_{i,i+1} + W_{i,i-1})n_i + W_{i,i-1}n_{i-1} + W_{i,i+1}n_{i+1} \\ \dots \\ \frac{dn_n}{dt} = -W_{n,n-1}n_n + W_{n,n-1}n_{n-1} \end{cases} \quad (5)$$

In our case, n_i is a particular energy state. $W_{i,j}$ is a transition parameter between particular n_j and n_k or in other words a transition rate. The task is to find $\frac{dn_0}{dt}$ in terms of n_0. In this way all elementary steps between n_0 and n_n can be included to give an overall rate and effective lifetime of one of the ground states. We take n_n to be the final state, i.e. if the system gets into that state it cannot go back until a long time has passed. The stationary state approximation means that we assume $\frac{dn_i}{dt}$ is equal to zero for all the intermediate states.

As we mentioned earlier, there are several equivalent transitions paths. We need to take that into account in the master equations for the system:

$$\begin{cases} \frac{dn_0}{dt} = -6W_{0,1}n_0 + 6W_{1,0}n_1 \\ \frac{dn_1}{dt} = -(2W_{1,2} + W_{10})n_1 + W_{1,0}n_0 + 2W_{1,2}n_2 \\ \frac{dn_i}{dt} = -(2W_{i,i+1} + 2W_{i,i-1})n_i + 2W_{i,i-1}n_{i-1} + 2W_{i,i+1}n_{i+1} \\ \dots \\ \frac{dn_5}{dt} = -W_{5,6}n_5 - 2W_{5,4}n_5 + 2W_{4,5}n_4 \end{cases} \quad (6)$$

When we solve equation (6) to obtain the $\frac{dn_0}{dt} = -n_0(G)$ where the G is a combination of the $W_{i,j}$ and gives the effective transition rate between the two ground states. The resulting rate can be compared with the Arrhenius expression, eq. (3) to extract the effective pre-exponential factor.

4. Results

By using the methodology described above, we have estimated the lifetime of a ground state of the kagome ring at 420K to be around 12 seconds. At 300 K, the lifetime is several days. The effective pre-exponential factor is about 10^8 s^{-1} and the effective activation energy is 0.74 eV. These rates are in excellent agreement with the experiment even though the pre-exponential factor is four orders of magnitude lower than the one assumed in ref. [4] for the interpretation of the experimental data. The next step is to investigate the influence of the internal magnetic structure of an islands during a spin transition. This can be important because the islands are relatively large, so the magnetization reversal process could involve the formation of a temporary domain wall rather than a coherent rotation of all the spins in the island. This would strongly affect the activation energy as well as the pre-exponential factor.

Acknowledgment

This work was financially supported by the Government of Russian Federation, Grant 074-U01, RFBR grant №14-02-00102 and the Icelandic Science Fund.

References

[1] C. Castelnovo, R. Moessner, S. L. Sondhi 2008 Nature 42-45

[2] Kapaklis V, Arnalds U, Harman-Clarke A, Papaioannou E Th, Karimipour M, Korelis P, Taroni A, Holdsworth P C W, Bramwell Steven T, Björgvin H 2012 New Journal of Physics **14** 035009

[3] L. Anghinolfi, H. Luetkens, J. Perron, M. G. Flokstra, O. Sendetskyi, A. Suter, T. Prokscha, P. M. Derlet, S. L. Lee & L. J. Heyderman Nature Communications 6 8278

[4] Farhan A, Derlet P M, Kleibert A, Balan A, Chopdekar R V, Wyss M, Anghinolfi L, Nolting F, Heyderman L J 2013 Nature Physics **9** 375
[5] Henkelman G, Uberuaga B P, Jónsson H 2000 Journal of Chemical Physics **113** 22
[6] Bessarab P F, Uzdin V M, Jónsson H 2013 PRL **110** 020604

Saint Petersburg OPEN 2016 IOP Publishing
Journal of Physics: Conference Series **741** (2016) 012183 doi:10.1088/1742-6596/741/1/012183

Crossover temperature for quantum tunnelling in spin systems

S M Vlasov[1,2], P F Bessarab[3,4], V M Uzdin[2,4], H Jónsson[1,5]
[1]Science Institute of the University of Iceland, Reykjavík, Iceland
[2]Department of Natural Sciences, University ITMO, St. Petersburg, 197101 Russia
[3]Department of Materials and Nanophysics, Electrum 229, Royal Institute of Technology (KTH), SE-16440 Kista, Sweden
[4]Department of Physics, St. Petersburg State University, St. Petersburg, 198504 Russia
[5]Department of Applied Physics, Aalto University, FIN-00076 Espoo, Finland

Abstract. We derive an analytical expression for the crossover temperature corresponding to the transition from classical activation mechanism to temperature assisted quantum tunnelling in spin systems. The crossover temperature depends on the magnetic configuration and Hessian at the first order saddle point on the energy surface of the system. The theory is applied to several single spin models, including a system with four-fold anisotropy. Good agreement is obtained with experimental results for a molecular magnet containing Mn_4.

1. Introduction

Investigations of the quantum tunnelling in spin systems have been the topic of theoretical and experimental studies over the past few decades. From the theoretical point of view, these problems are of great interest for fundamental basis of quantum theory. They are also practical, because small magnetic particles are good candidates for memory devices and the knowledge of their quantum behaviour helps us evaluate the lower limit of the size of a potential memory unit [1].

In case of spin systems, transitions between stable states separated by a potential barrier can occur via classical thermal activation over the barrier or by quantum tunnelling through the barrier. At sufficiently high temperature the transitions are governed by thermal activation and the rate, Γ, obeys the Arrhenius law, $\Gamma = \Gamma_0 e^{-U/k_B T}$, with U being the height of the energy barrier between the states. As the temperature is lowered, quantum tunnelling becomes the dominant mechanism and the rate eventually becomes temperature independent if the final state is equal or lower in energy than the initial state. The crossover temperature, T_c, indicates at what temperature the transition mechanism changes from thermal activation to thermally activated quantum tunnelling [2].

We propose a general approach for calculating the crossover temperature and give an analytical formula for T_c derived for a wide range of spin systems with arbitrary anisotropy in an external magnetic field. The formula reduces to known analytical solutions for simple spin systems [3,4], but also works for complex systems, for which only numerical results have been available so far.

2. Derivation of the formula

The thermally averaged escape rate is given by

$$\Gamma(T) = \frac{1}{Z_0}\sum_n \Gamma_n\, e^{\frac{-E_n}{k_B T}}, \tag{1}$$

where Z_0 is the partition function of the initial state, Γ_n are quantum transition probabilities from energy levels E_n, and T is the temperature of the bath. The transition probabilities can be estimated from the Euclidean (imaginary-time) action as functional integral from the $\exp(-S_E)$. Within the semi-classical approximation, the dominant contribution to the escape rate is given by the saddle point of the action and the escape rate can be estimated with an exponential accuracy as

$$\Gamma(T) \propto exp[-S(\boldsymbol{q}_{inst})], \tag{2}$$

where $\boldsymbol{q}_{inst}$ is the classical path that corresponds to a saddle point to the action S surface. Hereafter we use atomic units ($\hbar = 1, e = 1, m_e = 1$).

We consider a system with a quantum spin number, s, energy specified as a function of the angles θ and φ giving the direction of the magnetic vector. The Euclidean action is

$$S(\theta,\varphi) = \int_{-\beta/2}^{\beta/2} d\tau[-is(1- \cos\theta)\dot{\varphi} + E(\theta,\varphi)]. \tag{3}$$

where $\beta=1/k_BT$.

Our method is based on a quadratic expansion of the action around the saddle point of the potential energy surface. In order to find the saddle point, we write the first variation of the action (3):

$$\delta S = \int_{-\beta/2}^{\beta/2} d\tau\left[\left(-is\sin\theta\,\dot{\varphi} + \frac{\partial E(\theta,\varphi)}{\partial\theta}\right)\delta\theta + \left(is\sin\theta\,\dot{\theta} + \frac{\partial E(\theta,\varphi)}{\partial\varphi}\right)\delta\varphi\right], \tag{4}$$

and by setting it to zero, we obtain the equations of motion:

$$\dot{\varphi} = \frac{-i}{s\sin\theta}\frac{\partial E(\theta,\varphi)}{\partial\theta}, \tag{5}$$

$$\dot{\theta} = \frac{i}{s\sin\theta}\frac{\partial E(\theta,\varphi)}{\partial\varphi}. \tag{6}$$

These equations have two types of the solutions: The first is the trivial one, $\theta = \theta_0$ and $\varphi = \varphi_0$, corresponding to a point on the energy surface $E(\theta,\varphi)$. The second one is the instanton – a closed trajectory, for which the energy is conserved. When the temperature is close to the crossover temperature, the amplitude of the instanton is small and the trajectory involves only small oscillations in the vicinity of the saddle point, $\theta = \theta^\dagger$ and $\varphi = \varphi^\dagger$, on the potential surface

$$\theta = \theta^\dagger + \delta\theta, \quad \varphi = \varphi^\dagger + \delta\varphi \tag{7}$$

We expand the action (3) in a Taylor series around the saddle point in order to get the second variation

$$S(\theta,\varphi) = S(\theta^\dagger,\varphi^\dagger) + \delta S + \frac{1}{2}\delta^2 S, \tag{8}$$

where

$$\delta^2 S = \int_{-\beta/2}^{\beta/2} d\tau[-2is\sin\theta\,\delta\dot{\varphi}\delta\theta + (a\delta^2\theta + 2b\delta\theta\delta\varphi + c\delta^2\varphi)], \tag{9}$$

with

$$a = \frac{\partial^2 E}{\partial \theta^2}, \qquad b = \frac{\partial^2 E}{\partial \theta \partial \varphi}, \qquad c = \frac{\partial^2 E}{\partial \varphi^2} \tag{10}$$

Since the $\delta\theta$ and $\delta\varphi$ in (7) correspond to closed trajectories, they can be expanded in Fourier series:

$$\delta\theta = \sum_{n=-\infty}^{\infty} \theta_n e^{\frac{2\pi i \tau}{\beta}}, \tag{11}$$

$$\delta\varphi = \sum_{n=-\infty}^{\infty} \varphi_n e^{\frac{2\pi i \tau}{\beta}}, \tag{12}$$

and the second variation of the action in (9) written in terms of the Fourier coefficients θ_n and φ_n:

$$\frac{1}{2}\delta^2 S = \beta \sum_{n=0}^{\infty} \left[\frac{2\pi s \sin\theta^{\dagger}}{\beta} n(\varphi_n \theta_n^* - \theta_n \varphi_n^*) + a\theta_n\theta_n^* + b(\varphi_n\theta_n^* + \theta_n\varphi_n^*) + c\varphi_n\varphi_n^*\right]. \tag{13}$$

At sufficiently high temperature, $\delta^2 S$ has only one negative eigenvalue, giving correspondence between $\delta^2 S$ and the Hessian at the saddle point. However, as the temperature is lowered, a second negative eigenvalue appears, signifying the crossover from the classical regime to a quantum tunnelling regime. Thus, the crossover temperature is the highest temperature at which the $\delta^2 S$ has two negative eigenvalues. The expression for the crossover temperature can be written as:

$$T_c = \frac{\sqrt{b^2 - ac}}{2\pi k_B s \sin\theta^{\dagger}}. \tag{14}$$

3. Applications

3.1. Uniaxial spin system with a two-fold anisotropy

In order to test our approach, we choose the simple and well-studied spin system with axial symmetry and transverse applied magnetic field:

$$\hat{H} = -DS_z^2 - H_x S_x, \tag{15}$$

here D is an anisotropy and H_x is an applied field. The corresponding classical energy surface is:

$$E(\theta, \varphi) = -Ds^2 \cos\theta^2 - H_x s \sin\theta \cos\varphi, \tag{16}$$

with a saddle point at $(\theta^{\dagger}, \varphi^{\dagger}) = \left(\frac{\pi}{2}, 0\right)$. The crossover temperature obtained from eq. (14) is

$$T_c = \frac{\sqrt{H_x(2Ds - H_x)}}{2\pi k_B}. \tag{17}$$

This solution is equivalent to previously published results for this special case (see eq. 13 in ref. [3]).

3.2. Uniaxial spin system with a four-fold anisotropy

When eq. (14) is applied to a system with four-fold transverse anisotropy, described by the Hamiltonian:

$$\hat{H} = -DS_z^2 - BS_z^4 - C(S_+^2 + S_-^2) - H_x S_x, \tag{18}$$

the higher-order anisotropy terms are found to play a crucial role. Even for a small value of the parameter C (usually it is four orders of magnitude smaller than D), the crossover temperature becomes non-zero even in the absence of a field, as can be seen from figure 1. The crossover temperature in this case is

$$T_c = \frac{\sqrt{(H_x + 32Cs^3)(2Ds - 8Cs^3 - H_x)}}{2\pi k_B}. \tag{19}$$

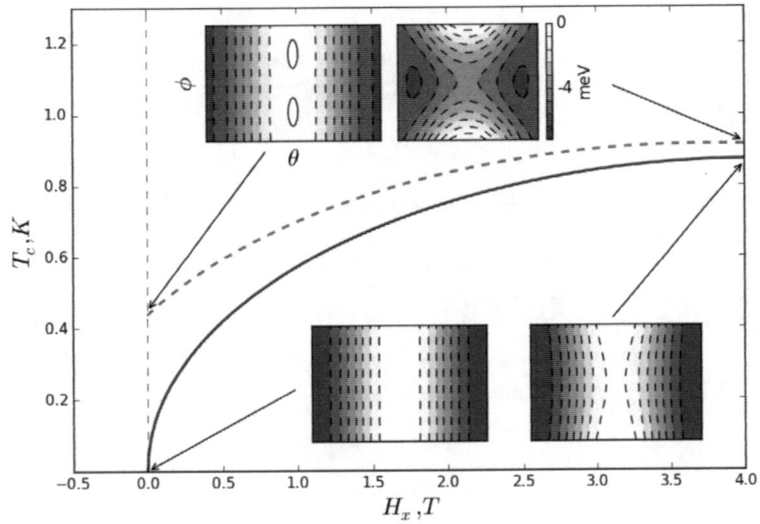

Figure 1. Dependence of T_c on applied magnetic field, H_x, for a systems with two-fold and four-fold anisotropy. Solid line shows results for a system with two-fold anisotropy, eq. (15). The dashed line shows results for a system with four-fold anisotropy, eq. (18). All parameters are taken from ref. [6]. By adding higher-order anisotropy, the crossover temperature becomes finite even in the absence of a magnetic field. The insets show contour graphs of the energy surfaces (two-fold anisotropy below, fourfold above) at zero field and a field of 4 T.

This example shows how the method works for more complex systems for which only numerical results have been reported previously.

3.3. Crossover temperature of Mn_4 complex

The molecular magnet $Mn_4O_3Cl(O_2CCH_3)_3(dbm)_3$ (Mn_4 for short) is a trigonal pyramidal complex with a spin ground state corresponding to $s = 9/2$. The Kramers degeneracy prohibits a molecule with a half-integer spin to tunnel. A transverse component in the Hamiltonian is needed to break the symmetry so as to allow for tunnelling [5]. The system is described by the Hamiltonian:

$$\hat{H} = D\left[S_z^2 - \frac{1}{3}s(s+1)\right] + B_4^0 O_4^0 + B_4^4 O_4^4, \qquad (20)$$

where $O_4^0 = 35S_z^4 - 30s(s+1)S_z^2 + 25S_z^2 + 6s(s+1)$ and $O_4^4 = \frac{1}{2}(S_+^4 + S_-^4)$. The application of our formula to estimate the crossover temperature gives $T_c = 0.6\ K$, which is in excellent agreement with experimental measurements [5].

4. Conclusions

We have derived an analytical expression for the crossover temperature in spin systems characterized by a single magnetic vector. The crossover temperature corresponds to the transition from classical activation mechanism to temperature assisted quantum tunnelling. Using our formula, T_c , can be estimated given the magnetic configuration and Hessian at the first order saddle point on the energy surface characterizing the system. The formula has been applied to several single spin models, including a system with four-fold anisotropy. It is in agreement with previously obtained analytical results for simple axial spin system. It was also applied to a system with four-fold symmetry for which only numerical simulations had been reported previously. Good agreement with experimental results were obtained for a molecular magnet containing Mn_4.

Acknowledgments
This work was supported by the Government of Russian Federation (Grant 074-U01), RFBR (Grant 14-02-00102), the Icelandic Research Fund and the Göran Gustafsson Foundation for Research in Natural Sciences and Medicine (Grant 1303A).

References
[1] Loth S Baumann S, Lutz C P, Eigler D M, Heinrich A J 2012 Science **335**, 196
[2] Benderskii V A, Goldanskii V I, Makarov D E 1993 Phys. Rep. **233**, 195
[3] Garanin D A, Hidalgo_X M, Chudnovsky EM 1998 Phys. Rev. B **57**, 13639
[4] Kim G-H 1999 Phys. Rev. B **59**, 11847
[5] Aubin S M J, Dilley N R, Pardi L, Krzystek J, Wemple M W, Brunel L-C, Maple M B, Christou G, Hendrickson D N 1998 J. Am. Chem. Soc. **120**, 4991
[6] Park C-S 2003 J. of Magnetism and Magnetic Materials **267**, 281

Tip-surface interaction and rate of magnetic transitions

A V Ivanov [1, 2], P F Bessarab [3], V M Uzdin [1, 4] and H Jónsson [2, 5]
[1]Saint Petersburg State University, 199034, Saint Petersburg, Russia.
[2]Science Institute of the University of Iceland, VR-III, 107 Reykjavík, Iceland
[3]Department of Materials and Nanophysics, Electrum 229, Royal Institute of Technology (KTH), SE-16440 Kista, Sweden
[4]Saint Petersburg National Research University of Information Technologies, Mechanics and Optics, 197101 Saint Petersburg, Russia.
[5]Department of Applied Physics, Aalto University, Espoo, Finland.

Email: alekcey92@inbox.ru

Abstract. Calculations of the interaction between a magnetic tip of an atomic force microscope with a magnetic surface using the non-collinear extension of the Alexander-Anderson model are described. The mechanism and rate of thermally activated magnetic transitions in a cluster of atoms at the tip is investigated. The results are compared with experimental data and found to be in good agreement with measured lifetimes [R. Schmidt *et. al.*, *Phys. Rev. B* 86, 174402 (2012).]. The results are also compared with previously reported density functional theory calculations.

1. Introduction

Magnetic exchange force microscopy (MExFM) makes it possible to detect the magnetic exchange force between a magnetic tip and a magnetic surface with atomic resolution [1-3]. In a recent experiment [2, 3], a ferromagnetic tip was scanned over an antiferromagnetic Fe layer on a W(001) substrate. The image of the surface shows an alternating pattern since the tip-surface exchange interaction is favorable for surface atoms with one direction of the magnetic moment and unfavorable for the other surface atoms. The strength of the interaction depends on the distance between the tip and the surface. Later, it was shown that the magnetic state of the small cluster supported at the tip can switch with time and that its average lifetime depends strongly on the tip-surface distance [4]. This illustrates the effect of the exchange interaction on the lifetime of the magnetic state and suggests a possible way to record and read data.

Theoretical calculations based on density functional theory (DFT) have been carried out to help interpret the results of such experiments [2,3,5]. The tip was modeled as a square pyramidal arrangement of 14 Fe, Cr or Mn atoms. The results of the calculations showed that an antiferromagnetic alignment of the tip and the underlying surface atom is of lower energy than ferromagnetic alignment. This was also found to be the case when the tip was represented by a single Fe, Cr or Mn atom. The calculations did not address the mechanism and rate of the magnetic transitions.

Journal of Physics: Conference Series **741** (2016) 012184 doi:10.1088/1742-6596/741/1/012184

2. Methodology

The simulations presented here made use of the non-collinear extension of the Alexander-Anderson (NCAA) model [6,7]. Two types of simulations of the system were carried out. In one case, the total number of d-electrons in the system was fixed. In the other case, the d-electron energy level was fixed with respect to the Fermi level and the number of d-electrons found from self-consistency calculations for each atom. Electroneutrality is ensured in that case by transfer of s(p) electrons to or from the bath of the extended metal. The effect of the exchange interaction and thereby the distance between the tip and the surface on the thermally activated transitions in the magnetic cluster at the tip is calculated.

The magnetic cluster at the tip was represented either by a single Fe, Mn or Cr atom or by a square pyramidal arrangement of 14 Fe atoms. The surface was represented by a layer of 13 Fe atoms when the tip consisted of a single atom, but by a layer of 25 atoms when the larger, 14 atom tip was simulated. The underlying substrate is effectively taken into account in the NCAA model through a hybridization parameter and the Fermi level. The larger tip cluster model was used in calculations of the lifetime of the magnetic states.

The rate of magnetic transitions in the tip was calculated by first finding the minimum energy path (MEP) between the ferromagnetic and antiferromagnetic states of the tip cluster with respect to the underlying surface atom. The geodesic nudged elastic band method [8] was used to find the MEPs. The maximum energy along the MEP corresponds to a first order saddle point on the energy surface and gives an estimate of the activation energy. The pre-exponential factor was calculated using harmonic transition state theory (HTST) for magnetic systems [9] which gives an Arrhenius expression for the transition rate. The corresponding expression for the average lifetime, τ, of a magnetic state is:

$$\tau = \tau_0 \exp\left(\frac{E_a}{k_B T}\right),$$

where E_a the activation energy, k_B the Boltzmann constant and T the temperature. The pre-exponential factor τ_0 is calculated from the curvature of the energy surface at the saddle point and at the initial state minimum [9].

3. Results

3.1. Tip-surface interaction

The tip-surface exchange energy is the energy difference between a parallel and antiparallel alignment of the magnetic vectors at the tip with respect to the magnetic vector of the underlying surface atom. When the exchange energy is positive, the ferromagnetic alignment is of lower energy than the antiferromagnetic one. Figure 1 shows the calculated exchange energy for a single atom representation of the tip as a function of the hopping parameter, V, in the NCAA model that represents the overlap of wave function of the tip atom with the wave function of the underlying surface atom. The closer the tip is to the surface the larger this overlap is and larger the hopping parameter. Results for Fe, Mn and Cr tip atom are shown. When the total number of d-electrons in the system is fixed, see figure 1(a), the exchange coupling is antiferromagnetic and the exchange interaction is weakest for Fe and strongest for Cr. This is similar to the results of the DFT calculations [5]. In this case, the maximum of absolute value of the exchange energy for the Cr tip is a factor of 20 larger than for Fe atom at the tip and a factor of 4 larger than for the Mn atom. However, when the Fermi level is fixed, see figure 1(b), and the number of d-electrons is found from self-consistency calculations for each value of the hopping parameter, the exchange energy is of opposite sign for a tip with an Fe and Mn atom, i.e. the interaction is ferromagnetic, and the exchange interaction for Fe is stronger than for Cr.

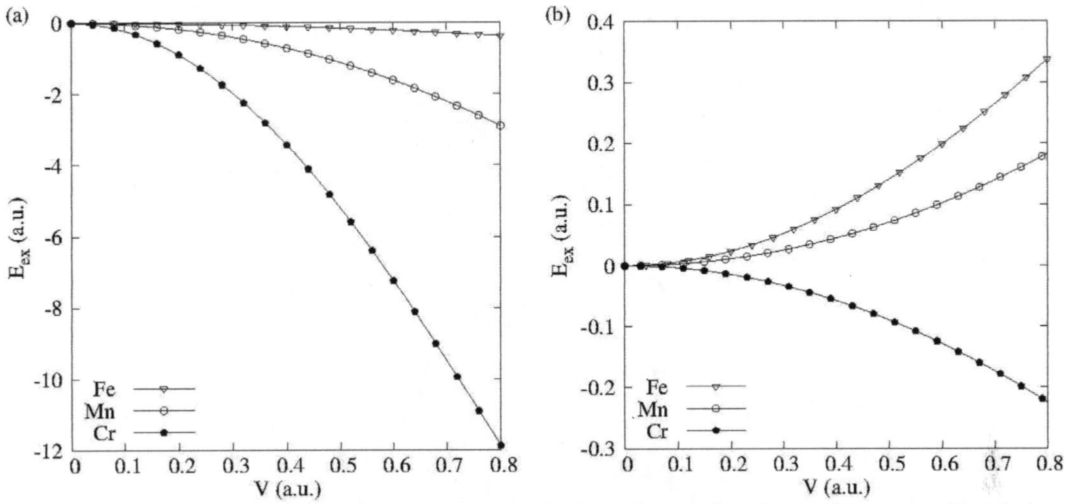

Figure 1. Exchange interaction between a magnetic tip and an underlying surface atom. The tip is represented by a single atom in this case, either an Fe, Mn, or Cr atom. The exchange energy is shown as a function of the hopping parameter, V, in the non-collinear Alexander Anderson model. A larger hopping parameter corresponds to shorter distance between the tip and the surface. (a) Calculated result for a fixed total number of d-electrons in the system. (b) Calculated results for a fixed Fermi level. The sign of the exchange interaction is different in the two cases for Fe and Mn tip.

The two different cases show a significant difference in the calculated exchange energy. When the total number of d-electrons in the system is fixed, the results are similar to the DFT results [2,3,5]. However, when the Fermi level is fixed, a different sign of the exchange interaction is obtained for the Fe and Mn tips. Similar reversal of the sign of the exchange interaction was found for larger models of the Fe and Mn tips. Fixing the Fermi level is a better representation of the experimental situation since the tip and substrate are macroscopic and contain a large number of itinerant s(p)-electrons. From the experiments it is not possible to tell whether the interaction is ferromagnetic or antiferromagnetic.

2.2. Thermally activated transitions

In these calculations the tip was modeled as a square pyramid of 14 Fe atoms and the Fermi level was fixed. The two magnetic states of the tip correspond to either parallel (P) or antiparallel (A) alignment with the magnetic vector of the underlying surface atom. If the tip does not interact with the surface, then the energy of these two states is the same. The MEP (figure 2, (1) solid line) has a maximum rise in energy of 23.3 meV. When the tip interacts with the surface, the energy of the P state becomes lower than that of the A state. The MEP is therefore no longer symmetric as shown in figure 2 ((2) dot-dashed line). The activation energy for the transition from the P state to the A state is 27.0 meV but 19.5 meV for the reverse transition. When an applied magnetic field in the direction of the tip is included, the energy of the two states is changed, as shown by the (3) dashed and (4) dotted lines in figure 2. If the magnetic field points in the direction of the P state, then the energy of this state is lowered while the energy of the A state is increased. The activation energy for the transition from P state to A state then increases but the activation energy for the back transition decreases. For a field of 1 T, the two energy barriers become 29.5 meV and 17.6 meV, respectively. If the magnetic field points in the opposite direction of the P state, then the change in the energy barrier is opposite, namely the energy barrier for the transition from P state to A state decreases and the energy barrier for the back transition increases. They become 24.5 meV and 21.7 meV, respectively.

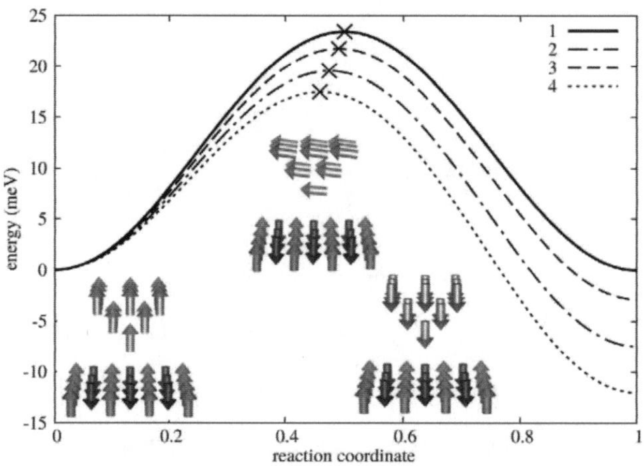

Figure 2. The minimum energy path for a transition between the magnetic states corresponding to antiparallel and parallel alignment of the 14 atoms tip with the underlying surface atom. The (1) full line corresponds to a tip that does not interact with a surface. The (2) dash-dotted line corresponds to a tip-surface separation of 450 pm. The (3) dashed and (4) dotted lines represent the system in a magnetic field of 1 T, pointing antiparallel and parallel to the tip, respectively. The maxima along the minimum energy paths are indicated with an x. They correspond to saddle points on the energy surface and give an estimate of the activation energy for the transition. The alignment of the magnetic vectors in the initial state, at the saddle points and in the final state is shown in the insets.

The pre-exponential factor was calculated and found to be $6 \cdot 10^{-11}$s for the isolated tip. The interaction with the surface changes this value insignificantly. Thus, at a temperature of 8.1 K, as in the experimental study, the lifetime of the magnetic state of the isolated tip is about 5 hours. When the tip is brought to a distance of about 450 pm from the surface, the average lifetime of the A state becomes 62 seconds and the lifetime of P state increases to 8 weeks. This means that a transition from the P state to A state will not occur on the time scale of the experiment, only transitions from A state to P state are fast enough. When a magnetic field of 1 T is included, the lifetimes of the metastable A state increases to 24 minutes and or decreases to 4 seconds depending on the direction of magnetic field. These results are consistent with the experimental observations of Schmidt *et. al.* [4]. They found the metastable tip state to have a lifetime of 27 ± 5 s and 17 ± 3 s. When the distance between tip and surface is taken to be 430 pm, the 14 atoms tip reproduces these times when a small magnetic field (60 mT) is included. The energy barrier for the A to P transition is then 18.9 meV and 18.6 meV, depending on the direction of the field.

4. Conclusion

The calculations described here largely support the interpretation of the experimental results of Schmidt *et. al.* [4]. Their assignment of the metastable and stable state in terms of the direction of the magnetic vectors of the tip, which was based on the results of the DFT calculations, is however different from our results because the sign of the exchange interaction is different when the Fermi level is fixed as compared with a fixed total number of d-electrons. In the NCAA model it is assumed that the magnetic atoms interact with free s(p)-electrons of a non-magnetic metal and the energy of the localized electrons with respect to the Fermi level is a parameter. The two approaches, fixing the Fermi level and fixing the number of d-electrons, give different sign of the exchange energy for Fe and Mn tips. Since the DFT calculations [2,3,5] were carried out for a small system with an unsupported tip, they are closer to the NCAA simulations with a fixed number of d-electrons rather than the NCAA calculations with fixed Fermi level.

In order to estimate the thermal stability of the magnetic state of the tip, the geodesic nudged elastic band method was employed to calculate minimum energy paths for the transition. This gave estimates of the activation energy for various situations. The pre-exponential factor was calculated using harmonic transition state theory. A model of the ferromagnetic Fe tip consisting of 14 atoms and a tip-surface distance of 430 – 450 pm gave a lifetime of the metastable, A state in close agreement with the experimental observations [4]. The energy barrier was 23.3 meV for an isolated tip but decreased as the tip is brought close to the surface. By changing the tip-surface distance, the lifetime of the metastable state can be varied from seconds to hours.

Acknowledgments

This work was supported by the Icelandic Research Fund, the Russian Foundation of Basic Research (Grant No. 14-02-00102). PB gratefully acknowledges support from the Göran Gustafsson Foundation for Research in Natural Sciences and Medicine (Grant No. 1303A)

References

[1] U. Kaiser, A. Schwarz, and R. Wiesendanger, *Nature* 446, 522 (2007).
[2] R. Schmidt, C. Lazo, H. Hölscher, U. H. Pi, V. Caciuc, A. Schwarz, R. Wiesendanger and S. Heinze, *Nano Lett.* 9, 200 (2009).
[3] R. Schmidt, C. Lazo, U. Kaiser, A. Schwarz, S. Heinze, and R. Wiesendanger *Phys. Rev. Letter* 106, 257202 (2011).
[4] R. Schmidt, A. Schwarz, and R. Wiesendanger, *Phys. Rev. B* 86, 174402 (2012).
[5] Cesar Lazo and Stefan Heinze *Phys. Rev. B* 84, 144428 (2011).
[6] S. Alexander and P.W. Anderson, *Phys. Rev.* 133, A1594 (1964).
[7] P.F. Bessarab, V. M. Uzdin and H. Jónsson, *Phys. Rev. B* 89, 214424 (2014).[8] P.F. Bessarab, V.M. Uzdin and H. Jónsson, *Comput. Phys. Commun.* 196 335-347 (2015).
[9] P.F. Bessarab, V. M. Uzdin and H. Jónsson, *Phys. Rev. B* 85, 184409 (2012).

Fractal analyses of porous sol-gel nanocomposites modified by fullerenol $C_{60}(OH)_n$ (n = 22–24)

E V Maraeva[1], I A Averin[2], M S Istomina[1], V A Moshnikov[1,3], S S Nalimova[1], I A Pronin[3], K N Semenov[4]

[1]Department of Micro-and Nanoelectronics, Saint Petersburg Electrotechnical University "LETI", ul. Prof. Popova, 5, St. Petersburg 197376, Russia
[2]Department of Nano- and Microelectronics, Penza State University, 40 Krasnaya St., 440026, Penza, Russia
[3]Department of Integral Electronics, St.Petersburg Polytechnic University, Polytechnicheskaya 29, St.Petersburg 195251, Russia
[4]Department of Chemical Thermodynamics and Kinetics, Saint-Petersburg State University, Universitetskaya nab., 7/9, St. Petersburg 199034, Russia

Abstract. In this study silicon dioxide – tin dioxide nanomaterials modified by fullerenol were obtained through sol-gel technology. The porousity changes caused by the fullerenols addition are discussed.

1. Introduction
Fullerenols are among the most important and promising fullerene derivatives which can be easily synthesized with readjusting properties by varying the number of hydroxyl groups [1-3]. The aim of this study was to investigate the influence of the fullerenols $C_{60}(OH)_n$ (n = 22–24) addition on features of $SiO_2 - SnO_2$ sol-gel system layers formation using the method of fractal dimension determining.

2. Experiment
In this study sol-gel method for porous layers formation was chosen [4-9]. An inorganic salt $SnCl_2 \cdot 2H_2O$ as the precursor of tin dioxide and tetraethoxysilane as the precursor of silicon dioxide were selected. Buthanol (C_4H_9OH) was used as a solvent. The modifying agent (fullerenol $C_{60}(OH)_n$ (n = 22–24) was introduced at the stage of preparing the solution, i.e. sol, and the concentration of fullerenols in the alcohol was 1 mg/l. To obtain powders, 10 % aqua ammonia (NH_4OH) was added to the solutions-sols. The powders were heat-treated for 30 minutes at the temperature of 600 °C.

AFM experiments were performed using NTEGRA-Therma nanolaboratory (NT-MDT, Zelenograd, Russia). Commercial etched silicon tips NSG 01 with typical resonance frequency of 150 kHz were used as AFM probes. X-ray diffraction experiments were carried out on «DRN Farad» (*Cr-Kα*). The original program for the analyses of the patterns fractal dimension by the method «perimeter - square» was used [10].

3. Results and discussion

During the experiments, a series of samples of the "silicon dioxide – tin dioxide" system was obtained using the sol-gel method which was both modified and not modified by fullerenols, with different ratios of components (mol. %): 90% SiO_2–10% SnO_2; 50% SiO_2–50% SnO_2. Some technological features are discussed in [11].

The microstructure of the synthesized samples was investigated by the atomic force microscopy method in the tapping mode in the NTEGRA nanolaboratory. Figure 1 illustrates the morphology of the sample surface containing 10 mol. % SnO_2 and 90 mol. % SiO_2 without and with the addition of the $C_{60}(OH)_n$ (n = 22–24) fullerenol. It has been found that the addition of the fullerenol greatly affects the morphology of the surface of the oxide composite: the density of grain distribution and their average size increases [11]. The atomic force microscopy data indicate that on the surface of the sample prepared with addition of the fullerenol aggregates of ~2 microns observed in figure 1 (a) are not formed.

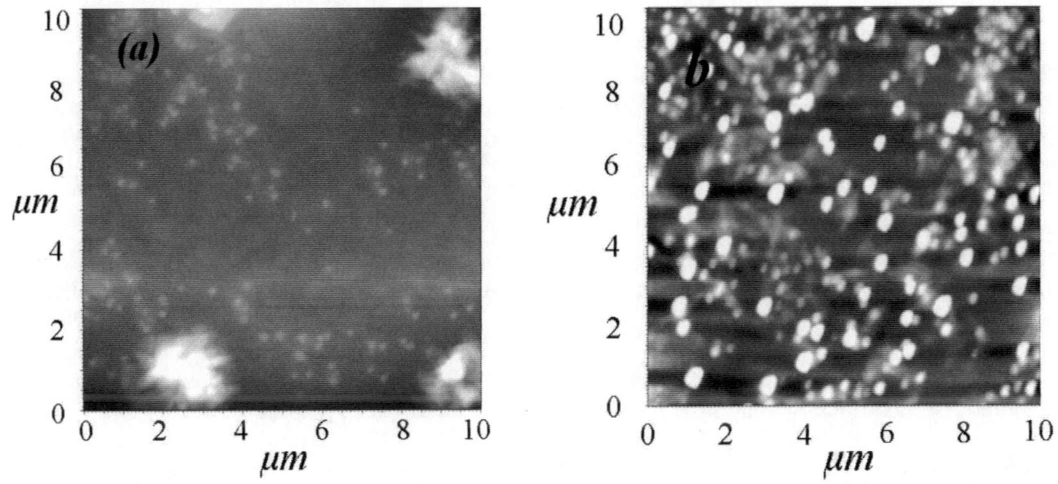

Figure 1. AFM images of the sample containing 90 SiO_2–10 SnO_2 (mol. %) synthesized without adding the fullerenol: (a) and synthesized with the addition of the fullerenol (b) (scan space size - 10 × 10 μm^2)

In the figures 2 (a, b) the depences $S_c(\delta) = f[P_c(\delta)]$ in logarithmic coordinates for the patternes shown in the figures 1 (a, b).

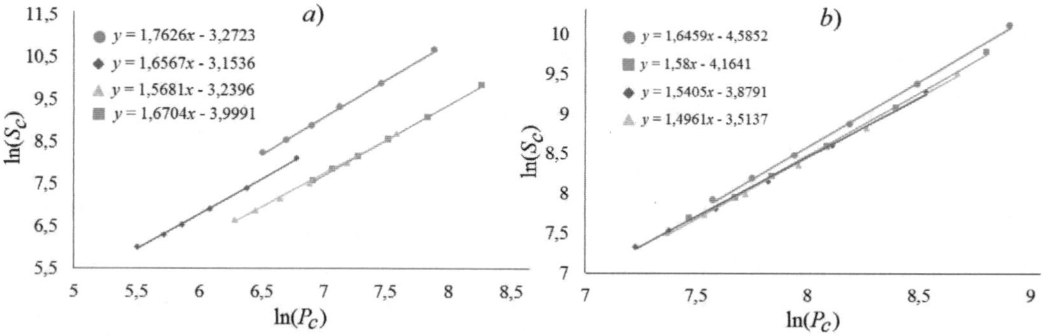

Figure 2. The depences $S_c(\delta) = f[P_c(\delta)]$ for 4 horizontal cross-sections of the patterns 90 SiO_2–10 SnO_2 (mol. %) synthesized without adding the fullerenol: (a) and synthesized with the addition of the fullerenol (b)

The slope coefficient reducing for all the lines corresponds to rising of the height at that the cross-section was made. It can be seen from the figures that the addition of fullerenols in initial solution results in the reducing of the line's slope coefficient (fractal dimension meanings are D_f=1.83 for the first image and D_f=1.78 for the second one).

4. Conclusions

Porous nanocomposites based on silicon dioxide – tin dioxide both modified and not modified by fullerenols were obtained using sol-gel method. It was determined that the introducing of fullerenol $C_{60}(OH)_n$ (n = 22–24) in silicon dioxide – tin dioxide nanocomposites results in reducing of fractal dimension, that argues for new porous system formation. It is shown that the method of fractal dimension determining «perimeter - square» can be used for analyses of porousity changes caused by the fullerenols addition.

Acknowledgments
This study was supported by the Ministry of Education and Science of the Russian Federation within the framework of the project part of the state assignment for the St. Petersburg Electrotechnical University "LETI" (project № 16.2112.2014/K) (a part of the work connecting with obtaining porous nanocomposites and AFM investigation), the basic part of the state assignment for the Penza State University (project №. 2014/151, project code 117) (a part of the work connecting with the analyses of the fractal dimension),

References

[1] Semenov K N, Charykov N A and Keskinov V N 2011 *Journal of Chemical & Engineering Data* **56** 230
[2] Semenov K N, Charykov N A, Murin I V and Pukharenko Y V 2015 *Journal of Molecular Liquids* **202** 1
[3] Sidorov L N, Yurovskaya M A, Borschevskii A Y, Trushkov I V and Ioffe I N 2005 *Fullerenes, in Russian* (Moscow: Ekzamen)
[4] Maximov A I, Moshnikov V A, Tairov Yu M and Shilova O A 2008 *Basics of sol-gel technology of nanocomposites, in Russian* (St. Petersburg: Elmor)
[5] Gracheva I E, Moshnikov V A, Karpova S S and Maraeva E V 2011 *Journal of Physics: Conference Series* **291** 012017

[6] Gracheva E I, Moshnikov V A, Maraeva E V, Karpova S S, Alexandrova O A, Alekseyev N I, Kuznetsov V V, Olchowik G, Semenov K N, Startseva A V, Sitnikov A V and Olchowik J M 2012 *Journal of Non-Crystalline Solids* **358** 433

[7] Maraeva E V, Bobkov A A, Maximov A I, Moshnikov V A, Nalimova S S 2015 *Journal of Physics: Conference Series* **643** 012116

[8] Pronin I A, Donkova B V, Dimitrov D Tz, Averin I A, Pencheva J A, Moshnikov V A 2014 *Semiconductors* **48(7)** 842

[9] Averin I A, Igoshina S E, Moshnikov V A, Karmanov A A, Pronin I A, Terukov E I 2015 *Technical Physics*, **60(6)** 928

[10] Averin I A, Pronin I A, Karmanov A A, Moshnikov V A 2015 *Fractal analyses of sol-gel layers* (ECM licence № 015614039)

[11] Maraeva E V, Istomina M S, Moshnikov V A, Nalimova S S, Maximov A I, Alexeev N I, Semenov K N 2016 *Journal of Physics: Conference Series* **690** 012031

Metal-carbon nanosystem IR-PVA/Fe-Co for catalysis in the Fischer-Tropsch synthesis

A A Vasilev[1,2], E L Dzidziguri[1], M I Ivantsov[2], M N Efimov[2]

[1] Department of Functional Nanosystems and High-Temperature Materials, National University of Science and Technology "MISiS", Moscow 119049, Russia
[2] Chemistry of Polyconjugated Systems Laboratory, A.V. Topchiev Institute of Petrochemical Synthesis RAS, Moscow 119991, Russia

Abstract. Metal-carbon nanosystems consisting of nanodimensional bimetallic particles of Fe-Co dispersed in a carbon matrix for the Fischer-Tropsch synthesis were studied. Prepared metal-carbon nanopowders samples were characterized by X-ray diffraction (XRD), transmission electron microscopy (TEM) and scanning electron microscopy (SEM). It was shown formation of FeCo nanoparticles with body-centered cubic structures started at 400 °C. FeCo nanoparticles have spherical form, the mean size is 7 – 12 nm and uniform distribution in a carbon matrix. The metal-carbon nanosystem demonstrates a catalytic activity in the Fischer-Tropsch synthesis. The maximum yield of liquid hydrocabons C_{5+} was 92 g/m^3 while the selectivity for the target product – 35 %.

1. Introduction

Associated petroleum gas (APG) utilization is one of the priority tasks facing the Russia's oil and gas complex. Imbalance of petroleum production and processing of APG has led to the fact that the volume of the gaseous hydrocarbons which are annually burned on torch installations exceeds 20 billion m^3 that is equivalent to losses to 500 billion roubles per year [1]. As the decree of the Government of the Russian Federation № 1148 dated 08.11.12 about the increase of payment for excessive APG flaring was approved most large companies have started paid more attention to the program based on the petroleum products synthesis, namely, APG processing into motor fuel, methanol and other valuable products via the "gas to liquid" technology. This technology is based on the chemical Fischer-Tropsch synthesis [2, 3].

Fischer-Tropsch (FT) synthesis is a catalytic process that converts syngas (CO and H_2) into components of liquid fuels [4]. Mainly cobalt and iron are applied as commercial catalysts [5]. At the present time these catalysts based on Co and Fe, used by Shell (Malaysia), Sasol (South Africa) and Qatar Petroleum (Qatar), require the recovery stage before FT process [6]. Synthesis of metal-carbon nanosystem containing a catalytic non-pyrophoric nanoscale particles which don't require recovery procedures is an important task, which will reduce excess expenses.

Thus, the goals of this work were preparation of metal-carbon nanosystems based on FeCo nanoparticles in infrared (IR) pyrolyzed polyvinyl alcohol support and investigation their structure and catalytic properties depend on synthesis conditions.

2. Materials and methods

Metal-carbon nanosystems were prepared under the conditions of IR pyrolysis of a precursor based on solution of polyvinyl alcohol (PVA) and iron and cobalt hydronitrates ($Fe(NO_3)_3 \cdot 9H_2O$ and $Co(NO_3)_2 \cdot 6H_2O$) at 5 different temperatures (300, 400, 500, 600, 700 °C), with a mole ratio of metal components 1 : 1. The concentration of PVA in water was 5 % wt. and the summary metal concentrations in polymer was 20 % wt.

Pyrolysis was conducted in the IR-chamber of pulse photon annealing device [7]. IR-annealing was accomplished at two-step mode: preliminary annealing in air at 150 and 200 °C for 15 minutes during which there is removal the solvent and initial structuring of PVA and then the main annealing in an argon atmosphere at a specified temperature. The main annealing time was 2 minutes.

The size and morphology of the metal nanoparticles were evaluated by TEM (JEM-1011) and SEM (JSM 6700F) analysis. Histograms of particle size distribution were obtained via estimation of particle size from TEM micrographs.

X-ray diffraction analysis was conducted at room temperature on a DIFREY-401 diffractometer using Cr-Kα radiation source with the Bragg-Brentano focusing geometry. Spherical crystallite diameter distribution was estimated from X-ray diffraction data.

The Fischer–Tropsch synthesis was carried out in catalytic flow system with a fixed catalyst bed under conditions of continuous operation at 20 bar and bulk velocity of the feed syngas 1000 h^{-1} (molar ratio CO : H_2 = 1 : 1) in the temperature range from 220 to 380 °C. The temperature rise was accomplished stepwise (20 °C every 12 hours). Gas sampling and fluid analysis is accomplished at the end of each isothermal regime.

Catalytic tests were conducted without the preliminary stage of recovery. Initial syngas and gaseous products of synthesis were analyzed by the method of gas-solid chromatography on the chromatograph "Cristallux-4000".

3. Results and discussion

Figure 1 shows the TEM images of samples synthesized at different temperatures. FeCo spherical nanoparticles dispersed in translucent carbon matrix are observed and have a uniform distribution.
Particle size distribution histograms (Figure 2) were obtained via estimation of particle size from TEM micrographs. Table 1 represents the average particle sizes D_{av}.

SEM images at high magnification in Figure 3 demonstrate a spherical form of nanoparticles localized on the carbon matrix surface.

Figure 4 shows XRD patterns of samples obtained at annealing temperatures of 300, 400, 500 and 600 °C. For a better visual perception the XRD patterns are shifted relative to each other by the same

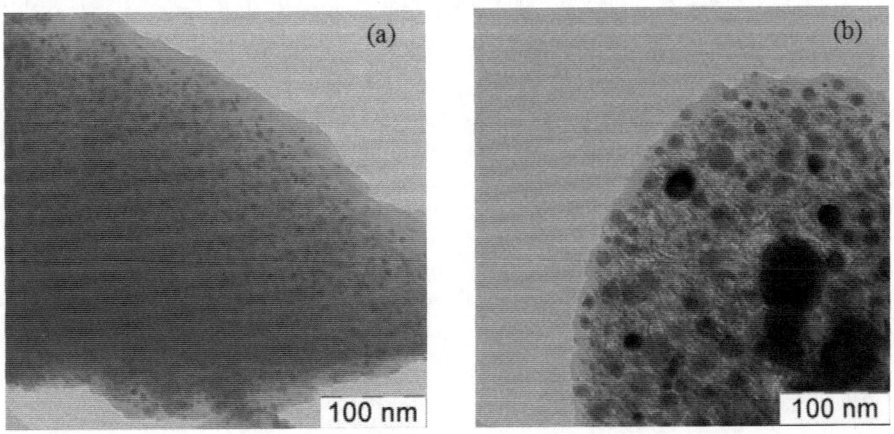

Figure 1. TEM micrographs of IR-PVA/Fe-Co samples obtained at: **(a)** 500 °C; **(b)** 700 °C.

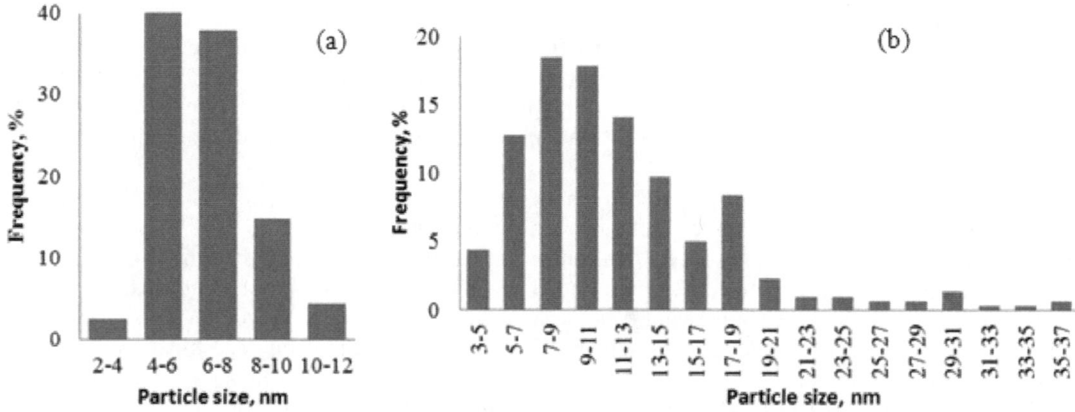

Figure 2. Particle size distribution histograms of FeCo nanoparticles obtained at: **(a)** 600 °C; **(b)** 700 °C.

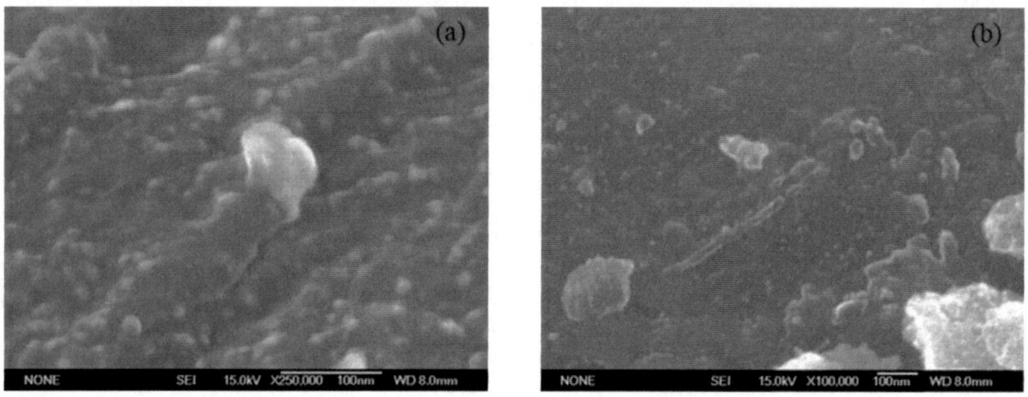

Figure 3. SEM micrographs of IR-PVA/Fe-Co samples obtained at: **(a)** 600 °C; **(b)** 700 °C.

scale along the intensity axis. X-ray phase analysis showed that FeCo nanoparticles with body-centered cubic structure start forming at 400 °C. In the temperature range of the IR-annealing from 300 to 500 °C the XRD patterns peaks corresponding to iron and cobalt oxides are observed, which are superimposed on each other.

X-ray diffraction analysis shows that FeCo reflection peaks at the range from 600 to 700 °C have a symmetrical shape and are not displaced relative to each other, whereas at 500 °C the diffraction maximum is shifted towards larger angles. This fact indicates FeCo solid solution formation with unequal ratio of components and with a smaller lattice constant at 500 °C. The observed iron oxide peaks at 500 °C indicate that FeCo solid solution formation is not completed.

Figure 5 represents the reflection lines corresponding to the maximum intensity peak of FeCo solid solution, the crystal plane of (110). The diffraction maximum corresponding to graphite on the diffraction pattern of 600 °C sample appears (Figure 4). The crystalline phase amount of carbon increases as temperature increases. At the same time there are structural changes in FeCo solid solution detected by the narrowing of the intensity peaks (Figure 5).

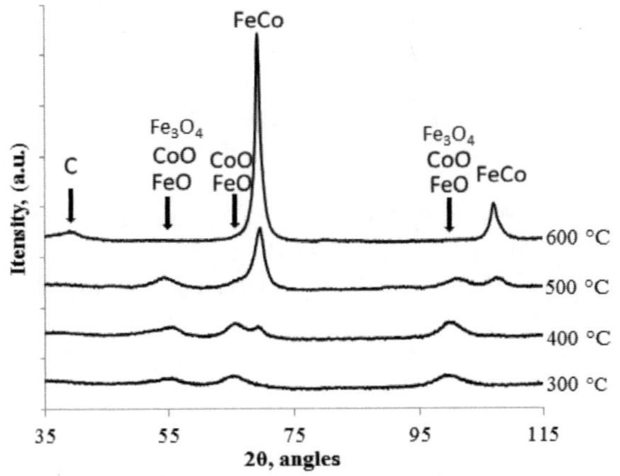

Figure 4. X-ray diffraction patterns of IR-PVA/Fe-Co samples.

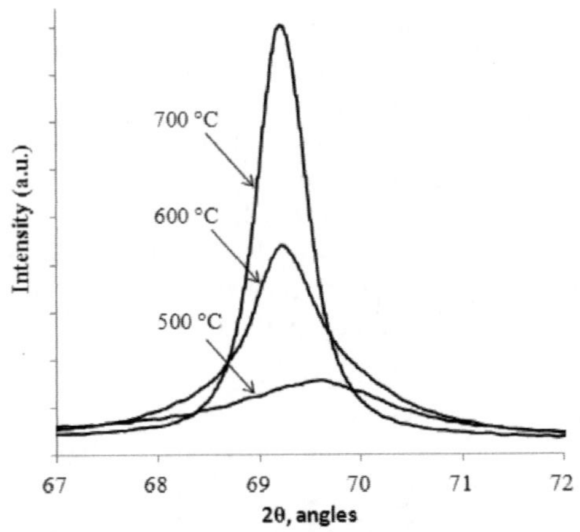

Figure 5. X-ray diffraction patterns of IR-PVA/Fe-Co samples obtained at T = 500 – 700 °C, the crystal plane of (110).

Table 1 shows the calculation results of the structural characteristics of FeCo phase according to X-ray analysis. The lattice constant *a* in the material is smaller than in a bulk analogue due to specific features of nanoscale metal condition.

Figure 6 shows the estimated lattice constant dependence on Fe content assisting in quantitative assessment of the composition of FeCo forming phases. The starting point of the expected dependence for nanoscale material was determined by extrapolation of lattice period values of FeCo solid solution with different compositions synthesized via the method described above. It is expected that the further graph course is similar to bulk material. The lattice period values of iron and FeCo solid solution for bulk samples were taken from reference data [8].

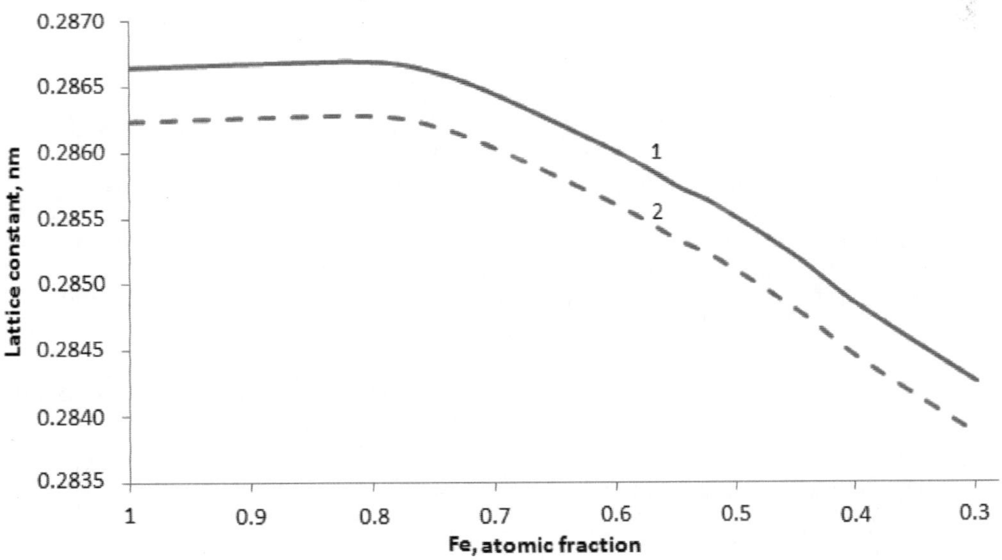

Figure 6. The curves of FeCo lattice constant dependence on iron content; 1 – bulk material, 2 – nanosize material.

Table 1 also represents the results of compositions determination of FeCo solid solutions by lattice periods experimentally obtained. In the sample synthesized at 500 °C FeCo composition is 30 : 70 due to an incomplete reduction of iron oxide.

Figure 7 shows the calculation result of the coherent scattering region (CSR) size distribution by the method of Selivanov-Smyslov [9]. The sample obtained at 600 °C has the most narrow distribution and smaller size of the CSR.

According to data of Table 1 particles consist of a single crystallite in the sample synthesized at 600 °C. The average crystallite size larger than the average particles size in the sample annealed at 700 °C. This can be explained by the locality of the TEM method, while X-ray analysis is an integral.

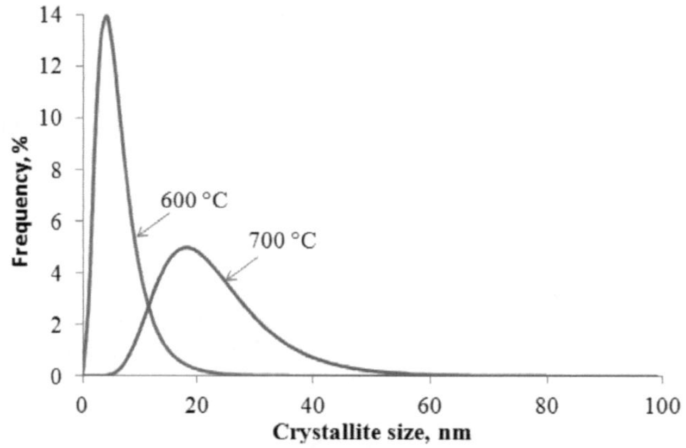

Figure 7. Crystallite size distribution.

Table 1. Dimensional and structural characteristics of FeCo phase

T, °C	*a*, nm	D_{el}, nm	D_{CSR}, nm	Fe : Co in mass percent
500	0.2838	-	-	30 : 70
600	0.2852	6.(6)	7	50 : 50
700	0.2852	12	23	50 : 50

It is known that the metal-carbon nanosystems obtained by the IR-annealing of polymer materials, exhibit a high catalytic activity in Fischer-Tropsch synthesis [10].

The catalytic activity study was conducted for two samples obtained at 600 and 700 °C. Table 2 shows the main indicators of the Fischer-Tropsch synthesis in the presence of the sample IR-PVA/FeCo (600 °C). The CO conversion reaches values close to 100 %. The yield of liquid hydrocarbons C_{5+} is 92 g/m^3 while selectivity was relatively low – 35 %. The thermodynamic yield of the group [–CH_2–] is 208.5 g/m^3 conceding a synthesis gas [11].

Table 2. The main indicators of the Fischer-Tropsch synthesis in the presence of catalyst IR-PVA/Fe-Co (600 °C)

Synthesis temperature, °C	CO conversion K_{CO}, %	Yield of liquid hydrocarbons C_{5+}, g/m^3	Productivity, g/kg FeCo h	Selectivity for liquid hydrocarbons C_{5+}, %
220	2.8	8.1	218	100.0
280	12.1	16.9	456	52.2
320	33.8	32.0	862	35.3
340	96.0	92.4	2488	35.7
360	98.7	50.4	1358	18.9

Increasing of the catalytic indicators during synthesis in the range of 220 – 300 °C are observed. In the range of 320 to 360 °C, the synthesis indicators reach their maximum values due to the optimum performance of the catalyst. Above 380 °C a rapid decrease of the synthesis indicators is observed that may be associated with the system deactivation, i.e. agglomeration of active phase or matrix destruction. The selectivity value of a liquid product is not high due to intensive secondary processes: methane generation, water-gas reaction).

Table 3 represents the main indicators of the Fischer-Tropsch synthesis in the presence of the sample IR-PVA/FeCo (700 °C). The CO conversion value is 90 - 95 % but the yield of liquid hydrocarbons, selectivity and productivity are lower compared with the previous sample.

Table 3. The main indicators of the Fischer-Tropsch synthesis in the presence of catalyst IR-PVA/Fe-Co (700 °C)

Synthesis temperature, °C	CO conversion K_{CO}, %	Yield of liquid hydrocarbons C_{5+}, g/m^3	Productivity, g/kg FeCo h	Selectivity for liquid hydrocarbons C_{5+}, %
220	4.7	10.1	221	78.7
280	10.0	9.4	206	34.3
320	29.7	32.6	716	39.8
340	91.6	74.9	1645	30.0
360	94.7	73.8	1621	28.5

4. Conclusions

The obtained research results show that FeCo nanoparticles with body-centered cubic structure start forming at 400 °C. The metallic phase of samples synthesized at 600 and 700 °C consist of FeCo solid solution only. Nanoparticles have spherical form, the mean size is 7 – 12 nm.

Catalytic tests in Fischer-Tropsch synthesis revealed activity of investigated samples (conversion of CO reached 95 – 100 %). The yield of synthetic hydrocarbons in the presence of nanosystems obtained at 600 and 700 °C was 92 and 75 g/m^3, respectively, while selectivity was relatively low (30 – 35 %).

References

[1] Kutepova E, Knizhnikov A, Kochi K 2011 *Associated Gas Utilization in Russia: Issues and Prospects* (Moscow: WWF-Russia-KPMG) p 43

[2] Solovyanov A A, Andreeva N N, Kryukov V A, Lyats K G 2008 *Strategy of Associated Petroleum Gas Utilization in the Russian Federation* (Moscow: CJSC Redaktsiya gazety Kvorum) p 320

[3] Kozyukov E A, Krylov A 2008 *Artificial flammable gas and liquid fuel* (Moscow: Moscow Aviation Institute Press) p 221

[4] Krylova A U, Kulikova M V 2014 *Solid Fuel Chemistry* **48** 230

[5] Sedighi B, Feyzi V, Joshaghani M 2015 *J. Taiwan Institute of Chemical Engineers* **50** 108

[6] Sineva L V, Mordkovich V Z, Ermolaev V S 2012 *Catalysis in Industry* 13

[7] Zemtsov L M, Karpacheva G P, Efimov M N 2006 *J. Polymer Science* **48** 977

[8] Barabash O M, Koval Y N 1986 *Crystalline Structure of Metals and Alloys* (Kiev: Naukova Dumka) p 598

[9] Selivanov V N, Smislov E F 1993 *Crystallography Reports* **38** 174

[10] Kulikova M V, Zemtsov L M, Sagitov S A, Efimov M N, Krylova A Y, Karpacheva G P, Khadzhiev S N 2014 *Solid Fuel Chemistry* **48** 32

[11] 1980 *Chemicals from Coal* ed U Falbe (Moscow: Chemistry) p 616

High-coercive garnet films for thermo-magnetic recording

V N Berzhansky[1], Y V Danishevskaya[1,2], A S Nedviga[1], H T Milyukova[1]

[1] Physico-Technical Institute of the V.I.Vernadsky Crimean Federal University, Simferopol 295007, Russia
[2] Crimean University of Culture, Art and Tourism, Simferopol 295017, Russia

Abstract. The possibility of using high-coercive of garnet films for thermo-magnetic recording is related with the presence of the metastable domain structure, which arises due to a significant mismatch of the lattice parameters of the film and the substrate. In the work the connection between facet crystal structure of elastically strained ferrite garnets films and the domain structure in them is established by methods of phase contrast and polarization microscopy.

1. Introduction

Uniaxial anisotropy high-coercive garnet films are an optimal media for thermo-magnetic recording by laser beams or contact printing. For example, contact printing magneto-optic films were used for criminalistics examinations of magnetic recordings [1]. For these purposes the films had a low Curie temperature ($T_C < 100$ C). The effective technological way to increase coercivity H_c is to make strains by increasing mismatch $\Delta a = (a_f - a_s)$ between crystalline lattices of film a_f and substrate a_s. Stresses are proportional to the mismatch and networks of misfit dislocations are formed usually to relief them. As in the case of polycrystalline films obtained by the deposition of the aerosol, the crystal structure of the films due to a strongly inhomogeneous internal stresses [2]. Strain-induced morphology has a great influence on parameters of mono-crystalline garnet films, grown by means of liquid phase epitaxy.

2. High-Coercive Garnet Films

2.1. Synthesis of High-Coercive Garnet Films

Garnet films $(Bi,Sm,Lu)_3(Fe,Al,Ga)_5O_{12}$ synthesized by liquid phase epitaxy, using isothermal dipping technique. As substrate monocrystalline wafers were taken $Gd_3Ga_5O_{12}$ (GGG) with a lattice parameter 12.383 Å and (111) crystallographic orientation. Mismatch of film-substrate crystal constants are positive Δa = 0.04-0.12 Å. After mechanical polishing by diamond abrasives (with optical surface quality) and ion-beam etching the thickness of garnet films is 1-2 microns. As a rule, this processing increases garnet films coercivity H_c [1,3]. As a result, the grown films are H_c = 80-550 Oe, angle of Faraday rotation θ = 2-2.3°/μm (λ_l = 532 nm), $4\pi M_s$ > 1.0 kGs, h = 4-5 μm and T_C = 60-210 °C. Uniaxial anisotropy field H_k = 1.3-1.7 kOe.

2.2. Magnetic and Morphological Properties of the Films

The initial morphology of films with a great Δa, which is due to the occurrence of the facet structure, is shown in **Figure 1 (a, b)**. Relief of the surface and internal stresses in the films particularly well revealed in the interferometer of Interfako microscope. All of the samples are characterized by block structure. It forms a mosaic. The size of its elements is about 1-5 microns. Block sizes reveal dependence on mismatch of film-substrate crystal constants Δa. In the films with smaller mismatch, the block size is less. These blocks are regions of relatively free from dislocations. The boundaries of the blocks consist of dislocation clusters. Typical film relief after epitaxy caused by the intersection dislocations with surface.

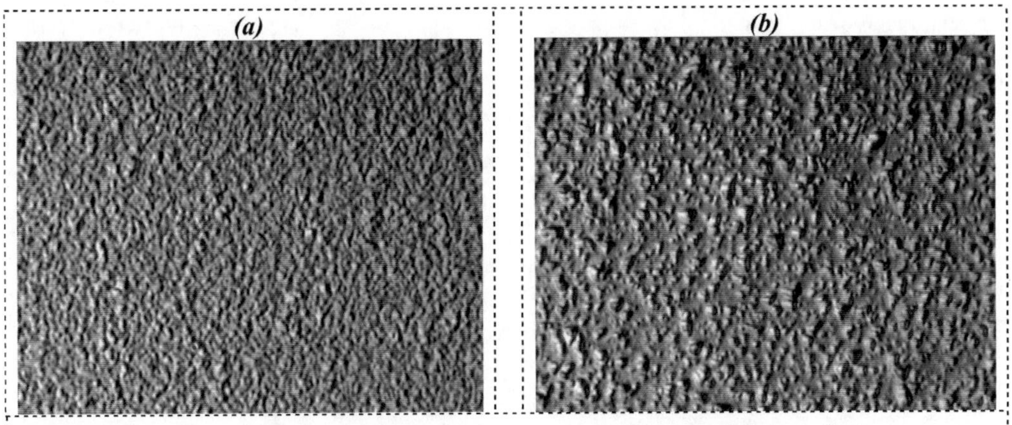

Figure 1 The facet structure of the films with Δa=0,074 Å (a) и Δa=0,103 Å (b). The block size is 2 (a) and 5 (b) microns (Nomarskii differential-interference contrast).

Figure 2 shows the magneto-optical images of two types of domain structures (DS), which are observed in such films. Domain structure on Figure 2 (a) obtained after demagnetization of the film by an alternating magnetic field. This domain structure is stable. It retains its form during remagnetization. Domain structure on Figure 2 (b) obtained by the method of thermo-demagnetization, during the heating above T_C. Such domain structure is metastable, it turns into a stable DS during remagnetization. Fractal analysis was used for a description of the type of domain structure. It is found that the fractal dimension of the meta-stable domain structure is higher (D=1,7) than the dimension of the stable domain structure (D=1,3). Both images are retaining a hexagonal symmetry, which is conditioned by cubic anisotropy in (111) plane.

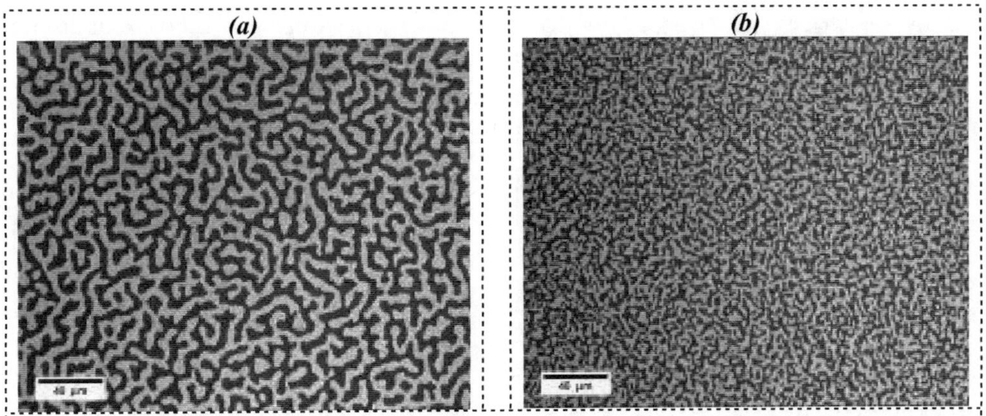

Figure 2 Stable (a) and meta-stable (b) domain structures.

2.3. Dislocations in the Films

It has been established that a network of misfit dislocations is located above the interface of substrate-film and its position is dependent on the value Δa [3]. This means that pseudomorphic layer, where the elastic stresses are mainly concentrated, has a definite value of thickness h_p. At large mismatches the line of appearance of the first dislocations above the interface tends to decrease h_p. Selective etching been used to identify the detection of defects in ferrite garnet films. Interface (ends of the film) and wedges are etched. Wedges were obtained by a special polishing, as a result of which the film thickness ranged from h to 0. The samples were studied using differential interference contrast (Figure 3) and scanning electron microscopy (Figure 4). Etch pits reveal the existence of dislocations near the interface (boundary) film-substrate, that observed that both the film-substrate interface and the wedges. The distribution of density over the thickness of the films was evaluated by observing dislocations in differential-interference microscope wedges in monochromatic light on the location of the interference fringes of equal thickness. The greater the mismatch, it corresponds to the lower height of appearance of the first the dislocation above the interface. In the image of the wedge Figure 3 it's clear the first dislocation in the film with constant mismatch film-substrate 0,045 Å appear with the achievement of the epitaxial layer thickness of about 0.2 microns. Half of the average dislocation density is located at a distance higher than 0.6 microns above the interface. The maximum density of dislocations is focused on the film surface.

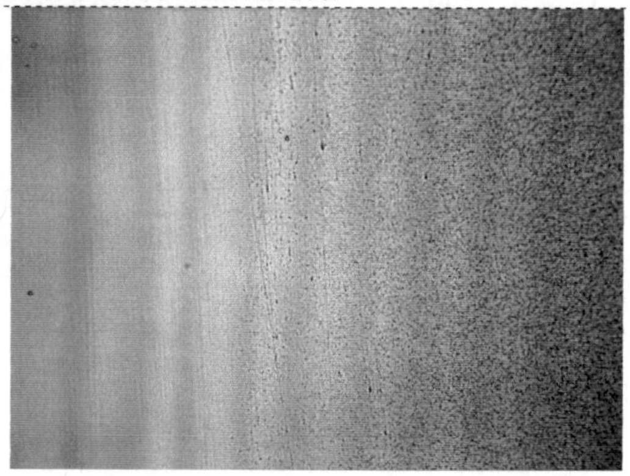

Figure 3 The distribution of dislocations in etched wedge of garnet film. Interference stripes of equal thickness (590 nm). (Differential-interference contrast)

Importantly, the images and the ends of wedges after the selective etching of films misfit dislocations are not located directly on the interface of substrate-film.

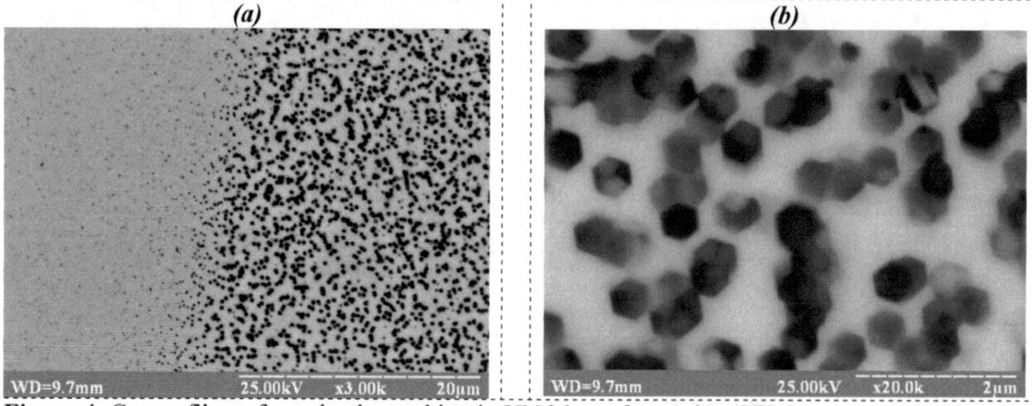

Figure 4 Garnet films after selective etching in HNO3: surface (a,b). (SEM).

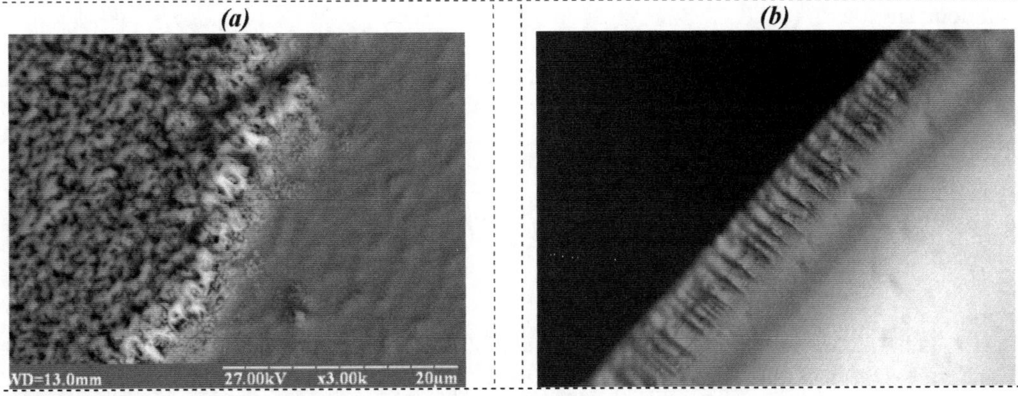

Figure 5 Garnet films after selective etching in HNO3: 45° views (a) (SEM). View of the film-substrate interface (b) (Nomarskii differential-interference contrast).

The relief of surface the film-substrate interface is visible when using differential interference contrast. On Figure 5 dark spots are etching pits of misfit dislocations, located above the film-substrate interface. The so-called "tracery creek" formed by a grid of screw dislocations is located still higher.

2.4. The relationship of the form of domain structure with morphological features

The research using methods of phase contrast and polarization microscopy has detected the relationship of the form of domain structure with morphological features faceted garnet films.

We used a standard polarizing microscope equipped with a surveillance system by phase contrast (having a revolving holder with phase plate). When the polarizer is off, we can see the internal structure of a transparent film of garnet (called facets). At the included polarizer, the image of the internal structure is superimposed with image of the domain structure (Figure 6).

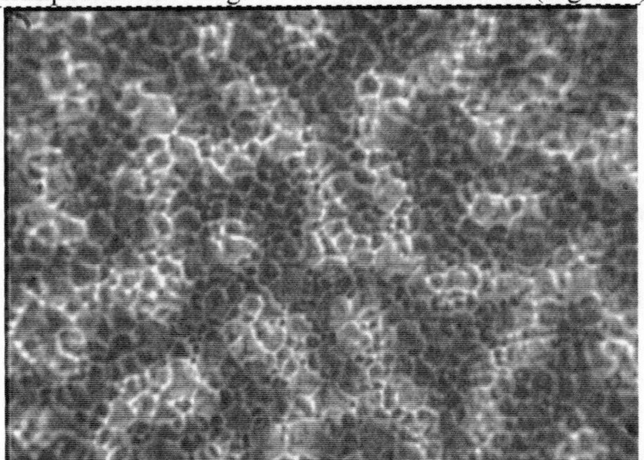

Figure 6 The domain structure observed by phase contrast at the included polarizer. The structural elements are in the form of grains. The grain size of 1.5 microns. The period of the domain structure of 6 microns.

The combination of these two methods of obtaining of contrast shows that the domain walls extend along the lines of equal brightness and encircle elements faceted structure of the film. Since the accumulations of dislocations are local weakly magnetic zones of the film, it is clear that domain walls will be distributed along their geometric borders.

3. Conclusion

In the high-coercivity ferrite garnet films with a large lattice mismatch film-substrate observed two types of domain structure - stable and metastable. It was found that the fractal dimension of a stable domain structure is lower than the metastable. It is shown that correlation of the domain structure of faceted crystal structure of the films due to lattice mismatch film-substrate.

Acknowledgments
This work is supported by the RF Ministry of Education and Science in the framework of the base part of the state task №2015 / 701 (project 3879).

References

[1] Vishnevskii V, Nesteruk A, Nedviga A, Dubinko S and Prokopov A 2007 *Sens. Lett.* **5** 29
[2] Shin K-H, Mizoguchi M, Inoue M 2007 *Journ. of Magnetics* **12** 129
[3] Berzhansky V, Nedviga A, Vishnevskii V, Prokopov A 2009 *Solid St. Phenomena* **152-153** 11
[4] Mikherskii R, Dubinko S, Vishnevskii V, Nedviga A and Prokopov A 2000 *Ukrainian J. Phys.* **45** 368

Laser-induced surface modification of metals and alloys in liquid argon medium

V S Kazakevich[1], P V Kazakevich[1], P S Yaresko[1], D A Kamynina[1,2]

[1] Samara branch of P.N. Lebedev Physical Institute of the Russian Academy of Sciences, Samara 443011, Russia
[2] Samara State Aerospace University, Samara, 443011, Russia

Abstract. Micro and nanostructuring of metals and alloys surfaces (Ti, Mo, Ni, T30K4) was considered by subnanocosecond laser radiation in stationary and dynamic mode in the liquid argon, ethanol and air. Depending of structures size on the samples surface from the energy density and the number of pulses were built. Non-periodic (NSS) and periodic (PSS) surface structures with periods about λ-λ/2 were obtained. PSS formation took place as at the target surface so at the NSS surface.

1. Introduction

Micro and nano-structuring of metal surfaces is of significant interest, because it allows to extend optical, mechanical and magnetic properties of the initial materials. Structuring of the sample surface is possible to realize using different methods: chemical etching, lithography, mechanical impact, laser technology, and etc. Laser methods are distinguish by relatively simple and allow to avoid using of toxic and explosive substances in comparison with chemical methods [1].

The objective of this work was to obtain micro and nanostructures on the metals surfaces (Ti, Mo, Ni) and T30K4 alloy surface by laser ablation in liquid argon and their comparison with structures obtained in the ethanol and air media.

The focus of the work done on structures obtaining in liquid argon because argon is an inert gas and practically insoluble in molten metals that helps to minimize chemical interactions. Structures obtained by laser ablation in liquids characteristic by larger aspect ratio. Moreover, most of the laser ablation in a liquid medium experimental data is got at room temperature. Go to the cryogenic liquid medium can expand the applicability of the laser ablation method.

2. Experimental techniques

Structuring of the samples (Ti, Ni, Mo and T30K4) surface was produced by Nd: YAG laser (RL-PQ-30/300) radiation (λ = 1064 nm, τ = 250 ps, ν = 20 Hz, Q = 0.3 mJ, Q_S = 0.1÷ 1.57 J / cm^2). The traditional optical scheme for the laser ablation in liquids method was used [2].

In the case of use a liquid argon medium, this scheme was supplemented by a cryogenic chamber, which avoids its boiling point (Figure 1).

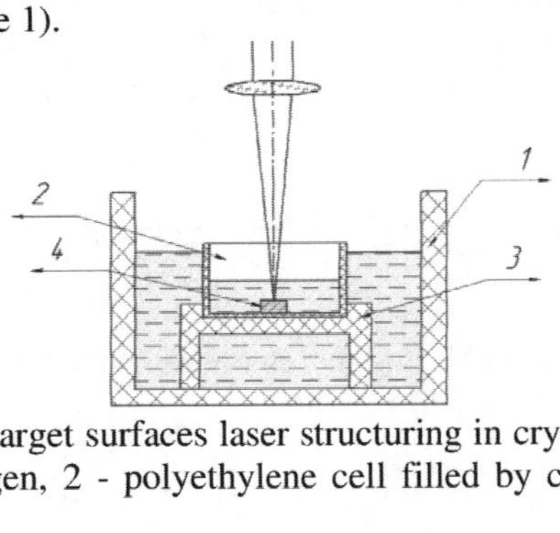

Figure 1. Sketch of the camera for target surfaces laser structuring in cryogenic liquids. 1 - thermally insulated cell filled by liquid nitrogen, 2 - polyethylene cell filled by cryogenic liquid (argon), 3 - plastic stand, 4 - sample.

Structuring of the surface was carried out in dynamic and stationary modes. In the case of a dynamic mode cell with the sample moved by Standa motorized tables relatively to the focal spot. Move options: speed of 500 μm/s, step offset by the other coordinate - 16 microns. In this mode, the irradiation of a titanium target in ethanol (95%) and liquid argon was realized.

In the case of a stationary mode cell with the sample remains fixed relatively to the focal spot. In this mode the dynamic of changes in the shape and size of the structures on the surface of Ti, Ni, Mo and T30K4 in liquid argon, ethanol and air media have been considered in the pulse number ranging from 50 to 20 000.

The overall view and element analysis of the obtained structures were received by using a scanning electron microscope Carl Zeiss Evo 50 with nitrogen-free energy dispersive X-Max detector 80 (EDX). The period of structures was determined by a fast fourier transform of surfaces SEM images. Spectrophotometric analysis of the samples was carried out using a spectrophotometer SF-56 with diffuse reflectance attachment.

3. Results and Discussion

As in the work [3] colour changes as a result of interaction of the subnanosecond IR laser pulses with a titanium surface in ethanol media in the dynamic mode with the difference energy density was observed. Key points: №1 - gray, the energy density is 0.112 J / cm^2; №3 - yellow, 0.125 J / cm^2; №5 - black, 0.145 J / cm^2; №8 - deep yellow, 0.203 J / cm^2; №14 - metallic gray, 1.570 J / cm^2 (Table 1).

Table 1. The dependence of the samples surface color from energy density.

№	1	2	3	4	5	6	7	8	9	10	11	12	13	14
Photography	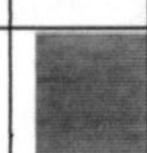	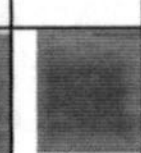	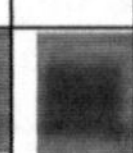		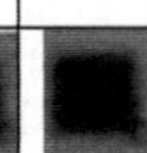			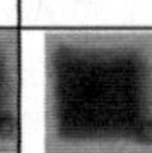	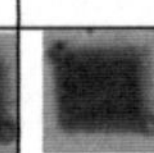		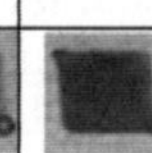	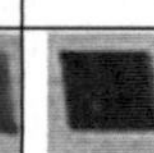	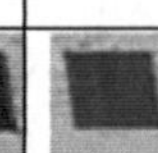	
Energy density, J / cm^2	0.112	0.119	0.125	0.134	0.145	0.164	0.182	0.203	0.244	0.319	0.438	0.641	1.219	1.570

For samples №5 (black) and №8 (deep yellow) absorption spectra were obtained and presented on figure 2.

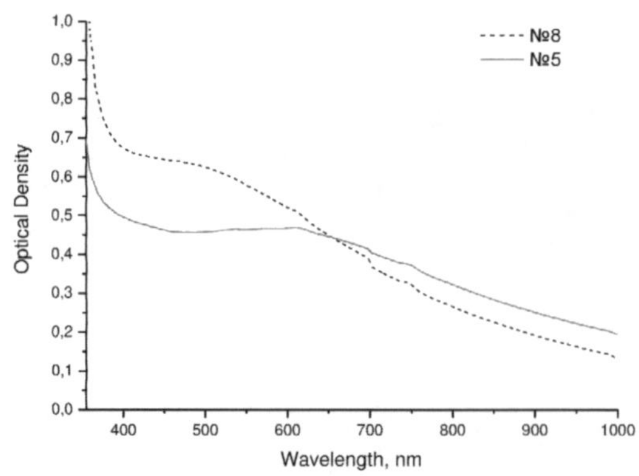

Figure 2. The absorption spectra of titanium surfaces irradiated in the ethanol environment (energy densities №8 - 0.203 J / cm^2, №5 - 0.145 J / cm^2).

In the transition from sample №5 to sample №8 shift of the maximum absorption band from 610 nm to 480 nm was observed. As shown in [4], such changes in the absorption spectra of the Ti surface may be associated with the different thickness of the oxide film on the sample surface.

Since the ethanol contains 5% of the water, in the laser ablation process the oxidative reactions are occurred, the speed of which will depend on the laser energy density. Elemental analysis showed that the maximum oxygen content was recorded on the sample surface №5 (black), its amount 23%. The surface sample №8 (deep yellow), in by turn, contains 15% of the oxygen. This can be explained by the fact that, depending on the oxygen concentration, the thickness of the oxide film is changed and, as a consequence, the interference absorption also changed [4].

Analysis of the surfaces by scanning electron microscopy irradiated in the ethanol environment showed cracking of the surface layer (figure 3a), which on the original sample was not observed. Nanoparticles / structures on the sample surface, which could affect on the sample color due to plasmon resonance phenomena, were not detected.

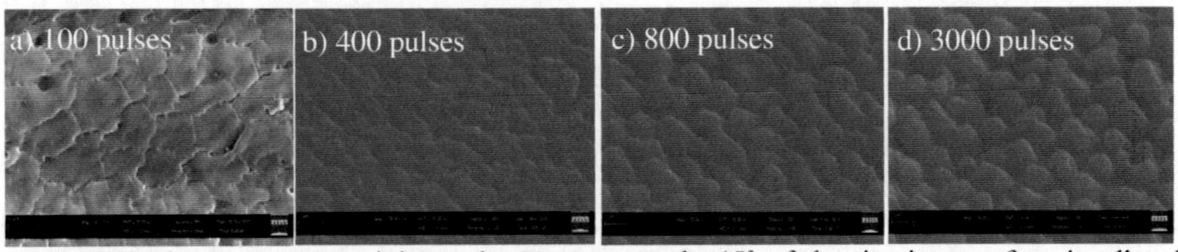

Figure 3. SEM images **a)** at a right angle, **b)** at an angle 45° of the titanium surface irradiated in ethanol medium. The energy density was 0.145 J/cm^2.

By increasing the number of laser pulses that act on a target, the cracking of the surface layer acts as a mask for developing structures (Figure 3 a-d), since the melting point of the oxide layer TiO = 1770 °C, TiO_2 = 1 870 °C, while the melting point of pure titanium - 1668 ° C.

In the case of laser ablation with the same parameters, but in the liquid argon so happens colouring of the sample surface and observed cracking of the surface layer acting as a mask for the structures growth. Energy density varied in the range of 0.2 to 1.2 J/cm^2 (Figure 4 a, b). The surface color after the impact of different energy density of radiation has a mostly yellowish tone, which may be due to a chemical reaction $Ti + TiO_2 \rightarrow 2TiO$, passing in argon at 1550 °C and leads to the transformation of TiO_2 film in TiO.

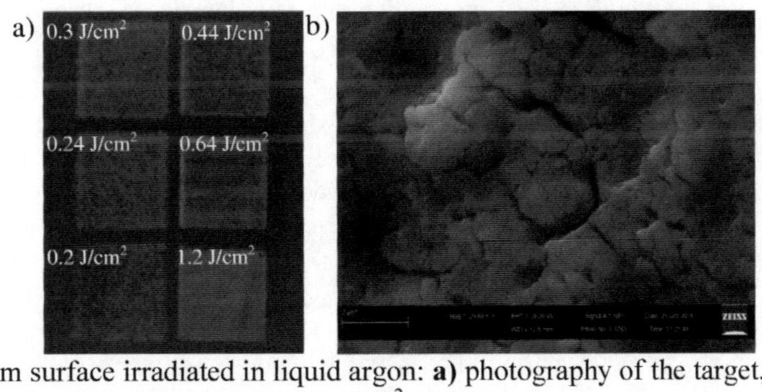

Figure 4. Titanium surface irradiated in liquid argon: **a)** photography of the target, **b)** SEM image of a sample irradiated at an energy density of 0.64 J/cm^2.

At stationary multi-pulse exposure due to the accumulative effect by displacing the melt by steam - gas cloud, the formation and growth of microcones with a base diameter from 1 to 6 microns in liquid argon was observed. Such structures receiving in room temperature liquids and air have been described in several papers [3, 5]. Displacing melt by the steam - gas cloud during laser ablation in liquids will be more active than during laser ablation in gases, that resulting to more developed structures (Figure 5).

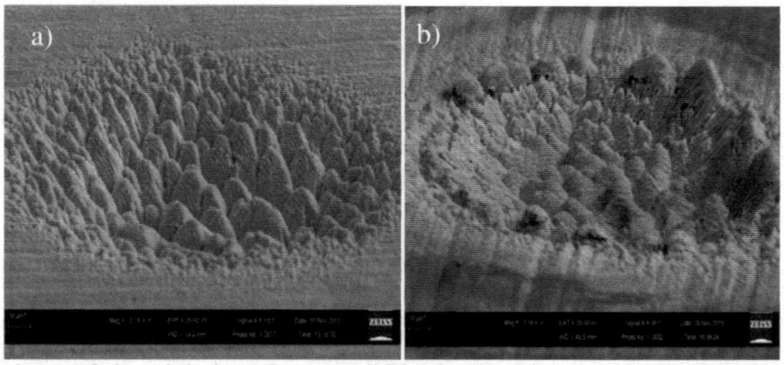

Figure 5. Comparison of the nickel surface modified in liquid argon **(a)** and air **(b)**; 20,000 pulses; 0.25 J/cm^2.

In the boundaries of the one laser spot resolved structures with different sizes (from 0.5 microns to 15 microns). It is also related to the inhomogeneity of the energy density distribution of the real laser beam on the target surface, and consequently, with various depth of the molten bath forming at laser spot center and at its periphery.

At an energy density of 0.3 J/cm^2 distinguish the different area sown by microcones (periphery), and the central "bowl" of the melt covered by drop-shaped microstructures (Figure 6). With the number of pulses growth, microcones, evolving towards the center on the periphery reborn in part-shaped structures, and a crater in the center is forming.

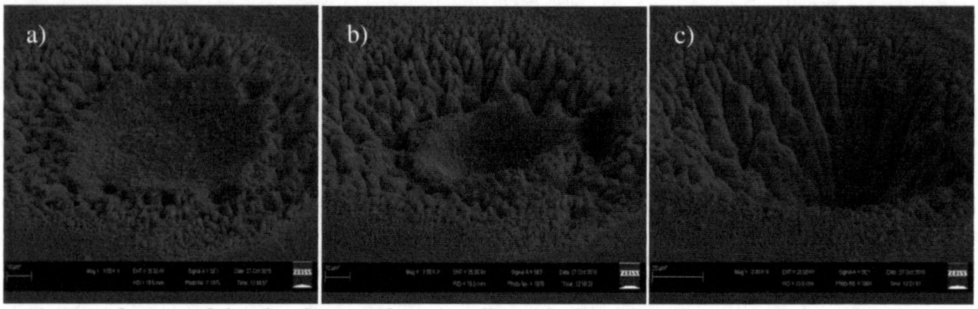

Figure 6. SEM image of the titanium surface, irradiated in liquid argon at an energy density of 0.3 J / cm^2: **a)** 100 pulses, **b)** pulses 300, **c)** 800 pulses.

On figure 7 the dependence of the structures average diameter from the number of pulses for different energy densities was represented. The graph shows that with an increase the number of pulses at a constant energy density, microcones consolidation was observed. The energy density change leads to the possibility of a control structures growth rate. The surface structures were reported on the metals and alloys such as a Ti, Mo, Ni, T30K4. Presented dependences are in good agreement with the results obtained in [3].

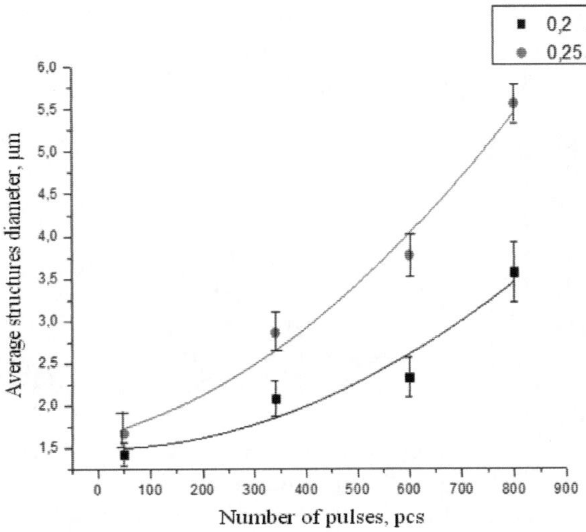

Figure 7. The dependence of the structures size from the number of pulses.

In the case of laser ablation of metals and alloys in the liquid argon and air, formation of surface periodic structures was observed (Figure 5). As shown in [6], this is due to the interference of the incident linearly polarized laser radiation with surface electromagnetic waves that arise at the interface between two media with dielectric constants of different sign at the scattering of light on the sample inhomogeneities. Orientation and period of structures depend on the target material, the environment surrounding the sample, as well as the wavelength, polarization, and intensity of the laser radiation.

At laser ablation in liquid argon in the center of the laser spot, where the energy density is maximum, on the microcones surface the nearwave PSS formation was observed (Figure 5,8a). On the periphery, in the formation conditions of the molten bath with substantially smaller depth, as well PSS formation was registered. The period of the such structures on metals as Ti, Mo, Ni is in the order of λ / 2. As an example, Figure 8b shows a typical image of such PSS obtained on the Mo surface at an energy density of 0.25 J / cm^2. The obtained PSS have two mutually orthogonal orientations that may be associated with the polarization rotation of radiation by an angle 90°, during the laser ablation process, by nonlinear optical shutter included in the laser construction.

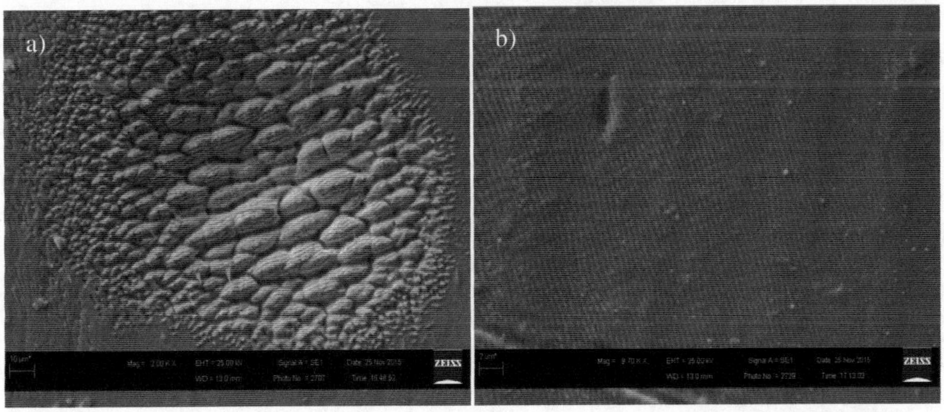

Figure 8(a, b) A typical SEM image of the structures obtained by laser ablation in liquid argon on the Mo surface. The energy density was 0.25 J / cm^2. **(a)** A general view, **(b)** an enlarged fragment of the sample surface at the periphery of the laser spot.

As a result of the T30K4 alloy laser ablation, apart the growth, the change in the ratio of the elemental composition of the surface structures was observed. This may be due to a difference of melting temperature of the alloy components (Figure 9).

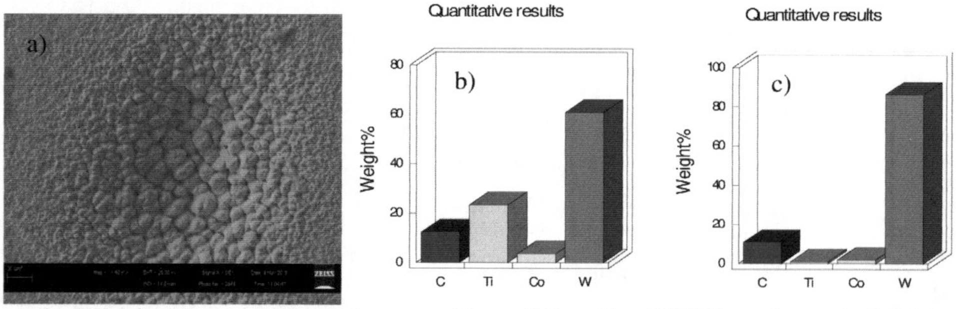

Figure 9. SEM image and elemental composition of the alloy T30K4 surface: **a)** SEM image of structures obtained during laser ablation in liquid argon, **b)** the elemental composition of the original alloy T30K4 surface, **c)** the elemental composition of the sample surface after laser ablation process in liquid argon.

4. Conclusion

Laser ablation method in liquid argon is enough promising because it makes impossible the chemical interaction both with remote from the surface material and with the target surface. This technique allows to obtain microcones with different sizes (from 0.5 microns to 15 microns) and surface periodic structures with periods of ~ λ, $\lambda / 2$. As well as in dynamic action it leads to changing in the spectral characteristics of a target.

Acknowledgments

This work was supported in the framework of the resolution № 244 of the Government of Samara region on May 7, 2015 on the approval for appointing of cash payments procedure for young researchers and designers working in the Samara region in 2016.

References

[1] Makarov G N 2013 *J. Physics-Uspekhi*. **183** 673

[2] Zijie Y, Douglas B Chrisey 2012 *J. of Photochemistry and Photobiology C: Photochemistry Reviews*. **13** 204

[3] Barmina E V, Stratacis E, Fotacis K, Shafeev G A 2010 *J. Quantum Electronics*. **40** 1012

[4] Stepanov A Y, Sotnikova L V, Vladimirov A A, Khaneft A V, Prosvirkina E V, Titov F V, Diaghilev D V 2014 *J. Polzunovsky messenger* **3** 53

[5] Makarov S V 2014 Nano and microstructuring of the surface of metals and semiconductors in the air by femtosecond laser pulses (Moscow: P.N. Lebedev Physical Institute of the Russian Academy of Sciences) p 149

[6] Libenson M V 1996 *J. International Soros Science Education Program*. **10** 92

Negative circular polarization dynamics in InP/InGaP quantum dots

S V Nekrasov[1,3], Yu G Kusrayev[1], I A Akimov[1,2], V L Korenev[1,2], L Langer[2] and M Salewski[2]
[1]Optics of semiconductors Lab., Ioffe Institute RAS, Saint Petersburg 194021, Russia
[2]Experimentelle Physik 2, Technische Universität Dortmund, Dortmund D-44227, Germany
[3]Author, to whom any correspondence should be addressed

Abstract. Photoluminescence (PL) negative circular polarization (NCP) dynamics of InP/InGaP quantum dots (QDs) was studied. Time resolved measurements of PL demonstrated that NCP vanishes, when transverse magnetic field is applied, while oscillations of polarization (that are typical for both low-dimensional and bulk materials) do not occur. Hole g-factor spread in the QD ensemble was supposed to be the most probable reason for such NCP magnetic field behavior. The dependence of NCP dynamics on the repetition period of excitation laser pulses was investigated. In case of fairly small repetition period (T = 13.3 ns) long living NCP (13.3 ns < t < 133 ns) was detected, what was ascribed to resident electron spin orientation, accumulated during many laser pulses. In that regime more than one luminescence polarization decay time exist.

1. Introduction

Quantum dots (QDs) are promising objects for quantum information processing devices, owing to long spin memory [1]. Operation of such devices will be based on spin orientation utilizing. The point is that one of the most important spin relaxation mechanisms of bulk materials, Dyakonov-Perel mechanism, is suppressed in QDs [2].

In X- trion (two electrons and hole) luminescence unusual phenomena of circular polarization negative degree (NCP) takes place [3,4,5,6]. This fact means that the sign of luminescence circular polarization is opposite to the sign of excitation light circular polarization. In current research we have studied dynamics of NCP with and without external magnetic field (applied in Voigt geometry) to clarify NCP formation details in InP quantum dots.

2. Sample

The sample contains one layer of self-assembled InP quantum dots in $In_{0.48}Ga_{0.52}P$ matrix, grown by metalorganic chemical vapour deposition (MOCVD). The distinctive feature of the sample is presence of two characteristic dot sizes, see figure 1. QDs are lens-shaped and mean lateral size of small dots is 100 nm, of big dots - 130 nm.

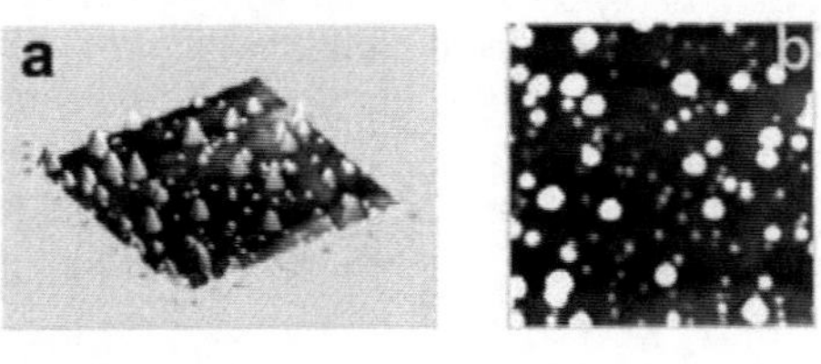

Figure 1. Atomic force microscopy images (a and b, size 2.5x2.5 μm^2) of the similar uncapped sample [7]. The $In_{0.48}Ga_{0.52}P$ cap thickness in the studied sample equals 40 nm.

3. Experimental setup

Both CW and time-resolved studies of optical orientation were conducted. The sample was placed in a helium cryostat (T = 6 K or 10 K), luminescence excitation was provided using circular polarized emission of a Ti:Al_2O_3 laser with 700 nm wavelength. In case of time-resolved measurements a pulse picosecond laser was used. Luminescence was collected in backscattering geometry and passed sequentially through a polarization analyzer, a monochromator and a streak camera (in case of time-resolved measurements) or an avalanche diode (for time averaged studies). After that the signal was processed by a pulse counting card and a PC. In some experiments magnetic field in Voight geometry was applied. For certain time averaged studies modulation of the excitation light circular polarization sign at 50 kHz frequency was conducted, using a photoelastic modulator, in order to prevent orientation of nuclear spins [8].

4. Results

Spectra of intensity and polarization are presented on figure 2. Two intensity peaks correspond to two characteristic dot sizes. We perform studies of long-wavelength line (big QDs), where photoluminescence (PL) circular polarization degree is negative. NCP presence in big QDs luminescence is referred to resident electrons accumulation in those dots (InGaP layers have residual n-doping), where the ground energy state is lower, in comparison to small QDs. It should be recalled that NCP is specific feature of X- trion emission.

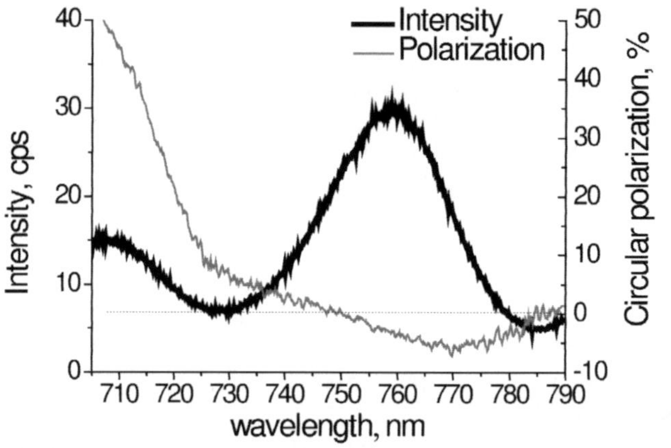

Figure 2. Intensity and circular polarization (optical orientation) PL spectra. Temperature 10 K, CW excitation, 697 nm excitation wavelength.

To clarify NCP formation details the PL dynamics with and without external magnetic field (applied in Voigt geometry) was studied. Time resolved measurements of the PL demonstrated that NCP vanishes with increasing of transverse magnetic field (figure 3 a,b), while oscillations of polarization (that are typical for both low-dimensional and bulk materials and are the consequence of the Larmor precession of electron spins) do not occur.

Three possible reasons for the polarization vanishing without oscillations have been proposed and examined:

(i) as far as NCP corresponds to X- trion emission, the orientation of resident electron spins is important for a trion formation process. If this spin orientation is accumulated with many laser pulses, the Larmor precession of resident electron spins, aligned by different pulses, starts at different times, and the coherence is lost;

(ii) the spread of nuclear magnetic fields acting on electron spins, localized in a QD, will result in the spread of Larmor precession frequencies of electron spins (inhomogeneous broadening), what may cause oscillations vanishing;

(iii) since the polarization of the X- trion emission is determined by hole spin orientation as well, the hole g-factor spread in the QD ensemble may result in vanishing of polarization degree in transverse magnetic field instead of oscillations.

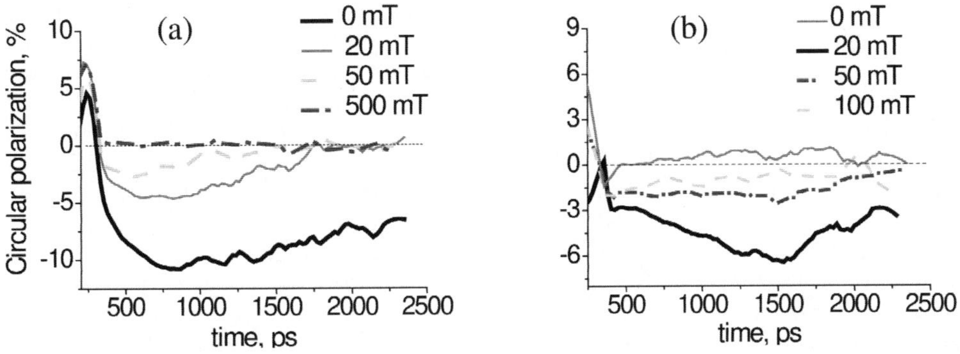

Figure 3. Circular polarization time dependences at different magnetic fields in Voight geometry. Temperature 6 K, 700 nm excitation wavelength, 766+/-17 nm detection wavelength, laser pulse repetition periods equal 13.3 ns (a) and 133 ns (b).

Scenario (i) - NCP accumulation with many laser pulses - was studied experimentally. Figure 4 shows time averaged NCP degree dependence on laser pulses repetition rate. The circular polarization degree doesn't depend on frequency of laser pumping, at periods from 133 ns and higher, while for 13.3 ns period the value of NCP is substantially bigger. We propose that in 13.3 ns laser pulse repetition period accumulation of resident electron spin orientation with many laser pulses takes place, thus mean spin orientation of resident electrons and consequently the degree of NCP increase. Spin relaxation time of resident electrons (τ_s) can be estimated as follows: 13.3 ns < τ_s < 133 ns. It was concluded that scenario (i) doesn't cause the effect, because oscillations are absent both in the NCP accumulation with many laser pulses regime (figure 3 a) and the separate laser pulses regime (figure 3 b).

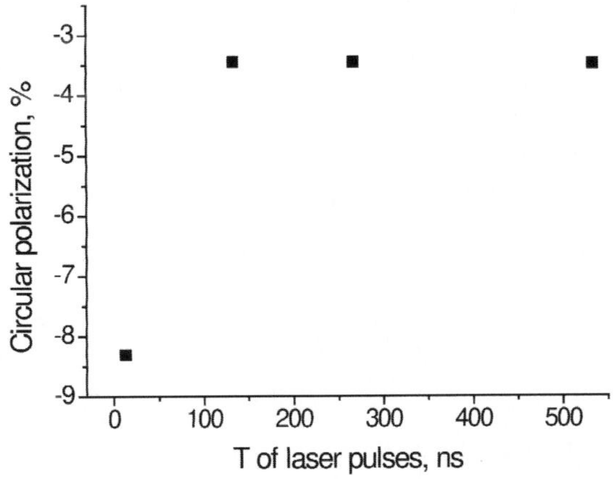

Figure 4. The time integrated circular polarization dependence on the period of excitation laser pulses. Temperature 6 K, 700 nm excitation wavelength, 766 nm detection wavelength

Our analysis has shown also that assumption (ii) – the nuclear fields spread in the QD ensemble - is unlikely, because in that case oscillations are expected to appear in big magnetic fields (which are much bigger than nuclear fields), what does not happen. First of all, presence of nuclear magnetic fields, acting on electron spins, will be demonstrated. Figure 5 shows (CW) NCP dependence on magnetic field in Faraday (5 a) and Voigt (5 b) configurations. The Hanle curve (5 b) in case of excitation with the constant polarization (black dots) is two times wider than in case of modulated polarization (red dots). The polarization modulation frequency equals 50 kHz and prevents nuclear spin orientation. NCP field dependence in Faraday geometry is moved relatively y axis by a value of 300 G. One can conclude that nuclear fields, acting on electrons, persist, and nuclear quadrupole interaction manifests itself [9]. Indeed, in the standard theory of optical orientation Knight field enabled nuclear fields increase the spin depolarization rate in transverse magnetic field, what leads to narrowing of the Hanle curve. That contradicts the obtained results. Presence of the nuclear electric strain-induced quadrupole interaction causes fixing of nuclear field in the structure growth direction, what explains the Hanle curve broadening (external magnetic field has to overcome nuclear field) and the longitudinal field shift. The value of nuclear magnetic fields of approximately 300 G can be estimated from NCP field dependences. A simulation has shown that oscillation vanishing takes place, only when the typical value of nuclear magnetic field fluctuations equals to the value of total magnetic field (which is the sum of nuclear and external fields). Thus in external magnetic fields much bigger than nuclear fields oscillations are expected to be recovered. But they are not recovered even in 5000 G external magnetic fields, what can be seen in figure 3 a.

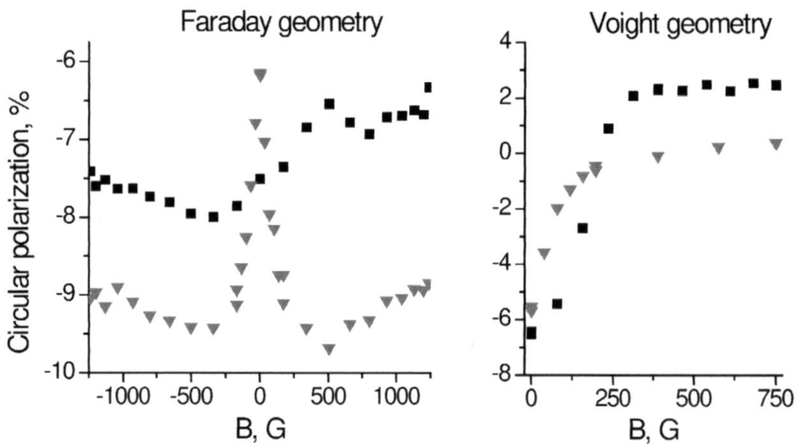

Figure 5. The circular polarization dependence on magnetic field. (a) Faraday geometry, (b) Voight geometry. Squares – excitation with constant polarization, triangles – the polarization was modulated at a frequency of 50 kHz. Temperature 11 K, 697 nm CW excitation wavelength, 766 nm detection wavelength.

The heavy hole g-factor scattering (scenario (iii)) is supposed to be the main candidate for explanation of the oscillations lack and requires further research. It will be shown why considerable g-factor spread in the QD ensemble may be realized. Heavy hole g-tensor is asymmetric, and in case of quantum dots with high enough symmetry (axial or D_{2d}) g-tensor components in the structure plane equal zero to a first approximation (actually they are nonzero, but very small [10],[11]). In-plane anisotropy of QDs will result in admixture of light and heavy hole states [12]. Thus g-factor in the structure plane appears. So in case of considerable in-plane anisotropy relatively high g-factor value can be achieved. Because of QDs forms scattering in the ensemble, transverse g-factor of heavy hole will scatter as well.

In order to figure out influence of NCP accumulation with many laser pulses on a negative polarization formation process we studied the NCP dynamics at different laser pulses repetition periods. Figure 6 shows the time dependence of PL circular polarization at two different pulse frequencies. Red curve (13.3 ns repetition period, corresponds to accumulation of negative polarization with many laser pulses) shows complicated dynamics of NCP – two polarization decay times. In case of 133 ns pumping repetition period (black curve), what corresponds to the separate pump pulses regime, long living NCP is absent. We conclude that such pulse frequency dependence is determined by presence of resident electron spin orientation in case of frequent pulses and absence, when pulses are rare. One can notice that the relatively high NCP degree is achieved without resident electron spin orientation, what is unusual. It should be noticed that observation of such long spin relaxation times (tens of nanoseconds) in luminescence requires long carrier lifetime as well. Appearance of nonradiative dark excitons due to hole spin relaxation during carriers thermalization process is proposed. Their radiative recombination can happen, if spin flip of a single carrier in a dark exciton takes place or, after forming a trion with a resident electron. Latter process also may require an electron spin flip for relaxation from triplet to singlet trion spins and further recombination. Characteristic times of these processes can be rather long.

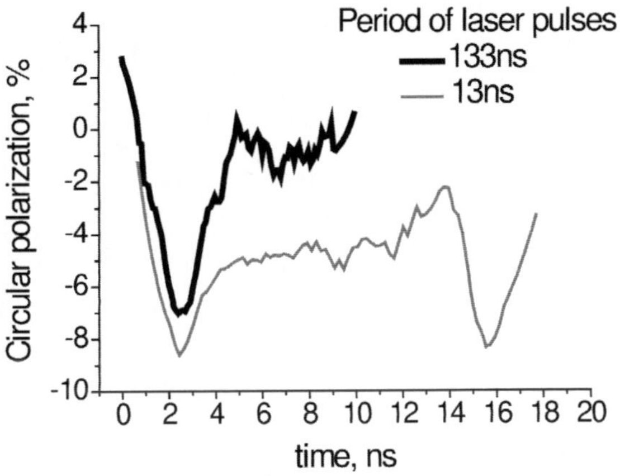

Figure 6. PL circular polarization time dependences at different laser pulse repetition periods. Temperature 6 K, 700 nm excitation wavelength, 766+/-17 nm detection wavelength.

5. Conclusions

To summarize, the photoluminescence negative circular polarization dynamics of InP quantum dots in InGaP matrix was studied. Time resolved measurements of PL demonstrated that NCP vanishes with increasing of external transverse magnetic field, while oscillations of polarization do not occur. Three reasons of the phenomenon were proposed and examined. The hole g-factor spread in the QD ensemble was supposed to be the most probable cause of such NCP magnetic field behavior. However, this scenario requires further studies. The dependence of NCP dynamics on the repetition period of excitation laser pulses was studied. In case of frequent laser excitation pulses (T = 13.3 ns) more than one polarization decay time exist. Presence of long living NCP (13.3 ns < t < 133 ns) in this excitation regime was ascribed to resident electron spin orientation, accumulated during many laser pulses.

Acknowledgments

Support of the Russian Basic Research Foundation (Grant No. 15-52-12017) is gratefully acknowledged.

References

[1] Kroutvar M, Ducommun Y, Heiss D, Bichler M, Schuh D, Abstreiter G and Finley J J 2004 *Nature* **432** 81
[2] Dyakonov M I and Perel V I 1971 *Sov. Phys. JETP* **33** 1053
[3] Dzhioev R I et al 1998 *Phys. Solid State* **40** 1587
[4] Cortez S et al 2002 *Phys. Rev. Lett.* **89** 207401
[5] Kalevich V K et al 2003 *Phys. Status Solidi B* **238** 250
[6] Ikezawa M et al 2005 *Phys. Rev. B* **72** 153302
[7] Mintairov A M, Merz James L, Blundell Steven A 2012 Molecular states of electrons: emission of Single Molecules in self-organized InP/GaInP quantum dots *Fingerprints in the optical and transport properties of quantum dots.* Ed Ameenah Al-Ahmadi (InTech)
[8] Merkulov I A and Fleisher V G 1984 *Optical Orientation* ed Meier F and Zakharchenya B P (North-Holland, New York) chapter 5
[9] Dzhioev R I and Korenev V L 2007 *Phys. Rev. Lett.* **99** 037401
[10] Tischler J G, Bracker A S, Gammon D and Park D 2002 *Phys. Rev. B* **66** 081310
[11] Marie X et al 1999 *Phys. Rev.B* **60** 5811
[12] Dyakonov M I and Perel V I 1972 *JETP Lett.* **16** 398

Magnetic domain structure investigation of Bi: YIG-thin films by combination of AFM and cantilever-based aperture SNOM

Yu E Vysokikh[1], A V Shelaev[2], A R Prokopov[3], V I Shevyakov[1] and S Yu Krasnoborodko[1,2]
[1]National Research University of Electronic Technology, Zelenograd, Moscow, 124498, Russia
[2]NT-MDT Co, Zelenograd, Moscow, 124482, Russia
[3]V.I. Vernadsky Crimean Federal University, Simferopol, 295007, Russia

Abstract. We present the results of magnetic domain structure investigation by combination of atomic force microscopy (AFM) and scanning near-field optical microscopy (SNOM). Special hollow-pyramid AFM cantilevers with aperture was used. This combination allows us use same probe for both topography and domain structure visualization of Bi –substituted ferrite garnet films of micro- and nano-meter thickness. Samples were excited through aperture by tightly focused linearly polarized laser beam. Magneto-optical effect rotates polarization of transmitted light depend on domain orientation. Visualization of magnetic domains was performed by detecting cross polarized component of transmitted light. SNOM allows to obtain high resolution magnetic domain image and prevent sample from any disturbance by magnetic probe. Same area SNOM and MFM images are presented.

1. Introduction

A specific place among magnetic thin film materials belongs to bismuth-substituted ferrite garnet (Bi: YIG) films due to their unique optical (high transparency) and magneto-optical (large intrinsic magneto-optical activity) in the visible and near-infrared wavelength region of optical spectrum [1-2]. It determines its efficiency in different areas of applications [3-6]. Investigation of magnetic domain structure and its properties in nano-scale level is important to understand magnetic structure behavior. There are a lot of tools designed to investigate magnetic structure and properties of these objects [7-9]. Atomic force microscopy (AFM) is a powerful tool for study different nanostructures. One of the most popular and widely used methods of AFM is magnetic force microscopy (MFM), enables us to study not only topography but magnetic structure of different kind of samples with high spatial resolution [10,11]. MFM is based on the principle of magnetic interaction between a micromechanical probe (cantilever) and sample surface.

There are a number of factors that negatively affect the quality of the MFM images and limits this technique: electrostatic tip-sample interaction, external magnetic field, cantilever magnetic coating material properties: its rigidity, the thickness of the magnetic coating, and others [12,13]. Meanwhile Kerr and Faraday optical effects are well known for a long time. Special microscopes for magnetic structure visualization were designed based on these effects [14]. Lateral resolution of these method is limited by diffraction limit and typically not exceed 0.5 μm. Sub-wavelength resolution could be achieved by scanning near field optical microscopy (SNOM). The combination of the best sides of these two methods (AFM and Faraday/Kerr SNOM) could allow us precisely investigate domain structure of thin films in nano-scale level.

SNOM itself is also well known as a method of magneto-optical visualization of magnetic samples [12-18]. Different types of experimental setup were presented (illumination through the hole [12,13], photon scanning tunneling magneto-optical microscopy [14,15] and apertureless scattering [13,19,21]). The most common SNOM configurations based on fiber probes with metal coating. The weakness of these probes is fragility and complicated behavior in shear-force mode [22]. Combination of AFM and cantilever-based SNOM allows use strongly reproducible silicon cantilevers instead of fibers probes. Feedback system in AFM mode is precise and reliable compare to shear-force feedback system [11,22].

2. Experiment

2.1. Experiment scheme

Special hollow-pyramid AFM cantilever with aperture was used. Cantilever aperture was performed by focused ion beam etching at the centre of pyramid with diameter of 50 – 150 nm. These probes are designed both for contact and tapping AFM modes. Cantilever probes demonstrate high thermal stability under strong excitation laser power. 20 mW of 473 nm laser beam was tightly focused in to the aperture. No degradation was observed. High durability of probes allows to operate in contact mode for rigid samples. Delicate samples could be measured by tapping mode.

AFM head with build-in 100x, 0.7 NA objective allows tightly focus laser beam and position it on the cantilever aperture. 1064 nm was used in AFM deflection system to avoid interference. Sample was illuminated through cantilever aperture by linearly polarized laser beam in near field. Probe holder was designed with hole at the centre for light transmission from laser to PMT. Sample holder was mounted onto piezo scanner as showed on figure 1.

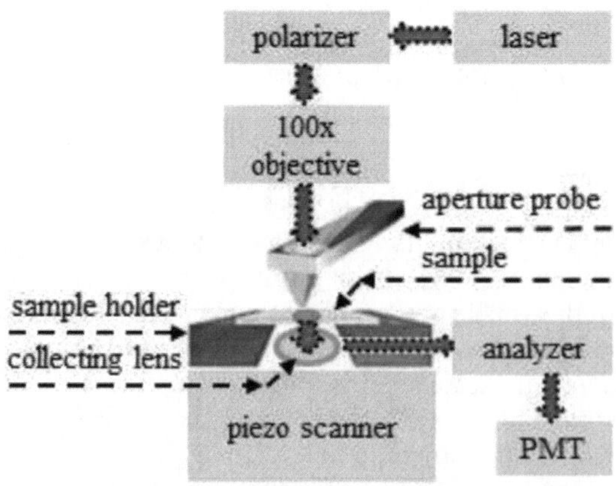

Figure 1. Scheme of AFM and cantilever-based aperture SNOM combination.

Cantilever approaches to the sample surface by step motor. After approaching tip-sample distance is controlled by AFM feedback system. Piezo scanner moves the sample under cantilever line by line (scan area up to 100 μm x 100 μm). "Objective-cantilever-collecting lens" system position is constant independent on scanner moving.

2.2. Samples description

Two types of Bi: YIG films were designed for the measurements.

2.2.1. LPE-film.

The films of nominal composition $(Bi,Gd,Lu,Ca)_3(Fe,Ga,Ge)_5O_{12}$ and "easy axis" (i.e. normal to the film plane) magnetic anisotropy were synthesized by means of liquid phase epitaxy (LPE) method on surfaces of mono-crystalline gadolinium gallium garnet $Gd_3Ga_5O_{12}$ substrates (LPE-films) [23]. Crystallographic orientation of all samples was (111); thickness of the films was 4-6 µm. The films characterized by period of equilibrium domain structure of 4 µm, effective uniaxial anisotropy field 2.5 kOe, saturation field H_S = 42 Oe, specific Faraday rotation 10°/µm at wavelength of light 630 nm.

2.2.2. RIBS-films.

120 nm thick Bi: YIG films of nominal composition $Bi_{2.3}Dy_{0.7}Fe_{4.2}Ga_{0.8}O_{12}$ were deposited on hot (350 °C) (111) GGG single-crystal substrates by reactive ion-beam sputtering (RIBS) technique in argon–oxygen mixture with following annealing at 680 °C in the air at atmospheric pressure as described earlier (RIBS-films) [9, 24]. To determine the magneto-optical properties of the films (Faraday rotation angle Θ_F, coercivity force Hc, squareness ratio K_S and others) the Faraday hysteresis loops (FHL) were measured using hand-made computer control magneto-polarimeter at a fixed radiation wavelength of 655 nm.

2.3. Topography and magneto-optical measurements of LPE-film

Firstly, the topography of Bi:YIG film was obtained by tapping AFM mode, figure 2. Then magneto-optical contrast was obtained by combination of AFM and cantilever-based aperture SNOM, figure 3. Polarization of transmitted light (wavelength – 473 nm) was rotated by magneto-optical effect depend on domain orientation. Visualization of magnetic domains was performed by detecting cross polarized component in transmitted light.

Near-field polarization microscopy allowed to obtain domain structure images with resolution down to ~100nm. Resolution mostly depends on diameter of aperture, distance between probe and sample. There are several factors which could impact on resolution such as film thickness, domain structure periodicity, mutual orientation of domain walls and light polarization, angle of incidence [14].

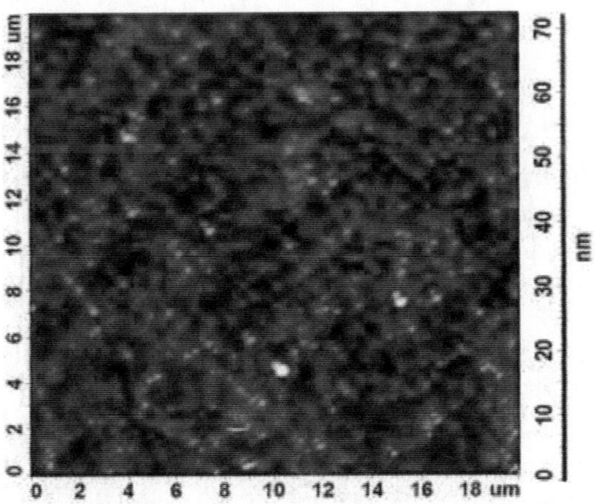

Figure 2. Topography of LPE-film obtained by semi-contact AFM with aperture cantilever.

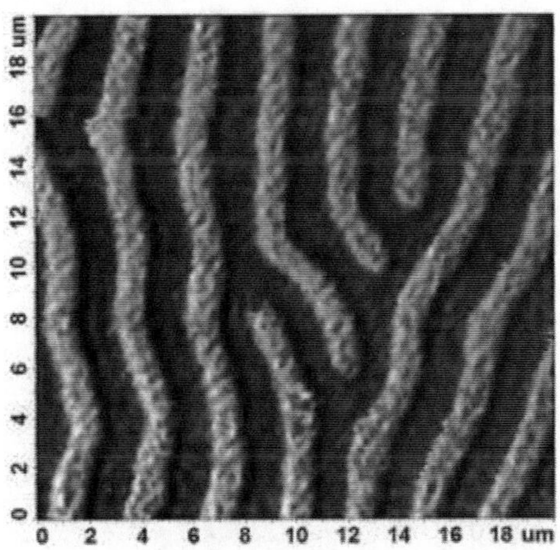

Figure 3. Magneto-optical contrast of LPE-film obtained by cantilever-based aperture SNOM simultaneously with topography from the same area.

2.4. Far-field, MFM and cantilever-based SNOM measurements comparison

Far-field and cantilever based SNOM magneto-optical contrast was obtained for LPE-film, figure 4(a) and 4(b) correspondently. Far-field image shows domain structure with standard optical resolution which is limited by diffraction. Then domain structure of the same was measured by conventional two-pass MFM (magnetic force microscopy) mode, figure 4(c). Silicon cantilever with CoCr magnetic coatings was used for MFM.

SNOM map of magnetic domains provides direct information of the magnetic domain structure unlike MFM image which provides distribution of magnetic force gradient. Moreover, unlike MFM probe non-magnetic hollow-pyramid probe prevents sample from possible disturbance by internal probe magnetic moment. After all same area was investigated in far-field magneto-optical mode again to show influence of magnetic cantilever on domain structure, figure 4(d).

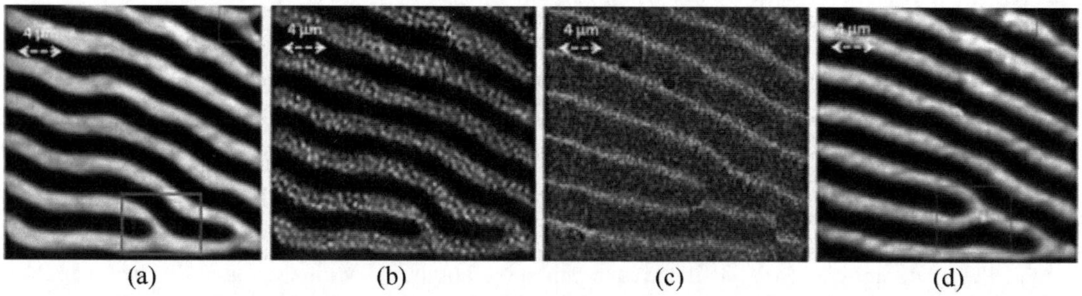

Figure 4. Far-field (a), cantilever based SNOM (b), MFM image (c) of magnetic domain structure of LPE-film. And far-field magneto-optical image (d) after remagnetization of remarkable areas due to magnetic cantilever and sample interaction during MFM measurements.

2.5. Sub-wavelength resolution in cantilever-based aperture SNOM

Special RIBS-film was prepared with a low time of crystallization annealing - 1.2min. This sample was created to determine minimal time when garnet phase created and domain structure appeared. Even with such a low annealing time magnetooptical FLH could be measured. Domain structure is also visible. Film parameters: Θ_F = 0.26°, H_C = 290 Oe, K_S = 0.96, and easy axis magnetic anisotropy showed on figure 5 (curve 1). FHL for another RIBS-film with annealing time 20 min is also showed on figure 5 (curve 2) for comparison.

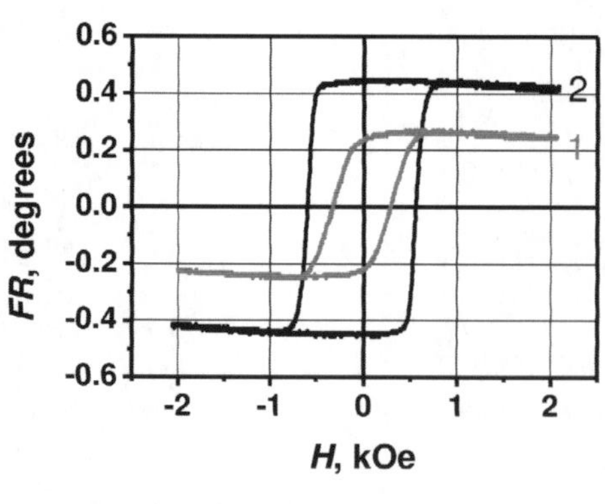

Figure 5. Magneto-optical Faraday hysteresis loops of the Bi: YIG RIBS-films annealed at 680 °C during 1.2 (1) and 20 (2) minutes.

Sputtered RIBS-film with a low time of crystallization annealing was investigated. Domain structure is clearly seen with optical resolution ~150 nm by the method of AFM and cantilever-based aperture SNOM combination. Domain size is ~300 nm. Domain structure and cross section are showed on figure 6 and figure 7 correspondently.

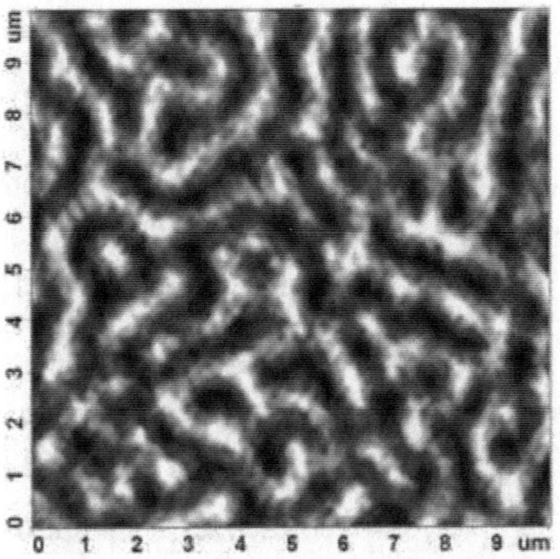

Figure 6. Domain structure of RIBS-film with small period obtained by combination of AFM and cantilever based SNOM, scan area 10x10 um, ~ optical resolution 150 nm.

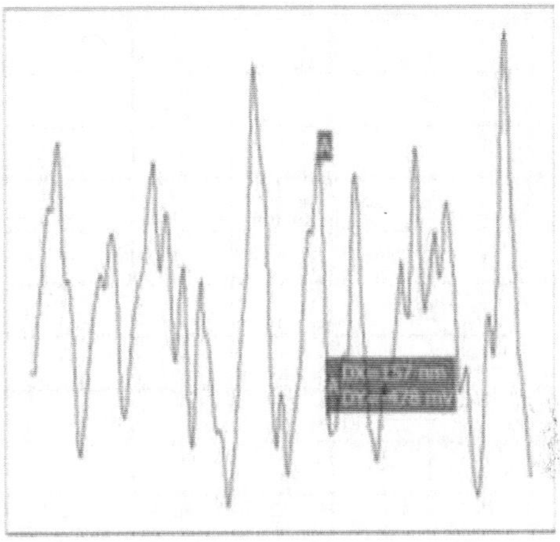

Figure 7. Cross section of domain structure showed on figure 5.

3. Conclusion

Efficiency of combination of AFM and cantilever-based SNOM method is demonstrated. Remagnetization and other disturbances of conventional MFM are eliminated by excluding magnetic coating of cantilever. Fragility of fibre probes in shear force mode in conventional SNOM are also excluded. Resolution of magneto-optical method improved by using near-field illuminating scheme with aperture cantilever. High resolution image of RIBS-film domain structure was obtained with optical resolution ~150 nm.

Acknowledgments
A R Prokopov acknowledges support by the RF Ministry of Education and Science in the framework of the base part of the state task No. 2015/701 (project 3879) of the Bi: YIG films synthesis.

References

[1] Okuda T, Koshizuka N, Hayashi K, Taniguchi H, Satoh K and Yamamoto H 1988 Synthesis of new magnetooptical material, bismuth iron garnet *J. Magnetics in Japan* **3(6)** 483-84
[2] Adachi N, Denysenkov V P, Khartsev S I, Grishin A M, Okuda T 2000 Epitaxial $Bi_3Fe_5O_{12}$ (001) films grown by pulsed laser deposition and reactive ion beam sputtering techniques *J. Appl. Phys* **88(5)** 2734-39
[3] Hansen P, Tolksdorf W, Witter K and Robertson J 1984 Recent advances of bismuth garnet materials research for bubble and magneto-optical applications *IEEE Trans. Magn.* **20(5)** 1099–1104
[4] Inoue M, Uchida H, Nishimura K and Lim P B 2006 Magnetophotonic crystals – a novel magneto-optic material with artificial periodic structures *J. Mater. Chem.* **16** 678–84

[5] Inoue M, Levy M and Baryshev A V 2010 Magnetophotonics: From Theory to Applications *Springer Series in Materials Science* 178

[6] Prokopov A R, Vetoshko P M, Shumilov A G, Shaposhnikov A N, Kuz'michev A N, Koshlyakova N N, Berzhansky V N, Zvezdin A K and Belotelov V I 2016 Epitaxial Bi-Gd-Sc iron-garnet films for magnetophotonic applications *Journal of Alloys and Compounds* 403-407

[7] Kobayashi H, Okuda T, Kobayashi N, Sakamoto I and Hayashi N 1991 Conversion electron Mossbauer study of the $(Bi,Y)_3Fe_5O_{12}$ single-crystal films *J. Magn. Magn. Mater.* **115(2-3)** 255–59

[8] Leitenmeier S, Koerner T, Griesbauer J and Herbort M 2008 Studies of the growth of epitaxial bismuth-substituted iron garnet on gadolinium gallium garnet single crystals by pulsed laser deposition *J. Cryst. Growth.* **310** 5392–5401.

[9] Berzhansky V, Mikhailova T, Shaposhnikov A, Prokopov A, Karavainikov A, Kotov V, Balabanov D and Burkov V 2013 Magneto-optics of nanoscale Bi:YIG films *Appl. Opt.* **52** 6599

[10] Shevyakov V I and Chaplygin Yu A 2013 Investigation into the effect of cantilever structural parameters on the sensitivity of magnetic force microscopy *J. Nanotechnologies in Russia* **8** 186-190

[11] Mironov V L 2004 *Fundamentals of Scanning Probe Microscopy* (Moscow: Technosphera)

[12] Alexeev A M, Komkov V N, Krasnoborod'ko S Yu, Shubin A B and Shevyakov V I 2010 *J. Proceedings of universities. Electronics* **6** 63-66

[13] Krasnoborod'ko S Yu, Shubin A B and Shevyakov V.I. 2011 *J. Russian Microelectronics* **93**

[14] Dickson W, Stashkevitch A, Yousse J B, Takahashi S and Zayats A V 2005 SNOM imaging of thick ferromagnetic films: Image formation mechanisms and limitation *J. Optics Communications* **250** 126-136

[15] Betzig E, Trautman J K, Wolfe R, Gyorgy E M, Finn P L, Kryder M H and Chang C H 1992 *Appl. Phys. Lett.* **61** 142

[16] Silva T J and Schultz S 1996 *Rev. Sci. Instrum.* **67** 715

[17] Matthes F, Bruckl H and Reiss G 1998 *Ultramicroscopy* **71** 24

[18] Bertrand P, Conin L, Hermann C, Lampel G, Peretti J and Safarov V I 1998 *J. Appl. Phys.* **83** 6834

[19] Bergossi O, Wioland H, Hudlet S, Deturche R and Royer P 1999 *Jpn. J. Appl. Phys.* **38** L655

[20] Dickson W, Takahashi S, Pollard R, Atkinson R and Zayats A V 2003 *J. Microsc.* **209** 194-198

[21] Wioland H, Bergossi O, Hudlet S, Mackay K and Royer P 1999 *Eur. Phys. J. Appl. Phys.* **5** 289

[22] Wei P K and Fann W S 1998 The probe dynamics under shear force in near-field scanning optical microscopy *J. Appl. Phys* **83(7)** 3461–68

[23] Berzhansky V N, Nedviga A S, Vishnevskii VG and Prokopov A R 2009 Garnet films for thermo-magnetic recording and nanotechnology purposes *Solid State Phenomena* **152-153** 11-14.

[24] Berzhansky V N, Mikhailova T V, Karavainikov A V 2012 Microcavity One-Dimensional Magnetophotonic Crystals with Double Layer Iron Garnet *J. Magn. Soc. Jpn.* **36(1-2)** 42-45

Investigation of dielectric properties of polymer composites reinforced with carbon nanotubes in the frequency band of 0.01 Hz - 10 MHz

A A Goshev, M K Eseev, S N Kapustin, L N Vinnik, A S Volkov

Northern Arctic Federal University named after M.V. Lomonosov, Severnaya Dvina Emb. 17, 163002, Arkhangelsk, Russia

Abstract. The goal of this work is experimental study of dielectric properties of polymer nanocomposites reinforced with multiwalled carbon nanotubes (MWCNTs) in alternating electric field in low frequency band of 0.01 Hz – 10 MHz. We investigated the influence, functionalization degree, aspect ratio, concentration of carbon nanotubes (CNTs) on dielectric properties of polymer sample. We also studied the dependence of dielectric properties on the polymerization temperature. The dependence of CNTs agglomeration on sample polymerization temperature and temperature's influence on conductivity has been shown. We conducted model calculation of percolation threshold and figured out its dependence on CNTs aspect ratio.

1. Introduction

Intensive research of composite materials which include nanostructures such as CNTs as filler are being conducted recently. Due to large surface area of CNT and significant differences in the physical properties between nanocomposite and macrostate material, properties of nanocomposites are not additive characteristics of each phase and can radically differ from those of each of its components [1]. The data on the study of electrical and other properties of nanocomposites with CNTs is provided in the review [2]. As is known, the functionalization of CNTs - is an effective way to reduce their tendency to agglomerate, increasing the affinity of the components to the matrix of polymer compositions. Methods of functionalization are presented in the review [3]. Early we conducted experimental study of dielectric properties of composites with a matrix of silica filled with multi-walled CNTs at different concentrations in the scattering and absorption of electromagnetic radiation [4,5]. The aim of this work is an experimental study of dielectric properties of polymer nanocomposites with a matrix of epoxy reinforced with multi-walled CNTs at different concentrations and influence of functionalization degree and the concentration of CNTs on the dielectric properties.

2. The influence of polymerization temperature on composite conductivity

As we assume, a significant dispersion of dielectric characteristics values noted in [2] is connected with the parameter unconsidered earlier (polymerization temperature). In this work we additionally studied the influence of this temperature on composites electric properties. We used epoxy ED-20 with solidifier of cold solidification PEPA as a composite matrix. MWCNTs Taunit-MD (0.1% by weight) were used as a filler. For uniform distribution in the polymer matrix in the sample tubes were dispersed by ultrasonic disperser (MEF 91.1). The solidifier was added at fixed temperature after this. We have investigated the frequency dependence of our nanocomposite electrical characteristics by dielectric relaxation spectroscopy method [2,4], according to which the sample is placed between the plates of the capacitor. The sample was exposed to an alternating electric field with a frequency varying in the band of 0.01 Hz - 10 MHz. The measured values were real and imaginary part of permittivity (ε', ε") and conductivity σ. Study of electrical properties of composites was accomplished using broadband dielectric spectrometer *Novocontrol concept 80*. We studied experimentally the dependence of polymer composite conductivity on temperature, at which the addition of solidifier had been made (figure 1).

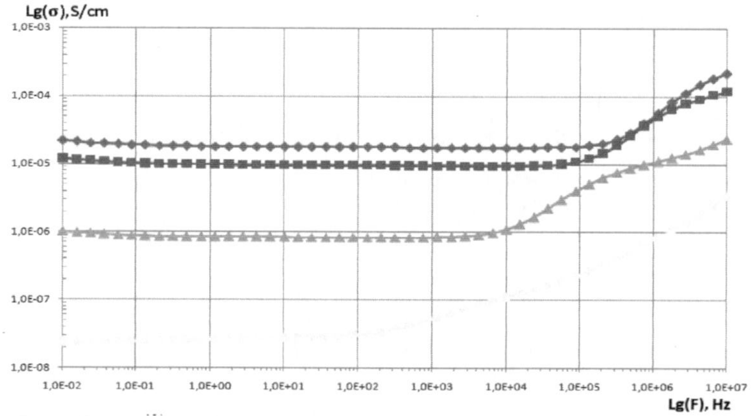

Figure 1. The dependence of conductivity of the composite with 0.1% of CNTs on the frequency of alternating electric field. Data for composites obtained by varying the polymerization temperature: curve – belong to the temperature T=20°C, curve ▲ – T=60°C, curve ■ –T=80°C, curve ♦ – T=100°C.

The result obtained agrees with the agglomeration mechanism studied by us earlier [4]. CNTs are prone to agglomeration because they have poor wettability by polymeric matrix in liquid state. That's why the bigger the solidification time, which directly depends on the temperature in the moment of addition of solidifier into dispersed medium, the bigger amount of CNTs transfer from the state of homogeneous distribution into agglomeration state (figure 2). Figure 2 also shows the model demonstrating behavior of the sample in measuring cell between plates of the capacitor.

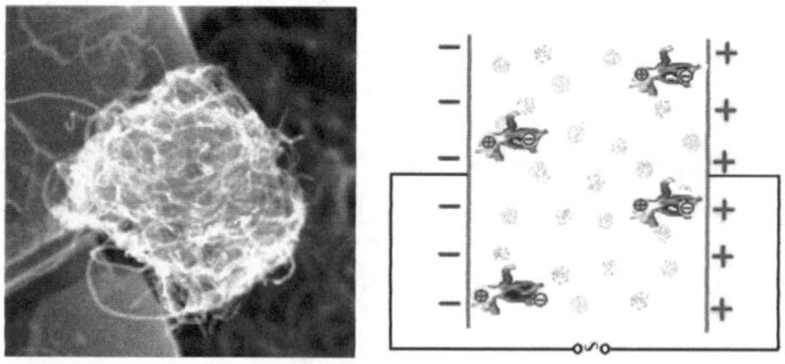

Figure 2. Image of the CNTs agglomerates (*Tescan 3SEM*). Model representation of the agglomerates in the measuring cell.

The effect of agglomeration is well observed on figure 1. In particular, we can see the displacement of percolation threshold into high frequency band from 10^2 Hz at T=20°C to $1.1 \cdot 10^5$ Hz at T=100°C. Table 1 shows the difference in conductivity of samples of same concentration at different frequencies.

Table 1.Increase of the conductivity at different frequencies depending on the sample temperature.

F(Hz)	$\frac{\sigma_{T=100°C}}{\sigma_{T=80°C}}$	$\frac{\sigma_{T=100°C}}{\sigma_{T=60°C}}$	$\frac{\sigma_{T=100°C}}{\sigma_{T=20°C}}$
10^6	1,13	5,08	67,6
10^3	1,86	20,8	319
1	1,84	21,1	675

3. Dielectric properties of composite with CNTs depending on functionalization degree

As known graphene sheet of CNTs can form only weak the van der Waals forces. For an uniform distribution in the polymer matrix and the formation of covalent bonds on the nanotubes surface they are functionalized.

Functionalization of MWCNTs is an effective way to decrease their tendency for agglomeration, to provide directed derivatization of the surface (figure 3), to increase affinity to a binder and other components of polymeric composition (figure 4). However, the oxidation process is accompanied with the destruction of MWCNTs external wall and with the decrease of aspect ratio, which is the consequence of shortening (burning of the ends) (figure 4).

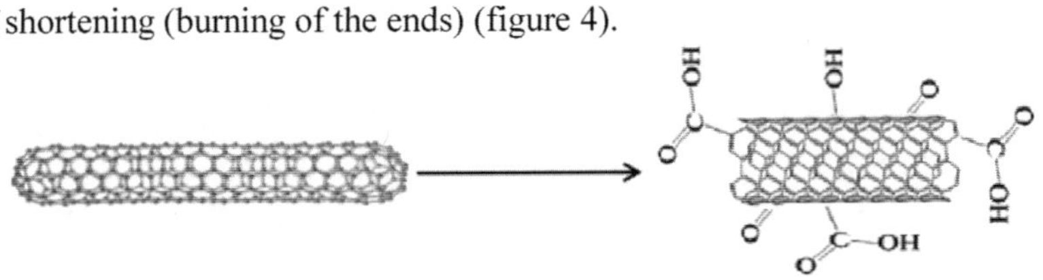

Figure 3. The process of CNTs functionalization in quasi liquid layer of hydrogen peroxide

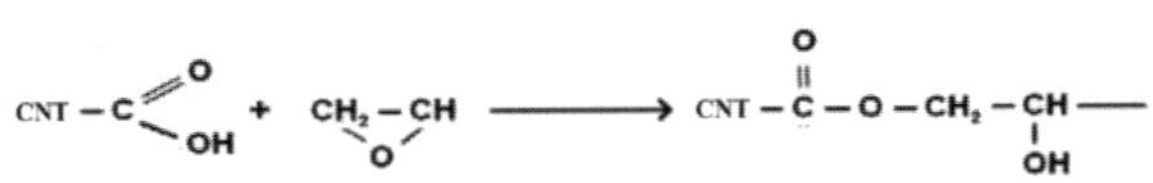

Figure 4. The connection scheme of CNTs functionalization to the polymer matrix through carboxyl groups.

In this experiment dielectric properties of samples of equal concentration of MWCNTs obtained by CVD method [3] were compared with ones of MWCNTs undergone the process of functionalization in quasi liquid layer of hydrogen peroxide during four hours at temperature of T=150°C and the velocity of addition of v=10(ml/h). According to the data obtained in section 1, the temperature of solidifier addition was fixed to be constant T=60°C. The results of real part of permittivity dependence on frequency are shown below.

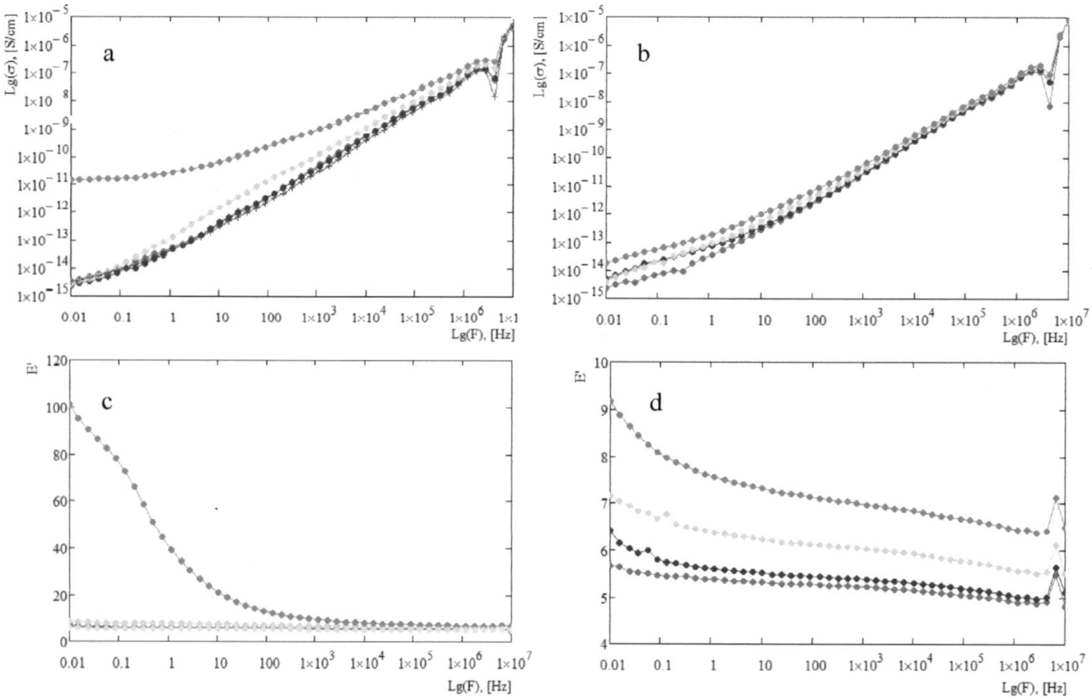

Figure 5. The dependence of the conductivity (a,b) and permittivity (c,d) of the sample on frequency of the alternating electric field.●●● - 0,3%; ●●● - 1%; ●●● - 3%;●●● - 6%; +++ - 0% (CNTs concentration in the composite); (a,c)- samples with CNTs, (b,d) - samples with functionalized CNTs.

The experiment shows that conductivity and permittivity of sample at low concentration changes insignificantly. Considerable change is observed at concentrations >3%. With the growth of CNTs amount in sample both absolute value of conductivity and value of the frequency, at which the transition from constant conductivity value to frequency-dependent value occurs, increases. The addition of functionalized CNTs leads to decrease of conductivity and increase of percolation threshold to more the 6%. The reason for that, as we assume, is in shortening of CNTs during the oxidation. The reason of such huge value of saturation threshold ~6% (figure 5a), in our opinion, is in small aspect ratio of CNTs.

4. Dependence of conductivity on aspect ratio

For the detection of polymeric composite dielectric properties dependence on the aspect ratio $\alpha = L/D$ (L-length, D – external diameter of MWCNTs) the CNTs of two types were used in the experiment. Multi-walled nanotubes "*Taunit-MD*" (*NANOTECH*, Tambov) are pictured on figure 6. MWCNTs obtained in our lab by CVD method (CNT_{Lab}) with nickel catalyst [4] are shown on figure 7. The polymeric composites based on epoxy with different percentage of CNTs were prepared using these tubes. The methodic of sample preparation is described in section 1.
Visual method of SEM was used for approximate estimation of the aspect ratio.

Journal of Physics: Conference Series **741** (2016) 012191 doi:10.1088/1742-6596/741/1/012191

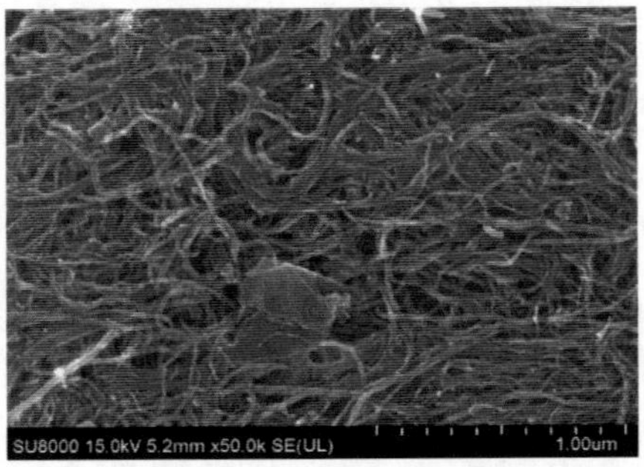

Figure 6. Image of the CNT "Taunit MD" (SEM *Hitachi SU8000*) α ≈ 800.

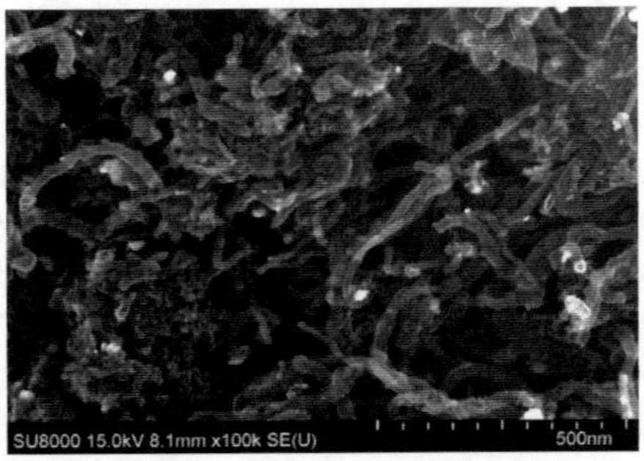

Figure 7. Image of the CNT_{Lab} (SEM *Hitachi SU8000*) α ≈ 14.

The results of experiments are shown on figure 8.

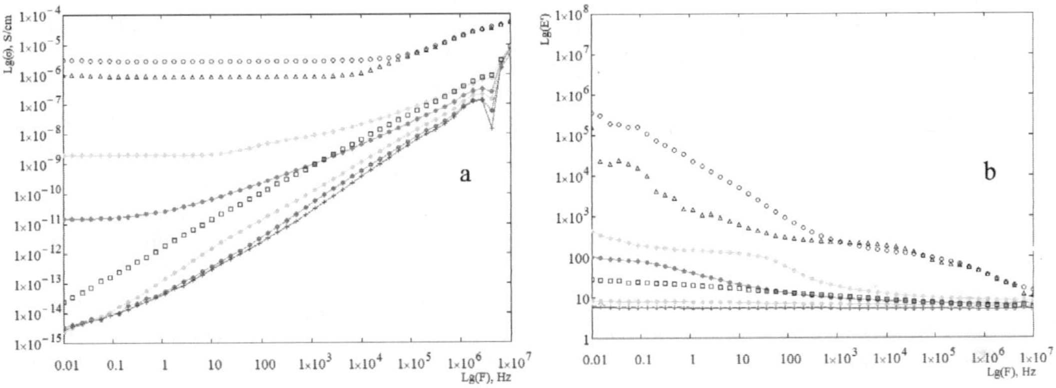

Figure 8.Graphs of the dielectric characteristics of the epoxy composite conductivity (a), the real part of the permittivity (b), on frequency of the electric field: ●●● -0,3%; ●●●- 1%; ●●●- 3%;●●●- 6%; ●●●- 9%, +++- 0% – samples prepared in the laboratory by CVD method. □- 0,06%, △- 0,1%, ○ - 0,3%, - samples with "Taunit-MD".

The measurement of pure tubes outside the polymeric matrix shown insignificant difference in their dielectric properties. Consequently, the difference in dielectric properties of samples could be connected with geometric characteristics of CNTs (aspect ratio). This process is shown on figure 9.

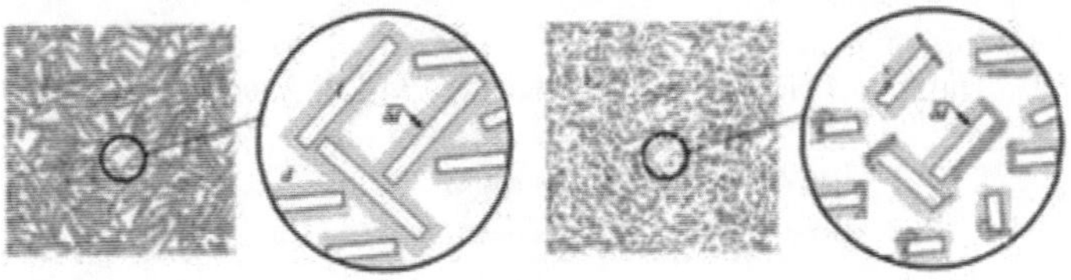

Figure 9. Model representation of CNTs conductivity percolation with different aspect ratio.

On this basis we conducted analytical calculation of percolation threshold on aspect ratio of CNTs by formula:

$$\phi_c = 1 - \exp\left[-\frac{1.4\cdot\left(\frac{\pi}{4}\right)D^2L+\left(\frac{\pi}{6}\right)D^3}{\left(\frac{4\pi}{3}\right)D^3+2\pi D^2L+\left(\frac{\pi}{2}\right)DL^2}\right], \quad (1)$$

where D and L are diameter and length of CNT. Expending (1) on small parameter $\frac{D}{L} \to 0$ we obtain:

$$\phi_c \approx 0.7 \cdot \frac{D}{L}.$$

In this model the percolation threshold for CNT_{Lab} (with aspect ratio of $\alpha \approx 14$) obtains value of $\phi_c = 5{,}25\%$, and for "Taunit-MD" with $\alpha \approx 800$, $\phi_c = 0{,}0875\%$, which is confirmed in experiment (figure 9). The functionalization of tubes by the acid leads to the change of their aspect ratio. Neglecting the contact resistance connected with the presence of carboxyl groups, we can define the upper boundary of the aspect ratio α less the 10.

5. Conclusions

We formulate main conclusions and results.

- The process of influence of temperature of the solidifier addition into the composite matrix (polymerization temperature) on the dielectric properties of the sample has been considered. The method of decrease of the degree of nanotubes agglomeration in polymeric composite via decrease of polymerization time has been proposed.
- It's been shown that with the growth of CNTs concentration in sample both absolute value of conductivity and value of the frequency, at which the transition from constant conductivity value to frequency-dependent value occurs, increases.
- The influence of CNTs functionalization (in hydrogen peroxide vapors) on dielectric properties of polymer reinforced with them was studied by the dielectric relaxation spectroscopy method. A significant decrease of composites with functionalized CNTs conductivity has been noted. This may be the result of two reasons: shortening (burning) of CNTs during functionalization and deterioration of contact between tubes via screening by carboxyl groups.
- The dependence of dielectric characteristics of composite on aspect ratio of CNTs has been studied. It's been shown, that sample dielectric properties strongly depend on aspect ratio. The dependence between aspect ratio and percolation threshold has been detected.

Acknowledgements

This work was supported by the project the Ministry of Education and Science of Russian Federation №3635 "Investigation of nanocomposites properties with controlled modification of the structure reinforced with carbon nanotubes".

We thank I. Chistyakov and our colleagues from Zelinsky Institute of Organic Chemistry (RAS) for help and assistance.

References

[1] Rakov E G 2013 *Russ. Chem. Rev.* **82** 538
[2] Eletskii A V, Knizhnik A A, Potapkin B V, Kenny J M 2015 *Phys. Usp.* **58** (3) 209
[3] Dyachkova T P, et al 2013 Nanosystems: Phys., Chem., Math. **4** (5) 605
[4] Goshev A A, et al 2015 J. Phys.: Conf. Series **643** 012126
[5] Eseev M K, et al 2016 Nanosystems: Phys., Chem., Math. **7** (1) 180

Thermal stability and hydrogenation behavior of Zr–1Nb alloy with TiN_x and Ti/TiN_x coatings

E B Kashkarov[1], O V Vilkhivskaya[2], S A Zakharchenko[1]
[1]General Physics Department, Tomsk Polytechnic University, Tomsk 634050, Russia
[2]Department of Atomic Energy Safety, Troitsk Institute for Innovation and Fusion Research, Moscow, Troitsk 142190, Russia

Abstract. Titanium nitride coatings were deposited by reactive dc magnetron sputtering (dcMS) to protect Zr–1Nb alloys from hydrogen embrittlement. Dense titanium (Ti) interlayer was prepared between TiN_x protection film and a Zr substrate to improve thermal stability and adhesion between the TiN_x and the substrate at high temperatures. Hydrogen absorption of Zr–1Nb with TiN_x and Ti/TiN_x at 623 K was reduced in comparison with uncoated Zr–1Nb. No peeling or cracks of Ti/TiN_x coatings is observed after thermal cycling up to 1073 K. The high temperature (1073 K) hydrogenation behaviour differs from the hydrogenation at lower temperature by increasing the amount of dissolved hydrogen in the β-phase of zirconium. The higher rate of hydrogen absorption by Zr–1Nb with TiN_x was observed due to the coating delamination as a result of differences in thermal expansion coefficients, while Ti/TiN_x demonstrates the lower hydrogen absorption at 1073 K and good adhesion strength.

1. Introduction

Hypothetical nuclear accidents can create real danger to the zirconium alloys and stability of parts made of these alloys, especially such as loss of coolant accident (LOCA) and reactivity initiated accidents (RIA). Hydrogen degradation can manifest itself in appearance of hydride phases resulting in substantial loss of plasticity, increase in ductile-brittle transition, and sometimes in decrease in mechanical strength [1–3]. In Russian Water-Water Energetic Reactor (WWER) and High Power Channel Reactor (RBMK) the Zr–1Nb alloy is used as cladding material. Rapid changes in cladding temperature and increase in the hydrogen absorption rate are the results of LOCA and RIA accidents. The magnetron sputtering and vacuum arc deposition are the main methods used for more than 60 years in various industries for coating deposition and surface modification of structural materials [4–6]. The main advantage of the VAD method is the higher degree of ionization and the higher energy of deposited particles, which increases the density and adhesion of coating by generating of the mixing zone contained deposited and substrate materials. Titanium nitride (TiN) deposited by magnetron sputtering and vacuum arc deposition is promising in protection of zirconium alloys from hydrogen embrittlement [7–9]. Furthermore, TiN coatings have excellent erosion resistance, which favourably affects the fretting resistance of the coatings. The differences between the temperature expansion coefficient of the TiN coating and zirconium alloy lead to the coating degradation. It is known that the coatings obtained by dcMS have low adhesion strength in comparison with FVAD coatings. Moreover, the titanium thermal expansion coefficient of 8.6×10^{-6} K^{-1} is in the range between zirconium (5.7×10^{-6} K^{-1}) and titanium nitride (9.35×10^{-6} K^{-1}). It suggests that it will sustain the balancing thermal expansion coefficient between the substrate and the coating and improve the adhesion strength. In this work, we investigate the operational stability characteristics of hydrogen-impermeable TiN and Ti/TiN protective coatings deposited on Zr–1Nb substrates by dcMS and FVAD/ dcMS methods under high-temperature hydrogenation and thermocycling conditions.

2. Experimental procedure

2.1 Coating Deposition

The Zr–1Nb alloy samples with fixed size of 0.5 mm in thickness and 20 mm in diameter were previously polished using sandpaper to the roughness of 0.1 μm. Initially, the samples were rinsed with alcohol and put into vacuum chamber. Then, the samples were subjected to ion bombardment in argon glow discharge to clean the surface from oxides and organic contamination at 1500 V for 5 min. Finally, TiN and Ti/TiN coatings were deposited onto Zr–1Nb substrates by dcMS and FVAD/dcMS methods, respectively. For magnetron sputtering, high-purity Ti target (99.99%) with dimensions of 100×8 mm was used. TiN coatings were deposited for 30 min by reactive sputtering with the flow rates ratio of Q_{Ar}/Q_{N_2}=4/3 at the constant pressure 0.15 Pa, target power 2.6 kW and current 4 A. The distance between the target and the samples was 150 mm. For FVAD – the cathodic vacuum arc evaporation system equipped with coaxial plasma filter [10] was used. Process parameters for FVAD of Ti (99.8 % purity) interlayer are: 0.2 Pa total pressure, 110 A arc current, 150 V negative pulsed substrate bias with 100 kHz pulse repetition frequency, 5 μs pulse length and 3 A peak pulse current.

2.2. Hydrogenation and Thermal Cycling

Hydrogenation of the samples was performed from hydrogen atmosphere on the automated complex Gas Reaction Controller (USA) at 623 K and 1073 K temperatures for 60 min. The heating rate and hydrogen pressure in the chamber was 6 K/min and 2 atm, respectively. The hydrogenation temperatures were chosen to simulate the working temperature of zirconium fuel cladding during stationary operation and in LOCA transient temperature conditions in the nuclear reactor core.
Thermal cycling of the samples was performed with the water cooled vacuum furnace. The residual pressure in the chamber was $(6–7)\times10^{-3}$ Pa. Heating and cooling rates were 200 and ~ 2000 K/min, respectively. Three cycles of heating and cooling the Zr–1Nb specimens coated with TiN and Ti/TiN were performed.

2.3. Characterization

The morphology of the coatings was studied using scanning electron microscope (Hitachi TM-3000, Japan). The phase identification and structural investigations of the deposited coatings were performed by X-ray diffraction with CuK_α radiation (1.5410 Å wavelength) using Shimadzu XRD-7000 (Japan) diffractometer in asymmetric mode at 40 kV, 30 mA and θ=3° fixed angle. The film thickness was measured by simple ball-cratering method performed with Calotest CAT-S. The measurement of adhesion strength was performed with MicroScratchTester MST-S-AX technique by scratch method using a diamond indenter. Scratching parameters are 0.01 N initial load, 20 N final load, 9.63 mm/min scratching speed and 10 mm scratch length.

3. Results and discussion

Figure 1 shows that deposited TiN_x coating has a cubic NaCl-type structure. It can be seen that the orientation in the plane (200) is the reflection of major intensity and it remains as the most intense for both TiN_x and Ti/TiN_x coatings. Similar results in preferred TiN orientation in plane (200) have also been observed in the report [11]. The less intensive (220) and (222) reflections have been observed. The reflections in plane (111) are difficult to resolve due to the merger with substrate (Zr) reflections. Ti interlayer is formed after vacuum arc deposition.

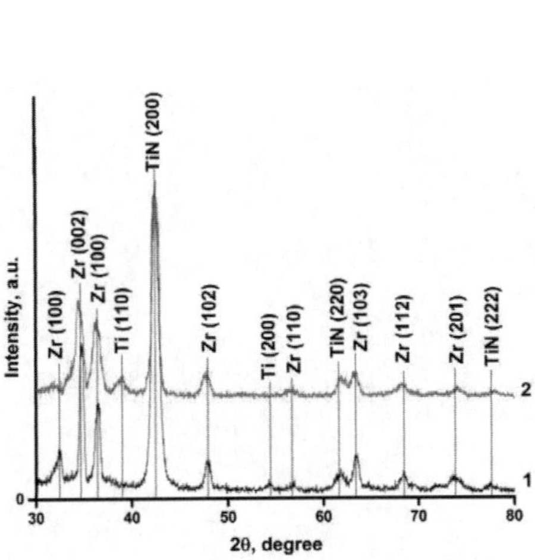

Figure 1. XRD pattern of Zr–1Nb with TiN_x (1) and Ti/TiN (2)

Figure 2 compares samples states after hydrogenation from hydrogen atmosphere at 623 K for an elongated period of time to the point where the initial Zr–1Nb specimen cracks up to pieces, while the specimen with TiN_x coating retains its shape. Surely, at the operational (normal) conditions of the reactor core, primary circuit coolant (for instance, for H_2O in WWER-1000: temperature ~600 K and pressure of 16 MPa) hydrogen absorption rate is not that intensive. At high burnups (>50 MW·days/kgU), developing of hydrides of mainly tangential orientation contains fraction of hydrogen that does not exceed 0.01 % [12].

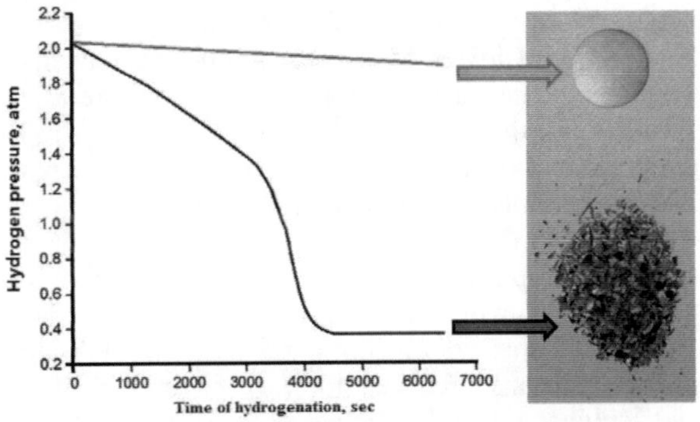

Figure 2. Hydrogen absorption curve for the initial (bottom) and coated with TiN_x (top) samples at 623 K for 2 hours

Figure 3 shows SEM surface morphologies of both TiN_x and Ti/TiN_x coatings before and after thermal cycling up to 1073 K (three cycles). The sputtered TiN_x coating shows excessive surface roughness and traces of grinding (Figure 3). After thermal cycling, the surface of the sample with TiN_x contained peeling of the coating and cracks formed during thermal expansion (Figure 3a and 3b), which was not observed in the sample with a Ti/TiN_x coating (Figure 3c and 3d).

The deposition of TiN_x coating reduces hydrogenation of Zr–1Nb alloy at 623 K. The sufficient reduction of the hydrogen absorption was observed for 1.5 μm TiN_x coating, while the thickness of coating was measured by ball-cratering method. However, the formation of cracks and peelings of coating after thermal cycling up to 1073 K has a strongly negative effect on hydrogenation behaviour and protective properties of the coating.

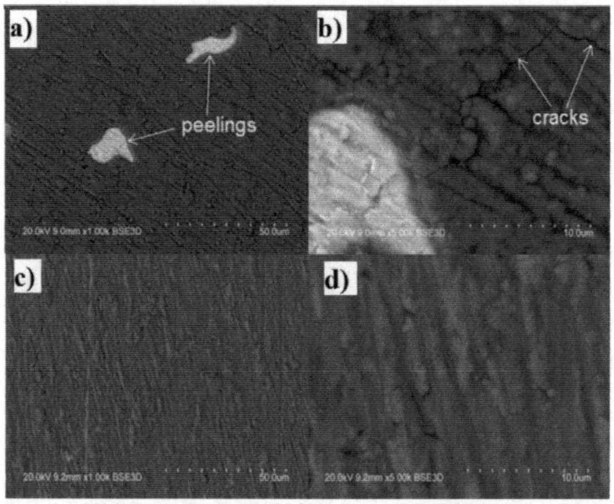

Figure 3. SEM images of TiN (a, b) and Ti/TiN (c, d) after thermal cycling up to 1073 K

The hydrogen absorption rate of Ti/TiN_x is on the same level with TiN_x without interlayer during hydrogenation at 623 K (Figure 4a). However, hydrogenation behavior at the temperature of 1073 K significantly differs for all the samples. The thermal solubility of hydrogen in α-Zr is very low, it is equal to ~ 6 at. % at a temperature of eutectoid transformation (820 K) and decreases rapidly with temperature. Nevertheless, the high-temperature β-Zr (bcc) dissolves up to ~ 50 at. % of hydrogen. Hydrogen is a very strong β-stabilizing element – with increasing hydrogen content the temperature of α→β transformation decreases from 1136 K for undoped Zr to ~ 820 K at a hydrogen concentration of ~ 6 at. % [13]. At room temperature, the thermal solubility of α-Zr hydrogen solubility is less than 1 ppm. Thus, only a small amount of hydrogen in the lattice of α-Zr can be dissolved, which causes the rapid precipitation of hydrides that increases hydrogen absorption rate of the initial sample by the cracking of hydrides. Nevertheless, the samples coated with TiN_x and Ti/TiN_x accumulate a small amount of hydrogen for 1-hour hydrogenation, which explains the linear dependence of hydrogen absorption determined by a diffusion barrier of titanium nitride coating. Hydrogen saturation of Zr–1Nb at 1073 K leads to the phase transformation (α+β)→β with increasing hydrogen content. In this case, hydrogen dissolves in the β-phase and the hydrogen absorption curve becomes saturated gradually when the concentration of hydrogen is determined by Sievert law. At high temperatures, zirconium hydride decomposes and does not transform the dissolved hydrogen in the bound state. The hydrogen absorption at 1073 K is also reduced for the coated with TiN_x and Ti/TiN_x samples. However, we can observe a higher rate of hydrogen absorption by Zr–1Nb with TiN_x due to the coating delamination as a result of differences in thermal expansion coefficients.

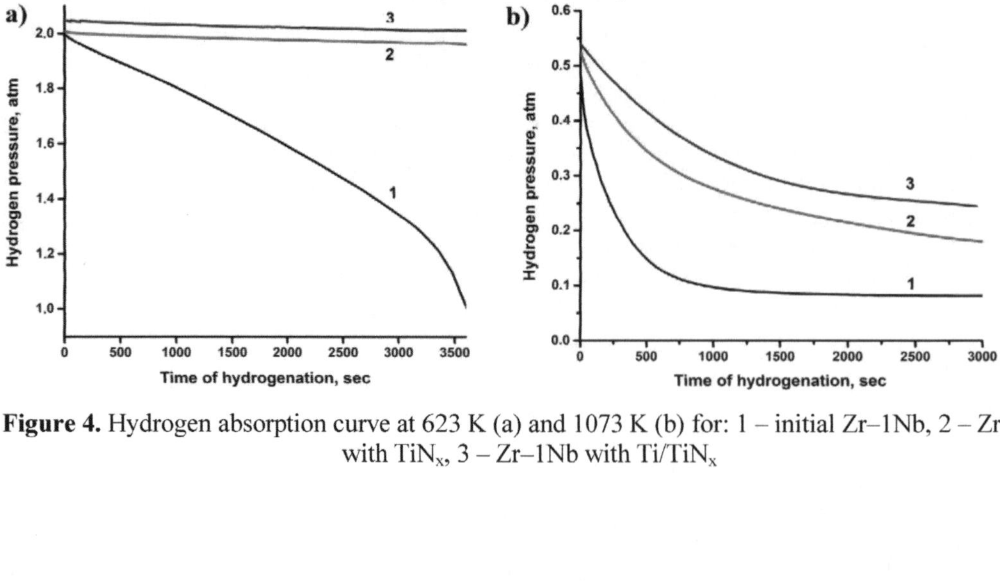

Figure 4. Hydrogen absorption curve at 623 K (a) and 1073 K (b) for: 1 – initial Zr–1Nb, 2 – Zr–1Nb with TiN_x, 3 – Zr–1Nb with Ti/TiN_x

The adhesion strength of TiN_x and Ti/TiN_x coatings was measured by scratch method. Typycal SEM image of the scratch after thermal cycling is shown in Figure 5, where coatings retained on the substrate up to 8 N normal load applied to indenter. This result demonstrates that Ti/TiN_x coating retains the adhesion properties when exposed to rapidly changing temperatures in the range of 293-1073 K.

Figure 5. Typical SEM image of scratch for Ti/TiN_x coating on Zr–1Nb substrate

4. Conclusion

TiN_x and Ti/TiN_x coatings were deposited onto Zr–1Nb by dcMS and combination of FVAD and dcMS methods, respectively. TiN_x coating significantly reduces hydrogen absorption of Zr–1Nb alloy. The cracking of TiN_x coating was observed after thermal cycling up to 1073 K. Nevertheless, Ti/TiN_x coatings are resistant during thermal cycling in the temperature range of 293–1073 K, which is associated with the balancing of thermal expansion coefficient between Zr–1Nb and TiN_x coating. The hydrogen sorption rate of Ti/TiN_x remains at the same level with TiN_x during hydrogenation at 623 K. The hydrogenation behaviour of Zr–1Nb significantly differs at 1073 K for Ti/TiN_x and TiN_x coatings; the latter has a higher rate of hydrogen absorption.

Acknowledgment

The reported study was funded by RFBR according to the research project No. 16-38-00709 and "UMNIK" program by Fund of assistance to development of small forms of the enterprises in the scientific and technical sphere.

References

[1] Chernyaeva T P, Ostapov A V 2013 *Problems of Atomic Science and Technology* **87** 16–31.
[2] Kashkarov E B, Nikitenkov N N, Syrtanov M S, Sutygina A N, Shulepov I A, Lider A M 2016 *Appl. Surf. Sci.* **370** (2016) 142–148.
[3] Zielinski A, Sobieszczyk S 2011 *Int. J. Hydrogen Energy* **36** 8619–8629.
[4] Kelly P J, Arnell R D 2000 *Vacuum* **56** 159–172.
[5] Sidelev D V, Yurjev Y N 2014 *Advanced Materials Research* **1040** 748–752.
[6] Anders A 2014 *Proc. Int. Symp. on Discharges and Electrical Insulation in Vacuum*, ISDEIV 201–204.
[7] Kashkarov E B, Nikitenkov N N, Tyurin Yu I, Syrtanov M S, Zhang L 2015 *IOP Conf. Ser.: Mater. Sci. Eng.* **81** (1) 012017.
[8] S.V. Ivanova E M, Glagovskiy I A, Khazov et al. 2008 *Proc. IV Int. Conf. on Interaction of Hydrogen Isotopes with Structural Materials* 51–75.
[9] Kim I, Khatkhatay F et al. 2012 *J. Nucl. Mater.* **429** 143–148.
[10] Ryabchikov A I and Stepanov I B 2009 *Surface and Coating Technology* **203** 2784.
[11] Dupin N, Ansara I, Servant C, Toffolon C, Lemaignan C, Brachet J C 1999 *J. Nucl. Mater.* **275** (3) 287–295.
[12] Markov D V et al, State of fuel rods spent in the VVER-1000 reactor up to a fuel burnup of 75 MWdays/kgU, JSC "SSC RIAR", IAEA, NCL Collection Store.
[13] Hong E, Dunand D C, Choe H 2010 *Int. J. Hydrogen Energy* **35** (11) 5708–5713.

Aluminum films deposition by magnetron sputtering systems: Influence of target state and pulsing unit

D V Sidelev[1], A V Yuryeva[1], V P Krivobokov[1], A S Shabunin[1], M S Syrtanov[1], Z Koishybayeva[1]
[1]Institute of Physics and Technology, Tomsk Polytechnic University, Tomsk 634050, Russia

Abstract. This article reports on technological possibilities of magnetron sputtering systems with solid-state and liquid-phase targets to deposition of aluminum films and its structure. The comparison of deposition rates of magnetron sputtering systems with direct current (DC), mid-frequency (MF) and high power pulsed (HiPIMS) supplies is shown. The optical emission spectroscopy indicates a high component of target material ions in discharge gap only to HiPIMS technique. Al films are a (111)-line oriented in DC and MF power supply cases, for high power pulsed unit – aluminum films also have intense (220)-line. The dependence of grain sizes and sputtering technique parameters is obtained.

1. Introduction

Al films have been widely employed as a reflective layer with high electrical conduction. Such coatings are interested to space applications as being a reflector element of satellite [1]. There is required a thick Al layer (1-3 μm) with radio-frequency transparent protective upper-layer (SiO_2 or Al_2O_3). Now, ion-plasma techniques are demand to thin films deposition [2-5]. These technologies are enabled to required thickness uniformity on large-scale substrates, structure homogeneity, high purity and adhesion, low heat flow on substrate. Thus, magnetron sputtering techniques are used to deposit of Al-based functional coatings on low-melting and carbon composite materials. Above, our colleagues made technology of thermal-control coatings deposition, based on ion-plasma techniques instead of resistive evaporation [6]. Such way has been helped on increase of output of technological process and part of nondefective units (up to 90%).

Today, the actual tasks in magnetron sputtering techniques are to increase a deposition rate[(a)] and adhesion of growing coatings to substrate[(b)], deposit films with predictable properties[(c)]. Therefore, HiPIMS and liquid-phase target magnetrons are attractive. The first technique has a low deposition rate [7], but form a short-term (30-200 μs) high-current discharge with plasma density 10^{13} cm^3, for comparing DC and MF magnetrons have only 10^{11} cm^3. The last investigations of HiPIMS techniques are indicated on film densification [8, increase adhesion [9] and others changing in film properties. The unique construction of the second magnetron type has a special pot to melting target [10] and evaporation initiation. Thereby, the significantly increase of target erosion due to combined mechanism behaviour was. In this case, the deposition rate of metal coatings can be enhanced to 10^2-10^3 nm/s. Moreover, the so-called "self-sputtering" process can initiates in liquid-phase target technique and magnetron discharge changes into a self-sustained mode. For this case, plasma discharge operates on metal vapor at 7×10^{-3}-2×10^{-2} Pa. These technologies are primarily interested for ultra-high deposition rates and film purity. Such magnetron sputtering techniques can be beneficial to Al films deposition for space applications. In this study the comparing of technological possibilities of different magnetron sputtering systems to Al films deposition and its structure investigation were done.

2. Experimental

Aluminum films were deposited in an ion-plasma vacuum installation by magnetron sputtering systems with Al target (99.995%). The base pressure was 5×10^{-3} Pa. Here, magnetrons with solid-state disk (dia. 90 mm) and liquid ring-shaped (inner dia. - 54 mm, outer dia. - 104 mm) targets were used. Three different power supply have been chosen. The pulsed DC-power supply was used to liquid-phase target magnetron. MF (1-100 kHz) and HiPIMS (0.1-15 kHz) pulsing units have been connected with solid-state target magnetron. The discharge parameters were measured by oscilloscope Tektronix 2022 B. All depositions were performed in the power stabilizationon mode. For substrates, 30×30 mm^2 borosilicate glasses were used. We use a specific deposition rate (v_s), which refers to thickness deposited per time and power of pulsing unit, in nm/(kW×s).

Table 1. Deposition modes of Al films

№	Target state	W_{av}, kW	Power unit parameter			P, Pa	h, μm	v_s, nm/(kW×s)
			v, kHz	t_0, μs	t_p, μs			
s_1	solid	2.5	100	8	2	0.18	0.5	0.49
s_2			1	800	200		0.5	0.45
s_3			DC	-	-		0.5	0.46
s_4			1	140	860		0.51	0.185
s_5			7	20	123		0.49	0.15
s_6			15	10	52		0.49	0.17
l_1	liquid	3	132	6	1.5	10^{-2} (*)	1.2	24.5
l_2		3.2				10^{-2} (*)	0.23	6.25
l_3		4				0.18	0.4	10
l_4		3				0.18	0.42	7.7

Note: W_{av} – average power; v – pulse frequency; t_0 – operating pulse; t_p – pause between pulses; P – pressure; h – film thickness; * - self-sputtering mode.

The effect of power supply on the magnetron discharge properties was studied by optical emission spectroscopy. The double-channel (225-920 nm) fiber-optic spectrometer AvaSpec was used. The integration time was 5 ms, the number of measurements for averaging – 100, focal distance – 300 mm. We use NIST atomic spectra database (ver. 5.3) for identification spectral lines of plasma discharge.

The film thickness was analyzed by spherical microsection method (Calotest CAT-S0000, CSEM). XRD measurements of Al films were performed using a Shimadzu XRD-7000S with accelerating tube Cu-Kα (40 kV and 30 mA) in 2·θ from 30° to 90° with exposure step 0.03°. The scanning angle was 1°.

3. Results and discussion

A pulsed DC and MF magnetron discharges strongly depend on the operating mode. There are three alternant processes: plasma build-up(a), stationary plasma (b) and decaying plasma process(c) [11]. Any recorded operating pulse parameters by oscilloscope are illustrated it. Figure 1 corresponds to specific deposition rates on pulse frequency to DC and MF solid-state target magnetron system. The stationary discharge mode is the longest-term at 1 kHz and current pulse has a practically rectangular shape. The operating voltage is an identical to DC mode, so v_s (for 1 kHz) is less to DC mode due to decaying plasma phase. The change of pulse frequency from 1 to 100 kHz at constant duty cycle results in decrease of t_p. So, there is an increasing of residual ionization from 0 to 3.36 A, deposition converts from pulse mode to "quasi-stationary" at high pulse frequency (plasma build-up phase is shorten). The specific deposition rate is higher to 60-100 kHz than for DC mode due to increase a discharge voltage. The low v_s in1-40 kHz is correlated with non-optimal time-ratio of plasma phases and discharge voltage.

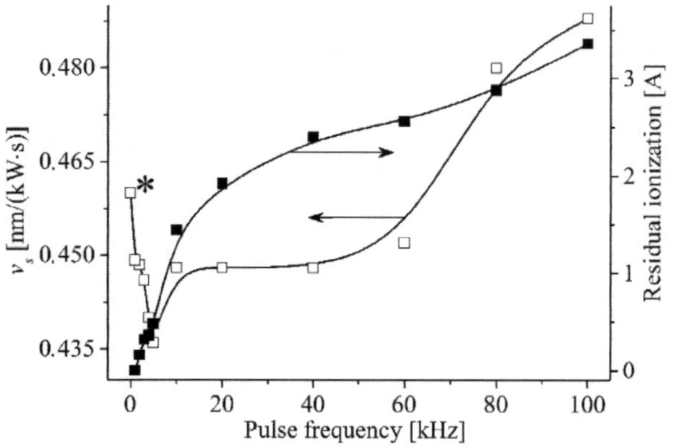

Figure 1. The specific deposition rate and residual ionization on pulse frequency of MF magnetron. The label (*) is aligned to v_s of DC technique with solid target.

The dependences of v_s on pulse frequency and duty cycle of HiPIMS technique are shown in Figure 2. The specific deposition rate is linearly decreased in 1-7 kHz. The increase of pulse frequency is caused to change of current pulse shape from rectangular to triangular. Firstly, the current increased and saturated to 24 A (for 1 kHz), then the discharge current stabilized. For 7-15 kHz, the pulse current increased at permanent average power 2.5 kW (33 A for 15 kHz).

The decrease of duty cycle is caused in fall of v_s. It is resulted in decrease of stationary plasma phase and increase pulse current. In this case, the self-sputtering mode is affected on productivity of magnetron sputtering system [12]. The difference between v_s for HiPIMS and MF discharge is non-stealthy.

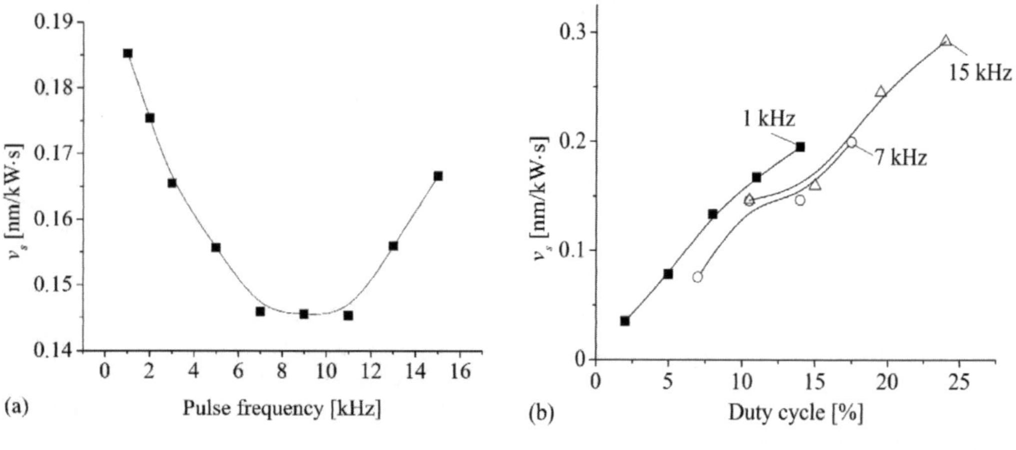

Figure 2. The specific deposition rate on pulse frequency (*a*) and duty cycle (*b*) of HiPIMS technique.

The main advantage of liquid-target magnetron is the high deposition rate. The specific deposition rate is 42-163 times higher than for high power impulse magnetron and 14-50 times - MF sputtering. The main mechanism of increase v_s is an additional evaporation of liquid-state target to sputtering [6].

Figure 3.a corresponds to optical emission spectra of magnetron discharge. The DC and MF sputtering systems have practically equal spectral patterns. It is a high content of Ar and Al atoms and relatively low concentration of Ar^+ and Al^+ ions. The emission spectra of HiPIMS technique have significant differences. There are increases of concentration of ions components (Ar^+, Al^+) and falling of neutral atoms intensity (Ar). Thus, the dense plasma forms in the HiPIMS case. Moreover, the decrease of Al intensity (394.4 nm) confirms of fall v_s comparing to the MF discharge.

Figure 3.b corresponds to emission lines of Ar atoms (842.47, 912.29 nm), Ar^+ (358.84, 465.79, 704.21 nm) and Al^+ (281.69, 623.18 nm) ions to the high power pulsed magnetron discharge. The obtained results have a good agreement with data of deposition rates and shapes of current pulses. The pulse time has a predominant influence of plasma composition. The generation mechanism of Al^+ ions is an ionization of Ar atoms[(a)], sputtering of Al target[(b)] and ionization of Al atoms[(c)]. The increase of Al^+ ions is inherent to long-time pulses and initiated by drop of pulse frequency at constant duty cycle. It results in effect of ion bombardment to substrate. On the contrary, the positive ions of aluminium can chance to accelerate to cathode and take a part in sputtering of Al target. There is a "self-sputtering" and its probability increases with t_0.

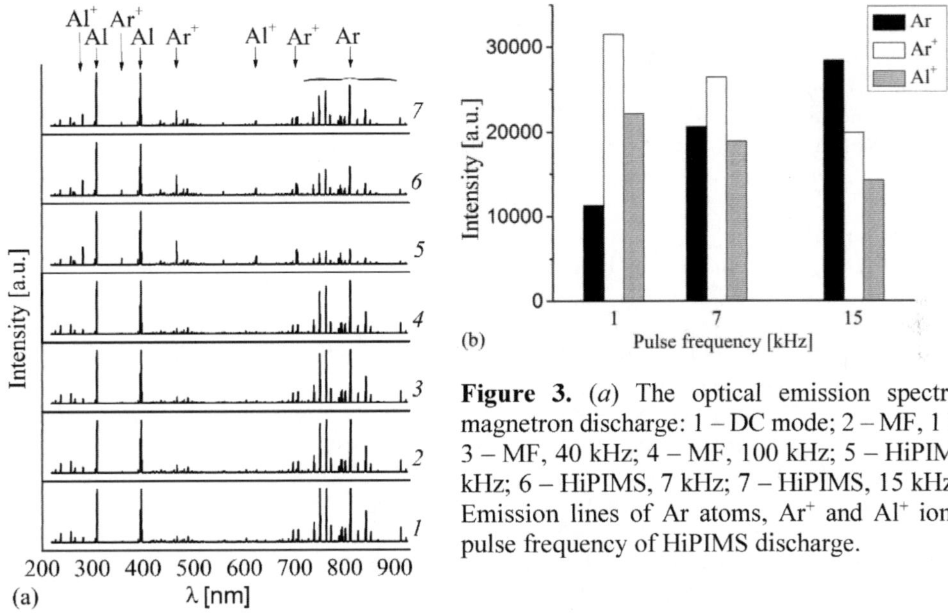

Figure 3. (*a*) The optical emission spectra of magnetron discharge: 1 – DC mode; 2 – MF, 1 kHz; 3 – MF, 40 kHz; 4 – MF, 100 kHz; 5 – HiPIMS, 1 kHz; 6 – HiPIMS, 7 kHz; 7 – HiPIMS, 15 kHz. (*b*) Emission lines of Ar atoms, Ar^+ and Al^+ ions on pulse frequency of HiPIMS discharge.

The diffraction patterns are shown in Figure 4. The textures of DC and MF coatings are equal. There are high-intensive (111) and low-intensive (200), (220), (311) crystallographic directions. The grain sizes are 20-24 nm calculated by Debye-Scherrer formula.

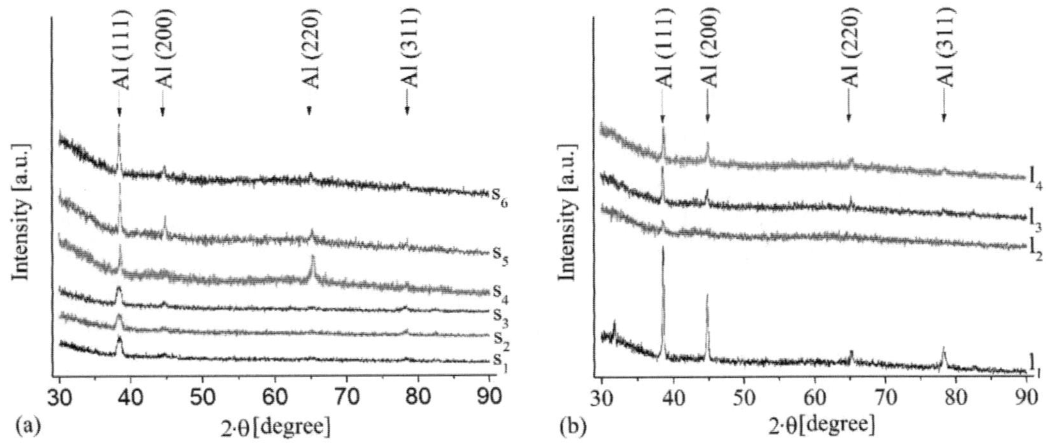

Figure 4. XRD structure of Al films: a – solid-state target magnetron; b - liquid target magnetron.

The HiPIMS coatings have more intensive (220) line. The decrease of pulse frequency is resulted in forming structure with dominant (111) and (220) intensities. Moreover, narrowing of (111) lines to 0.4-0.5 μm-thick films are indicated to form coarse-grained coatings. It is good illustrated to Al films deposited by HiPIMS and liquid-target techniques. For high power pulsed magnetron sputtering system, grain sizes are more 26 nm, for liquid target magnetron – 53 nm and higher. In the first case, it can be resulted in forming denser structure due to the higher concentration of Al^+ ions in deposition flow. For liquid target technique, the high-intensive evaporation is caused in grain growth.

4. Conclusions

The solid-state and liquid target magnetron sputtering systems with different pulsing units were used to Al film deposition. The comparing of technological possibilities of deposition techniques shows on a variance of specific deposition rates and plasma discharge composition. The liquid target magnetron has a more productive mode (6.25-24.5 nm/(kW×s)) and forms coatings with high grain sizes (higher 53 nm) due to intensive evaporation of liquid target material. The using of HiPIMS technique is caused in the increase of ion component Al^+ in deposition flow and grains growth of Al coatings to 26 nm and higher.

Acknowledgments
This investigation was supported by Russian Science Foundation (project № 15-19-00026).

References

[1] Asainov O H, Bainov D D, Krivobokov V P, Romanenko S E, Chernyatina A A 2011 *Russian Physics Journal* **54** 158
[2] Erofeev E V, Kagadei V A, Kazimirov A I, Fedin I V 2015 *Proc. Int. Sib. Conf. on Control and Communications SIBCON (Omsk)* 7147054 (St. Petersburg: Russia / Institute of Electrical and Electronics Engineers)
[3] Kashkarov E B, Nikitenkov N N, Syrtanov M S, Babihina M N 2016 *IOP Conf. Series: Materials Science and Engineering* **110** 012046 1
[4] Nikitenkov N N, Sutygina A N, Shulepov I A, Sivin D O, Kashkarov E B 2015 *IOP Conference Series: Materials Science and Engineering* **81** 012018 1
[5] Yurjev Y, Sidelev D 2013 *Journal of Physics: Conference Series* **479** 012018 1
[6] Kashkarov E B, Nikitenkov N N, Syrtanov M S, Sutygina A N, Shulepov I A, Lider A M 2016 *Applied Surface Science* **370** 142
[7] Asainov O H, Bainov D D, Krivobokov V P, Mihailov M N, Pachenko O V, Ydakov C V, Chernyatina A A 2011 *Russian Physics Journal* **54** 154
[8] Anders A 2010 *Journal of Vacuum Science & Technology A* **28** 2010
[9] Samuelsson M, Lundin D, Jensen J, Raadu M A, Gudmundsson J T, Helmersson U 2010 *Surface & Coatings Technology* **205** 591
[10] Bandorf R, Waschke S, Carreri F C, Vergöhl M, Grundmeier G, Bräuer G 2015 *Surface & Coatings Technology*, in press.
[11] Bleykher G A, Krivobokov V P, Yurjeva A V, Sadykova I 2016 *Vacuum* **124** 11
[12] Musil J, Lestina J, Vclek J, Tolg T 2001 *Journal of Vacuum Science and Technology A* **19** 420
[13] Sarakinos K, Alami J, Konstantinidis S 2010 *Surface & Coatings Technology* **204** 1661

Synthesis and characterization of nickel hydroxide nano-particles obtained by chemical deposition method under different precipitation conditions

Y P Vinichenko[1] and E N Sidorova[1]

[1]Department of functional nanosystems and high-temperature materials, The National University of Science and Technology "MISiS", Moscow 119991, Russia

Abstract. The aim of this investigation was to produce nickel hydroxide nanoparticles by chemical deposition method using different synthesis conditions such as temperature, pH and washing type. The phase composition was analyzed by X-ray diffractometer, which reveals that all the particles were β-$Ni(OH)_2$ phase with hexagonal lattice. XRD analysis also allowed determining the coherent scattering region size in order to compare it with particle size. Transmission electron microscopy showed that the $Ni(OH)_2$ particles have film forms with a thickness of about 1 nm. The surface area was determined using Brunauer-Emmett-Teller method and varies from 2 to 41 m^2/gr.

1. Introduction

Nowadays the most promising and high developing field of research is production [1 – 3] and using nanomaterials [4 – 6] due to their unique properties caused by small size of microconstituents. Nano-sized structures with different morphologies (particles, films, tubes) are in demand in such important spheres of human activities as medicine, energy, microelectronics, etc. [7 – 9].

Nickel nanoparticles are widely used in various fields due to its good catalytic, magnetic and tribological properties [10 – 12]. For example using suspension with nano-nickel as oil additives increases enduring quality of engine parts. Also attractive is implementation nanosized nickel in producing recording and storage information systems [13], permanent magnets [11], magnetic cooling systems [11], for directional medicine transport [14], for treatment of cardiovascular disease [15] etc. There are a lot of reactions which are capable of catalyzing by nickel, one of them is water dissociation on hydrogen and oxygen at 300 °C [16]. Moreover, nickel based compounds exhibits high electrochemical performance and corrosion which allow them to find its application in various fields such as electrochemistry and battery technology (nickel - cadmium and nickel - metal hydride batteries) [17].

There are different methods available for the nickel nanoparticles [18] that can in different ways influence on morphology and grain size of final product. Nevertheless the problem of controlling the dispersion and particle shape has not been resolved. Thus, the main aim of this paper is to study the influence of precipitation conditions on the particle size and morphology of nickel hydroxide.

2. Experiment

The deposition process is represented on Figure 1. Nickel hydroxide powders were prepared using chemical precipitation, in which sodium hydroxide (NaOH) was added into nickel nitrate ($Ni(NO_3)_2$)

solution under different conditions listed in Table 1. There were 8 powders obtained under three different conditions such as four different temperature (15, 20, 30 and 45 °C), three pH (8, 9, 10) and different types of washing (decantation, centrifugation and ultrasonic sound). After the different condition the green precipitate was dried at 50 °C in oven. After the drying process the powder was characterized using different characterization techniques. Phase analysis and sizes of coherent scattering region were determined using X-ray diffractometer Difray 401 (Cr Kα source). Images of nickel hydroxide nanoparticles were obtained on transmission electron microscope LEO 912AB. The values of surface area were carried out using NOVA 1200e.

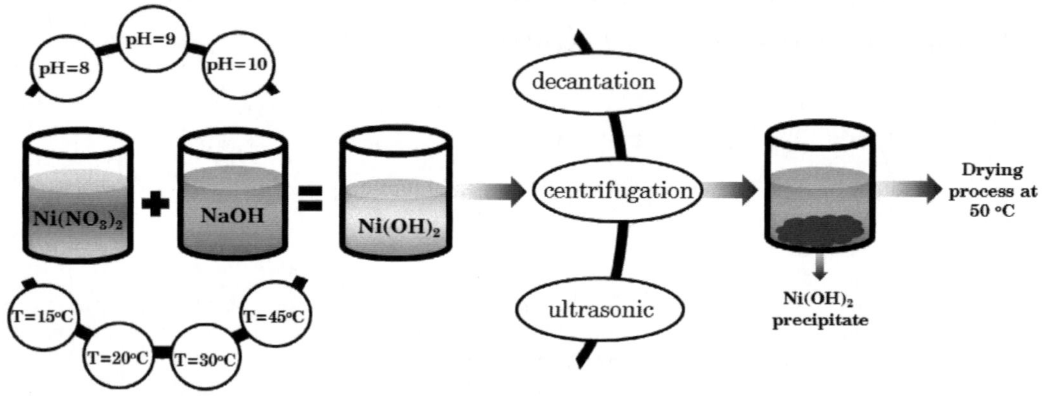

Figure 1. Schematic representation of precipitation reaction used for production of $Ni(OH)_2$ under different conditions.

Table 1. Different approaches used in chemical precipitation.

Sample	Temperature, °C	pH value	Washing type
S1	15	9	centrifugation
S2	20	9	centrifugation
S3	30	9	centrifugation
S4	45	9	centrifugation
S5	20	8	centrifugation
S6	20	10	centrifugation
S7	20	9	decantation
S8	20	9	ultra sonic

3. Results and discussion

The characteristic results reveal the information about the nanoparticles. XRD analysis confirms that the prepared nanoparticles are found to be β-$Ni(OH)_2$ phase with hexagonal lattice. From analyzing the broadening of reflections 20 °C was found to be the best temperature for obtaining the higher dispersed particles (Figure 2). Similarly the lowest grain sizes were found at pH 9 and centrifugation was found to be the best method of washing to obtain the nanoparticle.

The coherent scattering region sizes were determined by XRD analysis for lattice planes (100) and (101), and the obtained results are represented in Table 2.

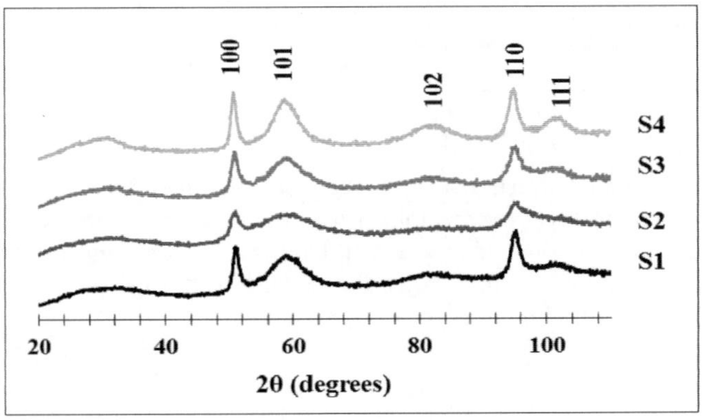

Figure 2. XRD patterns of samples obtained at different temperature.

Table 2. Sizes of CSR for different planes.

Sample		S1	S2	S3	S4	S5	S6	S7	S8
Size, nm	(100)	38.8	15.4	33.8	29.7	24.9	28.7	35.7	19.0
	(101)	4.0	5.2	7.9	5.6	4.5	4.8	5.2	4.0

Morphology and particle size of powders were studied using transmission electron microscopy. Figure 3 shows TEM images of $Ni(OH)_2$ samples under different pH value. The TEM analysis illustrates that particles are like aggregates of thin films with the average particle size of 20 nm, with thickness of 1 nm.

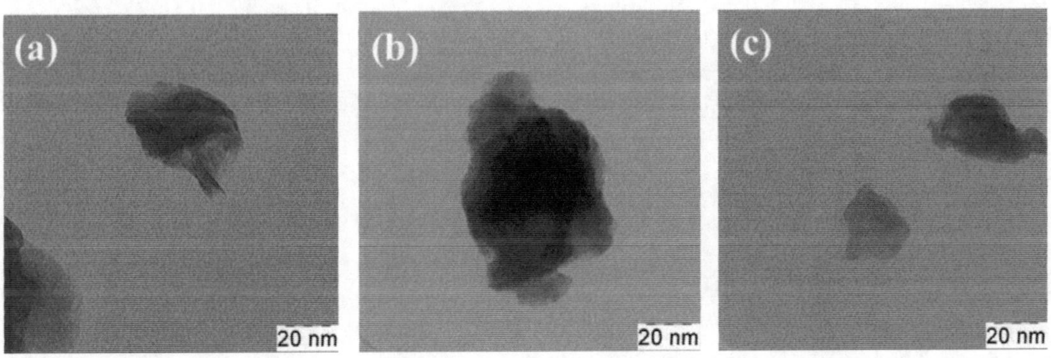

Figure 3. TEM image of $Ni(OH)_2$ nanoparticles at different pH **(a)** pH 8, **(b)** pH 9 and **(c)** pH 10.

Table 3 introduces the surface area that was determined using Brunauer-Emmett-Teller method. Taking account of cylindrical form of particles the surface area can be related to the average particle size by equation (1).

$$S = (2d + 4h) \cdot d^{-1} \cdot h^{-1} \cdot \rho^{-1} \qquad (1)$$

Where d – average diameter, m;
h – average altitude, m;
ρ – density of $Ni(OH)_2$, gr/m^2.

Table 3. Surface area of the prepared samples

Sample	S1	S2	S3	S4	S5	S6	S7	S8
S, m^2/gr	6	2	9	41	4	27	39	3

BET analysis reveals that the highest area surface of 41 m^2/gr was obtained for $Ni(OH)_2$ nanoparticles prepared at 45 °C with maintaining pH 9 for precipitation and collection of particles by centrifugation. Low values of surface area can be explained by the fact that particles form aggregates. So the Figure 4 shows the model of particle structure: thin films that consist of hexagonal lattices closely join to each other forming a bar structure.

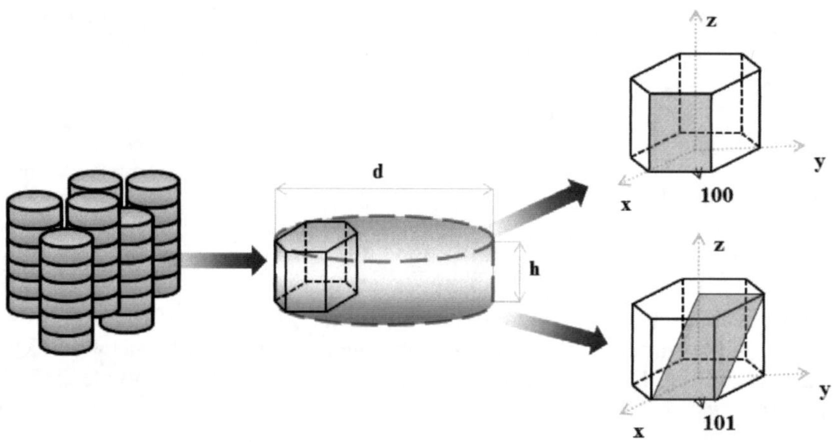

Figure 4. The model of particle structure

Using TEM images the particle size distribution was plotted in Figure 5 for samples obtained under different pH value. All powders have quit narrow size distribution, which indicates that particles are seemed to be monodispersed.

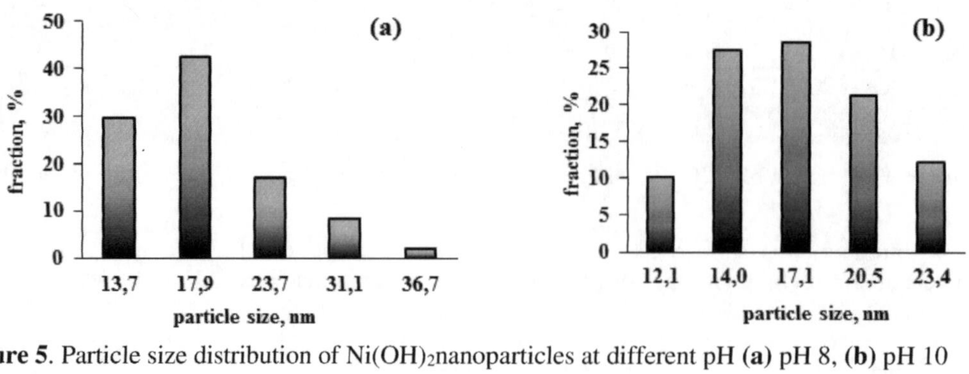

Figure 5. Particle size distribution of $Ni(OH)_2$nanoparticles at different pH **(a)** pH 8, **(b)** pH 10

Comparing the average particle size obtained from TEM images and sizes of coherent scattering region determined by XRD analysis it can be assumed that one particle contain one CSR.

4. Conclusions

Synthesis of nickel hydroxide nanoparticles was obtained using different precipitation conditions. After characterizations of powders with different techniques we determined that all samples consist of β-$Ni(OH)_2$ phase with hexagonal lattice, and particles look like thin films. The similarity of sizes from TEM images and CSR leads to the fact that one particle consists of one coherent scattering region. The equation between cylindrical particles sizes and surface area was also deducted. Low values of surface area can be explained by the fact that particles form aggregates.

References

[1] Buesser B and Pratsinis S 2012 *J. Rev. Chem. Biomol. Eng.* **3** 103 – 127
[2] Keshril A and Agarwal A 2012 *J. Nanoscience and Nanotechnology Letters* **4** 228 – 250
[3] Wankhede P, Sharma P and Jha A 2013 *J. of Engineering Research and Applications* **3** 1664 – 69
[4] Yang S and Dengteng G 2014 *J. Materials Science and Engineering* **3** 1 – 7
[5] Froudakis G 2011 *J. Materials Today* **14** 324 – 328
[6] Baati T, Bourasset F and Gharb N 2012 *J. Biomaterials* **33** 4936 – 346
[7] Zhang Y, Liu X, Ru C, Zhang Y, Dong L and Sun Y 2011 *J. of Microelectromechanical systems* **20** 959 – 967
[8] Bakry R, Vallant R, Najam-ul-Haq M, Rainer M, Szabo Z and Huck C 2007 *J. Nanomedicine* **2** 639 – 649
[9] Zhenan B, Vosgueritchian M, Wang H, Koleilat G, Liu N and Kim M 2013 *J. GCEP Report* **5** 1 – 16
[10] Singh S K and Xu Q 2010 *J. R. Soc.* **46** 6545 – 47
[11] Wang Z, Kuok M, Ng S, Lockwood D, Cottam M, Nielsch K, Wehrspohn R and Gosele U 2002 *J. Phys. Rev. Lett.* **89** 1
[12] Qiu S, Zhou Z, Dong J, Chen G 2001 *J. of Tribology* **123** 441 – 443
[13] Nielsch K, Wehrspohn R, Barthel J, Kirschner J and Gosele U 2001 *J. Applied Physics Letters* **73** 1360 – 62
[14] McHenry M and Laughlin D 2000 *J. Acta Materialia* **48** 223 – 238
[15] Al-Qubaisi M, Rasedee A, Flaifel M, Ahmad S, Hussein-Al-Ali S, Hussein M, Eid E, Zainal Z, Saeed M, Ilowefah M, Fakurazi S, Isa N and Zowalaty M 2013 *J. Nanomedicine* **8** 2597 – 508
[16] Kang G, Gillespie P, Gunnison a, Moreira A, Tchou-Wong K and Chen L 2010 *J. Environmental Health Perspectives* **119** 176 – 181
[17] Fan D and White E R 1998 *J. Electrochem. Soc.* **145** 34 – 39
[18] David S, Lockwood J, Bock C, Barry R 2015 *J. R. Soc.* A **471** 1 – 65

Luminescent properties of alumina ceramics doped with chromium oxide

V Kortov, A Kiryakov, V Pustovarov

Ural Federal University, Yekaterinburg, Russia

E-mail: arseny.kiriakov@urfu.ru

Abstract. Ceramics doped with chromium oxide were synthesized from alumina nanopowder at high heating and cooling rates. XRD analysis of the obtained samples shows that they consist mainly of Al_2O_3 α-phase. Photoluminescence (PL) spectra in the visible spectral region and thermoluminescence (TL) curves were measured. An effect of the dopant concentration on the intensity and shape of the PL bands as well as on the TL yield was found. Annealing of the quenching defects which emerged during the synthesis changed the PL spectra. The centers responsible for PL and TL in the synthesized ceramics were identified.

1. Introduction

The investigation of optical properties of oxygen-deficient alumina ceramics synthesized by solid-phase method from nanopowder showed the opportunities of their use as phosphors, which can arouse interest in optoelectronics. It is known that photoluminescence (PL) properties of such ceramics depend on the concentration of the oxygen vacancies, which form the traps of charge carriers, and take part in recombination processes. It is established that the following conditions of synthesis influence the concentration of anion vacancies in the oxygen sublattice of alumina: the temperature of synthesis, the time of isothermal annealing, the pressure in chamber, and the presence of a strong reducing agent, such as graphite [1]. However, during the synthesis in the highly reducing medium (deep vacuum, the presence of graphite), the loss of mass and changing of the sample sizes take place, which leads to some particular difficulties of the usage of luminescent ceramics. In this regard, development of a synthesis method of luminescent Al_2O_3-based ceramics which do not cause the abovementioned negative effects is an important task. One of the study directions is doping of ceramics with the introduction of different impurities. It is known that ions of Cr^{3+} can replace Al_i^{3+} in the crystal lattice of alumina. In this case one can expect the formation of new centers which can increase PL and TL without the use of strong reducing medium while synthesis. The purpose of this work is to study luminescence of oxygen-deficient alumina ceramics doped with chromium oxide.

2. Materials and Methods

α-Al_2O_3 nanopowder of high purity (99.99%) with the grain size of 50-70 nm and nominally pure microcrystalline powder of chromium oxide was used to produce blend. The powders were mixed and, after the addition of ethanol, the suspension was formed. Further mixing took place in an ultrasound bath. The mixture was dried at 343 K for 8 hours. After that the powder which consisted of oxides mixture was subject to uniaxial pressing at a pressure of 5 MPa. The obtained compacts 6.1 mm in diameter and 1 mm thick were annealed in the heating chamber of gas-analyzer EAO-220 (Balzers).

The device allowed us to get fast heating (200 K/s) of the samples put into carbon crucible due to transmission of high current. In this case the samples were heated to 1,773-1,873 K. The cooling rate of samples was 126 K/s. Quenching defects were annealed in the electrovacuum furnace SNVE 9/18 which heats the samples up to 2,073 K at a rate of 0.5 K/s. The annealed samples were cooled down to a room temperature with the furnace. The PL spectra were recorded by luminescence spectrophotometer Perkin Elmer LS-55 in the range of 200-900 nm. The excitation energy of quanta was 5.3 eV. Thermoluminescence was measured with an experimental unit at 2 K/s rate. XRD analysis was carried out on XPert PRO MPD (PANalitical) diffractometer in copper radiation with a beta-filter on the secondary beam. The voltage on the tube was 40 kV, the anode current was 30 mA. To exclude the effects of the sample texturing on XRD patterns, the samples were rotated in a horizontal plane at the angular rate of 0.25 revolution per second.

3. Results and Discussion

Figure 1 shows PL spectra of the synthesized alumina ceramics doped with chrome oxide. It can be seen that the intensity of the PL bands grows in the range of 2.2-3.4 eV with an increasing concentration of the impurity from 0.005 up to 0.01 weight %. Further increase in Cr_2O_3 concentration leads to decreasing emission in the given range due to concentration quenching. It is also noteworthy that PL yields grow in the 1.7-1.9 V band. Cr^{3+} ions which replace aluminum ions in the crystal lattice luminesce in the alumina single crystal in the reported range [2]. Increasing PL intensity in the given band indicates introduction of Cr^{3+} ions into Al_2O_3 crystalline lattice.

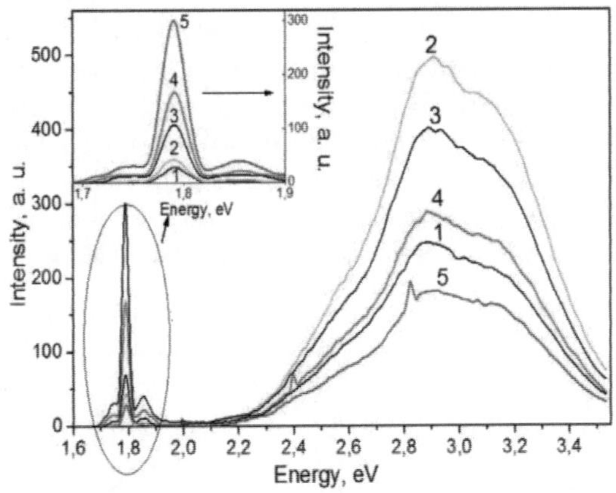

Figure 1. PL spectra of alumina ceramics (E_{ex}=5.3eV, T=300 K) with different Cr_2O_3 concentrations (wt.%): 1 – 0.005; 2 – 0.01; 3 – 0.02; 4 – 0.03; 5 – 0.1.

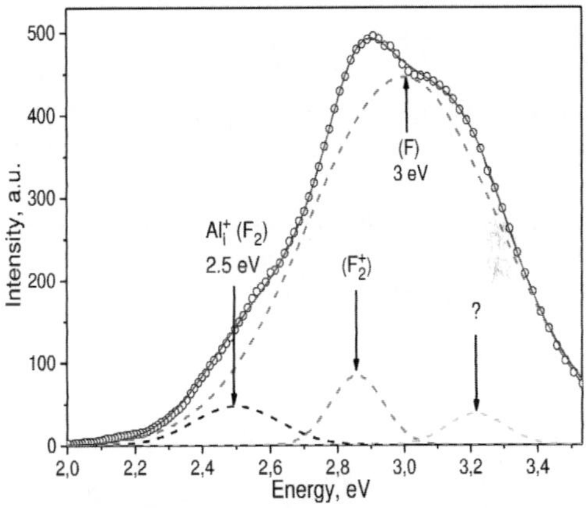

Figure 2. Gaussian decomposition of PL spectrum (curve 2, Figure 1).

Figure 2 features PL spectrum for the sample with the maximum intensity (curve 2, figure 1), which was deconvoluted into gaussians. It is known that in oxygen-deficient alumina the PL band with the maximum at 2.5 eV is caused by interstitial aluminum ions and F_2 aggregate centers (oxygen divacancies) [3]. Emission of F-centers (oxygen vacancies with two trapped electrons) is registered in the 3 eV band. The band with the maximum near 2.8 eV is associated with F_2^+ aggregate centers (oxygen divacancies with one trapped electron) [4]. The band with the maximum of 3.2 eV is presumably connected with deformation by (quenching) defects which emerged at high heating and cooling rates.

The analysis of the samples with a XRD method showed that they are α-Al_2O_3 solid solution with the lattice parameters: a(Å)=4,756(2); c(Å)=12,987(4). The XRD pattern features traces of another phase in the form of weak reflexes on the angles of 20,85°, 28,8°, 38,9° 42,6° (Figure 3). Additional weak reflexes in XRD may be caused by a presence of a small amount of chromium oxide which did not react with Al_2O_3 matrix while the samples were being synthesized.

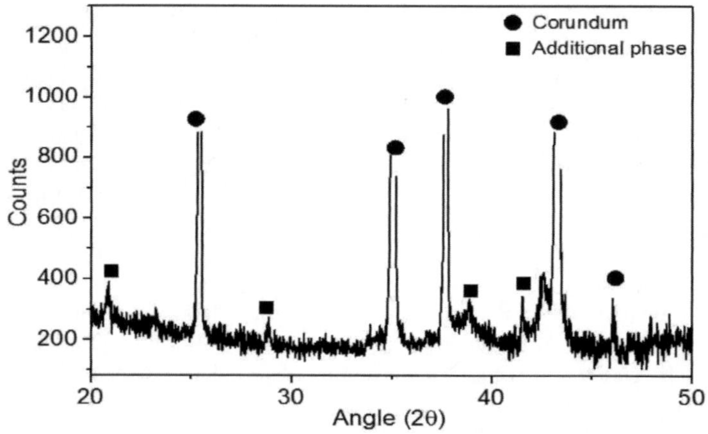

Figure 3. XRD pattern of the alumina ceramics doped with chromium oxide (0.02 wt.%).

To eliminate the effects of the tempering defects on the PL of the samples under study, thermovacuum annealing was carried out at T = 1,773 K for 180 min. The changes of PL spectra after annealing are shown in Figure 4. General decrease in PL yield and redistribution of the intensities of luminescence bands are observed both in the range of 1.7-1.9eV and in the range of 2.2-3.4eV. The maximum PL intensity is seen at the concentration of chromium oxide of 0.04-0.05 weight %. Deconvolution of the spectrum into gaussians in the range of 2.2 – 3.4 eV (Figure 5) allows singling out three components with the photon energies in the maximums 2.5 eV (Al_i^+ interstitial aluminium ions and F_2 – centers), 2.8 eV (F_2^+ aggregate centers) and 3 eV (F-centers). It is also noteworthy that there is no band with the maximum 3.2 eV after annealing of the samples, which confirms the assumption on its association with the deformation defects.

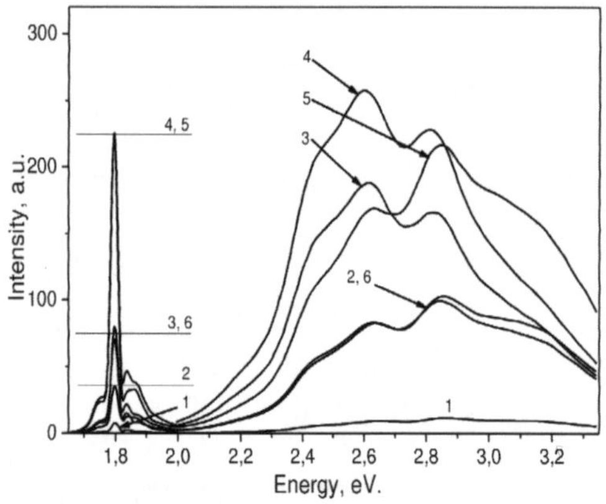

Figure 4. PL spectra of alumina ceramics (E_{ex}=5.3eV, T=300 K) after annealing with different Cr_2O_3 concentrations (wt.%): 1 – 0.005; 2 – 0.01; 3 – 0.02; 4 – 0.04; 5 – 0.05; 6 – 0.1.

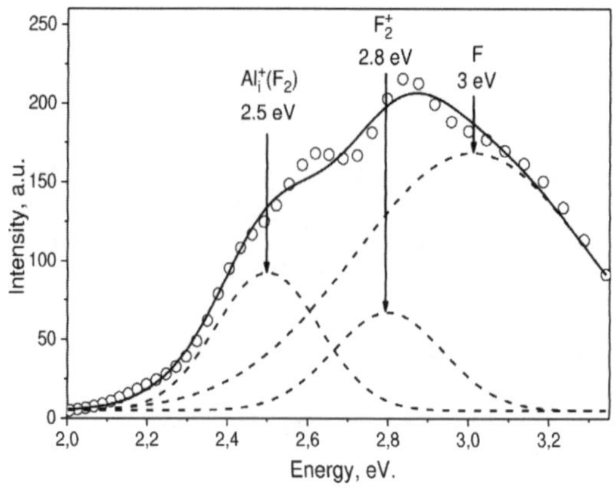

Figure 5. Gaussian decomposition of PL spectrum (curve 5, Figure 4)

A peak at 600 K is registered on the TL curve (Figure 6). It is known that impurity Cr^{3+} ions luminesce in alumina at this temperature. The intensity of TL at T=600 K correlates with spectroscopic PL measurements depending on the dopant concentration. In addition, on the TL curve one can single out a peak at 515 K associated with formation of F_2 centers, as well as a low-intensity peak at T=430 K which is caused by silicon impurity in the precursor nanopowder. The TL peak at 460 K is characteristic for oxygen-deficient alumina and is due to the presence of oxygen vacancies (F-centers) in the samples under study.

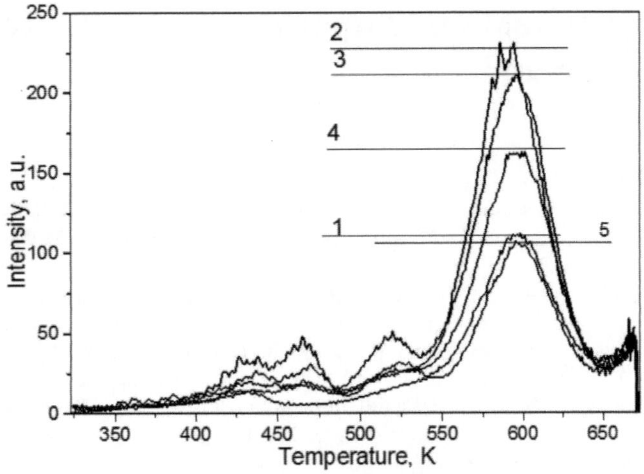

Figure 6. TL glow curves of alumina ceramics with different Cr_2O_3 concentrations (wt.%): 1 – 0.005; 2 – 0.02; 3 – 0.03; 4 – 0.05; 5 – 0.1.

4. Conclusions

Ceramics with photoluminescence in the wide region of visible light were synthesized from alumina nanopowder doped with chromium oxide. PL spectra and the curves of TL ceramics were measured; the main luminescence centers were identified. Concentration dependences of luminescence yield on Cr_2O_3 content were found. Functional materials for optoelectronic applications can be created on the basis of the obtained ceramics.

Acknowledgments

The work has been done as a part of the government task (№3.1016.2014/K) of the Ministry of Education and Science of the Russian Federation.

References

[1] Kortov, V., Zvonarev, S., Kiryakov, A., Ananchenko, D. Mat. Chem. Phys. 2016 **170**, p 168.
[2] Lapraz, D., at all. Phys. Stat. Sol. 1991 **126**, p 521.
[3] Solov'ev, S.V., Milman, I.I., Syurdo, A.I. Phys. Sol. Stat. 2012 **54**, p 726.
[4] Izerrouken, M., Benyahia, T. Nucl. Instrum. Meth. Phys. Res. B. 2010 **268**, p 2987.

Saint Petersburg OPEN 2016 IOP Publishing
Journal of Physics: Conference Series **741** (2016) 012196 doi:10.1088/1742-6596/741/1/012196

Infrared mediated production of C-Co-Pd nanocomposites under different annealing temperature and it's structural and phase composition analysis

A S Kurtsova[1]
[1]Department of functional nanosystems and high-temperature materials, The National University of Science and Technology "MISiS", Moscow 119991, Russia

Abstract. In the present study C-Co-Pd nanocomposites were prepared via Infrared annealing mediated method using precursor such as polyacrylonitrile (PAN), $CoCl_2 \cdot 6H_2O$ and $PdCl_2$. In addition the effect annealing temperatures such as 200, 250, 300, 400, 500, 600, 700, 800 and 900 °C on phase composition and structural characterization was analyzed using X-ray diffractometry (XRD) and transmission electron microscopy (TEM). Results reveals that the increase in temperature drastically influenced the phase and size of the C-Co-Pd nanocomposites.

1. Introduction

In recent years there is a tremendous interest in poly-conjugated material based systems due to its tremendous physic-chemical properties. The enhanced property of metal based carbon materials is not also depends on the poly-conjugated systems but it also depends on the nature of the incorporated materials. Based on this currently most of the research is focused on the development and its applications in various fields such as information systems, magnetic resonance imaging (MRI), fuel cell and as heterogeneous catalyst [1, 3].

Thus, in the present study we have focused on the preparation of C-Co-Pd nanocomposites by Infra-red annealing method using polyacrylonitrile (PAN), $CoCl_2 \cdot 6H_2O$ and $PdCl_2$ as precursors. In addition, the effects of IR pyrolysis temperature on the structural properties of C-Co-Pd nanocomposites are also analysed. [2].

2. Experiment

The precursor such as polyacrylonitrile (PAN), $CoCl_2 \cdot 6H_2O$ and $PdCl_2$ were procured from Reachem, Russia and were used without any further purification method. The scheme of obtaining of nanopowders C-Co-Pd is showed on Figure 1.

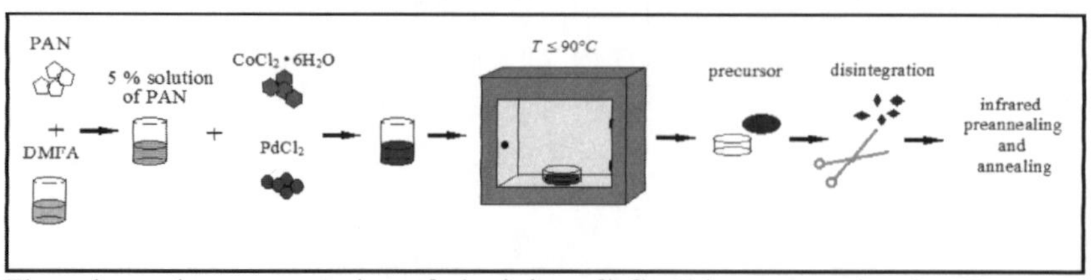

Figure 1. The schematic representation of obtaining C-Co-Pd nanopowders by IR annealing of precursor.

The schematic diagram clearly shows that the solution of PAN and metal compounds were mixed, solvent removal in the heating chamber and followed by preliminary and final IR-annealing [2]. The nanopowders were obtained at different annealing temperatures such as 200, 250, 300, 400, 500, 600, 700, 800 and 900 °C.
Preliminary annealing is carried out at a temperature of 150 °C for 15 minutes and then at 200 ° C for 15 minutes. After the pre-annealing the sample is crushed in a mortar [4].
Final annealing is then carried out to install Photon IR heating in an argon atmosphere. The argon pressure in the reactor was set slightly above atmospheric pressure, so that air does not enter the reaction space [4, 5]. The installation has 12 infrared lamps KG-220, with a total capacity of 12 kW and a maximum intensity of radiation in the range of 0.8 to 1.2 microns. The reactor is a quartz tube lamps mounted on the outside perimeter of the reactor, the reactor lengthwise. Thus, the lamp system and the electrical contacts are isolated from the reaction zone, which allows annealed samples in different environments (vacuum, inert or active gas). To ensure the safety of personnel, the infrared lamps are closed outside the reflective screen made of polished aluminum with water and air cooling system [6]. The latter is due to the work of three fans located in each end of the installation and designed to meet each other.
Within 15 minutes the samples were first annealed at 150 and 200 °C also for 15 minutes to remove water molecules adsorbed from the air after the preliminary annealing, and subsequent cooling [4 – 6]. Then the sample was heated to a fixed temperature (e.g. 700 °C), and the sample was kept at this temperature, the sample was cooled for two minutes. Temperature regime for preliminary annealing and final annealing were plotted in Figure 2.

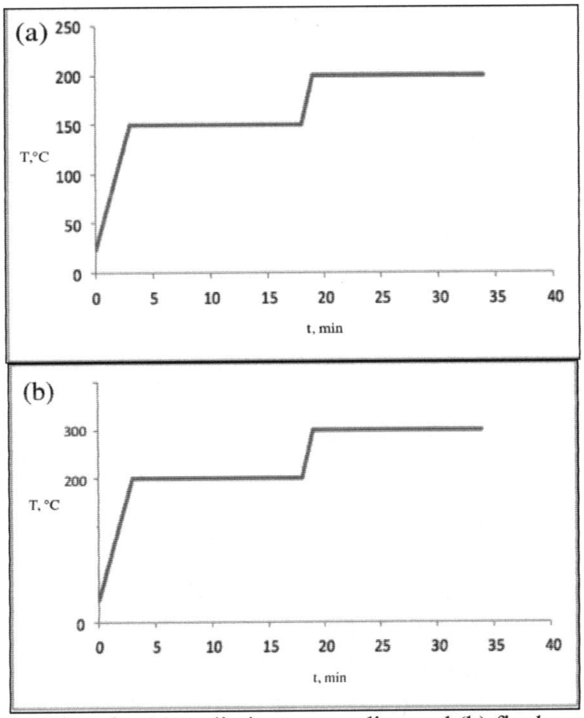

Figure 2. Temperature regime for (a) preliminary annealing and (b) final annealing.

The obtained powders under different heat treatment were further characterised using different techniques. The phase analysis of the nanocomposites is performed by X-ray diffraction (XRD) analysis using Difray-402 diffractometer (Scientific instruments, Russia) in the 2θ range from 20 to120 degree using chromium source (λ = 2.2909 Å) [7]. The structural analysis of the nanocomposites is performed by Transmission electron microscopy investigation using JEM-2010 (Jeol, Japan) with resolution up to 1.4 Å [7].

3. Results and discussion

The C-Co-Pd nanocomposites were prepared by Infrared annealing mediated method under different annealing temperatures. Qualitative phase analysis according to the X-ray diffraction (XRD) patterns was conducted, which showed that a solid solution Co based on Pd appear [3]. There is an amorphous carbon halo at all temperatures was observed which is illustrated in Figure 3.

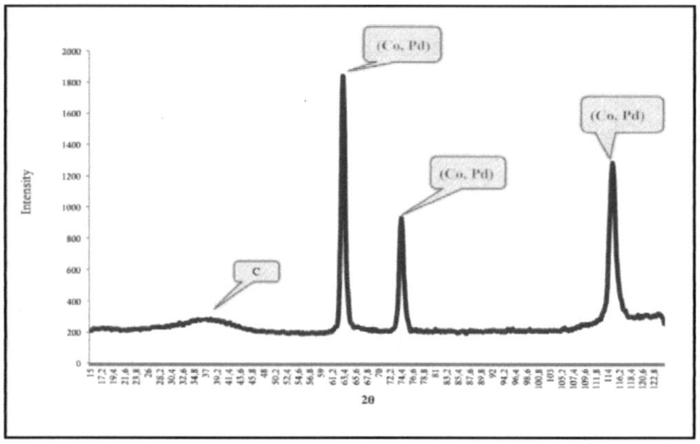

Figure 3. Diffractogram of C-Co-Pd nanopowder at 900 °C.

Figure 4 shows a shifting of the plane (111) of each samples synthesized at different temperatures. This shifting is due to the increase in the cobalt concentration in the solution and decrease in the period of grating [7, 8].

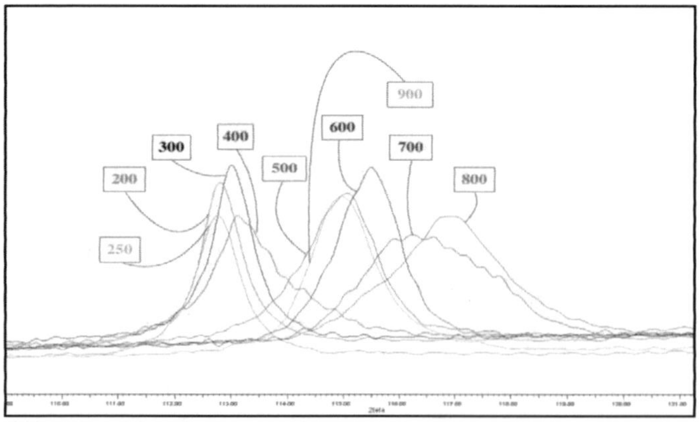

Figure 4. X-ray diffraction (XRD) pattern of C-Co-Pd nanocomposites synthesized at different temperatures 200, 250, 300, 400, 500, 600, 700, 800 and 900 °C.

The calculating of the periods of grating and sizes of coherent-scattering regions was also carried. The computation of sizes coherent-scattering regions was conducted on diffraction peak tailings by Selivanov-Smyslov method [8 – 11]. It is apparent that with the increase in temperature of pyrolysis the wide particle size distribution was observed. It is explained that particles start to make contact with each other more intensive with rising temperature, resulting in that particles size increase [10]. Figure 5 illustrates the calculating of CSR size distribution at different temperatures.

Journal of Physics: Conference Series **741** (2016) 012196 doi:10.1088/1742-6596/741/1/012196

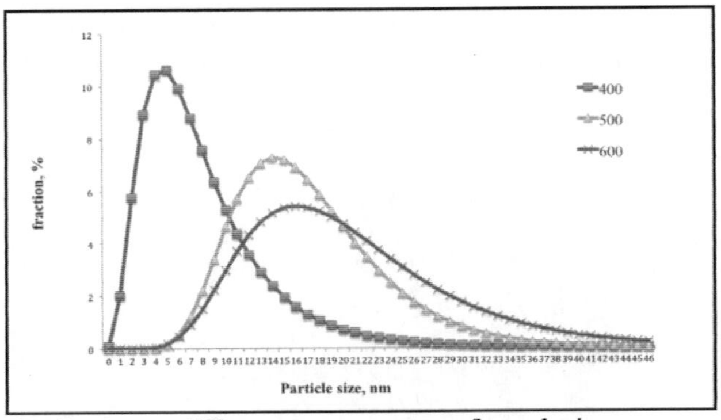

Figure 5. CSR size distribution at different temperatures of pyrolysis

Results of accounting of the periods of grating and the quantitative assessment of the composition of solid solution are represented in Table 1.

Table 1. Effect of heating and the quantitative assessment of the composition of solid solution

Temperature of IR annealing (°C)	Parameter a (nm) (to the maximum of peak)	Composition of solid solution	
		at. % Co	at. % Pd
200	0.3857	-	-
250	0.3885	-	-
300	0.3876	0.75	99.25
400	0.3875	0.9	99.1
500	0.3841	21	79
600	0.3829	27	73
700	0.3807	29	71
800	0.3789	34	66
900	0.3802	32	68

The quantitative assessment of the composition of the resulting solid solution phase performed by the difference in the palladium periods of grating in the coarse crystalline and nanocrystalline states. Figure 6 the TEM images of samples synthesized at different IR-annealing temperatures. At 300°C the particles were found to be pyramidal and pentagonal shaped with less than 100nm, whereas at 700°C the particles were spherical shaped with particles size more that 100nm [12 – 15].

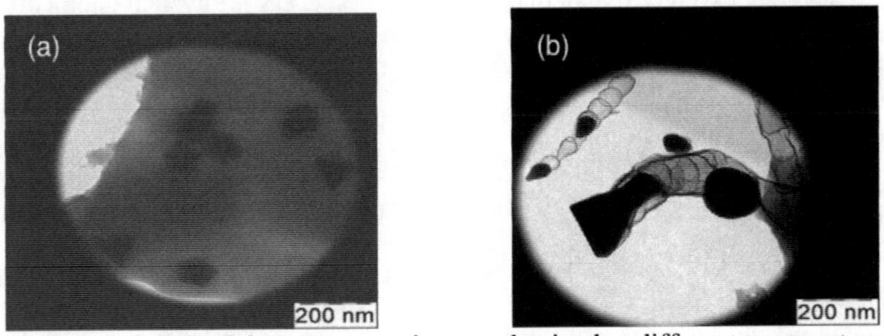

Figure 6. TEM images of C-Co-Pd nanocomposites synthesized at different temperatures (a) 300°C and (b) 700°C.

Using TEM images the particle sizes were calculated by fortuitous transverses method and the particle size distribution was plotted in Figure 7. Moreover, a function of lognormal distribution was found.

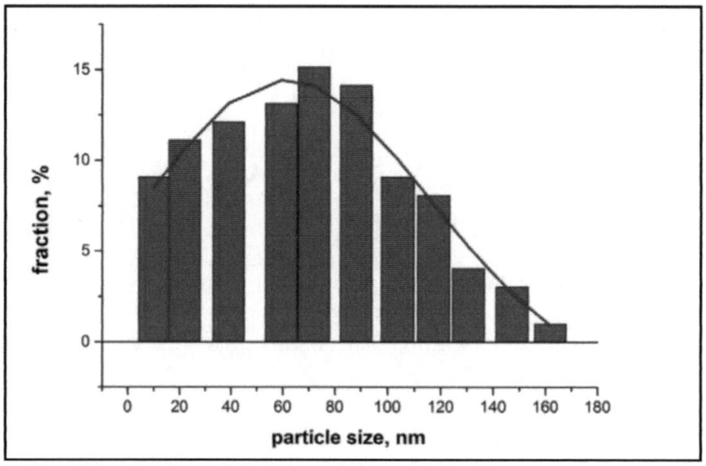

Figure 7. Particle Size Distribution of CoPd particles at 800 °C

4. Conclusions

In the present study C-Co-Pd nanocomposites were prepared via Infrared annealing mediated method using precursor such as polyacrylonitrile (PAN), $CoCl_2 \cdot 6H_2O$ and $PdCl_2$. In addition the effect annealing temperatures such as 200, 250, 300, 400, 500, 600, 700, 800 and 900 °C on phase composition and structural characterization using XRD and TEM. Results reveals that the increase in temperature drastically influenced the phase and size of the C-Co-Pd nanocomposites.

References

[1] Maltsev P P 2008 *Nanotechnology. Nanomaterials. Nanosystem engineering* (Moscow: Technosphere)
[2] Eliseev A A, Lukashin A V 2010 *Functional nanomaterials* (Moscow: Fizmatlit)
[3] Parmon V, Noskov V 2001 J. Catalysis in industry 1
[4] Kolodov V, Trineeva V, Vasilchikov U, Zakharov A 2011 J. Scientific and technological 3 24
[5] Radkevich N G, Ilin A P, Sitnikov A S 2001 *Physical chemistry of ultrafine systems* (Ekaterinburg: Avrora)
[6] Pomogailo A D, Rozenberg A S, Yflyand I E 2000 *Nanoparticles of metals in polymers* (Moscow: Himiya)
[7] Vekilova G V, Ivanov A N, Yagodkin U D 2009 *The diffraction and microscopic methods and instruments for the analysis of nanoparticles and nanomaterials* (Moscow: MISiS)
[8] Salova O, Mikhalenko N, Mikhalenko I, Gryaznov V 2008 J. Physical chemistry 72
[9] Zakharov U, Pugachev V, Popova A 2008 J. Advanced materials 6
[10] Gorelik S S, Skakov U A Rastorguev L N 2004 *X-ray and electron-optical analysis* (Moscow: MISiS)
[11] Matsuo Y 2000 J. Phys. Soc. Japan 4 32
[12] Mallat T, Petro J, Marczis L 2000 J. Electroanal. Chem. Iterfac. Electrochem 1 208
[13] Dyakonova N P, Ivanov A N 2001 *Crystallography, X-ray and electron microscopy* (Moscow: MISiS)
[14] Chernorykov N G, Nipryk O V 2012 *Theory and Practice of X-ray analysis* (Nigny Novgorod: NNGY)
[15] Barabash S M, Koval U N 2006 *The structure and properties of metals and alloys* (Kiev: Naykovaya Dymka)

Structure of metal-oxide Ti-Ta-$(Ti,Ta)_xO_y$ coatings during spark alloying and induction-thermal oxidation

V Koshuro[1], A Fomin[1], M Fomina[1], I Rodionov[1], B Brzhozovskii[1], V Martynov[1], A Zakharevich[2], A Aman[3,4], A Oseev[3,4], S Majcherek[3,4], S Hirsch[3,4]
[1] Yuri Gagarin State Technical University of Saratov, Saratov 410054, Russia
[2] Saratov State University, Saratov 410012, Russia
[3] University of Applied Sciences, Brandenburg, Germany
[4] Otto-von-Guericke University, Magdeburg, Germany

Abstract. The study focuses on combined spark alloying of titanium and titanium alloy surface and porous matrix structure oxidation. The metal-oxide coatings morphology is the result of melt drop transfer, heat treatment, and oxidation. The study establishes the influence of technological regimes of alloying and oxidation on morphological heterogeneity of metal-oxide system Ti-Ta-$(Ti,Ta)_xO_y$.

1. Introduction
Titanium and tantalum are applied in medicine due to their high biocompatibility. Previously, these metals were mainly used for the production of corrosion-resistant construction components operating at high temperatures. Particular attention was paid to the study of the oxidation kinetics, weight gain, chemical and phase transformations occurring in the microstructure [1]. The effect of prolonged (up to 16 hrs) heat treatment in oxygen, argon-oxygen and nitrogen-oxygen mixtures within the temperature range from 800 to 1400 °C was studied. It was concluded that a mixture of oxides of titania TiO_2, tantalum pentoxide Ta_2O_5, etc. are formed on the surface.

The application of oxide ceramics based on Ta_2O_5 in the form of coatings for endosseous implants, including the form of nanotubes, contributes significantly to the corrosion, biocompatible and osseoconductive qualities of these small metal items.

2. Methodology
Samples of commercially pure (cp) titanium VT1-0 and titanium alloy VT16 were subjected to electro-spark alloying with tantalum and subsequent oxidation with induction heat treatment (IHT).

Spark alloying is carried out in pulsed mode with the following parameters: pulse duration within 100 ms, current amplitude from 0.8 to 2.2 A, and pulse energy ranging from 0.01 to 0.1 J [2]. The technological regime is the current varying from 0.8 to 2.8 A. In this work the effect of 3 steps of current ranging was studied: 0.8, 1.5 и 2.2 A.

Main functional elements of the technological equipment for the production of metal heterogeneous coatings are shown on the screw-cutting lathe design and device for spark alloying surface modification "EFI-46A" (Figure 1).

Figure 1. Design of functional elements during spark alloying (top view).

The operating area includes 1 – slide bronze electrode; 2 – three-jaw chuck with the tool for sample attachment; 3 – titanium sample; 4 – tantalum electrode; 5 – electrode holder; 6 – cover of the electrode holder with low-frequency mechanical oscillator (100 Hz); 7 – copper conductor line.

Further, the surface of the prepared samples is oxidized in the course of treatment with HFC. Maximum power consumption P_{max} of the laboratory device for treatment with HFC during heating of the products with the geometry of this type is within 550 W, the frequency of the current in the inductor is 90±2 kHz; when it reaches a predetermined treatment temperature of 800 °C – power consumption P of the device equals 150 W. In the experiment the length of IHT was 30 and 300 s, which is associated with the probable appearance of nano-sized and submicron oxide crystals [3].

The surface morphology of the samples after the selected technological treatment is studied using scanning electron microscopy (SEM) in the micro- and nanoscale to understand the patterns of structure formation of the surface layer and coatings. SEM combined with energy-dispersive X-ray analysis (EDX) of chemical composition of samples is performed on "MIRA II LMU" with "INCA PentaFETx3" detector. Changes in tantalum and oxygen concentrations as well as other impurities measured in atomic percent (at.%) are studied in the areas of samples with various morphology.

3. Results

The deposition of the required conductive material (tantalum) on the surface of the metal substrate takes 4 main steps: 1 – convergence of the electrodes at a predetermined potential difference providing the interelectrode gap breakdown (IGB) in 5 to 300 μm; 2 – breakdown of the IGB and heat T required for melting of the contact parts of the electrodes; 3 – contact (short circuit mode) of the heated electrodes with the development of certain pressure P (impact); 4 – retraction of electrode-tool to form a single crystallized micro-ingot (drop) – coating splat (Figure 2).

The resulting tantalum coatings on titanium alloys are characterised by developed morphological heterogeneity of the surface, and their microstructure resembles the texture of plasma-sprayed thermal coatings [4]. When a current is 0.8 A the transfer of microdroplets occurs and a coating of splats with a large spread in size is formed as a result. This is associated with the fact that titanium surface has not enough time to warm up and contacting tantalum microdroplets start to splash across a colder titanium substrate.

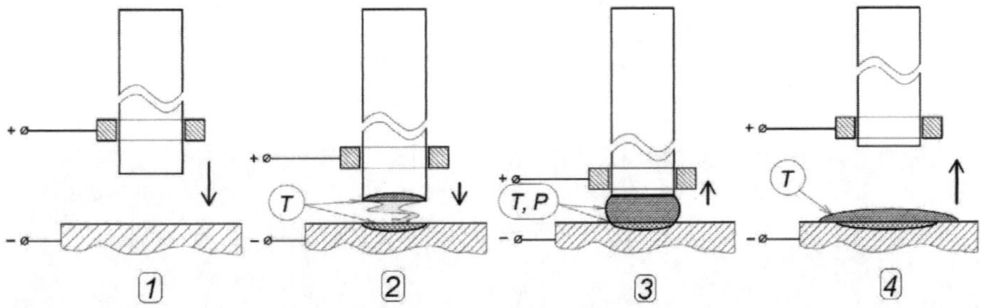

Figure 2. The scheme of the formation of metallic coating during electro-spark alloying.

The increase in the current contributes to more efficient transfer of tantalum on the surface of titanium and more efficient heating of the titanium substrate to the visco-plastic state. As a result, the share of large splats of about 150–300 μm increases in the coating. The best morphological parameters are formed at current of 2.2 A. During the interaction well heated materials of both electrodes are stirred intensively to form a homogeneous porous structure (Figure 3a).

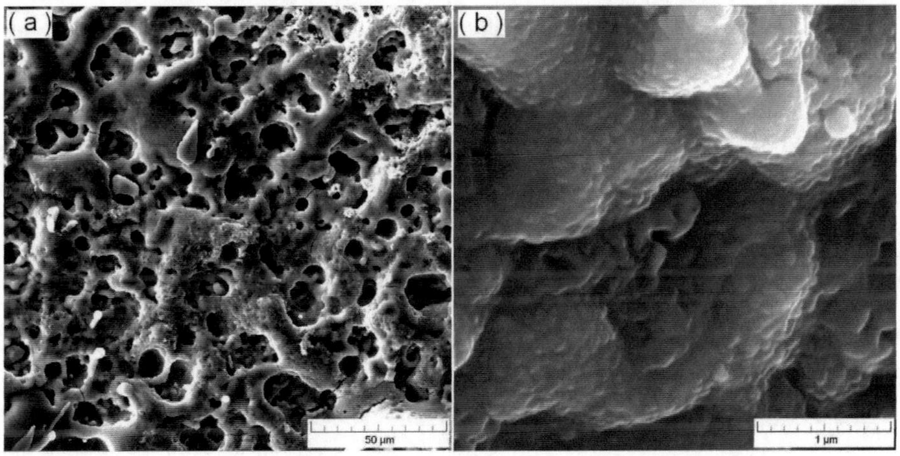

Figure 3(a, b). Morphology of a tantalum coating produced on titanium **(a)**; nanostructure **(b)**.

Crystallization processes of splats having different sizes are not the same. This contributes to significant differences in morphology in the nanoscale. Surface morphology of splats is characterized by the grains of 90 to 120 nm and their agglomeration of particles of about 200–300 nm (Figure 3b).

To change the structure and surface properties of titanium substrate alloyed with tantalum induction heat treatment (IHT) can be effectively used. Significant changes in the morphology of a titanium-tantalum system start at the temperature of 800 °C. During the short IHT of 30 s the splat surface is uniformly covered with oxide crystals (Figure 4a). Longer IHT within 300 s contributes to a significant increase in oxide crystals (Figure 4a).

To determine the probable phase state of the resulting metal-oxide Ti-Ta-$(Ti,Ta)_xO_y$ systems chemical elemental analysis has also been performed. Most detailed study has been given to the concentration of introduced alloying tantalum and oxygen saturation of the system with the formation of oxide crystals. The content of tantalum on the surface of titanium substrate ranges from 10 to 12 at.%, which corresponds to a monolayer of splats. When coatings of greater thickness (2–3 alloying cycles) are formed the tantalum concentration can reach 18–25 at.%. Further IHT contributes to intensification of oxidation and oxygen concentration increases to 57–65 at.% at the short treatment of 30 s. Longer IHT within 300 s contributes to a slight increase in oxygen concentration and equals 65–69 at.%.

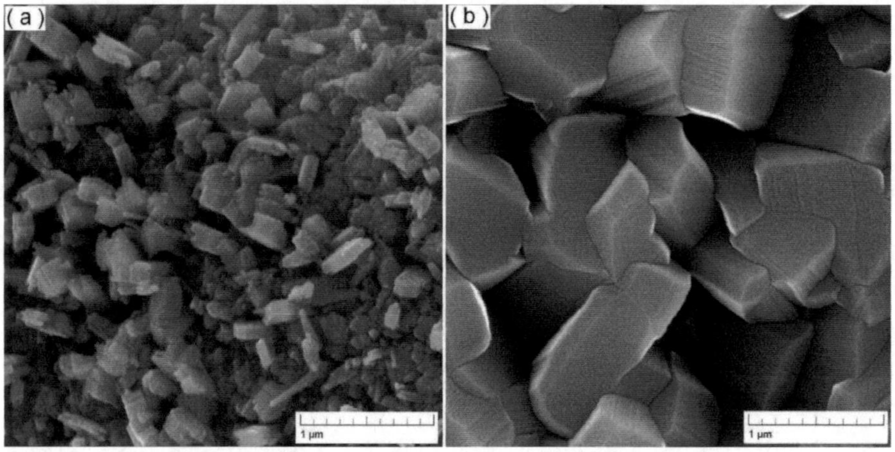

Figure 4(a, b). Nanostructure of Ti-Ta-$(Ti,Ta)_xO_y$ system obtained after IHT on cp-titanium at different durations of treatment: 30 s **(a)**; 300 s **(b)**.

4. Conclusions

Thus, the surface structure of samples of titanium after electric-spark alloying with tantalum and subsequent IHT is characterized by the formation of the coatings of Ti-Ta-(Ti,Ta)xOy system. These functional coatings are formed of nano-sized and submicron oxide crystals. The most pronounced nanostructure is formed at the short duration of IHT of about 30 s and is characterized by the size of nanograins of 60–90 nm, the concentration of tantalum is about 1.1–1.5 at.% and oxygen is at least 60–65 at.%.

Acknowledgments

The research was conducted with the support of the grant of the President of the Russian Federation MD-3156.2015.8 and project 1189 in the framework of the basic part of the state task for educational institutions of higher education under the jurisdiction of the Ministry of Education of the Russian Federation in the field of scientific research. The study of IHT effect on the oxidized surface morphology was performed by Fomin A.A. within the framework of grant 11.687.2016/DAAD and scholarship SP-617.2015.4.

References

[1] Park Y S and Butt D P 1999 *Oxidation of Metals* **51** 383
[2] Koshuro V A and Fomin A A 2014 Proc. 10th Int. Vacuum Electron Sources Conf. and 2nd Int. Conf. on Emission Electronics (Saint-Petersburg) (Saint-Petersburg: IEEE) p 145
[3] Fomin A, Dorozhkin S, Fomina M, Koshuro V, Rodionov I, Zakharevich A, Petrova N and Skaptsov A 2016 *Ceramics International* **42** 10838
[4] Fomin A A 2015 *Handbook of Nanoceramic and Nanocomposite Coatings and Materials* (Oxford: Butterworth-Heinemann) p 293

Production of C–Fe–Pd nanocomposites *via* Infra-red radiation and its structural characterization

A V Kovtun[1], E L Dzidziguri[1], D G Muratov[2]
[1]Department of Functional Nanosystems and High-Temperature Materials, National University of Science and Technology "MISiS", Moscow 119049, Russia
[2]Chemistry of Polyconjugated Systems Laboratory, A.V. Topchiev Institute of Petrochemical Synthesis RAS, Moscow 119991, Russia

E-mail: cvetaevaanja@rambler.ru

Abstract. The objective of the present study is to produce C–Fe–Pd nanocomposites using pyrolyzed infrared (IR) radiation. The structural characterization was analyzed using X-ray diffraction (XRD), transmission electron microscopy (TEM) and X-ray fluorescence analysis. In addition, the size of metal nanocomposites as a function of the pyrolysis temperature, and its distribution within the carbon matrix is also characterized.

1. Introduction
Currently, development of different process methods for the production of nanocomposites with tailored properties is one of the priority tasks of modern materials science. The peculiarity of such materials, structural characteristics and functional properties of nanosystems, depending on the conditions for obtaining is an important task. This due to this it opens a new research area that could possibility control the structure and consequently its properties [1].

Bimetalnanosized Fe-Pd particles systems possess excellent catalytic activity, which leads to their use at the present time. Furthermore, the C–Fe–Pd nanocomposites system is receiving increasing attention as it is being used to enhance magnetic media with high recording density and storing information [2].

In this regard, the aim of this work was to obtain the metal-carbon systems with bimetallic Fe-Pd particles, and to study their properties depending on the different temperatures (300, 400, 500, 600, 700 °C).This method for obtaining nanocomposites based on the pyrolysis of organic and inorganic substancesis the most efficient ways of obtaining nanocomposites for the perspective of industrial applications.

2. Experiment
Metal-carbon nanocomposite was obtained by thermal decomposition of the precursor under the influence of infrared heating to install Photon. The precursor was obtained by dissolving PAN, iron salts and palladium in a solution of DMF and subsequent drying. In addition the structural properties of C–Fe–Pd nanocomposites are analyzed under different temperatures (300, 400, 500, 600, 700 °C).

Journal of Physics: Conference Series **741** (2016) 012198 doi:10.1088/1742-6596/741/1/012198

3. Results and discussion

The diffraction patterns of the samples presented in Figure 1 was produced by a qualitative phase analysis. X-ray phase analysis showed the presence in the system IR-PAN/FePd phase of amorphous carbon. The radiographs of the samples determined that at all temperatures the production of nanopowder is formed a solid solution of Fe in Pd.

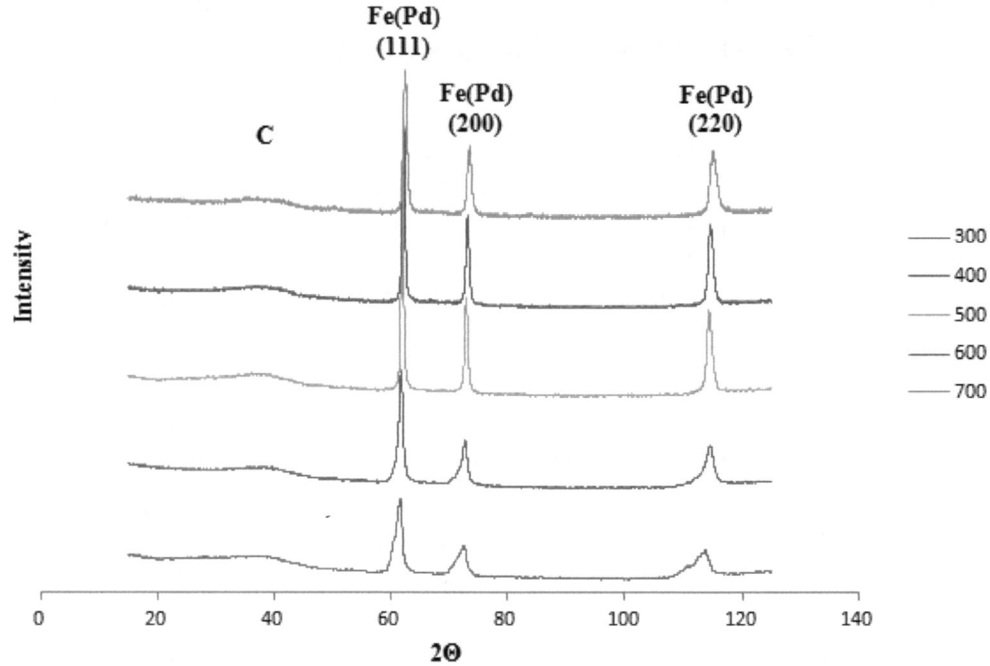

Figure 1. The diffraction pattern of the samples.

The obtained samples were further anneald from temperature 200 to 700 ° C . At temperature 700 ° C, a peak corresponding to palladium, acquires a symmetric form, becomes more intense and narrow, indicating an increase in the size of crystallites and enhancement of the structure (Figure 2). At temperatures 200, 250 and 300 ºC the X-ray lines observed with the "rush" from the side of smaller angles. Based on the above results it is assumed that the dissolution of hydrogen in palladium nanoparticles.

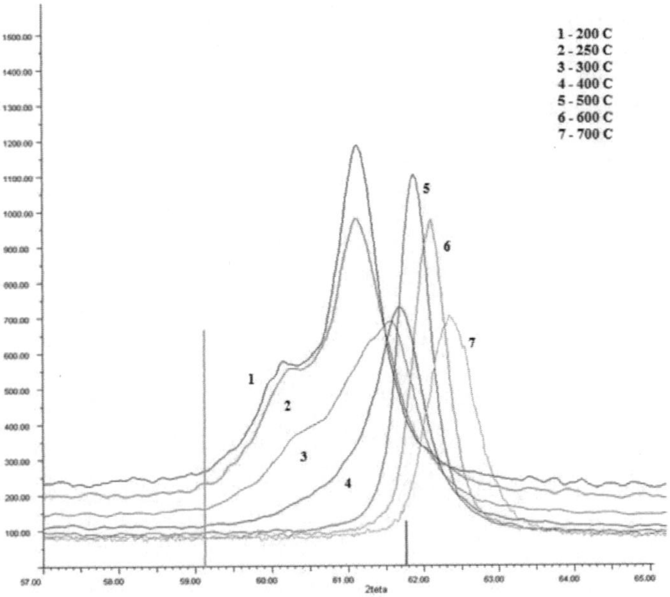

Figure 2. X-ray diffraction patterns of C–Fe–Pd nanocomposites.

The lattice constant of FCC-Pd in the nanocomposite IR-PAN/Fe/Pd is less than the values typical for coarse-grained FCC-Pd due to specific features of nano state of the metal (Table 1).

Table 1. The values of gratings Pd in a coarse-grained and nanocrystalline samples.

Sample	**Lattice constant a (nm)**
Pd (nano)	0.3881
Pd (array)	0.3890

According to the difference of periods of the lattice of palladium in coarse-grained and nanocrystalline states has been determined dependence for nanosized materials, which was performed quantitative estimation of the composition of the formed metallic phase.The concentration of iron in the solid solution increases with increasing pyrolysis temperature. This is evidenced by the shift in the peak intensity on X-ray film in the region of large angles and a decrease in grating period phase of palladium (Table 2).

Table 2. The lattice period for phase Pd under different temperature

Annealing temperature, °C	The parameter a (nm)
300	0.3885
400	0.3863
500	0.3855
600	0.3845
700	0.3833

To calculate the size of coherent scattering regions the Selivanov–Smyslov method was used for the broadened diffraction peaks. At 500 °C and 600 °C the peaks shifted strongly to higher values, hence the average size also increased drastically. In addition, it is clear from the Figure 3 that the increase in temperature leads to increase the crystallites size.

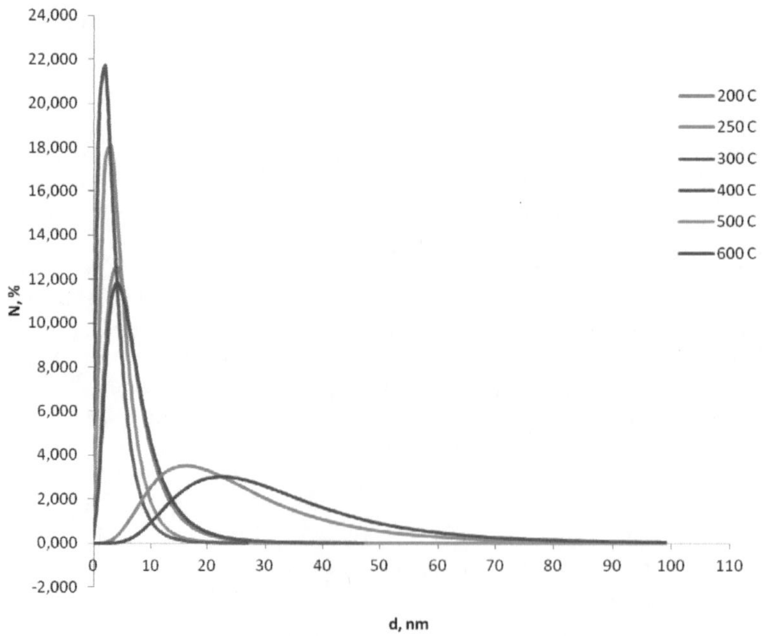

Figure 3. Distribution of crystallites size.

At 500 °C and especially at 600 °C graphics strongly blurred to higher values, the average size also increased drastically. As can be seen from the obtained data, the temperature increase leads to an increase in the size of crystallites.

Figure 4 shows the transmission electron microscopy micrograph of the obtained samples. It is clear from the images that the metal nanoparticles (dark areas on the micrograph) are rounded and encapsulated in a polymeric matrix (the gray area). In some areas there are coagulation of the particles also observed (Figure 4). The photomicrographs also reveals that the polymer matrix has a layered structure.

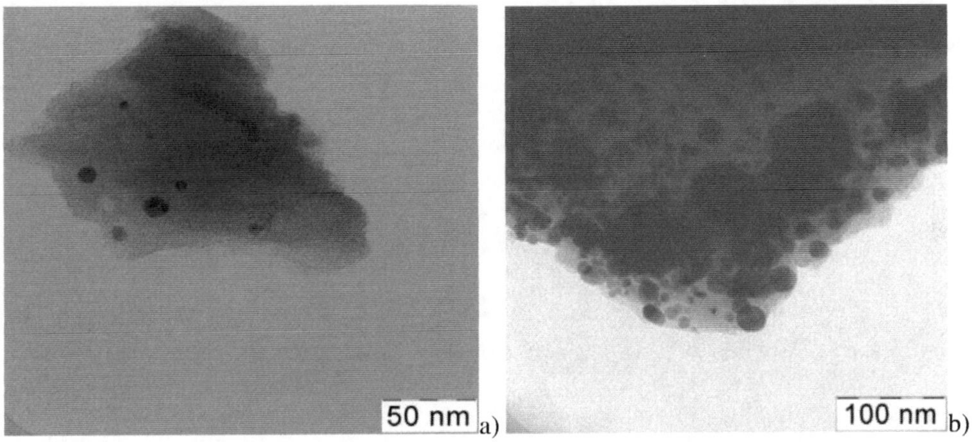

Figure 4. TEM images of C–Fe–Pd nanocomposites at different temperature (a) 400°C and (b) 500°C.

The image shows that there are large dark thickened backdown present. The large thickness are polymer layers or larger particles of the metallic phase. In the course of obtaining the nanocomposite undergoes various reactions and structural transformations, the nature and sequence of which may change with increasing temperatures, at which the gain material. As a consequence, in some areas there may be conditions for coagulation of the particles.

The photographs, obtained with the electron microscope, was used to measured the size of metal particles. Based on the data obtained histograms showing the distribution of nanoscale particles. The obtained results, reveals that with increasing pyrolysis temperature, the size of nanoparticles increases. This may be due to the coagulation of particles and the growth rate increases with increasing temperature of pyrolysis. The experimental curve (Figure 5) shows clearly that there is a long "tail" to larger particles, which is described by logarithmically normal law.

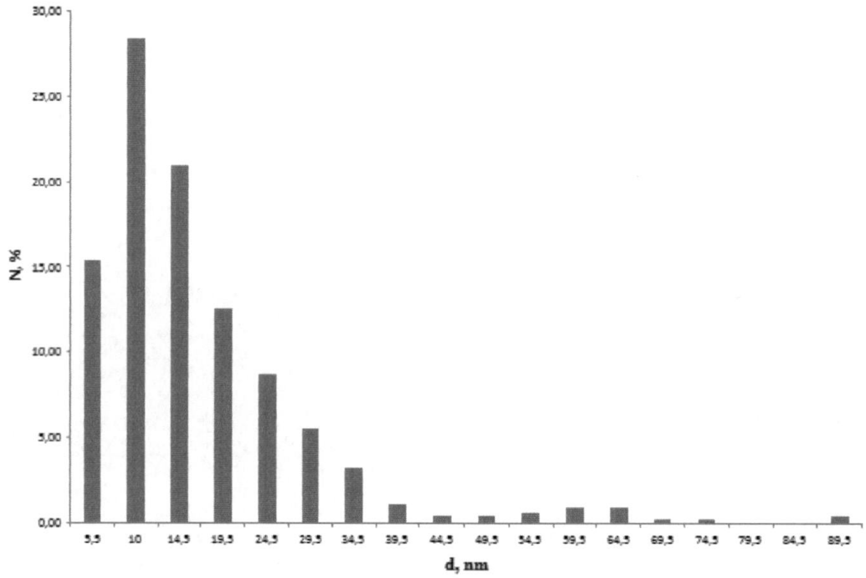

Figure 5. Histogram of size distribution of particles in the sample obtained at 500 °C.

References

[1] Ozkana S Zh, Dzidzigurib E L, Karpacheva G P, Bondarenko G N 2011 *J. Nanotechnologies in Russia* **6** 750

[2] Bagdasarova K A, Perov N S, Karpacheva G P, Pile S E, Dzidziguri E L 2011 *J. Advanced Materials* **168** 349

Approaches to numerical solution of 2D Ising model

K S Soldatov[1], K V Nefedev[1,2] , V Yu Kapitan[1] and P D Andriushchenko[1]

[1]School of Natural Sciences, Far Eastern Federal University
Vladivostok, 690950, Russia
[2]Institute of Applied Mathematics of Far Eastern Branch Russian Academy of Science, 7 Radio Str.,Vladivostok 690041, Russia

Abstract. Parallel algorithm of partition function calculation of two-dimensional Ising model for systems with a finite number of spins was developed. Within a method of complete enumeration by using MPI technology with subsequent optimization of a parallel code time of calculations was reduced considerably. Partition function was calculated for systems of 16, 25, 36 Ising spins. Based on the obtained results, main thermodynamic and magnetic values dependences (such as heat capacity, magnetic susceptibility, mean square magnetization) for ferromagnetic and antiferromagnetic interactions was investigated. The analysis of a different configurations contribution showed, that states with the minimum energy have essential influence on dependences of thermodynamic values. Comparison with the results obtained by the Wang Landau algorithm was performed.

1. Introduction

Studying of systems consisting a large number of interacting particles is one of the most fundamental and actual problems of the modern physics [1]. The task is interesting both from the practical and theoretical points of view. Special attention should paid to behavior of such systems near a stability boundary. First of all, there is a problem of definition of these borders, i.e. limiting values of energies for various systems, then studying a peculiar nature of thermodynamic behavior of the modelled substance in case when loss of stability of the system caused by emergence of particular type of the ordering called by phase transition. For the study of phase transitions and the knowledge of their nature is convenient to use classical lattice models such as the Ising model [2, 3]. In this regard, a comparison of the results of the analytical theory and numerical simulations, developed by the authors in [4–8], would be very useful and productive in terms of the development of our understanding of the processes of ordering and randomization.

2. Partition function

Difficulties in calculation of partition function, for a lattice consisting of a finite number of Ising spins generally consists in calculation of degeneracy multiplicity of spin excess M_i (a difference between number of spins down and number of spins up) by energy E_i (sum of energies of all pair interactions) see equation (1).

$$Z = \sum_{s_1=\pm1} \dots \sum_{s_N=\pm1} \exp(J\sum_{i,j}^{N} s_i s_j + \mu h \sum_{j=1}^{N} s_j)) \tag{1}$$

It's important to notice that sums of degeneracy multiplicities of energy within each fixed value of spin excess M_i, represent binomial coefficients. The developed algorithm based on representation of a spin lattice in the form of bit vectors:

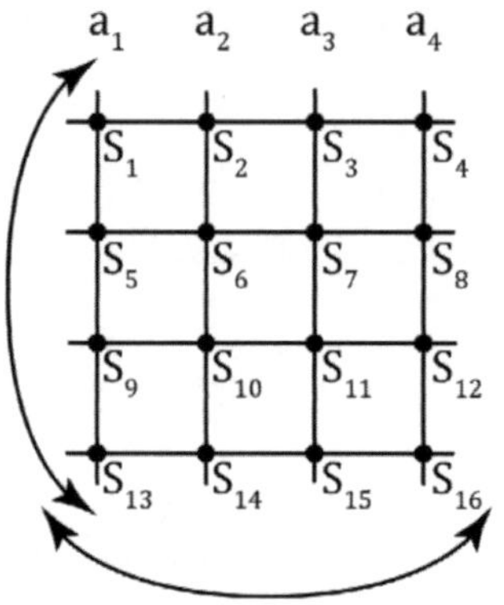

Figure 1. Representation of a lattice of Ising spins in the form of bit vectors.

- The number of rows and columns (and total number of spins) is set. The array containing integer variable $\{a_i\}$ is initialize. Every integer variable $\{a_i\}$ in binary notation represents corresponding lattice column (e.g. $a_i = 14$ that means "1110" in binary notation that means $S_1 = 1, S_5 = 1, S_9 = 1, S_{13} = -1$ on figure 1).
- By means of serial change of values of each integer variable $\{a_i\}$ all possible configurations of the system will passed.
- Spin excess (the sum of all one's) is calculated using the Kernighan's algorithm [9].
- Exchange pair interactions energy E_n between spins at lattice sites are calculated by logical XOR (^) between all columns and rows including periodic boundary conditions.
- The final value of E_n is equal to number of all pairs opposite spins. Energy of the system configuration in general:

$$E_{tot} = -2(N - E_n). \tag{2}$$

3. Parallel execution

The standard of Massage Passing Interface (MPI) expands possibilities of calculations and does possible parallel algorithmization that allows to solve problems, which could not be solved for acceptable time for single-threaded mode. Parallel algorithm enumerates all possible configurations and calculates energy and spin excess in multithread mode.

Calculation of small systems, such as 4×4, and 5×5, shows restriction to increase of efficiency with increase number of involved processes. This result is caused by the Amdahl's law according to which, the increase of effectiveness of calculations depends on algorithm of a task and is bounded above for any task [10].

The advanced methods of load distribution on the threads had to be used, as some threads received different number of subtasks. The number of p iterations, approximately identical in duration, which is required to be distributed between L processes, can be presented through integer numbers of a and b in the following form:

$$p = aL + b,\ 0 \le b < L. \tag{3}$$

Journal of Physics: Conference Series **741** (2016) 012199 doi:10.1088/1742-6596/741/1/012199

Width of an interval of iterations for each process p_k, $k = 0, \dots, L-1$ can be calculated as

$$p_k = \begin{cases} a+1, & k < b; \\ a, & k \geq b. \end{cases} \tag{4}$$

The first b intervals are wider than the others by one. Left border of each interval n_k^l is calculated based on the same ratios:

$$n_k^l = ak + \begin{cases} k, & k < b; \\ b, & k \geq b. \end{cases} \tag{5}$$

Thus, interval of iterations $[p_k^l; p_k^l + p_k)$ on each process is distributed. The described approach allows combine conveniences of block distribution and the acceptable balance.

Table 1. Run time of the balanced parallel software

Size 4 × 4		Size 5 × 5		Size 6 × 6	
N,proc	T,sec	N,proc	T,sec	N,proc	T,sec
2	0,018	2	13,17	8	5741,55
4	0,0078	4	4,35	16	2622,05
8	0,0057	8	1,80	32	1244,01
16	0,0065	16	0,82	64	595,48
20	0,0072	32	0,42	128	445,76
24	0,0086	48	0,46	192	320,63
32	0,0087	64	0,41	256	213,02

4. Wang-Landau algorithm

To describe a system in the state of thermodynamic equilibrium is necessary to know the partition function that contains all the information about the statistical properties of the system. The problem is that the number of members of the functional series, which is a partition function, increases exponentially with the increasing number of particles; that is, if a system and its parameters are researched by a brute-force method, such an approach would exponentially increase the computation time with the increasing number of particles. Therefore, statistical physics currently uses the Monte Carlo methods. The WL method is an efficient algorithm to calculate the energy density of states (DOS), g(E), with high accuracy, and was successfully applied to many problems [11-13].

a) b)

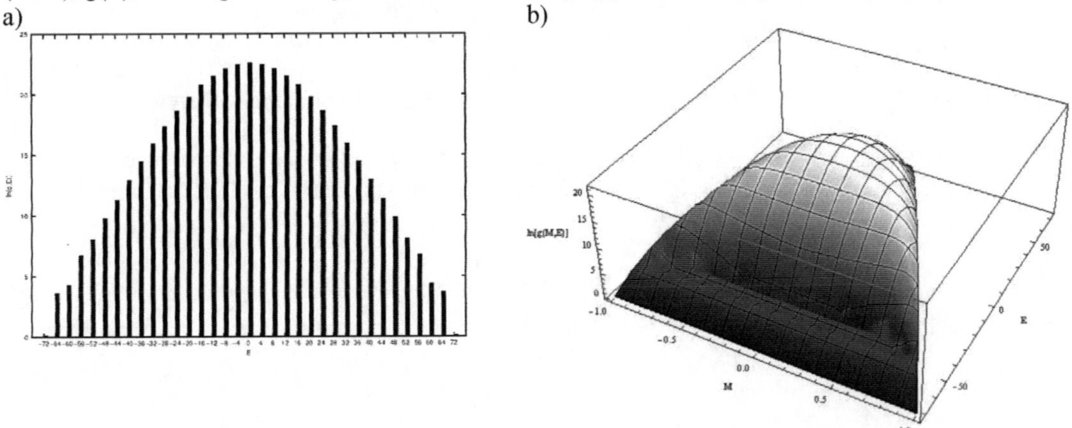

Figure 2. (a) Density of states $g(E)$ for a 6 × 6. (b) Joint density of states $g(M, E)$ for a 6 × 6.

Wang-Landau method is based on the representation of the partition function Z in the form of an expansion in the number of levels of density of states $g(E)$ with a given energy E [11]:

$$Z = \sum_E g(E)\exp\left[\frac{-E}{kT}\right] \tag{6}$$

If $g(E)$ is known, we can calculate the mean energy, heat capacity and other thermodynamic quantities at any temperature from the relation:

$$\langle E\rangle = \frac{1}{Z}\sum_E Eg(E)\exp\left[\frac{-E}{kT}\right]. \tag{7}$$

In the figure 2, as an example, are represented the density of states for a $6x6$ system of spins.

5. Discussion of results

By means of described algorithm, heat capacity, mean square magnetization, magnetic susceptibility for ferromagnetic and antiferromagnetic interactions were calculated. Heat capacity can be calculated as

$$C(h=0) = \frac{\delta U}{\delta T}. \tag{8}$$

Heat capacity reaches the maximum in points $T_1 = 2.434$, $T_2 = 2.408$, $T_3 = 2.391$ for systems 4×4, 5×5, 6×6 respectively. When the size of system increasing, the peak becomes sharper, more expressed and displaced to a point $T_C = 2.269$ — to the phase transition temperature calculated by Onsager [14].

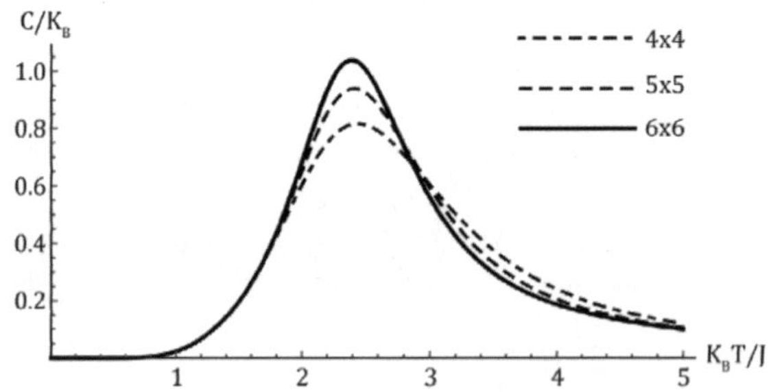

Figure 3. Dependence of a heat capacity for systems with finite number of Ising spins.

Mean square magnetization is calculated by formula (9). The sign of a constant of exchange interaction J of partition function is equal positive for ferromagnetic model, and negative for antiferromagnetic system.

$$\langle M^2\rangle(h=0) = \frac{1}{N}\frac{T^2}{Z^N}\frac{\delta^2 Z_N}{\delta h^2}. \tag{9}$$

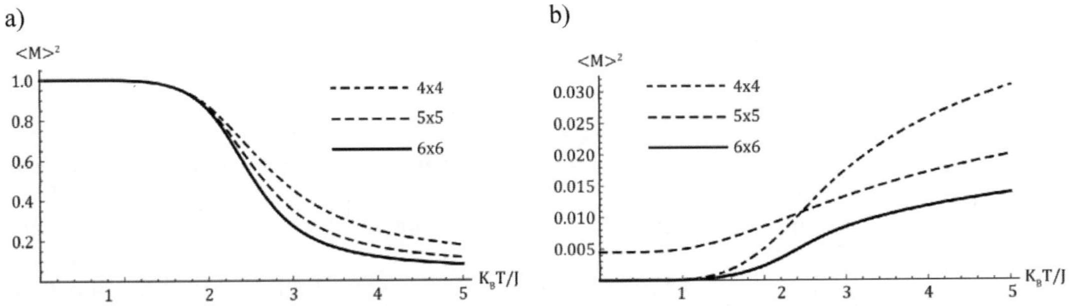

Figure 4. Dependences of mean square magnetization for a ferromagnetic exchange interaction $J > 0$ (a) and antiferromagnetic exchange interaction $J < 0$ (b).

Also, when number of particle increases, angle of the magnetization curve increases relatively to x-axis.

An abrupt tendency of magnetization to zero with infinite slope is observed at the limit of infinite number of particles in Curie point $T_C = 2.269$, characterizing the second-order phase transition [14]. Magnetic susceptibility χ is calculated as

$$\chi(T > 0) = \frac{\delta M}{\delta h}. \tag{10}$$

Increase of number of particles leads to higher peak of magnetic susceptibility, which becomes sharper at decreasing temperature.

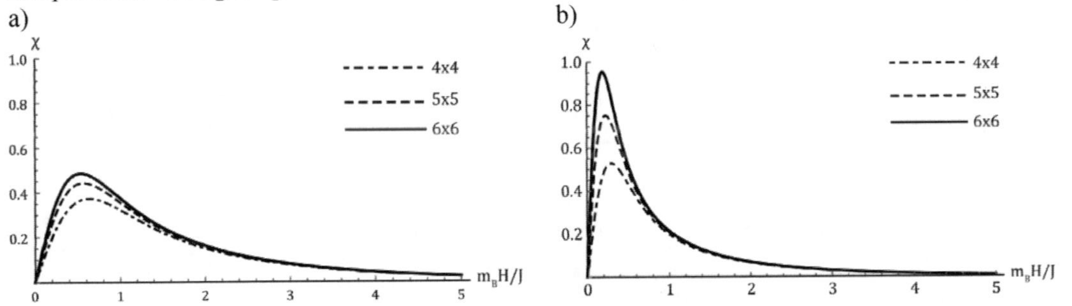

Figure 5. Dependences of magnetic susceptibility at $T = 5$ (a) and at $T = 2.269$ (b).

6. Conclusion

Realized parallel algorithm allows us to calculate coefficients of partition function within two-dimensional Ising model with a finite number of spins, using the minimum amount of resources. Existence of logical parallelism in the considered task gives wide opportunities for application of various approaches to use of parallel programming tools. By means of received results, the curves of main thermodynamic averages were investigated, contribution made in them was estimated too. States with minimum values of energy have the greatest impact, especially at increase number of particles in system. Such states rather easily can be calculated analytically, however complete enumeration of all configurations is the essential necessity for the most correct solution to the problem.

Acknowledgments
This work was supported by RFBR according to the research project No 16-32-00202 mol_a and the Ministry of Education and Science of the Russian Federation in frame of the scholarship of the President of the Russian Federation for young scientists and postgraduate students performing advanced research and development in priority areas of modernization of the Russian Economics for 2015-2017 years (SP-118.2015.5, order #184, 10/03/2015).

References

[1] Aharoni, Amikam, Introduction to the Theory of Ferromagnetism, Clarendon Press, pp. 315, 1996.
[2] W. Lenz, Physikalische Zeitschrift **21**, 613–615 (1920).
[3] E. Ising, Beitrag zur Theorie des Ferromagnetismus. Zeitschrift für Physik 31, 253–258 (1925).
[4] K. Nefedev, Y. Ivanov, A. Peretyatko, and V. Belokon, "Magnetic states of nanodot arrays. physical and numerical experiments," Solid State Phenomena, vol. 168, pp. 325–328, 2011.
[5] Belokon V., Kapitan V., Dyachenko, O., J. of Magnetism and Magnetic Materials, V 401 pp. 651-655, 2016.
[6] V. I. Belokon, K. V. Nefedev, V. Y. Kapitan, O. I. Dyachenko, "Magnetic States of Nanoparticles with RKKY Interaction", Advanced Materials Research, Vols. 774-776, pp. 523-527, 2013
[7] K V Nefedev, V I Belokon, V Yu Kapitan and O I Dyachenko, J. Phys.: Conf. Ser. 490 012163, 2014
[8] V I Belokon, V Yu Kapitan and O I Dyachenko, J. Phys.: Conf. Ser. 490 012165, 2014
[9] Kernighan, B.W., Ritchie, D.M., C Programming Language, Prentice-Hall, pp. 274, 1988).
[10] Amdahl, Gene M., Validity of the single processor approach to achieving large scale computing capabilities, A FIPS Computer conference, pp. 3-4, 1967).
[11] Wang F., Landau D. P.//Physical Review Letters. 2001. V. 86. P. 2050-2053.
[12] C. Yamaguchi and Y. Okabe, J. Phys. A: Math. Gen. 34, 8781 (2001).
[13] Y. Okabe and H. Otsuka, J. Phys. A: Math. Gen. 39, 9093 (2006).
[14] Onsager, L., Crystal Statictics A Two-Dimensional Model with and Order-Disorder Transition, Physical Review, pp. 117, 1944.

Electrical properties and the structure changing of polycrystalline materials from the system Cu-As-Ge-S at high pressure

V Zaikova, N Melnikova, A Tebenkov, Ya Volkova and E Chubareshko

Institute of Natural Science, Ural Federal University, Ekaterinburg 620000, Russia

Abstract. The results of experimental studies of magnetoresistance of polycrystalline materials $(GeS)_{1-x}(CuAsS_2)_x$ under high pressure (up to 50 GPa) are presented. The appearance of large negative magnetoresistance was found. The structure changing during applying high pressures was studied.

1. Introduction

Multicomponent semiconductors based on copper chalcogenids are well known as promising for fundamental and applied research materials, one of the interesting features of which is appearance of negative magnetoresistance [1,2]. Polycrystalline chalcogenides $(GeS)_{1-x}(CuAsS_2)_x$ $0.1\leq x\leq 1.0$ are promising materials for fundamental and applied research, because of its interesting electrical properties, in particular they combine semiconducting properties with ionic and mixed (electronic-ionic) electrotransfer and other interesting features [3-7]. The temperature, time and pressure dependences of the AC and DC electrical properties of these materials have been studied in [3-7]. This work is aimed to study of pressure dependences (up to 50 GPa) of magnetoresistance and to determine a correlation between the structure changing and detected negative magnetoresistance for materials $(GeS)_{1-x}(CuAsS_2)_x$ with x =0.3; 0.4; 0.6.

2. Materials and experimental details

Polycrystalline materials $(GeS)_{1-x}(CuAsS_2)_x$ were synthesized by melt crystallization processes. The initial high-purity (no worse than special purity grade) components were melted in the regime of step-by-step heating in quartz ampoules evacuated to a residual pressure of 10^{-4} Pa, filled with superhigh-pure helium or argon [3]. These materials crystallize in cubic or tetragonal syngonies depending on the composition and show mixed (electronic-ionic) conductance with the part of ionic conductivity from 0.15 up to 0.4 [3,4]. AC electrical properties of these materials at atmospheric and high pressures are already well studied [3,4]. Current work is devoted to investigation of magnetoresistance of $(GeS)_{1-x}(CuAsS_2)_x$ with x=0.3; 0.4; 0.6, which crystallize in tetragonal syngony [3].

The baric dependences of electric properties was studied in pressure range 15-50 GPa using an equipment with high pressure cells with a rounded coneplane-type anvils made of carbonado-type artificial diamonds. These anvils with good conductivity make it possible to examine the electrical properties of samples placed into high pressure cell [8]. The equipment allows high-pressure measurements in transverse magnetic field up to 1 T and at temperatures from 78 K to room temperature. High-pressure electrical measurements were made using samples with a diameter of ~0.2 mm and thickness from 10 to 30 μm. The samples were obtained by compression in high pressure cell of the initial powdered materials. Magnetoresistance measurements at ambient pressure and room temperature were carried out for bulk samples using the cell with cooper electrodes.

X-ray diffraction experiments at high pressures were carried out on a special diffractometer consisting of a high-flow X-ray generator FR-D, a focusing optical system Flux Max, and a BrukerAPEX CCD detector at the Institute of Geochemistry and Geophysics, University Bayreuth (Germany), in a high pressure cell with diamond anvils.

3. Results and discussions

It was found that electroresistance of studied materials decreases with increasing of pressure (*P)* (figure 1). Analyses of experimental results show that in magnetic field 0,75 T and more in pressure range 20-40 GPa materials exhibit negative magnetoresistance (MR), and there are one or several extremes at curves *MR*(*P*). The pressure dependences of magnetoresistance for $(GeS)_{1-x}(CuAsS_2)_x$, x=0.3 are shown in figure 2 and typical field dependences for these materials are presented in figure 3. Baric ranges of *MR*(*P*) curve extremes are correspond to baric ranges in which features of behaviour of electrical properties in AC electrical field were observed [4]. The field dependences of magnetoresistance of $(GeS)_{1-x}(CuAsS_2)_x$, x=0.3, at pressures 26 GPa and 30 GPa were studied additionally at temperature 78 K (in figure 4 results obtained under pressure of 26 GPa are shown). We can see that at low temperature effect of negative magnetoresistance is also observed.

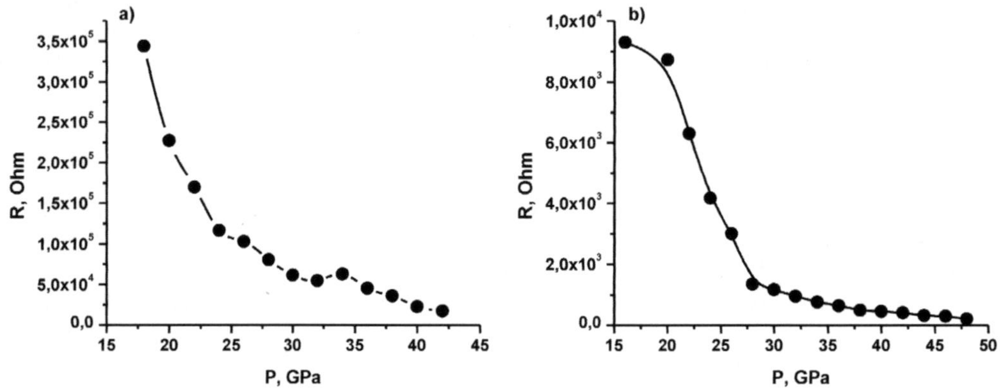

Figure 1. Pressure dependence of electroresistance for $(GeS)_{1-x}(CuAsS_2)_x$ a) x=0.3, b) x=0.4.

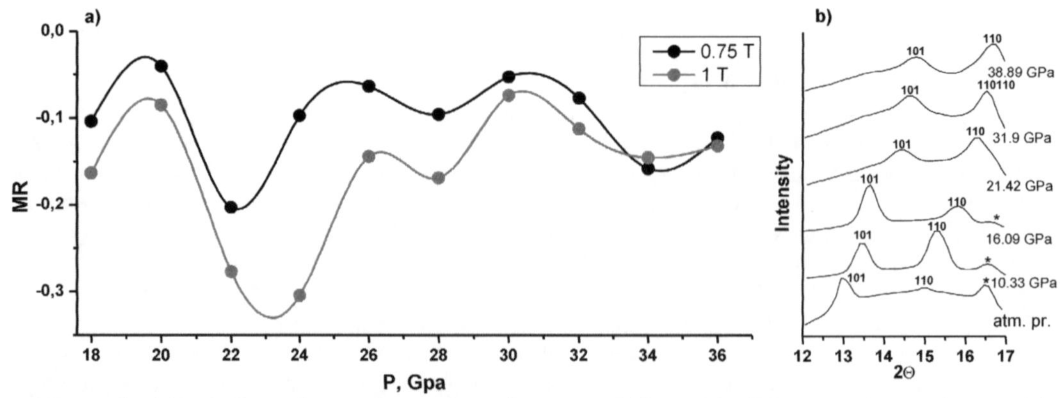

Figure 2. a) Baric dependence of magnetoresistance and b) results of X-ray study (peak tagged * corresponds to Re-gasket) for $(GeS)_{1-x}(CuAsS_2)_x$, x=0.3

The origin of negative magnetoresistance (NMR) in nonmagnetic semiconductors is still subject for discussion and studies. Some authors suggest that appearance of NMR is connected with splitting of impurity levels in a magnetic field [9], some other authors connect the NMR appearance with material loose structure and small carries mobility [11]. Other theory connects NMR appearance with baric or temperature structure phase changes or significant structural defects caused by changing of external parameters [1, 12-13]. Current work authors take the view that in case of NMR in polycrystalline chalcogenide semiconductors the latter theory is more probable explication. For this theory confirmation the X-ray experiments for compound $(GeS)_{1-x}(CuAsS_2)_x$, x=0.3 were run at pressures up to 45 GPa. This material crystallizes in tetragonal syngony and at atmospheric pressure has lattice parameters a = 0.376 nm, c = 0,521 nm. It was found that pressure ranges which correspond to features of behavior *MR*(*P*) (see figure 2) are in agreement with pressure intervals with significant changing of lattice parameters.

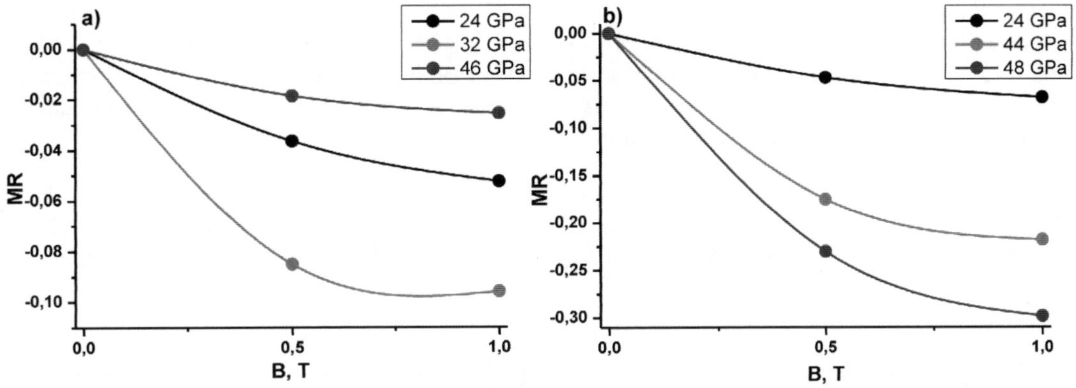

Figure 3. Field dependences of magnetoresistance at pressures up to 50 GPa for $(GeS)_{1-x}(CuAsS_2)_x$ a) x=0.4, b) x=0.6

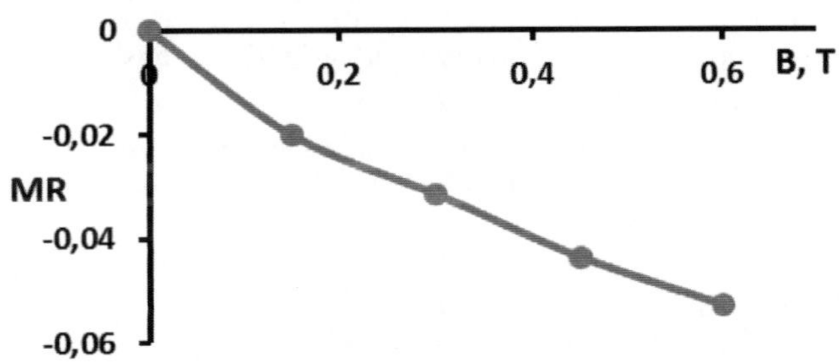

Figure 4. Field dependence of magnetoresistance for $(GeS)_{1-x}(CuAsS_2)_x$, x=0.3, at pressure 26 GPa at 78 K.

Table 1. The distances between lattice planes (101) *d* and the relative change *d* on the unit pressure at different pressures.

P, GPa	*d*, nm (101)	Δd/ΔP
atm. pr.	0.317	
10.33	0.307	0.0010
16.09	0.301	0.0010
21.42	0.286	0.0028
31.9	0.278	0.0008
38.89	0.276	0.0003

On the X-ray patterns (figure 2b) we can see the significant changes in position of lines, corresponding to reflections from crystallographic planes (101) and (110) at pressure range from 16 GPa to 22 GPa. In table 1 the data for the distances between lattice planes (101), which correspond to the most intensive peak on the X-ray patterns, are presented. It should be mentioned that at atmospheric pressure material shows small (near zero) MR then at pressures more than 18 GPa this material shows huge negative MR and there is maximum of MR absolute value near 22 -24 GPa (figure 2 a). Significant changes in impedance behavior for this material were observed at pressure range about 22-24 GPa [1]. All descried results can be interpreted as that at structure changing the formation of large number of defects cause changing in electronic structure of material which in turn cause appearance of visible NMR. Similar situation for polycrystalline cupper chalcogenide $CuInS_2$, $CuInSe_2$ were described in work [1], where extremes of MR(P) curve correspond to the structure phase transition.

4. Conclusion

Polycrystalline materials $(GeS)_{1-x}(CuAsS_2)_x$ show huge negative magnetoresistance at high pressure. The pressure and field dependences of magnetoresistance and results of X-ray diffraction study were analysed and the presence of negative magnetoresistance was suggested as consequence of changing of crystal and electronic structure.

Acknowledgments
This work was supported by the Russian Foundation for Basic Research, project No. 16-02-00857-a. The authors are grateful to Prof L.S. Dubrovinsky, University Bayreuth (Germany) for the possibility to conduct high pressure X-ray measurements.

References

[1] Melnikova N and et al 2014 *Bulletin of the Russian Academy of Sciences. Physics* **78** №4 299-303
[2] Melnikova N and et al 2015 *J. Phys.: Conf. Ser.* **653** 012098
[3] Melnikova N and et al 2007 *International Scientific Journal for Alternative Energy and Ecology ISJAEE* **49** №5 56-63.
[4] Melnikova N and et al 2015 *Physics of the Solid State* **57** №10 2025
[5] Baranova E and et al 1999 *Solid State Ionic* **124** 255
[6] Melnikova N and et al 2010 *Azerbaijan Journal of Physics* **XVI** №3-4 37
[7] Gorin Yu and et al 1997 *Tech. Phys. Lett.* **23 (7)** 550.
[8] Babushkin A N 1992 *High Press. Res.* **6** 349
[9] Toyozawa Y 1962 *J. Phys. Soc. Japan.* **17** №6 968-1024
[10] Seeger K 1982 *Semiconductor Physics* (Berlin: Springer)
[11] Shchennikov V.V.1996 *Physic of solid state* **38** №9 1470-1472.
[12] Alekperova Sh. et al 2007 *Physic of solid state* **49** №3 512-515

Influence of the linear magneto-electric effect on the lateral shift of light reflected from a magneto-electric film

Yu S Dadoenkova[1,2,3], F F L Bentivegna[4], N N Dadoenkova[2,3], R V Petrov[1], I L Lyubchanskii[3] and M I Bichurin[1]

[1]Institute of Electronic and Information Systems, Novgorod State University, Veliky Novgorod 173003, Russia
[2]Ulyanovsk State University, Ulyanovsk 432017, Russia
[3]Donetsk Physical and Technical Institute of the NAS of Ukraine, Donetsk 83114, Ukraine
[4]Lab-STICC (UMR 6285), CNRS, UBL, École Nationale d'Ingénieurs de Brest, 29238 Brest Cedex 3, France

Abstract. We present a theoretical investigation of the lateral shift of an infrared light beam reflected from a magnetic film deposited on a non-magnetic dielectric substrate, taking into account the linear magneto-electric interaction in the magnetic film. We use the stationary phase method to evaluate the lateral shift. It is shown that the magneto-electric coupling leads to a six-fold enhancement of the lateral shift amplitude of a *p*-(*s*-) polarized incident beam reflected into a *s*-(*p*-) polarized beam. A reversal of the magnetization in the film leads to a nonreciprocal sign change of the lateral shift.

1. Introduction

When a spatially limited optical beam impinges on an optical system, the reflected beam may be laterally shifted in the plane of incidence with respect to the behaviour predicted by geometric optics for idealized light rays. This effect, called the Goos-Hänchen (GH) effect [1], can be observed in many configurations and optical structures, and has recently found practical applications, e.g. for the design of an optical heterodyne sensor [2] or an optical waveguide switch [3], for biosensing technologies [4] or for the detection of chemical vapours [5]. The GH shift is very sensitive to the nature and geometrical characteristics of the constituents of the optical system and thus can be used to measure such parameters as refractive indices, beam apertures and incidence angles, temperature, displacement, or film thicknesses [2].

The development of magneto-photonic devices, however, has also raised the issue of the potential influence of magnetization on the GH shift. In magneto-optical materials, the lateral shift has been shown to reach values up to several tens of light wavelengths [6-8] and to be tunable *via* the application of an external magnetic field [9, 10]. Thus the GH shift ought to be taken into account for the design of integrated optical or magneto-optical devices.

It is also well known that some magnetic materials possess magneto-electric (ME) properties [11-13], which result in the induction of a magnetization by an electric field or induction of a dielectric polarization by a magnetic field. Since it has been shown that the linear ME interaction can greatly affect the complex reflectivity of a magnetic/dielectric bilayer [14-16], it can thus be expected that ME properties also influence the GH shift of a light beam upon reflection from a system including magnetic materials.

This paper purports to study the dependency of the GH shift on the ME interaction in a magnetic film deposited on a non-magnetic dielectric substrate, and to determine how this shift could be controlled through the application of an external magnetic field to that system.

2. General analysis

Let us consider the reflection of a finite-size light beam from a magnetic film of thickness d deposited on a semi-infinite dielectric isotropic substrate, as depicted in figure 1. The interfaces between the materials are parallel to the (xy) plane of a Cartesian system of coordinates. An electromagnetic harmonic wave of angular frequency ω impinges the upper surface of the film under oblique incidence angle θ from the vacuum. Without loss of generality, the plane of incidence can be chosen as (xz) and the incident and reflected electromagnetic waves can be decomposed into s- and p- components of their electric field strengths $E_{s,p}^{(i,r)}$ with respect to that plane, where superscripts (i) and (r) correspond to the incident and reflected fields, respectively. The magnetic film is magnetized at saturation in the polar magneto-optical configuration, i. e., perpendicularly to the interface, and its magnetization can be reversed *via* an externally-applied magnetic field. We assume that this magnetic film exhibits linear ME coupling.

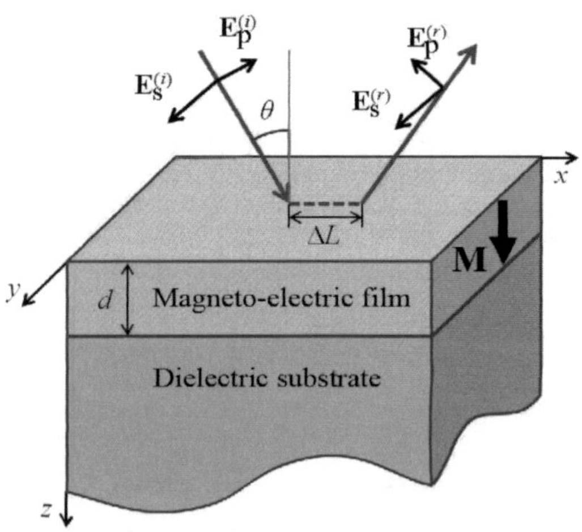

Figure 1. Schematic of the system under study: a magnetic film (thickness d) grown on a semi-infinite isotropic dielectric substrate. The saturation magnetization $\mathbf{M} = \{0, 0, M_s\}$ in the magnetic film is perpendicular to the plane of the film. The s- and p-components of the incident (i) and reflected (r) optical electric fields are denoted $E_{s,p}^{(i,r)}$. The GH shift in the plane of incidence upon reflection is ΔL.

When a wavepacket impinges on the uppermost surface of the magnetic film, its reflection is known to depart from the behaviour predicted by ray optics, in that it undergoes a non-negligible GH lateral shift ΔL in the plane of incidence (in our case, along the x-axis in figure 1). Note that this shift can assume positive as well as negative values. A shift towards increasing values of x, such as represented in figure 1, will thus be deemed positive. The amplitude ΔL of the lateral shift can be obtained with the stationary-phase approach initially proposed by Artmann [17]. Assuming a monochromatic incident Gaussian beam of waist w_0 whose transverse field distribution is given by

$$E^{(i)}(x) = E_0 \exp\left[-\frac{x^2}{2w_0^2}\right], \tag{1}$$

the lateral shift of the reflected wavepacket then simply writes [7]

$$\Delta L = -\frac{\partial \psi}{\partial k_x} + \frac{\partial \ln|R|}{\partial k_x}\frac{\partial^2 \psi}{\partial k_x^2}\left(w_0^2 + \frac{\partial^2 \ln|R|}{\partial k_x^2}\right)^{-1}, \tag{2}$$

where $R(k_x)$ is the complex reflection coefficient for any given k_x component of the incident wavevector $\boldsymbol{k}$ along the x-axis.

Taking into account the linear ME interaction, the constitutive material equations expressing the relationship in the frequency domain of the electric displacement vector **D** and the magnetic induction **B** in the magnetic film with the electric field **E** and the magnetic field **H** of the electromagnetic wave read [11]:

$$D_i = \varepsilon_0 \varepsilon_{ij} E_j + \alpha_{ij} H_j, \qquad B_i = \mu_0 \mu_{ij} H_j + \alpha_{ij} E_j, \tag{3}$$

where ε_0 and μ_0 are the vacuum permittivity and permeability, ε_{ij} and μ_{ij} are the relative permittivity and permeability tensor elements of the magnetic film, and α_{ij} are the elements of the ME tensor of the magnetic medium. The latter is diagonal in crystals with cubic symmetry ($\alpha_{ij} = \alpha\,\delta_{ij}$) [11].

The numerical simulations described in Section 3 will be carried out with a magnetic garnet as a magnetic medium. Indeed, magnetic garnets, which are widely used in magneto-photonic devices, exhibit good transparency properties in the near-infrared regime. They are also magnetically bigyrotropic in that regime, i.e., their relative permittivity and permeability tensors both depend on the local magnetization. Hence, tensors ε_{ij} and μ_{ij} can be expanded in power series of the magnetization vector **M** [18]. Neglecting terms above the first-order terms in such expansions, they write, in the polar magneto-optical configuration and in optically isotropic crystals:

$$\varepsilon_{ij} = \begin{pmatrix} \tilde{\varepsilon} & if^{(e)}m_z & 0 \\ -if^{(e)}m_z & \tilde{\varepsilon} & 0 \\ 0 & 0 & \tilde{\varepsilon} \end{pmatrix}, \qquad \mu_{ij} = \begin{pmatrix} \tilde{\mu} & if^{(m)}m_z & 0 \\ -if^{(m)}m_z & \tilde{\mu} & 0 \\ 0 & 0 & \tilde{\mu} \end{pmatrix}, \tag{4}$$

where $m_z = M_z/|\mathbf{M}|$, $\tilde{\varepsilon}$ and $\tilde{\mu}$ are the crystallographic components of the diagonal relative permittivity and permeability tensors of the medium, and $f^{(e)}$ and $f^{(m)}$ are the linear gyroelectric and gyromagnetic coefficients of the crystal.

Taking into account the boundary conditions at each interface of the system (in our geometry, continuity of the tangential x- and y-components of the electric and magnetic fields), one can then relate the amplitudes of the reflected $E_{s,p}^{(r)}$ and incident $E_{s,p}^{(i)}$ optical electric field components *via* the reflection matrix $\hat{\Re}$ as:

$$\begin{pmatrix} E_s^{(r)} \\ E_p^{(r)} \end{pmatrix} = \hat{\Re} \begin{pmatrix} E_s^{(i)} \\ E_p^{(i)} \end{pmatrix}, \quad \text{with} \quad \hat{\Re} = \begin{pmatrix} \Re_{ss} & \Re_{sp} \\ \Re_{ps} & \Re_{pp} \end{pmatrix}. \tag{5}$$

In general, the reflection matrix $\hat{\Re}$ has four non-zero components. Its off-diagonal components correspond to cross-polarization upon reflection of light and arise from the optical anisotropy of magneto-optical and ME couplings and thus, in turn, from the crystalline symmetry of the magnetic film.

Using Eq. (2) and taking into account Eqs. (3-5), one can finally obtain the value of the GH shift for each of the four polarimetric combinations of incident and reflected field components. In the following Section, these lateral shifts will be presented in the reduced form $\Delta X_{ij} = \Delta L_{ij}/\lambda_0$ $(i, j = s, p)$, where λ_0 is the incident light wavelength. The reduced shift ΔX_{ps} (resp., ΔX_{sp}) thus describes the cross-polarized contribution to the GH effect, i.e., the lateral shift of a *s*- (resp., *p*-) polarized incident wave reflected into a *p*- (resp., *s*-) polarized wave.

3. Results and discussion

The calculation of the lateral shift in the system described in Section 2 was performed in the case of a magnetic layer of yttrium iron garnet (YIG) $Y_3Fe_5O_{12}$ grown on a substrate of gadolinium gallium $Gd_3Ga_5O_{12}$ garnet. Symmetry rules forbid the linear ME effect in crystals with an inversion centre such as YIG. However, in thin epitaxial films (of thicknesses less than 4 μm) symmetry lowering takes place, and the inversion centre vanishes [19]. The linear ME constant in thin YIG films can reach large values up to 30 ps.m^{-1} [19]. For our numerical calculations, we considered a magnetic film thickness $d = 2$ μm and used the following material parameters (at $\lambda_0 = 1.15$ μm): $\tilde{\varepsilon} = 4.56$, $\tilde{\mu} = 1$, $f^{(e)} = -2.47 \times 10^{-4}$, $f^{(m)} = 8.65 \times 10^{-5}$ [20], and $\alpha = 30$ ps.m^{-1} [19].

In order to establish the influence of the ME coupling on the optical properties of the system, hence on the GH shift it imposes to an incident Gaussian beam, we have compared the dependency of the latter as a function of the beam incidence angle θ when the ME interaction is taken into account in our calculations ($\alpha = 30$ ps.m^{-1}) and when it is not ($\alpha = 0$). Our results first show that the GH shift observed upon the reflection of a *s*- (*p*-)polarized incident wave into a similarly *s*- (*p*-)polarized wave almost does not depend on the ME interaction. On the other hand, the ME interaction noticeably affects the cross-polarized lateral shift upon the reflection of a *p*-(*s*-) polarized incident wave into a *s*-(-*p*) polarized wave.

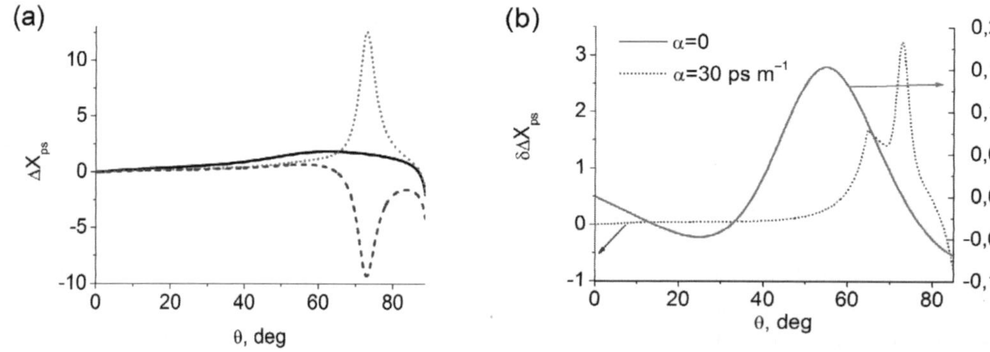

Figure 2. (a) Reduced GH shift ΔX_{ps} as a function of the incidence angle θ of light for $\alpha = 0$ (ME coupling neglected, black solid line) and $\alpha = 30$ ps.m^{-1} (ME coupling taken into account) for $M_s > 0$ ($\Delta X_{ps}(M \uparrow\uparrow z)$, red dotted line) and for $M_s < 0$ ($\Delta X_{ps}(M \uparrow\downarrow z)$, blue dashed line); **(b)** Difference in the GH shift $\delta\Delta X_{ps} = \Delta X_{ps}(M \uparrow\uparrow z) - |\Delta X_{ps}(M \uparrow\downarrow z)|$ upon magnetization reversal for $\alpha = 0$ (red solid line) and $\alpha = 30$ ps.m^{-1} (blue dotted line).

The reduced GH shift ΔX_{ps} is shown in figure 2(a) as function of the incidence angle θ. Comparing the black solid and red dotted lines, one can see that ΔX_{ps} exhibits up to a six-fold increase when the ME coupling is taken into account. This enhancement is maximal for incidence angles close to about 73°, where ΔX_{ps} reaches a peak value of about $12\lambda_0$. In this case, the magnetization **M** in the film points towards the overall direction of the incident beam ($M_s > 0$ and $k_z > 0$), as shown in figure 1, and the shift is positive. It should be noted that both cross-polarized contributions to the GH shift are equal ($\Delta X_{ps} = \Delta X_{sp}$) due to the crystalline symmetry of the magnetic crystal YIG.

Moreover, using an external magnetic field to reverse the initial magnetization **M** of the magnetic film inverts the GH shift direction [see red dotted and blue dashed lines in figure 2(a)] in the presence of ME interaction in the magnetic film. In this case the amplitude of the peak of the lateral shift is about $-9\lambda_0$, and its angular position experiences a slight displacement of about 1° towards larger values of θ. The difference in ΔX_{ps} upon magnetization reversal is shown in figure 2(b). One observes that the inversion of the GH shift is non-symmetrical when the ME coupling is taken into account ($\alpha = 30$ ps.m^{-1}), as the difference $\delta\Delta X_{ps} = \Delta X_{ps}(\mathrm{M}\uparrow\uparrow z) - |\Delta X_{ps}(\mathrm{M}\uparrow\downarrow z)|$ is non-zero [blue dotted line in figure 2(b)] and can reach up to a value of almost 3. Neglecting the ME coupling in the calculations ($\alpha = 0$), on the other hand, results only in a very slight change of the GH effect upon magnetization reversal [red solid line in figure 2(b)], without inversion of the direction of the lateral shift. This confirms that the ME effect does play an essential part in the GH shift in the system.

This conclusion is further substantiated with a study of the effect of the value of the ME constant α on the GH shift, as shown in figure 3 for both directions of saturation magnetization in the magnetic layer. As already seen in figure 2(b), there is almost no change in the GH shift upon magnetization reversal for $\alpha = 0$. When α gradually increases, however, the contrast in GH shift between the two opposite directions of saturation magnetization keeps increasing as well, and both the sign reversal of the shift and the asymmetry of that reversal (i.e., with $\delta\,\Delta X_{ps} \neq 0$) become more pronounced. Overall, the maximum value of the shift increases (in absolute value) with α. For small values of α, an increase of that constant leads to a slight shift of the maximum of ΔX_{ps} towards larger incidence angles for both directions of the magnetization [compare the black solid, red dotted, and blue dashed lines in figure 3]. For increasingly larger values of α, the angular position of that maximum remains almost unchanged [see blue dashed and magenta dash-dotted lines in figure 3], both upon magnetization reversal and upon an increase of the ME constant.

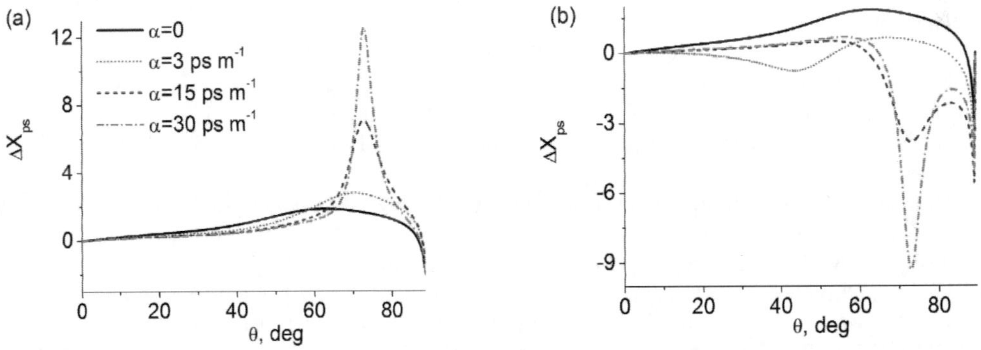

Figure 3. Reduced GH shift ΔX_{ps} as a function of the incidence angle θ of light for different values of the ME constant $\alpha = 0$, 3, 15, and 30 ps.m^{-1} (black solid, red dotted, blue dashed, and magenta dash-dotted lines, respectively) for **(a)** $\mathrm{M}\uparrow\uparrow z$ and **(b)** $\mathrm{M}\uparrow\downarrow z$.

Consequently, this sensitivity of the GH shift to α makes it possible to use GH shift measurements in order to estimate the ME constant of a magneto-electric medium (provided it is transparent at the wavelength used for measurements), in contrast to Faraday rotation measurements [15]. More generally, the value and sign of the GH shift are sensitive to all the materials properties mathematically summarized by Eqs. (3) and (4), upon which the reflectivity factors of Eq. (5) depend in a complex but entirely analytical way. Measuring the lateral shift of light for a set of well-chosen angles of incidence and states of polarization of the incident and reflected beams can be a way to detect small variations of the material properties under the influence of external conditions (strain, wave frequency, temperature, etc.). Moreover, the ability to control, enhance, reduce or reverse the GH shift in such a magneto-electric heterostructure provides a route towards magnetically-controlled photonic devices.

4. Conclusions

We have studied theoretically the lateral, or Goos-Hänchen, shift of an infrared light beam reflected from the upper surface of a magnetic film deposited on a non-magnetic dielectric substrate, taking into account the linear magneto-electric interaction in the magnetic film. It has been shown that the magneto-electric coupling leads to an about six-fold increase of the lateral shift upon reflection of a *p*-(*s*-) polarized incident wave to a *s*-(*p*-) polarized wave, up to about 10 times the incident wavelength. Magnetization reversal leads to a nonreciprocal sign change of the shift when magneto-electric coupling is present in the system.

Thus, the linear magneto-electric interaction, in combination with an external magnetic field, allows to controllably enhance the lateral shift of light to values that can be easily measured. This shows that the use of magneto-electric materials can find applications, through the measurement of the Goos-Hänchen shift they induce, for instance to the determination of very tiny variations of material properties, or to the design of magnetically-tunable optical switches and sensors.

Acknowledgments

This research is supported partly by a grant from the Russian Science Foundation (Project No. 15-19-10036); a grant from the Ministry of Education and Science of the Russian Federation (Project No. 14.Z50.31.0015); European Union's Horizon 2020 research and innovation programme under the Marie Skłodowska-Curie grant agreement (Project No. 644348); the MPNS COST Action (Project No. MP1403); grant No. 14MOESMOMO14 from École Nationale d'Ingénieurs de Brest (France).

References

[1] Goos F and Hänchen H 1947 *Ann. der Phys.* **1** 333
[2] Hashimoto T and Yoshino T 1989 *Opt. Lett.* **14** 913
[3] Sakata T, Togo H and Shimokawa F 2000 *Appl. Phys. Lett.* **76** 2841
[4] Sun J, Wang X, Yin C, Xiao P, Li H and Cao Z 2012 *J. Appl. Phys.* **112** 083104
[5] Nie Y, Li Y, Wu Z, Wang X, Yuan W and Sang M 2014 *Opt. Express* **22** 8943
[6] Borisov S B, Dadoenkova N N, Lyubchanskii I L and Lyubchanskii M I 1998 *Opt. Spectrosc.* **85** 225
[7] Dadoenkova Yu S, Bentivegna F F L, Dadoenkova N N, Lyubchanskii I L and Lee Y P 2016 *J. Opt. Soc. Am. B* **33** 393
[8] Tang T, Qin J, Xie J, Deng L and Bi L 2014 *Opt. Express* **22** 27042
[9] Lima F, Dumelow T, Albuquerque E L and da Costa J A P 2011 *J. Opt. Soc. Am. B* **28** 306
[10] Luo C, Dai X, Xiang Y and Wen S 2015 *IEEE Photon. J.* **7** 6100310
[11] O'Dell T H 1970 *The Electrodynamics of Magnetoelectric Media* (Amsterdam: NorthHolland)
[12] Fiebig M 2005 *J. Phys. D: Appl. Phys.* **38** R123
[13] Chupis I E 2010 Ferroelectromagnets. Fifty years after discovery (arxiv.org/abs/1012.2024)
[14] Dadoenkova Yu S, Lyubchanskii I L, Lee Y P and Rasing Th 2009 *Eur. Phys. J. B* **71** 401
[15] Dadoenkova Yu S, Lyubchanskii I L, Lee Y P and Rasing Th 2010 *Low Temperature Physics* **36** 538
[16] Dadoenkova Yu S, Lyubchanskii I L, Lee Y P and Rasing Th 2014 *J. Opt. Soc. Am. B* **31** 626
[17] Artmann K 1948 *Ann. Phys. (Leipzig)* **1-2** 87
[18] Zvezdin A K and Kotov V A 1997 *Modern Magnetooptics and Magnetooptical Materials* (Bristol: Institute of Physics Publishing)
[19] Krichevtsov B B, Pavlov V V, Pisarev R V and Selitsky A G 1994 *Ferroelectrics* **161** 65
[20] Torfeh M and Le Gall H 1981 *Phys. Status Solidi A* **63** 247

Production of tantalum nanopowders by electric explosion of wire

A V Pustovalov[1], M N Vlasyuk[1], A V Korshunov[1]
[1]Institute of High Technology Physics, Tomsk Polytechnic University, Tomsk 634028, Russia

Abstract. The possibility of producing metallic tantalum nanopowders using electric explosion of Ta-wire was explored. The regimes of the electric explosion were determined which provided maximum level of energy brought into the wire. The basic properties of the obtained powders were examined using low-temperature nitrogen adsorption, atomic-emission spectrometry, as well as by means of transmission and scanning electron microscopes.

1. Introduction

Tantalum and its alloys are widely used in metallurgy, instrumentation, aerospace engineering, medicine, and nuclear power. But the main consumer of metallic tantalum is electronics, in particular in high-capacity tantalum capacitors. The tantalum powders of particle size 3-30 μm are used in the initial stage of capacitor production [1]. It is known that metal properties may change with decreasing powder particles size from micron to nanometer scale. The method of electric explosion of a wire (EEW) [2] is one of the promising and inexpensive methods for producing metal powders with a particle size of about 100 nm. The possibility of obtaining metal Al, Cu, Fe, Mo, Ni, W powders using this method is described in the literature in detail [3-5]. There was determined a number of distinctive properties of electroexplosive powders. However, the data on the possibility of obtaining tantalum powders by EEW and their properties are insufficient.

Thus, the main objective of the present work is to determine the possibility of obtaining metallic tantalum powders using electric explosion of wire and to explore the basic properties of the obtained powders.

2. Materials and methods

The experimental setup developed in Tomsk Polytechnic University was used to obtain the powders by the wire exploding method (Figure 1).

The setup works as follows. The distance between high-voltage and ground electrodes in reactor 3 must be fixed. The setup is evacuated and filled with argon to pressure of $2 \cdot 10^5$ Pa. Then the high-voltage power supply charges the capacitor (C) of the impulse-current generator to the voltage level needed. After that the wire feeding system is activated. When the wire approaches the high-voltage electrode a switch goes off, the capacitor is discharged on the wire section between the high-voltage and grounded electrodes. As a result, the wire explodes. Then the process is repeated. The fan moves products of electric explosion first to the coarse particles filter and then to the fine particles filter. Coarse and fine powder particles settle in the containers of the respective filters, and the filtered argon is fed to the reactor again. After obtaining the necessary amount of powder the containers are removed and the powder is passivated in air-argon atmosphere.

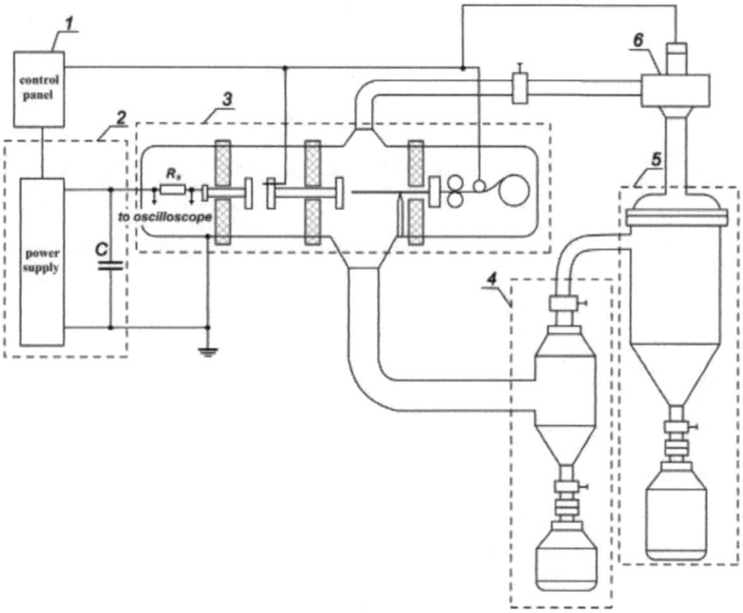

Figure 1. Setup for the production of nanopowders using electric explosion of wires. Control panel (1), impulse-current generator (2); reactor (3); coarse particles collecting filter (4); fine particles collecting filter (5); fan (6)

The electric explosion of tantalum wire was studied using generalized variables of the theory of similarity of EEW [6]:

$$\varepsilon = \frac{CU_0^2 \cdot 10^{-18}}{d^4 Z}, \left[\frac{J}{m^4 Ohm}\right]; \ \lambda = \frac{l \cdot 10^{-6}}{d^2 Z}, \left[\frac{1}{Ohm \cdot m}\right]; \ \nu = \frac{10^3 \cdot \sqrt{LC}}{d}, \left[\frac{s}{m}\right] \quad (1)$$

where C is capacitance, U_0 is voltage of capacitor charging, Z is impedance of the circuit, d is diameter of exploding wire, L is inductance of the circuit.

To perform the study the following parameters of the circuit have been chosen: capacitance 0.78–2.2 μF, charging voltage 16–32 kV, circuit inductance 0.7–0.85 μH, diameter of Ta-wire 0.2 mm, wire length 40–120 mm. The energy input into the conductor was calculated using oscillograms of current run through the current shunt (R_S, Figure 1) according to the equation:

$$W_{EEW} = \frac{CU_0^2}{2} - \frac{C\left(U_0 - \frac{1}{C}\int_0^{t_{EEW}} i(t)dt\right)^2}{2} - \frac{L \cdot i(t_{EEW})^2}{2} - R\int_0^{t_{EEW}} i(t)^2 dt \quad (2)$$

where i is EEW current, R is circuit resistance, t_{EEW} is duration of EEW current impulse.

To simplify the analysis of the obtained data the energy input into the conductor was expressed in the form of dimensionless values:

$$e/e_S = \frac{W_{EEW}}{V_W \cdot e_S} \quad (3)$$

where e is specific energy, V_W is volume of exploding wire, e_S is sublimation energy of tantalum (71.7 J/mm^3).

Additionally, the coefficient of energy transfer η from the energy power supply to the wire was determined according to equation (4):

$$\eta = \frac{W_{EEW}}{W_0} \cdot 100\% \quad (4)$$

Analysis of the EEW-regimes was performed using the following procedure: the modes of critical explosion [7] was studied first, and then the dependencies $\lambda_b = f(\varepsilon)$ and $\eta_b = f(\varepsilon)$ were determined. Secondly the EEW-regimes with conductor length less then critical were checked.
After studying EEW of Ta-wire under different initial conditions the optimal mode of EEW was found. Under this mode a sample of tantalum powder was obtained.
The following parameters were determined for all the obtained powders:
- specific surface area using low-temperature nitrogen adsorption (BET, "Sorbtometr-M");
- morphology and average particle size by means of transmission (TEM) and scanning (SEM) electron microscopes (Philips CM-12, SEM – 515);
- impurities by means of atomic-emission spectrometer "MAAS" andanalyzers TSN 600, TC 844 LECO.

3. Results and discussion

Figure 2 shows typical oscillograms of current for electric explosion of Ta-wires.

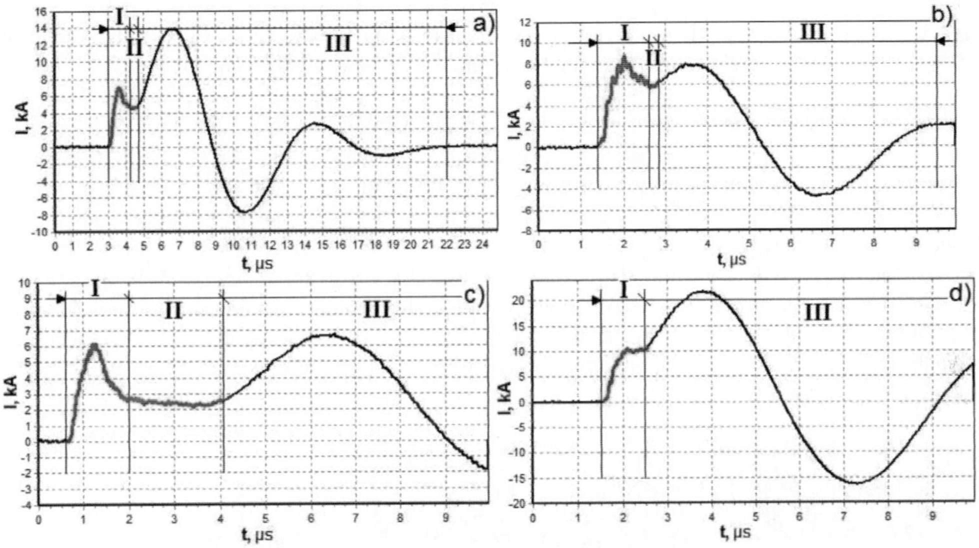

Figure 2. Oscillograms of current for electric explosion of Ta-wires:
(**a**) and (**b**) the critical mode; (**c**) wire length above critical, (**d**) wire length less then critical

The process of electric explosion of tantalum is similar to the process of electrical explosion of tungsten [8], and often is divided into three stages. First and second stages - EEW, the third – arc discharge. The arc discharge stage is present in all cases, but under some explosion regimes the moment occurs when EWW-current does not change (plateau current) (Fig. 1, a, b and c). The EEW-regimes with the lowest possible plateau current were chosen as a baseline – the critical modes of EEW.
Analysis of the waveform of the current allowed to determine the empirical relationships describing the critical modes of EEW:

$$\lambda_b = 4.5\,\varepsilon^{0.2};\ \eta_b = 27.9\,\varepsilon^{-0.6};\ e_b/e_s = 0.55\,\varepsilon^{0.2} \quad (5)$$

It should be noted that with increasing ε the energy, injected into the conductor in the critical regime of explosion, increased quite slowly. So for $\varepsilon = 0.15$ the input energy equals $0.4e_s$. The increase in ε up to 1.2 leads to an increase of energy up to $0.6e_s$. To increase the input energy when a wire exploded the regimes were studied when the wire length was less then critical under the other parameters were constant (C, U_0, d, L). This allowed us to determine the change in the energy properties of the explosion relative to the base – critical EEW.

Figure 3 shows the dependencies of the specific input energy and the coefficient of energy transfer on the wire length.

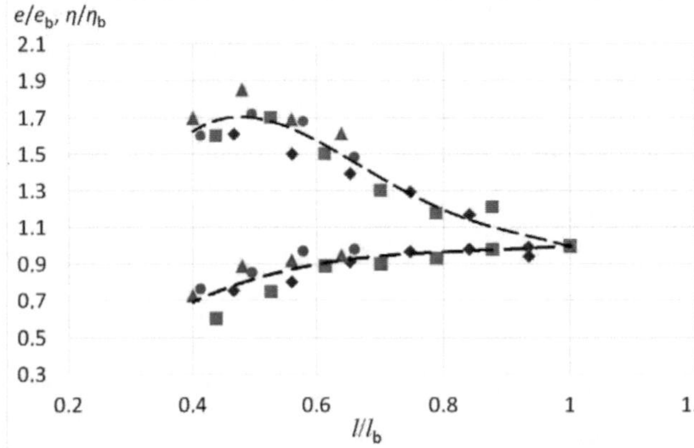

Figure 3. The dependence of the specific input energy and the coefficient of energy transfer from the power supply to the wire relatively the critical EEW

Decreasing the length of the exploding wire leads to decrease in η with simultaneous increase in the specific energy input into the wire. However, the value of e increases for all the studied wire lengths. When the length of wire was equal to $0.5l_b$ the maximum level of input energy was observed. Further decrease in the length of exploding wire leads to a decrease in specific energy which is due to the early development of the arc discharge stage.

To obtain tantalum powders the mode with the maximum level of input energy was chosen ($U_0 = 26$ kV, $C = 2.2$ μF, $l = 60$ mm). This EEW-mode provided the level of input energy $1.1e_s$. According to the data obtained using scanning and transmission microscopes the sample consists of two fractions. The first fraction (Figure 4 a - c) is of micron spherical particles with an average size of about 2 μm. The average size of the particles of the second fraction is about 200 nm (Figure 4 d,e).

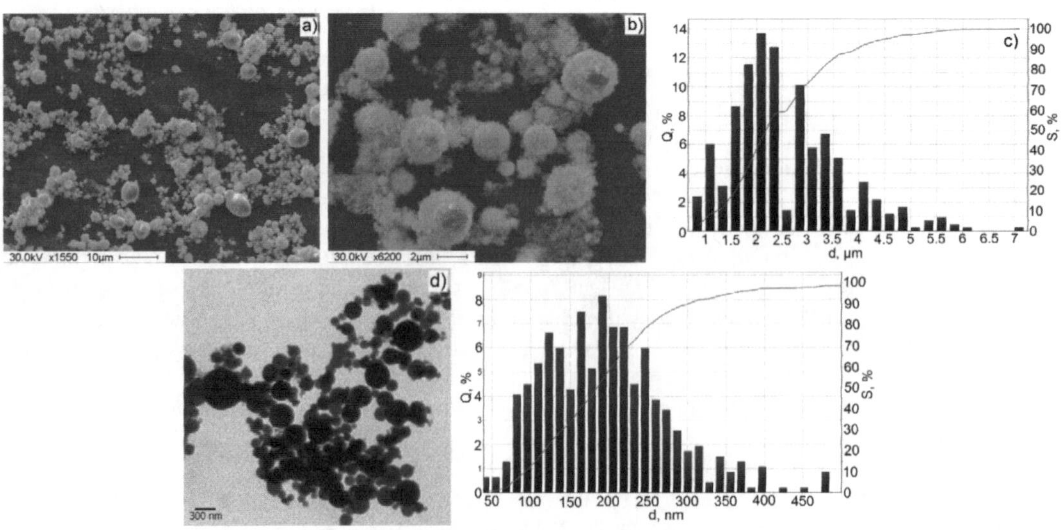

Figure 4. (a and b) SEM and **(d)** TEM photos of the obtained tantalum powder.

According to [2,3] formation of large micron particles caused by melting and splattering of a part of metal wire followed by crystallization of drops. Particles of nanometer size are formed by condensation of metal vapor. Specific surface area of the obtained powders was 2.3 m^2/g.
Spectral analysis of the samples showed the presence of metallic impurities not exceeding 0.15 %. The amount of adsorbed gas was 0.94 % - O_2, 0.11 % - N_2 and 0.46 % - H_2. The presence of metallic impurities is determined by the initial purity of the tantalum wire. The increase of the adsorbed gas is caused by high reactivity of the obtained powder.

4. Summary
Based on the obtained data there has been determined the possibility of Ta nanopowders production using electric explosion of Ta-wire. The obtained powders have a bimodal distribution of particle size, a small amount of extraneous metallic impurities and adsorbed gases.

Acknowledgments
This study was supported by the Russian Foundation for Basic Research (grant RFBR-mol-a 16-38-00051).

References
[1] Kolosov V.N., Miroshnichenko M.N., Orlov V.M., Prokhorova T.Y. 2011 Russian metallurgy (metally) 7 pp. 660-662
[2] Kotov Yu A 2003 Journal of Nanoparticle Research 5 pp. 539–550
[3] P Sen, Joyee Ghosh, Alqudami Abdullah, Prashant Kumar 2003 Proc. Indian Acad. Sci. (Chem. Sci.) 115 (5 & 6) pp. 499-508
[4] Geun Hee Lee, Joong Hak Park, Chang Kyu Rhee, and Whong Whoe Kim 2003 J. Ind. Eng. Chem. 9(1) pp 71-75
[5] F. Tepper, M. Lerner, D. Ginley 2004 Encyclopedia Entry. Dekker Encyclopedia of Nanoscience and Nanotechnology pp. 1921 – 1933.
[6] S. Kwon, Y.H. Jung, N. A.Yavorovsky, A.P. Illyn, J.S. Kim 2001 Scripta Mater 44 pp. 2247-2251.
[7] Yavorovskiy N.A. Pustovalov A.V. 2013 Russian Physics Journal 56 (7/2) pp. 164-169.
[8] A. Kloss, A. D. Rakhel and H. Hess 1998 International Journal of Thermophysics 19 (3) pp. 983-991

Investigation of thermal linear expansion for nanostructured $Si_{0.8}Ge_{0.2}P_{0.022}$ in wide temperature range

M S Rogachev[1], **L M Pavlova**[1], **Yu I Shtern** [1]

[1]National Research University of Electronic Technology, Moscow, Zelenograd, 124498, Russia

Abstract. The results of investigation of thermal linear expansion for high temperature thermoelectric material nanostructured $Si_{0.8}Ge_{0.2}P_{0.022}$ n-type with maximum thermoelectric figure of merit Z = 0.98 10^{-3} K are presented. Investigations were carried out by dilatometric method in the temperature range from 300 to 1220 K in dynamic heating and cooling regimes with using of infrared radiation source. Temperature dependence of thermal linear expansion coefficient (TLEC) was analyzed. The average value of TLEC for $Si_{0.8}Ge_{0.2}P_{0.022}$ was determined, which is equal to ~$5.9 \cdot 10^{-6}$ K^{-1}.

1. Introduction

Development of alternative sources of energy is one of the most challenging directions in the modern science and technology. Thermoelectric generators (TEG), which directly convert heat into electric energy, are one of these alternative sources. Different sources can be used for their work. Thermoelectric generators are used when extremely reliable power sources are required, which need no service and have long exploitation time. Thereby thermoelectricity can and should become one of the alternative technologies for generation of electric power.

TEG undergo multiple thermal cycling in the temperature range of 300-1200 K, when exist extremely high temperature gradient of 45 K/mm. The work of TEG is based on the semiconductor thermoelements. Due to the thermal expansion of materials during the TEG operation thermoelements must withstand high mechanical stresses. Mechanical strength of semiconductors used for the fabrication of thermoelements is much less than that of the other TEG construction units. So, mechanical strength of the device is limited by the strength of the thermoelectric materials.

Strength of any construction, which is exploited at varying temperature conditions, cannot be calculated without thermal expansion data for construction materials. This is especially important for thermoelectric generators.

This work is devoted to the experimental investigation of the thermal linear expansion for effective nanostructured $Si_{0.8}Ge_{0.2}P_{0.022}$ n-type, which have high value of Z in the range from $0.85 \cdot 10^{-3}$ K^{-1} (900 K) to $0.89 \cdot 10^{-3}$ K^{-1} (1200 K) with maximum of $0.98 \cdot 10^{-3}$ K^{-1} at 1050 K.

Experimental investigation of thermal expansion for SiGe solid solution is very insufficient and is limited by dilatometric [1, 2] and X-ray [3] data. Thermal expansion coefficient like other thermo- and electrophysical properties of SiGe is dependent not only on the composition, but on the structure, and so on the method of the thermoelectric material preparation [4].

One of the perspective fabrication methods of the effective nanostructured thermoelectric materials is hot pressing at sufficiently high temperatures. The main problems, which faces technologists producing solid nanostructured materials are connected with the need to maintain nanostructure as during the fabrication, so during the exploitation of thermoelements.

2. Experiment

In this work thermal linear expansion of $Si_{0.8}Ge_{0.2}P_{0.022}$ n-type fabricated by hot pressing of synthesized nanoparticles with given composition and nanoparticle size of 20 nm were investigated. Initial materials were monocrystalline Si p-type ingot with resistivity of 10 Ohm·cm, diameter of 63 mm and grown by zone melting in the direction of [100], and granulated Ge with 99.99 wt.% purity. Si and Ge were taken in the ratio of 4:1 with doping by 2.2 at.% of phosphorus. Preliminary initial materials were grinded in the knife mill XS-10, and the average particle size of 500 mcm was obtained. Then grinding was carried out in the planetary ball mill Activator 2S. The average particle size of 80-120 nm was obtained after the grinding during 30 min at the rate of 350 rev/min. After that mechano-chemical synthesis of $Si_{0.8}Ge_{0.2}P_{0.022}$ in the planetary high-energy mill Retsch PM400-MA was carried out. The average particle size of 20 nm was obtained after the grinding during 22 hours at the rate of 400 rev/min. The glove box with inert atmosphere (argon) was used for grinding to prevent oxidation. Hot pressing was carried out with using of hydraulic press IP-1000M at the pressure of 120 MPa, temperature of 1100 °C during 5 min.

Thermal linear expansion coefficient was studied by dilatometeric method with using of high-rate dilatometer DL-1500RH (ULVAC Sinku-Riko, Japan). Heating of sample was carried out in infrared oven RHL-E45P. Pt - (Pt10%Rh) thermocouples were used for the measuring of temperature.

Relative linear elongation of the sample of electrolytically pure copper was measure at different rates for the calibration of dilatometer. Results of our measurements of $\Delta L/L_0$ at different heating rates of 3, 5 and 50 K/min in comparison with the data for copper standard are presented in figure 1.

As can be seen from the figure 1, experimental results obtained at slow heating rate of 3 K/min are most close to the Cu standard in all temperature range.

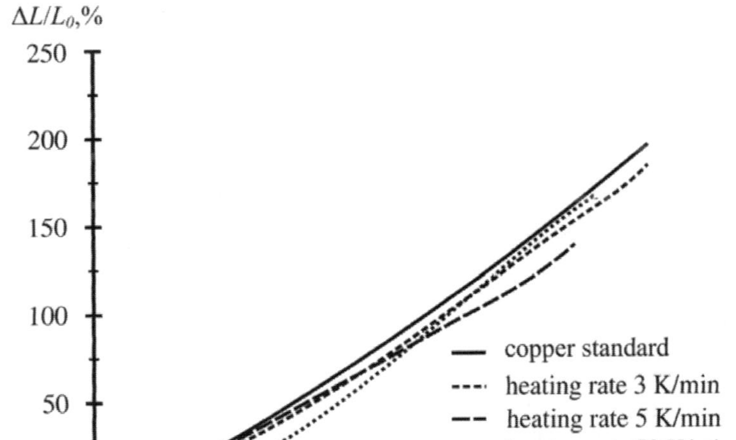

Figure 1. Comparison of the temperature dependences of relative linear expansion for electrolytically pure copper measured at heating rates of 3, 5 and 50 K/min with the data for copper standard.

3. Results

Dilatometric measurements of $Si_{0.8}Ge_{0.2}P_{0.022}$ n-type were carried out in the temperature range from 300 to 1220 K in the dynamic heating and cooling conditions. Samples with rectangular parallelepiped sizes of 4×4×20 mm were tested. Temperature dependences of the samples elongations were measured during the experiments. The accuracy of the samples elongation measurements was ±0.1 mcm.

Thermal cycling is a type of thermal treatment of the material. Due to the possible deformations of the sample after the first heating, dilatometric data for the following cycles can differs from that of the first measurement. The reasons for the deformation can be not only high plasticity [5] of the material, but also different structural and phase transformations observed after the infrared heating of the sample during a long period of time [6-8].

Results of the thermal expansion measurements of nanostructured $Si_{0.8}Ge_{0.2}P_{0.022}$ for two heating and cooling cycles at different rates (3 and 30 K/min) are presented in figure 2. Reliability of each cycle of measurement was controlled by the returning of the expansion curve to the initial point after cooling.

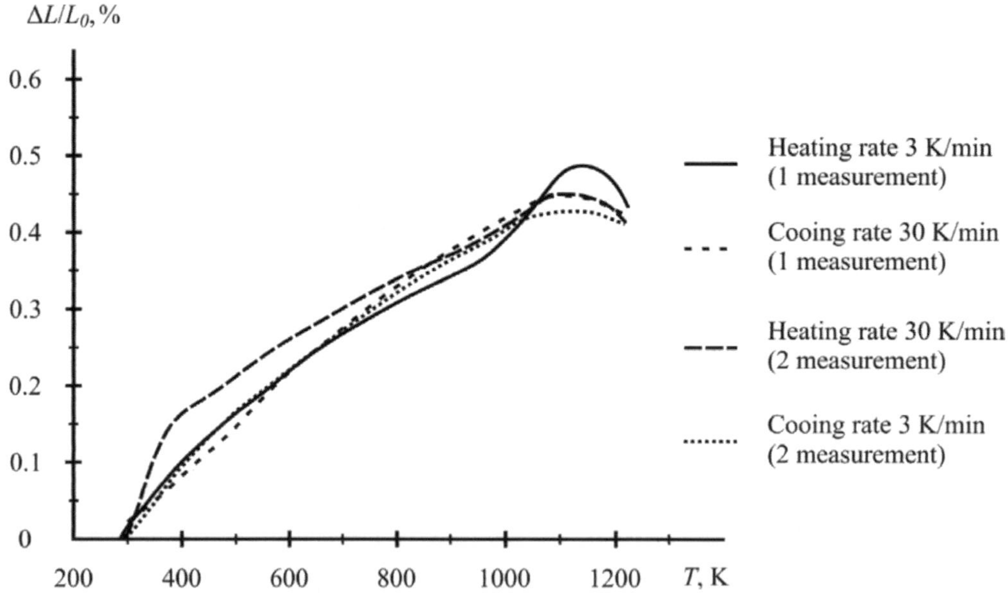

Figure 2. Temperature dependences of the relative linear elongation of $Si_{0.8}Ge_{0.2}P_{0.022}$ sample.

The data of first and second heating with the rates of 3 and 30 K/min are poorly reproduced, while cooling curves at different rates differs much less. This indirectly indicates on the structural transformations in the investigated alloy, which are possibly connected with the transformations in pure silicon [9] and agglomeration of nanoparticles during infrared heating.

It is seen from the figure 2 that all curves has maximums at the temperature of about 1050 K, especially pronounced for first heating. This is also indirect evidence of the structural transformations in the investigated material.

The data of relative linear elongation for $Si_{0.8}Ge_{0.2}P_{0.022}$ *n*-type obtained at different heating rates (3 and 30 K/min) on the dilatometer DL-1500 RH are compared with the data for nanostructured and mechanically alloyed SiGe *n*-type [2] obtained at the horizontal type dilatometer DIL-402C (NETZSCH) in figure 3. It should be noted that convective heating of samples is used in the dilatometer DIL-402C. Temperature dependence of relative linear elongation for pure silicon obtained with using of dilatometer AD-80 [9] is also shown in this figure.

Processing of the experimental data for $\Delta L/L_0$ was carried out with using of interpolation polynomial:

$$\frac{\Delta L}{L_0}(\%) = 0.86639 - 8.9197\cdot10^{-3}T + 3.2952\cdot10^{-5}T^2 - 5.26027\cdot10^{-8}T^3 + \\ +3.9062\cdot10^{-11}T^4 - 1.0952\cdot10^{-14}T^5 . \quad (1)$$

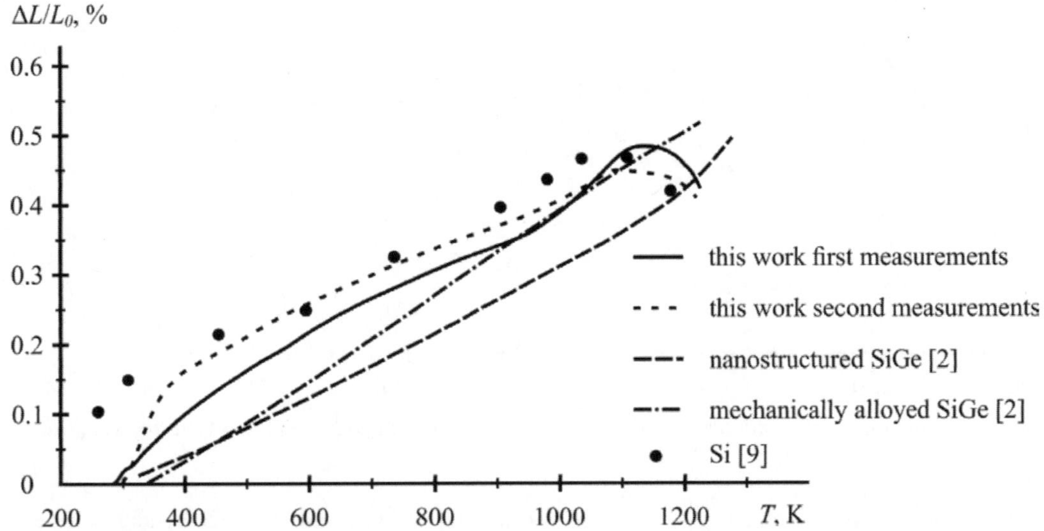

Figure 3. Temperature dependences of relative linear elongation for SiGe fabricated by different method, and Si.

It should be noted that data obtained for $Si_{0.8}Ge_{0.2}P_{0.022}$ at heating rate of 30 K/min are close to those for the silicon. At high temperatures for silicon can be also seen maximum for $\Delta L/L_0$ [9].

The values of relative linear elongation for $Si_{0.8}Ge_{0.2}P_{0.022}$ at heating rate of 3 K/min were used for the calculation of average value of thermal linear expansion coefficient (TLEC). Calculations were carried out with using of equation:

$$\alpha = \Delta L / L_0 \cdot (T - T_0), \quad (2)$$

where α is average value of thermal linear expansion coefficient in the temperature range of T-T_0; T is the temperature of TLEC estimation; L_0 is the length of the sample at the initial temperature T_0. The initial temperature of the measurements was 297.15 K.

The error of the determination of thermal linear expansion coefficient does not exceed 10 %.

Temperature dependence of the average TLEC for $Si_{0.8}Ge_{0.2}P_{0.022}$ is presented in figure 4. Experimental temperature dependences of the average TLEC for $Si_{0.8}Ge_{0.2}$ n-type [1-3] are also shown for comparison. Curve 1 correspond to the data of this work obtained with using of the equation (1); curves 2-4 correspond to the dilatometric data for $Si_{0.8}Ge_{0.2}$ obtained by the zone melting, and nunostructuring and mechanical alloying, respectively [1, 2]; dots correspond to the X-ray data [3].

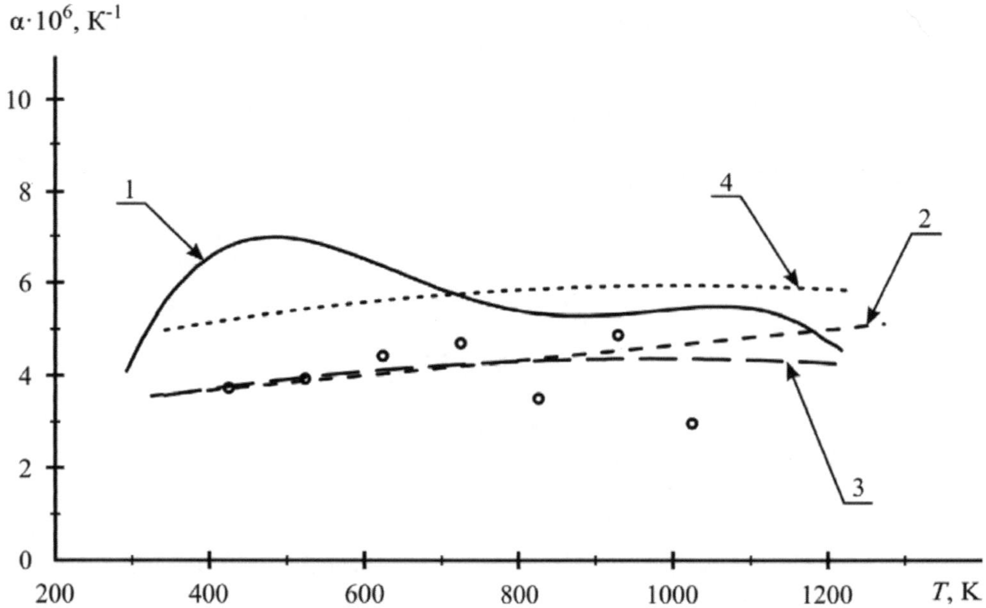

Figure 4. Temperature dependences of the average TLEC for $Si_{0.8}Ge_{0.2}P_{0.022}$ n-type in comparison with the data of works [1-3].

Obtained in this work average value of TLEC for investigated material in the temperature range of 300-1200 K was estimated as ~$5.9 \cdot 10^{-6}$ K^{-1} and correlates with that for the mechanically alloyed SiGe [2].

4. Conclusion

Analysis of the information showed that investigation of the thermal expansion of effective high temperature thermoelectric material $Si_{0.8}Ge_{0.2}$ is quite limited.

Investigation of thermal expansion of nanostructured $Si_{0.8}Ge_{0.2}P_{0.022}$ n-type was carried out by the dilatometric method in the temperature range of 300-1200 K in the dynamic heating and cooling conditions with using of infrared source of heating.

Non-monotonous character of TLEC temperature dependence for investigated material was established which can be due to the different processes in the material such as phase transformations.

The average value of thermal linear expansion coefficient for investigated material in the temperature range of 300-1200 K was estimated as ~$5.9 \cdot 10^{-6}$ K^{-1} and correlates with that for the mechanically alloyed SiGe.

Acknowledgments
This work was supported by the Ministry of Education and Science of Russian Federation project no. 14.578.21.0038 (RFMEFI57814X0038).

References

[1] RCA 1972 *Silicon Germanium Thermoelectric Materials and Module Development Program, 1st Quarterly Report* (Oak Ridge: OSTI)
[2] Ravi V, Firdosy S, Caillat T, Brandon E, Van Der Walde K, Maricic L, Sayir A 2009 *J. of Electronic Mat.* **38** 1433-42
[3] Dismukes J P, Ekstrom L, Paff R J 1964 *J. Phys. Chem.* **68** 3021-27
[4] Ni J E, Case E D, Schmidt R D, Wu C-I, Hogan T P, Trejo R M, Kirkham M J, Lara-Curzio E, Kanatzidis M G 2013 *J. of Mat. Science* **48** 6233-44

[5] Biront V S, Anikina V I, Kovalyova A A 2009 *J. of Siberian Federal University. Engineering & Technologies* **2** 384-93
[6] Fattakhov Y V, Galyautdinov M F, L'vova T N, Khaybullin I B 1997 *Zhurnal tekhnicheskoy fiziki* **67** 97-9
[7] Klimkov Yu M, Mayorov V S, Khoroshev M V 2014 *Vzaimodeystvie lazernogo izlucheniya s veshchestvom* (Moscow: MIIGAiK)
[8] Kudryashov S I, Emel'yanov V I 2002 *J.of Experimental and Theoretical Phys.* **94** 94-107
[9] Glazov V M, Kol'tsov V B, Kutsova V Z, Regel A R, Taran Yu N, Timoshina G G, Fal'kevich E S 1991 *Physics and technics of semiconductors* **25** 588-95

CF_3Br plasma cryo etching of low-k porous dielectric

I Clemente[1,2], N Koehler[3], A Miakonkikh[1,2], S Zimmermann[3,4], S E Schulz[3,4], K Rudenko[1,2]
[1]Institute of Physics and Technology, Moscow 117218, Russia
[2]Moscow Institute of Physics and Technology, Dolgoprudny 147001, Russia
[3]Technische Universität Chemnitz, Center for Microtechnologies, Chemnitz 09126, Germany
[4]Fraunhofer Institute for Electronic Nanosystems (ENAS), Chemnitz 09126, Germany

Abstract. Process of plasma etching of CVD low-k dielectric was studied. We used CF_3Br low pressure ICP plasma for etching at cryo temperatures (-20°C – - 100°C), pressures (5-20 mTorr) and RF bias with effective DC voltage 80-140 V. Refractive index of film and its thickness were measured by spectral ellipsometry. Ellipsometric porosimetry was employed to compare pore size distribution before and after etching of films. Measurements show increasing of etch rate increase with decreasing sample temperature.

1.Introduction

Etching of low-k porous dielectrics has critical applications in realization of advanced interconnect technologies of CMOS for VLSI ICs for technology node lower than 45 nm. Developing of low-k dielectrics allows improving RC delays in interconnections. Nowadays possibility of application low-k dielectrics with dielectric constant smaller than 2.5 for sub 10 nm technology nodes is under the research. With introduction of dielectrics with porosity up to 30% the etching issues become more challenging, because of plasma induced damage [1]. One of possible approach is to use cryogenic etching to reduce damage by condensation of plasma gases or etching products in pores which presume dielectric from damage. We propose to use CF_3Br plasma to perform etching. It is known that in ICP discharge the main way of ionization of this molecule is through detachment of Br^+ ion, which makes bromine main etching agent.

We are considering etching rates depending on temperatures, DC bias and gas pressure and also we study change in porosity of films by ellipsometric porosimetry (EP) technique [2] before and after etching.

2.Experimental

We used a porous low-k material with a initial thickness of 460 nm and a porosity of 20 %. The films consist of two pore size components with a average diameter of 1.1 nm for the smaller pores and 2.2 nm for the bigger pores. The porous films were deposited by a PECVD process, which uses a two injector head system for the matrix and for the porogen precursors. We used diethoxymethylsilane (DEMS) as matrix precursor and norbornadiene (BHD) as proven precursor.

The plasma etching was performed on 1x1 cm^2 samples on cryogenic ICP cluster plasma etching tool (PlasmaLab 100 Dual, OIPT). This tool allows cryo etching at temperatures from -150°C to room temperatures. Good thermal contact between chuck and sample are provided by helium backing at 7

Torr and mechanical clamps. The total flow of He to the chamber was no more than 8 sccm. The processed sample was transferred from one chamber to the other without vacuum break.
We perform etching procedure as follows. First sample was heated under vacuum ($1x10^{-6}$ mTorr) at 60°C for 2 minutes to remove adsorbed atmospheric water. Then sample was transferred to cryo chamber. Cryo process consists of following steps:
1) Temperature and pressure stabilization. The sample was placed on helium backing chuck and gas pressure was set to saturate pores with adsorbing gas; 2) etching for 30 seconds; 3) post etch pumping of chamber to base vacuum.
Etching were performed at three temperatures -20°C, -60°C and -100°C, at pressure from 5 to 20 mTorr. After etching the samples were heated to 60°C without vacuum break to remove etching byproducts which could interact with atmospheric water. After that, the refractive index and the film thickness were measured by ellipsometer (M2000X®, Woollam Co) after annealing for 20 minutes at 200°C.

3.Results and discussion

Etching rates measured by *ex situ* spectral ellipsometry presented in Figure 1. Refractive index of films did not changed taking into account measurement error 1.36±0.01.

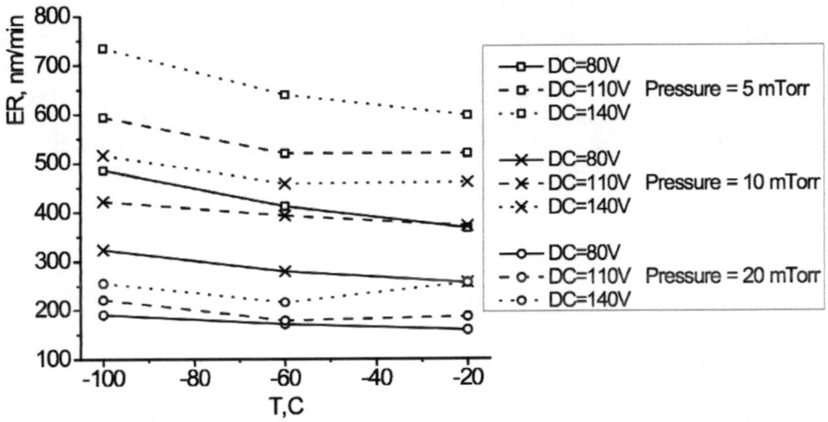

Figure 1. Etch rate of low-k CVD porous film in CF_3Br plasma depending on substrate temperature, pressure and effective bias voltage

The etch rate of film increases with DC bias voltage and with pressure, i. e. with energy and flux of etching particles. Dependence on temperature shows increase in etch rate with cooling down, with weak minima at -60°C which corresponds to behavior of spin-on low-k SiOCH dielectric etched in the same conditions [3].

Acknowledgments
The reported study was partially supported by RFBR, research project # 16-07-01233.

References

[1] M. R. Baklanov, J.-F. de Marneffe, D. Shamiryan et al J. Appl. Phys. **113**, (2013) 041101
[2] M.R. Baklanov, K.P. Mogilnikov, Microelectronic Engineering **64** (2002) 335–349
[3] A.Rezvanov, A. V. Miakonkikh, A. S. Vishnevskiy et al, Abstract book, Materials for advanced metallization, Grenoble, France, 20-23 March 2016 (in press).

Thermal and magnetotransport coefficients in doped $SmMnO_3$ manganites

M V Fetisova, O A Martynova, V E Gasumyants
Peter the Great St. Petersburg Polytechnic University, Polytechnicheskaya str., 29, St. Petersburg, 195251 Russia

Abstract. We present the results of a complex experimental study of the electron transport phenomena in doped manganites of the $SmMnO_3$ system. The temperature dependences of the resistivity, thermopower, and Nernst coefficient were measured in different applied magnetic fields. Specific features of these dependences in different temperature ranges were revealed and discussed. Besides, all the studied transport coefficients demonstrate a similar behavior near the temperature of the magneto-ordered transition, T_c. The effects of the colossal magnetoresistance, colossal thermopower, and colossal Nernst coefficient were observed to be qualitatively analogous; all of them show a sharp rise at $T \approx T_c$. It was also found that the value of the Nernst coefficient at temperatures below T_c depends strongly on the applied magnetic field thus demonstrating the anomalous behavior analogous to the one of the Hall coefficient.

1. Introduction

Materials demonstrating the colossal magnetoresistance (CMR) effect (the so-called manganites with a general formula $R_{1-x}A_xMnO_3$, where R is a rare-earth metal, A is an alkaline-earth metal) have attracted considerable interest from both practical and fundamental points of view. On the one hand, manganites are considered as perspective materials for applications in magnetic storage systems, magnetic sensors, optoelectronics, and spintronics (see, for example, [1] as a review). On the other hand, they are characterized by a very complicated magnetic phase diagram, a presence of different types of ordering, and a strong coupling between charge, spin, and orbital degrees of freedom associated with Mn ions. As a result, manganites are fascinating systems allowing one to study different intriguing material properties associated with the existence of strong correlation effects. This is why a huge number of papers devoted to both experimental and theoretical investigations of manganites have been published (reviews of the main results can be found in [2-6]). Nevertheless, despite a lot of models proposed and developed to describe unusual properties of manganites, the nature of the CMR effect remains unclear.

One of the widely used methods for studying manganite properties is the experimental investigation of the transport coefficients. First of all, some of them demonstrate an anomalous behavior near the temperature of the magneto-ordered transition, T_c, that is a direct consequence of the CMR effect. Second, their temperature dependences both above and below T_c are well known to be characterized by some specific features [2, 6–11] and the questions on their correct description and analysis, as well as on the conduction mechanisms realized in manganites in different temperature ranges call for further investigations. However, most of papers are concentrated, alongside with the magnetization study, on the resistivity, ρ, and magnetoresistance effect. The thermopower, S, which is sensitive to the type of the charge carriers contributing to the conduction process and the energy spectrum structure has also been widely investigated [6–11] including its modifications under applied magnetic field [7, 8, 12–15]. At the same time, only a few papers have been devoted to the experimental study of the Nernst effect. Note that the Nernst coefficient, Q, in both electron- and hole-doped manganites also demonstrates an anomalous behavior at $T \leq T_c$ [16–20]. According to the classical theory of the electron transport, this magnetotransport coefficient is sensitive not to the charge-carrier type, but to the scattering mechanism. This is obviously of interest to consider the Nernst data together with results on other transport coefficients obtained on the same samples of varied composition, since it is such a comparative analysis that could clarify the nature of the CMR effect.

For the above reasons, this paper aims to perform the complex experimental study of the transport coefficients (the resistivity, thermopower, and Nernst coefficient) in doped manganites of the $SmMnO_3$ system and to reveal the specific features of the modification of their temperature dependences near the temperature of the magneto-ordered transition when applying the magnetic field.

2. Samples and Experimental Details

Ceramic $Sm_{1-x-y}Ce_xSr_yMnO_3$ samples with different cerium and strontium contents ($x = 0.05$, 0.1, $y = 0.35$, 0.4) were used for investigations. All the samples were prepared by the standard solid-state processing technique from high-purity oxides. X-ray diffraction analysis has shown all the samples to be almost of single phase with amount of foreign impurities not exceeding 1–2 %. The sample homogeneity was controlled by measuring the local values of the thermopower in various points on the sample surface at room temperature.

The $\rho(T)$ dependences were measured by the standard four-probe low-frequency ac ($f = 20$ Hz) method. The thermopower was measured by a differential method relative to copper electrodes at the temperature difference between the two ends of a sample of about 2 K and then calculated by correcting for the absolute thermopower of copper. For studying the magnetic field influence on these coefficients, as well as for the Nernst coefficient measurements a constant magnetic field of varied magnitude (up to $H = 1.8$ T) was applied to a sample. To increase the Nernst signal and to exclude a contribution of even magnetic effects to the measured Nernst voltage we used thin samples (of about 1 mm in the ∇T direction) and a reversible magnetic field. The temperature difference between the two ends of a sample was about 10 K. Details of the Nernst coefficient measurements can be found elsewhere [21]. All the transport coefficients were measured in the temperature range of $T = 77$–310 K. The absolute error in determining the Nernst coefficient values when measuring the $Q(T)$ dependences did not exceed 10 %, the minimal securely fixed Q value at $T = 300$ K was about 0.1 nV/(K·T).

3. Experimental Results and Discussion

Figure 1 shows the temperature dependences of the resistivity for the $Sm_{0.55}Sr_{0.45}MnO_3$ and $Sm_{0.45}Ce_{0.1}Sr_{0.45}MnO_3$ samples at various magnetic fields H, for other samples they are qualitatively analogous. The resistivity increases strongly with decreasing temperature, has a sharp peak near the temperature of the magneto-ordered transition, and then falls down. Two points should be noted. First, a weak cerium doping leads to a strong decrease in the T_c value (from 143 K for cerium-free sample down to ≈90 K and ≈75 K for samples with $x = 0.05$ and 0.1 correspondingly, independently of the cerium position in the lattice). Second, the $\rho(T)$ curves in the temperature range of $T = T_c$–300 K fit well by simple exponential functions and despite the absolute values of the resistivity differ essentially for samples of various compositions, the resistivity activation energy remains almost unchanged being about 50 meV. This indicates the characteristics of the conduction process above T_c not to be influenced by the cerium doping. As magnetic field increases, the peak on the $\rho(T)$ curves shifts to higher temperatures and the maximal resistivity value decreases that is typical for manganites of different systems [1–3, 5].

The $S(T)$ dependences measured at zero magnetic field and at $H = 1.5$ T for samples of the $Sm_{0.55}Sr_{0.45}MnO_3$ and $Sm_{0.5}Ce_{0.05}Sr_{0.45}MnO_3$ compositions are presented in figure 2. The thermopower is negative over the whole measured temperature range. For all the studied samples cerium doping leads to a slight increase in the absolute value of the thermopower at $T = 300$ K. At temperatures far from the magneto-ordered transition, the $S(T)$ dependences are rather weak. Below T_c the thermopower has a very small value and demonstrates a metallic-like behavior. Above T_c, i.e., in the paramagnetic phase, the thermopower has a value of several tens of μV/K and changes almost linearly with temperature. However. in the cerium-free sample the absolute S value decreases with increasing temperature (see figure 2 (a)), while in all cerium-containing samples it increases with a varying slope. The latter indicates, contrary to the case of the resistivity, that the charge-carrier parameters are influenced by the cerium doping. It should be also noted that, as a whole, the thermopower temperature dependences in Sm-based manganites are not as complicated as in the case of La-based ones [8, 11, 13, 14]. At temperatures near T_c, the absolute value of the thermopower falls dramatically and the temperature of this drop increases with increasing magnetic field. This obviously points to a strong modification of the charge-carrier system properties occurring at the transition to the magneto-ordered state. Thus the thermopower behavior near T_c corresponds qualitatively to that of the resistivity.

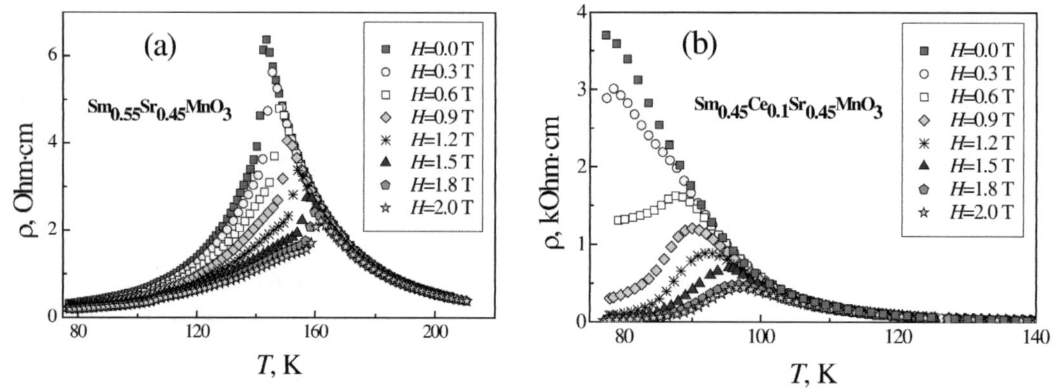

Figure 1. Temperature dependences of the resistivity at various magnetic fields for $Sm_{0.55}Sr_{0.45}MnO_3$ (a) and $Sm_{0.45}Ce_{0.1}Sr_{0.45}MnO_3$ (b) samples

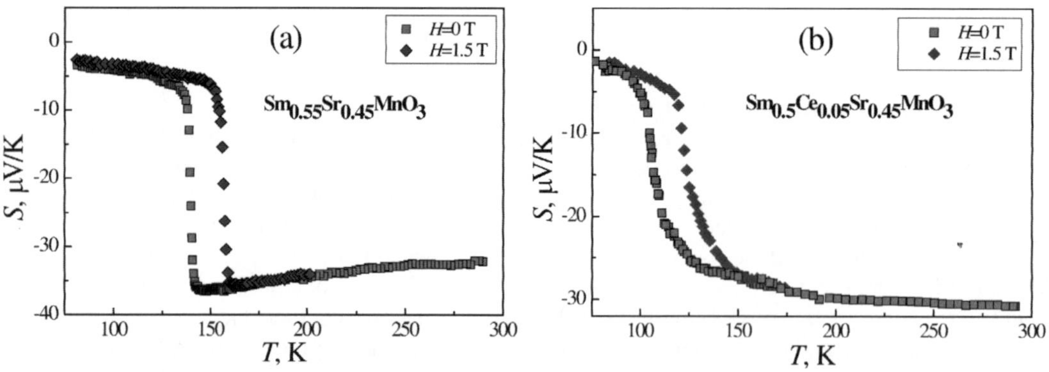

Figure 2. Temperature dependences of the thermopower at $H = 0$, 1.5 T for $Sm_{0.55}Sr_{0.45}MnO_3$ (a) and $Sm_{0.5}Ce_{0.05}Sr_{0.45}MnO_3$ (b) samples

The Nernst coefficient for all the investigated samples demonstrates a similar behavior. For example, the experimental data obtained on the $Sm_{0.55}Sr_{0.45}MnO_3$ and $Sm_{0.55}Ce_{0.05}S\ r_{0.4}MnO_3$ samples are presented in figure 3. The Nernst coefficient is positive over the whole temperature range studied (both above and below T_c) and its values at $T = 300$ K are extremely low (about 1–2 nV/(K·T) depending on the sample compositions). Besides, one can see that there are two temperature ranges on

Journal of Physics: Conference Series **741** (2016) 012205 doi:10.1088/1742-6596/741/1/012205

the $Q(T)$ dependences differing in both the temperature behavior of the Nernst coefficient and the character of the magnetic field influence on its value.

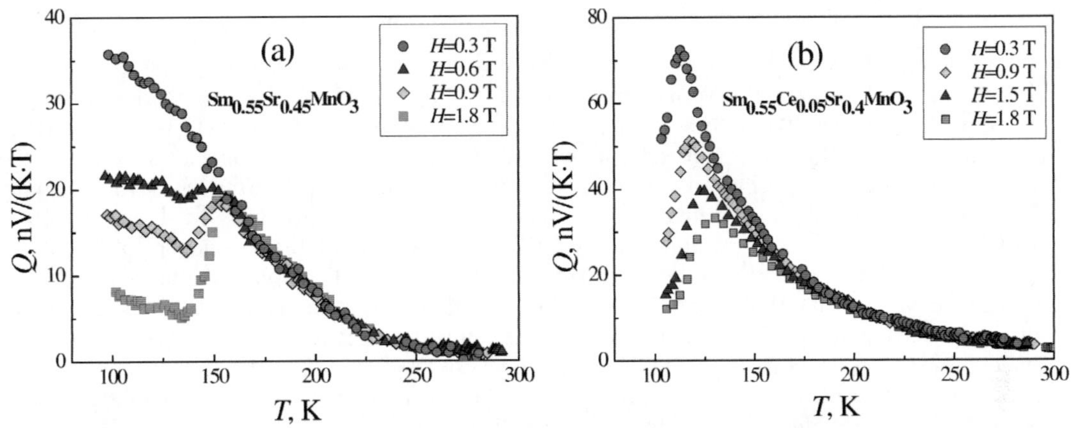

Figure 3. Temperature dependences of the Nernst coefficient at various magnetic fields for $Sm_{0.55}Sr_{0.45}MnO_3$ (a) and $Sm_{0.55}Ce_{0.05}Sr_{0.4}MnO_3$ (b) samples

Firstly, at $T > T_c$, the Q value increases very rapidly with decreasing temperature. Like the case of the resistivity, the $Q(T)$ curves can be fitted by an exponential law, but with a different activation energy. For the $Sm_{0.55}Sr_{0.45}MnO_3$ sample its value is 46 meV that corresponds well to the resistivity activation energy, while for cerium-containing samples it is equal to 15–20 meV, i.e., more than twice less in comparison with the resistivity case. At temperatures sufficiently lower T_c, the Nernst coefficient also increases with decreasing temperature, but rather weakly (see figure 3 (a), an analogous Q behavior at $T < T_c$ was observed in the $Sm_{0.5}Ce_{0.05}Sr_{0.45}MnO_3$ sample, for other samples the corresponding temperature range was not observed due to a limited measurement range). Note that for some samples a character of the $Q(T)$ dependences measured in low magnetic fields (up to 0.3 T) does not change at the transition to the magneto-ordered state, so that the Nernst coefficient at $T < T_c$ continues to increase rapidly (see figure 3 (a)). Understanding the physical reasons for the observed peculiarities of the temperature dependence of the Nernst coefficient both above and below T_c and choosing an adequate model allowing one to describe and to explain simultaneously all the specific features of the temperature dependences of the transport coefficients call for further investigations.

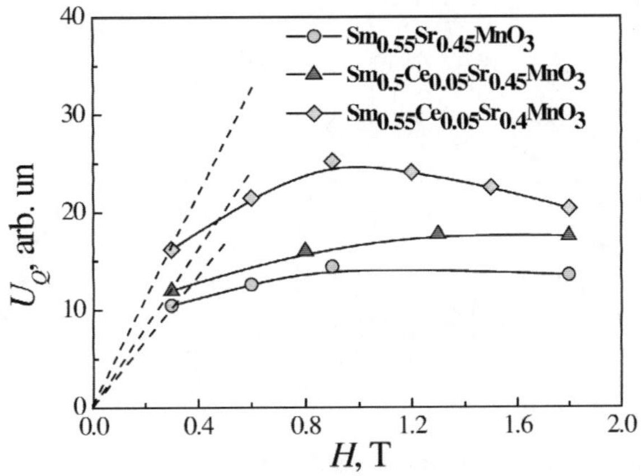

Figure 4. Dependence of the Nernst voltage measured at a temperature below T_c on applied magnetic field for selected samples. Dashed lines show expected classical dependences

Secondly, at $T > T_c$, the $Q(T)$ dependences measured in different magnetic fields coincide completely with each other that corresponds to the classical theory of the electron transport phenomena. Let us remind that the Nernst coefficient is determined as $Q = \frac{d \cdot U_Q}{H \cdot l \cdot \Delta T}$, where d is the sample thickness in the ΔT direction, U_Q is the measured Nernst voltage, H is an applied magnetic field, l is a distance between the Nernst leads, ΔT is the temperature difference applied to a sample. Thus, normally $Q = \mathrm{const}(H)$ or U_Q is proportional to H. However, in the magneto-ordered state (at $T < T_c$), the Nernst coefficient demonstrates an anomalous behavior. Its value depends strongly on the applied magnetic field decreasing with an increase in H (for instance, for the $Sm_{0.55}Sr_{0.45}MnO_3$ sample at $T = 100$ K the Q values measured in magnetic fields of $H = 0.3$ T and 1.8 T are equal to 35 nV/(K·T) and 8 nV/(K·T), correspondingly, see figure 3)). As a result, the dependence of the Nernst voltage on the applied magnetic field differs from the classical case (see figure 4). In this respect, it should be noted that the Hall coefficient in manganites is known to demonstrate an analogous anomalous behavior at $T < T_c$ depending also on the applied magnetic field [6, 8, 13]. Both these anomalous effects have obviously the same nature caused by a combined effect of the external (applied) and internal (induced by directionally oriented magnetic moments inside a sample) magnetic fields on the measured value of magnetotransport coefficients.

One can see from figure 4 that for all samples even in low magnetic fields the obtained $U_Q(H)$ dependences deviate from the expected classical ones smoothing out with increasing magnetic field, while in high fields the type of the $U_Q(H)$ curve is different depending on the sample composition that can be related to different magnetic structures of specific samples. In our opinion, the observed weakening of the $U_Q(H)$ dependence as compared to the classical linear one can be explained within the phase separation model repeatedly used to explain unusual manganite properties [1, 2, 5, 6]. We believe that the results obtained for the Nernst coefficient in the magneto-ordered state can be associated with the existence of an additional paramagnetic phase preventing the charge-carrier transfer.

At last, let us look at the Nernst coefficient behavior near the temperature of the magneto-ordered transition (see figure 3). For all the studied samples, if the applied magnetic field exceeds 0.3 T, the Nernst coefficient demonstrates a maximum at $T \approx T_c$ shifting slightly to high temperatures with increasing H (like the case of the $\rho(T)$ dependences) and falls with further decrease in temperature, so that the higher the applied magnetic field, the stronger the Nernst coefficient decrease. Like the resistivity case, increasing magnetic field suppresses the value of the Nernst coefficient at temperatures $T < T_c$.

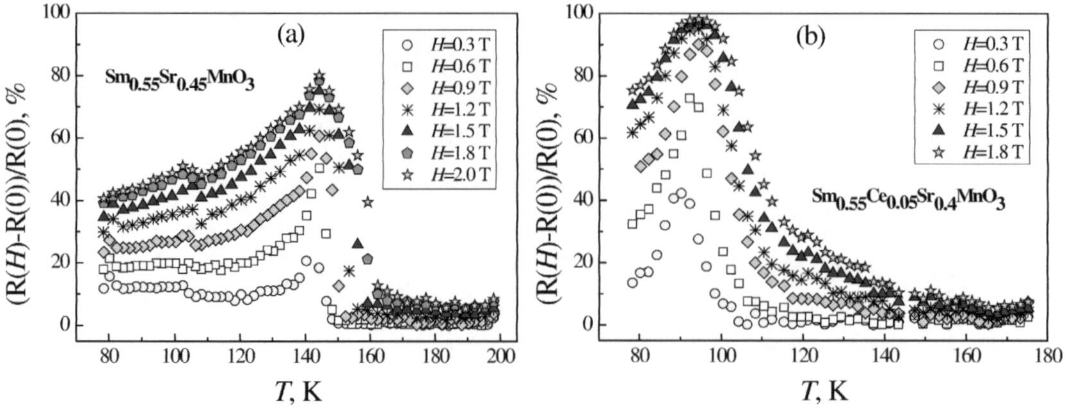

Figure 5. Magnetoresistance at various magnetic fields as a function of temperature for $Sm_{0.55}Sr_{0.45}MnO_3$ (a) and $Sm_{0.55}Ce_{0.05}Sr_{0.4}MnO_3$ (b) samples

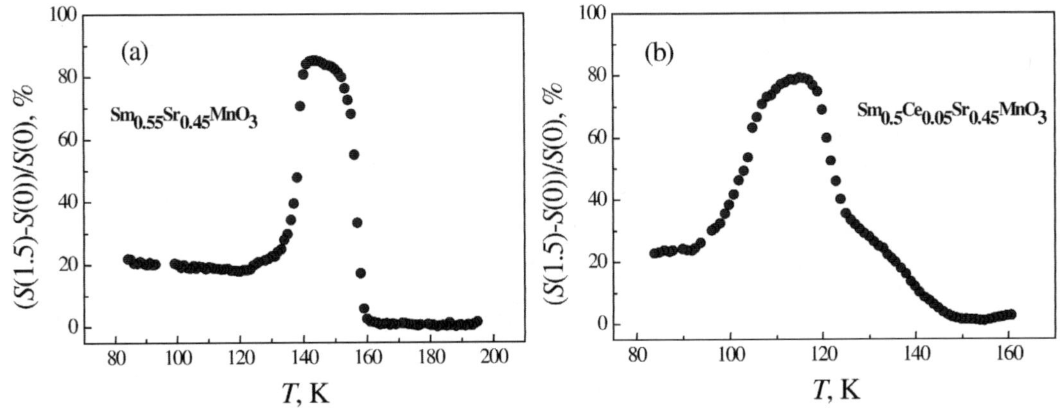

Figure 6. Magnetothermopower at H =1.5 T as a function of temperature for $Sm_{0.55}Sr_{0.45}MnO_3$ (a) and $Sm_{0.5}Ce_{0.05}Sr_{0.45}MnO_3$ (b) samples

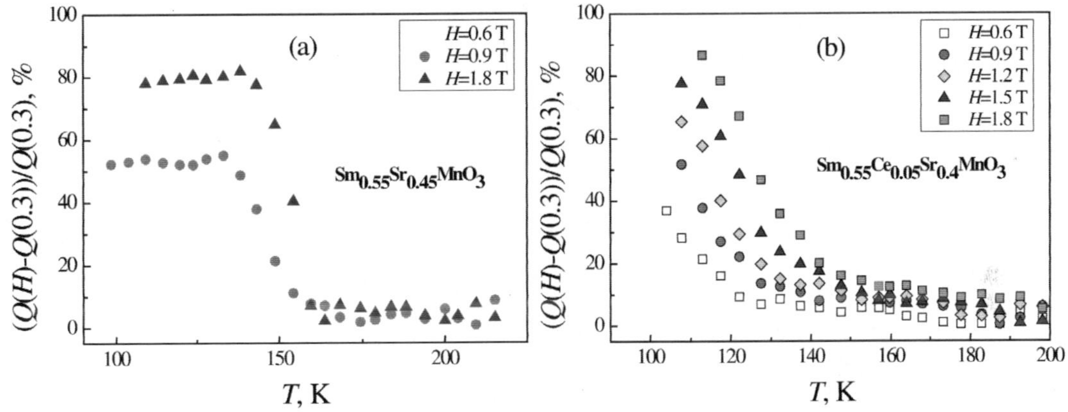

Figure 7. Fractional change of the Nernst coefficient at various magnetic fields as a function of temperature for $Sm_{0.55}Sr_{0.45}MnO_3$ (a) and $Sm_{0.55}Ce_{0.05}Sr_{0.4}MnO_3$ (b) samples

It can be thus stated that the temperature dependences of all the three studied transport coefficients are characterized by a similar anomalous behavior near the temperature of the magneto-ordered transition. This can be clearly demonstrated by a comparison of the temperature dependence of the CMR effect with the ones of the colossal magnetothermopower and the colossal Nernst effect defined as a fractional change of the Q value measured in a fixed H value relative to the one measured in a lowest magnetic field (H =0.3 T). The temperature dependences of these effects for some of the studied samples are presented in figures 5–7. As temperature decreases, all of them show a sharp rise at $T \approx T_c$, the higher the applied magnetic field, the larger the values of all these effects. Note that the CMR effect has a tendency to saturate in a high magnetic field whose value decreases with increasing cerium content, while in case of the colossal Nernst effect such saturation is not observed. Besides, contrary to the case of the magnetoresistance and magnetothermopower effects, the magnitude of the colossal Nernst effect does not demonstrate a sharp peak near $T \approx T_c$ remaining almost unchanged in the whole temperature range of $T \leq T_c$ (see figure 7 (a)).

In our opinion, the experimental results on the temperature dependences of the resistivity, thermopower, and Nernst coefficient and a character of their modification under applied magnetic field obviously indicate that the transition to the magneto-ordered state in manganites is accompanied by a significant modification in parameters of the charge-carrier system including changes in both the concentration and mobility of the charge carriers.

4. Conclusions

In summary, this paper presents the results of the complex experimental study of the temperature dependences of the resistivity, thermopower, and Nernst coefficient in manganites of the $Sm_{1-x-y}Ce_xSr_yMnO_3$ system. The common peculiarities of these dependences characteristic of different temperature ranges (both above and below the temperature of the magneto-ordered transition) were revealed. Besides, the $\rho(T)$, $S(T)$, and $Q(T)$ dependences were observed to be characterized by a qualitatively similar behavior near T_c that is related to a significant modification of the charge-carrier system parameters occurring at the transition to the magneto-ordered state. Special attention is paid to the Nernst coefficient demonstrating an anomalous behavior at $T \leq T_c$. The value of Q in this temperature range was observed to be strongly dependent on the applied magnetic field, so that the $U_Q(H)$ dependence is smoothed with increasing H. This is more probably due to the existence of an additional paramagnetic phase at temperatures below T_c and can be considered as an evidence in favor of the phase separation model.

References

[1] Haghiri-Gosnet A-M, Renard J-P 2003 *J. Phys. D: Appl. Phys.* **36** R127
[2] Nagaev E L 2001 *Phys. Rep.* **346** 387
[3] Izyumov Yu A, Skryabin Yu N 2001 *Physics – Uspekhi* **44** 109
[4] Dunaevskii S M 2004 *Phys. Solid State* **46** 193
[5] Tokura Y 2006 *Rep. Prog. Phys.* **69** 797
[6] Salamon M B, Jaime M 2001 *Rev. Mod. Phys.* **73** 583
[7] Bebenin N G, Zainullina R I, Mashkautsan V V, Burkhanov A M, Ustinov V V, Vasil'ev V G, Slobodin B V 1998 *J Exp. Theor. Phys.* **86** 534
[8] Bebenin N G, Zainullina R I, Mashkautsan V V, Ustinov V V, Vasil'ev V G, Slobodin B V 2001 *Phys. Solid State* **43** 501
[9] Reddy G L, Lakshmi Y K, Rao S M, Reddy P V 2014 *J. Magn. & Magn. Mat.* **362** 20
[10] Salazar D, Arias D, Dura O J, de la Torre M A L 2014 *J. Alloys Comp.* **583** 141
[11] Fetisova M V, Martynova O A, Gasumyants V E 2016 *J. Phys.: Conf. Ser.* **690** 012007
[12] Crespi V H, Lu L, Jia Y X, Khazeni K, Zettl A, Cohen M L 1996 *Phys. Rev. B* **53** 14303
[13] Zainullina R I, Bebenin N G, Mashkautsan V V, Burkhanov A M, Sukhorukov Yu P, Ustinov V V, Vasil'ev V G, Slobodin B V 2000 *Phys. Solid State* **42** 292
[14] Vergara J, Ortega-Hertogs R J, Madurga V, Sapina F, El-Fadli Z, Martınez E, Beltran A, Rao K V 1999 *Phys. Rev. B* **60** 1127
[15] Aliev A M., Abdulvagidov Sh B, Batdalov A B, Kamilov I K, Gorbenko O Yu, Amelichev V A, Kaul' A R, Kurbakov A I, Trunov V A 2003 *Phys. Solid State* **45** 130
[16] Suryanarayanan R, Gasumyants V, Ageev . 1999 *Phys. Rev. B* **59** R9019
[17] Suryanarayanan R, Gasumyants V, Ageev N 1999 *Solid State Commun.* **110** 309
[18] Suryanarayanan R, Gasumyants V, Ageev N 2000 *J. Magn. & Magn. Mat.* **211** 226
[19] Suryanarayanan R, Gasumyants V 2002 *J. Phys. D: Appl. Phys.* **35** 2077
[20] Suryanarayanan R, Gasumyants V 2005 *J. Phys. Chem. Solids* **66** 143
[21] Gasumyants V E, Ageev N V, Elizarova M V 2005 *Phys. Solid State* **47** 202

New system for manipulation of nanoobjects based on composite Ti_2NiCu/Pt nanotweezers with shape memory effect

A M Zhikharev[1], A V Irzhak[2,3], M Y Beresin[1,3], P V Lega[1], V V Koledov[1], N N Kasyanov[1,3] and G S Martynov [1,3]

[1] Laboratory magnetic phenomena in microelectronics, Kotelnikov Institute of Radioengineering and Electronics RAS, Moscow 125009, Russia.

[2]X-Ray acoustooptics laboratory, Institute of Microelectronics Technology and High Purity Materials RAS, Moscow Region, Noginsk, Chernogolovka 142432, Russia

[3] Department of the Material Science of Semiconductors and Dielectrics, National University of Science and Technology "MISiS", Moscow 119049, Russia

Abstract. We report the new system for manipulation of nanoobjects based on composite Ti2NiCu/Pt nanotweezers with shape memory effect. The design consists of the bimetallic Ti_2NiCu/Pt shape memory nanotweezers placed on a tip of electrochemically etched tungsten needle. The semiconductor diode placed on the tip of the needle plays both role of resistive element of the heater and temperature sensor for feedback control loop closing. The device is compatible with existing positioning systems like OmniProbe®, Kleindiek®, etc. and may find numerous practical applications in various tasks of nanotechnology connected with 3D manipulation.

1. Introduction

Nowadays the manipulation of the individual nanoscale objects in the vacuum chamber of scanning electron microscope (SEM) using existing manipulators (probe-type like OmniProbe® or Kleindiek®) performs a difficult and exhausting task. Despite this, such manipulations are demanded in different fields of experimental physics, biology, medicine, and nanoelectronics. The application of tweezers-type devices can make this work easier. Sizes of modern actuators are fundamentally limited by the corresponding physical effects. For example, relative deformation of piezoelectric actuators are about 10^{-4}, thus the sizes of device are greater than displacement by several orders of magnitude [1]. Alternatively, materials with shape memory effect (SME) demonstrate the large deformation (up to 10%) and strong force at shape recovery. For example, Ti_2NiCu shape memory alloy (SMA) can generate actuation stress up to 500 MPa at temperatures of phase transition between 30 °C and 70 °C [2-5] and withstands more than 2000 heating-cooling cycles [6]. High durability, small size and, respectively, small energy consumption makes SMA nanotweezers one of the best candidates to use in vacuum chambers of scanning electron and ion microscopes. Magnetic shape memory materials are known, but their use is hampered by the need to use strong magnetic fields (~ 8-10 T), which may induce interference on SEM systems [7-9]. Previous design of SME nanotweezers, presented in [10, 11], has proved itself as a useful device for performing 3D manipulation. In particular, the nanotweezers were applied for submicron and nanosized sample preparation for studying electrical properties of quasi-one dimensional conductors with a charge density waves – TaS_3 whiskers. Figure 1 shows the process of capturing TaS_3 whisker of about 500 nm thick. As a result the samples were

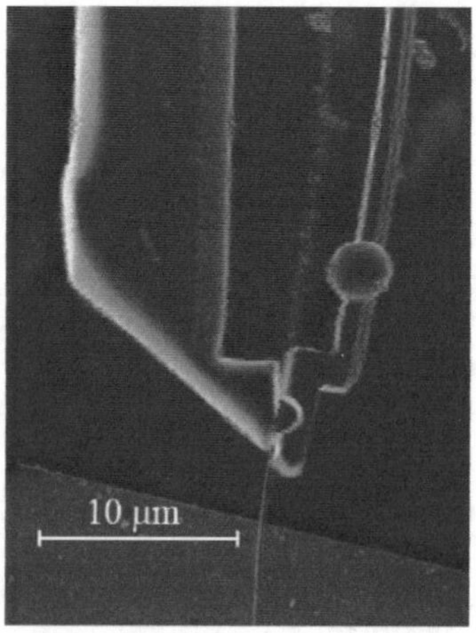

Figure 1. Grabbing of TaS_3 nanowhisker with SME nanotweezers.

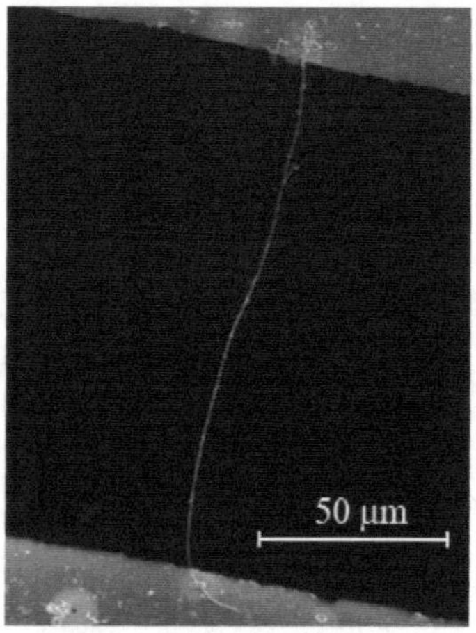

Figure 2. Sample with ohmic contacts for studying electrical properties of TaS_3 nanowhisker.

obtained with ohmic contacts, representing a thin bridge of TaS_3 whisker connected to two separate golden contacts by platinum layer, obtained by ion stimulated chemical vapour deposition (CVD) [12]. An image of one of the prepared sample is represented in Figure 2. A significant disadvantage of such design of the nanotweezers control system was the need for laser installation inside the vacuum chamber of SEM for nanotweezers control. In this work we present the robust SME nanotweezers system easy compatible with the existing manipulation systems, applicable in SEMs and FIB devices. The aim of the present paper is to describe the results of modelling, design and tests of the new nanomanipulation system.

2. Design and implementation of the control system.

2.1. Technical implementation

The new system of thermally controlled nanotweezers made of bimetal composite Ti_2NiCu/Pt with SME produced by FIB-milling and CVD. Nanotweezers were connected to tungsten needle with a heating element disposed. The tungsten needles are obtained from a wire by the method of cathode etching in 5% solution of KOH at a constant current. Manufactured tips of the needles have radii of curvature about 0.3÷1 μm. Typical size and form of the needles tip shown in the Figure 3.The SME nanotweezers attached to the end of the tungsten needle by the layers of CVD platinum as shown at Figure 4.The schematic representation of the installed in the SEM vacuum chamber system shown at Figure 5. The semiconductor diode mounted near the top of needle by thermally conductive glue. It plays both roles of resistive heating element and temperature sensor for feedback control loop closing. Brass cone used for mounting contact pads, is required for connecting of thin and fragile golden wires from semiconductor diode with the stout wires running to DAC-ADC module of control system, connected to a PC. To facilitate installation in vacuum chamber of SEM, tungsten needle were attached to a collet. The device is compatible with existing positioning systems like Zyvex®, Kleindiek®, etc. and may find applications in various tasks of nanotechnology connected with 3D manipulation of different kinds of micro- and nanoobjects. Automatically controlled heating system requires minimum modification to scanning electron microscope to be installed and can maintain the temperature of working body near martensitic-austenitic phase transition threshold, thus increasing the response speed of device. The new control system makes the nanotweezers much more robust and technological compared to existing ones [13].

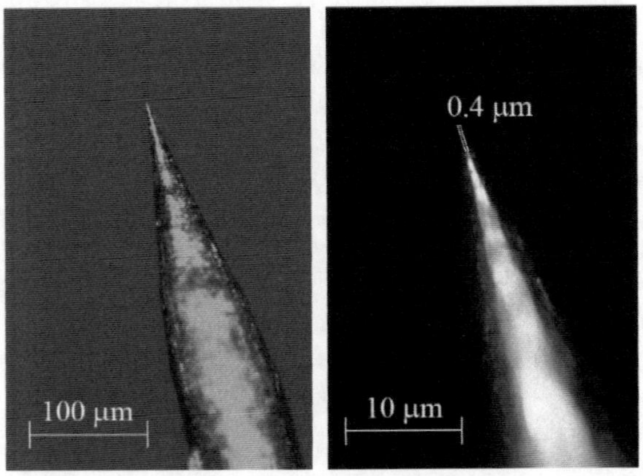

Figure 3. Needle tip obtained by electrochemical etching at different magnifications.

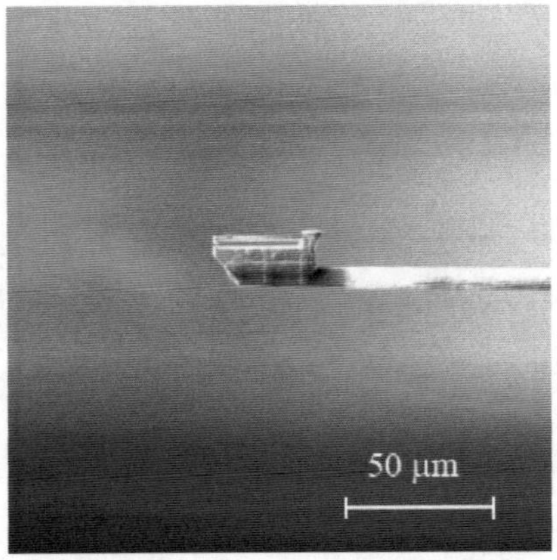

Figure 4. Tweezers on the tip of the tungsten needle.

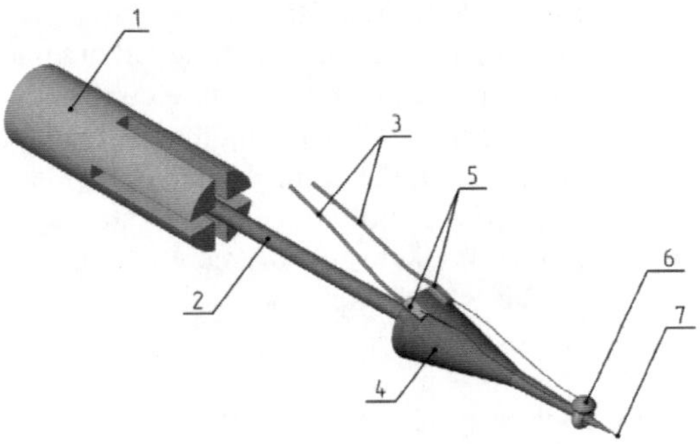

Figure 5. Schematic representation of nanotweezers handling and heating system. 1 – collet for installation in SEM vacuum chamber. 2 – tungsten needle. 3 – wires running to DAC-ADC module. 4 – brass cone used for contact plate mounting. 5 – isolated contact pads. 6 – semiconductor diode with 20 μm thin golden wires. 7 – SME nanotweezers on the tip of the tungsten needle.

2.2. Math modeling

As shown in [14], the nanotweezers on the tip of the tungsten needle can be treated as negligibly small. The heating/sensing diode is placed near the tip of the needle (Figure 5). One can write on the basis of energy balance the, heat conduction equation for temperature T, as function of coordinate x and time t, as follows:

$$\mathrm{C}\rho f(x)\frac{\partial T}{\partial t} = \frac{\partial}{\partial x}\left(\lambda\frac{\partial T}{\partial x}f(x)\right) + q(x) \tag{1}$$

$$f(x) = \pi\left(\frac{R-r}{L}x + R\right)^2 \tag{2}$$

where C – heat capacity, ρ – density, λ – heat conductivity of the tungsten needle. R and r correspond to cross-sectional radii of the tungsten wire and needle tip, respectively. L – distance between the diode centre and the tip of the needle, the coordinate of centre of the diode is set equal to zero. $f(x)$ refers to the tapered needle cross-section as a function of coordinate x, while $q(x)$ correspond to heating power density, produced by the diode. The boundary and initial conditions are defined, respectively, as:

$$\left.\frac{\partial T}{\partial x}\right|_{x=L} = 0\,, \; T|_{x=0} = T_0 \tag{3}$$

$$T\big|_{t=0} = T_0 \tag{4}$$

Using modeling software COMSOL Multiphysics® the analysis was done according to the described above mathematical model. Comparison of the results of math modeling and experimentally obtained data is shown in the Figure 6. As it can be seen the theoretical curve has good correlation with the experimental one. The successful modeling makes it possible to further carry out an analysis of an optimal structure of the installation setup, allowing to achieve a maximum operating speed of the SMA nanotweezers and minimal energy consumption.

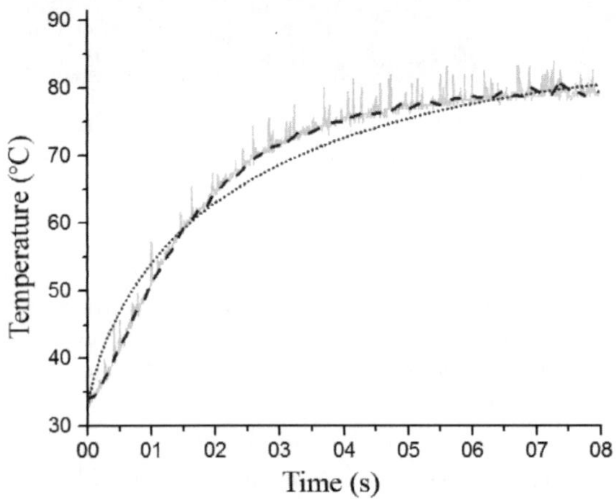

Figure 6. Comparison of the results of math modeling and experimental obtained data. Solid line on the picture corresponds to experimental data directly obtained from ADC module of control system. Dashed line – Fourier filtered experimental data. Dotted line – results of math modelling in COMSOL Multiphysics®.

3. Application of the SME nanotweezers.

To test the possibility of using nanotweezers with new control system in the SEM vacuum chamber and evaluation of the dislocation of its position, resulting from thermal expansion, a number of experiments on the manipulation of ZnO whiskers of different sizes have been carried out. For this purposes nanomanipulation system was installed instead of the Kleindiek® manipulator's probe and was connected to its electrical contacts. The experiments have demonstrated the ability to perform basic manipulation tasks such as capture, secure retention and movement of individual samples extracted from the forest of ZnO nanowhiskers. Key stages of the experiments manipulation of micro- and nanoscale structures are shown in Figures 7 and 8, respectively.

4. Summary

We report the new system for nanomanipulation based on layered Ti_2NiCu/Pt composite nanotweezers with SME, controlled automatically by resistive heating element. The system has demonstrated its compatibility with existing nanopositioning systems such as Kliendiek™ and allows to perform its installation avoiding SEM modifications. Proposed theoretical model of nanotweezers results are in good correlation with the experimentally obtained data. Actual manipulation of micro- and nanoscale objects is demonstrated.

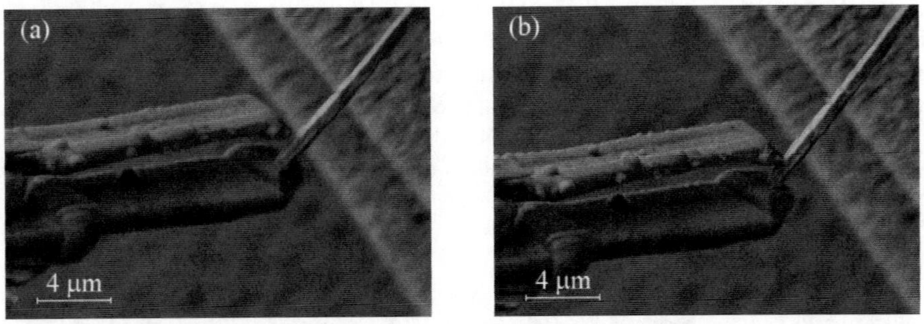

Figure 7. Manipulating with micro-sized ZnO whisker. **(a)** Nanotweezers in cold state. **(b)** Tweezers in hot state.

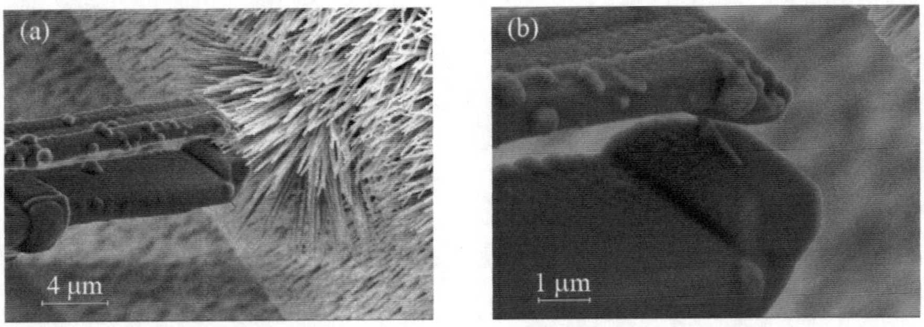

Figure 8. Manipulating with nano-sized ZnO whisker. **(a)** Nanotweezers in cold state. **(b)** Tweezers with the part of ZnO whisker in arms.

Acknowledgments
This work was supports by the Russian Science Foundation Grant No. 14-19-01644.

References

[1] Dikan V A, Mashirov A V, Zakharov D I, Mazaev P V, Zhikharev A M, Kalashnikov V S, Koledov V V, von Gratowsky S V, Sitnikov N N, Irzhak A V *et al.* 2016 *J. Commun. Technol. Electron.* **61(3)** 302-10

[2] Irzhak A V, Tabachkova N Y, Sagalova T B, Shelyakov A V and Koledov V V 2014 *Bull. Russ. Acad. Sci., Phys.* **78(12)** 1379-81

[3] Irzhak A V, Zakharov D I, Kalashnikov V S, Koledov V V, Kuchin D S, Lebedev G A, Lega P V, Perov E P, Pikhtin N A, Pushin V G *et al.* 2010 *J. Commun. Technol. Electron.* 55**(7)** 818-830

[4] Lee A P, Ciarlo D R, Krulevitch P A, Lehew S, Trevino J and Northrup A 1996 *Sens. Actuators, A* **54** 755

[5] Irzhak A V, Kalashnikov V S, Koledov V V, Kuchin D S, Lebedev G A, Lega P V, Pikhtin N A, Tarasov I S, Shavrov V G and Shelyakov A V 2010. *Tech. Phys. L.* **36(4)** 329-32

[6] Krulevitch P, Lee A P, Ramsey P B, Trevino J C, Hamilton J and Northrup A 1996 *J. Microelectromech. Syst.* **5(4)** 277

[7] Irzhak A, Koledov V, Zakharov D, Lebedev G, Mashirov A, Afonina V, Akatyeva K, Kalashnikov V, Sitnikov N, Tabachkova *et al.* 2014 *J. Alloy Compd.* **586 (Suppl. 1)** S464-8

[8] Kalimullina E, Kamantsev A, Koledov V, Shavrov V, Nizhankovskii V, Irzhak A, Albertini F, Fabbrici S, Ranzieri P and Ari-Gur P 2014 *P. Status Solidi* C **11(5-6)** 1023-25

[9] Akatyeva K, Afonina V, Albertini F, Von Gratowski S, Irzhak A, Fabbrici S, Khovaylo V, Koledov V, Krasnoperov E and Shavrov V 2012 *Sol. St. Phen.* **190** 295-8

[10] Zakharov D, Lebedev G, Koledov *et al.* 2010 *Phys. Procedia* **10** 58-64

[11] Zakharov D, Lebedev G, Irzhak A, Afonina V, Mashirov A, Kalashnikov V, Koledov V, Shelyakov A, Podgorny D, Tabachkova N and Shavrov V 2012 *VTT Symp.* **21(5)** 052001

[12] Zhikharev A, Zybtsev S, Pokrovskii V, Irzhak A *et al.* 2013 3M-NANO 2013. *Conf. Proc.* 2013 (Suzhou) (Piscataway, NJ : IEEE) 196

[13] Zhao H, Chang M, Liu X, Gabayno J L and Chen H T 2014 *J. Micromech. Microeng.* **24** 095012

[14] Lega P V, Koledov V V, Kuchin D S, Mazaev P V, Zhikharev A M, Mashirov A V, Kalashnikov V S, Zybtsev S A, Pokrovskii V Y, Shavrov V G *et al.* 2015 *J. Commun. Technol. Electron.* **60(10)** 1124-33

Precise mass detector based on "W needle – C nanowire" nanomechanical system

S Y Lukashenko[1,2,*], F E Komissarenko[1,3], I S Mukhin[3], V V Lysak[1], D A Averkiev[4], I D Sapozhnikov[2] and A O Golubok[1,2]

[1] Department of Nanotechnology and Material Science, ITMO University, Kronverskiy av. 49, 197101, St. Petersburg, Russia
[2] Institute for Analytical Instrumentation of RAS, Rizhsky pr 26, St. Petersburg, 190103, Russia
[3] Academic University, Russian Academy of Sciences, ul. Khlopina 8/3, 194021, St. Petersburg, Russia
[4] Department of computer photonics and digital video processing, ITMO University, Kronverskiy av. 49, 197101, St. Petersburg, Russia

Abstract. Nanomechanical system (NMS) based on amorphous carbon nanowhiskers localized on the top of tungsten tip were fabricated and investigated. The whiskers were grown in the scanning electron microscope (SEM) chamber using focused electron beam technique. The manipulation of SiO_2 and TiO_2 nanospheres was provided in SEM by means of dielectrophoretic force. Oscillation trajectories and amplitude-frequency characteristic of the oscillator were visualized at low pressure using a scanning electron microscope. The estimation of mass sensitivity of NMS was conducted.

1. Introduction

Nanomechanical oscillators (NMO) with small mass and volume are perspective sensitive elements for ultralow physical values sensors [1-3]. NMO's are particularly interesting as they show quantum properties at low temperatures and high resonance frequencies, though having macro- sizes in comparison with atoms [4].

NMO is a perspective mass sensor, which can be used for weightening of single cell or detection of biochemical reactions [5]. For using NMOs at atmospheric pressure, it is important to keep their Q-factor on high enough level. Works [6,7] were dedicated to studying NMO behaviour at atmospheric pressure.

State-of-the-art NMOs though having a number of advantages, have significantly lower Q-factor in comparison with macroscopic mechanical oscillators. It is important to find optimal construction of nanomechanical system (NMS).

This work presents the investigation of NMS "W needle – C nanowire". This system consists of single amorphous carbon nanowire, localized on a tungsten tip. This system can be also used for creation of specialized sensors based on local interaction probe sensors of scanning probe microscopes.

2. Experiment

We grow the amorphous carbon nanowire (CNW) on the top of sharp W tip with ~100 nm radius. CNW was made from precursor gas using focused electron beam induced deposition (FEBID) technique [8] inside the chamber of scanning electron microscope (SEM) Inspect S50 (FEI, USA). The oscillation were excited by applying of the sinusoidal signal from high-frequency generator to a piezoactuator 'Figure 1'. Measuring the amplitude-frequency characteristics (AFC) of as-grown structures was provided by using the SEM trajectory visualization method [6]. NMO with diameter (50-200) nm and length (1-3) μm 'Figure 2a' were created depend on growing parameters. W tips were fabricated from polycrystalline tungsten wire with base diameter of 150 μm using the electrochemical etching method in 5% alkali solution. Using W tip provides simple and convenient manipulation of the single CNW to excite and register the resonant frequencies.

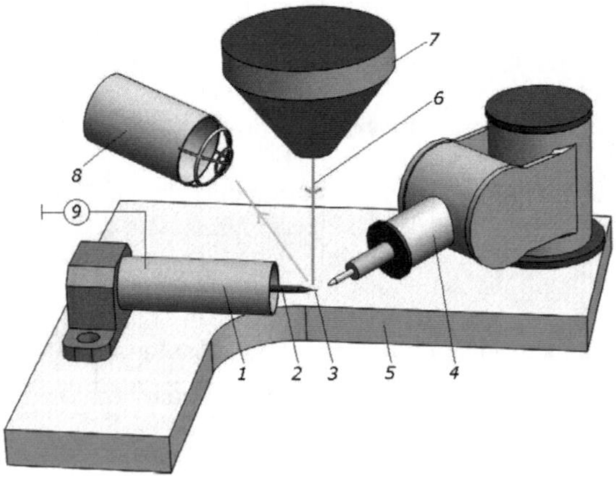

Figure 1. Experimental setup for the C nanowire growth and AFC measurement at low pressure. (1) piezotube, (2) W tip, (3) C nanowire, (4) micromanipulator, (5) base, (6) electron beam, (7) focused lens, (8) secondary electron detector, (9) high-frequency generator.

Ultrasonic wave propagates through W wire and excites CNW oscillations perpendicularly to electron beam axis. In this case, the sharp W needle concentrates the energy to the fixed end of the oscillator. SEM images of the CNW oscillations are obtained by means of secondary electron detector. The accelerating voltage of electron column was 20 kV. Mechanical fluctuations of the CNW blur the image on the SEM monitor. The amplitude of the blurred image has maximum on the resonance frequency of CNW.

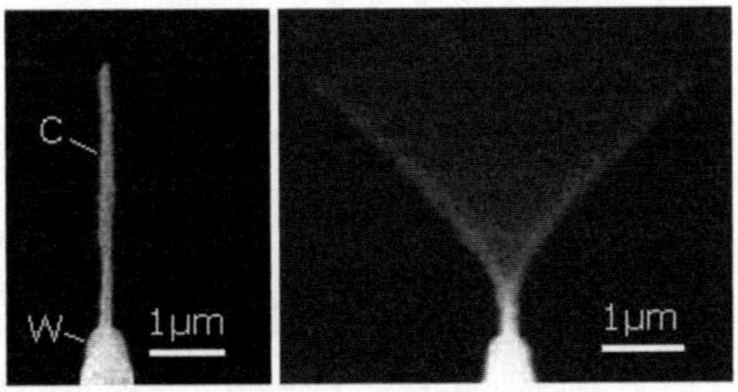

(a) (b)

Figure 2. SEM images of carbon nanowire (CNW) on tungsten tip (a); the resonance of this CNW at 3,68 MHz.

The resonant frequency of grown C nanowire was 3,68 MHz, on the 'Figure 2b' one can see the blurred image carbon nanowire on the tungsten tip at resonance.

SiO_2 and TiO_2 nanospheres (350 nm diameter and 100 nm, respectively) were manipulated with piezomanipulator (Kleindiek, Germany) in SEM chamber. First of all SiO_2 and TiO_2 nanospheres were landed on a silicon substrate. Then nanospheres were captured with the tip, installed on piezomanipulator with the help of dielectrophoretical force as described in [10]:

$$F_D = \int_V (\mathrm{D} - \varepsilon_0 E) \nabla \mathrm{E}\, \mathrm{dV},$$

where D – electrical induction, E – electric field, ε – the dialectric constant.

Influenced by the electron beam, the ungrounded metal tip was negatively charged. This metal tip produced the field, so dielectric nanosphere placed near this tip was polarized and attracted to the negatively charged tip. In particular, when the W needle approaches at a distance of ~ 300 nm to the nanosphere it jumps from substrate on the top of the needle. On the second stage, the nanospheres were moved to the apex of nanowire 'Figure 3a'. Attaching the nanospheres on the CNW was produced by a short (about 1 sec) local deposition of carbon at the point of contact between nanosphere and the C nanowire 'Figure 3b'. After that, the manipulator moved aside and nanosphere kept at the top of the NMO 'Figure 3c'.

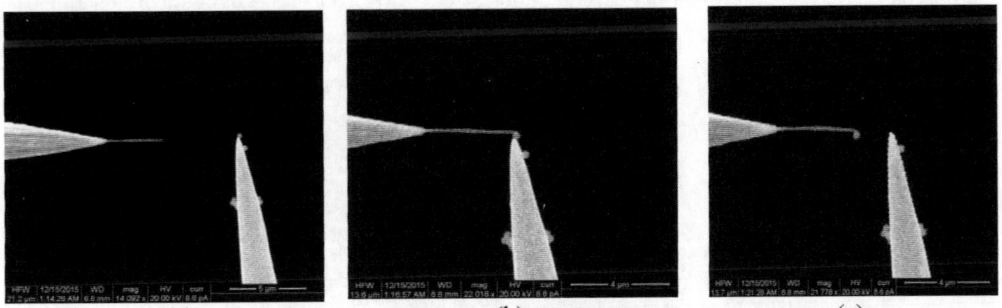

(a) (b) (c)

Figure 3. Electronic images of approaching the tip with a SiO_2 nanosphere to the carbon nanowire (a); the contact between SiO_2 nanosphere and nanowire, nanowelding (b); removal of the tip (c).

The resonant frequency of "W needle – C nanowire" nanomechanical system was measured again after fixing of the particle on the CNW 'Figure 4'.

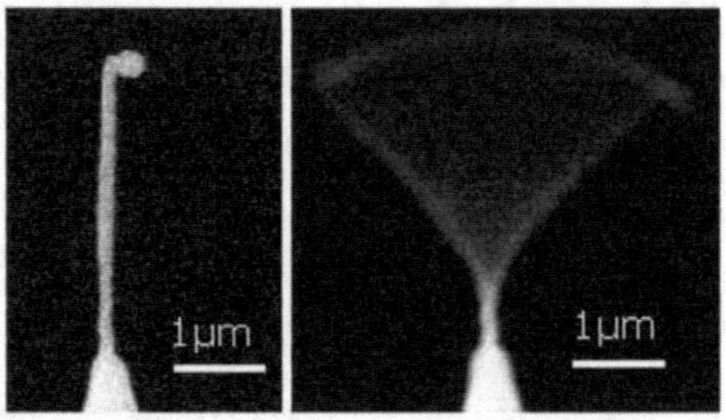

(a) (b)

Figure 4. Electronic images of "W needle – C nanowire – SiO_2 nanospere" system (a); the resonance of CNW wih attached mass at 2,58 MHz (b).

Estimation of Nanospheres Mass

In order to estimate the attached mass, we use a simple equation that describes the relation of the increased oscillator mass $\Delta M$and resonant frequency shift $\Delta\omega$[11]:
$\frac{\Delta\omega}{\omega} = \frac{\Delta M}{2M_{eff}}$, where M_{eff} - effective mass (for cantilevered beams $M_{eff} = \frac{33}{140}M_{NMO}$), ω- resonant frequency of initial nanowire. $M_{NMO} = \pi r_{NMO}^2 * l * \rho_C$, where $\rho_C = 2{,}2\, g/cm^3$ - density of amorphous carbon, r_{NMO}- radius of the NMO, l- length of the NMO. The second formula for estimation of the attached mass is general relation between density and volume $\Delta M = V\rho = \frac{4}{3}\pi r^3\rho$, where r – radius of nanospheres and the density is $\rho_{SiO_2} = 2{,}4\, g/cm^3$, $\rho_{TiO_2} = 3{,}9\, g/cm^3$ for SiO_2 and TiO_2 nanospheres accordingly.

All SiO_2 spheres were the same sizes with diameter of 350 nm, the mass was about $\Delta M = V * \rho = 31 * 10^{-15}(g)$using density and volume. The estimation via relation between frequency shift and attached mass is $M_{SiO_2} = 26fg$.

The mass of quartz nanospheres is too big for accurate estimation of mass sensitivity for our nanomechanical system. Therefore, we also manipulated with TiO_2 spheres with diameter of 100 – 120 nm. In this case, we measured resonant frequencies after fixing four TiO_2 nanospheres sequentially on the apex of CNW 'Figure 5'.

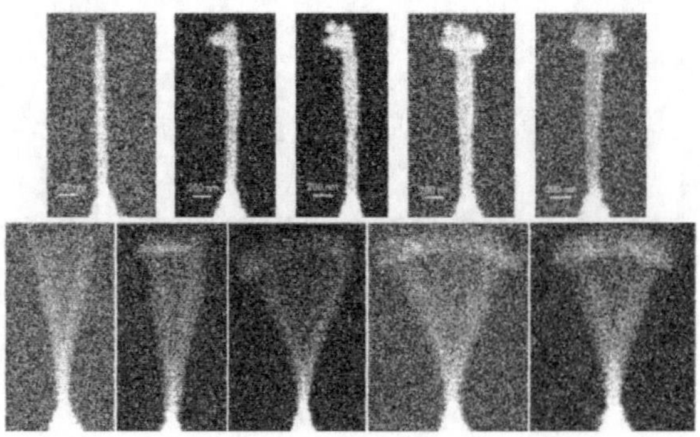

Figure 5. Series of electronic images of CNW resonances after sequentially addition of four TiO_2 nanoparticles.

The average mass of TiO_2 nanosheres $M_{TiO_2} = 3{,}5 \pm 0{,}5fg$. The approximation error of calculation is 5-15%. 'Figure 6' shows response of nanomechanical system on attaching four TiO_2 nanoparticles. The initial resonance frequency of the nanooscillator was 13,42 MHz.

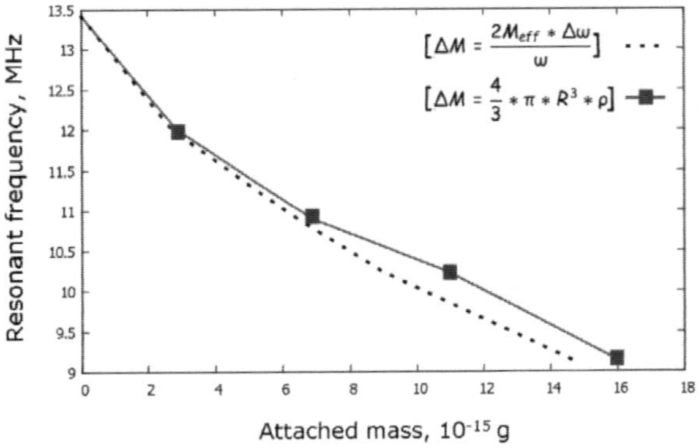

Figure 6. Graph of nanooscillator resonant frequency versus attached mass.

The results show good correlation between calculation of attached mass via classic linear approximation using the resonant frequency shift and calculation using volume and density of nanosheres. These results show a high sensitivity for the mass measurement. In recent work [9], we showed that at low pressure the Q-factor of our system is about 100. In particular, by setting for the smallest shift frequency ~ 50 kHz we get value about ~ 10^{-16} g that is the minimum detectable mass of concerned "W needle – C nanowire" nanomechanical system.

Acknowledgments
This work is supported by Government of the Russian Federation (074-U01), Russian Foundation for the Basic Research 16-32-60094 mol_a_dk, ITMO University grant 615893 and Program UMNIK.

References
[1] Wasisto H Merzschm S Waag A 2013 *Sensors and Actuators B* vol 180 p 77-89
[2] Jensen K Kwanpyo K 2008 *Nature Nanotechnology* vol 3 p 533-537
[3] Durand C Casset F 2008 *IEEE Electron device letters* vol 29 p 494-496
[4] O'Connell A Hofheinz M Ansmann M 2010 *Nature* vol 464 p 697
[5] Jensen K Kim K Zettl A 2008 *Nature Nanotechnology* vol 3 p 533
[6] Yum K Wang Z Suryavanshi A 2004 *Journal of applied physics* vol 96 p 3933
[7] Ekinci K L Karabacak D M Yakhot V 2008 *PRL* vol 101 p 264501
[8] Mukhin I Fadeev I Zhukov M Dubrovskii V Golubok A O 2015 *Ultramicroscopy* vol 148 p 151
[9] Lukashenko S Y Mukhin I S Veniaminov A V Sapozhnikov I D Lysak V V Golubok A O 2016 *Phys. Status Solidi A* p 1–5
[10] Denisyuk A Komissarenko F Mukhin I 2014 *Microelectronic Engineering* vol 121 p 15-18
[11] Bhiladvala R B Wang Z J 2004 *Phys. Rev. E* vol 69 p 036307

Influence of ion-plasma treatment on residual stress in the microcantilever

A S Babushkin, I V Uvarov and I I Amirov

Yaroslavl Branch of the Institute of Physics and Technology RAS,
21 Universitetskaya Street, Yaroslavl 150007, Russia

Abstract. The influence of ion-plasma treatment on residual stress in the microcantilever is investigated. The ability of treatment with energy below the sputtering threshold to affect the mechanical stress is shown. It is also demonstrated that a preliminary vacuum thermal annealing of samples reduces the influence of ion bombardment on the residual stress. With the increase of the annealing temperature the effect of ion bombardment disappears.

1. Introduction

Today the most of micromechanical thin-film structures are fabricated using standard technological processes of material deposition, such as vapor deposition or magnetron sputtering. These methods cause the residual mechanical stress in the film [1]. This stress can affect properties and reliability of a microelectromechanical device. Thus, the understanding of the mechanisms of relaxation and generation of stress is very important. The value and type of the residual stress depends on the deposited material and conditions of deposition. In the case of magnetron sputtering it depends on the ion energy, pressure inside the chamber, substrate temperature, etc.

In [2-4] the possibility of ion bombardment to influence the residual stress is shown. The energy of ions was in the order of several keV. Ion bombardment with such an energy leads to the sputtering of the material. It is a negative factor from the practical view. The aim of this work is to investigate the change of the mechanical stress in the microcantilever during the treatment with ions having the energy below the sputtering threshold.

2. Experiment details

The influence of ion-plasma treatment on the mechanical stress in the microcantilevers made of Cr and having the thickness of 500 nm, width from 2 to 10 µm and length from 10 to 100 µm was investigated. Cantilevers were fabricated as follows. At the first step the boron-doped silicon wafer with the diameter of 100 mm and the thickness of 460 µm was thermally oxidized (figure 1, step 1). SiO_2 layer had the thickness of 1 µm. Then the sacrificial polysilicon layer with the thickness of 1 µm (figure 1, step 2) and the chromium layer with the thickness of 0.5 µm (figure 1, step 3) were deposited by magnetron sputtering. The cantilevers were formed by contact photolithography technique and wet etching of Cr (figure 1, steps 4-7). At the last step the release of the cantilever was performed by plasma-chemical etching of the sacrificial layer in SF_6 plasma (figure 1, step 8). Fabricated cantilevers are shown in figure 2. Investigation of the cantilever details and geometrical measurements were performed by SEM Zeiss Supra 40.

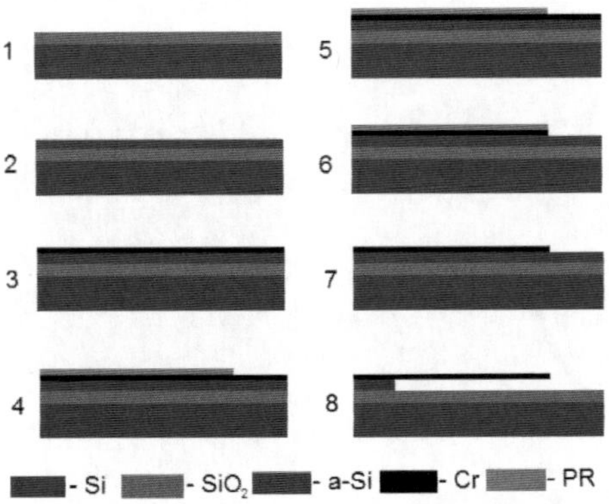

Figure 1. The main steps of the cantilever fabrication.

Figure 2. SEM image of the fabricated cantilevers.

Two batches of samples were fabricated. Cantilevers in both batches were bent up after the release. Due to the difference in the deposition conditions the cantilevers from the second batch curled more (curvature $\kappa = 8.5\times10^3$ m^{-1}) than the cantilevers from the first batch (curvature $\kappa = 5.3\times10^3$ m^{-1}).

Plasma treatment of the cantilever surface was performed using the radio-frequency high-density low-pressure inductively coupled plasma (RF ICP) [5]. Inductive power of the plasma reactor was 800 W, argon flow was 10 sccm, operating pressure was 0.08 Pa. RF bias power at the substrate was 30, 60 and 100 W, the average ion energy is determined by the self-bias potential. Maximum average energy of Ar+ ions was ~ 50 eV for the 100 W RF bias power. Duration of each treatment was 60 s. The ion current density during the experiment was 6.4 mA/cm^2. Stacked ion dose was calculated from the ion current density.

The field of mechanical stress in the microcantilever can be presented as a polynomial [6]:

$$\sigma_{\text{total}} = \sum_{k=0}^{\infty} \sigma_k \left(\frac{z}{h/2} \right)^k, \tag{1}$$

where z is the coordinate across the thickness h of the cantilever, it varies from $-h/2$ to $h/2$ with an origin chosen at the mid-plane of the film. In the first approximation, σ_{total} is the superposition of the constant mean stress σ_0 and the gradient stress σ_1, the former being symmetric and the latter anti-symmetric about the mid-plane. The effect of higher-order terms is neglected:

$$\sigma_{total} \approx \sigma_0 + \sigma_1 \left(\frac{z}{h/2} \right). \tag{2}$$

After the release the constant component of stress leads to a change of the cantilever length ΔL and gradient of the stress leads to a bending of the cantilever [7, 8]:

$$\sigma_0 = E\varepsilon = E\frac{\Delta L}{L}, \tag{3}$$

$$\sigma_1 = \frac{Eh}{R} = Eh\kappa, \tag{4}$$

where L is the length of the cantilever, R and κ are radius of curvature and the curvature of the cantilever respectively, E is the Young's modulus (300 GPa for Cr).

The gradient of mechanical stress σ_1 was determined after each treatment by the curvature of the cantilever. The curvature was calculated by the cantilever tip deflection from the straight position Δz (figure 3) using the following expressions:

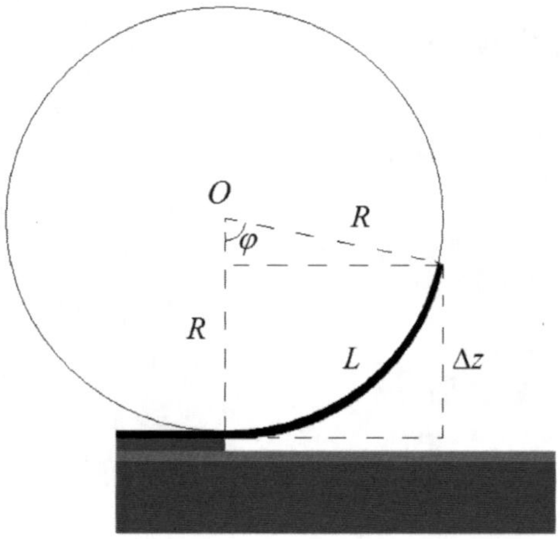

Figure 3. Geometry of the curled cantilever.

$$\Delta z = R - R\cos\varphi = 2R\sin^2\frac{\varphi}{2}, \tag{5}$$

$$\varphi = \frac{L}{R}, \tag{6}$$

$$\Delta z = \frac{2}{\kappa}\sin^2\frac{\kappa L}{2}. \tag{7}$$

3. Results and discussion

Two experiments were performed: in the first one the influence of the preliminary vacuum thermal annealing on the result of ion-plasma treatment was studied, and in the second one the effect of the heat sink on the result of ion-plasma treatment was investigated.

3.1. Effect of the preliminary vacuum thermal annealing on the result of ion-plasma treatment

Samples from the first batch were used for the experiment. Before the step of the cantilever release, the part of the samples was annealed in vacuum during 1 h at the temperature of 300, 450 and 500 °C, and another part of samples was not annealed. Plasma treatment was performed after the release. To avoid the heating of the cantilevers during the treatment the vacuum grease was used which provided better heat removal from the sample to the substrate holder.

The treatment with an average ion energy of 30 eV did not affect the bending of cantilevers (the total dose of ions was 26.3×10^{17} cm^{-2}). The treatment with an average energy of 50 eV affected the bending as follows. This treatment did not affect the curvature of cantilevers which were annealed at 450 °C and 500 °C. But it affected the curvature of cantilevers which were not annealed and were annealed at 300 °C. These cantilevers bent up even more. Results of the treatment of the cantilevers with lateral dimensions 100×10 μm^2 are shown in figure 4. Results obtained with the cantilevers of other dimensions are similar. A high temperature vacuum annealing makes cantilevers immune to the influence of ion-plasma treatment on the residual stress.

The increase of the cantilever curvature during the ion-plasma treatment means an increase of the stress gradient due to the generation of tensile stresses in the surface region of the cantilever (figure 5b).

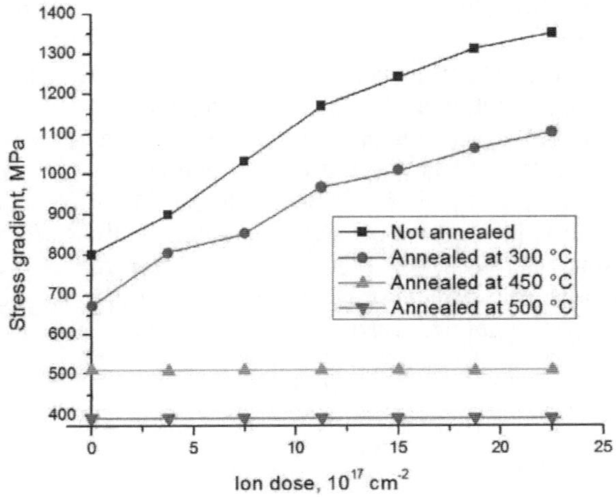

Figure 4. Dependence of the stress gradient in the cantilever on the dose of Ar ions with an average energy of 50 eV.

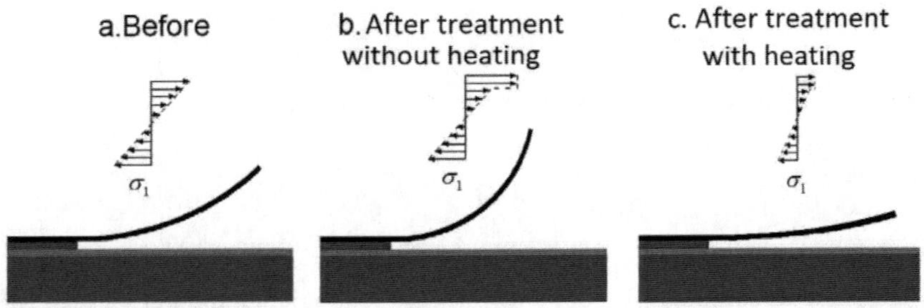

Figure 5. The stress distribution in the thickness of the cantilever before and after the ion-plasma treatment.

3.2. Effect of the heat sink and the order of the cantilever release on the result of the treatment
Samples from the second batch were used for the experiment. Treatments were performed in the following regimes: (a) the treatment of the unreleased cantilevers with the heat sink, the average ion energy of 15 eV («cold plasma annealing»); (b) the treatment of the released cantilevers with the heat sink, the average energy of 15 eV; (c) the treatment of the unreleased cantilevers without the heat sink, the average energy of 15 eV; (d) the treatment of the unreleased cantilevers without the heat sink, the average energy of 50 eV («hot plasma annealing»). After the treatments in the regimes (a), (c) and (d) the release of the cantilevers was performed to determine the stress gradient.

The results of the treatment of the cantilevers with lateral dimensions of 100×10 μm^2 are shown in figure 6. The treatment of the unreleased cantilevers with the heat sink was able to influence the stress even when the average ion energy was 15 eV (figure 6, line (a)). As in the first experiment (figure 4), the curvature of the cantilever increased, which meant the generation of the tensile stress in the surface region and the growth of the stress gradient.

The treatment of the released cantilevers with the ion energy of 15 eV with the heat sink does not affect the bending (figure 6, line (b)). This agrees with the results of the first experiment in which the energy of 30 eV was not enough to affect the stress in the released cantilevers. It means that for the released cantilevers the threshold energy exists. Below this energy the plasma treatment has no influence on the mechanical stress.

The plasma treatment of the unreleased cantilevers with no heat sink gave the opposite result in comparison with the case of the heat sink. The treatment under these conditions led to the straightening of the cantilevers (figure 6, lines (c) and (d)). The absence of the heat sink caused the heating of the cantilevers approximately to 250 °C. Probably, it had an effect similar to the thermal annealing and reduced the stress gradient (figure 5c).

The occurrence of the tensile stress during the ion bombardment was discussed in several papers [3, 4]. Chromium film bombarded with Ar ions having energy of 110 keV was studied in [3]. Authors attributed the occurrence of tensile stress to the decrease of interatomic distance with the increase of irradiation dose through the formation of defect clusters. Bombardment of Pt/Si cantilevers with ions of noble gases having energy of 0.5–4 keV was investigated in [4]. The occurrence of tensile stress was explained by the accumulation of vacancies due to the escape of interstitial atoms on the surface of the sample. In these studies the ion energy was of the order of a few keV. In our experiment the ions with energies of 15-50 eV were able to change the mechanical stress significantly. But we had the flow of ions three orders of magnitude higher than in mentioned papers, so the energy doses were the same.

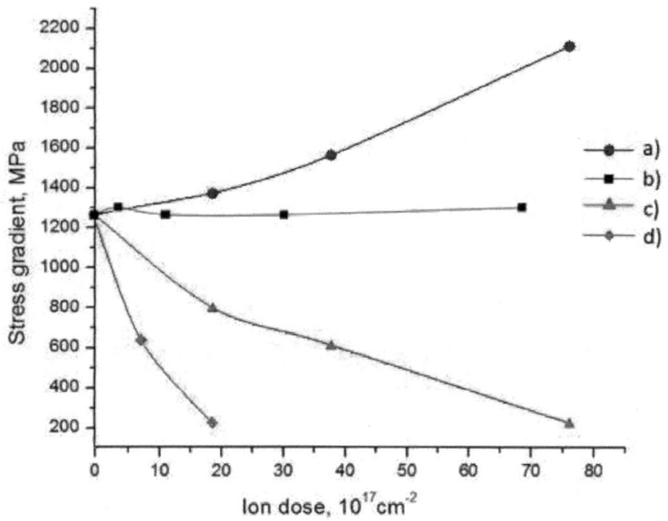

Figure 6. Dependence of the stress gradient in the cantilever on the dose of Ar ions obtained at the different regimes of the treatment.

4. Conclusions

Plasma treatment with the ion energy below the sputtering threshold is able to influence the mechanical stresses in the microcantilever. It can be used to control the shape of the cantilever. Choosing the treatment conditions it is possible to increase or decrease the bending of the cantilever.

Acknowledgments

The reported study was supported by RFBR, research project No. 16-37-60065 mol_a_dk.

References

[1] Knystautas E 2005 *Engineering Thin Films and Nanostructures with Ion Beams* (New York: University of Rochester)
[2] Lu P, Shen F, O Shea S J, Lee K H and Ng T Y 2001 *Mater. Phys. Mech.* **4** 51-5
[3] Fayeulle S, Misra A, Kung H and Nastasi M 1999 *Nucl. Instr. Meth. Phys. Res.* B **148** 227-31
[4] Chan W-L, Zhao K, Vo N, Ashkenazy Y, Cahill D G and Averback R S 2008 *Phys. Rev.* B **77** 205405
[5] Zimin S P, Amirov I I and Gorlachev E S 2011 *Semicond. Sci. Technol.* **26** 055018
[6] Fang W and Wickert J A 1996 *J. Micromech. Microeng.* **6** 301-9
[7] Mehner H, Leopold S and Hoffmann M 2013 *J. Micromech. Microeng.* **23** 095030
[8] Thuau D, Koutsos V and Cheung R 2013 *Soft Materials* **11** 414-20

The thermal characteristics of superlattice structures based on AlGaInN solid solution

A S Evseenkov, S A Tarasov, A V Solomonov, S M Altimime, A S Obukhova
Saint-Petersburg Electrotechnical University "LETI", Prof. Popova 5, St. Petersburg 197376, Russia

Abstract. The blue light-emitting structures based on solid solutions of the system AlGaInN that contained superlattices $In_{0.9}Ga_{0.1}N/In_{0.99}Ga_{0.01}N$ and $Al_{0.15}Ga_{0.85}N/GaN$ were investigated, and its basic parameters and characteristics were defined. The main difference between the samples was an "upper blocking layer" that formed by AlGaN solid solutions with a different stoichiometric composition or as a superlattice. Spectral and thermal investigations of samples were conducted, and the temperatures of the active region were calculated. The optimal types of structures for different operation modes were offered.

1. Introduction
Improving of the white and blue light sources based on gallium nitride and its solid solutions is one of the key tasks of the semiconductor industry. Low emitting power, insufficient quantum efficiency and fast degradation of the structures at high operating currents still limit using of light emitting semiconductor devices as effective light sources.

Electroluminescence efficiency is a main characteristic of LEDs. One of the main reasons of its decreasing is a low quality of epitaxial heterostructures [1-4]. Effect of Auger-recombination is making a significant influence at high-power mode. In addition, a significant influence on the efficiency has self-heating processes in the emitting structure. It is actual to study these processes and to investigate of the influence of light emitting structure construction on the efficiency of luminescence. Superlattice (SL) may improve the quality of the structure and enhance the efficiency of blocking layer above the active region. In the paper the impact of SL to self-heating and the efficiency of light-emitting structures was examined.

2. The samples
In this work were investigated four types of samples of structures based on solid solutions of AlGaInN [5,6] system and disposed on sapphire substrates of diameter 3 " (figure 1). These structures were created by the method of metalorganic vapour phase epitaxy (MOCVD). Samples in cross-section is a «sandwich structure» with the quantum well and superlattices.

An important feature of investigated structures was using superlattices based on solid solutions AlGaN or InGaN. First SL was between bottom n-layer and active layer. It consisted of 15 pairs $In_{0,9}Ga_{0,1}N/In_{0,99}Ga_{0,01}N$ (width of every layer – 1 nm). This SL was intended to reduce the concentration of defects in the active layer (AL). AL comprised single quantum well (SQW) $In_{0,15}Ga_{0,75}N$. It width was 2.5 nm. The barrier layer i-GaN blocked the process of diffusion between p-emitter and AL.

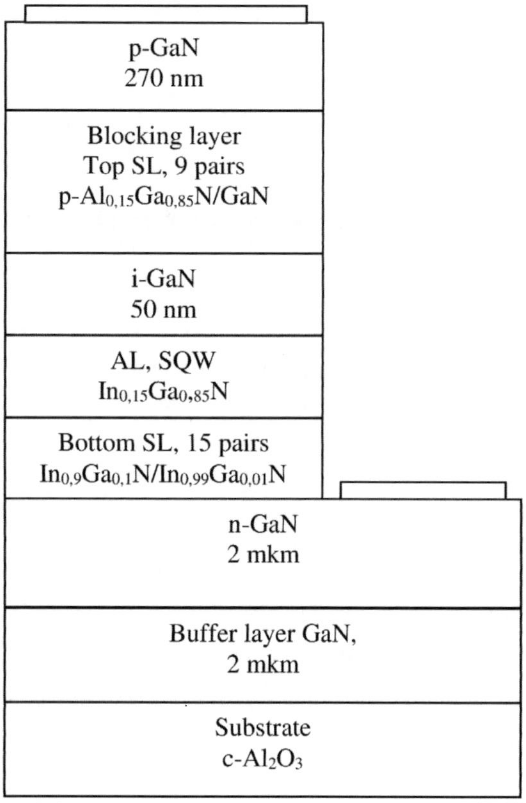

Figure 1. Light-emitting structures 114, based on solid solutions AlGaN or InGaN

Main attention was given for exploring the influence of upper blocking layer structure on the parameters of light emitting crystals, especially on the process of current distribution and thermal effects connected with it. First type of samples (114) contained blocking layer with a SL based on 9 pairs of p-$Al_{0.15}Ga_{0.85}N$/GaN with a thickness of every layer 1.5 nm. Samples with more simple structure of blocking layer had explored to determine all possible advantages of superlattice. Two types of crystals contained a simpler layer, based on a solid solution of aluminium gallium nitride: p-$Al_{0.15}Ga_{0.85}N$ (115) or p-$Al_{0.1}Ga_{0.9}N$ (116). In the last case, the blocking layer didn't contain AlN and was grown from a uniform p-GaN (117). In all samples, the thickness of the blocking layer was 27 nm and the concentration of the dopant was approx. $3 \cdot 10^{17}$ см$^{-3}$. A layer of p-GaN with a thickness of 270 nm completed the structure of all samples.

3. Experiment

The investigations were conducted using the developed hardware and software system. It includes thermostats, integrating spheres, quick scanning spectrometers produced by OceanOptics company and other measuring equipment [7-9]. Non-contact determination of the active layer temperature of the light-emitting structures is an important feature of the system. The experiment was conducted in accordance with international standard CIE 127:2007. The installation includes a probe station. This allowed investigating the temperature characteristics of the crystals in the unseparated substrates.

It was shown that an increase of the direct current leads to growth of the AL temperature to a value of 100 °C and above. This increasing was determined by the processes of charge transport and a change in the probability of quantum transitions in the structure. The relation between the efficiency of luminescence and thermal effects in the structures was studied. The figure 2 shows the luminescence spectra of the LED structure 116 at different currents. It can be seen that the shift of the maximum emission is a superposition of short-wave driving in connection with the Stark effect and temperature shift to longer wavelengths, which begins to dominate at a current rise above 150 mA. On a set of spectral characteristics was made a comparison between the sample spectral characteristics (Figure 3). The relative performance characteristics and the AL temperature are shown in Figure 4.

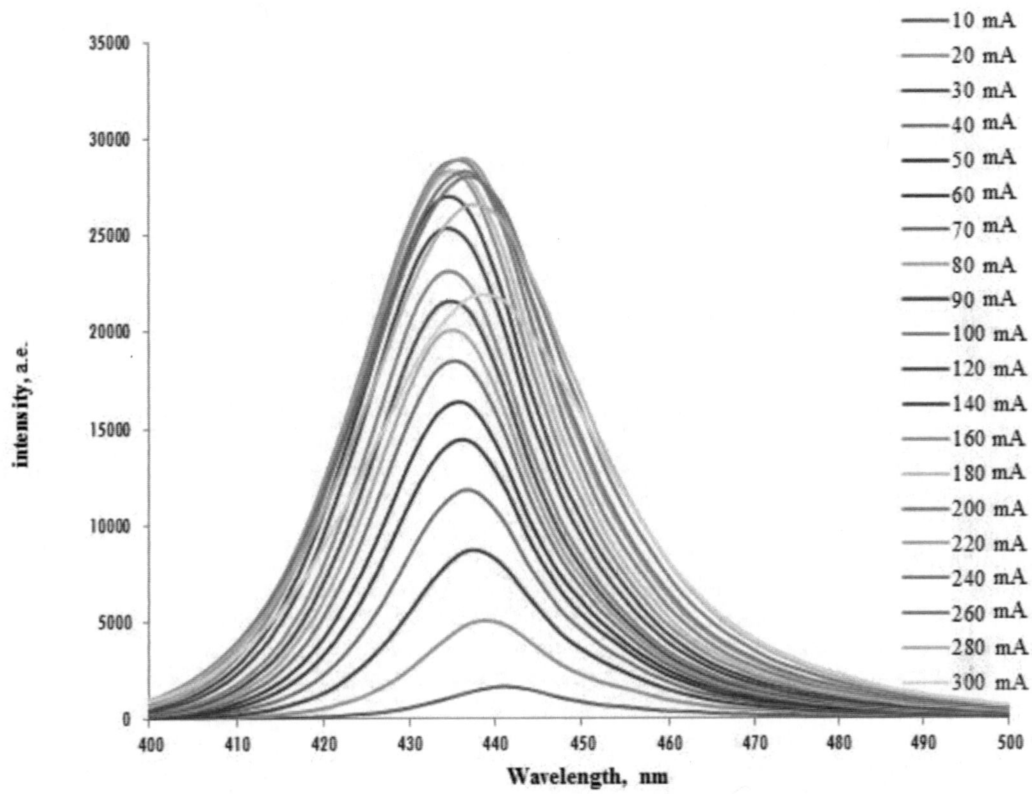

Figure 2. The luminescence spectra of the structure 116 at different currents

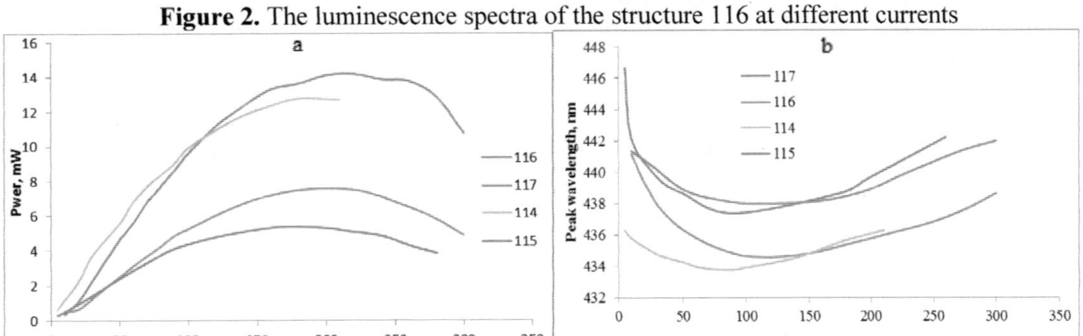

Figure 3. Power-current characteristics (a) and peak wavelength-current characteristics (b)

It was shown that the use of the SL in the blocking layer allows to provide the highest efficiency at low currents, due to improved barrier properties of the layer. The efficiency and power of all samples decreased when the current increased due to the heating of the structures and the increasing intensity of the Auger recombination (figures 3b and 4b).It is found that this reduction in efficiency for samples with 10 % AlN in the BL (116) is slower than for samples with SL (114). This leads to a smooth increase in the temperature of the active region of such samples (fig. 4b), so the use of such structures at high operating currents is preferable.

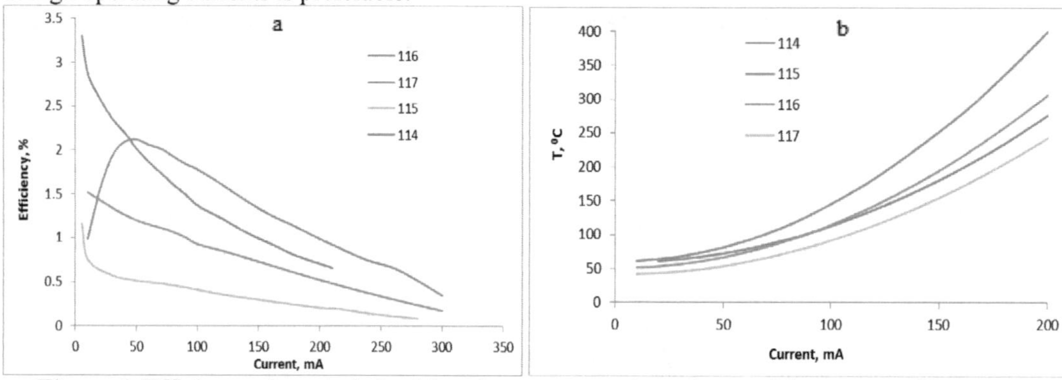

Figure 4. Efficiency characteristics (a) and temperature dependence of the currents (b)

This was associated with a sharp increase in the intensity of Auger recombination processes in the structures with SL. In this paper, the relation between the Auger recombination effects and thermal properties of structures is examined, as well as its effect on the power efficiency of the samples. Preliminary estimates show that the analysis of these effects allows for a quick assessment of the quality of structures, perspective for introduction in industrial enterprises.

4. Conclusions

The study of AlGaInN light-emitting structures containing superlattice $In_{0.9}Ga_{0.1}N$ / $In_{0.99}Ga_{0.01}N$ and $Al_{0.15}Ga_{0.85}N$ / GaN, was conducted, and its main parameters and characteristics were defined. It was shown that the highest efficiency at low currents was observed in the structures having a superlattice in the blocking layer. At large currents the efficiency was decreased for all samples. Nevertheless this decrease in efficiency was slower for samples with 10 % of AlN in the blocking layer than for samples with SL. This leads to a smooth increase in the temperature of the active region of such samples,. This is associated with a sharp increase in the intensity of Auger recombination processes in the structures with SL. Thus light-emitting structures containing $Al_{0.15}Ga_{0.85}N$ / GaN superlattice is preferable at low operating currents. At low operating currents structures with 10 % of AlN in the blocking layer are more preferable.

Acknowledgments

The work was supported by the Ministry of Education and Science of the Russian Federation in the framework of the project of the state task in the field of scientific activity, project number 16.1307.2014K.

References

[1] Kim J K Alhmed N, Frank W. 2008 *Appl Phys Lett* **93** 221111

[2] Evseenkov A S, Tarasov S A, Lamkin I A, Solomonov A V, Kurin S Y. 2015 IEEE North West Russia Section Young Researchers In Electrical And Electronic Engineering Conference, ELCONRUSNW 2015 27-29

[3] Kurin S, Antipov A, Barash I, Roenkov A, Makarov Y, Usikov A, Helava H, Solomonov A, Tarasov S, Evseenkov A, Lamkin I. 2015 *Physica Status Solidi (C) Current Topics in Solid State Physics* **12** № 4-5 369-371

[4] Evseenkov A S, Tarasov S A, et al 2015 *IOP Publishing Journal of Physics: Conference Series* **643** 012033

[5] Panfeng J, Naixin L, Tongboo W 2011 *Journal of Semiconductors* Vol. **32** 11

[6] Menkovich E A,Tarasov S A, Lamkin I A, Kurin S Yu, Antipov A A, Roenkov A D, Barash I S,

Helava H I, Makarov Yu N 2013 *Journal of Physics: Conference Series* **461** 012027
[7] Gu Y, Narendran N. 2004 *Proceedings of SPIE* **5187**: 107-114
[8] Solomonov A V, Tarasov S A, Men'kovich E A, Lamkin I A, Kurin S Yu, Antipov A A, Barash I S, Roenkov A D, Helava H I, Makarov Yu N 2014 *Semiconductors* **48** 245–250
[9] Menkovich E A, Solomonov A V, Tarasov S A, Yurgin P A 2014 *Functional materials* **21** 186-189

AUTHOR INDEX

AUTHOR INDEX

AUTHOR INDEX

AUTHOR INDEX

AUTHOR INDEX

AUTHOR INDEX

AUTHOR INDEX

AUTHOR INDEX

Institute of Physics
Dirac House, Temple Back
Bristol BS1 6BE UK

ISSN: 1742-6588
ISBN 978-1-5108-2995-4

CURRAN ASSOCIATES INC.
proceedings.com